EXPANDED EIGHTH EDITION

A Survey of MATHEMATICS with Applications

ALLEN R. ANGEL
Monroe Community College

CHRISTINE D. ABBOTT
Monroe Community College

DENNIS C. RUNDE
Manatee Community College

PEARSON
Addison Wesley

Boston San Francisco New York
London Toronto Sydney Tokyo Singapore Madrid
Mexico City Munich Paris Cape Town Hong Kong Montreal

Publisher: Greg Tobin
Executive Editor: Anne Kelly
Acquisitions Editor: Marnie Greenhut
Senior Project Editor: Rachel Reeve
Associate Project Editor: Elizabeth Bernardi
Editorial Assistant: Leah Goldberg
Senior Managing Editor: Karen Wernholm
Senior Production Supervisor: Peggy McMahon
Design Supervisor: Barbara T. Atkinson
Cover Designer: Barbara T. Atkinson
Cover Image: Rows of speakers, © Getty Images, Inc.
Cover Photo Photographer: Chad Baker
Photo Researcher: Beth Anderson
Media Producer: Ashley O'Shaughnessy
Software Development: Eric Gregg and Eileen Morse
Executive Marketing Manager: Becky Anderson
Marketing Coordinator: Bonnie Gill
Senior Author Support/Technology Specialist: Joe Vetere
Rights and Permissions Advisor: Dana Weightman
Manufacturing Manager: Evelyn Beaton
Text Design, Production Coordination, Illustrations, and Composition: Nesbitt Graphics, Inc.

Library of Congress Cataloging-in-Publication Data
Angel, Allen R., 1942–
A survey of mathematics with applications.—Expanded 8th ed. / Allen R. Angel, Christine D. Abbott, Dennis C. Runde.
p. cm.
Includes index.
High school binding ISBN 0-13-135481-7
1. Mathematics–Textbooks. I. Abbott, Christine D. II. Runde, Dennis C.
QA39.3.A54 2009
510—dc22 2007060062

ISBN-13: 978-0-13-135481-4 ISBN-10: 0-13-135481-7
1 2 3 4 5 6 7 8 9 10—VHP—11 10 09 08 07

To my wife, Kathy Angel (photo on page 60)
A.R.A.

To my husband Jason; and children, Matthew and Jake (photo on page 197)
C.D.A.

To my parents, Bud and Tina Runde
D.C.R.

CONTENTS

TO THE STUDENT

Mathematics is an exciting, living study. Its applications shape the world around you and influence your everyday life. We hope that as you read this book you will realize just how important mathematics is and gain an appreciation of both its usefulness and its beauty. We also hope to teach you some practical mathematics that you can use every day and that will prepare you for further mathematics courses.

The primary purpose of this text is to provide material that you can read, understand, and enjoy. To this end, we have used straightforward language and tried to relate the mathematical concepts to everyday experiences. We have also provided many detailed examples for you to follow.

The concepts, definitions, and formulas that deserve special attention are in boxes or are set in boldface type. Within each category, the exercises are graded, with more difficult problems appearing at the end. Exercises with numbers set in color are writing exercises. At the end of most exercise sets are Challenge Problem/Group Activity Exercises that contain challenging or exploratory exercises. At the end of each chapter are Group Projects that reinforce the material learned or provide related material.

Each chapter has a summary, review exercises, and a chapter test. Be sure to read the chapter summary, work the review exercises, and take the chapter test. The answers to the odd-numbered exercises, all review exercises, all chapter test exercises, and selected recreational mathematics exercises appear in the answer section in the back of the text. You should, however, use the answers only to check your work.

It is difficult to learn mathematics without becoming involved. To be successful, we suggest that you read the text carefully *and work each exercise in each assignment in detail.* Check with your instructor to determine which supplements are available for your use.

We welcome your suggestions and your comments. You may contact us at the following address:

Allen Angel
c/o Marketing
Mathematics and Statistics
Addison-Wesley
75 Arlington St., Suite 300
Boston, MA 02116

Or by email at:

math@awl.com
Subject: for Allen Angel

Good luck in your adventure in mathematics!

Allen R. Angel

Christine D. Abbott

Dennis C. Runde

MATH We Use It Every Day!

We present *A Survey of Mathematics with Applications*, eighth edition, with the vision in mind that we use math every day. In this edition, we stress how mathematics is used in our daily lives whenever possible. Our primary goal is to give students a text they can read, understand, and enjoy while learning how mathematics affects the world around them. Numerous real-life applied examples motivate topics. A variety of interesting exercises demonstrate the real-life nature of mathematics and its importance in students' lives.

The text is intended for students who require a broad-based general overview of mathematics, especially those considering majoring in the liberal arts, elementary education, the social sciences, business, nursing, and allied health fields. It is particularly suitable for those courses that satisfy the minimum competency requirement in mathematics for graduation or transfer.

The expanded version of *A Survey of Mathematics with Applications*, eighth edition, contains all the material covered in the basic text, with additional chapters on graph theory and on voting and apportionment. Although many wished inclusion of these topics, others did not. Therefore, we have two versions of the book.

CONTENT REVISION

In this edition, we have revised and expanded certain topics to introduce new material and to increase understanding.

Chapter 1 "Critical Thinking Skills," has been updated with exciting and current examples and exercises.

Chapter 2 "Sets," now includes new material on the difference of two sets and the Cartesian product of two sets.

Chapter 3 "Logic," has more exercises and a greater variety of exercises, such as in Section 3.1. Certain material has been rewritten for greater clarity. A new section (3.7) on switching circuits has been added.

Chapter 4 "Systems of Numeration," has increased use of the bases 2, 8, and 16, which have applications to modern computers and electronics. See Section 4.3.

Chapter 5 "Number Theory and the Real Number System," has the most current number theory information (largest prime number, most accurate value of pi). Updated examples and exercises include the most current economic numbers, such as federal debt and gross domestic product.

Chapter 6 "Algebra, Graphs, and Functions," has more and a greater variety of examples and exercises dealing with real-life situations. Additional exercises involving exponential equations have been included.

Chapter 7 "Systems of Linear Equations and Inequalities," now includes annotations and more detailed explanations of certain topics.

Chapter 8 "The Metric System," has many examples and interesting photographs of real-life (metric) situations taken from around the world.

Chapter 9 "Geometry," now includes information on finding surface area. Some information on non-Euclidean geometry has been rewritten.

Chapter 10 "Mathematical Systems," has additional exercises and examples. We have tied this material to real-life situations more.

Chapter 11 "Consumer Mathematics," contains current interest rates and updated information on items that may be of interest to students, including material on sources of credit and mutual funds. Additional information on stocks, bonds, and mutual funds is also included. Section 11.6, "Ordinary Annuities, Sinking Funds, and Retirement Investments" has been added.

Chapter 12 "Probability," has a greater variety of examples and exercises, and many that deal with real-life situations have been added. Certain material has been rewritten for greater clarity. Calculator keystrokes for solving permutation and combination exercises are included.

Chapter 13 "Statistics," now includes calculator keystrokes and commands for using Microsoft Excel to determine several statistical measures (see the Technology Tip on page 906). Material related to standard scores has been revised.

Chapter 14 "Graph Theory" includes many updated examples and exercises, and some material has been rewritten for greater clarity.

Chapter 15 "Voting and Apportionment" includes current information about voting and elections. Certain material has been rewritten for greater clarity.

NEW AND REVISED FEATURES

Several important improvements in presentation have also been made.

- Each section opening contains interesting and motivational applications to introduce each section and illustrate the real-world nature of the material in that section.
- Technology Tips that explain how a graphing calculator or a computer spreadsheet may be used to work certain problems have been added in selected sections.
- The number of examples has been increased throughout the text to promote student understanding.
- Sources have been added; up-to-date tables, graphs, and charts make the material more relevant and encourage students to read graphs and analyze data.
- The feature previously called Mathematics Everywhere is now called Mathematics Today. Current mathematical topics related to material presented in this book are included in the Mathematics Today boxes. The information provided in these boxes

relates mathematics to students' everyday lives. This material will help students see the need for mathematics and gain appreciation for it.

- Additional Did You Know? and Profiles in Mathematics features have been added throughout the book.
- In various exercise sets, the number and variety of exercises has been increased. Approximately 40% of the exercises have been revised or updated to reflect current data, new material in the text, and the needs and interests of today's students.

CONTINUING FEATURES

Several features appear throughout the book, adding interest and provoking thought.

- *Chapter openings* Interesting and motivational applications introduce each chapter and illustrate the real-world nature of the chapter topics.
- *Problem solving* Beginning in Chapter 1, students are introduced to problem solving and critical thinking. The theme of problem solving is then continued throughout the text, and special problem-solving exercises are presented in the exercise sets.
- *Critical thinking skills* In addition to a focus on problem solving, this book also features sections on inductive and deductive reasoning, estimation, and dimensional analysis.
- *Profiles in Mathematics* Brief historical sketches and vignettes present the stories of people who have advanced the discipline of mathematics.
- *Did You Know?* The colorful, engaging, and lively Did You Know? boxes highlight the connection of mathematics to history, the arts and sciences, technology, and a broad variety of disciplines.
- *Group Projects* At the end of each chapter are suggested projects that can be used to have students work together. These projects can also be assigned to individual students if desired.
- *Chapter Summaries, Review Exercises, and Chapter Tests* End-of-chapter summaries, exercises, and tests help students review material and prepare for exams.

TECHNOLOGY RESOURCES

MathXL®, www.mathxl.com

MathXL® is a powerful online homework, tutorial, and assessment system that accompanies Pearson Education textbooks in mathematics or statistics. With MathXL, instructors can create, edit, and assign online homework and tests using algorithmically generated exercises correlated at the objective level to the textbook. They can also create and assign their own online exercises and import TestGen tests for added flexibility. All student work is tracked in MathXL's online gradebook. Students can take chapter tests in MathXL and receive personalized study plans based on their test results. The study plan diagnoses weaknesses and links students directly to tutorial exercises for the objectives they need to study and retest. Students can also access supplemental animations and video clips directly from selected exercises. MathXL is available for purchase only. For more information, visit our website at www.mathxl.com or contact your sales representative.

InterAct Math Tutorial Website: www.interactmath.com

Get practice and tutorial help online! The interactive InterAct Math tutorial website provides algorithmically generated practice exercises that correlate directly to the exercises in the book. Students can retry an exercise as many times as they like with new values each time for unlimited practice and mastery. Each exercise is accompanied by an interactive guided solution that provides helpful feedback for incorrect answers. Students can also view a worked-out sample problem that guides them through an exercise similar to the one they're working on.

Video Lectures on CD with Optional Captioning

The video lectures for this text are also available on CD-ROM, making it easy and convenient for students to watch the videos from a computer at home or on campus. The complete digitized video set, affordable and portable for students, is ideal for distance learning or supplemental instruction (ISBN 13: 978-0-321-51091-4; ISBN 10: 0-321-51091-7). Available for purchase.

SUPPLEMENTS for Students

Student's Solutions Manual
ISBN 13: 978-0-321-51089-1; ISBN 10: 0-321-51089-5

This for-sale manual contains solutions to all odd-numbered exercises and to all review and chapter test exercises.

SUPPLEMENTS for Instructors

Annotated Instructor's Edition
ISBN 13: 978-0-321-50132-5; ISBN 10: 0-321-50132-2

This special edition of the text includes answers next to the exercises, when they fit. Answers that do not fit next to the exercise are placed in a separate section in the back of the book. Answers to all text exercises are included.

Instructor's Solutions Manual
ISBN 13: 978-0-321-51092-1; ISBN 10: 0-321-51092-5

This manual contains solutions to all exercises in the text and answers to Group Projects.

Instructor's Testing Manual
ISBN 13: 978-0-321-51093-8; ISBN 10: 0-321-51093-3

This manual includes three alternate tests per chapter.

New! Insider's Guide
ISBN 13: 978-0-321-52833-9; ISBN 10: 0-321-52833-6

This manual includes resources and helpful teaching tips designed to help teachers with course preparation and classroom management.

PowerPoint Lecture Presentation

Available through www.aw-bc.com/irc. This classroom presentation software covers all-important topics from sections in the text.

TestGen®

TestGen enables instructors to build, edit, print, and administer tests using a computerized bank of questions developed to cover all the objectives of the text. TestGen is algorithmically based, allowing instructors to create multiple but equivalent versions of the same question or test with the click of a button. Instructors can also modify test bank questions or add new questions. Tests can be printed or administered online. The software and testbank are available for download at www.aw-bc.com/irc.

ACKNOWLEDGMENTS

We thank our spouses, Kathy Angel, Jason Abbott, and Kris Runde, for their support and encouragement throughout the project. They helped us in a great many ways, including proofreading, typing, and offering valuable suggestions. We are grateful for their wonderful support and understanding while we worked on the book.

We also thank our children: Robert and Steven Angel; Matthew and Jake Abbott; and Alex, Nicholas, and Max Runde. They also gave us support and encouragement and were very understanding when we could not spend as much time with them as we wished because of book deadlines. Without the support and understanding of our families, this book would not be a reality.

We thank Patricia Nelson and Paul Lorczak for their conscientious job of checking the text and answers for accuracy. We also thank Sherry Tornwall of the University of Florida for reading the manuscript and suggesting improvements to its content. And thanks to Becky Troutman for preparing the Index of Applications.

Many people at Addison-Wesley deserve thanks, including all those listed on the copyright page. In particular, we thank Anne Kelly, executive editor; Marnie Greenhut, acquisitions editor; Rachel Reeve, senior project editor; Elizabeth Bernardi, associate project editor; Leah Goldberg, editorial assistant; Peggy McMahon, senior production supervisor; Becky Anderson, executive marketing manager; Bonnie Gill, marketing assistant; Ashley O'Shaughnessy, media producer; Barbara Atkinson, senior designer; Beth Anderson, photo researcher; and Karen Wernholm, managing editor. We also thank Maria McColligan of Nesbitt Graphics, Inc., for her assistance as project manager for this project.

Elka Block and Frank Purcell also deserve our thanks for the excellent work they did on the *Student's Solutions Manual* and the *Instructor's Solutions Manual* and Debra McGivney for her work on the *Instructor's Testing Manual*.

Finally, we thank the reviewers from all editions of the book and all the students who have offered suggestions for improving it. A list of reviewers for all editions of this book follows. Thanks to you all for helping make *A Survey of Mathematics with Applications* the most successful liberal arts mathematics textbook in the country.

Allen R. Angel

Christine D. Abbott

Dennis C. Runde

REVIEWERS FOR THIS AND PREVIOUS EDITIONS

*Mary Anne Anthony-Smith, Santa Ana College

Frank Asta, College of DuPage, IL

Robin L. Ayers, Western Kentucky University

Hughette Bach, California State University–Sacramento

Madeline Bates, Bronx Community College, NY

Rebecca Baum, Lincoln Land Community College, IL

Vivian Baxter, Fort Hays State University, KS

Una Bray, Skidmore College, NY
David H. Buckley, Polk Community College, FL
Robert C. Bueker, Western Kentucky University
Carl Carlson, Moorhead State University, MN
Kent Carlson, St. Cloud State University, MN
Donald Catheart, Salisbury State College, MD
Yungchen Cheng, Southwest Missouri State University
Joseph Cleary, Massasoit Community College, MA
Donald Cohen, SUNY Ag & Tech College at Cobleskill, NY
David Dean, Santa Fe Community College, FL
*John Diamantopoulos, Northeastern State University, OK
*Greg Dietrich, Florida Community College at Jacksonville
Charles Downey, University of Nebraska
Annie Droullard, Polk Community College, FL
Ruth Ediden, Morgan State University, MD
Lee Erker, Tri-County Community College, NC
*Nancy Eschen, Florida Community College at Jacksonville
Karen Estes, St. Petersburg College, FL
Teklay Fessanaye, Santa Fe Community College, FL
Kurtis Fink, Northwest Missouri State University
Raymond Flagg, McPherson College, KS
Penelope Fowler, Tennessee Wesleyan College
Gilberto Garza, El Paso Community College, TX
Judith L. Gersting, Indiana University–Purdue University at Indianapolis
Lucille Groenke, Mesa Community College, AZ
John Hornsby, University of New Orleans, LA
Nancy Johnson, Broward Community College, FL
*Phyllis H. Jore, Valencia Community College, FL
*Heidi Kiley, Suffolk County Community College, NY
Daniel Kimborowicz, Massasoit Community College, MA
Mary Lois King, Tallahassee Community College, FL
David Lehmann, Southwest Missouri State University
Peter Lindstrom, North Lake College, TX
James Magliano, Union College, NJ
Yash Manchanda, East Los Angeles College & Fullerton College, CA
Don Marsian, Hillsborough Community College, FL
Marilyn Mays, North Lake College, TX
Robert McGuigan, Westfield State College, MA
Wallace H. Memmer, Brookdale Community College, NJ
Maurice Monahan, South Dakota State University
Julie Monte, Daytona Beach Community College, FL

Karen Mosely, Alabama Southern Community College
*Kathleen Offenholley, Brookdale Community College, NY
Edwin Owens, Pennsylvania College of Technology
Wing Park, College of Lake County, IL
Bettye Parnham, Daytona Beach Community College, FL
Joanne Peeples, El Paso Community College, TX
Nelson Rich, Nazareth College, NY
Kenneth Ross, University of Oregon
Ronald Ruemmler, Middlesex County College, NJ
Rosa Rusinek, Queensborough Community College, NY
Len Ruth, Sinclair Community College, OH
John Samoylo, Delaware County Community College, PA
Sandra Savage, Orange Coast College, CA
Gerald Schultz, Southern Connecticut State University
Richard Schwartz, College of Staten Island, NY
Kara Shavo, Mercer County Community College, NJ
Minnie Shuler, Chipola Junior College, FL
Paula R. Stickles, University of Southern Indiana
Kristin Stoley, Blinn College–Bryan, TX
Steve Sworder, Saddleback College, CA
Shirley Thompson, Moorhead College, GA
Alvin D. Tinsley, Central Missouri State University
*Sherry Tornwall, University of Florida
William Trotter, University of South Carolina
Zia Uddin, Lock Haven University of Pennsylvania
Sandra Welch, Stephen F. Austin State University, TX
Joyce Wellington, Southeastern Community College, NC
Sue Welsch, Sierra Nevada College
Robert F. Wheeler, Northern Illinois University
Susan Wirth, Indian River Community College, FL
James Wooland, Florida State University
*Judith B. Wood, Central Florida Community College
Jean Woody, Tulsa Community College, OK
Michael A. Zwick, Monroe Community College, NY
**Denotes reviewers for eighth edition.*

CHAPTER 1

Critical Thinking Skills

▲ When creating a monthly budget, we may need to make problem-solving decisions to reduce our expenses.

WHAT YOU WILL LEARN

- Inductive and deductive reasoning processes
- Estimation
- Problem-solving techniques

WHY IT IS IMPORTANT

Life constantly presents new problems. The more sophisticated our society becomes, the more complex the problems. We as individuals are constantly solving problems. For example, when we consider ways to reduce our expenses or when we plan a trip, we make problem-solving decisions. Businesses are constantly trying to solve problems that involve making a profit for the company and keeping customers satisfied.

The goal of this chapter is to help you master the skills of reasoning, estimating, and problem solving. These skills will aid you in solving problems in the remainder of this book as well as problems you will encounter in everyday life.

1.1 INDUCTIVE REASONING

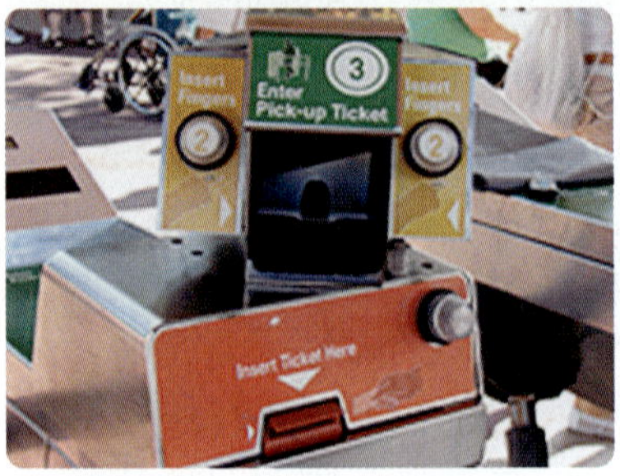

▲ A machine like this is used at Disney World to identify seasonal ticket holders.

Thus far, no two people have been found to have the same fingerprints or DNA, so fingerprints and DNA have become forms of identification. As technology improves, so do identification techniques. Today, when you purchase a seasonal pass at Disney World, you place your forefinger and middle finger in a device that measures and records your fingers' size, shape, and density. Then you are given a seasonal pass. Each time you enter Disney World, you put those two fingers in the device and it indicates whether you are or are not allowed entry under the seasonal pass that you show. The belief that no two people have the same finger size, shape, and density is based on a type of reasoning we will discuss in this section.

Inductive Reasoning

Before looking at some examples of inductive reasoning and problem solving, let us first review a few facts about certain numbers. The ***natural numbers*** or ***counting numbers*** are the numbers 1, 2, 3, 4, 5, 6, 7, 8, ... The three dots, called an ***ellipsis***, mean that 8 is not the last number but that the numbers continue in the same manner. A word that we sometimes use is "divisible." If $a \div b$ has a remainder of zero, then *a is divisible by b*. The counting numbers that are divisible by 2 are 2, 4, 6, 8, ... These numbers are called the *even counting numbers*. The numbers that are not divisible by 2 are 1, 3, 5, 7, 9, ... These numbers are the *odd counting numbers*. When we refer to *odd numbers* or *even numbers*, we mean odd or even counting numbers.

Recognizing patterns is sometimes helpful in solving problems, as Examples 1 and 2 illustrate.

EXAMPLE 1 *The Product of Two Even Numbers*

If two even numbers are multiplied together, will the product always be an even number?

SOLUTION To answer this question, we will examine the products of several pairs of even numbers to see if there is a pattern.

$2 \times 2 = 4$	$4 \times 6 = 24$	$6 \times 8 = 48$
$2 \times 4 = 8$	$4 \times 8 = 32$	$6 \times 10 = 60$
$2 \times 6 = 12$	$4 \times 10 = 40$	$6 \times 12 = 72$
$2 \times 8 = 16$	$4 \times 12 = 48$	$6 \times 14 = 84$

All the products are even numbers. Thus, we might predict from these examples that the product of any two even numbers is an even number. ●

EXAMPLE 2 *The Sum of an Odd Number and an Even Number*

If an odd number and an even number are added, will the sum be an odd or an even number?

SOLUTION Let's look at a few examples in which one number is odd and the other number is even.

$$3 + 4 = 7 \qquad 9 + 6 = 15 \qquad 23 + 18 = 41$$
$$5 + 12 = 17 \qquad 5 + 14 = 19 \qquad 81 + 32 = 113$$

All these sums are odd numbers. Therefore, we might predict that the sum of an odd number and an even number is an odd number. ●

In Examples 1 and 2, we cannot conclude that the results are true for all counting numbers. From the patterns developed, however, we can make predictions. This type of reasoning process, arriving at a general conclusion from specific observations or examples, is called ***inductive reasoning***, or ***induction***.

Inductive reasoning is the process of reasoning to a general conclusion through observations of specific cases.

Induction often involves observing a pattern and from that pattern predicting a conclusion. Imagine an endless row of dominoes. You knock down the first, which knocks down the second, which knocks down the third, and so on. Assuming the pattern will continue uninterrupted, you conclude that any one domino that you select in the row will eventually fall, even though you may not witness the event.

Inductive reasoning is often used by mathematicians and scientists to predict answers to complicated problems. For this reason, inductive reasoning is part of the ***scientific method***. When a scientist or mathematician makes a prediction based on specific observations, it is called a ***hypothesis*** or ***conjecture***. After looking at the products in Example 1, we might conjecture that the product of two even numbers will be an even number. After looking at the sums in Example 2, we might conjecture that the sum of an odd number and an even number is an odd number.

Examples 3 and 4 illustrate how we arrive at a conclusion using inductive reasoning.

DID YOU KNOW?

The Eyes Tell It All

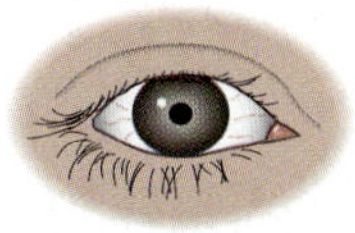

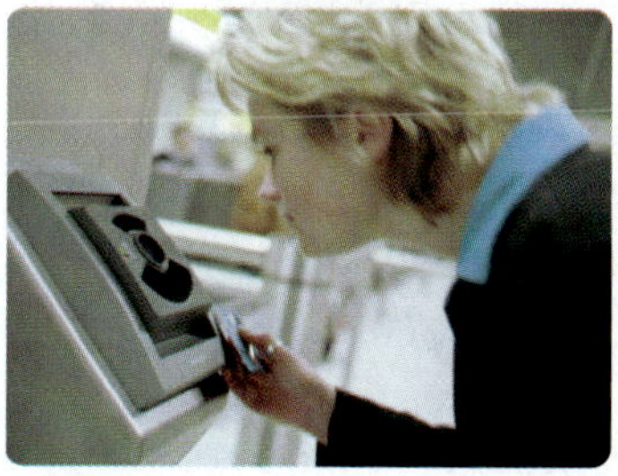

Automated teller machines (ATMs) are now experimenting with determining identity by scanning the iris of a person's eye. When you open a bank account, your iris is scanned and the image is entered into a computer. When you use an ATM, a powerful camera automatically (without you even knowing it) checks the veins in your iris against the computer's files. Shortly, we will no longer need personal identification numbers (PINs) to withdraw money from our bank accounts. Iris scanning is also used now by law enforcement to locate missing children. The Children's Identification and Location Database (CHILD) Project, begun in 2004, is a voluntary program that uses iris-scanning machines. As of August 2005, 1200 sheriff's departments in 46 states had signed up with the CHILD Project. By 2010, the project hopes to have 3 million to 5 million participants in its registry. In the future, iris scanning may also be used to track down Alzheimer's and mentally disabled patients. Because iris scanning can be done in a matter of seconds, it can be a much quicker procedure than fingerprinting and just as accurate. Because the iris-scanning method of identification relies on the observation of specific cases to form a general conclusion, it is based on inductive reasoning.

EXAMPLE 3 *Fingerprints and DNA*

What reasoning process has led to the conclusion that no two people have the same fingerprints or DNA? This conclusion has resulted in fingerprints and DNA being used in courts of law as evidence to convict persons of crimes.

SOLUTION In millions of tests, no two people have been found to have the same fingerprints or DNA. By induction, then, we believe that fingerprints and DNA provide a unique identification and can therefore be used in a court of law as evidence. Is it possible that sometime in the future two people will be found who do have exactly the same fingerprints or DNA? ●

EXAMPLE 4 *Divisibility by 9*

Consider the conjecture "If the sum of the digits of a number is divisible by 9, then the number itself is divisible by 9." We will test several numbers to see if the conjecture appears to be true or false.

SOLUTION Let's look at some numbers whose sum of the digits is divisible by 9.

Number	Sum of the Digits	Is the Sum of the Digits Divisible by 9?	Is the Number Divisible by 9?
576	$5 + 7 + 6 = 18$	yes; $18 \div 9 = 2$	yes; $576 \div 9 = 64$
2115	$2 + 1 + 1 + 5 = 9$	yes; $9 \div 9 = 1$	yes; $2115 \div 9 = 235$
6777	$6 + 7 + 7 + 7 = 27$	yes; $27 \div 9 = 3$	yes; $6777 \div 9 = 753$
49,302	$4 + 9 + 3 + 0 + 2 = 18$	yes; $18 \div 9 = 2$	yes; $49{,}032 \div 9 = 5448$

In each case, we find that if the sum of the digits of a number is divisible by 9, then the number itself is divisible by 9. From these examples, we might be tempted to generalize that the conjecture "If the sum of the digits of a number is divisible by 9, then the number itself is divisible by 9"* is true.

EXAMPLE 5 *Pick a Number, Any Number*

Pick any number, multiply the number by 6, add 9 to the product, divide the sum by 3, and subtract 3 from the quotient. Repeat this procedure for several different numbers and then make a conjecture about the relationship between the original number and the final number.

SOLUTION Let's go through this one together.

Pick a number:	say, 5
Multiply the number by 6:	$6 \times 5 = 30$
Add 9 to the product:	$30 + 9 = 39$
Divide the sum by 3:	$39 \div 3 = 13$
Subtract 3 from the quotient:	$13 - 3 = 10$

Note that we started with the number 5 and finished with the number 10. If you start with the number 2, you will end with the number 4. Starting with 3 would result in a final number of 6, 4 would result in 8, and so on. On the basis of these few examples, we may conjecture that when you follow the given procedure, the number you end with will always be twice the original number.

The result reached by inductive reasoning is often correct for the specific cases studied but not correct for all cases. History has shown that not all conclusions arrived at by inductive reasoning are correct. For example, Aristotle (384–322 B.C.) reasoned inductively that heavy objects fall at a faster rate than light objects. About 2000 years later, Galileo (1564–1642) dropped two pieces of metal—one 10 times heavier than the other—from the Leaning Tower of Pisa in Italy. He found that both hit the ground at exactly the same moment, so they must have traveled at the same rate.

When forming a general conclusion using inductive reasoning, you should test it with several special cases to see whether the conclusion appears correct. If a special case is found that satisfies the conditions of the conjecture but produces a different result, such a case is called a *counterexample*. A counterexample proves that the conjecture is false because only one exception is needed to show that a conjecture is not valid. Galileo's counterexample disproved Aristotle's conjecture. If a counterexample cannot be found, the conjecture is neither proven nor disproven.

DID YOU KNOW?

An Experiment Revisited

Apollo 15 astronaut David Scott used the moon as his laboratory to show that a heavy object (a hammer) does indeed fall at the same rate as a light object (a feather). Had Galileo dropped a hammer and feather from the Tower of Pisa, the hammer would have fallen more quickly to the ground and he still would have concluded that a heavy object falls faster than a lighter one. If it is not the object's mass that is affecting the outcome, then what is it? The answer is air resistance or friction: Earth has an atmosphere that creates friction on falling objects. The moon does not have an atmosphere; therefore, no friction is created.

*This statement is in fact true, as is discussed in Section 5.1.

Deductive Reasoning

A second type of reasoning process is called *deductive reasoning*, or *deduction*. Mathematicians use deductive reasoning to *prove* conjectures true or false.

Deductive reasoning is the process of reasoning to a specific conclusion from a general statement.

EXAMPLE 6 *Pick a Number, n*

Prove, using deductive reasoning, that the procedure in Example 5 will always result in twice the original number selected.

SOLUTION To use deductive reasoning, we begin with the *general* case rather than specific examples. In Example 5, specific cases were used. Let's select the letter n to represent *any number*.

Pick any number:	n
Multiply the number by 6:	$6n$ — 6n means 6 times n
Add 9 to the product:	$6n + 9$
Divide the sum by 3:	$\frac{6n+9}{3} = \frac{\overset{2}{\cancel{6}}n}{\underset{1}{\cancel{3}}} + \frac{\overset{3}{\cancel{9}}}{\underset{1}{\cancel{3}}} = 2n + 3$
Subtract 3 from the quotient:	$2n + 3 - 3 = 2n$

Note that for any number n selected, the result is $2n$, or twice the original number selected.

In Example 5, you may have *conjectured*, using specific examples and inductive reasoning, that the result would be twice the original number selected. In Example 6, we *proved*, using deductive reasoning, that the result will always be twice the original number selected.

SECTION 1.1 EXERCISES*

CONCEPT/WRITING EXERCISES

1. a) List the natural numbers.

   b) What is another name for the natural numbers?

2. a) What does it mean to say, "a is divisible by b," where a and b represent natural numbers?

   b) List three natural numbers that are divisible by 4.

   c) List three natural numbers that are divisible by 9.

In Exercises 3–6, explain your answer in one or two sentences.

3. What is inductive reasoning?

4. What is deductive reasoning?

5. What is a counterexample?

6. What is a conjecture?

7. Which type of reasoning is generally used to arrive at a conjecture?

*Exercise numbers set in color indicate writing exercises.

8. Which type of reasoning is used to prove a conjecture?

9. For many years, doctors noticed that many of their patients with heart disease also had high blood pressure. Doctors reasoned that high blood pressure increases a person's chance of developing heart disease. What type of reasoning did the doctors use? Explain.

10. You have purchased one lottery ticket each week for many months and have not won more than \$5.00. You decide, based on your past experience, that you are not going to win the grand prize and so you stop playing the lottery. What type of reasoning did you use? Explain.

PRACTICE THE SKILLS

In Exercises 11–14, use inductive reasoning to predict the next line in the pattern.

11.
$1 \times 7 = 7$
$2 \times 7 = 14$
$3 \times 7 = 21$
$4 \times 7 = 28$

12.
$13 \times 10 = 130$
$13 \times 11 = 143$
$13 \times 12 = 156$
$13 \times 13 = 169$

13.
1
1 1
1 2 1
1 3 3 1
↘↙↘↙↘↙
1 4 6 4 1

14.
$10 = 10^1$
$100 = 10^2$
$1000 = 10^3$
$10{,}000 = 10^4$

In Exercises 15–18, draw the next figure in the pattern (or sequence).

15. 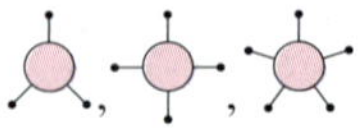

16. 

17. 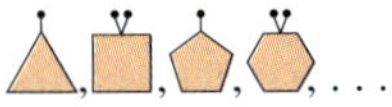

18. 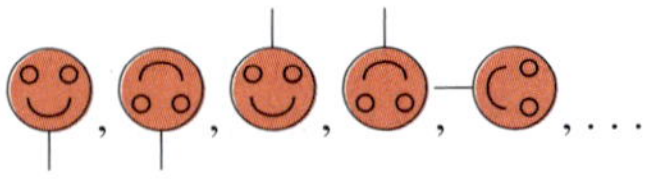

In Exercises 19–28, use inductive reasoning to predict the next three numbers in the pattern (or sequence).

19. 5, 10, 15, 20, . . .

20. 19, 16, 13, 10, . . .

21. $1, -1, 1, -1, 1, \ldots$

22. $5, 3, 1, -1, \ldots$

23. $16, 4, 1, \frac{1}{4}, \ldots$

24. $4, -20, 100, -500, \ldots$

25. 1, 4, 9, 16, 25, . . .

26. 0, 1, 3, 6, 10, 15, . . .

27. 1, 1, 2, 3, 5, 8, 13, 21, . . .

28. $3, -\frac{9}{4}, \frac{27}{16}, -\frac{81}{64}, \ldots$

PROBLEM SOLVING

29. Find the letter that is the 118th entry in the following sequence. Explain how you determined your answer.

Y, R, R, Y, R, R, Y, R, R, Y, R, R, Y, R, R, . . .

30. **a)** Select a variety of one- and two-digit numbers between 1 and 99 and multiply each by 9. Record your results.

b) Find the sum of the digits in each of your products in part (a). If the sum is not a one-digit number, find the sum of the digits again until you obtain a one-digit number.

c) Make a conjecture about the sum of the digits when a one- or two-digit number is multiplied by 9.

31. ***A Square Pattern*** The ancient Greeks labeled certain numbers as **square numbers**. The numbers 1, 4, 9, 16, 25, and so on are square numbers.

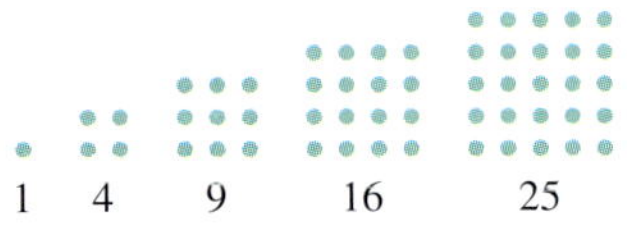

a) Determine the next three square numbers.

b) Describe a procedure to determine the next five square numbers without drawing the figures.

c) Is 72 a square number? Explain how you determined your answer.

32. ***A Triangular Pattern*** The ancient Greeks labeled certain numbers as **triangular numbers**. The numbers 1, 3, 6, 10, 15, 21, and so on are triangular numbers.

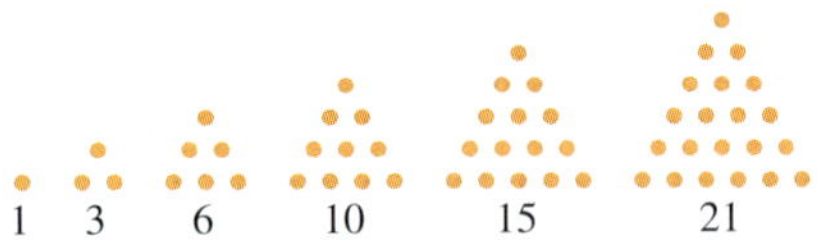

a) Can you determine the next two triangular numbers?

b) Describe a procedure to determine the next five triangular numbers without drawing the figures.

c) Is 72 a triangular number? Explain how you determined your answer.

33. ***Quilt Design*** The pattern shown is taken from a quilt design known as a triple Irish chain. Complete the color pattern by indicating the color assigned to each square.

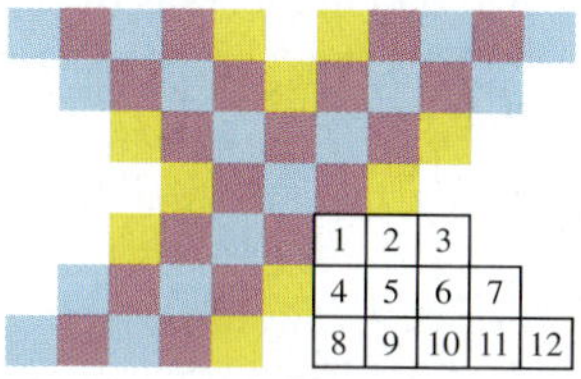

34. ***Triangles in a Triangle*** Four rows of a triangular figure are shown.

a) If you added six additional rows to the bottom of this triangle, using the same pattern displayed, how many triangles would appear in the 10th row?

b) If the triangles in all 10 rows were added, how many triangles would appear in the entire figure?

35. ***Television Advertising*** The graph shows the amount of money, in billions of dollars, advertisers spent or plan to spend on placing their products into television programs from 2000 to 2008.

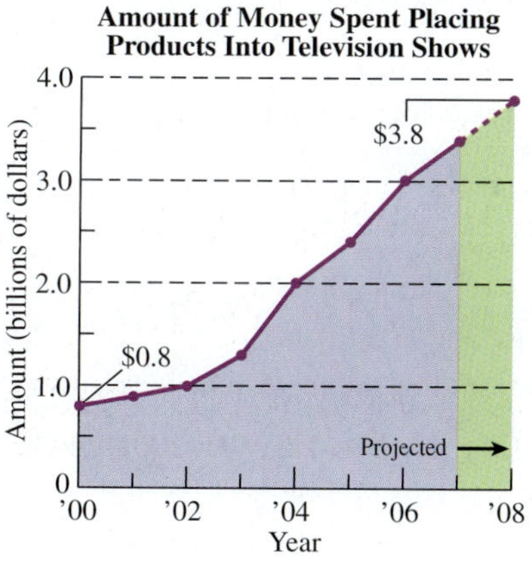

a) Assuming this trend continues, use the graph to predict the amount of money advertisers will spend on placing their products into television programs in 2012. Round your answer to the nearest billion dollars.

b) Explain how you are using inductive reasoning to determine your answer.

36. ***Cable TV Cost*** The graph above and to the right shows the average monthly cost for basic cable television for each year from 1994 through 2005.

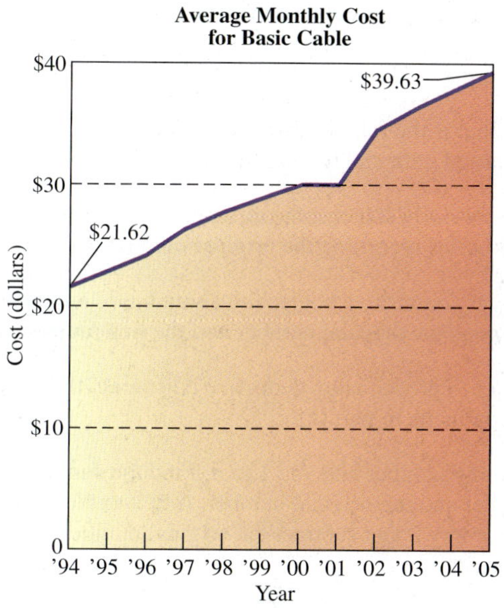

a) Assuming this trend continues, use the graph to predict the average monthly cost for basic cable TV in 2008. Round your answer to the nearest dollar.

b) Using the graph, predict the average monthly cost for basic cable TV in 2012. Round your answer to the nearest dollar.

c) Explain how you are using inductive reasoning to determine your answer.

In Exercises 37 and 38, draw the next diagram in the pattern (or sequence).

37. 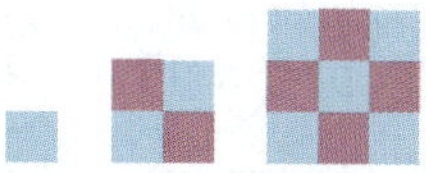

38. 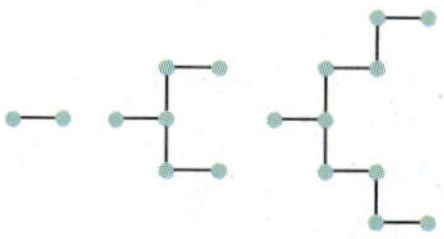

39. Pick any number, multiply the number by 8, add 16 to the product, divide the sum by 8, and subtract 2 from the quotient. See Example 5.

a) What is the relationship between the number you started with and the final number?

b) Arbitrarily select some different numbers and repeat the process, recording the original number and the result.

c) Can you make a conjecture about the relationship between the original number and the final number?

d) Try to prove, using deductive reasoning, the conjecture you made in part (c). See Example 6.

40. Pick any number and multiply the number by 4. Add 6 to the product. Divide the sum by 2 and subtract 3 from the quotient.

a) What is the relationship between the number you started with and the final answer?

b) Arbitrarily select some different numbers and repeat the process, recording the original number and the results.

c) Can you make a conjecture about the relationship between the original number and the final number?

d) Try to prove, using deductive reasoning, the conjecture you made in part (c).

41. Pick any number and add 1 to it. Find the sum of the new number and the original number. Add 9 to the sum. Divide the sum by 2 and subtract the original number from the quotient.

a) What is the final number?

b) Arbitrarily select some different numbers and repeat the process. Record the results.

c) Can you make a conjecture about the final number?

d) Try to prove, using deductive reasoning, the conjecture you made in part (c).

42. Pick any number and add 10 to the number. Divide the sum by 5. Multiply the quotient by 5. Subtract 10 from the product. Then subtract your original number.

a) What is the result?

b) Arbitrarily select some different numbers and repeat the process, recording the original number and the result.

c) Can you make a conjecture regarding the result when this process is followed?

d) Try to prove, using deductive reasoning, the conjecture you made in part (c).

In Exercises 43–48, find a counterexample to show that each statement is incorrect.

43. The sum of two odd numbers is an odd number.

44. The quotient of any two counting numbers is a counting number.

45. When a counting number is added to 3 and the sum is divided by 2, the quotient will be an even number.

46. The product of any two counting numbers is divisible by 2.

47. The difference of any two counting numbers will be a counting number.

48. The sum of any two odd numbers is divisible by 4.

49. ***Interior Angles of a Triangle***

a) Construct a triangle and measure the three interior angles with a protractor. What is the sum of the measures?

b) Construct three other triangles, measure the angles, and record the sums. Are your answers the same?

c) Make a conjecture about the sum of the measures of the three interior angles of a triangle.

50. ***Interior Angles of a Quadrilateral***

a) Construct a quadrilateral (a four-sided figure) and measure the four interior angles with a protractor. What is the sum of the measures?

b) Construct three other quadrilaterals, measure the angles, and record the sums. Are your answers the same?

c) Make a conjecture about the sum of the measures of the four interior angles of a quadrilateral.

CHALLENGE PROBLEMS/GROUP ACTIVITIES

51. Complete the following square of numbers. Explain how you determined your answer.

1	2	3	4
2	5	10	17
3	10	25	52
4	17	52	?

52. Find the next three numbers in the sequence.

1, 8, 11, 88, 101, 111, 181, 1001, 1111, . . .

RECREATIONAL MATHEMATICS

53. 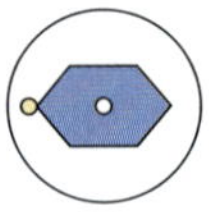 is to 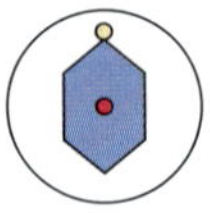 as 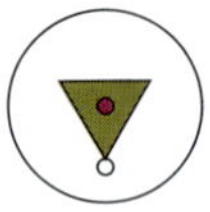 is to

a) 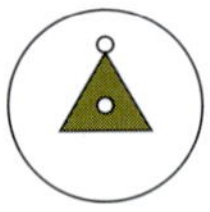 b) 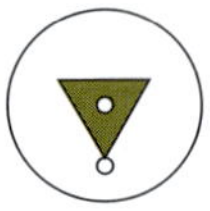 c) 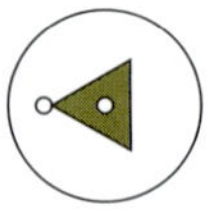

d) 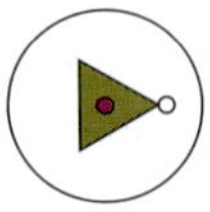 e) 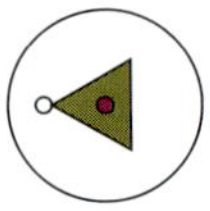

INTERNET/RESEARCH ACTIVITIES

54. a) Using newspapers, the Internet, magazines, and other sources, find examples of conclusions arrived at by inductive reasoning.

    b) Explain how inductive reasoning was used in arriving at the conclusion.

55. When a jury decides whether or not a defendant is guilty, do the jurors collectively use primarily inductive reasoning, deductive reasoning, or an equal amount of each? Write a brief report supporting your answer.

1.2 ESTIMATION

▲ Estimate the cost of the seventeen thirty-nine cent stamps shown in this photo.

What is the approximate sales tax on an $18,000 car if the sales tax is 8%? What is the approximate cost to purchase 52 forty-one cent stamps? In this section, we will introduce a technique for arriving at an approximate answer to a question.

An important step in solving mathematical problems—or, in fact, *any* problem—is to make sure that the answer you've arrived at makes sense. One technique for determining whether an answer is reasonable is to estimate. ***Estimation*** is the process of arriving at an approximate answer to a question. This section demonstrates several estimation methods.

To estimate, or approximate, an answer, we often round numbers as illustrated in the following examples. The symbol ≈ means *is approximately equal to.*

EXAMPLE 1 *Estimating the Cost of Library Books*

Jean Boehm decides to purchase new books for the Sherman Elementary School Library. Estimate her cost if she purchases 19 books at $3.99 each.

SOLUTION We may round the amounts as follows to obtain an estimate.

Number		Number Rounded
19	→	20
× $3.99	→	× $4.00
		$80.00

Thus, the 19 books would cost approximately $80.00 written ≈ $80. ●

In Example 1, the true cost is $3.99 × 19, or $75.81. *Estimates are not meant to give exact values for answers but are a means of determining whether your answer is reasonable.* If you calculated an answer of $75.81 and then did a quick estimate to check it, you would know that the answer is reasonable because it is close to your estimated answer.

EXAMPLE 2 *Two Ways to Estimate*

At a local supermarket, Kaitlyn purchased a cheesecake for \$5.89, lettuce for \$1.09, bread for \$1.98, laundry detergent for \$4.79, ground beef for \$4.26, steaks for \$15.37, and a green onion for \$0.92. The cashier said the total bill was \$44.08. Use estimation to determine whether this amount is reasonable.

SOLUTION The most expensive item is \$15.37, and the least expensive is \$0.92. How should we estimate? We will estimate two different ways. First, we will round the cost of each item to the nearest 10 cents. For the second method, we will round the cost of each item to the nearest dollar. Rounding to the nearest 10 cents is more accurate. To determine whether the total bill is reasonable, however, we may need to round only to the nearest dollar.

	Round to the Nearest 10 Cents		Round to the Nearest Dollar	
Cheesecake	\$5.89 →	\$5.90	\$5.89 →	\$6.00
Lettuce	1.09 →	1.10	1.09 →	1.00
Bread	1.98 →	2.00	1.98 →	2.00
Laundry detergent	4.79 →	4.80	4.79 →	5.00
Ground beef	4.26 →	4.30	4.26 →	4.00
Steaks	15.37 →	15.40	15.37 →	15.00
Onion	0.92 →	0.90	0.92 →	1.00
		\$34.40		\$34.00

Using either estimate, we find that the bill of \$44.08 is quite high. Therefore, Kaitlyn should check the bill carefully before paying it. Adding the prices of all seven items gives the true cost of \$34.30.

EXAMPLE 3 *Select the Best Estimate*

The number of bushels of grapes produced at a vineyard are 71,309 Cabernet Sauvignon, 123,879 French Colombard, 106,490 Chenin Blanc, 5960 Charbono, and 12,104 Chardonnay. Select the best estimate of the total number of bushels produced by the vineyard.

a) 500,000 b) 30,000 c) 300,000 d) 5,000,000

SOLUTION Following are suggested roundings. On the left, the numbers are rounded to thousands. For a less accurate estimate, round to ten thousands, as illustrated on the right.

Round to the Nearest Thousand		Round to the Nearest Ten Thousand	
71,309 →	71,000	71,309 →	70,000
123,879 →	124,000	123,879 →	120,000
106,490 →	106,000	106,490 →	110,000
5960 →	6000	5960 →	10,000
12,104 →	12,000	12,104 →	10,000
	319,000		320,000

Either rounding procedure indicates that the best estimate is (c), or 300,000.

EXAMPLE 4 *Using Estimation in Calculations*

The odometer of an automobile reads 58,289.6 miles.

a) If the automobile averaged 22.1 miles per gallon for that mileage, estimate the number of gallons of gasoline used.

b) If the cost of the gasoline averaged $2.39 per gallon, estimate the total cost of the gasoline.

SOLUTION

a) To estimate the number of gallons, divide the mileage by the number of miles per gallon.

$$\frac{58{,}289.6}{22.1}$$

Round these numbers to obtain an estimate.

$$\frac{60{,}000}{20} = 3000$$

Therefore, the car used approximately 3000 gallons of gasoline.

b) Rounding the price of the gasoline to $2.40 per gallon gives the cost of the gasoline as $3000 \times \$2.40$, or $7200.

Now let's look at some different types of estimation problems.

EXAMPLE 5 *Estimating Distances on Trails to the West*

In the mid-1800s, thousands of settlers followed trails to the West to gain cheap, fertile land and a chance to make a fortune. One of the main trails to the West was the Oregon Trail, which ran from Independence, Missouri, to the Oregon Territory. Another trail, the California Trail, ran from Fort Hall on the Oregon Trail to Sacramento, California. A map of the Oregon Trail and the California Trail is shown below.

Trails West, 1850

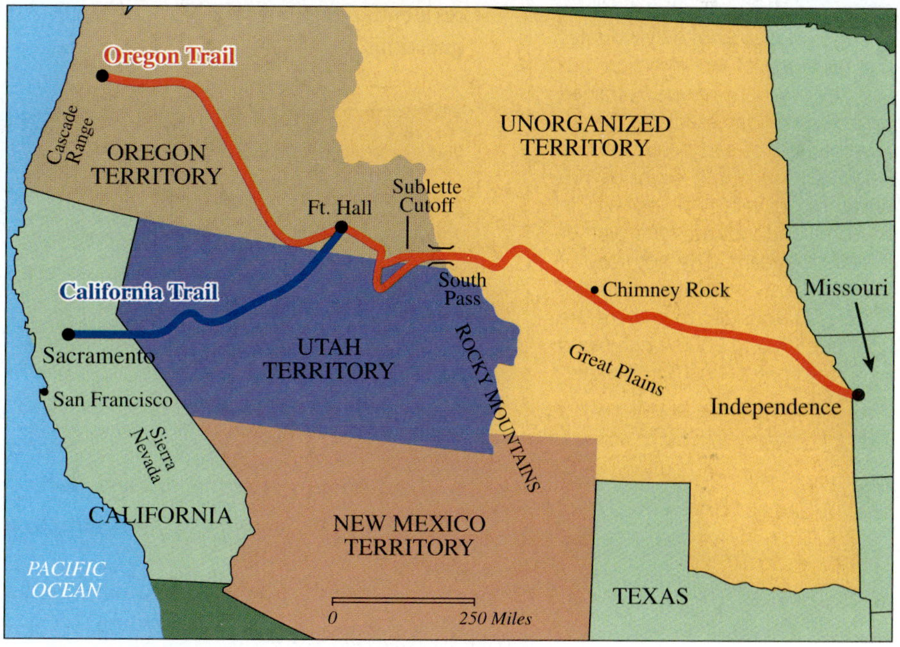

DID YOU KNOW?

Estimating Techniques in Medicine

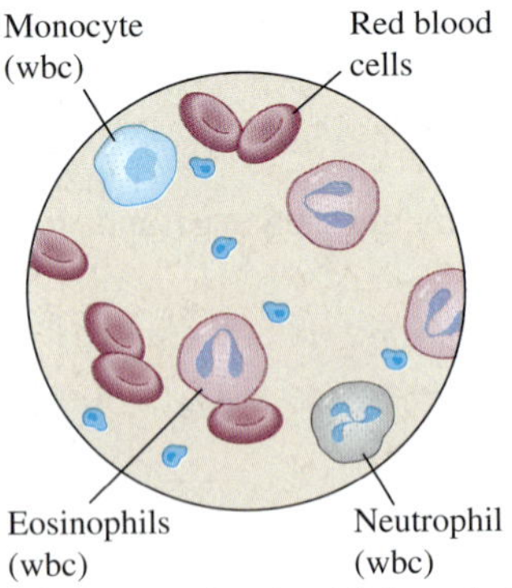

A high count of eosinophils (a particular kind of white blood cell, or wbc) can be an indicator of an allergic reaction.

Estimating is one of the diagnostic tools used by the medical profession. Physicians take a small sample of blood, tissue, or body fluids to be representative of the body as a whole. Human blood contains different types of white blood cells that fight infection. When a bacterium or virus gets into the blood, the body responds by producing more of the type of white blood cell whose job it is to destroy that particular invader. Thus, an increased level of white blood cells in a sample of blood not only indicates the presence of an infection but also helps identify its type. A trained medical technician estimates the relative number of each kind of white blood cell found in a count of 100 white blood cells. An increase in any one kind indicates the type of infection present. The accuracy of this diagnostic tool is impressive when you consider that there are normally 5000 to 9000 white blood cells in a dropful (1 cubic millimeter) of blood.

a) Using the Oregon Trail, estimate the distance from Independence, Missouri, to Ft. Hall in the Oregon Territory.

b) Using the California Trail, estimate the distance from Fort Hall in the Oregon Territory to Sacramento, California.

SOLUTION

a) Using a ruler and the scale given on the map, we can determine that approximately $\frac{3}{4}$ inch represents 250 miles (mi). One way to determine the distance from Independence, Missouri, to Ft. Hall in the Oregon Territory is to mark off $\frac{3}{4}$-inch intervals along the Oregon Trail. If you do so, you should obtain approximately 4 intervals. Thus, the distance is about 4×250 mi, or about 1000 mi.

Sometimes on a map like this one, it may be difficult to get an accurate estimate because of the curves on the map. To get a more accurate estimate, you may want to use a piece of string. Place the beginning of the string at Independence, Missouri, and, using tape or pins, align the string with the road. Indicate on the string the point represented by Ft. Hall. Then remove the string and make $\frac{3}{4}$-inch interval markings on the string (or measure the length of string you marked off). If, for example, your string from Independence, Missouri, to Ft. Hall measures $3\frac{1}{4}$ inch, then the distance is about 1083 mi.

b) Using the procedure discussed in part (a), we estimate that the distance from Fort Hall in the Oregon Territory to Sacramento, California, is about $1\frac{5}{8}$ inch, or about 2.2 three-quarter-inch intervals. Thus, the distance from Fort Hall in the Oregon Territory to Sacramento, California, is about 2.2×250 mi, or about 550 mi. ●

EXAMPLE 6 *Estimated Energy Use*

Some utility bills contain graphs illustrating the amount of electricity and natural gas used. The following graphs show gas and electric use at a specific residence for a period of 13 months, starting in November 2006 and going through November 2007 (the month of the current bill). Also shown, in red, is the bill for the average residential customer for November 2007. Using these graphs, answer the following questions.

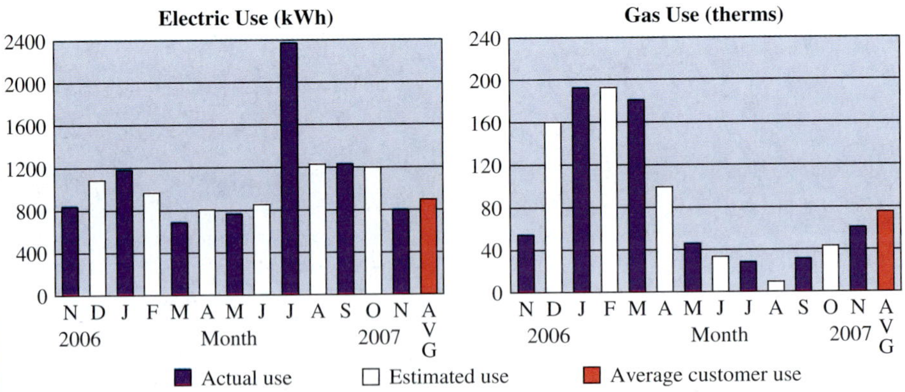

a) How often were actual gas and electric readings made?

b) Estimate the number of therms of gas used by the average residential customer in November 2007.

c) Estimate the amount of gas used by the resident in November 2007.

d) If the cost of gas is \$1.71295 per therm, estimate the gas bill in March 2007.

e) In which month was the most electricity used? How many kilowatt hours (kWh) of electricity were used in this month?

f) If the cost of electricity is 8.1716 cents per kWh, estimate the cost of the electricity in January 2007.

SOLUTION

a) Actual readings were made every other month, in November 2006, January 2007, March 2007, May 2007, July 2007, September 2007, and November 2007. Thus, actual readings were made 7 times.

b) In November 2007, approximately 75 therms were used by the average residential customer, as shown by the height of the red bar.

c) In November 2007, approximately 60 therms were used by the resident.

d) In March 2007, about 180 therms were used. The rate is $1.71295 per therm. To get a rough approximation, round the rate to $1.70 per therm.

$$1.70 \times 180 = \$306$$

Thus, the cost of gas used was about $306.

e) The most electricity was used in July 2007. Approximately 2400 kWh of electricity were used.

f) In January 2007, about 1200 kWh were used. Write 8.1716 cents as $0.081716. Rounding the rate to $0.08 per kWh and multiplying by 1200 yields an estimate of $96.

$$0.08 \times 1200 = \$96$$

Thus, the cost of electricity in January 2007 was about $96.

EXAMPLE 7 *Estimating the Number of Birds in a Photo*

Scientists who are concerned about dwindling animal populations often use aerial photography to make estimates. Estimate the number of birds in the accompanying photograph.

DID YOU KNOW?

Early Estimation

The ancient Egyptians and Greeks used sticks to estimate heights. To measure height, a person held a stick in his or her outstretched arm until it just covered the height of the object. The person then turned the stick through a right angle. Measuring the distance the stick appeared to cover on the ground provided an estimate of the object's height.

SOLUTION To estimate the number of birds, we can divide the photograph into rectangles with equal areas and then select one area that appears to be representative of all the areas. Estimate (or count) the number of birds in this single area and then multiply this number by the number of equal areas.

Let's divide the photo into 20 approximately equal areas. We will select the middle region in the bottom row as the representative region. We enlarge this region and count the birds in it. If half a bird is in the region, we count it (see enlargement). There are 13 birds in this region. Multiplying by 20 gives $13 \times 20 = 260$. Thus, there are about 260 birds in the photo.

In problems similar to that in Example 7, the number of regions or areas into which you choose to divide the total area is arbitrary. Generally, the more regions, the better the approximation, as long as the region selected is representative of the other regions in the map, diagram, or photo.

When you estimate an answer, the amount that your approximation differs from the actual answer will depend on how you round the numbers. Thus, in estimating the product of $196{,}000 \times 0.02520$, using the rounded values $195{,}000 \times 0.025$ would yield an estimate much closer to the true answer than using the rounded values $200{,}000 \times 0.03$. Without a calculator, however, the product of $195{,}000 \times 0.025$ might be more difficult to find than $200{,}000 \times 0.03$. When estimating, you need to determine the accuracy desired in your estimate and round the numbers accordingly.

SECTION 1.2 EXERCISES

PRACTICE THE SKILLS

In Exercises 1–57, your answers may vary from the answers given in the back of the text, depending on how you round your answers.

In Exercises 1–12, estimate the answer. There is no one correct estimate. Your answer, however, should be something near the answer given.

1. $523 + 47.8 + 821.6 + 733 + 92.7$

2. $3.76 + 76 + 821.7 + 654.93 + 321 + 0.89$

3. $297{,}700 \times 4087$

4. 1854×0.0096

5. $\dfrac{405}{0.049}$

6. $297.521 - 85.964$

7. 0.63×1523

8. $51{,}608 \times 6981$

9. 9% of 2164

10. 18% of 1576

11. $592 \times 2070 \times 992.62$

12. $296.3 \div 0.0096$

PROBLEM SOLVING

In Exercises 13–24, estimate the answer.

13. The income earned for 27 hours at $8.25 an hour

14. The cost of copying 13 pages at $0.15 a page

15. The cost of 6 grocery items if the items cost $4.23, $2.79, $0.79, $7.62, $12.38, and $4.99

16. A 7% sales tax on a refrigerator that sells for $599

17. The distance traveled when driving 57 miles per hour for 3.2 hours

18. One fourth of an annual salary of $47,600

19. The weight of the load of an 18-wheel truck if the weight of the truck when empty is 14,292 pounds and the weight of the truck when loaded is 27,453 pounds

20. The weight of one pork chop in a package of six pork chops if the weight of the package is 3.25 lb

21. A 15% tip on a meal that costs $26.32

22. The average daily distance traveled if Paul Simon traveled 173 miles in one week

23. The number of months needed to save $400 if you save $23 each month

24. The average monthly gasoline expense for the Stappenbacks if their annual gasoline expense is $2900

25. ***Utility Bill*** Joe Martinez pays $139.99 per month for the Time Warner All-n-One package. This package includes local telephone service, unlimited long-distance telephone service, cable TV, and Road Runner Internet service. His previous monthly costs were local telephone service $29.99, long-distance telephone service $36.99, cable TV $59.99 and Road Runner Internet service $49.99. Estimate the amount Joe saves per month with the All-n-One package.

26. ***Estimating Weights*** In a tug of war, the weight of the members of the two three-person teams is given below. Estimate the difference in the weights of the teams.

Team A	Team B
189	183
172	229
191	167

27. ***Picking Strawberries*** Chuck Chase hires 11 people to pick strawberries from his field. He agrees to pay them $1.50 for each quart they pick. Estimate the total amount of money he will have to pay if each person picks 8 quarts.

28. ***Estimating Area*** Mrs. Sanchez determines that her lawn contains an average of 3.8 grubs per square foot. If her rectangular lawn measures 60 ft by 80.2 ft, estimate the total number of grubs in her lawn.

29. ***Currency*** Estimate the difference in the value of 100 Mexican pesos and 50 U.S. dollars. Assume that one Mexican peso is about 0.089 U.S. dollar.

30. ***The Cost of a Vacation*** The Kleins are planning a vacation in the Great Smoky Mountains National Park. Their round-trip airfare from Houston, Texas, to Knoxville, Tennessee, totals $973. Car rental is $61 per day and lodging is a total of $97 per day, and they estimate a total of $150 per day for food, gas, and other miscellaneous items. If they are planning to stay six full days and nights, estimate their total expenses.

31. ***A Hike in Yellowstone National Park*** Below is a map of a trail in Yellowstone National Park. Using the scale on the map, estimate the distance of the route shown in red starting at Fishing Bridge and ending at West Thumb.

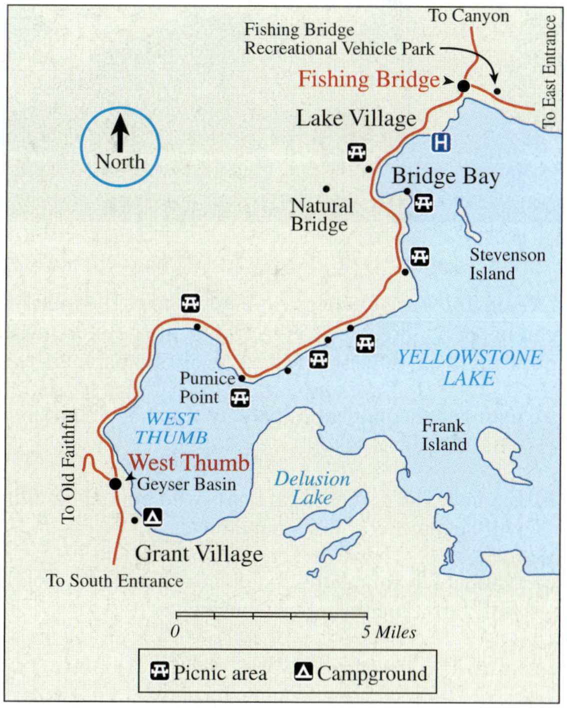

Source: National Park Service

32. ***A Trolley Ride in San Diego*** Below is a map of the trolley route in San Diego, California. Using the scale on the map, estimate the distance of the trolley route shown in red starting at Hazard Shopping Center and ending at Old Town.

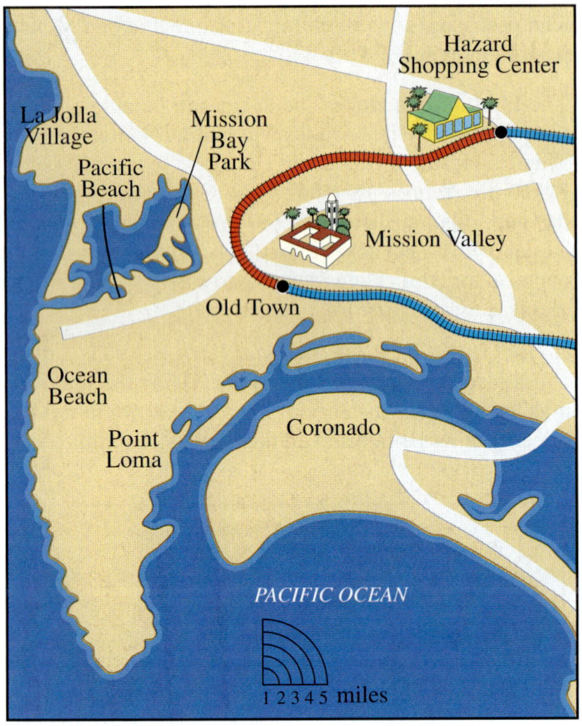

33. ***Dream Jobs*** The circle graph below shows the kind of job parents of children ages 5 to 17 hope their children will grow up to have. The survey involved 504 parents.

a) Estimate the number of parents who hope their child will become a doctor.

b) Estimate the number of parents who hope their child will become a professional athlete.

c) Estimate the number of parents who hope their child will start a successful new business.

Jobs That Parents Hope Their Children Will Have

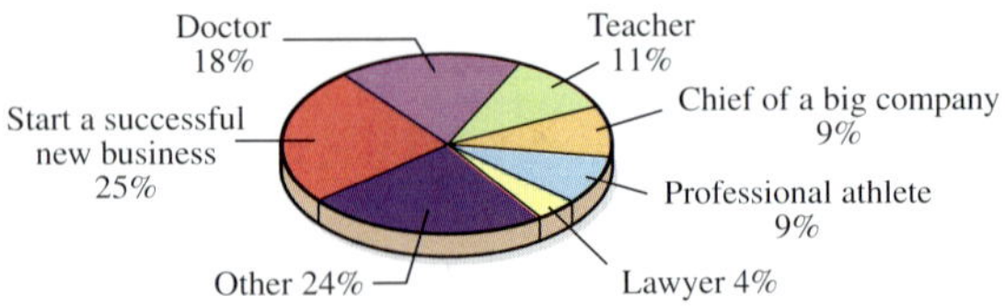

Source: Office Depot Back-to-School survey by Harris Interactive

34. ***Latino Population*** The circle graph shows the Latino population in the United States in 2005 by country of origin. Approximately 40 million Latinos were living in the United States in 2005.

a) Estimate the number of Latinos whose country of origin is Mexico.

b) Estimate the number of Latinos whose country of origin is El Salvador.

Latino Population in the U.S. by Place of Origin

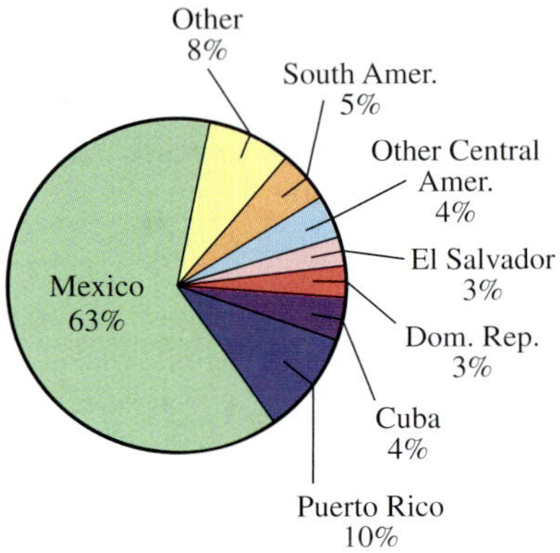

Source: U.S. Census Bureau

35. ***An Aging Population*** The bar graph shows population figures for 1900 and 2000 and estimated population figures for 2050.

a) Estimate the number of people 65 and over in 1900.

b) Estimate the number of people 65 and older in 2050.

c) Estimate the increase in the number of people 65 and older from 2000 to 2050.

d) Estimate the total U.S. population in 2000 by adding the five categories.

Population by Age

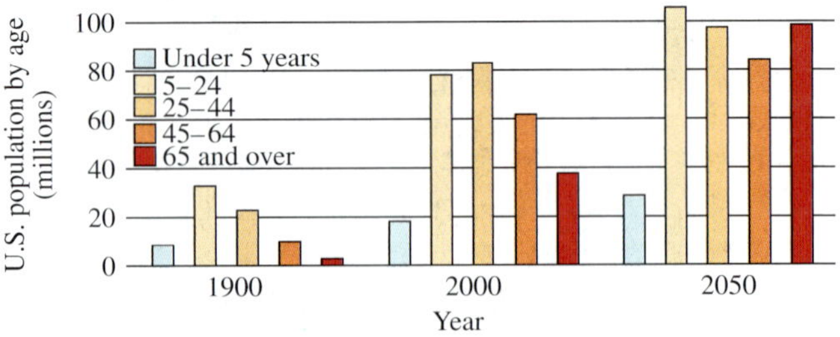

Source: U.S. Census Bureau

36. *Gaining Weight* As the graph shows, as a society we tend to get heavier as we get older. Also, with age, the amount of muscle tends to drop, and fat accounts for a greater percentage of weight.

a) Estimate the average percent of body fat for a woman, age 18 to 25.

b) Estimate the average percent of body fat for a man, age 56+.

c) Greg, an average 40-year-old man, weighs 179 lb. Estimate the number of pounds of body fat he has.

Getting Older Usually Means Getting Fatter

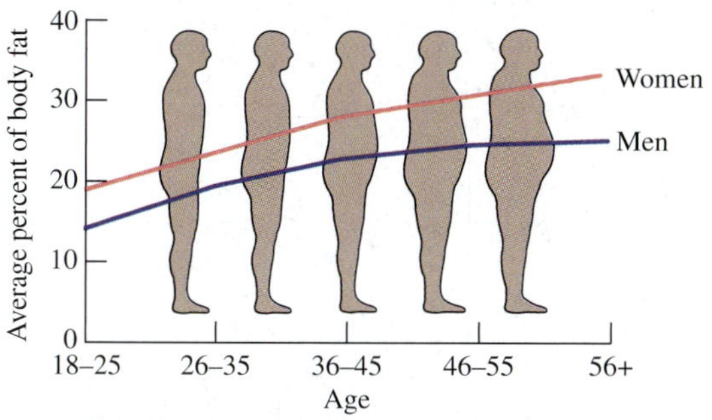

Source: Mayo Clinic Newsletter

37. *Land Ownership by the Federal Government* The federal government owns a great deal of land in the United States. The following graph shows the percent of land owned by the federal government in the 12 states in which the federal government owns the greatest percentage of a state's land.

Percent of Land Owned by the Federal Government

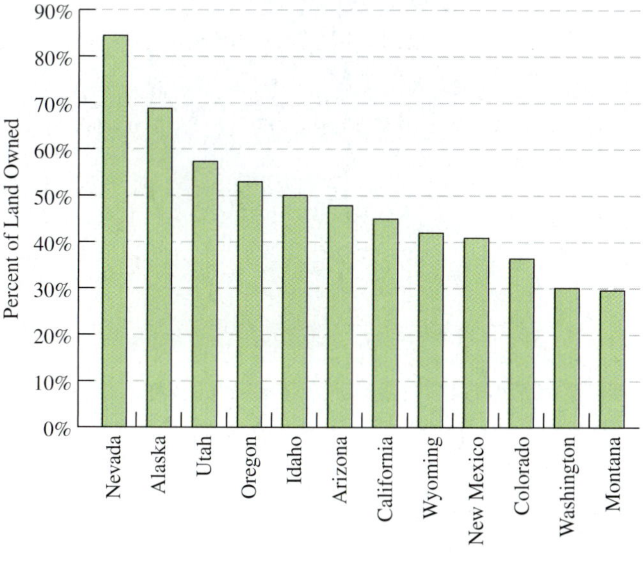

Source: Office of Governmentwide Policy, General Services Administration

a) Estimate the percent of land owned by the federal government in Nevada.

b) Estimate the difference in the percent of land owned by the federal government between Alaska and Oregon.

c) Nevada has a total area of 70,264,320 acres. Estimate the number of acres owned by the federal government in Nevada.

d) By just looking at the graph, is it possible to determine whether the federal government owns more land in Nevada or Utah? Explain.

38. *Calories and Exercise* The chart shows the calories burned per hour for an average person who weighs 150 lb.

a) Estimate the number of calories Phyllis Nye, who weighs 150 lb, burns in a week if she stair-climbs for 2 hours each week and jogs at 5 miles per hour for 4 hours each week.

b) Estimate the difference in the calories Phyllis will burn each week if she runs for 4 hours at 8 miles per hour rather than does casual bike riding for 4 hours.

c) Assume Phyllis jogs at 5 miles per hour for 3 hours and bicycles at 13 miles per hour for 3 hours each week. Estimate the number of calories she will burn in a year from these exercises.

Activity	Calories* per Hour
Running, 8 mph	920
Bicycling, 13 mph	545
Jogging, 5 mph	545
Air-walking	480
Stair-climbing	410
Weight-lifting	410
Walking, 4 mph	330
Casual bike riding	300

*For a 150 lb person.

In Exercises 39 and 40, estimate the maximum number of smaller figures (at left) that can be placed in the larger figure (at right) without the small figures overlapping.

39.

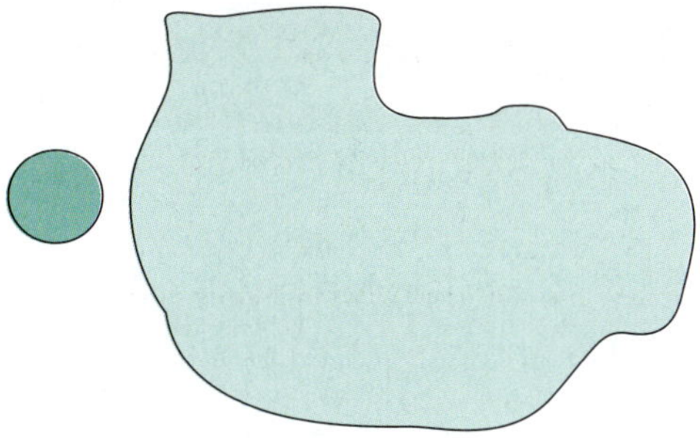

40.

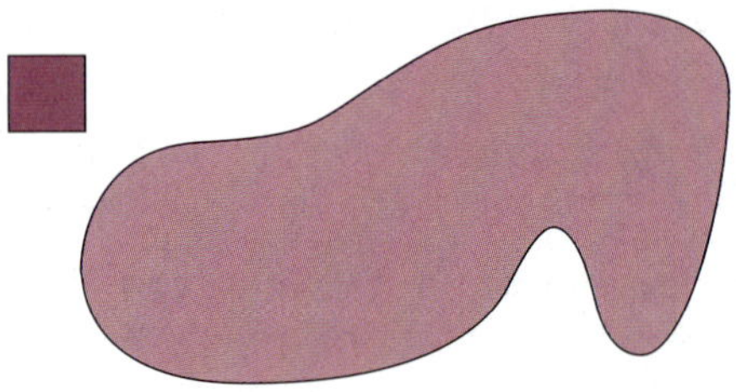

41. Estimate the number of bananas shown in the photo.

42. Estimate the number of berries shown in the photo.

In Exercises 43 and 44, estimate, in degrees, the measure of the angles depicted. For comparison purposes a right angle, ∟, measures 90°.

43.

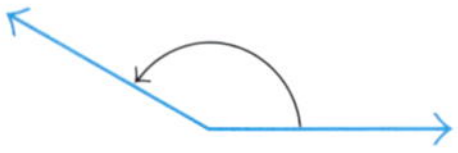

44.

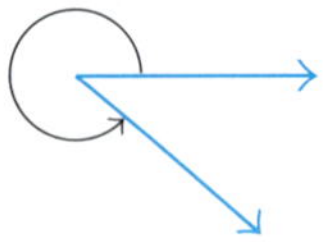

In Exercises 45 and 46, estimate the percent of area that is shaded in the following figures.

45.

46.

In Exercises 47 and 48, if each square represents one square unit, estimate the area of the shaded figure in square units.

47.

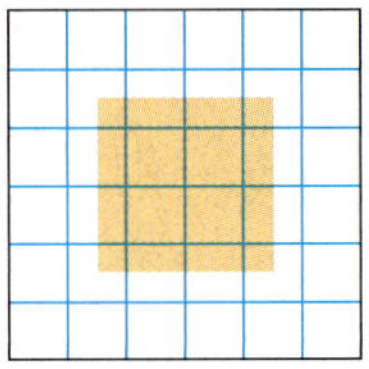

48.

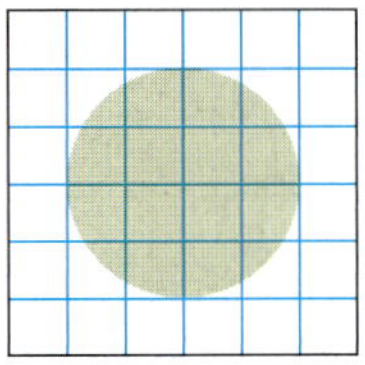

49. ***Statue of Liberty*** The length of the torch of the Statue of Liberty, from the tip of the flame to the bottom of the torch's baton, is 29 feet. Estimate the height of the Statue of Liberty from the top of the base to the top of the torch (the statue itself).

50. ***Estimating Heights*** If the height of the middle person in the photo is 62 in. tall, estimate the height of the tree.

51. ***Distance*** Estimate, without a ruler, a distance of 12 in. Measure the distance. How good was your estimate?

52. ***Weight*** In a bag, place objects that you believe have a total weight of 10 lb. Weigh the bag to determine the accuracy of your estimate.

53. ***Phone Call*** Estimate the number of times the phone will ring in 1 minute if unanswered. Have a classmate phone you so that you can count the rings and thus test your estimate.

54. ***Temperature*** Fill a glass with water and estimate the water's temperature. Then use a thermometer to measure the temperature and check your estimate.

55. ***Pennies*** Estimate the number of pennies that will fill a 3-ounce (oz) paper cup. Then actually fill a 3 oz paper cup with pennies, counting them to determine the accuracy of your estimate.

56. ***Height*** Estimate the ratio of your height to your waist size. Then have a friend measure your height and waist size. Determine the stated ratio and check the accuracy of your estimate.

57. ***Walking Speed*** Estimate how fast you can walk 60 ft. Then mark off a distance of 60 ft and use a watch with a second hand to time yourself walking it. Determine the accuracy of your estimate.

CHALLENGE PROBLEMS/GROUP ACTIVITIES

58. ***Shopping*** Make a shopping list of 20 items you use regularly that can be purchased at a supermarket. Beside each item write down what you estimate to be its price. Add these price guesses to estimate the total cost of the 20 items. Next, make a trip to your local supermarket and record the actual price of each item. Add these prices to determine the actual total cost. How close was your estimate? (Don't forget to add tax on the taxable items.)

59. ***A Ski Vacation*** Two friends, Tiffany Connolly and Ana Pott, are planning a skiing vacation in the Rockies. They plan to purchase round-trip airline tickets from Atlanta, Georgia, to Denver, Colorado. They will fly into Denver on a Friday morning, rent a midsize car, and drive to Aspen that same day. They will stay at the Holiday Inn in Aspen. They will begin skiing at the Buttermilk Ski Area on Saturday, ski up to and including Wednesday, drive back to Denver on Thursday, and fly out of Denver Thursday evening.

    a) Estimate the total cost of the vacation for the two friends. Do not forget items such as food, tips, gas, and other incidentals.

    b) Using informational sources, including the Internet, determine the airfare cost, hotel cost, cost of ski tickets, cost of a car rental, and so forth. You will need to make an estimate for food and other incidentals.

    c) How close was your estimate in part (a) to the amount you found in part (b)? Was your estimate in part (a) lower or higher than the amount obtained in part (b)?

RECREATIONAL MATHEMATICS

60. ***A Dime*** Look at a dime. Around the edge of a dime are many lines. Estimate the number of lines there are around the edge of a dime.

61. ***Golf Ball*** Look at a golf ball. Estimate the number of dimples (depressed areas) on a golf ball.

62. ***A Million Dollars***

    a) Estimate the time it would take, in days, to spend \$1 million if you spent \$1 a second until the \$1 million is used up.

    b) Calculate the actual time it would take, in days, to spend \$1 million if you spent \$1 a second. How close was your estimate?

INTERNET/RESEARCH ACTIVITIES

63. *Water Usage*

a) About how much water does your household use per day? Use the following data to estimate your household's daily water usage.

How much water do you use?

Activity	Typical Use
Running clothes washer	40 gal
Bath	35 gal
5-minute shower	25 gal
Doing dishes in sink, water running	20 gal
Running dishwasher	11 gal
Flushing toilet	4 gal
Brushing teeth, water running	2 gal

Source: U.S. Environmental Protection Agency

b) Determine from your water department (or company) your household's average daily usage by obtaining the total number of gallons used per year and dividing that amount by 365. How close was your estimate in part (a)?

c) Current records indicate that the average household uses about 300 gal of water per day (the average daily usage is 110 gal per person). Based on the number of people in your household, do you believe your household uses more or less than the average amount of water? Explain your answer.

64. Develop a monthly budget by estimating your monthly income and your monthly expenditures. Your monthly income should equal your monthly expenditures.

65. Identify three ways that you use estimation in your daily life. Discuss each of them briefly and give examples.

1.3 PROBLEM SOLVING

▲ Getting the best deal for office supplies, such as CDs, can help businesses maximize their profits.

To maximize their profits, businesses try to keep their expenses down. We, as individuals, also try to keep our expenses down, and we often look for "bargains" or the "best deal." For example, suppose we need to purchase a large number of CD-RW discs and only have $250 to spend. How can we determine the maximum number of CDs that we can purchase if CDs are sold in packs of 50 for $38 and in packs of 25 for $20? We'll answer this question in Example 1.

Solving mathematical puzzles and real-life mathematical problems can be enjoyable. You should work as many exercises in this section as possible. By doing so, you will sample a variety of problem-solving techniques.

You can approach any problem by using a general procedure developed by George Polya. Before learning Polya's problem-solving procedure, let's consider an example that illustrates the procedure.

EXAMPLE 1 *Saving Money When Purchasing CDs*

A law firm owned by Karen Morris plans to purchase a large number of CD-RW discs. One supplier, Staples, is selling packs of 50 CDs for $38 and packs of 25 CDs for $20. Only complete packs of CDs are sold.

a) Find the maximum number of CDs that can be purchased for $250 or less. Indicate how many packs of 50 CDs and how many packs of 25 CDs will be purchased.

b) If the maximum number of CDs determined in part (a) is purchased in the most economical way, how much will the CDs cost?

SOLUTION

a) The first thing to do is to read the problem carefully. Read it at least twice. Be sure you understand the facts given and what you are being asked to find. Next, make a list of the given facts and determine which facts are relevant to answering the question or questions asked.

GIVEN INFORMATION

Law firm owner: Karen Morris

Supplier: Staples

A pack of 50 CDs costs \$38.

A pack of 25 CDs costs \$20.

Only complete packs of CDs can be purchased.

We need to determine the maximum number of CDs the law firm can purchase for \$250 or less. To determine this number, we need to know the number of CDs in each of the packs and the cost of each pack. We also need to know that only complete packs of CDs can be purchased.

RELEVANT INFORMATION

A pack of 50 CDs costs \$38.

A pack of 25 CDs costs \$20.

Only complete packs of CDs can be purchased.

The next step is to determine the answer to the question. That is, we need to determine the maximum number of CDs that can be purchased for \$250 or less.

We now need a plan for solving the problem. One method is to set up a table or chart to compare costs of different combinations of packs of CDs. Start by using the maximum number of packs of 50 CDs. Then reduce the number of packs of 50 CDs and add more packs of 25 CDs. In each case, we need to keep the cost at \$250 or less.

Because 1 pack of 50 CDs costs \$38, we can determine the number of packs of 50 CDs that can be purchased by dividing 250 by 38. Because the quotient is about 6.58 and because only whole packs of CDs may be purchased, only 6 packs of 50 CDs may be purchased. Six packs would cost $6 \times 38 = \$228$. The remaining \$22 from the \$250 could be used to purchase packs of 25 CDs. Because each pack of 25 CDs costs \$20, only 1 pack of 25 CDs could be purchased. Thus, for \$250 or less, one option is 6 packs of 50 CDs and 1 pack of 25 CDs. This option is indicated in the first row of the table below. Also given in the table is the cost of this option, which is \$248. We complete the other rows of the table in a similar manner.

Packs of 50 and Packs of 25 CDs	Number of CDs	Cost
6 packs of 50 and 1 pack of 25	$(6 \times 50) + (1 \times 25) = 325$	\$248
5 packs of 50 and 3 packs of 25	$(5 \times 50) + (3 \times 25) = 325$	\$250
4 packs of 50 and 4 packs of 25	$(4 \times 50) + (4 \times 25) = 300$	\$232
3 packs of 50 and 6 packs of 25	$(3 \times 50) + (6 \times 25) = 300$	\$234
2 packs of 50 and 8 packs of 25	$(2 \times 50) + (8 \times 25) = 300$	\$236
1 pack of 50 and 10 packs of 25	$(1 \times 50) + (10 \times 25) = 300$	\$238
0 packs of 50 and 12 packs of 25	$(0 \times 50) + (12 \times 25) = 300$	\$240

PROFILE IN MATHEMATICS

George Polya

George Polya (1887–1985) was educated in Europe and taught at Stanford University. In his book *How to Solve It,* Polya outlines four steps in problem solving. We will use Polya's four steps as guidelines for problem solving.

The question asks us to find the maximum number of CDs that can be purchased for \$250 or less. From the second column of the table, we see that the answer is 325 CDs. This result can be done in two different ways: either 6 packs of 50 and 1 pack of 25 or 5 packs of 50 and 3 packs of 25.

b) When comparing the two possibilities for purchasing the 325 CDs discussed in part (a), we see that the most economical way to purchase the CDs is to purchase 6 packs of 50 CDs and 1 pack of 25 CDs. The cost is \$248. ●

Following is a general procedure for problem solving as given by George Polya. Note that Example 1 demonstrates many of these guidelines.

GUIDELINES FOR PROBLEM SOLVING

1. *Understand the problem.*
   - Read the problem *carefully* at least twice. In the first reading, get a general overview of the problem. In the second reading, determine (a) exactly what you are being asked to find and (b) what information the problem provides.
   - Try to make a sketch to illustrate the problem. Label the information given.
   - Make a list of the given facts that are pertinent to the problem.
   - Determine if the information you are given is sufficient to solve the problem.
2. *Devise a plan to solve the problem.*
   - Have you seen the problem or a similar problem before? Are the procedures you used to solve the similar problem applicable to the new problem?
   - Can you express the problem in terms of an algebraic equation? (We explain how to write algebraic equations in Chapter 6.)
   - Look for patterns or relationships in the problem that may help in solving it.
   - Can you express the problem more simply?
   - Can you substitute smaller or simpler numbers to make the problem more understandable?
   - Will listing the information in a table help in solving the problem?
   - Can you make an educated guess at the solution? Sometimes if you know an approximate solution, you can work backward and eventually determine the correct procedure to solve the problem.
3. *Carry out the plan.*
   Use the plan you devised in step 2 to solve the problem.
4. *Check the results.*
   - Ask yourself, "Does the answer make sense?" and "Is the answer reasonable?" If the answer is not reasonable, recheck your method for solving the problem and your calculations.
   - Can you check the solution using the original statement?
   - Is there an alternative method to arrive at the same conclusion?
   - Can the results of this problem be used to solve other problems?

The following examples show how to apply the guidelines for problem solving.

EXAMPLE 2 *Hotel Cost*

At the Courtyard by Marriot Hotel in Irving, Texas (near the Dallas/Fort Worth airport), the room rate is \$159 per day on weekdays and \$89 per day on weekends (Saturday and Sunday). In addition, a 13% sales tax is added to the cost of a room. In-room movies cost \$10.95 plus the 13% sales tax. Robin Ayers stays on the third floor of the Courtyard for four nights (Wednesday, Thursday, Friday, and Saturday) and watches three movies. Determine her hotel bill when she checks out.

SOLUTION We need to find the total cost of the hotel bill. Let's make a list of the information given and mark with an asterisk (*) the information that is pertinent to solving the problem.

*Cost of room per day on Wednesday, Thursday, and Friday = $159 + sales tax

*Cost of room on Saturday = $89 + sales tax

*Days at hotel: one Wednesday, one Thursday, one Friday, one Saturday

*Sales tax is 13%

*Cost per movie = $10.95 + sales tax

*Movies watched: 3

Room on third floor

All the information is needed to solve the problem except for the floor on which Robin stayed.

Let's first determine the cost of the room, before tax, for the four days.

Day	Cost (before tax)
Wednesday	$159
Thursday	$159
Friday	$159
Saturday	$89
	$566

The cost of three movies before tax is 3 × $10.95 = $32.85. Thus, the total amount of the bill before tax is $566 + $32.85 = $598.85. Let's determine the sales tax to be added. We could find the sales tax for each individual item and add it to the items separately. Because the same tax rate applies to each item, however, it is easier to just determine 13% of the total amount.

$$\begin{aligned} \text{Sales tax} &= 13\% \text{ of } \$598.85 \\ &= 0.13(598.85) = \$77.85 \end{aligned}$$

The total hotel bill is determined by adding the tax to the pretax amount. Thus, the total hotel bill is $598.85 + $77.85 = $676.70.

In Example 2, the total cost could have been determined in other ways. We presented the method we believe you would understand best.

EXAMPLE 3 *Retirement*

It is never too early to start planning for retirement. U.S. Census Bureau data indicate that at age 65 the average woman will live another 19.8 years and the average man will live another 16.8 years. The data also indicate that about 39% of the average person's retirement income will come from Social Security.

When discussing retirement planning, many investment firms and financial planners use the graph in Fig. 1.1,* which shows how long a typical retiree's assets (or "nest egg") will last based on the percentage of the assets withdrawn each year.

*The information in this graph is based on past performance of the stock market, with 50% invested in large company stocks and 50% invested in intermediate-term bonds. Past performance is not indicative of future results.

DID YOU KNOW?

Archimedes Treatise, On Floating Bodies

Scientists are using problem-solving techniques and modern technology to unlock ancient secrets. In May 2005, scientists at the Stanford Linear Accelerator Center used x-rays to decipher pages of the only known copy of a 2300-year-old Greek text by the mathematician Archimedes. Scholars believe that the 174-page treatise, *On Floating Bodies*, was copied by a scribe in the 10th century from Archimedes' original Greek scrolls, written in the 3rd century B.C. The treatise was erased 200 years later by a monk who reused the parchment for a prayer book, creating a twice-used parchment.

To reveal much of the hidden text, the scientists used digital cameras and processing techniques as well as ultraviolet and infrared filters developed for medicine and space research. Through the use of x-rays, scientists are now able to read the remaining pages of the treatise. The team, consisting of researchers from R. B. Toth Associates, Rochester Institute of Technology, Johns Hopkins University, Rutgers University, and ConocoPhillips, plan to create an interactive DVD once they've deciphered the entire text. The text and diagrams contain the roots of modern calculus and gravitational theory.

DID YOU KNOW?

Two Medieval Riddles

A boatman must transport a wolf, a goat, and a bundle of herbs across a river. If the boat is so small that he can carry only one of them at a time, how can he make the crossing so that the wolf will not devour the goat and the goat will not eat the herbs?

(Clue: The boatman must make multiple round trips.)

As I was going to St. Ives,
I met a man with seven wives;
Every wife had seven sacks,
Every sack had seven cats,
Every cat had seven kits.
Kits, cats, sacks, and wives,
How many were going to St. Ives?

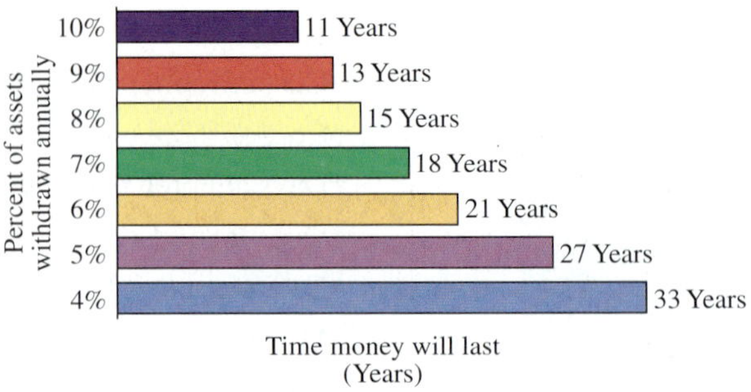

Source: Ned Davis Research

Figure 1.1

a) If a typical retiree has retirement assets of \$500,000, how much can he or she withdraw annually if he or she wishes the assets to last 21 years?

b) How much should a retiree have in assets if he or she wishes to withdraw \$25,000 annually and wishes his or her assets to last 18 years?

SOLUTION

a) Quite a bit of information is provided in the example. We will first need to determine what information is needed to answer the question. To answer the question, we only need to use the information provided in the graph. From the graph, we can see that for assets to last 21 years, about 6% of the assets can be withdrawn annually. The amount that can be withdrawn annually is found as follows.

$$\text{Amount} = 6\% \text{ of assets}$$
$$\text{Amount} = 0.06\,(500{,}000) = 30{,}000$$

Thus, about \$30,000 can be withdrawn annually.

b) Again, to solve this part of the example, we only need the information provided in the graph. From the graph, we can determine that if a retiree wishes for his or her assets to last 18 years, then 7% of the assets can be withdrawn annually. We need to determine the total assets such that 7% of the total assets is \$25,000.

$$7\% \text{ of assets} = \$25{,}000 \quad \text{or} \quad 0.07 \times \text{assets} = \$25{,}000$$

Because 0.07 is multiplied by the assets to obtain \$25,000, we can determine the assets by dividing the \$25,000 by 0.07 as follows.

$$\text{Assets} = \frac{25{,}000}{0.07} = 357{,}142.85$$

Therefore, if the retiree has assets of about \$357,142.85, he or she will be able to withdraw \$25,000 annually and the assets will last 18 years.

EXAMPLE 4 *Determining a Tip*

The cost of Sarah Decker's meal before tax is \$26.00.

a) If a $6\frac{1}{2}\%$ sales tax is added to her bill, determine the total cost of the meal including tax.

b) If Sarah wants to leave a 10% tip on the *pretax* cost of the meal, how much should she leave?

c) If she wants to leave a 15% tip on the *pretax* cost of the meal, how much should she leave?

SOLUTION

a) The sales tax is $6\frac{1}{2}\%$ of \$26.00. To determine the sales tax, first change the $6\frac{1}{2}\%$ to a decimal number. $6\frac{1}{2}\%$ when written as a decimal number is 0.065 (if you have forgotten how to change a percent to a decimal number, read Section 11.1). Next, multiply the decimal number, 0.065, by the amount, \$26.00.

$$\begin{aligned}\text{Sales tax} &= 6\tfrac{1}{2}\% \text{ of } 26.00\\ &= 0.065(26.00) = 1.69\end{aligned}$$

The sales tax is \$1.69. The total bill is the cost of the meal plus the sales tax.

$$\begin{aligned}\text{Total bill} &= \text{cost of meal} + \text{sales tax}\\ &= 26.00 + 1.69 = 27.69\end{aligned}$$

Thus, the bill, including sales tax, is \$27.69.

b) To find 10% of any number, we can multiply the number by 0.10.

$$\begin{aligned}10\% \text{ of pretax cost} &= 0.10(26.00)\\ &= 2.60\end{aligned}$$

A simple way to find 10% of any number is to simply move the decimal point in the number one place to the left. Moving the decimal point in \$26.00 one place to the left gives \$2.60, the same answer we obtained by our calculations.

c) To find 15% of \$26.00, multiply as follows.

$$15\% \text{ of } 26.00 = 0.15(26.00) = 3.90$$

Thus, 15% of \$26.00 is \$3.90. A second method to find a 15% tip is to find 10% of the cost, as in part (b), then add to it half that amount. Following this procedure, we get

$$\$2.60 + \frac{\$2.60}{2} = \$2.60 + \$1.30 = \$3.90$$

In most cases, tips are rounded. If the service is excellent, some people leave a 20% tip. Can you give two methods to determine a 20% tip on \$26.00? Determine the 20% tip now.

EXAMPLE 5 *A Recipe for 6*

The chart near the top of page 26 shows the amount of each ingredient recommended to make 2, 4, and 8 servings of Potato Buds. Determine the amount of each ingredient necessary to make 6 servings of Potato Buds by using the following procedures.

a) Multiply the amount for 2 servings by 3.*

b) Add the amounts for 2 servings to the amounts for 4 servings.

c) Find the average of the amounts for 4 servings and for 8 servings.

d) Subtract the amounts for 2 servings from the amounts for 8 servings.

*Addition, subtraction, multiplication, and division of fractions are discussed in detail in Section 5.3.

MATHEMATICS TODAY

Decisions, Decisions

Vail, Colorado

We deal with problem solving daily. There are aspects of problem solving to almost every decision we make. Can we afford to take that vacation in the mountains? Which vehicle should I purchase? What shall I major in in college? The list of questions we ask ourselves daily goes on and on. To make decisions, we often need to consider and weigh many factors.

Often, different branches of mathematics are involved in our decision-making process. For example, we often use statistical data and consider the probability (or chance) of an event occurring or not occurring when we make decisions. Probability and statistics are two branches of mathematics covered in later chapters of the book.

For many people, problem solving is a recreational activity, as is evident by the great number of crossword puzzles and puzzle books sold daily.

Today, as you go about your daily business, keep a record of all the problem-solving decisions you need to make! If you do this conscientiously, you will be amazed at the outcome.

e) Compare the answers for parts (a) through (d). Are they the same? If not, explain why not.

f) Which is the correct procedure for obtaining 6 servings?

Servings	**2**	**4**	**8**
Water	$\frac{2}{3}$ cup	$1\frac{1}{3}$ cups	$2\frac{2}{3}$ cups
Milk	2 tbsp	$\frac{1}{3}$ cup	$\frac{2}{3}$ cup
Butter or margarine	1 tbsp	2 tbsp	4 tbsp
Salt*	$\frac{1}{4}$ tsp	$\frac{1}{2}$ tsp	1 tsp
Potato Buds	$\frac{2}{3}$ cup	$1\frac{1}{3}$ cups	$2\frac{2}{3}$ cups

*Less salt can be used if desired.

SOLUTION

a) We multiply the amounts for 2 servings by 3.

Water: $3(\frac{2}{3}) = 2$ cups

Milk: $3(2) = 6$ tablespoons (tbsp)

Butter or margarine: $3(1) = 3$ tbsp

Salt: $3(\frac{1}{4}) = \frac{3}{4}$ teaspoon (tsp)

Potato Buds: $3(\frac{2}{3}) = 2$ cups

b) We find the amount of each ingredient by adding the amount for 2 and 4 servings.

Water: $\frac{2}{3}$ cup + $1\frac{1}{3}$ cup = 2 cups

Milk: 2 tbsp + $\frac{1}{3}$ cup

To add these two amounts, we must convert one of them so that both ingredients have the same units. By looking in a cookbook or a book of conversion factors, we see that 16 tbsp = 1 cup. The milk in part (a) was given in tablespoons, so we convert $\frac{1}{3}$ cup to tablespoons to compare answers. One-third cup equals $\frac{1}{3}(16) = \frac{16}{3}$ or $5\frac{1}{3}$ tbsp. Therefore,

Milk: 2 tbsp + $5\frac{1}{3}$ tbsp = $7\frac{1}{3}$ tbsp

Let's continue with the rest of the ingredients:

Butter: 1 tbsp + 2 tbsp = 3 tbsp

Salt: $\frac{1}{4}$ tsp + $\frac{1}{2}$ tsp = $\frac{3}{4}$ tsp

Potato Buds: $\frac{2}{3}$ cup + $1\frac{1}{3}$ cups = 2 cups

c) We compute the amounts of the ingredients by finding the average of the amounts for 4 and 8 servings. We do so by adding the amounts for each ingredient and dividing the sum by 2.

Water: $\dfrac{1\frac{1}{3}\text{ cups} + 2\frac{2}{3}\text{ cups}}{2} = \dfrac{4\text{ cups}}{2} = 2\text{ cups}$

Milk: $\dfrac{\frac{1}{3}\text{ cup} + \frac{2}{3}\text{ cup}}{2} = \dfrac{1\text{ cup}}{2} = \frac{1}{2}\text{ cup (or 8 tbsp)}$

Butter: $\dfrac{2\text{ tbsp} + 4\text{ tbsp}}{2} = \dfrac{6\text{ tbsp}}{2} = 3\text{ tbsp}$

Salt: $\dfrac{\frac{1}{2}\text{ tsp} + 1\text{ tsp}}{2} = \dfrac{\frac{3}{2}\text{ tsp}}{2} = \frac{3}{4}\text{ tsp}$

Potato Buds: $\dfrac{1\frac{1}{3}\text{ cups} + 2\frac{2}{3}\text{ cups}}{2} = \dfrac{4\text{ cups}}{2} = 2\text{ cups}$

d) We obtain the amounts of ingredients by subtracting the amounts for 2 servings from the amounts for 8 servings.

Water: $2\frac{2}{3}$ cups $-\ \frac{2}{3}$ cup $= 2$ cups

Milk: $\frac{2}{3}\text{ cup} - 2\text{ tbsp} = \frac{2}{3}(16)\text{ tbsp} - 2\text{ tbsp}$
$= \frac{32}{3}\text{ tbsp} - \frac{6}{3}\text{ tbsp}$
$= \frac{26}{3}\text{ tbsp, or } 8\frac{2}{3}\text{ tbsp}$

Butter: 4 tbsp − 1 tbsp = 3 tbsp

Salt: 1 tsp $-\ \frac{1}{4}$ tsp $= \frac{3}{4}$ tsp

Potato Buds: $2\frac{2}{3}$ cups $-\ \frac{2}{3}$ cup $= 2$ cups

e) Comparing the answers in parts (a) through (d), we find that the amounts of all ingredients, except milk, are the same. For milk, we get the following results.

Part (a): Milk = 6 tbsp | Part (c): Milk = 8 tbsp

Part (b): Milk = $7\frac{1}{3}$ tbsp | Part (d): Milk = $8\frac{2}{3}$ tbsp

Why are all these answers different? After rechecking, we find that all our calculations are correct, so we must look deeper. Note that milk is the only ingredient that has different units for 2 servings and 4 servings. Let's check the relationship between 2 tbsp and $\frac{1}{3}$ cup. In going from 2 servings to 4 servings, we would expect that $\frac{1}{3}$ cup should be twice 2 tbsp. We know that 1 cup = 16 tbsp, so

$$\tfrac{1}{3}\text{ cup} = \tfrac{1}{3}(16) = \tfrac{16}{3} = 5\tfrac{1}{3}\text{ tbsp}$$

Therefore, instead of the 4 tbsp of milk we expected for 4 servings, we get $5\frac{1}{3}$ tbsp. This change causes all our calculations for milk to be different.

f) Which is the correct answer? Because all our calculations for milk are correct, there is no single correct answer. All our answers are correct. Using 8 tbsp instead of $5\frac{1}{3}$ tbsp might make the Potato Buds a little thinner. When we cook, we generally do not add the *exact* amount recommended. We rely on experience to alter the recommended amounts according to individual taste.

Many real-life problems, such as the one in Example 6, can be solved by using proportions. A proportion is a statement of equality between two ratios (or fractions).*

EXAMPLE 6 *Spraying Weed Killer*

The instructions on the Ortho Weed-B-Gon lawn weed killer indicate that to cover 1000 square feet (ft^2) of lawn, 20 teaspoons (tsp) of the weed killer should be mixed in 5 gallons (gal) of water. Ron Haines wishes to spray his lawn with the weed killer using his pressurized sprayer.

a) How much weed killer should be mixed with 8 gal of water to get a solution of the proper strength?

b) How much weed killer is needed to cover an area of 2820 ft^2 of lawn?

*Proportions are discussed in greater detail in Section 6.2.

DID YOU KNOW?

East Meets West: Magic Squares

A Chinese myth says that in about 2200 B.C., a divine tortoise emerged from the Yellow River. On his back was a special diagram of numbers from which all mathematics was derived. The Chinese called this diagram Lo Shu. The Lo Shu diagram is the first known magic square.

Arab traders brought the Chinese magic square to Europe during the Middle Ages, when the plague was killing millions of people. Magic squares were considered strong talismans against evil, and possession of a magic square was thought to ensure health and wealth.

SOLUTION

a) Use the information that 20 teaspoons of weed killer is to be mixed with 5 gal of water.

$$\text{Given ratio} \left\{ \frac{20 \text{ tsp}}{5 \text{ gal water}} \right. = \frac{?\text{ tsp}}{8 \text{ gal}} \quad \begin{matrix} \leftarrow \text{Item to be found} \\ \leftarrow \text{Other information given} \end{matrix}$$

Notice in the proportion that teaspoons and gallons are placed in the same relative positions. Often, the unknown quantity is replaced with an x. The proportion may be written as follows and solved using cross multiplication.

$$\frac{20}{5} = \frac{x}{8}$$
$$20(8) = 5x$$
$$160 = 5x$$
$$\frac{160}{5} = \frac{\cancel{5}x}{\cancel{5}} \quad \text{Divide both sides of the equation by 5 to solve for } x.$$
$$32 = x$$

Thus, Ron must mix 32 tsp [or $10\frac{2}{3}$ tablespoons (tbsp) or $\frac{2}{3}$ cup] of the weed killer with 8 gal of water. This answer seems reasonable because we would expect to get an answer greater than 20 tsp.

b) To answer this question, we use the same procedure discussed in part (a). This time, we will use the information that 1000 ft^2 requires 20 tsp of weed killer. The areas may be placed either on the top or the bottom of the fraction, as long as they are placed in the same relative position.

$$\text{Given ratio} \left\{ \frac{1000 \text{ sq ft}}{20 \text{ tsp}} \right. = \frac{2820 \text{ sq ft}}{?\text{ tsp}} \quad \begin{matrix} \leftarrow \text{Other information given} \\ \leftarrow \text{Item to be found} \end{matrix}$$

Now replace the question mark with an x and solve the proportion.

$$\frac{1000}{20} = \frac{2820}{x}$$
$$1000(x) = 20(2820)$$
$$1000x = 56{,}400$$
$$\frac{\cancel{1000}x}{\cancel{1000}} = \frac{56{,}400}{1000} \quad \text{Divide both sides of the equation by 1000 to solve for } x.$$
$$x = 56.4$$

Thus, about 56.4 tsp are needed. This answer is reasonable because we would expect the answer to be more than twice the 20 tsp required for 1000 ft^2. •

Most of the problems solved so far have been practical ones. Many people, however, enjoy solving brainteasers. One example of such a puzzle follows.

EXAMPLE 7 *Magic Squares*

A magic square is a square array of numbers such that the numbers in all rows, columns, and diagonals have the same sum. Use the digits 1, 2, 3, 4, 5, 6, 7, 8, and 9 to construct a magic square.

SOLUTION The first step is to create a figure with nine cells as in Fig. 1.2(a). We must place the nine numbers in the cells so that the same sum is obtained in each row, column, and diagonal. Common sense tells us that 7, 8, and 9 cannot be in the same row, column, or diagonal. We need some small and large numbers in the same row, column, and diagonal. To see a relationship, we list the numbers in order:

1, 2, 3, 4, 5, 6, 7, 8, 9

Note that the middle number is 5 and the smallest and largest numbers are 1 and 9, respectively. The sum of 1, 5, and 9 is 15. If the sum of 2 and 8 is added to 5, the sum is 15. Likewise 3, 5, 7, and 4, 5, 6 have sums of 15. We see that in each group of three numbers the sum is 15 and 5 is a member of the group.

9	5	1

(a)

		8
9	5	1
2		

(b)

4		8
9	5	1
2		6

(c)

4	3	8
9	5	1
2	7	6

(d)

Figure 1.2

Because 5 is the middle number in the list of numbers, place 5 in the center square. Place 9 and 1 to the left and right of 5 as in Fig. 1.2(a). Now we place the 2 and the 8. The 8 cannot be placed next to 9 because $8 + 9 = 17$, which is greater than 15. Place the smaller number 2 next to the larger number 9. We elected to place the 2 in the lower left-hand cell and the 8 in the upper right-hand cell as in Fig. 1.2(b). The sum of 8 and 1 is 9. To arrive at a sum of 15, we place 6 in the lower right-hand cell as in Fig. 1.2(c). The sum of 9 and 2 is 11. To arrive at a sum of 15, we place 4 in the upper left-hand cell as in Fig. 1.2(c). Now the diagonals 2, 5, 8 and 4, 5, 6 have sums of 15. The numbers that remain to be placed in the empty cells are 3 and 7. Using arithmetic, we can see that 3 goes in the top middle cell and 7 goes in the bottom middle cell as in Fig. 1.2(d). A check shows that the sum in all the rows, columns, and diagonals is 15. •

The solution to Example 7 is not unique. Other arrangements of the nine numbers in the cells will produce a magic square. Also, other techniques of arriving at a solution for a magic square may be used. In fact, the process described will not work if the number of squares is even, for example, 16 instead of 9. Magic squares are not limited to the operation of addition or to the set of counting numbers.

SECTION 1.3 EXERCISES

PRACTICE THE SKILLS/PROBLEM SOLVING

1. ***Reading a Map*** The scale on a map is 1 inch = 18 miles. How long a distance is a route on the map if it measures 4.25 in.?

2. ***Blueprints*** Chalon Bridges, an architect, is designing a shopping mall. The scale of her plan is 1 in. = 12 ft. If one store in the mall is to have a frontage of 82 ft, how long will the line representing that store's frontage be on the blueprint?

3. ***Height of a Tree*** At a given time of day, the ratio of the height of an object to the length of its shadow is the same for all objects. If a 3-ft stick in the ground casts a shadow of 1.2 ft, find the length of the shadow of a 48.4 ft tree. See the diagram on page 30.

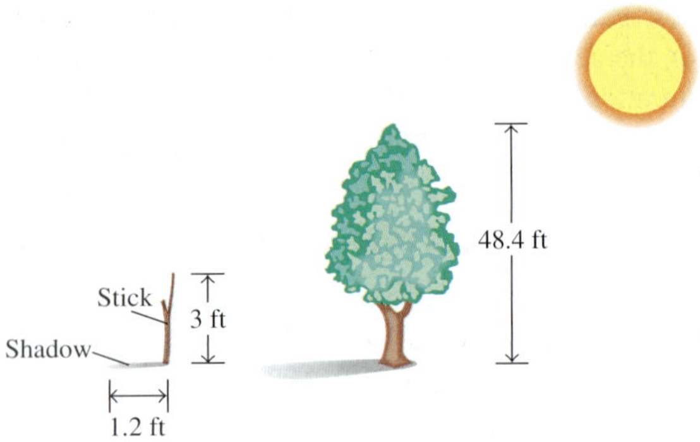

4. ***Use of Fertilizer*** A bag of fertilizer covers 6000 ft^2. How many bags are needed to cover an area of 32,000 ft^2?

5. ***College Tuition Cost*** According to the College Board, the average tuition at a private four-year college increased by 5.7% from 2004 to 2005. The average tuition at a private four-year college in 2004 was $27,461. Determine the average tuition at a private four-year college in 2005. Round your answer to the nearest dollar.

6. ***Fishing Charter*** Sam Stevens and four other friends want to schedule a 6-hour fishing trip on Lake Ontario with Reel Easy Sport Fishing. The cost for five people is $445. The cost for six people is $510. How much money per person will Sam and each of his friends save if they can find one more friend to go on their fishing trip? Assume that they will split the total cost equally.

7. ***Housing Market*** The following table shows where U.S. house prices have increased the most and the least, on average, for the 5-year period 2000–2005. The table also shows the increase from 2004–2005 for those areas.

   a) If a house in Nevada cost $150,000 in 2004, how much did a similar house cost in 2005?

   b) From 2004 to 2005, how much more did a $200,000 house in California increase in price than a $200,000 house in New Hampshire?

   c) If a house in Tennessee cost $180,000 in 2000, how much did a similar house cost in 2005?

Where house prices have gone up most	**% change 2000–2005**	**% change 2004–2005**
District of Columbia	108.1%	22.2%
California	103.0%	25.4%
Rhode Island	97.6%	17.1%
Nevada	84.7%	31.2%
Hawaii	82.9%	24.4%
Florida	80.5%	21.4%
Maryland	77.9%	21.0%
New Jersey	76.5%	15.8%
New Hampshire	72.3%	12.1%
Massachusetts	71.8%	11.6%
U.S.	**50.5%**	**12.5%**
Where prices have lagged		
Utah	17.5%	6.3%
Indiana	19.9%	4.1%
Nebraska	21.8%	5.4%
Mississippi	21.8%	4.9%
Tennessee	22.4%	5.5%

Source: Office of Federal Housing Enterprise Oversight

8. ***Who Receives Social Security?*** The circle graph below shows that Social Security benefits are not just for retirees. In 2005, about 48 million Americans received Social Security benefits. Use the circle graph below to answer the following questions.

   a) How many more retirees than disabled workers received Social Security benefits?

   b) How many fewer spouses and children of retired and disabled workers than survivors of deceased workers received Social Security benefits?

People Receiving Social Security, 2005

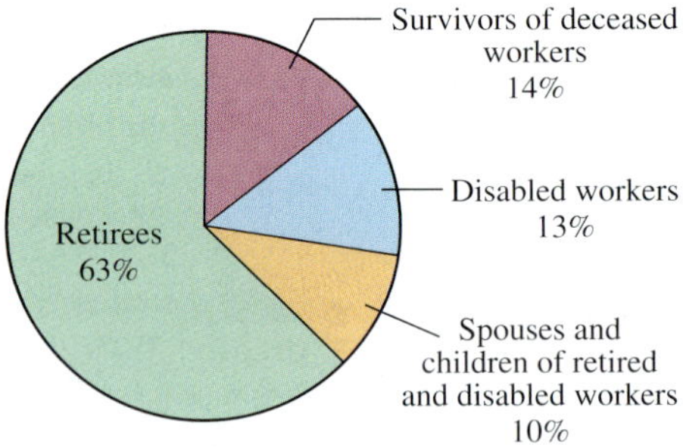

Source: Social Security Administration

9. ***Parking Costs*** The Main Street Garage charges $2.50 for the first hour of parking and $1.00 for each additional hour or part thereof. Denise Tomey parks her car in the garage from 9 A.M. to 5 P.M., 5 days a week. How much money does she save by paying a weekly parking rate of $35.00?

10. *School Pictures* Luann Alexander has two packages from which to choose to purchase her son's school picture. Package 1 costs $27 and includes 23 pictures. Package 2 costs $17 and includes 10 pictures. If the pictures are all the same size, which package offers the better price per picture?

11. *Buying a Computer* Emily Putnam wants to purchase a computer that sells for $1250. She can either pay the total amount at the time of purchase or she can agree to pay the store $120 down and $80 a month for 15 months. How much money can she save by paying the total amount at the time of purchase?

12. *Flu Shots* The price of a flu shot at Maxim Health Systems increased by 25% from 2005 to 2006. In 2005, Maxim Health Systems charged $20 for a flu shot. If Maxim Health sold flu shots to 2 million people in 2005 and to 2 million people in 2006, how much more money did they earn from selling flu shots in 2006 than in 2005?

13. *Buying a House* The Browns want to purchase a house that costs $140,000. They plan to take out a $100,000 mortgage on the house and put $40,000 as a down payment. The bank informs them that with a 15-year mortgage their monthly payment would be $840.62 and with a 30-year mortgage their monthly payment would be $620.28. Determine the amount they would save on the cost of the house if they selected the 15-year mortgage rather than the 30-year mortgage.

14. *Getting an 80 Average* On four exams, Wallace Memmer's grades were 79, 93, 91, and 68. What grade must he obtain on his fifth exam to have an 80 average?

15. *Japanese Sizes* The following chart shows men's jacket sizes as would be given in the United States and in Japan.

a) Justin Smith is in Japan and finds a sports jacket he wishes to buy. He is a size 48 in the United States. Determine the size of the jacket he should try on.

b) Determine a procedure (or a formula) for converting a jacket from a U.S. size to a size in Japan.

U.S.	chest	34	36	38	40	42	44	46	48
Japan		86.5	91.5	96.5	101.5	106.5	112	117	?

16. *Playing a Lottery* In one state lottery game, you must select a four-digit number (digits may be repeated). If your number matches exactly the four-digit number selected by the lottery commission, you win.

a) How many different numbers may be chosen?

b) If you purchase one lottery ticket, what is your chance of winning?

17. *Energy Value and Energy Consumption* The table gives the approximate energy values of some foods, in kilojoules (kJ) and the energy requirements of some activities.

a) How soon would you use up the energy from a fried egg by swimming?

b) How soon would you use up the energy from a hamburger by walking?

c) How soon would you use up the energy from a piece of strawberry shortcake by cycling?

d) How soon would you use up the energy from a hamburger and a chocolate milkshake by running?

Food	Energy Value (kJ)	Activity	Energy Consumption (kJ/min)
Chocolate milkshake	2200	Walking	25
Fried egg	460	Cycling	35
Hamburger	1550	Swimming	50
Strawberry shortcake	1400	Running	80
Glass of skim milk	350		

18. *Gas Mileage* Wendy Weisner fills her gas tank completely and makes a note that the odometer reads 38,451.4 miles. The next time she put gas in her car, filling the tank took 12.6 gal and the odometer read 38,687.0 miles. Determine the number of miles per gallon that Wendy's car got on this tank of gas.

19. *Household Credit Debt* The following graph shows the U.S. household credit debt, not including mortgages, from 1980 to 2005. Total debt includes revolving debt (the red area of the graph), such as credit card debt, and nonrevolving debt (the green area of the graph), such as auto and other installment loans.

a) Use the green area of the graph to estimate the U.S. household nonrevolving debt in 2005.

b) In 2005, there were 108,819,000 U.S. households. Estimate the average household total credit debt in 2005.

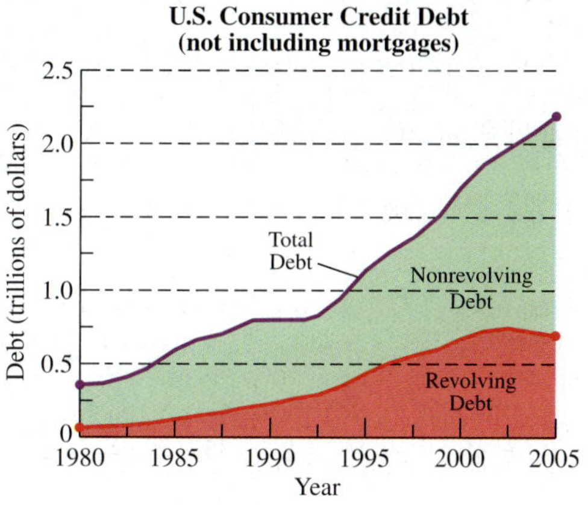

20. ***Saving for a Stereo*** Fernando Diez works 40 hours per week and makes $8.50 per hour.

a) How much money can he expect to earn in 1 year (52 weeks)?

b) If he saves all the money he earns, how long will he have to work to save for a stereo receiver that costs $1275?

21. ***Mail-Order Purchase*** Mary Liotta purchased 4 tires by mail order. She paid $52.80 per tire plus $5.60 per tire for shipping and handling. There is no sales tax on this purchase because the tires were purchased out of state. She also had to pay $8.56 per tire for mounting and balancing. At a local tire store, her total for the 4 tires with mounting and balancing would be $324 plus an 8% sales tax. How much did Mary save by purchasing the tires through the mail?

22. ***Sealing a Gym Floor*** A gymnasium floor has an area of 2400 square yards. Each gallon of floor sealant covers an area of 350 square feet. How many gallons of sealant are needed to cover the gymnasium floor?

23. ***Profit Margins*** The following chart shows retail stores' average percent profit margin on certain items.

Product Category	Average Profit Margin (%)
Video equipment	12
Audio components	14
Stereo speakers	20–25
Extended warranties	50–60

Source: *Consumer Reports*

a) Determine the average profit of a store that has the list price on a camcorder of $620.

b) Determine the average profit of a store that has a list price on a pair of speakers for $1200 (use a 22% profit margin).

c) If you negotiate with the salesperson and get him or her to sell the speakers for $1000, find the store's profit (use a 22% profit margin).

24. ***Income Taxes*** The federal income tax rate schedule for a joint return in 2006 is illustrated in the table. If Steve and Maureen Tomlin paid $13,365 in federal taxes, find the family's adjusted gross income.

Adjusted Gross Income	Taxes
$0–$15,100	10% of income
$15,100–$61,300	$1510 + 15% of amount over $15,100
$61,300–$123,700	$8440 + 25% of amount over $61,300
$123,700–$188,450	$24,040 + 28% of amount over $123,700
$188,450–$336,550	$42,170 + 33% of amount over $188,450
$336,550 and above	$91,043 + 35% of amount over $336,550

25. ***Leaking Faucet*** A faucet is leaking at a rate of one drop of water per second. Assume that the volume of one drop of water is 0.1 cubic centimeter (0.1 cm^3).

a) Determine the volume of water in cubic centimeters lost in 1 year.

b) How many days would it take to fill a rectangular basin 30 cm by 20 cm by 20 cm?

26. ***Wasted Water*** A faucet leaks 1 oz of water per minute.

a) How many gallons of water are wasted in a year? (A gallon contains 128 oz.)

b) If water costs $11.20 per 1000 gal, how much additional money is being spent on the water bill?

27. ***Airport Parking*** The chart shows parking rates at John F. Kennedy (JFK) Airport in New York, NY, as of August 1, 2006.

JFK Airport Parking Rates

SHORT-TERM RATES			
First half hour	$\frac{1}{2}$ hour–1 hour	Each Additional Hour After the First	Daily Maximum
$3	$6	$3	$30

LONG-TERM RATES	
Daily	Additional Day 0–8 Hours
$15	$5

a) Jeff Grace is going out of town for 5 full days. How much will he save by parking in the long-term lot rather than the short-term lot?

b) What is the cost of parking in the short-term lot for 4 hours?

c) If Jeff plans to park at the airport for 5 hours, is it cheaper to park in short-term or long-term parking? How much is the difference?

28. ***Tire Pressure*** When a car's tire pressure is 30 pounds per square inch (psi), it averages 20.8 mpg of gasoline. If the tire pressure is increased to 35 psi, the car averages 21.6 mpg of gasoline.

a) If Mr. Levy drives an average of 20,000 mi per year, how many gallons of gasoline will he save in a year by increasing his tire pressure from 30 to 35 psi?

b) If gasoline costs $3.00 per gallon, how much will he save in a year?

c) If we assume that there are about 140 million cars in the United States and that these changes are typical of each car, how many gallons of gasoline would be saved if all drivers increased their cars' tire pressure?

29. ***Air Pollution*** The table illustrates the 10 countries that produced the most carbon dioxide emissions in 2004. The table also illustrates the emissions per capita for each country listed. (The term *per capita* means per person.)

Producers of Carbon Dioxide

	Total Emissions (in millions of metric tons)	**Emissions per Capita** (in metric tons)
United States	5912.2	19.8
China	4707.3	3.6
Russia	1684.8	11.9
Japan	1262.1	9.9
India	1112.8	1.0
Germany	862.2	10.5
Canada	588.0	17.8
United Kingdom	579.7	9.6
South Korea	496.8	10.2
Italy	485.0	8.3

Source: U.S. Department of Energy

a) By looking at the data provided, is it possible to determine the population of each country? If so, explain how to do so.

b) Using the procedure you gave in part (a), determine the population of the United States.

c) Using the procedure you gave in part (a), determine the population of China.

30. ***Adjusting for Inflation*** Assume that the rate of inflation is 6% per year for the next 2 years. What will the price of a dishwasher be 2 years from now if the dishwasher costs $799 today?

31. ***Investing*** You place $1000 in a mutual fund. The first year, the value of the fund increases by 10%. The second year, the value of the fund decreases by 10%. Determine the value of the fund at the end of the second year. Is it greater than, less than, or equal to your initial investment?

32. ***X-rays*** With a certain medical insurance policy, the customer must first pay an annual $100 deductible, then the policy covers 80% of the cost of x-rays. The first insurance claims for a specific year submitted by Yungchen Cheng are for two x-rays. The first x-ray cost $640, and the second x-ray cost $920. How much, in total, will Yungchen need to pay for these x-rays?

33. ***A Photo Safari*** Kelli Hammer is planning a trip to Africa where she will participate in a photo safari. She is planning to bring a great deal of film. A photography store is selling 4 packs of film for $17 and 10 packs of the same film for $41.

a) If she wishes to purchase only the 4 packs and 10 packs and wishes to spend a maximum of $200 on film, what is the maximum number of rolls of film she can purchase?

b) What will be the cost of the film?

34. ***Buying Film*** Erika Gutierrez is planning a vacation to Australia and wishes to bring a large supply of film. At Wal-Mart, 4 packs of 24-exposure film costs $4.08 and 4 packs of the same film with 36 exposures costs $5.76.

a) If she wishes to spend a maximum of $50 on film and get the most exposures, how many 4 packs of 24 exposures and how many 4 packs of 36 exposures should she purchase?

b) How many exposures will she get?

c) What will be the cost of the film? If there is more than one choice in part (a), give the minimum cost.

35. ***Making Cream of Wheat*** The following amounts of ingredients are recommended to make various servings of Nabisco Instant Cream of Wheat. *Note:* 16 tbsp = 1 cup.

Ingredient	1 Serving	2 Servings	4 Servings
Mix water or milk	1 cup	2 cups	$3\frac{3}{4}$ cups
With salt (optional)	$\frac{1}{8}$ tsp	$\frac{1}{4}$ tsp	$\frac{1}{2}$ tsp
Add Cream of Wheat	3 tbsp	$\frac{1}{2}$ cup	$\frac{3}{4}$ cup

Determine the amount of each ingredient needed to make 3 servings using the following procedures.

a) Multiply the amounts for 1 serving by 3.

b) Find the average of the amounts for 2 and 4 servings.

c) Subtract the amounts for 1 serving from the amounts for 4 servings.

d) Compare the answers obtained in parts (a) through (c) and explain any differences.

36. ***Making Rice*** Following are the amounts of ingredients recommended to make various servings of Uncle Ben's Original Converted Rice. *Note:* 1 tbsp = 3 tsp.

Ingredient	2 Servings	4 Servings	6 Servings	12 Servings
Rice (cups)	$\frac{1}{2}$	1	$1\frac{1}{2}$	3
Water (cups)	$1\frac{1}{3}$	$2\frac{1}{4}$	$3\frac{1}{3}$	6
Salt (teaspoons)	$\frac{1}{4}$	$\frac{1}{2}$	$\frac{3}{4}$	$1\frac{1}{2}$
Butter	1 tsp	2 tsp	1 tbsp	2 tbsp

Determine the amount of each ingredient needed to make 8 servings using the following procedures.

a) Multiply the amount for 2 servings by 4.

b) Multiply the amount for 4 servings by 2.

c) Add the amounts for 2 and 6 servings.

d) Subtract the amount for 4 servings from the amount for 12 servings.

e) Compare the answers obtained in parts (a) through (d) and explain any differences.

Solve the following problems.

37. ***One Square Foot*** How many square inches, 1 in. by 1 in., fit in an area of 1 square foot, 1 ft by 1 ft?

38. ***Cubic Inches*** How many cubic inches fit in 1 cubic foot?

39. ***Rectangle*** If the length and width of a rectangle each double, what happens to the area of the rectangle?

40. ***Cube*** If the length, width, and height of a cube all double, what happens to the volume of the cube?

41. ***Pole in a Lake*** A pole is in the middle of a small lake. One half of the pole is in the ground. One third of the pole is covered by water. Eleven feet, or one-sixth, of the pole are out of water. What is the length of the pole?

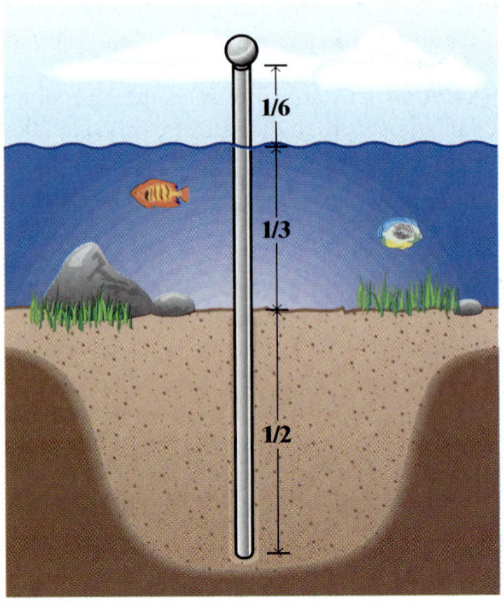

42. ***Buying Candy*** How much do 10 pieces of candy cost if 1000 pieces cost $10?

43. ***A Balance*** On the balance below, where should the one missing block ■ be placed so that the balance would balance on the triangle (the fulcrum). Assume that each block has the same weight.

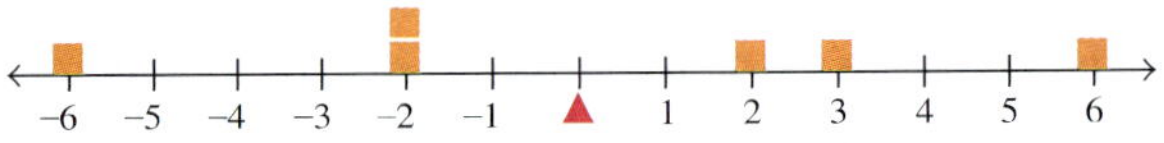

44. ***Ties, Ties, Ties*** All my ties are red except two. All my ties are blue except two. All my ties are brown except two. How many ties do I have?

45. ***Buying a Yacht*** Four partners decide to share the cost of a yacht equally. By bringing in an additional partner, they can reduce the cost to each of the five partners by $3000. What is the total cost of the yacht?

46. ***Palindromes*** A *palindrome* is a number (or word) that reads the same forward and backward. The numbers 1991 and 43234 are examples of palindromes. How many palindromes are there between the numbers 2000 and 3000? List them.

25. *Railroad Car Estimation* The scale of a model railroad is 1 in. = 12.5 ft. Estimate the size of an actual box car if this drawing is the same size as the model box car.

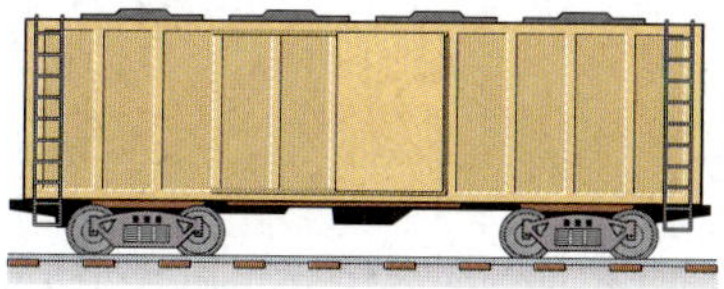

1.3

Solve the following problems.

26. *Change from a Twenty* Jeff Howard parked his car in a lot that charged \$2.00 for the first hour and \$1.50 for each additional hour. He left the car in the lot for 8 hr. How much change did he receive from a \$20 bill?

27. *Buying in Quantity* A six-pack of cola costs \$2.69. A carton of 4 six-packs costs \$9.60. How much will be saved by purchasing the carton rather than 4 individual six-packs?

28. *Jet Ski Rental* The rental cost of a jet ski from Nola Akala's Ski Rental is \$15 per 15 min, and the cost from Jill Berkman's Ski Rental is \$25 per half hour. If you plan to rent the jet ski for 2 hr, which is the better deal, and by how much?

29. *Oscars* In 2001, shortly before the Academy Awards show, many Oscars were lost by the shipping company. Fifty-two of the 55 Oscars were found, before the awards ceremony, in a dumpster by an Illinois man. The man was awarded \$50,000 (and two tickets to the ceremony). The actual cost to have each Oscar produced was \$327. How much more had the man been awarded than the actual cost to produce the 52 Oscars he found (disregarding the cost of the tickets)?

30. *Applying Fertilizer* Ron Williams needs to apply fertilizer to his lawn on his farm. A 30-pound bag of fertilizer will cover an area of 2500 square feet.

    a) How many pounds are needed to cover an area of 24,000 square feet?

    b) If Ron only has 150 pounds of fertilizer, how many square feet can he fertilize?

31. *Auto Insurance* Most insurance companies reduce premiums by 10% until age 25 for people who successfully pass a driver's education course. A particular driver's education course costs \$60. Patrick Flanigan, who just turned 18, has auto insurance that costs \$530 per year. By taking the driver's education course, how much would he save in auto insurance, including the cost of the course, from the age of 18 until the age of 25?

32. *Pediatric Dosage* If 1.5 milligrams of a medicine is to be given for 10 lb of body weight, how many milligrams should be given to a child who weighs 47 lb?

33. *Qualifying for a Mortgage* Banks will usually grant an applicant a mortgage if the monthly payments are not greater than 28% of the person's take-home pay. What is the maximum monthly mortgage payment you can make if your gross salary is \$4500 a month and your payroll deductions are 30% of your gross salary?

34. *Flying West* New York City is on eastern standard time, St. Louis is on central standard time (1 hr earlier than eastern standard time), and Las Vegas is on Pacific standard time (3 hr earlier than eastern standard time). A flight leaves New York City at 9 A.M. eastern standard time, stops for 50 min in St. Louis, and arrives in Las Vegas at 1:35 P.M. Pacific time. How long is the plane actually flying?

35. *Crossing Time Zones* The international date line is an imaginary line of longitude (from the North Pole to the South Pole) on Earth's surface between Japan and Hawaii in the Pacific Ocean. Crossing the line east to west adds a day to the present date. Crossing the line west to east subtracts a day. At 3:00 P.M. on July 25 in Hawaii, what is the time and date in Tokyo, Japan, which is four time zones to the west?

36. *Conversions* 1 in. = 2.54 cm.

    a) How many square centimeters are in a square inch?

    b) How many cubic centimeters are in a cubic inch?

    c) How long is a centimeter in terms of inches?

37. *Dot Pattern* If the following pattern is continued, how many dots will be in the hundredth figure?

38. *Magic Square* The following magic square uses each number from 6 to 21 exactly once. Complete the magic square by using the unused numbers from 6 through 21 exactly once.

21	7		18
10		15	
14	12	11	17
9	19		

39. *Magic Square* Create a magic square by using the numbers 13, 15, 17, 19, 21, 23, 25, 27, and 29. The sum of the numbers in every row, column, and diagonal must be 63.

40. ***Microbes in a Jar*** A colony of microbes doubles in number every second. A single microbe is placed in a jar, and in an hour the jar is full. When was the jar half full?

41. ***Brothers and Sisters*** Jim Carraway has four more brothers than sisters. How many more brothers than sisters does his sister Mary have?

42. ***A Missing Dollar*** Three friends check into a single room in a motel and pay \$10 apiece. The room costs \$25 instead of \$30, so a clerk is sent to the room to give \$5 back. The friends each take back \$1, and the clerk is given \$2 for his trouble. Now each of the friends paid \$9, a total of \$27, and the clerk received \$2. What happened to the missing dollar?

43. ***The Average Weight*** Four women in a room have an average weight of 130 lb. A fifth woman who weighs 180 lb enters the room. Find the average weight of all five women.

44. ***Change for a Dollar*** Could a person have \$1.15 worth of change in his pocket and still not be able to give someone change for a dollar bill? If so, what coins might he have?

45. ***Volume of a Cube*** Here is a flat pattern for a cube to be formed by folding. The sides of each square are 6 cm. Find the volume of the cube.

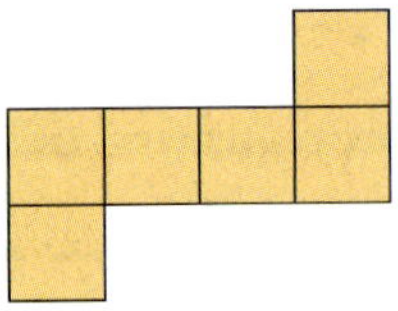

46. ***The Heavier Coin*** You have 13 coins, which all look alike. Twelve coins weigh exactly the same, but the other one is heavier. You have a pan balance. Tell how to find the heavier coin in just three weighings.

47. ***The Sum of Numbers*** Find the sum of the first 500 counting numbers. (*Hint:* Group in pairs.)

48. ***Balancing a Scale*** On a balance scale, three green balls balance six blue balls, two yellow balls balance five blue balls, and six blue balls balance four white balls. How many blue balls are needed to balance four green, two yellow, and two white balls?

49. ***Palindromes*** How many three-digit numbers greater than 100 are palindromes?

50. ***Figures*** Describe the fifth figure.

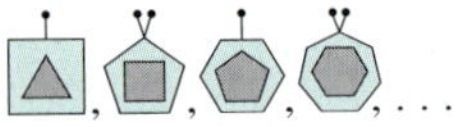

51. ***Patterns*** How many orange tiles will be required to build the sixth figure in this pattern?

52. ***Sum of Numbers*** Place the numbers 1 through 12 in the 12 circles so that the sum of the numbers in each of the six rows is 26. Use each number from 1 through 12 exactly once.

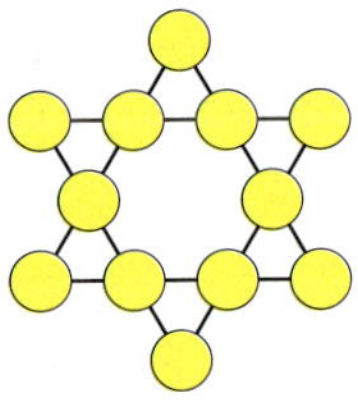

53. ***People in a Line*** People stand in a line at movies, the grocery store, and many other places.

a) In how many ways can two people stand in a line?

b) In how many ways can three people stand in a line?

c) In how many ways can four people stand in a line?

d) In how many ways can five people stand in a line?

e) Using the results from parts (a) through (d), make a conjecture about the number of ways in which n people can stand in a line.

CHAPTER 1 TEST

In Exercises 1 and 2, use inductive reasoning to determine the next three numbers in the pattern.

1. 3, 7, 11, 15, . . .

2. $1, \frac{1}{2}, \frac{1}{4}, \frac{1}{8}, \ldots$

3. Pick any number, multiply the number by 5, and add 10 to the number. Divide the sum by 5. Subtract 1 from the quotient.

a) What is the relationship between the number you started with and the final answer?

b) Arbitrarily select some different numbers and repeat the process. Record the original number and the results.

c) Make a conjecture about the relationship between the original number and the final answer.

d) Prove, using deductive reasoning, the conjecture made in part (c).

In Exercises 4 and 5, estimate the answers.

4. $0.18 \times 58{,}000$

5. $\dfrac{210{,}000}{0.12}$

6. ***Estimating Area*** If each square represents one square unit, estimate the area of the shaded figure.

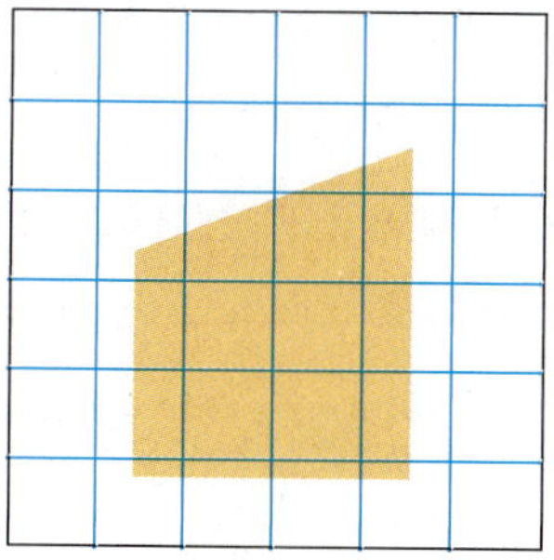

7. ***Body Mass Index*** The federal government gives a procedure to determine if a child is overweight. To make this decision, first determine the child's body mass index (BMI). Then compare the BMI with one of the two charts, one for boys and one for girls, provided by the government. On the right, we give the chart for boys up to age 20. To determine a child's BMI:

1) Divide the child's weight (in pounds) by the child's height (in inches).
2) Divide the results from part 1 by the child's height again.
3) Multiply the result from part 2 by 703.

Richard is a 14-year-old boy who weighs 130 lb and is 63 in. tall.

a) Determine his BMI.

b) Does he appear to be at risk for being overweight, or is he overweight? Explain.

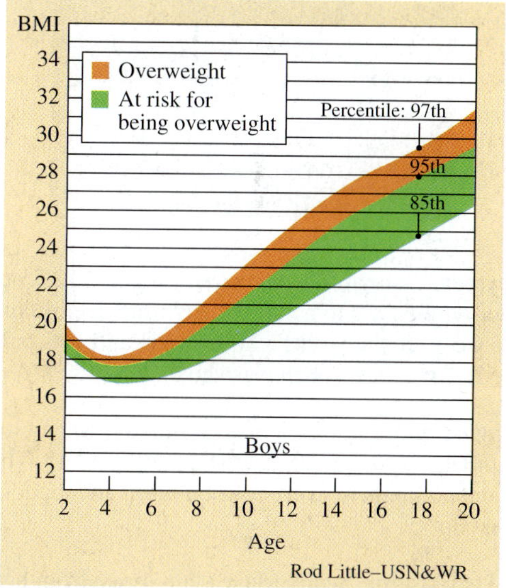

8. ***Cell Phone Cost*** AT & T charges \$59.99 for its Nation 900 with Rollover Plan for cell phones, which includes 900 minutes of call time. The company charges \$0.40 for each additional minute. If the Smiths' cell phone bill for December was \$74.39, how many additional minutes did they use during that month?

9. ***Cans of Soda*** At a local store a six-pack of soda costs \$2.59 and individual cans cost \$0.80. What is the maximum number of cans of soda that can be purchased for \$15?

10. ***Cutting Wood*** How much time does it take Carla Knab, a carpenter, to cut a 10 ft length of wood into four equal pieces, if each cut takes $2\frac{1}{2}$ min?

11. ***Determining Size*** In this photo of *Sunflowers*, a 1889 painting by Vincent van Gogh, 1 in. equals 15.8 in. on the actual painting. Find the dimensions of the actual painting.

12. ***Payment Shortfall*** Monica Wilson gets $12.75 per hour with time and a half for any time over 40 hours per week. If she works a 50 hr week and gets paid $652.25, by how much was she underpaid?

13. ***Magic Square*** Create a magic square by using the numbers 5, 10, 15, 20, 25, 30, 35, 40, and 45. The sum of the numbers in every row, column, and diagonal must be 75.

14. ***A Drive to the Beach*** Mary Chin drove from her home to the beach that is 30 mi from her house. The first 15 mi she drove at 60 mph, and the next 15 mi she drove at 30 mph. Would the trip take more, less, or the same time if she traveled the entire 30 mi at a steady 45 mph?

15. ***Pick Five Numbers*** From the six numbers 2, 6, 8, 9, 11, and 13, pick five that, when multiplied, give 11,232.

16. ***Jelly Bean Guess*** One guess is off by 9, another guess is off by 17, and yet another guess is off by 31. How many jelly beans are in the jar?

17. ***Buying Plants*** David Mackin wants to purchase nine herb plants. Countryside Nursery has herbs that are on sale at three for $3.99. David has a coupon for 25% off an unlimited number of herb plants at the original price of $1.75 per plant.

    a) Determine the cost of purchasing nine plants at the sale price.

    b) Determine the cost of purchasing nine plants if the coupon is used.

    c) Which is the least expensive way to purchase the nine plants, and by how much?

18. ***Arranging Letters*** In how many different ways can four letters, A, B, C, D, be arranged?

GROUP PROJECTS

HOLIDAY SHOPPING

1. It is December 1 and John needs to begin his holiday shopping. He intends to purchase gifts for three people: his girlfriend, Melissa; his mother, Ruth; and his father, Don. He doesn't want to spend more than a total of $325, including the 7% sales tax.

    a) If John were to spend the $325 equally among the three people, approximate the amount that would be spent on each person.

    b) If John were to spend the $325 equally among the three people, determine the maximum amount, *before tax*, that he could spend on each person and not exceed the maximum of $325, including tax.

c) John decides to get a new set of wrenches for his father. He sees the specific set he wants on sale at Sears. He calls four Sears stores to see if they have the set of wrenches in stock. They all reply that the set is out of stock. He decides that calling additional Sears stores is useless because he believes that they will also tell him that the set of wrenches is out of stock. What type of reasoning did John use in arriving at his conclusion? Explain.

d) John finds an equivalent set of wrenches at a True Value hardware store. The set he is considering is a combination set that contains both standard U.S. size and metric size wrenches. The store's regular price before tax is \$62 for the set, but it is selling them for 10% off the regular price. He can also purchase the same wrenches by purchasing two separate sets, one for standard U.S. size wrenches and the other for metric sizes. Each of these sets has a regular price, before tax, of \$36, but both are on sale for 20% off their regular prices. Can John purchase the combination set or the two individual sets less expensively?

e) How much will John save, *after tax*, by using the less expensive method?

GOING ON VACATION

2. Bob and Kristen Williams decide to go on a vacation. They live in San Francisco, California, and plan to drive to New Orleans, Louisiana.

a) Obtain a map that shows routes that they may take from San Francisco to New Orleans. Write directions for them from San Francisco to New Orleans via the shortest distance. Use major highways whenever possible.

b) Use the scale on the map to estimate the one-way distance to New Orleans.

c) If the Williams estimate that they will average 50 mph (including comfort stops), estimate the travel time, in hours, to New Orleans.

d) If the Williams want to travel about 400 miles per day, locate a town in the vicinity of where they will stop each evening.

e) If they begin each segment of the trip each day at 9 A.M., at about what time will they look for a hotel each evening?

f) Use the information provided in parts (a) through (e) to estimate the time of day they will arrive in New Orleans.

g) Estimate the mileage of a typical midsized car and the cost per gallon of a gallon of regular unleaded gasoline. Then estimate the cost of gasoline for the Williams' trip.

h) Estimate the cost of a typical breakfast, a typical lunch, and a typical dinner for two adults, and the cost of a typical motel room. Then estimate the total cost, including meals, gas, and lodging, for the Williams' trip from San Francisco to New Orleans (one way).

PROBLEM SOLVING

3. Four acrobats who bill themselves as the "Tumbling Tumbleweeds" finish up their act with the amazing "Human Pillar," in which the acrobats form a tower, each one standing on the shoulders of the one below. Each acrobat (Ernie, Jed, Tex, and Zeke Tumbleweed) wears a different distinctive item of western garb (chaps, holster, Stetson hat, or leather vest) in the act. Can you identify the members of the "Human Pillar," from top to bottom, by name and apparel?

a) Jed Tumbleweed is not on top, but he is somewhere above the man in the Stetson.

b) Zeke Tumbleweed does not wear the holster.

c) The man in the vest is not on top.

d) The man in the chaps is somewhere above Tex but somewhere below Zeke.

ORDER	NAME	APPAREL
______	______	______
______	______	______
______	______	______
______	______	______

CHAPTER 2

Sets

▲ Children learn how to classify sets, such as shapes and colors, at a very early age.

WHAT YOU WILL LEARN

- Methods to indicate sets, equal sets, and equivalent sets
- Subsets and proper subsets
- Venn diagrams
- Set operations such as complement, intersection, union, difference and Cartesian product
- Equality of sets
- Applications of sets
- Infinite sets

WHY IT IS IMPORTANT

A basic human impulse is to sort or classify things. As you will see in this chapter, putting elements into sets helps you order and arrange your world. It allows you to deal with large quantities of information. Set building is a learning tool that helps answer the question, What are the characteristics of this group? Studying sets is also important because sets underlie other mathematical topics such as logic and abstract algebra.

2.1 SET CONCEPTS

▲ Horses may be classified in many ways.

Can you think of a few different categories or groups to which you belong? One way you could categorize yourself is by your gender. Another way is by your academic major. A third way is by your state of residence. In this section, we will discuss ways to sort or classify objects. We will also discuss different methods that can be used to indicate collections of objects.

We encounter sets in many different ways every day of our lives. A ***set*** is a collection of objects, which are called ***elements*** or ***members*** of the set. For example, the United States is a collection or set of 50 states plus the District of Columbia. The 50 individual states plus the District of Columbia are the members or elements of the set that is called the United States.

A set is ***well defined*** if its contents can be clearly determined. The set of U.S. presidents is a well-defined set because its contents, the presidents, can be named. The set of the three best movies is not a well-defined set because the word *best* is interpreted differently by different people. In this text, we use only well-defined sets.

Three methods are commonly used to indicate a set: (1) description, (2) roster form, and (3) set-builder notation.

The method of indicating a set by ***description*** is illustrated in Example 1.

EXAMPLE 1 *Description of Sets*

Write a description of the set containing the elements Monday, Tuesday, Wednesday, Thursday, Friday, Saturday, Sunday.

SOLUTION The set is the days of the week. •

Listing the elements of a set inside a pair of ***braces***, { }, is called ***roster form***. The braces are an essential part of the notation because they identify the contents as a set. For example, $\{1, 2, 3\}$ is notation for the set whose elements are 1, 2, and 3, but (1, 2, 3) and [1, 2, 3] are not sets because parentheses and brackets do not indicate a set. For a set written in roster form, commas separate the elements of the set. The order in which the elements are listed is not important.

Sets are generally named with capital letters. For example, the name commonly selected for the set of ***natural numbers*** or ***counting numbers*** is N.

NATURAL NUMBERS

$$N = \{1, 2, 3, 4, 5, \dots\}$$

The three dots after the 5, called an *ellipsis,* indicate that the elements in the set continue in the same manner. An ellipsis followed by a last element indicates that the elements continue in the same manner up to and including the last element. This notation is illustrated in Example 2(b).

PROFILE IN MATHEMATICS

Georg Cantor

Georg Cantor (1845–1918), born in St. Petersburg, Russia, is recognized as the founder of set theory. Cantor's creative work in mathematics was nearly lost when his father insisted that he become an engineer rather than a mathematician. His two major books on set theory, *Foundations of General Theory of Aggregates* and *Contributions to the Founding of the Theory of Transfinite Numbers,* were published in 1883 and 1895, respectively. See the Profile in Mathematics on page 91 for more information on Cantor and Leopold Kronecker.

▲ The planets of Earth's solar system.

EXAMPLE 2 *Roster Form of Sets*

Express the following in roster form.

a) Set A is the set of natural numbers less than 5.

b) Set B is the set of natural numbers less than or equal to 75.

c) Set P is the set of planets in Earth's solar system.

SOLUTION

a) The natural numbers less than 5 are 1, 2, 3, and 4. Thus, set A in roster form is $A = \{1, 2, 3, 4\}$.

b) $B = \{1, 2, 3, 4, \ldots, 75\}$. The 75 after the ellipsis indicates that the elements continue in the same manner up to and including the number 75.

c) $P = \{$Mercury, Venus, Earth, Mars, Jupiter, Saturn, Uranus, Neptune$\}$* ●

EXAMPLE 3 *The Word* Inclusive

Express the following in roster form.

a) The set of natural numbers between 4 and 9.

b) The set of natural numbers between 4 and 9, inclusive.

SOLUTION

a) $A = \{5, 6, 7, 8\}$

b) $B = \{4, 5, 6, 7, 8, 9\}$. Note that the word *inclusive* indicates that the values of 4 and 9 are included in the set. ●

The symbol $\in$, read, "is an element of," is used to indicate membership in a set. In Example 3, because 6 is an element of set A, we write $6 \in A$. This may also be written $6 \in \{5, 6, 7, 8\}$. We may also write $9 \notin A$, meaning that 9 is not an element of set A.

Set-builder notation (sometimes called *set-generator notation*) may be used to symbolize a set. Set-builder notation is frequently used in algebra. The following example illustrates its form.

D	$=$	$\{$	x	$\mid$	Condition(s) $\}$
↑	↑	↑	↑	↑	↑
Set D	is	the set of	all elements x	such that	the condition(s) x must meet in order to be a member of the set.

Consider $E = \{x \mid x \in N \text{ and } x > 10\}$. The statement is read: "Set E is the set of all the elements x such that x is a natural number and x is greater than 10." The conditions that x must meet to be a member of the set are $x \in N$, which means that x must be a natural number, and $x > 10$, which means that x must be greater than 10. The numbers that meet both conditions are the set of natural numbers greater than 10. Set E in roster form is

$$E = \{11, 12, 13, 14, \ldots\}$$

*In August 2006, Pluto was reclassified as a dwarf planet.

EXAMPLE 4 *Using Set-Builder Notation*

a) Write set $B = \{1, 2, 3, 4, 5\}$ in set-builder notation.

b) Write, in words, how you would read set B in set-builder notation.

SOLUTION

a) Because set B consists of the natural numbers less than 6, we write

$$B = \{x \mid x \in N \text{ and } x < 6\}$$

Another acceptable answer is $B = \{x \mid x \in N \text{ and } x \leq 5\}$.

b) Set B is the set of all elements x such that x is a natural number and x is less than 6.

EXAMPLE 5 *Roster Form to Set-Builder Notation*

a) Write set C = {North America, South America, Europe, Asia, Australia, Africa, Antarctica} in set-builder notation.

b) Write in words how you would read set C in set-builder notation.

SOLUTION

a) $C = \{x \mid x \text{ is a continent}\}$.

b) Set C is the set of all elements x such that x is a continent.

EXAMPLE 6 *Set-Builder Notation to Roster Form*

Write set $A = \{x \mid x \in N \text{ and } 2 \leq x < 8\}$ in roster form.

SOLUTION $A = \{2, 3, 4, 5, 6, 7\}$

EXAMPLE 7 *Largest Cities*

The chart shows the 10 most populated U.S. cities in 2005. Let set C be the set of cities in California that are among the 10 most populated U.S. cities in 2005. Write set C in roster form.

Ten Most Populated Cities in the U.S., 2005	Population
New York, New York	8,143,197
Los Angeles, California	3,844,829
Chicago, Illinois	2,842,518
Houston, Texas	2,016,582
Philadelphia, Pennsylvania	1,463,281
Phoenix, Arizona	1,461,575
San Antonio, Texas	1,256,509
San Diego, California	1,255,540
Dallas, Texas	1,213,825
San Jose, California	912,332

Source: U.S. Census Bureau

SOLUTION By examining the chart we find that three California cities appear in the chart. They are Los Angeles, San Diego, and San Jose. Thus, set $C = \{$Los Angeles, San Diego, San Jose$\}$.

A set is said to be ***finite*** if it either contains no elements or the number of elements in the set is a natural number. The set $B = \{2, 4, 6, 8, 10\}$ is a finite set because the number of elements in the set is 5, and 5 is a natural number. A set that is not finite is said to be ***infinite***. The set of counting numbers is one example of an infinite set. Infinite sets are discussed in more detail in Section 2.6.

Another important concept is equality of sets.

Set A is **equal** to set B, symbolized by $A = B$, if and only if set A and set B contain exactly the same elements.

For example, if set $A = \{1, 2, 3\}$ and set $B = \{3, 1, 2\}$, then $A = B$ because they contain exactly the same elements. The order of the elements in the set is not important. If two sets are equal, both must contain the same number of elements. The number of elements in a set is called its *cardinal number.*

The **cardinal number** of set A, symbolized by $n(A)$, is the number of elements in set A.

Both set $A = \{1, 2, 3\}$ and set $B = \{$England, Brazil, Japan$\}$ have a cardinal number of 3; that is, $n(A) = 3$, and $n(B) = 3$. We can say that set A and set B both have a cardinality of 3.

Two sets are said to be *equivalent* if they contain the same number of elements.

Set A is **equivalent** to set B if and only if $n(A) = n(B)$.

Any sets that are equal must also be equivalent. Not all sets that are equivalent are equal, however. The sets $D = \{$a, b, c$\}$ and $E = \{$apple, orange, pear$\}$ are equivalent because both have the same cardinal number, 3. Because the elements differ, however, the sets are not equal.

Two sets that are equivalent or have the same cardinality can be placed in ***one-to-one correspondence***. Set A and set B can be placed in one-to-one correspondence if every element of set A can be matched with exactly one element of set B and every element of set B can be matched with exactly one element of set A. For example, there is a one-to-one correspondence between the student names on a class list and the student identification numbers because we can match each student with a student identification number.

Consider set S, states, and set C, state capitals.

$$S = \{\text{North Carolina, Georgia, South Carolina, Florida}\}$$
$$C = \{\text{Columbia, Raleigh, Tallahassee, Atlanta}\}$$

DID YOU KNOW?

Creature Cards

We learn to group objects according to what we see as the relevant distinguishing characteristics. One way used by educators to measure this ability is through visual cues. An example can be seen in this test, called "Creature Cards," offered by the Education Development Center. How would you describe membership in the set of Jexums?

Two different one-to-one correspondences for sets S and C follow.

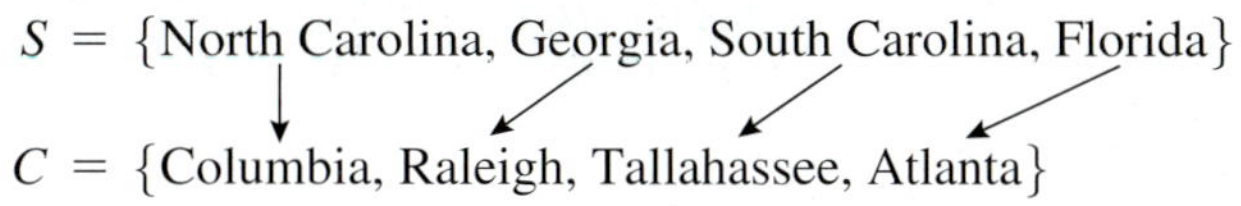

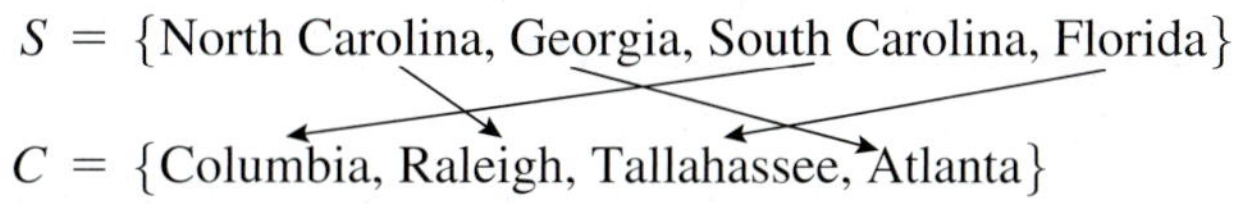

Other one-to-one correspondences between sets S and C are possible. Do you know which capital goes with which state?

Null or Empty Set

Some sets do not contain any elements, such as the set of zebras that live in your house.

> The set that contains no elements is called the **empty set** or **null set** and is symbolized by $\{\ \}$ or $\varnothing$.

Note that $\{\varnothing\}$ is not the empty set. This set contains the element $\varnothing$ and has a cardinality of 1. The set $\{0\}$ is also not the empty set because it contains the element 0. It also has a cardinality of 1.

EXAMPLE 8 *Natural Number Solutions*

Indicate the set of natural numbers that satisfies the equation $x + 2 = 0$.

SOLUTION The values that satisfy the equation are those that make the equation a true statement. Only the number -2 satisfies this equation. Because -2 is not a natural number, the solution set of this equation is $\{\ \}$ or $\varnothing$. ●

Universal Set

Another important set is a *universal set*.

> A **universal set,** symbolized by U, is a set that contains all the elements for any specific discussion.

When a universal set is given, only the elements in the universal set may be considered when working the problem. If, for example, the universal set for a particular problem is defined as $U = \{1, 2, 3, 4, \ldots, 10\}$, then only the natural numbers 1 through 10 may be used in that problem.

SECTION 2.1 EXERCISES

CONCEPT/WRITING EXERCISES

In Exercises 1–12, answer each question with a complete sentence.

1. What is a set?
2. What is an ellipsis, and how is it used?
3. What are the three ways that a set can be written? Give an example of each.
4. What is an infinite set?
5. What is a finite set?
6. What are equal sets?
7. What are equivalent sets?
8. What is the cardinal number of a set?
9. What is the empty set?
10. What are the two ways to indicate the empty set?
11. What is a universal set?
12. What does a one-to-one correspondence of two sets mean?

PRACTICE THE SKILLS

In Exercises 13–18, determine whether each set is well defined.

13. The set of the best books
14. The set of the easiest courses at your school
15. The set of states that have a common border with Kansas
16. The set of the four states in the United States having the largest areas
17. The set of astronauts who walked on the moon

18. The set of the most interesting teachers at your school

In Exercises 19–24, determine whether each set is finite or infinite.

19. $\{2, 4, 6, 8, \ldots\}$
20. The set of multiples of 6 between 0 and 90
21. The set of odd numbers greater than 25
22. The set of fractions between 1 and 2
23. The set of odd numbers greater than 15
24. The set of apple trees in Gro-More Farms Orchards

In Exercises 25–34, express each set in roster form. You may need to use a world almanac or some other reference source.

25. The set of states in the United States whose names begin with the letter M
26. The set of oceans in the world
27. The set of natural numbers between 10 and 178
28. $C = \{x \mid x + 6 = 10\}$
29. $B = \{x \mid x \in N \text{ and } x \text{ is even}\}$
30. The set of states west of the Mississippi River that have a common border with the state of Florida
31. The set of football players over the age of 70 who are still playing in the National Football League
32. The set of states in the United States that have no common border with any other state
33. $E = \{x \mid x \in N \text{ and } 14 \leq x < 85\}$
34. The set of states in the United States that are not in the contiguous 48 states

In Exercises 35–38, use the following table, which shows the average price for a Big Mac in selected countries as of March 20, 2005. Let the 15 selected countries represent the universal set.

Average Price for a Big Mac

Country	Price
Switzerland	\$5.46
Denmark	\$4.97
Sweden	\$4.46
United Kingdom	\$3.61
Germany	\$3.58
New Zealand	\$3.16
United States*	\$3.00
Turkey	\$2.80
Peru	\$2.74
Canada	\$2.60
Chile	\$2.56
Japan	\$2.50
Australia	\$2.46
Czech Republic	\$2.45
China	\$1.26

*Average price in four selected cities.
Source: *Economist*

Use the list to represent each set in roster form.

35. The set of countries in which the average price for a Big Mac is more than \$3.00

36. The set of countries in which the average price for a Big Mac is less than \$2.50

37. The set of countries in which the average price for a Big Mac is between \$3.25 and \$5.50

38. The set of countries in which the average price for a Big Mac is between \$2.00 and \$2.99

In Exercises 39–42, use the following graph, which shows the digital camera sales, in millions, for the years 1996–2005.

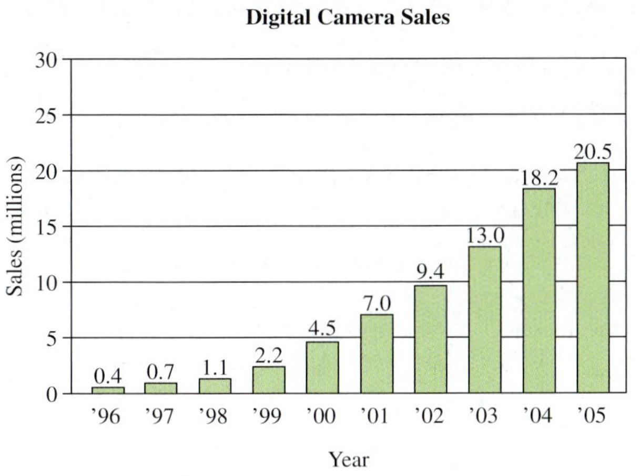

Source: PMA Marketing Research

Use the graph to represent each set in roster form.

39. The set of years included in the graph in which digital camera sales were more than 13 million

40. The set of years included in the graph in which digital camera sales were less than 6 million

41. The set of years included in the graph in which digital camera sales were between 1 million and 7 million

42. The set of years included in the graph in which digital camera sales were more than 25 million

In Exercises 43–50, express each set in set-builder notation.

43. $B = \{5, 6, 7, 8, 9, 10, 11, 12\}$

44. $A = \{1, 2, 3, 4, 5, 6, 7, 8, 9\}$

45. $C = \{3, 6, 9, 12, \dots\}$

46. $D = \{5, 10, 15, 20, \dots\}$

47. E is the set of odd natural numbers

48. A is the set of national holidays in the United States in July

49. C is the set of months that contain less than 30 days

50. $F = \{15, 16, 17, \dots, 100\}$

In Exercises 51–58, write a description of each set.

51. $A = \{1, 2, 3, 4, 5, 6, 7\}$

52. $D = \{3, 6, 9, 12, 15, 18, \dots\}$

53. $V = \{a, e, i, o, u\}$

54. $S = \{$Bashful, Doc, Dopey, Grumpy, Happy, Sleepy, Sneezy$\}$

55. $T = \{$oak, maple, elm, pine, ...$\}$

56. $E = \{x \mid x \in N \text{ and } 4 \le x < 11\}$

57. $S = \{$Spring, Summer, Fall, Winter$\}$

58. $B = \{$John Lennon, Ringo Starr, Paul McCartney, George Harrison$\}$

▲ The Beatles

In Exercises 59–62, use the following list, which shows the 10 corporations with the highest favorable consumer opinion ratings as of December 2005. Let the 10 corporations in the list represent the universal set.

Corporation	Percent of Consumers with a Favorable Opinion
1. Johnson & Johnson	91
2. Google	91
3. Home Depot	90
4. Target	85
5. Coca-Cola	85
6. Toyota	84
7. Microsoft	83
8. Southwest Airlines	83
9. United Airlines	78
10. McDonald's	74

Source: Pew Research Center

Use the list to represent each set in roster form.

59. $\{x \mid x$ is a corporation in which at least 90 percent of consumers reported having a favorable opinion$\}$

60. $\{x \mid x$ is a corporation in which fewer than 80 percent of consumers reported having a favorable opinion$\}$

61. $\{x \mid x$ is a corporation in which between 75 and 80 percent of consumers reported having a favorable opinion$\}$

62. $\{x \mid x$ is a corporation in which between 80 and 95 percent of consumers reported having a favorable opinion$\}$

In Exercises 63–66, use the following graph which shows the number of Internal Revenue Service (IRS) audits of individual taxpayers, in millions, for the years 1996–2004.

Audits on the Rise

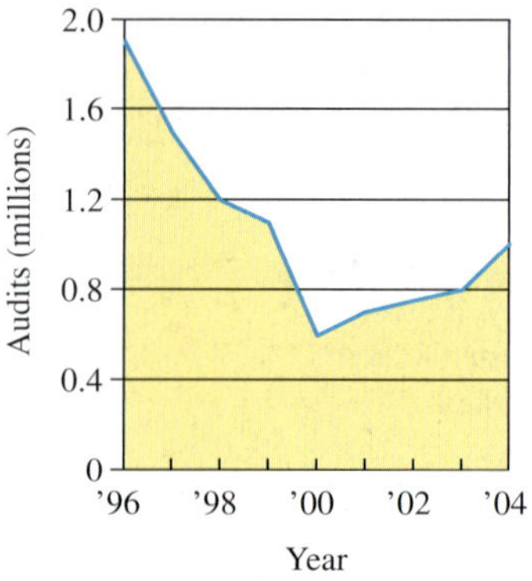

Source: Internal Revenue Service

Use the graph to represent each set in roster form.

63. The set of years in which the number of IRS audits of individual taxpayers exceeded 1 million

64. The set of years in which the number of IRS audits of individual taxpayers were between 0.5 million and 2 million

65. The set of years in which the number of IRS audits of individual taxpayers were fewer than 1.4 million

66. The set of years in which the number of IRS audits of individual taxpayers exceeded 2 million

In Exercises 67–74, state whether each statement is true or false. If false, give the reason.

67. $\{e\} \in \{a, e, i, o, u\}$

68. $b \in \{a, b, c, d, e, f\}$

69. $h \in \{a, b, c, d, e, f\}$

70. Mickey Mouse $\in$ {characters created by Walt Disney}

71. $3 \notin \{x \mid x \in N$ and x is odd$\}$

72. Maui $\in$ {capital cities in the United States}

73. *Titanic* $\in$ {top 10 motion pictures with the greatest revenues}

74. $2 \in \{x \mid x$ is an odd natural number$\}$

In Exercises 75–78, for the sets $A = \{2, 4, 6, 8\}$, $B = \{1, 3, 7, 9, 13, 21\}$, $C = \{\ \}$, and $D = \{\#, \&, \%, \square, *\}$.

75. Determine $n(A)$.

76. Determine $n(B)$.

77. Determine $n(C)$.

78. Determine $n(D)$.

In Exercises 79–84, determine whether the pairs of sets are equal, equivalent, both, or neither.

79. $A = \{\text{algebra, geometry, trigonometry}\}$,
$B = \{\text{geometry, trigonometry, algebra}\}$

80. $A = \{7, 9, 10\}$, $B = \{a, b, c\}$

81. $A = \{\text{grapes, apples, oranges}\}$,
$B = \{\text{grapes, peaches, apples, oranges}\}$

82. A is the set of Siamese cats.
B is the set of cats.

83. A is the set of letters in the word *tap*.
B is the set of letters in the word *ant*.

84. A is the set of states.
B is the set of state capitals.

PROBLEM SOLVING

85. Set-builder notation is often more versatile and efficient than listing a set in roster form. This versatility is illustrated with the two sets.

$$A = \{x \mid x \in N \text{ and } x > 2\}$$
$$B = \{x \mid x > 2\}$$

a) Write a description of set A and set B.

b) Explain the difference between set A and set B. (*Hint:* Is $4\frac{1}{2} \in A$? Is $4\frac{1}{2} \in B$?)

c) Write set A in roster form.

d) Can set B be written in roster form? Explain your answer.

86. Start with sets

$$A = \{x \mid 2 < x \le 5 \text{ and } x \in N\}$$

and

$$B = \{x \mid 2 < x \le 5\}$$

a) Write a description of set A and set B.

b) Explain the difference between set A and set B.

c) Write set A in roster form.

d) Can set B be written in roster form? Explain your answer.

*A cardinal number answers the question "How many?" An **ordinal number** describes the relative position that an element occupies. For example, Molly's desk is the third desk from the aisle.*

In Exercises 87–90, determine whether the number used is a cardinal number or an ordinal number.

87. J. K. Rowling has written 7 Harry Potter books.

▲ J. K. Rowling

88. Study the chart on page 25 in the book.

89. Lincoln was the sixteenth president of the United States.

90. Emily paid $35 for her new blouse.

91. Describe three sets of which you are a member.

92. Describe three sets that have no members.

93. Write a short paragraph explaining why the universal set and the empty set are necessary in the study of sets.

CHALLENGE PROBLEM/GROUP ACTIVITY

94. a) In a given exercise, a universal set is not specified, but we know that actor Orlando Bloom is a member of the universal set. Describe five different possible universal sets of which Orlando Bloom is a member.

b) Write a description of one set that includes all the universal sets in part (a).

INTERNET/RESEARCH ACTIVITY

95. Georg Cantor is recognized as the founder and a leader in the development of set theory. Do research and write a paper on his life and his contributions to set theory and to the field of mathematics. References include history of mathematics books, encyclopedias, and the Internet.

2.2 SUBSETS

▲ The set of occupations contains firefighters.

Consider the following sets. Set A = {architect, firefighter, mail carrier}. Set B = {architect, engineer, firefighter, mail carrier, teacher}. Note that each element of set A is also an element of set B. In this section, we will discuss how to illustrate the relationship between two sets, A and B, such that each element of set A is also an element of set B.

In our complex world, we often break larger sets into smaller more manageable sets, called *subsets*. For example, consider the set of people in your class. Suppose we categorize the set of people in your class according to the first letter of their last name (the A's, B's, C's, etc.). When we do so, each of these sets may be considered a subset of the original set. Each of these subsets can be separated further. For example, the set of people whose last name begins with the letter A can be categorized as either male or female or by their age. Each of these collections of people is also a subset. A given set may have many different subsets.

Set A is a **subset** of set B, symbolized by $A \subseteq B$, if and only if all the elements of set A are also elements of set B.

The symbol $A \subseteq B$ indicates that "set A is a subset of set B." The symbol $\nsubseteq$ is used to indicate "is not a subset." Thus, $A \nsubseteq B$ indicates that set A is not a subset of set B. *To show that set A is not a subset of set B, we must find at least one element of set A that is not an element of set B.*

EXAMPLE 1 *A Subset?*

Determine whether set A is a subset of set B.

a) A = {marigold, pansy, geranium}
   B = {marigold, pansy, begonia, geranium}

b) $A = \{2, 3, 4, 5\}$
   $B = \{2, 3\}$

c) $A = \{x \mid x \text{ is a yellow fruit}\}$
   $B = \{x \mid x \text{ is a red fruit}\}$

d) A = {vanilla, chocolate, rocky road}
   B = {chocolate, vanilla, rocky road}

SOLUTION

a) All the elements of set A are contained in set B, so $A \subseteq B$.

b) The elements 4 and 5 are in set A but not in set B, so $A \nsubseteq B$ (A is not a subset of B). In this example, however, all the elements of set B are contained in set A; therefore, $B \subseteq A$.

c) There are fruits, such as bananas, that are in set A that are not in set B, so $A \nsubseteq B$.

d) All the elements of set A are contained in set B, so $A \subseteq B$. Note also that $B \subseteq A$. In fact, set A = set B. ●

Proper Subsets

> Set A is a **proper subset** of set B, symbolized by $A \subset B$, if and only if all the elements of set A are elements of set B and set $A \neq$ set B (that is, set B must contain at least one element not in set A).

Consider the sets $A = \{\text{red, blue, yellow}\}$ and $B = \{\text{red, orange, yellow, green, blue, violet}\}$. Set A is a *subset* of set B, $A \subseteq B$, because every element of set A is also an element of set B. Set A is also a *proper subset* of set B, $A \subset B$, because set A and set B are not equal. Now consider $C = \{\text{car, bus, train}\}$ and $D = \{\text{train, car, bus}\}$. Set C is a subset of set D, $C \subseteq D$, because every element of set C is also an element of set D. Set C, however, is not a proper subset of set D, $C \not\subset D$, because set C and set D are equal sets.

EXAMPLE 2 *A Proper Subset?*

Determine whether set A is a proper subset of set B.

a) $A = \{\text{jazz, pop, hip hop}\}$
   $B = \{\text{classical, jazz, pop, rap, hip hop}\}$

b) $A = \{a, b, c, d\}$ $B = \{a, c, b, d\}$

SOLUTION

a) All the elements of set A are contained in set B, and sets A and B are not equal; thus, $A \subset B$.

b) Set A = set B, so $A \not\subset B$. (However, $A \subseteq B$.)

Every set is a subset of itself, but no set is a proper subset of itself. For all sets A, $A \subseteq A$, but $A \not\subset A$. For example, if $A = \{1, 2, 3\}$, then $A \subseteq A$ because every element of set A is contained in set A, but $A \not\subset A$ because set A = set A.

Let $A = \{\ \}$ and $B = \{1, 2, 3, 4\}$. Is $A \subseteq B$? To show $A \not\subseteq B$, you must find at least one element of set A that is not an element of set B. Because this cannot be done, $A \subseteq B$ must be true. Using the same reasoning, we can show that *the empty set is a subset of every set, including itself.*

EXAMPLE 3 *Element or Subset?*

Determine whether the following are true or false.

a) $3 \in \{3, 4, 5\}$

b) $\{3\} \in \{3, 4, 5\}$

c) $\{3\} \in \{\{3\}, \{4\}, \{5\}\}$

d) $\{3\} \subseteq \{3, 4, 5\}$

e) $3 \subseteq \{3, 4, 5\}$

f) $\{\ \} \subseteq \{3, 4, 5\}$

SOLUTION

a) $3 \in \{3, 4, 5\}$ is a true statement because 3 is a member of the set $\{3, 4, 5\}$.

b) $\{3\} \in \{3, 4, 5\}$ is a false statement because $\{3\}$ is a set, and the set $\{3\}$ is not an element of the set $\{3, 4, 5\}$.

DID YOU KNOW?

Rainbows

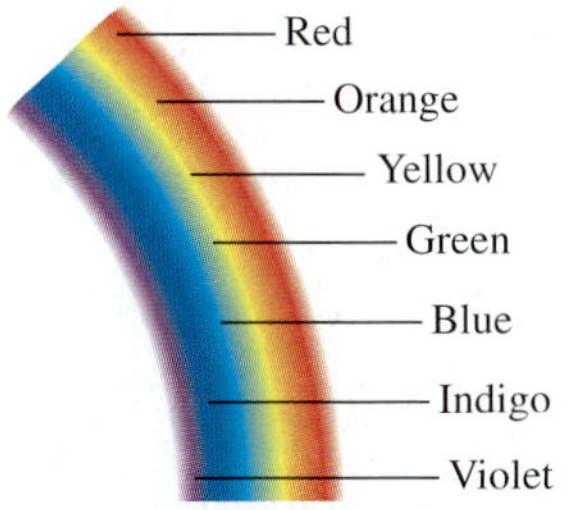

Colors of primary rainbow

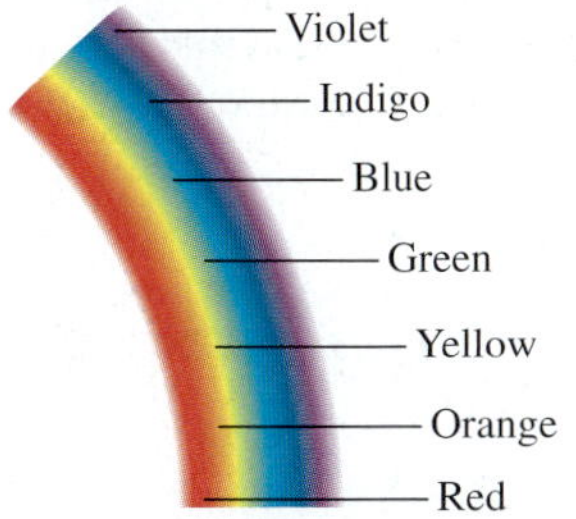

Colors of secondary rainbow

Most rainbows we see are primary rainbows, but there are rare moments when a second, fainter rainbow can be seen behind the first. In this secondary rainbow, the light pattern has been reversed. Both rainbows contain the same set of colors, so each set of colors is a subset of the other.

c) $\{3\} \in \{\{3\}, \{4\}, \{5\}\}$ is a true statement because $\{3\}$ is an element in the set. The elements of the set $\{\{3\}, \{4\}, \{5\}\}$ are themselves sets.

d) $\{3\} \subseteq \{3, 4, 5\}$ is a true statement because every element of the first set is an element of the second set.

e) $3 \subseteq \{3, 4, 5\}$ is a false statement because the 3 is not in braces, so it is not a set and thus cannot be a subset. The 3 is an element of the set as indicated in part (a).

f) $\{\ \} \subseteq \{3, 4, 5\}$ is a true statement because the empty set is a subset of every set.

Number of Subsets

How many distinct subsets can be made from a given set? The empty set has no elements and has exactly one subset, the empty set. A set with one element has two subsets. A set with two elements has four subsets. A set with three elements has eight subsets. This information is illustrated in Table 2.1. How many subsets will a set with four elements contain?

The **number of distinct subsets** of a finite set A is 2^n, where n is the number of elements in set A.

Table 2.1 Number of Subsets

Set	Subsets	Number of Subsets
$\{\ \}$	$\{\ \}$	$1 = 2^0$
$\{a\}$	$\{a\}$ $\{\ \}$	$2 = 2^1$
$\{a, b\}$	$\{a, b\}$ $\{a\}, \{b\}$ $\{\ \}$	$4 = 2 \times 2 = 2^2$
$\{a, b, c\}$	$\{a, b, c\}$ $\{a, b\}, \{a, c\}, \{b, c\}$ $\{a\}, \{b\}, \{c\}$ $\{\ \}$	$8 = 2 \times 2 \times 2 = 2^3$

By continuing this table with larger and larger sets, we can develop a general formula for finding the number of distinct subsets that can be made from any given set.

EXAMPLE 4 *Distinct Subsets*

a) Determine the number of distinct subsets for the set {S, L, E, D}.

b) List all the distinct subsets for the set {S, L, E, D}.

c) How many of the distinct subsets are proper subsets?

SOLUTION

a) Since the number of elements in the set is 4, the number of distinct subsets is $2^4 = 2 \times 2 \times 2 \times 2 = 16$.

b)

{S, L, E, D}	{S, L, E}	{S, L}	{S}	{ }
	{S, L, D}	{S, E}	{L}	
	{S, E, D}	{S, D}	{E}	
	{L, E, D}	{L, E}	{D}	
		{L, D}		
		{E, D}		

c) There are 15 proper subsets. Every subset except {S, L, E, D} is a proper subset.

EXAMPLE 5 *Variations of Pizza*

Sharon Bhatt is going to purchase a pizza at Pizza Hut. To her cheese pizza, she can add any of the following toppings: pepperoni, sausage, onions, mushrooms, anchovies, and ham. How many different variations of the pizza and toppings can be made?

SOLUTION Sharon can order the cheese pizza with no extra toppings, any one topping, any two toppings, any three toppings, and so on, up to six toppings. One technique used in problem solving is to consider similar problems that you have solved previously. If you think about this problem, you will realize that it is the same as asking how many distinct subsets can be made from a set with six elements. The number of different variations of the pizza is the same as the number of possible subsets of a set that has six elements. There are 2^6 or 64 possible subsets of a set with six elements. Therefore, there are 64 possible variations of the pizza and toppings.

DID YOU KNOW?

The Ladder of Life

Scientists use sets to classify and categorize knowledge. In biology, the science of classifying all living things is called *taxonomy*. More than 2000 years ago, Aristotle formalized animal classification with his "ladder of life": higher animals, lower animals, higher plants, lower plants. Today, living organisms are classified into six kingdoms (or sets) called animalia, plantae, archaea, eubacteria, fungi, and protista. Even more general groupings of living things are made according to shared characteristics. The groupings, from most general to most specific, are kingdom, phylum, class, order, family, genus, and species. For example, a zebra, *Equus burchelli*, is a member of the genus *Equus*, as is the horse, *Equus caballus*. Both the zebra and the horse are members of the universal set called the kingdom of animals and the same family, Equidae; they are members of different species (*E. burchelli* and *E. caballus*), however.

SECTION 2.2 EXERCISES

CONCEPT/WRITING EXERCISES

In Exercises 1–6, answer each question with a complete sentence.

1. What is a subset?
2. What is a proper subset?
3. Explain the difference between a subset and a proper subset.
4. Write the formula for determining the number of distinct subsets for a set with n distinct elements.
5. Write the formula for determining the number of distinct proper subsets for a set with n distinct elements.
6. Can any set be a proper subset of itself? Explain.

PRACTICE THE SKILLS

In Exercises 7–24, answer true or false. If false, give the reason.

7. Spanish $\subseteq$ {Spanish, French, Latin, German, Italian}
8. { } $\in$ {salt, pepper, basil, garlic powder}
9. { } $\subseteq$ {table, chair, sofa}
10. red $\subset$ {red, green, blue}
11. $5 \notin \{2, 4, 6\}$
12. {engineer, social worker} $\subseteq$ {physician, attorney, engineer, teacher}
13. $\{\ \} = \{\emptyset\}$
14. {motorboat, kayak} $\subseteq$ {kayak, fishing boat, sailboat, motorboat}
15. $\emptyset = \{\ \}$
16. $0 = \{\ \}$
17. $\{0\} = \emptyset$
18. $\{3, 8, 11\} \subseteq \{3, 8, 11\}$
19. {swimming} $\in$ {sailing, waterskiing, swimming}
20. $\{3, 5, 9\} \not\subseteq \{3, 9, 5\}$
21. $\{\ \} \subseteq \{\ \}$
22. $\{1\} \in \{\{1\}, \{2\}, \{3\}\}$
23. {Panasonic, Sharp, Pioneer} $\subset$ {Sharp, Panasonic, Pioneer}
24. $\{b, a, t\} \subseteq \{t, a, b\}$

In Exercises 25–32, determine whether $A = B$, $A \subseteq B$, $B \subseteq A$, $A \subset B$, $B \subset A$ or if none of these answers applies. (There may be more than one answer.)

25. $A =$ {Chevrolet, Pontiac, Honda, Lexus}
    $B =$ {Chevrolet, Lexus}
26. $A = \{x \mid x \in N \text{ and } x < 6\}$
    $B = \{x \mid x \in N \text{ and } 1 \leq x \leq 5\}$
27. Set A is the set of states that border the Atlantic Ocean. Set B is the set of states east of the Mississippi River.
28. $A = \{1, 3, 5, 7, 9\}$
    $B = \{3, 9, 5, 7, 6\}$
29. $A = \{x \mid x$ is a brand of ice cream$\}$
    $B =$ {Breyers, Ben & Jerry's, Häagen-Dazs}
30. $A = \{x \mid x$ is a sport that uses a ball$\}$
    $B =$ {basketball, soccer, tennis}
31. Set A is the set of natural numbers between 2 and 7. Set B is the set of natural numbers greater than 2 and less than 7.
32. Set A is the set of all exercise classes offered at Gold's Gym. Set B is the set of aerobic exercise classes offered at Gold's Gym.

In Exercises 33–38, list all the subsets of the sets given.

33. $D = \emptyset$

34. $A = \{\bigcirc\}$

35. $B = \{\text{pen, pencil}\}$

36. $C = \{\text{steak, pork, chicken}\}$

PROBLEM SOLVING

37. For set $A = \{a, b, c, d\}$,

   a) list all the subsets of set A.

   b) state which of the subsets in part (a) are not proper subsets of set A.

38. A set contains nine elements.

   a) How many subsets does it have?

   b) How many proper subsets does it have?

In Exercises 39–50, if the statement is true for all sets A and B, write "true." If it is not true for all sets A and B, write "false." Assume that $A \neq \emptyset$, $U \neq \emptyset$, and $A \subset U$.

39. If $A \subseteq B$, then $A \subset B$.

40. If $A \subset B$, then $A \subseteq B$.

41. $A \subseteq A$

42. $A \subset A$

43. $\emptyset \subset A$

44. $\emptyset \subseteq A$

45. $A \subseteq U$

46. $\emptyset \subset \emptyset$

47. $\emptyset \subset U$

48. $U \subseteq \emptyset$

49. $\emptyset \subseteq \emptyset$

50. $U \subset \emptyset$

51. ***Building a House*** The Jacobsens are planning to build a house in a new development. They can either build the base model offered by the builder or add any of the following options: deck, hot tub, security system, hardwood flooring. How many different variations of the house are possible?

52. ***Salad Toppings*** Donald Wheeler is ordering a salad at a Ruby Tuesday restaurant. He can purchase a salad consisting of just lettuce, or he can add any of the following items: cucumber, onion, tomato, carrot, green pepper, olive, mushroom. How many different variations of a salad are possible?

53. ***Telephone Features*** A customer with Verizon can order telephone service with some, all, or none of the following features: call waiting, call forwarding, caller identification, three-way calling, voice mail, fax line. How many different variations of the set of features are possible?

54. ***Hamburger Variations*** Customers ordering hamburgers at Vic and Irv's Hamburger stand are always asked, "What do you want on it?" The choices are ketchup, mustard, relish, hot sauce, onions, lettuce, tomato. How many different variations are there for ordering a hamburger?

55. If $E \subseteq F$ and $F \subseteq E$, what other relationship exists between E and F? Explain.

56. How can you determine whether the set of boys is equivalent to the set of girls at a roller-skating rink?

57. For the set $D = \{a, b, c\}$

   a) is a an element of set D? Explain.

   b) is c a subset of set D? Explain.

   c) is $\{a, b\}$ a subset of set D? Explain.

CHALLENGE PROBLEM/GROUP ACTIVITY

58. ***Hospital Expansion*** A hospital has four members on the board of directors: Arnold, Benitez, Cathy, and Dominique.

   a) When the members vote on whether to add a wing to the hospital, how many different ways can they vote (abstentions are not allowed)? For example, Arnold—yes, Benitez—no, Cathy—no, and Dominique—yes is one of the many possibilities.

   b) Make a listing of all the possible outcomes of the vote. For example, the vote described in part (a) could be represented as (YNNY).

   c) How many of the outcomes given in part (b) would result in a majority supporting the addition of a wing to the hospital? That is, how many of the outcomes have three or more Y's?

RECREATIONAL MATHEMATICS

59. How many elements must a set have if the number of proper subsets of the set is $\frac{1}{2}$ of the total number of subsets of the set?

60. If $A \subset B$ and $B \subset C$, must $A \subset C$?

61. If $A \subset B$ and $B \subseteq C$, must $A \subset C$?

62. If $A \subseteq B$ and $B \subseteq C$, must $A \subset C$?

INTERNET/RESEARCH ACTIVITY

63. On page 57, we discussed the ladder of life. Do research and indicate all the different classifications in the Linnaean system, from most general to the most specific, in which a koala belongs.

2.3 VENN DIAGRAMS AND SET OPERATIONS

▲ Some soft drinks contain caffeine, some soft drinks contain sugar, and some soft drinks contain caffeine *and* sugar.

Suppose your boss sends you to the store to purchase a soft drink that contains caffeine *and* sugar. You were a bit distracted when he made his request and thought he said to purchase a soft drink that contains caffeine *or* sugar. Which soft drinks are contained in the set of soft drinks that contain caffeine *and* sugar? Which soft drinks are contained in the set of soft drinks that contain caffeine *or* sugar? These two questions are quite different. The first involves soft drinks joined by the word *and*. The second involves soft drinks joined by the word *or*. In this section, you will learn how to illustrate these and other set relationships.

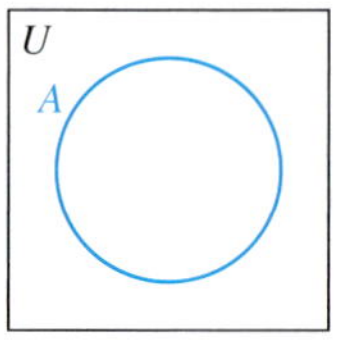

Figure 2.1

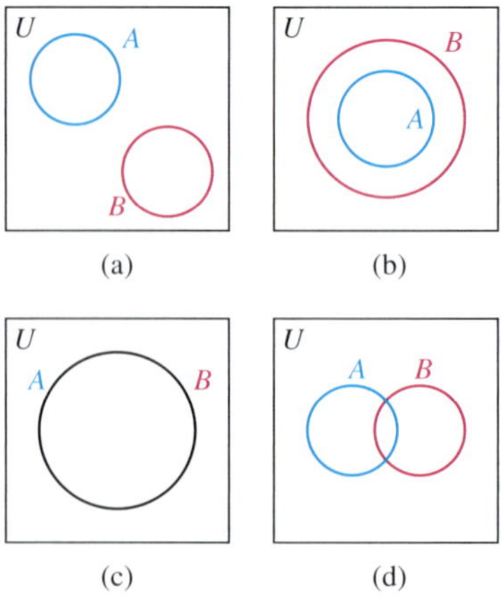

Figure 2.2

A useful technique for illustrating set relationships is the Venn diagram, named for English mathematician John Venn (1834–1923). Venn invented the diagrams and used them to illustrate ideas in his text on symbolic logic, published in 1881.

In a Venn diagram, a rectangle usually represents the universal set, U. The items inside the rectangle may be divided into subsets of the universal set. The subsets are usually represented by circles. In Fig. 2.1, the circle labeled A represents set A, which is a subset of the universal set.

Two sets may be represented in a Venn diagram in any of four different ways, as shown in Fig. 2.2. Two sets A and B are ***disjoint*** when they have no elements in common. Two disjoint sets A and B are illustrated in Fig. 2.2(a). If set A is a proper subset of set B, $A \subset B$, the two sets may be illustrated as in Fig. 2.2(b). If set A contains exactly the same elements as set B, that is, $A = B$, the two sets may be illustrated as in Fig. 2.2(c). Two sets A and B with some elements in common are shown in Fig. 2.2(d), which is regarded as the most general form of a Venn diagram.

If we label the regions of the diagram in Fig. 2.2(d) using I, II, III, and IV, we can illustrate the four possible cases with this one diagram, Fig. 2.3, on page 61.

CASE 1: DISJOINT SETS When sets A and B are disjoint, they have no elements in common. Therefore, region II of Fig. 2.3 is empty.

CASE 2: SUBSETS When $A \subseteq B$, every element of set A is also an element of set B. Thus, there can be no elements in region I of Fig. 2.3. If $B \subseteq A$, however, then region III of Fig. 2.3 is empty.

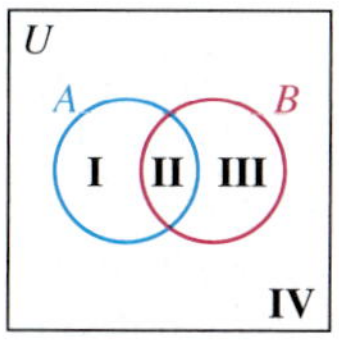

Figure 2.3

CASE 3: EQUAL SETS When set A = set B, all the elements of set A are elements of set B and all the elements of set B are elements of set A. Thus, regions I and III of Fig. 2.3 are empty.

CASE 4: OVERLAPPING SETS When sets A and B have elements in common, those elements are in region II of Fig. 2.3. The elements that belong to set A but not to set B are in region I. The elements that belong to set B but not to set A are in region III.

In each of the four cases, any element not belonging to set A or set B is placed in region IV.

Venn diagrams will be helpful in understanding set operations. The basic operations of arithmetic are $+$, $-$, $\times$, and $\div$. When we see these symbols, we know what procedure to follow to determine the answer. Some of the operations in set theory are $'$, $\cap$, and $\cup$. They represent complement, intersection, and union, respectively.

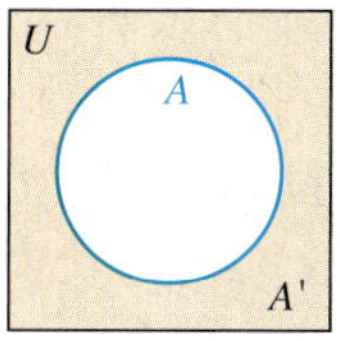

Figure 2.4

Complement

The **complement** of set A, symbolized by A', is the set of all the elements in the universal set that are not in set A.

In Fig. 2.4, the shaded region outside of set A within the universal set represents the complement of set A, or A'.

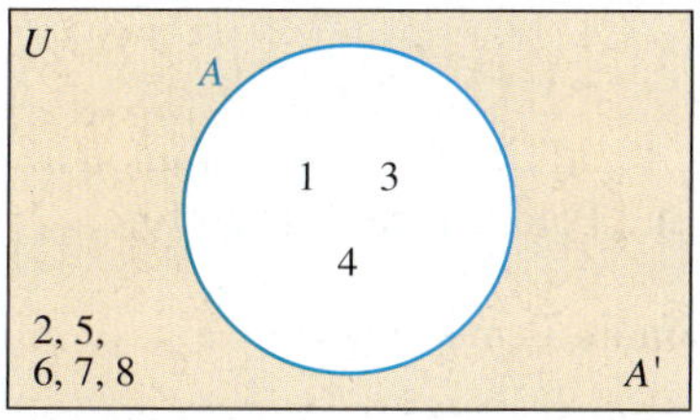

Figure 2.5

EXAMPLE 1 *A Set and Its Complement*

Given

$$U = \{1, 2, 3, 4, 5, 6, 7, 8\} \text{ and } A = \{1, 3, 4\}$$

find A' and illustrate the relationship among sets U, A, and A' in a Venn diagram.

SOLUTION The elements in U that are not in set A are 2, 5, 6, 7, 8. Thus, $A' = \{2, 5, 6, 7, 8\}$. The Venn diagram is illustrated in Fig. 2.5.

Intersection

The word *intersection* brings to mind the area common to two crossing streets. The red car in the figure is in the intersection of the two streets. The set operation intersection is defined as follows.

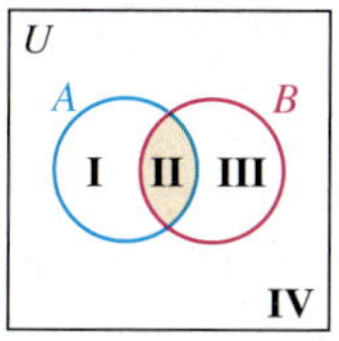

Figure 2.6

The **intersection** of sets A and B, symbolized by $A \cap B$, is the set containing all the elements that are common to both set A and set B.

The shaded region, region II, in Fig. 2.6 represents the intersection of sets A and B.

EXAMPLE 2 *Sets with Overlapping Regions*

Let the universal set, U, represent the 50 states in the United States. Let set A represent the set of states with a population of more than 10 million people as of 2005. Let set B represent the set of states that have at least one city with a population of more than 1 million people, as of 2005 (see the table). Draw a Venn diagram illustrating the relationship between set A and set B.

States With a Population of More Than 10 Million People	States with at Least One City with a Population of More Than 1 Million People
California	California
Texas	Texas
New York	New York
Florida	Illinois
Illinois	Pennsylvania
Pennsylvania	Arizona
Ohio	
Michigan	

Source: Bureau of the U.S. Census, U.S. Dept. of Commerce

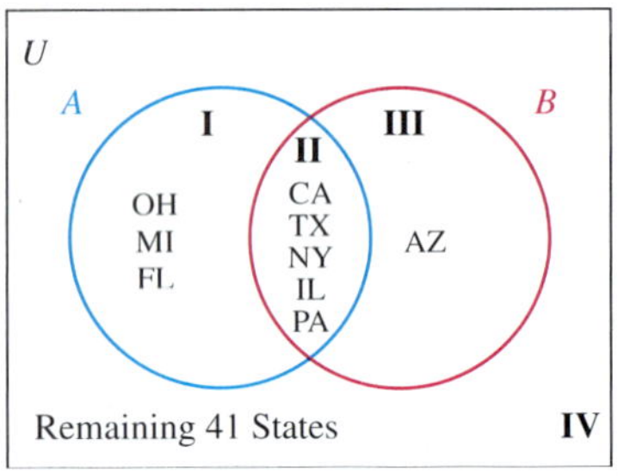

Figure 2.7

SOLUTION First determine the intersection of sets A and B. The states common to both sets are California, Texas, New York, Illinois, and Pennsylvania. Therefore,

$$A \cap B = \{\text{California, Texas, New York, Illinois, Pennsylvania}\}$$

Place these elements in region II of Fig. 2.7. Complete region I by determining the elements in set A that have not been placed in region II. Therefore, Ohio, Michigan, and Florida are placed in region I. Complete region III by determining the elements in set B that have not been placed in region II. Thus, Arizona is placed in region III. Finally, place those elements in U that are not in either set within the rectangle but are outside both circles. This group includes the remaining 41 states, which are placed in region IV.

EXAMPLE 3 *The Intersection of Sets*

Given

$$U = \{1, 2, 3, 4, 5, 6, 7, 8, 9, 10\}$$
$$A = \{1, 2, 3, 8\}$$
$$B = \{1, 3, 6, 7, 9\}$$
$$C = \{\ \}$$

find

a) $A \cap B$. b) $A \cap C$. c) $A' \cap B$. d) $(A \cap B)'$.

SOLUTION

a) $A \cap B = \{1, 2, 3, 8\} \cap \{1, 3, 6, 7, 9\} = \{1, 3\}$. The elements common to both set A and set B are 1 and 3.

b) $A \cap C = \{1, 2, 3, 8\} \cap \{\ \} = \{\ \}$. There are no elements common to both set A and set C.

c) $A' = \{4, 5, 6, 7, 9, 10\}$

$A' \cap B = \{4, 5, 6, 7, 9, 10\} \cap \{1, 3, 6, 7, 9\}$
$= \{6, 7, 9\}$

d) To find $(A \cap B)'$, first determine $A \cap B$.

$A \cap B = \{1, 3\}$ from part (a)

$(A \cap B)' = \{1, 3\}' = \{2, 4, 5, 6, 7, 8, 9, 10\}$

Union

The word *union* means to unite or join together, as in marriage, and that is exactly what is done when we perform the operation of union.

The **union** of set A and set B, symbolized by $A \cup B$, is the set containing all the elements that are members of set A or of set B (or of both sets).

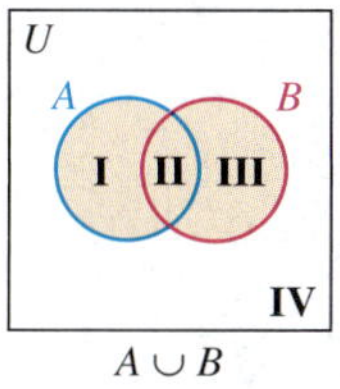

Figure 2.8

The three shaded regions of Fig. 2.8, regions I, II, and III, together represent the union of sets A and B. If an element is common to both sets, it is listed only once in the union of the sets.

EXAMPLE 4 *Determining Sets from a Venn Diagram*

Use the Venn diagram in Fig. 2.9 to determine the following sets.

a) U b) A c) B' d) $A \cap B$
e) $A \cup B$ f) $(A \cup B)'$ g) $n(A \cup B)$

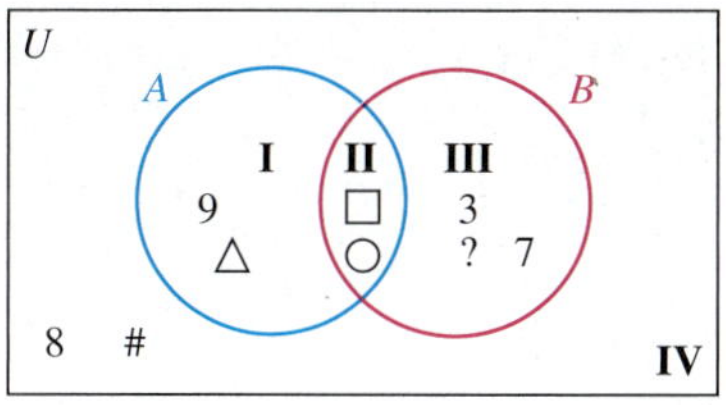

Figure 2.9

SOLUTION

a) The universal set consists of all the elements within the rectangle, that is, the elements in regions I, II, III, and IV. Thus, $U = \{9, \triangle, \square, \bigcirc, 3, 7, ?, \#, 8\}$.

b) Set A consists of the elements in regions I and II. Thus, $A = \{9, \triangle, \square, \bigcirc\}$.

c) B' consists of the elements outside set B, or the elements in regions I and IV. Thus, $B' = \{9, \triangle, \#, 8\}$.

d) $A \cap B$ consists of the elements that belong to both set A and set B (region II). Thus, $A \cap B = \{\square, \bigcirc\}$.

e) $A \cup B$ consists of the elements that belong to set A or set B (regions I, II, or III). Thus, $A \cup B = \{9, \triangle, \square, \bigcirc, 3, 7, ?\}$.

f) $(A \cup B)'$ consists of the elements in U that are not in $A \cup B$. Thus, $(A \cup B)' = \{\#, 8\}$.

g) $n(A \cup B)$ represents the *number of elements* in the union of sets A and B. Thus, $n(A \cup B) = 7$, as there are seven elements in the union of sets A and B.

EXAMPLE 5 ***The Union of Sets***

Given

$$\begin{aligned} U &= \{1, 2, 3, 4, 5, 6, 7, 8, 9, 10\} \\ A &= \{1, 2, 4, 6\} \\ B &= \{1, 3, 6, 7, 9\} \\ C &= \{\ \} \end{aligned}$$

determine each of the following.

a) $A \cup B$ b) $A \cup C$ c) $A' \cup B$ d) $(A \cup B)'$

SOLUTION

a) $A \cup B = \{1, 2, 4, 6\} \cup \{1, 3, 6, 7, 9\} = \{1, 2, 3, 4, 6, 7, 9\}$

b) $A \cup C = \{1, 2, 4, 6\} \cup \{\ \} = \{1, 2, 4, 6\}$. Note that $A \cup C = A$.

c) To determine $A' \cup B$, we must determine A'.

$$\begin{aligned} A' &= \{3, 5, 7, 8, 9, 10\} \\ A' \cup B &= \{3, 5, 7, 8, 9, 10\} \cup \{1, 3, 6, 7, 9\} \\ &= \{1, 3, 5, 6, 7, 8, 9, 10\} \end{aligned}$$

d) Find $(A \cup B)'$ by first determining $A \cup B$, and then find the complement of $A \cup B$.

$$\begin{aligned} A \cup B &= \{1, 2, 3, 4, 6, 7, 9\} \text{ from part (a)} \\ (A \cup B)' &= \{1, 2, 3, 4, 6, 7, 9\}' = \{5, 8, 10\} \end{aligned}$$

EXAMPLE 6 ***Union and Intersection***

Given

$$\begin{aligned} U &= \{\text{a, b, c, d, e, f, g}\} \\ A &= \{\text{a, b, e, g}\} \\ B &= \{\text{a, c, d, e}\} \\ C &= \{\text{b, e, f}\} \end{aligned}$$

determine each of the following.

a) $(A \cup B) \cap (A \cup C)$ b) $(A \cup B) \cap C'$ c) $A' \cap B'$

SOLUTION

a) $(A \cup B) \cap (A \cup C) = \{\text{a, b, c, d, e, g}\} \cap \{\text{a, b, e, f, g}\}$
$= \{\text{a, b, e, g}\}$

b) $(A \cup B) \cap C' = \{\text{a, b, c, d, e, g}\} \cap \{\text{a, c, d, g}\}$
$= \{\text{a, c, d, g}\}$

c) $A' \cap B' = \{\text{c, d, f}\} \cap \{\text{b, f, g}\}$
$= \{\text{f}\}$

The Meaning of *and* and *or*

The words *and* and *or* are very important in many areas of mathematics. We use these words in several chapters in this book, including the probability chapter. The word

and is generally interpreted to mean ***intersection***, whereas ***or*** is generally interpreted to mean ***union***. Suppose $A = \{1, 2, 3, 5, 6, 8\}$ and $B = \{1, 3, 4, 7, 9, 10\}$. The elements that belong to set A *and* set B are 1 and 3. These are the elements in the intersection of the sets. The elements that belong to set A *or* set B are 1, 2, 3, 4, 5, 6, 7, 8, 9, and 10. These are the elements in the union of the sets.

The Relationship Between $n(A \cup B)$, $n(A)$, $n(B)$, and $n(A \cap B)$

Having looked at unions and intersections, we can now determine a relationship between $n(A \cup B)$, $n(A)$, $n(B)$, and $n(A \cap B)$. Suppose set A has eight elements, set B has five elements, and $A \cap B$ has two elements. How many elements are in $A \cup B$? Let's make up some arbitrary sets that meet the criteria specified and draw a Venn diagram. If we let set $A = \{\text{a, b, c, d, e, f, g, h}\}$, then set B must contain five elements, two of which are also in set A. Let set $B = \{\text{g, h, i, j, k}\}$. We construct a Venn diagram by filling in the intersection first, as shown in Fig. 2.10. The number of elements in $A \cup B$ is 11. The elements g and h are in both sets, and if we add $n(A) + n(B)$, we are counting these elements twice.

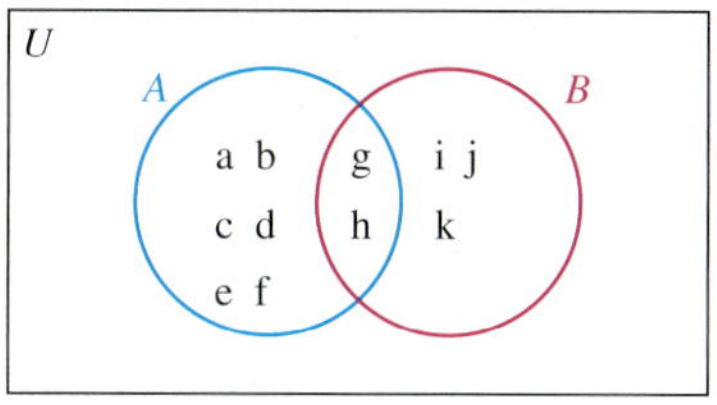

Figure 2.10

To find the number of elements in the union of sets A and B, we can add the number of elements in sets A and B and then subtract the number of elements common to both sets.

FOR ANY FINITE SETS *A* AND *B*,

$$n(A \cup B) = n(A) + n(B) - n(A \cap B)$$

EXAMPLE 7 *How Many Visitors Speak Spanish or French?*

The results of a survey of visitors at the Grand Canyon showed that 25 speak Spanish, 14 speak French, and 4 speak both Spanish and French. How many speak Spanish or French?

SOLUTION If we let set A be the set of visitors who speak Spanish and let set B be the set of visitors who speak French, then we need to determine $n(A \cup B)$. We can use the above formula to find $n(A \cup B)$.

$$\begin{aligned} n(A \cup B) &= n(A) + n(B) - n(A \cap B) \\ n(A \cup B) &= 25 + 14 - 4 \\ &= 35 \end{aligned}$$

Thus, 35 of the visitors surveyed speak either Spanish or French. ●

EXAMPLE 8 *The Number of Elements in Set A*

Of the homes listed for sale with RE/MAX, 39 have either a three-car garage or a fireplace, 31 have a fireplace, and 18 have both a three-car garage and a fireplace. How many of these homes have a three-car garage?

SOLUTION If we let set A be the set of homes with a three-car garage and set B be the set of homes with a fireplace, we need to determine $n(A)$. We are given the number of homes with either a three-car garage or a fireplace, which is $n(A \cup B)$. We are also given the number of homes with a fireplace, $n(B)$, and the number of homes that have both a three-car garage and a fireplace, $n(A \cap B)$. We can use the formula $n(A \cup B) = n(A) + n(B) - n(A \cap B)$ to solve for $n(A)$.

$$\begin{aligned} n(A \cup B) &= n(A) + n(B) - n(A \cap B) \\ 39 &= n(A) + 31 - 18 \\ 39 &= n(A) + 13 \\ 39 - 13 &= n(A) + 13 - 13 \\ 26 &= n(A) \end{aligned}$$

Thus, the number of homes listed for sale that have a three-car garage is 26.

Two other set operations are the difference of two sets and the Cartesian product. We will first discuss the difference of two sets.

Difference of Two Sets

The **difference** of two sets A and B, symbolized $A - B$, is the set of elements that belong to set A but not to set B.

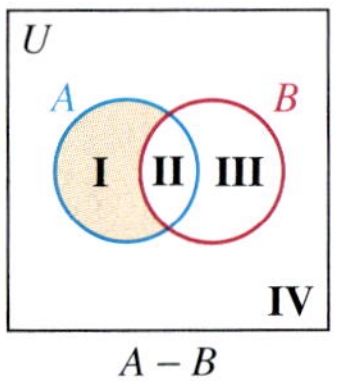

Figure 2.11

Using set-builder notation, the difference of two sets A and B is indicated by $A - B = \{x \mid x \in A \text{ and } x \notin B\}$. The shaded region, region I, in Fig. 2.11 represents the difference of two sets A and B, or $A - B$.

EXAMPLE 9 *The Difference of Two Sets*

Given

$$\begin{aligned} U &= \{a, b, c, d, e, f, g, h, i, j, k\} \\ A &= \{b, c, e, f, g, h\} \\ B &= \{a, b, c, g, i\} \\ C &= \{b, e, g\} \end{aligned}$$

determine

a) $A - B$ b) $A - C$ c) $A' - B$ d) $A - C'$

SOLUTION

a) $A - B$ is the set of elements that are in set A but not set B. The elements that are in set A but not set B are $e, f,$ and h. Therefore, $A - B = \{e, f, h\}$.

b) $A - C$ is the set of elements that are in set A but not set C. The elements that are in set A but not set C are $c, f,$ and h. Therefore, $A - C = \{c, f, h\}$.

c) To determine $A' - B$, we must first determine A'.

$$A' = \{a, d, i, j, k\}$$

$A' - B$ is the set of elements that are in set A' but not set B. The elements that are in set A' but not set B are d, j, and k. Therefore, $A' - B = \{d, j, k\}$.

d) To determine $A - C'$, we must first determine C'.

$$C' = \{a, c, d, f, h, i, j, k\}$$

$A - C'$ is the set of elements that are in set A but not set C'. The elements that are in set A but not set C' are b, e, and g. Therefore, $A - C' = \{b, e, g\}$. •

Next we discuss the Cartesian product.

Cartesian Product

The **Cartesian product** of set A and set B, symbolized by $A \times B$ and read "A cross B," is the set of all possible *ordered pairs* of the form (a, b), where $a \in A$ and $b \in B$.

To determine the ordered pairs in a Cartesian product, select the first element of set A and form an ordered pair with each element of set B. Then select the second element of set A and form an ordered pair with each element of set B. Continue in this manner until you have used each element of set A.

EXAMPLE 10 *The Cartesian Product of Two Sets*

Given $A = \{\text{orange, banana, apple}\}$ and $B = \{1, 2\}$, determine the following.

a) $A \times B$ b) $B \times A$ c) $A \times A$ d) $B \times B$

SOLUTION

a) $A \times B = \{$(orange, 1), (orange, 2) (banana, 1), (banana, 2), (apple, 1), (apple, 2)$\}$

b) $B \times A = \{$(1, orange), (1, banana), (1, apple), (2, orange), (2, banana), (2, apple)$\}$

c) $A \times A = \{$(orange, orange), (orange, banana), (orange, apple), (banana, orange), (banana, banana), (banana, apple), (apple, orange), (apple, banana), (apple, apple)$\}$

d) $B \times B = \{(1, 1), (1, 2), (2, 1), (2, 2)\}$ •

We can see from Example 10 that, in general, $A \times B \neq B \times A$. The ordered pairs in $A \times B$ are not the same as the ordered pairs in $B \times A$ because (orange, 1) $\neq$ (1, orange).

In general, if a set A has m elements and a set B has n elements, then the number of ordered pairs in $A \times B$ will be $m \times n$. In Example 10, set A contains 3 elements and set B contains 2 elements. Notice that $A \times B$ contains 3×2 or 6 ordered pairs.

SECTION 2.3 EXERCISES

CONCEPT/WRITING EXERCISES

In Exercises 1–5, use Fig. 2.2 as a guide to draw a Venn diagram that illustrates the situation described.

1. Set A and set B are disjoint sets.

2. $A \subset B$

3. $B \subset A$

4. $A = B$

5. Set A and set B are overlapping sets.

6. How do we obtain the union of two sets A and B, $A \cup B$?

7. If we are given set A, how do we obtain A complement, A'?

8. In a Venn diagram with two overlapping sets, which region(s) represents $A \cup B$?

9. How do we obtain the intersection of two sets A and B, $A \cap B$?

10. In a Venn diagram with two overlapping sets, which region(s) represents $A \cap B$?

11. a) Which set operation is the word *or* generally interpreted to mean?

    b) Which set operation is the word *and* generally interpreted to mean?

12. Give the relationship between $n(A \cup B)$, $n(A)$, $n(B)$, and $n(A \cap B)$.

13. What is the difference of two sets A and B?

14. a) What is the Cartesian product of set A and set B?

    b) If set A has m elements and set B has n elements, how many elements will be in $A \times B$?

PRACTICE THE SKILLS/PROBLEM SOLVING

15. ***National Parks*** For the sets U, A, and B, construct a Venn diagram and place the elements in the proper regions.

    U = {Badlands, Death Valley, Glacier, Grand Teton, Mammoth Cave, Mount Rainier, North Cascades, Shenandoah, Yellowstone, Yosemite}

    A = {Badlands, Glacier, Grand Teton, Mount Rainier, Yellowstone}

    B = {Death Valley, Glacier, Mammoth Cave, Mount Rainier, Yosemite}

▲ Yosemite National Park

16. ***Appliances and Electronics*** For the sets U, A, and B, construct a Venn diagram and place the elements in the proper regions.

    U = {microwave oven, washing machine, dryer, refrigerator, dishwasher, compact disc player, videocassette recorder, computer, camcorder, television}

    A = {microwave oven, washing machine, dishwasher, computer, television}

    B = {washing machine, dryer, refrigerator, compact disc player, computer, television}

17. ***Occupations*** The following table shows the fastest-growing occupations for college graduates based on employment in 2002 and the estimated employment in 2012. Let the occupations in the table represent the universal set.

Fastest-Growing Occupations for College Graduates, 2002–2012

	Employment (in thousands of jobs)	
Occupation	**2002**	**2012**
Systems analysts	186	292
Physician assistants	63	94
Medical records technicians	147	216
Software engineers, applications	394	573
Software engineers, software	281	409
Physical therapist assistants	50	73
Fitness and aerobics instructors	183	264
Database administrators	110	159
Veterinary technicians	53	76
Dental hygienists	148	212

Source: U.S. Bureau of Labor Statistics

Let A = the set of fastest-growing occupations for college graduates whose 2002 employment was at least 150,000.

Let B = the set of fastest-growing occupations for college graduates whose estimated employment in 2012 is at least 400,000.

Construct a Venn diagram illustrating the sets.

18. ***Golf Statistics*** The following table shows the number of wins at the Masters Golf Tournament and the U.S. Open for selected golfers as of March 1, 2007. Let these golfers represent the universal set.

Golfer	Masters Golf Tournament	U.S. Open
Tiger Woods	4	2
Phil Mickelson	2	0
Payne Stewart	0	2
Fred Couples	1	0
Ray Floyd	1	1
Jack Nicklaus	6	4
Arnold Palmer	4	1
Ben Hogan	2	4
V. J. Singh	1	0
Retief Goosen	0	2

Let A = the set of golfers who won the Masters Golf Tournament at least two times.

Let B = the set of golfers who won the U.S. Open at least two times.

Construct a Venn diagram illustrating the sets.

19. Let U represent the set of animals in U.S. zoos. Let A represent the set of animals in the San Diego zoo. Describe A'.

▲ San Diego Zoo

20. Let U represent the set of U.S. colleges and universities. Let A represent the set of U.S. colleges and universities in the state of Mississippi. Describe A'.

In Exercises 21–26,

U is the set of insurance companies in the U.S.

A is the set of insurance companies that offer life insurance.

B is the set of insurance companies that offer car insurance.

Describe each of the following sets in words.

21. A'

22. B'

23. $A \cup B$

24. $A \cap B$

25. $A \cap B'$

26. $A \cup B'$

In Exercises 27–32,

U is the set of furniture stores in the U.S.

A is the set of furniture stores that sell mattresses.

B is the set of furniture stores that sell outdoor furniture.

C is the set of furniture stores that sell leather furniture.

Describe the following sets.

27. $A \cap B$

28. $A \cup C$

29. $B' \cap C$

30. $A \cap B \cap C$

31. $A \cup B \cup C$

32. $A' \cup C'$

In Exercises 33–40, use the Venn diagram in Fig. 2.12 to list the set of elements in roster form.

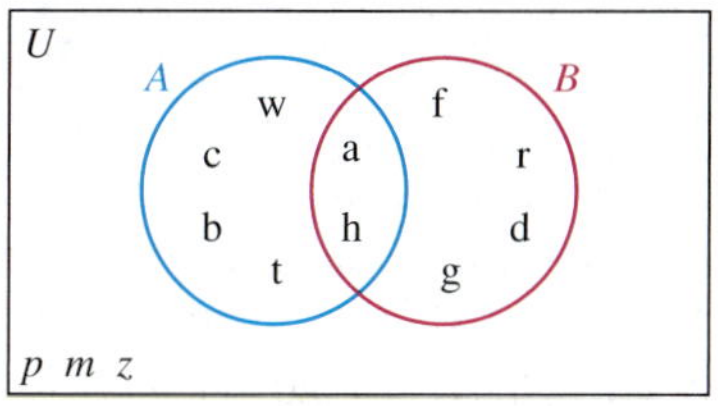

Figure 2.12

33. A

34. B

35. $A \cap B$

36. U

37. $A \cup B$

38. $(A \cup B)'$

39. $A' \cap B'$

40. $(A \cap B)'$

In Exercises 41–48, use the Venn diagram in Fig. 2.13 to list the set of elements in roster form.

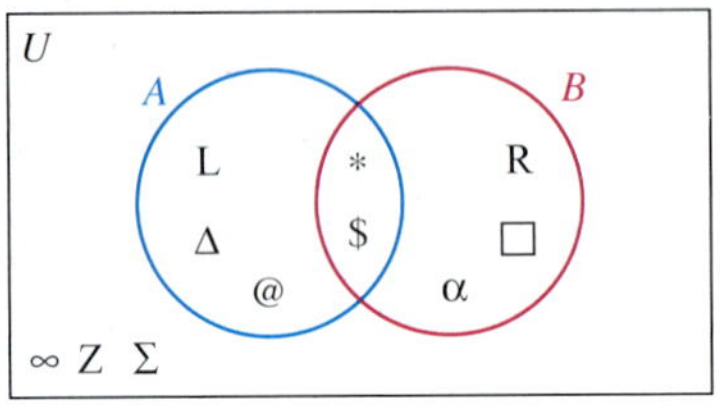

Figure 2.13

41. A

42. B

43. U

44. $A \cap B$

45. $A' \cup B$

46. $A \cup B'$

47. $A' \cap B$

48. $(A \cup B)'$

In Exercises 49–58, let

$$U = \{1, 2, 3, 4, 5, 6, 7, 8\}$$
$$A = \{1, 2, 4, 5, 8\}$$
$$B = \{2, 3, 4, 6\}$$

Determine the following.

49. $A \cup B$

50. $A \cap B$

51. B'

52. $A \cup B'$

53. $(A \cup B)'$

54. $A' \cap B'$

55. $(A \cup B)' \cap B$

56. $(A \cup B) \cap (A \cup B)'$

57. $(B \cup A)' \cap (B' \cup A')$

58. $A' \cup (A \cap B)$

In Exercises 59–68, let

$$U = \{a, b, c, d, e, f, g, h, i, j, k\}$$
$$A = \{a, c, d, f, g, i\}$$
$$B = \{b, c, d, f, g\}$$
$$C = \{a, b, f, i, j\}$$

Determine the following.

59. B'

60. $B \cup C$

61. $A \cap C$

62. $A' \cup B'$

63. $(A \cap C)'$

64. $(A \cap B) \cup C$

65. $A \cup (C \cap B)'$

66. $A \cup (C' \cup B')$

67. $(A' \cup C) \cup (A \cap B)$

68. $(C \cap B) \cap (A' \cap B)$

In Exercises 69–76, let

$$U = \{1, 2, 3, 4, 5, 6, 7, 8, 9, 10\}$$
$$A = \{1, 2, 4, 6, 9\}$$
$$B = \{1, 3, 4, 5, 8\}$$
$$C = \{4, 5, 9\}$$

Determine the following.

69. $A - B$

70. $A - C$

71. $A - B'$

72. $A' - C$

73. $(A - B)'$

74. $(A - B)' - C$

75. $C - A'$

76. $(C - A)' - B$

In Exercises 77–82, let

$$A = \{a, b, c\}$$
$$B = \{1, 2\}$$

77. Determine $A \times B$.

78. Determine $B \times A$.

79. Does $A \times B = B \times A$?

80. Determine $n(A \times B)$.

81. Determine $n(B \times A)$.

82. Does $n(A \times B) = n(B \times A)$?

PROBLEM SOLVING

In Exercises 83–96, let

$$U = \{x \mid x \in N \text{ and } x < 10\}$$
$$A = \{x \mid x \in N \text{ and } x \text{ is odd and } x < 10\}$$
$$B = \{x \mid x \in N \text{ and } x \text{ is even and } x < 10\}$$
$$C = \{x \mid x \in N \text{ and } x < 6\}$$

Determine the following.

83. $A \cap B$

84. $A \cup B$

85. $A' \cup B$

86. $(B \cup C)'$

87. $A \cap C'$

88. $A \cap B'$

89. $(B \cap C)'$

90. $(A \cup C) \cap B$

91. $(C' \cup A) \cap B$

92. $(C \cap B) \cup A$

93. $(A \cap B)' \cup C$

94. $(A' \cup C) \cap B$

95. $(A' \cup B') \cap C$

96. $(A' \cap C) \cup (A \cap B)$

97. When will a set and its complement be disjoint? Explain and give an example.

98. When will $n(A \cap B) = 0$? Explain and give an example.

99. ***Pet Ownership*** The results of a survey of customers at PetSmart showed that 27 owned dogs, 38 owned cats, and 16 owned both dogs and cats. How many people owned either a dog or a cat?

100. ***Chorus and Band*** At Henniger High School, 46 students sang in the chorus or played in the stage band, 30 students played in the stage band, and 4 students sang in the chorus and played in the stage band. How many students sang in the chorus?

101. Consider the formula

$$n(A \cup B) = n(A) + n(B) - n(A \cap B)$$

a) Show that this relation holds for $A = \{a, b, c, d\}$ and $B = \{b, d, e, f, g, h\}$.

b) Make up your own sets A and B, each consisting of at least six elements. Using these sets, show that the relation holds.

c) Use a Venn diagram and explain why the relation holds for any two sets A and B.

102. The Venn diagram in Fig. 2.14 shows a technique of labeling the regions to indicate membership of elements in a particular region. Define each of the four regions with a set statement. (*Hint:* $A \cap B'$ defines region I.)

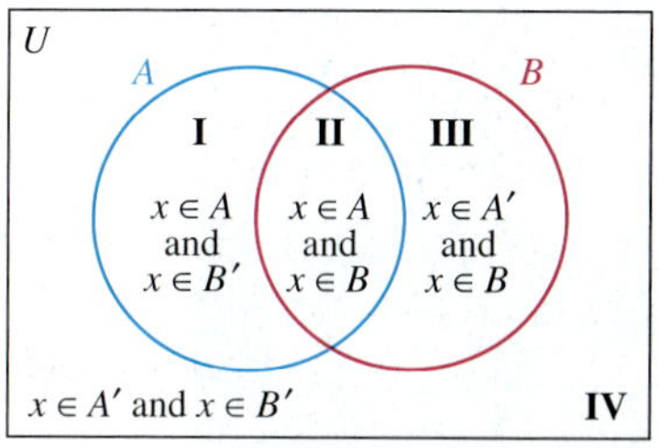

Figure 2.14

In Exercises 103–112, let $U = \{0, 1, 2, 3, 4, 5, \ldots\}$, $A = \{1, 2, 3, 4, \ldots\}$, $B = \{4, 8, 12, 16, \ldots\}$, *and* $C = \{2, 4, 6, 8, \ldots\}$. *Determine the following.*

103. $A \cup B$

104. $A \cap B$

105. $B \cap C$

106. $B \cup C$

107. $A \cap C$

108. $A' \cap C$

109. $B' \cap C$

110. $(B \cup C)' \cup C$

111. $(A \cap C) \cap B'$

112. $U' \cap (A \cup B)$

CHALLENGE EXERCISES/GROUP ACTIVITIES

In Exercises 113–120, determine whether the answer is $\emptyset$, *A, or U. (Assume* $A \neq \emptyset$, $A \neq U$.)

113. $A \cup A'$

114. $A \cap A'$

115. $A \cup \emptyset$

116. $A \cap \emptyset$

117. $A' \cup U$

118. $A \cap U$

119. $A \cup U$

120. $A \cup U'$

In Exercises 121–126, determine the relationship between set A and set B if

121. $A \cap B = B$.

122. $A \cup B = B$.

123. $A \cap B = \emptyset$.

124. $A \cup B = A$.

125. $A \cap B = A$.

126. $A \cup B = \emptyset$.

2.4 VENN DIAGRAMS WITH THREE SETS AND VERIFICATION OF EQUALITY OF SETS

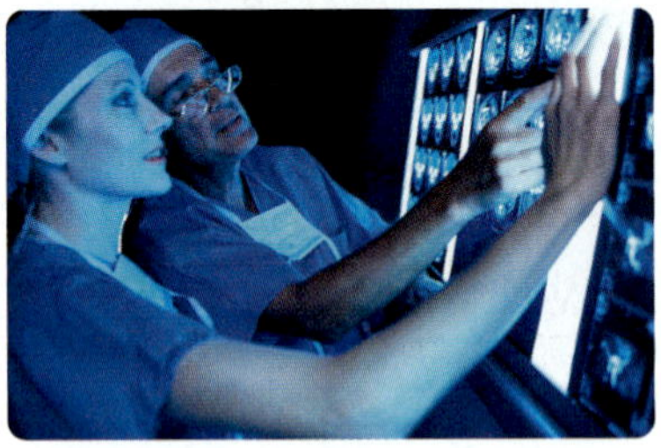

▲ Some of the highest-rated hospitals in the field of neurology are also among the highest-rated hospitals in the field of orthopedics.

Consider the set of the 10 highest-rated hospitals in the field of neurology. Next consider the set of the 10 highest-rated hospitals in the field of cardiology. Also, consider the set of the 10 highest-rated hospitals in the field of orthopedics. Are there any hospitals that are among the 10 highest rated in the fields of neurology, cardiology, and orthopedics? Are there any hospitals that are among the 10 highest rated in the fields of neurology and cardiology, but not orthopedics? In this section, we will learn how to use Venn diagrams to answer questions like these.

In Section 2.3, we learned how to use Venn diagrams to illustrate two sets. Venn diagrams can also be used to illustrate three sets.

For three sets, A, B, and C, the diagram is drawn so the three sets overlap (Fig. 2.15), creating eight regions. The diagrams in Fig. 2.16 emphasize selected regions of three intersecting sets. *When constructing Venn diagrams with three sets, we generally start with region V and work outward* as explained in the following procedure.

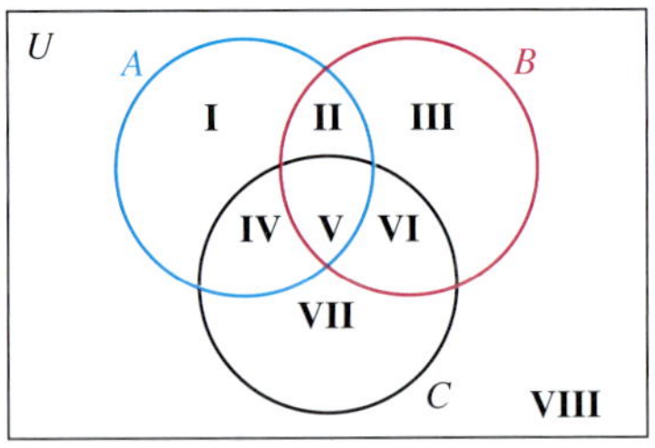

Figure 2.15

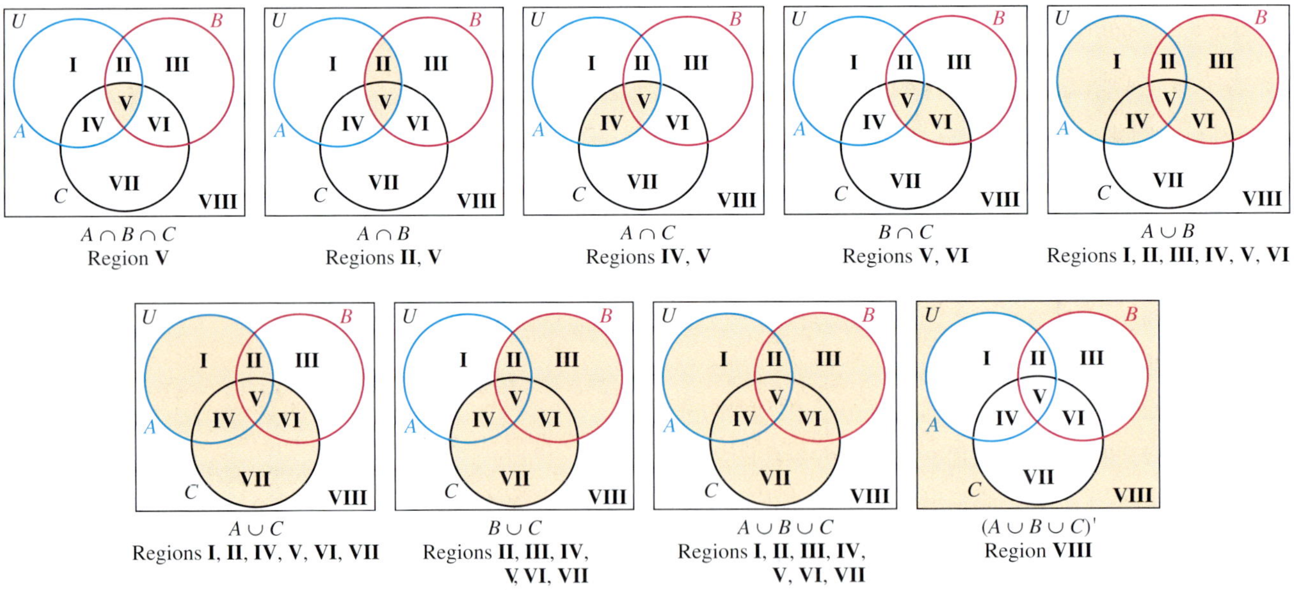

Figure 2.16

GENERAL PROCEDURE FOR CONSTRUCTING VENN DIAGRAMS WITH THREE SETS, A, B, AND C

1. Determine the elements to be placed in region V by finding the elements that are common to all three sets, $A \cap B \cap C$.
2. Determine the elements to be placed in region II. Find the elements in $A \cap B$. The elements in this set belong in regions II and V. Place the elements in the set $A \cap B$ that are not listed in region V in region II. The elements in regions IV and VI are found in a similar manner.
3. Determine the elements to be placed in region I by determining the elements in set A that are not in regions II, IV, and V. The elements in regions III and VII are found in a similar manner.
4. Determine the elements to be placed in region VIII by finding the elements in the universal set that are not in regions I through VII.

Example 1 illustrates the general procedure.

EXAMPLE 1 *Constructing a Venn Diagram for Three Sets*

Construct a Venn diagram illustrating the following sets.

$$U = \{1, 2, 3, 4, 5, 6, 7, 8, 9, 10, 11, 12, 13, 14\}$$
$$A = \{1, 5, 8, 9, 10, 12\}$$
$$B = \{2, 4, 5, 9, 10, 13\}$$
$$C = \{1, 3, 5, 8, 9, 11\}$$

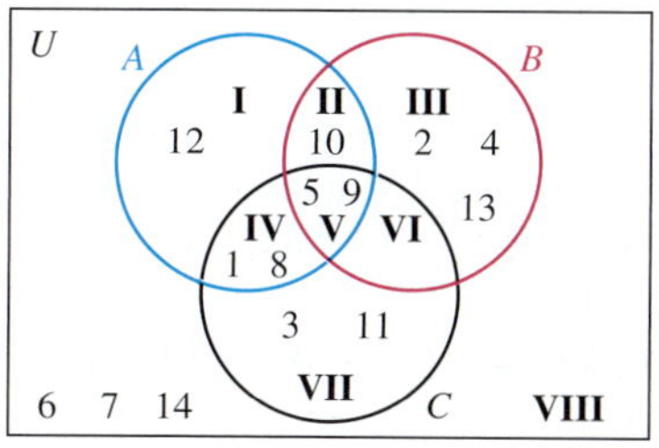

Figure 2.17

SOLUTION First find the intersection of all three sets. Because the elements 5 and 9 are in all three sets, $A \cap B \cap C = \{5, 9\}$. The elements 5 and 9 are placed in region V in Fig. 2.17. Next complete region II by determining the intersection of sets A and B.

$$A \cap B = \{5, 9, 10\}$$

$A \cap B$ consists of regions II and V. The elements 5 and 9 have already been placed in region V, so 10 must be placed in region II.

Now determine what numbers go in region IV.

$$A \cap C = \{1, 5, 8, 9\}$$

Since 5 and 9 have already been placed in region V, place the 1 and 8 in region IV. Now determine the numbers to go in region VI.

$$B \cap C = \{5, 9\}$$

Since both the 5 and 9 have been placed in region V, there are no numbers to be placed in region VI. Now complete set A. The only element of set A that has not previously been placed in regions II, IV, or V is 12. Therefore, place the element 12 in region I. The element 12 that is placed in region I is only in set A and not in set B or set C. Using set B, complete region III using the same general procedure used to determine the numbers in region I. Using set C, complete region VII by using the same procedure used to complete regions I and III. To determine the elements in region VIII, find the elements in U that have not been placed in regions I–VII. The elements 6, 7, and 14 have not been placed in regions I–VII, so place them in region VIII. ●

Venn diagrams can be used to illustrate and analyze many everyday problems. One example follows.

EXAMPLE 2 *Blood Types*

Human blood is classified (typed) according to the presence or absence of the specific antigens A, B, and Rh in the red blood cells. Antigens are highly specified proteins and carbohydrates that will trigger the production of antibodies in the blood to fight infection. Blood containing the Rh antigen is labeled positive, +, while blood lacking the Rh antigen is labeled negative, −. Blood lacking both A and B antigens is called type O. Sketch a Venn diagram with three sets A, B, and Rh and place each type of blood listed in the proper region. A person has only one type of blood.

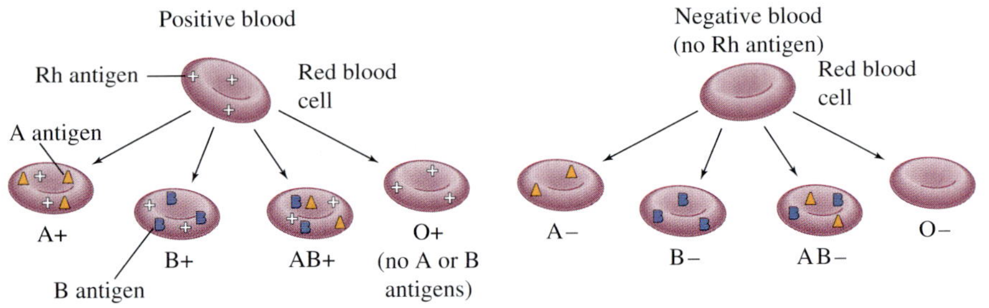

SOLUTION As illustrated in Chapter 1, the first thing to do is to read the question carefully and make sure you understand what is given and what you are asked to find. There are three antigens A, B, and Rh. Therefore, begin by naming the three circles in a Venn diagram with the three antigens; see Fig. 2.18.

Any blood containing the Rh antigen is positive, and any blood not containing the Rh antigen is negative. Therefore, all blood in the Rh circle is positive, and all blood outside the Rh circle is negative. The intersection of all three sets, region V, is AB+. Region II contains only antigens A and B and is therefore AB−. Region I is A− because it contains only antigen A. Region III is B−, region IV is A+, and region VI is B+. Region VII is O+, containing only the Rh antigen. Region VIII, which lacks all three antigens, is O−. ●

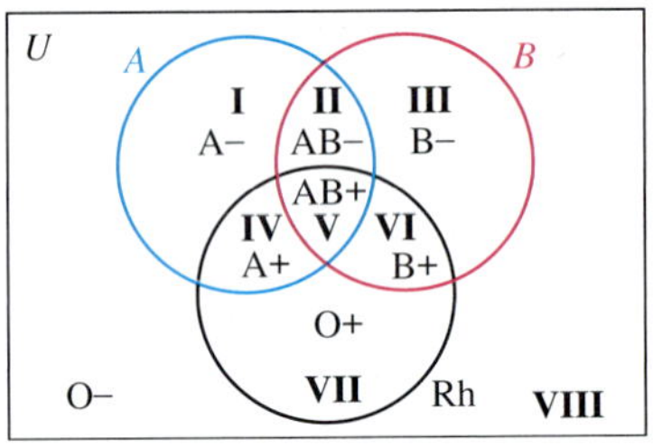

Figure 2.18

Verification of Equality of Sets

In this chapter, for clarity we may refer to operations on sets, such as $A \cup B'$ or $A \cap B \cap C$, as *statements involving sets* or simply as *statements*. Now we discuss how to determine if two statements involving sets are equal.

Consider the question: Is $A' \cup B = A' \cap B$ *for all sets A and B*? For the specific sets $U = \{1, 2, 3, 4, 5\}$, $A = \{1, 3\}$, and $B = \{2, 4, 5\}$, is $A' \cup B = A' \cap B$? To answer the question, we do the following.

Find $A' \cup B$	**Find $A' \cap B$**
$A' = \{2, 4, 5\}$	$A' = \{2, 4, 5\}$
$A' \cup B = \{2, 4, 5\}$	$A' \cap B = \{2, 4, 5\}$

For these sets, $A' \cup B = A' \cap B$, because both set statements are equal to $\{2, 4, 5\}$. At this point you may believe that $A' \cup B = A' \cap B$ for all sets A and B.

If we select the sets $U = \{1, 2, 3, 4, 5\}$, $A = \{1, 3, 5\}$, and $B = \{2, 3\}$, we see that $A' \cup B = \{2, 3, 4\}$ and $A' \cap B = \{2\}$. For this case, $A' \cup B \neq A' \cap B$. Thus, we have proved that $A' \cup B \neq A' \cap B$ for all sets A and B by using a *counterexample.* A counterexample, as explained in Chapter 1, is an example that shows a statement is not true.

In Chapter 1, we explained that proofs involve the use of deductive reasoning. Recall that deductive reasoning begins with a general statement and works to a specific conclusion. To verify, or determine whether set statements are equal for any two sets selected, we use deductive reasoning with Venn diagrams. Venn diagrams are used because they can illustrate general cases. To determine if statements that contain sets, such as $(A \cup B)'$ and $A' \cap B'$, are equal for all sets A and B, we use the regions of Venn diagrams. If both statements represent the same regions of the Venn diagram, then the statements are equal for all sets A and B. See Example 3.

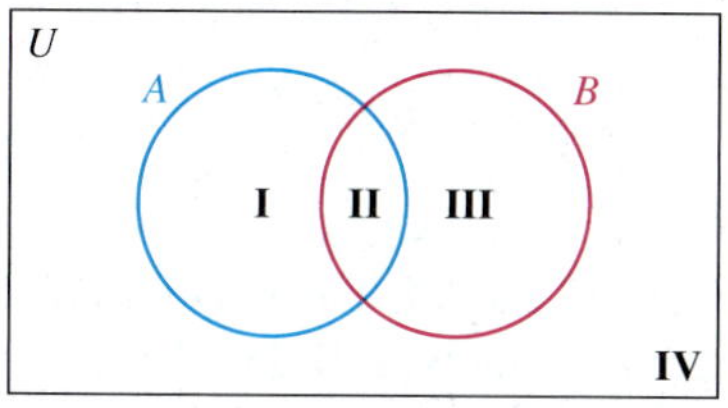

Figure 2.19

EXAMPLE 3 *Equality of Sets*

Determine whether $(A \cap B)' = A' \cup B'$ for all sets A and B.

SOLUTION Draw a Venn diagram with two sets A and B, as in Fig. 2.19. Label the regions as indicated.

Find $(A \cap B)'$

Set	Corresponding Regions
A	I, II
B	II, III
$A \cap B$	II
$(A \cap B)'$	I, III, IV

Find $A' \cup B'$

Set	Corresponding Regions
A'	III, IV
B'	I, IV
$A' \cup B'$	I, III, IV

Both statements are represented by the same regions, I, III, and IV, of the Venn diagram. Thus, $(A \cap B)' = A' \cup B'$ for all sets A and B. ●

In Example 3, when we proved that $(A \cap B)' = A' \cup B'$, we started with two general sets and worked to the specific conclusion that both statements represented the same regions of the Venn diagram. We showed that $(A \cap B)' = A' \cup B'$ *for all sets A and B.* No matter what sets we choose for A and B, this statement will be true. For example, let $U = \{1, 2, 3, 4, 5, 6, 7, 8, 9, 10\}$, $A = \{3, 4, 6, 10\}$, and $B = \{1, 2, 4, 5, 6, 8\}$.

$$\begin{aligned} (A \cap B)' &= A' \cup B' \\ \{4, 6\}' &= \{3, 4, 6, 10\}' \cup \{1, 2, 4, 5, 6, 8\}' \\ \{1, 2, 3, 5, 7, 8, 9, 10\} &= \{1, 2, 5, 7, 8, 9\} \cup \{3, 7, 9, 10\} \\ \{1, 2, 3, 5, 7, 8, 9, 10\} &= \{1, 2, 3, 5, 7, 8, 9, 10\} \end{aligned}$$

We can also use Venn diagrams to prove statements involving three sets.

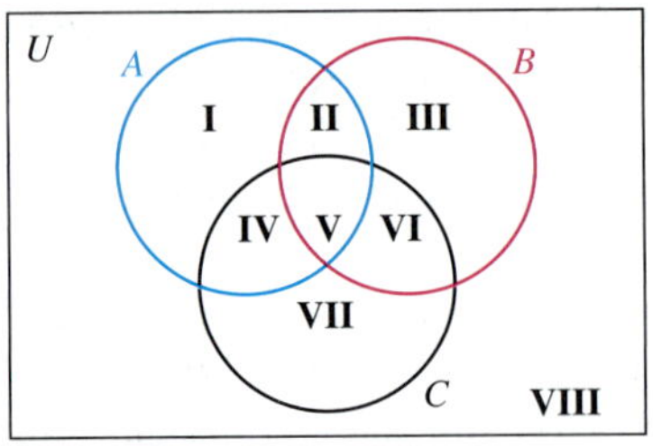

Figure 2.20

MATHEMATICS TODAY

Using Venn Diagrams

A financial planning company uses the following Venn diagram to illustrate the financial planning services it offers. From the diagram, we can see that this company offers financial planning in an "intersection" of the areas investment, retirement, and college planning, the intersection of all three sets.

We categorize items on a daily basis, from filing items to planning meals to planning social activities. Children are taught how to categorize items at an early age when they learn how to classify items according to color, shape, and size. Biologists categorize items when they classify organisms according to shared characteristics. A Venn diagram is a very useful tool to help order and arrange items and to picture the relationship between sets.

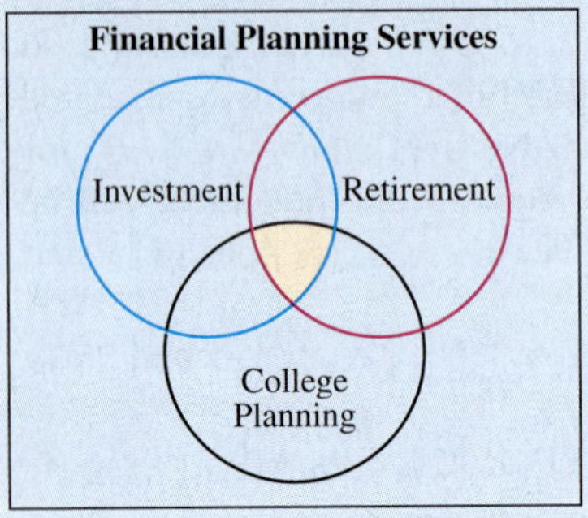

EXAMPLE 4 *Equality of Sets*

Determine whether $A \cup (B \cap C) = (A \cup B) \cap (A \cup C)$ for all sets, A, B, and C.

SOLUTION Because the statements include three sets, A, B, and C, three circles must be used. The Venn diagram illustrating the eight regions is shown in Fig. 2.20.

First we will find the regions that correspond to $A \cup (B \cap C)$, and then we will find the regions that correspond to $(A \cup B) \cap (A \cup C)$. If both answers are the same, the statements are equal.

Find $A \cup (B \cap C)$

Set	Corresponding Regions
A	I, II, IV, V
$B \cap C$	V, VI
$A \cup (B \cap C)$	I, II, IV, V, VI

Find $(A \cup B) \cap (A \cup C)$

Set	Corresponding Regions
$A \cup B$	I, II, III, IV, V, VI
$A \cup C$	I, II, IV, V, VI, VII
$(A \cup B) \cap (A \cup C)$	I, II, IV, V, VI

The regions that correspond to $A \cup (B \cap C)$ are I, II, IV, V, and VI, and the regions that correspond to $(A \cup B) \cap (A \cup C)$ are also I, II, IV, V, and VI. The results show that both statements are represented by the same regions, namely, I, II, IV, V, and VI, and therefore $A \cup (B \cap C) = (A \cup B) \cap (A \cup C)$ for all sets A, B, and C. ●

In Example 4, we proved that $A \cup (B \cap C) = (A \cup B) \cap (A \cup C)$ for all sets A, B, and C. Show that this statement is true for the specific sets $U = \{1, 2, 3, 4, 5, 6, 7, 8, 9, 10\}$, $A = \{1, 2, 3, 7\}$, $B = \{2, 3, 4, 5, 7, 9\}$, and $C = \{1, 4, 7, 8, 10\}$.

De Morgan's Laws

In set theory, logic, and other branches of mathematics, a pair of related theorems known as De Morgan's laws make it possible to transform statements and formulas into alternative and often more convenient forms. In set theory, ***De Morgan's laws*** are symbolized as follows.

DE MORGAN'S LAWS

1. $(A \cap B)' = A' \cup B'$
2. $(A \cup B)' = A' \cap B'$

Law 1 was verified in Example 3. We suggest that you verify law 2 at this time. The laws were expressed verbally by William of Ockham in the fourteenth century. In the nineteenth century, Augustus De Morgan expressed them mathematically. De Morgan's laws will be discussed more thoroughly in Chapter 3, Logic.

SECTION 2.4 EXERCISES

CONCEPT/WRITING EXERCISES

1. How many regions are created when constructing a Venn diagram with three overlapping sets?

2. When constructing a Venn diagram with three overlapping sets, which region do you generally complete first?

3. When constructing a Venn diagram with three overlapping sets, after completing region V, which regions do you generally complete next?

4. A Venn diagram contains three sets, A, B, and C, as in Fig. 2.15 on page 72. If region V contains 4 elements and there are 12 elements in $B \cap C$, how many elements belong in region VI? Explain.

5. A Venn diagram contains three sets, A, B, and C, as in Fig. 2.15 on page 72. If region V contains 4 elements and there are 9 elements in $A \cap B$, how many elements belong in region II? Explain.

6. Give De Morgan's laws.

7. a) For $U = \{1, 2, 3, 4, 5\}$, $A = \{1, 4, 5\}$, and $B = \{1, 4, 5\}$, does $A \cup B = A \cap B$?

   b) By observing the answer to part (a), can we conclude that $A \cup B = A \cap B$ for all sets A and B? Explain.

   c) Using a Venn diagram, determine if $A \cup B = A \cap B$ for all sets A and B.

8. What type of reasoning do we use when using Venn diagrams to verify or determine whether set statements are equal?

PRACTICE THE SKILLS

9. Construct a Venn diagram illustrating the following sets.

$$U = \{a, b, c, d, e, f, g, h, i, j\}$$
$$A = \{c, d, e, g, h, i\}$$
$$B = \{a, c, d, g\}$$
$$C = \{c, f, i, j\}$$

10. Construct a Venn diagram illustrating the following sets.

U = {Delaware, Pennsylvania, New Jersey, Georgia, Connecticut, Massachusetts, Maryland, South Carolina, New Hampshire, Virginia, New York, North Carolina, Rhode Island}

A = {New York, New Jersey, Pennsylvania, Massachusetts, New Hampshire}

B = {Delaware, Connecticut, Georgia, Maryland, New York, Rhode Island}

C = {New York, South Carolina, Rhode Island, Massachusetts}

11. Construct a Venn diagram illustrating the following sets.

U = {Circuit City, Best Buy, Wal-Mart, Kmart, Target, Sears, JCPenney, Costco, Kohl's, Gap, Gap Kids, Foot Locker, Old Navy, Macy's}

A = {Circuit City, Wal-Mart, Target, JCPenney, Old Navy}

B = {Best Buy, Target, Costco, Old Navy, Macy's}

C = {Target, Sears, Kohl's, Gap, JCPenney}

12. Construct a Venn diagram illustrating the following sets.

U = {*The Lion King, Aladdin, Cinderella, Beauty and the Beast, Snow White and the Seven Dwarfs, Toy Story, 101 Dalmatians, The Little Mermaid, The Incredibles*}

A = {*Aladdin, Toy Story, The Lion King, Snow White and the Seven Dwarfs*}

B = {*Snow White and the Seven Dwarfs, Toy Story, The Lion King, Beauty and the Beast*}

C = {*Snow White and the Seven Dwarfs, Toy Story, Beauty and the Beast, Cinderella, 101 Dalmatians*}

13. Construct a Venn diagram illustrating the following sets.

U = {peach, pear, banana, apple, grape, melon, carrot, corn, orange, spinach}

A = {pear, grape, melon, carrot}

B = {peach, pear, banana, spinach, corn}

C = {pear, banana, apple, grape, melon, spinach}

14. Construct a Venn diagram illustrating the following sets.

U = {Louis Armstrong, Glenn Miller, Stan Kenton, Charlie Parker, Duke Ellington, Benny Goodman, Count Basie, John Coltrane, Dizzy Gillespie, Miles Davis, Thelonius Monk}

A = {Stan Kenton, Count Basie, Dizzy Gillespie, Duke Ellington, Thelonius Monk}

B = {Louis Armstrong, Glenn Miller, Count Basie, Duke Ellington, Miles Davis}

C = {Count Basie, Miles Davis, Stan Kenton, Charlie Parker, Duke Ellington}

15. ***Olympic Medals*** Consider the following table, which shows countries that won at least 22 medals in the 2004 Summer Olympics. Let the countries shown in the table represent the universal set.

Country	Gold Medals	Silver Medals	Bronze Medals	Total Medals
United States	35	39	29	103
Russia	27	27	38	92
China	32	17	14	63
Australia	17	16	16	49
Germany	14	16	18	48
Japan	16	9	12	37
France	11	9	13	33
Italy	10	11	11	32
South Korea	9	12	9	30
Great Britain	9	9	12	30
Cuba	9	7	11	27
Ukraine	9	5	9	23
Netherlands	4	9	9	22

Source: *2006 Time Almanac*

Let A = set of teams that won at least 48 medals.

Let B = set of teams that won at least 20 gold medals.

Let C = set of teams that won at least 10 bronze medals.

Construct a Venn diagram that illustrates the sets A, B, and C.

16. ***Popular TV Shows*** Let U = {*60 Minutes*, *American Idol–Tues.*, *American Idol–Wed.*, *CSI*, *CSI:Miami*, *Desperate Housewives*, *Friends*, *Grey's Anatomy*, *Law & Order*, *The Apprentice*}. Sets A, B, and C that follow show the five most watched television shows for the years 2004, 2005, and 2006, respectively (according to Nielsen Media Research). Then

A = {*CSI*, *American Idol–Tues.*, *American Idol–Wed.*, *Friends*, *The Apprentice*}

B = {*CSI*, *American Idol–Tues.*, *American Idol–Wed.*, *Desperate Housewives*, *CSI:Miami*}

C = {*American Idol–Tues.*, *American Idol–Wed.*, *CSI*, *Desperate Housewives*, *Grey's Anatomy*}

Construct a Venn diagram illustrating the sets.

Best Hospitals For Exercises 17–22, use the chart below which shows the *U.S. News and World Report* top 10 rankings of hospitals in 2005 in the fields of neurology, cardiology, and orthopedics. The universal set is the set of all U.S. hospitals.

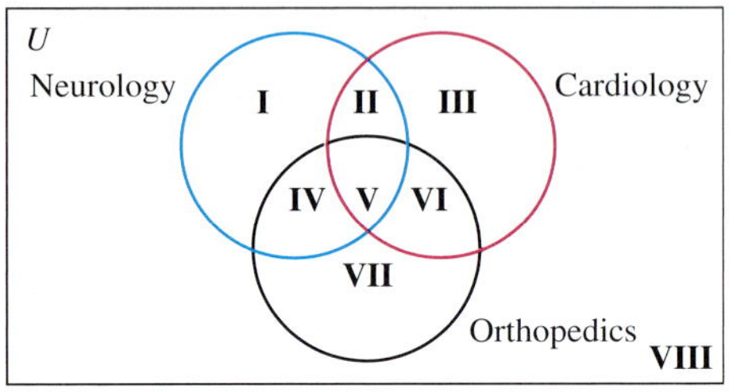

Figure 2.21

Top-Rated Hospitals

Neurology	Cardiology	Orthopedics
1. Mayo Clinic	1. Cleveland Clinic	1. Mayo Clinic
2. Johns Hopkins	2. Mayo Clinic	2. Hospital for Special Surgery
3. New York Presbyterian University Hospital	3. Johns Hopkins	3. Massachusetts General
4. Massachusetts General	4. Duke University Medical Center	4. Johns Hopkins
5. University of California, San Francisco	5. Massachusetts General	5. Cleveland Clinic
6. Cleveland Clinic	6. Brigham and Women's Hospital	6. UCLA Medical Center
7. St. Joseph's Hospital and Medical Center	7. New York–Presbyterian University Hospital	7. University of Iowa Hospitals and Clinics
8. Barnes-Jewish	8. Texas Heart Institute at St. Luke's Episcopal Hospital	8. Rush University Medical Center
9. UCLA Medical Center	9. Barnes-Jewish	9. University of Washington Medical Center
10. Methodist Hospital	10. University of Alabama	10. Duke University Medical Center

Source: *U.S. News and World Report*

Indicate in which region, I–VIII, in Fig. 2.21 each of the following hospitals belongs.

17. Mayo Clinic

18. UCLA Medical Center

19. Methodist Hospital

20. Barnes-Jewish

21. Brigham and Women's Hospital

22. Yale–New Haven Hospital

Rankings of Government Agencies For Exercises 23–28, use the following table, which shows the top 10 government agencies based on employee satisfaction in the areas of effective leadership, family-friendly culture, and pay and benefits. The universal set is the set of all government agencies.

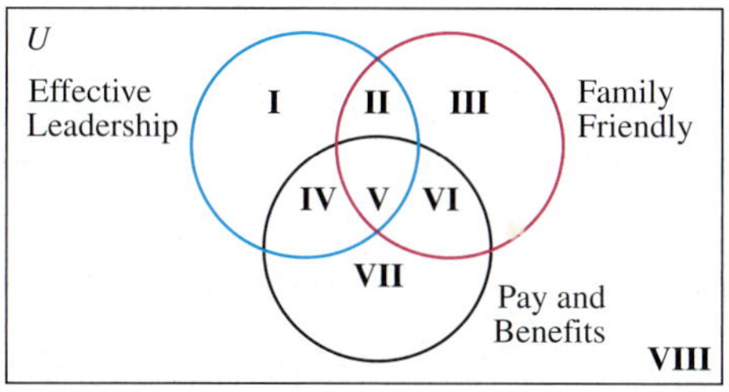

Figure 2.22

Rankings of Government Agencies

Effective Leadership	Family-Friendly Culture	Pay and Benefits
1. National Aeronautics and Space Administration	1. Federal Deposit Insurance Corp.	1. Federal Deposit Insurance Corp.
2. Nuclear Regulatory Commission	2. General Services Administration	2. Office of Management and Budget
3. Office of Management and Budget	3. Environmental Protection Agency	3. Securities and Exchange Commission
4. General Services Administration	4. Nuclear Regulatory Commission	4. National Aeronautics and Space Administration
5. National Science Foundation	5. National Aeronautics and Space Administration	5. Nuclear Regulatory Commission
6. Department of State	6. Office of Personnel Management	6. General Services Administration
7. Department of Energy	7. Department of Energy	7. Department of Commerce
8. Department of Defense	8. Equal Employment Opportunity Commission	8. Agency for International Development
8. Agency for International Development	9. Department of Agriculture	9. Office of Personnel Management
10. Securities and Exchange Commission	9. Department of Education	9. Small Business Administration

Source: Partnership for Public Service

Indicate in which region, I–VIII, in Fig. 2.22 each of the following agencies belongs.

23. Department of Energy

24. National Aeronautics and Space Administration

25. Department of Justice

26. Small Business Administration

27. Department of Agriculture

28. Office of Personnel Management

Figures *In Exercises 29–40, indicate in Fig. 2.23 the region in which each of the figures would be placed.*

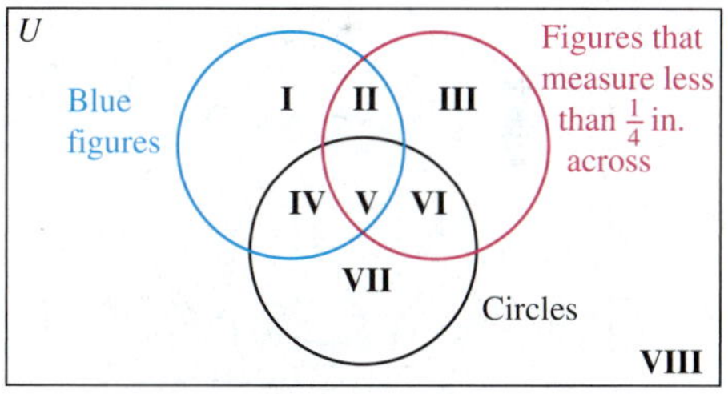

Figure 2.23

29. 

30. 

31. 

32. 

33. 

34. 

35. 

36. 

37. 

38.

39. 

40. 

Senate Bills *During a session of the U.S. Senate, three bills were voted on. The votes of six senators are shown below the figure. Determine in which region of Fig. 2.24 each senator would be placed. The set labeled Bill 1 represents the set of senators who voted yes on Bill 1, and so on.*

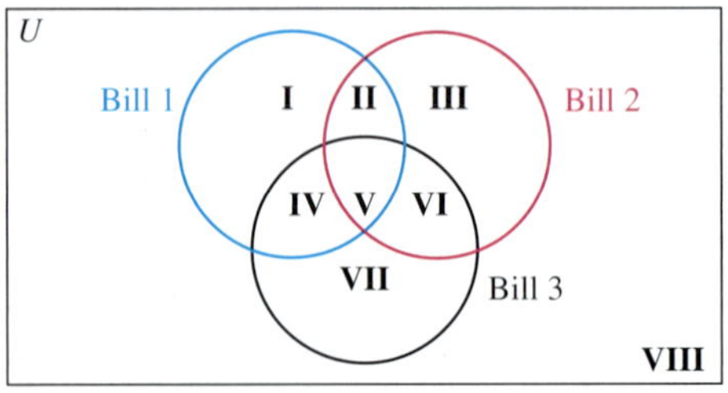

Figure 2.24

SENATOR	BILL 1	BILL 2	BILL 3
41. Feingold	yes	no	no
42. Lott	no	no	yes
43. McCain	no	no	no
44. Mikulski	yes	yes	yes
45. Obama	no	yes	yes
46. Specter	no	yes	no

In Exercises 47–60, use the Venn diagram in Fig. 2.25 to list the sets in roster form.

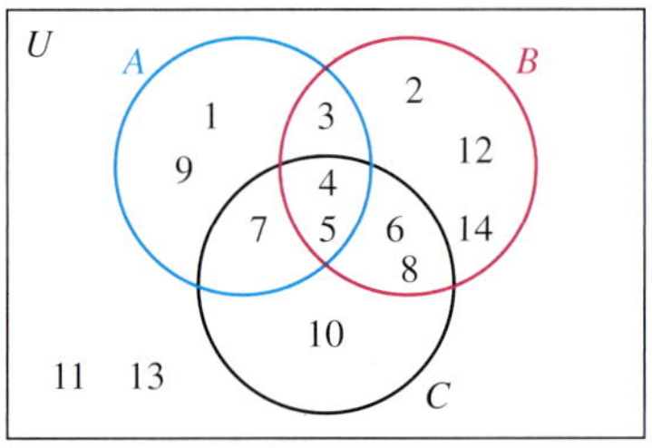

Figure 2.25

47. A

48. U

49. B

50. C

51. $A \cap B$

52. $A \cap C$

53. $(B \cap C)'$

54. $A \cap B \cap C$

55. $A \cup B$

56. $B \cup C$

57. $(A \cup C)'$

58. $A \cap (B \cup C)$

59. A'

60. $(A \cup B \cup C)'$

In Exercises 61–68, use Venn diagrams to determine whether the following statements are equal for all sets A and B.

61. $(A \cup B)'$, $A' \cap B'$

62. $(A \cap B)'$, $A \cup B'$

63. $A' \cup B'$, $A \cap B$

64. $(A \cup B)'$, $(A \cap B)'$

65. $A' \cap B'$, $(A \cap B)'$

66. $A' \cap B'$, $A \cup B'$

67. $(A' \cap B)'$, $A \cup B'$

68. $A' \cap B'$, $(A' \cap B')'$

In Exercises 69–78, use Venn diagrams to determine whether the following statements are equal for all sets A, B, and C.

69. $A \cap (B \cup C)$, $(A \cap B) \cup C$

70. $A \cup (B \cap C)$, $(B \cap C) \cup A$

71. $A \cap (B \cup C)$, $(B \cup C) \cap A$

72. $A \cup (B \cap C)'$, $A' \cap (B \cup C)$

73. $A \cap (B \cup C)$, $(A \cap B) \cup (A \cap C)$

74. $A \cup (B \cap C)$, $(A \cup B) \cap (A \cup C)$

75. $A \cap (B \cup C)'$, $A \cap (B' \cap C')$

76. $(A \cup B) \cap (B \cup C)$, $B \cup (A \cap C)$

77. $(A \cup B)' \cap C$, $(A' \cup C') \cap (B' \cup C)$

78. $(C \cap B)' \cup (A \cap B)'$, $A \cap (B \cap C)$

In Exercises 79–82, use set statements to write a description of the shaded area. Use union, intersection and complement as necessary. More than one answer may be possible.

79.

80.

81.

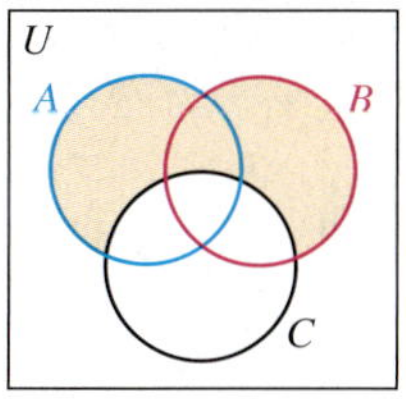

82.

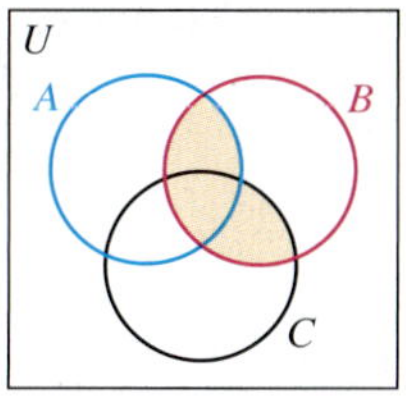

PROBLEM SOLVING

83. Let

$$U = \{1, 2, 3, 4, 5, 6, 7, 8, 9, 10\}$$
$$A = \{1, 2, 3, 4\}$$
$$B = \{3, 6, 7\}$$
$$C = \{6, 7, 9\}$$

a) Show that $(A \cup B) \cap C = (A \cap C) \cup (B \cap C)$ for these sets.

b) Make up your own sets A, B, and C. Verify that $(A \cup B) \cap C = (A \cap C) \cup (B \cap C)$ for your sets A, B, and C.

c) Use Venn diagrams to verify that $(A \cup B) \cap C = (A \cap C) \cup (B \cap C)$ for all sets A, B, and C.

84. Let

$$U = \{\text{a, b, c, d, e, f, g, h, i}\}$$
$$A = \{\text{a, c, d, e, f}\}$$
$$B = \{\text{c, d}\}$$
$$C = \{\text{a, b, c, d, e}\}$$

a) Determine whether $(A \cup C)' \cap B = (A \cap C)' \cap B$ for these sets.

b) Make up your own sets, A, B, and C. Determine whether $(A \cup C)' \cap B = (A \cap C)' \cap B$ for your sets.

c) Determine whether $(A \cup C)' \cap B = (A \cap C)' \cap B$ for all sets A, B, and C.

85. ***Blood Types*** A hematology text gives the following information on percentages of the different types of blood worldwide.

Type	Positive Blood, %	Negative Blood, %
A	37	6
O	32	6.5
B	11	2
AB	5	0.5

Construct a Venn diagram similar to the one in Example 2 and place the correct percent in each of the eight regions.

86. Define each of the eight regions in Fig. 2.26 using sets A, B, and C and a set operation. (*Hint:* $A \cap B' \cap C'$ defines region I.)

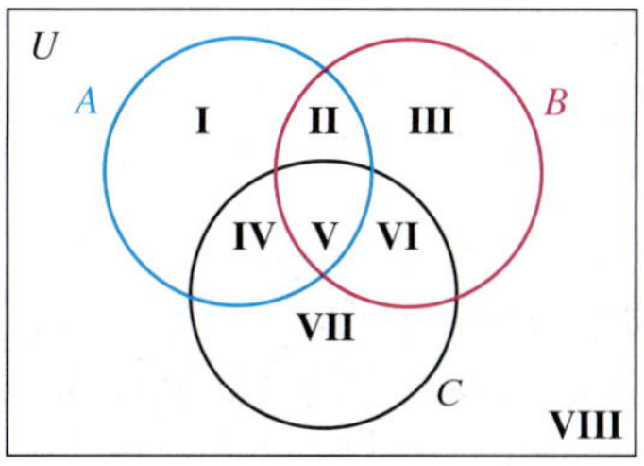

Figure 2.26

87. ***Categorizing Contracts*** J & C Mechanical Contractors wants to classify its projects. The contractors categorize set A as construction projects, set B as plumbing projects, and set C as projects with a budget greater than \$300,000.

a) Draw a Venn diagram that can be used to categorize the company projects according to the listed criteria.

b) Determine the region of the diagram that contains construction projects and plumbing projects with a budget greater than \$300,000. Describe the region using sets A, B, and C with set operations. Use union, intersection, and complement as necessary.

c) Determine the region of the diagram that contains plumbing projects with a budget greater than \$300,000 that are not construction projects. Describe the region using sets A, B and C with set operations. Use union, intersection, and complement as necessary.

d) Determine the region of the diagram that contains construction projects and nonplumbing projects whose budget is less than or equal to \$300,000. Describe the region using sets A, B, and C with set operations. Use union, intersection, and complement as necessary.

CHALLENGE PROBLEM/GROUP ACTIVITY

88. We were able to determine the number of elements in the union of two sets with the formula

$$n(A \cup B) = n(A) + n(B) - n(A \cap B).$$

Can you determine a formula for finding the number of elements in the union of three sets? In other words, write a formula to determine $n(A \cup B \cup C)$. [*Hint*: The formula will contain each of the following: $n(A)$, $n(B)$, $n(C)$, $n(A \cap B \cap C')$, $n(A \cap B' \cap C)$, $n(A' \cap B \cap C)$, and $2n(A \cap B \cap C)$.]

RECREATIONAL MATHEMATICS

89. a) Construct a Venn diagram illustrating four sets, *A*, *B*, *C*, and *D*. (*Hint*: Four circles cannot be used, and you should end up with 16 *distinct* regions.) Have fun!

b) Label each region with a set statement (see Exercise 86). Check all 16 regions to make sure that *each is distinct*.

INTERNET/RESEARCH ACTIVITY

90. The two Venn diagrams on the right illustrate what happens when colors are added or subtracted. Do research in an art text, an encyclopedia, the Internet, or another source and write a report explaining the creation of the colors in the Venn diagrams, using such terms as union of colors and subtraction (or difference) of colors.

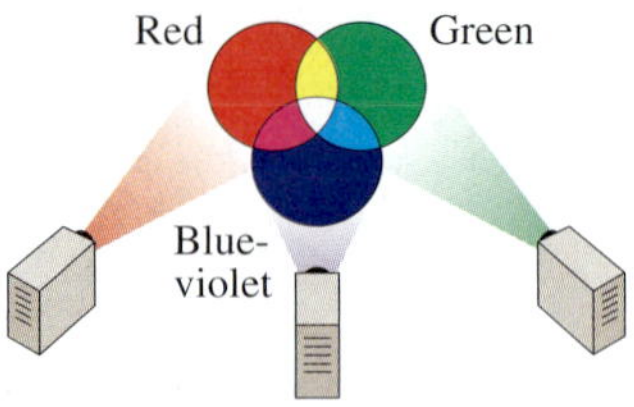

Additive color mixing

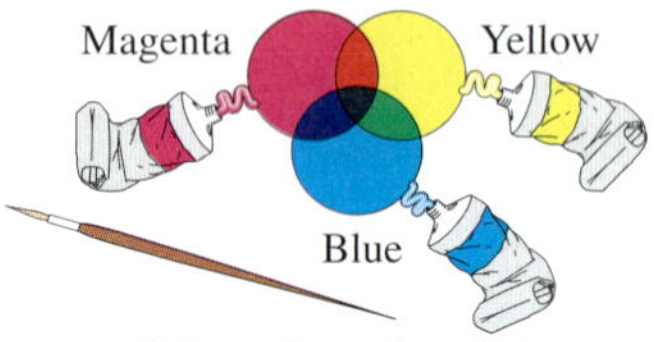

Subtractive color mixing

2.5 APPLICATIONS OF SETS

▲ We can use a Venn diagram to determine how many members of a health club use a particular piece of equiment.

The members of a health club were surveyed about their use of the equipment at the club. Suppose the results of the survey show how many members use the treadmill, how many members use the StairMaster, and how many members use both pieces of equipment. How can the manager of the club use this information to determine how many members use only the treadmill? In this section, we will learn how to use Venn diagrams to answer this type of question.

We can solve practical problems involving sets by using the problem-solving process discussed in Chapter 1: Understand the problem, devise a plan, carry out the plan, and then examine and check the results. First determine: What is the problem? or What am I looking for? To devise the plan, list all the facts that are given and how they are related. *Look for key words or phrases* such as "only set *A*," "set *A* and set *B*," "set *A* or set *B*," "set *A* and set *B* and not set *C*." Remember that *and* means intersection, *or* means union, and *not* means complement. The problems we solve in this section contain two or three sets of elements, which can be represented in a Venn diagram. Our plan will generally include drawing a Venn diagram, labeling the diagram, and filling in the regions of the diagram.

Whenever possible, follow the procedure in Section 2.4 for completing the Venn diagram and then answer the questions. *Remember, when drawing Venn diagrams, we generally start with the intersection of the sets and work outward.*

EXAMPLE 1 *Fitness Equipment*

Fitness for Life health club is considering adding additional cardiovascular equipment. It is considering two types of equipment, treadmills (T) and StairMasters (S). The health club surveyed a sample of members and asked which equipment they had used in the previous month. Of 150 members surveyed, it was determined that

102 used the treadmills.
71 used the StairMasters.
40 used both types.

Of those surveyed,

a) how many did not use either the treadmill or the StairMaster?
b) how many used the treadmill but not the StairMaster?
c) how many used the StairMaster but not the treadmill?
d) how many used either the treadmill or StairMaster?

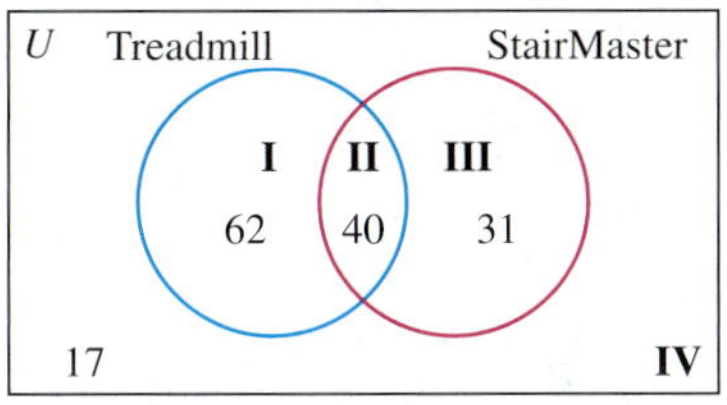

Figure 2.27

SOLUTION The problem provides the following information.

The number of members surveyed is 150: $n(U)$ is 150.

The number of members surveyed who used the treadmill is 102: $n(T) = 102$.

The number of members surveyed who used the StairMaster is 71: $n(S) = 71$.

The number of members surveyed who used both the treadmill and the StairMaster is 40: $n(T \cap S) = 40$.

We illustrate this information on the Venn diagram shown in Fig. 2.27. We already know that $T \cap S$ corresponds to region II. As $n(T \cap S) = 40$, we write 40 in region II. Set T consists of regions I and II. We know that set T, the members who used the treadmill, contains 102 members. Therefore, region I contains $102 - 40$, or 62 members. We write the number 62 in region I. Set S consists of regions II and III. As $n(S) = 71$, the total in these two regions must be 71. Region II contains 40, leaving $71 - 40$ or 31 for region III. We write 31 in region III.

The total number of members surveyed who used the treadmill or the StairMaster is found by adding the numbers in regions I, II, and III. Therefore, $n(T \cup S) = 62 + 40 + 31 = 133$. The number in region IV is the difference between $n(U)$ and $n(T \cup S)$. There are $150 - 133$, or 17 members in region IV.

a) The members surveyed who did not use either the treadmill or the StairMaster are those members of the universal set who are not contained in set T or set S. The 17 members in region IV did not use the treadmill or StairMaster.
b) The 62 members in region I are those members surveyed who used the treadmill but not the StairMaster.
c) The 31 members in region III are those members surveyed who used the StairMaster but not the treadmill.
d) The members in regions I, II, or III are those members surveyed who used either the treadmill or the StairMaster. Thus, $62 + 40 + 31$ or 133 members surveyed used either the treadmill or the StairMaster. Notice that the 40 members in region II who use both types of equipment are included in those members surveyed who used either the treadmill or the StairMaster. ●

Similar problems involving three sets can be solved, as illustrated in Example 2.

EXAMPLE 2 *Software Purchases*

CompUSA has recorded recent sales for three types of computer software: games, educational software, and utility programs. The following information regarding software purchases was obtained from a survey of 893 customers.

545 purchased games.

497 purchased educational software.

290 purchased utility programs.

297 purchased games and educational software.

196 purchased educational software and utility programs.

205 purchased games and utility programs.

157 purchased all three types of software.

Use a Venn diagram to answer the following questions. How many customers purchased

a) none of these types of software?
b) only games?
c) at least one of these types of software?
d) exactly two of these types of software?

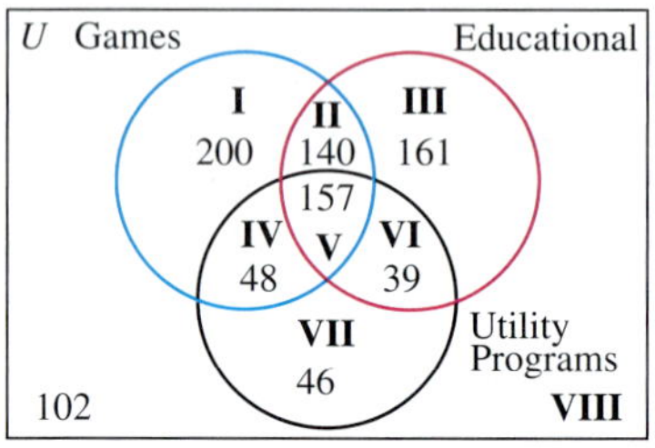

Figure 2.28

SOLUTION Begin by constructing a Venn diagram with three overlapping circles. One circle represents games, another educational software, and the third utilities. See Fig 2.28. Label the eight regions.

Whenever possible, work from the center of the diagram outwards. First fill in region V. Since 157 customers purchased all three types of software, we place 157 in region V. Next determine the number to be placed in region II. Regions II and V together represent the customers who purchased both games and educational software. Since 297 customers purchased both of these types of software, the sum of the numbers in these regions must be 297. Since 157 have already been placed in region V, $297 - 157 = 140$ must be placed in region II. Now we determine the number to be placed in region IV. Since 205 customers purchased both games and utility programs, the sum of the numbers in regions IV and V must be 205. Since 157 have already been placed in region V, $205 - 157 = 48$ must be placed in region IV. Now determine the number to be placed in region VI. A total of 196 customers purchased educational software and utility programs. The numbers in regions V and VI must total 196. Since 157 have already been placed in region V, the number to be placed in region VI is $196 - 157 = 39$.

Now that we have determined the numbers for regions V, II, IV, and VI, we can determine the numbers to be placed in regions I, III, and VII. We are given that 545 customers purchased games. The sum of the numbers in regions I, II, IV, and V must be 545. To determine the number to be placed in region I, subtract the amounts in regions II, IV, and V from 545. There must be $545 - 140 - 48 - 157 = 200$ in region I. Determine the numbers to be placed in regions III and VII in a similar manner.

$$\text{Region III} = 497 - 140 - 157 - 39 = 161$$
$$\text{Region VII} = 290 - 48 - 157 - 39 = 46$$

Now that we have determined the numbers in regions I through VII, we can determine the number to be placed in region VIII. Adding the numbers in regions I through VII yields a sum of 791. The difference between the total number of customers surveyed, 893, and the sum of the numbers in regions I through VII must be placed in region VIII.

$$\text{Region VIII} = 893 - 791 = 102$$

Now that we have completed the Venn diagram, we can answer the questions.

a) One hundred two customers did not purchase any of these types of software. These customers are indicated in region VIII.

b) Region I represents those customers who purchased only games. Thus, 200 customers purchased only games.

c) The words *at least one* mean "one or more." All those in regions I through VII purchased at least one of the types of software. The sum of the numbers in regions I through VII is 791, so 791 customers purchased at least one of the types of software.

d) The customers in regions II, IV, and VI purchased exactly two of the types of software. Summing the numbers in these regions 140 + 48 + 39 we find that 227 customers purchased exactly two of these types of software. Notice that we did not include the customers in region V. Those customers purchased all three types of software. ●

The procedure to work problems like those given in Example 2 is generally the same. Start by completing region V. Next complete regions II, IV, and VI. Then complete regions I, III, and VII. Finally, complete region VIII. When you are constructing Venn diagrams, be sure to check your work carefully.

TIMELY TIP When constructing a Venn diagram, the most common mistake made by students is forgetting to subtract the number in region V from the respective values in determining the numbers to be placed in regions II, IV, and VI.

EXAMPLE 3 *Travel Packages*

Liberty Travel surveyed 125 potential customers. The following information was obtained.

68 wished to travel to Hawaii.
53 wished to travel to Las Vegas.
47 wished to travel to Disney World.
34 wished to travel to Hawaii and Las Vegas.
26 wished to travel to Las Vegas and Disney World.
23 wished to travel to Hawaii and Disney World.
18 wished to travel to all three destinations.

▲ Hawaii

Use a Venn diagram to answer the following questions. How many of those surveyed

a) did not wish to travel to any of these destinations?
b) wished to travel only to Hawaii?
c) wished to travel to Disney World *and* Las Vegas, but not to Hawaii?
d) wished to travel to Disney World *or* Las Vegas, but not to Hawaii?
e) wished to travel to exactly one of these destinations?

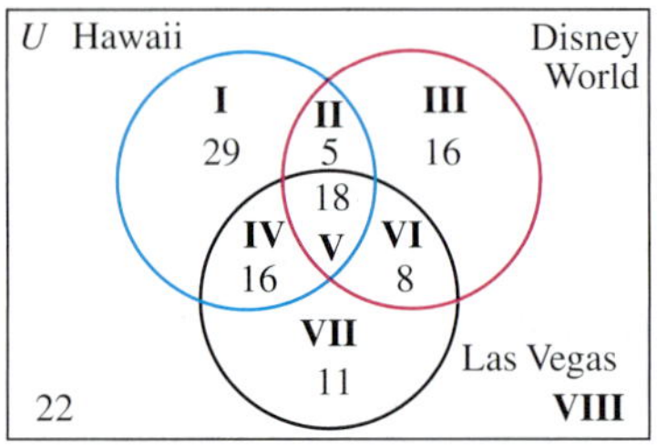

Figure 2.29

SOLUTION The Venn diagram is constructed using the procedures we outlined in Example 2. The diagram is illustrated in Fig. 2.29. We suggest you construct the diagram by yourself now and check your diagram with Fig. 2.29.

a) Twenty-two potential customers did not wish to travel to any of these destinations (see region VIII).

b) Twenty-nine potential customers wished to travel only to Hawaii (see region I).

c) Those potential customers in region VI wished to travel to Disney World *and* Las Vegas, but not to Hawaii. Therefore, eight customers satisfied the criteria.

d) The word *or* in this type of problem means one or the other or both. All the potential customers in regions II, III, IV, V, VI, and VII wished to travel to Disney World or Las Vegas. Those in regions II, IV, and V also wished to travel to Hawaii. The potential customers that wished to travel to Disney World or Las Vegas, but not to Hawaii, are found by adding the numbers in regions III, VI, and VII. There are $16 + 8 + 11 = 35$ potential customers who satisfy the criteria.

e) Those potential customers in regions I, III, and VII wished to travel to exactly one of the destinations. Therefore, $29 + 16 + 11 = 56$ customers wished to travel to exactly one of these destinations. ●

SECTION 2.5 EXERCISES

PRACTICE THE SKILLS/PROBLEM SOLVING

In Exercises 1–15, draw a Venn diagram to obtain the answers.

1. ***Campus Activities*** At a local college, a survey was taken to determine participation in different campus activities. Of 150 students surveyed, it was determined that
32 participated in clubs
27 participated in intramural sports
18 participated in both clubs and intramural sports

Of those surveyed,

a) how many participated only in clubs?

b) how many participated only in intramural sports?

c) how many did not participate in either clubs or intramural sports?

2. ***Landscape Purchases*** Agway Lawn and Garden collected the following information regarding purchases from 130 of its customers.
74 purchased shrubs.
70 purchased trees.
41 purchased both shrubs and trees.

Of those surveyed,

a) how many purchased only shrubs?

b) how many purchased only trees?

c) how many did not purchase either of these items?

3. ***Real Estate*** The Maiello's are moving to Wilmington, Delaware. Their real estate agent located 83 houses listed for sale, in the Wilmington area, in their price range. Of these houses listed for sale,
47 had a family room.
42 had a deck.
30 had a family room and a deck.

How many had

a) a family room but not a deck?

b) a deck but not a family room?

c) either a family room or a deck?

4. ***Soft Drink Taste Test*** A soft drink company wishes to introduce a new soft drink. The company is considering two flavors, lemon lime and cherry cola. In a sample of 125 people, it was found that

80 liked lemon lime.
68 liked cherry cola.
42 liked both types.

a) How many liked only lemon lime?

b) How many liked only cherry cola?

c) How many liked either one or the other or both?

d) How many did not like either flavor?

5. ***Cultural Activities*** Thirty-three U.S. cities were researched to determine whether they had a professional sports team, a symphony, or a children's museum. The following information was determined.
16 had a professional sports team.
17 had a symphony.
15 had a children's museum.
11 had a professional sports team and a symphony.
7 had a professional sports team and a children's museum.
9 had a symphony and children's museum.
5 had all three activities.

How many of the cities surveyed had

a) only a professional sports team?

b) a professional sports team and a symphony, but not a children's museum?

c) a professional sports team or a symphony?

d) a professional sports team or a symphony, but not a children's museum?

e) exactly two of the activities?

6. ***Amusement Parks*** In a survey of 85 amusement parks, it was found that
24 had a hotel on site.
55 had water slides.
38 had a wave pool.
13 had a hotel on site and water slides.
10 had a hotel on site and a wave pool.
19 had water slides and a wave pool.
7 had all three features.

How many of the amusement parks surveyed had

a) only water slides?

b) exactly one of these features?

c) at least one of these features?

d) exactly two of these features?

e) none of these features?

▲ See Exercise 6

7. ***Book Purchases*** A survey of 85 customers was taken at Barnes & Noble regarding the types of books purchased. The survey found that
44 purchased mysteries.
33 purchased science fiction.
29 purchased romance novels.
13 purchased mysteries and science fiction.
5 purchased science fiction and romance novels.
11 purchased mysteries and romance novels.
2 purchased all three types of books.

How many of the customers surveyed purchased

a) only mysteries?

b) mysteries and science fiction, but not romance novels?

c) mysteries or science fiction?

d) mysteries or science fiction, but not romance novels?

e) exactly two types?

8. ***Movies*** A survey of 350 customers was taken at Regal Cinemas in Austin, Texas, regarding the type of movies customers liked. The following information was determined.
196 liked dramas.
153 liked comedies.
88 liked science fiction.
59 liked dramas and comedies.
37 liked dramas and science fiction.
32 liked comedies and science fiction.
21 liked all three types of movies.

Of the customers surveyed, how many liked

a) none of these types of movies?

b) only dramas?

c) exactly one of these types of movies?

d) exactly two of these types of movies?

e) dramas or comedies?

9. *Transportation* Seventy businesspeople in Sacramento, California, were asked how they traveled to work during the previous month. The following information was determined.
28 used public transportation.
18 rode in a car pool.
37 drove alone.
5 used public transportation and rode in a car pool.
9 used public transportation and drove alone.
4 rode in a car pool and drove alone.
3 used all three forms of transportation.

How many of those surveyed

a) only used public transportation?

b) only drove alone?

c) used public transportation and rode in a car pool, but did not drive alone?

d) used public transportation or rode in a car pool, but did not drive alone?

e) used none of these forms of transportation?

10. *Colleges and Universities* In a survey of four-year colleges and universities, it was found that
356 offered a liberal arts degree.
293 offered a computer engineering degree.
285 offered a nursing degree.
193 offered a liberal arts degree and a computer engineering degree.
200 offered a liberal arts degree and a nursing degree.
139 offered a computer engineering degree and a nursing degree.
68 offered a liberal arts degree, a computer engineering degree, and a nursing degree.
26 offered none of these degrees.

a) How many four-year colleges and universities were surveyed?

Of the four-year colleges and universities surveyed, how many offered

b) a liberal arts degree and a nursing degree, but not a computer engineering degree?

c) a computer engineering degree, but neither a liberal arts degree nor a nursing degree?

d) exactly two of these degrees?

e) at least one of these degrees?

11. *Homeowners' Insurance Policies* A committee of the Florida legislature decided to analyze 350 homeowners' insurance policies to determine if the consumers' homes were covered for damage due to sinkholes, mold, and floods. The following results of the analysis were determined.
170 homes were covered for damage due to sinkholes.
172 homes were covered for damage due to mold.
234 homes were covered for damage due to floods.
105 homes were covered for damage due to sinkholes and mold.
115 homes were covered for damage due to mold and floods.
109 homes were covered for damage due to sinkholes and floods.
78 homes were covered for damage due to all three conditions.

How many of the homes

a) were covered for damage due to mold but were not covered for damage due to sinkholes?

b) were covered for damage due to sinkholes or mold?

c) were covered for damage due to mold and floods, but were not covered for damage due to sinkholes?

d) were not covered for damage due to any of the three conditions?

12. *Appetizers Survey* Da Tulio's Restaurant hired Dennis Goldstein to determine what kind of appetizers customers liked. He surveyed 100 people, with the following results: 78 liked shrimp cocktail, 56 liked mozzarella sticks, and 35 liked both shrimp cocktail and mozzarella sticks. Every person interviewed liked one or the other or both kinds of appetizers. Does this result seem correct? Explain your answer.

13. *Discovering an Error* An immigration agent sampled cars going from the United States into Canada. In his report, he indicated that of the 85 cars sampled,

35 cars were driven by women.
53 cars were driven by U.S. citizens.
43 cars had two or more passengers.
27 cars were driven by women who are U.S. citizens.
25 cars were driven by women and had two or more passengers.
20 cars were driven by U.S. citizens and had two or more passengers.
15 cars were driven by women who are U.S. citizens and had two or more passengers.

After his supervisor reads the report, she explains to the agent that he made a mistake. Explain how his supervisor knew that the agent's report contained an error.

CHALLENGE PROBLEMS/GROUP ACTIVITIES

14. *Parks* A survey of 300 parks showed the following.

15 had only camping.
20 had only hiking trails.
35 had only picnicking.
185 had camping.
140 had camping and hiking trails.
125 had camping and picnicking.
210 had hiking trails.

Find the number of parks that

a) had at least one of these features.

b) had all three features.

c) did not have any of these features.

d) had exactly two of these features.

15. *Surveying Farmers* A survey of 500 farmers in a midwestern state showed the following.

125 grew only wheat.
110 grew only corn.
90 grew only oats.
200 grew wheat.
60 grew wheat and corn.
50 grew wheat and oats.
180 grew corn.

Find the number of farmers who

a) grew at least one of the three.

b) grew all three.

c) did not grow any of the three.

d) grew exactly two of the three.

RECREATIONAL MATHEMATICS

16. *Number of Elements* A universal set U consists of 12 elements. If sets A, B, and C are proper subsets of U and $n(U) = 12$, $n(A \cap B) = n(A \cap C) = n(B \cap C) = 6$, $n(A \cap B \cap C) = 4$, and $n(A \cup B \cup C) = 10$, determine

a) $n(A \cup B)$ b) $n(A' \cup C)$ c) $n(A \cap B)'$

2.6 INFINITE SETS

▲ Georg Cantor, founder of set theory

Which set is larger, the set of integers or the set of even integers? One might argue that because the set of even integers is a subset of the set of integers, the set of integers must be larger than the set of even integers. Yet both sets are infinite sets, so how can we determine which set is larger? This question puzzled mathematicians for centuries until 1874, when Georg Cantor developed a method of determining the cardinal number of an infinite set. In this section, we will discuss infinite sets and how to determine the number of elements in an infinite set.

On page 48, we state that a finite set is a set in which the number of elements is zero or the number of elements can be expressed as a natural number. On page 48, we define a one-to-one correspondence. To determine the number of elements in a finite set, we can place the set in a one-to-one correspondence with a subset of the set of counting numbers. For example, the set $A = \{\#, ?, \$\}$ can be placed in one-to-one correspondence with set $B = \{1, 2, 3\}$, a subset of the set of counting numbers.

$$A = \{\#, ?, \$\}$$
$$\downarrow \ \downarrow \ \downarrow$$
$$B = \{1, 2, 3\}$$

Because the cardinal number of set B is 3, the cardinal number of set A is also 3. Any two sets such as set A and set B that can be placed in a one-to-one correspondence must have the same number of elements (therefore the same cardinality) and must be equivalent sets. Note that $n(A)$ and $n(B)$ both equal 3.

German mathematician Georg Cantor (1845–1918), known as the father of set theory, thought about sets that were not bounded. He called an unbounded set an *infinite set* and provided the following definition.

An **infinite set** is a set that can be placed in a one-to-one correspondence with a proper subset of itself.

In Example 1, we use Cantor's definition of an infinite set to show that the set of counting numbers is infinite.

EXAMPLE 1 *The Set of Natural Numbers*

Show that $N = \{1, 2, 3, 4, 5, \ldots, n, \ldots\}$ is an infinite set.

SOLUTION To show that the set N is infinite, we establish a one-to-one correspondence between the counting numbers and a proper subset of itself. By removing the first element from the set of counting numbers, we get the set $\{2, 3, 4, 5, \ldots\}$, which is a proper subset of the set of counting numbers. Now we establish the one-to-one correspondence.

$$\begin{array}{lcl} \text{Counting numbers} & = & \{1, 2, 3, 4, 5, \ldots, \quad n \quad, \ldots\} \\ & & \ \ \downarrow \ \downarrow \ \downarrow \ \downarrow \ \downarrow \qquad\quad \downarrow \\ \text{Proper subset} & = & \{2, 3, 4, 5, 6, \ldots, n + 1, \ldots\} \end{array}$$

Note that for any number, n, in the set of counting numbers, its corresponding number in the proper subset is one greater, or $n + 1$. We have now shown the desired one-to-one correspondence, and thus the set of counting numbers is infinite. ●

Note in Example 1 that we showed the pairing of the general terms $n \rightarrow (n + 1)$. Showing a one-to-one correspondence of infinite sets requires showing the pairing of the general terms in the two infinite sets.

In the set of counting numbers, n represents the general term. For any other set of numbers, the general term will be different. The general term in any set should be written in terms of n such that when 1 is substituted for n in the general term, we get the first number in the set; when 2 is substituted for n in the general term, we get the second number in the set; when 6 is substituted for n in the general term, we get the sixth number in the set; and so on.

Consider the set $\{4, 9, 14, 19, \ldots\}$. Suppose we want to write the general term for this set (or sequence) of numbers. What would the general term be? The numbers differ by 5, so the general term will be of the form $5n$ plus or minus some number. Substituting 1 for n yields 5(1), or 5. Because the first number in the set is 4, we need to subtract 1 from the 5. Thus, the general term is $5n - 1$. Note that when $n = 1$, the value is $5(1) - 1$ or 4; when $n = 2$, the value is $5(2) - 1$ or 9; when $n = 3$, the value is $5(3) - 1$ or 14; and so on. Therefore, we write the set of numbers with the general term as

$$\{4, 9, 14, 19, \ldots, 5n - 1, \ldots\}$$

Now that you are aware of how to determine the general term of a set of numbers, we can do some more problems involving sets.

EXAMPLE 2 *The Set of Even Numbers*

Show that the set of even counting numbers $\{2, 4, 6, 8, \ldots, 2n, \ldots\}$ is an infinite set.

SOLUTION First create a proper subset of the set of even counting numbers by removing the first number from the set. Then establish a one-to-one correspondence.

$$\begin{array}{ll} \text{Even counting numbers:} & \{2, 4, 6, 8, \ldots, 2n, \ldots\} \\ & \ \ \downarrow\ \downarrow\ \downarrow\ \downarrow \qquad\quad \downarrow \\ \text{Proper subset:} & \{4, 6, 8, 10, \ldots, 2n + 2, \ldots\} \end{array}$$

A one-to-one correspondence exists between the two sets, so the set of even counting numbers is infinite. •

EXAMPLE 3 *The Set of Multiples of Four*

Show that the set $\{4, 8, 12, 16, \ldots, 4n, \ldots\}$ is an infinite set.

SOLUTION

$$\begin{array}{ll} \text{Given set:} & \{4, 8, 12, 16, 20, \ldots, 4n, \ldots\} \\ & \ \ \downarrow\ \ \downarrow\ \ \downarrow\ \ \downarrow\ \ \downarrow \qquad\quad \downarrow \\ \text{Proper subset:} & \{8, 12, 16, 20, 24, \ldots, 4n + 4, \ldots\} \end{array}$$

Therefore, the given set is an infinite set. •

Countable Sets

In his work with infinite sets, Cantor developed ideas on how to determine the cardinal number of an infinite set. He called the cardinal number of infinite sets "transfinite cardinal numbers" or "transfinite powers." He defined a set as ***countable*** if it is finite or if it can be placed in a one-to-one correspondence with the set of counting numbers. All infinite sets that can be placed in a one-to-one correspondence with the set of counting numbers have cardinal number ***aleph-null***, symbolized $\aleph_0$ (the first Hebrew letter, aleph, with a zero subscript, read "null").

PROFILE IN MATHEMATICS

Leopold Kronecker

Mathematician Leopold Kronecker (1823–1891), Cantor's former mentor, ridiculed Cantor's theories and prevented Cantor from gaining a position at the University of Berlin. Although Cantor's work on infinite sets is now considered a masterpiece, it generated heated controversy when originally published. Cantor's claim that the infinite set was unbounded offended the religious views of the time that God had created a complete universe, which could not be wholly comprehended by humans. Eventually Cantor was given the recognition due to him, but by then the criticism had taken its toll on his health. He had several nervous breakdowns and spent his last days in a mental hospital. See the Profile in Mathematics on page 45 for more information on Cantor.

EXAMPLE 4 *The Cardinal Number of the Set of Even Numbers*

Show that the set of even counting numbers has cardinal number $\aleph_0$.

SOLUTION In Example 2, we showed that a set of even counting numbers is infinite by setting up a one-to-one correspondence between the set and a proper subset of itself.

Now we will show that it is countable and has cardinality $\aleph_0$ by setting up a one-to-one correspondence between the set of counting numbers and the set of even counting numbers.

$$\begin{array}{ll} \text{Counting numbers:} & N = \{1, 2, 3, 4, \ldots, \; n, \ldots\} \\ & \quad\;\;\; \downarrow\; \downarrow\; \downarrow\; \downarrow \qquad\;\; \downarrow \\ \text{Even counting numbers:} & E = \{2, 4, 6, 8, \ldots, 2n, \ldots\} \end{array}$$

For each number n in the set of counting numbers, its corresponding number is $2n$. Since we found a one-to-one correspondence between the set of counting numbers and the set of even counting numbers, the set of even counting numbers is countable. Thus, the cardinal number of the set of even counting numbers is $\aleph_0$; that is, $n(E) = \aleph_0$. As we mentioned earlier, the set of even counting numbers is an infinite set since it can be placed in a one-to-one correspondence with a proper subset of itself. Therefore, the set of even counting numbers is both infinite and countable.

Any set that can be placed in a one-to-one correspondence with the set of counting numbers has cardinality $\aleph_0$ and is infinite and is countable.

EXAMPLE 5 *The Cardinal Number of the Set of Odd Numbers*

Show that the set of odd counting numbers has cardinality $\aleph_0$.

SOLUTION To show that the set of odd counting numbers has cardinality $\aleph_0$, we need to show a one-to-one correspondence between the set of counting numbers and the set of odd counting numbers.

$$\begin{array}{ll} \text{Counting numbers:} & N = \{1, 2, 3, 4, 5, \ldots, \quad n \quad, \ldots\} \\ & \quad\;\;\; \downarrow\; \downarrow\; \downarrow\; \downarrow\; \downarrow \qquad\quad \downarrow \\ \text{Odd counting numbers:} & O = \{1, 3, 5, 7, 9, \ldots, 2n - 1, \ldots\} \end{array}$$

Since there is a one-to-one correspondence, the odd counting numbers have cardinality $\aleph_0$; that is, $n(O) = \aleph_0$.

We have shown that both the odd and even counting numbers have cardinality $\aleph_0$. Merging the odd counting numbers with the even counting numbers gives the set of counting numbers, and we may reason that

$$\aleph_0 + \aleph_0 = \aleph_0$$

This result may seem strange, but it is true. What could such a statement mean? Well, consider a hotel with infinitely many rooms. If all the rooms are occupied, the hotel is, of course, full. If more guests appear wanting accommodations, will they be turned away? The answer is *no,* for if the room clerk were to reassign each guest to a new room

▲ . . . where there's always room for one more. . .

with a room number twice that of the present room, all the odd-numbered rooms would become unoccupied and there would be space for more guests!

Cantor showed that there are different orders of infinity. Sets that are countable and have cardinal number $\aleph_0$ are the lowest order of infinity. Cantor showed that the set of integers and the set of rational numbers (fractions of the form p/q, where $q \neq 0$) are infinite sets with cardinality $\aleph_0$. He also showed that the set of real numbers (discussed in Chapter 5) could not be placed in a one-to-one correspondence with the set of counting numbers and that they have a higher order of infinity.

SECTION 2.6 EXERCISES

CONCEPT/WRITING EXERCISES

1. What is an infinite set as defined in this section?
2. a) What is a countable set?
   b) How can we determine if a given set has cardinality $\aleph_0$?

PRACTICE THE SKILLS

In Exercises 3–12, show that the set is infinite by placing it in a one-to-one correspondence with a proper subset of itself. Be sure to show the pairing of the general terms in the sets.

3. $\{5, 6, 7, 8, 9, \dots\}$
4. $\{20, 21, 22, 23, 24, \dots\}$
5. $\{3, 5, 7, 9, 11, \dots\}$
6. $\{20, 22, 24, 26, 28, \dots\}$
7. $\{3, 7, 11, 15, 19, \dots\}$
8. $\{4, 8, 12, 16, 20, \dots\}$
9. $\{6, 11, 16, 21, 26, \dots\}$
10. $\{1, \frac{1}{2}, \frac{1}{3}, \frac{1}{4}, \frac{1}{5}, \dots\}$
11. $\{\frac{1}{2}, \frac{1}{4}, \frac{1}{6}, \frac{1}{8}, \frac{1}{10}, \dots\}$
12. $\{\frac{6}{13}, \frac{7}{13}, \frac{8}{13}, \frac{9}{13}, \frac{10}{13}, \dots\}$

In Exercises 13–22, show that the set has cardinal number $\aleph_0$ by establishing a one-to-one correspondence between the set of counting numbers and the given set. Be sure to show the pairing of the general terms in the sets.

13. $\{3, 6, 9, 12, 15, \dots\}$
14. $\{50, 51, 52, 53, 54, \dots\}$
15. $\{4, 6, 8, 10, 12, \dots\}$
16. $\{0, 2, 4, 6, 8, \dots\}$
17. $\{2, 5, 8, 11, 14, \dots\}$
18. $\{6, 11, 16, 21, 26, \dots\}$
19. $\{5, 9, 13, 17, 21, \dots\}$
20. $\{\frac{1}{2}, \frac{1}{4}, \frac{1}{6}, \frac{1}{8}, \dots\}$
21. $\{\frac{1}{3}, \frac{1}{4}, \frac{1}{5}, \frac{1}{6}, \frac{1}{7}, \dots\}$
22. $\{\frac{1}{2}, \frac{2}{3}, \frac{3}{4}, \frac{4}{5}, \frac{5}{6}, \dots\}$

CHALLENGE PROBLEMS/GROUP ACTIVITIES

In Exercises 23–26, show that the set has cardinality $\aleph_0$ by establishing a one-to-one correspondence between the set of counting numbers and the given set.

23. $\{1, 4, 9, 16, 25, \dots\}$
24. $\{2, 4, 8, 16, 32, \dots\}$
25. $\{3, 9, 27, 81, 243, \dots\}$
26. $\{\frac{1}{3}, \frac{1}{6}, \frac{1}{12}, \frac{1}{24}, \frac{1}{48}, \dots\}$

RECREATIONAL MATHEMATICS

In Exercises 27–31, insert the symbol $<$, $>$, or $=$ in the shaded area to make a true statement.

27. $\aleph_0$ ■ $\aleph_0 + \aleph_0$
28. $2\aleph_0$ ■ $\aleph_0 + \aleph_0$
29. $2\aleph_0$ ■ $\aleph_0$

30. $\aleph_0 + 5$ ▢ $\aleph_0 - 3$

31. $n(N)$ ▢ $\aleph_0$

32. There are a number of paradoxes (a statement that appears to be true and false at the same time) associated with infinite sets and the concept of infinity. One of these, called *Zeno's Paradox,* is named after the mathematician Zeno, born about 496 B.C. in Italy. According to Zeno's paradox, suppose Achelles starts out 1 meter behind a tortoise. Also, suppose Achelles walks 10 times as fast as the tortoise crawls. When Achelles reaches the point where the tortoise started, the tortoise is 1/10 of a meter ahead of Achelles; when Achelles reaches the point where the tortoise was 1/10 of a meter ahead, the tortoise is now 1/100 of a meter ahead; and so on. According to Zeno's Paradox, Achelles gets closer and closer to the tortoise but never catches up to the tortoise.

a) Do you believe the reasoning process is sound? If not, explain why not.

b) In actuality, if this situation were real, would Achelles ever pass the tortoise?

INTERNET/RESEARCH ACTIVITIES

33. Do research to explain how Cantor proved that the set of rational numbers has cardinal number $\aleph_0$.

34. Do research to explain how it can be shown that the real numbers do not have cardinal number $\aleph_0$.

CHAPTER 2 SUMMARY

IMPORTANT FACTS

And is generally interpreted to mean *intersection*.

Or is generally interpreted to mean *union*.

DE MORGAN'S LAWS

$$(A \cap B)' = A' \cup B'$$
$$(A \cup B)' = A' \cap B'$$

For any sets A and B,

$$n(A \cup B) = n(A) + n(B) - n(A \cap B).$$

Number of distinct subsets of a finite set with n elements is 2^n.

Symbol	Meaning
$\in$	is an element of
$\notin$	is not an element of
$n(A)$	number of elements in set A
$\emptyset$ or { }	the empty set
U	the universal set
$\subseteq$	is a subset of
$\not\subseteq$	is not a subset of
$\subset$	is a proper subset of
$\not\subset$	is not a proper subset of
$'$	complement
$\cap$	intersection
$\cup$	union
$-$	difference of two sets
$\times$	cartesian product
$\aleph_0$	aleph-null

CHAPTER 2 REVIEW EXERCISES

2.1, 2.2, 2.3, 2.4, 2.6

In Exercises 1–14, state whether each is true or false. If false, give a reason.

1. The set of colleges located in the state of Oregon is a well-defined set.

2. The set of the three best cities in the United States is a well-defined set.

3. maple $\in$ {oak, elm, maple, sycamore}

4. $\{ \} \subset \emptyset$

5. {3, 6, 9, 12, ...} and {2, 4, 6, 8, ...} are disjoint sets.

6. {*a*, *b*, *c*, 1, 2} is an example of a set in roster form.

7. {plane, train, automobile} = {train, plane, motorcycle}

8. {apple, orange, banana, pear} is equivalent to {tomato, corn, spinach, radish}.

9. If A = {a, e, i, o, u}, then $n(A) = 5$.

10. $A = \{1, 4, 9, 16, \ldots\}$ is a countable set.

11. $A = \{1, 4, 7, 10, \ldots, 31\}$ is a finite set.

12. $\{2, 5, 7\} \subseteq \{2, 5, 7, 10\}$.

13. $\{x \mid x \in N \text{ and } 3 < x \leq 9\}$ is a set in set-builder notation.

14. $\{x \mid x \in N \text{ and } 2 < x \leq 12\} \subseteq \{1, 2, 3, 4, 5, \ldots, 20\}$

In Exercises 15–18, express each set in roster form.

15. Set A is the set of odd natural numbers between 5 and 16.

16. Set B is the set of states that border Kansas.

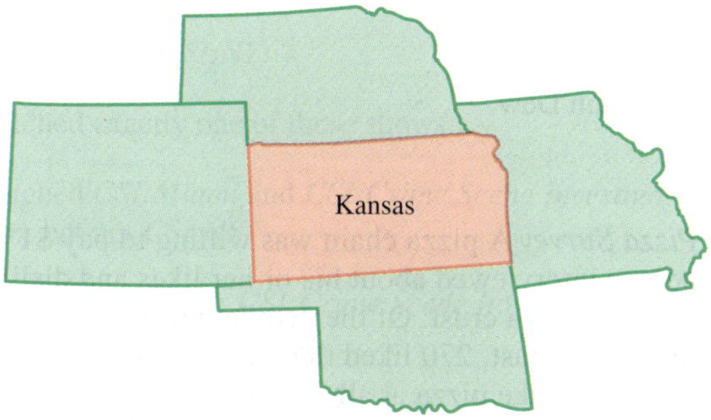

17. $C = \{x \mid x \in N \text{ and } x < 162\}$

18. $D = \{x \mid x \in N \text{ and } 8 < x \leq 96\}$

In Exercises 19–22, express each set in set-builder notation.

19. Set A is the set of natural numbers between 52 and 100.

20. Set B is the set of natural numbers greater than 42.

21. Set C is the set of natural numbers less than 5.

22. Set D is the set of natural numbers between 27 and 51, inclusive.

In Exercises 23–26, express each set with a written description.

23. $A = \{x \mid x$ is a capital letter of the English alphabet from E through M inclusive$\}$

24. $B = \{$penny, nickel, dime, quarter, half-dollar$\}$

25. $C = \{x, y, z\}$

26. $D = \{x \mid 3 \leq x < 9\}$

In Exercises 27–36, let

$$U = \{1, 2, 3, 4, \ldots, 10\}$$
$$A = \{1, 3, 5, 7\}$$
$$B = \{5, 7, 9, 10\}$$
$$C = \{1, 7, 10\}$$

Determine the following.

27. $A \cap B$

28. $A \cup B'$

29. $A' \cap B$

30. $(A \cup B)' \cup C$

31. $A - B$

32. $A - C'$

33. $A \times C$

34. $B \times A$

35. The number of subsets of set B

36. The number of proper subsets of set A

37. For the following sets, construct a Venn diagram and place the elements in the proper region.

$U = \{$lion, tiger, leopard, cheetah, puma, lynx, panther, jaguar$\}$

$A = \{$tiger, puma, lynx$\}$

$B = \{$lion, tiger, jaguar, panther$\}$

$C = \{$tiger, lynx, cheetah, panther$\}$

16. $A - B$

17. $A \times C$

18. Using the sets provided for Exercises 12–17, draw a Venn diagram illustrating the relationship among the sets.

19. Use a Venn diagram to determine whether

$$A \cap (B \cup C') = (A \cap B) \cup (A \cap C')$$

for all sets A, B, and C. Show your work.

20. ***Snacks at a Circus*** Of the 155 people who purchased snacks at a circus,

76 purchased cotton candy.
90 purchased peanuts.
107 purchased popcorn.
52 purchased cotton candy and peanuts.
54 purchased cotton candy and popcorn.
57 purchased peanuts and popcorn.
35 purchased all three snacks.

Construct a Venn diagram and then determine how many people purchased

a) exactly one of these snacks.

b) none of these snacks.

c) at least two of these snacks.

d) cotton candy and peanuts, but not popcorn.

e) cotton candy or peanuts, but not popcorn.

f) only popcorn.

21. Show that the following set is infinite by setting up a one-to-one correspondence between the set and a proper subset of itself.

$$\{7, 8, 9, 10, \dots\}$$

22. Show that the following set has cardinal number $\aleph_0$ by setting up a one-to-one correspondence between the set of counting numbers and the set.

$$\{1, 3, 5, 7, \dots\}$$

GROUP PROJECTS

SELECTING A FAMILY PET

1. The Wilcox family is considering buying a dog. They have established several criteria for the family dog: It must be one of the breeds listed in the table, must not shed, must be less than 16 in. tall, and must be good with children.

   a) Using the information in the table,* construct a Venn diagram in which the universal set is the dogs listed. Indicate the set of dogs to be placed in each region of the Venn diagram.

   b) From the Venn diagram constructed in part (a), determine which dogs will meet the criteria set by the Wilcox family. Explain.

Breed	Sheds	Less than 16 in.	Good with children
Airedale	no	no	no
Basset hound	yes	yes	yes
Beagle	yes	yes	yes
Border terrier	no	yes	yes
Cairn terrier	no	yes	no
Cocker spaniel	yes	yes	yes
Collie	yes	no	yes
Dachshund	yes	yes	no
Poodle, miniature	no	yes	no
Schnauzer, miniature	no	yes	no
Scottish terrier	no	yes	no
Wirehaired fox terrier	no	yes	no

*The information is a collection of the opinions of an animal psychologist, Dr. Daniel Tortora, and a group of veterinarians.

CLASSIFICATION OF THE DOMESTIC CAT

2. Read the Did You Know? feature on page 57. Do research and indicate the name of the following groupings to which the domestic cat belongs.

a) Kingdom

b) Phylum

c) Class

d) Order

e) Family

f) Genus

g) Species

WHO LIVES WHERE

3. On Diplomat Row, an area of Washington, D.C., there are five houses. Each owner is a different nationality, each has a different pet, each has a different favorite food, each has a different favorite drink, and each house is painted a different color.

The green house is directly to the right of the ivory house.
The Senegalese has the red house.
The dog belongs to the Spaniard.
The Afghanistani drinks tea.
The person who eats cheese lives next door to the fox.
The Japanese eats fish.
Milk is drunk in the middle house.
Apples are eaten in the house next to the horse.
Ale is drunk in the green house.
The Norwegian lives in the first house.
The peach eater drinks whiskey.
Apples are eaten in the yellow house.
The banana eater owns a snail.
The Norwegian lives next door to the blue house.

For each house find

a) the color.

b) the nationality of the occupant.

c) the owner's favorite food.

d) the owner's favorite drink.

e) the owner's pet.

f) Finally, the crucial question is: Does the zebra's owner drink vodka or ale?

CHAPTER 3

Logic

▲ Logic is used to communicate effectively and make convincing arguments.

WHAT YOU WILL LEARN

- Statements, quantifiers, and compound statements
- Statements involving the words *not, and, or, if . . . then . . .,* and *if and only if*
- Truth tables for negations, conjunctions, disjunctions, conditional statements, and biconditional statements
- Self-contradictions, tautologies, and implications
- Equivalent statements, De Morgan's laws, and variations of conditional statements
- Symbolic arguments and standard forms of arguments
- Euler diagrams and syllogistic arguments
- Using logic to analyze switching circuits

WHY IT IS IMPORTANT

The study of logic enables us to communicate effectively, make more convincing arguments, and develop patterns of reasoning for decision making. The study of logic also prepares an individual to better understand the thought processes involved in learning other areas of mathematics and other subjects.

3.1 STATEMENTS AND LOGICAL CONNECTIVES

▲ "When it rains it pours." Logic symbols can be used to help us analyze statements made by advertisers.

Almost everyday we see or hear advertisements that attempt to influence our buying habits. Advertisements often rely on spoken or written statements that are used to favorably portray the advertised product and form a convincing argument that will persuade us to purchase the product. Some familiar advertising statements are: *Don't leave home without it; It takes a licking and keeps on ticking; Sometimes you feel like a nut, sometimes you don't;* and *When it rains it pours*. In this section, we will learn how to represent statements using logic symbols that may help us better understand the nature of the statement. We will use these symbols throughout the chapter to analyze more complicated statements. Such statements appear in everyday life in, for example, legal documents, product instructions, and game rules in addition to advertising.

History

The ancient Greeks were the first people to systematically analyze the way humans think and arrive at conclusions. Aristotle (384–322 B.C.) organized the study of logic for the first time in a work called *Organon*. As a result of his work, Aristotle is called the father of logic. The logic from this period, called ***Aristotelian logic***, has been taught and studied for more than 2000 years.

Since Aristotle's time, the study of logic has been continued by other great philosophers and mathematicians. Gottfried Wilhelm Leibniz (1646–1716) had a deep conviction that all mathematical and scientific concepts could be derived from logic. As a result, he became the first serious student of ***symbolic logic***. One difference between symbolic logic and Aristotelian logic is that in symbolic logic, as its name implies, symbols (usually letters) represent written statements. A self-educated English mathematician, George Boole (1815–1864), is considered to be the founder of symbolic logic because of his impressive work in this area. Among Boole's publications are *The Mathematical Analysis of Logic* (1847) and *An Investigation of the Law of Thought* (1854). Mathematician Charles Dodgson, better known as Lewis Carroll, incorporated many interesting ideas from logic into his books *Alice's Adventures in Wonderland* and *Through the Looking Glass* and his other children's stories.

Logic has been studied through the ages to exercise the mind's ability to reason. Understanding logic will enable you to think clearly, communicate effectively, make more convincing arguments, and develop patterns of reasoning that will help you in making decisions. It will also help you to detect the fallacies in the reasoning or arguments of others such as advertisers and politicians. Studying logic has other practical applications, such as helping you to understand wills, contracts, and other legal documents.

The study of logic is also good preparation for other areas of mathematics. If you preview Chapter 12, on probability, you will see formulas for the probability of A or B and the probability of A and B, symbolized as $P(A \text{ or } B)$ and $P(A \text{ and } B)$, respectively. Special meanings of common words such as *or* and *and* apply to all areas of mathematics. The meaning of these and other special words is discussed in this chapter.

Logic and the English Language

In reading, writing, and speaking, we use many words such as *and, or,* and *if . . . then . . .* to connect thoughts. In logic we call these words ***connectives***. How are these words interpreted in daily communication? A judge announces to a convicted offender, "I hereby sentence you to five months of community service *and* a fine of $100." In this case, we normally interpret the word *and* to indicate that *both* events will take place. That is, the person must perform community service and must also pay a fine.

Now suppose a judge states, "I sentence you to six months in prison *or* 10 months of community service." In this case, we interpret the connective *or* as meaning the convicted person must either spend the time in jail or perform community service, but not both. The word *or* in this case is the ***exclusive or***. When the exclusive *or* is used, one or the other of the events can take place, but *not both*.

In a restaurant, a waiter asks, "May I interest you in a cup of soup or a sandwich?" This question offers three possibilities: You may order soup, you may order a sandwich, or you may order both soup and a sandwich. The *or* in this case is the ***inclusive or***. When the inclusive *or* is used, one or the other, *or both* events can take place. *In this chapter, when we use the word* or *in a logic statement, it will mean the* inclusive or *unless stated otherwise*.

If–then statements are often used to relate two ideas, as in the bank policy statement "If the average daily balance is greater than $500, then there will be no service charge." If–then statements are also used to emphasize a point or add humor, as in the statement "If the Cubs win, then I will be a monkey's uncle."

Now let's look at logic from a mathematical point of view.

Statements and Logical Connectives

▲ The Brooklyn Bridge in New York City

A sentence that can be judged either true or false is called a ***statement***. Labeling a statement true or false is called *assigning a truth value* to the statement. Here are some examples of statements.

1. The Brooklyn Bridge goes over San Francisco Bay.
2. Disney World is in Idaho.
3. The Mississippi River is the longest river in the United States.

In each case, we can say that the sentence is either true or false. Statement 1 is false because the Brooklyn Bridge does not go over San Francisco Bay. Statement 2 is false because Disney World is in Florida. By looking at a map or reading an almanac, we can determine that the Mississippi River is the longest river in the United States; therefore, statement 3 is true.

The three sentences discussed above are examples of ***simple statements*** because they convey one idea. Sentences combining two or more ideas that can be assigned a truth value are called ***compound statements***. Compound statements are discussed shortly.

Quantifiers

Sometimes it is necessary to change a statement to its opposite meaning. To do so, we use the ***negation*** of a statement. For example, the negation of the statement "Emily is at home" is "Emily is not at home." The negation of a true statement is always a false statement, and the negation of a false statement is always a true statement. We must use special caution when negating statements containing the words ***all***, ***none*** (or ***no***), and ***some***. These words are referred to as ***quantifiers***.

Consider the statement "All lakes contain fresh water." We know this statement is false because the Great Salt Lake in Utah contains salt water. Its negation must therefore be true. We may be tempted to write its negation as "No lake contains fresh water," but this statement is also false because Lake Superior contains fresh water. Therefore, "No lakes contain fresh water" is not the negation of "All lakes contain fresh water." The correct negation of "All lakes contain fresh water" is "Not all lakes contain fresh water" or "At least one lake does not contain fresh water" or "Some lakes do not contain fresh water." These statements all imply that at least one lake does not contain fresh water, which is a true statement.

Now consider the statement "No birds can swim." This statement is false because at least one bird, the penguin, can swim. Therefore, the negation of this statement must be true. We may be tempted to write the negation as "All birds can swim," but because this statement is also false it cannot be the negation. The correct negation of the statement is "Some birds can swim" or "At least one bird can swim," each of which is a true statement.

Now let's consider statements involving the quantifier *some,* as in "Some students have a driver's license." This statement is true, meaning that at least one student has a driver's license. The negation of this statement must therefore be false. The negation is "No student has a driver's license," which is a false statement.

Consider the statement "Some students do not ride motorcycles." This statement is true because it means "At least one student does not ride a motorcycle." The negation of this statement must therefore be false. The negation is "All students ride motorcycles," which is a false statement.

The negation of quantified statements is summarized as follows:

Form of statement	Form of negation
All are.	Some are not.
None are.	Some are.
Some are.	None are.
Some are not.	All are.

The following diagram might help you to remember the statements and their negations:

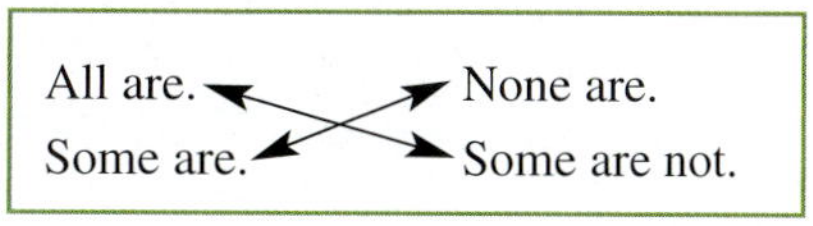

The quantifiers diagonally opposite each other are the negations of each other.

EXAMPLE 1 *Write Negations*

Write the negation of each statement.

a) Some snakes are poisonous.

b) All swimming pools are rectangular.

DID YOU KNOW?

Playing on Words

George Boole, Augustus De Morgan, and other mathematicians of the nineteenth century were anxious to make logic an abstract science that would operate like algebra but be applicable to all fields. One of the problems logicians faced was that verbal language could be ambiguous and could easily lead to confusion and contradiction. Comedians Bud Abbott and Lou Costello had fun with the ambiguity of language in their skit about the baseball players: "Who's on first, What's on second, I Don't Know is on third—Yeah, but who's on first?" (in the film *The Naughty Nineties*, 1945).

SOLUTION

a) Since *some* means "at least one," the statement "Some snakes are poisonous" is the same as "At least one snake is poisonous." Because it is a true statement, its negation must be false. The negation is "No snakes are poisonous," which is a false statement.

b) The statement "All swimming pools are rectangular" is a false statement since some pools are circular, some are oval, and some have other shapes. Its negation must therefore be true. The negation may be written as "Some swimming pools are not rectangular" or "Not all swimming pools are rectangular" or "At least one swimming pool is not rectangular." Each of these statements is true.

Compound Statements

Statements consisting of two or more simple statements are called **compound statements**. The connectives often used to join two simple statements are

and, **or,** **if,... then...,** **if and only if**

In addition, we consider a simple statement that has been negated to be a compound statement. The word *not* is generally used to negate a statement.

To reduce the amount of writing in logic, it is common to represent each simple statement with a lowercase letter. For example, suppose we are discussing the simple statement "Leland is a farmer." Instead of writing "Leland is a farmer" over and over again, we can let p represent the statement "Leland is a farmer." Thereafter we can simply refer to the statement with the letter p. It is customary to use the letters p, q, r, and s to represent simple statements, but other letters may be used instead. Let's now look at the connectives used to make compound statements.

Not Statements

The negation is symbolized by $\sim$ and read "not." For example, the negation of the statement "Steve is a college student" is "Steve is not a college student." If p represents the simple statement "Steve is a college student," then $\sim p$ represents the compound statement "Steve is not a college student." For any statement p, $\sim(\sim p) = p$. For example, the negation of the statement "Steve is not a college student" is "Steve is a college student."

Consider the statement "Inga is not at home." This statement contains the word *not,* which indicates that it is a negation. To write this statement symbolically, we let p represent "Inga *is* at home." Then $\sim p$ would be "Inga is not at home." *We will use this convention of letting letters such as p, q, or r represent statements that are not negated. We will represent negated statements with the negation symbol,* $\sim$.

And Statements

The *conjunction* is symbolized by $\wedge$ and read "and." The $\wedge$ looks like an A (for And) with the bar missing. Let p and q represent the simple statements.

p: You will perform 5 months of community service.
q: You will pay a $100 fine.

Then the following is the conjunction written in symbolic form.

You will perform 5 months of community service	and	you will pay a \$100 fine.
↑	↑	↑
p	$\wedge$	q.

The conjunction is generally expressed as *and*. Other words sometimes used to express a conjunction are *but, however,* and *nevertheless*.

▲ Billie Joe Armstrong of Green Day

EXAMPLE 2 *Write a Conjunction*

Write the following conjunction in symbolic form.

Green Day is not on tour, but Green Day is recording a new CD.

SOLUTION Let t and r represent the simple statements.

t: Green Day is on tour.
r: Green Day is recording a new CD.

In symbolic form, the compound statement is $\sim t \wedge r$. ●

In Example 2, the compound statement is "Green Day is not on tour, but Green Day is recording a new CD." This statement could also be represented as "Green Day is not on tour, but *they* are recording a new CD." In this problem, it should be clear the word *they* means *Green Day*. Therefore, the statement, "Green Day is not on tour, but they are recording a new CD" would also be symbolized as $\sim t \wedge r$.

Or Statements

The ***disjunction*** is symbolized by $\vee$ and read "or." The *or* we use in this book (except where indicated in the exercise sets) is the *inclusive or* described on page 102.

EXAMPLE 3 *Write a Disjunction*

Let

p: Maria will go to the circus.
q: Maria will go to the zoo.

Write the following statements in symbolic form.

a) Maria will go to the circus or Maria will go to the zoo.
b) Maria will go to the zoo or Maria will not go to the circus.
c) Maria will not go to the circus or Maria will not go to the zoo.

SOLUTION

a) $p \vee q$ b) $q \vee \sim p$ c) $\sim p \vee \sim q$ ●

> **PROFILE IN MATHEMATICS**
>
> ***George Boole***
>
> 
>
> The self-taught English mathematician George Boole (1815–1864) took the operations of algebra and used them to extend Aristotelian logic. He used symbols such as x and y to represent particular qualities or objects in question. For example, if x represents all butterflies, then $1 - x$ represents everything else except butterflies. If y represents the color yellow, then $(1 - x)(1 - y)$ represents everything except butterflies and things that are yellow, or yellow butterflies. This development added a computational dimension to logic that provided a basis for twentieth-century work in the field of computing.

Because *or* represents the *inclusive or,* the statement "Maria will go to the circus or Maria will go to the zoo" in Example 3(a) may mean that Maria will go to the circus, or that Maria will go to the zoo, or that Maria will go to both the circus *and* the zoo. The statement in Example 3(a) could also be written as "Maria will go to the circus or the zoo."

When a compound statement contains more than one connective, a comma can be used to indicate which simple statements are to be grouped together. When we write the compound statement symbolically, ***the simple statements on the same side of the comma are to be grouped together within parentheses.***

For example, "Pink is a singer (p) or Geena Davis is an actress (g), and Dallas is in Texas (d)" is written $(p \vee g) \wedge d$. Note that the p and g are both on the same side of the comma in the written statement. They are therefore grouped together within parentheses. The statement "Pink is a singer, or Geena Davis is an actress and Dallas is in Texas" is written $p \vee (g \wedge d)$. In this case, g and d are on the same side of the comma and are therefore grouped together within parentheses.

EXAMPLE 4 *Understand How Commas Are Used to Group Statements*

Let

p: Dinner includes soup.
q: Dinner includes salad.
r: Dinner includes the vegetable of the day.

Write the following statements in symbolic form.

a) Dinner includes soup, and salad or the vegetable of the day.
b) Dinner includes soup and salad, or the vegetable of the day.

SOLUTION

a) The comma tells us to group the statement "Dinner includes salad" with the statement "Dinner includes the vegetable of the day." Note that both statements are on the same side of the comma. The statement in symbolic form is $p \wedge (q \vee r)$.

In mathematics, we always evaluate the information within the parentheses first. Since the conjunction, $\wedge$, is outside the parentheses and is evaluated *last,* this statement is considered a *conjunction*.

b) The comma tells us to group the statement "Dinner includes soup" with the statement "Dinner includes salad." Note that both statements are on the same side of the comma. The statement in symbolic form is $(p \wedge q) \vee r$. Since the disjunction, $\vee$, is outside the parentheses and is evaluated *last,* this statement is considered a *disjunction*.

The information provided in Example 4 is summarized below.

Statement	Symbolic representation	Type of statement
Dinner includes soup, and salad or the vegetable of the day.	$p \wedge (q \vee r)$	conjunction
Dinner includes soup and salad, or the vegetable of the day.	$(p \wedge q) \vee r$	disjunction

A negation symbol has the effect of negating only the statement that directly follows it. To negate a compound statement, we must use parentheses. When a negation symbol is placed in front of a statement in parentheses, it negates the entire statement in parentheses. The negation symbol in this case is read, "It is not true that . . ." or "It is false that . . ."

EXAMPLE 5 ***Change Symbolic Statements to Words***

Let

p: Jozsef is making breakfast.
q: Arum is setting the table.

Write the following symbolic statements in words.

a) $p \wedge \sim q$ b) $\sim p \vee \sim q$ c) $\sim(p \wedge q)$

SOLUTION

a) Jozsef is making breakfast and Arum is not setting the table.
b) Jozsef is not making breakfast or Arum is not setting the table.
c) It is false that Jozsef is making breakfast and Arum is setting the table.

Recall that the word *but* may also be used in a conjunction. Therefore, Example 5(a) could also be written "Jozsef is making breakfast, *but* Arum is not setting the table."

Part (b) of Example 5 is a disjunction, since it can be written $(\sim p) \vee (\sim q)$. Part (c), which is $\sim(p \wedge q)$, is a negation since the negation symbol negates the entire statement within parentheses. The similarity of these two statements is discussed in Section 3.4.

Occasionally, we come across a ***neither–nor*** statement, such as "John is neither handsome nor rich." This statement means that John is not handsome *and* John is not rich. If p represents "John is handsome" and q represents "John is rich," this statement is symbolized by $\sim p \wedge \sim q$.

If–Then Statements

The ***conditional*** is symbolized by $\rightarrow$ and is read "if–then." The statement $p \rightarrow q$ is read "If p, then q."* The conditional statement consists of two parts: the part that precedes the arrow is the ***antecedent***, and the part that follows the arrow is the ***consequent***.† In the conditional statement $p \rightarrow q$, the p is the antecedent and the q is the consequent.

In the conditional statement $\sim(p \vee q) \rightarrow (p \wedge q)$, the antecedent is $\sim(p \vee q)$ and the consequent is $(p \wedge q)$. An example of a conditional statement is "If you drink your milk, then you will grow up to be healthy." A conditional symbol may be placed between any two statements even if the statements are not related.

*Some books indicate that $p \rightarrow q$ may also be read "p implies q." Many higher-level mathematics books, however, indicate that $p \rightarrow q$ may be read "p implies q" only under certain conditions. Implications are discussed in Section 3.3.

†Some books refer to the antecedent as the hypothesis or premise and the consequent as the conclusion.

Sometimes the word *then* in a conditional statement is not explicitly stated. For example, the statement "If you get an A, I will buy you a car" is a conditional statement because it actually means "If you get an A, then I will buy you a car."

▲ *Zoe* by Beth Anderson

EXAMPLE 6 *Write Conditional Statements*

Let

p: The portrait is a pastel.
q: The portrait is by Beth Anderson.

Write the following statements symbolically.

a) If the portrait is a pastel, then the portrait is by Beth Anderson.
b) If the portrait is by Beth Anderson, then the portrait is not a pastel.
c) It is false that if the portrait is by Beth Anderson then the portrait is a pastel.

SOLUTION

a) $p \rightarrow q$ b) $q \rightarrow \sim p$ c) $\sim(q \rightarrow p)$

EXAMPLE 7 *Use Commas When Writing a Symbolic Statement in Words*

Let

p: Jorge is enrolled in calculus.
q: Jorge's major is criminal justice.
r: Jorge's major is engineering.

Write the following symbolic statements in words and indicate whether the statement is a negation, conjunction, disjunction, or conditional.

a) $(q \rightarrow \sim p) \vee r$ b) $q \rightarrow (\sim p \vee r)$

SOLUTION The parentheses indicate where to place the commas in the sentences.

a) "If Jorge's major is criminal justice then Jorge is not enrolled in calculus, or Jorge's major is engineering." This statement is a disjunction because $\vee$ is outside the parentheses.
b) "If Jorge's major is criminal justice, then Jorge is not enrolled in calculus or Jorge's major is engineering." This statement is a conditional because $\rightarrow$ is outside the parentheses.

If and Only if Statements

The *biconditional* is symbolized by $\leftrightarrow$ and is read "if and only if." The phrase *if and only if* is sometimes abbreviated as "iff." The statement $p \leftrightarrow q$ is read "p if and only if q."

▲ Martin St. Louis

EXAMPLE 8 *Write Statements Using the Biconditional*

Let

p: The Tampa Bay Lightning win the Stanley Cup.
q: Martin St. Louis is the MVP.

Write the following symbolic statements in words.

a) $p \leftrightarrow \sim q$ b) $\sim(q \leftrightarrow \sim p)$

SOLUTION

a) The Tampa Bay Lightning win the Stanley Cup if and only if Martin St. Louis is not the MVP.

b) It is false that Martin St. Louis is the MVP if and only if the Tampa Bay Lightning do not win the Stanley Cup. ●

You will learn later that $p \leftrightarrow q$ means the same as $(p \rightarrow q) \wedge (q \rightarrow p)$. Therefore, the statement "I will go to college if and only if I can pay the tuition" has the same logical meaning as "If I go to college then I can pay the tuition, and if I can pay the tuition then I will go to college."

A summary of the connectives discussed in this section is given in Table 3.1.

Table 3.1 Logical Connectives

Formal Name	Symbol	Read	Symbolic Form
Negation	$\sim$	"Not"	$\sim p$
Conjunction	$\wedge$	"And"	$p \wedge q$
Disjunction	$\vee$	"Or"	$p \vee q$
Conditional	$\rightarrow$	"If-then"	$p \rightarrow q$
Biconditional	$\leftrightarrow$	"If and only if"	$p \leftrightarrow q$

SECTION 3.1 EXERCISES

CONCEPT/WRITING EXERCISES

1. a) What is a statement?

   b) What is a simple statement?

   c) What is a compound statement?

2. Fill in the blanks to make the following statements true.

   a) The negation of a true statement is always a ______ statement.

   b) The negation of a ______ statement is always a true statement.

3. Give three words that can be used as a quantifier in a statement.

4. Represent the statement, "The telephone does not have caller ID" symbolically. Explain your answer.

5. Write the general form of the negation for statements of the form

a) none are.

b) some are not.

c) all are.

d) some are.

6. Draw the symbol used to represent the

a) conditional.

b) disjunction.

c) conjunction.

d) negation.

e) biconditional.

7. a) When the *exclusive or* is used as a connective between two events, can both events take place? Explain.

b) When the *inclusive or* is used as a connective between two events, can both events take place? Explain.

c) Which *or*, the *inclusive or* or the *exclusive or*, is used in this chapter?

8. Explain how a comma is used to indicate the grouping of simple statements.

PRACTICE THE SKILLS/PROBLEM SOLVING

In Exercises 9–22, indicate whether the statement is a simple statement or a compound statement. If it is a compound statement, indicate whether it is a negation, conjunction, disjunction, conditional, or biconditional by using both the word and its appropriate symbol (for example, "a negation," ~).

9. Jana Bryant is teaching statistics or she is teaching calculus.

10. If you burn the chicken wings, then you can feed them to the dog.

11. Time will go backwards if and only if you travel faster than the speed of light.

12. Louis Armstrong did not play the drums.

▲ Louis Armstrong, see Exercise 12

13. Bobby Glewen joined the Army and he got married.

14. The book was neither a novel nor an autobiography.

15. The hurricane did $400,000 worth of damage to DeSoto County.

16. Inhibor Melendez will be admitted to law school if and only if he earns his bachelor's degree.

17. It is false that Jeffery Hilt is a high school teacher and a grade school teacher.

18. If Cathy Smith walks 4 miles today, then she will be sore tomorrow.

19. Mary Jo Woo ran 4 miles today and she lifted weights for 30 minutes.

20. Nancy Wallin went to the game, but she did not eat a hot dog.

21. It is false that if John Wubben fixes your car then you will need to pay him in cash.

22. If Buddy and Evelyn Cordova are residents of Budville, then they must vote for mayor on Tuesday.

In Exercises 23–34, write the negation of the statement.

23. All butterflies are insects.

24. All houses are wired using parallel circuits.

25. No aldermen are running for mayor.

26. Some diet sodas contain saccharin.

27. Some turtles do not have claws.

28. No teachers made the roster.

29. No bicycles have three wheels.

30. All horses have manes.

31. Some pine trees do not produce pinecones.

32. No one likes asparagus.

33. Some pedestrians are in the crosswalk.

34. Some dogs with long hair do not get cold.

In Exercises 35–40, write the statement in symbolic form. Let

p: The tent is pitched.
q: The bonfire is burning.

35. The tent is not pitched.

36. The tent is pitched and the bonfire is burning.

37. The bonfire is not burning or the tent is not pitched.

38. The bonfire is not burning if and only if the tent is not pitched.

39. If the tent is not pitched, then the bonfire is not burning.

40. The bonfire is not burning, however the tent is pitched.

In Exercises 41–46, write the statement in symbolic form. Let

p: The chili is spicy.
q: The sour cream is cold.

41. The chili is not spicy, but the sour cream is cold.

42. Neither is the chili spicy nor is the sour cream cold.

43. The sour cream is not cold if and only if the chili is spicy.

44. If the chili is spicy, then the sour cream is not cold.

45. It is false that the chili is spicy or the sour cream is cold.

46. It is false that if the sour cream is not cold then the chili is spicy.

In Exercises 47–56, write the compound statement in words. Let

p: Ken Jennings won 74 games of *Jeopardy!*
q: Ken Jennings won more than \$3 million.

▲ Alex Trebek and Ken Jennings

47. $\sim q$

48. $\sim p$

49. $p \wedge q$

50. $q \vee p$

51. $\sim p \rightarrow q$

52. $\sim p \leftrightarrow \sim q$

53. $\sim p \vee \sim q$

54. $\sim(q \vee p)$

55. $\sim(p \wedge q)$

56. $\sim p \wedge \sim q$

In Exercises 57–66, write the statements in symbolic form. Let

p: The temperature is 90°.
q: The air conditioner is working.
r: The apartment is hot.

57. The temperature is 90° and the air conditioner is not working, and the apartment is hot.

58. The temperature is not 90° and the air conditioner is working, but the apartment is hot.

59. The temperature is 90° and the air conditioner is working, or the apartment is hot.

60. If the apartment is hot and the air conditioner is working, then the temperature is 90°.

61. If the temperature is 90°, then the air conditioner is working or the apartment is not hot.

62. The temperature is not 90° if and only if the air conditioner is not working, or the apartment is not hot.

63. The apartment is hot if and only if the air conditioner is working, and the temperature is 90°.

64. It is false that if the apartment is hot then the air conditioner is not working.

65. If the air conditioner is working, then the temperature is 90° if and only if the apartment is hot.

66. The apartment is hot or the air conditioner is not working, if and only if the temperature is 90°.

In Exercises 67–76, write each symbolic statement in words. Let

p: The water is 70°.

q: The sun is shining.

r: We go swimming.

67. $(p \vee q) \wedge \sim r$

68. $(p \wedge q) \vee r$

69. $\sim p \wedge (q \vee r)$

70. $(q \rightarrow p) \vee r$

71. $\sim r \rightarrow (q \wedge p)$

72. $(q \wedge r) \rightarrow p$

73. $(q \rightarrow r) \wedge p$

74. $\sim p \rightarrow (q \vee r)$

75. $(q \leftrightarrow p) \wedge r$

76. $q \rightarrow (p \leftrightarrow r)$

Dinner Menu *In Exercises 77–80, use the following information to arrive at your answers. Many restaurant dinner menus include statements such as the following. All dinners are served with a choice of: Soup or Salad, and Potatoes or Pasta, and Carrots or Peas. Which of the following selections are permissible? If a selection is not permissible, explain why. See the discussion of the* exclusive or *on page 102.*

77. Soup, salad, and peas

78. Salad, pasta, and carrots

79. Soup, potatoes, pasta, and peas

80. Soup, pasta, and potatoes

In Exercises 81–89, (a) select letters to represent the simple statements and write each statement symbolically by using parentheses and (b) indicate whether the statement is a negation, conjunction, disjunction, conditional, or biconditional.

81. I bought the watch in Tijuana and I did not pay $100.

82. If the conference is in Las Vegas, then we can see Wayne Newton or we can play poker.

83. It is false that if your speed is below the speed limit then you will not get pulled over.

84. If dinner is ready then we can eat, or we cannot go to the restaurant.

85. If the food has fiber or the food has vitamins, then you will be healthy.

86. If Corliss is teaching then Faye is in the math lab, if and only if it is not a weekend.

87. You may take this course, if and only if you did not fail the previous course or you passed the placement test.

88. If the car has gas and the battery is charged, then the car will start.

89. The classroom is empty if and only if it is the weekend, or it is 7 A.M.

CHALLENGE PROBLEMS/GROUP ACTIVITIES

90. ***An Ancient Question*** If Zeus could do anything, could he build a wall that he could not jump over? Explain your answer.

91. **a)** Make up three simple statements and label them p, q, and r. Then write compound statements to represent $(p \vee q) \wedge r$ and $p \vee (q \wedge r)$.

b) Do you think that the statements for $(p \vee q) \wedge r$ and $p \vee (q \wedge r)$ mean the same thing? Explain.

RECREATIONAL MATHEMATICS

92. ***Sudoku*** Sudoku is a logic-based puzzle that originated in Japan. The goal is to enter numerical digits in each of the empty squares so that every digit from 1 to 9 appears exactly one time in each of the rows, in each of the columns, and in each of the nine 3 by 3 boxes. Completing the puzzle requires patience and logical ability. For more information, see www.websudoku.com. Complete the following Sudoku puzzle.

	1		4	8		5	6	
5					9	8		
	3				1	4		7
8	2			9		1		
6			1		4			9
		3		6			4	5
9		1	5				2	
		7	2					4
	5	2		7	8		3	

INTERNET/RESEARCH ACTIVITIES

93. ***Legal Documents*** Obtain a legal document such as a will or rental agreement and copy one page of the document. Circle every connective used. Then list the number of times each connective appeared. Be sure to include conditional statements from which the word *then* was omitted from the sentence. Give the page and your listing to your instructor.

94. Write a report on the life and accomplishments of George Boole, who was an important contributor to the development of logic. In your report, indicate how his work eventually led to the development of the computer. References include encyclopedias, history of mathematics books, and the Internet.

3.2 TRUTH TABLES FOR NEGATION, CONJUNCTION, AND DISJUNCTION

▲ Under what conditions is the statement *Idaho grows the most potatoes or Iowa grows the most corn* true?

Consider the following statement: Idaho grows the most potatoes or Iowa grows the most corn. Under what conditions can the statement be considered true? Under what conditions can the statement be considered false? In this section, we will introduce a tool used to help us analyze such statements.

A ***truth table*** is a device used to determine when a compound statement is true or false. Five basic truth tables are used in constructing other truth tables. Three are discussed in this section (Tables 3.2, 3.4, and 3.7), and two are discussed in the next section. Section 3.5 uses truth tables in determining whether a logical argument is valid or invalid.

Negation

The first truth table is for *negation*. If p is a true statement, then the negation of p, "not p," is a false statement. If p is a false statement, then "not p" is a true statement. For

Table 3.2 Negation

	p	$\sim p$
Case 1	T	F
Case 2	F	T

Table 3.3

	p	q
Case 1	T	T
Case 2	T	F
Case 3	F	T
Case 4	F	F

Table 3.4 Conjunction

	p	q	$p \wedge q$
Case 1	T	T	T
Case 2	T	F	F
Case 3	F	T	F
Case 4	F	F	F

example, if the statement "The shirt is blue" is true, then the statement "The shirt is not blue" is false. These relationships are summarized in Table 3.2. For a simple statement, there are exactly two true–false cases, as shown.

If a compound statement consists of two simple statements p and q, there are four possible cases, as illustrated in Table 3.3. Consider the statement "The test is today and the test covers Chapter 5." The simple statement "The test is today" has two possible truth values, true or false. The simple statement "The test covers Chapter 5" also has two truth values, true or false. Thus, for these two simple statements there are four distinct possible true–false arrangements. Whenever we construct a truth table for a compound statement that consists of two simple statements, we begin by listing the four true–false cases shown in Table 3.3.

Conjunction

To illustrate the conjunction, consider the following situation. You have recently purchased a new house. To decorate it, you ordered a new carpet and new furniture from the same store. You explain to the salesperson that the carpet must be delivered before the furniture. He promises that the carpet will be delivered on Thursday and that the furniture will be delivered on Friday.

To help determine whether the salesperson kept his promise, we assign letters to each simple statement. Let p be "The carpet will be delivered on Thursday" and q be "The furniture will be delivered on Friday." The salesperson's statement written in symbolic form is $p \wedge q$. There are four possible true–false situations to be considered (Table 3.4).

CASE 1: p is true and q is true. The carpet is delivered on Thursday and the furniture is delivered on Friday. The salesperson has kept his promise and the compound statement is true. Thus, we put a T in the $p \wedge q$ column.

CASE 2: p is true and q is false. The carpet is delivered on Thursday but the furniture is not delivered on Friday. Since the furniture was not delivered as promised, the compound statement is false. Thus, we put an F in the $p \wedge q$ column.

CASE 3: p is false and q is true. The carpet is not delivered on Thursday but the furniture is delivered on Friday. Since the carpet was not delivered on Thursday as promised, the compound statement is false. Thus, we put an F in the $p \wedge q$ column.

CASE 4: p is false and q is false. The carpet is not delivered on Thursday and the furniture is not delivered on Friday. Since the carpet and furniture were not delivered as promised, the compound statement is false. Thus, we put an F in the $p \wedge q$ column.

Examining the four cases, we see that in only one case did the salesperson keep his promise: in case 1. Therefore, case 1 (T, T) is true. In cases 2, 3, and 4, the salesperson did not keep his promise and the compound statement is false. The results are summarized in Table 3.4, the truth table for the conjunction.

The **conjunction** $p \wedge q$ is true only when both p and q are true.

EXAMPLE 1 *Construct a Truth Table*

Construct a truth table for $p \wedge \sim q$.

SOLUTION Because there are two statements, p and q, construct a truth table with four cases; see Table 3.5(a). Then write the truth values under the p in the compound statement and label this column 1, as in Table 3.5(b). Copy these truth values directly from the p column on the left. Write the corresponding truth values under the q in the compound statement and call this column 2, as in Table 3.5(c). Copy the truth values for column 2 directly from the q column on the left. Now find the truth values of $\sim q$ by negating the truth values in column 2 and call this column 3,

Table 3.5

(a)

	p	q	$p \wedge \sim q$
Case 1	T	T	
Case 2	T	F	
Case 3	F	T	
Case 4	F	F	

(b)

p	q	$p \wedge \sim q$
T	T	T
T	F	T
F	T	F
F	F	F
		1

(c)

p	q	p	$\wedge$	$\sim$	q
T	T	T			T
T	F	T			F
F	T	F			T
F	F	F			F
		1			2

(d)

p	q	p	$\wedge$	$\sim$	q
T	T	T		F	T
T	F	T		T	F
F	T	F		F	T
F	F	F		T	F
		1		3	2

(e)

p	q	p	$\wedge$	$\sim$	q
T	T	T	F	F	T
T	F	T	T	T	F
F	T	F	F	F	T
F	F	F	F	T	F
		1	4	3	2

as in Table 3.5(d). Use the conjunction table, Table 3.4, and the entries in the columns labeled 1 and 3 to complete the column labeled 4, as in Table 3.5(e). The results in column 4 are obtained as follows:

Row 1: T $\wedge$ F is F. Row 2: T $\wedge$ T is T.

Row 3: F $\wedge$ F is F. Row 4: F $\wedge$ T is F.

The answer is always the last column completed. The columns labeled 1, 2, and 3 are only aids in arriving at the answer labeled column 4. ●

The statement $p \wedge \sim q$ in Example 1 actually means $p \wedge (\sim q)$. In the future, instead of listing a column for q and a separate column for its negation, we will make one column for $\sim q$, which will have the opposite values of those in the q column on the left. Similarly, when we evaluate $\sim p$, we will use the opposite values of those in the p column on the left. This procedure is illustrated in Example 2.

In Example 1, we spoke about *cases* and also *columns*. Consider Table 3.5(e). This table has four cases indicated by the four different rows of the two left-hand (unnumbered) columns. The four *cases* are TT, TF, FT, and FF. In every truth table with two letters, we list the four cases (the first two columns) first. Then we complete the remaining columns in the truth table. In Table 3.5(e), after completing the two left-hand columns, we complete the remaining columns in the order indicated by the numbers below the columns. We will continue to place numbers below the columns to show the order in which the columns are completed.

In discussion of the truth table in Example 2, and all following truth tables, if we say column 1, it means the column labeled 1. Column 2 will mean the column labeled 2, and so on.

TIMELY TIP When constructing truth tables it is very important to keep your entries in neat columns and rows. If you are using lined paper, put only one row of the table on each line. If you are not using lined paper, using a straightedge may help you correctly enter the information into the truth table's rows and columns.

EXAMPLE 2 *Construct and Interpret a Truth Table*

a) Construct a truth table for the following statement: Jose is not an artist and Jose is not a musician.

b) Under which conditions will the compound statement be true?

c) Suppose "Jose is an artist" is a false statement and "Jose is a musician" is a true statement. Is the compound statement given in part (a) true or false?

SOLUTION

a) First write the simple statements in symbolic form by using simple nonnegated statements.

Let

p: Jose is an artist.
q: Jose is a musician.

Therefore, the compound statement may be written $\sim p \wedge \sim q$. Now construct a truth table with four cases, as shown in Table 3.6.

Fill in the column labeled 1 by negating the truth values under p on the far left. Fill in the column labeled 2 by negating the values under q in the second column from the left. Fill in the column labeled 3 by using the columns labeled 1 and 2 and the definition of conjunction.

In the first row, to determine the entry for column 3, we use false for $\sim p$ and false for $\sim q$. Since false $\wedge$ false is false (see case 4 of Table 3.4), we place an F in column 3, row 1. In the second row, we use false for $\sim p$ and true for $\sim q$. Since false $\wedge$ true is false (see case 3 of Table 3.4), we place an F in column 3, row 2. In the third row, we use true for $\sim p$ and false for $\sim q$. Since true $\wedge$ false is false (see case 2 of Table 3.4), we place an F in column 3, row 3. In the fourth row, we use true for $\sim p$ and true for $\sim q$. Since true $\wedge$ true is true (see case 1 of Table 3.4), we place a T in column 3, row 4.

Table 3.6

p	q	$\sim p$	$\wedge$	$\sim q$
T	T	F	F	F
T	F	F	F	T
F	T	T	F	F
F	F	T	T	T
		1	3	2

b) The compound statement in part (a) will be true only in case 4 (circled in blue) when both simple statements, p and q, are false, that is, when Jose is not an artist and Jose is not a musician.

c) We are told that p, "Jose is an artist," is a false statement and that q, "Jose is a musician," is a true statement. From the truth table (Table 3.6), we can determine that when p is false and q is true, the compound statement, case 3, (circled in red) is false.

Disjunction

Civil Technician

Municipal program for redevelopment seeks on-site technician. **The applicant must have a two-year college degree in civil technology or five years of related experience**. Interested candidates please call 555-1234.

Consider the job description in the margin that describes several job requirements. Who qualifies for the job? To help analyze the statement, translate it into symbolic form. Let p be "A requirement for the job is a two-year college degree in civil technology" and q be "A requirement for the job is five years of related experience." The statement in symbolic form is $p \vee q$. For the two simple statements, there are four distinct cases (see Table 3.7).

Table 3.7 Disjunction

p	q	$p \vee q$
T	T	T
T	F	T
F	T	T
F	F	F

CASE 1: p is true and q is true. A candidate has a two-year college degree in civil technology and five years of related experience. The candidate has both requirements and qualifies for the job. Consider qualifying for the job as a true statement and not qualifying as a false statement. Since the candidate qualifies for the job, we put a T in the $p \vee q$ column.

CASE 2: p is true and q is false. A candidate has a two-year college degree in civil technology but does not have five years of related experience. The candidate still qualifies for the job with the two-year college degree. Thus, we put a T in the $p \vee q$ column.

CASE 3: p is false and q is true. The candidate does not have a two-year college degree in civil technology but does have five years of related experience. The candidate qualifies for the job with the five years of related experience. Thus, we put a T in the $p \vee q$ column.

CASE 4: p is false and q is false. The candidate does not have a two-year college degree in civil technology and does not have five years of related experience. The candidate does not meet either of the two requirements and therefore does not qualify for the job. Thus, we put an F in the $p \vee q$ column.

In examining the four cases, we see that there is only one case in which the candidate does not qualify for the job: case 4. As this example indicates, an *or* statement will be true in every case, except when both simple statements are false. The results are summarized in Table 3.7, the truth table for the disjunction.

The **disjunction**, $p \vee q$, is true when either p is true, q is true, or both p and q are true.

The disjunction $p \vee q$ is false only when p and q are both false.

EXAMPLE 3 *Truth Table with a Negation*

Construct a truth table for $\sim(\sim q \wedge p)$.

SOLUTION First construct the standard truth table listing the four cases. Then work within parentheses. The order to be followed is indicated by the numbers below the columns (see Table 3.8). Under $\sim q$, column 1, write the negation of the q column. Then, in column 2, copy the values from the p column. Next, complete the *and* column, column 3, using columns 1 and 2 and the truth table for the conjunction. The *and* column is true only when both statements are true, as in case 2. Finally, negate the values in the *and* column, column 3, and place these negated values in column 4. By examining the truth table you can see that the compound statement $\sim(\sim q \wedge p)$ is false only in case 2, that is, when p is true and q is false.

Table 3.8

p	q	$\sim$	$(\sim q$	$\wedge$	$p)$
T	T	T	F	F	T
T	F	F	T	T	T
F	T	T	F	F	F
F	F	T	T	F	F
		4	1	3	2

A GENERAL PROCEDURE FOR CONSTRUCTING TRUTH TABLES

1. Study the compound statement and determine whether it is a negation, conjunction, disjunction, conditional, or biconditional statement, as was done in Section 3.1. The answer to the truth table will appear under $\sim$ if the statement is a negation, under $\wedge$ if the statement is a conjunction, under $\vee$ if the statement is a disjunction, under $\rightarrow$ if the statement is a conditional, and under $\leftrightarrow$ if the statement is a biconditional.
2. Complete the columns under the simple statements, p, q, r, and their negations, $\sim p$, $\sim q$, $\sim r$, within parentheses, if present. If there are nested parentheses (one pair of parentheses within another pair), work with the innermost pair first.
3. Complete the column under the connective within the parentheses, if present. You will use the truth values of the connective in determining the final answer in step 5.
4. Complete the column under any remaining statements and their negations.
5. Complete the column under any remaining connectives. Recall that the answer will appear under the column determined in step 1. If the statement is a conjunction, disjunction, conditional, or biconditional, you will obtain the truth values for the connective by using the last column completed on the left side and on the right side of the connective. If the statement is a negation, you will obtain the truth values by negating the truth values of the last column completed within the grouping symbols on the right side of the negation. Be sure to circle or highlight your answer column or number the columns in the order they were completed.

EXAMPLE 4 *Use the General Procedure to Construct a Truth Table*

Construct a truth table for the statement $(\sim p \vee q) \wedge \sim p$.

SOLUTION We will follow the general procedure outlined in the box. This statement is a conjunction, so the answer will be under the conjunction symbol. Complete columns under $\sim p$ and q within the parentheses and call these columns 1 and 2, respectively (see Table 3.9). Complete the column under the disjunction, $\vee$, using the truth values in columns 1 and 2, and call this column 3. Next complete the column under $\sim p$, and call this column 4. The answer, column 5, is determined from the definition of the conjunction and the truth values in column 3, the last column completed on the left side of the conjunction, and column 4.

Table 3.9

p	q	$(\sim p$	$\vee$	$q)$	$\wedge$	$\sim p$
T	T	F	T	T	F	F
T	F	F	F	F	F	F
F	T	T	T	T	T	T
F	F	T	T	F	T	T
		1	3	2	5	4

Table 3.10

	p	q	r
Case 1	T	T	T
Case 2	T	T	F
Case 3	T	F	T
Case 4	T	F	F
Case 5	F	T	T
Case 6	F	T	F
Case 7	F	F	T
Case 8	F	F	F

So far, all the truth tables we have constructed have contained at most two simple statements. Now we will explain how to construct a truth table that consists of three simple statements, such as $(p \wedge q) \wedge r$. When a compound statement consists of three simple statements, there are eight different true–false possibilities, as illustrated in Table 3.10. To begin such a truth table, write four Ts and four Fs in the column under p. Under the second statement, q, pairs of Ts alternate with pairs of Fs. Under the third statement, r, T alternates with F. This technique is not the only way of listing the cases, but it ensures that each case is unique and that no cases are omitted.

EXAMPLE 5 *Construct a Truth Table with Eight Cases*

a) Construct a truth table for the statement "Santana is home and he is not at his desk, or he is sleeping."

b) Suppose that "Santana is home" is a false statement, that "Santana is at his desk" is a true statement, and that "Santana is sleeping" is a true statement. Is the compound statement in part (a) true or false?

SOLUTION

a) First we will translate the statement into symbolic form.

Let

p: Santana is home.

q: Santana is at his desk.

r: Santana is sleeping.

In symbolic form, the statement is $(p \wedge \sim q) \vee r$.

Since the statement is composed of three simple statements, there are eight cases. Begin by listing the eight cases in the three left-hand columns; see Table 3.11. By examining the statement, you can see that it is a disjunction. Therefore, the answer will be in the $\vee$ column. Fill out the truth table by working in parentheses first. Place values under p, column 1, and $\sim q$, column 2. Then find the conjunctions of columns 1 and 2 to obtain column 3. Place the values of r in column 4. To obtain the answer, column 5, use columns 3 and 4 and the information for the disjunction contained in Table 3.7 on page 117.

Table 3.11

p	q	r	$(p$	$\wedge$	$\sim q)$	$\vee$	r
T	T	T	T	F	F	T	T
T	T	F	T	F	F	F	F
T	F	T	T	T	T	T	T
T	F	F	T	T	T	T	F
F	T	T	F	F	F	(T)	T
F	T	F	F	F	F	F	F
F	F	T	F	F	T	T	T
F	F	F	F	F	T	F	F
			1	3	2	5	4

b) We are given the following:

p: Santana is home—false.
q: Santana is at his desk—true.
r: Santana is sleeping—true.

We need to find the truth value of the following case: false, true, true. In case 5 of the truth table, p, q, and r are F, T, and T, respectively. Therefore, under these conditions, the original compound statement is true (as circled in the table).

We have learned that a truth table with one simple statement has two cases, a truth table with two simple statements has four cases, and a truth table with three simple statements has eight cases. In general, *the number of distinct cases in a truth table with n distinct simple statements is* 2^n. The compound statement $(p \vee q) \vee (r \wedge \sim s)$ has four simple statements, p, q, r, s. Thus, a truth table for this compound statement would have 2^4, or 16, distinct cases.

When we construct a truth table, we determine the truth values of a compound statement for every possible case. If we want to find the truth value of the compound statement for any specific case when we know the truth values of the simple statements, we do not have to develop the entire table. For example, to determine the truth value for the statement

$$2 + 3 = 5 \quad \text{and} \quad 1 + 1 = 3$$

we let p be $2 + 3 = 5$ and q be $1 + 1 = 3$. Now we can write the compound statement as $p \wedge q$. We know that p is a true statement and q is a false statement. Thus, we can substitute T for p and F for q and evaluate the statement:

$$\begin{array}{c} p \wedge q \\ \mathrm{T} \wedge \mathrm{F} \\ \mathrm{F} \end{array}$$

Therefore, the compound statement $2 + 3 = 5$ and $1 + 1 = 3$ is a false statement.

EXAMPLE 6 *Determine the Truth Value of a Compound Statement*

Determine the truth value for each simple statement. Then, using these truth values, determine the truth value of the compound statement.

a) 15 is less than or equal to 9.

b) George Washington was the first U.S. president or Abraham Lincoln was the second U.S. president, but there has not been a U.S. president born in Antarctica.

SOLUTION

a) Let

p: 15 is less than 9.
q: 15 is equal to 9.

The statement "15 is less than or equal to 9" means that 15 is less than 9 or 15 is equal to 9. The compound statement can be expressed as $p \vee q$. We know that both p and q are false statements since 15 is greater than 9, so we substitute F for p and F for q and evaluate the statement:

$$p \vee q$$
$$\mathrm{F} \vee \mathrm{F}$$
$$\mathrm{F}$$

Therefore, the compound statement "15 is less than or equal to 9" is a false statement.

b) Let

p: George Washington was the first U.S. president.

q: Abraham Lincoln was the second U.S. president.

r: There has been a U.S. president who was born in Antarctica .

The compound statement can be written in symbolic form as $(p \vee q) \wedge \sim r$. Recall that *but* is used to express a conjunction. We know that p is a true statement and that q is a false statement. We also know that r is a false statement since all U.S. presidents must be born in the United States. Thus, since r is a false statement, the negation, $\sim r$, is a true statement. So we will substitute T for p, F for q, and T for $\sim r$ and then evaluate the statement:

$$(p \vee q) \wedge \sim r$$
$$(\mathrm{T} \vee \mathrm{F}) \wedge \mathrm{T}$$
$$\mathrm{T} \wedge \mathrm{T}$$
$$\mathrm{T}$$

Therefore, the original compound statement is a true statement.

EXAMPLE 7 *Pet Ownership in the United States*

The number of pets owned in the United States in 2005 is shown in Fig. 3.1. Use this graph to determine the truth value of the following statement: There are more dogs owned than cats and there are fewer reptiles owned than birds, or the most numerous pets owned are not fish.

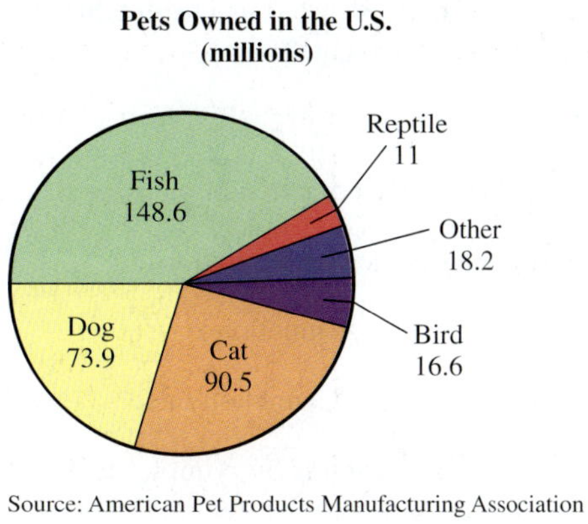

Figure 3.1

SOLUTION Let

p: There are more dogs owned than cats.
q: There are fewer reptiles owned than birds.
r: The most numerous pets owned are fish.

The given compound statement can be written in symbolic form as $(p \wedge q) \vee \sim r$. From Fig. 3.1, we see that statement p is false: there are actually more cats owned than dogs. We also see that statement q is true: there are fewer reptiles owned than birds. We also see that statement r is true: the most numerous pets owned are fish. Since r is true, its negation, $\sim r$, is false. Therefore, we substitute F for p, T for q, and F for $\sim r$, which gives

$$(p \wedge q) \vee \sim r$$
$$(\mathrm{F} \wedge \mathrm{T}) \vee \mathrm{F}$$
$$\mathrm{F} \vee \mathrm{F}$$
$$\mathrm{F}$$

Thus, the original compound statement is a false statement.

SECTION 3.2 EXERCISES

CONCEPT/WRITING EXERCISES

1. a) How many distinct cases must be listed in a truth table that contains two simple statements?

   b) List all the cases.

2. a) How many distinct cases must be listed in a truth table that contains three simple statements?

   b) List all the cases.

3. a) Construct the truth table for the conjunction $p \wedge q$.

   b) Under what circumstances is the conjunction true?

   c) Construct the truth table for the disjunction $p \vee q$.

   d) Under what circumstances is the disjunction false?

4. Suppose a compound statement contains five simple statements. How many distinct cases would the truth table for this compound statement have?

PRACTICE THE SKILLS/PROBLEM SOLVING

In Exercises 5–20, construct a truth table for the statement.

5. $p \wedge \sim p$
6. $p \vee \sim p$
7. $q \vee \sim p$
8. $p \wedge \sim q$
9. $\sim p \vee \sim q$
10. $\sim(p \vee \sim q)$
11. $\sim(p \wedge \sim q)$
12. $\sim(\sim p \wedge \sim q)$
13. $\sim q \vee (p \wedge r)$
14. $(p \vee \sim q) \wedge r$
15. $r \vee (p \wedge \sim q)$
16. $(r \wedge q) \wedge \sim p$
17. $(r \vee \sim p) \wedge \sim q$
18. $\sim p \wedge (q \vee r)$
19. $(\sim q \wedge r) \vee p$
20. $\sim r \vee (\sim p \wedge q)$

In Exercises 21–30, write the statement in symbolic form and construct a truth table.

21. The cookies are warm and the milk is cold.

22. The zoo is open, but it is not a nice day.

23. I have a new cell phone, but I do not have a new battery.

24. It is false that Wanda Garner is the president or that Judy Ackerman is the treasurer.

25. It is false that Jasper Adams is a tutor and Mark Russo is a secretary.

26. Mike made pizza and Dennis made a chef salad, but Gil burned the lemon squares.

27. The copier is out of toner, or the lens is dirty or the corona wires are broken.

28. I am hungry, and I want to eat a healthy lunch and I want to eat in a hurry.

29. The Congress must act on the bill, and the president must sign the bill or not sign the bill.

30. Gordon Langeneger likes the Mac Pro and he likes the MacBook Pro, but he does not like the Pentium IV.

In Exercises 31–42, determine the truth value of the statement if

a) p is true, q is false, and r is true.

b) p is false, q is true, and r is true.

31. $(\sim p \wedge r) \wedge q$

32. $\sim p \vee (q \wedge r)$

33. $(\sim p \vee \sim q) \vee \sim r$

34. $(\sim q \wedge \sim p) \vee \sim r$

35. $(p \vee \sim q) \wedge \sim(p \wedge \sim r)$

36. $(p \wedge \sim q) \vee r$

37. $(\sim r \wedge p) \vee q$

38. $\sim q \vee (r \wedge p)$

39. $(\sim q \vee \sim p) \wedge r$

40. $(\sim r \vee \sim p) \vee \sim q$

41. $(\sim p \vee \sim q) \vee (\sim r \vee q)$

42. $(\sim r \wedge \sim q) \wedge (\sim r \vee \sim p)$

In Exercises 43–50, determine the truth value for each simple statement. Then use these truth values to determine the truth value of the compound statement. (You may have to use a reference source such as the Internet or an encyclopedia.)

43. $18 \div 3 = 9$ or $56 \div 8 = 7$

44. $17 \geq 17$ and $-3 > -2$

45. Virginia borders the Atlantic Ocean or California borders the Indian Ocean.

46. Hawaii is the 50th state or Alaska lies on the equator, and Birmingham is the capital of Missouri.

47. Steven Spielberg is a movie director and Tom Hanks is an actor, but John Madden is not a sports announcer.

▲ John Madden and Melissa Stark

48. Quebec is in Texas or Toronto is in California, and Cedar Rapids is in Iowa.

49. Iraq is in Africa or Iran is in South America, and Syria is in the Middle East.

50. Holstein is a breed of cattle and collie is a breed of dogs, or beagle is not a breed of cats.

Carbon Dioxide Emissions *In Exercises 51–54, use the table to determine the truth value of each simple statement. Then determine the truth value of the compound statement.*

Countries with the Highest Carbon Dioxide (CO_2) Emissions in 2004

Country	CO_2 Emissions (in megatons)	Per Capita CO_2 Emissions (in tons)
United States	5652	19.66
China	3271	2.55
Russia	1503	10.43
Japan	1207	9.47
India	1016	0.97

Source: International Energy Agency

51. The United States had the lowest per capita CO_2 emissions and China had the highest per capita CO_2 emissions.

52. The United States had more CO_2 emissions than China and Russia had combined or the United States had more CO_2 emissions than Russia, Japan, and India had combined.

53. India had lower per capita CO_2 emissions than Japan had or China had lower per capita CO_2 emissions than India had.

54. Russia had lower per capita CO_2 emissions than the United States had, but Russia had higher per capita CO_2 emissions than Japan had.

Sleep Time *In Exercises 55–58, use the graph, which shows the number of hours Americans sleep, to determine the truth value of each simple statement. Then determine the truth value of the compound statement.*

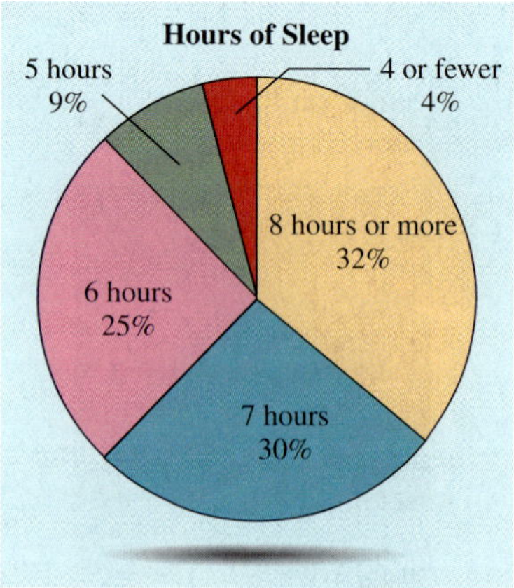

55. It is false that 30% of Americans get 6 hours of sleep each night and 9% get 5 hours of sleep each night.

56. Twenty-five percent of Americans get 6 hours of sleep each night, and 30% get 7 hours of sleep each night or 9% do not get 5 hours of sleep each night.

57. Thirteen percent of Americans get 5 or fewer hours of sleep each night or 32% get 6 or more hours of sleep each night, and 30% get 8 or more hours of sleep each night.

58. Over one-half of all Americans get 7 or fewer hours of sleep each night, and over one-quarter get 6 or fewer hours of sleep each night.

In Exercises 59–62, let

p: Tanisha owns a convertible.
q: Joan owns a Volvo.

Translate each statement into symbols. Then construct a truth table for each and indicate under what conditions the compound statement is true.

59. Tanisha owns a convertible and Joan does not own a Volvo.

60. Tanisha does not own a convertible, but Joan owns a Volvo.

61. Tanisha owns a convertible or Joan does not own a Volvo.

62. Tanisha does not own a convertible or Joan does not own a Volvo.

In Exercises 63–66, let

p: The house is owned by an engineer.
q: The heat is solar generated.
r: The car is run by electric power.

Translate each statement into symbols. Then construct a truth table for each and indicate under what conditions the compound statement is true.

63. The car is run by electric power or the heat is solar generated, but the house is owned by an engineer.

64. The house is owned by an engineer and the heat is solar generated, or the car is run by electric power.

65. The heat is solar generated, or the house is owned by an engineer and the car is not run by electric power.

66. The house is not owned by an engineer, and the car is not run by electric power and the heat is solar generated.

Obtaining a Loan *In Exercises 67 and 68, read the requirements and each applicant's qualifications for obtaining a loan.*

a) Identify which of the applicants would qualify for the loan.

b) For the applicants who do not qualify for the loan, explain why.

67. To qualify for a loan of $40,000, an applicant must have a gross income of $28,000 if single, $46,000 combined income if married, and assets of at least $6,000.

Mrs. Rusinek, married with three children, earns $42,000. Mr. Rusinek does not have an income. The Rusineks have assets of $42,000.

Mr. Duncan is not married, works in sales, and earns $31,000. He has assets of $9000.

Mrs. Tuttle and her husband have total assets of $43,000. One earns $35,000, and the other earns $23,500.

68. To qualify for a loan of $45,000, an applicant must have a gross income of $30,000 if single, $50,000 combined income if married, and assets of at least $10,000.

Mr. Argento, married with two children, earns $37,000. Mrs. Argento earns $15,000 at a part-time job. The Argentos have assets of $25,000.

Ms. McVey, single, has assets of $19,000. She works in a store and earns $25,000.

Mr. Siewert earns $24,000 and Ms. Fox, his wife, earns $28,000. Their assets total $8000.

69. ***Airline Special Fares*** An airline advertisement states, "To get the special fare you must purchase your tickets between January 1 and February 15 and fly round trip between March 1 and April 1. You must depart on a Monday, Tuesday, or Wednesday, and return on a Tuesday, Wednesday, or Thursday, and stay over at least one Saturday."

a) Determine which of the following individuals will qualify for the special fare.

b) If the person does not qualify for the special fare, explain why.

Wing Park plans to purchase his ticket on January 15, depart on Monday, March 3, and return on Tuesday, March 18.

Gina Vela plans to purchase her ticket on February 1, depart on Wednesday, March 12, and return on Thursday, April 3.

Kara Shavo plans to purchase her ticket on February 14, depart on Tuesday, March 4, and return on Monday, March 19.

Christos Supernaw plans to purchase his ticket on January 4, depart on Monday, March 10, and return on Thursday, March 13.

Alex Chang plans to purchase his ticket on January 1, depart on Monday, March 3, and return on Monday, March 10.

▲ See Exercise 69

PROBLEM SOLVING/GROUP ACTIVITIES

In Exercises 70 and 71, construct a truth table for the symbolic statement.

70. $\sim[(\sim(p \vee q)) \vee (q \wedge r)]$

71. $[(q \wedge \sim r) \wedge (\sim p \vee \sim q)] \vee (p \vee \sim r)$

72. On page 120, we indicated that a compound statement consisting of n simple statements had 2^n distinct true–false cases.

a) How many distinct true–false cases does a truth table containing simple statements p, q, r, and s have?

b) List all possible true–false cases for a truth table containing the simple statements p, q, r, and s.

c) Use the list in part (b) to construct a truth table for $(q \wedge p) \vee (\sim r \wedge s)$.

d) Construct a truth table for $(\sim r \wedge \sim s) \wedge (\sim p \vee q)$.

73. Must $(p \wedge \sim q) \vee r$ and $(q \wedge \sim r) \vee p$ have the same number of trues in their answer columns? Explain.

INTERNET/RESEARCH ACTIVITIES

74. Do research and write a report on each of the following.

a) The relationship between *negation* in logic and *complement* in set theory.

b) The relationship between *conjunction* in logic and *intersection* in set theory.

c) The relationship between *disjunction* in logic and *union* in set theory.

3.3 TRUTH TABLES FOR THE CONDITIONAL AND BICONDITIONAL

▲ Under what conditions is the statement *If you get an A, then I will buy you a car* true?

Suppose I said to you, "If you get an A, then I will buy you a car." As we discussed in Section 3.1, this statement is called a *conditional* statement. In this section, we will discuss under what conditions a conditional statement is true and under what conditions a conditional statement is false.

Conditional

In Section 3.1, we mentioned that the statement preceding the conditional symbol is called the *antecedent* and that the statement following the conditional symbol is called the *consequent*. For example, consider $(p \vee q) \rightarrow [\sim(q \wedge r)]$. In this statement, $(p \vee q)$ is the antecedent and $[\sim(q \wedge r)]$ is the consequent.

To develop a truth table for the conditional statement, consider the statement "If you get an A, then I will buy you a car." Assume this statement is true except when I have actually broken my promise to you.

Let

p: You get an A.

q: I buy you a car.

Translated into symbolic form, the statement becomes $p \rightarrow q$. Let's examine the four cases shown in Table 3.12.

Table 3.12 Conditional

p	q	$p \rightarrow q$
T	T	T
T	F	F
F	T	T
F	F	T

CASE 1: (T, T) You get an A, and I buy a car for you. I have met my commitment, and the statement is true.

CASE 2: (T, F) You get an A, and I do not buy a car for you. I have broken my promise, and the statement is false.

What happens if you don't get an A? If you don't get an A, I no longer have a commitment to you, and therefore I cannot break my promise.

CASE 3: (F, T) You do not get an A, and I buy you a car. I have not broken my promise, and therefore the statement is true.

CASE 4: (F, F) You do not get an A, and I don't buy you a car. I have not broken my promise, and therefore the statement is true.

The conditional statement is false when the antecedent is true and the consequent is false. In every other case the conditional statement is true.

The **conditional statement** $p \rightarrow q$ is true in every case except when p is a true statement and q is a false statement.

EXAMPLE 1 *A Truth Table with a Conditional*

Construct a truth table for the statement $\sim p \rightarrow \sim q$.

Table 3.13

p	q	$\sim p$	$\rightarrow$	$\sim q$
T	T	F	T	F
T	F	F	T	T
F	T	T	F	F
F	F	T	T	T
		1	3	2

SOLUTION Because this statement is a conditional, the answer will lie under the $\rightarrow$. Fill out the truth table by placing the appropriate truth values under $\sim p$, column 1, and under $\sim q$, column 2 (see Table 3.13). Then, using the information given in the truth table for the conditional and the truth values in columns 1 and 2, determine the solution, column 3. In row 1, the antecedent, $\sim p$, is false and the consequent, $\sim q$, is also false. Row 1 is F $\rightarrow$ F, which according to row 4 of Table 3.12, is T. Likewise, row 2 of Table 3.13 is F $\rightarrow$ T, which is T. Row 3 is T $\rightarrow$ F, which is F. Row 4 is T $\rightarrow$ T, which is T.

EXAMPLE 2 *A Conditional Truth Table with Three Simple Statements*

Construct a truth table for the statement $p \rightarrow (\sim q \wedge r)$.

Table 3.14

p	q	r	p	$\rightarrow$	$(\sim q$	$\wedge$	$r)$
T	T	T	T	F	F	F	T
T	T	F	T	F	F	F	F
T	F	T	T	T	T	T	T
T	F	F	T	F	T	F	F
F	T	T	F	T	F	F	T
F	T	F	F	T	F	F	F
F	F	T	F	T	T	T	T
F	F	F	F	T	T	F	F
			4	5	1	3	2

SOLUTION Because this statement is a conditional, the answer will lie under the $\rightarrow$. Work within the parentheses first. Place the truth values under $\sim q$, column 1, and r, column 2 (Table 3.14). Then take the conjunction of columns 1 and 2 to obtain column 3. Next, place the truth values under p in column 4. To determine the answer, column 5, use columns 3 and 4 and the information of the conditional statement given in Table 3.12. Column 4 represents the truth values of the antecedent, and column 3 represents the truth values of the consequent. Remember that the conditional is false only when the antecedent is true and the consequent is false, as in cases (rows) 1, 2, and 4 of column 5.

EXAMPLE 3 *Examining an Advertisement*

An advertisement for Perky Morning coffee makes the following claim: "If you drink Perky Morning coffee, then you will not be sluggish and you will have a great day." Translate the statement into symbolic form and construct a truth table.

SOLUTION Let

p: You drink Perky Morning coffee.
q: You will be sluggish.
r: You will have a great day.

In symbolic form, the claim is

$$p \rightarrow (\sim q \wedge r)$$

This symbolic statement is identical to the statement in Table 3.14, and the truth tables are the same. Column 3 represents the truth values of $(\sim q \wedge r)$, which corresponds to the statement "You will not be sluggish and you will have a great day." Note that column 3 is true in cases (rows) 3 and 7. In case 3, since p is true, you drank Perky Morning coffee. In case 7, however, since p is false, you did not drink Perky Morning coffee. From this information we can conclude that it is possible for you to not be sluggish and for you to have a great day without drinking Perky Morning coffee.

A truth table alone cannot tell us whether a statement is true or false. It can, however, be used to examine the various possibilities.

Biconditional

The *biconditional statement*, $p \leftrightarrow q$, means that $p \rightarrow q$ and $q \rightarrow p$, or, symbolically, $(p \rightarrow q) \wedge (q \rightarrow p)$. To determine the truth table for $p \leftrightarrow q$, we will construct the truth table for $(p \rightarrow q) \wedge (q \rightarrow p)$ (Table 3.15). Table 3.16 shows the truth values for the biconditional statement.

Table 3.15

p	q	$(p$	$\rightarrow$	$q)$	$\wedge$	$(q$	$\rightarrow$	$p)$
T	T	T	T	T	T	T	T	T
T	F	T	F	F	F	F	T	T
F	T	F	T	T	F	T	F	F
F	F	F	T	F	T	F	T	F
		1	3	2	7	4	6	5

Table 3.16 Biconditional

p	q	$p \leftrightarrow q$
T	T	T
T	F	F
F	T	F
F	F	T

From Table 3.16 we see that the biconditional statement is true when the antecedent and the consequent have the same truth value and false when the antecedent and consequent have different truth values.

The **biconditional statement,** $p \leftrightarrow q$, is true only when p and q have the same truth value, that is, when both are true or both are false.

EXAMPLE 4 *A Truth Table Using a Biconditional*

Construct a truth table for the statement $p \leftrightarrow (q \rightarrow \sim r)$.

SOLUTION Since there are three letters, there must be eight cases. The parentheses indicate that the answer must be under the biconditional (Table 3.17). Use columns 3 and 4 to obtain the answer in column 5. When columns 3 and 4 have the same truth values, place a T in column 5. When columns 3 and 4 have different truth values, place an F in column 5.

Table 3.17

p	q	r	p	$\leftrightarrow$	$(q$	$\rightarrow$	$\sim r)$
T	T	T	T	F	T	F	F
T	T	F	T	T	T	T	T
T	F	T	T	T	F	T	F
T	F	F	T	T	F	T	T
F	T	T	F	T	T	F	F
F	T	F	F	F	T	T	T
F	F	T	F	F	F	T	F
F	F	F	F	F	F	T	T
			4	5	1	3	2

DID YOU KNOW?

Satisfiability Problems

Suppose you are hosting a dinner party for seven people: Yasumasa, Marie, Albert, Stephen, Leonhard, Karl, and Emmy. You need to develop a seating plan around your circular dining room table that would satisfy all your guests. Albert and Emmy are great friends and must sit together. Yasumasa and Karl haven't spoken to each other in years and cannot sit by each other. Leonhard must sit by Marie or by Albert, but he cannot sit by Karl. Stephen insists he sit by Albert. Can you come up with a plan that would satisfy all your guests? Now imagine the difficulty of such a problem as the list of guests, and their demands, grows.

Problems such as this are known as *satisfiability* problems. The symbolic logic you are studying in this chapter allows computer scientists to represent these problems with symbols. Once in symbolic form, a problem can be studied with computers using algorithms. An *algorithm* is a general procedure for accomplishing a task. Even with the fastest computers, some satisfiability problems take an enormous amount of time to solve. Other problems may not even have a solution. Computer scientists and mathematicians are continually working to improve the algorithms used to solve satisfiability problems.

One solution to the problem posed above is as follows:

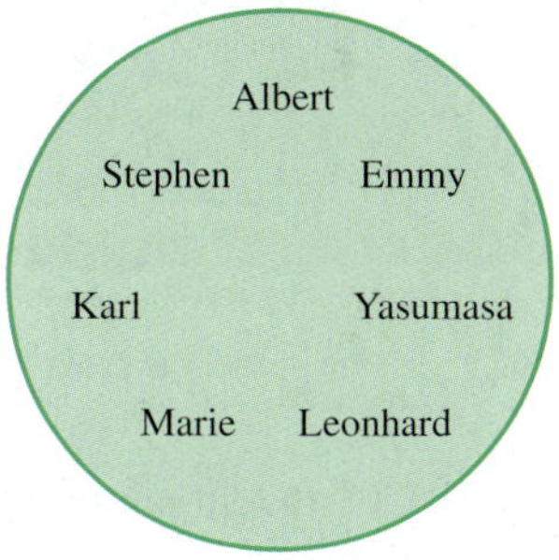

In Section 3.2, we showed that finding the truth value of a compound statement for a specific case does not require constructing an entire truth table. Examples 5 and 6 illustrate this technique for the conditional and the biconditional.

EXAMPLE 5 *Determine the Truth Value of a Compound Statement*

Determine the truth value of the statement $(\sim p \leftrightarrow q) \rightarrow (\sim q \leftrightarrow r)$ when p is false, q is true, and r is false.

SOLUTION Substitute the truth value for each simple statement:

$$\begin{array}{c}(\sim p \leftrightarrow q) \rightarrow (\sim q \leftrightarrow r)\\ (\mathrm{T} \leftrightarrow \mathrm{T}) \rightarrow (\mathrm{F} \leftrightarrow \mathrm{F})\\ \mathrm{T} \quad \rightarrow \quad \mathrm{T}\\ \mathrm{T}\end{array}$$

For this specific case, the statement is true.

EXAMPLE 6 *Determine the Truth Value of a Compound Statement*

Determine the truth value for each simple statement. Then use the truth values to determine the truth value of the compound statement.

a) If 15 is an even number, then 29 is an even number.

b) Northwestern University is in Illinois and Marquette University is in Alaska, if and only if Purdue University is in Alabama.

SOLUTION

a) Let

p: 15 is an even number.
q: 29 is an even number.

Then the statement "If 15 is an even number, then 29 is an even number" can be written $p \rightarrow q$. Since 15 is not an even number, p is a false statement. Also, since 29 is not an even number, q is a false statement. We substitute F for p and F for q and evaluate the statement:

$$\begin{array}{c}p \rightarrow q\\ \mathrm{F} \rightarrow \mathrm{F}\\ \mathrm{T}\end{array}$$

Therefore, "If 15 is an even number, then 29 is an even number" is a true statement.

b) Let

p: Northwestern University is in Illinois.
q: Marquette University is in Alaska.
r: Purdue University is in Alabama.

The original compound statement can be written $(p \wedge q) \leftrightarrow r$. By checking the Internet or other references we can find that Northwestern University is in Illinois, Marquette University is in Wisconsin, and Purdue University is in Indiana. Therefore, p is a true statement, but q and r are false statements. We will substitute T for p, F for q, and F for r and evaluate the compound statement:

$$\begin{array}{c} (p \wedge q) \leftrightarrow r \\ (\mathrm{T} \wedge \mathrm{F}) \leftrightarrow \mathrm{F} \\ \mathrm{F} \quad \leftrightarrow \mathrm{F} \\ \mathrm{T} \end{array}$$

Therefore, the original compound statement is true.

EXAMPLE 7 *Using Real Data in Compound Statements*

The graph in Fig. 3.2 represents the student population by age group in 2005 for Everett Community College (EvCC). Use this graph to determine the truth value of the following compound statements.

a) If 18- to 21-year-old students account for 25% of the EvCC student population, then 22- to 25-year-old students account for 12% of the EvCC student population and 31- to 35-year-old students account for 16% of the EvCC student population.

b) Students under age 18 years old account for 10% of the EvCC student population, if and only if 26- to 30-year-old students account for 25% of the EvCC student population or 36- to 40-year-old students account for 9% of the EvCC student population.

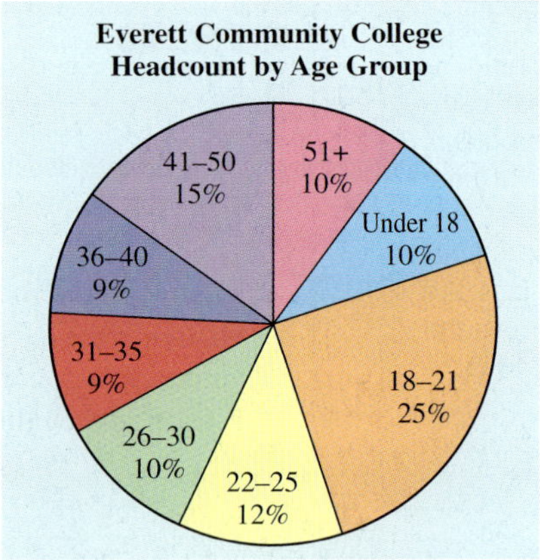

Source: Everett Community College Web Site

Figure 3.2

SOLUTION

a) Let

p: 18- to 21-year-old students account for 25% of the EvCC student population.

q: 22- to 25-year-old students account for 12% of the EvCC student population.

r: 31- to 35-year-old students account for 16% of the EvCC student population.

Then the original statement can be written $p \rightarrow (q \wedge r)$. We can see from Fig. 3.2 that both p and q are true statements and that r is a false statement. Substitute T for p, T for q, and F for r and evaluate the statement:

$$\begin{array}{l} p \rightarrow (q \wedge r) \\ \mathrm{T} \rightarrow (\mathrm{T} \wedge \mathrm{F}) \\ \mathrm{T} \rightarrow \quad \mathrm{F} \\ \quad \mathrm{F} \end{array}$$

Therefore, "If 18- to 21-year-old students account for 25% of the EvCC student population, then 22- to 25-year-old students account for 12% of the EvCC student population and 31- to 35-year-old students account for 16% of the EvCC student population" is a false statement.

b) Let

p: Students under age 18 years old account for 10% of the EvCC student population.

q: 26- to 30-year-old students account for 25% of the EvCC student population.

r: 36- to 40-year-old students account for 9% of the EvCC student population.

Then the original statement can be written $p \leftrightarrow (q \vee r)$. We can see from Fig. 3.2 that p is true, q is false, and r is true. Substitute T for p, F for q, and T for r and evaluate the statement:

$$\begin{array}{l} p \leftrightarrow (q \vee r) \\ \mathrm{T} \leftrightarrow (\mathrm{F} \vee \mathrm{T}) \\ \mathrm{T} \leftrightarrow \quad \mathrm{T} \\ \quad \mathrm{T} \end{array}$$

Therefore, the original statement, "Students under age 18 years old account for 10% of the EvCC student population, if and only if 26- to 30-year-old students account for 25% of the EvCC student population or 36- to 40-year-old students account for 9% of the EvCC student population" is a true statement. ●

Self-Contradictions, Tautologies, and Implications

Two special situations can occur in the truth table of a compound statement: The statement may always be false, or the statement may always be true. We give such statements special names.

A **self-contradiction** is a compound statement that is always false.

When every truth value in the answer column of the truth table is false, then the statement is a self-contradiction.

EXAMPLE 8 *All Falses, a Self-Contradiction*

Construct a truth table for the statement $(p \leftrightarrow q) \wedge (p \leftrightarrow \sim q)$.

SOLUTION See Table 3.18. In this example, the truth values are false in each case of column 5. This statement is an example of a self-contradiction or a *logically false statement.* ●

Table 3.18

p	q	$(p \leftrightarrow q)$	$\wedge$	$(p$	$\leftrightarrow$	$\sim q)$
T	T	T	F	T	F	F
T	F	F	F	T	T	T
F	T	F	F	F	T	F
F	F	T	F	F	F	T
		1	5	2	4	3

A **tautology** is a compound statement that is always true.

When every truth value in the answer column of the truth table is true, the statement is a tautology.

▲ "Heads I win, tails you lose." Do you think that this statement is a tautology, self-contradiction, or neither? See Problem-Solving Exercise 81.

EXAMPLE 9 *All Trues, a Tautology*

Construct a truth table for the statement $(p \wedge q) \rightarrow (p \vee r)$.

SOLUTION The answer is given in column 3 of Table 3.19. The truth values are true in every case. Thus, the statement is an example of a tautology or a *logically true statement*.

Table 3.19

p	q	r	$(p \wedge q)$	$\rightarrow$	$(p \vee r)$
T	T	T	T	T	T
T	T	F	T	T	T
T	F	T	F	T	T
T	F	F	F	T	T
F	T	T	F	T	T
F	T	F	F	T	F
F	F	T	F	T	T
F	F	F	F	T	F
			1	3	2

The conditional statement $(p \wedge q) \rightarrow (p \vee r)$ is a tautology. Conditional statements that are tautologies are called *implications*. In Example 9, we can say that $p \wedge q$ implies $p \vee r$.

An **implication** is a conditional statement that is a tautology.

In any implication the antecedent of the conditional statement implies the consequent. In other words, if the antecedent is true, then the consequent must also be true. That is, the consequent will be true whenever the antecedent is true.

EXAMPLE 10 *An Implication?*

Determine whether the conditional statement $[(p \wedge q) \wedge p] \rightarrow q$ is an implication.

SOLUTION If the conditional statement is a tautology, the conditional statement is an implication. Because the conditional statement is a tautology (see Table 3.20), the conditional statement is an implication. The antecedent $[(p \wedge q) \wedge p]$ implies the consequent q. Note that the antecedent is true only in case 1 and that the consequent is also true in case 1.

Table 3.20

p	q	$[(p \wedge q)$	$\wedge$	$p]$	$\rightarrow$	q
T	T	T	T	T	T	T
T	F	F	F	T	T	F
F	T	F	F	F	T	T
F	F	F	F	F	T	F
		1	3	2	5	4

SECTION 3.3 EXERCISES

CONCEPT/WRITING EXERCISES

1. a) Construct the truth table for the conditional statement $p \rightarrow q$.

   b) Explain when the conditional statement is true and when it is false.

   c) Construct the truth table for the biconditional statement $p \leftrightarrow q$.

   d) Explain when the biconditional statement is true and when it is false.

2. a) What is the statement preceding the conditional symbol called?

   b) What is the statement following the conditional symbol called?

3. a) Explain the procedure to determine the truth value of a compound statement when specific truth values are provided for the simple statements.

   b) Follow the procedure in part (a) and determine the truth value of the symbolic statement

$$[(p \leftrightarrow q) \vee (\sim r \rightarrow q)] \rightarrow \sim r$$

   when p is true, q is true, and r is false.

4. What is a tautology?

5. What is a self-contradiction?

6. What is an implication?

PRACTICE THE SKILLS

In Exercises 7–16, construct a truth table for the statement.

7. $\sim p \rightarrow q$

8. $\sim p \rightarrow \sim q$

9. $\sim(p \rightarrow \sim q)$

10. $\sim(\sim p \leftrightarrow q)$

11. $\sim q \leftrightarrow p$

12. $(p \leftrightarrow q) \rightarrow p$

13. $p \leftrightarrow (q \vee p)$

14. $(\sim q \wedge p) \rightarrow \sim q$

15. $q \rightarrow (p \rightarrow \sim q)$

16. $(p \vee q) \leftrightarrow (p \wedge q)$

In Exercises 17–26, construct a truth table for the statement.

17. $\sim p \rightarrow (q \wedge r)$

18. $q \vee (p \rightarrow \sim r)$

19. $p \leftrightarrow (\sim q \rightarrow r)$

20. $(\sim p \rightarrow \sim r) \rightarrow q$

21. $(q \vee \sim r) \leftrightarrow \sim p$

22. $(p \wedge r) \rightarrow (q \vee r)$

23. $(\sim r \vee \sim q) \rightarrow p$

24. $[r \wedge (q \vee \sim p)] \leftrightarrow \sim p$

25. $(p \rightarrow q) \leftrightarrow (\sim q \rightarrow \sim r)$

26. $(\sim p \leftrightarrow \sim q) \rightarrow (\sim q \leftrightarrow r)$

In Exercises 27–32, write the statement in symbolic form. Then construct a truth table for the symbolic statement.

27. If I take niacin, then I will stay healthy and I will have lower cholesterol.

28. The goalie will make the save if and only if the stopper is in position, or the forward cannot handle the ball.

29. The election was fair if and only if the polling station stayed open until 8 P.M., or we will request a recount.

30. If the dam holds then we can go fishing, if and only if the pole is not broken.

31. If Mary Andrews does not send me an e-mail then we can call her, or we can write to Mom.

32. It is false that if Eileen Jones went to lunch, then she cannot take a message and we will have to go home.

In Exercises 33–38, determine whether the statement is a tautology, self-contradiction, or neither.

33. $p \rightarrow \sim p$

34. $(p \wedge \sim q) \leftrightarrow \sim p$

35. $p \wedge (q \wedge \sim p)$

36. $(p \wedge \sim q) \rightarrow q$

37. $(\sim q \rightarrow p) \vee \sim q$

38. $[(p \rightarrow q) \vee r] \leftrightarrow [(p \wedge q) \rightarrow r]$

In Exercises 39–44, determine whether the statement is an implication.

39. $\sim p \rightarrow (p \vee q)$

40. $(p \wedge q) \rightarrow (\sim p \vee q)$

41. $(q \wedge p) \rightarrow (p \wedge q)$

42. $(p \vee q) \rightarrow (p \vee \sim r)$

43. $[(p \rightarrow q) \wedge (q \rightarrow p)] \rightarrow (p \leftrightarrow q)$

44. $[(p \vee q) \wedge r] \rightarrow (p \vee q)$

In Exercises 45–56, if p is true, q is false, and r is true, find the truth value of the statement.

45. $\sim p \rightarrow (q \rightarrow r)$

46. $(p \vee q) \rightarrow \sim r$

47. $q \leftrightarrow (\sim p \vee r)$

48. $r \rightarrow (\sim p \leftrightarrow \sim q)$

49. $(\sim p \wedge \sim q) \vee \sim r$

50. $\sim[p \rightarrow (q \wedge r)]$

51. $(p \wedge r) \leftrightarrow (p \vee \sim q)$

52. $(\sim p \vee q) \rightarrow \sim r$

53. $(\sim p \leftrightarrow r) \vee (\sim q \leftrightarrow r)$

54. $(r \rightarrow \sim p) \wedge (q \rightarrow \sim r)$

55. $\sim[(p \vee q) \leftrightarrow (p \rightarrow \sim r)]$

56. $[(\sim r \rightarrow \sim q) \vee (p \wedge \sim r)] \rightarrow q$

PROBLEM SOLVING

In Exercises 57–64, determine the truth value for each simple statement. Then, using the truth values, determine the truth value of the compound statement.

57. If $2 + 7 = 9$, then $15 - 3 = 12$.

58. If $\frac{3}{4} < 1$ and $\frac{7}{8} > 1$, then $\frac{15}{16} < 1$.

59. A cat has whiskers or a fish can swim, and a chicken lays eggs.

60. Tallahassee is in Florida and Atlanta is in Georgia, or Chicago is in Mississippi.

61. Apple makes computers, if and only if Nike makes sports shoes or Rolex makes watches.

62. Spike Lee is a movie director, or if Halle Berry is a schoolteacher then George Clooney is a circus clown.

63. Valentine's Day is in February or President's Day is in March, and Thanksgiving is in November.

64. Honda makes automobiles or Honda makes motorcycles, if and only if Toyota makes cereal.

In Exercises 65–68, use the information provided about the moons for the planets Jupiter and Saturn to determine the truth values of the simple statements. Then determine the truth value of the compound statement.

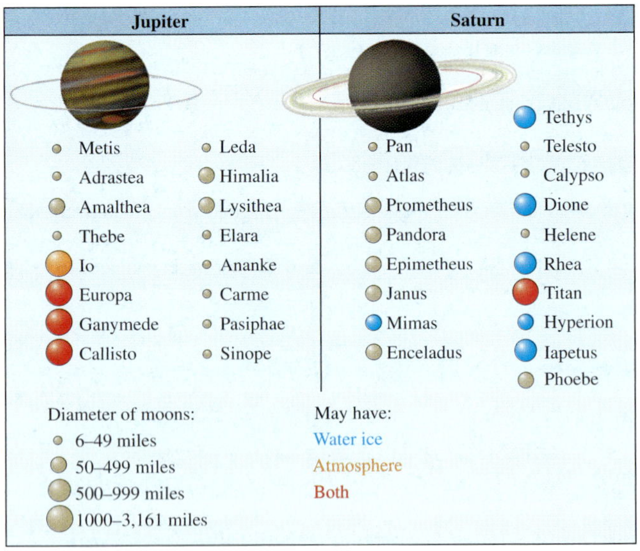

Source: Data from *Time* Magazine

65. ***Jupiter's Moons*** Io has a diameter of 1000–3161 miles or Thebe may have water, and Io may have atmosphere.

66. ***Moons of Saturn*** Titan may have water and Titan may have atmosphere, if and only if Janus may have water.

67. ***Moon Comparisons*** Phoebe has a larger diameter than Rhea if and only if Callisto may have water ice, and Calypso has a diameter of 6–49 miles.

68. ***Moon Comparisons*** If Jupiter has 16 moons or Saturn does not have 18 moons, then Saturn has 7 moons that may have water ice.

In Exercises 69 and 70, use the graphs to determine the truth values of each simple statement. Then determine the truth value of the compound statement.

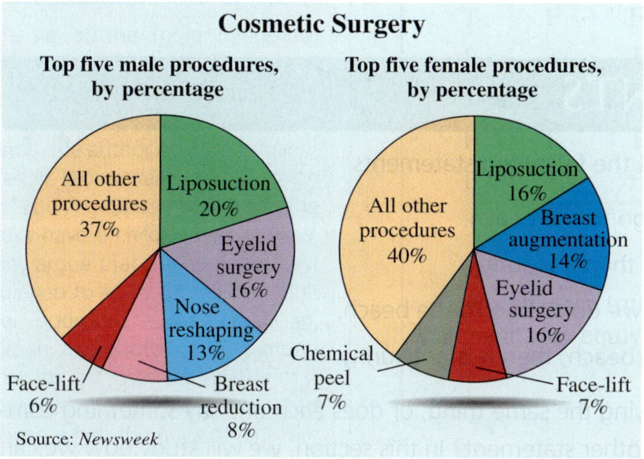

69. ***Most Common Cosmetic Surgery*** The most common cosmetic surgery procedure for females is liposuction or the most common procedure for males is eyelid surgery, and 20% of male cosmetic surgery is for nose reshaping.

70. ***Face-lifts and Eyelid Surgeries*** 7% of female cosmetic surgeries are for face-lifts and 10% of male cosmetic surgeries are for face-lifts, if and only if males have a higher percent of eyelid surgeries than females.

In Exercises 71–76, suppose both of the following statements are true.

p: Muhundan spoke at the teachers' conference.

q: Muhundan received the outstanding teacher award.

Find the truth values of each compound statement.

71. If Muhundan received the outstanding teacher award, then Muhundan spoke at the teachers' conference.

72. If Muhundan spoke at the teachers' conference, then Muhundan did not receive the outstanding teacher award.

73. If Muhundan did not receive the outstanding teacher award, then Muhundan spoke at the teachers' conference.

74. Muhundan did not receive the outstanding teacher award if and only if Muhundan spoke at the teachers' conference.

75. Muhundan received the outstanding teacher award if and only if Muhundan spoke at the teachers' conference.

76. If Muhundan did not receive the outstanding teacher award, then Muhundan did not speak at the teachers' conference.

77. ***A New Computer*** Your parents make the following statement to your sister, "If you get straight A's this semester, then we will buy you a new computer." At the end of the semester your parents buy your sister a new computer. Can you conclude that your sister got straight A's? Explain.

78. ***Job Interview*** Consider the statement "If your interview goes well, then you will be offered the job." If you are interviewed and then offered the job, can you conclude that your interview went well? Explain.

CHALLENGE PROBLEMS/GROUP ACTIVITIES

In Exercises 79 and 80, construct truth tables for the symbolic statement.

79. $[p \vee (q \rightarrow \sim r)] \leftrightarrow (p \wedge \sim q)$

80. $[(r \rightarrow \sim q) \rightarrow \sim p] \vee (q \leftrightarrow \sim r)$

81. Is the statement "Heads I win, tails you lose" a tautology, a self-contradiction, or neither? Explain your answer.

82. Construct a truth table for

a) $(p \vee q) \rightarrow (r \wedge s)$.

b) $(q \rightarrow \sim p) \vee (r \leftrightarrow s)$.

RECREATIONAL MATHEMATICS

83. ***Cat Puzzle*** Solve the following puzzle. The Joneses have four cats. The parents are Tiger and Boots, and the kittens are Sam and Sue. Each cat insists on eating out of its own bowl. To complicate matters, each cat will eat only its own brand of cat food. The colors of the bowls are red, yellow, green, and blue. The different types of cat food are Whiskas, Friskies, Nine Lives, and Meow Mix. Tiger will eat Meow Mix if and only if it is in a yellow bowl. If Boots

The truth tables for these statements are given in Tables 3.22 and 3.23, respectively. The answers in the columns labeled 5 are not identical, so the statements are not equivalent.

Table 3.22

p	q	r	$(p$	$\wedge$	$q)$	$\rightarrow$	r
T	T	T	T	T	T	T	T
T	T	F	T	T	T	F	F
T	F	T	T	F	F	T	T
T	F	F	T	F	F	T	F
F	T	T	F	F	T	T	T
F	T	F	F	F	T	T	F
F	F	T	F	F	F	T	T
F	F	F	F	F	F	T	F
			1	3	2	5	4

Table 3.23

p	q	r	$(\sim p$	$\vee$	$\sim q)$	$\rightarrow$	$\sim r$
T	T	T	F	F	F	T	F
T	T	F	F	F	F	T	T
T	F	T	F	T	T	F	F
T	F	F	F	T	T	T	T
F	T	T	T	T	F	F	F
F	T	F	T	T	F	T	T
F	F	T	T	T	T	F	F
F	F	F	T	T	T	T	T
			1	3	2	5	4

EXAMPLE 3 *Which Statements Are Logically Equivalent?*

Determine which statement is logically equivalent to "It is not true that the tire is both out of balance and flat."

a) If the tire is not flat, then the tire is not out of balance.
b) The tire is not out of balance or the tire is not flat.
c) The tire is not flat and the tire is not out of balance.
d) If the tire is not out of balance, then the tire is not flat.

SOLUTION To determine whether any of the choices are equivalent to the given statement, first write the given statements and the choices in symbolic form. Then construct truth tables and compare the answer columns of the truth tables.

Let

p: The tire is out of balance.
q: The tire is flat.

The given statement may be written "It is not true that the tire is out of balance and the tire is flat." The statement is expressed in symbolic form as $\sim(p \wedge q)$. Using p and q as indicated, choices (a) through (d) may be expressed symbolically as

a) $\sim q \rightarrow \sim p$. b) $\sim p \vee \sim q$. c) $\sim q \wedge \sim p$. d) $\sim p \rightarrow \sim q$.

Now construct a truth table for the given statement (Table 3.24) and for each statement (a) through (d), given in Table 3.25(a) through (d). By examining the truth tables, we see that the given statement, $\sim(p \wedge q)$, is logically equivalent to choice (b), $\sim p \vee \sim q$. Therefore, the correct answer is "The tire is not out of balance or the tire is not flat." This statement is logically equivalent to the statement "It is not true that the tire is both out of balance and flat."

Table 3.24

p	q	$\sim$	$(p$	$\wedge$	$q)$
T	T	F	T	T	T
T	F	T	T	F	F
F	T	T	F	F	T
F	F	T	F	F	F
		4	1	3	2

Table 3.25

		(a)			(b)			(c)			(d)		
p	q	$\sim q$	$\rightarrow$	$\sim p$	$\sim p$	$\vee$	$\sim q$	$\sim q$	$\wedge$	$\sim p$	$\sim p$	$\rightarrow$	$\sim q$
T	T	F	T	F	F	F	F	F	F	F	F	T	F
T	F	T	F	F	F	T	T	T	F	F	F	T	T
F	T	F	T	T	T	T	F	F	F	T	T	F	F
F	F	T	T	T	T	T	T	T	T	T	T	T	T

De Morgan's Laws

Example 3 showed that a statement of the form $\sim(p \wedge q)$ is equivalent to a statement of the form $\sim p \vee \sim q$. Thus, we may write $\sim(p \wedge q) \Leftrightarrow \sim p \vee \sim q$. This equivalent statement is one of two special laws called De Morgan's laws. The laws, named after Augustus De Morgan, an English mathematician, were first introduced in Section 2.4, where they applied to sets.

DE MORGAN'S LAWS

1. $\sim(p \wedge q) \Leftrightarrow \sim p \vee \sim q$
2. $\sim(p \vee q) \Leftrightarrow \sim p \wedge \sim q$

You can demonstrate that De Morgan's second law is true by constructing and comparing truth tables for $\sim(p \vee q)$ and $\sim p \wedge \sim q$. Do so now.

When using De Morgan's laws, if it becomes necessary to negate an already negated statement, use the fact that $\sim(\sim p)$ is equivalent to p. For example, the negation of the statement "Today is not Monday" is "Today is Monday."

EXAMPLE 4 *Use De Morgan's Laws*

Select the statement that is logically equivalent to "I do not have investments, but I do not have debts."

a) I do not have investments or I do not have debts.

b) It is false that I have investments and I have debts.

c) It is false that I have investments or I have debts.

d) I have investments or I have debts.

SOLUTION To determine which statement is equivalent, write each statement in symbolic form.

Let

p: I have investments.

q: I have debts.

The statement, "I do not have investments, but I do not have debts" written symbolically is $\sim p \wedge \sim q$. Recall that the word *but* means the same thing as *and*. Now, write parts (a) through (d) symbolically.

a) $\sim p \vee \sim q$ b) $\sim(p \wedge q)$ c) $\sim(p \vee q)$ d) $p \vee q$

De Morgan's law shows that $\sim p \wedge \sim q$ is equivalent to $\sim(p \vee q)$. Therefore, the answer is (c): "It is false that I have investments or I have debts."

EXAMPLE 5 *Using De Morgan's Laws to Write an Equivalent Statement*

Write a statement that is logically equivalent to "It is not true that tomatoes are poisonous or eating peppers cures the common cold."

SOLUTION Let

p: Tomatoes are poisonous.
q: Eating peppers cures the common cold.

The given statement is of the form $\sim(p \vee q)$. Using the second of De Morgan's laws, we see that an equivalent statement in symbols is $\sim p \wedge \sim q$. Therefore, an equivalent statement in words is "Tomatoes are not poisonous and eating peppers does not cure the common cold."

Consider $\sim(p \wedge q) \Leftrightarrow \sim p \vee \sim q$, one of De Morgan's laws. To go from $\sim(p \wedge q)$ to $\sim p \vee \sim q$, we negate both the p and the q within parentheses; change the conjunction, $\wedge$, to a disjunction, $\vee$; and remove the negation symbol preceding the left parentheses and the parentheses themselves. We can use a similar procedure to obtain equivalent statements. For example,

$$\sim(\sim p \wedge q) \Leftrightarrow p \vee \sim q$$
$$\sim(p \wedge \sim q) \Leftrightarrow \sim p \vee q$$

We can use a similar procedure to obtain equivalent statements when a disjunction is within parentheses. Note that

$$\sim(\sim p \vee q) \Leftrightarrow p \wedge \sim q$$
$$\sim(p \vee \sim q) \Leftrightarrow \sim p \wedge q$$

EXAMPLE 6 *Using De Morgan's Laws to Write an Equivalent Statement*

Use De Morgan's laws to write a statement logically equivalent to "Benjamin Franklin was not a U.S. president, but he signed the Declaration of Independence."

SOLUTION Let

p: Benjamin Franklin was a U.S. president.
q: Benjamin Franklin signed the Declaration of Independence.

PROFILE IN MATHEMATICS

Augustus De Morgan

Augustus De Morgan (1806–1871), the son of a member of the East India Company, was born in India and educated at Trinity College, Cambridge. One of the great reformers of logic in the nineteenth century, De Morgan made his greatest contribution to the subject by realizing that logic as it had come down from Aristotle was narrow in scope and could be applied to a wider range of arguments. His work laid the foundation for modern, symbolic logic.

The statement written symbolically is $\sim p \wedge q$. Earlier we showed that

$$\sim p \wedge q \Leftrightarrow \sim(p \vee \sim q)$$

Therefore, the statement "It is false that Benjamin Franklin was a U.S. president or Benjamin Franklin did not sign the Declaration of Independence" is logically equivalent to the given statement.

There are strong similarities between the topics of sets and logic. We can see them by examining De Morgan's laws for sets and logic.

De Morgan's laws: set theory	**De Morgan's laws: logic**
$(A \cap B)' = A' \cup B'$	$\sim(p \wedge q) \Leftrightarrow \sim p \vee \sim q$
$(A \cup B)' = A' \cap B'$	$\sim(p \vee q) \Leftrightarrow \sim p \wedge \sim q$

The complement in set theory, $'$, is similar to the negation, $\sim$, in logic. The intersection, $\cap$, is similar to the conjunction, $\wedge$; and the union, $\cup$, is similar to the disjunction, $\vee$. If we were to interchange the set symbols with the logic symbols, De Morgan's laws would remain, but in a different form.

Both $'$ and $\sim$ can be interpreted as *not*.

Both $\cap$ and $\wedge$ can be interpreted as *and*.

Both $\cup$ and $\vee$ can be interpreted as *or*.

For example, the set statement $A' \cup B$ can be written as a statement in logic as $\sim a \vee b$.

Statements containing connectives other than *and* and *or* may have equivalent statements. To illustrate this point, construct truth tables for $p \rightarrow q$ and for $\sim p \vee q$. The truth tables will have the same answer columns and therefore the statements are equivalent. That is,

$$p \rightarrow q \Leftrightarrow \sim p \vee q$$

With these equivalent statements, we can write a conditional statement as a disjunction or a disjunction as a conditional statement. For example, the statement "If the game is polo, then you ride a horse" can be equivalently stated as "The game is not polo or you ride a horse."

To change a conditional statement to a disjunction, negate the antecedent, change the conditional symbol to a disjunction symbol, and keep the consequent the same. To change a disjunction statement to a conditional statement, negate the first statement, change the disjunction symbol to a conditional symbol, and keep the second statement the same.

EXAMPLE 7 *Rewriting a Disjunction as a Conditional Statement*

Write a conditional statement that is logically equivalent to "The Oregon Ducks will win or the Oregon State Beavers will lose." Assume that the negation of winning is losing.

SOLUTION Let

p: The Oregon Ducks will win.
q: The Oregon State Beavers will win.

The original statement may be written symbolically as $p \vee \sim q$. To write an equivalent conditional statement, negate the first statement, p, change the disjunction symbol to a conditional symbol, and keep the second statement the same. Symbolically, the equivalent statement is $\sim p \rightarrow \sim q$. The equivalent statement in words is "If the Oregon Ducks lose, then the Oregon State Beavers will lose." •

Negation of the Conditional Statement

Now we will discuss how to negate a conditional statement. To negate a statement we use the fact that $p \rightarrow q \Leftrightarrow \sim p \vee q$ and De Morgan's laws. Examples 8 and 9 show the process.

EXAMPLE 8 *The Negation of a Conditional Statement*

Determine a statement equivalent to $\sim(p \rightarrow q)$.

SOLUTION Begin with $p \rightarrow q \Leftrightarrow \sim p \vee q$, negate both statements, and use De Morgan's laws.

$$
\begin{aligned}
p \rightarrow q &\Leftrightarrow \sim p \vee q \\
\sim(p \rightarrow q) &\Leftrightarrow \sim(\sim p \vee q) && \text{Negate both statements} \\
&\Leftrightarrow p \wedge \sim q && \text{De Morgan's laws}
\end{aligned}
$$

Therefore, $\sim(p \rightarrow q)$ is equivalent to $p \wedge \sim q$. •

EXAMPLE 9 *Write an Equivalent Statement*

Write a statement that is equivalent to "It is false that if it is snowing then we cannot go to the basketball game."

SOLUTION Let

p: It is snowing.
q: We can go to the basketball game.

The given statement can then be represented symbolically as $\sim(p \rightarrow \sim q)$. Using the procedure illustrated in Example 8 we can determine that $\sim(p \rightarrow \sim q)$ is equivalent to $p \wedge q$. You should verify that the two statements are equivalent now. Therefore, an equivalent statement is "It is snowing and we can go to the basketball game." •

Variations of the Conditional Statement

We know that $p \rightarrow q$ is equivalent to $\sim p \vee q$. Are any other statements equivalent to $p \rightarrow q$? Yes, there are many. Now let's look at the variations of the conditional statement to determine whether any are equivalent to the conditional statement. *The variations of the conditional statement are made by switching and/or negating the antecedent and the consequent of a conditional statement.* The variations of the conditional statement are the ***converse*** of the conditional, the ***inverse*** of the conditional, and the ***contrapositive*** of the conditional.

Listed here are the variations of the conditional with their symbolic form and the words we say to read each one.

VARIATIONS OF THE CONDITIONAL STATEMENT

Name	Symbolic form	Read
Conditional	$p \rightarrow q$	"If p, then q"
Converse of the conditional	$q \rightarrow p$	"If q, then p"
Inverse of the conditional	$\sim p \rightarrow \sim q$	"If not p, then not q"
Contrapositive of the conditional	$\sim q \rightarrow \sim p$	"If not q, then not p"

To write the converse of the conditional statement, switch the order of the antecedent and the consequent. To write the inverse, negate both the antecedent and the consequent. To write the contrapositive, switch the order of the antecedent and the consequent and then negate both of them.

Are any of the variations of the conditional statement equivalent? To determine the answer, we can construct a truth table for each variation, as shown in Table 3.26. It reveals that the conditional statement is equivalent to the contrapositive statement and that the converse statement is equivalent to the inverse statement.

Table 3.26

p	q	Conditional $p \rightarrow q$	Contrapositive $\sim q \rightarrow \sim p$	Converse $q \rightarrow p$	Inverse $\sim p \rightarrow \sim q$
T	T	T	T	T	T
T	F	F	F	T	T
F	T	T	T	F	F
F	F	T	T	T	T

EXAMPLE 10 *The Converse, Inverse and Contrapositive*

For the conditional statement "If the song contains sitar music, then the song was written by George Harrison," write the

a) converse. b) inverse. c) contrapositive.

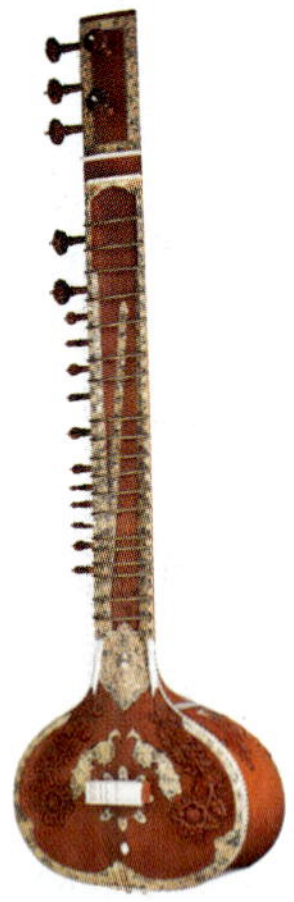

▲ A sitar

SOLUTION

a) Let

p: The song contains sitar music.

q: The song was written by George Harrison.

The conditional statement is of the form $p \rightarrow q$, so the converse must be of the form $q \rightarrow p$. Therefore, the converse is "If the song was written by George Harrison, then the song contains sitar music."

b) The inverse is of the form $\sim p \rightarrow \sim q$. Therefore, the inverse is "If the song does not contain sitar music, then the song was not written by George Harrison."

c) The contrapositive is of the form $\sim q \rightarrow \sim p$. Therefore, the contrapositive is "If the song was not written by George Harrison, then the song does not contain sitar music."

EXAMPLE 11 *Determine the Truth Values*

Let

p: The number is divisible by 9.

q: The number is divisible by 3.

Write the following statements and determine which are true.

a) The conditional statement, $p \rightarrow q$

b) The converse of $p \rightarrow q$

c) The inverse of $p \rightarrow q$

d) The contrapositive of $p \rightarrow q$

SOLUTION

a) *Conditional statement:* $(p \rightarrow q)$
If the number is divisible by 9, then the number is divisible by 3. This statement is true. A number divisible by 9 must also be divisible by 3, since 3 is a divisor of 9.

b) *Converse of the conditional:* $(q \rightarrow p)$
If the number is divisible by 3, then the number is divisible by 9. This statement is false. For instance, 6 is divisible by 3, but 6 is not divisible by 9.

c) *Inverse of the conditional:* $(\sim p \rightarrow \sim q)$
If the number is not divisible by 9, then the number is not divisible by 3. This statement is false. For instance, 6 is not divisible by 9, but 6 is divisible by 3.

d) *Contrapositive of the conditional:* $(\sim q \rightarrow \sim p)$
If the number is not divisible by 3, then the number is not divisible by 9. The statement is true, since any number that is divisible by 9 must be divisible by 3.

EXAMPLE 12 *Use the Contrapositive*

Use the contrapositive to write a statement logically equivalent to "If you don't eat your meat, then you can't have pudding."

SOLUTION Let

p: You do eat your meat.
q: You can have pudding.

The given statement written symbolically is

$$\sim p \rightarrow \sim q$$

The contrapositive of the statement is

$$q \rightarrow p$$

Therefore, an equivalent statement is "If you can have pudding, then you do eat your meat." ●

The contrapositive of the conditional is very important in mathematics. Consider the statement "If a^2 is not a whole number, then a is not a whole number." Is this statement true? You may find this question difficult to answer. Writing the statement's contrapositive may enable you to answer the question. The contrapositive is "If a is a whole number, then a^2 is a whole number." Since the contrapositive is a true statement, the original statement must also be true.

EXAMPLE 13 *Which Are Equivalent?*

Determine which, if any, of the following statements are equivalent. You may use De Morgan's laws, the fact that $p \rightarrow q \Leftrightarrow \sim p \vee q$, information from the variations of the conditional, or truth tables.

a) If you leave by 9 A.M., then you will get to your destination on time.
b) You do not leave by 9 A.M. or you will get to your destination on time.
c) It is false that you will get to your destination on time or you did not leave by 9 A.M.
d) If you do not get to your destination on time, then you did not leave by 9 A.M.

SOLUTION Let

p: You leave by 9 A.M.
q: You will get to your destination on time.

In symbolic form, the four statements are

a) $p \rightarrow q$. b) $\sim p \vee q$. c) $\sim(q \vee \sim p)$. d) $\sim q \rightarrow \sim p$.

Which of these statements are equivalent? Earlier in this section, you learned that $p \rightarrow q$ is equivalent to $\sim p \vee q$. Therefore, statements (a) and (b) are equivalent. Statement (d) is the contrapositive of statement (a). Therefore, statement (d) is also equivalent to statement (a) and statement (b). All these statements have the same truth table (Table 3.27 on page 146).

DID YOU KNOW?

Fuzzy Logic

Many modern computers work solely with two values, 1 or 0. This constraint makes it difficult for a computer to evaluate vague concepts that human beings deal with on a regular basis, such as *bright*, *slow*, and *light*. More and more computers are becoming more capable of handling such vague concepts thanks to the introduction of *fuzzy logic*. Unlike the traditional computer logic, fuzzy logic is based on the assignment of a value between 0 and 1, inclusively, that can vary from setting to setting. For example, a camera using fuzzy logic may assign *bright* a value of 0.9 on a sunny day, 0.4 on a cloudy day, and 0.1 at night. Fuzzy logic also makes use of logical statements like those studied in this chapter. One such statement is "If X and Y, then Z." For example, a computer chip in a camera programmed with fuzzy logic may use the rule "If the day is *bright* and the film speed is *slow*, then let in less *light*." Fuzzy logic is discussed further in Exercises 89 and 90.

Table 3.27

		(a)	(b)	(d)
p	q	$p \rightarrow q$	$\sim p \vee q$	$\sim q \rightarrow \sim p$
T	T	T	T	T
T	F	F	F	F
F	T	T	T	T
F	F	T	T	T

Now let's look at statement (c). To determine whether $\sim(q \vee \sim p)$ is equivalent to the other statements, we will construct its truth table (Table 3.28) and compare the answer column with the answer columns in Table 3.27.

Table 3.28

		(c)			
p	q	$\sim$	$(q$	$\vee$	$\sim p)$
T	T	F	T	T	F
T	F	T	F	F	F
F	T	F	T	T	T
F	F	F	F	T	T
		4	1	3	2

None of the three answer columns of the truth table in Table 3.27 is the same as the answer column of the truth table in Table 3.28. Therefore $\sim(q \vee \sim p)$ is not equivalent to any of the other statements. Therefore, statements (a), (b), and (d) are equivalent to each other.

SECTION 3.4 EXERCISES

CONCEPT/WRITING EXERCISES

1. a) What are equivalent statements?

   b) Explain how you can determine whether two statements are equivalent.

2. Suppose two statements are connected with the biconditional and the truth table is constructed. If the answer column of the truth table has all trues, what must be true about these two statements? Explain.

3. Write De Morgan's laws for logic.

4. For a statement of the form $p \rightarrow q$, symbolically indicate the form of the

   a) converse.

   b) inverse.

   c) contrapositive.

5. Which of the following are equivalent statements?
   a) The converse
   b) The contrapositive
   c) The inverse
   d) The conditional
6. Write a disjunction that is logically equivalent to $p \rightarrow q$.
7. If $p \rightarrow q$ is equivalent to $\sim p \vee q$, write a conjunction that is equivalent to $\sim(p \rightarrow q)$.
8. Write a conjunction involving two conditional statements that is logically equivalent to $p \leftrightarrow q$. See page 128.

PRACTICE THE SKILLS

In Exercises 9–18, use De Morgan's laws to determine whether the two statements are equivalent.

9. $\sim(p \wedge q), \sim p \wedge q$
10. $\sim(p \vee \sim q), \sim p \wedge q$
11. $\sim(p \wedge q), \sim(q \vee \sim p)$
12. $\sim p \vee \sim q, \sim(p \wedge q)$
13. $\sim(p \vee q), \sim p \wedge \sim q$
14. $\sim(p \wedge q), \sim p \wedge \sim q$
15. $(\sim p \vee \sim q) \rightarrow r, \sim(p \wedge q) \rightarrow r$
16. $q \rightarrow \sim(p \wedge \sim r), q \rightarrow \sim p \vee r$
17. $\sim(p \rightarrow \sim q), p \wedge q$
18. $\sim(\sim p \rightarrow q), \sim p \wedge \sim q$

In Exercises 19–30, use a truth table to determine whether the two statements are equivalent.

19. $p \rightarrow q, \sim p \vee q$
20. $\sim(p \rightarrow q), p \wedge \sim q$
21. $\sim q \rightarrow \sim p, p \rightarrow q$
22. $q \rightarrow p, \sim p \rightarrow \sim q$
23. $(p \vee q) \vee r, p \vee (q \vee r)$
24. $p \vee (q \wedge r), \sim p \rightarrow (q \wedge r)$
25. $p \wedge (q \vee r), (p \wedge q) \vee r$
26. $\sim(q \rightarrow p) \vee r, (p \vee q) \wedge \sim r$
27. $(p \rightarrow q) \wedge (q \rightarrow r), (p \rightarrow q) \rightarrow r$
28. $\sim q \rightarrow (p \wedge r), \sim(p \vee r) \rightarrow q$
29. $(p \rightarrow q) \wedge (q \rightarrow p), (p \leftrightarrow q)$
30. $[\sim(p \rightarrow q)] \wedge [\sim(q \rightarrow p)], \sim(p \leftrightarrow q)$

PROBLEM SOLVING

In Exercises 31–38, use De Morgan's laws to write an equivalent statement for the sentence.

31. It is false that the Rocky Mountains are in the East and the Appalachian Mountains are in the West.
32. It is false that Johanna Chan is the secretary or Davidson Pierre is the treasurer.
33. The watch was neither a Swatch watch nor was the watch a Swiss Army watch.

34. The pot roast is hot, but it is not well done.
35. The hotel does not have a weight room or the conference center does not have an auditorium.
36. Robert Farinelli is an authorized WedgCor dealer or he is not going to work for Prism Construction Company.
37. If Ashley Tabai takes the new job, then she will not move or she will buy a new house in town.
38. If Phil Murphy buys us dinner, then we will not go to the top of the CN Tower but we will be able to walk to the Red Bistro Restaurant.

In Exercises 39–46, use the fact that $p \rightarrow q$ is equivalent to $\sim p \vee q$ to write an equivalent form of the given statement.

39. If Ena Salter selects a new textbook, then she will have to write a new syllabus.

40. Maxwell's favorite TV show is *Jimmy Neutron* or his favorite TV show is *Danny Phantom.*

▲ Jimmy Neutron

41. Bob the Tomato visited the nursing home or he did not visit the Cub Scout meeting.

42. If Joanne Ernst goes to the Lightning game, then she will not go to the Devil Rays game.

43. If the plumbers meet in Kansas City then the Rainmakers will provide the entertainment.

44. Mary Beth Headlee organized the conference or John Waters does not work at Sinclair Community College.

45. Chase is not hiding or the pitcher is broken.

46. If Weezer is not on the radio, then Tim Ollendick is working.

In Exercises 47–54, use the fact that $\sim(p \rightarrow q)$ *is equivalent to* $p \wedge \sim q$ *to write the statement in an equivalent form.*

47. It is false that if we go to Cincinnati, then we will go to the zoo.

48. It is false that if General Electric makes the telephone, then the telephone is made in the United States.

49. I am cold and the heater is not working.

50. The Badgers beat the Nittany Lions and the Bucks beat the 76ers.

51. It is not true that if Borders has a sale then we will buy $100 worth of books.

52. Thompson is sick today but Allen didn't go to school.

53. John Deere will hire new workers and the city of Dubuque will retrain the workers.

54. My cell phone is not made by Motorola and my carrier is Alltel.

In Exercises 55–60, write the converse, inverse, and contrapositive of the statement. (For Exercise 60, use De Morgan's laws.)

55. If we work every night, then we can finish the quilt in 1 week.

56. If your cell phone is beeping, then you need to charge the battery.

57. If I go to Mexico, then I buy silver jewelry.

58. If Bob Dylan records a new CD, then he will go on tour.

59. If that annoying paper clip shows up on my computer screen, then I will scream.

60. If the sun is shining, then we will go down to the marina and we will take out the sailboat.

In Exercises 61–66, write the contrapositive of the statement. Use the contrapositive to determine whether the conditional statement is true or false.

61. If a natural number is not divisible by 5, then the natural number is not divisible by 10.

62. If the opposite sides of the quadrilateral are not parallel, then the quadrilateral is not a parallelogram.

63. If a natural number is divisible by 3, then the natural number is divisible by 6.

64. If $1/n$ is not a natural number, then n is not a natural number.

65. If two lines do not intersect in at least one point, then the two lines are parallel.

66. If $\dfrac{m \cdot a}{m \cdot b} \neq \dfrac{a}{b}$, then m is not a counting number.

In Exercises 67–82, determine which, if any, of the three statements are equivalent (see Example 13).

67. a) Bill Rush is not the editor or Bill Rush is the vice president.

b) If Bill Rush is the vice president, then Bill Rush is not the editor.

c) If Bill Rush is the editor, then Bill Rush is not the vice president.

68. a) If Fido is our dog's name, then Rex is not our dog's name.

b) It is false that Fido is our dog's name and Rex is not our dog's name.

c) Fido is not our dog's name or Rex is our dog's name.

69. a) The office is not cool and the copier is jammed.

b) If the office is not cool, then the copier is not jammed.

c) It is false that the office is cool or the copier is not jammed.

70. a) The test is not written or the review sheet is not ready.

b) If the test is written, then the review sheet is ready.

c) It is false that the review sheet is ready and the test is not written.

71. a) Today is not Sunday or the library is open.

b) If today is Sunday, then the library is not open.

c) If the library is open, then today is not Sunday.

72. a) If you are fishing at 1 P.M., then you are driving a car at 1 P.M.

b) You are not fishing at 1 P.M. or you are driving a car at 1 P.M.

c) It is false that you are fishing at 1 P.M. and you are not driving a car at 1 P.M.

▲ See Exercise 72

73. a) The grass grows and the trees are blooming.

b) If the trees are blooming, then the grass does not grow.

c) The trees are not blooming or the grass does not grow.

74. a) Johnny Patrick is chosen as department chair if and only if he is the only candidate.

b) If Johnny Patrick is chosen as department chair then he is the only candidate, and if Johnny Patrick is the only candidate then he is chosen as department chair.

c) Johnny Patrick is not chosen as department chair and he is not the only candidate.

75. a) It is false that if you do not drink milk then your cholesterol count will be lower.

b) Your cholesterol count will be lower if and only if you drink milk.

c) It is false that if you drink milk then your cholesterol count will not be lower.

76. a) Bruce Springsteen will not go on tour if and only if Clarence Clemmons does not play the saxophone in his band.

b) It is false that Bruce Springsteen will go on tour if and only if Clarence Clemmons does not play the saxophone in his band.

c) If Bruce Springsteen goes on tour, then Clarence Clemmons plays saxophone in his band.

77. a) If the pay is good and today is Monday, then I will take the job.

b) If I do not take the job, then it is false that the pay is good or today is Monday.

c) The pay is good and today is Monday, or I will take the job.

78. a) If you are 18 years old and a citizen of the United States, then you can vote in the presidential election.

b) You can vote in the presidential election, if and only if you are a citizen of the United States and you are 18 years old.

c) You cannot vote in the presidential election, or you are 18 years old and you are not a citizen of the United States.

79. a) The package was sent by Federal Express, or the package was not sent by United Parcel Service but the package arrived on time.

b) The package arrived on time, if and only if it was sent by Federal Express or it was not sent by United Parcel Service.

c) If the package was not sent by Federal Express, then the package was not sent by United Parcel Service but the package arrived on time.

80. a) If we put the dog outside or we feed the dog, then the dog will not bark.

b) If the dog barks, then we did not put the dog outside and we did not feed the dog.

c) If the dog barks, then it is false that we put the dog outside or we feed the dog.

81. a) The car needs oil, and the car needs gas or the car is new.

b) The car needs oil, and it is false that the car does not need gas and the car is not new.

c) If the car needs oil, then the car needs gas or the car is not new.

82. a) The mortgage rate went down, if and only if Tim purchased the house and the down payment was 10%.

b) The down payment was 10%, and if Tim purchased the house then the mortgage rate went down.

c) If Tim purchased the house, then the mortgage rate went down and the down payment was not 10%.

83. If p and q represent two simple statements, and if $p \rightarrow q$ is a false statement, what must be the truth value of the converse, $q \rightarrow p$? Explain.

84. If p and q represent two simple statements, and if $p \rightarrow q$ is a false statement, what must be the truth value of the inverse, $\sim p \rightarrow \sim q$? Explain.

85. If p and q represent two simple statements, and if $p \rightarrow q$ is a false statement, what must be the truth value of the contrapositive, $\sim q \rightarrow \sim p$? Explain.

86. If p and q represent two simple statements, and if $p \rightarrow q$ is a true statement, what must be the truth value of the contrapositive, $\sim q \rightarrow \sim p$? Explain.

CHALLENGE PROBLEMS/GROUP ACTIVITIES

87. We learned that $p \rightarrow q \Leftrightarrow \sim p \vee q$. Determine a conjunction that is equivalent to $p \rightarrow q$. (*Hint:* There are many answers.)

88. Determine whether $\sim[\sim(p \vee \sim q)] \Leftrightarrow p \vee \sim q$. Explain the method(s) you used to determine your answer.

89. In an appliance or device that uses fuzzy logic, a change in one condition causes a change in a second condition. For example, in a camera, if the brightness increases, the lens aperture automatically decreases to get the proper exposure on the film. Name at least 10 appliances or devices that make use of fuzzy logic and explain how fuzzy logic is used in each appliance or device. See the Did You Know? on page 146.

90. In symbolic logic, a statement is either true or false (consider true to have a value of 1 and false a value of 0). In fuzzy logic, nothing is true or false, but everything is a matter of degree. For example, consider the statement "The sun is shining." In fuzzy logic, this statement may have a value between 0 and 1 and may be constantly changing. For example, if the sun is partially blocked by clouds, the value of this statement may be 0.25. In fuzzy logic, the values of connective statements are found as follows for statements p and q.

Not p has a truth value of $1 - p$.
$p \wedge q$ has a truth value equal to the lesser of p and q.
$p \vee q$ has a truth value equal to the greater of p and q.
$p \rightarrow q$ has a truth value equal to the lesser of 1 and $1 - p + q$.
$p \leftrightarrow q$ has a truth value equal to $1 - |p - q|$, that is, 1 minus the absolute value* of p minus q.

Suppose the statement "p: The sun is shining" has a truth value of 0.25 and the statement "q: Mary is getting a tan" has a truth value of 0.20. Find the truth value of

a) $\sim p$.
b) $\sim q$.
c) $p \wedge q$.
d) $p \vee q$.
e) $p \rightarrow q$.
f) $p \leftrightarrow q$.

INTERNET/RESEARCH ACTIVITIES

91. Do research and write a report on fuzzy logic.

92. Read one of Lewis Carroll's books and write a report on how he used logic in the book. Give at least five specific examples.

93. Do research and write a report on the life and achievements of Augustus De Morgan. Indicate in your report his contributions to sets and logic.

3.5 SYMBOLIC ARGUMENTS

▲ John Mellencamp and his band sing the national anthem.

Consider the following statements.

If John Mellencamp sings the national anthem, then U2 will play at halftime.

John Mellencamp sings the national anthem.

If you accept these two statements as true, then what logical conclusion could you draw? Do you agree that you can logically conclude that U2 will play at halftime? In this section, we will use our knowledge of logic to study the structure of such statements to draw logical conclusions.

Previously in this chapter, we used symbolic logic to determine the truth value of a compound statement. We now extend those basic ideas to determine whether we can draw logical conclusions from a set of given statements. Consider once again the two statements.

If John Mellencamp sings the national anthem, then U2 will play at halftime.

John Mellencamp sings the national anthem.

*Absolute values are discussed in Section 13.8.

DID YOU KNOW?

The Dog Did Nothing

In the case of the disappearance of the racehorse Silver Blaze, Sherlock Holmes demonstrated that sometimes the absence of a clue is itself a clue. The local police inspector asked him, "Is there any point to which you would wish to draw my attention?" Holmes replied, "To the curious incident of the dog in the nighttime." The inspector, confused, asked: "The dog did nothing in the nighttime." "That was the curious incident," remarked Sherlock Holmes. From the lack of the dog's bark, Holmes concluded that the horse had been "stolen" by a stablehand. How did Holmes reach his conclusion?

These statements in the following form constitute what we will call a *symbolic argument*.

Premise 1: If John Mellencamp sings the national anthem, then U2 will play at halftime.
Premise 2: John Mellencamp sings the national anthem.
Conclusion: U2 will play at halftime.

A ***symbolic argument*** consists of a set of ***premises*** and a ***conclusion***. It is called a symbolic argument because we generally write it in symbolic form to determine its validity.

An **argument is valid** when its conclusion necessarily follows from a given set of premises.

An **argument is invalid** or a **fallacy** when the conclusion does not necessarily follow from the given set of premises.

An argument that is not valid is invalid. The argument just presented is an example of a valid argument, as the conclusion necessarily follows from the premises. Now we will discuss a procedure to determine whether an argument is valid or invalid. We begin by writing the argument in symbolic form. To write the argument in symbolic form, we let p and q be

p: John Mellencamp sings the national anthem.
q: U2 will play at halftime.

Symbolically, the argument is written

Premise 1: $p \rightarrow q$
Premise 2: p
Conclusion: $\therefore q$ (The three-dot triangle is read "therefore.")

Write the argument in the following form.

If [*premise 1* **and** *premise 2*] **then** *conclusion*

$$[(p \rightarrow q) \quad \wedge \quad p] \quad \rightarrow \quad q$$

Then construct a truth table for the statement $[(p \rightarrow q) \wedge p] \rightarrow q$ (Table 3.29). *If the truth table answer column is true in every case, then the statement is a tautology, and the argument is valid. If the truth table is not a tautology, then the argument is invalid.* Since the statement is a tautology (see column 5), the conclusion necessarily follows from the premises and the argument is valid.

Table 3.29

p	q	$[(p \rightarrow q)$	$\wedge$	$p]$	$\rightarrow$	q
T	T	T	T	T	T	T
T	F	F	F	T	T	F
F	T	T	F	F	T	T
F	F	T	F	F	T	F
		1	3	2	5	4

Once we have demonstrated that an argument in a particular form is valid, all arguments with exactly the same form will also be valid. In fact, many of these forms have been assigned names. The argument form just discussed,

$$\begin{array}{l} p \rightarrow q \\ \underline{p \quad\quad} \\ \therefore\ q \end{array}$$

is called the ***law of detachment***, or ***modus ponens***.

▲ Alaska

EXAMPLE 1 *Determining Validity without a Truth Table*

Determine whether the following argument is valid or invalid.

If Canada is north of the United States, then Alaska is in Mexico.
Canada is north of the United States.
∴ Alaska is in Mexico.

SOLUTION Translate the argument into symbolic form.

Let

c: Canada is north of the United States.
a: Alaska is in Mexico.

In symbolic form, the argument is

$$\begin{array}{l} c \rightarrow a \\ \underline{c \quad\quad} \\ \therefore\ a \end{array}$$

This argument is also the law of detachment. Therefore, it is a valid argument. ●

Note that the argument in Example 1 is valid even though the conclusion, "Alaska is in Mexico," is a false statement. It is also possible to have an invalid argument in which the conclusion is a true statement. When an argument is valid, the conclusion necessarily follows from the premises. It is not necessary for the premises or the conclusion to be true statements in an argument.

PROCEDURE TO DETERMINE WHETHER AN ARGUMENT IS VALID

1. Write the argument in symbolic form.
2. Compare the form of the argument with forms that are known to be valid or invalid. If there are no known forms to compare it with, or you do not remember the forms, go to step 3.
3. If the argument contains two premises, write a conditional statement of the form

$$[(\text{premise } 1) \wedge (\text{premise } 2)] \rightarrow \text{conclusion}$$

4. Construct a truth table for the statement in step 3.
5. If the answer column of the truth table has all trues, the statement is a tautology, and the argument is valid. If the answer column does not have all trues, the argument is invalid.

Examples 1 through 4 contain two premises. When an argument contains more than two premises, step 3 of the procedure will change slightly, as will be explained shortly.

EXAMPLE 2 *Determining Validity with a Truth Table*

Determine whether the following argument is valid or invalid.

If you score 90% on the final exam, then you will get an A in the course.
You will not get an A in the course.
∴ You do not score 90% on the final exam.

SOLUTION We first write the argument in symbolic form.

Let

p: You score 90% on the final exam.
q: You will get an A in the course.

In symbolic form, the argument is

$$\begin{array}{l} p \rightarrow q \\ \underline{\sim q} \\ \therefore \sim p \end{array}$$

As we have not tested an argument in this form, we will construct a truth table to determine whether the argument is valid or invalid. We write the argument in the form $[(p \rightarrow q) \wedge \sim q] \rightarrow \sim p$, and construct a truth table (Table 3.30). Since the answer, column 5, has all T's, the argument is valid.

Table 3.30

p	q	$[(p \rightarrow q)$	$\wedge$	$\sim q]$	$\rightarrow$	$\sim p$
T	T	T	F	F	T	F
T	F	F	F	T	T	F
F	T	T	F	F	T	T
F	F	T	T	T	T	T
		1	3	2	5	4

The argument form in Example 2 is an example of the *law of contraposition*, or *modus tollens*.

EXAMPLE 3 *Another Symbolic Argument*

Determine whether the following argument is valid or invalid.

The grass is green or the grass is full of weeds.
The grass is not green.
∴ The grass is full of weeds.

SOLUTION Let

p: The grass is green.
q: The grass is full of weeds.

In symbolic form, the argument is

$$\begin{array}{l} p \vee q \\ \underline{\sim p} \\ \therefore q \end{array}$$

As this form is not one of those we are familiar with, we will construct a truth table. We write the argument in the form $[(p \vee q) \wedge \sim p] \rightarrow q$. Next we construct a truth table, as shown in Table 3.31. The answer to the truth table, column 5, is true in *every case*. Therefore, the statement is a tautology, and the argument is valid.

Table 3.31

p	q	$[(p \vee q)$	$\wedge$	$\sim p]$	$\rightarrow$	q
T	T	T	F	F	T	T
T	F	T	F	F	T	F
F	T	T	T	T	T	T
F	F	F	F	T	T	F
		1	3	2	5	4

The argument form in Example 3 is an example of ***disjunctive syllogism***. Other standard forms of arguments are given in the following chart.

STANDARD FORMS OF ARGUMENTS

Valid Arguments	*Law of Detachment*	*Law of Contraposition*	*Law of Syllogism*	*Disjunctive Syllogism*
	$\begin{array}{l} p \rightarrow q \\ \underline{p} \\ \therefore q \end{array}$	$\begin{array}{l} p \rightarrow q \\ \underline{\sim q} \\ \therefore \sim p \end{array}$	$\begin{array}{l} p \rightarrow q \\ \underline{q \rightarrow r} \\ \therefore p \rightarrow r \end{array}$	$\begin{array}{l} p \vee q \\ \underline{\sim p} \\ \therefore q \end{array}$
Invalid Arguments	*Fallacy of the Converse*	*Fallacy of the Inverse*		
	$\begin{array}{l} p \rightarrow q \\ \underline{q} \\ \therefore p \end{array}$	$\begin{array}{l} p \rightarrow q \\ \underline{\sim p} \\ \therefore \sim q \end{array}$		

As we saw in Example 1, it is not always necessary to construct a truth table to determine whether or not an argument is valid. The next two examples will show how we can identify an argument as one of the standard arguments given in the chart above.

MATHEMATICS TODAY

Freedom of Speech or Misleading Advertising?

The first amendment to the United States Constitution guarantees that Americans have the freedom of speech. There are, however, limits to what we may say or write. For example, companies are limited in what they can say or write to advertise their products. The Federal Trade Commission (FTC) states that "ads must be truthful and not misleading; that advertisers must have evidence to back up their assertions; and that ads cannot be unfair." Although most advertisements are truthful and fair, some enter into a "gray area" of truthfulness. Some of these breaches of fairness may be found in logical fallacies either made directly or implicitly by the context of the ad's wording or artwork. The FTC has recently taken action against companies marketing a variety of products on the Internet aimed at treating or curing a variety of diseases. The FTC refers to this action as Operation Cure-All. Go to www.ftc.gov/healthclaims/ to see examples of products removed from the market by this FTC action.

EXAMPLE 4 *Identifying the Law of Syllogism in an Argument*

Determine whether the following argument is valid or invalid.

If my laptop battery is dead, then I use my home computer.
If I use my home computer, then my kids will play outside.
∴ If my laptop battery is dead, then my kids will play outside.

SOLUTION Let

p: My laptop battery is dead.
q: I use my home computer.
r: My kids will play outside.

In symbolic form, the argument is

$$\begin{array}{l} p \rightarrow q \\ \underline{q \rightarrow r} \\ \therefore p \rightarrow r \end{array}$$

The argument is in the form of the law of syllogism. Therefore, the argument is valid, and there is no need to construct a truth table.

EXAMPLE 5 *Identifying Common Fallacies in Arguments*

Determine whether the following arguments are valid or invalid.

a)

If it is snowing, then we put salt on the driveway.
We put salt on the driveway.
∴ It is snowing.

b)

If it is snowing, then we put salt on the driveway.
It is not snowing.
∴ We do not put salt on the driveway.

SOLUTION

a) Let

p: It is snowing.
q: We put salt on the driveway.

In symbolic form, the argument is

$$\begin{array}{l} p \rightarrow q \\ \underline{q} \\ \therefore p \end{array}$$

This argument is in the form of the fallacy of the converse. Therefore, the argument is a fallacy, or invalid.

b) Using the same symbols defined in the solution to part (a), in symbolic form, the argument is

$$\begin{array}{l} p \rightarrow q \\ \underline{\sim p} \\ \therefore \sim q \end{array}$$

This argument is in the form of the fallacy of the inverse. Therefore, the argument is a fallacy, or invalid.

TIMELY TIP If you are not sure whether an argument with two premises is one of the standard forms or if you do not remember the standard forms, you can always determine whether a given argument is valid or invalid by using a truth table. To do so, follow the boxed procedure on page 153.

In Example 5(b), if you did not recognize that this argument was of the same form as the fallacy of the inverse you could construct the truth table for the conditional statement

$$[(p \rightarrow q) \wedge \sim p] \rightarrow \sim q$$

The true–false values under the conditional column, $\rightarrow$, would be T, T, F, T. Since the statement is not a tautology, the argument is invalid.

Now we consider an argument that has more than two premises. When an argument contains more than two premises, the statement we test, using a truth table, is formed by taking the conjunction of all the premises as the antecedent of a conditional statement and the conclusion as the consequent of the conditional statement. One example is an argument of the form

$$\begin{array}{l} p_1 \\ p_2 \\ \underline{p_3} \\ \therefore c \end{array}$$

We evaluate the truth table for $[p_1 \wedge p_2 \wedge p_3] \rightarrow c$. When we evaluate $[p_1 \wedge p_2 \wedge p_3]$, it makes no difference whether we evaluate $[(p_1 \wedge p_2) \wedge p_3]$ or $[p_1 \wedge (p_2 \wedge p_3)]$ because both give the same answer. In Example 6, we evaluate $[p_1 \wedge p_2 \wedge p_3]$ from left to right; that is, $[(p_1 \wedge p_2) \wedge p_3]$.

EXAMPLE 6 *An Argument with Three Premises*

Use a truth table to determine whether the following argument is valid or invalid.

If my cell phone company is Verizon, then I can call you free of charge.
I can call you free of charge or I can send you an e-mail.
I can send you an e-mail or my cell phone company is Verizon.
∴ My cell phone company is Verizon.

SOLUTION This argument contains three simple statements.

Let

p: My cell phone company is Verizon.
q: I can call you free of charge.
r: I can send you an e-mail.

In symbolic form, the argument is

$$\begin{array}{l} p \rightarrow q \\ q \vee r \\ \underline{r \vee p} \\ \therefore\ p \end{array}$$

Write the argument in the form

$$[(p \rightarrow q) \wedge (q \vee r) \wedge (r \vee p)] \rightarrow p.$$

Now construct the truth table (Table 3.32). The answer, column 7, is not true in every case. Thus, the argument is a fallacy, or invalid.

Table 3.32

p	q	r	$[(p \rightarrow q)$	$\wedge$	$(q \vee r)$	$\wedge$	$(r \vee p)]$	$\rightarrow$	p
T	T	T	T	T	T	T	T	T	T
T	T	F	T	T	T	T	T	T	T
T	F	T	F	F	T	F	T	T	T
T	F	F	F	F	F	F	T	T	T
F	T	T	T	T	T	T	T	F	F
F	T	F	T	T	T	F	F	T	F
F	F	T	T	T	T	T	T	F	F
F	F	F	T	F	F	F	F	T	F
			1	3	2	5	4	7	6

Let's now investigate how we can arrive at a valid conclusion from a given set of premises.

EXAMPLE 7 *Determine a Logical Conclusion*

Determine a logical conclusion that follows from the given statements. "If the price of gas is below \$3.00 per gallon, then we will drive to the Offspring concert. We will not drive to the Offspring concert. Therefore . . ."

SOLUTION If you recognize a specific form of an argument, you can use your knowledge of that form to draw a logical conclusion.

Let

p: The price of gas is below \$3.00 per gallon.
q: We will drive to the Offspring concert.

The argument is of the following form.

$$\begin{array}{l} p \to q \\ \underline{\sim q} \\ \therefore\ ? \end{array}$$

If the question mark is replaced with a $\sim p$, this argument is of the form of the law of contraposition. Thus, a logical conclusion is "Therefore, the price of gas is not below \$3.00 per gallon."

SECTION 3.5 EXERCISES

CONCEPT/WRITING EXERCISES

1. a) What does it mean when an argument is a valid argument?

   b) What does it mean when an argument is a fallacy?

2. Explain how to determine whether an argument with premises p_1 and p_2 and conclusion c is a valid argument or an invalid argument.

3. Is it possible for an argument to be invalid if its conclusion is true? Explain your answer.

4. Is it possible for an argument to be valid if its conclusion is false? Explain your answer.

5. Is it possible for an argument to be invalid if the premises are all true? Explain your answer.

6. Is it possible for an argument to be valid if the premises are all false? Explain your answer.

In Exercises 7–10, (a) indicate the form of the valid argument and (b) write an original argument in words for each form.

7. Law of detachment
8. Law of syllogism
9. Law of contraposition
10. Disjunctive syllogism

In Exercises 11 and 12, (a) indicate the form of the fallacy, and (b) write an original argument in words for each form.

11. Fallacy of the converse
12. Fallacy of the inverse

PRACTICE THE SKILLS

In Exercises 13–32, determine whether the argument is valid or invalid. You may compare the argument to a standard form, given on page 155, or use a truth table.

13. $\begin{array}{l} a \to b \\ \underline{\sim a} \\ \therefore \sim b \end{array}$

14. $\begin{array}{l} c \vee d \\ \underline{\sim c} \\ \therefore d \end{array}$

15. $\begin{array}{l} e \to f \\ \underline{e} \\ \therefore f \end{array}$

16. $\begin{array}{l} g \to h \\ \underline{h \to i} \\ \therefore g \to i \end{array}$

17. $\begin{array}{l} j \vee k \\ \underline{\sim k} \\ \therefore j \end{array}$

18. $\begin{array}{l} l \to m \\ \underline{\sim m} \\ \therefore \sim l \end{array}$

19. $\begin{array}{l} n \to o \\ \underline{o} \\ \therefore n \end{array}$

20. $\begin{array}{l} r \to s \\ \underline{r} \\ \therefore s \end{array}$

21. $\begin{array}{l} t \to u \\ \underline{\sim u} \\ \therefore \sim t \end{array}$

22. $\begin{array}{l} v \to w \\ \underline{w} \\ \therefore v \end{array}$

23. $\begin{array}{l} x \to y \\ \underline{y \to z} \\ \therefore x \to z \end{array}$

24. $\begin{array}{l} x \to y \\ \underline{\sim x} \\ \therefore \sim y \end{array}$

25. $\begin{array}{l} p \leftrightarrow q \\ \underline{q \wedge r} \\ \therefore p \vee r \end{array}$

26. $\begin{array}{l} p \leftrightarrow q \\ \underline{q \to r} \\ \therefore \sim r \to \sim p \end{array}$

27. $\begin{array}{l} r \leftrightarrow p \\ \underline{\sim p \wedge q} \\ \therefore p \wedge r \end{array}$

28. $\begin{array}{l} p \vee q \\ \underline{r \wedge p} \\ \therefore q \end{array}$

29. $\begin{array}{l} p \to q \\ q \vee r \\ \underline{r \vee p} \\ \therefore p \end{array}$

30. $\begin{array}{l} p \to q \\ q \to r \\ \underline{r \to p} \\ \therefore q \to p \end{array}$

31. $p \rightarrow q$
$r \rightarrow \sim p$
$p \vee r$
$\therefore q \vee \sim p$

32. $p \leftrightarrow q$
$p \vee r$
$q \rightarrow r$
$\therefore q \vee r$

PROBLEM SOLVING

In Exercises 33–50, (a) translate the argument into symbolic form and (b) determine if the argument is valid or invalid. You may compare the argument to a standard form or use a truth table.

33. If Will Smith wins an Academy Award, then he will retire from acting.
Will Smith did not win an Academy Award.
∴ Will Smith will not retire from acting.

34. If the car is a Road Runner, then the car is fast.
The car is fast.
∴ The car is a Road Runner.

35. If the baby is a boy, then we will name him Alexander Martin.
The baby is a boy.
∴ We will name him Alexander Martin.

36. If I can get my child to preschool by 8:45 A.M., then I can take the 9:00 A.M. class.
If I can take the 9:00 A.M. class, then I can be done by 2:00 P.M.
∴ If I can get my child to preschool by 8:45 A.M., then I can be done by 2:00 P.M.

37. If the guitar is a Les Paul model, then the guitar is made by Gibson.
The guitar is not made by Gibson.
∴ The guitar is not a Les Paul model.

▲ Les Paul

38. We will go for a bike ride or we will go shopping.
We will not go shopping.
∴ We will go for a bike ride.

39. If we planted the garden by the first Friday in April, then we will have potatoes by the Fourth of July.
We will have potatoes by the Fourth of July.
∴ We planted the garden by the first Friday in April.

40. If you pass general chemistry, then you can take organic chemistry.
You pass general chemistry.
∴ You can take organic chemistry.

41. Sarah Hughes will win an Olympic gold medal in figure skating or Joey Cheek will win an Olympic gold medal in speed skating.
Sarah Hughes will not win an Olympic gold medal in figure skating.
∴ Joey Cheek will win an Olympic gold medal in speed skating.

42. If Nicholas Thompson teaches this course, then I will get a passing grade.
I did not get a passing grade.
∴ Nicholas Thompson did not teach the course.

43. If it is cold, then graduation will be held indoors.
If graduation is held indoors, then the fireworks will be postponed.
∴ If it is cold, then the fireworks will be postponed.

44. If the canteen is full, then we can go for a walk.
We can go for a walk and we will not get thirsty.
∴ If we go for a walk, then the canteen is not full.

45. Marie works at the post office and Jim works for Target.
If Jim works for Target, then Tommy gets an internship.
∴ If Tommy gets an internship, then Marie works at the post office.

46. Vitamin C helps your immune system or niacin helps reduce cholesterol.
If niacin helps reduce cholesterol, then vitamin E enhances your skin.
∴ Vitamin C helps your immune system and vitamin E enhances your skin.

47. It is snowing and I am going skiing.
If I am going skiing, then I will wear a coat.
∴ If it is snowing, then I will wear a coat.

48. The garden has vegetables or the garden has flowers.
If the garden does not have flowers, then the garden has vegetables.
∴ The garden has flowers or the garden has vegetables.

49. If the house has electric heat, then the Flynns will buy the house.
If the price is not less than $100,000, then the Flynns will not buy the house.
∴ If the house has electric heat, then the price is less than $100,000.

50. If there is an atmosphere, then there is gravity.
If an object has weight, then there is gravity.
∴ If there is an atmosphere, then an object has weight.

In Exercises 51–60, translate the argument into symbolic form. Then determine whether the argument is valid or invalid.

51. If the prescription was called in to Walgreen's, then you can pick it up by 4:00 P.M. You cannot pick it up by 4:00 P.M. Therefore, the prescription was not called in to Walgreen's.

52. The printer has a clogged nozzle or the printer does not have toner. The printer has toner. Therefore, the printer has a clogged nozzle.

53. Max is playing Game Boy with the sound off or Max is wearing headphones. Max is not playing Game Boy with the sound off. Therefore, Max is wearing headphones.

54. If the cat is in the room, then the mice are hiding. The mice are not hiding. Therefore, the cat is not in the room.

55. The test was easy and I received a good grade. The test was not easy or I did not receive a good grade. Therefore, the test was not easy.

56. If Bonnie passes the bar exam, then she will practice law. Bonnie will not practice law. Therefore, Bonnie did not pass the bar exam.

57. The baby is crying but the baby is not hungry. If the baby is hungry then the baby is crying. Therefore, the baby is hungry.

58. If the car is new, then the car has air conditioning. The car is not new and the car has air conditioning. Therefore, the car is not new.

59. If the football team wins the game, then Dave played quarterback. If Dave played quarterback, then the team is not in second place. Therefore, if the football team wins the game, then the team is in second place.

60. The engineering courses are difficult and the chemistry labs are long. If the chemistry labs are long, then the art tests are easy. Therefore, the engineering courses are difficult and the art tests are not easy.

In Exercises 61–67, using the standard forms of arguments and other information you have learned, supply what you believe is a logical conclusion to the argument. Verify that the argument is valid for the conclusion you supplied.

61. If you eat an entire bag of M & M's, then your face will break out.
You eat an entire bag of M & M's.
Therefore, . . .

62. If the temperature hits 100°, then we will go swimming.
We did not go swimming.
Therefore, . . .

63. I am stressed out or I have the flu.
I do not have the flu.
Therefore, . . .

64. If I can get Nick to his piano lesson by 3:30 P.M., then I can do my shopping.
If I can do my shopping, then we do not need to order pizza again.
Therefore, . . .

65. If you close the deal, then you will get a commission.
You did not get a commission.
Therefore, . . .

66. If Katherine is at her ballet lesson, then Allyson will take a nap.
Katherine is at her ballet lesson.
Therefore, . . .

67. If you do not pay off your credit card bill, then you will have to pay interest.
If you have to pay interest, then the bank makes money.
Therefore, . . .

CHALLENGE PROBLEMS/GROUP ACTIVITIES

68. Determine whether the argument is valid or invalid.

If Lynn wins the contest or strikes oil, then she will be rich.
If Lynn is rich, then she will stop working.
______
∴ If Lynn does not stop working, she did not win the contest.

69. Is it possible for an argument to be invalid if the conjunction of the premises is false in every case of the truth table? Explain your answer.

RECREATIONAL MATHEMATICS

70. René Descartes was a seventeenth-century French mathematician and philosopher. One of his most memorable statements is, "I think, therefore, I am." This statement is the basis for the following joke.

Descartes walks into an inn. The innkeeper asks Descartes if he would like something to drink. Descartes replies, "I think not," and promptly vanishes into thin air!

This joke can be summarized in the following argument: If I think, then I am. I think not. Therefore, I am not.

a) Represent this argument symbolically.

b) Is it a valid argument?

c) Explain your answer using either a standard form of argument or using a truth table.

INTERNET/RESEARCH ACTIVITIES

71. Show how logic is used in advertising. Discuss several advertisements and show how logic is used to persuade the reader.

72. Find examples of valid (or invalid) arguments in printed matter such as newspaper or magazine articles. Explain why the arguments are valid (or invalid).

3.6 EULER DIAGRAMS AND SYLLOGISTIC ARGUMENTS

▲ Adam Brody and Meg Ryan in the movie *In the Land of Women.*

While trying to choose a movie to watch on Valentine's Day, your spouse says to you, "All Meg Ryan movies are romantic comedies. Meg Ryan is in the movie *In the Land of Women*. Therefore, *In the Land of Women* is a romantic comedy." This argument uses a type of argument that we will discuss and analyze in this section.

In Section 3.5, we showed how to determine the validity of *symbolic arguments* using truth tables and comparing the arguments to standard forms. This section presents another form of argument called a ***syllogistic argument***, better known by the shorter name ***syllogism***. The validity of a syllogistic argument is determined by using Euler (pronounced "oiler") diagrams, as is explained shortly.

Syllogistic logic, a deductive process of arriving at a conclusion, was developed by Aristotle in about 350 B.C. Aristotle considered the relationships among the four types of statements that follow.

All ______ are ______.
No ______ are ______.
Some ______ are ______.
Some ______ are not ______.

PROFILE IN MATHEMATICS

Leonhard Euler

Swiss mathematician Leonhard Euler (pronounced "oiler"; 1707–1783) produced such a large volume of work that not all of it has yet been compiled. It has been estimated that a complete edition of his work would make up over 60 volumes. Although his greatest contribution was made in the field of modern analysis, his work also explored physics, astronomy, trigonometry, and calculus. Euler also had a great influence on the notation and symbolism of mathematics, and it was through his work that the symbols e and π came into common use. His facility for mental calculation and his uncommon memory allowed him to continue his prolific career even after he lost his vision in both eyes.

Examples of these statements are: *All doctors are tall. No doctors are tall. Some doctors are tall. Some doctors are not tall.* Since Aristotle's time, other types of statements have been added to the study of syllogistic logic, two of which are

______ is a ______.

______ is not a ______.

Examples of these statements are: *Maria is a doctor. Maria is not a doctor.*

The difference between a symbolic argument and a syllogistic argument can be seen in the following chart. Symbolic arguments use the connectives *and, or, not, if–then,* and *if and only if.* Syllogistic arguments use the quantifiers *all, some,* and *none,* which were discussed in Section 3.1.

SYMBOLIC ARGUMENTS VERSUS SYLLOGISTIC ARGUMENTS

	Words or phrases used	*Method of determining validity*
Symbolic argument	and, or, not, if–then, if and only if	Truth tables or by comparison with standard forms of arguments
Syllogistic argument	all are, some are, none are, some are not	Euler diagrams

As with symbolic logic, the premises and the conclusion together form an argument. An example of a syllogistic argument is

All German shepherds are dogs.
All dogs bark.
―――――――――――――――
∴ All German shepherds bark.

This is an example of a valid argument. Recall from Section 3.5 that an argument is *valid* when its conclusion necessarily follows from a given set of premises. Recall that an argument in which the conclusion does not necessarily follow from the given premises is said to be an *invalid argument* or a *fallacy.*

Before we give another example of a syllogism, let's review the Venn diagrams discussed in Section 2.3 in relationship with Aristotle's four statements.

All *A*s are *B*s	No *A*s are *B*s	Some *A*s are *B*s	Some *A*s are not *B*s

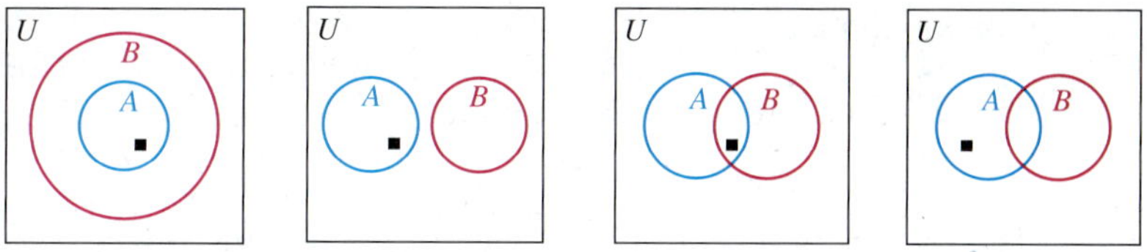

If an element is in set *A*, then it is in set *B*.	If an element is in set *A*, then it is not in set *B*.	There is at least one element that is in both set *A* and set *B*.	There is at least one element that is in set *A* that is not in set *B*.

One method used to determine whether an argument is valid or is a fallacy is by means of an ***Euler diagram***, named after Leonhard Euler, who used circles to represent sets in syllogistic arguments. The technique of using Euler diagrams is illustrated in Example 1.

EXAMPLE 1 *Using an Euler Diagram*

Determine whether the following syllogism is valid or invalid.

All keys are made of brass.
All things made of brass are valuable.
∴ All keys are valuable.

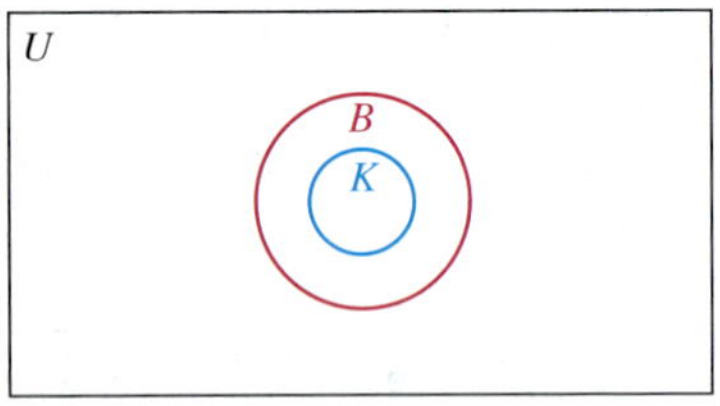

Figure 3.3

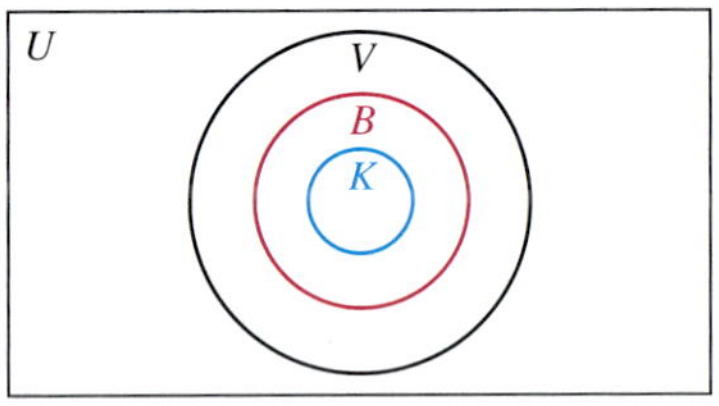

Figure 3.4

SOLUTION To determine whether this syllogism is valid or not valid, we will construct an Euler diagram. We begin with the first premise, "All keys are made of brass." As shown in Fig. 3.3, the inner blue circle labeled K represents the set of all keys and the outer red circle labeled B represents the set of all brass objects. The first premise requires that the inner blue circle must be entirely contained within the outer red circle. Next, we will represent the second premise, "All things made of brass are valuable." As shown in Fig. 3.4, the outermost black circle labeled V represents the set of all valuable objects. The second premise dictates that the red circle, representing the set of brass objects, must be entirely contained within the black circle, representing the set of valuable objects. Now, examine the completed Euler diagram in Fig. 3.4. Note that the premises force the set of keys to be within the set of valuable objects. Therefore, the argument is valid since the conclusion, "All keys are valuable," necessarily follows from the set of premises. ●

The syllogism in Example 1 is valid even though the conclusion, "All keys are valuable," is not a true statement. Similarly, a syllogism can be invalid, or a fallacy, even if the conclusion is a true statement.

When we determine the validity of an argument, we are determining whether the conclusion necessarily follows from the premises. When we say that an argument is valid, we are saying that if all the premises are true statements, then the conclusion must also be a true statement.

The form of the argument determines its validity, not the particular statements. For example, consider the syllogism

All Earth people have two heads.
All people with two heads can fly.
∴ All Earth people can fly.

The form of this argument is the same as that of the previous valid argument in Example 1. Therefore, this argument is also valid.

EXAMPLE 2 *Analyzing a Syllogism*

Determine whether the following syllogism is valid or invalid.

All judges are lawyers.
William is a judge.
∴ William is a lawyer.

SOLUTION The statement "All judges are lawyers" is illustrated in Fig. 3.5. Note that the *J* circle must be completely inside the *L* circle. The second premise, "William is a judge," tells us that William must be placed in the inner circle labeled *J* (Fig. 3.6). The Euler diagram illustrates that by placing William inside the *J* circle William must also be inside the *L* circle. Therefore, the conclusion "William is a lawyer" necessarily follows from the premises and the argument is valid. ●

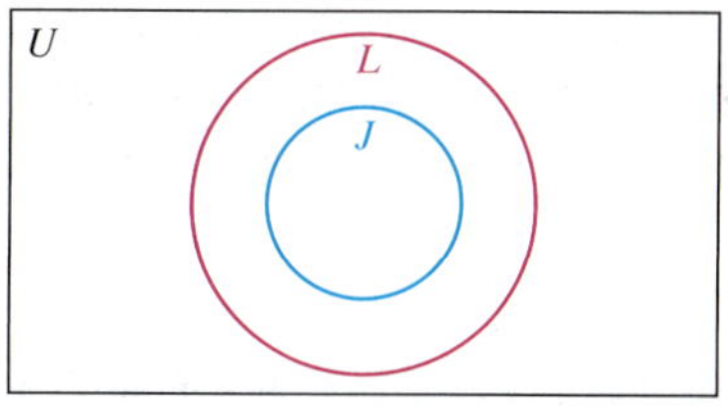

Figure 3.5

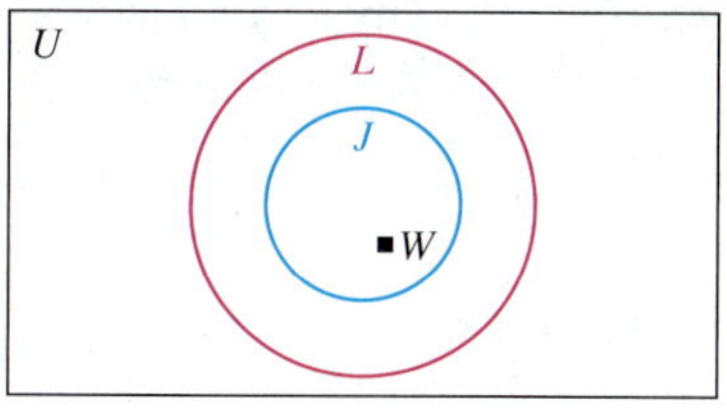

Figure 3.6

In both Example 1 and Example 2, we had no choice as to where the second premise was to be placed in the Euler diagram. In Example 1, the set of brass objects had to be placed inside the set of valuable objects. In Example 2, William had to be placed inside the set of judges. Often when determining the truth value of a syllogism, a premise can be placed in more than one area in the diagram. *We always try to draw the Euler diagram so that the conclusion does not necessarily follow from the premises. If that can be done, then the conclusion does not necessarily follow from the premises and the argument is invalid.* If we cannot show that the argument is invalid, only then do we accept the argument as valid. We illustrate this process in Example 3.

EXAMPLE 3 *Ballerinas and Athletes*

Determine whether the following syllogism is valid or is invalid.

All ballerinas are athletic.
Keyshawn is athletic.
∴ Keyshawn is a ballerina.

SOLUTION The premise "All ballerinas are athletic" is illustrated in Fig. 3.7(a). The next premise, "Keyshawn is athletic," tells us that Keyshawn must be placed in the set of athletic people. Two diagrams in which both premises are satisfied are shown in Fig. 3.7(b) and (c). By examining Fig. 3.7(b), however, we see that Keyshawn is not a ballerina. Therefore, the conclusion "Keyshawn is a ballerina" does not necessarily follow from the set of premises. Thus, the argument is invalid, or a fallacy.

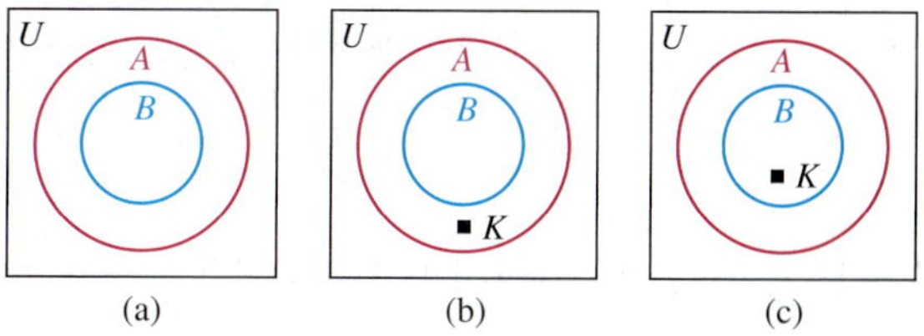

Figure 3.7

●

EXAMPLE 4 *Parrots and Chickens*

Determine whether the following syllogism is valid or invalid.

No parrots eat chicken.
Fletch does not eat chicken.
∴ Fletch is a parrot.

DID YOU KNOW?

Boolean Algebra

Logic, or Boolean algebra, affects each of us daily. Nowadays, many aspects of our lives are computer related. There are computer chips in our cars, watches, radios, cameras, supermarket checkouts, and most appliances, to name just a few uses. The outcomes of these computer chips are based on the *not, and,* and *or* truth tables we have discussed. Sometimes abstract mathematics is not applied until many years after its development, which is exactly what happened with the abstract Boolean algebra concepts considered in the seventeenth and eighteenth centuries.

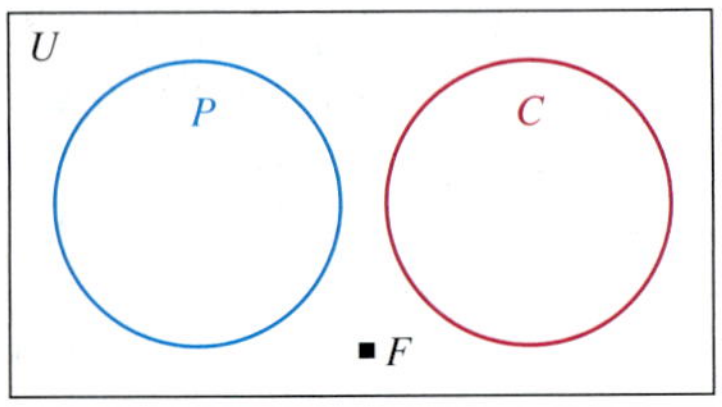

Figure 3.8

SOLUTION The first premise tells us that parrots and things that eat chicken are disjoint sets—that is, sets that do not intersect. The diagram in Fig. 3.8 satisfies the two given premises and also shows that Fletch is not a parrot. Therefore, the argument is invalid, or is a fallacy.

TIMELY TIP Note that in Example 4 if we placed Fletch in circle P, the argument would appear to be valid. Remember that *whenever testing the validity of an argument, always try to show that the argument is invalid.* If there is any way of showing that the conclusion does not necessarily follow from the premises, then the argument is invalid.

EXAMPLE 5 *A Syllogism Involving the Word Some*

Determine whether the following syllogism is valid or invalid.

All As are Bs.
Some Bs are Cs.

$\therefore$ Some As are Cs.

SOLUTION The premise "All As are Bs" is illustrated in Fig. 3.9. The premise "Some Bs are Cs" means that there is at least one B that is a C. We can illustrate this set of premises in four ways, as illustrated in Fig. 3.10.

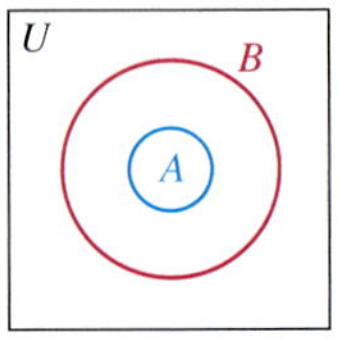

Figure 3.9

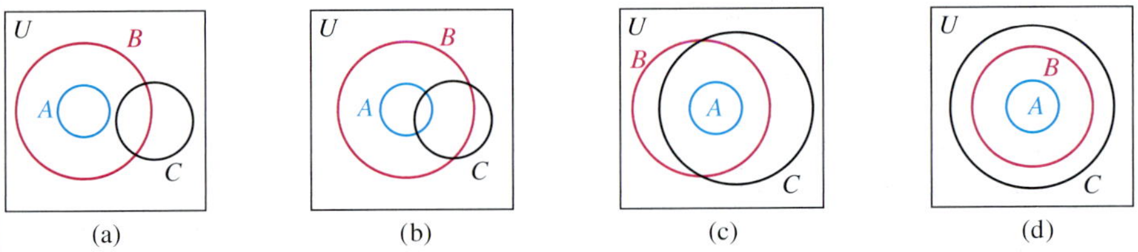

Figure 3.10

In all four illustrations, we see that (1) all As are Bs and (2) some Bs are Cs. The conclusion is "Some As are Cs." Since at least one of the illustrations, Fig. 3.10(a), shows that the conclusion does not necessarily follow from the given premises, the argument is invalid.

EXAMPLE 6 *Frogs and Cats*

Determine whether the following syllogism is valid or invalid.

No frogs eat tuna.
All cats eat tuna.

$\therefore$ No frogs are cats.

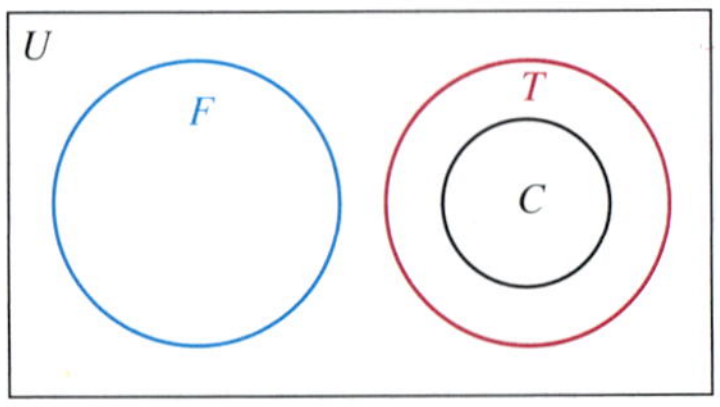

Figure 3.11

SOLUTION The first premise tells us that frogs and the things that eat tuna are disjoint sets as shown in Fig. 3.11. The second premise tells us that the set of cats is a subset of the things that eat tuna. Therefore, the circle representing the set of cats must go within the circle representing the set of the things that eat tuna.

Note that the set of frogs and the set of cats cannot be made to intersect without violating a premise. Thus, the conclusion "No frogs are cats" necessarily follows from the premises and the argument is valid. Note that we did not say that this conclusion is true, only that the argument is valid. •

SECTION 3.6 EXERCISES

CONCEPT/WRITING EXERCISES

1. a) If an Euler diagram can only be drawn in a way in which the conclusion necessarily follows from the premises, what can be said about the syllogistic argument under consideration?

   b) If an Euler diagram can be drawn in a way in which the conclusion does not necessarily follow from the premises, what can be said about the syllogistic argument under consideration?

2. Explain the difference between a symbolic argument and a syllogistic argument.

3. Represent the given statement with two circles as you would when constructing an Euler diagram.

   a) All *A*s are *B*s.

   b) Some *C*s are *D*s.

   c) No *E*s are *F*s.

4. a) What does it mean when we determine that an argument is a valid argument?

   b) What does it mean when we determine that an argument is invalid?

5. Can an argument be valid if the conclusion is a false statement? Explain your answer.

6. Can an argument be invalid if the conclusion is a true statement? Explain.

PRACTICE THE SKILLS/PROBLEM SOLVING

In Exercises 7–30, use an Euler diagram to determine whether the syllogism is valid or invalid.

7. All gingerbread men can sing.
   Gingy is a gingerbread man.
   ∴ Gingy can sing.

▲ See Exercise 7

8. All cats have whiskers.
   All things with whiskers are dogs.
   ∴ All cats are dogs.

9. No candy bars are health foods.
   All Snickers are candy bars.
   ∴ No Snickers are health foods.

10. All dolphins are mammals.
    All mammals are vertebrates.
    ∴ All dolphins are vertebrates.

11. All theme parks have walkways.
    Metropolitan Community College has walkways.
    ∴ Metropolitan Community College is a theme park.

12. All golfers have rain gear.
    John Pearse has rain gear.
    ∴ John Pearse is a golfer.

13. No horses buck.
    Palominos are horses.
    ∴ Palominos do not buck.

14. No jockeys weigh more than 200 pounds.
    Deb Otto is not a jockey.
    ∴ Deb Otto weighs more than 200 pounds.

15. Some mushrooms are poisonous.
A morel is a mushroom.
∴ A morel is poisonous.

16. Some policemen are polite.
Jarod Harshbarger is a policeman.
∴ Jarod Harshbarger is not polite.

17. Some farmers are politicians.
Some politicians are senators.
∴ Some farmers are senators.

18. Some professional golfers give golf lessons.
All people who belong to the PGA are professional golfers.
∴ All people who belong to the PGA give golf lessons.

19. No lawn weeds are flowers.
Sedge is not a flower.
∴ Sedge is a lawn weed.

20. Some caterpillars are furry.
All furry things are mammals.
∴ Some caterpillars are mammals.

21. Some flowers love sunlight.
All things that love sunlight love water.
∴ Some flowers love water.

22. Some CD players are MP3 players.
All iPods are MP3 players.
∴ Some CD players are iPods.

23. No scarecrows are tin men.
No tin men are lions.
∴ No scarecrows are lions.

24. All pilots can fly.
All astronauts can fly.
∴ Some pilots are astronauts.

25. Some dogs wear glasses.
Fido wears glasses.
∴ Fido is a dog.

26. All rainy days are cloudy.
Today it is cloudy.
∴ Today is a rainy day.

27. All sweet things taste good.
All things that taste good are fattening.
All things that are fattening put on pounds.
∴ All sweet things put on pounds.

28. All books have red covers.
All books that have red covers contain 200 pages.
Some books that contain 200 pages are novels.
∴ All books that contain 200 pages are novels.

29. All country singers play the guitar.
All country singers play the drums.
Some people who play the guitar are rock singers.
∴ Some country singers are rock singers.

30. Some hot dogs are made of turkey.
All things made of turkey are edible.
Some things that are made of beef are edible.
∴ Some hot dogs are made of beef.

CHALLENGE PROBLEM/GROUP ACTIVITY

31. *Sets and Logic* Statements in logic can be translated into set statements: for example, $p \wedge q$ is similar to $P \cap Q$; $p \vee q$ is similar to $P \cup Q$; and $p \rightarrow q$ is equivalent to $\sim p \vee q$, which is similar to $P' \cup Q$. Euler diagrams can also be used to show that arguments similar to those discussed in Section 3.5 are valid or invalid. Use Euler diagrams to show that the following symbolic argument is invalid.

$$\begin{array}{l} p \rightarrow q \\ \underline{p \vee q} \\ \therefore \sim p \end{array}$$

INTERNET/RESEARCH ACTIVITY

32. Leonhard Euler is considered one of the greatest mathematicians of all time. Do research and write a report on Euler's life. Include information on his contributions to sets and to logic. Also indicate other areas of mathematics in which he made important contributions. References include encyclopedias, history of mathematics books, and the Internet.

3.7 SWITCHING CIRCUITS

▲ What must be true for a lamp to produce light?

Suppose you are sitting at your desk and want to turn on a lamp so you can read a book. If a wall switch controls the power to the outlet that the lamp is plugged into, and the lamp has an on/off switch, what conditions must be true for the lamp's bulb to light? The wiring in our homes can be explained and described using logic. In this section, we will learn how to analyze electrical circuits using logic.

Using Symbolic Statements to Represent Circuits

A common application of logic is switching circuits. To understand the basic concepts of switching circuits, let us examine a few simple circuits that are common in most homes. The typical lamp has a cord, which is plugged into a wall outlet. Somewhere between the bulb in the lamp and the outlet is a switch to turn the lamp on and off. A switch is often referred to as being on or off. When the switch is in the *on* position, the current flows through the switch and the bulb lights up. When the switch is in the *on* position, we can say that the switch is *closed* and that current will flow through the switch. When the switch is in the *off* position, the current does not flow through the switch and the bulb does not light. When the switch is in the *off* position, we can say that the switch is *open*, and the current does not flow through the switch. The basic configuration of a switch is shown in Fig. 3.12.

Figure 3.12

Electric circuits can be expressed as logical statements. We represent switches as letters, using T to represent a closed switch (or current flow) and F to represent an open switch (or no current flow). This relationship is indicated in Table 3.33.

Table 3.33

Switch	Lightbulb
T	on (switch closed)
F	off (switch open)

Occasionally, we have a wall switch connected to a wall outlet and a lamp plugged into the wall outlet (Fig. 3.13). We then say that the wall switch and the switch on the lamp are in *series*, meaning that for the bulb in the lamp to light, both switches must be on at the same time (Fig. 3.14 on page 170). On the other hand, the bulb will not light if either switch is off or if both switches are off. In either of these conditions, the electricity will not flow through the circuit. In a ***series circuit***, the current can take only one path. If any switch in the path is open, the current cannot flow.

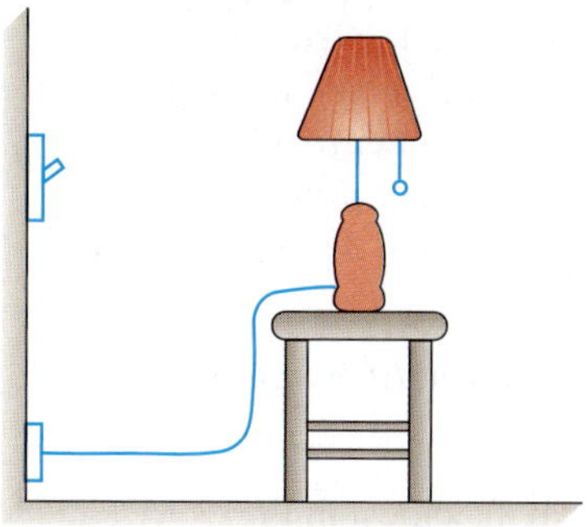

Figure 3.13

To illustrate this situation symbolically, let p represent the wall switch and q the lamp switch. The letter T will be used to represent both a closed switch and the bulb lighting. The letter F will represent an open switch and the bulb not lighting. Thus, we have four possible cases.

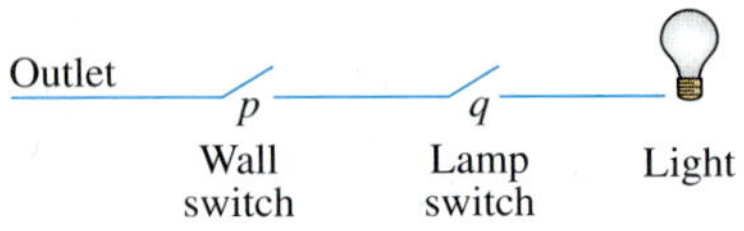

Figure 3.14

CASE 1: Both switches are closed; that is, p is T and q is T. The light is on, T.

CASE 2: Switch p is closed and switch q is open; that is, p is T and q is F. The light is off, F.

CASE 3: Switch p is open and switch q is closed; that is, p is F and q is T. The light is off, F.

CASE 4: Both switches are open; that is, p is F and q is F. The light is off, F.

Table 3.34

p	q	Light	$p \wedge q$
T	T	on (T)	T
T	F	off (F)	F
F	T	off (F)	F
F	F	off (F)	F

Table 3.34 summarizes the results.

The on–off results are the same as the truth table for the conjunction $p \wedge q$ if we think of "on" as true and "off" as false.

Switches in series will always be represented with a conjunction, $\wedge$.

Another type of electric circuit used in the home is the ***parallel circuit***, in which there are two or more paths that the current can take. If the current can pass through either path or both (see Fig. 3.15), the light will go on. The letter T will be used to represent both a closed switch and the bulb lighting. The letter F will represent an open switch and the bulb not lighting. Thus, we have four possible cases.

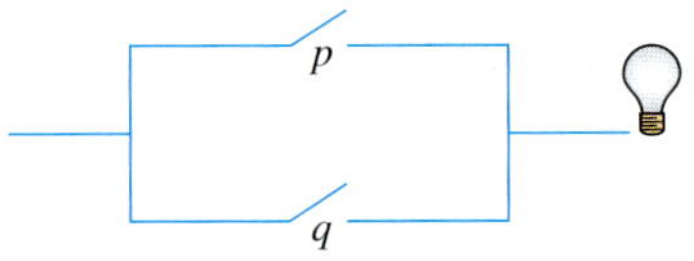

Figure 3.15

CASE 1: Both switches are closed; that is, p is T and q is T. The light is on, T.

CASE 2: Switch p is closed and switch q is open; that is, p is T and q is F. The light is on, T.

CASE 3: Switch p is open and switch q is closed; that is, p is F and q is T. The light is on, T.

CASE 4: Both switches are open; that is, p is F and q is F. The light is off, F.

Table 3.35

p	q	Light	$p \vee q$
T	T	on (T)	T
T	F	on (T)	T
F	T	on (T)	T
F	F	off (F)	F

Table 3.35 summarizes the results. The on–off results are the same as the $p \vee q$ truth table if we think of "on" as true and "off" as false.

Switches in parallel will always be represented with a disjunction, $\vee$.

Sometimes it is necessary to have two or more switches in the same circuit that will both be open at the same time and both be closed at the same time. In such circuits, we will use the same letter to represent both switches. For example, in the circuit shown in Fig. 3.16, there are two switches labeled p. Therefore, both of these switches must be open at the same time and both must be closed at the same time. One of the p switches cannot be open at the same time the other p switch is closed.

We can now combine some of these basic concepts to analyze more circuits.

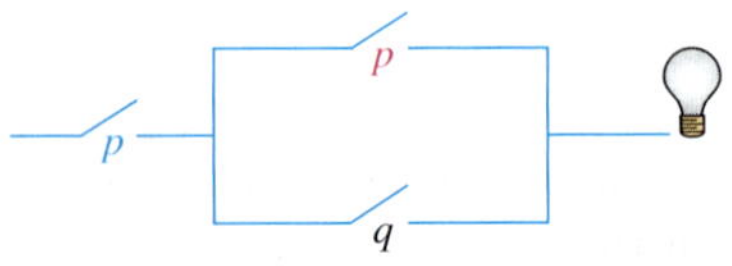

Figure 3.16

EXAMPLE 1 ***Representing a Switching Circuit with Symbolic Statements***

a) Write a symbolic statement that represents the circuit shown in Fig. 3.16.

b) Construct a truth table to determine when the light will be on.

SOLUTION

a) In Fig. 3.16, there is a blue switch p on the left, and to the right there is a branch containing a red switch p and a switch q. The current must flow through the blue switch p into the branch on its right. Therefore, the blue switch p is in series with the branch containing the red switch p and switch q. After the current flows through the blue switch p it reaches the branch on its right. At this point, the current has the option of flowing into the red switch p, switch q, or both of these switches. Therefore, the branch containing the red switch p and switch q is a parallel branch. We say these two switches are in parallel. The entire circuit in symbolic form is $p \wedge (p \vee q)$. Note that the parentheses are very important. Without parentheses, the symbolic statement could be interpreted as $(p \wedge p) \vee q$. The diagram for $(p \wedge p) \vee q$ is illustrated in Fig. 3.17. We shall see later in this section that the circuits in Fig. 3.16 and Fig. 3.17 are not the same.

b) The truth table for the statement (Table 3.36) indicates that the light will be on only in the cases in which p is true or when switch p is closed. ●

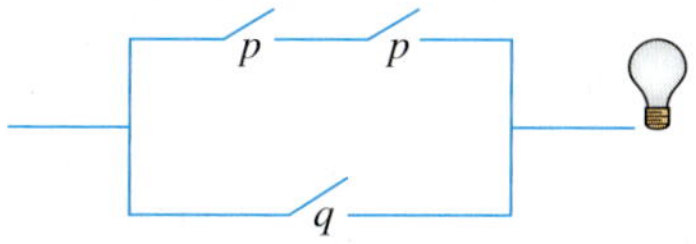

Figure 3.17

Table 3.36

p	q	p	$\wedge$	$(p \vee q)$
T	T	T	T	T
T	F	T	T	T
F	T	F	F	T
F	F	F	F	F
		1	3	2

EXAMPLE 2 *Representing a Switching Circuit with Symbolic Statements*

a) Write a symbolic statement that represents the circuit in Fig. 3.18.

b) Construct a truth table to determine when the light will be on.

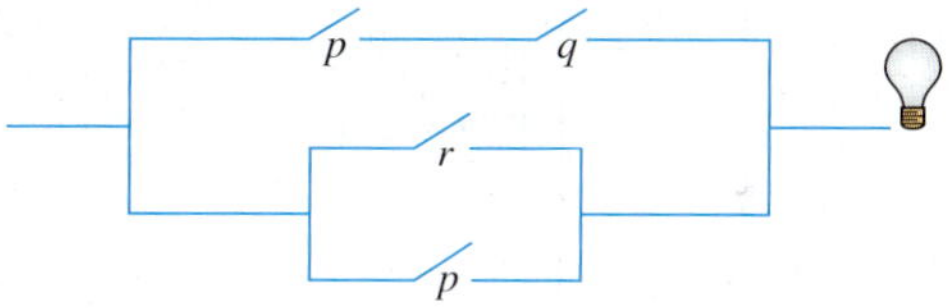

Figure 3.18

SOLUTION

a) The upper branch of the circuit contains two switches, p and q, in series. We can represent that branch with the statement $p \wedge q$. The lower branch of the circuit has two switches, r and p, in parallel. We can represent that branch with the statement $r \vee p$. The upper branch is in parallel with the lower branch. Putting the two branches together, we get the statement $(p \wedge q) \vee (r \vee p)$.

b) The truth table for the statement (Table 3.37) shows that cases 6 and 8 are false. Thus, the light will be off in these cases and on in all the other cases. In examining the truth values, we see that the statement is false only in the cases in which both p and r are false or when both switches are open. By examining the diagram or truth table, we can see that the current will flow if switch p is closed or switch r is closed. ●

Table 3.37

p	q	r	$(p \wedge q)$	$\vee$	$(r \vee p)$
T	T	T	T	T	T
T	T	F	T	T	T
T	F	T	F	T	T
T	F	F	F	T	T
F	T	T	F	T	T
F	T	F	F	F	F
F	F	T	F	T	T
F	F	F	F	F	F
			1	3	2

Drawing Switching Circuits That Represent Symbolic Statements

We will next study how to draw a switching circuit that represents a given symbolic statement. Suppose we are given the statement $(p \wedge q) \vee r$ and are asked to construct a circuit corresponding to it. Remember that $\wedge$ indicates a series branch

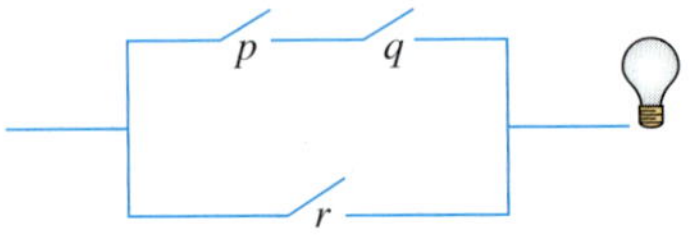

Figure 3.19

Figure 3.20

and ∨ indicates a parallel branch. Working first within parentheses, we see that switches p and q are in series. This series branch is in parallel with switch r, as indicated in Fig. 3.19.

Occasionally, it is necessary to have two switches in the same circuit such that when one switch is open, the other switch is closed; and when one switch is closed, the other switch is open. Therefore, the two switches will never both be open together and never be closed together. This situation can be represented by using p for one of the switches and $\overline{p}$ for the other switch. The switch labeled $\overline{p}$ corresponds to $\sim p$ in a logic statement. For example, in a series circuit, $p \wedge \sim p$ would be represented by Fig. 3.20. In this case, the light would never go on; the switches would counteract each other. When switch p is closed, switch $\overline{p}$ is open; and when p is open, $\overline{p}$ is closed.

EXAMPLE 3 *Representing a Symbolic Statement as a Switching Circuit*

Draw a switching circuit that represents $[(p \wedge \sim q) \vee (r \vee q)] \wedge s$.

SOLUTION In the statement, p and $\sim q$ have $\wedge$ between them, so switches p and $\overline{q}$ are in series, as represented in Fig. 3.21. Also in the statement, r and q have $\vee$ between them, so switches r and q are in parallel, as represented in Fig. 3.22.

Because $(p \wedge \sim q)$ and $(r \vee q)$ are connected with $\vee$, the two branches are in parallel with each other. The parallel branches are represented in Fig. 3.23. Finally, s in the statement is connected to the rest of the statement with $\wedge$. Therefore, switch s is in series with the entire rest of the circuit, as illustrated in Fig. 3.24.

Figure 3.21

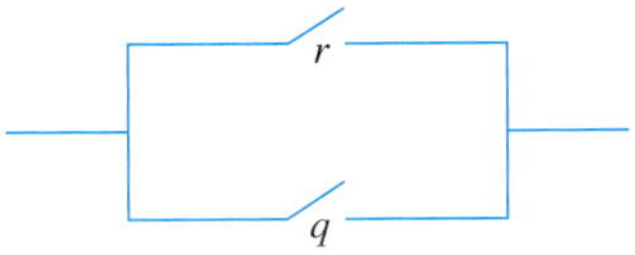

Figure 3.22

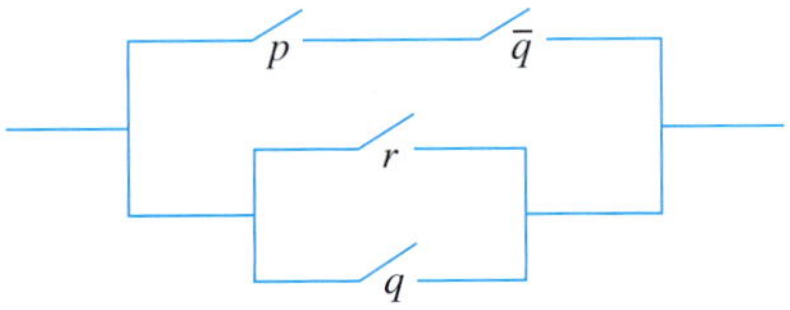

Figure 3.23

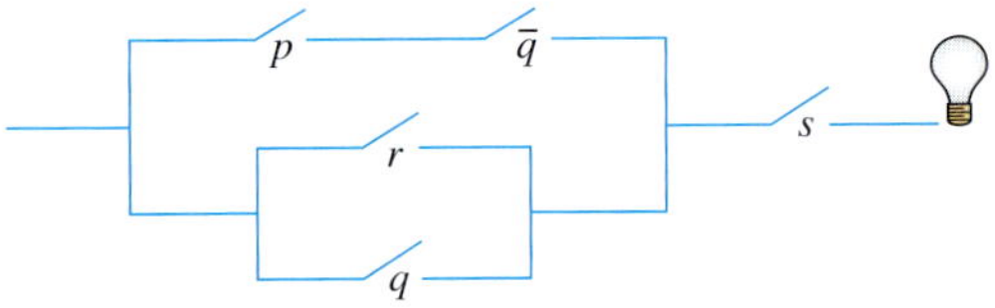

Figure 3.24

Equivalent Circuits

Sometimes two circuits that look very different will actually have the exact same conditions under which the light will be on. If we were to analyze the truth tables for the corresponding symbolic statements for such circuits, we would find that they have identical answer columns. In other words, the corresponding symbolic statements are equivalent.

Equivalent circuits are two circuits that have equivalent corresponding symbolic statements.

DID YOU KNOW?

Applications of Logic

George Boole provided the key that would unlock the door to modern computing. Not until 1938, however, did Claude Shannon, in his master's thesis at MIT, propose uniting the on–off capability of electrical switches with Boole's two-value system of 0's and 1's. The operations AND, OR, and NOT and the rules of logic laid the foundation for computer gates. Such gates determine whether the current will pass. The closed switch (current flow) is represented as 1, and the open switch (no current flow) is represented as 0. For more information, see the group projects on page 180.

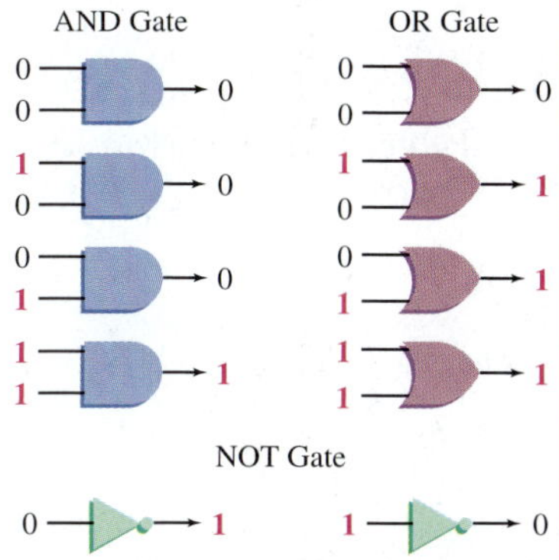

As you can see from the accompanying diagrams, the gates function in essentially the same way as a truth table. The gates shown here are the simplest ones, representing simple statements. There are other gates, such as the NAND and NOR gates, that are combinations of NOT, AND, and OR gates. The microprocessing unit of a computer uses thousands of these switches.

To determine whether two circuits are equivalent, we will analyze the answer columns of the truth tables of their corresponding symbolic statements. If the answer columns from their corresponding symbolic statements are identical, then the circuits are equivalent.

EXAMPLE 4 *Are the Circuits Equivalent?*

Determine whether the two circuits are equivalent.

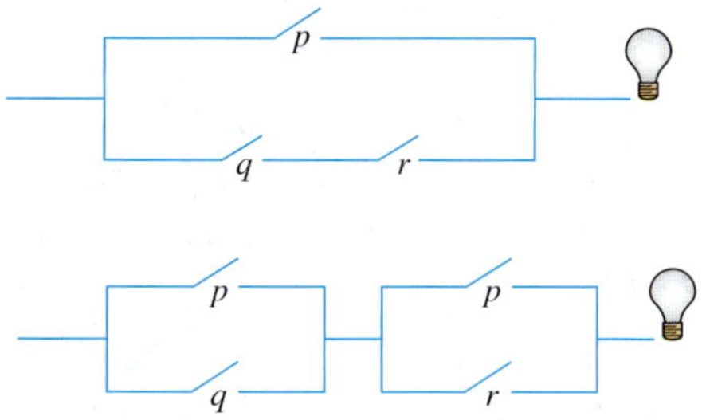

SOLUTION The symbolic statement that represents the first circuit is $p \vee (q \wedge r)$. The symbolic statement that represents the second circuit is $(p \vee q) \wedge (p \vee r)$. The truth tables for these statements are shown in Table 3.38.

Table 3.38

p	q	r	p	$\vee$	$(q \wedge r)$	$(p \vee q)$	$\wedge$	$(p \vee r)$
T	T	T	T	T	T	T	T	T
T	T	F	T	T	F	T	T	T
T	F	T	T	T	F	T	T	T
T	F	F	T	T	F	T	T	T
F	T	T	F	T	T	T	T	T
F	T	F	F	F	F	T	F	F
F	F	T	F	F	F	F	F	T
F	F	F	F	F	F	F	F	F

Note that the answer columns for the two statements are identical. Therefore, $p \vee (q \wedge r)$ is equivalent to $(p \vee q) \wedge (p \vee r)$ and the two circuits are equivalent.

SECTION 3.7 EXERCISES

CONCEPT/WRITING EXERCISES

1. a) In your own words, describe a series circuit.

   b) Which type of logical connective can be used to represent a series circuit?

2. a) In your own words, describe a parallel circuit.

   b) Which type of logical connective can be used to represent a parallel circuit?

3. Explain why the lightbulb will never go on in the following circuit.

4. Explain why the lightbulb will always be on in the following circuit.

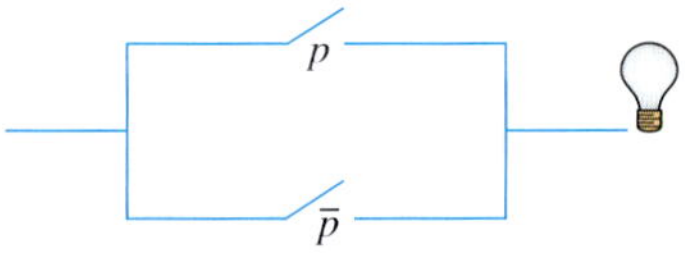

PRACTICE THE SKILLS/PROBLEM SOLVING

In Exercises 5–12, (a) write a symbolic statement that represents the circuit and (b) construct a truth table to determine when the lightbulb will be on. That is, determine which switches must be open and which switches must be closed for the lightbulb to be on.

5.

6.

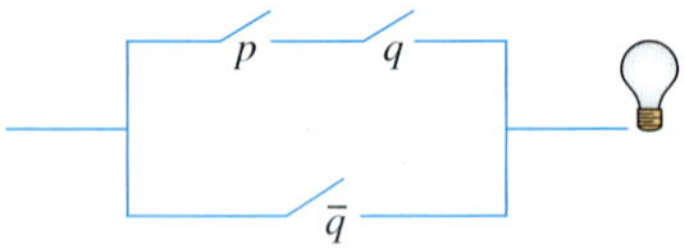

7.

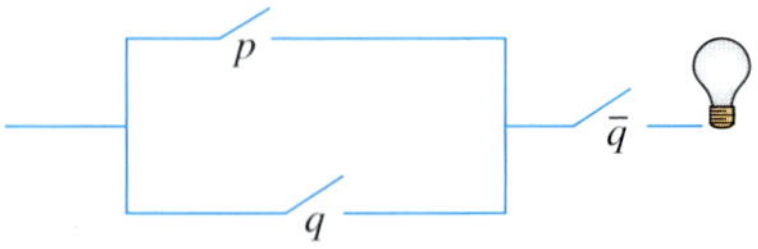

8.

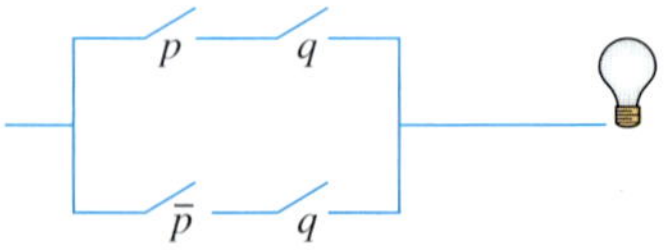

9.

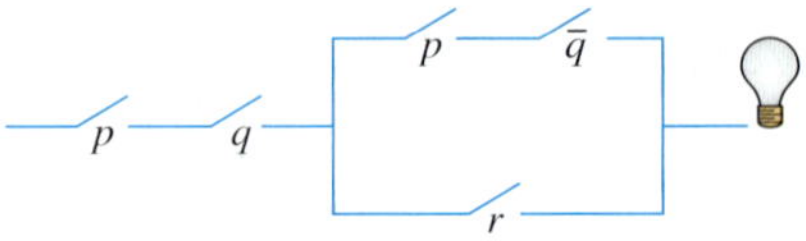

10.

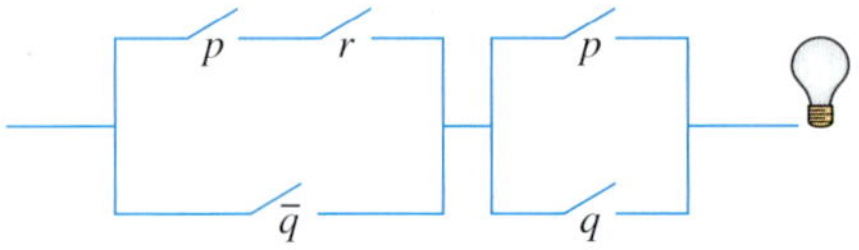

11.

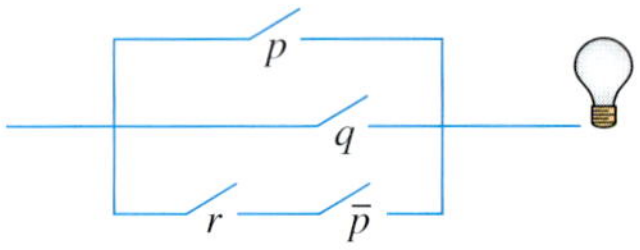

12.

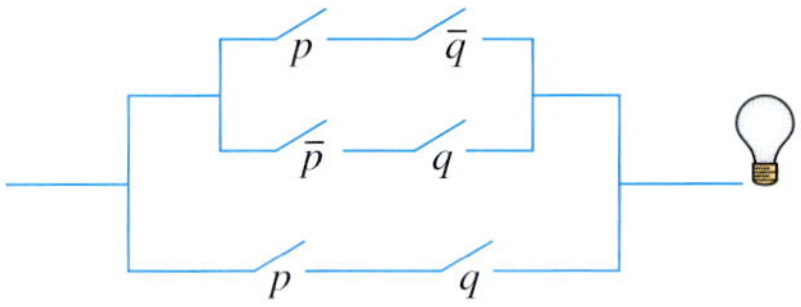

In Exercises 13–20, draw a switching circuit that represents the symbolic statement.

13. $p \vee q$

14. $p \vee (q \wedge r)$

15. $(p \vee q) \wedge r$

16. $\sim p \wedge (q \vee r)$

17. $(p \vee q) \wedge (r \vee s)$

18. $(p \wedge q) \vee (r \wedge s)$

19. $[(p \vee q) \vee (r \wedge q)] \wedge (\sim p)$

20. $[(p \vee q) \wedge r] \vee (\sim p \wedge q)$

In Exercises 21–26, represent each circuit with a symbolic statement. Then use a truth table to determine if the circuits are equivalent.

21.

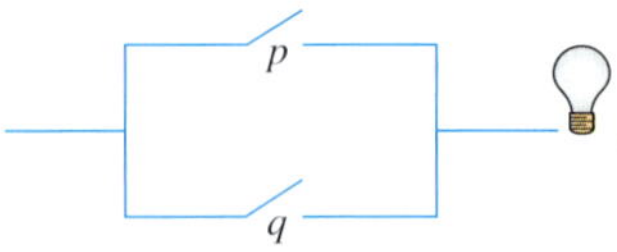

,

22.

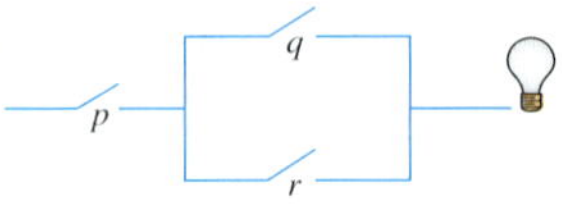

,
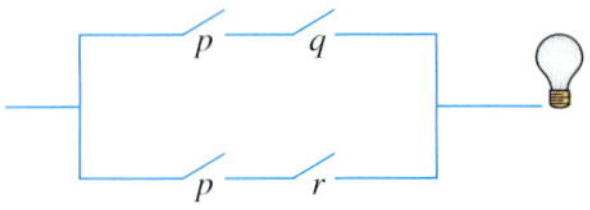

23. 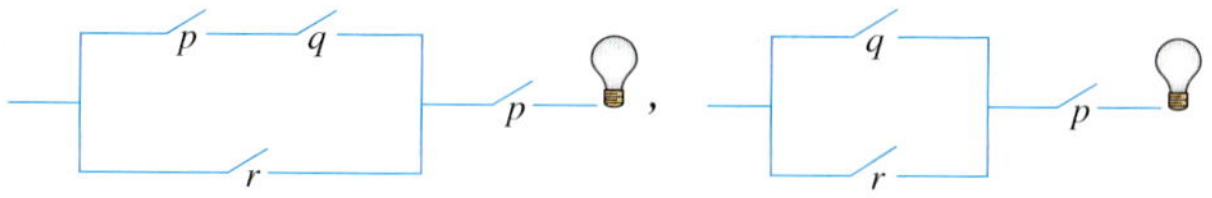

24. 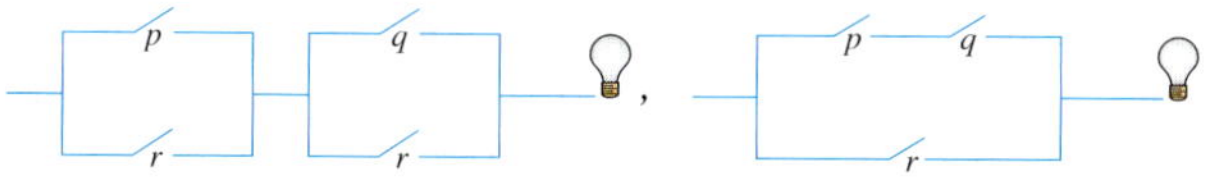

25. 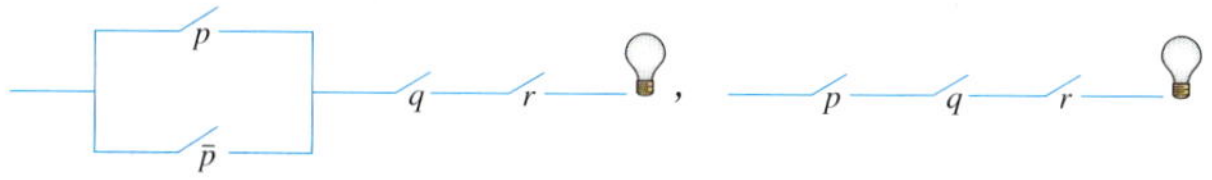

26. 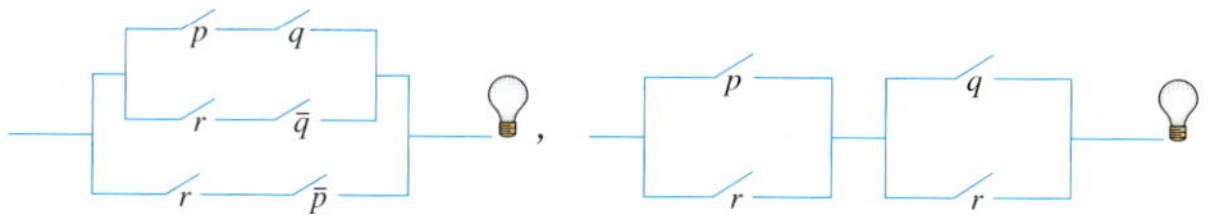

CHALLENGE PROBLEMS/GROUP ACTIVITIES

27. Design a circuit that can be represented by

a) $p \to q$

b) $\sim(p \to q)$

(*Hint:* See Section 3.4.)

28. Design two circuits that each involve switches labeled p, q, r, and s that appear to be different but are actually equivalent circuits.

INTERNET/RESEARCH ACTIVITIES

29. Digital computers use gates that work like switches to perform calculations. (See Did You Know? on page 173.) Information is fed into the gates and information leaves the gates, according to the type of gate. The three basic gates used in computers are the NOT gate, the AND gate, and the OR gate. Do research on the three types of gates.

a) Explain how each gate works.

b) Explain the relationship between each gate and the corresponding logic connectives *not, and,* and *or*.

c) Illustrate how two or more gates can be combined to form a more complex gate.

CHAPTER 3 SUMMARY

Important Facts

Quantifiers

Form of Statement	Form of Negation
All are.	Some are not.
None are.	Some are.
Some are.	None are.
Some are not.	All are.

Summary of connectives

Formal Name	Symbol	Read	Symbolic Form
Negation	$\sim$	not	$\sim p$
Conjunction	$\wedge$	and	$p \wedge q$
Disjunction	$\vee$	or	$p \vee q$
Conditional	$\to$	if–then	$p \to q$
Biconditional	$\leftrightarrow$	if and only if	$p \leftrightarrow q$

Basic truth tables

Negation

p	$\sim p$
T	F
F	T

p	q	Conjunction $p \wedge q$	Disjunction $p \vee q$	Conditional $p \to q$	Biconditional $p \leftrightarrow q$
T	T	T	T	T	T
T	F	F	T	F	F
F	T	F	T	T	F
F	F	F	F	T	T

De Morgan's laws

$$\sim(p \wedge q) \Leftrightarrow \sim p \vee \sim q$$
$$\sim(p \vee q) \Leftrightarrow \sim p \wedge \sim q$$

In Exercises 45–48, determine which, if any, of the three statements are equivalent.

45. a) If you read 10 books in the summer, then you reach your goal.

b) You do not read 10 books in the summer or you reach your goal.

c) It is false that you read 10 books in the summer and you do not reach your goal.

46. a) The screwdriver is on the workbench if and only if the screwdriver is not on the counter.

b) If the screwdriver is not on the counter, then the screwdriver is not on the workbench.

c) It is false that the screwdriver is on the counter and the screwdriver is not on the workbench.

47. a) If $2 + 3 = 6$, then $3 + 1 = 5$.

b) $2 + 3 = 6$ if and only if $3 + 1 \neq 5$.

c) If $3 + 1 \neq 5$, then $2 + 3 \neq 6$.

48. a) If the sale is on Tuesday and I have money, then I will go to the sale.

b) If I go to the sale, then the sale is on Tuesday and I have money.

c) I go to the sale, or the sale is on Tuesday and I have money.

3.5

In Exercises 49–52, determine whether the argument is valid or invalid. You may compare the argument to a standard form or use a truth table.

49. $p \rightarrow q$
$\sim p$
$\therefore\ q$

50. $p \wedge q$
$q \rightarrow r$
$\therefore\ p \rightarrow r$

51. If Jose Macias is the manager, then Kevin Geis is the coach.
If Kevin Geis is the coach, then Tim Weisman is the umpire.
∴ If Jose Macias is the manager, then Tim Weisman is the umpire.

52. If we eat at Joe's, then we will get heartburn. We get heartburn or we take Tums. Therefore, we do not get heartburn.

3.6

In Exercises 53–56, use an Euler diagram to determine whether the argument is valid or invalid.

53. All plumbers wear overalls.
Some electricians wear overalls.
∴ Some electricians are plumbers.

54. Some submarines are yellow.
All dandelions are yellow.
∴ Some dandelions are submarines.

55. No Beatles are Rolling Stones.
Some Rolling Stones are Small Faces.
∴ No Beatles are Small Faces.

56. All bears are furry.
Teddy is furry.
∴ Teddy is a bear.

3.7

57. a) Write the corresponding symbolic statement of the circuit shown.

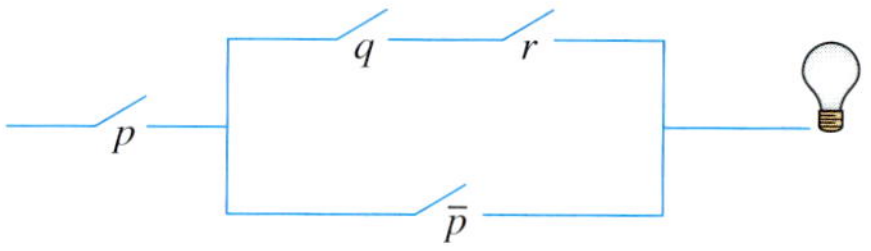

b) Construct a truth table to determine when the lightbulb will be on.

58. Construct a diagram of a circuit that corresponds to the symbolic statement $(p \vee q) \vee (p \wedge q)$.

59. Determine whether the circuits shown are equivalent.

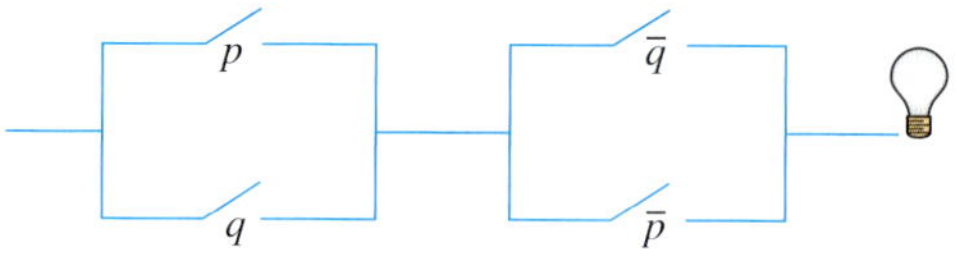

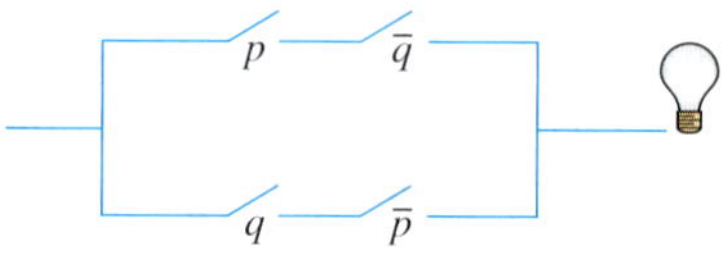

CHAPTER 3 TEST

In Exercises 1–3, write the statement in symbolic form.

p: Ann is the secretary.

q: Dick is the vice president.

r: Elaine is the president.

1. Ann is the secretary but Elaine is the president, or Dick is not the vice president.

2. If Elaine is the president then Dick is the vice president, or Ann is not the secretary.

3. It is false that Elaine is the president if and only if Dick is not the vice president.

In Exercises 4 and 5, use p, q, and r as above to write each symbolic statement in words.

4. $(\sim p \wedge r) \leftrightarrow \sim q$

5. $(p \vee \sim q) \rightarrow r$

In Exercises 6 and 7, construct a truth table for the given statement.

6. $[\sim(p \rightarrow r)] \wedge q$

7. $(q \leftrightarrow \sim r) \vee p$

In Exercises 8 and 9, find the truth value of the statement.

8. $2 + 6 = 8$ or $7 - 12 = 5$.

9. Harrison Ford is an actor or Gerald Ford was a president, if and only if the Mississippi is a river.

In Exercises 10 and 11, given that p is true, q is false, and r is true, determine the truth value of the statement.

10. $(r \vee q) \leftrightarrow (p \wedge \sim q)$

11. $[\sim(r \rightarrow \sim p)] \wedge (q \rightarrow p)$

12. Determine whether the pair of statements are equivalent.

$$\sim p \vee q, \qquad \sim(p \wedge \sim q)$$

In Exercises 13 and 14, determine which, if any, of the three statements are equivalent.

13. **a)** If the bird is red, then it is a cardinal.

 b) The bird is not red or it is a cardinal.

 c) If the bird is not red, then it is not a cardinal.

14. **a)** It is not true that the test is today or the concert is tonight.

 b) The test is not today and the concert is not tonight.

 c) If the test is not today, then the concert is not tonight.

15. *Translate the following argument into symbolic form. Determine whether the argument is valid or invalid by comparing the argument to a recognized form or by using a truth table.*

 If the soccer team wins the game, then Sue played fullback. If Sue played fullback, then the team is in second place. Therefore, if the soccer team wins the game, then the team is in second place.

16. *Use an Euler diagram to determine whether the syllogism is valid or is a fallacy.*

 All living things contain carbon.
 Roger contains carbon.
 ___
 ∴ Therefore, Roger is a living thing.

In Exercises 17 and 18, write the negation of the statement.

17. All highways are roads.

18. Nick played football and Max played baseball.

19. Write the converse, inverse, and contrapositive of the conditional statement, "If the garbage truck comes, then today is Saturday."

20. Construct a diagram of a circuit that corresponds to $(p \wedge q) \vee (\sim p \vee \sim q)$

GROUP PROJECTS

COMPUTER GATES

1. Gates in computers work on the same principles as switching circuits. The three basic types of gate are the NOT gate, the AND gate, and the OR gate. Each is illustrated along with a table that indicates current flow entering and exiting the gate. If current flows into a NOT gate, then no current exits, and vice versa. Current exits an AND gate only when both inputs have a current flow. Current exits an OR gate if current flows through either, or both, inputs. In the table, a 1 represents a current flow and a 0 indicates no current flow. For example, in the AND gate, if there is a current flow in input *A* (I_a has a value of 1) and no current flow in input *B* (I_b has a value of 0), there is no current flow in the output (O has a value of 0); see row 2 of the AND Gate table.

NOT gate

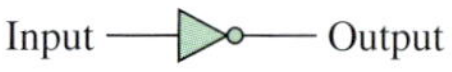

NOT gate

I	**O**
1	0
0	1

AND gate

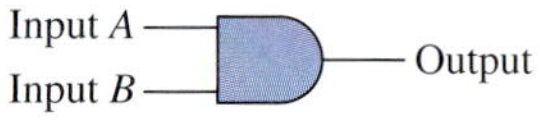

OR gate

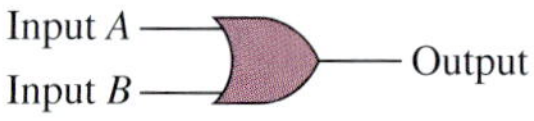

AND gate

I_a	I_b	**O**
1	1	1
1	0	0
0	1	0
0	0	0

OR gate

I_a	I_b	**O**
1	1	1
1	0	1
0	1	1
0	0	0

a) If 1 is considered true and 0 is considered false, explain how these tables are similar to the *not, and,* and *or* truth tables.

For the inputs indicated in the following figures determine whether the output is 1 or 0.

b)

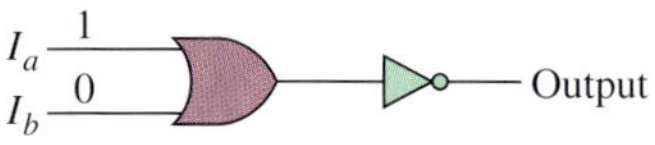

c)

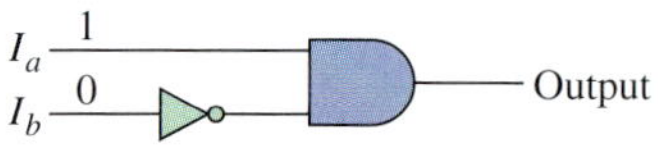

d)

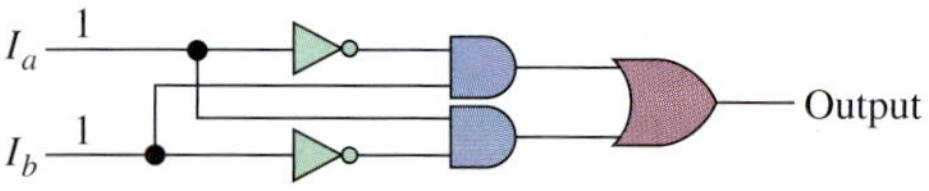

e) What values for I_a and I_b will give an output of 1 in the figure in part (d)? Explain how you determined your answer.

f) Construct a truth table using 1's and 0's for the following gate. Your truth table should have columns I_a, I_b, and O and should indicate the four possible cases for the inputs and each corresponding output.

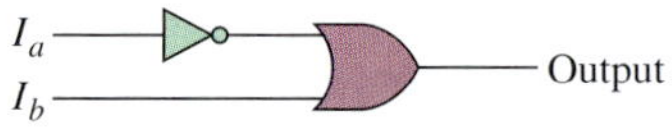

LOGIC GAME

2. **a)** Shown is a photograph of a logic game at the Ontario Science Centre. There are 12 balls on top of the game board, numbered from left to right, with ball 1 on the extreme left and ball 12 on the extreme right. On the platform in front of the players are 12 buttons. Button 1 corresponds to ball 1, button 2 corresponds to ball 2, and so on. When 6 buttons are pushed, the 6 respective balls are released. When 1 or 2 balls reach an *and* gate or an *or* gate, a single ball may or may not pass through the gate. The object of the game is to select a proper combination of 6 buttons that will allow 1 ball to reach the bottom. Using your knowledge of *and* and *or*, select a combination of 6 buttons that will result in a win. (There is more than one answer.) Explain how you determined your answer.

b) Construct a game similar to this one where 15 balls are at the top and 8 balls must be selected to allow 1 ball to reach the bottom.

c) Indicate all solutions to the game you constructed in part (c).

CHAPTER 4

Systems of Numeration

▲ There are many ways that we can communicate numerically.

WHAT YOU WILL LEARN

- Additive, multiplicative, and ciphered systems of numeration
- Place-value systems of numeration
- Egyptian, Hindu–Arabic, Roman, Chinese, Ionic Greek, Babylonian, and Mayan numerals
- Converting base 10 numerals to numerals in other bases
- Converting numerals in other bases to base 10 numerals
- Performing addition, subtraction, multiplication, and division in other bases
- Other computational methods such as duplation and mediation, the lattice method, and Napier's rods

WHY IT IS IMPORTANT

The number system most of the world uses today, called the Hindu–Arabic system, is only one way to communicate numerically. In this chapter, we will study other systems of numeration. We will also study how to perform basic arithmetic in other bases and with methods other than those most of us were taught when we were children.

4.1 ADDITIVE, MULTIPLICATIVE, AND CIPHERED SYSTEMS OF NUMERATION

▲ Hines Ward in front of the 2006 Super Bowl XL Logo

We are all familiar with the numerals we grew up with: 1, 2, 3, and so forth. Now think of another set of numerals. You probably thought of Roman numerals: I, II, III, and so forth. Some common modern-day uses of Roman numerals include clock and watch numbers, the copyright year on movie and television credits, outlines, and page numbers at the beginning of books. In fact, if you look at the pages in this textbook that precede page 1, you will see Roman numerals in use. Perhaps an even more famous use of Roman numerals is their use in the logo for the Super Bowl each year. In this section, we will study several sets of numerals besides our familiar 1, 2, 3, ...

Just as the first attempts to write were made long after the development of speech, the first representation of numbers by symbols came long after people had learned to count. A tally system using physical objects, such as scratch marks in the soil or on a stone, notches on a stick, pebbles, or knots on a vine, was probably the earliest method of recording numbers.

In primitive societies, such a tally system adequately served the limited need for recording livestock, agriculture, or whatever was counted. As civilization developed, however, more efficient and accurate methods of calculating and keeping records were needed. Because tally systems are impractical and inefficient, societies developed symbols to replace them. The Egyptians, for example, used the symbol ∩, and the Babylonians used the symbol < to represent the number we symbolize by 10.

A ***number*** is a quantity, and it answers the question "How many?" A ***numeral*** is a symbol such as ∩, <, or 10 used to represent the number. We think a number but write a numeral. The distinction between number and numeral will be made here only if it is helpful to the discussion.

In language, relatively few letters of the alphabet are used to construct a large number of words. Similarly, in arithmetic, a small variety of numerals can be used to represent all numbers. In general, when representing a number, we use as few numerals as possible. One of the greatest accomplishments of humankind has been the development of systems of numeration, whereby all numbers are "created" from a few symbols. Without such systems, mathematics would not have developed to its present level.

> A **system of numeration** consists of a set of numerals and a scheme or rule for combining the numerals to represent numbers.

Four types of numeration systems used by different cultures are the topic of this chapter. They are additive (or repetitive), multiplicative, ciphered, and place-value systems. You do not need to memorize all the symbols, but you should understand the principles behind each system. By the end of this chapter, we hope that you better understand the system we use, the ***Hindu–Arabic system***, and its relationship to other types of systems.

DID YOU KNOW?

An Important Discovery

Archaeologist Denise Schmandt-Besserat made a breakthrough discovery about early systems of numeration. She realized that the little clay geometric objects that had been found in many archaeological sites had actually been used by people to account for their goods. Later in history, these tokens were impressed on a clay tablet to represent quantities, the beginning of writing.

DID YOU KNOW?

The Rhind Papyrus

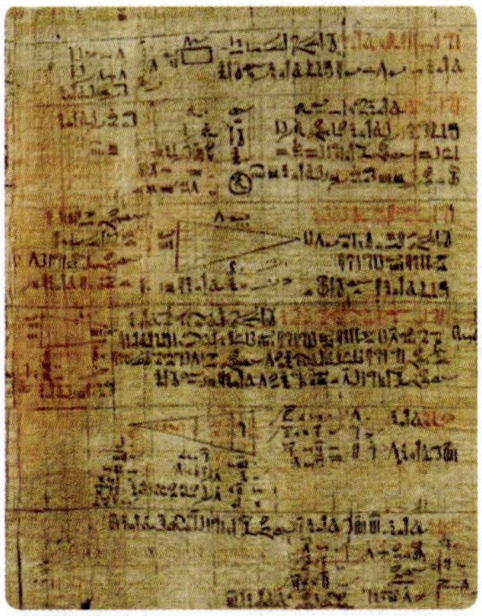

Much of our knowledge of Egyptian mathematics and numeration comes from a roll of papyrus measuring 18 ft by 13 in. It was discovered in a shop in Luxor, Egypt, in 1858 by Henry Rhind, a Scottish lawyer turned archaeologist. Unlike the straightforward accounting of property and events common to Egyptian tombs, the Rhind Papyrus has inscribed on it 85 mathematical problems and solutions involving addition, subtraction, multiplication, division, and geometry.

The key to translation of the papyrus was the Rosetta Stone, which had been discovered some 60 years earlier by one of Napoleon's officers. Both of these treasures now reside at the British Museum in London.

Additive Systems

An additive system is one in which the number represented by a particular set of numerals is simply the sum of the values of the numerals. The additive system of numeration is one of the oldest and most primitive types of numeration systems. One of the first additive systems, the Egyptian hieroglyphic system, dates back to about 3000 B.C. The Egyptians used symbols for the powers of 10: 10^0 or 1, 10^1 or 10, 10^2 or $10 \cdot 10$, 10^3 or $10 \cdot 10 \cdot 10$, and so on. Table 4.1 lists the Egyptian hieroglyphic numerals with the equivalent Hindu–Arabic numerals.

To write the number 600 in Egyptian hieroglyphics, we write the numeral for 100 six times: 𓍢𓍢𓍢𓍢𓍢𓍢.

Table 4.1 Egyptian Hieroglyphics

Hindu–Arabic Numerals	Egyptian Numerals	Description
1	𓏺	Staff (vertical stroke)
10	𓎆	Heel bone (arch)
100	𓍢	Scroll (coiled rope)
1,000	𓆼	Lotus flower
10,000	𓂭	Pointing finger
100,000	𓆐	Tadpole (or whale)
1,000,000	𓁨	Astonished person

EXAMPLE 1 *From Egyptian to Hindu–Arabic Numerals*

Write the following numeral as a Hindu–Arabic numeral.

𓂭𓂭𓂭𓍢𓍢𓍢𓎆𓏺𓏺𓏺𓏺

SOLUTION

$$10{,}000 + 10{,}000 + 10{,}000 + 100 + 100 + 100 + 10 + 1 + 1 + 1 + 1$$
$$= 30{,}314$$

EXAMPLE 2 *From Hindu–Arabic to Egyptian Numerals*

Write 43,628 as an Egyptian numeral.

SOLUTION

$$43{,}628 = 40{,}000 + 3000 + 600 + 20 + 8$$

𓂭𓂭𓂭𓂭𓆼𓆼𓆼𓍢𓍢𓍢𓍢𓍢𓍢𓎆𓎆𓏺𓏺𓏺𓏺𓏺𓏺𓏺𓏺

In the Egyptian hieroglyphic system, the order of the symbols is not important. For example, 𓆐𓏺𓏺𓍢𓍢𓎆 and 𓏺𓏺𓍢𓍢𓆐𓎆 both represent 100,212.

Users of additive systems easily accomplished addition and subtraction by combining or removing symbols. Multiplication and division were more difficult; they were performed by a process called *duplation and mediation* (see Section 4.5). The Egyptians had no symbol for zero, but they did have an understanding of fractions. The symbol 𓂋 was used to take the reciprocal of a number; thus, 𓂋𓏺𓏺𓏺 meant $\frac{1}{3}$ and 𓂋𓎆𓏺

was $\frac{1}{11}$. Writing large numbers in the Egyptian hieroglyphic system takes longer than in other systems because so many symbols have to be listed. For example, 45 symbols are needed to represent 99,999.

The Roman numeration system, a second example of an additive system, was developed later than the Egyptian system. Roman numerals (Table 4.2) were used in most European countries until the eighteenth century. They are still commonly seen on buildings, on clocks, and in books. Roman numerals are selected letters of the Roman alphabet.

Table 4.2 Roman Numerals

Roman numerals	I	V	X	L	C	D	M
Hindu–Arabic numerals	1	5	10	50	100	500	1000

The Roman system has two advantages over the Egyptian system. The first is that it uses the subtraction principle as well as the addition principle. Starting from the left, we add each numeral unless its value is smaller than the value of the numeral to its right. In that case, we subtract its value from the value of the numeral to its right. Only the numbers 1, 10, 100, 1000, . . . can be subtracted, and they can only be subtracted from the next two higher numbers. For example, C (100) can be subtracted only from D (500) or M (1000). The symbol DC represents 500 + 100, or 600, and CD represents 500 − 100, or 400. Similarly, MC represents 1000 + 100, or 1100, and CM represents 1000 − 100, or 900.

DID YOU KNOW?

Roman Numerals

Roman numerals remained popular on large clock faces long after their disappearance from daily transactions because they are easier to read from a distance than Hindu–Arabic numerals. The clock shown in this photo is in Vancouver, Canada.

EXAMPLE 3 *From Roman Numerals to Hindu–Arabic Numerals*

Write MMCCCLXII as a Hindu–Arabic numeral.

SOLUTION Since each numeral is larger than the one on its right, no subtraction is necessary.

$$\begin{aligned}\text{MMCCCLXII} &= 1000 + 1000 + 100 + 100 + 100 + 50 + 10 + 1 + 1\\ &= 2362\end{aligned}$$

EXAMPLE 4 *From Roman Numerals to Hindu–Arabic Numerals*

Write CMLXIV as a Hindu–Arabic numeral.

SOLUTION As you read the numerals from left to right, you will note that C (100) has a smaller value than M (1000) and that C is to the left of M. Therefore, CM represents 1000 − 100, or 900. Also note that I (1) has a smaller value than V (5) and that I is to the left of V. Therefore, IV represents 5 − 1, or 4. The rest of the numerals can be added from left to right.

$$\text{CMLXIV} = (1000 - 100) + 50 + 10 + (5 - 1) = 964$$

EXAMPLE 5 *Writing a Roman Numeral*

Write 289 as a Roman numeral.

SOLUTION

$$289 = 200 + 80 + 9 = 100 + 100 + 50 + 10 + 10 + 10 + 9$$

(Nine is treated as 10 − 1.)

$$289 = \text{CCLXXXIX}$$

In the Roman numeration system, a symbol is not repeated more than three consecutive times. For example, the number 646 would be written DCXLVI instead of DCXXXXVI.

The second advantage of the Roman numeration system over the Egyptian numeration system is that it makes use of the multiplication principle for numerals greater than 1000. A bar above a symbol or group of symbols indicates that the symbol or symbols are to be multiplied by 1000. Thus, $\overline{\text{V}} = 5 \times 1000 = 5000$, $\overline{\text{X}} = 10 \times 1000 = 10{,}000$, and $\overline{\text{CD}} = 400 \times 1000 = 400{,}000$. Other examples are $\overline{\text{VI}} = 6 \times 1000 = 6000$, $\overline{\text{XIX}} = 19 \times 1000 = 19{,}000$, and $\overline{\text{XCIV}} = 94 \times 1000 = 94{,}000$. This practice greatly reduces the number of symbols needed to write large numbers.

EXAMPLE 6 *Writing a Large Roman Numeral*

Write 12,345 as a Roman numeral.

SOLUTION

$$
\begin{aligned}
12{,}345 &= (12 \times 1000) + 300 + 40 + 5 \\
&= [(10 + 1 + 1) \times 1000] + (100 + 100 + 100) + (50 - 10) + 5 \\
&= \overline{\text{XII}}\text{CCCXLV}
\end{aligned}
$$

Multiplicative Systems

Multiplicative numeration systems are more similar to our Hindu–Arabic system than are additive systems. In a multiplicative system, 642 might be written (6) (100) (4) (10) (2) or

6
100
4
10
2

Note that no addition signs are needed in the representation. From this illustration, try to formulate a rule explaining how multiplicative systems work.

The principle example of a multiplicative system is the traditional Chinese system. The numerals used in this system are given in Table 4.3.

Table 4.3 Traditional Chinese Numerals

Traditional Chinese numerals	零	一	二	三	四	五	六	七	八	九	十	百	千
Hindu–Arabic numerals	0	1	2	3	4	5	6	7	8	9	10	100	1000

Chinese numerals are always written vertically. The numeral on top will be from 1 to 9 inclusive. This numeral is to be multiplied by the power of 10 below it. Thus, 20 is written

二
十 } $2 \times 10 = 20$

and 400 is written

四
百 } $4 \times 100 = 400$

DID YOU KNOW?

Hindu–Arabic Numerals

The Hindu–Arabic system of numeration is used throughout the world. In many countries, numbers are displayed in Hindu–Arabic numerals as well as the numerals used in the country itself.

EXAMPLE 7 *A Traditional Chinese Numeral*

Write 538 as a Chinese numeral.

SOLUTION

538 =	500 =	5	五
		100	百
	30 =	3	三
		10	十
	8 =	8	八

Note that in Example 7 the units digit, the 8, is not multiplied by a power of the base.

When writing Chinese numerals, there are some special cases that need to be considered. When writing a numeral between 11 and 19 it is not necessary to include the 1 before the 10. Thus, 18 would be written 十八 rather than 一十八. Another special case involves the use of zero.

When more than one consecutive zero occurs (except at the end of a numeral) you need to write a zero, but only once for two or more consecutive zeros. Zeros are not included at the end of numerals. The top two illustrations that follow show how zeros are used within a numeral, and the bottom two show that zeros are not used at the end of a numeral.

406 =	
四百	$4 \times 100 = 400$
零	$0 \times 10 = 0$
六	$6 = 6$

4006 =	
四千	$4 \times 1000 = 4000$
零	$0 \times 100,\ 0 \times 10 = 0$
六	$6 = 6$

460 =	
四百	$4 \times 100 = 400$
六十	$6 \times 10 = 60$

4600 =	
四千	$4 \times 1000 = 4000$
六百	$6 \times 100 = 600$

EXAMPLE 8 *Traditional Chinese Numerals*

Write the following as traditional Chinese numerals.

a) 7080 b) 7008

SOLUTION In part (a), there is one zero between the 7 and the 8. In part (b), there are two zeros between the 7 and the 8. As just mentioned, the symbol for zero is used only once in each of these numerals.

a) 7080 =	
七千	7×1000
零	0×100
八十	8×10

b) 7008 =	
七千	7×1000
零	$0 \times 100,\ 0 \times 10$
八	8

TIMELY TIP Notice the difference between our Hindu–Arabic numeration system, which is a positional numeration system, and the Chinese system, which is a multiplicative numeration system. Below we show how to write 5678 as a Chinese numeral if the Chinese system was a positional value system similar to ours.

Multiplicative

五	5
千	1000
六	6
百	100
七	7
十	10
八	8

Positional Value

五	5
六	6
七	7
八	8

Note that the multiples of base 10 are removed when writing positional value numerals. We will discuss positional value systems in more detail shortly.

Ciphered Systems

A ciphered numeration system is one in which there are numerals for numbers up to and including the base and for multiples of the base. The numbers represented by a particular set of numerals is the sum of the values of the numerals.

Ciphered numeration systems require the memorization of many different symbols but have the advantage that numerals can be written in a compact form. The ciphered numeration system that we discuss is the Ionic Greek system (Table 4.4). The Ionic Greek system was developed in about 3000 B.C., and it used letters of the Greek alphabet for numerals. Other ciphered systems include the Hebrew, Coptic, Hindu, Brahmin, Syrian, Egyptian Hieratic, and early Arabic systems.

Table 4.4 Ionic Greek Numerals

1	α	alpha	60	ξ	xi
2	β	beta	70	o	omicron
3	γ	gamma	80	π	pi
4	δ	delta	90	ϙ*	koppa
5	ε	epsilon	100	ρ	rho
6	f*	digamma	200	σ	sigma
7	ζ	zeta	300	τ	tau
8	η	eta	400	υ	upsilon
9	θ	theta	500	ϕ	phi
10	ι	iota	600	χ	chi
20	κ	kappa	700	ψ	psi
30	λ	lambda	800	ω	omega
40	μ	mu	900	ϡ*	sampi
50	ν	nu			

*Ancient Greek letters not used in the classic or modern Greek language.

The classic Greek alphabet contains only 24 letters, however 27 symbols were needed. Thus, the Greeks used three obsolete Greek letters—f, ϙ, and ϡ—that are not part of the classic Greek alphabet. To distinguish words from numerals, the Greeks would place a mark similar to our apostrophe to the right and above each letter that was used as a numeral. In this text, we will not use this mark because we will not be using Greek words and numerals together.

We can write 45 as 40 + 5. When 45 is written as a Greek numeral, the plus sign is omitted:

$$45 = \mu\varepsilon$$

Similarly, 768 written as a Greek numeral is $\psi\xi\eta$.

When the Greek letter iota, ι, is placed to the left and above a numeral, it represents the numeral multiplied by 1000. For example,

$$^{\iota}\theta = 9 \times 1000 = 9000$$
$$^{\iota}\zeta = 7 \times 1000 = 7000$$

EXAMPLE 9 *The Ionic Greek System: A Ciphered System*

Write $\phi\ \nu\ \gamma$ as a Hindu–Arabic numeral.

SOLUTION $\phi = 500$, $\nu = 50$, and $\gamma = 3$. The sum is 553. ●

EXAMPLE 10 *Writing an Ionic Greek Numeral*

Write 1654 as an Ionic Greek numeral.

SOLUTION

$$\begin{aligned} 1654 &= 1000 + 600 + 50 + 4 \\ &= (1 \times 1000) + 600 + 50 + 4 \\ &= {}^{\iota}\alpha \qquad\qquad \chi \qquad \nu \qquad \delta \\ &= {}^{\iota}\alpha\ \chi\ \nu\ \delta \end{aligned}$$

●

SECTION 4.1 EXERCISES

CONCEPT/WRITING EXERCISES

1. What is the difference between a number and a numeral?
2. List four numerals given in this section that may be used to represent the number ten.
3. What is a system of numeration?
4. List four numerals given in this section that may be used to represent the number one hundred.
5. What is the name of the system of numeration that we presently use?
6. Explain how numbers are represented in an additive numeration system.
7. Explain how numbers are represented in a multiplicative numeration system.
8. Explain how numbers are represented in a ciphered numeration system.

PRACTICE THE SKILLS

In Exercises 9–14, write the Egyptian numeral as a Hindu–Arabic numeral.

9. 𓍢𓍢𓍢𓎆𓎆𓎆𓎆𓏺𓏺𓏺𓏺𓏺𓏺
10. 𓆼𓍢𓍢𓎆𓏺𓏺𓏺
11. 𓆼𓆼𓍢𓍢𓍢𓍢𓎆𓎆𓏺𓏺𓏺
12. 𓂭𓂭𓂭𓆼𓍢𓍢𓎆
13. 𓆐𓆐𓆐𓆐𓂭𓂭𓂭𓆼𓆼𓆼𓆼𓍢𓍢𓎆𓏺𓏺𓏺𓏺
14. 𓁨𓁨𓁨𓆐𓆐𓆐𓍢𓍢𓍢𓍢𓎆𓎆𓎆𓏺

In Exercises 15–20, write the numeral as an Egyptian numeral.

15. 53
16. 224
17. 2045
18. 1812
19. 173,845
20. 3,235,614

In Exercises 21–32, write the Roman numeral as a Hindu–Arabic numeral.

21. VIII
22. XXIX
23. XLIII
24. CDLXXIV
25. MCCXXXVI
26. MCMLXIV
27. MMCMXLVI
28. MDCCXLVI
29. $\overline{\text{X}}$MMDCLXVI
30. $\overline{\text{L}}$MCMXLIV
31. $\overline{\text{IX}}$CDLXIV
32. $\overline{\text{V}}$MCCCXXXIII

In Exercises 33–44, write the numeral as a Roman numeral.

33. 27
34. 89
35. 341
36. 477
37. 2005
38. 4285
39. 4793
40. 6274
41. 9999
42. 14,315
43. 20,644
44. 99,999

In Exercises 45–52, write the Chinese numeral as a Hindu–Arabic numeral.

45. 七十四
46. 六十二
47. 四千零八十一
48. 三千零二十九
49. 八千五百五十
50. 三千四百八十七
51. 四千零三
52. 五千六百零二

In Exercises 53–60, write the numeral as a traditional Chinese numeral.

53. 53
54. 178
55. 378
56. 2001
57. 4260
58. 6905
59. 7056
60. 3009

In Exercises 61–66, write the numeral as a Hindu–Arabic numeral.

61. κ ϛ
62. ρ λ η
63. σ ο θ
64. ϡ ε
65. ͵β ω π γ
66. ͵ζ υ ξ ζ

In Exercises 67–72, write the numeral as an Ionic Greek numeral.

67. 59
68. 178
69. 726
70. 2008
71. 5005
72. 9999

In Exercises 73–76, write the numeral as numerals in the indicated systems of numeration.

73. 𓆼𓎆𓎆𓏺 in Hindu–Arabic, Roman, traditional Chinese, and Greek
74. MCMXXXVI in Hindu–Arabic, Egyptian, Greek, and traditional Chinese
75. 五百二十七 in Hindu–Arabic, Egyptian, Roman, and Greek
76. υκβ in Hindu–Arabic, Egyptian, Roman, and traditional Chinese

CHALLENGE PROBLEMS/GROUP ACTIVITIES

77. Write the Roman numeral for 999,999.

78. Make up your own additive system of numeration and indicate the symbols and rules used to represent numbers. Using your system of numeration, write

a) your age.

b) the year you were born.

c) the current year.

In Exercises 79–81, compare the advantages and disadvantages of a ciphered system of numeration with those of the named system.

79. An additive system

80. A multiplicative system

81. The Hindu–Arabic system

RECREATIONAL MATHEMATICS

82. Without using any type of writing instrument, what can you do to make the following incorrect statement a correct statement?

$$\text{XI} + \text{I} = \text{X}$$

83. Words and numerals that read the same both backward and forward are called *palindromes.* Some examples are the words CIVIC and RACECAR, and the numerals 121 and 32523. Using Roman numerals, list the last year that was a palindrome.

84. Which year in the past 2000 years required the most Roman numerals to write? Write out the year in Roman numerals.

INTERNET/RESEARCH ACTIVITIES

85. ***Egypt, China, and Greece*** In this section, we discussed the numeration systems of Egypt, China, and Greece. Choose one of these countries and write a paper that does the following.

a) State the current numerals used in your chosen country.

b) Explain how that country's current system of numeration works. If more than one system is used, discuss the system used most commonly.

▲ Great Pyramid of Khufu in Giza, Egypt

4.2 PLACE-VALUE OR POSITIONAL-VALUE NUMERATION SYSTEMS

▲ Place value is used in many games on *The Price is Right.*

Have you ever watched the television game show *The Price is Right*? Chances are that you have seen one of the many games that require contestants to place digits in the correct order to guess the price of a car, appliance, or other fabulous prize. Such games require the contestant to have an understanding of the concept of *place value* that is used in our modern number system. In this section, we will study two other number systems that also rely on place value.

Today the most common type of numeration system is the place-value system. The Hindu–Arabic numeration system, used in the United States and many other countries, is an example of a place-value system. In a ***place-value system***, which is also

Eighteenth-century mathematician Pierre Simon, Marquis de Laplace, speaking of the positional principle, said: "The idea is so simple that this very simplicity is the reason for our not being sufficiently aware of how much attention it deserves."

called a ***positional value system***, the value of the symbol depends on its position in the representation of the number. For example, the 2 in 20 represents 2 tens, and the 2 in 200 represents 2 hundreds. A true positional-value system requires a ***base*** and a set of symbols, including a symbol for zero and one for each counting number less than the base. Although any number can be written in any base, the most common positional system is the base 10 system which is called the ***decimal number system***.

The Hindus in India are credited with the invention of zero and the other symbols used in our system. The Arabs, who traded regularly with the Hindus, also adopted the system, thus the name Hindu–Arabic. Not until the middle of the fifteenth century, however, did the Hindu–Arabic numerals take the form we know today.

The Hindu–Arabic numerals and the positional system of numeration revolutionized mathematics by making addition, subtraction, multiplication, and division much easier to learn and very practical to use. Merchants and traders no longer had to depend on the counting board or abacus. The first group of mathematicians, who computed with the Hindu–Arabic system rather than with pebbles or beads on a wire, were known as the "algorists."

In the Hindu–Arabic system, the symbols 0, 1, 2, 3, 4, 5, 6, 7, 8, and 9 are called ***digits***. The base 10 system was developed from counting on fingers, and the word *digit* comes from the Latin word for fingers.

The positional values in the Hindu–Arabic system are

$$\ldots, (10)^5, (10)^4, (10)^3, (10)^2, 10, 1$$

To evaluate a numeral in the Hindu–Arabic system, we multiply the first digit on the right by 1. We multiply the second digit from the right by the base, 10. We multiply the third digit from the right by the base squared, 10^2 or 100. We multiply the fourth digit from the right by the base cubed, 10^3 or 1000, and so on. In general, we multiply the digit n places from the right by 10^{n-1}. Therefore, we multiply the digit eight places from the right by 10^7. Using the place-value rule, we can write a number in ***expanded form***. In expanded form, 1234 is written

$$1234 = (1 \times 10^3) + (2 \times 10^2) + (3 \times 10) + (4 \times 1)$$

or

$$(1 \times 1000) + (2 \times 100) + (3 \times 10) + 4$$

> **DID YOU KNOW?**
>
> ***Babylonian Numerals***
>
> 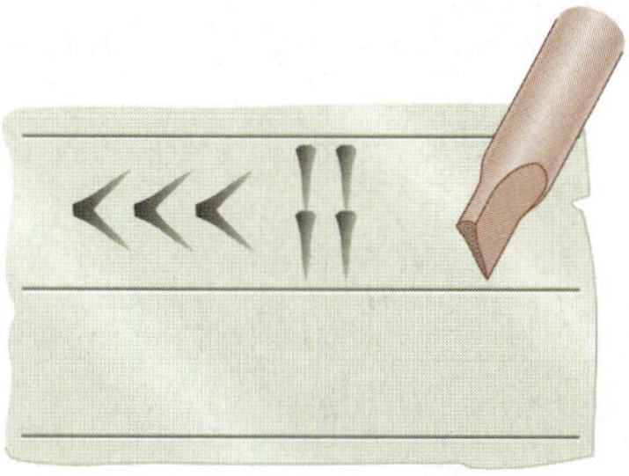
>
> The form Babylonian numerals took is directly related to their writing materials. Babylonians used a reed (later a stylus) to make their marks in wet clay. The end could be used to make a thin wedge, 𒁹, which represents a unit, or a wider wedge, 𒌋, which represents 10 units. The clay dried quickly, so the writings tended to be short but extremely durable.

Babylonian Numerals

The oldest known numeration system that resembled a place-value system was developed by the Babylonians in about 2500 B.C. Their system resembled a place-value system with a base of 60, a sexagesimal system. It was not a true place-value system because it lacked a symbol for zero. The lack of a symbol for zero led to a great deal of ambiguity and confusion. Table 4.5 gives the Babylonian numerals.

The positional values in the Babylonian system are

$$\ldots, (60)^3, (60)^2, 60, 1$$

Table 4.5 Babylonian Numerals

Babylonian Numerals	𒁹	𒌋
Hindu–Arabic numerals	1	10

In a Babylonian numeral, a gap is left between the characters to distinguish between the various place values. From right to left, the sum of the first group of numerals is multiplied by 1. The sum of the second group is multiplied by 60. The sum of the third group is multiplied by $(60)^2$, and so on.

DID YOU KNOW?

Counting Boards

One of the earliest counting devices, used in most ancient civilizations, was the counting board. On such a board, each column represents a positional value. The number of times a value occurs is represented by markers (beads, stones, sticks) in the column. An empty column signifies "no value." The widespread use of counting boards meant that Europeans were already long accustomed to working with positional values when they were introduced to Hindu–Arabic numerals in the fifteenth century. Some people in China, Japan, Russia, Eastern Europe, and the United States still commonly use a type of counting board known as the abacus to perform routine computations.

EXAMPLE 1 *The Babylonian System: A Positional Value System*

Write  as a Hindu–Arabic numeral.

SOLUTION

60's units

$$\underbrace{10 + 10 + 10}_{60\text{'s}} \qquad \underbrace{10 + 10 + 1 + 1}_{\text{units}}$$

$$(30 \times 60) + (22 \times 1)$$
$$1800 + 22 = 1822$$

The Babylonians used the symbol  to indicate subtraction. The numeral  represents 10 − 2, or 8. The numeral  represents 40 − 3, or 37 in base 10 or decimal notation.

EXAMPLE 2 *From Babylonian to Hindu–Arabic Numerals*

Write  as a Hindu–Arabic numeral.

SOLUTION The place value of these three groups of numerals from left to right is

$$(60)^2, \quad 60, \quad 1$$

or

$$3600, \quad 60, \quad 1$$

The numeral in the group on the right has a value of 20 − 2, or 18. The numeral in the center group has a value of 10 + 1, or 11. The numeral on the left represents 1 + 1, or 2. Multiplying each group by its positional value gives

$$\begin{aligned} &(2 \times 60^2) + (11 \times 60) + (18 \times 1) \\ &= (2 \times 3600) + (11 \times 60) + (18 \times 1) \\ &= 7200 + 660 + 18 \\ &= 7878 \end{aligned}$$

To explain the procedure used to convert from a Hindu–Arabic numeral to a Babylonian numeral, we will consider a length of time. How can we change 9820 seconds into hours, minutes, and seconds? Since there are 3600 seconds in an hour (60 seconds to a minute and 60 minutes to an hour), we can find the number of hours in 9820 seconds by dividing 9820 by 60^2, or 3600.

$$\begin{array}{r l} 2 & \leftarrow \text{Hours} \\ 3600\overline{)9820} & \\ \underline{7200} & \\ 2620 & \leftarrow \text{Remaining seconds} \end{array}$$

DID YOU KNOW?

On the High Seas

Vestiges of the Babylonian sexagesimal system are still with us today, especially in navigation. Navigators use degrees, minutes, and seconds of longitude and latitude and global positioning systems (GPS) to chart their course. If you look at a globe or world map, you will see that Earth is divided into 360° of latitude and 360° of longitude. Each degree can be divided into 60 minutes, and each minute can be divided into 60 seconds.

Early explorers had to have an easy means of computing angles as they guided their ships by the stars. Base 60 easily divides into halves, thirds, fourths, fifths, sixths, tenths, twelfths, fifteenths, twentieths, and thirtieths, making such computations easier. Hence, the use of a base 60 system became popular.

Now we can determine the number of minutes by dividing the remaining seconds by 60, the number of seconds in a minute.

$$\begin{array}{r l} 43 & \leftarrow \text{Minutes} \\ 60\overline{)2620} & \\ \underline{2400} & \\ 220 & \\ \underline{180} & \\ 40 & \leftarrow \text{Remaining seconds} \end{array}$$

Since the remaining number of seconds, 40, is less than the number of seconds in a minute, our task is complete.

$$9820 \text{ sec} = 2 \text{ hr, } 43 \text{ min, and } 40 \text{ sec}$$

The same procedure is used to convert a decimal (base 10) numeral to a Babylonian numeral or any numeral in a different base.

EXAMPLE 3 *From Hindu–Arabic to Babylonian Numerals*

Write 2519 as a Babylonian numeral.

SOLUTION The Babylonian numeration system has positional values of

$$\ldots, 60^3, 60^2, 60, 1$$

which can be expressed as

$$\ldots, 216000, 3600, 60, 1$$

The largest positional value less than or equal to 2519 is 60. To determine how many groups of 60 are in 2519, divide 2519 by 60.

$$\begin{array}{r l} 41 & \leftarrow \text{Groups of 60} \\ 60\overline{)2519} & \\ \underline{240} & \\ 119 & \\ \underline{60} & \\ 59 & \leftarrow \text{Units remaining} \end{array}$$

Thus, $2519 \div 60 = 41$ with remainder 59. There are 41 groups of 60 and 59 units remaining. Because the remainder, 59, is less than the base, 60, no further division is necessary. The remainder represents the number of units when the numeral is written in expanded form. Therefore, $2519 = (41 \times 60) + (59 \times 1)$. When written as a Babylonian numeral, 2519 is

EXAMPLE 4 *Using Division to Determine a Babylonian Numeral*

Write 6270 as a Babylonian numeral.

SOLUTION Divide 6270 by the largest positional value less than or equal to 6270. That value is 3600.

$$6270 \div 3600 = 1 \text{ with remainder } 2670$$

There is one group of 3600 in 6270. Next divide the remainder 2670 by 60 to determine the number of groups of 60 in 2670.

$$2670 \div 60 = 44 \text{ with remainder } 30$$

There are 44 groups of 60 and 30 units remaining.

$$6270 = (1 \times 60^2) + (44 \times 60) + (30 \times 1)$$

Thus, 6270 written as a Babylonian numeral is

𒁹 𒌋𒌋𒌋𒌋𒁹𒁹𒁹𒁹 𒌋𒌋𒌋

Mayan Numerals

Another place-value system is the Mayan numeration system. The Mayans, who lived on the Yucatan Peninsula in present day Mexico, developed a sophisticated numeration system based on their religious and agricultural calendar. The numerals in this system are written vertically rather than horizontally, with the units position on the bottom. In the Mayan system, the numeral in the bottom row is to be multiplied by 1. The numeral in the second row from the bottom is to be multiplied by 20. The numeral in the third row is to be multiplied by 18×20, or 360. You probably expected the numeral in the third row to be multiplied by 20^2 rather than 18×20. It is believed that the Mayans used 18×20 so that their numeration system would conform to their calendar of 360 days. The positional values above 18×20 are 18×20^2, 18×20^3, and so on.

Positional Values in the Mayan System

$\ldots 18 \times (20)^3$,	$18 \times (20)^2$,	18×20,	20,	1
or... 144,000,	7200,	360,	20,	1

The digits 0, 1, 2, 3, . . . , 19 of the Mayan systems are formed by a simple grouping of dots and lines, as shown in Table 4.6.

Table 4.6 Mayan Numerals

0	1	2	3	4	5	6	7	8	9
⬭	•	••	•••	••••	—	• —	•• —	••• —	•••• —
10	**11**	**12**	**13**	**14**	**15**	**16**	**17**	**18**	**19**
=	• =	•• =	••• =	•••• =	≡	• ≡	•• ≡	••• ≡	•••• ≡

EXAMPLE 5 *The Mayan System: A Positional Value System*

Write

•
=
•••
•
≡

as a Hindu–Arabic numeral.

SOLUTION In the Mayan numeration system, the first three positional values are

$$18 \times 20$$
$$20$$
$$1$$

• =	$= 11 \times (18 \times 20)$	$= 3960$
•••	$= 3 \times 20$	$= 60$
• ≡	$= 16 \times 1$	$= 16$
		4036

DID YOU KNOW?

Sacred Mayan Glyphs

3

4

In addition to their base 20 numerals, the Mayans had a holy numeration system used by priests to create and maintain calendars. They used a special set of hieroglyphs that consisted of pictograms of Mayan gods. For example, the number 3 was represented by the god of wind and rain, the number 4 by the god of sun.

DID YOU KNOW?

Numerals of the World

Although most countries presently use a place value (or positional value) numeration system with base 10, the numerals used for the digits differ by country. The photo of the artwork entitled *Numbers* by Jan Fleck shows numerals currently used in many countries of the world. For example, in Burmese, the numeral ၃ has a value of 3.

EXAMPLE 6 *From Mayan to Hindu–Arabic Numerals*

Write the following as a Hindu–Arabic numeral.

••
•••
―
•
═
••••

SOLUTION

••	=	$2 \times [18 \times (20)^2]$	=	14,400
••• over one bar	=	$8 \times (18 \times 20)$	=	2880
• over two bars	=	11×20	=	220
••••	=	4×1	=	4
				17,504

EXAMPLE 7 *From Hindu–Arabic to Mayan Numerals*

Write 4025 as a Mayan numeral.

SOLUTION To convert from a Hindu–Arabic to a Mayan numeral, we use a procedure similar to the one used to convert to a Babylonian numeral. The Mayan positional values are . . . , 7200, 360, 20, 1. The greatest positional value less than or equal to 4025 is 360. Divide 4025 by 360.

$$4025 \div 360 = 11 \text{ with remainder } 65$$

There are 11 groups of 360 in 4025. Next, divide the remainder, 65, by 20.

$$65 \div 20 = 3 \text{ with remainder } 5$$

There are 3 groups of 20 with five units remaining.

$$4025 = (11 \times 360) + (3 \times 20) + (5 \times 1)$$

4025 written as a Mayan numeral is

11×360	=	• over two bars
3×20		•••
5×1		―

TIMELY TIP Notice that changing a numeral from the Babylonian or Mayan numeration system *to the Hindu–Arabic* (or decimal or base 10) system involves *multiplication*. Changing a numeral *from the Hindu–Arabic system* to the Babylonian or Mayan numeration system involves *division*.

SECTION 4.2 EXERCISES

CONCEPT/WRITING EXERCISES

1. What is another name for a place-value system?

2. What is the most common type of numeration system used in the world today?

3. Consider the numerals 40 and 400 in the Hindu–Arabic numeration system. What does the 4 represent in each numeral?

4. What is the most common base used for a positional value system? Explain why you believe the base you named is the most common base.

5. a) What is the base in the Hindu–Arabic numeration system?

   b) What are the digits in the Hindu–Arabic numeration system?

6. In a true positional-value system, what symbols are required?

7. Explain how to represent a number in expanded form in a positional-value numeration system.

8. Why was the Babylonian system not a true place-value system?

9. a) The Babylonian system did not have a symbol for zero. Why did this lead to some confusion?

   b) Write 133 and 7980 as Babylonian numerals.

10. Consider the Babylonian numeral . Give two Hindu–Arabic numerals this numeral may represent. Explain your answer.

11. List the first five positional values, starting with the units position, for the Mayan numeration system.

12. Describe two ways that the Mayan place-value system differs from the Hindu–Arabic place-value system.

PRACTICE THE SKILLS

In Exercises 13–24, write the Hindu–Arabic numeral in expanded form.

13. 23 **14.** 84 **15.** 359

16. 562 **17.** 897 **18.** 3769

19. 4387 **20.** 23,468 **21.** 16,402

22. 125,678 **23.** 346,861 **24.** 3,765,934

In Exercises 25–30, write the Babylonian numeral as a Hindu–Arabic numeral.

25.  **26.** 

27.  **28.** 

29.  **30.** 

In Exercises 31–36, write the numeral as a Babylonian numeral.

31. 35 **32.** 129 **33.** 471

34. 512 **35.** 3685 **36.** 12,435

In Exercises 37–42, write the Mayan numeral as a Hindu–Arabic numeral.

37.  **38.**  **39.** 

40.  **41.**  **42.** 

In Exercises 43–48, write the numeral as a Mayan numeral.

43. 17 **44.** 257 **45.** 297

46. 406 **47.** 2163 **48.** 17,708

In Exercises 49 and 50, write the numeral in the indicated systems of numeration.

49.  in Hindu–Arabic and Babylonian

50.  in Hindu–Arabic and Mayan

In Exercises 51 and 52, suppose a place-value numeration system has base *, with digits represented by the symbols* , , , *and* . *Write each expression in expanded form.*

51.  **52.** 

CHALLENGE PROBLEMS/GROUP ACTIVITIES

53. a) Is there a largest numeral in the Mayan numeration system? Explain.

b) Write the Mayan numeral for 999,999.

54. a) Is there a largest numeral in the Babylonian numeration system? Explain.

b) Write the Babylonian numeral for 999,999.

In Exercises 55–58, first convert each numeral to a Hindu–Arabic numeral and then perform the indicated operation. Finally, convert the answer back to a numeral in the original numeration system.

55. 

56. 

57. 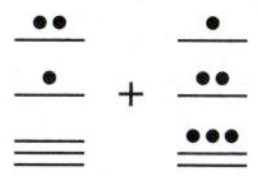

58. 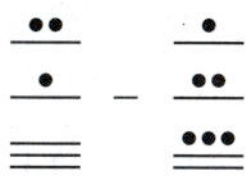

59. ***Comparisons of Systems*** Compare the advantages and disadvantages of a place-value system with those of

a) additive numeration systems.

b) multiplicative numeration systems.

c) ciphered numeration systems.

60. ***Your Own System*** Create your own place-value system. Write 2009 in your system.

INTERNET/RESEARCH ACTIVITIES

61. Investigate and write a report on the development of the Hindu–Arabic system of numeration. Start with the earliest records of this system in India.

62. The Arabic numeration system currently in use is a base 10 positional-value system, which uses different symbols than the Hindu–Arabic numeration system. Write the symbols used in the Arabic system of numeration and their equivalent symbols in the Hindu–Arabic numeration system. Write 54, 607, and 2000 in Arabic numerals.

4.3 OTHER BASES

▲ Electronic devices have computer chips that use numeral bases other than base 10.

Have you done any of the following activities today: read your e-mail, browsed the Internet, watched television, listened to a music CD or MP3, driven a car, or gone grocery shopping? Chances are that while doing any of these or many other daily activities you used a device with a computer chip that uses numerals other than our familiar base 10 (decimal) numerals. In this section, we will study base 2 (binary), base 8 (octal), and base 16 (hexadecimal) numerals, all of which are used in most modern computers as well as in many other modern electronic devices.

The positional values in the Hindu–Arabic numeration system are

$$\ldots, (10)^4, (10)^3, (10)^2, 10, 1$$

The positional values in the Babylonian numeration system are

$$\ldots, (60)^4, (60)^3, (60)^2, 60, 1$$

Thus, 10 and 60 are called the ***bases*** of the Hindu–Arabic and Babylonian systems, respectively.

Any counting number greater than 1 may be used as a base for a positional-value numeration system. If a positional-value system has a base b, then its positional values will be

$$\ldots, b^4, b^3, b^2, b, 1$$

The positional values in a base 8 system are

$$\ldots, 8^4, 8^3, 8^2, 8, 1$$

and the positional values in a base 2 system are

$$\ldots, 2^4, 2^3, 2^2, 2, 1$$

As we indicated in Section 4.2, the Mayan numeration system is based on the number 20. It is not, however, a true base 20 positional-value system. Why not?

The reason for the almost universal acceptance of base 10 numeration systems is that most human beings have 10 fingers. Even so, there are still some positional-value numeration systems that use bases other than 10. Some societies are still using a base 2 numeration system. They include some groups of people in Australia, New Guinea, Africa, and South America. Bases 3 and 4 are also used in some areas of South America. Base 5 systems were used by some primitive tribes in Bolivia, but the tribes are now extinct. The pure base 6 system occurs only sparsely in Northwest Africa. Base 6 also occurs in other systems in combination with base 12, the *duodecimal system*.

▲ Computers make use of the binary, octal, and hexadecimal number systems.

Our society still contains remnants of other base systems from other cultures. For example, there are 12 inches in a foot and 12 months in a year. Base 12 is also evident in the dozen, the 24-hour day, and the gross (12×12). Remains of base 60 are found in measurements of time (60 seconds to 1 minute, 60 minutes to 1 hour) and angles (60 seconds to 1 minute, 60 minutes to 1 degree).

Computers and many other electronic devices make use of three numeration systems: the ***binary*** (base 2), the ***octal*** (base 8), and the ***hexadecimal*** (base 16) numeration systems. One familiar example of the binary numeration system is the bar codes found on most items purchased in stores today. Computers can use the binary numeration system because the binary number system consists of only the digits 0 and 1. These digits can be represented with electronic switches that are either off (0) or on (1). All data that we enter into a computer can be converted into a series of single binary digits. Each binary digit is known as a ***bit***. The octal numeration system is used when eight bits of data are grouped together to form a ***byte***. In the American Standard Code for Information Interchange (ASCII) code, the byte 01000001 represents the character A, and 01100001 represents the character a. Other characters along with their decimal, binary, octal, and hexadecimal representation can be found at the web site www.asciitable.com. The hexadecimal numeration system is used to create computer languages. Two such examples of computer languages that rely on the hexadecimal system are HTML and CSS, both of which are used heavily in creating Internet web pages. Computers can easily convert between binary (base 2), octal (base 8), and hexadecimal (base 16) numbers.

Bases Less Than 10

A place-value system with base b must have b distinct symbols, one for zero and one for each numeral less than the base. A base 6 system must have symbols for 0, 1, 2, 3, 4, and 5. All numerals in base 6 are constructed from these 6 symbols. A base 8 system must have symbols for 0, 1, 2, 3, 4, 5, 6, and 7. All numerals in base 8 are constructed from these 8 symbols, and so on.

A numeral in a base other than base 10 will be indicated by a subscript to the right of the numeral. Thus, 123_5 represents a base 5 numeral. The numeral 123_6 represents a base 6 numeral. The value of 123_5 is not the same as the value of 123_{10}, and the value of 123_6 is not the same as the value of 123_{10}. A base 10 numeral may be written without a subscript. For example, 123 means 123_{10} and 456 means 456_{10}. For clarity in certain problems, we will use the subscript 10 to indicate a numeral in base 10.

We can represent 273 as one group of 243, no groups of 81, one group of 27, no groups of 9, one group of 3, and no units.

$$\begin{aligned} 273 &= (1 \times 243) + (0 \times 81) + (1 \times 27) + (0 \times 9) + (1 \times 3) + (0 \times 1) \\ &= (1 \times 3^5) + (0 \times 3^4) + (1 \times 3^3) + (0 \times 3^2) + (1 \times 3) + (0 \times 1) \\ &= 101010_3 \end{aligned}$$

Bases Greater Than 10

Recall that a place-value system with base b must have symbols for the digits from 0 up to one less than the base. For example, base 6 uses the digits 0, 1, 2, 3, 4, and 5; and base 8 uses the digits 0, 1, 2, 3, 4, 5, 6, and 7. What happens if the base is larger than ten? We will need single digit symbols to represent the numbers ten, eleven, twelve, . . . up to one less than the base. *In this textbook, whenever a base larger than ten is used we will use the capital letter A to represent ten, the capital letter B to represent eleven, the capital letter C to represent twelve, and so on.* For example, for base 12, known as the ***duodecimal system***, we use the symbols 0, 1, 2, 3, 4, 5, 6, 7, 8, 9, A, and B, where A represents ten and B represents eleven. For base 16, known as the hexadecimal system, we use the symbols 0, 1, 2, 3, 4, 5, 6, 7, 8, 9, A, B, C, D, E, and F.

EXAMPLE 6 *Converting to and from Base 12*

a) Convert $39BA_{12}$ to base 10. b) Convert 6893 to base 12.

SOLUTION

a) In base 12, the positional values are . . . 12^3, 12^2, 12, 1 or . . . 1728, 144, 12, 1. Since B has the value of eleven and A has the value of ten, we perform the following calculation.

$$\begin{aligned} 39BA_{12} &= (3 \times 12^3) + (9 \times 12^2) + (11 \times 12) + (10 \times 1) \\ &= (3 \times 1728) + (9 \times 144) + 132 + 10 \\ &= 5184 + 1296 + 132 + 10 = 6622 \end{aligned}$$

b) The highest power of base 12 that is less than or equal to 6893 is 12^3, or 1728. To convert 6893 to base 12, we will use a process similar to the one we used in Examples 4 and 5. Remember that in base 12 we represent ten with the numeral A and eleven with the numeral B.

$$\begin{aligned} 6893 \div 1728 &= 3 \text{ with remainder } 1709 \\ 1709 \div 144 &= \text{B with remainder } 125 \\ 125 \div 12 &= \text{A with remainder } 5 \end{aligned}$$

Note that B has a value of eleven

Note that A has a value of ten

The remainder, 5, is less than the base, 12, so no further division is necessary. Thus, the answer is $3BA5_{12}$. To check this answer, perform the calculation $(3 \times 12^3) + (11 \times 12^2) + (10 \times 12) + (5 \times 1)$ to verify that you obtain 6893 in base 10.

DID YOU KNOW?

Dozenal Society

The Dozenal Society of America, formerly the Duodecimal Society of America, is a voluntary nonprofit organization that conducts research and educates the public in the use of the base 12 numeration system. Throughout history, 12 has been used for many groupings of items. Some familiar examples are 12 inches in a foot, 12 hours on a clock, 12 months in a year, 12 items in a dozen, and 12 dozen in a gross. For more information about the Dozenal Society you can visit their web site at www.polar.sunynassau.edu/~dozenal/index.html. The site is hosted by Nassau Community College, in Garden City, New York.

EXAMPLE 7 *Converting to and from Base 16*

a) Convert $7DE_{16}$ to base 10. b) Convert 6713 to base 16.

SOLUTION

a) In a base 16 system, the positional values are . . . , 16^3, 16^2, 16, 1 or . . . 4096, 256, 16, 1. Since D has a value of thirteen and E has a value of fourteen, we perform the following calculation.

$$\begin{aligned} 7DE_{16} &= (7 \times 16^2) + (D \times 16) + (E \times 1) \\ &= (7 \times 256) + (13 \times 16) + (14 \times 1) \\ &= 1792 + 208 + 14 \\ &= 2014 \end{aligned}$$

b) The highest power of base 16 less than or equal to 6713 is 16^3, or 4096. If we obtain a quotient greater than nine but less than sixteen, we will use the corresponding letter A through F.

$$\begin{aligned} 6713 \div 4096 &= 1 \text{ with remainder } 2617 \\ 2617 \div 256 &= A \text{ with remainder } 57 \\ 57 \div 16 &= 3 \text{ with remainder } 9 \end{aligned}$$

Note that A has a value of ten

Thus, $6713 = 1A39_{16}$.

TIMELY TIP It is important to remember the following items presented in this section.

- If a numeral is shown without a base, we assume the numeral is a base 10 numeral.
- When converting a base 10 numeral to a different base, your answer should never contain a digit greater than or equal to that different base.
- When changing a numeral given in a base other than 10 to a numeral in base 10, use multiplication. When changing a base 10 numeral to a numeral in a different base, use division.

SECTION 4.3 EXERCISES

CONCEPT/WRITING EXERCISES

1. a) In your own words, explain how to convert a numeral in a base other than 10 to a base 10 numeral.

   b) In your own words, explain how to convert a base 10 numeral to a numeral in a base other than 10.

2. List the digits that are used when writing numerals in each of the following bases.

   a) 2 b) 8

   c) 12 d) 16

3. For each of the following, explain why the numeral is written incorrectly.

   a) 234_4 b) 1030001_2

   c) $4A8LB4A_{12}$ d) $BCFGA_{16}$

4. a) What is 10_2 equal to in base 10?

   b) What is 10_8 equal to in base 10?

   c) What is 10_{16} equal to in base 10?

   d) What is 10_{32} equal to in base 10?

PRACTICE THE SKILLS

In Exercises 5–22, convert the given numeral to a numeral in base 10.

5. 1_2
6. 10_2
7. 23_5
8. 41_5
9. 270_8
10. 306_8
11. 309_{12}
12. 1005_{12}
13. 573_{16}
14. 423_{16}
15. 110101_2
16. 1110111_2
17. 7654_8
18. 5032_8
19. $A91_{12}$
20. $B52_{12}$
21. $C679_{16}$
22. $D20E_{16}$

In Exercises 23–40, convert each of the following to a numeral in the base indicated.

23. 6 to base 2
24. 25 to base 2
25. 347 to base 3
26. 513 to base 3
27. 53 to base 4
28. 591 to base 4
29. 102 to base 5
30. 549 to base 7
31. 1098 to base 8
32. 2921 to base 8
33. 1432 to base 9
34. 2170 to base 9
35. 9004 to base 12
36. 13,312 to base 12
37. 493 to base 16
38. 3243 to base 16
39. 9455 to base 16
40. 39,885 to base 16

In Exercises 41–50, convert 2009 to a numeral in the base indicated.

41. 3
42. 4
43. 5
44. 6
45. 7
46. 9
47. 11
48. 13
49. 15
50. 17

PROBLEM SOLVING

In Exercises 51–54, assume the numerals given are in a base 5 numeration system. The numerals in this system and their equivalent Hindu–Arabic numerals are

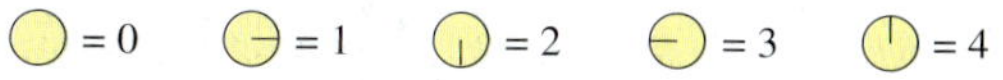

Write the Hindu–Arabic numerals equivalent to each of the following.

51. 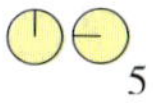
52. 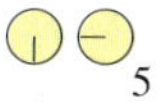
53. 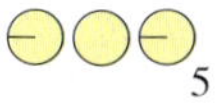
54. 

In Exercises 55–58, write the Hindu Arabic numerals in the numeration system discussed in Exercises 51–54.

55. 19
56. 23
57. 74
58. 85

In Exercises 59–62, suppose colors as indicated below represent numerals in a base 4 numeration system.

Write the Hindu–Arabic numerals equivalent to each of the following.

59. 
60. 
61. 
62. 

In Exercises 63–66, write the Hindu–Arabic numerals in the base 4 numeration system discussed in Exercises 59–62. You will need to use the colors indicated above to write the answer.

63. 10
64. 15
65. 60
66. 56

67. ***Another Conversion Method*** There is an alternative method for changing a base 10 numeral to a different base. This method will be used to convert 328 to base 5. Dividing 328 by 5 gives a quotient of 65 and a remainder of 3. Write the quotient below the dividend and the remainder on the right, as shown.

5	328	remainder
	65	3

Continue this process of division by 5.

5	328	remainder
5	65	3
5	13	0
5	2	3
	0	2

(Answer read upward along the arrow.)

(In the last division, since the dividend, 2, is smaller than the divisor, 5, the quotient is 0 and the remainder is 2.)

Note that the division continues until the quotient is zero. The answer is read from the bottom number to the top number in the remainder column. Thus, $328 = 2303_5$.

a) Explain why this procedure results in the proper answer.

b) Convert 683 to base 5 by this method.

c) Convert 763 to base 8 by this method.

CHALLENGE PROBLEMS/GROUP ACTIVITIES

68. a) Use the numerals 0, 1, and 2 to write the first 20 numbers in the base 3 numeration system.

b) What is the next numeral after 222_3?

69. a) *Your Own Numeration System* Make up your own base 20 positional-value numeration system. Indicate the 20 numerals you will use to represent the 20 numbers less than the base.

b) Write 523 and 5293 in your base 20 numeration system.

70. *Computer Code* The ASCII code used by most computers uses the last seven positions of an eight-bit byte to represent all the characters on a standard keyboard. How many different orderings of 0's and 1's (or how many different characters) can be made by using the last seven positions of an eight-bit byte?

RECREATIONAL MATHEMATICS

71. Find b if $111_b = 43$.

72. Find d if $ddd_5 = 124$.

73. Suppose a base 4 place-value system has its digits represented by colors as follows:

a) Determine the value of  in base 10.

b) Write 177 in the base 4 system using only the four colors given in the exercise.

INTERNET/RESEARCH ACTIVITIES

74. Investigate and write a report on the use of the duodecimal (base 12) system of numeration. You may wish to contact the Dozenal Society (see the Did You Know? on page 201) for more information.

75. We mentioned at the beginning of this section that some societies still use a base 2 and base 3 numeration system. These societies are in Australia, New Guinea, Africa, and South America. Write a report on these societies, covering the symbols they use and how they combine these symbols to represent numbers in their numeration system.

4.4 COMPUTATION IN OTHER BASES

▲ Students do arithmetic by "carrying" and "borrowing."

You may recall learning how to add, subtract, multiply, and divide when you were in grade school. You may recall "carrying" when adding and "borrowing" when subtracting. For example, when performing 75 + 18, you would add 5 and 8 to get 13. You would write down the 3 and "carry the 1." When performing 43 − 19 you would "borrow from the 4" so you could perform 13 − 9 to get 4. In this section, we will use similar procedures when working with numerals that have bases other than base 10. By working in other bases, we will gain a better understanding of our own base 10 place-value system.

Addition

When computers perform calculations, they do so in base 2, the binary system. In this section, we will explain how to perform calculations in base 2 and other bases.

In a base 2 system, the only digits are 0 and 1, and the place values are

$$\ldots,\ 2^4,\ 2^3,\ 2^2,\ 2,\ 1$$

or

$$\ldots,\ 16,\ 8,\ 4,\ 2,\ 1$$

Suppose we want to add $1_2 + 1_2$. The subscript 2 indicates that we are adding in base 2. Remember that the answer to $1_2 + 1_2$ must be written using only the digits 0 and 1. The sum of $1_2 + 1_2$ is 10_2, which represents 1 group of two and 0 units in base 2. Recall that 10_2 means $1(2) + 0(1)$.

If we wanted to find the sum of $10_2 + 1_2$, we would add the digits in the right-hand, or units, column. Since $0_2 + 1_2 = 1_2$, the sum of $10_2 + 1_2 = 11_2$.

We are going to work additional examples and exercises in base 2, so rather than performing individual calculations in every problem, we can construct and use an addition table, Table 4.7, for base 2 (just as we used an addition table in base 10 when we first learned to add in base 10).

Table 4.7 Base 2 Addition Table

+	0	1
0	0	1
1	1	10

EXAMPLE 1 *Adding in Base 2*

Add

$$\begin{array}{r} 1101_2 \\ +\ 111_2 \\ \hline \end{array}$$

SOLUTION Begin by adding the digits in the right-hand, or units, column. From previous discussion, and as can be seen in Table 4.7, $1_2 + 1_2 = 10_2$. Place the 0 under the units column and carry the 1 to the 2's column, the second column from the right.

$$\begin{array}{rcccc} & 2^3 & 2^2 & 2 & 1 \\ & \downarrow & \downarrow & \downarrow & \downarrow \\ & 1 & 1 & {}^{1}0 & 1 \\ + & & 1 & 1 & 1 \\ \hline & & & & 0_2 \end{array}$$

Place value of columns

Now add the three digits in the 2's column, $1_2 + 0_2 + 1_2$. Treat it as $(1_2 + 0_2) + 1_2$. Therefore, add $1_2 + 0_2$ to get 1_2, then add $1_2 + 1_2$ to get 10_2. Place the 0 under the 2's column and carry the 1 to the 2^2 column (the third column from the right).

$$\begin{array}{rcccc} & 1 & {}^{1}1 & {}^{1}0 & 1 \\ + & & 1 & 1 & 1 \\ \hline & & & 0 & 0_2 \end{array}$$

Now add the three 1's in the 2^2 column to get $(1_2 + 1_2) + 1_2 = 10_2 + 1_2 = 11_2$. Place the 1 under the 2^2 column and carry the 1 to the 2^3 column (the fourth column from the right).

$$\begin{array}{rcccc} & {}^{1}1 & {}^{1}1 & {}^{1}0 & 1 \\ + & & 1 & 1 & 1 \\ \hline & & 1 & 0 & 0_2 \end{array}$$

Now add the two 1's in the 2^3 column, $1_2 + 1_2 = 10_2$. Place the 10 as follows.

$$\begin{array}{rccccc} & & {}^{1}1 & {}^{1}1 & {}^{1}0 & 1 \\ + & & & 1 & 1 & 1 \\ \hline & 1 & 0 & 1 & 0 & 0_2 \end{array}$$

Therefore, the sum is 10100_2.

Table 4.8 Base 5 Addition Table

+	0	1	2	3	4
0	0	1	2	3	4
1	1	2	3	4	10
2	2	3	4	10	11
3	3	4	10	11	12
4	4	10	11	(12)	13

DID YOU KNOW?

Speaking to Machines

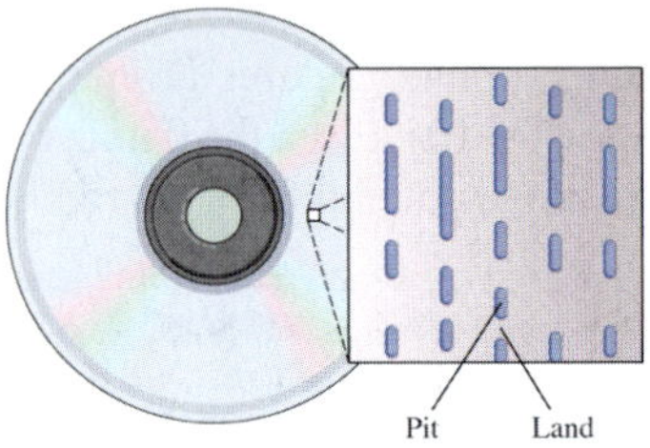

On digital video discs (DVDs) and compact discs (CDs), video and sound are digitally encoded on the underside of the disc in a binary system of pits and "lands" (nonpits). To play the disc, a laser beam tracks along the spiral and is reflected when it hits a land (signal sent = 1), but it is not reflected by the pits (no signal = 0). The binary sequence is then converted into video images and music.

For the past 600 years, we have used the Hindu–Arabic system of numeration without change. Our base 10 numeration system seems so obvious to us, perhaps because of our 10 fingers and 10 toes, but it would be rash to think that numbers in other bases are not useful. In fact, one of the most significant numeration systems is the binary system, or base 2. This system, with its elemental simplicity, is what is used by computers to process information and "talk" to one another. When a computer receives a command or data, every character in the command or data must first be converted into a binary numeral for the computer to understand and use it.

Let's now look at addition in a base 5 system. In base 5, the only digits are 0, 1, 2, 3, and 4, and the positional values are

$$\begin{array}{rl} & \ldots,\ 5^4,\ 5^3,\ 5^2,\ 5,\ 1 \\ \text{or} & \ldots,\ 625,\ 125,\ 25,\ 5,\ 1 \end{array}$$

What is the sum of $4_5 + 3_5$? We can consider this to mean $(1 + 1 + 1 + 1) + (1 + 1 + 1)$. We can regroup the seven 1's into one group of five and two units as $(1 + 1 + 1 + 1 + 1) + (1 + 1)$. Thus, the sum of $4_5 + 3_5 = 12_5$ (circled in Table 4.8). Recall that 12_5 means $1(5) + 2(1)$. We can use this same procedure in obtaining the remaining values in the base 5 addition table.

EXAMPLE 2 *Use the Base 5 Addition Table*

Add $\begin{array}{r} 44_5 \\ +\ 43_5 \\ \hline \end{array}$

SOLUTION First determine from Table 4.8 that $4_5 + 3_5$ is 12_5. Record the 2 and carry the 1 to the 5's column.

$$\begin{array}{rrr} & {}^{1}4 & 4_5 \\ + & 4 & 3_5 \\ \hline & & 2_5 \end{array}$$

Add the numbers in the second column, $(1_5 + 4_5) + 4_5 = 10_5 + 4_5 = 14_5$. Record the 14.

$$\begin{array}{rrr} & {}^{1}4 & 4_5 \\ + & 4 & 3_5 \\ \hline 1 & 4 & 2_5 \end{array}$$

The sum is 142_5. ●

EXAMPLE 3 *Add in Base 5*

Add $\begin{array}{r} 1234_5 \\ +\ 2042_5 \\ \hline \end{array}$

SOLUTION

$$\begin{array}{rrrr} 1 & {}^{1}2 & {}^{1}3 & 4_5 \\ +2 & 0 & 4 & 2_5 \\ \hline 3 & 3 & 3 & 1_5 \end{array}$$
●

You can develop an addition table for any base and use it to add in that base. As you get more comfortable with addition in other bases, however, you may prefer to add in other bases by using mental arithmetic. To do so, convert the sum from the given base to base 10 and then convert the base 10 numeral back into the given base. You must clearly understand how to convert from base 10 to the given base, as discussed in Section 4.3. For example, to add $7_9 + 8_9$, add $7 + 8$ in base 10 to get 15_{10} and then mentally convert 15_{10} to 16_9 using the procedure given earlier. Remember,

16_9 when converted to base 10 becomes $1(9) + 6(1)$, or 15. Addition using this procedure is illustrated in Examples 4 and 5.

EXAMPLE 4 *Adding in Base 10; Converting to Base 3*

Add

$$\begin{array}{r} 1022_3 \\ +\ 2121_3 \\ \hline \end{array}$$

SOLUTION To solve this problem, make the necessary conversions by using mental arithmetic. $2 + 1 = 3_{10} = 10_3$. Record the 0 and carry the 1.

$$\begin{array}{rrrr} 1 & 0 & {}^{1}2 & 2_3 \\ +2 & 1 & 2 & 1_3 \\ \hline & & & 0_3 \end{array}$$

$1 + 2 + 2 = 5_{10} = 12_3$. Record the 2 and carry the 1.

$$\begin{array}{rrrr} 1 & {}^{1}0 & {}^{1}2 & 2_3 \\ +2 & 1 & 2 & 1_3 \\ \hline & & 2 & 0_3 \end{array}$$

$1 + 0 + 1 = 2_{10} = 2_3$. Record the 2.

$$\begin{array}{rrrr} 1 & {}^{1}0 & {}^{1}2 & 2_3 \\ +2 & 1 & 2 & 1_3 \\ \hline & 2 & 2 & 0_3 \end{array}$$

$1 + 2 = 3_{10} = 10_3$. Record the 10.

$$\begin{array}{r} 1022_3 \\ +\ 2121_3 \\ \hline 10220_3 \end{array}$$

The sum is 10220_3.

EXAMPLE 5 *Adding in Base 10; Converting to Base 4*

Add

$$\begin{array}{r} 322_4 \\ 223_4 \\ 230_4 \\ +\ 132_4 \\ \hline \end{array}$$

SOLUTION Adding the digits in the right-hand column gives $2 + 3 + 0 + 2 = 7_{10} = 13_4$. Record the 3 and carry the 1. Adding the 1 with the digits in the next column gives $1 + 2 + 2 + 3 + 3 = 11_{10} = 23_4$. Record the 3 and carry the 2. Adding the 2 with the digits in the next column gives $2 + 3 + 2 + 2 + 1 = 10_{10} = 22_4$. Record both digits. The sum is 2233_4.

$$\begin{array}{rrrr} & {}^{2}3 & {}^{1}2 & 2_4 \\ & 2 & 2 & 3_4 \\ & 2 & 3 & 0_4 \\ + & 1 & 3 & 2_4 \\ \hline 2 & 2 & 3 & 3_4 \end{array}$$

Subtraction

Subtraction can also be performed in bases other than base 10. Always remember that when you "borrow," you borrow the amount of the base given in the subtraction problem. For example, if subtracting in base 5, when you borrow, you borrow 5. If subtracting in base 12, when you borrow, you borrow 12.

EXAMPLE 6 *Subtracting in Base 5*

Subtract
$$\begin{array}{r} 3032_5 \\ -1004_5 \\ \hline \end{array}$$

SOLUTION We will perform the subtraction in base 10 and convert the results to base 5. Since 4 is greater than 2, we must borrow one group of 5 from the preceding column. This action gives a sum of $5 + 2$, or 7 in base 10. Now we subtract 4 from 7; the difference is 3. We complete the problem in the usual manner. The 3 in the second column becomes a 2, $2 - 0 = 2$. In the third column, $0 - 0 = 0$. Finally, in the fourth column, $3 - 1 = 2$.

$$\begin{array}{r} 3032_5 \\ -1004_5 \\ \hline 2023_5 \end{array}$$

In the next example, we will subtract base 12 numerals. Recall from Section 4.3 that we use the capital letter A to represent the number ten and the capital letter B to represent the number eleven in base 12.

EXAMPLE 7 *Subtracting in Base 12*

Subtract
$$\begin{array}{r} 8\ 2\ \text{B}_{12} \\ -3\ \text{A}\ 6_{12} \\ \hline \end{array}$$

SOLUTION Recall that B represents eleven. Therefore, in the units column we have $11 - 6 = 5$. Next, recall that A represents ten. Therefore, in the next column we must subtract 10 from 2. Since 10 is greater than 2, borrowing is necessary. We must borrow one group of 12 from the preceding column. We then have a sum of $12 + 2$, or 14. We can now subtract 10 from 14 and the difference is 4. The 8 in the third column becomes 7, and $7 - 3 = 4$.

$$\begin{array}{r} 8\ 2\ \text{B}_{12} \\ -3\ \text{A}\ 6_{12} \\ \hline 4\ 4\ 5_{12} \end{array}$$

Multiplication

Multiplication can also be performed in bases other than base 10. Doing so is helped by forming a multiplication table for the base desired. Suppose we want to determine the product of $4_5 \times 3_5$. In base 10, 4×3 means there are four groups of three units. Similarly, in a base 5 system, $4_5 \times 3_5$ means there are four groups of three units, or

$$(1 + 1 + 1) + (1 + 1 + 1) + (1 + 1 + 1) + (1 + 1 + 1)$$

Table 4.9 Base 5 Multiplication Table

×	0	1	2	3	4
0	0	0	0	0	0
1	0	1	2	3	4
2	0	2	4	11	13
3	0	3	11	14	22
4	0	4	13	(22)	31

Regrouping the 12 units above into groups of five gives

$$(1 + 1 + 1 + 1 + 1) + (1 + 1 + 1 + 1 + 1) + (1 + 1)$$

or two groups of five, and two units. Thus, $4_5 \times 3_5 = 22_5$.

We can construct other values in the base 5 multiplication table in the same way. You may, however, find it easier to multiply the values in the base 10 system and then change the product to base 5 by using the procedure discussed in Section 4.3. Multiplying 4×3 in base 10 gives 12, and converting 12 from base 10 to base 5 gives 22_5.

The product of $4_5 \times 3_5$ is circled in Table 4.9, the base 5 multiplication table. The other values in the table may be found by either method discussed.

EXAMPLE 8 *Using the Base 5 Multiplication Table*

Multiply

$$\begin{array}{r} 13_5 \\ \times\ 3_5 \\ \hline \end{array}$$

SOLUTION Use the base 5 multiplication table to find the products. When the product consists of two digits, record the right digit and carry the left digit. Multiplying gives $3_5 \times 3_5 = 14_5$. Record the 4 and carry the 1.

$$\begin{array}{r} {}^{1}13_5 \\ \times\ \ 3_5 \\ \hline 4\ \ \end{array}$$

$(3_5 \times 1_5) + 1_5 = 4_5$. Record the 4.

$$\begin{array}{r} {}^{1}13_5 \\ \times\ \ 3_5 \\ \hline 44_5 \end{array}$$

The product is 44_5.

Constructing a multiplication table is often tedious, especially when the base is large. To multiply in a given base without the use of a table, multiply in base 10 and convert the products to the appropriate base before recording them. This procedure is illustrated in Example 9.

EXAMPLE 9 *Multiplying in Base 7*

Multiply

$$\begin{array}{r} 43_7 \\ \times\ 25_7 \\ \hline \end{array}$$

SOLUTION $5 \times 3 = 15_{10} = 2(7) + 1(1) = 21_7$. Record the 1 and carry the 2.

$$\begin{array}{r} {}^{2}43_7 \\ \times\ 25_7 \\ \hline 1\ \ \end{array}$$

$(5 \times 4) + 2 = 20 + 2 = 22_{10} = 3(7) + 1(1) = 31_7$. Record the 31.

$$\begin{array}{r} {}^{2}43_7 \\ \times\ 25_7 \\ \hline 311\ \ \end{array}$$

> **DID YOU KNOW?**
>
> *Ever-Changing Numerals*
>
> 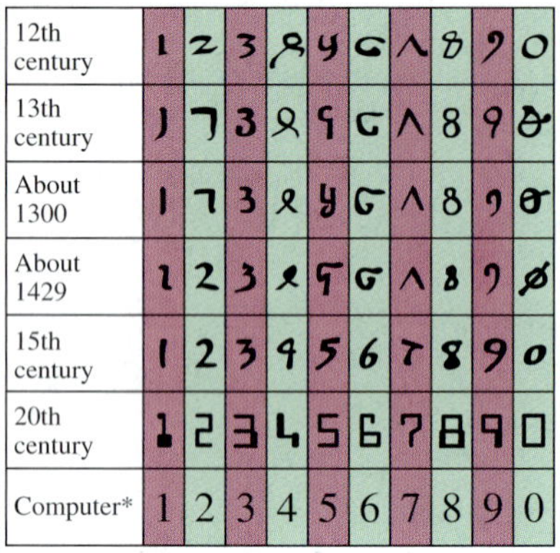
>
>
> * Times New Roman font
>
> During the Middle Ages, Western Europeans were reluctant to give up Roman numerals in favor of Hindu–Arabic numerals. The rapid expansion of trade and commerce during the fifteenth century, however, caused the need for quicker systems of calculation. The invention of movable type in 1450 also ensured a certain consistency in the way numerals were depicted, yet we still find ways to alter them.

$2 \times 3 = 6_{10} = 6_7$. Record the 6.

$$\begin{array}{r} {}^{2}43_7 \\ \times\ 25_7 \\ \hline 311 \\ 6 \end{array}$$

$2 \times 4 = 8_{10} = 1(7) + 1(1) = 11_7$. Record the 11. Now add in base 7 to determine the answer. Remember, in base 7, there are no digits greater than 6.

$$\begin{array}{r} {}^{2}43_7 \\ \times\ 25_7 \\ \hline 311 \\ 116 \\ \hline 1501_7 \end{array}$$

Division

Division is performed in much the same manner as long division in base 10. A detailed example of a division in base 5 is illustrated in Example 10. The same procedure is used for division in any other base.

EXAMPLE 10 *Dividing in Base 5*

Divide $2_5\overline{)143_5}$.

SOLUTION Using the multiplication table for base 5, Table 4.9 on page 209, we list the multiples of the divisor, 2.

$$\begin{aligned} 2_5 \times 1_5 &= 2_5 \\ 2_5 \times 2_5 &= 4_5 \\ 2_5 \times 3_5 &= 11_5 \\ 2_5 \times 4_5 &= 13_5 \end{aligned}$$

Since $2_5 \times 4_5 = 13_5$, which is the largest product less than 14_5, 2_5 divides into 14_5 four times.

$$\begin{array}{r} 4 \\ 2_5\overline{)143_5} \\ 13 \\ \hline 1 \end{array}$$

Subtract 13_5 from 14_5. The difference is 1_5. Record the 1. Now bring down the 3 as when dividing in base 10.

$$\begin{array}{r} 4 \\ 2_5\overline{)143_5} \\ 13 \\ \hline 13_{} \end{array}$$

We see that $2_5 \times 4_5 = 13_5$. Use this information to complete the problem.

$$
\begin{array}{r}
44_5 \\
2_5 \overline{)143_5} \\
\underline{13} \\
13 \\
\underline{13} \\
0
\end{array}
$$

Therefore, $143_5 \div 2_5 = 44_5$ with remainder 0_5.

A division problem can be checked by multiplication. If the division was performed correctly, (quotient × divisor) + remainder = dividend. We can check Example 10 as follows.

$$(44_5 \times 2_5) + 0_5 = 143_5$$

$$
\begin{array}{r}
44_5 \\
\times \quad 2_5 \\
\hline
143_5
\end{array}
\quad \text{Check}
$$

EXAMPLE 11 *Dividing in Base 8*

Divide $6_8\overline{)4071_8}$.

SOLUTION The multiples of 6 in base 8 are

$6_8 \times 1_8 = 6_8$ $\quad 6_8 \times 2_8 = 14_8$ $\quad 6_8 \times 3_8 = 22_8$ $\quad 6_8 \times 4_8 = 30_8$

$6_8 \times 5_8 = 36_8$ $\quad 6_8 \times 6_8 = 44_8$ $\quad 6_8 \times 7_8 = 52_8$

$$
\begin{array}{r}
536_8 \\
6_8 \overline{)4071_8} \\
\underline{36} \\
27 \\
\underline{22} \\
51 \\
\underline{44} \\
5
\end{array}
$$

Thus, the quotient is 536_8, with remainder 5_8.

Be careful when subtracting! In the above division problem, we had to borrow when subtracting 6 from 0 and when subtracting 4 from 1. Remember that we borrow 10_8, which is the same as 8 in base 10.

CHECK: Does $(536_8 \times 6_8) + 5_8 = 4071_8$?

$$
\begin{array}{r}
536_8 \\
\times \quad 6_8 \\
\hline
4064_8
\end{array}
+ 5_8 = 4071_8 \quad \text{True}
$$

SECTION 4.4 EXERCISES

CONCEPT/WRITING EXERCISES

1. a) What are the first five positional values, from right to left, in base b?

   b) What are the first five positional values, from right to left, in base 2?

2. In the addition

$$\begin{array}{r} 367_8 \\ +24_8 \\ \hline \end{array}$$

what are the positional values of the first column on the right, the second column from the right, and the third column from the right? Explain how you determined your answer.

3. Suppose you add two base 3 numerals and you obtain an answer of 20323_3. Can your answer be correct? Explain.

4. Suppose you add two base 5 numerals and you obtain an answer of 463_5. Can your answer be correct? Explain.

5. In your own words, explain how to subtract two numerals in a given base. Include in your explanation what you do when, in one column, you must subtract a larger numeral from a smaller numeral.

6. In your own words, explain how to add two numerals in a given base. In your explanation, answer the question, "What happens when the sum of the numerals in a column is greater than the base?"

PRACTICE THE SKILLS

In Exercises 7–18, add in the indicated base.

7. $\begin{array}{r} 21_3 \\ +\ 20_3 \\ \hline \end{array}$ 8. $\begin{array}{r} 23_4 \\ +\ 13_4 \\ \hline \end{array}$ 9. $\begin{array}{r} 1234_5 \\ +\ 341_5 \\ \hline \end{array}$

10. $\begin{array}{r} 1101_2 \\ +\ 111_2 \\ \hline \end{array}$ 11. $\begin{array}{r} 799_{12} \\ +\ 218_{12} \\ \hline \end{array}$ 12. $\begin{array}{r} 222_3 \\ +\ 22_3 \\ \hline \end{array}$

13. $\begin{array}{r} 1112_3 \\ +\ 1011_3 \\ \hline \end{array}$ 14. $\begin{array}{r} 470_{12} \\ +\ 347_{12} \\ \hline \end{array}$ 15. $\begin{array}{r} 14631_7 \\ +\ 6040_7 \\ \hline \end{array}$

16. $\begin{array}{r} 1341_8 \\ +\ 341_8 \\ \hline \end{array}$ 17. $\begin{array}{r} 1110_2 \\ +\ 110_2 \\ \hline \end{array}$ 18. $\begin{array}{r} 43A_{16} \\ +\ 496_{16} \\ \hline \end{array}$

In Exercises 19–30, subtract in the indicated base.

19. $\begin{array}{r} 201_3 \\ -120_3 \\ \hline \end{array}$ 20. $\begin{array}{r} 512_6 \\ -421_6 \\ \hline \end{array}$ 21. $\begin{array}{r} 2138_9 \\ -1207_9 \\ \hline \end{array}$

22. $\begin{array}{r} AB32_{12} \\ -\ 207_{12} \\ \hline \end{array}$ 23. $\begin{array}{r} 1101_2 \\ -\ 111_2 \\ \hline \end{array}$ 24. $\begin{array}{r} 1221_3 \\ -\ 202_3 \\ \hline \end{array}$

25. $\begin{array}{r} 1001_2 \\ -\ 110_2 \\ \hline \end{array}$ 26. $\begin{array}{r} 2173_8 \\ -1654_8 \\ \hline \end{array}$ 27. $\begin{array}{r} 4223_7 \\ -\ 304_7 \\ \hline \end{array}$

28. $\begin{array}{r} 4232_5 \\ -2341_5 \\ \hline \end{array}$ 29. $\begin{array}{r} 2100_3 \\ -1012_3 \\ \hline \end{array}$ 30. $\begin{array}{r} 4E7_{16} \\ -189_{16} \\ \hline \end{array}$

In Exercises 31–42, multiply in the indicated base.

31. $\begin{array}{r} 22_3 \\ \times\ 2_3 \\ \hline \end{array}$ 32. $\begin{array}{r} 124_5 \\ \times\ 3_5 \\ \hline \end{array}$ 33. $\begin{array}{r} 647_8 \\ \times\ 5_8 \\ \hline \end{array}$

34. $\begin{array}{r} 12_4 \\ \times\ 30_4 \\ \hline \end{array}$ 35. $\begin{array}{r} 512_6 \\ \times\ 23_6 \\ \hline \end{array}$ 36. $\begin{array}{r} 124_{12} \\ \times\ 6_{12} \\ \hline \end{array}$

37. $\begin{array}{r} B12_{12} \\ \times\ 83_{12} \\ \hline \end{array}$ 38. $\begin{array}{r} 6A3_{12} \\ \times\ 24_{12} \\ \hline \end{array}$ 39. $\begin{array}{r} 111_2 \\ \times\ 101_2 \\ \hline \end{array}$

40. $\begin{array}{r} 584_9 \\ \times\ 24_9 \\ \hline \end{array}$ 41. $\begin{array}{r} 316_7 \\ \times\ 16_7 \\ \hline \end{array}$ 42. $\begin{array}{r} 7A9_{12} \\ \times\ 12_{12} \\ \hline \end{array}$

In Exercises 43–54, divide in the indicated base.

43. $1_2\overline{)101_2}$ 44. $2_3\overline{)120_3}$ 45. $2_6\overline{)451_6}$

46. $7_9\overline{)123_9}$ 47. $2_4\overline{)312_4}$ 48. $6_{12}\overline{)431_{12}}$

49. $2_4\overline{)213_4}$ 50. $5_6\overline{)214_6}$ 51. $3_5\overline{)224_5}$

52. $4_6\overline{)210_6}$ 53. $6_7\overline{)404_7}$ 54. $3_7\overline{)2101_7}$

PROBLEM SOLVING

In Exercises 55–58, the numerals in a base 5 numeration system are as illustrated with their equivalent Hindu–Arabic numerals.

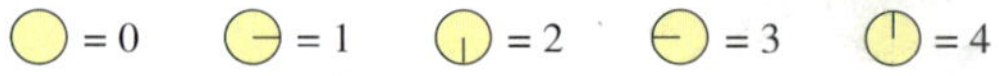

Add the following base 5 numerals.

55.
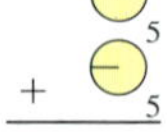

56.
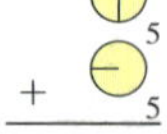

57.
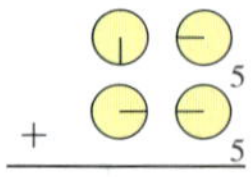

58.
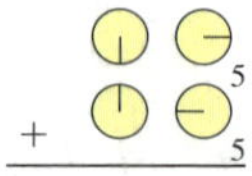

In Exercises 59–66, assume the numerals given are in a base 4 numeration system. In this system, suppose colors are used as numerals, as indicated below.

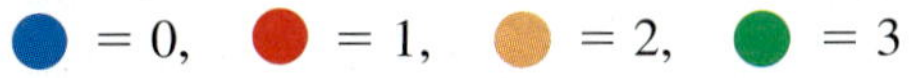

Add the following base 4 numerals. Your answers will contain a variety of the colors indicated.

59. 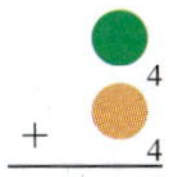

60. 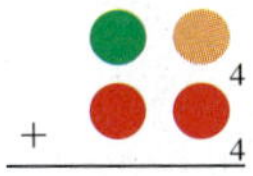

61. 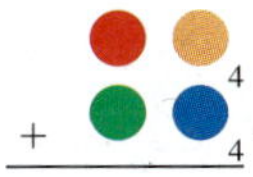

62. 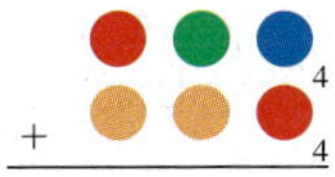

Subtract the following in base 4. Your answer will contain a variety of the colors indicated.

63. 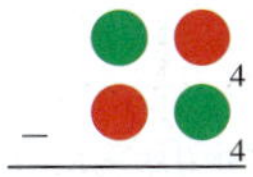

64. 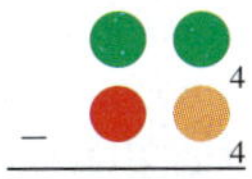

65. 

66. 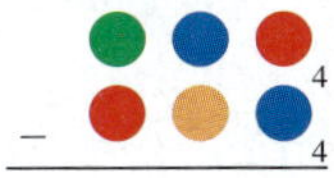

For Exercises 67 and 68, study the pattern in the boxes. The numeral in the bottom row of each box represents the value of each dot in the box directly above it. For example, the following box represents $(3 \times 7^2) + (2 \times 7) + (4 \times 1)$, or the numeral 324_7. This numeral in base 10 is 165.

• • •	• •	• • • •
7^2	7	1

67. Determine the base 5 numeral represented by the dots in the top row of the boxes. Then convert the base 5 numeral to a numeral in base 10.

• •	• • •		• •
5^3	5^2	5	1

68. Fill in the correct amount of dots in the columns above the base values if the numeral represented by the dots is to equal 327 in base 10.

9^2	9	1

CHALLENGE PROBLEMS/GROUP ACTIVITIES

In Exercises 69–72, perform the indicated operation.

69. $\begin{array}{r} FAB_{16} \\ \times \quad 4_{16} \\ \hline \end{array}$

70. $D_{16}\overline{)FACE_{16}}$

71. $20_4\overline{)223_4}$

72. $14_5\overline{)242_5}$

73. Consider the multiplication

$$\begin{array}{r} 462_8 \\ \times\ 35_8 \\ \hline \end{array}$$

a) Multiply the numerals in base 8.

b) Convert 462_8 and 35_8 to base 10.

c) Multiply the base 10 numerals determined in part (b).

d) Convert the answer obtained in base 8 in part (a) to base 10.

e) Are the answers obtained in parts (c) and (d) the same? Why or why not?

RECREATIONAL MATHEMATICS

74. Determine b, by trial and error, if $1304_b = 204$.

75. In a base 4 system, each of the four numerals is represented by one of the following colors:

Determine the value of each color if the following addition is true in base 4.

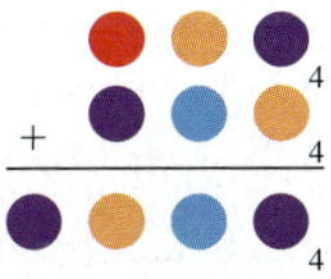

INTERNET/RESEARCH ACTIVITIES

76. Write a report on how computers use the binary (base 2), octal (base 8), and hexadecimal (base 16) numeration systems.

77. One method used by computers to perform subtraction is the "end around carry method." Do research and write a report explaining, with specific examples, how a computer performs subtraction by using the end around carry method.

4.5 EARLY COMPUTATIONAL METHODS

▲ Multiplication can be used to calculate the amount of your paycheck.

Suppose that last week you worked at your job for 27 hours at a rate of $8 per hour. How much money did you make? You may find the answer by multiplying 27 × 8 "by hand."

$$\begin{array}{r} {}^{5}27 \\ \times \;\; 8 \\ \hline 216 \end{array}$$

Although most of us would use this method, it is not the only method for multiplying two numbers together. Early civilizations used various other methods. In this section, we will study three other methods of multiplication.

Duplation and Mediation

The first method of multiplication we will study is *duplation and mediation*. This method, also known as *Russian peasant multiplication,* is still used in parts of Russia today. Duplation refers to doubling one number, and mediation refers to halving one number. The duplation and mediation method is similar to a method used by the ancient Egyptians as described on the Rhind Papyrus. Example 1 illustrates this method of multiplication.

EXAMPLE 1 *Using Duplation and Mediation*

Multiply 21 × 32 using duplation and mediation.

SOLUTION Write 21 and 32 with a dash between the two numbers. Divide the number on the left, 21, by 2, drop the remainder, and place the quotient, 10, under the 21. Double the number on the right, 32, obtaining 64, and place it under the 32. You will then have the following number pairs.

21 — 32
10 — 64

Continue this process, dividing the number in the left-hand column by 2, disregarding the remainder, and doubling the number in the right-hand column, as shown below. When a 1 appears in the left-hand column, stop.

21 — 32
10 — 64
5 — 128
2 — 256
1 — 512

Next, cross out all the even numbers in the left-hand column and the corresponding numbers in the right-hand column.

21 — 32
~~10 — 64~~
5 — 128
~~2 — 256~~
1 — 512

Now add the remaining numbers in the right-hand column to obtain 32 + 128 + 512 = 672. The number obtained, 672, is the product of 21 and 32. Thus, $21 \times 32 = 672$.

Lattice Multiplication

Another method of multiplication is *lattice multiplication*. This method's name comes from the use of a grid, or lattice, when multiplying two numbers. It is also known as the *gelosia* method. The lattice method was brought to Europe by Fibonacci, in the 1200s who likely learned it from the Egyptians, Arabs, or Hindus (see the *Profile in Mathematics* on page 298).

EXAMPLE 2 *Using Lattice Multiplication*

Multiply 312×75 using lattice multiplication.

SOLUTION To multiply 312×75 using lattice multiplication, first construct a rectangle consisting of three columns (one for each digit of 312) and two rows (one for each digit of 75).

Place the digits 3, 1, 2 above the boxes and the digits 7, 5 on the right of the boxes, as shown in Fig. 4.1. Then place a diagonal in each box.

Complete each box by multiplying the number on top of the box by the number to the right of the box (Fig. 4.2). Place the units digit of the product below the diagonal and the tens digit of the product above the diagonal.

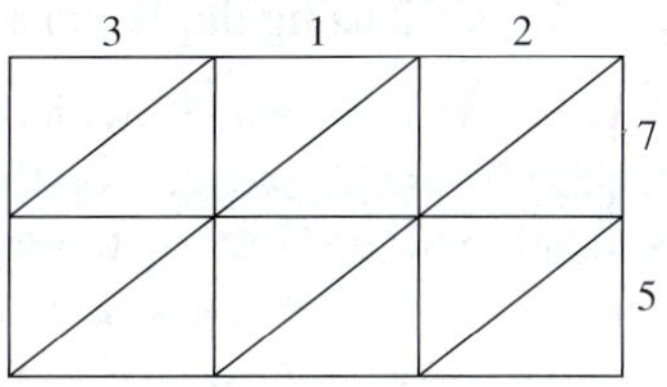

Figure 4.1

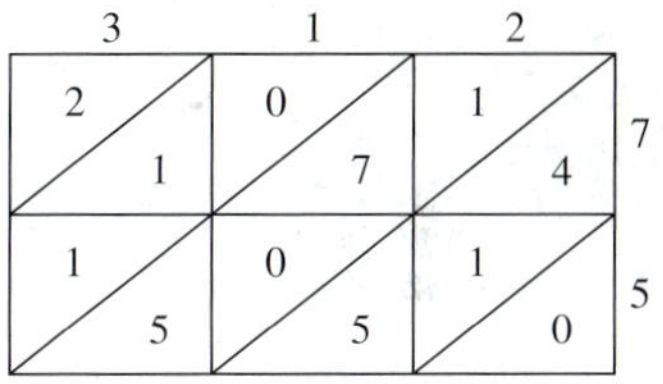

Figure 4.2

Add the numbers along the diagonals, as shown with the blue shaded arrows in Fig. 4.3, starting with the bottom right diagonal. If the sum in a diagonal is 10 or greater, record the units digit below the rectangle and carry the tens digit to the next diagonal to the left.

For example, when adding 4, 1, and 5 (along the second blue diagonal from the right), the sum is 10. Record the 0 below the rectangle and carry the 1 to the next blue diagonal. The sum of 1 + 1 + 7 + 0 + 5 is 14. Record the 4 and carry the 1 to the next blue diagonal. The sum of the numbers in the next blue diagonal is 1 + 0 + 1 + 1 or 3.

The answer is read down the left-hand column and along the bottom, as shown by the purple arrow in Fig. 4.3. Therefore, $312 \times 75 = 23{,}400$.

Napier's Rods

The third method used to multiply numbers was developed from the lattice method by John Napier in the early 1600s and is known as *Napier's rods* or *Napier's bones*. This method is similar to that used in modern computers. Napier developed a system of

MATHEMATICS TODAY

The New Old Math

Sometimes the *new* way to solve a problem is to use an *old* method. Such is the case in some elementary schools with regard to multiplying whole numbers. In addition to teaching the traditional multiplication algorithm, some schools are also teaching the lattice method of multiplication. Lattice multiplication, along with the Hindu–Arabic numerals, was introduced to Europe in the thirteenth century by Fibonacci. (See the *Profile in Mathematics?* on page 298.) Lattice multiplication is recommended by the National Council of Teachers of Mathematics as a way to help students increase their conceptual understanding as they are developing computational ability. This method gained in popularity when it was included in several elementary school textbooks as a way to help students understand how the traditional multiplication algorithm works.

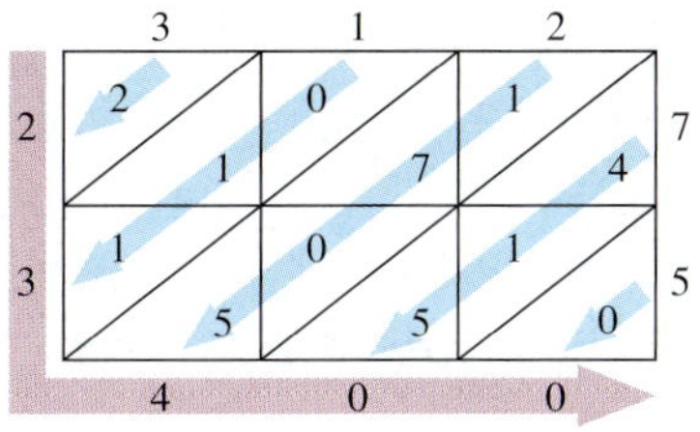

Figure 4.3

PROFILE IN MATHEMATICS

John Napier

During the seventeenth century, the growth of scientific fields such as astronomy required the ability to perform often unwieldy calculations. Scottish mathematician John Napier (1550–1617) made great contributions toward solving the problem of computing these numbers. His inventions include simple calculating machines and a device for performing multiplication and division known as Napier's rods. Napier also developed the theory of logarithms.

separate rods (which were often made out of bones) numbered 0 through 9 and an additional rod for an index, numbered vertically 1 through 9 (Fig. 4.4). Each rod is divided into 10 blocks. Each block below the first block contains a multiple of the number in the first block, with a diagonal separating the digits. The units digits are placed to the right of the diagonals and the tens digits to the left. Example 3 explains how Napier's rods are used to multiply numbers.

INDEX	0	1	2	3	4	5	6	7	8	9
1	0/0	0/1	0/2	0/3	0/4	0/5	0/6	0/7	0/8	0/9
2	0/0	0/2	0/4	0/6	0/8	1/0	1/2	1/4	1/6	1/8
3	0/0	0/3	0/6	0/9	1/2	1/5	1/8	2/1	2/4	2/7
4	0/0	0/4	0/8	1/2	1/6	2/0	2/4	2/8	3/2	3/6
5	0/0	0/5	1/0	1/5	2/0	2/5	3/0	3/5	4/0	4/5
6	0/0	0/6	1/2	1/8	2/4	3/0	3/6	4/2	4/8	5/4
7	0/0	0/7	1/4	2/1	2/8	3/5	4/2	4/9	5/6	6/3
8	0/0	0/8	1/6	2/4	3/2	4/0	4/8	5/6	6/4	7/2
9	0/0	0/9	1/8	2/7	3/6	4/5	5/4	6/3	7/2	8/1

Figure 4.4

EXAMPLE 3 *Using Napier's Rods*

Multiply 8 × 365, using Napier's rods.

SOLUTION To multiply 8 × 365, line up the rods 3, 6, and 5 to the right of the index 8, as shown in Fig. 4.5. Below the 3, 6, and 5 place the blocks that contain the products of 8 × 3, 8 × 6, and 8 × 5, respectively. To obtain the answer, add along the diagonals as in the lattice method.

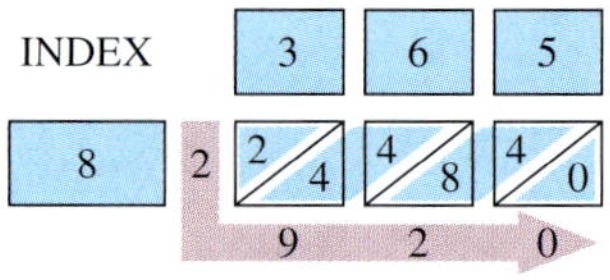

Thus, 8 × 365 = 2920.

INDEX	3	6	5
1	0/3	0/6	0/5
2	0/6	1/2	1/0
3	0/9	1/8	1/5
4	1/2	2/4	2/0
5	1/5	3/0	2/5
6	1/8	3/6	3/0
7	2/1	4/2	3/5
8	2/4	4/8	4/0
9	2/7	5/4	4/5

Figure 4.5

Example 4 illustrates the procedure to follow to multiply numbers containing more than one digit, using Napier's rods.

EXAMPLE 4 *Using Napier's Rods to Multiply Two- and Three-Digit Numbers*

Multiply 48 × 365, using Napier's rods.

SOLUTION 48 × 365 = (40 + 8) × 365

Write (40 + 8) × 365 = (40 × 365) + (8 × 365). To find 40 × 365, determine 4 × 365 and multiply the product by 10. To evaluate 4 × 365, set up

Napier's rods for 3, 6, and 5 with index 4, and then evaluate along the diagonals, as indicated.

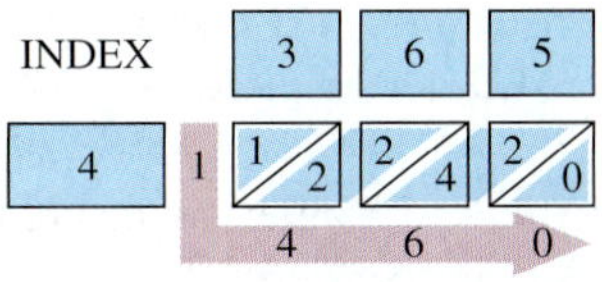

Therefore, $4 \times 365 = 1460$. Then $40 \times 365 = 1460 \times 10 = 14{,}600$.

$$\begin{aligned} 48 \times 365 &= (40 \times 365) + (8 \times 365) \\ &= 14{,}600 + 2920 \\ &= 17{,}520 \end{aligned}$$

$8 \times 365 = 2920$ from Example 3

SECTION 4.5 EXERCISES

CONCEPT/WRITING EXERCISES

1. What are the three early computational methods discussed in this section?

2. a) Explain in your own words how multiplication by duplation and mediation is performed.

   b) Using the procedure given in part (a), multiply 73×21.

3. a) Explain in your own words how lattice multiplication is performed.

   b) Using the procedure given in part (a), multiply 143×26.

4. a) Explain in your own words how multiplication using Napier's rods is performed.

   b) Using the procedure given in part (a), multiply 34×7.

PRACTICE THE SKILLS

In Exercises 5–12, multiply using duplation and mediation.

5. 9×171
6. 39×57
7. 27×53
8. 138×41
9. 35×236
10. 96×53
11. 93×93
12. 49×124

In Exercises 13–20, use lattice multiplication to determine the product.

13. 5×191
14. 6×227
15. 8×469
16. 9×509
17. 75×12
18. 47×259
19. 314×652
20. 634×832

In Exercises 21–28, multiply using Napier's rods.

21. 3×49
22. 4×54
23. 5×63
24. 6×171
25. 5×125
26. 75×125
27. 9×6742
28. 7×3456

PROBLEM SOLVING

In Exercises 29 and 30, we show lattice multiplications. (a) Determine the numbers being multiplied. Explain how you determined your answer. (b) Determine the product.

29.

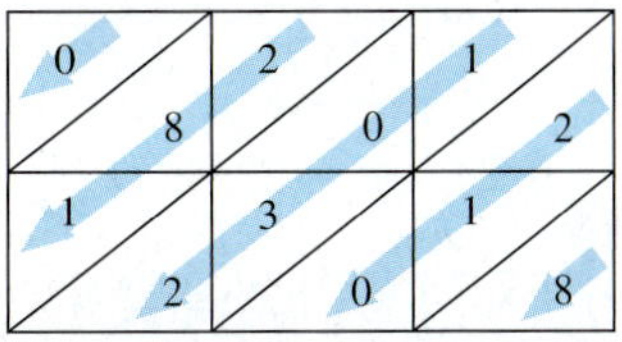

30.

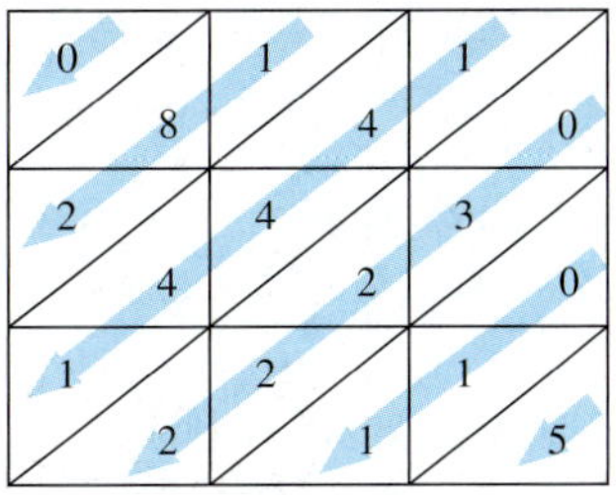

In Exercises 31 and 32, we solve a multiplication problem using Napier's rods. (a) Determine the numbers being multiplied. Each empty box contains a single digit. Explain how you determined your answer. (b) Determine the product.

31.

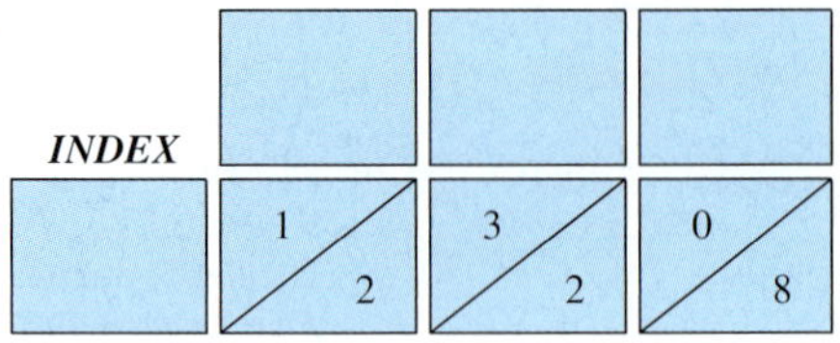

32.

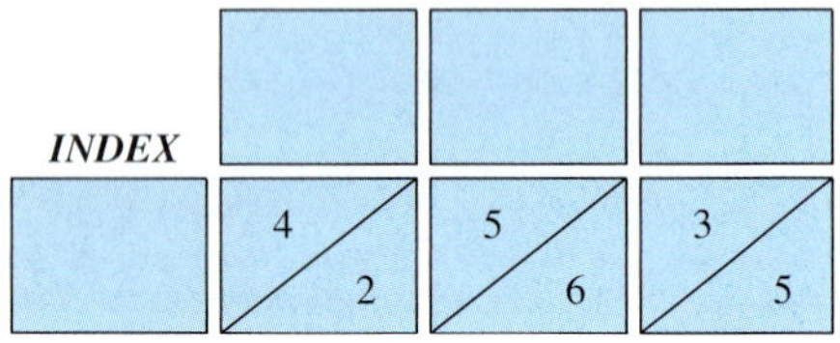

CHALLENGE PROBLEMS/GROUP ACTIVITIES

In Exercises 33 and 34, use the method of duplation and mediation to perform the multiplication. See Section 4.1 for Egyptian and Roman numerals. Write the answer in the numeration system in which the exercise is given.

33. (∩|||) • (∩∩||) **34.** (XXVI) · (LXVII)

In Exercises 35 and 36, (a) use lattice multiplication to perform the multiplication. (Hint: Be sure not to list any number greater than or equal to the base within the box.) Write the answer in the base in which the exercise is given. (b) Multiply the numbers as explained in Section 4.4. If you do not obtain the results obtained in part (a), explain why.

35. $21_3 \times 21_3$ **36.** $24_5 \times 234_5$

RECREATIONAL MATHEMATICS

37. Obtain a clean U.S. $1 bill. On the back side of the bill is a circle containing a pyramid. On the base of the pyramid is a Roman numeral. a) Determine the value of that Roman numeral in our Hindu–Arabic system.

While you have that bill out, let's consider something else. For many people the number 13 is considered an unlucky number. In fact, many hotels do not have a thirteenth floor because many guests refuse to stay on the thirteenth floor of a hotel. Yet, if you look at the back of a $1 bill, you will find

13 steps on the pyramid.
13 letters in the Latin words *Annuit Coeptis.*
13 letters in "E Pluribus Unum."
13 stars above the eagle.
13 plum feathers on each span of the eagle's wing.
13 bars on the eagle's shield.
13 leaves on the olive branch.
13 fruits.
13 arrows.

In addition, the U.S. flag has 13 stripes, there were 13 original colonies, and don't forget the important Thirteenth Amendment that abolished slavery. b) So why, in your opinion, do we as a society fear the number 13? By the way, the official name for the fear of the number 13 is *triskaidekaphobia.*

INTERNET/RESEARCH ACTIVITIES

38. In addition to Napier's rods, John Napier is credited with making other important contributions to mathematics. Write a report on John Napier and his contributions to mathematics.

39. Write a paper explaining why the duplation and mediation method works.

CHAPTER 4 SUMMARY

IMPORTANT FACTS

TYPES OF NUMERATION SYSTEMS

Additive (Egyptian hieroglyphics, Roman)
Multiplicative (traditional Chinese)
Ciphered (Ionic Greek)
Place-value (Babylonian, Mayan, Hindu–Arabic)

EARLY COMPUTATIONAL METHODS

Duplation and mediation
Lattice multiplication
Napier's rods

POSITIONAL VALUES

Hindu–Arabic (base 10) . . . , $(10)^4$, $(10)^3$, $(10)^2$, 10, 1
Base b . . . , b^4, b^3, b^2, b, 1

CHAPTER 4 REVIEW EXERCISES

4.1, 4.2

In Exercises 1–6, assume an additive numeration system in which a = 1, b = 10, c = 100, and d = 1000. Find the value of the numeral.

1. *ddcbba* **2.** *dccbaaaa* **3.** *bcccad*

4. *cbdadaaa* **5.** *ddcccbaaaa* **6.** *ccbaddac*

In Exercises 7–12, assume the same additive numeration system as in Exercises 1–6. Write the numeral in terms of a, b, c, and d.

7. 31 **8.** 125 **9.** 293

10. 2008 **11.** 6851 **12.** 2314

In Exercises 13–18, assume a multiplicative numeration system in which a = 1, b = 2, c = 3, d = 4, e = 5, f = 6, g = 7, h = 8, i = 9, x = 10, y = 100, and z = 1000. Find the value of the numeral.

13. *cxb* **14.** *hxe*

15. *gydxi* **16.** *dzfxh*

17. *ezfydxh* **18.** *fziye*

In Exercises 19–24, assume the same multiplicative numeration system as in Exercises 13–18. Write the Hindu–Arabic numeral in that system.

19. 37 **20.** 157 **21.** 862

22. 3094 **23.** 6004 **24.** 2001

In Exercises 25–36, use the following ciphered numeration system.

Decimal	1	2	3	4	5	6	7	8	9
Units	a	b	c	d	e	f	g	h	i
Tens	j	k	l	m	n	o	p	q	r
Hundreds	s	t	u	v	w	x	y	z	A
Thousands	B	C	D	E	F	G	H	I	J
Ten thousands	K	L	M	N	O	P	Q	R	S

Convert the numeral to a Hindu–Arabic numeral.

25. rc **26.** tc **27.** woh

28. NGzqc **29.** PEvqa **30.** Pwki

Write each of the following in the ciphered numeration system.

31. 31 **32.** 274 **33.** 493

34. 1997 **35.** 53,467 **36.** 75,496

In Exercises 37–42, convert 1776 to a numeral in the indicated numeration system.

37. Egyptian **38.** Roman **39.** Chinese

40. Ionic Greek **41.** Babylonian **42.** Mayan

In Exercises 43–48, convert the numeral to a Hindu–Arabic numeral.

43. 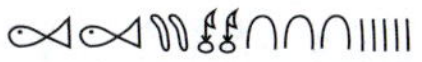

44. 八千二百五十四

45. $\chi\pi\varepsilon$

46. MCMXCI

47. 

48. 

4.3

In Exercises 49–54, convert the numeral to a Hindu–Arabic numeral.

49. 47_8 **50.** 101_2 **51.** 130_4

52. 3425_7 **53.** $A94_{12}$ **54.** 20220_3

In Exercises 55–60, convert 463 to a numeral in the base indicated.

55. base 2 **56.** base 3 **57.** base 4

58. base 8 **59.** base 12 **60.** base 16

4.4

In Exercises 61–66, add in the base indicated.

61. $52_7 + 55_7$ **62.** $10110_2 + 11001_2$ **63.** $9B_{12} + 87_{12}$

64. $2B9_{16} + 456_{16}$

65. $3024_5 + 4023_5$

66. $3407_8 + 7014_8$

In Exercises 67–72, subtract in the base indicated.

67. $4032_7 - 321_7$

68. $1001_2 - 101_2$

69. $A7B_{12} - 95_{12}$

70. $4321_5 - 442_5$

71. $1713_8 - 1243_8$

72. $F64_{16} - 2A3_{16}$

In Exercises 73–78, multiply in the base indicated.

73. $1011_2 \times 101_2$

74. $221_3 \times 22_3$

75. $34_5 \times 21_5$

76. $476_8 \times 23_8$

77. $126_{12} \times 47_{12}$

78. $1A3_{16} \times 12_{16}$

In Exercises 79–84, divide in the base indicated.

79. $2_3\overline{)120_3}$

80. $2_4\overline{)320_4}$

81. $3_5\overline{)130_5}$

82. $4_6\overline{)3020_6}$

83. $3_6\overline{)2034_6}$

84. $6_8\overline{)5072_8}$

4.5

85. Multiply 142 × 24, using the duplation and mediation method.

86. Multiply 142 × 24, using lattice multiplication.

87. Multiply 142 × 24, using Napier's rods.

CHAPTER 4 TEST

1. Explain the difference between a numeral and a number.

In Exercises 2–7, convert the numeral to a Hindu–Arabic numeral.

2. MMCDLXXXIV

3. 

4. 八千零九十

5. 

6. 

7. ʹβψμε

In Exercises 8–12, convert the base 10 numbered to a numeral in the numeration system indicated.

8. 2124 to Egyptian

9. 2476 to Ionic Greek

10. 1434 to Mayan

11. 1596 to Babylonian

12. 3706 to Roman

In Exercises 13–16, describe briefly each of the systems of numeration. Explain how each type of numeration system is used to represent numbers.

13. Additive system

14. Multiplicative system

15. Ciphered system

16. Place-value system

In Exercises 17–20, convert the numeral to a Hindu–Arabic numeral

17. 23_4

18. 403_5

19. 101101_2

20. $3A7_{12}$

In Exercises 21–24, convert each of the following to a numeral in the base indicated.

21. 36 to base 2

22. 93 to base 8

23. 2356 to base 12

24. 2938 to base 16

In Exercises 25–28, perform the indicated operations.

25. $1101_2 + 1011_2$

26. $324_6 - 142_6$

27. $45_6 \times 23_6$

28. $3_5\overline{)1210_5}$

29. Multiply 35 × 28, using duplation and mediation.

30. Multiply 43 × 196, using lattice multiplication.

GROUP PROJECTS

U.S. POSTAL SERVICE BAR CODES

Wherever we look nowadays, we see bar codes. We find them on items we buy at grocery stores and department stores and on many pieces of mail we receive. There are various types of bar codes, but each can be considered a type of numeration system. Although bar codes may vary in design, most are made up of a series of long and short bars. (New bar codes now being developed use a variety of shapes.) In this group project, we explain how postal bar codes are used.

The U.S. Postal Service introduced a bar coding system for zip codes in 1976. The system became known as Postnet (*post*al *n*umeric *e*ncoding *t*echnique), and it has been refined over the years. Our basic zip code consists of five digits. The post office would like us to use the basic zip code followed by a hyphen and four additional digits. The post office refers to this nine-digit zip code as "zip + 4."

The Postnet bar code uses a series of long and short bars. A bar code may contain either 52 or 62 bars. The code designates the location to which the letter is being sent. The following bar code, with 52 bars, is for an address in Pittsburgh, Pennsylvania.

15250-7406 (Pittsburgh, PA)

In bar codes, each short bar represents 0 and each long bar represents 1. Each code starts and ends with a long bar that is *not* used in determining the zip + 4. If the code contains 52 bars, the code represents the zip + 4 and an extra digit referred to as a check digit. If the code contains 62 bars, it contains the zip + 4, the last two digits of the address number, and a check digit. If the code contains 52 bars, the sum of the zip + 4 and the check digit must equal a number that is divisible by 10. If the code contains 62 bars, the sum of the zip + 4, the last two digits of the address number, and the check digit must equal a number that is divisible by 10. The check digit is added to make each sum divisible by 10.

In a postal bar code, each of the digits 0 through 9 is represented by a series of five digits containing zeros and ones:

11000 (0) 00011 (1) 00101 (2) 00110 (3) 01001 (4)
01010 (5) 01100 (6) 10001 (7) 10010 (8) 10100 (9)

Consider the postal code from Pittsburgh given earlier. If you disregard the bar on the left, the next five bars are ıııll. Since each small bar represents a 0 and each large bar represents a 1, these five bars can be represented as 00011. From the chart, we see that this represents the number 1. The first five bars (after the bar on the far left has been excluded) tell the region of the country in which the address is located on the map shown below. Notice that Pennsylvania is located, along with New York and Delaware, in the region marked 1 on the map.

National zip code areas

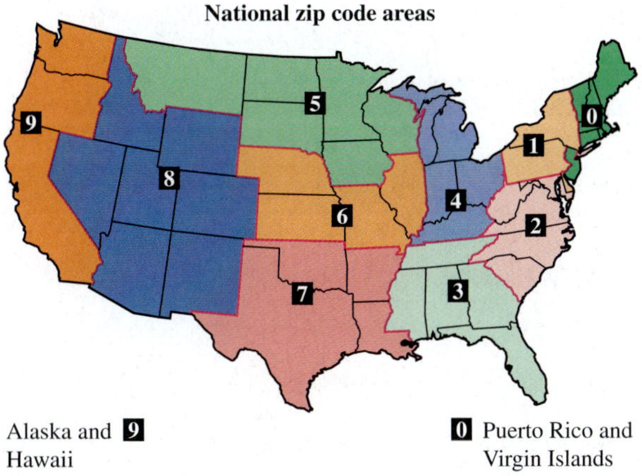

The second five lines in the bar code, ılılı, represent 01010 and have a value of 5. The other digits in the zip + 4 are determined in a similar manner. This code has 52 bars. The 45 bars, after the first bar, give the zip + 4. If you add the digits in the zip + 4, you get $1 + 5 + 2 + 5 + 0 + 7 + 4 + 0 + 6 = 30$. Since 30 is divisible by 10, the five bars to the left of the bar on the very right should be 0. Note that llııı is represented as 11000 and has a value of 0. If, for example, the sum of the nine digits in the zip + 4 were 36, then the last five digits would need to represent the number 4, to make the sum of the digits divisible by 10. The five bars to the left of the last bar on the right are always used as a check.

Now let's work some problems.

a) For the Postnet code

determine the zip + 4 and the check digit. Then check by adding the zip + 4 and the check digit. Is the sum divisible by 10?

b) For each of the following Postnet codes, determine the zip + 4, the last two numbers of the address number (if applicable), and the check digit.

i)

ii)

c) Construct the Postnet code of long and short bars for each of the numbers. The numbers represent the zip + 4 and the last two digits of the address number. Do not forget the check digit.

i) 32226-8600-34 **ii)** 20794-1063-50

d) Construct the 52-bar Postnet code for your college's zip + 4. Don't forget to include the check digit.

CHAPTER 5

Number Theory and the Real Number System

▲ Number theory plays an important role in cell phone technology.

WHAT YOU WILL LEARN

- An introduction to number theory
- Prime numbers
- Integers, rational numbers, irrational numbers, and real numbers
- Properties of real numbers
- Rules of exponents and scientific notation
- Arithmetic and geometric sequences
- The Fibonacci sequence

WHY IT IS IMPORTANT

Every time we use a computer, make a telephone call, or watch television we use a device that relies on numbers to operate. In addition to playing many roles in our everyday lives, numbers are also used to describe the natural world, communicate information, and model problems facing scientists and researchers. In this chapter, we will see how number theory, the study of numbers and their properties, makes all these roles possible.

5.1 NUMBER THEORY

▲ There are many ways to display 48 pieces of chocolate.

Suppose you decide to start your own chocolate candy company. You would like to sell boxes containing 48 pieces of candy. You must decide how to arrange the pieces within the box. You could have six rows with eight pieces in each row, or four rows with twelve pieces in each row, or three layers with four rows each with four pieces in each row. Each of these possibilities involves writing the number 48 as the product of two or more smaller numbers. In this section, we will discuss many other similar problems related to the writing of numbers as the product of smaller numbers.

This chapter introduces ***number theory***, the study of numbers and their properties. The numbers we use to count are called the ***counting numbers*** or ***natural numbers***. Because we begin counting with the number 1, the set of natural numbers begins with 1. The set of natural numbers is frequently denoted by N:

$$N = \{1, 2, 3, 4, 5, \ldots\}$$

Any natural number can be expressed as a product of two or more natural numbers. For example, $8 = 2 \times 4$, $16 = 4 \times 4$, and $19 = 1 \times 19$. The natural numbers that are multiplied together are called factors of the product. For example,

$$\underset{\text{Factors}}{\underbrace{\overset{\uparrow}{2} \times \overset{\uparrow}{4}}} = 8$$

A natural number may have many factors. For example, what pairs of numbers have a product of 18?

$$1 \cdot 18 = 18$$
$$2 \cdot 9 = 18$$
$$3 \cdot 6 = 18$$

The numbers 1, 2, 3, 6, 9, and 18 are all factors of 18. Each of these numbers divides 18 without a remainder.

If a and b are natural numbers, we say that a is a ***divisor*** of b or a ***divides*** b, symbolized $a \mid b$, if the quotient of b divided by a has a remainder of 0. If a divides b, then b is ***divisible*** by a. For example, 4 divides 12, symbolized $4 \mid 12$, since the quotient of 12 divided by 4 has a remainder of 0. Note that 12 is divisible by 4. The notation $7 \nmid 12$ means that 7 does not divide 12. Note that every factor of a natural number is also a divisor of the natural number. *Caution:* Do not confuse the symbols $a \mid b$ and a / b; $a \mid b$ means "a divides b" and a / b means "a divided by b" $(a \div b)$. The symbols a / b and $a \div b$ indicate that the operation of division is to be performed, and b may or may not be a divisor of a.

Prime and Composite Numbers

Every natural number greater than 1 can be classified as either a prime number or a composite number.

PROFILE IN MATHEMATICS

Eratosthenes of Cyrene

Eratosthenes of Cyrene (275–195 B.C.) was born in northern Africa near the present-day city of Shahhat, Libya. Eratosthenes is best known for being the first to estimate accurately the diameter of Earth. He is also credited for developing a method of finding prime numbers known as the **sieve of Eratosthenes**. Although he is most known for his work in mathematics, Eratosthenes also was influential in the fields of history, geography, and astronomy. In addition, Eratosthenes served for many years as the director of the famous library in Alexandria, Egypt. Although Eratosthenes was a highly regarded scholar throughout the ancient world, only fragments of his writing remain today. Eratosthenes was near 80 years old when, after going blind, he died from voluntary starvation.

A **prime number** is a natural number greater than 1 that has exactly two factors (or divisors), itself and 1.

The number 5 is a prime number because it is divisible only by the factors 1 and 5. The first eight prime numbers are 2, 3, 5, 7, 11, 13, 17, and 19. The number 2 is the only even prime number. All other even numbers have at least three divisors: 1, 2, and the number itself.

A **composite number** is a natural number that is divisible by a number other than itself and 1.

Any natural number greater than 1 that is not prime is composite. The first eight composite numbers are 4, 6, 8, 9, 10, 12, 14, and 15.

The number 1 is neither prime nor composite; it is called a ***unit***. The number 38 has at least three divisors, 1, 2, and 38, and hence is a composite number. In contrast, the number 23 is a prime number since its only divisors are 1 and 23.

More than 2000 years ago, the ancient Greeks developed a technique for determining which numbers are prime numbers and which are not. This technique is known as the ***sieve of Eratosthenes***, for the Greek mathematician Eratosthenes of Cyrene who first used it.

To find the prime numbers less than or equal to any natural number, say, 50, using this method, list the first 50 counting numbers (Fig. 5.1). Cross out 1 since it is not a prime number. Circle 2, the first prime number. Then cross out all the multiples of 2: 4, 6, 8, . . . , 50. Circle the next prime number, 3. Cross out all multiples of 3 that are not already crossed out. Continue this process of crossing out multiples of prime numbers until you reach the prime number p, such that $p \cdot p$, or p^2, is greater than the last number listed, in this case 50. Therefore, we next circle 5 and cross out its multiples. Then circle 7 and cross out its multiples. The next prime number is 11, and $11 \cdot 11$, or 121, is greater than 50, so you are done. At this point, circle all the remaining numbers to obtain the prime numbers less than or equal to 50. The prime numbers less than or equal to 50 are 2, 3, 5, 7, 11, 13, 17, 19, 23, 29, 31, 37, 41, 43, and 47.

Figure 5.1

Now we turn our attention to composite numbers and their factors. The rules of divisibility given in the chart on page 226 are helpful in finding divisors (or factors) of composite numbers.

Rules of Divisibility

Divisible by	Test	Example
2	The number is even.	924 is divisible by 2 since 924 is even.
3	The sum of the digits of the number is divisible by 3.	924 is divisible by 3 since the sum of the digits, $9 + 2 + 4 = 15$, and 15 is divisible by 3.
4	The number formed by the last two digits of the number is divisible by 4.	924 is divisible by 4 since the number formed by the last two digits, 24, is divisible by 4.
5	The number ends in 0 or 5.	265 is divisible by 5 since the number ends in 5.
6	The number is divisible by both 2 and 3.	924 is divisible by 6 since it is divisible by both 2 and 3.
8	The number formed by the last three digits of the number is divisible by 8.	5824 is divisible by 8 since the number formed by the last three digits, 824, is divisible by 8.
9	The sum of the digits of the number is divisible by 9.	837 is divisible by 9 since the sum of the digits, 18, is divisible by 9.
10	The number ends in 0.	290 is divisible by 10 since the number ends in 0.

The test for divisibility by 6 is a particular case of the general statement that the product of two prime divisors of a number is a divisor of the number. Thus, for example, if both 3 and 7 divide a number, then 21 will also divide the number.

Note that the chart does not list rules for the divisibility for the number 7. The rule for 7 is given in Exercise 92 at the end of this section. You may find after using the rule that the easiest way to check divisibility by 7 is simply to perform the division.

EXAMPLE 1 *Using the Divisibility Rules*

Determine whether 384,540 is divisible by

a) 2 b) 3 c) 4 d) 5 e) 6 f) 8 g) 9 h) 10

SOLUTION

a) Since 384,540 is even, it is divisible by 2.

b) The sum of the digits of 384,540 is $3 + 8 + 4 + 5 + 4 + 0 = 24$. Since 24 is divisible by 3, the number 384,540 is divisible by 3.

c) The number formed by the last two digits is 40. Since 40 is divisible by 4, the number 384,540 is divisible by 4.

d) Since 384,540 has 0 as the last digit, the number 384,540 is divisible by 5.

e) Since 384,540 is divisible by both 2 and 3, the number 384,540 is divisible by 6.

f) The number formed by the last three digits is 540. Since 540 is not divisible by 8, the number 384,540 is not divisible by 8.

g) The sum of the digits of 384,540 is 24. Since 24 is not divisible by 9, the number 384,540 is not divisible by 9.

h) Since 384,540 has 0 as the last digit, the number 384,540 is divisible by 10. •

Every composite number can be expressed as a product of prime numbers. The process of breaking a given composite number down into a product of prime numbers is called ***prime factorization***. The prime factorization of 18 is $3 \times 3 \times 2$. No other natural number listed as a product of primes will have the same prime factorization as 18. The *fundamental theorem of arithmetic* states this concept formally. (A ***theorem*** is a statement or proposition that can be proven true.)

THE FUNDAMENTAL THEOREM OF ARITHMETIC
Every composite number can be expressed as a *unique* product of prime numbers.

In writing the prime factorization of a number, the order of the factors is immaterial. For example, we may write the prime factors of 18 as $3 \times 3 \times 2$ or $2 \times 3 \times 3$ or $3 \times 2 \times 3$.

A number of techniques can be used to find the prime factorization of a number. Two methods are illustrated.

Method 1: Branching

To find the prime factorization of a number, select any two numbers whose product is the number to be factored. If the factors are not prime numbers, continue factoring each composite number until all factors are prime.

EXAMPLE 2 *Prime Factorization by Branching*

Write 2100 as a product of primes.

SOLUTION Select any two numbers whose product is 2100. Among the many choices, two possibilities are $21 \cdot 100$ and $30 \cdot 70$. Let us consider $21 \cdot 100$. Since neither 21 nor 100 are prime numbers, find any two numbers whose product is 21 and any two numbers whose product is 100. Continue branching as shown in Fig. 5.2 until the numbers in the last row are all prime numbers. To determine the answer, write the product of all the prime factors. The branching diagram is sometimes called a *factor tree.*

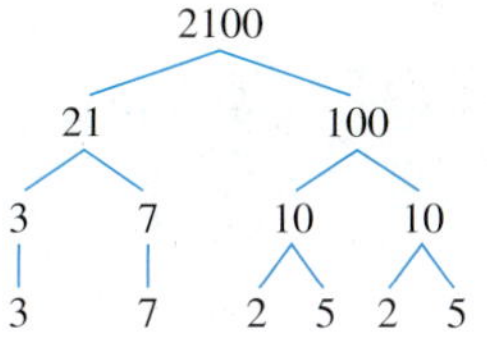

Figure 5.2

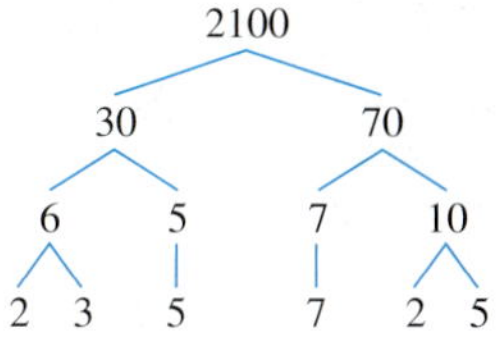

Figure 5.3

We see that the numbers in the last row of factors in Fig. 5.2 are all prime numbers. Thus, the prime factorization of 2100 is $3 \cdot 7 \cdot 2 \cdot 5 \cdot 2 \cdot 5 = 2 \cdot 2 \cdot 3 \cdot 5 \cdot 5 \cdot 7 = 2^2 \cdot 3 \cdot 5^2 \cdot 7$. Note from Fig. 5.3 that had we chosen 30 and 70 as the first pair of factors, the prime factorization would still be $2^2 \cdot 3 \cdot 5^2 \cdot 7$. •

DID YOU KNOW?

The Long-Lost Factoring Machine

In 1989, Jeffrey Shallit of the University of Waterloo came across an article in a 1920 French journal regarding a machine that was built in 1914 for factoring numbers. Eugene Oliver Carissan, an amateur mathematician who had invented the machine, wrote the article. After reading the article, Shallit wondered whatever became of the factoring machine.

After considerable searching, Shallit found the machine in a drawer of an astronomical observatory in Floirac, France. The machine was in good condition, and it still worked. By rotating the machine by a hand crank at two revolutions per minute, an operator could process 35 to 40 numbers per second. Carissan needed just 10 minutes to prove that 708,158,977 is a prime number, an amazing feat in precomputer times. The machine is now housed at the Conservatoire National des Arts et Métiers in Paris.

Method 2: Division

To obtain the prime factorization of a number by this method, divide the given number by the smallest prime number by which it is divisible. Place the quotient under the given number. Then divide the quotient by the smallest prime number by which it is divisible and again record the quotient. Repeat this process until the quotient is a prime number. The prime factorization is the product of all the prime divisors and the prime (or last) quotient. This procedure is illustrated in Example 3.

EXAMPLE 3 *Prime Factorization by Division*

Write 2100 as a product of prime numbers.

SOLUTION Because 2100 is an even number, the smallest prime number that divides it is 2. Divide 2100 by 2. Place the quotient, 1050, below the 2100. Repeat this process of dividing each quotient by the smallest prime number that divides it.

2	2100
2	1050
3	525
5	175
5	35
	7

The final quotient, 7, is a prime number, so we stop. The prime factorization of 2100 is

$$2 \cdot 2 \cdot 3 \cdot 5 \cdot 5 \cdot 7 = 2^2 \cdot 3 \cdot 5^2 \cdot 7$$

Note that despite the different methods used in Examples 2 and 3, the answer is the same.

PROFILE IN MATHEMATICS

Srinivasa Ramanujan

One of the most interesting mathematicians of modern times is Srinivasa Ramanujan (1887–1920). Born to an impoverished middle-class family in India, he virtually taught himself higher mathematics. He went to England to study with the number theorist G. H. Hardy. Hardy tells the story of a taxicab ride he took to visit Ramanujan. The cab had the license plate number 1729, and he challenged the young Indian to find anything interesting in that. Without hesitating, Ramanujan pointed out that it was the smallest positive integer that could be represented in two different ways as the sum of two cubes: $1^3 + 12^3$ and $9^3 + 10^3$.

Greatest Common Divisor

The discussion in Section 5.3 of how to reduce fractions makes use of the greatest common divisor (GCD). We will now discuss how to determine the GCD of a set of numbers. One technique of finding the GCD is to use prime factorization.

The **greatest common divisor (GCD)** of a set of natural numbers is the largest natural number that divides (without remainder) every number in that set.

What is the GCD of 12 and 18? One way to determine it is to list the divisors (or factors) of 12 and 18:

Divisors of 12 {**1, 2, 3**, 4, **6**, 12}
Divisors of 18 {**1, 2, 3, 6**, 9, 18}

The common divisors are 1, 2, 3, and 6. Therefore, the greatest common divisor is 6.

If the numbers are large, this method of finding the GCD is not practical. The GCD can be found more efficiently by using prime factorization.

TO DETERMINE THE GREATEST COMMON DIVISOR OF TWO OR MORE NUMBERS

1. Determine the prime factorization of each number.
2. List each prime factor with the smallest exponent that appears in each of the prime factorizations.
3. Determine the product of the factors found in step 2.

Example 4 illustrates this procedure.

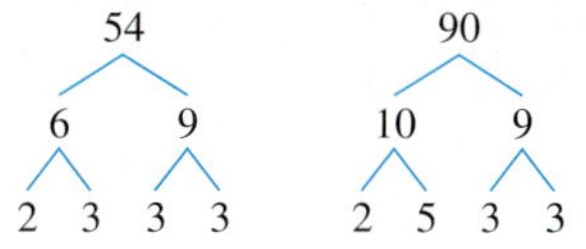

Figure 5.4

EXAMPLE 4 *Using Prime Factorization to Determine the GCD*

Determine the GCD of 54 and 90.

SOLUTION The branching method of finding the prime factors of 54 and 90 is illustrated in Fig. 5.4.

a) The prime factorization of 54 is $2 \cdot 3^3$, and the prime factorization of 90 is $2 \cdot 3^2 \cdot 5$.

b) The prime factors with the smallest exponents that appear in each of the factorizations of 54 and 90 are 2 and 3^2.

c) The product of the factors found in step 2 is $2 \cdot 3^2 = 2 \cdot 9 = 18$. The GCD of 54 and 90 is 18. Eighteen is the largest natural number that divides both 54 and 90.

EXAMPLE 5 *Determining the GCD*

Determine the GCD of 315 and 450.

SOLUTION

a) The prime factorization of 315 is $3^2 \cdot 5 \cdot 7$, and the prime factorization of 450 is $2 \cdot 3^2 \cdot 5^2$. You should verify these answers using either the branching method or the division method.

b) The prime factors with the smallest exponents that appear in each of the factorizations of 315 and 450 are 3^2 and 5.

c) The product of the factors found in step 2 is $3^2 \cdot 5 = 9 \cdot 5 = 45$. The greatest common divisor of 315 and 450 is 45. It is the largest natural number that will evenly divide both 315 and 450.

DID YOU KNOW?

Friendly Numbers

The ancient Greeks often thought of numbers as having human qualities. For example, the numbers 220 and 284 were considered "friendly" or "amicable" numbers because each number was the sum of the other number's proper factors. (A proper factor is any factor of a number other than the number itself.) If you sum all the proper factors of 284 (1 + 2 + 4 + 71 + 142), you get the number 220, and if you sum all the proper factors of 220 (1 + 2 + 4 + 5 + 10 + 11 + 20 + 22 + 44 + 55 + 110), you get 284.

Least Common Multiple

To perform addition and subtraction of fractions (Section 5.3), we use the least common multiple (LCM). We will now discuss how to determine the LCM of a set of numbers.

The **least common multiple (LCM)** of a set of natural numbers is the smallest natural number that is divisible (without remainder) by each element of the set.

DID YOU KNOW?

Safer Secrets

Prime numbers have for many years played an essential role in protecting the databases of government and business. From 1977 through 2000, data were *encrypted* to allow access only to legitimate users. The key to entering the database was a large composite number more than 300 digits long. This number, n, was the product of two very large prime numbers, $p \cdot q$. The number n was publicly available, so information could be entered into the database. The numbers p and q, however, were given only to those users who had the right to access the database.

By 1997, realizing a better protection system was needed, the National Institute for Standardization and Technology began a global contest to choose a better protection system. In October 2000, the Rijndael Block Cipher, developed by two Belgian cryptographers, was selected as the winner. The new system involves various areas of mathematics, including many that are discussed in this book such as permutations, modular arithmetic, polynomials, matrices, and group theory.

What is the least common multiple of 12 and 18? One way to determine the LCM is to list the multiples of each number:

Multiples of 12 {12, 24, **36**, 48, 60, **72**, 84, 96, **108**, 120, 132, **144**, . . . }
Multiples of 18 {18, **36**, 54, **72**, 90, **108**, 126, **144**, 162, . . . }

Some common multiples of 12 and 18 are 36, 72, 108, and 144. The least common multiple, 36, is the smallest number that is divisible by both 12 and 18. Usually, the most efficient method of finding the LCM is to use prime factorization.

TO DETERMINE THE LEAST COMMON MULTIPLE OF TWO OR MORE NUMBERS

1. Determine the prime factorization of each number.
2. List each prime factor with the greatest exponent that appears in any of the prime factorizations.
3. Determine the product of the factors found in step 2.

Example 6 illustrates this procedure.

EXAMPLE 6 *Using Prime Factorization to Determine the LCM*

Determine the LCM of 54 and 90.

SOLUTION

a) Determine the prime factors of each number. In Example 4, we determined that

$$54 = 2 \cdot 3^3 \qquad \text{and} \qquad 90 = 2 \cdot 3^2 \cdot 5$$

b) List each prime factor with the greatest exponent that appears in either of the prime factorizations: 2, 3^3, 5.

c) Determine the product of the factors found in step 2:

$$2 \cdot 3^3 \cdot 5 = 2 \cdot 27 \cdot 5 = 270$$

Thus, 270 is the LCM of 54 and 90. It is the smallest natural number that is divisible by both 54 and 90.

EXAMPLE 7 *Determining the LCM*

Determine the LCM of 315 and 450.

SOLUTION

a) Determine the prime factorization of each number. In Example 5, we determined that

$$315 = 3^2 \cdot 5 \cdot 7 \quad \text{and} \quad 450 = 2 \cdot 3^2 \cdot 5^2.$$

b) List each prime factor with the greatest exponent that appears in either of the prime factorizations: 2, 3^2, 5^2, 7.

c) Determine the product of the factors from step (b)

$$2 \cdot 3^2 \cdot 5^2 \cdot 7 = 2 \cdot 9 \cdot 25 \cdot 7 = 3150$$

Thus, 3150 is the least common multiple of 315 and 450. It is the smallest natural number that is evenly divisible by both 315 and 450.

MATHEMATICS TODAY

GIMPS

Curtis Cooper and Stephen Boone

The Great Internet Mersenne Prime Search (GIMPS) was started by George Woltman to provide free software and access to a large database to coordinate the efforts of thousands of people interested in seeking new Mersenne prime numbers. Scott Kurowski provided the technology and services needed to make it simple for anyone to join the project via the Internet. GIMPS allows participants to use their computers to search for prime numbers while their computers are idle. A computer communicates, via the specialized software, with the database. The largest prime number found so far was found by Curtis Cooper and Stephen Boone on September 4, 2006 in conjunction with GIMPS. The number, $2^{32,582,657} - 1$, is 9,808,358 digits long, is the 44th Mersenne prime number found and is the 10th discovered by GIMPS participants.

The Search For Larger Prime Numbers

More than 2000 years ago, the Greek mathematician Euclid proved that there is no largest prime number. Mathematicians, however, continue to strive to find larger and larger prime numbers.

Marin Mersenne (1588–1648), a seventeenth-century monk, found that numbers of the form $2^n - 1$ are often prime numbers when n is a prime number. For example,

$$2^2 - 1 = 4 - 1 = 3 \qquad 2^3 - 1 = 8 - 1 = 7$$
$$2^5 - 1 = 32 - 1 = 31 \qquad 2^7 - 1 = 128 - 1 = 127$$

Numbers of the form $2^n - 1$ that are prime are referred to as ***Mersenne primes***. The first 10 Mersenne primes occur when $n = 2, 3, 5, 7, 13, 17, 19, 31, 61, 89$. The first time the expression $2^n - 1$ does not generate a prime number, for prime number n, is when n is 11. The number $2^{11} - 1$ is a composite number (see Exercise 90).

Scientists frequently use Mersenne primes in their search for larger and larger primes. The largest prime number found to date was discovered on September 4, 2006, by Curtis Cooper and Stephen Boone, professors at Central Missouri State University. Cooper and Boone had worked in conjunction with the Great Internet Mersenne Prime Search (GIMPS; see Mathematics Today at left) to use the university's computers during down time to conduct the search. The number is the Mersenne prime $2^{32,582,657} - 1$. This record prime number is the 44th known Mersenne prime, and when written out it is 9,808,358 digits long, which is more than 12 miles long if written using a standard 12-point font.

More About Prime Numbers

Another mathematician who studied prime numbers was Pierre de Fermat (1601–1665). A lawyer by profession, Fermat became interested in mathematics as a hobby. He became one of the finest mathematicians of the seventeenth century. Fermat conjectured that each number of the form $2^{2^n} + 1$, now referred to as a ***Fermat number***, was prime for each natural number n. Recall that a *conjecture* is a supposition that has not been proved nor disproved. In 1732, Leonhard Euler proved that for $n = 5$, $2^{32} + 1$ was a composite number, thus disproving Fermat's conjecture.

Since Euler's time, mathematicians have only been able to evaluate the sixth, seventh, eighth, ninth, tenth, and eleventh Fermat numbers to determine whether they are prime or composite. Each of these numbers has been shown to be composite. The eleventh Fermat number was factored by Richard Brent and François Morain in 1988. The sheer magnitude of the numbers involved makes it difficult to test these numbers, even with supercomputers.

In 1742, Christian Goldbach conjectured in a letter to Euler that every even number greater than or equal to 4 can be represented as the sum of two (not necessarily distinct) prime numbers (for example, $4 = 2 + 2$, $6 = 3 + 3$, $8 = 3 + 5$, $10 = 5 + 5$, $12 = 5 + 7$). This conjecture became known as ***Goldbach's conjecture***, and it remains unproven to this day. The ***twin prime conjecture*** is another famous long-standing conjecture. ***Twin primes*** are primes of the form p and $p + 2$ (for example, 3 and 5, 5 and 7, 11 and 13). This conjecture states that there are an infinite number of pairs of twin primes. At the time of this writing, the largest twin primes are of the form $16,869,987,339,975 \cdot 2^{171,960} \pm 1$, which was found by a group of Hungarian mathematicians in September 2005.

SECTION 5.1 EXERCISES

CONCEPT/WRITING EXERCISES

1. What is number theory?

2. What does "a and b are factors of c" mean?

3. a) What does "a divides b" mean?

   b) What does "a is divisible by b" mean?

4. What is a prime number?

5. What is a composite number?

6. What does the fundamental theorem of arithmetic state?

7. a) What is the least common multiple (LCM) of a set of natural numbers?

   b) In your own words, explain how to find the LCM of a set of natural numbers by using prime factorization.

   c) Find the LCM of 16 and 40 by using the procedure given in part (b).

8. a) What is the greatest common divisor (GCD) of a set of natural numbers?

   b) In your own words, explain how to find the GCD of a set of natural numbers by using prime factorization.

   c) Find the GCD of 16 and 40 by using the procedure given in part (b).

9. What are Mersenne primes?

10. What is a conjecture?

11. What is Goldbach's conjecture?

12. What are twin primes?

PRACTICE THE SKILLS

13. Use the sieve of Eratosthenes to find the prime numbers up to 100.

14. Use the sieve of Eratosthenes to find the prime numbers up to 150.

In Exercises 15–26, determine whether the statement is true or false. Modify each false statement to make it a true statement.

15. 7 is a factor of 28.

16. $7 \mid 24$.

17. 13 is a multiple of 26.

18. 6 is a divisor of 18.

19. 8 is divisible by 56.

20. 15 is a factor of 45.

21. If a number is not divisible by 5, then it is not divisible by 10.

22. If a number is not divisible by 10, then it is not divisible by 5.

23. If a number is divisible by 3, then every digit of the number is divisible by 3.

24. If every digit of a number is divisible by 3, then the number itself is divisible by 3.

25. If a number is divisible by 2 and 3, then the number is divisible by 6.

26. If a number is divisible by 3 and 5, then the number is divisible by 15.

In Exercises 27–32, determine whether the number is divisible by each of the following numbers: 2, 3, 4, 5, 6, 8, 9, and 10.

27. 11,115

28. 33,813

29. 474,138

30. 170,820

31. 1,882,320

32. 3,941,221

33. Determine a number that is divisible by 2, 3, 4, 5, and 6.

34. Determine a number that is divisible by 3, 4, 5, 9, and 10.

In Exercises 35–46, find the prime factorization of the number.

35. 48

36. 58

37. 168

38. 315

39. 332

40. 399

41. 513

42. 663

43. 1336

44. 1313

45. 2001

46. 3190

In Exercises 47–56, find (a) the greatest common divisor (GCD) and (b) the least common multiple (LCM).

47. 6 and 21

48. 15 and 24

49. 20 and 35

50. 32 and 224

51. 40 and 900

52. 120 and 240

53. 96 and 212

54. 240 and 285

55. 24, 48, and 128

56. 18, 78, and 198

PROBLEM SOLVING

57. Find the next two sets of twin primes that follow the set 11, 13.

58. The primes 2 and 3 are consecutive natural numbers. Is there another pair of consecutive natural numbers both of which are prime? Explain.

59. For each pair of numbers, determine whether the numbers are relatively prime. Write yes or no as your answer.

a) 10, 21

b) 22, 26

c) 27, 28

d) 85, 119

60. Find the first five Mersenne prime numbers.

61. Find the first three Fermat numbers and determine whether they are prime or composite.

62. Show that Goldbach's conjecture is true for the even numbers 4 through 20.

63. ***Setting Up Chairs*** Jerrett Dumouchel is setting up chairs for his school band concert. He needs to put 120 chairs on the gymnasium floor in rows of equal size, and there must be at least 2 chairs in each row. List the number of rows and the number of chairs in each row that are possible.

64. ***U.S. Senate Committees*** The U.S. Senate consists of 100 members. Senate committees are to be formed so that each of the committees contains the same number of senators and each senator is a member of exactly one committee. The committees are to have more than 2 members but fewer than 50 members. There are various ways that these committees can be formed.

a) What size committees are possible?

b) How many committees are there for each size?

65. ***Taking Medicine*** Sue Chen-Harper takes the medicine bisphosphonate once every 30 days and the medicine pegaspargase once every 14 days. If Sue took both medicines on June 1, how many days would it be before she has to take both medicines on the same day again?

66. ***Barbie and Ken*** Mary Lois King collects Barbie dolls and Ken dolls. She has 390 Barbie dolls and 468 Ken dolls. Mary Lois wishes to display the dolls in groups so that the same number of dolls are in each group and that each doll belongs to one group. If each group is to consist only of Barbie dolls or only of Ken dolls, what is the largest number of dolls Mary Lois can have in each group?

67. ***Toy Car Collection*** Martha Goshaw collects Matchbox® and HotWheels® toy cars. She has 70 red cars and 175 blue cars. She wants to line up her cars in groups so that each group has the same number of cars and each group contains only red cars or only blue cars. What is the largest number of cars she can have in a group?

68. ***Stacking Trading Cards*** Desmond Freeman collects trading cards. He has 432 baseball cards and 360 football cards. He wants to make stacks of cards on a table so that each stack contains the same number of cards and each card belongs to one stack. If the baseball and football cards must not be mixed in the stacks, what is the largest number of cards that he can have in a stack?

69. ***Tree Rows*** Elizabeth Dwyer is the manager at Queen Palm Nursery and is in charge of displaying potted trees in rows.

Elizabeth has 150 citrus trees and 180 palm trees. She wants to make rows of trees so that each row has the same number of trees and each tree is in a row. If the citrus trees and the palm trees must not be mixed in the rows, what is the largest number of trees that she can have in a row?

70. ***Car Maintenance*** For many sport utility vehicles, it is recommended that the oil be changed every 3500 miles and that the tires be rotated every 6000 miles. If Carmella Gonzalez just had the oil changed and tires rotated on her SUV during the same visit to her mechanic, how many miles will she drive before she has the oil changed and tires rotated again during the same visit?

71. ***Work Schedules*** Sara Pappas and Harry Kinnan both work the 3:00 P.M. to 11:00 P.M. shift. Sara has every fifth night off and Harry has every sixth night off. If they both have tonight off, how many days will pass before they have the same night off again?

72. ***Prime Numbers*** Consider the first eight prime numbers greater than 3. The numbers are 5, 7, 11, 13, 17, 19, 23, and 29.

a) Determine which of these prime numbers differs by 1 from a multiple of the number 6.

b) Use inductive reasoning and the results obtained in part (a) to make a conjecture regarding prime numbers.

c) Select a few more prime numbers and determine whether your conjecture appears to be correct.

73. State a procedure that defines a divisibility test for 15.

74. State a procedure that defines a divisibility test for 22.

Euclidean Algorithm *Another method that can be used to find the greatest common divisor is known as the Euclidean algorithm. We illustrate this procedure by finding the GCD of 60 and 220.*

First divide 220 by 60 as shown below. Disregard the quotient 3 and then divide 60 by the remainder 40. Continue this process of dividing the divisors by the remainders until you obtain a remainder of 0. The divisor in the last division, in which the remainder is 0, is the GCD.

$$\begin{array}{r} 3 \\ 60\overline{)220} \\ \underline{180} \\ 40 \end{array} \qquad \begin{array}{r} 1 \\ 40\overline{)60} \\ \underline{40} \\ 20 \end{array} \qquad \begin{array}{r} 2 \\ 20\overline{)40} \\ \underline{40} \\ 0 \end{array}$$

Since 40/20 had a remainder of 0, the GCD is 20.

In Exercises 75–80, use the Euclidean algorithm to find the GCD.

75. 15, 40

76. 12, 28

77. 35, 105

78. 78, 104

79. 150, 180

80. 210, 560

Perfect Numbers *A number whose* ***proper factors*** *(factors other than the number itself) add up to the number is called a* ***perfect number****. For example, 6 is a perfect number because its proper factors are 1, 2, and 3, and $1 + 2 + 3 = 6$. Determine which, if any, of the following numbers are perfect.*

81. 12

82. 28

83. 48

84. 496

CHALLENGE PROBLEMS/GROUP ACTIVITIES

85. ***Number of Factors*** The following procedure can be used to determine the *number of factors* (or *divisors*) of a composite number. Write the number in prime factorization form. Examine the exponents on the prime numbers in the prime factorization. Add 1 to each exponent and then find the product of these numbers. This product gives the number of positive divisors of the composite number.

a) Use this procedure to determine the number of divisors of 60.

b) To check your answer, list all the divisors of 60. You should obtain the same number of divisors found in part (a).

86. Recall that if a number is divisible by both 2 and 3, then the number is divisible by 6. If a number is divisible by both 2 and 4, is the number necessarily divisible by 8? Explain your answer.

87. The product of any three consecutive natural numbers is divisible by 6. Explain why.

88. A number in which each digit except 0 appears exactly three times is divisible by 3. For example, 888,444,555 and 714,714,714 are both divisible by 3. Explain why this outcome must be true.

89. Use the fact that if $a|b$ and $a|c$, then $a|(b + c)$ to determine whether 54,036 is divisible by 18. (*Hint:* Write 54,036 as $54{,}000 + 36$.)

90. Show that the $2^n - 1$ is a (Mersenne) prime for $n = 2, 3, 5$, and 7 but composite for $n = 11$.

91. Goldbach also conjectured in his letter to Euler that *every integer greater than 5 is the sum of three prime numbers.* For example, $6 = 2 + 2 + 2$ and $7 = 2 + 2 + 3$. Show that this conjecture is true for integers 8 through 20.

92. ***Divisibility by Seven*** The following describes a procedure to determine whether a number is divisible by 7. We will demonstrate the procedure with the number 203.

i) Remove the units digit from the number, double the units digit, and subtract it from the remaining number. For 203, we will remove the 3, double it to get 6, and then subtract $20 - 6$ to get 14.

ii) If this new number is divisible by 7, then so is the original number. In our case, since 14 is divisible by 7, then 203 is also divisible by 7.

iii) If you are not sure if the new number is divisible by 7, you can repeat the process described in the first step.

Use the procedure described above to determine whether the following numbers are divisible by 7.

a) 329 **b)** 553 **c)** 583 **d)** 4823

RECREATIONAL MATHEMATICS

93. ***Country, Animal, Fruit*** Select a number from 1 to 10. Multiply your selected number by 9. Add the digits of the product together (if the product has more than one digit). Now subtract 5. Determine which letter of the alphabet corresponds to the number you ended with (for example, $1 = A$, $2 = B$, $3 = C$, and so on). Think of a country whose name begins with that letter. Remember the last letter of the name of that country. Think of an animal whose name begins with that letter. Remember the last letter in the name of that animal. Think of a fruit whose name begins with that letter. a) What country, animal, and fruit did you select? Turn to the answer section to see if your response matches the responses of over 90% of the people who attempt this activity. b) Can you explain why most people select the given answer?

RESEARCH ACTIVITIES

94. Conduct an Internet search on the GIMPS project. (See Mathematics Today on page 231.) Write a report describing the history and development of the project. Include a current update of the project's findings.

95. Do research and explain what *deficient numbers* and *abundant numbers* are. Give an example of each type of number. References include history of mathematics books, encyclopedias, and the Internet.

5.2 THE INTEGERS

▲ Fahrenheit temperatures can be positive, negative, or zero.

Answer the question, What is the temperature, to the nearest degree Fahrenheit, outside right now? Depending on where you are and what time of year it is, you may have answered with a positive number, a negative number, or zero. These numbers are examples of the group of numbers called the *integers* that we will study in this section. We will also study the operations of addition, subtraction, multiplication, and division using the integers.

In Section 5.1, we introduced the natural or counting numbers:

$$N = \{1, 2, 3, 4, \dots\}$$

Another important set of numbers, the ***whole numbers***, help to answer the question, How many?

$$\text{Whole numbers} = \{0, 1, 2, 3, 4, \dots\}$$

Note that the set of whole numbers contains the number 0 but that the set of counting numbers does not. If a farmer were asked how many chickens were in a coop, the answer would be a whole number. If the farmer had no chickens, he or she would answer zero. Although we use the number 0 daily and take it for granted, the number 0 as we know it was not used and accepted until the sixteenth century.

DID YOU KNOW?

Triangular Numbers

One very simple reason that the ancient Greek mathematicians thought of mathematics in terms of whole numbers and their ratios (the rational numbers) was that they were still working with numbers that were represented by objects, such as squares and triangles, not number symbols. Still, that did not prevent them from drawing some conclusions about number theory. Consider two consecutive triangular numbers (introduced in Section 1.1), for instance 3 and 6. You can see from the diagram that the sum of the two triangular numbers is equal to the square number 9, the square of a side of the larger triangle.

If the temperature is 12°F and drops 20°, the resulting temperature is −8°F. This type of problem shows the need for negative numbers. The set of ***integers*** consists of the negative integers, 0, and the positive integers.

$$\text{Integers} = \{\underbrace{\ldots, -4, -3, -2, -1}_{\text{Negative integers}}, 0, \underbrace{1, 2, 3, \ldots}_{\text{Positive integers}}\}$$

The term *positive integers* is yet another name for the natural numbers or counting numbers.

An understanding of addition, subtraction, multiplication, and division of the integers is essential in understanding algebra (Chapter 6). To aid in our explanation of addition and subtraction of integers, we introduce the real number line (Fig. 5.5). To construct the real number line, arbitrarily select a point for zero to serve as the starting point. Place the positive integers to the right of 0, equally spaced from one another. Place the negative integers to the left of 0, using the same spacing. The real number line contains the integers and all the other real numbers that are not integers. Some examples of real numbers that are not integers are indicated in Fig. 5.5, namely $-\frac{5}{2}$, $\frac{1}{2}$, $\sqrt{2}$, and π. We discuss real numbers that are not integers in the next two sections.

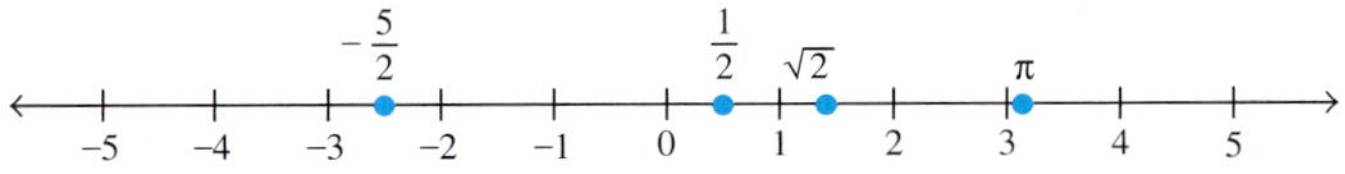

Figure 5.5

The arrows at the ends of the real number line indicate that the line continues indefinitely in both directions. Note that for any natural number, n, on the number line, the *opposite of* that number, $-n$, is also on the number line. This real number line was drawn horizontally, but it could just as well have been drawn vertically. In fact, in the next chapter, we show that the axes of a graph are the union of two number lines, one horizontal and the other vertical.

The number line can be used to determine the greater (or lesser) of two integers. Two *inequality symbols* that we will use in this chapter are $>$ and $<$. The symbol $>$ is read "is greater than," and the symbol $<$ is read "is less than." Expressions that contain an inequality symbol are called *inequalities*. On the number line, the numbers increase from left to right. The number 3 is greater than 2, written $3 > 2$. Observe that 3 is to the right of 2. Similarly, we can see that $0 > -1$ by observing that 0 is to the right of -1 on the number line.

Instead of stating that 3 is greater than 2, we could state that 2 is less than 3, written $2 < 3$. Note that 2 is to the left of 3 on the number line. We can also see that $-1 < 0$ by observing that -1 is to the left of 0. The inequality symbol always points to the smaller of the two numbers when the inequality is true.

EXAMPLE 1 *Writing an Inequality*

Insert either $>$ or $<$ in the shaded area between the paired numbers to make the statement correct.

a) $-3 \quad 1$ b) $-3 \quad -5$ c) $-6 \quad -4$ d) $0 \quad -7$

SOLUTION

a) $-3 < 1$ since -3 is to the left of 1 on the number line.

b) $-3 > -5$ since -3 is to the right of -5 on the number line.

c) $-6 < -4$ since -6 is to the left of -4 on the number line.

d) $0 > -7$ since 0 is to the right of -7 on the number line.

Addition of Integers

Addition of integers can be represented geometrically using a number line. To do so, begin at 0 on the number line. Represent the first addend (the first number to be added) by an arrow starting at 0. Draw the arrow to the right if the addend is positive. If the addend is negative, draw the arrow to the left. From the tip of the first arrow, draw a second arrow to represent the second addend. Draw the second arrow to the right or left, as just explained. The sum of the two integers is found at the tip of the second arrow.

EXAMPLE 2 ***Adding Integers***

Evaluate the following using the number line.

a) $3 + (-5)$ b) $-1 + (-4)$ c) $-6 + 4$ d) $3 + (-3)$

SOLUTION

a)

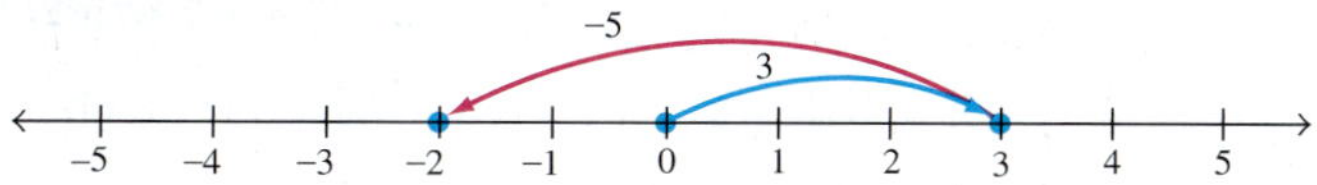

Thus, $3 + (-5) = -2$.

b)

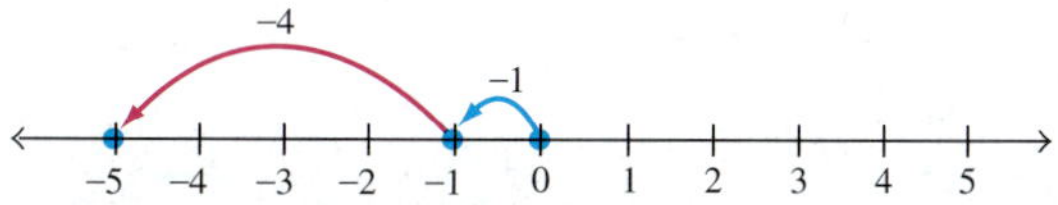

Thus, $-1 + (-4) = -5$.

c)

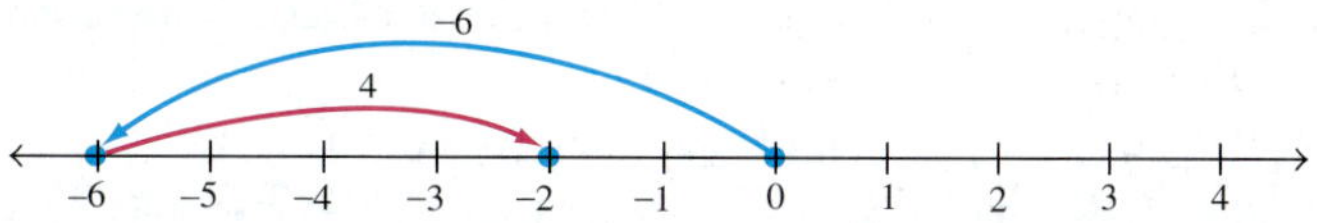

Thus, $-6 + 4 = -2$.

d)

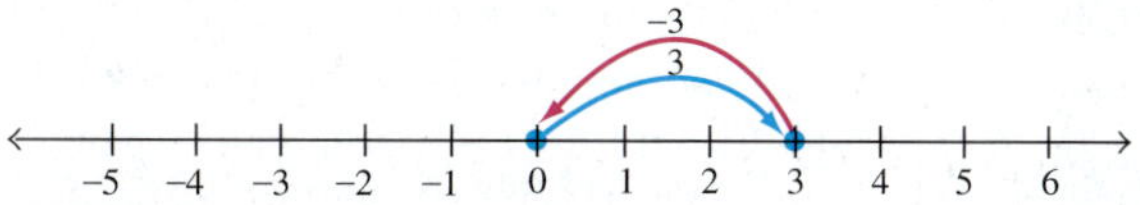

Thus, $3 + (-3) = 0$.

In Example 2(d), the number -3 is said to be the additive inverse of 3 and 3 is the additive inverse of -3 because their sum is 0. In general, the ***additive inverse*** of the number n is $-n$ since $n + (-n) = 0$. Inverses are discussed more formally in Chapter 10.

Subtraction of Integers

Any subtraction problem can be rewritten as an addition problem. To do so, we use the following definition of subtraction.

SUBTRACTION

$$a - b = a + (-b)$$

The rule for subtraction indicates that to subtract b from a, *add* the additive inverse of b to a. For example,

$$3 - 5 = 3 + (-5)$$

Subtraction ↑ Addition ↑ Additive inverse of 5 ↑

Now we can determine the value of $3 + (-5)$.

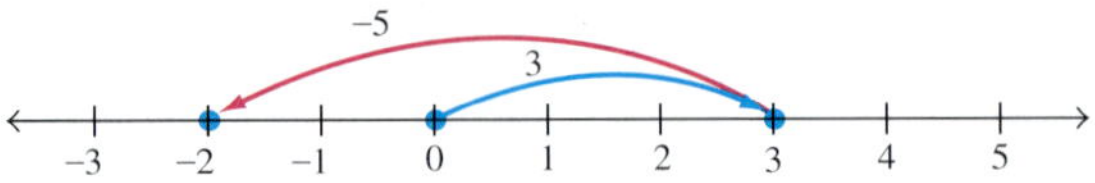

Thus, $3 - 5 = 3 + (-5) = -2$.

EXAMPLE 3 *Subtracting Integers*

Evaluate $-4 - (-3)$ using the number line.

SOLUTION We are subtracting -3 from -4. The additive inverse of -3 is 3; therefore, we add 3 to -4. We now add $-4 + 3$ on the number line to obtain the answer -1.

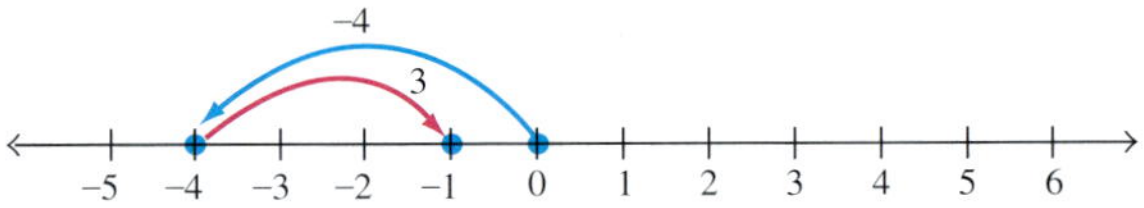

Thus, $-4 - (-3) = -4 + 3 = -1$.

In Example 3, we found that $-4 - (-3) = -4 + 3$. In general, $a - (-b) = a + b$. As you get more proficient in working with integers, you should be able to answer questions involving them without drawing a number line.

EXAMPLE 4 *Subtracting: Adding the Inverse*

Evaluate.

a) $-7 - 3$ b) $-7 - (-3)$ c) $7 - (-3)$ d) $7 - 3$

SOLUTION

a) $-7 - 3 = -7 + (-3) = -10$ b) $-7 - (-3) = -7 + 3 = -4$

c) $7 - (-3) = 7 + 3 = 10$ d) $7 - 3 = 7 + (-3) = 4$

EXAMPLE 5 *Elevation Difference*

The highest point on Earth is Mount Everest, in the Himalayas, at a height of 29,035 ft above sea level. The lowest point on Earth is the Mariana Trench, in the Pacific Ocean, at a depth of 36,198 ft below sea level (−36,198 ft). Find the vertical height difference between Mount Everest and the Mariana Trench.

DID YOU KNOW?

Earthly Extremes

More than 1400 people have stood on the top of Nepal's 29,035-foot-high Mount Everest. Very few people have visited the Pacific Ocean's 36,198-foot-deep Mariana Trench. In 1960, Jacques Piccard and Don Walsh were the first to make the journey in a vessel designed to withstand the immense pressure: 7 tons per square inch. The diving time was $8\frac{1}{2}$ hours, of which only 20 minutes were spent on the ocean floor.

SOLUTION We obtain the vertical difference by subtracting the lower elevation from the higher elevation.

$$29{,}035 - (-36{,}198) = 29{,}035 + 36{,}198 = 65{,}233$$

The vertical difference is 65,233 ft.

Multiplication of Integers

The multiplication property of zero is important in our discussion of multiplication of integers. It indicates that the product of 0 and any number is 0.

MULTIPLICATION PROPERTY OF ZERO

$$a \cdot 0 = 0 \cdot a = 0$$

We will develop the rules for multiplication of integers using number patterns. The four possible cases are

1. positive integer × positive integer,
2. positive integer × negative integer,
3. negative integer × positive integer, and
4. negative integer × negative integer.

CASE 1: *POSITIVE INTEGER × POSITIVE INTEGER* The product of two positive integers can be defined as repeated addition of a positive integer. Thus, $3 \cdot 2$ means $2 + 2 + 2$. This sum will always be positive. Thus, a positive integer times a positive integer is a positive integer.

CASE 2: *POSITIVE INTEGER × NEGATIVE INTEGER* Consider the following patterns:

$$3(3) = 9$$
$$3(2) = 6$$
$$3(1) = 3$$

Note that each time the second factor is reduced by 1, the product is reduced by 3. Continuing the process gives

$$3(0) = 0$$

What comes next?

$$3(-1) = -3$$
$$3(-2) = -6$$

The pattern indicates that a positive integer times a negative integer is a negative integer.

We can confirm this result by using the number line. The expression $3(-2)$ means $(-2) + (-2) + (-2)$. Adding $(-2) + (-2) + (-2)$ on the number line, we obtain a sum of -6.

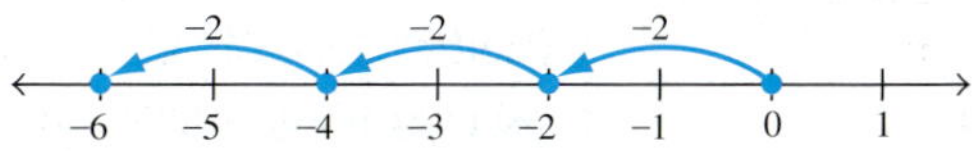

CASE 3: ***NEGATIVE INTEGER × POSITIVE INTEGER*** A procedure similar to that used in case 2 will indicate that a negative integer times a positive integer is a negative integer.

CASE 4: ***NEGATIVE INTEGER × NEGATIVE INTEGER*** We have illustrated that a positive integer times a negative integer is a negative integer. We make use of this fact in the following pattern:

$$\begin{aligned} 4(-4) &= -16 \\ 3(-4) &= -12 \\ 2(-4) &= -8 \\ 1(-4) &= -4 \end{aligned}$$

In this pattern, each time the first term is decreased by 1, the product is increased by 4. Continuing this process gives

$$\begin{aligned} 0(-4) &= 0 \\ (-1)(-4) &= 4 \\ (-2)(-4) &= 8 \end{aligned}$$

This pattern illustrates that a negative integer times a negative integer is a positive integer.

The examples were restricted to integers. The rules for multiplication, however, can be used for any numbers. We summarize them as follows.

RULES FOR MULTIPLICATION

1. The product of two numbers with *like signs* (positive × positive or negative × negative) is a *positive number.*
2. The product of two numbers with *unlike signs* (positive × negative or negative × positive) is a *negative number.*

EXAMPLE 6 *Multiplying Integers*

Evaluate.

a) $4 \cdot 8$ b) $4 \cdot (-8)$ c) $(-4) \cdot 8$ d) $(-4) \cdot (-8)$

SOLUTION

a) $4 \cdot 8 = 32$ b) $4 \cdot (-8) = -32$
c) $(-4) \cdot 8 = -32$ d) $(-4) \cdot (-8) = 32$

Division of Integers

You may already realize that a relationship exists between multiplication and division.

$$6 \div 2 = 3 \quad \text{means that} \quad 3 \cdot 2 = 6$$

$$\frac{20}{10} = 2 \quad \text{means that} \quad 2 \cdot 10 = 20$$

These examples demonstrate that division is the reverse process of multiplication.

DIVISION

For any a, b, and c where $b \neq 0$, $\frac{a}{b} = c$ means that $c \cdot b = a$.

We discuss the four possible cases for division of integers, which are similar to those for multiplication.

CASE 1: ***POSITIVE INTEGER ÷ POSITIVE INTEGER*** A positive integer divided by a positive integer is positive.

$$\frac{6}{2} = 3 \qquad \text{since} \qquad 3(2) = 6$$

CASE 2: ***POSITIVE INTEGER ÷ NEGATIVE INTEGER*** A positive integer divided by a negative integer is negative.

$$\frac{6}{-2} = -3 \qquad \text{since} \qquad (-3)(-2) = 6$$

CASE 3: ***NEGATIVE INTEGER ÷ POSITIVE INTEGER*** A negative integer divided by a positive integer is negative.

$$\frac{-6}{2} = -3 \qquad \text{since} \qquad (-3)(2) = -6$$

CASE 4: ***NEGATIVE INTEGER ÷ NEGATIVE INTEGER*** A negative integer divided by a negative integer is positive.

$$\frac{-6}{-2} = 3 \qquad \text{since} \qquad 3(-2) = -6$$

The examples were restricted to integers. The rules for division, however, can be used for any numbers. You should realize that division of integers does not always result in an integer. The rules for division are summarized as follows.

RULES FOR DIVISION

1. The quotient of two numbers with *like signs* (positive ÷ positive or negative ÷ negative) is a *positive number.*
2. The quotient of two numbers with *unlike signs* (positive ÷ negative or negative ÷ positive) is a *negative number.*

EXAMPLE 7 *Dividing Integers*

Evaluate.

a) $\frac{42}{7}$ b) $\frac{-42}{7}$ c) $\frac{42}{-7}$ d) $\frac{-42}{-7}$

SOLUTION

a) $\frac{42}{7} = 6$ b) $\frac{-42}{7} = -6$ c) $\frac{42}{-7} = -6$ d) $\frac{-42}{-7} = 6$

In the definition of division, we stated that the denominator could not be 0. Why not? Suppose we are trying to find the quotient $\frac{5}{0}$. Let's say that this quotient is equal to some number x. Then we would have $\frac{5}{0} = x$. If true, it would mean that $5 = x \cdot 0$. The right side of the equation is $x \cdot 0$, which is equal to 0 for any real value of x. That leads us to conclude that $5 = 0$, which is false. Thus, there is no number that can replace x that makes the equation $\frac{5}{0} = x$ true. Therefore, in mathematics, division by 0 is not allowed and we say that a quotient of any number divided by zero is *undefined*.

SECTION 5.2 EXERCISES

CONCEPT/WRITING EXERCISES

1. Explain how to add numbers using a number line.
2. What is the additive inverse of a number n?
3. Explain how to rewrite a subtraction problem as an addition problem.
4. Explain the rules for division of real numbers.
5. Explain the rules for multiplication of real numbers.
6. Explain why the quotient of a number divided by 0 is undefined.

PRACTICE THE SKILLS

In Exercises 7–16, evaluate the expression.

7. $-4 + 7$
8. $3 + (-7)$
9. $-9 + 5$
10. $-2 + (-2)$
11. $[6 + (-11)] + 0$
12. $(2 + 5) + (-4)$
13. $[(-3) + (-4)] + 9$
14. $[8 + (-3)] + (-2)$
15. $[(-23) + (-9)] + 11$
16. $[5 + (-13)] + 18$

In Exercises 17–26, evaluate the expression.

17. $1 - 7$
18. $-4 - 8$
19. $-5 - 4$
20. $6 - (-3)$
21. $-5 - (-3)$
22. $-4 - 4$
23. $14 - 20$
24. $8 - (-3)$
25. $[5 + (-3)] - 4$
26. $6 - (8 + 6)$

In Exercises 27–36, evaluate the expression.

27. $-5 \cdot 6$
28. $9(-2)$
29. $(-8)(-8)$
30. $-4(11)$
31. $[(-8)(-2)] \cdot 6$
32. $4(-5)(-6)$
33. $(5 \cdot 6)(-2)$
34. $(-9)(-1)(-2)$
35. $[(-3)(-6)] \cdot [(-5)(8)]$
36. $[(-8 \cdot 4) \cdot 5](-2)$

In Exercises 37–46, evaluate the expression.

37. $-28 \div (-4)$
38. $-72 \div 9$
39. $11 \div (-11)$
40. $-100 \div 20$
41. $56/-8$
42. $-75/15$
43. $-210/14$
44. $186/-6$
45. $144 \div (-3)$
46. $(-900) \div (-4)$

In Exercises 47–56, determine whether the statement is true or false. Modify each false statement to make it a true statement.

47. Every whole number is an integer.
48. Every integer is a whole number.
49. The difference of any two negative integers is a negative integer.
50. The sum of any two negative integers is a negative integer.
51. The product of any two positive integers is a positive integer.
52. The difference of a positive integer and a negative integer is always a negative integer.
53. The quotient of a negative integer and a positive integer is always a negative number.
54. The quotient of any two negative integers is a negative number.

55. The sum of a positive integer and a negative integer is always a positive integer.

56. The product of a positive integer and a negative integer is always a positive integer.

In Exercises 57–66, evaluate the expression.

57. $(7 + 11) \div 3$

58. $(-8) \div [64 \div (-8)]$

59. $[(-7)(-6)] - 22$

60. $[8(-2)] - 15$

61. $(4 - 8)(3)$

62. $[18 \div (-2)](-3)$

63. $[2 + (-17)] \div 3$

64. $(5 - 9) \div (-4)$

65. $[(-22)(-3)] \div (2 - 13)$

66. $[15(-4)] \div (-6)$

In Exercises 67–70, write the numbers in increasing order from left to right.

67. 0, −3, 3, −6, 6, −9

68. 100, −10, 1, 0, −1, 10

69. −5, −2, −3, −1, −4, −6

70. 106, 33, −47, −108, 72, −76

PROBLEM SOLVING

71. ***Dow Jones Industrial Average*** On March 28, 2006, the Dow Jones Industrial Average (DJIA) opened at 11,250 points. During that day it lost 95 points. On March 29, 2006, it gained 61 points. On March 30, 2006, it lost 65 points. On March 31, 2006, it lost 42 points. Determine the closing DJIA on March 31, 2006.

72. ***Pit Score*** While playing the game of Pit, John Pearse began with a score of zero points. He then gained 100 points, lost 40 points, gained 90 points, lost 20 points, and gained 80 points on his next five rounds. What is John's score after five rounds?

73. ***Elevation Difference*** Mount Whitney, in the Sierra Nevada mountains of California, is the highest point in the contiguous United States. It is 14,495 ft above sea level. Death Valley, in California and Nevada, is the lowest point in the United States, 282 ft below sea level. Find the vertical height difference between Mount Whitney and Death Valley.

74. ***Submarine Depth*** The *USS Toledo* submarine is moving at an underwater depth of 1000 feet. It then rises 600 feet, dives 700 feet, and finally dives 400 feet. What is its final depth?

75. ***Football Yardage*** In the first four plays of a football game, the New Orleans Saints lost 6 yards, gained 4 yards, lost 1 yard, and gained 12 yards. What is the total number of yards gained in the first four plays? Did the Saints make a first down? (Ten yards are needed for a first down.)

▲ Reggie Bush of the New Orleans Saints

76. ***Extreme Temperatures*** The hottest temperature ever recorded in the United States was 134°F, which occurred at Greenland Ranch, California, in Death Valley on July 10, 1913. The coldest temperature ever recorded in the United States was −79.8°F, which occurred at Prospect Creek Camp, Alaska, in the Endicott Mountains on January 23, 1971. Determine the difference between these two temperatures.

77. ***Time Zone Calculations*** Part of a World Standard Time Zones chart used by airlines and the United States Navy is shown. The scale along the bottom is just like a number line with the integers $-12, -11, \ldots, 11, 12$ on it.

a) Find the difference in time between Amsterdam (zone $+1$) and Los Angeles (zone -8).

b) Find the difference in time between Boston (zone -5) and Puerto Vallarta (zone -7).

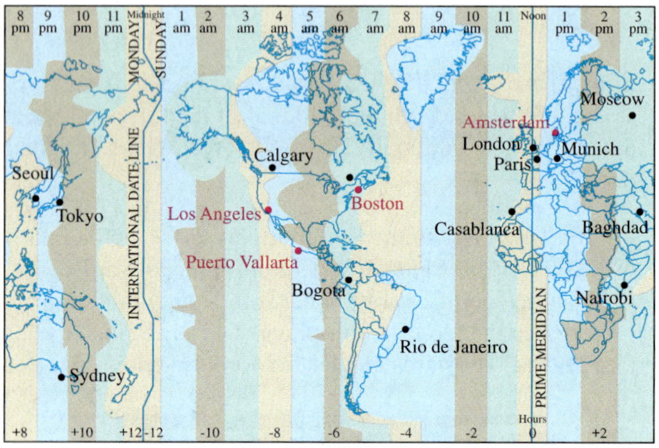

78. Explain why $\dfrac{a}{b} = \dfrac{-a}{-b}$.

CHALLENGE PROBLEMS/GROUP ACTIVITIES

79. Find the quotient:

$$\frac{-1 + 2 - 3 + 4 - 5 + \cdots - 99 + 100}{1 - 2 + 3 - 4 + 5 - \cdots + 99 - 100}$$

80. ***Pentagonal Numbers*** Triangular numbers and square numbers were introduced in the Section 1.1 Exercises. There are also **pentagonal numbers**, which were also studied by the Greeks. Four pentagonal numbers are 1, 5, 12, and 22.

1 5 12 22

a) Determine the next three pentagonal numbers.

b) Describe a procedure to determine the next five pentagonal numbers without drawing the figures.

c) Is 72 a pentagonal number? Explain how you determined your answer.

81. Place the appropriate plus or minus signs between each digit so that the total will equal 1.

$$0 \quad 1 \quad 2 \quad 3 \quad 4 \quad 5 \quad 6 \quad 7 \quad 8 \quad 9 = 1$$

RECREATIONAL MATHEMATICS

82. ***Four 4's*** The game of *Four 4's* is a challenging way to learn about some of the operations on integers. In this game, you must use exactly four 4's, and no other digits, along with one or more of the operations of addition, subtraction, multiplication, and division† to write the expressions. The object of the game is to write an expression that is equal to 0. Then write a second expression that is equal to 1, write a third that is equal to 2, and so on up to 9. You may use as many grouping symbols (that is, parentheses and brackets) as you wish, but you must use exactly four 4's. For example, one way to obtain 1 is as follows: $(4 + 4) \div (4 + 4) = 8 \div 8 = 1$. One way to obtain 2 is as follows: $(4 \div 4) + (4 \div 4) = 1 + 1 = 2$.

a) Use the rules as defined above to obtain each whole number 0 through 9.

b) Use the rules as defined above to obtain the following whole numbers: 12, 15, 16, 17, 20.

c) We will now change our rules to allow the number 44 to count as two of the four 4's. Use the number 44 and two other fours to obtain the whole number 10.

RESEARCH ACTIVITY

83. Do research and write a report on the history of the number 0 in the Hindu–Arabic numeration system.

†This game will be expanded in future exercise sets to include other operations such as exponents and square roots.

5.3 THE RATIONAL NUMBERS

▲ Rational numbers are frequently used in recipes.

Among the ingredients called for in a recipe for carrot cake are $1\frac{1}{3}$ cups flour, $\frac{1}{2}$ teaspoon salt, $1\frac{1}{3}$ teaspoons baking powder, $1\frac{1}{3}$ teaspoons baking soda, $1\frac{1}{3}$ teaspoons cinnamon, $\frac{1}{2}$ teaspoon cloves, and $\frac{1}{2}$ teaspoon ginger. How much of each of these ingredients would you need to use if you wish to cut the recipe in half? In this section, we will review the use of fractions such as those in the carrot cake recipe. We will also learn about the operations of addition, subtraction, multiplication, and division using fractions.

We introduced the number line in Section 5.1 and discussed the integers in Section 5.2. The numbers that fall between the integers on the number line are either rational or irrational numbers. In this section, we discuss the rational numbers, and in Section 5.4, we discuss the irrational numbers.

Any number that can be expressed as a quotient of two integers (denominator not 0) is a rational number.

The set of **rational numbers**, denoted by Q, is the set of all numbers of the form p/q, where p and q are integers and $q \neq 0$.

The following numbers are examples of rational numbers:

$$\frac{1}{3}, \quad \frac{3}{4}, \quad -\frac{7}{8}, \quad 1\frac{2}{3}, \quad 2, \quad 0, \quad \frac{15}{7}$$

The integers 2 and 0 are rational numbers because each can be expressed as the quotient of two integers: $2 = \frac{2}{1}$ and $0 = \frac{0}{1}$. In fact, every integer n is a rational number because it can be written in the form of $\frac{n}{1}$.

Numbers such as $\frac{1}{3}$ and $-\frac{7}{8}$ are also called ***fractions***. The number above the fraction line is called the ***numerator***, and the number below the fraction line is called the ***denominator***.

Reducing Fractions

Sometimes the numerator and denominator in a fraction have a common divisor (or common factor). For example, both the numerator and denominator of the fraction $\frac{6}{10}$ have the common divisor 2. When a numerator and denominator have a common divisor, we can *reduce the fraction to its lowest terms.*

A fraction is said to be in its lowest terms (or reduced) when the numerator and denominator are relatively prime (that is, have no common divisors other than 1). To reduce a fraction to its lowest terms, divide both the numerator and the denominator by the greatest common divisor. Recall that a procedure for finding the greatest common divisor was discussed in Section 5.1.

The fraction $\frac{6}{10}$ is reduced to its lowest terms as follows.

$$\frac{6}{10} = \frac{6 \div 2}{10 \div 2} = \frac{3}{5}$$

EXAMPLE 1 *Reducing a Fraction to Lowest Terms*

Reduce $\dfrac{54}{90}$ to its lowest terms.

SOLUTION On page 229 in Example 4 of Section 5.1, we determined that the GCD of 54 and 90 is 18. Divide the numerator and the denominator by GCD, 18.

$$\frac{54}{90} = \frac{54 \div 18}{90 \div 18} = \frac{3}{5}$$

Since there are no common divisors of 3 and 5 other than 1, the fraction $\frac{3}{5}$ is in its lowest terms.

Mixed Numbers and Improper Fractions

Consider the number $2\frac{3}{4}$. It is an example of a ***mixed number***. It is called a mixed number because it consists of an integer, 2, and a fraction, $\frac{3}{4}$. The mixed number $2\frac{3}{4}$ means $2 + \frac{3}{4}$. The mixed number $-4\frac{1}{4}$ means $-(4 + \frac{1}{4})$. Rational numbers greater than 1 or less than -1 that are not integers may be represented as mixed numbers, or as ***improper fractions***. An improper fraction is a fraction whose numerator is greater than its denominator. An example of an improper fraction is $\frac{8}{5}$. Figure 5.6 shows both mixed numbers and improper fractions indicated on a number line. In this section, we show how to convert mixed numbers to improper fractions and vice versa.

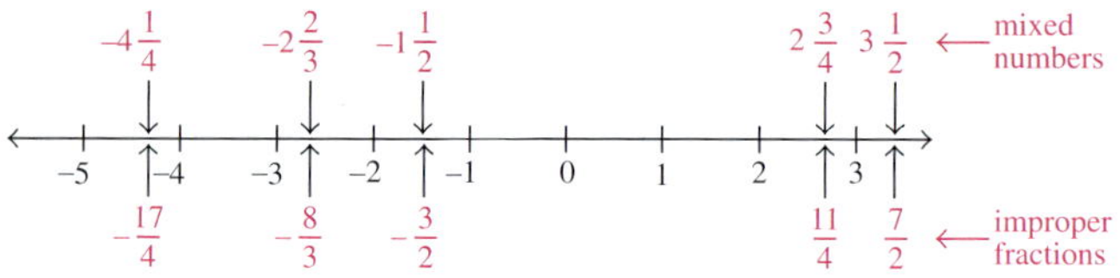

Figure 5.6

We begin by limiting our discussion to positive mixed numbers and positive improper fractions.

CONVERTING A POSITIVE MIXED NUMBER TO AN IMPROPER FRACTION

1. Multiply the denominator of the fraction in the mixed number by the integer preceding it.
2. Add the product obtained in step 1 to the numerator of the fraction in the mixed number. This sum is the numerator of the improper fraction we are seeking. The denominator of the improper fraction we are seeking is the same as the denominator of the fraction in the mixed number.

EXAMPLE 2 *Converting Mixed Numbers to Improper Fractions*

Convert the following mixed numbers to improper fractions.

a) $1\dfrac{2}{3}$ b) $4\dfrac{5}{8}$

SOLUTION

a) $1\frac{2}{3} = \frac{3 \cdot 1 + 2}{3} = \frac{3 + 2}{3} = \frac{5}{3}$

b) $4\frac{5}{8} = \frac{8 \cdot 4 + 5}{8} = \frac{32 + 5}{8} = \frac{37}{8}$

Notice that both $\frac{5}{3}$ and $\frac{37}{8}$ have numerators larger than their denominators; therefore, both numbers are improper fractions.

Now let's discuss converting an improper fraction to a mixed number.

CONVERTING A POSITIVE IMPROPER FRACTION TO A MIXED NUMBER

1. Divide the numerator by the denominator. Identify the quotient and the remainder.
2. The quotient obtained in step 1 is the integer part of the mixed number. The remainder is the numerator of the fraction in the mixed number. The denominator in the fraction of the mixed number will be the same as the denominator in the original fraction.

EXAMPLE 3 *From Improper Fraction to Mixed Number*

Convert the following improper fractions to mixed numbers.

a) $\frac{8}{5}$ b) $\frac{225}{8}$

SOLUTION

a) Divide the numerator, 8, by the denominator, 5.

$$\begin{array}{r} 1 \\ 5\overline{)8} \\ \underline{5} \\ 3 \end{array}$$

Divisor → 5; 1 ← Quotient; 8 ← Dividend; 3 ← Remainder

Therefore,

$$\frac{8}{5} = 1\frac{3}{5}$$

Quotient → 1; 3 ← Remainder; 5 ← Divisor

The mixed number is $1\frac{3}{5}$.

b) Divide the numerator, 225, by the denominator, 8.

$$\begin{array}{r} 28 \\ 8\overline{)225} \\ \underline{16} \\ 65 \\ \underline{64} \\ 1 \end{array}$$

Divisor → 8; 28 ← Quotient; 225 ← Dividend; 1 ← Remainder

Therefore,

$$\frac{225}{8} = \underset{\text{Quotient}}{28}\frac{1}{8} \begin{matrix} \leftarrow \text{Remainder} \\ \leftarrow \text{Divisor} \end{matrix}$$

The mixed number is $28\frac{1}{8}$.

Up to this point, we have only worked with positive mixed numbers and positive improper fractions. When converting a negative mixed number to an improper fraction, or a negative improper fraction to a mixed number, it is best to ignore the negative sign temporarily. Perform the calculation as described earlier and then reattach the negative sign.

EXAMPLE 4 *Negative Mixed Numbers and Improper Fractions*

a) Convert $-1\frac{2}{3}$ to an improper fraction.

b) Convert $-\frac{8}{5}$ to a mixed number.

SOLUTION

a) First, ignore the negative sign and examine $1\frac{2}{3}$. We learned in Example 2(a) that $1\frac{2}{3} = \frac{5}{3}$. To convert $-1\frac{2}{3}$ to an improper fraction, we reattach the negative sign. Thus, $-1\frac{2}{3} = -\frac{5}{3}$.

b) We learned in Example 3(a) that $\frac{8}{5} = 1\frac{3}{5}$. Therefore, $-\frac{8}{5} = -1\frac{3}{5}$.

Terminating or Repeating Decimal Numbers

Note the following important property of the rational numbers.

> Every *rational number* when expressed as a decimal number will be either a terminating or a repeating decimal number.

Examples of terminating decimal numbers are 0.5, 0.75, and 4.65. Examples of repeating decimal numbers are 0.333 . . . , 0.2323 . . . , and 8.13456456 . . . . One way to indicate that a number or group of numbers repeat is to place a bar above the number or group of numbers that repeat. Thus, 0.333 . . . may be written $0.\overline{3}$, 0.2323 . . . may be written $0.\overline{23}$, and 8.13456456 . . . may be written $8.13\overline{456}$.

EXAMPLE 5 *Terminating Decimal Numbers*

Show that the following rational numbers can be expressed as terminating decimal numbers.

a) $\frac{3}{4}$ b) $-\frac{7}{20}$ c) $\frac{13}{8}$

SOLUTION To express the rational number in decimal form, divide the numerator by the denominator. If you use a calculator or long division, you will obtain the following results.

a) $\frac{3}{4} = 0.75$ b) $-\frac{7}{20} = -0.35$ c) $\frac{13}{8} = 1.625$

EXAMPLE 6 *Repeating Decimal Numbers*

Show that the following rational numbers can be expressed as repeating decimal numbers.

a) $\frac{2}{3}$ b) $\frac{14}{99}$ c) $1\frac{5}{36}$

SOLUTION If you use a calculator or long division, you will see that each fraction results in a repeating decimal number.

a) $2 \div 3 = 0.6666\ldots$ or $0.\overline{6}$

b) $14 \div 99 = 0.141414\ldots$ or $0.\overline{14}$

c) $1\frac{5}{36} = \frac{41}{36} = 1.138888\ldots$ or $1.13\overline{8}$

Note that in each part of Example 6, the quotient when expressed as a decimal number has no final digit and continues indefinitely. Each number is a repeating decimal number.

When a fraction is converted to a decimal number, the maximum number of digits that can repeat is $n - 1$, where n is the denominator of the fraction. For example, when $\frac{2}{7}$ is converted to a decimal number, the maximum number of digits that can repeat is $7 - 1$, or 6.

Converting Decimal Numbers to Fractions

We can convert a terminating or repeating decimal number into a quotient of integers. The explanation of the procedure will refer to the positional values to the right of the decimal point, as illustrated here:

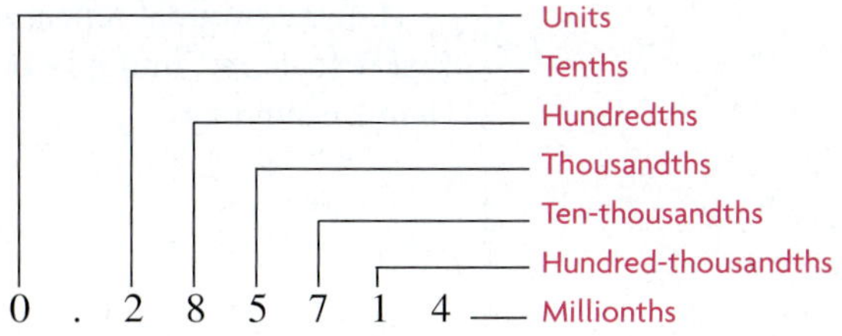

Example 7 demonstrates how to convert from a decimal number to a fraction.

EXAMPLE 7 *Converting Decimal Numbers to Fractions*

Convert the following terminating decimal numbers to a quotient of integers. If necessary, reduce the quotient to lowest terms.

a) 0.7 b) 0.35 c) 0.016 d) 3.41

SOLUTION When converting a terminating decimal number to a quotient of integers, we observe the last digit to the right of the decimal point. The position of this digit will indicate the denominator of the quotient of integers.

a) $0.7 = \frac{7}{10}$ because the 7 is in the tenths position.

b) $0.35 = \frac{35}{100}$ because the right-most digit, 5, is in the hundredths position. We finish by reducing the fraction to lowest terms:

$$\frac{35}{100} = \frac{35 \div 5}{100 \div 5} = \frac{7}{20}$$

c) $0.016 = \frac{16}{1000}$ because the right-most digit, 6, is in the thousandths position. We finish by reducing the fraction to lowest terms:

$$\frac{16}{1000} = \frac{16 \div 8}{1000 \div 8} = \frac{2}{125}$$

d) We begin by writing 3.41 as the mixed number, $3\frac{41}{100}$ because the right-most digit, 1, is in the hundredths position. To write the mixed number as a ratio of two integers, we follow the procedure shown in Example 2, on page 247:

$$3\frac{41}{100} = \frac{100 \cdot 3 + 41}{100} = \frac{341}{100}$$

The fraction $\frac{341}{100}$ cannot be reduced further.

Converting a repeating decimal number to a quotient of integers is more difficult than converting a terminating decimal number to a quotient of integers. To do so, we must "create" another repeating decimal number with the same repeating digits so that when one repeating decimal number is subtracted from the other repeating decimal number, the difference will be a whole number. To create a number with the same repeating digits, multiply the original repeating decimal number by 10 if one digit repeats, by 100 if two digits repeat, by 1000 if three digits repeat, and so on. Examples 8 through 10 demonstrate this procedure.

EXAMPLE 8 *Converting a Repeating Decimal Number to a Fraction*

Convert $0.\overline{3}$ to a quotient of integers.

SOLUTION $0.\overline{3} = 0.3\overline{3} = 0.33\overline{3}$, and so on.

Let the original repeating decimal number be n; thus, $n = 0.\overline{3}$. Because one digit repeats, we multiply both sides of the equation by 10, which gives $10n = 3.\overline{3}$. Then we subtract.

$$\begin{aligned} 10n &= 3.\overline{3} \\ -\quad n &= 0.\overline{3} \\ \hline 9n &= 3.0 \end{aligned}$$

Note that $10n - n = 9n$ and $3.\overline{3} - 0.\overline{3} = 3.0$.

Next, we solve for n by dividing both sides of the equation by 9.

$$\frac{9n}{9} = \frac{3.0}{9}$$

$$n = \frac{3}{9} = \frac{1}{3}$$

Therefore, $0.\overline{3} = \frac{1}{3}$. Evaluate $1 \div 3$ on a calculator now and see what value you get.

EXAMPLE 9 *Converting a Repeating Decimal Number to a Fraction*

Convert $0.\overline{35}$ to a quotient of integers.

SOLUTION Let $n = 0.\overline{35}$. Since two digits repeat, multiply both sides of the equation by 100. Thus, $100n = 35.\overline{35}$. Now we subtract n from $100n$.

$$\begin{aligned} 100n &= 35.\overline{35} \\ -\quad n &= 0.\overline{35} \\ \hline 99n &= 35 \end{aligned}$$

Finally, we divide both sides of the equation by 99.

$$\frac{99n}{99} = \frac{35}{99}$$

$$n = \frac{35}{99}$$

Therefore, $0.\overline{35} = \frac{35}{99}$. Evaluate $35 \div 99$ on a calculator now and see what value you get.

EXAMPLE 10 *Converting a Repeating Decimal Number to a Fraction*

Convert $12.14\overline{2}$ to a quotient of integers.

SOLUTION This example is different from the two preceding examples in that the repeating digit, 2, is not directly to the right of the decimal point. When this situation arises, move the decimal point to the right until the repeating terms are directly to its right. For each place the decimal point is moved, the number is multiplied by 10. In this example, the decimal point must be moved two places to the right. Thus, the number must be multiplied by 100.

$$\begin{aligned} n &= 12.14\overline{2} \\ 100n &= 100 \times 12.14\overline{2} = 1214.\overline{2} \end{aligned}$$

Now proceed as in the previous two examples. Since one digit repeats, multiply both sides by 10.

$$\begin{aligned} 100n &= 1214.\overline{2} \\ 10 \times 100n &= 10 \times 1214.\overline{2} \\ 1000n &= 12142.\overline{2} \end{aligned}$$

Now subtract $100n$ from $1000n$ so that the repeating part will drop out.

$$\begin{aligned} 1000n &= 12142.\overline{2} \\ -\ 100n &= 1214.\overline{2} \\ \hline 900n &= 10928 \\ n &= \frac{10,928}{900} = \frac{2732}{225} \end{aligned}$$

Therefore, $12.14\overline{2} = \frac{2732}{225}$. Evaluate $2732 \div 225$ on a calculator now and see what value you get.

Multiplication and Division of Fractions

The product of two fractions is found by multiplying the numerators together and multiplying the denominators together.

MULTIPLICATION OF FRACTIONS

$$\frac{a}{b} \cdot \frac{c}{d} = \frac{a \cdot c}{b \cdot d} = \frac{ac}{bd}, \quad b \neq 0, \quad d \neq 0$$

EXAMPLE 11 *Multiplying Fractions*

Evaluate.

a) $\frac{3}{5} \cdot \frac{7}{8}$ b) $\left(\frac{-2}{3}\right)\left(\frac{-4}{9}\right)$ c) $\left(1\frac{7}{8}\right)\left(2\frac{1}{4}\right)$

SOLUTION

a) $\frac{3}{5} \cdot \frac{7}{8} = \frac{3 \cdot 7}{5 \cdot 8} = \frac{21}{40}$

b) $\left(\frac{-2}{3}\right)\left(\frac{-4}{9}\right) = \frac{(-2)(-4)}{(3)(9)} = \frac{8}{27}$

c) $\left(1\frac{7}{8}\right)\left(2\frac{1}{4}\right) = \frac{15}{8} \cdot \frac{9}{4} = \frac{135}{32} = 4\frac{7}{32}$

The *reciprocal* of any number is 1 divided by that number. The product of a number and its reciprocal must equal 1. Examples of some numbers and their reciprocals follow.

Number		Reciprocal		Product
3	$\cdot$	$\frac{1}{3}$	=	1
$\frac{3}{5}$	$\cdot$	$\frac{5}{3}$	=	1
-6	$\cdot$	$-\frac{1}{6}$	=	1

To find the quotient of two fractions, multiply the first fraction by the reciprocal of the second fraction.

DIVISION OF FRACTIONS

$$\frac{a}{b} \div \frac{c}{d} = \frac{a}{b} \cdot \frac{d}{c} = \frac{ad}{bc}, \quad b \neq 0, \quad d \neq 0, \quad c \neq 0$$

EXAMPLE 12 *Dividing Fractions*

Evaluate.

a) $\frac{5}{7} \div \frac{3}{4}$ b) $\left(-\frac{3}{5}\right) \div \frac{7}{8}$

SOLUTION

a) $\frac{5}{7} \div \frac{3}{4} = \frac{5}{7} \cdot \frac{4}{3} = \frac{5 \cdot 4}{7 \cdot 3} = \frac{20}{21}$

b) $\left(-\frac{3}{5}\right) \div \frac{7}{8} = \frac{-3}{5} \cdot \frac{8}{7} = \frac{-3 \cdot 8}{5 \cdot 7} = \frac{-24}{35} = -\frac{24}{35}$

DID YOU KNOW?

Mathematical Music

When the string of a musical instrument is plucked or bowed, it moves in a wavelike pattern like the strings shown here. The vibration this creates in the surrounding air is what your eardrums detect as sound.

The ancient Greeks believed that in nature, all harmony and everything of beauty could be explained with rational numbers. This belief was reinforced by the discovery that the sound of plucked strings could be quite pleasing if the strings plucked were in the ratio of 1 to 2 (an octave), 2 to 3 (a fifth), 3 to 4 (a fourth), and so on. Thus, the secret of harmony lies in the rational numbers such as $\frac{1}{2}$, $\frac{2}{3}$, and $\frac{3}{4}$.

The theory of vibrating strings has applications today that go well beyond music. How materials vibrate, and hence the stress they can absorb, is a vital matter in the construction of rockets, buildings, and bridges.

Addition and Subtraction of Fractions

Before we can add or subtract fractions, the fractions must have a common denominator. A common denominator is another name for a common multiple of the denominators. The *lowest common denominator (LCD)* is the least common multiple of the denominators.

To add or subtract two fractions with a common denominator, we add or subtract their numerators and retain the common denominator.

ADDITION AND SUBTRACTION OF FRACTIONS

$$\frac{a}{c} + \frac{b}{c} = \frac{a+b}{c}, \quad c \neq 0; \qquad \frac{a}{c} - \frac{b}{c} = \frac{a-b}{c}, \quad c \neq 0$$

EXAMPLE 13 *Adding and Subtracting Fractions*

Evaluate.

a) $\frac{5}{16} + \frac{7}{16}$ b) $\frac{13}{21} - \frac{8}{21}$

SOLUTION

a) $\frac{5}{16} + \frac{7}{16} = \frac{5+7}{16} = \frac{12}{16} = \frac{3}{4}$ b) $\frac{13}{21} - \frac{8}{21} = \frac{13-8}{21} = \frac{5}{21}$

Note that in Example 13, the denominators of the fractions being added or subtracted were the same; that is, they have a common denominator. *When adding or subtracting two fractions with unlike denominators, first rewrite each fraction with a common denominator. Then add or subtract the fractions.*

Writing fractions with a common denominator is accomplished with the *fundamental law of rational numbers.*

FUNDAMENTAL LAW OF RATIONAL NUMBERS

If a, b, and c are integers, with $b \neq 0$ and $c \neq 0$, then

$$\frac{a}{b} = \frac{a}{b} \cdot \frac{c}{c} = \frac{a \cdot c}{b \cdot c}$$

The terms $\frac{a}{b}$ and $\frac{a \cdot c}{b \cdot c}$ are called ***equivalent fractions***. For example, since $\frac{5}{12} = \frac{5 \cdot 5}{12 \cdot 5} = \frac{25}{60}$, the fractions $\frac{5}{12}$ and $\frac{25}{60}$ are equivalent fractions. We will see the importance of equivalent fractions in the next two examples.

EXAMPLE 14 *Subtracting Fractions with Unlike Denominators*

Evaluate $\frac{5}{12} - \frac{3}{10}$.

SOLUTION Using prime factorization (Section 5.1), we find that the LCM of 12 and 10 is 60. We will therefore express each fraction as an equivalent fraction with a denominator of 60. Sixty divided by 12 is 5. Therefore, the denominator, 12, must be multiplied by 5 to get 60. If the denominator is multiplied by 5, the numerator must also be multiplied by 5 so that the value of the fraction remains unchanged. Multiplying both numerator and denominator by 5 is the same as multiplying by 1.

We follow the same procedure for the other fraction, $\frac{3}{10}$. Sixty divided by 10 is 6. Therefore, we multiply both the denominator, 10, and the numerator, 3, by 6 to obtain an equivalent fraction with a denominator of 60.

$$\begin{aligned}\frac{5}{12} - \frac{3}{10} &= \left(\frac{5}{12} \cdot \frac{5}{5}\right) - \left(\frac{3}{10} \cdot \frac{6}{6}\right)\\ &= \frac{25}{60} - \frac{18}{60}\\ &= \frac{7}{60}\end{aligned}$$

EXAMPLE 15 *Adding Fractions with Unlike Denominators*

Evaluate $\frac{1}{54} + \frac{1}{90}$.

SOLUTION On page 230, in Example 6 of Section 5.1, we determined that the LCM of 54 and 90 is 270. Rewrite each fraction as an equivalent fraction using the LCM as the common denominator.

$$\begin{aligned}\frac{1}{54} + \frac{1}{90} &= \left(\frac{1}{54} \cdot \frac{5}{5}\right) + \left(\frac{1}{90} \cdot \frac{3}{3}\right)\\ &= \frac{5}{270} + \frac{3}{270}\\ &= \frac{8}{270}\end{aligned}$$

Now we reduce $\frac{8}{270}$ by dividing both 8 and 270 by 2, their greatest common factor.

$$\frac{8}{270} = \frac{8 \div 2}{270 \div 2} = \frac{4}{135}$$

EXAMPLE 16 *Rice Preparation*

Following are the instructions given on a box of Minute Rice. Determine the amount of (a) rice and water, (b) salt, and (c) butter or margarine needed to make 3 servings of rice.

DIRECTIONS

1. Bring water, salt, and butter (or margarine) to a boil.
2. Stir in rice. Cover; remove from heat. Let stand 5 minutes. Fluff with fork.

To Make	Rice & Water (Equal Measures)	Salt	Butter or Margarine (If Desired)
2 servings	$\frac{2}{3}$ cup	$\frac{1}{4}$ tsp	1 tsp
4 servings	$1\frac{1}{3}$ cups	$\frac{1}{2}$ tsp	2 tsp

SOLUTION Since 3 is halfway between 2 and 4, we can find the amount of each ingredient by finding the average of the amount for 2 and 4 servings. To do so, we add the amounts for 2 servings and 4 servings and divide the sum by 2.

a) Rice and water: $\dfrac{\frac{2}{3} + 1\frac{1}{3}}{2} = \dfrac{\frac{2}{3} + \frac{4}{3}}{2} = \dfrac{\frac{6}{3}}{2} = \dfrac{2}{2} = 1$ cup

b) Salt: $\dfrac{\frac{1}{4} + \frac{1}{2}}{2} = \dfrac{\frac{1}{4} + \frac{2}{4}}{2} = \dfrac{\frac{3}{4}}{2} = \dfrac{3}{4} \cdot \dfrac{1}{2} = \dfrac{3}{8}$ tsp

c) Butter or margarine: $\dfrac{1 + 2}{2} = \dfrac{3}{2}$, or $1\frac{1}{2}$ tsp

The solution to Example 16 can be found in other ways. Suggest two other procedures for solving the same problem.

SECTION 5.3 EXERCISES

CONCEPT/WRITING EXERCISES

1. Describe the set of rational numbers.

2. a) Explain how to write a terminating decimal number as a fraction.

   b) Write 0.213 as a fraction.

3. a) Explain how to reduce a fraction to lowest terms.

   b) Reduce $\frac{28}{35}$ to lowest terms using the procedure in part (a).

4. Explain how to convert an improper fraction to a mixed number.

5. Explain how to convert a mixed number to an improper fraction.

6. a) Explain how to multiply two fractions.

   b) Multiply $\frac{14}{15} \cdot \frac{25}{28}$ using the procedure in part (a).

7. a) Explain how to determine the reciprocal of a number.

   b) Determine the reciprocal of -2 by using the procedure in part(a).

8. a) Explain how to divide two fractions.

   b) Divide $\frac{14}{33} \div \frac{10}{21}$ by using the procedure in part (a).

9. a) Explain how to add or subtract two fractions having a common denominator.

b) Add $\frac{11}{24} + \frac{3}{24}$ using the procedure in part (a).

c) Subtract $\frac{37}{48} - \frac{13}{48}$ using the procedure in part (a).

10. a) Explain how to add or subtract two fractions having unlike denominators.

b) Add $\frac{2}{3} + \frac{5}{8}$ using the procedure in part (a).

c) Subtract $\frac{5}{6} - \frac{2}{15}$ using the procedure in part (a).

11. In your own words, state the fundamental law of rational numbers.

12. Are $\frac{7}{11}$ and $\frac{21}{33}$ equivalent fractions? Explain your answer.

PRACTICE THE SKILLS

In Exercises 13–22, reduce each fraction to lowest terms.

13. $\frac{5}{10}$ 14. $\frac{12}{16}$ 15. $\frac{24}{54}$

16. $\frac{36}{56}$ 17. $\frac{95}{125}$ 18. $\frac{13}{221}$

19. $\frac{112}{176}$ 20. $\frac{120}{135}$ 21. $\frac{45}{495}$

22. $\frac{124}{148}$

In Exercises 23–28, convert each mixed number to an improper fraction.

23. $3\frac{5}{8}$ 24. $2\frac{1}{4}$ 25. $-1\frac{15}{16}$

26. $-7\frac{1}{5}$ 27. $-4\frac{15}{16}$ 28. $11\frac{9}{16}$

In Exercises 29–32, write the number of inches indicated by the arrows as an improper fraction.

29.
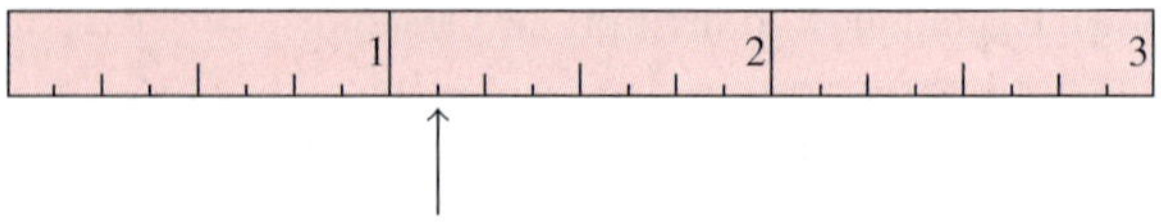

30.
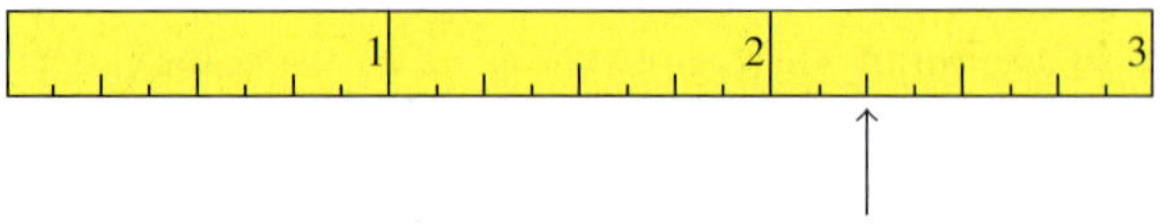

31.
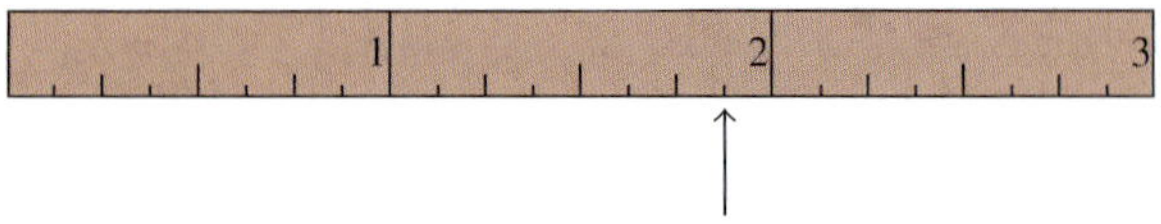

32.
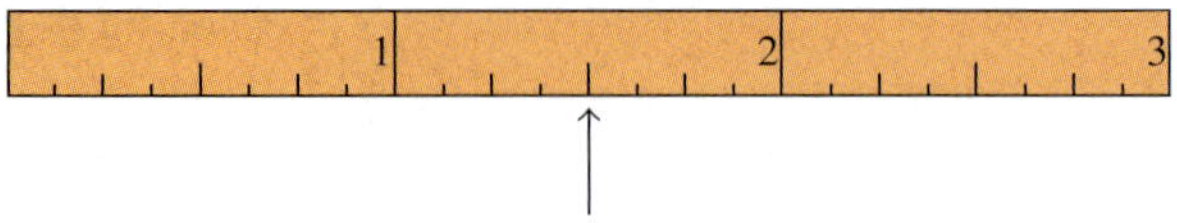

In Exercises 33–38, convert each improper fraction to a mixed number.

33. $\frac{13}{5}$ 34. $\frac{23}{4}$ 35. $-\frac{73}{6}$

36. $-\frac{157}{12}$ 37. $-\frac{878}{15}$ 38. $\frac{1028}{21}$

In Exercises 39–48, express each rational number as terminating or repeating decimal number.

39. $\frac{7}{10}$ 40. $\frac{15}{16}$ 41. $\frac{2}{9}$

42. $\frac{1}{6}$ 43. $\frac{3}{8}$ 44. $\frac{23}{7}$

45. $\frac{13}{6}$ 46. $\frac{115}{15}$ 47. $\frac{85}{15}$

48. $\frac{100}{9}$

In Exercises 49–58, express each terminating decimal number as a quotient of two integers. If possible, reduce the quotient to lowest terms.

49. 0.75 50. 0.29 51. 0.045

52. 0.375 53. 0.2 54. 0.251

55. 0.0131 56. 0.2345 57. 0.0001

58. 0.0053

In Exercises 59–68, express each repeating decimal number as a quotient of two integers. If possible, reduce the quotient to lowest terms.

59. $0.\overline{1}$ 60. $0.\overline{5}$ 61. $1.\overline{9}$

62. $0.\overline{23}$ 63. $1.\overline{36}$ 64. $0.\overline{135}$

65. $2.0\overline{5}$ 66. $2.4\overline{9}$ 67. $3.4\overline{78}$

68. $4.1\overline{19}$

In Exercises 69–78, perform the indicated operation and reduce your answer to lowest terms.

69. $\frac{1}{2} \cdot \frac{4}{5}$ 70. $\frac{2}{9} \div \frac{4}{7}$

71. $\left(\frac{-3}{8}\right)\left(\frac{-16}{15}\right)$

72. $\left(-\frac{3}{5}\right) \div \frac{10}{21}$

73. $\frac{7}{8} \div \frac{8}{7}$

74. $\frac{3}{7} \div \frac{3}{7}$

75. $\left(\frac{3}{5} \cdot \frac{4}{7}\right) \div \frac{1}{3}$

76. $\left(\frac{4}{7} \div \frac{4}{5}\right) \cdot \frac{1}{7}$

77. $\left[\left(-\frac{2}{3}\right)\left(\frac{5}{8}\right)\right] \div \left(-\frac{7}{16}\right)$

78. $\left(\frac{7}{15} \cdot \frac{5}{8}\right) \div \left(\frac{7}{9} \cdot \frac{5}{2}\right)$

In Exercises 79–88, perform the indicated operation and reduce your answer to lowest terms.

79. $\frac{1}{4} + \frac{2}{3}$

80. $\frac{7}{8} - \frac{1}{6}$

81. $\frac{2}{11} + \frac{5}{22}$

82. $\frac{5}{12} + \frac{7}{36}$

83. $\frac{5}{9} - \frac{7}{54}$

84. $\frac{17}{25} - \frac{43}{100}$

85. $\frac{1}{12} + \frac{1}{48} + \frac{1}{72}$

86. $\frac{3}{5} + \frac{7}{15} + \frac{9}{75}$

87. $\frac{1}{30} - \frac{3}{40} - \frac{7}{50}$

88. $\frac{4}{25} - \frac{9}{100} - \frac{7}{40}$

PROBLEM SOLVING

Alternative methods for adding and subtracting two fractions are shown. These methods may not result in a solution in its lowest terms.

$$\frac{a}{b} + \frac{c}{d} = \frac{ad + bc}{bd} \quad \text{and} \quad \frac{a}{b} - \frac{c}{d} = \frac{ad - bc}{bd}$$

In Exercises 89–94, use the appropriate formula to evaluate the expression.

89. $\frac{2}{3} + \frac{1}{8}$

90. $\frac{1}{5} + \frac{4}{9}$

91. $\frac{5}{6} - \frac{7}{8}$

92. $\frac{7}{3} - \frac{5}{12}$

93. $\frac{3}{8} + \frac{5}{12}$

94. $\left(\frac{2}{3} + \frac{1}{4}\right) - \frac{3}{5}$

In Exercises 95–100, evaluate each expression.

95. $\left(\frac{2}{3} \cdot \frac{9}{10}\right) + \frac{2}{5}$

96. $\left(\frac{7}{6} \div \frac{4}{3}\right) - \frac{11}{12}$

97. $\left(\frac{3}{4} + \frac{1}{6}\right) \div \left(2 - \frac{7}{6}\right)$

98. $\left(\frac{1}{3} \cdot \frac{3}{7}\right) + \left(\frac{3}{5} \cdot \frac{10}{11}\right)$

99. $\left(3 - \frac{4}{9}\right) \div \left(4 + \frac{2}{3}\right)$

100. $\left(\frac{2}{5} \div \frac{4}{9}\right)\left(\frac{3}{5} \cdot 6\right)$

In Exercises 101–112, write an expression that will solve the problem and then evaluate the expression.

101. ***Height Increase*** When David Conway finished his first year of high school, his height was $69\frac{7}{8}$ inches. When he returned to school after the summer, David's height was $71\frac{5}{8}$ inches. How much did David's height increase over the summer?

102. ***Thistles*** Diane Helbing has four different varieties of thistles invading her pasture. She estimates that of these thistles, $\frac{1}{2}$ are Canada thistles, $\frac{1}{4}$ are bull thistles, $\frac{1}{6}$ are plumeless thistles, and the rest are musk thistles. What fraction of the thistles are musk thistles?

103. ***Stairway Height*** A stairway consists of 14 stairs, each $8\frac{5}{8}$ inches high. What is the vertical height of the stairway?

104. ***Alphabet Soup*** Margaret Cannata's recipe for alphabet soup calls for (among other items) $\frac{1}{4}$ cup snipped parsley, $\frac{1}{8}$ teaspoon pepper, and $\frac{1}{2}$ cup sliced carrots. Margaret is expecting company and needs to multiply the amounts of the ingredients by $1\frac{1}{2}$ times. Determine the amount of (a) snipped parsley, (b) pepper, and (c) sliced carrots she needs for the soup.

105. ***Sprinkler System*** To repair his sprinkler system, Tony Gambino needs a total of $20\frac{5}{16}$ inches of PVC pipe. He has on hand pieces that measure $2\frac{1}{4}$ inches, $3\frac{7}{8}$ inches, and $4\frac{1}{4}$ inches in length. If he can combine these pieces and use them in the repair, how long of a piece of PVC pipe will Tony need to purchase to repair his sprinkler system?

106. ***Crop Storage*** Todd Schroeder has a silo on his farm in which he can store silage made from his various crops. He currently has a silo that is $\frac{1}{4}$ full of corn silage, $\frac{2}{5}$ full of hay silage, and $\frac{1}{3}$ full of oats silage. What fraction of Todd's silo is currently in use?

107. ***Department Budget*** Jaime Bailey is chair of the humanities department at Santa Fe Community College. Jaime has a budget in which $\frac{1}{2}$ of the money is for photocopying, $\frac{2}{5}$ of the money is for computer-related expenses, and the rest of the money is for student tutors in the foreign languages lab. What fraction of Jaime's budget is for student tutors?

108. ***Art Supplies*** Denise Viale teaches kindergarten and is buying supplies for her class to make papier-mâché piggy banks. Each piggy bank to be made requires $1\frac{1}{4}$ cups of flour. If Denise has 15 students who are going to make piggy banks, how much flour does Denise need to purchase?

109. ***Height of a Computer Stand*** The instructions for assembling a computer stand include a diagram illustrating its dimensions. Find the total height of the stand.

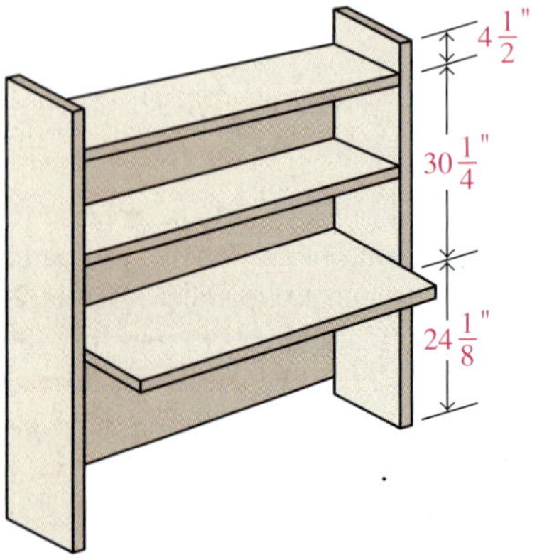

110. ***Width of a Picture*** The width of a picture is $24\frac{7}{8}$ in., as shown in the diagram. Find x, the distance from the edge of the frame to the center.

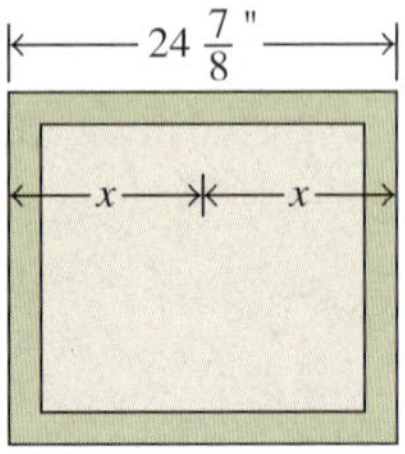

111. ***Traveling Interstate 5*** While on vacation, Janet Mazerella traveled the full length of Interstate Highway 5 from the U.S.-Canadian border to the U.S.-Mexican border. She recorded the following travel times between cities: Blaine, Washington, to Seattle: 1 hour 49 minutes; Seattle to Portland: 2 hours 48 minutes; Portland to Sacramento: 9 hours 6 minutes; Sacramento to Los Angeles: 6 hours 3 minutes; Los Angeles to San Diego: 2 hours 9 minutes; San Diego to San Ysidro, California: 22 minutes.

a) Write each of these times as a mixed number with minutes represented as a fraction with a denominator of 60. Do not reduce the fractional part of the mixed number.

b) What is Janet's total driving time from Blaine, Washington, to San Ysidro, California? Give your answer as a mixed number and in terms of hours and minutes.

▲ Interstate Highway 5

112. ***Traveling Across Texas*** While on vacation, Omar Rodriguez traveled the entire length of Interstate Highway 10 in Texas. Omar recorded the following travel times between cities on his trip. Anthony to El Paso: 25 minutes; El Paso to San Antonio: 7 hours 54 minutes; San Antonio to Houston: 3 hours 1 minute; Houston to Beaumont: 1 hour 23 minutes; Beaumont to Orange: 28 minutes.

a) Write each of these times as a mixed number with minutes represented as a fraction with a denominator of 60. Do not reduce the fractional part of the mixed number.

b) What is Omar's total driving time from Anthony to Orange? Give your answer as a mixed number and in terms of hours and minutes.

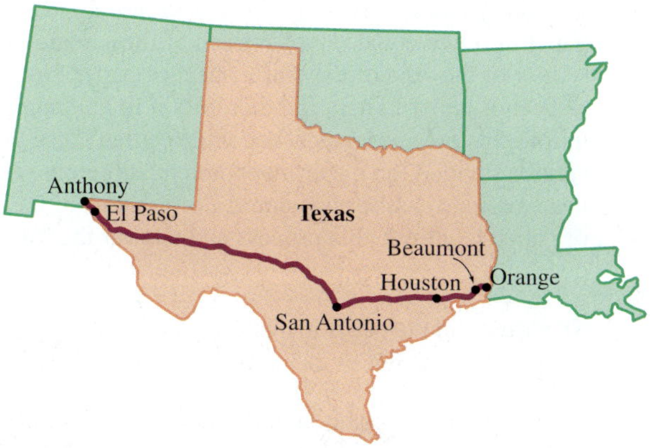

▲ Interstate Highway 10

▲ The Alamo, San Antonio, Texas

CHALLENGE PROBLEMS/GROUP ACTIVITIES

113. ***Cutting Lumber*** If a piece of wood $8\frac{3}{4}$ ft long is to be cut into four equal pieces, find the length of each piece. (Allow $\frac{1}{8}$ in. for each saw cut.)

114. ***Dimensions of a Room*** A rectangular room measures 8 ft 3 in. by 10 ft 8 in. by 9 ft 2 in. high.

a) Determine the perimeter of the room in feet. Write your answer as a mixed number.

b) Calculate the area of the floor of the room in square feet.

c) Calculate the volume of the room in cubic feet. Use volume = length × width × height. Write your answer as a mixed number.

115. ***Hanging a Picture*** The back of a framed picture that is to be hung is shown. A nail is to be hammered into the wall, and the picture will be hung by the wire on the nail.

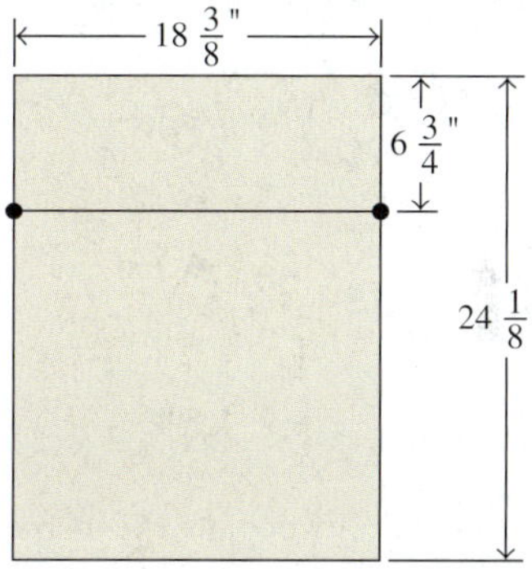

a) If the center of the wire is to rest on the nail and a side of the picture is to be 20 in. from the window, how far from the window should the nail be placed?

b) If the top of the frame is to be $26\frac{1}{4}$ in. from the ceiling, how far from the ceiling should the nail be placed? (Assume the wire will not stretch.)

c) Repeat part (b) if the wire will stretch $\frac{1}{4}$ in. when the picture is hung.

116. ***Increasing a Book Size*** The dimensions of the cover of a book have been increased from $8\frac{1}{2}$ in. by $9\frac{1}{4}$ in. to $8\frac{1}{2}$ in. by $10\frac{1}{4}$ in. By how many square inches has the surface area increased? Use area = length × width.

Dense Set of Numbers *A set of numbers is said to be a* ***dense set*** *if between any two distinct members of the set there exists a third distinct member of the set. The set of integers is not dense since between any two consecutive integers there is not another integer. For example, between 1 and 2 there are no other integers. The set of rational numbers is dense because between any two distinct rational numbers there exists a third distinct rational number. For example, we can find a rational number between 0.243 and 0.244. The number 0.243 can be written as 0.2430, and 0.244 can be written as 0.2440. There are many numbers between these two.*

Some of them are 0.2431, 0.2435, and 0.243912. In Exercises 117–122, find a rational number between the two numbers in each pair.

117. 0.10 and 0.11

118. 5.03 and 5.003

119. −2.176 and −2.175

120. 1.3457 and 1.34571

121. 4.872 and 4.873

122. −3.7896 and −3.7895

Halfway Between Two Numbers *To find a rational number halfway between any two rational numbers given in fraction form, add the two numbers together and divide their sum by 2. In Exercises 123–128, find a rational number halfway between the two fractions in each pair.*

123. $\frac{1}{3}$ and $\frac{2}{3}$

124. $\frac{2}{7}$ and $\frac{3}{7}$

125. $\frac{1}{100}$ and $\frac{1}{10}$

126. $\frac{7}{13}$ and $\frac{8}{13}$

127. $\frac{1}{10}$ and $\frac{1}{100}$

128. $\frac{1}{2}$ and $\frac{2}{3}$

129. ***Cooking Oatmeal*** Following are the instructions given on a box of oatmeal. Determine the amount of water (or milk) and oats needed to make $1\frac{1}{2}$ servings by:

a) Adding the amount of each ingredient needed for 1 serving to the amount needed for 2 servings and dividing by 2.

b) Adding the amount of each ingredient needed for 1 serving to half the amount needed for 1 serving.

DIRECTIONS

1. Boil water or milk and salt (if desired).
2. Stir in oats.
3. Stirring occasionally, cook over medium heat for 5 minutes.

Servings	**1**	**2**
Water (or milk)	1 cup	$1\frac{3}{4}$ cup
Oats	$\frac{1}{2}$ cup	1 cup
Salt (optional)	dash	$\frac{1}{8}$ tsp

130. Consider the rational number $0.\overline{9}$.

a) Use the method from Example 8 on page 250 to convert $0.\overline{9}$ to a quotient of integers.

b) Find a number halfway between $0.\overline{9}$ and 1 by adding the two numbers and dividing by 2.

c) Find $\frac{1}{3} + \frac{2}{3}$. Express $\frac{1}{3}$ and $\frac{2}{3}$ as repeating decimals. Now find the same sum using the repeating decimal representation of $\frac{1}{3}$ and $\frac{2}{3}$.

d) What conclusion can you draw from parts (a), (b), and (c)?

RECREATIONAL MATHEMATICS

131. ***Paper Folding*** Fold a sheet of paper in half. Now unfold the paper. You will see that this one crease divided the paper into two equal regions. Each of these regions represents $\frac{1}{2}$ of the area of the entire sheet of paper. Next, fold a sheet of paper in half and then fold it in half again. Now unfold this piece of paper. You will see that these two creases divided the paper into four equal regions. Each of these regions can be considered $\frac{1}{4}$ of the area of the sheet of paper. Continue this process and answer the following questions.

a) If you fold the paper in half three times, each region will be what fraction of the area of the sheet of paper?

b) If you fold the paper in half four times, each region will be what fraction of the area of the sheet of paper?

c) How many creases will you need to form regions that are $\frac{1}{32}$ of the area of the sheet of paper?

d) How many creases will you need to form regions that are $\frac{1}{64}$ of the area of the sheet of paper?

INTERNET/RESEARCH ACTIVITY

132. The ancient Greeks are often considered the first true mathematicians. Write a report summarizing the ancient Greeks' contributions to rational numbers. Include in your report what they learned and believed about the rational numbers. References include encyclopedias, history of mathematics books, and Internet web sites.

5.4 THE IRRATIONAL NUMBERS AND THE REAL NUMBER SYSTEM

▲ The distance from the northwest corner of a football field to the southeast corner is an irrational number.

A football field is 160 ft wide and 360 ft long. A football player catches the football from a kickoff in the northwest corner of one end zone and runs across the field to the southeast corner of the other end zone, as shown by the dashed line in the diagram.

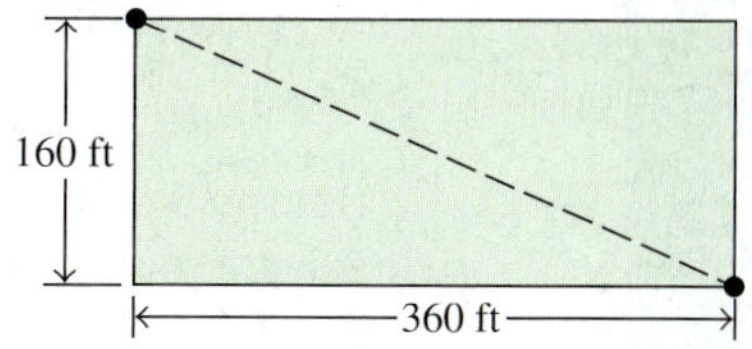

How far did the football player run? The answer to this question is an example of an *irrational* number. In this section, we will introduce irrational numbers along with the operations of addition, subtraction, multiplication, and division of irrational numbers. The answer to the above question is $40\sqrt{97}$ feet, or about 394 feet.

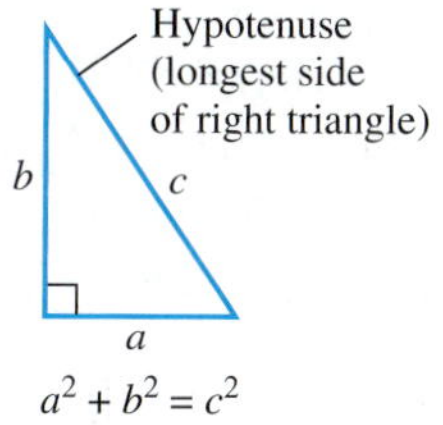

Figure 5.7

Pythagoras (ca. 585–500 B.C.), a Greek mathematician, is credited with providing a written proof that in any *right triangle* (a triangle with a 90° angle; see Fig. 5.7), the square of the length of one side (a^2) added to the square of the length of the other side (b^2) equals the square of the length of the hypotenuse (c^2). The formula $a^2 + b^2 = c^2$ is now known as the ***Pythagorean theorem***.* Pythagoras found that the solution of the formula, where $a = 1$ and $b = 1$, is not a rational number.

$$\begin{aligned} a^2 + b^2 &= c^2 \\ 1^2 + 1^2 &= c^2 \\ 1 + 1 &= c^2 \\ 2 &= c^2 \end{aligned}$$

There is no rational number that when squared will equal 2. This fact prompted a need for a new set of numbers, the irrational numbers.

In Section 5.2, we introduced the real number line. The points on the real number line that are not rational numbers are referred to as irrational numbers. Recall that every rational number is either a terminating or a repeating decimal number. Therefore, irrational numbers, when represented as decimal numbers, will be nonterminating, nonrepeating decimal numbers.

An **irrational number** is a real number whose decimal representation is a nonterminating, nonrepeating decimal number.

A nonrepeating decimal number such as 5.12639537. . . can be used to indicate an irrational number. Notice that no number or set of numbers repeat on a continuous basis, and the three dots at the end of the number indicate that the number continues

*The Pythagorean theorem is discussed in more detail in Section 9.3.

PROFILE IN MATHEMATICS

Pythagoras of Samos

Pythagoras of Samos founded a philosophical and religious school in southern Italy in the sixth century B.C. The scholars at the school, known as Pythagoreans, produced important works of mathematics, astronomy, and theory of music. Although the Pythagoreans are credited with proving the Pythagorean theorem, it was known to the ancient Babylonians 1000 years earlier. The Pythagoreans were a secret society that formed a model for many secret societies in existence today. One practice was that students were to spend their first three years of study in silence, while their master, Pythagoras, spoke to them from behind a curtain. Among other philosophical beliefs of the Pythagoreans was "that at its deepest level, reality is mathematical in nature."

indefinitely. Nonrepeating number patterns can be used to indicate irrational numbers. For example, 6.1011011101111 . . . and 0.525225222 . . . are both irrational numbers.

The expression $\sqrt{2}$ is read "the square root of 2" or "radical 2." The symbol $\sqrt{}$ is called the ***radical sign***, and the number or expression inside the radical sign is called the ***radicand***. In $\sqrt{2}$, 2 is the radicand.

The square roots of some numbers are rational, whereas the square roots of other numbers are irrational. The ***principal*** (or ***positive***) ***square root*** of a number n, written $\sqrt{n}$, is the positive number that when multiplied by itself, gives n. Whenever we mention the term "square root" in this text, we mean the principal square root. For example,

$$\sqrt{9} = 3 \quad \text{since} \quad 3 \cdot 3 = 9$$
$$\sqrt{36} = 6 \quad \text{since} \quad 6 \cdot 6 = 36$$

Both $\sqrt{9}$ and $\sqrt{36}$ are examples of numbers that are rational numbers because their square roots, 3 and 6 respectively, are terminating decimal numbers.

Returning to the problem faced by Pythagoras: If $c^2 = 2$, then c has a value of $\sqrt{2}$, but what is $\sqrt{2}$ equal to? The $\sqrt{2}$ is an irrational number, and it cannot be expressed as a terminating or repeating decimal number. It can only be approximated by a decimal number: $\sqrt{2}$ is approximately 1.4142135 (to seven decimal places). Later in this section, we will discuss using a calculator to approximate irrational numbers.

Other irrational numbers include $\sqrt{3}$, $\sqrt{5}$, and $\sqrt{37}$. Another important irrational number used to represent the ratio of a circle's circumference to its diameter is pi, symbolized π. Pi is approximately 3.1415926.

We have discussed procedures for performing the arithmetic operations of addition, subtraction, multiplication, and division with rational numbers. We can perform the same operations with the irrational numbers. Before we can proceed, however, we must understand the numbers called perfect squares. Any number that is the square of a natural number is said to be a ***perfect square***. Some perfect squares are shown in the following chart.

Natural numbers	1,	2,	3,	4,	5,	6, . . .
Squares of the natural numbers	1^2,	2^2,	3^2,	4^2,	5^2,	6^2, . . .
or perfect squares	1,	4,	9,	16,	25,	36, . . .

The numbers 1, 4, 9, 16, 25, and 36 are some of the perfect square numbers. Can you determine the next two perfect square numbers? How many perfect square numbers are there? The square root of a perfect square number will be a natural number. For example, $\sqrt{1} = 1$, $\sqrt{4} = 2$, $\sqrt{9} = 3$, $\sqrt{16} = 4$, $\sqrt{25} = 5$, and so on.

The number that multiplies a radical is called the radical's ***coefficient***. For example, in $3\sqrt{5}$, the 3 is the coefficient of the radical.

Some irrational numbers can be simplified by determining whether there are any perfect square factors in the radicand. If there are, the following rule can be used to simplify the radical.

PRODUCT RULE FOR RADICALS

$$\sqrt{a \cdot b} = \sqrt{a} \cdot \sqrt{b}, \qquad a \geq 0, \qquad b \geq 0$$

To simplify a radical, write the radical as a product of two radicals. One of the radicals should contain the greatest perfect square that is a factor of the radicand in the original expression. Then simplify the radical containing the perfect square factor.

> **DID YOU KNOW?**
>
> ***A Piece of the Pi***
>
> 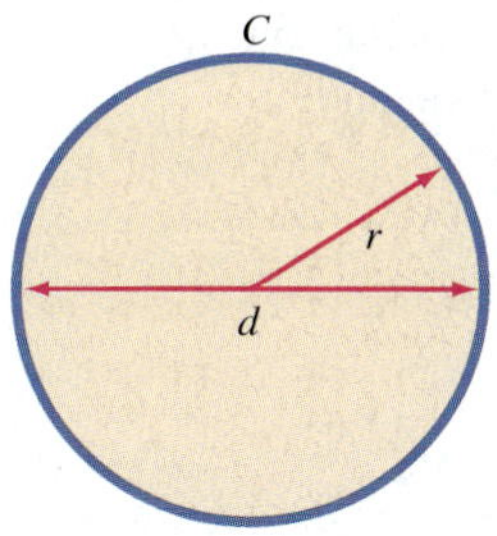
>
>
> $C = \pi d$
> or $\pi = \frac{C}{d}$
>
> American philosopher and psychologist William James wrote in 1909: "The thousandth decimal of π sleeps there though no one may ever try to compute it." It sleeps no more. We now know π, the ratio of a circle's circumference to its diameter, to over 1.2 trillion digits.
>
> In December 2002, Yasumasa Kanada and others at the University of Tokyo announced that they had calculated π to 1,241,100,000,000 decimal places, beating their previous record set in 1999. Their computation of π consumed more than 600 hours of time on a Hitachi SR8000 supercomputer.
>
> This record, like the record for the largest prime number, will most likely be broken in the near future (it might already be broken as you read this). Mathematicians and computer scientists continue to improve both their computers and their methods used to find numbers like the most accurate approximation for π or the largest prime number.

For example,

$$\sqrt{18} = \sqrt{9 \cdot 2} = \sqrt{9} \cdot \sqrt{2} = 3 \cdot \sqrt{2} = 3\sqrt{2}$$

and

$$\sqrt{75} = \sqrt{25 \cdot 3} = \sqrt{25} \cdot \sqrt{3} = 5 \cdot \sqrt{3} = 5\sqrt{3}$$

EXAMPLE 1 *Simplifying Radicals*

Simplify.

a) $\sqrt{18}$ b) $\sqrt{80}$

SOLUTION

a) Since 9 is a perfect square factor of 18, we write

$$\sqrt{18} = \sqrt{9 \cdot 2} = \sqrt{9} \cdot \sqrt{2} = 3 \cdot \sqrt{2} = 3\sqrt{2}$$

Since 2 has no perfect square factors, $\sqrt{2}$ cannot be simplified.

b) Since 16 is a perfect square factor of 80, we write

$$\sqrt{80} = \sqrt{16 \cdot 5} = \sqrt{16} \cdot \sqrt{5} = 4 \cdot \sqrt{5} = 4\sqrt{5}$$

In Example 1(b), you can obtain the correct answer if you start out factoring differently:

$$\sqrt{80} = \sqrt{4 \cdot 20} = \sqrt{4} \cdot \sqrt{20} = 2 \cdot \sqrt{20} = 2\sqrt{20}$$

Note that 20 has 4 as a perfect square factor.

$$2\sqrt{20} = 2\sqrt{4 \cdot 5} = 2 \cdot \sqrt{4} \cdot \sqrt{5} = 2 \cdot 2 \cdot \sqrt{5} = 4\sqrt{5}$$

The second method will eventually give the same answer, but it requires more work. It is best to try to factor out the *largest* perfect square factor from the radicand.

Addition and Subtraction of Irrational Numbers

To add or subtract two or more square roots with the same radicand, add or subtract their coefficients while keeping the common radicand. The answer is the sum or difference of the coefficients multiplied by the common radical.

EXAMPLE 2 *Adding and Subtracting Radicals with the Same Radicand*

Simplify.

a) $4\sqrt{3} + 5\sqrt{3}$ b) $2\sqrt{5} - 7\sqrt{5} + \sqrt{5}$

SOLUTION

a) $4\sqrt{3} + 5\sqrt{3} = (4 + 5)\sqrt{3} = 9\sqrt{3}$

b) $2\sqrt{5} - 7\sqrt{5} + \sqrt{5} = (2 - 7 + 1)\sqrt{5} = -4\sqrt{5}$

Note that $\sqrt{5} = 1\sqrt{5}$.

DID YOU KNOW?

Is the Computer Always Right?

We often assume that the computer is always right. We base this on the history and performance of the computer language and programs we use. Obviously, we cannot check calculations that are being made in billionths of a second. We could run a computation twice or on two different systems to be sure we get the same result to a specified number of decimal places. That's what is done with the big supercomputers. Their computation of the value of π often serves as a test of the program's ability to compute accurately.

EXAMPLE 3 *Subtracting Radicals with Different Radicands*

Simplify $5\sqrt{3} - \sqrt{12}$.

SOLUTION These radicals cannot be subtracted in their present form because they contain different radicands. When that occurs, determine whether one or more of the radicals can be simplified so that they have the same radicand.

$$\begin{aligned} 5\sqrt{3} - \sqrt{12} &= 5\sqrt{3} - \sqrt{4 \cdot 3} \\ &= 5\sqrt{3} - \sqrt{4} \cdot \sqrt{3} \\ &= 5\sqrt{3} - 2\sqrt{3} \\ &= (5 - 2)\sqrt{3} = 3\sqrt{3} \end{aligned}$$

Multiplication of Irrational Numbers

When multiplying irrational numbers, we again make use of the product rule for radicals. After the radicands are multiplied, simplify the remaining radical when possible.

EXAMPLE 4 *Multiplying Radicals*

Simplify.

a) $\sqrt{3} \cdot \sqrt{27}$ b) $\sqrt{3} \cdot \sqrt{7}$ c) $\sqrt{6} \cdot \sqrt{10}$

SOLUTION

a) $\sqrt{3} \cdot \sqrt{27} = \sqrt{3 \cdot 27} = \sqrt{81} = 9$

b) $\sqrt{3} \cdot \sqrt{7} = \sqrt{3 \cdot 7} = \sqrt{21}$

c) $\sqrt{6} \cdot \sqrt{10} = \sqrt{6 \cdot 10} = \sqrt{60} = \sqrt{4 \cdot 15} = \sqrt{4} \cdot \sqrt{15} = 2\sqrt{15}$

Division of Irrational Numbers

To divide irrational numbers, use the following rule. After performing the division, simplify when possible.

QUOTIENT RULE FOR RADICALS

$$\frac{\sqrt{a}}{\sqrt{b}} = \sqrt{\frac{a}{b}}, \qquad a \geq 0, \qquad b > 0$$

EXAMPLE 5 *Dividing Radicals*

Divide.

a) $\dfrac{\sqrt{8}}{\sqrt{2}}$ b) $\dfrac{\sqrt{96}}{\sqrt{2}}$

SOLUTION

a) $\dfrac{\sqrt{8}}{\sqrt{2}} = \sqrt{\dfrac{8}{2}} = \sqrt{4} = 2$

b) $\dfrac{\sqrt{96}}{\sqrt{2}} = \sqrt{\dfrac{96}{2}} = \sqrt{48} = \sqrt{16 \cdot 3} = \sqrt{16} \cdot \sqrt{3} = 4\sqrt{3}$

Rationalizing the Denominator

A denominator of a fraction is *rationalized* when it contains no radical expressions. To rationalize a denominator that contains only a square root, multiply both the numerator and denominator of the fraction by a number that will result in the radicand in the denominator becoming a perfect square. (This action is the equivalent of multiplying the fraction by 1 because the value of the fraction does not change.) Then simplify the fractions when possible.

EXAMPLE 6 *Rationalizing the Denominator*

Rationalize the denominator of the following.

a) $\dfrac{5}{\sqrt{2}}$ b) $\dfrac{5}{\sqrt{12}}$ c) $\dfrac{\sqrt{5}}{\sqrt{10}}$

SOLUTION

a) Multiply the numerator and denominator by a number that will make the radicand in the denominator a perfect square.

$$\frac{5}{\sqrt{2}} = \frac{5}{\sqrt{2}} \cdot \frac{\sqrt{2}}{\sqrt{2}} = \frac{5\sqrt{2}}{\sqrt{4}} = \frac{5\sqrt{2}}{2}$$

Note that the 2's in the answer cannot be divided out because one 2 is a radicand and the other is not.

b) $\dfrac{5}{\sqrt{12}} = \dfrac{5}{\sqrt{12}} \cdot \dfrac{\sqrt{3}}{\sqrt{3}} = \dfrac{5\sqrt{3}}{\sqrt{36}} = \dfrac{5\sqrt{3}}{6}$

You could have obtained the same answer to this problem by multiplying both the numerator and denominator by $\sqrt{12}$ and then simplifying. Try to do so now.

c) Write $\dfrac{\sqrt{5}}{\sqrt{10}}$ as $\sqrt{\dfrac{5}{10}}$ and reduce the fraction to obtain $\sqrt{\dfrac{1}{2}}$. By the quotient rule for radicals, $\sqrt{\dfrac{1}{2}} = \dfrac{\sqrt{1}}{\sqrt{2}}$ or $\dfrac{1}{\sqrt{2}}$. Now rationalize the denominator of $\dfrac{1}{\sqrt{2}}$.

$$\frac{1}{\sqrt{2}} = \frac{1}{\sqrt{2}} \cdot \frac{\sqrt{2}}{\sqrt{2}} = \frac{\sqrt{2}}{2}$$

Approximating Square Roots on a Scientific Calculator

Consider the irrational number the square root of two. We use the symbol $\sqrt{2}$ to represent the *exact value* of this number. Although exact values are important, approximations are also important, especially when working with application problems. We can use a scientific calculator to obtain approximations for square roots. Scientific calculators generally have one of the following square root keys:*

$\boxed{\sqrt{\ }}$ or $\boxed{\sqrt{x}}$

*If your calculator has the $\sqrt{\ }$ symbol printed *above* the key instead of on the face of the key, you can access the square root function by first pressing the "2nd" or the "inverse" key.

For simplicity, we will refer to the square root key with the [$\sqrt{\ }$] symbol.

To approximate $\sqrt{2}$, perform the following keystrokes:

[2] [$\sqrt{\ }$]

or, depending on your model of calculator, you may have to do the following:

[$\sqrt{\ }$] [2] [ENTER]

The display on your calculator may read 1.414213562. Your calculator may display more or fewer digits. It is important to realize that 1.414213562 is a rational number *approximation* for the irrational number $\sqrt{2}$. The symbol $\approx$ means *is approximately equal to,* and we write

$$\sqrt{2} \approx 1.414213562$$

Exact value (irrational number) ↗ ↖ Approximation (rational number)

EXAMPLE 7 *Approximating Square Roots*

Use a scientific calculator to approximate the following square roots. Round each answer to two decimal places.

a) $\sqrt{3}$ b) $\sqrt{19}$ c) $\sqrt{79}$ d) $\sqrt{371}$

SOLUTION

a) $\sqrt{3} \approx 1.73$ b) $\sqrt{19} \approx 4.36$

c) $\sqrt{79} \approx 8.89$ d) $\sqrt{371} \approx 19.26$

SECTION 5.4 EXERCISES

CONCEPT/WRITING EXERCISES

1. Explain the difference between a rational number and an irrational number.

2. What is the principal square root of a number?

3. What is a perfect square?

4. a) State the product rule for radicals.

   b) State the quotient rule for radicals.

5. a) Explain how to add or subtract square roots that have the same radicand.

   b) Using the procedure in part (a), add $3\sqrt{5} + 8\sqrt{5} - 6\sqrt{5}$.

6. What does it mean to rationalize the denominator?

7. a) Explain how to rationalize a denominator that contains a square root.

   b) Using the procedure in part (a), rationalize the denominator of $\frac{2}{\sqrt{3}}$.

8. a) Explain how to approximate square roots on your calculator.

   b) Using the procedure in part (a), approximate $\sqrt{11}$. Round your answer to the nearest hundredth.

PRACTICE THE SKILLS

In Exercises 9–18, determine whether the number is rational or irrational.

9. $\sqrt{25}$

10. $\sqrt{50}$

11. $\frac{3}{5}$

12. 0.212112111...

13. 3.575775777...

14. π

15. $\frac{22}{7}$

16. 3.14159

17. 3.14159...

18. $\frac{\sqrt{5}}{\sqrt{5}}$

In Exercises 19–28, evaluate the expression.

19. $\sqrt{16}$ **20.** $\sqrt{144}$ **21.** $\sqrt{100}$

22. $-\sqrt{121}$ **23.** $-\sqrt{169}$ **24.** $\sqrt{25}$

25. $-\sqrt{81}$ **26.** $-\sqrt{36}$ **27.** $-\sqrt{100}$

28. $\sqrt{289}$

In Exercises 29–38, classify the number as a member of one or more of the following sets: the rational numbers, the integers, the natural numbers, the irrational numbers.

29. 2 **30.** -3

31. $\sqrt{25}$ **32.** $\frac{3}{8}$

33. 0.040040004 **34.** 2.718

35. $-\frac{7}{8}$ **36.** 0.123123123

37. $0.\overline{123}$ **38.** 0.123112311123...

In Exercises 39–48, simplify the radical.

39. $\sqrt{12}$ **40.** $\sqrt{20}$ **41.** $\sqrt{48}$

42. $\sqrt{50}$ **43.** $\sqrt{63}$ **44.** $\sqrt{75}$

45. $\sqrt{84}$ **46.** $\sqrt{90}$ **47.** $\sqrt{162}$

48. $\sqrt{200}$

In Exercises 49–58, perform the indicated operation.

49. $3\sqrt{5} + 2\sqrt{5}$ **50.** $4\sqrt{6} - 7\sqrt{6}$

51. $5\sqrt{18} - 7\sqrt{8}$ **52.** $2\sqrt{18} - 3\sqrt{50}$

53. $4\sqrt{12} - 7\sqrt{27}$ **54.** $2\sqrt{7} + 5\sqrt{28}$

55. $5\sqrt{3} + 7\sqrt{12} - 3\sqrt{75}$

56. $13\sqrt{2} + 2\sqrt{18} - 5\sqrt{32}$

57. $\sqrt{8} - 3\sqrt{50} + 9\sqrt{32}$

58. $\sqrt{63} + 13\sqrt{98} - 5\sqrt{112}$

In Exercises 59–68, perform the indicated operation. Simplify the answer when possible.

59. $\sqrt{3}\sqrt{27}$ **60.** $\sqrt{5}\sqrt{15}$ **61.** $\sqrt{6}\sqrt{10}$

62. $\sqrt{2}\sqrt{10}$ **63.** $\sqrt{10}\sqrt{20}$ **64.** $\sqrt{11}\sqrt{33}$

65. $\frac{\sqrt{20}}{\sqrt{5}}$ **66.** $\frac{\sqrt{125}}{\sqrt{5}}$ **67.** $\frac{\sqrt{72}}{\sqrt{8}}$

68. $\frac{\sqrt{145}}{\sqrt{5}}$

In Exercises 69–78, rationalize the denominator.

69. $\frac{1}{\sqrt{5}}$ **70.** $\frac{3}{\sqrt{3}}$ **71.** $\frac{\sqrt{3}}{\sqrt{7}}$

72. $\frac{\sqrt{2}}{\sqrt{7}}$ **73.** $\frac{\sqrt{20}}{\sqrt{3}}$ **74.** $\frac{\sqrt{50}}{\sqrt{14}}$

75. $\frac{\sqrt{5}}{\sqrt{3}}$ **76.** $\frac{\sqrt{15}}{\sqrt{3}}$ **77.** $\frac{\sqrt{10}}{\sqrt{6}}$

78. $\frac{\sqrt{2}}{\sqrt{27}}$

PROBLEM SOLVING

Approximating Radicals *The following diagram is a sketch of a 16 in. ruler marked using $\frac{1}{2}$ inches.*

In Exercises 79–84, without using a calculator, indicate between which two adjacent ruler marks each of the following irrational numbers will fall. Explain how you obtained your answer. Support your answer by obtaining an approximation with a calculator.

79. $\sqrt{5}$ in. **80.** $\sqrt{17}$ in.

81. $\sqrt{107}$ in. **82.** $\sqrt{135}$ in.

83. $\sqrt{170}$ in. **84.** $\sqrt{200}$ in.

In Exercises 85–90, determine whether the statement is true or false. Rewrite each false statement to make it a true statement. A false statement can be modified in more than one way to be made a true statement.

85. $\sqrt{c}$ is a rational number for any composite number c.

86. $\sqrt{p}$ is a rational number for any prime number p.

87. The sum of any two rational numbers is always a rational number.

88. The product of any two rational numbers is always a rational number.

89. The product of an irrational and a rational number is always an irrational number.

90. The product of any two irrational numbers is always an irrational number.

In Exercises 91–94, give an example to show that the stated case can occur.

91. The sum of two irrational numbers may be an irrational number.

92. The sum of two irrational numbers may be a rational number.

93. The product of two irrational numbers may be an irrational number.

94. The product of two irrational numbers may be a rational number.

95. Without doing any calculations, determine whether $\sqrt{2} = 1.414$. Explain your answer.

96. Without doing any calculations, determine whether $\sqrt{17} = 4.123$. Explain your answer.

97. The number π is an irrational number. Often the values 3.14 or $\frac{22}{7}$ are used for π. Does π equal either 3.14 or $\frac{22}{7}$? Explain your answer.

98. Give an example to show that $\sqrt{a+b} \neq \sqrt{a} + \sqrt{b}$.

99. Give an example to show that $\sqrt{a \cdot b} = \sqrt{a} \cdot \sqrt{b}$.

100. ***A Swinging Pendulum*** The time T required for a pendulum to swing back and forth may be found by the formula

$$T = 2\pi\sqrt{\frac{l}{g}}$$

where l is the length of the pendulum and g is the acceleration of gravity. Find the time in seconds for the pendulum to swing back and forth if $l = 35$ cm and $g = 980\text{ cm/sec}^2$. Round your answer to the nearest tenth of a second.

101. ***Estimating Speed of a Vehicle*** The speed that a vehicle was traveling, s, in miles per hour, when the brakes were first applied, can be estimated using the formula $s = \sqrt{\dfrac{d}{0.04}}$ where d is the length of the vehicle's skid marks, in feet.

a) Determine the speed of a car that made skid marks 4 ft long.

b) Determine the speed of a car that made skid marks 16 ft long.

c) Determine the speed of a car that made skid marks 64 ft long.

d) Determine the speed of a car that made skid marks 256 ft long.

102. ***Dropping an Object*** The formula $t = \dfrac{\sqrt{d}}{4}$ can be used to estimate the time, t, in seconds it takes for an object dropped to travel d feet.

a) Determine the time it takes for an object to drop 100 ft.

b) Determine the time it takes for an object to drop 400 ft.

c) Determine the time it takes for an object to drop 900 ft.

d) Determine the time it takes for an object to drop 1600 ft.

CHALLENGE PROBLEMS/GROUP ACTIVITIES

103. a) Is $\sqrt{0.04}$ rational or irrational? Explain.

b) Is $\sqrt{0.7}$ rational or irrational? Explain.

104. One way to find a rational number between two distinct rational numbers is to add the two distinct rational numbers and divide by 2. Do you think that this method will work for finding an irrational number between two distinct irrational numbers? Explain.

RECREATIONAL MATHEMATICS

105. ***More Four 4's*** In Exercise 82 on page 244, we introduced some of the basic rules of the game Four 4's. We now expand our operations to include square roots. For example, one way to obtain the whole number 8 is $\sqrt{4} + \sqrt{4} + \sqrt{4} + \sqrt{4} = 2 + 2 + 2 + 2 = 8$. Using the rules given on page 244, and using at least one square root of 4, $\sqrt{4}$, play Four 4's to obtain the following whole numbers:

a) 11

b) 13

c) 14

d) 18

INTERNET/RESEARCH ACTIVITIES

In Exercises 106 and 107, references include history of mathematics books, encyclopedias, and Internet web sites.

106. Write a report on the history of the development of the irrational numbers.

107. Write a report on the history of pi. In your report, indicate when the symbol π was first used and list the first 10 digits of π.

5.5 REAL NUMBERS AND THEIR PROPERTIES

▲ Real number properties are often learned as we learn arithmetic facts.

Evaluate $10 + 5$ and then evaluate $5 + 10$. Evaluate 10×5 and then evaluate 5×10. Evaluate $10 - 5$ and then evaluate $5 - 10$. Finally, evaluate $10 \div 5$ and then evaluate $5 \div 10$. You will notice that $10 + 5 = 5 + 10$ and that $10 \times 5 = 5 \times 10$. On the other hand, $10 - 5 \neq 5 - 10$ and $10 \div 5 \neq 5 \div 10$. These simple examples demonstrate a certain property that holds for addition and multiplication, but not for subtraction and division. In this section, we will study the set of *real* numbers and their properties.

Now that we have discussed both the rational and irrational numbers, we can discuss the real numbers and the properties of the real number system. The union of the rational numbers and the irrational numbers is the ***set of real numbers***, symbolized by $\mathbb{R}$.

Figure 5.8 on page 270 illustrates the relationship among various sets of numbers. It shows that the natural numbers are a subset of the whole numbers, the integers, the rational numbers, and the real numbers. For example, since the number 3 is a natural or counting number, it is also a whole number, an integer, a rational number, and a real number. Since the rational number $\frac{1}{4}$ is outside the set of integers, it is not an integer, a whole number, or a natural number. The number $\frac{1}{4}$ is a real number, however, as is the irrational number $\sqrt{2}$. Note that the real numbers are the union of the rational numbers and the irrational numbers.

DID YOU KNOW?

An Important Number

Some years ago, a group of French mathematicians who worked under the collective pseudonym "Monsieur Nicholas Bourbaki" embarked on the development of an encyclopedic description of all mathematics. They devoted 200 pages simply to introduce the innocent-looking concept, the number 1.

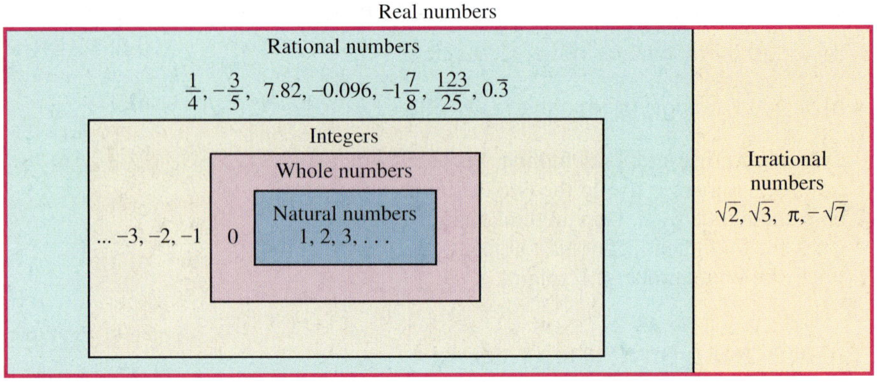

Figure 5.8

The relationship between the various sets of numbers in the real number system can also be illustrated with a tree diagram, as in Fig. 5.9.

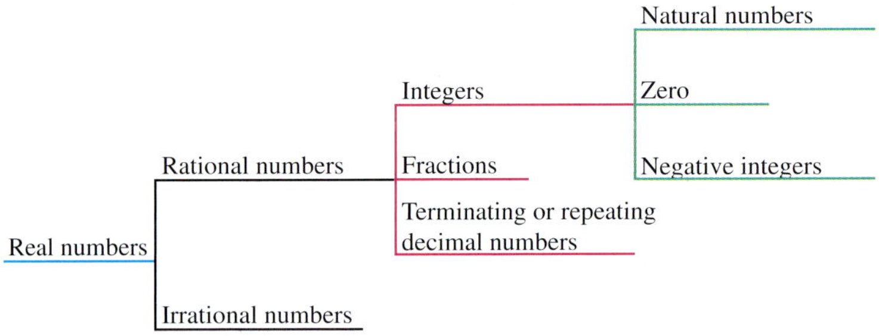

Figure 5.9

Figure 5.9 shows that, for example, the natural numbers are a subset of the integers, the rational numbers, and the real numbers. We can also see, for example, the natural numbers, zero, and the negative integers together form the integers.

Properties of the Real Number System

We are now prepared to consider the properties of the real number system. The first property we will discuss is *closure*.

If an operation is performed on any two elements of a set and the result is an element of the set, we say that the set is **closed** under that given operation.

Is the sum of any two natural numbers a natural number? The answer is yes. Thus, we say that the natural numbers are closed under the operation of addition.

Are the natural numbers closed under the operation of subtraction? If we subtract one natural number from another natural number, must the difference always be a natural number? The answer is no. For example, $3 - 5 = -2$, and -2 is not a natural number. Therefore, the natural numbers are not closed under the operation of subtraction.

MATHEMATICS TODAY

Brain Man

Daniel Tammet

In a segment of *60 Minutes* on January 29, 2007, Morley Safer interviewed 21-year-old Daniel Tammet, who has a mysterious disorder of the brain called savant syndrome. As a child, Tammet had a massive epileptic seizure, which may have contributed to his condition. Tammet is called "Brain Man" in his native England. Unlike most savants, however, he has no obvious mental disability, and most important to scientists, he can describe his own thought process. He may very well be a scientific Rosetta stone, a key to understanding the brain. In a period of a few weeks before he appeared on *60 Minutes,* Tammet memorized 23,514 decimal digits of pi. He recited them at Oxford University; it took more than 5 hours, and he made no mistakes. Tammet can also perform complicated calculations, such as 31^4, instantly. He has developed a rare crossing of the senses called synesthesia. Tammet said, "I see numbers in my head as colors, shapes, and textures." For example, when Daniel says he sees pi and does instant computations, he says he is not calculating, but, rather, says the answer appears to him as a landscape of colorful shapes. Tammet is working with scientists at the California Center for Brain Study, who are trying to unlock the secrets of the brain. Tammet runs his own online educational business, has written a book about his experiences, and does motivational speaking for parents of children with autism.

EXAMPLE 1 *Closure of Sets*

Determine whether the set of whole numbers are closed under the operations of (a) multiplication and (b) division.

SOLUTION

a) If we multiply any two whole numbers, will the product always be a whole number? You may want to try multiplying several whole numbers. You will see that yes, the product of any two whole numbers is a whole number. Thus, the set of whole numbers are closed under the operation of multiplication.

b) If we divide any two whole numbers, will the quotient always be a whole number? The answer is no. For example $1 \div 2 = \frac{1}{2}$, and $\frac{1}{2}$ is not a whole number. Therefore, the set of whole numbers are not closed under the operation of division.

Next we will discuss three important properties: the commutative property, the associative property, and the distributive property. A knowledge of these properties is essential for the understanding of algebra. We begin with the commutative property.

COMMUTATIVE PROPERTY

Addition	**Multiplication**
$a + b = b + a$	$a \cdot b = b \cdot a$

for any real numbers a and b.

The commutative property states that the *order* in which two numbers are added or multiplied is not important. For example, $4 + 5 = 5 + 4 = 9$ and $3 \cdot 6 = 6 \cdot 3 = 18$. Note that the commutative property does not hold for the operations of subtraction or division. For example,

$$4 - 7 \neq 7 - 4 \qquad \text{and} \qquad 9 \div 3 \neq 3 \div 9$$

Now we introduce the associative property.

ASSOCIATIVE PROPERTY

Addition	**Multiplication**
$(a + b) + c = a + (b + c)$	$(a \cdot b) \cdot c = a \cdot (b \cdot c)$

for any real numbers a, b, and c.

The associative property states that when adding or multiplying three real numbers, we may place parentheses around any two adjacent numbers. The result is the same regardless of the placement of parentheses. For example,

$$\begin{aligned} (3 + 4) + 5 &= 3 + (4 + 5) \\ 7 + 5 &= 3 + 9 \\ 12 &= 12 \end{aligned} \qquad \begin{aligned} (3 \cdot 4) \cdot 5 &= 3 \cdot (4 \cdot 5) \\ 12 \cdot 5 &= 3 \cdot 20 \\ 60 &= 60 \end{aligned}$$

The associative property does not hold for the operations of subtraction and division. For example,

$$(10 - 6) - 2 \neq 10 - (6 - 2) \qquad \text{and} \qquad (27 \div 9) \div 3 \neq 27 \div (9 \div 3)$$

Note the difference between the commutative property and the associative property. The commutative property involves a change in *order,* whereas the associative property involves a change in *grouping* (or the *association* of numbers that are grouped together).

Another property of the real numbers is the distributive property of multiplication over addition.

DISTRIBUTIVE PROPERTY OF MULTIPLICATION OVER ADDITION

$$a \cdot (b + c) = a \cdot b + a \cdot c$$

for any real numbers a, b, and c.

For example, if $a = 3$, $b = 4$, and $c = 5$, then

$$3 \cdot (4 + 5) = (3 \cdot 4) + (3 \cdot 5)$$
$$3 \cdot 9 = 12 + 15$$
$$27 = 27$$

This result indicates that, when using the distributive property, you may either add first and then multiply or multiply first and then add. Note that the distributive property involves two operations, addition and multiplication. Although positive integers were used in the example, any real numbers could have been used.

We frequently use the commutative, associative, and distributive properties without realizing that we are doing so. To add $13 + 4 + 6$, we may add the $4 + 6$ first to get 10. To this sum we then add 13 to get 23. Here we have done the equivalent of placing parentheses around the $4 + 6$. We can do so because of the associative property of addition.

To multiply 102×11 in our heads, we might multiply $100 \times 11 = 1100$ and $2 \times 11 = 22$ and add these two products to get 1122. We are permitted to do so because of the distributive property.

$$102 \times 11 = (100 + 2) \times 11 = (100 \times 11) + (2 \times 11)$$
$$= 1100 + 22 = 1122$$

EXAMPLE 2 *Identifying Properties of Real Numbers*

Name the property illustrated.

a) $2 + 5 = 5 + 2$

b) $(x + 3) + 5 = x + (3 + 5)$

c) $4 \cdot (3 \cdot y) = (4 \cdot 3) \cdot y$

d) $9(w + 3) = 9 \cdot w + 9 \cdot 3$

e) $5 + (z + 3) = 5 + (3 + z)$

f) $(2p) \cdot 5 = 5 \cdot (2p)$

SOLUTION

a) Commutative property of addition

b) Associative property of addition

c) Associative property of multiplication

d) Distributive property of multiplication over addition

e) The only change between the left and right sides of the equal sign is the order of the z and 3 within the parentheses. The order is changed from $z + 3$ to $3 + z$ using the commutative property of addition.

f) The order of 5 and $(2p)$ is changed by using the commutative property of multiplication.

EXAMPLE 3 *Simplifying by Using the Distributive Property*

Use the distributive property to simplify

a) $2(3 + \sqrt{5})$ b) $\sqrt{2}(7 + \sqrt{3})$

SOLUTION

a) $2(3 + \sqrt{5}) = (2 \cdot 3) + (2 \cdot \sqrt{5})$
$= 6 + 2\sqrt{5}$

b) $\sqrt{2}(7 + \sqrt{3}) = (\sqrt{2} \cdot 7) + (\sqrt{2} \cdot \sqrt{3})$
$= 7\sqrt{2} + \sqrt{6}$

Note that $\sqrt{2} \cdot 7$ is written $7\sqrt{2}$.

EXAMPLE 4 *Distributive Property*

Use the distributive property to multiply $5(z + 3)$. Then simplify the result.

SOLUTION

$$5(z + 3) = 5 \cdot z + 5 \cdot 3$$
$$= 5z + 15$$

We summarize the properties mentioned in this section as follows, where a, b, and c are any real numbers.

Commutative property of addition	$a + b = b + a$
Commutative property of multiplication	$a \cdot b = b \cdot a$
Associative property of addition	$(a + b) + c = a + (b + c)$
Associative property of multiplication	$(a \cdot b) \cdot c = a \cdot (b \cdot c)$
Distributive property of multiplication over addition	$a \cdot (b + c) = a \cdot b + a \cdot c$

SECTION 5.5 EXERCISES

CONCEPT/WRITING EXERCISES

1. What are the real numbers?

2. What symbol is used to represent the set of real numbers?

3. What does it mean if a set is closed under a given operation?

4. Give the commutative property of addition, explain what it means, and give an example illustrating it.

5. Give the commutative property of multiplication, explain what it means, and give an example illustrating it.

6. Give the associative property of multiplication, explain what it means, and give an example illustrating it.

7. Give the associative property of addition, explain what it means, and give an example illustrating it.

8. Give the distributive property of multiplication over addition, explain what it means, and give an example illustrating it.

PRACTICE THE SKILLS

In Exercises 9–12, determine whether the natural numbers are closed under the given operation.

9. Subtraction

10. Addition

11. Multiplication

12. Division

In Exercises 13–16, determine whether the integers are closed under the given operation.

13. Addition

14. Subtraction

15. Multiplication

16. Division

In Exercises 17–20, determine whether the rational numbers are closed under the given operation.

17. Subtraction

18. Addition

19. Division

20. Multiplication

In Exercises 21–24, determine whether the irrational numbers are closed under the given operation.

21. Addition

22. Subtraction

23. Division

24. Multiplication

In Exercises 25–28, determine whether the real numbers are closed under the given operation.

25. Subtraction

26. Addition

27. Multiplication

28. Division

29. Does $(5 + x) + 6 = (x + 5) + 6$ illustrate the commutative property or the associative property? Explain your answer.

30. Does $(x + 5) + 6 = x + (5 + 6)$ illustrate the commutative property or the associative property? Explain your answer.

31. Give an example to show that the commutative property of addition may be true for the negative integers.

32. Give an example to show that the commutative property of multiplication may be true for the negative integers.

33. Does the commutative property hold for the integers under the operation of subtraction? Give an example to support your answer.

34. Does the commutative property hold for the rational numbers under the operation of division? Give an example to support your answer.

35. Give an example to show that the associative property of addition may be true for the negative integers.

36. Give an example to show that the associative property of multiplication may be true for the negative integers.

37. Does the associative property hold for the integers under the operation of division? Give an example to support your answer.

38. Does the associative property hold for the integers under the operation of subtraction? Give an example to support your answer.

39. Does the associative property hold for the real numbers under the operation of division? Give an example to support your answer.

40. Does $a + (b \cdot c) = (a + b) \cdot (a + c)$? Give an example to support your answer.

In Exercises 41–56, state the name of the property illustrated.

41. $3(y + 5) = 3 \cdot y + 3 \cdot 5$

42. $6 + 7 = 7 + 6$

43. $(7 \cdot 8) \cdot 9 = 7 \cdot (8 \cdot 9)$

44. $c + d = d + c$

45. $(24 + 7) + 3 = 24 + (7 + 3)$

46. $4 \cdot (11 \cdot x) = (4 \cdot 11) \cdot x$

47. $\sqrt{3} \cdot 7 = 7 \cdot \sqrt{3}$

48. $\frac{3}{8} + \left(\frac{1}{8} + \frac{3}{2}\right) = \left(\frac{3}{8} + \frac{1}{8}\right) + \frac{3}{2}$

49. $-1(x + 4) = (-1) \cdot x + (-1) \cdot 4$

50. $(p + q) + r = p + (q + r)$

51. $\sqrt{5} \cdot 2 = 2 \cdot \sqrt{5}$

52. $(r + s) + t = t + (r + s)$

53. $(r + s) \cdot t = (r \cdot t) + (s \cdot t)$

54. $g \cdot (h + i) = (h + i) \cdot g$

55. $(f \cdot g) + (j \cdot h) = (g \cdot f) + (j \cdot h)$

56. $(k + l) \cdot (m + n) = (l + k) \cdot (m + n)$

In Exercises 57–68, use the distributive property to multiply. Then, if possible, simplify the resulting expression.

57. $4(z + 1)$

58. $-1(a + b)$

59. $-\frac{3}{4}(x - 12)$

60. $-x(-y + z)$

61. $6\left(\frac{x}{2} + \frac{2}{3}\right)$

62. $24\left(\frac{x}{3} - \frac{1}{8}\right)$

63. $32\left(\frac{1}{16}x - \frac{1}{32}\right)$

64. $15\left(\frac{2}{3}x - \frac{4}{5}\right)$

65. $\sqrt{2}(\sqrt{8} - \sqrt{2})$

66. $-3(2 - \sqrt{3})$

67. $5(\sqrt{2} + \sqrt{3})$

68. $\sqrt{5}(\sqrt{15} - \sqrt{20})$

In Exercises 69–74, name the property used to go from step to step. You only need to name the properties indicated by an a), b), c), or d). For example, in Exercise 69, you need to supply an answer for part a), and an answer to go from the second step of part a) to part b).

69. a) $3(a + 2) + 5 = (3 \cdot a + 3 \cdot 2) + 5$
$= (3a + 6) + 5$

b) $= 3a + (6 + 5)$
$= 3a + 11$

70. a) $2(b + 4) + 1 = (2 \cdot b + 2 \cdot 4) + 1$
$= (2b + 8) + 1$

b) $= 2b + (8 + 1)$
$= 2b + 9$

71. a) $7(k + 1) + 2k = (7 \cdot k + 7 \cdot 1) + 2k$
$= (7k + 7) + 2k$

b) $= 7k + (7 + 2k)$

c) $= 7k + (2k + 7)$

d) $= (7k + 2k) + 7$
$= 9k + 7$

72. a) $11(h + 6) + 5h = (11 \cdot h + 11 \cdot 6) + 5h$
$= (11h + 66) + 5h$

b) $= 11h + (66 + 5h)$

c) $= 11h + (5h + 66)$

d) $= (11h + 5h) + 66$
$= 16h + 66$

73. a) $9 + 2(t + 3) + 7t = 9 + (2 \cdot t + 2 \cdot 3) + 7t$
$= 9 + (2t + 6) + 7t$

b) $= 9 + (6 + 2t) + 7t$

c) $= (9 + 6) + 2t + 7t$
$= 15 + 9t$

d) $= 9t + 15$

74. a) $7 + 5(s + 4) + 3s = 7 + (5 \cdot s + 5 \cdot 4) + 3s$
$= 7 + (5s + 20) + 3s$

b) $= 7 + (20 + 5s) + 3s$

c) $= (7 + 20) + 5s + 3s$
$= 27 + 8s$

d) $= 8s + 27$

In Exercises 75–80, determine whether the activity can be used to illustrate the commutative property. For the property to hold, the end result must be identical, regardless of the order in which the actions are performed.

75. Feeding your dog and giving your dog water

76. Putting your wallet in your back pocket and putting your keys in your front pocket

77. Washing clothes and drying clothes

78. Turning on a computer and typing a term paper on the computer

79. Filling your car with gasoline and washing the windshield

80. Turning on a lamp and reading a book

In Exercises 81–88, determine whether the activity can be used to illustrate the associative property. For the property to hold, doing the first two actions followed by the third would produce the same end result as doing the second and third actions followed by the first.

81. Putting batteries in your calculator, pressing the ON key, and pressing the 7 key

82. Filling your car with gas, washing the windshield, and checking the tire pressure

83. Sending a holiday card to your grandmother, sending one to your parents, and sending one to your teacher

84. Mowing the lawn, trimming the bushes, and removing dead limbs from trees

85. Brushing your teeth, washing your face, and combing your hair

86. Cracking an egg, pouring out the egg, and cooking the egg

87. Taking a bath, brushing your teeth, and taking your vitamins

88. While making meatloaf, mixing in the milk, mixing in the spices, and mixing in the bread crumbs

CHALLENGE PROBLEMS/GROUP ACTIVITIES

89. Describe two other activities that can be used to illustrate the commutative property (see Exercises 75–80).

90. Describe three other activities that can be used to illustrate the associative property (see Exercises 81–88).

91. Does $0 \div a = a \div 0$ (assume $a \neq 0$)? Explain.

RECREATIONAL MATHEMATICS

92. a) Consider the three words *man eating tiger.* Does (*man eating*) *tiger* mean the same as *man* (*eating tiger*)?

b) Does (*horse riding*) *monkey* mean the same as *horse* (*riding monkey*)?

c) Can you find three other nonassociative word triples?

INTERNET/RESEARCH ACTIVITY

93. A set of numbers that was not discussed in this chapter is the set of *complex numbers.* Write a report on complex numbers. Include their relationship to the real numbers.

5.6 RULES OF EXPONENTS AND SCIENTIFIC NOTATION

▲ The diameter of this galaxy may be given in scientific notation.

What is the current world population? What is the diameter of an atom? What is the diameter of a galaxy? What is the wavelength of an x-ray? What is the current national federal debt? All these questions have answers that may involve either very large numbers or very small numbers. One way to accurately represent such numbers is with *scientific notation*. In this section, we will study how to use scientific notation when answering questions such as those posed above.

An understanding of exponents is important in solving problems in algebra. In the expression 5^2, the 2 is referred to as the ***exponent*** and the 5 is referred to as the ***base***. We read 5^2 as 5 to the second power, or 5 squared, which means

$$5^2 = \underbrace{5 \cdot 5}_{\text{2 factors of 5}}$$

DID YOU KNOW?

A Very Large Number

170,141,183,460,469,231,731,687,303,715,884,105,727 is a very large number. How would you read this number? Take a breath.

170 undecillion, 141 decillion, 183 nonillion, 460 octillion, 469 septillion, 231 sextillion, 731 quintillion, 687 quadrillion, 303 trillion, 715 billion, 884 million, 105 thousand, 727.

Often numbers this large can be represented with an approximation involving scientific notation. This number, however, is a prime number, and its *exact* representation is very important. Mathematicians frequently represent such large numbers by using exponents. Here, another exact, and more efficient, representation of this number is $2^{127} - 1$.

The number 5 to the third power, or 5 cubed, written 5^3, means

$$5^3 = \underbrace{5 \cdot 5 \cdot 5}_{\text{3 factors of 5}}$$

In general, the number b to the nth power, written b^n, means

$$b^n = \underbrace{b \cdot b \cdot b \cdot \cdots \cdot b}_{n \text{ factors of } b}$$

EXAMPLE 1 *Evaluating the Power of a Number*

Evaluate.

a) 3^2 b) $(-7)^2$ c) 4^3 d) 1^{100} e) 100^1

SOLUTION

a) $3^2 = 3 \cdot 3 = 9$

b) $(-7)^2 = (-7) \cdot (-7) = 49$

c) $4^3 = 4 \cdot 4 \cdot 4 = 64$

d) $1^{100} = 1$

The number 1 multiplied by itself any number of times equals 1.

e) $100^1 = 100$

Any number with an exponent of 1 equals the number itself.

EXAMPLE 2 *The Importance of Parentheses*

Evaluate.

a) $(-2)^4$ b) -2^4 c) $(-2)^5$ d) -2^5

SOLUTION

a) $(-2)^4 = (-2)(-2)(-2)(-2) = 4(-2)(-2) = -8(-2) = 16$

b) -2^4 means take the opposite of 2^4 or $-1 \cdot 2^4$.

$-1 \cdot 2^4 = -1 \cdot 2 \cdot 2 \cdot 2 \cdot 2 = -1 \cdot 16 = -16$

c) $(-2)^5 = (-2)(-2)(-2)(-2)(-2) = 4(-2)(-2)(-2) = -8(-2)(-2)$
$= 16(-2) = -32$

d) $-2^5 = -1 \cdot 2^5 = -1 \cdot 2 \cdot 2 \cdot 2 \cdot 2 \cdot 2 = -1 \cdot 32 = -32$

From Example 2, we can see that $(-x)^n \neq -x^n$, where n is an even natural number.

Rules of Exponents

Now that we know how to evaluate powers of numbers we can discuss the rules of exponents. Consider

$$2^2 \cdot 2^3 = \underbrace{2 \cdot 2}_{\text{2 factors}} \cdot \underbrace{2 \cdot 2 \cdot 2}_{\text{3 factors}} = 2^5$$

This example illustrates the product rule for exponents.

PRODUCT RULE FOR EXPONENTS

$$a^m \cdot a^n = a^{m+n}$$

Therefore, by using the product rule, $2^2 \cdot 2^3 = 2^{2+3} = 2^5$.

EXAMPLE 3 ***Using the Product Rule for Exponents***

Use the product rule to simplify.

a) $2^3 \cdot 2^7$ b) $5^2 \cdot 5^6$

SOLUTION

a) $2^3 \cdot 2^7 = 2^{3+7} = 2^{10}$ b) $5^2 \cdot 5^6 = 5^{2+6} = 5^8$

Consider

$$\frac{2^5}{2^2} = \frac{2 \cdot 2 \cdot 2 \cdot \cancel{2} \cdot \cancel{2}}{\cancel{2} \cdot \cancel{2}} = 2 \cdot 2 \cdot 2 = 2^3$$

This example illustrates the quotient rule for exponents.

QUOTIENT RULE FOR EXPONENTS

$$\frac{a^m}{a^n} = a^{m-n}, \qquad a \neq 0$$

Therefore, $\frac{2^5}{2^2} = 2^{5-2} = 2^3$.

EXAMPLE 4 ***Using the Quotient Rule for Exponents***

Use the quotient rule to simplify.

a) $\frac{5^8}{5^5}$ b) $\frac{8^{12}}{8^5}$

SOLUTION

a) $\frac{5^8}{5^5} = 5^{8-5} = 5^3$ b) $\frac{8^{12}}{8^5} = 8^{12-5} = 8^7$

Consider $2^3 \div 2^3$. The quotient rule gives

$$\frac{2^3}{2^3} = 2^{3-3} = 2^0$$

But $\frac{2^3}{2^3} = \frac{8}{8} = 1$. Therefore, 2^0 must equal 1. This example illustrates the zero exponent rule.

ZERO EXPONENT RULE

$$a^0 = 1, \qquad a \neq 0$$

Note that 0^0 is not defined by the zero exponent rule.

EXAMPLE 5 ***The Zero Power***

Use the zero exponent rule to simplify.

a) 2^0 b) $(-2)^0$ c) -2^0 d) $(5x)^0$ e) $5x^0$

SOLUTION

a) $2^0 = 1$ b) $(-2)^0 = 1$ c) $-2^0 = -1 \cdot 2^0 = -1 \cdot 1 = -1$

d) $(5x)^0 = 1$ e) $5x^0 = 5 \cdot x^0 = 5 \cdot 1 = 5$

Consider $2^3 \div 2^5$. The quotient rule yields

$$\frac{2^3}{2^5} = 2^{3-5} = 2^{-2}$$

But $\frac{2^3}{2^5} = \frac{\cancel{2} \cdot \cancel{2} \cdot \cancel{2}}{\cancel{2} \cdot \cancel{2} \cdot \cancel{2} \cdot 2 \cdot 2} = \frac{1}{2^2}$. Since $\frac{2^3}{2^5}$ equals both 2^{-2} and $\frac{1}{2^2}$, then 2^{-2} must equal $\frac{1}{2^2}$. This example illustrates the negative exponent rule.

NEGATIVE EXPONENT RULE

$$a^{-m} = \frac{1}{a^m}, \qquad a \neq 0$$

EXAMPLE 6 ***Using the Negative Exponent Rule***

Use the negative exponent rule to simplify.

a) 2^{-3} b) 5^{-1}

SOLUTION

a) $2^{-3} = \frac{1}{2^3} = \frac{1}{8}$ b) $5^{-1} = \frac{1}{5^1} = \frac{1}{5}$

Consider $(2^3)^2$.

$$(2^3)^2 = (2^3)(2^3) = 2^{3+3} = 2^6$$

This example illustrates the power rule for exponents.

POWER RULE FOR EXPONENTS

$$(a^m)^n = a^{m \cdot n}$$

Thus, $(2^3)^2 = 2^{3 \cdot 2} = 2^6$.

EXAMPLE 7 ***Evaluating a Power Raised to Another Power***

Use the power rule to simplify.

a) $(5^4)^3$ b) $(7^2)^5$

SOLUTION

a) $(5^4)^3 = 5^{4 \cdot 3} = 5^{12}$ b) $(7^2)^5 = 7^{2 \cdot 5} = 7^{10}$

DID YOU KNOW?

Large and Small Numbers

The diameter of a galaxy may be 1×10^5 light-years

The diameter of an atom may be 1×10^{-10} meter

Our everyday activities don't require us to deal with quantities much above those in the thousands: $6.95 for lunch, a 100 lap race, a $15,000 car loan, and so on. Yet as modern technology has developed, so has our ability to study all aspects of the universe we live in, from the very large to the very small. Modern technology can be used with the rules of exponents and scientific notation to study everything from the diameter of a galaxy to the diameter of an atom.

SUMMARY OF THE RULES OF EXPONENTS

$a^m \cdot a^n = a^{m+n}$	Product rule for exponents
$\dfrac{a^m}{a^n} = a^{m-n}, \quad a \neq 0$	Quotient rule for exponents
$a^0 = 1, \quad a \neq 0$	Zero exponent rule
$a^{-m} = \dfrac{1}{a^m}, \quad a \neq 0$	Negative exponent rule
$(a^m)^n = a^{m \cdot n}$	Power rule for exponents

Scientific Notation

Often scientific problems deal with very large and very small numbers. For example, the distance from Earth to the sun is about 93,000,000 miles. The wavelength of a yellow color of light is about 0.0000006 meter. Because working with many zeros is difficult, scientists developed a notation that expresses such numbers with exponents. For example, consider the distance from Earth to the sun, 93,000,000 miles.

$$\begin{aligned} 93{,}000{,}000 &= 9.3 \times 10{,}000{,}000 \\ &= 9.3 \times 10^7 \end{aligned}$$

The wavelength of a yellow color of light is about 0.0000006 meter.

$$\begin{aligned} 0.0000006 &= 6.0 \times 0.0000001 \\ &= 6.0 \times 10^{-7} \end{aligned}$$

The numbers 9.3×10^7 and 6.0×10^{-7} are written in a form called ***scientific notation***. Each number written in scientific notation is written as a number greater than or equal to 1 and less than 10 multiplied by some power of 10.

Some examples of numbers in scientific notation are

$$3.7 \times 10^3, \qquad 2.05 \times 10^{-3}, \qquad 5.6 \times 10^8, \qquad \text{and} \qquad 1.00 \times 10^{-5}$$

The following is a procedure for writing a number in scientific notation.

TO WRITE A NUMBER IN SCIENTIFIC NOTATION

1. Move the decimal point in the original number to the right or left until you obtain a number greater than or equal to 1 and less than 10.
2. Count the number of places you have moved the decimal point to obtain the number in step 1. If the decimal point was moved to the left, the count is to be considered positive. If the decimal point was moved to the right, the count is to be considered negative.
3. Multiply the number obtained in step 1 by 10 raised to the count found in step 2. (Note that the count determined in step 2 is the exponent on the base 10.)

DID YOU KNOW?

Catalan's Conjecture

Eugène Catalan

Preda Mihailescu

Consider all the squares of positive integers:

$1^2 = 1$, $2^2 = 4$, $3^2 = 9$, $4^2 = 16$, and so on

Now consider all the cubes of positive integers:

$1^3 = 1$, $2^3 = 8$, $3^3 = 27$, $4^3 = 64$, and so on

Next, consider all the fourth powers of positive integers:

$1^4 = 1$, $2^4 = 16$, $3^4 = 81$, $4^4 = 256$, and so on

If we were to put *all* powers of the positive integers into one set, the set would begin as follows: $\{1, 4, 8, 9, 16, 25, 27, 32, 36, 49, 64, \ldots\}$. In 1844, Belgian mathematician Eugène Catalan proposed that of all the numbers in this infinite set, only 8 and 9 are consecutive integers. Although this conjecture was easily stated, the formal proof of it eluded mathematicians for more than 150 years. In April 2002, Preda Mihailescu, a Romanian-born mathematician teaching at the University of Paderborn, Germany, announced that he had proved the Catalan conjecture.

EXAMPLE 8 *Converting from Decimal Notation to Scientific Notation*

Write each number in scientific notation.

a) In 2006, the population of the United States was about 299,000,000.
b) In 2006, the population of China was about 1,306,000,000.
c) In 2006, the population of the world was about 6,522,000,000
d) The diameter of a hydrogen atom nucleus is about 0.0000000000011 millimeter.
e) The wavelength of an x-ray is about 0.000000000492 meter.

SOLUTION

a) $299{,}000{,}000 = 2.99 \times 10^8$
b) $1{,}306{,}000{,}000 = 1.306 \times 10^9$
c) $6{,}522{,}000{,}000 = 6.522 \times 10^9$
d) $0.0000000000011 = 1.1 \times 10^{-12}$
e) $0.000000000492 = 4.92 \times 10^{-10}$

To convert from a number given in scientific notation to decimal notation we reverse the procedure.

TO CHANGE A NUMBER IN SCIENTIFIC NOTATION TO DECIMAL NOTATION

1. Observe the exponent on the 10.
2. a) If the exponent is positive, move the decimal point in the number to the right the same number of places as the exponent. Adding zeros to the number might be necessary.
   b) If the exponent is negative, move the decimal point in the number to the left the same number of places as the exponent. Adding zeros might be necessary.

EXAMPLE 9 *Converting from Scientific Notation to Decimal Notation*

Write each number in decimal notation.

a) The average distance from Mars to the sun is about 1.4×10^8 miles.
b) The half-life of uranium-235 is about 4.5×10^9 years.
c) The average grain size in siltstone is 1.35×10^{-3} inch.
d) A *millimicron* is a unit of measure used for very small distances. One millimicron is about 3.94×10^{-8} inch.

SOLUTION

a) $1.4 \times 10^8 = 140{,}000{,}000$
b) $4.5 \times 10^9 = 4{,}500{,}000{,}000$
c) $1.35 \times 10^{-3} = 0.00135$
d) $3.94 \times 10^{-8} = 0.0000000394$

In scientific journals and books, we occasionally see numbers like 10^{15} and 10^{-6}. We interpret these numbers as 1×10^{15} and 1×10^{-6}, respectively, when converting the numbers to decimal form.

EXAMPLE 10 ***Multiplying Numbers in Scientific Notation***

Multiply $(2.1 \times 10^5)(9 \times 10^{-3})$. Write the answer in scientific notation and in decimal notation.

SOLUTION

$$\begin{aligned}(2.1 \times 10^5)(9 \times 10^{-3}) &= (2.1 \times 9)(10^5 \times 10^{-3})\\ &= 18.9 \times 10^2\\ &= 1.89 \times 10^3 \quad \text{Scientific notation}\\ &= 1{,}890 \quad \text{Decimal notation}\end{aligned}$$

EXAMPLE 11 ***Dividing Numbers Using Scientific Notation***

Divide $\frac{0.000000000048}{24{,}000{,}000{,}000}$. Write the answer in scientific notation.

SOLUTION First write each number in scientific notation.

$$\begin{aligned}\frac{0.000000000048}{24{,}000{,}000{,}000} = \frac{4.8 \times 10^{-11}}{2.4 \times 10^{10}} &= \left(\frac{4.8}{2.4}\right)\left(\frac{10^{-11}}{10^{10}}\right)\\ &= 2.0 \times 10^{-11-10}\\ &= 2.0 \times 10^{-21}\end{aligned}$$

Scientific Notation on the Scientific Calculator

One advantage of using scientific notation when working with very large and very small numbers is the ease with which you can perform operations. Performing these operations is even easier with the use of a scientific calculator. Most scientific calculators have a scientific notation key labeled “Exp,” “EXP,” or “EE.” We will refer to the scientific notation key as [EXP]. The following keystrokes reading downwards can be used to enter the number 4.3×10^6.

Keystroke(s)	Calculator display
4.3	4.3
[EXP]	4.3^{00}
6	4.3^{06}

Your calculator may have some slight variations to the display shown here. The display 4.3^{06} means 4.3×10^6. We now will use our calculators to perform some computations using scientific notation.

EXAMPLE 12 ***Use Scientific Notation on a Calculator to Find a Product***

Multiply $(4.3 \times 10^6)(2 \times 10^{-4})$ using a scientific calculator. Write the answer in decimal notation.

SOLUTION Our sequence of keystrokes is as follows:

Keystroke(s)	Display	
4.3	4.3	
EXP	4.3^{00}	
6	4.3^{06}	
×	4300000	4.3×10^6 is now entered in the calculator. Most calculators will convert to decimal notation (if the display can show the number)
2	2.	
EXP	$2.^{00}$	
4	$2.^{04}$	Enter the positive form of the exponent
+/−	$2.^{-04}$	Make the exponent negative*
=	860.	Press = to obtain the answer of 860†

EXAMPLE 13 *Use Scientific Notation on a Calculator to Find a Quotient*

Divide $\dfrac{0.000000000048}{24{,}000{,}000{,}000}$ using a scientific calculator. Write the answer in scientific notation.

SOLUTION We first rewrite the numerator and denominator using scientific notation. See Example 11.

$$\frac{0.000000000048}{24{,}000{,}000{,}000} = \frac{4.8 \times 10^{-11}}{2.4 \times 10^{10}}$$

Next we use a scientific calculator to perform the computation. The keystrokes are as follows:

4.8 EXP 11 +/− ÷ 2.4 EXP 10 =

The display on the calculator is $2.^{-21}$, which means 2.0×10^{-21}.

EXAMPLE 14 *U.S. Debt per Person*

On June 15, 2006, the U.S. Department of the Treasury estimated the U.S. federal debt to be about \$8.38 trillion. On this same day, the U.S. Bureau of the Census estimated the U.S. population to be about 299 million people. Determine the average debt, per person, by dividing the U.S. federal debt by the U.S. population.

DID YOU KNOW?

What's the Difference Between Debt and Deficit?

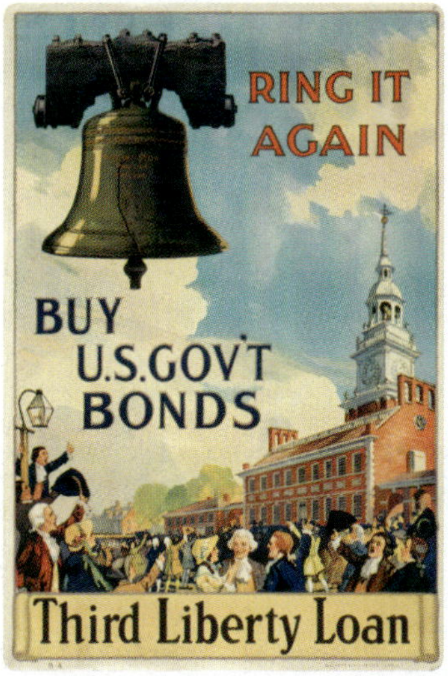

A World War I–era poster encouraging investment in U.S. government bonds

We often hear economists and politicians talk about such things as revenue, expenditures, deficit, surplus, and national debt. Revenue is the money a government collects annually, mostly through taxes. Expenditures are the money a government spends annually. If revenue exceeds expenditures, a surplus occurs; if expenditures exceed revenue, a deficit occurs. The national debt is the total of all the budget deficits and (the few) surpluses encountered by the U.S. federal government for more than 200 years. How big is the national debt? Search the Internet for the latest figures. As of June 15, 2006, the national debt was \$8,381,261,219,125.38. This amount is owed to investors worldwide—perhaps including you—who own U.S. government bonds.

*Some calculators will require you to enter the negative sign before entering the exponent.

†Some calculators will display the answer in scientific notation. In this case, the display will show $8.6^{\,02}$, which means 8.6×10^2 and equals 860.

SOLUTION We will first write the numbers involved using decimal notation. Then we will convert them to scientific notation.

$$8.38 \text{ trillion} = 8{,}380{,}000{,}000{,}000 = 8.38 \times 10^{12}$$
$$299 \text{ million} = 299{,}000{,}000 = 2.99 \times 10^{8}$$

Now we will divide 8.38×10^{12} by 2.99×10^{8} using a scientific calculator. The keystrokes are

8.38 [EXP] 12 [÷] 2.99 [EXP] 8 [=]

The display then shows 28026.7558. This number indicates that on June 15, 2006, the U.S. government owed about $28,026.76 per man, woman, and child living in the United States.

SECTION 5.6 EXERCISES

CONCEPT/WRITING EXERCISES

1. In the expression 2^3, what is the name given to the 2, and what is the name given to the 3?

2. Explain the meaning of b^n.

3. a) Explain the product rule for exponents.

   b) Use the product rule to simplify $2^3 \cdot 2^4$.

4. a) Explain the quotient rule for exponents.

   b) Use the quotient rule to simplify $\frac{5^6}{5^4}$.

5. a) Explain the zero exponent rule.

   b) Use the zero exponent rule to simplify 7^0.

6. a) Explain the negative exponent rule.

   b) Use the negative exponent rule to simplify 2^{-3}.

7. a) Explain the power rule for exponents.

   b) Use the power rule to simplify $(3^2)^4$.

8. Explain how you can simplify the expression 1^{500}.

9. Explain how you can simplify the following expressions and then simplify the expression.

   a) -1^{500} b) $(-1)^{500}$

   c) -1^{501} d) $(-1)^{501}$

10. a) In your own words, explain how to change a number in decimal notation to scientific notation.

    b) Using the procedure in part (a), change 0.000739 to scientific notation.

    c) Using the procedure in part (a), change 1,940,000,000 to scientific notation.

11. a) In your own words, explain how to change a number in scientific notation to decimal notation.

    b) Using the procedure in part (a), change 2.91×10^{-5} to decimal notation.

    c) Using the procedure in part (a), change 7.02×10^{6} to decimal notation.

12. A number is given in scientific notation. What does it indicate about the number when the exponent on the 10 is (a) positive, (b) zero, and (c) negative?

PRACTICE THE SKILLS

In Exercises 13–40, evaluate the expression.

13. a) 3^2 b) 2^3

14. a) 2^5 b) 5^2

15. a) $(-5)^2$ b) -5^2

16. a) -3^2 b) $(-3)^2$

17. a) -2^4 b) $(-2)^4$

18. a) $(-3)^4$ b) -3^4

19. a) -4^3 b) $(-4)^3$

20. a) $(-2)^7$ b) -2^7

21. a) $\left(\frac{1}{8}\right)^2$ b) $\left(-\frac{3}{4}\right)^2$

22. a) $-\left(\frac{2}{3}\right)^2$ b) $\left(\frac{1}{6}\right)^3$

23. a) 1000^1 b) 1^{1000} **24.** a) 2008^1 b) 1^{2008}

25. a) $3^2 \cdot 3^3$ b) $(-3)^2 \cdot (-3)^3$

26. a) $2^3 \cdot 2$ b) $(-2)^3 \cdot (-2)$

27. a) $\frac{5^7}{5^5}$ b) $\frac{(-5)^7}{(-5)^5}$ **28.** a) $\frac{4^5}{4^2}$ b) $\frac{(-4)^5}{(-4)^2}$

29. a) 6^0 b) -6^0 **30.** a) $(-6)^0$ b) $-(-6)^0$

31. a) $(6x)^0$ b) $6x^0$ **32.** a) $-6x^0$ b) $(-6x)^0$

33. a) 3^{-3} b) 7^{-2} **34.** a) 5^{-1} b) 2^{-4}

35. a) -9^{-2} b) $(-9)^{-2}$

36. a) -5^{-2} b) $(-5)^{-2}$

37. a) $(2^2)^3$ b) $(2^3)^2$

38. a) $\left[\left(\frac{1}{2}\right)^2\right]^3$ b) $\left[\left(-\frac{1}{2}\right)^2\right]^3$

39. a) $4^3 \cdot 4^{-2}$ b) $2^{-2} \cdot 2^{-2}$

40. a) $(-5)^{-2}(-5)^{-2}$ b) $(-1)^{-5}(-1)^{-5}$

In Exercises 41–56, express the number in scientific notation.

41. 175,000 **42.** 496,000,000 **43.** 0.00023

44. 0.000034 **45.** 0.56 **46.** 0.00467

47. 19,000 **48.** 1,260,000,000 **49.** 0.000186

50. 0.0003 **51.** 0.00000423 **52.** 54,000

53. 711 **54.** 0.02 **55.** 0.153

56. 416,000

In Exercises 57–72, express the number in decimal notation.

57. 1.7×10^2 **58.** 2.03×10^4 **59.** 1.097×10^{-4}

60. 4.003×10^7 **61.** 8.62×10^{-5} **62.** 2.19×10^{-4}

63. 3.12×10^{-1} **64.** 4.6×10^1 **65.** 9×10^6

66. 7.3×10^4 **67.** 2.31×10^2 **68.** 1.04×10^{-2}

69. 3.5×10^4 **70.** 2.17×10^{-6}

71. 1×10^4 **72.** 1×10^{-3}

In Exercises 73–82, (a) perform the indicated operation without the use of a calculator and express each answer in decimal notation. (b) Confirm your answer from part (a) by using a scientific calculator to perform the operations. If the calculator displays the answer in scientific notation, convert the answer to decimal notation.

73. $(3 \times 10^2)(2.5 \times 10^3)$

74. $(2 \times 10^8)(3.9 \times 10^{-5})$

75. $(5.1 \times 10^1)(3 \times 10^{-4})$

76. $(1.6 \times 10^{-2})(4 \times 10^{-3})$

77. $\frac{7.5 \times 10^6}{3 \times 10^4}$ **78.** $\frac{5.2 \times 10^7}{4 \times 10^9}$

79. $\frac{8.4 \times 10^{-6}}{4 \times 10^{-3}}$ **80.** $\frac{25 \times 10^3}{5 \times 10^{-2}}$

81. $\frac{4 \times 10^5}{2 \times 10^4}$ **82.** $\frac{16 \times 10^3}{8 \times 10^{-3}}$

In Exercises 83–92, (a) perform the indicated operation without the use of a calculator and express each answer in scientific notation. (b) Confirm your answer from part (a) by using a scientific calculator to perform the operations. If the calculator displays the answer in decimal notation, convert the answer to scientific notation.

83. (5,000,000)(20,000) **84.** (0.0015)(4,000,000)

85. (0.003)(0.00015) **86.** (230,000)(3000)

87. $\frac{5{,}600{,}000}{80{,}000}$ **88.** $\frac{28{,}000}{0.004}$

89. $\frac{0.00004}{200}$ **90.** $\frac{0.0012}{0.000006}$

91. $\frac{150{,}000}{0.0005}$ **92.** $\frac{24{,}000}{8{,}000{,}000}$

PROBLEM SOLVING

In Exercises 93–96, list the numbers from smallest to largest.

93. 1.03×10^4; 1.7; 3.6×10^{-3}; 9.8×10^2

94. 8.5×10^{-5}; 8.2×10^3; 1.3×10^{-1}; 6.2×10^4

95. 40,000; 4.1×10^3; 0.00079; 8.3×10^{-5}

96. 267,000,000; 3.14×10^7; 1,962,000; 4.79×10^6

In Exercises 97–118, you may use a scientific calculator to perform the necessary operations.

97. ***U.S. Population*** On June 16, 2006, the U.S. Bureau of the Census estimated that the population of the United States was about 2.990×10^8 people and the population of the world was about 6.522×10^9 people. Determine the ratio of the U.S. population to the world's population by dividing the U.S. population by the world's population. Write your answer as a decimal number rounded to the nearest thousandth.

98. ***China's Population*** On June 16, 2006, the U.S. Bureau of the Census estimated that the population of China was about 1.307×10^9 people and the population of the world was about 6.522×10^9 people. Determine the ratio of China's population to the world's population by dividing China's population by the world's population. Write your answer as a decimal number rounded to the nearest thousandth.

99. ***India's Population*** On June 16, 2006, the U.S. Bureau of the Census estimated that the population of India was about 1.095×10^9 people and the population of the world was about 6.522×10^9 people. Determine the ratio of India's population to the world's population by dividing India's population by the world's population. Write your answer as a decimal number rounded to the nearest thousandth.

100. ***U.S. Debt per Person: 2000 versus 2006*** In Example 14, the U.S. government debt was discussed. It was found that the debt in 2006 was about \$28,026.76 per person. In 2000, the U.S. government debt was about \$5.67 trillion and the population of the United States was about 273 million.

a) Determine the amount of debt per person in the United States in 2000.

b) How much more per person did the U.S. government owe in 2006 than in 2000?

101. ***Computer Speed*** As of December 26, 2006, the fastest computer in the world was IBM's BlueGene/L System located at Lawrence Livermore National Laboratory. This computer was capable of performing about 280.6 trillion calculations per second. At this rate, how long would it take to perform a task requiring 7.6×10^{65} calculations? Write your answer in scientific notation.

▲ IBM's BlueGene/L System Computer, see Exercise 101

102. ***Gross Domestic Product*** The gross domestic product (GDP) of a country is the total national output of goods and services produced within that country. In 2005, the GDP of the United States was about \$12.4 trillion and the U.S. population was about 299 million people. Determine the U.S. GDP per person by dividing the GDP by the population. Round your answer to the nearest whole dollar.

103. ***China's GDP*** In 2005, the GDP (see Exercise 102) of China was about \$8.86 trillion and the population of China was about 1.307 billion people. Determine China's GDP per person by dividing the GDP by the population. Round your answer to the nearest whole dollar.

104. ***Traveling to Jupiter*** The distance from Earth to the planet Jupiter is approximately 4.5×10^8 mi. If a spacecraft traveled at a speed of 25,000 mph, how many hours would the spacecraft need to travel from Earth to Jupiter? Use distance = rate $\times$ time.

105. ***Traveling to the Moon*** The distance from Earth to the moon is approximately 239,000 mi. If a spacecraft travels at a speed of 20,000 mph, how many hours would the spacecraft need to travel from Earth to the moon? Use distance = rate $\times$ time.

106. ***Bucket Full of Molecules*** A drop of water contains about 40 billion molecules. If a bucket has half a million drops of water in it, how many molecules of water are in the bucket? Write your answer in scientific notation.

107. ***Blood Cells in a Cubic Millimeter*** If a cubic millimeter of blood contains 5,800,000 red blood cells, how many red blood cells are contained in 50 cubic millimeters of blood? Write your answer in scientific notation.

108. ***Radioactive Isotopes*** The half-life of a radioactive isotope is the time required for half the quantity of the isotope to decompose. The half-life of uranium 238 is 4.5×10^9 years, and the half-life of uranium 234 is 2.5×10^5 years. How many times greater is the half-life of uranium 238 than uranium 234?

109. ***1950 Niagara Treaty*** The 1950 Niagara Treaty between the United States and Canada requires that during the tourist season a minimum of 100,000 cubic feet of water per second (ft^3/sec) flow over Niagara Falls (another 130,000–160,000 ft^3/sec are diverted for power generation). Find the minimum amount of water that will flow over the falls in a 24-hour period during the tourist season. Write your answer in scientific notation.

110. ***Disposable Diaper Quantity*** Laid end to end, the 18 billion disposable diapers thrown away in the United States each year would reach the moon and back seven times. If the distance from Earth to the moon is 2.38×10^5 miles, what is the length of all these diapers placed end to end? Write your answer in decimal notation.

111. ***Mutual Fund Manager*** Lauri Mackey is the fund manager for the Mackey Mutual Fund. This mutual fund has total assets of \$1.2 billion. Lauri wants to maintain the investments in this fund according to the following pie chart.

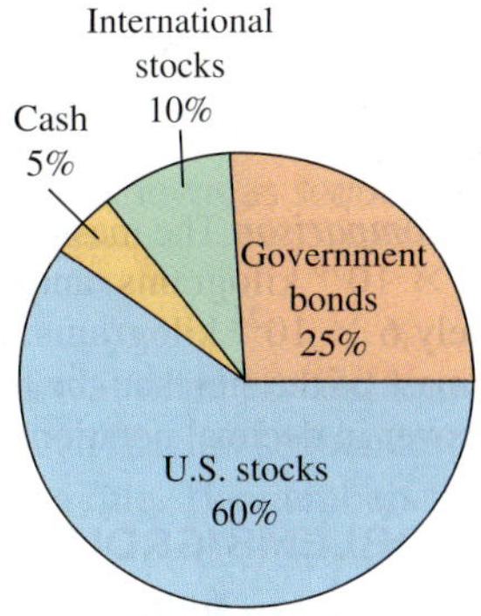

a) How much of the \$1.2 billion should be invested in U.S. stocks?

b) How much should be invested in government bonds?

c) How much should be invested in international stocks?

d) How much should remain in cash?

112. ***Another Mutual Fund Manager*** Susan Dratch is the fund manager for the Dratch Mutual Fund. This mutual fund has total assets of \$3.4 billion. Susan wants to maintain the investments in this fund according to the following pie chart.

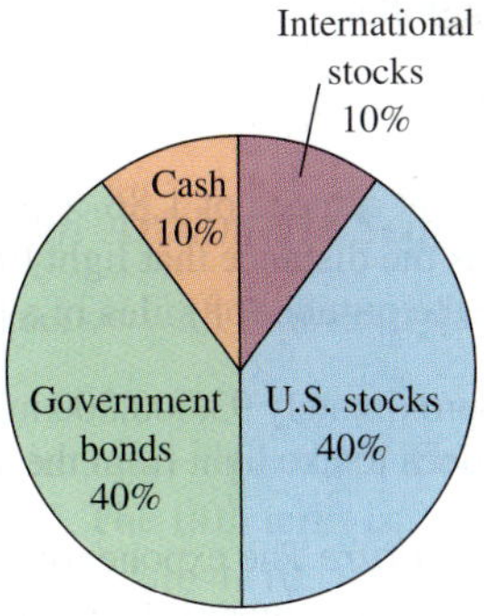

a) How much of the \$3.4 billion should be invested in U.S. stocks?

b) How much should be invested in government bonds?

c) How much should be invested in international stocks?

d) How much should remain in cash?

PROFILE IN MATHEMATICS

Carl Friedrich Gauss

Carl Friedrich Gauss (1777–1855), often called the "Prince of Mathematicians," made significant contributions to the fields of algebra, geometry, and number theory. Gauss was only 22 years old when he proved the fundamental theorem of algebra for his doctoral dissertation.

When Gauss was only 10, his mathematics teacher gave him the problem of finding the sum of the first 100 natural numbers, thinking that this task would keep Gauss busy for a while. Gauss recognized a pattern in the sequence of numbers when he considered the sum of the following numbers.

$$\begin{array}{ccccccccc} 1 & + & 2 & + & 3 & + \cdots + & 99 & + & 100 \\ 100 & + & 99 & + & 98 & + \cdots + & 2 & + & 1 \\ \hline 101 & + & 101 & + & 101 & + \cdots + & 101 & + & 101 \end{array}$$

He had the required answer in no time at all. When he added, he had one hundred 101's. Therefore, the sum of the first 100 natural numbers is $\frac{1}{2}(100)(101) = 5050$.

to represent the general or nth term of a sequence. Thus, a sequence may be symbolized as

$$a_1, a_2, a_3, a_4, \ldots, a_n, \ldots$$

For example, in the sequence 2, 5, 8, 11, 14, . . . , we have

$$a_1 = 2, a_2 = 5, a_3 = 8, a_4 = 11, a_5 = 14, \ldots.$$

When we know the first term of an arithmetic sequence and the common difference, we can use the following formula to find the value of any specific term.

GENERAL OR *N*TH TERM OF AN ARITHMETIC SEQUENCE

$$a_n = a_1 + (n - 1)d$$

EXAMPLE 3 *Determining the Twelfth Term of an Arithmetic Sequence*

Determine the twelfth term of the arithmetic sequence whose first term is -5 and whose common difference is 3.

SOLUTION To determine the twelfth term, or a_{12}, replace n in the formula with 12, a with -5, and d with 3.

$$\begin{aligned} a_n &= a_1 + (n - 1)d \\ a_{12} &= -5 + (12 - 1)3 \\ &= -5 + (11)3 \\ &= -5 + 33 \\ &= 28 \end{aligned}$$

The twelfth term is 28. As a check, we have listed the first 12 terms of the sequence: $-5, -2, 1, 4, 7, 10, 13, 16, 19, 22, 25, 28$.

EXAMPLE 4 *Determining an Expression for the nth Term*

Write an expression for the general or nth term, a_n, for the sequence 1, 6, 11, 16, ...

SOLUTION In this sequence, the first term, a_1, is 1 and the common difference, d, is 5. We substitute these values into $a_n = a_1 + (n - 1)d$ to obtain an expression for the nth term, a_n.

$$\begin{aligned} a_n &= a_1 + (n - 1)d \\ &= 1 + (n - 1)5 \\ &= 1 + 5n - 5 \\ &= 5n - 4 \end{aligned}$$

The expression is $a_n = 5n - 4$. Note that when $n = 1$, the first term is $5(1) - 4 = 1$. When $n = 2$, the second term is $5(2) - 4 = 6$, and so on.

We can use the following formula to find the sum of the first n terms in an arithmetic sequence.

SUM OF THE FIRST *N* TERMS IN AN ARITHMETIC SEQUENCE

$$s_n = \frac{n(a_1 + a_n)}{2}$$

In this formula, s_n represents the sum of the first n terms, a_1 is the first term, a_n is the nth term, and n is the number of terms in the sequence from a_1 to a_n.

EXAMPLE 5 ***Determining the Sum of an Arithmetic Sequence***

Determine the sum of the first 25 even natural numbers.

SOLUTION The sequence we are discussing is

$$2, 4, 6, 8, 10, \ldots, 50$$

In this sequence the first term is 1, the 25th term is 50, and there are 25 terms; therefore, $a_1 = 1$, $a_{25} = 50$, and $n = 25$. Thus, the sum of the first 25 terms is

$$s_n = \frac{n(a_1 + a_n)}{2}$$

$$s_{25} = \frac{25(2 + 50)}{2}$$

$$= \frac{25(52)}{2}$$

$$= \frac{1300}{2}$$

$$= 650$$

The sum of the first 25 even natural numbers, $2 + 4 + 6 + 8 + \cdots + 50$, is 650.

Geometric Sequences

The next type of sequence we will discuss is the geometric sequence. A ***geometric sequence*** is one in which the ratio of any term to the term that directly precedes it is a constant. This constant is called the ***common ratio***. The common ratio, r, can be found by taking any term except the first and dividing that term by the preceding term.

Examples of geometric sequences	Common ratios
$2, 4, 8, 16, 32, \ldots$	$r = 4 \div 2 = 2$
$-3, 6, -12, 24, -48, \ldots$	$r = 6 \div (-3) = -2$
$\frac{2}{3}, \frac{2}{9}, \frac{2}{27}, \frac{2}{81}, \ldots$	$r = \frac{2}{9} \div \frac{2}{3} = \left(\frac{2}{9}\right)\left(\frac{3}{2}\right) = \frac{1}{3}$

DID YOU KNOW?

A Heavenly Sequence

The sequence 0, 3, 6, 12, 24, 48, 96, 192, . . . , which resembles a geometric sequence, is known collectively as the **Titius–Bode law**. The sequence, discovered in 1766 by two German astronomers, was of great importance to astronomy in the eighteenth and nineteenth centuries. When 4 was added to each term and the result was divided by 10, the sequence closely corresponded with the observed mean distance from the sun to the known principal planets of the solar system (in astronomical units, or au): Mercury (0.4), Venus (0.7), Earth (1.0), Mars (1.6), missing planet (2.8), Jupiter (5.2), Saturn (10.0), missing planet (19.6). The exciting discovery of Uranus in 1781 with a mean distance of 19.2 astronomical units was in near agreement with the Titius–Bode law and stimulated the search for an undiscovered planet at a predicted 2.8 astronomical units. This effort led to the discovery of Ceres and other members of the asteroid belt. Although the Titius–Bode law broke down after discoveries of Neptune (30.1 au) and Pluto (39.5 au), many scientists still believe that other applications of the Titius–Bode law will emerge in the future.

To construct a geometric sequence when the first term, a_1, and common ratio are known, multiply the first term by the common ratio to get the second term. Then multiply the second term by the common ratio to get the third term, and so on.

EXAMPLE 6 *The First Five Terms of a Geometric Sequence*

Write the first five terms of the geometric sequence whose first term, a_1, is 3 and whose common ratio, r, is 4.

SOLUTION The first term is 3. The second term, found by multiplying the first term by 4, is $3 \cdot 4$, or 12. The third term is $12 \cdot 4$, or 48. The fourth term is $48 \cdot 4$, or 192. The fifth term is $192 \cdot 4$, or 768. Thus, the first five terms of the sequence are 3, 12, 48, 192, 768. ●

When we know the first term of a geometric sequence and the common ratio, we can use the following formula to find the value of the general or nth term, a_n.

GENERAL OR *N*TH TERM OF A GEOMETRIC SEQUENCE

$$a_n = a_1 r^{n-1}$$

EXAMPLE 7 *Determining the Twelfth Term of a Geometric Sequence*

Determine the twelfth term of the geometric sequence whose first term is −4 and whose common ratio is 2.

SOLUTION In this sequence, $a_1 = -4$, $r = 2$, and $n = 12$. Substituting the values into the formula, we obtain

$$\begin{aligned} a_n &= a_1 r^{n-1} \\ a_{12} &= -4 \cdot 2^{12-1} \\ &= -4 \cdot 2^{11} \\ &= -4 \cdot 2048 \\ &= -8192 \end{aligned}$$

As a check, we have listed the first twelve terms of the sequence:

−4, −8, −16, −32, −64, −128, −256, −512, −1024, −2048, −4096, −8192 ●

EXAMPLE 8 *Determining an Expression for the nth Term*

Write an expression for the general or nth term, a_n, of the sequence 2, 6, 18, 54, ...

SOLUTION In this sequence, the first term is 2, so $a_1 = 2$. Note that $\frac{6}{2} = 3$, $\frac{18}{6} = 3$, $\frac{54}{18} = 3$, and so on; therefore, $r = 3$. We substitute $a_1 = 2$ and $r = 3$ into $a_n = a_1 r^{n-1}$ to obtain an expression for the nth term, a_n.

$$\begin{aligned} a_n &= a_1 r^{n-1} \\ &= 2(3)^{n-1} \end{aligned}$$

The expression is $a_n = 2(3)^{n-1}$. Note than when $n = 1$, $a_1 = 2(3)^0 = 2(1) = 2$. When $n = 2$, $a_2 = 2(3)^1 = 6$, and so on. •

We can use the following formula to find the sum of the first n terms of a geometric sequence.

SUM OF THE FIRST *N* TERMS OF A GEOMETRIC SEQUENCE

$$s_n = \frac{a_1(1 - r^n)}{1 - r}, \qquad r \neq 1$$

EXAMPLE 9 *Adding the First n Terms of a Geometric Sequence*

Determine the sum of the first five terms in the geometric sequence whose first term is 4 and whose common ratio is 2.

SOLUTION In this sequence, $a_1 = 4$, $r = 2$, and $n = 5$. Substituting these values into the formula, we get

$$s_n = \frac{a_1(1 - r^n)}{1 - r}$$

$$s_5 = \frac{4[1 - (2)^5]}{1 - 2}$$

$$= \frac{4(1 - 32)}{-1}$$

$$= \frac{4(-31)}{-1} = \frac{-124}{-1} = 124$$

The sum of the first five terms of the sequence is 124. The first five terms of the sequence are 4, 8, 16, 32, 64. If you add these five numbers, you will obtain the sum 124. •

EXAMPLE 10 *Pounds and Pounds of Silver*

As a reward for saving his kingdom from a band of thieves, a king offered a knight one of two options. The knight's first option was to be paid 100,000 pounds of silver all at once. The second option was to be paid over the course of a month. On the first day, he would receive one pound of silver. On the second day, he would receive two pounds of silver. On the third day, he would receive four pounds of silver, and so on, each day receiving double the amount given on the previous day. Assuming the month has 30 days, which option would provide the knight with more silver?

SOLUTION The first option pays the knight 100,000 pounds of silver. The second option pays according to the geometric sequence 1, 2, 4, 8, 16, ... In this sequence,

$a_1 = 1, r = 2$, and $n = 30$. The sum of this sequence can be found by substituting these values into the formula to obtain

$$
\begin{aligned}
s_n &= \frac{a_1(1 - r^n)}{1 - r} \\
s_{30} &= \frac{1(1 - 2^{30})}{1 - 2} \\
&= \frac{1 - 1{,}073{,}741{,}824}{-1} \\
&= \frac{-1{,}073{,}741{,}823}{-1} \\
&= 1{,}073{,}741{,}823
\end{aligned}
$$

Thus, the knight would get 1,073,741,823 pounds of silver with the second option. The second option would result in an additional 1,073,641,823 pounds of silver than with the first option.

SECTION 5.7 EXERCISES

CONCEPT/WRITING EXERCISES

1. State the definition of *sequence* and give an example.

2. What are the numbers that make up a sequence called?

3. a) State the definition of *arithmetic sequence* and give an example.

   b) State the definition of *geometric sequence* and give an example.

4. a) In the arithmetic sequence 1, 6, 11, 16, 21, . . . , state the common difference, d.

   b) In the geometric sequence $-1, 3, -9, 27, -81, \ldots$, state the common ratio, r.

5. For an arithmetic sequence, state the meaning of each of the following symbols.

   a) a_n b) a_1 c) d d) s_n

6. For a geometric sequence, state the meaning of each of the following symbols.

   a) a_n b) a_1 c) r d) s_n

PRACTICE THE SKILLS

In Exercises 7–14, write the first five terms of the arithmetic sequence with the first term, a_1, and common difference, d.

7. $a_1 = 5, d = 1$
8. $a_1 = 4, d = 2$
9. $a_1 = 12, d = -2$
10. $a_1 = -11, d = 5$
11. $a_1 = 5, d = -2$
12. $a_1 = -3, d = -4$
13. $a_1 = \frac{3}{4}, d = \frac{1}{4}$
14. $a_1 = \frac{5}{2}, d = -\frac{3}{2}$

In Exercises 15–22, determine the indicated term for the arithmetic sequence with the first term, a_1, and common difference, d.

15. Determine a_5 when $a_1 = 4, d = 1$.
16. Determine a_8 when $a_1 = -5, d = 4$.
17. Determine a_{10} when $a_1 = -5, d = 2$.
18. Determine a_{12} when $a_1 = 7, d = -3$.
19. Determine a_{20} when $a_1 = \frac{4}{5}, d = -1$.

20. Determine a_{14} when $a_1 = \frac{1}{3}, d = \frac{2}{3}$.

21. Determine a_{22} when $a_1 = -23, d = \frac{5}{7}$.

22. Determine a_{15} when $a_1 = \frac{4}{3}, d = \frac{1}{3}$.

In Exercises 23–30, write an expression for the general or nth term, a_n, for the arithmetic sequence.

23. $1, 2, 3, 4, \ldots$

24. $1, 3, 5, 7, \ldots$

25. $2, 4, 6, 8, \ldots$

26. $-3, 0, 3, 6, \ldots$

27. $\frac{1}{4}, 1, \frac{7}{4}, \frac{5}{2}, \ldots$

28. $-\frac{1}{4}, 0, \frac{1}{4}, \frac{1}{2}, \ldots$

29. $-3, -\frac{3}{2}, 0, \frac{3}{2}, \ldots$

30. $-5, -2, 1, 4, \ldots$

In Exercises 31–38, determine the sum of the terms of the arithmetic sequence. The number of terms, n, is given.

31. $1, 2, 3, 4, \ldots, 50; n = 50$

32. $2, 4, 6, 8, \ldots, 100; n = 50$

33. $1, 3, 5, 7, \ldots, 99; n = 50$

34. $-4, -7, -10, -13, \ldots, -28; n = 9$

35. $11, 6, 1, -4, \ldots, -24; n = 8$

36. $-\frac{1}{16}, 0, \frac{1}{16}, \frac{1}{8}, \ldots, \frac{17}{16}; n = 19$

37. $\frac{1}{5}, \frac{2}{5}, \frac{3}{5}, \frac{4}{5}, \ldots, \frac{24}{5}; n = 24$

38. $-\frac{3}{4}, -\frac{1}{2}, -\frac{1}{4}, 0, \ldots, \frac{17}{4}; n = 21$

In Exercises 39–46, write the first five terms of the geometric sequence with the first term, a_1, and common ratio, r.

39. $a_1 = 1, r = 5$

40. $a_1 = 1, r = 2$

41. $a_1 = 2, r = -2$

42. $a_1 = 8, r = \frac{1}{2}$

43. $a_1 = -3, r = -1$

44. $a_1 = -6, r = -2$

45. $a_1 = 81, r = -\frac{1}{3}$

46. $a_1 = \frac{16}{15}, r = \frac{1}{2}$

In Exercises 47–54, determine the indicated term for the geometric sequence with the first term, a_1, and common ratio, r.

47. Determine a_6 when $a_1 = 5, r = 2$.

48. Determine a_6 when $a_1 = 1, r = 3$.

49. Determine a_3 when $a_1 = 3, r = \frac{1}{2}$.

50. Determine a_7 when $a_1 = -3, r = -3$.

51. Determine a_7 when $a_1 = -5, r = 3$.

52. Determine a_8 when $a_1 = \frac{1}{2}, r = -2$.

53. Determine a_{10} when $a_1 = -2, r = 3$.

54. Determine a_{18} when $a_1 = -5, r = -2$.

In Exercises 55–62, write an expression for the general or nth term, a_n, for the geometric sequence.

55. $1, 2, 4, 8, \ldots$

56. $1, 3, 9, 27, \ldots$

57. $-1, 1, -1, 1, \ldots$

58. $-16, -8, -4, -2, \ldots$

59. $2, 1, \frac{1}{2}, \frac{1}{4}, \ldots$

60. $-3, 6, -12, 24, \ldots$

61. $9, 3, 1, \frac{1}{3}, \frac{1}{9}, \ldots$

62. $-4, -\frac{8}{3}, -\frac{16}{9}, -\frac{32}{27}, \ldots$

In Exercises 63–70, determine the sum of the first n terms of the geometric sequence for the values of a_1 and r.

63. $n = 5, a_1 = 6, r = 2$

64. $n = 4, a_1 = 4, r = 3$

65. $n = 6, a_1 = -3, r = 4$

66. $n = 9, a_1 = -3, r = 5$

67. $n = 11, a_1 = -7, r = 3$

68. $n = 15, a_1 = -1, r = 2$

69. $n = 15, a_1 = -1, r = -2$

70. $n = 10, a_1 = 512, r = \frac{1}{2}$

PROBLEM SOLVING

71. Determine the sum of the first 100 natural numbers.

72. Determine the sum of the first 100 odd natural numbers.

73. Determine the sum of the first 100 even natural numbers.

74. Determine the sum of the first 50 multiples of 3.

75. ***Pendulum Movement*** Each swing of a pendulum (from far left to far right) is 3 in. shorter than the preceding swing. The first swing is 8 ft.

a) Find the length of the twelfth swing.

b) Determine the total distance traveled by the pendulum during the first 12 swings.

76. ***Clock Strikes*** A clock strikes once at 1 o'clock, twice at 2 o'clock, three times at 3 o'clock, and so on. How many times does it strike over a 12 hr period?

77. ***Annual Pay Raises*** Rita Fernandez is given a starting salary of \$35,000 and promised a \$1400 raise per year after each of the next 8 years.

    a) Determine her salary during her eighth year of work.

    b) Determine the total salary she received over the 8 years.

78. ***A Bouncing Ball*** Each time a ball bounces, the height attained by the ball is 6 in. less than the previous height attained. If on the first bounce the ball reaches a height of 6 ft, find the height attained on the eleventh bounce.

79. ***Squirrels and Pinecones*** A tree squirrel cut down 1 pinecone on the first day of October, 2 pinecones on the second day, 3 pinecones on the third day, and so on. How many pinecones did this squirrel cut down during the month of October, which contains 31 days?

80. ***Enrollment Increase*** The current enrollment at Loras College is 8000 students. If the enrollment increases by 8% per year, determine the enrollment 10 years later.

81. ***Decomposing Substance*** A certain substance decomposes and loses 20% of its weight each hour. If there are originally 200 g of the substance, how much remains after 6 hr?

82. ***Samurai Sword Construction*** While making a traditional Japanese samurai sword, the master sword maker prepares the blade by heating a bar of iron until it is white hot. He then folds it over and pounds it smooth. Therefore, after each folding, the number of layers of steel is doubled. Assuming the sword maker starts with a bar of one layer and folds it 15 times, how many layers of steel will the finished sword contain?

83. ***Salary Increase*** If your salary were to increase at a rate of 6% per year, find your salary during your 15th year if your original salary is \$31,000.

84. ***A Bouncing Ball*** When dropped, a ball rebounds to four-fifths of its original height. How high will the ball rebound after the fourth bounce if it is dropped from a height of 30 ft?

85. ***Value of a Stock*** Ten years ago, Nancy Hart purchased \$2000 worth of shares in RCF, Inc. Since then, the price of the stock has roughly tripled every 2 years. Approximately how much are Nancy's shares worth today?

86. ***A Baseball Game*** During a baseball game, the visiting team scored 1 run in the first inning, 2 runs in the second inning, 3 runs in the third inning, 4 runs in the fourth inning, and so on. The home team scored 1 run in the first inning, 2 runs in the second inning, 4 runs in the third inning, 8 runs in the fourth inning, and so on. What is the score of the game after eight innings?

Inning	1	2	3	4	5	6	7	8	9
Visitors	1	2	3	4					
Home	1	2	4	8					

CHALLENGE PROBLEMS/GROUP ACTIVITIES

87. A geometric sequence has $a_1 = 82$ and $r = \frac{1}{2}$; find s_6.

88. ***Sums of Interior Angles*** The sums of the interior angles of a triangle, a quadrilateral, a pentagon, and a hexagon are 180°, 360°, 540°, and 720°, respectively. Use this pattern to find a formula for the general term, a_n, where a_n represents the sum of the interior angles of an n-sided polygon.

89. ***Divisibility by 6*** Determine how many numbers between 7 and 1610 are divisible by 6.

90. Determine r and a_1 for the geometric sequence with $a_2 = 24$ and $a_5 = 648$.

91. ***Total Distance Traveled by a Bouncing Ball*** A ball is dropped from a height of 30 ft. On each bounce, it attains a height four-fifths of its original height (or of the previous bounce). Find the total vertical distance traveled by the ball after it has completed its fifth bounce (has therefore hit the ground six times).

RECREATIONAL MATHEMATICS

92. ***A Wagering Strategy*** The following is a strategy used by some people involved in games of chance. A player begins by betting a standard bet, say \$1. If the player wins, the player again bets \$1 in the next round. If the player loses, the player bets \$2 in the next round. Next, if the player wins, the player again bets \$1; if the player loses, the player now bets \$4 in the next round. The process continues as long as the player keeps playing, betting \$1 after a win or doubling the previous bet after a loss.

a) Assume a player is using a \$1 standard bet and loses five times in a row. How much money should the player bet in the sixth round? How much money has the player lost at the end of the fifth round?

b) Assume a player is using a \$10 standard bet and loses five times in a row. How much money should the player bet in the sixth round? How much money has the player lost at the end of the fifth round?

c) Assume a player is using a \$1 standard bet and loses 10 times in a row. How much money should the player bet in the 11th round? How much money has the player lost at the end of the 10th round?

d) Assume a player is using a \$10 standard bet and loses 10 times in a row. How much money should the player bet in the 11th round? How much money has the player lost at the end of the 10th round?

e) Why is this strategy dangerous?

INTERNET/RESEARCH ACTIVITY

93. A topic generally associated with sequences is *series.*

a) Research *series* and explain what a series is and how it differs from a sequence. Also write a formal definition of series. Give examples of different kinds of series.

b) Write the arithmetic series associated with the arithmetic sequence 1, 4, 7, 10, 13, ...

c) Write the geometric series associated with the geometric sequence 3, 6, 12, 24, 48, ...

d) What is an infinite geometric series?

e) Determine the sum of the terms of the infinite geometric series $1 + \frac{1}{2} + \frac{1}{4} + \frac{1}{8} + \frac{1}{16} + \cdots$.

5.8 FIBONACCI SEQUENCE

▲ The number of ancestors of a male bee is among the numbers found in the Fibonacci sequence.

Consider the following numbers found in nature: the number of petals on a daisy, the number of sunflower seeds in a sunflower, the number of scales on a pineapple, the number of ancestors of a male bee. In this section, we will find that all these numbers are part of a special sequence of numbers called the Fibonacci sequence.

Our discussion of sequences would not be complete without mentioning a sequence known as the ***Fibonacci sequence***. The sequence is named after Leonardo of Pisa, also known as Fibonacci. He was one of the most distinguished mathematicians of the Middle Ages. This sequence is first mentioned in his book *Liber Abacci* (Book of the Abacus), which contained many interesting problems, such as: "A certain man put a pair of rabbits in a place surrounded on all sides by a wall. How many pairs of rabbits can be produced from that pair in a year if it is assumed that every month each pair begets a new pair which from the second month becomes productive?"

The solution to this problem (Fig. 5.10) led to the development of the sequence that bears its author's name: the Fibonacci sequence. The sequence is shown in Table 5.1. The numbers in the columns titled *Pairs of Adults* form the Fibonacci sequence.

PROFILE IN MATHEMATICS

Fibonacci

Leonardo of Pisa (1170–1250) is considered one of the most distinguished mathematicians of the Middle Ages. He was born in Italy and was sent by his father to study mathematics with an Arab master. When he began writing, he referred to himself as Fibonacci, or "son of Bonacci," the name by which he is known today. In addition to the famous sequence bearing his name, Fibonacci is also credited with introducing the Hindu–Arabic number system into Europe. His 1202 book, *Liber Abacci* (Book of the Abacus), explained the use of this number system and emphasized the importance of the number zero.

FIBONACCI SEQUENCE

1, 1, 2, 3, 5, 8, 13, 21, . . .

Table 5.1

Month	Pairs of Adults	Pairs of Babies	Total Pairs
1	1	0	1
2	1	1	2
3	2	1	3
4	3	2	5
5	5	3	8
6	8	5	13
7	13	8	21
8	21	13	34
9	34	21	55
10	55	34	89
11	89	55	144
12	144	89	233

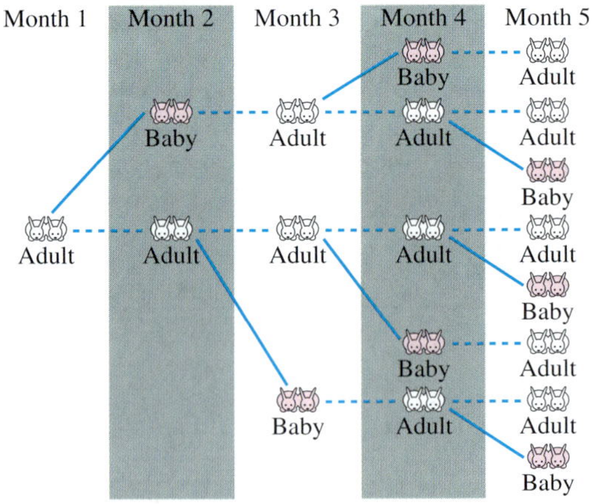

Figure 5.10

▲ The head of a sunflower

In the Fibonacci sequence, the first and second terms are 1. The sum of these two terms is the third term. The sum of the second and third terms is the fourth term, and so on.

In the middle of the nineteenth century, mathematicians made a serious study of this sequence and found strong similarities between it and many natural phenomena. Fibonacci numbers appear in the seed arrangement of many species of plants and in the petal counts of various flowers. For example, when the flowering head of the sunflower matures to seed, the seeds' spiral arrangement becomes clearly visible. A typical count of these spirals may give 89 steeply curving to the right, 55 curving more shallowly to the left, and 34 again shallowly to the right. The largest known specimen to be examined had spiral counts of 144 right, 89 left, and 55 right. These numbers, like the other three mentioned, are consecutive terms of the Fibonacci sequence.

On the heads of many flowers, petals surrounding the central disk generally yield a Fibonacci number. For example, some daisies contain 21 petals, and others contain 34, 55, or 89 petals. (People who use a daisy to play the "love me, love me not" game will likely pluck 21, 34, 55, or 89 petals before arriving at an answer.)

Fibonacci numbers are also observed in the structure of pinecones and pineapples. The tablike or scalelike structures called bracts that make up the main body of the pinecone form a set of spirals that start from the cone's attachment to the branch. Two sets of oppositely directed spirals can be observed, one steep and the other more gradual. A count on the steep spiral will reveal a Fibonacci number, and a count on the gradual one will be the adjacent smaller Fibonacci number, or if not, the next smaller Fibonacci number. One investigation of 4290 pinecones from 10 species of pine trees found in California revealed that only 74 cones, or 1.7%, deviated from this Fibonacci pattern.

Like pinecone bracts, pineapple scales are patterned into spirals, and because they are roughly hexagonal in shape, three distinct sets of spirals can be counted. Generally, the number of pineapple scales in each spiral are Fibonacci numbers.

Table 5.2

Numbers	Ratio
1, 1	$\frac{1}{1} = 1$
1, 2	$\frac{2}{1} = 2$
2, 3	$\frac{3}{2} = 1.5$
3, 5	$\frac{5}{3} = 1.666\ldots$
5, 8	$\frac{8}{5} = 1.6$
8, 13	$\frac{13}{8} = 1.625$
13, 21	$\frac{21}{13} \approx 1.615$
21, 34	$\frac{34}{21} \approx 1.619$
34, 55	$\frac{55}{34} \approx 1.618$
55, 89	$\frac{89}{55} \approx 1.618$

Fibonacci Numbers and Divine Proportions

In 1753, while studying the Fibonacci sequence, Robert Simson, a mathematician at the University of Glasgow, noticed that when he took the ratio of any term to the term that immediately preceded it, the value he obtained remained in the vicinity of one specific number. To illustrate this, we indicate in Table 5.2 the ratio of various pairs of sequential Fibonacci numbers.

The ratio of the 50th term to the 49th term is 1.6180. Simson proved that the ratio of the $(n + 1)$ term to the nth term as n gets larger and larger is the irrational number $(\sqrt{5} + 1)/2$, which begins 1.61803. . . . This number was already well known to mathematicians at that time as the ***golden number***.

Many years earlier, Bavarian astronomer and mathematician Johannes Kepler wrote that for him the golden number symbolized the Creator's intention "to create like from like." The golden number $(\sqrt{5} + 1)/2$ is frequently referred to as "phi," symbolized by the Greek letter Φ.

Figure 5.11

The ancient Greeks, in about the sixth century B.C., sought unifying principles of beauty and perfection, which they believed could be described by using mathematics. In their study of beauty, the Greeks used the term *golden ratio*. To understand the golden ratio, let's consider the line segment AB in Fig. 5.11. When this line segment is

▲ The Great Pyramid of Gizeh

divided at a point C such that the ratio of the whole, AB, to the larger part, AC, is equal to the ratio of the larger part, AC, to the smaller part, CB, each ratio AB/AC and AC/CB is referred to as a ***golden ratio***. The proportion these ratios form, $AB/AC = AC/CB$, is called the ***golden proportion***. Furthermore, each ratio in the proportion will have a value equal to the golden number, $(\sqrt{5} + 1)/2$.

$$\frac{AB}{AC} = \frac{AC}{CB} = \frac{\sqrt{5} + 1}{2} \approx 1.618$$

The Great Pyramid of Gizeh in Egypt, built about 2600 B.C., is the earliest known example of use of the golden ratio in architecture. The ratio of any of its sides of the square base (775.75 ft) to its altitude (481.4 ft) is about 1.611. Other evidence of the use of the golden ratio appears in other Egyptian buildings and tombs.

In medieval times, people referred to the golden proportion as the ***divine proportion***, reflecting their belief in its relationship to the will of God.

Twentieth-century architect Le Corbusier developed a scale of proportions for the human body that he called the Modulor (Fig. 5.12). Note that the navel separates the entire body into golden proportions, as do the neck and knee.

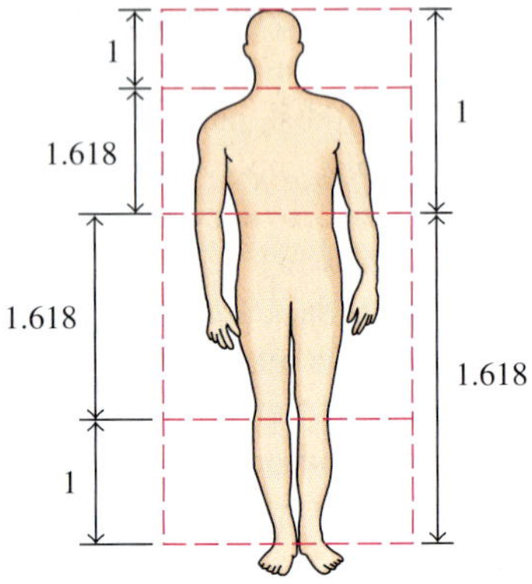

Figure 5.12

From the golden proportion, the ***golden rectangle*** can be formed, as shown in Fig. 5.13.

$$\frac{\text{Length}}{\text{Width}} = \frac{a + b}{a} = \frac{a}{b} = \frac{\sqrt{5} + 1}{2}$$

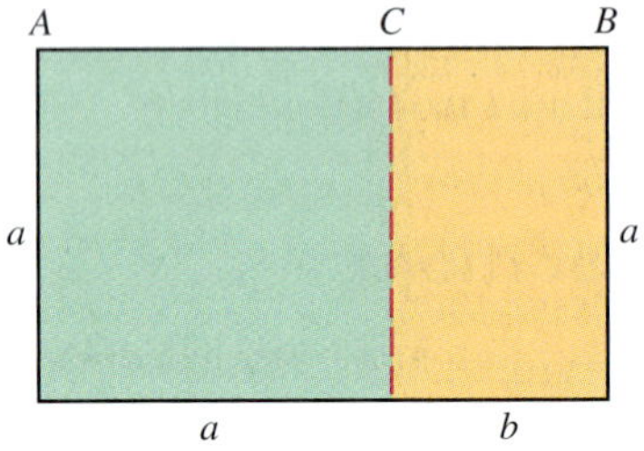

Figure 5.13

Note that when a square is cut off one end of a golden rectangle, as in Fig. 5.13, the remaining rectangle has the same properties as the original golden rectangle (creating "like from like" as Johannes Kepler had written) and is therefore itself a golden rectangle. Interestingly, the curve derived from a succession of diminishing golden rectangles, as shown in Fig. 5.14, is the same as the spiral curve of the chambered nautilus. The same curve appears on the horns of rams and some other animals. It is the same curve that is observed in the plant structures mentioned earlier: sunflowers, other flower heads, pinecones, and pineapples. You will recall that Fibonacci numbers were observed in each of these plant structures. The curve shown in Fig. 5.14 closely approximates what mathematicians call a *logarithmic spiral*.

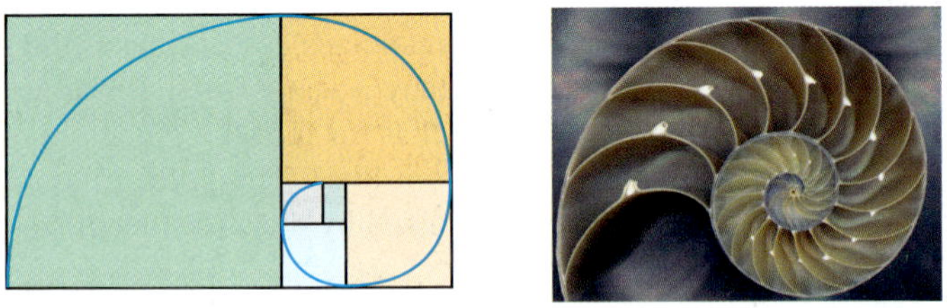

Figure 5.14

▲ The Parthenon

Ancient Greek civilization used the golden rectangle in art and architecture. The main measurements of many buildings of antiquity, including the Parthenon in Athens, are governed by golden ratios and rectangles. Greek statues, vases, urns, and so on also exhibit characteristics of the golden ratio. It is for Phidas, considered the greatest of Greek sculptors, that the golden ratio was named "phi." The proportions can be found abundantly in his work.

The proportions of the golden rectangle can be found in the work of many artists, from the old masters to the moderns. For example, the golden rectangle can be seen in the painting *Invitation to the Sideshow (La Parade de Cirque),* 1887, by George Seurat, a French neoimpressionist artist.

▲ *Invitation to the Sideshow (La Parade de Cirque)*, 1887, by George Seurat

▲ *Fibonacci's Garden* by Caryl Bryer Fallert

In addition to using the golden rectangle in art, several artists have used Fibonacci numbers in art. One contemporary example is the 1995 work by Caryl Bryer Fallert called *Fibonacci's Garden*. This artwork is a quilt constructed from two separate fabrics that are put together in a pattern based on the Fibonacci sequence.

Fibonacci numbers are also found in another form of art, namely music. Perhaps the most obvious link between Fibonacci numbers and music can be found on the piano keyboard. An octave (Fig. 5.15) on a keyboard has 13 keys: 8 white keys and 5 black keys (the 5 black keys are in one group of 2 and one group of 3).

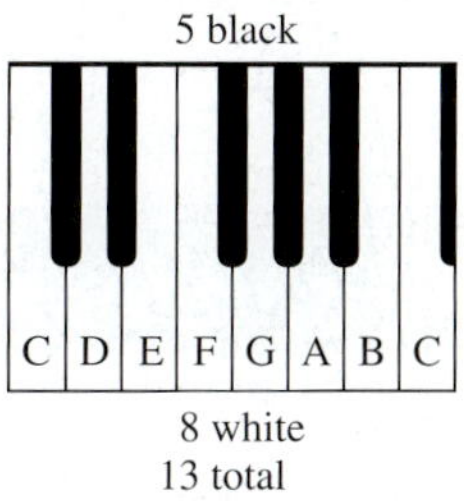

Figure 5.15

In Western music, the most complete scale, the chromatic scale, consists of 13 notes (from C to the next higher C). Its predecessor, the diatonic scale, contains 8 notes (an octave). The diatonic scale was preceded by a 5-note pentatonic scale (*penta* is Greek for "five"). Each number is a Fibonacci number.

The visual arts deal with what is pleasing to the eye, whereas musical composition deals with what is pleasing to the ear. Whereas art achieves some of its goals by using division of planes and area, music achieves some of its goals by a similar division of time, using notes of various duration and spacing. The musical intervals considered by many to be the most pleasing to the ear are the major sixth and minor sixth. A major sixth, for example, consists of the note C, vibrating at about 264† vibrations per second, and note A, vibrating at about 440 vibrations per second. The ratio of 440 to 264 reduces to 5 to 3, or $\frac{5}{3}$, a ratio of two consecutive Fibonacci numbers. An example of a minor sixth is E (about 330 vibrations per second) and C (about 528 vibrations per second). The ratio 528 to 330 reduces to 8 to 5, or $\frac{8}{5}$, the next ratio of two consecutive Fibonacci numbers. The vibrations of any sixth interval reduce to a similar ratio.

Patterns that can be expressed mathematically in terms of Fibonacci relationships have been found in Gregorian chants and works of many composers, including Bach, Beethoven, and Bartók. A number of twentieth-century musical works, including Ernst Krenek's *Fibonacci Mobile*, were deliberately structured by using Fibonacci proportions.

A number of studies have tried to explain why the Fibonacci sequence and related items are linked to so many real-life situations. It appears that the Fibonacci numbers are a part of natural harmony that is pleasing to both the eye and the ear. In the nineteenth century, German physicist and psychologist Gustav Fechner tried to determine which dimensions were most pleasing to the eye. Fechner, along with psychologist Wilhelm Wundt, found that most people do unconsciously favor golden dimensions when purchasing greeting cards, mirrors, and other rectangular objects. This discovery has been widely used by commercial manufacturers in their packaging and labeling designs, by retailers in their store displays, and in other areas of business and advertising.

†Frequencies of notes vary in different parts of the world and change over time.

DID YOU KNOW?

Fibonacci and the Male Bee's Ancestors

The most frequent example given to introduce the Fibonacci sequence involves rabbits producing offspring, two at a time. Although this example makes for a nice introduction to Fibonacci sequences, it is not at all realistic. A much better example comes from the breeding practices of bees. Female or worker bees are produced when the queen bee mates with a male bee. Male bees are produced from the queen's unfertilized eggs. In essence, then, female bees have two parents, whereas male bees only have one parent. The family tree of a male bee would look like this:

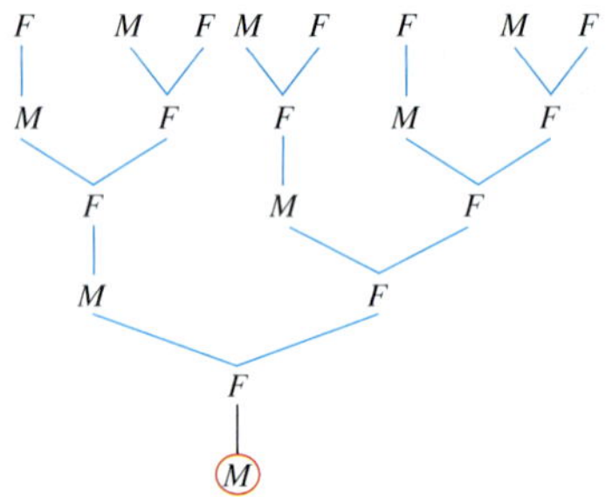

From this tree, we can see that the **1** male bee (circled) has **1** parent, **2** grandparents, **3** great-grandparents, **5** great-great-grandparents, **8** great-great-great-grandparents, and so on. We see the Fibonacci sequence as we move back through the male bees' generations.

SECTION 5.8 EXERCISES

CONCEPT/WRITING EXERCISES

1. Explain how to construct the Fibonacci sequence.

2. a) Write out the first ten terms of the Fibonacci sequence.

   b) Divide the ninth term by the eighth term, rounding to the nearest thousandth.

   c) Divide the tenth term by the ninth term, rounding to the nearest thousandth.

   d) Try a few more divisions and then make a conjecture about the result.

3. a) What is the value of the golden number?

b) What is the golden ratio?

c) What is the golden proportion?

d) What is the golden rectangle?

4. Describe three examples of where the golden ratio can be found

a) in nature.

b) in manufactured items.

5. In your own words, explain the relationship between the golden number, golden ratio, golden proportion, and golden rectangle.

6. Describe three examples of where Fibonacci numbers can be found

a) in nature.

b) in manufactured items.

PRACTICE THE SKILLS/PROBLEM SOLVING

7. a) To what decimal value is $(\sqrt{5} + 1)/2$ aproximately equal?

b) To what decimal value is $(\sqrt{5} - 1)/2$ approximately equal?

c) By how much do the results in parts (a) and (b) differ?

8. The eleventh Fibonacci number is 89. Examine the first six digits in the decimal expression of its reciprocal, $\frac{1}{89}$. What do you find?

9. Find the ratio of the second to the first term of the Fibonacci sequence. Then find the ratio of the third to the second term of the sequence and determine whether this ratio was an increase or decrease from the first ratio. Continue this process for 10 ratios and then make a conjecture regarding the increasing or decreasing values in consecutive ratios.

10. A musical composition is described as follows. Explain why this piece is based on the golden ratio.

Entire Composition

34 measures	55 measures	21 measures	34 measures
Theme	Fast, Loud	Slow	Repeat of theme

11. The sum of any 10 consecutive Fibonacci numbers is always divisible by 11. Select any 10 consecutive Fibonacci numbers and show that for your selection this statement is true.

12. The greatest common factor of any two consecutive Fibonacci numbers is 1. Show that this statement is true for the first 15 Fibonacci numbers.

13. For any four consecutive Fibonacci numbers, the difference of the squares of the middle two numbers equals the product of the smallest and largest numbers. Select four consecutive Fibonacci numbers and show that this pattern holds for the numbers you selected.

14. Twice any Fibonacci number minus the next Fibonacci number equals the second number preceding the original number. Select a number in the Fibonacci sequence and show that this pattern holds for the number selected.

15. Determine the ratio of the length to width of various photographs and compare these ratios to Φ.

16. Determine the ratio of the length to the width of a 6 inch by 4 inch standard index card, and compare the ratio to Φ.

17. Determine the ratio of the length to width of several picture frames and compare these ratios to Φ.

18. Determine the ratio of the length to the width of your television screen and compare this ratio to Φ.

19. Determine the ratio of the length to the width of a computer screen and compare this ratio to Φ.

20. Determine the ratio of the length to width of a desktop in your classroom and compare this ratio to Φ.

21. Determine the ratio of the length to the width of this textbook and compare this ratio to Φ.

22. Find three physical objects whose dimensions are very close to a golden rectangle.

a) List the articles and record the dimensions.

b) Compute the ratios of their lengths to their widths.

c) Find the difference between the golden ratio and the ratio you obtain in part (b)—to the nearest tenth—for each object.

In Exercises 23–30, determine whether the sequence is a Fibonacci-type sequence (each term is the sum of the two preceding terms). If it is, determine the next two terms of the sequence.

23. 1, 2, 3, 5, 6, 7, . . .

24. 1, 3, 4, 7, 11, 18, . . .

25. $-1, 1, 0, 1, 1, 2, \ldots$

26. $1, 4, 9, 16, 25, 36, \ldots$

27. $5, 10, 15, 25, 40, 65, \ldots$

28. $\frac{1}{4}, \frac{1}{4}, \frac{1}{2}, \frac{3}{4}, 1\frac{1}{4}, 2, \ldots$

29. $-5, 3, -2, 1, -1, 0, \ldots$

30. $-4, 5, 1, 6, 7, 13, \ldots$

31. a) Select any two nonzero digits and add them to obtain a third digit. Continue adding the two previous terms to get a Fibonacci-type sequence.

b) Form ratios of successive terms to show how they will eventually approach the golden number.

32. Repeat Exercise 31 for two different nonzero numbers.

33. a) Select any three consecutive terms of a Fibonacci sequence. Subtract the product of the terms on each side of the middle term from the square of the middle term. What is the difference?

b) Repeat part (a) with three different consecutive terms of the sequence.

c) Make a conjecture about what will happen when you repeat this process for any three consecutive terms of a Fibonacci sequence.

34. ***Pascal's Triangle*** One of the most famous number patterns involves ***Pascal's triangle***. The Fibonacci sequence can be found by using Pascal's triangle. Can you explain how that can be done? A hint is shown.

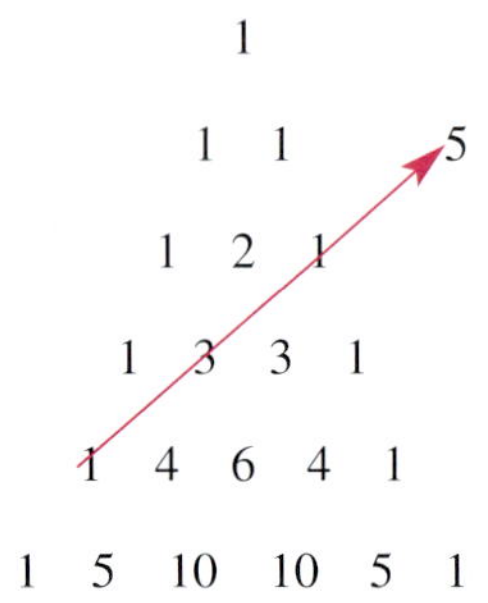

35. ***Lucas Sequence*** **a)** A sequence related to the Fibonacci sequence is the ***Lucas sequence***. The Lucas sequence is formed in a manner similar to the Fibonacci sequence. The first two numbers of the Lucas sequence are 1 and 3. Write the first eight terms of the Lucas sequence.

b) Complete the next two lines of the following chart.

$$
\begin{aligned}
1 + 2 &= 3\\
1 + 3 &= 4\\
2 + 5 &= 7\\
3 + 8 &= 11\\
5 + 13 &= 18
\end{aligned}
$$

c) What do you observe about the first column in the chart in part (b)?

CHALLENGE PROBLEMS/GROUP ACTIVITIES

36. ***Fibonacci-Type Sequence*** The following sequence represents a Fibonacci-type sequence (each term is the sum of the two preceding terms). Here x represents any natural number from 1 to 10:

$$-10, x, -10 + x, -10 + 2x, -20 + 3x, -30 + 5x, \ldots$$

For example, if $x = 2$, the first 10 terms of the sequence would be $-10, 2, -8, -6, -14, -20, -34, -54, -88, -142$.

Write out the first 10 terms of this Fibonacci-type sequence for x equal to the following.

a) 4 **b)** 5 **c)** 6 **d)** 7 **e)** 8

f) For values of x of 4, 5, and 6, you should have found that each term after the seventh term in the sequence was a negative number. For values of x of 7 and 8, you should have found that each term after the seventh term in the sequence was a positive number. Do you believe that for any value of x greater than or equal to 7, each term after the seventh term of the sequence will always be a positive number? Explain.

37. Draw a line of length 5 in. Determine and mark the point on the line that will create the golden ratio. Explain how you determined your answer.

38. The divine proportion is $(a + b)/a = a/b$ (see Fig. 5.13), which can be written $1 + (b/a) = a/b$. Now let $x = a/b$, which gives $1 + (1/x) = x$. Multiply both sides of this equation by x to get a quadratic equation and then use the quadratic formula (Section 6.8) to show that one answer is $x = (1 + \sqrt{5})/2$ (the golden ratio).

39. ***Pythagorean Triples*** A Pythagorean triple is a set of three whole numbers, $\{a, b, c\}$, such that $a^2 + b^2 = c^2$. For example, since $6^2 + 8^2 = (10)^2$, $\{6, 8, 10\}$ is a Pythagorean triple. The following steps show how to find Pythagorean triples using any four consecutive Fibonacci numbers. Here we will demonstrate the process with the Fibonacci numbers 3, 5, 8, and 13.

1. Determine the product of 2 and the two inner Fibonacci numbers. We have $2(5)(8) = 80$, which is the first number in the Pythagorean triple. So $a = 80$.
2. Determine the product of the two outer numbers. We have $3(13) = 39$, which is the second number in the Pythagorean triple. So $b = 39$.
3. Determine the sum of the squares of the inner two numbers. We have $5^2 + 8^2 = 25 + 64 = 89$, which is the third number in the Pythagorean triple. So $c = 89$.

This process has produced the Pythagorean triple, $\{80, 39, 89\}$. To verify,

$$\begin{aligned}(80)^2 + (39)^2 &= (89)^2\\ 6400 + 1521 &= 7921\\ 7921 &= 7921\end{aligned}$$

Use this process to produce four other Pythagorean triples.

40. ***Reflections*** When two panes of glass are placed face to face, four interior reflective surfaces exist labeled 1, 2, 3, and 4. If light is not reflected, it has just one path through the glass (see the figure below). If it has one reflection, it can be reflected in two ways. If it has two reflections, it can be reflected in three ways. Use this information to answer parts (a) through (c).

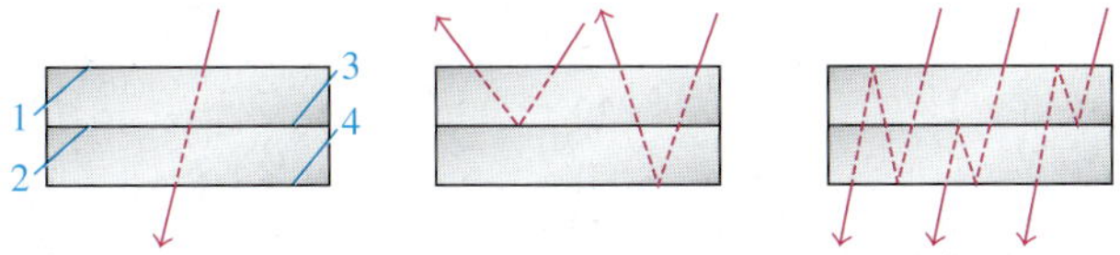

a) If a ray is reflected three times, there are five paths it can follow. Show the paths.

b) If a ray is reflected four times, there are eight paths it can follow. Show the paths.

c) How many paths can a ray follow if it is reflected five times? Explain how you determined your answer.

INTERNET/RESEARCH ACTIVITIES

41. Write a report on the history and mathematical contributions of Fibonacci.

42. The digits 1 through 9 have evolved considerably since they appeared in Fibonacci's book *Liber Abacci.* Write a report tracing the history of the evolution of the digits 1 through 9 since Fibonacci's time.

43. Write a report indicating where the golden ratio and golden rectangle have been used in art and architecture. You may wish to include information on art and architecture related to the golden ratio and Fibonacci sequences.

CHAPTER 5 SUMMARY

IMPORTANT FACTS

FUNDAMENTAL THEOREM OF ARITHMETIC

Every composite number can be expressed as a unique product of prime numbers.

SETS OF NUMBERS

Natural or counting numbers: $\{1, 2, 3, 4, \dots\}$

Whole numbers: $\{0, 1, 2, 3, 4, \dots\}$

Integers: $\{\dots, -3, -2, -1, 0, 1, 2, 3, \dots\}$

Rational numbers: Numbers of the form p/q, where p and q are integers, $q \neq 0$. Every rational number when expressed as a decimal number will be either a terminating or repeating decimal number.

Irrational number: A real number whose representation is a nonterminating, nonrepeating decimal number (not a rational number).

DEFINITION OF SUBTRACTION

$$a - b = a + (-b)$$

FUNDAMENTAL LAW OF RATIONAL NUMBERS

$$\frac{a}{b} = \frac{a}{b}\cdot\frac{c}{c} = \frac{ac}{bc}, \qquad b \neq 0, \qquad c \neq 0$$

RULES OF RADICALS

Product rule for radicals:

$$\sqrt{a \cdot b} = \sqrt{a}\cdot\sqrt{b}, \qquad a \geq 0, \qquad b \geq 0$$

Quotient rule for radicals:

$$\frac{\sqrt{a}}{\sqrt{b}} = \sqrt{\frac{a}{b}}, \qquad a \geq 0, \qquad b > 0$$

PROPERTIES OF REAL NUMBERS

Commutative property of addition: $a + b = b + a$

Commutative property of multiplication: $a \cdot b = b \cdot a$

Associative property of addition:

$$(a + b) + c = a + (b + c)$$

Associative property of multiplication:

$$(a \cdot b) \cdot c = a \cdot (b \cdot c)$$

Distributive property: $a \cdot (b + c) = ab + ac$

RULES OF EXPONENTS

Product rule for exponents: $a^m \cdot a^n = a^{m+n}$

Quotient rule for exponents: $\dfrac{a^m}{a^n} = a^{m-n}, \qquad a \neq 0$

Zero exponent rule: $a^0 = 1, \quad a \neq 0$

Negative exponent rule: $a^{-m} = \frac{1}{a^m}, \quad a \neq 0$

Power rule: $(a^m)^n = a^{m \cdot n}$

ARITHMETIC SEQUENCE

General or nth term: $a_n = a_1 + (n - 1)d$

Sum of the first n terms: $s_n = \frac{n(a_1 + a_n)}{2}$

GEOMETRIC SEQUENCE

General or nth term: $a_n = a_1 r^{n-1}$

Sum of the first n terms: $s_n = \frac{a_1(1 - r^n)}{1 - r}, \quad r \neq 1$

FIBONACCI SEQUENCE

$$1, 1, 2, 3, 5, 8, 13, 21, \ldots$$

GOLDEN NUMBER

$$\frac{\sqrt{5} + 1}{2} \approx 1.618$$

GOLDEN PROPORTION

$$\frac{a + b}{a} = \frac{a}{b}$$

CHAPTER 5 REVIEW EXERCISES

5.1

In Exercises 1 and 2, determine whether the number is divisible by each of the following numbers: 2, 3, 4, 5, 6, 8, 9, and 10.

1. 158,340 **2.** 400,644

In Exercises 3–7, determine the prime factorization of the number.

3. 540 **4.** 693 **5.** 840

6. 882 **7.** 1452

In Exercises 8–13, determine the GCD and LCM of the numbers.

8. 30, 105 **9.** 63, 108 **10.** 45, 250

11. 90, 300 **12.** 60, 40, 96 **13.** 36, 108, 144

14. ***Train Stops*** From 1912 to 1971, the Milwaukee Road Railroad Company had a train stop every 15 days in Dubuque, Iowa. During this same period, the same train also stopped in Des Moines, Iowa, every 9 days. If on April 18, 1964, the train made a stop in Dubuque and a stop in Des Moines, how many days was it until the train again stopped in both cities on the same day?

▲ See Exercise 14

5.2

In Exercises 15–22, use a number line to evaluate the expression.

15. $4 + (-7)$ **16.** $-2 + 5$

17. $-2 + (-4)$ **18.** $4 - 8$

19. $-5 - 4$ **20.** $-3 - (-6)$

21. $(-3 + 7) - 4$ **22.** $-1 + (9 - 4)$

In Exercises 23–30, evaluate the expression.

23. $-5 \cdot 3$ **24.** $(-2)(-12)$

25. $(14)(-4)$

26. $\frac{-35}{-7}$

27. $\frac{12}{-6}$

28. $[8 \div (-4)](-3)$

29. $-20 \div (-4 - 1)$

30. $[(-30) \div (10)] \div (-1)$

5.3

In Exercises 31–39, express the fraction as a terminating or repeating decimal.

31. $\frac{3}{10}$ **32.** $\frac{11}{25}$ **33.** $\frac{15}{40}$

34. $\frac{13}{4}$ **35.** $\frac{6}{7}$ **36.** $\frac{7}{12}$

37. $\frac{3}{8}$ **38.** $\frac{11}{16}$ **39.** $\frac{5}{7}$

In Exercises 40–46, express the decimal number as a quotient of two integers.

40. 0.225 **41.** 1.4 **42.** $0.\overline{6}$ **43.** $0.\overline{51}$

44. 0.083 **45.** 0.0073 **46.** $2.3\overline{4}$

In Exercises 47–50, express each mixed number as an improper fraction.

47. $1\frac{3}{4}$ **48.** $4\frac{1}{6}$ **49.** $-3\frac{1}{4}$ **50.** $-35\frac{3}{8}$

In Exercises 51–54, express each improper fraction as a mixed number.

51. $\frac{11}{5}$ **52.** $\frac{75}{8}$ **53.** $-\frac{12}{7}$ **54.** $-\frac{136}{5}$

In Exercises 55–63, perform the indicated operation and reduce your answer to lowest terms.

55. $\frac{1}{3} + \frac{3}{4}$

56. $\frac{11}{12} - \frac{2}{3}$

57. $\frac{1}{6} + \frac{5}{4}$

58. $\frac{7}{16} \cdot \frac{12}{21}$

59. $\frac{5}{9} \div \frac{6}{7}$

60. $\left(\frac{4}{5} + \frac{5}{7}\right) \div \frac{4}{5}$

61. $\left(\frac{2}{3} \cdot \frac{1}{7}\right) \div \frac{4}{7}$

62. $\left(\frac{1}{5} + \frac{2}{3}\right)\left(\frac{3}{8}\right)$

63. $\left(\frac{1}{5} \cdot \frac{2}{3}\right) + \left(\frac{1}{5} \div \frac{1}{2}\right)$

64. ***Cajun Turkey*** A recipe for Roasted Cajun Turkey calls for $\frac{1}{8}$ teaspoon of cayenne pepper per pound of turkey. If Jennifer Sabo is preparing a turkey that weighs $17\frac{3}{4}$ pounds, how much cayenne pepper does she need?

5.4

In Exercises 65–80, simplify the expression. Rationalize the denominator when necessary.

65. $\sqrt{45}$ **66.** $\sqrt{200}$ **67.** $\sqrt{5} + 7\sqrt{5}$

68. $\sqrt{2} - 4\sqrt{2}$ **69.** $\sqrt{8} + 6\sqrt{2}$ **70.** $\sqrt{3} - 7\sqrt{27}$

71. $\sqrt{28} + \sqrt{63}$ **72.** $\sqrt{3} \cdot \sqrt{6}$ **73.** $\sqrt{8} \cdot \sqrt{6}$

74. $\frac{\sqrt{300}}{\sqrt{3}}$ **75.** $\frac{\sqrt{56}}{\sqrt{2}}$ **76.** $\frac{4}{\sqrt{3}}$

77. $\frac{\sqrt{7}}{\sqrt{5}}$ **78.** $3(2 + \sqrt{7})$

79. $\sqrt{3}(4 + \sqrt{6})$ **80.** $\sqrt{3}(\sqrt{6} + \sqrt{15})$

5.5

In Exercises 81–90, state the name of the property illustrated.

81. $5 + 7 = 7 + 5$

82. $5 \cdot m = m \cdot 5$

83. $(1 + 2) + 3 = 1 + (2 + 3)$

84. $9(x + 1) = 9 \cdot x + 9 \cdot 1$

85. $4 + (5 + 6) = (4 + 5) + 6$

86. $(3 + 5) + (4 + 3) = (4 + 3) + (3 + 5)$

87. $(3 \cdot a) \cdot b = 3 \cdot (a \cdot b)$

88. $a \cdot (2 + 3) = (2 + 3) \cdot a$

89. $2(x + 3) = (2 \cdot x) + (2 \cdot 3)$

90. $x \cdot 2 + 6 = 2 \cdot x + 6$

In Exercises 91–96, determine whether the set of numbers is closed under the given operation.

91. Natural numbers, subtraction

92. Whole numbers, multiplication

93. Integers, division

94. Real numbers, subtraction

95. Irrational numbers, multiplication

96. Rational numbers, division

5.6

In Exercises 97–104, evaluate each expression.

97. 5^2 **98.** 5^{-2} **99.** $\frac{9^5}{9^3}$ **100.** $5^2 \cdot 5$

101. 7^0 **102.** 4^{-3} **103.** $(2^3)^2$ **104.** $(3^2)^2$

In Exercises 105–108, write each number in scientific notation.

105. 8,200,000,000 **106.** 0.0000158

107. 0.02309 **108.** 4,950,000

In Exercises 109–112, express each number in decimal notation.

109. 2.8×10^5 **110.** 1.39×10^{-4}

111. 1.75×10^{-4} **112.** 1×10^7

In Exercises 113–116, (a) perform the indicated operation and write your answer in scientific notation. (b) Confirm the result found in part (a) by performing the calculation on a scientific calculator.

113. $(3 \times 10^4)(2 \times 10^{-9})$

114. $(5 \times 10^6)(7.5 \times 10^5)$

115. $\frac{8.4 \times 10^3}{4 \times 10^2}$

116. $\frac{1.5 \times 10^{-3}}{5 \times 10^{-4}}$

In Exercises 117–121, (a) perform the indicated calculation by first converting each number to scientific notation. Write your answer in decimal notation. (b) Confirm the result found in part (a) by performing the calculation on a scientific calculator.

117. (550,000)(2,000,000) **118.** (35,000)(0.00002)

119. $\frac{8{,}400{,}000}{70{,}000}$ **120.** $\frac{0.000002}{0.0000004}$

121. ***Space Distances*** The distance from Earth to the sun is about 1.49×10^{11} meters. The distance from Earth to the moon is about 3.84×10^8 meters. The distance from Earth to the sun is how many times larger than the distance from Earth to the moon? Use a scientific calculator and round your answer to the nearest whole number.

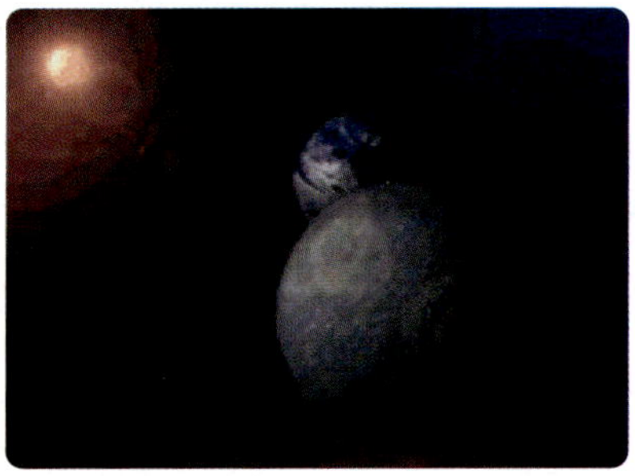

122. ***Outstanding Debt*** As a result of a recent water and sewer system improvement, the city of Galena, Illinois, has an outstanding debt of \$20,000,000. If the population of Galena is 3600 people, how much would each person have to contribute to pay off the outstanding debt?

5.7

In Exercises 123–128, determine whether the sequence is arithmetic or geometric. Then determine the next two terms of the sequence.

123. $1, 5, 9, 13, \ldots$ **124.** $\frac{1}{2}, 1, 2, 4, \ldots$

125. $-3, -6, -9, -12, \ldots$ **126.** $\frac{1}{2}, \frac{1}{4}, \frac{1}{8}, \frac{1}{16}, \ldots$

127. $1, 4, 7, 10, 13, \ldots$ **128.** $\frac{1}{2}, -\frac{1}{2}, \frac{1}{2}, -\frac{1}{2}, \ldots$

In Exercises 129–134, determine the indicated term of the sequence with the given first term, a_1, and common difference, d, or common ratio, r.

129. Determine a_6 when $a_1 = 2, d = 5$.

130. Determine a_9 when $a_1 = -6, d = 2$.

131. Determine a_{10} when $a_1 = -20, d = 5$.

132. Determine a_5 when $a_1 = 3, r = 2$.

133. Determine a_5 when $a_1 = 4, r = \frac{1}{2}$.

134. Determine a_4 when $a_1 = -6, r = 2$.

In Exercises 135–138, determine the sum of the arithmetic sequence. The number of terms, n, is given.

135. $3, 6, 9, 12, \ldots, 150; n = 50$

136. $-4, -3\frac{3}{4}, -3\frac{1}{2}, -3\frac{1}{4}, \ldots, -2\frac{1}{4}; n = 8$

137. $100, 94, 88, 82, \ldots, 58; n = 8$

138. $0.5, 0.75, 1.00, 1.25, \ldots, 5.25; n = 20$

In Exercises 139–142, determine the sum of the first n terms of the geometric sequence for the values of a_1 and r.

139. $n = 5, a_1 = 2, r = 4$

140. $n = 4, a_1 = 3, r = 2$

141. $n = 5, a_1 = 3, r = -2$

142. $n = 6, a_1 = 1, r = -2$

In Exercises 143–148, first determine whether the sequence is arithmetic or geometric; then write an expression for the general or nth term, a_n.

143. $3, 6, 9, 12, \ldots$

144. $1, 4, 7, 10, \ldots$

145. $4, \frac{5}{2}, 1, -\frac{1}{2}, \ldots$

146. $3, 6, 12, 24, \ldots$

147. $2, -2, 2, -2, \ldots$

148. $5, \frac{5}{3}, \frac{5}{9}, \frac{5}{27}, \ldots$

5.8

In Exercises 149–152, determine whether the sequence is a Fibonacci-type sequence. If so, determine the next two terms.

149. $0, 1, 1, 2, 2, 3, 3, 4, 4, \ldots$

150. $-1, 0, -1, -1, -2, -3, -5, \ldots$

151. $1, 4, 3, -1, -4, -5, \ldots$

152. $-10, 10, 0, 10, 20, \ldots$

CHAPTER 5 TEST

1. Which of the numbers 2, 3, 4, 5, 6, 8, 9, and 10 divide 48,395?
2. Determine the prime factorization of 414.
3. Evaluate $[(-3) + 7] - (-4)$.
4. Evaluate $-7 - 13$.
5. Evaluate $[(-70)(-5)] \div (8 - 10)$.
6. Convert $4\frac{5}{8}$ to an improper fraction.
7. Convert $\frac{176}{9}$ to a mixed number.
8. Write $\frac{5}{8}$ as a terminating or repeating decimal.
9. Express 6.45 as a quotient of two integers.
10. Evalute $(\frac{5}{16} \div 3) + (\frac{4}{5} \cdot \frac{1}{2})$.
11. Perform the operation and reduce the answer to lowest terms: $\frac{17}{24} - \frac{7}{12}$.
12. Simplify $\sqrt{75} + \sqrt{48}$.
13. Rationalize the denominator $\frac{\sqrt{2}}{\sqrt{7}}$.
14. Determine whether the integers are closed under the operation of multiplication. Explain your answer.

Name the property illustrated.

15. $(x + 3) + 5 = x + (3 + 5)$
16. $4(x + 3) = 4x + 12$

Evaluate.

17. $\frac{4^5}{4^2}$ **18.** $4^3 \cdot 4^2$ **19.** 3^{-4}

20. Perform the operation by first converting the numerator and denominator to scientific notation. Write the answer in scientific notation.

$$\frac{7,200,000}{0.000009}$$

21. Write an expression for the general or nth term, a_n, of the sequence $-2, -6, -10, -14, \ldots$.

22. Determine the sum of the terms of the arithmetic sequence. The number of terms, n, is given.

$$-2, -5, -8, -11, \ldots, -32; n = 11$$

23. Determine a_6 when $a_1 = 2$ and $r = 3$.

24. Determine the sum of the first eight terms of the sequence when $a_1 = -2$ and $r = 2$.

25. Write an expression for the general or nth term, a_n, of the sequence 3, 6, 12, 24, ...

26. Write the first 10 terms of the Fibonacci sequence.

GROUP PROJECTS

1. ***Making Rice*** The amount of ingredients needed to make 3 and 5 servings of rice are:

To Make	Rice and Water	Salt	Butter
3 servings	1 cup	$\frac{3}{8}$ tsp	$1\frac{1}{2}$ tsp
5 servings	$1\frac{2}{3}$ cup	$\frac{5}{8}$ tsp	$2\frac{1}{2}$ tsp

Find the amount of each ingredient needed to make (a) 2 servings, (b) 1 serving, and (c) 29 servings. Explain how you determined your answers.

2. ***Finding Areas***

a) Determine the area of the trapezoid shown by finding the area of the three parts indicated and finding the sum of the three areas. The necessary geometric formulas are given in Chapter 9.

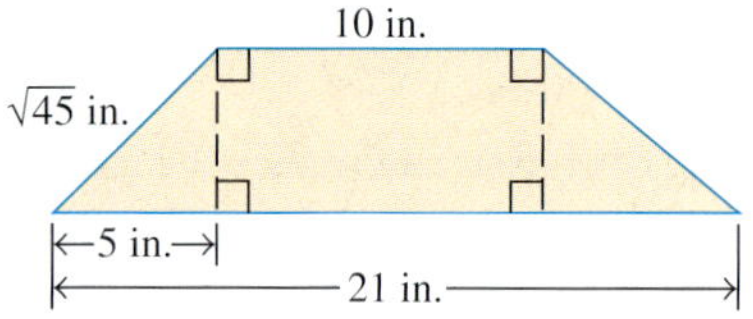

b) Determine the area of the trapezoid by using the formula for the area of a trapezoid given in Chapter 9.

c) Compare your answers from parts (a) and (b). Are they the same? If not, explain why they are different.

3. ***Medical Insurance*** On a medical insurance policy (such as Blue Cross/Blue Shield), the policyholder may need to make copayments for prescription drugs, office visits, and procedures until the total of all copayments reaches a specified amount. Suppose that the copayment on the Gattelaro's medical policy for prescription drugs is 50% of the cost; the copayment for office visits is $10; and the copayment for all medical tests, x-rays, and other procedures is 20% of the cost. After the family's copayment totals $500 in a calendar year, all medical and prescription bills are paid in full by the insurance company. The Gattelaros had the following medical expenses from January 1 through April 30.

Date	Reason	Cost before Copayment
January 10	Office visit	$40
	Prescription	$44
February 27	Office visit	$40
	Medical tests	$188
April 19	Office visit	$40
	X-rays	$348
	Prescription	$76

a) How much had the Gattelaros paid in copayments from January 1 through April 30?

b) How much had the medical insurance company paid?

c) What is the remaining copayment that must be paid by the Gattelaros before the $500 copayment limit is reached?

4. ***A Branching Plant*** A plant grows for two months and then adds a new branch. Each new branch grows for two months and then adds another branch. After the second month, each branch adds a new branch every month. Assume the growth begins in January.

a) How many branches will there be in February?

b) How many branches will there be in May?

c) How many branches will there be after 12 months?

d) How is this problem similar to the problem involving rabbits that appeared in Fibonacci's book *Liber Abacci* (see page 298)?

CHAPTER 6

Algebra, Graphs, and Functions

▲ Algebra is a tool for solving everyday problems such as changing a recipe to increase or decrease the number of servings.

WHAT YOU WILL LEARN

- Order of operations
- Solving linear and quadratic equations, and linear inequalities in one variable
- Evaluating a formula
- Solving application problems involving linear, quadratic, and exponential equations
- Solving application problems dealing with variation
- Graphing equations and functions, including linear, quadratic, and exponential equations

WHY IT IS IMPORTANT

Algebra is one of the most practical tools for solving everyday problems. You probably use algebra in your daily life without realizing it. For example, you use a coordinate system when you consult a map to find directions. You solve simple equations when you change a recipe to increase or decrease the number of servings. To evaluate how much interest you will earn in a savings account or to figure out how long it will take you to travel a given distance, you use common formulas that are algebraic equations. The symbolic language of algebra makes it an excellent tool for solving problems because it allows us to write lengthy expressions in compact form. English philosopher Alfred North Whitehead explained the power of algebra when he stated, "By relieving the brain of unnecessary work, a good notation sets the mind free to concentrate on more advanced problems."

6.1 ORDER OF OPERATIONS

▲ You can use algebra to determine the sales tax rate when purchasing items such as tires.

Suppose you see an advertisement for tires that states: "Purchase Michelin All-Season Tires for \$135 each. This price includes mounting." You purchase four tires and have them mounted on your car. Your total price, including sales tax, comes to \$583.50. What is the sales tax rate? You may be able to solve this problem and other problems in everyday life by using arithmetic or trial and error. With a knowledge of algebra, however, you can often find the solution with much less effort. In this section, we will discuss how to use algebra to answer questions like the one above.

Algebra is a generalized form of arithmetic. The word *algebra* is derived from the Arabic word *al-jabr* (meaning "reunion of broken parts"), which was the title of a book written by the mathematician Muhammed ibn-Musa al Khwarizmi in about A.D. 825.

Algebra uses letters of the alphabet called ***variables*** to represent numbers. Often the letters x and y are used to represent variables. However, any letter may be used as a variable. A symbol that represents a specific quantity is called a ***constant***.

Multiplication of numbers and variables may be represented in several different ways in algebra. Because the "times" sign might be confused with the variable x, a dot between two numbers or variables indicates multiplication. Thus, $3 \cdot 4$ means 3 times 4, and $x \cdot y$ means x times y. Placing two letters or a number and a letter next to one another, with or without parentheses, also indicates multiplication. Thus, $3x$ means 3 times x, xy means x times y, and $(x)(y)$ means x times y.

An ***algebraic expression*** (or simply an ***expression***) is a collection of variables, numbers, parentheses, and operation symbols. Some examples of algebraic expressions are

$$x, \qquad x + 2, \qquad 3(2x + 3), \qquad \frac{3x + 1}{2x - 3}, \qquad \text{and} \qquad x^2 + 7x + 3$$

Two algebraic expressions joined by an equal sign form an ***equation***. Some examples of equations are

$$x + 2 = 4, \qquad 3x + 4 = 1, \qquad \text{and} \qquad x + 3 = 2x$$

The ***solution to an equation*** is the number or numbers that replace the variable to make the equation a true statement. For example, the solution to the equation $x + 3 = 4$ is 1. When we find the solution to an equation, we ***solve the equation***.

We can determine if any number is a solution to an equation by ***checking the solution***. To check the solution, we substitute the number for the variable in the equation. If the resulting statement is a true statement, that number is a solution to the equation. If the resulting statement is a false statement, the number is not a solution to the equation. To check the number 1 in the equation $x + 3 = 4$, we do the following.

$$x + 3 = 4$$
$$1 + 3 = 4 \quad \text{Substitute 1 for } x$$
$$4 = 4 \quad \text{True}$$

The same number is obtained on both sides of the equal sign, so 1 is the solution. For the equation $x + 3 = 4$, the only solution is 1. Any other value of x would result in the check being a false statement.

DID YOU KNOW?

Broken Bones

Imagine yourself walking down a street in Spain during the Middle Ages. You see a sign over a door: "*Algebrista y Sangrador.*" Inside, you would find a person more ready to give you a haircut than help you with your algebra. The sign translates into "Bonesetter and Bloodletter," relatively simple medical treatments administered in barbershops of the day.

The root word *al-jabr*, which the Muslims (Moors) brought to Spain along with some concepts of algebra, suggests the restoring of broken parts. The parts might be bones, or they might be mathematical expressions that are broken into separate parts and the parts moved from one side of an equation to the other and reunited in such a way as to make a solution more obvious.

DID YOU KNOW?

Google or a Googol

Google Headquarters in Mountain View, California

When many of us hear the word *google,* we think of the name of one of the most popular search engines on the Internet. To others, the word *googol* (pronounced the same as *google*) means something entirely different. A googol is the name of a very large number: 10^{100}. In 1938, Edward Kasner, an American mathematician, named the number 10^{100} a googol. He was trying to define a name for this very large number when his 9-year-old nephew suggested that Kasner call this number a googol. If you are looking for a number that is even larger than a googol, try a googolplex, which is 10^{googol}, or $10^{10^{100}}$.

To ***evaluate an expression*** means to find the value of the expression for a given value of the variable. To evaluate expressions and solve equations, you must have an understanding of exponents. Exponents (Section 5.6) are used to abbreviate repeated multiplication. For example, the expression 5^2 means $5 \cdot 5$. The 2 in the expression 5^2 is the ***exponent***, and the 5 is the ***base***. We read 5^2 as "5 to the second power" or "5 squared," and 5^2 means $5 \cdot 5$ or 25.

In general, the number b to the nth power, written b^n, means

Exponent

$$b^n = \underbrace{b \cdot b \cdot b \cdot \cdots \cdot b}_{n \text{ factors of } b}$$

Base

An exponent refers only to its base. In the expression -5^2, the base is 5. In the expression $(-5)^2$, the base is -5.

$$-5^2 = -(5)^2 = -1(5)^2 = -1(5)(5) = -25$$
$$(-5)^2 = (-5)(-5) = 25$$

Note that $-5^2 \neq (-5)^2$ since $-25 \neq 25$.

Order of Operations

To evaluate an expression or to check the solution to an equation, we need to know the ***order of operations*** to follow. For example, suppose we want to evaluate the expression $2 + 3x$ when $x = 4$. Substituting 4 for x, we obtain $2 + 3 \cdot 4$. What is the value of $2 + 3 \cdot 4$? Does it equal 20, or does it equal 14? Some standard rules, called the order of operations, have been developed to ensure that there is only one correct answer. In mathematics, unless parentheses indicate otherwise, always perform multiplication before addition. Thus, the correct answer is 14.

$$2 + 3 \cdot 4 = 2 + (3 \cdot 4) = 2 + 12 = 14$$

The order of operations for evaluating an expression is as follows.

ORDER OF OPERATIONS

1. First, perform all operations within parentheses or other grouping symbols (according to the following order).
2. Next, perform all exponential operations (that is, raising to powers or finding roots).
3. Next, perform all multiplications and divisions from left to right.
4. Finally, perform all additions and subtractions from left to right.

DID YOU KNOW?

An Important Breakthrough

This painting by Charles Demuth, called *I Saw the Figure 5 in Gold,* depicts the abstract nature of numbers. Mathematician and philosopher Bertrand Russell observed in 1919 that it must have required "many ages to discover that a brace of pheasants and a couple of days were both instances of the number 2." The discovery that numbers could be used not only to count objects such as the number of birds but also to represent abstract quantities represented a breakthrough in the development of algebra.

TIMELY TIP Some students use the phrase "*P*lease *E*xcuse *M*y *D*ear *A*unt *S*ally" or the word "PEMDAS" (*P*arentheses, *E*xponents, *M*ultiplication, *D*ivision, *A*ddition, *S*ubtraction) to remind them of the order of operations. Remember: Multiplication and division are of the same order, and addition and subtraction are of the same order.

EXAMPLE 1 *Evaluating an Expression*

Evaluate the expression $-x^2 + 4x + 16$ for $x = 3$.

SOLUTION Substitute 3 for each x and use the order of operations to evaluate the expression.

$$\begin{aligned} &-x^2 + 4x + 16 \\ &= -(3)^2 + 4(3) + 16 \\ &= -9 + 12 + 16 \\ &= 3 + 16 \\ &= 19 \end{aligned}$$

EXAMPLE 2 *Finding the Height*

A ball is thrown upward off a bridge 40 feet above ground. Its height, in feet, above ground, t seconds after it is thrown, can be determined by the expression $-16t^2 + 30t + 40$. Find the height of the ball, above ground, 2 seconds after it is thrown.

SOLUTION Substitute 2 for each t.

$$\begin{aligned} &-16t^2 + 30t + 40 \\ &= -16(2)^2 + 30(2) + 40 \\ &= -16(4) + 30(2) + 40 \\ &= -64 + 60 + 40 \\ &= -4 + 40 \\ &= 36 \end{aligned}$$

The ball is 36 feet above ground 2 seconds after it is thrown.

EXAMPLE 3 *Substituting for Two Variables*

Evaluate $-3x^2 + 2xy - 2y^2$ when $x = 2$ and $y = 3$.

SOLUTION Substitute 2 for each x and 3 for each y; then evaluate using the order of operations.

$$\begin{aligned} &-3x^2 + 2xy - 2y^2 \\ &= -3(2)^2 + 2(2)(3) - 2(3)^2 \\ &= -3(4) + 2(2)(3) - 2(9) \\ &= -12 + 12 - 18 \\ &= 0 - 18 \\ &= -18 \end{aligned}$$

EXAMPLE 4 *Is 2 a Solution?*

Determine whether 2 is a solution to the equation $3x^2 + 7x - 11 = 15$.

SOLUTION To determine whether 2 is a solution to the equation, substitute 2 for each x in the equation. Then evaluate the left-hand side of the equation using the order of operations. If the left-hand side of the equation equals 15, then both sides of the equation have the same value and 2 is a solution. When checking the solution, we use $\stackrel{?}{=}$, which means we are not sure if the statement is true.

$$3(2)^2 + 7(2) - 11 \stackrel{?}{=} 15$$
$$3(4) + 14 - 11 \stackrel{?}{=} 15$$
$$12 + 14 - 11 \stackrel{?}{=} 15$$
$$26 - 11 \stackrel{?}{=} 15$$
$$15 = 15 \quad \text{True}$$

Because 2 makes the equation a true statement, 2 is a solution to the equation.

SECTION 6.1 EXERCISES

CONCEPT/WRITING EXERCISES

1. What is a *variable*?

2. What is a *constant*?

3. What does it mean when we state that *a number is a solution to an equation*?

4. What is an algebraic expression? Illustrate an algebraic expression with an example.

5. a) For the term 4^5, identify the base and the exponent.

   b) In your own words, explain how to evaluate 4^5.

6. In your own words, explain the order of operations.

7. a) Evaluate $8 + 16 \div 4$ using the order of operations.

   b) Evaluate $9 + 6 \cdot 3$ using the order of operations.

8. a) Explain why $-x^2$ will always be a negative number for any nonzero real number selected for x.

   b) Explain why $(-x)^2$ will always be a positive number for any nonzero real number selected for x.

PRACTICE THE SKILLS

In Exercises 9–28, evaluate the expression for the given value(s) of the variable(s).

9. $x^2, \quad x = -5$

10. $x^2, \quad x = 6$

11. $-x^2, \quad x = -2$

12. $-x^2, \quad x = -3$

13. $-2x^3, \quad x = -7$

14. $-x^3, \quad x = -4$

15. $x - 7, \quad x = 4$

16. $8x - 3, \quad x = \frac{5}{2}$

17. $-4x + 4, \quad x = -2$

18. $x^2 - 3x + 8, \quad x = 5$

19. $-x^2 + 3x - 10, \quad x = -2$

20. $7x^2 + 5x - 11, \quad x = -1$

21. $\frac{1}{2}x^2 - 5x + 2, \quad x = \frac{2}{3}$

22. $\frac{2}{3}x^2 + x - 1, \quad x = \frac{1}{2}$

23. $8x^3 - 4x^2 + 7, \quad x = \frac{1}{2}$

24. $-x^2 + 5xy, \quad x = 2, \quad y = -3$

25. $2x^2 + xy + 3y^2, \quad x = -2, \quad y = 1$

26. $3x^2 + \frac{2}{5}xy - \frac{1}{5}y^2, \quad x = 2, \quad y = 5$

27. $4x^2 - 10xy + 3y^2, \quad x = 2, \quad y = -1$

28. $(x - 2y)^2, \quad x = 4, \quad y = -2$

In Exercises 29–38, determine whether the value(s) is (are) a solution to the equation.

29. $8x + 3 = 23, \quad x = 3$

30. $6x - 7 = -31, \quad x = -4$

31. $x - 3y = 0, \quad x = 6, \quad y = 3$

32. $4x + 2y = -2, \quad x = -2, \quad y = 3$

33. $x^2 - 3x + 6 = 5, \quad x = 2$

34. $-2x^2 + x + 5 = 20, \quad x = 3$

35. $2x^2 + x = 28, \quad x = -4$

36. $y = x^2 + 4x - 5, \quad x = -1, \quad y = -8$

37. $y = -x^2 + 3x - 1, \quad x = 3, \quad y = -1$

38. $y = x^3 - 3x^2 + 1, \quad x = 2, \quad y = -3$

PROBLEM SOLVING

39. ***Sales Tax*** If the sales tax on an item is 7%, the sales tax, in dollars, on an item costing d dollars can be found by using the expression $0.07d$. Determine the sales tax on a refrigerator costing \$899.

40. ***8 Trillion Calculations*** If a computer can do a calculation in 0.000002 sec, the time required to do n calculations can be determined by the expression $0.000002n$. Determine the number of seconds needed for the computer to do 8 trillion (8,000,000,000,000) calculations.

41. ***Electronic Filing*** The graph shows the number of taxpayers, in millions, filing their taxes electronically for the years 2000 through 2005. The number of taxpayers, in millions, filing their taxes electronically per year can be approximated by the expression $6.2x + 34.8$, where x is the number of years since 2000. For example, the year 2001 would correspond to $x = 1$. Assuming this trend continues, use the expression to approximate the number of taxpayers who will file their taxes electronically in 2010.

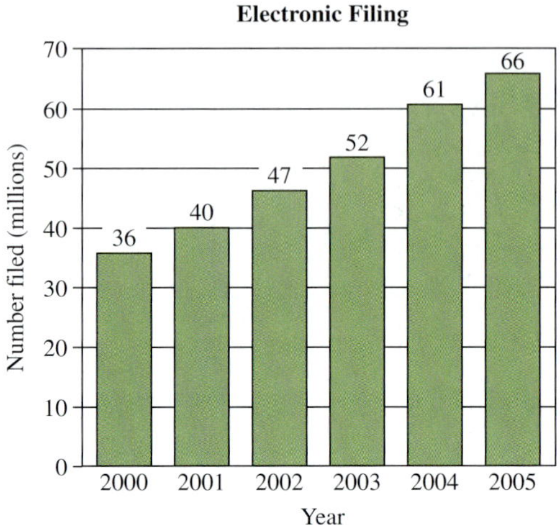

Source: Internal Revenue Service

42. ***Cost of a Tour*** The cost, in dollars, for Crescent City Tours to provide a tour for x people can be determined by the expression $220 + 2.75x$. Determine the cost for Crescent City Tours to provide a tour for 75 people.

43. ***Drying Time*** The time, in minutes, needed for clothes hanging on a line outdoors to dry, at a specific temperature and wind speed, depends on the humidity, h. The time can be approximated by the expression $2h^2 + 80h + 40$, where h is the percent humidity expressed as a decimal number. Find the length of time required for clothing to dry if there is 60% humidity.

44. ***Stopping Distance*** A typical car's stopping distance, in feet, on wet pavement can be approximated by the expression $0.08s^2 + 0.24s - 7.10$, where s is the speed of the car, in miles per hour (mph), before braking and $60 \leq s \leq 80$ miles per hour. Use the expression to approximate the car's stopping distance on wet pavement if the speed of the car before braking is 72 mph.

45. ***Grass Growth*** The rate of growth of grass, in inches per week, depends on a number of factors, including rainfall and temperature. For a certain area, this rate can be approximated by the expression $0.2R^2 + 0.003RT + 0.0001T^2$, where R is the weekly rainfall, in inches, and T is the average weekly temperature, in degrees Fahrenheit. Find the amount of growth of grass for a week in which the rainfall is 2 in. and the average temperature is 70°F.

CHALLENGE PROBLEMS/GROUP ACTIVITIES

46. Explain why $(-1)^n = 1$ for any even number n.

47. Does $(x + y)^2 = x^2 + y^2$? Complete the table and state your conclusion.

x	y	$(x + y)^2$	$x^2 + y^2$
2	3		
−2	−3		
−2	3		
2	−3		

48. Suppose n represents any natural number. Explain why 1^n equals 1.

INTERNET/RESEARCH ACTIVITY

49. When were exponents first used? Write a paper explaining how exponents were first used and when mathematicians began writing them in the present form.

6.2 LINEAR EQUATIONS IN ONE VARIABLE

▲ A linear equation can be used to approximate the number of adults in the United States age 65 or older.

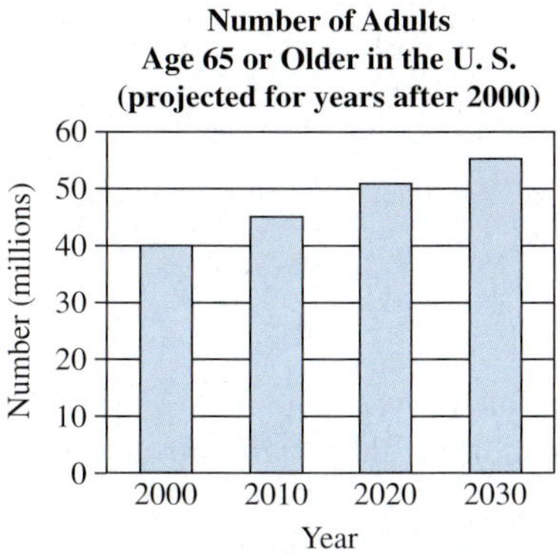

Source: U.S. Census Bureau

Figure 6.1

The graph in Figure 6.1 shows that the number of adults in the United States age 65 or older is projected to increase steadily from 2000 through 2030. The number of adults in the United States age 65 or older can be approximated by the equation $A = 0.51x + 40.1$, where A represents the number of adults in the United States age 65 or older, in millions, and x represents the number of years after 2000. Suppose we wish to determine the year when there will be 60 million adults in the United States age 65 or older. How would we do so? We could substitute 60 for A in the equation to obtain the equation $60 = 0.51x + 40.1$. We would then need to solve this equation for x. In this section, we explain how to solve equations such as $60 = 0.51x + 40.1$.

In Section 6.1, we stated that two algebraic expressions joined by an equal sign form an equation. The solution to some equations, such as $x + 3 = 4$, can be easily found by trial and error. Solving more difficult equations such as $60 = 0.51x + 40.1$ or $2x - 3 = 4(x + 3)$, however, requires us to understand some basic properties of equations.

The parts that are added or subtracted in an algebraic expression are called ***terms***. The expression $0.51x + 40.1$ contains two terms: $0.51x$ and $+ 40.1$. The expression $4x - 3y - 5$ contains three terms: $4x$, $-3y$, and -5. The + and − signs that break the expression into terms are part of the terms. When listing the terms of an expression, however, it is not necessary to include the + sign at the beginning of the term.

The numerical part of a term is called its ***numerical coefficient*** or, simply, its ***coefficient***. In the term $4x$, the 4 is the numerical coefficient. In the term $-4y$, the -4 is the numerical coefficient.

Like terms are terms that have the same variables with the same exponents on the variables. ***Unlike terms*** have different variables or different exponents on the variables.

Like Terms	Unlike Terms
$2x$, $7x$ (same variable, x)	$2x$, 9 (only first term has a variable)
$-8y$, $3y$ (same variable, y)	$5x$, $6y$ (different variables)
-4, 10 (both constants)	x, 8 (only first term has a variable)
$-5x^2$, $6x^2$ (same variable with same exponent)	$2x^3$, $3x^2$ (different exponents)

To ***simplify an expression*** means to combine like terms by using the commutative, associative, and distributive properties discussed in Chapter 5. For convenience, we list these properties below.

PROPERTIES OF THE REAL NUMBERS

$a(b + c) = ab + ac$	Distributive property
$a + b = b + a$	Commutative property of addition
$ab = ba$	Commutative property of multiplication
$(a + b) + c = a + (b + c)$	Associative property of addition
$(ab)c = a(bc)$	Associative property of multiplication

EXAMPLE 1 *Combining Like Terms*

Combine like terms in each expression.

a) $7x + 3x$ b) $6y - 2y$

c) $x + 12 - 3x + 7$ d) $-2x + 4 - 6y - 11 - 5y + 3x$

SOLUTION

a) We use the distributive property (in reverse) to combine like terms.

$$\begin{aligned} 7x + 3x &= (7 + 3)x && \text{Distributive property} \\ &= 10x \end{aligned}$$

b) $6y - 2y = (6 - 2)y = 4y$

c) $$\begin{aligned} x + 12 - 3x + 7 &= x - 3x + 12 + 7 && \text{Rearrange terms, place like terms together.} \\ &= -2x + 19 && \text{Combine like terms.} \end{aligned}$$

d) $$\begin{aligned} &-2x + 4 - 6y - 11 - 5y + 3x \\ &= -2x + 3x - 6y - 5y + 4 - 11 && \text{Rearrange terms, place like terms together.} \\ &= x - 11y - 7 && \text{Combine like terms.} \end{aligned}$$

We are able to rearrange the terms of an expression, as was done in Example 1(c) and (d) by the commutative and associative properties that were discussed in Section 5.5.

The order of the terms in an expression is not crucial. However, when listing the terms of an expression we generally list the terms in alphabetical order with the constant, the term without a variable, last.

DID YOU KNOW?

In the Beginning

François Viète

Greek philosopher Diophantus of Alexandria (A.D. 250), who invented notations for powers of a number and for multiplication and division of simple quantities, is thought to have made the first attempts at algebra. Not until the sixteenth century, however, did French mathematician François Viète (1540–1603) use symbols to represent numbers, the foundation of symbolic algebra. The work of René Descartes (1596–1660), though, is considered to be the starting point of modern-day algebra. In 1707, Sir Isaac Newton (1643–1727) gave symbolic mathematics the name *universal arithmetic*.

Solving Equations

Recall that to solve an equation means to find the value or values for the variable that make(s) the equation true. In this section, we discuss solving ***linear (or first-degree) equations***. A linear equation in one variable is one in which the exponent on the variable is 1. Examples of linear equations are $5x - 1 = 3$ and $2x + 4 = 6x - 5$.

Equivalent equations are equations that have the same solution. The equations $2x - 5 = 1$, $2x = 6$, and $x = 3$ are all equivalent equations since they all have the same solution, 3. When we solve an equation, we write the given equation as a series of simpler equivalent equations until we obtain an equation of the form $x = c$, where c is some real number.

To solve any equation, we have to ***isolate the variable***. That means getting the variable by itself on one side of the equal sign. The four properties of equality that we are about to discuss are used to isolate the variable. The first is the addition property.

ADDITION PROPERTY OF EQUALITY
If $a = b$, then $a + c = b + c$ for all real numbers a, b, and c.

The addition property of equality indicates that the same number can be added to both sides of an equation without changing the solution.

EXAMPLE 2 *Using the Addition Property of Equality*

Find the solution to the equation $x - 9 = 15$.

SOLUTION To isolate the variable, add 9 to both sides of the equation.

$$x - 9 = 15$$
$$x - 9 + 9 = 15 + 9$$
$$x + 0 = 24$$
$$x = 24$$

CHECK:

$$x - 9 = 15$$
$$24 - 9 \stackrel{?}{=} 15 \quad \text{Substitute 24 for } x.$$
$$15 = 15 \quad \text{True}$$

In Example 2, we showed the step $x + 0 = 24$. This step is usually done mentally, and the step is usually not listed.

SUBTRACTION PROPERTY OF EQUALITY
If $a = b$, then $a - c = b - c$ for all real numbers a, b, and c.

The subtraction property of equality indicates that the same number can be subtracted from both sides of an equation without changing the solution.

EXAMPLE 3 *Using the Subtraction Property of Equality*

Find the solution to the equation $x + 11 = 19$.

SOLUTION To isolate the variable, subtract 11 from both sides of the equation.

$$x + 11 = 19$$
$$x + 11 - 11 = 19 - 11$$
$$x = 8$$

Note in Example 3 that we did not subtract 19 from both sides of the equation, because doing so would not result in getting x on one side of the equal sign by itself.

Now we discuss the multiplication property.

MULTIPLICATION PROPERTY OF EQUALITY
If $a = b$, then $a \cdot c = b \cdot c$ for all real numbers a, b, and c, where $c \neq 0$.

The multiplication property of equality indicates that both sides of the equation can be multiplied by the same nonzero number without changing the solution.

EXAMPLE 4 *Using the Multiplication Property of Equality*

Find the solution to $\frac{x}{6} = 3$.

SOLUTION To solve this equation, multiply both sides of the equation by 6.

$$\frac{x}{6} = 3$$
$$6\left(\frac{x}{6}\right) = 6(3)$$
$$\frac{\overset{1}{\cancel{6}}x}{\underset{1}{\cancel{6}}} = 18$$
$$1x = 18$$
$$x = 18$$

In Example 4, we showed the steps $\frac{6x}{6} = 18$ and $1x = 18$. Usually, we will not illustrate these steps.

DIVISION PROPERTY OF EQUALITY
If $a = b$, then $\frac{a}{c} = \frac{b}{c}$ for all real numbers a, b, and c, $c \neq 0$.

The division property of equality indicates that both sides of an equation can be divided by the same nonzero number without changing the solution. Note that the divisor, c, cannot be 0 because division by 0 is not permitted.

EXAMPLE 5 *Using the Division Property of Equality*

Solve the equation $4x = 16$.

SOLUTION To solve this equation, divide both sides of the equation by 4.

$$4x = 16$$
$$\frac{4x}{4} = \frac{16}{4}$$
$$x = 4$$

An *algorithm* is a general procedure for accomplishing a task. The following general procedure is an algorithm for solving linear (or first-degree) equations. Sometimes the solution to an equation may be found more easily by using a variation of this general procedure. Remember that the primary objective in solving any equation is to isolate the variable.

A GENERAL PROCEDURE FOR SOLVING LINEAR EQUATIONS

1. If the equation contains fractions, multiply both sides of the equation by the lowest common denominator (or least common multiple). This step will eliminate all fractions from the equation.
2. Use the distributive property to remove parentheses when necessary.
3. Combine like terms on the same side of the equal sign when possible.
4. Use the addition or subtraction property to collect all terms with a variable on one side of the equal sign and all constants on the other side of the equal sign. It may be necessary to use the addition or subtraction property more than once. This process will eventually result in an equation of the form $ax = b$, where a and b are real numbers.
5. Solve for the variable using the division or multiplication property. The result will be an answer in the form $x = c$, where c is a real number.

EXAMPLE 6 *Using the General Procedure*

Solve the equation $2x - 9 = 19$.

SOLUTION Our goal is to isolate the variable; therefore, we start by getting the term $2x$ by itself on one side of the equation.

$$2x - 9 = 19$$
$$2x - 9 + 9 = 19 + 9 \quad \text{Add 9 to both sides of the equation (addition property) (step 4).}$$
$$2x = 28$$
$$\frac{2x}{2} = \frac{28}{2} \quad \text{Divide both sides of the equation by 2 (division property) (step 5).}$$
$$x = 14$$

A check will show that 14 is the solution to $2x - 9 = 19$.

EXAMPLE 7 *Solving a Linear Equation*

Solve the equation $4 = 5 + 2(t + 1)$ for t.

SOLUTION Our goal is to isolate the variable t. To do so, follow the general procedure for solving equations.

$$
\begin{aligned}
4 &= 5 + 2(t + 1) \\
4 &= 5 + 2t + 2 && \text{Distributive property (step 2)} \\
4 &= 2t + 7 && \text{Combine like terms (step 3).} \\
4 - 7 &= 2t + 7 - 7 && \text{Subtraction property (step 4)} \\
-3 &= 2t \\
\frac{-3}{2} &= \frac{2t}{2} && \text{Division property (step 5)} \\
-\frac{3}{2} &= t
\end{aligned}
$$

EXAMPLE 8 *Solving an Equation Containing Fractions*

Solve the equation $\frac{2x}{3} + \frac{1}{3} = \frac{3}{4}$.

SOLUTION When an equation contains fractions, we generally begin by multiplying each term of the equation by the lowest common denominator, LCD (see Chapter 5). In this example, the LCD is 12 because 12 is the smallest number that is divisible by both 3 and 4.

$$
\begin{aligned}
12\left(\frac{2x}{3} + \frac{1}{3}\right) &= 12\left(\frac{3}{4}\right) && \text{Multiply both sides of the equation by the LCD (step 1).} \\
12\left(\frac{2x}{3}\right) + 12\left(\frac{1}{3}\right) &= 12\left(\frac{3}{4}\right) && \text{Distributive property (step 2)} \\
\overset{4}{\cancel{12}}\left(\frac{2x}{\underset{1}{\cancel{3}}}\right) + \overset{4}{\cancel{12}}\left(\frac{1}{\underset{1}{\cancel{3}}}\right) &= \overset{3}{\cancel{12}}\left(\frac{3}{\underset{1}{\cancel{4}}}\right) && \text{Divide out common factors.} \\
8x + 4 &= 9 \\
8x + 4 - 4 &= 9 - 4 && \text{Subtraction property (step 4)} \\
8x &= 5 \\
\frac{8x}{8} &= \frac{5}{8} && \text{Division property (step 5)} \\
x &= \frac{5}{8}
\end{aligned}
$$

A check will show that $\frac{5}{8}$ is the solution to the equation. You could have worked the problem without first multiplying both sides of the equation by the LCD. Try it!

EXAMPLE 9 *Variables on Both Sides of the Equation*

Solve the equation $6x + 8 = 10x + 12$.

SOLUTION Note that the equation has an x on both sides of the equal sign. In equations of this type, you might wonder what to do first. It actually does not

DID YOU KNOW?

A New Concept

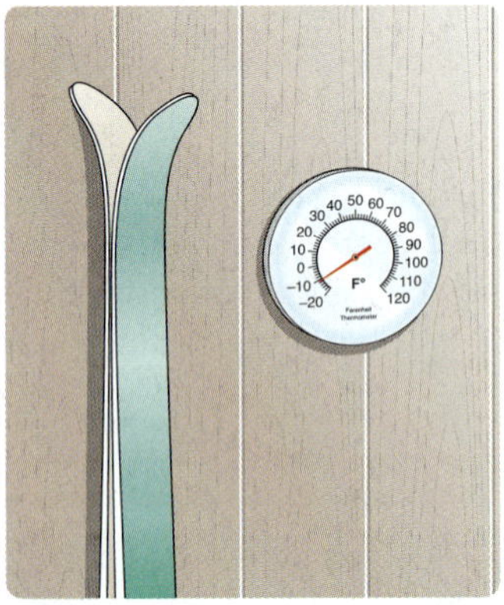

Well into the sixteenth century, mathematicians found it difficult to accept the idea that the solution to a problem (such as Example 7) could be a negative number because negative numbers could not be accepted as physically real. In the early days of algebra, someone working a problem did not isolate a variable by subtracting like terms. Instead, a problem would be put into a form that allowed only positive coefficients and answers. Albert Girard (1595–1637) contributed to the evolution of a correct understanding of negative quantities.

matter as long as you do not forget the goal of isolating the variable x. Let's collect the terms containing a variable on the left-hand side of the equation.

$$\begin{aligned} 6x + 8 &= 10x + 12 \\ 6x + 8 - 8 &= 10x + 12 - 8 && \text{Subtraction property (step 4)} \\ 6x &= 10x + 4 \\ 6x - 10x &= 10x - 10x + 4 && \text{Subtraction property (step 4)} \\ -4x &= 4 \\ \frac{-4x}{-4} &= \frac{4}{-4} && \text{Division property (step 5)} \\ x &= -1 \end{aligned}$$

In the solution to Example 9, the terms containing the variable were collected on the left-hand side of the equal sign. Now work Example 9, collecting the terms with the variable on the right-hand side of the equal sign. If you do so correctly, you will get the same result.

EXAMPLE 10 *Solving an Equation Containing Decimals*

Solve the equation $4x - 0.48 = 0.8x + 4$ and check your solution.

SOLUTION This equation may be solved with the decimals, or you may multiply each term by 100 and eliminate the decimals. We will solve the equation with the decimals.

$$\begin{aligned} 4x - 0.48 &= 0.8x + 4 \\ 4x - 0.48 + 0.48 &= 0.8x + 4 + 0.48 && \text{Addition property} \\ 4x &= 0.8x + 4.48 \\ 4x - 0.8x &= 0.8x - 0.8x + 4.48 && \text{Subtraction property} \\ 3.2x &= 4.48 \\ \frac{3.2x}{3.2} &= \frac{4.48}{3.2} && \text{Division property} \\ x &= 1.4 \end{aligned}$$

CHECK:

$$\begin{aligned} 4x - 0.48 &= 0.8x + 4 \\ 4(1.4) - 0.48 &\stackrel{?}{=} 0.8(1.4) + 4 && \text{Substitute 1.4 for each } x \text{ in the equation.} \\ 5.6 - 0.48 &\stackrel{?}{=} 1.12 + 4 \\ 5.12 &= 5.12 && \text{True} \end{aligned}$$

In Chapter 5, we explained that $a - b$ can be expressed as $a + (-b)$. We use this principle in Example 11.

EXAMPLE 11 *Using the Definition of Subtraction*

Solve $12 = 2(3r - 4) - 4r$ for r.

SOLUTION Our goal is to isolate the variable r. To do so, follow the general procedure for solving equations.

$$
\begin{aligned}
12 &= 2(3r - 4) - 4r \\
12 &= 2[3r + (-4)] - 4r && \text{Definition of subtraction} \\
12 &= 2(3r) + 2(-4) - 4r && \text{Distributive property} \\
12 &= 6r - 8 - 4r \\
12 &= 2r - 8 && \text{Combine like terms.} \\
12 + 8 &= 2r - 8 + 8 && \text{Addition property} \\
20 &= 2r \\
\frac{20}{2} &= \frac{2r}{2} && \text{Division property} \\
10 &= r
\end{aligned}
$$

TIMELY TIP Remember that the goal in solving an equation is to get the variable alone on one side of the equal sign by using the general procedure for solving equations.

So far, every equation has had exactly one solution. Some equations, however, have no solution, and others have more than one solution. Example 12 illustrates an equation that has no solution, and Example 13 illustrates an equation that has an infinite number of solutions.

EXAMPLE 12 *An Equation with No Solution*

Solve $3(x - 4) + x + 2 = 6x - 2(x + 3)$.

SOLUTION

$$
\begin{aligned}
3(x - 4) + x + 2 &= 6x - 2(x + 3) \\
3x - 12 + x + 2 &= 6x - 2x - 6 && \text{Distributive property} \\
4x - 10 &= 4x - 6 && \text{Combine like terms.} \\
4x - 4x - 10 &= 4x - 4x - 6 && \text{Subtraction property} \\
-10 &= -6 && \text{False}
\end{aligned}
$$

During the process of solving an equation, if you obtain a false statement like $-10 = -6$, or $-4 = 0$, the equation has ***no solution***. An equation that has no solution is called a ***contradiction***. The equation $3(x - 4) + x + 2 = 6x - 2(x + 3)$ is a contradiction and thus has no solution.

EXAMPLE 13 *An Equation with Infinitely Many Solutions*

Solve $2(x + 4) - 3(x - 5) = -x + 23$.

SOLUTION

$$
\begin{aligned}
2(x + 4) - 3(x - 5) &= -x + 23 \\
2x + 8 - 3x + 15 &= -x + 23 && \text{Distributive property} \\
-x + 23 &= -x + 23 && \text{Combine like terms.}
\end{aligned}
$$

Note that at this point both sides of the equation are the same. Every real number will satisfy this equation. This equation has an infinite number of solutions. An equation of this type is called an *identity*. When solving an equation, you can tell that the equation is an identity if you notice that the same expression appears on both sides of the equal sign. The solution to any linear equation with one variable that is an identity is *all real numbers*. If you continue to solve an equation that is an identity, you will end up with $0 = 0$, as follows.

$$\begin{aligned} -x + 23 &= -x + 23 \\ -x + x + 23 &= -x + x + 23 && \text{Addition property} \\ 23 &= 23 && \text{Combine like terms.} \\ 23 - 23 &= 23 - 23 && \text{Subtraction property} \\ 0 &= 0 && \text{True for any value of } x \end{aligned}$$

Proportions

A *ratio* is a quotient of two quantities. An example is the ratio of 2 to 5, which can be written 2 : 5 or $\frac{2}{5}$ or 2/5. Ratios are used in proportions.

A **proportion** is a statement of equality between two ratios.

An example of a proportion is $\frac{a}{b} = \frac{c}{d}$, where $b \neq 0$ and $d \neq 0$. Consider the proportion

$$\frac{x + 2}{5} = \frac{x + 5}{8}$$

We can solve this proportion by first multiplying both sides of the equation by the LCD, 40.

$$\begin{aligned} \frac{x + 2}{5} &= \frac{x + 5}{8} \\ \overset{8}{\cancel{40}}\left(\frac{x + 2}{\cancel{5}}\right) &= \overset{5}{\cancel{40}}\left(\frac{x + 5}{\cancel{8}}\right) && \text{Multiplication property} \\ 8(x + 2) &= 5(x + 5) \\ 8x + 16 &= 5x + 25 \\ 3x + 16 &= 25 \\ 3x &= 9 \\ x &= 3 \end{aligned}$$

A check will show that 3 is the solution.

Proportions can often be solved more easily by using cross multiplication.

CROSS MULTIPLICATION

If $\frac{a}{b} = \frac{c}{d}$, then $ad = bc$, where $b \neq 0$, $d \neq 0$.

Let's use cross multiplication to solve the proportion $\frac{x+2}{5} = \frac{x+5}{8}$.

$$\frac{x+2}{5} = \frac{x+5}{8}$$

$$8(x+2) = 5(x+5) \quad \text{Cross multiplication}$$

$$8x + 16 = 5x + 25$$

$$3x + 16 = 25$$

$$3x = 9$$

$$x = 3$$

Many practical application problems can be solved using proportions.

TO SOLVE APPLICATION PROBLEMS USING PROPORTIONS

1. Represent the unknown quantity by a variable.
2. Set up the proportion by listing the given ratio on the left-hand side of the equal sign and the unknown and other given quantity on the right-hand side of the equal sign. When setting up the right-hand side of the proportion, the same respective quantities should occupy the same respective positions on the left and right. For example, an acceptable proportion might be

$$\frac{\text{miles}}{\text{hour}} = \frac{\text{miles}}{\text{hour}}$$

3. Once the proportion is properly written, drop the units and use cross multiplication to solve the equation.
4. Answer the question or questions asked using the appropriate units.

EXAMPLE 14 *Water Usage*

The cost for water in Orange County is \$1.64 per 750 gallons (gal) of water used. What is the water bill if 30,000 gallons are used?

SOLUTION This problem may be solved by setting up a proportion. One proportion that can be used is

$$\frac{\text{cost of 750 gal}}{\text{750 gal}} = \frac{\text{cost of 30,000 gal}}{\text{30,000 gal}}$$

The unknown quantity is the cost for 30,000 gallons of water, so we will call this quantity x. The proportion then becomes

$$\text{Given ratio} \left\{ \frac{1.64}{750} \right. = \frac{x}{30{,}000}$$

Now we solve for x by using cross multiplication.

$$(1.64)(30{,}000) = 750x$$

$$49{,}200 = 750x$$

$$\frac{49{,}200}{750} = \frac{750x}{750}$$

$$\$65.60 = x$$

The cost of 30,000 gallons of water is \$65.60.

EXAMPLE 15 ***Determining the Amount of Insulin***

Insulin comes in 10 cubic centimeter (cc) vials labeled in the number of units of insulin per cubic centimeter of fluid. A vial of insulin marked U40 has 40 units of insulin per cubic centimeter of fluid. If a patient needs 30 units of insulin, how much fluid should be drawn into the syringe from the U40 vial?

SOLUTION The unknown quantity, x, is the number of cubic centimeters of fluid to be drawn into the syringe. Following is one proportion that can be used to find that quantity.

Given ratio
$$\frac{40 \text{ units}}{1 \text{ cc}} = \frac{30 \text{ units}}{x \text{ cc}}$$
$$40x = 30(1)$$
$$40x = 30$$
$$x = \frac{30}{40} = 0.75$$

The nurse or doctor putting the insulin in the syringe should draw 0.75 cc of the fluid.

SECTION 6.2 EXERCISES

CONCEPT/WRITING EXERCISES

1. Define and give an example of a *term.*
2. Define and give an example of *like terms.*
3. Define and give an example of a *numerical coefficient.*
4. Define and give an example of a *linear equation.*
5. Explain how to simplify an expression. Give an example.
6. State the addition property of equality. Give an example.
7. State the multiplication property of equality. Give an example.
8. State the subtraction property of equality. Give an example.
9. State the division property of equality. Give an example.
10. Define *algorithm.*
11. Define and give an example of a *ratio.*
12. Define and give an example of a *proportion.*
13. Are $3x$ and $\frac{1}{2}x$ like terms? Explain.
14. Are $4x$ and $4y$ like terms? Explain.

PRACTICE THE SKILLS

In Exercises 15–38, combine like terms.

15. $4x + 6x$
16. $-7x - 5x$
17. $5x - 3x + 12$
18. $9x - 6x + 21$
19. $7x + 3y - 4x + 8y$
20. $x - 4x + 3$
21. $-3x + 2 - 5x$
22. $-3x + 4x - 2 + 5$
23. $2 - 3x - 2x + 1$
24. $-0.2x + 1.7x - 4$
25. $6.2x - 8.3 + 7.1x$
26. $\frac{2}{3}x + \frac{1}{6}x - 5$
27. $\frac{1}{5}x - \frac{1}{3}x - 4$
28. $7s + 4t + 9 - 2t - 2s - 15$
29. $6x - 4y - 5y + 4x + 3$

30. $2p - 4q - 3p + 4q - 15$

31. $2(s + 3) + 6(s - 4) + 1$

32. $6(r - 3) - 2(r + 5) + 10$

33. $0.2(x + 4) + 1.2(x - 3)$

34. $\frac{3}{5}(x + 1) - \frac{3}{10}x$

35. $\frac{1}{4}x + \frac{4}{5} - \frac{2}{3}x$

36. $n - \frac{3}{4} + \frac{5}{9}n - \frac{1}{6}$

37. $0.5(2.6x - 4) + 2.3(1.4x - 5)$

38. $\frac{2}{3}(3x + 9) - \frac{1}{4}(2x + 5)$

In Exercises 39–64, solve the equation.

39. $y - 7 = 10$

40. $3y + 6 = 24$

41. $15 = 9 - 6x$

42. $18 = 8 + 2x$

43. $\frac{3}{x} = \frac{7}{8}$

44. $\frac{x - 1}{5} = \frac{x + 5}{15}$

45. $\frac{1}{2}x + \frac{1}{3} = \frac{2}{3}$

46. $\frac{1}{2}y + \frac{1}{3} = \frac{1}{4}$

47. $0.9x - 1.2 = 2.4$

48. $5x + 0.050 = -0.732$

49. $6t - 8 = 4t - 2$

50. $\frac{x}{4} + 2x = \frac{1}{3}$

51. $\frac{x - 3}{2} = \frac{x + 4}{3}$

52. $\frac{x - 5}{4} = \frac{x - 9}{3}$

53. $6t - 7 = 8t + 9$

54. $12x - 1.2 = 3x + 1.5$

55. $2(x + 3) - 4 = 2(x - 4)$

56. $3(x + 2) + 2(x - 1) = 5x - 7$

57. $4(x - 4) + 12 = 4(x - 1)$

58. $6(t + 2) - 14 = 6t - 2$

59. $\frac{1}{3}(x + 3) = \frac{2}{5}(x + 2)$

60. $\frac{2}{3}(x - 4) = \frac{1}{4}(x + 1)$

61. $3x + 2 - 6x = -x - 15 + 8 - 5x$

62. $6x + 8 - 22x = 28 + 14x - 10 + 12x$

63. $4(t - 3) + 8 = 4(2t - 6)$

64. $8.3y - 3.1(y - 4) = 29.04$

PROBLEM SOLVING

In Exercises 65 and 66, use the water rate for Newton, New Jersey, which is \$7.75 per 1000 gallons of water.

65. ***Water Bill*** What is the water bill if a resident of Newton uses 27,000 gal?

66. ***Limiting the Cost*** How many gallons of water can a customer use if the water bill is not to exceed \$200?

67. ***Concrete Sealer*** A gallon of concrete sealer covers 360 ft^2. How much sealer is needed to cover a basement with a surface area of 1440 ft^2?

68. ***Fajitas*** A recipe for six servings of beef fajitas requires 16 oz of beef sirloin.

a) If the recipe were to be made for nine servings, how many ounces of beef sirloin would be needed?

b) How many servings of beef fajitas can be made with 32 oz of beef sirloin?

69. ***Watching Television*** Nielson Media Research determines the number of people who watch a television show. One rating point means that about 1,102,000 households watched the show. The top-rated television show for the week of October 16, 2006, was *Grey's Anatomy* with a rating of 14.4. About how many households watched *Grey's Anatomy* that week?

70. ***Topsoil*** A 40-pound bag of topsoil will cover a surface area of 12 ft^2.

a) How many pounds of topsoil are needed to cover a surface area of 480 ft^2?

b) How many bags of topsoil must be purchased to cover a surface area of 480 ft^2?

71. ***Speed Limit*** When Jacob Abbott crossed over from Niagara Falls, New York, to Niagara Falls, Canada, he saw a sign that said 50 miles per hour (mph) is equal to 80 kilometers per hour (kph).

▲ Niagara Falls, Canada

a) How many kilometers per hour are equal to 1 mph?

b) On a stretch of the Queen Elizabeth Way, the speed limit is 90 kph. What is the speed limit in miles per hour?

72. ***The Proper Dosage*** A doctor asks a nurse to give a patient 250 milligrams (mg) of the drug Simethicone. The drug is available only in a solution whose concentration is 40 mg Simethicone per 0.6 millimeter (mm) of solution. How many millimeters of solution should the nurse give the patient?

Amount of Insulin *In Exercises 73 and 74, how much insulin (in cc) would be given for the following doses? (Refer to Example 15 on page 327.)*

73. 15 units of insulin from a vial marked U40

74. 35 units of insulin from a vial marked U40

75. a) In your own words, summarize the procedure to use to solve an equation.

b) Solve the equation $2(x + 3) = 4x + 3 - 5x$ with the procedure you outlined in part (a).

76. a) What is an identity?

b) When solving an equation, how will you know if the equation is an identity?

77. a) What is a contradiction?

b) When solving an equation, how will you know if the equation is a contradiction?

CHALLENGE PROBLEMS/GROUP ACTIVITIES

78. ***Depth of a Submarine*** The pressure, P, in pounds per square inch (psi) exerted on an object x ft below the sea is given by the formula $P = 14.70 + 0.43x$. The 14.70 represents the weight in pounds of the column of air (from sea level to the top of the atmosphere) standing over a 1 in. by 1 in. square of seawater. The $0.43x$ represents the weight in pounds of a column of water 1 in. by 1 in. by x ft (see Fig. 6.2).

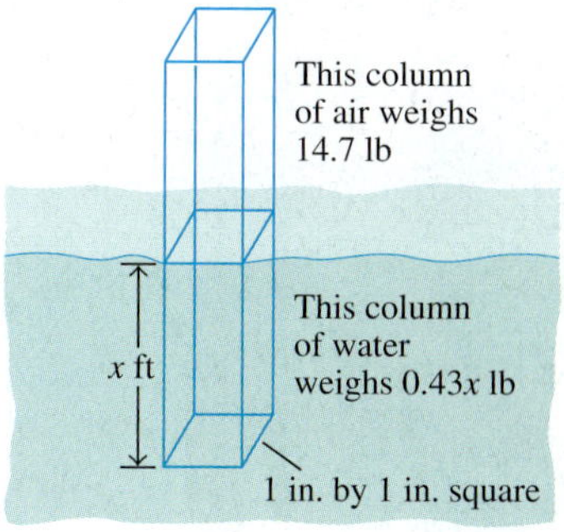

Figure 6.2

a) A submarine is built to withstand a pressure of 148 psi. How deep can that submarine go?

b) If the pressure gauge in the submarine registers a pressure of 128.65 psi, how deep is the submarine?

79. a) ***Gender Ratios*** If the ratio of males to females in a class is 2 : 3, what is the ratio of males to all the students in the class? Explain your answer.

b) If the ratio of males to females in a class is $m : n$, what is the ratio of males to all the students in the class?

80. Determine whether the given equations are equivalent.

a) $\dfrac{x}{3} = \dfrac{x + 2}{6}$, $6x = 3x + 6$

b) $2x - 9 = 19$, $\dfrac{2x - 9}{4} = \dfrac{19}{4}$

INTERNET/RESEARCH ACTIVITIES

81. Ratio and proportion are used in many different ways in everyday life. Submit two articles from newspapers, magazines, or the Internet in which ratios and/or proportions are used. Write a brief summary of each article explaining how ratio and/or proportion were used.

82. Write a report explaining how the ancient Egyptians used equations. Include in your discussion the forms of the equations used.

6.3 FORMULAS

▲ A formula is used to determine the amount of carpet needed in a room.

Suppose you need to purchase new carpet for your family room, which is in the shape of a rectangle. To determine the amount of carpet you need, you would need to determine the area of the room by multiplying the length of the room by the width of the room. In this section, we will discuss many applications that can be solved by using a special kind of equation called a *formula*.

A ***formula*** is an equation that typically has a real-life application. To ***evaluate a formula***, substitute the given values for their respective variables and then evaluate using the order of operations given in Section 6.1. Many of the formulas given in this section are discussed in greater detail in other parts of the book.

EXAMPLE 1 *Simple Interest*

The simple interest formula,* interest = principal × rate × time, or $i = prt$, is used to find the interest you must pay on a simple interest loan when you borrow principal, p, at simple interest rate, r, in decimal form, for time, t. Chris Campbell borrows \$4000 at a simple interest rate of 7% for 3 years.

a) How much will Chris Campbell pay in interest at the end of 3 years?

b) What is the total amount he will repay the bank at the end of 3 years?

SOLUTION

a) Substitute the values of p, r, and t into the formula; then evaluate.

$$\begin{aligned} i &= prt \\ &= 4000(0.07)(3) \\ &= 840 \end{aligned}$$

Thus, Chris must pay \$840 interest.

b) The total he must pay at the end of 3 years is the principal, \$4000, plus the \$840 interest, for a total of \$4840. ●

EXAMPLE 2 *Volume of an Ice-Cream Box*

The formula for the volume of a rectangular box† is volume = length × width × height, or $V = lwh$. Use the formula $V = lwh$ to find the width of a rectangular box of ice cream if $l = 7$ in., $h = 3.5$ in., and $V = 122.5$ in.^{3}.

SOLUTION We substitute the appropriate values into the volume formula and solve for the desired quantity, w.

$$\begin{aligned} V &= lwh \\ 122.5 &= (7)w(3.5) \\ 122.5 &= 24.5w \end{aligned}$$

---

*The simple interest formula is discussed in Section 11.2.

†The volume formula is discussed in Section 9.4.

$$\frac{122.5}{24.5} = \frac{24.5w}{24.5}$$
$$5 = w$$

Therefore, the width of the ice-cream box is 5 in.

In Example 1, we used the formula $i = prt$. In Example 2, we used the formula $V = lwh$. In these examples, we used a mathematical equation to represent real phenomena. When we represent real phenomena, such as finding simple interest, mathematically we say we have created a ***mathematical model*** or simply a ***model*** to represent the situation. A model may be a single formula, or equation, or a system of many equations. By using models we gain insight into real-life situations, such as how much interest you will accumulate in your savings account. We will use mathematical models throughout this chapter and elsewhere in the text. In some exercises in this and the next chapter, when you are asked to determine an equation to represent a real-life situation, we will sometimes write the word *modeling* in the instructions.

Many formulas contain Greek letters, such as μ (mu), σ (sigma), Σ (capital sigma), δ (delta), ϵ (epsilon), π (pi), θ (theta), and λ (lambda). Example 3 makes use of Greek letters.

EXAMPLE 3 *A Statistics Formula*

A formula used in the study of statistics to find the standard score (or z-score) is

$$z = \frac{\overline{x} - \mu}{\frac{\sigma}{\sqrt{n}}}$$

Find the value of z when $\overline{x}$ (read "x bar") $= 120$, $\mu = 100$, $\sigma = 16$, and $n = 4$.

SOLUTION

$$z = \frac{\overline{x} - \mu}{\frac{\sigma}{\sqrt{n}}} = \frac{120 - 100}{\frac{16}{\sqrt{4}}} = \frac{20}{\frac{16}{2}} = \frac{20}{8} = 2.5$$

Some formulas contain ***subscripts***. Subscripts are numbers (or letters) placed below and to the right of variables. They are used to help clarify a formula. For example, if two different amounts are used in a problem, they may be symbolized as A and A_0, or A_1 and A_2. Subscripts are read using the word *sub*; for example, A_0 is read "A sub zero" and A_1 is read "A sub one."

Exponential Equations

Many real-life problems, including population growth, growth of bacteria, and decay of radioactive substances, increase or decrease at a very rapid rate. For example, in Fig. 6.3 on page 332, which shows the amount of money borrowed by college students through private loans, in billions of dollars, from 1999–2000 to 2004–2005, the graph is increasing rapidly. This is an example in which the

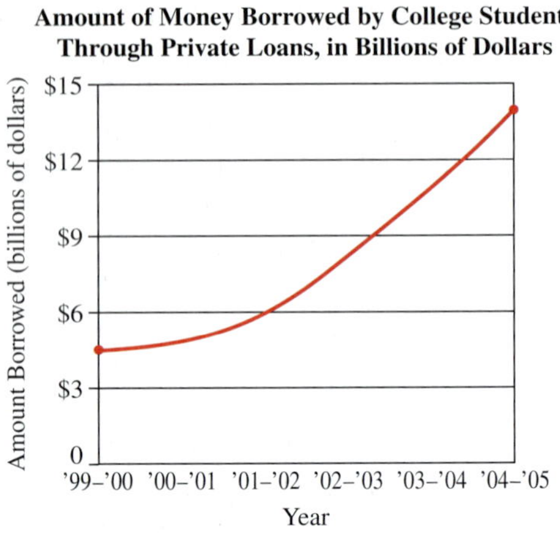

Source: College Board

Figure 6.3

graph is increasing ***exponentially***. The equation of a graph that increases or decreases exponentially is called an ***exponential equation***. An exponential equation is of the form $y = a^x, a > 0, a \neq 1$. We often use exponential equations to model real-life problems. In Section 6.10, we will discuss exponential equations in more detail.

In an exponential equation, letters other than x and y may be used to represent the variables. The following are examples of exponential equations: $y = 2^x$, $A = (\frac{1}{2})^x$, and $P = 2.3^t$. Note in exponential equations that the variable is the exponent of some positive constant that is not equal to 1. In many real-life applications, the variable t will be used to represent time. Problems involving exponential equations can be evaluated much more easily if you use a calculator containing a y^x, x^y, or $\wedge$ key.

The following equation, referred to as the ***exponential growth*** or ***decay formula***, is used to solve many real-life problems.

$$P = P_0 a^{kt}, \qquad a > 0, \quad a \neq 1$$

In the formula, P_0 represents the original amount present, P represents the amount present after t years, and a and k are constants.

When $k > 0$, P increases as t increases and we have exponential growth. When $k < 0$, P decreases as t increases and we have exponential decay.

EXAMPLE 4 *Using an Exponential Decay Formula*

Carbon dating is used by scientists to find the age of fossils, bones, and other items. The formula used in carbon dating is

$$P = P_0 2^{-t/5600}$$

where P_0 represents the original amount of carbon 14 (C_{14}) present and P represents the amount of C_{14} present after t years. If 10 mg of C_{14} is present in an animal bone recently excavated, how many milligrams will be present in 3000 years?

DID YOU KNOW?

Determining the Age of Bones and Fossils

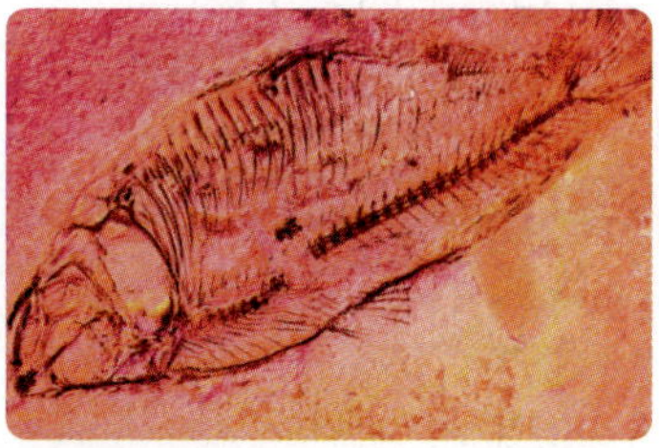

Carbon-14 dating is a method through which scientists estimate the age of organic materials such as bones, cloth, wood, and plant fibers. This dating technique was discovered by Willard Frank Libby and his colleagues in 1949. In 1960, Libby received the Nobel Prize in Chemistry for his work with carbon-14 dating. Carbon-14 dating is reliable for dating objects up to approximately 60,000 years old.

SOLUTION Substituting the values in the formula gives

$$P = P_0 2^{-t/5600}$$
$$P = 10(2)^{-3000/5600}$$
$$P \approx 10(2)^{-0.54} \qquad \text{(Recall that } \approx \text{ means "is approximately equal to.")}$$
$$P \approx 10(0.69)$$
$$P \approx 6.9 \text{ mg}$$

Thus, in 3000 years, approximately 6.9 mg of the original 10 mg of C_{14} will remain. •

In Example 4, we used a calculator to evaluate $(2)^{-3000/5600}$. The steps used to find this quantity on a calculator with a [y^x] key are

2 [y^x] [( ] 3000 [+/−] [÷] 5600 [ )] [=] .6898170602

After the [=] key is pressed, the calculator displays the answer 0.6898170602. To evaluate $10(2)^{-3000/5600}$ on a scientific calculator, we can press the following keys.

10 [×] 2 [y^x] [( ] 3000 [+/−] [÷] 5600 [ )] [=] 6.898170602

Notice that the answer obtained using the calculator steps shown above is a little more accurate than the answer we gave when we rounded the values before the final answer in Example 4.

When the a in the formula $P = P_0 a^{kt}$ is replaced with the very special letter e, we get the ***natural exponential growth or decay formula***

$$P = P_0 e^{kt}$$

The letter e represents an irrational number whose value is approximately 2.7183. The number e plays an important role in mathematics and is used in finding the solution to many application problems.

To evaluate $e^{(0.04)5}$ on a calculator, as will be needed in Example 5, press*

[( ] .04 [×] 5 [ )] [inv] [ln] = 1.221402758

(inv: Inverse key; ln: Natural logarithm key)

After the [ln] key is pressed, the calculator displays the answer 1.221402758.

To evaluate $10{,}000e^{(0.04)5}$ on a calculator, press

10000 [×] [( ] .04 [×] 5 [ )] [inv] [ln] [=] 12214.02758

In this calculation, after the [=] key is pressed, the calculator displays the answer 12214.02758.

*Keys to press may vary on some calculators.

EXAMPLE 5 *Using an Exponential Growth Formula*

Banks often credit compound interest continuously. When that is done, the principal amount in the account, P, at any time t can be calculated by the natural exponential formula $P = P_0e^{kt}$, where P_0 is the initial principal invested, k is the interest rate in decimal form, and t is the time.

Suppose \$10,000 is invested in a savings account at a 4% interest rate compounded continuously. What will be the balance (or principal) in the account in 5 years?

SOLUTION

$$\begin{aligned} P &= P_0e^{kt} \\ &= 10{,}000e^{(0.04)5} \\ &= 10{,}000e^{(0.20)} \\ &\approx 10{,}000(1.221402758) \\ &\approx 12{,}214 \end{aligned}$$

Thus, after 5 years, the account's value will have grown from \$10,000 to about \$12,214, an increase of about \$2214.

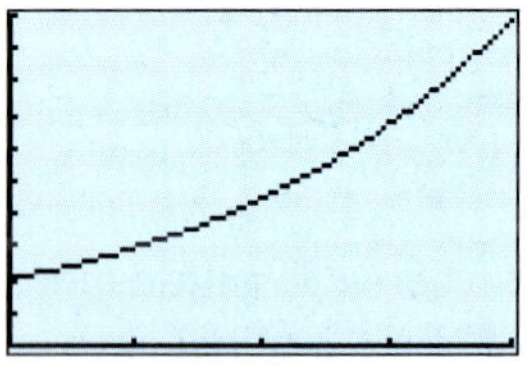

$P = 10{,}000e^{0.04t}$

Figure 6.4

Graphing calculators are a tool that can be used to graph equations. Figure 6.4 shows the graph of $P = 10{,}000e^{0.04t}$ as it appears on the screen (or window) of a Texas Instrument TI-84 Plus graphing calculator. To obtain this screen, the domain (or the x-values) and range (or the y-values) of the window need to be set to selected values. We will speak a little more about graphing calculators shortly. In Section 6.10, we will explain how to graph exponential equations by plotting points.

EXAMPLE 6 *Population of Arizona*

The population of Arizona is growing exponentially at the rate of about 3.5% per year. In 2005, the population of Arizona was about 5,939,292 people. Arizona's expected population, t years after 2005, is given by the formula $P = 5{,}939{,}292e^{0.035t}$. Assuming that the population of Arizona continues to grow at the same rate, determine the expected population of Arizona in 2012.

SOLUTION Since 2012 is 7 years after 2005, $t = 7$ years.

$$\begin{aligned} P &= 5{,}939{,}292e^{0.035t} \\ &= 5{,}939{,}292e^{0.035(7)} && \text{Substitute 7 for } t. \\ &= 5{,}939{,}292e^{0.245} \\ &\approx 5{,}939{,}292(1.277621313) \\ &\approx 7{,}588{,}166.045 \end{aligned}$$

Thus, in 2012, the population of Arizona is expected to be about 7,588,166 people.

TIMELY TIP When doing calculations on the calculator, do not round any value before obtaining the final answer. By not rounding, you will obtain a more accurate answer. For example, if we work Example 6 on a calculator and rounded $e^{0.245}$ to 1.28, we would determine that the population of Arizona in 2012 is expected to be about 7,602,294, which is a less accurate answer.

DID YOU KNOW?

$L = mc^2$
$E = mc^2$

Many of you recognize the formula $E = mc^2$ used by Albert Einstein in 1912 to describe his groundbreaking theory of relativity. In the formula, E is energy, m is mass, and c is the speed of light. In his theory, Einstein hypothesized that time was not absolute and that mass and energy were related.

Einstein's 72-page handwritten manuscript, which went on display in 1999 at Jerusalem's Israel Museum, shows that Einstein toyed with using an L rather than an E to represent energy. He wrote $L = mc^2$, and then scratched out the L and replaced it with an E to produce the equation $E = mc^2$. Yisrail Galili at Jerusalem's Hebrew University said the L might have been a lambda, a Greek letter sometimes used in functions related to energy. He said it might also be related to Joseph-Louis Lagrange, an eighteenth-century Italian mathematician.

Solving for a Variable in a Formula or Equation

Often in mathematics and science courses, you are given a formula or an equation expressed in terms of one variable and asked to express it in terms of a different variable. For example, you may be given the formula $P = i^2r$ and asked to solve the formula for r. To do so, treat each of the variables, except the one you are solving for, as if it were a constant. Then solve for the variable desired, using the properties previously discussed. Examples 7 through 9 show how to do this task.

When graphing equations in Section 6.7, you will sometimes have to solve the equation for the variable y as is done in Example 7.

EXAMPLE 7 *Solving for a Variable in an Equation*

Solve the equation $2x + 5y - 10 = 0$ for y.

SOLUTION We need to isolate the term containing the variable y. Begin by moving the constant, -10, and the term $2x$ to the right-hand side of the equation.

$$2x + 5y - 10 = 0$$

$$2x + 5y - 10 + 10 = 0 + 10 \quad \text{Addition property}$$

$$2x + 5y = 10$$

$$-2x + 2x + 5y = -2x + 10 \quad \text{Subtraction property}$$

$$5y = -2x + 10$$

$$\frac{5y}{5} = \frac{-2x + 10}{5} \quad \text{Division property}$$

$$y = \frac{-2x + 10}{5}$$

$$y = -\frac{2x}{5} + \frac{10}{5}$$

$$y = -\frac{2}{5}x + 2$$

Note that once you have found $y = \frac{-2x + 10}{5}$, you have solved the equation for y. The solution can also be expressed in the form $y = -\frac{2}{5}x + 2$. This form of the equation is convenient for graphing equations, as will be explained in Section 6.7. Example 7 can also be solved by moving the y term to the right-hand side of the equal sign. Do so now and note that you obtain the same answer.

EXAMPLE 8 *Solving for a Variable in a Formula*

An important formula used in statistics is

$$z = \frac{x - \mu}{\sigma}$$

Solve this formula for x.

PROFILE IN MATHEMATICS

Sophie Germain

Because she was a woman, Sophie Germain (1776–1831) was denied admission to the École Polytechnic, the French academy of mathematics and science. Not to be stopped, she obtained lecture notes from courses in which she had an interest, including one taught by Joseph-Louis Lagrange. Under the pen name M. LeBlanc, she submitted a paper on analysis to Lagrange, who was so impressed with the report that he wanted to meet the author and personally congratulate "him." When he found out that the author was a woman, he became a great help and encouragement to her. Lagrange introduced Germain to many of the French scientists of the time.

In 1801, Germain wrote to the great German mathematician Carl Friedrich Gauss to discuss Fermat's equation, $x^n + y^n = z^n$. He commended her for showing "the noblest courage, quite extraordinary talents and a superior genius."

SOLUTION To isolate the term x, use the general procedure for solving linear equations given in Section 6.2. Treat each letter, except x, as if it were a constant.

$$z = \frac{x - \mu}{\sigma}$$

$$z \cdot \sigma = \frac{x - \mu}{\not\sigma} \cdot \not\sigma \qquad \text{Multiply both sides of the equation by } \sigma$$

$$z\sigma = x - \mu$$

$$z\sigma + \mu = x - \mu + \mu \qquad \text{Add } \mu \text{ to both sides of the equation.}$$

$$z\sigma + \mu = x$$

$$\text{or} \quad x = z\sigma + \mu$$

EXAMPLE 9 *The Tax-Free Yield Formula*

A formula that may be important to you now or sometime in the future is the tax-free yield formula, $T_f = T_a(1 - F)$. This formula can be used to convert a taxable yield, T_a, into its equivalent tax-free yield, T_f, where F is the federal income tax bracket of the individual. A taxable yield is an interest rate for which income tax is paid on the interest made. A tax-free yield is an interest rate for which income tax does not have to be paid on the interest made.

a) For someone in a 25% tax bracket, find the equivalent tax-free yield of a 4% taxable investment.

b) Solve this formula for T_a. That is, write a formula for taxable yield in terms of tax-free yield.

SOLUTION

a) $T_f = T_a(1 - F)$

$$= 0.04(1 - 0.25) = 0.04(0.75) = 0.03, \quad \text{or} \quad 3\%$$

Thus, a taxable investment of 4% is equivalent to a tax-free investment of 3% for a person in a 25% income tax bracket.

b)

$$T_f = T_a(1 - F)$$

$$\frac{T_f}{1 - F} = \frac{T_a\cancel{(1 - F)}}{\cancel{1 - F}} \qquad \text{Divide both sides of the equation by } 1 - F.$$

$$\frac{T_f}{1 - F} = T_a, \quad \text{or} \quad T_a = \frac{T_f}{1 - F}$$

SECTION 6.3 EXERCISES

CONCEPT/WRITING EXERCISES

1. What is a formula?

2. Explain how to evaluate a formula.

3. What are subscripts?

4. What is the simple interest formula?

5. What is an exponential equation?

6. a) In an exponential equation of the form $y = a^x$, what are the restrictions on a?

   b) In an exponential equation of the form $y = P_0 a^{kt}$, what does P_0 represent?

PRACTICE THE SKILLS

In Exercises 7–40, use the formula to find the value of the indicated variable for the values given. Use a calculator when one is needed. When necessary, round answers to the nearest hundredth.

7. $A = lw$; determine A when $l = 4$ and $w = 14$ (geometry).

8. $P = a + b + c$; determine P when $a = 25$, $b = 53$, and $c = 32$ (geometry).

9. $P = 2l + 2w$; determine P when $l = 12$ and $w = 16$ (geometry).

10. $F = ma$; determine m when $F = 40$ and $a = 5$ (physics).

11. $K = \frac{1}{2}mv^2$; determine m when $v = 30$ and $K = 4500$ (physics).

12. $p = i^2 r$; determine r when $p = 62{,}500$ and $i = 5$ (electronics).

13. $S = \pi r(r + h)$; determine S when $r = 8$, $\pi = 3.14$, and $h = 2$ (geometry).

14. $B = \frac{703w}{h^2}$; determine B when $w = 130$ and $h = 67$ (body mass index).

15. $z = \frac{x - \mu}{\sigma}$; determine μ when $z = 2.5$, $x = 42.1$, and $\sigma = 2$ (statistics).

16. $S = 2hw + 2lw + 2lh$; determine l when $S = 122$, $h = 4$, and $w = 3$ (geometry).

17. $T = \frac{PV}{k}$; determine P when $T = 80$, $V = 20$, and $k = 0.5$ (physics).

18. $m = \frac{a + b + c}{3}$; determine a when $m = 70$, $b = 60$, and $c = 90$ (statistics).

19. $A = P(1 + rt)$; determine P when $A = 3600$, $r = 0.04$, and $t = 5$ (economics).

20. $m = \frac{a + b}{2}$; determine a when $m = 70$ and $b = 77$ (statistics).

21. $V = \pi r^2 h$; determine h when $V = 942$, $\pi = 3.14$, and $r = 5$ (geometry).

22. $\text{F} = \frac{9}{5}\text{C} + 32$; determine F when C = 7 (temperature conversion).

23. $\text{C} = \frac{5}{9}(\text{F} - 32)$; determine C when F = 77 (temperature conversion).

24. $\text{K} = \frac{\text{F} - 32}{1.8} + 273.1$; determine K when F = 100 (chemistry).

25. $m = \frac{y_2 - y_1}{x_2 - x_1}$; determine m when $y_2 = 8$, $y_1 = -4$, $x_2 = -3$, and $x_1 = -5$ (mathematics).

26. $z = \dfrac{\bar{x} - \mu}{\frac{\sigma}{\sqrt{n}}}$; determine z when $\bar{x} = 66$, $\mu = 60$, $\sigma = 15$, and $n = 25$ (statistics).

27. $S = R - rR$; determine R when $S = 186$ and $r = 0.07$ (for determining sale price when an item is discounted).

28. $S = C + rC$; determine C when $S = 115$ and $r = 0.15$ (for determining selling price when an item is marked up).

29. $E = a_1 p_1 + a_2 p_2 + a_3 p_3$; determine E when $a_1 = 5$, $p_1 = 0.2$, $a_2 = 7$, $p_2 = 0.6$, $a_3 = 10$, and $p_3 = 0.2$ (probability).

30. $x = \frac{-b + \sqrt{b^2 - 4ac}}{2a}$; determine x when $a = 2$, $b = -5$, and $c = -12$ (mathematics).

31. $s = -16t^2 + v_0t + s_0$; determine s when $t = 4$, $v_0 = 30$, and $s_0 = 150$ (physics).

32. $R = O + (V - D)r$; determine O when $R = 670$, $V = 100$, $D = 10$, and $r = 4$ (economics).

33. $P = \dfrac{f}{1 + i}$; determine f when $i = 0.08$ and $P = 3000$ (investment banking).

34. $Q_w = m_wc_w(T_f - T_w)$; determine Q_w when $m_w = 0.5$, $c_w = 4186$, $T_f = 22$, and $T_w = 19$ (physics).

35. $R_T = \dfrac{R_1R_2}{R_1 + R_2}$; determine R_T when $R_1 = 100$ and $R_2 = 200$ (electronics).

36. $P = \dfrac{nRT}{V}$; determine V when $P = 12$, $n = 10$, $R = 60$, and $T = 8$ (chemistry).

37. $a_n = a_1 + (n - 1)d$; determine a_n when $a_1 = 15$, $n = 4$, and $d = 8$ (mathematics).

38. $A = P\left(1 + \dfrac{r}{n}\right)^{nt}$; determine A when $P = 100$, $r = 6\%$, $n = 1$, and $t = 3$ (banking).

39. $P = P_0e^{kt}$; determine P when $P_0 = 5000$, $k = 0.06$, and $t = 7$ (economics).

40. $S = S_0e^{-0.028t}$; determine S when $S_0 = 1000$ and $t = 35$ (chemistry).

In Exercises 41–50, solve the equation for y.

41. $4x - 9y = 14$

42. $11x - 15y = 23$

43. $8x + 7y = 21$

44. $-9x + 4y = 11$

45. $2x - 3y + 6 = 0$

46. $3x + 4y = 0$

47. $-2x + 3y + z = 15$

48. $5x + 3y - 2z = 22$

49. $9x + 4z = 7 + 8y$

50. $2x - 3y + 5z = 0$

In Exercises 51–70, solve for the variable indicated.

51. $d = rt$ for r

52. $V = lwh$ for w

53. $p = a + b + c$ for a

54. $p = a + b + s_1 + s_2$ for s_1

55. $V = \frac{1}{3}Bh$ for B

56. $V = \pi r^2h$ for h

57. $C = 2\pi r$ for r

58. $y = mx + b$ for m

59. $y = mx + b$ for b

60. $V = \dfrac{1}{3}\pi r^2h$ for h

61. $P = 2l + 2w$ for w

62. $A = \dfrac{d_1d_2}{2}$ for d_2

63. $A = \dfrac{a + b + c}{3}$ for c

64. $\dfrac{w_1}{w_2} = \dfrac{f_2 - f}{f - f_1}$ for w_1

65. $P = \dfrac{KT}{V}$ for T

66. $\dfrac{P_1V_1}{T_1} = \dfrac{P_2V_2}{T_2}$ for V_2

67. $F = \frac{9}{5}C + 32$ for C

68. $C = \frac{5}{9}(F - 32)$ for F

69. $S = 2\pi rh + 2\pi r^2$ for h

70. $A = \dfrac{1}{2}h(b_1 + b_2)$ for b_2

PROBLEM SOLVING

In Exercises 71–78, when appropriate, round answers to the nearest hundredth.

71. ***Savings Account*** Christine Northrup borrowed \$4500 from a bank at a simple interest rate of 2.5% for one year.

a) Determine how much interest Christine paid at the end of 1 year.

b) Determine the total amount Christine will repay the bank at the end of 1 year.

72. ***Interest on a Loan*** Jeff Hubbard borrowed \$800 from his brother for 2 years. At the end of 2 years, he repaid the \$800 plus \$128 in interest. What simple interest rate did he pay?

73. ***Volume of an Ice-Cream Cone*** An ice-cream cone is filled with ice cream to the top of the cone. Determine the volume, in cubic inches, of ice cream in the cone if the cone's radius is 1 in. and the height is 4 in. The formula for the volume of a cone is $V = \dfrac{1}{3}\pi r^2h$. Use the pi key, $\boxed{\pi}$, on your calculator, or 3.14 for π if your calculator does not have a pi key.

74. ***Body Mass Index*** A person's body mass index (BMI) is found by the formula $B = \frac{703w}{h^2}$, where w is the person's weight, in pounds, and h is the person's height, in inches. Lance Bass is 6 ft tall and weighs 200 lb.

a) Determine his BMI.

b) If Lance would like to have a BMI of 26, how much weight would he need to gain or lose?

75. ***Bacteria*** The number of a certain type of bacteria, y, present in a culture is determined by the equation $y = 2000(3)^x$, where x is the number of days the culture has been growing. Find the number of bacteria present after 5 days.

76. ***Adjusting for Inflation*** If P is the price of an item today, the price of the same item n years from today, P_n, is $P_n = P(1 + r)^n$, where r is the constant rate of inflation. Determine the price of a movie ticket 10 years from today if the price today is \$8.00 and the annual rate of inflation is constant at 3%.

77. ***Value of New York City*** Assume the value of the island of Manhattan has grown at an exponential rate of 8% per year since 1626 when Peter Minuit of the Dutch West India Company purchased the island for \$24. The value of the island, V, at any time, t, in years after 1626, can be found by the formula $V = 24e^{0.08t}$. What is the value of the island in 2008, 382 years after Minuit purchased it? Use a scientific calculator and give your answer in scientific notation as provided by your calculator.

78. ***Radioactive Decay*** Radium-226 is a radioactive isotope that decays exponentially at a rate of 0.0428% per year. The amount of radium-226, R, remaining after t years can be found by the formula $R = R_0e^{-0.000428t}$, where R_0 is the original amount present. If there are originally 10 grams of radium-226, determine the amount of radium-226 remaining after 1000 years.

CHALLENGE PROBLEM/GROUP ACTIVITY

79. Determine the volume of the block shown in Fig. 6.5, excluding the hole. The formula for the volume of a rectangular solid is $V = lwh$. The formula for the volume of a cylinder is $V = \pi r^2h$.

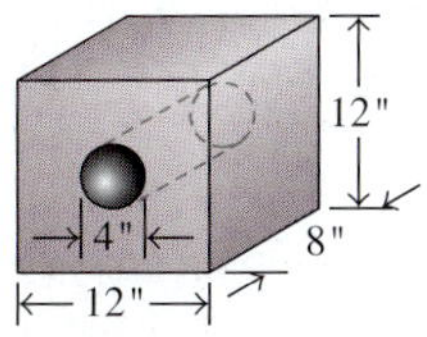

Figure 6.5

6.4 APPLICATIONS OF LINEAR EQUATIONS IN ONE VARIABLE

▲ A mathematical equation can be used to determine a sales representative's weekly salary.

Suppose your weekly salary as a sales representative consists of a weekly base salary of \$400 and a 6% commission on the total dollar amount of all sales you make. How many dollars in sales would you need to make in a week to earn a total of \$1000 a week? To answer this question you must first express the written problem as a mathematical equation. Then you can use the general procedure for solving linear equations discussed in Section 6.2 to obtain the answer. In this section, we will explain how to write mathematical equations to represent word problems.

One reason to study algebra is that it can be used to solve everyday problems. In this section, we will do two things: (1) show how to translate a written problem into a mathematical equation and (2) show how linear equations can be used in solving everyday problems. We begin by illustrating how English phrases can be written as mathematical expressions. When writing a mathematical expression, we may use any letter to represent the variable. In the following illustrations, we use the letter x.

DID YOU KNOW?

Survival of the Fittest

Symbols come and symbols go; the ones that find the greatest acceptance are the ones that survive. The Egyptians used pictorial symbols: a pair of legs walking forward for addition or backward for subtraction. Robert Recorde (1510–1558) used two parallel lines to represent "equals" because "no 2 thynges can be moore equalle." Some symbols evolved from abbreviations, such as the "+" sign, which comes from the Latin *et* meaning "and." The evolution of others is less clear. The invention of the printing press in the fifteenth century led to a greater standardization of symbols already in use.

Phrase	Mathematical expression
Six more than a number	$x + 6$
A number increased by 3	$x + 3$
Four less than a number	$x - 4$
A number decreased by 9	$x - 9$
Twice a number	$2x$
Four times a number	$4x$
3 decreased by a number	$3 - x$
The difference between a number and 5	$x - 5$

Sometimes the phrase that must be converted to a mathematical expression involves more than one operation.

Phrase	Mathematical expression
Four less than 3 times a number	$3x - 4$
Ten more than twice a number	$2x + 10$
The sum of 5 times a number and 3	$5x + 3$
Eight times a number, decreased by 7	$8x - 7$

The word *is* often represents the equal sign.

Phrase	Mathematical equation
Six more than a number is 10.	$x + 6 = 10$
Five less than a number is 20.	$x - 5 = 20$
Twice a number, decreased by 6 is 12.	$2x - 6 = 12$
A number decreased by 13 is 6 times the number.	$x - 13 = 6x$

The following is a general procedure for solving word problems.

TO SOLVE A WORD PROBLEM

1. Read the problem carefully at least twice to be sure that you understand it.
2. If possible, draw a sketch to help visualize the problem.
3. Determine which quantity you are being asked to find. Choose a letter to represent this unknown quantity. Write down exactly what this letter represents.
4. Write the word problem as an equation.
5. Solve the equation for the unknown quantity.
6. Answer the question or questions asked.
7. Check the solution.

This general procedure for solving word problems is illustrated in Examples 1 through 4. In these examples, the equations we obtain are mathematical models of the given situations.

DID YOU KNOW?

Aha!

Aha:

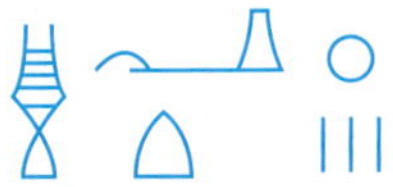

or

Heap:

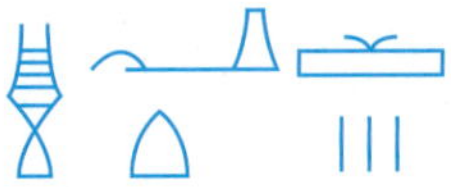

The Egyptians as far back as 1650 B.C. had a knowledge of linear equations. They used the words *aha* or *heap* in place of the variable. Problems involving linear equations can be found in the Rhind Papyrus (see Chapter 4, p. 183).

EXAMPLE 1 *How Much Can You Purchase?*

Roberto Raynor spent \$339.95 on textbooks at the bookstore. In addition to textbooks, he wanted to purchase as many notebooks as possible, but he only has a total of \$350 to spend. If a notebook, including tax, costs \$0.99, how many notebooks can he purchase?

SOLUTION In this problem, the unknown quantity is the number of notebooks Roberto can purchase. Let's select n to represent the number of notebooks he can purchase. Then we construct an equation using the given information that will allow us to solve for n.

Let

$$n = \text{number of notebooks Roberto can purchase}$$

Then

$$\$0.99n = \text{cost for } n \text{ notebooks at } \$0.99 \text{ per notebook}$$

$$\begin{array}{ccccc} \text{Cost of textbooks} & + & \text{cost of notebooks} & = & \text{total amount to spend} \\ \$339.95 & + & \$0.99n & = & \$350 \end{array}$$

Now solve the equation.

$$\begin{aligned} 339.95 + 0.99n &= 350 \\ 339.95 - 339.95 + 0.99n &= 350 - 339.95 \\ 0.99n &= 10.05 \\ \frac{0.99n}{0.99} &= \frac{10.05}{0.99} \\ n &\approx 10.15 \end{aligned}$$

Therefore, Roberto can purchase 10 notebooks. When we solve the equation, we obtain $n = 10.15$ (to the nearest hundredth). Since he cannot purchase a part of a notebook, only 10 notebooks can be purchased.

CHECK: The check is made with the information given in the original problem.

$$\begin{aligned} \text{Total amount to spend} &= \text{cost of textbooks} + \text{cost of notebooks} \\ &= 339.95 + 0.99(10) \\ &= 339.95 + 9.90 \\ &= 349.85 \end{aligned}$$

This result would leave 15 cents change from the \$350 he has to spend, which is not enough to purchase another notebook. Therefore, this answer checks. ●

EXAMPLE 2 *Dividing Land*

Mark IV Construction Company purchased 100 acres of land to be split into three parcels of land on which to build houses. One parcel of land will be three times as large as each of the other two. How many acres of land will each parcel contain?

SOLUTION Two parcels will contain the same number of acres, and the third parcel will contain three times that amount.

Let

$$x = \text{number of acres in the first parcel}$$
$$x = \text{number of acres in the second parcel}$$
$$3x = \text{number of acres in the third parcel}$$

Then

$$\begin{aligned} x + x + 3x &= \text{total amount of land} \\ x + x + 3x &= 100 \\ 5x &= 100 \\ x &= 20 \end{aligned}$$

Thus, two parcels will contain 20 acres of land. The third parcel will contain 3(20), or 60, acres of land. A check in the original problem will verify that this answer is correct. ●

EXAMPLE 3 *Dimensions of an Exercise Area*

Dr. Christine Seidel, a veterinarian, wants to fence in a large rectangular region in the yard behind her office for exercising dogs that are boarded overnight. She has 130 ft of fencing to use for the perimeter of the region. What should the dimensions of the region be if she wants the length to be 15 ft greater than the width?

SOLUTION The formula for finding the perimeter of a rectangle is $P = 2l + 2w$, where P is the perimeter, l is the length, and w is the width. A diagram, such as the one shown below, is often helpful in solving problems of this type.

Let w equal the width of the region. The length is 15 ft more than the width, so $l = w + 15$. The total distance around the region P, is 130 ft.

Substitute the known quantities in the formula.

$$\begin{aligned} P &= 2l + 2w \\ 130 &= 2(w + 15) + 2w \\ 130 &= 2w + 30 + 2w \\ 130 &= 4w + 30 \\ 100 &= 4w \\ 25 &= w \end{aligned}$$

The width of the region is 25 ft and the length of the region is $25 + 15 = 40$ ft. ●

In shopping and other daily activities, we are occasionally asked to solve problems using percents. The word *percent* means "per hundred." Thus, for example, 7%

means 7 per hundred, or $\frac{7}{100}$. When $\frac{7}{100}$ is converted to a decimal number, we obtain 0.07. Thus, $7\% = 0.07$.

Let's look at one example involving percent. (See Section 11.1 for a more detailed discussion of percent.)

EXAMPLE 4 ***Art Show***

Kim Stappenbeck is selling her original paintings at an art show. Determine the cost of a painting before tax if the total cost of a painting, including an 8% sales tax, is to be $145.80.

SOLUTION We are asked to find the cost of a painting before sales tax.

Let

$$x = \text{cost of a painting before sales tax.}$$

Then

$$0.08x = 8\% \text{ of the cost of the painting (the sales tax)}$$

$$\begin{aligned}
\text{Cost of a painting before tax} + \text{tax on a painting} &= 145.80 \\
x + 0.08x &= 145.80 \\
1.08x &= 145.80 \\
\frac{1.08x}{1.08} &= \frac{145.80}{1.08} \\
x &= \frac{145.80}{1.08} \\
x &= 135
\end{aligned}$$

Thus, the cost of a painting before tax is $135.

SECTION 6.4 EXERCISES

CONCEPT/WRITING EXERCISES

1. What is the difference between a mathematical expression and an equation?
2. Give an example of a mathematical expression and an example of a mathematical equation.

In Exercises 3–6, write the mathematical expression as a phrase. (There are several acceptable answers.)

3. $x + 4$
4. $x - 7$
5. $2x - 3$
6. $3x + 5$

PRACTICE THE SKILLS

In Exercises 7–18, write the phrase as a mathematical expression.

7. 8 more than x
8. 5 less than y
9. 3 increased by 2 times z
10. 5 times x decreased by 8
11. The sum of 6 times w and 9

12. 4 more than twice a number

13. 6 more than 4 times x

14. 8 increased by 5 times x

15. 18 decreased by s, divided by 4

16. The sum of 8 and t, divided by 2

17. Three times the sum of a number and 7

18. The quotient of 8 and y, decreased by 3 times x

In Exercises 19–30, write an equation and solve.

19. 5 more than a number is 11.

20. A number decreased by 4 is 9.

21. The difference between a number and 4 is 20.

22. Seven multiplied by a number is 56.

23. Ten less than four times a number is 42.

24. Sixteen increased by 8 times a number is 88.

25. Twelve more than 4 times a number is 32.

26. Eight less than 3 times a number is 6 times the number increased by 10.

27. A number increased by 6 is 3 less than twice the number.

28. A number divided by 3 is 4 less than the number.

29. A number increased by 10 is 2 times the sum of the number and 3.

30. Twice a number decreased by 3 is 4 more than the number.

PROBLEM SOLVING

In Exercises 31–50, set up an equation that can be used to solve the problem. Solve the equation and determine the desired value(s).

31. **MODELING** - *Reimbursed Expenses* When sales representatives for Pfizer Pharmaceuticals drive to out-of-town meetings that require an overnight stay, they receive $150 for lodging plus $0.42 per mile driven. How many miles did Joe Kotaska drive if Pfizer reimbursed him $207.54 for an overnight trip?

32. **MODELING** - *Sales Commission* Tito Sanchez receives a weekly salary of $400 at Anderson's Appliances. He also receives a 6% commission on the total dollar amount of all sales he makes. What must his total sales be in a week if he is to make a total of $790?

33. **MODELING** - *Jet Skiing* At Action Water Sports of Ocean City, Maryland, the cost of renting a Jet Ski is $42 per half hour, which includes a 5% sales tax. Determine the cost of a half hour Jet Ski rental before tax.

34. **MODELING** - *Pet Supplies* PetSmart has a sale offering 10% off of all pet supplies. If Amanda Miller spent $15.72 on pet supplies before tax, what was the price of the pet supplies she purchased before the discount?

35. **MODELING** - *Copying* Ronnie McNeil pays 8¢ to make a copy of a page at a copy shop. She is considering purchasing a photocopy machine that is on sale for $250, including tax. How many copies would Ronnie have to make in the copy shop for her cost to equal the purchase price of the photocopy machine she is considering buying?

36. **MODELING** - *Number of CDs* Samantha Silverstone and Josie Appleton together receive 12 free compact discs by joining a compact disc club. How many CDs will each receive if Josie is to have three times as many as Samantha?

37. **MODELING** - *Scholarship Donation* Each year, Andrea Choi donates a total of $1000 for scholarships at Mercer County Community College. This year, she wants the amount she donates for scholarships for liberal arts to be three times the amount she donates for scholarships for business. Determine the amount she will donate for each type of scholarship.

38. **MODELING** - *Homeowners Association* Cross Creek Townhouses Homeowners Association needs to charge each homeowner a supplemental assessment to help pay for some unexpected repairs to the townhouses. The association has $2000 in its reserve fund that it will use to help pay for the repairs. How much must the association charge each of the 50 homeowners if the total cost for the repairs is $13,350?

39. **MODELING** - *Dimensions of a Deck* Jim Yuhas is building a rectangular deck and wants the length to be 3 ft greater than the width. What will be the dimensions of the deck if the perimeter is to be 54 ft?

▲ See Exercise 39 on page 344

40. **MODELING** - ***Floor Area*** The total floor space in three barns is 45,000 ft^2. The two smaller ones have the same area, and the largest one has three times the area of the smaller ones.

a) Determine the floor space for each barn.

b) Can merchandise that takes up 8500 ft^2 of floor space fit into either of the smaller barns?

41. **MODELING** - ***Wild Mustangs*** According to the Bureau of Land Management, the wild mustang population of Nevada in 2004 was 2515 more than six times the wild mustang population in Utah. The total number of wild mustangs in Nevada and Utah was 21,730. Determine the number of wild mustangs in each state in 2004.

42. **MODELING** - ***Counting Calories*** According to *The Step Diet Book*, the number of steps it takes for a 150 lb person to burn off the calories from eating a cheeseburger and drinking a 12 oz soda is 11,040. The number of steps it takes to burn off the calories from eating a cheeseburger is 690 more than twice the number of steps it takes to burn off the calories from drinking a 12 oz soda. Determine the number of steps needed for a 150 lb person to burn off the calories from eating a cheeseburger and from drinking a 12 oz soda.

43. **MODELING** - ***Tornados*** The typical number of tornados in the United States in May is 11 more than 14 times the typical number of tornados in December. The sum of the typical number of tornados in December and in May is 341. Determine the typical number of tornados in the month of December and also in the month of May.

44. **MODELING** - ***Car Purchase*** The Gilberts purchased a car. If the total cost, including a 5% sales tax, was $14,512, find the cost of the car before tax.

45. **MODELING** - ***Enclosing Two Pens*** Chuck Salvador has 140 ft of fencing in which he wants to build two connecting, adjacent square pens (see the figure). What will be the dimensions if the length of the entire enclosed region is to be twice the width?

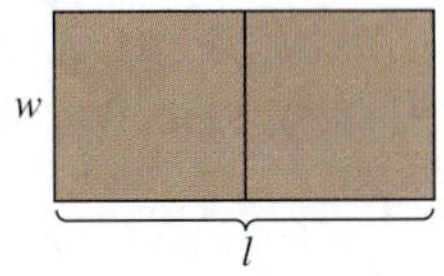

46. **MODELING** - ***Dimensions of a Bookcase*** A bookcase with three shelves, as shown in the diagram, is to be built by a woodworking student. If the height of the bookcase is to be 2 ft longer than the length of a shelf and the total amount of wood to be used is 32 ft, find the dimensions of the bookcase.

47. **MODELING** - ***Laundry Cost*** The cost of doing the family laundry for a month at a local laundromat is $70. A new washer and dryer cost a total of $760. How many months would it take for the cost of doing the laundry at the laundromat to equal the cost of a new washer and dryer?

48. **MODELING** - ***Renting Videos*** Hollywood Video offers customers two options to rent DVDs. The Unlimited Option costs $28 a month for unlimited DVD rentals. The Single Option costs $3.50 per DVD rental. How many DVDs must be rented in a month for the cost of the Unlimited Option to equal the cost of the Single Option?

49. **MODELING** - ***Airfare*** Rachel James has been told that with her half-off airfare coupon, her airfare from New York to San Diego will be $257.00. The $257.00 includes a 7% tax *on the regular fare.* On the way to the airport, Rachel realizes that she has lost her coupon. What will her regular fare be before tax?

▲ Hotel del Coronado, San Diego, California, see Exercise 49 on page 345

50. MODELING - *Truck Rentals* The cost of renting a small truck at the U-Haul rental agency is \$35 per day plus 20¢ a mile. The cost of renting the same truck at the Ryder rental agency is \$25 per day plus 32¢ a mile. How far would you have to drive in one day for the cost of renting from U-Haul to equal the cost of renting from Ryder?

CHALLENGE PROBLEMS/GROUP ACTIVITIES

51. *Income Tax* Some states allow a husband and wife to file individual tax returns (on a single form) even though they have filed a joint federal tax return. It is usually to the taxpayers' advantage to do so when both husband and wife work. The smallest amount of tax owed (or the largest refund) will occur when the husband's and wife's taxable incomes are the same.

Mr. McAdams's 2007 taxable income was \$34,200, and Mrs. McAdams's taxable income for that year was \$36,400. The McAdams's total tax deduction for the year was \$3640. This deduction can be divided between Mr. and Mrs. McAdams any way they wish. How should the \$3640 be divided between them to result in each individual having the same taxable income and therefore the greatest tax refund?

52. Write each equation as a sentence. There are many correct answers.

a) $x + 3 = 13$ b) $3x + 5 = 8$

c) $3x - 8 = 7$

53. Show that the sum of any three consecutive integers is 3 less than 3 times the largest.

54. *Auto Insurance* A driver education course at the East Lake School of Driving costs \$45 but saves those under 25 years of age 10% of their annual insurance premiums until they are 25. Dan has just turned 18, and his insurance costs \$100.00 per month.

a) When will the amount saved from insurance equal the price of the course?

b) Including the cost of the course, when Dan turns 25, how much will he have saved?

RECREATIONAL MATHEMATICS

55. The relationship between Fahrenheit temperature (F) and Celsius temperature (C) is shown by the formula $F = \frac{9}{5}C + 32$. At what temperature will a Fahrenheit thermometer read the same as a Celsius thermometer?

6.5 VARIATION

▲ Variation can be used to estimate the property tax on a home.

Your real estate agent tells you the property tax on a \$170,000 home in a certain town is \$2550. Suppose you are about to purchase a home in this town for \$190,000. We can use variation to estimate the property tax on the home you are about to purchase. In this section, we will discuss four different types of variation and show how variation can be used to solve real-life problems.

Direct Variation

Many scientific formulas are expressed as variations. A *variation* is an equation that relates one variable to one or more other variables through the operations of multiplication or division (or both operations). There are essentially four types of variation problems: direct, inverse, joint, and combined variation.

In *direct variation*, the values of the two related variables increase together or decrease together; that is, as one increases so does the other, and as one decreases so does the other.

Consider a car traveling at 40 miles an hour. The car travels 40 miles in 1 hour, 80 miles in 2 hours, and 120 miles in 3 hours. Note that, as the time increases, the distance traveled increases, and, as the time decreases, the distance traveled decreases.

The formula used to calculate distance traveled is

$$\text{Distance} = \text{rate} \cdot \text{time}$$

Since the rate is a constant 40 miles per hour, the formula can be written

$$d = 40t$$

We say that distance *varies directly* as time or that distance is *directly proportional* to time.

The preceding equation is an example of direct variation.

DIRECT VARIATION

If a variable y varies directly with a variable x, then

$$y = kx$$

where k is the **constant of proportionality** (or the variation constant).

Examples 1 through 4 illustrate direct variation.

EXAMPLE 1 *Direct Variation in Physics*

The length that a spring will stretch, S, varies directly with the force (or weight) F, attached to the spring. Write the equation for the length that a spring will stretch, S, if the constant of proportionality is 0.07.

SOLUTION

$S = kF$ S varies directly as F.

$S = 0.07F$ Constant of proportionality is 0.07.

EXAMPLE 2 *Direct Variation in Medicine*

The recommended dosage, d, of the antibiotic drug vancomycin is directly proportional to a person's weight, w.

a) Write this variation as an equation.

b) Find the recommended dosage, in milligrams, for Doug Kulzer, who weighs 192 lb. Assume the constant of proportionality for the dosage is 18.

SOLUTION

a) $d = kw$

b) $d = 18(192) = 3456$

The recommended dosage for Doug Kulzer is 3456 mg.

In certain variation problems, the constant of proportionality, k, may not be known. In such cases, we can often find it by substituting the given values in the variation formula and solving for k.

EXAMPLE 3 *Finding the Constant of Proportionality*

Suppose w varies directly as the square of y. If w is 60 when y is 20, find the constant of proportionality.

SOLUTION Since w varies directly as the *square of y*, we begin with the formula $w = ky^2$. Since the constant of proportionality is not given, we must find k using the given information. Substitute 60 for w and 20 for y.

$$\begin{aligned} w &= ky^2 \\ 60 &= k(20)^2 \\ 60 &= 400k \\ \frac{60}{400} &= \frac{400k}{400} \\ 0.15 &= k \end{aligned}$$

Thus, the constant of proportionality is 0.15.

EXAMPLE 4 *Using the Constant of Proportionality*

The area, a, of a picture projected on a movie screen varies directly as the square of the distance, d, from the projector to the screen. If a projector at a distance of 25 feet projects a picture with an area of 100 square feet, what is the area of the projected picture when the projector is at a distance of 40 feet?

SOLUTION We begin with the formula $a = kd^2$. Since the constant of proportionality is not given, we must determine k, using the given information.

$$\begin{aligned} a &= kd^2 \\ 100 &= k(25)^2 \\ 100 &= k(625) \\ \frac{100}{625} &= k \\ 0.16 &= k \end{aligned}$$

We now use $k = 0.16$ to determine a when $d = 40$.

$$\begin{aligned} a &= kd^2 \\ a &= 0.16d^2 \\ a &= 0.16(40)^2 \\ a &= 0.16(1600) \\ a &= 256 \text{ ft}^2 \end{aligned}$$

Thus, the area of a projected picture is 256 ft^2 when the projector is at a distance of 40 ft.

Inverse Variation

A second type of variation is *inverse variation*. When two quantities vary inversely, as one quantity increases the other quantity decreases, and vice versa.

To explain inverse variation, we use the formula, distance = rate · time. If we solve for time, we get time = distance/rate. Assume the distance is fixed at 100 miles; then

$$\text{Time} = \frac{100}{\text{rate}}$$

At 100 miles per hour, it would take 1 hour to cover this distance. At a rate of 50 miles an hour, it would take 2 hours. At a rate of 25 miles an hour, it would take 4 hours. Note that as the rate (or speed) decreases, the time increases and vice versa.

The preceding equation can be written

$$t = \frac{100}{r}$$

This equation is an example of an inverse variation. The time and rate are inversely proportional. The constant of proportionality in this case is 100.

INVERSE VARIATION

If a variable y varies inversely with a variable x, then

$$y = \frac{k}{x}$$

where k is the constant of proportionality.

Two quantities *vary inversely*, or are *inversely proportional*, when as one quantity increases the other quantity decreases and vice versa. Examples 5 and 6 illustrate inverse variation.

EXAMPLE 5 *Inverse Variation in Astronomy*

The velocity, v, of a meteor approaching Earth varies inversely as the square root of its distance from the center of Earth. Assuming the velocity is 2 miles per second at a distance, d, of 6400 miles from the center of Earth, determine the equation that expresses the relationship between the velocity of a meteor and its distance from the center of Earth.

SOLUTION Since the velocity of the meteor varies inversely as the *square root* of its distance from the center of Earth, the general form of the equation is

$$v = \frac{k}{\sqrt{d}}$$

To determine k, we substitute the given values for v and d.

$$2 = \frac{k}{\sqrt{6400}}$$

$$2 = \frac{k}{80}$$

$$(2)(80) = k$$

$$160 = k$$

Thus, the formula is $v = \dfrac{160}{\sqrt{d}}$.

EXAMPLE 6 ***Using the Constant of Proportionality***

Suppose y varies inversely as x. If $y = 8$ when $x = 15$, find y when $x = 18$.

SOLUTION First write the inverse variation, then solve for k.

$$y = \frac{k}{x}$$

$$8 = \frac{k}{15}$$

$$120 = k$$

Now substitute 120 for k in $y = \frac{k}{x}$ and find y when $x = 18$.

$$y = \frac{120}{x} = \frac{120}{18} = 6.7 \quad \text{(to the nearest tenth)}$$

Joint Variation

One quantity may vary directly as a product of two or more other quantities. This type of variation is called *joint variation*.

JOINT VARIATION

The general form of a joint variation, where y varies directly as x and z, is

$$y = kxz$$

where k is the constant of proportionality.

EXAMPLE 7 ***Joint Variation in Geometry***

The area, A, of a triangle varies jointly as its base, b, and height, h. If the area of a triangle is 48 in.^{2} when its base is 12 in. and its height is 8 in., find the area of a triangle whose base is 15 in. and whose height is 20 in.

SOLUTION First write the joint variation, then substitute the known values and solve for k.

$$A = kbh$$

$$48 = k(12)(8)$$

$$48 = k(96)$$

$$\frac{48}{96} = k$$

$$\tfrac{1}{2} = k$$

Now solve for the area of the given triangle.

$$A = kbh$$

$$= \tfrac{1}{2}(15)(20)$$

$$= 150 \text{ in.}^2$$

SUMMARY OF VARIATIONS

Direct	Inverse	Joint
$y = kx$	$y = \frac{k}{x}$	$y = kxz$

Combined Variation

Often in real-life situations, one variable varies as a combination of variables. The following examples illustrate the use of *combined variations*.

EXAMPLE 8 *Combined Variation in Engineering*

The load, L, that a horizontal beam can safely support varies jointly as the width, w, and the square of the depth, d, and inversely as the length, l. Express L in terms of w, d, l, and the constant of proportionality, k.

SOLUTION

$$L = \frac{kwd^2}{l}$$

EXAMPLE 9 *Hot Dog Price, Combined Variation*

The owners of Henrietta Hots find their weekly sales of hot dogs, S, vary directly with their advertising budget, A, and inversely with their hot dog price, P. When their advertising budget is \$600 and the price of a hot dog is \$1.50, they sell 5600 hot dogs a week.

a) Write a variation expressing S in terms of A and P. Include the value of the constant of proportionality.

b) Find the expected sales if the advertising budget is \$800 and the hot dog price is \$1.75.

SOLUTION

a) Since S varies directly as A and inversely as P, we begin with the equation

$$S = \frac{kA}{P}$$

We now find k using the known values.

$$5600 = \frac{k(600)}{1.50}$$

$$5600 = 400k$$

$$14 = k$$

Therefore, the equation for the weekly sales of hot dogs is $S = \frac{14A}{P}$.

b)
$$S = \frac{14A}{P} = \frac{14(800)}{1.75} = 6400$$

Henrietta Hots can expect to sell 6400 hot dogs a week if the advertising budget is \$800 and the hot dog price is \$1.75.

EXAMPLE 10 ***Combined Variation***

A varies jointly as B and C and inversely as the square of D. If $A = 1$ when $B = 9$, $C = 4$, and $D = 6$, find A when $B = 8$, $C = 12$, and $D = 5$.

SOLUTION We begin with the equation

$$A = \frac{kBC}{D^2}$$

We must first find the constant of proportionality, k, by substituting the known values for A, B, C, and D and solving for k.

$$1 = \frac{k(9)(4)}{6^2}$$
$$1 = \frac{36k}{36}$$
$$1 = k$$

Thus, the constant of proportionality equals 1. Now we find A for the corresponding values of B, C, and D.

$$A = \frac{kBC}{D^2}$$
$$A = \frac{(1)(8)(12)}{5^2} = \frac{96}{25} = 3.84$$

SECTION 6.5 EXERCISES

CONCEPT/WRITING EXERCISES

In Exercises 1–4, use complete sentences to answer the question.

1. Describe direct variation.
2. Describe inverse variation.
3. Describe joint variation.
4. Describe combined variation.

In Exercises 5–20, use your intuition to determine whether the variation between the indicated quantities is direct or inverse.

5. The distance between two cities on a map and the actual distance between the two cities

6. The time required to fill a pool with a hose and the volume of water coming from the hose

7. The time required to boil water on a burner and the temperature of the burner

8. A person's salary and the amount of money withheld from his or her salary for federal income taxes

9. The interest earned on an investment and the interest rate

10. The volume of a balloon and its radius

11. A person's speed and the time needed for the person to complete the race

12. The time required to cool a room and the temperature of the room

13. The number of workers hired to install a fence and the time required to install the fence

14. The number of calories in a slice of pizza and the size of the slice

15. The time required to defrost a frozen hamburger in a room and the temperature of the room

16. On Earth, the weight and mass of an object

17. The number of people in line at a bank and the amount of time required for the last person to reach the teller

18. The number of books that can be placed upright on a shelf 3 ft long and the width of the books

19. The amount of fertilizer needed to fertilize a lawn and the area of the lawn

20. The percent of light that filters through water and the depth of the water

In Exercises 21 and 22, use Exercises 5–20 as a guide.

21. Name two items that have not been mentioned in this section that have a direct variation.

22. Name two items that have not been mentioned in this section that have an inverse variation.

PRACTICE THE SKILLS

In Exercises 23–40, (a) write the variation and (b) determine the quantity indicated.

23. y varies directly as x. Determine y when $x = 15$ and $k = 8$.

24. x varies inversely as y. Determine x when $y = 7$ and $k = 14$.

25. m varies inversely as the square of n. Determine m when $n = 8$ and $k = 16$.

26. r varies directly as the square of s. Determine r when $s = 2$ and $k = 13$.

27. A varies directly as B and inversely as C. Determine A when $B = 5$, $C = 10$, and $k = 5$.

28. M varies directly as J and inversely as C. Determine M when $J = 10$, $C = 20$, and $k = 8$.

29. F varies jointly as D and E. Determine F when $D = 3$, $E = 10$, and $k = 7$.

30. A varies jointly as R_1 and R_2 and inversely as the square of L. Determine A when $R_1 = 120$, $R_2 = 8$, $L = 5$, and $k = \frac{3}{2}$.

31. t varies directly as the square of d and inversely as f. If $t = 192$ when $d = 8$ and $f = 4$, determine t when $d = 10$ and $f = 6$.

32. y varies directly as the square root of t and inversely as s. If $y = 12$ when $t = 36$ and $s = 2$, determine y when $t = 81$ and $s = 4$.

33. Z varies jointly as W and Y. If $Z = 12$ when $W = 9$ and $Y = 4$, determine Z when $W = 50$ and $Y = 6$.

34. y varies directly as the square of R. If $y = 4$ when $R = 4$, determine y when $R = 8$.

35. H varies directly as L. If $H = 15$ when $L = 50$, determine H when $L = 10$.

36. C varies inversely as J. If $C = 7$ when $J = 0.7$, determine C when $J = 12$.

37. A varies directly as the square of B. If $A = 245$ when $B = 7$, determine A when $B = 12$.

38. F varies jointly as M_1 and M_2 and inversely as the square of d. If $F = 20$ when $M_1 = 5$, $M_2 = 10$, and $d = 0.2$, determine F when $M_1 = 10$, $M_2 = 20$, and $d = 0.4$.

39. F varies jointly as q_1 and q_2 and inversely as the square of d. If $F = 80$ when $q_1 = 4$, $q_2 = 16$, and $d = 0.4$, determine F when $q_1 = 12$, $q_2 = 20$, and $d = 0.2$.

40. S varies jointly as I and the square of T. If $S = 4$ when $I = 10$ and $T = 4$, determine S when $I = 4$ and $T = 6$.

PROBLEM SOLVING

In Exercises 41–49, (a) write the variation and (b) determine the quantity indicated.

41. ***Property Tax*** The property tax, t, on a home is directly proportional to the assessed value, v, of the home. If the property tax on a home with an assessed value of \$140,000 is \$2100, what is the property tax on a home with an assessed value of \$180,000?

42. ***Resistance*** The resistance, R, of a wire varies directly as its length, L. If the resistance of a 30 ft length of wire is 0.24 ohm, determine the resistance of a 40 ft length of wire.

43. ***Speaker Loudness*** The loudness of a stereo speaker, l, measured in decibels (dB), is inversely proportional to the square of the distance, d, of the listener from the speaker. If the loudness is 20 dB when the listener is 6 ft from the speaker, what is the loudness when the listener is 3 ft from the speaker?

44. ***Melting an Ice Cube*** The time, t, for an ice cube to melt is inversely proportional to the temperature, T, of the water in which the ice cube is placed. If it takes an ice cube 2 minutes to melt in 75°F water, how long will it take an ice cube of the same size to melt in 80°F water?

▲ See Exercise 44

45. ***Video Rentals*** The number of weekly videotape rentals, R, at Busterblock Video varies directly with the advertising budget, A, and inversely with the daily rental price, P. When the video store's advertising budget is \$600 and the rental price is \$3 per day, it rents 4800 tapes per week. How many tapes would it rent per week if the store increased its advertising budget to \$700 and raised its rental price to \$3.50?

46. ***Stopping Distance of a Car*** The stopping distance, d, of a car after the brakes are applied varies directly as the square of the speed, s, of the car. If a car traveling at a speed of 40 mph can stop in 80 ft, what is the stopping distance of a car traveling at 65 mph?

47. ***Guitar Strings*** The number of vibrations per second, v, of a guitar string varies directly as the square root of the tension, t, and inversely as the length of the string, l. If the number of vibrations per second is 5 when the tension is 225 kg and the length of the string is 0.60 m, determine the number of vibrations per second when the tension is 196 kg and the length of the string is 0.70 m.

48. ***Electrical Resistance*** The electrical resistance of a wire, R, varies directly as its length, L, and inversely as its cross-sectional area, A. If the resistance of a wire is 0.2 ohm when the length is 200 ft and its cross-sectional area is 0.05 in.^{2}, what is the resistance of a wire whose length is 5000 ft with a cross-sectional area of 0.01 in.^{2}?

49. ***Phone Calls*** The number of phone calls between two cities during a given time period, N, varies directly as the populations p_1 and p_2 of the two cities and inversely to the distance, d, between them. If 100,000 calls are made between two cities 300 mi apart and the populations of the cities are 60,000 and 200,000, how many calls are made between two cities with populations of 125,000 and 175,000 that are 450 mi apart?

50. **a)** If y varies directly as x and the constant of proportionality is 2, does x vary directly or inversely as y? Explain.

b) Give the new constant of proportionality for x as a variation of y.

51. **a)** If y varies inversely as x and the constant of proportionality is 0.3, does x vary directly or inversely as y? Explain.

b) Give the new constant of proportionality for x as a variation of y.

CHALLENGE PROBLEMS/GROUP ACTIVITIES

52. ***Photography*** An article in the magazine *Outdoor and Travel Photography* states, "If a surface is illuminated by a point-source of light, the intensity of illumination produced is inversely proportional to the square of the distance separating them. In practical terms, this means that foreground objects will be grossly overexposed if your background subject is properly exposed with a flash. Thus direct flash will not offer pleasing results if there are any intervening objects between the foreground and the subject."

If the subject you are photographing is 4 ft from the flash and the illumination on this subject is $\frac{1}{16}$ of the light of the flash, what is the intensity of illumination on an intervening object that is 3 ft from the flash?

53. ***Water Cost*** In a specific region of the country, the amount of a customer's water bill, W, is directly proportional to the average daily temperature for the month, T, the lawn area, A, and the square root of F, where F is the family size, and inversely proportional to the number of inches of rain, R.

In one month, the average daily temperature is 78°F and the number of inches of rain is 5.6. If the average family of four who has a thousand square feet of lawn pays \$72.00 for water for that month, estimate the water bill in the same month for the average family of six who has 1500 ft^2 of lawn.

6.6 LINEAR INEQUALITIES

▲ An inequality can be used to determine how many textbooks must be sold so that a bookstore's revenue is greater than its cost.

Suppose your campus bookstore needs to determine how many textbooks for a particular course must be sold so that the bookstore's textbook revenue is greater than its cost. To determine this number, an inequality would be set up. In this section, we will discuss how to set up and how to solve an inequality.

The symbols of inequality are as follows.

SYMBOLS OF INEQUALITY

$a < b$ means that a is less than b.
$a \leq b$ means that a is less than or equal to b.
$a > b$ means that a is greater than b.
$a \geq b$ means that a is greater than or equal to b.

An ***inequality*** consists of two (or more) expressions joined by an inequality sign.

Examples of inequalities

$$3 < 5, \qquad x < 2, \qquad 3x - 2 \geq 5$$

A statement of inequality can be used to indicate a set of real numbers. For example, $x < 2$ represents the set of all real numbers less than 2. Listing all these numbers is impossible, but some are -2, -1.234, -1, $-\frac{1}{2}$, 0, $\frac{97}{163}$, 1.

EXAMPLE 6 *A Solution of Only Integers*

Solve the inequality $x + 5 < 9$, where x is an integer, and graph the solution set on the number line.

SOLUTION To find the solution set, isolate x on one side of the inequality symbol.

$$x + 5 < 9$$

$$x + 5 - 5 < 9 - 5 \quad \text{Subtract 5 from both sides of the inequality.}$$

$$x < 4$$

Since x is an integer and is less than 4, the solution set is the set of integers less than 4, or $\{\ldots, -2, -1, 0, 1, 2, 3\}$. To graph the solution set, we make solid dots at the corresponding points on the number line. The ellipsis (the three dots) to the left of -3 indicate that all the integers to the left of -3 are included.

TIMELY TIP Some linear inequalities have no solution. When solving an inequality, if you obtain a statement that is always false, such as $3 > 5$, the answer is *no solution*. There is no real number that makes the statement true.

Some linear inequalities have all real numbers as their solution. When solving an inequality, if you obtain a statement that is always true, such as $2 > 1$, the solution is *all real numbers*.

Solving Compound Inequalities

An inequality of the form $a < x < b$ is called a *compound inequality*. Consider the compound inequality $-3 < x \leq 2$, which means that $-3 < x$ *and* $x \leq 2$.

EXAMPLE 7 *A Compound Inequality*

Graph the solution set of the inequality $-3 < x \leq 2$

a) where x is an integer. b) where x is a real number.

SOLUTION

a) The solution set is all the integers between -3 and 2, including the 2 but not including the -3, or $\{-2, -1, 0, 1, 2\}$.

b) The solution set consists of all the real numbers between -3 and 2, including the 2 but not including the -3.

EXAMPLE 8 *Solving a Compound Inequality*

Solve the compound inequality for x and graph the solution set.

$$-4 < \frac{x + 3}{2} \leq 5$$

"In mathematics the art of posing problems is easier than that of solving them."

Georg Cantor

SOLUTION To solve this compound inequality, we must isolate the x in the middle term. To do so, we use the same principles used to solve inequalities.

$$-4 < \frac{x+3}{2} \leq 5$$

$$2(-4) < 2\left(\frac{x+3}{2}\right) \leq 2(5) \qquad \text{Multiply each part of the inequality by 2.}$$

$$-8 < x + 3 \leq 10$$

$$-8 - 3 < x + 3 - 3 \leq 10 - 3 \qquad \text{Subtract 3 from each part of the inequality.}$$

$$-11 < x \leq 7$$

The solution set is graphed on the number line as follows.

$-11 < x \leq 7$

EXAMPLE 9 *Average Grade*

A student must have an average (the mean) on five tests that is greater than or equal to 80% but less than 90% to receive a final grade of B. Devon's grades on the first four tests were 98%, 76%, 86%, and 92%. What range of grades on the fifth test would give him a B in the course?

SOLUTION The unknown quantity is the range of grades on the fifth test. First construct an inequality that can be used to find the range of grades on the fifth test. The average (mean) is found by adding the grades and dividing the sum by the number of tests.

Let x = the fifth grade. Then

$$\text{Average} = \frac{98 + 76 + 86 + 92 + x}{5}$$

For Devon to obtain a B in this course, his average must be greater than or equal to 80 but less than 90.

$$80 \leq \frac{98 + 76 + 86 + 92 + x}{5} < 90$$

$$80 \leq \frac{352 + x}{5} < 90$$

$$5(80) \leq 5\left(\frac{352 + x}{5}\right) < 5(90) \qquad \text{Multiply the three terms of the inequality by 5.}$$

$$400 \leq 352 + x < 450$$

$$400 - 352 \leq 352 - 352 + x < 450 - 352 \qquad \text{Subtract 352 from all three terms.}$$

$$48 \leq x < 98$$

Thus, a grade of 48% up to but not including a grade of 98% on the fifth test will result in a grade of B in this course.

TIMELY TIP Remember to change the direction of the inequality symbol when multiplying or dividing both sides of an inequality by a negative number.

SECTION 6.6 EXERCISES

CONCEPT/WRITING EXERCISES

1. Give the four inequality symbols we use in this section and indicate how each is read.

2. a) What is an inequality?

   b) Give an example of three inequalities.

3. When solving an inequality, under what conditions do you need to change the direction of the inequality symbol?

4. Does $x < 2$ have the same meaning as $2 > x$? Explain.

5. Does $x > -3$ have the same meaning as $-3 < x$? Explain.

6. When graphing the solution set to an inequality on the number line, when should you use an open circle and when should you use a closed circle?

7. a) What is a compound inequality?

   b) Give an example of a compound inequality.

8. a) Give an example of an inequality whose solution on a number line would contain a closed circle.

   b) Give an example of an inequality whose solution on a number line would contain an open circle.

PRACTICE THE SKILLS

In Exercises 9–26, graph the solution set of the inequality, where x is a real number, on the number line.

9. $x \geq 4$

10. $x < 7$

11. $x - 8 \geq 3$

12. $x + 5 > -2$

13. $-3x \leq 18$

14. $-4x < 12$

15. $\frac{x}{5} < 4$

16. $\frac{x}{6} > 2$

17. $\frac{-x}{3} \geq 3$

18. $\frac{x}{2} \geq -4$

19. $2x + 6 \geq 14$

20. $3x + 12 < 5x + 14$

21. $4(2x - 1) < 2(4x - 3)$

22. $-2(x - 3) \leq 4x + 6$

23. $-1 \leq x \leq 3$

24. $-4 < x \leq 1$

25. $3 < x - 7 \leq 6$

26. $\frac{1}{2} < \frac{x + 4}{2} \leq 4$

In Exercises 27–46, graph the solution set of the inequality, where x is an integer, on the number line.

27. $x > 1$

28. $-2 < x$

29. $-3x \leq 27$

30. $3x \geq 27$

31. $x - 2 < 4$

32. $-5x \leq 15$

33. $\frac{x}{3} \leq -2$

34. $\frac{x}{4} \geq -3$

35. $-\frac{x}{4} \geq 2$

36. $\frac{3x}{5} \leq 3$

37. $-15 < -4x - 3$

38. $-5x - 1 < 17 - 2x$

39. $3(x + 4) \geq 4x + 13$

40. $-2(x - 1) < 3(x - 4) + 5$

41. $5(x + 4) - 6 \leq 2x + 8$

42. $-3 \leq x < 5$

43. $1 > -x > -5$

44. $-2 < 2x + 3 < 6$

45. $0.3 \leq \frac{x + 2}{10} \leq 0.5$

46. $-\frac{1}{3} < \frac{x - 2}{12} \leq \frac{1}{4}$

PROBLEM SOLVING

47. *Federal Drug Costs* The following bar graph shows the government's projected cost, in billions of dollars, for the Medicare prescription drug benefit for the years 2005–2015.

Projected Cost of Medicare Prescription Drug Benefit

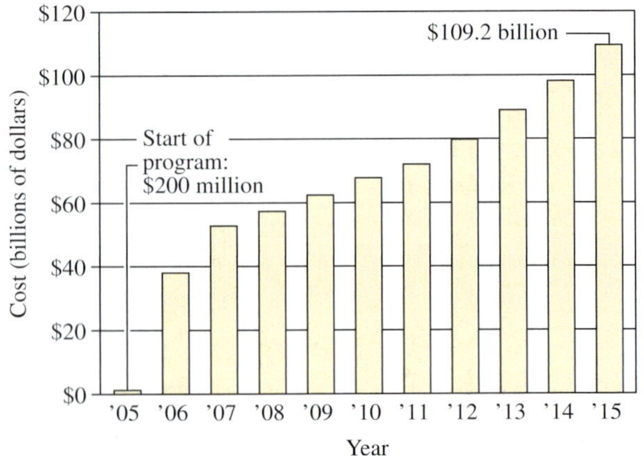

Source: CMS

In which years is the projected cost

a) $> \$60$ billion? **b)** $\leq \$60$ billion?

c) $> \$20$ billion? **d)** $\leq \$40$ billion?

48. ***Work Stoppages*** The following bar graph shows the number of work stoppages involving at least 1000 workers, including strikes and lockouts, in the United States for the years 1999–2004.

Source: Bureau of Labor Statistics, U.S. Department of Labor

In which years was the number of work stoppages in the United States

a) <25? **b)** >30? **c)** <40? **d)** >20?

49. ***Video Rental*** Movie Mania offers two rental plans. One has an annual fee, and the other has no annual fee. The annual membership fee and the daily charge per video for each plan are shown in the table. Determine the maximum number of videos that can be rented for the no fee plan to cost less than the annual fee plan.

Rental Plan	Yearly Fee	Daily Charge per Video
Annual fee	\$30	\$1.49
No fee	None	\$2.99

50. ***Salary Plans*** Bobby Exler recently accepted a sales position in Portland, Oregon. He can select between the two salary plans shown in the table. Determine the dollar amount of weekly sales that would result in Bobby earning more with Plan B than with Plan A.

Salary Plan	Weekly Salary	Commission on Sales
Plan A	\$500	6%
Plan B	\$400	8%

51. ***College Employment*** To be eligible for financial assistance for college, Samantha Lavin can earn no more than \$3000 during her summer employment. She will earn \$725 cutting lawns. She is also earning \$6.50 per hour at an evening job as a cashier at a grocery store. What is the maximum number of hours Samantha can work at the grocery store without jeopardizing her financial assistance?

52. ***Moving Boxes*** A janitor must move a large shipment of books from the first floor to the fifth floor. Each box of books weighs 60 lb, and the janitor weighs 180 lb. The sign on the elevator reads, "Maximum weight 1200 lb."

a) Write a statement of inequality to determine the maximum number of boxes of books the janitor can place on the elevator at one time. (The janitor must ride in the elevator with the books.)

b) Determine the maximum number of boxes that can be moved in one trip.

53. ***Parking Costs*** A parking garage in Richmond, Virginia, charges \$1.75 for the first hour and \$0.50 for each additional hour or part of an hour. What is the maximum length of time Tom Shalhoub can park in the garage if he wishes to pay no more than \$4.25?

54. ***Making a Profit*** For a business to realize a profit, its revenue, R, must be greater than its costs, C; that is, a profit will only result if $R > C$ (the company breaks even when $R = C$). A book publishing company has a weekly cost equation of $C = 2x + 2000$ and a weekly revenue equation of $R = 12x$, where x is the number of books produced and sold in a week. How many books must be sold weekly for the company to make a profit?

55. ***Finding Velocity*** The velocity, v, in feet per second, t sec after a tennis ball is projected directly upward is given by the formula $v = 84 - 32t$. How many seconds after being projected upward will the velocity be between 36 ft/sec and 68 ft/sec?

56. ***Speed Limit*** The minimum speed for vehicles on a highway is 40 mph, and the maximum speed is 55 mph. If Philip Rowe has been driving nonstop along the highway for 4 hr, what range in miles could he have legally traveled?

57. ***A Grade of B*** In Example 9 on page 359, what range of grades on the fifth test would result in Devon receiving a grade of B if his grades on the first four tests were 78%, 64%, 88%, and 76%?

58. ***Tent Rental*** The Cuhayoga Community College Planning Committee wants to rent tents for the spring job fair. Rent-a-Tent charges \$325 for setup and delivery of its tents. This fee is charged regardless of the number of tents delivered and set up. In addition, Rent-a-Tent charges \$125 for each tent rented. If the minimum amount the planning committee wishes to spend is \$950 and the maximum amount they wish to spend is \$1200, determine the minimum and maximum number of tents the committee can rent.

CHALLENGE PROBLEMS/GROUP ACTIVITIES

59. ***Painting a House*** Donovan Davis is painting the exterior of his house. The instructions on the paint can indicate that 1 gal covers from 250 to 400 ft^2. The total surface of the house to be painted is 2750 ft^2. Determine the number of gallons of paint he could use and express the answer as an inequality.

60. ***Final Exam*** Teresa's five test grades for the semester are 86%, 78%, 68%, 92%, and 72%. Her final exam counts one-third of her final grade. What range of grades on her final exam would result in Teresa receiving a final grade of B in the course? (See Example 9.)

61. A student multiplied both sides of the inequality $-\frac{1}{3}x \le 4$ by -3 and forgot to reverse the direction of the inequality symbol. What is the relation between the student's incorrect solution set and the correct solution set? Is there any number in both the correct solution set and the student's incorrect solution set? If so, what is it?

INTERNET/RESEARCH ACTIVITY

62. Find a newspaper or a magazine article that contains the mathematical concept of inequality.

a) From the information in the article write a statement of inequality.

b) Summarize the article and explain how you arrived at the inequality statement in part (a).

6.7 GRAPHING LINEAR EQUATIONS

▲ The cost of mailing a package may depend on the weight of the package.

The profit, *p*, of a company may depend on the amount of sales, *s*. The cost, *c*, of mailing a package may depend on the weight, *w*, of the package. Real-life problems, such as these two examples, often involve two or more variables. In this section, we will discuss how to graph equations with two variables.

To be able to work with equations with two variables requires understanding the ***Cartesian*** (or ***rectangular***) ***coordinate system***, named after the French mathematician René Descartes. The rectangular coordinate system consists of two perpendicular number lines (Fig. 6.6 on page 363). The horizontal line is the ***x-axis***, and the vertical line is the ***y-axis***. The point of intersection of the *x*-axis and *y*-axis is called the ***origin***. The numbers on the axes to the right and above the origin are positive. The numbers on the axes on the left and below the origin are negative. The axes divide the plane into four parts: the first, second, third, and fourth ***quadrants***.

We indicate the location of a point in the rectangular coordinate system by means of an ***ordered pair*** of the form (x, y). The x-coordinate is always placed first, and the y-coordinate is always placed second in the ordered pair. Consider the point illustrated in Fig. 6.7. Since the x-coordinate of the point is 5 and the y-coordinate is 3, the ordered pair that represents this point is (5, 3).

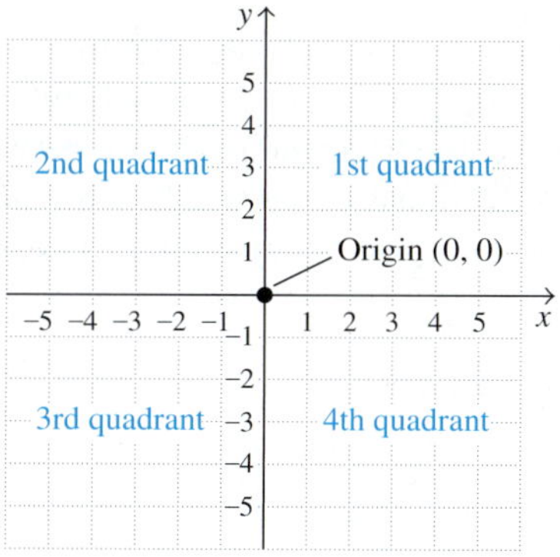

Figure 6.6

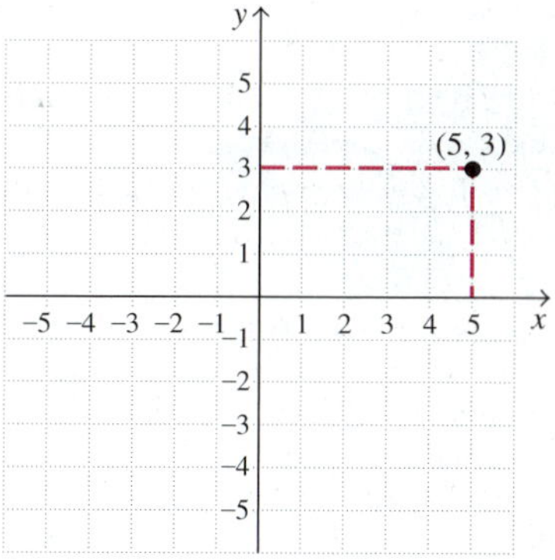

Figure 6.7

The origin is represented by the ordered pair (0, 0). Every point on the plane can be represented by one and only one ordered pair (x, y), and every ordered pair (x, y) represents one and only one point on the plane.

EXAMPLE 1 *Plotting Points*

Plot the points $A(-2, 4)$, $B(-3, -1)$, $C(4, 0)$, $D(2, 1)$, and $E(0, -3)$.

SOLUTION Point A has an x-coordinate of -2 and a y-coordinate of 4. Project a vertical line up from -2 on the x-axis and a horizontal line to the left from 4 on the y-axis. The two lines intersect at the point denoted A (Fig. 6.8). The other points are plotted in a similar manner.

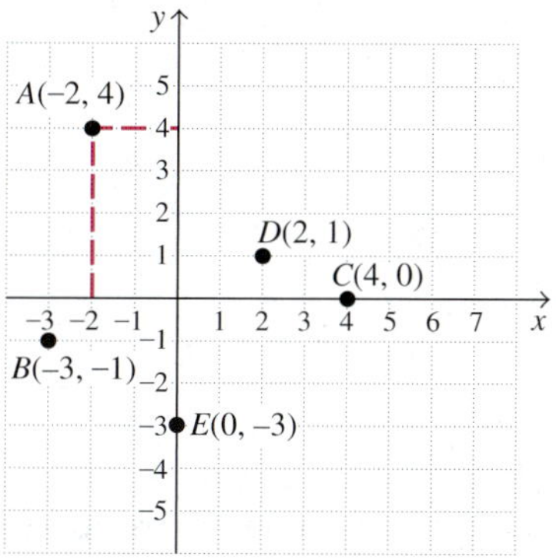

Figure 6.8

MATHEMATICS TODAY

Hollywood's Four Quadrants

Mike Myers, the voice of *Shrek*

Hollywood studio executives use four "quadrants" to divide up the moviegoing audience. Men 25 years and older, men younger than 25, women 25 years and older, and women younger than 25 are the age groups represented by Hollywood's four quadrants. If a studio produces a movie that appeals to all four quadrants, it is sure to have a hit movie. If the movie appeals to none of the four quadrants, the movie is sure to fail. A challenge for studio executives is to determine the core quadrant and then try to make sure no other movie geared toward the same quadrant debuts at the same time. One of the biggest hit movies of the summer of 2004 was *Shrek 2*, a movie that appealed to all four quadrants.

PROFILE IN MATHEMATICS

René Descartes

The Mathematician and the Fly

According to legend, French mathematician and philosopher René Descartes (1596–1650) did some of his best thinking in bed. He was a sickly child, and so the Jesuits who undertook his education allowed him to stay in bed each morning as long as he liked. This practice he carried into adulthood, seldom getting up before noon. One morning as he watched a fly crawl about the ceiling, near the corner of his room, he was struck with the idea that the fly's position could best be described by the connecting distances from it to the two adjacent walls. These became the coordinates of his rectangular coordinate system and were appropriately named after him (Cartesian coordinates) and not the fly.

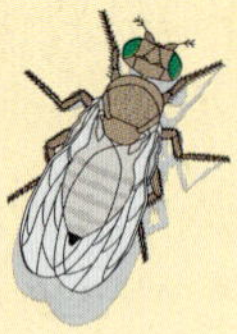

EXAMPLE 2 *A Parallelogram*

The points, A, B, and C are three vertices of a parallelogram with two sides parallel to the x-axis. Plot the three points below and determine the coordinates of the fourth vertex, D.

$$A(1, 2) \qquad B(2, 4) \qquad C(7, 4)$$

SOLUTION A parallelogram is a figure that has opposite sides that are of equal length and are parallel. (Parallel lines are two lines in the same plane that do not intersect.) The horizontal distance between points B and C is 5 units (see Fig. 6.9). Therefore, the horizontal distance between points A and D must also be 5 units. This problem has two possible solutions, as illustrated in Fig. 6.9. In each figure, we have indicated the given points in red.

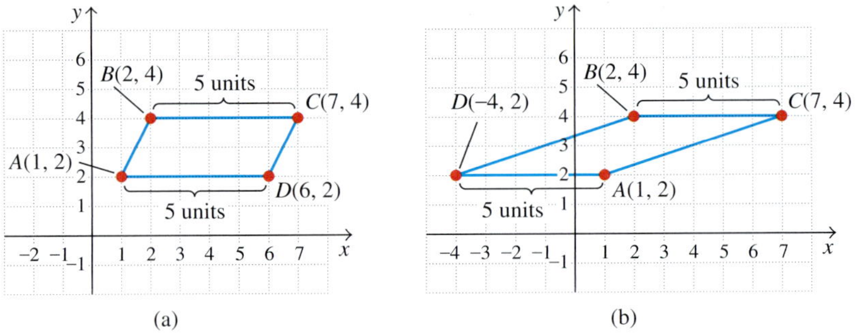

Figure 6.9

The solutions are the points $(6, 2)$ and $(-4, 2)$.

Graphing Linear Equations by Plotting Points

Consider the following equation in two variables: $y = x + 1$. Every ordered pair that makes the equation a true statement is a solution to, or satisfies, the equation. We can mentally find some ordered pairs that satisfy the equation $y = x + 1$ by picking some values of x and solving the equation for y. For example, suppose we let $x = 1$; then $y = 1 + 1 = 2$. The ordered pair $(1, 2)$ is a solution to the equation $y = x + 1$. We can make a chart of other ordered pairs that are solutions to the equation.

x	y	**Ordered Pair**
1	2	(1, 2)
2	3	(2, 3)
3	4	(3, 4)
4.5	5.5	(4.5, 5.5)
−3	−2	(−3, −2)

How many other ordered pairs satisfy the equation? Infinitely many ordered pairs satisfy the equation. Since we cannot list all the solutions, we show them by means of a graph. A ***graph*** is an illustration of all the points whose coordinates satisfy an equation.

DID YOU KNOW?

Graphing Calculators

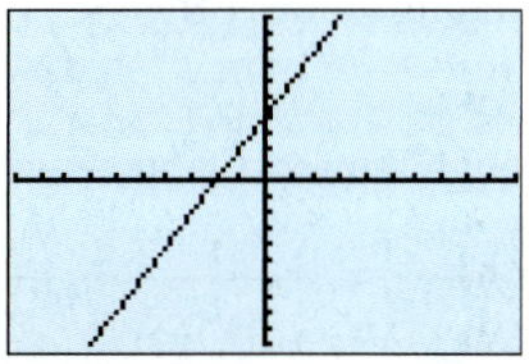

Figure 6.12

On page 334, we illustrated the graph of an exponential equation we obtained from the screen (or window) of a TI-84 Plus graphing calculator. Figure 6.12 shows the graph of the equation $y = 2x + 4$ as illustrated on the screen of that calculator.

Graphing calculators are being used more and more in many mathematics courses. Other makers of graphing calculators include Casio, Sharp, and Hewlett-Packard. Graphing calculators can do a great many more things than a scientific calculator can. If you plan on taking additional mathematics courses, you may find that a graphing calculator is required for those courses.

The points (1, 2), (2, 3), (3, 4), (4.5, 5.5), and $(-3, -2)$ are plotted in Fig. 6.10. With a straightedge we can draw one line that contains all these points. This line, when extended indefinitely in either direction, passes through all the points in the plane that satisfy the equation $y = x + 1$. The arrows on the ends of the line indicate that the line extends indefinitely.

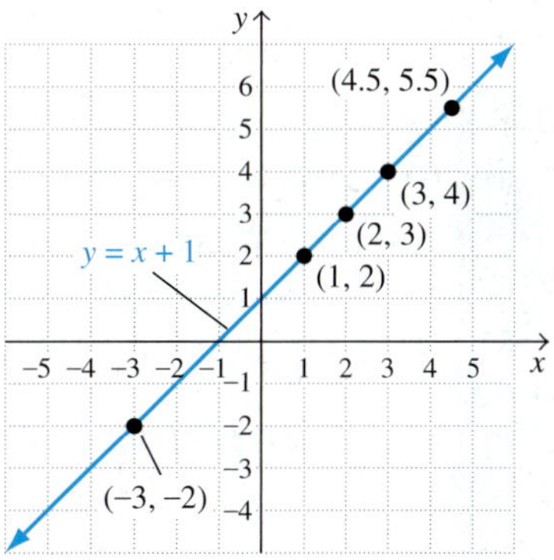

Figure 6.10

All equations of the form $ax + by = c$, $a \neq 0$ and $b \neq 0$, will be straight lines when graphed. Thus, such equations are called ***linear equations in two variables***. The exponents on the variables x and y must be 1 for the equation to be linear. Since only two points are needed to draw a line, only two points are needed to graph a linear equation. It is always a good idea to plot a third point as a checkpoint. If no error has been made, all three points will be in a line, or ***collinear***. One method that can be used to obtain points is to solve the equation for y, substitute values for x, and find the corresponding values of y.

EXAMPLE 3 *Graphing an Equation by Plotting Points*

Graph $y = 2x + 2$.

SOLUTION Since the equation is already solved for y, select values for x and find the corresponding values for y. The table indicates values arbitrarily selected for x and the corresponding values for y. The ordered pairs are (0, 2), (1, 4), and $(-1, 0)$. The graph is shown in Fig. 6.11.

x	y
0	2
1	4
−1	0

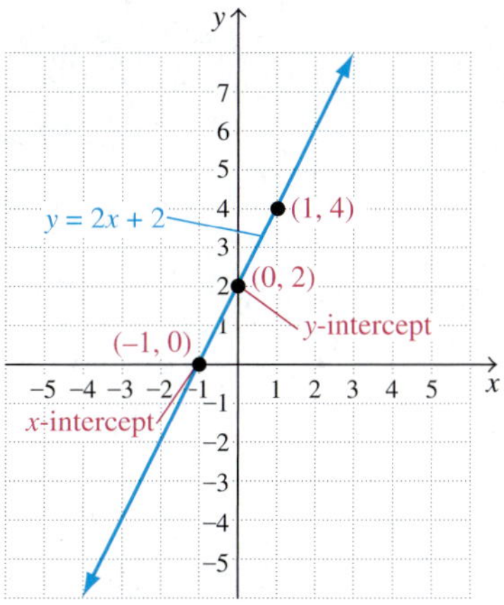

Figure 6.11

DID YOU KNOW?

Use of Grids

Grids have long been used in mapping. In archaeological digs, a rectangular coordinate system may be used to chart the location of each find.

TO GRAPH LINEAR EQUATIONS BY PLOTTING POINTS

1. Solve the equation for y.
2. Select at least three values for x and find their corresponding values of y.
3. Plot the points.
4. The points should be in a straight line. Draw a line through the set of points and place arrow tips at both ends of the line.

In step 4 of the procedure, if the points are not in a straight line, recheck your calculations and find your error.

Graphing by Using Intercepts

Example 3 contained two special points on the graph, $(-1, 0)$ and $(0, 2)$. At these points, the line crosses the x-axis and the y-axis, respectively. The ordered pairs $(-1, 0)$ and $(0, 2)$ represent the ***x*-intercept** and the ***y*-intercept**, respectively. Another method that can be used to graph linear equations is to find the x- and y-intercepts of the graph.

FINDING THE X- AND Y-INTERCEPTS

To find the x-intercept, set $y = 0$ and solve the equation for x.
To find the y-intercept, set $x = 0$ and solve the equation for y.

A linear equation may be graphed by finding the x- and y-intercepts, plotting the intercepts, and drawing a straight line through the intercepts. When graphing by this method, you should always plot a checkpoint before drawing your graph. To obtain a checkpoint, select a nonzero value for x and find the corresponding value of y. The checkpoint should be collinear with the x- and y-intercepts.

EXAMPLE 4 *Graphing by Using Intercepts*

Graph $2x - 4y = 8$ by using the x- and y-intercepts.

SOLUTION To find the x-intercept, set $y = 0$ and solve for x.

$$\begin{aligned} 2x - 4y &= 8 \\ 2x - 4(0) &= 8 \\ 2x &= 8 \\ x &= 4 \end{aligned}$$

The x-intercept is $(4, 0)$. To find the y-intercept, set $x = 0$ and solve for y.

$$\begin{aligned} 2x - 4y &= 8 \\ 2(0) - 4y &= 8 \\ -4y &= 8 \\ y &= -2 \end{aligned}$$

The y-intercept is $(0, -2)$. As a checkpoint, try $x = 2$ and find the corresponding value for y.

$$\begin{aligned} 2x - 4y &= 8 \\ 2(2) - 4y &= 8 \\ 4 - 4y &= 8 \\ -4y &= 4 \\ y &= -1 \end{aligned}$$

The checkpoint is the ordered pair $(2, -1)$.

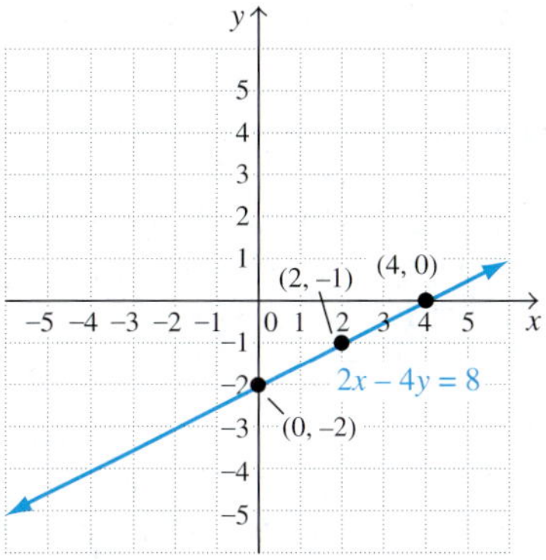

Figure 6.13

Since all three points in Fig. 6.13 are collinear, draw a line through the three points to obtain the graph.

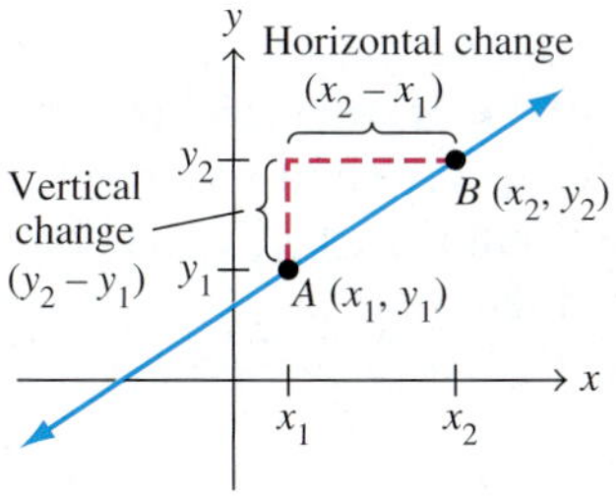

Figure 6.14

Slope

Another useful concept when you are working with straight lines is slope, which is a measure of the "steepness" of a line. The ***slope of a line*** is a ratio of the vertical change to the horizontal change for any two points on the line. Consider Fig. 6.14. Point A has coordinates (x_1, y_1), and point B has coordinates (x_2, y_2). The vertical change between points A and B is $y_2 - y_1$. The horizontal change between points A and B is $x_2 - x_1$. Thus, the slope, which is often symbolized with the letter m, can be found as follows.

$$\textbf{Slope} = \frac{\text{vertical change}}{\text{horizontal change}}$$

$$m = \frac{y_2 - y_1}{x_2 - x_1}$$

The Greek capital letter delta, Δ, is often used to represent the words "the change in." Therefore, slope may be defined as

$$m = \frac{\Delta y}{\Delta x}$$

DID YOU KNOW?

Up Up and Away

Thomas Morgenstern

Although we may not think much about it, the slope of a line is something we are altogether familiar with. You confront it every time you run up the stairs, late for class, moving 8 inches horizontally for every 6 inches up. The 2006 Olympic gold medalist Thomas Morgenstern of Austria is familiar with the concept of slope. He speeds down a steep 140 meters of a ski ramp at speeds of more than 60 mph before he takes flight.

A line may have a positive slope, a negative slope, or a zero slope, or the slope may be undefined, as indicated in Fig. 6.15. A line with a positive slope rises from left to right, as shown in Fig. 6.15(a). A line with a negative slope falls from left to right, as shown in Fig. 6.15(b). A horizontal line, which neither rises nor falls, has a slope of zero, as shown in Fig. 6.15(c). Since a vertical line does not have any horizontal change (the x value remains constant) and since we cannot divide by 0, the slope of a vertical line is undefined, as shown in Fig. 6.15(d).

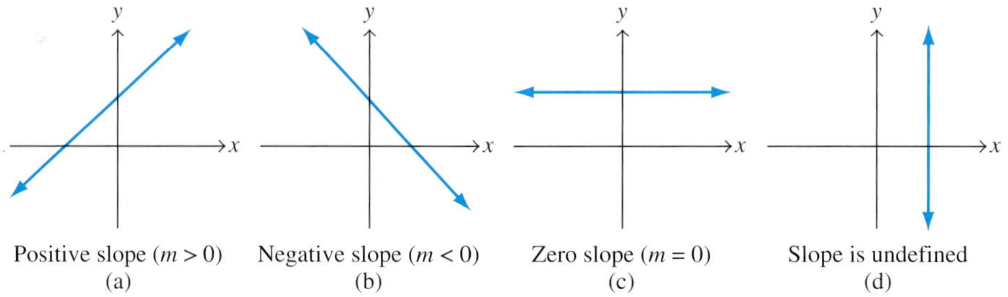

Figure 6.15

EXAMPLE 5 *Finding the Slope of a Line*

Determine the slope of the line that passes through the points $(-1, -3)$ and $(1, 5)$.

SOLUTION Let's begin by drawing a sketch, illustrating the points and the line. See Fig. 6.16(a).

We will let (x_1, y_1) be $(-1, -3)$ and (x_2, y_2) be $(1, 5)$. Then

$$\text{Slope} = \frac{y_2 - y_1}{x_2 - x_1} = \frac{5 - (-3)}{1 - (-1)} = \frac{5 + 3}{1 + 1} = \frac{8}{2} = \frac{4}{1} = 4$$

The slope of 4 means that there is a vertical change of 4 units for each horizontal change of 1 unit; see Fig. 6.16(b). The slope is positive, and the line rises from left to right. Note that we would have obtained the same result if we let (x_1, y_1) be $(1, 5)$ and (x_2, y_2) be $(-1, -3)$. Try this method now and see.

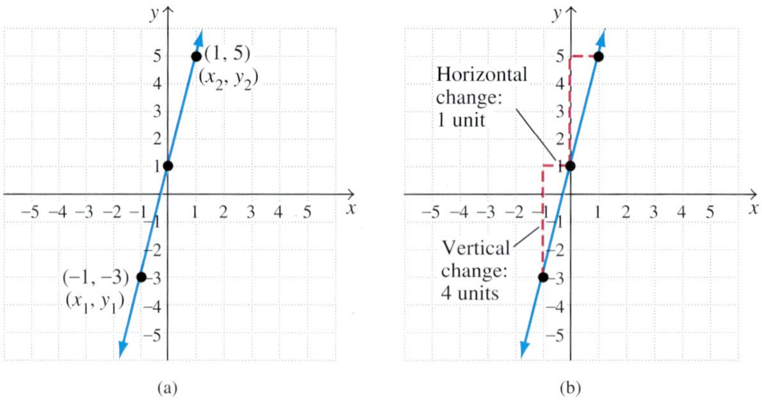

Figure 6.16

Graphing Linear Equations by Using the Slope and y-Intercept

A linear equation given in the form $y = mx + b$ is said to be in slope–intercept form.

SLOPE–INTERCEPT FORM OF THE EQUATION OF A LINE

$$y = mx + b$$

where m is the slope of the line and $(0, b)$ is the y-intercept of the line.

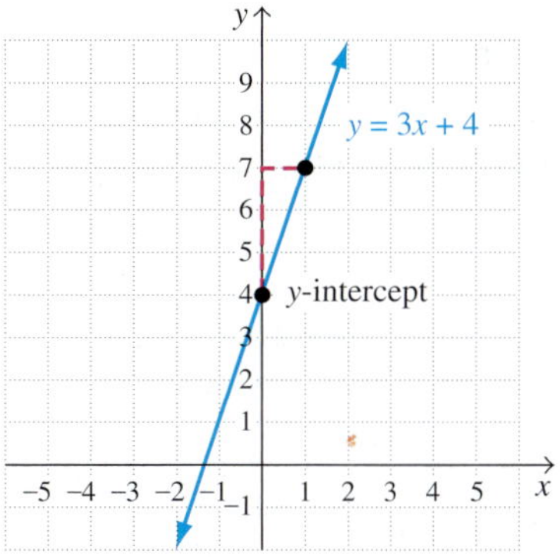

Figure 6.17

In the equation $y = mx + b$, b represents the value of y where the graph of the equation $y = mx + b$ crosses the y-axis.

Consider the graph of the equation $y = 3x + 4$, which appears in Fig. 6.17. By examining the graph, we can see that the y-intercept is (0, 4). We can also see that the graph has a positive slope because it rises from left to right. Because the vertical change is 3 units for every 1 unit of horizontal change, the slope must be $\frac{3}{1}$ or 3.

We could graph this equation by marking the y-intercept at (0, 4) and then moving *up* 3 units and to the *right* 1 unit to get another point. If the slope were -3, which means $\frac{-3}{1}$, we could start at the y-intercept and move *down* 3 units and to the *right* 1 unit. Thus, if we know the slope and y-intercept of a line, we can graph the line.

TO GRAPH LINEAR EQUATIONS BY USING THE SLOPE AND Y-INTERCEPT

1. Solve the equation for y to place the equation in slope–intercept form.
2. Determine the slope and y-intercept from the equation.
3. Plot the y-intercept.
4. Obtain a second point using the slope.
5. Draw a straight line through the points.

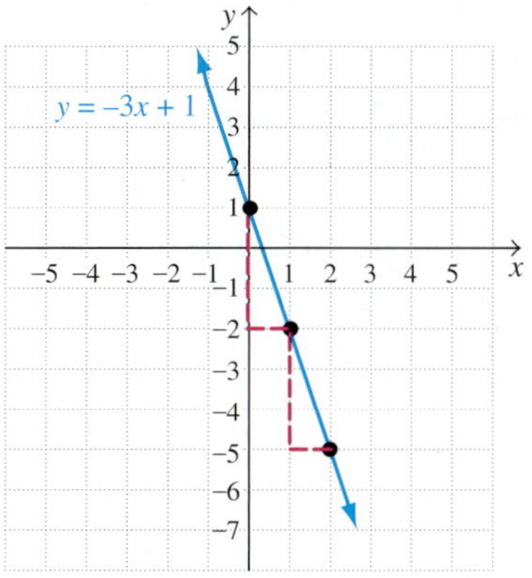

Figure 6.18

EXAMPLE 6 *Graphing an Equation by Using the Slope and y-Intercept*

Graph $y = -3x + 1$ by using the slope and y-intercept.

SOLUTION The slope is -3 or $\dfrac{-3}{1}$ and the y-intercept is (0, 1) (see Fig. 6.18). Plot (0, 1) on the y-axis. Then plot the next point by moving *down* 3 units and to the *right* 1 unit. A third point has been plotted in the same way. The graph of $y = -3x + 1$ is the line drawn through these three points. ●

EXAMPLE 7 *Writing an Equation in Slope–Intercept Form*

a) Write $4x - 3y = 9$ in slope–intercept form.

b) Graph the equation.

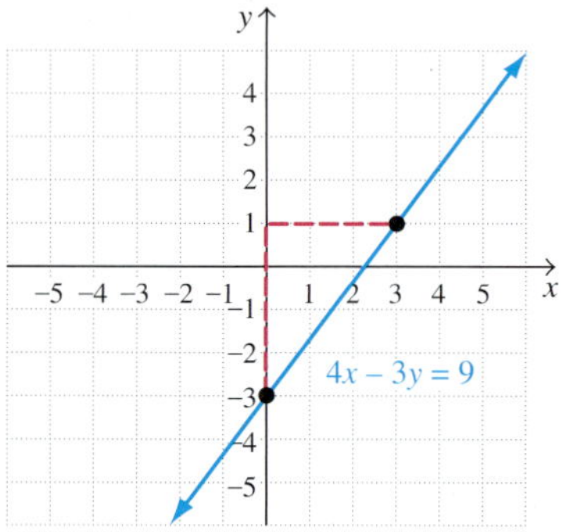

Figure 6.19

SOLUTION

a) To write $4x - 3y = 9$ in slope–intercept from, we solve the given equation for y.

$$4x - 3y = 9$$
$$4x - 4x - 3y = -4x + 9$$
$$-3y = -4x + 9$$
$$\frac{-3y}{-3} = \frac{-4x + 9}{-3}$$
$$y = \frac{-4x}{-3} + \frac{9}{-3} \quad \text{or} \quad y = \frac{4}{3}x - 3$$

Thus, in slope–intercept form, the equation is $y = \frac{4}{3}x - 3$.

b) The y-intercept is $(0, -3)$, and the slope is $\frac{4}{3}$. Plot a point at $(0, -3)$ on the y-axis, then move *up* 4 units and to the *right* 3 units to obtain the second point (see Fig. 6.19). Draw a line through the two points. ●

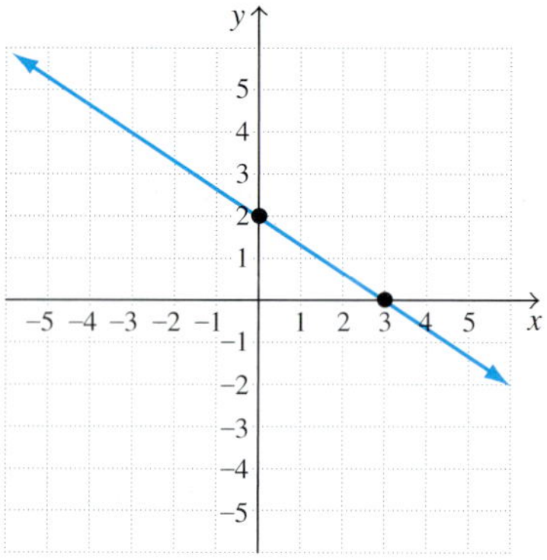

Figure 6.20

EXAMPLE 8 *Determining the Equation of a Line from Its Graph*

Determine the equation of the line in Fig. 6.20.

SOLUTION If we determine the slope and the y-intercept of the line, we can write the equation using slope–intercept form, $y = mx + b$. We see from the graph that the y-intercept is (0, 2); thus, $b = 2$. The slope of the line is negative because the graph falls from left to right. The change in y is 2 units for every 3-unit change in x. Thus, m, the slope of the line, is $-\frac{2}{3}$.

$$y = mx + b$$
$$y = -\frac{2}{3}x + 2$$

The equation of the line is $y = -\frac{2}{3}x + 2$. ●

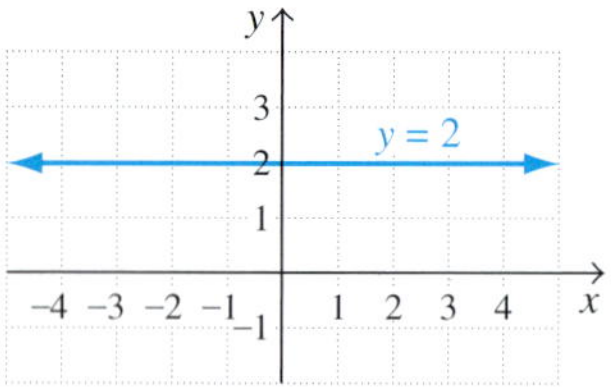

Figure 6.21

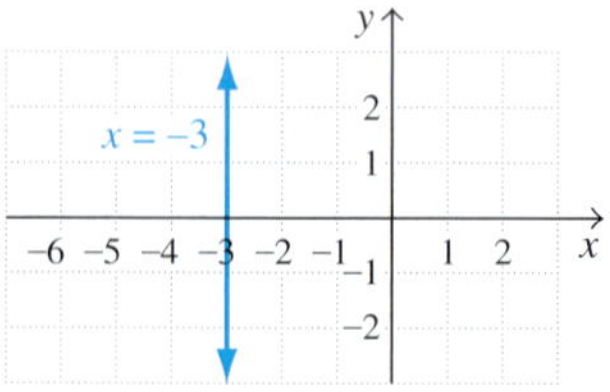

Figure 6.22

EXAMPLE 9 *Horizontal and Vertical Lines*

In the Cartesian coordinate system, graph (a) $y = 2$ and (b) $x = -3$.

SOLUTION

a) For any value of x, the value of y is 2. Therefore, the graph will be a horizontal line through $y = 2$ (Fig. 6.21).

b) For any value of y, the value of x is -3. Therefore, the graph will be a vertical line through $x = -3$ (Fig. 6.22).

Note that the graph of $y = 2$ has a slope of 0. The slope of the graph of $x = -3$ is undefined. ●

In graphing the equations in this section, we labeled the horizontal axis the x-axis and the vertical axis the y-axis. For each equation, we can determine values for y by substituting values for x. Since the value of y depends on the value of x, we refer to y as the ***dependent variable*** and x as the ***independent variable***. We label the *vertical axis* with the *dependent variable* and the *horizontal axis* with the *independent variable*. For the equation $C = 3n + 5$, the C is the dependent variable and n is the independent variable. Thus, to graph this equation, we label the vertical axis C and the horizontal axis n.

In many graphs, the values to be plotted on one axis are much greater than the values to be plotted on the other axis. When that occurs, we can use different scales on the horizontal and the vertical axes, as illustrated in Examples 10 and 11. The next two examples illustrate applications of graphing.

EXAMPLE 10 *Using a Graph to Determine Area*

The Professional Patio Company installs brick patios. The area of brick, a, in square feet, the company can install in t hours can be approximated by the formula $a = 5t$.

a) Graph $a = 5t$, for $t \leq 6$.

b) Use the graph to estimate the area of brick the company can install in 4 hours.

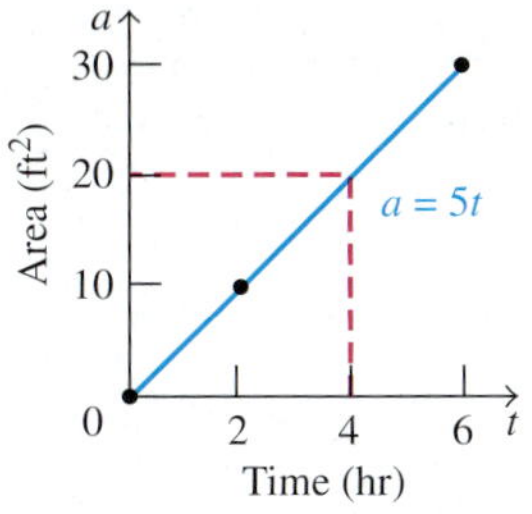

Figure 6.23

SOLUTION

a) Since $a = 5t$ is a linear equation, its graph will be a straight line. Select three values for t, find the corresponding values for a, and then draw the graph (Fig. 6.23).

$$a = 5t$$

Let $t = 0$, $a = 5(0) = 0$

Let $t = 2$, $a = 5(2) = 10$

Let $t = 6$, $a = 5(6) = 30$

t	***a***
0	0
2	10
6	30

b) By drawing a vertical line from $t = 4$ on the time axis up to the graph and then drawing a horizontal line across to the area axis, we can determine that the area installed in 4 hours is 20 ft^2.

EXAMPLE 11 *Using a Graph to Determine Profits*

Javier Mungro owns a business that sells mountain bicycles. He believes the profit (or loss) from the bicycles sold can be estimated by the formula $P = 60S - 90{,}000$, where S is the number of bicycles sold.

a) Graph $P = 60S - 90{,}000$, for $S \leq 5000$ bicycles.

b) From the graph, estimate the number of bicycles that must be sold for the business to break even (Fig. 6.24).

c) From the graph, estimate the number of bicycles sold if the profit from selling bicycles is \$150,000.

SOLUTION

a) Select values for S and find the corresponding values of P.

S	***P***
0	−90,000
2000	30,000
5000	210,000

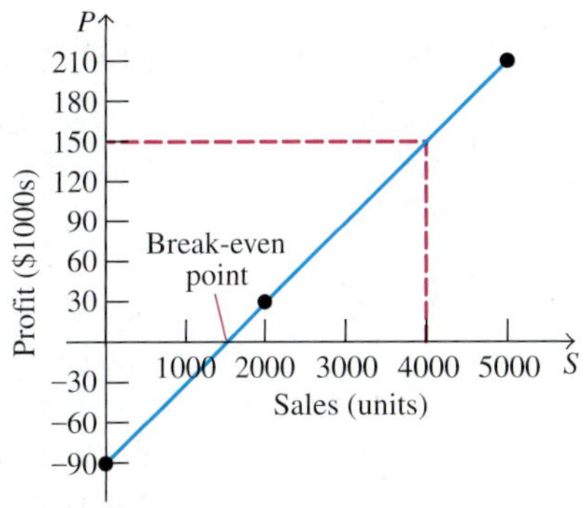

Figure 6.24

b) The break-even point is the number of bicycles that must be sold for the company to have neither a profit nor a loss. The break-even point is where the graph intercepts the horizontal, or S-axis, for that is where the profit, P, is 0. To break even, 1500 bicycles must be sold.

c) We can obtain the answer by drawing a horizontal line from 150 on the profit axis. Since the horizontal line cuts the graph at 4 on the S axis, 4000 bicycles were sold.

SECTION 6.7 EXERCISES

CONCEPT/WRITING EXERCISES

1. What is a graph?
2. Explain how to find the x-intercept of a linear equation.
3. Explain how to find the y-intercept of a linear equation.
4. What is the slope of a line?
5. **a)** Explain in your own words how to find the slope of a line between two points.

   b) Based on your explanation in part (a), find the slope of the line through the points (6, 2) and (−3, 5).
6. Describe the three methods used to graph a linear equation given in this section.
7. **a)** In which quadrant is the point (1, 4) located?

   b) In which quadrant is the point (−2, −5) located?
8. What is the minimum number of points needed to graph a linear equation?

PRACTICE THE SKILLS

In Exercises 9–16, plot all the points on the same axes.

9. (2, 4) **10.** (−4, 2) **11.** (−5, −1)

12. (4, 0) **13.** (0, 2) **14.** (0, 0)

15. (0, −5) **16.** $(2\frac{1}{2}, 5\frac{1}{2})$

In Exercises 17–24, plot all the points on the same axes.

17. (1, 4) **18.** (−3, 0) **19.** (−2, −5)

20. (0, −1) **21.** (0, 2) **22.** (−4, 5)

23. (4, −1) **24.** (4.5, 3.5)

In Exercises 25–34 (indicated on Fig. 6.25), write the coordinates of the corresponding point.

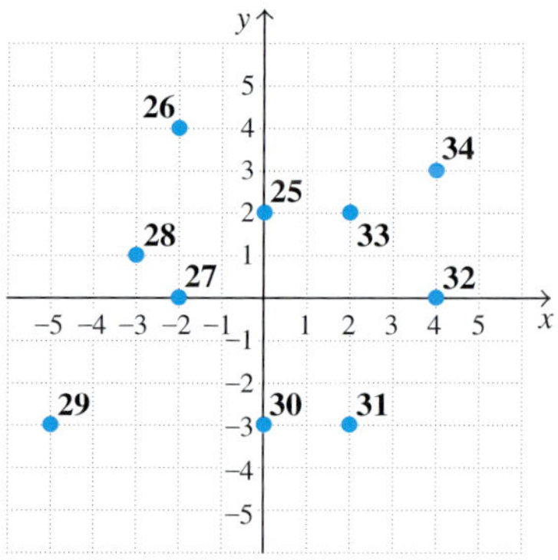

Figure 6.25

In Exercises 35–42, determine which ordered pairs satisfy the given equation.

35. $x + 2y = 9$ (5, 2) (1, 4) (0, 5)

36. $4x - y = 4$ (0, −4) (1, 0) (2, −3)

37. $2x - 3y = 10$ (8, 2) (−1, 4) $(0, -\frac{10}{3})$

38. $3y = 4x + 2$ (2, 1) (1, 2) $(0, \frac{2}{3})$

39. $7y = 3x - 5$ (1, −1), (−3, −2), (2, 5)

40. $3x - 5y = -12$ (1, 3), (−1, 2), (−4, 0)

41. $\frac{x}{4} + \frac{2y}{3} = 2$ (8, 0), $(1, \frac{1}{2})$, (0, 3)

42. $\frac{x}{2} + 3y = 4$ $(0, \frac{4}{3})$, (8, 0), (10, −2)

In Exercises 43–46, graph the equation and state the slope of the line if the slope exists (see Example 9).

43. $x = -1$ **44.** $x = 3$ **45.** $y = 2$ **46.** $y = -4$

In Exercises 47–56, graph the equation by plotting points, as in Example 3.

47. $y = x - 1$

48. $y = x + 4$

49. $y = 3x - 1$

50. $y = -x + 4$

51. $y + 3x = 6$

52. $y - 4x = 8$

53. $y = \frac{1}{2}x + 4$

54. $3y = 2x - 3$

55. $3y = x + 6$

56. $y = -\frac{2}{3}x$

In Exercises 57–66, graph the equation, using the x- and y-intercepts, as in Example 4.

57. $x + y = 4$

58. $x - y = 5$

59. $2x + y = 6$

60. $3x - y = -3$

61. $2x = -4y - 8$

62. $y = 4x + 4$

63. $y = 3x + 5$

64. $3x + 6y = 9$

65. $3x - y = 5$

66. $6x = 3y - 9$

In Exercises 67–76, determine the slope of the line through the given points. If the slope is undefined, so state.

67. (6, 2) and (8, 6)

68. (5, 3) and (3, 5)

69. $(-5, 6)$ and $(7, -9)$

70. $(-5, 6)$ and (8, 8)

71. (5, 2) and $(-3, 2)$

72. $(-3, -5)$ and $(-1, -2)$

73. $(8, -3)$ and (8, 3)

74. (2, 6) and $(2, -3)$

75. $(-3, -1)$ and $(8, -9)$

76. (2, 1) and $(-5, -9)$

In Exercises 77–86, graph the equation using the slope and y-intercept, as in Examples 6 and 7.

77. $y = x - 3$

78. $y = -x - 4$

79. $y = 2x + 3$

80. $y = 3x - 2$

81. $y = -\frac{1}{3}x + 2$

82. $y = -x - 2$

83. $7y = 4x - 7$

84. $3x + 2y = 6$

85. $3x - 2y + 6 = 0$

86. $3x + 4y - 8 = 0$

In Exercises 87–90, determine the equation of the graph.

87.

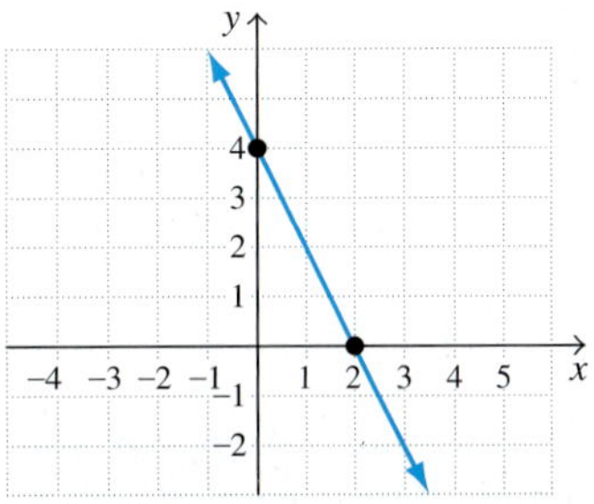

88.

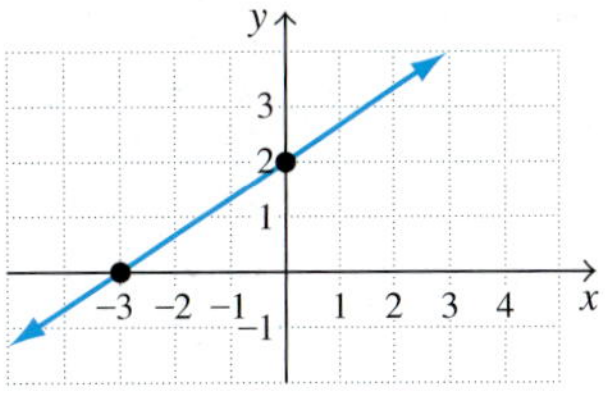

89.

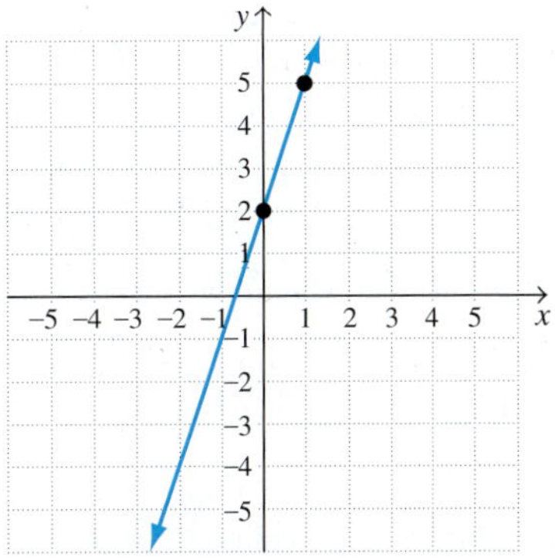

90.

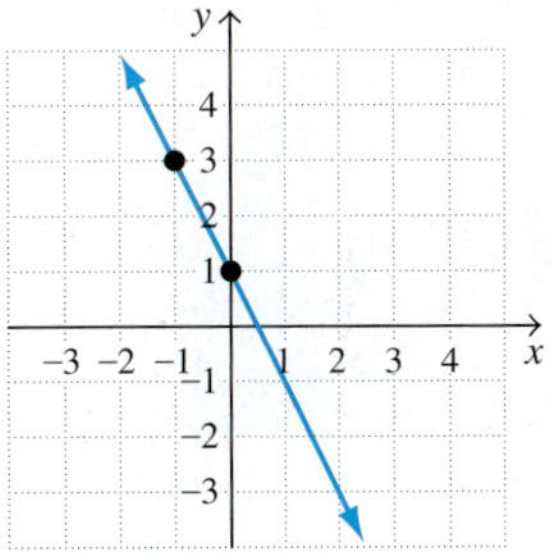

PROBLEM SOLVING

In Exercises 91 and 92, points A, B, and C are three vertices of a rectangle (the points where two sides meet). Plot the three points. (a) Find the coordinates of the fourth point, D, to complete the rectangle. (b) Find the area of the rectangle; use A = lw.

91. $A(-1, 3)$, $B(3, 3)$, $C(-1, -2)$

92. $A(-4, 2)$, $B(7, 2)$, $C(7, 8)$

In Exercises 93 and 94, points A, B, and C are three vertices of a parallelogram with sides parallel to the x-axis. Plot the three points. Find the coordinates of the fourth point, D, to complete the parallelogram. Note: There are two possible answers for point D.

93. $A(3, 2)$, $B(5, 5)$, $C(9, 5)$

94. $A(-2, 2)$, $B(3, 2)$, $C(6, -1)$

In Exercises 95–98, for what value of b will the line joining the points P and Q be parallel to the indicated axis?

95. $P(2, -3)$, $Q(6, b)$; x-axis

96. $P(4, 7)$, $Q(b, -2)$; y-axis

97. $P(3b - 1, 5)$, $Q(8, 4)$; y-axis

98. $P(-6, 2b + 3)$, $Q(7, -1)$; x-axis

99. ***Hanging Wallpaper*** Tanisha Vizquez owns a wallpaper hanging business. Her charge, C, for hanging wallpaper is \$40 plus \$0.30 per square foot of wallpaper she hangs, or $C = 40 + 0.30s$, where s is the number of square feet of wallpaper she hangs.

a) Graph $C = 40 + 0.30s$, for $s \leq 500$.

b) From the graph, estimate her charge if she hangs 300 square feet of wallpaper.

c) If her charge is \$70, use the equation for C to determine how many square feet of wallpaper she hung.

100. ***Selling Chocolates*** Ryan Stewart sells chocolate on the Internet. His monthly profit, p, in dollars, can be estimated by $p = 15n - 300$, where n is the number of dozens of chocolates he sells in a month.

a) Graph $p = 15n - 300$, for $n \leq 60$.

b) From the graph, estimate his profit if he sells 40 dozen chocolates in a month.

c) How many dozens of chocolates must he sell in a month to break even?

101. ***Cable Cost*** The Schwimmer's cost for cable television is given by $C = 75 + 25x$, where x is the number of months of service.

a) Graph $C = 75 + 25x$, for $x \leq 50$.

b) From the graph, estimate the Schwimmer's cost for 30 months of cable television.

c) If the Schwimmer's total cost for cable television is \$975, estimate the number of months of service.

102. ***Earning Simple Interest*** When \$1000 is invested in a savings account paying simple interest for a year, the interest, i, in dollars, earned can be found by the formula $i = 1000r$, where r is the rate in decimal form.

a) Graph $i = 1000r$, for r up to and including a rate of 15%.

b) If the rate is 4%, what is the simple interest?

c) If the rate is 6%, what is the simple interest?

In Exercises 103 and 104, a set of points is plotted. Also shown is a straight line through the set of points that is called the line of best fit *(or a* regression line, *as will be discussed in Chapter 13, Statistics.)*

103. ***Determining the Number of Defects*** The graph on the top of page 375 shows the daily number of workers absent from the assembly line at J. B. Davis Corporation and the number of defects coming off the assembly line for 8 days. (The two points indicated on the line do not represent any of the 8 days.) The line of best fit, the blue line on the graph, can be used to approximate the number of defects coming off the assembly line per day for a given number of workers absent.

a) Determine the slope of the line of best fit using the two points indicated.

b) Using the slope determined in part (a) and the y-intercept, $(0, 9)$, determine the equation of the line of best fit.

c) Using the equation you determined in part (b), determine the approximate number of defects for a day if 3 workers are absent.

d) Using the equation you determined in part (b), approximate the number of workers absent for a day if there are 17 defects that day.

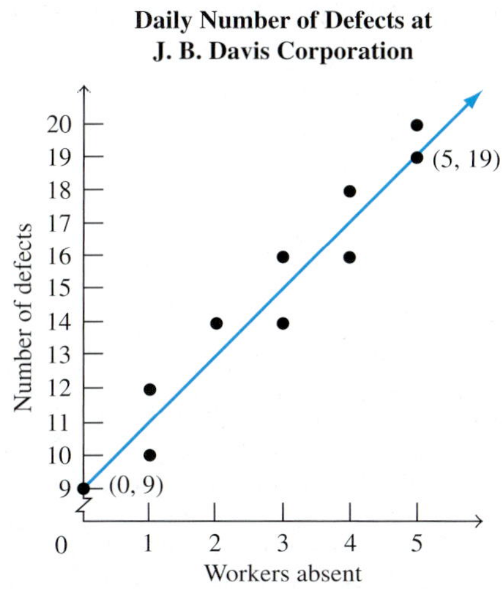

104. ***Determining a Test Grade*** The graph shows the hours studied and the test grades on a biology test for six students. (The two points indicated on the line do not represent any of the six students.) The line of best fit, the red line on the graph, can be used to approximate the test grade the average student receives for the number of hours he or she studies.

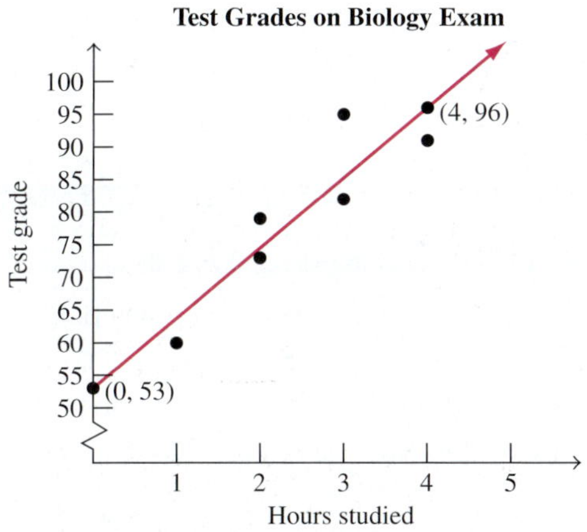

a) Determine the slope of the line of best fit using the two points indicated.

b) Using the slope determined in part (a) and the y-intercept, (0, 53), determine the equation of the line of best fit.

c) Using the equation you determined in part (b), determine the approximate test grade for a student who studied for 3 hours.

d) Using the equation you determined in part (b), determine the amount of time a student would need to study to receive a grade of 80 on the biology test.

105. ***Drug Sales*** The green graph shows U.S. sales of pain relief drugs, in billions of dollars, for the years 2000–2004. The red dashed line can be used to approximate the U.S. sales of these pain relief drugs. If we let 0 represent 2000, 1 represent 2001, and so on, then 2004 would be represented by 4. Using the ordered pairs (0, 2.60) and (4, 2.41),

a) determine the slope of the red dashed line.

b) determine the equation of the red dashed line using (0, 2.60) as the y-intercept.

c) Using the equation you determined in part (b), estimate the U.S. sales of these pain relief drugs in 2002.

d) Using the equation you determined in part (b), determine the year in which U.S. sales of these pain relief drugs was \$2.4 billion.

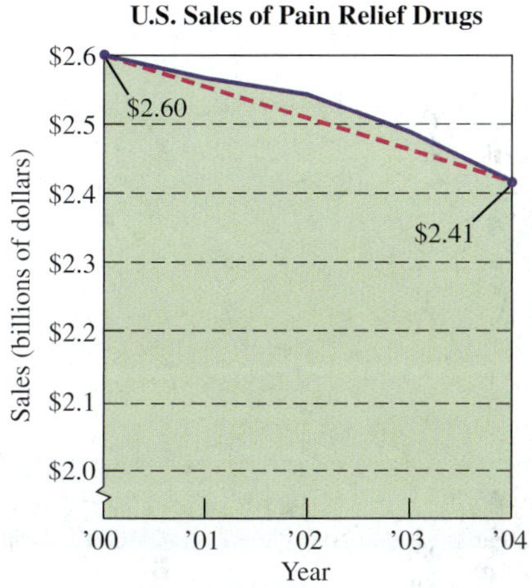

Source: AC Nielsen

106. ***Airline Passengers*** The blue graph on page 376 shows the actual number of passengers, in millions, who departed from Greater Rochester International Airport (GRIA) for the years 2001–2005. The red dashed line can be used to approximate the number of passengers, in millions, who departed from GRIA. If we let 0 represent 2001 and 1 represent 2002, and so on, then 2005 would be represented by 4. Using the ordered pairs (0, 1.15) and (4, 1.45),

a) determine the slope of the red dashed line.

b) determine the equation of the red dashed line using (0, 1.15) as the y-intercept of the graph.

c) Using the equation you determined in part (b), estimate the number of passengers who departed from GRIA in 2003.

d) Assuming this trend continues, use the equation you determined in part (b) to determine the year in which there will be 1.55 million passengers departing from GRIA.

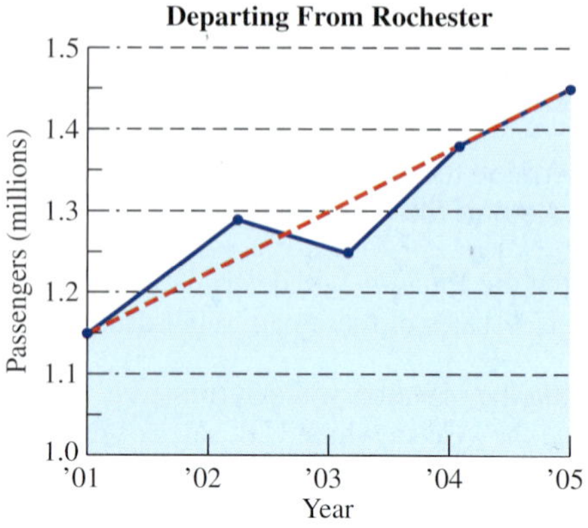

Source: Monroe County

CHALLENGE PROBLEMS/GROUP ACTIVITIES

107. a) Two lines are parallel when they do not intersect no matter how far they are extended. Explain how you can determine, without graphing the equations, whether two equations will be parallel lines when graphed.

b) Determine whether the graphs of the equations $2x - 3y = 6$ and $4x = 6y + 6$ are parallel lines.

108. In which quadrants will the set of points that satisfy the equation $x + y = 1$ lie? Explain.

INTERNET/RESEARCH ACTIVITY

109. René Descartes is known for his contributions to algebra. Write a paper on his life and his contributions to algebra.

6.8 LINEAR INEQUALITIES IN TWO VARIABLES

▲ A linear inequality in two variables may be used to represent the number of grills that can be shipped to each of two stores.

Suppose a manufacturer of gas grills needs to determine how many grills to ship to each of two stores. Also suppose the manufacturer can ship at most 300 grills in total. A linear inequality in two variables can be used to represent the number of grills that can be shipped to each store. In this section, we will discuss how to solve linear inequalities in two variables.

In Section 6.6, we introduced linear inequalities in one variable. Now we will introduce linear inequalities in two variables. Some examples of linear inequalities in two variables are $2x + 3y \leq 7$, $x + 7y \geq 5$, and $x - 3y < 6$.

The solution set of a linear inequality in one variable may be indicated on a number line. The solution set of a linear inequality in two variables is indicated on a coordinate plane.

An inequality that is strictly less than ($<$) or greater than ($>$) will have as its solution set a ***half-plane***. A half-plane is the set of all the points on one side of a line. An inequality that is less than or equal to ($\leq$) or greater than or equal to ($\geq$) will have as its solution set the set of points that consists of a half-plane and a line. To indicate that the line is part of the solution set, we draw a solid line. To indicate that the line is not part of the solution set, we draw a dashed line.

TO GRAPH LINEAR INEQUALITIES IN TWO VARIABLES

1. Mentally substitute the equal sign for the inequality sign and plot points as if you were graphing the equation.
2. If the inequality is $<$ or $>$, draw a dashed line through the points. If the inequality is $\leq$ or $\geq$, draw a solid line through the points.
3. Select a test point not on the line and substitute the x- and y-coordinates into the inequality. If the substitution results in a true statement, shade in the area on the same side of the line as the test point. If the test point results in a false statement, shade in the area on the opposite side of the line as the test point.

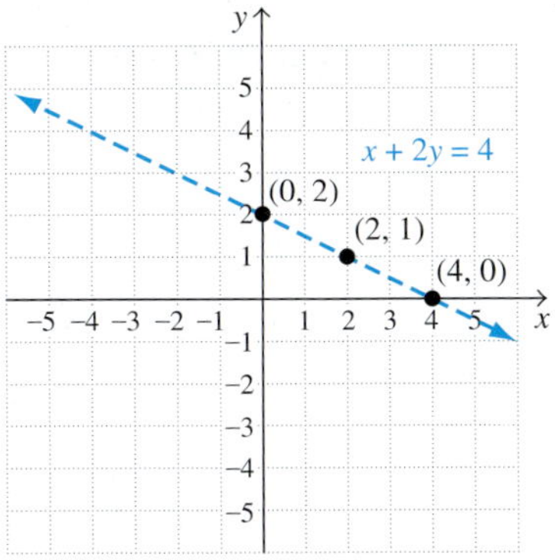

Figure 6.26

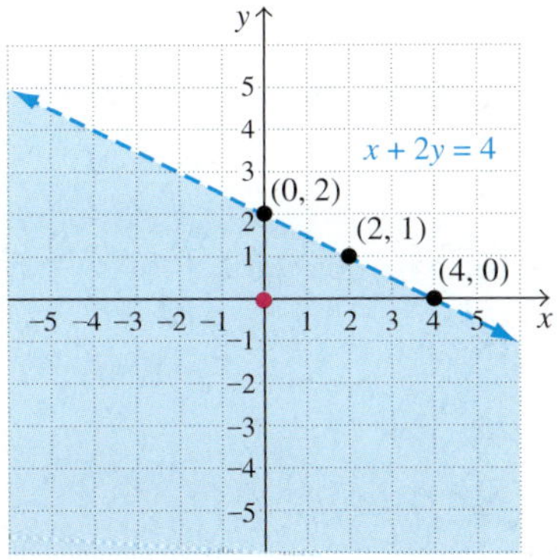

Figure 6.27

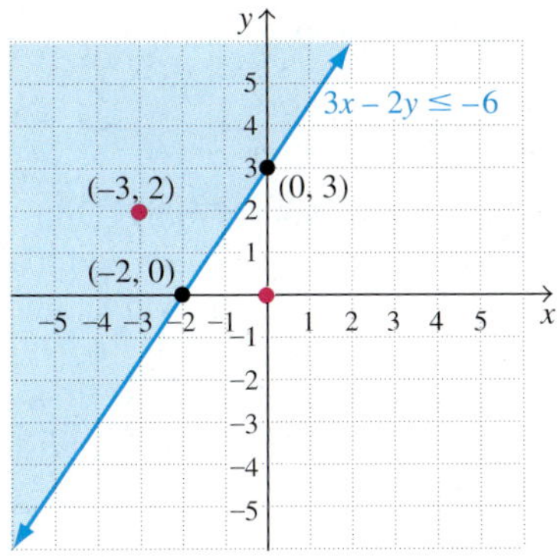

Figure 6.28

EXAMPLE 1 *Graphing an Inequality*

Draw the graph of $x + 2y < 4$.

SOLUTION To obtain the solution set, start by graphing $x + 2y = 4$. Since the original inequality is strictly "less than," draw a dashed line (Fig 6.26). The dashed line indicates that the points on the line are not part of the solution set.

The line $x + 2y = 4$ divides the plane into three parts, the line itself and two *half-planes*. The line is the boundary between the two half-planes. The points in one half-plane will satisfy the inequality $x + 2y < 4$. The points in the other half-plane will satisfy the inequality $x + 2y > 4$.

To determine the solution set to the inequality $x + 2y < 4$, pick any point on the plane that is not on the line. The simplest point to work with is the origin, (0, 0). Substitute $x = 0$ and $y = 0$ into $x + 2y < 4$.

$$x + 2y < 4$$
$$\text{Is } 0 + 2(0) < 4?$$
$$0 + 0 < 4$$
$$0 < 4 \quad \text{True}$$

Since 0 is less than 4, the point (0, 0) is part of the solution set. All the points on the same side of the graph of $x + 2y = 4$ as the point (0, 0) are members of the solution set. We indicate that by shading the half-plane that contains (0, 0). The graph is shown in Fig. 6.27.

EXAMPLE 2 *Graphing an Inequality*

Draw the graph of $3x - 2y \leq -6$.

SOLUTION First draw the graph of the equation $3x - 2y = -6$. Use a solid line because the points on the boundary line are included in the solution set (see Fig. 6.28). Now pick a point that is not on the line. Take (0, 0) as the test point.

$$3x - 2y \leq -6$$
$$\text{Is } 3(0) - 2(0) \leq -6?$$
$$0 \leq -6 \quad \text{False}$$

Since 0 is not less than or equal to $-6 (0 \nleq -6)$, the solution set is the line and the half-plane that does not contain the point (0, 0).

If you had arbitrarily selected the test point $(-3, 2)$ from the other half-plane, you would have found that the inequality would be true: $3(-3) - 2(2) \leq -6$, or $-13 \leq -6$. Because -13 is less than or equal to -6, the point $(-3, 2)$ would be in the half-plane containing the solution set. ●

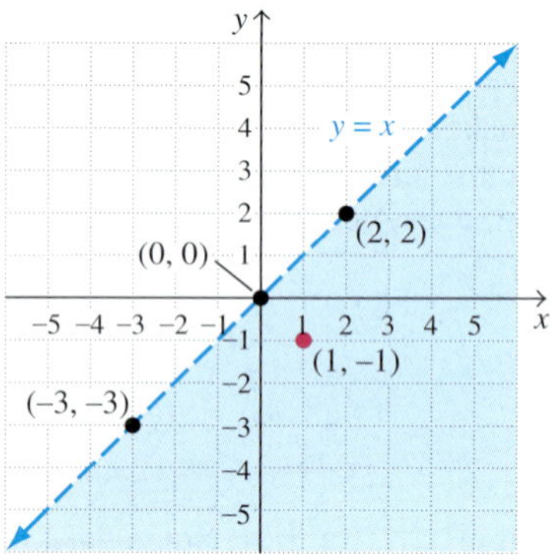

Figure 6.29

EXAMPLE 3 *Graphing an Inequality*

Draw the graph of $y < x$.

SOLUTION The inequality is strictly "less than," so the boundary line is not part of the solution set. In graphing the equation $y = x$, draw a dashed line (Fig. 6.29). Since (0, 0) is *on* the line, it cannot serve as a test point. Let's pick the point $(1, -1)$.

$$y < x$$
$$-1 < 1 \quad \text{True}$$

Since $-1 < 1$ is true, the solution set is the half-plane containing the point $(1, -1)$. ●

SECTION 6.8 EXERCISES

CONCEPT/WRITING EXERCISES

1. Outline the procedure used to graph inequalities in two variables.

2. Explain why we use a solid line when graphing an inequality containing $\leq$ or $\geq$ and we use a dashed line when graphing an inequality containing $<$ or $>$.

3. What is a half-plane?

4. a) How do the graphs of $x < 1$ and $x > 1$ differ?

   b) How do the graphs of $y > x + 3$ and $y \geq x + 3$ differ?

5. For each inequality, determine whether (1, 3) is a solution to the inequality.

   a) $2x + 3y < 11$ b) $2x + 3y \leq 11$

   c) $2x + 3y \geq 11$ d) $2x + 3y > 11$

6. For each inequality, determine whether $(-2, 4)$ is a solution to the inequality.

   a) $4x - 2y < -10$ b) $4x - 2y \leq -10$

   c) $4x - 2y > -10$ d) $4x - 2y \geq -10$

PRACTICE THE SKILLS

In Exercises 7–28, draw the graph of the inequality.

7. $x \leq 3$
8. $y \geq -3$
9. $y > x - 4$
10. $y < x + 1$
11. $y \geq 2x - 6$
12. $y < -2x + 2$
13. $3x - 4y > 12$
14. $x + 2y > 4$
15. $3x - 4y \leq 9$
16. $4y - 3x \geq 9$
17. $3x + 2y < 6$
18. $-x + 2y < 2$
19. $x + y > 0$
20. $x + 2y \leq 0$
21. $5x + 2y \geq 10$
22. $y \geq -2x + 1$
23. $3x - 2y < 12$
24. $y \leq 3x - 4$
25. $\frac{2}{5}x - \frac{1}{2}y \leq 1$
26. $\frac{1}{2}x + \frac{2}{3}y \geq 1$
27. $0.1x + 0.3y \leq 0.4$
28. $0.2x + 0.5y \leq 0.3$

PROBLEM SOLVING

29. *Gas Grills* A manufacturer of gas grills must produce and ship x gas grills to one store and y gas grills to a second store. The maximum number of gas grills the manufacturer can produce and ship is 300.

    a) Write an inequality in two variables that represents this problem.

    b) Graph the inequality.

30. ***Buying CDs and DVDs*** Reggie Wayne is allowed to spend a maximum of \$100 for purchasing x DVDs and y CDs. Each DVD costs \$20, and each CD costs \$10.

a) State this problem as an inequality in two variables.

b) Graph the inequality.

CHALLENGE PROBLEMS/GROUP ACTIVITIES

31. ***Men's Shirts*** The Tommy Hilfiger Company must ship x men's shirts to one outlet and y men's shirts to a second outlet. The maximum number of shirts the manufacturer can produce and ship is 250. We can represent this situation with the inequality $x + y \leq 250$.

a) Can x or y be negative? Explain.

b) Graph the inequality.

c) Write one or two paragraphs interpreting the information that the graph provides.

32. Which of the following inequalities have the same graph? Explain how you determined your answer.

a) $2x - y < 8$ **b)** $-2x + y > -8$

c) $2x - 4y < 16$ **d)** $y > 2x - 8$

6.9 SOLVING QUADRATIC EQUATIONS BY USING FACTORING AND BY USING THE QUADRATIC FORMULA

▲ A quadratic equation can be used to determine the width of a border around a rectangular swimming pool.

You are adding a brick border of uniform width around your rectangular swimming pool. If you know the surface area of the pool and the area you would like the brick border to cover, you can determine the width of the border by solving a quadratic equation. In this section, we will learn two methods that can be used to solve quadratic equations.

We begin this section by discussing binomials.

A ***binomial*** is an expression that contains two terms in which each exponent that appears on a variable is a whole number.

Examples of binomials

$x + 3$ $x - 5$

$3x + 5$ $4x - 2$

To multiply two binomials, we can use the ***FOIL method***. The name of the method, FOIL, is an acronym to help its users remember it as a method that obtains the products of the *F*irst, *O*uter, *I*nner, and *L*ast terms of the binomials.

First Outer Inner Last

$$(a + b)(c + d) = a \cdot c + a \cdot d + b \cdot c + b \cdot d$$

After multiplying the first, outer, inner, and last terms, combine all like terms.

EXAMPLE 1 *Multiplying Binomials*

Multiply $(x + 4)(x + 7)$.

SOLUTION The FOIL method of multiplication yields

F O I L

$$\begin{aligned}(x + 4)(x + 7) &= x \cdot x + x \cdot 7 + 4 \cdot x + 4 \cdot 7 \\ &= x^2 + 7x + 4x + 28 \\ &= x^2 + 11x + 28\end{aligned}$$

Note that the like terms $7x$ and $4x$ were combined to get $11x$. ●

EXAMPLE 2 *Multiplying Binomials*

Multiply $(2x - 1)(x + 4)$.

SOLUTION

F O I L

$$\begin{aligned}(2x - 1)(x + 4) &= 2x \cdot x + 2x \cdot 4 + (-1) \cdot x + (-1) \cdot 4 \\ &= 2x^2 + 8x - x - 4 \\ &= 2x^2 + 7x - 4\end{aligned}$$ ●

Factoring Trinomials of the Form $x^2 + bx + c$

The expression $x^2 + 8x + 15$ is an example of a trinomial. A ***trinomial*** is an expression containing three terms in which each exponent that appears on a variable is a whole number.

Example 1 showed that

$$(x + 4)(x + 7) = x^2 + 11x + 28$$

Since the product of $x + 4$ and $x + 7$ is $x^2 + 11x + 28$, we say that $x + 4$ and $x + 7$ are ***factors*** of $x^2 + 11x + 28$. To ***factor*** an expression means to write the expression as a product of its factors. For example, to factor $x^2 + 11x + 28$ we write

$$x^2 + 11x + 28 = (x + 4)(x + 7)$$

Let's look at the factors more closely.

$$4 + 7 = 11$$

$$4 \cdot 7 = 28$$

$$x^2 + 11x + 28 = (x + 4)(x + 7)$$

Note that the sum of the two numbers in the factors is $4 + 7$, or 11. The 11 is the coefficient of the x term. Also note that the product of the numbers in the two factors is $4 \cdot 7$, or 28. The 28 is the constant in the trinomial. In general, when factoring an expression of the form $x^2 + bx + c$, we need to find two numbers whose product is c and whose sum is b. When we determine the two numbers, the factors will be of the form

$$(x + \square)\ (x + \square)$$

↑ One number ↑ Other number

EXAMPLE 3 *Factoring a Trinomial*

Factor $x^2 + 5x + 6$.

SOLUTION We need to find two numbers whose product is 6 and whose sum is 5. Since the product is $+6$, the two numbers must both be positive or both be negative. Because the coefficient of the x term is positive, only the positive factors of 6 need to be considered. Can you explain why? We begin by listing the positive numbers whose product is 6.

Factors of 6	Sum of Factors
1(6)	$1 + 6 = 7$
2(3)	$2 + 3 = 5$

Since $2 \cdot 3 = 6$ and $2 + 3 = 5$, the numbers we are seeking are 2 and 3. Thus, we write

$$x^2 + 5x + 6 = (x + 2)(x + 3)$$

Note that $(x + 3)(x + 2)$ is also an acceptable answer.

TO FACTOR TRINOMIAL EXPRESSIONS OF THE FORM $x^2 + bx + c$

1. Find two numbers whose product is c and whose sum is b.
2. Write factors in the form

$$(x + \square)\ (x + \square)$$

↑ One number from step 1 ↑ Other number from step 1

3. Check your answer by multiplying the factors using the FOIL method.

If, for example, the numbers found in step 1 of the above procedure were 6 and -4, the factors would be written $(x + 6)(x - 4)$.

EXAMPLE 4 *Factoring a Trinomial*

Factor $x^2 - 6x - 16$.

SOLUTION We must find two numbers whose product is -16 and whose sum is -6. Begin by listing the factors of -16.

Factors of -16	Sum of Factors
$-16(1)$	$-16 + 1 = -15$
$-8(2)$	$-8 + 2 = -6$
$-4(4)$	$-4 + 4 = 0$
$-2(8)$	$-2 + 8 = 6$
$-1(16)$	$-1 + 16 = 15$

The table lists all the factors of -16. The only factors listed whose product is -16 and whose sum is -6 are -8 and 2. We listed all factors in this example so that you could see, for example, that $-8(2)$ is a different set of factors than $-2(8)$. Once you find the factors you are looking for, there is no need to go any further. The trinomial can be written in factored form as

$$x^2 - 6x - 16 = (x - 8)(x + 2)$$

Factoring Trinomials of the Form $ax^2 + bx + c, a \neq 1$

Now we discuss how to factor an expression of the form $ax^2 + bx + c$, where a, the coefficient of the squared term, is not equal to 1.

Consider the multiplication problem $(2x + 1)(x + 3)$.

$$\begin{aligned}(2x + 1)(x + 3) &= 2x \cdot x + 2x \cdot 3 + 1 \cdot x + 1 \cdot 3 \\ &= 2x^2 + 6x + x + 3 \\ &= 2x^2 + 7x + 3\end{aligned}$$

Since $(2x + 1)(x + 3) = 2x^2 + 7x + 3$, the factors of $2x^2 + 7x + 3$ are $2x + 1$ and $x + 3$.

Let's study the coefficients more closely.

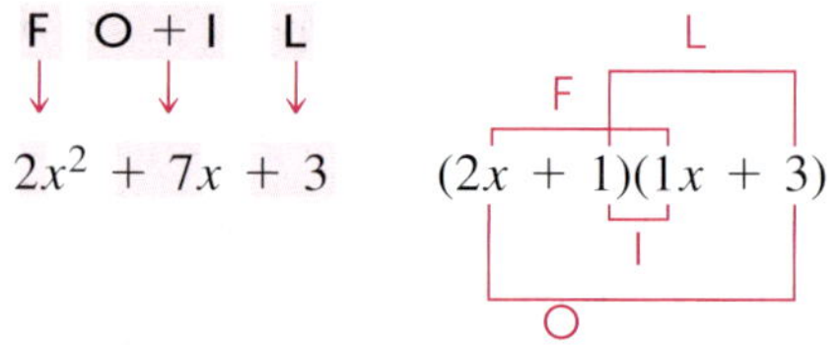

$\mathbf{F} = 2 \cdot 1 = 2$ $\quad$ $\mathbf{O + I} = (2 \cdot 3) + (1 \cdot 1) = 7$ $\quad$ $\mathbf{L} = 1 \cdot 3 = 3$

Note that the product of the coefficient of the first terms in the multiplication of the binomials equals 2, the coefficient of the squared term. The sum of the products of the coefficients of the outer and inner terms equals 7, the coefficient of the x-term. The product of the last terms equals 3, the constant.

A procedure to factor expressions of the form $ax^2 + bx + c, a \neq 1$, follows.

TO FACTOR TRINOMIAL EXPRESSIONS OF THE FORM $ax^2 + bx + c, a \neq 1$

1. Write all pairs of factors of the coefficient of the squared term, a.
2. Write all pairs of factors of the constant, c.
3. Try various combinations of these factors until the sum of the products of the outer and inner terms is bx.
4. Check your answer by multiplying the factors using the FOIL method.

EXAMPLE 5 *Factoring a Trinomial, $a \neq 1$*

Factor $3x^2 + 17x + 10$.

SOLUTION The only positive factors of 3 are 3 and 1. Therefore, we write

$$3x^2 + 17x + 10 = (3x \qquad)(x \qquad)$$

The number 10 has both positive and negative factors. However, since both the constant, 10, and the sum of the products of the outer and inner terms, 17, are positive, the two factors must be positive. Why? The positive factors of 10 are 1(10) and 2(5). The following is a list of the possible factors.

Possible Factors	Sum of Products of Outer and Inner Terms
$(3x + 1)(x + 10)$	$31x$
$(3x + 10)(x + 1)$	$13x$
$(3x + 2)(x + 5)$	$17x$ ← Correct middle term
$(3x + 5)(x + 2)$	$11x$

Thus, $3x^2 + 17x + 10 = (3x + 2)(x + 5)$.

Note that factoring problems of this type may be checked by using the FOIL method of multiplication. We will check the results to Example 5:

$$\begin{aligned}(3x + 2)(x + 5) &= 3x \cdot x + 3x \cdot 5 + 2 \cdot x + 2 \cdot 5\\ &= 3x^2 + 15x + 2x + 10\\ &= 3x^2 + 17x + 10\end{aligned}$$

Since we obtained the expression we started with, our factoring is correct.

DID YOU KNOW?

Algorithm

Recall that the series of steps taken to solve a certain type of equation is called an *algorithm*. Because mathematical procedures to solve equations can be generalized, it is possible to program computers to solve problems. For example, the equation $4x^2 + 7x + 3 = 0$ is an equation of the form $ax^2 + bx + c = 0$. If you can find a procedure to solve an equation of the general form, it can be used to solve all equations of that form.

EXAMPLE 6 *Factoring a Trinomial, $a \neq 1$*

Factor $6x^2 - 11x - 10$.

SOLUTION The factors of 6 will be either $6 \cdot 1$ or $2 \cdot 3$. Therefore, the factors may be of the form $(6x \quad)(x \quad)$ or $(2x \quad)(3x \quad)$. When there is more than one set of factors for the first term, we generally try the medium-sized factors first. If that does not work, we try the other factors. Thus, we write

$$6x^2 - 11x - 10 = (2x \quad)(3x \quad)$$

The factors of -10 are $(-1)(10)$, $(1)(-10)$, $(-2)(5)$, and $(2)(-5)$. There will be eight different pairs of possible factors of the trinomial $6x^2 - 11x - 10$. Can you list them?

The correct factoring is $6x^2 - 11x - 10 = (2x - 5)(3x + 2)$. ●

Note that in Example 6 we first tried factors of the form $(2x \quad)(3x \quad)$. If we had not found the correct factors using them, we would have tried $(6x \quad)(x \quad)$.

Solving Quadratic Equations by Factoring

In Section 6.2, we solved linear, or first-degree, equations. In those equations, the exponent on all variables was 1. Now we deal with the *quadratic equation*. The standard form of a quadratic equation in one variable is shown in the box.

STANDARD FORM OF A QUADRATIC EQUATION

$$ax^2 + bx + c = 0, \qquad a \neq 0$$

Note that in the standard form of a quadratic equation, the greatest exponent on x is 2 and the right side of the equation is equal to zero. *To solve a quadratic equation* means to find the value or values that make the equation true. In this section, we will solve quadratic equations by factoring and by the quadratic formula.

To solve a quadratic equation by factoring, set one side of the equation equal to 0 and then use the *zero-factor* property.

ZERO-FACTOR PROPERTY

If $a \cdot b = 0$, then $a = 0$ or $b = 0$.

The zero-factor property indicates that if the product of two factors is 0, then one (or both) of the factors must have a value of 0.

EXAMPLE 7 *Using the Zero-Factor Property*

Solve the equation $(x + 2)(x - 5) = 0$.

SOLUTION When we use the zero factor property, either $(x + 2)$ or $(x - 5)$ must equal 0 for the product to equal 0. Thus, we set each individual factor equal to 0 and solve each resulting equation for x.

$$(x + 2)(x - 5) = 0$$

$$x + 2 = 0 \quad \text{or} \quad x - 5 = 0$$

$$x = -2 \qquad\qquad x = 5$$

Thus, the solutions are -2 and 5.

CHECK: $x = -2$ $\qquad$ $x = 5$

$$(x + 2)(x - 5) = 0 \qquad (x + 2)(x - 5) = 0$$

$$(-2 + 2)(-2 - 5) = 0 \qquad (5 + 2)(5 - 5) = 0$$

$$0(-7) = 0 \qquad 7(0) = 0$$

$$0 = 0 \quad \text{True} \qquad 0 = 0 \quad \text{True}$$

TO SOLVE A QUADRATIC EQUATION BY FACTORING

1. Use the addition or subtraction property to make one side of the equation equal to 0.
2. Factor the side of the equation not equal to 0.
3. Use the zero-factor property to solve the equation.

Examples 8 and 9 illustrate this procedure.

EXAMPLE 8 *Solving a Quadratic Equation by Factoring*

Solve the equation $x^2 - 8x = -15$.

SOLUTION First add 15 to both sides of the equation to make the right-hand side of the equation equal to 0.

$$x^2 - 8x = -15$$

$$x^2 - 8x + 15 = -15 + 15$$

$$x^2 - 8x + 15 = 0$$

Factor the left-hand side of the equation. The object is to find two numbers whose product is 15 and whose sum is -8. Since the product of the numbers is positive and the sum of the numbers is negative, the two numbers must both be negative. The numbers are -3 and -5. Note that $(-3)(-5) = 15$ and $-3 + (-5) = -8$.

$$x^2 - 8x + 15 = 0$$

$$(x - 3)(x - 5) = 0$$

Now use the zero-factor property to find the solution.

$$x - 3 = 0 \quad \text{or} \quad x - 5 = 0$$

$$x = 3 \qquad\qquad x = 5$$

The solutions are 3 and 5.

EXAMPLE 9 *Solving a Quadratic Equation by Factoring*

Solve the equation $3x^2 - 13x + 4 = 0$.

SOLUTION $3x^2 - 13x + 4$ factors into $(3x - 1)(x - 4)$. Thus, we write

$$3x^2 - 13x + 4 = 0$$
$$(3x - 1)(x - 4) = 0$$
$$3x - 1 = 0 \quad \text{or} \quad x - 4 = 0$$
$$3x = 1 \qquad\qquad x = 4$$
$$x = \frac{1}{3}$$

The solutions are $\frac{1}{3}$ and 4.

TIMELY TIP As stated on page 383, every factoring problem can be checked by multiplying the factors. If you have factored correctly, the product of the factors will be identical to the original expression that was factored. If we wished to check the factoring of Example 9, we would multiply $(3x - 1)(x - 4)$. Since the product of the factors is $3x^2 - 13x + 4$, which is the expression we started with, our factoring is correct.

Solving Quadratic Equations by Using the Quadratic Formula

Not all quadratic equations can be solved by factoring. When a quadratic equation cannot be easily solved by factoring, we can solve the equation with the *quadratic formula*. The quadratic formula can be used to solve any quadratic equation.

QUADRATIC FORMULA

For a quadratic equation in standard form, $ax^2 + bx + c = 0$, $a \neq 0$, the quadratic formula is

$$x = \frac{-b \pm \sqrt{b^2 - 4ac}}{2a}$$

In the quadratic formula, the plus or minus symbol, $\pm$, is used. If, for example, $x = 2 \pm 3$, then $x = 2 + 3 = 5$ or $x = 2 - 3 = -1$.

It is possible for a quadratic equation to have no real solution. In solving an equation, if the radicand (the expression inside the square root) is a negative number, the quadratic equation has *no real solution*.

To use the quadratic formula, first write the quadratic equation in standard form. Then determine the values for a (the coefficient of the squared term), b (the coefficient of the x term), and c (the constant). Finally, substitute the values of a, b, and c into the quadratic formula and evaluate the expression.

DID YOU KNOW?

The Mathematics of Motion

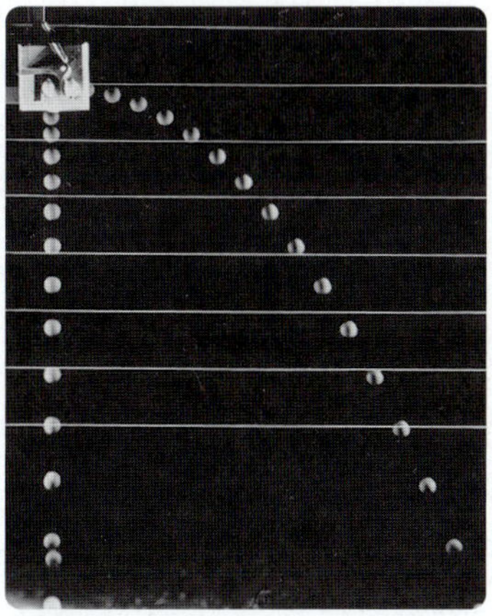

The free fall of an object is something that has interested scientists and mathematicians for centuries. It is described by a quadratic equation. Shown here is a time-lapse photo that shows the free fall of a ball in equal-time intervals. What you see can be described verbally this way: The rate of change in velocity in each interval is the same; therefore, velocity is continuously increasing and acceleration is constant.

EXAMPLE 10 *Solving a Quadratic Equation by Using the Quadratic Formula*

Solve the equation $x^2 - 5x - 14 = 0$ by using the quadratic formula.

SOLUTION In this equation, $a = 1$, $b = -5$, and $c = -14$.

$$x = \frac{-b \pm \sqrt{b^2 - 4ac}}{2a} = \frac{-(-5) \pm \sqrt{(-5)^2 - 4(1)(-14)}}{2(1)}$$

$$= \frac{5 \pm \sqrt{25 + 56}}{2}$$

$$= \frac{5 \pm \sqrt{81}}{2}$$

$$= \frac{5 \pm 9}{2}$$

$$\frac{5 + 9}{2} = \frac{14}{2} = 7 \quad \text{or} \quad \frac{5 - 9}{2} = \frac{-4}{2} = -2$$

The solutions are 7 and -2.

Note that Example 10 can also be solved by factoring. We suggest you do so now.

EXAMPLE 11 *Irrational Solutions to a Quadratic Equation*

Solve $4x^2 - 8x = -1$ by using the quadratic formula.

SOLUTION Begin by writing the equation in standard form by adding 1 to both sides of the equation, which gives the following.

$$4x^2 - 8x + 1 = 0$$

$$a = 4, \qquad b = -8, \qquad c = 1$$

$$x = \frac{-b \pm \sqrt{b^2 - 4ac}}{2a} = \frac{-(-8) \pm \sqrt{(-8)^2 - 4(4)(1)}}{2(4)}$$

$$= \frac{8 \pm \sqrt{64 - 16}}{8}$$

$$= \frac{8 \pm \sqrt{48}}{8}$$

Since $\sqrt{48} = \sqrt{16}\sqrt{3} = 4\sqrt{3}$ (see Section 5.4), we write

$$\frac{8 \pm \sqrt{48}}{8} = \frac{8 \pm 4\sqrt{3}}{8} = \frac{\overset{1}{\cancel{4}}(2 \pm \sqrt{3})}{\underset{2}{\cancel{8}}} = \frac{2 \pm \sqrt{3}}{2}$$

The solutions are $\frac{2 + \sqrt{3}}{2}$ and $\frac{2 - \sqrt{3}}{2}$.

Note that the solutions to Example 11 are irrational numbers.

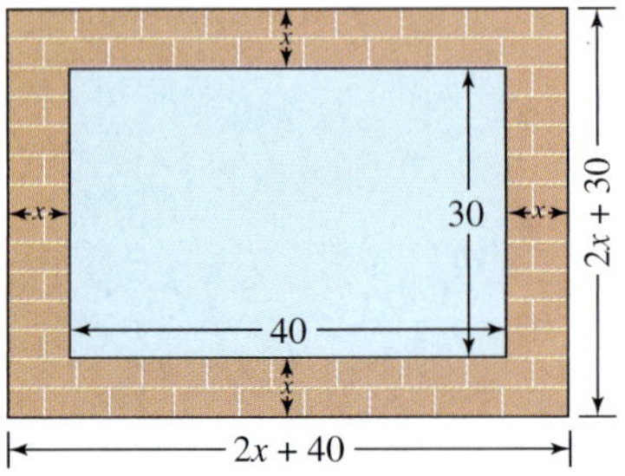

Figure 6.30

EXAMPLE 12 *Brick Border*

Diane Cecero and her husband recently installed an inground rectangular swimming pool measuring 40 ft by 30 ft. They want to add a brick border of uniform width around all sides of the pool. How wide can they make the brick border if they purchased enough brick to cover an area of 296 ft^2?

SOLUTION Let's make a diagram of the pool and the brick border (Fig. 6.30). Let $x =$ the uniform width of the brick border. Then the total length of the larger rectangular area, the pool plus the border, is $2x + 40$. The total width of the larger rectangular area is $2x + 30$.

The area of the brick border can be found by subtracting the area of the pool from the area of the pool plus the brick border.

$$\text{Area of pool} = l \cdot w = (40)(30) = 1200 \text{ ft}^2$$
$$\text{Area of pool plus brick border} = l \cdot w = (2x + 40)(2x + 30)$$
$$= 4x^2 + 140x + 1200$$
$$\text{Area of the brick border} = \text{area of pool plus brick border} - \text{area of pool}$$
$$= (4x^2 + 140x + 1200) - 1200$$
$$= 4x^2 + 140x$$

The area of the brick border must be 296 ft^2. Therefore,

$$296 = 4x^2 + 140x$$

or

$$4x^2 + 140x - 296 = 0$$
$$4(x^2 + 35x - 74) = 0 \quad \text{Factor out 4 from each term.}$$
$$\frac{4}{4}(x^2 + 35x - 74) = \frac{0}{4} \quad \text{Divide both sides of the equation by 4.}$$
$$x^2 + 35x - 74 = 0$$
$$(x + 37)(x - 2) = 0 \quad \text{Factor trinomial.}$$
$$x + 37 = 0 \quad \text{or} \quad x - 2 = 0$$
$$x = -37 \qquad x = 2$$

Since lengths are positive, the only possible answer is $x = 2$. Thus, they can make a brick border 2 ft wide all around the pool.

SECTION 6.9 EXERCISES

CONCEPT/WRITING EXERCISES

1. What is a *binomial*? Give three examples of binomials.
2. What is a *trinomial*? Give three examples of trinomials.
3. In your own words, explain the FOIL method used to multiply two binomials.
4. In your own words, state the zero-factor property.
5. Give the standard form of a quadratic equation.
6. Have you memorized the quadratic formula? If not, you need to do so. Without looking at the book, write the quadratic formula.

PRACTICE THE SKILLS

In Exercises 7–22, factor the trinomial. If the trinomial cannot be factored, so state.

7. $x^2 + 8x + 15$ **8.** $x^2 + 9x + 14$

9. $x^2 - x - 2$ **10.** $x^2 + 3x - 4$

11. $x^2 + 2x - 24$ **12.** $x^2 - 6x + 8$

13. $x^2 - 2x - 3$ **14.** $x^2 - 5x - 6$

15. $x^2 - 10x + 21$ **16.** $x^2 - x - 20$

17. $x^2 - 49$ **18.** $x^2 - 64$

19. $x^2 + 3x - 28$ **20.** $x^2 + 4x - 32$

21. $x^2 + 2x - 63$ **22.** $x^2 - 2x - 48$

In Exercises 23–34, factor the trinomial. If the trinomial cannot be factored, so state.

23. $2x^2 - x - 6$ **24.** $3x^2 - 7x - 6$

25. $5x^2 + 16x + 3$ **26.** $2x^2 + 3x - 20$

27. $5x^2 + 12x + 4$ **28.** $2x^2 - 9x + 10$

29. $4x^2 + 11x + 6$ **30.** $4x^2 + 20x + 21$

31. $4x^2 - 11x + 6$ **32.** $6x^2 - 11x + 4$

33. $8x^2 - 10x - 3$ **34.** $6x^2 - 19x + 3$

In Exercises 35–38, solve each equation, using the zero-factor property.

35. $(x - 3)(x + 6) = 0$ **36.** $(2x + 1)(x - 5) = 0$

37. $(3x + 4)(2x - 1) = 0$ **38.** $(x - 6)(5x - 4) = 0$

In Exercises 39–58, solve each equation by factoring.

39. $x^2 + 7x + 10 = 0$ **40.** $x^2 + x - 12 = 0$

41. $x^2 - 7x + 6 = 0$ **42.** $x^2 - 10x + 21 = 0$

43. $x^2 - 15 = 2x$ **44.** $x^2 - 7x = -6$

45. $x^2 = 4x - 3$ **46.** $x^2 - 13x + 40 = 0$

47. $x^2 - 81 = 0$ **48.** $x^2 - 64 = 0$

49. $x^2 + 5x - 36 = 0$ **50.** $x^2 + 12x + 20 = 0$

51. $3x^2 + 10x = 8$ **52.** $3x^2 - 5x = 2$

53. $5x^2 + 11x = -2$ **54.** $2x^2 = -5x + 3$

55. $3x^2 - 4x = -1$ **56.** $5x^2 + 16x + 12 = 0$

57. $6x^2 - 11x + 3 = 0$ **58.** $4x^2 + 11x - 3 = 0$

In Exercises 59–78, solve the equation, using the quadratic formula. If the equation has no real solution, so state.

59. $x^2 + 2x - 15 = 0$ **60.** $x^2 + 12x + 27 = 0$

61. $x^2 - 3x - 18 = 0$ **62.** $x^2 - 6x - 16 = 0$

63. $x^2 - 8x = 9$ **64.** $x^2 = -8x + 15$

65. $x^2 - 2x + 3 = 0$ **66.** $2x^2 - x - 3 = 0$

67. $x^2 - 4x + 2 = 0$ **68.** $2x^2 - 5x - 2 = 0$

69. $3x^2 - 8x + 1 = 0$ **70.** $2x^2 + 4x + 1 = 0$

71. $3x^2 - 6x - 5 = 0$ **72.** $4x^2 - 5x - 3 = 0$

73. $2x^2 + 7x + 5 = 0$ **74.** $3x^2 = 9x - 5$

75. $3x^2 - 10x + 7 = 0$ **76.** $4x^2 + 7x - 1 = 0$

77. $4x^2 + 6x = -5$ **78.** $5x^2 - 4x = -2$

CHALLENGE PROBLEMS/GROUP ACTIVITIES

79. ***Flower Garden*** Karen and Kurt Ohliger's backyard has a width of 20 meters and a length of 30 meters. Karen and Kurt want to put a rectangular flower garden in the middle of the backyard leaving a strip of grass of uniform width around all sides of the flower garden. If they want to have 336 square meters of grass, what will be the width and length of the garden?

80. ***Lawn Maintenance Business*** The weekly profit, p, of a lawn maintenance business, in dollars, can be approximated by the equation

$$p = x^2 + 40x - 110$$

where x is the number of lawns maintained in a week. How many lawns must be maintained to have a weekly profit of $1090?

81. **a)** Explain why solving $(x - 4)(x - 7) = 6$ by setting each factor equal to 6 is not correct.

b) Determine the correct solution to $(x - 4)(x - 7) = 6$.

82. The radicand in the quadratic formula, $b^2 - 4ac$, is called the **discriminant**. How many real number solutions will the quadratic equation have if the discriminant is (a) greater than 0, (b) equal to 0, or (c) less than zero? Explain your answer.

83. Write an equation that has solutions -1 and 3.

INTERNET/RESEARCH ACTIVITIES

84. Italian mathematician Girolamo Cardano (1501–1576) is recognized for his skill in solving equations. Write a paper about his life and his contributions to mathematics, in particular his contribution to solving equations.

85. Chinese mathematician Foo Ling Awong, who lived during the Pong dynasty, developed a technique, other than trial and error, to factor trinomials of the form $ax^2 + bx + c$, $a \neq 1$. Write a paper about his life and his contributions to mathematics, in particular his technique for factoring trinomials in the form $ax^2 + bx + c$, $a \neq 1$. (References include history of mathematics books, encyclopedias, and the Internet.)

6.10 FUNCTIONS AND THEIR GRAPHS

▲ The more oranges you purchase, the greater the cost.

At a supermarket, oranges cost \$0.20 each. Thus, one orange costs $1 \times \$0.20 = \$0.20$, two oranges cost $2 \times \$0.20 = \$0.40$, three oranges cost $3 \times \$0.20 = \$0.60$, and so on. In this section, we will discuss relationships between two quantities such as the number of oranges purchased and the cost of the oranges purchased.

Relations and *functions* are both extremely important mathematical concepts that we will introduce in this section. A ***relation*** is any set of ordered pairs. Every graph is a set of ordered pairs. Therefore, any graph will be a relation. We can indicate the relation between the number of oranges purchased and the cost of the oranges discussed in the section opening paragraph in a table of values.

Number of Oranges	Cost
0	0.00
1	0.20
2	0.40
3	0.60
⋮	⋮
10	2.00
⋮	⋮

In general, the cost for purchasing n oranges will be 20 cents times the number of oranges, or $0.20n$ dollars. We can represent the cost, c, of n items by the equation $c = 0.20n$. Since the value of c depends on the value of n, we refer to c as the *dependent variable* and n as the *independent variable. Note that for each value of the independent variable, n, there is one and only one value of the dependent variable, c.*

An equation such as $c = 0.20n$ is called a *function*. In the equation $c = 0.20n$, the value of c depends on the value of n, so we say that "c is a function of n."

A **function** is a special type of relation where each value of the independent variable corresponds to a unique value of the dependent variable.

The set of values that can be used for the independent variable is called the *domain* of the function, and the resulting set of values obtained for the dependent variable is called the *range*. The domain and range for the function $c = 0.20n$ are illustrated in Fig. 6.31.

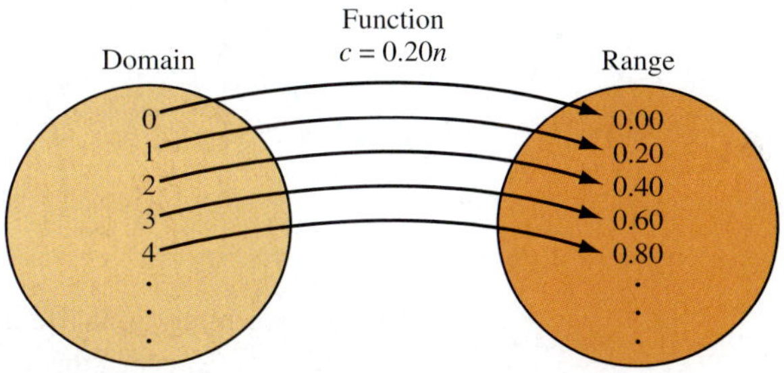

Figure 6.31

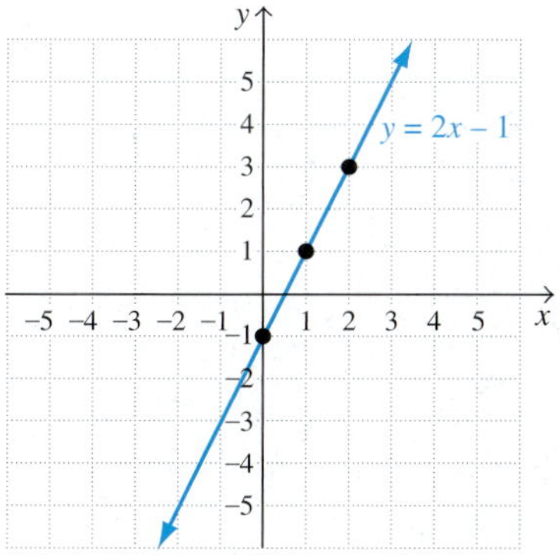

Figure 6.32

When we graphed equations of the form $ax + by = c$ in Section 6.7, we found that they were straight lines. For example, the graph of $y = 2x - 1$ is illustrated in Fig. 6.32.

Is the equation $y = 2x - 1$ a function? To answer this question, we must ask, Does each value of x correspond to a unique value of y? The answer is yes; therefore, this equation is a function.

For the equation $y = 2x - 1$, we say that "y is a function of x" and write $y = f(x)$. The notation $f(x)$ is read "f of x." When we are given an equation that is a function, we may replace the y in the equation with $f(x)$ because $f(x)$ represents y. Thus, $y = 2x - 1$ may be written $f(x) = 2x - 1$.

To evaluate a function for a specific value of x, replace each x in the function with the given value, then evaluate. For example, to evaluate $f(x) = 2x - 1$ when $x = 8$, we do the following.

$$f(x) = 2x - 1$$
$$f(8) = 2(8) - 1 = 16 - 1 = 15$$

Thus, $f(8) = 15$. Since $f(x) = y$, when $x = 8$, $y = 15$. What is the domain and range of $f(x) = 2x - 1$? Because x can be any real number, the domain is the set of real numbers, symbolized $\mathbb{R}$. The range is also $\mathbb{R}$.

We can determine whether a graph represents a function by using the *vertical line test*: If a vertical line can be drawn so that it intersects the graph at more than one point, then each value of x does not have a unique value of y and the graph does not represent a function. If a vertical line cannot be made to intersect the graph in at least two different places, then the graph represents a function.

EXAMPLE 1 *Using the Vertical Line Test*

Use the vertical line test to determine which of the graphs in Figure 6.33 represent functions.

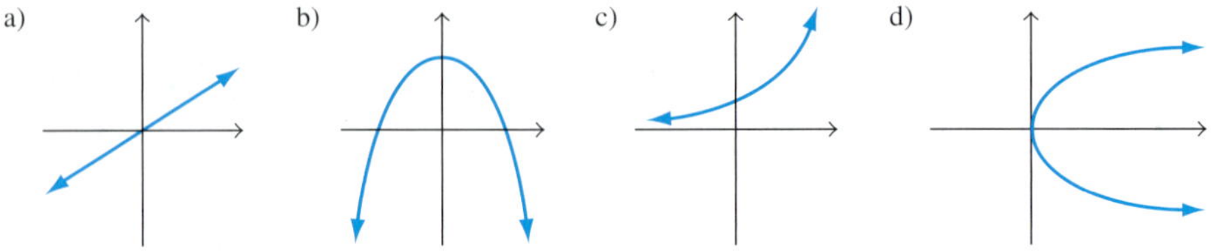

Figure 6.33

SOLUTION (a), (b), and (c) represent functions, but (d) does not.

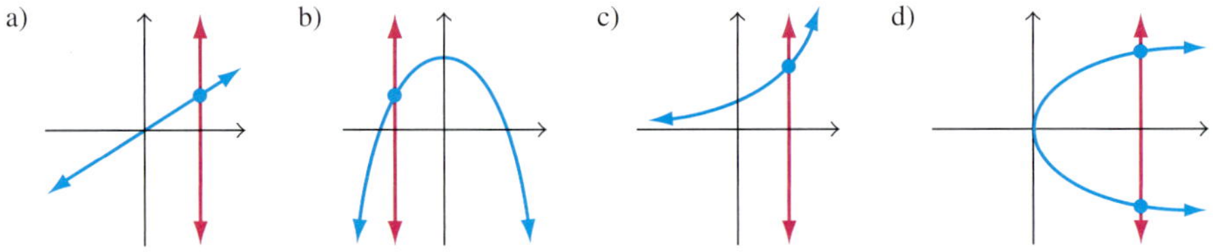

There are many real-life applications of functions. In fact, all the applications illustrated in Sections 6.2 through 6.4 are functions.

In this section, we will discuss three types of functions: linear functions, quadratic functions, and exponential functions.

Linear Functions

In Section 6.7, we graphed linear equations. The graph of any linear equation of the form $y = ax + b$ will pass the vertical line test, and so equations of the form $y = ax + b$ are ***linear functions***. If we wished, we could write the linear function as $f(x) = ax + b$ since $f(x)$ means the same as y.

EXAMPLE 2 *Cost as a Linear Function*

Adam Finiteri's weekly cost of operating a taxi, c, is given by the function $c(m) = 52 + 0.23m$, where m is the number of miles driven per week. What is his weekly cost if he drives 200 miles in a week?

SOLUTION Substitute 200 for m in the function.

$$c(m) = 52 + 0.23m$$
$$c(200) = 52 + 0.23(200)$$
$$c(200) = 52 + 46 = 98$$

Thus, if Adam drives 200 miles in a week, his weekly cost will be \$98.

Graphs of Linear Functions

The graphs of linear functions are straight lines that will pass the vertical line test. In Section 6.7, we discussed how to graph linear equations. Linear functions can be graphed by plotting points, by using intercepts, or by using the slope and y-intercept.

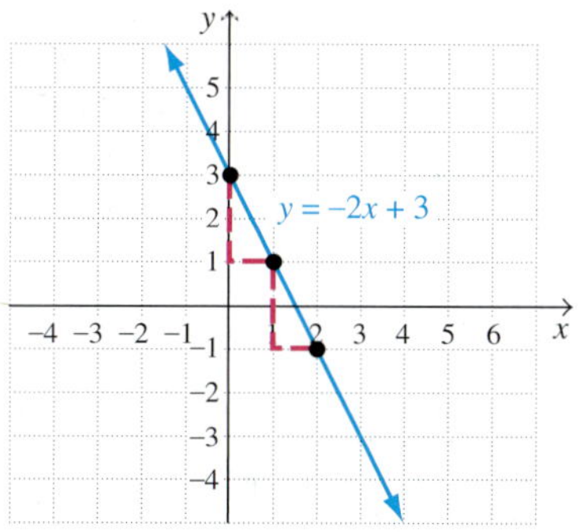

Figure 6.34

EXAMPLE 3 *Graphing a Linear Function*

Graph $f(x) = -2x + 3$ by using the slope and y-intercept.

SOLUTION Since $f(x)$ means the same as y, we can rewrite this function as $y = -2x + 3$. From Section 6.7, we know that the slope is -2 and the y-intercept is (0, 3). Plot (0, 3) on the y-axis. Then plot the next point by moving *down* 2 units and to the *right* 1 unit (see Fig. 6.34). A third point has been plotted in the same way. The graph of $f(x) = -2x + 3$ is the line drawn through these three points. ●

Quadratic Functions

The standard form of a quadratic equation is $y = ax^2 + bx + c, a \neq 0$. We will learn shortly that graphs of equations of this form always pass the vertical line test and are functions. Therefore, equations of the form $y = ax^2 + bx + c, a \neq 0$, may be referred to as ***quadratic functions***. We may express quadratic functions using function notation as $f(x) = ax^2 + bx + c$. Two examples of quadratic functions are $y = 2x^2 + 5x - 7$ and $y = -\frac{1}{2}x^2 + 4$.

EXAMPLE 4 *Landing on the Moon*

On July 20, 1969, Neil Armstrong became the first person to walk on the moon. The velocity, v, of his spacecraft, the *Eagle*, in meters per second, was a function of time before touchdown, t, given by

$$v = f(t) = 3.2t + 0.45$$

The height of the spacecraft, h, above the moon's surface, in meters, was also a function of time before touchdown, given by

$$h = g(t) = 1.6t^2 + 0.45t$$

What was the velocity of the spacecraft and its distance from the surface of the moon

a) at 3 seconds before touchdown? b) at touchdown (0 seconds)?

SOLUTION

a)
$$\begin{aligned} v = f(t) &= 3.2t + 0.45, \\ f(3) &= 3.2(3) + 0.45 \\ &= 9.6 + 0.45 \\ &= 10.05 \end{aligned} \qquad \begin{aligned} h = g(t) &= 1.6t^2 + 0.45t \\ g(3) &= 1.6(3)^2 + 0.45(3) \\ &= 1.6(9) + 1.35 \\ &= 14.4 + 1.35 \\ &= 15.75 \end{aligned}$$

The velocity 3 seconds before touchdown was 10.05 meters per second, and the height 3 seconds before touchdown was 15.75 meters.

b)
$$\begin{aligned} v = f(t) &= 3.2t + 0.45, \\ f(0) &= 3.2(0) + 0.45 \\ &= 0 + 0.45 \\ &= 0.45 \end{aligned} \qquad \begin{aligned} h = g(t) &= 1.6t^2 + 0.45t \\ g(0) &= 1.6(0)^2 + 0.45(0) \\ &= 0 + 0 \\ &= 0 \end{aligned}$$

The touchdown velocity was 0.45 meter per second. At touchdown, the *Eagle* is on the moon, and the distance from the surface of the moon is therefore 0 meter. ●

DID YOU KNOW?

Sea of Tranquillity

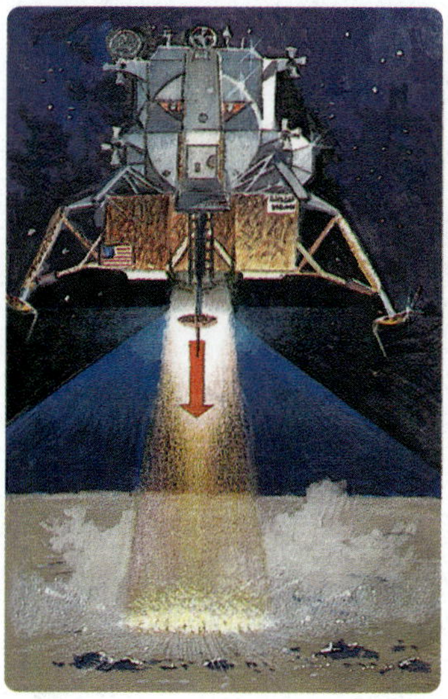

Apollo II touched down at Mare Tranqillitatis, the Sea of Tranquillity. The rock samples taken there placed the age of the rocks at 3.5 billion years old, as old as the oldest known Earth rocks.

DID YOU KNOW?

Computing Reality with Mathematical Models

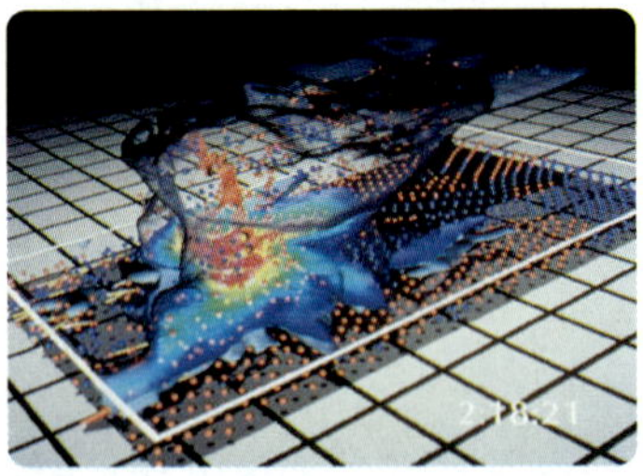

It is a relatively easy matter for scientists and mathematicians to describe and predict a simple motion like that of a falling object. When the phenomenon is complicated, such as making an accurate prediction of the weather, the mathematics becomes much more difficult. The National Weather Service has devised an algorithm that takes the temperature, pressure, moisture content, and wind velocity of more than 250,000 points in Earth's atmosphere and applies a set of equations that, it believes, will reasonably predict what will happen at each point over time. Researchers at the National Center for Supercomputing Applications are working on a computer model that simulates thunderstorms. They want to know why some thunderstorms can turn severe, even deadly.

Graphs of Quadratic Functions

The graph of every quadratic function is a parabola. Two parabolas are illustrated in Fig. 6.35. Note that both graphs represent functions since they pass the vertical line test. A parabola opens upward when the coefficient of the squared term, a, is greater than 0, as shown in Figure 6.35(a). A parabola opens downward when the coefficient of the squared term, a, is less than 0, as shown in Fig. 6.35(b).

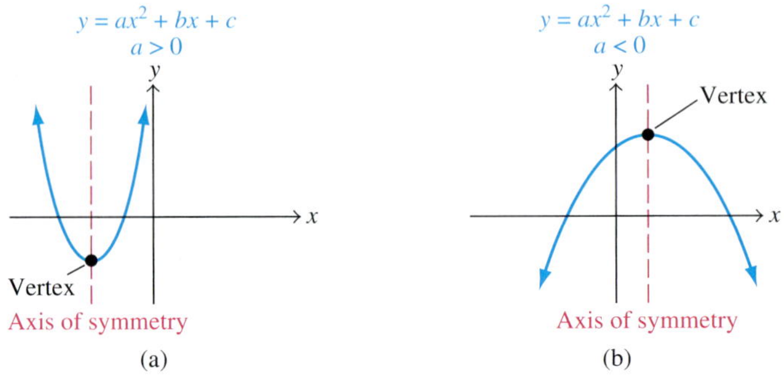

Figure 6.35

The *vertex* of a parabola is the lowest point on a parabola that opens upward and the highest point on a parabola that opens downward. Every parabola is *symmetric* with respect to a vertical line through its vertex. This line is called the *axis of symmetry* of the parabola. The x-coordinate of the vertex and the equation of the axis of symmetry can be found by using the following formula.

AXIS OF SYMMETRY OF A PARABOLA

$$x = \frac{-b}{2a}$$

This formula also gives the x-coordinate of the vertex of a parabola.

Once the x-coordinate of the vertex has been determined, the y-coordinate can be found by substituting the value found for the x-coordinate into the quadratic equation and evaluating the equation. This procedure is illustrated in Example 5.

EXAMPLE 5 *Describing the Graph of a Quadratic Equation*

Consider the equation $y = -2x^2 + 12x - 10$.

a) Determine whether the graph will be a parabola that opens upward or downward.

b) Determine the equation of the axis of symmetry of the parabola.

c) Determine the vertex of the parabola.

SOLUTION

a) Since $a = -2$, which is less than 0, the parabola opens downward.

b) To find the axis of symmetry, we use the equation $x = -\frac{b}{2a}$. In the equation $y = -2x^2 + 12x - 10$, $a = -2$, $b = 12$, and $c = -10$, so

$$x = \frac{-b}{2a} = \frac{-(12)}{2(-2)} = \frac{-12}{-4} = 3$$

The equation of the axis of symmetry is $x = 3$.

c) The x-coordinate of the vertex is 3 from part (b). To find the y-coordinate, we substitute 3 for x in the equation and then evaluate.

$$\begin{aligned} y &= -2x^2 + 12x - 10 \\ &= -2(3)^2 + 12(3) - 10 \\ &= -2(9) + 36 - 10 \\ &= -18 + 36 - 10 \\ &= 8 \end{aligned}$$

Therefore, the vertex of the parabola is located at the ordered pair (3, 8) on the graph.

GENERAL PROCEDURE TO SKETCH THE GRAPH OF A QUADRATIC EQUATION

1. Determine whether the parabola opens upward or downward.
2. Determine the equation of the axis of symmetry.
3. Determine the vertex of the parabola.
4. Determine the y-intercept by substituting $x = 0$ into the equation.
5. Determine the x-intercepts (if they exist) by substituting $y = 0$ into the equation and solving for x.
6. Draw the graph, making use of the information gained in steps 1 through 5. Remember that the parabola will be symmetric with respect to the axis of symmetry.

In step 5, you may use either factoring or the quadratic formula to determine the x-intercepts.

EXAMPLE 6 *Graphing a Quadratic Equation*

Sketch the graph of the equation $y = x^2 - 6x + 8$.

SOLUTION We follow the steps outlined in the general procedure.

1. Since $a = 1$, which is greater than 0, the parabola opens upward.
2. Axis of symmetry: $x = \dfrac{-b}{2a} = \dfrac{-(-6)}{2(1)} = \dfrac{6}{2} = 3$

   Thus, the axis of symmetry is $x = 3$.
3. y-coordinate of vertex:

   $y = x^2 - 6x + 8$

   $= (3)^2 - 6(3) + 8 = 9 - 18 + 8 = -1$

   Thus, the vertex is at $(3, -1)$.
4. y-intercept: $y = x^2 - 6x + 8$

   $y = 0^2 - 6(0) + 8 = 8$

   Thus, the y-intercept is at (0, 8).
5. x-intercepts: $0 = x^2 - 6x + 8$, or $x^2 - 6x + 8 = 0$

   We can solve this equation by factoring.

$$\begin{aligned} x^2 - 6x + 8 &= 0 \\ (x - 4)(x - 2) &= 0 \end{aligned}$$

$$x - 4 = 0 \quad \text{or} \quad x - 2 = 0$$
$$x = 4 \qquad\qquad x = 2$$

Thus, the x-intercepts are (4, 0) and (2, 0).

DID YOU KNOW?

Gravity and the Parabola

Any golfer knows that a golf ball will arch in the same path going down as it did going up. Early gunners knew it, too. To hit a distant target, the cannon barrel was pointed skyward, not directly at the target. A cannonball fired at a 45° angle will travel the greatest horizontal distance. What the golfer and gunner alike were allowing for is the effect of gravity on a projectile. Projectile motion follows a parabolic path. Galileo was neither a gunner nor a golfer, but he gave us the formula that effectively describes that motion and the distance traveled by an object if it is projected at a specific angle with a specific initial velocity.

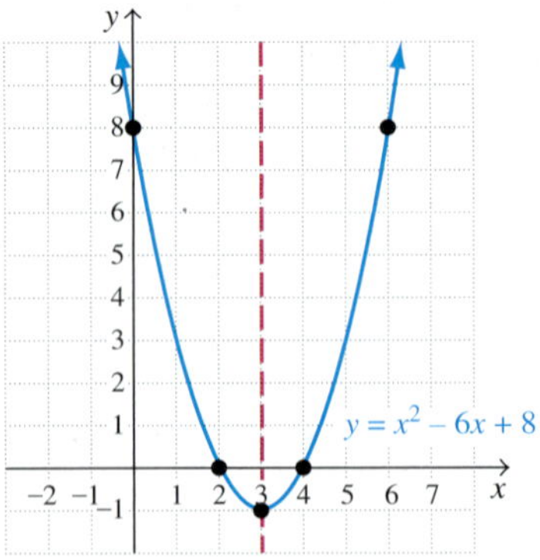

Figure 6.36

6. Plot the vertex $(3, -1)$, the y-intercept $(0, 8)$, and the x-intercepts $(4, 0)$ and $(2, 0)$. Then sketch the graph (Fig. 6.36).

Note that the domain of the graph in Example 6, the possible x values, is the set of all real numbers, $\mathbb{R}$. The range, the possible y values, is the set of all real numbers greater than or equal to -1. When graphing parabolas, if you think you need additional points to graph the equation, you can always substitute values for x and find the corresponding values of y and plot those points. For example, if you substituted 1 for x, the corresponding value of y is 3. Thus, you could plot the point $(1, 3)$.

EXAMPLE 7 *Domain and Range of a Quadratic Function*

a) Sketch the graph of the function $f(x) = -2x^2 + 3x + 4$.

b) Determine the domain and range of the function.

SOLUTION

a) Since $f(x)$ means y, we can replace $f(x)$ with y to obtain $y = -2x^2 + 3x + 4$. Now graph $y = -2x^2 + 3x + 4$ using the steps outlined in the general procedure.

1. Since $a = -2$, which is less than 0, the parabola opens downward.

2. Axis of symmetry: $x = \dfrac{-b}{2a} = \dfrac{-(3)}{2(-2)} = \dfrac{-3}{-4} = \dfrac{3}{4}$

   Thus, the axis of symmetry is $x = \frac{3}{4}$.

3. y-coordinate of vertex:

$$\begin{aligned} y &= -2x^2 + 3x + 4 \\ &= -2\left(\frac{3}{4}\right)^2 + 3\left(\frac{3}{4}\right) + 4 \\ &= -2\left(\frac{9}{16}\right) + \frac{9}{4} + 4 \\ &= -\frac{9}{8} + \frac{9}{4} + 4 \\ &= -\frac{9}{8} + \frac{18}{8} + \frac{32}{8} = \frac{41}{8} \quad \text{or} \quad 5\frac{1}{8} \end{aligned}$$

   Thus, the vertex is at $\left(\frac{3}{4}, 5\frac{1}{8}\right)$.

4. y-intercept: $y = -2x^2 + 3x + 4$

$$= -2(0)^2 + 3(0) + 4 = 4$$

   Thus, the y-intercept is $(0, 4)$.

5. x-intercepts: $y = -2x^2 + 3x + 4$

$$0 = -2x^2 + 3x + 4 \quad \text{or} \quad -2x^2 + 3x + 4 = 0$$

This equation cannot be factored, so we will use the quadratic formula to solve it.

$$a = -2, \qquad b = 3, \qquad c = 4$$

$$\begin{aligned} x &= \frac{-b \pm \sqrt{b^2 - 4ac}}{2a} \\ &= \frac{-3 \pm \sqrt{3^2 - 4(-2)(4)}}{2(-2)} \\ &= \frac{-3 \pm \sqrt{9 + 32}}{-4} \\ &= \frac{-3 \pm \sqrt{41}}{-4} \end{aligned}$$

Since $\sqrt{41} \approx 6.4$,

$$x \approx \frac{-3 + 6.4}{-4} \approx \frac{3.4}{-4} \approx -0.85 \qquad \text{or} \qquad x \approx \frac{-3 - 6.4}{-4} \approx \frac{-9.4}{-4} \approx 2.35$$

Thus, the x-intercepts are about $(-0.85, 0)$ and $(2.35, 0)$.

6. Plot the vertex $(\frac{3}{4}, 5\frac{1}{8})$, the y-intercept $(0, 4)$, and the x-intercepts $(-0.85, 0)$ and $(2.35, 0)$. Then sketch the graph (Fig. 6.37).

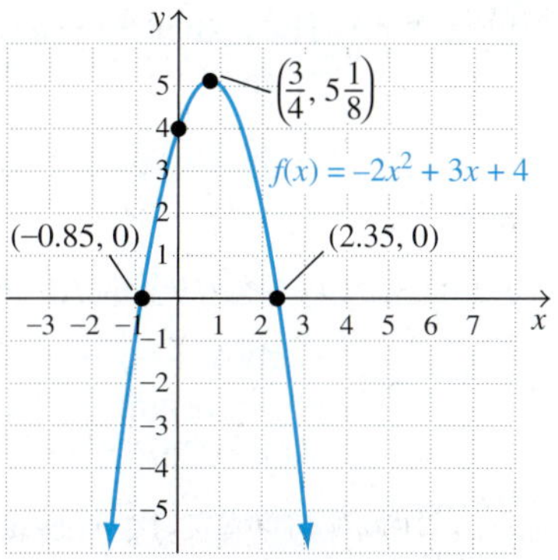

Figure 6.37

b) The domain, the values that can be used for x, is the set of all real numbers, $\mathbb{R}$. The range, the values of y, is $y \leq 5\frac{1}{8}$.

When we use the quadratic formula to find the x-intercepts of a graph, if the radicand, $b^2 - 4ac$, is a negative number, the graph has no x-intercepts. The graph will lie totally above or below the x-axis.

Exponential Functions

In Section 6.3, we discussed exponential equations. Recall that exponential equations are of the form $y = a^x, a > 0, a \neq 1$. The graph of every exponential equation will pass the vertical line test, and so every exponential equation is also an exponential function. Exponential functions may be written as $f(x) = a^x, a > 0, a \neq 1$. When the base a is replaced with the base e, which was discussed on page 333, we get the ***natural exponential function*** $f(x) = e^x$.

In Section 6.3, we also introduced the natural exponential growth or decay formula $P = P_0e^{kt}$. We can write this formula in function notation as $P(t) = P_0e^{kt}$. This expression is referred to as the ***natural exponential growth or decay function***. In Examples 8 and 9, we use the natural exponential growth or decay function.

EXAMPLE 8 *Evaluating an Exponential Decay Function*

Plutonium, a radioactive material used in most nuclear reactors, decays exponentially at a rate of 0.003% per year. If there are originally 2000 grams of plutonium, the amount of plutonium, P, remaining after t years is $P(t) = 2000e^{-0.00003t}$. How much plutonium will remain after 50 years?

SOLUTION Substitute 50 years for t in the function, then evaluate using a calculator as described in Section 6.3.

$$\begin{aligned} P(t) &= 2000e^{-0.00003t} \\ P(50) &= 2000e^{-0.00003(50)} \\ &= 2000e^{-0.0015} \\ &\approx 2000(0.9985011244) \\ &\approx 1997.0 \text{ grams} \end{aligned}$$

Thus, after 50 years, the amount of plutonium remaining will be about 1997 grams.

EXAMPLE 9 *Evaluating an Exponential Decay Function*

Carbon-14 is used by scientists to find the age of fossils and other artifacts. If an object originally had 25 grams of carbon-14, the amount present after t years is $f(t) = 25e^{-0.00012010t}$. How much carbon-14 will be found after 350 years?

SOLUTION Substitute 350 for t in the function, then evaluate using a calculator as described in Section 6.3.

$$\begin{aligned} f(t) &= 25e^{-0.00012010t} \\ f(350) &= 25e^{-0.00012010(350)} \\ &= 25e^{-0.042035} \\ &\approx 25(0.9588362207) \\ &\approx 23.97090552 \\ &\approx 24 \text{ grams} \end{aligned}$$

Thus, after 350 years, about 24 grams of carbon-14 will remain.

Graphs of Exponential Functions

What does the graph of an ***exponential function*** of the form $y = a^x, a > 0, a \neq 1$, look like? Examples 10 and 11 illustrate graphs of exponential functions.

DID YOU KNOW?

Population Growth

Population growth during certain time periods can be described by an exponential function. Whether it is a population of bacteria, fish, flowers, or people, the same general trend emerges: a period of rapid (exponential) growth, which is then followed by a leveling-off period.

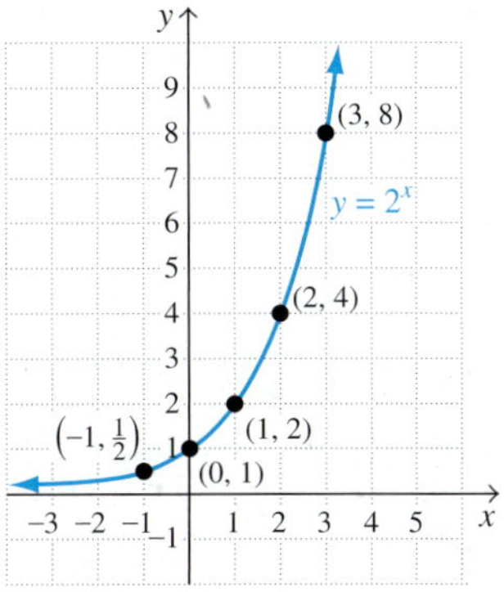

Figure 6.38

EXAMPLE 10 *Graphing an Exponential Function with a Base Greater Than 1*

a) Graph $y = 2^x$.

b) Determine the domain and range of the function.

SOLUTION

a) Substitute values for x and find the corresponding values of y. The graph is shown in Fig. 6.38.

$$y = 2^x$$

$$x = -3, \quad y = 2^{-3} = \frac{1}{2^3} = \frac{1}{8}$$

$$x = -2, \quad y = 2^{-2} = \frac{1}{2^2} = \frac{1}{4}$$

$$x = -1, \quad y = 2^{-1} = \frac{1}{2^1} = \frac{1}{2}$$

$$x = 0, \quad y = 2^0 = 1$$

$$x = 1, \quad y = 2^1 = 2$$

$$x = 2, \quad y = 2^2 = 4$$

$$x = 3, \quad y = 2^3 = 8$$

x	y
-3	$\frac{1}{8}$
-2	$\frac{1}{4}$
-1	$\frac{1}{2}$
0	1
1	2
2	4
3	8

b) The domain is all real numbers, $\mathbb{R}$. The range is $y > 0$. Note that y can never have a value of 0.

All exponential functions of the form $y = a^x$, $a > 1$, will have the general shape of the graph illustrated in Fig. 6.38. Since $f(x)$ is the same as y, the graphs of functions of the form $f(x) = a^x$, $a > 1$, will also have the general shape of the graph illustrated in Fig. 6.38. Can you now predict the shape of the graph of $y = e^x$? Remember: e has a value of about 2.7183.

EXAMPLE 11 *Graphing an Exponential Function with a Base Between 0 and 1*

a) Graph $y = (\frac{1}{2})^x$.

b) Determine the domain and range of the function.

SOLUTION

a) We begin by substituting values for x and calculating values for y. We then plot the ordered pairs and use these points to sketch the graph. To evaluate a fraction with a negative exponent, we use the fact that

$$\left(\frac{a}{b}\right)^{-x} = \left(\frac{b}{a}\right)^x$$

For example,

$$\left(\frac{1}{2}\right)^{-3} = \left(\frac{2}{1}\right)^3 = 8$$

Then

$$y = \left(\frac{1}{2}\right)^x$$

$$x = -3, \quad y = \left(\frac{1}{2}\right)^{-3} = 2^3 = 8$$

$$x = -2, \quad y = \left(\frac{1}{2}\right)^{-2} = 2^2 = 4$$

$$x = -1, \quad y = \left(\frac{1}{2}\right)^{-1} = 2^1 = 2$$

$$x = 0, \quad y = \left(\frac{1}{2}\right)^{0} = 1$$

$$x = 1, \quad y = \left(\frac{1}{2}\right)^{1} = \frac{1}{2}$$

$$x = 2, \quad y = \left(\frac{1}{2}\right)^{2} = \frac{1}{4}$$

$$x = 3, \quad y = \left(\frac{1}{2}\right)^{3} = \frac{1}{8}$$

x	y
-3	8
-2	4
-1	2
0	1
1	$\frac{1}{2}$
2	$\frac{1}{4}$
3	$\frac{1}{8}$

The graph is illustrated in Fig. 6.39.

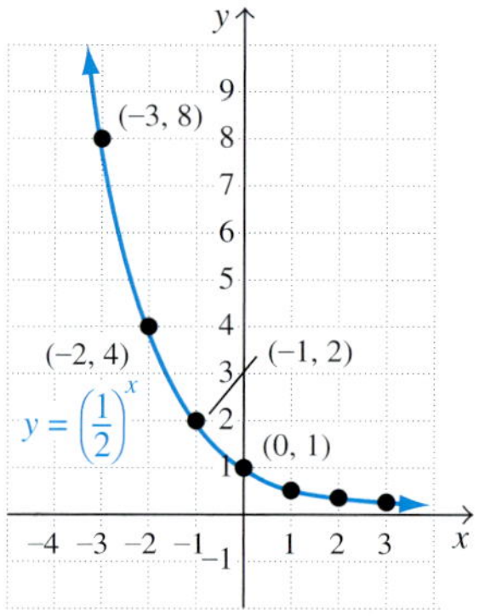

Figure 6.39

b) The domain is the set of all real numbers, $\mathbb{R}$. The range is $y > 0$.

All exponential functions of the form $y = a^x$ or $f(x) = a^x$, $0 < a < 1$, will have the general shape of the graph illustrated in Fig. 6.39.

EXAMPLE 12 *Is the Growth Exponential?*

The number of new stores opened by Starbucks in the United States has increased greatly since 1994. The graph on the next page shows the number of Starbucks stores in the United States for selected years and projected to 2010.

a) Does the graph approximate the graph of an exponential function?

b) Estimate the number of Starbucks stores in the United States in 2006.

SOLUTION

a) Yes, the graph has the approximate shape of an exponential function. A function that increases rapidly with this general shape is, or approximates, an exponential function.

b) From the graph, we see that Starbucks had about 7500 U.S. stores in 2006.

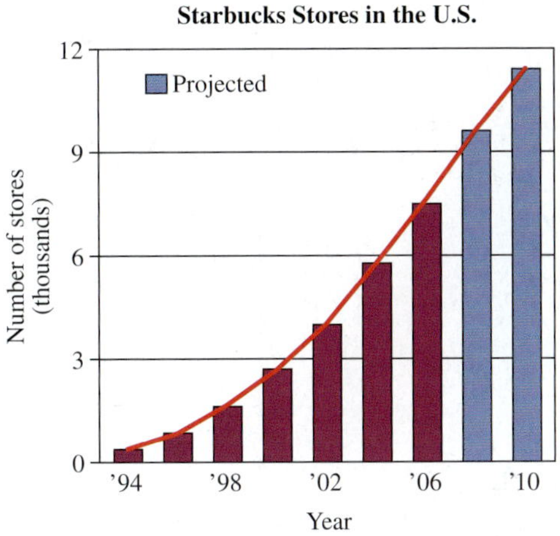

Source: Starbucks, CIBC World Markets

SECTION 6.10 EXERCISES

CONCEPT/WRITING EXERCISES

1. What is a function?
2. What is a relation?
3. What is the domain of a function?
4. What is the range of a function?
5. Explain how and why the vertical line test can be used to determine whether a graph is that of a function.
6. Give three examples of one quantity being a function of another quantity.
7. What is the formula used to determine the equation for the axis of symmetry of a parabola?
8. Explain how to determine whether the graph of a quadratic equation of the form $y = ax^2 + bx + c$, $a \neq 0$ opens upward or downward.

PRACTICE THE SKILLS

In Exercises 9–26, determine whether the graph represents a function. If it does represent a function, give its domain and range.

9.

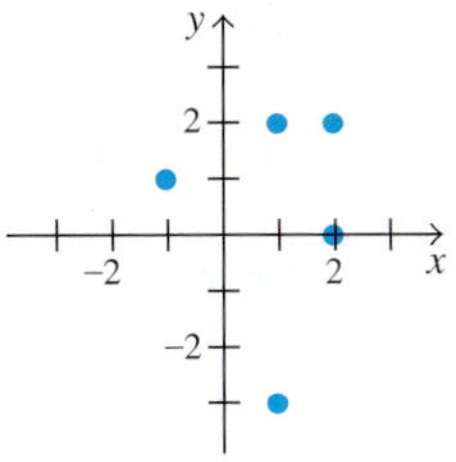

10.

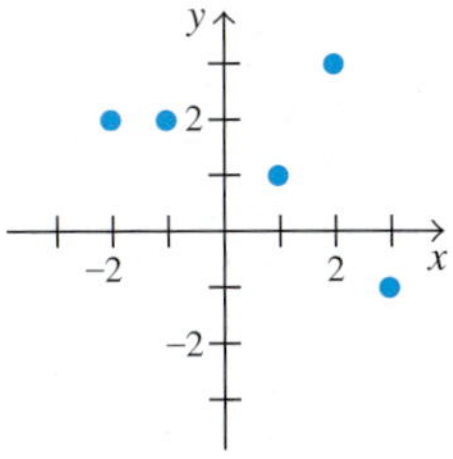

11.

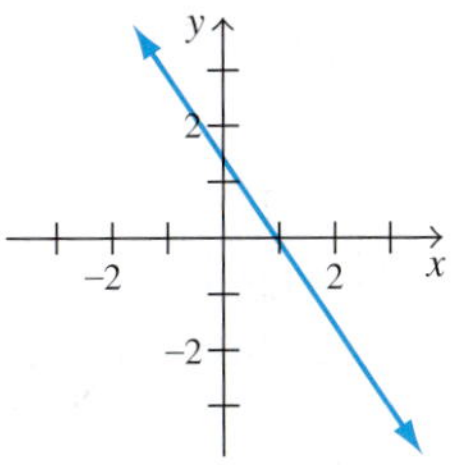

12.

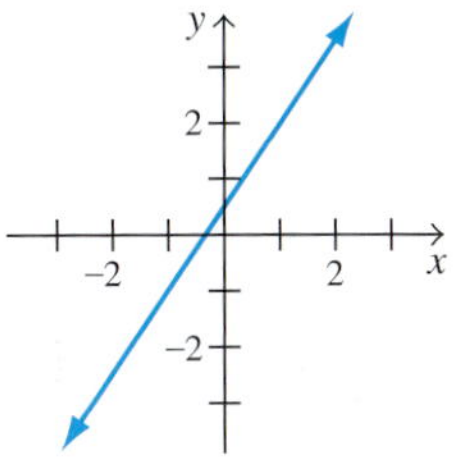

13.

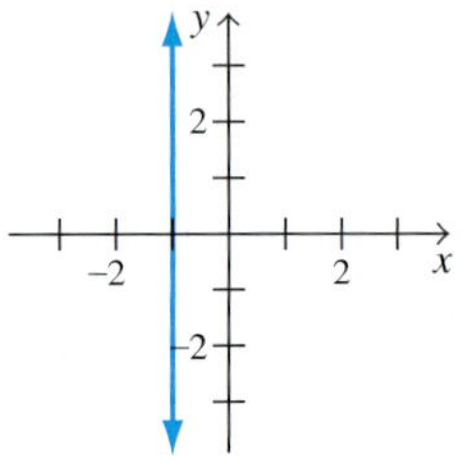

14.

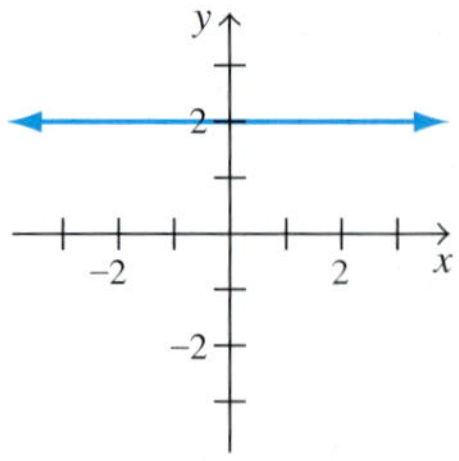

15.

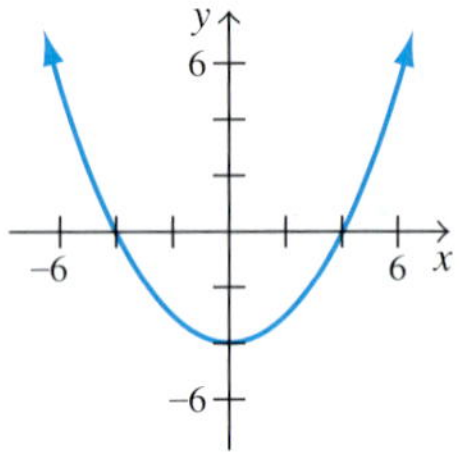

16.

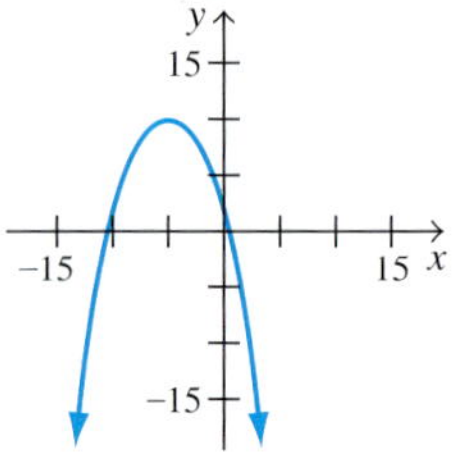

17.

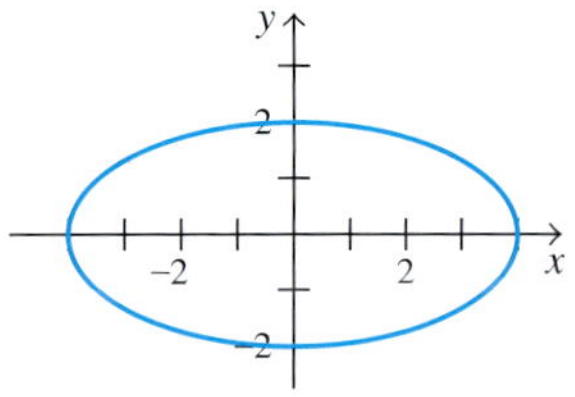

18.

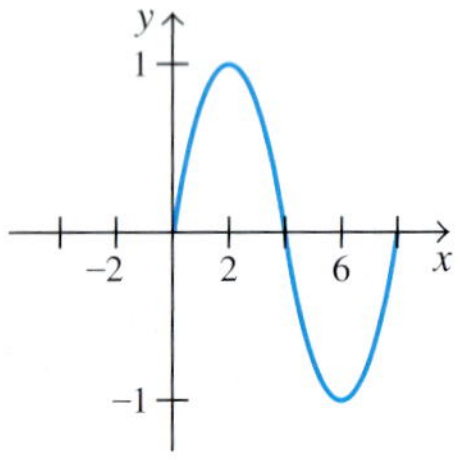

19.

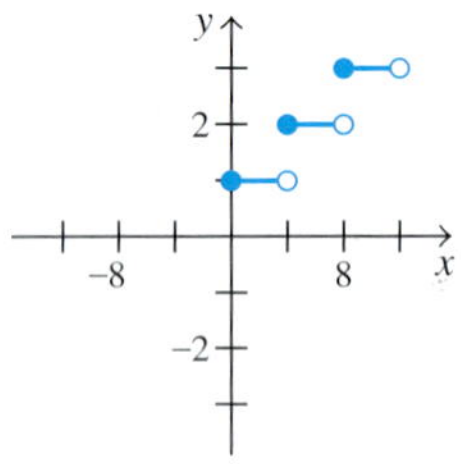

20.

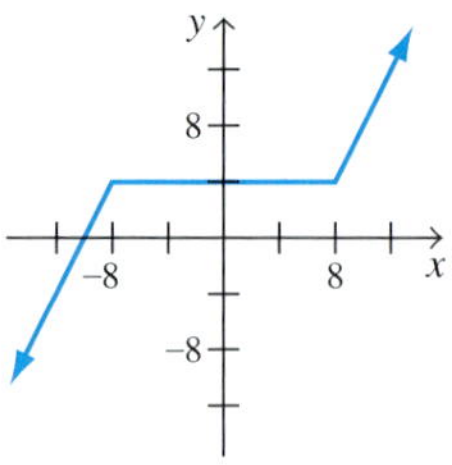

21.

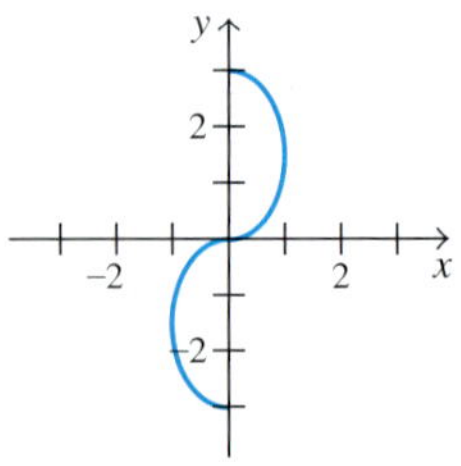

22.

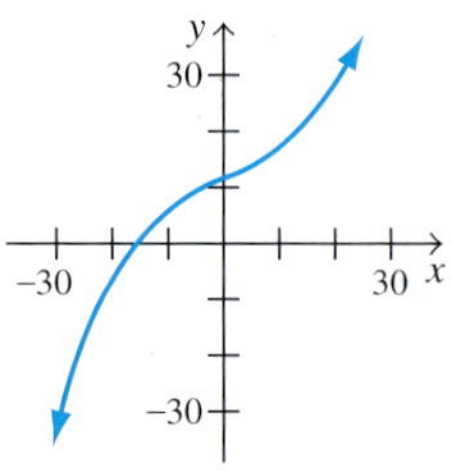

23.

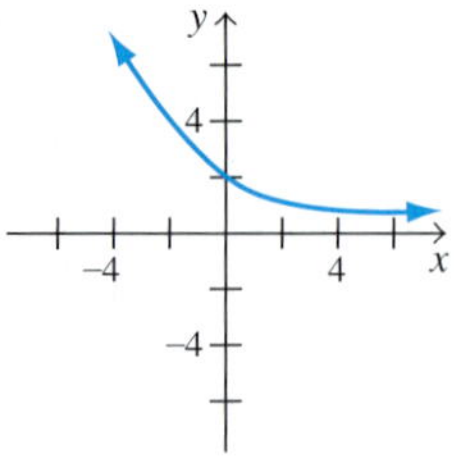

24.

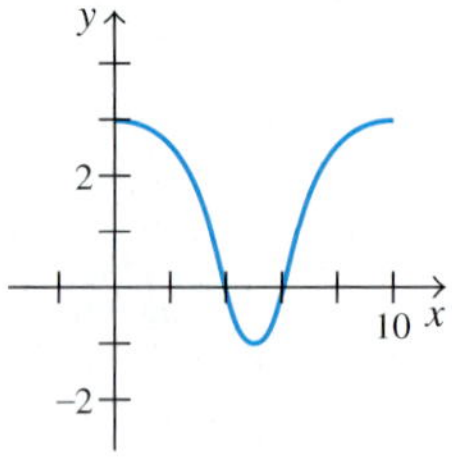

25.

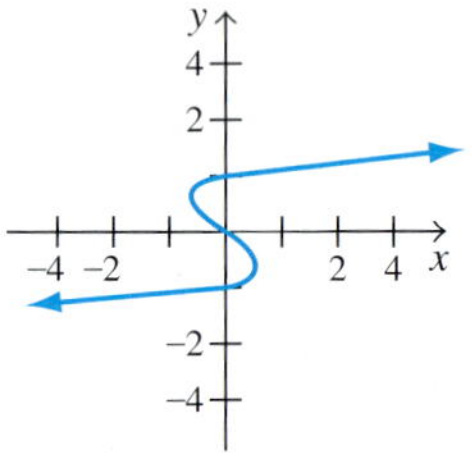

26.

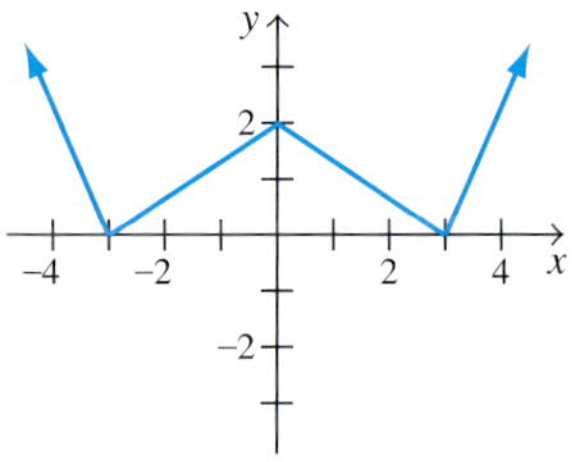

In Exercises 27–32, determine whether the set of ordered pairs is a function.

27. $\{(5, 9), (3, 2), (8, 3), (9, 15)\}$

28. $\{(2, -3), (4, -5), (5, -7), (6, -9)\}$

29. $\{(6, 4), (6, 1), (4, 0)\}$

30. $\{(4, 1), (4, 3), (4, 5)\}$

31. $\{(8, 3), (2, 3), (-1, 3)\}$

32. $\{(1, -1), (1, 0), (1, 1)\}$

In Exercises 33–46, evaluate the function for the given value of x.

33. $f(x) = x + 5, \quad x = 3$

34. $f(x) = 2x + 4, \quad x = 1$

35. $f(x) = -2x - 7, \quad x = -4$

36. $f(x) = -5x + 3, \quad x = -1$

37. $f(x) = 10x - 6, \quad x = 0$

38. $f(x) = 7x - 6, \quad x = 4$

39. $f(x) = x^2 - 3x + 1, \quad x = 4$

40. $f(x) = x^2 - 5, \quad x = 7$

41. $f(x) = -x^2 - 2x + 1, \quad x = 4$

42. $f(x) = 2x^2 + 3x - 1, \quad x = 3$

43. $f(x) = 4x^2 - 6x - 9, \quad x = -3$

44. $f(x) = 3x^2 + 2x - 5, \quad x = -4$

45. $f(x) = -5x^2 + 3x - 9, \quad x = -1$

46. $f(x) = -3x^2 - 6x + 10, \quad x = -2$

In Exercises 47–52, graph the function by using the slope and y-intercept.

47. $f(x) = 3x - 2$

48. $f(x) = -x - 1$

49. $f(x) = -4x + 2$

50. $f(x) = 2x + 5$

51. $f(x) = \frac{3}{2}x - 1$

52. $f(x) = -\frac{1}{2}x + 3$

In Exercises 53–68,

a) *determine whether the parabola will open upward or downward.*

b) *find the equation of the axis of symmetry.*

c) *find the vertex.*

d) *find the y-intercept.*

e) *find the x-intercepts if they exist.*

f) *sketch the graph.*

g) *find the domain and range of the function.*

53. $y = x^2 - 9$

54. $y = x^2 - 16$

55. $y = -x^2 + 4$

56. $y = -x^2 + 16$

57. $y = -2x^2 - 8$

58. $f(x) = -x^2 - 4$

59. $y = 2x^2 - 3$

60. $f(x) = -3x^2 - 6$

61. $f(x) = x^2 + 2x + 6$

62. $y = x^2 - 8x + 1$

63. $y = x^2 + 5x + 6$

64. $y = x^2 - 7x - 8$

65. $y = -x^2 + 4x - 6$

66. $y = -x^2 + 8x - 8$

67. $y = -2x^2 + 3x - 2$

68. $y = -4x^2 - 6x + 4$

In Exercises 69–80, draw the graph of the function and state the domain and range.

69. $y = 3^x$

70. $f(x) = 4^x$

71. $y = (\frac{1}{3})^x$

72. $y = (\frac{1}{4})^x$

73. $f(x) = 2^x + 1$

74. $y = 3^x - 1$

75. $y = 4^x + 1$

76. $y = 2^x - 1$

77. $y = 3^{x-1}$

78. $y = 3^{x+1}$

79. $f(x) = 4^{x+1}$

80. $y = 4^{x-1}$

PROBLEM SOLVING

81. ***Yearly Profit*** Marta Rivera is a part owner of a newly opened bagel company. Marta's yearly profit, in dollars, is given by the function $p(x) = 0.3x - 4000$, where x is the number of bagels sold per year. If Marta sells 150,000 bagels a year, determine her yearly profit.

82. ***Finding Distances*** The distance a car travels, $d(t)$, at a constant 60 mph is given by the function $d(t) = 60t$, where t is the time in hours. Find the distance traveled in

a) 3 hours.

b) 7 hours.

83. ***U.S. Foreign-Born Population*** The following graph indicates the percentage of the U.S. population that was foreign-born for each decade between 1930 and 2000, and projected percentage for 2010. The function $f(x) = 0.005x^2 - 0.36x + 11.8$ can be used to estimate the percentage of the U.S. population that was foreign-born, where x is the number of years since 1930 and $0 \leq x \leq 80$.

a) Use the function $f(x)$ to estimate the percentage of the U.S. population that was foreign-born in 2004. Round your answer to the nearest percent.

b) Of the years illustrated on the graph, determine the year in which the percentage of the U.S. population that was foreign-born was a minimum.

c) Determine the x-coordinate of the vertex of the graph of the function $f(x)$. Then use this value in the function $f(x)$ to estimate the minimum percentage of the U.S. population that was foreign-born. Round your answer for the vertex to the nearest whole number. Round your answer for the population to the nearest percent.

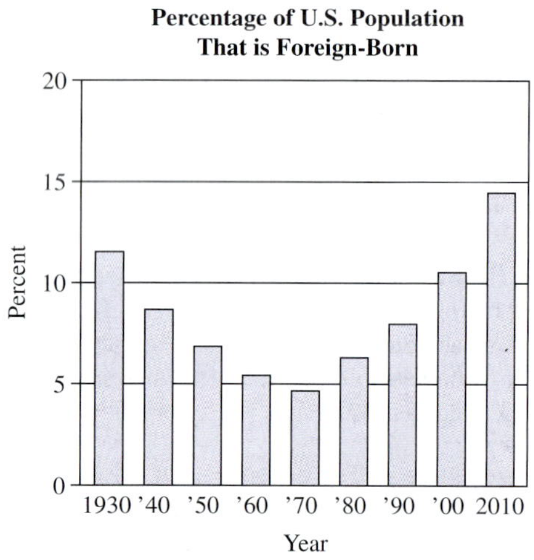

Source: Bureau of the Census, U.S. Dept. of Commerce

84. ***Hourly Temperature*** The following graph indicates the hourly temperature in Rochester, New York, from 11 P.M. on Thursday, February 16, 2006, through 7 A.M. on Friday, February 17, 2006. The function $t(x) = -1.11x^2 + 6.60x + 51.28$, can be used to estimate the hourly temperature in Rochester where x is the number of hours after midnight on February 16, 2006, and $-1 \leq x \leq 7$.

a) Use the function $t(x)$ to estimate the temperature in Rochester at 6 A.M. on February 17, 2006. Round your answer to the nearest degree.

b) Use the graph to determine the hour(s) that the temperature was a maximum.

c) Determine the x-coordinate of the vertex of the graph of the function $t(x)$. Then use this value in the function $t(x)$ to estimate the maximum temperature. Round your answer for the vertex to the nearest hour. Round your answer for the temperature to the nearest degree.

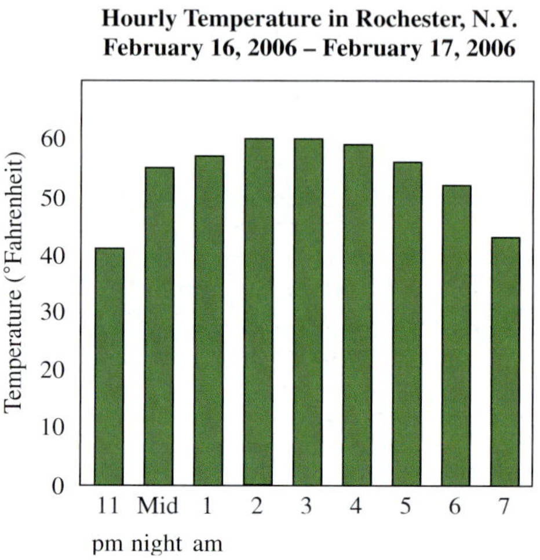

Source: 13 WHAM News Weather Team

85. ***Expected Growth*** The town of Lockport currently has 4000 residents. The expected future population can be approximated by the function $P(x) = 4000(1.3)^{0.1x}$, where x is the number of years in the future. Find the expected population of Lockport in

a) 10 years.

b) 50 years.

86. ***Radioactive Decay*** Strontium-90 is a radioactive isotope that decays exponentially at a rate of 2.8% per year. The amount of strontium-90, S, remaining after t years can be found by the function $S(t) = S_0e^{-0.028t}$, where S_0 is the original amount present. If there are originally 1000 grams of strontium-90, determine the amount of strontium-90 remaining after 40 years. Round your answer to the nearest gram.

87. ***Medicare Premiums*** The graph at the top of page 405 shows the monthly Medicare Part B premiums for Medicare beneficiaries for the years 2000 through 2005.

a) Does the graph approximate the graph of an exponential function?

b) Estimate the monthly Medicare Part B premium for Medicare beneficiaries in 2003.

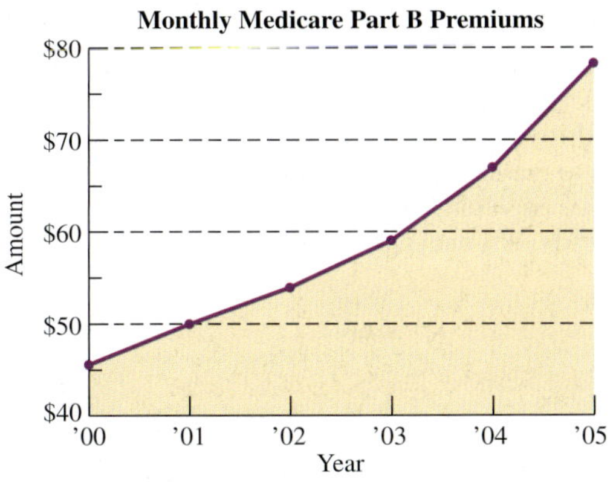

Source: Centers for Medicare and Medicaid Services

88. ***U.S. Cellular Telephone Subscribers*** The following graph shows the number of U.S. cellular subscribers, in thousands, for the years 1985 through 2004.

a) Does the graph approximate the graph of an exponential function?

b) Estimate the number of U.S. cellular subscribers in 2004.

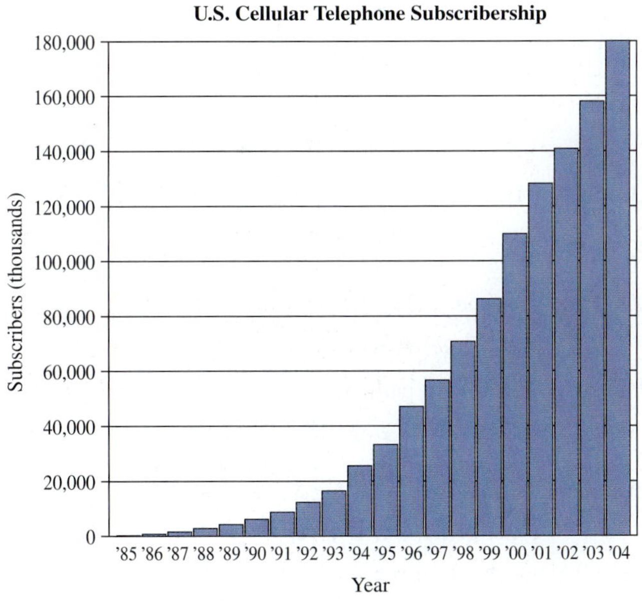

Source: The CTIA Semi-Annual Wireless Industry Survey

89. The spacing of the frets on the neck of a classical guitar is determined from the equation $d = (21.9)(2)^{(20-x)/12}$, where $x =$ the fret number and $d =$ the distance in centimeters of the xth fret from the bridge.

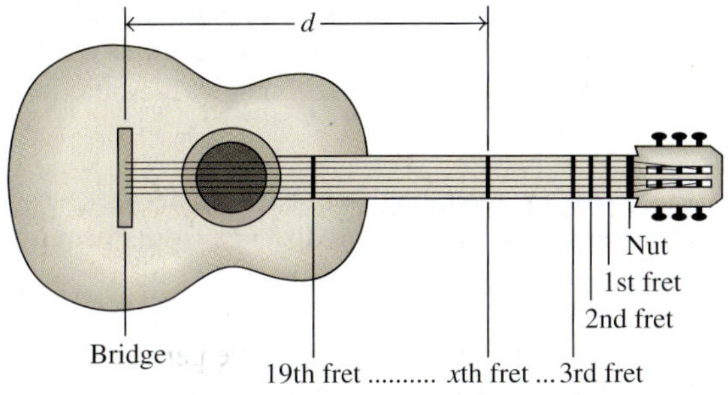

a) Determine how far the 19th fret should be from the bridge (rounded to one decimal place).

b) Determine how far the 4th fret should be from the bridge (rounded to one decimal place).

c) The distance of the nut from the bridge can be found by letting $x = 0$ in the given exponential equation. Find the distance from the nut to the bridge (rounded to one decimal place).

CHALLENGE PROBLEMS/GROUP ACTIVITIES

90. ***Appreciation of a House*** A house initially cost \$150,000. The value, V, of the house after n years if it appreciates at a constant rate of 4% per year can be determined by the function $V = f(n) = \$150{,}000(1.04)^n$.

a) Determine $f(8)$ and explain its meaning.

b) After how many years is the value of the house greater than \$250,000? (Find by trial and error.)

91. ***Target Heart Rate*** While exercising, a person's recommended target heart rate is a function of age. The recommended number of beats per minute, y, is given by the function $y = f(x) = -0.85x + 187$, where x represents a person's age in years. Determine the number of recommended heart beats per minute for the following ages and explain the results.

a) 20

b) 30

c) 50

d) 60

e) What is the age of a person with a recommended target heart rate of 85?

92. *Speed of Light* Light travels at about 186,000 miles per second through space. The distance, d, in miles that light travels in t seconds can be determined by the function $d(t) = 186{,}000t$.

a) Light reaches the moon from Earth in about 1.3 sec. Determine the approximate distance from Earth to the moon.

b) Express the distance in miles, d, traveled by light in t minutes as a function of time, t.

c) Light travels from the sun to Earth in about 8.3 min. Determine the approximate distance from the sun to Earth.

INTERNET/RESEARCH ACTIVITY

93. The idea of using variables in algebraic equations was introduced by the French mathematician François Viète (1540–1603). Write a paper about his life and his contributions to mathematics. In particular, discuss his work with algebra equations. References include history of mathematics books, encyclopedias, and the Internet.

CHAPTER 6 SUMMARY

IMPORTANT FACTS

PROPERTIES USED TO SOLVE EQUATIONS

Addition property of equality

If $a = b$, then $a + c = b + c$.

Subtraction property of equality

If $a = b$, then $a - c = b - c$.

Multiplication property of equality

If $a = b$, then $ac = bc$.

Division property of equality

If $a = b$, then $a/c = b/c$, $c \neq 0$.

VARIATION

Direct: $y = kx$

Inverse: $y = \dfrac{k}{x}$

Joint: $y = kxz$

INEQUALITY SYMBOLS

$a < b$ means that a is less than b.

$a \leq b$ means that a is less than or equal to b.

$a > b$ means that a is greater then b.

$a \geq b$ means that a is greater than or equal to b.

INTERCEPTS

To find the x-intercept, set $y = 0$ and solve the resulting equation for x.

To find the y-intercept, set $x = 0$ and solve the resulting equation for y.

SLOPE

Slope (m): $m = \dfrac{y_2 - y_1}{x_2 - x_1}$

EQUATIONS AND FORMULAS

Linear equation in two variables:

$$ax + by = c, \qquad a \neq 0 \quad \text{and} \quad b \neq 0$$

Exponential growth or decay formula:

$$P = P_0 a^{kt}, \qquad a \neq 1, \qquad a > 0$$

Slope–intercept form of a line:

$$y = mx + b$$

Quadratic equation in one variable:

$$ax^2 + bx + c = 0, \qquad a \neq 0$$

Quadratic formula:

$$x = \frac{-b \pm \sqrt{b^2 - 4ac}}{2a}$$

Quadratic equation (or function) in two variables:

$$y = ax^2 + bx + c, \qquad a \neq 0$$

Axis of symmetry of a parabola:

$$x = \frac{-b}{2a}$$

Exponential function:

$$f(x) = a^x, \qquad a \neq 1, \qquad a > 0$$

Natural exponential function:

$$f(x) = e^x$$

Natural exponential growth or decay function:

$$P(t) = P_0 e^{kt}$$

ZERO-FACTOR PROPERTY

If $a \cdot b = 0$, then $a = 0$ or $b = 0$.

CHAPTER 6 REVIEW EXERCISES

6.1

In Exercises 1–6, evaluate the expression for the given value(s) of the variable.

1. $x^2 + 15, \quad x = 2$

2. $-x^2 - 7, \quad x = -1$

3. $4x^2 - 2x + 5, \quad x = 2$

4. $-x^2 + 7x - 3, \quad x = \frac{1}{2}$

5. $4x^3 - 7x^2 + 3x + 1, \quad x = -1$

6. $3x^2 - xy + 2y^2, \quad x = 1, \quad y = -2$

6.2

In Exercises 7–9, combine like terms.

7. $3x - 2 + x + 7$

8. $2x + 5(x - 2) + 8x$

9. $4(x - 1) + \frac{1}{3}(9x + 3)$

In Exercises 10–14, solve the equation for the given variable.

10. $4s + 10 = -30$

11. $3t + 14 = -6t - 13$

12. $\dfrac{x + 5}{2} = \dfrac{x - 3}{4}$

13. $4(x - 2) = 3 + 5(x + 4)$

14. $\dfrac{x}{4} + \dfrac{3}{5} = 7$

15. ***Making Oatmeal*** A recipe for Hot Oats Cereal calls for 2 cups of water and for $\frac{1}{3}$ cup of dry oats. How many cups of dry oats would be used with 3 cups of water?

16. ***Laying Blocks*** A mason lays 120 blocks in 1 hr 40 min. How long will it take her to lay 300 blocks?

6.3

In Exercises 17–20, use the formula to find the value of the indicated variable for the values given.

17. $A = bh$
Find A when $b = 12$ and $h = 4$ (geometry).

18. $V = 2\pi R^2 r^2$
Find V when $R = 3$, $r = 1\frac{3}{4}$, and $\pi = 3.14$ (geometry).

19. $z = \dfrac{\bar{x} - \mu}{\dfrac{\sigma}{\sqrt{n}}}$
Find $\bar{x}$ when $z = 2$, $\mu = 100$, $\sigma = 3$, and $n = 16$ (statistics).

20. $E = mc^2$
Find m when $E = 400$ and $c = 4$ (physics).

In Exercises 21–24, solve for y.

21. $8x - 4y = 24$

22. $2x + 7y = 15$

23. $2x - 3y + 52 = 30$

24. $-3x - 4y + 5z = 4$

In Exercises 25–28, solve for the variable indicated.

25. $A = lw$, for w

26. $P = 2l + 2w$, for w

27. $L = 2(wh + lh)$, for l

28. $a_n = a_1 + (n - 1)d$, for d

6.4

In Exercises 29–32, write the phrase in mathematical terms.

29. 7 decreased by 4 times x

30. 2 times y increased by 7

31. 10 increased by 3 times r

32. The difference between 9 divided by q and 15

In Exercises 33–36, write an equation that can be used to solve the problems. Solve the equation and find the desired value(s).

33. Three increased by 7 times a number is 17.

34. The product of 3 and a number, increased by 8, is 6 less than the number.

35. Five times the difference of a number and 4 is 45.

36. Fourteen more than 10 times a number is 8 times the sum of the number and 12.

In Exercises 37–40, write the equation and then find the solution.

37. MODELING - ***Investing*** Jim Lawton received an inheritance of \$15,000. If he wants to invest twice as much money in mutual funds as in bonds, how much should he invest in mutual funds?

38. MODELING - ***Lawn Chairs*** Larry's Lawn Chair Company has fixed costs of \$15,000 per month and variable costs of \$9.50 per lawn chair manufactured. The company has \$95,000 available to meet its total monthly expenditures. What is the maximum number of lawn chairs the company can manufacture in a month? (Fixed costs, such as rent and insurance, are those that occur regardless of the level of production. Variable costs, such as those for materials, depend on the level of production.)

39. MODELING - ***Endangered and Threatened Mammals*** According to the U.S. Fish and Wildlife Service, the number of species of endangered mammals in the United States in 2005 was 2 more than 6 times the number of species of threatened mammals. The sum of the number of species of endangered mammals and the number of species of threatened mammals in the United States in 2005 was 79. Determine the number of species of endangered mammals and the number of species of threatened mammals.

▲ The Florida panther is an endangered mammal.

40. MODELING - ***Restaurant Profit*** John Smith owns two restaurants. His profit for a year at restaurant A is \$12,000 greater than his profit at restaurant B. The total profit from both restaurants is \$68,000. Determine the profit at each restaurant.

6.5

In Exercises 41–44, find the quantity indicated.

41. s is directly proportional to t. If $s = 40$ when $t = 5$, find s when $t = 8$.

42. J is inversely proportional to the square of A. If $J = 8$ when $A = 3$, find J when $A = 6$.

43. W is directly proportional to L and inversely proportional to A. If $W = 80$ when $L = 100$ and $A = 20$, find W when $L = 50$ and $A = 40$.

44. z is jointly proportional to x and y and inversely proportional to the square of r. If $z = 12$ when $x = 20$, $y = 8$, and $r = 8$, find z when $x = 10$, $y = 80$, and $r = 3$.

45. ***Buying Fertilizer***

a) A 30 lb bag of fertilizer will cover an area of 2500 ft^2. How many pounds of fertilizer are needed to cover an area of 12,500 ft^2?

b) How many bags of fertilizer are needed?

46. ***Map Reading*** The scale of a map is 1 in. to 30 mi. What distance on the map represents 120 mi?

47. ***Electric Bill*** An electric company charges \$0.162 per kilowatt-hour (kWh). What is the electric bill if 740 kWh are used in a month?

48. ***Stretching a Spring*** The length that a spring will stretch, S, varies directly with the weight, w, attached to the spring. If a spring stretches 4.2 in. when a 60 lb weight is attached, how far will the spring stretch when a 25 lb weight is attached?

6.6

In Exercises 49–52, graph the solution set for the set of real numbers.

49. $8 + 7x \leq 5x - 4$

50. $2x + 8 \geq 5x + 11$

51. $3(x + 9) \leq 4x + 11$

52. $-3 \leq x + 1 < 7$

In Exercises 53–56, graph the solution set for the set of integers.

53. $2 + 4x > -10$

54. $5x + 13 \geq -22$

55. $-1 < x \leq 9$

56. $-8 \leq x + 2 \leq 7$

6.7

In Exercises 57–60, graph the ordered pair in the Cartesian coordinate system.

57. $(1, -3)$

58. $(-2, 2)$

59. $(-4, 3)$

60. $(5, 3)$

In Exercises 61 and 62, points A, B, and C are vertices of a rectangle. Plot the points. Find the coordinates of the fourth point, D, to complete the rectangle. Find the area of the rectangle.

61. $A(-3, 3)$, $B(2, 3)$, $C(2, -1)$

62. $A(-3, 1)$, $B(-3, -2)$, $C(4, -2)$

In Exercises 63–66, graph the equation by plotting points.

63. $x - y = 2$

64. $2x + 3y = 12$

65. $x = y$

66. $x = -3$

In Exercises 67–70, graph the equation, using the x- and y-intercepts.

67. $x - 2y = 6$

68. $x + 3y = 6$

69. $4x - 3y = 12$

70. $2x + 3y = 9$

In Exercises 71–74, find the slope of the line through the given points.

71. $(1, 3), (6, 5)$

72. $(3, -1), (5, -4)$

73. $(-1, -4), (2, 3)$

74. $(6, 2), (6, -2)$

In Exercises 75–78, graph the equation by plotting the y-intercept and then plotting a second point by making use of the slope.

75. $y = 2x - 5$

76. $2y - 4 = 3x$

77. $2y + x = 8$

78. $y = -x - 1$

In Exercises 79 and 80, determine the equation of the graph.

79.

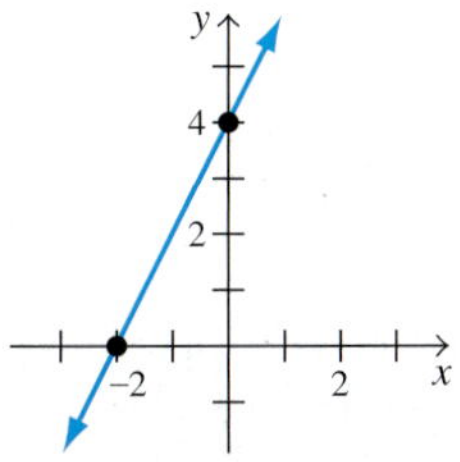

80.

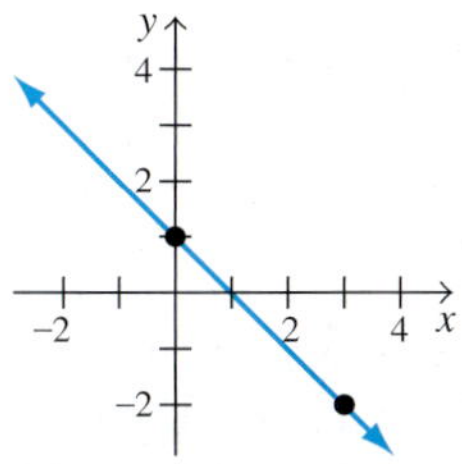

81. ***Car Wash Profit*** The monthly profit, p, in dollars, at Delta Sonic Car Wash can be estimated by $p = 20n - 400$, where n is the number of cars washed in 1 week.

a) Draw a graph of weekly profit versus number of cars washed for up to and including 100 cars.

b) Use the function to determine the weekly profit if 50 cars are washed.

c) If the weekly profit was \$1000, use the function to determine how many cars were washed.

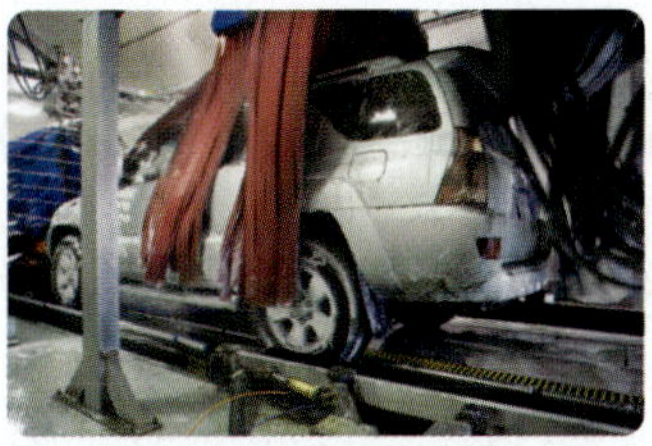

82. ***Business Space Rental*** The monthly rental cost, C, in dollars, for space in the Galleria Mall can be approximated by the equation $C = 1.70A + 3000$, where A is the area, in square feet, of space rented.

a) Draw a graph of monthly rental cost versus square feet for up to and including 12,000 ft^2.

b) Determine the monthly rental cost if 2000 ft^2 are rented.

c) If the rental cost is \$10,000 per month, how many square feet are rented?

6.8

In Exercises 83–86, graph the inequality.

83. $2x + 3y \leq 12$

84. $4x + 2y \geq 12$

85. $2x - 3y > 12$

86. $-7x - 2y < 14$

6.9

In Exercises 87–92, factor the trinomial. If the trinomial cannot be factored, so state.

87. $x^2 + 11x + 18$

88. $x^2 + x - 30$

89. $x^2 - 10x + 24$

90. $x^2 - 9x + 20$

91. $6x^2 + 7x - 3$

92. $2x^2 + 13x - 7$

In Exercises 93–96, solve the equation by factoring.

93. $x^2 + 4x + 3 = 0$

94. $x^2 + 3x = 18$

95. $3x^2 - 17x + 10 = 0$

96. $3x^2 = -7x - 2$

In Exercises 97–100, solve the equation using the quadratic formula. If the equation has no real solution, so state.

97. $x^2 - 4x - 1 = 0$

98. $x^2 - 6x - 16 = 0$

99. $2x^2 - 3x + 4 = 0$

100. $2x^2 - x - 3 = 0$

6.10

In Exercises 101–104, determine whether the graph represents a function. If it does represent a function, give its domain and range.

101.

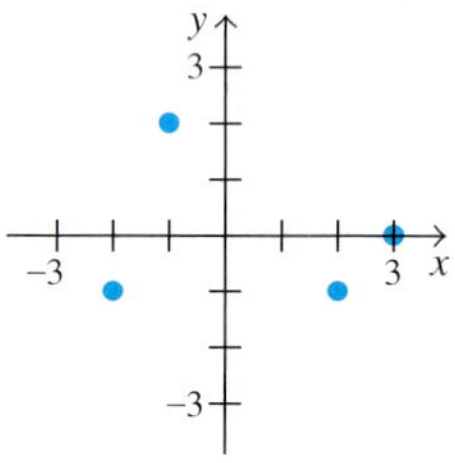

102.

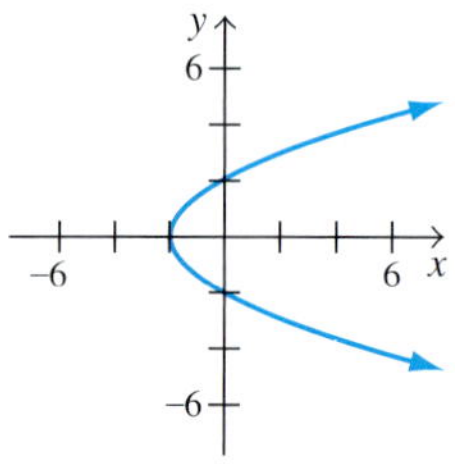

103.

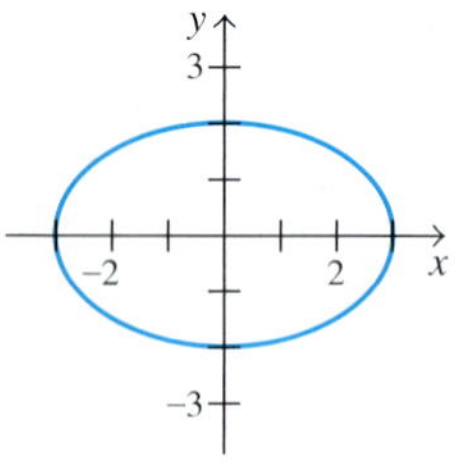

104.

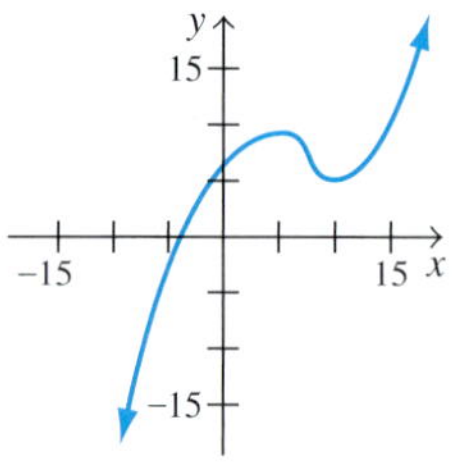

In Exercises 105–108, evaluate f(x) for the given value of x.

105. $f(x) = 3x - 7, \quad x = 4$

106. $f(x) = -2x + 9, \quad x = -3$

107. $f(x) = 2x^2 - 3x + 4, \quad x = 5$

108. $f(x) = -4x^2 + 7x + 9, \quad x = -2$

In Exercises 109 and 110, for each function

a) *determine whether the parabola will open upward or downward.*

b) *determine the equation of the axis of symmetry.*

c) *determine the vertex.*

d) *determine the y-intercept.*

e) *determine the x-intercepts if they exist.*

f) *sketch the graph.*

g) *determine the domain and range.*

109. $y = -x^2 - 4x + 21$

110. $f(x) = 2x^2 + 8x + 6$

In Exercises 111 and 112, draw the graph of the function and state the domain and range.

111. $y = 2^{2x}$

112. $y = (\frac{1}{2})^x$

6.2, 6.3, 6.10

113. ***Gas Mileage*** The gas mileage, m, of a specific car can be estimated by the equation (or function)

$$m = 30 - 0.002n^2, \qquad 20 \le n \le 80$$

where n is the speed of the car in miles per hour. Estimate the gas mileage when the car travels at 60 mph.

114. ***Auto Accidents*** The approximate number of accidents in one month, n, involving drivers between 16 and 30 years of age inclusive can be approximated by the equation

$$n = 2a^2 - 80a + 5000, \qquad 16 \le a \le 30$$

where a is the age of the driver. Approximate the number of accidents in one month that involved

a) 18-year-olds

b) 25-year-olds.

115. ***Filtered Light*** The percent of light filtering through Swan Lake, P, can be approximated by the function $P(x) = 100(0.92)^x$, where x is the depth in feet. Find the percent of light filtering through at a depth of 4.5 ft.

CHAPTER 6 TEST

1. Evaluate $4x^2 + 3x - 1$, when $x = -2$.

In Exercises 2 and 3, solve the equation.

2. $3x + 5 = 2(4x - 7)$

3. $-2(x - 3) + 6x = 2x + 3(x - 4)$

In Exercises 4 and 5, write an equation to represent the problem. Then solve the equation.

4. The product of 3 and a number, increased by 5 is 17.

5. ***Salary*** Mary Gilligan's salary is \$350 per week plus 6% commission of sales. How much in sales must Mary make to earn a total of \$710 per week?

6. Evaluate $L = ah + bh + ch$ when $a = 2, b = 5, c = 4$, and $h = 7$.

7. Solve $3x + 5y = 11$ for y.

8. L varies jointly as M and N and inversely as P. If $L = 12$ when $M = 8$, $N = 3$, and $P = 2$, find L when $M = 10$, $N = 5$, and $P = 15$.

9. For a constant area, the length, l, of a rectangle varies inversely as the width, w. If $l = 15$ ft when $w = 9$ ft, find the length of a rectangle with the same area if the width is 20 ft.

10. Graph the solution set of $-3x + 11 \leq 5x + 35$ on the real number line.

11. Determine the slope of the line through the points $(-2, 8)$ and $(1, 14)$.

In Exercises 12 and 13, graph the equation.

12. $y = 2x - 4$

13. $2x - 3y = 15$

14. Graph the inequality $3y \geq 5x - 12$.

15. Solve the equation $x^2 + 5x = -4$ by factoring.

16. Solve the equation $3x^2 + 2x = 8$ by using the quadratic formula.

17. Determine whether the graph is a function. Explain your answer.

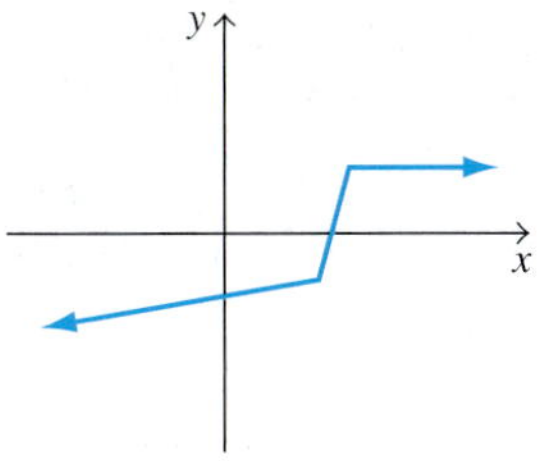

18. Evaluate $f(x) = -3x^2 - 12x + 5$ when $x = -2$.

19. For the equation $y = x^2 - 2x + 4$,

a) determine whether the parabola will open upward or downward.

b) determine the equation of the axis of symmetry.

c) determine the vertex.

d) determine the y-intercept.

e) determine the x-intercepts if they exist.

f) sketch the graph.

g) determine the domain and range of the function.

GROUP PROJECTS

ARCHEOLOGY: GAINING INFORMATION FROM BONES

1. Archeologists have developed formulas to predict the height and, in some cases, the age at death of the deceased by knowing the lengths of certain bones in the body. The long bones of the body grow at approximately the same rate. Thus, a linear relationship exists between the length of the bones and the person's height. If the length of one of these major bones—the femur (F), the tibia (T), the humerus (H), and the radius (R)— is known, the height, h, of a person can be calculated with one of the following formulas. The relationship between bone length and height is different for males and females.

Male	**Female**
$h = 2.24F + 69.09$	$h = 2.23F + 61.41$
$h = 2.39T + 81.68$	$h = 2.53T + 72.57$
$h = 2.97H + 73.57$	$h = 3.14H + 64.98$
$h = 3.65R + 80.41$	$h = 3.88R + 73.51$

All measurements are in centimeters.

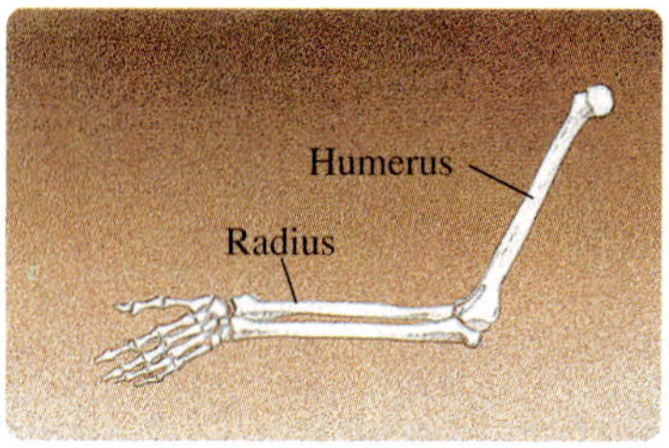

a) Measure your humerus and use the appropriate formula to predict your height in centimeters. How close is this predicted height to your actual height? (The result is an approximation because measuring a bone covered with flesh and muscle is difficult.)

b) Determine and describe where the femur and tibia bones are located.

c) Dr. Juarez, an archeologist, had one female humerus that was 29.42 cm in length. He concluded that the height of the entire skeleton would have been 157.36 cm. Was his conclusion correct?

d) If a 21-year-old woman is 167.64 cm tall, about how long should her tibia be?

e) Sometimes the age of a person may be determined by using the fact that the height of a person, and the length of his or her long bones, decreases at the rate of 0.06 cm per year after the age of 30.

i. At age 30, Jolene is 168 cm tall. Estimate the length of her humerus.

ii. Estimate the length of Jolene's humerus when she is 60 years old.

f) Select six people of the same gender and measure their height and one of the bones for which an equation is given (the same bone on each person). Each measurement should be made to the nearest 0.5 cm. For each person, you will have two measurements, which can be considered an ordered pair (bone length, height). Plot the ordered pairs on a piece of graph paper, with the bone length on the horizontal axis and the height on the vertical axis. Start the scale on both axes at zero. Draw a straight line that you feel is the best approximation, or best fit, through these points. Determine where the line crosses the y-axis and the slope of the line. Your y-intercept and slope should be close to the values in the given equation for that bone. (Reference: M. Trotter and G. C. Gleser, "Estimation of Stature from Long Bones of American Whites and Negroes," *American Journal of Physical Anthropology,* 1952, 10:463–514.)

GRAPHING CALCULATOR

2. The functions that we graphed in this chapter can be easily graphed with a graphing calculator (or grapher). If you do not have a graphing calculator, borrow one from your instructor or a friend.

a) Explain how you would set the domain and range. The calculator key to set the domain and range may be labeled *range* or *window*. Set the grapher with the following range or window settings:

Xmin = −12, Xmax = 12, Xscl = 1,
Ymin = −13, Ymax = 6, and Yscl = 1.

b) Explain how to enter a function in the graphing calculator. Enter the function $y = 3x^2 - 7x - 8$ in the calculator.

c) Graph the function you entered in part (b).

d) Learn how to use the *trace feature.* Then use it to estimate the x-intercepts. Record the estimated values for the x-intercepts.

e) Learn how to use the *zoom feature* to obtain a better approximation of the x-intercepts. Use the zoom feature twice and record the x-intercepts each time.

CHAPTER 7

Systems of Linear Equations and Inequalities

▲ Business owners may solve systems of equations to determine under what conditions their business will be profitable.

WHAT YOU WILL LEARN

- Solving systems of linear equations by graphing, by using the substitution method, by using the addition method, and by using matrices
- Solving systems of linear inequalities
- Adding, subtracting, and multiplying matrices
- Solving application problems using linear programming

WHY IT IS IMPORTANT

Imagine that you and a friend enter the T-shirt business. To make a profit in your business, you need to keep track of the cost of your materials, your overhead, the quantity of T-shirts sold, the price at which you sell the T-shirts, and many other items. To do so, you may need to develop and solve systems of equations.

7.1 SYSTEMS OF LINEAR EQUATIONS

▲ Solving a system of equations can help determine which of two landscaping companies offers the least expensive service.

Suppose you needed to have some landscaping work done at your home. In your local newspaper, two different landscaping services each have advertised specials. One company charges a lower consultation fee than the other company but charges a higher hourly rate for labor. How do you determine the number of hours of service needed for both services to have the same cost? How do you decide which company is less expensive for the services you need? In this section, we will introduce one way to use algebra to answer these questions.

In algebra, it is often necessary to find the common solution to two or more equations. We refer to the equations in this type of problem as a *system of linear equations* or as *simultaneous linear equations*. The solution to a system of linear equations may be found by a number of techniques. In this section, we illustrate how to solve a system of linear equations by graphing.

Solutions to Systems of Linear Equations

A *solution to a system of equations* is the ordered pair or ordered pairs that satisfy *all* equations in the system. A system of linear equations may have exactly one solution, no solution, or infinitely many solutions.

EXAMPLE 1 *Is the Ordered Pair a Solution?*

Determine which of the ordered pairs is a solution to the following system of linear equations.

$$x + 2y = 8$$
$$2x - 3y = 2$$

a) $(6, 1)$ b) $(4, 2)$ c) $(1, 0)$

SOLUTION For the ordered pair to be a solution to the system, it must satisfy each equation in the system. Substitute the values of x and y into each equation.

a) $x + 2y = 8$ $\qquad$ $2x - 3y = 2$

$6 + 2(1) = 8$ $\qquad$ $2(6) - 3(1) = 2$

$8 = 8$ True $\qquad$ $9 = 2$ False

Since $(6, 1)$ does not satisfy *both* equations, it is not a solution to the system.

b) $x + 2y = 8$ $\qquad$ $2x - 3y = 2$

$4 + 2(2) = 8$ $\qquad$ $2(4) - 3(2) = 2$

$8 = 8$ True $\qquad$ $2 = 2$ True

Since $(4, 2)$ satisfies *both* equations, it is a solution to the system.

c) $x + 2y = 8$ $\qquad$ $2x - 3y = 2$

$1 + 2(0) = 8$ $\qquad$ $2(1) - 3(0) = 2$

$1 = 8$ False $\qquad$ $2 = 2$ True

Since $(1, 0)$ does not satisfy *both* equations, it is not a solution to the system. ●

Solving a System of Linear Equations by Graphing

To find the solution to a system of linear equations graphically, we graph both of the equations on the same axes. The coordinates of the point or points of intersection of the graphs are the solution or solutions to the system of equations.

$x + y = 4$

x	y
0	4
1	3
4	0

$2x - y = -1$

x	y
0	1
1	3
−2	−3

PROCEDURE FOR SOLVING A SYSTEM OF EQUATIONS BY GRAPHING

1. Determine three ordered pairs that satisfy each equation.
2. Plot the points that correspond to the ordered pairs and sketch the graphs of *both* equations on the same axes.
3. The coordinates of the point or points of intersection of the graphs are the solution or solutions to the system of equations.

When two linear equations are graphed, three situations are possible. The two lines may intersect at one point, as in Example 2; or the two lines may be parallel and not intersect, as in Example 3; or the two equations may represent the same line, as in Example 4.

Since the solution to a system of equations may not be integer values, you may not be able to obtain the exact solution by graphing.

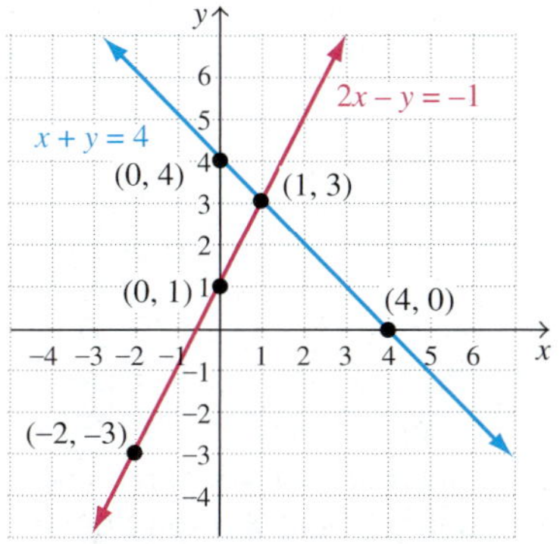

Figure 7.1

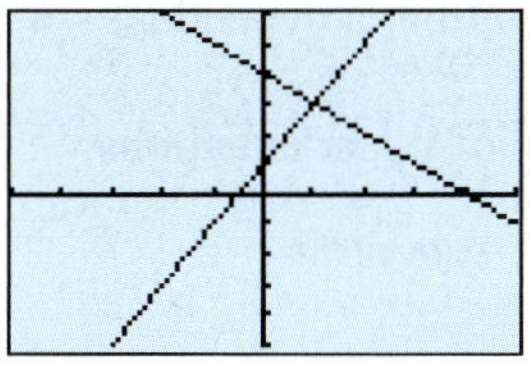

Figure 7.2

EXAMPLE 2 *A System with One Solution*

Determine the solution to the following system of equations graphically.

$$x + y = 4$$
$$2x - y = -1$$

SOLUTION To determine the solution, graph both $x + y = 4$ and $2x - y = -1$ on the same axes (Fig. 7.1). Three points that satisfy each equation are shown in the tables above Fig. 7.1. Figure 7.2 shows the system $x + y = 4$ and $2x - y = -1$ graphed on a Texas Instrument TI-84 Plus graphing calculator.

The graphs intersect at (1, 3), which is the solution to the system of equations. This point is the only point that satisfies *both* equations.

CHECK:

$$x + y = 4 \qquad 2x - y = -1$$
$$1 + 3 = 4 \qquad 2(1) - 3 = -1$$
$$4 = 4 \quad \text{True} \qquad 2 - 3 = -1$$
$$-1 = -1 \quad \text{True}$$

TIMELY TIP When checking the solution to a system of equations, it is important to check the solution in both equations. The ordered pair solution must satisfy *both* equations of the system.

The system of equations in Example 2 is an example of a *consistent system of equations*. A consistent system of equations is one that has a solution.

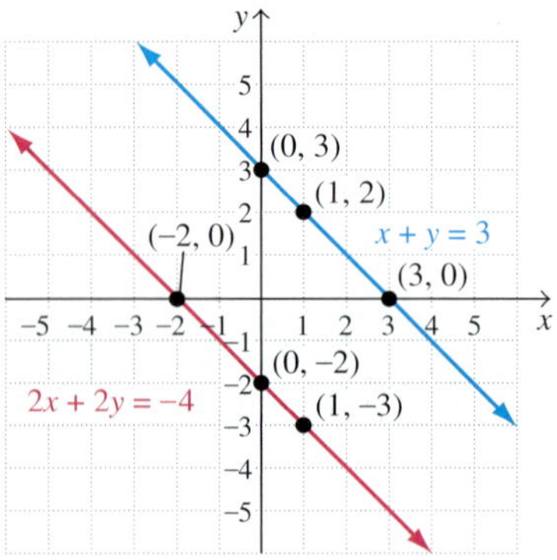

Figure 7.3

EXAMPLE 3 *A System with No Solution*

Determine the solution to the following system of equations graphically.

$$x + y = 3$$
$$2x + 2y = -4$$

SOLUTION Three ordered pairs that satisfy the equation $x + y = 3$ are (0, 3), (3, 0), and (1, 2). Three ordered pairs that satisfy the equation $2x + 2y = -4$ are $(-2, 0)$, $(0, -2)$, and $(1, -3)$. The graphs of both equations are given in Fig. 7.3. Since the two lines are parallel, they do not intersect; therefore, the system has *no solution.*

The system of equations in Example 3 has no solution. A system of equations that has no solution is called an ***inconsistent system***.

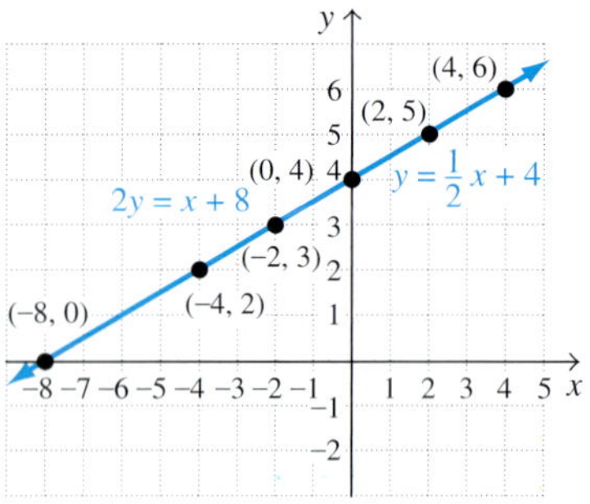

Figure 7.4

EXAMPLE 4 *A System with an Infinite Number of Solutions*

Determine the solution to the following system of equations graphically.

$$y = \frac{1}{2}x + 4$$
$$2y = x + 8$$

SOLUTION Three ordered pairs that satisfy the equation $y = \frac{1}{2}x + 4$ are (0, 4), (2, 5), and $(-2, 3)$. Three ordered pairs that satisfy the equation $2y = x + 8$ are $(-8, 0)$, (4, 6), and $(-4, 2)$. Graph the equations on the same axes (Fig. 7.4). Because all six points are on the same line, the two equations represent the same line. Therefore, every ordered pair that is a solution to one equation is also a solution to the other equation. Every point on the line satisfies both equations; thus, this system has an *infinite number of solutions*. Solving the second equation for y reveals that the equations are equivalent.

When a system of equations has an infinite number of solutions, as in Example 4, it is called a ***dependent system***. Note that because it has a solution, a dependent system is also a consistent system.

Figure 7.5 summarizes the three possibilities for a system of linear equations.

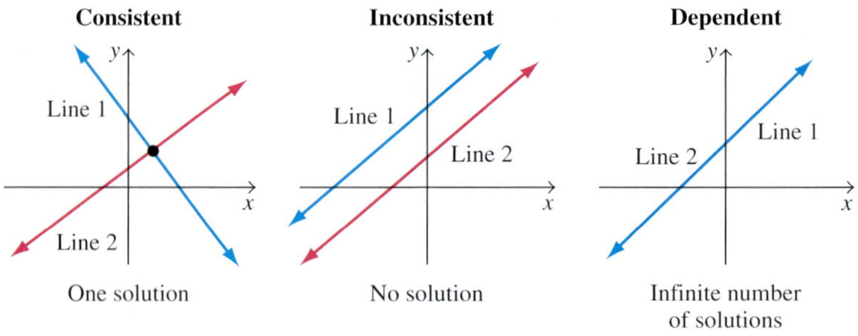

Figure 7.5

In Chapter 6, we introduced *modeling*. Recall that a *mathematical model* is an equation or system of equations that represents a real-life situation. In Examples 5 and 6, we develop equations that model a real-life situation.

EXAMPLE 5 **MODELING** - *A Landscape Service Application*

Tom's Tree and Landscape Service charges a consultation fee of \$200 plus \$50 per hour for labor for landscaping. Lawn Perfect Landscape Service charges a consultation fee of \$300 plus \$25 per hour for labor for landscaping.

a) Write a system of equations to represent the cost, C, of the two landscaping services, each with h hours of labor.

b) Graph both equations on the same axes and determine the number of hours needed for both services to have the same cost.

c) If the Johnsons need 7 hours of landscaping service done at their home, which service is less expensive?

SOLUTION Let h = the number of hours of labor. The total cost of each service is the consultation fee plus the cost of the labor.

a) Tom's Tree and Landscape Service: $C = 200 + 50h$

Lawn Perfect Landscape Service: $C = 300 + 25h$

b) We graphed the cost, C, versus the number of hours of labor, h, for 0 to 10 hours (Fig. 7.6). On the graph, the lines intersect at the point (4, 400). Thus, for 4 hours of service, both services would have the same cost, \$400.

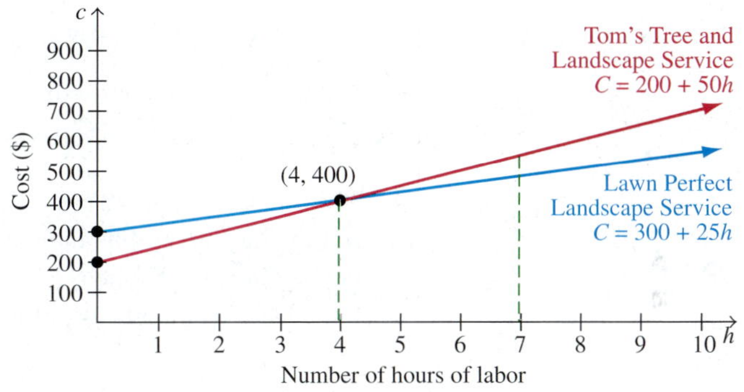

Figure 7.6

c) The graph shows that for more than 4 hours, Lawn Perfect is the least expensive service. Thus, for 7 hours, Lawn Perfect is less expensive than Tom's Tree and Landscape Service. ●

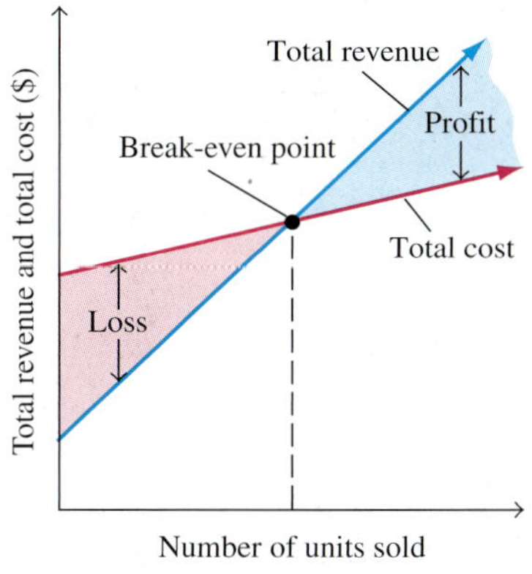

Figure 7.7

Break-Even Analysis

Manufacturers use a technique called ***break-even analysis*** to determine how many units of an item must be sold for the business to "break even," that is, for its total revenue to equal its total cost. Suppose we let the horizontal axis represent the number of units manufactured and sold and the vertical axis represent dollars. Then linear equations for cost, C, and revenue, R, can both be sketched on the same axes (Fig. 7.7). Both C and R are expressed in dollars, and both are a function of the number of units. Profit, P, is the difference between revenue, R, and cost, C. Thus, $P = R - C$. If revenue is greater than cost, the company makes a profit. If cost is greater than revenue, the company has a loss.

Initially, the cost graph is higher than the revenue graph because of fixed (overhead) costs such as rent and utilities. During low levels of sales, the manufacturer suffers a loss (the cost graph is greater). During higher levels of sales, the manufacturer realizes a profit (the revenue graph is greater). The point at which the two graphs intersect is called the ***break-even point***. At that number of units sold, revenue equals cost and the manufacturer breaks even.

EXAMPLE 6 MODELING - *Profit and Loss in Business*

At a collectibles show, Richard Lane can sell model cars for \$25. The costs for making the cars are a fixed cost of \$150 and a production cost of \$10 apiece.

a) Write an equation that represents Richard's revenue. Write an equation that represents Richard's cost.
b) How many model cars must Richard sell to break even?
c) Write an equation for the profit formula. Use the formula to determine Richard's profit if he sells 14 model cars.
d) How many model cars must Richard sell to make a profit of \$450?

SOLUTION

a) Let x denote the number of model cars made and sold. The revenue is given by the equation

$$R = 25x \quad (\$25 \text{ times the number of units})$$

and the cost is given by the equation

$$C = 150 + 10x \quad (\$150 \text{ plus } \$10 \text{ times the number of units})$$

b) The break-even point is the point at which the revenue and cost graphs intersect. In Fig. 7.8, the graphs intersect at the point (10, 250), which is the break-even point. Thus, for Richard to break even, he must sell 10 model cars. When 10 model cars are made and sold, the cost and revenue are both \$250.

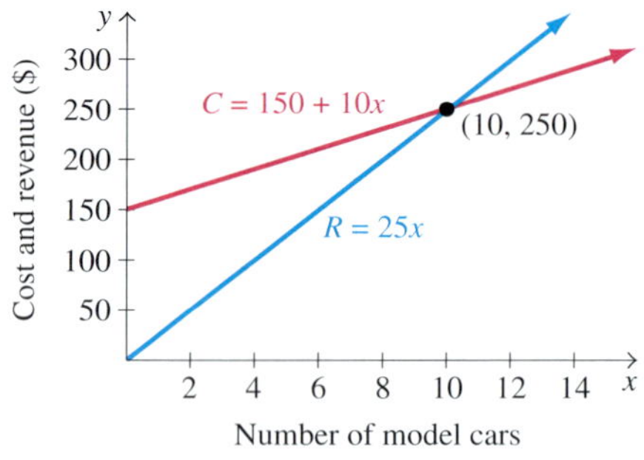

Figure 7.8

c) Profit is equal to the revenue minus the cost. Therefore, the profit formula is

$$\begin{aligned} P &= R - C \\ &= 25x - (150 + 10x) \\ &= 25x - 150 - 10x \\ &= 15x - 150 \end{aligned}$$

For 14 cars, the profit is found as follows.

$$P = 15x - 150$$
$$= 15(14) - 150 = 60$$

Richard has a profit of \$60 if he sells 14 model cars. By observing the graph, we can see that if Richard sells 14 model cars, he will have a profit since at 14 cars the revenue line is greater than the cost line.

d) We can determine the number of model cars that Richard must sell to have a profit of \$450 by using the profit formula. Substituting 450 for P we have

$$P = 15x - 150$$
$$450 = 15x - 150$$
$$600 = 15x$$
$$40 = x$$

Thus, Richard must sell 40 model cars to make a profit of \$450.

TIMELY TIP Following is a summary of the different types of systems of linear equations.

- A *consistent system of equations* is one that has a solution.
- An *inconsistent system of equations* is one that has no solution.
- A *dependent system of equations* is one that has an infinite number of solutions.

SECTION 7.1 EXERCISES

CONCEPT/WRITING EXERCISES

1. What is a system of linear equations?
2. What is the solution to a system of linear equations?
3. Define an *inconsistent system of equations*.
4. Define a *consistent system of equations*.
5. Define a *dependent system of equations*.
6. a) Outline the procedure for solving a system of linear equations by graphing.

   b) What is a disadvantage of solving a system of linear equations by graphing?
7. If a system of linear equations has no solution, what does that mean about the graphs of the equations in the system?
8. If a system of linear equations has one solution, what does that mean about the graphs of the equations in the system?
9. If a system of linear equations has an infinite number of solutions, what does that mean about the graphs of the equations in the system?
10. Can a system of linear equations have exactly two solutions? Explain.

PRACTICE THE SKILLS

In Exercises 11 and 12, determine which ordered pairs are solutions to the given system.

11. $y = 3x - 4$ $\quad$ $y = -x + 8$ $\qquad$ (3, 5) $\quad$ (2, 2) $\quad$ (1, 7)

12. $x + 2y = 6$ $\quad$ $x - y = -6$ $\qquad$ (−2, 4) $\quad$ (2, 2) $\quad$ (3, −9)

In Exercises 13–16, solve the system of equations graphically.

13. $x = 2$
$y = 4$

14. $x = -1$
$y = 3$

15. $x = 4$
$y = -3$

16. $x = -5$
$y = -3$

In Exercises 17–32, solve the system of equations graphically. If the system does not have a single ordered pair as a solution, state whether the system is inconsistent or dependent.

17. $x = 3$
$y = -x - 2$

18. $y = 2$
$y = x - 1$

19. $y = 4x - 8$
$y = -x + 7$

20. $x + y = 6$
$-x + y = 4$

21. $x + 2y = 0$
$2x - 3y = -14$

22. $3x - y = 1$
$4x - 3y = 3$

23. $2x + y = 3$
$2y = 6 - 4x$

24. $y = 2x - 4$
$2x + y = 0$

25. $y = x + 3$
$y = -1$

26. $x = 1$
$x + y + 3 = 0$

27. $2x - y = -3$
$2x + y = -9$

28. $3x + 2y = 6$
$6x + 4y = 12$

29. $2x - 3y = 12$
$3y - 2x = 9$

30. $y = \frac{1}{3}x - 4$
$3y - x = 4$

31. $y = \frac{4}{3}x - 2$
$2x + 2y = 10$

32. $2(x - 1) + 2y = 0$
$3x + 2(y + 2) = 0$

33. a) If the two lines in a system of equations have different slopes, how many solutions will the system have? Explain your answer.

b) If the two lines in a system of equations have the same slope but different y-intercepts, how many solutions will the system have? Explain.

c) If the two lines in a system of equations have the same slope and the same y-intercept, how many solutions will the system have? Explain.

34. Indicate whether the graph shown represents a consistent, inconsistent, or dependent system. Explain your answer.

a)

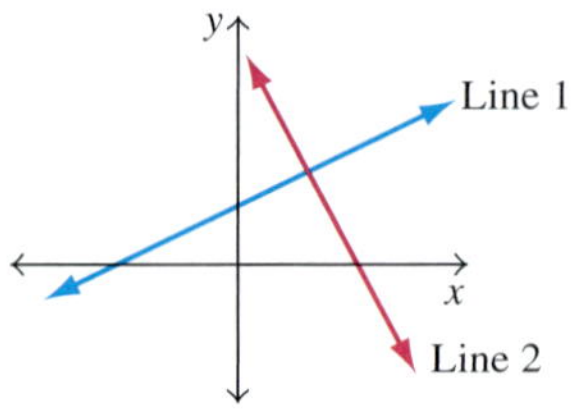

b)

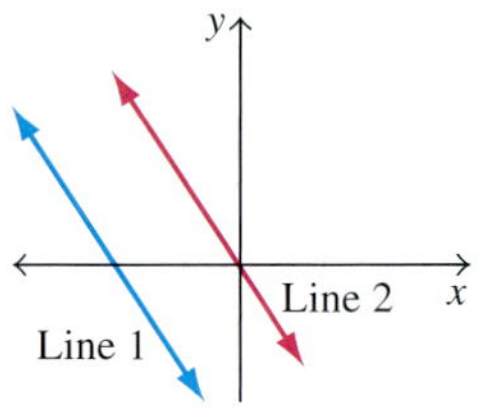

c)

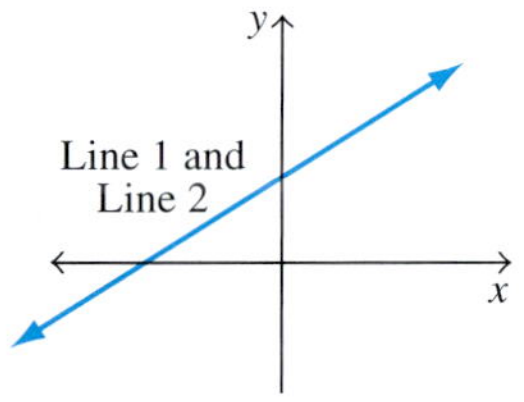

In Exercises 35–46, determine without graphing whether the system of equations has exactly one solution, no solution, or an infinite number of solutions. (Consider your answers to Exercise 33.)

35. $3x + y = 9$
$y = -3x + 9$

36. $4x + 3y = 8$
$6y = -8x + 4$

37. $3x + y = -6$
$4x - 2y = -8$

38. $x - 3y = 8$
$3x - y = 6$

39. $3x + y = 7$
$y = -3x + 9$

40. $x + 4y = 12$
$x = 4y + 3$

41. $2x - 3y = 6$
$x - \frac{3}{2}y = 3$

42. $x - 2y = 6$
$x + 2y = 4$

43. $3x = 6y + 5$
$y = \frac{1}{2}x - 3$

44. $3y = 6x + 4$
$-2x + y = \frac{4}{3}$

45. $4x + 7y = 2$
$4x = 6 + 7y$

46. $12x - 5y = 4$
$3x + 4y = 6$

PROBLEM SOLVING

Two lines are **perpendicular** *to each other when they meet at a right angle (a 90° angle). Two lines are perpendicular when their slopes are* **negative reciprocals**. *The negative reciprocal of 2 is $-\frac{1}{2}$, the negative reciprocal of $\frac{3}{5}$ is $-1/(\frac{3}{5})$ or $-\frac{5}{3}$, and so on. If a represents any real number, except 0, its negative reciprocal is $-1/a$. Note that the product of a number and its negative reciprocal is -1. In Exercises 47–50, determine, by finding the slope of each line, whether the lines will be perpendicular to each other when graphed.*

47. $4y - 2x = 15$
$3y - 5x = 9$

48. $4y - x = 6$
$y = x + 8$

49. $2x + y = 3$
$2y - x = 5$

50. $6x + 5y = 3$
$-10x = 2 + 12y$

In Exercises 51–55, part of the question involves determining a system of equations that models the situation.

51. MODELING - ***Hot-Water Heater Repair*** Phillip Hoffman's plumber says it will cost Phillip $250 to repair his water heater. Phillip can purchase a new, more efficient heater for $700. His current heater costs $375 per year for energy costs, and the new heater would cost $225 per year.

a) Write a system of equations with one equation representing the total cost of repairing the heater plus annual energy costs and the other equation representing the total cost of replacing the heater plus annual energy costs.

b) Graph both equations (for up to and including 6 years) on the same axes.

c) Determine the number of years it would take for the total cost of repair to equal the total cost of replacement.

52. MODELING - ***Landscaping Revisited*** In Example 5, assume that Tom's Tree and Landscape Service charges $200 for a consultation fee plus $60 per hour for labor and that Lawn Perfect Landscape Service charges $305 for a consultation fee plus $25 per hour for labor.

a) Write the system of equations to represent the cost of the two landscaping services.

b) Graph both equations for 0 to 10 hours on the same axes.

c) Determine the number of hours of landscaping that must be used for both services to have the same cost.

53. MODELING - ***Selling Backpacks*** Benjamin's Backpacks can sell backpacks for $25 per backpack. The costs for making the backpacks are a fixed cost of $400 and a production cost of $15 per backpack (see Example 6 for an example of cost and revenue equations).

a) Write the cost and revenue equations.

b) Graph both equations, for 0 to 50 backpacks, on the same axes.

c) Use the graph to determine the number of backpacks Benjamin's Backpacks must sell to break even.

d) Write the profit formula.

e) Use the profit formula to determine whether Benjamin's Backpacks makes a profit or loss if it sells 30 backpacks. What is the profit or loss?

f) How many backpacks must Benjamin's Backpacks sell to realize a profit of $1000?

54. **MODELING** - ***Purchasing Stocks*** When buying or selling stock for a customer, the Mark Demo Agency charges $40 plus 8 cents per share of stock purchased or sold. Andy Harris and Associates charges $15 plus 18 cents per share of stock purchased or sold.

   a) Write a system of equations to represent the cost of purchasing or selling stock with each company.

   b) Graph both equations (for up to and including 350 shares of stock) on the same axes.

   c) Determine the number of shares of stock that must be purchased or sold for the total cost to be the same.

   d) If 300 shares of stock are to be purchased, which firm would be less expensive?

55. **MODELING** - ***Manufacturing PDAs*** A manufacturer sells a certain personal digital assistant, or PDA, for $300 per unit. Manufacturing costs consist of a fixed cost of $8400 and a production cost of $230 per unit.

   a) Write the cost and revenue equations.

   b) Graph both equations (for up to and including 150 units) on the same axes.

   c) Use the graph to determine the number of units the manufacturer must sell to break even.

   d) Write the profit formula.

   e) Use the profit formula to determine the manufacturer's profit or loss if 100 units are sold.

   f) How many units must the manufacturer sell to make a profit of $1260?

56. Explain how you can determine whether a system of two linear equations will be consistent, dependent, or inconsistent without graphing the equations.

CHALLENGE PROBLEMS/GROUP ACTIVITIES

57. **MODELING** - ***Job Offers*** Hubert Hotchkiss had two job offers for sales positions. One pays a salary of $500 per week plus a 15% commission on his dollar sales volume. The second position pays a salary of $650 per week with no commission.

   a) For each offer, write an equation that expresses the weekly pay.

   b) Graph the system of equations and determine the solution.

   c) For what dollar sales volume will the two offers result in the same pay?

58. **MODELING** - ***Long-Distance Calling***

   a) In September 2006, an AT&T One Rate Plan charged 7 cents per minute for long-distance calls with a monthly fee of $3.95. The Sprint Nickel Anytime Plan charged 5 cents per minute for long-distance calls with a monthly fee of $8.95. Write an equation to determine the monthly cost for long-distance service with the AT&T One Rate Plan and write an equation to determine the monthly cost for long-distance service with the Sprint Nickel Anytime Plan.

   b) Graph the system of equations and determine the solution.

   c) After how many minutes will the cost for the two long-distance service plans be the same?

59. ***Points of Intersection*** a) If two lines have different slopes, what is the maximum possible number of points of intersection?

   b) If three lines all have different slopes, what is the maximum possible number of points of intersection of any two lines?

   c) If four lines all have different slopes, what is the maximum possible number of points of intersection of any two lines?

   d) If five lines all have different slopes, what is the maximum possible number of points of intersection of any two lines?

   e) Is there a pattern in the number of points of intersection? If so, explain the pattern. Use the pattern to determine the maximum possible number of points of intersection for six lines.

RECREATIONAL MATHEMATICS

60. Connect all the following points using exactly four straight-line segments. Do not lift your pencil off the paper.

INTERNET/RESEARCH ACTIVITY

61. ***The Rhind Papyrus*** The Rhind Papyrus indicates that the early Egyptians used linear equations. Do research and write a paper on the symbols used in linear equations and the use of the linear equations by the early Egyptians. (References include history of mathematics books, encyclopedias, and the Internet.)

7.2 SOLVING SYSTEMS OF LINEAR EQUATIONS BY THE SUBSTITUTION AND ADDITION METHODS

▲ Solving a system of equations may help determine which of two cell phone plans offers the least expensive service.

Bill Ramierez is considering two cell phone plans. Both plans offer 300 free minutes each month. One plan charges \$30 per month plus 45 cents for each additional minute after 300 minutes. The other plan charges \$35 per month plus 20 cents for each additional minute after 300 minutes. How would Bill determine how long he would have to talk on the phone, in 1 month, for the two plans to have the same cost? In this section, we will illustrate two different algebraic methods, the substitution method and the addition method, for answering this question.

We first discuss the substitution method.

Substitution Method

PROCEDURE FOR SOLVING A SYSTEM OF EQUATIONS USING THE SUBSTITUTION METHOD

1. Solve one of the equations for one of the variables. If possible, solve for a variable with a numerical coefficient of 1. By doing so, you may avoid working with fractions.
2. Substitute the expression found in step 1 into the other equation. This step yields an equation in terms of a single variable.
3. Solve the equation found in step 2 for the variable.
4. Substitute the value found in step 3 into the equation you rewrote in step 1 and solve for the remaining variable.

Examples 1, 2, and 3 illustrate the *substitution method*. These systems of equations are the same as in Examples 2, 3, and 4 in Section 7.1.

EXAMPLE 1 *A Single Solution, by the Substitution Method*

Solve the following system of equations by substitution.

$$x + y = 4$$
$$2x - y = -1$$

SOLUTION The numerical coefficients of the x and y terms in the equation $x + y = 4$ are both 1. Thus, we can solve this equation for either x or y. Let's solve for x in the first equation.

STEP 1.

$$x + y = 4$$
$$x + y - y = 4 - y \quad \text{Subtract } y \text{ from both sides of the equation.}$$
$$x = 4 - y$$

STEP 2. Substitute $4 - y$ for x in the second equation.

$$2x - y = -1$$
$$2(4 - y) - y = -1$$

STEP 3. Now solve the equation for y.

$$8 - 2y - y = -1 \quad \text{Distributive property}$$
$$8 - 3y = -1$$
$$8 - 8 - 3y = -1 - 8 \quad \text{Subtract 8 from both sides of the equation.}$$
$$-3y = -9$$
$$\frac{-3y}{-3} = \frac{-9}{-3} \quad \text{Divide both sides of the equation by } -3.$$
$$y = 3$$

STEP 4. Substitute $y = 3$ in the equation solved for x and determine the value of x.

$$x = 4 - y$$
$$x = 4 - 3$$
$$x = 1$$

Thus, the solution is the ordered pair (1, 3). This answer checks with the solution obtained graphically in Section 7.1, Example 2. ●

TIMELY TIP When solving a system of equations, once you successfully solve for one of the variables, make sure you solve for the other variable. Remember that a solution to a system of equations must contain a numerical value for each variable in the system.

EXAMPLE 2 *No Solution, by the Substitution Method*

Solve the following system of equations by substitution.

$$x + y = 3$$
$$2x + 2y = -4$$

SOLUTION The numerical coefficients of the x and y terms in the equation $x + y = 3$ are both 1. Thus, we can solve this equation for either x or y. Let us solve for y in the first equation.

$$x + y = 3$$
$$x - x + y = 3 - x \qquad \text{Subtract } x \text{ from both sides of the equation.}$$
$$y = 3 - x$$

Now substitute $3 - x$ for y in the second equation.

$$2x + 2y = -4$$
$$2x + 2(3 - x) = -4$$
$$2x + 6 - 2x = -4 \qquad \text{Distributive property}$$
$$6 = -4 \qquad \text{False}$$

Since 6 cannot be equal to -4, there is no solution to the system of equations. Thus, the system of equations is inconsistent. This answer checks with the solution obtained graphically in Section 7.1, Example 3. ●

When solving the system in Example 2, we obtained $6 = -4$ and indicated that the system was inconsistent and that there was no solution. When solving a system of equations, if you obtain a false statement, such as $4 = 0$ or $-2 = 0$, the system is *inconsistent* and has *no solution.*

EXAMPLE 3 ***An Infinite Number of Solutions, by the Substitution Method***

Solve the following system of equations by substitution.

$$y = \frac{1}{2}x + 4$$
$$2y = x + 8$$

SOLUTION The first equation $y = \frac{1}{2}x + 4$ is already solved for y, so we will substitute $\frac{1}{2}x + 4$ for y in the second equation.

$$2y = x + 8$$
$$2\left(\frac{1}{2}x + 4\right) = x + 8$$
$$x + 8 = x + 8 \qquad \text{Distributive property}$$
$$x - x + 8 = x - x + 8 \qquad \text{Subtract } x \text{ from both sides of the equation.}$$
$$8 = 8 \qquad \text{True}$$

Since 8 equals 8, the system has an infinite number of solutions. Thus, the system of equations is dependent. This answer checks with the solution obtained in Section 7.1, Example 4. ●

When solving Example 3, we obtained $8 = 8$ and indicated that the system was dependent and had an infinite number of solutions. When solving a system of equations, if you obtain a true statement, such as $0 = 0$ or $8 = 8$, the system is *dependent* and has an *infinite number of solutions.*

DID YOU KNOW?

How to Succeed in Business

Economics, a science dependent on mathematics, dates back to just before the Industrial Revolution of the eighteenth century. Technologies were being invented and applied to the manufacture of cloth, iron, transportation, and agriculture. These new technologies led to the development of mathematically based economic models that often included systems of equations. French economist Jules Dupuit (1804–1866) suggested a method to calculate the value of railroad bridges; Irish economist Dionysis Larder (1793–1859) showed railroad companies how to structure their rates so as to increase their profits.

Addition Method

If neither of the equations in a system of linear equations has a variable with a coefficient of 1, it is generally easier to solve the system by using the ***addition*** (or ***elimination***) ***method***.

To solve a system of linear equations by the addition method, it is necessary to obtain two equations whose sum will be a single equation containing only one variable. To achieve this goal, we rewrite the system of equations as two equations where the coefficients of one of the variables are opposites (or additive inverses) of each other. For example, if one equation has a term of $2x$, we might rewrite the other equation so that its x term will be $-2x$. To obtain the desired equations, it might be necessary to multiply one or both equations in the original system by a number. When an equation is to be multiplied by a number, we will place brackets around the equation and place the number that is to multiply the equation before the brackets. For example, $4[2x + 3y = 6]$ means that each term on both sides of the equal sign in the equation $2x + 3y = 6$ is to be multiplied by 4:

$$4[2x + 3y = 6] \quad \text{gives} \quad 8x + 12y = 24$$

This notation will make our explanations much more efficient and easier for you to follow.

PROCEDURE FOR SOLVING A SYSTEM OF EQUATIONS BY THE ADDITION METHOD

1. If necessary, rewrite the equations so that the terms containing the variables appear on one side of the equal sign and the constants appear on the other side of the equal sign.
2. If necessary, multiply one or both equations by a constant(s) so that when you add the equations, the sum will be an equation containing only one variable.
3. Add the equations to obtain a single equation in one variable.
4. Solve for the variable in the equation obtained in step 3.
5. Substitute the value found in step 4 into either of the original equations and solve for the other variable.

EXAMPLE 4 *Eliminating a Variable by the Addition Method*

Solve the following system of equations by the addition method.

$$x + y = 5$$
$$2x - y = 7$$

SOLUTION Since the coefficients of the y terms, 1 and -1, are additive inverses, the sum of the y terms will be zero when the equations are added. Thus, the sum of the two equations will contain only one variable, x. Add the two equations to obtain one equation in one variable. Then solve for the remaining variable.

$$\begin{aligned} x + y &= 5 \\ \underline{2x - y} &\underline{= 7} \\ 3x \quad\;\; &= 12 \\ x &= 4 \end{aligned}$$

Now substitute 4 for x in either of the original equations to find the value of y.

$$\begin{aligned} x + y &= 5 \\ 4 + y &= 5 \\ y &= 1 \end{aligned}$$

The solution to the system is (4, 1).

EXAMPLE 5 *Multiplying by −1 in the Addition Method*

Solve the following system of equations by the addition method.

$$\begin{aligned} x + 4y &= 10 \\ x + 2y &= 6 \end{aligned}$$

SOLUTION We want the sum of the two equations to have only one variable. We can eliminate the variable x by multiplying either equation by -1 and then adding the two equations. We will multiply the first equation by -1.

$$\begin{aligned} -1[x + 4y &= 10] &\quad \text{gives} \quad -x - 4y &= -10 \\ x + 2y &= 6 & x + 2y &= 6 \end{aligned}$$

We now have a system of equations equivalent to the original system.

Now add the two equations.

$$\begin{aligned} -x - 4y &= -10 \\ \underline{x + 2y} &= \underline{\ \ 6} \\ -2y &= -4 \\ y &= 2 \end{aligned}$$

Now we solve for x by substituting 2 for y in either of the original equations.

$$\begin{aligned} x + 4y &= 10 \\ x + 4(2) &= 10 \\ x + 8 &= 10 \\ x &= 2 \end{aligned}$$

The solution is (2, 2).

EXAMPLE 6 *Multiplying One Equation in the Addition Method*

Solve the following system of equations by the addition method.

$$\begin{aligned} 4x + y &= 6 \\ 3x + 2y &= 7 \end{aligned}$$

SOLUTION We can multiply the top equation by -2 and then add the two equations to eliminate the variable y.

$$\begin{aligned} -2[4x + y = 6] \quad &\text{gives} \quad -8x - 2y = -12 \\ 3x + 2y = 7 \quad & \quad\quad\quad\; 3x + 2y = 7 \end{aligned}$$

$$\begin{array}{r} -8x - 2y = -12 \\ 3x + 2y = 7 \\ \hline -5x \quad\quad = -5 \\ x = 1 \end{array}$$

Now we find y by substituting 1 for x in either of the original equations.

$$\begin{aligned} 4x + y &= 6 \\ 4(1) + y &= 6 \\ 4 + y &= 6 \\ y &= 2 \end{aligned}$$

The solution is (1, 2).

Note that in Example 6 we could have eliminated the variable x by multiplying the top equation by 3 and the bottom equation by -4, then adding the two equations. Try this method now.

EXAMPLE 7 *Multiplying Both Equations*

Solve the following system of equations by the addition method.

$$\begin{aligned} 3x - 4y &= 8 \\ 2x + 3y &= 9 \end{aligned}$$

SOLUTION In this system, we cannot eliminate a variable by multiplying only one equation by an integer value and then adding. To eliminate a variable, we can multiply each equation by a different number. To eliminate the variable x, we can multiply the top equation by 2 and the bottom by -3 (or the top by -2 and the bottom by 3) and then add the two equations. If we want, we can instead eliminate the variable y by multiplying the top equation by 3 and the bottom by 4 and then adding the two equations. Let's eliminate the variable x.

$$\begin{aligned} 2[3x - 4y = 8] \quad &\text{gives} \quad 6x - 8y = 16 \\ -3[2x + 3y = 9] \quad &\text{gives} \quad -6x - 9y = -27 \end{aligned}$$

$$\begin{array}{r} 6x - 8y = 16 \\ -6x - 9y = -27 \\ \hline -17y = -11 \\ y = \dfrac{11}{17} \end{array}$$

We could now find x by substituting $\frac{11}{17}$ for y in either of the original equations. Although it can be done, it gets messy. Instead, let's solve for x by eliminating the variable y from the two original equations. To do so, we multiply the first equation by 3 and the second equation by 4.

$$3[3x - 4y = 8] \quad \text{gives} \quad 9x - 12y = 24$$
$$4[2x + 3y = 9] \quad \text{gives} \quad 8x + 12y = 36$$

$$\begin{aligned} 9x - 12y &= 24 \\ 8x + 12y &= 36 \\ \hline 17x \qquad &= 60 \\ x &= \frac{60}{17} \end{aligned}$$

The solution to the system is $\left(\frac{60}{17}, \frac{11}{17}\right)$.

TIMELY TIP If you obtain an equation such as $0 = 6$, or any other equation that is false when solving a system of linear equations, the system is *inconsistent* (the two equations represent parallel lines; see Fig. 7.3 on page 416) and there is no solution.

If you obtain the equation $0 = 0$ when solving a system of linear equations by either the substitution or the addition method, the system is *dependent* (both equations represent the same line; see Fig. 7.4 on page 416) and there are an infinite number of solutions.

EXAMPLE 8 MODELING - *When Are Repair Costs the Same?*

Melinda Melendez needs to purchase a new radiator for her car and have it installed by a mechanic. She is considering two garages: Steve's Repair and Greg's Garage. At Steve's Repair, the parts cost \$200 and the labor cost is \$50 per hour. At Greg's Garage, the parts cost \$375 and the labor cost is \$25 per hour. How many hours would the repair need to take for the total cost at each garage to be the same?

SOLUTION We are asked to find the number of hours the repair would need to take for each garage to have the same total cost, C. First write a system of equations to represent the total cost for each of the garages. The total cost consists of the cost of the parts and the labor cost. The labor cost depends on the number of hours of labor.

Let $x =$ the number of hours of labor.

$$\text{Total cost} = \text{cost of parts} + \text{labor cost}$$
$$\text{Steve's Repair: } C = 200 + 50x$$
$$\text{Greg's Garage: } C = 375 + 25x$$

We want to determine when the cost will be the same, so we set the two costs equal to each other (substitution method) and solve the resulting equation.

$$200 + 50x = 375 + 25x$$
$$200 - 200 + 50x = 375 - 200 + 25x \qquad \text{Subtract 200 from both sides of the equation.}$$
$$50x = 175 + 25x$$
$$50x - 25x = 175 + 25x - 25x \qquad \text{Subtract } 25x \text{ from both sides of the equation.}$$
$$25x - 175$$
$$\frac{25x}{25} = \frac{175}{25} \qquad \text{Divide both sides of the equation by 25.}$$
$$x = 7$$

Thus, for 7 hours of labor, the cost at both garages would be the same. If we construct a graph (Fig. 7.9) of the two cost equations, the point of intersection is (7, 550). If the repair were to require 7 hours of labor, the total cost at either garage would be \$550.

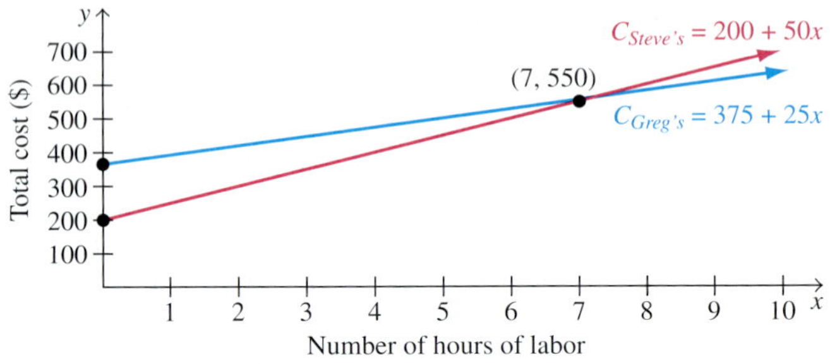

Figure 7.9

EXAMPLE 9 MODELING - *A Mixture Problem*

Pat Kuby, a pharmacist, needs 500 milliliters (mℓ) of a 10% phenobarbital solution. She has only a 5% phenobarbital solution and a 25% phenobarbital solution available. How many milliliters of each solution should she mix to obtain the desired solution?

SOLUTION First we set up a system of equations. The unknown quantities are the amount of 5% solution and the amount of the 25% solution that must be used. Let

$$x = \text{number of m}\ell\text{ of 5\% solution}$$
$$y = \text{number of m}\ell\text{ of 25\% solution}$$

We know that 500 mℓ of solution are needed. Thus,

$$x + y = 500$$

The total amount of phenobarbital in a solution is determined by multiplying the percent of phenobarbital by the number of milliliters of solution. The second equation comes from the fact that

$$\begin{pmatrix}\text{Total amount of}\\\text{phenobarbital in}\\\text{5\% solution}\end{pmatrix} + \begin{pmatrix}\text{total amount of}\\\text{phenobarbital in}\\\text{25\% solution}\end{pmatrix} = \begin{pmatrix}\text{total amount of}\\\text{phenobarbital}\\\text{in 10\% mixture}\end{pmatrix}$$

$$0.05x \quad + \quad 0.25y \quad = \quad 0.10(500)$$

$$\text{or} \qquad 0.05x + 0.25y = 50$$

The system of equations is

$$\begin{aligned} x + y &= 500 \\ 0.05x + 0.25y &= 50 \end{aligned}$$

Let's solve this system of equations by using the addition method. There are various ways of eliminating one variable. To obtain integer values in the second equation, we can multiply both sides of the equation by 100. The result will be an x term of $5x$. If we multiply both sides of the first equation by -5, that will result in an x term of $-5x$. By following this process, we can eliminate the x terms from the system.

$$-5[x + y = 500] \quad \text{gives} \quad -5x - 5y = -2500$$
$$100[0.05x + 0.25y = 50] \quad \text{gives} \quad 5x + 25y = 5000$$

$$\begin{aligned} -5x - 5y &= -2500 \\ 5x + 25y &= 5000 \\ \hline 20y &= 2500 \\ \frac{20y}{20} &= \frac{2500}{20} \\ y &= 125 \end{aligned}$$

Now we determine x.

$$\begin{aligned} x + y &= 500 \\ x + 125 &= 500 \\ x &= 375 \end{aligned}$$

Therefore, 375 mℓ of a 5% phenobarbital solution must be mixed with 125 mℓ of a 25% phenobarbital solution to obtain 500 mℓ of a 10% phenobarbital solution. ●

Example 9 can also be solved by using substitution. Try to do so now.

SECTION 7.2 EXERCISES

CONCEPT/WRITING EXERCISES

1. In your own words, explain how to solve a system of linear equations by using the substitution method.

2. In your own words, explain how to solve a system of linear equations by using the addition method.

3. How will you know, when solving a system of linear equations by either the substitution or the addition method, whether the system is dependent?

4. How will you know, when solving a system of linear equations by either the substitution or the addition method, whether the system is inconsistent?

5. When solving the following system of equations by the substitution method, which variable, in which equation, would you choose to solve for in order to make the solution easier to obtain? Explain your answer. Do not solve the system.

$$x + 3y = 3$$
$$3x + 4y = -1$$

6. When solving the following system of equations by the addition method, what will your first step be in solving the system? Explain your answer. Do not solve the system.

$$2x + y = 6$$
$$3x + 3y = 9$$

PRACTICE THE SKILLS

In Exercises 7–24, solve the system of equations by the substitution method. If the system does not have a single ordered pair as a solution, state whether the system is inconsistent or dependent.

7. $y = x + 8$
   $y = -x + 4$

8. $y = 4x - 3$
   $y = 3x - 1$

9. $6x + 5y = 1$
   $x - 3y = 4$

10. $4x - y = 3$
    $3x - y = 1$

11. $y - x = 4$
    $x - y = 3$

12. $x + y = 3$
    $y + x = 5$

13. $3y + 2x = 4$
$3y = 6 - x$

14. $x = 5y - 12$
$x - y = 0$

15. $y - 2x = 3$
$2y = 4x + 6$

16. $y = 2$
$y + x + 3 = 0$

17. $x = y + 3$
$x = -3$

18. $x + 2y = 6$
$y = 2x + 3$

19. $y + 3x - 4 = 0$
$2x - y = 7$

20. $x + 4y = 7$
$2x + 3y = 5$

21. $x = 2y + 3$
$y = 3x - 1$

22. $x + 4y = 9$
$2x - y - 6 = 0$

23. $y = -2x + 3$
$4x + 2y = 12$

24. $2x + y = 12$
$x = -\frac{1}{2}y + 6$

In Exercises 25–40, solve the system of equations by the addition method. If the system does not have a single ordered pair as a solution, state whether the system is inconsistent or dependent.

25. $3x + y = 9$
$2x - y = 6$

26. $x + 3y = 9$
$x - 3y = -3$

27. $x + y = 12$
$x - 2y = -3$

28. $2x + y = 10$
$-2x + 2y = -16$

29. $2x - y = -4$
$-3x - y = 6$

30. $x + y = 6$
$-2x + y = -3$

31. $4x + 3y = -1$
$2x - y = -13$

32. $2x + y = 6$
$3x + y = 5$

33. $2x + y = 11$
$x + 3y = 18$

34. $5x - 2y = 11$
$-3x + 2y = 1$

35. $3x - 4y = 11$
$3x + 5y = -7$

36. $4x - 2y = 6$
$4y = 8x - 12$

37. $4x + y = 6$
$-8x - 2y = 13$

38. $2x + 3y = 6$
$5x - 4y = -8$

39. $3x - 4y = 2$
$4x + 3y = 11$

40. $6x + 6y = 1$
$4x + 9y = 4$

PROBLEM SOLVING

In Exercises 41–52, write a system of equations that can be used to solve the problem. Then solve the system and determine the answer.

41. **MODELING - *Owning a Business*** Sosena Milion can join a small business as a full partner and receive a salary of \$12,000 per year plus 15% of the year's profit, or she can join as a sales manager with a salary of \$27,000 per year plus 5% of the year's profit. What must the year's profit be for her total earnings to be the same whether she joins as a full partner or as a sales manager?

42. **MODELING - *Investments*** David Stewart invested \$25,000 in two different corporate bonds for 1 year. One bond pays a 4.5% simple interest rate, and the other pays a 6% simple interest rate. The total annual interest David received from both bonds was \$1380. Find the amount he invested in each bond. For 1 year, the simple interest on a specific corporate bond is found by multiplying the amount invested by the simple interest rate.

43. **MODELING - *Pizza Orders*** Pizza Corner sells medium and large specialty pizzas. A medium Meat Lovers pizza costs \$10.95, and a large Meat Lovers pizza costs \$14.95. One Saturday a total of 50 Meat Lovers pizzas were sold, and the receipts from the Meat Lovers pizzas were \$663.50. How many medium and how many large Meat Lovers pizzas were sold?

44. **MODELING - *Basketball Game*** The University of Maryland women's basketball team made 45 field goals in a recent game; some were 2-pointers and some were 3-pointers. How many 2-point baskets were made and how many 3-point baskets were made if Maryland scored 101 points?

45. **MODELING - *Chemical Mixture*** Antonio Gonzalez is a chemist and needs 10 liters (ℓ) of a 40% hydrochloric acid solution. He discovers he is out of the 40% hydrochloric acid solution and does not have sufficient time to reorder. He checks his supply shelf and finds he has a large supply of both 25% and 50% hydrochloric acid solutions. He decides to use the 25% and 50% solutions to make 10 ℓ of a 40% solution. How many liters of the 25% solution and of the 50% solution should he mix?

46. **MODELING - *Sets of Dishes*** A restaurant manager purchased 100 sets of dishes. One design cost \$30 per set, and another design cost \$40 per set. If the manager spent \$3200 on the dishes, how many sets of each design did she purchase?

47. **MODELING** - *Choosing a Copy Service* Lori Lanier recently purchased a high-speed copier for her home office and wants to purchase a service contract on the copier. She is considering two sources for the contract. The Economy Sales and Service Company charges \$18 a month plus 2 cents per copy. Office Superstore charges \$24 a month but only 1.5 cents per copy. How many copies would Lori need to make for the monthly costs of both plans to be the same?

48. **MODELING** - *Hardwood Floor Installation* The cost to purchase a particular type of hardwood flooring at Home Depot is \$2.65 per square foot. In addition, the installation cost is \$468.75. The cost to purchase the same flooring at Hardwood Guys is \$3.10 per square foot. In addition, the installation cost is \$412.50.

a) Determine the number of square feet of this flooring that Roberto Cruz must purchase for the total cost of the flooring and installation to be the same from both stores.

b) If Roberto needs to purchase and have installed 196 square feet of this flooring, which store would be less expensive?

49. **MODELING** - *Nut and Pretzel Mix* Dave Chwalik wants to purchase 20 pounds of party mix for a total of \$30. To obtain the mixture, he will mix nuts that cost \$3 per pound with pretzels that cost \$1 per pound. How many pounds of each type of mix should he use?

50. **MODELING** - *Laboratory Research* Animals in an experiment are to be kept on a strict diet. Each animal is to receive, among other things, 20 g of protein and 6 g of carbohydrates. The scientist has only two food mixes of the following compositions available.

	Protein (%)	Carbohydrates (%)
Mix *A*	10	6
Mix *B*	20	2

How many grams of each mix should she use to obtain the right diet for a single animal?

51. **MODELING** - *Concert Ticket Prices* Finger Lakes Performing Arts Center sold 4600 tickets for a summer jazz concert. Covered amphitheatre tickets cost \$27 and lawn seats cost \$14. If \$104,700 in ticket sales was collected for the concert, how many tickets of each type were sold?

52. **MODELING** - *Golf Club Membership* Membership in Oakwood Country Club costs \$3000 per year and entitles a member to play a round of golf for a greens fee of \$18. At Pinecrest Country Club, membership costs \$2500 per year and the greens fee is \$20.

a) How many rounds must a golfer play in a year for the costs at the two clubs to be the same?

b) If Tamika Johnson planned to play 30 rounds of golf in a year, which club would be the least expensive?

53. **MODELING** - *Car Sales* The following graph shows that General Motors' (GM's) share of the U.S. car and truck market has declined from 2001 through 2005 and is expected to continue to decline through 2010. The graph also shows that Toyota cars and trucks sales are increasing and are expected to continue to increase through 2010. GM's percentage of the U.S. market share can be modeled by the equation $y = -0.83x + 28.33$, and Toyota's market share can be modeled by the equation $y = 0.89x + 9.11$, where x represents the number of years since 2000 and $x \geq 1$. Assuming this trend continues, use the substitution method to approximate when GM's percentage of market share will equal Toyota's.

Collision Course

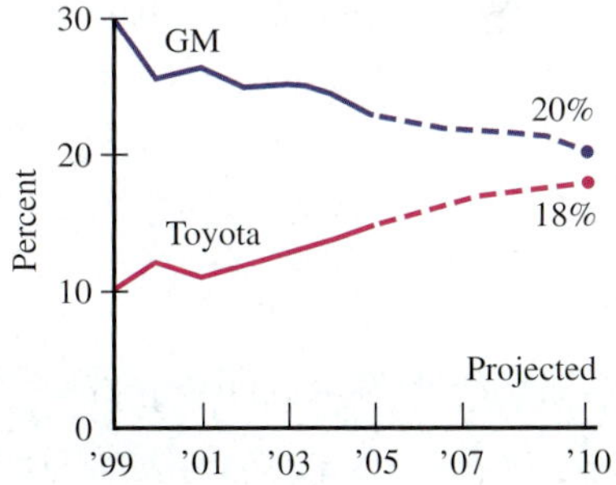

Source: Center for Automotive Research

54. **MODELING** - *Printing Photos* The graph on the top of page 434 shows that the number of traditional photos printed at retail photography labs has declined from 2000 through 2005. The graph also shows that the number of digital photos printed at retail photography labs has increased over this time period. The number of traditional photos printed at labs can be modeled by the equation $y = -2.3x + 29.9$, and the number of digital photos printed at labs can be modeled by the equation $y = 1.5x + 0.4$, where x represents the number of years since 2000. Assuming this trend

continues, use the substitution method to approximate when the number of traditional photos printed will equal the number of digital photos printed at retail photography labs.

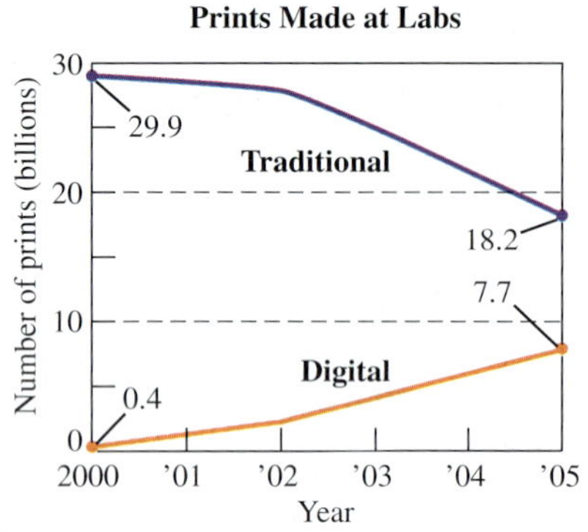

Source: Data from Photo Marketing Association International

CHALLENGE PROBLEMS/GROUP ACTIVITIES

55. Solve the following system of equations for u and v by first substituting x for $\frac{1}{u}$ and y for $\frac{1}{v}$.

$$\frac{1}{u} + \frac{2}{v} = 8$$
$$\frac{3}{u} - \frac{1}{v} = 3$$

56. Develop a system of equations that has (6, 5) as its solution. Explain how you developed your system of equations.

57. The substitution or addition methods can also be used to solve a system of three equations in three variables. Consider the following system.

$$x + y + z = 7$$
$$x - y + 2z = 9$$
$$-x + 2y + z = 4$$

The *ordered triple* (x, y, z) is the solution to the system if it satisfies all three equations.

a) Show that the ordered triple (2, 1, 4) is a solution to the system.

b) Use the substitution or addition method to determine the solution to the system. (*Hint:* Eliminate one variable by using two equations. Then eliminate the same variable by using two different equations.)

58. Construct a system of two equations that has no solution. Explain how you know the system has no solution.

59. Construct a system of two equations that has an infinite number of solutions. Explain how you know the system has an infinite number of solutions.

60. When solving a system of equations by the substitution method, a student obtained the equation $0 = 0$ and gave the solution as (0, 0). What is the student's error?

61. In parts (a)–(d), make up a system of linear equations whose solution will be the ordered pair given. *Hint:* It may be helpful to visualize possible graphs that have the given solution. There are many possible answers for each part.

a) (0, 0) **b)** (1, 0) **c)** (0, 1) **d)** (1, 1)

7.3 MATRICES

▲ In this section, we introduce a method to display information from a survey such as opinions of college students.

Five hundred students at the University of Delaware were asked if they were in favor of or opposed to an increase in their student fees to pay for building a new meeting room for student clubs. Each student was also asked to indicate whether he or she was a freshman, sophomore, junior, or senior. How can the responses from the survey be displayed? In this section, we will introduce a method used to display information such as responses from a survey.

A ***matrix*** is a rectangular array of elements. An array is a systematic arrangement of numbers or symbols in rows and columns. Matrices (the plural of matrix) may be used

> **DID YOU KNOW?**
>
> ***Matrices Are Everywhere***
>
> 
>
>
> You are already familiar with matrices, although you may not be aware of it. The matrix is a good way to display numerical data, as illustrated on this trail sign in Yosemite National Park.

to display information and to solve systems of linear equations. The following matrix displays the responses from the survey of 500 students at the University of Delaware regarding an increase in their student fees.

Rows \ Columns	Freshmen	Sophomores	Juniors	Seniors
In favor	102	93	22	35
Opposed	82	94	23	49

The numbers in the rows and columns of a matrix are called the ***elements*** of the matrix. The matrix given above contains eight elements. The ***dimensions*** of a matrix may be indicated with the notation $r \times s$, where r is the number of rows and s is the number of columns in the matrix. Because the matrix given above has 2 rows and 4 columns, it is a 2 by 4, written 2×4, matrix. In this text, from this point onward, we use brackets, [ ], to indicate a matrix. Consider the two matrices below. A matrix that contains the same number of rows and columns is called a ***square matrix***. Following is an example of a 2×2 square matrix and a 3×3 square matrix.

$$\begin{bmatrix} 2 & 3 \\ 5 & 2 \end{bmatrix} \qquad \begin{bmatrix} 4 & 6 & -1 \\ 2 & 3 & 0 \\ 5 & 2 & 1 \end{bmatrix}$$

Two matrices are equal if and only if they have the same elements in the same relative positions.

EXAMPLE 1 *Equal Matrices*

Given $A = B$, determine x and y.

$$A = \begin{bmatrix} 3 & 7 \\ 4 & 9 \end{bmatrix}, \qquad B = \begin{bmatrix} x & 7 \\ 4 & y \end{bmatrix}$$

SOLUTION Since the matrices are equal, the corresponding elements must be the same, so $x = 3$ and $y = 9$.

Addition of Matrices

Two matrices can be added only if they have the same dimensions (same number of rows and same number of columns). To obtain the sum of two matrices with the same dimensions, add the corresponding elements of the two matrices.

EXAMPLE 2 *Adding Matrices*

Determine $A + B$ if

$$A = \begin{bmatrix} 3 & 4 \\ -1 & 7 \end{bmatrix} \quad \text{and} \quad B = \begin{bmatrix} 2 & 8 \\ 4 & 0 \end{bmatrix}$$

SOLUTION

$$A + B = \begin{bmatrix} 3 & 4 \\ -1 & 7 \end{bmatrix} + \begin{bmatrix} 2 & 8 \\ 4 & 0 \end{bmatrix}$$

$$= \begin{bmatrix} 3 + 2 & 4 + 8 \\ -1 + 4 & 7 + 0 \end{bmatrix} = \begin{bmatrix} 5 & 12 \\ 3 & 7 \end{bmatrix}$$

EXAMPLE 3 MODELING - *Sales of Bicycles*

Peddler's Bicycle Corporation owns and operates two stores, one in Pennsylvania and one in New Jersey. The number of mountain bicycles (MB) and racing bicycles (RB) sold in each store during January through June and during July through December are indicated in the matrices that follow. We will call the matrices A and B.

$$\begin{array}{l} \\ \\ \text{Jan.–June} \\ \text{July–Dec.} \end{array} \overset{\textit{Pennsylvania}}{\overset{\begin{matrix} MB & RB \end{matrix}}{\begin{bmatrix} 515 & 425 \\ 290 & 250 \end{bmatrix}}} = A \qquad \overset{\textit{New Jersey}}{\overset{\begin{matrix} MB & RB \end{matrix}}{\begin{bmatrix} 520 & 350 \\ 180 & 271 \end{bmatrix}}} = B$$

Determine the total number of each type of bicycle sold by the corporation during each time period.

SOLUTION To solve the problem, we add matrices A and B.

$$\begin{array}{l} \\ \text{Jan.–June} \\ \text{July–Dec.} \end{array} \overset{\begin{matrix} MB \qquad\qquad RB \end{matrix}}{\begin{bmatrix} 515 + 520 & 425 + 350 \\ 290 + 180 & 250 + 271 \end{bmatrix}} = \overset{\begin{matrix} MB & RB \end{matrix}}{\begin{bmatrix} 1035 & 775 \\ 470 & 521 \end{bmatrix}}$$

We can see from the sum matrix that during the period from January through June, a total of 1035 mountain bicycles and 775 racing bicycles were sold. During the period from July through December, a total of 470 mountain bicycles and 521 racing bicycles were sold.

Subtraction of Matrices

Only matrices with the same dimension may be subtracted. To do so, we subtract each entry in one matrix from the corresponding entry in the other matrix.

EXAMPLE 4 *Subtracting Matrices*

Determine $A - B$ if

$$A = \begin{bmatrix} 3 & 6 \\ 5 & -1 \end{bmatrix} \quad \text{and} \quad B = \begin{bmatrix} 2 & -4 \\ 8 & -3 \end{bmatrix}$$

SOLUTION

$$\begin{aligned} A - B &= \begin{bmatrix} 3 & 6 \\ 5 & -1 \end{bmatrix} - \begin{bmatrix} 2 & -4 \\ 8 & -3 \end{bmatrix} \\ &= \begin{bmatrix} 3 - 2 & 6 - (-4) \\ 5 - 8 & -1 - (-3) \end{bmatrix} = \begin{bmatrix} 1 & 10 \\ -3 & 2 \end{bmatrix} \end{aligned}$$

Multiplying a Matrix by a Real Number

A matrix may be multiplied by a real number by multiplying each entry in the matrix by the real number. Sometimes when we multiply a matrix by a real number, we call that real number a *scalar*.

DID YOU KNOW?

Spreadsheets

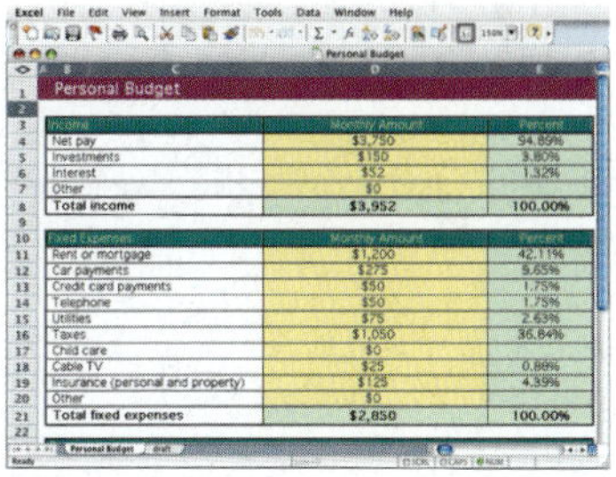

One of the most useful applications of computer software is the spreadsheet. On a computer screen, it looks very much like an accountant's ledger, but there the similarities end. A spreadsheet uses a matrix that assigns an identifying number and letter, in maplike fashion, to each "cell" in the matrix. The data may be dates, numbers, or sums of money. You instruct the computer what operation to perform in each column and row. You can change an entry to see what effect that has on the rest of the spreadsheet. The computer will automatically recompute all affected cells. The computer has the capacity to store information in thousands of cells and can reformat the data in graph form.

EXAMPLE 5 *Multiplying a Matrix by a Scalar*

For matrices A and B, determine (a) $3A$ and (b) $3A - 2B$.

$$A = \begin{bmatrix} 1 & 4 \\ -3 & 5 \end{bmatrix}, \qquad B = \begin{bmatrix} -1 & 3 \\ 5 & 6 \end{bmatrix}$$

SOLUTION

a) $3A = 3\begin{bmatrix} 1 & 4 \\ -3 & 5 \end{bmatrix} = \begin{bmatrix} 3(1) & 3(4) \\ 3(-3) & 3(5) \end{bmatrix} = \begin{bmatrix} 3 & 12 \\ -9 & 15 \end{bmatrix}$

b) We determined $3A$ in part (a). Now we find $2B$.

$$2B = 2\begin{bmatrix} -1 & 3 \\ 5 & 6 \end{bmatrix} = \begin{bmatrix} 2(-1) & 2(3) \\ 2(5) & 2(6) \end{bmatrix} = \begin{bmatrix} -2 & 6 \\ 10 & 12 \end{bmatrix}$$

$$3A - 2B = \begin{bmatrix} 3 & 12 \\ -9 & 15 \end{bmatrix} - \begin{bmatrix} -2 & 6 \\ 10 & 12 \end{bmatrix}$$

$$= \begin{bmatrix} 3-(-2) & 12-6 \\ -9-10 & 15-12 \end{bmatrix} = \begin{bmatrix} 5 & 6 \\ -19 & 3 \end{bmatrix}$$

Multiplication of Matrices

Multiplication of matrices is slightly more difficult than addition of matrices. Multiplication of matrices is possible only when the number of *columns* of the first matrix, A, is the same as the number of *rows* of the second matrix, B. We use the notation

$$\underset{3 \times 4}{A}$$

to indicate that matrix A has three rows and four columns. Suppose matrix A is a 3×4 matrix and matrix B is a 4×5 matrix. Then

$$\underset{3 \times 4}{A} \qquad \underset{4 \times 5}{B}$$

Same

Product matrix 3×5

This notation indicates that matrix A has four columns and matrix B has four rows. Therefore, we can multiply these two matrices. The product matrix will have the same number of rows as matrix A and the same number of columns as matrix B. Thus, the dimensions of the product matrix are 3×5.

EXAMPLE 6 *Can These Matrices Be Multiplied?*

Determine which of the following pairs of matrices can be multiplied.

a) $A = \begin{bmatrix} 3 & 2 \\ 5 & 7 \end{bmatrix}, \quad B = \begin{bmatrix} 0 & 6 \\ 4 & 1 \end{bmatrix}$

b) $A = \begin{bmatrix} 2 & 3 \\ 5 & 6 \end{bmatrix}, \quad B = \begin{bmatrix} 2 & 4 & -1 \\ 6 & 8 & 0 \end{bmatrix}$

c) $A = \begin{bmatrix} 2 & 1 & 4 \\ 3 & 2 & 8 \end{bmatrix}, \quad B = \begin{bmatrix} 2 & 1 & 3 \\ 1 & 0 & -2 \end{bmatrix}$

SOLUTION

a)

$$\underset{2 \times 2}{A} \qquad \underset{2 \times 2}{B}$$

Same

Because matrix A has two columns and matrix B has two rows, the two matrices can be multiplied. The product is a 2×2 matrix.

b)

$$\underset{2 \times 2}{A} \qquad \underset{2 \times 3}{B}$$

Same

Because matrix A has two columns and matrix B has two rows, the two matrices can be multiplied. The product is a 2×3 matrix.

c)

$$\underset{2 \times 3}{A} \qquad \underset{2 \times 3}{B}$$

Not Same

Because matrix A has three columns and matrix B has two rows, the two matrices cannot be multiplied.

To explain matrix multiplication, let's use matrices A and B that follow.

$$A = \begin{bmatrix} 3 & 2 \\ 5 & 7 \end{bmatrix} \quad \text{and} \quad B = \begin{bmatrix} 0 & 6 \\ 4 & 1 \end{bmatrix}$$

Since A contains two rows and B contains two columns, the product matrix will contain two rows and two columns. To multiply two matrices, we use a row–column scheme of multiplying. The numbers in the *first row* of matrix A are multiplied by the numbers in the *first column* of matrix B. These products are then added to determine the entry in the product matrix.

$$A \times B = \begin{bmatrix} 3 & 2 \\ 5 & 7 \end{bmatrix}\begin{bmatrix} 0 & 6 \\ 4 & 1 \end{bmatrix}$$

First row $\begin{bmatrix} 3 & 2 \\ 5 & 7 \end{bmatrix}$ First column $\begin{bmatrix} 0 & 6 \\ 4 & 1 \end{bmatrix}$

$$\begin{aligned}(3 \times 0) + (2 \times 4) &= 0 + 8 \\ &= 8\end{aligned}$$

The 8 is placed in the first-row, first-column position of the product matrix. The other numbers in the product matrix are obtained similarly, as illustrated in the matrix that follows.

First row, First column

$$\begin{bmatrix} 3 & 2 \\ 5 & 7 \end{bmatrix} \quad \begin{bmatrix} 0 & 6 \\ 4 & 1 \end{bmatrix}$$

$(3 \times 0) + (2 \times 4) = 8$

First row, Second column

$$\begin{bmatrix} 3 & 2 \\ 5 & 7 \end{bmatrix} \quad \begin{bmatrix} 0 & 6 \\ 4 & 1 \end{bmatrix}$$

$(3 \times 6) + (2 \times 1) = 20$

$$A \times B = \begin{bmatrix} 8 & 20 \\ 28 & 37 \end{bmatrix}$$

Second row, First column

$$\begin{bmatrix} 3 & 2 \\ 5 & 7 \end{bmatrix} \quad \begin{bmatrix} 0 & 6 \\ 4 & 1 \end{bmatrix}$$

$(5 \times 0) + (7 \times 4) = 28$

Second row, Second column

$$\begin{bmatrix} 3 & 2 \\ 5 & 7 \end{bmatrix} \quad \begin{bmatrix} 0 & 6 \\ 4 & 1 \end{bmatrix}$$

$(5 \times 6) + (7 \times 1) = 37$

We can shorten the procedure as follows.

$$\begin{aligned} A \times B &= \begin{bmatrix} 3 & 2 \\ 5 & 7 \end{bmatrix}\begin{bmatrix} 0 & 6 \\ 4 & 1 \end{bmatrix} \\ &= \begin{bmatrix} 3(0) + 2(4) & 3(6) + 2(1) \\ 5(0) + 7(4) & 5(6) + 7(1) \end{bmatrix} \\ &= \begin{bmatrix} 8 & 20 \\ 28 & 37 \end{bmatrix} \end{aligned}$$

In general, if

$$A = \begin{bmatrix} a & b \\ c & d \end{bmatrix} \quad \text{and} \quad B = \begin{bmatrix} e & f \\ g & h \end{bmatrix}$$

then

$$A \times B = \begin{bmatrix} a & b \\ c & d \end{bmatrix}\begin{bmatrix} e & f \\ g & h \end{bmatrix} = \begin{bmatrix} ae + bg & af + bh \\ ce + dg & cf + dh \end{bmatrix}$$

Let's do one more multiplication of matrices.

EXAMPLE 7 *Multiplying Matrices*

Determine $A \times B$ if

$$A = \begin{bmatrix} 1 & 5 \\ 4 & 7 \end{bmatrix} \quad \text{and} \quad B = \begin{bmatrix} 5 & -1 & 3 \\ 2 & 8 & 0 \end{bmatrix}$$

SOLUTION Matrix A contains two columns, and matrix B contains two rows. Thus, the matrices can be multiplied. Since matrix A contains two rows and matrix B contains three columns, the product matrix will contain two rows and three columns.

$$\begin{aligned} A \times B &= \begin{bmatrix} 1 & 5 \\ 4 & 7 \end{bmatrix}\begin{bmatrix} 5 & -1 & 3 \\ 2 & 8 & 0 \end{bmatrix} \\ &= \begin{bmatrix} 1(5) + 5(2) & 1(-1) + 5(8) & 1(3) + 5(0) \\ 4(5) + 7(2) & 4(-1) + 7(8) & 4(3) + 7(0) \end{bmatrix} \\ &= \begin{bmatrix} 15 & 39 & 3 \\ 34 & 52 & 12 \end{bmatrix} \end{aligned}$$

PROFILE IN MATHEMATICS

James Sylvester, William Rowan Hamilton, and Arthur Cayley

James Sylvester

William Rowan Hamilton

Arthur Cayley

Three mathematicians played important roles in the development of matrix theory: James Sylvester (1814–1897), William Rowan Hamilton (1805–1865), and Arthur Cayley (1821–1895). Sylvester and Cayley cooperated in developing matrix theory. Sylvester was the first to use the term *matrix*. Hamilton, a noted physicist, astronomer, and mathematician, also used what was essentially the algebra of matrices under the name *linear and vector functions*. The mathematical concept of *vector space* grew out of Hamilton's work on the algebra of vectors.

DID YOU KNOW?

The Prisoner's Dilemma

When two parties pursue conflicting interests, the situation can sometimes be described and modeled in a matrix under a branch of mathematics known as *game theory*. Consider a famous problem called "the prisoner's dilemma." A pair of criminal suspects, A and B, are being held in separate jail cells and cannot communicate with each other. Each one is told that there are four possible outcomes: If both confess, each receives a 3-year sentence. If A confesses and B does not, A receives a 1-year sentence, whereas B receives a 10-year sentence. If B confesses and A does not, B receives a 1-year sentence and A receives a 10-year sentence. Finally, if neither confesses, each will be imprisoned for 2 years. (Try arranging this situation in a matrix.)

If neither prisoner knows whether the other will confess, what should each prisoner do? A study of game theory shows that it is in each prisoner's best interest to confess to the crime.

It should be noted that multiplication of matrices *is not* commutative; that is, $A \times B \neq B \times A$, except in special instances.

Square matrices have a ***multiplicative identity matrix***. The multiplicative identity matrices for a 2×2 and a 3×3 matrix, denoted I, follow. Note that in any multiplicative identity matrix, 1's go diagonally from top left to bottom right and all other elements in the matrix are 0's.

$$I = \begin{bmatrix} 1 & 0 \\ 0 & 1 \end{bmatrix} \qquad I = \begin{bmatrix} 1 & 0 & 0 \\ 0 & 1 & 0 \\ 0 & 0 & 1 \end{bmatrix}$$

For any square matrix, A, $A \times I = I \times A = A$.

EXAMPLE 8 *Using the Identity Matrix in Multiplication*

Use the multiplicative identity matrix for a 2×2 matrix and matrix A to show that $A \times I = A$.

$$A = \begin{bmatrix} 4 & 3 \\ 2 & 1 \end{bmatrix}$$

SOLUTION The identity matrix is $I = \begin{bmatrix} 1 & 0 \\ 0 & 1 \end{bmatrix}$.

$$\begin{aligned} A \times I &= \begin{bmatrix} 4 & 3 \\ 2 & 1 \end{bmatrix}\begin{bmatrix} 1 & 0 \\ 0 & 1 \end{bmatrix} \\ &= \begin{bmatrix} 4(1) + 3(0) & 4(0) + 3(1) \\ 2(1) + 1(0) & 2(0) + 1(1) \end{bmatrix} \\ &= \begin{bmatrix} 4 & 3 \\ 2 & 1 \end{bmatrix} = A \end{aligned}$$

Verify for yourself that $I \times A = A$ as well. •

Example 9 illustrates an application of multiplication of matrices.

EXAMPLE 9 *A Manufacturing Application*

The Fancy Frock Company manufactures three types of women's outfits: a dress, a two-piece suit (skirt and jacket), and a three-piece suit (skirt, jacket, and a vest). On a particular day, the firm produces 20 dresses, 30 two-piece suits, and 50 three-piece suits. Each dress requires 4 units of material and 1 hour of work to produce, each two-piece suit requires 5 units of material and 2 hours of work to produce, and each three-piece suit requires 6 units of material and 3 hours to produce. Use matrix multiplication to determine the total number of units of material and the total number of hours needed for that day's production.

SOLUTION Let matrix A represent the number of each type of women's outfits produced.

	Dress	Two piece	Three piece
$A =$	[20	30	50]

The units of material and time requirements for each type are indicated in matrix B.

$$B = \begin{bmatrix} 4 & 1 \\ 5 & 2 \\ 6 & 3 \end{bmatrix} \begin{matrix} \text{Dress} \\ \text{Two piece} \\ \text{Three piece} \end{matrix}$$

(Columns: Material, Hours)

The product of A and B, or $A \times B$, will give the total number of units of material and the total number of hours of work needed for that day's production.

$$\begin{aligned} A \times B &= [20 \quad 30 \quad 50]\begin{bmatrix} 4 & 1 \\ 5 & 2 \\ 6 & 3 \end{bmatrix} \\ &= [20(4) + 30(5) + 50(6) \qquad 20(1) + 30(2) + 50(3)] \\ &= [530 \quad 230] \end{aligned}$$

Thus, a total of 530 units of material and a total of 230 hours of work are needed that day.

TIMELY TIP Matrices can only be added or subtracted if they have the same dimensions.

Matrices can only be multiplied if the number of *columns* in the first matrix is the same as the number of *rows* in the second matrix.

SECTION 7.3 EXERCISES

CONCEPT/WRITING EXERCISES

1. What is a matrix?

2. Explain how to determine the dimensions of a matrix.

3. What is a square matrix?

4. How many columns does a 3×2 matrix have?

5. How many rows does a 4×3 matrix have?

6. To add or subtract two matrices, what must be true about the dimensions of those matrices?

7. **a)** In your own words, explain the procedure used to add matrices.

   b) Use the procedure given in part (a) to add

$$\begin{bmatrix} 5 & 4 & -1 \\ 3 & 9 & 5 \end{bmatrix} \quad \text{and} \quad \begin{bmatrix} 2 & 5 & -6 \\ -1 & 7 & 4 \end{bmatrix}$$

8. **a)** In your own words, explain the procedure used to subtract matrices.

   b) Use the procedure given in part (a) to subtract

$$\begin{bmatrix} 8 & 4 & 2 \\ 0 & -2 & 4 \end{bmatrix} \quad \text{from} \quad \begin{bmatrix} 3 & -5 & 6 \\ -2 & 3 & 4 \end{bmatrix}$$

9. **a)** To multiply two matrices, what must be true about the dimensions of those matrices?

   b) What will be the dimensions of the product matrix when multiplying a 2×2 matrix with a 2×3 matrix?

10. **a)** In your own words, explain the procedure used to multiply matrices.

    b) Use the procedure given in part (a) to multiply

$$\begin{bmatrix} 8 & -1 \\ 6 & 0 \end{bmatrix} \quad \text{by} \quad \begin{bmatrix} 2 & -3 \\ 5 & -4 \end{bmatrix}$$

11. a) What is the multiplicative identity matrix for a 2×2 matrix?

b) What is the multiplicative identity matrix for a 3×3 matrix?

12. A company has three offices: East, West, and Central. Each office has five divisions. The number of employees in each division of the three offices is as follows:
East: 110, 232, 103, 190, 212
West: 107, 250, 135, 203, 189
Central: 115, 218, 122, 192, 210

Express this information in the form of a 3×5 matrix.

PRACTICE THE SKILLS

In Exercises 13–16, determine $A + B$.

13. $A = \begin{bmatrix} 1 & 8 \\ 2 & 7 \end{bmatrix}$, $B = \begin{bmatrix} -4 & 1 \\ 7 & 2 \end{bmatrix}$

14. $A = \begin{bmatrix} 5 & 6 & -7 \\ 0 & 1 & -1 \end{bmatrix}$, $B = \begin{bmatrix} -4 & 2 & -8 \\ 7 & -3 & 0 \end{bmatrix}$

15. $A = \begin{bmatrix} 2 & 1 \\ -1 & 4 \\ 6 & 0 \end{bmatrix}$, $B = \begin{bmatrix} -3 & 3 \\ -4 & 0 \\ 1 & 6 \end{bmatrix}$

16. $A = \begin{bmatrix} 2 & 6 & 3 \\ -1 & -6 & 4 \\ 3 & 0 & 5 \end{bmatrix}$, $B = \begin{bmatrix} -1 & 3 & 1 \\ 7 & -2 & 1 \\ 2 & 3 & 8 \end{bmatrix}$

In Exercises 17–20, determine $A - B$.

17. $A = \begin{bmatrix} 4 & -2 \\ -3 & 5 \end{bmatrix}$, $B = \begin{bmatrix} -2 & 5 \\ 9 & 1 \end{bmatrix}$

18. $A = \begin{bmatrix} 10 & 1 \\ 12 & 2 \\ -3 & -9 \end{bmatrix}$, $B = \begin{bmatrix} -3 & 3 \\ 4 & 7 \\ -2 & 6 \end{bmatrix}$

19. $A = \begin{bmatrix} -5 & 1 \\ 8 & 6 \\ 1 & -5 \end{bmatrix}$, $B = \begin{bmatrix} -6 & -8 \\ -10 & -11 \\ 3 & -7 \end{bmatrix}$

20. $A = \begin{bmatrix} 5 & 3 & -1 \\ 7 & 4 & 2 \\ 6 & -1 & -5 \end{bmatrix}$, $B = \begin{bmatrix} 4 & 3 & 6 \\ -2 & -4 & 9 \\ 0 & -2 & 4 \end{bmatrix}$

In Exercises 21–26,

$$A = \begin{bmatrix} 1 & 2 \\ 0 & 5 \end{bmatrix}, \quad B = \begin{bmatrix} 3 & 2 \\ 5 & 0 \end{bmatrix}, \quad \text{and} \quad C = \begin{bmatrix} -2 & 3 \\ 4 & 0 \end{bmatrix}.$$

Determine the following.

21. $2B$

22. $-3B$

23. $2B + 4C$

24. $2B + 3A$

25. $4B - 2C$

26. $3C - 2A$

In Exercises 27–32, determine $A \times B$.

27. $A = \begin{bmatrix} 1 & 3 \\ 0 & 6 \end{bmatrix}$, $B = \begin{bmatrix} 2 & 6 \\ 8 & 4 \end{bmatrix}$

28. $A = \begin{bmatrix} 1 & -1 \\ 2 & 6 \end{bmatrix}$, $B = \begin{bmatrix} 4 & -2 \\ -3 & -2 \end{bmatrix}$

29. $A = \begin{bmatrix} 2 & 3 & -1 \\ 0 & 4 & 6 \end{bmatrix}$, $B = \begin{bmatrix} 2 \\ 4 \\ 1 \end{bmatrix}$

30. $A = \begin{bmatrix} 1 & 1 \\ 1 & 1 \end{bmatrix}$, $B = \begin{bmatrix} 1 & -1 \\ -1 & 2 \end{bmatrix}$

31. $A = \begin{bmatrix} 4 & 7 & 6 \\ -2 & 3 & 1 \\ 5 & 1 & 2 \end{bmatrix}$, $B = \begin{bmatrix} 1 & 0 & 0 \\ 0 & 1 & 0 \\ 0 & 0 & 1 \end{bmatrix}$

32. $A = \begin{bmatrix} 1 & -3 \\ 7 & 2 \end{bmatrix}$, $B = \begin{bmatrix} 0 & 4 \\ 6 & 1 \end{bmatrix}$

In Exercises 33–38, determine $A + B$ and $A \times B$. If an operation cannot be performed, explain why.

33. $A = \begin{bmatrix} 2 & 3 & 5 \\ 4 & 0 & 3 \end{bmatrix}$, $B = \begin{bmatrix} 7 & -2 & 3 \\ 2 & -1 & 1 \end{bmatrix}$

34. $A = \begin{bmatrix} 6 & 4 & -1 \\ 2 & 3 & 4 \end{bmatrix}$, $B = \begin{bmatrix} 1 & 0 \\ 4 & -1 \end{bmatrix}$

35. $A = \begin{bmatrix} 4 & 5 & 3 \\ 6 & 2 & 1 \end{bmatrix}$, $B = \begin{bmatrix} 3 & 2 \\ 4 & 6 \\ -2 & 0 \end{bmatrix}$

36. $A = \begin{bmatrix} 6 & 5 \\ 4 & 3 \\ 2 & 1 \end{bmatrix}$, $B = \begin{bmatrix} 6 & 5 \\ 4 & 3 \\ 2 & 1 \end{bmatrix}$

37. $A = \begin{bmatrix} 1 & 2 \\ 3 & 4 \end{bmatrix}$, $B = \begin{bmatrix} -3 \\ 2 \end{bmatrix}$

38. $A = \begin{bmatrix} 5 & -1 \\ 6 & -2 \end{bmatrix}$, $B = \begin{bmatrix} 1 & 2 \\ 3 & 4 \end{bmatrix}$

In Exercises 39–41, show the commutative property of addition, $A + B = B + A$, holds for matrices A and B.

39. $A = \begin{bmatrix} 3 & 5 \\ -2 & -3 \end{bmatrix}, \quad B = \begin{bmatrix} 4 & 5 \\ 6 & 7 \end{bmatrix}$

40. $A = \begin{bmatrix} 9 & 4 \\ 1 & 7 \end{bmatrix}, \quad B = \begin{bmatrix} 2 & 0 \\ -1 & 6 \end{bmatrix}$

41. $A = \begin{bmatrix} 0 & -1 \\ 3 & -4 \end{bmatrix}, \quad B = \begin{bmatrix} 8 & 1 \\ 3 & -4 \end{bmatrix}$

42. Create two matrices with the same dimensions, A and B, and show that $A + B = B + A$.

In Exercises 43–45, show that the associative property of addition, $(A + B) + C = A + (B + C)$, holds for the matrices given.

43. $A = \begin{bmatrix} 5 & 2 \\ 3 & 6 \end{bmatrix}, \quad B = \begin{bmatrix} 3 & 4 \\ -2 & 7 \end{bmatrix}, \quad C = \begin{bmatrix} -1 & 4 \\ 5 & 0 \end{bmatrix}$

44. $A = \begin{bmatrix} 4 & 1 \\ 6 & 7 \end{bmatrix}, \quad B = \begin{bmatrix} -9 & 1 \\ -7 & 2 \end{bmatrix}, \quad C = \begin{bmatrix} -6 & -3 \\ 3 & 6 \end{bmatrix}$

45. $A = \begin{bmatrix} 7 & 4 \\ 9 & -36 \end{bmatrix}, \quad B = \begin{bmatrix} 5 & 6 \\ -1 & -4 \end{bmatrix}, \quad C = \begin{bmatrix} -7 & -5 \\ -1 & 3 \end{bmatrix}$

46. Create three matrices with the same dimensions, A, B, and C, and show that $(A + B) + C = A + (B + C)$.

In Exercises 47–51, determine whether the commutative property of multiplication, $A \times B = B \times A$, holds for the matrices given.

47. $A = \begin{bmatrix} 1 & -2 \\ 4 & -3 \end{bmatrix}, \quad B = \begin{bmatrix} -1 & -3 \\ 2 & 4 \end{bmatrix}$

48. $A = \begin{bmatrix} 3 & 1 \\ 6 & 6 \end{bmatrix}, \quad B = \begin{bmatrix} 1 & 0 \\ 0 & 1 \end{bmatrix}$

49. $A = \begin{bmatrix} 4 & 2 \\ 1 & -3 \end{bmatrix}, \quad B = \begin{bmatrix} 2 & 4 \\ -3 & 1 \end{bmatrix}$

50. $A = \begin{bmatrix} -3 & 2 \\ 6 & -5 \end{bmatrix}, \quad B = \begin{bmatrix} -\frac{5}{3} & -\frac{2}{3} \\ -2 & -1 \end{bmatrix}$

51. $A = \begin{bmatrix} 3 & 2 & 1 \\ 4 & 2 & 0 \\ 0 & -2 & 5 \end{bmatrix}, \quad B = \begin{bmatrix} 1 & 0 & 0 \\ 0 & 1 & 0 \\ 0 & 0 & 1 \end{bmatrix}$

52. Create two square matrices A and B with the same dimensions, and determine whether $A \times B = B \times A$.

In Exercises 53–57, show that the associative property of multiplication, $(A \times B) \times C = A \times (B \times C)$, holds for the matrices given.

53. $A = \begin{bmatrix} 1 & 3 \\ 4 & 0 \end{bmatrix}, \quad B = \begin{bmatrix} 4 & 2 \\ 3 & 1 \end{bmatrix}, \quad C = \begin{bmatrix} 2 & 1 \\ 3 & 0 \end{bmatrix}$

54. $A = \begin{bmatrix} -2 & 3 \\ 0 & 4 \end{bmatrix}, \quad B = \begin{bmatrix} 4 & 0 \\ 3 & 5 \end{bmatrix}, \quad C = \begin{bmatrix} 3 & 4 \\ -2 & 5 \end{bmatrix}$

55. $A = \begin{bmatrix} 4 & 3 \\ -6 & 2 \end{bmatrix}, \quad B = \begin{bmatrix} 1 & 2 \\ 0 & 1 \end{bmatrix}, \quad C = \begin{bmatrix} 4 & 3 \\ 0 & -2 \end{bmatrix}$

56. $A = \begin{bmatrix} -1 & -2 \\ -3 & -4 \end{bmatrix}, \quad B = \begin{bmatrix} 1 & 0 \\ 0 & 1 \end{bmatrix}, \quad C = \begin{bmatrix} 0 & 0 \\ 0 & 0 \end{bmatrix}$

57. $A = \begin{bmatrix} 3 & 4 \\ -1 & -2 \end{bmatrix}, \quad B = \begin{bmatrix} 0 & 1 \\ 1 & 0 \end{bmatrix}, \quad C = \begin{bmatrix} 2 & 0 \\ 3 & 0 \end{bmatrix}$

58. Create three matrices, A, B, and C, and show that $(A \times B) \times C = A \times (B \times C)$.

PROBLEM SOLVING

59. MODELING - *Purchasing Produce* Matrix A represents the weight, in pounds, of tomatoes, onions, and carrots that Wegmans Food Markets purchased from two different local farmers in 1 week. Matrix B represents the weight, in pounds, of tomatoes, onions, and carrots that Wegmans Food Markets purchased from the same farmers the following week.

$$\begin{matrix} & \text{Tomatoes} & \text{Onions} & \text{Carrots} & \\ A = & \left[\begin{matrix} 40 \\ 38 \end{matrix}\right. & \begin{matrix} 22 \\ 25 \end{matrix} & \left.\begin{matrix} 31 \\ 34 \end{matrix}\right] & \begin{matrix} \text{Chase's Farm} \\ \text{Gro-More Farms} \end{matrix} \end{matrix}$$

$$\begin{matrix} & \text{Tomatoes} & \text{Onions} & \text{Carrots} & \\ B = & \left[\begin{matrix} 48 \\ 40 \end{matrix}\right. & \begin{matrix} 36 \\ 29 \end{matrix} & \left.\begin{matrix} 39 \\ 37 \end{matrix}\right] & \begin{matrix} \text{Chase's Farm} \\ \text{Gro-More Farms} \end{matrix} \end{matrix}$$

Use matrix addition to determine the total weight, in pounds, Wegmans purchased over the 2-week period for each item from each farm.

60. MODELING - *Sweatshirt Inventory* Dick's Sporting Goods sells sweatshirts for adults and youth. Each type of sweatshirt comes in four sizes: small, medium, large, and extra large. Matrix A represents the stock on hand at the beginning of a given week for each type of sweatshirt. Matrix B represents the stock on hand at the end of the same week for the same three types of sweatshirts.

$$A = \begin{bmatrix} 31 & 18 \\ 39 & 16 \\ 41 & 22 \\ 34 & 21 \end{bmatrix} \begin{matrix} \text{Small} \\ \text{Medium} \\ \text{Large} \\ \text{Extra large} \end{matrix} \quad (\text{columns: Adult, Youth})$$

$$B = \begin{bmatrix} 14 & 9 \\ 18 & 9 \\ 19 & 15 \\ 15 & 9 \end{bmatrix} \begin{matrix} \text{Small} \\ \text{Medium} \\ \text{Large} \\ \text{Extra large} \end{matrix} \quad (\text{columns: Adult, Youth})$$

Use matrix subtraction to determine the number of sweatshirts sold in the given week for each size of each type.

61. MODELING - *Supplying Muffins* A bakery supplies chocolate chip, blueberry, and coffee cake muffins to Java's Coffee Shop and to Spot Coffee Shop. Matrix A shows the number of each type of muffin, in dozens, supplied to each coffee shop in 1 week. The bakery's cost for ingredients, per dozen, for chocolate chip, blueberry, and coffee cake muffins is \$3, \$2, and \$1.5, respectively. Matrix B represents the bakery's cost for ingredients, per dozen, for each type of muffin.

$$A = \begin{bmatrix} 7 & 8.5 & 10 \\ 7.5 & 8 & 11 \end{bmatrix} \begin{matrix} \text{Java's Coffee Shop} \\ \text{Spot Coffee Shop} \end{matrix} \quad (\text{columns: Chocolate chip, Blueberry, Coffee cake})$$

$$B = \begin{bmatrix} 3 \\ 2 \\ 1.5 \end{bmatrix} \begin{matrix} \text{Chocolate chip} \\ \text{Blueberry} \\ \text{Coffee cake} \end{matrix}$$

Use matrix multiplication to determine the total weekly cost of ingredients for each shop.

62. MODELING - *Menu Choices* The cafeteria manager at the University of Virginia recently added three new meals that each consist of a meat, a vegetable, a salad, and a dessert to the lunch menu. The first day these new meals were offered, 35 students purchased meal 1, 28 students purchased meal 2, and 23 students purchased meal 3. Matrix A represents the number of students who purchased each type of meal. Matrix B represents the cost per serving, in cents, for each item from the three meals.

$$A = \begin{bmatrix} 35 & 28 & 23 \end{bmatrix} \quad (\text{columns: Meal 1, Meal 2, Meal 3})$$

$$B = \begin{bmatrix} 45 & 12 & 8 & 10 \\ 52 & 8 & 6 & 15 \\ 49 & 7 & 9 & 11 \end{bmatrix} \begin{matrix} \text{Meal 1} \\ \text{Meal 2} \\ \text{Meal 3} \end{matrix} \quad (\text{columns: Meat, Vegetable, Salad, Dessert})$$

Use matrix multiplication to determine the total cost for each item.

63. MODELING - *Cookie Company Costs* The Original Cookie Factory bakes and sells four types of cookies: chocolate chip, sugar, molasses, and peanut butter. Matrix A shows the number of units of various ingredients used in baking a dozen of each type of cookie.

$$A = \begin{bmatrix} 2 & 2 & \frac{1}{2} & 1 \\ 3 & 2 & 1 & 2 \\ 0 & 1 & 0 & 3 \\ \frac{1}{2} & 1 & 0 & 0 \end{bmatrix} \begin{matrix} \text{Chocolate chip} \\ \text{Sugar} \\ \text{Molasses} \\ \text{Peanut butter} \end{matrix} \quad (\text{columns: Sugar, Flour, Milk, Eggs})$$

The cost, in cents per cup or per egg, for each ingredient when purchased in small quantities and in large quantities is given in matrix B.

$$B = \begin{bmatrix} 10 & 12 \\ 5 & 8 \\ 8 & 8 \\ 4 & 6 \end{bmatrix} \begin{matrix} \text{Sugar} \\ \text{Flour} \\ \text{Milk} \\ \text{Eggs} \end{matrix} \quad (\text{columns: Large quantities, Small quantities})$$

Use matrix multiplication to find a matrix representing the comparative cost per item for small and large quantities purchased.

In Exercises 64 and 65, use the information given in Exercise 63. Suppose a typical day's order consists of 40 dozen chocolate chip cookies, 30 dozen sugar cookies, 12 dozen molasses cookies, and 20 dozen peanut butter cookies.

64. a) Express these orders as a 1×4 matrix.

b) Use matrix multiplication to determine the amount of each ingredient needed to fill the day's order.

65. Use matrix multiplication to determine the cost under the two purchase options (small and large quantities) to fill the day's order.

66. MODELING - *Food Prices* To raise money for a local charity, the Spanish Club at Montclair High School sold hot dogs, soft drinks, and candy bars for 3 days in the student lounge. The sales for the 3 days are summarized in matrix A.

	Hot dogs	Soft drinks	Candy bars
Day 1	52	50	75
Day 2	48	43	60
Day 3	62	57	81

$$A = \begin{bmatrix} 52 & 50 & 75 \\ 48 & 43 & 60 \\ 62 & 57 & 81 \end{bmatrix}$$

The cost and revenue (in dollars) for hot dogs, soft drinks, and candy are summarized in matrix B.

	Cost	Revenue
Hot dogs	0.30	0.75
Soft drinks	0.25	0.50
Candy bars	0.15	0.45

$$B = \begin{bmatrix} 0.30 & 0.75 \\ 0.25 & 0.50 \\ 0.15 & 0.45 \end{bmatrix}$$

Multiply the two matrices to form a 3×2 matrix that shows the total cost and revenue for each item.

In Exercises 67 and 68, there are many acceptable answers.

67. a) Construct two matrices A and B whose product is a 3×1 matrix. Explain how you determined your answer.

b) For your matrices, determine $A \times B$.

68. a) Construct two matrices A and B whose product is a 4×1 matrix. Explain how you determined your answer.

b) For your matrices, determine $A \times B$.

Two matrices whose product is the multiplicative identity matrix are said to be ***multiplicative inverses****. That is, if $A \times B = B \times A = I$, where I is the multiplicative identity matrix, then A and B are multiplicative inverses.*

In Exercises 69 and 70, determine whether A and B are multiplicative inverses.

69. $A = \begin{bmatrix} 5 & -2 \\ -2 & 1 \end{bmatrix}, \quad B = \begin{bmatrix} 1 & 2 \\ 2 & 5 \end{bmatrix}$

70. $A = \begin{bmatrix} 7 & 3 \\ 2 & 1 \end{bmatrix}, \quad B = \begin{bmatrix} 1 & -3 \\ -2 & 7 \end{bmatrix}$

CHALLENGE PROBLEMS/GROUP ACTIVITIES

In Exercises 71 and 72, determine whether the statement is true or false. Give an example to support your answer.

71. $A - B = B - A$, where A and B are any matrices.

72. For scalar a and matrices B and C, $a(B + C) = aB + aC$.

73. MODELING - *Sofa Manufacturing Costs* The number of hours of labor required to manufacture one sofa of various sizes is summarized in matrix L.

Sofa size	Cutting	Assembly	Packing
Small	1.4 hr	0.7 hr	0.3 hr
Medium	1.8 hr	1.4 hr	0.3 hr
Large	2.7 hr	2.8 hr	0.5 hr

(Columns: Department)

$$L = \begin{bmatrix} 1.4 \text{ hr} & 0.7 \text{ hr} & 0.3 \text{ hr} \\ 1.8 \text{ hr} & 1.4 \text{ hr} & 0.3 \text{ hr} \\ 2.7 \text{ hr} & 2.8 \text{ hr} & 0.5 \text{ hr} \end{bmatrix}$$

The hourly labor rates for cutting, assembly, and packing at the Ames City Plant and at the Bay City Plant are given in matrix C.

Department	Ames City	Bay City
Cutting	$14	$12
Assembly	$10	$9
Packaging	$7	$5

(Columns: Plant)

$$C = \begin{bmatrix} \$14 & \$12 \\ \$10 & \$9 \\ \$7 & \$5 \end{bmatrix}$$

a) What is the total labor cost for manufacturing a small-sized sofa at the Ames City plant?

b) What is the total cost for manufacturing a large-sized sofa at the Bay City plant?

c) Determine the product $L \times C$ and explain the meaning of the results.

74. Is it possible that two matrices could be added but not multiplied? If so, give an example.

75. Is it possible that two matrices could be multiplied but not added? If so, give an example.

RECREATIONAL MATHEMATICS

76. Make up two matrices A and B such that

$$A + B = \begin{bmatrix} 1 & 0 \\ 0 & 1 \end{bmatrix} \text{ and } A \times B = \begin{bmatrix} 0 & 0 \\ 0 & 0 \end{bmatrix}.$$

INTERNET/RESEARCH ACTIVITIES

77. Find an article that shows information illustrated in matrix form. Write a short paper explaining how to interpret the information provided by the matrix. Include the article with your report.

78. ***Messages*** The study of encoding and decoding messages is called *cryptography*. Do research on current real-life uses of cryptography and write a paper on how matrix multiplication is used to encode and decode messages. In your paper, include current real-life uses of cryptography.

7.4 SOLVING SYSTEMS OF LINEAR EQUATIONS BY USING MATRICES

▲ If you know the total cost of two different types of candy in a box and the weight of each type of candy, you can use a system of equations to determine the cost of 1 pound of each item.

Consider the following situation. One package of chocolate-covered mints and chocolate-covered cherries sells for \$23. A different package of chocolate-covered mints and chocolate-covered cherries sells for \$14. If you know the weight of the chocolate-covered mints and the weight of the chocolate-covered cherries in each package, would you be able to determine the cost of 1 pound of the chocolate-covered mints and the cost of 1 pound of the chocolate-covered cherries? There are several ways to determine these costs. In this section, we will illustrate a method for answering this question by using matrices.

In Section 7.3, we introduced matrices. Now we will discuss the procedure to solve a system of linear equations using matrices. We will illustrate how to solve a system of two equations and two unknowns. Systems of equations containing three equations and three unknowns (called third-order systems) and higher-order systems can also be solved by using matrices, but we will not discuss third or higher order systems in this book.

The first step in solving a system of equations using matrices is to represent the system of equations with an ***augmented matrix***. An augmented matrix consists of two smaller matrices, one for the coefficients of the variables in the equations and one for the constants in the equations. To determine the augmented matrix, first write each equation in standard form, $ax + by = c$. For the system of equations below, its augmented matrix is shown to its right.

System of equations	**Augmented matrix**
$a_1x + b_1y = c_1$ $a_2x + b_2y = c_2$	$\left[\begin{array}{cc\|c} a_1 & b_1 & c_1 \\ a_2 & b_2 & c_2 \end{array}\right]$

DID YOU KNOW?

Early Matrices

The earliest known use of a matrix to solve linear equations appeared in the ancient Chinese mathematical classic *Jiuzhang Suanshu (Nine Chapters on the Mathematical Art)* in about 200 B.C. The use of matrices to solve problems did not appear in the West until the nineteenth century. Perhaps the fact that the Chinese used a counting board that took the form of a grid made it easier for them to make the leap to the development and use of matrices. The image above shows the same type of counting board as appeared in a book dated 1795.

Following is another example.

System of equations	**Augmented matrix**
$x + 2y = 8$ $3x - y = 7$	$\left[\begin{array}{cc\|c} 1 & 2 & 8 \\ 3 & -1 & 7 \end{array}\right]$

Note that the vertical bar in the augmented matrix separates the numerical coefficients from the constants. The matrix is just a shortened way of writing the system of equations. Thus, we can solve a system of equations by using matrices in a manner very similar to solving a system of equations with the addition method.

To solve a system of equations by using matrices, we use *row transformations* to obtain new matrices that have the same solution as the original system. We will discuss three row transformation procedures.

PROCEDURES FOR ROW TRANSFORMATIONS

1. Any two rows of a matrix may be interchanged (which is the same as interchanging any two equations in the system of equations).
2. All the numbers in any row may be multiplied by any nonzero real number (which is the same as multiplying both sides of an equation by any nonzero real number).
3. All the numbers in any row may be multiplied by any nonzero real number, and these products may be added to the corresponding numbers in any other row of numbers.

We use row transformations to obtain an augmented matrix whose numbers to the left of the vertical bar are the same as in the *multiplicative identity matrix*. From this type of augmented matrix, we can determine the solution to the system of equations. For example, if we get

$$\left[\begin{array}{cc|c} 1 & 0 & 3 \\ 0 & 1 & -2 \end{array}\right]$$

it tells us that $1x + 0y = 3$ or $x = 3$, and $0x + 1y = -2$ or $y = -2$. Thus, the solution to the system of equations that yielded this augmented matrix is $(3, -2)$. Now let's work an example.

EXAMPLE 1 *Using Row Transformations*

Solve the following system of equations by using matrices.

$$x + 2y = 5$$
$$3x - y = 8$$

SOLUTION First we write the augmented matrix.

$$\left[\begin{array}{cc|c} 1 & 2 & 5 \\ 3 & -1 & 8 \end{array}\right]$$

MATHEMATICS TODAY

Matrices and Computer Graphics

Doc Hudson and Lightning from *Cars*

Did you ever wonder how the characters in computer-generated movies such as *Shrek* and *Cars* are created? To create the graphics for computer-generated films, an object is first represented as a collection of geometric figures. Each geometric figure is then represented as points in the Cartesian coordinate plane. The coordinates of the points representing the object are written in columns in a matrix. Matrix operations are used if an object needs to be transformed by scaling, stretching, reflecting, or translating (see transformational geometry in Section 9.5). For example, in the movie *Cars*, matrix multiplication is used to translate the car called Lightning McQueen from one position to another.

To add a shadow to an object, graphic artists perform several transformations on the matrix representing the object. Each transformation is created using matrix multiplication. The computer graphics used to create special effects in video games and used in areas such as medical imaging, architectural engineering, and weather forecasting are all based on matrices.

Our goal is to obtain a matrix of the form

$$\left[\begin{array}{cc|c} 1 & 0 & c_1 \\ 0 & 1 & c_2 \end{array}\right]$$

where c_1 and c_2 may represent any real numbers. It is generally easier to work by columns. Therefore, we will try to get the first column of the augmented matrix to be $\begin{smallmatrix}1\\0\end{smallmatrix}$ and the second column to be $\begin{smallmatrix}0\\1\end{smallmatrix}$. Since the element in the top left position is already a 1, we must work to change the 3 in the first column, second row, into a 0. We use row transformation procedure 3 to change the 3 into a 0. If we multiply the top row of numbers by -3 and add these products to the second row of numbers, the element in the first column, second row will become a 0:

$$\left[\begin{array}{cc|c} 1 & 2 & 5 \\ 3 & -1 & 8 \end{array}\right] \quad \text{Original augmented matrix}$$

The top row of numbers multiplied by -3 gives

$$1(-3), \qquad 2(-3), \qquad \text{and} \qquad 5(-3)$$

Now add these products to their respective numbers in row 2.

$$\left[\begin{array}{cc|c} 1 & 2 & 5 \\ 3 + 1(-3) & -1 + 2(-3) & 8 + 5(-3) \end{array}\right] = \left[\begin{array}{cc|c} 1 & 2 & 5 \\ 0 & -7 & -7 \end{array}\right]$$

The next step is to obtain a 1 in the second column, second row. At present, -7 is in that position. To change the -7 to a 1, we use row transformation procedure 2. If we multiply -7 by $-\frac{1}{7}$, the product will be 1. Therefore, we multiply all the numbers in the second row by $-\frac{1}{7}$ to get

$$\left[\begin{array}{cc|c} 1 & 2 & 5 \\ 0(-\frac{1}{7}) & -7(-\frac{1}{7}) & -7(-\frac{1}{7}) \end{array}\right] = \left[\begin{array}{cc|c} 1 & 2 & 5 \\ 0 & 1 & 1 \end{array}\right]$$

The next step is to obtain a 0 in the second column, first row. At present, a 2 is in that position. Multiplying the numbers in the second row by -2 and adding the products to the corresponding numbers in the first row gives a 0 in the desired position.

$$\left[\begin{array}{cc|c} 1 + 0(-2) & 2 + 1(-2) & 5 + 1(-2) \\ 0 & 1 & 1 \end{array}\right] = \left[\begin{array}{cc|c} 1 & 0 & 3 \\ 0 & 1 & 1 \end{array}\right]$$

We now have the desired augmented matrix:

$$\left[\begin{array}{cc|c} 1 & 0 & 3 \\ 0 & 1 & 1 \end{array}\right]$$

With this matrix, we see that $1x + 0y = 3$, or $x = 3$, and $0x + 1y = 1$, or $y = 1$. The solution to the system is (3, 1).

CHECK:

$$x + 2y = 5 \qquad\qquad 3x - y = 8$$
$$3 + 2(1) = 5 \qquad\qquad 3(3) - 1 = 8$$
$$5 = 5 \quad \text{True} \qquad\qquad 8 = 8 \quad \text{True}$$

Now we give a general procedure to change an augmented matrix to the desired form.

TO CHANGE AN AUGMENTED MATRIX TO THE FORM

$$\left[\begin{array}{cc|c} 1 & 0 & c_1 \\ 0 & 1 & c_2 \end{array}\right]$$

Use row transformations to:

1. Change the element in the first column, first row, to a 1.
2. Change the element in the first column, second row, to a 0.
3. Change the element in the second column, second row, to a 1.
4. Change the element in the second column, first row, to a 0.

Generally, when changing an element in the augmented matrix to a 1, *we use step* 2 *in the row transformation box on page* 447. *When changing an element to a* 0, *we use step* 3 *in the row transformation box.*

EXAMPLE 2 *Using Matrices to Solve a System of Equations*

Solve the following system of equations using matrices.

$$2x + 4y = 6$$
$$4x - 2y = -8$$

SOLUTION First write the augmented matrix.

$$\left[\begin{array}{cc|c} 2 & 4 & 6 \\ 4 & -2 & -8 \end{array}\right]$$

To obtain a 1 in the first column, first row, multiply the numbers in the first row by $\frac{1}{2}$.

$$\left[\begin{array}{cc|c} 1 & 2 & 3 \\ 4 & -2 & -8 \end{array}\right]$$

To obtain a 0 in the first column, second row, multiply the numbers in the first row by -4 and add the products to the corresponding numbers in the second row.

$$\left[\begin{array}{cc|c} 1 & 2 & 3 \\ 4 + 1(-4) & -2 + 2(-4) & -8 + 3(-4) \end{array}\right] = \left[\begin{array}{cc|c} 1 & 2 & 3 \\ 0 & -10 & -20 \end{array}\right]$$

To obtain a 1 in the second column, second row, multiply the numbers in the second row by $-\frac{1}{10}$.

$$\left[\begin{array}{cc|c} 1 & 2 & 3 \\ 0(-\frac{1}{10}) & -10(-\frac{1}{10}) & -20(-\frac{1}{10}) \end{array}\right] = \left[\begin{array}{cc|c} 1 & 2 & 3 \\ 0 & 1 & 2 \end{array}\right]$$

To obtain a 0 in the second column, first row, multiply the numbers in the second row by -2 and add the products to the corresponding numbers in the first row.

$$\left[\begin{array}{cc|c} 1 + 0(-2) & 2 + 1(-2) & 3 + 2(-2) \\ 0 & 1 & 2 \end{array}\right] = \left[\begin{array}{cc|c} 1 & 0 & -1 \\ 0 & 1 & 2 \end{array}\right]$$

The solution to the system of equations is $(-1, 2)$. •

Inconsistent and Dependent Systems

Assume that you solve a system of two equations and obtain an augmented matrix in which one row of numbers on the left side of the vertical line are all zeroes but a zero does not appear in the same row on the right side of the vertical line. This situation indicates that the system is inconsistent and has no solution. For example, a system of equations that yields the following augmented matrix is an inconsistent system.

$$\left[\begin{array}{cc|c} 1 & 2 & 5 \\ 0 & 0 & 4 \end{array}\right]$$ Inconsistent system

The second row of the matrix represents the equation

$$0x + 0y = 4 \qquad \text{or} \qquad 0 = 4$$

which is never true. This matrix represents a system of equations that has no solution.

If you obtain a matrix in which a 0 appears across an entire row, the system of equations is dependent. For example, a system of equations that yields the following matrix is a dependent system.

$$\left[\begin{array}{cc|c} 1 & 5 & -6 \\ 0 & 0 & 0 \end{array}\right]$$ Dependent system

The second row of the matrix represents the equation

$$0x + 0y = 0 \qquad \text{or} \qquad 0 = 0$$

which is always true. This matrix represents a system of equations that has an infinite number of solutions.

Triangularization Method

Another procedure to solve a system of two equations is to use row transformation procedures to obtain an augmented matrix of the form

$$\left[\begin{array}{cc|c} 1 & a & b \\ 0 & 1 & c \end{array}\right]$$

where a, b, and c represent real numbers. This procedure is called the ***triangularization method*** because the ones and zeroes form a triangle.

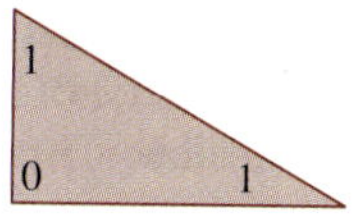

When the matrix is in this form, we can write the following system of equations.

$$\begin{aligned} 1x + ay &= b \\ 0x + 1y &= c \end{aligned} \quad \text{or} \quad \begin{aligned} x + ay &= b \\ y &= c \end{aligned}$$

Using substitution, we can solve the system.

EXAMPLE 3 *Solving a System of Equations Using the Triangularization Method*

Solve the following system of equations using the triangularization method.

$$\begin{aligned} 2x + 4y &= 6 \\ 4x - 2y &= -8 \end{aligned}$$

SOLUTION In Example 2, in the process of solving the system we obtained the augmented matrix

$$\left[\begin{array}{cc|c} 1 & 2 & 3 \\ 0 & 1 & 2 \end{array}\right]$$

This matrix represents the following system of equations.

$$\begin{aligned} x + 2y &= 3 \\ y &= 2 \end{aligned}$$

To solve for x, we substitute 2 for y in the equation

$$\begin{aligned} x + 2y &= 3 \\ x + 2(2) &= 3 \\ x + 4 &= 3 \\ x &= -1 \end{aligned}$$

Thus, the solution to the system is $(-1, 2)$, as was obtained in Example 2. You may use either method when solving a system of equations with matrices unless your instructor specifies otherwise.

EXAMPLE 1 *Solving a System of Inequalities*

Graph the following system of inequalities and indicate the solution set.

$$x + y < 5$$
$$x - y < 3$$

SOLUTION Graph both inequalities on the same axes. First draw the graph of $x + y < 5$. When drawing the graph, remember to use a dashed line, since the inequality is "less than" (see Fig. 7.10a). If you have forgotten how to graph inequalities, review Section 6.8. Shade the half plane that satisfies the inequality $x + y < 5$.

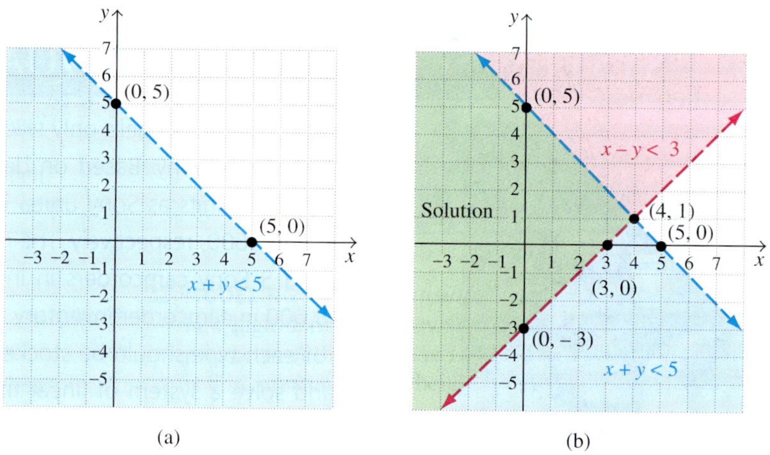

Figure 7.10

Now, on the same axes, shade the half plane that satisfies the inequality $x - y < 3$ (see Fig. 7.10b). The solution set consists of all the points common to the two shaded half planes. These are the points in the region on the graph containing both color shadings. In Figure 7.10(b), we have indicated this region in green. Figure 7.10(b) shows that the two lines intersect at (4, 1). This ordered pair can also be found by any of the algebraic methods discussed in Sections 7.2 and 7.3.

EXAMPLE 2 *Solving a System of Linear Inequalities*

Graph the following system of inequalities and indicate the solution set.

$$4x - 2y \geq 8$$
$$2x + 3y < 6$$

SOLUTION Graph the inequality $4x - 2y \geq 8$. Remember to use a solid line because the inequality is "greater than or equal to"; see Fig. 7.11(a).

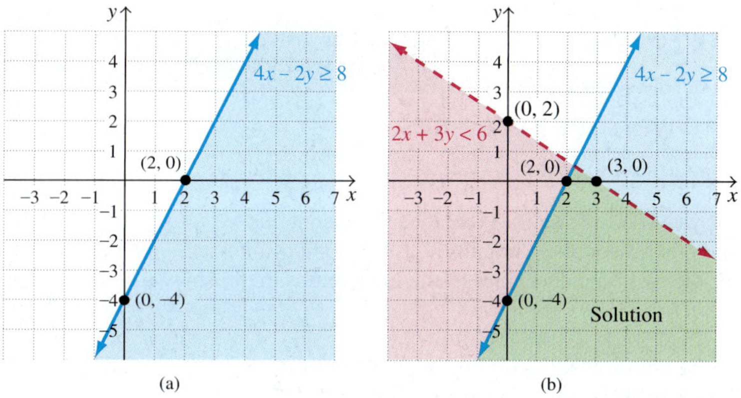

Figure 7.11

On the same set of axes, draw the graph of $2x + 3y < 6$. Use a dashed line since the inequality is "less than"; see Fig. 7.11(b). The solution is the region of the graph that contains both color shadings and the part of the solid line that satisfies the inequality $2x + 3y < 6$. Note that the point of intersection of the two lines is not a part of the solution set.

EXAMPLE 3 *Another System of Inequalities*

Graph the following system of inequalities and indicate the solution set.

$$x \geq -2$$
$$y < 3$$

SOLUTION Graph the inequality $x \geq -2$; see Fig. 7.12(a). On the same axes, graph the inequality $y < 3$; see Fig. 7.12(b). The solution set is that region of the graph that is shaded in both colors and the part of the solid line that satisfies the inequality $y < 3$. The point of intersection of the two lines, $(-2, 3)$, is not part of the solution because it does not satisfy the inequality $y < 3$.

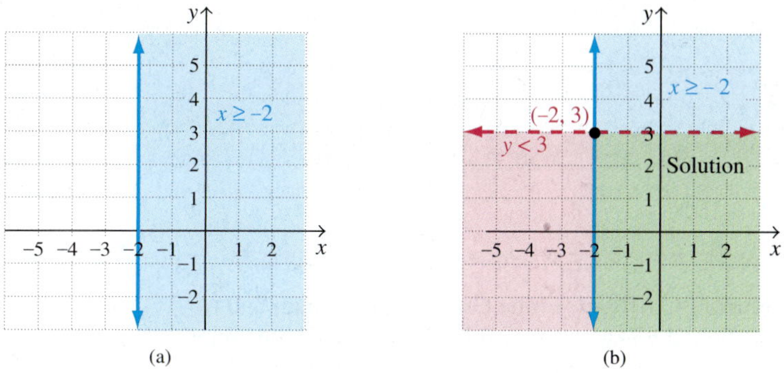

Figure 7.12

SECTION 7.5 EXERCISES

CONCEPT/WRITING EXERCISES

1. What is the solution set of a system of linear inequalities?

2. If in a system of inequalities one inequality contains $\leq$ and the other inequality contains $>$, is the point of intersection of the two boundary lines of the inequalities in the solution set? Explain.

3. If in a system of inequalities one inequality contains $\leq$ and the other inequality contains $\geq$, is the point of intersection of the two boundary lines of the inequalities in the solution set? Explain.

4. If in a system of inequalities one inequality contains $<$ and the other inequality contains $>$, is the point of intersection of the two boundary lines of the inequalities in the solution set? Explain.

PRACTICE THE SKILLS

In Exercises 5–20, graph the system of linear inequalities and indicate the solution set.

5. $y > x + 4$
   $y > 3x$

6. $y > -3x$
   $y \leq -2x + 1$

7. $x + y \geq 1$
   $x - y > 3$

8. $x + y < 4$
   $3x + 2y \geq 6$

9. $x - y < 4$
   $x + y < 5$

10. $3x - y \leq 6$
    $x - y > 4$

11. $x + 2y \geq 4$
    $3x - y \geq -6$

12. $x - 3y \leq 3$
    $x + 2y \geq 4$

13. $y \leq 3x$
    $x \geq 3y$

14. $y \leq 4$
    $x - y < 1$

15. $x \leq 0$
    $y \leq 0$

16. $x \geq 1$
    $y \leq 1$

17. $4x + 2y > 8$
    $x \geq y - 1$

18. $5y > 3x + 10$
    $3y < -2x - 3$

19. $3x + 2y > 8$
    $x < 5y - 5$

20. $3x \geq 2y + 10$
    $x \leq y + 8$

CHALLENGE PROBLEMS/GROUP ACTIVITIES

21. **MODELING** - *Craft Sales* Julie Gratien makes small and large decorated bowls. It takes Julie 20 minutes to decorate a small bowl and 30 minutes to decorate a large bowl. She can spend no more than a total of 600 minutes decorating the bowls. The number of small bowls made needs to be at least twice the number of large bowls made. Julie must also make at least 10 small bowls and 5 large bowls.

    a) Using x to represent the number of small bowls made and y to represent the number of large bowls made, translate the problem into a system of linear inequalities.

    b) Solve the system graphically. Graph sales for small bowls on the horizontal axis and sales for large bowls on the vertical axis.

    c) Select a point in the solution set. Determine the sales for the two types of bowls that corresponds to the point selected.

22. **MODELING** - *Special Diet* Ruben Gonzalez is on a special diet. He must consume fewer than 500 calories at a meal that consists of one serving of chicken and one serving of rice. The meal must contain at least 150 calories from each source.

    a) Using x to represent the number of calories from chicken and y to represent the number of calories from rice, translate the problem into a system of linear inequalities.

    b) Solve the system graphically. Graph calories from chicken on the horizontal axis and calories from rice on the vertical axis.

    c) There are about 180 calories in 3 oz of chicken and about 200 calories in 8 oz of rice. Select a point in the solution set. For the point selected, determine the number of ounces of chicken and the number of ounces of rice to be served.

23. Write a system of linear inequalities whose solution is the second quadrant, including the axes.

24. a) Do all systems of linear inequalities have solutions? Explain.

b) Write a system of inequalities that has no solution.

25. Can a system of linear inequalities have a solution set consisting of a single point? Explain.

26. Can a system of linear inequalities have as its solution set all the points on the coordinate plane? Explain your answer, giving an example to support it.

27. Write a system of linear inequalities that has the ordered pair (0, 0) as its only solution. There are many possible answers.

28. Write a system of linear inequalities that has the following ordered pairs as some of its solutions. There are many possible answers.

$\ldots(-3, -3), (-2, -2), (-1, -1), (0, 0), (1, 1), (2, 2), (3, 3), \ldots$

7.6 LINEAR PROGRAMMING

▲ Linear programming can be used to determine how many skateboards and in-line skates a company should make in order to maximize profit.

Consider a small company that manufactures skateboards and in-line skates. The maximum number of skateboards and pairs of in-line skates that can be produced per day is 20. To meet demand, the company must make at least 2 pairs of in-line skates per day. The company wants to make at least 3 skateboards but no more than 6 skateboards per day. The company makes a profit of $25 on a skateboard and a profit of $20 on a pair of in-line skates. How many skateboards and pairs of in-line skates should be made to maximize profit while satisfying all requirements? In this section, we will discuss how to solve questions such as this one by using a method called linear programming.

Government, business, and industry often require decision makers to find cost-effective solutions to a variety of problems. Linear programming provides businesses and governments with a mathematical form of decision making that makes the most efficient use of time and resources. Linear programming often serves as a method of expressing the relationships in many of these problems and uses systems of linear inequalities.

The typical linear programming problem has many variables and is generally so lengthy that it is solved on a computer by a technique called the ***simplex method***. The simplex method was developed in the 1940s by George B. Dantzig (see the Profile in Mathematics on page 459). Linear programming is used to solve problems in the social sciences, health care, land development, nutrition, military, and many other fields.

We will not discuss the simplex method in this textbook. We will merely give a brief introduction to how linear programming works. You can find a detailed explanation in books on finite mathematics.

In a linear programming problem, there are restrictions called ***constraints***. Each constraint is represented as a linear inequality. The list of constraints forms a system of linear inequalities. When the system of inequalities is graphed, we often obtain a region bounded on all sides by line segments (Fig. 7.13). This region is called the ***feasible region***. The points where two or more boundaries intersect are called the ***vertices*** of the feasible region. The points on the boundary of the region and the points inside the feasible region are the solution set for the system of inequalities.

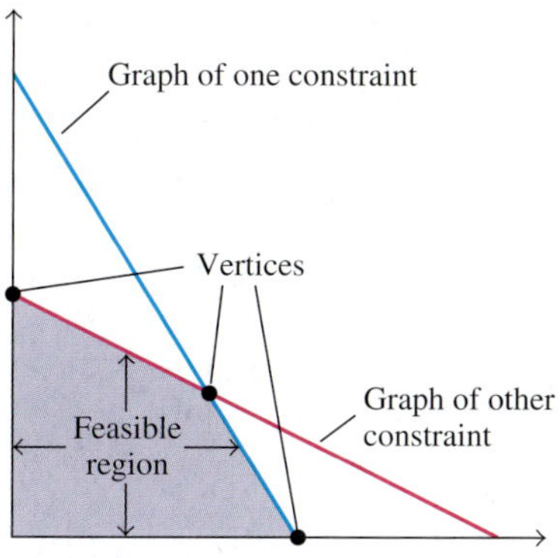

Figure 7.13

DID YOU KNOW?

The Logistics of D-Day

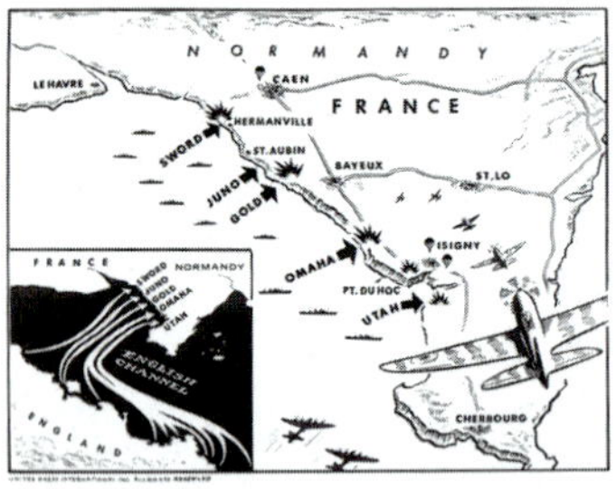

Winston Churchill called the invasion of Normandy "the most difficult and complicated operation that has ever taken place."

Linear programming was first used to deal with the age-old military problem of logistics: obtaining, maintaining, and transporting military equipment and personnel. George B. Dantzig developed the simplex method for the Allies of World War II to do just that. Consider the logistics of the Allied invasion of Normandy. Meteorologic experts had settled on three possible dates in June 1944. It had to be a day when low tide and first light would coincide, when the winds should not exceed 8 to 13 mph, and when visibility was not less than 3 miles. A force of 170,000 assault troops was to be assembled and moved to 22 airfields in England where 1200 air transports and 700 gliders would then take them to the coast of France to converge with 5000 ships of the D-day armada. The code name for the invasion was Operation Overlord, but it is known to most as D-day.

For each linear programming problem, we will obtain a formula of the form $K = Ax + By$, called the ***objective function***. The objective function is the formula for the quantity K (or some other variable) that we want to maximize or minimize. The values we substitute for x and y determine the value of K. From the information given in the problem, we determine the real number constants A and B. In a particular linear programming problem, a typical equation that might be used to find the maximum profit, P, is $P = 3x + 7y$. We would find the maximum profit by substituting the ordered pairs (x, y) of the vertices of the feasible region into the formula $P = 3x + 7y$ to see which ordered pair yields the greatest value of P and therefore the maximum profit. The ordered pair that yields the smallest value of P determines the minimum profit.

Linear programming is used to determine which ordered pair will yield the maximum (or minimum) value of the variable that is being maximized (or minimized). The fundamental principle of linear programming provides a rule for finding those maximum and minimum values.

FUNDAMENTAL PRINCIPLE OF LINEAR PROGRAMMING

If the objective function, $K = Ax + By$, is evaluated at each point in a feasible region, the maximum and minimum values of the equation occur at vertices of the region.

Linear programming is a powerful tool for finding the maximum and minimum values of an objective function. Using the fundamental principle of linear programming, we are quickly able to determine the maximum and minimum values of an objective function by using just a few of the infinitely many points in the feasible region.

Example 1 illustrates how the fundamental principle is used to solve a linear programming problem.

EXAMPLE 1 MODELING - *Using the Fundamental Principle of Linear Programming*

The Ric Shaw Chair company makes two types of rocking chairs, a plain chair and a fancy chair. Each rocking chair must be assembled and then finished. The plain chair takes 4 hours to assemble and 4 hours to finish. The fancy chair takes 8 hours to assemble and 12 hours to finish. The company can provide at most 160 worker-hours of assembling and 180 worker-hours of finishing a day. If the profit on a plain chair is \$40 and the profit on a fancy chair is \$65, how many rocking chairs of each type should the company make per day to maximize profits? What is the maximum profit?

SOLUTION From the information given, we know the following facts.

	Assembly Time (hr)	Finishing Time (hr)	Profit (\$)
Plain chair	4	4	40.00
Fancy chair	8	12	65.00

PROFILE IN MATHEMATICS

George B. Dantzig

George B. Dantzig

The simplex method, a technique used to solve linear programming problems, was developed by George B. Dantzig (1914–2005) in the 1940s. Dantzig's career included implementing the simplex method on computers for the RAND Corporation and teaching operations research and computer science at Stanford University. His book, *Linear Programming and Extensions*, is considered one of the classic books on linear programming. As a graduate student at the University of California, Berkeley, Dantzig became legendary in the mathematics field for solving several problems he mistakenly thought were homework exercises. It turned out the problems were famous unsolved problems in mathematics.

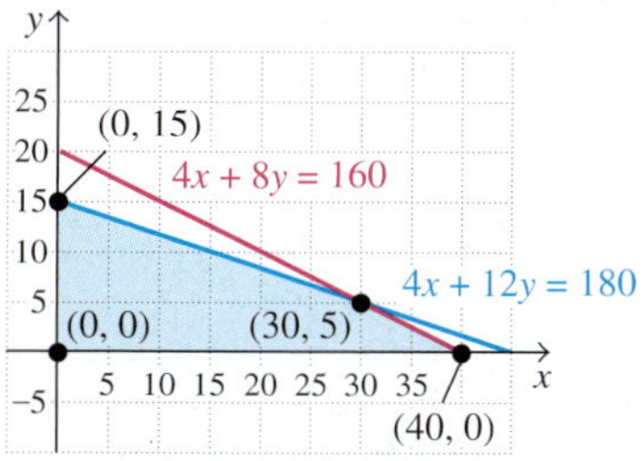

Figure 7.14

Let

$$x = \text{the number of plain chairs made per day}$$
$$y = \text{the number of fancy chairs made per day}$$
$$40x = \text{profit on the plain chairs}$$
$$65y = \text{profit on the fancy chairs}$$
$$P = \text{the total profit}$$

The total profit is the sum of the profit on the plain chairs and the profit on the fancy chairs. Since $40x$ is the profit on the plain chairs and $65y$ is the profit on the fancy chairs, the profit formula is $P = 40x + 65y$.

The maximum profit, P, is dependent on several conditions, called *constraints*. The number of chairs manufactured each day cannot be a negative amount. This condition gives us the constraints $x \geq 0$ and $y \geq 0$. Another constraint is determined by the total number of hours allocated for assembling. Four hours are needed to assemble the plain chair, so the total number of hours per day to assemble x plain chairs is $4x$. Eight hours are required to assemble a fancy chair, so the total number of hours needed to assemble y fancy chairs is $8y$. The maximum number of hours allocated for assembling is 160 per day. Thus, the third constraint is $4x + 8y \leq 160$. The final constraint is determined by the number of hours allotted for finishing. Finishing a plain chair takes 4 hours, or $4x$ hours to finish x plain chairs. Finishing a fancy chair takes 12 hours, or $12y$ hours to finish y fancy chairs. The total number of hours allotted for finishing is 180 per day. Therefore, the fourth constraint is $4x + 12y \leq 180$. Thus, the four constraints are

$$x \geq 0$$
$$y \geq 0$$
$$4x + 8y \leq 160$$
$$4x + 12y \leq 180$$

The list of constraints is a system of linear inequalities in two variables. The solution to the system of inequalities is the set of ordered pairs that satisfy all the constraints. These points are plotted in Fig. 7.14. Note that the solution to the system consists of the colored region and the solid boundaries. The points (0, 0), (0, 15), (30, 5), and (40, 0) are the points at which the boundaries intersect. These points can also be found by the addition or substitution method described in Section 7.2.

The goal in this example is to maximize the profit. The objective function is given by the profit formula $P = 40x + 65y$. According to the fundamental principle, the maximum profit will be found at one of the vertices of the feasible region.

Calculate P for each one of the vertices.

	$P = 40x + 65y$
At (0, 0),	$P = 40(0) + 65(0) = 0$
At (0, 15),	$P = 40(0) + 65(15) = 975$
At (30, 5),	$P = 40(30) + 65(5) = 1525$
At (40, 0),	$P = 40(40) + 65(0) = 1600$

The maximum profit is at (40, 0), which means that the company should manufacture 40 plain rocking chairs and no fancy rocking chairs. The maximum profit would be \$1600. The minimum profit would be at (0, 0), when no rocking chairs of either style were manufactured.

A variation of the problem in Example 1 could be that the company knows that it cannot make more than 15 plain rocking chairs per day. With this additional constraint, we now have the following set of constraints.

$$\begin{aligned} x &\geq 0 \\ x &\leq 15 \\ y &\geq 0 \\ 4x + 8y &\leq 160 \\ 4x + 12y &\leq 180 \end{aligned}$$

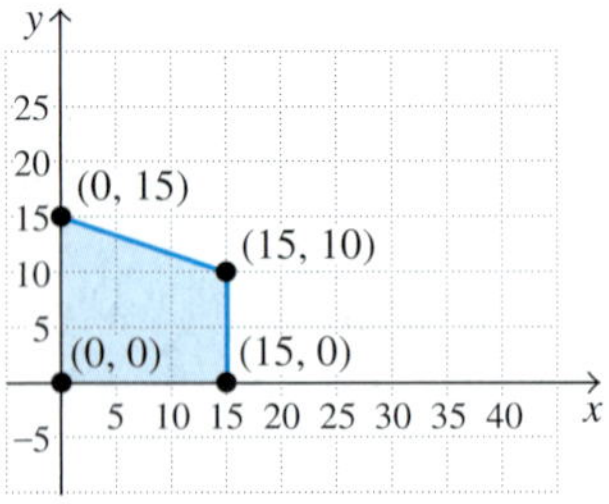

Figure 7.15

The graph of the feasible region of these constraints is shown in Fig. 7.15.

The vertices of the feasible region are (0, 0), (0, 15), (15, 10), and (15, 0). To determine the maximum profit, we calculate P for each of these vertices:

$$P = 40x + 65y$$

$$\begin{aligned} &\text{At } (0, 0), && P = 40(0) + 65(0) = 0 \\ &\text{At } (0, 15), && P = 40(0) + 65(15) = 975 \\ &\text{At } (15, 10), && P = 40(15) + 65(10) = 1250 \\ &\text{At } (15, 0), && P = 40(15) + 65(0) = 600 \end{aligned}$$

This set of constraints gives the maximum profit of \$1250 when the company manufactures 15 plain rocking chairs and 10 fancy rocking chairs.

EXAMPLE 2 MODELING - *Washers and Dryers, Maximizing Profit*

The Admiral Appliance Company makes washers and dryers. The company must manufacture at least one washer per day to ship to one of its customers. No more than 6 washers can be manufactured due to production restrictions. The number of dryers manufactured cannot exceed 7 per day. Also, the number of washers manufactured cannot exceed the number of dryers manufactured per day. If the profit on each washer is \$20 and the profit on each dryer is \$30, how many of each appliance should the company make per day to maximize profits? What is the maximum profit?

SOLUTION Let

$$\begin{aligned} x &= \text{the number of washers manufactured per day} \\ y &= \text{the number of dryers manufactured per day} \\ 20x &= \text{the profit on washers} \\ 30y &= \text{the profit on dryers} \\ P &= \text{the total profit} \end{aligned}$$

The maximum profit is dependent on several constraints. The number of appliances manufactured each day cannot be a negative amount. This condition gives us the constraints $x \geq 0$ and $y \geq 0$. The company must manufacture at least one washer

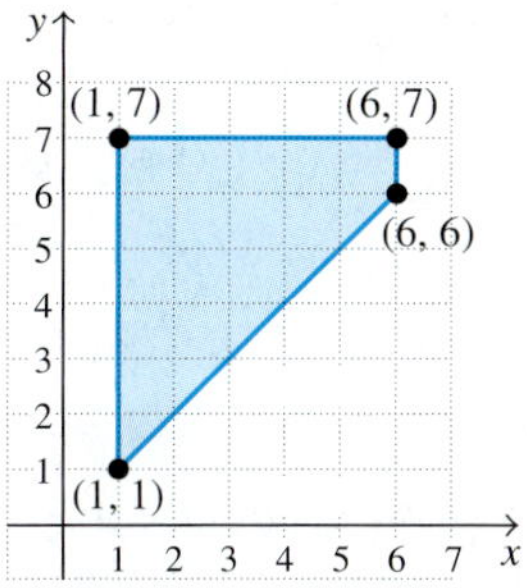

Figure 7.16

per day; therefore, $x \geq 1$. No more than 6 washers can be manufactured per day; therefore, $x \leq 6$. No more than 7 dryers can be manufactured per day; therefore, $y \leq 7$. The number of washers cannot exceed the number of dryers manufactured per day; therefore, $x \leq y$. Thus, the six constraints are

$$x \geq 0, y \geq 0, x \geq 1, x \leq 6, y \leq 7, x \leq y$$

In this example, the objective function is the profit formula. Since $20x$ is the profit on x washers and $30y$ is the profit on y dryers, the profit formula is $P = 20x + 30y$. Figure 7.16 shows the feasible region. The feasible region consists of the shaded region and the boundaries. The vertices of the feasible region are at (1, 1), (1, 7), (6, 7), and (6, 6).

Next we calculate the value of the objective function, P, at each one of the vertices.

$$P = 20x + 30y$$

At (1, 1), $P = 20(1) + 30(1) = 50$

At (1, 7), $P = 20(1) + 30(7) = 230$

At (6, 7), $P = 20(6) + 30(7) = 330$

At (6, 6), $P = 20(6) + 30(6) = 300$

The maximum profit is at (6, 7). Therefore, the company should manufacture 6 washers and 7 dryers to maximize its profit. The maximum profit is \$330.

Use the following steps to solve a linear programming problem.

SOLVING A LINEAR PROGRAMMING PROBLEM

1. Determine all necessary constraints.
2. Determine the objective function.
3. Graph the constraints and determine the feasible region.
4. Determine the vertices of the feasible region.
5. Determine the value of the objective function at each vertex.

The solution is determined by the values in the ordered pair of the vertex that yields the maximum or minimum value of the objective function.

SECTION 7.6 EXERCISES

CONCEPT/WRITING EXERCISES

1. What are constraints in a linear programming problem? How are they represented?

2. In a linear programming problem, how is a feasible region formed?

3. What are the points of intersection of the boundaries of the feasible region called?

4. a) What is the general form of the objective function?

   b) What is the purpose of the objective function in a linear programming problem?

5. In your own words, state the fundamental principle of linear programming.

6. A profit function is $P = 4x + 6y$ and the vertices of the feasible region are (1, 1), (1, 4), (5, 1), and (7, 1). Determine the maximum profit. Explain how you determined your answer.

PRACTICE THE SKILLS

Exercises 7–10 show a feasible region and its vertices. Find the maximum and minimum values of the given objective function.

7. $K = 6x + 4y$

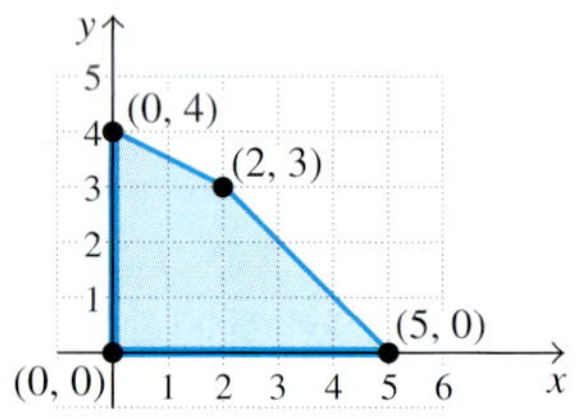

8. $K = 10x + 8y$

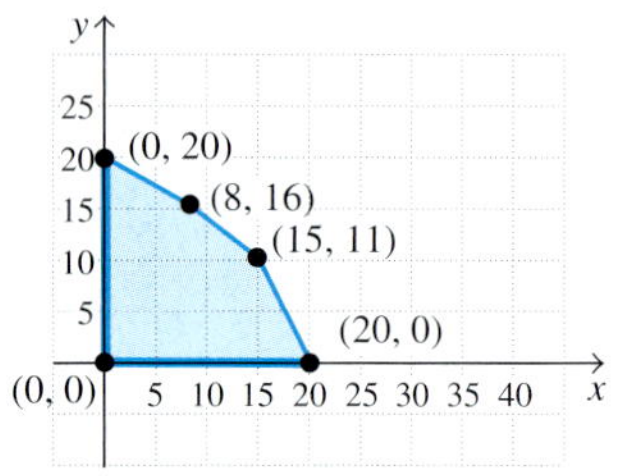

9. $K = 2x + 3y$

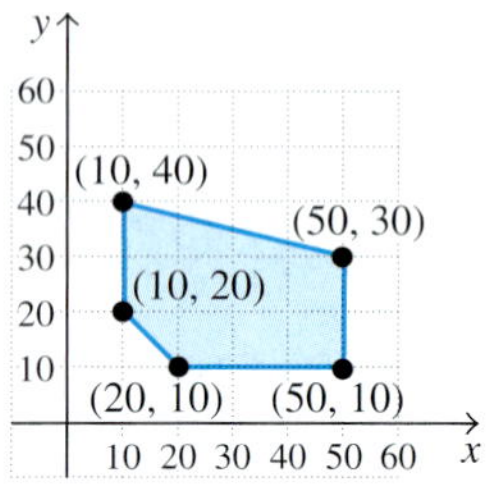

10. $K = 40x + 50y$

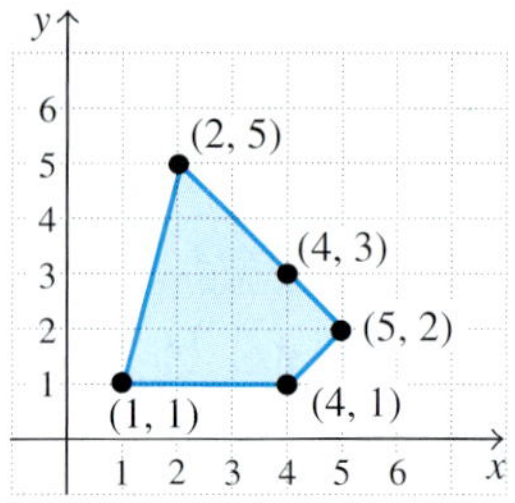

In Exercises 11–16, a set of constraints and a profit formula are given.

a) *Draw the graph of the constraints and find the vertices of the feasible region.*

b) *Use the vertices as obtained in part (a) to determine the maximum and minimum profit.*

11. $x + y \le 6$
$2x + y \le 8$
$x \ge 0$
$y \ge 0$
$P = 4x + 5y$

12. $2x + 3y \le 8$
$4x + 2y \le 8$
$x \ge 0$
$y \ge 0$
$P = 2x + 6y$

13. $x + y \le 4$
$x + 3y \le 6$
$x \ge 0$
$y \ge 0$
$P = 7x + 6y$

14. $x + y \le 50$
$x + 3y \le 90$
$x \ge 0$
$y \ge 0$
$P = 20x + 40y$

15. $2x + 3y \ge 18$
$4x + 2y \le 20$
$x \ge 1$
$y \ge 4$
$P = 2.20x + 1.65y$

16. $x + 2y \le 14$
$7x + 4y \ge 28$
$x \ge 2$
$x \le 10$
$y \ge 1$
$P = 15.13x + 9.35y$

PROBLEM SOLVING

17. **MODELING** - ***Stocking Cameras*** A small electronics store stocks two brands of digital cameras, Kodak and Canon. The manager does not want to keep more than 24 cameras on hand. The number of Kodak cameras stocked needs to be at least twice the number of Canon cameras stocked. She also wants to stock at least 4 Canon cameras. Assume that the store makes a profit of \$40 on a Kodak camera and a profit of \$55 on a Canon camera.

a) List the constraints.

b) Determine the objective function for maximizing profit.

c) Graph the set of constraints.

d) Determine the vertices of the feasible region.

e) How many cameras of each brand should the manager stock to maximize the store's profit?

f) Determine the maximum profit.

18. **MODELING** - ***On Wheels*** The Boards and Blades Company manufactures skateboards and in-line skates. The company can produce a maximum of 20 skateboards and pairs of in-line skates per day. It makes a profit of \$25 on a skateboard and a profit of \$20 on a pair of in-line skates. The company's planners want to make at least 3 skateboards but not more than 6 skateboards per day. To keep customers happy, they must make at least 2 pairs of in-line skates per day.

a) List the constraints.

b) Determine the objective function.

c) Graph the set of constraints.

d) Determine the vertices of the feasible region.

e) How many skateboards and pairs of in-line skates should be made to maximize the profit?

f) Find the maximum profit.

▲ Central Park, New York City, see Exercise 18

19. **MODELING - *Paint Production*** A paint supplier has two machines that produce both indoor paint and outdoor paint. To meet one of its contractual obligations, the company must produce at least 60 gal of indoor paint and 100 gal of outdoor paint. Machine I makes 3 gal of indoor paint and 10 gal of outdoor paint per hour. Machine II makes 4 gal of indoor paint and 5 gal of outdoor paint per hour. It costs $28 per hour to run machine I and $33 per hour to run machine II.

a) List the constraints.

b) Determine the objective function.

c) Graph the set of constraints.

d) Determine the vertices of the feasible region.

e) How many hours should each machine be operated to fulfill the contract at a minimum cost?

f) Determine the minimum cost.

CHALLENGE PROBLEMS/GROUP ACTIVITIES

20. **MODELING - *Hot Dog Profits*** To make one package of all-beef hot dogs, a manufacturer uses 1 lb of beef; to make one package of regular hot dogs, the manufacturer uses $\frac{1}{2}$ lb each of beef and pork. The profit on the all-beef hot dogs is 40 cents per pack and the profit on regular hot dogs is 30 cents per pack. If there are 200 lb of beef and 150 lb of pork available, how many packs of all-beef and regular hot dogs should the manufacturer make to maximize the profit? What is the profit?

21. **MODELING - *Car Seats and Strollers*** A company makes car seats and strollers. Each car seat and stroller passes through three processes: assembly, safety testing, and packaging. A car seat requires 1 hr in assembly, 2 hr in safety testing, and 1 hr in packaging. A stroller requires 3 hr in assembly, 1 hr in safety testing, and 1 hr in packaging. Employee work schedules allow for 24 hr per day for assembly, 16 hr per day for safety testing, and 10 hr per day for packaging. The profit for each car seat is $25, and the profit for each stroller is $35. How many units of each type should the company make per day to maximize the profit? What is the maximum profit?

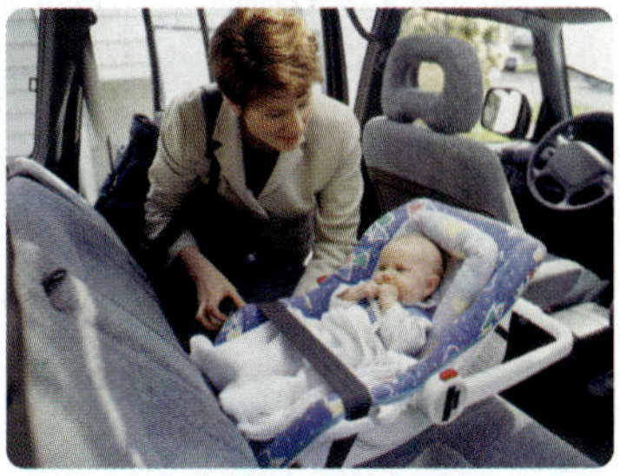

22. **MODELING - *Special Diet*** A dietitian prepares a special diet using two food groups, *A* and *B*. Each ounce of food group *A* contains 3 units of vitamin C and 1 unit of vitamin D. Each ounce of food group *B* contains 1 unit of vitamin C and 2 units of vitamin D. The minimum daily requirements with this diet are at least 9 units of vitamin C and at least 8 units of vitamin D. Each ounce of food group *A* costs 50 cents, and each ounce of food group *B* costs 30 cents.

a) List the constraints.

b) Determine the objective function for minimizing cost.

c) Graph the set of constraints.

d) Determine the vertices of the feasible region.

e) How many ounces of each food group should be used to meet the daily requirements and minimize the cost?

f) Determine the minimum cost.

INTERNET/RESEARCH ACTIVITY

23. *Operations research* draws on several disciplines, including mathematics, probability theory, statistics, and economics. George B. Dantzig (see the Profile in Mathematics on page 459) was one of the key people in developing operations research. Write a paper on Dantzig and his contributions to operations research and linear programming.

CHAPTER 7 SUMMARY

IMPORTANT FACTS

SYSTEMS OF EQUATIONS

Consistent

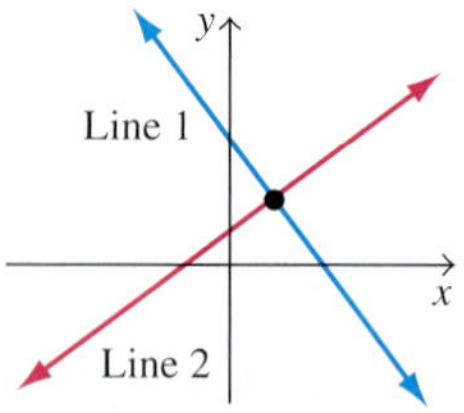

One solution

Inconsistent

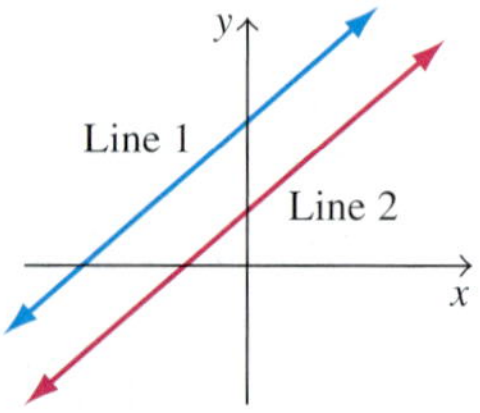

No solution

Dependent

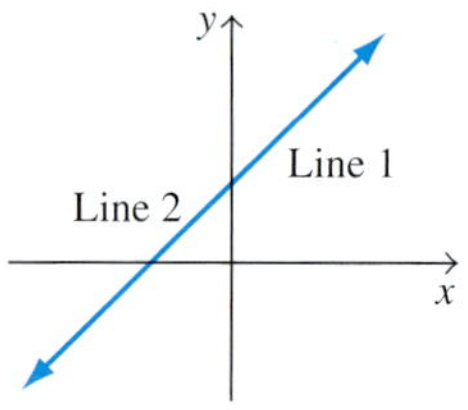

Infinite number of solutions

METHODS OF SOLVING SYSTEMS OF EQUATIONS

1. Graphing
2. Substitution
3. Addition (or elimination) method

MULTIPLICATIVE IDENTITY MATRIX

$$\begin{bmatrix} 1 & 0 \\ 0 & 1 \end{bmatrix}, \begin{bmatrix} 1 & 0 & 0 \\ 0 & 1 & 0 \\ 0 & 0 & 1 \end{bmatrix}, \ldots$$

FUNDAMENTAL PRINCIPLE OF LINEAR PROGRAMMING

If the objective function $K = Ax + By$ is evaluated at each point in a feasible region, the maximum and minimum values of the equation occur at vertices of the region.

CHAPTER 7 REVIEW EXERCISES

7.1

In Exercises 1–4, solve the system of equations graphically. If the system does not have a single ordered pair as a solution, state whether the system is inconsistent or dependent.

1. $x = 2$
$y = -1$

2. $x + 2y = 4$
$2x - 2y = 2$

3. $x = 3$
$x + y = 5$

4. $x + 2y = 5$
$2x + 4y = 4$

In Exercises 5–8, determine without graphing whether the system of equations has exactly one solution, no solution, or an infinite number of solutions.

5. $y = \frac{1}{3}x + 5$
$3y - x = 15$

6. $2x + y = 4$
$y = -2x + 6$

7. $6y - 2x = 20$
$4y + 2x = 10$

8. $2x - 4y = 8$
$-2x + y = 6$

7.2

In Exercises 9–12, solve the system of equations by the substitution method. If the system does not have a single ordered pair as a solution, state whether the system is inconsistent or dependent.

9. $-x + y = -2$
$x + 2y = 5$

10. $x - 2y = 9$
$y = 2x - 3$

11. $2x - y = 4$
$3x - y = 2$

12. $3x + y = 1$
$3y = -9x - 4$

In Exercises 13–18, solve the system of equations by the addition method. If the system does not have a single ordered pair as a solution, state whether the system is inconsistent or dependent.

13. $x - 2y = 8$
$2x + y = 6$

14. $2x + y = 2$
$-3x - y = 5$

15. $x + y = 2$
$x + 3y = -2$

16. $4x - 8y = 16$
$x - 2y = 4$

17. $3x - 4y = 10$
$5x + 3y = 7$

18. $3x + 4y = 6$
$2x - 3y = 4$

7.3

Given $A = \begin{bmatrix} 1 & -3 \\ 2 & 4 \end{bmatrix}$ *and* $B = \begin{bmatrix} -2 & -5 \\ 6 & 3 \end{bmatrix}$, *determine the following.*

19. $A + B$

20. $A - B$

21. $2A$

22. $2A - 3B$

23. $A \times B$

24. $B \times A$

7.4

In Exercises 25–30, use an augmented matrix to solve the system of equations.

25. $x + 3y = 8$
$x + y = 4$

26. $-x + y = 4$
$x + 3y = 4$

27. $2x + y = 3$
$3x - y = 12$

28. $2x + 3y = 2$
$4x - 9y = 4$

29. $x + 3y = 3$
$3x - 2y = 2$

30. $3x - 6y = -9$
$4x + 5y = 14$

7.1–7.4

31. MODELING - ***Borrowing Money*** A company borrows \$600,000 for 1 year to expand its product line. Some of the money was borrowed at a 4% simple interest rate, and the rest of the money was borrowed at a 6% simple interest rate. How much money was borrowed at each rate if the total annual interest was \$29,000?

32. MODELING - ***Chemistry*** In chemistry class, Tom Northrup has an 80% acid solution and a 50% acid solution. How much of each solution should he mix to get 100 liters of a 75% acid solution?

33. MODELING - ***Landscaping*** The Garden Factory purchased 4 tons of topsoil and 3 tons of mulch for \$1529. The next week the company purchased 2 tons of topsoil and 5 tons of mulch for \$1405. Determine the price per ton for topsoil and the price per ton for mulch.

34. MODELING - ***Cool Air*** Emily Richelieu needs to purchase a new air conditioner for the office. Model 1600A costs \$950 to purchase and \$32 per month to operate. Model 6070B, a more efficient unit, costs \$1275 to purchase and \$22 per month to operate.

a) After how many months will the total cost of both units be equal?

b) Which model will be the more cost effective if the life of both units is guaranteed for 10 years?

35. MODELING - ***Minimizing Parking Costs*** The cost of parking in All-Day parking lot is \$5 for the first hour and \$0.50 for each additional hour. Sav-a-Lot parking lot costs \$4.25 for the first hour and \$0.75 for each additional hour.

a) In how many hours after the first hour would the total cost of parking at All-Day and Sav-a-Lot be the same?

b) If Mark McMahon needed to park his car for 5 hr, which parking lot would be less expensive?

7.5

In Exercises 36–39, graph the system of linear inequalities and indicate the solution set.

36. $y \le 3x - 1$
$y > -2x + 1$

37. $2x + y < 8$
$y \ge 2x - 1$

38. $x + 3y \le 6$
$2x - 7y \ge 14$

39. $x - y > 5$
$6x + 5y \le 30$

7.6

40. The set of constraints and profit formula for a linear programming problem are

$$\begin{aligned} 2x + 3y &\le 12 \\ 2x + y &\le 8 \\ x &\ge 0 \\ y &\ge 0 \\ P &= 5x + 3y \end{aligned}$$

a) Draw the graph of the constraints and determine the vertices of the feasible region.

b) Use the vertices to determine the maximum and minimum profit.

CHAPTER 7 TEST

1. From a graph, explain how you would identify a consistent system of equations, an inconsistent system of equations, and a dependent system of equations.

2. Solve the system of equations graphically.

$$y = 2x - 12$$
$$2x + 2y = -6$$

3. Determine without graphing whether the system of equations has exactly one solution, no solution, or an infinite number of solutions.

$$4x + 5y = 6$$
$$-3x + 5y = 13$$

Solve the system of equations by the method indicated.

4. $x + y = -1$
   $2x + 3y = -5$
   (substitution)

5. $y = 3x - 7$
   $y = 5x - 3$
   (substitution)

6. $x - y = 4$
   $2x + y = -10$
   (addition)

7. $4x + 3y = 5$
   $2x + 4y = 10$
   (addition)

8. $3x + 4y = 6$
   $2x - 3y = 4$
   (addition)

9. $x + 3y = 4$
   $5x + 7y = 4$
   (matrices)

In Exercises 10–12, for $A = \begin{bmatrix} 2 & -5 \\ 1 & 3 \end{bmatrix}$ *and* $B = \begin{bmatrix} -1 & -3 \\ 5 & 2 \end{bmatrix}$, *determine the following.*

10. $A + B$ 11. $A - 2B$ 12. $A \times B$

13. Graph the system of linear inequalities and indicate the solution set.

$$y < -2x + 2$$
$$y > 3x + 2$$

Solve Exercises 14 and 15 by using a system of equations.

14. **MODELING - *Truck Rental*** U-Haul charges a daily fee plus a mileage charge to rent a truck. Dorothy DiMento rented a truck with U-Haul and was charged \$132 for 3 days rental and 150 miles driven. Elena Dilai rented the same truck and was charged \$142 for 2 days rental and 400 miles driven. Determine the daily fee and the mileage charge for renting this truck.

15. **MODELING - *Checking Accounts*** The charge for maintaining a checking account at Union Bank is \$6 per month plus 10 cents for each check that is written. The charge at Citrus Bank is \$2 per month and 20 cents per check.

   a) How many checks would a customer have to write in a month for the total charges to be the same at both banks?

   b) If Brent Pickett planned to write 14 checks per month, which bank would be the least expensive?

16. The set of constraints and profit formula for a linear programming problem are

$$x + 3y \leq 6$$
$$4x + 3y \leq 15$$
$$x \geq 0$$
$$y \geq 0$$
$$P = 6x + 4y$$

   a) Draw the graph of the constraints and determine the vertices of the feasible region.

   b) Use the vertices to determine the maximum and minimum profit.

GROUP PROJECTS

1. Make up three different systems of equations that have (1, 4) as a solution. Explain how you determined your systems.

LINEAR PROGRAMMING

2. **MODELING** - *Profit from Bookcases* The Bookholder Company manufactures two types of bookcases out of both oak and walnut. Model 01 requires 5 board feet of oak and 2 board feet of walnut. Model 02 requires 4 board feet of oak and 3 board feet of walnut. A profit of \$75 is made on each Model 01 bookcase and a profit of \$125 is made on each Model 02 bookcase. The company has a supply of 1000 board feet of oak and 600 board feet of walnut. The company has orders for 40 Model 01 bookcases and 50 Model 02 bookcases. These orders indicate the minimum number the company must manufacture of each model.

   a) Write the set of constraints.

   b) Write the objective function.

   c) Graph the set of constraints.

   d) Determine the number of bookcases of each type the company should manufacture in order to maximize profits.

   e) Determine the maximum profit.

CREATE YOUR OWN WORD PROBLEM

3. **a)** Write a word problem that can be solved by using a system of two equations with two unknowns.

   b) For the problem in part (a), write the system of equations and find the answer.

   c) Explain how you developed the problem in part (a).

CHAPTER 8

The Metric System

▲ Everywhere outside the United States you may see traffic signs given in metric units.

WHAT YOU WILL LEARN

- The advantages of using the metric system
- The basic units used in the metric system
- Conversions within the metric system
- Determining length, area, volume, mass, and temperature in the metric system
- Dimensional analysis and converting to and from the metric system

WHY IT IS IMPORTANT

When you leave the United States, whether you travel to Canada, Mexico, or most other places in the world, you may see metric measurements being used. Clothing sizes may be given in centimeters, gasoline may be sold in liters, and speed limit signs may be given in kilometers per hour. Each day in the United States we see and use metric measurements. For example, soda is sold in liters, medicines are measured in milligrams, and tire sizes are given in millimeters. An understanding of the metric system will help you both at home and when you travel outside the United States.

8.1 BASIC TERMS AND CONVERSIONS WITHIN THE METRIC SYSTEM

▲ The sciences use metric measurements.

Have you ever taken a science course? If so, you most likely worked with metric measurements. For example, in chemistry you may have worked with liter containers, and in physics you may have worked with a mass in kilograms rather than a weight in pounds. Have you ever asked yourself why the sciences use metric measurement? In this section, we will explain some benefits of using the metric system.

Most countries of the world use the ***Système international d'unités*** or ***SI system***. The SI system is generally referred to as the ***metric system*** in the United States. The metric system was named for the Greek word *metron,* meaning "measure." The standard units in the metric system have gone through many changes since the system was first developed in France during the French Revolution. For example, one unit of measure, the meter, was first defined as one ten-millionth of the distance between the North Pole and the equator. Later, the meter was defined as 1,650,763.73 wavelengths of the orange–red line of krypton 86. Since 1893, the meter has been defined as the distance traveled by light in a vacuum in $\frac{1}{299,792,458}$ of a second.

Two systems of weights and measures exist side by side in the United States today, the ***U.S. customary system*** and the metric system. The metric system is used predominantly in the automotive, construction, farm equipment, computer, and bottling industries and in health-related professions. Furthermore, almost every industry that ships internationally uses at least some metric measures.

In this chapter, we will discuss the metric measurements of length, area, volume, mass, and temperature. Using the metric system has many advantages. Some of them are summarized here.

1. The metric system is the worldwide accepted standard measurement system. All industrial nations that trade internationally, except the United States, use the metric system as the official system of measurement.
2. There is only one basic unit of measurement for each physical quantity. In the U.S. customary system, many units are often used to represent the same physical quantity. For example, when discussing length, we use inches, feet, yards, miles, and so on. Converting from one of these units to the other is often a tedious task (consider changing 12 miles to inches). In the metric system, we can make many conversions by simply moving the decimal point.
3. The SI system is based on the number 10, and there is less need for fractions because most quantities can be expressed as decimals.

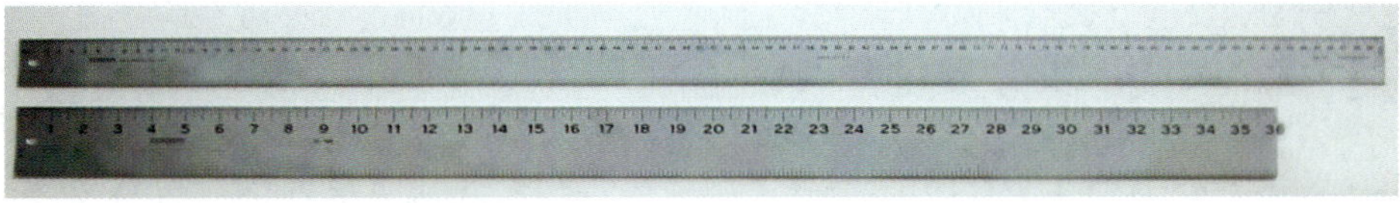
▲ A meter (top figure) is a little longer than a yard (bottom figure).

DID YOU KNOW?

Lost in Space

The missing *Mars Climate Orbiter*

In September 1999, the United States lost a $125 million spacecraft, the *Mars Climate Orbiter,* as it approached Mars. Two spacecraft teams, one at NASA's Jet Propulsion Laboratory (JPL) and the other at a Lockheed Martin facility, were unknowingly exchanging some vital information in different measurement units. The spacecraft team at Lockheed sent some measurements to the spacecraft team at JPL using U.S. customary units. The JPL team assumed the information it received was in metric units. The mix-up in units led to the JPL scientists giving the spacecraft's computer the wrong information, which led to the spacecraft entering the Martian atmosphere, where it burned up. NASA has taken steps to prevent this error from ever happening again.

▲ One- and two-liter bottles

Basic Terms

Because the official definitions of many metric terms are quite technical, we present them informally.

The ***meter*** (m) is commonly used to measure *length* in the metric system. One meter is a little more than a yard. A door is about 2 meters high.

The ***kilogram*** (kg) is commonly used to measure *mass*. (The difference between mass and weight is discussed in Section 8.3.) One kilogram is about 2.2 pounds. A newborn baby may have a mass of about 3 kilograms. The gram (g), a unit of mass derived from the kilogram, is used to measure small amounts. A nickel has a mass of about 5 grams.

The ***liter*** (ℓ) is commonly used to measure *volume*. One liter is a little more than a quart. The gas tank of a compact car may hold 50 liters of gasoline.

Thus,

$$1 \text{ m} \approx 1 \text{ yd}$$
$$1 \text{ kg} \approx 2.2 \text{ lb}$$
$$1 \ \ell \approx 1 \text{ qt}$$

The term ***degree Celsius*** (°C) is used to measure temperature. The freezing point of water is 0°C, and the boiling point of water is 100°C. The temperature on a warm day may be 30°C.

$$0°\text{C} = 32°\text{F} \quad \text{Water freezes}$$
$$22°\text{C} = 71.6°\text{F} \quad \text{Comfortable room temperature}$$
$$37°\text{C} = 98.6°\text{F} \quad \text{Body temperature}$$
$$100°\text{C} = 212°\text{F} \quad \text{Water boils}$$

Prefixes

The metric system is based on the number 10 and therefore is a decimal system. Prefixes are used to denote a multiple or part of a base unit. Table 8.1 summarizes the more commonly used prefixes and their meanings. In the table, where we mention "base units" we mean metric units without prefixes, such as meter, gram, or liter. From Table 8.1, we can determine that a *deka*meter represents 10 meters and a *centi*meter represents $\frac{1}{100}$ of a meter. Also, 1 kiloliter = 1000 liters, 1 kilogram = 1000 grams, and 1 milliliter = $\frac{1}{1000}$ liter, and so on.

Table 8.1 Metric Prefixes

Prefix	Symbol	Meaning
kilo	k	1000 × base unit
hecto	h	100 × base unit
deka	da	10 × base unit
—	—	base unit
deci	d	$\frac{1}{10}$ of base unit
centi	c	$\frac{1}{100}$ of base unit
milli	m	$\frac{1}{1000}$ of base unit

In the metric system, a comma is commonly used the way we use a decimal point (see the photo on the left, taken in Prague, Czech Republic), and a decimal point is used the way we use a comma (see the photo on the right, taken in Vienna, Austria).

In the metric system, groups of three digits in large numbers are often separated by a space, not a comma. For example, thirty thousand is written 30 000 and nine million is written 9 000 000. Groups of three digits to the right of the decimal point are also separated by spaces. For example, 16 millionths is written 0,000 016 in many countries of the world. We will use the decimal point in this book and write 0.000 016. In this section, we will separate groups of three digits using spaces. Note, however, that the space between groups of three digits is usually omitted if there are only four digits to the left or right of the decimal point. Thus, we will write three thousand as 3000 and five ten-thousandths as 0.0005.

▲ The computer's hard drive can store 400 gigabytes (400 GB) of information.

For scientific work that involves very large and very small quantities, the following prefixes are also used: *mega* (M) is one million times the base unit, *giga* (G) is one billion times the base unit, *tera* (T) is one trillion times the base unit, *micro* (μ, the Greek letter mu) is one millionth of the base unit, *nano* (n) is one billionth of the base unit, and *pico* (p) is one trillionth of the base unit.

In this book, the abbreviations or symbols for units of measure are not pluralized, but full names are. For example, 5 milliliters is symbolized as 5 mℓ, not 5 mℓs. Some countries that use the metric system do not use an "s" in their abbreviations, whereas others do.

Conversions within the Metric System

We will use Table 8.2 to help demonstrate how to change from one metric unit to another metric unit using the meter as our base unit (for example, meters to kilometers, and so on).

The meters in Table 8.2 can be replaced by grams, liters, or any other base unit of the metric system. Regardless of which unit we choose, the procedure is the same. For purposes of explanation, we have used the meter.

Table 8.2 Changing Metric Units

Measure of length	kilometer	hectometer	dekameter	meter	decimeter	centimeter	millimeter
Symbol	km	hm	dam	m	dm	cm	mm
Number of meters	1000 m	100 m	10 m	1 m	0.1 m	0.01 m	0.001 m

Table 8.2 shows that 1 hectometer equals 100 meters and 1 millimeter is 0.001 (or $\frac{1}{1000}$) meter. The millimeter is the smallest unit in the table. A centimeter is 10 times

▲ Our neighbors in Canada (and also Mexico) use the metric system. As you will learn shortly, the distance to the Botanical Gardens is about 0.6 mile and the distance to Niagara-on-the-Lake is about 9 miles from the sign.

as large as a millimeter, a decimeter is 10 times as large as a centimeter, a meter is 10 times as large as a decimeter, and so on. Because each unit is 10 times as large as the unit on its right, converting from one unit to another is simply a matter of multiplying or dividing by powers of 10.

CHANGING UNITS WITHIN THE METRIC SYSTEM

1. To change from a smaller unit to a larger unit (for example, from meters to kilometers), move the decimal point in the original quantity one place to the left for each larger unit of measurement until you obtain the desired unit of measurement.
2. To change from a larger unit to a smaller unit (for example, from kilometers to meters), move the decimal point in the original quantity one place to the right for each smaller unit of measurement until you obtain the desired unit of measurement.

DID YOU KNOW?

Additional Metric Prefaces

Many publications about the metric system are available free from the U.S. government. You may contact the National Institute of Standards and Technology (under the U.S. Department of Customs), through the Web site at www.nist.gov/, or you may write to the institute's office (Gaithersburg, MD 20899). Two worthwhile publications are *Metric Style Guide for the News Media* and *A Brief History of Measurement Systems*. The following interesting chart was selected and modified from the latter.

METRIC PREFIXES

Multiples and Submultiples	Prefixes	Symbols
1 000 000 000 000 000 000 000 000 = 10^{24}	yotta	Y
1 000 000 000 000 000 000 000 = 10^{21}	zetta	Z
1 000 000 000 000 000 000 = 10^{18}	exa	E
1 000 000 000 000 000 = 10^{15}	peta	P
1 000 000 000 000 = 10^{12}	tera	T
1 000 000 000 = 10^{9}	giga	G
1 000 000 = 10^{6}	mega	M
1000 = 10^{3}	kilo	k
100 = 10^{2}	hecto	h
10 = 10^{1}	deka	da*
1 = 10^{0}		
0.1 = 10^{-1}	deci	d
0.01 = 10^{-2}	centi	c
0.001 = 10^{-3}	milli	m
0.000 001 = 10^{-6}	micro	μ
0.000 000 001 = 10^{-9}	nano	n
0.000 000 000 001 = 10^{-12}	pico	p
0.000 000 000 000 001 = 10^{-15}	femto	f
0.000 000 000 000 000 001 = 10^{-18}	atto	a
0.000 000 000 000 000 000 001 = 10^{-21}	zepto	z
0.000 000 000 000 000 000 000 001 = 10^{-24}	yocto	y

*Some countries use D for deka.

EXAMPLE 1 *Changing Units*

a) Convert 375.6 m to km.

b) Convert 14 g to cg.

c) Convert 0.76 ℓ to mℓ.

d) Convert 240 daℓ to kℓ.

SOLUTION

a) Table 8.2 shows that dekameters, hectometers, and kilometers are all larger units of measurements than meters. Kilometers appear three places to the left of meters in the table. Therefore, to change a measure from meters to kilometers, we must move the decimal point in the given number three places to the left, or

$$375.6 \text{ m} = 0.3756 \text{ km}$$

Note that since we are changing from a smaller unit of measurement (meter) to a larger unit of measurement (kilometer), the answer will be a smaller number of units.

b) Grams are a larger unit of measurement than centigrams. To convert grams to centigrams, we move the decimal point two places to the right, or

$$14 \text{ g} = 1400 \text{ cg}$$

Note that since we are changing from a larger unit of measurement (gram) to a smaller unit of measurement (centigram), the answer will be a larger number of units.

c) $0.76\ \ell = 760\ \text{m}\ell$

d) $240\ \text{da}\ell = 2.40\ \text{k}\ell$

EXAMPLE 2 *Two More Conversions*

a) Convert 305 mm to hectometers.

b) Convert 6.34 dam to decimeters.

SOLUTION

a) Table 8.2 shows that hectometers are five places to the left of millimeters. Therefore, to make the conversion, we must move the decimal point in the given number five places to the left, or

$$305 \text{ mm} = 0.003\ 05 \text{ hm}$$

b) Table 8.2 shows that decimeters are two places to the right of dekameters. Therefore, to make the conversion, we must move the decimal point in the given number two places to the right, or

$$6.34 \text{ dam} = 634 \text{ dm}$$

EXAMPLE 3 *A Metric Road Sign*

The sign in the photo, from Prague, Czech Republic, shows that there is a Subway sandwich shop 750 meters to the right.

a) Determine the distance in kilometers.

b) Determine the distance in centimeters.

SOLUTION

a) We must move the decimal point three places to the left to change from meters to kilometers. Therefore,

$$750 \text{ m} = 0.750 \text{ km}$$

b) We must move the decimal point two places to the right to change from meters to centimeters. Therefore,

$$750 \text{ m} = 75\,000 \text{ cm}$$

EXAMPLE 4 *Comparing Lengths*

Arrange in order from smallest to largest length: 3.4 m, 3421 mm, and 104 cm.

SOLUTION

To be compared, these lengths should all be in the same units of measure. Let's convert all the measures to millimeters, the smallest units of the lengths being compared.

$$3.4 \text{ m} = 3400 \text{ mm} \qquad 3421 \text{ mm} \qquad 104 \text{ cm} = 1040 \text{ mm}$$

Since the lengths, in millimeters, from smallest to largest are 1040, 3400, 3421, the lengths arranged in order from smallest to largest are 104 cm, 3.4 m, and 3421 mm.

SECTION 8.1 EXERCISES

CONCEPT/WRITING EXERCISES

1. What is the name commonly used for the Système international d'unités in the United States?

2. What is the name of the system of measurement primarily used in the United States today?

3. List three advantages of the metric system.

4. What metric unit is commonly used to measure

   a) length? **b)** mass?

   c) volume? **d)** temperature?

5. **a)** Explain how to convert from one metric unit of length to a different metric unit of length. Then use this procedure in parts (b) and (c).

   b) Convert 214.6 cm to kilometers.

   c) Convert 60.8 hm to decimeters.

6. What is the name of the prefix that is

   a) a million times the basic unit?

   b) one millionth of the base unit?

7. Without referring to any table, name as many of the metric system prefixes as you can and give their meanings. If you don't already know all the prefixes in Table 8.1, memorize them now.

8. **a)** How many times greater is 1 dam than 1 dm?

   b) Convert 1 dam to decimeters.

   c) Convert 1 dm to dekameters.

9. **a)** How many times greater is 1 hectometer than 1 centimeter?

   b) Convert 1 hm to centimeters.

   c) Convert 1 cm to hectometers.

10. **a)** What is the freezing temperature of water in the metric system?

    b) What is the boiling point of water in the metric system?

    c) What is normal human body temperature in the metric system?

PRACTICE THE SKILLS

In Exercises 11–16, fill in the blank.

11. One meter is a little longer than a _____.

12. One kilogram is a little more than _____ pounds.

13. One nickel has a mass of about _____ grams.

14. The temperature on a warm day may be _____ °C.

15. A comfortable room temperature may be _____ °C.

16. A door may be _____ meters high.

In Exercises 17–22, match the prefix with the one letter, a–f, that gives the meaning of the prefix.

17. Milli	**a)** $\frac{1}{100}$ of base unit
18. Kilo	**b)** $\frac{1}{1000}$ of base unit
19. Hecto	**c)** 100 times base unit
20. Deka	**d)** 1000 times base unit
21. Deci	**e)** 10 times base unit
22. Centi	**f)** $\frac{1}{10}$ of base unit

23. Complete the following.

a) 1 hectogram = ____ grams

b) 1 milligram = ____ gram

c) 1 kilogram = ____ grams

d) 1 centigram = ____ gram

e) 1 dekagram = ____ grams

f) 1 decigram = ____ gram

24. Complete the following.

a) 1 dekaliter = ____ liters

b) 1 centiliter = ____ liter

c) 1 milliliter = ____ liter

d) 1 deciliter = ____ liter

e) 1 kiloliter = ____ liters

f) 1 hectoliter = ____ liters

In Exercises 25–30, without referring to any of the tables or your notes, give the symbol and the equivalent in grams for the unit.

25. Centigram **26.** Milligram **27.** Decigram

28. Dekagram **29.** Hectogram **30.** Kilogram

Maximum Mass *In Exercises 31 and 32, use the photo taken in Canada, which shows the maximum mass for vehicles allowed on the street.*

31. What is the maximum mass in grams?

32. What is the maximum mass in milligrams?

In Exercises 33–42, fill in the missing values.

33. 5 m = ____ mm

34. 35.7 hg = ____ g

35. 0.085 hℓ = ____ kℓ

36. 8 dam = ____ m

37. 242.6 cm = ____ hm

38. 1.34 mℓ = ____ ℓ

39. 2435 mg = ____ hg

40. 14.27 kℓ = ____ ℓ

41. 1.34 hm = ____ cm

42. 0.000 062 kg = ____ mg

In Exercises 43–50, convert the given unit to the unit indicated.

43. 32.5 kg to hectograms

44. 7.3 m to millimeters

45. 895 ℓ to milliliters

46. 24 dm to kilometers

47. 140 cg to grams

48. 6049 mm to meters

49. 40 302 mℓ to dekaliters

50. 0.034 mℓ to liters

In Exercises 51–56, arrange the quantities in order from smallest to largest.

51. 2.3 dam, 0.47 km, 590 cm

52. 514 hm, 62 km, 680 m

53. 1.4 kg, 1600 g, 16 300 dg

54. 4.3 ℓ, 420 cℓ, 0.045 kℓ

55. 2.6 km, 203 000 mm, 52.6 hm

56. 0.032 kℓ, 460 dℓ, 48 000 cℓ

PROBLEM SOLVING

57. ***Who Ran Faster*** Jim ran 100 m, and Bob ran 100 yd in the same length of time. Who ran faster? Explain.

58. ***Walking*** Would you be walking faster if you walked 1 dam in 10 min or 1 hm in 10 min? Explain.

59. ***Water Removal*** One pump removes 1 daℓ of water in 1 min, and another pump removes 1 dℓ of water in 1 min. Which pump removes water faster? Explain.

60. ***Balance*** If 5 kg are placed on one side of a balance and a 15 lb weight is placed on the other side, which way would the balance tip? Explain.

61. ***Framing a Masterpiece*** The painting by Picasso, including the frame, measures 74 cm by 99 cm.

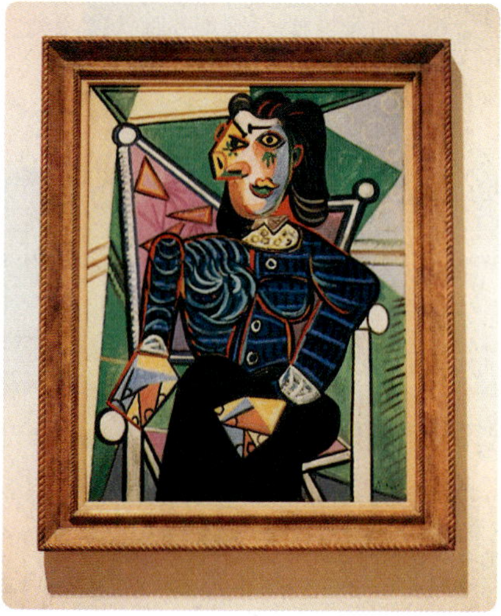

a) How many centimeters of framing were needed to frame the painting?

b) How many millimeters of framing were needed to frame the painting?

62. ***Calcium Tablets*** Sean takes two 250 mg chewable calcium tablets each day.

a) How many milligrams of calcium will Sean take in a week?

b) How many grams of calcium will Sean take in a week?

63. ***Gas Consumption*** Dale Ewen drove 1200 km and used 187 ℓ of gasoline. What was his average rate of gas use for the trip

a) in kilometers per liter?

b) in meters per liter?

64. ***Track and Field*** The high school has a 400-m oval track. If Patty Burgess runs around the track eight times, how many kilometers has she traveled?

65. ***Liters of Soda*** A bottle of soda contains 360 mℓ.

a) How many milliliters are contained in a six-bottle carton?

b) How many liters does the amount in part (a) equal?

c) At $2.45 for the carton of soda, what is its cost per liter?

66. ***Turkey Dinner*** After a turkey is cooked it weighs 6.9 kg.

a) What is its weight in grams?

b) If Marie Sinclair cuts off one-third of the turkey and places it in the freezer, how many decigrams of turkey has she placed in the freezer?

67. ***A Home Run*** A baseball diamond is a square whose sides are about 27 m in length.

a) How many meters does a batter run if he hits a home run?

b) How many kilometers?

c) How many millimeters?

68. ***Fill 'er Up*** In Europe, gas may cost the equivalent of about $1.63 (American) per liter. What will be the cost of filling the gas tank of a car that has a capacity of 37.7 ℓ?

69. ***Tennis Stadium*** This photo taken at the Roland Garros Tennis Stadium in Paris shows the distances from the Roland Garros Stadium to tennis stadiums where the other three grand slams of tennis are played.

a) How much further is Melbourne Park (in Australia) than Flushing Meadows (in New York)?

b) What is the distance determined in part (a) in meters?

70. ***Elevator*** A sign in an elevator in Mexico City indicates that the maximum load is 375 kg. If the mass of the five people in the elevator are 92 kg, 100 kg, 62 kg, 96 kg, and 128 kg, by how much is the maximum load exceeded?

CHALLENGE EXERCISES/GROUP ACTIVITIES

In Exercises 71–74, fill in the blank to make a true statement.

71. 1 gigameter = ____ megameters

72. 1 nanogram = ____ micrograms

73. 1 teraliter = ____ picoliters

74. 1 megagram = ____ nanograms

Calcium *The recommended daily amount of calcium for an American adult is 0.8 g. In Exercises 75–78, how much of the food indicated must an adult eat to satisfy the entire daily allowance using only that food?*

75. Eggs: 1 egg contains 27 mg calcium.

76. Milk: 1 cup contains 288 mg calcium.

77. Broccoli: 1 cup (cooked) contains 195 mg calcium.

78. Raisin bran: 49 g contains 1.6 mg calcium.

Large and Small Numbers *One advantage of the metric system is that by using the proper prefix, you can write large and small numbers without large groups of zeroes. In Exercises 79–84, write an equivalent metric measurement without using any zeroes. For example, you can write 3000 m without zeroes as 3 km and 0.0003 hm as 3 cm.*

79. 7000 cm

80. 4000 mm

81. 0.000 06 hg

82. 3000 dm

83. 0.02 kℓ

84. 500 cm

RECREATIONAL MATHEMATICS

In Exercises 85–94, unscramble the word to make a metric unit of measurement.

85. magr

86. migradec

87. rteli

88. raktileed

89. terem

90. leritililm

91. reketolim

92. timenceret

93. greseed sulesic

94. togmeharc

INTERNET/RESEARCH ACTIVITIES

95. Write a report on the development of the metric system in Europe. Indicate which individual people had the most influence in its development.

96. Write a report on why you believe many Americans oppose switching to the metric system. Give your opinion about whether the United States will eventually switch to the metric system and, if so, when it might do so.

8.2 LENGTH, AREA, AND VOLUME

▲ When shipped overseas, U.S. cars will give gas mileage, the car's weight, and other measurements in metric units.

When U.S. companies send their cars overseas, the gas mileage given on the car's window sticker is usually given in liters per hundred kilometers ($\frac{1}{100}$ km) rather than miles per gallon. Other measurements, such as the car's length and weight, are also given in metric measurements. When clothing manufacturers ship clothing overseas, the clothing sizes will be given in metric units, such as centimeters. Can you convert a pants waist size of 38 inches to centimeters? In this chapter, you will learn how to make conversions to and from the metric system.

This section and the next section are designed to help you *think metric*, that is, to become acquainted with day-to-day usage of metric units. In this section, we consider length, area, and volume.

Length

The basic unit of length in the metric system is the meter. In all English-speaking countries except the United States, *meter* is spelled "metre." Until 1960, the meter was officially defined by the length of a platinum bar kept in a vault in France. The modern definition of the meter is based on the speed of light, a constant that has been defined with great precision. Other commonly used units of length are the kilometer, centimeter, and millimeter. The meter, which is a little longer than 1 yard, is used to measure things that we normally measure in yards and feet. A man whose height is about 2 meters is a tall man. A tractor trailer unit (an 18-wheeler) is about 18 meters long.

The kilometer is used to measure what we normally measure in miles. For example, the distance from New York to Seattle is about 5120 kilometers. One kilometer is about 0.6 mile, and 1 mile is about 1.6 kilometers.

Centimeters and millimeters are used to measure what we normally measure in inches. The centimeter is a little less than $\frac{1}{2}$ inch (see Fig. 8.1), and the millimeter is a little less than $\frac{1}{20}$ inch. A millimeter is about the thickness of a dime. A book may measure 20 cm by 25 cm with a thickness of about 3 cm. Millimeters are often used in scientific work and other areas in which small quantities must be measured. The length of a small insect may be measured in millimeters.

DID YOU KNOW?

1 Yard = 3 Feet = 36 Inches = . . .

In the U.S. Customary system, 27 different units of length are used. How many of them can you name? Don't forget rod, mil, paris line, toise, cubit, and light-year. The different units can be found in the *CRC Handbook of Chemistry and Physics* or at the website www.hbcpnetbase.com.

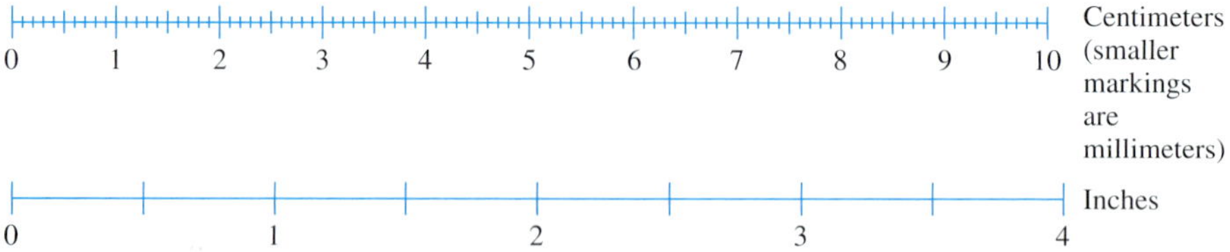

Figure 8.1

▲ The Ted Williams statue outside Fenway Park, Boston

EXAMPLE 1 *Choosing an Appropriate Unit of Length*

Determine which metric unit of length you would use to express the following.

a) The height of the Ted Williams statue shown in the photo.
b) The length of your arm
c) The length of a flea
d) The height of the Sears Tower in Chicago
e) The diameter of a half-dollar
f) The distance between Amarillo, Texas, and Detroit, Michigan.
g) The diameter of a round wastepaper basket
h) The diameter of a pencil
i) Your waist size
j) Your height

SOLUTION

a) Meters or centimeters
b) Centimeters
c) Millimeters
d) Meters
e) Centimeters or millimeters
f) Kilometers
g) Centimeters
h) Millimeters
i) Centimeters
j) Meters or centimeters

In some parts of this solution, more than one possible answer is listed. Measurements can often be made by using more than one unit. For example, if someone asks your height, you might answer $5\frac{1}{2}$ feet or 66 inches. Both answers are correct. ●

Area

In Chapter 9, we provide formulas and discuss procedures for finding the area and volume of many geometric figures. The procedures and formulas for finding area and volume are the same regardless of whether the units are metric units or customary units. When finding areas and volumes, each side of the figure must be given in (or converted to) the same unit.

The area enclosed in a square with 1-centimeter sides (Fig. 8.2) is $1 \text{ cm} \times 1 \text{ cm} = 1 \text{ cm}^2$. A square whose sides are 2 cm (Fig. 8.3) has an area of $2 \text{ cm} \times 2 \text{ cm} = 2^2 \text{ cm}^2 = 4 \text{ cm}^2$.

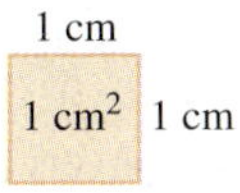

Figure 8.2

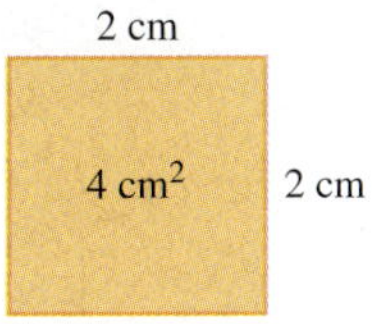

Figure 8.3

Areas are always expressed in square units, such as square centimeters, square kilometers, or square meters. When finding areas, be careful that all the numbers being multiplied are expressed in the same units.

In the metric system, the square centimeter replaces the square inch. The square meter replaces the square foot and square yard. In the future, you might purchase carpet or other floor covering by the square meter instead of by the square foot.

For measuring large land areas, the metric system uses a square unit 100 meters on each side (a square hectometer). This unit is called a ***hectare*** (pronounced "hectair" and symbolized ha). A hectare is about 2.5 acres. One square mile of land contains about 260 hectares. Very large units of area are measured in square kilometers. One square kilometer is about $\frac{4}{10}$ square mile.

▲ Yellowstone National Park, Wyoming, see Example 2(a)

EXAMPLE 2 *Choosing an Appropriate Unit of Area*

Determine which metric unit of area you would use to measure the area of the following.

a) Yellowstone National Park (see photo)
b) The top of a kitchen table
c) The floor of the classroom
d) A person's property with an average-sized lot
e) A newspaper page
f) A baseball field
g) An ice-skating rink
h) A dime
i) A lens in eyeglasses
j) A dollar bill

SOLUTION

a) Square kilometers or hectares
b) Square meters
c) Square meters
d) Square meters or hectares
e) Square centimeters
f) Hectares or square meters
g) Square meters
h) Square millimeters or square centimeters
i) Square centimeters
j) Square centimeters

To find the area, A, of a square, we use the formula $A = s^2$, where s is the length of a side of the square. The area can also be found by the formula area = side × side. We use this information in Example 3.

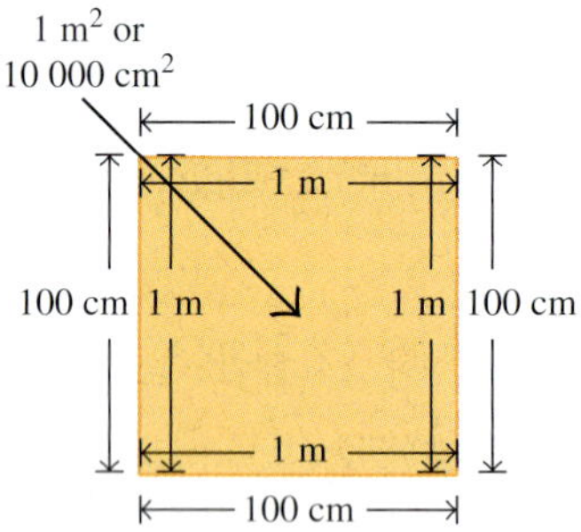

Figure 8.4

EXAMPLE 3 *Converting Square Meters to Square Centimeters*

A square meter is how many times as large as a square centimeter?

SOLUTION A square meter is a square whose sides are 1 meter long. Since 1 m equals 100 cm, we can replace 1 m with 100 cm (see Fig. 8.4). The area of $1\text{ m}^2 = 1\text{ m} \times 1\text{ m} = 100\text{ cm} \times 100\text{ cm} = 10\,000\text{ cm}^2$. Thus, the area of one square meter is 10,000 times the area of one square centimeter. This technique can be used to convert from any square unit to a different square unit.

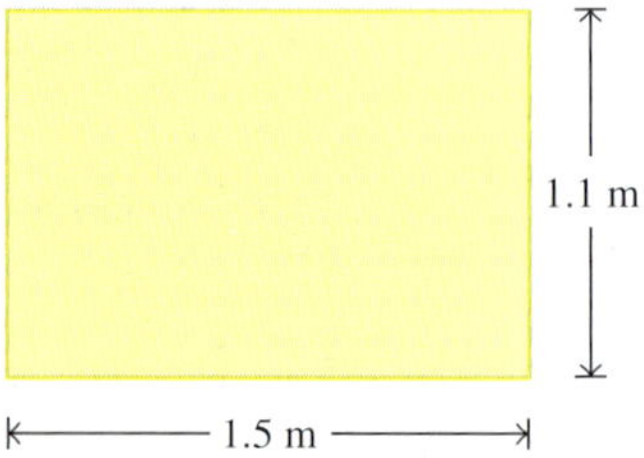

Figure 8.5

EXAMPLE 4 *Table Top*

Find the area of a rectangular table top if its length is 1.5 m and its width is 1.1 m (see Fig. 8.5).

SOLUTION To find the area, we use the formula

$$\text{Area} = \text{length} \times \text{width}$$

or

$$A = l \times w$$

Substituting values for l and w, we have

$$\begin{aligned} A &= 1.5 \text{ m} \times 1.1 \text{ m} \\ &= 1.65 \text{ m}^2 \end{aligned}$$

Notice that the area is measured in square meters. ●

Figure 8.6

EXAMPLE 5 ***A Quarter***

A quarter has a diameter of about 2.4 cm (Fig. 8.6). Find the surface area of one side of a quarter.

SOLUTION The formula for the area of a circle is $A = \pi r^2$, where π is *approximately* 3.14. The radius, r, is one-half the diameter. Since the diameter is about 2.4 cm, the radius is about 1.2 cm. Substituting values for π and r, we get the following.

$$\begin{aligned} A &= \pi r^2 \\ &\approx 3.14(1.2 \text{ cm})^2 \\ &\approx 4.52 \text{ cm}^2 \end{aligned}$$

Thus, the area is approximately 4.52 square centimeters. Recall from earlier chapters that the symbol $\approx$ means "is approximately equal to." ●

TIMELY TIP Many calculators contain a $\boxed{\pi}$ key. If your calculator contains a $\boxed{\pi}$ key, you should use that key to input the value of π. If you do so, you will get a more accurate answer than if you used 3.14 for pi.

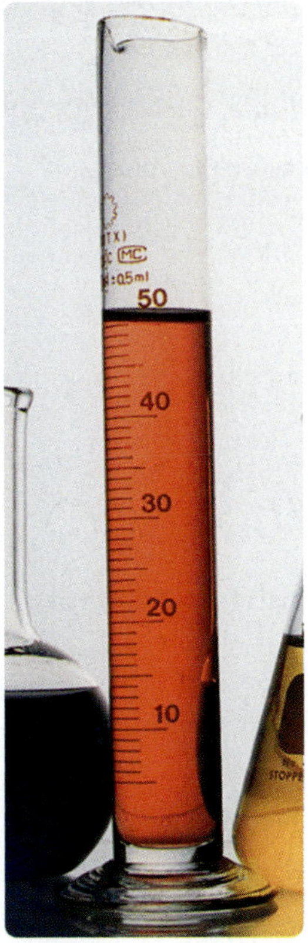

Figure 8.7

Volume

When a figure has only two dimensions—length and width—we can find its area. When a figure has three dimensions—length, width, and height—we can find its volume. The volume of an item can be considered the space occupied by the item.

In the metric system, volume may be expressed in terms of liters or cubic meters, depending on what is being measured. In all English-speaking countries except the United States, *liter* is spelled "litre."

The volume of liquids is expressed in liters. A liter is a little larger than a quart. Liters are used in place of pints, quarts, and gallons. A liter can be divided into 1000 equal parts, each of which is called a milliliter. Figure 8.7 illustrates a 50 mℓ graduated cylinder. In chemistry, 1000 mℓ and other metric graduated cylinders are often used. Milliliters are used to express the volume of very small amounts of liquid. Drug dosages are often expressed in milliliters. An 8 oz cup will hold about 240 mℓ of liquid.

The kiloliter, 1000 liters, is used to represent the volume of large amounts of liquid. Tank trucks carrying gasoline to service stations hold about 10.5 kℓ of gasoline.

Cubic meters are used to express the volume of large amounts of solid and gaseous material. The volume of a dump truck's load of topsoil is measured in cubic meters. The volume of natural gas used to heat a house may soon be measured in cubic meters instead of cubic feet.

The liquid in a liter container will fit exactly in a cubic decimeter (Fig. 8.8). Note that 1 ℓ = 1000 mℓ and that $1\ dm^3 = 1000\ cm^3$. Because $1\ \ell = 1\ dm^3$, *1 mℓ must equal 1 cm^3*. Other useful facts are illustrated in Table 8.3. Thus, within the metric system, conversions are much simpler than in the U.S. customary system. For example, how would you change cubic feet of water into gallons of water?

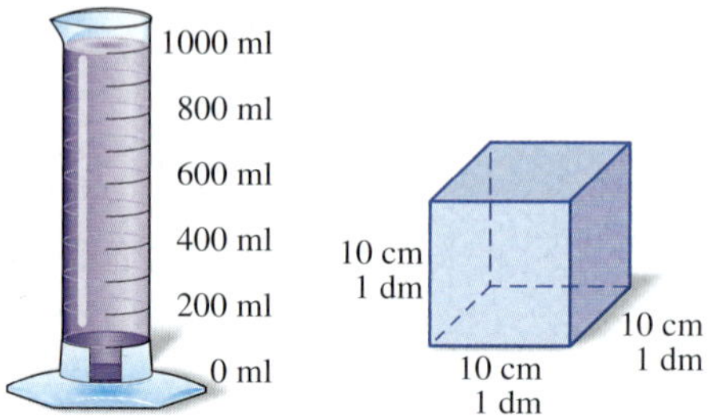

Figure 8.8

Table 8.3

Volume in Cubic Units		Volume in Liters
$1\ cm^3$	=	1 mℓ
$1\ dm^3$	=	1 ℓ
$1\ m^3$	=	1 kℓ

EXAMPLE 6 *Choosing an Appropriate Unit of Volume*

▲ See Example 6(a)

Determine which metric unit of volume you would use to measure the volume of the following.

a) The water in a swimming pool
b) A carton of milk
c) A truckload of topsoil
d) A drug dosage
e) Sand in a paper cup
f) A dime
g) Water in a drinking glass
h) Water in a full bath tub
i) The storage area of a sports utility vehicle with the back seats folded down or removed.
j) Concrete used to lay the foundation for a basement

SOLUTION

a) Kiloliters or liters
b) Liters
c) Cubic meters
d) Milliliters
e) Cubic centimeters
f) Cubic millimeters
g) Milliliters
h) Liters
i) Cubic meters
j) Cubic meters

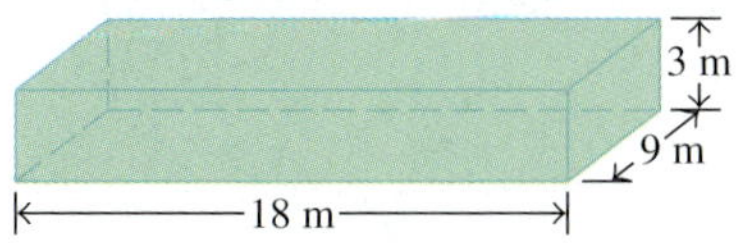

Figure 8.9

EXAMPLE 7 *Swimming Pool Volume*

A swimming pool is 18 m long and 9 m wide, and it has a uniform depth of 3 m (Fig. 8.9). Find (a) the volume of the pool in cubic meters and (b) the volume of water in the pool in kiloliters.

SOLUTION

a) To find the volume in cubic meters, we use the formula

$$V = l \times w \times h$$

Substituting values for l, w, and h we have

$$V = 18 \text{ m} \times 9 \text{ m} \times 3 \text{ m}$$
$$= 486 \text{ m}^3$$

b) Since $1 \text{ m}^3 = 1 \text{ k}\ell$, the pool will hold 486 kℓ of water.

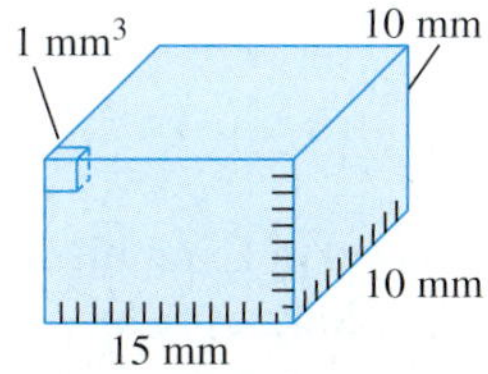

Figure 8.10

EXAMPLE 8 *Choose an Appropriate Unit*

Select the most appropriate answer. The volume of a shoe box is approximately

a) 1500 mm^3. b) 6500 ℓ. c) 6500 cm^3.

SOLUTION A shoe box is not a liquid, so its volume is not expressed in liters. Thus, (b) is not the answer. The volume of the rectangular solid in Fig. 8.10 is approximately 1500 mm^3, so (a) is not an appropriate answer. A shoe box may measure about 33 cm × 18 cm × 11 cm, or 6534 cm^3. Therefore, 6500 cm^3 or (c) is the most appropriate answer.

When the volume of a liquid is measured, the abbreviation cc is often used instead of cm^3 to represent cubic centimeters. For example, a nurse may give a patient an injection of 3 cc or 3 mℓ of the drug ampicillin.

EXAMPLE 9 *Measuring Medicine*

A nurse must give a patient 3 cc of the drug gentamicin mixed in 100 cc of a normal saline solution.

a) How many milliliters of the drug will the nurse administer?

b) What is the total volume of the drug and saline solution in milliliters?

SOLUTION

a) Because 1 cc is equal in volume to 1 mℓ, the nurse will administer 3 mℓ of the drug.

b) The total volume is 3 + 100 or 103 cc, which is equal to 103 mℓ.

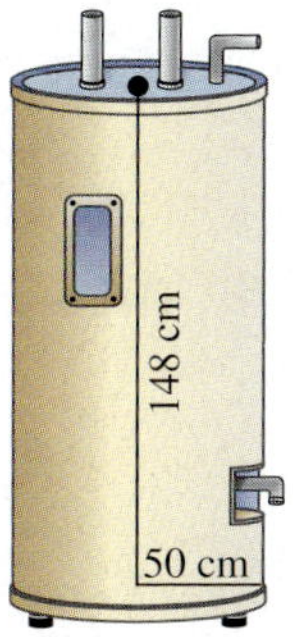

Figure 8.11

EXAMPLE 10 *A Hot-Water Heater*

A hot-water heater, in the shape of a right circular cylinder, has a radius of 50 cm and a height of 148 cm. What is the capacity, in liters, of the hot-water heater?

SOLUTION The hot-water heater is illustrated in Fig. 8.11. The formula for the volume of a right circular cylinder is $V = \pi r^2 h$, where π is approximately 3.14.

DID YOU KNOW?

Metrics and Medicine

Both milliliters and cubic centimeters are commonly used in medicine. In the United States, cubic centimeters are commonly denoted cc rather than the cm^3 used in the metric system. A patient's intake and output of fluids and intravenous injections are commonly measured in cubic centimeters. Drug dosage is measured in milliliters.

The following question is from a nursing exam. Can you determine the correct answer?

In caring for a patient after delivery, you are to give 12 units of Pitocin (in 1000 cc of intravenous fluid). The ampule is labeled 10 units per 0.5 mℓ. How much of the solution would you draw and give?

a) 0.6 cc
b) 1.2 cc
c) 6.0 cc
d) 9.6 cc

Answer: (a)

Since we want the capacity in liters, we will express all the measurements in meters. The volume will then be given in cubic meters, which can be easily converted to liters. Thus, 50 cm = 0.5 m, and 148 cm = 1.48 m.

$$\begin{aligned} V &= \pi r^2 h \\ &\approx 3.14(0.5)^2(1.48) \\ &\approx 3.14(0.25)(1.48) \approx 1.1618 \text{ m}^3 \end{aligned}$$

We want the volume in liters, so we must change the answer from cubic meters to liters.

$$1 \text{ m}^3 = 1000 \ \ell$$

So,

$$1.1618 \text{ m}^3 = 1.1618 \times 1000 = 1161.8 \ \ell$$

EXAMPLE 11 *Comparing Volume Units*

a) How many times larger is a cubic meter than a cubic centimeter?
b) How many times larger is a cubic dekameter than a cubic meter?

SOLUTION

a) The procedure used to determine the answer is similar to that used in Example 3 in this section. First we draw a cubic meter, which is a cube 1 m long by 1 m wide by 1 m high. In Fig. 8.12, we represent each meter as 100 centimeters. The volume of the cube is its length times its width times its height, or

$$\begin{aligned} V &= l \times w \times h \\ &= 100 \text{ cm} \times 100 \text{ cm} \times 100 \text{ cm} = 1\,000\,000 \text{ cm}^3 \end{aligned}$$

Since $1 \text{ m}^3 = 1\,000\,000 \text{ cm}^3$, a cubic meter is one million times larger than a cubic centimeter.

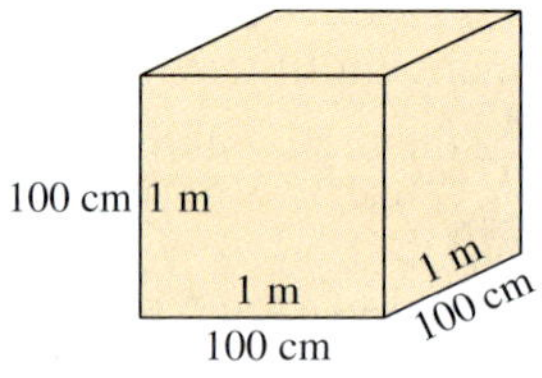

Figure 8.12

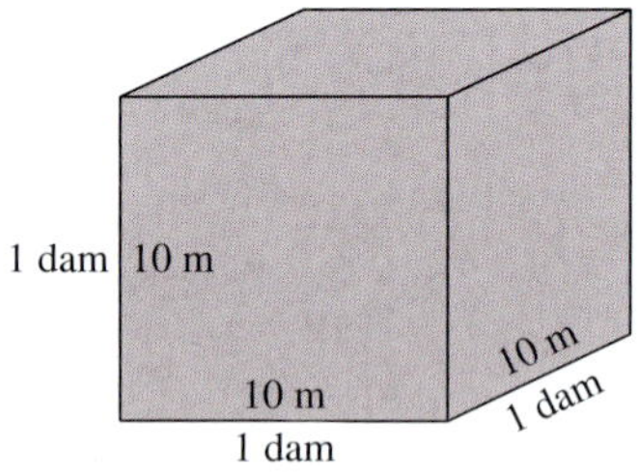

Figure 8.13

b) Work part (b) in a similar manner (Fig. 8.13). A dekameter is 10 meters. Thus,

$$\begin{aligned} V &= l \times w \times h \\ &= 10 \text{ m} \times 10 \text{ m} \times 10 \text{ m} = 1000 \text{ m}^3 \end{aligned}$$

Since $1 \text{ dam}^3 = 1000 \text{ m}^3$, a cubic dekameter is one thousand times larger than a cubic meter.

SECTION 8.2 EXERCISES

CONCEPT/WRITING EXERCISES

In Exercises 1–12, an object has been measured and the measurement has been written with the unit indicated. Indicate what was measured: length, area, or volume.

1. m^3 **2.** mm **3.** ha **4.** cm

5. cc **6.** ℓ **7.** cm^3 **8.** kℓ

9. m^2 **10.** dℓ **11.** km **12.** cm^2

13. Estimate, in centimeters, the length of this book.

14. Estimate your height in (a) centimeters and (b) meters.

15. Estimate, in meters, the length of the classroom in which your mathematics course is held.

16. Estimate, in square centimeters, the surface area of this book.

17. Estimate, in square centimeters, the surface area of a dollar bill.

18. Estimate, in centimeters, the length of your arm.

19. One liter of liquid has the equivalent volume of which of the following: a cubic centimeter, a cubic decimeter, or a cubic meter?

20. One cubic meter has the equivalent volume of which of the following liquid measures: a liter, a milliliter, or a kiloliter?

21. One milliliter of liquid has the equivalent volume of which of the following: a cubic centimeter, a cubic decimeter, or a cubic meter?

22. Which metric measurement is used to measure very large areas of land?

23. Is the hectare a measure of length, area, or volume?

24. A hectare has an area of about how many acres: 2.5, 25, or 250?

PRACTICE THE SKILLS

In Exercises 25–36, indicate the metric unit of measurement that you would use to express the following.

25. The length of a pencil

26. The distance between cities

27. The diameter of a dime

28. The diameter of a Frisbee

29. The length of a newborn infant

30. The diameter of a helicopter landing pad

31. The distance between freeway exits

32. The width of an Olympic-size swimming pool

33. The height of a cereal box

34. The length of a butterfly

35. The distance to the moon

36. The height of an adult male

In Exercises 37–44, choose the best answer.

37. The length of a tennis court is about how long?

a) 36 cm **b)** 36 km **c)** 36 m

38. A U.S. postage stamp is about how wide and how long?

a) 2 cm × 3 cm **b)** 2 mm × 3 mm

c) 2 hm × 3 hm

39. The distance between Los Angeles and San Diego could be how far?

a) 130 cm **b)** 130 m **c)** 130 km

40. A grown woman is about how tall?

a) 160 cm **b)** 160 mm **c)** 160 dm

41. The width of a ruler is about how wide?

a) 4 mm **b)** 4 cm **c)** 4 dm

42. The diameter of a coffee cup is about which of the following?

a) 8 mm **b)** 8 cm **c)** 8 dm

43. The Sears Tower in Chicago is about how tall?

a) 375 cm **b)** 375 km **c)** 375 m

44. The length of the New River Gorge Bridge near Fayetteville, West Virginia is about how long?

a) 1000 dam **b)** 1000 m **c)** 1000 cm

▲ New River Gorge Bridge

In Exercises 45–50, (a) estimate the item in metric units and (b) measure it with a metric ruler. Record your result.

45. The width of a card from a deck of cards

46. The width of a classroom door

47. The length of a car

48. The diameter of a can of soda

49. The height of a milk carton

50. The thickness of 10 sheets of paper.

In Exercises 51–56, replace the customary measure (shown in parentheses) with the appropriate metric measure.

51. Give him a _____ (inch), and he will take a _____ (mile).

52. There was a crooked man and he walked a crooked _____ (mile).

53. One hundred _____ (yard) dash.

54. I wouldn't touch a skunk with a 10-_____ (foot) pole.

55. I found a _____ (inch) worm.

56. This is a _____ (mile)stone in my life.

In Exercises 57–66, indicate the metric unit of measurement you would use to express the area of the following.

57. A small television screen

58. The floor of your classroom

59. A kitchen table

60. A building lot for a house

61. A baseball field

62. A postage stamp

63. A ceiling tile

64. Disney World

65. A professional basketball court

66. The city of San Antonio, Texas

▲ Riverwalk, San Antonio, Texas

In Exercises 67–74, choose the best answer.

67. A U.S. postage stamp has an area of about

a) 5 cm^2. **b)** 5 mm^2. **c)** 5 dm^2.

68. The area of a city lot is about

a) 800 m^2. **b)** 800 hm^2. **c)** 800 cm^2.

69. The area of a city lot is about

a) $\frac{1}{8}$ m^2. **b)** $\frac{1}{8}$ ha. **c)** $\frac{1}{8}$ km^2.

70. The area of a floor tile is about

a) 930 m^2. **b)** 930 km^2. **c)** 930 cm^2.

71. The area of one side of a \$1 bill is about

a) 100 cm^2. **b)** 100 m^2. **c)** 100 mm^2.

72. The area of the screen of a tabletop TV is about

a) 1200 dm^2. **b)** 1200 mm^2. **c)** 1200 cm^2.

73. The area of Grand Canyon National Park is about

a) 4900 m^2 **b)** 4900 cm^2 **c)** 4900 km^2

74. The area of a U.S. flag is about

a) 2.2 cm^2. **b)** 2.2 m^2. **c)** 2.2 km^2.

In Exercises 75–80, (a) estimate the area of the item in metric units and (b) measure it in metric units and compute its area.

75. A typical photograph

76. The cover of this book

77. A $20 bill

78. The top of your teacher's desk

79. The bottom of a 12 oz soda can

80. The face of a penny

In Exercises 81–90, determine the metric unit that would best be used to measure the volume of the following.

81. Water in a hot-water heater

82. Liquid in an eye dropper

83. Water flowing over Niagara Falls per minute

84. Oil needed to change the oil in your car

85. A bag of topsoil

86. A truckload of ready-mix concrete

87. Water coming out of a drinking fountain in 1 minute

88. Soda in a bottle of soda

89. Air in a hot air balloon

90. Air in a soccer ball

In Exercises 91–98, choose the best answer to indicate the volume of the following.

91. A shoe box

a) 7780 mm^3 **b)** 7780 dm^3 **c)** 7780 cm^3

92. Water that could fit in a thimble

a) 3 mℓ **b)** 3 ℓ **c)** 3 dℓ

93. Water in a 24-ft-diameter above-ground circular swimming pool

a) 55 ℓ **b)** 55 mℓ **c)** 55 kℓ

94. Soda in a can of soda

a) 355 ℓ **b)** 355 mℓ **c)** 355 m^3

95. A carry-on suitcase

a) 0.04 cm^3 **b)** 0.04 mm^3 **c)** 0.04 m^3

96. Juice that can be squeezed out of an orange

a) 120 kℓ **b)** 120 mℓ **c)** 120 ℓ

97. Air in a balloon with a diameter of 4 meters

a) 30 m^3 **b)** 30 cm^3 **c)** 30 km^3

98. Air in a soccer ball

a) 5 000 m^3 **b)** 5 000 cm^3 **c)** 5 000 mm^3

In Exercises 99–102, (a) estimate the volume in metric units and (b) compute the actual volume of the item.

99. Air in a cardboard box that is 61 cm long, 61 cm wide, and 41 cm tall (Use $V = lwh$.)

100. Water in a water bed that is 2 m long, 1.5 m wide, and 25 cm deep

101. Oil in a barrel that has a height of 1 m and a diameter of 0.5 m (Use $V = \pi r^2 h$.)

102. Water in a cylindrical tank that is 40 cm in diameter and 2 m high

PROBLEM SOLVING

103. *Area* Use a metric ruler to measure the length and width of the sides of the rectangle. Then compute the area of the rectangle. Give your answers in metric units.

104. *Area* Use a metric ruler to find the radius of the circle. Then compute the area of the circle. Give your answers in metric units.

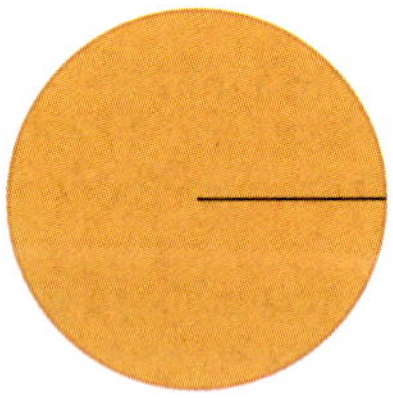

105. *Helipad* A helicopter landing area, generally in the shape of a circle, is called a helipad. A specific helipad is a circle of radius 10.2 meters. Determine the area of the helipad. Use the formula $A = \pi r^2$.

106. *Painting* The total area of a framed painting including the matting is 2540 cm^2 (see photo). If the length and width of the actual painting are 37 cm and 28 cm, respectively, determine the area of the matting.

107. *A Walkway* A rectangular building 50 m by 70 m is surrounded by a walk 1.5 m wide.

a) Find the area of the region covered by the building and the walk.

b) Find the area of the walk.

108. *Farmland* Mrs. Manecki has purchased a farm that is in the shape of a rectangle. The dimensions of the piece of land are 1.4 km by 3.75 km.

a) How many square kilometers of land did she purchase?

b) If 1 km^2 equals 100 ha, determine the amount of land she purchased in hectares.

109. *Gymnasium* The Boys and Girls Club is planning on having a large rectangular gymnasium built. The dimensions of the gymnasium are to be 62.4 m by 50.5 m.

a) Determine the area of the gymnasium in square meters.

b) If 1 m^2 equals 0.0001 ha, determine the area of the gymnasium in hectares.

110. *Volume of Water* **a)** What is the volume of water in a rectangular swimming pool that is 18 m long and 10 m wide and has an average depth of 2.5 m? Give your answer in cubic meters.

b) How many kiloliters of water will the pool hold?

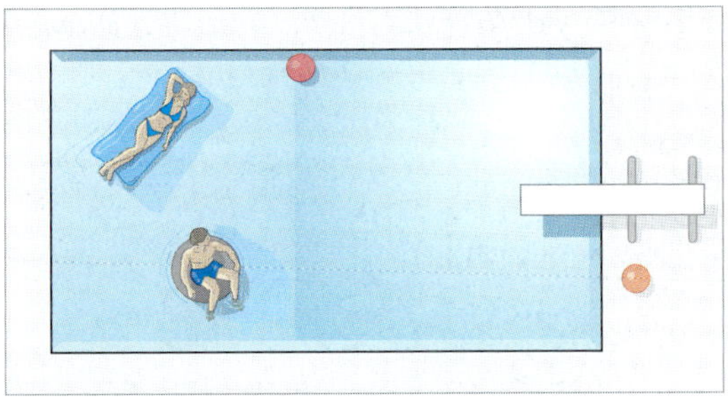

111. *Cost of Paint* The first coat of paint for the outside of a building requires 1 ℓ of paint for each 10 m^2. The second coat requires 1 ℓ for every 15 m^2. If the paint costs \$4.75 per liter, what will be the cost of two coats of paint for the four outside walls of a building 20 m long, 12 m wide, and 6 m high?

112. ***Volume of a Can*** A can of soup has a diameter of 8.0 cm and a height of 12.5 cm. Determine the volume of the can in cubic centimeters.

113. ***Fish Tank Volume*** A rectangular fish tank is 70 cm long, 40 cm wide, and 20 cm high.

a) How many cubic centimeters of water will the tank hold?

b) How many milliliters of water will the tank hold?

c) How many liters of water will the tank hold?

114. ***Hot-Water Heater*** A cylindrical shaped hot-water heater has a diameter of 0.56 m and a height of 1.17 m. If the hot-water heater is filled, determine the volume of water in the hot-water heater in

a) cubic meters.

b) liters.

115. How many times larger is a square dekameter than a square meter?

116. How many times larger is a square kilometer than a square dekameter?

117. How many times larger is a cubic meter than a cubic decimeter?

118. How many times larger is a cubic centimeter than a cubic millimeter?

In Exercises 119–126, replace the question mark with the appropriate value.

119. 1 cm^2 = ? mm^2 **120.** 1 hm^2 = ? cm^2

121. 1 km^2 = ? hm^2 **122.** 1 cm^2 = ? m^2

123. 1 mm^3 = ? dm^3 **124.** 1 dm^3 = ? mm^3

125. 1 m^3 = ? cm^3 **126.** 1 hm^3 = ? km^3

In Exercises 127–130, fill in the blank.

127. 218 cm^3 = ____ mℓ **128.** 435 cm^3 = ____ ℓ

129. 76 kℓ = ____ m^3 **130.** 4.2 ℓ = ____ cm^3

Glacier *In Exercises 131 and 132, assume that a part of a rectangular glacier that contains 60 cubic meters of ice calves (or breaks) off and falls into the ocean.*

▲ A glacier in Alaska

131. When the ice that has fallen into the ocean melts, determine the approximate amount of water, in kiloliters, obtained from the ice.

132. When the ice melts, determine the approximate amount of water, in cubic centimeters, obtained from the ice.

CHALLENGE PROBLEMS/GROUP ACTIVITIES

133. Starting with a straight piece of wood of sufficient size, construct a meterstick. Indicate decimeters, centimeters, and millimeters on the meterstick. Use the centimeter measure in Fig. 8.1 as a guide.

134. Construct a metric tape measure from a piece of tape or rope and then determine your waist measurement.

In Exercises 135 and 136, fill in the blank to make a true statement.

135. 6.7 kℓ = ____ dm^3 **136.** 1.4 ha = ____ cm^2

137. ***Conversions*** In Example 3, we illustrated how to change an area in a metric unit to an area measured with a different metric unit.

a) Using Example 3 as a guide, change 1 square mile to square inches.

b) Is converting from one unit of area to a different unit of area generally easier in the metric system or the U.S. customary system? Explain.

138. ***Conversions*** In Example 11, we illustrated how to change a volume in one metric unit to a volume measured with a different metric unit.

a) Using Example 11 as a guide, change 6 yd^3 (a volume 1 yard by 2 yards by 3 yards) into cubic inches.

b) Is converting from one unit of volume to a different unit of volume generally easier in the metric system or the U.S. customary system? Explain.

RECREATIONAL MATHEMATICS

139. ***Water Usage*** **a)** How much water do we use daily? On the average, people in the United States use more water than people anywhere else in the world. Take a guess at the number of liters of water used per day per person in the United States.

b) Now take a guess at the number of liters used per day per person in the United Kingdom.

Compare your answers to those given in the answer section.

INTERNET/RESEARCH ACTIVITIES

140. ***The Meter*** The definition of the meter has changed several times throughout history. Write a one- to two-page report on the history of the meter, from when it was first named to the present.

8.3 MASS AND TEMPERATURE

▲ This astronaut, while floating in space, has mass but no weight.

Suppose you go on a trip to Canada. When you turn on the television, the weatherperson says the temperature today will be sunny and 28°Celsius. How do you dress to go outside? Should you wear shorts or a sweater or a jacket? What is the temperature in degrees Fahrenheit?

On your television, you are watching U.S. astronauts floating in space. Do these floating astronauts have any weight? Do they have any mass? In this section, we will explore these two important metric measurements, temperature and mass.

Mass

Weight and mass are not the same. ***Mass*** is a measure of the amount of matter in an object. It is determined by the molecular structure of the object, and it will not change from place to place. Weight is a measure of the gravitational pull on an object. For example, the gravitational pull of Earth is about six times as great as the gravitational pull of the moon. Thus, a person on the moon weighs about $\frac{1}{6}$ as much as on Earth, even though the person's mass remains the same. In space, where there is no gravity, a person has no weight.

Even on Earth, the gravitational pull varies from point to point. The closer you are to Earth's center, the greater the gravitational pull. Thus, a person weighs very slightly less on a mountain than in a nearby valley. Because the mass of an object does not vary with location, scientists generally use mass rather than weight.

Although weight and mass are not the same, on Earth they are proportional to each other (the greater the weight, the greater the mass). Therefore, for our purposes, we can treat weight and mass as the same.

The *kilogram* is the basic unit of mass in the metric system. It is about 2.2 pounds. The official kilogram is a cylinder of platinum–iridium alloy kept by the International Bureau of Weights and Measures, located in Sèvres, near Paris. (See the Did You Know? in the margin on the next page.)

Items that we normally measure in pounds are usually measured in kilograms in other parts of the world. For example, an average-sized man has a mass of about 75 kg.

The *gram* (a unit that is 0.001 kg) is relatively small and is used for items normally measured in ounces. A nickel has a mass of about 5 g, a cube of sugar has a mass of about 2 g, and a large paper clip has a mass of about 1 g.

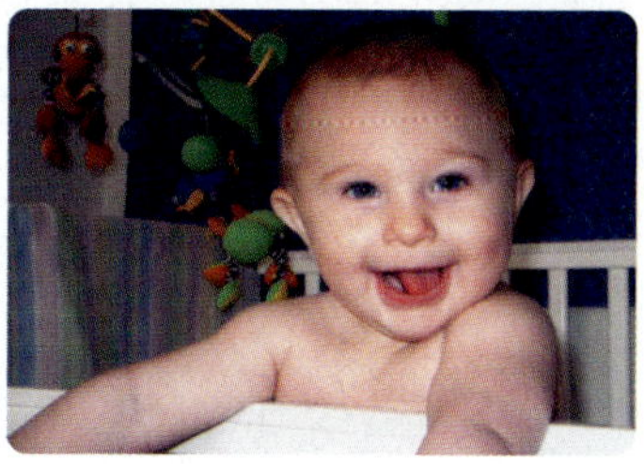

▲ See Exercise 1 (a)

The *milligram* is used extensively in the medical and scientific fields as well as in the pharmaceutical industry. Nearly all bottles of tablets are now labeled in either milligrams or grams.

The *metric tonne* (t) is used to express the mass of heavy items. One metric tonne equals 1000 kg. A metric ton is a little larger than our customary ton of 2000 lb. The mass of a large truck may be expressed in metric tonnes.

EXAMPLE 1 *Choosing the Appropriate Unit*

Determine which metric unit you would use to express the mass of the following.

a) A 1-year-old child
b) An orca (or killer whale)
c) A teaspoon
d) A box of cereal
e) A laptop computer
f) A fly
g) A frog
h) A refrigerator

SOLUTION

a) Kilograms
b) Metric tonnes
c) Grams
d) Grams
e) Grams or kilograms
f) Milligrams
g) Grams
h) Kilograms

One kilogram of water has a volume of exactly 1 liter. In fact, 1 liter is defined to be the volume of 1 kilogram of water at a specified temperature and pressure. Thus, mass and volume are easily interchangeable in the metric system. Converting from weight to volume is not nearly as convenient in the U.S. customary system. For example, how would you change pounds of water to cubic feet or gallons of water in our customary system?

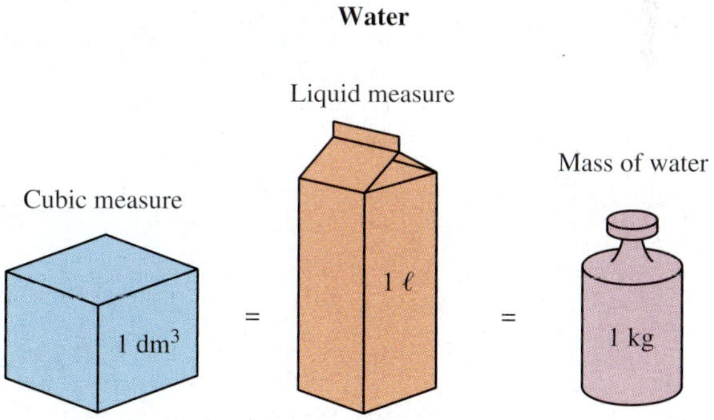

Figure 8.14 One cubic decimeter of water has the volume of one liter and the mass of one kilogram.

Since $1\text{ dm}^3 = 1000\text{ cm}^3$, $1\ell = 1000\text{ m}\ell$, and $1\text{kg} = 1000\text{g}$, we have the following relationship.

$$\begin{aligned} 1000\text{ cm}^3 &= 1000\text{ m}\ell = 1000\text{ g} \\ \text{or,}\quad 1\text{ cm}^3 &= 1\text{ m}\ell = 1\text{ g} \end{aligned}$$

Figure 8.14 illustrates the relationship between volume of water in cubic decimeters, in liters, and mass in kilograms. Table 8.4 expands on this relationship between the volume and mass of water.

DID YOU KNOW?

The Kilogram

Since 1889, a single platinum–iridium bar has been sealed in an airtight jar in the International Bureau of Weights and Measures in Sèvres, France.

Nicknamed "Le Grand K," this bar constitutes the one and only true kilogram. Of all the standard international units of measure, the kilogram remains the only one whose definition relies on a physical artifact. All the other units have their definitions rooted in constants of nature, such as the speed of light or atomic vibrations.

The kilogram has varied by as much as 0.05 part per million in the last 100 years. The cause of that variance remains unknown.

Some scientists want to redefine the kilogram in a way that will make the standard absolute, unchanging, and accessible to anyone, anywhere.

DID YOU KNOW?

Penny-wise, Kilo-foolish

Steven Thoburn

Steven Thoburn, a British grocer, had been charged in England with violating a British law. His crime? He used a non-metric scale to sell bananas to undercover officers. Thoburn pleaded not guilty but was convicted of the crime. In April 2002, a district judge handed Thoburn a 6-month conditional suspended sentence. Thoburn appealed the decision but died of natural causes at age 39 before the appeal was completed. Thoburn has since become known as the metric martyr.

Table 8.4 Volume and Mass of Water

Volume in Cubic Units		Volume in Liters		Mass of Water
1 cm^3	=	1 mℓ	=	1 g
1 dm^3	=	1 ℓ	=	1 kg
1 m^3	=	1 kℓ	=	1 t (1000 kg)

EXAMPLE 2 *Volume of a Fish Tank*

A fish tank is 1 m long, 50 cm high, and 250 mm wide (Fig. 8.15).

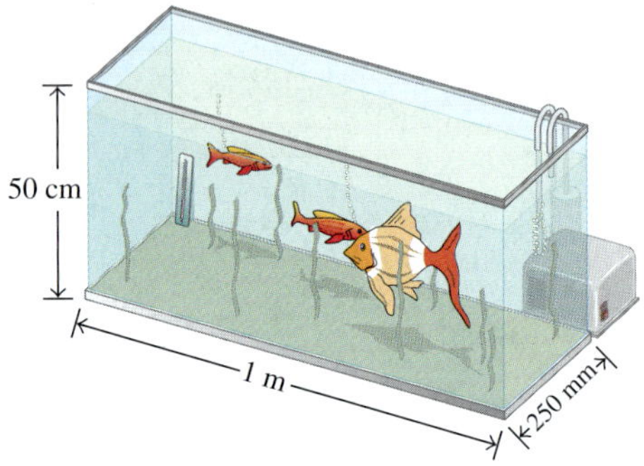

Figure 8.15

a) Determine the number of liters of water the tank holds.

b) What is the mass of the water in kilograms?

SOLUTION

a) We must convert all the measurements to the same units. Let's convert them all to meters: 50 cm is 0.5 m, and 250 mm is 0.25 m.

$$\begin{aligned} V &= l \times w \times h \\ &= 1 \times 0.25 \times 0.5 \\ &= 0.125 \text{ m}^3 \end{aligned}$$

Since 1 m^3 of water = 1 kℓ of water,

$$0.125 \text{ m}^3 = 0.125 \text{ k}\ell\text{, or } 125\ \ell \text{ of water}$$

b) Since 1 ℓ has a mass of 1 kg, 125 ℓ has a mass of 125 kg of water. ●

To convince yourself of the advantages of the metric system, do a similar problem involving the U.S. customary system of measurement, such as Challenge Problems/Group Activities Exercise 75 at the end of this section.

Temperature

The Celsius scale is used to measure temperatures in the metric system. Figure 8.16 on page 493 shows a thermometer with the Fahrenheit scale on the left and the Celsius scale on the right.

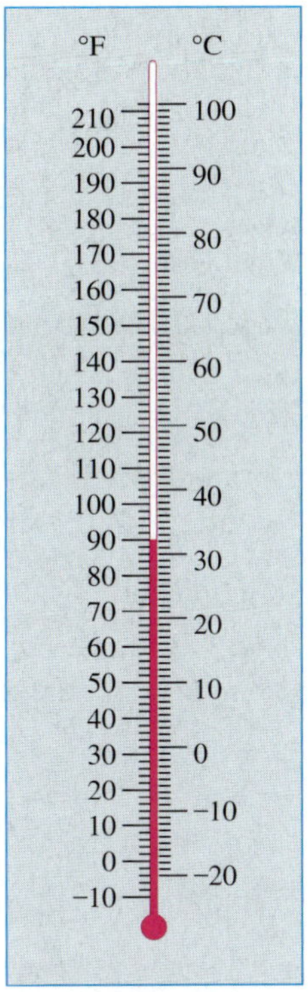

Figure 8.16

The Celsius scale was named for Swedish astronomer Anders Celsius (1701–1744), who first devised it in 1742. On the Celsius scale, water freezes at 0°C and boils at 100°C. In the past, the Celsius thermometer was called a "centigrade thermometer." Recall that *centi* means $\frac{1}{100}$, and there are 100 degrees between the freezing point of water and the boiling point of water. Thus, 1°C is $\frac{1}{100}$ of this interval. Table 8.5 gives some common temperatures in both degrees Celsius (°C) and degrees Fahrenheit (°F).

Table 8.5

Celsius Temperature		Fahrenheit Temperature
−18°C	A very cold day	0°F
0°C	Freezing point of water	32°F
10°C	A warm winter day	50°F
20°C	A mild spring day	68°F
30°C	A warm summer day	86°F
37°C	Body temperature	98.6°F
100°C	Boiling point of water	212°F
177°C	Oven temperature for baking	351°F

At a temperature of −40° the Celsius and Fahrenheit temperatures are the same. That is, −40°C = −40°F. See Exercise 74.

EXAMPLE 3 *Metric Temperatures*

Choose the best answer. (Refer to the dual-scale thermometer in Fig. 8.16.)

a) Chicago, Illinois, on New Year's Day might have a temperature of

i) −10°C. ii) 20°C. iii) 45°C.

b) Washington, D.C., on July 4 might have a temperature of

i) 15°C. ii) 30°C. iii) 40°C.

c) The oven temperature for baking a cake might be

i) 60°C. ii) 100°C. iii) 175°C.

SOLUTION

a) A temperature of 20°C is possible if it is a very mild winter, but 45°C is much too hot. The best answer for a normal winter is −10°C.

b) The best estimate is 30°C. A temperature of 15°C is too chilly, and 40°C is too hot for July 4.

c) A cake bakes at temperatures well above boiling, so the only reasonable answer is 175°C.

Comparing the temperature in Table 8.5, we see that the Celsius scale has 100° from the boiling point of water to the freezing point of water and the Fahrenheit scale has 180° from the boiling point of water to the freezing point of water. Therefore, one Celsius degree represents a greater change in temperature than one

Fahrenheit degree does. In fact, one Celsius degree is the same as $\frac{180}{100}$, or $\frac{9}{5}$ Fahrenheit degrees. When converting from one system to the other system, use the following formulas.

FROM CELSIUS TO FAHRENHEIT

$$F = \frac{9}{5}C + 32$$

FROM FAHRENHEIT TO CELSIUS

$$C = \frac{5}{9}(F - 32)$$

DID YOU KNOW?

It's a Metric World

Figure 8.17

The United States is the only westernized country not currently using the metric system as its primary system of measurement. The only countries in the world besides the United States not committed to using the metric system are Liberia and Myanmar (although Liberia and Myanmar use it in practice); see Fig. 8.17.

The European Union (EU) adopted a directive that requires all exporters to EU nations to indicate the dimensions of their products in metric units. Currently, U.S. manufacturers who export goods are doing so. Little by little, the United States is using more metric measurements. For example, soft drinks come in liter bottles and prescription drug dosages are given in metric units. Maybe in the not too distant future gasoline will be measured in liters, not gallons, as it is in Canada and Mexico.

EXAMPLE 4 *Convert to °C*

A typical setting for home thermostats is 72°F. What is the equivalent temperature on the Celsius thermometer?

SOLUTION We use the formula $C = \frac{5}{9}(F - 32)$ to convert from °F to °C. Substituting $F = 72$ gives

$$\begin{aligned} C &= \frac{5}{9}(72 - 32) \\ &= \frac{5}{9}(40) \\ &\approx 22.2 \end{aligned}$$

Thus, the equivalent temperature of 72°F is about 22.2°C.

EXAMPLE 5 *Convert to °F*

If the temperature outdoors is 28°C, will you need to wear a sweater if going outdoors?

SOLUTION We use the formula $F = \frac{9}{5}C + 32$ to convert from °C to °F. Substituting $C = 28$ yields

$$\begin{aligned} F &= \frac{9}{5}(28) + 32 \\ &= 50.4 + 32 \\ &= 82.4 \end{aligned}$$

Since the temperature is about 82.4°F, you will not need to wear a sweater.

SECTION 8.3 EXERCISES

CONCEPT/WRITING EXERCISES

1. What is the basic unit of mass in the metric system?
2. The mass of a nickel is about how many grams?
3. One kilogram is a little more than how many pounds?
4. What unit of mass is used to express the mass of very heavy items?
5. Give an estimate of the average temperature, in degrees Celsius, in Florida in August.
6. Give an estimate of the average temperature, in degrees Celsius, in North Dakota in February.
7. Give an estimate, in degrees Celsius, of what you would consider an ideal outdoor temperature.
8. **a)** Is a person's mass the same in space as on Earth? Explain.

   b) Is a person's weight the same in space as on Earth? Explain.

PRACTICE THE SKILLS

In Exercises 9–18, indicate the metric unit of measurement that would best express the mass of the following.

9. A woman
10. A dime
11. A pair of eyeglasses
12. A box of cereal
13. A new pencil
14. A sports utility vehicle
15. A refrigerator
16. A mosquito
17. A full-grown whale
18. A calculator

In Exercises 19–24, select the best answer.

19. The mass of a 5 lb bag of flour is about how much?

a) 2.27 g **b)** 2.27 kg **c)** 2.27 dag

20. The mass of a tea bag is about how much?

a) 4 mg **b)** 4 kg **c)** 4 g

21. The mass of a coffee pot filled with coffee is about how much?

a) 1.4 mg **b)** 1.4 kg **c)** 1.4 g

22. The mass of a box of cornflakes is about how much?

a) 0.45 t **b)** 0.45 g **c)** 0.45 kg

23. The mass of a full-grown elephant is about how much?

a) 2800 g **b)** 2800 kg **c)** 2800 dag

24. The mass of a full-size car is about how much?

a) 1 962 000 hg **b)** 380 kg **c)** 1.6 t

In Exercises 25–28, estimate the mass of the item. If a scale with metric measure is available, find the mass.

25. Your body
26. A telephone book
27. A gallon of water
28. A tomato

In Exercises 29–38, choose the best answer. Use Table 8.5 and Fig. 8.16 to help select your answers.

29. Freezing rain is most likely to occur at a temperature of

a) −25°C. **b)** 32°C. **c)** 0°C.

▲ Rochester, New York

30. The thermostat for an air conditioner was set for 80°F. This setting is closest to

a) 2°C. **b)** 27°C. **c)** 57°C.

31. The temperature of the water in a certain lake is 5°C. You could

a) ice fish.

b) dress warmly and walk along the lake.

c) swim in the lake.

32. What might be the temperature at which a refrigerator is set?

a) 30°C **b)** 5°C **c)** 0°C

33. The temperature in Phoenix, Arizona on a summer day might be

a) 15°C **b)** 20°C **c)** 40°C

34. The weather forecast calls for a high of 32°C. You should plan to wear

a) a down-lined jacket.

b) a sweater.

c) a bathing suit.

35. What might be the temperature of an apple pie baking in the oven?

a) 90°C **b)** 100°C **c)** 177°C

36. The temperature of the water in a car's radiator when the car's engine is operating at its normal temperature might be

a) 70°C. **b)** 300°C. **c)** 110°C.

37. The temperature of water in a hot tub might be

a) 30°C. **b)** 50°C. **c)** 40°C.

38. The temperature of the snow in a snowman might be

a) −15°C **b)** −5°C **c)** 5°C

In Exercises 39–52, convert each temperature as indicated. When appropriate, give your answer to the nearest tenth of a degree.

39. 25°C = ____°F **40.** −5°C = ____°F

41. 92°F = ____°C **42.** −10°F = ____°C

43. 0°F = ____°C **44.** 98°F = ____°C

45. 37°C = ____°F **46.** −4°C = ____°F

47. 13°F = ____°C **48.** 75°F = ____°C

49. 0°C = ____°F **50.** 50°C = ____°F

51. −20°F = ____°C **52.** 425°F = ____°C

In Exercises 53 and 54, use the photo of a seismographic pool in Yellowstone National Park in Wyoming. The pool is heated by volcanic activity under Earth's surface.

53. Convert 165°F to degrees Celsius to verify that it is approximately 74°C.

54. Convert 74°C to degrees Fahrenheit to verify that it is approximately 165°F.

In Exercises 55–60, use the following graph, which shows the daily low and high temperatures, in degrees Celsius, for the week in the Outback in Australia. The week illustrated was unseasonably warm. Determine the following temperatures in degrees Fahrenheit.

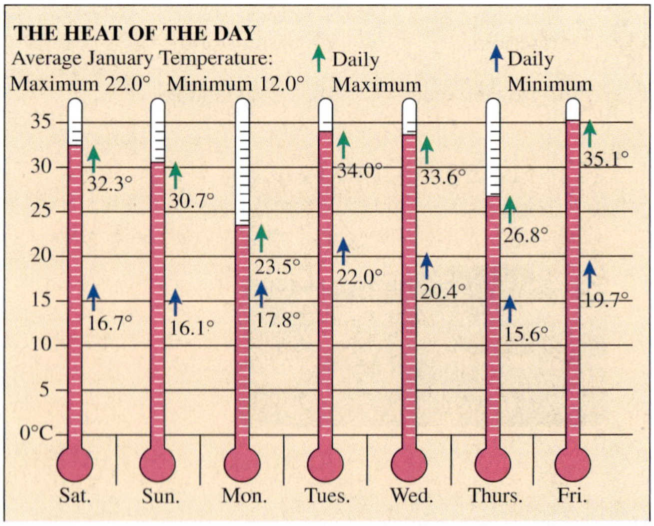

55. The average January maximum temperature

56. The maximum temperature for the week

57. The maximum temperature on Friday

58. The maximum temperature on Saturday

59. The range of temperatures on Monday

60. The range of temperatures on Tuesday

PROBLEM SOLVING

The photo shows the cost of popcorn and sabu dhana at a farm market in the country of Fiji. Use the information in the photo to answer Exercises 61 and 62.

61. ***Popcorn*** Determine the cost, in dollars, of 8.1 kg of popcorn if popcorn cost \$2.50 per kg.

62. ***Sabu Dhana*** Determine the cost, in dollars, of 5.2 kg of sabu dhana if sabu cost \$2.60 per kg. (Sabu dhana is a by-product of the tapioca root.)

63. ***Salt and Soda*** A mixture of 45 g of salt and 29 g of baking soda is poured into 370 mℓ of water. What is the total mass of the mixture in grams?

64. ***Jet Fuel*** A jet can travel about 1 km on 17 kg of fuel. How many metric tonnes of fuel will the jet use flying nonstop between Baltimore and Los Angeles, a distance of about 4320 km?

65. ***A Storage Tank*** The dimensions of a storage tank are length 16 m, width 12 m, and height 12 m. If the tank is filled with water, determine

a) the volume of water in the tank in cubic meters.

b) the number of kiloliters of water the tank will hold.

c) the mass of the water in metric tonnes.

66. ***A Water Heater*** A hot-water heater in the shape of a right circular cylinder has a radius of 50 cm and a height of 150 cm. If the tank is filled with water, determine

a) the volume of water in the tank in cubic meters.

b) the number of liters of water the tank will hold.

c) the mass of the water in kilograms.

67. ***Is There a Problem?*** A temperature display at a bank flashes the temperature in degrees Fahrenheit and then flashes the temperature in degrees Celsius. If it flashes 78°F, then 20°C, is there a problem with the display? Explain.

68. ***Fever or Chills?*** Maria's body temperature is 38.2°C. Should she take an aspirin or put on a sweater? Explain.

In Exercises 69–72, convert as indicated.

69. 4.2 kg = ____ t

70. 9.52 t = ____ kg

71. 17.4 t = ____ g

72. 1 460 000 mg = ____ t

CHALLENGE PROBLEMS/GROUP ACTIVITIES

73. ***Gatorade*** Gatorade is poured into a plastic bottle that holds 1.2 ℓ of liquid. The bottle is then placed in a freezer. When the bottle is removed from the freezer, the plastic is cut away, leaving just the frozen Gatorade.

a) What is the approximate mass of the frozen Gatorade in grams?

b) What is the approximate volume of the frozen Gatorade in cubic centimeters?

74. Show that $-40°C = -40°F$.

In Example 2, we showed how to find the volume and mass of water in a fish tank. Exercise 75 demonstrates how much more complicated solving a similar problem is in the U.S. customary system.

75. ***Fish Tank*** A fish tank is 1 yd long by 1.5 ft high by 15 in. wide.

a) Determine the volume of water in the fish tank in cubic feet.

b) Determine the weight of the water in pounds. One cubic foot of water weighs about 62.5 lb.

c) If 1 gal of water weighs about 8.3 lb, how many gallons will the tank hold?

RECREATIONAL MATHEMATICS

76. ***Balance the Scale*** Determine the quantity needed to replace the question mark to make the scale balance. The weight times the distance on both sides of the fulcrum (the triangle) must be the same to make the scale balance.

77. ***Interesting Facts*** The 2006 *Guinness Book of World Records* provides some interesting facts.

a) The lowest temperature ever recorded in the United States was −62.11°C on January 23, 1971, in Prospect Creek, Alaska. What is this temperature in degrees Fahrenheit?

b) International Falls, Minnesota, has the lowest annual mean (average) temperature in the United States (including Alaska). Its mean annual temperature is about 2.5°C. What is this temperature in degrees Fahrenheit?

c) The highest temperature produced in a laboratory was about 918,000,000°F. What is this temperature in degrees Celsius?

▲ Ice Box Day, International Falls, Minnesota, see Exercise 77(b)

INTERNET/RESEARCH ACTIVITY

78. Do industries in your area export goods? If so, are they training employees to use and understand the metric system? Contact local industries that export goods and write a report on your findings.

8.4 DIMENSIONAL ANALYSIS AND CONVERSIONS TO AND FROM THE METRIC SYSTEM

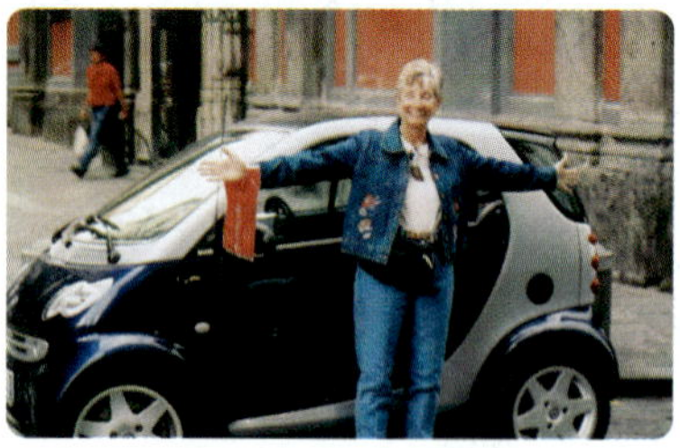

▲ Can you determine the length, in feet, of the Smart Car?

Pictured here is a Smart Car, commonly driven throughout Europe and soon to be sold in the United States. Some of the specifications of the Smart Car shown are length, 2500 mm; width, 1515 mm; maximum speed, 135 km/hr; gross weight, 990 kg; 0–100 km/hr, 18.3 sec; engine size, 698 cc: and fuel economy, 4.8 ℓ/100 km. Can you change these specifications to U.S. customary system measurements? To determine the Smart Car's specifications, you will need to know how to convert measurements from the metric system to the U.S. customary system. We will explain how to do so in this section.

You may sometimes need to change units of measurement in the metric system to equivalent units in the U.S. customary system. To do so, use ***dimensional analysis***, which is a procedure used to convert from one unit of measurement to a different unit of measurement. To perform dimensional analysis, you must first understand what is meant by a unit fraction. A ***unit fraction*** is any fraction in which the numerator and

DID YOU KNOW?

The Importance of Measurements

John Quincy Adams, 6th president of the United States

We all realize how important the abilities to read and to write are. Yet do we realize how important an understanding of measurements is in our daily lives? As John Quincy Adams wrote in his report to the U.S. Congress in 1821: "Weights and measures may be ranked among the necessities of life to every individual of human society. They enter into the economical arrangements and daily concerns of every family. They are necessary to every occupation of human industry."

denominator contain different units and the value of the fraction is 1. From Table 8.6, we can obtain many unit fractions involving U.S. customary units.

Table 8.6 U.S. Customary Units

1 foot = 12 inches
1 yard = 3 feet
1 mile = 5280 feet
1 pound = 16 ounces
1 ton = 2000 pounds
1 cup (liquid) = 8 fluid ounces
1 pint = 2 cups
1 quart = 2 pints
1 gallon = 4 quarts
1 minute = 60 seconds
1 hour = 60 minutes
1 day = 24 hours
1 year = 365 days

Examples of Unit Fractions

$$\frac{12 \text{ in.}}{1 \text{ ft}} \qquad \frac{1 \text{ ft}}{12 \text{ in.}} \qquad \frac{16 \text{ oz}}{1 \text{ lb}} \qquad \frac{1 \text{ lb}}{16 \text{ oz}} \qquad \frac{60 \text{ min}}{1 \text{ hr}} \qquad \frac{1 \text{ hr}}{60 \text{ min}}$$

In each of these examples, the numerator equals the denominator, so the value of the fraction is 1.

To convert an expression from one unit of measurement to a different unit, multiply the given expression by the unit fraction (or fractions) that will result in the answer having the units you are seeking. When two fractions are being multiplied and the same unit appears in the numerator of one fraction and in the denominator of the other fraction, that common unit may be divided out and eliminated. For example, suppose we want to convert 30 inches to feet. We consider the following:

$$30 \text{ in.} = \text{? ft}$$

Since inches are given, we will need to eliminate them. Thus, inches will need to appear in the denominator of the unit fraction. We need to convert to feet, so feet will need to appear in the numerator of the unit fraction. If we multiply a quantity in inches by a unit fraction containing feet/inches, the inches will divide out as follows, leaving feet. In the following illustration we have omitted the numbers in the unit fraction so we can concentrate on the units.

$$(\cancel{\text{in.}})\left(\frac{\text{ft}}{\cancel{\text{in.}}}\right) = \text{ft}$$

Thus, to convert 30 inches to feet, we do the following.

$$30 \text{ in.} = (30 \cancel{\text{in.}})\left(\frac{1 \text{ ft}}{12 \cancel{\text{in.}}}\right) = \frac{30}{12}\text{ft} = 2.5 \text{ ft}$$

In Examples 1 through 3, we will give examples that do not involve the metric system. After that, we will use dimensional analysis to make conversions to and from the metric system.

▲ Mexican currency

EXAMPLE 1 ***Using Dimensional Analysis***

A container contains 26 ounces of salt. Convert 26 ounces to pounds.

SOLUTION One pound is 16 ounces. Therefore, we write

$$26 \text{ oz} = (26 \cancel{\text{oz}})\left(\frac{1 \text{ lb}}{16 \cancel{\text{oz}}}\right) = \frac{26}{16}\text{lb} = 1.625 \text{ lb}$$

Thus, 26 oz equals 1.625 lb. •

EXAMPLE 2 ***Mexican Pesos***

On November 4, 2006, $1 U.S. could be exchanged for about 10.87 Mexican pesos. What was the amount in U.S. dollars of 2500 pesos?

SOLUTION

$$2500 \text{ pesos} = 2500 \cancel{\text{pesos}}\left(\frac{\$1.00}{10.87 \cancel{\text{pesos}}}\right) = \$\frac{2500}{10.87} \approx \$229.99$$

Thus, 2500 pesos had a value of about $229.99. •

If more than one unit needs to be changed, more than one multiplication may be needed, as illustrated in Example 3.

EXAMPLE 3 ***Using Several Unit Fractions***

Convert 60 miles per hour to feet per second.

SOLUTION Let's consider the units given and where we want to end up. We are given $\frac{\text{mi}}{\text{hr}}$ and wish to end with $\frac{\text{ft}}{\text{sec}}$. Thus, we need to change miles into feet and hours into seconds. Because two units need to be changed, we will need to multiply the given quantity by two unit fractions, one for each conversion. First we show how to convert the units of measurement from miles per hour to feet per second:

$$\left(\frac{\cancel{\text{mi}}}{\cancel{\text{hr}}}\right)\left(\frac{\text{ft}}{\cancel{\text{mi}}}\right)\left(\frac{\cancel{\text{hr}}}{\text{sec}}\right) \quad \text{gives an answer in} \quad \frac{\text{ft}}{\text{sec}}$$

Now we multiply the given quantity by the appropriate unit fractions to obtain the answer:

$$60\frac{\text{mi}}{\text{hr}} = \left(60\frac{\cancel{\text{mi}}}{\cancel{\text{hr}}}\right)\left(\frac{5280 \text{ ft}}{1 \cancel{\text{mi}}}\right)\left(\frac{1 \cancel{\text{hr}}}{3600 \text{ sec}}\right) = \frac{(60)(5280)}{(1)(3600)}\frac{\text{ft}}{\text{sec}}$$

$$= 88\frac{\text{ft}}{\text{sec}}$$

Note that $\left(60\frac{\text{mi}}{\text{hr}}\right)\left(\frac{1 \text{ hr}}{3600 \text{ sec}}\right)\left(\frac{5280 \text{ ft}}{1 \text{ mi}}\right)$ will give the same answer. •

Conversions to and from the Metric System

Now we will apply dimensional analysis to the metric system.

Table 8.7 is used in making conversions to and from the metric system. The values given in Table 8.7 on page 501 are often approximations. A more exact table of conversion factors may be found in many science books at your college's library or on the Internet. However, we can use this table to obtain many unit fractions.

Table 8.7 shows that 1 in. = 2.54 cm. From this equality, we can write the two unit fractions

$$\frac{1 \text{ in.}}{2.54 \text{ cm}} \quad \text{or} \quad \frac{2.54 \text{ cm}}{1 \text{ in.}}$$

Examples of other unit fractions from Table 8.7 are

$$\frac{1 \text{ yd}}{0.9 \text{ m}}, \quad \frac{0.9 \text{ m}}{1 \text{ yd}}, \quad \frac{1 \text{ gal}}{3.8 \ \ell}, \quad \frac{3.8 \ \ell}{1 \text{ gal}}, \quad \frac{1 \text{ lb}}{0.45 \text{ kg}}, \quad \text{and} \quad \frac{0.45 \text{ kg}}{1 \text{ lb}}$$

To change from a metric unit to a customary unit or vice versa, multiply the given quantity by the unit fraction whose product will result in the units you are seeking. For example, to convert 5 in. to centimeters, multiply 5 in. by a unit fraction with centimeters in the numerator and inches in the denominator.

$$\begin{aligned} 5 \text{ in.} &= (5 \not{\text{in.}})\left(\frac{2.54 \text{ cm}}{1 \not{\text{in.}}}\right) \\ &= 5(2.54) \text{ cm} \\ &= 12.7 \text{ cm} \end{aligned}$$

Table 8.7 Conversions Table

Length	
1 inch (in.) = 2.54	centimeters (cm)
1 foot (ft) = 30	centimeters (cm)
1 yard (yd) = 0.9	meter (m)
1 mile (mi) = 1.6	kilometers (km)
Area	
1 square inch (in.^2) = 6.5	square centimeters (cm^2)
1 square foot (ft^2) = 0.09	square meter (m^2)
1 square yard (yd^2) = 0.8	square meter (m^2)
1 square mile (mi^2) = 2.6	square kilometers (km^2)
1 acre = 0.4	hectare (ha)
Volume	
1 teaspoon (tsp) = 5	milliliters (mℓ)
1 tablespoon (tbsp) = 15	milliliters (mℓ)
1 fluid ounce (fl oz) = 30	milliliters (mℓ)
1 cup (c) = 0.24	liter (ℓ)
1 pint (pt) = 0.47	liter (ℓ)
1 quart (qt) = 0.95	liter (ℓ)
1 gallon (gal) = 3.8	liters (ℓ)
1 cubic foot (ft^3) = 0.03	cubic meter (m^3)
1 cubic yard (yd^3) = 0.76	cubic meter (m^3)
Weight (Mass)	
1 ounce (oz) = 28	grams (g)
1 pound (lb) = 0.45	kilogram (kg)
1 ton (T) = 0.9	tonne (t)

▲ See Example 4(b)

EXAMPLE 4 *Volume and Length Conversions*

a) A recipe requires $3\frac{1}{2}$ cups of chicken broth. How many liters does this amount equal?

b) A man measures 1.86 m (see photo). What is his height in feet?

SOLUTION

a) In Table 8.7, under the heading of volume, we see that 1 cup = 0.24 ℓ. Thus, the unit fractions involving cups and liters are

$$\frac{1 \text{ cup}}{0.24\ \ell} \quad \text{or} \quad \frac{0.24\ \ell}{1 \text{ cup}}$$

We need to convert from cups to liters. Since $3\frac{1}{2}$ is 3.5 in decimal form, we write

$$3.5 \text{ cups} = 3.5\ \cancel{\text{cups}}\left(\frac{0.24\ \ell}{1\ \cancel{\text{cup}}}\right) = (3.5)(0.24)\ \ell = 0.84\ \ell$$

b) In Table 8.7, under the heading of length, we see that there is no conversion given from meters to feet. There are a number of ways this example could be worked. One method is to convert meters to yards and then convert yards to feet. The procedure is shown below.

$$\begin{aligned} 1.86 \text{ m} &= (1.86\ \cancel{\text{m}})\left(\frac{1\ \cancel{\text{yd}}}{0.9\ \cancel{\text{m}}}\right)\left(\frac{3 \text{ ft}}{1\ \cancel{\text{yd}}}\right) \\ &= \frac{(1.86)(3)}{0.9} \text{ ft} \\ &= 6.2 \text{ ft} \end{aligned}$$

Thus, the man is 6.2 ft.

▲ Land for sale in Fiji, measured in hectares, see Example 5

EXAMPLE 5 *Area Conversion*

The photo shows an area of 31.46 hectares for sale. Find the area in acres.

SOLUTION From Table 8.7, we determine that 1 acre = 0.4 ha. Thus,

$$31.46 \text{ ha} = (31.46\ \cancel{\text{ha}})\left(\frac{1 \text{ acre}}{0.4\ \cancel{\text{ha}}}\right) = \frac{31.46}{0.4} \text{acres} = 78.65 \text{ acres}$$

▲ See Example 6

EXAMPLE 6 *Weight (Mass) Conversion*

The photo shows that peppers (in New Zealand) cost \$3.75 per kilogram. Determine the cost per pound for the peppers.

SOLUTION First we will determine the number of pounds that is equivalent to 1 kilogram. From Table 8.7, we obtain the unit fraction

$$\frac{0.45 \text{ kg}}{1 \text{ lb}}$$

Next, we use this unit fraction to determine the pepper's cost per pound.

$$\frac{\$3.75}{1\text{ kg}} = \left(\frac{\$3.75}{1\ \cancel{\text{kg}}}\right)\left(\frac{0.45\ \cancel{\text{kg}}}{1\text{ lb}}\right)$$
$$= \$3.75(0.45)\text{ per pound}$$
$$\approx \$1.69\text{ per pound}$$

Therefore, the peppers cost about $1.69 per pound.

EXAMPLE 7 *Administering a Medicine*

A nurse must administer 4 cc of codeine elixir to a patient.

a) How many milliliters of the drug will be administered?

b) How many ounces is this dosage equivalent to?

SOLUTION

a) Since 1 cc = 1 mℓ, the nurse will administer 4 mℓ of the drug.

b) Since 1 fl oz = 30 mℓ,

$$4\text{ m}\ell = (4\ \cancel{\text{m}\ell})\left(\frac{1\text{ fl oz}}{30\ \cancel{\text{m}\ell}}\right) = \frac{4}{30}\text{fl oz} \approx 0.13\text{ fl oz}$$

Suppose we want to convert 150 millimeters to inches. Table 8.7 does not have a conversion factor from millimeters to inches, but it does have one for inches to centimeters. Because 1 inch = 2.54 centimeters and 1 centimeter = 10 millimeters, we can reason that 1 inch = 25.4 millimeters. Therefore, unit fractions we may use are as follows.

$$\frac{1\text{ in.}}{25.4\text{ mm}} \quad \text{or} \quad \frac{25.4\text{ mm}}{1\text{ in.}}$$

We can solve the problem as follows.

$$150\text{ mm} = (150\ \cancel{\text{mm}})\left(\frac{1\text{ in.}}{25.4\ \cancel{\text{mm}}}\right) = \frac{150}{25.4}\text{in.}$$
$$\approx 5.91\text{ in.}$$

If we wish, we can use dimensional analysis using two unit fractions to make the conversion. The procedure follows:

$$150\text{ mm} = (150\ \cancel{\text{mm}})\left(\frac{1\ \cancel{\text{cm}}}{10\ \cancel{\text{mm}}}\right)\left(\frac{1\text{ in.}}{2.54\ \cancel{\text{cm}}}\right) = \frac{150}{(10)(2.54)}\text{ in.}$$
$$\approx 5.91\text{ in.}$$

▲ Key West, Florida

EXAMPLE 8 *Southernmost Point*

The photo shows that the southernmost point in the continental United States, in Key West, Florida, is 90 miles away from Cuba. Determine this distance in kilometers.

SOLUTION We need to convert miles to kilometers. From Table 8.7, we find the unit fractions.

$$\frac{1 \text{ mi}}{1.6 \text{ km}} \quad \text{or} \quad \frac{1.6 \text{ km}}{1 \text{ mi}}$$

Since we are converting 90 miles to kilometers, the miles in the unit fraction must go in the denominator. Thus, we use the unit fraction 1.6 km/l mi.

$$90 \text{ mi} = 90 \cancel{\text{mi}}\left(\frac{1.6 \text{ km}}{1 \cancel{\text{mi}}}\right) = 144 \text{ km}$$

Therefore, 90 miles is equal to 144 kilometers.

EXAMPLE 9 *Understanding the Label*

The label on a bottle of Vicks Formula 44D Cough Syrup indicates that the active ingredient is dextromethorphan hydrobromide and that 5 mℓ (or 1 teaspoon) contains 10 mg of this ingredient. If the recommended dosage for adults is 3 teaspoons, determine the following.

a) How many milliliters of cough medicine should be taken?

b) How many milligrams of the active ingredient should be taken?

c) If the bottle contains 8 fluid ounces of medicine, how many milligrams of the active ingredient are in the bottle?

SOLUTION

a) Since each teaspoon contains 5 mℓ and 3 teaspoons should be taken, 15 mℓ of the cough medicine should be taken.

$$3 \text{ tsp} = (3 \cancel{\text{tsp}})\left(\frac{5 \text{ m}\ell}{1 \cancel{\text{tsp}}}\right) = 15 \text{ m}\ell$$

b) Since each teaspoon contains 10 mg of the active ingredient, 30 mg of the active ingredient should be taken.

$$3 \text{ tsp} = (3 \cancel{\text{tsp}})\left(\frac{10 \text{ mg}}{1 \cancel{\text{tsp}}}\right) = 30 \text{ mg}$$

c) Table 8.7 shows that each fluid ounce contains 30 mℓ. Since each 5 mℓ contains 10 mg of the active ingredient, we can work the problem as follows.

$$8 \text{ fl oz} = (8 \cancel{\text{fl oz}})\left(\frac{30 \cancel{\text{m}\ell}}{1 \cancel{\text{fl oz}}}\right)\left(\frac{10 \text{ mg}}{5 \cancel{\text{m}\ell}}\right) = \frac{8(30)(10)}{5} \text{ mg} = 480 \text{ mg}$$

Therefore, there are 480 mg (or 0.48 g) of the active ingredient in the bottle of cough syrup.

EXAMPLE 10 *Determining Dosage by Weight*

Drug dosage is often administered according to a patient's weight. For example, 30 mg of the drug vancomicin is to be given for each kilogram of a person's weight. If Martha Greene, who weighs 136 lb, is to be given the drug, what dosage should she be given?

SOLUTION First we need to convert Martha's weight into kilograms. From Table 8.7, we see that 1 lb = 0.45 kg. We obtain our unit fraction from this information. Then we need to determine the number of milligrams of the drug for Martha's weight in kilograms. To do so, write the given ratio of 30 mg of the drug for each kilogram as $\frac{30\text{ mg}}{1\text{ kg}}$. Note that this ratio is not a unit fraction since the numerator and denominator are not equivalent. The answer may be found as follows.

$$136\text{ lb} = (136\text{ }\cancel{\text{lb}})\left(\frac{0.45\text{ }\cancel{\text{kg}}}{1\text{ }\cancel{\text{lb}}}\right)\left(\frac{30\text{ mg}}{1\text{ }\cancel{\text{kg}}}\right) = (136)(0.45)(30)\text{ mg} = 1836\text{ mg}$$

Thus, 1836 mg, or 1.836 g, of the drug should be given. •

MATHEMATICS TODAY

Metrics on the Moon

On Earth, only the United States, Liberia, and Myanmar (formerly Burma) still use U.S. customary units instead of metric units. And now the moon will be metric, too. On January 8, 2007, NASA and 13 other space agencies decided to use the metric system for all lunar exploration programs. All the agencies agreed that using a single measurement system would make the human habitants and vehicles placed on the moon by different space agencies more compatible with one another. That could come in handy if, say, one agency's moon base needs emergency spare parts from another agency's base. In addition, metric standardization will make it easier for countries to form new partnerships and collaborations after their lunar operations are already in place.

SECTION 8.4 EXERCISES

CONCEPT/WRITING EXERCISES

1. What is dimensional analysis?

2. What is a unit fraction?

3. Give a unit fraction that relates seconds and minutes. Explain how you determined the unit fraction.

4. Give a unit fraction that relates feet and yards. Explain how you determined the unit fraction.

5. When converting from kilograms to pounds, which unit fraction would you use? Explain.

$$\frac{1\text{ lb}}{0.45\text{ kg}} \quad \text{or} \quad \frac{0.45\text{ kg}}{1\text{ lb}}$$

6. When converting from centimeters to feet, which unit fraction would you use? Explain.

$$\frac{1\text{ ft}}{30\text{ cm}} \quad \text{or} \quad \frac{30\text{ cm}}{1\text{ ft}}$$

7. When converting from gallons to liters, which unit fraction would you use? Explain.

$$\frac{1\text{ gal}}{3.8\ \ell} \quad \text{or} \quad \frac{3.8\ \ell}{1\text{ gal}}$$

8. When converting from square yards to square meters, which unit fraction would you use? Explain.

$$\frac{1\text{ yd}^2}{0.8\text{ m}^2} \quad \text{or} \quad \frac{0.8\text{ m}^2}{1\text{ yd}^2}$$

PRACTICE THE SKILLS

In Exercises 9–24, convert the quantity to the indicated units. When appropriate, give your answer to the nearest hundredth.

9. 62 in. to centimeters

10. 9 lb to kilograms

11. 4.2 ft to meters

12. 427 g to ounces

13. 120 kg to pounds

14. 20 yd^2 to square meters

15. 39 mi to kilometers

16. 765 mm to inches

17. 675 ha to acres

18. 192 oz to grams

19. 15.6 ℓ to pints

20. 4 T to tonnes

21. 3.8 km^2 to square miles

22. 25.6 mℓ to fluid ounces

23. 120 lb to kilograms

24. 6.2 acres to hectares

In Exercises 25–32, replace the measurement(s) indicated in blue with an equivalent metric measure(s). For example, a foot could be replaced with 30 cm.

25. More bounce to the ***ounce***.

26. An ***ounce*** of prevention is worth a ***pound*** of cure.

27. He demanded his ***pound*** of flesh.

28. ***Five foot two*** and eyes of blue.

29. Give him an ***inch*** and he'll take a ***mile***.

30. A miss is as good as a ***mile***.

31. The longest ***yard***.

32. First down and ***10 yards*** to go.

In Exercises 33–36, use the part of the scorecard that shows the distance in meters for the first four holes of the Millbrook Resort Golf Course in Queenstown, New Zealand. Determine the distances indicated.

HOLE	BLACK Tees	BLUE Tees	HANDICAP	PAR					WHITE Tees	RED Tees
1	505	505	3	5					466	414
2	185	175	15	3					137	91
3	366	357	11	4					344	287
4	396	376	7	4					376	303

33. Hole 1, black tees, in yards

34. Hole 2, blue tees, in yards

35. Hole 3, white tees, in feet

36. Hole 4, red tees, in feet

PROBLEM SOLVING

37. ***Speed Limit*** The photo shows that the speed limit is 60 kph. Determine the speed in miles per hour.

38. ***How Far?*** Carol Ann Harle's new car traveled 105 mi on 5 gal of gasoline. How many kilometers can Carol Ann's car travel with the same amount of gasoline?

39. ***Buying Carpet*** Victoria Montoya is buying outdoor carpet for her lanai, which is 6 yd by 9 yd. The carpeting is sold in square meters. How many square meters of carpeting will she need?

40. ***Cincinnati to Columbus*** The distance from Cincinnati, Ohio, to Columbus, Ohio, is about 110 mi. What is the distance in kilometers?

41. ***The QEW*** Part of the Queen Elizabeth Way in Canada has a speed limit of 80 kph. What is the speed in miles per hour?

42. ***Milliliters in a Glass*** A glass holds 8 fl oz. How many milliliters will it hold?

43. ***Poison Dart Frog*** A full-grown strawberry poison dart frog (see photo) has a weight of about 6 g. What is its weight in ounces?

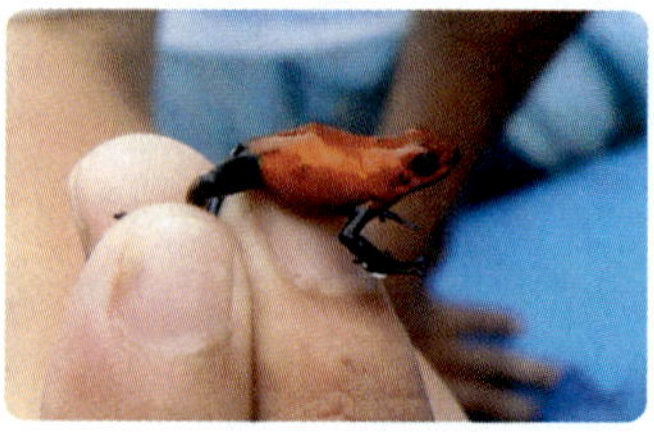

44. ***Swimming Pool*** A swimming pool holds 12,500 gal of water. What is this volume in kiloliters?

45. ***Building a Basement*** A basement is to be 50 ft long, 30 ft wide, and 8 ft high. How much dirt will have to be removed when this basement is built? Answer in cubic meters.

46. ***Area of Yosemite National Park*** Yosemite National Park has an area of 1189 mi^2. What is its area in square kilometers?

47. ***Cost of Rice*** If rice costs \$1.10 per kilogram, determine the cost of 1 pound of rice.

48. ***Weight Restriction*** The weight restriction on a road in France is 3.5 t.

a) How many tons does this weight equal?

b) How many pounds?

49. ***Capacity of a Tank Truck*** A tank truck holds 34.5 kℓ of gasoline. How many gallons does it hold?

50. ***Cost per Gram*** A 0.25 oz bottle of Chanel perfume costs \$80. What is the cost per gram?

51. ***Car's Engine*** A specific car's engine has a capacity of 5.7 ℓ of oil. How many quarts of oil does the engine have?

52. ***A Precious Stone*** One gram is the same as 5 carats. David Erich's new ring contains a precious stone that is $\frac{1}{8}$ carat. Find the weight of the stone in grams.

53. ***Death Valley Elevation*** The lowest elevation in the United States is −282 ft at Badwater in Death Valley, California. Determine this elevation in

a) centimeters.

b) meters.

54. ***Sign in Costa Rica*** Using the sign in the photo at the right, determine the distance to Lecheria #2 in

a) kilometers.

b) feet.

55. ***Square Meters to Square Feet*** One meter is about 3.3 ft. Use this information to determine

a) the equivalent of one square meter in square feet.

b) the equivalent of one cubic meter in cubic feet.

56. ***Acres to Hectares*** One acre is about 0.4 hectare.

a) Use this information to determine the equivalent of 15.3 acres in hectares.

b) If 1 hectare is 10 000 m^2, determine the area of the 15.3 acres in square meters.

57. ***Dosage for a Child*** The recommended dosage of the drug codeine for pediatric patients is 1 mg per kilogram of a child's weight. What dosage of codeine should be given to April Adam, who weighs 56 lb?

58. ***Dosage for a Man*** For each kilogram of a person's weight, 1.5 mg of the antibiotic drug gentamicin is to be administered. If Ron Gigliotti weighs 170 lb, how much of the drug should he receive?

59. ***Ampicillin*** The recommended dosage of the drug ampicillin for pediatric patients is 200 mg per kilogram of a patient's weight. If Janine Baker weighs 76 lb, how much ampicillin should she receive?

60. ***Medicine for a Dog*** For each kilogram of weight of a dog, 5 mg of the drug bretylium is to be given. If Blaster, an Irish setter, weighs 82 lb, how much of the drug should be given?

61. ***Active Ingredients*** The label on the bottle of Triaminic expectorant indicates that each teaspoon (5 mℓ) contains 12.5 mg of the active ingredient phenylpropanolamine hydrochloride.

a) Determine the amount of the active ingredient in the recommended adult dosage of 2 teaspoons.

b) Determine the quantity of the active ingredient in a 12 oz bottle.

62. ***Stomach Ache Remedy*** The label on the bottle of Maximum Strength Pepto-Bismol indicates that each tablespoon contains 236 mg of the active ingredient bismuth subsalicyate.

a) Determine the amount of the active ingredient in the recommended dosage of 2 tablespoons.

b) If the bottle contains 8 fl oz, determine the quantity of the active ingredient in the bottle.

63. ***Disney Magic*** The Disney Magic Cruise Ship is 964 feet long, has a weight of 85,000 tons, and can travel 28 mph.

a) Determine the length of the ship in meters.

b) Determine the weight in tonnes.

c) Determine the speed in kilometers per hour.

64. ***Making Cookies*** Change all the measurements in the cookie recipe to metric units. Do not forget pan size, temperature, and size of cookies.

Magic Cookie Bar
$\frac{1}{2}$c graham cracker crumbs
12 oz nuts
8 oz chocolate pieces
$1\frac{1}{3}$c flaked coconut
$1\frac{1}{3}$c condensed milk

Coat the bottom of a 9 in. × 13 in. pan with melted margarine. Add rest of ingredients one by one: crumbs, nuts, chocolate, and coconut. Pour condensed milk over all. Bake at 350°F for 25 minutes. Allow to cool 15 minutes before cutting. Makes about two dozen $1\frac{1}{2}$in. by 3 in. bars.

65. ***Peppers*** The photo on the top right shows that in Rome, Italy, peppers cost 2 euros (€) per kilogram.

a) Determine the cost per pound, in euros, for the peppers.

b) If 1 € can be converted to $1.30 U.S., determine the cost, in U.S. dollars, of 1 pound of peppers.

▲ See Exercise 65

66. ***Curry*** The photo shows Indian curry for sale at a street market in Albi, France.

a) Indian curry cost 7 € per 100 grams. Determine the price, in euros, of 1 kg of curry.

b) Determine the cost in euros of 1 pound of curry.

c) If 1 € can be converted to $1.30 U.S., determine the cost, in U.S. dollars, of 1 pound of curry.

CHALLENGE PROBLEMS/GROUP ACTIVITIES

67. ***Nursing Question*** The following question was selected from a nursing exam. Can you answer it?

In caring for a patient after delivery, you are to give 0.2 mg Ergotrate Maleate. The ampule is labeled $\frac{1}{300}$ grain/mℓ. How much would you draw and give? (60 mg = 1 grain)

a) 15 cc **b)** 1.0 cc **c)** 0.5 cc **d)** 0.01 cc

68. ***How Much Beef*** Paul Gosse is planning a picnic and plans on purchasing 0.18 kg of ground beef for each 100 lb of weight of guests who will be in attendance. If he expects 15 people whose average weight is 130 lb, how many pounds of beef should he purchase?

69. ***An Auto Engine*** The displacement of automobile engines is measured in liters. A 2007 Ford Explorer has a 4.0 ℓ engine.

a) Determine the displacement of the engine in cubic centimeters.

b) Determine the displacement of the engine in cubic inches.

RECREATIONAL MATHEMATICS

In Exercises 70–75, answer the question, What metric unit am I?

70. I am a length greater than a yard, but less than a kitchen tabletop.

71. I am a weight greater than a calculator, but less than a large bottle of ketchup.

72. I am an area greater than an acre, but less than a square kilometer.

73. I am a liquid volume greater than a quart, but less than a gallon.

74. I am a weight greater than a ton, but less than a full-grown elephant.

75. I am a length greater than an inch, but less than a yard.

In Exercises 76–85, try to solve the puzzle. What is

76. 1 millionth of a mouthwash?

77. 2000 pounds of Chinese soup?

78. 448 grams of cake?

79. 1000 aches?

80. 1 million bicycles?

81. 1 million phones?

82. 10 cards?

83. 2000 mockingbirds?

84. 1 millionth of a fish?

85. 10 rations?

CHAPTER 8 SUMMARY

IMPORTANT FACTS

Metric Units

Prefix	Symbol	Meaning
kilo	k	1000 × base unit
hecto	h	100 × base unit
deka	da	10 × base unit
—	—	base unit
deci	d	$\frac{1}{10}$ of base unit
centi	c	$\frac{1}{100}$ of base unit
milli	m	$\frac{1}{1000}$ of base unit

Water

Volume in Cubic Units		Volume in Liters		Mass of Water
1 cm^3	=	1 mℓ	=	1g
1 dm^3	=	1 ℓ	=	1 kg
1 m^3	=	1 kℓ	=	1 t (1000 kg)

Temperature

$$°C = \frac{5}{9}(°F - 32)$$

$$°F = \frac{9}{5}°C + 32$$

CHAPTER 8 REVIEW EXERCISES

8.1

In Exercises 1–6, indicate the meaning of the prefix.

1. Centi **2.** Kilo **3.** Milli

4. Hecto **5.** Deka **6.** Deci

In Exercises 7–12, change the given quantity to that indicated.

7. 40 mg to grams

8. 3.2 ℓ to centiliters

9. 0.0016 cm to millimeters

10. 1 000 000 mg to kilograms

11. 4.62 kℓ to liters

12. 192.6 dag to decigrams

In Exercises 13 and 14, arrange the quantities from smallest to largest.

13. 2.67 kℓ, 3000 mℓ, 14 630 cℓ

14. 0.047 km, 4700 m, 47 000 cm

8.2, 8.3

In Exercises 15–24, indicate the metric unit of measurement that would best express the following.

15. The diameter of a pizza

16. The mass of a cellular telephone

17. The temperature of the sun's surface

18. The diameter of a quarter

19. The area of a room of a house

20. The volume of a glass of milk

21. The length of an ant

22. The mass of a car

23. The distance from Miami, Florida, to Los Angeles, California

24. The height a dolphin can jump

In Exercises 25 and 26, (a) first estimate the following in metric units and then (b) measure with a metric ruler. Record your results.

25. Your height

26. The length of a new pencil

In Exercises 27–32, select the best answer.

27. The length of the distance between New York City and Boston, Massachusetts, is about

a) 3100 m. **b)** 3100 km. **c)** 310 km.

28. The mass of a full-grown border collie is about

a) 600 g. **b)** 20 kg. **c)** 100 kg.

29. The volume of a gallon of gasoline is about

a) 0.1 kℓ. **b)** 0.5 ℓ. **c)** 4 ℓ.

30. The area of a large vegetable garden in a person's yard may be

a) 200 m^2. **b)** 0.5 ha. **c)** 0.02 km^2.

31. The temperature on a hot summer day in Georgia may be

a) 34°C. **b)** 55°C. **c)** 25°C.

32. The height of a giant sequoia tree is about

a) 300 m. **b)** 3000 cm. **c)** 0.3 m.

33. Convert 3600 kg to tonnes.

34. Convert 4.3 t to grams.

35. The temperature on the thermostat shown is 24°C. What is the Fahrenheit temperature?

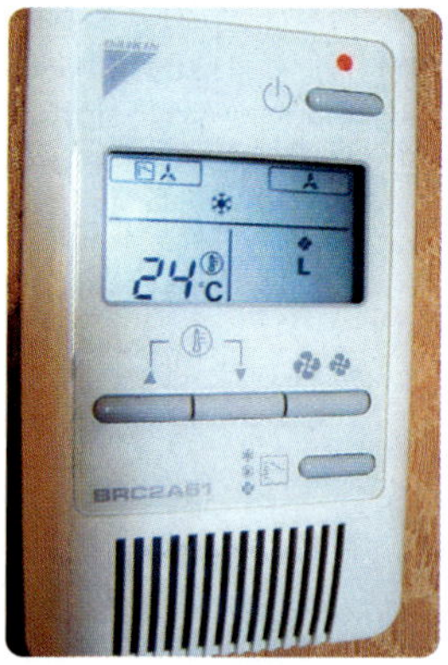

36. If the room temperature is 68°F, what is the Celsius temperature?

37. If your outdoor thermometer shows a temperature of −6°F, what is the Celsius temperature?

38. If Lynn Colgin's body temperature is 39°C, what is her Fahrenheit temperature?

39. Measure, in centimeters, each of the line segments, then compute the area of the figure.

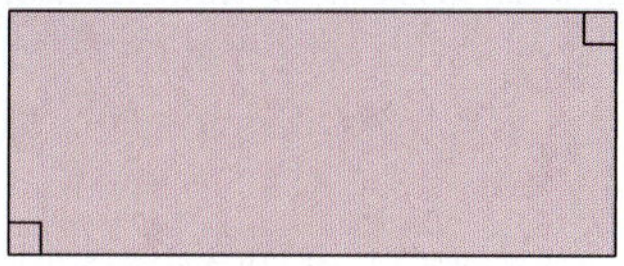

40. Measure, in centimeters, the radius of the circle, then compute the area of the circle.

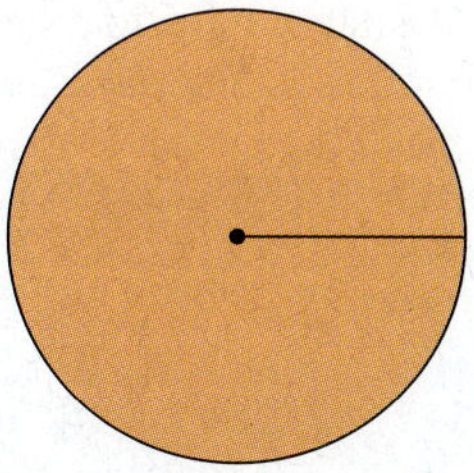

41. a) *A Swimming Pool's Volume* What is the volume of water in a full rectangular swimming pool that is 10 m long and 4 m wide and has an average depth of 2 m? Answer in cubic meters.

b) What is the mass of the water in kilograms?

42. *Computer Monitor* A rectangular computer monitor's screen measures 33.7 cm by 26.7 cm. Determine the area in

a) square centimeters.

b) square meters.

43. *Volume of a Fish Tank* A small fish tank measures 80 cm long, 40 cm wide, and 30 cm high.

a) What is its volume in cubic centimeters?

b) What is its volume in cubic meters?

c) How many milliliters of water will the tank hold?

d) How many kiloliters of water will the tank hold?

44. A square kilometer is a square with length and width both 1 km. How many times larger is a square kilometer than a square dekameter?

8.4

In Exercises 45–58, change the given quantity to the indicated quantity. When appropriate, round answers to the nearest hundredth.

45. 25 cm = ____ in.

46. 105 kg = ____ lb

47. 83 yd = ____ m

48. 100 m = ____ yd

49. 45 mph = ____ kph

50. 60 ℓ = ____ qt

51. 20 gal = ____ ℓ

52. 60 m^3 = ____ yd^3

53. 83 cm^2 = ____ $in.^2$

54. 4 qt = ____ ℓ

55. 15 yd^3 = ____ m^3

56. 62 mi = ____ km

57. 27 cm = ____ ft

58. $3\frac{1}{4}$ in. = ____ mm

59. *Building a Chimney* Anne Kelly bought 700 bricks to build a chimney. Each brick has a mass of 1.5 kg.

a) What is the total mass of the bricks in kilograms?

b) What is the total weight of the bricks in pounds?

60. *Carpeting a Room* Patricia Burgess is buying new carpet for her family room. The room is 15 ft wide and 24 ft long. The carpeting is sold only in square meters. How many square meters of carpeting will she need? Round your answer to the nearest tenth of a square meter.

61. *Milk Tank* A cylindrical milk tank can store 50,000 gal of milk.

a) Determine the volume in kiloliters.

b) Estimate the weight of the milk in kilograms. Assume that milk has the same weight as water.

62. *The Speed Limit* The speed limit on a certain road is 65 mph. What is the speed limit in

a) kilometers per hour?

b) meters per hour?

63. *A Water Tank* A rectangular tank used to test leaks in tires is 90 cm long by 70 cm wide by 40 cm deep.

a) Determine the number of liters of water the tank holds.

b) What is the mass of the water in kilograms?

64. *Oranges* If the cost of oranges is \$3.50 per kilogram, determine the cost of 1 lb of oranges.

CHAPTER 8 TEST

1. Change 497 cℓ to daℓ.

2. Change 273 hm to cm.

3. How many times greater is a kilometer than a dekameter?

4. ***Jogging*** A high school track is an oval that measures 400 m around. If Dave Camp jogs around the track six times, how many kilometers has he gone?

In Exercises 5–9, choose the best answer.

5. The length of this page is about

a) 10 cm.

b) 25 cm.

c) 60 cm.

6. The surface area of the top of a kitchen table is about

a) 2 m^2.

b) 200 cm^2.

c) 2000 cm^2.

7. The amount of gasoline that an automobile's gas tank can hold is about

a) 200 ℓ.

b) 20 ℓ.

c) 75 ℓ.

8. The mass of a cell phone is about

a) 0.1 t.

b) 2 kg.

c) 150 g.

9. The outside temperature on a snowy day is about

a) 18°C.

b) −2°C.

c) −40°C.

10. How many times greater is a square meter than a square centimeter?

11. How many times greater is a cubic meter than a cubic millimeter?

12. Convert 225 oz. to grams.

13. ***Distance*** The photo of the sign shows distances, in kilometers, from Positano, Italy, to other nearby Italian cities. How far, in miles, is Salerno from the sign?

14. Change −15°F to degrees Celsius.

15. Change 20°C to degrees Fahrenheit.

16. ***Elevator*** A sign in an elevator in Nice, France, indicates that its maximum capacity is 4 persons or 300 kg. Determine the maximum capacity in

a) grams.

b) pounds.

17. ***At the Aquarium*** A fish tank at an aquarium is 20 m long by 20 m wide by 8 m deep.

a) Determine the volume of the tank in cubic meters.

b) Determine the number of liters of water the tank holds.

c) Determine the weight of the water in kilograms.

18. ***Cost of Paint*** The first coat of paint for the outside walls of a building requires 1 ℓ of paint for each 10 m^2 of wall surface. The second coat requires 1 ℓ for every 15 m^2. If the paint costs \$3.50 per liter, what will be the cost of two coats of paint for the four outside walls of a building 20 m long, 15 m wide, and 6 m high?

GROUP PROJECTS

HEALTH AND MEDICINE

Throughout this chapter, we have shown the importance of the metric system in the medical professions. The following two questions involve applications of the metric system to medicine.

1. a) Twenty milligrams of the drug lincomycin is to be given for each kilogram of a person's weight. The drug is to be mixed with 250 cc of a normal saline solution, and the mixture is to be administered intravenously over a 1 hr period. Clyde Dexter, who weighs 196 lb, is to be given the drug. Determine the dosage of the drug he will be given.

b) At what rate per minute should the 250 cc solution be administered?

2. a) At a pharmacy, a parent asks a pharmacist why her child needs such a small dosage of a certain medicine. The pharmacist explains that a general formula may be used to estimate a child's dosage of certain medicines. The formula is

$$\text{Child's dose*} = \frac{\begin{pmatrix}\text{child's weight} \\ \text{in kilograms}\end{pmatrix}}{67.5 \text{ kg}} \times \text{adult dose}$$

What is the amount of medicine you would give a 60 lb child if the adult dosage of the medicine is 70 mg?

b) At what weight, in pounds, would the child receive an adult dose?

TRAVELING TO OTHER COUNTRIES

3. Dale Pollinger is a buyer at General Motors and travels frequently on business to foreign countries. He always plans ahead and does his holiday shopping overseas, where he can purchase items not easily found in the United States.

a) On a trip to Tokyo, he decides to buy a kimono for his sister, Kathy. To determine the length of a kimono, one measures, in centimeters, the distance from the bottom of a person's neck to 5 cm above the floor. If the distance from the bottom of Kathy's neck to the floor is 5 ft 2 in., calculate the length of the kimono that Dale should purchase.

b) If the conversion rate at the time is 1 U.S. dollar = 117.25 yen and the kimono cost 8695.5 yen, determine the cost of the kimono in U.S. dollars.

c) On a trip to Mexico City, Mexico, Dale finds a small replica of a Mayan castle that he wants to purchase for his wife, Sue. He is going directly from Mexico to Rome, so he wants to mail the castle back to the United States. The mailing rate from Mexico to the United States is 10 pesos per hundred grams. Determine the mailing cost, in U.S. dollars, if the castle weighs 6 lb and the exchange rate is 1 peso = 0.092 U.S. dollar.

d) This question has three parts. While traveling to Italy, Dale finds that unleaded gasoline cost 1.238 € per liter.

▲ Positano, Italy on the Amalfi Coast

1) How much will it cost him, in euros, to fill the 53 ℓ gas tank of his rented car? **2)** If the exchange rate is 1 euro = \$1.17 U.S., what will it cost in U.S. dollars to fill the tank? **3)** What is the cost, in U.S. dollars, of a gallon of gasoline at this gas station?

*Consult a pediatrician before giving a child medicine.

CHAPTER 9

Geometry

▲ Baseballs and baseball diamonds are examples of geometric objects.

WHAT YOU WILL LEARN

- Points, lines, planes, and angles
- Polygons, similar figures, and congruent figures
- Perimeter and area
- Pythagorean theorem
- Circles
- Volume
- Transformational geometry, symmetry, and tessellations
- The Möbius strip, Klein bottle, and maps
- Non-Euclidean geometry and fractal geometry

WHY IT IS IMPORTANT

Many objects we encounter each day can be described in terms of geometry. Spherical baseballs hit on square baseball "diamonds," cylindrical cans of soda, and rectangular computer screens are all examples of geometric objects we frequently encounter. In addition to these items, scientists also use geometry to conduct research and create items that improve our lives. For example, the antenna on your cell phone is an example of a geometric object known as a *fractal*. In this chapter, we will study many geometric objects that we see in our everyday lives.

9.1 POINTS, LINES, PLANES, AND ANGLES

▲ Billiard games involve many geometric concepts.

Playing billiards involves many geometric concepts. In this section, we will introduce the geometric terms *points, lines, planes*, and *angles*. In a typical billiard game, each of these terms is evident. The billiard balls rest on specific points on the table, which is part of a plane. After being struck with a cue, the ball travels along a path that is part of a line. Angles are involved in the path the ball takes when the ball hits a table bumper or another ball. The concepts we will discuss in this section form an important basis for the study of geometry.

Human beings recognized shapes, sizes, and physical forms long before geometry was developed. Geometry as a science is said to have begun in the Nile Valley of ancient Egypt. The Egyptians used geometry to measure land and to build pyramids and other structures.

The word *geometry* is derived from two Greek words, *ge*, meaning earth, and *metron*, meaning measure. Thus, geometry means "earth measure" or "measurement of the earth."

Unlike the Egyptians, the Greeks were interested in more than just the applied aspects of geometry. The Greeks attempted to apply their knowledge of logic to geometry. In about 600 B.C., Thales of Miletus was the first to be credited with using deductive methods to develop geometric concepts. Another outstanding Greek mathematician, Pythagoras, continued the systematic development of geometry that Thales had begun.

In about 300 B.C., Euclid (see Profile in Mathematics on page 516) collected and summarized much of the Greek mathematics of his time. In a set of 13 books called *Elements*, Euclid laid the foundation for plane geometry, which is also called ***Euclidean geometry***.

Euclid is credited with being the first mathematician to use the ***axiomatic method*** in developing a branch of mathematics. First, Euclid introduced ***undefined terms*** such as point, line, plane, and angle. He related these to physical space by such statements as "A line is length without breadth" so that we may intuitively understand them. Because such statements play no further role in his system, they constitute primitive or undefined terms.

Second, Euclid introduced certain ***definitions***. The definitions are introduced when needed and are often based on the undefined terms. Some terms that Euclid introduced and defined include triangle, right angle, and hypotenuse.

Third, Euclid stated certain primitive propositions called ***postulates*** (now called ***axioms****) about the undefined terms and definitions. The reader is asked to accept these statements as true on the basis of their "obviousness" and their relationship with the physical world. For example, the Greeks accepted all right angles as being equal, which is Euclid's fourth postulate.

Fourth, Euclid proved, using deductive reasoning (see Section 1.1), other propositions called ***theorems***. One theorem that Euclid proved is known as the Pythagorean theorem: "The sum of the areas of the squares constructed on the arms of a right triangle

*The concept of the axiom has changed significantly since Euclid's time. Now any statement may be designated as an axiom, whether it is self-evident or not. All axioms are *accepted* as true. A set of axioms forms the foundation for a mathematical system.

PROFILE IN MATHEMATICS

Euclid

Euclid (320–275 B.C.) lived in Alexandria, Egypt, and was a teacher and scholar at Alexandria's school called the *Museum*. It was here that Euclid collected and arranged many of the mathematical results known at the time. This collection of works became his 13-volume masterpiece known as *Elements*. Beginning with a list of definitions, postulates, and axioms, Euclid proved one theorem after another, using only previously proven results. This method of proof became a model of mathematical and scientific investigation that survives today. Remarkably, the geometry in *Elements* does not rely on making exact geometric measurements using a ruler or protractor. Rather, the work is developed using only an unmarked straightedge and a drawing compass. Next to the Bible, Euclid's *Elements* may be the most translated, published, and studied of all the books produced in the Western world.

is equal to the area of the square constructed on the hypotenuse." He also proved that the sum of the angles of a triangle is 180°.

Using only 10 axioms, Euclid deduced 465 propositions (or theorems) in plane and solid geometry, number theory, and Greek geometric algebra.

Point and Line

Three basic terms in geometry are ***point***, ***line***, and ***plane***. These three terms are not given a formal definition, but we recognize points, lines, and planes when we see them.

Let's consider some properties of a line. Assume that a line means a straight line unless otherwise stated.

1. A line is a set of points. Each point is on the line and the line passes through each point. When we wish to refer to a specific point, we will label it with a single capital letter. For example, in Figure 9.1(a) three points are labled A, B, and C, respectively.
2. Any two distinct points determine a unique line. Figure 9.1(a) illustrates a line. The arrows at both ends of the line indicate that the line continues in each direction. The line in Fig. 9.1(a) may be symbolized with any two points on the line by placing a line with a double-sided arrow above the letters that correspond to the points, such as $\overleftrightarrow{AB}$, $\overleftrightarrow{BA}$, $\overleftrightarrow{AC}$, $\overleftrightarrow{CA}$, $\overleftrightarrow{BC}$, or $\overleftrightarrow{CB}$.
3. Any point on a line separates the line into three parts: the point itself and two ***half lines*** (neither of which includes the point). For example, in Fig. 9.1(a) point B separates the line into the point B and two half lines. Half line BA, symbolized $\overset{\circ\!\rightarrow}{BA}$, is illustrated in Fig. 9.1(b). The open circle above the B indicates that point B is not included in the half line. Figure 9.1(c) illustrates half line BC, symbolized $\overset{\circ\!\rightarrow}{BC}$.

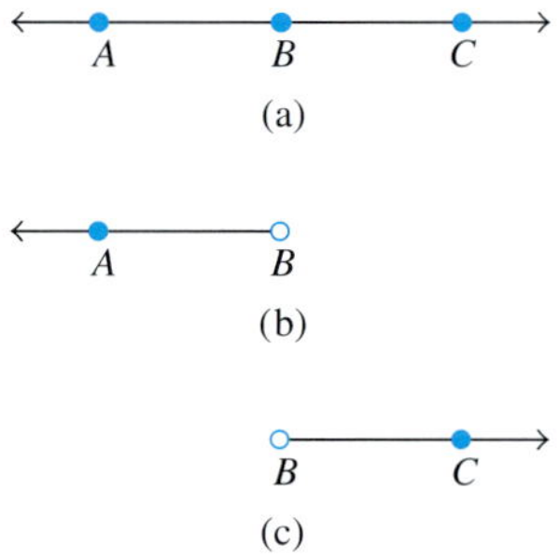

Figure 9.1

Look at the half line $\overset{\circ\!\rightarrow}{AB}$ in Fig. 9.2(b) on page 517. If the ***end point***, A, is included with the set of points on the half line, the result is called a ***ray***. Ray AB, symbolized $\overrightarrow{AB}$, is illustrated in Fig. 9.2(c). Ray BA, symbolized $\overrightarrow{BA}$, is illustrated in Fig. 9.2(d).

A ***line segment*** is that part of a line between two points, including the end points. Line segment AB, symbolized $\overline{AB}$, is illustrated in Fig. 9.2(e).

An open line segment is the set of points on a line between two points, excluding the end points. Open line segment AB, symbolized $\overset{\circ\!-\!\circ}{AB}$, is illustrated in Fig. 9.2(f).

Figure 9.2(g) illustrates two half open line segments, symbolized $\overset{\circ\!-}{AB}$ and $\overset{-\!\circ}{AB}$.

DID YOU KNOW?

Compass and Straightedge Constructions

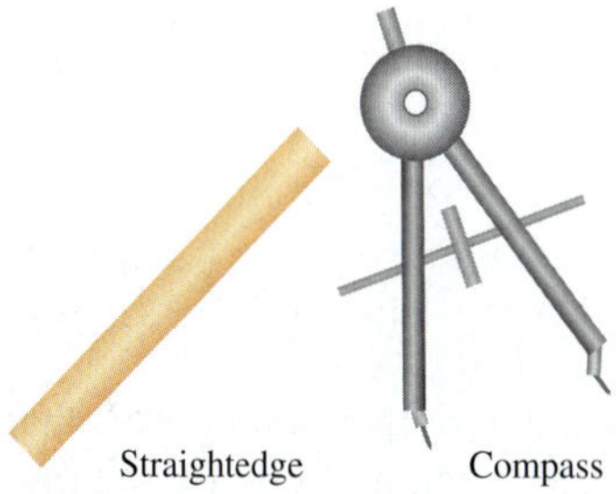

Geometric constructions were central to ancient Greek mathematics. Although these constructions are often referred to as *Euclidean constructions*, they were used centuries before Euclid wrote his classic work, *Elements*. The tools *allowed* in geometric constructions are a pencil, an unmarked straightedge, and a drawing compass. The straightedge is used to draw line segments, and the compass is used to draw circles and arcs. The Internet has many sites devoted to classic geometric constructions. Exercise 105 shows you how to construct a triangle with sides of equal length (an equilateral triangle).

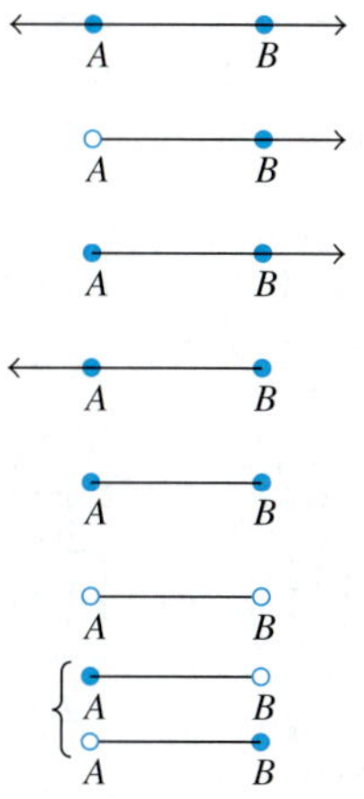

Description	Symbol
(a) Line AB	$\overleftrightarrow{AB}$
(b) Half line AB	$\overset{\circ\!\rightarrow}{AB}$
(c) Ray AB	$\overrightarrow{AB}$
(d) Ray BA	$\overrightarrow{BA}$
(e) Line segment AB	$\overline{AB}$
(f) Open line segment AB	$\overset{\circ\!-\!\circ}{AB}$
(g) Half open line segments AB	$\overset{-\!\circ}{AB}$, $\overset{\circ\!-}{AB}$

Figure 9.2

In Chapter 2, we discussed intersection of sets. Recall that the intersection (symbolized $\cap$) of two sets is the set of elements (points in this case) common to both sets.

Consider the rays $\overrightarrow{AB}$ and $\overrightarrow{BA}$ in Fig. 9.3(a). The intersection of $\overrightarrow{AB}$ and $\overrightarrow{BA}$ is $\overline{AB}$. Thus, $\overrightarrow{AB} \cap \overrightarrow{BA} = \overline{AB}$.

We also discussed the union of two sets in Chapter 2. The union (symbolized $\cup$) of two sets is the set of elements (points in this case) that belong to either of the sets or both sets. The union of $\overrightarrow{AB}$ and $\overrightarrow{BA}$ is $\overleftrightarrow{AB}$ (Fig. 9.3b). Thus, $\overrightarrow{AB} \cup \overrightarrow{BA} = \overleftrightarrow{AB}$.

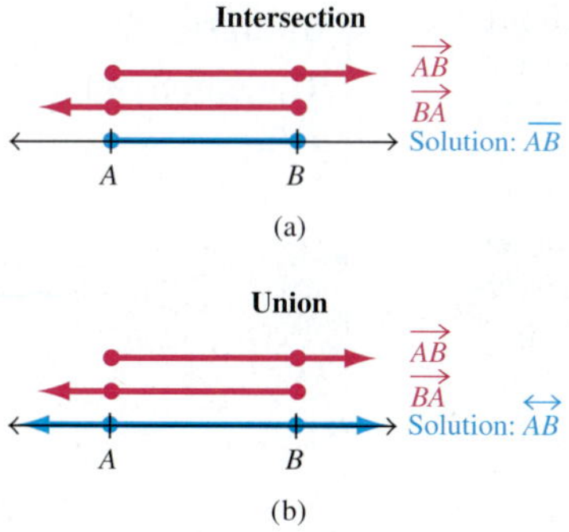

Figure 9.3

EXAMPLE 1 *Unions and Intersections of Parts of a Line*

Using line AD, determine the solution to each part.

a) $\overrightarrow{AB} \cap \overrightarrow{DC}$ b) $\overrightarrow{AB} \cup \overrightarrow{DC}$ c) $\overline{AB} \cap \overrightarrow{CD}$ d) $\overline{AD} \cup \overset{\circ\!\rightarrow}{CA}$

DID YOU KNOW?

Let There Be Light

A Kodak Camera advertisement appeared in the first issue of the *Photographic Herald and Amateur Sportsman*, in November 1889. The slogan "You press the button, we do the rest" summed up George Eastman's groundbreaking snapshot camera system.

Sunlight travels in straight lines, or rays. When these rays pass through a small hole of a darkened room, they converge and then spread out to form an inverted image of the sun on the opposite wall. In the sixteenth century, artists used lenses to focus images on paper, which could then be traced to create a reproduction. In the eighteenth century, chemists discovered that certain salts of silver would darken when exposed to light, effectively creating a negative image. For the population at large, it all came together in 1888 when 23-year-old George Eastman, using basic geometric concepts, developed the first modern camera intended for the "casual millions," which sold for $25 apiece.

SOLUTION

a) $\overrightarrow{AB} \cap \overrightarrow{DC}$

Ray *AB* and ray *DC* are shown below. The intersection of these two rays is that part of line *AD* that is a part of *both* ray *AB and* ray *DC*. The intersection of ray *AB* and ray *DC* is line segment *AD*.

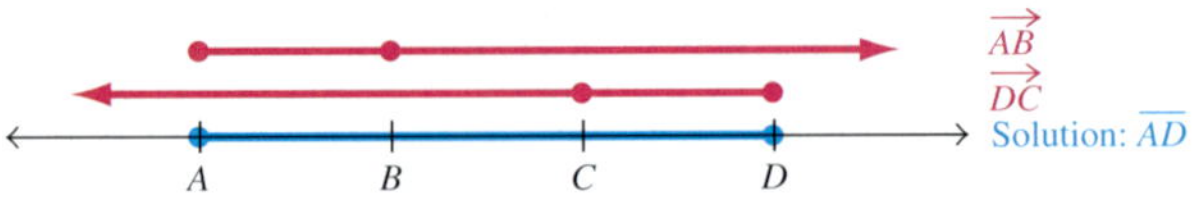

$$\overrightarrow{AB} \cap \overrightarrow{DC} = \overline{AD}$$

b) $\overrightarrow{AB} \cup \overrightarrow{DC}$

Once again ray *AB* and ray *DC* are shown below. The union of these two rays is that part of line *AD* that is part of *either* ray *AB or* ray *DC*. The union of ray *AB* and ray *DC* is the entire line *AD*.

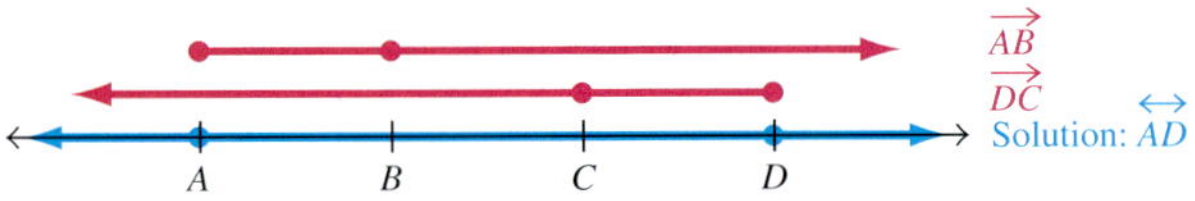

$$\overrightarrow{AB} \cup \overrightarrow{DC} = \overleftrightarrow{AD}$$

c) $\overline{AB} \cap \overrightarrow{CD}$

Line segment *AB* and ray *CD* have no points in common, so their intersection is empty.

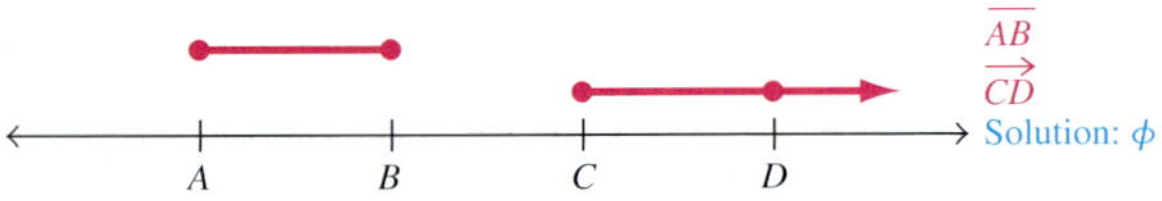

$$\overline{AB} \cap \overrightarrow{CD} = \emptyset$$

d) $\overline{AD} \cup \overset{\circ\rightarrow}{CA}$

The union of line segment *AD* and half line *CA* is ray *DA* (or equivalently, $\overrightarrow{DB}$ or $\overrightarrow{DC}$).

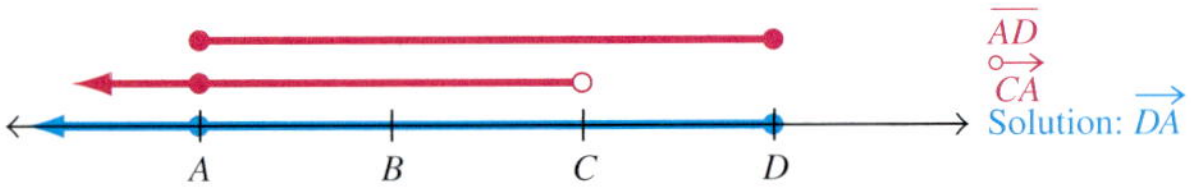

$$\overline{AD} \cup \overset{\circ\rightarrow}{CA} = \overrightarrow{DA}$$

Plane

The term *plane* is one of Euclid's undefined terms. For our purposes, we can think of a plane as a two-dimensional surface that extends infinitely in both directions, like an infinitely large blackboard. Euclidean geometry is called ***plane geometry*** because it is the study of two-dimensional figures in a plane.

Two lines in the same plane that do not intersect are called ***parallel lines***. Figure 9.4(a) on page 519 illustrates two parallel lines in a plane ($\overleftrightarrow{AB}$ is parallel to $\overleftrightarrow{CD}$).

Properties of planes include the following:

1. Any three points that are not on the same line (noncollinear points) determine a unique plane (Fig. 9.4b).
2. A line in a plane divides the plane into three parts, the line and two half planes (Fig. 9.4c).

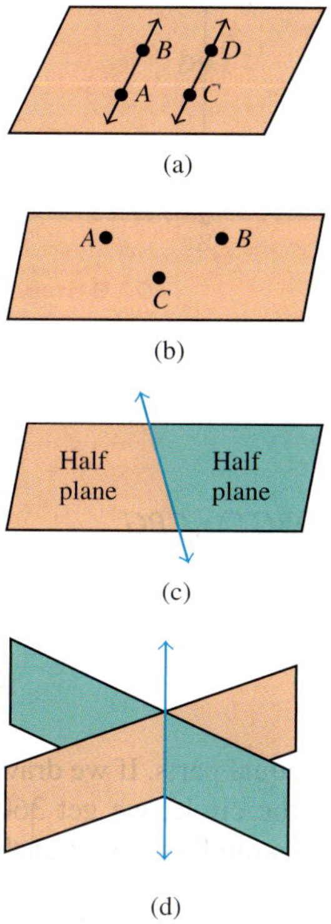

Figure 9.4

3. Any line and a point not on the line determine a unique plane.
4. The intersection of two distinct planes is a line (Fig. 9.4d).

Two planes that do not intersect are said to be ***parallel planes***. For example, in Fig. 9.5 plane ABE is parallel to plane GHF.

Two lines that do not lie in the same plane and do not intersect are called ***skew lines***. Figure 9.5 illustrates many skew lines (for example, $\overleftrightarrow{AB}$ and $\overleftrightarrow{CD}$).

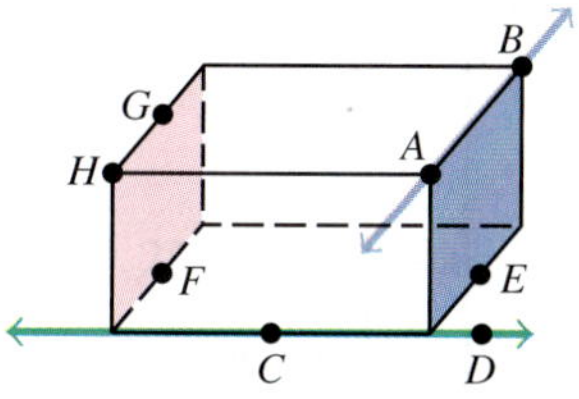

Figure 9.5

Angles

An ***angle***, denoted $\measuredangle$, is the union of two rays with a common end point (Fig. 9.6):

$$\overrightarrow{BA} \cup \overrightarrow{BC} = \measuredangle ABC \text{ (or } \measuredangle CBA\text{)}$$

An angle can be formed by the rotation of a ray about a point. An angle has an initial side and a terminal side. The initial side indicates the position of the ray prior to rotation; the terminal side indicates the position of the ray after rotation. The point common to both rays is called the ***vertex*** of the angle. The letter designating the vertex is always the middle one of the three letters designating an angle. The rays that make up the angle are called its ***sides***.

There are several ways to name an angle. The angle in Fig. 9.6 may be denoted

$$\measuredangle ABC, \qquad \measuredangle CBA, \qquad \text{or} \qquad \measuredangle B$$

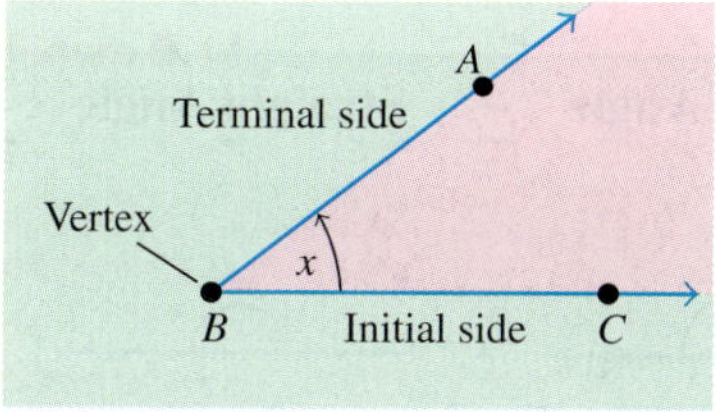

Figure 9.6

An angle divides a plane into three distinct parts: the angle itself, its interior, and its exterior. In Fig. 9.6, the angle is represented by the blue lines, the interior of the angle is shaded pink, and the exterior is shaded green.

The ***measure of an angle***, symbolized m, is the amount of rotation from its initial side to its terminal side. In Fig. 9.6, the letter x represents the measure of $\measuredangle ABC$; therefore, we may write $m\measuredangle ABC = x$.

Angles can be measured in ***degrees***, radians, or gradients. In this text, we will discuss only the degree unit of measurement. The symbol for degrees is the same as the symbol for temperature degrees. An angle of 45 degrees is written 45°. A *protractor* is used to measure angles. The angle shown being measured by the protractor in Fig. 9.7 is 50°.

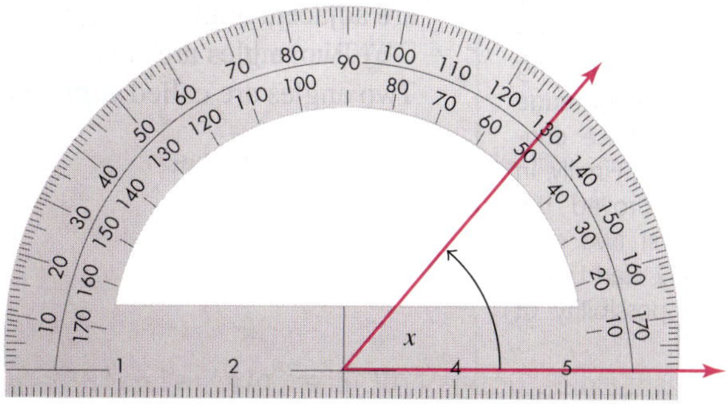
Figure 9.7

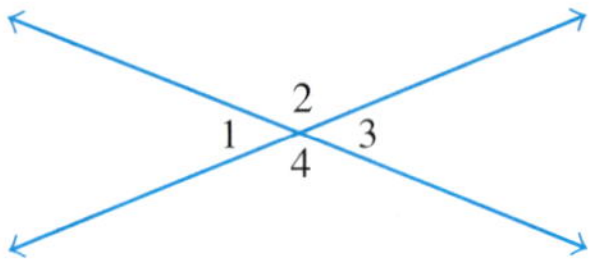

Figure 9.13

When two straight lines intersect, the nonadjacent angles formed are called ***vertical angles***. In Fig. 9.13, ∡1 and ∡3 are vertical angles, and ∡2 and ∡4 are vertical angles. We can show that vertical angles have the same measure, that is, they are equal. For example, Fig. 9.13 shows that

$$m\measuredangle 1 + m\measuredangle 2 = 180^\circ. \qquad \text{Why?}$$

$$m\measuredangle 2 + m\measuredangle 3 = 180^\circ. \qquad \text{Why?}$$

Since ∡2 has the same measure in both cases, $m\measuredangle 1$ must equal $m\measuredangle 3$.

Vertical angles have the same measure.

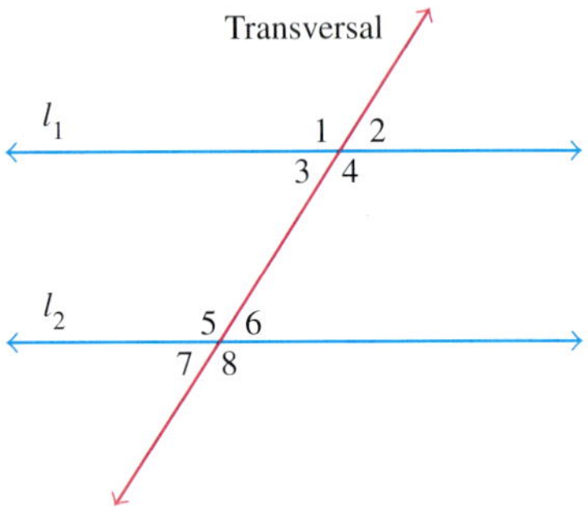

Figure 9.14

A line that intersects two different lines, l_1 and l_2, at two different points is called a ***transversal***. Figure 9.14 illustrates that when two parallel lines are cut by a transversal, eight angles are formed. Angles 3, 4, 5, and 6 are called ***interior angles***, and angles 1, 2, 7, and 8 are called ***exterior angles***. Eight pairs of supplementary angles are formed. Can you list them?

Special names are given to the angles formed by a transversal crossing two parallel lines. We describe these angles below.

Name	Description	Illustration	Pairs of Angles Meeting Criteria
Alternate interior angles	Interior angles on opposite sides of the transversal	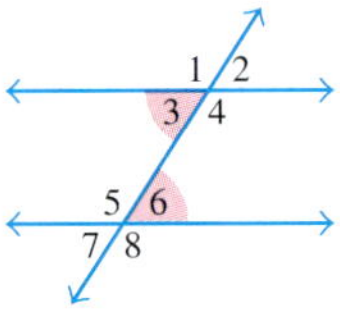	∡3 and ∡6 ∡4 and ∡5
Alternate exterior angles	Exterior angles on opposite sides of the transversal	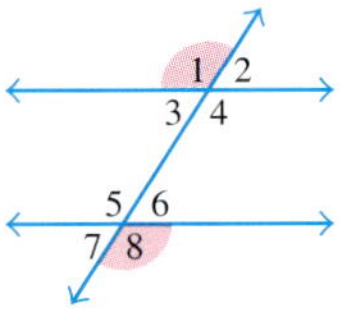	∡1 and ∡8 ∡2 and ∡7
Corresponding angles	One interior and one exterior angle on the same side of the transversal	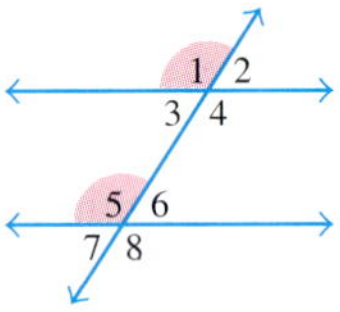	∡1 and ∡5 ∡2 and ∡6 ∡3 and ∡7 ∡4 and ∡8

When two parallel lines are cut by a transversal,

1. alternate interior angles have the same measure.
2. alternate exterior angles have the same measure.
3. corresponding angles have the same measure.

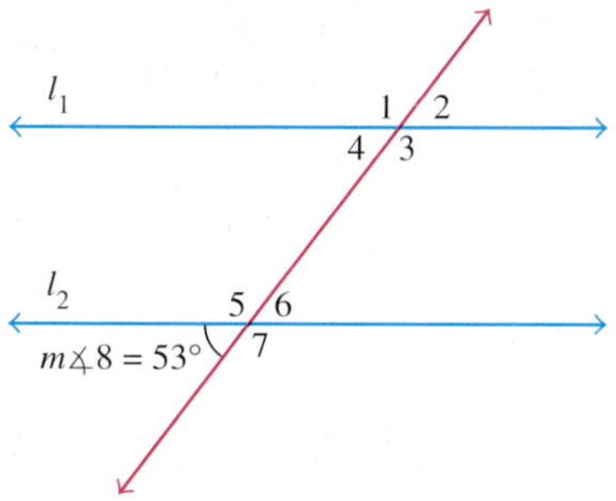

Figure 9.15

EXAMPLE 6 *Determining Angle Measures*

Figure 9.15 shows two parallel lines cut by a transversal. Determine the measure of $\measuredangle 1$ through $\measuredangle 7$.

SOLUTION

$m\measuredangle 6 = 53°$	$\measuredangle 8$ and $\measuredangle 6$ are vertical angles.
$m\measuredangle 5 = 127°$	$\measuredangle 8$ and $\measuredangle 5$ are supplementary angles.
$m\measuredangle 7 = 127°$	$\measuredangle 5$ and $\measuredangle 7$ are vertical angles.
$m\measuredangle 1 = 127°$	$\measuredangle 1$ and $\measuredangle 7$ are alternate exterior angles.
$m\measuredangle 4 = 53°$	$\measuredangle 4$ and $\measuredangle 6$ are alternate interior angles.
$m\measuredangle 2 = 53°$	$\measuredangle 6$ and $\measuredangle 2$ are corresponding angles.
$m\measuredangle 3 = 127°$	$\measuredangle 3$ and $\measuredangle 1$ are vertical angles.

In Example 6, the angles could have been determined in alternate ways. For example, we mentioned $m\measuredangle 1 = 127°$ because $\measuredangle 1$ and $\measuredangle 7$ are alternate exterior angles. We could have also stated that $m\measuredangle 1 = 127°$ because $\measuredangle 1$ and $\measuredangle 5$ are corresponding angles.

SECTION 9.1 EXERCISES

CONCEPT/WRITING EXERCISES

1. a) What are the four key parts in the axiomatic method used by Euclid?
   b) Discuss each of the four parts.
2. What is the difference between an axiom and a theorem?
3. What are skew lines?
4. What are parallel lines?
5. What are adjacent angles?
6. What are complementary angles?
7. What are supplementary angles?
8. What is a straight angle?
9. What is an obtuse angle?
10. What is a right angle?
11. What is an acute angle?
12. Draw two intersecting lines. Identify the two pairs of vertical angles.

PRACTICE THE SKILLS

In Exercises 13–20, identify the figure as a line, half line, ray, line segment, open line segment, or half open line segment. Denote the figure by its appropriate symbol.

13. 

14. 

15. 

16. 

17. 

18. 

19. 

20. 

In Exercises 21–32, use the figure to find the following:

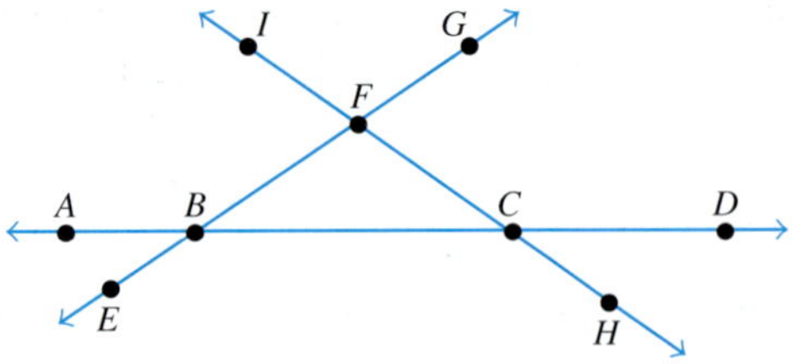

21. $\angle ABF \cap \angle GBC$
22. $\overset{\circ\rightarrow}{FE} \cup \overrightarrow{FG}$
23. $\overset{-\circ}{BC} \cup \overset{-\circ}{CD}$
24. $\overline{AB} \cup \overline{BD}$
25. $\angle ICA \cap \overleftrightarrow{EG}$
26. $\angle IFG \cap \angle EFH$
27. $\overset{\circ\rightarrow}{AB} \cap \overrightarrow{HC}$
28. $\overset{\circ\rightarrow}{BD} \cap \overset{\circ\rightarrow}{CB}$
29. $\overline{FG} \cup \overrightarrow{FC}$
30. $\overline{BC} \cup \overline{CF} \cup \overline{FB}$
31. $\overset{\circ\rightarrow}{BD} \cup \overset{\circ\rightarrow}{CB}$
32. $\{C\} \cap \overset{\circ\rightarrow}{CH}$

In Exercises 33–44, use the figure to find each of the following.

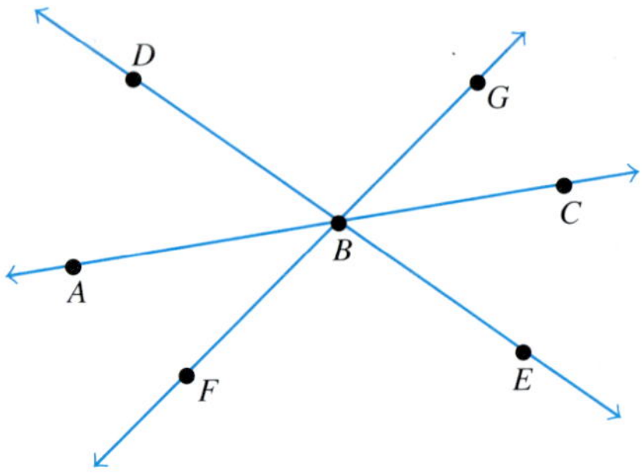

33. $\overleftrightarrow{DE} \cap \overleftrightarrow{FG}$
34. $\overrightarrow{BD} \cup \overset{-\circ}{BE}$
35. $\angle GBC \cap \angle CBE$
36. $\overleftrightarrow{DE} \cup \overset{\circ\rightarrow}{BE}$
37. $\angle ABE \cup \overset{\circ-\circ}{AB}$
38. $\overset{\circ\rightarrow}{BF} \cup \overrightarrow{BE}$
39. $\overrightarrow{FG} \cap \overrightarrow{BF}$
40. $\overrightarrow{BE} \cup \overline{BD}$
41. $\overset{\circ-\circ}{AC} \cap \overline{AC}$
42. $\overset{\circ-\circ}{AC} \cap \overset{\circ-\circ}{BE}$
43. $\overrightarrow{EB} \cap \overset{-\circ}{BE}$
44. $\overline{FB} \cup \overline{BG}$

In Exercises 45–52, classify the angle as acute, right, straight, obtuse, or none of these angles.

45. 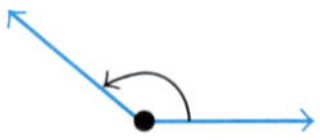

46. 

47. 

48. 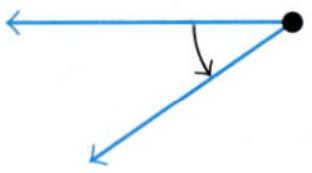

49. 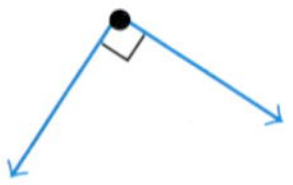

50. 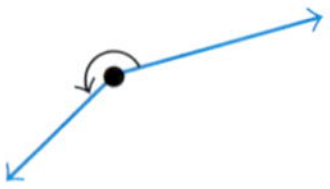

51. 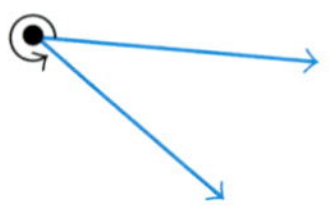

52. 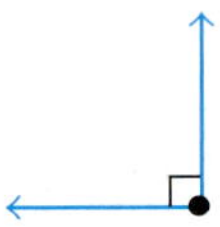

In Exercises 53–58, find the complementary angle of the given angle.

53. $26°$
54. $89°$
55. $32\frac{3}{4}°$
56. $31\frac{2}{5}°$
57. $64.7°$
58. $0.01°$

In Exercises 59–64, find the supplementary angle of the given angle.

59. $89°$
60. $8°$
61. $20.5°$
62. $148.7°$
63. $43\frac{5}{7}°$
64. $64\frac{7}{16}°$

In Exercises 65–70, match the names of the angles with the corresponding figure in parts (a)–(f).

65. Vertical angles
66. Corresponding angles
67. Complementary angles
68. Supplementary angles
69. Alternate exterior angles
70. Alternate interior angles

a) 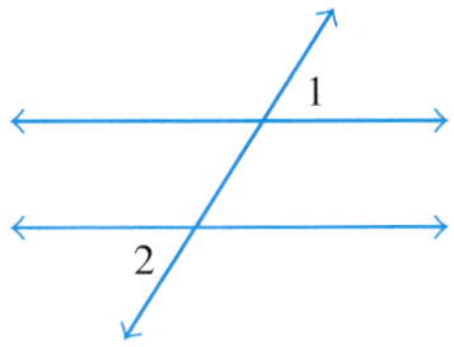

b) 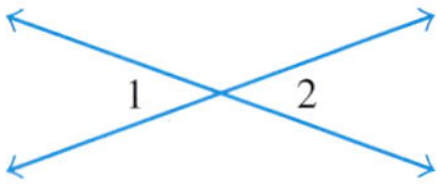

c) 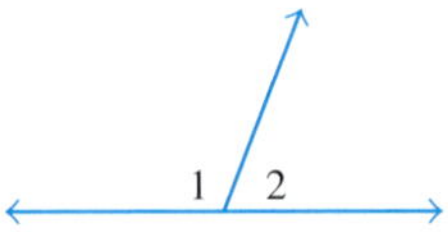

d) 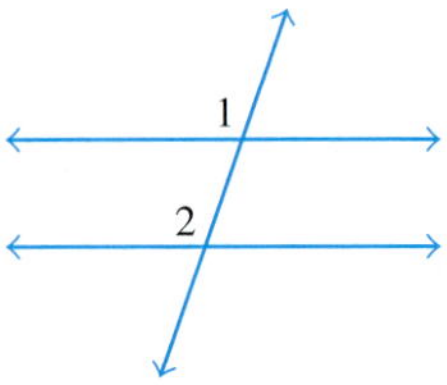

e)

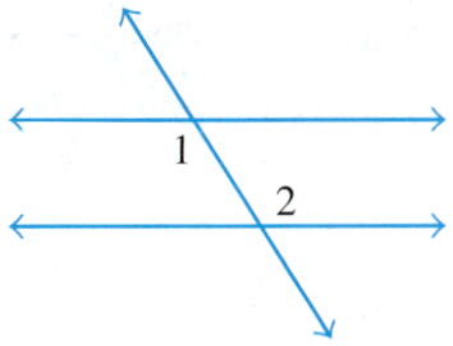

f)

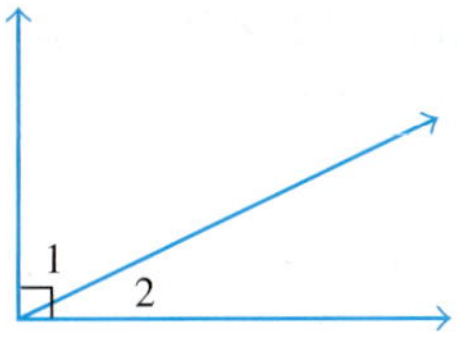

PROBLEM SOLVING

71. MODELING - *Complementary Angles* If $\measuredangle 1$ and $\measuredangle 2$ are complementary angles and if the measure of $\measuredangle 1$ is eight more than the measure of $\measuredangle 2$, determine the measures of $\measuredangle 1$ and $\measuredangle 2$.

72. MODELING - *Complementary Angles* The difference between the measures of two complementary angles is 16°. Determine the measures of the two angles.

73. MODELING - *Supplementary Angles* The difference between the measures of two supplementary angles is 88°. Determine the measures of the two angles.

74. MODELING - *Supplementary Angles* If $\measuredangle 1$ and $\measuredangle 2$ are supplementary angles and if the measure of $\measuredangle 2$ is 17 times the measure of $\measuredangle 1$, determine the measures of the two angles.

In Exercises 75–78, parallel lines are cut by the transversal shown. Determine the measures of $\measuredangle 1$ through $\measuredangle 7$.

75.

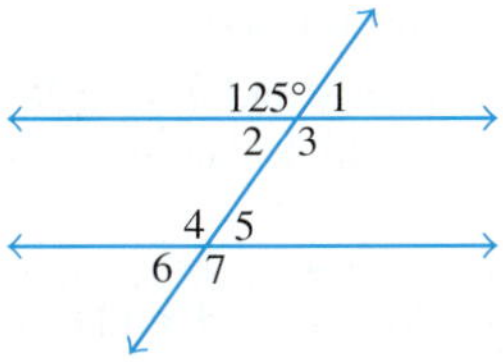

76.

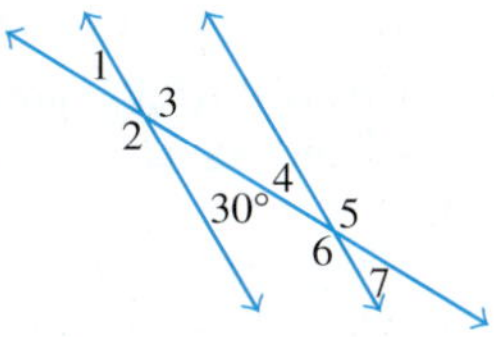

77.

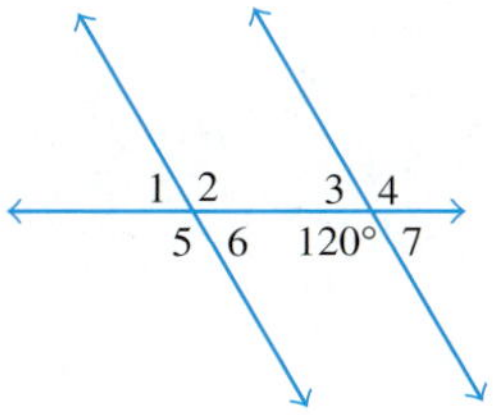

78.

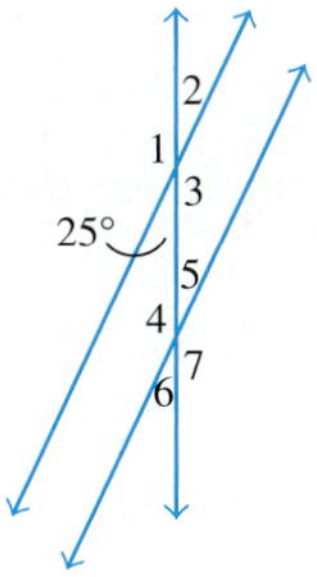

In Exercises 79–82, the angles are complementary angles. Determine the measures of $\measuredangle 1$ and $\measuredangle 2$.

79.

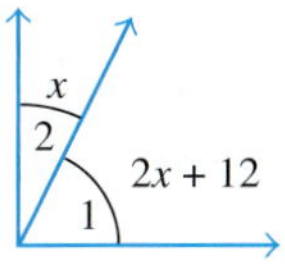

80.

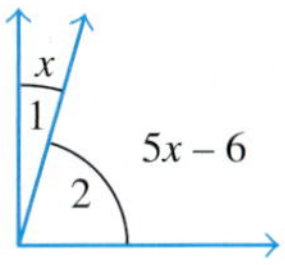

81.

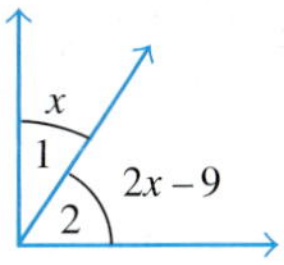

82.

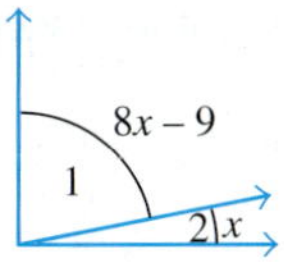

In Exercises 83–86, the angles are supplementary angles. Determine the measures of $\measuredangle 1$ and $\measuredangle 2$.

83.

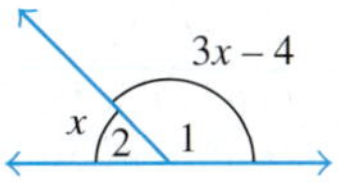

84.

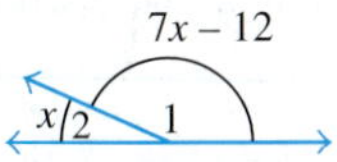

85.

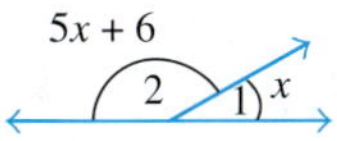

86.

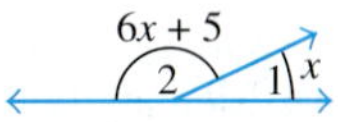

87. a) How many lines can be drawn through a given point?

b) How many planes can be drawn through a given point?

88. What is the intersection of two distinct nonparallel planes?

89. How many planes can be drawn through a given line?

90. a) Will three noncollinear points A, B, and C always determine a plane? Explain.

b) Is it possible to determine more than one plane with three noncollinear points? Explain.

c) How many planes can be constructed through three collinear points?

The figure suggests a number of lines and planes. The lines may be described by naming two points, and the planes may be described by naming three points. In Exercises 91–98, use the figure to name the following.

91. Two parallel planes

92. Two parallel lines

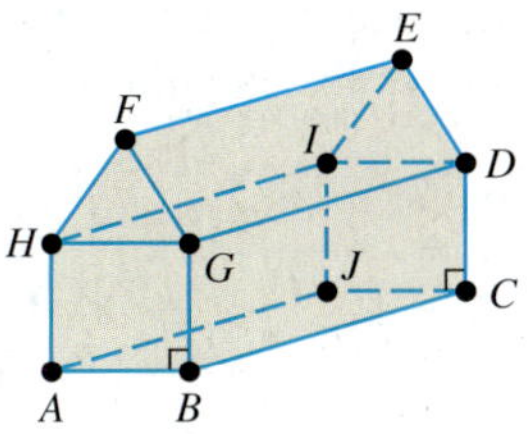

93. Two lines that intersect at right angles

94. Two planes that intersect at right angles

95. Three planes whose intersection is a single point

96. Three planes whose intersection is a line

97. A line and a plane whose intersection is a point

98. A line and a plane whose intersection is a line

In Exercises 99–104, determine whether the statement is always true, sometimes true, or never true. Explain your answer.

99. Two lines that are both parallel to a third line must be parallel to each other.

100. A triangle contains exactly two acute angles.

101. Vertical angles are complementary angles.

102. Alternate exterior angles are supplementary angles.

103. Alternate interior angles are complementary angles.

104. A triangle contains two obtuse angles.

CHALLENGE PROBLEMS/GROUP ACTIVITIES

105. Use a straightedge and a compass to construct a triangle with sides of equal length (an equilateral triangle) by doing the following:

a) Use the straightedge to draw a line segment of any length and label the end points A and B (Fig. a).

b) Place one end of the compass at point A and the other end at point B and draw an arc as shown (Fig. b).

c) Now turn the compass around and draw another arc as shown. Label the point of intersection of the two arcs C (Fig. c).

d) Draw line segments AC and BC. This completes the construction of equilateral triangle ABC (Fig. d).

Figure a

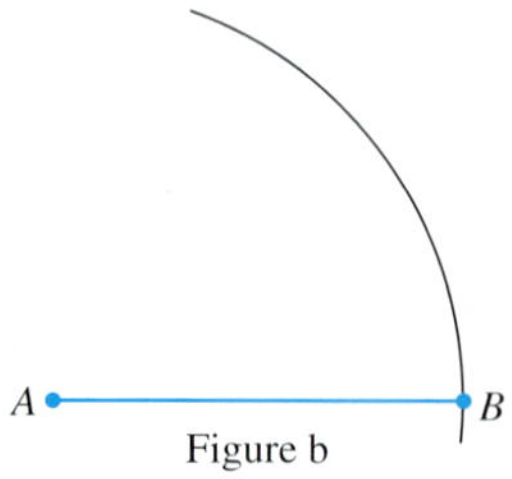

Figure b

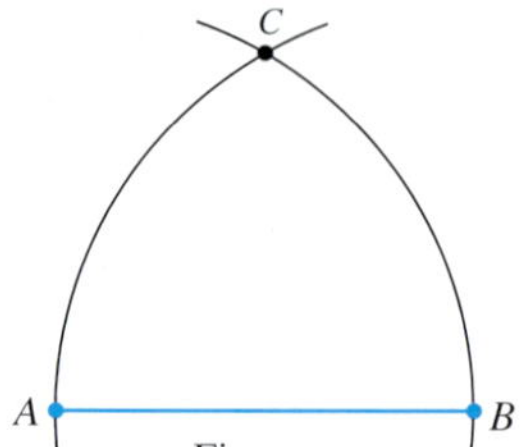

Figure c

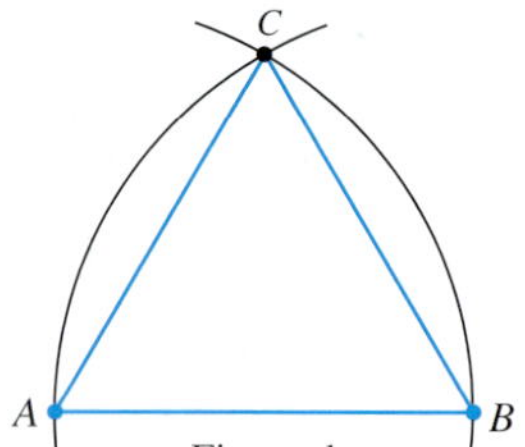

Figure d

106. If lines l and m are parallel lines and if lines l and n are skew lines, is it true that lines m and n must also be skew? (*Hint:* Look at Fig. 9.5 on page 519.) Explain your answer and include a sketch to support your answer.

107. Two lines are *perpendicular* if they intersect at right angles. If lines l and m are perpendicular and if lines m and n are perpendicular, is it true that lines l and n must also be perpendicular? Explain your answer and include a sketch to support your answer.

108. Suppose you have three distinct lines, all lying in the same plane. Find all the possible ways in which the three lines can be related. Sketch each case (four cases).

RECREATIONAL MATHEMATICS

109. If two straight lines intersect at a point, determine the sum of the measures of the 4 angles formed.

110. $\measuredangle ABC$ and $\measuredangle CBD$ are complementary and $m\measuredangle CBD$ is twice the $m\measuredangle ABC$. $\measuredangle ABD$ and $\measuredangle DBE$ are supplementary angles.

a) Draw a sketch illustrating $\measuredangle ABC$, $\measuredangle CBD$, and $\measuredangle DBE$.

b) Determine $m\measuredangle ABC$.

c) Determine $m\measuredangle CBD$.

d) Determine $m\measuredangle DBE$.

INTERNET/RESEARCH ACTIVITIES

111. Using the Internet and other sources, write a research paper on Euclid's contributions to geometry.

112. Using the Internet and other sources, write a research paper on the three classic geometry problems of Greek antiquity (see the Did You Know? on page 520).

113. Search the Internet or other sources such as a geometry textbook to study the geometric constructions that use a straightedge and a compass only. Prepare a poster demonstrating five of these basic constructions.

9.2 POLYGONS

▲ The shapes of these road signs are examples of polygons.

What shape would you use to best describe the following road signs: a stop sign, a yield sign, a speed limit sign? In this section, we will study the shapes of these and other geometric figures that can be classified as *polygons*.

A ***polygon*** is a closed figure in a plane determined by three or more straight line segments. Examples of polygons are given in Fig. 9.16.

The straight line segments that form the polygon are called its ***sides***, and a point where two sides meet is called a ***vertex*** (plural ***vertices***). The union of the sides of a polygon and its interior is called a ***polygonal region***. A ***regular polygon*** is one whose sides are all the same length and whose interior angles all have the same measure. Figures 9.16(b) and (d) are regular polygons.

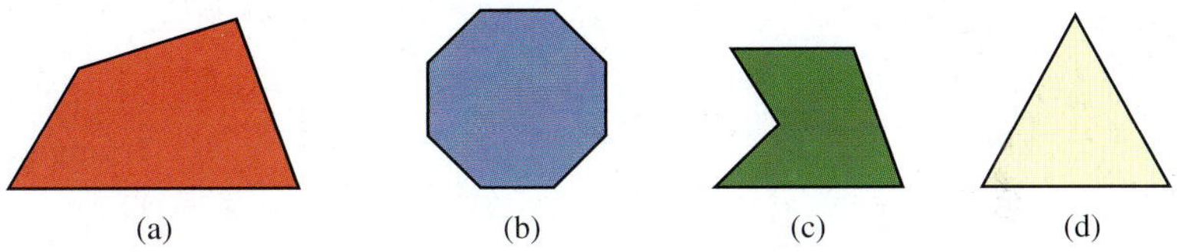

Figure 9.16

Polygons are named according to their number of sides. The names of some polygons are given in Table 9.1.

Table 9.1

Number of Sides	Name	Number of Sides	Name
3	Triangle	8	Octagon
4	Quadrilateral	9	Nonagon
5	Pentagon	10	Decagon
6	Hexagon	12	Dodecagon
7	Heptagon	20	Icosagon

One of the most important polygons is the triangle. The sum of the measures of the interior angles of a triangle is 180°. To illustrate, consider triangle *ABC* given in Fig. 9.17. The triangle is formed by drawing two transversals through two parallel lines l_1 and l_2 with the two transversals intersecting at a point on l_1.

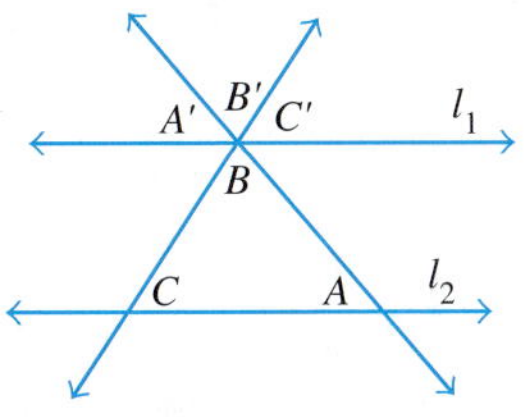

Figure 9.17

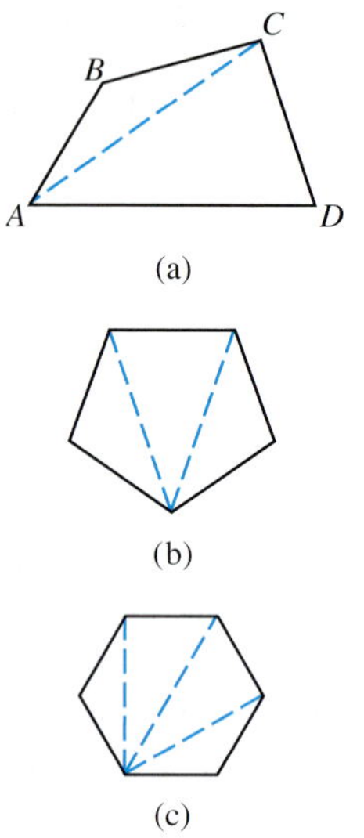

Figure 9.18

In Fig. 9.17, notice that $\measuredangle A$ and $\measuredangle A'$ are corresponding angles. Recall from Section 9.1 that corresponding angles are equal, so $m\measuredangle A = m\measuredangle A'$. Also, $\measuredangle C$ and $\measuredangle C'$ are corresponding angles; therefore, $m\measuredangle C = m\measuredangle C'$. Next, we notice that $\measuredangle B$ and $\measuredangle B'$ are vertical angles. In Section 9.1, we learned that vertical angles are equal; therefore, $m\measuredangle B = m\measuredangle B'$. Figure 9.17 shows that $\measuredangle A', \measuredangle B'$, and $\measuredangle C'$ form a straight angle; therefore, $m\measuredangle A' + m\measuredangle B' + m\measuredangle C' = 180°$. Since $m\measuredangle A = m\measuredangle A', m\measuredangle B = m\measuredangle B'$, and $m\measuredangle C = m\measuredangle C'$, we can reason that $m\measuredangle A + m\measuredangle B + m\measuredangle C = 180°$. This example illustrates that the sum of the interior angles of a triangle is 180°.

Consider the quadrilateral *ABCD* (Fig. 9.18a). Drawing a straight line segment between any two vertices forms two triangles. Since the sum of the measures of the angles of a triangle is 180°, the sum of the measures of the interior angles of a quadrilateral is $2 \cdot 180°$, or 360°.

Now let's examine a pentagon (Fig. 9.18b). We can draw two straight line segments to form three triangles. Thus, the sum of the measures of the interior angles of a five-sided figure is $3 \cdot 180°$, or 540°. Figure 9.18(c) shows that four triangles can be drawn in a six-sided figure. Table 9.2 summarizes this information.

Table 9.2

Sides	Triangles	Sum of the Measures of the Interior Angles
3	1	$1(180°) = 180°$
4	2	$2(180°) = 360°$
5	3	$3(180°) = 540°$
6	4	$4(180°) = 720°$

If we continue this procedure, we can see that for an *n*-sided polygon the sum of the measures of the interior angles is $(n - 2)180°$.

The **sum** of the measures of the interior angles of an *n*-sided polygon is $(n - 2)180°$.

EXAMPLE 1 *Angles of a Hexagon*

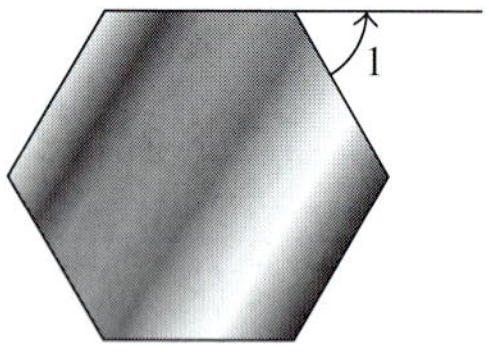

Figure 9.19

The surfaces of the heads of many bolts are in the shape of regular hexagons. A regular hexagon is a six-sided figure with all the sides the same length and all interior angles with the same measure. See Fig. 9.19. Determine

a) the measure of an interior angle.

b) the measure of exterior $\measuredangle 1$.

SOLUTION

a) Using the formula $(n - 2)180°$, we can determine the sum of the measures of the interior angles of a hexagon as follows.

$$\begin{aligned} \text{Sum} &= (6 - 2)180° \\ &= 4(180°) \\ &= 720° \end{aligned}$$

The measure of an interior angle of a regular polygon can be determined by dividing the sum of the interior angles by the number of angles.

MATHEMATICS TODAY

Making Movies Come Alive

Mr. Tumnus from *The Chronicles of Narnia*

Mathematics plays a key role in the animation you see in movies such as those in *The Chronicles of Narnia.* The images you see on the movie screen are created using software that combines pixels (the smallest piece of a screen image) into geometric shapes including polygons. These shapes are then stored in a computer and manipulated using various mathematical techniques so that the new shapes formed (from the original geometric shapes) approximate curves. Each movie frame has over 2 million pixels and can have over 40 million polygons. With such a huge amount of data, computers are used to carry out the mathematics needed to create animation. One computer animation specialist stated that "it's all controlled by math. . . . All those little *X*'s, *Y*'s, and *Z*'s that you had in school—oh my gosh, suddenly they all apply."

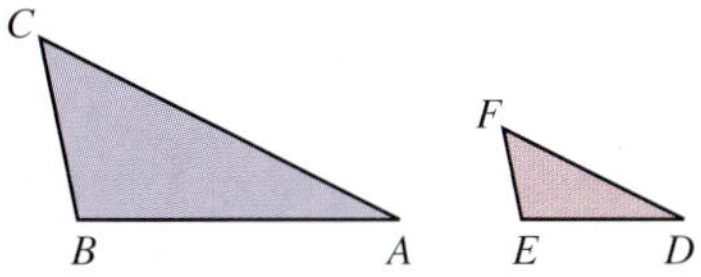

Figure 9.20

The measure of an interior angle of a regular hexagon is determined as follows:

$$\text{Measure} = \frac{720^\circ}{6} = 120^\circ$$

b) Since $\measuredangle 1$ is the supplement of an interior angle,

$$m\measuredangle 1 = 180^\circ - 120^\circ = 60^\circ$$

To discuss area in the next section, we must be able to identify various types of triangles and quadrilaterals. The following is a summary of certain types of triangles and their characteristics.

TRIANGLES

Acute Triangle

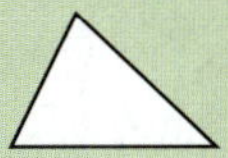

All angles are acute angles.

Obtuse Triangle

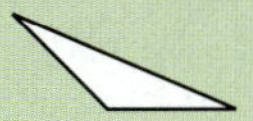

One angle is obtuse.

Right Triangle

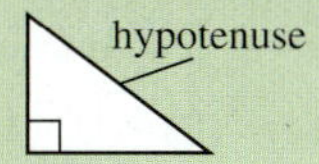

One angle is a right angle.

Isosceles Triangle

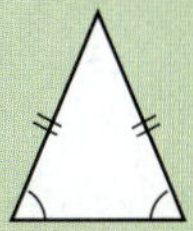

Two equal sides
Two equal angles

Equilateral Triangle

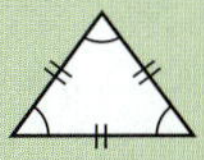

Three equal sides
Three equal angles
(60° each)

Scalene Triangle

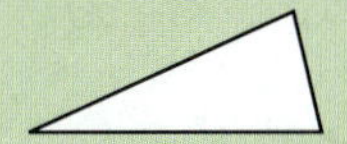

No two sides are equal in length.

Similar Figures

In everyday living, we often have to deal with geometric figures that have the "same shape" but are of different sizes. For example, an architect will make a small-scale drawing of a floor plan or a photographer will make an enlargement of a photograph. Figures that have the same shape but may be of different sizes are called ***similar figures***. Two similar figures are illustrated in Fig. 9.20.

Similar figures have *corresponding angles* and *corresponding sides*. In Fig. 9.20, triangle *ABC* has angles *A*, *B*, and *C*. Their respective corresponding angles in triangle *DEF* are angles *D*, *E*, and *F*. Sides $\overline{AB}$, $\overline{BC}$, and $\overline{AC}$ in triangle *ABC* have corresponding sides $\overline{DE}$, $\overline{EF}$, and $\overline{DF}$, respectively, in triangle *DEF*.

Two polygons are **similar** if their corresponding angles have the same measure and the lengths of their corresponding sides are in proportion.

In Figure 9.20, $\measuredangle A$ and $\measuredangle D$ have the same measure, $\measuredangle B$ and $\measuredangle E$ have the same measure, and $\measuredangle C$ and $\measuredangle F$ have the same measure. Also, the lengths of corresponding sides of similar triangles are in proportion.

When we refer to the line segment *AB*, we place a line over the *AB* and write $\overline{AB}$. **When we refer to the *length* of a line segment, we do not place a bar above the two letters**. Thus, when we refer to the *length* of line segment $\overline{AB}$, we write *AB* (without the bar on top of the two letters). For example, if we write $AB = 12$, we are indicating the length of line segment $\overline{AB}$ is 12. When setting up proportions, as is shown below, we will be comparing the lengths of corresponding sides of similar figures, and so we will not be using the bar above the letters of the line segments. The proportion below shows that the lengths of the corresponding sides of the similar triangles in Figure 9.20 are in proportion.

$$\frac{AB}{DE} = \frac{BC}{EF} = \frac{AC}{DF}$$

EXAMPLE 2 *Similar Figures*

Consider the similar figures in Fig. 9.21.

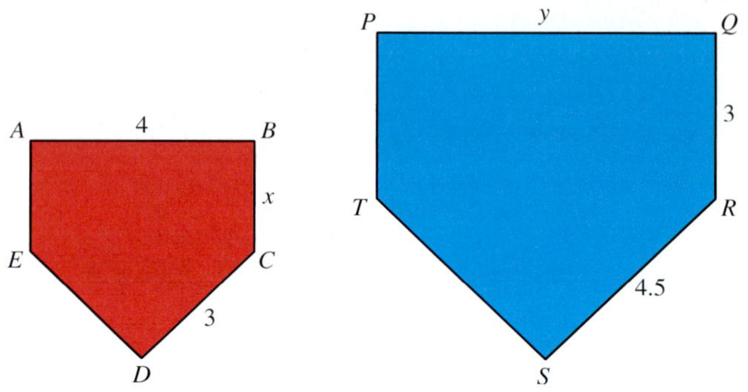

Figure 9.21

Determine

a) the length of side $\overline{BC}$. b) the length of side $\overline{PQ}$.

SOLUTION

a) We will represent the length of side $\overline{BC}$ with the variable x. Because the corresponding sides of similar figures must be in proportion, we can write a proportion (as explained in Section 6.2) to determine the length of side $\overline{BC}$. Corresponding sides $\overline{CD}$ and $\overline{RS}$ are known, so we use them as one ratio in the proportion. The side corresponding to $\overline{BC}$ is $\overline{QR}$.

$$\frac{BC}{QR} = \frac{CD}{RS}$$

$$\frac{x}{3} = \frac{3}{4.5}$$

Now we solve for x.

$$x \cdot 4.5 = 3 \cdot 3$$

$$4.5x = 9$$

$$x = 2$$

Thus, the length of side $\overline{BC}$ is 2 units.

b) We will represent the length of side $\overline{PQ}$ with the variable y. The side corresponding to $\overline{PQ}$ is $\overline{AB}$. We will work part (b) in a manner similar to part (a).

$$\frac{PQ}{AB} = \frac{RS}{CD}$$

$$\frac{y}{4} = \frac{4.5}{3}$$

$$y \cdot 3 = 4 \cdot 4.5$$
$$3y = 18$$
$$y = 6$$

Thus, the length of side $\overline{PQ}$ is 6 units.

EXAMPLE 3 ***Using Similar Triangles to Find the Height of a Tree***

Saraniti Walker plans to remove a tree from her back yard. She needs to know the height of the tree. Saraniti is 5 ft tall and determines that when her shadow is 8 ft long, the shadow of the tree is 50 ft long (see Fig. 9.22). How tall is the tree?

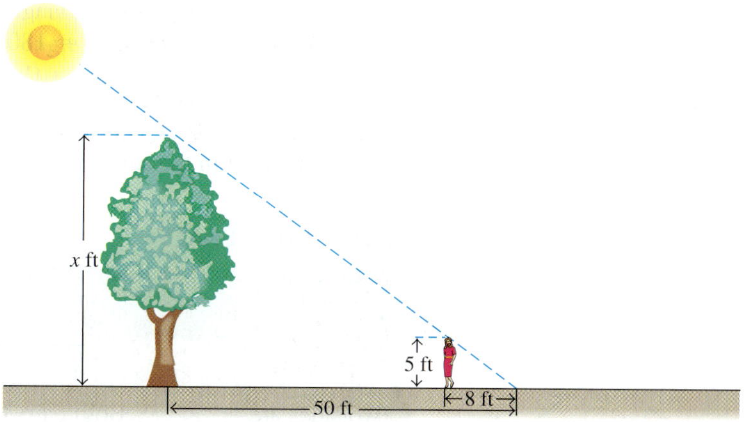

Figure 9.22

SOLUTION We will let x represent the height of the tree. From Fig. 9.22, we can see that the triangle formed by the sun's rays, Saraniti, and her shadow is similar to the triangle formed by the sun's rays, the tree, and its shadow. To find the height of the tree, we will set up and solve the following proportion:

$$\frac{\text{Height of the tree}}{\text{Height of Saraniti}} = \frac{\text{length of tree's shadow}}{\text{length of Saraniti's shadow}}$$
$$\frac{x}{5} = \frac{50}{8}$$
$$8x = 250$$
$$x = 31.25$$

Therefore, the tree is 31.25 ft tall.

Congruent Figures

If the corresponding sides of two similar figures are the same length, the figures are called *congruent figures*. Corresponding angles of congruent figures have the same measure, and the corresponding sides are equal in length. Two congruent figures coincide when placed one upon the other.

DID YOU KNOW?

Buckyballs

The molecular structure of C_{60} resembles the patterns found on a soccer ball.

Buckminsterfullerenes, known as buckyballs, are a class of pure carbon molecules. Along with graphite and diamond, buckyballs are the only naturally occurring forms of pure carbon. Named after American architect-engineer F. Buckminster Fuller, who designed hemispherical geodesic domes from hexagonal and pentagonal faces, fullerenes are the most spherical molecules known. Discovered in 1985 at Rice University, buckyballs are used primarily as microscopic lubricant and have potential applications in molecular medical engineering, electrical superconductivity, and computer chip design. The most common form of buckminsterfullerene contains 60 carbon atoms and has the chemical symbol C_{60}. The molecular structure of C_{60} contains 12 pentagons and 20 hexagons arranged in a pattern similar to that found on a soccer ball.

17.

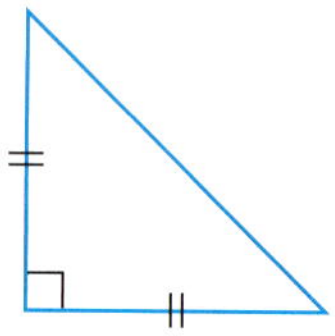

18.

19.

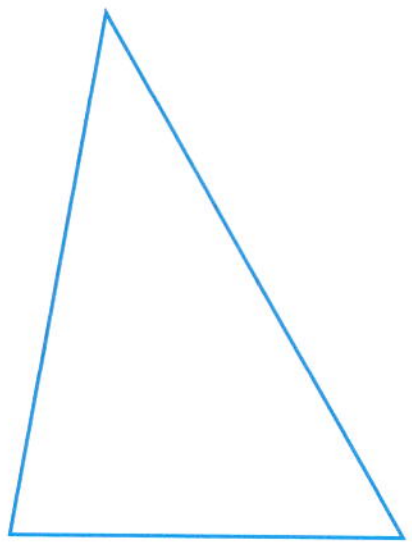

20.

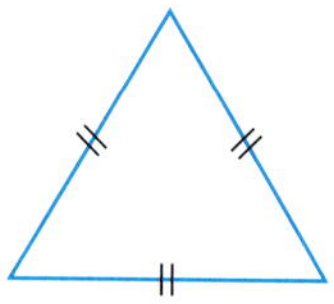

21.

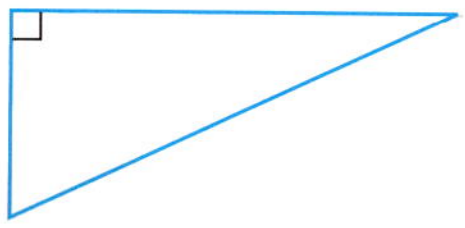

22.

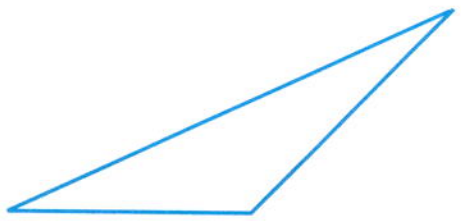

In Exercises 23–28, identify the quadrilateral.

23.

24.

25.

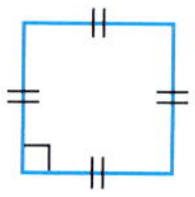

26.

27.

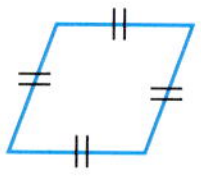

28.

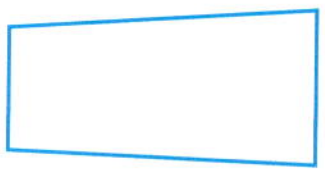

In Exercises 29–32, find the measure of ∡x.

29.

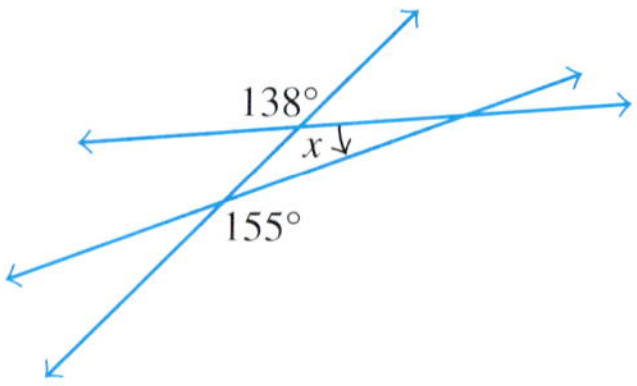

30.

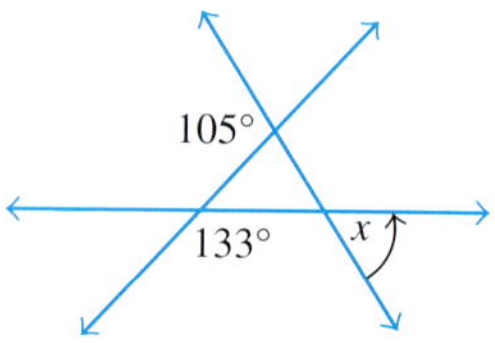

31.

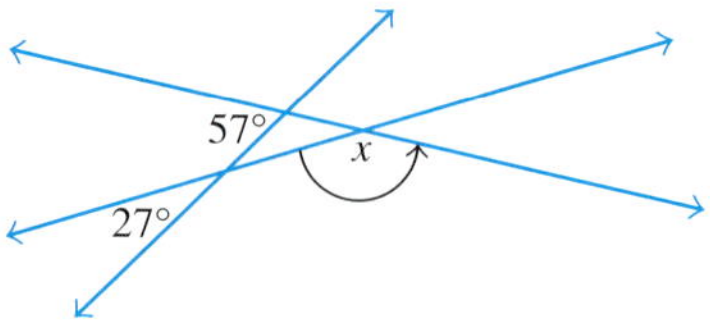

32.

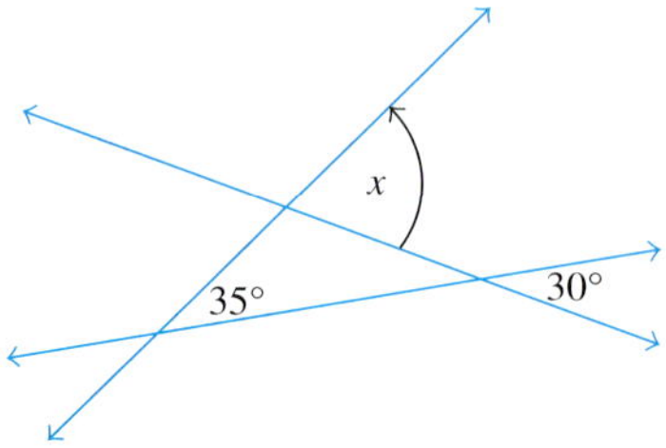

In Exercises 33–34, lines l_1 and l_2 are parallel. Determine the measures of ∡1 through ∡12.

33.

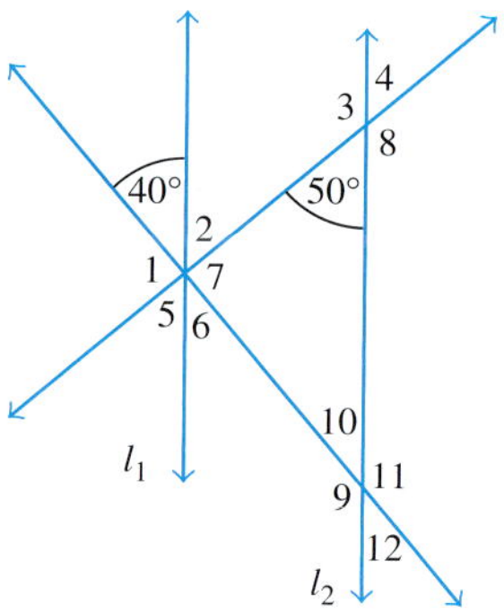

34.

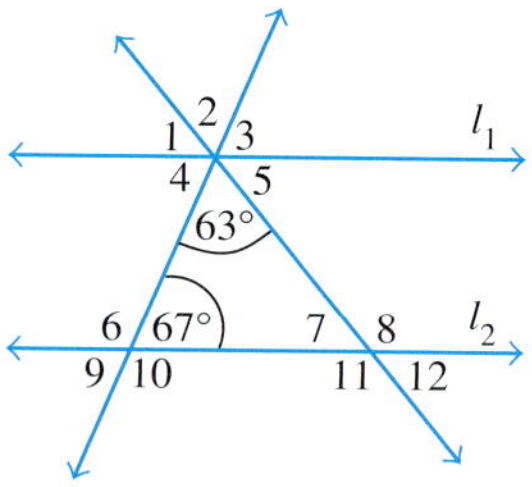

In Exercises 35–40, determine the sum of the measures of the interior angles of the indicated polygon.

35. Hexagon

36. Heptagon

37. Octagon

38. Decagon

39. Icosagon

40. Dodecagon

$$y \cdot 3 = 4 \cdot 4.5$$
$$3y = 18$$
$$y = 6$$

Thus, the length of side $\overline{PQ}$ is 6 units.

EXAMPLE 3 *Using Similar Triangles to Find the Height of a Tree*

Saraniti Walker plans to remove a tree from her back yard. She needs to know the height of the tree. Saraniti is 5 ft tall and determines that when her shadow is 8 ft long, the shadow of the tree is 50 ft long (see Fig. 9.22). How tall is the tree?

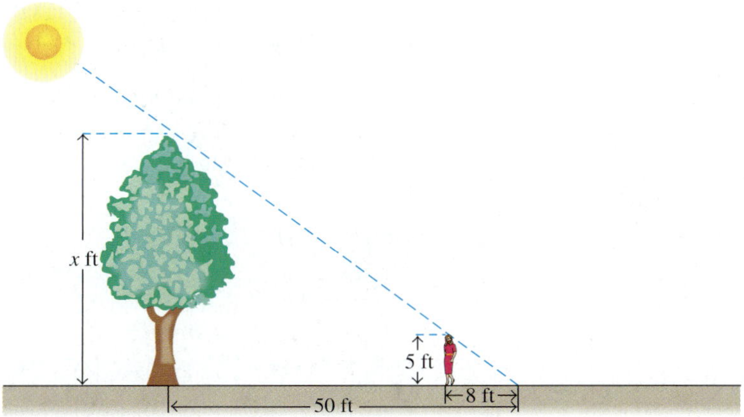

Figure 9.22

SOLUTION We will let x represent the height of the tree. From Fig. 9.22, we can see that the triangle formed by the sun's rays, Saraniti, and her shadow is similar to the triangle formed by the sun's rays, the tree, and its shadow. To find the height of the tree, we will set up and solve the following proportion:

$$\frac{\text{Height of the tree}}{\text{Height of Saraniti}} = \frac{\text{length of tree's shadow}}{\text{length of Saraniti's shadow}}$$
$$\frac{x}{5} = \frac{50}{8}$$
$$8x = 250$$
$$x = 31.25$$

Therefore, the tree is 31.25 ft tall.

Congruent Figures

If the corresponding sides of two similar figures are the same length, the figures are called *congruent figures*. Corresponding angles of congruent figures have the same measure, and the corresponding sides are equal in length. Two congruent figures coincide when placed one upon the other.

DID YOU KNOW?

Buckyballs

The molecular structure of C_{60} resembles the patterns found on a soccer ball.

Buckminsterfullerenes, known as buckyballs, are a class of pure carbon molecules. Along with graphite and diamond, buckyballs are the only naturally occurring forms of pure carbon. Named after American architect-engineer F. Buckminster Fuller, who designed hemispherical geodesic domes from hexagonal and pentagonal faces, fullerenes are the most spherical molecules known. Discovered in 1985 at Rice University, buckyballs are used primarily as microscopic lubricant and have potential applications in molecular medical engineering, electrical superconductivity, and computer chip design. The most common form of buckminsterfullerene contains 60 carbon atoms and has the chemical symbol C_{60}. The molecular structure of C_{60} contains 12 pentagons and 20 hexagons arranged in a pattern similar to that found on a soccer ball.

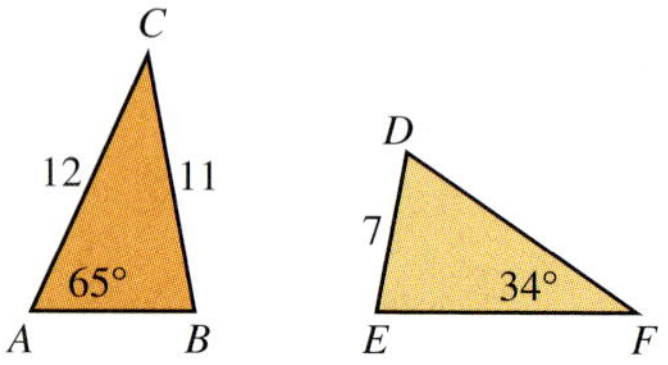

Figure 9.23

EXAMPLE 4 *Congruent Triangles*

Triangles ABC and DEF in Fig. 9.23 are congruent. Determine

a) the length of side $\overline{DF}$.
b) the length of side $\overline{AB}$.
c) $m\measuredangle FDE$.
d) $m\measuredangle ACB$.
e) $m\measuredangle ABC$.

SOLUTION Because ΔABC is congruent to ΔDEF, we know that the corresponding side lengths are equal and corresponding angle measures are equal.

a) $DF = AC = 12$
b) $AB = DE = 7$
c) $m\measuredangle FDE = m\measuredangle CAB = 65°$
d) $m\measuredangle ACB = m\measuredangle DFE = 34°$
e) The sum of the angles of a triangle is 180°. Since $m\measuredangle BAC = 65°$ and $m\measuredangle ACB = 34°$, $m\measuredangle ABC = 180° - 65° - 34° = 81°$.

Earlier we learned that ***quadrilaterals*** are four-sided polygons, the sum of whose interior angles is 360°. Quadrilaterals may be classified according to their characteristics, as illustrated in the summary box below.

QUADRILATERALS

Trapezoid

Two sides are parallel.

Parallelogram

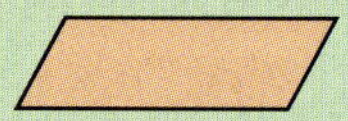

Both pairs of opposite sides are parallel. Both pairs of opposite sides are equal in length.

Rhombus

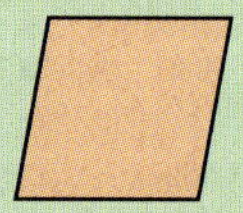

Both pairs of opposite sides are parallel. The four sides are equal in length.

Rectangle

Both pairs of opposite sides are parallel. Both pairs of opposite sides are equal in length. The angles are right angles.

Square

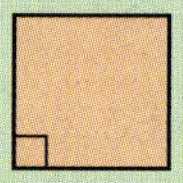

Both pairs of opposite sides are parallel. The four sides are equal in length. The angles are right angles.

EXAMPLE 5 *Angles of a Trapezoid*

Trapezoid $ABCD$ is shown in Fig. 9.24.

a) Determine the measure of the interior angle, x.
b) Determine the measure of the exterior angle, y.

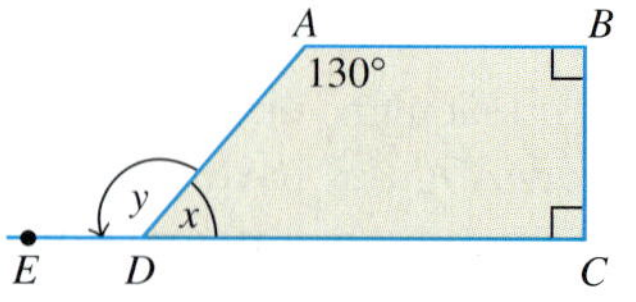

Figure 9.24

SOLUTION

a) We know that each of the two right angles in trapezoid $ABCD$ has a measure of 90°. We also know that the sum of the interior angles in any quadrilateral is 360°. Therefore, we have

$$
\begin{aligned}
m\measuredangle DAB + m\measuredangle ABC + m\measuredangle BCD + m\measuredangle x &= 360^\circ \\
130^\circ + 90^\circ + 90^\circ + m\measuredangle x &= 360^\circ \\
310^\circ + m\measuredangle x &= 360^\circ \\
m\measuredangle x &= 50^\circ
\end{aligned}
$$

Thus, the measure of the interior angle, x, is 50°.

b) Angle x and angle y are supplementary angles. Therefore, $m\measuredangle x + m\measuredangle y = 180^\circ$ and $m\measuredangle y = 180^\circ - m\measuredangle x = 180^\circ - 50^\circ = 130^\circ$. Thus, the measure of the exterior angle, y, is 130°.

Note that in Example 5 part (b) we could also have determined the measure of angle y as follows. By the definition of a trapezoid, sides $\overline{AB}$ and $\overline{CD}$ must be parallel. Therefore, side $\overline{AD}$ may be considered a transversal and $\measuredangle BAD$ and $\measuredangle ADE$ are alternate interior angles. Recall from Section 9.1 that alternate interior angles are equal. Thus, $m\measuredangle BAD = m\measuredangle ADE$ and $m\measuredangle y = 130^\circ$.

SECTION 9.2 EXERCISES

CONCEPT/WRITING EXERCISES

1. What is a polygon?
2. What distinguishes regular polygons from other polygons?
3. List six different types of triangles and in your own words describe the characteristics of each.
4. List five different types of quadrilaterals and in your own words describe the characteristics of each.
5. What are congruent figures?
6. What are similar figures?

In Exercises 7–14, (a) name the polygon. If the polygon is a quadrilateral, give its specific name. (b) State whether or not the polygon is a regular polygon.

7.
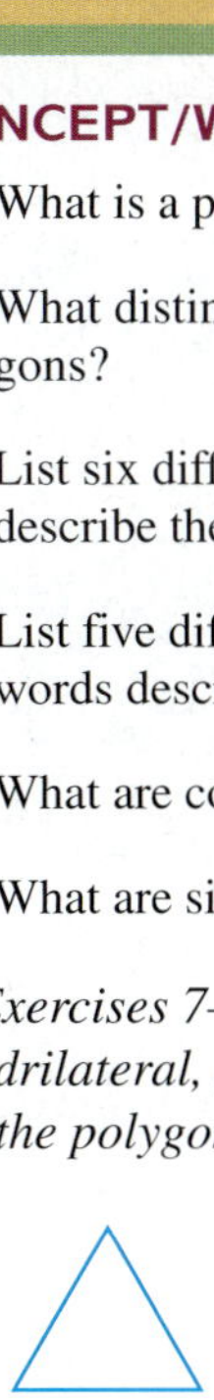

8.
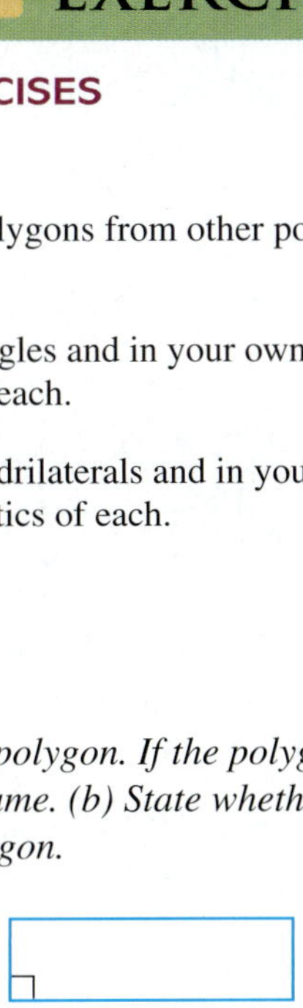

9.
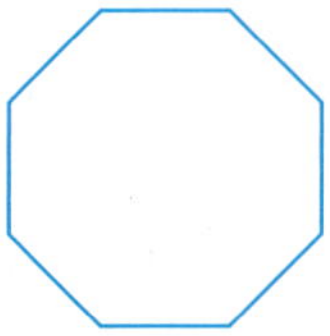

10.
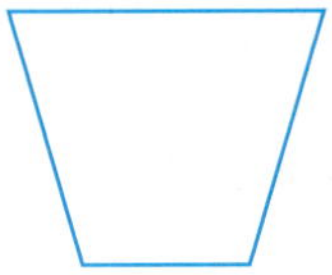

11.
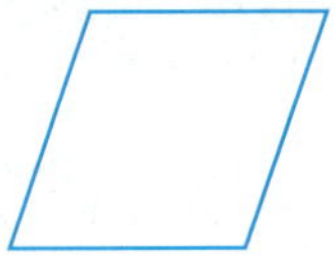

12.
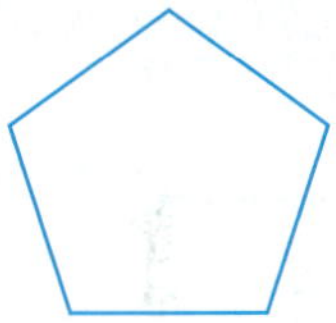

13.
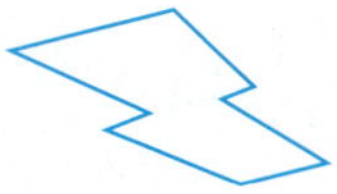

14.

In Exercises 15–22, identify the triangle as (a) scalene, isosceles, or equilateral and as (b) acute, obtuse, or right. The parallel markings (the two small parallel lines) on two or more sides indicate that the marked sides are of equal length.

15.
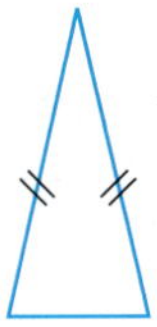

16.

17.

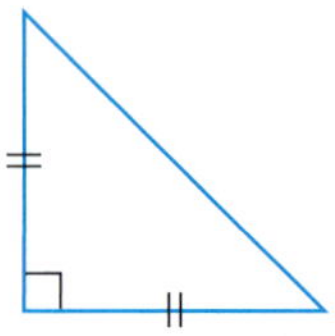

18.

19.

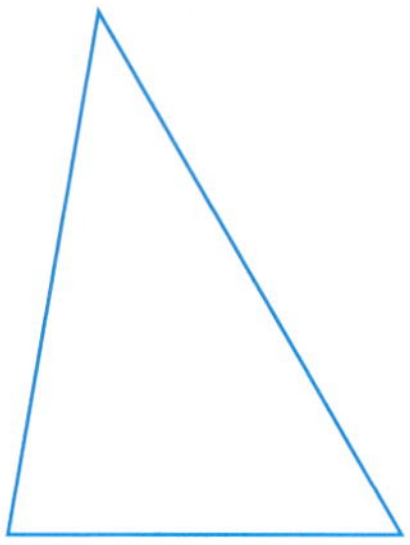

20.

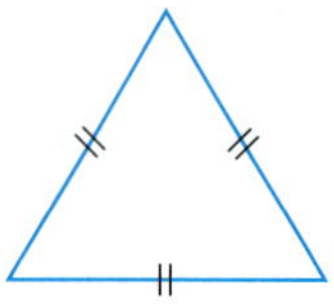

21.

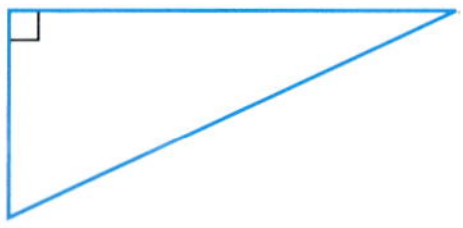

22.

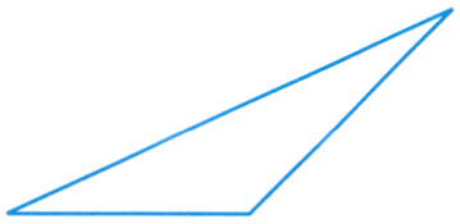

In Exercises 23–28, identify the quadrilateral.

23.

24.

25.

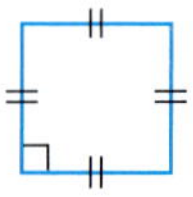

26.

27.

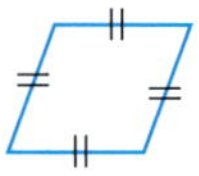

28.

In Exercises 29–32, find the measure of ∡x.

29.

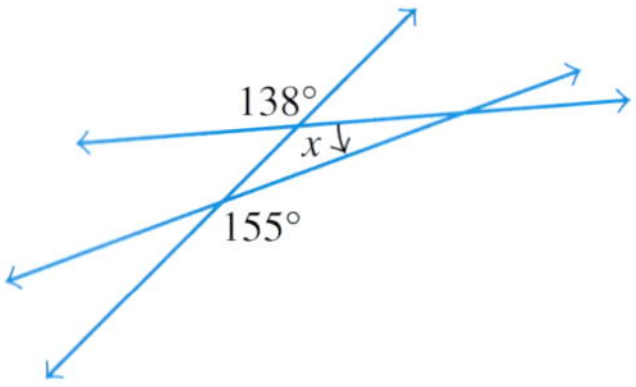

30.

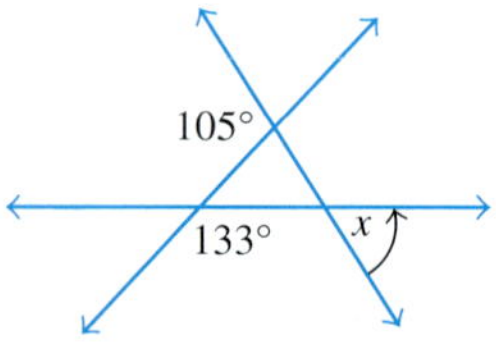

31.

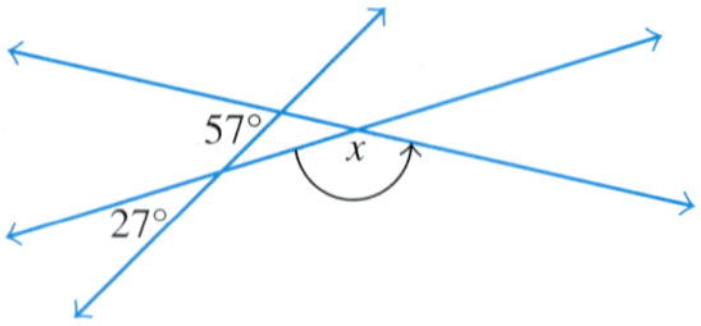

32.

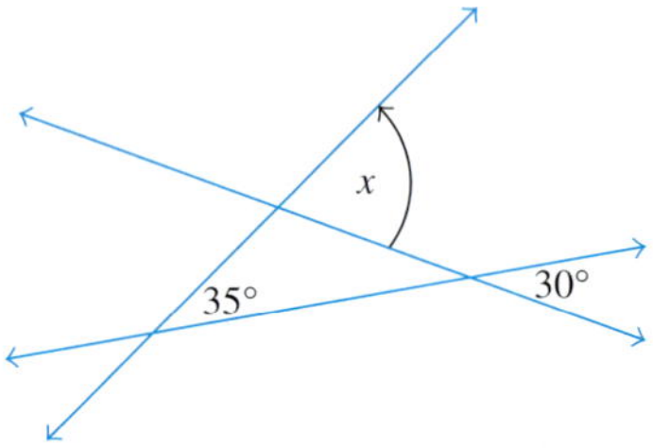

In Exercises 33–34, lines l_1 and l_2 are parallel. Determine the measures of ∡1 through ∡12.

33.

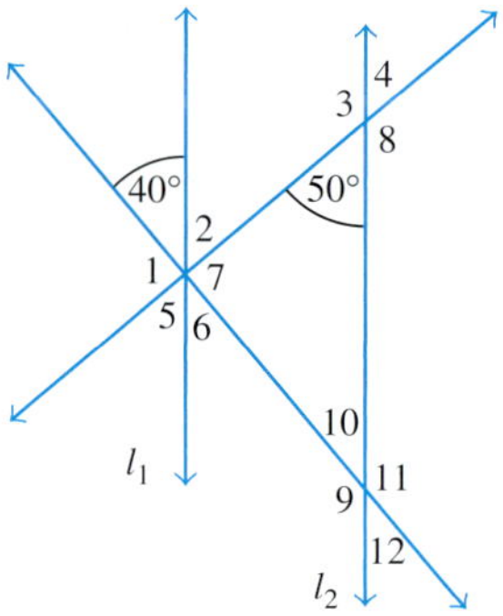

34.

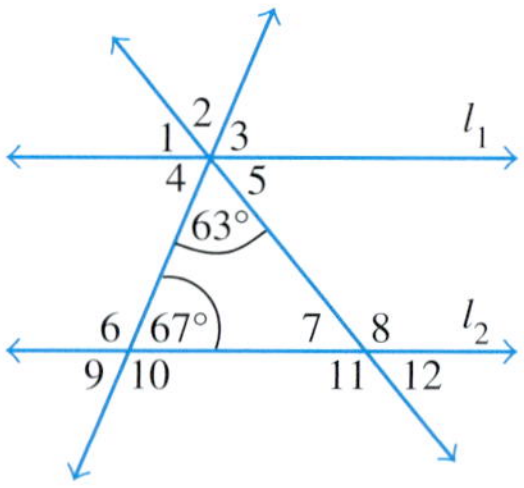

In Exercises 35–40, determine the sum of the measures of the interior angles of the indicated polygon.

35. Hexagon

36. Heptagon

37. Octagon

38. Decagon

39. Icosagon

40. Dodecagon

In Exercises 41–46, (a) determine the measure of an interior angle of the named regular polygon. (b) If a side of the polygon is extended, determine the supplementary angle of an interior angle. See Example 1.

41. Triangle

42. Quadrilateral

43. Pentagon

44. Nonagon

45. Decagon

46. Icosagon

In Exercises 47–52, the figures are similar. Find the length of side x and side y.

47.

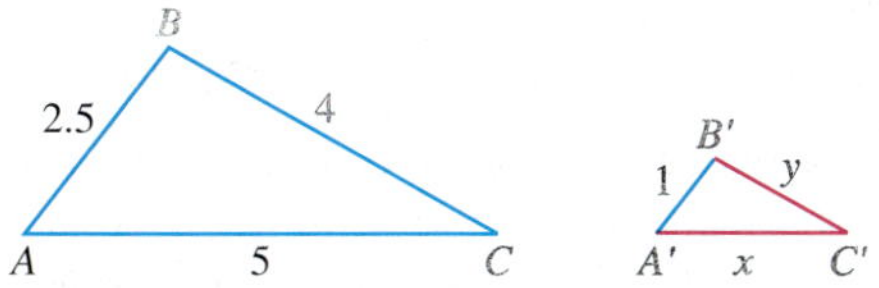

48.

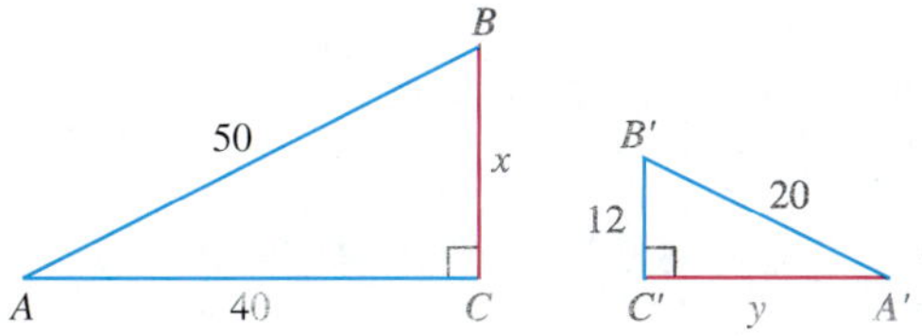

49.

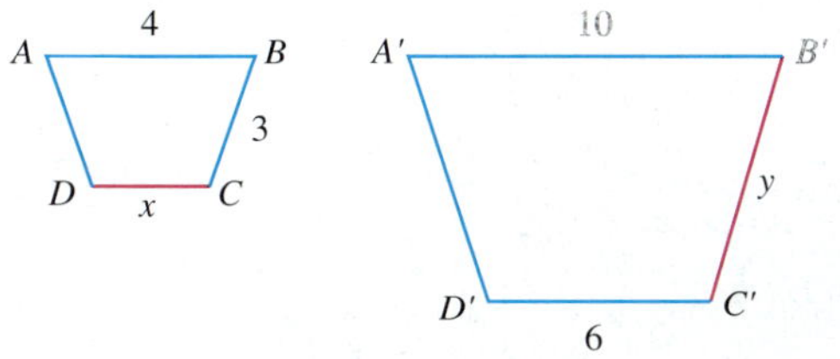

50.

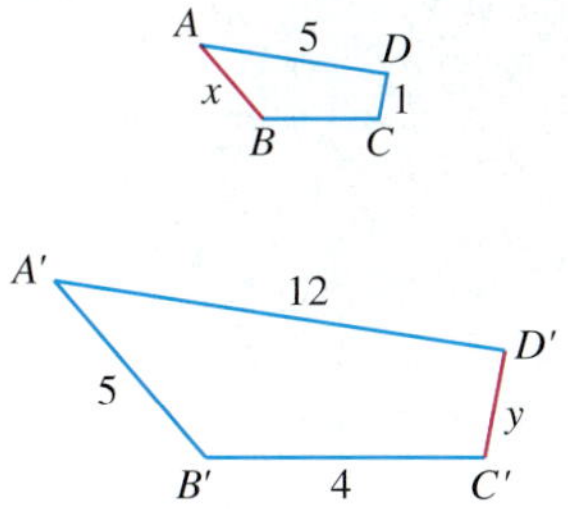

51.

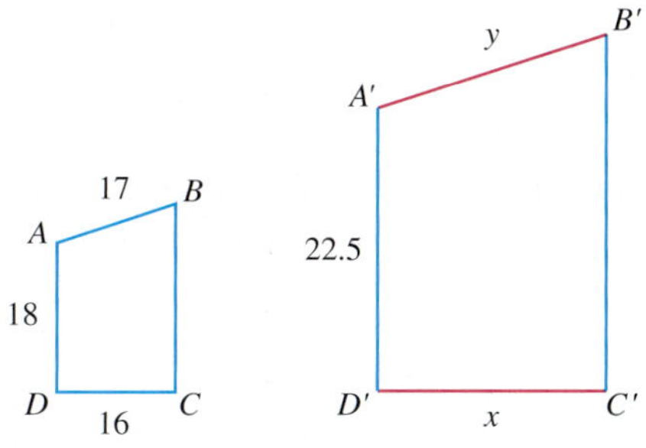

52.

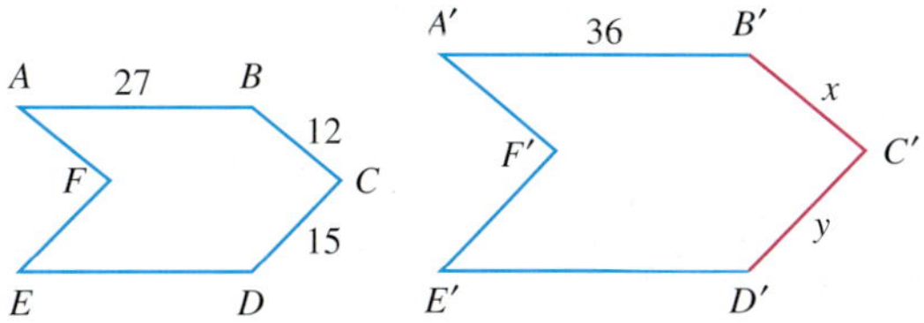

In Exercises 53–56, triangles ABC and DEC are similar figures. Determine the length of

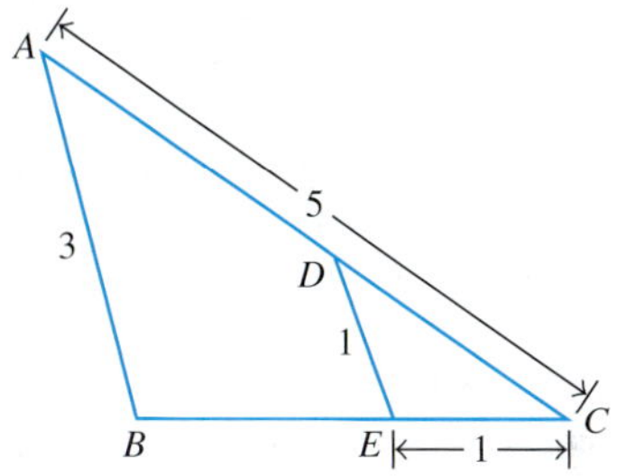

53. side $\overline{BC}$.

54. side $\overline{DC}$.

55. side $\overline{AD}$.

56. side $\overline{BE}$.

In Exercises 57–62, find the length of the sides and the measures of the angles for the congruent triangles ABC and A′B′C′.

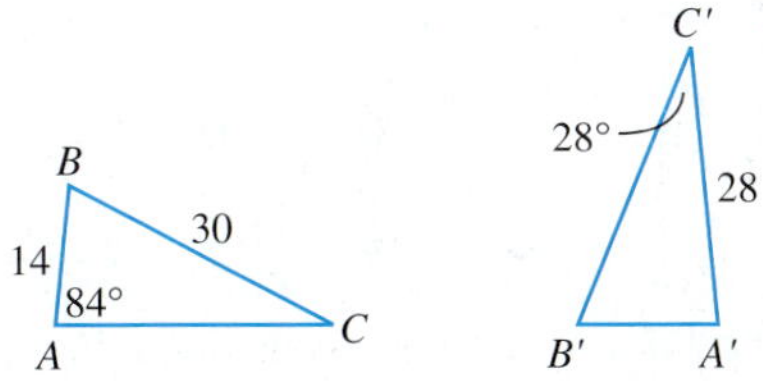

57. The length of side $\overline{AC}$

58. The length of side $\overline{A'B'}$

59. The length of side $\overline{B'C'}$

60. $\measuredangle B'A'C'$

61. $\measuredangle ACB$

62. $\measuredangle ABC$

In Exercises 63–68, determine the length of the sides and the measures of the angles for the congruent quadrilaterals ABCD and A′B′C′D′.

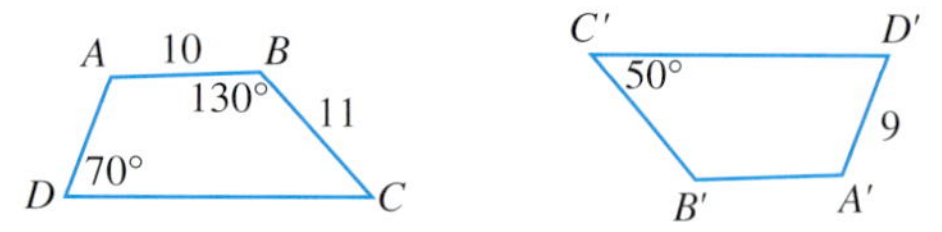

63. The length of side $\overline{AD}$

64. The length of side $\overline{B'C'}$

65. The length of side $\overline{A'B'}$

66. $\measuredangle BCD$

67. $\measuredangle A'D'C'$

68. $\measuredangle DAB$

PROBLEM SOLVING

In Exercises 69–72, determine the measure of the angle. In the figure, $\measuredangle ABC$ makes an angle of 125° with the floor and l_1 and l_2 are parallel.

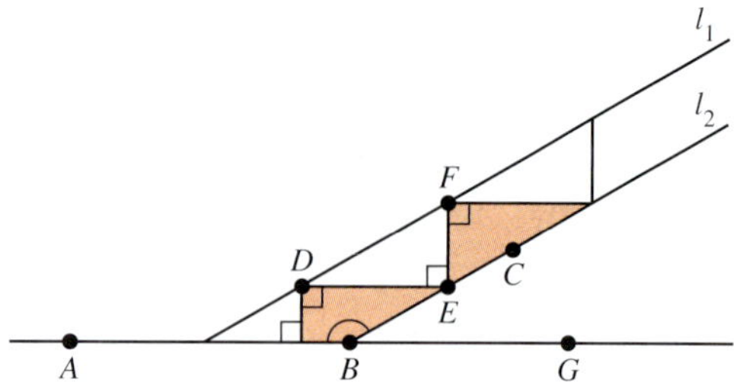

69. $\measuredangle GBC$

70. $\measuredangle EDF$

71. $\measuredangle DFE$

72. $\measuredangle DEC$

73. ***Height of a Silo*** Steve Runde is buying a farm and needs to determine the height of a silo on the farm. Steve, who is 6 ft tall, notices that when his shadow is 9 ft long, the shadow of the silo is 105 ft long (see diagram). How tall is the silo? Note that the diagram is not to scale.

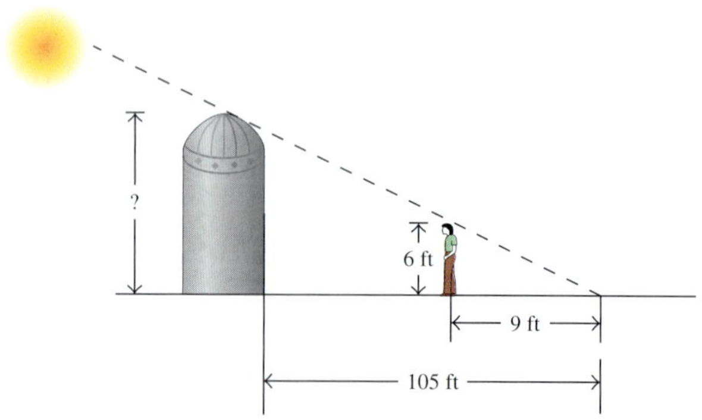

74. ***Angles on a Picnic Table*** The legs of a picnic table form an isosceles triangle as indicated in the figure. If $\measuredangle ABC = 80°$, determine $m\measuredangle x$ and $m\measuredangle y$ so that the top of the table will be parallel to the ground.

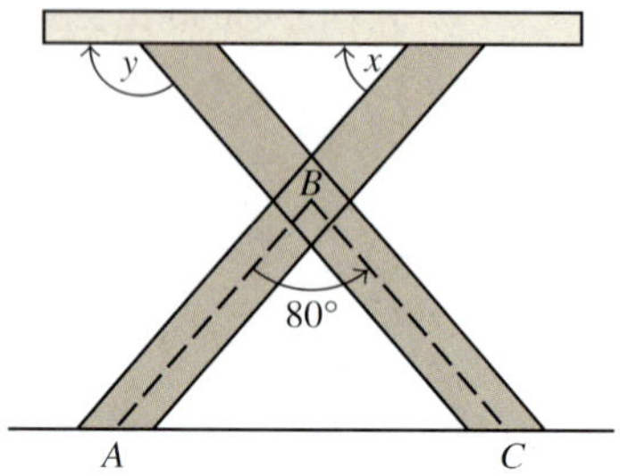

75. ***Distances in Minnesota*** A triangle can be formed by drawing line segments on a map of Minnesota connecting the cities of Austin, Rochester, and St. Paul (see figure). If the actual distance from Austin to Rochester is approximately 44 miles, use the lengths of the line segments indicated in the figure along with similar triangles to approximate

a) the actual distance from St. Paul to Austin.

b) the actual distance from St. Paul to Rochester.

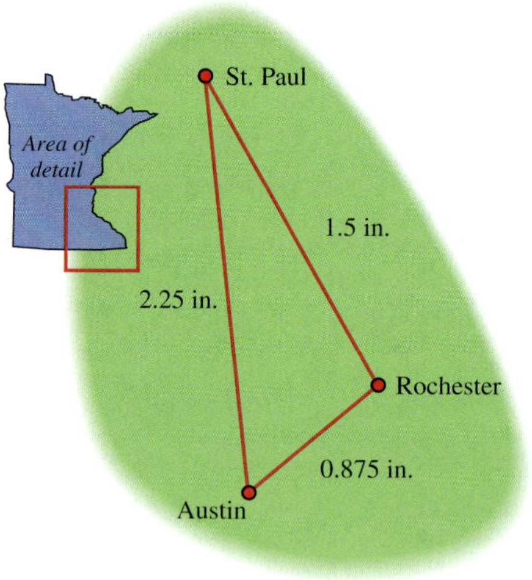

76. ***Distances in Illinois*** A triangle can be formed by drawing line segments on a map of Illinois connecting the cities of Rockford, Chicago, and Bloomington (see figure on top of page 537). If the actual distance from Chicago to Rockford is approximately 90 miles, use the lengths of the line segments indicated in the figure along with similar triangles to approximate

a) the actual distance from Chicago to Bloomington.

b) the actual distance from Bloomington to Rockford.

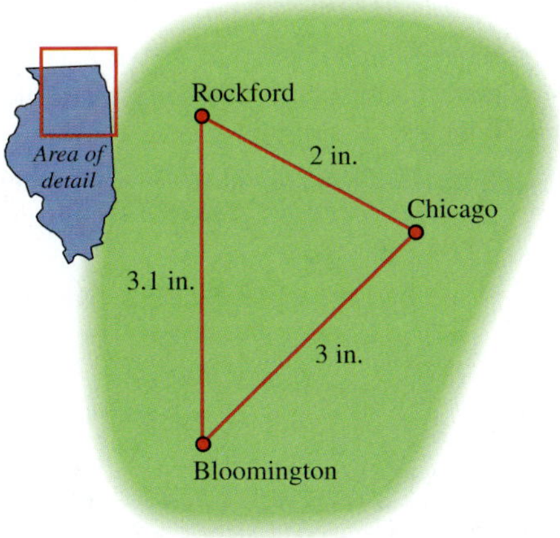

CHALLENGE PROBLEMS/GROUP ACTIVITIES

Scaling Factor *Examine the similar triangles ABC and $A'B'C'$ in the figure below.*

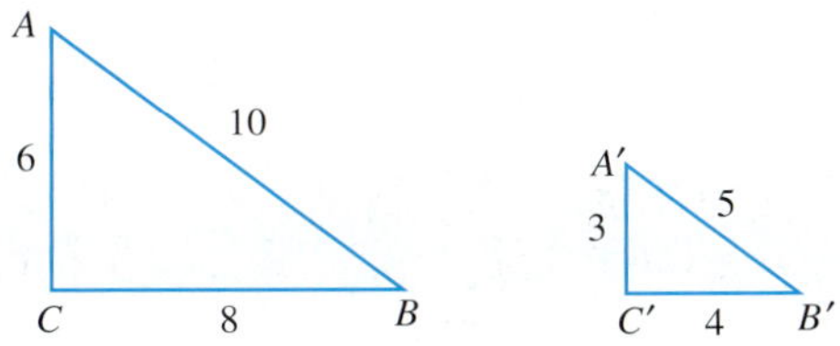

If we calculate the ratios $\dfrac{AB}{A'B'}, \dfrac{BC}{B'C'}$, *and* $\dfrac{CA}{C'A'}$, *we see that each of these ratios is equal to 2. We call this common ratio the* ***scaling factor*** *of ΔABC with respect to $\Delta A'B'C'$. If we calculate the reciprocal ratios* $\dfrac{A'B'}{AB}, \dfrac{B'C'}{BC}$, *and* $\dfrac{C'A'}{CA}$, *we see that each of these ratios is equal to $\frac{1}{2}$. We call this common ratio the scaling factor of $\Delta A'B'C'$ with respect to ΔABC. Every pair of similar figures has two scaling factors that show the relationship between the corresponding side lengths. Notice that the length of each side of ΔABC is two times the length of the corresponding side in $\Delta A'B'C'$. We can also state that the length of each side of $\Delta A'B'C'$ is one-half the length of the corresponding side of ΔABC.*

77. In the figure, ΔDEF is similar to $\Delta D'E'F'$. The length of the sides of ΔDEF is shown in the figure. If the scaling factor of ΔDEF with respect to $\Delta D'E'F'$ is 3, determine the length of the sides of triangle $\Delta D'E'F'$.

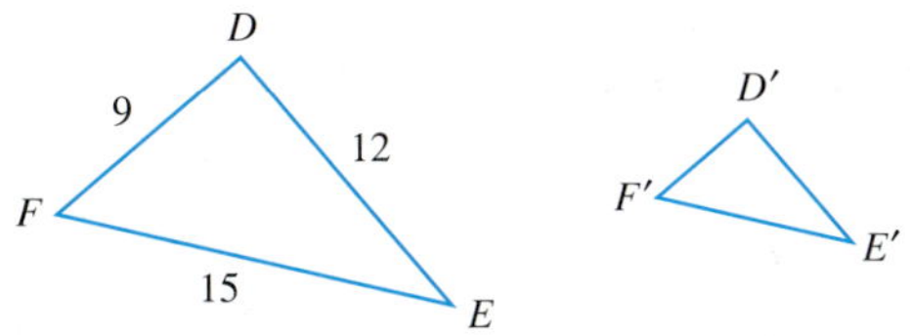

78. In the figure, quadrilateral *EFGH* is similar to quadrilateral $E'F'G'H'$. The length of the sides of quadrilateral *EFGH* is shown in the figure. If the scaling factor of quadrilateral $E'F'G'H'$ with respect to quadrilateral *EFGH* is $\frac{1}{3}$, determine the length of the sides of quadrilateral $E'F'G'H'$.

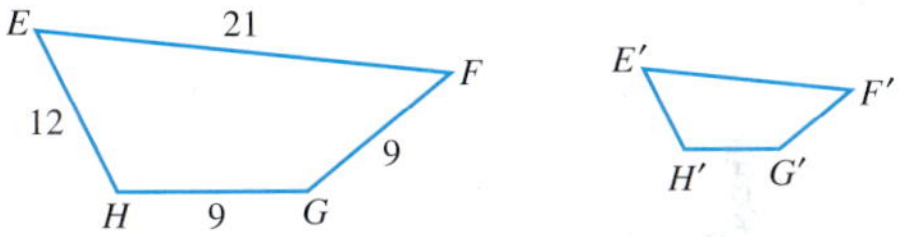

79. ***Height of a Wall*** You are asked to measure the height of an inside wall of a warehouse. No ladder tall enough to measure the height is available. You borrow a mirror from a salesclerk and place it on the floor. You then move away from the mirror until you can see the reflection of the top of the wall in it, as shown in the figure.

a) Explain why triangle *HFM* is similar to triangle *TBM*. (*Hint:* In the reflection of light the angle of incidence equals the angle of reflection. Thus, $\measuredangle HMF = \measuredangle TMB$.)

b) If your eyes are $5\frac{1}{2}$ ft above the floor and you are $2\frac{1}{2}$ ft from the mirror and the mirror is 20 ft from the wall, how high is the wall?

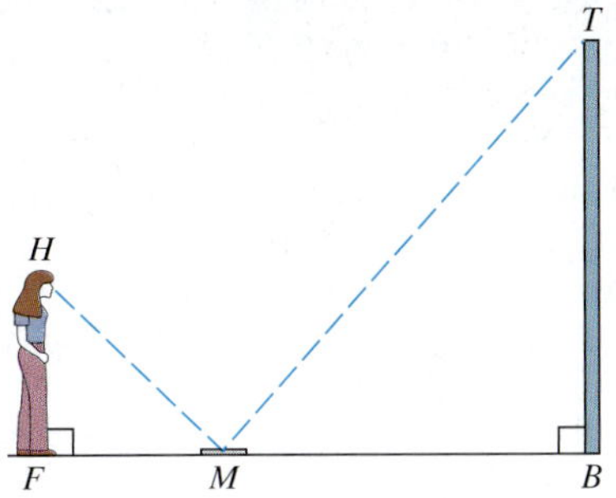

RECREATIONAL MATHEMATICS

80. *Distance Across a Lake*

a) In the figure, $m\measuredangle CED = m\measuredangle ABC$. Explain why triangles ABC and DEC must be similar.

b) Determine the distance across the lake, DE.

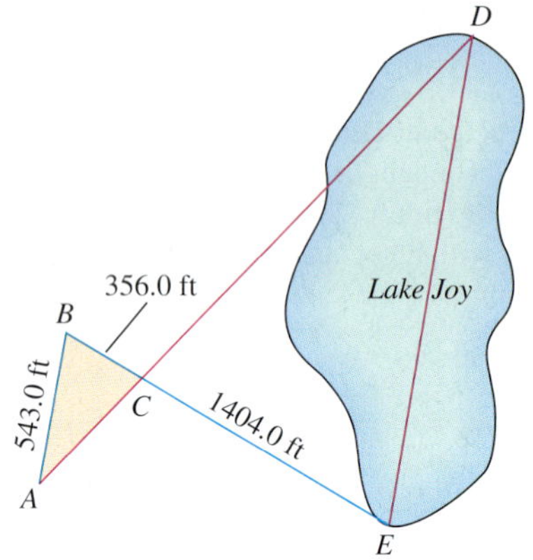

INTERNET/RESEARCH ACTIVITIES

81. Using the Internet and history of mathematics texts, write a paper on the history and use of the theodolite, a surveying instrument.

82. Using the Internet and other sources, write a paper on the use of geometry in the photographic process. Include discussions on the use of similar figures.

9.3 PERIMETER AND AREA

▲ We will study the perimeter and area of rectangles, such as those on which basketball games are played.

If you were to walk around the outside edge of a basketball court, how far would you walk? If you had to place floor tiles that each measured 1 foot by 1 foot on a basketball court, how many tiles would you need? In this section, we will study the geometric concepts of perimeter and area that are used to answer these and other questions.

Perimeter and Area

The ***perimeter***, P, of a two-dimensional figure is the sum of the lengths of the sides of the figure. In Figs. 9.25 and 9.26, the sums of the lengths of the red line segments are the perimeters. Perimeters are measured in the same units as the sides. For example, if the sides of a figure are measured in feet, the perimeter will be measured in feet.

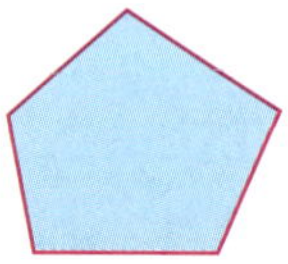

Figure 9.25

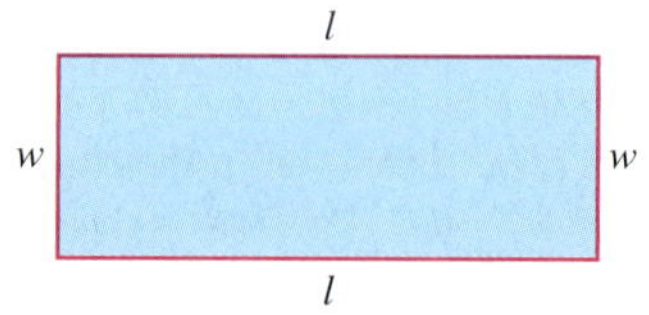

Figure 9.26

The *area*, A, is the region within the boundaries of the figure. The blue color in Figs. 9.25 and 9.26 indicates the areas of the figures. Area is measured in square units. For example, if the sides of a figure are measured in inches, the area of the figure will be measured in square inches (in.^{2}). (See Table 8.7 on page 501 for common units of area in the U.S. customary and metric systems.)

Consider the rectangle in Fig. 9.26. Two sides of the rectangle have length l, and two sides of the rectangle have width w. Thus, if we add the lengths of the four sides to get the perimeter, we find $P = l + w + l + w = 2l + 2w$.

PERIMETER OF A RECTANGLE

$$P = 2l + 2w$$

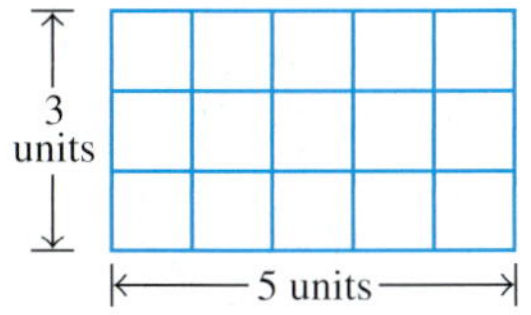

Figure 9.27

Consider a rectangle of length 5 units and width 3 units (Fig. 9.27). Counting the number of 1-unit by 1-unit squares within the figure we obtain the area of the rectangle, 15 square units. The area can also be obtained by multiplying the number of units of length by the number of units of width, or 5 units $\times$ 3 units $=$ 15 square units. We can find the area of a rectangle by the formula area $=$ length $\times$ width.

AREA OF A RECTANGLE

$$A = l \times w$$

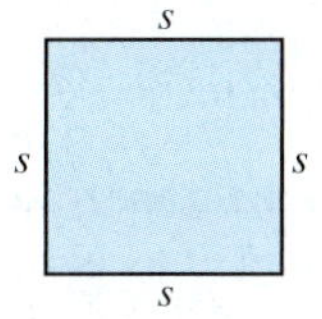

Figure 9.28

Using the formula for the area of a rectangle, we can determine the formulas for the areas of other figures.

A square (Fig. 9.28) is a rectangle that contains four equal sides. Therefore, the length equals the width. If we call both the length and the width of the square s, then

$$A = l \times w, \quad \text{so} \quad A = s \times s = s^2$$

AREA OF A SQUARE

$$A = s^2$$

A parallelogram with height h and base b is shown in Fig. 9.29(a).

Figure 9.29

If we were to cut off the red portion of the parallelogram on the left, Fig. 9.29(a), and attach it to the right side of the figure, the resulting figure would be a rectangle, Fig. 9.29(b). Since the area of the rectangle is $b \times h$, the area of the parallelogram is also $b \times h$.

DID YOU KNOW?

Area versus Perimeter

New Mexico

Which geographic area, Italy or New Mexico, do you think has the larger land area? As a matter of fact, the land areas of Italy and New Mexico are very similar: Italy's area is 116,304 mi², and New Mexico's area is 121,593 mi². Italy's perimeter, 5812 mi, is almost five times greater than New Mexico's 1200 mi, however.

AREA OF A PARALLELOGRAM

$$A = b \times h$$

Consider the triangle with height, h, and base, b, shown in Fig. 9.30(a). Using this triangle and a second identical triangle, we can construct a parallelogram, Fig. 9.30(b). The area of the parallelogram is bh. The area of the triangle is one-half that of the parallelogram. Therefore, the area of the triangle is $\frac{1}{2}$(base)(height).

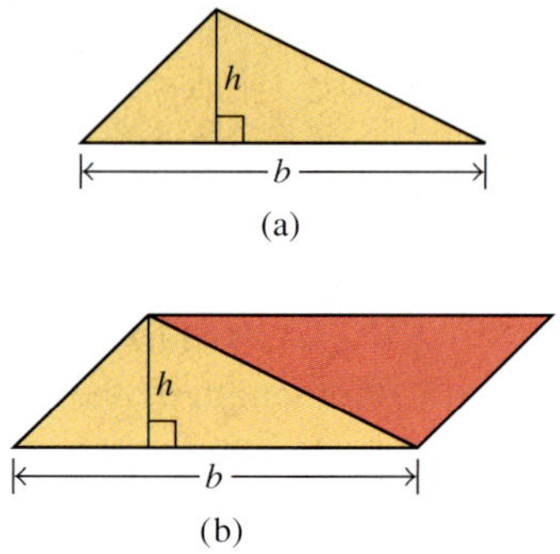

Figure 9.30

AREA OF A TRIANGLE

$$A = \frac{1}{2}bh$$

Now consider the trapezoid shown in Fig. 9.31(a). We can partition the trapezoid into two triangles by drawing diagonal $\overline{DB}$, as in Fig. 9.31(b). One triangle has base $\overline{AB}$ (called b_2) with height $\overline{DE}$, and the other triangle has base $\overline{DC}$ (called b_1) with height $\overline{FB}$. Note that the line used to measure the height of the triangle need not be inside the triangle. Because heights $\overline{DE}$ and $\overline{FB}$ are equal, both triangles have the same height, h. The area of triangles DCB and ADB are $\frac{1}{2}b_1h$ and $\frac{1}{2}b_2h$, respectively. The area of the trapezoid is the sum of the areas of the triangles, $\frac{1}{2}b_1h + \frac{1}{2}b_2h$, which can be written $\frac{1}{2}h(b_1 + b_2)$.

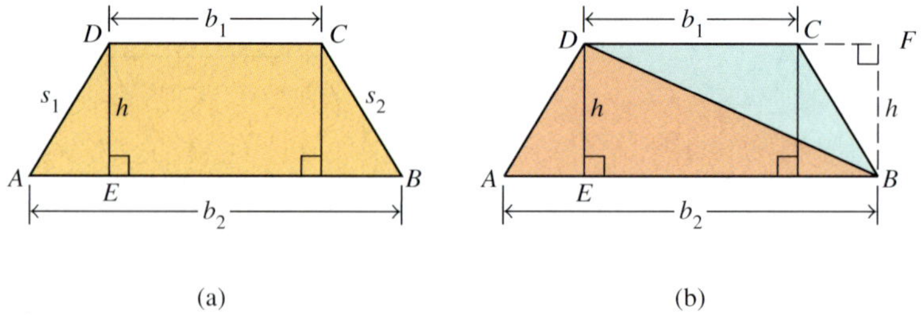

Figure 9.31

AREA OF A TRAPEZOID

$$A = \frac{1}{2}h(b_1 + b_2)$$

Following is a summary of the perimeters and areas of selected figures.

PERIMETERS AND AREAS

Triangle

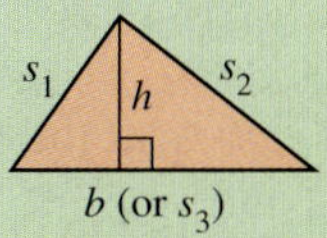

$P = s_1 + s_2 + s_3 \ (s_3 = b)$

$A = \frac{1}{2}bh$

Square

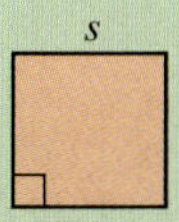

$P = 4s$

$A = s^2$

Rectangle

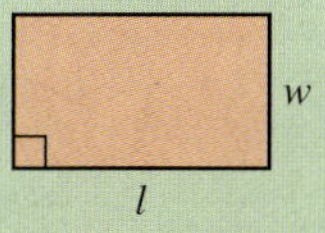

$P = 2l + 2w$

$A = lw$

Parallelogram

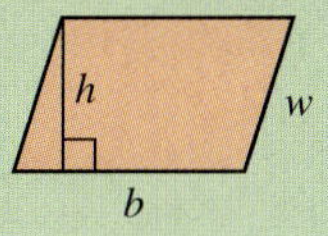

$P = 2b + 2w$

$A = bh$

Trapezoid

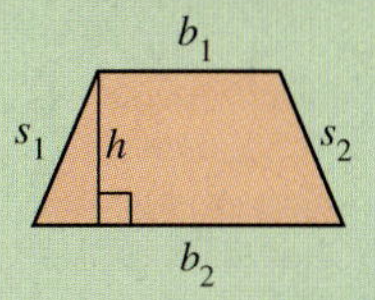

$P = s_1 + s_2 + b_1 + b_2$

$A = \frac{1}{2}h(b_1 + b_2)$

EXAMPLE 1 *Sodding a Backyard*

Kevin Geis wishes to replace the grass (sod) in his backyard. One pallet of Milberger's sod costs \$130 and covers 450 square feet. If Kevin's backyard is rectangular in shape with a length of 90 feet and width of 60 feet, determine

a) the area of the backyard.

b) how many pallets of sod Kevin needs to purchase.

c) the cost of the sod purchased.

SOLUTION

a) The area of the backyard is

$$A = l \cdot w = 90 \cdot 60 = 5400 \text{ ft}^2$$

The area of the backyard is in square feet because both the length and width are measured in feet.

b) To determine the number of pallets of sod Kevin needs, divide the area of the backyard by the area covered by one pallet of sod.

$$\frac{\text{Area of backyard}}{\text{Area covered by one pallet}} = \frac{5400}{450} = 12$$

Kevin needs to purchase 12 pallets of sod.

c) The cost of 12 pallets of sod is $12 \times \$130$, or \$1560.

Pythagorean Theorem

We introduced the Pythagorean theorem in Chapter 5. Because this theorem is an important tool for finding the perimeter and area of triangles, we restate it here.

MATHEMATICS TODAY

Pythagorean Theorem

The Pythagorean theorem is one of the most famous theorems of all time. One book, *Pythagorean Propositions*, contains 370 different proofs of the Pythagorean theorem. U.S. President James A. Garfield gave one notable proof. The Pythagorean theorem has found its way into popular culture as well. In the movie *The Wizard of Oz*, the scarecrow incorrectly recites the Pythagorean theorem once the wizard grants him his diploma. In the play *The Pirates of Penzance*, the Major General refers to the Pythagorean theorem when he sings "I'm teeming with a lot o' news, with many cheerful facts about the square of the hypotenuse." Lewis Carroll, author of *Through the Looking Glass* and *Alice's Adventures in Wonderland*, stated that the Pythagorean theorem "is as dazzlingly beautiful now as it was in the day when Pythagoras discovered it." Actually, the Pythagorean theorem was known to the ancient Babylonians in about 1600 B.C., 1000 years before Pythagoras, and it continues to play a huge role in mathematics.

PYTHAGOREAN THEOREM

The sum of the squares of the lengths of the legs of a right triangle equals the square of the length of the hypotenuse.

$$\text{leg}^2 + \text{leg}^2 = \text{hypotenuse}^2$$

Symbolically, if a and b represent the lengths of the legs and c represents the length of the hypotenuse (the side opposite the right angle), then

$$a^2 + b^2 = c^2$$

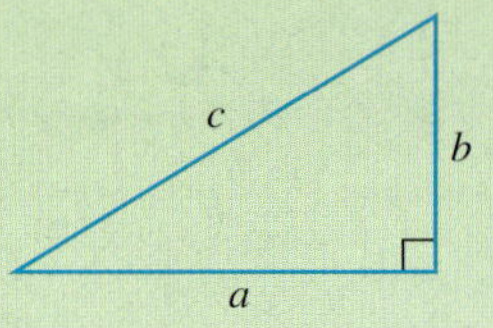

EXAMPLE 2 *Crossing a Moat*

The moat surrounding a castle is 18 ft wide and the wall by the moat of the castle is 24 ft high (see Fig. 9.32). If an invading army wishes to use a ladder to cross the moat and reach the top of the wall, how long must the ladder be?

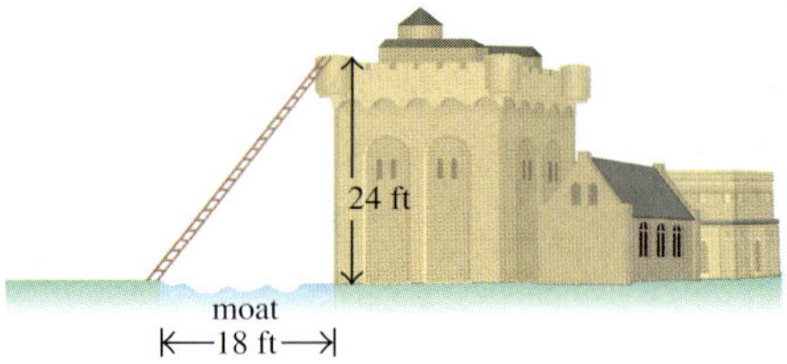

Figure 9.32

SOLUTION The moat, the castle wall, and the ladder form a right triangle. The moat and the castle wall form the legs of the triangle (sides a and b), and the ladder forms the hypotenuse (side c). By the Pythagorean theorem,

$$c^2 = a^2 + b^2$$
$$c^2 = (18)^2 + (24)^2$$
$$c^2 = 324 + 576$$
$$c^2 = 900$$
$$\sqrt{c^2} = \sqrt{900} \quad \text{Take the square root of both sides of the equation.}$$
$$c = 30$$

Therefore, the ladder would need to be at least 30 ft long.

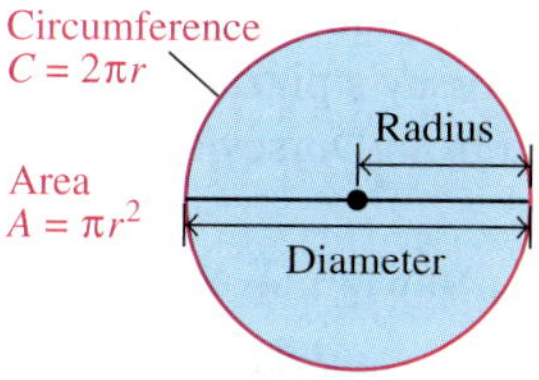

Figure 9.33

Circles

A commonly used plane figure that is not a polygon is a *circle*. A ***circle*** is a set of points equidistant from a fixed point called the center. A ***radius***, r, of a circle is a line segment from the center of the circle to any point on the circle (Fig. 9.33). A ***diameter***, d, of a circle is a line segment through the center of a circle with both end points on the circle. Note that the diameter of the circle is twice its radius. The ***circumference*** is the length of the simple closed curve that forms the circle. The formulas for the area and circumference of a circle are given in Fig. 9.33. The symbol pi, π, was introduced in Chapter 5. Recall that π is approximately 3.14. If your calculator contains a $\boxed{\pi}$ key, you should use that key when working calculations involving pi.

DID YOU KNOW?

City Planning

The Roman poet Virgil tells the story of Queen Dido, who fled to Africa after her brother murdered her husband. There, she begged for some land from King Iarbus, telling him she only needed as much land as the hide of an ox would enclose. Being very clever, she decided that the greatest area would be enclosed if she tore the hide into thin strips and formed the strips into a circle. On this land she founded the city of Byrsa (the Greek word for "hide"), later known as Carthage in present-day Tunisia.

EXAMPLE 3 *Comparing Pizzas*

Victoria Montoya wishes to order a large cheese pizza. She can choose among three pizza parlors in town: Antonio's, Brett's, and Dorsey's. Antonio's large cheese pizza is a round 16-in.-diameter pizza that sells for \$15. Brett's large cheese pizza is a round 14-in.-diameter pizza that sells for \$12. Dorsey's large cheese pizza is a square 12-in. by 12-in. pizza that sells for \$10. All three pizzas have the same thickness. To get the most for her money, from which pizza parlor should Victoria order her pizza?

SOLUTION To determine the best value, we will calculate the cost per square inch of pizza for each of the three pizzas. To do so, we will divide the cost of each pizza by its area. The areas of the two round pizzas can be determined using the formula for the area of a circle, $A = \pi r^2$. Since the radius is half the diameter, we will use $r = 8$ and $r = 7$ for Antonio's and Brett's large pizzas, respectively. The area for the square pizza can be determined using the formula for the area of a square, $A = s^2$. We will use $s = 12$.

$$\text{Area of Antonio's pizza} = \pi r^2 \approx (3.14)(8)^2 \approx 3.14(64) \approx 200.96 \text{ in.}^2{*}$$
$$\text{Area of Brett's pizza} = \pi r^2 \approx (3.14)(7)^2 \approx 3.14(49) \approx 153.86 \text{ in.}^2{*}$$
$$\text{Area of Dorsey's pizza} = s^2 = (12)^2 = 144 \text{ in.}^2$$

Now, to find the cost per square inch of pizza, we will divide the cost of the pizza by the area of the pizza.

$$\text{Cost per square inch of Antonio's pizza} \approx \frac{\$15}{200.96 \text{ in.}^2} \approx \$0.0746$$

Thus, Antonio's pizza costs about \$0.0746, or about 7.5 cents, per square inch.

$$\text{Cost per square inch of Brett's pizza} \approx \frac{\$12}{153.86 \text{ in.}^2} \approx \$0.0780$$

Thus, Brett's pizza costs about \$0.0780, or about 7.8 cents, per square inch.

$$\text{Cost per square inch of Dorsey's pizza} = \frac{\$10}{144 \text{ in.}^2} \approx \$0.0694$$

*If you use the $\boxed{\pi}$ key on your calculator, your answers will be slightly more accurate.

3. a) How do you convert an area from square yards into square feet?

   b) How do you convert an area from square feet into square yards?

4. a) How do you convert an area from square feet into square inches?

   b) How do you convert an area from square inches into square feet?

PRACTICE THE SKILLS

In Exercises 5–8, find the area of the triangle.

5.
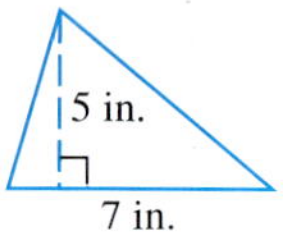

6.
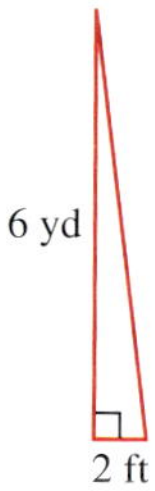

7.
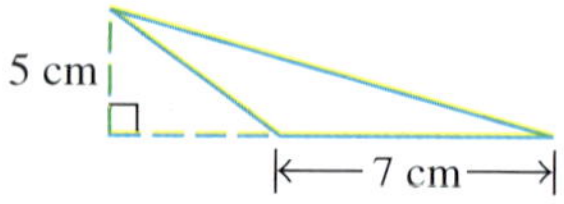

8.
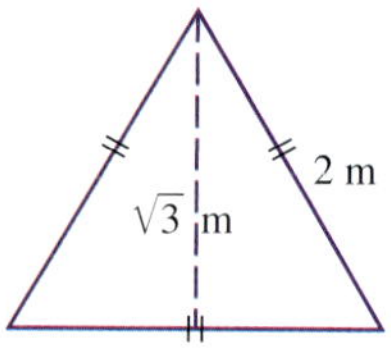

In Exercises 9–14, determine (a) the area and (b) the perimeter of the quadrilateral.

9.
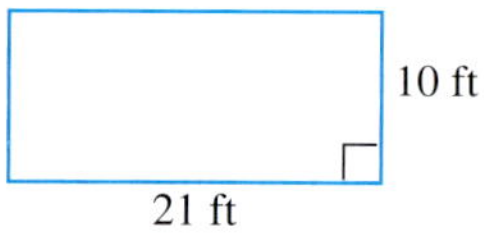

10.
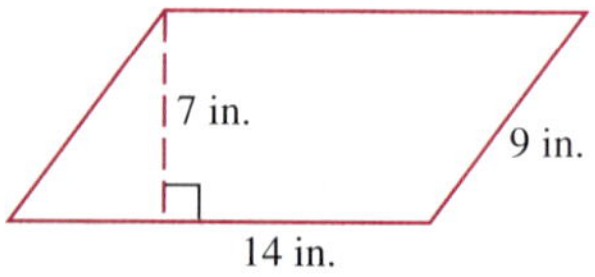

11.
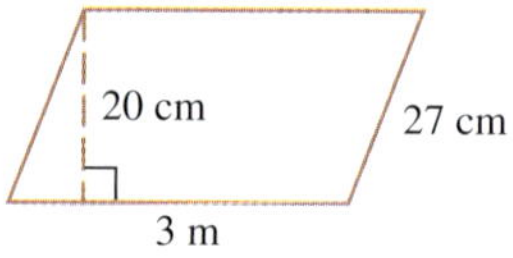

12.
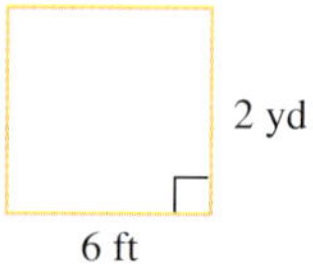

13.
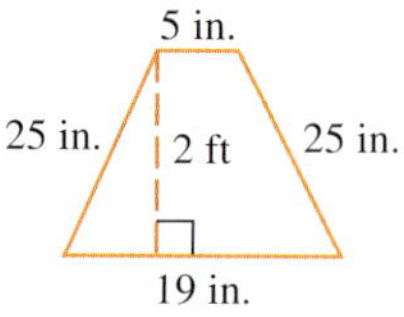

14.
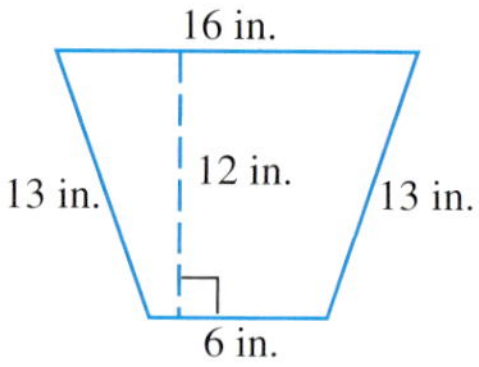

In Exercises 15–18, determine (a) the area and (b) the circumference of the circle. Use the [π] *key on your calculator and round your answer to the nearest hundredth.*

15.
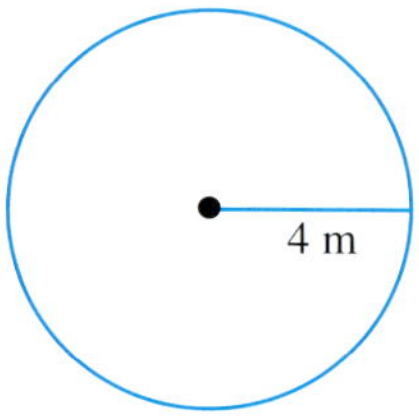

16.
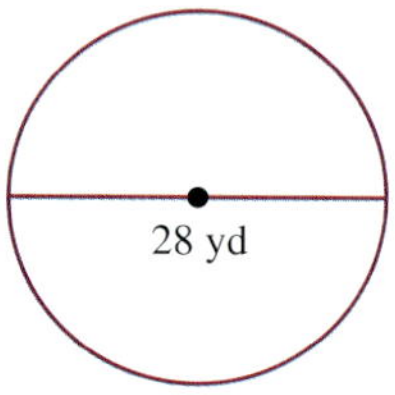

17.
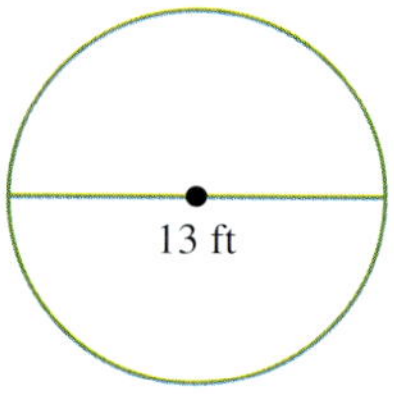

18.
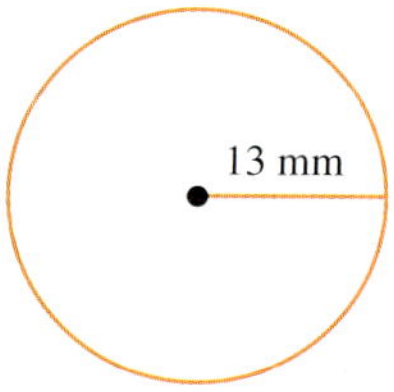

In Exercises 19–22, (a) use the Pythagorean theorem to determine the length of the unknown side of the triangle, (b) determine the perimeter of the triangle, and (c) determine the area of the triangle.

19.
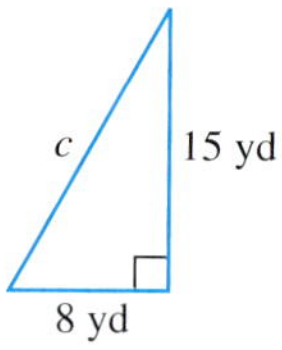

20.
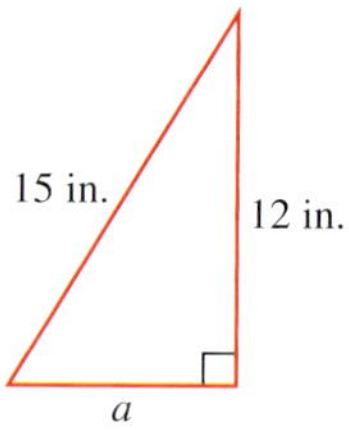

21.
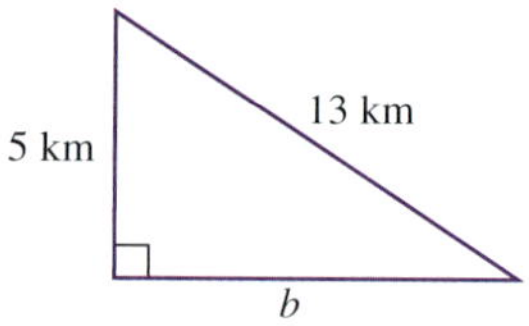

22.
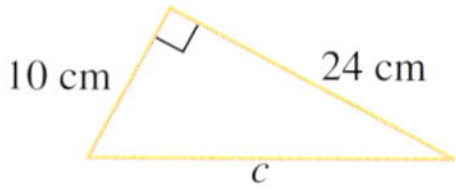

PROBLEM SOLVING

In Exercises 23–32, find the shaded area. When appropriate, use the $\boxed{\pi}$ *key on your calculator and round your answer to the nearest hundredth.*

23.

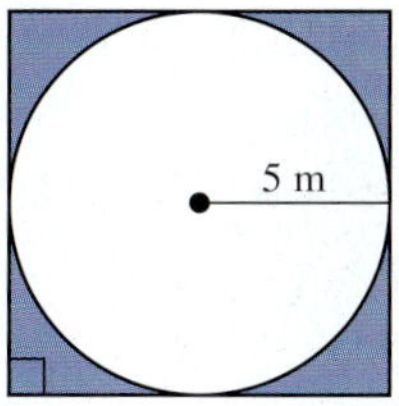

24.

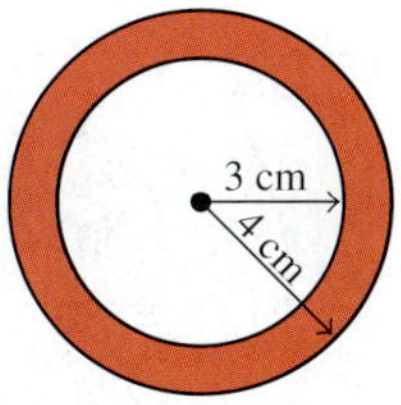

25.

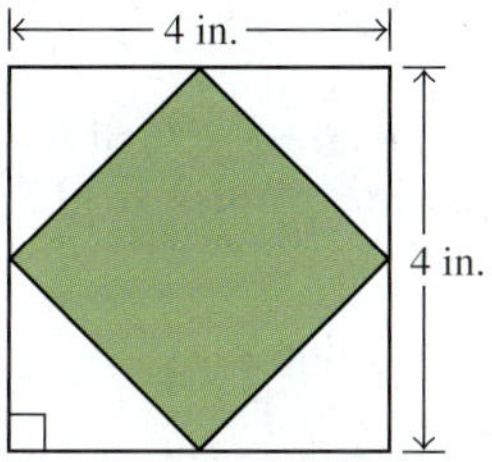

26.

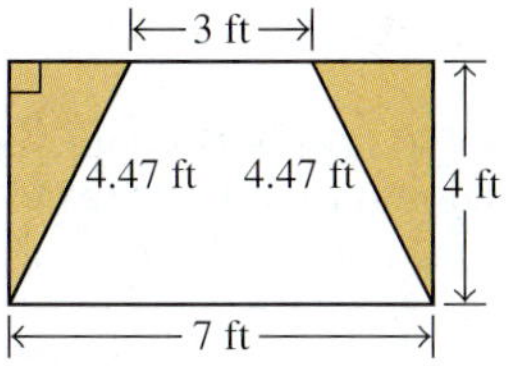

27.

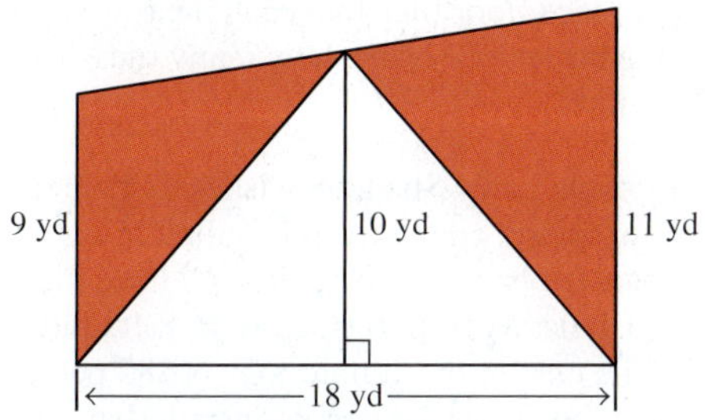

28.

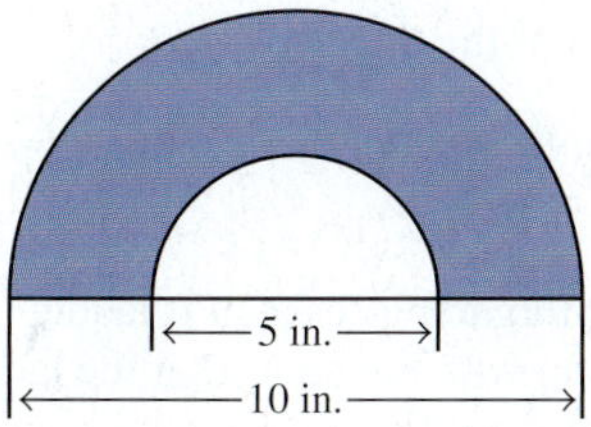

29.

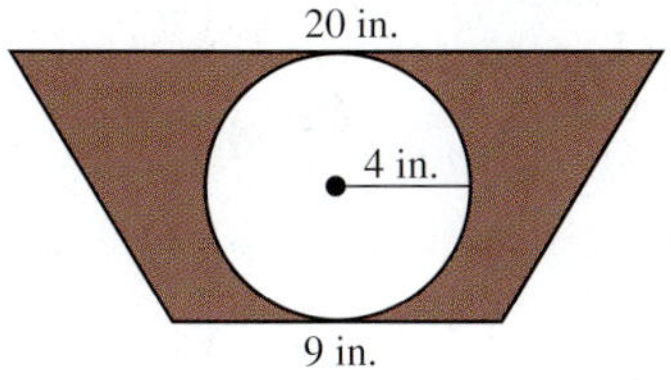

30.

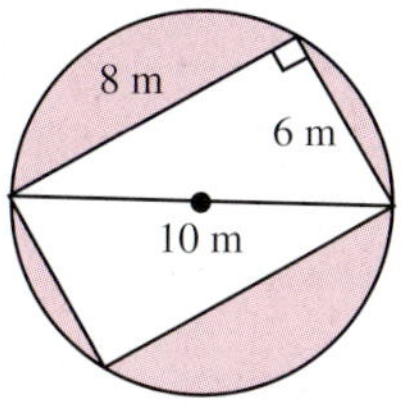

31.

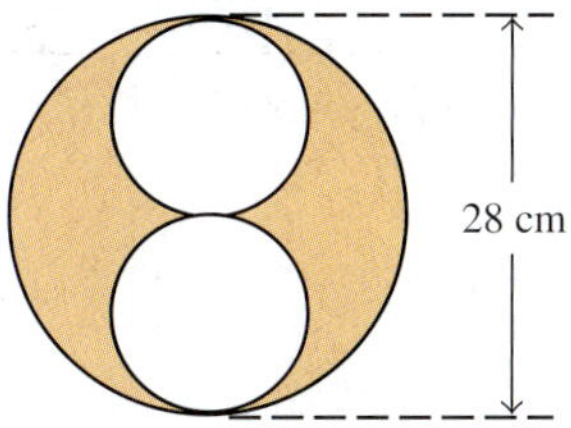

32.

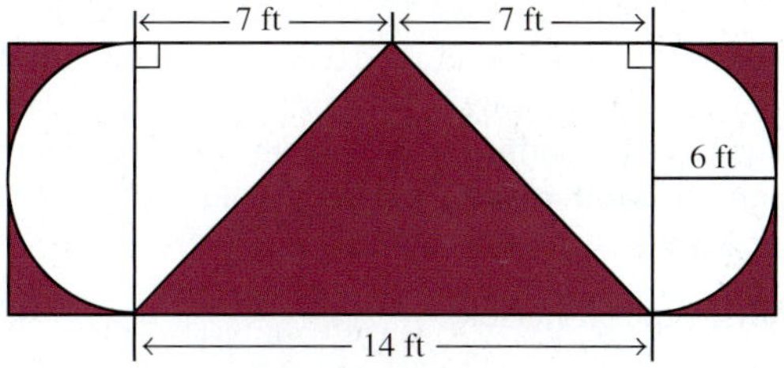

One square yard equals 9 ft². Use this information to convert the following.

33. 207 ft² to square yards

34. 15.2 ft² to square yards

35. 14.7 yd² to square feet

36. 15.2 yd² to square feet

One square meter equals 10,000 cm². Use this information to convert the following.

37. 23.4 m² to square centimeters

38. 0.375 m² to square centimeters

39. 8625 cm² to square meters

40. 608 cm² to square meters

Nancy Wallin has just purchased a new house that is in need of new flooring. In Exercises 41–46, use the measurements given on the floor plans of Nancy's house to obtain the answer.

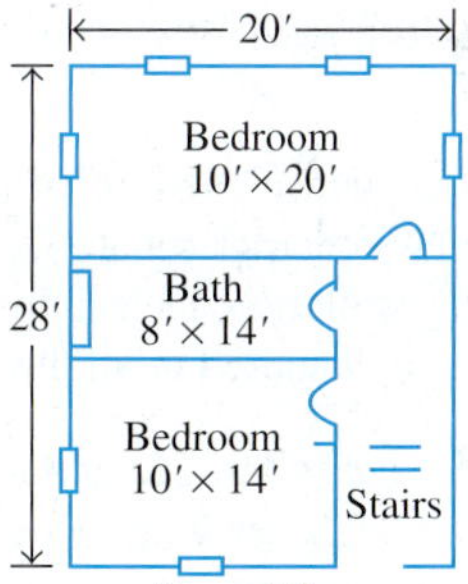

Second floor

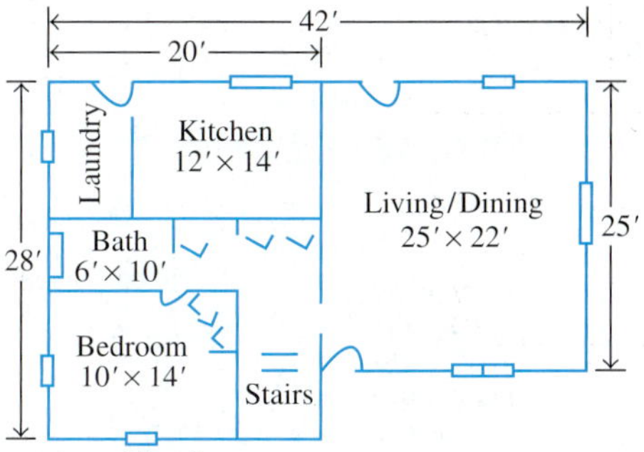

First floor

41. *Cost of Hardwood Flooring* The cost of Mannington Chestnut hardwood flooring is \$10.86 per square foot if Nancy installs the flooring herself or \$13.86 per square foot if she has the flooring installed by the flooring company. Determine the cost for hardwood flooring in the living/dining room if

a) Nancy installs it herself.

b) Nancy has it installed by the flooring company.

42. *Cost of Laminate Flooring* The cost of Pergo Select Helsinki Birch laminate flooring is \$5.89 per square foot if Nancy installs the flooring herself or \$8.89 per square foot if she has the flooring installed by the flooring company. Determine the cost for the flooring in the living/dining room if

a) Nancy installs it herself.

b) Nancy has it installed by the flooring company.

43. *Cost of Ceramic Tile* The cost of Mohawk Porcelain ceramic tile is \$8.50 per square foot. This price includes the cost of installation. Determine the cost for Nancy to have this ceramic tile installed in the kitchen and in both bathrooms.

44. *Cost of Linoleum* The cost of Armstrong Solarian Woodcut linoleum is \$5.00 per square foot. This price includes the cost of installation. Determine the cost for Nancy to have this linoleum installed in the kitchen and in both bathrooms.

45. *Cost of Berber Carpeting* The cost of Bigelow Commodore Berber carpeting is \$6.06 per square foot. This price includes the cost of installation. Determine the cost for Nancy to have this carpeting installed in all three bedrooms.

46. *Cost of Saxony Carpeting* The cost of DuPont Stainmaster Saxony carpeting is \$5.56 per square foot. This price includes the cost of installation. Determine the cost for Nancy to have this carpeting installed in all three bedrooms.

47. *Cost of a Lawn Service* Clarence and Rose Cohen's home lot is illustrated here. Clarence and Rose wish to hire Picture Perfect Lawn Service to cut their lawn. How much will it cost Clarence and Rose to have their lawn cut if Picture Perfect charges \$0.02 per square yard?

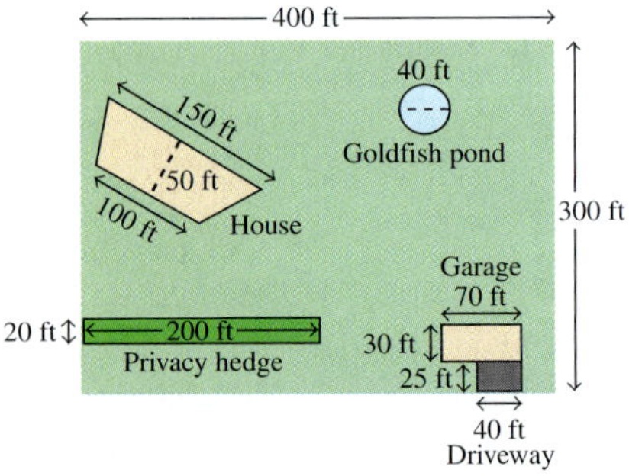

48. *Cost of a Lawn Service* Jim and Wendy Scott's home lot is illustrated here. The Scotts wish to hire a lawn service to cut their lawn. M&M Lawn Service charges \$0.02 per square yard of lawn. How much will it cost the Scotts to have their lawn cut?

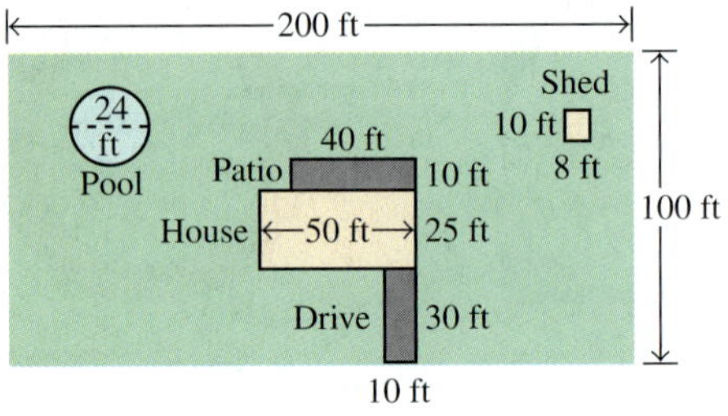

49. *Area of a Basketball Court* A National Basketball Association (NBA) basketball court is a rectangle that is 94 ft long and 50 ft wide.

a) If you were to walk around the outside edge of a basketball court, how far would you walk?

b) If you had to place floor tiles that each measured 1 foot by 1 foot on a basketball court, how many tiles would you need?

50. *Quartz Countertops* Larry Shedden wishes to have three Cambria Windsor quartz countertops installed in his new kitchen. The countertops are rectangular and have the following dimensions: $3\frac{1}{2}$ ft × 6 ft, $2\frac{1}{2}$ ft × 8 ft, and 3 ft × $11\frac{1}{2}$ ft. The cost of the countertops is \$87 per square foot, which includes the cost of installation.

a) Determine the total area of the three countertops.

b) Determine the total cost to have all three countertops installed.

51. *Ladder on a Wall* Lorrie Morgan places a 29 ft ladder against the side of a building with the bottom of the ladder 20 ft away from the building (see figure). How high up on the wall does the ladder reach?

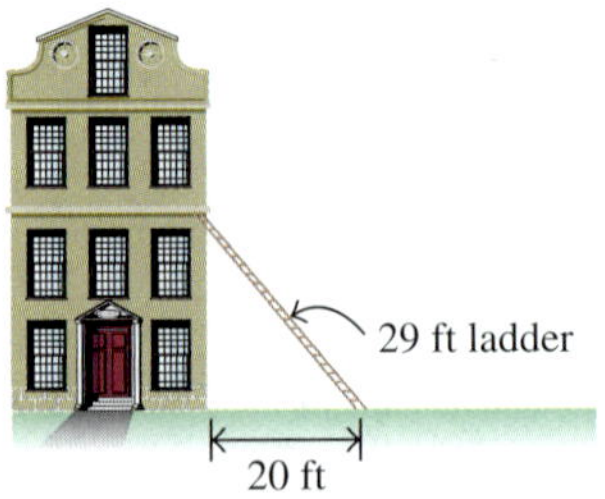

52. ***Docking a Boat*** Brian Murphy is bringing his boat into a dock that is 9 ft above the water level (see figure). If a 41 ft rope is attached to the dock on one side and to the boat on the other side, determine the horizontal distance from the dock to the boat.

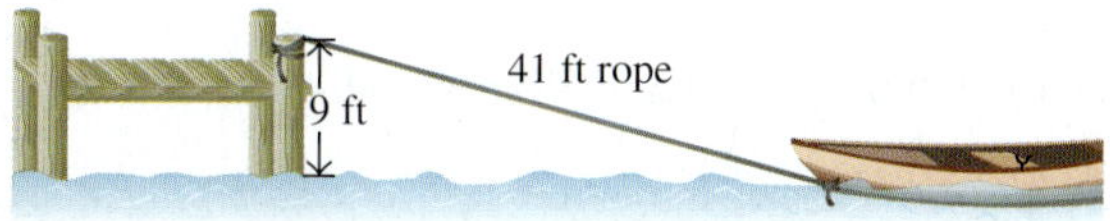

53. ***The Green Monster*** In Fenway Park, home of baseball's Boston Red Sox, the left field wall is known as the *Green Monster*. The distance from home plate down the third baseline to the bottom of the wall is 310 feet (see photo). In left field, at the end of the baseline, the Green Monster is perpendicular to the ground and is 37 feet tall. Determine the distance from home plate to the top of the Green Monster along the third baseline. Round your answer to the nearest foot.

CHALLENGE PROBLEMS/GROUP ACTIVITIES

54. ***Plasma Television*** The screen of a plasma television is in the shape of a rectangle with a diagonal of length 43 in. If the height of the screen is 21 in., determine the width of the screen.

55. ***Doubling the Sides of a Square*** In the figure below, an original square with sides of length s is shown. Also shown is a larger square with sides double in length, or $2s$.

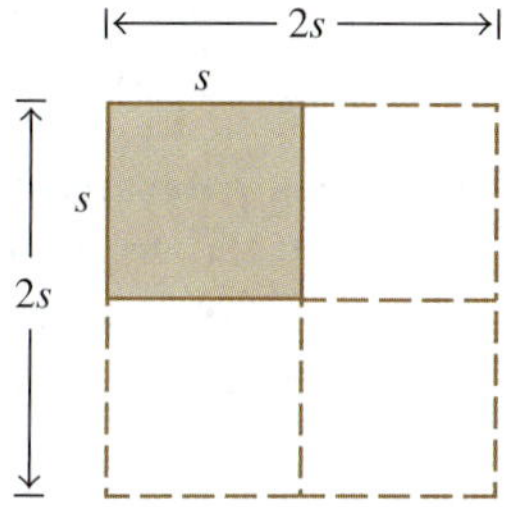

a) Express the area of the original square in terms of s.

b) Express the area of the larger square in terms of s.

c) How many times larger is the area of the square in part (b) than the area of the square in part (a)?

56. ***Doubling the Sides of a Parallelogram*** In the figure below, an original parallelogram with base b and height h is shown. Also shown is a larger parallelogram with base and height double in length, or $2b$ and $2h$, respectively.

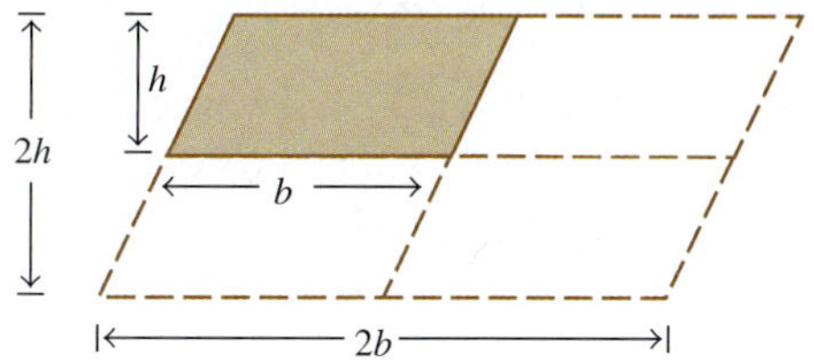

a) Express the area of the original parallelogram in terms of b and h.

b) Express the area of the larger parallelogram in terms of b and h.

c) How many times larger is the area of the parallelogram in part (b) than the area of the parallelogram in part (a)?

57. ***Heron's Formula*** A second formula for determining the area of a triangle (called Heron's formula) is

$$A = \sqrt{s(s-a)(s-b)(s-c)}$$

where $s = \frac{1}{2}(a + b + c)$ and a, b, and c are the lengths of the sides of the triangle. Use Heron's formula to determine the area of right triangle ABC and check your answer using the formula $A = \frac{1}{2}ab$.

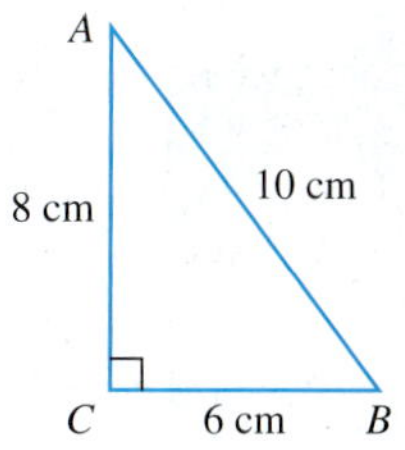

58. ***Expansion of*** $(a + b)^2$ In the figure, one side of the largest square has length $a + b$. Therefore, the area of the largest square is $(a + b)^2$. Answer the following questions to find a formula for the expansion of $(a + b)^2$.

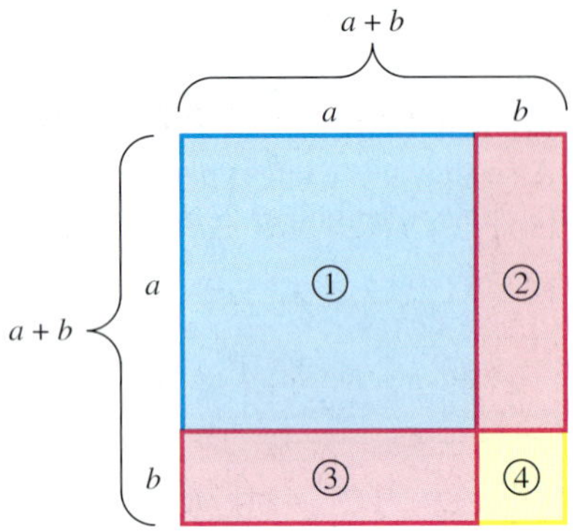

a) What is the area of the square marked ①?

b) What is the area of the rectangle marked ②?

c) What is the area of the rectangle marked ③?

d) What is the area of the square marked ④?

e) Add the four areas found in parts (a) through (d) to write a formula for the expansion of $(a + b)^2$.

RECREATIONAL MATHEMATICS

59. ***Sports Areas*** Use the Internet to research the official dimensions of the playing surface of your favorite sport and determine the area of the playing surface. Choices may include a basketball court, volleyball court, tennis court, racquetball court, hockey rink, football field, soccer field, lacrosse field, and a baseball diamond.

60. ***Scarecrow's Error*** In the movie *The Wizard of Oz*, once the scarecrow gets his diploma he states the following: "In an isosceles triangle, the sum of the square roots of the two equal sides is equal to the square root of the third side." Discuss why this statement is incorrect.

INTERNET/RESEARCH ACTIVITIES

For Exercises 61–63, references include the Internet, history of mathematics textbooks, and encyclopedias.

61. Research the proof of the Pythagorean theorem provided by President James Garfield. Write a brief paper and make a poster of this proof and the associated diagrams.

62. The early Babylonians and Egyptians did not know about π and had to devise techniques to approximate the area of a circle. Do research and write a paper on the techniques these societies used to approximate the area of a circle.

63. Write a paper on the contributions of Heron of Alexandria to geometry.

9.4 VOLUME AND SURFACE AREA

▲ The amount of paint in a can refers to volume, and the amount of wall space that the paint covers refers to surface area.

On the label of a can of Color Place paint is the following sentence: "One gallon will cover about 400 sq. ft." This sentence refers to the two main geometric topics that we will cover in this section. *One gallon* refers to volume of the paint in the can, and *400 sq. ft* refers to the surface area that the paint will cover.

When discussing a one-dimensional figure such as a line, we can find its length. When discussing a two-dimensional figure such as a rectangle, we can find its area and its perimeter. When discussing a three-dimensional figure such as a cube, we can find its volume and its surface area. ***Volume*** is a measure of the capacity of a three-dimensional figure. ***Surface area*** is the sum of the areas of the surfaces of a three-dimensional figure. Volume refers to the amount of material that you can put *inside* a three-dimensional figure, and surface area refers to the total area that is on the *outside* surface of the figure.

Solid geometry is the study of three-dimensional solid figures, also called ***space figures***. Volumes of three-dimensional figures are measured in cubic units such as cubic feet or cubic meters. Surface areas of three-dimensional figures are measured in square units such as square feet or square meters.

Rectangular Solids, Cylinders, Cones, and Spheres

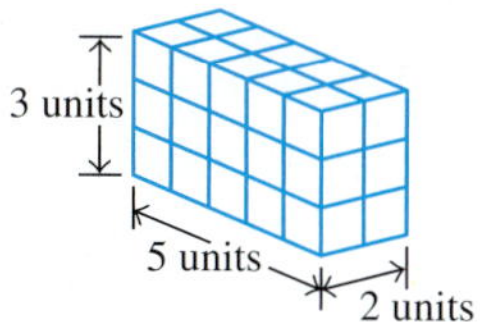

Figure 9.35

We will begin our discussion with the *rectangular solid*. If the length of the solid is 5 units, the width is 2 units, and the height is 3 units, the total number of cubes is 30 (Fig. 9.35). Thus, the volume is 30 cubic units. The volume of a rectangular solid can also be found by multiplying its length times width times height; in this case, 5 units $\times$ 2 units $\times$ 3 units = 30 cubic units. In general, the volume of any rectangular solid is $V = l \times w \times h$.

The surface area of the rectangular solid in Fig. 9.35 is the sum of the area of the surfaces of the rectangular solid. Notice that each surface of the rectangular solid is a rectangle. The left and right side of the rectangular solid each has an area of 5 units $\times$ 3 units, or 15 square units. The front and back sides of the rectangular solid each has an area of 2 units $\times$ 3 units, or 6 square units. The top and bottom sides of the rectangular solid each has an area of 5 units $\times$ 2 units, or 10 square units. Therefore, the surface area of the rectangular solid is $2(5 \times 3) + 2(2 \times 3) + 2(5 \times 2) = 2(15) + 2(6) + 2(10) = 30 + 12 + 20 = 62$ square units. In general, the surface area of any rectangular solid is $SA = 2lw + 2wh + 2lh$.

VOLUME AND SURFACE AREA OF A RECTANGULAR SOLID

$$V = lwh \qquad SA = 2lw + 2wh + 2lh$$

A *cube* is a rectangular solid with the same length, width, and height. If we call the length of the side of the cube s and use the volume and surface area formulas for a rectangular solid, substituting s in for l, w, and h, we obtain $V = s \cdot s \cdot s = s^3$ and $SA = 2 \cdot s \cdot s + 2 \cdot s \cdot s + 2 \cdot s \cdot s = 2s^2 + 2s^2 + 2s^2 = 6s^2$.

VOLUME AND SURFACE AREA OF A CUBE

$$V = s^3 \qquad SA = 6s^2$$

Now consider the right circular cylinder shown in Fig. 9.36(a) on page 552. When we use the term *cylinder* in this book, we mean a right circular cylinder. The volume of the cylinder is found by multiplying the area of the circular base, πr^2, by the height, h, to get $V = \pi r^2 h$.

The surface area of the cylinder is the sum of the area of the top, the area of the bottom, and the area of the side of the cylinder. Both the top and bottom of the cylinder are circles with an area of πr^2. To determine the area of the side of the cylinder, examine Fig. 9.36(b) and (c).

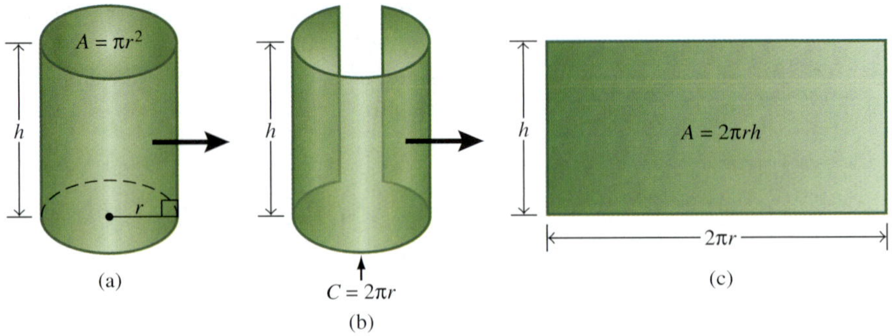

Figure 9.36

Notice how when flattened, the side of the cylinder is a rectangle whose length is the circumference of the base of the cylinder and whose width is the height of the cylinder. Thus, the side of the cylinder has an area of $2\pi r \cdot h$, or $2\pi rh$. Therefore, the surface area of the cylinder is $\pi r^2 + \pi r^2 + 2\pi rh$, or $SA = 2\pi rh + 2\pi r^2$.

VOLUME AND SURFACE AREA OF A CYLINDER

$$V = \pi r^2 h \qquad SA = 2\pi rh + 2\pi r^2$$

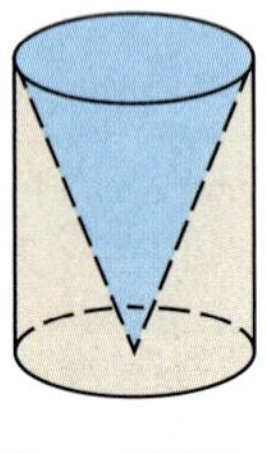

Figure 9.37

Now consider the right circular cone illustrated in Fig. 9.37. When we use the term *cone* in this book, we mean a right circular cone. The volume of a cone is less than the volume of a cylinder that has the same base and the same height. In fact, the volume of the cone is one-third the volume of the cylinder. The formula for the surface area of a cone is the sum of the area of the circular base of the cone, πr^2, and the area of the side of the cone, $\pi r\sqrt{r^2 + h^2}$, or $SA = \pi r^2 + \pi r\sqrt{r^2 + h^2}$. The derivation of the area of the side of the cone is beyond the scope of this book.

VOLUME AND SURFACE AREA OF A CONE

$$V = \tfrac{1}{3}\pi r^2 h \qquad SA = \pi r^2 + \pi r\sqrt{r^2 + h^2}$$

The next shape we will discuss in this section is the *sphere*. Baseballs, tennis balls, and so on have the shape of a sphere. The formulas for the volume and surface area of a sphere are as follows. The derivation of the volume and surface area of a sphere are beyond the scope of this book.

VOLUME AND SURFACE AREA OF A SPHERE

$$V = \tfrac{4}{3}\pi r^3 \qquad SA = 4\pi r^2$$

Following is a summary of the formulas for the volumes and surface areas of the three-dimensional figures we have discussed thus far.

VOLUMES AND SURFACE AREAS

Rectangular Solid

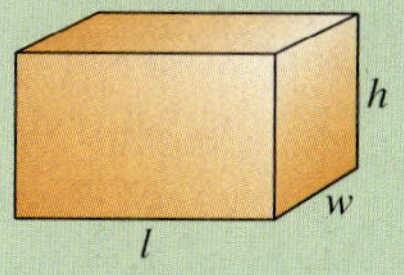

$V = lwh$

$SA = 2lw + 2wh + 2lh$

Cube

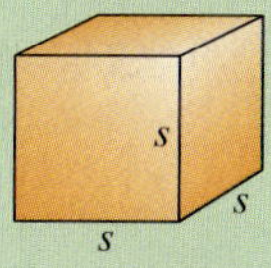

$V = s^3$

$SA = 6s^2$

Cylinder

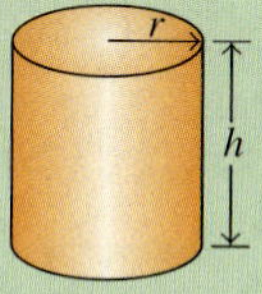

$V = \pi r^2 h$

$SA = 2\pi rh + 2\pi r^2$

Cone

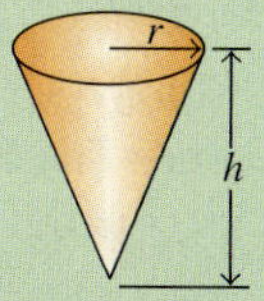

$V = \frac{1}{3}\pi r^2 h$

$SA = \pi r^2 + \pi r\sqrt{r^2 + h^2}$

Sphere

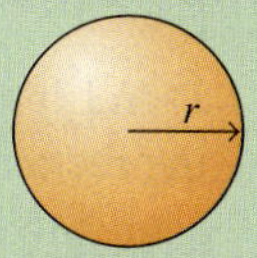

$V = \frac{4}{3}\pi r^3$

$SA = 4\pi r^2$

EXAMPLE 1 *Volume and Surface Area*

Determine the volume and surface area of each of the following three-dimensional figures. When appropriate, use the $\boxed{\pi}$ key on your calculator and round your answer to the nearest hundredths.

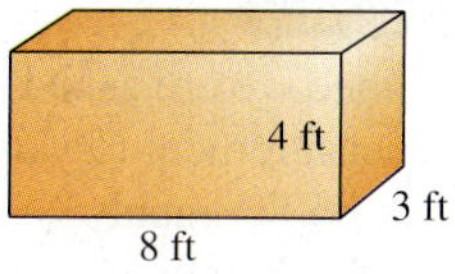

(a)

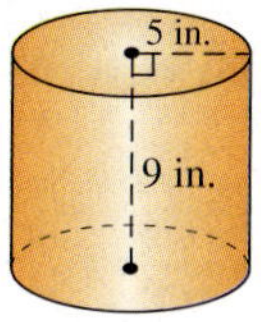

(b)

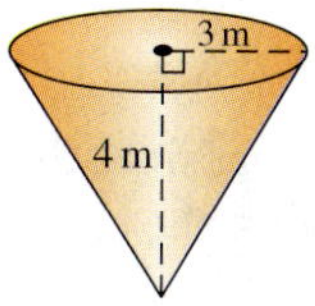

(c)

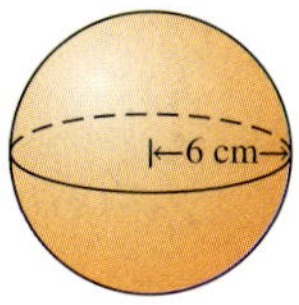

(d)

SOLUTION

a) $V = lwh = 8 \cdot 3 \cdot 4 = 96 \text{ ft}^3$

$SA = 2lw + 2wh + 2lh = 2 \cdot 8 \cdot 3 + 2 \cdot 3 \cdot 4 + 2 \cdot 8 \cdot 4$
$= 48 + 24 + 64 = 136 \text{ ft}^2$

b) $V = \pi r^2 h = \pi \cdot 5^2 \cdot 9 = 225\pi \approx 706.86 \text{ in.}^3$

$SA = 2\pi rh + 2\pi r^2 = 2\pi \cdot 5 \cdot 9 + 2\pi \cdot 5^2$
$= 90\pi + 50\pi = 140\pi \approx 439.82 \text{ in.}^2$

c) $V = \frac{1}{3}\pi r^2 h = \frac{1}{3}\pi \cdot 3^2 \cdot 4 = \frac{1}{3}\pi \cdot 9 \cdot 4 = 12\pi \text{ m}^3 \approx 37.70 \text{ m}^3$

$SA = \pi r^2 + \pi r\sqrt{r^2 + h^2} = \pi \cdot 3^2 + \pi \cdot 3 \cdot \sqrt{3^2 + 4^2}$
$= \pi \cdot 9 + \pi \cdot 3 \cdot \sqrt{25}$
$= \pi \cdot 9 + \pi \cdot 3 \cdot 5 = 9\pi + 15\pi$
$= 24\pi \approx 75.40 \text{ m}^2$

d) $V = \frac{4}{3}\pi r^3 = \frac{4}{3} \cdot \pi \cdot 6^3 = \frac{4}{3} \cdot \pi \cdot 216 = 288\pi \approx 904.78 \text{ cm}^3$

$SA = 4\pi r^2 = 4 \cdot \pi \cdot 6^2 = 4 \cdot \pi \cdot 36 = 144\pi \approx 452.39 \text{ cm}^2$

EXAMPLE 2 *Replacing a Sand Volleyball Court*

Joan Lind is the manager at the Colony Apartments and needs to replace the sand in the rectangular sand volleyball court. The court is 30 ft wide by 60 ft long, and the sand has a uniform depth of 18 in. (see figure below). Sand sells for \$15 per cubic yard.

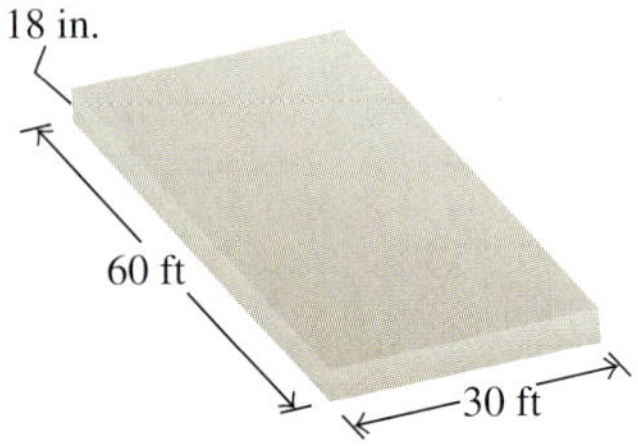

a) How many cubic yards of sand does Joan need?

b) How much will the sand cost?

SOLUTION

a) Since we are asked to find the volume in cubic yards, we will convert each measurement to yards. There are 3 ft in a yard. Thus, 30 ft equals $\frac{30}{3}$ or 10 yd, and 60 ft equals $\frac{60}{3}$ or 20 yd. There are 36 in. in a yard, so 18 in. equals $\frac{18}{36}$, or $\frac{1}{2}$ yd. The amount of sand needed is determined using the formula for the volume of a rectangular solid, $V = l \cdot w \cdot h$. In this case, the height of the rectangular solid can be considered the depth of the sand.

$$V = l \cdot w \cdot h = 10 \cdot 20 \cdot \tfrac{1}{2} = 100 \text{ yd}^3$$

Note that since the measurements for length, width, and height are each in terms of yards, the answer is in terms of cubic yards.

b) One cubic yard of sand costs \$15, so 100 yd^3 will cost $100 \times \$15$, or \$1500.

EXAMPLE 3 *Homecoming Float*

The basketball team at Southwestern High School is building a float for the homecoming parade. On the float is a large papier-maché basketball that has a radius of 4.5 ft. Team members need to know the surface area of the basketball so that they can determine how much paint they will need to buy to paint the basketball.

a) Determine the surface area of the basketball.

b) If 1 quart of paint covers approximately 100 sq ft, how many quarts of paint will team members need to buy?

SOLUTION

a) The basketball has the shape of a sphere. We will use the formula for the surface area of a sphere: $SA = 4\pi r^2 = 4 \cdot \pi \cdot (4.5)^2 = 81\pi \approx 254.47$ sq ft.

b) Since each quart of paint will cover about 100 sq ft, they will need about $\frac{254.47}{100} = 2.5447$ quarts. Since you cannot buy a portion of a quart of paint, the team will need to buy 3 quarts of paint to paint the basketball. ●

EXAMPLE 4 *Silage Storage*

Gordon Langeneger has three silos on his farm. The silos are each in the shape of a right circular cylinder (see Fig. 9.38). One silo has a 12 ft diameter and is 40 ft tall. The second silo has a 14 ft diameter and is 50 ft tall. The third silo has an 18 ft diameter and is 60 ft tall.

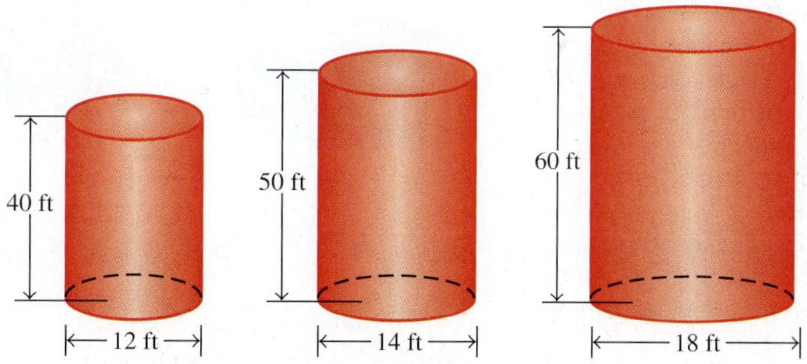

Figure 9.38

a) What is the total capacity of the three silos in cubic feet?

b) If Gordon fills all three of his silos and then feeds his cattle 150 ft^3 of silage per day, in how many days will all three silos be empty?

SOLUTION

a) The capacity of each silo can be determined using the formula for the volume of a right circular cylinder, $V = \pi r^2 h$. Since the radius is half the diameter, the radii for the three silos are 6 ft, 7 ft, and 9 ft, respectively. Now let's determine the volumes.

$$\text{Volume of the first silo} = \pi r^2 h = \pi \cdot 6^2 \cdot 40$$
$$\approx 3.14 \cdot 36 \cdot 40 \approx 4521.6 \text{ ft}^3$$

DID YOU KNOW?

Fractured Volumes

Picasso, Pablo. *Les Demoiselles d'Avignon*. Paris (June–July 1907). Oil on canvas, 8 ft × 7 ft 8 in. (243.9 cm × 233.7 cm). The Museum of Modern Art, New York. Acquired through the Lillie P. Bliss Bequest. Photograph © 1996 The Museum of Modern Art, New York. © Estate of Pablo Picasso.

The cubist painters of the early twentieth century sought to analyze forms as geometric shapes. They were greatly influenced by Paul Cézanne who, in a famous letter, wrote of treating nature in terms of the cylinder, sphere, and cone. In one of the first paintings of the cubist period, *Les Demoiselles d'Avignon*, Pablo Picasso dismissed the idea of the human figure as a dynamic unity and instead fractured volumes and planes to present them from multiple angles of vision at one time.

$$\text{Volume of the second silo} = \pi r^2 h = \pi \cdot 7^2 \cdot 50 \approx 3.14 \cdot 49 \cdot 50 \approx 7693.0 \text{ ft}^3$$

$$\text{Volume of the third silo} = \pi r^2 h = \pi \cdot 9^2 \cdot 60 \approx 3.14 \cdot 81 \cdot 60 \approx 15{,}260.4 \text{ ft}^3$$

Therefore, the total capacity of all three silos is about

$$4521.6 + 7693.0 + 15{,}260.4 \approx 27{,}475.0 \text{ ft}^3.$$

b) To find how long it takes to empty all three silos, we will divide the total capacity by 150 ft^3, the amount fed to Gordon's cattle every day.

$$\frac{27{,}475}{150} \approx 183.17$$

Thus, the silos will be empty in about 183 days.

Polyhedra, Prisms, and Pyramids

Now let's discuss polyhedra. A *polyhedron* (plural is *polyhedra*) is a closed surface formed by the union of polygonal regions. Figure 9.39 illustrates some polyhedra.

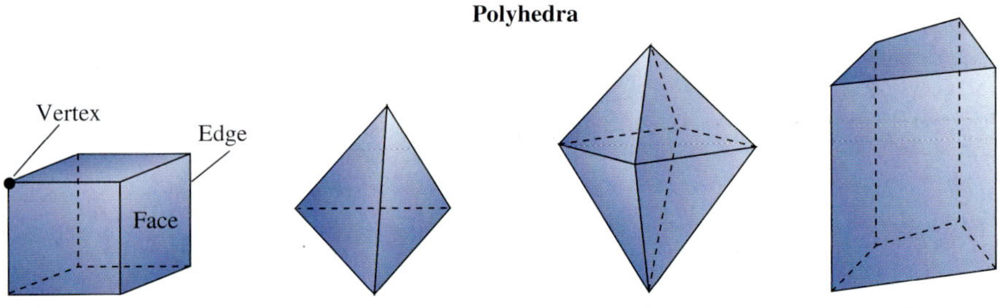

Figure 9.39

Each polygonal region is called a *face* of the polyhedron. The line segment formed by the intersection of two faces is called an *edge*. The point at which two or more edges intersect is called a *vertex*. In Fig. 9.39(a), there are 6 faces, 12 edges, and 8 vertices. Note that

Number of vertices	−	number of edges	+	number of faces	= 2
8	−	12	+	6	= 2

This formula, credited to Leonhard Euler, is true for any polyhedron.

EULER'S POLYHEDRON FORMULA

Number of vertices − number of edges + number of faces = 2

We suggest that you verify that this formula holds for Fig. 9.39(b), (c), and (d).

DID YOU KNOW?

Platonic Solids

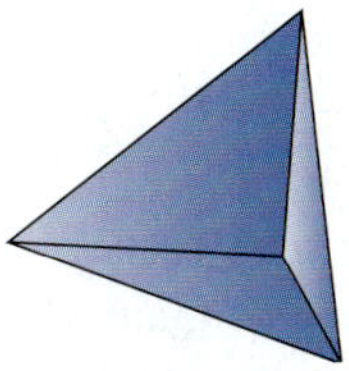

Tetrahedron: 4 faces, 4 vertices, 6 edges

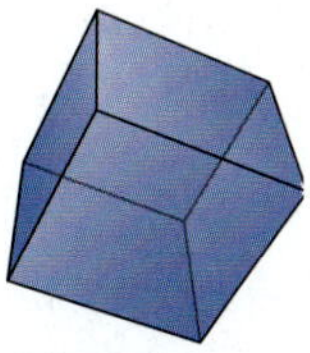

Cube: 6 faces, 8 vertices, 12 edges

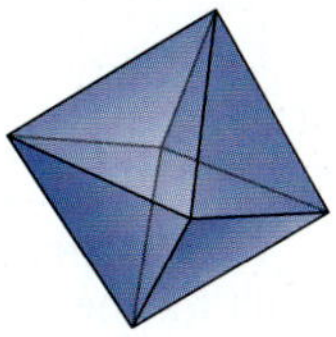

Octahedron: 8 faces, 6 vertices, 12 edges

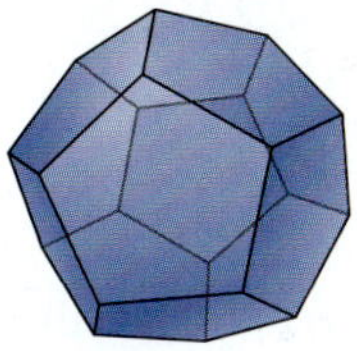

Dodecahedron: 12 faces, 20 vertices, 30 edges

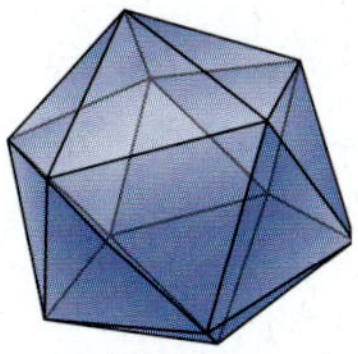

Icosahedron: 20 faces, 12 vertices, 30 edges

A platonic solid is a polyhedron whose faces are all regular polygons of the same size and shape. There are exactly five platonic solids, as shown above. Platonic solids, also called regular polyhedra, are named after the ancient Greek philosopher Plato and are included in Euclid's *Elements* (see the Profile in Mathematics on page 516). The figures above show each of the platonic solids, along with the number of faces, vertices, and edges for each.

EXAMPLE 5 *Using Euler's Polyhedron Formula*

A certain polyhedron has 13 vertices and 8 faces. Determine the number of edges on the polyhedron.

SOLUTION Since we are seeking the number of edges, we will let x represent the number of edges on the polyhedron. Next, we will use Euler's polyhedron formula to set up an equation:

$$\text{Number of vertices} - \text{number of edges} + \text{number of faces} = 2$$

$$13 - x + 8 = 2$$

$$21 - x = 2$$

$$-x = -19$$

$$x = 19$$

Therefore, the polyhedron has 19 edges. ●

A ***platonic solid***, also known as a ***regular polyhedron***, is a polyhedron whose faces are all regular polygons of the same size and shape. There are exactly five platonic solids. All five platonic solids are illustrated in the Did You Know? at left.

A ***prism*** is a special type of polyhedron whose bases are congruent polygons and whose sides are parallelograms. These parallelogram regions are called the *lateral faces* of the prism. If all the lateral faces are rectangles, the prism is said to be a ***right prism***. The prisms illustrated in Fig. 9.40 are all right prisms. When we use the word *prism* in this book, we are referring to a right prism.

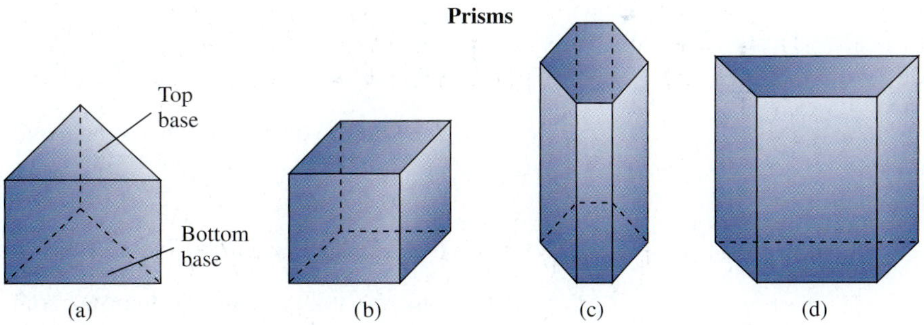

Figure 9.40

The volume of any prism can be found by multiplying the area of the base, B, by the height, h, of the prism.

VOLUME OF A PRISM

$$V = Bh$$

where B is the area of a base and h is the height.

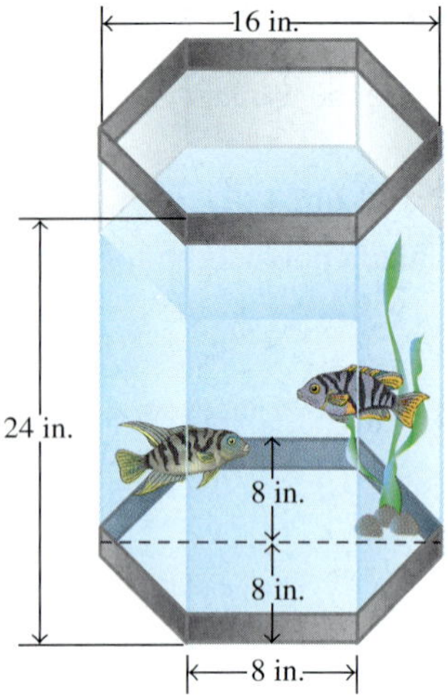

Figure 9.41

EXAMPLE 6 *Volume of a Hexagonal Prism Fish Tank*

Frank Nicolzaao's fish tank is in the shape of a hexagonal prism as shown in Fig. 9.41. Use the dimensions shown in the figure and the fact that 1 gal $=$ 231 in.^{3} to

a) determine the volume of the fish tank in cubic inches.

b) determine the volume of the fish tank in gallons (round your answer to the nearest gallon).

SOLUTION

a) First we will need to calculate the area of the hexagonal base of the fish tank. Notice from Fig. 9.41 that by drawing a diagonal as indicated, the base can be divided into two identical trapezoids. To find the area of the hexagonal base, we will calculate the area of one of these trapezoids and then multiply by 2.

$$\text{Area of one trapezoid} = \tfrac{1}{2}h(b_1 + b_2)$$
$$= \tfrac{1}{2}(8)(16 + 8) = 96 \text{ in.}^2$$
$$\text{Area of the hexagonal base} = 2(96) = 192 \text{ in.}^2$$

Now to determine the volume of the fish tank, we will use the formula for the volume of a prism, $V = Bh$. We already determined that the area of the base, B, is 192 in.^{2}

$$V = B \cdot h = 192 \cdot 24 = 4608 \text{ in.}^3$$

In the above calculation, the area of the base, B, was measured in square inches and the height was measured in inches. The product of square inches and inches is cubic inches, or in.^{3}.

b) To determine the volume of the fish tank in gallons, we will divide the volume of the fish tank in cubic inches by 231.

$$V = \frac{4608}{231} \approx 19.95 \text{ gal}$$

Thus, the volume of the fish tank is approximately 20 gal.

EXAMPLE 7 *Volumes Involving Prisms*

Determine the volume of the remaining solid after the cylinder, triangular prism, and square prism have been cut from the solid (Fig. 9.42).

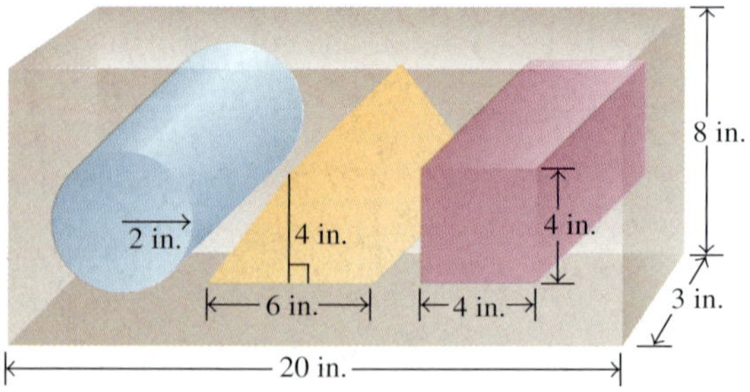

Figure 9.42

SOLUTION To determine the volume of the remaining solid, first determine the volume of the rectangular solid. Then subtract the volume of the two prisms and the cylinder that were cut out.

$$\begin{aligned}
\text{Volume of rectangular solid} &= l \cdot w \cdot h \\
&= 20 \cdot 3 \cdot 8 = 480 \text{ in.}^3 \\
\text{Volume of circular cylinder} &= \pi r^2 h \\
&\approx (3.14)(2^2)(3) \\
&\approx (3.14)(4)(3) \approx 37.68 \text{ in.}^3 \\
\text{Volume of triangular prism} &= \text{area of the base} \cdot \text{height} \\
&= \tfrac{1}{2}(6)(4)(3) = 36 \text{ in.}^3 \\
\text{Volume of square prism} &= s^2 \cdot h \\
&= 4^2 \cdot 3 = 48 \text{ in.}^3 \\
\text{Volume of solid} &\approx 480 - 37.68 - 36 - 48 \\
&\approx 358.32 \text{ in.}^3
\end{aligned}$$

Another special category of polyhedra is the *pyramid*. Unlike prisms, pyramids have only one base. The figures illustrated in Fig. 9.43 are pyramids. Note that all but one face of a pyramid intersect at a common vertex.

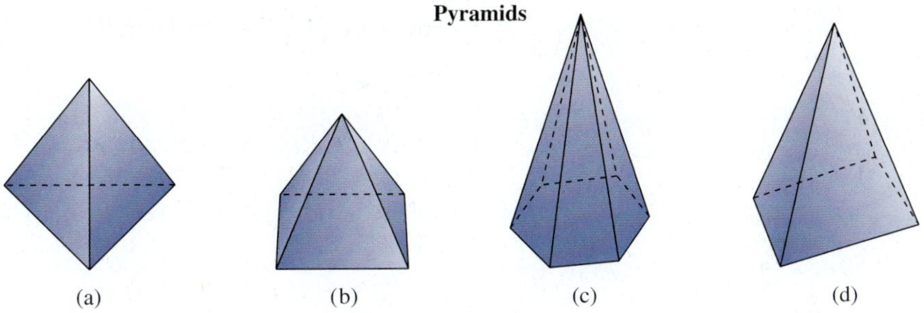

Figure 9.43

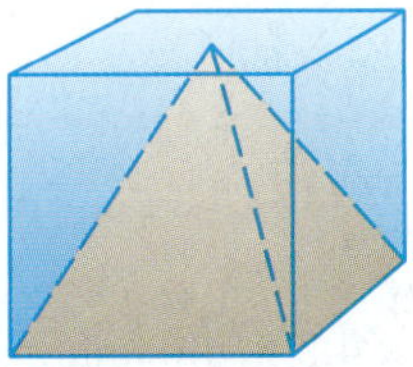

Figure 9.44

If a pyramid is drawn inside a prism, as shown in Fig. 9.44, the volume of the pyramid is less than that of the prism. In fact, the volume of the pyramid is one-third the volume of the prism.

VOLUME OF A PYRAMID

$$V = \tfrac{1}{3}Bh$$

where B is the area of the base and h is the height.

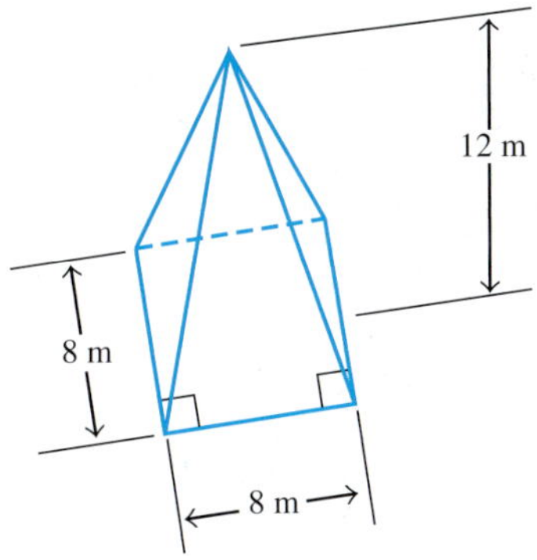

Figure 9.45

EXAMPLE 8 *Volume of a Pyramid*

Find the volume of the pyramid shown in Fig. 9.45.

SOLUTION First find the area of the base of the pyramid. Since the base of the pyramid is a square,

$$\text{Area of base} = s^2 = 8^2 = 64 \text{ m}^2$$

Now use this information to find the volume of the pyramid.

$$V = \frac{1}{3} \cdot B \cdot h$$
$$= \frac{1}{3} \cdot 64 \cdot 12$$
$$= 256 \text{ m}^3$$

Thus, the volume of the pyramid is 256 m^3.

Cubic Unit Conversions

In certain situations, converting volume from one cubic unit to a different cubic unit might be necessary. For example, when purchasing topsoil you might have to change the amount of topsoil from cubic feet to cubic yards prior to placing your order. Example 9 shows how that may be done.

EXAMPLE 9 *Cubic Yards and Cubic Feet*

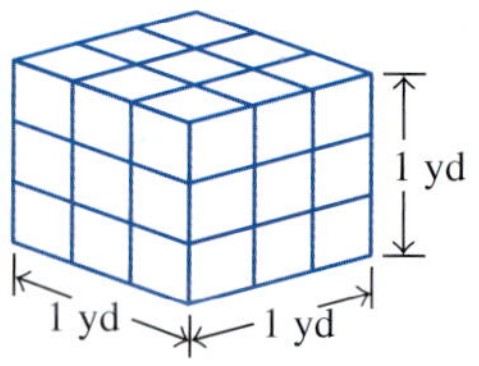

Figure 9.46

a) Convert 1 yd^3 to cubic feet. (See Fig. 9.46.)
b) Convert 18.5 yd^3 to cubic feet.
c) Convert 1302.75 ft^3 to cubic yards.

SOLUTION

a) We know that 1 yd = 3ft. Thus, 1 $\text{yd}^3 = 3 \text{ ft} \times 3 \text{ ft} \times 3 \text{ ft} = 27 \text{ ft}^3$.
b) In part (a), we learned that 1 $\text{yd}^3 = 27\text{ft}^3$. Thus, 18.5 $\text{yd}^3 = 18.5 \times 27 = 499.5 \text{ ft}^3$.
c) In part (b), we converted from cubic yards to cubic feet by *multiplying* the number of cubic yards by 27. Now, to convert from cubic feet to cubic yards we will *divide* the number of cubic feet by 27. Therefore, 1302.75 $\text{ft}^3 = \frac{1302.75}{27} \text{ yd}^3 = 48.25 \text{ yd}^3$.

EXAMPLE 10 *Filling in a Swimming Pool*

Julianne Peterson recently purchased a home with a rectangular swimming pool. The pool is 30 ft long and 15 ft wide, and it has a uniform depth of 4.5 ft. Julianne lives in a cold climate, so she plans to fill the pool in with dirt to make a flower garden. How many cubic yards of dirt will Julianne have to purchase to fill in the swimming pool?

SOLUTION To find the amount of dirt, we will use the formula for the volume of a rectangular solid:

$$V = lwh$$
$$= (30)(15)(4.5)$$
$$= 2025 \text{ ft}^3$$

Now we must convert this volume from cubic feet to cubic yards. In Example 9, we learned that 1 $\text{yd}^3 = 27 \text{ ft}^3$. Therefore, 2025 $\text{ft}^3 = \frac{2025}{27} = 75 \text{ yd}^3$. Thus, Julianne needs to purchase 75 yd^3 of dirt to fill in her swimming pool.

SECTION 9.4 EXERCISES

CONCEPT/WRITING EXERCISES

1. a) Define *volume*.

   b) Define *surface area*.

2. What is solid geometry?

3. What is the difference between a polyhedron and a regular polyhedron?

4. What is the difference between a prism and a right prism?

5. In your own words, explain the difference between a prism and a pyramid.

6. In your own words, state Euler's polyhedron formula.

PRACTICE THE SKILLS

In Exercises 7–16, determine (a) the volume and (b) the surface area of the three-dimensional figure. When appropriate, use the $\boxed{\pi}$ *key on your calculator and round your answer to the nearest hundredth.*

7.

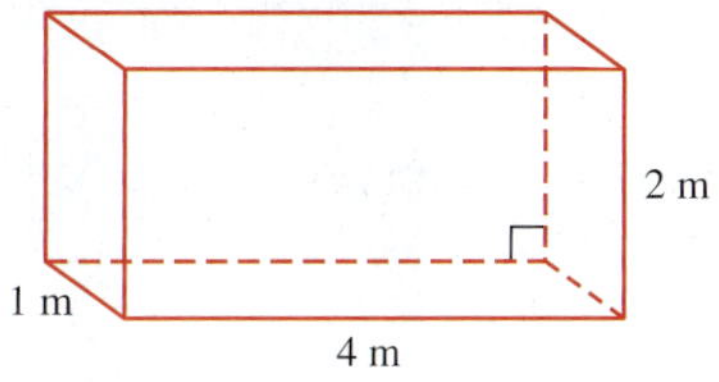

8.

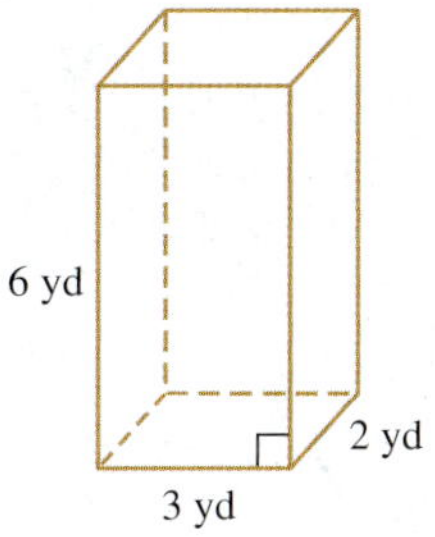

9.

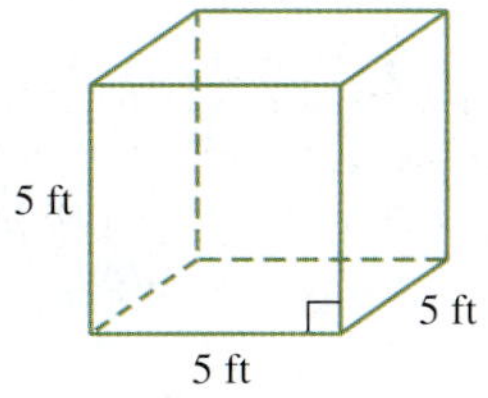

10.

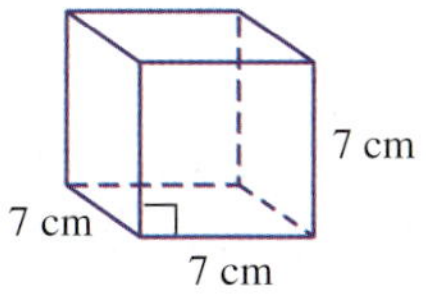

11.

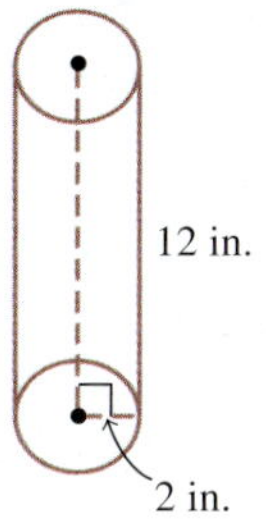

12.

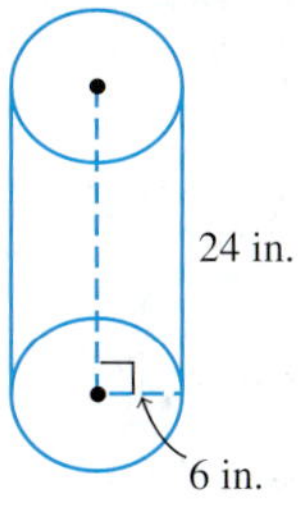

13.

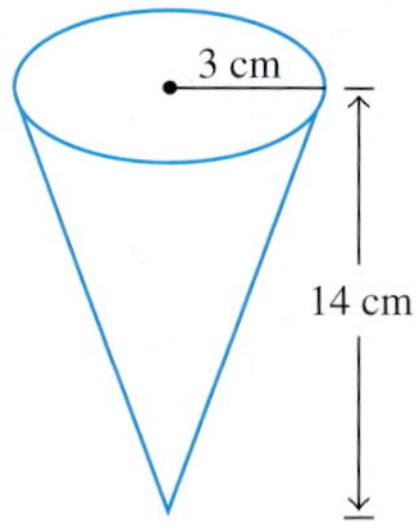

14.

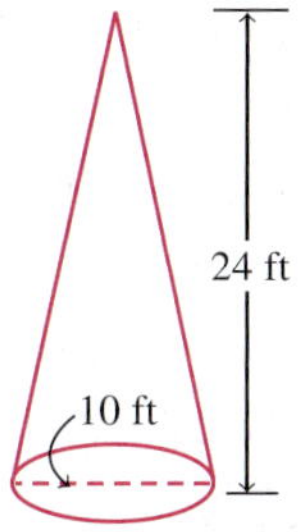

15.

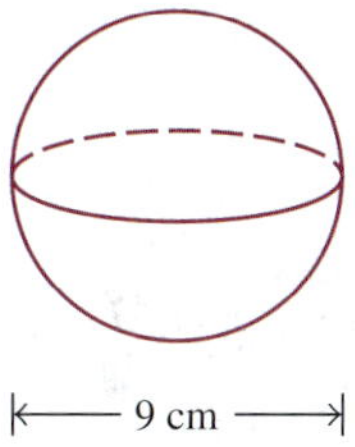

16.

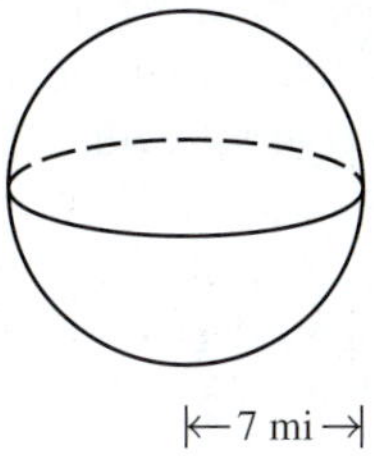

In Exercises 17–20, determine the volume of the three-dimensional figure. When appropriate, round your answer to the nearest hundredth.

17.

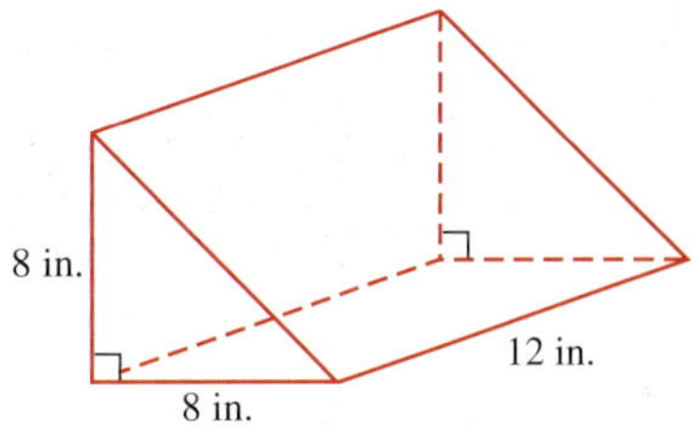

18.

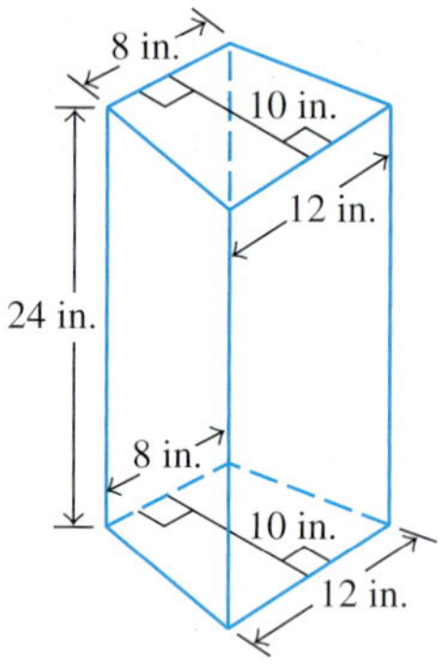

19.

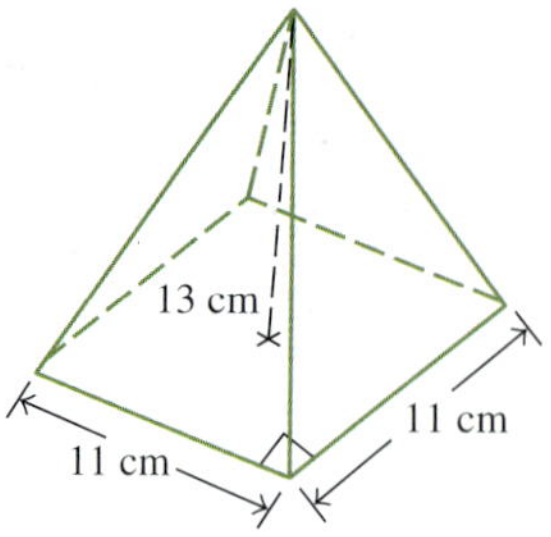

20.

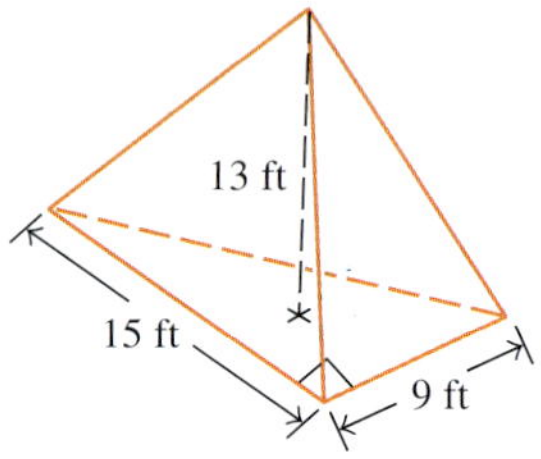

PROBLEM SOLVING

In Exercises 21–28, determine the volume of the shaded region. When appropriate, use the [π] *key on your calculator and round your answer to the nearest hundredth.*

21.

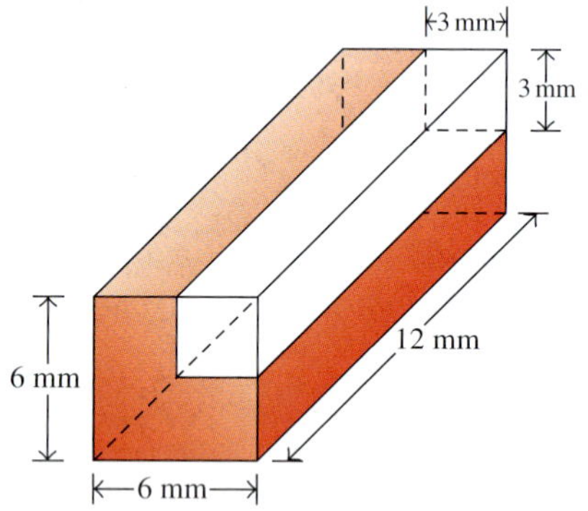

22.

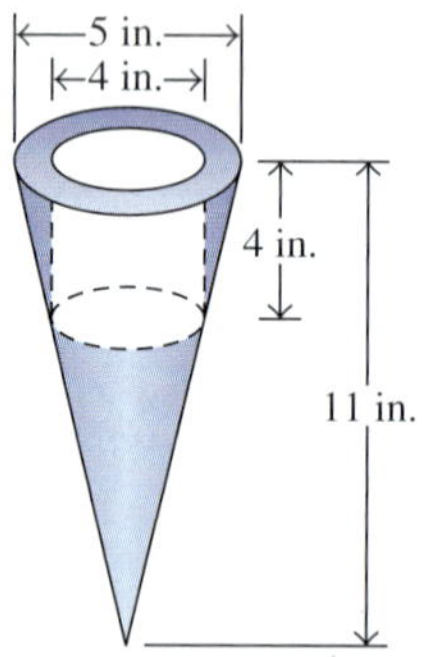

23.

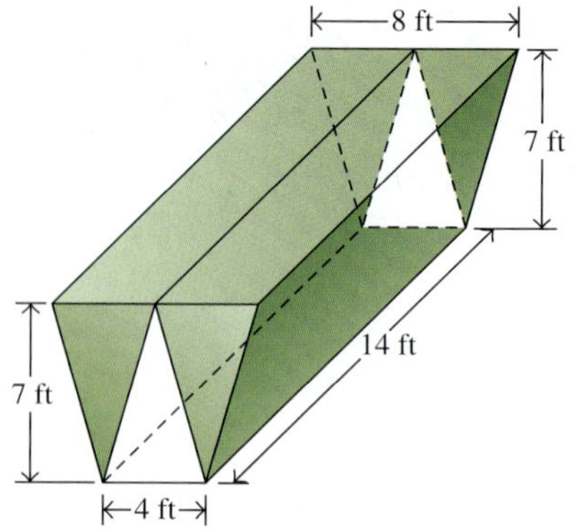

24.

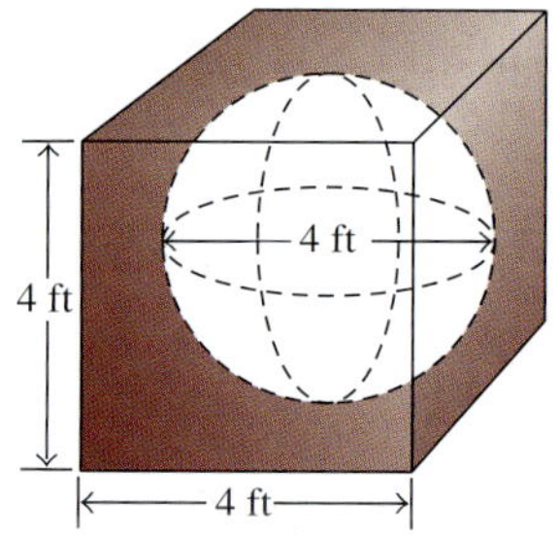

25.

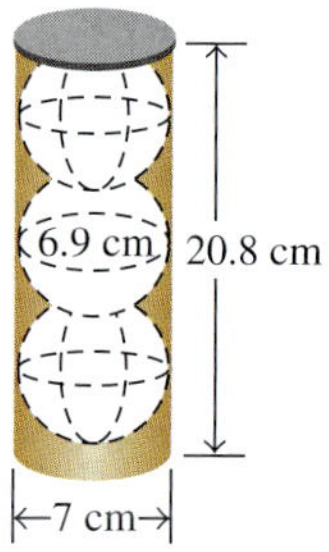

26.

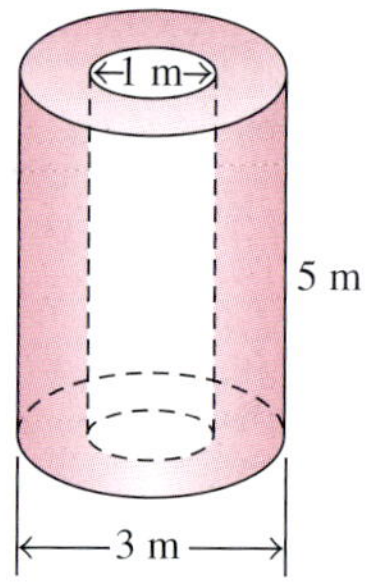

27.

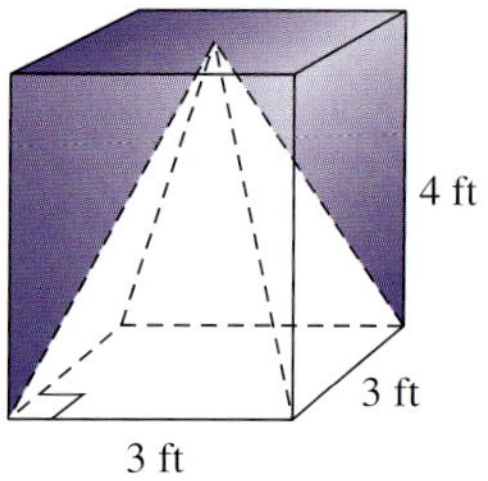

28.

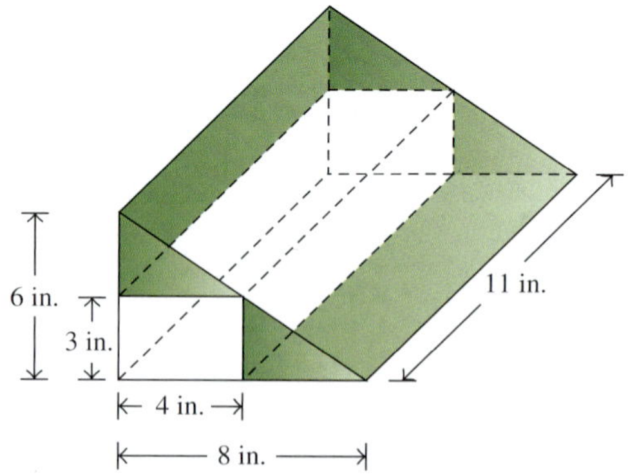

In Exercises 29–32, use the fact that 1 yd^3 equals 27 ft^3 to make the conversion.

29. 9 yd^3 to cubic feet

30. 7.25 yd^3 to cubic feet

31. 153 ft^3 to cubic yards

32. 2457 ft^3 to cubic yards

In Exercises 33–36, use the fact that 1 m^3 equals 1,000,000 cm^3 to make the conversion.

33. 3.7 m^3 to cubic centimeters

34. 17.6 m^3 to cubic centimeters

35. 7,500,000 cm^3 to cubic meters

36. 7,300,000 cm^3 to cubic meters

37. ***Ice-Cream Comparison*** The Louisburg Creamery packages its homemade ice cream in tubs and in boxes. The tubs are in the shape of a right cylinder with a radius of 3 in. and height of 5 in. The boxes are in the shape of a cube with each side measuring 5 in. Determine the volume of each container.

38. ***Volume of a Freezer*** The dimensions of the interior of an upright freezer are height 46 in., width 25 in., and depth 25 in. Determine the volume of the freezer

a) in cubic inches. **b)** in cubic feet.

39. ***CD Case*** A compact disc case is a rectangular solid that is 142 mm long, 125 mm wide, and 10 mm high. Determine its surface area.

40. ***Globe Surface Area*** The Everest model globe has a diameter of 20 in. Determine the surface area of this globe. Round your answer to the nearest hundredth.

41. ***Volume of a Bread Pan*** A bread pan is 12 in. × 4 in. × 3 in. How many quarts does it hold, if 1 $in.^3 \approx 0.01736$ qt?

42. ***Gasoline Containers*** Mark Russo has two right cylindrical containers for storing gasoline. One has a diameter of 10 in. and a height of 12 in. The other has a diameter of 12 in. and a height of 10 in.

a) Which container holds the greater amount of gasoline, the taller one or the one with the greater diameter?

b) What is the difference in volume?

43. ***A Fish Tank***

a) How many cubic centimeters of water will a rectangular fish tank hold if the tank is 80 cm long, 50 cm wide, and 30 cm high?

b) If 1 cm^3 holds 1 mℓ of liquid, how many milliliters will the tank hold?

c) If 1ℓ = 1000 mℓ, how many liters will the tank hold?

44. ***The Pyramid of Cheops*** The Pyramid of Cheops in Egypt has a square base measuring 720 ft on a side. Its height is 480 ft. What is its volume?

45. ***Engine Capacity*** The engine in a 1957 Chevrolet Corvette has eight cylinders. Each cylinder is a right cylinder with a bore (diameter) of 3.875 in. and a stroke (height) of 3 in. Determine the total displacement (volume) of this engine.

9.5 TRANSFORMATIONAL GEOMETRY, SYMMETRY, AND TESSELLATIONS

▲ Capital letters can be used to display symmetry.

Consider the capital letters of the alphabet. Now consider a kindergarten-age child who is practicing to write capital letters. Usually, children will write some of the letters "backwards." However, some letters cannot be made backwards. For example, the capital letter A would look the same backwards as it does forwards, but the capital letters B and C would look different backwards than they do forwards. The capital letters that appear the same forwards as they do backwards have a property called *symmetry* that we will define and study in this section. Symmetry is part of another branch of geometry known as *transformational geometry*.

In our study of geometry, we have thus far focused on definitions, axioms, and theorems that are used in the study of *Euclidean geometry*. We will now introduce a second type of geometry called *transformational geometry*. In ***transformational geometry***, we study various ways to move a geometric figure without altering the shape or size of the figure. When discussing transformational geometry, we often use the term *rigid motion*.

The act of moving a geometric figure from some starting position to some ending position without altering its shape or size is called a **rigid motion** (or **transformation**).

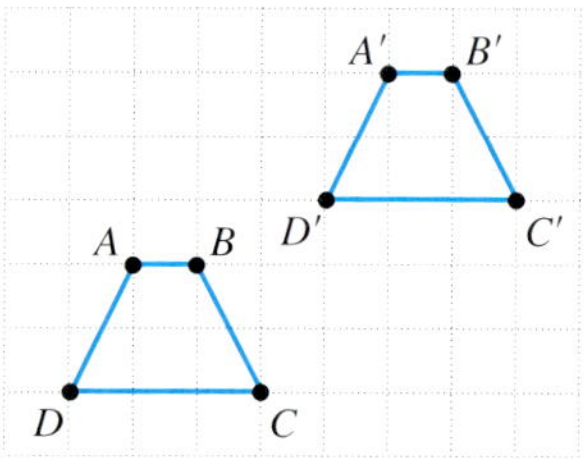

Figure 9.47

Consider trapezoid $ABCD$ in Figure 9.47. If we move each point on this trapezoid 4 units to the right and 3 units up, the trapezoid is in the location specified by trapezoid $A'B'C'D'$. This figure illustrates one type of rigid motion. When studying rigid motions, we are only concerned about the starting and ending positions of the figure and not what happens in between. When discussing rigid motions of two-dimensional figures, there are four basic types of rigid motions: reflections, rotations, translations, and glide reflections. We call these four types of rigid motions the ***basic rigid motions in a plane***. After we discuss the four rigid motions we will discuss symmetry of geometric figures and tessellations.

Reflections

The first rigid motion we will study is *reflection*. In our everyday life, we are quite familiar with the concept of reflection. In transformational geometry, a reflection is an image of a geometric figure that appears on the opposite side of a designated line.

A **reflection** is a rigid motion that moves a geometric figure to a new position such that the figure in the new position is a mirror image of the figure in the starting position. In two dimensions, the figure and its mirror image are equidistant from a line called the **reflection line** or the **axis of reflection**.

Figure 9.48 shows trapezoid $ABCD$, a reflection line l, and the reflected trapezoid $A'B'C'D'$. Notice that vertex A is 6 units to the *left* of reflection line l and that vertex A' is 6 units to the *right* of reflection line l. Next notice that vertex B is 2 units to the *left* of l and that vertex B' is 2 units to the *right* of l. A similar relationship holds true for vertices C and C' and for vertices D and D'. It is important to see that the trapezoid is not simply *moved* to the other side of the reflection line, but instead it is *reflected*. Notice in the trapezoid $ABCD$ that the longer base $\overline{BC}$ is on the *right* side of the trapezoid, but in the reflected trapezoid $A'B'C'D'$ the longer base $\overline{B'C'}$ is on the *left* side of the trapezoid. Finally, notice the colors of the sides of the two trapezoids. Side $\overline{AB}$ in trapezoid $ABCD$ and side $\overline{A'B'}$ in the reflected trapezoid are both blue. Side $\overline{BC}$ and side $\overline{B'C'}$ are both red, sides $\overline{CD}$ and $\overline{C'D'}$ are both gold, and sides $\overline{DA}$ and $\overline{D'A'}$ are both green. In this section, we will occasionally use such color coding to help you visualize the effect of a rigid transformation on a figure.

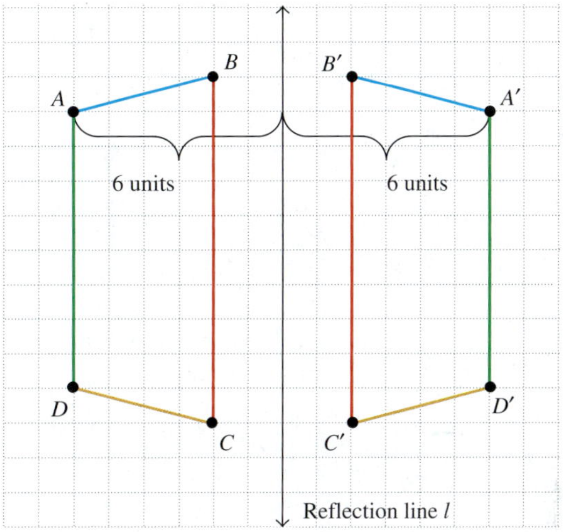

Figure 9.48

EXAMPLE 1 *Reflection of a Square*

Construct the reflection of square $ABCD$, shown in Fig. 9.49, about reflection line l.

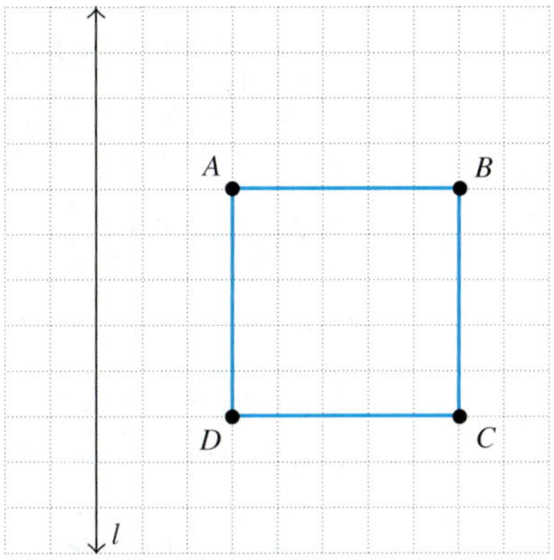

Figure 9.49

SOLUTION The reflection of square $ABCD$ will be called $A'B'C'D'$. To determine the position of the reflection, we first examine vertex A in Fig. 9.49. Notice that vertex A is 3 units to the *right* of line l. Thus, in the reflected square $A'B'C'D'$, vertex A' must also be 3 units away from, but to the *left* of, reflection line l. Next, notice that vertex B is 8 units to the right of l. Thus, in the reflected square $A'B'C'D'$, vertex B' must also be 8 units away from, but to the *left* of, reflection line l. Next, notice that vertices C and D are 8 units and 3 units, respectively, to the *right* of l. Thus, vertices C' and D' must be 8 units and 3 units, respectively, to the *left* of l. Figure 9.50 shows vertices A', B', C', and D'. Finally, we draw line segments between vertices A' and B', B' and C', C' and D', and D' and A' to form the sides of the reflection, square $A'B'C'D'$, as illustrated in Fig. 9.50.

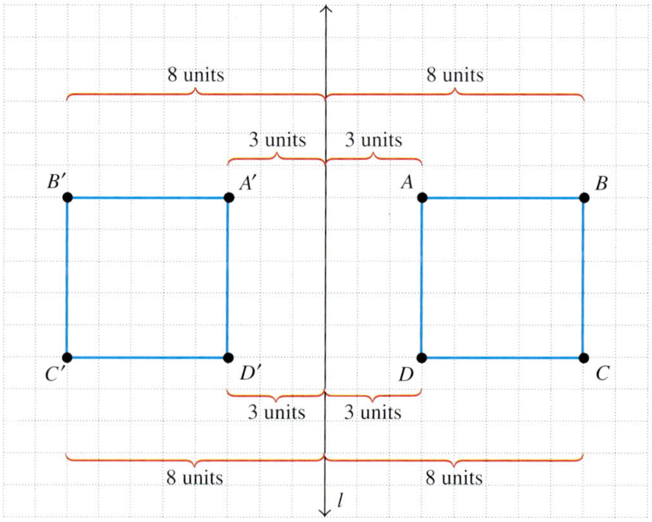

Figure 9.50

In Example 1, the reflection line did not intersect the figure being reflected. We will now study an example in which the reflection line goes directly through the figure to be reflected.

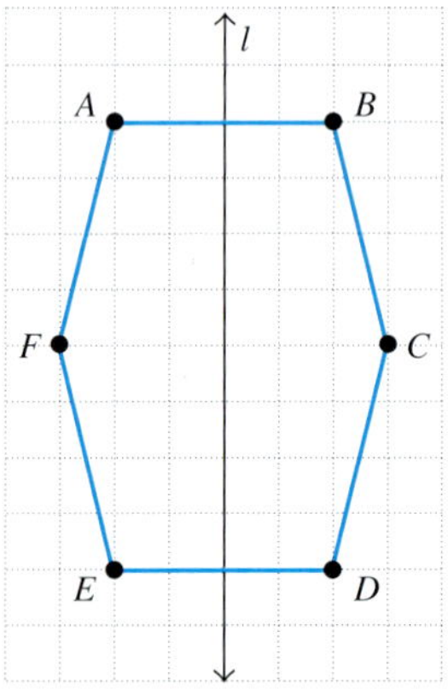

Figure 9.51

EXAMPLE 2 *Reflection of a Hexagon*

Construct the reflection of hexagon $ABCDEF$, shown in Fig. 9.51, about reflection line l.

SOLUTION From Fig. 9.51, we see that vertex A in hexagon $ABCDEF$ is 2 units to the left of reflection line l. Thus, vertex A' in the reflected hexagon will be 2 units to the right of l (see Fig. 9.52). Notice that vertex A' of the reflected hexagon is in the same location as vertex B of hexagon $ABCDEF$ in Fig. 9.51.

We next see that vertex B in hexagon $ABCDEF$ is 2 units to the right of l. Thus, vertex B' in the reflected hexagon will be 2 units to the left of l. Notice that vertex B' of the reflected hexagon is in the same location as vertex A of hexagon $ABCDEF$. We continue this process to determine the locations of vertices C', D', E', and F' of the reflected hexagon. Notice once again that each vertex of

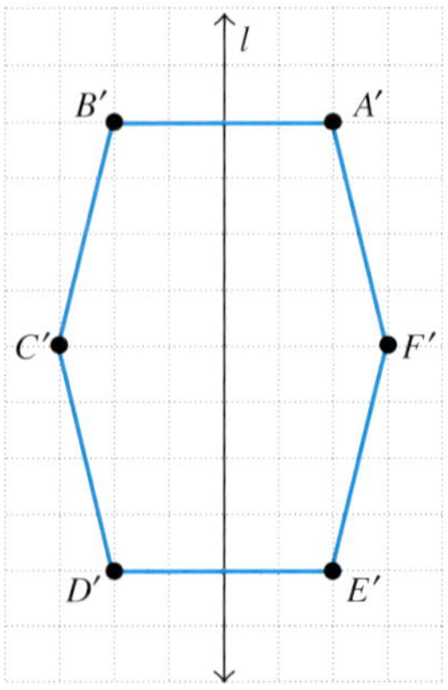

Figure 9.52

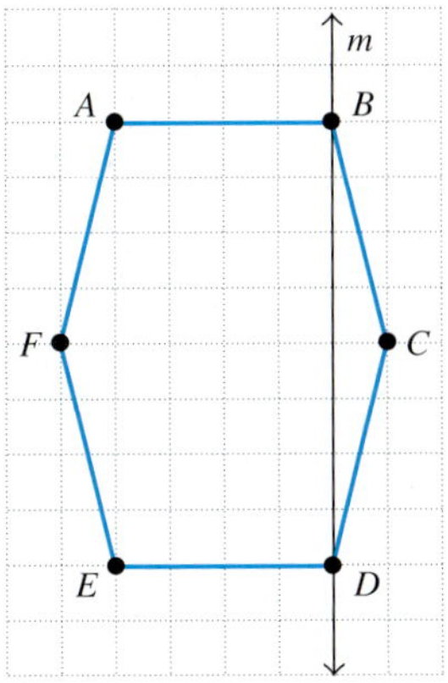

Figure 9.53

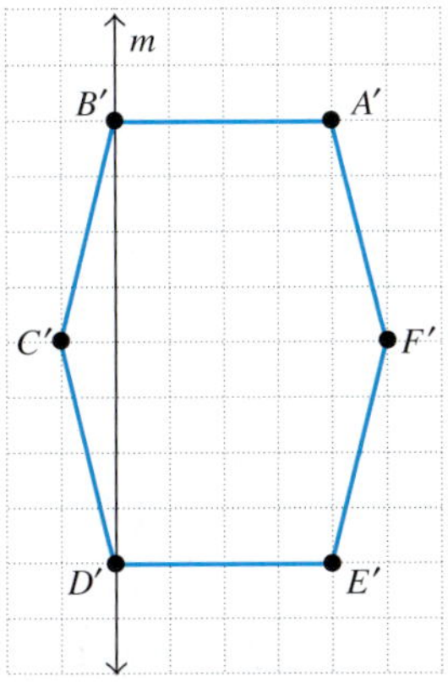

Figure 9.54

the reflected hexagon is in the same location as a vertex of hexagon $ABCDEF$. Finally, we draw the line segments to complete the reflected hexagon $A'B'C'D'E'F'$ (see Fig. 9.52). For this example, we see that other than the vertex labels, the positions of the hexagon before and after the reflection are identical. ●

In Example 2, the reflection line was in the center of the hexagon in the original position. As a result, the reflection line was also in the center of the reflected hexagon. In this particular case the reflected hexagon lies directly on top of the hexagon in its original position. We will revisit reflections such as that in Example 2 again when we discuss *reflective symmetry* later in this section.

Now consider hexagon $ABCDEF$ in Fig. 9.53 and its reflection about line m, hexagon $A'B'C'D'E'F'$ in Fig. 9.54. Notice that the positions of the hexagon before and after the reflection, relative to line m, are not the same. Furthermore, if we line up reflection line m in Fig. 9.53 and Fig. 9.54, we would see that hexagon $ABCDEF$ and hexagon $A'B'C'D'E'F'$ are in different positions.

Translations

The next rigid motion we will discuss is the *translation*. In a translation, we simply move a figure along a straight line to a new position.

A **translation** (or **glide**) is a rigid motion that moves a geometric figure by sliding it along a straight line segment in the plane. The direction and length of the line segment completely determine the translation.

After conducting a translation, we say the figure was *translated* to a new position.

A concise way to indicate the direction and the distance that a figure is moved during a translation is with a ***translation vector.*** In mathematics, vectors are typically represented with boldface letters. For example, in Fig. 9.55 we see trapezoid $ABCD$ and a translation vector, **v**, which is pointing to the right and upward. This translation vector indicates a translation of 9 units to the right and 4 units upward. Notice that in Fig. 9.55 the translated vector appears on the right side of the polygon. The placement of the translation vector does not matter. Therefore, the translation vector could have been placed to the left, above, or below the polygon, and the translation would not change. When trapezoid $ABCD$ is translated using **v**, every point on trapezoid $ABCD$ is moved 9 units to the right and 4 units upward. This movement is demonstrated for vertex A in Fig. 9.56(a) on page 570. Figure 9.56(b) shows trapezoid $ABCD$ and the translated trapezoid $A'B'C'D'$. Notice in Fig. 9.56(b) that every point on trapezoid $A'B'C'D'$ is 9 units to the right and 4 units up from its corresponding point on trapezoid $ABCD$.

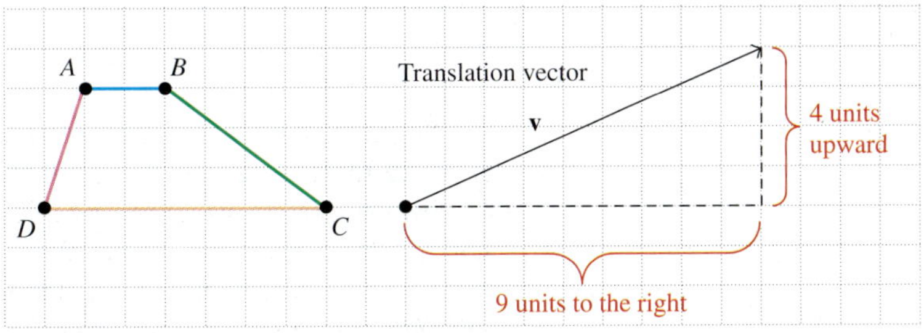

Figure 9.55

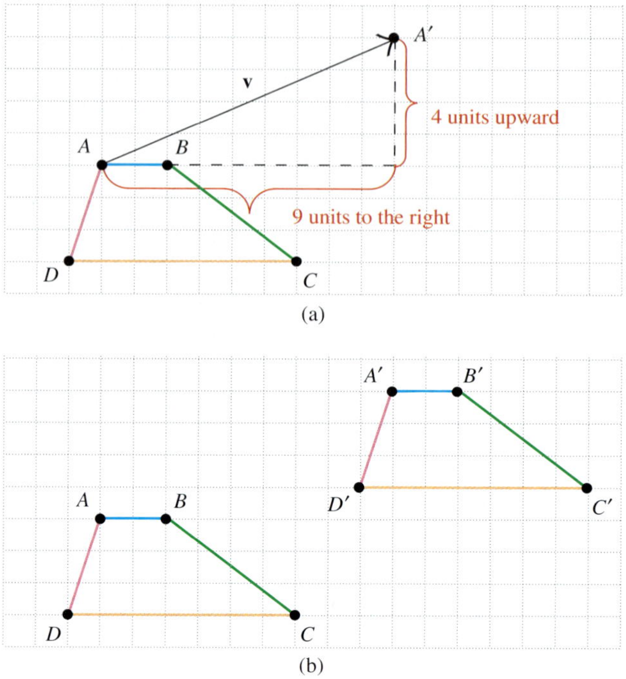

Figure 9.56

EXAMPLE 3 *A Translated Parallelogram*

Given parallelogram $ABCD$ and translation vector, **v**, shown in Fig. 9.57, construct the translated parallelogram $A'B'C'D'$.

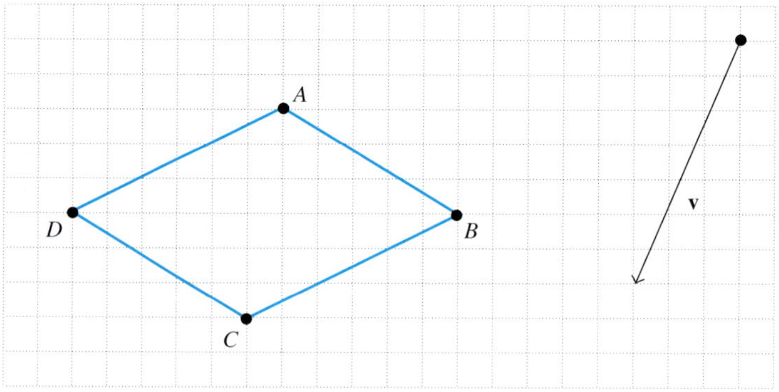

Figure 9.57

SOLUTION The translated figure will be a parallelogram of the same size and shape as parallelogram $ABCD$. We notice that the translation vector, **v**, points 7 units downward and 3 units to the left. We next examine vertex A. To determine the location of vertex A' of the translated parallelogram, start at vertex A of parallelogram $ABCD$ and move down 7 units and to the left 3 units. We label this vertex A' (see Fig. 9.58a on page 571). We determine vertices B', C', and D' in a similar manner by moving down 7 units and to the left 3 units from vertices B, C, and D, respectively.

Figure 9.58(b) shows parallelogram *ABCD* and the translated parallelogram $A'B'C'D'$. Notice in Fig. 9.58(b) that every point on parallelogram $A'B'C'D'$ is 7 units down and 3 units to the left of its corresponding point on parallelogram *ABCD*.

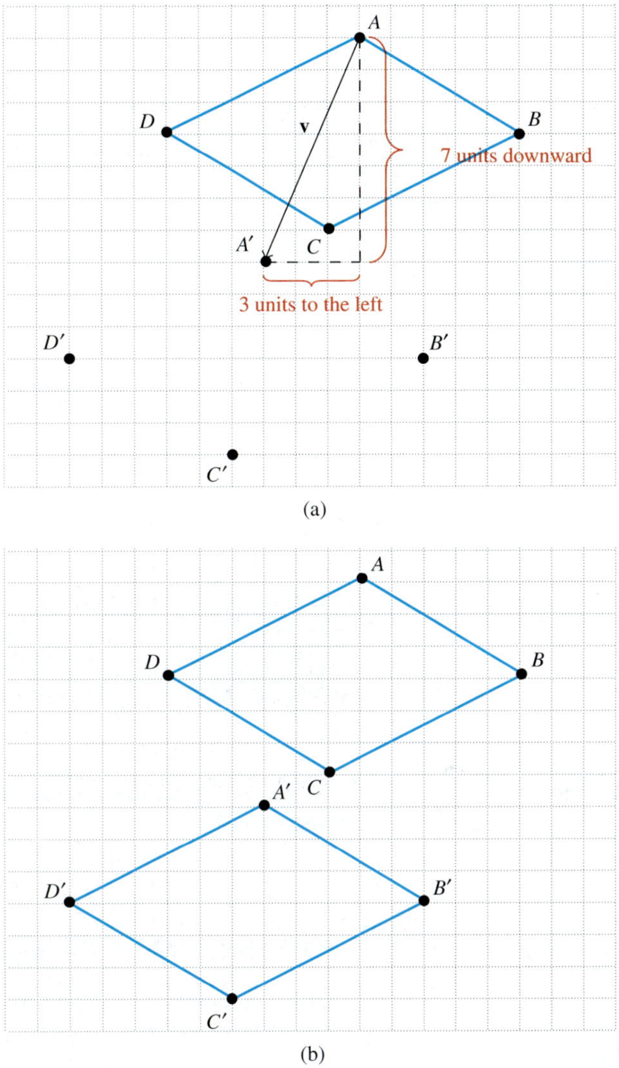

Figure 9.58

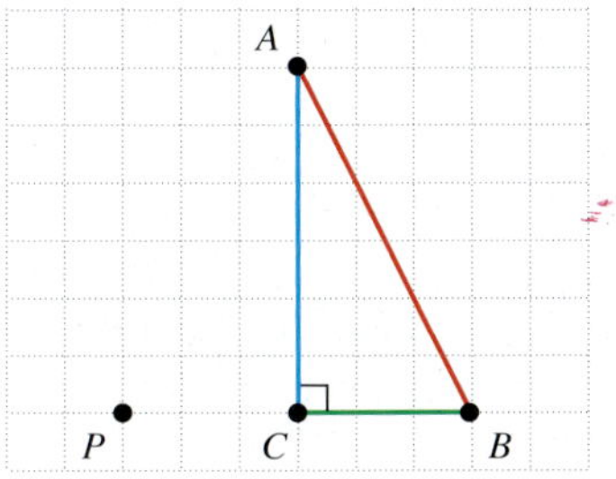

Figure 9.59

Rotations

The next rigid motion we will discuss is *rotation*. To help visualize a rotation, examine Fig. 9.59, which shows right triangle *ABC* and point *P* about which right triangle *ABC* is to be rotated.

Imagine that this page was removed from this book and attached to a bulletin board with a single pin through point *P*. Next imagine rotating the page 90° in the *counterclockwise* direction. The triangle would now appear as triangle $A'B'C'$ shown in Fig. 9.60. Next, imagine rotating the original triangle 180° in a counterclockwise direction. The triangle would now appear as triangle $A''B''C''$ shown in Fig. 9.61.

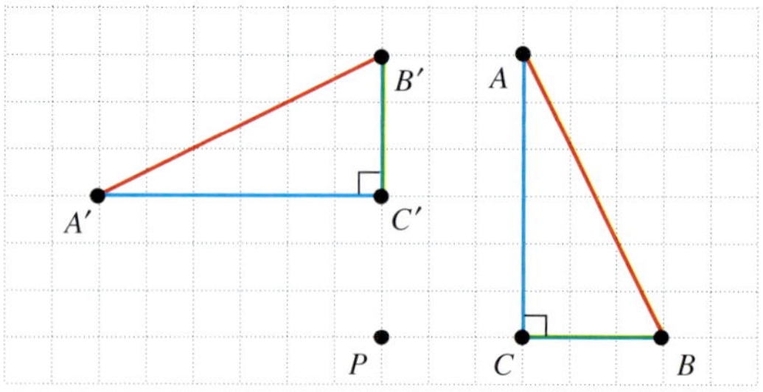

Figure 9.60

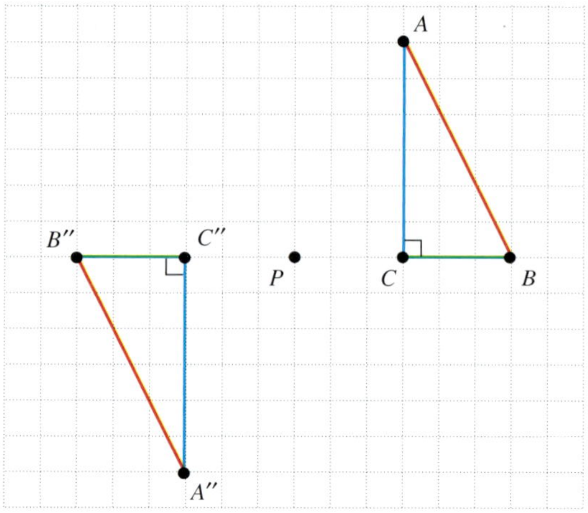

Figure 9.61

Now that we have an intuitive idea of how to determine a rotation, we give the definition of rotation.

A **rotation** is a rigid motion performed by rotating a geometric figure in the plane about a specific point, called the **rotation point** or the **center of rotation**. The angle through which the object is rotated is called the **angle of rotation**.

We will measure angles of rotation using degrees. In mathematics, generally, *counterclockwise angles have positive degree measures and clockwise angles have negative degree measures.*

EXAMPLE 4 *A Rotated Rectangle*

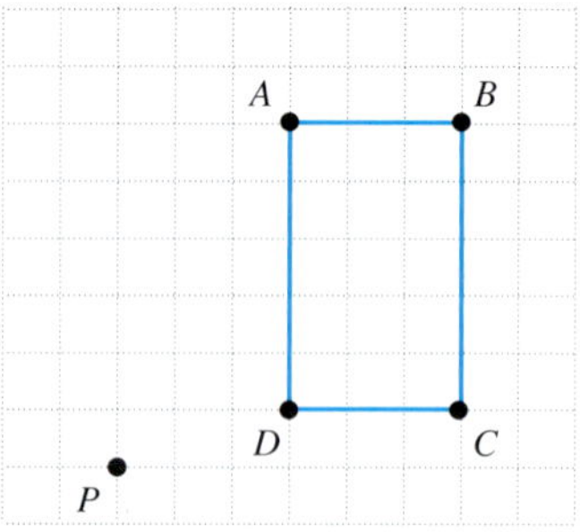

Figure 9.62

Given rectangle $ABCD$ and rotation point P, shown in Fig. 9.62, construct rectangles that result from rotations through

a) 90°. b) 180°. c) 270°.

SOLUTION

a) First, since 90 is a *positive* number, we will rotate the figure in a counterclockwise direction. We also note that the rotated rectangle will be the same size and shape as rectangle $ABCD$. To get an idea of what the rotated rectangle will look like, pick up this book and rotate it counterclockwise 90°. Figure 9.63 on page 573 shows rectangle $ABCD$ and rectangle $A'B'C'D'$ which is rectangle $ABCD$ rotated 90°about point P. Notice how line segment AB in rectangle $ABCD$ is horizontal, but in the rotated rectangle in Fig. 9.63 line segment $A'B'$ is vertical. Also notice that in rectangle $ABCD$ vertex D is 3 units to the *right* and 1 unit *above* rotation point P, but in the rotated rectangle, vertex D' is 3 units *above* and 1 unit to the *left* of rotation point P.

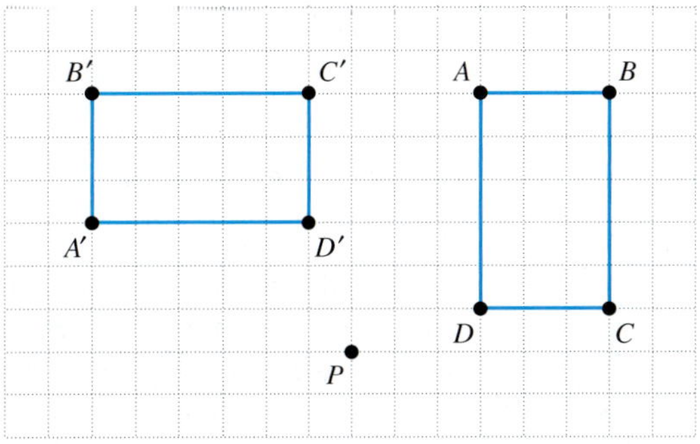

Figure 9.63

b) To gain some perspective on a 180° rotation, again pick up this book, but this time rotate the book 180° in the counterclockwise direction. The rotated rectangle $A''B''C''D''$ is shown along with the rectangle $ABCD$ in Fig. 9.64.

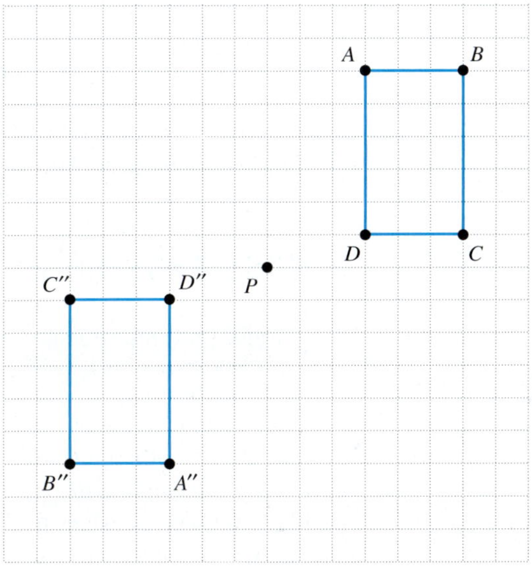

Figure 9.64

c) To gain some perspective on a 270° rotation, rotate this book 270° in the counterclockwise direction. The rotated rectangle $A'''B'''C'''D'''$ is shown along with rectangle $ABCD$ in Fig. 9.65.

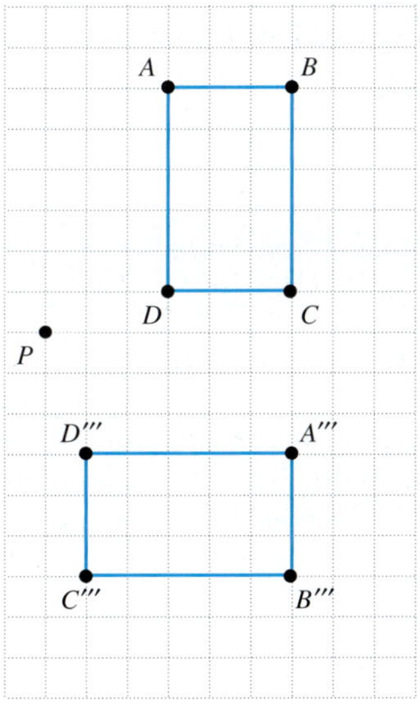

Figure 9.65

Thus far in our examples of rotations, the rotation point was outside the figure being rotated. We now will study an example where the rotation point is inside the figure to be rotated.

EXAMPLE 5 *A Rotation Point Inside a Polygon*

Given polygon $ABCDEFGH$ and rotation point P, shown in Fig. 9.66, construct polygons that result from rotations through

a) 90°. b) 180°.

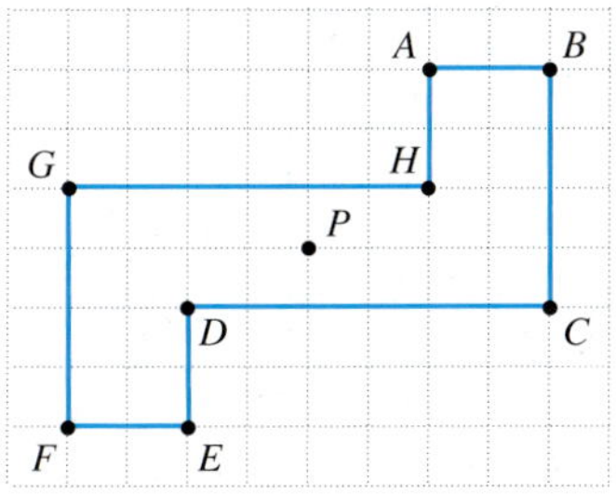

Figure 9.66

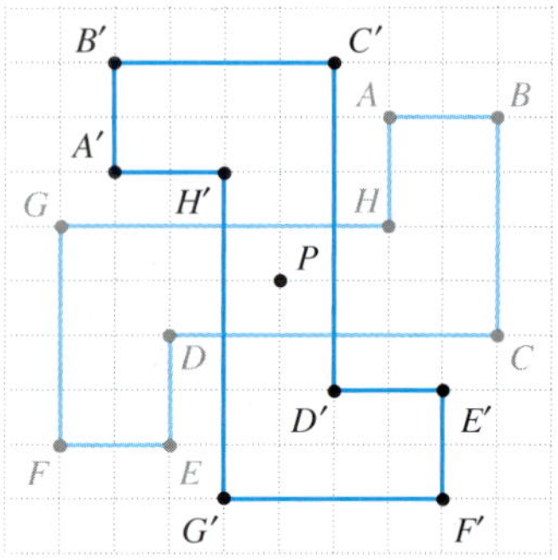

Figure 9.67

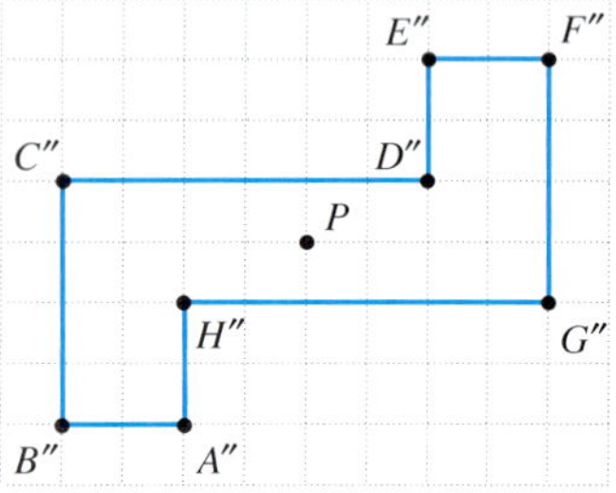

Figure 9.68

SOLUTION

a) We will rotate the polygon 90° in a counterclockwise direction. The resulting polygon will be the same size and shape as polygon *ABCDEFGH*. To visualize what the rotated polygon will look like, pick up this book and rotate it counterclockwise 90°. Figure 9.67 shows the polygon *ABCDEFGH,* in pale blue, and the rotated polygon $A'B'C'D'E'F'G'H'$, in deeper blue. Notice how line segments $\overline{AB}$, $\overline{CD}$, $\overline{EF}$, and $\overline{GH}$ in polygon *ABCDEFGH* are *horizontal*, but in the rotated polygon $A'B'C'D'E'F'G'H'$, line segments $\overline{A'B'}$, $\overline{C'D'}$, $\overline{E'F'}$, and $\overline{G'H'}$ are *vertical*. Also notice in polygon *ABCDEFGH* that line segment $\overline{GH}$ is 1 unit *above* rotation point *P*, but in the rotated polygon, line segment $\overline{G'H'}$ is 1 unit to the *left* of rotation point *P*.

b) To visualize the polygon obtained through a 180° rotation, we can pick up this book and rotate it 180° in the counterclockwise direction. Notice from Fig. 9.68 that vertex A'' of the rotated polygon is in the same position as vertex *E* of polygon *ABCDEFGH*. Also notice from Fig. 9.68 that vertex B'' of the rotated polygon is in the same position as vertex *F* of polygon *ABCDEFGH*. In fact, each of the vertices in the rotated polygon is in the same position as a different vertex in polygon *ABCDEFGH*. From Fig. 9.68 we see that, other than vertex labels, the position of rotated polygon $A''B''C''D''E''F''G''H''$ is the same as the position of polygon *ABCDEFGH*.

The polygon used in Example 5 will be discussed again later when we discuss *rotational symmetry*. The three rigid motions we have discussed thus far are reflection, translation, and rotation. Now we will discuss the fourth rigid motion, *glide reflection*.

Glide Reflections

A **glide reflection** is a rigid motion formed by performing a *translation* (or *glide*) followed by a *reflection*.

A glide reflection, as its name suggests, is a translation (or glide) followed by a reflection. Both translations and reflections were discussed earlier in this section. Consider triangle *ABC* (shown in blue), translation vector **v**, and reflection line *l* in Fig. 9.69. The translation of triangle *ABC,* obtained using translation vector **v**, is triangle $A'B'C'$ (shown in red). The reflection of triangle $A'B'C'$ about reflection line *l* is triangle $A''B''C''$ (shown in green). Thus, triangle $A''B''C''$ is the glide reflection of triangle *ABC* using translation vector **v** and reflection line *l*.

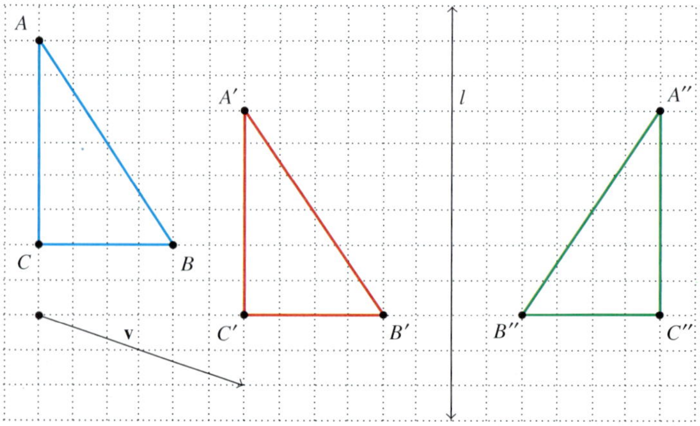

Figure 9.69

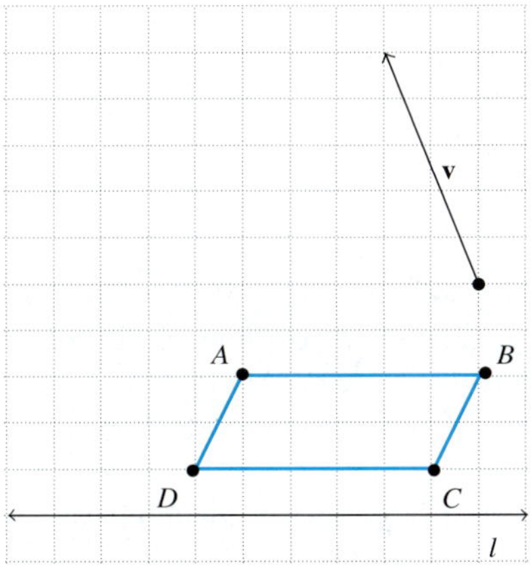

Figure 9.70

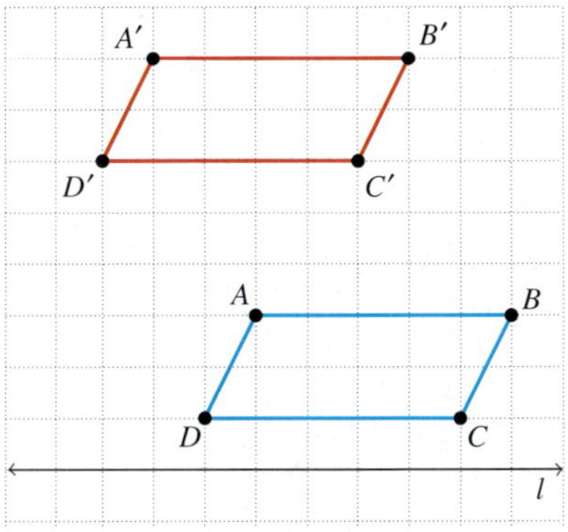

(a)

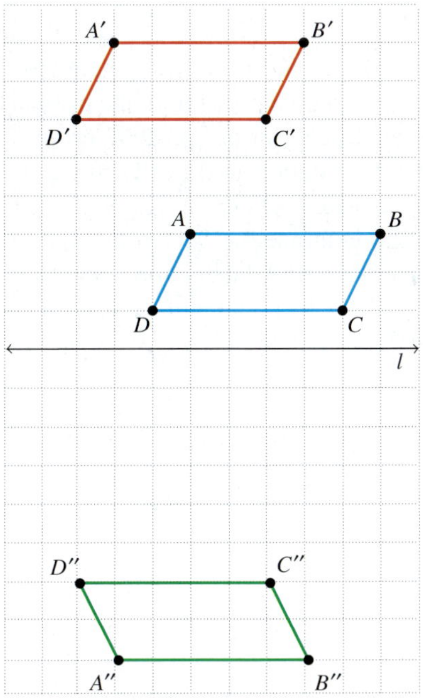

(b)

Figure 9.71

EXAMPLE 6 *A Glide Reflection of a Parallelogram*

Construct a glide reflection of parallelogram *ABCD*, shown in Fig. 9.70, using translation vector **v** and reflection line *l*.

SOLUTION To construct the glide reflection of parallelogram *ABCD*, first translate the parallelogram 2 units to the left and 5 units up as indicated by translation vector **v**. This translated parallelogram is labeled $A'B'C'D'$, shown in red in Fig. 9.71(a). Next, we will reflect parallelogram $A'B'C'D'$ about reflection line *l*. Parallelogram $A'B'C'D'$, shown in red, and the reflected parallelogram, labeled $A''B''C''D''$, shown in green, are shown in Fig. 9.71(b). The glide reflection of the parallelogram *ABCD* is parallelogram $A''B''C''D''$.

Symmetry

We are now ready to discuss symmetry. Our discussion of symmetry involves a rigid motion of an object.

A **symmetry** of a geometric figure is a rigid motion that moves the figure back onto itself. That is, the beginning position and ending position of the figure must be identical.

Suppose we start with a figure in a specific position and perform a rigid motion on this figure. If the position of the figure after the rigid motion is identical to the position of the figure before the rigid motion (if the beginning and ending positions of the figure coincide), then the rigid motion is a symmetry and we say that the figure has symmetry. For a two-dimensional figure, there are four types of symmetries: reflective symmetry, rotational symmetry, translational symmetry, and glide reflective symmetry. In this textbook, however, we will only discuss reflective symmetry and rotational symmetry.

Consider the polygon and reflection line *l* shown in Fig. 9.72(a). If we use the rigid motion of reflection and reflect the polygon *ABCDEFGH* about line *l*, we get polygon $A'B'C'D'E'F'G'H'$. Notice that the ending position of the polygon is identical to the starting position, as shown in Fig. 9.72(b). Compare Fig. 9.72(a) with Fig. 9.72(b). Although the vertex labels are different, the reflected polygon is in the same position as the polygon in the original position. Thus, we say that the polygon has ***reflective symmetry*** about line *l*. We refer to line *l* as a ***line of symmetry***.

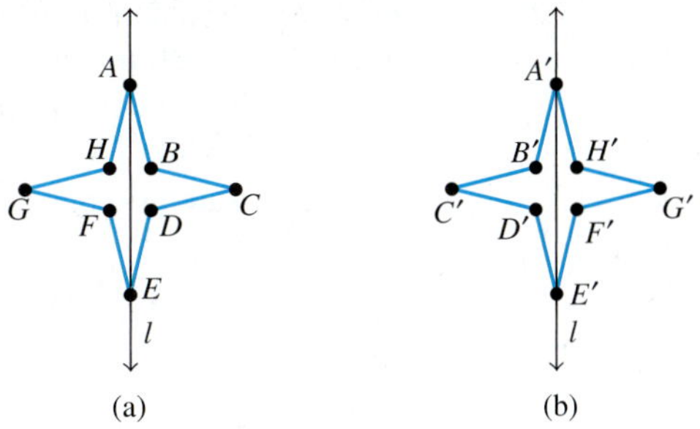

Figure 9.72

Recall Example 2 on page 568 in which hexagon *ABCDEF* was reflected about reflection line *l*. Examine the hexagon in the original position (Fig. 9.51) and the hexagon in the final position after being reflected about line *l* (Fig. 9.52). Other than the labels of the vertices, the beginning and ending positions of the hexagon are identical. Therefore, hexagon *ABCDEF* has reflective symmetry about line *l*.

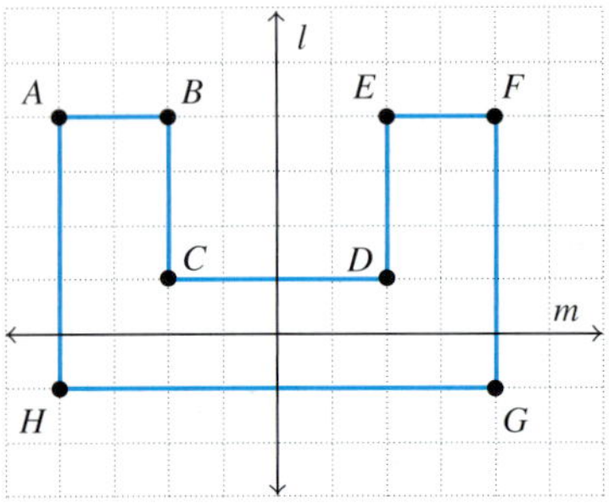

Figure 9.73

EXAMPLE 7 *Reflective Symmetries of Polygons*

Determine whether the polygon shown in Fig. 9.73 has reflective symmetry about each of the following lines.

a) Line l b) Line m

SOLUTION

a) Examine the reflection of the polygon about line l as seen in Fig. 9.74(a). Notice that other than the vertex labels, the beginning and ending positions of the polygon are identical. Thus, the polygon has reflective symmetry about line l.

b) Examine the reflection of the polygon about line m as seen in Fig. 9.74(b). Notice that the position of the reflected polygon is different from the original position of the polygon. Thus, the polygon does not have reflective symmetry about line m. ●

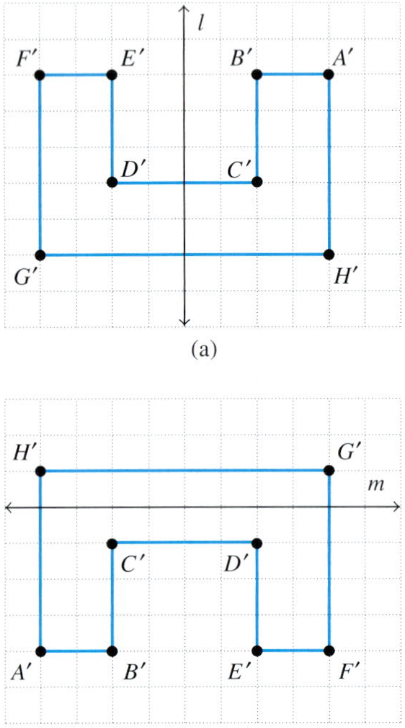

(a)

(b)

Figure 9.74

We will now discuss a second type of symmetry, rotational symmetry. Consider the polygon and rotation point P shown in Fig. 9.75(a). The rigid motion of rotation of polygon $ABCDEFGH$ through a 90° angle about point P gives polygon $A'B'C'D'E'F'G'H'$ shown in Fig. 9.75(b). Compare Fig. 9.75(a) with Fig. 9.75(b). Although the vertex labels are different, the position of the polygon before and after the rotation is identical. Thus, we say that the polygon has 90° ***rotational symmetry*** about point P. We refer to point P as the ***point of symmetry***.

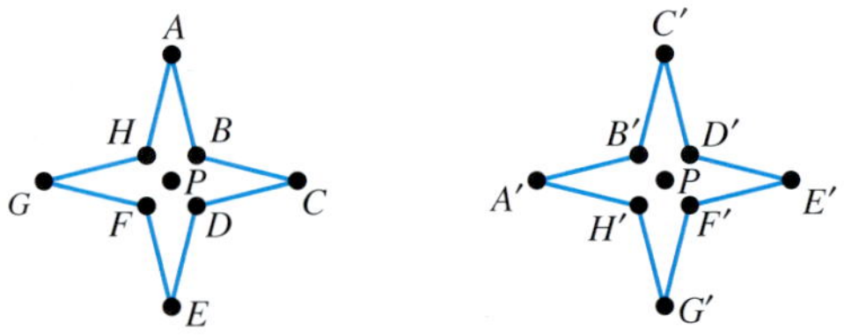

(a) (b)

Figure 9.75

Recall Example 5 on page 573 in which polygon $ABCDEFGH$ was rotated 90° about point P in part (a) and 180° in part (b). First examine the polygon in the original position in Fig. 9.66 on page 573 and the 90° rotated polygon in Fig. 9.67 on page 574. Notice the position of the polygon after the 90° rotation is different from the original position of the polygon. Therefore, polygon $ABCDEFGH$ in Fig. 9.67 does not have 90° rotational symmetry about point P. Now examine the 180° rotated polygon in Fig. 9.68 on page 574. Notice that other than the vertex labels, the position of the two polygons $ABCDEFGH$ and $A'B'C'D'E'F'G'H'$ is identical with respect to rotation about point P. Therefore, polygon $ABCDEFGH$ in Fig. 9.66 has 180° rotational symmetry about point P.

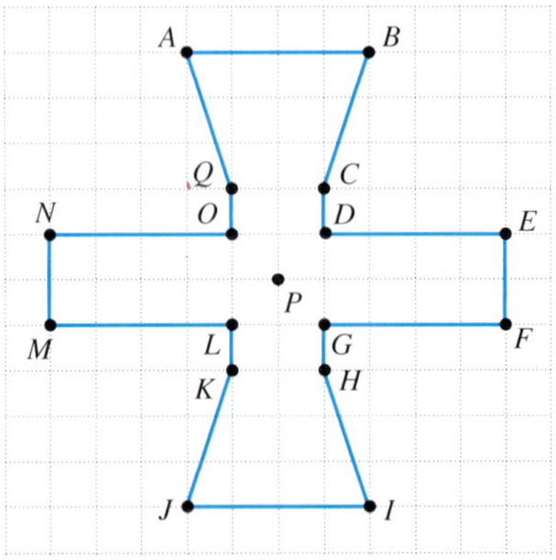

Figure 9.76

EXAMPLE 8 *Rotational Symmetries*

Determine whether the polygon shown in Fig. 9.76 has rotational symmetry about point P for rotations through each of the following angles.

a) 90° b) 180°

SOLUTION

a) To determine whether the polygon has 90° counterclockwise rotational symmetry about point P we rotate the polygon 90° as shown in Fig. 9.77(a). Compare Fig. 9.77(a) with Fig. 9.76. Notice that the position of the polygon after

PROFILE IN MATHEMATICS

Maurits Cornelius Escher

In addition to being wonderfully engaging art, the work of M. C. Escher (1898–1972) also displays some of the more beautiful and intricate aspects of mathematics. Escher's work involves Euclidean, non-Euclidean (to be studied shortly), and transformational geometries. Amazingly, Escher had no formal training in mathematics.

In 1936, Escher became obsessed with tessellations, that is, with creating art that used objects to cover a plane so as to leave no gaps. Symmetry became a cornerstone of Escher's famous tessellations. In 1995, the mathematician Donald Coxeter published a paper in which he proved that the mathematics Escher displayed in the etching *Circle Limit III* (see page 601) was indeed perfectly consistent with mathematical theory.

Escher kept a notebook in which he kept background information for his artwork. In this notebook, Escher characterized all possible combinations of shapes, colors, and symmetrical properties of polygons in the plane. By doing so, Escher had unwittingly developed areas of a branch of mathematics known as *crystallography* years before any mathematician had done so!

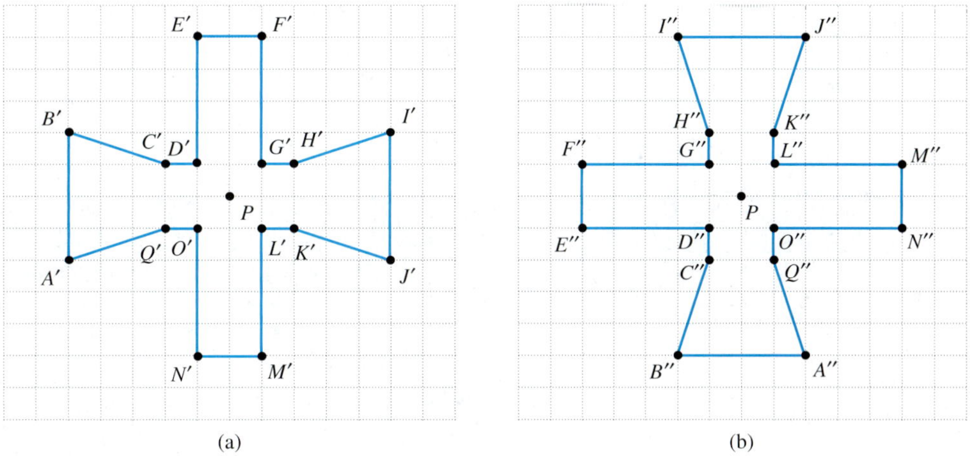

Figure 9.77

the rotation in Fig. 9.77(a) is different than the original position of the polygon (Fig. 9.76). Therefore, the polygon does not have 90° rotational symmetry.

b) To determine whether the polygon has 180° counterclockwise rotational symmetry about the point P, we rotate the polygon 180° as shown in Fig. 9.77(b). Compare Fig. 9.77(b) with Fig. 9.76. Notice that other than vertex labels, the position of the polygon after the rotation in Fig. 9.77(b) is identical to the position of the polygon before the rotation (Fig. 9.76). Therefore, the polygon has 180° rotational symmetry.

Tessellations

A fascinating application of transformational geometry is the creation of *tessellations*.

A **tessellation** (or **tiling**) is a pattern consisting of the repeated use of the same geometric figures to entirely cover a plane, leaving no gaps. The geometric figures used are called the **tessellating shapes** of the tessellation.

Figure 9.78 shows an example of a tessellation from ancient Egypt. Perhaps the most famous person to incorporate tessellations into his work is M. C. Escher (see Profiles in Mathematics at left).

Figure 9.78

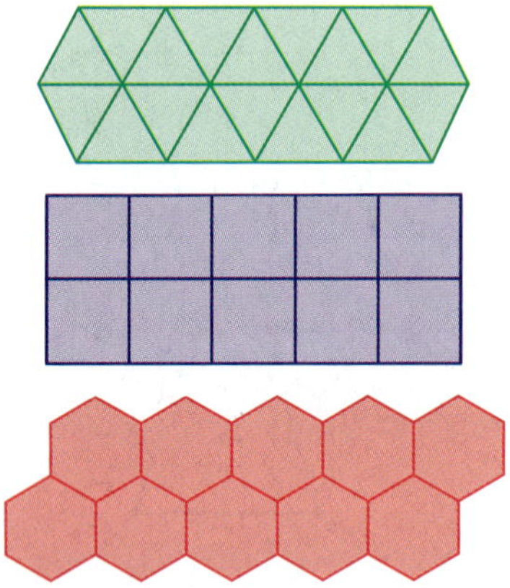

Figure 9.79

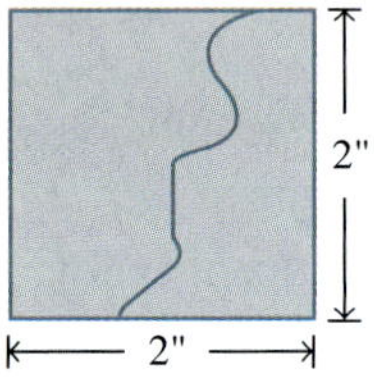

Figure 9.80

The simplest tessellations use one single regular polygon as the tessellating shape. Recall that a *regular polygon* is one whose sides are all the same length and whose interior angles all have the same measure. A tessellation that uses one single regular polygon as the tessellating shape is called a *regular tessellation*. It can be shown that only three regular tessellations exist: those that use an equilateral triangle, a square, or a regular hexagon as the tessellating shape. Figure 9.79 shows each of these regular tessellations. Notice that each tessellation can be obtained from a single tessellating shape through the use of reflections, translations, or rotations.

We will now learn how to create unique tessellations. We will do so by constructing a unique tessellating shape from a square. We could also construct other tessellating shapes using an equilateral triangle or a regular hexagon. If you wish to follow along with our construction, you will need some lightweight cardboard, a ruler, cellophane tape, and a pair of scissors. We will start by measuring and cutting out a square 2 in. by 2 in. from the cardboard. We next cut the square into two parts by cutting it from top to bottom using any kind of cut. One example is shown in Fig. 9.80. We then rearrange the pieces and tape the two vertical edges together as shown in Fig. 9.81. Next we cut this new shape into two parts by cutting it from left to right using any kind of cut as shown in Fig. 9.82. We then rearrange the pieces and tape the two horizontal edges together as shown in Fig. 9.83. This completes our tessellating shape.

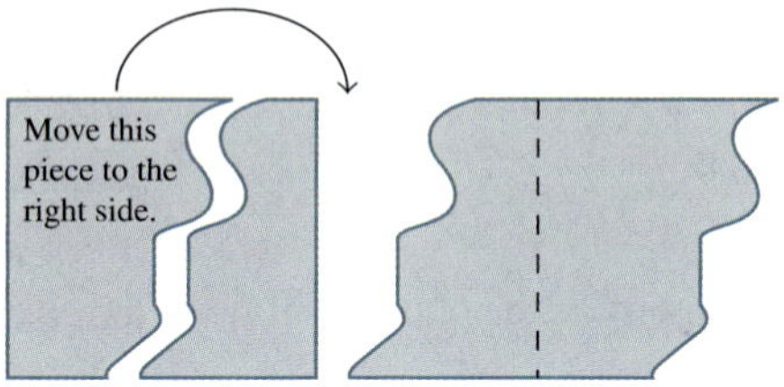

Figure 9.81

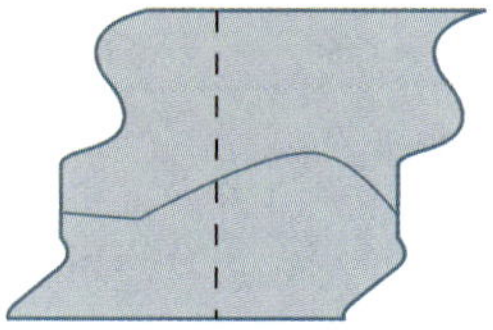

Figure 9.82

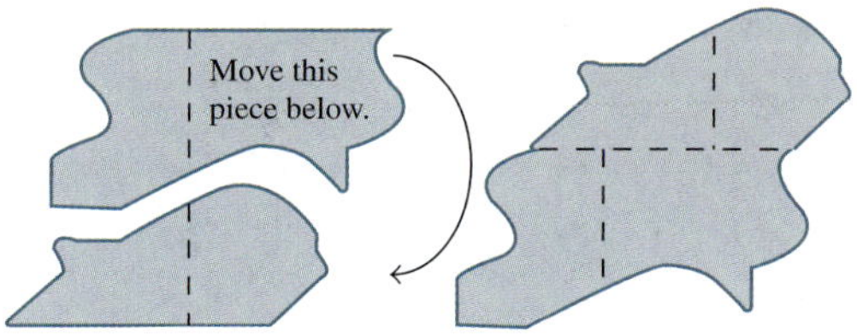

Figure 9.83

Figure 9.84

We now set the cardboard tessellating shape in the middle of a blank piece of paper (the tessellating shape can be rotated to any position as a starting point) and trace the outline of the shape onto the paper. Next move the tessellating shape so that it lines up with the figure already drawn and trace the outline again. Continue to do that until the page is completely covered. Once the page is covered with the tessellation, we can add some interesting colors or even some unique sketches to the tessellation. Figure 9.84 shows one tessellation created using the tessellation shape in Fig. 9.83. In Fig. 9.84, the tessellation shape was rotated about 45° counterclockwise.

An infinite number of different tessellations can be created using the method described by altering the cuts made. We could also create different tessellations using an equilateral triangle, a regular hexagon, or other types of polygons. There are also other, more complicated ways to create the tessellating shape. The Internet has many sites devoted to the creation of tessellations by hand. Many computer programs that generate tessellations are also available.

SECTION 9.5 EXERCISES

CONCEPT/WRITING EXERCISES

1. In the study of transformational geometry, what is a rigid motion? List the four rigid motions studied in this section.

2. What is transformational geometry?

3. In terms of transformational geometry, describe a reflection.

4. Describe how to construct a reflection of a given figure about a given line.

5. In terms of transformational geometry, describe a rotation.

6. Describe how to construct a rotation of a given figure, about a given point, through a given angle.

7. In terms of transformational geometry, describe a translation.

8. Describe how to construct a translation of a given figure using a translation vector.

9. In terms of transformational geometry, describe a glide reflection.

10. Describe how to construct a glide reflection of a given figure using a given translation vector and a given reflection line.

11. Describe what it means for a figure to have reflective symmetry about a given line.

12. Describe what it means for a figure to have rotational symmetry about a given point.

13. What is a tessellation?

14. Describe one way to make a unique tessellation from a 2-in. by 2-in. cardboard square.

PRACTICE THE SKILLS/PROBLEM SOLVING

In Exercises 15–22, use the given figure and lines of reflection to construct the indicated reflections. Show the figure in the positions both before and after the reflection.

In Exercises 15 and 16, use the following figure. Construct

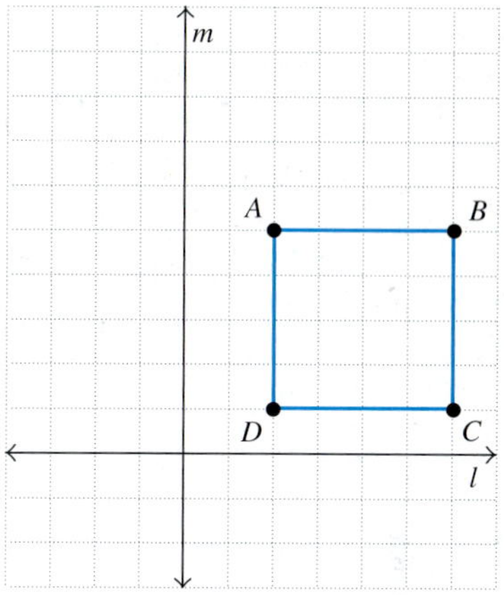

15. the reflection of square $ABCD$ about line m.

16. the reflection of square $ABCD$ about line l.

In Exercises 17 and 18, use the following figure. Construct

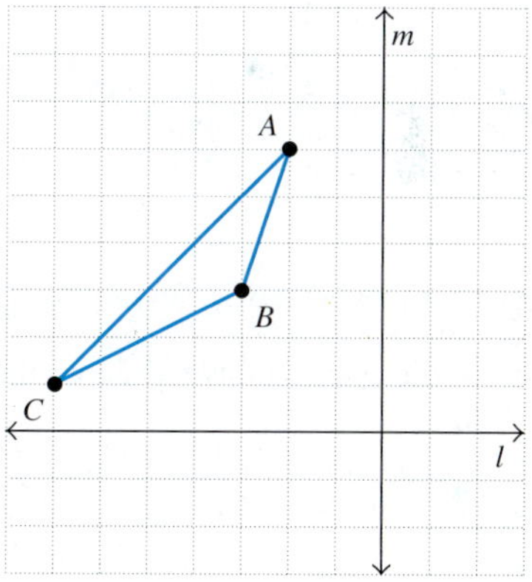

17. the reflection of triangle ABC about line l.

18. the reflection of triangle ABC about line m.

In Exercises 19 and 20, use the following figure. Construct

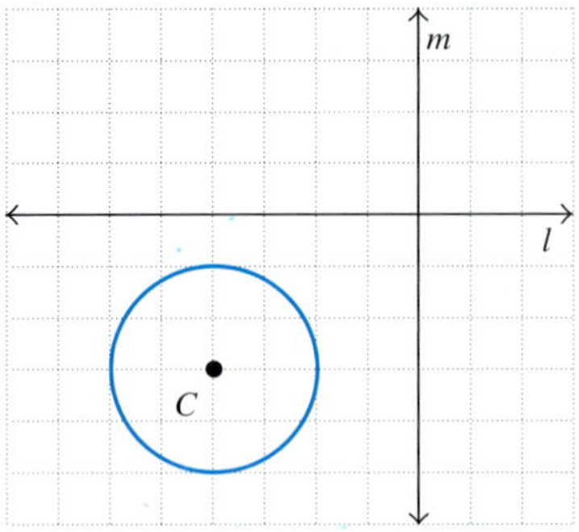

19. the reflection of circle C about line l.

20. the reflection of circle C about line m.

In Exercises 21 and 22, use the following figure. Construct

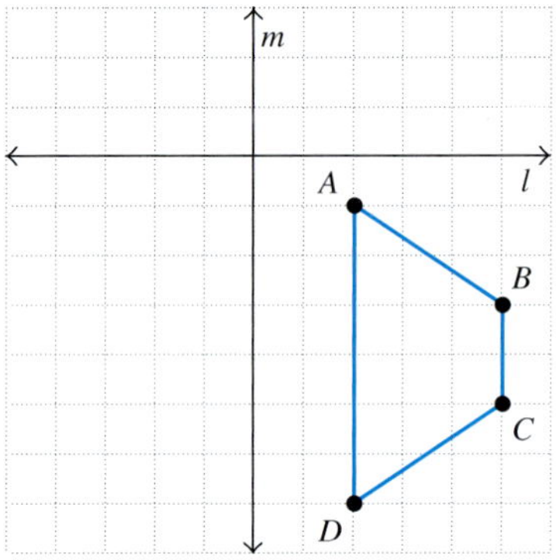

21. the reflection of trapezoid $ABCD$ about line m.

22. the reflection of the trapezoid $ABCD$ about line l.

In Exercises 23–30, use the translation vectors, **v** *and* **w** *shown below, to construct the translations indicated in the exercises. Show the figure in the positions both before and after the translation.*

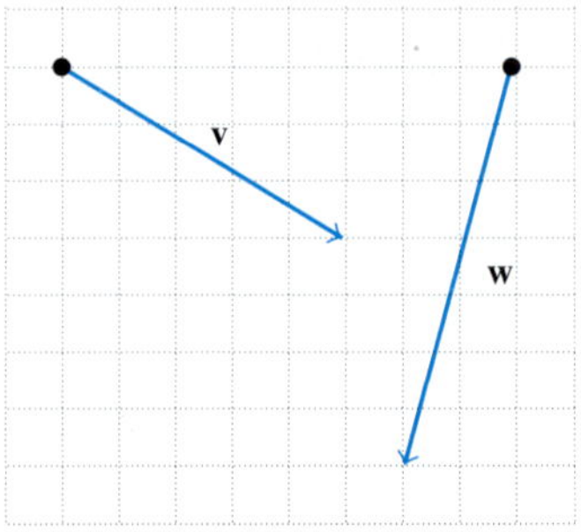

In Exercises 23 and 24, use the following figure. Construct

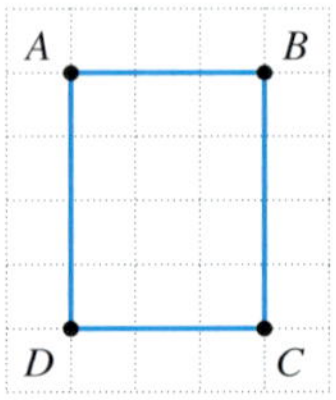

23. the translation of rectangle $ABCD$ using translation vector **v**.

24. the translation of rectangle $ABCD$ using translation vector **w**.

In Exercises 25 and 26, use the following figure. Construct

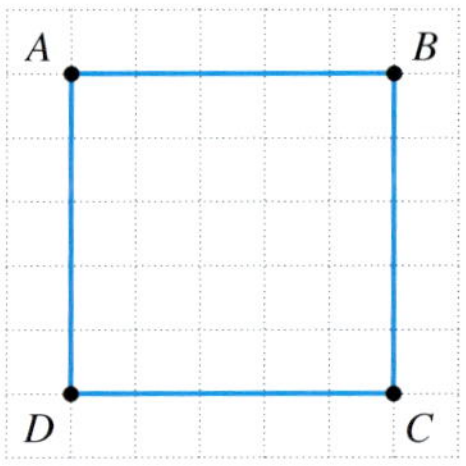

25. the translation of square $ABCD$ using translation vector **w**.

26. the translation of square $ABCD$ using translation vector **v**.

In Exercises 27 and 28, use the following figure. Construct

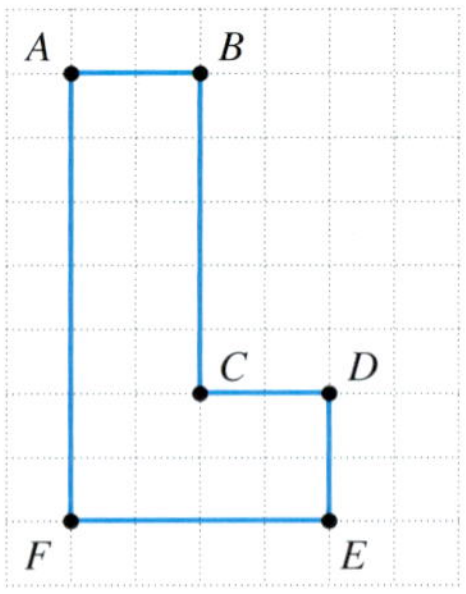

27. the translation of polygon $ABCDEF$ using translation vector **v**.

28. the translation of polygon $ABCDEF$ using translation vector **w**.

In Exercises 29 and 30, use the following figure. Construct

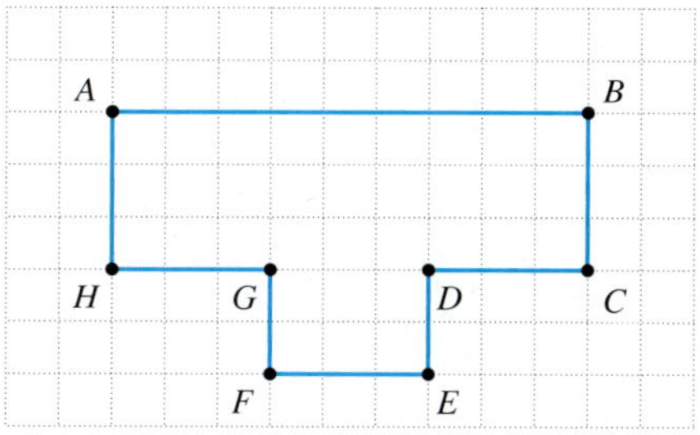

29. the translation of polygon *ABCDEFGH* using translation vector **w** (shown on page 580).

30. the translation of polygon *ABCDEFGH* using translation vector **v** (shown on page 580).

In Exercises 31–38, use the given figure and rotation point P to construct the indicated rotations. Show the figure in the positions both before and after the rotation.

In Exercises 31 and 32, use the following figure. Construct

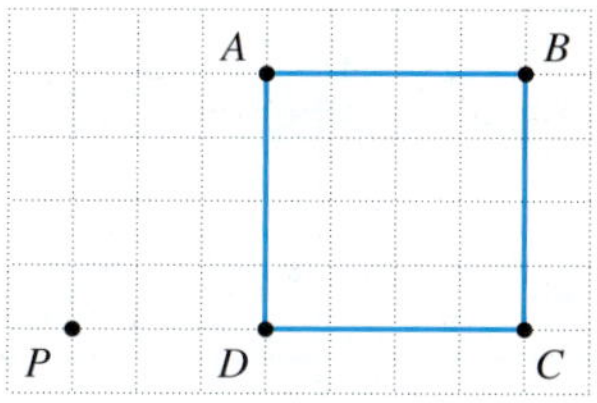

31. a 90° rotation of square *ABCD* about point *P*.

32. a 180° rotation of square *ABCD* about point *P*.

In Exercises 33 and 34, use the following figure. Construct

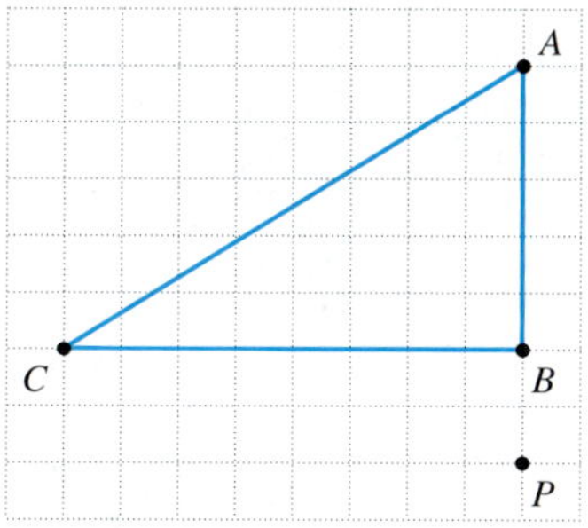

33. a 180° rotation of triangle *ABC* about point *P*.

34. a 270° rotation of triangle *ABC* about point *P*.

In Exercises 35 and 36, use the following figure. Construct

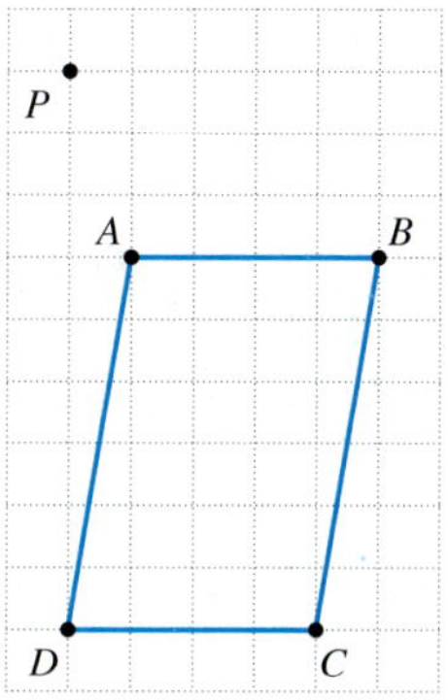

35. a 270° rotation of parallelogram *ABCD* about point *P*.

36. a 180° rotation of parallelogram *ABCD* about point *P*.

In Exercises 37 and 38, use the following figure. Construct

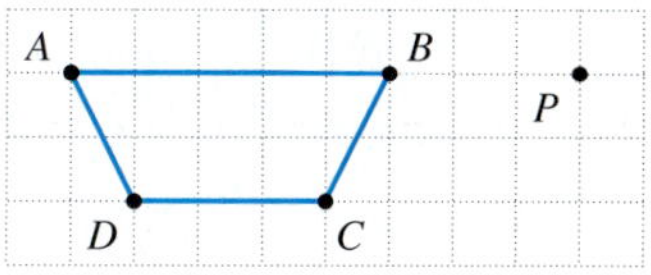

37. a 90° rotation of trapezoid *ABCD* about point *P*.

38. a 270° rotation of trapezoid *ABCD* about point *P*.

In Exercises 39–46, use the given figure, translation vectors **v** *and* **w**, *and reflection lines l and m to construct the indicated glide reflections. Show the figure in the positions before and after the glide reflection.*

In Exercises 39 and 40, use the following figure. Construct

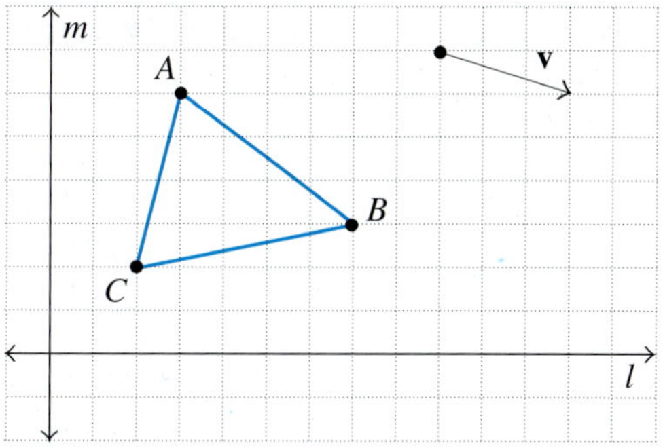

39. a glide reflection of triangle *ABC* using vector **v** and reflection line *l*.

40. a glide reflection of triangle *ABC* using vector **v** (shown on page 581) and reflection line *m*.

In Exercises 41 and 42, use the following figure. Construct

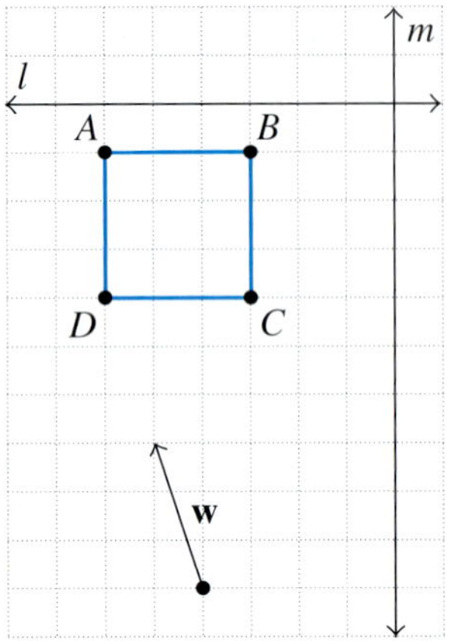

41. a glide reflection of square *ABCD* using vector **w** and reflection line *l*.

42. a glide reflection of square *ABCD* using vector **w** and reflection line *m*.

In Exercises 43 and 44, use the following figure. Construct

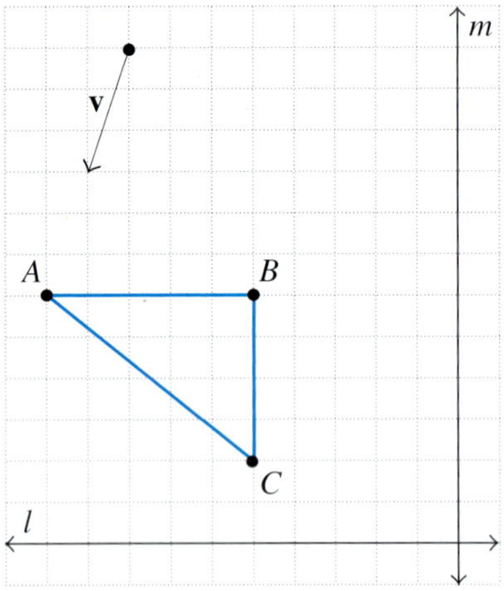

43. a glide reflection of triangle *ABC* using vector **v** and reflection line *l*.

44. a glide reflection of triangle *ABC* using vector **v** and reflection line *m*.

In Exercises 45 and 46, use the following figure. Construct

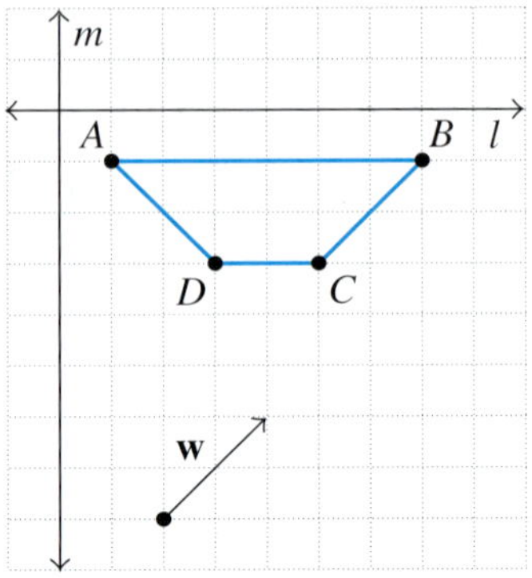

45. a glide reflection of trapezoid *ABCD* using vector **w** and reflection line *l*.

46. a glide reflection of trapezoid *ABCD* using vector **w** and reflection line *m*.

47. **a)** Reflect triangle *ABC*, shown below, about line *l*. Label the reflected triangle $A'B'C'$.

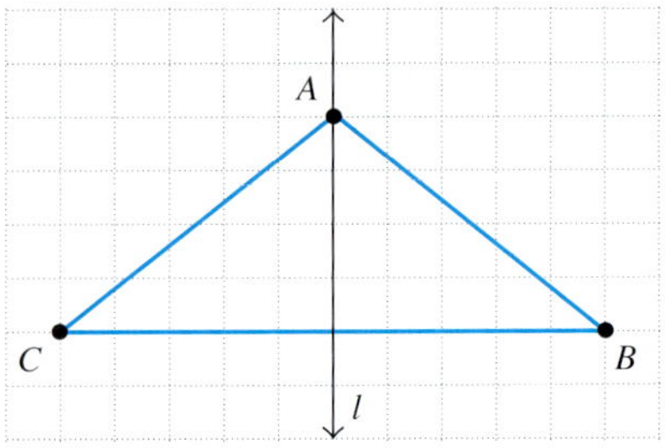

b) Other than vertex labels, is the position of triangle $A'B'C'$ identical to the position of triangle *ABC*?

c) Does triangle *ABC* have reflective symmetry about line *l*?

48. **a)** Reflect rectangle *ABCD*, shown below, about line *l*. Label the reflected rectangle $A'B'C'D'$.

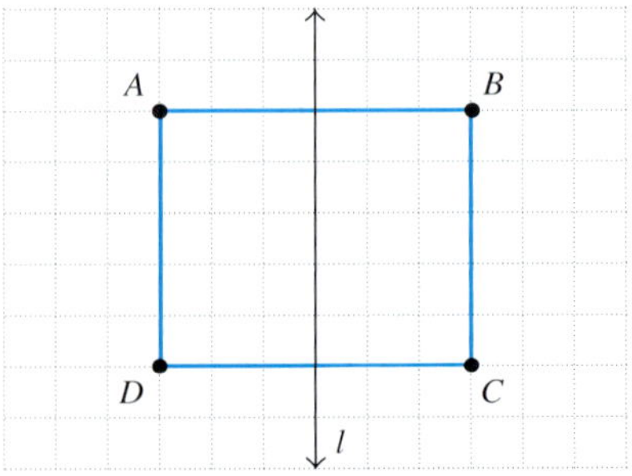

b) Other than vertex labels, is the position of rectangle $A'B'C'D'$ identical to the position of rectangle *ABCD*?

c) Does rectangle *ABCD* have reflective symmetry about line *l*?

49. a) Reflect parallelogram *ABCD*, shown below, about line *l*. Label the reflected parallelogram $A'B'C'D'$.

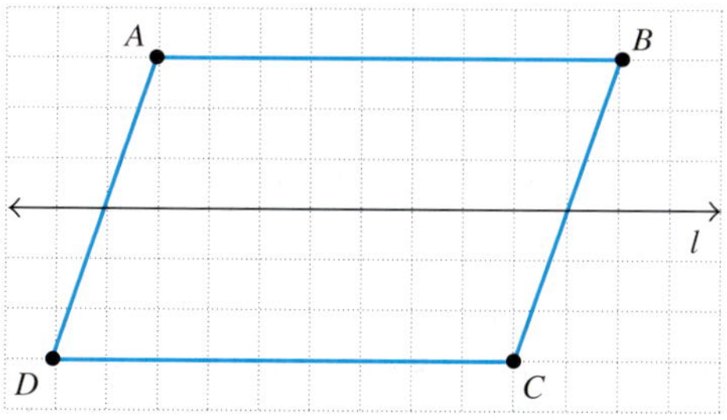

b) Other than vertex labels, is the position of parallelogram $A'B'C'D'$ identical to the position of parallelogram *ABCD*?

c) Does parallelogram *ABCD* have reflective symmetry about line *l*?

50. a) Reflect square *ABCD*, shown below, about line *l*. Label the reflected square $A'B'C'D'$.

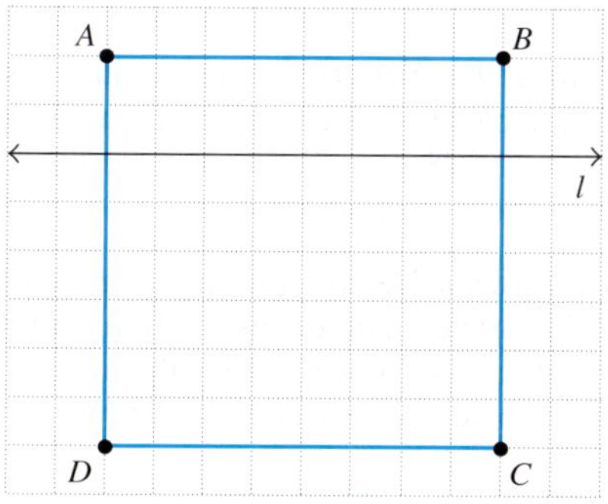

b) Other than vertex labels, is the position of square $A'B'C'D'$ identical to the position of square *ABCD*?

c) Does square *ABCD* have reflective symmetry about line *l*?

51. a) Rotate rectangle *ABCD*, shown below, 90° about point *P*. Label the rotated rectangle $A'B'C'D'$.

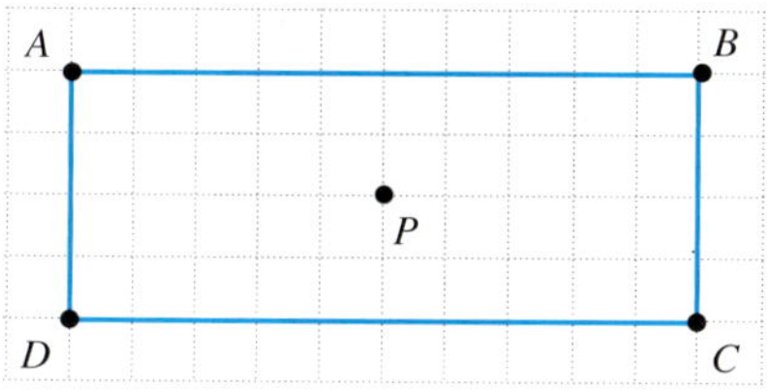

b) Other than vertex labels, is the position of rectangle $A'B'C'D'$ identical to the position of rectangle *ABCD*?

c) Does rectangle *ABCD* have 90° rotational symmetry about point *P*?

d) Now rotate the rectangle in the original position, rectangle *ABCD*, 180° about point *P*. Label the rotated rectangle $A''B''C''D''$.

e) Other than vertex labels, is the position of rectangle $A''B''C''D''$ identical to the position of rectangle *ABCD*?

f) Does rectangle *ABCD* have 180° rotational symmetry about point *P*?

52. a) Rotate parallelogram *ABCD*, shown below, 90° about point *P*. Label the rotated parallelogram $A'B'C'D'$.

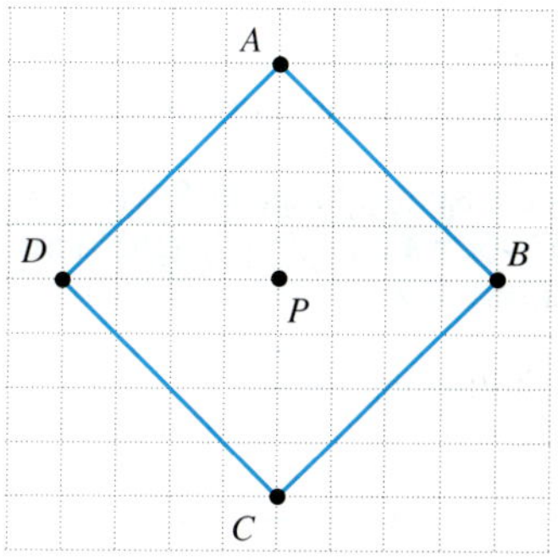

b) Other than vertex labels, is the position of parallelogram $A'B'C'D'$ identical to the position of parallelogram *ABCD*?

c) Does parallelogram *ABCD* have 90° rotational symmetry about point *P*?

d) Now rotate the parallelogram in the original position, parallelogram *ABCD*, shown above, 180° about point *P*. Label the rotated parallelogram $A''B''C''D''$.

e) Other than vertex labels, is the position of parallelogram $A''B''C''D''$ identical to the position of parallelogram *ABCD*?

f) Does parallelogram *ABCD* have 180° rotational symmetry about point *P*?

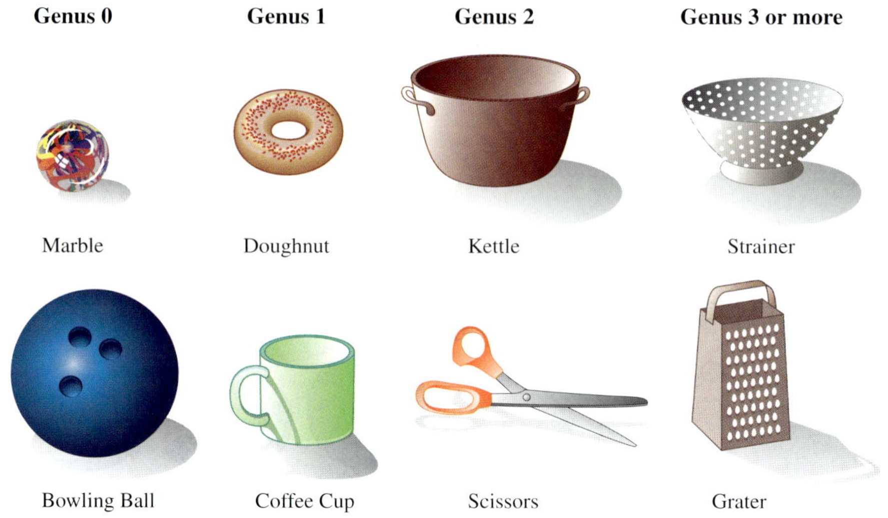

Figure 9.97

SECTION 9.6 EXERCISES

CONCEPT/WRITING EXERCISES

1. Explain why topology is sometimes referred to as "rubber sheet geometry."

2. What is a Möbius strip?

3. Explain how to make a Möbius strip.

4. What is a Klein bottle?

5. What is the maximum number of colors needed to create a map on a flat surface if no two regions colored the same are to share a common border?

6. What is the maximum number of colors needed to create a map if no two regions colored the same are to share a common border if the surface is a

   a) Möbius strip?

   b) torus?

7. What is a Jordan curve?

8. When testing to determine whether a point is inside or outside a Jordan curve, explain why if you count an odd number of lines, the point is inside the curve, and if you count an even number of lines, the point is outside the curve.

9. How is the genus of a figure determined?

10. When are two figures topologically equivalent?

PRACTICE THE SKILLS

In Exercises 11–16, color the map by using a maximum of four colors so that no two regions with a common border have the same color.

11.

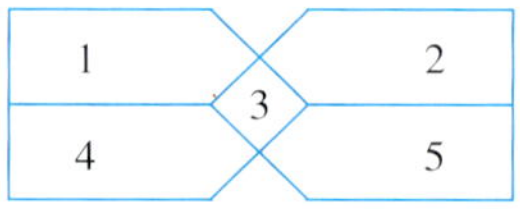

12.

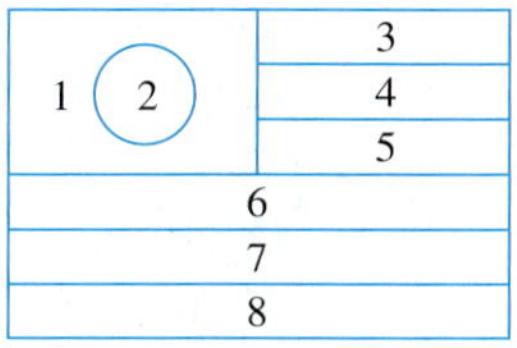

13.

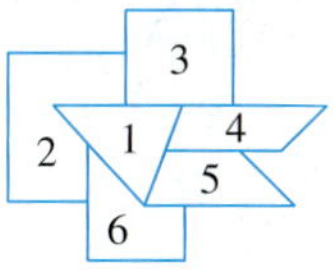

14.

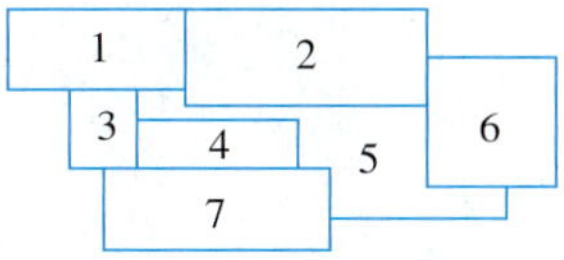

15.

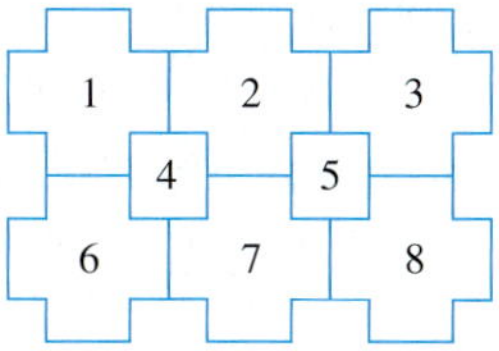

16.

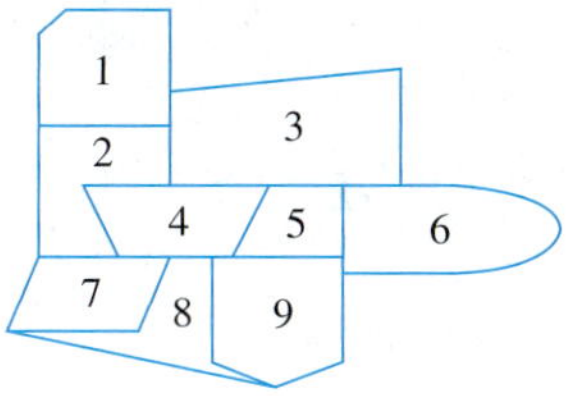

Using the Four-Color Theorem *In Exercises 17–20, maps show certain areas of the United States, Canada, and Mexico. Shade in the states (or provinces) using a maximum of four colors so that no two states (or provinces) with a common border have the same color.*

17.

18.

19.

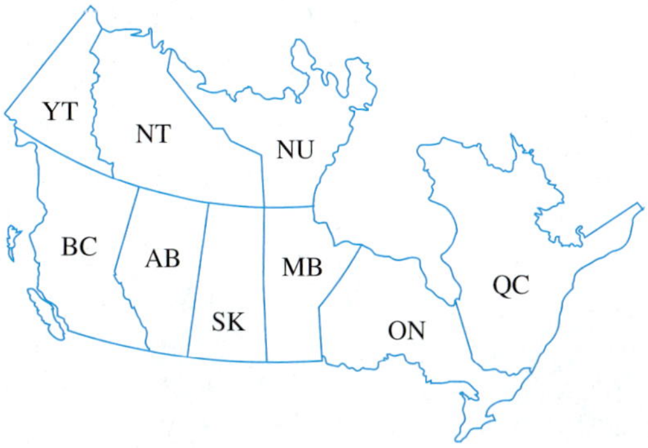

20.

21. Determine whether point A in Fig. 9.95(d) on page 589 is inside or outside the Jordan curve.

22. Determine whether point B in Fig. 9.95(d) is inside or outside the Jordan curve.

At right is a Jordan curve. In Exercises 23–28, determine if the point is inside or outside the curve.

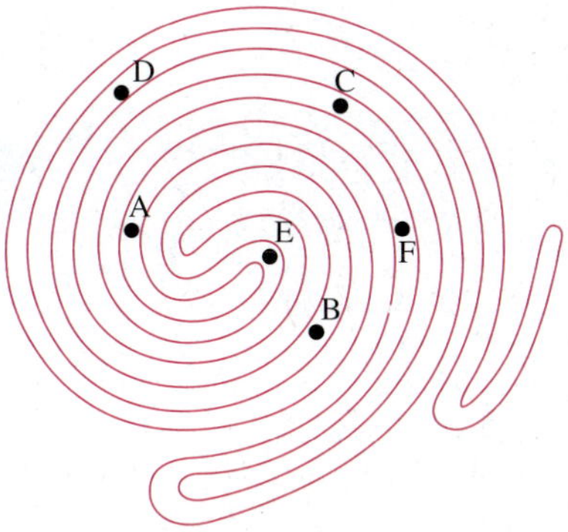

23. Point F

24. Point E

25. Point D

26. Point C

27. Point B

28. Point A

In Exercises 29–40, give the genus of the object. If the object has a genus larger than 5, write "larger than 5."

29. 

30. 

31. 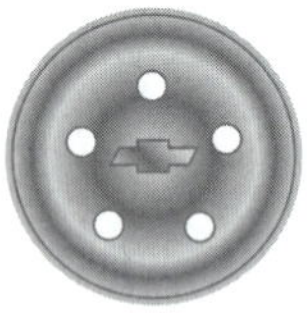

32. 

33. 

34. 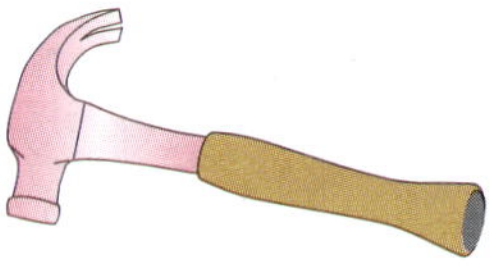

35. 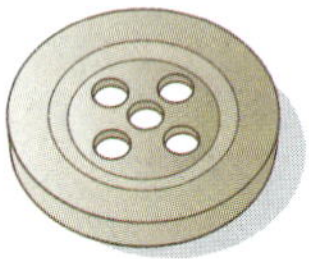

36. 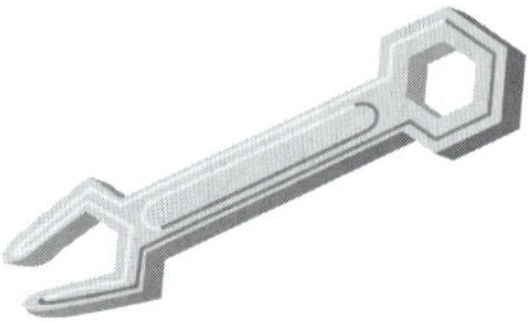

37. 

38. 

39. 

40. 

41. Name at least three objects not mentioned in this section that have

a) genus 0.

b) genus 1.

c) genus 2.

d) genus 3 or more.

42. Use the result of Experiment 1 on page 586 to find the number of edges on a Möbius strip.

43. Use the result of Experiment 2 on page 586 to find the number of surfaces on a Möbius strip.

44. How many separate strips are obtained in Experiment 3 on page 586?

45. How many separate strips are obtained in Experiment 4 on page 586?

46. Make a Möbius strip. Cut it one-third of the way from the edge, as in Experiment 4. You should get two loops, one going through the other. Determine whether either (or both) of these loops is itself a Möbius strip.

47. a) Take a strip of paper, give it one full twist, and connect the ends. Is the result a Möbius strip with only one side? Explain.

b) Determine the number of edges, as in Experiment 1.

c) Determine the number of surfaces, as in Experiment 2.

d) Cut the strip down the middle. What is the result?

48. Take a strip of paper, make one whole twist and another half twist, and then tape the ends together. Test by a method of your choice to determine whether this has the same properties as a Möbius strip.

CHALLENGE PROBLEMS/GROUP ACTIVITIES

49. Using clay (or glazing compound), make a doughnut. Without puncturing or tearing the doughnut, reshape it into a topologically equivalent figure, a cup with a handle.

50. Using at most four colors, color the map of South America. Do not use the same color for any two countries that share a common border.

Source: www.mapsofworld.com <http://www.mapsofworld.com> used with permission.

51. Using at most four colors, color the following map of the counties of New Mexico. Do not use the same color for any two counties that share a common border.

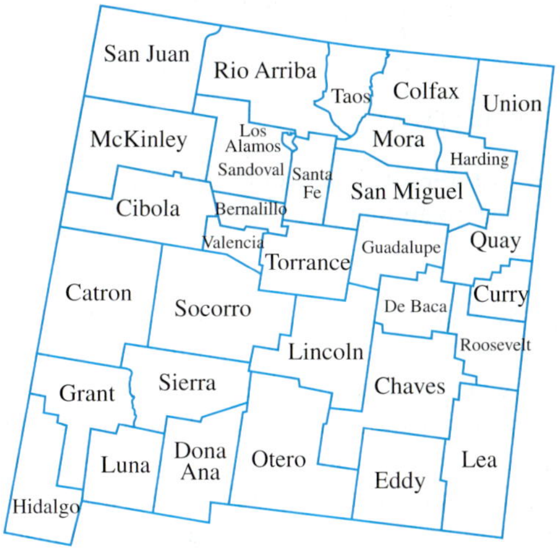

RECREATIONAL MATHEMATICS

52. ***An Interesting Surface*** Construct the following surface using two strips of paper, scissors, and tape. Then answer the following questions.

a) How many sides does this surface have?

b) How many edges does this surface have?

c) Attempt to cut the surface "in half" by making a small slit in the middle of the paper surface. Then cut along the surface (see dashed line in the figure), keeping the scissors the same distance from the edge. In your own words, describe what happens.

Although the surface shown in the figure above shares some of the same traits as a Möbius strip, this surface is not topologically equivalent to the Möbius strip.

INTERNET/RESEARCH ACTIVITIES

53. Use the Internet to find a map of your state that shows the outline of all the counties within your state. Print this map and, using at most four colors, color it. Do not use the same color for any two counties that share a common border.

54. The short story *Paul Bunyan versus the Conveyor Belt* (1947) by William Hazlett Upson focuses on a conveyor belt in the shape of a Möbius strip. The story can be found in several books that include mathematical essays. Read Upson's short story and write a 200-word description of what Paul Bunyan does to the conveyor belt. Confirm the outcome of the story by repeating Paul's actions with a paper Möbius strip.

PYTHAGOREAN THEOREM

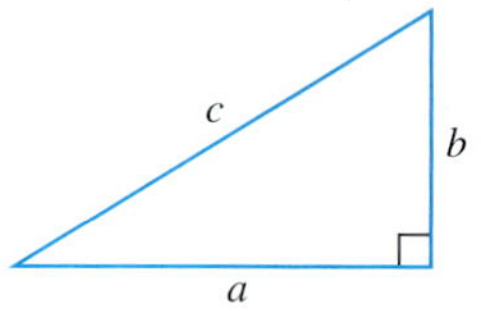

$a^2 + b^2 = c^2$

CIRCLE

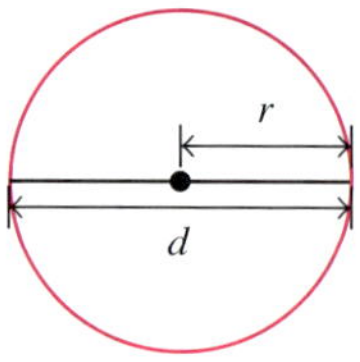

$A = \pi r^2; C = 2\pi r$ or $C = \pi d$

CUBE

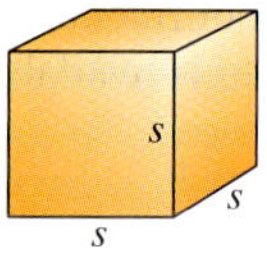

$V = s^3$
$SA = 6s^2$

RECTANGULAR SOLID

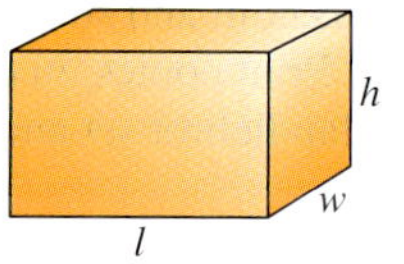

$V = lwh$
$SA = 2lw + 2wh + 2lh$

CYLINDER

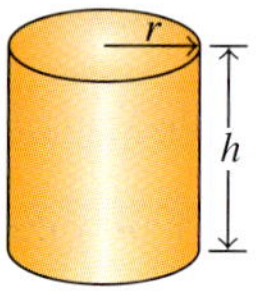

$V = \pi r^2 h$
$SA = 2\pi rh + 2\pi r^2$

CONE

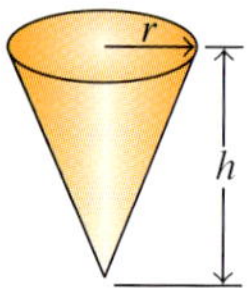

$V = \frac{1}{3}\pi r^2 h$
$SA = \pi r^2 + \pi r\sqrt{r^2 + h^2}$

SPHERE

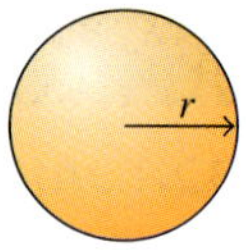

$V = \frac{4}{3}\pi r^3$
$SA = 4\pi r^2$

PRISM

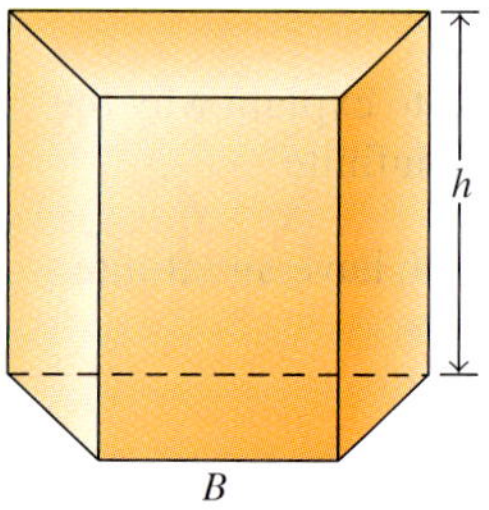

$V = Bh$, where B is the area of the base

PYRAMID

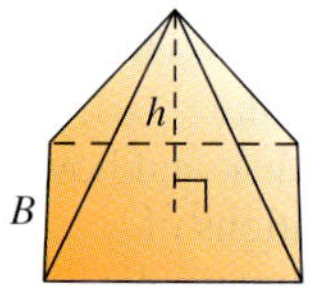

$V = \frac{1}{3}Bh$, where B is the area of the base

FIFTH POSTULATE IN EUCLIDEAN GEOMETRY

Given a line and a point not on the line, only one line can be drawn through the given point parallel to the given line.

FIFTH POSTULATE IN ELLIPTICAL GEOMETRY

Given a line and a point not on the line, no line can be drawn through the given point parallel to the given line.

FIFTH POSTULATE IN HYPERBOLIC GEOMETRY

Given a line and a point not on the line, two or more lines can be drawn through the given point parallel to the given line.

CHAPTER 9 REVIEW EXERCISES

9.1

In Exercises 1–6, use the figure shown to determine the following.

1. $\measuredangle ABF \cap \measuredangle CBH$
2. $\overrightarrow{AB} \cap \overrightarrow{DC}$
3. $\overline{BF} \cup \overline{FC} \cup \overline{BC}$
4. $\mathring{\overrightarrow{BH}} \cup \overrightarrow{HB}$
5. $\overleftrightarrow{HI} \cap \overleftrightarrow{EG}$
6. $\mathring{\overrightarrow{CF}} \cap \mathring{\overrightarrow{CG}}$

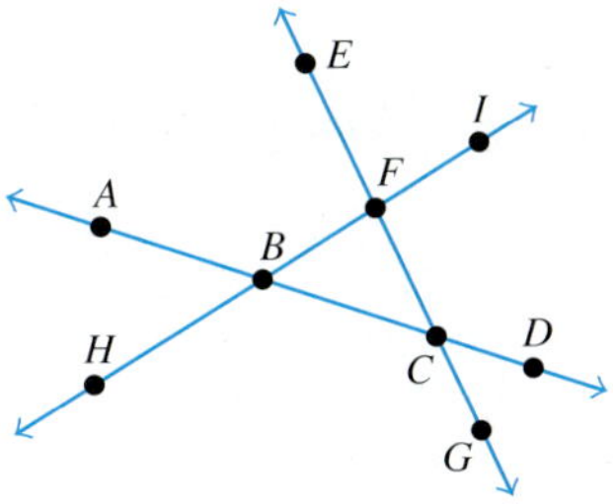

7. $m\measuredangle A = 23.7°$. Determine the measure of the complement of $\measuredangle A$.

8. $m\measuredangle B = 124.7°$. Determine the measure of the supplement $\measuredangle B$.

9.2

In Exercises 9–12, use the similar triangles ABC and A′B′C shown to determine the following.

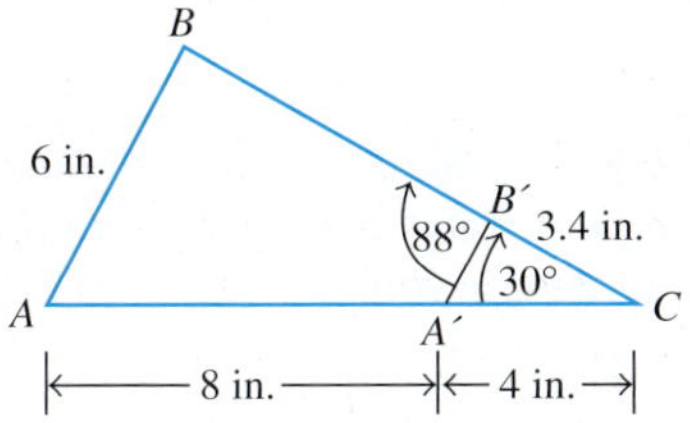

9. The length of $\overline{BC}$
10. The length of $\overline{A'B'}$
11. $m\measuredangle BAC$
12. $m\measuredangle ABC$

13. In the following figure, l_1 and l_2 are parallel lines. Determine $m\measuredangle 1$ through $m\measuredangle 6$.

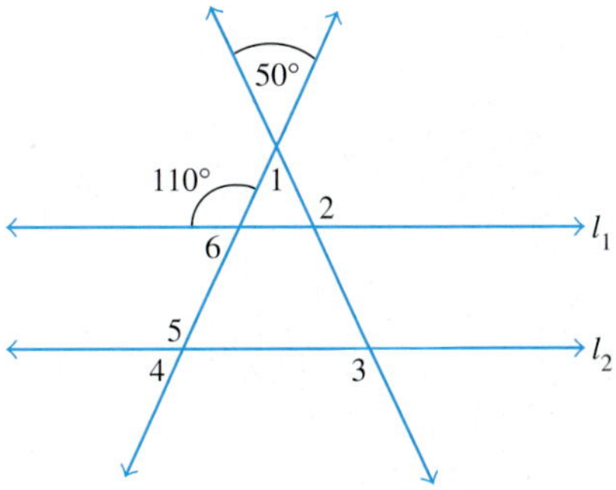

14. Determine the sum of the measures of the interior angles of a pentagon.

9.3

In Exercises 15–18, determine (a) the area and (b) the perimeter of the figure.

15.

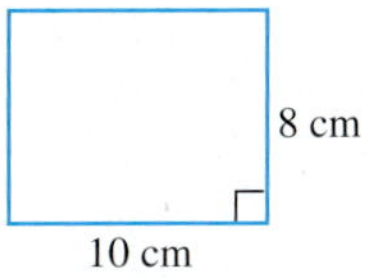

16.

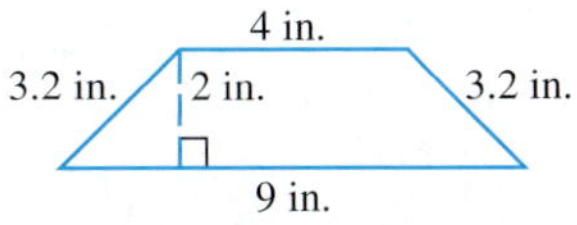

17.

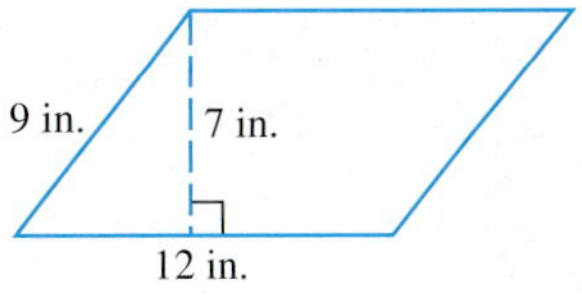

18.

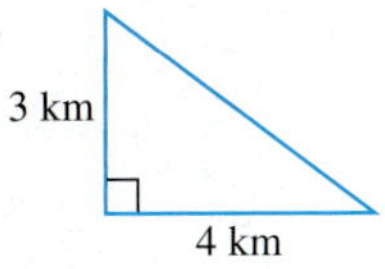

19. Determine (a) the area and (b) the circumference of the circle. Use the π key on a calculator and round your answer to the nearest hundredth.

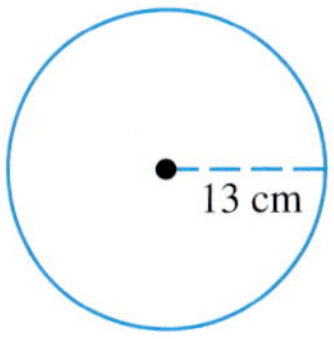

In Exercises 20 and 21, determine the shaded area. When appropriate, use the [π] *key on your calculator and round your answer to the nearest hundredth.*

20.

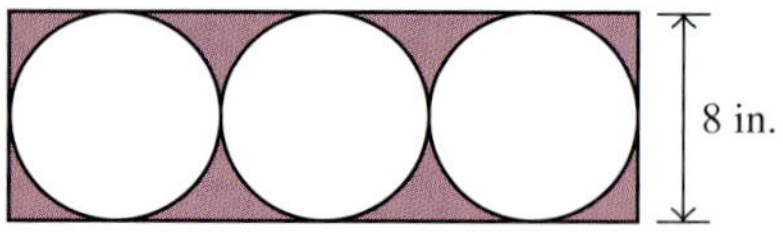

21.

22. ***Cost of Kitchen Tile*** Determine the total cost of covering a 14 ft by 16 ft kitchen floor with ceramic tile. The cost of the tile selected is $5.25 per square foot.

9.4

In Exercises 23–26, determine (a) the volume and (b) the surface area of the figure. When appropriate, use the [π] *key on your calculator and round your answer to the nearest hundredth.*

23.

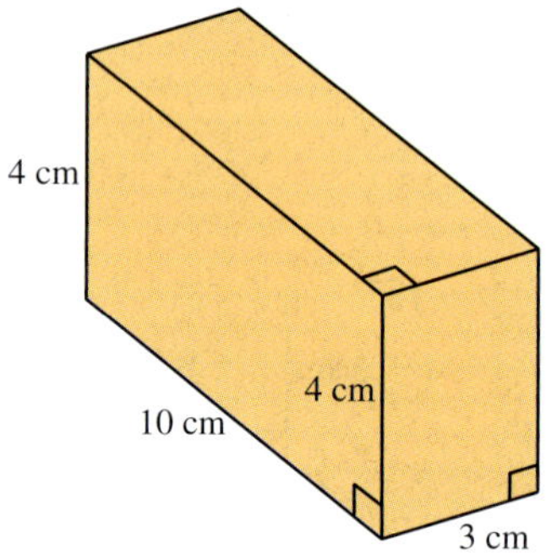

24.

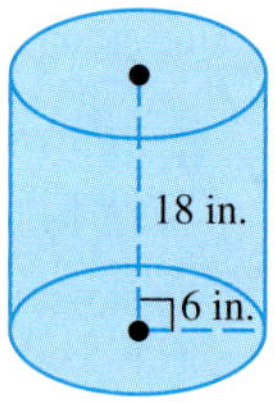

25.

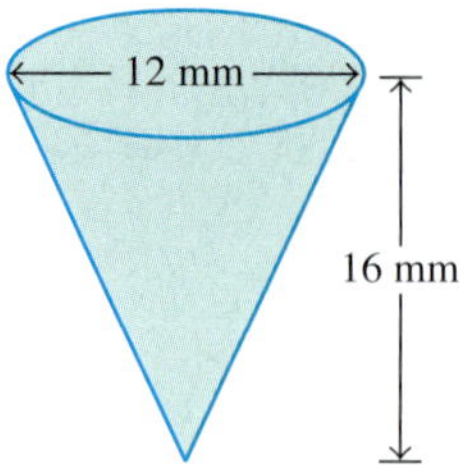

26.

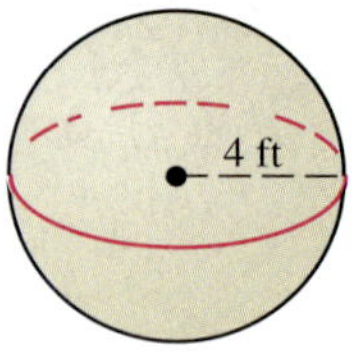

In Exercises 27 and 28, determine the volume of the figure.

27.

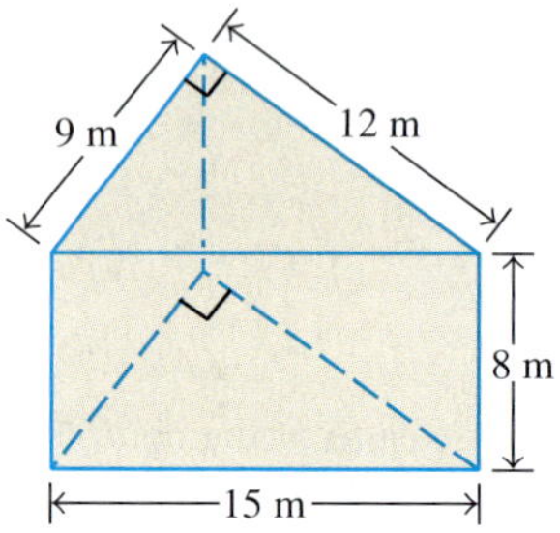

28.

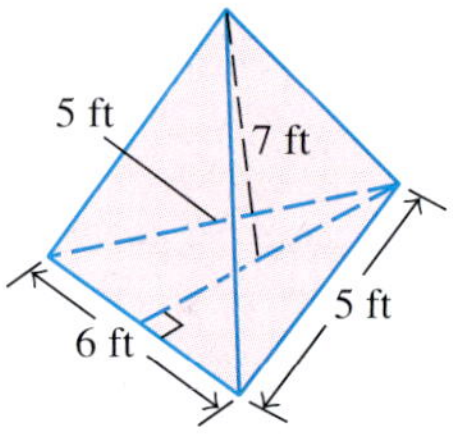

In Exercises 29 and 30, determine the volume of the shaded area. When appropriate, use the [π] *key and round your answer to the nearest hundredth.*

29.

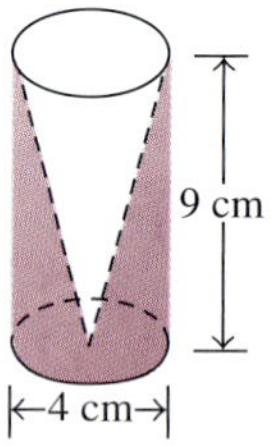

30.

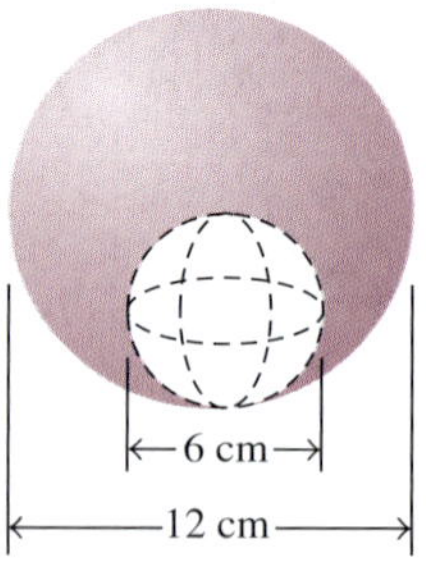

31. ***Water Trough*** Steven Dale has a water trough whose ends are trapezoids and whose sides are rectangles, as illustrated. He is afraid that the base it is sitting on will not support the weight of the trough when it is filled with water. He knows that the base will support 4800 lb.

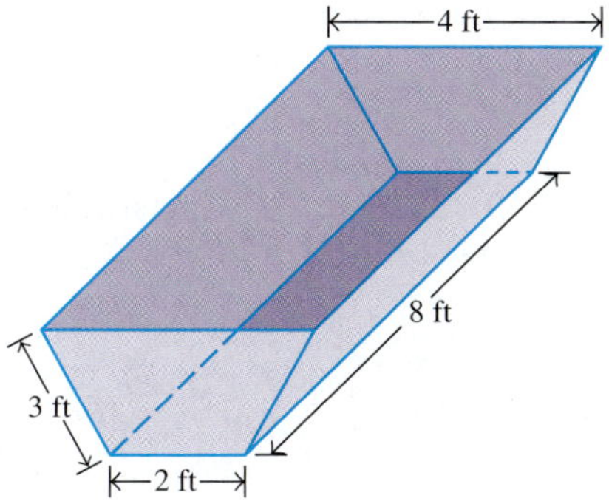

a) If the trough is filled with water, determine the number of cubic feet of water contained in the trough.

b) Determine the total weight, assuming that the trough weighs 375 lb and the water weighs 62.4 lb per cubic foot. Is the base strong enough to support the trough filled with water?

c) If 1 gal of water weighs 8.3 lb, how many gallons of water will the trough hold?

9.5

In Exercises 32 and 33, use the given triangle and reflection lines to construct the indicated reflections. Show the triangle in the positions both before and after the reflection.

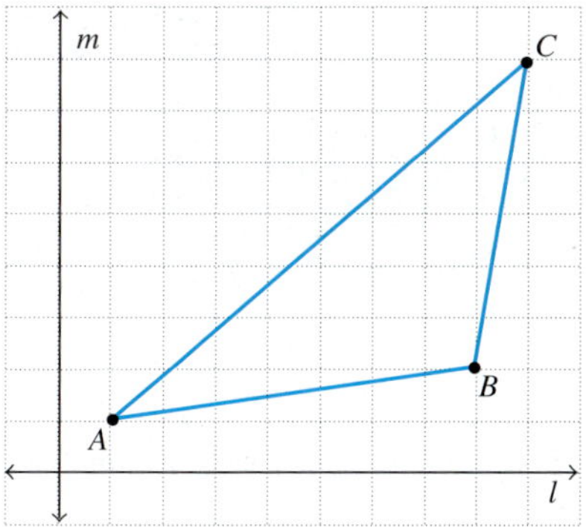

32. Construct the reflection of triangle ABC about line l.

33. Construct the reflection of triangle ABC about line m.

In Exercises 34 and 35, use translation vectors **v** *and* **w** *to construct the indicated translations. Show the parallelogram in the positions both before and after the translation.*

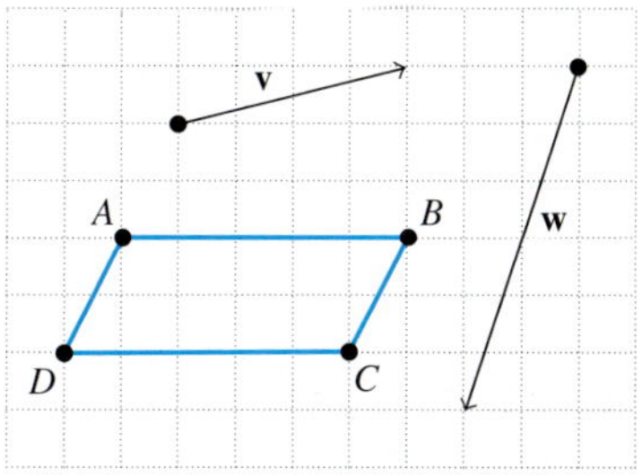

34. Construct the translation of parallelogram $ABCD$ using translation vector **v**.

35. Construct the translation of parallelogram $ABCD$ using translation vector **w**.

In Exercises 36–38, use the given figure and rotation point P to construct the indicated rotations. Show the trapezoid in the positions both before and after the rotation.

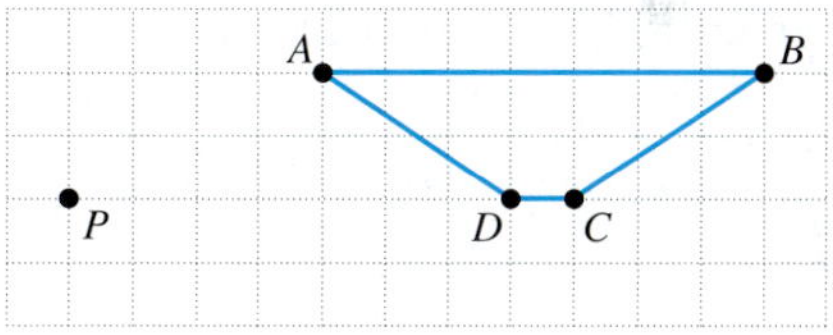

36. Construct a 90° rotation of trapezoid $ABCD$ about point P.

37. Construct a 180° rotation of trapezoid $ABCD$ about point P.

38. Construct a 270° rotation of trapezoid $ABCD$ about point P.

In Exercises 39 and 40, use the given figure, translation vector **v***, and reflection lines l and m to construct the indicated glide reflections. Show the triangle in the positions both before and after the glide reflection.*

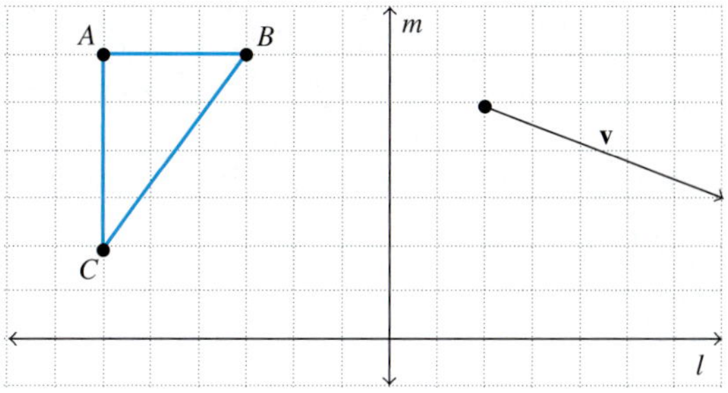

39. Construct a glide reflection of triangle ABC using vector **v** and reflection line l.

40. Construct a glide reflection of triangle ABC using vector **v** and reflection line m.

In Exercises 41 and 42, use the following figure to answer the following questions.

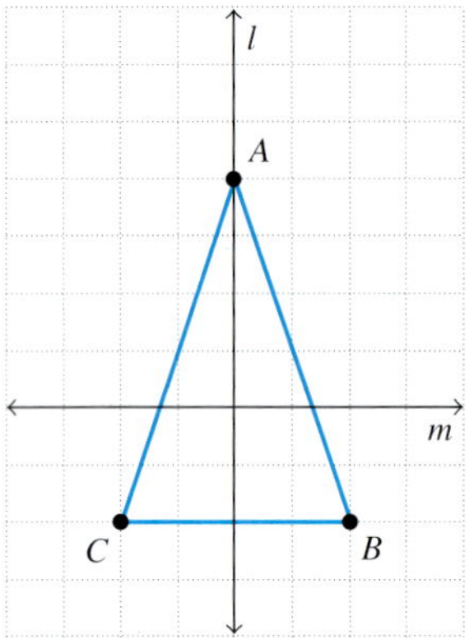

41. Does triangle ABC have reflective symmetry about line l? Explain.

42. Does triangle ABC have reflective symmetry about line m? Explain.

In Exercises 43 and 44, use the following figure to answer the following questions.

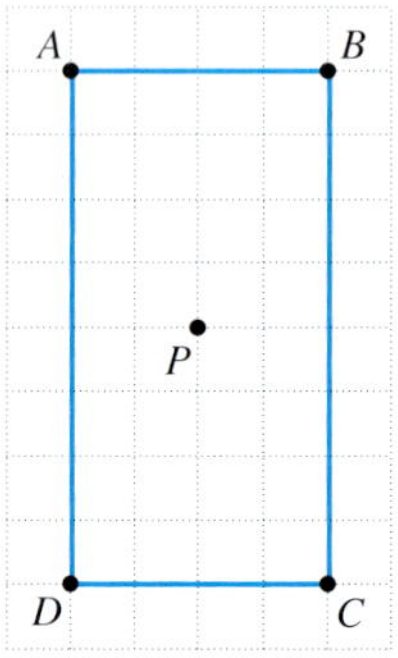

43. Does rectangle $ABCD$ have 90° rotational symmetry about point P? Explain.

44. Does rectangle $ABCD$ have 180° rotational symmetry about point P? Explain.

9.6

45. Give an example of an object that has

a) genus 0.

b) genus 1.

c) genus 2.

d) genus 3 or more.

46. The map shows the states of Germany. Shade in the states using a maximum of four colors so that no two states with a common border have the same color.

47. Determine whether point A is inside or outside the Jordan curve.

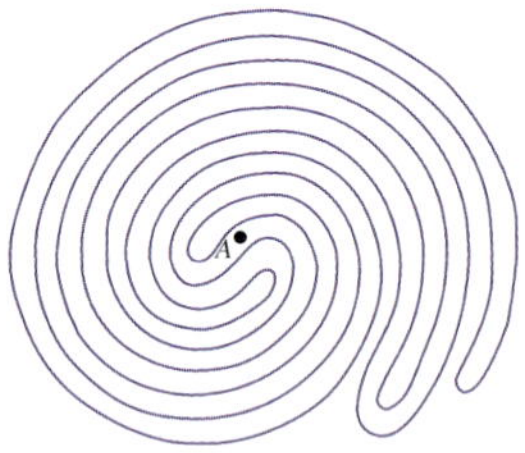

9.7

48. State the fifth axiom of Euclidean, elliptical, and hyperbolic geometry.

49. Develop a fractal by beginning with a square and replacing each side —— with a ⎍. Repeat this process twice.

50. Construct a Koch snowflake by beginning with an equilateral triangle and replacing each side with a _/_. Repeat this process twice.

CHAPTER 9 TEST

In Exercises 1–4, use the figure to describe the following sets of points.

1. $\overleftrightarrow{AF} \cap \overleftrightarrow{EF}$
2. $\overline{BC} \cup \overline{CD} \cup \overline{BD}$
3. $\measuredangle EDF \cap \measuredangle BDC$
4. $\overrightarrow{AC} \cup \overrightarrow{BA}$

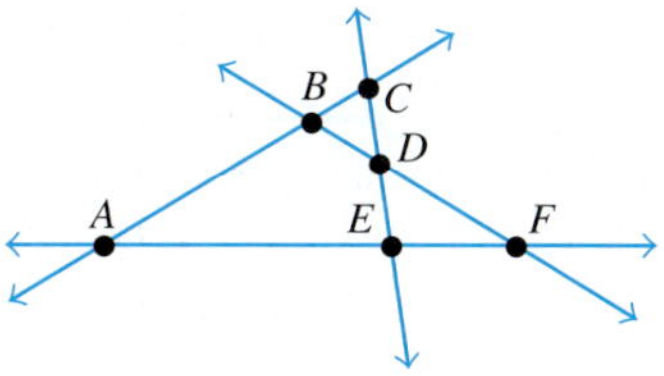

5. $m\measuredangle A = 12.4°$. Determine the measure of the complement of $\measuredangle A$.

6. $m\measuredangle B = 51.7°$. Determine the measure of the supplement of $\measuredangle B$.

7. In the figure, determine the measure of $\measuredangle x$.

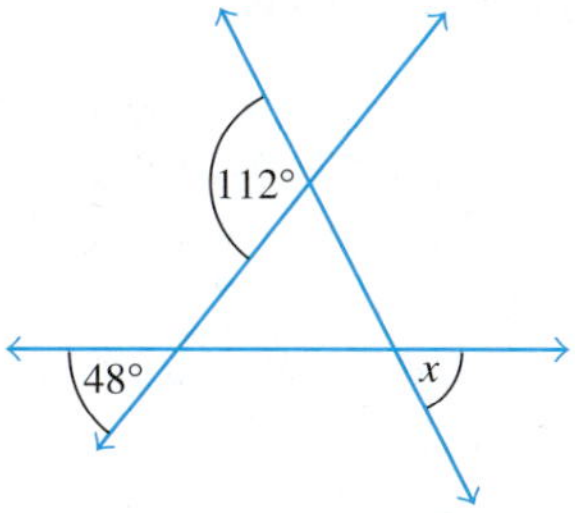

8. Determine the sum of the measures of the interior angles of a decagon.

9. Triangles ABC and $A'B'C'$ are similar figures. Determine the length of side $\overline{B'C'}$.

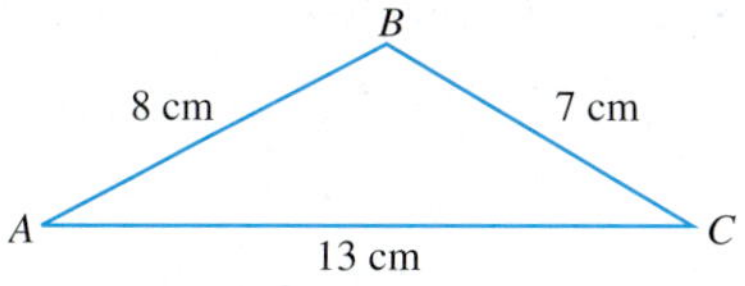

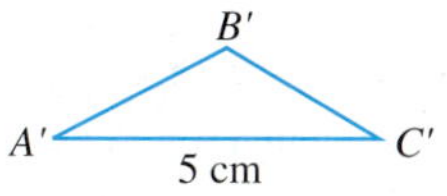

10. Right triangle ABC (see top of right-hand column) has one leg of length 5 in. and a hypotenuse of length 13 in.

    a) Determine the length of the other leg.

    b) Determine the perimeter of the triangle.

    c) Determine the area of the triangle.

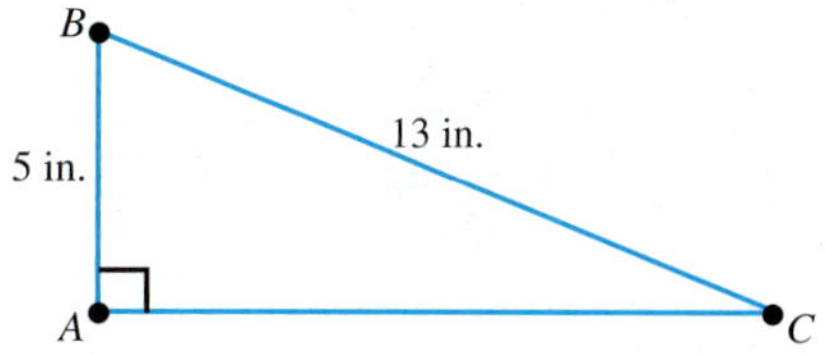

11. Determine (a) the volume and (b) the surface area of a sphere of diameter 14 cm.

12. Determine the volume of the shaded area. Use the π key on your calculator and round your answer to the nearest hundredth.

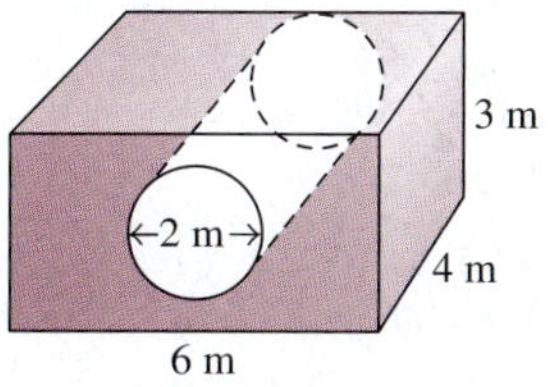

13. Determine the volume of the pyramid.

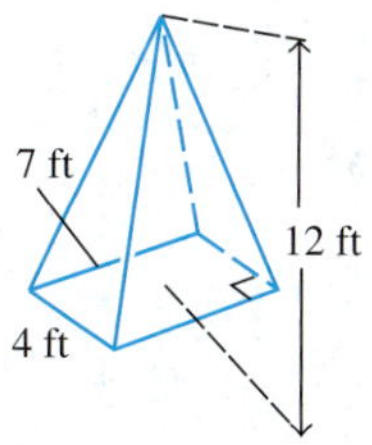

14. Construct a reflection of square $ABCD$, shown below, about line l. Show the square in the positions both before and after the reflection.

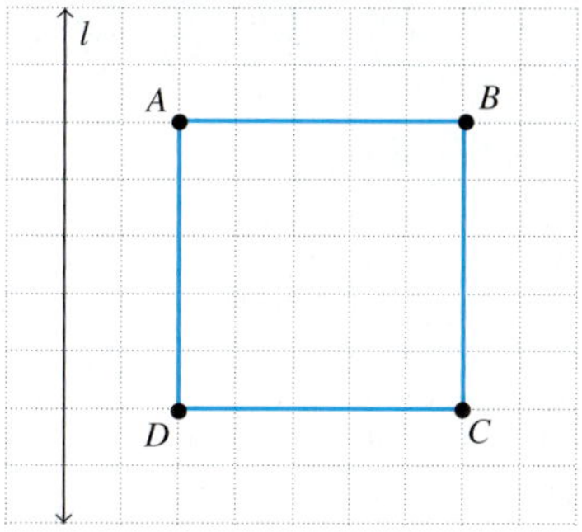

15. Construct a translation of quadrilateral $ABCD$, shown below, using translation vector **v**. Show the quadrilateral in the positions both before and after the translation.

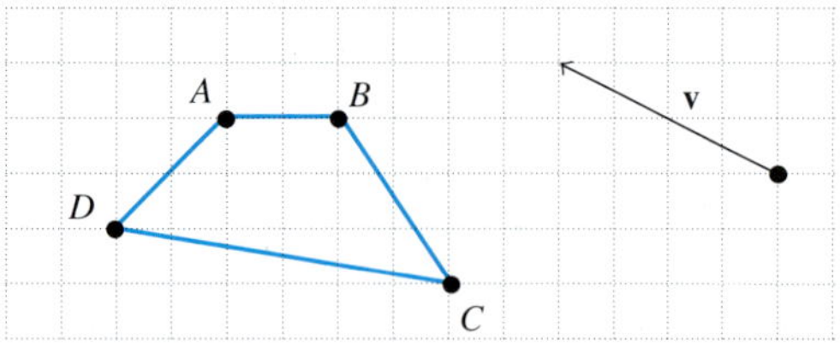

16. Construct a 180° rotation of triangle ABC, shown below, about rotation point P. Show the triangle in the positions both before and after the rotation.

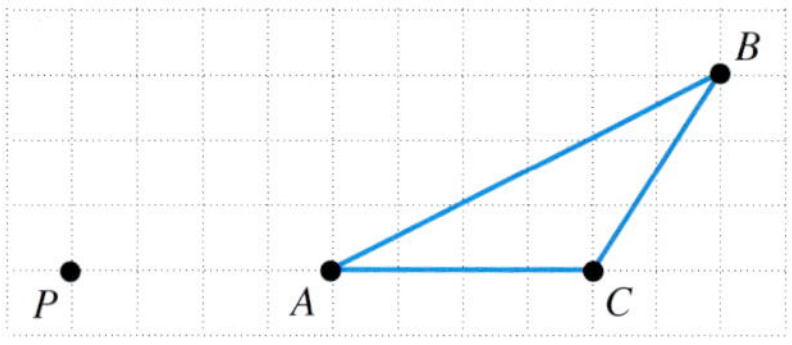

17. Construct a glide reflection of rectangle $ABCD$, shown below, using translation vector **v** and reflection line l. Show the rectangle in the positions both before and after the glide reflection.

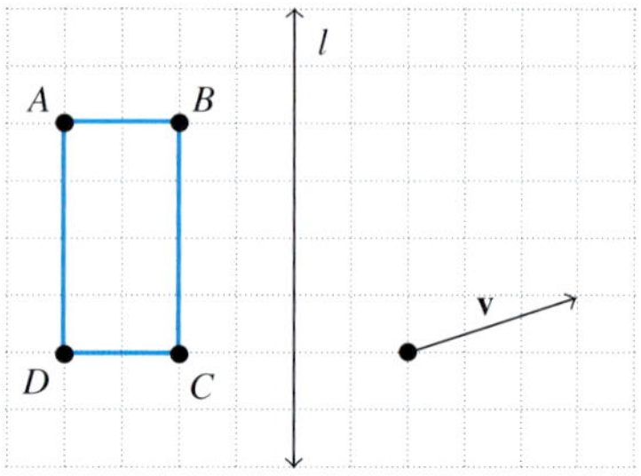

18. Use the figure below to answer the following questions.

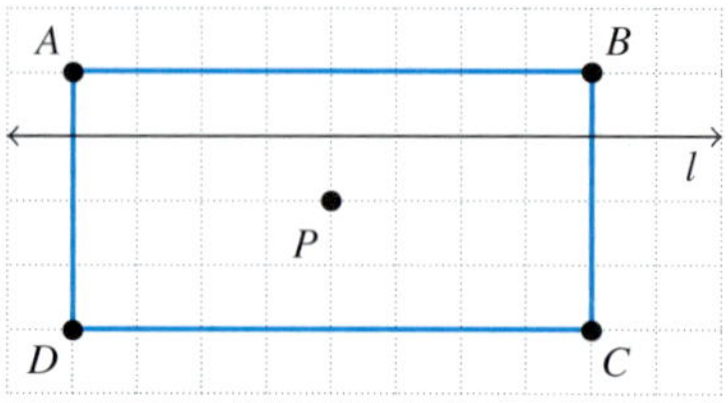

a) Does rectangle $ABCD$ have reflective symmetry about line l? Explain.

b) Does rectangle $ABCD$ have 180° rotational symmetry about point P? Explain.

19. What is a Möbius strip?

20. a) Sketch an object of genus 1.

b) Sketch an object of genus 2.

21. Explain how the fifth axiom in Euclidean geometry, elliptical geometry, and hyperbolic geometry differ.

GROUP PROJECTS

SUPPORTING A HOT TUB

1. Samantha Saraniti is thinking of buying a circular hot tub 12 ft in diameter, 4 ft deep, and weighing 475 lb. She wants to place the hot tub on a deck built to support 30,000 lb.

a) Determine the volume of the water in the hot tub in cubic feet.

b) Determine the number of gallons of water the hot tub will hold. (*Note:* 1 $ft^3 \approx 7.5$ gal.)

c) Determine the weight of the water in the hot tub. (*Hint:* Fresh water weighs about 62.4 lb/ft^3.)

d) Will the deck support the weight of the hot tub and water?

e) Will the deck support the weight of the hot tub, water, and four people, whose average weight is 115 lb?

DESIGNING A RAMP

2. David and Sandra Jessee are planning to build a ramp so that the front entrance of their home is wheelchair accessible. The ramp will be 3 feet wide. It will rise 2 in. for each foot of length of horizontal distance. Where the ramp meets the porch, the ramp must be 2 ft high. To provide stability for the ramp, the Jessees will install a slab of concrete 4 in. thick and 6 in. longer and wider than the ramp (see accompanying figure). The top of the slab will be level with the ground. The ramp may be constructed of concrete or pressure-treated lumber. You are to estimate the cost of materials for constructing the slab, the ramp of concrete, and the ramp of pressure-treated lumber.

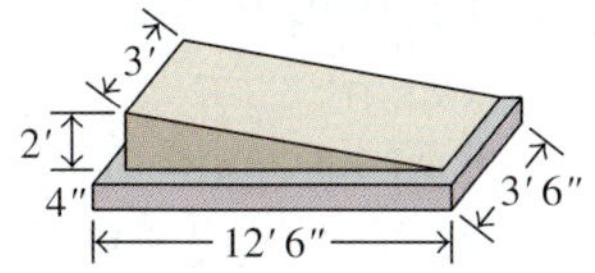

Slab

a) Determine the length of the base of the ramp.

b) Determine the dimensions of the concrete slab on which the ramp will set.

c) Determine the volume of the concrete in cubic yards needed to construct the slab.

d) If ready-mix concrete costs \$45 per cubic yard, determine the cost of the concrete needed to construct the slab.

Concrete Ramp

e) To build the ramp of concrete, a form in the shape of the ramp must be framed. The two sides of the form are triangular, and the shape of the end, which is against the porch, is rectangular. The form will be framed from $\frac{3}{4}$ in. plywood, which comes in 4 ft $\times$ 8 ft sheets. Determine the number of sheets of plywood needed. Assume that the entire sheet(s) will be used to make the sides and the end of the form and that there is no waste.

f) If the plywood costs \$18.95 for a 4 ft $\times$ 8 ft sheet, determine the cost of the plywood.

g) To brace the form, the Jessees will need two boards 2 in. $\times$ 4 in. $\times$ 8 ft (referred to as 8 ft 2 $\times$ 4's) and six pieces of lumber 2 in. $\times$ 4 in. $\times$ 3 ft. These six pieces of lumber will be cut from 8 ft 2 $\times$ 4 boards. Determine the number of 8 ft 2 $\times$ 4 boards needed.

h) Determine the cost of the 8 ft 2 $\times$ 4 boards needed in part (g) if one board costs \$2.14.

i) Determine the volume, in cubic yards, of concrete needed to fill the form.

j) Determine the cost of the concrete needed to fill the form.

k) Determine the total cost of materials for building the ramp of concrete by adding the results in parts (d), (f), (h), and (j).

Wooden Ramp

l) Determine the length of the top of the ramp.

m) The top of the ramp will be constructed of $\frac{5}{4}$ in. $\times$ 6 in. $\times$ 10 ft pressure-treated lumber. The boards will be butted end to end to make the necessary length and will be supported from underneath by a wooden frame. Determine the number of boards needed to cover the top of the ramp. The boards are laid lengthwise on the ramp.

n) Determine the cost of the boards to cover the top of the ramp if the price of a 10 ft length is \$6.47.

o) To support the top of the ramp, the Jessees will need 10 pieces of 8 ft 2 $\times$ 4's. The price of a pressure-treated 8 ft 2 $\times$ 4 is \$2.44. Determine the cost of the supports.

p) Determine the cost of the materials for building a wooden ramp by adding the amounts from parts (d), (n), and (o).

q) Are the materials for constructing a concrete ramp or a wooden ramp less expensive?

CHAPTER 10

Mathematical Systems

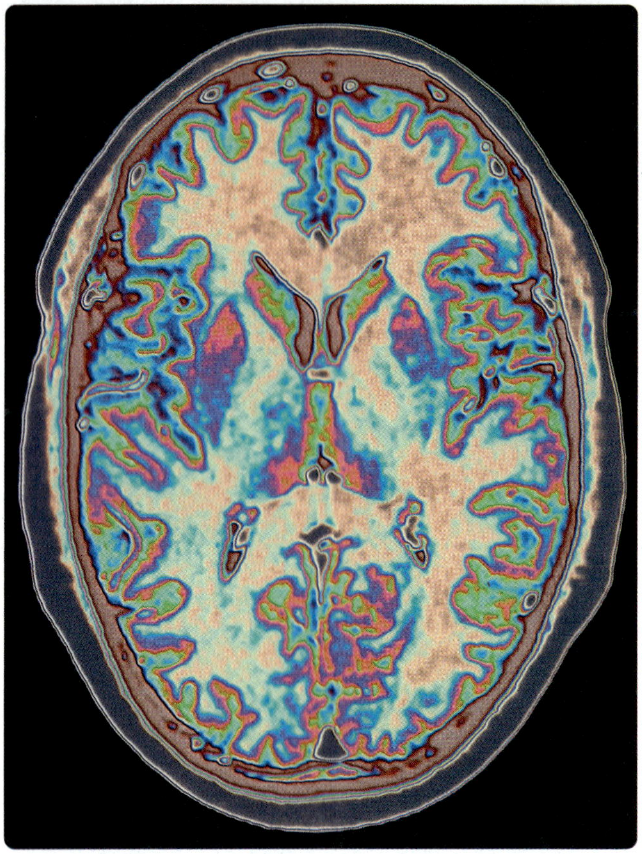

▲ Group theory is used in the field of medical image analysis including magnetic resonance imaging.

WHAT YOU WILL LEARN

- Mathematical systems
- Commutative and associative properties
- Closure, identity element, and inverses
- Groups and commutative groups
- Clock arithmetic
- Mathematical systems without numbers
- Modular arithmetic

WHY IT IS IMPORTANT

Mathematics plays a crucial role in the study of the interaction between many objects of interest to scientists. One branch of mathematics of particular importance to this study is group theory. Physicists and chemists rely on group theory to study the behavior of the tiniest particles that make up an atom in a scientific field known as *quantum mechanics*. Group theory can be used to predict the interactive behavior of subatomic particles within an atom, atoms within a molecule, and molecules within a chemical reaction. Group theory also plays an important role in robotics, computer graphics, weather forecasting, musical theory, and medical image analysis.

10.1 GROUPS

▲ The big bang theory uses groups to address questions relating to the size, structure, and future of the universe.

How big is the universe? Is there an end to the universe? Does the universe continue to expand? Such questions are addressed by the scientific theory known as the *big bang theory*. In this section, we will introduce a mathematical structure known as a *group*. Many modern scientific theories, including the big bang theory, make use of groups and other mathematical topics discussed in this book.

We begin our discussion by introducing a mathematical system. As you will learn shortly, you already know and use many mathematical systems.

A **mathematical system** consists of a set of elements and at least one binary operation.

In the above definition, we mention binary operation. A ***binary operation*** is an operation, or rule, that can be performed on two and only two elements of a set. The result is a single element. When we add *two* integers, the sum is *one* integer. When we multiply *two* integers, the product is *one* integer. Thus, addition and multiplication are both binary operations. Is finding the reciprocal of a number a binary operation? No, it is an operation on a single element of a set.

TIMELY TIP Throughout this section, we will work with many different sets of numbers, such as natural numbers, whole numbers, integers, rational numbers, and real numbers. A review of the numbers that make up each of these sets can be found on page 270 in Section 5.5.

When you learned how to add integers, you were introduced to a mathematical system. The set of elements is the set of integers, and the binary operation is addition. When you learned how to multiply integers, you became familiar with a second mathematical system. The set of integers with the operation of subtraction and the set of integers with the operation of division are two other examples of mathematical systems since subtraction and division are also binary operations.

Some mathematical systems are used in solving everyday problems, such as planning work schedules. Others are more abstract and are used primarily in research, chemistry, physical structure, matter, the nature of genes, and other scientific fields.

Commutative and Associative Properties

Once a mathematical system is defined, its structure may display certain properties. Consider the set of integers

$$I = \{\ldots, -3, -2, -1, 0, 1, 2, 3, \ldots\}$$

Recall that the ellipsis, the three dots at each end of the set, indicates that the set continues in the same manner.

The set of integers can be studied with the operations of addition, subtraction, multiplication, and division as separate mathematical systems. For example, when we study the set of integers under the operations of addition or multiplication, we see that the commutative and associative properties hold. The general forms of the properties are shown here.

FOR ANY ELEMENTS *a*, *b*, AND *c*	*Addition*	*Multiplication*
Commutative property	$a + b = b + a$	$a \cdot b = b \cdot a$
Associative property	$(a + b) + c = a + (b + c)$	$(a \cdot b) \cdot c = a \cdot (b \cdot c)$

The integers *are commutative* under the operations of *addition and multiplication*. For example,

Addition		*Multiplication*
$2 + 4 = 4 + 2$	and	$2 \cdot 4 = 4 \cdot 2$
$6 = 6$		$8 = 8$

The integers, however, *are not commutative* under the operations of *subtraction and division*. For example,

Subtraction		*Division*
$4 - 2 \neq 2 - 4$	and	$4 \div 2 \neq 2 \div 4$
$2 \neq -2$		$2 \neq \frac{1}{2}$

The integers *are associative* under the operations of *addition and multiplication*. For example,

Addition		*Multiplication*
$(1 + 2) + 3 = 1 + (2 + 3)$	and	$(1 \cdot 2) \cdot 3 = 1 \cdot (2 \cdot 3)$
$3 + 3 = 1 + 5$		$2 \cdot 3 = 1 \cdot 6$
$6 = 6$		$6 = 6$

The integers, however, *are not associative* under the operations of *subtraction and division*. See Exercises 21 and 22 at the end of this section.

To say that a set of elements is commutative under a given operation means that the commutative property holds for *any* elements a and b in the set. Similarly, to say that a set of elements is associative under a given operation means that the associative property holds for *any* elements a, b, and c in the set.

Consider the mathematical system consisting of the set of integers under the operation of addition. Because the set of integers is infinite, this mathematical system is an example of an *infinite mathematical system*. We will study certain properties of this mathematical system. The first property we will examine is closure.

Closure

The sum of any two integers is an integer. Therefore, the set of integers is said to be *closed*, or to satisfy the *closure property*, under the operation of addition.

If a binary operation is performed on any two elements of a set and the result is an element of the set, then that set is **closed** (or has **closure**) under the given binary operation.

Is the set of integers closed under the operation of multiplication? The answer is yes. When any two integers are multiplied, the product will be an integer.

Is the set of integers closed under the operation of subtraction? Again, the answer is yes. The difference of any two integers is an integer.

Is the set of integers closed under the operation of division? The answer is no because two integers may have a quotient that is not an integer. For example, if we select the integers 2 and 3, the quotient of 2 divided by 3 is $\frac{2}{3}$, which is not an integer. Thus, the integers are not closed under the operation of division.

We showed that the set of integers was not closed under the operation of division by finding two integers whose quotient was not an integer. A specific example illustrating that a specific property is not true is called a *counterexample*. Mathematicians and scientists often try to find a counterexample to confirm that a specific property is not always true.

Identity Element

Now we will discuss the identity element for the set of integers under the operation of addition. Is there an element in the set that, when added to any given integer, results in a sum that is the given integer? The answer is yes. The sum of 0 and any integer is the given integer. For example, $1 + 0 = 0 + 1 = 1$, $-4 + 0 = 0 + (-4) = -4$, and so on. For this reason, we call 0 the *additive identity element* for the set of integers. Note that for any integer a, $a + 0 = 0 + a = a$.

An **identity element** is an element in a set such that when a binary operation is performed on it and any given element in the set, the result is the given element.

Is there an identity element for the set of integers under the operation of multiplication? The answer is yes; it is the number 1. Note that $2 \cdot 1 = 1 \cdot 2 = 2$, $3 \cdot 1 = 1 \cdot 3 = 3$, and so on. For any integer a, $a \cdot 1 = 1 \cdot a = a$. For this reason, 1 is called the *multiplicative identity element* for the set of integers.

Inverses

What integer, when added to 4, gives a sum of 0; that is, $4 + \square = 0$? The shaded area is to be filled in with the integer -4: $4 + (-4) = 0$. We say that -4 is the additive inverse of 4 and that 4 is the additive inverse of -4. Note that the sum of the element and its additive inverse gives the additive identity element 0. What is the additive inverse of 12? Since $12 + (-12) = 0$, -12 is the additive inverse of 12.

Other examples of integers and their additive inverses are

Element	+	Additive Inverse	=	Identity Element
0	+	0	=	0
2	+	(−2)	=	0
−5	+	5	=	0

Note that for the operation of addition, every integer a has a unique inverse, $-a$, such that $a + (-a) = -a + a = 0$.

When a binary operation is performed on two elements in a set and the result is the identity element for the binary operation, each element is said to be the **inverse** of the other.

Does every integer have an inverse under the operation of multiplication? For multiplication, the product of an integer and its inverse must yield the multiplicative identity element, 1. What is the multiplicative inverse of 2? That is, 2 times what number gives 1?

$$2 \cdot \square = 1 \qquad 2 \cdot \tfrac{1}{2} = 1$$

However, since $\frac{1}{2}$ is not an integer, 2 does not have a multiplicative inverse in the set of integers.

Group

Let's review what we have learned about the mathematical system consisting of the set of integers under the operation of addition.

1. The set of integers is *closed* under the operation of addition.
2. The set of integers has an *identity element* under the operation of addition.
3. Each element in the set of integers has an *inverse* under the operation of addition.
4. The *associative property* holds for the set of integers under the operation of addition.

The set of integers under the operation of addition is an example of a *group*. The properties of a group can be summarized as follows.

PROPERTIES OF A GROUP

Any mathematical system that meets the following four requirements is called a **group**.

1. The set of elements is *closed* under the given operation.
2. An *identity element* exists for the set under the given operation.
3. Every element in the set has an *inverse* under the given operation.
4. The set of elements is *associative* under the given operation.

TIMELY TIP Note that the set of elements need *not* be commutative for the mathematical system to be a group. Also note that *every* element in the set must have an inverse for the mathematical system to be a group.

It is often very time consuming to show that the associative property holds for all cases. In many of the examples that follow, we will state that the associative property holds for the given set of elements under the given operation.

PROFILES IN MATHEMATICS

Niels Abel (1802–1829), Évariste Galois (1811–1832)

Niels Abel

Évariste Galois

Untimely ends: Important contributions to the development of group theory were made by two young men who would not live to see their work gain acceptance: Niels Abel, a Norwegian, and Évariste Galois, a Frenchman. Although their work showed brilliance, neither was noticed in his lifetime by the mathematics community. Abel lived in poverty and died of malnutrition and tuberculosis at the age of 26, just 2 days before a letter arrived with the offer of a teaching position. Galois died of complications from a gunshot wound he received in a duel over a love affair. The 20-year-old man had spent the night before summarizing his theories in the hope that eventually someone would find it "profitable to decipher this mess." After their deaths, both men had important elements of group theory named after them.

DID YOU KNOW?

The Enormous Theorem

The largest theorem known to the mathematical world involves the classification of a special type of group. The official name of the theorem is the *classification of finite, simple groups,* but it is commonly referred to by mathematicians as the *enormous theorem.* The full proof consists of more than 15,000 pages in about 500 separate publications written by more than 100 mathematicians. In addition, many parts of the proof were completed by computer and some parts are so complex or massive that no particular mathematician can claim to have a complete understanding of the entire proof. Instead, mathematicians must rely on the reputed expertise of the individual contributors to the theorem, including computer programmers and chip designers.

Commutative Group

The commutative property does not need to hold for a mathematical system to be a group. However, if a mathematical system meets the four requirements of a group and is also commutative under the given operation, the mathematical system is a *commutative* (or *abelian*) *group.* The abelian group is named after Niels Abel (see the Profiles in Mathematics on page 614).

A group that satisfies the commutative property is called a **commutative group** (or **abelian group**).

Because the commutative property holds for the set of integers under the operation of addition, the set of integers under the operation of addition is not only a group, but it is a commutative group.

PROPERTIES OF A COMMUTATIVE GROUP

A mathematical system is a **commutative group** if all five of the following conditions hold.

1. The set of elements is *closed* under the given operation.
2. An *identity element* exists for the set under the given operation.
3. Every element in the set has an *inverse* under the given operation.
4. The set of elements is *associative* under the given operation.
5. The set of elements is *commutative* under the given operation.

To determine whether a mathematical system is a group under a given operation, *check, in the following order,* whether (1) the system is closed under the given operation, (2) there is an identity element in the set for the given operation, (3) every element in the set has an inverse under the given operation, and (4) the associative property holds under the given operation. If *any* of these four requirements is *not* met, stop and state that the mathematical system is not a group. If asked to determine whether the mathematical system is a commutative group, you also need to determine whether the commutative property holds for the given operation.

EXAMPLE 1 *Is It a Group?*

Determine whether the set of rational numbers under the operation of multiplication forms a group.

SOLUTION Recall from Chapter 5 that the rational numbers are the set of numbers of the form $\frac{p}{q}$ where p and q are integers and $q \neq 0$. The set of rational numbers includes all fractions and integers.

1. *Closure*: The product of any two rational numbers is a rational number. Therefore, the rational numbers are closed under the operation of multiplication.
2. *Identity element*: The multiplicative identity element for the set of rational numbers is 1. Note, for example, that $3 \cdot 1 = 1 \cdot 3 = 3$, and $\frac{3}{8} \cdot 1 = 1 \cdot \frac{3}{8} = \frac{3}{8}$. For any rational number a, $a \cdot 1 = 1 \cdot a = a$.
3. *Inverse elements*: For the mathematical system to be a group under the operation of multiplication, *each and every* rational number must have a multiplicative inverse in the set of rational numbers. Remember that for the operation of

MATHEMATICS TODAY

The Expanding Universe

Woody Allen and Diane Keaton in *Annie Hall*

How and when will the universe end? In the Woody Allen movie *Annie Hall*, a boy (Woody Allen), having just read that the universe is expanding, becomes so worried that he can't do his homework. "Someday it will break apart," he tells his psychiatrist, "and that will be the end of everything." But his mother snaps, "You're here in Brooklyn! Brooklyn is not expanding!"

The universe could break apart. That, however, will not be for trillions of years, and life as we know it on Earth will cease long before that. Our sun, halfway through its estimated 10-billion-year lifetime, is slowly brightening. In about 1 billion years, its energy output will have increased by at least 10%, turning Earth into a Venus-like hothouse incapable of supporting life as we know it. Long before that, humans may make plans to escape to Mars, which is farther from the sun and hence cooler. Although quite some time away, scientists are already considering various scenarios to make the surface of Mars a livable habitat.

multiplication, the product of a number and its inverse must give the multiplicative identity element, 1. Let's check a few rational numbers:

Rational Number	·	Inverse	=	Identity Element
3	·	$\frac{1}{3}$	=	1
$\frac{2}{3}$	·	$\frac{3}{2}$	=	1
$-\frac{1}{5}$	·	-5	=	1

Looking at these examples you might deduce that each rational number does have an inverse. However, one rational number, 0, does not have an inverse.

$$0 \cdot ? = 1$$

Because there is no rational number that when multiplied by 0 gives 1, 0 does not have a multiplicative inverse. Since *every* rational number does not have an inverse, this mathematical system is not a group.

There is no need at this point to check the associative property because we have already shown that the mathematical system of rational numbers under the operation of multiplication is not a group.

EXAMPLE 2 *Real Numbers under Addition*

Determine whether the set of real numbers under the operation of addition forms a commutative group.

SOLUTION Recall from Chapter 5 that the real numbers can be thought of as the numbers that correspond to *each* point on the real number line. The set of real numbers include all rational numbers and all irrational numbers.

1. *Closure*: The sum of any two real numbers is a real number. Therefore, the real numbers are closed under the operation of addition.
2. *Identity element*: The additive identity element for the set of real numbers is 0. For example, $\frac{2}{3} + 0 = 0 + \frac{2}{3} = \frac{2}{3}$ and $\sqrt{5} + 0 = 0 + \sqrt{5} = \sqrt{5}$. For any real number a, $a + 0 = 0 + a = a$.
3. *Inverse elements*: For the set of real numbers under the operation of addition to be a commutative group, each and every real number must have an additive inverse in the set of real numbers. For the operation of addition, the sum of a real number and its additive inverse, or *opposite*, must be the additive identity element, 0. From Chapter 5, we saw that every real number a has an opposite $-a$ such that $a + (-a) = -a + a = 0$. Recall also that $0 + 0 = 0$; therefore, 0 is its own additive inverse. Thus, all real numbers have an additive inverse.
4. *Associative property*: From Chapter 5, we know that the real numbers are associative under the operation of addition. That is, for any real numbers a, b, and c, $(a + b) + c = a + (b + c)$.
5. *Commutative property*: From Chapter 5, we know that the real numbers are commutative under the operation of addition. That is, for any real numbers a and b, $a + b = b + a$.

Since all five properties (closure, identity element, inverse elements, associative property, and commutative property) hold, the set of real numbers under the operation of addition form a commutative group.

SECTION 10.1 EXERCISES

CONCEPT/WRITING EXERCISES

1. What is a binary operation?

2. What does a mathematical system consist of?

3. Explain why each of the following is a binary operation. Give an example to illustrate each binary operation.

   a) Addition **b)** Subtraction

   c) Multiplication **d)** Division

4. What properties are required for a mathematical system to be a group?

5. What properties are required for a mathematical system to be a commutative group?

6. What is another name for a commutative group?

7. What is an identity element? Give the additive and multiplicative identity elements for the set of integers.

8. Explain the closure property. Give an example of the property.

9. What is an inverse element? Give the additive and multiplicative inverse of the number 2 for the set of rational numbers.

10. What is a counterexample?

11. Is it possible that a mathematical system is a group but not a commutative group? Explain.

12. Is it possible that a mathematical system is a commutative group but not a group? Explain.

13. Which of the following properties is not required for a mathematical system to be a group?
   a) Closure
   b) An identity element
   c) Every element must have an inverse.
   d) The commutative property must apply.
   e) The associative property must apply.

14. For the set of integers, list two operations that are not binary. Explain.

PRACTICE THE SKILLS

15. Give the commutative property of addition and illustrate the property with an example.

16. Give the commutative property of multiplication and illustrate the property with an example.

17. Give the associative property of multiplication and illustrate the property with an example.

18. Give the associative property of addition and illustrate the property with an example.

19. Give an example to show that the commutative property does not hold for the set of integers under the operation of division.

20. Give an example to show that the commutative property does not hold for the set of integers under the operation of subtraction.

21. Give an example to show that the associative property does not hold for the set of integers under the operation of subtraction.

22. Give an example to show that the associative property does not hold for the set of integers under the operation of division.

PROBLEM SOLVING

In Exercises 23–38, explain your answer.

23. Is the set of integers a commutative group under the operation of addition?

24. Is the set of integers a group under the operation of addition?

25. Is the set of positive integers a group under the operation of addition?

26. Is the set of positive integers a commutative group under the operation of addition?

27. Is the set of positive integers a group under the operation of subtraction?

28. Is the set of negative integers a group under the operation of division?

29. Is the set of negative integers a commutative group under the operation of addition?

30. Is the set of integers a group under the operation of multiplication?

31. Is the set of positive integers a commutative group under the operation of multiplication?

32. Is the set of negative integers a group under the operation of multiplication?

33. Is the set of rational numbers a group under the operation of addition?

34. Is the set of rational numbers a commutative group under the operation of multiplication?

35. Is the set of rational numbers a commutative group under the operation of division?

36. Is the set of rational numbers a group under the operation of subtraction?

37. Is the set of rational numbers a commutative group under the operation of subtraction?

38. Is the set of positive integers a commutative group under the operation of division?

CHALLENGE PROBLEMS/GROUP ACTIVITIES

In Exercises 39–42, explain your answer.

39. Is the set of real numbers a group under the operation of addition?

40. Is the set of real numbers a group under the operation of multiplication?

41. Is the set of irrational numbers a group under the operation of addition?

42. Is the set of irrational numbers a group under the operation of multiplication?

43. Create a mathematical system with two binary operations. Select a set of elements and two binary operations so that one binary operation with the set of elements meets the requirements for a group and the other binary operation does not. Explain why the one binary operation with the set of elements is a group. For the other binary operation and the set of elements, find counterexamples to show that it is not a group.

INTERNET/RESEARCH ACTIVITY

44. There are other classifications of mathematical systems besides groups. For example, there are *rings* and *fields*. Do research to determine the requirements that must be met for a mathematical system to be (a) a ring and (b) a field. (c) Is the set of real numbers, under the operations of addition and multiplication, a field? Ask your instructor for references to use.

10.2 FINITE MATHEMATICAL SYSTEMS

▲ Timekeeping makes use of mathematical systems.

Suppose you decide to use your slow cooker to cook a roast for your family's dinner. You would like to eat at 6:30 P.M., and it will take 10 hours for the roast to cook. What time should you begin to cook the roast in the slow cooker? Or, suppose you need to take medicine every eight hours. You last took the medicine at 7:45 A.M. When should you take the medicine again? Questions such as these involve the use of mathematical systems that we will study in this section.

In the preceding section, we presented infinite mathematical systems. In this section, we present some finite mathematical systems. A ***finite mathematical system*** is one whose set contains a finite number of elements.

Clock Arithmetic

Let's develop a finite mathematical system called ***clock 12 arithmetic***. The set of elements in this system will be the hours on a clock: {1, 2, 3, 4, 5, 6, 7, 8, 9, 10, 11, 12}.

Figure 10.1

The binary operation that we will use is addition, which we define as movement of the hour hand in a clockwise direction. Assume that it is 4 o'clock. What time will it be in 9 hours? (See Fig. 10.1.) If we add 9 hours to 4 o'clock, the clock will read 1 o'clock. Thus, $4 + 9 = 1$ in clock arithmetic. Would $9 + 4$ be the same as $4 + 9$? Yes it will because $4 + 9 = 9 + 4 = 1$.

Table 10.1 is the addition table for clock arithmetic. Its elements are based on the definition of addition as previously illustrated. For example, the sum of 4 and 9 is 1, so we put a 1 in the table where the row to the right of the 4 intersects the column below the 9. Likewise, the sum of 11 and 10 is 9, so we put a 9 in the table where the row to the right of the 11 intersects the column below the 10.

The binary operation of this system is defined by the table. It is denoted by the symbol +. To determine the value of $a + b$, where a and b are any two numbers in the set, find a in the left-hand column and find b along the top row. Assume that there is a horizontal line through a and a vertical line through b; the point of intersection of these two lines is where you find the value of $a + b$. For example, $10 + 4 = 2$ has been circled in Table 10.1. Note that $4 + 10$ also equals 2, but this result will not necessarily hold for all examples in this chapter.

Table 10.1 Clock 12 Arithmetic

+	1	2	3	4	5	6	7	8	9	10	11	12
1	2	3	4	5	6	7	8	9	10	11	12	1
2	3	4	5	6	7	8	9	10	11	12	1	2
3	4	5	6	7	8	9	10	11	12	1	2	3
4	5	6	7	8	9	10	11	12	1	2	3	4
5	6	7	8	9	10	11	12	1	2	3	4	5
6	7	8	9	10	11	12	1	2	3	4	5	6
7	8	9	10	11	12	1	2	3	4	5	6	7
8	9	10	11	12	1	2	3	4	5	6	7	8
9	10	11	12	1	2	3	4	5	6	7	8	9
10	11	12	1	(2)	3	4	5	6	7	8	9	10
11	12	1	2	3	4	5	6	7	8	9	10	11
12	1	2	3	4	5	6	7	8	9	10	11	12

EXAMPLE 1 *A Commutative Group?*

Determine whether the clock arithmetic system under the operation of addition is a commutative group.

SOLUTION Check the five requirements that must be satisfied for a commutative group.

1. *Closure*: Is the set of elements in clock arithmetic closed under the operation of addition? Yes it is since Table 10.1 contains only the elements in the set $\{1, 2, 3, 4, 5, 6, 7, 8, 9, 10, 11, 12\}$. If Table 10.1 had contained an element other than the numbers 1 through 12, the set would not have been closed under addition.
2. *Identity element*: Is there an identity element for clock arithmetic? If the time is currently 4 o'clock, how many hours have to pass before it is 4 o'clock again? Twelve hours: $4 + 12 = 12 + 4 = 4$. In fact, given any hour, in 12 hours the clock will return to the starting point. Therefore, 12 is the additive identity element in clock arithmetic.

DID YOU KNOW?

The Changing Times

This photo shows the famous clock at the Cathedral of St. Marks (called the Basilica) in Piazza San Marco (St. Mark's Plaza) in Venice, Italy. Construction of the basilica began in 830, but it has been rebuilt a number of times due to fires. The clock tower was built in the fifteenth century. Note that the clock displays all 24 hours in Roman numerals. Also note the position of the hours. Noon, or XII, is where we expect 9 o'clock to be.

In Florence, Italy, the oldest building is probably the Duomo (Cathedral House) in the Piazza del Duomo in the city's center. Florentines believe it was constructed in about 1000. The large clock in the cathedral also displays 24 hours using Roman numerals. What would be the additive identity of clock arithmetic if we used the 24-hour clock shown? What would be the additive inverse of VI?

In examining Table 10.1, we see that the row of numbers next to the 12 in the left-hand column is identical to the row of numbers along the top. We also see that the column of numbers under the 12 in the top row is identical to the column of numbers on the left. The search for such a column and row is one technique for determining whether an identity element exists for a system defined by a table.

3. *Inverse elements*: Is there an inverse for the number 4 in clock arithmetic for the operation of addition? Recall that the identity element in clock arithmetic is 12. What number when added to 4 gives 12; that is, $4 + \square = 12$? Table 10.1 shows that $4 + 8 = 12$ and also that $8 + 4 = 12$. Thus, 8 is the additive inverse of 4 and 4 is the additive inverse of 8.

   To find the additive inverse of 7, find 7 in the left-hand column of Table 10.1. Look to the right of the 7 until you come to the identity element 12. Then determine the number at the top of this column. The number is 5. Since $7 + 5 = 5 + 7 = 12$, 5 is the inverse of 7 and 7 is the inverse of 5. The other inverses can be found in the same way. Table 10.2 shows each element in clock 12 arithmetic and its inverse. Note that each element in the set has an *inverse*.

Table 10.2 Clock 12 Inverses

Element	+	Inverse	=	Identity Element
1	+	11	=	12
2	+	10	=	12
3	+	9	=	12
4	+	8	=	12
5	+	7	=	12
6	+	6	=	12
7	+	5	=	12
8	+	4	=	12
9	+	3	=	12
10	+	2	=	12
11	+	1	=	12
12	+	12	=	12

4. *Associative property*: Now consider the associative property. Does $(a + b) + c = a + (b + c)$ for all values a, b, and c of the set? Remember to always evaluate the values within the parentheses first. Let's select some values for a, b, and c. Let $a = 2$, $b = 6$, and $c = 8$. Then

$$\begin{aligned}(2 + 6) + 8 &= 2 + (6 + 8)\\ 8 + 8 &= 2 + 2\\ 4 &= 4 \quad \text{True}\end{aligned}$$

   Let $a = 5$, $b = 12$, and $c = 9$. Then

$$\begin{aligned}(5 + 12) + 9 &= 5 + (12 + 9)\\ 5 + 9 &= 5 + 9\\ 2 &= 2 \quad \text{True}\end{aligned}$$

Randomly selecting *any* elements a, b, and c of the set reveals $(a + b) + c = a + (b + c)$. Thus, the system of clock arithmetic is associative under the operation of addition. Note that if there is just one set of values a, b, and c such that $(a + b) + c \neq a + (b + c)$, the system is not associative. Usually, you will not be asked to check every case to determine whether the associative property holds. *If every element in the set does not appear in every row and column of the table, however, you need to check the associative property carefully.*

5. *Commutative property*: Does the commutative property hold under the given operation? Does $a + b = b + a$ for all elements a and b of the set? Let's randomly select some values for a and b to determine whether the commutative property appears to hold. Let $a = 5$ and $b = 8$; then Table 10.1 shows that

$$5 + 8 = 8 + 5$$
$$1 = 1 \quad \text{True}$$

Let $a = 9$ and $b = 6$; then

$$9 + 6 = 6 + 9$$
$$3 = 3 \quad \text{True}$$

The commutative property holds for these two specific cases. In fact, if we were to select *any* values for a and b, we would find that $a + b = b + a$. Thus, the commutative property of addition is true in clock arithmetic. Note that if there is just one set of values a and b such that $a + b \neq b + a$, the system is not commutative.

This system satisfies the five properties required for a mathematical system to be a commutative group. Thus, clock arithmetic under the operation of addition is a commutative or abelian group.

Table 10.3 Symmetry about the Main Diagonal

+	0	1	2	3	4
0	0	1	2	3	4
1	1	2	3	4	0
2	2	3	4	0	1
3	3	4	0	1	2
4	4	0	1	2	3

One method that can be used to determine whether a system defined by a table is commutative under the given operation is to determine whether the elements in the table are symmetric about the ***main diagonal***. The main diagonal is the diagonal from the upper left-hand corner to the lower right-hand corner of the table. In Table 10.3, the main diagonal is shaded in color.

If the elements are symmetric about the main diagonal, then the system is commutative. If the elements are not symmetric about the main diagonal, then the system is not commutative. If you examine the system in Table 10.3, you see that its elements are symmetric about the main diagonal because the same numbers appear in the same relative positions on opposite sides of the main diagonal. Therefore, this mathematical system is commutative.

It is possible to have groups that are not commutative. Such groups are called ***noncommutative*** or ***nonabelian groups***. However, a *noncommutative group defined by a table must be at least a six-element by six-element table.* Nonabelian groups are illustrated in Exercises 83 through 85 at the end of this section.

Now we will look at another finite mathematical system.

Table 10.4 Four-Element System

☼	6	7	8	9
6	8	9	6	7
7	9	6	7	8
8	6	7	8	9
9	7	8	9	6

EXAMPLE 2 *A Finite System*

Consider the mathematical system defined by Table 10.4. Assume that the associative property holds for the given operation.

a) List the elements in the set of this mathematical system.

b) Identify the binary operation.

c) Determine whether this mathematical system is a commutative group.

SOLUTION

a) The set of elements for this mathematical system consists of the elements found on the top (or left-hand side) of the table: $\{6, 7, 8, 9\}$.

b) The binary operation is ☼.

c) We must determine whether the five requirements for a commutative group are satisfied.

1. *Closure*: All the elements in the table are in the original set of elements $\{6, 7, 8, 9\}$, so the mathematical system is closed.
2. *Identity element*: The identity element is 8. Note that the row of elements to the right of the 8 is identical to the top row *and* that the column of elements under the 8 is identical to the left-hand column.
3. *Inverse elements*: When an element operates on its inverse element, the result is the identity element. For this example, the identity element is 8. To determine the inverse of 6, determine which element (6, 7, 8, or 9) replaces the shaded area in the following equation and gives a true statement:

$$6 ☼ \square = 8$$

From Table 10.4, we see that 6 ☼ 6 = 8. Thus, 6 is its own inverse.

To determine the inverse of 7, determine which element replaces the shaded area in the following equation and gives a true statement:

$$7 ☼ \square = 8$$

From Table 10.4, we see that 7 ☼ 9 = 9 ☼ 7 = 8. Therefore, 7 is the inverse of 9 and 9 is the inverse of 7. The elements and their inverses are shown in Table 10.5. Every element has an inverse.

Table 10.5 Inverses under ☼

Element	☼	Inverse	=	Identity Element
6	☼	6	=	8
7	☼	9	=	8
8	☼	8	=	8
9	☼	7	=	8

4. *Associative property*: It is given that the associative property holds for this binary operation, ☼. One example of the associative property is

$$(7 ☼ 6) ☼ 9 = 7 ☼ (6 ☼ 9)$$
$$9 ☼ 9 = 7 ☼ 7$$
$$6 = 6 \quad \text{True}$$

5. *Commutative property*: The elements in Table 10.4 are symmetric about the main diagonal, so the commutative property holds for the operation of ☼. One example of the commutative property is

$$6 ☼ 7 = 7 ☼ 6$$
$$9 = 9 \quad \text{True}$$

The five necessary properties hold. Thus, the mathematical system is a commutative group.

Mathematical Systems Without Numbers

Thus far, all the systems we have discussed have been based on sets of numbers. Example 3 illustrates a mathematical system of symbols rather than numbers.

DID YOU KNOW?

Rubik's Cube

Group theory may be viewed as the study of the algebra of symmetry and transformations (see Section 9.5 for a discussion of transformational geometry). One of the best, and perhaps most entertaining, examples of this view is *Rubik's cube*. In 1974, Erno Rubik, a Hungarian teacher of architecture and design, presented the world with his popular puzzle. Each face of the cube is divided into nine squares, and each row and column of each face can rotate. The result is approximately 43 quintillion different arrangements of the colors on the cube. Rubik's cube is the most popular and best-selling puzzle in human history. Although the popularity of the puzzle peaked during the 1980s, clubs and annual contests are still devoted to solving the puzzle. For more information, visit the web site www.rubiks.com.

EXAMPLE 3 *Investigating a System of Symbols*

Use the mathematical system defined by Table 10.6 and determine

Table 10.6 A System of Symbols

$\boxdot$	@	*P*	*W*
@	*P*	*W*	@
P	*W*	@	*P*
W	@	*P*	*W*

a) the set of elements.
b) the binary operation.
c) closure or nonclosure of the system.
d) the identity element.
e) the inverse of @.
f) $(@ \boxdot P) \boxdot P$ and $@ \boxdot (P \boxdot P)$.
g) $W \boxdot P$ and $P \boxdot W$.

SOLUTION

a) The set of elements of this mathematical system is $\{@, P, W\}$.
b) The binary operation is $\boxdot$.
c) Because the table does not contain any symbols other than @, *P*, and *W*, the system is closed under $\boxdot$.
d) The identity element is *W*. Note that the row next to *W* in the left-hand column is the same as the top row and that the column under *W* is identical to the left-hand column. We see that

$$@ \boxdot W = W \boxdot @ = @$$
$$P \boxdot W = W \boxdot P = P$$
$$W \boxdot W = W$$

e) We know that

$$\text{element} \boxdot \text{inverse element} = \text{identity element},$$

and since *W* is the identity element, to find the inverse of @ we write

$$@ \boxdot \square = W$$

To find the inverse of @, we must determine the element to replace the shaded area. Since $@ \boxdot P = W$ and $P \boxdot @ = W$, *P* is the inverse of @.

f) We first evaluate the information within parentheses.

$$(@ \boxdot P) \boxdot P = W \boxdot P = P \quad \text{and} \quad @ \boxdot (P \boxdot P) = @ \boxdot @ = P$$

g) $W \boxdot P = P$ and $P \boxdot W = P$

Table 10.7

♫	α	β	γ	δ
α	δ	α	β	γ
β	α	β	γ	δ
γ	β	γ	δ	α
δ	γ	δ	α	β

Table 10.8 Inverses under ♫

Element	♫	Inverse	=	Identity Element
α	♫	γ	=	β
β	♫	β	=	β
γ	♫	α	=	β
δ	♫	δ	=	β

EXAMPLE 4 *Is the System a Commutative Group?*

Determine whether the mathematical system defined by Table 10.7 is a commutative group. Assume that the associative property holds for the given operation.

SOLUTION

1. *Closure*: The system is closed.
2. *Identity element*: The identity element is β.
3. *Inverse elements*: Each element has an inverse as illustrated in Table 10.8.
4. *Associative property*: It is given that the associative property holds. An example illustrating the associative property is

$$(\gamma \text{ ♫ } \delta) \text{ ♫ } \alpha = \gamma \text{ ♫ } (\delta \text{ ♫ } \alpha)$$
$$\alpha \text{ ♫ } \alpha = \gamma \text{ ♫ } \gamma$$
$$\delta = \delta \quad \text{True}$$

5. *Commutative property*: By examining Table 10.7, we can see that the elements are symmetric about the main diagonal. Thus, the system is commutative under the given operation, ♫. One example of the commutative property is

$$\alpha \text{ ♫ } \delta = \delta \text{ ♫ } \alpha$$
$$\gamma = \gamma \quad \text{True}$$

All five properties are satisfied. Thus, the system is a commutative group.

EXAMPLE 5 *Another System to Study*

Determine whether the mathematical system in Table 10.9 is a commutative group under the operation of ☺.

Table 10.9

☺	x	y	z
x	x	z	y
y	z	y	x
z	y	x	z

SOLUTION

1. The system is closed.
2. No row is identical to the top row, so there is no identity element. Therefore, this mathematical system is *not a group*. There is no need to go any further, but for practice, let's look at a few more items.
3. Since there is no identity element, there can be no inverses.
4. The associative property does not hold. The following counterexample illustrates the associative property does not hold for every case.

$$(x \text{ ☺ } y) \text{ ☺ } z \neq x \text{ ☺ } (y \text{ ☺ } z)$$
$$z \text{ ☺ } z \neq x \text{ ☺ } x$$
$$z \neq x$$

5. The table is symmetric about the main diagonal. Therefore, the commutative property does hold for the operation of ☺.

Note that the associative property does not hold even though the commutative property does hold. This outcome can occur when there is no identity element and every element does not have an inverse, as in this example.

EXAMPLE 6 *Is the System a Commutative Group?*

Determine whether the mathematical system in Table 10.10 is a commutative group under the operation of $*$.

Table 10.10

$*$	$\square$	b	c
$\square$	$\square$	b	c
b	b	b	$\square$
c	c	$\square$	c

SOLUTION

1. The system is closed.
2. There is an identity element, $\square$.
3. Each element has an inverse; $\square$ is the inverse of $\square$, b is the inverse of c, and c is the inverse of b.
4. Every element in the set does not appear in every row and every column of the table, so we need to check the associative property carefully. There are many specific cases in which the associative property does hold. However, the following counterexample illustrates that the associative property does not hold for every case.

$$(b*b)*c \neq b*(b*c)$$
$$b*c \neq b*\square$$
$$\square \neq b$$

5. The commutative property holds because there is symmetry about the main diagonal.

Since we have shown that the associative property does not hold under the operation of $*$, this system is not a group. Therefore, it cannot be a commutative group.

DID YOU KNOW?

Creating Patterns by Design

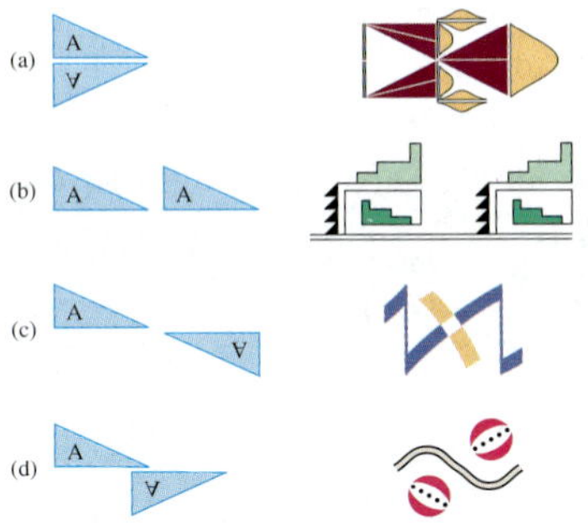

Patterns from *Symmetries of Culture* by Dorothy K. Washburn and Donald W. Crowe (University of Washington Press, 1988)

What makes group theory such a powerful tool is that it can be used to reveal the underlying structure of just about any physical phenomenon that involves symmetry and patterning, such as wallpaper or quilt patterns. Interest in the formal study of symmetry in design came out of the Industrial Revolution in the late nineteenth century. The new machines of the Industrial Revolution could vary any given pattern almost indefinitely. Designers needed a way to describe and manipulate patterns systematically. Shown here are the four geometric motions that generate all two-dimensional patterns: (a) reflection, (b) translation, (c) rotation, and (d) glide reflection. How these motions are applied, or not applied, is the basis of pattern analysis. The geometric motions, called *rigid motions*, were discussed in Section 9.5.

SECTION 10.2 EXERCISES

CONCEPT/WRITING EXERCISES

1. Explain how the clock 12 addition table is formed.
2. What is 12 + 12 in clock 12 arithmetic? Explain how you obtained your answer.
3. **a)** Explain how to add the numbers (2 + 11) + 5 in clock 12 arithmetic using the addition table.

   b) What is (2 + 11) + 5 in clock 12 arithmetic?

4. a) Explain how to determine a difference of two numbers in clock arithmetic by using the face of a clock.

   b) Determine 3 − 5 in clock 12 arithmetic using the method explained in part (a).

5. a) Explain how to find 2 − 6 in clock 12 arithmetic by adding the number 12 to one of the numbers.

   b) Determine 2 − 6 in clock 12 arithmetic using the method explained in part (a).

   c) Explain why the procedure you give in part (a) works.

6. Explain one method of determining whether a system defined by a table is commutative under the given operation.

7. Is there an identity element for addition in clock 12 arithmetic? If so, what is it?

8. Is clock 12 arithmetic closed under the operation of addition? Explain.

9. Does each element in clock 12 arithmetic have an inverse? If so, give each element and its corresponding inverse.

10. Give an example to illustrate the associative property of addition in clock 12 arithmetic.

11. Is clock 12 arithmetic commutative? Give an example to verify your answer.

12. Is clock 12 arithmetic under the operation of addition a commutative group? Explain.

13. Consider clock 7 arithmetic under the operation of addition. The set of elements in such a mathematical system is {1, 2, 3, 4, 5, 6, 7}.

   a) Which of these elements is the additive identity element?

   b) What is the additive inverse of 2? Explain your answer.

14. Consider clock 6 arithmetic under the operation of addition. The set of elements in such a mathematical system is {1, 2, 3, 4, 5, 6}.

   a) Which of these elements is the additive identity element?

   b) What is the additive inverse of 3? Explain your answer.

In Exercises 15 and 16, determine if the system is commutative. Explain how you determined your answer.

15.

I	*P*	*A*	*L*
P	*L*	*P*	*A*
A	*P*	*L*	*A*
L	*A*	*L*	*P*

16.

⊟	*A*	⊗	*W*
A	⊗	*W*	*A*
⊗	*W*	*A*	⊗
W	*A*	⊗	⊗

In Exercises 17 and 18, determine if the system has an identity element. If so, list the identity element. Explain how you determined your answer.

17.

W	*A*	*B*	*C*
A	*C*	*B*	*A*
B	*B*	*C*	*B*
C	*A*	*B*	*C*

18.

⊖	□	⊙	△
□	△	□	⊙
⊙	⊙	△	□
△	□	⊙	△

In Exercises 19 and 20, the identity element is C. Determine the inverse of (a) A, (b) B, and (c) C.

19.

☮	*A*	*B*	*C*
A	*B*	*C*	*A*
B	*C*	*A*	*B*
C	*A*	*B*	*C*

20.

⋕	*A*	*B*	*C*
A	*C*	*B*	*A*
B	*B*	*C*	*B*
C	*A*	*B*	*C*

PRACTICE THE SKILLS

In Exercises 21–32, use Table 10.1 on page 619 to determine the sum in clock 12 arithmetic.

21. 2 + 5
22. 6 + 10
23. 8 + 8
24. 11 + 7
25. 4 + 12
26. 12 + 12
27. 3 + (8 + 9)
28. (8 + 7) + 6
29. (6 + 4) + 8
30. (6 + 10) + 12
31. (7 + 8) + (9 + 6)
32. (7 + 11) + (9 + 5)

In Exercises 33–44, determine the difference in clock 12 arithmetic by starting at the first number and counting counterclockwise on the clock the number of units given by the second number.

33. 11 − 6
34. 12 − 5
35. 6 − 7
36. 8 − 11
37. 11 − 8
38. 4 − 10
39. 1 − 12
40. 6 − 10
41. 5 − 5
42. 8 − 8
43. 12 − 12
44. 5 − 8

45. Use the following figure to develop an addition table for clock 6 arithmetic. The figure will also be used in Exercises 46–54.

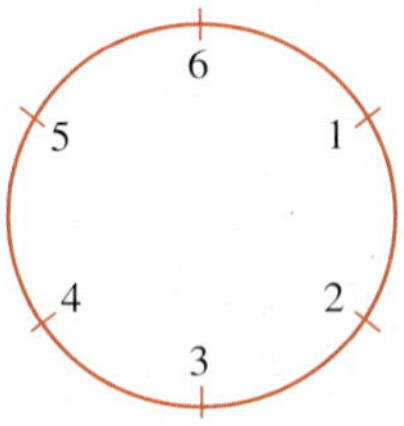

In Exercises 46–54, determine the sum or difference in clock 6 arithmetic.

46. 3 + 4 **47.** 4 + 5 **48.** 5 + 6

49. 5 − 2 **50.** 4 − 5 **51.** 2 − 6

52. 3 − 4 **53.** (3 − 5) − 6 **54.** 2 + (1 − 3)

55. Use the following figure to develop an addition table for clock 7 arithmetic. The figure will also be used in Exercises 56–64.

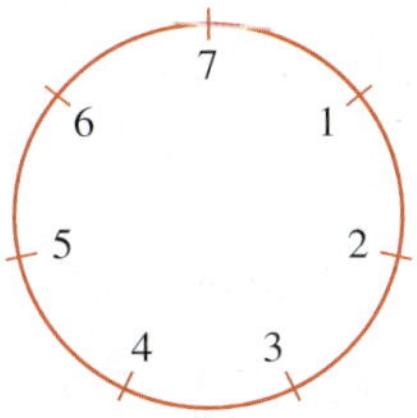

In Exercises 56–64, determine the sum or difference in clock 7 arithmetic.

56. 6 + 6 **57.** 4 + 7

58. 3 + 6 **59.** 7 + 6

60. 2 − 3 **61.** 3 − 6

62. 2 − 4 **63.** (4 − 5) − 6

64. 3 − (2 − 6)

65. Determine whether clock 7 arithmetic under the operation of addition is a commutative group. Explain.

66. A mathematical system is defined by a three-element by three-element table where every element in the set appears in each row and each column. Must the mathematical system be a commutative group? Explain.

67. Consider the mathematical system defined by the following table. Assume that the associative property holds for the given operation.

£	**0**	**2**	**4**	**6**
0	0	2	4	6
2	2	4	6	0
4	4	6	0	2
6	6	0	2	4

a) What are the elements of the set in this mathematical system?

b) What is the binary operation?

c) Is the system closed? Explain.

d) Is there an identity element for the system under the given operation? If so, what is it?

e) Does every element in the system have an inverse? If so, give each element and its corresponding inverse.

f) Give an example to illustrate the associative property.

g) Is the system commutative? Give an example to verify your answer.

h) Is the mathematical system a commutative group? Explain.

In Exercises 68–70, repeat parts (a)–(h) of Exercise 67 for the mathematical system defined by the following table. Assume that the associative property holds for the given operation.

68.

®	♣	♦	♥	♠
♣	♥	♠	♣	♦
♦	♠	♣	♦	♥
♥	♣	♦	♥	♠
♠	♦	♥	♠	♣

69.

	*	**5**	***L***
*	5	*L*	*
5	*L*	*	5
L	*	5	*L*

70.

	**3**	**5**	**8**	**4**
3	5	8	4	3
5	8	4	3	5
8	4	3	5	8
4	3	5	8	4

71. For the mathematical system

𓅭	f	r	o	m
f	f	r	o	m
r	r	o	m	f
o	o	m	f	r
m	m	f	r	o

determine

a) the elements in the set.

b) the binary operation.

c) closure or nonclosure of the system.

d) $(r$ 𓅭 $o)$ 𓅭 f

e) $(f$ 𓅭 $r)$ 𓅭 m

f) the identity element.

g) the inverse of r.

h) the inverse of m.

72. a) Is the following mathematical system a group? Explain your answer.

b) Find an example showing that the associative property does not hold for the given set of elements.

〰	1	2	3	4
1	2	3	4	1
2	3	4	1	2
3	4	1	2	3
4	1	4	3	2

In Exercises 73–78, for the mathematical system given, determine which of the five properties of a commutative group do not hold.

73.

‡	Δ	Φ	Γ
Δ	Δ	Φ	Γ
Φ	Φ	Γ	Δ
Γ	Γ	Δ	Π

74.

⊗	⊡	M	🔔
⊡	⊡	M	🔔
M	M	⊡	🔔
🔔	🔔	M	⊡

75.

⊡	⊙	*	?	T	P
⊙	T	P	⊙	*	*
*****	P	⊙	*	⊙	T
?	⊙	*	?	T	P
T	*	⊙	T	P	?
P	*	T	P	?	⊙

76.

☺	a	b	π	0	Δ
a	Δ	a	b	π	0
b	a	b	π	0	Δ
π	b	π	0	Δ	a
0	π	0	Δ	a	b
Δ	0	Δ	π	b	a

77.

⇔	a	b	c	d	e
a	c	d	e	a	b
b	d	e	a	b	c
c	e	a	b	c	d
d	a	b	c	e	d
e	b	c	d	e	a

78.

⍒	0	1	2	3	4	5
0	0	0	0	0	0	0
1	0	1	2	3	4	5
2	0	2	4	0	2	4
3	0	3	0	3	0	3
4	0	4	2	0	4	2
5	0	5	4	3	2	1

PROBLEM SOLVING

79. a) Consider the set consisting of two elements $\{E, O\}$, where E stands for an even number and O stands for an odd number. For the operation of addition, complete the table.

+	E	O
E		
O		

b) Determine whether this mathematical system forms a commutative group under the operation of addition. Explain your answer.

80. a) Let E and O represent even numbers and odd numbers, respectively, as in Exercise 79. Complete the table for the operation of multiplication.

×	E	O
E		
O		

b) Determine whether this mathematical system forms a commutative group under the operation of multiplication. Explain your answer.

In Exercises 81 and 82, make up your own mathematical system that is a group. List the identity element and the inverses of each element. Do so with sets containing

81. three elements.

82. four elements.

In Exercises 83 and 84 the tables shown are examples of noncommutative or nonabelian groups. For each exercise, do the following.

a) *Show that the system under the given operation is a group. (It would be very time consuming to prove that the associative property holds, but you can give some examples to show that it appears to hold.)*

b) *Determine a counterexample to show that the commutative property does not hold.*

83.

∞	1	2	3	4	5	6
1	5	3	4	2	6	1
2	4	6	5	1	3	2
3	2	1	6	5	4	3
4	3	5	1	6	2	4
5	6	4	2	3	1	5
6	1	2	3	4	5	6

84.

⊖	*A*	*B*	*C*	*D*	*E*	*F*
A	*E*	*C*	*D*	*B*	*F*	*A*
B	*D*	*F*	*E*	*A*	*C*	*B*
C	*B*	*A*	*F*	*E*	*D*	*C*
D	*C*	*E*	*A*	*F*	*B*	*D*
E	*F*	*D*	*B*	*C*	*A*	*E*
F	*A*	*B*	*C*	*D*	*E*	*F*

CHALLENGE PROBLEMS/GROUP ACTIVITIES

85. ***Book Arrangements*** Suppose that three books numbered 1, 2, and 3 are placed next to one another on a shelf. If we remove volume 3 and place it before volume 1, the new order of books is 3, 1, 2. Let's call this replacement *R*. We can write

$$R = \begin{pmatrix} 1 & 2 & 3 \\ 3 & 1 & 2 \end{pmatrix}$$

which indicates the books were switched in order from 1, 2, 3 to 3, 1, 2. Other possible replacements are *S*, *T*, *U*, *V*, and *I*, as indicated.

$$S = \begin{pmatrix} 1 & 2 & 3 \\ 2 & 1 & 3 \end{pmatrix} \quad T = \begin{pmatrix} 1 & 2 & 3 \\ 3 & 2 & 1 \end{pmatrix} \quad U = \begin{pmatrix} 1 & 2 & 3 \\ 1 & 3 & 2 \end{pmatrix}$$

$$V = \begin{pmatrix} 1 & 2 & 3 \\ 2 & 3 & 1 \end{pmatrix} \quad I = \begin{pmatrix} 1 & 2 & 3 \\ 1 & 2 & 3 \end{pmatrix}$$

Replacement set *I* indicates that the books were removed from the shelves and placed back in their original order. Consider the mathematical system with the set of elements, $\{R, S, T, U, V, I\}$ with the operation of $*$.

To evaluate $R * S$, write

$$R * S = \overset{R}{\begin{pmatrix} 1 & 2 & 3 \\ 3 & 1 & 2 \end{pmatrix}} * \overset{S}{\begin{pmatrix} 1 & 2 & 3 \\ 2 & 1 & 3 \end{pmatrix}}.$$

As shown in Fig. 10.2, *R* replaces 1 with 3 and *S* replaces 3 with 3 (no change), so $R * S$ replaces 1 with 3. *R* replaces 2 with 1 and *S* replaces 1 with 2, so $R * S$ replaces 2 with 2 (no change). *R* replaces 3 with 2 and *S* replaces 2 with 1, so $R * S$ replaces 3 with 1. $R * S$ replaces 1 with 3, 2 with 2, and 3 with 1.

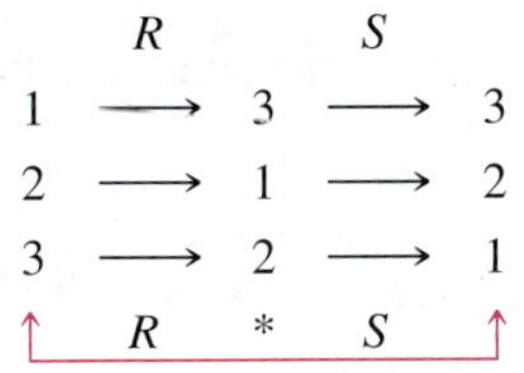

$$R * S = \begin{pmatrix} 1 & 2 & 3 \\ 3 & 2 & 1 \end{pmatrix} = T$$

Figure 10.2

Since this result is the same as replacement set *T*, we write $R * S = T$.

a) Complete the table for the operation using the procedure outlined.

*	*R*	*S*	*T*	*U*	*V*	*I*
R		*T*				
S						
T						
U						
V						
I						

b) Is this mathematical system a group? Explain.

c) Is this mathematical system a commutative group? Explain.

86. If a mathematical system is defined by a four-element by four-element table, how many specific cases must be illustrated to prove the set of elements is associative under the given operation?

Exercises 87–89 will help prepare you for the next section.

87. A table is shown below.

+	0	1	2	3	4
0	0		2		4
1	1		3		
2			4		1
3	3		0	1	2
4		0	1		

Fill in the blank areas of the table by performing the following operations.

1. Add the number in the left-hand column and the number in the top row.
2. Divide this sum by 5.
3. Place the *remainder* in the cell where the number in the left-hand column and the number in the top row would meet.

   For example, in the table, the number 2 is in red. This 2 is found as follows:

$$(3 + 4) \div 5 = 1, \text{ remainder } 2$$

88. Using Exercise 87 as a guide, complete the following table by dividing the sum of the numbers by 6. Place the remainders in the table as explained in Exercise 87.

+	0	1	2	3	4	5
0						
1						
2						
3						
4						
5						

89. Consider the mathematical system defined by the following table.

+	0	1	2	3
0	0	1	2	3
1	1	2	3	0
2	2	3	0	1
3	3	0	1	2

After working Exercises 87 and 88, can you explain how this table was formed?

RECREATIONAL MATHEMATICS

90. ***A Tiny Group*** Consider the mathematical system with the single element {1} under the operation of multiplication.

a) Is this system closed?

b) Does the system have an identity element?

c) Does 1 have an inverse?

d) Does the associative property hold?

e) Does the commutative property hold?

f) Is {1} under the operation of multiplication a commutative group?

INTERNET/RESEARCH ACTIVITIES

91. a) ***Matrices*** In Section 7.3, we introduced matrices. Show that 2×2 matrices under the operation of addition form a commutative group.

b) Show that 2×2 matrices under the operation of multiplication do not form a commutative group.

10.3 MODULAR ARITHMETIC

▲ When will your birthday next land on a Saturday?

If your birthday is on a Monday this year, on what day of the week will it fall next year? What if next year is a leap year? When will your birthday next fall on a Saturday? These questions can be addressed with the type of arithmetic that we will study in this section.

The clock arithmetic we discussed in the previous section is similar to modular arithmetic. The set of elements $\{0, 1, 2, 3, 4, 5, 6, 7, 8, 9, 10, 11\}$ together with the operation of addition is called a modulo 12 or mod 12 system. There is one difference in notation between clock 12 arithmetic and modulo 12 arithmetic. In the modulo 12 system, the symbol 12 is replaced with the symbol 0.

Modulo *m* Systems

A ***modulo m system*** consists of m elements, 0 through $m - 1$, and a binary operation. In this section, we will discuss modular arithmetic systems and their properties.

If today is Sunday, what day of the week will it be in 23 days? The answer, Tuesday, is arrived at by dividing 23 by 7. The quotient is 3 and the remainder is 2. Twenty-three days represent 3 weeks plus 2 days. Since we are interested only in the day of the week on which the twenty-third day will fall, the 3-week segment is unimportant to the answer. The remainder of 2 indicates the answer will be 2 days later than Sunday, which is Tuesday.

If we place the days of the week on a clock face as shown in Fig. 10.3, then in 23 days the hand would have made three complete revolutions and end on Tuesday. If we replace the days of the week with numbers, then a modulo 7 arithmetic system will result. See Fig. 10.4: Sunday = 0, Monday = 1, Tuesday = 2, and so on. If we start at 0 and move the hand 23 places, we will end at 2. Table 10.11 shows a modulo 7 addition table.

Table 10.11 Modular 7 Addition

+	0	1	2	3	4	5	6
0	0	1	2	3	4	5	6
1	1	2	3	4	5	6	0
2	2	3	4	5	6	0	1
3	3	4	5	6	0	1	2
4	4	5	6	0	1	2	③
5	5	6	0	1	2	3	4
6	6	0	1	2	3	4	5

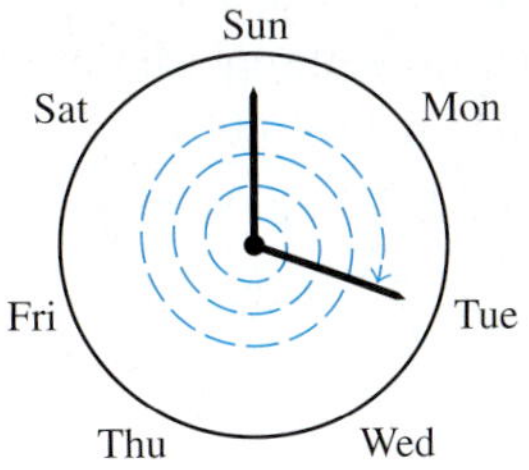

Figure 10.3

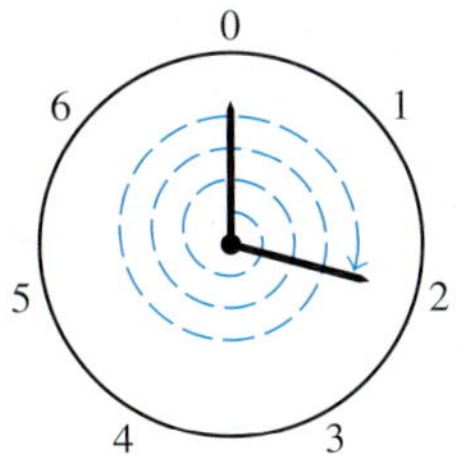

Figure 10.4

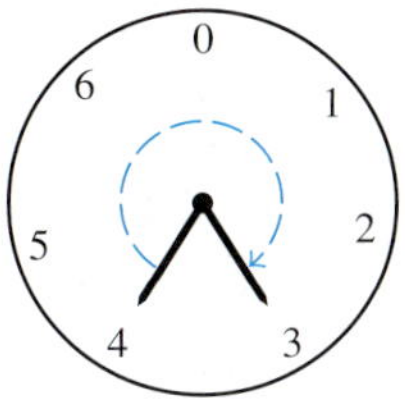

Figure 10.5

If we start at 4 and add 6, we end at 3 on the clock in Fig. 10.5. This number is circled in Table 10.11. The other numbers can be obtained in the same way.

A second method of determining the sum of $4 + 6$ in modulo 7 arithmetic is to divide the sum, 10, by 7 and observe the remainder.

$$10 \div 7 = 1, \quad \text{remainder } 3$$

The remainder, 3, is the sum of $4 + 6$ in a modulo 7 arithmetic system.

The concept of congruence is important in modular arithmetic.

> a is **congruent** to b modulo m, written, $a \equiv b \pmod{m}$, if a and b have the same remainder when divided by m.

We can show, for example, that $10 \equiv 3 \pmod 7$ by dividing both 10 and 3 by 7 and observing that we obtain the same remainder in each case.

$$10 \div 7 = 1, \quad \text{remainder } 3 \qquad \text{and} \qquad 3 \div 7 = 0, \quad \text{remainder } 3$$

Since the remainders are the same, 3 in each case, 10 is congruent to 3 modulo 7, and we may write $10 \equiv 3 \pmod 7$.

Now consider $37 \equiv 5 \pmod 8$. If we divide both 37 and 5 by 8, each has the same remainder, 5.

In any modulo system, we can develop a set of ***modulo classes*** by placing all numbers with the same remainder in the appropriate modulo class. In a modulo 7 system, every number must have a remainder of either 0, 1, 2, 3, 4, 5, or 6. Thus, a modulo 7 system has seven modulo classes. The seven classes are presented in Table 10.12.

Table 10.12 Modulo 7 Classes

0	1	2	3	4	5	6
0	1	2	3	4	5	6
7	8	9	10	11	12	13
14	15	16	17	18	19	20
21	22	23	24	25	26	27
28	29	30	31	32	33	34
⋮	⋮	⋮	⋮	⋮	⋮	⋮

Every number is congruent to a number from 0 to 6 in modulo 7. For example, $24 \equiv 3 \pmod 7$ because 24 is in the same modulo class as 3.

The solution to a problem in modular arithmetic, if it exists, will always be a number from 0 through $m - 1$, where m is the ***modulus*** of the system. For example, in a modulo 7 system, because 7 is the modulus, the solution will be a number from 0 through 6.

EXAMPLE 1 *Congruence Modulo 7*

Determine which number, from 0 through 6, the following numbers are congruent to in modulo 7.

a) 62 b) 53 c) 105

SOLUTION We could determine the answer by listing more entries in Table 10.12. Another method of finding the answer is to divide the given number by 7 and observe the remainder. In the solutions, we will use a question mark, ?, as a place holder. The ? will represent a number from 0 through 6. When we determine the answer, we will replace the ? with the answer.

a) $62 \equiv ? \pmod 7$

To determine the value that 62 is congruent to in mod 7, divide 62 by 7 and find the remainder.

$$\begin{array}{r} 8 \\ 7\overline{)62} \\ \underline{56} \\ 6 \end{array} \leftarrow \text{remainder}$$

$$62 \div 7 = 8, \quad \text{remainder } 6$$

Thus, $62 \equiv 6 \pmod 7$.

b) $53 \equiv ? \pmod 7$

$$53 \div 7 = 7, \quad \text{remainder } 4$$

Thus, $53 \equiv 4 \pmod 7$.

c) $105 \equiv ? \pmod 7$

$$105 \div 7 = 15, \quad \text{remainder } 0$$

Thus, $105 \equiv 0 \pmod 7$. ●

EXAMPLE 2 *Congruence Modulo 6*

Evaluate each of the following in modulo 6.

a) $3 + 5$ b) $5 - 2$ c) $3 \cdot 5$

SOLUTION In each part, because we are working in modulo 6, the answer will be a number from 0 through 5.

a)
$$\begin{aligned} 3 + 5 &\equiv ? \pmod 6 \\ 8 &\equiv ? \pmod 6 \\ 8 \div 6 = 1, &\quad \text{remainder } 2 \end{aligned}$$

Therefore, $3 + 5 \equiv 2 \pmod 6$.

b)
$$\begin{aligned} 5 - 2 &\equiv ? \pmod 6 \\ 3 &\equiv ? \pmod 6 \\ 3 &\equiv 3 \pmod 6 \end{aligned}$$

Remember that we want to replace the question mark with a number between 0 and 5, inclusive. Thus, $5 - 2 \equiv 3 \pmod 6$.

c)
$$\begin{aligned} 3 \cdot 5 &\equiv ? \pmod 6 \\ 15 &\equiv ? \pmod 6 \end{aligned}$$

Since $15 \div 6 = 2$, remainder 3, $15 \equiv 3 \pmod 6$. Thus, $3 \cdot 5 \equiv 3 \pmod 6$. ●

Note in Table 10.12 on page 632 that every number in the same modulo class differs by a multiple of the modulus, in this case a multiple of 7. Adding (or subtracting) a multiple of the modulus to (or from) a given number does not change the modulo class or congruence of the given number. For example, $3, 3 + 1(7), 3 + 2(7), 3 + 3(7), \ldots, 3 + n(7)$ are all in the same modulo class, namely, 3. We use this fact in the solution to Example 3.

EXAMPLE 3 *Using Modulo Classes in Subtraction*

Determine the positive number replacement (less than the modulus) for the question mark that makes the statement true.

a) $3 - 5 \equiv ? \pmod 7$ b) $? - 4 \equiv 3 \pmod 5$ c) $5 - ? \equiv 7 \pmod 8$

SOLUTION In each part, we wish to replace the question mark with a positive number less than the modulus. Therefore, in part (a) we wish to replace the ? with a positive number less than 7. In part (b), we wish to replace the ? with a positive number less than 5. In part (c), we wish to replace the ? with a positive number less than 8.

a) In mod 7, adding 7, or a multiple of 7, to a number results in a sum that is in the same modulo class. Thus, if we add 7, 14, 21, ... to 3, the result will be a number in the same modulo class. We want to replace 3 with an equivalent mod 7

DID YOU KNOW?

The Enigma

A cipher, or secret code, actually has group properties: A well-defined operation turns plain text into a cipher, and an inverse operation allows it to be deciphered. During World War II, the Germans used an encrypting device based on a modulo 26 system (for the 26 letters of the alphabet) known as the enigma. It has four rotors, each with the 26 letters. The rotors were wired to one another so that if an A were typed into the machine, the first rotor would contact a different letter on the second rotor, which contacted a different letter on the third rotor, and so on. The French secret service obtained the wiring instructions, and Polish and British cryptoanalysts were able to crack the ciphers using group theory. This breakthrough helped to keep the Allies abreast of the deployment of the German navy.

number that is greater than 5. Adding 7 to 3 yields a sum of 10, which is greater than 5.

$$\begin{aligned} 3 - 5 &\equiv ?\ (\text{mod } 7) \\ (3+7) - 5 &\equiv ?\ (\text{mod } 7) \\ 10 - 5 &\equiv ?\ (\text{mod } 7) \\ 5 &\equiv ?\ (\text{mod } 7) \\ 5 &\equiv 5\ (\text{mod } 7) \end{aligned}$$

Therefore, ? = 5 and $3 - 5 \equiv 5\ (\text{mod } 7)$.

b) We wish to replace the ? with a number less than 5. We know that $7 - 4 \equiv 3$ mod 5 because $3 \equiv 3$ mod 5. Therefore, we need to determine what number, less than 5, the number 7 is congruent to in mod 5. If we subtract the modulus, 5, from 7, we obtain 2. Thus, 2 and 7 are in the same modular class. Therefore, ? = 2.

$$\begin{aligned} ? - 4 &\equiv 3\ (\text{mod } 5) \\ 7 - 4 &\equiv 3\ (\text{mod } 5) \\ 2 - 4 &\equiv 3\ (\text{mod } 5) \end{aligned}$$

Notice in the last equivalence that if we add 5 to 2, we get $7 - 4 \equiv 3\ (\text{mod } 5)$, which is a true statement.

c) $5 - ? \equiv 7\ (\text{mod } 8)$

In mod 8, adding 8, or a multiple of 8, to a number results in a sum that is in the same modulo class. Thus, we can add 8 to 5 so that the statement becomes

$$\begin{aligned} (8+5) - ? &\equiv 7\ (\text{mod } 8) \\ 13 - ? &\equiv 7\ (\text{mod } 8) \end{aligned}$$

We can see that $13 - 6 = 7$. Therefore, ? = 6 and $5 - 6 \equiv 7\ (\text{mod } 8)$. ●

EXAMPLE 4 *Using Modulo Classes in Multiplication*

Determine all positive number replacements (less than the modulus) for the question mark that make the statement true.

a) $4 \cdot ? \equiv 2\ (\text{mod } 5)$ b) $3 \cdot ? \equiv 3\ (\text{mod } 6)$ c) $3 \cdot ? \equiv 4\ (\text{mod } 6)$

SOLUTION

a) One method of determining the solution is to replace the question mark with the numbers 0–4 and then determine the equivalent modulo class of the product. We use the numbers 0–4 because we are working in modulo 5.

$$\begin{aligned} 4 \cdot ? &\equiv 2\ (\text{mod } 5) \\ 4 \cdot 0 &\equiv 0\ (\text{mod } 5) \\ 4 \cdot 1 &\equiv 4\ (\text{mod } 5) \\ 4 \cdot 2 &\equiv 3\ (\text{mod } 5) \\ 4 \cdot 3 &\equiv 2\ (\text{mod } 5) \\ 4 \cdot 4 &\equiv 1\ (\text{mod } 5) \end{aligned}$$

Therefore, ? = 3 since $4 \cdot 3 \equiv 2\ (\text{mod } 5)$.

b) Since we are working in modulo 6, replace the question mark with the numbers 0–5 and follow the procedure used in part (a).

$$3 \cdot ? \equiv 3 \pmod 6$$
$$3 \cdot 0 \equiv 0 \pmod 6$$
$$3 \cdot 1 \equiv 3 \pmod 6$$
$$3 \cdot 2 \equiv 0 \pmod 6$$
$$3 \cdot 3 \equiv 3 \pmod 6$$
$$3 \cdot 4 \equiv 0 \pmod 6$$
$$3 \cdot 5 \equiv 3 \pmod 6$$

Therefore, replacing the question mark with 1, 3, or 5 results in true statements. The answers are 1, 3, and 5.

c) $3 \cdot ? \equiv 4 \pmod 6$

Examining the products in part (b) shows there are no values that satisfy the statement. The answer is "no solution."

Whenever a process is repetitive, modular arithmetic may be helpful in answering some questions about the process. Now let's look at an application of modular arithmetic.

EXAMPLE 5 *Work Schedule*

Ellis drives a bus for a living. His working schedule is to drive for 6 days, and then he gets 2 days off. If today is the third day that Ellis has been driving, determine the following.

a) Will he be driving 60 days from today?

b) Will he be driving 82 days from today?

c) Was he driving 124 days ago?

SOLUTION

a) Since Ellis drives for 6 days and then gets 2 days off, his working schedule may be considered a modular 8 system. That is, 8, 16, 24, . . . days from today will be just like today, the third day of the 8-day cycle.

If we divide 60 by 8, we obtain

$$\begin{array}{r} 7 \\ 8\overline{)60} \\ \underline{56} \\ 4 \leftarrow \text{remainder} \end{array}$$

Therefore, in 60 days Ellis will go through 7 complete cycles and be 4 days further into the next cycle. If we let D represent a driving day and N represent a not driving day, then Ellis's cycle may be represented as follows:

$$D\ D\ D\ D\ D\ D\ N\ N$$

↑ today — ↑ 4 days from today

Notice that 4 days from today will be his first not driving day. Therefore, Ellis will not be driving 60 days from today.

DID YOU KNOW?

Secrets of the Universe

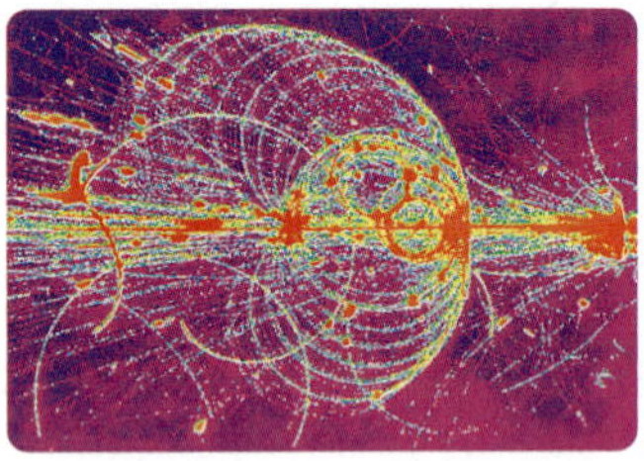

A computer simulation of protons and neutrons being smashed together.

Physicists use group theory in their search for the elementary particles that make up all matter. The operations are *transformations*: the coming together or splitting apart of elementary particles as they form larger particles such as protons and neutrons. Some transformations bring two particles together to form a different particle (binary operation), some leave a particle intact (identity), and some follow the rules of associativity and commutativity. In fact, understanding there were distinct "families" of particles related to one another as groups enabled physicists to predict the existence of certain particles before any experimental evidence was available.

b) We work this part in the same way we worked part (a). Divide 82 by 8 and determine the remainder.

```
   10
8)82
  80
   2 ← remainder
```

Thus, in 82 days it will be 2 days later in the cycle than it is today. Because he is currently in day 3 of his cycle, Ellis will be in day 5 of his cycle and will be driving the bus.

c) This part is worked in the same way as parts (a) and (b), but once we find the remainder we must move backward in the cycle.

```
    15
8)124
  120
    4 ← remainder
```

Thus, 124 days ago was 4 days earlier in the cycle. Marking day 3 of the cycle (indicated by the word *today*) and then moving 4 days backwards brings us to the first nondriving day. The two N's shown at the beginning of the letters below are actually the end days of the previous cycle.

N N D D D D D D N N

↑ 4 days earlier than today ↑ today

Therefore, 124 days ago Ellis was not driving. ●

Modular Arithmetic and Groups

Modular arithmetic systems under the operation of addition are commutative groups, as illustrated in Example 6.

EXAMPLE 6 *A Commutative Group*

Construct a mod 5 addition table and show that the mathematical system is a commutative group. Assume that the associative property holds for the given operation.

SOLUTION The set of elements in modulo 5 arithmetic is {0, 1, 2, 3, 4}; the binary operation is +.

+	0	1	2	3	4
0	0	1	2	3	4
1	1	2	3	4	0
2	2	3	4	0	1
3	3	4	0	1	2
4	4	0	1	2	3

PROFILE IN MATHEMATICS

Amalie Emmy Noether (1882–1935)

Professor Amalie Emmy Noether was described by Albert Einstein as the "most significant creative mathematical genius thus far produced since the higher education of women began." Because she was a woman, she was never to be granted a regular professorship in her native Germany. In 1933, with the rise of the Nazis to power, she and many other mathematicians and scientists emigrated to the United States. That brought her to Bryn Mawr College, at the age of 51, to her first full-time faculty appointment. She was to die just 2 years later.

Her special interest lay in abstract algebra, which includes such topics as groups, rings, and fields. She developed a number of theories that helped draw together many important mathematical concepts. Her innovations in higher algebra gained her recognition as one of the most creative abstract algebraists of modern times.

For this system to be a commutative group, it must satisfy the five properties of a commutative group.

1. *Closure*: Every entry in the table is a member of the set $\{0, 1, 2, 3, 4\}$, so the system is closed under addition.
2. *Identity element*: An easy way to determine whether there is an identity element is to look for a row in the table that is identical to the elements at the top of the table. Note that the row next to 0 is identical to the top of the table, which indicates that 0 *might be* the identity element. Now look at the column under the 0 at the top of the table. If this column is identical to the left-hand column, then 0 is the identity element. Since the column under 0 is the same as the left-hand column, 0 is the additive identity element in modulo 5 arithmetic.

Element	+	Identity	=	Element
0	+	0	=	0
1	+	0	=	1
2	+	0	=	2
3	+	0	=	3
4	+	0	=	4

3. *Inverse elements*: Does every element have an inverse? Recall that an element plus its inverse must equal the identity element. In this example, the identity element is 0. Therefore, for each of the given elements 0, 1, 2, 3, and 4, we must find the element that when added to it results in a sum of zero. These elements will be the inverses.

Element	+	Inverse	=	Identity	
0	+	?	=	0	Since $0 + 0 = 0$, 0 is its own inverse.
1	+	?	=	0	Since $1 + 4 = 0$, 4 is the inverse of 1.
2	+	?	=	0	Since $2 + 3 = 0$, 3 is the inverse of 2.
3	+	?	=	0	Since $3 + 2 = 0$, 2 is the inverse of 3.
4	+	?	=	0	Since $4 + 1 = 0$, 1 is the inverse of 4.

Note that each element has an inverse.

4. *Associative property*: It is given that the associative property holds. One example that illustrates the associative property is

$$(2 + 3) + 4 = 2 + (3 + 4)$$
$$0 + 4 = 2 + 2$$
$$4 = 4 \quad \text{True}$$

5. *Commutative property*: Is $a + b = b + a$ for *all* elements a and b of the given set? The table shows that the system is commutative because the elements are symmetric about the main diagonal. We will give one example to illustrate the commutative property.

$$4 + 2 = 2 + 4$$
$$1 = 1 \quad \text{True}$$

All five properties are satisfied. Thus, modulo 5 arithmetic under the operation of addition is a commutative group.

SECTION 10.3 EXERCISES

CONCEPT/WRITING EXERCISES

1. What does a *modulo m* system consist of?

2. a) Explain the meaning of the statement, "*a* is congruent to *b* modulo *m*."

   b) Explain the meaning of $27 \equiv 2 \pmod 5$.

3. In a modulo 3 system, how many modulo classes will there be? Construct a table similar to Table 10.12 on page 632 showing elements from each class.

4. In general, for a modulo *m* system, how are modulo classes developed?

5. In a modulo 16 system, how many modulo classes will there be? Explain.

6. In a modulo *n* system, how many modulo classes will there be? Explain.

7. Consider $29 \equiv ? \pmod 5$. Which of the following values could replace the question mark and result in a true statement? Explain.

   a) 20 b) 4 c) 9 d) 204

8. Consider $50 \equiv ? \pmod 7$. Which of the following values could replace the question mark and result in a true statement? Explain.

   a) 78 b) 71 c) 7 d) 17

PRACTICE THE SKILLS

In Exercises 9–16, assume that Sunday is represented as day 0, Monday is represented by day 1, and so on. If today is Thursday (day 4), determine the day of the week it will be in the specified number of days. Assume no leap years.

9. 10 days **10.** 161 days **11.** 365 days

12. 3 years **13.** 3 years, 34 days **14.** 463 days

15. 400 days **16.** 3 years, 27 days

In Exercises 17–24, consider 12 months to be a modulo 12 system. If it is currently October, determine the month it will be in the specified number of months.

17. 16 months **18.** 48 months

19. 2 years, 7 months **20.** 4 years, 8 months

21. 83 months **22.** 7 years

23. 105 months **24.** 5 years, 9 months

In Exercises 25–36, determine what number the sum, difference, or product is congruent to in mod 5.

25. $4 + 7$ **26.** $5 + 10$

27. $2 + 5 + 5$ **28.** $9 - 3$

29. $5 - 12$ **30.** $8 \cdot 7$

31. $8 \cdot 9$ **32.** $1 - 4$

33. $4 - 8$ **34.** $3 - 7$

35. $(15 \cdot 4) - 8$ **36.** $(4 - 9) \cdot 7$

In Exercises 37–50, find the modulo class to which each number belongs for the indicated modulo system.

37. 10, mod 5 **38.** 23, mod 7 **39.** 48, mod 12

40. 43, mod 6 **41.** 41, mod 9 **42.** 75, mod 8

43. 30, mod 7 **44.** 53, mod 4 **45.** −1, mod 7

46. −7, mod 4 **47.** −13, mod 11 **48.** −11, mod 13

49. 205, mod 10 **50.** −12, mod 4

In Exercises 51–66, determine all positive number replacements (less than the modulus) for the question mark that make the statement true.

51. $2 + 3 \equiv ? \pmod 6$ **52.** $? + 5 \equiv 3 \pmod 8$

53. $1 + ? \equiv 4 \pmod 5$ **54.** $4 + ? \equiv 3 \pmod 6$

55. $4 - ? \equiv 5 \pmod 6$ **56.** $2 \cdot 5 \equiv ? \pmod 7$

57. $5 \cdot ? \equiv 7 \pmod 9$ **58.** $3 \cdot ? \equiv 5 \pmod 6$

59. $2 \cdot ? \equiv 1 \pmod 6$ **60.** $3 \cdot ? \equiv 3 \pmod{12}$

61. $4 \cdot ? \equiv 4 \pmod{10}$ **62.** $? - 3 \equiv 4 \pmod 8$

63. $? - 8 \equiv 9 \pmod{12}$ **64.** $6 - ? \equiv 8 \pmod 9$

65. $3 \cdot ? \equiv 0 \pmod{10}$ **66.** $4 \cdot ? \equiv 5 \pmod 8$

PROBLEM SOLVING

67. ***Presidential Elections*** The upcoming presidential election years are 2008, 2012, 2016, ...

a) List the next five presidential election years after 2016.

b) What will be the first election year after the year 3000?

c) List the election years between the years 2550 and 2575.

68. ***Flight Schedules*** A pilot is scheduled to fly for 5 consecutive days and rest for 3 consecutive days. If today is the second day of her rest shift, determine whether she will be flying or resting

a) 60 days from today.

b) 90 days from today.

c) 240 days from today.

d) Was she flying 6 days ago?

e) Was she flying 20 days ago?

69. ***Workout Schedule*** A tennis pro's workout schedule is to give both morning and afternoon lessons for 3 days, rest for 1 day, give only morning lessons for 2 days, rest for 2 days, and then start the cycle again. If the tennis pro is on his rest for 1 day part of the schedule, determine what he will be doing

a) 28 days from today.

b) 60 days from today.

c) 127 days from today.

d) Will the tennis pro have a day off 82 days from today?

70. ***Physical Therapy*** A man has an Achilles' tendon injury and is receiving physical therapy. He must have physical therapy twice a day for 5 days, physical therapy once a day for 3 days, and then 2 days off; then the cycle begins again. If he is in his second day of his twice-a-day therapy cycle, determine what he will be doing

a) 20 days from today.

b) 49 days from today.

c) 103 days from today.

d) Will the man have a day off 78 days from today?

71. ***The Weekend Off*** A manager of a theater has both Saturday and Sunday off every 7 weeks. It is week 2 of the 7 weeks.

a) Determine the number of weeks before he will have both Saturday and Sunday off.

b) Will he have both Saturday and Sunday off 25 weeks from this week?

c) What is the first week, after 50 weeks from this week, that he will have both Saturday and Sunday off?

72. ***Nursing Shifts*** A nurse's work pattern at Community Hospital consists of working the 7 A.M.–3 P.M. shift for 3 weeks and then the 3 P.M.–11 P.M. shift for 2 weeks.

a) If it is the third week of the pattern, what shift will the nurse be working 6 weeks from now?

b) If it is the fourth week of the pattern, what shift will the nurse be working 7 weeks from now?

c) If it is the first week of the pattern, what shift will the nurse be working 11 weeks from now?

73. ***Restaurant Rotation*** A waiter at a restaurant works both daytime and evening shifts. He works daytime for 5 consecutive days, then evenings for 3 consecutive days, then daytime for 4 consecutive days, then evenings for 2 consecutive days. Then the rotation starts again. If it is day 2 of the 5-day consecutive daytime shift, determine whether he will be working the daytime or evening shift

a) 20 days from today.

b) 52 days from today.

c) 365 days from today.

Determine the sum or difference in clock 12 arithmetic.

5. 8 + 7 **6.** 5 + 12 **7.** 4 − 7

8. 4 + 7 + 9 **9.** 7 − 4 + 6 **10.** 2 − 8 − 7

11. List the properties of a group and explain what each property means.

12. What is another name for an abelian group?

In Exercises 13 and 14, explain your answer.

13. Determine whether the set of integers under the operation of addition forms a group.

14. Determine whether the set of integers under the operation of multiplication forms a group.

15. Determine whether the set of natural numbers under the operation of addition forms a group.

16. Determine whether the set of rational numbers under the operation of multiplication forms a group.

In Exercises 17–19, for the mathematical system, determine which of the five properties of a commutative group do not hold.

17.

●	*R*	*S*	⊗
R	*R*	⊗	*S*
S	⊗	*S*	*R*
⊗	*S*	*R*	⊗

18.

?	4	#	L	P
4	P	4	#	L
#	4	#	L	P
L	#	L	P	4
P	L	P	4	L

19.

□	!	?	Δ	*p*
!	?	Δ	!	*p*
?	Δ	*p*	?	!
Δ	!	?	Δ	*p*
p	*p*	!	*p*	Δ

20. Consider the following mathematical system. Assume the associative property holds for the given operation.

⌂	☺	●	♀	♂
☺	☺	●	♀	♂
●	●	♀	♂	☺
♀	♀	♂	☺	●
♂	♂	☺	●	♀

a) What are the elements of the set in this mathematical system?

b) What is the binary operation? ⌂

c) Is the system closed? Explain.

d) Is there an identity element for the system under the given operation?

e) Does every element in the system have an inverse? If so, give each element and its corresponding inverse.

f) Give an example to illustrate the associative property.

⌂ ⌂ ⌂ ⌂

g) Is the system commutative? Give an example.

⌂ ⌂

h) Is this mathematical system a commutative group? Explain.

10.3

In Exercises 21–30, find the modulo class to which the number belongs for the indicated modulo system.

21. 15, mod 5 **22.** 26, mod 7

23. 31, mod 6 **24.** 59, mod 8

25. 71, mod 12 **26.** 54, mod 4

27. 52, mod 12 **28.** 54, mod 14

29. 97, mod 11 **30.** 31, mod 20

In Exercises 31–40, determine all positive number replacements (less than the modulus) for the question mark that make the statement true.

31. $3 + 7 \equiv ? \pmod 9$ **32.** $? - 4 \equiv 2 \pmod 5$

33. $4 \cdot ? \equiv 3 \pmod 7$ **34.** $6 - ? \equiv 5 \pmod 7$

35. $? \cdot 4 \equiv 0 \pmod 8$ **36.** $10 \cdot 7 \equiv ? \pmod{11}$

37. $3 - 5 \equiv ? \pmod 7$ **38.** $? \cdot 7 \equiv 3 \pmod{10}$

39. $5 \cdot ? \equiv 3 \pmod 8$ **40.** $7 \cdot ? \equiv 2 \pmod 9$

41. Construct a modulo 6 addition table. Then determine whether the modulo 6 system forms a commutative group under the operation of addition.

42. Construct a modulo 4 multiplication table. Then determine whether the modulo 4 system forms a commutative group under the operation of multiplication.

43. ***Firefighter Shifts*** Linda Zientek, a firefighter, has the following work pattern. She works 3 days, has 4 days off, then works 4 days, then has 3 days off, and then the pattern repeats. Assume that today is the first day of the work pattern.

a) Will Linda be working 30 days from today?

b) Linda's sister is getting married 45 days from today. Will Linda have the day off?

CHAPTER 10 TEST

1. What is a mathematical system?

2. List the requirements needed for a mathematical system to be a commutative group.

3. Is the set of integers a commutative group under the operation of addition? Explain your answer completely.

4. Develop a clock 5 arithmetic addition table.

5. Is clock 5 arithmetic under the operation of addition a commutative group? Assume that the associative property holds. Explain your answer completely.

Determine the following in clock 5 arithmetic.

6. $1 + 3 + 4$

7. $2 - 5$

8. Consider the mathematical system

□	*W*	*S*	*T*	*R*
W	*T*	*R*	*W*	*S*
S	*R*	*W*	*S*	*T*
T	*W*	*S*	*T*	*R*
R	*S*	*T*	*R*	*W*

a) What is the binary operation?

b) Is this system closed? Explain.

c) Is there an identity element for this system under the given operation? Explain.

d) What is the inverse of the element *R*?

e) What is $(T \, \square \, R) \, \square \, W$?

In Exercises 9 and 10, determine whether the mathematical system is a commutative group. Explain your answer completely.

9.

*	*a*	*b*	*c*
a	*a*	*b*	*c*
b	*b*	*b*	*a*
c	*c*	*a*	*d*

10.

?	1	2	3
1	3	1	2
2	1	2	3
3	2	3	1

11. Determine whether the mathematical system is a commutative group. Assume that the associative property holds. Explain your answer.

○	@	$	&	%
@	%	@	$	&
$	@	$	&	%
&	$	&	%	@
%	&	%	@	$

In Exercises 12 and 13, determine the modulo class to which the number belongs for the indicated modulo system.

12. 27, mod 5

13. 91, mod 13

In Exercises 14–18, determine all positive number replacements for the question mark, less than the modulus, that make the statement true.

14. $7 + 7 \equiv ? \pmod 8$

15. $? - 3 \equiv 4 \pmod 5$

16. $3 - ? \equiv 7 \pmod 9$

17. $4 \cdot 2 \equiv ? \pmod 6$

18. $3 \cdot ? \equiv 2 \pmod 6$

19. To what number is 239 congruent to in modulo 8?

20. a) Construct a modulo 5 multiplication table.

b) Is this mathematical system a commutative group? Explain your answer completely.

GROUP PROJECTS

1. ***Rotating a Square*** The square *ABCD* below has a pin through it at point *P* so that the square can be rotated in a clockwise direction. The mathematical system consists of the set $\{A, B, C, D\}$ and the operation ♣. The elements of the set represent different rotations of the square clockwise, as follows.

A = rotate about the point P clockwise 90°
B = rotate about the point P clockwise 180°
C = rotate about the point P clockwise 270°
D = rotate about the point P clockwise 360°

The operation ♣ means *followed by*. For example, A ♣ B means a clockwise rotation of 90° followed by a clockwise rotation of 180° or a clockwise rotation of 270°. Since C represents a clockwise rotation of 270°, we write A ♣ $B = C$. This result and two other results are shown in the following table.

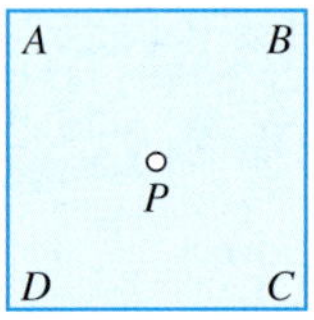

♣	A	B	C	D
A		C		
B			A	
C	D			
D				

Complete the table and determine if the mathematical system is a commutative group. Explain.

2. ***Product of Zero*** In arithmetic and algebra, the statement, "If $a \cdot b = 0$, then $a = 0$ or $b = 0$" is true. That is, for the product of two numbers to be 0, at least one of the factors must be 0. Can the product of two nonzero numbers equal 0 in a specific modulo system? If so, in what type of modulo systems can this result occur?

a) Construct multiplication tables for modulo systems 3–9.

b) Which, if any, of the multiplication tables in part (a) have products equal to 0 when neither factor is 0?

c) Which, if any, of the multiplication tables in part (a) have products equal to 0 only when at least one factor is 0?

d) Using the results in parts (b) and (c), can you write a conjecture as to which modulo systems have a product of 0 when neither factor is 0?

3. ***Conjecture about Multiplicative Inverses*** Are there certain modulo systems where all numbers have multiplicative inverses?

a) If you have not worked Group Projects Exercise 2, construct multiplication tables for modulo systems 3–9.

b) Which of the multiplication tables in part (a) contain multiplicative inverses for all nonzero numbers?

c) Which, if any, of the multiplication tables in part (a) do not contain multiplicative inverses for all nonzero numbers?

d) Using the results in parts (b) and (c), can you write a conjecture as to which modulo systems contain multiplicative inverses for all nonzero numbers?

CHAPTER 11

Consumer Mathematics

▲ Knowledge of consumer mathematics may help you reach your long-term goals.

WHAT YOU WILL LEARN

- Percents, fractions, and decimals
- Percent increase, percent decrease, and percent markup and markdown
- Simple interest
- Compound interest
- Present value
- Fixed and open-ended installment loans
- Mortgages
- Annuities, sinking funds, and retirement investments

WHY IT IS IMPORTANT

Managing your money takes much thought and planning. All your daily needs must be met, but at the same time you should consider your long-term goals such as purchasing a house, saving for college expenses, and investing for retirement. This chapter will provide information to help you reach these important goals.

11.1 PERCENT

▲ Daily weather reports include percents.

Your clock radio turns on at 6 A.M. and you hear a weather report that says there is a 75% chance of rain. Based on this information, you decide to carry your umbrella to your class. You used your knowledge of percents to make an everyday decision. In this section, we will investigate the meaning of percent and learn how percents relate to mathematics and to daily life.

Percent, Fractions, and Decimals

The study of mathematics is crucial to understanding how to make better financial decisions. A basic topic necessary for understanding the material in this chapter is percent.

The word *percent* comes from the Latin *per centum*, meaning "per hundred." A ***percent*** is simply a ratio of some number to 100. Thus, $\frac{15}{100} = 15\%$, and $\frac{x}{100} = x\%$.

Percents are useful in making comparisons. Consider Ross, who took two psychology tests. On the first test Ross answered 18 of the 20 questions correctly, and on the second test he answered 23 of 25 questions correctly. On which test did he have the higher score? One way to compare the results is to write ratios of the number of correct answers to the number of questions on the test and then convert the ratios to percents. We can find the grades in percent for each test by (a) writing a ratio of the number of correct answers to the total number of questions, (b) rewriting these ratios with a denominator of 100, and (c) expressing the ratios as percents.

Test 1 — (a), (b), (c)

$$\frac{\text{Number of correct answers}}{\text{Number of questions on the test}} = \frac{18}{20} = \frac{18 \times 5}{20 \times 5} = \frac{90}{100} = 90\%$$

Test 2 — (a), (b), (c)

$$\frac{\text{Number of correct answers}}{\text{Number of questions on the test}} = \frac{23}{25} = \frac{23 \times 4}{25 \times 4} = \frac{92}{100} = 92\%$$

By changing the results of both tests to percents, we have a common standard for comparison. The results show that Ross scored 90% on the first test and 92% on the second test. Thus, he had a higher score on the second test.

Another procedure to change a fraction to a percent follows.

PROCEDURE TO CHANGE A FRACTION TO A PERCENT

1. Divide the numerator by the denominator to obtain a decimal number.
2. Multiply the decimal number by 100 (which has the effect of moving the decimal point two places to the right).
3. Add a percent sign.

The above procedure has the effect of moving the decimal point in the original number two places to the right and adding a percent sign.

Note that steps 2 and 3 together are the equivalent of multiplying by 100%. Since $100\% = 100/100 = 1$, we are not changing the *value* of the number, we are simply changing the number to a percent.

EXAMPLE 1 *Converting Fractions to Percents*

Change each of the following fractions to a percent.

a) $\frac{67}{100}$ b) $\frac{11}{20}$ c) $\frac{15}{16}$

SOLUTION For each fraction, follow the three steps in the procedure box on page 646.

a) 1. $67 \div 100 = 0.67$ 2. $0.67 \times 100 = 67$ 3. 67%

Thus, $\frac{67}{100} = 67\%$.

b) 1. $11 \div 20 = 0.55$ 2. $0.55 \times 100 = 55$ 3. 55%

Thus, $\frac{11}{20} = 55\%$.

c) 1. $15 \div 16 = 0.9375$ 2. $0.9375 \times 100 = 93.75$ 3. 93.75%

Thus, $\frac{15}{16} = 93.75\%$.

In future examples, when changing a fraction to a percent, we will follow the three-step procedure given on page 646 without listing the individual steps. We will do this in Examples 4 and 5.

PROCEDURE TO CHANGE A DECIMAL NUMBER TO A PERCENT

1. Multiply the decimal number by 100.
2. Add a percent sign.

The procedure for changing a decimal number to a percent is equivalent to moving the decimal point two places to the right and adding a percent sign.

EXAMPLE 2 *Converting Decimal Numbers to Percents*

Change each of the following decimal numbers to a percent.

a) 0.14 b) 0.893 c) 0.7625

SOLUTION

a) $0.14 = (0.14 \times 100)\% = 14\%$. Thus, $0.14 = 14\%$.

b) $0.893 = (0.893 \times 100)\% = 89.3\%$. Thus, $0.893 = 89.3\%$.

c) $0.7625 = (0.7625 \times 100)\% = 76.25\%$. Thus, $0.7625 = 76.25\%$.

To change a number given as a percent to a decimal number, follow this procedure.

PROCEDURE TO CHANGE A PERCENT TO A DECIMAL NUMBER

1. Divide the number by 100.
2. Remove the percent sign.

The procedure for changing a number given as a percent to a decimal number is equivalent to moving the decimal point two places to the left and removing the percent sign. Another way to remember that is: percent means per hundred (or to *divide by 100*).

EXAMPLE 3 *Converting a Percent to a Decimal Number*

a) Change 35% to a decimal number.

b) Change 69.8% to a decimal number.

c) Change $\frac{1}{2}\%$ to a decimal number.

SOLUTION

a) $35\% = \frac{35}{100} = 0.35$. Thus, $35\% = 0.35$.

b) $69.8\% = \frac{69.8}{100} = 0.698$. Thus, $69.8\% = 0.698$.

c) $\frac{1}{2}\% = 0.5\% = \frac{0.5}{100} = 0.005$. Thus, $\frac{1}{2}\% = 0.005$. •

EXAMPLE 4 *How Old Is Old?*

A Yahoo! Question of the Week asked, "At what age do you consider someone old?" Out of the 3496 people who responded, 1017 said age 80 is old. What percent of respondents believe that age 80 is old?

SOLUTION To find the percent that believe age 80 is old, divide the number of those who responded that age 80 is old by the total number of respondents. Then move the decimal point two places to the right and add a percent sign:

$$\text{Percent who believe age 80 is old} = \frac{1017}{3496} \approx 0.2909 \approx 29.09\%$$

Thus, about 29.1% (to the nearest tenth of a percent) believe that age 80 is old. •

All answers in this section will be given to the nearest tenth of a percent. When rounding answers to the nearest tenth of a percent, we carry the division to four places after the decimal point (ten-thousandths) and then round to the nearest thousandths position.

EXAMPLE 5 *Cherry Production*

According to the U.S. Department of Agriculture, in 2005 the United States produced about 244 million pounds of cherries. Of these cherries, 190 million pounds were produced in Michigan, 19 million pounds were produced in Utah, 18 million pounds were produced in Washington, and 17 million pounds were produced in other states. Determine the percent of cherries produced in each of the following: Michigan, Utah, Washington, and other states.

SOLUTION To determine each percent, divide the number of cherries produced in each state by the total number of cherries produced in the United States. Because all the numbers given are in millions of pounds, we will calculate the percents using the number of millions of pounds.

$$\text{Percent of cherries produced in Michigan} = \frac{190}{244} \approx 0.7786\ldots \approx 77.9\%$$

$$\text{Percent of cherries produced in Utah} = \frac{19}{244} \approx 0.0778\ldots \approx 7.8\%$$

$$\text{Percent of cherries produced in Washington} = \frac{18}{244} \approx 0.0737\ldots \approx 7.4\%$$

$$\text{Percent of cherries produced in other states} = \frac{17}{244} \approx 0.0696\ldots \approx 7.0\%$$

The sum of the percents, 77.9% + 7.8% + 7.4% + 7.0%, equals 100.1%. Since we rounded each percent, the sum is slightly different from 100%.

Percent Change

The percent increase or decrease, or percent change, over a period of time is determined by the following formula:

$$\textbf{Percent change} = \frac{\begin{pmatrix}\text{amount in}\\ \text{latest period}\end{pmatrix} - \begin{pmatrix}\text{amount in}\\ \text{previous period}\end{pmatrix}}{\text{amount in previous period}} \times 100$$

If the amount in the latest period is greater than the amount in the previous period, the answer will be positive and will indicate a percent increase. If the amount in the latest period is smaller than the amount in the previous period, the answer will be negative and will indicate a percent decrease.

EXAMPLE 6 *Baseball Record*

In 2005, the major league baseball team with the most improved record for winning games was the Arizona Diamondbacks. In 2004, the Diamondbacks won 51 games. In 2005, the Diamondbacks won 77 games. Find the percent increase in the number of games won by the Diamondbacks from 2004 to 2005.

SOLUTION The previous period is 2004, and the latest period is 2005.

$$\begin{aligned}\text{Percent change} &= \frac{\begin{pmatrix}\text{amount in}\\ \text{latest period}\end{pmatrix} - \begin{pmatrix}\text{amount in}\\ \text{previous period}\end{pmatrix}}{\text{amount in previous period}} \times 100\\ &= \frac{77 - 51}{51} \times 100\\ &= \frac{26}{51} \times 100\\ &\approx 0.5098\ldots \times 100\\ &\approx 51.0\end{aligned}$$

Therefore, there was about a 51.0% increase in the number of games won by the Diamondbacks from 2004 to 2005.

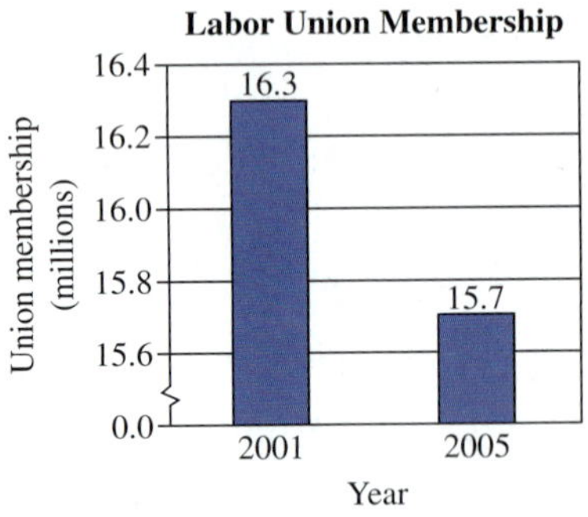

EXAMPLE 7 *Labor Union Membership*

There were approximately 15.7 million labor union members in the United States in 2005. This number is down from approximately 16.3 million labor union members in 2001 (see the bar graph in the margin). Determine the percent change in labor union membership from 2001 to 2005.

SOLUTION The previous period is 2001, and the latest period is 2005.

$$\text{Percent change} = \frac{\begin{pmatrix}\text{amount in}\\ \text{latest period}\end{pmatrix} - \begin{pmatrix}\text{amount in}\\ \text{previous period}\end{pmatrix}}{\text{amount in previous period}} \times 100$$

$$= \frac{15.7 - 16.3}{16.3} \times 100$$

$$= \frac{-0.6}{16.3} \times 100$$

$$\approx -0.0368 \times 100$$

$$\approx -3.7\%$$

Thus, union membership decreased by about 3.7% from 2001 to 2005.

Percent Markup and Markdown

A similar formula is used to calculate percent markup or markdown on cost. A positive answer indicates a markup, and a negative answer indicates a markdown.

$$\textbf{Percent markup (or markdown) on cost} = \frac{\text{selling price} - \text{dealer's cost}}{\text{dealer's cost}} \times 100$$

EXAMPLE 8 *Determining Percent Markup*

Holdren Hardware stores pay $48.76 for glass fireplace screens. They regularly sell them for $79.88. At a sale they sell them for $69.99. Determine

a) the percent markup on the regular price.

b) the percent markup on the sale price.

SOLUTION

a) We determine the percent markup on the regular price as follows.

$$\text{Percent markup} = \frac{\text{selling price} - \text{dealer's cost}}{\text{dealer's cost}} \times 100$$

$$= \frac{79.88 - 48.76}{48.76} \times 100$$

$$\approx 0.6382 \times 100$$

$$\approx 63.8$$

Thus, the percent markup on the regular price was about 63.8%.

b) We determine the percent markup on the sale price as follows.

$$\text{Percent markup} = \frac{69.99 - 48.76}{48.76} \times 100$$
$$\approx 0.4354 \times 100$$
$$\approx 43.5$$

Thus, the percent markup on the sale price was about 43.5%.

EXAMPLE 9 *Determining Percent Markdown*

To help get rid of its remaining 2008 calendars, Cardin's Pharmacy reduces the price to \$4.00. If Cardin's cost per calendar was \$5.00, determine the percent markdown on the calendars.

SOLUTION We will use the same formula that we used for percent markup in Example 8.

$$\text{Percent markdown} = \frac{\text{selling price} - \text{dealer's cost}}{\text{dealer's cost}} \times 100$$
$$= \frac{4.00 - 5.00}{5.00} \times 100$$
$$= -0.20 \times 100$$
$$= -20$$

Thus, the percent markdown on the calendars was 20%.

Other Percent Problems

In daily life, we may need to know how to solve any one of the following three types of problems involving percent:

1. What is a 15% tip on a restaurant bill of \$24.66? The problem can be stated as

   15% of \$24.66 is what number?

2. If Nancy Johnson made a sale of \$500 and received a commission of \$25, what percent of the sale is the commission? The problem can be stated as

   What percent of \$500 is \$25?

3. If the price of a jacket was reduced by 25% or \$12.50, what was the original price of the jacket? The problem can be stated as

   25% of what number is 12.50?

To answer these questions, we will write each problem as an equation. The word *is* means "is equal to," or =. In each problem, we will represent the unknown quantity with the letter x. Therefore, the preceding problems can be represented as

1. 15% of \$24.66 = x 2. x% of \$500 = \$25 3. 25% of x = \$12.50

The word *of* in such problems indicates multiplication. To solve each problem, change the percent to a decimal number and express the problem as an equation; then solve the equation for the variable x. The solutions follow.

DID YOU KNOW?

To Insure Promptness

The practice of tipping began in an English coffeehouse in the eighteenth century. Legend has it that a hurried businessman dropped a few pence on the table as a friendly incentive to the waiter. In fact, the coffeehouses eventually installed small collection boxes marked **T**o **I**nsure **P**romptness. The tip, as it became known, was paid before, not after, the meal. Today, there are many books and Internet web sites devoted to the guidelines of tipping. For example, according to the Original Tipping Page (*www.tipping.org*), it is proper to tip waiters and waitresses 15–20% of the total restaurant bill. Hair stylists should be tipped 15% of the total bill, and pizza delivery persons should receive \$1–\$5, depending on the distance driven. Skycaps at the airport should be tipped at least \$1 per bag, and newspaper delivery persons should receive \$15–\$25 at holiday time.

1. 15% of $\$24.66 = x$

$$0.15(24.66) = x \quad \text{15\% is written as 0.15 in decimal form.}$$
$$3.699 = x$$

Since 15% of \$24.66 is \$3.699, the tip would be \$3.70.

2. $x\%$ of $\$500 = \$25$

$$(0.01x)500 = 25 \quad x\% \text{ is written as } 0.01x \text{ in decimal form.}$$
$$5x = 25$$
$$\frac{5x}{5} = \frac{25}{5}$$
$$x = 5$$

Since 5% of \$500 is \$25, the commission is 5% of the sale.

3. 25% of $x = \$12.50$

$$0.25(x) = 12.50 \quad \text{25\% is written as 0.25 in decimal form.}$$
$$\frac{0.25x}{0.25} = \frac{12.50}{0.25}$$
$$x = 50$$

Since 25% of \$50 is \$12.50, the original price of the jacket was \$50.00.

EXAMPLE 10 *Down Payment on a House*

Melissa Bell wishes to buy a house for \$287,000. To obtain a mortgage, she needs to pay 20% of the selling price as a down payment. Determine the amount of Melissa's down payment.

SOLUTION We want to find the amount of the down payment. Let $x =$ the down payment. Then

$$\begin{aligned} x &= 20\% \text{ of the selling price} \\ &= 20\% \text{ of } \$287{,}000 \\ &= 0.20(287{,}000) \\ &= 57{,}400. \end{aligned}$$

Melissa will have a down payment of \$57,400.

EXAMPLE 11 *Chess Membership*

In 2006, the U.S. Chess Federation (USCF) had about 84,000 members. Of these members, about 11,000 were considered *youth* members (under the age of 18). What percent of USCF members were youth members?

SOLUTION We need to determine what percent of the 84,000 is 11,000. Let $x =$ percent of USCF members who are youth members. Then

$$\begin{aligned} x\% \text{ of } 84{,}000 &= 11{,}000 \\ 0.01x(84{,}000) &= 11{,}000 \\ 840x &= 11{,}000 \\ x &= \frac{11{,}000}{840} \\ x &\approx 13.1 \end{aligned}$$

Therefore, about 13.1% of the USCF members are youth members.

▲ The Taj Mahal in Agra, India

EXAMPLE 12 *Population of India*

About 337,567,000, or about 31%, of India's population is younger than 15 years old. What is the approximate population of India?

SOLUTION This problem can be stated as

31% of what number is 337,567,000?

Let x = the population of India. Then

$$\begin{aligned} 31\% \text{ of } x &= 337{,}567{,}000 \\ 0.31x &= 337{,}567{,}000 \\ x &= \frac{337{,}567{,}000}{0.31} \\ x &\approx 1{,}088{,}925{,}806 \end{aligned}$$

Thus, the population of India is about 1,088,925,806 people. ●

SECTION 11.1 EXERCISES

CONCEPT/WRITING EXERCISES

1. What is a percent?
2. Explain how to change a percent to a decimal number.
3. Explain how to change a fraction to a percent.
4. Explain how to change a decimal number to a percent.
5. Explain how to determine percent change.
6. Explain how to determine percent markup on cost.

PRACTICE THE SKILLS

In Exercises 7–14, change the number to a percent. Express your answer to the nearest tenth of a percent.

7. $\frac{3}{4}$ **8.** $\frac{3}{5}$ **9.** $\frac{11}{20}$

10. $\frac{7}{8}$ **11.** 0.007654 **12.** 0.5688

13. 3.78 **14.** 13.678

In Exercises 15–24, change the percent to a decimal number.

15. 5% **16.** 3.2% **17.** 5.15% **18.** 0.75%

19. $\frac{1}{4}\%$ **20.** $\frac{3}{8}\%$ **21.** $\frac{1}{5}\%$ **22.** 135.9%

23. 1% **24.** 0.50%

PROBLEM SOLVING

For Exercises 25–67, where appropriate, round answers to the nearest tenth of a percent.

25. ***Niacin*** The U.S. Department of Agriculture's recommended daily allowance (USRDA) of niacin for adults is 20 mg. One serving of Cheerios provides 5 mg of niacin. What percent of the USRDA of niacin does one serving of Cheerios provide?

26. ***Potassium*** The U.S. Department of Agriculture's recommended daily allowance (USRDA) of potassium for adults is 3500 mg. One serving of Cheerios provides 95 mg of potassium. What percent of the USRDA of potassium does one serving of Cheerios provide?

27. ***Reduced Fat Milk*** One serving of whole milk contains 8 g of fat. Reduced-fat milk contains 41.25% less fat per serving than whole milk. How many grams of fat does one serving of reduced-fat milk contain?

28. ***River Pollution*** In a study of U.S. rivers, 693,905 miles of river were studied. According to the U.S. Environmental Protection Agency, it was found that 36% of the total miles of rivers studied had impaired water quality. How many miles of rivers had impaired quality?

Florida Lottery *In Exercises 29–32, use the circle graph to answer the questions. In April 2006, the Florida Lottery had total education appropriations of $2,604,264. The areas where this money was appropriated along with the corresponding percents are represented in the circle graph below.*

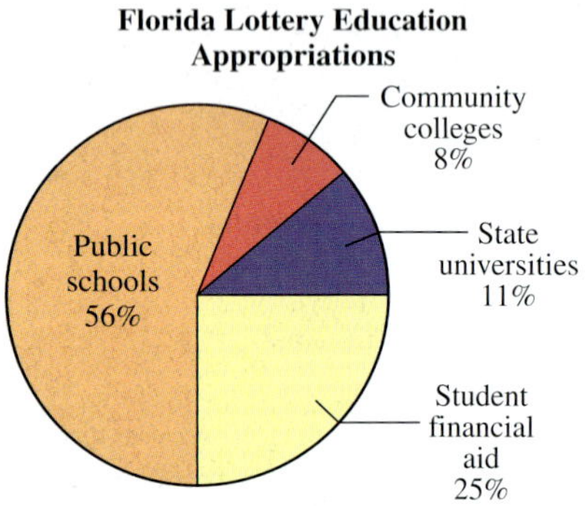

Source: Florida Department of Education

29. Determine the amount of money appropriated to community colleges.

30. Determine the amount of money appropriated to state universities.

31. Determine the amount of money appropriated to student financial aid.

32. Determine the amount of money appropriated to public schools.

Gasoline Cost *In Exercises 33–36, use the circle graph to answer the questions. The circle graph shows the breakdown of the cost, for a gallon of gasoline in the United States, in 2006.*

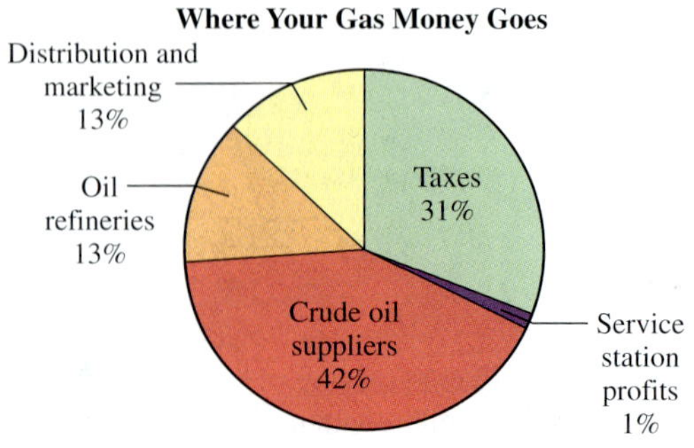

Source: American Petroleum Institute

Suppose in 2006 your family spent $5000 on gasoline.

33. Determine the amount of money that went to taxes.

34. Determine the amount of money that went to service station profits.

35. Determine the amount of money that went to crude oil suppliers.

36. Determine the amount of money that went to oil refineries.

Super Bowl Advertisers *In Exercises 37–40, use the circle graph to answer the questions. During the years 1986–2005, the top five Super Bowl advertisers spent about $559 million on advertising.*

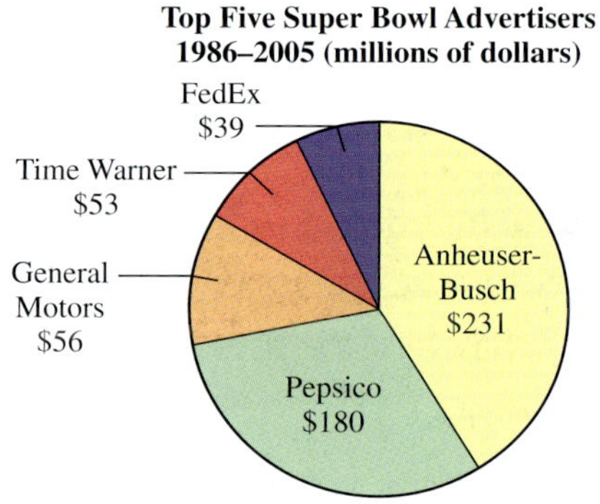

37. What percent of the total did Anheuser-Busch spend?

38. What percent of the total did Pepsico spend?

39. What percent of the total did General Motors spend?

40. What percent of the total did Time Warner spend?

41. ***Decreasing Population*** In 2000, the population of North Dakota was 642,200. In 2005, the population of North Dakota had decreased to 636,677. Determine the percent decrease in North Dakota's population from 2000 to 2005.

42. ***Increasing Population*** The population of the United States rose from approximately 248.7 million in 1990 to approximately 298.3 million in 2006. Determine the percent increase in the U.S. population from 1990 to 2006.

43. ***U.S. Debt*** The graph on top of page 655 shows the federal U.S. debt, in trillions of U.S. dollars, for the years 1966, 1976, 1986, 1996, and 2006.

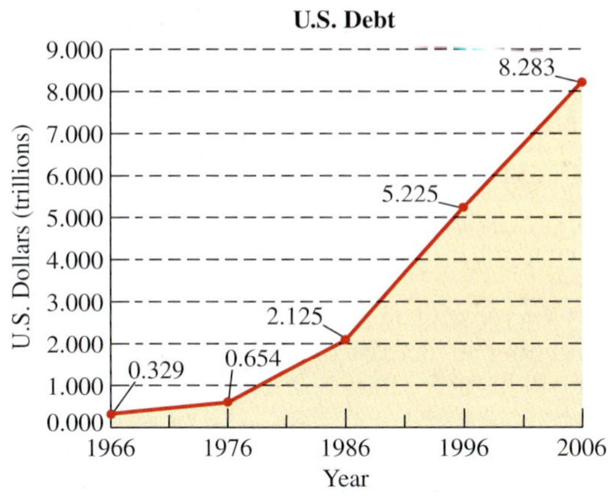

Source: Bureau of the Public Debt

a) Determine the percent increase in the U.S. debt from 1966 to 1976.

b) Determine the percent increase in the U.S. debt from 1976 to 1986.

c) Determine the percent increase in the U.S. debt from 1986 to 1996.

d) Determine the percent increase in the U.S. debt from 1996 to 2006.

44. ***American Smokers*** The graph shows the number of cigarette smokers, in millions, in the United States for the years 1990, 1994, 1998, 2002, and 2006.

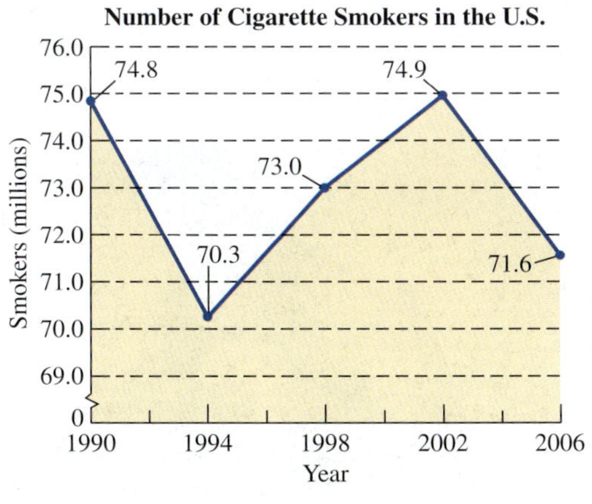

Source: Action on Smoking and Health

a) Determine the percent decrease in the number of smokers from 1990 to 1994.

b) Determine the percent increase in the number of smokers from 1994 to 1998.

c) Determine the percent increase in the number of smokers from 1998 to 2002.

d) Determine the percent decrease in the number of smokers from 2002 to 2006.

45. ***Dow Jones*** The following graph shows the Dow Jones Industrial Average (DJIA) for 1996–2005 at closing on the last trading day of the year.

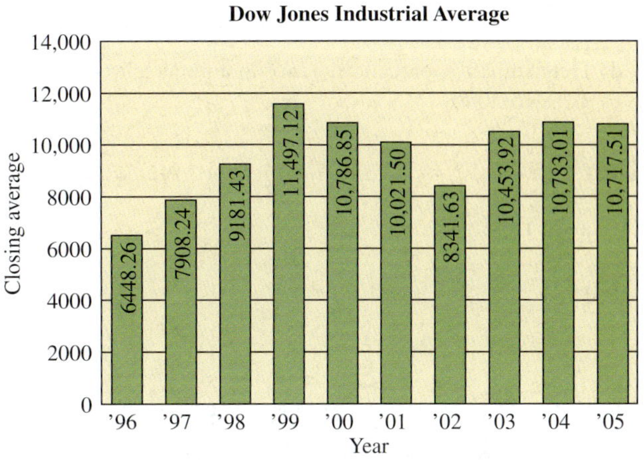

Source: Dow Jones indexes

a) Determine the percent increase in the DJIA from 1996 to 1999.

b) Determine the percent decrease in the DJIA from 1999 to 2002.

c) Determine the percent increase in the DJIA from 2002 to 2005.

d) Determine the percent increase in the DJIA from 1996 to 2005.

46. ***Presidential Salary*** The following graph shows the annual salary for the president of the United States for selected years from 1789 to 2006.

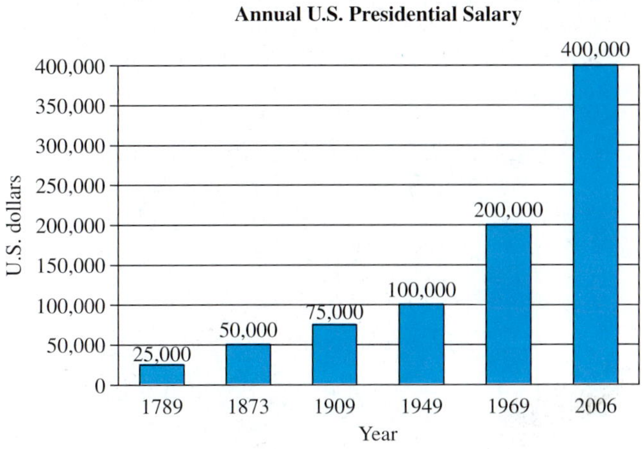

Source: *Congressional Quarterly*

a) Determine the percent increase in annual salary from 1789 to 1873.

b) Determine the percent increase in annual salary from 1909 to 1949.

c) Determine the percent increase in annual salary from 1969 to 2006.

d) Determine the percent increase in annual salary from 1789 to 2006.

In Exercises 47–52, determine the answer to the question.

47. What is 15% of $45.00?

48. What is 6.5% of $150.00?

49. What percent of 96 is 24?

50. What percent of 75 is 15?

51. Five percent of what number is 15?

52. Ten percent of what number is 75?

53. ***Tax and Tip*** According to the Original Tipping Page (*www.tipping.org*), it is appropriate to tip waiters and waitresses 15–20% of the total restaurant bill, *including the tax.* Mary and Keith's dinner costs $43.50 before tax, and the tax rate is 6%.

a) What is the tax on Mary and Keith's dinner?

b) What is the total bill, including tax, before the tip?

c) If Mary and Keith decide to tip 15% of the total bill, how much is the tip?

d) What is the total cost of the dinner including tax and tip?

54. ***Fishing*** The Gordon and Stallard Charter Fishing Boat Company has recently increased the number of crewmembers by 25%, or 10 crewmembers. What was the original number of crewmembers?

55. ***Percentage of A's*** In a mathematics class, eighteen students received an A on the third test, which is 150% of the students who received an A on the second test. How many students received an A on the second test?

56. ***Employee Increase*** The Fastlock Company hired 57 new employees, which increased its staff by 30%. What was the original number of employees?

57. ***Salary Increase*** Qami Brown's present salary is $36,500. He is getting an increase of 7% in his salary next year. What will his new salary be?

58. ***Salad Dressing Preference*** In a survey of 300 people, 17% prefer ranch dressing on their salad. How many people in the surveyed group prefer ranch dressing?

59. ***Vacuum Cleaner Sales*** A Kirby vacuum cleaner dealership sold 430 units in 2006 and 407 units in 2007. Find the percent increase or decrease in the number of units sold.

60. ***Vacuum Cleaner Markup*** If the Kirby dealership in Exercise 59 pays $320 for each unit and sells it for $699, what is the percent markup?

61. ***U.S. Poverty*** In the year 2001, the United States had 32.9 million people living in poverty. In the year 2004, there were 37.0 million people living in poverty. Determine the percent increase in the number of people in the United States living in poverty from 2001 to 2004.

62. ***Weight Loss*** On January 1, Max Saindoux weighed 235 pounds and decided to diet and exercise. On June 30, Max weighed 210 pounds. Determine the percent decrease in Max's weight from January 1 to June 30.

63. ***Television Sale*** The regular price of a Phillips color TV is $539.62. During a sale, Hill TV is selling the TV for $439. Find the percent decrease in the price of this TV.

64. ***Restaurant Markup*** The cost of a fish dinner to the owner of the Golden Wharf restaurant is $7.95. The fish dinner is sold for $11.95. Find the percent markup.

65. ***Truck Sale Profit*** Bonnie James sold a truck and made a profit of \$675. Her profit was 18% of the sale price. What was the sale price?

66. ***Furniture Sale*** Kane's Furniture Store advertised a table at a 15% discount. The original price was \$115, and the sale price was \$100. Was the sale price consistent with the ad? Explain.

67. ***Reselling a Car*** Quincy Carter purchased a used car for \$1000. He decided to sell the car for 10% above his purchase price. Quincy could not sell the car so he reduced his asking price by 10%. If he sells the car at the reduced price, will he have a profit or a loss or will he break even? Explain how you arrived at your answer.

CHALLENGE PROBLEMS/GROUP ACTIVITIES

68. ***Comparing Markdowns***

a) A coat is marked down 10%, and the customer is given a second discount of 15%. Is that the same as a single discount of 25%? Explain.

b) The regular price of a chair is \$189.99. Determine the sale price of the chair if the regular price is reduced by 10% and this price is then reduced another 15%.

c) Determine the sale price of the chair if the regular price of \$189.99 is reduced by 25%.

d) Examine the answers obtained in parts (b) and (c). Does your answer to part (a) appear to be correct? Explain.

69. ***Selling Ties*** The Tie Shoppe paid \$5901.79 for a shipment of 500 ties and wants to make a profit of 40% of the cost on the whole shipment. The store is having two special sales. At the first sale it plans to sell 100 ties for \$9.00 each, and at the second sale it plans to sell 150 ties for \$12.50 each. What should be the selling price of the other 250 ties for the Tie Shoppe to make a 40% profit on the whole shipment?

RECREATIONAL MATHEMATICS

70. In parts (a) through (c), determine which is the greater amount and by how much.

a) \$100 increased by 25% or \$200 decreased by 25%.

b) \$100 increased by 50% or \$200 decreased by 50%.

c) \$100 increased by 100% or \$200 decreased by 100%.

INTERNET/RESEARCH ACTIVITY

71. Find two circle graphs in newspapers, magazines, or on the Internet whose data are not given in percents. Redraw the graphs and label them with percents.

11.2 PERSONAL LOANS AND SIMPLE INTEREST

▲ A personal loan may be used to pay for emergency expenses.

Consider the following dilemma. Your car breaks down and the mechanic tells you that the repairs will cost \$450. You currently do not have the \$450, but your family desperately needs to have a car. One option for you would be to borrow the money from a friend, a family member, a bank, or other lending institution. In this section, we will discuss personal loans and the cost of obtaining such loans.

The money a bank or other lender is willing to lend you is called the amount of ***credit*** extended or the ***principal of the loan***. The amount of credit and the interest rate that you may obtain depend on the assurance you can give the lender that you will be able to repay the loan. Your credit is determined by your business reputation for honesty, by your earning power, and by what you can pledge as security to cover the loan. ***Security*** (or *collateral*) is anything of value pledged by the borrower that the lender may sell or keep if the borrower does not repay the loan. Acceptable security may be a business, a mortgage on a property, the title to an automobile, savings accounts, or stocks or bonds. The more marketable the security, the easier it is to obtain the loan, and in some cases marketability may help in getting a lower interest rate.

Bankers sometimes grant loans without security, but they require the signature of one or more other persons, called ***cosigners***, who guarantee the loan will be repaid. For either of the two types of loans, the secured loan or the cosigner loan, the borrower (and cosigner, if there is one) must sign an agreement called a ***personal note***. This document states the terms and conditions of the loan.

The most common way for individuals to borrow money is through an installment loan or using a credit card. (Installment loans and credit cards are discussed in Section 11.4.)

The concept of simple interest is essential to the understanding of installment buying. ***Interest*** is the money the borrower pays for the use of the lender's money. One type of interest is called simple interest. ***Simple interest*** is based on the entire amount of the loan for the total period of the loan. The formula used to find simple interest follows.

SIMPLE INTEREST FORMULA

$$\text{Interest} = \text{principal} \times \text{rate} \times \text{time}$$
$$i = prt$$

In the simple interest formula, the ***principal***, p, is the amount of money lent, the ***rate***, r, is the rate of interest expressed as a percent, and the ***time***, t, is the number of days, months, or years for which the money will be lent. *Time is expressed in the same period as the rate*. For example, if the rate is 2% per month, the time must be expressed in months. Typically, rate means the annual rate unless otherwise stated. Principal and interest are expressed in dollars in the United States.

Ordinary Interest

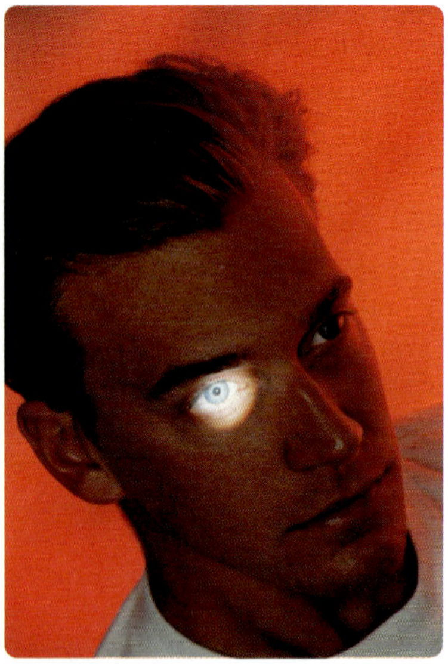

The most common type of simple interest is called *ordinary interest.* For computing ordinary interest, each month has 30 days and a year has 12 months or 360 days. On the *due date* of a *simple interest note* the borrower must repay the principal plus the interest. (*Note*: Simple interest will mean ordinary interest unless stated otherwise.)

EXAMPLE 1 *Eye Surgery Loan*

Chad Cohen needs to borrow \$2200 to have corrective eye surgery. From his credit union, he obtains an 18-month loan with an annual simple interest rate of 7.5%.

a) Calculate the simple interest on the loan.

b) Determine the amount (principal + interest) Chad will pay the credit union at the end of the 18 months.

SOLUTION

a) To determine the interest on the loan, we use the formula $i = prt$. We know that $p = \$2200$, $r = 7.5\%$, which when written as a decimal is 0.075, and t in years $= \frac{18}{12} = 1.5$. We substitute the appropriate values in the simple interest formula

$$\begin{aligned} i &= p \times r \times t \\ &= \$2200 \times 0.075 \times 1.5 \\ &= \$247.5 \end{aligned}$$

The simple interest on \$2200 at 7.5% for 18 months is \$247.50.

b) The amount to be repaid is equal to the principal plus interest, or

$$\begin{aligned} A &= p + i \\ &= \$2200 + \$247.50 \\ &= \$2447.50 \end{aligned}$$

To pay off his loan, Chad will pay the credit union \$2447.50 at the end of 18 months.

EXAMPLE 2 *Determining the Annual Rate of Interest*

Philip Mahler agrees to lend \$850 to his friend Joe Mahoney to help Joe travel to Cincinnati to attend a family wedding. Nine months later, Joe repaid the original \$850 plus \$51 interest. What annual rate of interest did Joe pay to Philip?

SOLUTION We need to solve for the interest rate, r. Since the time is 9 months, the time in years is $\frac{9}{12}$, or 0.75. Using the formula $i = prt$, we get

$$\begin{aligned} \$51 &= \$850 \times r \times 0.75 \\ 51 &= 637.5r \\ \frac{51}{637.5} &= \frac{637.5r}{637.5} \\ 0.08 &= r \end{aligned}$$

Divide both sides of the equation by 637.5.

Thus, the annual rate of interest paid by Joe to Philip is 0.08, or 8%.

EXAMPLE 3 *A Pawn Loan*

To obtain money for new eyeglasses, Gilbert French decides to pawn his trumpet. Gilbert borrows \$240 and after 30 days gets his trumpet back by paying the pawnbroker \$288. What annual rate of interest did Gilbert pay?

SOLUTION Gilbert paid \$288 − \$240 = \$48 in interest, and the length of the loan is for one month, or $\frac{1}{12}$ of a year.

Using the formula $i = prt$, we get

$$\begin{aligned} \$48 &= \$240 \times r \times \frac{1}{12} \\ 48 &= 20r \\ \frac{48}{20} &= r \\ 2.4 &= r \end{aligned}$$

DID YOU KNOW?

Pawn Loans

Since ancient times, pawnbroking has been a source of credit for people who cannot obtain loans from other financial institutions. A typical pawn transaction involves the customer presenting an item of value to a pawnbroker. The pawnbroker makes a loan to the customer based on a percentage of the value of the item. The pawnbroker keeps the item as collateral until the customer repays the loan plus the interest. If the customer defaults on the loan, the pawnbroker keeps the item and usually offers it for sale in his or her pawnshop. Because of the 15 to 20% default rate and the need to pay for retail space, pawn loans usually have a significantly higher rate of interest than other financial institutions (see Example 3).

DID YOU KNOW?

Origin of Banks

In ancient Babylonia, as early as 2000 B.C., temples were considered safe depositories for assets. It was believed that these sacred places enjoyed the special protection of the gods and were not likely to be robbed. It is no coincidence that many of the bank buildings built during the early twentieth century resembled ancient temples. In medieval times, monies were kept in vaults protected by armies of the ruling nobility. The word *bank* is derived from the Italian word *banca*, meaning "board." It refers to the counting boards used by merchants. Dishonest moneychangers in the marketplace had their boards smashed to prevent them from continuing in business. The word *bankrupt* is literally "a broken (ruptured) board."

The annual rate of interest as a decimal number is 2.4. To change this number to a percent, multiply the number by 100 and add a percent sign. Thus, the annual rate of interest paid is 240%. •

In Examples 1, 2, and 3, we illustrated a simple interest loan, for which the interest and principal are paid on the due date of the note. There is another type of loan, the ***discount note***, for which the interest is paid at the time the borrower receives the loan. The interest charged in advance is called the ***bank discount***. A Federal Reserve Treasury bill is a bank discount note issued by the U.S. government. Example 4 illustrates a discount note.

EXAMPLE 4 *True Interest Rate of a Discount Note*

Siegrid Cook took out a $500 loan using a 10% discount note for a period of 3 months. Determine

a) the interest she must pay to the bank on the date she receives the loan.

b) the net amount of money she receives from the bank.

c) the actual rate of interest for the loan.

SOLUTION

a) To determine the interest, use the simple interest formula.

$$\begin{aligned} i &= prt \\ &= \$500 \times 0.10 \times \frac{3}{12} \\ &= \$12.50 \end{aligned}$$

b) Since Siegrid must pay $12.50 interest when she first receives the loan, the net amount she receives is $500 − $12.50, or $487.50.

c) We calculate the actual rate of interest charged using the simple interest formula. In the formula for the principal, we use the amount Siegrid actually received from the bank, $487.50. For the interest, we use the interest calculated in part (a).

$$\begin{aligned} i &= prt \\ \$12.50 &= \$487.50 \times r \times \frac{3}{12} \\ 12.50 &= 121.875 \times r \\ \frac{12.50}{121.875} &= r \\ 0.1026 &\approx r \end{aligned}$$

Thus, the actual rate of interest is about 10.3% rather than the quoted 10%. •

EXAMPLE 5 *Partial Payments*

Ken Hurley wishes to purchase a new racing bicycle but does not have the $2000 purchase price. Luckily, the bike shop has two payment options. With option 1, Ken can pay $1000 as a down payment and then pay $1150 in 6 months. With option 2, Ken can pay $500 as a down payment and then pay $1700 in 6 months. Which payment option has a higher annual simple interest rate?

SOLUTION Option 1: To determine the principal of the loan, subtract the down payment from the purchase price of the bicycle. Therefore, $p = \$2000 - \$1000 = \$1000$.

DID YOU KNOW?

Tallying Accounts

Until 1826, British banks used tally sticks to record the amounts of loans and deposits. The tallies were flat pieces of wood about 8 in. long. The amount was recorded as notches, different values represented by different widths. A hand's width signified £1000, a thumb's width £100. The stick was split in two; the borrower got "the short end of the stick" and the money lender held the larger stock ("the stockholder"). Transactions could be verified by putting the sticks together to show the accounts "tallied."

In 1834, as the tallies were being burned in the furnaces of the House of Lords, the fire got out of control and burned the old House of Parliament.

To determine the interest charged, subtract the purchase price of the bicycle from the total amount paid. Therefore, $i = (\$1000 + \$1150) - \$2000 = \$150$. Then

$$
\begin{aligned}
i &= p \times r \times t \\
150 &= 1000 \times r \times \frac{1}{2} \\
150 &= 500r \\
\frac{150}{500} &= r \\
0.3 &= r
\end{aligned}
$$

Option 1 has a 30% annual simple interest rate.

Option 2: The principal is $p = \$2000 - \$500 = \$1500$, and interest is $i = (\$500 + \$1700) - \$2000 = \$200$. Then

$$
\begin{aligned}
i &= p \times r \times t \\
200 &= 1500 \times r \times \frac{1}{2} \\
200 &= 750r \\
\frac{200}{750} &= r \\
0.2667 &\approx r
\end{aligned}
$$

Option 2 has about a 26.7% annual simple interest rate. Therefore, option 1 charges a higher annual simple interest rate than option 2.

The United States Rule

A loan has a date of maturity, at which time the principal and interest are due. It is possible to make payments on a loan before the date of maturity. A payment that is less than the full amount owed and made prior to the due date is known as a ***partial payment***. A Supreme Court decision specified the method by which these payments are credited. The procedure is called the ***United States rule***.

The United States rule states that if a partial payment is made on the loan, interest is computed on the principal from the first day of the loan until the date of the partial payment. The partial payment is used to pay the interest first; then the rest of the payment is used to reduce the principal. The next time a partial payment is made, interest is calculated on the unpaid principal from the date of the previous date of payment. Again, the payment goes first to pay the interest, with the rest of the payment used to reduce the principal. An individual can make as many partial payments as he or she wishes; the procedure is repeated for each payment. The balance due on the date of maturity is found by computing interest due since the last partial payment and adding this interest to the unpaid principal.

The ***Banker's rule*** is used to calculate simple interest when applying the United States rule. The Banker's rule considers a year to have 360 days, and any fractional part of a year is the exact number of days of the loan.

To determine the exact number of days in a period, we can use Table 11.1 on page 662. Example 6 illustrates how to use the table.

DID YOU KNOW?

Selling Time

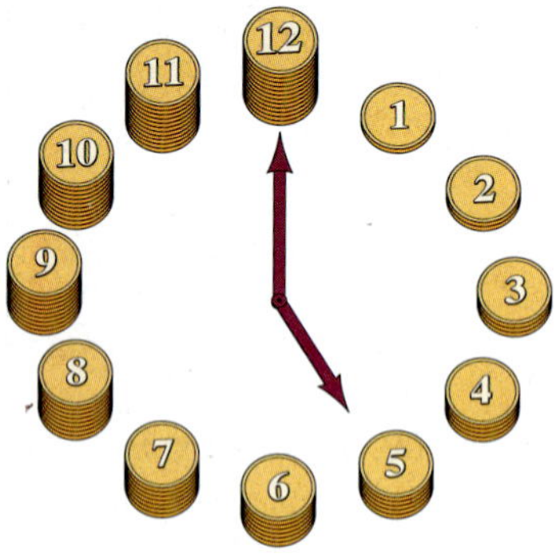

For most of human history, the practice of charging interest on money borrowed, or *usury*, was considered not only immoral, but a crime. This attitude arose in ancient times, in part because a peasant or farmer could literally be enslaved if payments could not be made. Charging interest over time was considered "selling time," something not ours to sell. What was especially objectionable was the idea of "pure interest," a fixed payment set at the time of the loan, even when there was no risk to the lender. Much has changed in the past 200 years. Borrowing money has become a fact of life for individuals, businesses, and governments alike. In 2005, the U.S. government paid about $352.4 billion in interest alone on a debt of more than $7.9 trillion.

Today the word *usury* means lending money at a rate of interest that is excessive or unlawfully high. A person who engages in such loans is often referred to as a "loan shark."

Table 11.1

	Days in Each Month											
	31	28	31	30	31	30	31	31	30	31	30	31
Day of Month	Jan.	Feb.	Mar.	Apr.	May	June	July	Aug.	Sept.	Oct.	Nov.	Dec.
Day 1	1	32	60	91	121	152	182	213	244	274	305	335
Day 2	2	33	61	92	122	153	183	214	245	275	306	336
Day 3	3	34	62	93	123	154	184	215	246	276	307	337
Day 4	4	35	63	94	124	155	185	216	247	277	308	338
Day 5	5	36	64	95	125	156	186	217	248	278	309	339
Day 6	6	37	65	96	126	157	187	218	249	279	310	340
Day 7	7	38	66	97	127	158	188	219	250	280	311	341
Day 8	8	39	67	98	128	159	189	220	251	281	312	342
Day 9	9	40	68	99	129	160	190	221	252	282	313	343
Day 10	10	41	69	100	130	161	191	222	253	283	314	344
Day 11	11	42	70	101	131	162	192	223	254	284	315	345
Day 12	12	43	71	102	132	163	193	224	255	285	316	346
Day 13	13	44	72	103	133	164	194	225	256	286	317	347
Day 14	14	45	73	104	134	165	195	226	257	287	318	348
Day 15	15	46	74	105	135	166	196	227	258	288	319	349
Day 16	16	47	75	106	136	167	197	228	259	289	320	350
Day 17	17	48	76	107	137	168	198	229	260	290	321	351
Day 18	18	49	77	108	138	169	199	230	261	291	322	352
Day 19	19	50	78	109	139	170	200	231	262	292	323	353
Day 20	20	51	79	110	140	171	201	232	263	293	324	354
Day 21	21	52	80	111	141	172	202	233	264	294	325	355
Day 22	22	53	81	112	142	173	203	234	265	295	326	356
Day 23	23	54	82	113	143	174	204	235	266	296	327	357
Day 24	24	55	83	114	144	175	205	236	267	297	328	358
Day 25	25	56	84	115	145	176	206	237	268	298	329	359
Day 26	26	57	85	116	146	177	207	238	269	299	330	360
Day 27	27	58	86	117	147	178	208	239	270	300	331	361
Day 28	28	59	87	118	148	179	209	240	271	301	332	362
Day 29	29		88	119	149	180	210	241	272	302	333	363
Day 30	30		89	120	150	181	211	242	273	303	334	364
Day 31	31		90		151		212	243		304		365

Add 1 day for leap year if February 29 falls between the two dates under consideration.

EXAMPLE 6 *Determining the Due Date of a Note*

Use Table 11.1 to determine (a) the due date of a loan made on March 15 for 120 days and (b) the number of days from April 18 to July 31.

SOLUTION

a) To determine the due date of the loan, do the following. In Table 11.1, find Day 15 in the left column (with heading Day of Month) and then move three columns to the right (heading at the top of the column is March) to find the number 74 (circled in red). Thus, March 15 is the 74th day of the year. Add 120 to 74 since the loan will be due 120 days after March 15:

$$74 + 120 = 194$$

Thus, the due date of the note is the 194th day of the year. Find 194 (circled in blue) in Table 11.1 in the column headed July. The number in the same row as 194 in the left column is Day 13. Thus, the due date of the note is July 13.

b) To determine the number of days from April 18 to July 31, use Table 11.1 to find that April 18 is the 108th day of the year and that July 31 is the 212th day of the year. Then find the difference: $212 - 108 = 104$. Thus, the number of days from April 18 to July 31 is 104 days.

EXAMPLE 7 *Using the Banker's Rule*

Determine the simple interest that will be paid on a \$300 loan at an interest rate of 5% for the period March 3 to May 3 using the Banker's rule.

SOLUTION The exact number of days from March 3 to May 3 is 61. The period of time in years is 61/360. Substituting in the simple interest formula gives

$$\begin{aligned} i &= prt \\ &= \$300 \times 0.05 \times \frac{61}{360} \\ &\approx \$2.54 \end{aligned}$$

The interest is \$2.54.

A borrower may wish to pay back part of the loan prior to the due date. The next example illustrates how a partial payment is credited under the United States rule. Making partial payments reduces the amount of interest paid and thus the cost of the loan.

EXAMPLE 8 *Using the United States Rule*

Cathy Panik is a mathematics teacher, and she plans to attend a national conference. To pay for her airfare, on November 1, 2006, Cathy takes out a 120-day loan for \$400 at an interest rate of 12.5%. Cathy uses some birthday gift money to make a partial payment of \$150 on January 5, 2007. She makes a second partial payment of \$100 on February 2, 2007.

a) Determine the due date of the loan.

b) Determine the interest and the amount credited to the principal on January 5.

c) Determine the interest and the amount credited to the principal on February 2.

d) Determine the amount that Cathy must pay on the due date.

SOLUTION

a) Using Table 11.1, we see that November 1 is the 305th day of the year. Next, we note that the sum of 305 and 120 is 425. Since this due date will extend into the next year, we subtract 365 from 425 to get 60. From Table 11.1, we see that the 60th day of the year is March 1. Therefore, the loan due date is March 1, 2007. Had 2007 been a leap year, the due date would have been February 29, 2007.

b) Using Table 11.1, January 5 is the 5th day of the year and November 1 is the 305th day of the year. The number of days from November 1 to January 5 can be computed as follows: $(365 - 305) + 5 = 65$. Then, using $i = prt$ and the Banker's rule, we get

$$\begin{aligned} i &= \$400 \times 0.125 \times \frac{65}{360} \\ &\approx \$9.03 \end{aligned}$$

The interest of \$9.03 that is due January 5, 2007, is deducted from the payment of \$150. The remaining payment of \$150 − \$9.03, or \$140.97, is then credited to the principal. Therefore, the adjusted principal is now \$400 − \$140.97, or \$259.03.

Note that if that payment had been the only partial payment made, then the balance due on March 1, 2007, would be calculated by determining the interest on the balance, \$259.03, for the remainder of the loan, 55 days, and adding this interest to the principal of \$259.03. If this payment had been the only partial payment made, then the balance due on March 1 would be

$$\begin{aligned} \text{Balance due} &= \text{principal} + \text{interest} \\ &\approx 259.03 + \left(259.03 \times 0.125 \times \frac{55}{360}\right) \\ &\approx 259.03 + 4.95 \\ &\approx 263.98 \end{aligned}$$

c) Since there was a second partial payment made, we use the Banker's rule to calculate the interest on the unpaid principal for the period from January 5 to February 2. According to Table 11.1, the number of days from January 5 to February 2 is $33 - 5$, or 28 days.

$$\begin{aligned} i &= \$259.03 \times 0.125 \times \frac{28}{360} \\ &\approx \$2.52 \end{aligned}$$

The interest of \$2.52 that is due February 2, 2007, is deducted from the payment of \$100. The remaining payment of \$100 − \$2.52, or \$97.48, is then credited to the principal. Therefore, the new adjusted principal is now \$259.03 − \$97.48, or \$161.55.

d) The due date of the loan is March 1. Using Table 11.1, we see that there are $60 - 33$, or 27, days from February 2 to March 1. The interest is computed on the remaining balance of \$161.51 by using the simple interest formula.

$$\begin{aligned} i &= \$161.55 \times 0.125 \times \frac{27}{360} \\ &\approx \$1.51 \end{aligned}$$

Therefore, the balance due on the maturity date of the loan is the sum of the principal and the interest, \$161.55 + \$1.51, or \$163.06. *Note*: The sum of the days in the three calculations, $65 + 28 + 27$, equals the total number of days in the loan, 120. ●

SECTION 11.2 EXERCISES

CONCEPT/WRITING EXERCISES

1. What is interest?
2. What is credit?
3. What is security (or collateral)?
4. What is a cosigner?
5. Explain what each letter in the simple interest formula, $i = prt$, represents.
6. What is a personal note?
7. Explain the United States rule.
8. Explain the difference between ordinary interest and interest under the Banker's rule.

In Exercises 9–18, determine the simple interest. (The rate is an annual rate unless otherwise noted. Assume 360 days in a year.)

PRACTICE THE SKILLS

9. $p = \$250$, $r = 3\%$, $t = 2$ years
10. $p = \$375$, $r = 4.5\%$, $t = 5$ years
11. $p = \$1100$, $r = 8.75\%$, $t = 90$ days
12. $p = \$2500$, $r = 3\frac{7}{8}\%$, $t = 9$ months
13. $p = \$587$, $r = 0.045\%$ per day, $t = 2$ months
14. $p = \$6742.75$, $r = 6.05\%$, $t = 90$ days
15. $p = \$2756.78$, $r = 10.15\%$, $t = 103$ days
16. $p = \$550.31$, $r = 8.9\%$, $t = 67$ days
17. $p = \$1372.11$, $r = 1\frac{3}{8}\%$ per month, $t = 6$ months
18. $p = \$41{,}864$, $r = 0.0375\%$ per day, $t = 60$ days

In Exercises 19–24, use the simple interest formula to determine the missing value.

19. $p = \$2000$, $r = ?$, $t = 4$ years, $i = \$400$
20. $p = ?$, $r = 6\%$, $t = 30$ days, $i = \$26.90$
21. $p = ?$, $r = 8\%$, $t = 3$ months, $i = \$12.00$
22. $p = \$800.00$, $r = 6\%$, $t = ?$, $i = \$64.00$
23. $p = \$957.62$, $r = 6.5\%$, $t = ?$, $i = \$124.49$
24. $p = \$1650.00$, $r = ?$, $t = 6.5$ years, $i = \$343.20$

PROBLEM SOLVING

25. ***Business Loan*** The city of Bradenton is offering simple interest loans to start-up businesses at a rate of 3.5%. Gaetano Cannata obtains one of these loans of \$6000 for 3 years to help pay for start-up costs of his restaurant, Ortygia. Determine the amount of money Gaetano must repay the city after 3 years.

26. ***Credit Union Loan*** Steve and Laurie Carah borrowed \$4500 from their credit union to remodel their kitchen. The simple interest rate was 4.75%, and the length of the loan was 3 years.

    a) How much did the Carahs pay for the use of the money?

    b) Determine the amount the Carahs paid to the credit union on the date of maturity of the loan.

27. ***Bank Personal Note*** Kelly Droessler borrowed \$3500 from a bank for 6 months. Her friend Ms. Harris was cosigner of Kelly's personal note. The bank collected $7\frac{1}{2}\%$ simple interest on the date of maturity.

    a) How much did Kelly pay for the use of the money?

    b) Determine the amount she repaid to the bank on the due date of the note.

28. ***Bank Discount Note*** Kwame Adebele borrowed \$2500 for 5 months from a bank, using U.S. government bonds as security. The bank discounted the loan at 8%.

    a) How much interest did Kwame pay the bank for the use of its money?

b) How much did he receive from the bank?

c) What was the actual rate of interest he paid?

29. ***Bank Discount Note*** Julie Jansen borrowed \$3650 from her bank for 8 months. The bank discounted the loan at 7.5%.

a) How much interest did Julie pay the bank for the use of its money?

b) How much did she receive from the bank?

c) What was the actual rate of interest she paid?

30. ***Credit Union Loan*** Enrico Montoyo wants to borrow \$350 for 6 months from his credit union, using his savings account as security. The credit union's policy is that the maximum amount a person can borrow is 80% of the amount in the person's savings account. The interest rate is 2 percentage points higher than the interest rate being paid on the savings account. The current rate on the savings account is $3\frac{1}{4}\%$.

a) How much money must Enrico have in his account to borrow \$350?

b) What is the rate of interest the credit union will charge for the loan?

c) Find the amount Enrico must repay in 6 months.

31. ***Investing Tuition Payments*** Sand Ridge School is requiring parents to pay half of the yearly tuition at the time of registration and half on the date classes begin. Registration is held 5 months prior to the beginning of school, and administrators expect 470 students to register. If annual tuition is \$4500 and if the money paid at the time of registration is placed in an account paying 5.4% simple interest, how much interest will Sand Ridge School earn by the time school begins?

32. ***A Pawn Loan*** Jeffrey Kowalski wants to take his mother out for dinner on her birthday, but he doesn't get paid until the following week. To borrow money, Jeffrey pawns his watch. Based on the value of the watch, the pawnbroker loans Jeffrey \$75. Fourteen days later, Jeffrey gets his watch back by paying the pawnbroker \$80.25. What annual simple interest rate did the pawnbroker charge Jeffrey?

In Exercises 33–38, determine the exact time from the first date to the second date. Use Table 11.1. Assume the year is not a leap year unless otherwise indicated.

33. February 14 to November 25

34. January 17 to October 10

35. February 2 to October 31 (the loan is due in a leap year)

36. June 14 to January 24

37. August 24 to May 15

38. December 21 to April 28

In Exercises 39–42, determine the due date of the loan, using the exact time, if the loan is made on the given date for the given number of days.

39. March 15 for 90 days

40. May 11 for 120 days

41. November 25 for 120 days (the loan is due in a leap year)

42. July 5 for 210 days

In Exercises 43–52, a partial payment is made on the date(s) indicated. Use the United States rule to determine the balance due on the note at the date of maturity. (The Effective Date is the date the note was written.) Assume the year is not a leap year.

	Principal	Rate	Effective Date	Partial Payment(s) Amount	Partial Payment(s) Date(s)	Maturity Date
43.	\$3000	4%	May 1	\$1000	June 1	July 1
44.	\$6000	7%	Jan. 15	\$2500	Mar. 1	Apr. 15
45.	\$2400	5.5%	Feb. 1	\$1000	May 1	Aug. 31
46.	\$8500	11.5%	Sep. 1	\$4250	Oct. 31	Dec. 31
47.	\$9000	6%	July 15	\$4000	Dec. 27	Feb. 1
48.	\$1000	12.5%	Jan. 1	\$300	Jan. 15	Feb. 15
49.	\$1800	15%	Aug. 1	\$500	Sept. 1	Nov. 1
				\$500	Oct. 1	
50.	\$5000	14%	Oct. 15	\$800	Nov. 15	Jan. 1
				\$800	Dec. 15	
51.	\$11,600	6%	Mar. 1	\$2000	Aug. 1	Dec. 1
				\$4000	Nov. 15	
52.	\$21,000	$4\frac{3}{8}\%$	July 12	\$8000	Oct. 10	Jan. 30
				\$6000	Dec. 8	

53. ***Company Loan*** On March 1, the Zwick Balloon Company signed a \$6500 note with simple interest of $10\frac{1}{2}\%$ for 180 days. The company made payments of \$1750 on May 1 and \$2350 on July 1. How much will the company owe on the date of maturity?

54. ***Restaurant Loan*** The Sweet Tooth Restaurant borrowed \$3000 on a note dated May 15 with simple interest of 11%. The maturity date of the loan is September 1. The restaurant made partial payments of \$875 on June 15 and \$940 on August 1. Find the amount due on the maturity date of the loan.

55. ***U.S. Treasury Bills*** The U.S. government borrows money by selling Treasury bills. Treasury bills are discounted notes issued by the U.S. government. On May 5, 2007, Kris Greenhalgh purchased a 182-day, \$1000 U.S. Treasury bill at a 4.34% discount. On the date of maturity, Kris will receive \$1000.

a) What is the date of maturity of the Treasury bill?

b) How much did Kris actually pay for the Treasury bill?

c) How much interest did the U.S. goverment pay Kris on the date of maturity?

d) What is the actual rate of interest of the Treasury bill? (Round the answer to the nearest hundredth of a percent.)

▲ United States Treasury building, Washington, D.C.

56. ***U.S. Treasury Bills*** On August 31, 2007, Trinity Lopez purchased a 364-day, \$6000 U.S. Treasury bill at a 4.4% discount. (See Exercise 55.)

a) What is the date of maturity of the Treasury bill (2008 is a leap year)?

b) How much did Trinity actually pay for the Treasury bill?

c) How much interest did the U.S. government pay Trinity on the date of maturity?

d) What is the actual rate of interest of the Treasury bill?

57. ***Tax Preparation Loan*** Many tax preparation organizations will prepay customers' tax refunds if they pay a one-time finance charge. In essence, the customer is borrowing the money (the refund minus the finance charge) from the tax preparer, prepaying the interest (as in a discount note), and then repaying the loan with the tax refund. This procedure allows customers access to their tax refund money without having to wait. Joy Stallard had a tax refund of \$743.21 due. She was able to get her tax refund immediately by paying a finance charge of \$39.95. What annual simple interest rate is Joy paying for this loan assuming

a) the tax refund check would be available in 5 days?

b) the tax refund check would be available in 10 days?

c) the tax refund check would be available in 20 days?

58. ***Prime Interest Rate*** Nick St. Louis borrowed \$600 for 3 months. The banker said that Nick must repay the loan at the rate of \$200 per month plus interest. The bank was charging a rate of 2% above the prime interest rate. The *prime interest rate* is the rate charged to preferred customers of the bank. During the first month the prime interest rate was 4.75%, during the second month it was 5%, and during the third month it was 5.25%.

a) Find the amount Nick paid the bank at the end of the first month, at the end of the second month, and at the end of the third month.

b) What was the total amount of interest Nick paid the bank?

CHALLENGE PROBLEM/GROUP ACTIVITY

59. ***U.S. Treasury Bills*** Mark Beiley purchased a 52-week U.S. Treasury bill (see Exercises 55 and 56) for \$93,337. The par value (the value of the bill upon maturity) was \$100,000.

a) What was the discount rate?

b) What was the actual interest rate Mark received?

c) Since U.S. Treasury bills are sold through auctions at Federal Reserve Banks, Mark did not know the purchase price of his Treasury bill until after he was notified by mail. If Mark had sent the Federal Reserve

Bank a check for \$100,000 to purchase the Treasury bill, how much would the Federal Reserve Bank have to rebate him upon notice of his purchase?

d) On the day Mark received the rebate that was discussed in part (c), he invested it in a 1-year certificate of deposit yielding 5% interest. What is the total amount of interest he will receive from both investments?

RECREATIONAL MATHEMATICS

60. ***Columbus Investment*** On August 3, 1492, Christopher Columbus set sail on a voyage that would eventually lead him to the Americas. If on this day Columbus had invested \$1 in a 5% simple interest account, determine the amount of interest the account would have earned by the following dates. Use a scientific calculator and disregard leap years in your calculations.

a) December 11, 1620 (Pilgrims land on Plymouth Rock)

b) July 4, 1776 (Declaration of Independence)

c) December 7, 1941 (U.S. enters World War II)

d) Today's date

INTERNET/RESEARCH ACTIVITY

61. ***Loan Sources*** Consider the following places where a loan may be obtained: banks, savings and loans, credit unions, and pawnshops. Write a report that includes the following information:

a) Describe the ownership of each.

b) Historically, what need is each fulfilling?

c) What are the advantages and disadvantages of obtaining a loan from each of those places listed?

11.3 COMPOUND INTEREST

▲ Albert Einstein

Albert Einstein said of compound interest, "The most powerful force in the universe is compound interest." What is this marvelous concept that warrants such a strong statement from one of the most intelligent human beings of all time? Another very intelligent person, Benjamin Franklin, described compound interest quite well when he said, "Money makes money and the money that money makes makes more money." In this section, we will introduce compound interest and show how it can be used to help you with investing for some of your long-term goals.

Compound Interest

An ***investment*** is the use of money or capital for income or profit. We can divide investments into two classes: fixed investments and variable investments. In a ***fixed investment***, the amount invested as principal is guaranteed and the interest is computed at a fixed rate. *Guaranteed* means that the exact amount invested will be paid back together with any accumulated interest. Examples of a fixed investment are savings accounts, money market deposit accounts, and certificates of deposit. Another fixed investment is a government savings bond. In a ***variable investment***, neither the principal nor the interest is guaranteed. Examples of variable investments are stocks, mutual funds, and commercial bonds.

Simple interest, introduced earlier in the chapter, is calculated once for the period of a loan or investment using the formula $i = prt$. The interest paid on savings accounts at most banks is compound interest. A bank computes the interest periodically (for example, daily or quarterly) and adds this interest to the original principal. The interest for the following period is computed by using the new principal (original principal plus interest). In effect, the bank is computing interest on interest, which is called compound interest.

Interest that is computed on the principal and any accumulated interest is called **compound interest**.

EXAMPLE 1 *Computing Compound Interest*

Marjorie Thrall recently won the \$1000 first prize in a raffle contest. Marjorie deposits the \$1000 in a 1-year certificate of deposit paying 2.0% compounded quarterly. Find the amount, A, to which the \$1000 will grow in 1 year.

SOLUTION Compute the interest for the first quarter using the simple interest formula. Add this interest to the principal to find the amount at the end of the first quarter. In our calculation, time is $\frac{1}{4}$ of a year, or $t = 0.25$.

$$\begin{aligned} i &= prt \\ &= \$1000 \times 0.02 \times 0.25 = \$5.00 \\ A &= \$1000 + \$5 = \$1005 \end{aligned}$$

Now repeat the process for the second quarter, this time using a principal of \$1005.

$$\begin{aligned} i &= \$1005 \times 0.02 \times 0.25 \approx \$5.03 \\ A &= \$1005 + \$5.03 = \$1010.03 \end{aligned}$$

For the third quarter, use a principal of \$1010.03.

$$\begin{aligned} i &= \$1010.03 \times 0.02 \times 0.25 \approx \$5.05 \\ A &= \$1010.03 + \$5.05 = \$1015.08 \end{aligned}$$

For the fourth quarter, use a principal of \$1015.08.

$$\begin{aligned} i &= \$1015.08 \times 0.02 \times 0.25 \approx \$5.08 \\ A &= \$1015.08 + \$5.08 = \$1020.16 \end{aligned}$$

Hence, the \$1000 grows to a final value of \$1020.16 over the 1-year period.

This example shows the effect of earning interest on interest, or compounding interest. In 1 year, the amount of \$1000 has grown to \$1020.16, compared with \$1020 that would have been obtained with a simple interest rate of 2%. Thus, in 1 year alone the gain was \$0.16 more with compound interest than with simple interest.

A simpler and less time-consuming way to calculate compound interest is to use the compound interest formula and a calculator.

COMPOUND INTEREST FORMULA

$$A = p\left(1 + \frac{r}{n}\right)^{nt}$$

where A is the amount that accumulates in the account, p is the principal, r is the annual interest rate as a decimal, n is the number of compounding periods per year, and t is the time in years.

We will now use this formula to show how an investment can grow using compound interest.

DID YOU KNOW?

A Penny Saved . . .

Benjamin Franklin left 1000 pounds sterling (£) in his will to the city of Boston, where he was born. The money was to be lent to young apprentices who would pay $\frac{1}{10}$ of the principal each year plus 5% interest compounded yearly. The money earned was to be used to benefit the inhabitants of Boston. By 1894, the fund's value had increased to 90,000 £. Part of this legacy became the Benjamin Franklin Institute of Technology, a 2-year technical college founded in 1908.

DID YOU KNOW?

Cost of Manhattan

In 1626, Peter Du Minuit, representing the Dutch West India Company, traded beads and blankets to the native American inhabitants of Manhattan Island for the island. The trade was valued at 60 Dutch guilders, about \$24. If at that time the \$24 had been invested at 6% interest compounded annually, the investment would be worth about \$111,442,737,812 in 2008. As a comparison, had the \$24 been invested for the 382 years in a 6% simple interest account, the investment would be worth only \$574.08. Had the \$24 been invested in a continuously compounded interest account at 6%, the investment would be worth about \$215,894,096,485. These results dramatically demonstrate the power of compounding interest over time.

EXAMPLE 2 *Using the Compound Interest Formula*

Kathy Mowers invested \$5600 in a savings account with an interest rate of 7.5% compounded monthly. If Kathy makes no other deposits into this account, determine the amount in the account after 10 years.

SOLUTION We will use the formula for compound interest, $A = p\left(1 + \frac{r}{n}\right)^{nt}$. The principal, p, is the amount of money invested, so $p = 5600$. The interest rate, r, is 7.5%, so $r = 0.075$. Because the interest is compounded monthly, there are 12 periods per year and $n = 12$. Because the money is invested for 10 years, $t = 10$.

$$\begin{aligned} A &= 5600\left(1 + \frac{0.075}{12}\right)^{12\cdot 10} \\ &= 5600(1 + 0.00625)^{120} \\ &= 5600(1.00625)^{120} \\ &\approx 5600(2.1120646) \\ &\approx 11{,}827.56 \end{aligned}$$

Thus, the amount in the account after 10 years would be \$11,827.56.

TECHNOLOGY TIP

Scientific Calculator

There are various ways to determine compound interest using a scientific calculator. In Example 2, we had to evaluate

$$5600\left(1 + \frac{0.075}{12}\right)^{12\cdot 10}$$

To evaluate this expression on a scientific calculator, enter the following keys.

5600 [×] [(] 1 [+] 0.075 [÷] 12 [)] [y^x] [(] 12 [×] 10 [)] [=]

After the [=] key is pressed, the answer 11827.56197 is displayed. This answer checks with the answer obtained in Example 2.

Microsoft Excel

The software program Excel can also be used to determine compound interest. Excel is a spreadsheet, so once you insert a formula, you can change the value of the variables and the new answer will be automatically displayed. Before you read the following procedure, read Appendix A on Excel at the back of the book.

To use Excel to determine the compound interest, set up the following columns.

	A	B	C	D	E
1	p	r	n	t	A
2	5600	0.075	12	10	
3					

Sheet1 / Sheet2 / Sheet3

Ready

Then, to set up the formula in cell E2, go to the formula bar on top and enter

= A2*(1+B2/C2)^(C2*D2)

Then press the Enter key. The answer, 11827.56, will appear in cell E2.

Suppose you now wanted to determine the compound interest for $p = 6000$, $r = 0.05$, $n = 24$, and $t = 6$. If you were to change the values in cells A2 through D2 to these values respectively, the new answer, 8096.626, will now be displayed in cell E2.

EXAMPLE 3 *Calculating Compound Interest*

Calculate the interest on \$650 at 8% compounded semiannually for 3 years, using the compound interest formula.

SOLUTION Since interest is compounded semiannually, there are two periods per year. Thus, $n = 2$, $r = 0.08$, and $t = 3$. Substituting into the formula, we find the amount, A.

$$\begin{aligned} A &= p\left(1 + \frac{r}{n}\right)^{nt} \\ &= 650\left(1 + \frac{0.08}{2}\right)^{(2)(3)} \\ &= 650(1.04)^6 \\ &\approx 650(1.2653190) \\ &\approx \$822.46 \end{aligned}$$

Since the total amount is \$822.46 and the original principal is \$650, the interest must be \$822.46 − \$650, or \$172.46. ●

In Example 3, the interest rate is stated as an annual rate of 8%, but the number of compounding periods per year is two. In applying the compound interest formula, the rate for one period is $r \div n$, which was $8\% \div 2$, or 4% per period.

Now we will calculate the amount, and interest, on \$1 invested at 8% compounded semiannually for 1 year. The result is

$$A = 1\left(1 + \frac{0.08}{2}\right)^{(2)(1)} = 1.0816$$

$$\begin{aligned} \text{Interest} &= \text{amount} - \text{principal} \\ i &= 1.0816 - 1 \\ &= 0.0816 \end{aligned}$$

The interest for 1 year is 0.0816. This amount written as a percent, 8.16%, is called the ***effective annual yield***. *Most financial institutions refer to the effective annual yield as the* ***annual percentage yield (APY)***. If \$1 was invested at a simple interest rate of 8.16% and \$1 was invested at 8% interest compounded semiannually (equivalent to an effective yield of 8.16%), the interest from both investments would be the same.

The **effective annual yield** or **annual percentage yield (APY)** is the simple interest rate that gives the same amount of interest as a compound rate over the same period of time.

Suncoast Schools Credit Union

CD Rates

Type	Rate	APY*
12 mo	4.50%	4.60%
24 mo	4.88%	5.00%
36 mo	4.97%	5.10%
48 mo	5.12%	5.25%

* Annual Percentage Yield

Many banks compound interest daily. When computing the effective annual yield, they use 360 for the number of periods in a year. To determine the effective annual yield for any interest rate, calculate the amount using the compound interest formula where p is \$1. Then subtract \$1 from that amount. The difference, written as a percent, is the effective annual yield, as illustrated in Example 4.

The sign in the margin shows interest rates and the corresponding annual percentage yields (APY) that were available for certificates of deposit (CDs) from Suncoast Schools Credit Union on April 7, 2006. To determine the effective annual yield (or the APY) for the 48-month CD, calculate the amount of interest earned on \$1 for 1 year.

$$A = 1\left(1 + \frac{0.0512}{360}\right)^{360} \approx 1.0525295 \approx 1.0525$$

From this result, we subtract 1 to get $1.0525 - 1 = 0.0525$, or 5.25%. Confirm that the other annual percentage yields shown on the sign are correct.

EXAMPLE 4 *Determining Annual Percentage Yield*

Determine the annual percentage yield or the effective annual yield for \$1 invested for 1 year at

a) 8% compounded daily.

b) 6% compounded quarterly.

SOLUTION

a) With daily compounding, $n = 360$.

$$\begin{aligned} A &= p\left(1 + \frac{r}{n}\right)^{nt} \\ &= 1\left(1 + \frac{0.08}{360}\right)^{(360)(1)} \\ &\approx 1.0832774 \approx 1.0833 \\ i &= A - 1 \\ &\approx 1.0833 - 1 \\ &\approx 0.0833 \end{aligned}$$

Thus, when the interest is 8% compounded daily, the annual percentage yield or the effective annual yield is about 8.33%.

b) With quarterly compounding, $n = 4$.

$$A = p\left(1 + \frac{r}{n}\right)^{nt}$$

$$= 1\left(1 + \frac{0.06}{4}\right)^{(4)(1)}$$

$$\approx 1.0613636 \approx 1.0614$$

$$i = A - 1$$

$$\approx 1.0614 - 1$$

$$\approx 0.0614$$

Thus, when the interest is 6% compounded quarterly, the annual percentage yield or the effective annual yield is about 6.14%.

There are numerous types of savings accounts. Many savings institutions compound interest daily. Some pay interest from the day of deposit to the day of withdrawal, and others pay interest from the first of the month on all deposits made before the tenth of the month. In each of these accounts in which interest is compounded daily, interest is entered into the depositor's account only once each quarter. Some savings institutions will not pay any interest on a day-to-day account if the balance falls below a set amount.

Present Value

You may wonder about what amount of money you must deposit in an account today to have a certain amount of money in the future. For example, how much must you deposit in an account today at a given rate of interest so that it will accumulate to \$25,000 to pay your child's college costs in 4 years? The principal, p, that would have to be invested now is called the ***present value***. Following is a formula for determining the present value.

PRESENT VALUE FORMULA

$$p = \frac{A}{\left(1 + \frac{r}{n}\right)^{nt}}$$

where p is the present value, or the principal to invest now, A is the amount to be accumulated in the account, r is the annual interest rate as a decimal, n is the number of compounding periods per year, and t is the time in years.

EXAMPLE 5 *Savings for College*

Will Hunting would like his daughter to attend college in 6 years when she finishes high school. Will would like to invest enough money in a certificate of deposit (CD) now to pay for his daughter's college expenses. If Will estimates that he will need \$30,000 in 6 years, how much should he invest now in a CD that has a rate of 4.72% compounded quarterly?

DID YOU KNOW?

Federal Reserve Banks

Federal Reserve Banks are institutions owned by commercial banks in the district they serve. The many functions performed by these banks include the following: clearing checks, issuing new currency, retiring damaged currency, making discount loans to banks within their districts, examining member state banks, collecting data on local business conditions, and conducting research to help establish monetary policy, including the setting of the discount interest rate used when banks borrow money from each other. Currently, there are Federal Reserve Banks in Boston, New York, Philadelphia, Richmond, Cleveland, Atlanta, Chicago, St. Louis, Minneapolis, Kansas City, Dallas, and San Francisco. Take a look at any dollar bill and you will see a circular symbol with the location of the Federal Reserve Bank that printed the bill.

▲ Bunker Hill Community College

SOLUTION To answer this question, we will use the present value formula with $A = \$30{,}000$, $r = 0.0472$, $n = 4$, and $t = 6$.

$$\begin{aligned} p &= \frac{A}{\left(1 + \frac{r}{n}\right)^{nt}} \\ &= \frac{30{,}000}{\left(1 + \frac{0.0472}{4}\right)^{4 \cdot 6}} \\ &= \frac{30{,}000}{(1.0118)^{24}} \\ &\approx \frac{30{,}000}{1.3251718} \\ &\approx 22{,}638.57 \end{aligned}$$

Will Hunting needs to invest approximately \$22,638.57 now to have \$30,000 in 6 years.

TECHNOLOGY TIP

Scientific Calculator

To determine the present value in Example 5, we had to evaluate the expression

$$\frac{30{,}000}{\left(1 + \frac{0.0472}{4}\right)^{4 \cdot 6}}$$

This expression can be evaluated on a scientific calculator as follows.

30000 [÷] [(] 1 [+] 0.0472 [÷] 4 [)] [y^x] [(] 4 [×] 6 [)] [=]

After the [=] key is pressed, the answer 22638.57326 is displayed. This answer agrees with the answer obtained in Example 5.

Excel

Read the Technology Tip on compound interest on page 670. To determine the present value using Excel, enter A, r, n, t, and p in cells A1 through E1, respectively. In cells A2 through D2, enter the values of A, r, n, and t, respectively. (See the table in the Technology Tip on page 670 for an example of the row and column setup.) When cell E2 is highlighted, in the formula box at the top, enter

= A2/(1 + B2/C2)^(C2*D2)

After the Enter key is pressed, the answer, 22638.57, is displayed in cell E2.

SECTION 11.3 EXERCISES

CONCEPT/WRITING EXERCISES

1. What is an investment?
2. What is a fixed investment?
3. What is a variable investment?
4. What is compound interest?
5. **a)** What is effective annual yield?

   b) What is another name for effective annual yield?
6. What is meant by present value in the present value formula?

PRACTICE THE SKILLS

In Exercises 7–14, use the compound interest formula to compute the total amount accumulated and the interest earned.

7. \$1000 for 5 years at 4% compounded annually
8. \$3000 for 2 years at 5% compounded semiannually
9. \$2000 for 4 years at 3% compounded quarterly
10. \$5000 for 3 years at 7% compounded quarterly
11. \$7000 for 3 years at 5.5% compounded monthly
12. \$4000 for 2 years at 6% compounded semiannually
13. \$8000 for 2 years at 4% compounded daily (use $n = 360$)
14. \$8500 for 5 years at 4.5% compounded monthly

In Exercises 15–18, use the present value formula to determine the amount to be invested now, or the present value needed.

15. The desired accumulated amount is \$50,000 after 10 years invested in an account with 5% interest compounded annually.
16. The desired accumulated amount is \$75,000 after 5 years invested in an account with 4% interest compounded semiannually.
17. The desired accumulated amount is \$100,000 after 4 years invested in an account with 4% interest compounded quarterly.
18. The desired accumulated amount is \$25,000 after 20 years invested in an account with 12% interest compounded quarterly.

PROBLEM SOLVING

19. ***Little League*** Braden River Little League receives a \$50,000 donation for building a new snack bar and office building. The league decides to invest this money in a money market account that pays 4% interest compounded quarterly. How much will the league have in this account after 2 years?

20. ***Contest Winnings*** Mary Robinson wins \$2500 in a singing contest and invests the money in a 4-year CD that pays 3% interest compounded monthly. How much money will Mary receive when she redeems the CD at the end of the 4 years?
21. ***Class Trip*** To help pay for a class trip at the end of their senior year, the sophomore class at Cortez High School invests \$1500 from fund-raisers in a 30-month CD paying 3.9% interest compounded monthly. Determine the amount the class will receive when it cashes in the CD after 30 months.
22. ***Investing Prize Winnings*** Marcella Laddon wins third prize in the Clearinghouse Sweepstakes and receives a check for \$250,000. After spending \$10,000 on a vacation, she decides to invest the rest in a money market account that pays 1.5% interest compounded monthly. How much money will be in the account after 10 years?
23. ***Investing Gifts and Scholarships*** Cliff Morris just graduated from high school and has received \$800 in gifts of cash from friends and relatives. In addition, Cliff received three scholarships in the amounts of \$150, \$300, and \$1000. If Cliff takes all his gift and scholarship money and invests it in a 24-month CD paying 2% interest compounded daily, how much will he have when he cashes in the CD at the end of the 24 months?

dollars financed is $\frac{1500}{100} = 15$. To determine the total finance charge, multiply the finance charge per \$100 by the number of hundreds of dollars financed.

$$\text{Total finance charge} = \$7.45 \times 15 = \$111.75$$

Therefore, Kristin will pay a total finance charge of \$111.75.

b) To determine the monthly payments, first calculate the total installment price by adding the finance charge to the purchase price.

$$\text{Total installment price} = \$1500 + \$111.75 = \$1611.75$$

Next divide the total installment price by the number of payments.

$$\text{Monthly payment} = \frac{\$1611.75}{24} \approx \$67.16$$

Kristin will have 24 monthly payments of \$67.16.

Table 11.2 Annual Percentage Rate Table for Monthly Payment Plans

Number of Payments	Annual Percentage Rate (Finance charge per \$100 of amount financed) 4.0%	4.5%	5.0%	5.5%	6.0%	6.5%	7.0%	7.5%	8.0%	8.5%	9.0%	9.5%	10.0%
6	1.17	1.32	1.46	1.61	1.76	1.90	2.05	2.20	2.35	2.49	2.64	2.79	2.93
12	2.18	2.45	2.73	3.00	3.28	3.56	3.83	4.11	4.39	4.66	4.94	5.22	5.50
18	3.20	3.60	4.00	4.41	4.82	5.22	5.63	6.04	6.45	6.86	7.28	7.69	8.10
24	4.22	4.75	5.29	5.83	6.37	6.91	7.45	8.00	8.54	9.09	9.64	10.19	10.75
30	5.25	5.92	6.59	7.26	7.94	8.61	9.30	9.98	10.66	11.35	12.04	12.74	13.43
36	6.29	7.09	7.90	8.71	9.52	10.34	11.16	11.98	12.81	13.64	14.48	15.32	16.16
48	8.38	9.46	10.54	11.63	12.73	13.83	14.94	16.06	17.18	18.31	19.45	20.59	21.74
60	10.50	11.86	13.23	14.61	16.00	17.40	18.81	20.23	21.66	23.10	24.55	26.01	27.48

Example 2 shows how to determine the annual percentage rate on a loan that has a schedule for repaying a fixed amount each period.

EXAMPLE 2 *Determining the APR*

Jan Ford is purchasing a new boat for \$25,000, including taxes. Jan decides to make a \$5000 down payment and finance the balance, \$20,000, through her bank. The loan officer informs her that her monthly payment will be \$410.33 for 60 months.

a) Determine the finance charge. b) Determine the APR.

SOLUTION

a) The total installment price is the down payment plus the total monthly installment payments.

$$\begin{aligned}\text{Total installment price} &= \$5000 + (60 \times \$410.33)\\ &= \$5000 + \$24{,}619.80 = \$29{,}619.80\end{aligned}$$

SECTION 11.3 EXERCISES

CONCEPT/WRITING EXERCISES

1. What is an investment?
2. What is a fixed investment?
3. What is a variable investment?
4. What is compound interest?
5. a) What is effective annual yield?
   b) What is another name for effective annual yield?
6. What is meant by present value in the present value formula?

PRACTICE THE SKILLS

In Exercises 7–14, use the compound interest formula to compute the total amount accumulated and the interest earned.

7. \$1000 for 5 years at 4% compounded annually
8. \$3000 for 2 years at 5% compounded semiannually
9. \$2000 for 4 years at 3% compounded quarterly
10. \$5000 for 3 years at 7% compounded quarterly
11. \$7000 for 3 years at 5.5% compounded monthly
12. \$4000 for 2 years at 6% compounded semiannually
13. \$8000 for 2 years at 4% compounded daily (use $n = 360$)
14. \$8500 for 5 years at 4.5% compounded monthly

In Exercises 15–18, use the present value formula to determine the amount to be invested now, or the present value needed.

15. The desired accumulated amount is \$50,000 after 10 years invested in an account with 5% interest compounded annually.
16. The desired accumulated amount is \$75,000 after 5 years invested in an account with 4% interest compounded semiannually.
17. The desired accumulated amount is \$100,000 after 4 years invested in an account with 4% interest compounded quarterly.
18. The desired accumulated amount is \$25,000 after 20 years invested in an account with 12% interest compounded quarterly.

PROBLEM SOLVING

19. ***Little League*** Braden River Little League receives a \$50,000 donation for building a new snack bar and office building. The league decides to invest this money in a money market account that pays 4% interest compounded quarterly. How much will the league have in this account after 2 years?

20. ***Contest Winnings*** Mary Robinson wins \$2500 in a singing contest and invests the money in a 4-year CD that pays 3% interest compounded monthly. How much money will Mary receive when she redeems the CD at the end of the 4 years?
21. ***Class Trip*** To help pay for a class trip at the end of their senior year, the sophomore class at Cortez High School invests \$1500 from fund-raisers in a 30-month CD paying 3.9% interest compounded monthly. Determine the amount the class will receive when it cashes in the CD after 30 months.
22. ***Investing Prize Winnings*** Marcella Laddon wins third prize in the Clearinghouse Sweepstakes and receives a check for \$250,000. After spending \$10,000 on a vacation, she decides to invest the rest in a money market account that pays 1.5% interest compounded monthly. How much money will be in the account after 10 years?
23. ***Investing Gifts and Scholarships*** Cliff Morris just graduated from high school and has received \$800 in gifts of cash from friends and relatives. In addition, Cliff received three scholarships in the amounts of \$150, \$300, and \$1000. If Cliff takes all his gift and scholarship money and invests it in a 24-month CD paying 2% interest compounded daily, how much will he have when he cashes in the CD at the end of the 24 months?

b) On June 1, instead of making the minimum payment, Mary Beth makes a payment of $200. Assuming there are no additional charges or cash advances, determine Mary Beth's minimum payment due on July 1.

SOLUTION

a) Since there is no new interest due, Mary Beth's minimum monthly payment is determined by taking 1.5% of the outstanding principal, or $0.015 \times \$1500 = \$22.50$. Rounding up to the nearest whole dollar, we determine that Mary Beth's minimum monthly payment due on June 1 is $23.

b) Mary Beth's minimum monthly payment will be the sum of the new interest charges plus 1.5% of her outstanding principal. To calculate her new interest charges, we will use the simple interest formula, $i = prt$. The principal p is $\$1500 - \$200 = \$1300$, the rate r is 0.04792% per day or 0.0004792 per day, and the time t is 30 days because June has 30 days.

$$\begin{aligned} i &= prt \\ &= (\$1300)(0.0004792)(30) \\ &\approx \$18.69 \end{aligned}$$

To determine 1.5% of the outstanding principal, multiply 0.015 by $1300 to get $19.50. Now we add the new interest to 1.5% of the outstanding principal to get $\$19.50 + \$18.69 = \$38.19$. Rounding up to the nearest whole dollar, we determine that Mary Beth's minimum monthly payment due on July 1 is $39.

In Example 5, Mary Beth made no additional charges during the month. When additional charges are made during the month, the finance charges on open-end installment loans or credit cards are generally calculated in one of two ways: the *unpaid balance method* or the *average daily balance method.* Example 6 illustrates the unpaid balance method, and Example 7 illustrates the average daily balance method.

With the ***unpaid balance method***, the borrower is charged interest or a finance charge on the unpaid balance from the previous charge period.

EXAMPLE 6 *Finance Charges Using the Unpaid Balance Method*

In October, Ed Laughbaum charged all the supplies for his Halloween party to his Visa card. On November 5, the billing date, Ed had a balance due of $275. From November 5 through December 4, he did some shopping and charged items totaling $320, and he also made a payment of $145.

a) Determine the finance charge due on December 5, Ed's next billing date, using the unpaid balance method. Assume that the interest rate charged is 1.3% per month.

b) Determine the new account balance on December 5.

SOLUTION

a) The finance charge is based on the $275 balance due on November 5. To find the finance charge due on December 5, we used the simple interest formula with a time of 1 month.

$$i = \$275 \times 0.013 \times 1 \approx \$3.58$$

The finance charge due on December 5 is $3.58.

SECTION 11.3 EXERCISES

CONCEPT/WRITING EXERCISES

1. What is an investment?
2. What is a fixed investment?
3. What is a variable investment?
4. What is compound interest?
5. **a)** What is effective annual yield?

   b) What is another name for effective annual yield?
6. What is meant by present value in the present value formula?

PRACTICE THE SKILLS

In Exercises 7–14, use the compound interest formula to compute the total amount accumulated and the interest earned.

7. \$1000 for 5 years at 4% compounded annually
8. \$3000 for 2 years at 5% compounded semiannually
9. \$2000 for 4 years at 3% compounded quarterly
10. \$5000 for 3 years at 7% compounded quarterly
11. \$7000 for 3 years at 5.5% compounded monthly
12. \$4000 for 2 years at 6% compounded semiannually
13. \$8000 for 2 years at 4% compounded daily (use $n = 360$)
14. \$8500 for 5 years at 4.5% compounded monthly

In Exercises 15–18, use the present value formula to determine the amount to be invested now, or the present value needed.

15. The desired accumulated amount is \$50,000 after 10 years invested in an account with 5% interest compounded annually.
16. The desired accumulated amount is \$75,000 after 5 years invested in an account with 4% interest compounded semiannually.
17. The desired accumulated amount is \$100,000 after 4 years invested in an account with 4% interest compounded quarterly.
18. The desired accumulated amount is \$25,000 after 20 years invested in an account with 12% interest compounded quarterly.

PROBLEM SOLVING

19. ***Little League*** Braden River Little League receives a \$50,000 donation for building a new snack bar and office building. The league decides to invest this money in a money market account that pays 4% interest compounded quarterly. How much will the league have in this account after 2 years?

20. ***Contest Winnings*** Mary Robinson wins \$2500 in a singing contest and invests the money in a 4-year CD that pays 3% interest compounded monthly. How much money will Mary receive when she redeems the CD at the end of the 4 years?
21. ***Class Trip*** To help pay for a class trip at the end of their senior year, the sophomore class at Cortez High School invests \$1500 from fund-raisers in a 30-month CD paying 3.9% interest compounded monthly. Determine the amount the class will receive when it cashes in the CD after 30 months.
22. ***Investing Prize Winnings*** Marcella Laddon wins third prize in the Clearinghouse Sweepstakes and receives a check for \$250,000. After spending \$10,000 on a vacation, she decides to invest the rest in a money market account that pays 1.5% interest compounded monthly. How much money will be in the account after 10 years?
23. ***Investing Gifts and Scholarships*** Cliff Morris just graduated from high school and has received \$800 in gifts of cash from friends and relatives. In addition, Cliff received three scholarships in the amounts of \$150, \$300, and \$1000. If Cliff takes all his gift and scholarship money and invests it in a 24-month CD paying 2% interest compounded daily, how much will he have when he cashes in the CD at the end of the 24 months?

dollars financed is $\frac{1500}{100} = 15$. To determine the total finance charge, multiply the finance charge per \$100 by the number of hundreds of dollars financed.

$$\text{Total finance charge} = \$7.45 \times 15 = \$111.75$$

Therefore, Kristin will pay a total finance charge of \$111.75.

b) To determine the monthly payments, first calculate the total installment price by adding the finance charge to the purchase price.

$$\text{Total installment price} = \$1500 + \$111.75 = \$1611.75$$

Next divide the total installment price by the number of payments.

$$\text{Monthly payment} = \frac{\$1611.75}{24} \approx \$67.16$$

Kristin will have 24 monthly payments of \$67.16.

Table 11.2 Annual Percentage Rate Table for Monthly Payment Plans

Number of Payments	Annual Percentage Rate												
	4.0%	**4.5%**	**5.0%**	**5.5%**	**6.0%**	**6.5%**	**7.0%**	**7.5%**	**8.0%**	**8.5%**	**9.0%**	**9.5%**	**10.0%**
	(Finance charge per \$100 of amount financed)												
6	1.17	1.32	1.46	1.61	1.76	1.90	2.05	2.20	2.35	2.49	2.64	2.79	2.93
12	2.18	2.45	2.73	3.00	3.28	3.56	3.83	4.11	4.39	4.66	4.94	5.22	5.50
18	3.20	3.60	4.00	4.41	4.82	5.22	5.63	6.04	6.45	6.86	7.28	7.69	8.10
24	4.22	4.75	5.29	5.83	6.37	6.91	7.45	8.00	8.54	9.09	9.64	10.19	10.75
30	5.25	5.92	6.59	7.26	7.94	8.61	9.30	9.98	10.66	11.35	12.04	12.74	13.43
36	6.29	7.09	7.90	8.71	9.52	10.34	11.16	11.98	12.81	13.64	14.48	15.32	16.16
48	8.38	9.46	10.54	11.63	12.73	13.83	14.94	16.06	17.18	18.31	19.45	20.59	21.74
60	10.50	11.86	13.23	14.61	16.00	17.40	18.81	20.23	21.66	23.10	24.55	26.01	27.48

Example 2 shows how to determine the annual percentage rate on a loan that has a schedule for repaying a fixed amount each period.

EXAMPLE 2 *Determining the APR*

Jan Ford is purchasing a new boat for \$25,000, including taxes. Jan decides to make a \$5000 down payment and finance the balance, \$20,000, through her bank. The loan officer informs her that her monthly payment will be \$410.33 for 60 months.

a) Determine the finance charge. b) Determine the APR.

SOLUTION

a) The total installment price is the down payment plus the total monthly installment payments.

$$\begin{aligned}\text{Total installment price} &= \$5000 + (60 \times \$410.33)\\ &= \$5000 + \$24{,}619.80 = \$29{,}619.80\end{aligned}$$

The finance charge is the total installment price minus the cash price.

$$\text{Finance charge} = \$29{,}619.80 - \$25{,}000 = \$4619.80$$

Therefore, Jan will pay a finance charge of $4619.80.

b) To determine the annual percentage rate, use Table 11.2 on page 680. First divide the finance charge by the amount financed and multiply the quotient by 100. The result is the finance charge per $100 of the amount financed.

$$\frac{\text{Finance charge}}{\text{Amount financed}} \times 100 = \frac{4619.80}{20{,}000} \times 100 = 0.23099 \times 100 \approx 23.10$$

Thus, Jan pays about $23.10 for each $100 being financed. To use Table 11.2, look for 60 in the far left-hand column under the heading Number of Payments. Then move across to the right until you find the value closest to $23.10. In this case, $23.10 is in the table (circled in blue). At the top of this column is the value of 8.5%. Therefore, the annual percentage rate is approximately 8.5%.

Much more complete APR tables similar to Table 11.2 are available at your local lending institution or on the Internet. Also available on the Internet are loan "calculators" that will calculate your payment for you if you provide the loan amount, the length of the loan, and the APR. One such loan calculator can be found at www.bankrate.com.

EXAMPLE 3 *Financing a Restored Car*

Tino Garcia borrowed $9800 to purchase a classic 1966 Ford Mustang. He does not recall the APR of the loan but remembers that there are 48 payments of $237. If he did not make a down payment on the car, determine the APR.

SOLUTION First determine the finance charge by subtracting the cash price from the total amount paid.

$$\begin{aligned}\text{Finance charge} &= (237 \times 48) - 9800\\ &= 11{,}376 - 9800\\ &= \$1576\end{aligned}$$

Next divide the finance charge by the amount of the loan and multiply this quotient by 100.

$$\frac{1576}{9800} \times 100 \approx 16.08$$

Next find 48 payments in the left column of Table 11.2. Move to the right until you find the value that is closest to 16.08. The value closest to 16.08 is 16.06 (circled in green). At the top of the column is the APR of 7.5%. Thus, the APR for Tino's loan is approximately 7.5%.

In Example 3, if Tino made all 48 payments, his finance charge would be $1576. If he decides to repay the loan after making 30 payments, must he pay the total finance charge? The answer is no. By paying off his loan early, Tino is not obligated to pay the entire finance charge. The amount of the reduction of the finance charge from

DID YOU KNOW?

Actuarial Method versus Rule of 78s

When making automobile or other installment loans, consumers should be aware of the "fine print" described in the loan agreement. If a consumer repays a loan early, the lender is required by law to rebate a portion of the original finance charge or interest. Two methods exist for determining this rebate. The actuarial method is the primary method used by financial institutions today. This method provides the consumer with an accurate interest rebate for repaying the loan early. The second method, the rule of 78s, however, is still used by some lenders. The rule of 78s charges the consumer more interest during the first few months of the loan. The result is a lower interest rebate for the consumer. The rule of 78s originated as an easy way for lenders to compute interest rebates. It has been phased out gradually as calculators and computers become more available. Today the rule of 78s is most likely to be applied by used car dealers who specialize in offering loans to borrowers with a poor credit history.

In 1992, the U.S. Congress outlawed the use of the rule of 78s for loans longer than 61 months. In 2001, federal legislation was introduced to outlaw it altogether. As of 2006, 17 states outlaw the use of the rule of 78s for all loans.

paying off a loan early is called the ***unearned interest***. We will learn about the most common way of calculating unearned interest, the *actuarial method*. In the past, another method, called the *rule of 78s*, was also used. The rule of 78s, however, is less favorable to the consumer and is outlawed in much of the country (See the Did You Know? at left and Exercises 39 and 40). Now we give the formula for finding the unearned interest using the actuarial method.

ACTUARIAL METHOD FOR UNEARNED INTEREST

$$u = \frac{n \cdot P \cdot V}{100 + V}$$

where u is the unearned interest, n is the number of remaining monthly payments (excluding the current payment), P is the monthly payment, and V is the value from the APR table that corresponds to the annual percentage rate for the number of remaining payments (excluding the current payment).

Example 4 illustrates the actuarial method for calculating unearned interest when an installment loan is paid off early.

EXAMPLE 4 *Using the Actuarial Method*

In Example 3, we determined the APR of Tino's loan to be 7.5%. Instead of making his 30th payment of his 48-payment loan, Tino wishes to pay his remaining balance and terminate the loan.

a) Use the actuarial method to determine how much interest Tino will save (the unearned interest, u) by repaying the loan early.

b) What is the total amount due to pay off the loan early on the day he makes his final payment?

SOLUTION

a) Recall from Example 3 that Tino's monthly payments are \$237.00. After 30 payments have been made, 18 payments remain. Thus, $n = 18$ and $P = \$237$. To determine V, use the APR table (Table 11.2). In the Number of Payments column, find the number of remaining payments, 18, and then look to the right until you reach the column headed by 7.5%, the APR. This row and column intersect at 6.04. Thus, $V = 6.04$. Now use the actuarial method formula to determine the unearned interest, u.

$$u = \frac{n \cdot P \cdot V}{100 + V} = \frac{(18)(237)(6.04)}{100 + 6.04} \approx 242.99$$

Tino will save \$242.99 in interest by the actuarial method.

b) Because the remaining payments total $18(\$237) = \$4266$, Tino's remaining balance excluding his 30th monthly payment, is

\$4266.00	Total of remaining payments (which includes interest)
− 242.99	Interest saved (unearned interest)
\$4023.01	Balance due (excluding the 30th monthly payment)

MATHEMATICS TODAY

Credit Card Minimum Payments

In January 2003, several government regulatory agencies, concerned with rising credit card debt, required credit card companies to raise the minimum monthly payments made by their cardholders. Under the old requirement for calculating minimum payments, it was quite possible for consumers to actually owe more money each month than they owed the previous month if they made only their minimum payments! The aim of the new requirement was to encourage more Americans to pay off their credit card debt quicker and thereby reduce the amount of interest owed by consumers. However, many credit card companies waited until January 2006—immediately following the holiday shopping season—to begin to raise their monthly minimum payments. The result was that many consumers could not pay the newer minimum monthly payments, which in some cases were more than double the previous minimum monthly payments. Realizing that it was in their own interest to help these consumers pay back the principal and interest owed, credit card companies were generally willing to work with such consumers.

A payment of \$4023.01 plus the 30th monthly payment of \$237 will terminate Tino's installment loan.

\$4023.01	Balance due (excluding the 30th monthly payment)
+ 237.00	30th monthly payment
\$4260.01	Total amount due

The total amount due is \$4260.01.

Open-End Installment Loan

A credit card is a popular way of making purchases or borrowing money. Use of a credit card is an example of an open-end installment loan. A typical credit card charge account with a bank or store may have the terms in Table 11.3.

Table 11.3 Credit Card Terms

Type of Charge	Daily Periodic Rate*	Annual Percentage Rate*
Purchases	0.04792%	17.49%
Cash advances	0.05477%	19.99%

*These rates vary with different credit card accounts and localities.

Typically, credit card monthly statements contain the following information: balance at the beginning of the period, balance at the end of the period (or new balance), the transactions for the period, statement closing date (or billing date), payment due date, and the minimum payment due. For *purchases*, there is no finance or interest charge if there is no previous balance due and you pay the entire new balance by the payment due date. The period between when a purchase is made and when the credit card company begins charging interest is called the *grace period* and is usually 20 to 25 days. However, if you use a credit card to borrow money, called a *cash advance*, there generally is no grace period and a finance charge is applied from the date you borrowed the money until the date you repay the money. When you make purchases or obtain cash advances, the minimum monthly payment will generally be any new fees and interest plus at least 1% of the outstanding principal (see Mathematics Today at left). This sum is then rounded up to the nearest whole dollar. Credit card companies often have \$20 as the lowest possible minimum monthly payment. These guidelines will vary among the different credit card companies. If you currently have a credit card, you can find out how your minimum monthly payment is calculated by reading the back of your monthly statement or by reading the literature given to you when you obtained the credit card.

EXAMPLE 5 *Calculating Minimum Monthly Payments*

Mary Beth Orrange's credit card company determines her minimum monthly payment by adding any new interest owed to 1.5% of the outstanding principal. The credit card company charges an interest rate of 0.04792% per day. For Mother's Day, May 17, Mary Beth used her credit card to purchase a \$1500 plasma television set for her mother. She made no other purchases in May.

a) Assuming Mary Beth owed no interest, determine Mary Beth's minimum payment due on June 1, her billing date.

Date	(i) Balance Due	(ii) Number of Days Balance Did Not Change	(iii) (Balance)(Days)
July 1	\$375.80	4	(\$375.80)(4) = \$1503.20
July 5	\$225.80	5	(\$225.80)(5) = \$1129.00
July 10	\$300.15	8	(\$300.15)(8) = \$2401.20
July 18	\$423.65	10	(\$423.65)(10) = \$4236.50
July 28	\$466.15	4	(\$466.15)(4) = \$1864.60
		31	(iv) Sum = \$11,134.50

(v) Divide this sum by the number of days in the billing cycle (in the month). The number of days may be found by adding the days in column (ii).

$$\frac{\$11{,}134.50}{31} = \$359.18$$

Thus, the average daily balance is \$359.18.

b) The finance charge for the month is found using the simple interest formula with the average daily balance as the principal.

$$i = \$359.18 \times 0.013 \times 1 = \$4.67$$

c) Since the finance charge for the month is \$4.67, the balance owed on August 1 is \$466.15 + \$4.67, or \$470.82.

The calculations in Example 7 are tedious. These calculations, however, are made almost instantaneously with computers.

Example 8 illustrates how a credit card may be used to borrow money.

EXAMPLE 8 *Using a Credit Card for a Cash Advance*

To obtain money to buy a stereo system, Bobby Bueker obtained a cash advance of \$1500 from his credit card. He borrowed the money on July 10 and repaid it on July 31. If Bobby is charged an interest rate of 0.05477% per day, how much did Bobby pay the credit card company on July 31?

SOLUTION The amount Bobby pays is the original principal plus any accrued interest. Interest on cash advances is generally calculated for the exact number of days of the loan, starting with the day the money is obtained. The time of the loan in this case is 21 days. Using the simple interest formula, we get the following:

$$\begin{aligned} i &= prt \\ &= \$1500 \times 0.0005477 \times 21 \\ &= \$17.25 \end{aligned}$$

Therefore, on July 31, Bobby must repay the credit card company \$1500 + \$17.25, or \$1517.25.

Anyone purchasing a car or other costly items should consider a number of different sources for a loan. Example 9 illustrates one method of making a comparison.

EXAMPLE 9 *Comparing Loan Sources*

Franz Helfenstein purchased carpeting costing \$2400 with his credit card. When the bill comes due on February 1, Franz realizes that he can pay \$350 per month until the debt is paid off. His credit card charges 1.5% interest per month.

a) Assuming Franz makes no other purchases with this credit card, how many payments are necessary to retire this debt?

b) What is the total interest Franz will pay?

c) How much money could Franz have saved by obtaining a fixed installment loan of \$2400 with an annual percentage rate of 6% interest with six equal monthly payments?

SOLUTION

a) Franz would make his first monthly payment of \$350 on February 1, resulting in a new balance of \$2400 − \$350 = \$2050. His next bill reflects the \$2050 balance plus the monthly interest. He continues to make payments until the debt is retired. For each date indicated, the amount on the far right represents the amount due on that date.

February 1	$\$2400 - \$350 = \$2050$
March 1	$\$2050 + 0.015(\$2050) = \$2080.75; \$2080.75 - \$350 = \$1730.75$
April 1	$\$1730.75 + 0.015(\$1730.75) = \$1756.71; \$1756.71 - \$350 = \$1406.71$
May 1	$\$1406.71 + 0.015(\$1406.71) = \$1427.81; \$1427.81 - \$350 = \$1077.81$
June 1	$\$1077.81 + 0.015(\$1077.81) = \$1093.98; \$1093.98 - \$350 = \$743.98$
July 1	$\$743.98 + 0.015(\$743.98) = \$755.14; \$755.14 - \$350 = \$405.14$
August 1	$\$405.14 + 0.015(\$405.14) = \$411.22; \$411.22 - \$350 = \$61.22$
September 1	$\$61.22 + 0.015(\$61.22) = \$62.14$

After eight payments—seven for \$350 and one for \$62.14—Franz has paid off his credit card bill for his carpeting.

b) To calculate the total interest paid, we add up all his payments and then subtract the cost of the carpeting:

$$\begin{aligned}\text{Total of all payments} &= 7(\$350) + \$62.14 \\ &= \$2512.14 \\ \text{Total interest} &= \$2512.14 - \$2400 \\ &= \$112.14\end{aligned}$$

c) To determine the interest that Franz pays with the installment loan, we will use Table 11.2 on page 680. From Table 11.2, we see that a fixed installment loan with an APR of 6% for 6 months corresponds to a finance charge of \$1.76 per \$100 of the amount financed. So the interest, or finance charge, is

$$\begin{aligned}\text{Finance charge} &= 1.76\left(\frac{2400}{100}\right) \\ &= 1.76(24) \\ &= \$42.24\end{aligned}$$

Therefore, Franz would save \$112.14 − \$42.24 = \$69.90 in interest by using an installment loan instead of a credit card.

SECTION 11.4 EXERCISES

CONCEPT/WRITING EXERCISES

1. Explain the difference between an open-end installment loan and a fixed installment loan.

2. Explain how an installment plan differs from a personal note.

3. What is an annual percentage rate (APR)?

4. What is a finance charge?

5. What is a total installment price?

6. a) What is unearned interest?

   b) What is the most common method used to calculate unearned interest?

7. Name the two methods used to determine the finance charge on an open-end installment loan.

8. What is a cash advance?

PRACTICE THE SKILLS/PROBLEM SOLVING

9. ***A New Roof*** Nancy Sattler paid \$10,000 for a new roof for her house. She paid 15% as a down payment and financed the balance with a 60-month fixed installment loan with an APR of 7.5%.

   a) Determine Nancy's finance charge.

   b) Determine Nancy's monthly payment.

10. ***A New Air Conditioner*** Juan Avalos paid \$7000 for a new central air-conditioning unit for his house. He paid 20% as a down payment and financed the balance with a 36-month fixed installment loan with an APR of 5%.

    a) Determine Juan's finance charge.

    b) Determine Juan's monthly payment.

11. ***New Appliances*** Becky Kubiac wishes to purchase all new kitchen appliances for her home for \$9900. To finance the purchase, Becky makes a 10% down payment and finances the balance with a 48-month fixed installment loan with an APR of 9.5%.

    a) Determine Becky's total finance charge.

    b) Determine Becky's monthly payment.

12. ***PT Cruiser GT*** Sam Bazzi purchased a 2006 PT Cruiser GT for \$25,000. Sam obtained a 60-month, no-money-down installment loan with an APR of 8%.

    a) Determine Sam's total finance charge.

    b) Determine Sam's monthly payment.

13. ***Financing a New Business*** Cheryl Sisson is a hair designer and wishes to convert her garage into a hair salon to use for her own business. The entire project would have a cash price of \$3200. She decides to finance the project by paying 20% down, with the balance paid in 60 monthly payments of \$53.14.

    a) What finance charge will Cheryl pay?

    b) What is the APR to the nearest half percent?

14. ***Financing a Computer*** Ilga Ross purchased a new laptop computer on a monthly purchase plan. The computer sold for \$1495. Ilga paid 5% down and \$64 a month for 24 months.

    a) What finance charge did Ilga pay?

    b) What is the APR to the nearest half percent?

15. ***Financing a Used Car*** Jack Fitzgerald wishes to purchase a used car that has a cash price of \$12,000. The installment terms include a down payment of \$3000 and 48 monthly payments of \$224.

    a) What finance charge will Jack pay?

    b) What is the APR to the nearest half percent?

16. *Financing Furniture* Mr. and Mrs. Chan want to buy furniture that has a cash price of $3450. On the installment plan, they must pay 25% of the cash price as a down payment and make six monthly payments of $437.

 a) What finance charge will the Chans pay?

 b) What is the APR to the nearest half percent?

17. *Early Repayment of a Loan* Ray Flagg took out a 60-month fixed installment loan of $12,000 to open a new pet store. He paid no money down and began making monthly payments of $232. Ray's business does better than expected and instead of making his 24th payment, Ray wishes to repay his loan in full.

 a) Determine the APR of the installment loan.

 b) How much interest will Ray save by paying off his loan early?

 c) What is the total amount due to pay off the loan?

18. *Early Repayment of a Loan* Jeslie Ann Hernandez has a 48-month installment loan with a fixed monthly payment of $83.81. The amount she borrowed was $3500. Instead of making her 18th payment, Jeslie Ann is paying the remaining balance on the loan.

 a) Determine the APR of the installment loan.

 b) How much interest will Jeslie Ann save by paying off the loan early?

 c) What is the total amount due to pay off the loan?

19. *Early Repayment of a Loan* Nina Abu buys a new sport utility vehicle for $32,000. She trades in her old truck and receives $10,000, which she uses as a down payment. She finances the balance at 8% APR over 36 months. Before making her 24th payment, she decides to pay off the loan.

 a) Use Table 11.2 to determine the total interest Nina would pay if all 36 payments were made.

 b) What were Nina's monthly payments?

 c) How much interest will Nina save by paying off the loan early?

 d) What is the total amount due to pay off the loan?

20. *Early Repayment of a Loan* The cash price for a new washer and dryer for Toshio Nakamura's new apartment was $1250. Toshio made a $100 down payment and financed the balance on a 24-month fixed payment installment loan. The monthly payments are $50.71. Instead of making his 12th payment, Toshio decides to pay off the loan.

 a) Determine the APR on the installment loan.

 b) How much interest will Toshio save by paying off the loan early?

 c) What is the total amount due to pay off the loan?

21. *Credit Card Minimum Monthly Payment* Shiing Shen Chern's credit card company determines his minimum monthly payment by adding all new interest to 1% of the outstanding principal. The credit card company charges an interest rate of 0.039698% per day. On March 17, Shiing uses his credit card to purchase airline tickets for his family for $2600. He makes no other purchases during March.

 a) Assuming Shiing had no new interest, determine Shiing's minimum payment due on April 1, his billing date.

 b) On April 1, instead of making the minimum payment, Shiing makes a payment of $500. Assuming there are no additional charges or cash advances, determine Shiing's minimum payment due on May 1.

22. *Credit Card Minimum Monthly Payment* Evelyn Boyd Granville's credit card company determines her minimum monthly payment by adding all new interest to 2% of the outstanding principal. The credit card company charges an interest rate of 0.04238% per day. On September 15, Evelyn uses her credit card to purchase a new picture window for her house for $3200. She makes no other purchases during September.

 a) Assuming Evelyn had no new interest, determine Evelyn's minimum payment due on October 1, her billing date.

 b) On October 1, instead of making the minimum payment, Evelyn makes a payment of $1200. Assuming there are no additional charges or cash advances, determine Evelyn's minimum payment due on November 1.

23. *Business Expenses* Kevin Devlin's credit card company determines his minimum monthly payment by adding all new interest to 1.5% of the outstanding principal. The credit card company charges an interest rate of 0.05163% per day. On November 12, Kevin used his credit card to pay for the following business expenses: van repairs ($677), equipment maintenance ($452), office supplies ($139), and dinner with clients ($141).

 a) Assuming Kevin had no new interest, determine his minimum payment due on December 1, his billing date.

 b) On December 1, instead of making his minimum payment, Kevin makes a payment of $300. Assuming there are no additional charges or cash advances, determine Kevin's minimum payment on January 1.

24. ***Vacation Expenses*** Harry Waldman's credit card company determines his minimum monthly payment by adding all new interest to 2.5% of the outstanding principal. The credit card company charges an interest rate of 0.03164% per day. On August 21, while on vacation, Harry used his credit card to pay for the following expenses: airfare ($359), car rental ($273), hotel ($653), meals ($315), and surfboard rental ($225).

a) Assuming Harry had no new interest, determine Harry's minimum payment due on September 1, his billing date.

b) On September 1, instead of making his minimum payment, Harry makes a payment of $750. Assuming there are no additional charges or cash advances, determine Harry's minimum payment on October 1.

25. ***Unpaid Balance Method*** On the April 5 billing date, Michaelle Chappell had a balance due of $1097.86 on her credit card. From April 5 through May 4, Michaelle charged an additional $425.79 and made a payment of $800.

a) Find the finance charge on May 5, using the unpaid balance method. Assume that the interest rate is 1.8% per month.

b) Find the new balance on May 5.

26. ***Unpaid Balance Method*** On September 5, the billing date, Verna Brown had a balance due of $567.20 on her credit card. The transactions during the following month were

September 8	Payment	$275.00
September 21	Charge: Airline ticket	330.00
September 27	Charge: Hotel bill	190.80
October 2	Charge: Clothing	84.75

a) Find the finance charge on October 5, using the unpaid balance method. Assume that the interest rate is 1.1% per month.

b) Find the new balance on October 5.

27. ***Unpaid Balance Method*** On February 3, the billing date, Carol Ann Bluesky had a balance due of $124.78 on her credit card. Her bank charges an interest rate of 1.25% per month. She made the following transactions during the month.

February 8	Charge: Art supplies	$25.64
February 12	Payment	100.00
February 14	Charge: Flowers delivered	67.23
February 25	Charge: Music CD	13.90

a) Find the finance charge on March 3, using the unpaid balance method.

b) Find the new balance on March 3.

28. ***Unpaid Balance Method*** On April 15, the billing date, Gabrielle Michaelis had a balance due of $57.88 on her credit card. She is redecorating her apartment and has the following transactions.

April 16	Charge: Paint	$64.75
April 20	Payment	45.00
May 3	Charge: Curtains	72.85
May 10	Charge: Chair	135.50

a) Find the finance charge on May 15, using the unpaid balance method. Assume that the interest rate is 1.35% per month.

b) Find the new balance on May 15.

29. ***Average Daily Balance Method*** The balance on the Razazadas' credit card on May 12, their billing date, was $378.50. For the period ending June 12, they had the following transactions.

May 13	Charge: Toys	$129.79
May 15	Payment	50.00
June 1	Charge: Clothing	135.85
June 8	Charge: Housewares	37.63

a) Find the average daily balance for the billing period.

b) Find the finance charge to be paid on June 12. Assume an interest rate of 1.3% per month.

c) Find the balance due on June 12.

30. ***Average Daily Balance Method*** The Levys' credit card statement shows a balance due of $1578.25 on March 23, the billing date. For the period ending April 23, they had the following transactions.

March 26	Charge: Party supplies	$79.98
March 30	Charge: Restaurant meal	52.76
April 3	Payment	250.00
April 15	Charge: Clothing	190.52
April 22	Charge: Car repairs	190.85

a) Find the average daily balance for the billing period.

b) Find the finance charge to be paid on April 23. Assume an interest rate of 1.3% per month.

c) Find the balance due on April 23.

31. ***Average Daily Balance Method*** Refer to Exercise 27. Instead of the unpaid balance method, suppose that Carol Ann's bank uses the average daily balance method.

a) Find Carol Ann's average daily balance for the billing period from February 3 to March 3. Assume it is not a leap year.

b) Find the finance charge to be paid on March 3.

c) Find the balance due on March 3.

d) Compare these answers with those in Exercise 27.

32. ***Average Daily Balance Method*** Refer to Exercise 26. Instead of the unpaid balance method, suppose Verna's bank uses the average daily balance method.

a) Find Verna's average daily balance for the billing period from September 5 to October 5.

b) Find the finance charge to be paid on October 5.

c) Find the balance due on October 5.

d) Compare these answers with those in Exercise 26.

33. ***A Cash Advance*** John Richards borrowed $875 against his charge account on September 12 and repaid the loan on October 14 (32 days later). Assume that the interest rate is 0.04273% per day.

a) How much interest did John pay on the loan?

b) What amount did he pay the bank when he repaid the loan?

34. ***A Cash Advance*** Travis Thompson uses his credit card to obtain a cash advance of $600 to pay for his textbooks in medical school. The interest rate charged for the loan is 0.05477% per day. Travis repays the money plus the interest after 27 days.

a) Determine the interest charged for the cash advance.

b) When he repaid the loan, how much did he pay the credit card company?

35. ***Comparing Loan Sources*** Grisha Stewart needs to borrow $1000 for an automobile repair. She finds that State National Bank charges 5% simple interest on the amount borrowed for the duration of the loan and requires the loan to be repaid in six equal monthly payments. Consumer's Credit Union offers loans of $1000 to be repaid in 12 monthly payments of $86.30.

a) How much interest is charged by the State National Bank?

b) How much interest is charged by the Consumer's Credit Union?

c) What is the APR, to the nearest half percent, on the State National Bank loan?

d) What is the APR, to the nearest half percent, on the Consumer Credit Union loan?

36. ***Comparing Loan Options*** Sara Lin wants to purchase a new television set. The purchase price is $890. If she purchases the set today and pays cash, she must take money out of her savings account. Another option is to charge the TV on her credit card, take the set home today, and pay next month. Next month she will have cash and can pay her credit card balance without paying any interest. The simple interest rate on her savings account is $5\frac{1}{4}\%$. How much is she saving by using the credit card instead of taking the money out of her savings account?

CHALLENGE PROBLEMS/GROUP ACTIVITY

37. ***Determining Purchase Price*** Ken Tucker bought a new car, but now he cannot remember the original purchase price. His payments are $379.50 per month for 36 months. He remembers that the salesperson said the simple interest rate for the period of the loan was 6%. He also recalls he was allowed $2500 on his old car. Find the original purchase price.

38. ***Comparing Loans*** Suppose the Chans in Exercise 16 use a credit card rather than an installment plan. Assume that they make the same down payment, have no finance charge the first month, make no additional purchases on their credit card, and pay $432 per month plus the finance charge starting with the second month. The interest rate is 1.3% per month.

a) How many months will it take them to repay the loan?

b) How much interest will they pay on the loan?

c) Which method of borrowing will cost the Chans the least amount of interest, the installment loan in Exercise 16 or the credit card?

APR involves a number of steps, including adding the amount paid for points to the total interest paid. The APR can then be determined using an APR table, a calculator, or an Internet web site. Often, when shopping for a mortgage, a buyer may have two choices: pay points and get a lower interest rate, or not pay points and get a higher interest rate. Web sites such as www.hsh.com allow potential buyers to research whether paying points is to their advantage.

Conventional Loans

Example 1 illustrates purchasing a house with a conventional mortgage loan.

EXAMPLE 1 *Down Payment and Points*

Patricia and Marshall Martin wish to purchase a house selling for \$249,000. They plan to obtain a loan from their bank. The bank requires a 15% down payment, payable to the seller, and a payment of 2 points, payable to the bank, at the time of closing.

a) Determine the Martins' down payment.

b) Determine the amount of the Martins' mortgage.

c) Determine the cost of the 2 points paid by the Martins on their mortgage.

SOLUTION

a) The down payment is 15% of \$249,000, or

$$0.15 \times \$249{,}000 = \$37{,}350$$

b) The mortgage on the Martin's new home is the selling price minus the down payment.

$$\$249{,}000 - \$37{,}350 = \$211{,}650$$

c) Each point equals 1% of the mortgage amount, so 2 points equals 2% of the mortgage amount.

$$0.02 \times \$211{,}650 = \$4233$$

At the closing, the Martins will pay the down payment of \$37,350 to the seller and the 2 points, or \$4233, to the bank.

Banks use a formula to determine the maximum monthly payment that they believe is within the purchaser's ability to pay. A mortgage loan officer first determines the buyer's ***adjusted monthly income*** by subtracting from the gross monthly income (total income before any deductions) any fixed monthly payments with more than 10 months remaining (such as for a student loan, a car, furniture, or a television). The loan officer then multiplies the adjusted monthly income by 28%. (This percent, and the maximum number of payments remaining on other fixed loans, may vary in different locations.) In general, this product is the maximum monthly house payment the lending institution believes that the purchaser can afford to pay. This payment must cover principal, interest, property taxes, and insurance. Taxes and insurance are not necessarily paid to the bank; they may be paid directly to the tax collector and the insurance company. Example 2 shows how a bank uses the formula to determine whether a prospective buyer qualifies for a mortgage.

EXAMPLE 2 *Qualifying for a Mortgage*

Suppose the Martins' (see Example 1) gross monthly income is \$7250 and they have 23 remaining monthly payments of \$225 on their car loan, 17 remaining monthly payments of \$175 on their daughter's orthodontic braces, and 11 remaining monthly payments of \$45 on a loan used to purchase new furniture. The property taxes and homeowners' insurance on the house they wish to buy are \$165 and \$115 per month, respectively. Their bank will approve a loan that has a total monthly mortgage payment of principal, interest, property taxes, and homeowners' insurance that is less than or equal to 28% of their adjusted monthly income.

a) Determine 28% of the Martins' adjusted monthly income.

b) The Martins want a 30-year, \$211,650 mortgage. If the interest rate is 7.0%, determine the total monthly mortgage payment (including principal, interest, property taxes, and homeowners' insurance) for this mortgage.

c) Determine whether the Martins qualify for this mortgage.

SOLUTION

a) To determine the Martins' adjusted monthly income, subtract the sum of their monthly payments, \$225 + \$175 + \$45 = \$445, from their gross monthly income, \$7250.

\$7250	Gross monthly income
− 445	Monthly payments
\$6805	Adjusted monthly income

Next, take 28% of the adjusted monthly income.

$$0.28 \times \$6805 = \$1905.40$$

Thus, 28% of the Martins' adjusted monthly income is \$1905.40.

b) To determine the total monthly mortgage payment, we first need to determine the monthly principal and interest payments. Then we add the monthly property taxes and homeowners' insurance. We will use a table to calculate the monthly principal and interest payments the Martins will need to pay. Table 11.4 on page 696 gives monthly principal and interest payments per \$1000 of mortgage. With an interest rate of 7.0% and a 30-year mortgage, the Martins would have a monthly principal and interest payment of \$6.65 (circled in blue) per thousand dollars of mortgage.

To determine the Martins' monthly principal and interest payment on their mortgage, first divide the mortgage amount by \$1000, which will give the number of thousands of dollars of the mortgage.

$$\frac{\$211{,}650}{\$1000} = 211.65$$

Then, to determine the monthly principal and interest payment, multiply the number of thousands of dollars of mortgage, 211.65, by the value found in Table 11.4, \$6.65.

$$211.65 \times \$6.65 \approx \$1407.47$$

Thus, the monthly principal and interest payment is \$1407.47. To this amount, we add the monthly property taxes, \$165, and the monthly homeowner's insurance, \$115.

$$\$1407.47 + \$165 + \$115 = \$1687.47$$

Therefore, the Martins' total monthly mortgage payment is \$1687.47.

DID YOU KNOW?

The Benefits of Home Ownership

Most of us are aware that a big part of the American dream is to own your own home. What we may not realize is that in addition to having a place to call your own, several important financial benefits occur when you own your own home. First, instead of paying rent to someone else, you make a mortgage payment that builds the equity in your home. *Equity* is the difference between the appraised value of your home and your loan balance, and it usually increases with each payment you make. As years go by, this equity may also help you qualify for other loans such as college and car loans. Second, the interest and real estate taxes you pay (in most cases) are deductible on your federal income tax returns. These deductions can add up to significant savings each year and may result in a larger tax refund. Finally, over time you can typically expect your home to increase in value. Thus, your home not only becomes your place of dwelling; in most cases, it also serves as a wise financial investment.

Thus, the first monthly principal and interest payment consists of \$1234.63 in interest and \$172.84 in principal. The \$172.84 is applied to reduce the outstanding balance due on the loan. Thus, the balance due on the loan after the first monthly payment is made is \$211,650.00 − \$172.84, or \$211,477.16.

By repeatedly using the simple interest formula month to month on the unpaid balance, you could calculate the principal and the interest for all the payments, which is a tedious task. However, a list containing the payment number, payment on the interest, payment on the principal, and balance of the loan, can be prepared using a computer. Such a list is called a loan ***amortization schedule***. One way to obtain an amortization schedule is by using a computer spreadsheet program. Another way is to access an amortization "calculator" program on the Internet. A part of the amortization schedule for the Martins' loan in Example 3 is given in Table 11.5. This schedule was generated from an Internet site called *Monthly Mortgage Payment Calculator* (*www.hsh.com*). Note that the monthly payment in Table 11.5 (\$1408.11) is slightly more than the monthly payment we calculated using Table 11.4 (\$1407.47). This difference is due to a rounding error that occurs when estimating the monthly principal and interest payment from Table 11.4.

Table 11.5 Amortization Schedule

Annual % Rate: 7.0 **Loan: \$211,650** **Periods: 360**		**Monthly Payment: \$1408.11** **Term: Years 30, Months 0**	
Payment Number	**Interest**	**Principal**	**Balance of Loan**
1	\$1234.62	\$173.49	\$211,476.51
2	\$1233.61	\$174.50	\$211,302.01
3	\$1232.60	\$175.52	\$211,126.49
4	\$1231.57	\$176.54	\$210,949.95
11	\$1224.24	\$183.88	\$209,684.99
12	\$1223.16	\$184.95	\$209,500.04
119	\$1063.49	\$344.62	\$181,968.54
120	\$1061.48	\$346.63	\$181,621.91
239	\$715.54	\$692.57	\$121,972.22
240	\$711.50	\$696.61	\$121,275.62
359	\$16.29	\$1391.83	\$1,399.95
360	\$8.17	\$1399.95	\$0.00

Adjustable-Rate Mortgages

Now let's consider ***adjustable-rate mortgages***, ARMs (also called *variable-rate mortgages*). The rules for ARMs vary from state to state and from bank to bank, so the material presented on ARMs may not apply in your state or at your local lending institution. Generally, with adjustable-rate mortgages, the monthly mortgage payment remains the same for a 1-, 2-, or 5-year period even though the interest rate of the mortgage may change every 3 months, 6 months, or some other predetermined period. The interest rate for an adjustable-rate mortgage may be based on an index that is determined by the Federal Home Loan Bank Association or on the interest rate of a 3-month, 6-month, or 1-year Treasury bill. The interest rate of a 3-month Treasury bill may change every 3 months, the interest rate on a 6-month Treasury bill may change every 6 months, and so on. When the base is a Treasury bill, the actual interest rate charged for the mortgage is often determined by adding 3% to $3\frac{1}{2}$%, called the ***add on rate*** or ***margin***, to the rate of the Treasury bill. Thus, if the rate of the Treasury bill is 6% and the add on rate is 3%, the interest rate charged is 9%.

EXAMPLE 4 *An Adjustable-Rate Mortgage*

Tony and Keisha Torrence purchased a house for \$115,000 with a down payment of \$23,100. They obtained a 30-year adjustable-rate mortgage with the following terms. The interest rate is based on a 6-month Treasury bill. The interest rate charged is 3% above the interest rate of the 6-month Treasury bill (3% is the add on rate). The interest rate is adjusted every 6 months on the date of adjustment. The interest rate will not change more than 1% (up or down) when it is adjusted. The maximum interest rate for the duration of the loan is 12%. There is no lower limit on the interest rate. The initial mortgage interest rate is 5.5%, and the monthly payments (including principal and interest) are adjusted every 5 years.

a) Determine the initial monthly payment.

b) Determine the adjusted interest rate in 6 months if the interest rate on the Treasury bill at that time is 2%.

SOLUTION

a) To determine the initial monthly payment of interest and principal, divide the amount of the loan, \$115,000 − \$23,100 = \$91,900, by \$1000. The result is 91.9. Now multiply the number of thousands of dollars of mortgage, 91.9, by the value found in Table 11.4 with $r = 5.5\%$ for 30 years. The value found in the table is \$5.68.

$$\$5.68 \times 91.9 \approx \$521.99$$

Thus, the initial monthly payment for principal and interest is \$521.99. This amount will not change for the first 5 years of the mortgage.

b) The adjusted interest rate in 6 months will be the Treasury bill rate plus the add on rate.

$$2\% + 3\% = 5\%$$

In Example 4(b), note that the rate after 6 months, 5%, is lower than the initial rate of 5.5%. Since the monthly payment remains the same, the additional money paid the bank is applied to reduce the principal. The monthly interest and principal payment of \$521.99 would pay off the loan in 30 years if the interest remained constant at 5.5%. What happens if the interest rate drops and stays lower than the initial 5.5% for the length of the loan? In this case, at the end of each 5-year period the bank reduces the monthly payment so that the loan will be paid off in 30 years. What happens if the interest rate increases above the initial 5.5% rate? In that case, part of, or if necessary, all the mortgage payment that would normally go toward repaying the principal would be used to meet the interest obligation. At the end of the 5-year period, the bank will increase the monthly payment so that the loan can be repaid by the end of the 30-year period. Or the bank may increase the time period of the loan beyond 30 years so that the monthly payment is affordable.

To prevent rapid increases in interest rates, some banks have a rate cap. A ***rate cap*** limits the maximum amount the interest rate may change. A ***periodic rate cap*** limits the amount the interest rate may increase in any one period. For example, your mortgage could provide that even if the index increases by 2% in 1 year, your rate can only go up 1% per year. An ***aggregate rate cap*** limits the interest rate increase and decrease over the entire life of the loan. If the initial interest rate is 6% and the aggregate rate cap is 2%, the interest rate could go no higher than 8% and no lower than 4% over the life of the mortgage. A ***payment cap*** limits the amount the monthly payment may change but does not limit changes in interest rates. If interest rates increase rapidly on a loan with a payment cap, the monthly payment may not be large enough to pay the monthly principal and interest on the loan. If that happens, the borrower could end up paying interest on interest.

Other Types of Mortgages

Conventional mortgages and adjustable-rate mortgages are not the only methods of financing the purchase of a house. Next, we briefly describe four other methods and then briefly discuss home equity loans.

FHA Mortgage A house may be purchased with a smaller down payment than with a conventional mortgage if the buyer qualifies for a Federal Housing Authority (FHA) loan. The loan application is made through a lender such as a bank, credit union, or mortgage broker, but the FHA guarantees the loan against default by the borrower. This guarantee by the FHA, along with the smaller down payment, makes it possible for many people who might not qualify for a conventional mortgage to buy a home. The down payment for an FHA loan may be as low as 3% of the purchase price, rather than the standard 5% or higher. FHA loans also allow the buyer to have slightly higher mortgage payments than with a conventional loan. Typically, lenders will approve an FHA mortgage with a monthly mortgage payment of 29% of the lenders' adjusted monthly income as opposed to 28% with a conventional loan. Furthermore, many of the costs that are paid by the buyer at the closing of a conventional mortgage may be added to the amount of money borrowed, again making it easier for the buyer to obtain a mortgage to buy a house.

Another advantage is that FHA loans can be assumed at the original rate of interest on the loan. For example, if you purchase a home today that already has a 6% FHA mortgage, you, the new buyer, can assume that 6% mortgage regardless of the current interest rates. However, to be able to assume a mortgage, the purchaser must be able to make a down payment equal to the difference between the purchase price and the balance due on the original mortgage. There are some drawbacks to an FHA loan. Each county in the United States has a maximum house purchase price that can be financed through an FHA loan. Also, because the buyer pays such a small down payment, the borrower must pay for loan insurance. FHA borrowers must pay a one-time upfront insurance premium that is 1.5% of the loan amount. The borrower also pays an annual premium of 0.5% of the balance due on the loan amount that can be added to the monthly mortgage amount. Even with these drawbacks, FHA loans are the only mortgage option for many people seeking to buy a home and are very popular with first-time home buyers.

MATHEMATICS TODAY

Credit Scores

When consumers borrow money by obtaining car loans, home mortgages, or other loans or through the use of credit cards, the lender takes a risk by giving money to the borrower in exchange for the borrower's promise to repay the loan plus interest. To minimize their risk, lenders do research on borrowers to determine the likelihood the borrowers will *default*, or fail to repay the loans. Lenders often obtain this research from a *consumer credit bureau*. Credit bureaus assign a credit score to an individual after researching the individual's credit history, income, age, and other factors. A high credit score suggests that the individual is more likely to repay a loan and often enables the borrower to get lower interest rates on loans. In March 2006, the three major credit bureaus in the United States—Equifax, Experian, and TransUnion—announced that they would all use a new system to establish a credit score called a *VantageScore*. You can find your own credit score online at www.creditscoring.com. The following table shows the range of credit scores, grades, and credit outlook for these scores.

Score	Grade	Credit Outlook
901–990	A	Excellent rating; low interest rate
801–900	B	Good rating but short of excellent
701–800	C	Fair rating; moderate risk
601–700	D	Poor rating; higher risk and high interest rate
501–600	F	Lowest rating; highest risk; may not qualify for loan

VA Mortgage

A veteran certified by the Department of Veterans Affairs (VA) who wants to purchase a house can apply for a mortgage with a bank or lending institution. The individual must meet the requirements set by the bank or lending institution for a mortgage. The VA guarantees repayment of a certain percentage of the loan obtained by the individual should the individual default on the loan. For example, if the loan is less than $45,000, the VA guarantees 50% of the loan. For loans from $45,000 to $144,000, the VA guarantees the lesser of 40% of the loan or $36,000. Since the bank has this guarantee from the VA, qualified veterans often do not have to make a down payment. A veteran who can make the monthly mortgage payments may therefore obtain a certain mortgage without a down payment.

The government sets the maximum interest rate that the lender may charge. A VA loan is always assumable. With a VA loan, the seller may be asked to pay points, but there is no monthly insurance premium.

Graduated Payment Mortgage (GPM)

A GPM mortgage is designed so that for the first 5 to 10 years the size of the mortgage payment is smaller than the payments for the remaining time of the mortgage. After the first 5- to 10-year period, the mortgage payments are increased. The mortgage payments then remain constant for the duration of the mortgage. Depending on the size of the mortgage, the lender might find that the monthly payments made for the first few years may actually be less than the interest owed on the loan for those few years. The interest not paid during the first few years of the mortgage is then added to the original loan. This type of loan is strictly for those who are confident their annual incomes will increase as rapidly as the mortgage payments do.

Balloon-Payment Mortgage (BPM)

The BPM type of loan could be for the person who needs time to find a permanent loan. It may work this way: The individual pays the interest for 3 to 8 years, the period of the loan. At the end of the 3- to 8-year period, the buyer must repay the entire remaining principal unless the lender agrees to a loan extension. Balloon-payment loans generally offer lower rates than conventional mortgages. This type of loan may be advantageous if the buyer plans to sell the house before the maturity date of the balloon-payment mortgage.

Home Equity Loans

As you make monthly payments and pay off the principal you owe on your home, you are said to be gaining *equity* in your home. **Equity** is the difference between the appraised value of your home and your loan balance. This equity can be used as collateral in obtaining a loan. Such a loan is referred to as a **home equity loan** or a **second mortgage**. One advantage of home equity loans over other types of loans (such as installment loans) is that the interest charged on a home equity loan is often tax deductible on federal income taxes. Home equity loans are commonly used for home improvements, for bill consolidation, or to pay for college education expenses.

For further information about the types of mortgages or loans discussed in this section, consult a loan officer at a local bank, a savings and loan, or a credit union. Additional information on buying a house may be obtained from the U.S. Government Printing Office in Pueblo, CO 81009. This information is also available on the Internet at www.access.gpo.gov.

SECTION 11.5 EXERCISES

CONCEPT/WRITING EXERCISES

1. What is a mortgage?

2. What is a down payment?

3. What is the difference between a variable-rate mortgage and a conventional mortgage?

4. a) What are points in a mortgage agreement?

   b) Explain how to determine the cost of x points.

5. Explain how to determine a buyer's adjusted monthly income.

6. What is an add on rate, or margin?

7. What is an amortization schedule?

8. Who insures an FHA loan? Who provides the money for an FHA loan?

9. What is equity?

10. What is a home equity loan?

PROBLEM SOLVING

11. ***Buying a House*** Greg Tobin is buying a house selling for \$350,000. His bank requires him to make a 20% down payment. The current mortgage rate is 6.5%.

    a) Determine the amount of the required down payment.

    b) Determine the monthly principal and interest payment for a 15-year mortgage with a 20% down payment.

12. ***Buying a Condo*** Lauren Morse is buying a condominium for \$289,000. Her credit union is requiring a down payment of 10%. The current mortgage rate is 8.0%.

    a) Determine the amount of the required down payment.

    b) Determine the monthly principal and interest payment for a 30-year mortgage with the 10% down payment.

13. ***Buying a Townhouse*** Karen Guardino is purchasing a brownstone townhouse in Brooklyn for \$1,750,000. The mortgage broker she is working with is requiring her to make a 15% down payment. The current mortgage rate is 6.0%.

    a) Determine the amount of the required down payment.

    b) Determine the monthly principal and interest payment for a 25-year loan with a 15% down payment.

▲ See Exercise 13

14. ***Buying a First Home*** Stacie Williams is purchasing her first home for \$275,000. She is obtaining an FHA mortgage through her credit union and is required to make a 3% down payment. The current mortgage rate is 7.5%.

    a) Determine the amount of the required down payment.

    b) Determine the monthly principal and interest payment for a 30-year loan with a 3% down payment.

15. ***Paying Points*** Martha Cutler is buying a house selling for \$195,000. The bank is requiring a minimum down payment of 20%. To obtain a 20-year mortgage at 6% interest, she must pay 2 points at the time of closing.

    a) What is the required down payment?

    b) With the 20% down payment, what is the amount of the mortgage?

    c) What is the cost of the 2 points?

16. ***Down Payment and Points*** The Nicols are buying a house selling for \$245,000. They pay a down payment of \$45,000 from the sale of their current house. To obtain a 15-year mortgage at 4.5% interest, the Nicols must pay 1.5 points at the time of closing.

    a) What is the amount of the mortgage?

    b) What is the cost of the 1.5 points?

17. ***Qualifying for a Mortgage*** Pietr and Helga Guenther's gross monthly income is $3200. They have 25 remaining car payments of $335. The Guenthers are applying for a 15-year, $150,000 mortgage at 5% interest to buy a new house. The taxes and insurance on the house are $225 per month. Their credit union will approve a loan that has a total monthly mortgage payment of principal, interest, property taxes, and homeowners' insurance that is less than or equal to 28% of their adjusted monthly income.

    a) Determine 28% of the Guenther's adjusted monthly income.

    b) Determine the Guenther's total monthly mortgage payment, including principal, interest, taxes, and homeowners' insurance.

    c) Do the Guenthers qualify for this mortgage?

18. ***Qualifying for a Mortgage*** Ting-Fang and Su-hua Zheng's gross monthly income is $4100. They have 18 remaining boat payments of $505. The Zhengs are applying for a 20-year, $275,000 mortgage at 9% interest to buy a new house. The taxes and insurance on the house are $425 per month. Their bank will approve a loan that has a total monthly mortgage payment of principal, interest, property taxes, and homeowners' insurance that is less than or equal to 28% of their adjusted monthly income.

    a) Determine 28% of the Zheng's adjusted monthly income.

    b) Determine the Zheng's total monthly mortgage payment, including principal, interest, taxes, and insurance.

    c) Do the Zhengs qualify for this mortgage?

19. ***A 30-Year Conventional Mortgage*** Ingrid Holzner obtains a 30-year, $63,750 conventional mortgage at 8.5% on a house selling for $75,000. Her monthly payment, including principal and interest, is $490.24. She pays 0 points.

    a) Determine the total amount Ingrid will pay for her house.

    b) How much of the cost will be interest?

    c) How much of the first payment on the mortgage is applied to the principal?

20. ***A 25-Year Conventional Mortgage*** Mr. and Mrs. Alan Bell obtain a 25-year, $110,000 conventional mortgage at 10.5% on a house selling for $160,000. Their monthly mortgage payment, including principal and interest, is $1038.40. They also pay 2 points at closing.

    a) Determine the total amount the Bells will pay for their house.

    b) How much of the cost will be interest (including the 2 points)?

    c) How much of the first payment on the mortgage is applied to the principal?

21. ***Evaluating a Loan Request*** The Rosens found a house selling for $113,500. The taxes on the house are $1200 per year, and insurance is $320 per year. They are requesting a conventional loan from the local bank. The bank is currently requiring a 15% down payment and 3 points, and the interest rate is 10%. The Rosen's gross monthly income is $4750. They have more than 10 monthly payments remaining on a car, a boat, and furniture. The total monthly payments for these items is $420. Their bank will approve a loan that has a total monthly mortgage payment of principal, interest, property taxes, and homeowners' insurance that is less than or equal to 28% of their adjusted monthly income.

    a) Determine the required down payment.

    b) Determine the cost of the 3 points.

    c) Determine 28% of their adjusted monthly income.

    d) Determine the monthly payments of principal and interest for a 20-year loan.

    e) Determine their total monthly payment, including homeowners' insurance and taxes.

    f) Determine whether the Rosens qualify for the 20-year loan.

    g) Determine how much of the first payment on the loan is applied to the principal.

22. ***Evaluating a Loan Request*** Kathy Fields wants to buy a condominium selling for $95,000. The taxes on the property are $1500 per year, and homeowners' insurance is $336 per year. Kathy's gross monthly income is $4000. She has 15 monthly payments of $135 remaining on her van. The bank is requiring 20% down and is charging 9.5% interest with no points. Her bank will approve a loan that has a total monthly mortgage payment of principal, interest, property taxes, and homeowners' insurance that is less than or equal to 28% of her adjusted monthly income.

    a) Determine the required down payment.

b) Determine 28% of her adjusted monthly income.

c) Determine the monthly payment of principal and interest for a 25-year loan.

d) Determine her total monthly payment, including homeowners' insurance and taxes.

e) Does Kathy qualify for the loan?

f) Determine how much of the first payment on the mortgage is applied to the principal.

g) Determine the total amount she pays for the condominium with a 25-year conventional loan. (Do not include taxes or homeowners' insurance.)

h) Determine the total interest paid for the 25-year loan.

23. ***Comparing Loans*** The Riveras are negotiating with two banks for a mortgage to buy a house selling for \$105,000. The terms at bank A are a 10% down payment, an interest rate of 10%, a 30-year conventional mortgage, and 3 points to be paid at the time of closing. The terms at bank B are a 20% down payment, an interest rate of 11.5%, a 25-year conventional mortgage, and no points. Which loan should the Riveras select for the total cost of the house to be less?

24. ***Comparing Loans*** Paul Westerberg is negotiating with two credit unions for a mortgage to buy a condominium selling for \$525,000. The terms at Grant County Teacher's Credit Union are a 20% down payment, an interest rate of 7.5%, a 15-year mortgage, and 1 point to be paid at the time of closing. The terms at Sinnipee Consumer's Credit Union are a 15% down payment, an interest rate of 8.5%, a 20-year mortgage, and no points. Which loan should Paul select for the total cost of the down payment, points, and total mortgage payments of the house to be less?

CHALLENGE PROBLEMS/GROUP ACTIVITIES

25. ***Changing Lengths of Mortgages*** Rose Hulman and her husband, George Mason, are purchasing a home for \$450,000. Their bank requires them to pay a down payment of 20%. The current mortgage rate is 10%, and they are required to pay 1 point at the time of closing. Determine the total amount Rose and George will pay for their house, including principal, interest, down payment, and points (do not include taxes and homeowners' insurance) if the length of their mortgage is

a) 10 years.

b) 20 years.

c) 30 years.

26. ***Principal and Interest Formula*** Rather than using Table 11.4, the following formula can be used to calculate principal and interest payments on a mortgage.

$$m = \frac{P\left(\frac{r}{n}\right)}{1 - \left(1 + \frac{r}{n}\right)^{-n \cdot t}}$$

In the formula, m is the principal and interest payment, P is the amount of the mortgage, r is mortgage rate in decimal form, n is the number of payments per year, and t is the time in years. This formula allows you to calculate principal and interest payments that could not be calculated using Table 11.4. For a mortgage of \$200,000, calculate the principal and interest payment with each of the following conditions.

a) $r = 6.75\%$, monthly payments for 30 years

b) $r = 13\%$, weekly payments for 15 years

c) $r = 9\%$, monthly payments for 40 years

d) $r = 3.9\%$, semimonthly (use $n = 24$) payments for 50 years

27. ***An Adjustable-Rate Mortgage*** The Simpsons purchased a house for \$105,000 with a down payment of \$5000. They obtained a 30-year adjustable-rate mortgage. The terms of the mortgage are as follows: The interest rate is based on a 3-month Treasury bill, the interest rate charged is 3.25% above the rate of the Treasury bill on the date of adjustment, the interest rate is adjusted every 3 months, the interest rate will not change more than 1% (up or down) when the interest rate is adjusted, the maximum interest rate that can be charged for the duration of the loan is 16%, there is no lower limit on the interest rate, the initial mortgage interest rate is 9%, and the monthly payment of interest and principal is adjusted semiannually.

a) Determine the initial monthly payment for interest and principal.

b) Determine an amortization schedule for months 1–3.

c) Determine the interest rate for months 4–6 if the interest rate on the Treasury bill at the time is 6.13%.

d) Determine an amortization schedule for months 4–6.

e) Determine the interest rate for months 7–9 if the interest rate on the Treasury bill at the time is 6.21%.

28. ***An Adjustable-Rate Mortgage*** The Bretz family purchased a house for $95,000 with a down payment of $13,000. They obtained a 30-year adjustable-rate mortgage. The terms of the mortgage are as follows: The interest rate is based on a 3-month Treasury bill, the interest rate charged is 3.25% above the rate of the Treasury bill on the date of adjustment, the interest rate is adjusted every 3 months, the interest rate will not change more than 1% (up or down) when the interest rate is adjusted, the maximum interest rate that can be charged for the duration of the loan is 16%, there is no lower limit on the interest rate, the initial mortgage interest rate is 8.5%, and the monthly payment of interest and principal is adjusted annually.

a) Determine the initial monthly payment for interest and principal.

b) Determine the interest rate in 3 months if the interest rate on the Treasury bill at the time is 5.65%.

c) Determine the interest rate in 6 months if the interest rate on the Treasury bill at the time is 4.85%.

29. ***Comparing Mortgages*** The Hassads are applying for a $90,000 mortgage. They can choose between a 30-year conventional mortgage and a 30-year variable-rate mortgage. The interest rate on a 30-year conventional mortgage is 9.5%. The terms of the variable-rate mortgage are 6.5% interest rate the first year, an annual cap of 1%, and an aggregate cap of 6%. The interest rates and the mortgage payments are adjusted annually. Assume that the interest rates for the variable-rate mortgage increase by the maximum amount each year. Then the monthly mortgage payments for the variable-rate mortgage for years 1–6 are $568.80, $629.10, $692.10, $756.90, $823.50, and $891.00, respectively.

a) Knowing that they will be in the house for only 6 years, which mortgage, the conventional mortgage or the variable-rate mortgage, will be the least expensive for that period?

b) How much will they save by choosing the less expensive mortgage?

INTERNET/RESEARCH ACTIVITIES

30. ***Finding Your Dream Home*** Examine a local newspaper to find your "dream home" and note the asking price. Next contact a loan officer from your local bank, savings and loan, or credit union. Assuming you can make a 20% down payment, determine the interest rates for a 15-year and a 30-year mortgage. Use an amortization "calculator" (see page 698) with the data you obtained to print amortizations schedules. Compare the monthly payments with the 15-year mortgage to those of the 30-year mortgage. Compare the total interest costs of the 15-year mortgage with those of the 30-year mortgage. Write a report summarizing your findings.

31. ***Closing Costs*** An important part of buying a house is the closing. The exact procedures for the closing differ with individual cases and in different parts of the country. In any closing, however, both the buyer and the seller have certain expenses. To determine what is involved in the closing of a property in your community, contact a lawyer, a real estate agent, or a banker. Explain that you are a student and that your objective is to understand the procedure for closing a real estate purchase and the costs to both buyer and seller. Select a specific piece of property that is for sale. Use the asking price to determine the total closing costs to both buyer and seller. The following is a partial list of the most common closing costs. Consider them in your research.

a) Fee for title search and title insurance
b) Credit report on buyer
c) Fees to the lender for services in granting the loan
d) Fee for property survey
e) Fee for recording of the deed
f) Appraisal fee
g) Lawyer's fee
h) Escrow accounts (taxes, insurance)
i) Mortgage assumption fee

11.6 ORDINARY ANNUITIES, SINKING FUNDS, AND RETIREMENT INVESTMENTS

▲ Saving money on a regular basis may lead to a comfortable retirement.

In Sections 11.2 and 11.3, we studied simple and compound interest, respectively. In both sections, we answered questions involving the investment of one lump sum of money. Although such questions may be relevant to our everyday financial matters, more appropriate questions might involve investing smaller amounts of money on a regular basis over a longer period of time. For example, the Brabsons, who are planning for retirement, might ask the question: If we deposit \$100 a month at 5% interest compounded monthly, how much money will we accumulate after 30 years when we're ready to retire? Also consider the Weismans, who are saving for their child's college education. They might ask the question: If we know we will need \$50,000 to send our child to college in 10 years, how much should we invest each month beginning now in an account paying 6% interest compounded monthly? To answer these and similar questions, we will study two investments, the *annuity* and the *sinking fund*.

Annuities

An **annuity** is an account into which, or out of which, a sequence of scheduled payments is made.

There are many different types of annuities. An annuity may be an investment account that you have with a bank, insurance company, or financial management firm. Annuities may contain investments in stocks, bonds, mutual funds, money market accounts, and other types of investments. Annuities are often used to save for long-term goals such as saving money for college or for retirement. An annuity can also be used to provide long-term regular payments to individuals. Lottery jackpots and professional athletes' salaries are often paid out over time from annuities. Retirees may invest some of their retirement savings into an annuity and then receive monthly payments that come from that annuity. We will focus primarily on two basic types of annuities that are used as investment accounts, *ordinary annuities* and *sinking funds*. Later in this section, we will discuss other types of annuities.

An annuity into which equal payments are made at regular intervals, with the interest compounded at the end of each interval and with a fixed interest rate for each compounding period, is called an **ordinary annuity** or a **fixed annuity**.

For example, the Brabsons, mentioned in the opening paragraph of this section, are asking a question that may involve an ordinary annuity. They plan to make \$100 payments each month into an account that pays 5% interest compounded monthly. With an ordinary annuity, the payment period and frequency of the compounding are the same, so the Brabsons make *monthly* payments *and* the interest is compounded *monthly*. They would like to know how much money would accumulate in this annuity after 30 years. The amount of money that is present in an ordinary annuity after t years is known as the ***accumulated amount*** or the ***future value*** of an annuity.

To determine the accumulated amount of an ordinary annuity, we could make many individual calculations using the compound interest formula discussed in Section 11.3. Such calculations would be quite tedious and time consuming even with the assistance of a scientific calculator. A better way is to use a computer to generate a spreadsheet. A portion of such a spreadsheet is shown below. This spreadsheet was generated using the data from the example involving the Brabsons. Note that the Brabsons would make monthly payments for 30 years for a total of 360 payments. Also note that had the Brabsons simply put $100 in their cookie jar each month, their total amount at the end of 30 years would have been $100 × 360, or $36,000. Instead, by investing in an interest-bearing annuity, they will have accumulated $83,225.86 after 30 years.

DID YOU KNOW?

Investing in Stocks

When the owners of a company wish to raise money to expand their company, they often decide to sell part of the company to investors. When an investor purchases a portion of a company, the investor is said to own *stock* in the company. The unit of measure of the stock is called a *share*. By selling shares of stock, a company is selling ownership of the company. By buying shares of stock, an investor is becoming a part owner, or *shareholder*, of the company. The shares of many companies' stock are often bought and sold, or *traded*, by investors through a *stock exchange*.

Large companies may have many millions or even billions of shares of stock available for trading to the general public. For example, in 2006, there were about 2.41 billion shares of the Coca-Cola Company owned by about 340,000 investors. On March 10, 2006, each of these shares was worth $42.77.

Because companies do not always distribute dividends (profits to shareholders) and the price of stocks may go down as well as go up, investing in stocks involves some risk of losing all or part of your investment. Investing in stocks over long periods of time, however, is usually a good investment. Since 1926, shareholders of large established companies have seen an average annual increase in their investment of more than 11%. This return exceeds the average annual increase in bonds, savings accounts, or certificates of deposit.

Accumulated Amount in an Ordinary Annuity with $100 Monthly Investments with Interest Rate 5% Compounded Monthly

Payment Number	Monthly Investment	Interest	Accumulated Amount
1	$100.00	$ —	$ 100.00
2	$100.00	$ 0.42	$ 200.42
3	$100.00	$ 0.84	$ 301.25
4	$100.00	$ 1.26	$ 402.51
5	$100.00	$ 1.68	$ 504.18
6	$100.00	$ 2.10	$ 606.28
7	$100.00	$ 2.53	$ 708.81
8	$100.00	$ 2.95	$ 811.76
9	$100.00	$ 3.38	$ 915.15
10	$100.00	$ 3.81	$ 1,018.96
⋮	⋮	⋮	⋮
351	$100.00	$328.58	$79,287.40
352	$100.00	$330.36	$79,717.76
353	$100.00	$332.16	$80,149.92
354	$100.00	$333.96	$80,583.88
355	$100.00	$335.77	$81,019.65
356	$100.00	$337.58	$81,457.23
357	$100.00	$339.41	$81,896.63
358	$100.00	$341.24	$82,337.87
359	$100.00	$343.07	$82,780.94
360	$100.00	$344.92	$83,225.86

DID YOU KNOW?

Investing in Bonds

A *bond* is a type of loan. When government agencies or corporations need money, they often borrow money from investors by selling, or *issuing*, bonds. When an investor purchases a bond, the investor is actually lending money to the bond's issuer. The issuer agrees to pay the investor a certain interest rate over a stated period of time, usually from 1 to 30 years. The date on which the issuer repays the loan is called the *date of maturity*. Although bonds are generally considered safer investments than stocks, they do have some risks. On rare occasions, issuers may fail to make interest payments or may fail to return the investment entirely. A more common risk is that the value of a bond may decrease if interest rates increase. Such a decrease may cause investors to lose some of their investment if they decide to sell their bond before the date of maturity. In general, though, bonds offer a very stable investment that usually provide a higher return on investment than savings accounts or certificates of deposit without many of the risks associated with investing in stocks.

Another way, and perhaps a more efficient way, to calculate the accumulated amount in an ordinary annuity is to use the ordinary annuity formula.

ORDINARY ANNUITY FORMULA

The accumulated amount, A, of an ordinary annuity with payments of p dollars made n times per year, for t years, at interest rate, r, compounded at the end of each payment period is given by the formula

$$A = \frac{p\left[\left(1+\frac{r}{n}\right)^{nt}-1\right]}{\frac{r}{n}}$$

EXAMPLE 1 *Using the Ordinary Annuity Formula*

Bill and Megan Lutes are depositing $250 each quarter in an ordinary annuity that pays 4% interest compounded quarterly. Determine the accumulated amount in this annuity after 35 years.

SOLUTION We will use the ordinary annuity formula. The payment, p, is $250, the interest rate, r, is 4% or 0.04, the account is compounded quarterly and the payments are made quarterly, so n is 4, and the number of years, t, is 35. We substitute these values into the formula to obtain the following.

$$A = \frac{p\left[\left(1+\frac{r}{n}\right)^{nt}-1\right]}{\frac{r}{n}}$$

$$= \frac{250\left[\left(1+\frac{0.04}{4}\right)^{(4\cdot 35)}-1\right]}{\frac{0.04}{4}}$$

Substitute the given values into the formula.

$$= \frac{250[1.01^{140}-1]}{0.01}$$

To evaluate 1.01^{140}, we use the y^x or the $\wedge$ key on a scientific calculator.

$$\approx \frac{250[4.027099-1]}{0.01}$$

We round 1.01^{140} to six decimal places.

$$\approx \frac{250(3.027099)}{0.01}$$

$$\approx \frac{756.77475}{0.01} \approx 75{,}677.475$$

Thus, there will be about $75,677.48 in Bill and Megan's annuity after 35 years.

Example 1 illustrated the power of establishing and maintaining a consistent investment plan and how using an ordinary annuity can help us save money.

TECHNOLOGY TIP

Scientific Calculator

In Example 1, we determined the accumulated amount by evaluating the expression

$$\frac{250\left[\left(1+\frac{0.04}{4}\right)^{(4\cdot 35)}-1\right]}{\frac{0.04}{4}}$$

There are various ways to evaluate this expression on a scientific calculator. One method is to press the following sequence of keys.

$$\boxed{(}\ 250\ \boxed{\times}\ \boxed{(}\ \boxed{(}\ 1\ \boxed{+}\ 0.04\ \boxed{\div}\ 4\ \boxed{)}\ \boxed{y^x}\ \boxed{(}\ 4\ \boxed{\times}$$

$$35\ \boxed{)}\ \boxed{-}\ 1\ \boxed{)}\ \boxed{)}\ \boxed{\div}\ \boxed{(}\ 0.04\ \boxed{\div}\ 4\ \boxed{)}\ \boxed{=}$$

After the $\boxed{=}$ key is pressed, the answer 75677.48042 is displayed.

Another way to perform the calculations is to use the store and display keys (or memory keys). Read the manual that came with your calculator to learn how to use those keys.

Microsoft Excel

Read the Technology Tip on page 670. To determine the accumulated interest using Excel, place the letters p, r, n, t and A in cells A1 through E1, respectively. In cells A2 through D2, place the values for p, r, n, and t, respectively. In cell E2, use the formula box to add the formula

= (A2*((1 + B2/C2)^(C2*D2) − 1))/(B2/C2)

After the Enter key is pressed, the result is as follows.

	A	B	C	D	E
1	*p*	*r*	*n*	*t*	*A*
2	250	0.04	4	35	75677.48
3					

Sheet1 / Sheet2 / Sheet3 /

Ready

Next we will study a specific type of annuity called a *sinking fund.*

Sinking Funds

We will return to the second question asked at the beginning of this section. Recall that the Weismans have a goal of saving \$50,000 in 10 years for their child's college education and want to know how much they should begin to invest each month in an account paying 6% interest compounded monthly. The Weismans can help reach this goal by investing in a special kind of annuity called a *sinking fund.*

A **sinking fund** is a type of annuity in which the goal is to save a specific amount of money in a specific amount of time.

There are many types of sinking funds that are used for many different purposes by individuals, corporations, and governments. Historically, sinking funds were used by governments to set aside money to pay off bonds that were used to borrow large sums of money for major civic projects such as sewers, roads, or bridges. The term *sinking fund* referred to the *sinking* of the debt that was owed to repay the bonds. Our discussion of sinking funds will focus on answering questions similar to the question facing the Weismans. Because they are interested in saving a specific amount of money, \$50,000, in a specific amount of time, 10 years, their investment can be considered a sinking fund.

TIMELY TIP An ordinary annuity is used when you wish to determine the accumulated amount obtained over n years when you contribute a fixed amount each period. A sinking fund is used when you wish to determine how much money an investor must invest each period to reach an accumulated amount at a specific time.

To determine the amount of money to be invested in a sinking fund, we can solve the ordinary annuity formula, given on page 708, for the payment, p. Solving this formula for p gives us the following formula.

SINKING FUND PAYMENT FORMULA

$$p = \frac{A\left(\frac{r}{n}\right)}{\left(1 + \frac{r}{n}\right)^{nt} - 1}$$

In the formula, p is the payment needed to reach the accumulated amount, A. Payments are made n times per year, for t years, into a sinking fund with interest rate r, compounded n times per year.

EXAMPLE 2 *Using the Sinking Fund Payment Formula*

Richard Ries would like to have \$30,000 in 5 years to buy a new grand piano. Richard decides to invest monthly in a sinking fund that pays 4.5% interest compounded monthly. How much should Richard invest in the sinking fund each month to accumulate \$30,000 in 5 years?

SOLUTION Since we are looking for how much Richard should invest each month to obtain an accumulated amount, we will use the sinking fund payment formula. The accumulated amount, A, is \$30,000, the interest rate, r, is 4.5%, the payments are made monthly and the interest is compounded monthly, so n is 12, and the number of years, t, is 5.

$$\begin{aligned} p &= \frac{A\left(\frac{r}{n}\right)}{\left(1+\frac{r}{n}\right)^{nt}-1} \\ &= \frac{30{,}000\left(\frac{0.045}{12}\right)}{\left(1+\frac{0.045}{12}\right)^{(12\cdot 5)}-1} \\ &\approx \frac{112.50}{0.2517958} \\ &\approx 446.79061 \end{aligned}$$

We often round to the nearest cent, or in this case, to \$446.79. However, because \$446.79 is actually slightly *less* than the answer we obtained, we will round our answer *up* to \$446.80 to ensure that Richard reaches his goal of \$30,000. *In all our exercises involving sinking fund calculations, we will round our final answers up to the nearest cent to ensure that we reach the desired amount in the sinking fund.* ●

TECHNOLOGY TIP

Scientific Calculator

To determine the sinking fund payment in Example 2, we evaluated the expression

$$\frac{30{,}000\left(\frac{0.045}{12}\right)}{\left(1+\frac{0.045}{12}\right)^{12\cdot 5}-1}$$

This expression can be evaluated on a scientific calculator as follows.

30000 [×] [(] 0.045 [÷] 12 [)] [÷] [(] [(] 1 [+] 0.045 [÷] 12 [)] [y^x] [(] 12 [×] 5 [)] [−] 1 [)]

After the [=] key is pressed, the answer 446.7905772 is displayed.

Microsoft Excel

Read the Technology Tip on page 709 and follow a similar procedure using the variables A, r, n, t, and p, respectively. To evaluate the sinking fund payment, in cell E2 enter the formula

= A2*(B2/C2)/((1 + B2/C2)^(C2*D2) − 1)

After the Enter key is pressed, the answer 446.7906 is displayed in cell E2.

In Example 2, we can check our answer by substituting \$446.80 back into the ordinary annuity formula. Doing this calculation with a scientific calculator gives us an accumulated amount of about \$30,000.63. If we substitute \$446.79 for A into the ordinary annuity formula, the accumulated amount would be about \$29,999.96, which is slightly short of the desired goal of \$30,000.

DID YOU KNOW?

Investing in Mutual Funds

A *mutual fund* is an investment tool that enables investors to indirectly own a wide variety of stocks, bonds, or other investments. When investors purchase shares in a mutual fund, they are actually placing their money in a pool along with many other investors. An investment company that invests in many different stocks, bonds, or other investments manages this pool of money. The investments within a mutual fund are called the mutual fund's *portfolio*. The investors of a mutual fund share the gains and losses from the investments within the portfolio. There are some distinct advantages to investing in mutual funds rather than investing in individual stocks and bonds. First, investors in mutual funds have their money managed by full-time professionals who research and evaluate hundreds of stocks and bonds every day. Second, because large sums of money are managed within a mutual fund, costs related to investing, known as *commissions*, are generally lower than they are for purchasing individual stocks and bonds. Third, when investors purchase shares in a mutual fund, they are indirectly purchasing shares in a multitude of stocks or bonds. This diversification can greatly help to reduce some of the risks of investing.

One disadvantage of mutual fund investing is the potential to miss out on a large return on investment. In general, though, investing in mutual funds is considered an excellent way to begin investing and to maintain diverse ownership in a variety of investments.

Other Annuities

Ordinary annuities and sinking funds are not the only types of annuities. We will now briefly discuss two other common types of annuities, *variable annuities* and *immediate annuities*.

Variable Annuities A *variable annuity* is an annuity that is invested in stocks, bonds, mutual funds, or other investments that do not provide a guaranteed interest rate. The term *variable* is used because the value of the annuity will vary depending on the performance of the investment options chosen by the investor. Like an ordinary annuity, an investor usually invests in a variable annuity by making regular periodic payments. Unlike an ordinary annuity, however, a variable annuity does not have a fixed rate of interest.

Most variable annuities are investments made to save for retirement. Upon retirement, the variable annuity investor generally receives periodic, usually monthly, payments for the rest of the investor's life. Often, variable annuity investors may name a designated person who will continue receiving the monthly payments after the investor's death. Such a person, usually a spouse or other family member, is called a *beneficiary*. A very attractive feature of variable annuities is that during the investment period, variable annuities are *tax-deferred*, meaning that no taxes are paid on the income and investment gains from the annuity until the investor begins receiving payments. Investors may also transfer money from one investment option to another within a variable annuity without paying taxes at the time of the transfer. Note, however, that payments received from a variable annuity after a person retires are generally subject to income taxes. One major drawback to investing in variable annuities is that the fees charged by the investment company are often higher than those charged for other investments. Another possible drawback of a variable annuity is that it is sometimes possible for an investor to lose money in a variable annuity. For example, if an investor invests in stocks that drop in price, the value of the annuity may decrease.

As mentioned, variable annuities are usually used to save for retirement. Next we will discuss another type of annuity that is commonly used once an investor nears or reaches retirement.

Immediate Annuities An *immediate annuity* is an annuity that is established with a lump sum of money for the purpose of providing the investor with regular, usually monthly, payments for the rest of the investor's life. In exchange for giving the investment company a lump sum of money, the investor is guaranteed to receive a monthly income for the duration of the investor's life.

An investor has many options when investing in an immediate annuity. One option is to receive monthly payments that increase a certain percentage each year. Another option is the ability to name a beneficiary who will continue receiving the monthly payments after the investor dies. A third option is a guarantee by the investment company that the investor's heirs will receive the remaining balance of the annuity upon the deaths of both the investor and the beneficiary. Although most of these options appear attractive to the investor, including any of these options in an immediate annuity contract may decrease the amount of the monthly payments.

Both variable annuities and immediate annuities have an array of options that carry several advantages and disadvantages. All these options are described in an informational brochure from the investment company called a ***prospectus***. Furthermore, many Internet web sites such as www.morningstar.com and www.smartmoney.com have independent information about annuities and other investments. Before investing in any annuity, it is very important to do independent research. Thoroughly read the prospectus, and ask any questions you may have.

Other Retirement Savings Options

▲ It is never too early to start saving for retirement.

Annuities can be used to save for retirement, but depending on the investor's circumstances, other investment options for retirement savings may be more advantageous. We will briefly discuss several options here.

Individual Retirement Accounts ***Individual retirement accounts***, also known as ***IRAs***, are savings accounts that allow individuals to invest up to a certain amount of money each year for the purpose of saving for retirement. IRAs have distinct tax advantages over ordinary savings accounts. There are several different kinds of IRAs, but most can be classified as either a *traditional IRA* or a *Roth IRA*. A ***traditional IRA*** is an IRA into which *pretax* money is invested, meaning that any money invested into a traditional IRA is not subject to income taxes. However, when money is withdrawn from a traditional IRA, the money *is* subject to income taxes. A ***Roth IRA*** is an IRA into which *posttax* money is invested, meaning that the investor has already paid income taxes on the money invested. When money is withdrawn from a Roth IRA after the investor reaches retirement age, the money *is not* subject to income taxes. Also, as of 2007, married couples with an adjusted gross income of more than $166,000 per year and single taxpayers with an adjusted gross income of more than $114,000 cannot invest in Roth IRAs.

Both types of IRAs have annual limits on the amount that can be invested. In 2008, investors age 49 or younger could invest an annual maximum of $5000 into a traditional or Roth IRA, whereas investors age 50 or older could invest an annual maximum of $6000. Both types of IRAs can contain a variety of investments, including, but not limited to, stocks, bonds, mutual funds, and money market funds. Depending on an investor's financial situation, there may be advantages of one type of IRA over the other. Therefore, before investing, it is important to do research or talk to a financial advisor.

401ks and 403bs A ***401k plan*** is a retirement savings plan that allows employees of private companies to make contributions of pre–income tax dollars that are then pooled with other employees' money. These funds are then invested in a variety of stocks, bonds, money markets, and mutual funds. Although 401k plans share many of the same tax advantages as IRAs, they also have several distinct advantages. First, employees are generally allowed to invest more money annually in a 401k plan than in an IRA. Often, employees are eligible to invest up to 15% of their annual income into a 401k plan. A second advantage is that some employers will match employee contributions up to a certain limit. A ***Roth 401k plan*** is a retirement savings plan that differs from a 401k plan in that an investor first pays income tax on the money used to invest in a Roth 401k plan. Like a Roth IRA, money withdrawn from a Roth 401k plan after the individual reaches retirement age is not subject to income tax.

A *403b plan* is a retirement plan similar to a 401k plan, but it is available only to employees of schools, civil governments, hospitals, charities, and other not-for-profit organizations. A 403b plan shares many of the same advantages as 401k plans. One major difference, though, is that a 403b plan is not necessarily established by the employer as is a 401k plan. Rather, a 403b plan is frequently established independently by the employee with a representative of an investment or insurance company.

As you can see from this brief discussion of annuities, IRAs, 401k plans, and 403b plans, many important details can affect your decisions as you begin to invest for retirement. For more information, and for other types of retirement accounts not mentioned here, consult a financial advisor or visit a library or Internet web site. Regardless of the source of your information, it is very important to do your own research and ask many questions prior to investing.

SECTION 11.6 EXERCISES

CONCEPT/WRITING EXERCISES

1. What is an *annuity*?

2. In the ordinary annuity formula

$$A = \frac{p\left[\left(1 + \frac{r}{n}\right)^{nt} - 1\right]}{\frac{r}{n}}$$

state what each of the following variables represents: A, p, r, n, and t.

3. **a)** What is an *ordinary* annuity?

 b) What is a *sinking fund*?

4. What is a variable annuity?

5. What is an immediate annuity?

6. Explain the difference between a *traditional IRA* and a *Roth IRA*.

7. Explain the difference between a *401k* and a *Roth 401k*.

8. What is a *403b* plan?

PRACTICE THE SKILLS

In Exercises 9–12, use the ordinary annuity formula

$$A = \frac{p\left[\left(1 + \frac{r}{n}\right)^{nt} - 1\right]}{\frac{r}{n}}$$

to determine the accumulated amount in each annuity. Round all answers to the nearest cent.

9. \$3000 invested annually for 30 years at 5% compounded annually

10. \$800 invested semiannually for 25 years at 7% compounded semiannually

11. \$400 invested quarterly for 35 years at 8% compounded quarterly

12. \$200 invested monthly for 40 years at 6% compounded monthly

In Exercises 13–16, use the sinking fund formula

$$p = \frac{A\left(\frac{r}{n}\right)}{\left(1 + \frac{r}{n}\right)^{nt} - 1}$$

to determine the payment needed to reach the accumulated amount. To ensure that enough is invested each period, round each answer up to the next cent.

13. Semiannual payments with 4% interest compounded semiannually for 15 years to accumulate \$20,000

14. Annual payments with 5% interest compounded annually for 20 years to accumulate \$35,000

15. Monthly payments with 6% interest compounded monthly for 35 years to accumulate \$250,000

16. Quarterly payments with 10% interest compounded quarterly for 40 years to accumulate \$1,000,000

PROBLEM SOLVING

In Exercises 17–20, round all answers to the nearest cent.

17. ***Retirement Savings*** To save for retirement, Jana Bryant invests \$50 each month in an ordinary annuity with 3% interest compounded monthly. Determine the accumulated amount in Jana's annuity after 35 years.

18. ***Starting a Business*** To save money to start a new business, Sandra Jessee invests \$200 each quarter in an ordinary annuity with 5% interest compounded quarterly. Determine the accumulated amount in Sandra's annuity after 20 years.

19. ***Saving for Graduate School*** To save for graduate school, Ena Salter invests \$2000 semiannually in an ordinary annuity with 7% interest compounded semiannually. Determine the accumulated amount in Ena's annuity after 10 years.

20. ***Grandchild's College Expenses*** To save for their grandchildren's college expenses, Ken and Laura Machol invest \$500 each month in an ordinary annuity with 6% interest compounded monthly. Determine the accumulated amount in the Machol's annuity after 18 years.

In Exercises 21–24, round all answers up to the next cent.

21. ***Saving for a New Car*** Erica Atalla would like to have \$25,000 to buy a new car in 5 years. To accumulate \$25,000 in 5 years, how much should she invest monthly in a sinking fund with 9% interest compounded monthly?

22. ***Campaign Fund-Raising*** Ashley Smith would like to run for Congress in 10 years and would like to have \$500,000 to spend on campaign expenses. How much should she invest monthly in a sinking fund with 6% interest compounded monthly to accumulate \$500,000 in 10 years?

▲ See Exercise 22

23. ***Becoming a Millionaire*** Becky Anderson would like to be a millionaire in 20 years. How much would she need to invest quarterly in a sinking fund paying 8% interest compounded quarterly to accumulate \$1,000,000 in 20 years?

24. ***Repaying a Debt*** The city of Corpus Christi wishes to repay a debt of \$25,000,000 in 50 years. How much should the city invest semiannually in a sinking fund paying 7.5% interest compounded semiannually to accumulate \$25,000,000 in 50 years?

CHALLENGE PROBLEMS/GROUP ACTIVITIES

25. ***Don't Wait to Invest*** The purpose of this exercise is to demonstrate that it is much more beneficial to begin to invest for retirement early than it is to wait. We will compare the investments of two investors, Alberto and Zachary. Alberto begins to invest immediately after graduating from college, whereas Zachary waits 10 years to begin investing. For the purposes of this demonstration, we will assume that the interest rate remains the same over the entire period of time.

    a) From age 25 to 35, Alberto invests \$100 per month in an annuity that pays 12% interest compounded monthly. Determine the accumulated amount in Alberto's annuity after 10 years.

    b) At age 35, Alberto stops making monthly payments into his annuity, but he takes the money he accumulated in his annuity and invests it in a savings account with an interest rate of 12% compounded monthly. Using the compound interest formula $A = p\left(1 + \frac{r}{n}\right)^{n \cdot t}$, determine the amount Alberto has in his savings account 30 years later when he retires at age 65.

    c) From age 35 to 65, Zachary invests \$100 per month in an annuity that pays 12% interest compounded monthly. Determine the accumulated amount in Zachary's annuity after 30 years.

    d) Determine the total amount Alberto invested during the 10 years from age 25 to 35.

e) Determine the total amount Zachary invested during the 30 years from age 35 to 65.

f) Which investor has more money from the investments described here upon retirement?

INTERNET/RESEARCH ACTIVITIES

26. ***Investing in Stocks*** Using the Internet, financial magazines, books, and other resources, write a paper on investing in stocks (see the Did You Know? on page 707). Include in your paper a description of the following terms: *New York Stock Exchange*, *National Association of Securities Dealers Automated Quotations (NASDAQ)*, *dividends*, and *capital gains*.

27. ***Investing in Bonds*** Using the Internet, financial magazines, books, and other resources, write a paper on investing in bonds (see the Did You Know? on page 708). Include in your paper a description of the following terms: *Treasury bonds*, *corporate bonds*, and *municipal bonds*.

28. ***Investing in Mutual Funds*** Using the Internet, financial magazines, books, and other resources, write a paper on investing in mutual funds (see the Did You Know? on page 712). Include in your paper a description of the following terms: *net asset value (NAV)*, *portfolio manager*, and *expense ratios*.

29. ***Financial Advisor Interview*** Annuities, IRAs, 401k plans, 403b plans, SEP accounts, and Keogh accounts are all types of investments that can be used to save for retirement. Interview a financial advisor to discuss each of these investment types. Write a paper in which you explain the differences among the different investment types and describe the advantages of each type of investment over the others.

CHAPTER 11 SUMMARY

IMPORTANT FACTS

ORDINARY INTEREST

When computing ordinary interest, each month is considered to have 30 days and a year is considered to have 360 days.

UNITED STATES RULE

If a partial payment is made on a loan, interest is computed on the principal from the first day of the loan until the date of the partial payment. The partial payment is used to pay the interest first; then the rest of the payment is used to reduce the principal.

BANKER'S RULE

When computing interest with the Banker's rule, a year is considered to have 360 days and any fractional part of a year is the exact number of days.

SIMPLE INTEREST FORMULA

Interest = principal × rate × time or $i = prt$

COMPOUND INTEREST FORMULA

$$A = p\left(1 + \frac{r}{n}\right)^{nt}$$

PRESENT VALUE FORMULA

$$p = \frac{A}{\left(1 + \frac{r}{n}\right)^{nt}}$$

ACTUARIAL METHOD

$$u = \frac{n \cdot P \cdot V}{100 + V}$$

ORDINARY ANNUITY FORMULA

$$A = \frac{p\left[\left(1 + \frac{r}{n}\right)^{nt} - 1\right]}{\frac{r}{n}}$$

SINKING FUND PAYMENT FORMULA

$$p = \frac{A\left(\frac{r}{n}\right)}{\left(1 + \frac{r}{n}\right)^{nt} - 1}$$

PERCENT CHANGE

$$\textbf{Percent change} = \frac{\text{amount in latest period} - \text{amount in previous period}}{\text{amount in previous period}} \times 100$$

PERCENT MARKUP

$$\textbf{Percent markup on cost} = \frac{\text{selling price} - \text{dealer's cost}}{\text{dealer's cost}} \times 100$$

CHAPTER 11 REVIEW EXERCISES

11.1

Change the number to a percent. Express your answer to the nearest tenth of a percent.

1. $\frac{3}{4}$ 2. $\frac{5}{6}$ 3. $\frac{5}{8}$

4. 0.041 5. 0.0098 6. 3.141

Change the percent to a decimal number.

7. 9% 8. 14.1% 9. 123%

10. $\frac{1}{4}\%$ 11. $\frac{5}{6}\%$ 12. 0.00045%

13. ***Lambeau Field*** Before undergoing renovations, the seating capacity at Lambeau Field (in Green Bay, Wisconsin) was 60,790. After the renovations were complete the seating capacity was 71,500. Determine the percent increase (to the nearest tenth of a percent) in the seating capacity of Lambeau Field.

▲ Lambeau Field

14. ***Salary Increase*** Charlotte Newsom had a salary of $46,200 in 2006 and a salary of $51,300 in 2007. Determine the percent increase in Charlotte's salary from 2006 to 2007.

In Exercises 15–17, solve for the unknown quantity.

15. What percent of 80 is 15?

16. Forty-four is 55% of what number?

17. What is 17% of 540?

18. ***Tipping*** At Empress Garden Restaurant, Vishnu and Krishna's bill comes to $42.79, including tax. If they wish to leave a 15% tip on the total bill, how much should they tip the waiter?

19. ***Increased Membership*** If the number of people in your chess club increased by 20%, or 8 people, what was the original number of people in the club?

20. ***Increased Membership*** The Sarasota Wheelers skateboard club had 75 members and increased the number of members to 95. What is the percent increase in the number of members?

11.2

In Exercises 21–24, determine the missing quantity by using the simple interest formula.

21. $p = \$6000, r = 2\%, t = 30 \text{ days}, i = ?$

22. $p = \$2700, r = ?, t = 100 \text{ days}, i = \$37.50$

23. $p = ?, r = 8\frac{1}{2}\%, t = 3 \text{ years}, i = \$114.75$

24. $p = \$5500, r = 11\frac{1}{2}\%, t = ?, i = \$316.25$

25. ***Roof Replacement Loan*** Chris Sharek borrowed $5300 from his father to replace the roof on his house. The loan was for 36 months and had a simple interest rate of 5.75%. Determine the amount Chris paid his father on the date of maturity of the loan.

26. ***A Bank Loan*** Lori Holdren borrowed $3000 from her bank for 240 days at a simple interest rate of 8.1%.

   a) How much interest did she pay for the use of the money?

   b) How much did she pay the bank on the date of maturity?

27. ***A Bank Loan*** Nikos Pappas borrowed \$6000 for 24 months from his bank, using stock as security. The bank discounted the loan at $11\frac{1}{2}\%$.

   a) How much interest did Nikos pay the bank for the use of the money?

   b) How much did he receive from the bank?

   c) What was the actual rate of interest?

28. ***Savings as Security*** Golda Frankl borrowed \$800 for 6 months from her bank, using her savings account as security. A bank rule limits the amount that can be borrowed in this manner to 85% of the amount in the borrower's savings account. The rate of interest is 2% higher than the interest rate being paid on the savings account. The current rate on the savings account is $5\frac{1}{2}\%$.

   a) What rate of interest will the bank charge for the loan?

   b) Determine the amount that Golda must repay in 6 months.

   c) How much money must she have in her account to borrow \$800?

11.3

29. ***Comparing Compounding Periods*** Determine the amount and the interest when \$5000 is invested for 5 years at 6%.

   a) compounded annually.

   b) compounded semi-annually.

   c) compounded quarterly.

   d) compounded monthly.

   e) compounded daily (use $n = 360$).

30. ***Total Amount*** Choi deposited \$2500 in a savings account that pays 4.75% interest compounded quarterly. What will be the total amount of money in the account 15 years from the day of deposit?

31. ***Effective Annual Yield*** Determine the effective annual yield of an investment if the interest is compounded daily at an annual rate of 5.6%.

32. ***Present Value*** How many dollars must you invest today to have \$40,000 in 20 years? Assume that the money earns 5.5% interest compounded quarterly.

11.4

33. ***Repaying a Loan Early*** Bill Jordan has a 48-month installment loan with a fixed monthly payment of \$176.14. The amount borrowed was \$7500. Instead of making his 24th payment, Bill is paying the remaining balance on the loan.

   a) Determine the APR of the installment loan.

   b) How much interest will Bill save, computed by the actuarial method?

   c) What is the total amount due to pay off the loan?

34. ***Repaying a Loan Early*** Carter Fenton is buying a book collection that costs \$4000. He is making a down payment of \$500 and 24 monthly payments of \$155.91.

   a) Determine the APR.

   b) Instead of making his 12th payment, Carter decides to pay the total remaining balance and terminate the loan. How much will Carter save by paying off the loan early?

   c) What is the total amount due to pay off the loan?

35. ***Installment Loan*** Dara Holliday's cost for a new wardrobe was \$3420. She made a down payment of \$860 and financed the balance on a 24-month fixed payment installment loan. The monthly payments are \$111.73. Instead of making her 12th payment, Dara decides to pay the total remaining balance and terminate the loan.

   a) Determine the APR of the installment loan.

   b) How much interest will Dara save, computed by the actuarial method?

   c) What is the total amount due to pay off the loan?

36. ***Finance Charge Comparison*** On June 1, the billing date, Tim McGarry had a balance due of \$485.75 on his credit card. The transactions during the month of June were

June 4	Payment	\$375.00
June 8	Charge: Car repair	370.00
June 21	Charge: Airline ticket	175.80
June 28	Charge: Clothing	184.75

a) Determine the finance charge on July 1 by using the unpaid balance method. Assume that the interest rate is 1.3% per month.

b) Determine the new account balance on July 1 using the finance charge found in part (a).

c) Determine the average daily balance for the period.

d) Determine the finance charge on July 1 by using the average daily balance method. Assume that the interest rate is 1.3% per month.

e) Determine the new account balance on July 1 using the finance charge found in part (d).

37. *Finance Charge Comparison* On August 5, the billing date, Pat Schaefer had a balance due of $185.72 on her credit card. The transactions during the month of August were

August 8	Charge: Shoes	$85.75
August 10	Payment	75.00
August 15	Charge: Dry cleaning	72.85
August 21	Charge: Textbooks	275.00

a) Determine the finance charge on September 5 by using the unpaid balance method. Assume that the interest rate is 1.4% per month.

b) Determine the new account balance on September 5 using the finance charge found in part (a).

c) Determine the average daily balance for the period.

d) Determine the finance charge on September 5 by using the average daily balance method. Assume that the interest rate is 1.4% per month.

e) Determine the new account balance on September 5 using the finance charge found in part (d).

38. *Financing a Corvette* David Snodgress bought a new Chevrolet Corvette for $72,000. He made a 20% down payment and financed the balance with the dealer on a 60-month payment plan. The monthly payments were $1100.26. Determine the

a) down payment.

b) amount to be financed.

c) total finance charge.

d) APR.

39. *Financing a Ski Outfit* Lucille Groenke can buy a cross-country skiing outfit for $275. The store is offering the following terms: $50 down and 12 monthly payments of $19.62.

a) Determine the interest Lucille would pay.

b) Determine the APR.

11.5

40. *Building a House* The Freemans have decided to build a new house. The contractor quoted them a price of $135,700. The taxes on the house will be $3450 per year, and homeowners' insurance will be $350 per year. They have applied for a conventional loan from a local bank. The bank is requiring a 25% down payment, and the interest rate on the loan is 9.5%. The Freemans' annual income is $64,000. They have more than 10 monthly payments remaining on each of the following: $218 on a car, $120 on new furniture, and $190 on a camper. Their bank will approve a loan that has a total monthly mortgage payment of principal, interest, property taxes, and homeowners' insurance that is less than or equal to 28% of their adjusted monthly income. Determine

a) the required down payment.

b) 28% of their adjusted monthly income.

c) the monthly payment of principal and interest for a 30-year loan.

d) their total monthly payment, including insurance and taxes.

e) Do the Freemans qualify for the mortgage?

41. *Thirty-Year Mortgage* James Whitehead purchased a home selling for $89,900 with a 15% down payment. The period of the mortgage is 30 years, and the interest rate is 11.5% with no points. Determine the

a) amount of the down payment.

b) monthly mortgage payment.

c) amount of the first payment applied to the principal.

d) total cost of the house.

e) total interest paid.

42. *Adjustable-Rate Mortgage* The Nguyens purchased a house for $105,000 with a down payment of $26,250. They obtained a 30-year adjustable-rate mortgage. The terms of the mortgage are as follows: The interest rate is based on the 6-month Treasury bill, the interest rate charged is 3.00% above the rate of the Treasury bill on the date of adjustment, the interest rate is adjusted every 6 months, the interest rate will not change more than 1% (up or down) when the interest rate is adjusted, the maximum interest rate that can be charged for the duration of the loan is 16%, there is no lower limit on the interest rate,

the initial mortgage interest rate is 7.5%, and the monthly payment of interest and principal is adjusted annually. Determine the

a) initial monthly payment for principal and interest.

b) interest rate in 6 months if the interest rate on the Treasury bill at the time is 5.00%.

c) interest rate in 6 months if the interest rate on the Treasury bill at the time is 4.75%.

11.6

43. ***Vacation Savings*** To save money to take an around-the-world vacation, Kim Ghiselin invests $250 monthly in an ordinary annuity with 9% interest compounded monthly. Determine the accumulated amount in Kim's annuity after 10 years.

44. ***Saving for a New Combine*** Steve Carah would like to save $100,000 to buy a used combine (a farm machine used to harvest crops). To accumulate $100,000 in 15 years, how much should he invest quarterly in a sinking fund with 7% interest compounded quarterly?

CHAPTER 11 TEST

1. Find the missing quantity by using the simple interest formula.

   a) $i = ?, p = \$2000, r = 4\%$ per year, $t = 6$ months

   b) $i = \$288, p = \$1200, r = 8\%$ per year, $t = ?$

In Exercises 2 and 3, Rob Wilson borrowed $5700 from a bank for 30 months. The rate of simple interest charged is 5.75%.

2. How much interest did he pay for the use of the money?

3. What is the amount he repaid to the bank on the due date of the loan?

In Exercises 4 and 5, Yolanda Fernandez received a $5400 loan with interest at 12.5% for 90 days on August 1. Yolanda made a payment of $3000 on September 15.

4. How much did she owe the bank on the date of maturity?

5. What total amount of interest did she pay on the loan?

6. Compute the amount and the compound interest.

	Principal	Time	Rate	Compounded
a)	$7500	2 years	3%	Quarterly
b)	$2500	3 years	6.5%	Monthly

*A **New Bicycle** In Exercises 7–9, a new touring bicycle sells for \$2350. To finance it through a bank the bank will require a down payment of 15% and monthly payments of \$90.79 for 24 months.*

7. How much money will the purchaser borrow from the bank?

8. What finance charge will the individual pay the bank?

9. What is the APR?

10. ***Actuarial Method*** Gino Sedillo borrowed \$7500. To repay the loan, he was scheduled to make 36 monthly installment payments of \$223.10. Instead of making his 24th payment, Gino decides to pay off the loan.

a) Determine the APR of the installment loan.

b) How much interest will Gino save (use the actuarial method)?

c) What is the total amount due to pay off the loan?

11. ***Unpaid Balance Method*** Michael Murphy's credit card statement shows a balance due of \$878.25 on March 23, the billing date. For the period ending on April 23, he had the following transactions.

March 26	Charge: Groceries	\$ 95.89
March 30	Charge: Restaurant bill	68.76
April 3	Payment	450.00
April 15	Charge: Clothing	90.52
April 22	Charge: Eyeglasses	450.85

a) Determine the finance charge on March 23 by using the unpaid balance method. Assume that the interest rate is 1.4% per month.

b) Determine the new account balance on April 23 using the finance charge found in part (a).

c) Determine the average daily balance for the period.

d) Determine the finance charge on March 23 by using the average daily balance method. Assume that the interest rate is 1.4% per month.

e) Determine the new account balance on April 23 using the finance charge found in part (d).

Building a House In Exercises 12–18, the Leungs decided to build a new house. The contractor quoted them a price of \$144,500, including the lot. The taxes on the house would be \$3200 per year, and homeowners' insurance would cost \$450 per year. They have applied for a conventional loan from a bank. The bank is requiring a 15% down payment, and the interest rate is $10\frac{1}{2}$% with 2 points. The Leung's annual income is \$86,500. They have more than 10 monthly payments remaining on each of the following: \$220 for a car, \$175 for new furniture, and \$210 on a college education loan. Their bank will approve a loan that has a total monthly mortgage payment of principal, interest, property taxes, and homeowners' insurance that is less than or equal to 28% of their adjusted monthly income.*

12. What is the required down payment?

13. Determine the amount paid for points.

14. Determine 28% of their adjusted monthly income.

15. Determine the monthly payments of principal and interest for a 30-year loan.

16. Determine their total monthly payments, including homeowners' insurance and taxes.

17. Do the Leungs meet the requirements for the mortgage?

18. a) Determine the total cost of the house (excluding homeowners' insurance and taxes) after 30 years.

b) How much of the total cost is interest, including points?

19. ***Opening a Restaurant*** To raise money to open her own restaurant, Carol Kuper invests \$500 monthly in an ordinary annuity with 6% interest compounded monthly. Determine the accumulated amount in Carol's annuity after 20 years.

20. ***Saving for a New Car*** Elisa Bolotin is a freshman in high school and would like to have \$15,000 in 4 years to buy a car for college. How much should she invest monthly in a sinking fund with 6% interest compounded monthly to accumulate \$15,000 in 4 years?

GROUP PROJECTS

MORTGAGE LOAN

1. The Young family is purchasing a $130,000 house with a VA mortgage. The bank is offering them a 25-year mortgage with an interest rate of 9.5%. They have $20,000 invested that could be used for a down payment. Since they do not need a down payment, Mr. Young wants to keep the money invested. Mrs. Young believes that they should make a down payment of $20,000.

   a) Determine the total cost of the house with no down payment.

   b) Determine the total cost of the house if they make a down payment of $20,000.

   c) Mr. Young believes that the $20,000 investment will have an annual rate of return of 10% compounded quarterly. Assuming that Mr. Young is right, calculate the value of the investment in 25 years. (Use the Compound Interest Formula on page 669.)

   d) If the Youngs use the $20,000 as a down payment, their monthly payments will decrease. Determine the difference of the monthly payments in parts (a) and (b).

   e) Assume that the difference in monthly payments, part (d), is invested each month at a rate of 6% compounded monthly for 25 years. Determine the value of the investment in 25 years. (Use the Ordinary Annuity Formula on page 708.)

   f) Use the information from parts (a)–(e) to analyze the problem. Would you recommend that the Youngs make the down payment of $20,000 and invest the difference in their monthly payments as in part (e) or that they do not make the down payment and keep the $20,000 invested as in part (c)? Explain.

CREDIT CARD TERMS

2. With each credit card comes a credit agreement (or security agreement) the cardholder must sign. Select two members of your group who have a major credit card (MasterCard, Visa, Discover, or American Express). If possible, one of the credit cards should be a gold or platinum card. Obtain a copy of the credit agreement signed by each cardholder and answer questions (a)–(p).

   a) What are the cardholder's responsibilities?

   b) What is the cardholder's maximum line of credit?

   c) What restrictions apply to the use of the credit card?

   d) How many days after the billing date does the cardholder have to make a payment without being charged interest?

   e) What is the minimum monthly payment required, and how is it determined?

   f) What is the interest rate charged on purchases?

   g) How does the bank determine when to start charging interest on purchases?

   h) Is there an annual fee for the credit card? If so, what is it?

   i) What late charge applies if payments are not made on time?

   j) What information is given on the monthly statement?

   k) How is the finance charge computed?

   l) What other fees, if any, may the bank charge you?

   m) If the card is lost, what is the responsibility of the cardholder?

   n) If the card is lost, what is the liability of the cardholder?

   o) What are the advantages of a gold or platinum card over a regular card?

   p) In your opinion, which of the two cards is more desirable? Explain.

CHAPTER 12

Probability

▲ Probability is involved in all games of chance, including lotteries.

WHAT YOU WILL LEARN

- Empirical probability and theoretical probability
- Compound probability, conditional probability, and binomial probability
- Odds against an event and odds in favor of an event
- Expected value
- Tree diagrams
- Mutually exclusive events and independent events
- The counting principle, permutations, and combinations

WHY IT IS IMPORTANT

Each year, millions of Americans play lotteries. If you play a lottery, your hope is to beat the odds and be the person with the winning numbers. Not satisfied with leaving things to chance, mathematicians of the sixteenth, seventeenth, and eighteenth centuries invented the study of probability to determine the likelihood of an event such as winning the lottery.

Although the rules of probability were first applied to gaming, they have many other applications. The cost of your car insurance, the weather forecast, the expected number of people who will attend an outdoor concert if it is raining all involve probability. In this chapter, we will learn many of the important concepts of probability that can help us make informed decisions in our lives.

12.1 THE NATURE OF PROBABILITY

▲ We can determine the probability of an event, such as the probability of someone purchasing a rock CD, based on actual observations.

On Monday morning before a music store opens, the manager reviews sales from the previous week. She notices that more rock compact discs (CDs) were sold than any other type of CD. Based on sales from the previous week, she can determine the probability that the first CD sold on Monday morning will be a rock CD. In this section, we will introduce several definitions that are important to the understanding of the probability of an event. We will also learn how to calculate the probability of an event based on actual observations from an experiment.

History

Probability is used in many areas, including public finance, medicine, insurance, elections, manufacturing, educational tests and measurements, genetics, weather forecasting, investments, opinion polls, the natural sciences, and games of chance. The study of probability originated from the study of games of chance. Archaeologists have found artifacts used in games of chance in Egypt dating from about 3000 B.C.

Mathematical problems relating to games of chance were studied by a number of mathematicians of the Renaissance. Italy's Girolamo Cardano (1501–1576) in his *Liber de Ludo Aleae* (book on the games of chance) presents one of the first systematic computations of probabilities. Although it is basically a gambler's manual, many consider it the first book ever written on probability. A short time later, two French mathematicians, Blaise Pascal (1623–1662) and Pierre de Fermat (1601–1665), worked together studying "the geometry of the die." In 1657, Dutch mathematician Christian Huygens (1629–1695) published *De Ratiociniis in Luno Aleae* (on ratiocination in dice games), which contained the first documented reference to the concept of mathematical expectation (see Section 12.4). Swiss mathematician Jacob Bernoulli (1654–1705), whom many consider the founder of probability theory, is said to have fused pure mathematics with the empirical methods used in statistical experiments. The works of Pierre-Simon de Laplace (1749–1827) dominated probability throughout the nineteenth century.

PROFILE IN MATHEMATICS

Jacob Bernoulli (1654–1705)

Swiss mathematician Jacob Bernoulli was considered a pioneer of probability theory. He was part of a famous family of mathematicians and scientists. In *Ars Conjectand*, published posthumously in 1713, he proposed that an increased degree of accuracy can be obtained by increasing the number of trials of an experiment. This theorem, called Bernoulli's theorem (of probability), is also known as the law of large numbers. Bernoulli's theorem of fluid dynamics, used in aircraft wing design, was developed by Daniel Bernoulli (1700–1782). Daniel was one of Jacob's younger brother Johann's three sons.

The Nature of Probability

Before we discuss the meaning of the word *probability* and learn how to calculate probabilities, we must introduce a few definitions.

An **experiment** is a controlled operation that yields a set of results.

The process by which medical researchers administer experimental drugs to patients to determine their reaction is one type of experiment.

The possible results of an experiment are called its **outcomes**.

For example, the possible outcomes from administering an experimental drug may be a favorable reaction, no reaction, or an adverse reaction.

A die (one of a pair of dice) contains six surfaces, called faces. Each face contains a unique number of dots, from 1 to 6. The sum of the dots on opposite surfaces is 7.

An **event** is a subcollection of the outcomes of an experiment.

For example, when a die is rolled, the event of rolling a number greater than 2 can be satisfied by any one of four outcomes: 3, 4, 5, or 6. The event of rolling a 5 can be satisfied by only one outcome, the 5 itself. The event of rolling an even number can be satisfied by any of three outcomes: 2, 4, or 6.

Probability is classified as either *empirical* (experimental) or *theoretical* (mathematical). ***Empirical probability*** is the relative frequency of occurrence of an event and is determined by actual observations of an experiment. ***Theoretical probability*** is determined through a study of the possible *outcomes* that can occur for the given experiment. We will indicate the probability of an event E by $P(E)$, which is read "P of E."

Empirical Probability

In this section, we will briefly discuss empirical probability. The emphasis in the remaining sections is on theoretical probability. Following is the formula for computing empirical probability, or relative frequency.

EMPIRICAL PROBABILITY (RELATIVE FREQUENCY)

$$P(E) = \frac{\text{number of times event } E \text{ has occurred}}{\text{total number of times the experiment has been performed}}$$

The probability of an event, whether empirical or theoretical, is always a number between 0 and 1, inclusive, and may be expressed as a decimal number or a fraction. An empirical probability of 0 indicates that the event has never occurred. An empirical probability of 1 indicates that the event has always occurred.

EXAMPLE 1 *Heads Up!*

In 100 tosses of a fair coin, 44 landed heads up. Find the empirical probability of the coin landing heads up.

SOLUTION Let E be the event that the coin lands heads up. Then

$$P(E) = \frac{44}{100} = 0.44$$

EXAMPLE 2 *Weight Reduction*

A pharmaceutical company is testing a drug that is supposed to help with weight reduction. The drug is given to 500 individuals with the following outcomes.

Weight reduced	Weight unchanged	Weight increased
379	62	59

If this drug is given to an individual, find the empirical probability that the person's weight is (a) reduced, (b) unchanged, (c) increased.

SOLUTION

a) Let E be the event that the weight is reduced.

$$P(E) = \frac{379}{500} = 0.758$$

b) Let E be the event that the weight is unchanged.

$$P(E) = \frac{62}{500} = 0.124$$

c) Let E be the event that the weight is increased.

$$P(E) = \frac{59}{500} = 0.118$$

Empirical probability is used when probabilities cannot be theoretically calculated. For example, life insurance companies use empirical probabilities to determine the chance of an individual in a certain profession, with certain risk factors, living to age 65.

DID YOU KNOW?

One Blue Eye, One Brown Eye

Seeing a dog or a person with different-colored eyes is unusual, but not that unusual. Features such as eye color are determined genetically, as are certain illnesses. Below is a chart listing some human genetic disorders and their rates of occurrence. Genetics, which has its foundation in probability, is used in explaining the cause of these illnesses, and it plays an important role in research to find cures for genetic disorders.

Disorder/Incidence

Cystic fibrosis:
1 out of 2500 Caucasian births

Down syndrome:
1 out of 800–1000 births

Duchenne muscular dystrophy:
1 out of 3300 male births

Fragile X syndrome:
1 out of 1500 male births
1 out of 2500 female births

Hemophilia A:
1 out of 8500 male births

Polycystic kidney disease:
1 out of 3000 births

Prader–Willi syndrome:
1 out of 12,000 births

Sickle-cell anemia:
1 out of 400–600 African-American births

Tay-Sachs disease:
1 out of 3600 Ashkenazi Jewish births

Empirical Probability in Genetics

Using empirical probability, Gregor Mendel (1822–1884) developed the laws of heredity by crossbreeding different types of "pure" pea plants and observing the relative frequencies of the resulting offspring. These laws became the foundation for the study of genetics. For example, when he crossbred a pure yellow pea plant and a pure green pea plant, the resulting offspring (the first generation) were always yellow; see Fig. 12.1(a). When he crossbred a pure round-seeded pea plant and a pure wrinkled-seeded pea plant, the resulting offspring (the first generation) were always round; see Fig. 12.1(b).

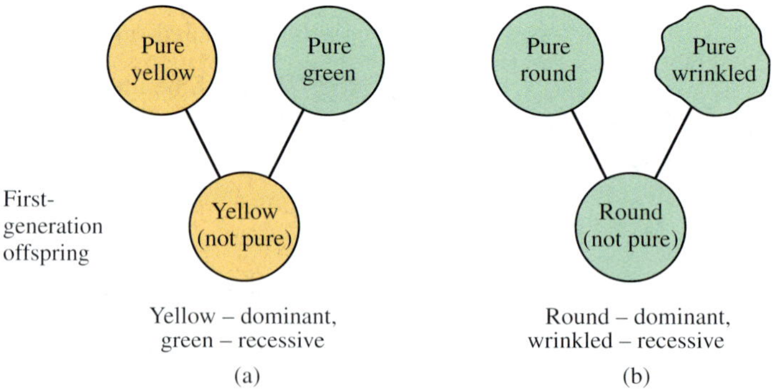

Figure 12.1

Mendel called traits such as yellow color and round seeds ***dominant*** because they overcame or "dominated" the other trait. He labeled the green color and the wrinkled traits ***recessive***.

> **DID YOU KNOW?**
>
> ### *The Royal Disease*
>
> 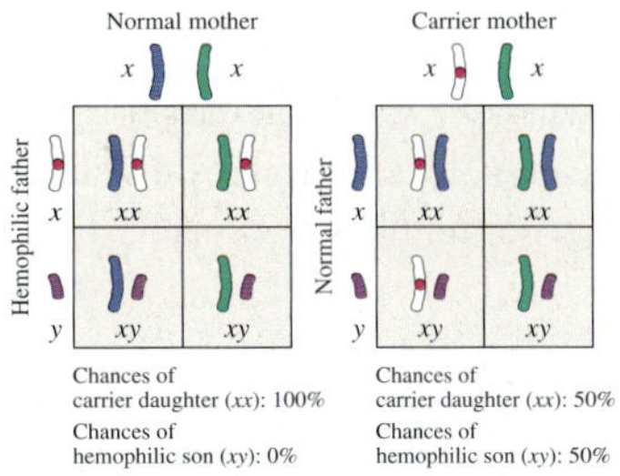
>
>
> The effect of genetic inheritance is dramatically demonstrated by the occurrence of hemophilia in some of the royal families of Europe. Great Britain's Queen Victoria (1819–1901) was the initial carrier. The disease was subsequently introduced into the royal lines of Prussia, Russia, and Spain through the marriages of her children.
>
> Hemophilia, a recessive sex-linked disease, keeps blood from clotting. As a result, even a minor bruise or cut can be dangerous. Females have a second gene that enables the blood to clot, which blocks the effects of the recessive carrier gene; males do not. So even though both males and females may be carriers, the disease afflicts only males.
>
> In humans, genes are located on 23 pairs of *chromosomes*. Each parent contributes one member of each pair to a child. The gene that affects blood clotting is carried on the *x* chromosome. Females have two *x* chromosomes; males have one *x* chromosome and one *y* chromosome. Thus, on the diagram above, *xx* represents a female and *xy* represents a male. A female with the defective gene, symbolized by , will be a carrier, whereas a male with the defective gene will have hemophilia.

Mendel then crossbred the offspring of the first generation. The resulting second-generation offspring had both the dominant and the recessive traits of their grandparents; see Fig. 12.2(a) and (b). What's more, these traits always appeared in approximately a 3 to 1 ratio of dominant to recessive.

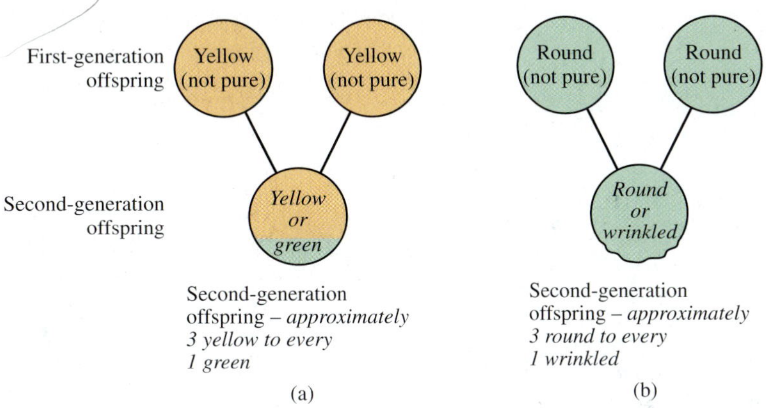

Figure 12.2

Table 12.1 lists some of the actual results of Mendel's experiments with pea plants. Note that the ratio of dominant trait to recessive trait in the second-generation offspring is about 3 to 1 for each experiment. The empirical probability of the dominant trait has also been calculated. How would you find the empirical probability of the recessive trait?

Table 12.1 Second-Generation Offspring

Dominant Trait	Number with Dominant Trait	Recessive Trait	Number with Recessive Trait	Ratio of Dominant to Recessive	P (Dominant Trait)
Yellow seeds	6022	Green seeds	2001	3.01 to 1	$\frac{6022}{8023} \approx 0.75$
Round seeds	5474	Wrinkled seeds	1850	2.96 to 1	$\frac{5474}{7324} \approx 0.75$

From his work, Mendel concluded that the sex cells (now called gametes) of the pure yellow (dominant) pea plant carried some factor that caused the offspring to be yellow and that the gametes of the green variety had a variant factor that "induced the development of green plants." In 1909, Danish geneticist W. Johannsen called these factors "genes." Mendel's work led to the understanding that each pea plant contains two genes for color, one that comes from the mother and the other from the father. If the two genes are alike—for instance, both for yellow plants or both for green plants—the plant will be that color. If the genes for color are different, the plant will grow the color of the dominant gene. Thus, if one parent contributes a gene for the plant to be yellow (dominant) and the other parent contributes a gene for the plant to be green (recessive), the plant will be yellow.

"The laws of probability, so true in general, so fallacious in particular."
Edward Gibbon, 1796

The Law of Large Numbers

Most of us accept that if a "fair coin" is tossed many, many times, it will land heads up approximately half of the time. Intuitively, we can guess that the probability that a fair coin will land heads up is $\frac{1}{2}$. Does that mean that if a coin is tossed twice, it will land heads up exactly once? If a fair coin is tossed 10 times, will there necessarily be five heads? The answer is clearly no. What, then, does it mean when we state that the probability that a fair coin will land heads up is $\frac{1}{2}$? To answer this question, let's examine Table 12.2, which shows what may occur when a fair coin is tossed a given number of times.

Table 12.2

Number of Tosses	Expected Number of Heads	Actual Number of Heads Observed	Relative Frequency of Heads
10	5	4	$\frac{4}{10} = 0.4$
100	50	45	$\frac{45}{100} = 0.45$
1000	500	546	$\frac{546}{1000} = 0.546$
10,000	5000	4852	$\frac{4852}{10{,}000} = 0.4852$
100,000	50,000	49,770	$\frac{49{,}770}{100{,}000} = 0.49770$

The far right column of Table 12.2, the relative frequency of heads, is a ratio of the number of heads observed to the total number of tosses of the coin. The relative frequency is the empirical probability, as defined earlier. Note that as the number of tosses increases, the relative frequency of heads gets closer and closer to $\frac{1}{2}$, or 0.5, which is what we expect.

The nature of probability is summarized by the law of large numbers.

The **law of large numbers** states that probability statements apply in practice to a large number of trials, not to a single trial. It is the relative frequency over the long run that is accurately predictable, not individual events or precise totals.

What does it mean to say that the probability of rolling a 2 on a die is $\frac{1}{6}$? It means that over the long run, on the average, one of every six rolls will result in a 2.

DID YOU KNOW?

Batting Averages

If Joe Mauer of the Minnesota Twins gets three hits in his first three at bats of the season, he is batting a thousand (1.000). Over the course of the 162 games of the season (with three or four at bats per game), however, his batting average will fall closer to 0.347 (his 2006 major league leading batting average). In 2006, out of 521 at bats, Mauer had 181 hits, an average above all other players' but much less than 1.000. His batting average is a relative frequency (or empirical probability) of hits to at bats. It is only the long-term average that we take seriously because it is based on the law of large numbers.

SECTION 12.1 EXERCISES

CONCEPT/WRITING EXERCISES

1. What is an experiment?

2. a) What are outcomes of an experiment?

   b) What is an event?

3. What is empirical probability, and how is empirical probability determined?

4. What are theoretical probabilities based on?

5. Explain in your own words the law of large numbers.

6. Explain in your own words why empirical probabilities are used in determining premiums for life insurance policies.

7. The theoretical probability of a coin landing heads up is $\frac{1}{2}$. Does this probability mean that if a coin is flipped two times, one flip will land heads up? If not, what does it mean?

8. To determine premiums, life insurance companies must compute the probable date of death. On the basis of a great deal of research, Mr. Duncan, age 36, is expected to live another 43.21 years. Does this determination mean that Mr. Duncan will live until he is 79.21 years old? If not, what does it mean?

9. a) Explain how you would find the empirical probability of rolling a 5 on a die.

   b) What do you believe is the empirical probability of rolling a 5?

   c) Determine the empirical probability of rolling a 5 by rolling a die 40 times.

10. The theoretical probability of rolling a 4 on a die is $\frac{1}{6}$. Does this probability mean that if a die is rolled six times one 4 will appear? If not, what does it mean?

PRACTICE THE SKILLS

11. ***Flip a Coin*** Flip a coin 50 times and record the results. Determine the empirical probability of flipping

    a) a head.

    b) a tail.

    c) Does the probability of flipping a head appear to be the same as flipping a tail?

12. ***Pair of Dice*** Roll a pair of dice 60 times and record the sums. Determine the empirical probability of rolling a sum of

    a) 2.

    b) 7.

    c) Does the probability of rolling a sum of 2 appear to be the same as the probability of rolling a sum of 7?

13. ***Roll a Die*** Roll a die 50 times and record the results. Determine the empirical probability of rolling

    a) a 1.

    b) a 4.

    c) Does the probability of rolling a 1 appear to be the same as the probability of rolling a 4? Explain.

14. ***Two Coins*** Flip two coins 50 times and record the number of times exactly one head was obtained. Determine the empirical probability of flipping exactly one head.

PROBLEM SOLVING

15. ***Birds at a Feeder*** The last 30 birds that fed at the Haines' bird feeder were 14 finches, 10 cardinals, and 6 blue jays. Use this information to determine the empirical probability that the next bird to feed from the feeder is

    a) a finch. b) a cardinal. c) a blue jay.

16. ***Music Purchases*** At the Virgin Music store in Times Square, 60 people entering the store were selected at random and were asked to choose their favorite type of music. Of the 60, 12 chose rock, 16 chose country, 8 chose classical, and 24 chose something other than rock, country, or classical. Determine the empirical probability that the next person entering the store favors

a) rock music.

b) country music.

c) something other than rock, country, or classical music.

17. ***Veterinarian*** In a given week, a veterinarian treated the following animals.

Animal	Number Treated
Dog	45
Cat	40
Bird	15
Rabbit	5

Determine the empirical probability that the next animal she treats is

a) a dog.

b) a cat.

c) a rabbit.

18. ***Prader–Willi Syndrome*** In a sample of 50,000 first-born babies, 5 were found to have Prader–Willi syndrome. Find the empirical probability that a family's first child will be born with this syndrome.

19. ***Studying Abroad*** The table at the top of the right-hand column shows the 10 most popular destinations for U.S. college students studying abroad for the 2003–2004 school year.

Destination	Number of Students
United Kingdom	32,237
Italy	21,922
Spain	20,080
France	13,718
Australia	11,418
Mexico	9293
Germany	5985
Ireland	5198
China	4737
Costa Rica	4510
Total	129,098

Source: Institute of International Education

If a student were selected at random from those who studied abroad in 2003–2004 at one of the destinations in the table, determine the empirical probability that the student studied in

a) China. **b)** Italy. **c)** Australia.

20. ***Travel Web Sites*** The following table shows the number of visitors, in millions, to the five most frequently visited travel web sites in November 2005.

Web Site	Number of Visitors
MapQuest	31,200,000
Expedia	14,200,000
Travelocity	11,200,000
Orbitz	10,900,000
Southwest Airlines	7,100,000
Total	74,600,000

Source: Nielsen/NetRatings

Assuming this trend continues, if a person chooses to visit only one of the listed web sites, determine the empirical probability the person will visit

a) Expedia.

b) Travelocity.

c) Southwest Airlines.

21. ***Bird Flu*** As of March 24, 2006, a total of 186 people had contracted the bird flu worldwide. The graph below shows the number of bird flu cases by country. If a person contracted the bird flu, determine the empirical probability that the person contracted it in

a) Indonesia.

b) Vietnam.

c) Turkey.

Bird Flu Cases by Country

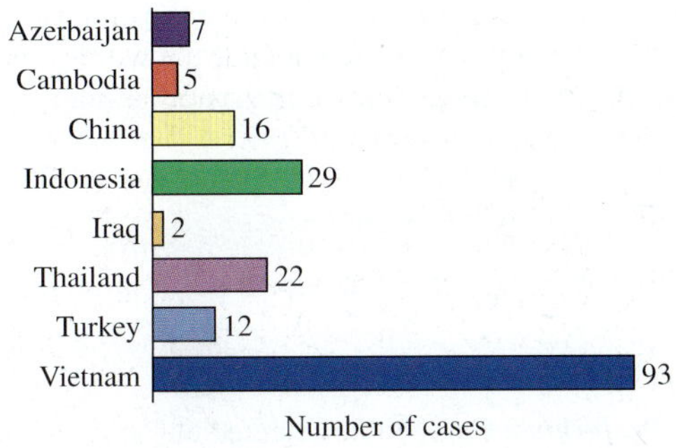

Sources: World Health Organization and Gannett News Service

22. ***Grade Distribution*** Mr. Doole's grade distribution over the past 3 years for a course in college algebra is shown in the chart below.

Grade	Number
A	43
B	182
C	260
D	90
F	62
I	8

If Sue Gilligan plans to take college algebra with Mr. Doole, determine the empirical probability that she receives a grade of

a) A. b) C. c) D or higher.

23. ***Election*** In an election for student council president at Russell Sage College, 80 students were polled and asked for whom they planned to vote. The table shows the results of the poll.

Candidate	Votes
Allison	22
Emily	18
Kimberly	20
Johanna	14
Other	6

If one student from Russell Sage College is selected at random, determine the empirical probability that the person planned to vote for

a) Allison.

b) Emily.

c) Kimberly.

d) Johanna.

e) Someone other than the four people listed above.

24. ***U.S. Foreign-Born Workers*** In 2004, there were 20 million foreign-born workers in the United States. The circle graph below shows the occupations of those workers.

Occupations of Foreign-Born Workers in the United States

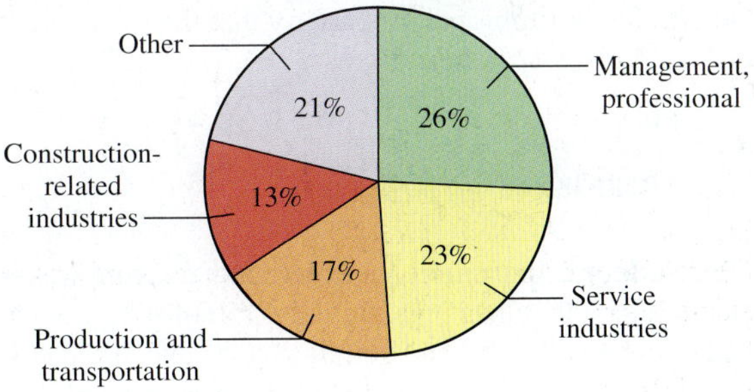

Source: Knight-Ridder

If one foreign-born worker is selected at random, determine the empirical probability the person's occupation is in

a) the service industries.

b) construction-related industries.

c) management or is a professional.

25. ***Hitting a Bull's-Eye*** The pattern of hits shown on the target resulted from a marksman firing 20 rounds. For a single shot,

a) determine the empirical probability that the marksman hits the 50-point bull's-eye (the center of the target).

b) determine the empirical probability that the marksman does not hit the bull's-eye.

c) determine the empirical probability that the marksman scores at least 20 points.

d) determine the empirical probability that the marksman does not score any points (the area outside the large circle).

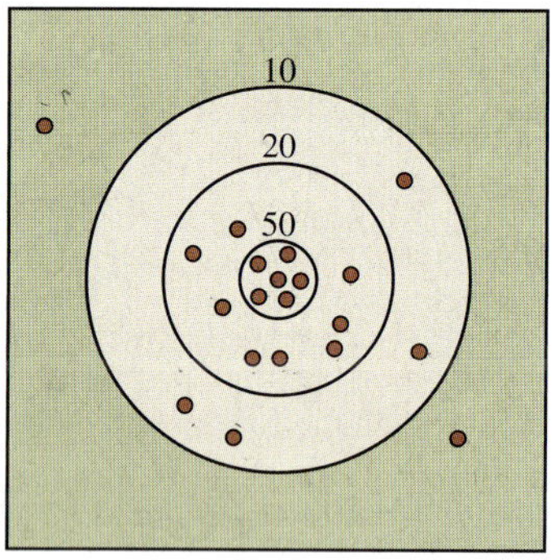

26. ***Rock Toss*** Jim Handy finds an irregularly shaped five-sided rock. He labels each side and tosses the rock 100 times. The results of his tosses are shown in the table. Determine the empirical probability that the rock will land on side 4 if tossed again.

Side	1	2	3	4	5
Frequency	32	18	15	13	22

27. ***Cell Biology Experiment*** An experimental serum was injected into 500 guinea pigs. Initially, 150 of the guinea pigs had circular cells, 250 had elliptical cells, and 100 had irregularly shaped cells. After the serum was injected, none of the guinea pigs with circular cells were affected, 50 with elliptical cells were affected, and all those with irregular cells were affected. Determine the empirical probability that a guinea pig with (a) circular cells, (b) elliptical cells, and (c) irregular cells will be affected by injection of the serum.

28. ***Baby Gender*** In the United States, more male babies are born than female. In 2004, 2,104,661 males were born and 2,007,391 females were born. Determine the empirical probability of an individual being born

a) male. b) female.

29. ***Mendel's Experiment*** In one of his experiments (see pages 726–727), Mendel crossbred nonpure purple flower pea plants. These purple pea plants had two traits for flowers, purple (dominant) and white (recessive). The result of this crossbreeding was 705 second-generation plants with purple flowers and 224 second-generation plants with white flowers. Determine the empirical probability of a second-generation plant having

a) white flowers. b) purple flowers.

30. ***Second-Generation Offspring*** In another experiment, Mendel crossbred nonpure tall pea plants. As a result, the second-generation offspring were 787 tall plants and 277 short plants. Determine the empirical probability of a second-generation plant being

a) tall. b) short.

CHALLENGE PROBLEM/GROUP ACTIVITY

31. a) ***Design an Experiment*** Which do you believe is used more frequently in a magazine or newspaper article, the word *a* or the word *the*?

b) Design an experiment to determine the empirical probabilities (or relative frequencies) of the words *a* and *the* appearing in a magazine or newspaper article.

c) Perform the experiment in part (b) and determine the empirical probabilities.

d) Which word, *a* or *the*, appears to occur more frequently?

RECREATIONAL MATHEMATICS

32. ***Cola Preference*** Can people selected at random distinguish Coke from Pepsi? Which do they prefer?

a) Design an experiment to determine the empirical probability that a person selected at random can select Coke when given samples of both Coke and Pepsi.

b) Perform the experiment in part (a) and determine the empirical probability.

c) Determine the empirical probability that a person selected at random will prefer Coke over Pepsi.

INTERNET/RESEARCH ACTIVITIES

33. Write a paper on how insurance companies use empirical probabilities in determining insurance premiums. An insurance agent may be able to direct you to a source of information.

34. Write a paper on how Gregor Mendel's use of empirical probability led to the development of the science of genetics. You may want to check with a biology professor to determine references to use.

12.2 THEORETICAL PROBABILITY

▲ We use probability to determine the likelihood of winning money at a casino.

Should you spend the money for a stamp to return a sweepstakes ticket? What are your chances of winning a lottery? If you go to a carnival, bazaar, or casino, which games provide the greatest chance of winning? These and similar questions can be answered once you have an understanding of theoretical probability that we will discuss in this section.

In the remainder of this chapter, the word probability will refer to theoretical probability.

Recall from Section 12.1 that the results of an experiment are called outcomes. When you roll a die and observe the number of points that face up, the possible outcomes are 1, 2, 3, 4, 5, and 6. It is equally likely that you will roll any one of the possible numbers.

If each outcome of an experiment has the same chance of occurring as any other outcome, we say that the outcomes are **equally likely outcomes**.

Can you think of a second set of equally likely outcomes when a die is rolled? An odd number is as likely to be rolled as an even number. Therefore, odd and even numbers are another set of equally likely outcomes.

TIMELY TIP To be able to do the problems in this section and the remainder of the chapter, you must have a thorough understanding of fractions. If you have forgotten how to work with fractions, we strongly suggest that you review Section 5.3 before beginning this section.

If an event E has *equally likely outcomes*, the probability of event E, symbolized by $P(E)$, may be calculated with the following formula.

PROBABILITY

$$P(E) = \frac{\text{number of outcomes favorable to } E}{\text{total number of possible outcomes}}$$

Example 1 illustrates how to use this formula.

EXAMPLE 1 ***Finding Probabilities***

A die is rolled. Find the probability of rolling

a) a 3. b) an even number. c) a number greater than 2.

d) a 7. e) a number less than 7.

SOLUTION

a) There are six possible equally likely outcomes: 1, 2, 3, 4, 5, and 6. The event of rolling a 3 can occur in only one way.

$$P(3) = \frac{\text{number of outcomes that will result in a 3}}{\text{total number of possible outcomes}} = \frac{1}{6}$$

b) The event of rolling an even number can occur in three ways: 2, 4, or 6.

$$P(\text{even number}) = \frac{\text{number of outcomes that result in an even number}}{\text{total number of possible outcomes}} = \frac{3}{6} = \frac{1}{2}$$

c) Four numbers are greater than 2, namely, 3, 4, 5 and 6.

$$P(\text{number greater than 2}) = \frac{4}{6} = \frac{2}{3}$$

d) No outcomes will result in a 7. Thus, the event cannot occur and the probability is 0.

$$P(7) = \frac{0}{6} = 0$$

e) All the outcomes 1 through 6 are less than 7. Thus, the event must occur and the probability is 1.

$$P(\text{number less than 7}) = \frac{6}{6} = 1$$

Four important facts about probability follow.

IMPORTANT FACTS

1. The probability of an event that cannot occur is 0.
2. The probability of an event that must occur is 1.
3. Every probability is a number between 0 and 1 inclusive; that is, $0 \leq P(E) \leq 1$.
4. The sum of the probabilities of all possible outcomes of an experiment is 1.

▲ Northern cardinal

EXAMPLE 2 *Choosing One Bird from a List*

The names of 15 birds and their food preferences are listed in Table 12.3 on page 735. Each of the 15 birds' names is listed on a slip of paper, and the 15 slips are placed in a bag. One slip is to be selected at random from the bag. Find the probability that the slip contains the name of

a) a finch (any type listed).

b) a bird that has a high attractiveness to cracked corn.

c) a bird that has a low attractiveness to peanut kernels, *and* a low attractiveness to cracked corn, *and* a high attractiveness to black-striped sunflower seeds.

d) a bird that has a high attractiveness to either peanut kernels *or* cracked corn (or both).

Table 12.3 Birds and Their Food Preferences

Bird	Peanut Kernels	Cracked Corn	Black-Striped Sunflower Seeds
American goldfinch	L	L	H
Blue jay	H	M	H
Chickadee	M	L	H
Common grackle	M	H	H
Evening grosbeak	L	L	H
House finch	M	L	H
House sparrow	L	M	M
Mourning dove	L	M	M
Northern cardinal	L	L	H
Purple finch	L	L	H
Scrub jay	H	L	H
Song sparrow	L	L	M
Tufted titmouse	H	L	H
White-crowned sparrow	H	M	H
White-throated sparrow	H	H	H

Source: *How to Attract Birds* (Ortho Books)

Note: H = high attractiveness; M = medium attractiveness; L = low attractiveness.

SOLUTION

a) Three of the 15 birds listed are finches (American goldfinch, house finch, and purple finch).

$$P(\text{finch}) = \frac{3}{15} = \frac{1}{5}$$

b) Two of the 15 birds listed have a high attractiveness to cracked corn (common grackle and white-throated sparrow).

$$P(\text{high attractiveness to cracked corn}) = \frac{2}{15}$$

c) Reading across the rows reveals that 4 birds have a low attractiveness to peanut kernels, a low attractiveness to cracked corn, and a high attractiveness to black-striped sunflower seeds (American goldfinch, evening grosbeak, northern cardinal, and purple finch).

$$P\left(\begin{array}{c}\text{low attractiveness to peanuts, and low to}\\ \text{corn, and high to black-striped sunflower seeds}\end{array}\right) = \frac{4}{15}$$

d) Six birds have a high attractiveness to either peanut kernels or to cracked corn (or both). They are the blue jay, common grackle, scrub jay, tufted titmouse, white-crowned sparrow, and white-throated sparrow.

$$P(\text{high attractiveness to peanut kernels or cracked corn}) = \frac{6}{15} = \frac{2}{5}$$

10. Using the definition of probability, explain in your own words why the probability of an event that cannot occur is 0.

11. Between what two numbers (inclusively) will all probabilities lie?

12. What is the sum of all the probabilities of all possible outcomes of an experiment?

PRACTICE THE SKILLS

13. ***Multiple-Choice Test*** A multiple-choice test has five possible answers for each question.

a) If you guess at an answer, what is the probability that you select the correct answer for one particular question?

b) If you eliminate one of the five possible answers and guess from the remaining possibilities, what is the probability that you select the correct answer to that question?

14. ***Remote Control*** A TV remote control has keys for channels 0 through 9. If you select one key at random,

a) what is the probability that you press channel 6?

b) what is the probability that you press a key for an even number?

c) what is the probability that you press a key for a number less than 7?

15. ***Raffle*** In a raffle where one number is chosen, determine the probability that you would win if you have a choice of 40 numbers to choose from. Explain your answer.

16. ***Raffle*** In a raffle where one number is chosen, determine the probability that you would win if you have a choice of 52 numbers to choose from. Explain your answer.

Select a Card *In Exercises 17–26, one card is selected at random from a deck of cards. Determine the probability that the card selected is*

17. a 5.

18. a 5 or a 7.

19. not a 5.

20. the five of diamonds.

21. a black card.

22. a diamond.

23. a red card or a black card.

24. a red card and a black card.

25. a card greater than 4 and less than 9.

26. a king and a club.

Spin the Spinner *In Exercises 27–30, assume that the spinner cannot land on a line. Determine the probability that the spinner lands on (a) red, (b) green, (c) yellow, (d) blue.*

27.

28.

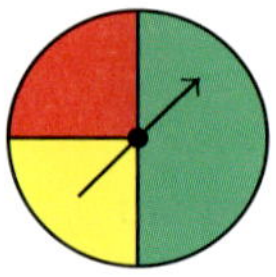

29.

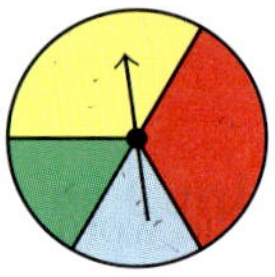

30.

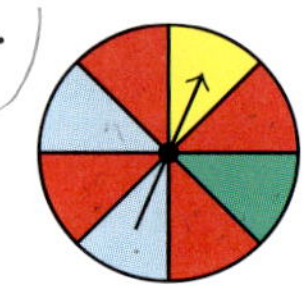

Picnic *In Exercises 31–34, a cooler at a picnic contains 100 cans of soda covered by ice. There are 30 cans of cola, 40 cans of orange soda, 10 cans of ginger ale, and 20 cans of root beer. The cans are all the same size and shape. If one can is selected at random from the cooler, determine the probability that the soda selected is*

31. root beer.

32. cola or orange soda.

33. cola, root beer, or orange soda.

34. ginger ale.

Wheel of Fortune *In Exercises 35–38, use the small replica of the Wheel of Fortune.*

If the wheel is spun at random, determine the probability of the sector indicated stopping under the pointer.

35. $600

36. A number greater than $700

b) $P(\text{not a } 5) = 1 - P(5) = 1 - \frac{1}{13} = \frac{12}{13}$

This probability could also have been found by noting that there are 48 cards that are not 5's in a deck of 52 cards.

$$P(\text{not a } 5) = \frac{48}{52} = \frac{12}{13}$$

c) There are 13 diamonds in the deck.

$$P(\text{diamond}) = \frac{13}{52} = \frac{1}{4}$$

d) There are 4 jacks, 4 queens, and 4 kings, or a total of 12 picture cards.

$$P(\text{jack } or \text{ queen } or \text{ king}) = \frac{12}{52} = \frac{3}{13}$$

e) The word *and* means that *both* events must occur. Since it is not possible to select one card that is both a heart and a club, the probability is 0.

$$P(\text{heart and club}) = \frac{0}{52} = 0$$

f) The cards that are both greater than 6 and less than 9 are 7's and 8's. There are four 7's and four 8's, or a total of eight cards.

$$P(\text{greater than 6 } and \text{ less than 9}) = \frac{8}{52} = \frac{2}{13}$$

SECTION 12.2 EXERCISES

CONCEPT/WRITING EXERCISES

1. What are equally likely outcomes?

2. Explain in your own words how to find the theoretical probability of an event.

3. State the relationship that exists for $P(A)$ and $P(\text{not } A)$.

4. If the probability that an event occurs is $\frac{3}{7}$, determine the probability that the event does not occur.

5. If the probability that an event occurs is 0.7, determine the probability that the event does not occur.

6. If the probability that an event does not occur is 0.65, determine the probability that the event occurs.

7. If the probability that an event does not occur is $\frac{5}{12}$, determine the probability that the event occurs.

8. How many of each of the following are there in a standard deck of cards?

   a) Total cards

   b) Hearts

   c) Red cards

   d) Fives

   e) Black cards

   f) Picture cards

   g) Aces

   h) Queens

9. Using the definition of probability, explain in your own words why the probability of an event that must occur is 1.

10. Using the definition of probability, explain in your own words why the probability of an event that cannot occur is 0.

11. Between what two numbers (inclusively) will all probabilities lie?

12. What is the sum of all the probabilities of all possible outcomes of an experiment?

PRACTICE THE SKILLS

13. ***Multiple-Choice Test*** A multiple-choice test has five possible answers for each question.

   a) If you guess at an answer, what is the probability that you select the correct answer for one particular question?

   b) If you eliminate one of the five possible answers and guess from the remaining possibilities, what is the probability that you select the correct answer to that question?

14. ***Remote Control*** A TV remote control has keys for channels 0 through 9. If you select one key at random,

   a) what is the probability that you press channel 6?

   b) what is the probability that you press a key for an even number?

   c) what is the probability that you press a key for a number less than 7?

15. ***Raffle*** In a raffle where one number is chosen, determine the probability that you would win if you have a choice of 40 numbers to choose from. Explain your answer.

16. ***Raffle*** In a raffle where one number is chosen, determine the probability that you would win if you have a choice of 52 numbers to choose from. Explain your answer.

Select a Card *In Exercises 17–26, one card is selected at random from a deck of cards. Determine the probability that the card selected is*

17. a 5.

18. a 5 or a 7.

19. not a 5.

20. the five of diamonds.

21. a black card.

22. a diamond.

23. a red card or a black card.

24. a red card and a black card.

25. a card greater than 4 and less than 9.

26. a king and a club.

Spin the Spinner *In Exercises 27–30, assume that the spinner cannot land on a line. Determine the probability that the spinner lands on (a) red, (b) green, (c) yellow, (d) blue.*

27.

28.

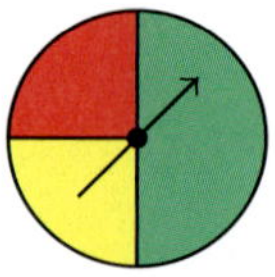

29.

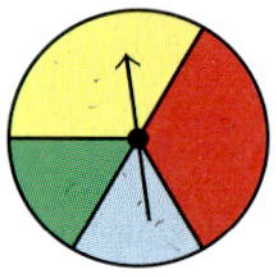

30.

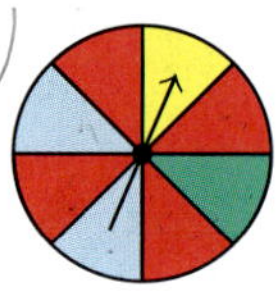

Picnic *In Exercises 31–34, a cooler at a picnic contains 100 cans of soda covered by ice. There are 30 cans of cola, 40 cans of orange soda, 10 cans of ginger ale, and 20 cans of root beer. The cans are all the same size and shape. If one can is selected at random from the cooler, determine the probability that the soda selected is*

31. root beer.

32. cola or orange soda.

33. cola, root beer, or orange soda.

34. ginger ale.

Wheel of Fortune *In Exercises 35–38, use the small replica of the Wheel of Fortune.*

If the wheel is spun at random, determine the probability of the sector indicated stopping under the pointer.

35. $600

36. A number greater than $700

37. Lose a turn or Bankrupt

38. \$2500 or Surprise

Basketballs *In Exercises 39–42, 30 basketballs (15 Spalding, 10 Wilson, and 5 other brand-name balls) are on a basketball court. Barry Wood closes his eyes and arbitrarily picks up a ball from the court. Determine the probability that the ball selected is*

39. a Spalding.

40. a Wilson.

41. not a Wilson.

42. a Wilson or a Spalding.

Traffic Light *In Exercises 43–46, a traffic light is red for 25 sec, yellow for 5 sec, and green for 55 sec. What is the probability that when you reach the light,*

43. the light is green.

44. the light is yellow.

45. the light is not red.

46. the light is not green.

TENNESSEE *In Exercises 47–52, each individual letter of the word TENNESSEE is placed on a piece of paper and all 9 pieces of paper are placed in a hat. If one letter is selected at random from the hat, determine the probability that*

47. the letter *S* is selected.

48. the letter *S* is not selected.

49. a consonant is selected.

50. the letter *T* or *N* is selected.

51. the letter *W* is selected.

52. the letter *V* is not selected.

Hurricane Evacuees *Following Hurricanes Katrina and Rita, many residents had to evacuate their homes. Some relocated in the same state, whereas others relocated to different states. In Exercises 53–56, use the following map, which shows the number of evacuees residing in each state and in the District of Columbia as of February 27, 2006.*

After the Storm:
The Spread of Evacuees

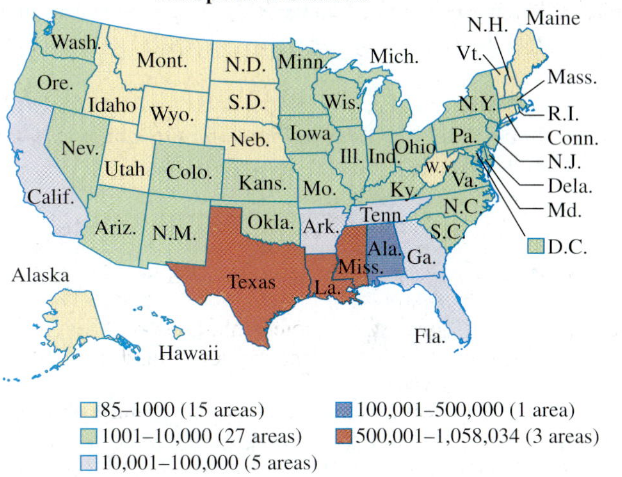

Source: Federal Emergency Management Agency

If 1 of the 51 areas illustrated on the map is selected at random, determine the probability that the area contained

53. between 500,001 and 1,058,034 residents who evacuated their homes.

54. between 10,001 and 100,000 residents who evacuated their homes.

55. between 1001 and 10,000 residents who evacuated their homes.

56. between 85 and 1000 residents who evacuated their homes.

Dart Board *In Exercises 57–60, a dart is thrown randomly and sticks on the circular dart board with 26 partitions, as shown.*

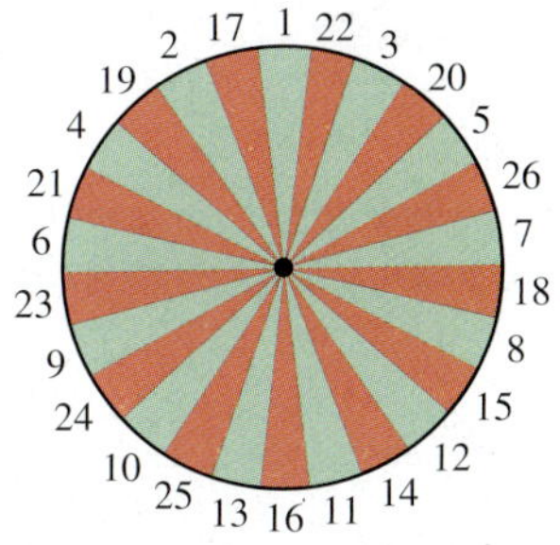

Assuming the dart cannot land on the black area or on a border between colors, determine the probability that the dart lands on

57. the area marked 15.

58. an orange area.

59. an area marked with a number greater than or equal to 22.

60. an area marked with a number greater than 6 and less than or equal to 9.

Car Dealership *In Exercises 61–66, refer to the following table, which shows the type and manufacturer of vehicles at a specific car dealership.*

	General Motors	Toyota	Total
Car	55	30	85
SUV	28	17	45
Total	83	47	130

If one person selects a vehicle at random from the dealership, determine the probability that the person selects

61. a car.

62. an SUV.

63. a vehicle manufactured by General Motors.

64. a vehicle manufactured by Toyota.

65. a car manufactured by General Motors.

66. an SUV manufactured by General Motors.

Stocking Peanut Butter *In Exercises 67–72, refer to the following table, which contains information about a shopping cart full of peanut butter jars that must be stocked on a shelf.*

Brand	Smooth	Chunky	Total
Peter Pan	10	6	16
Jif	7	5	12
Skippy	4	3	7
Other	2	1	3
Total	23	15	38

If a stock clerk selects one jar at random to place on the shelf, determine the probability he selects a jar of

67. Jif.

68. Skippy.

69. a chunky peanut butter.

70. a smooth peanut butter.

71. Peter Pan smooth peanut butter.

72. Jif chunky peanut butter.

Bean Bag Toss *In Exercises 73–77, a bean bag is randomly thrown onto the square table top shown below and does not touch a line.*

Determine the probability that the bean bag lands on

73. a red area.

74. a green area.

75. a yellow area.

76. a red or green area.

77. a yellow or green area.

78. a red or yellow area.

CHALLENGE PROBLEMS/GROUP ACTIVITIES

Before working Exercises 79 and 80, reread the material on genetics in Section 12.1.

79. ***Genetics*** Cystic fibrosis is an inherited disease that occurs in about 1 in every 2500 Caucasian births in North America and in about 1 in every 250,000 non-Caucasian births in North America. Let's denote the cystic fibrosis gene as *c* and a disease-free gene as *C*. Since the disease-free gene is dominant, only a person with *cc* genes will have the disease. A person who has *Cc* genes is a carrier of cystic fibrosis but does not actually have the disease. If one parent has *CC* genes and the other parent has *cc* genes, determine the probability that

a) an offspring will inherit cystic fibrosis, that is, *cc* genes.

b) an offspring will be a carrier of cystic fibrosis but not contract the disease.

80. ***Genetics*** Sickle-cell anemia is an inherited disease that occurs in about 1 in every 500 African-American births and about 1 in every 160,000 non-African-American births. Unlike cystic fibrosis, in which the cystic fibrosis gene is recessive, sickle-cell anemia is *codominant.* In other words, a person inheriting two sickle-cell genes will have sickle-cell anemia, whereas a person inheriting only one of the sickle-cell genes will have a mild version of sickle-cell anemia, called *sickle-cell trait.* Let's call the disease-free

Table 12.3 Birds and Their Food Preferences

Bird	Peanut Kernels	Cracked Corn	Black-Striped Sunflower Seeds
American goldfinch	L	L	H
Blue jay	H	M	H
Chickadee	M	L	H
Common grackle	M	H	H
Evening grosbeak	L	L	H
House finch	M	L	H
House sparrow	L	M	M
Mourning dove	L	M	M
Northern cardinal	L	L	H
Purple finch	L	L	H
Scrub jay	H	L	H
Song sparrow	L	L	M
Tufted titmouse	H	L	H
White-crowned sparrow	H	M	H
White-throated sparrow	H	H	H

Source: *How to Attract Birds* (Ortho Books)

Note: H = high attractiveness; M = medium attractiveness; L = low attractiveness.

SOLUTION

a) Three of the 15 birds listed are finches (American goldfinch, house finch, and purple finch).

$$P(\text{finch}) = \frac{3}{15} = \frac{1}{5}$$

b) Two of the 15 birds listed have a high attractiveness to cracked corn (common grackle and white-throated sparrow).

$$P(\text{high attractiveness to cracked corn}) = \frac{2}{15}$$

c) Reading across the rows reveals that 4 birds have a low attractiveness to peanut kernels, a low attractiveness to cracked corn, and a high attractiveness to black-striped sunflower seeds (American goldfinch, evening grosbeak, northern cardinal, and purple finch).

$$P\left(\begin{array}{c}\text{low attractiveness to peanuts, and low to}\\ \text{corn, and high to black-striped sunflower seeds}\end{array}\right) = \frac{4}{15}$$

d) Six birds have a high attractiveness to either peanut kernels or to cracked corn (or both). They are the blue jay, common grackle, scrub jay, tufted titmouse, white-crowned sparrow, and white-throated sparrow.

$$P(\text{high attractiveness to peanut kernels or cracked corn}) = \frac{6}{15} = \frac{2}{5}$$

In any experiment, an event must either occur or not occur. *The sum of the probability that an event will occur and the probability that it will not occur is 1.* Thus, for any event A we conclude that

$$P(A) + P(\text{not } A) = 1$$

or

$$P(\text{not } A) = 1 - P(A)$$

For example, if the probability that event A will occur is $\frac{5}{12}$, the probability that event A will not occur is $1 - \frac{5}{12}$, or $\frac{7}{12}$. Similarly, if the probability that event A will not occur is 0.3, the probability that event A will occur is $1 - 0.3 = 0.7$, or $\frac{7}{10}$. We make use of this concept in Example 3.

EXAMPLE 3 *Selecting One Card from a Deck*

A standard deck of 52 playing cards is shown in Figure 12.3. The deck consists of four suits: hearts, clubs, diamonds, and spades. Each suit has 13 cards, including numbered cards ace (1) through 10 and three picture (or face) cards, the jack, the queen, and the king. Hearts and diamonds are red cards; clubs and spades are black cards. There are 12 picture cards, consisting of 4 jacks, 4 queens, and 4 kings. One card is to be selected at random from the deck of cards. Find the probability that the card selected is

a) a 5.

b) not a 5.

c) a diamond.

d) a jack *or* queen *or* king (a picture card).

e) a heart *and* a club.

f) a card greater than 6 *and* less than 9.

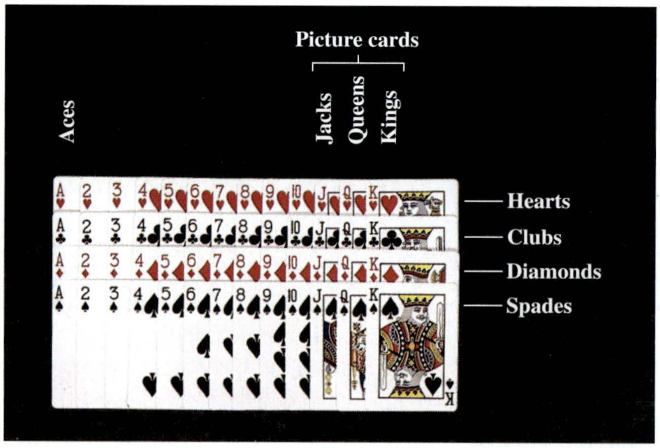

Figure 12.3

SOLUTION

a) There are four 5's in a deck of 52 cards.

$$P(5) = \frac{4}{52} = \frac{1}{13}$$

genes s_1 and the sickle cell gene s_2. If both parents have s_1s_2 genes, determine the probability that

a) an offspring will have sickle-cell anemia.

b) an offspring will have the sickle-cell trait.

c) an offspring will have neither sickle-cell anemia nor the sickle-cell trait.

In Exercises 81 and 82, the solutions involve material that we will discuss in later sections of the chapter. Try to solve them before reading ahead.

81. ***Marbles*** A bottle contains two red and two green marbles, and a second bottle also contains two red and two green marbles. If you select one marble at random from each bottle, determine the probability (to be discussed in Section 12.6) that you obtain

a) two red marbles.

b) two green marbles.

c) a red marble from the first bottle and a green marble from the second bottle.

82. ***Birds*** Consider Table 12.3 on page 735. Suppose you are told that one bird's name was selected at random from the birds listed and the bird selected has a low attractiveness to peanut kernels. Determine the probability (to be discussed in Section 12.7) that

a) the bird is a sparrow.

b) the bird has a high attractiveness to cracked corn.

c) the bird has a high attractiveness to black-striped sunflower seeds.

RECREATIONAL MATHEMATICS

83. ***Dice*** On a die, the sum of the dots on the opposite faces is seven. Two six-sided dice are placed together on top of one another, on a table, as shown in the figure below. The top and bottom faces of the bottom die and the bottom face of the top die cannot be seen. If you walk around the table, what is the sum of all the dots on all the visible faces of the dice?

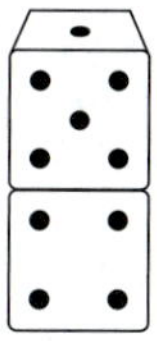

INTERNET/RESEARCH ACTIVITY

84. On page 724, we briefly discuss Jacob Bernoulli. The Bernoulli family produced several prominent mathematicians, including Jacob I, Johann I, and Daniel. Write a paper on the Bernoulli family, indicating some of the accomplishments of each of the three Bernoullis named and their relationship to one another. Indicate which Bernoulli the Bernoulli numbers are named after, which Bernoulli the Bernoulli theorem in statistics is named after, and which Bernoulli the Bernoulli theorem of fluid dynamics is named after.

12.3 ODDS

▲ What are the odds against the Chicago White Sox winning the World Series this year?

The odds against winning the Fabulous Fortune lottery are about 2.6 million to 1. The odds against being audited by the IRS this year are about 47 to 1. The odds against the Chicago White Sox winning the World Series this year may be 6 to 1. We see the word *odds* daily in newspapers and magazines and often use it ourselves. Yet there is widespread misunderstanding of its meaning. In this section, we will explain the meaning of odds. We will also discuss how to determine odds against an event and how to determine odds in favor of an event.

Odds Against an Event

The odds given at horse races, at craps, and at all gambling games in Las Vegas and other casinos throughout the world are always *odds against* unless they are otherwise specified. The ***odds against*** an event is a ratio of the probability that the event will fail to occur (failure) to the probability that the event will occur (success). Thus, *to find odds you must first know or determine the probability of success and the probability of failure.*

DID YOU KNOW?

Gambling in Ancient Times

Archaeologists have found evidence of gambling in all cultures, from the Stone Age Australian aborigines to the ancient Egyptians, and across cultures touched by the Roman Empire. There is an equally long history of moral and legal opposition to gambling.

$$\text{Odds against event} = \frac{P(\text{event fails to occur})}{P(\text{event occurs})} = \frac{P(\text{failure})}{P(\text{success})}$$

EXAMPLE 1 *Rolling a 4*

Determine the odds against rolling a 4 on one roll of a die.

SOLUTION Before we can determine the odds, we must first determine the probability of rolling a 4 (success) and the probability of not rolling a 4 (failure). When a die is rolled there are six possible outcomes: 1, 2, 3, 4, 5, and 6.

$$P(\text{rolling a 4}) = \frac{1}{6} \qquad P(\text{failure to roll a 4}) = \frac{5}{6}$$

Now that we know the probabilities of success and failure, we can determine the odds against rolling a 4.

$$\text{Odds against rolling a 4} = \frac{P(\text{failure to roll a 4})}{P(\text{rolling a 4})}$$

$$= \frac{\frac{5}{6}}{\frac{1}{6}} = \frac{5}{\cancel{6}_1} \cdot \frac{\cancel{6}^1}{1} = \frac{5}{1}$$

The ratio $\frac{5}{1}$ is commonly written as 5:1 and is read "5 to 1." Thus, the odds against rolling a 4 are 5 to 1.

TIMELY TIP The denominators of the probabilities in an odds problem will always divide out, as was shown in Example 1.

In Example 1, we considered the possible outcomes of the die: 1, 2, 3, 4, 5, 6. Over the long run, one of every six rolls will result in a 4, and five of every six rolls will result in a number other than 4. Therefore, if a person is gambling, for each dollar bet in favor of the rolling of a 4, $5 should be bet against the rolling of a 4 if the person is to break even. The person betting in favor of the rolling of a 4 will either lose $1 (if a number other than a 4 is rolled) or win $5 (if a 4 is rolled). The person betting against the rolling of a 4 will either win $1 (if a number other than a 4 is rolled) or lose $5 (if a 4 is rolled). If this game is played for a long enough period, each player theoretically will break even.

Example 2 involves a circle graph that contains percents; see Fig. 12.4 on page 743. Before we discuss Example 2, let us briefly discuss percents. Recall that probabilities are numbers between 0 and 1, inclusive. We can change a percent between 0% and 100% to a probability by writing the percent as a fraction or a decimal number. In Fig. 12.4, we see 51% in one of the sectors (or areas) of the circle. To change 51% to a probability, we can write $\frac{51}{100}$ or 0.51. Note that both the fraction and the decimal number are numbers between 0 and 1, inclusive.

EXAMPLE 2 *Hours Worked per Week*

The circle graph in Figure 12.4 shows the percent of U.S. workers who work various hours per week. If one U.S. worker is selected at random, use the graph to determine the odds against the person working 35–39 hours per week.

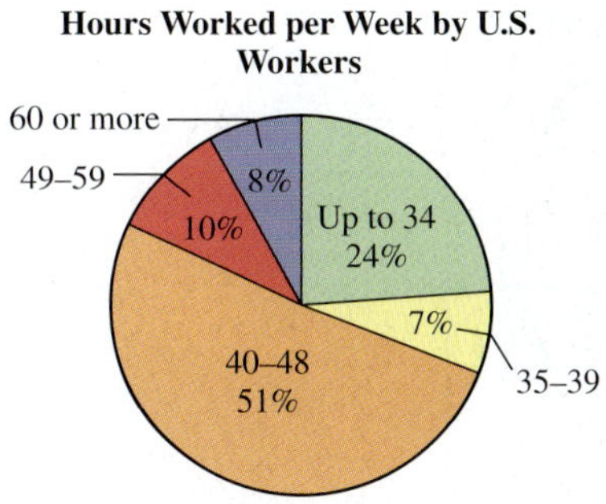

Source: *Statistical Abstract of the United States*

Figure 12.4

SOLUTION From the graph, we can determine that 7% of U.S. workers work 35–39 hours per week. When written as a probability, 7% is $\frac{7}{100}$. Therefore, the probability that a U.S. worker works 35–39 hours per week is $\frac{7}{100}$. The probability that a U.S. worker *does not* work 35–39 hours per week is $1 - \frac{7}{100} = \frac{93}{100}$.

$$\begin{aligned}\text{Odds against the person working 35–39 hours per week} &= \frac{P(\text{person does not work 35–39 hours per week})}{P(\text{person works 35–39 hours per week})}\\ &= \frac{93/100}{7/100}\\ &= \frac{93}{100}\cdot\frac{100}{7} = \frac{93}{7}, \text{ or } 93:7\end{aligned}$$

Thus, the odds against the person working 35–39 hours per week are 93:7

DID YOU KNOW?

Flying Is Safer

Many people fear the risks of air travel, but the odds in favor of dying in an airplane crash are relatively low: approximately 1 to 440,951. The odds in favor of dying in a car accident are approximately 1 to 17,625. Thus, the odds in favor of dying in a car accident are much greater than the odds in favor of dying in an airplane accident.

Odds in Favor of an Event

Although odds are generally given against an event, at times they may be given in favor of an event. The *odds in favor of* an event are expressed as a ratio of the probability that the event will occur to the probability that the event will fail to occur.

$$\text{Odds in favor of event} = \frac{P(\text{event occurs})}{P(\text{event fails to occur})} = \frac{P(\text{success})}{P(\text{failure})}$$

If the odds *against* an event are $a:b$, the odds *in favor of* the event are $b:a$.

EXAMPLE 3 *Parents and Children Communicating*

A group of parents whose children attended Middlebury College were asked to identify the topic on which their child requested the most advice during the past year. The circle graph in Figure 12.5 on page 744 shows the parents' responses.

MATHEMATICS TODAY

Deal or No Deal

A popular television show is the NBC game show *Deal or No Deal*. The rules are simple. A contestant chooses a briefcase from 26 briefcases. Inside each briefcase is a card indicating a specific, but unique, amount of money. As each round progresses, the contestant must either stay with the briefcase selected or make a "deal" with the "banker" to accept a cash offer in exchange for whatever dollar amount is in the chosen case. How does the banker know how much to offer, and when should the contestant accept or reject the banker's offer? *Deal or No Deal* is a repeated exercise in a branch of mathematics known as *decision theory*, in which the value of the *expected winnings* at any point in the game is the *mean* of the values contained in the unopened briefcases. To maximize the expected return without regard to risk, a contestant's optimal strategy is to reject the banker's offer if it is below the mean value of the amounts in the unrevealed briefcases. Typically, the banker's offer undervalues the remaining briefcases, especially at the start of the game. For further information, visit en.wikipedia.org/wiki/Deal_or_No_Deal. The Wikipedia article shows that many of the concepts presented in the probability and statistics chapters of this book may be used in deciding whether to accept or reject the banker's offer.

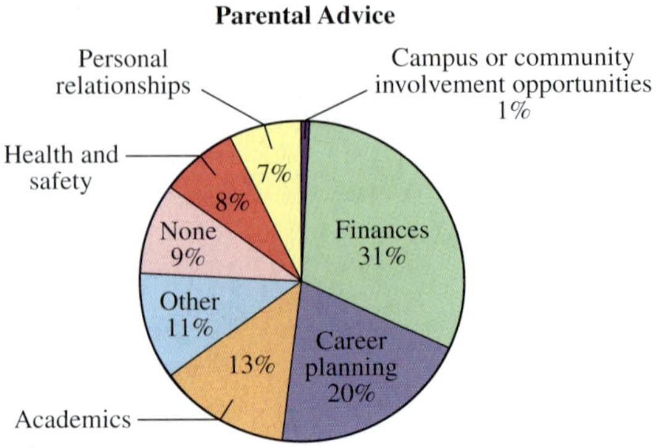

Figure 12.5

If one parent from those surveyed is selected at random, use the graph to determine

a) the odds against the parent saying that the child requested career planning advice.

b) the odds in favor of the parent saying that the child requested career planning advice.

SOLUTION

a) The graph shows that 20%, or $\frac{20}{100}$, of parents said that their child requested career planning advice. Thus, the probability that a child requested career planning advice is $\frac{20}{100}$. The probability that a child did not request career planning advice is therefore $1 - \frac{20}{100} = \frac{80}{100}$.

$$\text{Odds against a child having requested career planning advice} = \frac{P(\text{a child did not request career planning advice})}{P(\text{a child requested career planning advice})}$$

$$= \frac{80/100}{20/100} = \frac{80}{100} \cdot \frac{100}{20} = \frac{80}{20} = \frac{4}{1} \text{ or } 4:1$$

Thus, the odds against parents saying that their child requested career planning advice are 4:1.

b) The odds in favor of parents saying that their child requested career planning advice are 1:4.

Finding Probabilities from Odds

When odds are given, either in favor of or against a particular event, it is possible to determine the probability that the event occurs and the probability that the event does not occur. The denominators of the probabilities are found by adding the numbers in the odds statement. The numerators of the probabilities are the numbers given in the odds statements.

EXAMPLE 4 *Determining Probabilities from Odds*

The odds against Robin Murphy being admitted to the college of her choice are 9:2. Determine the probability that (a) Robin is admitted and (b) Robin is not admitted.

DID YOU KNOW?

What Were the Odds on This One?

In the year 1905, only two automobiles were registered in the entire state of Missouri.

In the year 1905, Missouri's only two registered automobiles were involved in a head-on collision!

SOLUTION

a) We have been given odds against and have been asked to find probabilities.

$$\text{Odds against being admitted} = \frac{P(\text{fails to be admitted})}{P(\text{is admitted})}$$

Since the odds statement is 9:2, the denominators of both the probability of success and the probability of failure must be 9 + 2 or 11. To get the odds ratio of 9:2, the probabilities must be $\frac{9}{11}$ and $\frac{2}{11}$. Since odds against is a ratio of failure to success, the $\frac{9}{11}$ and $\frac{2}{11}$ represent the probabilities of failure and success, respectively. Thus, the probability that Robin is admitted (success) is $\frac{2}{11}$.

b) The probability that Robin is not admitted (failure) is $\frac{9}{11}$. ●

Odds and probability statements are sometimes stated incorrectly. For example, consider the statement, "The odds of being selected to represent the district are 1 in 5." Odds are given using the word *to*, not *in*. Thus, there is a mistake in this statement. The correct statement might be, "The odds of being selected to represent the district are 1 to 5" or "The probability of being selected to represent the district is 1 in 5." Without additional information, it is not possible to tell which statement is the correct interpretation.

SECTION 12.3 EXERCISES

CONCEPT/WRITING EXERCISES

1. a) Explain how to determine the odds against an event.

   b) Explain how to determine the odds in favor of an event.

2. Explain the difference between the probability of an event and the odds in favor of an event.

3. Which odds are generally quoted, odds against or odds in favor?

4. Explain how to determine probabilities when you are given an odds statement.

5. The odds in favor of winning at Monopoly are 2:7. Determine the odds against winning at Monopoly.

6. The odds against Fancy Frank winning the horse race are 8:3. Determine the odds in favor of Fancy Frank winning.

7. If the odds against an event are 1:1, what is the probability that the event will

   a) occur.

   b) fail to occur.

   Explain your answer.

8. If the probability an event will occur is $\frac{1}{2}$, determine

   a) the probability that the event will fail to occur.

   b) the odds against the event occurring.

   c) the odds in favor of the event occurring.

   Explain your answer.

PRACTICE THE SKILLS/PROBLEM SOLVING

9. ***Dressing Up*** Lalo Jaquez is going to wear a blue sportcoat and is trying to decide what tie he should wear with it. In his closet, he has 15 ties, 8 of which go well with the sportcoat. If Lalo selects one tie at random, determine

   a) the probability that it goes well with the sportcoat.

   b) the probability that it does not go well with the sportcoat.

   c) the odds against it going well with the sportcoat.

   d) the odds in favor of it going well with the sportcoat.

10. ***Making a Donation*** In her wallet, Anne Kelly has 12 bills. Six are \$1 bills, two are \$5 bills, three are \$10 bills, and one is a \$20 bill. She passes a volunteer seeking donations for the Salvation Army and decides to select one bill at random from her wallet and give it to the Salvation Army. Determine

    a) the probability that she selects a \$5 bill.

    b) the probability that she does not select a \$5 bill.

    c) the odds in favor of her selecting a \$5 bill.

    d) the odds against her selecting a \$5 bill.

Deal or No Deal *In Exercises 11 and 12, consider the TV show Deal or No Deal. (See the Mathematics Today on page 744.) In the game show, there are 26 numbered cases, each indicating a cash prize ranging from 1 cent to \$1,000,000. Contestants on the show have a series of choices to make, and each time they can either accept the cash offer made by the banker or select a case from the remaining cases. Assume that one of the remaining cases contains a \$1,000,000 cash prize. Determine the odds against and the odds in favor of selecting the case containing the \$1,000,000 cash prize if there are*

11. seven cases remaining.

12. four cases remaining.

Toss a Die *In Exercises 13–16, a die is tossed. Determine the odds against rolling*

13. a 4.

14. an odd number.

15. a number less than 3.

16. a number greater than 4.

Deck of Cards *In Exercises 17–20, a card is picked from a standard deck of cards. Determine the odds against and the odds in favor of selecting*

17. a queen.

18. a heart.

19. a picture card.

20. a card greater than 5 (ace is low).

Spin the Spinner *In Exercises 21–24, assume that the spinner cannot land on a line. Determine the odds against the spinner landing on the color red.*

21. 

22. 

23. 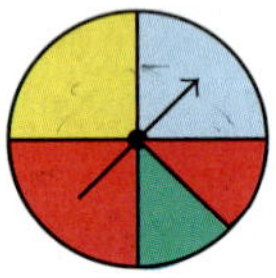

24. 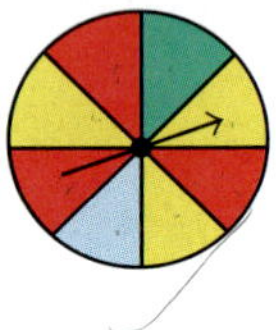

25. ***Students*** One person is selected at random from a class of 16 men and 14 women. Determine the odds against selecting

    a) a woman.

    b) a man.

26. ***Lottery*** One million tickets are sold for a lottery in which a single prize will be awarded.

    a) If you purchase a ticket, determine your odds against winning.

    b) If you purchase 10 tickets, determine your odds against winning.

Billiard Balls *In Exercises 27–32, use the rack of 15 billiard balls shown.*

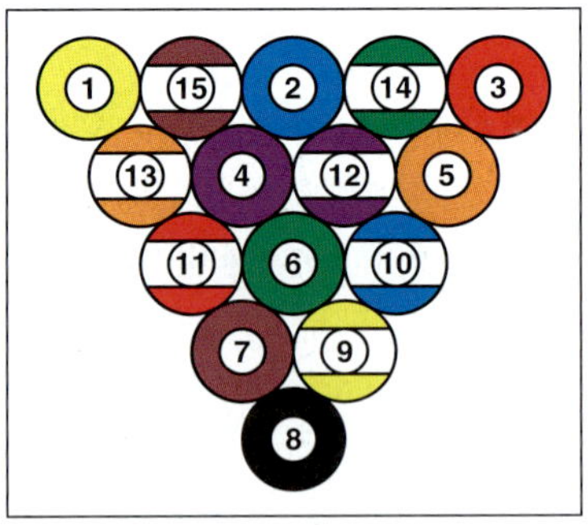

27. If one ball is selected at random, determine the odds against it containing a stripe. (Balls numbered 9 through 15 contain stripes.)

28. If one ball is selected at random, determine the odds in favor of it being a ball other than the 8 ball.

29. If one ball is selected at random, determine the odds in favor of it being an even-numbered ball.

30. If one ball is selected at random, determine the odds against it containing any red coloring (solid or striped).

31. If one ball is selected at random, determine the odds against it containing a number greater than or equal to 9.

32. If one ball is selected at random, determine the odds in favor of it containing two digits.

33. ***NASCAR Winnings*** The chart shows the winnings, in dollars, for the 10 highest-rated NASCAR drivers (according to point standings) for the 2005 season.

Driver	2005 Winnings
Tony Stewart	\$6,987,530
Greg Biffle	\$5,729,930
Carl Edwards	\$4,889,990
Mark Martin	\$5,994,350
Jimmie Johnson	\$6,796,660
Ryan Newman	\$5,578,110
Matt Kenseth	\$5,790,770
Rusty Wallace	\$4,868,980
Jeremy Mayfield	\$4,566,910
Kurt Busch	\$6,516,320

Source: www.NASCAR.com

If one of the drivers listed in the chart is selected at random, determine

a) the probability that the driver earned more than \$6 million in 2005.

b) the odds against the driver earning more than \$6 million in 2005.

▲ Tony Stewart

34. ***Rolling a Special Die*** A special die used in a game contains one dot on one side, two dots on two sides, and three dots on three sides. If the die is rolled, determine

a) the probability of rolling two dots.

b) the odds against rolling two dots.

35. ***Medical Tests*** The results of a medical test show that of 76 people selected at random who were given the test, 72 tested negative and 4 tested positive. Determine the odds against a person selected at random testing negative on the test. Explain how you determined your answer.

36. ***A Red Marble*** A box contains 9 red and 2 blue marbles. If you select one marble at random from the box, determine the odds against selecting a red marble. Explain how you determined your answer.

37. ***Scholarship Award*** The odds in favor of Wendy White winning a scholarship are 7 : 4. Determine the probability that

a) Wendy wins.

b) Wendy does not win.

38. ***Chicken Wing Contest*** The odds in favor of Boris Penzed winning the chicken wing eating contest are 3 : 8. Determine the probability that Boris will

a) win the contest.

b) not win the contest.

39. ***Getting Promoted*** The odds against Jason Judd getting promoted are 4 : 11. Determine the probability that Jason gets promoted.

40. ***Winning a Race*** The odds against Paul Phillips winning the 100 yard dash are 7 : 2. Determine the probability that

a) Paul wins. b) Paul does not win.

Playing Bingo *When playing bingo, 75 balls are placed in a bin and balls are selected at random. Each ball is marked with a letter and number as indicated in the following chart.*

B	I	N	G	O
1–15	16–30	31–45	46–60	61–75

For example, there are balls marked B1, B2, up to B15; I16, I17, up to I30; and so on (see photo). In Exercises 41–46, assuming one bingo ball is selected at random, determine

41. the probability that it contains the letter *G*.

42. the probability that it does not contain the letter *G*.

43. the odds in favor of it containing the letter *G*.

44. the odds against it containing the letter *G*.

45. the odds against it being *B*9.

46. the odds in favor of it being *B*9.

Blood Types *In Exercises 47–52, the following circle graph shows the percent of Americans with the various types of blood.*

Blood Types of Americans

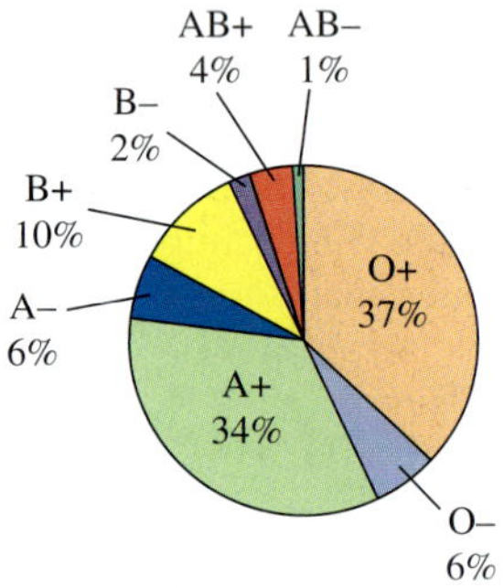

Source: 2003 Time Almanac

If one American is selected at random, use the graph to determine

47. the probability that the person has A+ blood.

48. the probability that the person has B− blood.

49. the odds against the person having A+ blood.

50. the odds in favor of the person having B− blood.

51. the odds in favor of the person having either O+ or O− blood.

52. the odds against the person having either A+ or O+ blood.

53. ***Rock Concert*** Suppose that the probability that a rock concert sells out is 0.9. Determine the odds against the concert selling out.

54. ***High Blood Pressure*** According to the U.S. Department of Health and Human Services, one in four Americans age 20 and older has high blood pressure. If an American who is age 20 or older is selected at random, determine the odds in favor of this person having high blood pressure.

55. ***Bookcase Assembly*** Suppose that the probability that all the parts needed to assemble a bookcase are included in the carton is $\frac{7}{8}$. Determine the odds in favor of the carton including all the needed parts.

56. ***On-Time Flight Arrivals*** A study of 175 airline flights in January 2006 showed that 138 of them arrived on time. If one of these flights was selected at random, determine

a) the probability that the flight arrived on time.

b) the odds against the flight arriving on time.

57. ***Birth Defects*** Birth defects affect 1 in 33 babies born in the United States each year.

a) What is the probability that a baby born in the United States will have a birth defect?

b) What are the odds against a baby born in the United States having a birth defect?

CHALLENGE PROBLEMS/GROUP ACTIVITIES

58. ***Odds Against*** Determine the odds against an even number or a number greater than 3 being rolled on a die.

59. ***Horse Racing*** Racetracks quote the approximate odds against each horse winning on a large board called a *tote board.* The odds quoted on a tote board for a race with five horses is as follows.

Horse Number	Odds
2	7:2
3	2:1
4	15:1
5	7:5
6	1:1

Determine the probability of each horse winning the race. (Do not be concerned that the sum of the probabilities is not 1.)

60. ***Roulette*** Turn to the roulette wheel illustrated on page 760. If the wheel is spun, determine

a) the probability that the ball lands on red.

b) the odds against the ball landing on red.

c) the probability that the ball lands on 0 or 00.

d) the odds in favor of the ball landing on 0 or 00.

RECREATIONAL MATHEMATICS

61. ***Multiple Births*** Multiple births make up about 3% of births per year in the United States. The following illustrates the number and type of multiple births in 2004.

Multiple Births in the United States in 2004

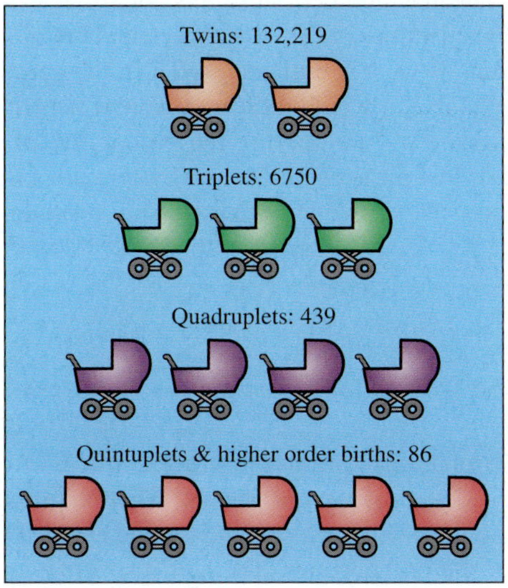

Source: National Center for Health Statistics

Using the above information, determine an estimate for the odds against a birth being a multiple birth in 2004.

INTERNET/RESEARCH ACTIVITIES

62. ***State Lottery*** Determine whether your state has a lottery. If so, do research and write a paper indicating

a) the probability of winning the grand prize.

b) the odds against winning the grand prize.

c) Explain, using real objects such as pennies or table tennis balls, what these odds actually mean.

63. ***Casino Advantages*** There are many types of games of chance to choose from at casinos. The house has the advantage in each game, but the advantages differ according to the game.

a) List the games available at a typical casino.

b) List those for which the house has the smallest advantage of winning.

c) List those for which the house has the greatest advantage of winning.

12.4 EXPECTED VALUE (EXPECTATION)

▲ Expected value can be used to determine the expected results of an experiment repeated many times.

Consider the following situation. Tim tells Barbara that he will give her \$1 if she can roll an even number on a single die. If she fails to roll an even number, she must give Tim \$1. Who would win money in the long run if this game were played many times? In this section, we will learn how to determine the expected results of an experiment over the long term.

Expected value, also called ***expectation***, is often used to determine the expected results of an experiment or business venture *over the long term.* People use expectation to make important decisions in many different areas. For example, expectation is used in business to predict future profits of a new product. In the insurance industry, expectation is used to determine how much each insurance policy should cost for the company to make an overall profit. Expectation is also used to predict the expected gain or loss in games of chance such as the lottery, roulette, craps, and slot machines.

In the situation with Tim and Barbara, we would expect in the long run that half the time Tim would win \$1 and half the time he would lose \$1; therefore, Tim would break even. Mathematically, we could find Tim's expected gain or loss by the following procedure:

$$\text{Tim's expected gain or loss} = P\binom{\text{Tim}}{\text{wins}} \cdot \binom{\text{amount}}{\text{Tim wins}} + P\binom{\text{Tim}}{\text{loses}} \cdot \binom{\text{amount}}{\text{Tim loses}}$$
$$= \frac{1}{2}(\$1) + \frac{1}{2}(-\$1) = \$0$$

Note that the loss is written as a negative number. This procedure indicates that Tim has an expected gain or loss (or expected value) of \$0. The expected value of zero indicates that he would indeed break even, as we had anticipated. Thus, the game is a *fair game*. If Tim's expected value were positive, it would indicate a gain; if negative, a loss.

The expected value, E, is calculated by multiplying the probability of an event occurring by the *net* amount gained or lost if the event occurs. If there are a number of different events and amounts to be considered, we use the following formula.

EXPECTED VALUE

$$E = P_1 \cdot A_1 + P_2 \cdot A_2 + P_3 \cdot A_3 + \cdots + P_n \cdot A_n$$

The symbol P_1 represents the probability that the first event will occur, and A_1 represents the net amount won or lost if the first event occurs. P_2 is the probability of the second event, and A_2 is the net amount won or lost if the second event occurs, and so on. The sum of these products of the probabilities and their respective amounts is the expected value. The expected value is the average (or mean) result that would be obtained if the experiment were performed a great many times.

EXAMPLE 1 *A New Business Venture*

JetBlue Airways is considering adding a route to the city of Minneapolis, Minnesota. Before the company makes its decision as to whether or not to service Minneapolis, it needs to consider many factors, including potential profits and losses. Factors that may affect the company's profits and losses include the number of competing airlines, the potential number of customers, the overhead costs, and fees it must pay. After considerable research, the company estimates that if it serves Minneapolis, there is a 60% chance of making a \$900,000 profit, a 10% chance of breaking even, and a 30% chance of losing \$1,400,000. How much can JetBlue Airways "expect" to make on this new route?

SOLUTION The three amounts to be considered are a gain of \$900,000, breaking even at \$0, and a loss of \$1,400,000. The probability of gaining \$900,000 is 0.6, the probability of breaking even is 0.1, and the probability of losing \$1,400,000 is 0.3.

$$\text{JetBlue's expectation} = \underbrace{P_1 \cdot A_1}_{\text{Gain}} + \underbrace{P_2 \cdot A_2}_{\text{Break even}} + \underbrace{P_3 \cdot A_3}_{\text{Loss}}$$
$$= (0.6)(\$900{,}000) + (0.1)(\$0) + (0.3)(-\$1{,}400{,}000)$$
$$= \$540{,}000 + \$0 - \$420{,}000$$
$$= \$120{,}000$$

DID YOU KNOW?

Tough Decision

The concept of expected value can be used to help evaluate the consequences of many decisions. You use this concept when you consider whether to double-park your car for a few minutes. The probability of being caught may be low, but the penalty, a parking ticket, may be high. You weigh these factors when you decide whether to spend time looking for a legal parking spot.

JetBlue Airways has an expectation, or expected average gain, of \$120,000 for adding this particular service. Thus, if the company opened routes like this one, with these particular probabilities and amounts, in the long run it would have an average gain of \$120,000 per route. However, you must remember that there is a 30% chance that JetBlue will lose \$1,400,000 on this *particular* route (or any particular route with these probabilities and amounts.)

EXAMPLE 2 *Test-Taking Strategy*

Maria is taking a multiple-choice exam in which there are five possible answers for each question. The instructions indicate that she will be awarded 2 points for each correct response, that she will lose $\frac{1}{2}$ point for each incorrect response, and that no points will be added or subtracted for answers left blank.

a) If Maria does not know the correct answer to a question, is it to her advantage or disadvantage to guess at an answer?

b) If she can eliminate one of the possible choices, is it to her advantage or disadvantage to guess at the answer?

SOLUTION

a) Let's determine the expected value if Maria guesses at an answer. Only one of five possible answers is correct.

$$P(\text{guesses correctly}) = \frac{1}{5} \qquad P(\text{guesses incorrectly}) = \frac{4}{5}$$

$$\begin{aligned}\text{Maria's expectation} &= \overbrace{P_1 \cdot A_1}^{\text{Guesses correctly}} + \overbrace{P_2 \cdot A_2}^{\text{Guesses incorrectly}} \\ &= \frac{1}{5}(2) + \frac{4}{5}\left(-\frac{1}{2}\right) \\ &= \frac{2}{5} - \frac{2}{5} = 0\end{aligned}$$

Thus, Maria's expectation is zero when she guesses. Therefore, over the long run she will neither gain nor lose points by guessing.

b) If Maria can eliminate one possible choice, one of four answers will be correct.

$$P(\text{guesses correctly}) = \frac{1}{4} \qquad P(\text{guesses incorrectly}) = \frac{3}{4}$$

$$\begin{aligned}\text{Maria's expectation} &= \overbrace{P_1 \cdot A_1}^{\text{Guesses correctly}} + \overbrace{P_2 \cdot A_2}^{\text{Guesses incorrectly}} \\ &= \frac{1}{4}(2) + \frac{3}{4}\left(-\frac{1}{2}\right) \\ &= \frac{2}{4} - \frac{3}{8} = \frac{4}{8} - \frac{3}{8} = \frac{1}{8}\end{aligned}$$

Since the expectation is a positive $\frac{1}{8}$, Maria will, on average, gain $\frac{1}{8}$ point each time she guesses when she can eliminate one possible choice.

EXAMPLE 3 *Selling Hot Dogs*

An outdoor hot dog vendor sells an average of 50 hot dogs per day in dry weather and an average of 15 per day in wet weather. If the weather in this area is wet 25% of the time, determine the expected (average) number of hot dogs sold per day.

SOLUTION The amounts in this example are the number of hot dogs sold. Since the weather is wet 25% of the time, it will be dry 100% − 25% = 75% of the time. When written as probabilities, 25% and 75% are 0.25 and 0.75, respectively.

$$\begin{aligned} E &= P(\text{dry}) \cdot (\text{number sold}) + P(\text{wet}) \cdot (\text{number sold}) \\ &= 0.75(50) + 0.25(15) = 37.5 + 3.75 = 41.25 \end{aligned}$$

Thus, the average, or expected, number of hot dogs sold per day is 41.25. ●

When we gave the expectation formula, we indicated that the amounts were the **net amounts**, which are the actual amounts gained or lost. Examples 4 and 5 illustrate how net amounts are used in two applications of expected value.

EXAMPLE 4 *Winning a Door Prize*

When Josh Rosenberg attends a charity event, he is given a free ticket for the $50 door prize. A total of 100 tickets will be given out. Determine his expectation of winning the door prize.

SOLUTION The probability of winning the door prize is $\frac{1}{100}$ since Josh has 1 of 100 tickets. If he wins, his net or actual winnings will be $50 since he did not pay for the ticket. The probability that Josh loses is $\frac{99}{100}$. If Josh loses, the amount he loses is $0 because he did not pay for the ticket.

$$\begin{aligned} \text{Expectation} &= P(\text{Josh wins}) \cdot (\text{amount won}) + P(\text{Josh loses}) \cdot (\text{amount lost}) \\ &= \frac{1}{100}(50) + \frac{99}{100}(0) = \frac{50}{100} = 0.50 \end{aligned}$$

Thus, Josh's expectation is $0.50, or 50 cents. ●

Now we will consider a problem similar to Example 4, but this time we will assume that Josh must purchase the ticket for the door prize.

EXAMPLE 5 *Winning a Door Prize*

When Josh Rosenberg attends a charity event, he is given the opportunity to purchase a ticket for the $50 door prize. The cost of the ticket is $2, and 100 tickets will be sold. Determine Josh's expectation if he purchases one ticket.

SOLUTION As in Example 4, Josh's probability of winning is $\frac{1}{100}$. However, if he does win, his actual or net winnings will be $48. The $48 is obtained by subtracting the cost of the ticket, $2, from the amount of the door prize, $50. There is also a probability of $\frac{99}{100}$ that Josh will not win the door prize. If he does not win the door prize, he has lost the $2 that he paid for the ticket. Therefore, we must consider two amounts when we determine Josh's expectation, winning $48 and losing $2.

$$\begin{aligned} \text{Expectation} &= P(\text{Josh wins}) \cdot (\text{amount won}) + P(\text{Josh loses}) \cdot (\text{amount lost}) \\ &= \frac{1}{100}(48) + \frac{99}{100}(-2) \end{aligned}$$

$$= \frac{48}{100} - \frac{198}{100} - \frac{150}{100} = -1.50$$

Josh's expectation is $-\$1.50$ when he purchases one ticket. ●

In Example 5, we determined that Josh's expectation was $-\$1.50$ when he purchased one ticket. If he purchased two tickets, his expectation would be $2(-\$1.50)$, or $-\$3.00$. We could also compute Josh's expectation if he purchased two tickets as follows:

$$E = \frac{2}{100}(46) + \frac{98}{100}(-4) = -3.00$$

This answer, $-\$3.00$, checks with the answer obtained by multiplying the expectation for a single ticket by 2.

Let's look at one more example in which a person must pay for a chance to win a prize. In the following example, there will be more than two amounts to consider.

EXAMPLE 6 *Raffle Tickets*

One thousand raffle tickets are sold for \$1 each. One grand prize of \$500 and two consolation prizes of \$100 will be awarded. The tickets are placed in a bin. The winning tickets will be selected from the bin. Assuming that the probability that any given ticket selected for the grand prize is $\frac{1}{1000}$ and the probability that any given ticket selected for a consolation prize is $\frac{2}{1000}$, determine

a) Irene Drew's expectation if she purchases one ticket.

b) Irene's expectation if she purchases five tickets.

SOLUTION

a) Three amounts are to be considered: the net gain in winning the grand prize, the net gain in winning one of the consolation prizes, and the loss of the cost of the ticket. If Irene wins the grand prize, her net gain is \$499 (\$500 minus \$1 spent for the ticket). If Irene wins one of the consolation prizes, her net gain is \$99 (\$100 minus \$1). The probability that Irene wins the grand prize is $\frac{1}{1000}$ and the probability that she wins a consolation prize is $\frac{2}{1000}$. The probability that she does not win a prize is $1 - \frac{1}{1000} - \frac{2}{1000} = \frac{997}{1000}$.

$$E = P_1 \cdot A_1 + P_2 \cdot A_2 + P_3 \cdot A_3$$
$$= \frac{1}{1000}(\$499) + \frac{2}{1000}(\$99) + \frac{997}{1000}(-\$1)$$
$$= \frac{499}{1000} + \frac{198}{1000} - \frac{997}{1000} = -\frac{300}{1000} = -0.30$$

Thus, Irene's expectation is $-\$0.30$ per ticket purchased.

b) On average, Irene loses 30 cents on each ticket purchased. On five tickets, her expectation is $(-\$0.30)(5)$, or $-\$1.50$. ●

TIMELY TIP In any expectation problem, the sum of the probabilities of all the events should always be 1. Note in Example 6 that the sum of the probabilities is $\frac{1}{1000} + \frac{2}{1000} + \frac{997}{1000} = \frac{1000}{1000} = 1$.

DID YOU KNOW?

It Pays to Be Original

Commonly Selected Numbers

7-14-21-28-35-42
1-2-3-4-5-6
5-10-15-20-25-30
3-8-13-18-23-28

The numbers you select when picking lottery numbers have no effect on your probability of winning. However, your expectation (expected winnings) varies greatly with the numbers you select. Because the jackpot is divided among all the winners, the fewer the number of winners the more each winner receives. Some groups of numbers are commonly selected. Some are illustrated above. In a lottery with a large jackpot, there may be as many as 10,000 people who select the numbers 7-14-21-28-35-42. If the jackpot was \$40 million and these numbers were selected, each winner would receive about \$4000. That is quite a difference from the \$40 million a single winner would receive. Many people use birthdays or other dates when selecting lottery numbers. Therefore, there may be fewer people selecting numbers greater than 31.

Fair Price

In Example 5, we determined that Josh's expectation was $-\$1.50$. Now let's determine how to find out how much should have been charged for a ticket so that his expectation would be \$0. If Josh's expectation were to be \$0, he could be expected to break even over the long run. Suppose that Josh paid 50 cents, or \$0.50, for the ticket. His expectation, if paying \$0.50 for the ticket, would be calculated as follows.

$$\begin{aligned}\text{Expectation} &= P(\text{Josh wins}) \cdot (\text{amount won}) + P(\text{Josh loses}) \cdot (\text{amount lost})\\ &= \frac{1}{100}(49.50) + \frac{99}{100}(-\$0.50)\\ &= \frac{49.50}{100} - \frac{49.50}{100} = 0\end{aligned}$$

Thus, if Josh paid 50 cents per ticket, his expectation would be \$0. The 50 cents, in this case, is called the fair price of the ticket. The ***fair price*** is the amount to be paid that will result in an expected value of \$0. The fair price may be found by adding the *cost to play* to the *expected value*.

Fair price = expected value + cost to play

In Example 5, the cost to play was \$2 and the expected value was determined to be $-\$1.50$. The fair price for a ticket in Example 5 may be found as follows.

$$\begin{aligned}\text{Fair price} &= \text{expected value} + \text{cost to play}\\ &= -1.50 + 2.00 = 0.50\end{aligned}$$

We obtained a fair price of \$0.50. If the tickets were sold for the fair price of \$0.50 each, Josh's expectation would be \$0, as shown above. Can you now find the fair price that Irene should pay for a raffle ticket in Example 6? In Example 6, the cost of a ticket was \$1 and we determined that the expected value was $-\$0.30$.

$$\begin{aligned}\text{Fair price} &= \text{expected value} + \text{cost to play}\\ &= -\$0.30 + \$1.00 = \$0.70\end{aligned}$$

Thus, the fair price for a ticket in Example 6 is \$0.70, or 70 cents. Verify for yourself now that if the tickets were sold for \$0.70, the expectation would be \$0.00.

EXAMPLE 7 *Expectation and Fair Price*

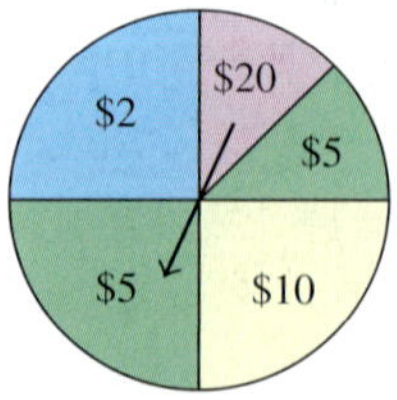

Suppose that you are playing a game in which you spin the pointer shown in the figure in the margin and you are awarded the amount shown under the pointer. If it costs \$8 to play the game, determine

a) the expectation of a person who plays the game.

b) the fair price to play the game.

SOLUTION

a) There are four numbers on which the pointer can land: 2, 5, 10, and 20. The following chart shows the probability of the pointer landing on each number and the actual amount won or lost if the pointer lands on that number. The probabilities are obtained using the areas of the circle. The amounts won or lost are determined by subtracting the cost to play, \$8, from each indicated amount.

Amount Shown on Wheel	**\$2**	**\$5**	**\$10**	**\$20**
Probability	$\frac{1}{4}$	$\frac{3}{8}$	$\frac{1}{4}$	$\frac{1}{8}$
Amount Won or Lost	−\$6	−\$3	\$2	\$12

Notice that the sum of the probabilities is $\frac{1}{4} + \frac{3}{8} + \frac{1}{4} + \frac{1}{8} = 1$, which shows that all possible outcomes have been considered.

Now let's find the expectation. There are four amounts to consider.

$$\begin{aligned}\text{Expectation} &= P(\text{lands on \$2}) \cdot (\text{amount}) + P(\text{lands on \$5}) \cdot (\text{amount}) \\ &\quad + P(\text{lands on \$10}) \cdot (\text{amount}) + P(\text{lands on \$20}) \cdot (\text{amount}) \\ &= \frac{1}{4}(-6) + \frac{3}{8}(-3) + \frac{1}{4}(2) + \frac{1}{8}(12) \\ &= -\frac{6}{4} - \frac{9}{8} + \frac{2}{4} + \frac{12}{8} \\ &= -\frac{12}{8} - \frac{9}{8} + \frac{4}{8} + \frac{12}{8} = -\frac{5}{8} = -\$0.625\end{aligned}$$

Thus, the expectation is −\$0.625.

b) Fair price = expectation + cost to play

$$= -\$0.625 + \$8 = \$7.375$$

Thus, the fair price to play the game is about \$7.38.

SECTION 12.4 EXERCISES

CONCEPT/WRITING EXERCISES

1. What does the expected value of an experiment or business venture represent?

2. a) What does an expected value of 0 mean?

   b) What does a negative expected value mean?

   c) What does a positive expected value mean?

3. What is meant by the fair price of a game of chance?

4. Write the formula used to find the expected value of an experiment with

   a) two possible outcomes.

   b) three possible outcomes.

5. If the expected value and cost to play are known for a particular game of chance, explain how you can determine the fair price to pay to play that game of chance.

6. Is the fair price to pay for a game of chance the same as the expected value of that game of chance? Explain your answer.

7. If a particular game costs \$1.50 to play and the expectation for the game is −\$1.00, what is the fair price to pay to play the game? Explain how you determined your answer.

8. If a particular game cost \$3.00 to play and the expectation for the game is −\$2.00, what is the fair price to pay to play the game? Explain how you determined your answer.

PRACTICE THE SKILLS/PROBLEM SOLVING

9. ***Three Tickets*** On a \$1 lottery ticket, Marty Smith's expected value is −\$0.40. What is Marty's expected value if he purchases three lottery tickets?

10. ***Expected Value*** If on a \$1 bet, Paul Goldstein's expected value is \$0.30, what is Paul's expected value on a \$5 bet?

11. ***Expected Attendance*** For an outdoor concert of the Los Angeles Philharmonic Orchestra at the Hollywood Bowl, concert organizers estimate that 14,000 people will attend if it is not raining. If it is raining, concert organizers estimate that 8400 people will attend. On the day of the concert, meteorologists predict a 70% chance of rain. Determine the expected number of people who will attend this concert.

▲ The Hollywood Bowl

12. ***A New Business*** In a proposed business venture, Stephanie Morrison estimates that there is a 60% chance she will make \$80,000 and a 40% chance she will lose \$20,000. Determine Stephanie's expected value.

13. ***Basketball*** Candace Parker is a star player for the University of Tennessee Volunteers women's basketball team. She has injured her ankle, and it is doubtful that she will be able to play in an upcoming game. If Candace can play, the coach estimates that the Volunteers will score 78 points. If Candace is not able to play, the coach estimates that they will score 62 points. The team doctor estimates that there is a 50% chance Candace will play. Determine the number of points the team can expect to score.

▲ Candace Parker (left) and Brittany Hunter

14. ***Career Fair Attendance*** For a Nursing and Allied Health Care Career Fair, organizers estimate that 50 people will attend if it does not rain and 65 will attend if it rains. The weather forecast indicates that there is a 40% chance it will not rain and a 60% chance it will rain on the day of the career fair. Determine the expected number of people who will attend the fair.

15. ***TV Shows*** The NBC television network is scheduling its fall lineup of shows. For the Thursday night 8 P.M. slot, NBC has selected the show *Heroes*. If its rival network CBS schedules the show *CSI: Crime Scene Investigation* during the same time slot, NBC estimates that *Heroes* will get 1.2 million viewers. However, if CBS schedules the show *The Unit* during that time slot, NBC estimates that *Heroes* will get 1.6 million viewers. NBC believes that the probability that CBS will show *CSI* is 0.4 and the probability that CBS will show *The Unit* is 0.6. Determine the expected number of viewers for the show *Heroes*.

16. ***Seattle Greenery*** In July in Seattle, the grass grows $\frac{1}{2}$ in. a day on a sunny day and $\frac{1}{4}$ in. a day on a cloudy day. In Seattle in July, 75% of the days are sunny and 25% are cloudy.

    a) Determine the expected amount of grass growth on a typical day in July in Seattle.

    b) Determine the expected total grass growth in the month of July in Seattle.

17. ***Investment Club*** The Triple L investment club is considering purchasing a certain stock. After considerable research, the club members determine that there is a 60% chance of making \$10,000, a 10% chance of breaking even, and a 30% chance of losing \$7200. Determine the expectation of this purchase.

18. ***Clothing Sale*** At a special clothing sale at the Crescent Oaks Country Club, after the cashier rings up your purchase, you select a slip of paper from a box. The slip of paper indicates the dollar amount, either \$5 or \$10, that is deducted from your purchase price. The probability of selecting a slip indicating \$5 is $\frac{7}{10}$, and the probability of selecting a slip indicating \$10 is $\frac{3}{10}$. If your original purchase before you select the slip of paper is \$200, determine

    a) the expected dollar amount to be deducted from your purchase.

    b) the expected dollar amount you will pay for your purchase.

19. ***Fortune Cookies*** At the Royal Dragon Chinese restaurant, a slip in the fortune cookies indicates a dollar amount that will be subtracted from your total bill. A bag of 10 fortune cookies is given to you from which you will select one. If seven fortune cookies contain "\$1 off," two contain "\$2 off," and one contains "\$5 off," determine the expectation of a selection.

20. ***Pick a Card*** Mike and Dave play the following game. Mike picks a card from a deck of cards. If he selects a heart, Dave gives him $5. If not, he gives Dave $2.

  a) Determine Mike's expectation.

  b) Determine Dave's expectation.

21. ***Roll a Die*** Alyssa and Gabriel play the following game. Alyssa rolls a die. If she rolls a 1, 2, or 3, Gabriel gives Alyssa $3. If Alyssa rolls a 4 or 5, Gabriel gives Alyssa $2. However, if Alyssa rolls a 6, she gives Gabriel $14.

  a) Determine Alyssa's expectation.

  b) Determine Gabriel's expectation.

22. ***Blue Chips and Red Chips*** A bag contains 3 blue chips and 2 red chips. Chi and Dolly play the following game. Chi selects one chip at random from the bag. If Chi selects a blue chip, Dolly gives Chi $5. If Chi selects a red chip, Chi gives Dolly $8.

  a) Determine Chi's expectation.

  b) Determine Dolly's expectation.

23. ***Multiple-Choice Test*** A multiple-choice exam has five possible answers for each question. For each correct answer, you are awarded 5 points. For each incorrect answer, 1 point is subtracted from your score. For answers left blank, no points are added or subtracted.

  a) If you do not know the correct answer to a particular question, is it to your advantage to guess? Explain.

  b) If you do not know the correct answer but can eliminate one possible choice, is it to your advantage to guess? Explain.

24. ***Multiple-Choice Test*** A multiple-choice exam has four possible answers for each question. For each correct answer, you are awarded 5 points. For each incorrect answer, 2 points are subtracted from your score. For answers left blank, no points are added or subtracted.

  a) If you do not know the correct answer to a particular question, is it to your advantage to guess? Explain.

  b) If you do not know the correct answer but can eliminate one possible choice, is it to your advantage to guess? Explain.

25. ***Raffle Tickets*** Five hundred raffle tickets are sold for $2 each. One prize of $400 is to be awarded.

  a) Raul Mondesi purchases one ticket. Determine his expected value.

  b) Determine the fair price of a ticket.

26. ***Raffle Tickets*** One thousand raffle tickets are sold for $1 each. One prize of $800 is to be awarded.

  a) Rena Condos purchases one ticket. Determine her expected value.

  b) Determine the fair price of a ticket.

27. ***Raffle Tickets*** Two thousand raffle tickets are sold for $3.00 each. Three prizes will be awarded: one for $1000 and two for $500. Assume that the probability that any given ticket is selected for the $1000 prize is $\frac{1}{2000}$ and the probability that any given ticket is selected for the $500 prize is $\frac{2}{2000}$. Jeremy Sharp purchases one of these tickets.

  a) Determine his expected value.

  b) Determine the fair price of a ticket.

28. ***Raffle Tickets*** Ten thousand raffle tickets are sold for $5 each. Four prizes will be awarded: one for $5000, one for $2500, and two for $1000. Assume that the probability that any given ticket is selected for the $5000 prize is $\frac{1}{10,000}$, the probability that any given ticket is selected for the $2500 prize is $\frac{1}{10,000}$, and the probability that any given ticket is selected for a $1000 prize is $\frac{2}{10,000}$. Sidhardt purchases one of these tickets.

  a) Determine his expected value.

  b) Determine the fair price of a ticket.

Spinners *In Exercises 29 and 30, assume that a person spins the pointer and is awarded the amount indicated by the pointer. Determine the person's expectation.*

29.

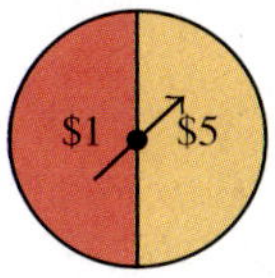

30.

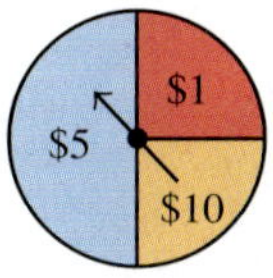

Spinners *In Exercises 31 and 32, assume that a person spins the pointer and is awarded the amount indicated if the pointer points to a positive number but must pay the amount indicated if the pointer points to a negative number. Determine the person's expectation if the person plays the game.*

31.

32.

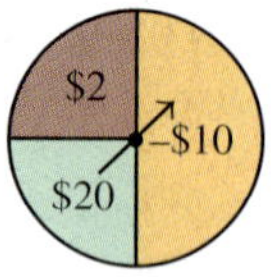

Selecting an Envelope *In Exercises 33–36, a person randomly selects one of the five envelopes shown below. Each envelope contains a check that the person gets to keep. Determine the person's expectation if the checks in the envelopes are as follows.*

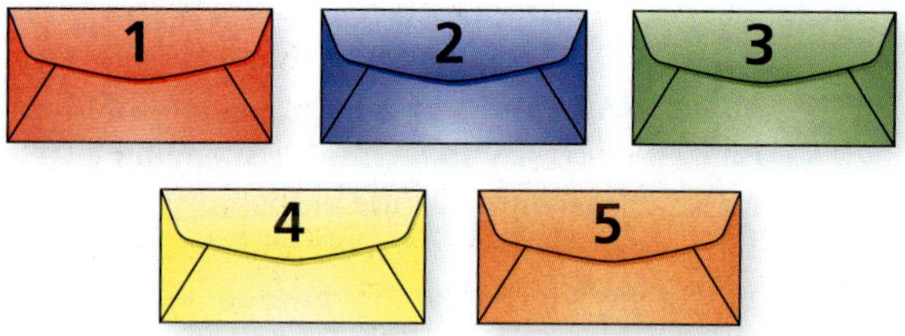

33. Two envelopes contain a $500 check, and three envelopes contain a $1000 check.

34. Four envelopes contain a $1000 check, and one envelope contains a $5000 check.

35. The five envelopes contain checks for $100, $200, $300, $400, and $1000, respectively.

36. Two envelopes contain a check for $600, two envelopes contain a check for $2000, and one envelope contains a check for $5000.

Spinners *In Exercises 37–40, assume that a person spins the pointer and is awarded the amount indicated by the pointer. If it costs $2 to play the game, determine*

a) *the expectation of a person who plays the game.*

b) *the fair price to play the game.*

37.

38.

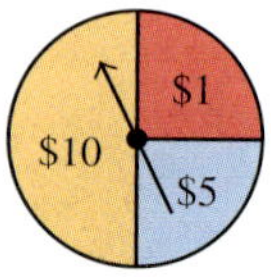

39.

40.

Selecting an Envelope *In Exercises 41–44, a person randomly selects one of the four envelopes shown below. Each envelope contains a check that the person gets to keep. However, before the person can select an envelope, he or she must pay $10 to play. For the value of the checks indicated in the exercises, determine*

a) *the expectation for a person who plays.*

b) *the fair price to play.*

41. Two envelopes contain a $5 check, and two envelopes contain a $20 check.

42. Three envelopes contain a $15 check, and one envelope contains a $0 check.

43. The checks in the envelopes are for $0, $1, $5, and $10, respectively.

44. The checks in the envelopes are for $2, $2, $10, and $400, respectively.

45. ***Reaching Base Safely*** Based on his past baseball history, Jim Devias has a 17% chance of reaching first base safely, a 10% chance of hitting a double, a 2% chance of hitting a triple, an 8% chance of hitting a home run, and a 63% chance of making an out at his next at bat. Determine Jim's expected number of bases for his next at bat.

46. ***Life Insurance*** According to Bristol Mutual Life Insurance's mortality table, the probability that a 20-year-old woman will survive 1 year is 0.994 and the probability that she will die within 1 year is 0.006. If a 20-year-old woman buys a $10,000 1-year policy for $100, what is Bristol Mutual's expected gain or loss?

47. ***Choosing a Colored Chip*** In a box, there are a total of 10 chips. The chips are orange, green, and yellow, as shown below.

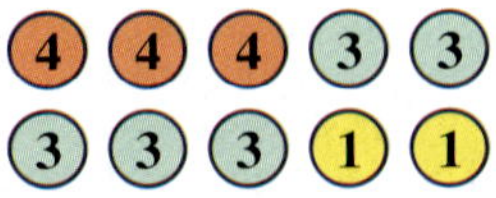

If you select an orange chip, you get 4 points, a green chip 3 points, and a yellow chip 1 point. If you select one chip at random, determine the expected number of points you will get.

48. ***Choosing a Colored Chip*** Repeat Exercise 47 but assume that an orange chip is worth 5 points, a green chip 2 points, and a yellow chip -3 points (3 points are taken away).

49. ***Employee Hiring*** The academic vice president at Brookdale Community College has requested that new academic programs be added to the college curriculum. If the college's Board of Trustees approves the new programs, the college will hire 75 new employees. If the new programs are not approved, the college will hire only 20 new employees. If the probability that the new programs will be approved is 0.65, what is the expected number of new employees to be hired by Brookdale Community College? Round your answer to the nearest whole number of employees.

50. ***Completing a Project*** A mechanical contractor is preparing for a construction project. He determines that if he completes the project on schedule, his net profit will be $450,000. If he completes the project between 0 and 3 months late, his net profit decreases to $120,000. If he completes the project more than 3 months late, his net loss is $275,000. The probability that he completes the project on schedule is 0.6, the probability that he completes the project between 0 and 3 months late is 0.3, and the probability that he completes the project more than 3 months late is 0.1. Determine his expected gain or loss for this project.

51. ***New Store*** Dunkin' Donuts is opening a new store. The company estimates that there is a 75% chance the store will have a profit of $10,000, a 10% chance the store will break even, and a 15% chance the store will lose $2000. Determine the expected gain or loss for this store.

52. ***China Cabinet*** The owner of an antique store estimates that there is a 40% chance she will make $2000 when she sells an antique china cabinet, a 50% chance she will make $750 when she sells the cabinet, and a 10% chance she will break even when she sells the cabinet. Determine the expected amount she will make when she sells the cabinet.

53. ***Rolling a Die*** A die is rolled many times, and the points facing up are recorded. Determine the expected (average) number of points facing up over the long run.

54. ***Lawsuit*** Don Vello is considering bringing a lawsuit against the Dummote Chemical Company. His lawyer estimates that there is a 70% chance Don will make $40,000, a 10% chance Don will break even, and a 20% chance they will lose the case and Don will need to pay $30,000 in legal fees. Estimate Don's expected gain or loss if he proceeds with the lawsuit.

55. ***Road Service*** On a clear day in Boston, the Automobile Association of American (AAA) makes an average of 110 service calls for motorist assistance, on a rainy day it makes an average of 160 service calls, and on a snowy day it makes an average of 210 service calls. If the weather in Boston is clear 200 days of the year, rainy 100 days of the year, and snowy 65 days of the year, determine the expected number of service calls made by the AAA in a given day.

56. ***Real Estate*** The expenses for Jorge Estrada, a real estate agent, to list, advertise, and attempt to sell a house are $1000. If Jorge succeeds in selling the house, he will receive a commission of 6% of the sales price. If an agent with a different company sells the house, Jorge still receives 3% of the sales price. If the house is unsold after 3 months, Jorge loses the listing and receives nothing. Suppose that the probability that he sells a $100,000 house is 0.2, the probability that another agent sells the house is 0.5, and the probability that the house is unsold after 3 months is 0.3. Determine Jorge's expectation if he accepts this house for listing. Should Jorge list the house? Explain.

Dart Board *In Exercises 57 and 58, assume that you are blindfolded and throw a dart at the dart board shown on the top of page 760. Assume that your dart sticks in the dart board, and not on a line.*

a) *Determine the probabilities that the dart lands on $1, $10, $20, and $100, respectively.*

b) *If you win the amount of money indicated by the section of the board where the dart lands, determine your expectation when you throw the dart.*

c) *If the game is to be fair, how much should you pay to play?*

57.

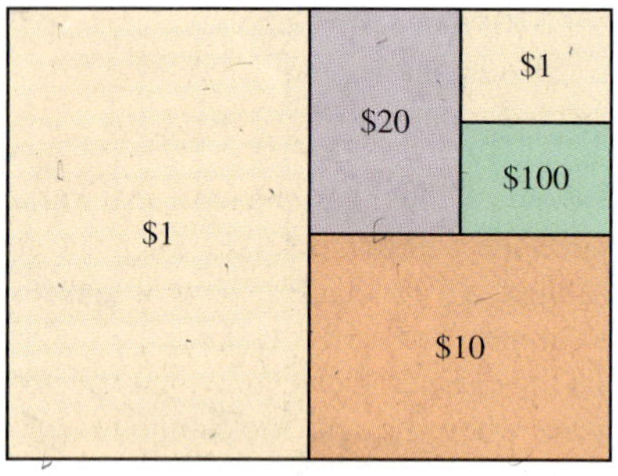

58.

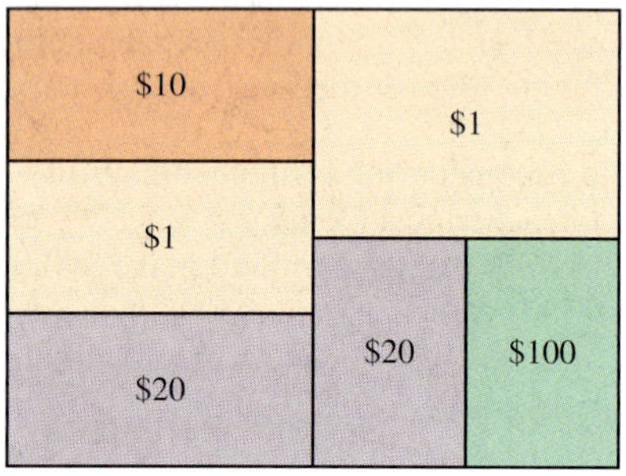

CHALLENGE PROBLEMS/GROUP ACTIVITIES

59. ***Term Life Insurance*** An insurance company will pay the face value of a term life insurance policy if the insured person dies during the term of the policy. For how much should an insurance company sell a 10-year term policy with a face value of $40,000 to a 30-year-old man for the company to make a profit? The probability of a 30-year-old man living to age 40 is 0.97. Explain your answer. Remember that the customer pays for the insurance before the policy becomes effective.

60. ***Lottery Ticket*** Is it possible to determine your expectation when you purchase a lottery ticket? Explain.

Roulette In Exercises 61 and 62, use the roulette wheel illustrated. A roulette wheel typically contains slots with numbers 1–36 and slots marked 0 and 00. A ball is spun on the wheel and comes to rest in one of the 38 slots. Eighteen numbers are colored red, and 18 numbers are colored black. The 0 and 00 are colored green. If you bet on one particular number and the ball lands on that number, the house pays off odds of 35 to 1. If you bet on a red number or black number and win, the house pays 1 to 1 (even money).

61. Determine the expected value of betting $1 on a particular number.

62. Determine the expected value of betting $1 on red.

RECREATIONAL MATHEMATICS

63. ***Wheel of Fortune*** The following is a miniature version of the Wheel of Fortune. When Dave Salem spins the wheel, he is awarded the amount on the wheel indicated by the pointer. If the wheel points to Bankrupt, he loses the total amount he has accumulated and also loses his turn. Assume that the wheel stops on a position at random and that each position is equally likely to occur.

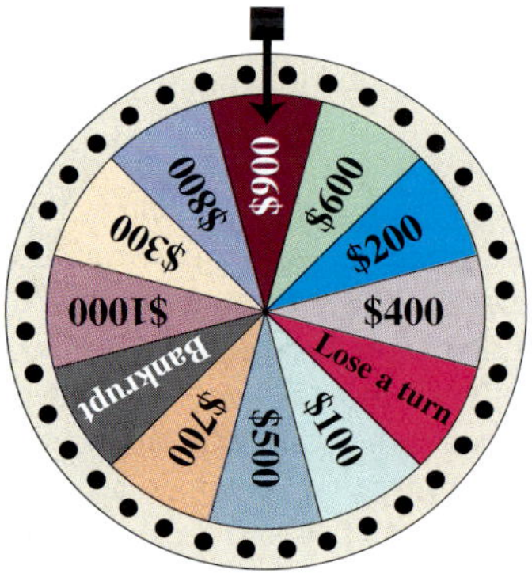

a) Determine Mr. Salem's expectation when he spins the wheel at the start of the game (he has no money to lose if he lands on Bankrupt).

b) If Mr. Salem presently has a balance of $1800, determine his expectation when he spins the wheel.

12.5 TREE DIAGRAMS

▲ We can use a tree diagram to illustrate the possible arrangements of boys and girls in a family with three children.

Suppose that a couple plans to have three children. What is the probability that the couple will have exactly two boys? What is the probability that the couple will have at least one girl? We stated earlier that the possible results of an experiment are called its outcomes. To solve probability problems such as the ones above, it is helpful if we are able to determine all the possible outcomes of an experiment. In this section, we will illustrate how the counting principle can be used to determine the number of outcomes of an experiment. We will also illustrate how tree diagrams can be used to determine all the possible outcomes of an experiment.

Now we will introduce the counting principle.

COUNTING PRINCIPLE

If a first experiment can be performed in M distinct ways and a second experiment can be performed in N distinct ways, then the two experiments in that specific order can be performed in $M \cdot N$ distinct ways.

If we wanted to find the number of possible outcomes when a coin is tossed and a die is rolled, we could reason that the coin has two possible outcomes, heads and tails. The die has six possible outcomes: 1, 2, 3, 4, 5, and 6. Thus, the two experiments together have $2 \cdot 6$, or 12, possible outcomes.

A list of all the possible outcomes of an experiment is called a ***sample space***. Each individual outcome in the sample space is called a ***sample point***. ***Tree diagrams*** are helpful in determining sample spaces.

A tree diagram illustrating all the possible outcomes when a coin is tossed and a die is rolled (see Fig. 12.6) has two initial branches, one for each possible outcome of the coin. Each of these branches will have six branches emerging from them, one for each possible outcome of the die. That will give a total of 12 branches, the same number of possible outcomes found by using the counting principle. We can obtain the sample space by listing all the possible combinations of branches. Note that this sample space consists of 12 sample points.

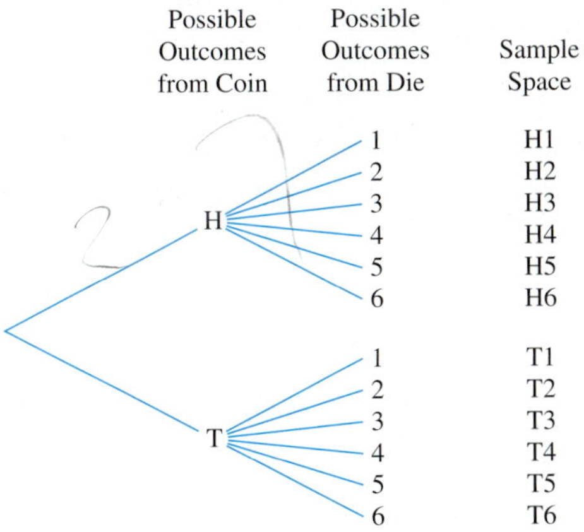

Figure 12.6

Example 1 uses the phrase "without replacement." This phrase tells us that once an item is selected, it cannot be selected again, making it impossible to select the same item twice.

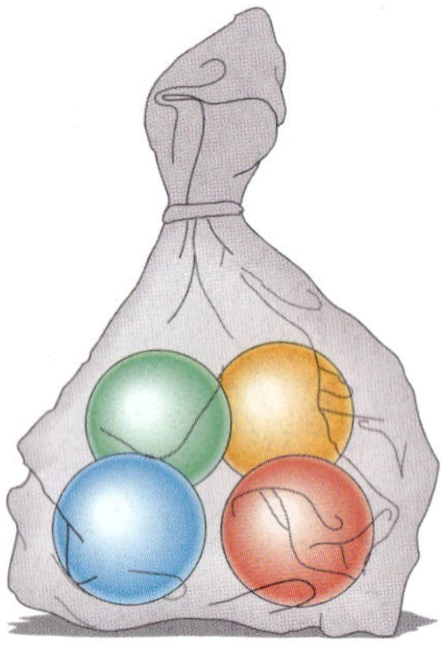

Figure 12.7

EXAMPLE 1 *Selecting Balls without Replacement*

Two balls are to be selected *without replacement* from a bag that contains one red, one blue, one green, and one orange ball (see Fig. 12.7).

a) Use the counting principle to determine the number of points in the sample space.
b) Construct a tree diagram and list the sample space.
c) Determine the probability that one orange ball is selected.
d) Determine the probability that a green ball followed by a red ball is selected.

SOLUTION

a) The first selection may be any one of the four balls. Once the first ball is selected, only three balls remain for the second selection. Thus, there are $4 \cdot 3$, or 12, sample points in the sample space.
b) The first ball selected can be red, blue, green, or orange. Since this experiment is done without replacement, the same colored ball cannot be selected twice. For example, if the first ball selected is red, the second ball selected must be either blue, green, or orange. The tree diagram and sample space are shown in Fig. 12.8. The sample space contains 12 points. That result checks with the answer obtained in part (a) using the counting principle.

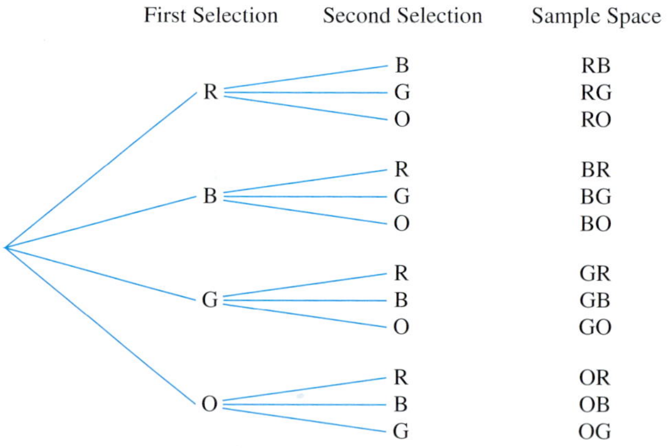

Figure 12.8

c) If we know the sample space, we can compute probabilities using the formula

$$P(E) = \frac{\text{number of outcomes favorable to } E}{\text{total number of outcomes}}$$

The total number of outcomes will be the number of points in the sample space. From Fig. 12.8, we determine that there are 12 possible outcomes. Six outcomes have one orange ball: RO, BO, GO, OR, OB, and OG.

$$P(\text{one orange ball is selected}) = \frac{6}{12} = \frac{1}{2}$$

d) One possible outcome meets the criteria of a green ball followed by a red ball: GR.

$$P(\text{green followed by red}) = \frac{1}{12}$$

The counting principle can be extended to any number of experiments, as illustrated in Example 2.

EXAMPLE 2 *Lunch Choices*

At Theresa's Restaurant, each lunch special consists of a sandwich, a beverage, and a dessert. The sandwich choices are roast beef (r) or ham (h). The beverage choices are coffee (c), tea (t), or soda (s). The dessert choices are ice cream (i) or apple pie (p).

a) Use the counting principle to determine the number of different lunch specials offered by the restaurant.

b) Construct a tree diagram and list the sample space.

c) If a customer randomly selects one of the lunch specials, determine the probability that both a roast beef sandwich and ice cream are selected.

d) If a customer randomly selects one of the lunch specials, determine the probability that neither tea nor apple pie is selected.

SOLUTION

a) There are 2 choices for a sandwich, 3 choices for a beverage and 2 choices for dessert. Using the counting principle, we can determine that there are $2 \cdot 3 \cdot 2$ or 12 different lunch specials.

b) The tree diagram illustrating the 12 lunch specials is given in Fig. 12.9.

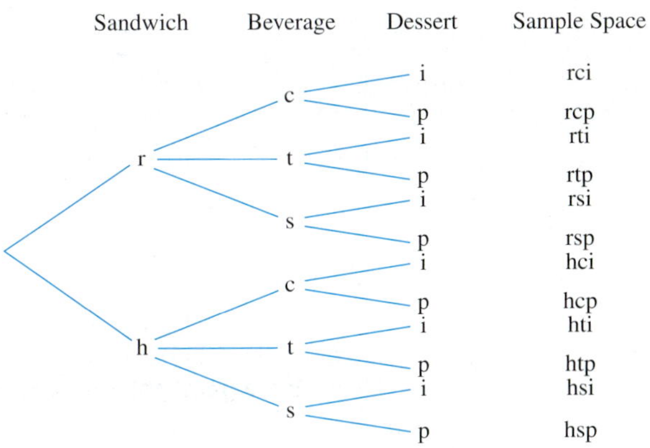

Figure 12.9

c) Of the 12 lunch specials, 3 contain both a roast beef sandwich and ice cream (rci, rti, rsi).

$$P(\text{roast beef and ice cream are selected}) = \frac{3}{12} = \frac{1}{4}$$

d) Of the 12 lunch specials, 4 contain neither tea nor apple pie (rci, rsi, hci, hsi).

$$P(\text{neither tea nor ice cream are selected}) = \frac{4}{12} = \frac{1}{3}$$

▲ Beyoncé

EXAMPLE 3 *Selecting Ticket Winners*

A radio station has two tickets to give away to a Beyoncé concert. It held a contest and narrowed the possible recipients down to four people: Christine (C), Mike Hammer (MH), Mike Levine (ML), and Phyllis (P). The names of two of these four people will be selected at random from a hat and the two people selected will be awarded the tickets.

a) Use the counting principle to determine the number of points in the sample space.

b) Construct a tree diagram and list the sample space.

c) Determine the probability that Christine is selected.

d) Determine the probability that neither Mike Hammer nor Mike Levine is selected.

e) Determine the probability that at least one Mike is selected.

SOLUTION

a) The first selection may be any one of the four people; see Fig. 12.10. Once the first person is selected, only three people remain for the second selection. Thus, there are $4 \cdot 3$ or 12 sample points in the sample space.

b)

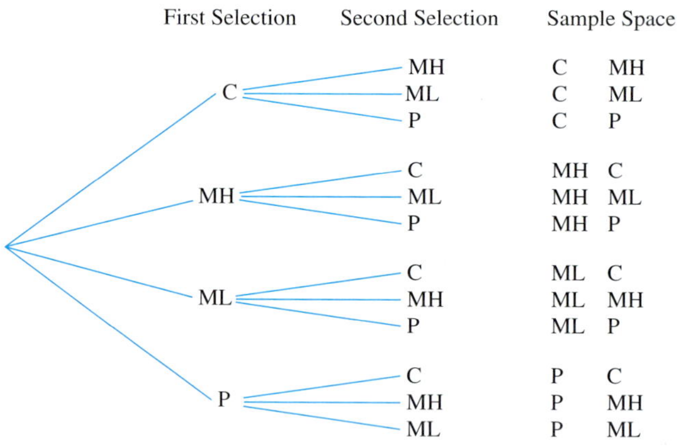

Figure 12.10

c) Of the 12 points in the sample space, 6 have Christine. They are C MH, C ML, C P, MH C, ML C, and P C.

$$P(\text{Christine is selected}) = \frac{6}{12} = \frac{1}{2}$$

d) Of the 12 points in the sample space, two have neither Mike. They are C P and P C.

$$P(\text{neither Mike selected}) = \frac{2}{12} = \frac{1}{6}$$

e) At least one Mike means that one or more Mikes are selected. There are 10 points in the sample space with at least one Mike (all those except C P and P C).

$$P(\text{at least one Mike is selected}) = \frac{10}{12} = \frac{5}{6}$$

In Example 3, if you add the probability of no Mike being selected with the probability of at least one Mike being selected, you get $\frac{1}{6} + \frac{5}{6}$, or 1. In any probability problem, if E is a specific event, then either E happens at least one time or it does not happen at all. Thus, $P(E$ happening at least once$) + P(E$ does not happen$) = 1$, which leads to the following rule.

$$P(\text{event happening at least once}) = 1 - P(\text{event does not happen})$$

For example, suppose that the probability of not getting any red flowers from the seeds that are planted is $\frac{2}{7}$. Then the probability of getting at least one red flower from the seeds that are planted is $1 - \frac{2}{7} = \frac{5}{7}$. We will use this rule in later sections.

Figure 12.11

In all the tree diagrams in this section, the outcomes were always equally likely; that is, each outcome had the same probability of occurrence. Consider a rock that has 4 faces such that each face has a different surface area and the rock is not uniform in density (see Fig. 12.11). When the rock is dropped, the probability that the rock lands on face 1 will not be the same as the probability that the rock lands on face 2. In fact, the probabilities that the rock lands on face 1, face 2, face 3, and face 4 may all be different. Therefore, the outcomes of the rock landing on face 1, face 2, face 3, and face 4 are not equally likely outcomes. Because the outcomes are not equally likely and we are not given additional information, we cannot determine the theoretical probability of the rock landing on each individual face. However, we can still determine the sample space indicating the faces that the rock may land on when the rock is dropped twice. The tree diagram and sample space are shown in Fig. 12.12.

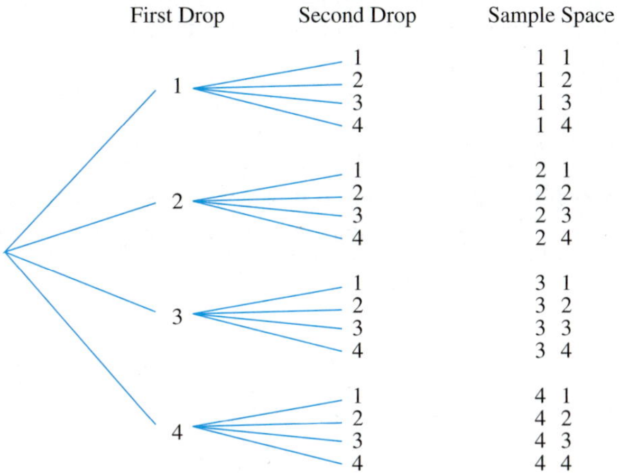

Figure 12.12

Since the outcomes are not equally likely, the probability of each of the 16 sample points in the sample space occurring cannot be determined. If the outcomes were equally likely, then each of the 16 points in the sample space would have a probability of $\frac{1}{16}$. See Exercises 29 and 30, which deal with outcomes that are not equally likely.

SECTION 12.5 EXERCISES

CONCEPT/WRITING EXERCISES

1. Explain the counting principle.

2. a) What is a sample space?

   b) What is a sample point?

3. If a first experiment can be performed in two distinct ways and a second experiment can be performed in seven distinct ways, how many possible ways can the two experiments be performed? Explain your answer.

4. In your own words, describe how to construct a tree diagram.

5. A problem states that two selections are made "without replacement." Explain what that means.

6. One experiment has five equally likely outcomes, and a second experiment has three equally likely outcomes. How many sample points will be in the sample space when the two experiments are performed one after the other?

PRACTICE THE SKILLS

7. ***Selecting States*** If two states are selected at random from the 50 U.S. states, use the counting principle to determine the number of possible outcomes if the states are selected

   a) with replacement.

   b) without replacement.

8. ***Selecting Dates*** If two dates are selected at random from the 365 days of the year, use the counting principle to determine the number of possible outcomes if the dates are selected

   a) with replacement.

   b) without replacement.

9. ***Selecting Lightbulbs*** A bag contains seven lightbulbs, all the same size and equally likely to be selected. Each lightbulb is a different brand. If you select three lightbulbs at random, use the counting principle to determine how many points will be in the sample space if the lightbulbs are selected

   a) with replacement.

   b) without replacement.

10. ***Remote Control*** Your television remote control has buttons for digits 0–9. If you press two buttons, how many numbers are possible if

    a) the same button may be pressed twice.

    b) the same button may not be pressed twice.

In Exercises 11–28, use the counting principle to determine the answer to part (a). Assume that each event is equally likely to occur.

11. ***Coin Toss*** Two coins are tossed.

    a) Determine the number of points in the sample space.

    b) Construct a tree diagram and list the sample space.

    Determine the probability that

    c) no heads are tossed.

    d) exactly one head is tossed.

    e) two heads are tossed.

12. ***Boys and Girls*** A couple plans to have two children.

    a) Determine the number of points in the sample space of the possible arrangements of boys and girls.

    b) Construct a tree diagram and list the sample space.

Assuming that boys and girls are equally likely, determine the probability that the couple has

c) two girls.

d) at least one girl.

e) a girl and then a boy.

13. *Cards* A box contains three cards. On one card there is a sun, on another card there is a question mark, and on the third card there is an apple.

Two cards are to be selected at random with replacement.

a) Determine the number of points in the sample space.

b) Construct a tree diagram and list the sample space.

Determine the probability that

c) two apples are selected.

d) a sun and then a question mark are selected.

e) at least one apple is selected.

14. *Cards* Repeat Exercise 13 but assume that the cards are drawn without replacement.

15. *Marble Selection* A hat contains four marbles: 1 yellow, 1 red, 1 blue, and 1 green. Two marbles are to be selected at random without replacement from the hat.

a) Determine the number of points in the sample space.

b) Construct a tree diagram and list the sample space.

Determine the probability of selecting

c) exactly 1 red marble.

d) at least 1 marble that is not red.

e) no green marbles.

16. *Three Coins* Three coins are tossed.

a) Determine the number of points in the sample space.

b) Construct a tree diagram and list the sample space.

Determine the probability that

c) no heads are tossed.

d) exactly one head is tossed.

e) three heads are tossed.

17. *Paint Choices* The Flotterons plan to purchase paint for the walls and paint for the trim in their living room. They will select one color for the walls and one color for the trim from the following colors.

Walls	Trim
Sable	Alabaster
Java	White
Chocolate	Oyster

a) Determine the number of points in the sample space.

b) Construct a tree diagram and list the sample space.

Determine the probability that they select

c) Java.

d) Java and oyster.

e) paint other than java.

18. *Pet Shop* A pet shop is selling a calico cat, a Siamese cat, a Persian cat, and a Himalayan cat. The Chens are going to select two cats to bring home as pets.

a) Determine the number of points in the sample space.

b) Construct a tree diagram and list the sample space.

Determine the probability that they select

c) the Persian cat.

d) the Persian cat and the calico cat.

e) cats other than the Persian cat.

PROBLEM SOLVING

19. ***Rolling Dice*** Two dice are rolled.

a) Determine the number of points in the sample space.

b) Construct a tree diagram and list the sample space.

Determine the probability that

c) a double (a 1, 1 or 2, 2, etc.) is rolled.

d) a sum of 7 is rolled.

e) a sum of 2 is rolled.

f) Are you as likely to roll a sum of 2 as you are of rolling a sum of 7? Explain your answer.

20. ***Voting*** At a homeowners' association meeting, a board member can vote yes, vote no, or abstain on a motion. There are three motions on which a board member must vote.

a) Determine the number of points in the sample space.

b) Construct a tree diagram and determine the sample space.

Determine the probability that a board member votes

c) no, yes, no in that order.

d) yes on exactly two of the motions.

e) yes on at least one motion.

21. ***Gift Cards*** Three different people are to be selected at random, and each will be given one gift card. There is one card from Home Depot, one from Best Buy, and one from Red Lobster. The first person selected gets to choose one of the cards. The second person selected gets to choose between the two remaining cards. The third person selected gets the third card.

a) Determine the number of points in the sample space.

b) Construct a tree diagram and determine the sample space.

Determine the probability that

c) the Best Buy card is selected first.

d) the Home Depot card is selected first and the Red Lobster card is selected last.

e) The cards are selected in this order: Best Buy, Red Lobster, Home Depot.

22. ***Shopping*** Susan Forman has to purchase a box of cereal, a bottle of soda, and a can of vegetables. The types of cereal, soda, and vegetables she is considering are shown below.

Cereal: Rice Krispies, Frosted Flakes, Honeycomb
Soda: orange, black cherry, ginger ale
Vegetables: peas, carrots

a) Determine the number of points in the sample space.

b) Construct a tree diagram and determine the sample space.

Determine the probability that she selects

c) Honeycomb for the cereal.

d) Rice Krispies and ginger ale.

e) a soda other than black cherry.

23. ***Apartment Options*** Don Cater plans to rent an apartment from Rustic Village Apartments. The following chart displays information regarding the possible number of bedrooms, bathrooms, and other features from which he can choose. For his apartment, Don will select the number of bedrooms, number of bathrooms, and one other feature.

Bedrooms	Bathrooms	Other Features
1	1	Fireplace
2	2	Hardwood floors
3		Balcony

a) Determine the number of points in the sample space.

b) Construct a tree diagram and determine the sample space.

Determine the probability that Don selects

c) a two-bedroom apartment.

d) a two-bedroom apartment with a fireplace.

e) an apartment without a balcony.

24. ***A New Computer*** You visit Computer City to purchase a new computer system. You are going to purchase a computer, printer, and monitor from among the following brands.

Computer	Printer	Monitor
Compaq	Hewlett-Packard	Omega
IBM	Epson	Toshiba
Gateway		
Dell		

a) Determine the number of points in the sample space.

b) Construct a tree diagram and determine the sample space.

Determine the probability of selecting

c) a Gateway computer.

d) a Hewlett-Packard printer.

e) a Gateway computer and a Hewlett-Packard printer.

25. ***Buying Appliances*** Mr. and Mrs. Miller just moved into a new home and need to purchase kitchen appliances. The chart below shows the brands of appliances they are considering.

Refrigerator	Stove	Dishwasher
General Electric	General Electric	General Electric
Kenmore	Frigidaire	KitchenAid
Maytag	Roper	Whirlpool

The Millers are going to purchase a refrigerator, a stove, and a dishwasher.

a) Determine the number of points in the sample space.

b) Construct a tree diagram and determine the sample space.

Determine the probability that they select

c) General Electric for all three appliances.

d) no General Electric appliances.

e) at least one General Electric appliance.

26. ***Literature Choices*** You decide to take a Literature course. A requirement for the course is that you must read one classic book, one nonfiction book, and one science fiction book from the list below.

Classic	Nonfiction	Science Fiction
A Farewell to Arms (F)	*John Adams* (J)	*War of the Worlds* (W)
The Grapes of Wrath (G)	*Band of Brothers* (B)	*Dune* (D)
Tom Sawyer (T)		
Moby-Dick (M)		

a) Determine the number of points in the sample space.

b) Construct a tree diagram and determine the sample space.

Determine the probability that

c) *John Adams* is selected.

d) either *A Farewell to Arms* or *Moby-Dick* is selected.

e) *Moby-Dick* is not selected.

27. ***Personal Characteristics*** An individual can be classified as male or female with red, brown, black, or blond hair and with brown, blue, or green eyes.

a) How many different classifications are possible (for example, male, red-headed, blue-eyed)?

b) Construct a tree diagram to determine the sample space.

c) If each outcome is equally likely, determine the probability that the individual will be a male with black hair and blue eyes.

d) Determine the probability that the individual will be a female with blond hair.

28. ***Mendel Revisited*** A pea plant must have exactly one of each of the following pairs of traits: short (s) or tall (t); round (r) or wrinkled (w) seeds; yellow (y) or green (g) peas; and white (wh) or purple (p) flowers (for example, short, wrinkled, green pea with white flowers).

a) How many different classifications of pea plants are possible?

b) Use a tree diagram to determine all the classifications possible.

c) If each characteristic is equally likely, find the probability that the pea plant will have round peas.

d) Determine the probability that the pea plant will be short, have wrinkled seeds, have yellow seeds, and have purple flowers.

CHALLENGE PROBLEMS/GROUP ACTIVITIES

29. ***Three Chips*** Suppose that a bag contains one white chip and two red chips. Two chips are going to be selected at random from the bag *with replacement*.

a) What is the probability of selecting a white chip from the bag on the first selection?

b) What is the probability of selecting a red chip from the bag on the first selection?

c) Are the outcomes of selecting a white chip and selecting a red chip on the first selection equally likely? Explain.

d) The sample space when two chips are selected from the bag with replacement is ww, wr, rw, rr. Do you believe that the probability of selecting ww is greater than, equal to, or less than the probability of selecting rr? Explain.

30. ***Thumbtacks*** A thumbtack is dropped on a concrete floor. Assume that the thumbtack can only land point up (u) or point down (d), as shown in the figure below.

If two thumbtacks are dropped, one after the other, the tree diagram below can be used to show the possible outcomes.

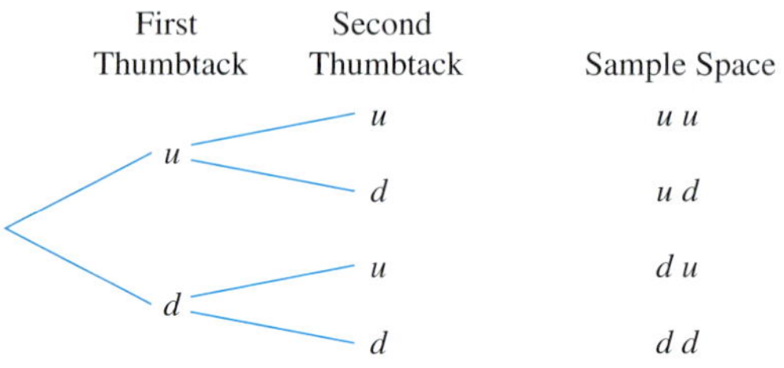

a) Do you believe that the outcomes of the thumbtack landing point up and the thumbtack landing point down are equally likely? Explain.

b) List the sample points in the sample space of this experiment.

c) Do you believe that the probability that both thumbtacks land point up (uu) is the same as the probability that both thumbtacks land point down (dd)? Explain.

d) Can you compute the theoretical probability of a thumbtack landing point up and the theoretical probability of a thumbtack landing point down? Explain.

e) Obtain a box of thumbtacks and drop the thumbtacks out of the box with care. Determine the empirical probability of a thumbtack landing point up when dropped and the empirical probability of a thumbtack landing point down when dropped.

RECREATIONAL MATHEMATICS

31. ***Ties*** All my ties are red except two. All my ties are blue except two. All my ties are brown except two. How many ties do I have?

32. ***Rock Faces*** An experiment consists of 3 parts: flipping a coin, tossing a rock, and rolling a die. If the sample space consists of 60 sample points, determine the number of faces on the rock.

12.6 *OR* AND *AND* PROBLEMS

▲ What is the probability that you select a movie that is a comedy or a movie rated PG?

Suppose you go to Blockbuster Video to rent a movie. You are overwhelmed with the number of movie choices, so you decide to randomly select a movie from the list of the top 100 rentals. What is the probability that you select a comedy *or* a movie rated PG? If you decide to randomly select two movies, what is the probability that the first movie selected is a comedy *and* the second movie selected is a drama? In this section, we will learn how to solve probability problems containing the words *or* and probability problems containing the word *and*.

In Section 12.5, we showed how to work probability problems by constructing sample spaces. Often it is inconvenient or too time consuming to solve a problem by first constructing a sample space. For example, if an experiment consists of selecting two cards with replacement from a deck of 52 cards, there would be $52 \cdot 52$ or 2704 points in the sample space. Trying to list all these sample points could take hours. In this section, we learn how to solve ***compound probability*** problems that contain the words *and* or *or* without constructing a sample space.

Or Problems

The ***or probability problem*** requires obtaining a "successful" outcome for *at least one* of the given events. For example, suppose that we roll one die and we are interested in finding the probability of rolling an even number *or* a number greater than 4. For this situation, rolling either a 2, 4, or 6 (an even number) or a 5 or 6 (a number greater than 4) would be considered successful. Note that the number 6 satisfies both criteria. Since 4 of the 6 numbers meet the criteria (the 2, 4, 5, and 6), the probability of rolling an even number *or* a number greater than 4 is $\frac{4}{6}$ or $\frac{2}{3}$.

A formula for finding the probability of event A or event B, symbolized $P(A \text{ or } B)$, follows.

$$P(A \text{ or } B) = P(A) + P(B) - P(A \text{ and } B)$$

Since we add (and subtract) probabilities to find $P(A \text{ or } B)$, this formula is sometimes referred to as the *addition formula*. We explain the use of the *or* formula in Example 1.

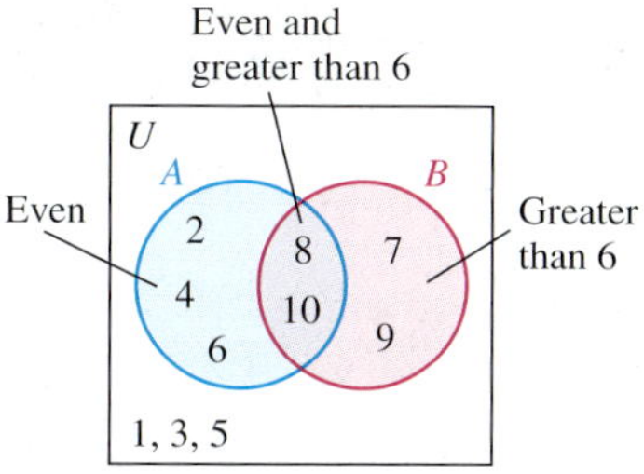

Figure 12.13

EXAMPLE 1 *Using the Addition Formula*

Each of the numbers 1, 2, 3, 4, 5, 6, 7, 8, 9, and 10 is written on a separate piece of paper. The 10 pieces of paper are then placed in a hat, and one piece is randomly selected. Determine the probability that the piece of paper selected contains an even number or a number greater than 6.

SOLUTION We are asked to find the probability that the number selected is *even* or is *greater than 6*. Let's use set A to represent the statement "the number is even" and set B to represent the statement "the number is greater than 6." Figure 12.13 is a Venn diagram, as introduced in Chapter 2, with sets A (even) and B (greater than 6). There are a total of 10 numbers, of which five are even (2, 4, 6, 8, and 10). Thus, the probability of selecting an even number is $\frac{5}{10}$. Four numbers are greater than 6: the 7, 8, 9, and 10. Thus, the probability of selecting a number greater than 6 is $\frac{4}{10}$. Two numbers are both even and greater than 6: the 8 and 10. Thus, the probability of selecting a number that is both even and greater than 6 is $\frac{2}{10}$.

If we substitute the appropriate statements for A and B in the formula, we obtain

$$P(A \text{ or } B) = P(A) + P(B) - P(A \text{ and } B)$$

$$P\begin{pmatrix}\text{even or}\\ \text{greater than 6}\end{pmatrix} = P(\text{even}) + P\begin{pmatrix}\text{greater}\\ \text{than 6}\end{pmatrix} - P\begin{pmatrix}\text{even and}\\ \text{greater than 6}\end{pmatrix}$$

$$= \frac{5}{10} + \frac{4}{10} - \frac{2}{10}$$

$$= \frac{7}{10}$$

Thus, the probability of selecting an even number or a number greater than 6 is $\frac{7}{10}$. The seven numbers that are even or greater than 6 are 2, 4, 6, 7, 8, 9, and 10. •

Example 1 illustrates that when finding the probability of A or B, we add the probabilities of events A and B and then subtract the probability of both events occurring simultaneously.

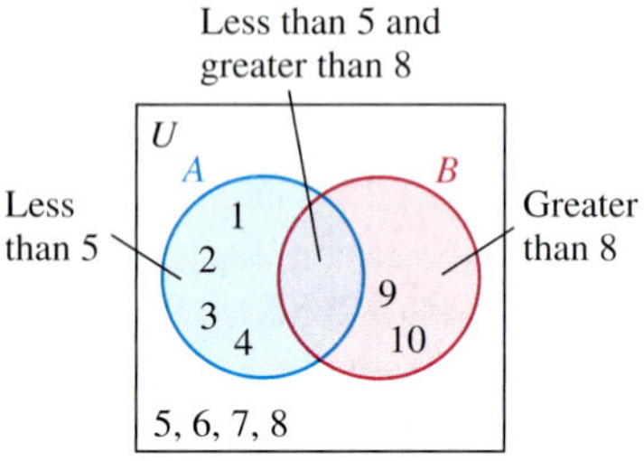

Figure 12.14

EXAMPLE 2 *Using the Addition Formula*

Consider the same sample space, the numbers 1 through 10, as in Example 1. If one piece of paper is selected, determine the probability that it contains a number less than 5 or a number greater than 8.

SOLUTION Let A represent the statement "the number is less than 5" and B represent the statement "the number is greater than 8." A Venn diagram illustrating these statements is shown in Fig. 12.14.

$$P(\text{number is less than 5}) = \frac{4}{10}$$

$$P(\text{number is greater than 8}) = \frac{2}{10}$$

Since there are no numbers that are *both* less than 5 and greater than 8, $P(\text{number is less than 5 and greater than 8}) = 0$. Therefore,

$$P\begin{pmatrix}\text{number is}\\ \text{less than 5}\\ \text{or greater}\\ \text{than 8}\end{pmatrix} = P\begin{pmatrix}\text{number is}\\ \text{less than 5}\end{pmatrix} + P\begin{pmatrix}\text{number is}\\ \text{greater than 8}\end{pmatrix} - P\begin{pmatrix}\text{number is}\\ \text{less than 5}\\ \text{and greater}\\ \text{than 8}\end{pmatrix}$$

$$= \frac{4}{10} + \frac{2}{10} - 0 = \frac{6}{10} = \frac{3}{5}$$

Thus, the probability of selecting a number less than 5 or greater than 8 is $\frac{3}{5}$. The six numbers that are less than 5 or greater than 8 are 1, 2, 3, 4, 9, and 10. •

In Example 2, it is impossible to select a number that is both less than 5 *and* greater than 8 when only one number is to be selected. Events such as these are said to be *mutually exclusive.*

Two events A and B are **mutually exclusive** if it is impossible for both events to occur simultaneously.

If events A and B are mutually exclusive, then $P(A \text{ and } B) = 0$ and the addition formula simplifies to $P(A \text{ or } B) = P(A) + P(B)$.

EXAMPLE 3 *Probability of A or B*

One card is selected from a standard deck of playing cards. Determine whether the following pairs of events are mutually exclusive and find $P(A \text{ or } B)$.

a) A = an ace, B = a jack

b) A = an ace, B = a heart

c) A = a red card, B = a black card

d) A = a picture card, B = a red card

SOLUTION

a) There are four aces and four jacks in a standard deck of 52 cards. It is impossible to select both an ace and a jack when only one card is selected. Therefore, these events are mutually exclusive.

$$P(\text{ace or jack}) = P(\text{ace}) + P(\text{jack}) = \frac{4}{52} + \frac{4}{52} = \frac{8}{52} = \frac{2}{13}$$

The ace of hearts is both an ace and a heart.

b) There are 4 aces and 13 hearts in a standard deck of 52 cards. One card, the ace of hearts, is both an ace and a heart. Therefore, these events are not mutually exclusive.

$$P(\text{ace}) = \frac{4}{52} \qquad P(\text{heart}) = \frac{13}{52} \qquad P(\text{ace and heart}) = \frac{1}{52}$$

$$\begin{aligned} P(\text{ace or heart}) &= P(\text{ace}) + P(\text{heart}) - P(\text{ace and heart}) \\ &= \frac{4}{52} + \frac{13}{52} - \frac{1}{52} \\ &= \frac{16}{52} = \frac{4}{13} \end{aligned}$$

c) There are 26 red cards and 26 black cards in a standard deck of 52 cards. It is impossible to select one card that is both a red card and a black card. Therefore, the events are mutually exclusive.

$$\begin{aligned} P(\text{red or black}) &= P(\text{red}) + P(\text{black}) \\ &= \frac{26}{52} + \frac{26}{52} = \frac{52}{52} = 1 \end{aligned}$$

Since $P(\text{red or black}) = 1$, a red card or a black card must be selected.

DID YOU KNOW?

They're All Girls

The probability that a husband and wife have nine children of which all are girls is $(\frac{1}{2})^9$, or $\frac{1}{512}$.

The odds against a family having nine girls in a row is 511 to 1. Thus, for every 512 families with nine children, on average, one family will have all girls and 511 will not. However, should another child be born to that one family, the probability of the tenth child being a girl is still $\frac{1}{2}$.

d) There are 12 picture cards in a standard deck of 52 cards. Six of the 12 picture cards are red (the jacks, queens, and kings of hearts and diamonds). Thus, selecting a picture card and a red card are not mutually exclusive.

$$P\begin{pmatrix}\text{picture card}\\ \text{or red card}\end{pmatrix} = P\begin{pmatrix}\text{picture}\\ \text{card}\end{pmatrix} + P\begin{pmatrix}\text{red}\\ \text{card}\end{pmatrix} - P\begin{pmatrix}\text{picture card}\\ \text{and red card}\end{pmatrix}$$

$$= \frac{12}{52} + \frac{26}{52} - \frac{6}{52}$$

$$= \frac{32}{52} = \frac{8}{13}$$

And Problems

A second type of probability problem is the ***and probability problem***, which requires obtaining a favorable outcome in *each* of the given events. For example, suppose that *two* cards are to be selected from a deck of cards and we are interested in the probability of selecting two aces (one ace *and* then a second ace). Only if *both* cards selected are aces would this experiment be considered successful. A formula for finding the probability of events A and B, symbolized $P(A \text{ and } B)$, follows.

$$P(A \text{ and } B) = P(A) \cdot P(B)$$

Since we multiply to find $P(A \text{ and } B)$, this formula is sometimes referred to as the *multiplication formula*. When using the multiplication formula, we **always assume that event A has occurred when calculating $P(B)$** because we are determining the probability of obtaining a favorable outcome in both of the given events.*

Unless we specify otherwise, $P(A \text{ and } B)$ indicates that we are determining the probability that event A occurs *and then* event B occurs (in that order). Consider a bag that contains three chips: 1 red chip, 1 blue chip, and 1 green chip. Suppose that two chips are selected from the bag with replacement. The tree diagram and sample space for the experiment are shown in Fig. 12.15. There are nine possible outcomes for the two selections, as indicated in the sample space.

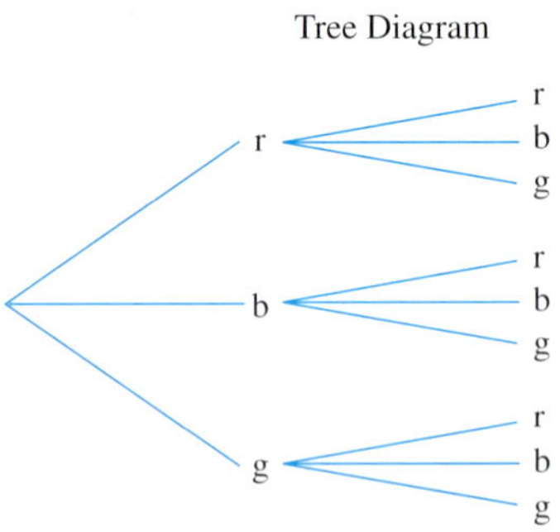

Sample Space
r r
r b
r g
b r
b b
b g
g r
g b
g g

Figure 12.15

*$P(B)$, assuming that event A has occurred, may be denoted $P(B \mid A)$, which is read "the probability of B, given A." We will discuss this type of probability (conditional probability) further in Section 12.7.

Note that the probability of selecting a red chip followed by a blue chip (rb), indicated by P(red and blue), is $\frac{1}{9}$. The probability of selecting a red chip and a blue chip, in any order (rb or br), is $\frac{2}{9}$. In this section, when we ask for $P(A \text{ and } B)$, it means the probability of event A occurring *and then* event B occurring, in that order.

EXAMPLE 4 *An Experiment with Replacement*

Two cards are to be selected *with replacement* from a deck of cards. Determine the probability that two queens will be selected.

SOLUTION Since the deck of 52 cards contains four queens, the probability of selecting a queen on the first draw is $\frac{4}{52}$. The card selected is then returned to the deck. Therefore, the probability of selecting a queen on the second draw remains $\frac{4}{52}$.

If we let A represent the selection of the first queen and B represent the selection of the second queen, the formula may be written as follows.

$$P(A \text{ and } B) = P(A) \cdot P(B)$$

$$P(2 \text{ queens}) = P(\text{queen 1 } \textit{and} \text{ queen 2}) = P(\text{queen 1}) \cdot P(\text{queen 2})$$

$$= \frac{4}{52} \cdot \frac{4}{52}$$

$$= \frac{1}{13} \cdot \frac{1}{13} = \frac{1}{169}$$

EXAMPLE 5 *An Experiment without Replacement*

Two cards are to be selected *without replacement* from a deck of cards. Determine the probability that two queens will be selected.

SOLUTION This example is similar to Example 4. However, this time we are doing the experiment without replacing the first card selected to the deck before selecting the second card.

The probability of selecting a queen on the first draw is $\frac{4}{52}$. When calculating the probability of selecting the second queen, we must assume that the first queen has been selected. Once this first queen has been selected, only 51 cards, including 3 queens, remain in the deck. The probability of selecting a queen on the second draw becomes $\frac{3}{51}$. The probability of selecting two queens without replacement is

$$P(2 \text{ queens}) = P(\text{queen 1}) \cdot P(\text{queen 2})$$

$$= \frac{4}{52} \cdot \frac{3}{51}$$

$$= \frac{1}{13} \cdot \frac{1}{17} = \frac{1}{221}$$

Now we introduce *independent events*.

Event A and event B are **independent events** if the occurrence of either event in no way affects the probability of occurrence of the other event.

DID YOU KNOW?

The Birthday Problem

Among 24 people chosen at random, what would you guess is the probability that at least 2 of them have the same birthday? It might surprise you to learn that it is greater than 0.5. There are 365 days on which the first person selected can have a birthday. That person has a 365/365 chance of having a birthday on one of those days. The probability that the second person's birthday is on any other day is 364/365. The probability that the third person's birthday is on a day different from the first two is 363/365, and so on. The probability that the 24th person has a birthday on any other day than the first 23 people is 342/365. Thus, the probability, P, that of 24 people, no 2 have the same birthday is $(365/365) \times (364/365) \times (363/365) \times \cdots \times (342/365) = 0.462$. Then the probability of at least 2 people of 24 having the same birthday is $1 - P = 1 - 0.462 = 0.538$, or slightly larger than $\frac{1}{2}$.

Rolling dice and tossing coins are examples of independent events. In Example 4, the events are independent since the first card was returned to the deck. The probability of selecting a queen on the second draw was not affected by the first selection. The events in Example 5 are not independent since the probability of the selection of the second queen was affected by removing the first queen selected from the deck. Such events are called ***dependent events***. *Experiments done with replacement will result in independent events, and those done without replacement will result in dependent events.*

EXAMPLE 6 *Independent or Dependent Events?*

One hundred people attended a charity benefit to raise money for cancer research. Three people in attendance will be selected at random without replacement, and each will be awarded one door prize. Are the events of selecting the three people who will be awarded the door prize independent or dependent events?

SOLUTION The events are dependent since each time one person is selected, it changes the probability of the next person being selected. In the first selection, the probability that a specific individual is selected is $\frac{1}{100}$. If that person is not selected first, the probability that the specific person is selected second changes to $\frac{1}{99}$. In general, in any experiment in which two or more items are selected *without replacement*, the events will be dependent.

The multiplication formula may be extended to more than two events, as illustrated in Example 7.

EXAMPLE 7 *Drug Reaction*

A new medicine was given to a sample of 25 of Dr. Cleary's patients with flu symptoms. Of the total, 19 patients reacted favorably, 2 reacted unfavorably, and 4 were unaffected. Three of these patients are selected at random. Determine the probability of each of the following.

a) All three reacted favorably.

b) The first patient reacted favorably, the second patient reacted unfavorably, and the third patient was unaffected.

c) No patient reacted favorably.

d) At least one patient reacted favorably.

SOLUTION Each time a patient is selected, the number of patients remaining decreases by one.

a) The probability that the first patient reacted favorably is $\frac{19}{25}$. If the first patient reacted favorably, of the 24 remaining patients only 18 patients are left who reacted favorably. The probability of selecting a second patient who reacted favorably is $\frac{18}{24}$. If the second patient reacted favorably, only 17 patients are left who reacted favorably. The probability of selecting a third patient who reacted favorably is $\frac{17}{23}$.

$$P\begin{pmatrix}\text{three patients}\\ \text{reacted}\\ \text{favorably}\end{pmatrix} = P\begin{pmatrix}\text{first patient}\\ \text{reacted}\\ \text{favorably}\end{pmatrix} \cdot P\begin{pmatrix}\text{second patient}\\ \text{reacted}\\ \text{favorably}\end{pmatrix} \cdot P\begin{pmatrix}\text{third patient}\\ \text{reacted}\\ \text{favorably}\end{pmatrix}$$

$$= \frac{19}{25} \cdot \frac{18}{24} \cdot \frac{17}{23} = \frac{969}{2300}$$

b) The probability that the first patient reacted favorably is $\frac{19}{25}$. Once a patient is selected, there are only 24 patients remaining. Two of the remaining 24 patients reacted unfavorably. Thus, the probability that the second patient reacted unfavorably is $\frac{2}{24}$. After the second patient is selected, there are 23 remaining patients, of which 4 were unaffected. The probability that the third patient was unaffected is therefore $\frac{4}{23}$.

$$P\left(\begin{array}{c}\text{first patient reacted favorably, the}\\ \text{second patient reacted unfavorably, and}\\ \text{the third patient was unaffected}\end{array}\right)$$

$$= P\left(\begin{array}{c}\text{first patient}\\ \text{reacted}\\ \text{favorably}\end{array}\right) \cdot P\left(\begin{array}{c}\text{second patient}\\ \text{reacted}\\ \text{unfavorably}\end{array}\right) \cdot P\left(\begin{array}{c}\text{third patient}\\ \text{was}\\ \text{unaffected}\end{array}\right)$$

$$= \frac{19}{25} \cdot \frac{2}{24} \cdot \frac{4}{23} = \frac{19}{1725}$$

c) If none of the patients reacted favorably, the patients either reacted unfavorably or were unaffected. Six patients did not react favorably (2 reacted unfavorably and 4 were unaffected). The probability that the first patient selected did not react favorably is $\frac{6}{25}$. After the first patient is selected, 5 of the remaining 24 patients did not react favorably. After the second patient is selected, 4 of the remaining 23 patients did not react favorably.

$$P\left(\begin{array}{c}\text{none}\\ \text{reacted}\\ \text{favorably}\end{array}\right) = P\left(\begin{array}{c}\text{first patient}\\ \text{did not react}\\ \text{favorably}\end{array}\right) \cdot P\left(\begin{array}{c}\text{second patient}\\ \text{did not react}\\ \text{favorably}\end{array}\right) \cdot P\left(\begin{array}{c}\text{third patient}\\ \text{did not react}\\ \text{favorably}\end{array}\right)$$

$$= \frac{6}{25} \cdot \frac{5}{24} \cdot \frac{4}{23} = \frac{1}{115}$$

d) In Section 12.5, we learned that

$$P(\text{event happening at least once}) = 1 - P(\text{event does not happen})$$

In part (c), we found that the probability of selecting three patients none of whom reacted favorably was $\frac{1}{115}$. Therefore, the probability that at least one of the patients selected reacted favorably can be found as follows.

$$P\left(\begin{array}{c}\text{at least one of the three}\\ \text{patients reacted favorably}\end{array}\right) = 1 - P\left(\begin{array}{c}\text{none of the three}\\ \text{patients reacted favorably}\end{array}\right)$$

$$= 1 - \frac{1}{115} = \frac{115}{115} - \frac{1}{115} = \frac{114}{115}$$

TIMELY TIP

Which formula to use

It is sometimes difficult to determine when to use the *or* formula and when to use the *and* formula. The following information may be helpful in deciding which formula to use.

Continued on next page

DID YOU KNOW?

Slot Machines

You pull the handle, or push the button, to activate the slot machine. You are hoping that each of the three reels stops on Jackpot. The first reel shows Jackpot, the second reel shows Jackpot, and you hold your breath hoping the third reel shows Jackpot. When the third reel comes to rest, though, the Jackpot appears either one row above or one row below the row containing the other two Jackpots. You say to yourself, "I came so close" and try again. In reality, however, you probably did not come close to winning the Jackpot at all. The instant you pull the slot machine's arm, or push the button, the outcome is decided by a computer inside the slot machine. The computer uses stop motors to turn each reel and stop it at predetermined points. A random number generator within the computer is used to determine the outcomes. Although each reel generally has 22 stops, each stop is not equally likely. The computer assigns only a few random numbers to the stop that places the Jackpots on all three reels at the same time and many more random numbers to stop the reels on stops that are not the Jackpot. If you do happen to hit the Jackpot, it is simply because the random number generator happened to generate the right sequence of numbers the instant you activated the machine. For more detailed information, check *www.howstuffworks.com*. Also see the Did You Know? on page 792.

Continued from previous page

Or formula

Or problems will almost always contain the word *or* in the statement of the problem. For example, determine the probability of selecting a heart *or* a 6. *Or* problems in this book generally involve only *one* selection. For example, "one card is selected" or "one die is rolled."

And formula

And problems often do *not* use the word *and* in the statement of the problem. For example, "determine the probability that both cards selected are red" or "determine the probability that none of those selected is a banana" are both *and*-type problems. *And* problems in this book will generally involve *more than one* selection. For example, the problem may read "two cards are selected" or "three coins are flipped."

SECTION 12.6 EXERCISES

CONCEPT/WRITING EXERCISES

1. a) In $P(A \text{ or } B)$, what does the word *or* indicate?

   b) In $P(A \text{ and } B)$, what does the word *and* indicate?

2. a) Give the formula for $P(A \text{ or } B)$.

   b) In your own words, explain how to determine $P(A \text{ or } B)$ with the formula.

3. a) What are mutually exclusive events? Give an example.

   b) How do you calculate $P(A \text{ or } B)$ when A and B are mutually exclusive?

4. a) Give the formula for $P(A \text{ and } B)$.

   b) In your own words, explain how to determine $P(A \text{ and } B)$ with the formula.

5. When finding $P(B)$ using the formula $P(A \text{ and } B)$, what do we always assume?

6. a) What are independent events? Give an example.

   b) What are dependent events? Give an example.

7. A family is selected at random. Let event A be the mother likes classical music. Let event B be the daughter likes classical music.

   a) Are events A and B mutually exclusive? Explain.

   b) Are they independent events? Explain.

8. A family is selected at random. Let event A be the father likes to cook. Let event B be the mother likes to golf.

   a) Are events A and B mutually exclusive? Explain.

   b) Are they independent events? Explain.

9. If events A and B are mutually exclusive, explain why the formula $P(A \text{ or } B) = P(A) + P(B) - P(A \text{ and } B)$ can be simplified to $P(A \text{ or } B) = P(A) + P(B)$.

10. a) Write a problem that you would use the *or formula* to solve. Solve the problem and give the answer.

    b) Write a problem that you would use the *and formula* to solve. Solve the problem and give the answer.

PRACTICE THE SKILLS

In Exercises 11–14, determine the indicated probability.

11. If $P(A) = 0.6$, $P(B) = 0.4$, and $P(A \text{ and } B) = 0.3$, determine $P(A \text{ or } B)$.

12. If $P(A \text{ or } B) = 0.9$, $P(A) = 0.7$, and $P(B) = 0.5$, determine $P(A \text{ and } B)$.

13. If $P(A \text{ or } B) = 0.7$, $P(A) = 0.6$, and $P(A \text{ and } B) = 0.3$, determine $P(B)$.

14. If $P(A \text{ or } B) = 0.6$, $P(B) = 0.3$, and $P(A \text{ and } B) = 0.1$, determine $P(A)$.

15. ***Exam Preparation*** Professor Connell is in charge of a program to prepare students for a high school equivalency exam. Records show that the probability that a student in the program needs help in mathematics is 0.7, the probability that a student needs help in English is 0.6, and the probability that a student needs help in both mathematics and English is 0.55. Determine the probability that a student in the program needs help in mathematics or English.

16. ***Car Repair*** The manager at Arango Automotive has found that the probability that a car brought into the shop requires an oil change is 0.6, the probability that a car brought into the shop requires brake repair is 0.4, and the probability that a car requires both an oil change and brake repair is 0.2. For a car brought into the shop, determine the probability that the car will require an oil change or brake repair.

Roll a Die *In Exercises 17–20, a single die is rolled one time. Determine the probability of rolling*

17. a 2 or 5.

18. an odd number or a number greater than 4.

19. a number greater than 4 or less than 2.

20. a number greater than 3 or less than 5.

Select One Card *In Exercises 21–26, one card is selected from a deck of playing cards. Determine the probability of selecting*

21. an ace or a 2.

22. a jack or a club.

23. a picture card or a red card.

24. a club or a red card.

25. a card less than 8 or a club. (*Note:* The ace is considered a low card.)

26. a card greater than 9 or a black card.

PROBLEM SOLVING

Select Two Cards *In Exercises 27–34, a board game uses the deck of 20 cards shown.*

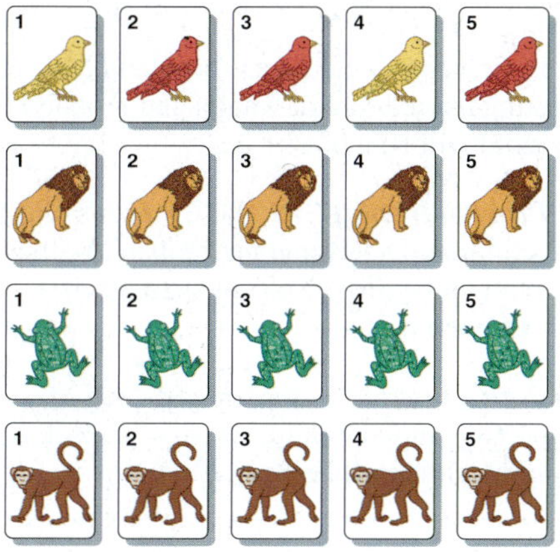

Two cards are selected at random from this deck. Determine the probability of the following

a) *with replacement.*

b) *without replacement.*

27. They both show monkeys.

28. They both show the number 3.

29. The first shows a lion, and the second shows a bird.

30. The first shows a 2, and the second shows a 4.

31. The first shows a red bird, and the second shows a monkey.

32. They both show even numbers.

33. Neither shows an even number.

34. The first shows a lion, and the second shows a red bird.

Select One Card *In the deck of cards used in Exercises 27–34, if one card is drawn, determine the probability that the card shows*

35. a frog or an even number.

36. a yellow bird or a number greater than 4.

37. a lion or a 5.

38. a red bird or an even number.

Two Spins *In Exercises 39–48, assume that the pointer cannot land on the line and that each spin is independent. If the pointer in Fig. 12.16 is spun twice, determine the probability that the pointer lands on*

39. red on both spins.

40. red and then yellow.

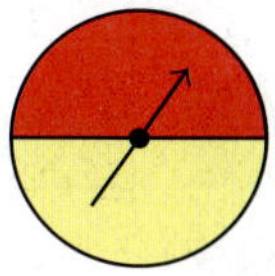

Figure 12.16

Two Spins *If the pointer in Fig. 12.17 is spun twice, determine the probability that the pointer lands on*

41. green and then red.

42. red on both spins.

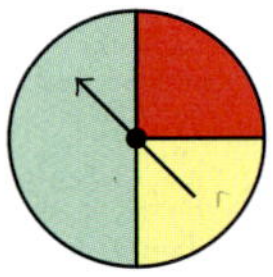

Figure 12.17

Two Spins *If the pointer in Fig. 12.18 is spun twice, determine the probability that the pointer lands on*

43. red on both spins.

44. a color other than green on both spins.

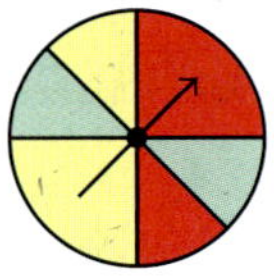

Figure 12.18

Two Spins *In Exercises 45–48, assume that the pointer in Fig. 12.16 on page 779 is spun and then the pointer in Fig. 12.17 is spun. Determine the probability of the pointers landing on*

45. red on both spins.

46. yellow on the first spin and red on the second spin.

47. a color other than red on both spins.

48. yellow on the first spin and a color other than yellow on the second spin.

Selecting an Envelope *In Exercises 49–56, consider the colored envelopes shown below.*

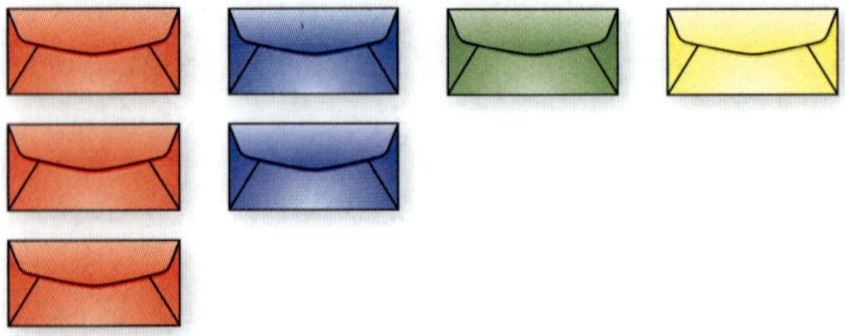

If one of the envelopes is selected at random, determine the probability that

49. a green or a red envelope is selected.

50. an envelope other than a blue envelope is selected.

If two envelopes are selected at random, with replacement, determine the probability that

51. both are red envelopes.

52. the first is a blue envelope and the second is a yellow envelope.

If three envelopes are selected at random, without replacement, determine the probability that

53. they are all red envelopes.

54. none is a red envelope.

55. the first is a red envelope, the second is a blue envelope, and the third is a blue envelope.

56. the first is a red envelope, the second is a green envelope, and the third is a red envelope.

Having a Family *In Exercises 57–60, a couple has three children. Assuming independence and that the probability of a boy is $\frac{1}{2}$, determine the probability that*

57. all three children are girls.

58. all three children are boys.

59. the youngest child is a boy and the two older children are girls.

60. the youngest child is a girl, the middle child is a boy, and the oldest child is a girl.

61. a) ***Five Children*** The Martinos plan to have five children. Determine the probability that all their children will be boys. (Assume that $P(\text{boy}) = \frac{1}{2}$ and assume independence.)

b) If their first four children are boys and Mrs. Martino is expecting another child, what is the probability that the fifth child will be a boy?

62. a) ***The Probability of a Girl*** The Bronsons plan to have eight children. Determine the probability that all their children will be girls. (Assume that $P(\text{girl}) = \frac{1}{2}$ and assume independence.)

b) If their first seven children are girls and Mrs. Bronson is expecting another child, what is the probability that the eighth child will be a girl?

Golf Balls *Angel Sanchez has seven golf balls in one pocket of his golf bag: 4 Titleist balls, 2 Top Flite balls, and 1 Pinnacle ball. In Exercises 63–66, two balls will be selected at random. Determine the probability of selecting each of the following*

a) *with replacement.*

b) *without replacement.*

63. a Titleist ball and then a Pinnacle ball

64. no Top Flite balls

65. at least one Top Flite ball

66. two Pinnacle balls

Health Insurance *A sample of 50 people yielded the following information about their health insurance.*

Number of People	Type of Insurance
24	Managed care plan
19	Traditional insurance
7	No insurance

Two people who provided information for the table were selected at random, without replacement. Determine the probability that

67. neither had traditional insurance.

68. they both had a managed care plan.

69. at least one had traditional insurance.

70. the first had traditional insurance and the second had a managed care plan.

Landscaping *A sample of 40 homeowners who recently hired a landscaping service yielded the following information about their landscaper.*

Number of Homeowners	Would You Recommend Your Landscaper to a Friend
23	Yes
7	No
10	Not sure

Three homeowners who provided information for the table were selected at random. Determine the probability that

71. they would all recommend their landscaper.

72. the first would not recommend the landscaper, but the second and third would recommend their landscapers.

73. the first two would not recommend their landscapers, and the third is not sure if he or she would recommend the landscaper.

74. the first would recommend his or her landscaper, but the second and third would not recommend their landscaper.

A New Medicine *In Exercises 75–78, a new medicine was given to a sample of 100 hospital patients. Of the total, 70 patients reacted favorably, 10 reacted unfavorably, and 20 were unaffected by the drug. Assume that this sample is representative of the entire population. If this medicine is given to Mr. and Mrs. Rivera and their son Carlos, what is the probability of each of the following? (Assume independence.)*

75. Mrs. Rivera reacts favorably.

76. Mr. and Mrs. Rivera react favorably, and Carlos is unaffected.

77. All three react favorably.

78. No one reacts favorably.

Multiple-Choice Exam *In Exercises 79–84, each question of a five-question multiple-choice exam has four possible answers. Gurshawn Salk picks an answer at random for each question. Determine the probability that he selects the correct answer on*

79. any one question.

80. only the first question.

81. only the third and fourth questions.

82. all five questions.

83. none of the questions.

84. at least one of the questions.

A Slot Machine *In Exercises 85–88, consider a slot machine.*

Most people who play slot machines end up losing money because the machines are designed to favor the casino (the house). There are 22 positions on each reel. Assume that the following is a list of the number of symbols of each type on each of the three reels and each symbol has the same chance of occurring (which is not the case; see the Did You Know? on page 777).

Pictures on Reels	Reels 1	2	3
Cherries	2	5	4
Oranges	5	4	5
Plums	6	4	4
Bells	3	4	4
Melons	3	2	3
Bars BAR	2	2	1
7s 7	1	1	1

For this slot machine, assuming that the wheels are independent, determine the probability of obtaining

85. an orange on the first reel.

86. bells on all three reels.

87. no bars.

88. three 7's.

Two Wheels *In Exercises 89–92, suppose that you spin the following double wheel.*

Assuming that the wheels are independent and each outcome is equally likely, determine the probability that you get

89. blue on both wheels.

90. red on the outer wheel and blue on the inner wheel.

91. red on neither wheel.

92. red on at least one wheel.

Hitting a Target *In Exercises 93–96, the probability that a heat-seeking torpedo will hit its target is 0.4. If the first torpedo hits its target, the probability that the second torpedo will hit the target increases to 0.9 because of the extra heat generated by the first explosion. If two heat-seeking torpedoes are fired at a target, determine the probability that*

93. neither hits the target.

94. the first hits the target and the second misses the target.

95. both hit the target.

96. the first misses the target and the second hits the target.

97. ***Polygenetic Afflictions*** Certain birth defects and syndromes are *polygenetic* in nature. Typically, the chance that an offspring will be born with a polygenetic affliction is small. However, once an offspring is born with the affliction, the probability that future offspring of the same parents will be born with the same affliction increases. Let's assume that the probability of a child being born with affliction A is 0.001. If a child is born with this affliction, the probability of a future child being born with the same affliction becomes 0.04.

a) Are the events of the births of two children in the same family with affliction *A* independent? Explain.

b) A couple plans to have one child. Determine the probability that the child will be born with this affliction.

A couple plans to have two children. Use the information provided to determine the probability that

c) both children will be born with the affliction.

d) the first will be born with the affliction and the second will not.

e) the first will not be born with the affliction and the second will.

f) neither will be born with the affliction.

98. ***Lottery Ticket*** In a bin are an equal number of balls marked with the digits 0, 1, 2, 3, . . . , 9. Three balls are to be selected from the bin, one after the other, at random with replacement to make the winning three-digit lottery number. Ms. Jones has a lottery ticket with a three-digit number in the range 000 to 999. Determine the probability that Ms. Jones's number is the winning number.

Chance of an Audit *In Exercises 99–102, assume that 36 in every 1000 people in the $31,850–$77,100 income bracket are audited yearly. Assuming that the returns to be audited are selected at random and each year's selections are independent of the previous year's selections, determine the probability that a person in this income bracket will be audited*

99. this year.

100. the next two years in succession.

101. this year but not next year.

102. neither this year nor next year.

CHALLENGE PROBLEMS/GROUP ACTIVITIES

103. ***Picking Chips*** A bag contains five red chips, three blue chips, and two yellow chips. Two chips are selected from the bag without replacement. Determine the probability that two chips of the same color are selected.

104. ***Peso Coins*** Ron has ten coins from Mexico: three 1-peso coins, one 2-peso coin, two 5-peso coins, one 10-peso coin, and three 20-peso coins. He selects two coins at random without replacement. Assuming that each coin is equally likely to be selected, determine the probability that Ron selects at least one 1-peso coin.

105. ***A Fair Game?*** Two playing cards are dealt to you from a well-shuffled standard deck of 52 cards. If either card is a diamond or if both are diamonds, you win; otherwise, you lose. Determine whether this game favors you, is fair, or favors the dealer. Explain your answer.

106. ***Picture Card Probability*** You have three cards: an ace, a king, and a queen. A friend shuffles the cards, selects two of them at random, and discards the third. You ask your friend to show you a picture card, and she turns over the king. What is the probability that she also has the queen?

RECREATIONAL MATHEMATICS

A Different Die *For Exercises 107–110, consider a six-sided die that has 1 dot on one side, 2 dots on two sides, and 3 dots on three sides.*

If the die is rolled twice, determine the probability of rolling

107. two 2's.

108. two 3's.

If the die is rolled only once, determine the probability of rolling

109. an even number or a number less than 3.

110. an odd number or a number greater than 1.

INTERNET/RESEARCH ACTIVITY

111. Girolamo Cardano (1501–1576) wrote *Liber de Ludo Aleae*, which is considered to be the first book on probability. Cardano had a number of different vocations. Do research and write a paper on the life and accomplishments of Girolamo Cardano.

12.7 CONDITIONAL PROBABILITY

▲ We use conditional probability to determine the probability of selecting a second rotten apple from a basket, given that the first apple selected is rotten.

A basket of 125 apples contains six rotten apples. Suppose that you are asked to randomly select two apples, without replacement, from the basket. What is the probability that both apples you select are rotten? The probability of selecting a rotten apple on the first selection is $\frac{6}{125}$. The probability of selecting a second rotten apple is $\frac{5}{124}$ since we assume that a rotten apple was removed from the basket with the first selection. Since the probability of selecting a second rotten apple is affected by the first rotten apple being selected, these two events are *dependent*. Probability problems involving dependent events can be solved by using conditional probability. In this section, we will discuss conditional probability problems.

In Section 12.6, we learned when two events are *dependent*, the occurrence of the first event, A, affected the probability of the second event, B, occurring. When we calculated $P(A \text{ and } B)$, when we determined the probability of event B, we assumed that event A occurred. That is, we calculated the probability of event B, given event A. The probability of event B, given event A is called a ***conditional probability***. The definition of conditional probability follows.

CONDITIONAL PROBABILITY

In general, the probability of event E_2 occurring, given that an event E_1 has happened (or will happen; the time relationship does not matter), is called a **conditional probability** and is written $P(E_2 \mid E_1)$.

The symbol $P(E_2 \mid E_1)$, read "the probability of E_2, given E_1," represents the probability of E_2 occurring, assuming that E_1 has already occurred (or will occur).

EXAMPLE 1 *Using Conditional Probability*

A single card is selected from a deck of cards. Determine the probability it is a club, given that it is black.

SOLUTION We are told that the card is black. Thus, only 26 cards are possible, of which 13 are clubs. Therefore,

$$P(\text{club} \mid \text{black}) \text{ or } P(\text{C} \mid \text{B}) = \frac{13}{26} = \frac{1}{2}$$

EXAMPLE 2 *Girls in a Family*

A family has two children. Assuming that boys and girls are equally likely, determine the probability that the family has

a) two girls.

b) two girls if you know that at least one of the children is a girl.

c) two girls given that the older child is a girl.

DID YOU KNOW?

Chance of Showers

At one time or another, you probably have been caught in a downpour on a day the weather forecaster had predicted sunny skies. Short-term (24-hour) weather forecasts are correct nearly 85% of the time, a level of accuracy achieved through the use of conditional probability. Computer models are used to analyze data taken on the ground and in the air and make predictions of atmospheric pressures at some future time, say 10 minutes ahead. Based on these predicted conditions, another forecast is then computed. This process is repeated until a weather map has been generated for the next 12, 24, 36, and 48 hours. Since each new prediction relies on the previous prediction being correct, the margin of error increases as the forecast extends farther into the future.

SOLUTION

a) To determine the probability that the family has two girls, we can determine the sample space of a family with two children. Then, from the sample space we can determine the probability that both children are girls. The sample space of two children can be determined by a tree diagram (see Fig. 12.19).

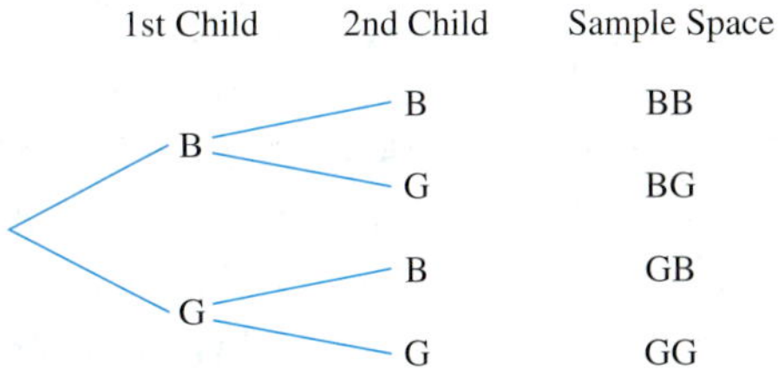

Figure 12.19

There are four possible equally likely outcomes: BB, BG, GB, and GG. Only one of the outcomes has two girls, GG. Thus,

$$P(2 \text{ girls}) = \frac{1}{4}$$

b) We are given that at least one of the children is a girl. Therefore, for this example the sample space is BG, GB, GG. Since there are three possibilities, of which only one has two girls, GG,

$$P(\text{both girls} \mid \text{at least one is a girl}) = \frac{1}{3}$$

c) If the older child is a girl, the sample space reduces to GB, GG. Thus,

$$P(\text{both girls} \mid \text{older child is a girl}) = \frac{1}{2}$$

A number of formulas can be used to find conditional probabilities. The one we will use follows.

CONDITIONAL PROBABILITY

For any two events, E_1 and E_2,

$$P(E_2 \mid E_1) = \frac{n(E_1 \text{ and } E_2)}{n(E_1)}$$

In the formula, $n(E_1 \text{ and } E_2)$ represents the number of sample points common to both event 1 and event 2, and $n(E_1)$ is the number of sample points in event E_1, the given event. Since the intersection of E_1 and E_2, symbolized $E_1 \cap E_2$, represents the sample points common to both E_1 and E_2, the formula can also be expressed as

$$P(E_2 \mid E_1) = \frac{n(E_1 \cap E_2)}{n(E_1)}$$

Figure 12.20 on page 786 is helpful in explaining conditional probability.

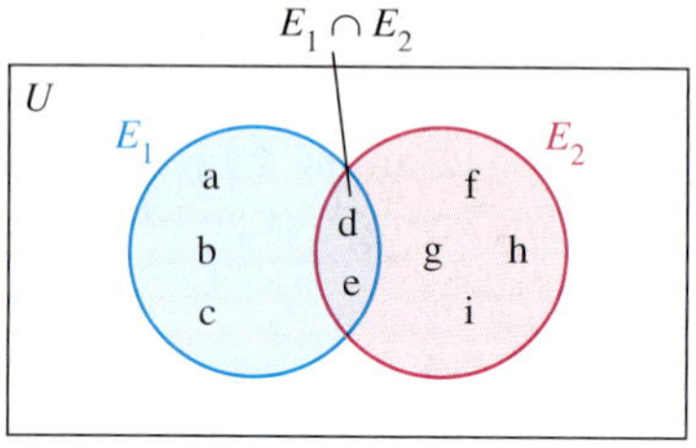

Figure 12.20

Here, the number of elements in E_1 is five, the number of elements in E_2 is six, and the number of elements in both E_1 and E_2, or $E_1 \cap E_2$, is two.

$$P(E_2 \mid E_1) = \frac{n(E_1 \text{ and } E_2)}{n(E_1)} = \frac{2}{5}$$

Thus, for this situation, the probability of selecting an element from E_2, given that the element is in E_1, is $\frac{2}{5}$.

EXAMPLE 3 *Using the Conditional Probability Formula*

Two hundred patients who either had hip surgery or knee surgery were asked whether they were satisfied, dissatisfied, or neutral regarding the results of their surgery. The responses are given in the table below.

Surgery	Satisfied	Dissatisfied	Total
Knee	70	25	95
Hip	90	15	105
Total	160	40	200

If one person from the 200 patients surveyed is selected at random, determine the probability that the person

a) was satisfied with the results of the surgery.

b) was satisfied with the results of the surgery, given that the person had knee surgery.

c) was dissatisfied with the results of the surgery, given that the person had hip surgery.

d) had hip surgery, given that the person was dissatisfied with the results of the surgery.

SOLUTION

a) The total number of patients is 200, of which 160 were satisfied with the results of the surgery.

$$P(\text{satisfied with the results of the surgery}) = \frac{160}{200} = \frac{4}{5}$$

b) We are given that the person had knee surgery. Thus, we have a conditional probability problem. Let E_1 be the given information "the person had knee surgery." Let E_2 be "the person was satisfied with the results of the surgery." We are being asked to determine $P(E_2 \mid E_1)$. The number of people who had knee surgery, $n(E_1)$, is 95. The number of people who had knee surgery and were satisfied with the results of the surgery, $n(E_1 \text{ and } E_2)$, is 70. Thus,

$$P(E_2 \mid E_1) = \frac{n(E_1 \text{ and } E_2)}{n(E_1)} = \frac{70}{95} = \frac{14}{19}$$

c) We are given that the person had hip surgery. Thus, this example is a conditional probability problem. Let E_1 be the given information "the person had hip surgery." Let E_2 be "the person was dissatisfied with the results of the surgery." We are asked to find $P(E_2 \mid E_1)$. The number of people who had hip surgery, $n(E_1)$, is 105.

DID YOU KNOW?

Dice-Y Music

During the eighteenth century, composers were fascinated with the idea of creating compositions by rolling dice. A musical piece attributed by some to Wolfgang Amadeus Mozart, titled "Musical Dice Game," consists of 176 numbered musical fragments of three-quarter beats each. Two charts show the performer how to order the fragments given a particular roll of two dice. Modern composers can create music by using the same random technique and a computer; as in Mozart's time, however, the music created by random selection of notes or bars is still considered by musicologists to be aesthetically inferior to the work of a creative composer.

The number of people who had hip surgery and were dissatisfied with the results of the surgery, $n(E_1 \text{ and } E_2)$, is 15. Thus,

$$P(E_2 \mid E_1) = \frac{n(E_1 \text{ and } E_2)}{n(E_1)} = \frac{15}{105} = \frac{1}{7}$$

d) We are given that the person was dissatisfied with the results of the surgery. Thus, we have a conditional probability problem. Let E_1 be the given information "the person was dissatisfied with the results of the surgery." Let E_2 be "the person had hip surgery." We are asked to determine $P(E_2 \mid E_1)$. The number of people who were dissatisfied with the results of the surgery, $n(E_1)$, is 40. The number of people who were dissatisfied with the results of the surgery and had hip surgery, $n(E_1 \text{ and } E_2)$, is 15. Thus,

$$P(E_2 \mid E_1) = \frac{n(E_1 \text{ and } E_2)}{n(E_1)} = \frac{15}{40} = \frac{3}{8}$$

In many of the examples, we used the words *given that*. Other words may be used instead. For example, in Example 3(b), the question could have been worded "was satisfied with their surgery *if* the person had knee surgery."

SECTION 12.7 EXERCISES

CONCEPT/WRITING EXERCISES

1. What does the notation $P(E_2 \mid E_1)$ mean?

2. Give the formula for $P(E_2 \mid E_1)$.

3. If $n(E_1 \cap E_2) = 4$ and $n(E_1) = 12$, determine $P(E_2 \mid E_1)$.

4. If $n(E_1 \cap E_2) = 5$ and $n(E_1) = 22$, determine $P(E_2 \mid E_1)$.

PRACTICE THE SKILLS

Select a Circle *In Exercises 5–10, consider the circles shown.*

Assume that one circle is selected at random and each circle is equally likely to be selected. Determine the probability of selecting

5. a 5, given that the circle is orange.

6. a 3, given that the circle is yellow.

7. an even number, given that the circle is not orange.

8. a number less than 2, given that the number is less than 5.

9. a red number, given that the circle is orange.

10. a number greater than 3, given that the circle is yellow.

Select a Number *In Exercises 11–16, consider the following figures.*

Assume that one number from 1 to 7 is equally likely to be selected at random. Each number corresponds to one of the seven figures shown. Determine the probability of selecting

11. a circle, given that an odd number is selected.

12. a circle, given that a number greater than or equal to 5 is selected.

13. a red figure, given that an even number is selected.

14. a red or a blue figure, given that an even number is selected.

15. a circle or square, given that a number less than 4 is selected.

16. a circle, given that an even number is selected.

Spin the Wheel *In Exercises 17–24, consider the following wheel.*

If the wheel is spun and each section is equally likely to stop under the pointer, determine the probability that the pointer lands on

17. a four, given that the color is purple.

18. an even number, given that the color is red.

19. purple, given that the number is odd.

20. a number greater than 6, given that the color is red.

21. a number greater than 4, given that the color is purple.

22. an even number, given that the color is red or purple.

23. gold, given that the number is greater than 5.

24. gold, given that the number is greater than 10.

Money from a Hat *In Exercises 25–28, assume that a hat contains four bills: a \$1 bill, a \$5 bill, a \$10 bill, and a \$20 bill. Two bills are to be selected at random with replacement. Construct a sample space as was done in Example 2 and determine the probability that*

25. both bills are \$1 bills.

26. both bills are \$1 bills if the first selected is a \$1 bill.

27. both bills are \$5 bills if at least one of the bills is a \$5 bill.

28. both bills have a value greater than a \$5 bill if the second bill is a \$10 bill.

▲ See Exercises 25–28

Two Dice *In Exercises 29–34, two dice are rolled one after the other. Construct a sample space and determine the probability that the sum of the dots on the dice total*

29. 6.

30. 6 if the first die is a 1.

31. 6 if the first die is a 3.

32. an even number if the second die is a 2.

33. a number greater than 7 if the second die is a 5.

34. a 7 or 11 if the first die is a 5.

PROBLEM SOLVING

Costliest Hurricanes *In Exercises 35–40, use the following information concerning the nine costliest hurricanes to strike the U.S. mainland.*

Hurricane	Category	Damage (billions of dollars)
Katrina (2005)	4	80.0
Andrew (1992)	5	26.5
Charlie (2004)	4	15.0
Wilma (2005)	3	14.4
Ivan (2004)	3	14.2
Rita (2005)	3	9.4
Frances (2004)	2	8.9
Hugo (1989)	4	7.0
Jeanne (2004)	3	6.9

Source: National Oceanic and Atmospheric Administration

If one hurricane from the list is selected at random, determine the probability that it

35. was a category 4.

36. had damages of at least \$16 billion.

37. had damages of at least $20 billion, given that it was a category 4.

38. had damages of at least $10 billion, given that it was a category 3.

39. was a category 5, given that it had damages of at least $25 billion.

40. was a category 3, given that it had damages of at least $10 billion.

E-Z Pass *In Exercises 41–46, use the following table, which shows the number of cars and trucks that used the Pennsylvania Turnpike on a particular day. The number of cars and trucks that used, and did not use, the E-Z Pass on that same day was also recorded.*

E-Z Pass	Cars	Trucks	Total
Used	527	316	843
Did not use	935	683	1618
Total	1462	999	2461

If one of these vehicles is selected at random, determine the probability (as a decimal number rounded to four decimal places) that the

41. vehicle was a car.

42. vehicle used the E-Z Pass.

43. vehicle used the E-Z Pass, given that the vehicle was a car.

44. vehicle used the E-Z Pass, given that the vehicle was a truck.

45. vehicle was a car, given that the vehicle used the E-Z Pass.

46. vehicle was a truck, given that the vehicle used the E-Z Pass.

Sales Effectiveness *In Exercises 47–52 use the following information. Sales representatives at a car dealership were split into two groups. One group used an aggressive approach to sell a customer a new automobile. The other group used a passive approach. The following table summarizes the records for 650 customers.*

Approach	Sale	No Sale	Total
Aggressive	100	250	350
Passive	220	80	300
Total	320	330	650

If one of these customers is selected at random, determine the probability

47. that the aggressive approach was used.

48. of a sale.

49. of no sale, given that the passive approach was used.

50. of a sale, given that the aggressive approach was used.

51. of a sale, given that the passive approach was used.

52. of no sale, given that the aggressive approach was used.

Age Distribution *For Exercises 53–58, use the following information concerning the age distribution of U.S. residents, based on 2005 population data. The data are rounded to the nearest million people.*

Age	Male	Female	Total
0–14	31	30	61
15–64	99	99	198
65 years or over	15	21	36
Total	145	150	295

Source: *The World Factbook*, January 2006.

If one of these individuals is selected at random, determine the probability that the person is

53. male.

54. 15–64 years old.

55. 15–64 years old, given that the person is female.

56. 65 years or over, given that the person is female.

57. female, given that the person is 0–14 years old.

58. male, given that the person is 15–64 years old.

Quality Control *In Exercises 59–64, Sally Horsefall, a quality control inspector, is checking a sample of light bulbs for defects. The following table summarizes her findings.*

Wattage	Good	Defective	Total
20	80	15	95
50	100	5	105
100	120	10	130
Total	300	30	330

If one of these lightbulbs is selected at random, determine the probability that the lightbulb is

59. good.

60. good, given that it is 50 watts.

61. defective, given that it is 20 watts.

62. good, given that it is 100 watts.

63. good, given that it is 50 or 100 watts.

64. defective, given that it is not 50 watts.

News Survey *In Exercises 65–70, 270 individuals are asked which evening news they watch most often. The results are summarized as follows.*

Viewers	ABC	NBC	CBS	Other	Total
Men	30	20	40	55	145
Women	50	10	20	45	125
Total	80	30	60	100	270

If one of these individuals is selected at random, determine the probability that the person watches

65. ABC or NBC.

66. ABC, given that the individual is a woman.

67. ABC or NBC, given that the individual is a man.

68. a station other than CBS, given that the individual is a woman.

69. a station other than ABC, NBC, or CBS, given that the individual is a man.

70. NBC or CBS, given that the individual is a woman.

▲ Katie Couric, *CBS Evening News*

CHALLENGE PROBLEMS/GROUP ACTIVITIES

Mutual Fund Holdings *Use the following information in Exercises 71–74. Mutual funds often hold many stocks. Each stock may be classified as a value stock, a growth stock, or a blend of the two. The stock may also be categorized by how large the company is. It may be classified as a large company stock, medium company stock, or small company stock. A selected mutual fund contains 200 stocks as illustrated in the following chart.*

	Value	Blend	Growth
Large	28	23	42
Medium	19	15	18
Small	26	12	17

Equity Investment Style

If one stock is selected at random from the mutual fund, determine the probability that it is

71. a large company stock.

72. a value stock.

73. a blend, given that it is a medium company stock.

74. a large company stock, given that it is a blend stock.

75. Consider the Venn diagram below. The numbers in the regions of the circle indicate the number of items that belong to that region. For example, 60 items are in set A but not in set B.

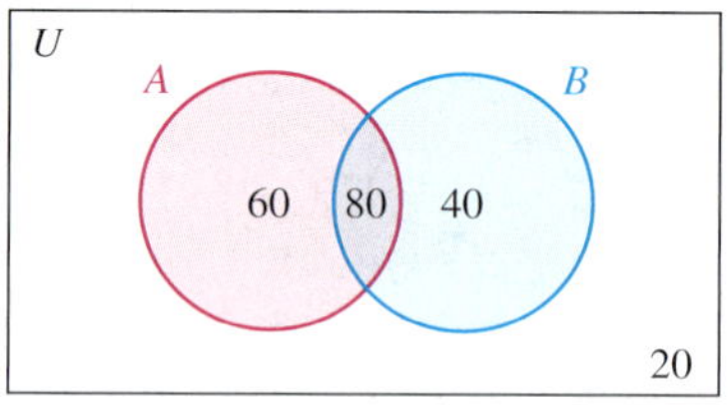

Determine **a)** $n(A)$ **b)** $n(B)$ **c)** $P(A)$ **d)** $P(B)$

Use the formula on page 785 to determine

e) $P(A \mid B)$.

f) $P(B \mid A)$.

g) Explain why $P(A \mid B) \neq P(A) \cdot P(B)$.

76. A formula we gave for conditional probability is

$$P(E_2 \mid E_1) = \frac{n(E_1 \text{ and } E_2)}{n(E_1)}$$

This formula may be derived from the formula

$$P(E_2 \mid E_1) = \frac{P(E_1 \text{ and } E_2)}{P(E_1)}$$

Can you explain why? [*Hint:* Consider what happens to the denominators of $P(E_1 \text{ and } E_2)$ and $P(E_1)$ when they are expressed as fractions and the fractions are divided out.]

77. Given that $P(A) = 0.3$, $P(B) = 0.5$, and $P(A \text{ and } B) = 0.15$, use the formula

$$P(E_2 \mid E_1) = \frac{P(E_1 \text{ and } E_2)}{P(E_1)}$$

to determine

a) $P(A \mid B)$.

b) $P(B \mid A)$.

c) Are A and B independent? Explain.

RECREATIONAL MATHEMATICS

In Exercises 78–83, suppose that each circle is equally likely to be selected. One circle is selected at random.

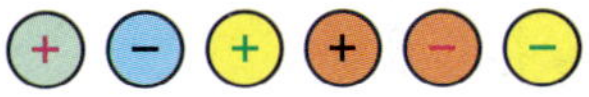

Determine the probability indicated.

78. $P(\text{green circle} \mid + \text{ obtained})$

79. $P(+ \mid \text{orange circle obtained})$

80. $P(\text{yellow circle} \mid - \text{ obtained})$

81. $P(\text{green } + \mid + \text{ obtained})$

82. $P(\text{green or orange circle} \mid \text{green } + \text{ obtained})$

83. $P(\text{orange circle with green } + \mid + \text{ obtained})$

12.8 THE COUNTING PRINCIPLE AND PERMUTATIONS

▲ In how many different ways can five televisions be displayed?

You are a manager at an electronics store and want to display five different plasma televisions on the showroom floor. In how many different ways can you display the five televisions? In this section, we will learn how to determine the number of different ordered arrangements of a set of objects. We will also discuss how we can use the counting principle to determine the number of distinct ways two or more experiments can be performed.

The Counting Principle

In Section 12.5, we introduced the counting principle, which is repeated here for your convenience.

> **COUNTING PRINCIPLE**
> If a first experiment can be performed in M distinct ways and a second experiment can be performed in N distinct ways, then the two experiments in that specific order can be performed in $M \cdot N$ distinct ways.

DID YOU KNOW?

Slot Machines

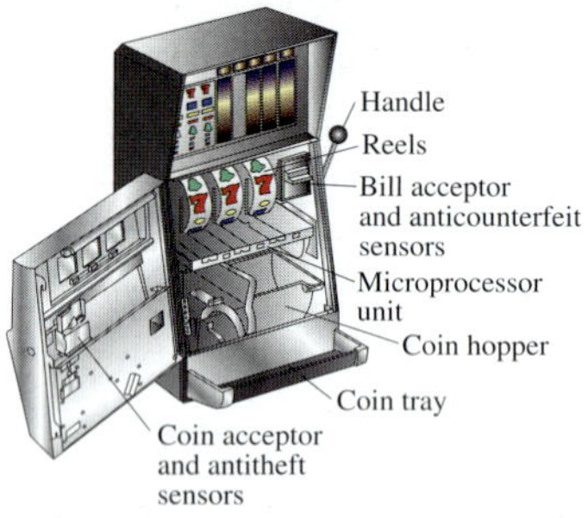

Source: *Scientific American*

Slot machines may qualify as the world's most user-friendly computer. Nowadays, slot machines are a conglomeration of microprocessors and memory chips. Each of three reels can stop at 22 positions (each one displaying various symbols or a blank space) for a total of 10,648 different arrangements ($22 \times 22 \times 22 = 10{,}648$). The heart of a contemporary slot machine is a microprocessor programmed to generate random numbers that can be assigned to any arrangement of the reels. In effect, the microprocessor generates random numbers that dictate the machine's display and payout. Because many or few random numbers can be assigned to any given arrangement, makers of slot machines can vary the odds as desired. For the top jackpot, the odds of obtaining the single correct arrangement may be set at 10 million to 1. See the Did You Know? on page 777 for additional information.

The counting principle is illustrated in Examples 1 and 2.

EXAMPLE 1 *Counting Principle: Passwords*

A password used to gain access to a computer account is to consist of two lowercase letters followed by four digits. Determine how many different passwords are possible if

a) repetition of letters and digits is permitted.

b) repetition of letters and digits is not permitted.

c) the first letter must be a vowel (*a*, *e*, *i*, *o*, *u*) and the first digit cannot be a 0, and repetition of letters and digits is not permitted.

SOLUTION There are 26 letters and 10 digits (0–9). We have six positions to fill, as indicated.

L	L	D	D	D	D

a) Since repetition is permitted, there are 26 possible choices for both the first and second positions. There are 10 possible choices for the third, fourth, fifth, and sixth positions.

26	26	10	10	10	10
L	L	D	D	D	D

Since $26 \cdot 26 \cdot 10 \cdot 10 \cdot 10 \cdot 10 = 6{,}760{,}000$, there are 6,760,000 different possible arrangements.

b) There are 26 possibilities for the first position. Since repetition of letters is not permitted, there are only 25 possibilities for the second position. The same reasoning is used when determining the number of digits for positions 3 through 6.

26	25	10	9	8	7
L	L	D	D	D	D

Since $26 \cdot 25 \cdot 10 \cdot 9 \cdot 8 \cdot 7 = 3{,}276{,}000$, there are 3,276,000 different possible arrangements.

c) Since the first letter must be an *a*, *e*, *i*, *o*, or *u*, there are five possible choices for the first position. The second position can be filled by any of the letters except for the vowel selected for the first position. Therefore, there are 25 possibilities for the second position.

Since the first digit cannot be a 0, there are nine possibilities for the third position. The fourth position can be filled by any digit except the one selected for the third position. Thus, there are nine possibilities for the fourth position. Since the fifth position cannot be filled by any of the two digits previously used, there are eight possibilities for the fifth position. The last position can be filled by any of the seven remaining digits.

5	25	9	9	8	7
L	L	D	D	D	D

Since $5 \cdot 25 \cdot 9 \cdot 9 \cdot 8 \cdot 7 = 567{,}000$, there are 567,000 different arrangements that meet the conditions specified.

EXAMPLE 2 *Counting Principle: T-Shirt Colors*

At Old Navy, a supply of solid-colored T-shirts has just been received. The T-shirts come in the following colors: green, blue, white, yellow, and red. Billy Bragg, the floor manager, decides to display one of each color T-shirt in a row on a shelf.

a) In how many different ways can he display the five different color T-shirts on a shelf?

b) If he wants to place the blue T-shirt in the middle, in how many different ways can he arrange the T-shirts?

c) If Billy wants the white T-shirt to be the first T-shirt and the blue T-shirt to be the last T-shirt, in how many different ways can he arrange the T-shirts?

SOLUTION

a) There are five positions to fill, using the five colors. In the first position, on the left, he can use any one of the five colors. In the second position, he can use any of the four remaining colors. In the third position, he can use any of the three remaining colors, and so on. The number of distinct possible arrangements is

$$\underline{5} \cdot \underline{4} \cdot \underline{3} \cdot \underline{2} \cdot \underline{1} = 120$$

b) We begin by satisfying the specified requirements stated. In this case, the blue T-shirt must be placed in the middle. Therefore, there is only one possibility for the middle position.

$$\underline{}\ \underline{}\ \underline{1}\ \underline{}\ \underline{}$$

For the first position, there are now four possibilities. For the second position, there will be three possibilities. For the fourth position, there will be two possibilities. Finally, in the last position, there is only one possibility.

$$\underline{4} \cdot \underline{3} \cdot \underline{1} \cdot \underline{2} \cdot \underline{1} = 24$$

Thus, under the condition stated, there are 24 different possible arrangements.

c) For the first T-shirt, there is only one possibility, the white T-shirt. For the last T-shirt, there is only one possibility, the blue T-shirt.

$$\underline{1}\ \underline{}\ \underline{}\ \underline{}\ \underline{1}$$

The second position can be filled by any of the three remaining T-shirts. The third position can be filled by any of the two remaining T-shirts. There is only one T-shirt left for the fourth position. Thus, the number of possible arrangements is

$$\underline{1} \cdot \underline{3} \cdot \underline{2} \cdot \underline{1} \cdot \underline{1} = 6$$

There are only six possible arrangements that satisfy the given conditions.

Permutations

Now we introduce the definition of a permutation.

A **permutation** is any *ordered arrangement* of a given set of objects.

Curly Moe Larry

"Larry, Curly, Moe" and "Curly, Moe, Larry" represent two different ordered arrangements or two different permutations of the same three names. In Example 2(a), there are 120 different ordered arrangements, or permutations, of the five colored T-shirts. In Example 2(b), there are 24 different ordered arrangements, or permutations possible, if the blue T-shirt must be displayed in the middle.

When determining the number of permutations possible, we assume that repetition of an item is not permitted. To help you understand and visualize permutations, we illustrate the various permutations possible when a triangle, rectangle, and circle are to be placed in a line, see Figure 12.21.

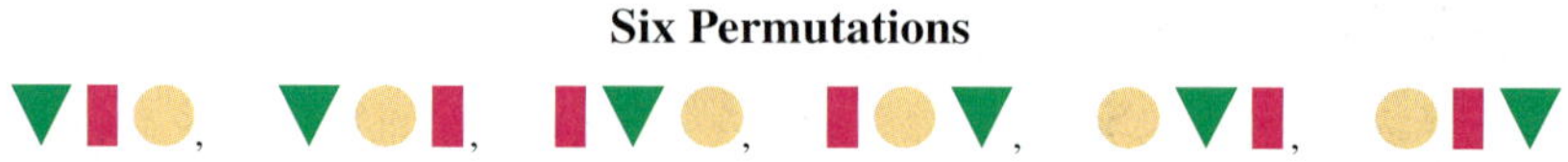

Figure 12.21

For this set of three shapes, six different arrangements, or six permutations, are possible. We can obtain the number of permutations by using the counting principle. For the first position, there are three choices. There are then two choices for the second position, and only one choice is left for the third position.

$$\text{Number of permutations} = 3 \cdot 2 \cdot 1 = 6$$

The product $3 \cdot 2 \cdot 1$ is referred to as 3 factorial, and is written 3!. Thus,

$$3! = 3 \cdot 2 \cdot 1 = 6$$

NUMBER OF PERMUTATIONS

The number of permutations of n distinct items is n factorial, symbolized $n!$, where

$$n! = n(n-1)(n-2)\cdots(3)(2)(1)$$

It is important to note that 0! is defined to be 1. Many calculators have the ability to determine factorials. Often to determine factorials you need to press the 2nd or INV key. Read your calculator manual to determine how to find factorials on your calculator.

EXAMPLE 3 *Children in Line*

In how many different ways can seven children be arranged in a line?

SOLUTION Since there are seven children, the number of permutations is 7!.

$$7! = 7 \cdot 6 \cdot 5 \cdot 4 \cdot 3 \cdot 2 \cdot 1 = 5040$$

The seven children can be arranged in 5040 different ways.

Example 4 illustrates how to use the counting principle to determine the number of permutations possible when only a part of the total number of items is to be selected and arranged.

Some of the many permutations of 3 of the 5 letters

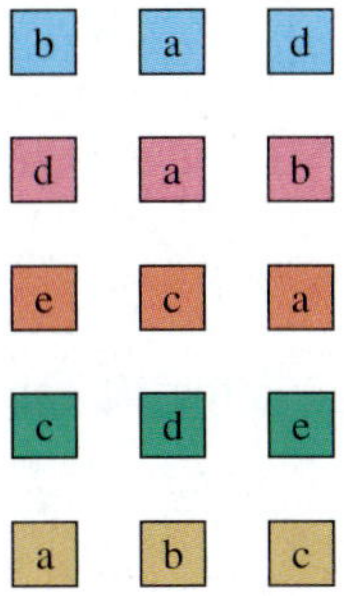

Figure 12.22

EXAMPLE 4 *Permutations of Three Out of Five Letters*

Consider the five letters a, b, c, d, e. In how many distinct ways can three letters be selected and arranged if repetition is not allowed?

SOLUTION We are asked to select and arrange only three of the five possible letters. Figure 12.22 shows some possibilities. Using the counting principle, we find that there are five possible letters for the first choice, four possible letters for the second choice, and three possible letters for the third choice:

$$5 \cdot 4 \cdot 3 = 60$$

Thus, there are 60 different possible ordered arrangements, or permutations. On the left, we show 5 of the 60 possible permutations. ●

In Example 4, we determined the number of different ways in which we could select and arrange three of the five items. We can indicate that result by using the notation ${}_5P_3$. The notation ${}_5P_3$ is read "the number of permutations of five items taken three at a time." The notation ${}_nP_r$ is read "the number of permutations of n items taken r at a time."

We use the counting principle below to evaluate ${}_8P_4$, ${}_9P_3$, and ${}_{10}P_5$. Note the relationship between the number preceding the P, the number following the P, and the last number in the product.

One more than $8 - 4$

$${}_8P_4 = 8 \cdot 7 \cdot 6 \cdot 5$$

One more than $9 - 3$

$${}_9P_3 = 9 \cdot 8 \cdot 7$$

One more than $10 - 5$

$${}_{10}P_5 = 10 \cdot 9 \cdot 8 \cdot 7 \cdot 6$$

To evaluate ${}_nP_r$, we begin with n and form a product of r consecutive descending factors. For example, to evaluate ${}_{10}P_5$, we start with 10 and form a product of five consecutive descending factors (see the preceding illustration).

In general, the number of permutations of n items taken r at a time, ${}_nP_r$, may be found by the formula

One more than $n - r$

$${}_nP_r = n(n-1)(n-2)\cdots(n-r+1)$$

Therefore, when evaluating ${}_{20}P_{15}$, we would find the product of consecutive decreasing integers from 20 to $(20 - 15 + 1)$ or 6, which is written as $20 \cdot 19 \cdot 18 \cdot 17 \cdot \cdots \cdot 6$.

Now let's develop an alternative formula that we can use to find the number of permutations possible when r objects are selected from n objects:

$${}_nP_r = n(n-1)(n-2)\cdots(n-r+1)$$

Now multiply the expression on the right side of the equals sign by $\dfrac{(n-r)!}{(n-r)!}$, which is equivalent to multiplying the expression by 1.

$${}_nP_r = n(n-1)(n-2)\cdots(n-r+1) \times \frac{(n-r)!}{(n-r)!}$$

For example,

$$_{10}P_5 = 10 \cdot 9 \cdot \cdots \cdot 6 \times \frac{5!}{5!}$$

or

$$_{10}P_5 = \frac{10 \cdot 9 \cdot \cdots \cdot 6 \times 5!}{5!}$$

Since $(n - r)!$ means $(n - r)(n - r - 1)\cdots(3)(2)(1)$, the expression for $_nP_r$ can be rewritten as

$$_nP_r = \frac{n(n-1)(n-2)\cdots(n-r+1)\overbrace{(n-r)(n-r-1)\cdots(3)(2)(1)}^{(n-r)!}}{(n-r)!}$$

Since the numerator of this expression is $n!$, we can write

$$_nP_r = \frac{n!}{(n-r)!}$$

For example,

$$_{10}P_5 = \frac{10!}{(10-5)!}$$

The number of permutations possible when r objects are selected from n objects is found by the **permutation formula**

$$_nP_r = \frac{n!}{(n-r)!}$$

In Example 4, we found that when selecting three of five letters, there were 60 permutations. We can obtain the same result using the permutation formula:

$$_5P_3 = \frac{5!}{(5-3)!} = \frac{5!}{2!} = \frac{5 \cdot 4 \cdot 3 \cdot \cancel{2 \cdot 1}}{\cancel{2 \cdot 1}} = 60$$

EXAMPLE 5 *Using the Permutation Formula*

You are among eight people forming a skiing club. Collectively, you decide to put each person's name in a hat and to randomly select a president, a vice president, and a secretary. How many different arrangements or permutations of officers are possible?

SOLUTION There are eight people, $n = 8$, of which three are to be selected; thus, $r = 3$.

$$_8P_3 = \frac{8!}{(8-3)!} = \frac{8!}{5!} = \frac{8 \cdot 7 \cdot 6 \cdot \cancel{5 \cdot 4 \cdot 3 \cdot 2 \cdot 1}}{\cancel{5 \cdot 4 \cdot 3 \cdot 2 \cdot 1}} = 336$$

Thus, with eight people there can be 336 different arrangements for president, vice president, and secretary.

In Example 5, the fraction

$$\frac{8 \cdot 7 \cdot 6 \cdot \cancel{5 \cdot 4 \cdot 3 \cdot 2 \cdot 1}}{\cancel{5 \cdot 4 \cdot 3 \cdot 2 \cdot 1}} = 336$$

can be also expressed as

$$\frac{8 \cdot 7 \cdot 6 \cdot 5!}{5!} = 336$$

The solution to Example 5, like other permutation problems, can also be obtained using the counting principle.

EXAMPLE 6

The Prince George County bicycle club has 10 different routes members wish to travel exactly once, but they only have 6 specific dates for their trips. In how many ways can the different routes be assigned to the dates scheduled for their trips?

SOLUTION There are 10 possible routes but only 6 specific dates scheduled for the trips. Since traveling route A on day 1 and traveling route B on day 2 is different than traveling route B on day 1 and traveling route A on day 2, we have a permutation problem. There are 10 possible routes; thus, $n = 10$. There are 6 routes that are going to be selected and assigned to different days; thus, $r = 6$. Now we calculate the number of different permutations of selecting and arranging the dates for 6 out of 10 possible routes.

$${}_{10}P_6 = \frac{10!}{(10-6)!} = \frac{10!}{4!} = \frac{10 \cdot 9 \cdot 8 \cdot 7 \cdot 6 \cdot 5 \cdot \cancel{4!}}{\cancel{4!}} = 151{,}200$$

There are 151,200 different ways that 6 routes can be selected and scheduled from the 10 possible routes.

Example 6 could also be worked using the counting principle because we are discussing an *ordered arrangement* (a permutation) that is done *without replacement*. For the first date scheduled, there are 10 possible outcomes. For the second date selected, there are 9 possible outcomes. By continuing this process we would determine that the number of possible outcomes for the 6 different trips is $10 \cdot 9 \cdot 8 \cdot 7 \cdot 6 \cdot 5 = 151{,}200$.

We have worked permutation problems (selecting and arranging, without replacement, r items out of n *distinct* items) by using the counting principle and using the permutation formula. When you are given a permutation problem, unless specified by your instructor, you may use either technique to determine its solution.

Permutations of Duplicate Items

So far, all the examples we have discussed in this section have involved arrangements with distinct items. Now we will consider permutation problems in which some of the items to be arranged are duplicates. For example, the name BOB contains three letters, of which the two Bs are duplicates. How many permutations of the letters in the name BOB are possible? If the two Bs were distinguishable (one red and the other blue), there would be six permutations.

BOB	BBO	OBB
BOB	BBO	OBB

However, if the Bs are not distinguishable (replacing all colored Bs with black print), we see that there are only three permutations.

BOB BBO OBB

The number of permutations of the letters in BOB can be computed as

$$\frac{3!}{2!} = \frac{3 \cdot \cancel{2 \cdot 1}}{\cancel{2 \cdot 1}} = 3$$

where 3! represents the number of permutations of three letters, assuming that none are duplicates, and 2! represents the number of ways the two items that are duplicates can be arranged (**BB** or **BB**). In general, we have the following rule.

PERMUTATIONS OF DUPLICATE OBJECTS

The number of distinct permutations of n objects where n_1 of the objects are identical, n_2 of the objects are identical, . . . , n_r of the objects are identical is found by the formula

$$\frac{n!}{n_1!n_2!\cdots n_r!}$$

EXAMPLE 7 *Duplicate Letters*

In how many different ways can the letters of the word "TALLAHASSEE" be arranged?

SOLUTION Of the 11 letters, three are A's, two are S's, two are L's and two are E's. The number of possible arrangements is

$$\frac{11!}{3!2!2!2!} = \frac{11 \cdot 10 \cdot 9 \cdot 8 \cdot 7 \cdot 6 \cdot 5 \cdot 4 \cdot \cancel{3 \cdot 2 \cdot 1}}{\cancel{3 \cdot 2 \cdot 1} \cdot \cancel{2} \cdot 1 \cdot \cancel{2} \cdot 1 \cdot \cancel{2} \cdot 1} = 11 \cdot 10 \cdot 9 \cdot 7 \cdot 6 \cdot 5 \cdot 4 = 831{,}600$$

There are 831,600 different possible arrangements of the letters in the word "TALLAHASSEE."

TECHNOLOGY TIP Many scientific calculators and all graphing calculators have the ability to evaluate permutations. Read your calculator's instruction manual to determine the procedure to follow to evaluate permutations.

To evaluate ${}_{10}P_6$ on a TI-83 Plus or a TI-84 Plus calculator, enter the number 10. Then press the MATH key. Use the right arrow key to scroll over to PRB, which stands for probability. Then scroll down to ${}_nP_r$. Press the ENTER key. You should now see 10 ${}_nP_r$ on your screen. Next press 6. Press the ENTER key again. The answer will be 151,200, which agrees with our answer in Example 6.

SECTION 12.8 EXERCISES

CONCEPT/WRITING EXERCISES

1. In your own words, state the counting principle.

2. In your own words, describe a permutation.

3. a) Give the formula for the number of permutations of n distinct items.

   b) Give the formula for the number of permutations of n objects when $n_1, n_2, \ldots, n_r$ of the objects are identical.

4. In your own words, explain how to find $n!$ for any positive integer n.

5. How do you read ${}_nP_r$? When you evaluate ${}_nP_r$, what does the outcome represent?

6. Does ${}_1P_1 = {}_1P_0$? Explain.

7. Give the formula for the number of permutations when r objects are selected from n objects.

8. a) Explain how to calculate $\frac{500!}{499!}$ without a calculator.

   b) Evaluate $\frac{500!}{499!}$.

PRACTICE THE SKILLS

In Exercises 9–20, evaluate the expression.

9. $4!$ 10. $7!$ 11. ${}_7P_2$ 12. ${}_5P_2$

13. $0!$ 14. ${}_6P_4$ 15. ${}_8P_0$ 16. ${}_6P_0$

17. ${}_9P_4$ 18. ${}_3P_3$ 19. ${}_8P_3$ 20. ${}_{10}P_6$

PROBLEM SOLVING

21. ***ATM Codes*** To use an automated teller machine, you generally must enter a four-digit code, using the digits 0–9. How many four-digit codes are possible if repetition of digits is permitted?

22. ***Vice Presidents*** The board of a pharmaceutical company has nine members. One board member will be selected as vice president of marketing, and a different board member will be selected as vice president of research. How many different arrangements of the two vice presidents are possible?

23. ***Passwords*** Assume that a password to log onto a computer account is to consist of three letters followed by two digits. Determine the number of possible passwords if

   a) repetition is not permitted.

   b) repetition is permitted.

24. ***Passwords*** Assume that a password to log onto a computer account is to consist of any four digits or letters (repetition is permitted). Determine the number of passwords possible if

   a) the letters are not case sensitive (that is, a lowercase letter is treated the same as an uppercase letter).

   b) the letters are case sensitive (that is, an uppercase letter is considered different than the same lowercase letter).

25. ***Car Door Locks*** Some doors on cars can be opened by pressing the correct sequence of buttons. A display of the five buttons by the door handle of a car follows.*

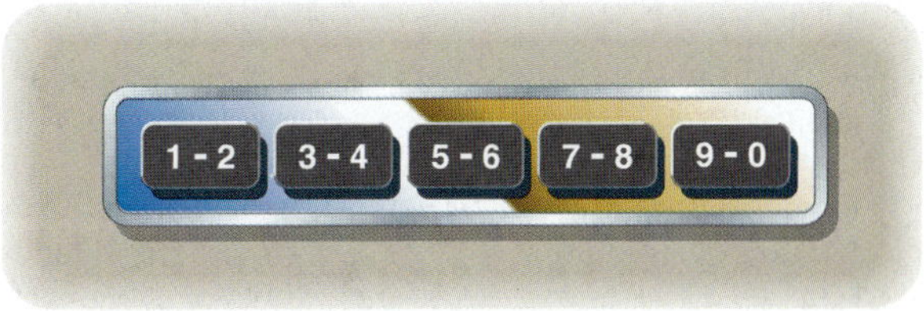

The correct sequence of five buttons must be pressed to unlock the door.

   a) If the same button may be pressed consecutively, how many possible ways can the five buttons be pressed (repetition is permitted)?

   b) If five buttons are pressed at random, determine the probability that a sequence that unlocks the door will be entered.

26. ***Social Security Numbers*** A social security number consists of nine digits. How many different social security numbers are possible if repetition of digits is permitted?

27. ***License Plate*** A license plate is to have five uppercase letters or digits. Determine the number of license plates possible if repetition is permitted and if any position can contain either a letter or digit with the exception that the first position cannot contain the letter O or the number 0.

*On most cars, although each key lists two numbers, the key acts as a single number. Therefore, if your code is 1, 6, 8, 5, 3, the code 2, 5, 7, 6, 4 will also open the lock.

28. ***Winning the Trifecta*** The trifecta at most racetracks consists of selecting the first-, second-, and third-place finishers in a particular race in their proper order. If there are seven entries in the trifecta race, how many tickets must you purchase to guarantee a win?

29. ***Choosing Classes*** Kai Lu plans to enroll in four classes: sociology, chemistry, economics, and humanities. There are seven sociology classes, four chemistry classes, three economics classes, and four humanities classes that fit his schedule. How many different ways can he select his four classes?

30. ***Geometric Shapes*** Consider the five figures shown.

In how many different ways can the figures be arranged

a) from left to right?

b) from top to bottom if placed one under the other?

c) from left to right if the triangle is to be placed on the far right?

d) from left to right if the circle is to be placed on the far left and the triangle is to be placed on the far right?

31. ***Arranging Pictures*** The six pictures shown are to be placed side by side along a wall.

In how many ways can they be arranged from left to right if

a) they can be arranged in any order?

b) the bird must be on the far left?

c) the bird must be on the far left and the giraffe must be next to the bird?

d) a four-legged animal must be on the far right?

32. ***Nursing Positions*** There are 15 candidates for three nursing positions available at Community General Hospital. One position is for the day supervisor. The second position is for the evening supervisor. The third position is for the night supervisor. If all 15 candidates are equally qualified for the three positions, in how many different ways can the three positions be filled?

▲ See Exercise 32

33. ***Club Officers*** If a club consists of 10 members, how many different arrangements of president, vice-president, and secretary are possible?

34. ***ISBN Codes*** Each book registered in the Library of Congress must have an ISBN code number. For an ISBN number of the form D-DD-DDDDDD-D, where D represents a digit from 0–9, how many different ISBN numbers are possible if repetition of digits is allowed? (See research activity Exercise 72 on page 802.)

35. ***Wedding Reception*** At the reception line of a wedding, the bride, the groom, the best man, the maid of honor, the four ushers, and the four bridesmaids must line up to receive the guests.

a) If these individuals can line up in any order, how many arrangements are possible?

b) If the groom must be the last in line and the bride must be next to the groom, and the others can line up in any order, how many arrangements are possible?

c) If the groom is to be last in line, the bride next to the groom, and males and females are to alternate, how many arrangements are possible?

36. ***Disc Jockey*** A disc jockey has 9 songs to play. Five are slow songs, and 4 are fast songs. Each song is to be played only once. In how many ways can the disc jockey play the 9 songs if

a) the songs can be played in any order.

b) the first song must be a slow song and the last song must be a slow song.

c) the first two songs must be fast songs.

Letter Codes *In Exercises 37–40, an identification code is to consist of three letters followed by four digits. How many different codes are possible if*

37. repetition is not permitted?

38. repetition is permitted?

39. repetition of letters is permitted, repetition of numbers is not permitted, and the first three entries must all be the same letter?

40. the first letter must be an *A*, *B*, *C*, or *D* and repetition is not permitted?

License Plates *In Exercises 41–44, a license plate is to consist of three digits followed by two uppercase letters. Determine the number of different license plates possible if*

41. repetition of numbers and letters is permitted.

42. repetition of numbers and letters is not permitted.

43. the first and second digits must be odd, and repetition is not permitted.

44. the first digit cannot be zero, and repetition is not permitted.

45. ***Possible Phone Numbers*** A telephone number consists of seven digits with the restriction that the first digit cannot be 0 or 1.

a) How many distinct telephone numbers are possible?

b) How many distinct telephone numbers are possible with three-digit area codes preceding the seven-digit number, where the first digit of the area code is not 0 or 1?

c) With the increasing use of cell phones and paging systems, our society is beginning to run out of usable phone numbers. Various phone companies are developing phone numbers that use 11 digits instead of 7. How many distinct phone numbers can be made with 11 digits assuming that the area code remains three digits and the first digit of the area code and the phone number cannot be 0 or 1?

46. ***Granola Bars*** Mrs. Williams and her four children go shopping at a local grocery store. Each of the children will be allowed to select one box of granola bars. On the store's shelf there are 12 boxes of granola bars, and each box contains a different type of bar. In how many ways can the selections be made?

47. ***Track Meet*** A track meet has 15 participants for the 100-meter event. The 6 participants with the fastest speeds will be listed, in the order of their speed, on the leader board. How many different ways are there for the names to be listed?

48. ***History Test*** In one question on a history test, a student is asked to match 10 dates with 10 events; each date can only be matched with 1 event. In how many different ways can this question be answered?

49. ***Color Permutations*** Determine the number of permutations of the colors in the spectrum that follows.

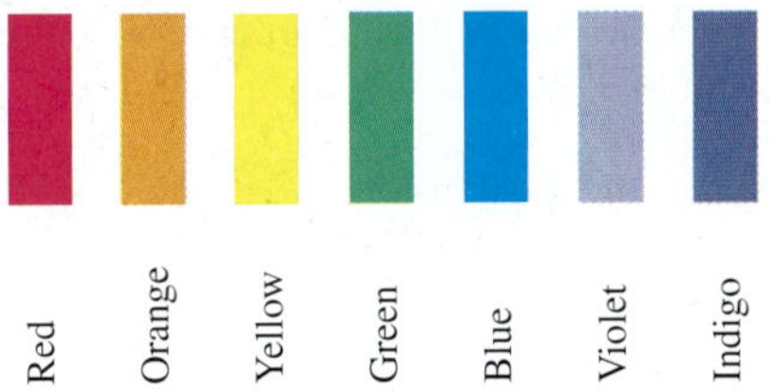

50. ***Drive-Through at a Bank*** A bank has three drive-through stations. Assuming that each is equally likely to be selected by customers, in how many different ways can the next six drivers select a station?

51. ***Computer Systems*** At a computer store, a customer is considering 5 different computers, 4 different monitors, 7 different printers, and 2 different scanners. Assuming that each of the components is compatible with one another and that one of each is to be selected, determine the number of different computer systems possible.

52. ***Selecting Furniture*** The Johnsons just moved into their new home and are selecting furniture for the family room. They are considering 5 different sofas, 2 different chairs, and 6 different tables. They plan to select one item from each category. Determine the number of different ways they can select the furniture.

53. Determine the number of permutations of the letters of the word "EDUCATION."

54. Determine the number of permutations of the letters of the word "SELECTION."

55. In how many ways can the letters in the word "DIFFERENCE" be arranged?

56. In how many ways can the letters in the word "MISSISSIPPI" be arranged?

57. In how many ways can the digits in the number 9,876,678 be arranged?

58. In how many ways can the digits of the number 2,142,332 be arranged?

59. ***Flag Messages*** Five different colored flags will be placed on a pole, one beneath another. The arrangement of the colors indicates the message. How many messages are possible if five flags are to be selected from nine different colored flags?

60. ***Multiple-Choice Test*** Keri Kershaw is taking a 12-question multiple-choice exam. Each question has three possible answers, (a), (b), and (c). In how many possible ways can Keri answer the questions?

61. ***Batting Order*** In how many ways can the manager of a National League baseball team arrange his batting order of nine players if

    a) the pitcher must bat last?

    b) there are no restrictions?

62. ***Painting Exhibit*** Five Monet paintings are to be displayed in a museum.

    a) In how many different ways can they be arranged if they must be next to one another?

    b) In how many different ways can they be displayed if a specific one is to be in the middle?

▲ *The Water Lily Pond, Pink Harmony,* 1900 by Claude Monet

CHALLENGE PROBLEMS/GROUP ACTIVITIES

63. ***Car Keys*** Door keys for a certain automobile are made from a blank key on which five cuts are made. Each cut may be one of five different depths.

    a) How many different keys can be made?

    b) If 400,000 of these automobiles are made such that each of the keys determined in part (a) opens the same number of cars, find how many cars can be opened by a specific key.

    c) If one of these cars is selected at random, what is the probability that the key selected at random will unlock the door?

64. ***Voting*** On a ballot, each committee member is asked to rank three of eight candidates for recommendation for promotion, giving first, second, and third choices (no ties). What is the minimum number of ballots that must be cast to guarantee that at least two ballots are the same?

65. ***Scrabble*** Nancy Lin, who is playing Scrabble with Dale Grey, has seven different letters. She decides to test each five-letter permutation before her next move. If each permutation takes 5 sec, how long will it take Nancy to check all the permutations?

66. ***Scrabble*** In Exercise 65, assume, of Nancy's seven letters, that three are identical and two are identical. How long will it take Nancy to try all different permutations of her seven letters?

67. Does ${}_nP_r = {}_nP_{(n-r)}$ for all whole numbers, where $n \geq r$? Explain.

RECREATIONAL MATHEMATICS

68. ***Stations*** There are eight bus stations from town A to town B. How many different single tickets must be printed so that a passenger may purchase a ticket from any station to any other station?

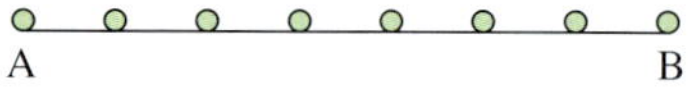

69. ***Bus Loop*** How many tickets with different points of origin and destination can be sold on a bus line that travels a loop with 25 stops?

Jumbles *Many newspapers contain Jumble puzzles, where the letters of a word are given out of order. In Exercises 70 and 71, determine*

a) *the number of possible arrangements of the letters given.*

b) *the word.*

70. HEICOC

71. ROSEGOC

INTERNET/RESEARCH ACTIVITY

72. When a book is published, it is assigned a 10-digit code number called the International Standard Book Number (ISBN). Do research and write a report on how this coding system works.

12.9 COMBINATIONS

▲ We can use combinations to determine how many different committees of three people can be selected from 25 people.

The members of the student senate at Austin Community College want to select three of their 25 members to attend a meeting with the faculty senate. In how many different ways can three members of the student senate be selected? If Frank, Sue, and Joe are the three students selected, does it make a difference which student was selected first, which student was selected second, or which student was selected third? For this example, the order of the selection of the students does not matter because the same set of students would attend the meeting. In this section, we will learn how to determine the number of combinations when the order of the selection of the items is not important to the final outcome.

When the order of the selection of the items is important to the final outcome, the problem is a permutation problem. When the order of the selection of the items is unimportant to the final outcome, the problem is a ***combination*** problem.

Recall from Section 12.8 that permutations are *ordered* arrangements. For example, *a*, *b*, *c* and *b*, *c*, *a* are two different permutations because the ordering of the three letters is different. The letters *a*, *b*, *c* and *b*, *c*, *a* represent the same combination of letters because the *same letters* are used in each set. However, the letters *a*, *b*, *c* and *a*, *b*, *d* represent two different combinations of letters because the letters contained in each set are different.

A **combination** is a distinct group (or set) of objects without regard to their arrangement.

EXAMPLE 1 *Permutation or Combination*

Determine whether the situation represents a permutation or combination problem.

a) A group of five friends, Arline, Inez, Judy, Dan, and Eunice, are forming a club. The group will elect a president and a treasurer. In how many different ways can the president and treasurer be selected?

b) Of the five individuals named, two will be attending a meeting together. In how many different ways can they do so?

SOLUTION

a) Since the president's position is different from the treasurer's position, we have a permutation problem. Judy as president with Dan as treasurer is different from Dan as president with Judy as treasurer. The order of the selection is important.

b) Since the order in which the two individuals selected to attend the meeting is not important, we have a combination problem. There is no difference if Judy is selected and then Dan is selected, or if Dan is selected and then Judy is selected. ●

In Section 12.8, you learned that $_nP_r$ represents the number of permutations when r items are selected from n distinct items. *Similarly, $_nC_r$ represents the number of combinations when r items are selected from n distinct items.*

Table 12.4 Summary

Counting Principle: If
ment can be performed
ways and a second expe
performed in N distinct
two experiments in that
can be performed in M
ways.

The counting principle
with or without repetiti
is used when determinin
of different ways that tw
periments can occur. It i
when there are specific
quirements, such as the
must be a 0 or 1.

SECTION

CONCEPT/WRITIN

1. In your own words,
2. What does $_nC_r$ mea
3. Give the formula fo
4. What is the relation
5. In your own words, mutation and a com
6. Assume that you ha are going to select f more combinations items? Explain.

PRACTICE THE SK

In Exercises 7–20, eva

7. $_4C_2$

9. a) $_6C_4$ b) $_6P_4$

11. a) $_8C_0$ b) $_8P_0$

13. a) $_{10}C_3$ b) $_{10}P_3$

DID YOU KNOW?

The Dead Man's Hand

James Butler "Wild Bill" Hickok (1837–1876), known as "the fastest gun in the West," was a famous scout and federal marshal. Hickok was shot in the back and mortally wounded by Jack McCall while playing poker in a saloon in Deadwood, Dakota Territory (now South Dakota). McCall was hanged for his deed. The cards Hickok was holding when shot are shown here. Since that time, the two pair, aces and eights, have become known as "the dead man's hand" (see Exercise 43). Hickok is buried next to Martha Canary, better known as "Calamity Jane," in a cemetery overlooking the town of Deadwood.

The number of ways in which two cards can be selected from a deck of 52 cards is $_{52}C_2$, or

$$_{52}C_2 = \frac{52!}{(52-2)!2!} = \frac{\overset{26}{\cancel{52}} \cdot 51 \cdot \cancel{50!}}{\cancel{50!} \cdot \cancel{2} \cdot 1} = 1326$$

Thus,

$$P(\text{selecting 2 picture cards}) = \frac{_{12}C_2}{_{52}C_2} = \frac{66}{1326} = \frac{11}{221}$$

Note that the same answer is obtained with either method. To give you more exposure to counting techniques, we will work the problems in this section using combinations.

EXAMPLE 1 *Committee of Three Women*

A club consists of four men and five women. Three members are to be selected at random to form a committee. What is the probability that the committee will consist of three women?

SOLUTION The order in which the three members are selected is not important. Therefore, we may work this problem using combinations.

$$P\begin{pmatrix}\text{committee consists}\\ \text{of 3 women}\end{pmatrix} = \frac{\text{number of possible committees with 3 women}}{\text{total number of possible 3-member committees}}$$

Since there is a total of 5 women, the number of possible committees with three women is $_5C_3 = 10$. Since there is a total of 9 people, the total number of possible three-member committees is $_9C_3 = 84$.

$$P(\text{committee consists of 3 women}) = \frac{10}{84} = \frac{5}{42}$$

The probability of randomly selecting a committee with three women is $\frac{5}{42}$. ●

EXAMPLE 2 *A Heart Flush*

A flush in the game of poker is five cards of the same suit (5 hearts, 5 diamonds, 5 clubs, or 5 spades). If you are dealt a five-card hand, find the probability that you will be dealt a heart flush.

SOLUTION The order in which the five hearts are dealt is not important. Therefore, we may work this problem using combinations.

$$P(\text{heart flush}) = \frac{\text{number of possible 5-card heart flushes}}{\text{total number of possible 5-card hands}}$$

Since there are 13 hearts in a deck of cards, the number of possible five-card heart flush hands is $_{13}C_5 = 1287$. The total number of possible five-card hands in a deck of 52 cards is $_{52}C_5 = 2{,}598{,}960$.

$$P(\text{heart flush}) = \frac{_{13}C_5}{_{52}C_5} = \frac{1287}{2{,}598{,}960} = \frac{33}{66{,}640}$$

The probability of being dealt a heart flush is $\frac{33}{66{,}640}$, or ≈ 0.000495. ●

22. ***Banana Split*** An ice-cream parlor has 20 different flavors. Cynthia orders a banana split and has to select 3 different flavors. How many different selections are possible?

23. ***Test Essays*** A student must select and answer four of five essay questions on a test. In how many ways can she do so?

24. ***Software Packages*** During a special promotion at CompUSA, a customer purchasing a computer and a printer is given the choice of 2 free software packages. If there are 9 different software packages from which to select, how many different ways can the 2 packages be selected?

25. ***Scholarships*** A scholarship committee has received 8 applications for a \$500 scholarship. The committee has decided to select 3 of the 8 candidates for further consideration. In how ways can the committee do so?

26. ***Attending Plays*** While visiting New York City, the Nygens want to attend 3 plays out of 10 plays they would like to see. In how many ways can they do so?

▲ Theater district, New York City

27. ***Taxi Ride*** A group of 7 people wants to use taxis to go to a local restaurant. When the first taxi arrives, the group decides that 4 people should get into the taxi. In how many ways can that be done?

28. ***Plants*** Mary Robinson purchased a package of 24 different plants, but she only needed 20 plants for planting. In how many ways can she select the 20 plants from the package to be planted?

29. ***Entertainers*** Ruth Eckerd Hall must select 8 of 12 possible entertainers for its summer schedule. In how many ways can that be done?

30. ***CD Purchase*** Neo Anderson wants to purchase six different CDs but only has enough money to purchase four. In how many ways can he select four of six CDs for purchase?

31. ***Posters*** Matthew Abbott has eight posters he would like to hang on the wall of his bedroom, but his wall is only wide enough to hang four posters. In how many ways can Matthew select the four posters to hang on his bedroom wall?

32. ***Washers and Dryers*** Sears has nine different washing machines in stock and six different clothes dryers in stock. The manager wants to place three of the nine washing machines and two of the six clothes dryers on sale. In how many ways can the manager select the items to be listed as sale items?

33. ***Quinella Bet*** A quinella bet consists of selecting the first- and second-place winners, in any order, in a particular event. For example, suppose you select a 2–5 quinella. If 2 wins and 5 finishes second, or if 5 wins and 2 finishes second, you win. Mr. Smith goes to a jai alai match. In the match, 8 jai alai teams compete. How many quinella tickets must Mr. Smith purchase to guarantee a win?

▲ Jai alai game

34. ***Test Question*** On an English test, Tito Ramirez must write an essay for three of the five questions in Part 1 and four of the six questions in Part 2. How many different combinations of questions can he answer?

35. ***Plasma and LCD TVs*** A television/stereo store has 12 different plasma televisions and 8 different LCD televisions in stock. The store's manager wishes to place 3 plasma televisions and 2 LCD televisions on sale. In how many ways can that be done?

36. ***Medical Research*** At a medical research center, an experimental drug is to be given to 16 people, 8 men and 8 women. If 14 men and 11 women have volunteered to be given the drug, in how many ways can the researcher choose the 16 people to be given the drug?

37. ***Dinner Party*** Sue Less is having a dinner party. She has 10 different bottles of red wine and 8 different bottles of white wine on her wine rack. She wants to select 4 bottles of red wine and 2 bottles of white wine to serve at her party. In how many ways can she do so?

38. ***Forming a Committee*** The Webster Town Board is forming a committee to explore ways to improve public safety in the town. The committee will consist of 4 representatives from the town board and 3 representatives from a citizens advisory board. If there are 7 town board members and 5 citizens advisory board members from which to choose, how many different ways can the committee be formed?

39. *Selecting Soda* Angel Ramirez is sent to the store to get 5 different bottles of regular soda and 3 different bottles of diet soda. If there are 10 different types of regular sodas and 7 different types of diet sodas to choose from, how many different choices does Angel have?

40. *Constructing a Test* A teacher is constructing a mathematics test consisting of 10 questions. She has a pool of 28 questions, which are classified by level of difficulty as follows: 6 difficult questions, 10 average questions, and 12 easy questions. How many different 10-question tests can she construct from the pool of 28 questions if her test is to have 3 difficult, 4 average, and 3 easy questions?

41. *Selecting Mutual Funds* Joe Chang recently graduated from college and now has a job that provides a retirement investment plan. Joe wants to diversify his investments, so he wants to invest in four stock mutual funds and two bond mutual funds. If he has a choice of eight stock mutual funds and five bond mutual funds, how many different selections of mutual funds does he have?

42. *Door Prize* As part of a door prize, Mary McCarty won three tickets to a baseball game and three tickets to a theater performance. She decided to give all the tickets to friends. For the baseball game she is considering six different friends, and for the theater she is considering eight different friends. In how many ways can she distribute the tickets?

43. *New Breakfast Cereals* General Mills is testing 6 oat cereals, 5 wheat cereals, and 4 rice cereals. If it plans to market 3 of the oat cereals, 2 of the wheat cereals, and 2 of the rice cereals, how many different combinations are possible?

44. *Catering Service* A catering service is making up trays of hors d'oeuvres. The hors d'oeuvres are categorized as inexpensive, average, and expensive. If the client must select three of the eight inexpensive, five of the nine average, and two of the four expensive hors d'oeuvres, how many different choices are possible?

CHALLENGE PROBLEMS/GROUP ACTIVITIES

45. *Test Answers* Consider a 10-question test in which each question can be answered either correctly or incorrectly.

a) How many different ways are there to answer the questions so that eight are correct and two are incorrect?

b) How many different ways are there to answer the questions so that at least eight are correct?

46. a) *A Dinner Toast* Four people at dinner make a toast. If each person is to tap glasses with each other person one at a time, how many taps will take place?

b) Repeat part (a) with five people.

c) How many taps will there be if there are n people at the dinner table?

47. *Pascal's Triangle* The notation ${}_nC_r$ may be written $\binom{n}{r}$.

a) Use this notation to evaluate each of the combinations in the following array. Form a triangle of the results, similar to the one given, by placing the answer to each combination in the same relative position in the triangle.

$$\binom{0}{0}$$

$$\binom{1}{0} \quad \binom{1}{1}$$

$$\binom{2}{0} \quad \binom{2}{1} \quad \binom{2}{2}$$

$$\binom{3}{0} \quad \binom{3}{1} \quad \binom{3}{2} \quad \binom{3}{3}$$

$$\binom{4}{0} \quad \binom{4}{1} \quad \binom{4}{2} \quad \binom{4}{3} \quad \binom{4}{4}$$

b) Using the number pattern in part (a), find the next row of numbers of the triangle (known as *Pascal's triangle*).

48. *Lottery Combinations* Determine the number of combinations possible in a state lottery where you must select

a) 6 of 46 numbers.

b) 6 of 47 numbers.

c) 6 of 48 numbers.

d) 6 of 49 numbers.

e) Does the number of combinations increase by the same amount going from part (a) to part (b) as from part (b) to part (c)?

49. a) *Table Seating Arrangements* How many distinct ways can four people be seated in a row?

b) How many distinct ways can four people be seated at a circular table?

50. Show that ${}_nC_r = {}_nC_{(n-r)}$.

51. *Forming a Committee* A group of 15 people wants to form a committee consisting of a chair, vice chair, and three additional members. How many different committees can be formed?

RECREATIONAL MATHEMATICS

52. a) *Combination Lock* To open a combination lock, you must know the lock's three-number sequence in its proper order. Repetition of numbers is permitted. Why is this lock more like a permutation lock than a combination lock? Why is it not a true permutation problem?

b) Assuming that a combination lock has 40 numbers, determine how many different three-number arrangements are possible if repetition of numbers is allowed.

c) Answer the question in part (b) if repetition is not allowed.

INTERNET/RESEARCH ACTIVITY

53. The area of mathematics called combinatorics is the science of counting. Do research and write a paper on combinatorics and its many applications.

12.10 SOLVING PROBABILITY PROBLEMS BY USING COMBINATIONS

▲ What is the probability of selecting two picture cards from a standard deck of 52 cards when the two cards are selected without replacement?

Suppose that we want to find the probability of selecting two picture cards (jacks, queens, or kings) when two cards are selected, without replacement, from a standard deck of cards. In Section 12.6, we used the *and* probability formula to solve this type of probability problem. In this section, we will learn another way to solve this type of probability problem by using combinations.

Using the *and* formula for the above example, we could reason as follows:

$$P(2 \text{ picture cards}) = P(\text{first picture card}) \cdot P(\text{second picture card})$$
$$= \frac{12}{52} \cdot \frac{11}{51} = \frac{132}{2652} \quad \text{or} \quad \frac{11}{221}$$

Since the order of the two picture cards selected is not important to the final answer, this problem can be considered a combination probability problem.

We can also find the probability of selecting two picture cards, using combinations, by finding the number of possible successful outcomes (selecting two picture cards) and dividing that answer by the total number of possible outcomes (selecting any two cards).

The number of ways in which two picture cards can be selected from the 12 picture cards in a deck is ${}_{12}C_2$, or

$${}_{12}C_2 = \frac{12!}{(12-2)!2!} = \frac{\overset{6}{\cancel{12}} \cdot 11 \cdot \cancel{10!}}{\cancel{10!} \cdot \cancel{2} \cdot 1} = 66$$

DID YOU KNOW?

The Dead Man's Hand

James Butler "Wild Bill" Hickok (1837–1876), known as "the fastest gun in the West," was a famous scout and federal marshal. Hickok was shot in the back and mortally wounded by Jack McCall while playing poker in a saloon in Deadwood, Dakota Territory (now South Dakota). McCall was hanged for his deed. The cards Hickok was holding when shot are shown here. Since that time, the two pair, aces and eights, have become known as "the dead man's hand" (see Exercise 43). Hickok is buried next to Martha Canary, better known as "Calamity Jane," in a cemetery overlooking the town of Deadwood.

The number of ways in which two cards can be selected from a deck of 52 cards is ${}_{52}C_2$, or

$$ {}_{52}C_2 = \frac{52!}{(52-2)!2!} = \frac{\overset{26}{\cancel{52}}\cdot 51\cdot \cancel{50!}}{\cancel{50!}\cdot \cancel{2}\cdot 1} = 1326 $$

Thus,

$$ P(\text{selecting 2 picture cards}) = \frac{{}_{12}C_2}{{}_{52}C_2} = \frac{66}{1326} = \frac{11}{221} $$

Note that the same answer is obtained with either method. To give you more exposure to counting techniques, we will work the problems in this section using combinations.

EXAMPLE 1 *Committee of Three Women*

A club consists of four men and five women. Three members are to be selected at random to form a committee. What is the probability that the committee will consist of three women?

SOLUTION The order in which the three members are selected is not important. Therefore, we may work this problem using combinations.

$$ P\left(\begin{array}{c}\text{committee consists}\\ \text{of 3 women}\end{array}\right) = \frac{\text{number of possible committees with 3 women}}{\text{total number of possible 3-member committees}} $$

Since there is a total of 5 women, the number of possible committees with three women is ${}_5C_3 = 10$. Since there is a total of 9 people, the total number of possible three-member committees is ${}_9C_3 = 84$.

$$ P(\text{committee consists of 3 women}) = \frac{10}{84} = \frac{5}{42} $$

The probability of randomly selecting a committee with three women is $\frac{5}{42}$.

EXAMPLE 2 *A Heart Flush*

A flush in the game of poker is five cards of the same suit (5 hearts, 5 diamonds, 5 clubs, or 5 spades). If you are dealt a five-card hand, find the probability that you will be dealt a heart flush.

SOLUTION The order in which the five hearts are dealt is not important. Therefore, we may work this problem using combinations.

$$ P(\text{heart flush}) = \frac{\text{number of possible 5-card heart flushes}}{\text{total number of possible 5-card hands}} $$

Since there are 13 hearts in a deck of cards, the number of possible five-card heart flush hands is ${}_{13}C_5 = 1287$. The total number of possible five-card hands in a deck of 52 cards is ${}_{52}C_5 = 2{,}598{,}960$.

$$ P(\text{heart flush}) = \frac{{}_{13}C_5}{{}_{52}C_5} = \frac{1287}{2{,}598{,}960} = \frac{33}{66{,}640} $$

The probability of being dealt a heart flush is $\frac{33}{66{,}640}$, or ≈ 0.000495.

EXAMPLE 3 *Employment Assignments*

A temporary employment agency has six men and five women who wish to be assigned for the day. One employer has requested four employees for security guard positions, and the second employer has requested three employees for moving furniture in an office building. If we assume that each of the potential employees has the same chance of being selected and being assigned at random and that only seven employees will be assigned, find the probability that

a) three men will be selected for moving furniture.

b) three men will be selected for moving furniture and four women will be selected for security guard positions.

SOLUTION

a) $$P\begin{pmatrix}\text{3 men selected}\\ \text{for moving furniture}\end{pmatrix} = \frac{\begin{pmatrix}\text{number of possible combinations}\\ \text{of 3 men selected}\end{pmatrix}}{\begin{pmatrix}\text{total number of possible combinations}\\ \text{for selecting 3 people}\end{pmatrix}}$$

The number of possible combinations with 3 men is ${}_6C_3$. The total number of possible selections of 3 people is ${}_{11}C_3$.

$$P\begin{pmatrix}\text{3 men selected}\\ \text{for moving furniture}\end{pmatrix} = \frac{{}_6C_3}{{}_{11}C_3} = \frac{20}{165} = \frac{4}{33}$$

Thus, the probability that 3 men are selected is $\frac{4}{33}$.

b) The number of ways of selecting 3 men out of 6 is ${}_6C_3$, and the number of ways of selecting 4 women out of 5 is ${}_5C_4$. The total number of possible selections when 7 people are selected from 11 is ${}_{11}C_7$. Since both the 3 men *and* the 4 women must be selected, the probability is calculated as follows:

$$P\begin{pmatrix}\text{3 men and}\\ \text{4 women selected}\end{pmatrix} = \frac{\begin{pmatrix}\text{number of combinations}\\ \text{of 3 men selected}\end{pmatrix}\cdot\begin{pmatrix}\text{number of combinations}\\ \text{of 4 women selected}\end{pmatrix}}{\begin{pmatrix}\text{total number of possible combinations}\\ \text{for selecting 7 people}\end{pmatrix}}$$

$$= \frac{{}_6C_3 \cdot {}_5C_4}{{}_{11}C_7} = \frac{20 \cdot 5}{330} = \frac{100}{330} = \frac{10}{33}$$

Thus, the probability is $\frac{10}{33}$.

EXAMPLE 4 *New Breakfast Cereals*

Kellogg's is testing 12 new cereals for possible production. They are testing 3 oat cereals, 4 wheat cereals, and 5 rice cereals. If we assume that each of the 12 cereals has the same chance of being selected and 4 new cereals will be produced, find the probability that

a) no wheat cereals are selected.

b) at least 1 wheat cereal is selected.

c) 2 wheat cereals and 2 rice cereals are selected.

SOLUTION

a) If no wheat cereals are to be selected, then only oat and rice cereals must be selected. Eight cereals are oat or rice. Thus, the number of ways that 4 oat or rice cereals may be selected from the 8 possible oat or rice cereals is ${}_8C_4$. The total number of possible selections is ${}_{12}C_4$.

$$P(\text{no wheat cereals}) = \frac{{}_8C_4}{{}_{12}C_4} = \frac{70}{495} = \frac{14}{99}$$

b) When 4 cereals are selected, the choice must contain either no wheat cereal or at least 1 wheat cereal. Since one of these outcomes must occur, the sum of the probabilities must be 1, or

$$P(\text{no wheat cereal}) + P(\text{at least 1 wheat cereal}) = 1$$

Therefore,

$$\begin{aligned} P(\text{at least 1 wheat cereal}) &= 1 - P(\text{no wheat cereal}) \\ &= 1 - \frac{14}{99} = \frac{99}{99} - \frac{14}{99} = \frac{85}{99} \end{aligned}$$

Note that the probability of selecting no wheat cereals, $\frac{14}{99}$, was found in part (a).

c) The number of ways of selecting 2 wheat cereals out of 4 wheat cereals is ${}_4C_2$, which equals 6. The number of ways of selecting 2 rice cereals out of 5 rice cereals is ${}_5C_2$, which equals 10. The total number of possible selections when 4 cereals are selected from the 12 choices is ${}_{12}C_4$ which equals 495. Since both the 2 wheat *and* the 2 rice cereals must be selected, the probability is calculated as follows.

$$P(\text{2 wheat and 2 rice}) = \frac{{}_4C_2 \cdot {}_5C_2}{{}_{12}C_4} = \frac{6 \cdot 10}{495} = \frac{60}{495} = \frac{4}{33}$$

EXAMPLE 5 *Rare Coins*

Conner Shanahan's rare coin collection is made up of 8 silver dollars, 7 quarters, and 5 dimes. Conner plans to sell 8 of his 20 coins to finance part of his college education. If he selects the coins at random, what is the probability that 3 silver dollars, 2 quarters, and 3 dimes are selected?

SOLUTION The number of ways that Conner can select 3 out of 8 silver dollars is ${}_8C_3$. The number of ways he can select 2 out of 7 quarters is ${}_7C_2$. The number of ways he can select 3 out of 5 dimes is ${}_5C_3$. He will select 8 coins from a total of 20 coins. The number of ways he can do so is ${}_{20}C_8$. The probability that Conner selects 3 silver dollars, 2 quarters, and 3 dimes is calculated as follows.

$$\begin{aligned} P(\text{3 silver dollars, 2 quarters, and 3 dimes}) &= \frac{{}_8C_3 \cdot {}_7C_2 \cdot {}_5C_3}{{}_{20}C_8} \\ &= \frac{56 \cdot 21 \cdot 10}{125{,}970} = \frac{11{,}760}{125{,}970} = \frac{392}{4199} \end{aligned}$$

SECTION 12.10 EXERCISES

CONCEPT/WRITING EXERCISES

In Exercises 1–8, set up the problem as if it were to be solved, but do not solve. Assume that each problem is to be done without replacement. Explain why you set up the exercises as you did.

1. Eight red balls and four blue balls are in a bag. If four balls from the bag are to be selected at random, determine the probability of selecting four red balls.

2. A class consists of 19 girls and 15 boys. If 12 of the students are to be selected at random, determine the probability that they are all girls.

3. Three letters are to be selected at random from the English alphabet of 26 letters. Determine the probability that 3 vowels (a, e, i, o, u) are selected.

4. Determine the probability of being dealt 3 aces from a standard deck of 52 cards when 3 cards are dealt.

5. A dog breeder has 15 puppies for sale, of which 8 are yellow Labrador retrievers. If 5 puppies are selected at random, determine the probability they are all yellow Labrador retrievers.

6. Of 80 people attending a dance, 28 have a college degree. If 4 people at the dance are selected at random, determine the probability that each of the 4 has a college degree.

7. A basket of 50 golf balls contains 22 Nike balls. If 8 balls are selected at random from the basket, determine the probability that *none* of those selected is a Nike ball.

8. A class of 16 people contains 4 people whose birthday is in October. If 3 people from the class are selected at random, determine the probability that *none* of those selected has an October birthday.

PRACTICE THE SKILLS/PROBLEM SOLVING

In Exercises 9–18, the problems are to be done without replacement. Use combinations to determine probabilities.

9. ***Green and Red Balls*** A bag contains four red balls and five green balls. You plan to select three balls at random. Determine the probability of selecting three green balls.

10. ***Selecting Lightbulbs*** A box contains 4 defective and 6 good lightbulbs. If two lightbulbs are to be selected at random from the box, determine the probability that both lightbulbs are defective.

11. ***Flu Serum*** A doctor has five doses of flu protection serum left. He has six women and eight men who want the medication. If the names of five of these people are selected at random, determine the probability that five men's names are selected.

12. ***Bills of Four Denominations*** Duc Tran's wallet contains 8 bills of the following denominations: four \$5 bills, two \$10 bills, one \$20 bill, and one \$50 bill. If Duc selects two bills at random, determine the probability that he selects two \$5 bills.

13. ***Selecting Digits*** Each of the digits 0–9 is written on a slip of paper, and the slips are placed in a hat. If three slips of paper are selected at random, determine the probability that the three numbers selected are greater than 4.

14. ***Selecting Books*** Barnes and Noble has 20 different books listed on its clearance list. Twelve books are listed as mystery, and 8 are listed as romance. If 4 books are selected at random from the list, determine the probability that they are all mystery books.

15. ***Gift Certificates*** The sales department at Atwell Studios consists of three people, the manufacturing department consists of six people, and the accounting department consists of two people. Three people will be selected at random from these people and will be given gift certificates to Sweet Tomatoes, a local restaurant. Determine the probability that two of those selected will be from the manufacturing department and one will be from the accounting department.

16. ***Faculty-Student Committee*** A committee of four is to be randomly selected from a group of seven teachers and eight students. Determine the probability that the committee will consist of two teachers and two students.

17. ***Winning the Grand Prize*** A lottery consists of 46 numbers. You select 6 numbers, and if they match the 6 numbers selected by the lottery commission, you win the grand prize. Determine the probability of winning the grand prize.

18. ***Red Cards*** You are dealt 5 cards from a standard deck of 52 cards. Determine the probability that you are dealt 5 red cards.

TV Game Show *In Exercises 19–22, a television game show has five doors, of which the contestant must pick two. Behind two of the doors are expensive cars, and behind the other three doors are consolation prizes. The contestant gets to keep the items behind the two doors she selects. Determine the probability that the contestant wins*

19. no cars.

20. both cars.

21. at least one car.

22. exactly one car.

Baseball *In Exercises 23–26, assume that a particular professional baseball team has 10 pitchers, 6 infielders, and 9 other players. If 3 players' names are selected at random, determine the probability that*

23. all 3 are infielders.

24. none of the three is a pitcher.

25. 2 are pitchers and 1 is an infielder.

26. 1 is a pitcher and 2 are players other than pitchers and infielders.

▲ See Exercises 23–26

Car Rental *In Exercises 27–30, a car rental agency has 10 midsized and 15 compact cars on its lot, from which 6 will be selected. Assuming that each car is equally likely to be selected and the cars are selected at random, determine the probability (as a decimal number rounded to four decimal places) that the cars selected consist of*

27. all midsized cars.

28. 2 midsized cars and 4 compact cars.

29. 3 midsized cars and 3 compact cars.

30. at least 1 compact car.

Airline Routes *In Exercises 31–34, an airline is given permission to fly 4 new routes of its choice. The airline is considering 12 new routes: 4 routes in Florida, 5 routes in California, and 3 routes in Texas. If the airline selects the 4 new routes at random from the 12 possibilities, determine the probability that*

31. 2 are in Florida and 2 are in Texas.

32. 3 are in California and 1 is in Florida.

33. 1 is in Florida, 1 is in California, and 2 are in Texas.

34. at least one is in Texas.

Theater *In Exercises 35–38, five men and six women are going to be assigned to a specific row of seats in a theater. If the 11 tickets for the numbered seats are given out at random, determine the probability that*

35. five women are given the first five seats next to the center aisle.

36. at least one woman is in one of the first five seats.

37. exactly one woman is in one of the first five seats.

38. three women are seated in the first three seats and two men are seated in the next two seats.

39. ***Working Overtime*** Of 24 employees at Lowe's home improvement store, 10 work as cashiers and 14 stock shelves. If 3 of the 24 employees are selected at random to work overtime, determine the probability that all 3 are cashiers.

40. ***Poker Probability*** A full house in poker consists of three of one kind and two of another kind in a five-card hand. For example, if a hand contains three kings and two 5's, it is a full house. If 5 cards are dealt at random from a standard deck of 52 cards, without replacement, determine the probability of getting three kings and two 5's.

41. ***A Royal Flush*** A royal flush consists of an ace, king, queen, jack, and 10 all in the same suit. If 7 cards are dealt at random from a standard deck of 52 cards, determine the probability of getting a

a) royal flush in spades.

b) royal flush in any suit.

42. ***Restaurant Staff*** The staff of a restaurant consists of 25 people, including 8 waiters, 12 waitresses, and 5 cooks. For Mother's Day, a total of 9 people will need to be selected to work. If the selections are made at random, determine the probability that 3 waiters, 4 waitresses, and 2 cooks will be selected.

43. ***"Dead Man's Hand"*** A pair of aces and a pair of 8's is often known as a "dead man's hand." (See the Did You Know? on page 810.)

a) Determine the probability of being dealt a dead man's hand (any two aces, any two 8's, and one other card that is not an ace or an 8) when 5 cards are dealt, without replacement, from a standard deck of 52 cards.

b) The actual cards "Wild Bill" Hickok was holding when he was shot were the aces of spades and clubs, the 8's of spades and clubs, and the 9 of diamonds. If you are dealt five cards without replacement, determine the probability of being dealt this exact hand.

CHALLENGE PROBLEMS/GROUP ACTIVITIES

44. ***Alternate Seating*** If three men and three women are to be assigned at random to six seats in a row at a theater, determine the probability that they will alternate by gender.

45. ***Selecting Officers*** A club consists of 15 people including Ali, Kendra, Ted, Alice, Marie, Dan, Linda, and Frank. From the 15 members, a president, vice president, and treasurer will be selected at random. An advisory committee of 5 other individuals will also be selected at random.

a) Determine the probability that Ali is selected president, Kendra is selected vice president, Ted is selected treasurer, and the other 5 individuals named form the advisory committee.

b) Determine the probability that 3 of the 8 individuals named are selected for the three officers' positions and the other 5 are selected for the advisory board.

46. ***A Marked Deck*** A number is written with a magic marker on each card of a deck of 52 cards. The number 1 is put on the first card, 2 on the second, and so on. The cards are then shuffled and cut. What is the probability that the top 4 cards will be in ascending order? (For example, the top card is 12, the second 22, the third 41, and the fourth 51.)

RECREATIONAL MATHEMATICS

47. ***Hair*** When the Isle of Flume took its most recent census, the population was 100,002 people. Nobody on the isle has more than 100,001 hairs on his or her head. Determine the probability that at least two people have exactly the same number of hairs on their head.

12.11 BINOMIAL PROBABILITY FORMULA

▲ We can use the binomial formula to determine the probability that selected customers waited on will leave a tip.

Suppose that you are a waiter at a restaurant and have learned from past experience that 80% of your customers leave a tip. If you wait on 6 customers, what is the probability that all 6 customers will leave a tip? If you wait on 8 customers, what is the probability that at least 5 of them will leave you a tip? In this section, we will learn how to use the binomial probability formula to answer these and similar questions.

Suppose that a basket contains three identical balls, except for their color. One is red, one is blue, and one is yellow (Fig. 12.23). Suppose further that we are going to select three balls *with replacement* from the basket. We can determine specific probabilities by examining the tree diagram shown in Fig. 12.24. Note that 27 different selections are possible, as indicated in the sample space.

Figure 12.23

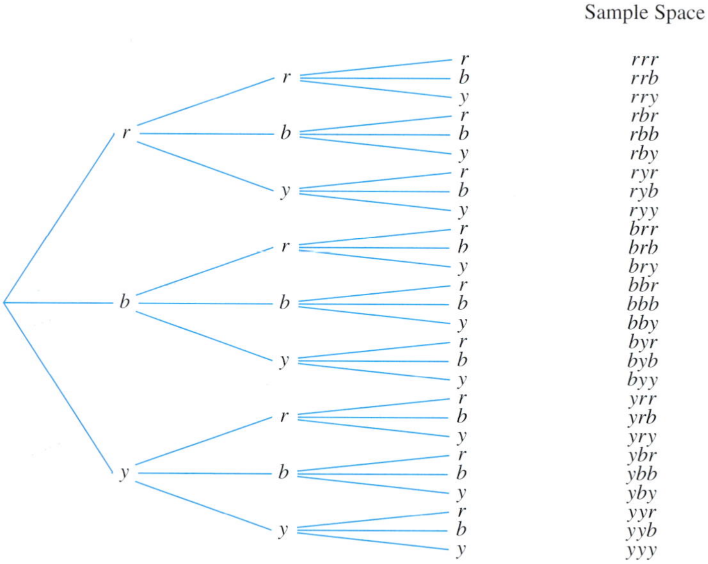

Figure 12.24

Our three selections may yield 0, 1, 2, or 3 red balls. We can determine the probability of selecting exactly 0, 1, 2, or 3 red balls by using the sample space. To determine the probability of selecting 0 red balls, we count those outcomes that do not contain a red ball. There are 8 of them (*bbb*, *bby*, *byb*, *byy*, *ybb*, *yby*, *yyb*, *yyy*). Thus, the probability of obtaining exactly 0 red balls is 8/27. We determine the probability of selecting exactly 1 red ball by counting the sample points that contain exactly 1 red ball. There are 12 of them. Thus, the probability is $\frac{12}{27}$, or $\frac{4}{9}$.

We can determine the probability of selecting exactly 2 red balls and exactly 3 red balls in a similar manner. The probabilities of selecting exactly 0, 1, 2, and 3 red balls are illustrated in Table 12.5 on page 817.

Table 12.5 A Probability Distribution for Three Balls Selected with Replacement

Number of Red Balls Selected, (x)	**Probability of Selecting the Number of Red Balls, $P(x)$**
0	$\frac{8}{27}$
1	$\frac{12}{27}$
2	$\frac{6}{27}$
3	$\frac{1}{27}$
	Sum $= \frac{27}{27} = 1$

Note that the sum of the probabilities is 1. This table is an example of a ***probability distribution***, which shows the probabilities associated with each specific outcome of an experiment. *In a probability distribution, every possible outcome must be listed and the sum of the probabilities must be 1.*

Let us specifically consider the probability of selecting 1 red ball in 3 selections. We see from Table 12.5 that this probability is $\frac{12}{27}$, or $\frac{4}{9}$. Can we determine this probability without developing a tree diagram? The answer is yes.

Suppose that we consider selecting a red ball success, S, and selecting a non–red ball failure, F. Furthermore, suppose that we let p represent the probability of success and q the probability of failure on any trial. Then $p = \frac{1}{3}$ and $q = \frac{2}{3}$. We can obtain 1 success in three selections in the following ways:

SFF FSF FFS

We can compute the probabilities of each of these outcomes using the multiplication formula because each of the selections is independent.

$$P(\text{SFF}) = P(\text{S}) \cdot P(\text{F}) \cdot P(\text{F}) = p \cdot q \cdot q = pq^2 = \frac{1}{3}\left(\frac{2}{3}\right)^2 = \frac{4}{27}$$

$$P(\text{FSF}) = P(\text{F}) \cdot P(\text{S}) \cdot P(\text{F}) = q \cdot p \cdot q = pq^2 = \frac{1}{3}\left(\frac{2}{3}\right)^2 = \frac{4}{27}$$

$$P(\text{FFS}) = P(\text{F}) \cdot P(\text{F}) \cdot P(\text{S}) = q \cdot q \cdot p = pq^2 = \frac{1}{3}\left(\frac{2}{3}\right)^2 = \frac{4}{27}$$

$$\text{Sum} = \frac{12}{27} = \frac{4}{9}$$

We obtained an answer of $\frac{4}{9}$, the same answer that was obtained using the tree diagram. Note that each of the 3 sets of outcomes above has 1 success and 2 failures. Rather than listing all the possibilities containing 1 success and 2 failures, we can use the combination formula to determine the number of possible combinations of 1 success in 3 trials. To do so, evaluate ${}_3C_1$.

$$ {}_3C_1 = \frac{3!}{(3-1)!1!} = \frac{3 \cdot 2 \cdot 1}{2 \cdot 1 \cdot 1} = 3 $$

Number of trials (3) ↗ ↖ Number of successes (1)

Thus, we see that there are 3 ways the 1 success could occur in 3 trials. To compute the probability of 1 success in 3 trials, we can multiply the probability of success in any one trial, $p \cdot q^2$, by the number of ways the 1 success can be arranged among the 3 trials, ${}_3C_1$. Thus, the probability of selecting 1 red ball, $P(1)$, in 3 trials may be found as follows.

$$ P(1) = ({}_3C_1)p^1q^2 = 3\left(\frac{1}{3}\right)\left(\frac{2}{3}\right)^2 = \frac{12}{27} = \frac{4}{9} $$

The binomial probability formula, which we introduce shortly, explains how to obtain expressions like $P(1) = ({}_3C_1)p^1q^2$ and is very useful in finding certain types of probabilities.

To use the binomial probability formula, the following three conditions must hold.

TO USE THE BINOMIAL PROBABILITY FORMULA

1. There are n repeated independent trials.
2. Each trial has two possible outcomes, *success* and *failure.*
3. For each trial, the probability of success (and failure) remains the same.

Before going further, let's discuss why we can use the binomial probability formula to find the probability of selecting a specific number of red balls when three balls are selected with replacement. First, since each trial is performed *with replacement*, the three trials are independent of each other. Second, we may consider selecting a red ball as success and selecting any ball of another color as failure. Third, for each selection, the probability of success (selecting a red ball) is $\frac{1}{3}$ and the probability of failure (selecting a ball of another color) is $\frac{2}{3}$. Now let's discuss the binomial probability formula.

BINOMIAL PROBABILITY FORMULA

The probability of obtaining exactly x successes, $P(x)$, in n independent trials is given by

$$ P(x) = ({}_nC_x)p^xq^{n-x} $$

where p is the probability of success on a single trial and q is the probability of failure on a single trial.

In the formula, p will be a number between 0 and 1, inclusive, and $q = 1 - p$. Therefore, if $p = 0.2$, then $q = 1 - 0.2 = 0.8$. If $p = \frac{3}{5}$, then $q = 1 - \frac{3}{5} = \frac{2}{5}$. Note that $p + q = 1$ and the values of p and q remain the same for each independent trial. The combination ${}_nC_x$ is called the *binomial coefficient.*

In Example 1, we use the binomial probability formula to solve the same problem we recently solved by using a tree diagram.

EXAMPLE 1 *Selecting Colored Balls with Replacement*

A basket contains 3 balls: 1 red, 1 blue, and 1 yellow. Three balls are going to be selected with replacement from the basket. Find the probability that

a) no red balls are selected.

b) exactly 1 red ball is selected.

c) exactly 2 red balls are selected.

d) exactly 3 red balls are selected.

SOLUTION

a) We will consider selecting a red ball a success and selecting a ball of any other color a failure. Since only 1 of the 3 balls is red, the probability of success on any single trial, p, is $\frac{1}{3}$. The probability of failure on any single trial, q, is $1 - \frac{1}{3} = \frac{2}{3}$. We are finding the probability of selecting 0 red balls, or 0 successes. Since x represents the number of successes, we let $x = 0$. There are 3 independent selections (or trials), so $n = 3$. In our calculations, we will need to evaluate $(\frac{1}{3})^0$. Note that any nonzero number raised to a power of 0 is 1. Thus, $(\frac{1}{3})^0 = 1$. We determine the probability of 0 successes, or $P(0)$, as follows.

$$\begin{aligned} P(x) &= ({}_nC_x)p^x q^{n-x} \\ P(0) &= ({}_3C_0)\left(\frac{1}{3}\right)^0\left(\frac{2}{3}\right)^{3-0} \\ &= (1)(1)\left(\frac{2}{3}\right)^3 \\ &= \left(\frac{2}{3}\right)^3 = \frac{8}{27} \end{aligned}$$

b) We are finding the probability of obtaining exactly 1 red ball or exactly 1 success in 3 independent selections. Thus, $x = 1$ and $n = 3$. We find the probability of exactly 1 success, or $P(1)$, as follows.

$$\begin{aligned} P(x) &= ({}_nC_x)p^x q^{n-x} \\ P(1) &= ({}_3C_1)\left(\frac{1}{3}\right)^1\left(\frac{2}{3}\right)^{3-1} \\ &= 3\left(\frac{1}{3}\right)\left(\frac{2}{3}\right)^2 \\ &= 3\left(\frac{1}{3}\right)\left(\frac{4}{9}\right) = \frac{4}{9} \end{aligned}$$

c) We are finding the probability of selecting exactly 2 red balls in 3 independent trials. Thus, $x = 2$ and $n = 3$. We find $P(2)$ as follows.

$$\begin{aligned} P(x) &= ({}_nC_x)p^x q^{n-x} \\ P(2) &= {}_3C_2\left(\frac{1}{3}\right)^2\left(\frac{2}{3}\right)^{3-2} \\ &= 3\left(\frac{1}{3}\right)^2\left(\frac{2}{3}\right)^1 \\ &= 3\left(\frac{1}{9}\right)\left(\frac{2}{3}\right) = \frac{2}{9} \end{aligned}$$

SOLUTION

a) Success is a tree survives. Thus, $p = 0.8$ and $q = 1 - p = 1 - 0.8 = 0.2$. We want to find the probability of 0 successes in 4 trials. Thus, $x = 0$ and $n = 4$. We find the probability of 0 successes, or $P(0)$, as follows.

$$\begin{aligned} P(x) &= ({}_nC_x)p^x q^{n-x} \\ P(0) &= ({}_4C_0)(0.8)^0(0.2)^{4-0} \\ &= 1(1)(0.2)^4 \\ &= 1(1)(0.0016) \\ &= 0.0016 \end{aligned}$$

Thus, the probability that none of the four trees planted will survive is 0.0016.

b) The probability that at least one tree of the four trees planted will survive can be found by subtracting from 1 the probability that none of the four trees survives. We worked problems of this type in earlier sections of the chapter, including Sections 12.6 and 12.10.

In part (a), we determined the probability that none of the four trees planted survives is 0.0016. Thus,

$$\begin{aligned} P(\text{at least one tree planted will survive}) &= 1 - P\begin{pmatrix}\text{none of the four trees} \\ \text{planted will survive}\end{pmatrix} \\ &= 1 - 0.0016 \\ &= 0.9984 \end{aligned}$$

SECTION 12.11 EXERCISES

CONCEPT/WRITING EXERCISES

1. What is a probability distribution?
2. What are the three requirements that must be met to use the binomial probability formula?
3. Write the binomial probability formula.
4. a) In the binomial probability formula, what do p and q represent?

   b) If $p = 0.7$, what is the value of q?

   c) If $p = 0.25$, what is the value of q?

PRACTICE THE SKILLS

In Exercises 5–10, assume that each of the n trials is independent and that p is the probability of success on a given trial. Use the binomial probability formula to find P(x).

5. $n = 5, x = 3, p = 0.2$
6. $n = 3, x = 2, p = 0.6$
7. $n = 5, x = 2, p = 0.4$
8. $n = 3, x = 3, p = 0.8$
9. $n = 6, x = 0, p = 0.5$
10. $n = 5, x = 3, p = 0.4$

PROBLEM SOLVING

11. *A Dozen Eggs* An egg distributor determines that the probability that any individual egg has a crack is 0.14.

    a) Write the binomial probability formula to determine the probability that exactly x eggs of n eggs are cracked.

    b) Write the binomial probability formula to determine the probability that exactly 2 eggs in a one-dozen egg carton are cracked. Do not evaluate.

12. *Getting Audited* The probability that a self-employed taxpayer will be audited by the Internal Revenue Service (IRS) is 0.0077.

    a) Write the binomial probability formula to determine the probability that exactly x out of n self-employed taxpayers selected at random will be audited by the IRS.

    b) Write the binomial probability formula to determine the probability that exactly 5 out of 20 self-employed taxpayers selected at random will be audited by the IRS. Do not evaluate.

In Exercises 13–21, use the binomial probability formula to answer the question. Round answers to five decimal places.

13. *Leaving a Tip* Thomas Zellner works as a waiter at Outback Steakhouse. He has learned from experience that 80% of customers who dine alone leave a tip. If Thomas waits on 6 customers dining alone, determine the probability that exactly 4 of them leave a tip.

14. *Traffic Tickets* In Georgia, the probability that a driver is actually given a ticket when he or she is pulled over for a traffic infraction by a member of the Georgia State Police is 0.6. If eight people who were pulled over are selected at random, determine the probability that exactly five of them were given tickets.

15. *Bank Loans* Records from a specific bank show that 70% of car loan applications are approved. If eight car loan applications from this bank are selected at random, determine the probability that exactly five of the applications are approved.

16. *Basketball* Jean Woody makes 80% of her free throws in a basketball game. Determine the probability that she makes exactly four of the next seven free throws.

17. *Dolphin Drug Care* When treated with the antibiotic resonocyllin, 92% of all dolphins are cured of a particular bacterial infection. If six dolphins with the particular bacterial infection are treated with resonocyllin, determine the probability that exactly four are cured.

18. *Manufacturing Lightbulbs* A quality control engineer at a GE lightbulb plant finds that 1% of its bulbs are defective. Determine the probability that exactly two of the next six bulbs made are defective.

19. *Water Heaters* The probability that a specific brand of water heater produces the water temperature it is set to produce is $\frac{4}{5}$. Determine the probability that if five of these water heaters are selected at random, exactly four of them will produce the water temperature they are set to produce.

20. *TV Purchases* At a Circuit City store, $\frac{1}{4}$ of those purchasing color televisions purchase a large-screen TV. Determine the probability that

    a) none of the next four people who purchase a color television at Circuit City purchases a large-screen TV.

    b) at least one of the next four people who purchase a color television at Circuit City purchases a large-screen TV.

21. *Multiple-Choice Quiz* Edward Dunn has to take a five-question multiple-choice quiz in his sociology class. Each question has four choices for answers, of which only one is correct. Assuming that Edward guesses on all five questions, what is the probability that he will answer

    a) all five questions correctly.

    b) exactly three questions correctly.

    c) at least three questions correctly.

CHALLENGE PROBLEMS/GROUP ACTIVITIES

22. *Transportation to Work* In a random sample of 80 working mothers in Duluth, Minnesota, the following data indicating how they get to work were obtained.

Mode of Transportation	Number of Mothers
Car	40
Bus	20
Bike	16
Other	4

    If this sample is representative of all working mothers in Duluth, determine the probability that exactly three of five working mothers selected at random

    a) take a car to work.

    b) take a bus to work.

23. *Selecting 6 Cards* Six cards are selected from a standard deck of playing cards with replacement. Determine the probability that

    a) exactly three picture cards are obtained.

    b) exactly two spades are obtained.

24. *Office Visit* The probability that a person visiting Dr. Guillermo Suarez's office is more than 60 years old is 0.7. Determine the probability that

a) exactly three of the next five people visiting the office are more than 60 years old.

b) at least three of the next five people visiting the office are more than 60 years old.

RECREATIONAL MATHEMATICS

25. ***Aruba*** The island of Aruba is well known for its beaches and predictable warm, sunny weather. In fact, Aruba's weather is so predictable that the daily newspapers don't even bother to print a forecast. Strangely enough, however, on New Year's Eve, as the islanders were counting down the last 10 sec of 2007, it began to rain. What is the probability, from 0 to 1, that 72 hr later the sun will be shining?

CHAPTER 12 SUMMARY

IMPORTANT FACTS

EMPIRICAL PROBABILITY

$$P(E) = \frac{\text{number of times event } E \text{ has occurred}}{\left(\begin{array}{c}\text{total number of times the}\\ \text{experiment has been performed}\end{array}\right)}$$

THE LAW OF LARGE NUMBERS

Probability statements apply in practice to a large number of trials, not to a single trial. It is the relative frequency over the long run that is accurately predictable, not individual events or precise totals.

THEORETICAL PROBABILITY

$$P(E) = \frac{\text{number of outcomes favorable to } E}{\text{total number of possible outcomes}}$$

The probability of an event that cannot occur is 0. The probability of an event that must occur is 1. Every probability must be a number between 0 and 1 inclusively; that is

$$0 \le P(E) \le 1$$

The sum of the probabilities of all possible outcomes of an event is 1.

$$P(A) + P(\text{not } A) = 1$$

ODDS AGAINST AN EVENT

$$\text{Odds against} = \frac{P(\text{event fails to occur})}{P(\text{event occurs})} = \frac{P(\text{failure})}{P(\text{success})}$$

ODDS IN FAVOR OF AN EVENT

$$\text{Odds in favor} = \frac{P(\text{event occurs})}{P(\text{event fails to occur})} = \frac{P(\text{success})}{P(\text{failure})}$$

EXPECTED VALUE

$$E = P_1A_1 + P_2A_2 + P_3A_3 + \cdots + P_nA_n$$

FAIR PRICE

$$\text{Fair price} = \text{expected value} + \text{cost to play}$$

COUNTING PRINCIPLE

If a first experiment can be performed in M distinct ways and a second experiment can be performed in N distinct ways, then the two experiments in that specific order can be performed in $M \cdot N$ distinct ways.

OR AND *AND* PROBLEMS

$$P(A \text{ or } B) = P(A) + P(B) - P(A \text{ and } B)$$
$$P(A \text{ and } B) = P(A) \cdot P(B)$$

CONDITIONAL PROBABILITY

$$P(E_2 \mid E_1) = \frac{n(E_1 \text{ and } E_2)}{n(E_1)}$$

The **number of permutations** of n items is $n!$.

$$n! = n(n-1)(n-2)\cdots(3)(2)(1)$$

PERMUTATION FORMULA

$$_nP_r = \frac{n!}{(n-r)!}$$

The number of different permutations of n objects where $n_1, n_2, \ldots, n_r$ of the objects are identical is

$$\frac{n!}{n_1!n_2!\cdots n_r!}$$

COMBINATION FORMULA

$$_nC_r = \frac{n!}{(n-r)!r!}$$

BINOMIAL PROBABILITY FORMULA

$$P(x) = (_nC_x)p^xq^{n-x}$$

CHAPTER 12 REVIEW EXERCISES

12.1–12.11

1. In your own words, explain the law of large numbers.

2. Explain how empirical probability can be used to determine whether a die is "loaded" (not a fair die).

3. ***Cars*** Of 40 people who purchase a vehicle at a car dealership, 8 purchased an SUV. Determine the empirical probability that the next person who purchases a vehicle from that car dealership purchases an SUV.

4. ***Cards*** Select a card from a deck of cards 40 times with replacement and compute the empirical probability of selecting a heart.

5. ***Television News*** In a small town, 200 people were asked whether they watched ABC, CBS, NBC, or Fox news. The results are indicated below.

Network	Number of People
ABC	80
CBS	30
NBC	55
Fox	35

Find the empirical probability that the next person selected at random from the town watches ABC news.

Digits *In Exercises 6–9, each of the digits 0, 1, 2, 3, 4, 5, 6, 7, 8, 9 is written on a piece of paper and all the pieces of paper are placed in a hat. One number is selected at random. Determine the probability that the number selected is*

6. even.

7. odd or greater than 5.

8. greater than 3 or less than 6.

9. even and greater than 6.

Cheese Preference *In Exercises 10–13, a taste test is given to 50 customers at a supermarket. The customers are asked to taste 4 types of cheese and to list their favorite. The results are summarized at the top of the right column.*

Type	Number of People
Cheddar	18
Colby jack	14
Muenster	11
Swiss	7

If one person who participated in the taste test is selected at random, determine the probability that the person's favorite was

10. colby jack.

11. muenster.

12. either cheddar or colby jack.

13. a cheese other than Swiss.

14. ***Obesity*** According to the U.S. Centers for Disease Control and Prevention, 31% of men in the United States are obese. If a man was selected at random, determine the odds

    a) against him being obese.

    b) in favor of him being obese.

15. ***Vegetable Mix-up*** Nicholas Delaney, a mischievous little boy, has removed labels on the eight cans of vegetables in the cabinet. Nicholas's father knows that there are three cans of corn, three cans of beans, and two cans of carrots. If the father selects and opens one can at random, determine the odds against his selecting a can of corn.

16. ***Horseracing*** The odds against Buttermilk winning the Triple Crown in horse racing are 82:3. Determine the probability that Buttermilk wins the Triple Crown.

17. *Coffee Shop Success* The probability that a new coffee shop will succeed at a given location is 0.7. Determine the odds in favor of the coffee shop succeeding.

18. *Raffle Tickets* One thousand raffle tickets are sold at $2 each. Three prizes of $200 and two prizes of $100 will be awarded. Assume that the probability that any given ticket is selected for a $200 prize is $\frac{3}{1000}$ and the probability that any given ticket is selected for a $100 prize is $\frac{2}{1000}$.

a) Determine the expectation of a person who purchases a ticket.

b) Determine the expectation of a person who purchases three tickets.

c) Determine the fair price to pay for a ticket.

19. *Expectation of a Card* If Cameron selects a picture card from a standard deck of 52 cards, Lindsey will give him $9. If Cameron does not select a picture card, he must give Lindsey $3.

a) Determine Cameron's expectation.

b) Determine Lindsey's expectation.

c) If Cameron plays this game 100 times, how much can he expect to lose or gain?

20. *Expected Attendance* If the day is sunny, 1000 people will attend the baseball game. If the day is cloudy, only 500 people will attend. If it rains, only 100 people will attend. The local meteorologist states that the probability of a sunny day is 0.4, of a cloudy day is 0.5, and of a rainy day is 0.1. Determine the number of people that are expected to attend.

21. *Club Officers* Tina, Jake, Gina, and Carla form a club. They plan to select a president and a vice president.

a) Construct a tree diagram showing all the possible outcomes.

b) List the sample space.

c) Determine the probability that Gina is selected president and Jake is selected vice president.

22. *A Coin and a Number* A coin is flipped and then a number from 1 through 4 is selected at random from a bag.

a) Construct a tree diagram showing all the possible outcomes.

b) List the sample space.

c) Determine the probability that a head is flipped and an odd number is selected.

d) Determine the probability that a head is flipped or an odd number is selected.

Spinning Two Wheels *In Exercises 23–28, the outer and inner wheels are spun.*

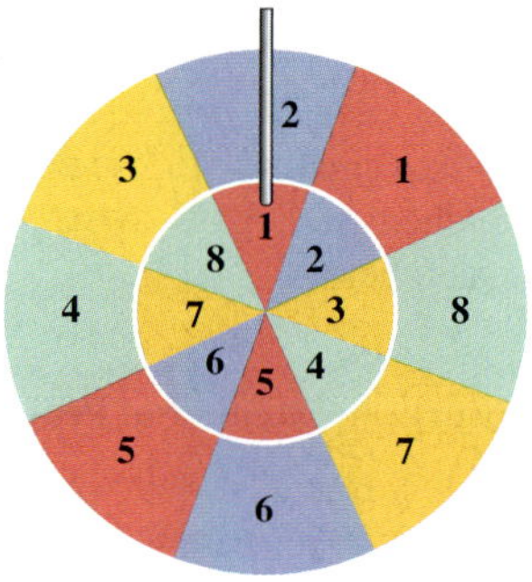

Assuming that the wheels are independent and the outcomes are equally likely, determine the probability of obtaining

23. even numbers on both wheels.

24. numbers greater than 5 on both wheels.

25. an odd number on the outer wheel and a number less than 6 on the inner wheel.

26. an even number or a number less than 6 on the outer wheel.

27. an even number or a color other than green on the inner wheel.

28. gold on the outer wheel and a color other than gold on the inner wheel.

Cereal Selection *In Exercises 29–32, assume that Mrs. Alexander is going to purchase 3 boxes of cereal at a convenience store. The store has 12 different boxes of cereal, including 5 different boxes made by General Mills, 4 different boxes made by Kellogg's, and 3 different boxes made by Post. If Mrs. Alexander selects 3 of these 12 boxes at random, determine the probability that she selects*

29. 3 boxes of General Mills.

30. no boxes of Kellogg's.

31. at least 1 box of Kellogg's.

32. boxes of General Mills, General Mills, and Post, in that order.

Spinner Probabilities *In Exercises 33–36, assume that the spinner cannot land on a line.*

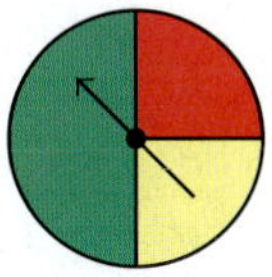

If spun once, determine

33. the probability that the spinner lands on yellow.

34. the odds against and the odds in favor of the spinner landing on yellow.

35. You are awarded $5 if the spinner lands on red, $10 if it lands on yellow, and $20 if it lands on green. Determine your expected value.

36. If the spinner is spun twice, determine the probability that it lands on red and then green (assume independence).

Spinner Probabilities *In Exercises 37–40, assume that the spinner cannot land on a line.*

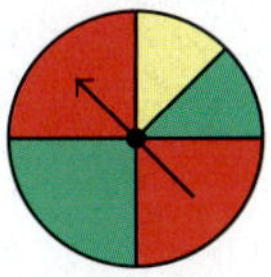

If spun once, determine

37. the probability that the spinner does not land on green.

38. the odds in favor of and the odds against the spinner landing on green.

39. A person wins $10 if the spinner lands on green, wins $5 if the spinner lands on red, and loses $20 if the spinner lands on yellow. Find the expectation of a person who plays this game.

40. If the spinner is spun three times, determine the probability that at least one spin lands on red.

Restaurant Service *In Exercises 41–44, use the results of a survey regarding the service at the Lobster House restaurant, which is summarized as follows.*

Meal	Service Rated Good	Service Rated Poor	Total
Lunch	65	10	75
Dinner	85	10	95
Total	150	20	170

If one person who completed the survey is selected at random, determine the probability that the person indicated that the

41. service was rated good.

42. service was rated good, given that the meal was dinner.

43. service was rated poor, given that the meal was lunch.

44. meal was dinner, given that the service was rated poor.

Neuroscience *In Exercises 45–48, assume that in a neuroscience course the students perform an experiment. Tests are given to determine if people are right brained, are left brained, or have no predominance. It is also recorded whether they are right handed or left handed. The following chart shows the results obtained.*

	Right Brained	Left Brained	No Predominance	Total
Right handed	40	130	60	230
Left handed	120	30	20	170
Total	160	160	80	400

If one person who completed the survey is selected at random, determine the probability the person selected is

45. right handed.

46. left brained, given that the person is left handed.

47. right handed, given that the person has no predominance.

48. right brained, given that the person is left handed.

49. ***Television Show*** Four contestants are on a television show. There are four different-colored rubber balls in a box, and each contestant gets to pick one from the box. Inside each ball is a slip of paper indicating the amount the contestant has won. The amounts are $10,000, $5000, $2000, and $1000.

a) In how many different ways can the contestants select the balls?

b) What is the expectation of a contestant?

50. *Spelling Bee* Five finalists remain in a high school spelling bee. Two will receive \$50 each, two will receive \$100 each, and one will receive \$500. How many different arrangements of prizes are possible?

51. *Candy Selection* Mrs. Williams takes her 3 children shopping. Each of her children gets to select a different type of candy that only that child will eat. At the store, there are only 10 boxes of candy left and each is a different type. In how many ways can the 3 children select the candy?

52. *Astronaut Selection* Three of nine astronauts must be selected for a mission. One will be the captain, one will be the navigator, and one will perform scientific experiments. Assuming each of the nine astronauts can perform any of the tasks, in how many ways can a three-person crew be selected so that each person has a different assignment?

53. *Medicine* Dr. Goldberg has three doses of serum for influenza type A. Six patients in the office require the serum. In how many different ways could Dr. Goldberg dispense the serum?

54. *Dogsled* Ten of 15 huskies are to be selected to pull a dogsled.

a) Assuming each of the nine astronauts can perform any of the tasks, in how many ways can this selection be made?

b) How many different arrangements of the 10 huskies on a dogsled are possible?

55. *Mega Millions* The Big Game Mega Millions is a multi-state lottery game offered in California, Georgia, Illinois, Maryland, Massachusetts, Michigan, New Jersey, New York, Ohio, Texas, Virginia, and Washington. To play, you select 5 numbers from 1 through 56 and 1 Big Money Ball number from 1 through 46. If you win the Big Game by matching all 6 numbers, your guaranteed minimum payoff is \$10 million. If you match the 5 numbers but do not match the Big Money number, your guaranteed minimum payoff is \$250,000.

a) What is the probability that you match the 5 numbers?

b) What is the probability that you have a Big Game win?

56. *Parent-Teacher Committee* A committee of 7 is to be formed from 10 parents and 12 teachers. If the committee is to consist of 2 parents and 5 teachers, how many different committees are possible?

57. *Selecting Test Subjects* In a psychology research laboratory, one room contains eight men and another room contains five women. Three men and two women are to be selected at random to be given a psychological test. How many different combinations of these people are possible?

58. *Choosing Two Aces* Two cards are selected at random, without replacement, from a standard deck of 52 cards. Determine the probability that two aces are selected (use combinations).

Color Chips In Exercises 59–62, a bag contains five red chips, three white chips, and two blue chips. Three chips are to be selected at random, without replacement. Determine the probability that

59. all are red.

60. the first two are red and the third is blue.

61. the first is red, the second is white, and the third is blue.

62. at least one is red.

Magazines *In Exercises 63–66, on a table in a doctor's office are six* Newsweek *magazines, five* U.S. News and World Report *magazines, and three* Time *magazines. If Ramona Cleary randomly selects three magazines, determine the probability that*

63. three *U.S. News and World Report* magazines were selected.

64. two *Newsweek* magazines and one *Time* magazine were selected.

65. no *Newsweek* magazines were selected.

66. at least one *Newsweek* magazine was selected.

67. ***New Homes*** In the community of Spring Hill, 60% of the homes purchased cost more than \$160,000.

a) Write the binomial probability formula to determine the probability that exactly x of the next n homes purchased in Spring Hill cost more than \$160,000.

b) Write the binomial probability formula to determine the probability that exactly 75 of the next 100 home purchases cost more than \$160,000.

68. ***Long-Stemmed Roses*** At the Floyd's Flower Shop, $\frac{1}{5}$ of people ordering flowers select long-stemmed roses. Determine the probability that exactly 3 of the next 5 customers ordering flowers select long-stemmed roses.

69. ***Taking a Math Course*** During any semester at City College, 60% of the students are taking a mathematics course. Determine the probability that of five students selected at random,

a) none is taking a mathematics course this semester.

b) at least one is taking a mathematics course this semester.

CHAPTER 12 TEST

1. ***Deep-Sea Fishing*** Of 40 people who went deep-sea fishing off the coast of Miami, 22 were fishing for tuna. Determine the empirical probability that the next person who goes deep-sea fishing off the coast of Miami will be fishing for tuna.

One Sheet of Paper *In Exercises 2–5, each of the numbers 1–9 is written on a sheet of paper and the nine sheets of paper are placed in a hat. If one sheet of paper is selected at random from the hat, determine the probability that the number selected is*

2. greater than 7.

3. odd.

4. even or greater than 4.

5. odd and greater than 4.

Two Sheets of Paper *In Exercises 6–9, if two of the same nine sheets of paper mentioned at the left are selected, without replacement, from the hat, determine the probability that*

6. both numbers are greater than 5.

7. both numbers are even.

8. the first number is odd and the second number is even.

9. neither of the numbers is greater than 6.

DID YOU KNOW?

Tune in Tomorrow

The A. C. Nielsen Company, which has been measuring the viewing population of TV shows for more than 50 years, uses a sample of about 25,000 metered households in the United States to draw conclusions about more than 110 million American households that have a television. An electronic measurement system, called the People Meter, is placed on each TV in the sample household. Nielsen uses the People Meter to measure what program is being tuned in and who is watching. Each household member is assigned a personal viewing button on the People Meter to keep track of which channels he or she watches and for how long. Nielsen then computes the rating of the show, using the data obtained from the sample. The People Meter is used to collect audience information for broadcast and cable networks, nationally syndicated programs, Spanish-language networks, and satellite distributors.

You can learn more about the sampling techniques used by A. C. Nielsen at Nielsenmedia.com.

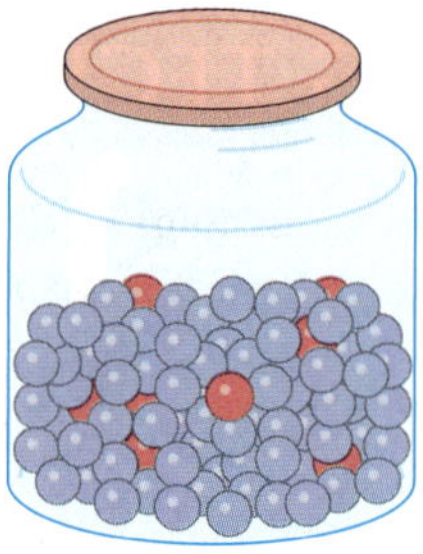

Figure 13.1

wrongly conclude that the jar contains all blue marbles. If the statistician takes a larger sample, say, 15 marbles, he or she is likely to select some red marbles. At that point, the statistician may make a prediction about the contents of the jar based on the sample selected. Of course, the most accurate result would occur if every object in the jar, the entire population, were observed. However, in most statistical experiments, observing the entire population is not practical.

Statisticians use samples instead of the entire population for two reasons: (a) it is often impossible to obtain data on an entire population, and (b) sampling is less expensive because collecting the data takes less time and effort. For example, suppose that you wanted to determine the number of each species of all the fish in a lake. To do so would be almost impossible without using a sample. If you did try to obtain this information from the entire population, the cost would be astronomical. Or suppose that you wanted to test soup cans for spoilage. If every can produced by the company was opened and tested, the company wouldn't have any product left to sell. Instead of testing the entire population of soup cans, a sample is selected. The results obtained from the sample of soup cans selected are used to make conclusions about the entire population of soup cans.

Later in this chapter we will discuss statistical measures such as the *mean* and the *standard deviation*. When statisticians calculate the mean and the standard deviation of the entire population, they use different symbols and formulas than when they calculate the mean and standard deviation of a sample. The following chart shows the symbols used to represent the mean and standard deviation of a sample and of a population. Note that the mean and standard deviation of a population are symbolized by Greek letters.

Measure	Sample	Population
Mean	$\bar{x}$ (read "x bar")	μ (mu)
Standard deviation	s	σ (sigma)

Unless otherwise indicated, in this book we will always assume that we are working with a sample and so we will use $\bar{x}$ *and* s. If you take a course in statistics, you will use all four symbols and different formulas for a sample and for a population.

Consider the task of determining the political strength of a certain candidate running in a national election. It is not possible for pollsters to ask each of the approximately 207 million eligible voters his or her preference of a candidate. Thus, pollsters must select and use a sample of the population to obtain their information.

> **DID YOU KNOW?**
>
> ***The Birth of Inferential Statistics***

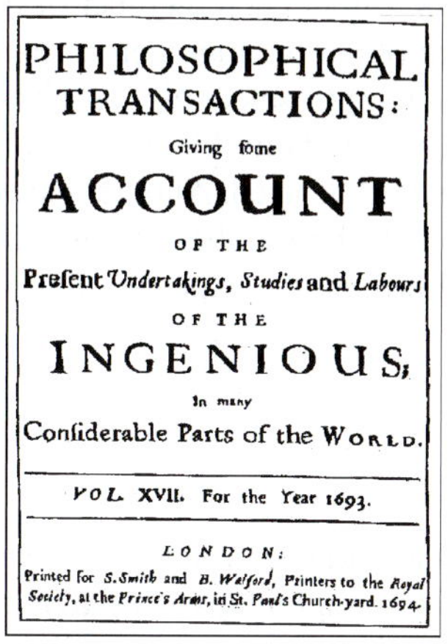

PHILOSOPHICAL TRANSACTIONS:

Giving fome

ACCOUNT

OF THE

Prefent *Undertakings, Studies* and *Labours*

OF THE

INGENIOUS,

In many

Confiderable Parts of the WORLD.

VOL. XVII. For the Year 1693.

LONDON:

Printed for *S. Smith* and *B. Walford*, Printers to the *Royal Society*, at the *Prince's Arms*, in St. *Paul's* Church-yard. 1694.

> John Gaunt, a London merchant, is credited with being the first person to make statistical predictions, or inferences, from a set of data rather than basing the predictions simply on the laws of chance. He studied the vital statistics (births, deaths, marriages) contained in the Bills of Mortality published during the years of the Great Plague. He observed that more males were born than females and that women lived longer than men. From these observations, he made predictions about life expectancies. The keeping of mortality statistics was stimulated considerably by the growth of the insurance industry.

How large a sample do you think they use to make predictions about an upcoming national election? You might be surprised to learn that pollsters use only about 1600 registered voters in their national sample. How can a pollster using such a small percentage of the population make an accurate prediction?

The answer is that when pollsters select a sample, they use sophisticated statistical techniques to obtain an unbiased sample. An ***unbiased sample*** is one that is a small replica of the entire population with regard to income, education, gender, race, religion, political affiliation, age, and so on. The procedures statisticians use to obtain unbiased samples are quite complex. The following sampling techniques will give you a brief idea of how statisticians obtain unbiased samples.

Random Sampling

If a sample is drawn in such a way that each time an item is selected each item in the population has an equal chance of being drawn, the sample is said to be a ***random sample***. When using a random sample, one combination of a specified number of items has the same probability of being selected as any other combination. When all the items in the population are similar with regard to the specific characteristic we are interested in, a random sample can be expected to produce satisfactory results. For example, consider a large container holding 300 tennis balls that are identical except for color. One-third of the balls are yellow, one-third are white, and one-third are green. If the balls can be thoroughly mixed between each draw of a tennis ball so that each ball has an equally likely chance of being selected, randomness is not difficult to achieve. However, if the objects or items are not all the same size, shape, or texture, it might be impossible to obtain a random sample by reaching into a container and selecting an object.

The best procedure for selecting a random sample is to use a random number generator or a table of random numbers. A random number generator is a device, usually a calculator or computer program, that produces a list of random numbers. A random number table is a collection of random digits in which each digit has an equal chance of appearing. To select a random sample, first assign a number to each element in the population. Numbers are usually assigned in order. Then select the number of random numbers needed, which is determined by the sample size. Each numbered element from the population that corresponds to a selected random number becomes part of the sample.

Systematic Sampling

When a sample is obtained by drawing every *n*th item on a list or production line, the sample is a ***systematic sample***. The first item should be determined by using a random number.

It is important that the list from which a systematic sample is chosen include the entire population being studied. See the Did You Know? called "Don't Count Your Votes Until They're Cast" on page 836. Another problem that must be avoided when this method of sampling is used is the constantly recurring characteristic. For example, on an assembly line, every 10th item could be the work of robot X. If only every 10th item is checked for defects, the work of other robots doing the same job may not be checked and may be defective.

PRACTICE THE SKILLS

Sampling Techniques *In Exercises 15–24, identify the sampling technique used to obtain a sample. Explain your answer.*

15. Faculty members at Cayuga Community College are classified according to the department in which they teach, and then random samples from each department are taken.

16. Every 10th iPod coming off an assembly line is checked for defects.

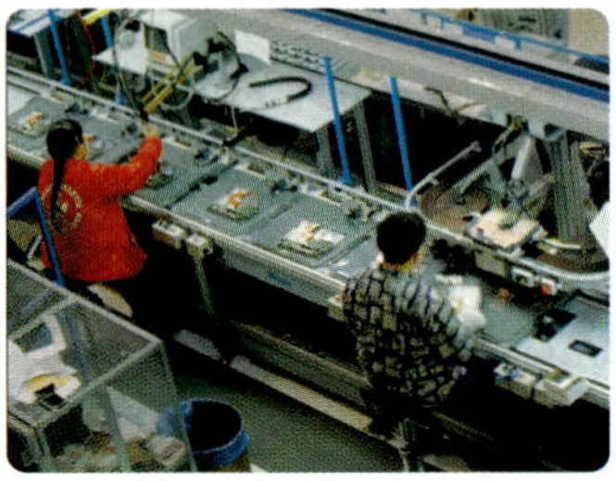

17. A state is divided into counties. A random sample of 12 counties is selected. A random sample from each of the 12 selected counties is selected.

18. A door prize is given away at a home improvement seminar. Tickets are placed in a bin, and the tickets are mixed up. Then a ticket is selected by a blindfolded person.

19. Every 17th person in line at a grocery store is asked his or her age.

20. The businesses in Iowa City are grouped according to type: medical, service, retail, manufacturing, financial, construction, restaurant, hotel, tourism, and other. A random sample of 10 businesses from each type is selected.

21. The first 25 students leaving the cafeteria are asked how many hours per week they work.

22. The Food and Drug Administration randomly selects five stores from each of four randomly selected sections of a large city and checks food items for freshness. These stores are used as a representative sample of the entire city.

23. Bingo balls in a bin are shaken, and then balls are selected from the bin.

24. The Student Senate at the University of North Carolina is electing a new president. The first 25 people leaving the library are asked for whom they will vote.

CHALLENGE PROBLEMS/GROUP ACTIVITIES

25. a) ***Random Sampling*** Select a topic and population of interest to which a random sampling technique can be applied to obtain data.

    b) Explain how you or your group would obtain a random sample for your population of interest.

    c) Actually obtain the sample by the procedure stated in part (b).

26. ***Data from Questionnaire*** Some subscribers of *Consumer Reports* respond to an annual questionnaire regarding their satisfaction with new appliances, cars, and other items. The information obtained from these questionnaires is then used as a sample from which frequency of repairs and other ratings are made by the magazine. Are the data obtained from these returned questionnaires representative of the entire population, or are they biased? Explain your answer.

RECREATIONAL MATHEMATICS

27. Statistically speaking, what is the most dangerous job in the United States?

28. Refer to the Did You Know? on page 775. Select a random sample of 30 people and see how many of the 30 people have the same birthday. (*Hint:* The probability of at least 2 of the 30 sharing the same birthday is greater than 0.5).

INTERNET/RESEARCH ACTIVITY

29. We have briefly introduced sampling techniques. Using statistics books and Internet websites as references, select one type of sampling technique (it may be one that we have not discussed in this section) and write a report on how statisticians obtain that type of sample. Also indicate when that type of sampling technique may be preferred. List two examples of when the sampling technique may be used.

13.2 THE MISUSES OF STATISTICS

▲ It is important to examine statistical statements about products before accepting the statements as fact.

Many of us may have seen an advertisement stating, "Four out of five dentists recommend sugarless gum for their patients who chew gum." Seeing an advertisement like this one may cause some of us to be a bit skeptical. Should we believe that sugarless chewing gum will not harm our teeth? Or should we investigate a bit further before buying that next pack of chewing gum? In this section, we will learn how to examine statistical statements before accepting them as fact.

Statistics, when used properly, is a valuable tool to society. However, many individuals, businesses, and advertising firms misuse statistics to their own advantage. You should examine statistical statements very carefully before accepting them as fact. You should ask yourself two questions: Was the sample used to gather the statistical data unbiased and of sufficient size? Is the statistical statement ambiguous; that is, can it be interpreted in more than one way?

Let's examine two advertisements. "Four out of five dentists recommend sugarless gum for their patients who chew gum." In this advertisement, we do not know the sample size and the number of times the experiment was performed to obtain the desired results. The advertisement does not mention that possibly only 1 out of 100 dentists recommended gum at all.

In a golf ball commercial, a "type A" ball is hit and a second ball is hit in the same manner. The type A ball travels farther. We are supposed to conclude that the type A is the better ball. The advertisement does not mention the number of times the experiment was previously performed or the results of the earlier experiments. Possible sources of bias include (1) wind speed and direction, (2) that no two swings are identical, and (3) that the ball may land on a rough or smooth surface.

Vague or ambiguous words also lead to statistical misuses or misinterpretations. The word *average* is one such culprit. There are at least four different "averages," some of which are discussed in Section 13.5. Each is calculated differently, and each may have a different value for the same sample. During contract negotiations, it is not uncommon for an employer to state publicly that the average salary of its employees is $45,000, whereas the employees' union states that the average is $40,000. Who is lying? Actually, both sides may be telling the truth. Each side will use the average that best suits its needs to present its case. Advertisers also use the average that most enhances their products. Consumers often misinterpret this average as the one with which they are most familiar.

Another vague word is *largest*. For example, ABC claims that it is the largest department store in the United States. Does that mean largest profit, largest sales, largest building, largest staff, largest acreage, or largest number of outlets?

Still another deceptive technique used in advertising is to state a claim from which the public may draw irrelevant conclusions. For example, a disinfectant manufacturer claims that its product killed 40,760 germs in a laboratory in 5 seconds. "To prevent colds, use disinfectant A." It may well be that the germs killed in the laboratory were not related to any type of cold germ. In another example, company C claims that its paper towels are heavier than its competition's towels. Therefore, they will hold more water. Is weight a measure of absorbency? A rock is heavier than a sponge, yet a sponge is more absorbent.

7. Morgan's is the largest department store in New York. So shop at Morgan's and save money.

8. Eighty percent of all automobile accidents occur within 10 miles of the driver's home. Therefore, it is safer to take long trips.

9. Arizona has the highest death rate for asthma in the United States. Therefore, it is unsafe to go to Arizona if you have asthma.

10. Thirty students said that they would recommend Professor Malone to a friend. Twenty students said that they would recommend Professor Wagner to a friend. Therefore, Professor Malone is a better teacher than Professor Wagner.

11. A steak is more expensive at Dino's Steak House than at Rick's Prime Rib House. Therefore, the quality of a steak at Dino's Steak House is better than the quality of a steak at Rick's Prime Rib House.

12. John Deere lawn tractors cost more than Toro lawn tractors. Therefore, John Deere lawn tractors will last longer than Toro lawn tractors.

13. The average depth of the pond is only 3 ft, so it is safe to go wading.

14. More men than women are involved in automobile accidents. Therefore, women are better drivers.

15. At West High School, half the students are below average in mathematics. Therefore, the school should receive more federal aid to raise student scores.

16. In 2007, more men than women applied for sales positions at Dick's Sporting Goods. Therefore, in 2007, more men than women were hired for sales positions at Dick's Sporting Goods.

17. ***Number of Males to Females*** The following table shows the number of males per 100 females in the U.S. for selected years.

Year	Number of males per 100 females
1980	94.5
1990	95.1
2000	96.3
2004	96.9

Source: U.S. Census Bureau

Draw a line graph that makes the increase in the number of males per 100 females for the years shown appear to be

a) small. **b)** large.

18. ***Four or More Years of College*** The following table shows the percent of the U.S. population with a bachelor's degree or higher for selected years.

Year	Percent
1998	24.4
1999	25.2
2000	25.6
2001	26.1
2002	26.7
2003	27.2
2004	27.7

Source: U.S. Department of Education

Draw a line graph that makes the increase in the percent of U.S. population with a bachelor's degree or higher appear to be

a) small. **b)** large.

First Marriage *In Exercises 19 and 20, use the following table.*

Median Age at First Marriage

Male		Female	
Year	**Age**	**Year**	**Age**
1998	26.7	1998	25.0
2000	26.8	2000	25.1
2002	26.9	2002	25.3
2004	27.1	2004	25.8

Source: U.S. Census Bureau

19. **a)** Draw a bar graph that appears to show a small increase in the median age at first marriage for males.

b) Draw a bar graph that appears to show a large increase in the median age at first marriage for males.

20. **a)** Draw a bar graph that appears to show a small increase in the median age at first marriage for females.

b) Draw a bar graph that appears to show a large increase in the median age at first marriage for females.

21. ***Online Purchasing*** The graph on the top of page 843 shows the percent of males and the percent of females surveyed who purchased clothing accessories online during the months from November 2006 to January 2007.

a) Draw a bar graph that shows the entire scale from 0 to 10.

b) Does the new graph give a different impression? Explain.

13.2 THE MISUSES OF STATISTICS

▲ It is important to examine statistical statements about products before accepting the statements as fact.

Many of us may have seen an advertisement stating, "Four out of five dentists recommend sugarless gum for their patients who chew gum." Seeing an advertisement like this one may cause some of us to be a bit skeptical. Should we believe that sugarless chewing gum will not harm our teeth? Or should we investigate a bit further before buying that next pack of chewing gum? In this section, we will learn how to examine statistical statements before accepting them as fact.

Statistics, when used properly, is a valuable tool to society. However, many individuals, businesses, and advertising firms misuse statistics to their own advantage. You should examine statistical statements very carefully before accepting them as fact. You should ask yourself two questions: Was the sample used to gather the statistical data unbiased and of sufficient size? Is the statistical statement ambiguous; that is, can it be interpreted in more than one way?

Let's examine two advertisements. "Four out of five dentists recommend sugarless gum for their patients who chew gum." In this advertisement, we do not know the sample size and the number of times the experiment was performed to obtain the desired results. The advertisement does not mention that possibly only 1 out of 100 dentists recommended gum at all.

In a golf ball commercial, a "type A" ball is hit and a second ball is hit in the same manner. The type A ball travels farther. We are supposed to conclude that the type A is the better ball. The advertisement does not mention the number of times the experiment was previously performed or the results of the earlier experiments. Possible sources of bias include (1) wind speed and direction, (2) that no two swings are identical, and (3) that the ball may land on a rough or smooth surface.

Vague or ambiguous words also lead to statistical misuses or misinterpretations. The word *average* is one such culprit. There are at least four different "averages," some of which are discussed in Section 13.5. Each is calculated differently, and each may have a different value for the same sample. During contract negotiations, it is not uncommon for an employer to state publicly that the average salary of its employees is \$45,000, whereas the employees' union states that the average is \$40,000. Who is lying? Actually, both sides may be telling the truth. Each side will use the average that best suits its needs to present its case. Advertisers also use the average that most enhances their products. Consumers often misinterpret this average as the one with which they are most familiar.

Another vague word is *largest*. For example, ABC claims that it is the largest department store in the United States. Does that mean largest profit, largest sales, largest building, largest staff, largest acreage, or largest number of outlets?

Still another deceptive technique used in advertising is to state a claim from which the public may draw irrelevant conclusions. For example, a disinfectant manufacturer claims that its product killed 40,760 germs in a laboratory in 5 seconds. "To prevent colds, use disinfectant A." It may well be that the germs killed in the laboratory were not related to any type of cold germ. In another example, company C claims that its paper towels are heavier than its competition's towels. Therefore, they will hold more water. Is weight a measure of absorbency? A rock is heavier than a sponge, yet a sponge is more absorbent.

An insurance advertisement claims that in Duluth, Minnesota, 212 people switched to insurance company Z. One may conclude that this company is offering something special to attract these people. What may have been omitted from the advertisement is that 415 people in Duluth, Minnesota, dropped insurance company Z during the same period.

A foreign car manufacturer claims that 9 of every 10 of a popular-model car it sold in the United States during the previous 10 years were still on the road. From this statement, the public is to conclude that this foreign car is well manufactured and would last for many years. The commercial neglects to state that this model has been selling in the United States for only a few years. The manufacturer could just as well have stated that 9 of every 10 of these cars sold in the United States in the previous 100 years were still on the road.

Charts and graphs can also be misleading or deceptive. In Fig. 13.2, the two graphs show the performance of two stocks over a 6-month period. Based on the graphs, which stock would you purchase? Actually, the two graphs present identical information; the only difference is that the vertical scale of the graph for stock B has been exaggerated.

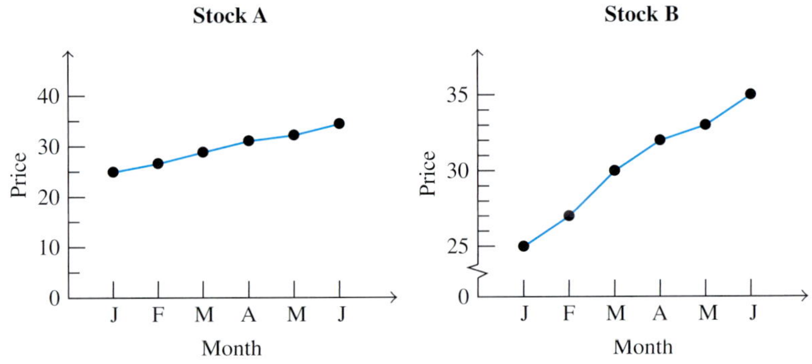

Figure 13.2

The two graphs in Fig. 13.3 show the same change. However, the graph in part (a) appears to show a greater increase than the graph in part (b), again because of a different scale.

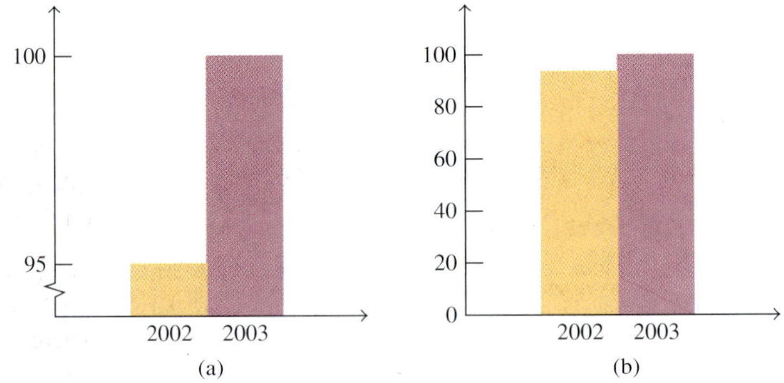

Figure 13.3

DID YOU KNOW?

Creative Displays

How Employers Make Workers Happy

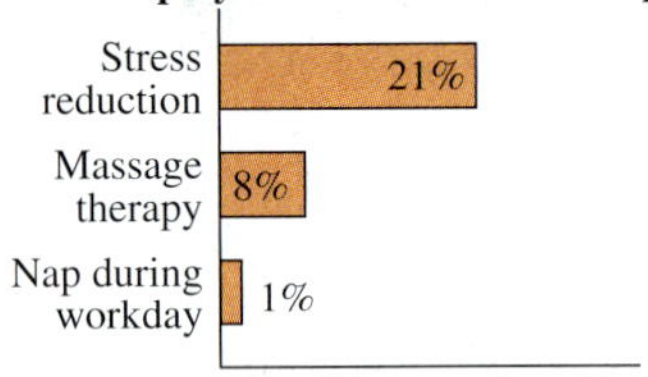

Visual graphics are often used to "dress up" what might otherwise be considered boring statistics. Although visually appealing, such creative displays of numerical data can be misleading. The graph above shows the percentage of employers that offer "perks" such as stress reduction, massage therapy, or a nap during the workday to make workers happy. This graph is misleading because the lengths of the bars are not proportional to one another as they should be to accurately reflect the percent of employers offering each of the named perks. For example, the bar for massage therapy should be eight times as long as the bar for nap during workday instead of being approximately four times as long, as the graph shows.

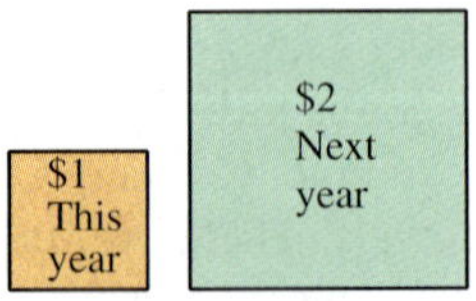

Figure 13.4

Consider a claim that if you invest $1, by next year you will have $2. This type of claim is sometimes misrepresented, as in Fig. 13.4. Actually, your investment has only doubled, but the area of the square on the right is four times that of the square on the left. By expressing the amounts as cubes (Fig. 13.5), you increase the volume eightfold.

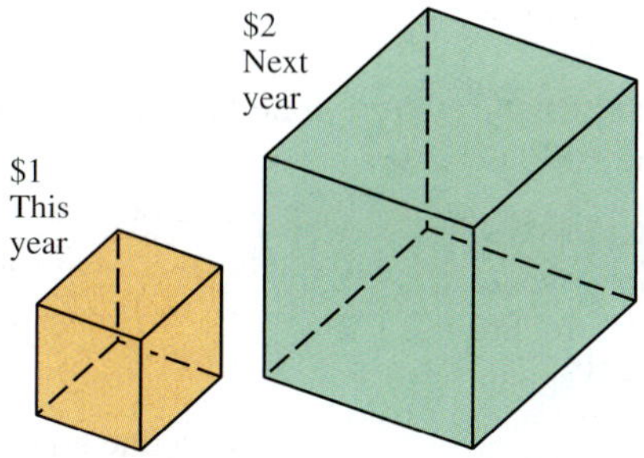

Figure 13.5

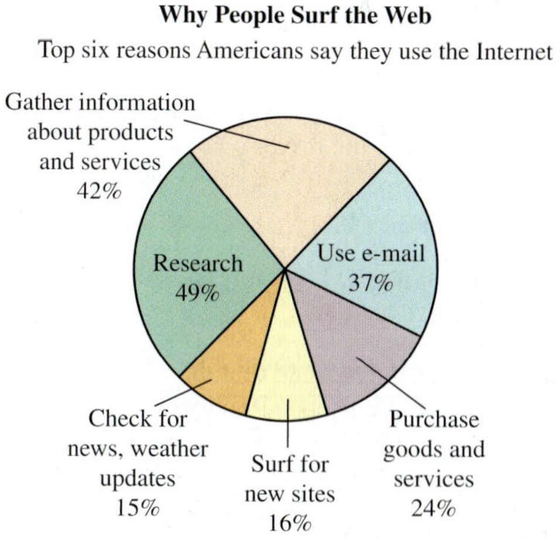

Figure 13.6

The graph in Fig. 13.6 is an example of a circle graph. We will discuss how to construct circle graphs in Section 13.4. In a circle graph, the total circle represents 100%. Therefore, the sum of the parts should add up to 100%. This graph is misleading since the sum of its parts is 183%. A graph other than a circle graph should have been used to display the top six reasons Americans say they use the Internet.

Despite the examples presented in this section, you should not be left with the impression that statistics is used solely for the purpose of misleading or cheating the consumer. As stated earlier, there are many important and necessary uses of statistics. Most statistical reports are accurate and useful. You should realize, however, the importance of being an aware consumer.

SECTION 13.2 EXERCISES

CONCEPT/WRITING EXERCISES

1. Find five advertisements or commercials that may be statistically misleading. Explain why each may be misleading.

2. The following circle graph appeared in a popular magazine in 2005. The graph shows various types of identity theft and the percent of victims of identity theft in 2004 that experienced these types of thefts. Is the graph misleading? Explain.

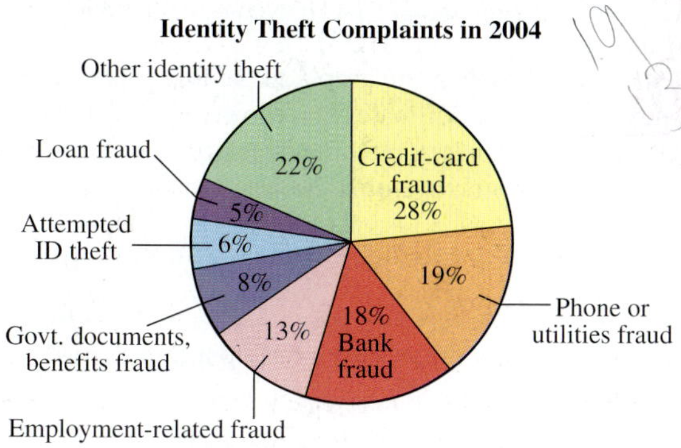

Source: Data from Federal Trade Commission. Image from *Newsweek*, July 4, 2005. © 2005.

PRACTICE THE SKILLS

Misinterpretations of Statistics *In Exercises 3–16, discuss the statement or graph and tell what possible misuses or misinterpretations may exist.*

3. In 2007, Liberty Travel received more requests for travel brochures to Hawaii than to Las Vegas. Therefore, in 2008, Liberty Travel sold more travel packages to Hawaii than to Las Vegas.

4. There are more empty spaces in the parking lot of Mama Mia's Italian restaurant than at Shanghi Chinese restaurant. Therefore, more people prefer Chinese food than Italian food.

5. Healthy Snacks cookies are fat free. So eat as many as you like and you will not gain weight.

6. Most accidents occur on Saturday night. That means that people do not drive carefully on Saturday night.

7. Morgan's is the largest department store in New York. So shop at Morgan's and save money.

8. Eighty percent of all automobile accidents occur within 10 miles of the driver's home. Therefore, it is safer to take long trips.

9. Arizona has the highest death rate for asthma in the United States. Therefore, it is unsafe to go to Arizona if you have asthma.

10. Thirty students said that they would recommend Professor Malone to a friend. Twenty students said that they would recommend Professor Wagner to a friend. Therefore, Professor Malone is a better teacher than Professor Wagner.

11. A steak is more expensive at Dino's Steak House than at Rick's Prime Rib House. Therefore, the quality of a steak at Dino's Steak House is better than the quality of a steak at Rick's Prime Rib House.

12. John Deere lawn tractors cost more than Toro lawn tractors. Therefore, John Deere lawn tractors will last longer than Toro lawn tractors.

13. The average depth of the pond is only 3 ft, so it is safe to go wading.

14. More men than women are involved in automobile accidents. Therefore, women are better drivers.

15. At West High School, half the students are below average in mathematics. Therefore, the school should receive more federal aid to raise student scores.

16. In 2007, more men than women applied for sales positions at Dick's Sporting Goods. Therefore, in 2007, more men than women were hired for sales positions at Dick's Sporting Goods.

17. ***Number of Males to Females*** The following table shows the number of males per 100 females in the U.S. for selected years.

Year	Number of males per 100 females
1980	94.5
1990	95.1
2000	96.3
2004	96.9

Source: U.S. Census Bureau

Draw a line graph that makes the increase in the number of males per 100 females for the years shown appear to be

a) small. **b)** large.

18. ***Four or More Years of College*** The following table shows the percent of the U.S. population with a bachelor's degree or higher for selected years.

Year	Percent
1998	24.4
1999	25.2
2000	25.6
2001	26.1
2002	26.7
2003	27.2
2004	27.7

Source: U.S. Department of Education

Draw a line graph that makes the increase in the percent of U.S. population with a bachelor's degree or higher appear to be

a) small. **b)** large.

First Marriage *In Exercises 19 and 20, use the following table.*

Median Age at First Marriage

Male		Female	
Year	**Age**	**Year**	**Age**
1998	26.7	1998	25.0
2000	26.8	2000	25.1
2002	26.9	2002	25.3
2004	27.1	2004	25.8

Source: U.S. Census Bureau

19. **a)** Draw a bar graph that appears to show a small increase in the median age at first marriage for males.

b) Draw a bar graph that appears to show a large increase in the median age at first marriage for males.

20. **a)** Draw a bar graph that appears to show a small increase in the median age at first marriage for females.

b) Draw a bar graph that appears to show a large increase in the median age at first marriage for females.

21. ***Online Purchasing*** The graph on the top of page 843 shows the percent of males and the percent of females surveyed who purchased clothing accessories online during the months from November 2006 to January 2007.

a) Draw a bar graph that shows the entire scale from 0 to 10.

b) Does the new graph give a different impression? Explain.

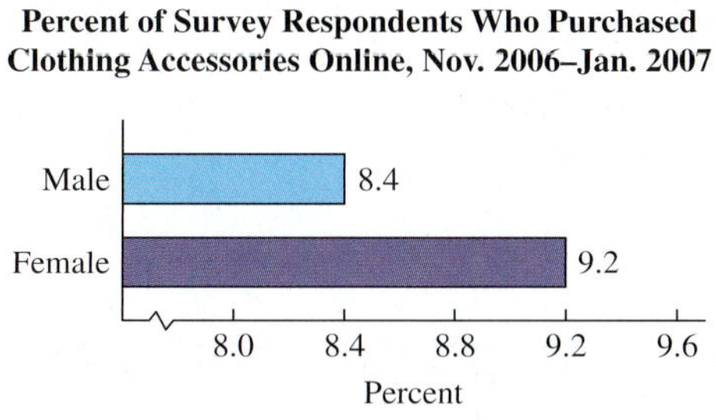

CHALLENGE PROBLEM/GROUP ACTIVITY

22. Consider the following graph, which shows the U.S. population in 2000 and the projected U.S. population in 2050.

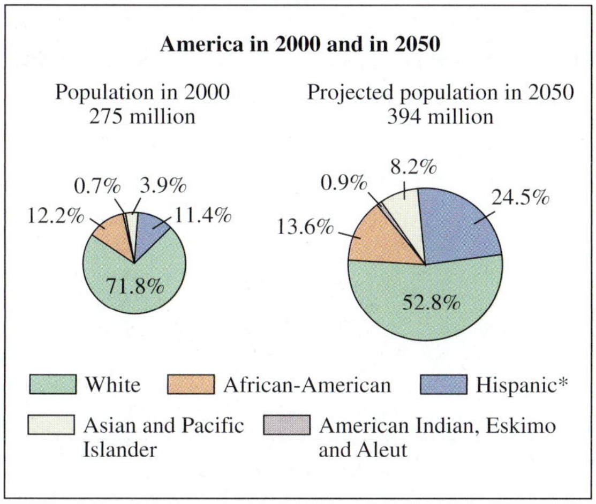

Source: U.S. Census Bureau

a) Compute the projected percent increase in population from 2000 to 2050 by using the formula given on page 649.

b) Measure the radius and then compute the area of the circle representing 2000. Use $A = \pi r^2$.

c) Repeat part (b) for the circle representing 2050.

d) Compute the percent increase in the size of the area of the circle from 2000 to 2050.

e) Are the circle graphs misleading? Explain your answer.

RECREATIONAL MATHEMATICS

23. What mathematical symbol can you place between 1 and 2 to obtain a number greater than 1 but less than 2?

INTERNET/RESEARCH ACTIVITY

24. Read the book *How to Lie with Statistics* by Darrell Huff and write a book report on it. Select three illustrations from the book that show how people manipulate statistics.

13.3 FREQUENCY DISTRIBUTIONS

▲ Organizing and summarizing data can make large sets of information more useful.

Suppose that you have a job waiting on tables in a restaurant. You decide to keep track of how much money you make each day from tips. After keeping a list of your tips each day for several months, you realize that you have a large set of data and want to condense your data into a more manageable form. In this section, we will learn a method to use to organize and summarize data.

It is not uncommon for statisticians and others to have to analyze thousands of pieces of data. A ***piece of data*** is a single response to an experiment. When the amount of data is large, it is usually advantageous to construct a frequency distribution. A ***frequency distribution*** is a listing of the observed values and the corresponding frequency of occurrence of each value. Example 1 shows how we construct a frequency distribution.

"Statistical thinking will one day be as necessary for efficient citizenship as the ability to read and write."

H. G. Wells

EXAMPLE 1 *Frequency Distribution*

The number of children per family is recorded for 64 families surveyed. Construct a frequency distribution of the following data:

0	1	1	2	2	3	4	5
0	1	1	2	2	3	4	5
0	1	1	2	2	3	4	6
0	1	2	2	2	3	4	6
0	1	2	2	2	3	4	7
0	1	2	2	3	3	4	8
0	1	2	2	3	3	5	8
0	1	2	2	3	3	5	9

SOLUTION Listing the number of children (observed values) and the number of families (frequency) gives the following frequency distribution.

Number of Children (Observed Values)	Number of Families (Frequency)
0	8
1	11
2	18
3	11
4	6
5	4
6	2
7	1
8	2
9	1
	64

Eight families had no children, 11 families had one child, 18 families had two children, and so on. Note that the sum of the frequencies is equal to the original number of pieces of data, 64.

Often data are grouped in classes to provide information about the distribution that would be difficult to observe if the data were ungrouped. Graphs called *histograms* and *frequency polygons* can be made of grouped data, as will be explained in Section 13.4. These graphs also provide a great deal of useful information.

When data are grouped in classes, certain rules should be followed.

RULES FOR DATA GROUPED BY CLASSES

1. The classes should be of the same "width."
2. The classes should not overlap.
3. Each piece of data should belong to only one class.

> **DID YOU KNOW?**
>
> ***Can You Count the F's?***
>
> Statistical errors often result from careless observations. To see how such errors can occur, consider the statement below. How many F's do you count in the statement? You can find the answer in the answer section (Section 13.3, Exercise 32).
>
> *FINISHED FILES ARE THE RESULT OF YEARS OF SCIENTIFIC STUDY COMBINED WITH THE EXPERIENCE OF YEARS.*

In addition, it is often suggested that a frequency distribution should be constructed with 5 to 12 classes. If there are too few or too many classes, the distribution may become difficult to interpret. For example, if you use fewer than 5 classes, you risk losing too much information. If you use more than 12 classes, you may gain more detail but you risk losing clarity. Let the spread of the data be a guide in deciding the number of classes to use.

To understand these rules, let's consider a set of observed values that go from a low of 0 to a high of 26. Let's assume that the first class is arbitrarily selected to go from 0 through 4. Thus, any of the data with values of 0, 1, 2, 3, 4 would belong in this class. We say that the ***class width*** is 5 since there are five integral values that belong to the class. This first class ended with 4, so the second class must start with 5. If this class is to have a width of 5, at what value must it end? The answer is 9 (5, 6, 7, 8, 9). The second class is 5–9. Continuing in the same manner, we obtain the following set of classes.

Lower class limits	*Classes*	Upper class limits
0	0–4	4
5	5–9	9
10	10–14	14
15	15–19	19
20	20–24	24
25	25–29	29

We need not go beyond the 25–29 class because the largest value we are considering is 26. The classes meet our three criteria: They have the same width, there is no overlap among the classes, and each of the values from a low of 0 to a high of 26 belongs to one and only one class.

The choice of the first class, 0–4, was arbitrary. If we wanted to have more classes or fewer classes, we would make the class widths smaller or larger, respectively.

The numbers 0, 5, 10, 15, 20, 25 are called the ***lower class limits***, and the numbers 4, 9, 14, 19, 24, 29 are called the ***upper class limits***. Each class has a width of 5. Note that the class width, 5, can be obtained by subtracting the first lower class limit from the second lower class limit: $5 - 0 = 5$. The difference between any two consecutive lower class or upper class limits is also 5.

EXAMPLE 2 *A Frequency Distribution of Consumer Magazines*

Table 13.1 on the next page shows the 2005 circulation for the 48 leading U.S. consumer magazines (excluding *AARP Magazine* and *AARP Bulletin*, which are far ahead of the other magazines). The circulation is rounded to the nearest ten thousand. Construct a frequency distribution of the data, letting the first class be 157–241.

SOLUTION Forty-eight pieces of data are given in *descending order* from highest to lowest. We are given that the first class is 157–241. The second class must therefore start at 242. To find the class width, we subtract 157 (the lower class limit of the first class) from 242 (the lower class limit of the second class) to obtain a class width of 85. The upper class limit of the second class is found by

Table 13.1

Magazine	Circulation (ten thousands)
Reader's Digest	1001
Better Homes & Gardens	761
TV Guide	735
National Geographic	538
Good Housekeeping	466
Family Circle	429
Ladies' Home Journal	411
Woman's Day	409
TIME	403
People	369
AAA Westways	368
Prevention	335
Sports Illustrated	324
Newsweek	312
Cosmopolitan	301
Playboy	301
Via Magazine	275
Southern Living	274
Guideposts	263
American Legion	253
Maxim	250
AAA Going Places	248
Redbook	243
O, the Oprah Magazine	240
Glamour	240
AAA Living	240
Parents	205
Smithsonian	205
U.S. News & World Report	203
Seventeen	203
Money	199
Martha Stewart Living	197
Parenting	193
ESPN the Magazine	189
Real Simple	186
Game Informer	183
Family Fun	180
Men's Health	178
In Style	177
Entertainment Weekly	176
Country Living	174
Cooking Light	172
Endless Vacation	172
VFW Magazine	169
US Weekly	166
Shape	165
Golf Digest	160
Home & Away	157

Source: Audit Bureau of Circulations

adding the class width, 85, to the upper class limit of the first class, 241. Therefore, the upper class limit of the second class is $241 + 85 = 326$. Thus,

157–241	first class
242–326	second class

The other classes are found using a similar technique. They are 327–411, 412–496, 497–581, 582–666, 667–751, 752–836, 837–921, 922–1006. Since the highest value in the data is 1001, there is no need to go any further. Note that each two consecutive lower class limits differ by 85, as does each two consecutive upper class limits. There are 25 pieces of data in the 157–241 class. There are 11 pieces of data in the 242–326 class, 6 in the 327–411 class, 2 in the 412–496 class, 1 in the 497–581 class, 0 in the 582–666 class, 1 in the 667–751 class, 1 in the 752–836 class, 0 in the 837–921 class, and 1 in the 922–1006 class. The complete frequency distribution of the 10 classes is given below. The number of magazines totals 48, so we have included each piece of data.

Circulation	Number of Magazines (ten thousands)
157–241	25
242–326	11
327–411	6
412–496	2
497–581	1
582–666	0
667–751	1
752–836	1
837–921	0
922–1006	1
	48

The ***modal class*** of a frequency distribution is the class with the greatest frequency. In Example 2, the modal class is 157–241. The ***midpoint of a class***, also called the ***class mark***, is found by adding the lower and upper class limits and dividing the sum by 2. The midpoint of the first class in Example 2 is

$$\frac{157 + 241}{2} = \frac{398}{2} = 199$$

Note that the difference between successive class marks is the class width. The class mark of the second class can therefore be obtained by adding the class width, 85, to the class mark of the first class, 199. The sum is $199 + 85 = 284$. Note that $\frac{242 + 326}{2} = 284$, which checks with the class mark obtained by adding the class width to the first class mark.

DID YOU KNOW?

Cyberspace Is the Place to Be

Seattle, Washington

Do you remember the days when you were only able to access the Internet from your office, school, or home? With wireless Internet access, we can now connect to the Internet to share information and enjoy entertainment while we are on the go. Today, wireless hotspots, locations offering wireless access to the Internet, show up in diverse places such as coffee shops, parks, gas stations, bowling alleys, airports, and golf courses in addition to more traditional places such as colleges and hotels. According to a survey conducted by Intel in April 2005, the Seattle-Bellevue-Everett-Tacoma area was the most accessible wireless Internet area in the United States. Many websites, such as www.wififreespot.com, list worldwide locations that offer free wireless Internet access.

EXAMPLE 3 *A Frequency Distribution of Family Income*

The following set of data represents the family income (in thousands of dollars, rounded to the nearest hundred) of 15 randomly selected families.

31.5	16.8	30.8	29.7	25.9
50.2	37.4	29.6	38.7	33.8
20.5	25.3	24.8	41.3	35.7

Construct a frequency distribution with a first class of 16.5–22.6.

SOLUTION First rearrange the data from lowest to highest so that the data will be easier to categorize.

16.8	25.3	29.7	33.8	38.7
20.5	25.9	30.8	35.7	41.3
24.8	29.6	31.5	37.4	50.2

The first class goes from 16.5 to 22.6. Since the data are in tenths, the class limits will also be given in tenths. The first class ends with 22.6; therefore, the second class must start with 22.7. The class width of the first class is $22.7 - 16.5$, or 6.2. The upper class limit of the second class must therefore be $22.6 + 6.2$, or 28.8. The frequency distribution is as follows.

Income ($1000)	Number of Families
16.5–22.6	2
22.7–28.8	3
28.9–35.0	5
35.1–41.2	3
41.3–47.4	1
47.5–53.6	1
	15

Note in Example 3 that the class width is 6.2, the modal class is $28.9 - 35.0$, and the class mark of the first class is $(16.5 + 22.6)/2$, or 19.55.

SECTION 13.3 EXERCISES

CONCEPT/WRITING EXERCISES

1. What is a frequency distribution?

2. How can a class width be determined using class limits?

3. Suppose that the first class of a frequency distribution is 9–15.

   a) What is the width of this class?

   b) What is the second class?

   c) What is the lower class limit of the second class?

   d) What is the upper class limit of the second class?

4. Repeat Exercise 3 for a frequency distribution whose first class is 12–20.

5. What is the modal class of a frequency distribution?

6. What is another name for the midpoint of a class? How is the midpoint of a class determined?

PRACTICE THE SKILLS/PROBLEM SOLVING

In Exercises 7 and 8, use the frequency distribution to determine

a) *the total number of observations.*
b) *the width of each class.*
c) *the midpoint of the second class.*
d) *the modal class (or classes).*
e) *the class limits of the next class if an additional class were to be added.*

7.

Class	Frequency
9–15	4
16–22	7
23–29	1
30–36	0
37–43	3
44–50	5

8.

Class	Frequency
40–49	7
50–59	5
60–69	3
70–79	2
80–89	7
90–99	1

9. ***Visits to the Library*** El Paso Community College is planning to expand its library. Forty students were asked how many times they visited the library during the previous semester. Their responses are given below. Construct a frequency distribution, letting each class have a width of 1 (as in Example 1).

0	1	1	3	4	5	7	8
0	1	2	3	5	5	7	8
0	1	2	3	5	5	7	9
1	1	2	3	5	6	8	10
1	1	3	4	5	6	8	10

10. ***Hot Dog Sales*** A hot dog vendor is interested in the number of hot dogs he sells each day at his hot dog cart. The number of hot dogs sold is indicated below for 32 consecutive days. Construct a frequency distribution, letting each class have a width of 1.

15	16	19	20	21	22	24	27
15	18	19	20	21	22	25	27
15	18	19	20	21	23	25	28
16	18	19	21	21	23	26	29

Note that there were no days in which the vendor sold 17 hot dogs. However, it is customary to include a missing value as an observed value and assign to it a frequency of 0.

▲ See Exercise 10

IQ Scores *In Exercises 11–14, use the following data, which show the result of 50 sixth-grade I.Q. scores.*

80	89	92	95	97	100	102	106	110	120
81	89	93	95	98	100	103	108	113	120
87	90	94	97	99	100	103	108	114	122
88	91	94	97	100	100	103	108	114	128
89	92	94	97	100	101	104	109	119	135

Use this data to construct a frequency distribution with a first class of

11. 78–86. **12.** 80–88.

13. 80–90. **14.** 80–92.

Placement Test Scores *In Exercises 15–18, use the following data, which represent the English placement test scores of a sample of 30 students.*

559	482	490	520	514
498	472	490	523	491
480	490	562	486	491
498	543	506	539	576
508	509	499	515	501
593	512	510	577	533

Use this data to construct a frequency distribution with a first class of

15. 472–492. **16.** 470–486.

17. 472–487. **18.** 472–496.

Magazine Circulation *In Exercises 19–22, use the data given in Table 13.1 on page 846 to construct a frequency distribution with a first class (in ten thousands) of*

19. 157–306. **20.** 150–296.

21. 157–256. **22.** 157–234.

City Population *In Exercises 23–26, use the following data, which represent the population of the 20 most populous cities in the world in 2006, in millions of people (rounded to the nearest 100,000).*

11.9	10.1	8.7	8.1	7.2
11.0	9.8	8.6	7.8	7.0
10.9	9.3	8.5	7.6	6.3
10.3	8.8	8.4	7.4	6.1

▲ Mumbai, India is the world's most populated city.

Use these data to construct a frequency distribution with a first class of

23. 6.0–6.9.

24. 5.5–6.4.

25. 5.5–6.5.

26. 6.0–6.4.

Residents in Poverty *In Exercises 27–30, use the data in the following table.*

Percent of U.S. Residents Living in Poverty in 2004, by State

State	Percent	State	Percent
AL	16.0	AR	16.4
AK	9.4	CA	13.2
AZ	13.9	CO	9.9
CT	9.1	NH	5.6
DE	8.2	NJ	8.3
FL	12.2	NM	17.3
GA	12.5	NY	14.6
HI	8.9	NC	15.1
ID	10.0	ND	9.7
IL	12.4	OH	11.3
IN	10.8	OK	11.8
IA	9.9	OR	12.1
KS	11.1	PA	10.9
KY	16.0	RI	11.5
LA	16.8	SC	13.8
ME	11.6	SD	13.0
MD	9.2	TN	15.0
MA	9.7	TX	16.7
MI	12.3	UT	9.5
MN	7.2	VT	8.2
MS	17.3	VA	9.7
MO	11.5	WA	12.0
MT	14.6	WV	15.8
NE	9.6	WI	11.0
NV	10.9	WY	9.9

Source: Bureau of the Census

Construct a frequency distribution with a first class of

27. 5.6–7.5.

28. 5.6–8.2.

29. 5.6–7.0.

30. 5.6–8.0.

RECREATIONAL MATHEMATICS

31. In what month do people take the least number of daily vitamins?

32. a) Count the number of F's in the sentence at the bottom of the Did You Know? on page 845.

b) Can you explain why so many people count the number of F's incorrectly?

13.4 STATISTICAL GRAPHS

▲ Graphs can be used to display numerical information.

For a class research project, you are required to collect information and write a report about the percent of college students who work in retail, service, or other areas. How could you display this information in a way that is helpful to the reader of your report? In this section, we will introduce four types of graphs that can be used to display information.

We will consider four types of graphs: the circle graph, the histogram, the frequency polygon, and the stem-and-leaf display.

Circle Graphs

Circle graphs (also known as pie charts) are often used to compare parts of one or more components of the whole to the whole. The circle graph in Fig. 13.7 shows what moviegoers say is the most annoying distraction during a movie. Since the total circle represents 100%, the sum of the percents of the sectors should be 100%, and it is.

In the next example, we will discuss how to construct a circle graph given a set of data.

Distractions at the Movies

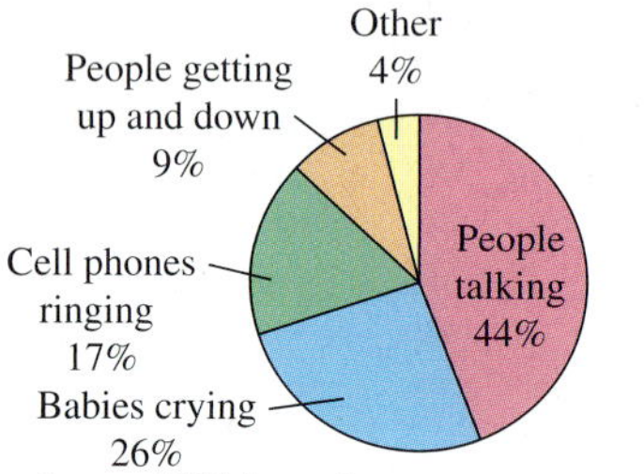

Source: AMC Entertainment

Figure 13.7

EXAMPLE 1 *Circus Performances*

Six hundred people who attended a Ringling Bros. and Barnum & Bailey circus were asked to indicate their favorite performance. The results are illustrated below.

Performance	Number of People
Tigers	230
Elephants	154
Acrobats	99
Jugglers	91
Other	26
	600

Use this information to construct a circle graph illustrating the percent of people whose favorite circus performance was tigers, elephants, acrobats, jugglers, and other performances.

SOLUTION Determine the measure of the central angle, as illustrated in the following table.

Performance	Number of People	Percent of Total (to the nearest tenth percent)	Measure of Central Angle
Tigers	230	$\frac{230}{600} \times 100 \approx 38.3\%$	$0.383 \times 360 = 137.9°$
Elephants	154	$\frac{154}{600} \times 100 \approx 25.7\%$	$0.257 \times 360 = 92.5°$
Acrobats	99	$\frac{99}{600} \times 100 \approx 16.5\%$	$0.165 \times 360 = 59.4°$
Jugglers	91	$\frac{91}{600} \times 100 \approx 15.2\%$	$0.152 \times 360 = 54.7°$
Other	26	$\frac{26}{600} \times 100 \approx 4.3\%$	$0.043 \times 360 = 15.5°$
Total	600	100.0%	360.0°

Now use a protractor (see Section 9.1, page 519) to construct a circle graph and label it properly, as illustrated in Fig. 13.8. The measure of the central angle for tigers is about 137.9°, for elephants it is about 92.5°, for acrobats it is about 59.4°, for jugglers it is about 54.7°, and for other performances it is about 15.5°.

Favorite Performance at the Circus

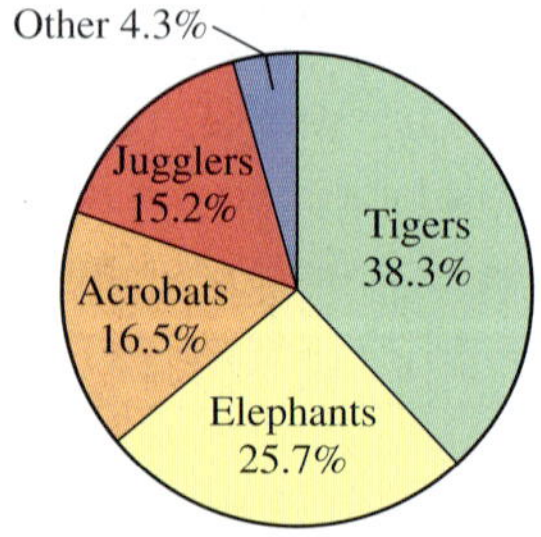

Figure 13.8

TIMELY TIP When computing values to construct a circle graph, if you round your values, the sum of the percents of the total may not exactly equal 100%, and the sum of the measures of the central angle may not exactly equal 360°.

Histograms and Frequency Polygons

Histograms and frequency polygons are statistical graphs used to illustrate frequency distributions. A ***histogram*** is a graph with observed values on its horizontal scale and frequencies on its vertical scale. A bar is constructed above each observed value (or class when classes are used), indicating the frequency of that value (or class). The horizontal scale need not start at zero, and the calibrations on the horizontal and vertical scales do not have to be the same. The vertical scale must start at zero. To accommodate large frequencies on the vertical scale, it may be necessary to break the scale. Because histograms and other bar graphs are easy to interpret visually, they are used a great deal in newspapers and magazines.

EXAMPLE 2 *Construct a Histogram*

The frequency distribution developed in Example 1, Section 13.3 on page 844, is repeated here. Construct a histogram of this frequency distribution.

Number of Children (Observed Values)	Number of Families (Frequency)
0	8
1	11
2	18
3	11
4	6
5	4
6	2
7	1
8	2
9	1

SOLUTION The vertical scale must extend at least to the number 18 since that is the greatest recorded frequency. The horizontal scale must include the numbers 0–9, the number of children observed. Eight families have no children. We indicate that by constructing a bar above the number 0, centered at 0, on the horizontal scale extended up to 8 on the vertical scale (see Fig. 13.9 on page 852). Eleven families have one child, so we construct a bar extending to 11 above the number 1, centered at 1, on the horizontal scale. We continue this procedure for each observed value. Both the horizontal and vertical scales should be labeled, the bars should be the same width and centered at the observed value, and the histogram should have a title. In a histogram, the bars should always touch.

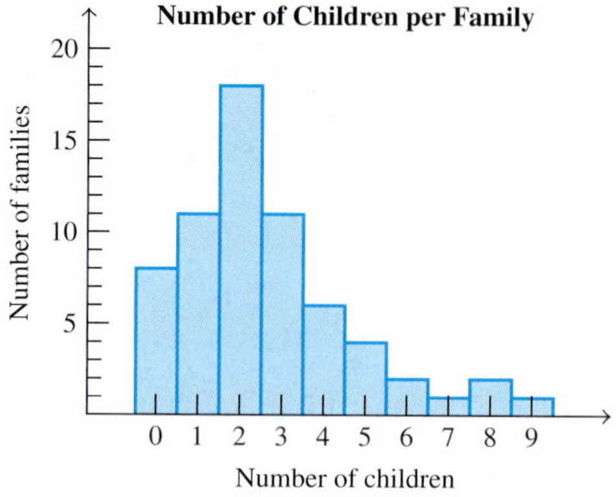

Figure 13.9

Frequency polygons are line graphs with scales the same as those of the histogram; that is, the horizontal scale indicates observed values and the vertical scale indicates frequency. To construct a frequency polygon, place a dot at the corresponding frequency above each of the observed values. Then connect the dots with straight-line segments. When constructing frequency polygons, always put in two additional class marks, one at the lower end and one at the upper end on the horizontal scale (values for these added class marks are not needed on the frequency polygon). Since the frequency at these added class marks is 0, the end points of the frequency polygon will always be on the horizontal scale.

EXAMPLE 3 *Construct a Frequency Polygon*

Construct a frequency polygon of the frequency distribution in Example 2.

SOLUTION Since eight families have no children, place a mark above the 0 at 8 on the vertical scale, as shown in Fig. 13.10. Because there are 11 families with one child, place a mark above the 1 on the horizontal scale at the 11 on the vertical scale, and so on. Connect the dots with straight-line segments and bring the end points of the graph down to the horizontal scale, as shown.

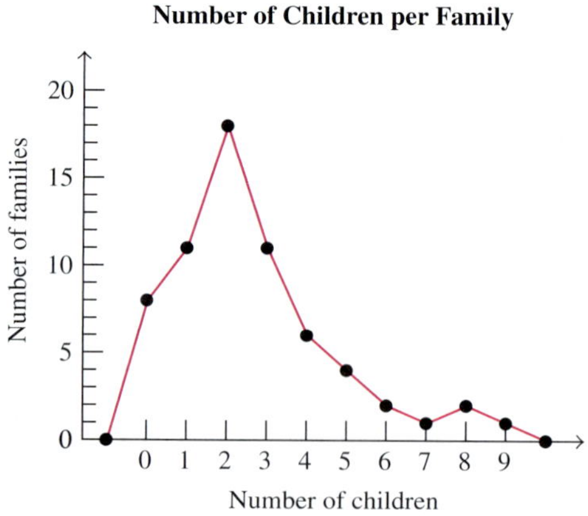

Figure 13.10

TIMELY TIP When constructing a histogram or frequency polygon, be sure to label both scales of the graph.

EXAMPLE 4 *Commuting Distances*

The frequency distribution of the one-way commuting distances for 70 workers is listed in Table 13.2. Construct a histogram and then construct a frequency polygon.

Table 13.2

Distance (miles)	Number of Workers
1–7	15
8–14	24
15–21	13
22–28	8
29–35	5
36–42	5

SOLUTION The histogram can be constructed with either class limits or class marks (class midpoints) on the horizontal scale. Frequency polygons are constructed with class marks on the horizontal scale. Since we will construct a frequency polygon on the histogram, we will use class marks. Recall that class marks are found by adding the lower class limit and upper class limit and dividing the sum by 2. For the first class, the class mark is $\frac{1+7}{2}$, or 4. Since the class widths are seven units, the class marks will also differ by seven units (see Fig. 13.11).

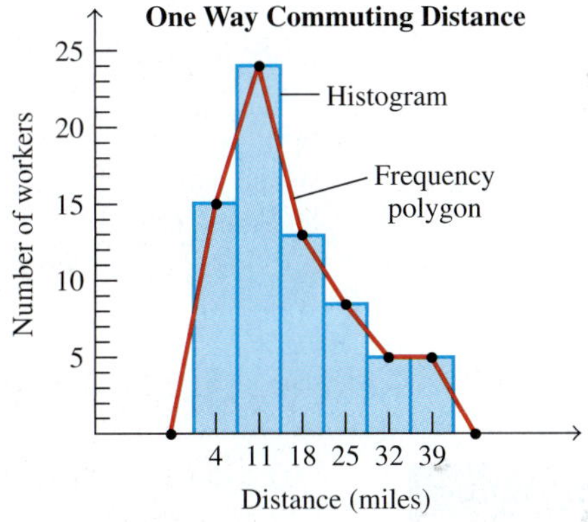

Figure 13.11

EXAMPLE 5 *Carry-on Luggage Weights*

The histogram in Fig. 13.12 shows the weights of selected pieces of carry-on luggage at an airport. Construct the frequency distribution from the histogram in Fig. 13.12.

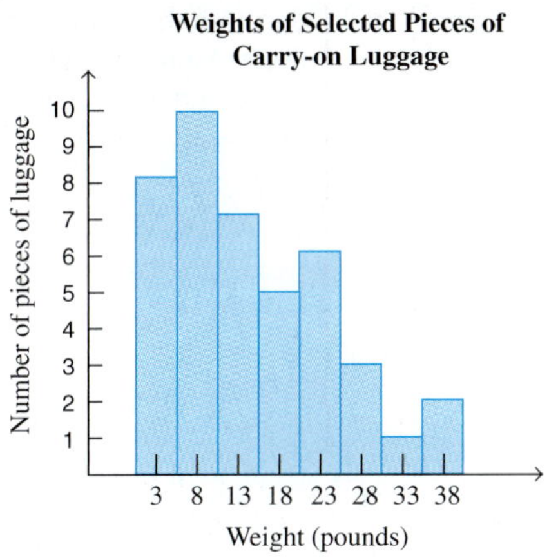

Figure 13.12

Table 13.3

Weight (pounds)	Number of Pieces of Luggage
1–5	8
6–10	10
11–15	7
16–20	5
21–25	6
26–30	3
31–35	1
36–40	2

SOLUTION There are five units between class midpoints, so each class width must also be five units. Since three is the midpoint of the first class, there must be two units below and two units above it. The first class must be 1–5. The second class must therefore be 6–10. The frequency distribution is given in Table 13.3. •

Stem-and-Leaf Displays

Frequency distributions and histograms provide very useful tools to organize and summarize data. However, if the data are grouped, we cannot identify specific data values in a frequency distribution and in a histogram. For example, in Example 5, we know that there are eight pieces of luggage in the class of 1 to 5 pounds, but we don't know the specific weights of those eight pieces of luggage.

A *stem-and-leaf display* is a tool that organizes and groups the data while allowing us to see the actual values that make up the data. To construct a stem-and-leaf display each value is represented with two different groups of digits. The left group of digits is called the *stem*. The remaining group of digits on the right is called the *leaf*. There is no rule for the number of digits to be included in the stem. Usually the units digit is the leaf and the remaining digits are the stem. For example, the number 53 would be broken up into 5 and 3. The 5 would be the stem and the 3 would be the leaf. The number 417 would be broken up into 41 and 7. The 41 would be the stem and the 7 would be the leaf. The number 6, which can be represented as 06, would be broken up into 0 and 6. The stem would be the 0 and the leaf would be the 6. With a stem-and-leaf display, the stems are listed, in ascending order, to the left of a vertical line. Then we place each leaf to the right of its corresponding stem, to the right of the vertical line.* Example 6 illustrates this procedure.

EXAMPLE 6 *Stem-and-Leaf Display*

The table below indicates the ages of a sample of 20 guests who stayed at Captain Fairfield House Bed and Breakfast. Construct a stem-and-leaf display.

29	31	39	43	56
60	62	59	58	32
47	27	50	28	71
72	44	45	44	68

SOLUTION By quickly glancing at the data, we can see the ages consist of two-digit numbers. Let's use the first digit, the tens digit, as our stem and the second digit, the units digit, as the leaf. For example, for an age of 62, the stem is 6 and the leaf is 2. Our values are numbers in the 20s, 30s, 40s, 50s, 60s, and 70s. Therefore, the stems will be 2, 3, 4, 5, 6, 7 as shown below.

2 |
3 |
4 |
5 |
6 |
7 |

*In stem-and-leaf displays, the leaves are sometimes listed from lowest digit to greatest digit, but that is not necessary.

Next we place each leaf on its stem. We will do so by placing the second digit of each value next to its stem, to the right of the vertical line. Our first value is 29. The 2 is the stem and the 9 is the leaf. Therefore, we place a 9 next to the stem of 2 and to the right of the vertical line.

2 | 9

The next value is 31. We will place a leaf of 1 next to the stem of 3.

2 | 9
3 | 1

The next value is 39. Therefore, we will place a leaf of 9 after the leaf of 1 that is next to the stem of 3.

2 | 9
3 | 1 9

We continue this process until we have listed all the leaves on the display. The diagram below shows the stem-and-leaf display for the ages of the guests. In our display, we will also include a legend to indicate the values represented by the stems and leaves. For example, 5 | 6 represents 56.

5 | 6 represents 56

Stem	Leaves
2	9 7 8
3	1 9 2
4	3 7 4 5 4
5	6 9 8 0
6	0 2 8
7	1 2

Every piece of the original data can be seen in a stem-and-leaf display. From the above diagram, we can see that five of the guests' ages were in the 40s. Only two guests were older than 70. Note that the stem-and-leaf display gives the same visual impression as a sideways histogram.

SECTION 13.4 EXERCISES

CONCEPT/WRITING EXERCISES

1. In your own words, explain how to construct a circle graph from a table of values.

2. a) What is listed on the horizontal axis of a histogram and frequency polygon?

   b) What is listed on the vertical axis of a histogram and frequency polygon?

3. In your own words, explain how to construct a frequency polygon from a set of data.

4. In your own words, explain how to construct a histogram from a set of data.

5. a) In your own words, explain how to construct a frequency polygon from a histogram.

   b) Construct a frequency polygon from the histogram at the top of the next page.

Number of Textbooks Required

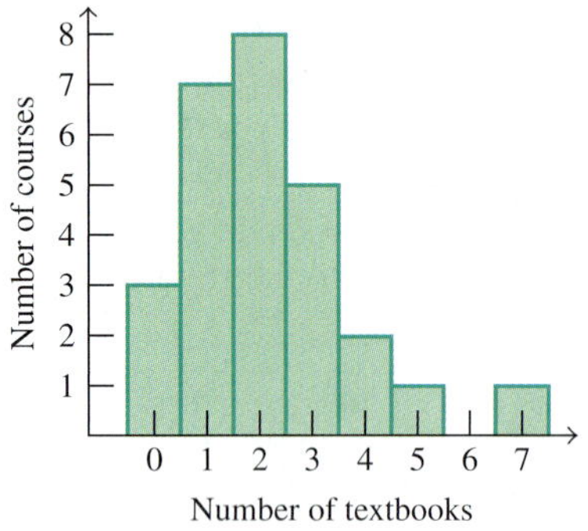

6. a) In your own words, explain how to construct a histogram from a frequency polygon.

   b) Construct a histogram from the frequency polygon below.

Number of Sick Days Taken Last Year

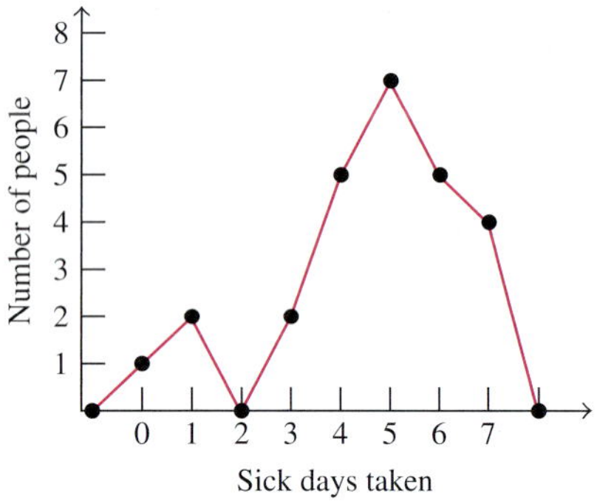

7. a) In your own words, explain how to construct a stem-and-leaf display.

   b) Construct a frequency distribution, letting each class have a width of 1, from the following stem-and-leaf display.

4 | 5 represents 45

Stem	Leaf
4	5 5 5 7 9
5	0 1 1

8. Construct a frequency distribution, letting each class have a width of 1, from the following stem-and-leaf display.

2 | 3 represents 23

Stem	Leaf
1	7 8 7 9 6
2	3 1 2 2 5 5 4

PRACTICE THE SKILLS

9. *College Costs* The cost to attend Rochester Institute of Technology (RIT) for the 2005–2006 school year was \$32,235. The following circle graph shows the percent of that cost for tuition, room, board, and student fees. Determine the cost, in dollars, for each category. Round answers to the nearest cent.

Cost to Attend Rochester Institute of Technology for 2005–2006

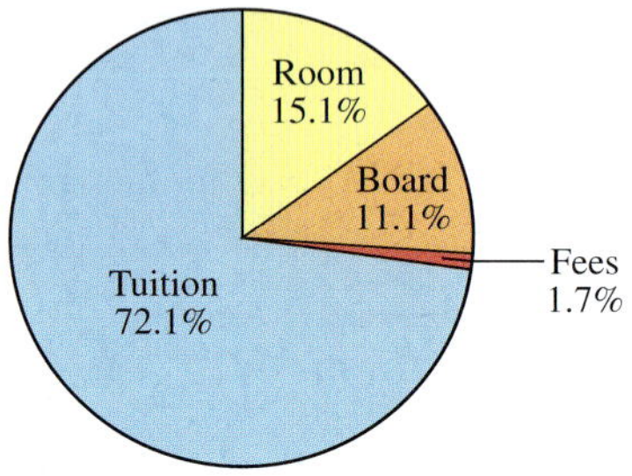

Source: RIT

10. *Bausch & Lomb Sales* The following circle graph shows the percent of Bausch & Lomb 2004 sales for the following categories: contact lenses, pharmaceuticals, lens care products, cataract/vitreoretinal, and refractive products. If the company's 2004 sales were \$2.2 billion, determine their sales from each category.

Bausch & Lomb Sales, 2004

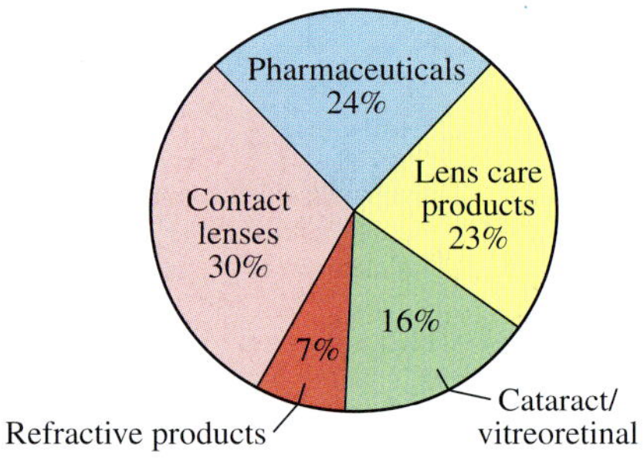

Source: Bausch & Lomb

11. *Eating at Fast-Food Restaurants* A sample of 600 people were asked which meal (breakfast, lunch, dinner, or a snack) they were most likely to eat at a fast-food restaurant. Their responses are given in the table below. Use this information to construct a circle graph illustrating the percent of people who gave each response. Round percents to the nearest tenth.

Meal	Number of Respondents
Breakfast	46
Lunch	293
Dinner	190
Snack	71

12. *Online Travel Websites* A sample of 500 travelers who used the Internet to book a trip were asked which travel website they used. Their responses are given in the table below. Use this information to construct a circle graph illustrating the percent of people who gave each response.

Online Travel Website	Number of Bookings
Travelocity	175
Expedia	125
Orbitz	85
Other	115

13. *Audition* The frequency distribution indicates the ages of a group of 50 dancers attending an audition.

Age	Number of People
17	2
18	6
19	8
20	9
21	5
22	10
23	7
24	3

a) Construct a histogram of the frequency distribution.

b) Construct a frequency polygon of the frequency distribution.

14. *Heights* The frequency distribution indicates the heights of 45 male high school seniors.

Height (in.)	Number of Males
64	2
65	6
66	7
67	9
68	10
69	6
70	3
71	0
72	2

a) Construct a histogram of the frequency distribution.

b) Construct a frequency polygon of the frequency distribution.

15. *DVDs* The frequency distribution indicates the number of DVDs owned by a sample of 40 people.

Number of DVDs	Number of People
6–13	4
14–21	5
22–29	10
30–37	11
38–45	6
46–53	3
54–61	1

a) Construct a histogram of the frequency distribution.

b) Construct a frequency polygon of the frequency distribution.

16. *Annual Salaries* The frequency distribution illustrates the annual salaries, in thousands of dollars, of the people in management positions at the X-Chek Corporation.

Salary (in $1000)	Number of People
35–40	3
41–46	6
47–52	9
53–58	8
59–64	7
65–70	6
71–76	4

a) Construct a histogram of the frequency distribution.

b) Construct a frequency polygon of the frequency distribution.

PROBLEM SOLVING

17. *Number of Televisions per Home* Use the histogram below to answer the following questions.

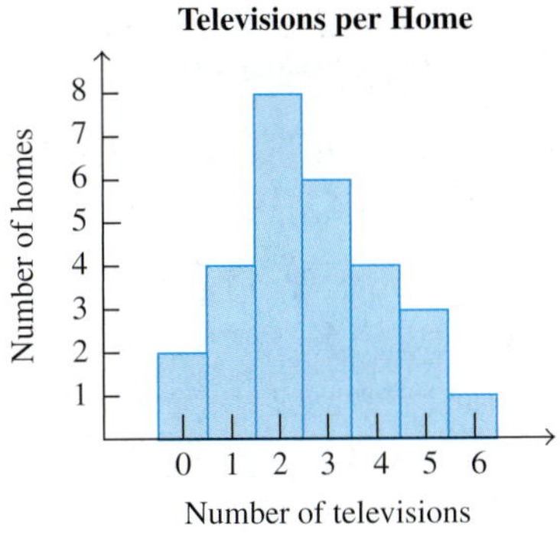

a) How many homes were included in the survey?

b) In how many homes were four televisions observed?

c) What is the modal class?

d) How many televisions were observed?

e) Construct a frequency distribution from this histogram.

18. *Car Insurance* Use the histogram below to answer the following questions.

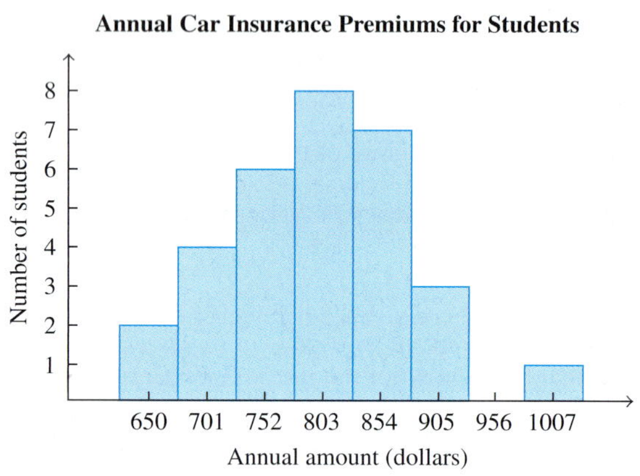

a) How many students were surveyed?

b) What are the lower and upper class limits of the first and second classes?

c) How many students have an annual car insurance premium in the class with a class mark of $752?

d) What is the class mark of the modal class?

e) Construct a frequency distribution from this histogram. Use a first class of 625–675.

19. *Text Messages* Use the frequency polygon below to answer the following questions.

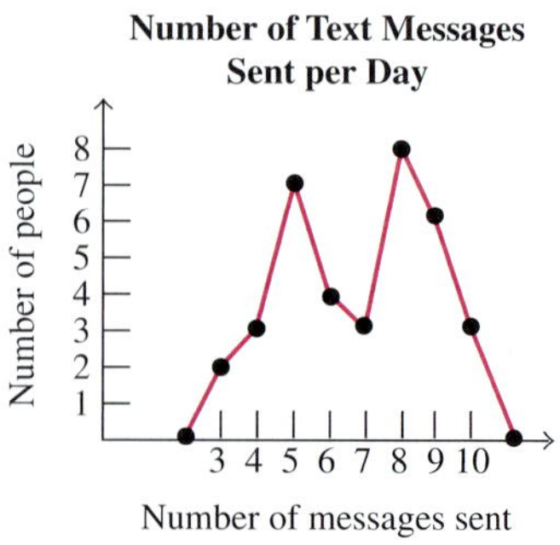

a) How many people sent five text messages?

b) How many people sent six or fewer text messages?

c) How many people were included in the survey?

d) Construct a frequency distribution from the frequency polygon.

e) Construct a histogram from the frequency distribution in part (d).

20. *San Diego Zoo* Use the frequency polygon below to answer the following questions.

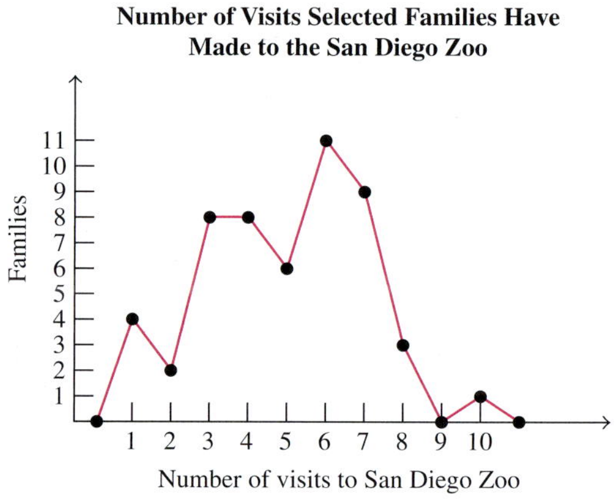

a) How many families visited the San Diego Zoo four times?

b) How many families visited the San Diego Zoo at least six times?

c) How many families were surveyed?

d) Construct a frequency distribution from the frequency polygon.

e) Construct a histogram from the frequency distribution in part (d).

21. Construct a histogram and a frequency polygon from the frequency distribution given in Exercise 7 of Section 13.3. See page 848.

22. Construct a histogram and a frequency polygon from the frequency distribution given in Exercise 8 of Section 13.3. See page 848.

23. ***Jogging Distances*** Twenty members of a health club who jog were asked how many miles they jog per week. The responses are as follows. Construct a stem-and-leaf display. For single digit data, use a stem of 0.

12	15	4	7	12	25	21
33	18	6	8	27	40	22
19	13	23	34	17	16	

24. ***College Credits*** Eighteen students in a geology class were asked how many college credits they had earned. The responses are as follows. Construct a stem-and-leaf display.

10	15	24	36	48	45
42	53	60	17	24	30
33	45	48	62	54	60

25. ***Starting Salaries*** Starting salaries (in thousands of dollars) for social workers with a bachelor of science degree and no experience are shown for a random sample of 25 different social workers.

27	28	29	31	33
28	28	29	31	33
28	28	30	32	33
28	29	30	32	34
28	29	30	32	34

a) Construct a frequency distribution. Let each class have a width of one.

b) Construct a histogram.

c) Construct a frequency polygon.

d) Construct a stem-and-leaf display.

26. ***Visiting a Museum*** The ages of a random sample of people visiting a museum are

20	23	25	30	32	35	39	44
21	23	26	30	33	35	40	45
21	24	27	30	34	35	40	45
22	24	28	31	34	37	40	46
23	25	28	31	34	38	42	47

a) Construct a frequency distribution with a first class of 20–24.

b) Construct a histogram.

c) Construct a frequency polygon.

d) Construct a stem-and-leaf display.

27. ***Broadway Shows*** The following table shows the number of performances for the 50 longest-running Broadway shows as of September 1, 2006.

Play	Performances
The Phantom of the Opera	7755*
Cats	7485
Les Miserables	6680
A Chorus Line	6137
Oh! Calcutta (revival)	5959
Beauty and the Beast	5080*
Rent	4301*
Miss Saigon	4092
Chicago (revival)	4080*
The Lion King	3633*
42nd Street	3486
Grease	3388
Fiddler on the Roof	3242
Life With Father	3224
Tobacco Road	3182
Hello, Dolly!	2844
My Fair Lady	2717
Annie	2377
Cabaret (revival)	2377
Man of La Mancha	2328
Abie's Irish Rose	2327
The Producers	2235*
Oklahoma!	2212
Smokey Joe's Café	2036
Mamma Mia!	2026*
Pippin	1944
South Pacific	1925
The Magic Show	1920
Aida	1852
Gemini	1819
Deathtrap	1793
Harvey	1775
Dancin'	1774

Table continued on next page

Table continued from privious page

Play	Performances
La Cage aux Folles	1761
Hair	1750
Hairspray	1685*
The Wiz	1672
Born Yesterday	1642
Crazy For You	1622
Ain't Misbehavin'	1604
The Best Little Whorehouse in Texas	1584
Mary, Mary	1572
Evita	1567
The Voice of the Turtle	1557
Jekyll & Hyde	1543
Barefoot in the Park	1530
Brighton Beach Memoirs	1530
42nd Street (revival)	1524
Dreamgirls	1521
Mame	1508

Source: League of American Theatres and Producers, Inc
*Still running as of September 1, 2006.

a) Construct a frequency distribution with a first class of 1508–2548.

b) Construct a histogram.

c) Construct a frequency polygon.

28. ***U.S. Presidents*** The ages of the 43 U.S. presidents at their first inauguration (as of 2007) are

57	57	49	52	50	42	54	55	64
61	61	64	56	47	51	51	56	46
57	54	50	46	55	56	60	61	54
57	68	48	54	55	55	62	52	
58	51	65	49	54	51	43	69	

a) Construct a frequency distribution with a first class of 42–47.

b) Construct a histogram.

c) Construct a frequency polygon.

CHALLENGE PROBLEMS/GROUP ACTIVITIES

29. a) ***Birthdays*** What do you believe a histogram of the months in which the students in your class were born (January is month 1 and December is month 12) would look like? Explain.

b) By asking, determine the month in which the students in your class were born (include yourself).

c) Construct a frequency distribution containing 12 classes.

d) Construct a histogram from the frequency distribution in part (c).

e) Construct a frequency polygon of the frequency distribution in part (c).

30. ***Social Security Numbers*** Repeat Exercise 29 for the last digit of the students' social security numbers. Include classes for the digits 0–9.

INTERNET/RESEARCH ACTIVITY

31. Over the years many changes have been made in the U.S. Social Security System.

a) Do research and determine the number of people receiving social security benefits for the years 1945, 1950, 1955, 1960, . . . , 2005. Then construct a frequency distribution and histogram of the data.

b) Determine the maximum amount that self-employed individuals had to pay into social security (the FICA tax) for the years 1945, 1950, 1955, 1960, . . . , 2005. Then construct a frequency distribution and a histogram of the data.

13.5 MEASURES OF CENTRAL TENDENCY

▲ An average is used to describe the fuel efficiency of a car.

Most people have an intuitive idea of what is meant by an "average." The term is used daily in many familiar ways. "This car averages 19 miles per gallon," "The average test grade was 82," and "The average height of adult males is 5 feet 9 inches" are three examples. In this section, we will introduce four different averages and discuss the circumstances in which each average is used.

Measures of Central Tendency

An *average* is a number that is representative of a group of data. There are at least four different averages: the mean, the median, the mode, and the midrange. Each is calculated differently and may yield different results for the same set of data. Each will result in a number near the center of the data; for this reason, averages are commonly referred to as *measures of central tendency*.

The *arithmetic mean*, or simply the *mean*, is symbolized either by $\overline{x}$ (read "x bar") or by the Greek letter mu, μ. The symbol $\overline{x}$ is used when the mean of a *sample* of the population is calculated. The symbol μ is used when the mean of the *entire population* is calculated. Unless otherwise indicated, we will assume that the data featured in this book represent samples; therefore, we will use $\overline{x}$ for the mean.

The Greek letter sigma, Σ, is used to indicate "summation." The notation Σx, read "the sum of x," is used to indicate the sum of all the data. For example, if there are five pieces of data, 4, 6, 1, 0, 5, then $\Sigma x = 4 + 6 + 1 + 0 + 5 = 16$.

Now we can discuss the procedure for determining the mean of a set of data.

The **mean**, $\overline{x}$, is the sum of the data divided by the number of pieces of data. The formula for calculating the mean is

$$\overline{x} = \frac{\Sigma x}{n}$$

where Σx represents the sum of all the data and n represents the number of pieces of data.

The most common use of the word *average* is the mean.

EXAMPLE 1 *Determine the Mean*

Determine the mean age of a group of patients at a doctor's office if the ages of the individuals are 27, 18, 48, 34, and 48.

SOLUTION

$$\overline{x} = \frac{\Sigma x}{n} = \frac{27 + 18 + 48 + 34 + 48}{5} = \frac{175}{5} = 35$$

Therefore, the mean, $\overline{x}$, is 35 years.

The mean represents "the balancing point" of a set of data. For example, if a seesaw were pivoted at the mean and uniform weights were placed at points corresponding to the ages in Example 1, the seesaw would balance. Figure 13.13 shows the five ages given in Example 1 and the calculated mean.

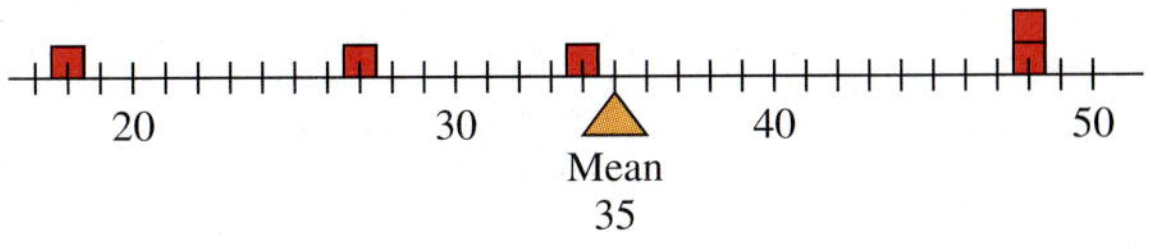

Figure 13.13

DID YOU KNOW?

Buying the American Dream

San Jose, California

One of the biggest dreams for most people is to own their own home. Yet depending on where you live, the American dream may be hard to achieve. In 2006, San Jose/Sunnyvale/Santa Clara, California, had the highest median home price for major metropolitan housing markets, \$775,000. Youngstown/Warren/Boardman, Ohio, had the lowest median home price for major metropolitan housing markets, \$81,500.

Youngstown, Ohio

Source: National Association of Realtors

A second average is the *median*. To find the median of a set of data, *rank the data* from smallest to largest, or largest to smallest, and determine the value in the middle of the set of *ranked data*. This value will be the median.

The **median** is the value in the middle of a set of *ranked data*.

EXAMPLE 2 *Determine the Median*

Determine the median of the patients' ages in Example 1 on page 861.

SOLUTION Ranking the data from smallest to largest gives 18, 27, 34, 48, and 48. Since 34 is the value in the middle of this set of ranked data (two pieces of data above it and two pieces below it), 34 years is the median. ●

When there are an even number of pieces of data, the median is halfway between the two middle pieces. In this case, to find the median, add the two middle pieces and divide this sum by 2.

EXAMPLE 3 *Determine the Median of an Even Number of Pieces of Data*

Determine the median of the following sets of data.

a) 9, 14, 16, 17, 11, 16, 11, 12

b) 7, 8, 8, 8, 9, 10

SOLUTION

a) Ranking the data gives 9, 11, 11, 12, 14, 16, 16, 17. There are eight pieces of data. Therefore, the median will lie halfway between the two middle pieces, the 12 and the 14. The median is $\frac{12 + 14}{2}$, or $\frac{26}{2}$, or 13.

b) There are six pieces of data, and they are already ranked. Therefore, the median lies halfway between the two middle pieces. Both middle pieces are 8's. The median is $\frac{8 + 8}{2}$, or $\frac{16}{2}$, or 8. ●

TIMELY TIP Data must be ranked before determining the median. A common error made when determining the median is neglecting to arrange the data in ascending (increasing) or in descending (decreasing) order.

A third average is the ***mode***.

The **mode** is the piece of data that occurs most frequently.

EXAMPLE 4 *Determine the Mode*

Determine the mode of the patients' ages in Example 1 on page 861.

SOLUTION The ages are 27, 18, 48, 34, and 48. The age 48 is the mode because it occurs twice and the other values occur only once. ●

If each piece of data occurs only once, the set of data has no mode. For example, the set of data 1, 2, 3, 4, 5 has no mode. If two values in a set of data occur more often than all the other data, we consider both these values as modes and say that the data are **bimodal*** (which means two modes). For example, the set of data 1, 1, 2, 3, 3, 5 has two modes, 1 and 3.

The last average we will discuss is the midrange. The *midrange* is the value halfway between the lowest (L) and highest (H) values in a set of data. It is found by adding the lowest and highest values and dividing the sum by 2. A formula for finding the midrange follows.

$$\textbf{Midrange} = \frac{\text{lowest value} + \text{highest value}}{2}$$

EXAMPLE 5 *Determine the Midrange*

Determine the midrange of the patients' ages given in Example 1 on page 861.

SOLUTION The ages of the patients are 27, 18, 48, 34, and 48. The lowest age is 18, and the highest age is 48.

$$\text{Midrange} = \frac{\text{lowest} + \text{highest}}{2} = \frac{18 + 48}{2} = \frac{66}{2} = 33 \text{ years}$$

The "average" of the ages 27, 18, 48, 34, 48 can be considered any one of the following values: 35 (mean), 34 (median), 48 (mode), or 33 (midrange). Which average do you feel is most representative of the ages? We will discuss this question later in this section.

EXAMPLE 6 *Measures of Central Tendency*

The salaries of eight selected social workers rounded to the nearest thousand dollars are 40, 25, 28, 35, 42, 60, 60, and 73. For this set of data, determine the (a) mean, (b) median, (c) mode, and (d) midrange. Then (e) list the measures of central tendency from lowest to highest.

SOLUTION

a) $\bar{x} = \frac{\Sigma x}{n} = \frac{40 + 25 + 28 + 35 + 42 + 60 + 60 + 73}{8} = \frac{363}{8} = 45.375$

b) Ranking the data from the smallest to largest gives

25, 28, 35, 40, 42, 60, 60, 73

Since there are an even number of pieces of data, the median is halfway between 40 and 42. The median = (40 + 42)/2 = 82/2 = 41.

c) The mode is the piece of data that occurs most frequently. The mode is 60.

d) The midrange = (L + H)/2 = (25 + 73)/2 = 98/2 = 49.

e) The averages from lowest to highest are the median, mean, midrange, and mode. Their values are 41, 45.375, 49, and 60, respectively.

At this point, you should be able to calculate the four measures of central tendency: mean, median, mode, and midrange. Now let's examine the circumstances in which each is used.

*Some textbooks say that sets of data such as 1, 1, 2, 3, 3, 5 have no mode.

DID YOU KNOW?

When Babies' Eyes Are Smiling

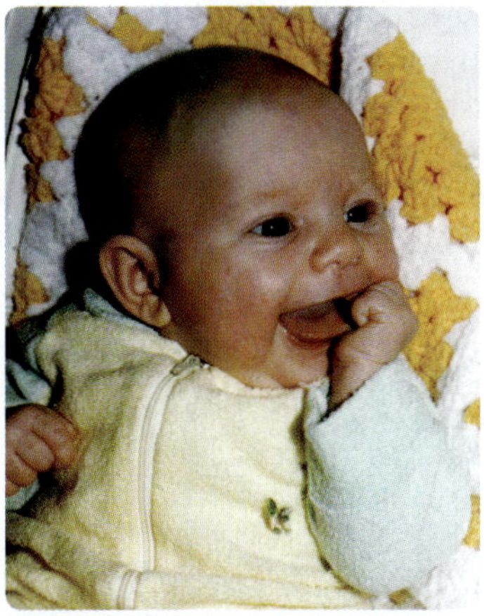

Experimental psychologists formulate hypotheses about human behavior, design experiments to test them, make observations, and draw conclusions from their data. They use statistical concepts at each stage to help ensure that their conclusions are valid. In one experiment, researchers observed that 2-month-old infants who learned to move their heads so as to make a mobile turn began to smile as soon as the mobile turned. Babies in the control group did not smile as often when the mobile moved independently of their head turning. The researchers concluded that it was not the movement of the mobile that made the infant smile; rather, the infants smiled at their own achievement.

The mean is used when each piece of data is to be considered and "weighed" equally. It is the most commonly used average. It is the only average that can be affected by *any* change in the set of data; for this reason, it is the most sensitive of all the measures of central tendency (see Exercise 23).

Occasionally, one or more pieces of data may be much greater or much smaller than the rest of the data. When this situation occurs, these "extreme" values have the effect of increasing or decreasing the mean significantly so that the mean will not be representative of the set of data. Under these circumstances, the median should be used instead of the mean. The median is often used in describing average family incomes because a relatively small number of families have extremely large incomes. These few incomes would inflate the mean income, making it nonrepresentative of the millions of families in the population.

Consider a set of exam scores from a mathematics class: 0, 16, 19, 65, 65, 65, 68, 69, 70, 72, 73, 73, 75, 78, 80, 85, 88, 92. Which average would best represent these grades? The mean is 64.06. The median is 71. Since only 3 of the 18 scores fall below the mean, the mean would not be considered a good representative score. The median of 71 probably would be the better average to use.

The mode is the piece of data, if any, that occurs most frequently. Builders planning houses are interested in the most common family size. Retailers ordering shirts are interested in the most common shirt size. An individual purchasing a thermometer might choose one, from those on display, whose temperature reading is the most common reading among those on display. These examples illustrate how the mode may be used.

The midrange is sometimes used as the average when the item being studied is constantly fluctuating. Average daily temperature, used to compare temperatures in different areas, is calculated by adding the lowest and highest temperatures for the day and dividing the sum by 2. The midrange is actually the mean of the high value and the low value of a set of data. Occasionally, the midrange is used to estimate the mean since it is much easier to calculate.

Sometimes an average itself is of little value, and care must be taken in interpreting its meaning. For example, Jim is told that the average depth of Willow Pond is only 3 feet. He is not a good swimmer but decides that it is safe to go out a short distance in this shallow pond. After he is rescued, he exclaims, "I thought this pond was only 3 feet deep." Jim didn't realize that an average does not indicate extreme values or the spread of the values. The spread of data is discussed in Section 13.6.

Measures of Position

Measures of position are used to describe the position of a piece of data in relation to the rest of the data. If you took the Scholastic Aptitude Test (SAT) before applying to college, your score was described as a measure of position rather than a measure of central tendency. ***Measures of position*** are often used to make comparisons, such as comparing the scores of individuals from different populations, and are generally used when the amount of data is large.

Two measures of position are ***percentiles*** and ***quartiles***. There are 99 percentiles dividing a set of data into 100 equal parts; see Fig. 13.14. For example, suppose that

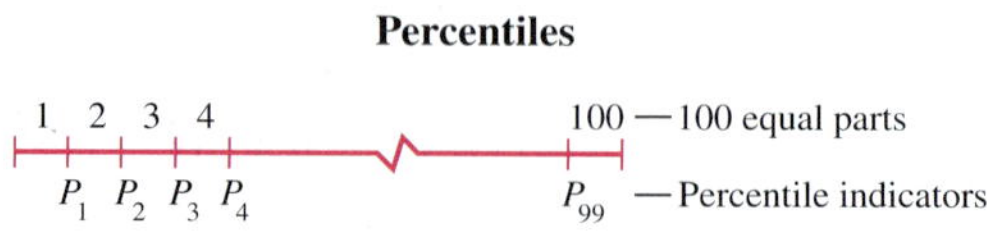

Figure 13.14

you scored 520 on the math portion of the SAT, and the score of 520 was reported to be in the 78th percentile of high school students. This wording *does not* mean that 78% of your answers were correct; it *does* mean that you outperformed about 78% of all those taking the exam. In general, a score in the nth percentile means that you outperformed about $n\%$ of the population who took the test and that $(100 - n)\%$ of the people taking the test performed better than you did.

EXAMPLE 7 *English Achievement Test*

Kara Hopkins took an English achievement test to obtain college credit by exam for freshman English. Her score was at the 81st percentile. Explain what that means.

SOLUTION If a score is at the 81st percentile, it means that about 81% of the scores are below that score. Therefore, Kara scored better than about 81% of the students taking the exam. Also, about 19% of all students taking the exam scored higher than she did.

Quartiles are another measure of position. Quartiles divide data into four equal parts: The first quartile is the value that is higher than about $\frac{1}{4}$ or 25% of the population. It is the same as the 25th percentile. The second quartile is the value that is higher than about $\frac{1}{2}$ the population and is the same as the 50th percentile, or the median. The third quartile is the value that is higher than about $\frac{3}{4}$ of the population and is the same as the 75th percentile; see Fig. 13.15.

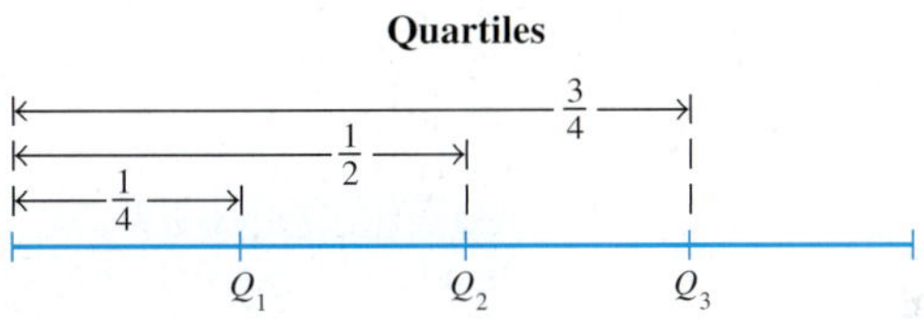

Figure 13.15

TO DETERMINE THE QUARTILES OF A SET OF DATA

1. List the data from smallest to largest.
2. Determine the median, or the 2nd quartile, of the set of data. If there is an odd number of pieces of data, the median is the middle value. If there is an even number of pieces of data, the median will be halfway between the two middle pieces of data.
3. The first quartile, Q_1, is the median of the lower half of the data; that is, Q_1 is the median of the data less than Q_2.
4. The third quartile, Q_3, is the median of the upper half of the data; that is, Q_3 is the median of the data greater than Q_2.

EXAMPLE 8 *Finding Quartiles*

Electronics World is concerned about the high turnover of its sales staff. A survey was done to determine how long (in months) the sales staff had been in their current positions. The responses of 27 sales staff follow. Determine Q_1, Q_2, and Q_3.

25	3	7	15	31	36	17	21	2
11	42	16	23	19	21	9	20	5
8	12	27	14	39	24	18	6	10

SOLUTION First we list the data from smallest to largest.

2	3	5	6	7	8	9	10	11
12	14	15	16	17	18	19	20	21
21	23	24	25	27	31	36	39	42

Next we determine the median. Since there are 27 pieces of data, an odd number, the median will be the middle value. The middle value is 17, with 13 pieces of data less than 17 and 13 pieces of data greater than 17. Therefore, the median, Q_2, is 17, shown in red.

To find Q_1, the median of the lower half of the data, we need to find the median of the 13 pieces of data that are less than Q_2. The middle value of the lower half of the data is 9. There are 6 pieces of data less than 9 and 6 pieces of data greater than 9. Therefore, Q_1 is 9, shown in blue.

To find Q_3, the median of the upper half of the data, we need to find the median of the 13 pieces of data that are greater than 17, or Q_2. The middle value of the upper half of the data is 24. There are 6 pieces of data greater than 17 but less than 24 and 6 pieces of data greater than 24. Therefore, Q_3 is 24, shown in blue.

TECHNOLOGY TIP Several computer software programs and calculators can be used to determine the mean of a set of data. These programs and calculators can also provide other types of statistical information that we will discuss in this chapter. We will provide the instructions for using a software program called Microsoft Excel as well as information on how to use Texas Instruments TI-83 Plus and TI-84 Plus graphing calculators. In the example, we will use the data from Example 1 on page 861, which represent the ages of patients in a doctor's office.

EXCEL

If you are new to Excel, first read the appendix on Excel in the back of the book. In our discussion, we will use the symbol > to indicate the next item to be selected from the list or menu of items.

Begin by entering the five pieces of data in column A. Press the Enter key after each piece of data is entered. Next select

Insert > Function . . . > Statistical > AVERAGE

Then click the OK box at the bottom. The program will then generate a gray box, where you need to enter the data. In the area to the right of **Number1**, you need to enter the data for which you want to find the mean. Since you have already entered the data in column A, rows 1 to 5, if A1:A5 is not already listed, you can enter A1:A5 in the area to the right of **Number1**. Then, at the bottom of the gray box, *Formula Results = 35* is displayed. The 35 is the mean of the set of data. If you then press the OK box, the mean will be displayed in the cell where the cursor was located.

24 9

If you do not want the mean displayed in a cell, press CANCEL. To find the median or mode you follow a similar procedure, except that instead of selecting AVERAGE you would select MEDIAN or MODE, respectively.

TI-83 PLUS AND TI-84 PLUS GRAPHING CALCULATORS

To enter the data, press STAT. Then highlight **1: Edit** and press the ENTER key. If any data are currently listed in column **L1**, move the cursor to the **L1** and press CLEAR and then the ENTER key. This step will eliminate all data from column **L1**. Now enter the data in column **L1**. After you enter each piece of data, press the ENTER key. After all the data has been entered, press STAT. Then highlight **CALC**. At this point, **1: 1−Var Stats** should be highlighted. Press the ENTER key twice. You will now see $\bar{x} = 35$. Thus, the mean is 35. Several other descriptive statistics that we will discuss shortly are also shown. If you scroll down, you will eventually see the values of Q_1, the median, and Q_3.

SECTION 13.5 EXERCISES

CONCEPT/WRITING EXERCISES

1. a) Describe the mean of a set of data and explain how to find it.

   b) Describe the median of a set of data and explain how to find it.

2. a) Describe the midrange of a set of data and explain how to find it.

   b) Describe the mode of a set of data and explain how to find it.

3. When might the median be the preferred average to use? Give an example.

4. When might the mode be the preferred average to use? Give an example.

5. When might the midrange be the preferred average to use? Give an example.

6. When might the mean be the preferred average to use? Give an example.

7. a) What symbol is used for the sample mean?

   b) What symbol is used for the population mean?

8. What is a set of *ranked data*?

9. Explain how to find the quartiles of a set of data.

10. a) What is another name for the 25th percentile?

    b) What is another name for the 50th percentile?

    c) What is another name for the 75th percentile?

PRACTICE THE SKILLS

In Exercises 11–20, determine the mean, median, mode, and midrange of the set of data. Round your answer to the nearest tenth.

11. 7, 8, 8, 10, 12, 12, 12, 23, 25

12. 10, 8, 11, 11, 11, 13, 15

13. 76, 82, 94, 55, 100, 52, 96

14. 4, 6, 10, 12, 10, 9, 365, 40, 37, 8

15. 1, 3, 5, 7, 9, 11, 13, 15

16. 40, 50, 30, 60, 90, 100, 140

17. 1, 7, 11, 27, 36, 14, 12, 9, 1

18. 1, 1, 1, 1, 4, 4, 4, 4, 6, 8, 10, 12, 15, 21

19. 6, 8, 12, 13, 11, 13, 15, 17

20. 5, 15, 5, 15, 5, 15

21. ***Best-Seller List*** The number of weeks the top 10 hardcover fiction novels were on the best-seller list as of July 8, 2007, is 1, 5, 2, 2, 3, 4, 5, 3, 3, 7. Determine the mean, median, mode, and midrange.

22. ***Daily Commission*** The amount of money Steve Kilner collected in sales commission in each of seven days is $48, $67, $51, $25, $102, $61, $80. Determine the mean, median, mode, and midrange.

PROBLEM SOLVING

23. ***Change in the Data*** The mean is the "most sensitive" average because it is affected by any change in the data.

a) Determine the mean, median, mode, and midrange for 1, 2, 3, 5, 5, 7, 11.

b) Change the 7 to a 10 in part (a). Determine the mean, median, mode, and midrange.

c) Which averages were affected by changing the 7 to a 10?

d) Which averages will be affected by changing the 11 to a 10 in part (a)?

24. ***Life Expectancy*** In 2005, the National Center for Health Statistics indicated a record "average life expectancy" of 77.6 years for the total U.S. population. The average life expectancy for men was 74.8 years, and for women it was 80.1 years. Which "average" do you think the National Center for Health is using? Explain your answer.

25. ***A Grade of B*** To get a grade of B, a student must have a mean average of 80 or greater. Jim Condor has a mean average of 79 for 10 quizzes. He approaches his teacher and asks for a B, reasoning that he missed a B by only one point. What is wrong with Jim's reasoning?

26. ***Employee Salaries*** The salaries of 10 employees of a small company follow.

$28,000	$64,000
25,000	24,000
31,000	27,000
26,000	81,000
26,000	29,000

Determine the

a) mean.

b) median.

c) mode.

d) midrange.

e) If the employees wanted to demonstrate the need for a raise, which average would they use to show they are being underpaid: the mean or the median? Explain.

f) If the management did not want to give the employees a raise, which average would they use: the mean or the median? Explain.

27. ***Passenger Traffic*** The 10 U.S. airports with the most passenger arrivals and departures in 2005 are listed below.

Airport	Passenger Arrivals and Departures (millions of people)
Hartsfield Atlanta	85.9
Chicago O'Hare	75.5
Los Angeles	61.5
Dallas–Fort Worth	59.1
Las Vegas	44.3
Denver	43.3
Phoenix Sky Harbor	41.2
JFK–New York	40.6
Houston	39.7
Minneapolis–St. Paul	37.6

Source: Infoplease.com

Determine to the nearest tenth the

a) mean.

b) median.

c) mode.

d) midrange.

28. ***Living Expenses*** Bob Exler's monthly living expenses for 1 year are as follows:

$1000	$850	$1370	$1400
1900	850	1350	1250
1600	900	1110	1230

When appropriate, round your answer to the nearest cent. Determine the

a) mean.

b) median.

c) mode.

d) midrange.

29. ***Amusement Park Attendance*** The 10 amusement parks with the highest attendance in 2004 are listed below.

Park	Attendance (millions of people)
Magic Kingdom at Walt Disney World	15.1
Disneyland	13.4
Tokyo Disneyland	13.2
Tokyo Disney Sea	12.2
Disneyland Paris	10.2
Universal Studios Japan	9.9
Epcot at Walt Disney World	9.4
MGM Studios at Walt Disney World	8.3
Lotte World	8.0
Animal Kingdom at Walt Disney World	7.8

Source: Amusement Business

Determine to the nearest tenth the

a) mean.

b) median.

c) mode.

d) midrange.

30. ***Exam Average*** Malcolm Sander's mean average on five exams is 81. Determine the sum of his scores.

31. ***Exam Average*** Jeremy Urban's mean average on six exams is 92. Determine the sum of his scores.

32. ***Creating a Data Set*** Construct a set of five pieces of data in which the mode has a lower value than the median and the median has a lower value than the mean.

33. ***Creating a Data Set*** Construct a set of six pieces of data with a mean, median, and midrange of 75 and where no two pieces of data are the same.

34. ***Creating a Data Set*** Construct a set of six pieces of data with a mean of 84 and where no two pieces of data are the same.

35. ***Water Park*** For the 2007 season, 24,000 people visited the Blue Lagoon Water Park. The park was open 120 days for water activities. The highest number of visitors on a single day was 500. The lowest number of visitors on a single day was 50. Determine whether it is possible to find the following with the given information. Explain your answer.

a) the mean number of visitors per day.

b) the median number of visitors per day.

c) the mode number of visitors per day.

d) the midrange number of visitors per day.

36. ***Determine a Necessary Grade*** A mean average of 80 or greater for five exams is needed for a final grade of B in a course. Jorge Rivera's first four exam grades are 73, 69, 85, and 80. What grade does Jorge need on the fifth exam to get a B in the course?

37. ***Grading Methods*** A mean average of 60 on seven exams is needed to pass a course. On her first six exams, Sheryl Ward received grades of 51, 72, 80, 62, 57, and 69.

a) What grade must she receive on her last exam to pass the course?

b) An average of 70 is needed to get a C in the course. Is it possible for Sheryl to get a C? If so, what grade must she receive on the seventh exam?

c) If her lowest grade of the exams already taken is to be dropped, what grade must she receive on her last exam to pass the course?

d) If her lowest grade of the exams already taken is to be dropped, what grade must she receive on her last exam to get a C in the course?

38. *Central Tendencies* Which of the measures of central tendency *must* be an actual piece of data in the distribution? Explain.

39. *Creating a Data Set* Construct a set of six pieces of data such that if only one piece of data is changed, the mean, median, and mode will all change.

40. *Changing One Piece of Data* Consider the set of data 1, 1, 1, 2, 2, 2. If one 2 is changed to a 3, which of the following will change: mean, median, mode, midrange? Explain.

41. *Changing One Piece of Data* Is it possible to construct a set of six different pieces of data such that by changing only one piece of data you cause the mean, median, mode, and midrange to change? Explain.

42. *Grocery Expenses* The Taylor's have recorded their weekly grocery expenses for the past 12 weeks and determined that the mean weekly expense was $85.20. Later Mrs. Taylor discovered that 1 week's expense of $74 was incorrectly recorded as $47. What is the correct mean?

43. *Percentiles* For any set of data, what must be done to the data before percentiles can be determined?

44. *Percentiles* Josie Waverly scored in the 73rd percentile on the verbal part of her College Board test. What does that mean?

45. *Percentiles* When a national sample of heights of kindergarten children was taken, Kevin Geis was told that he was in the 35th percentile. Explain what that means.

46. *Percentiles* A union leader is told that, when all workers' salaries are considered, the first quartile is $20,750. Explain what that means.

47. *Quartiles* The prices of a gallon of the 21 top-rated exterior paints, as rated in the June 2006 issue of *Consumer Reports*, are as follows:

$15	$19	$19	$20	$22	$22	$24
$24	$24	$25	$25	$27	$29	$30
$32	$34	$34	$35	$36	$39	$42

Determine

a) Q_2. b) Q_1. c) Q_3.

48. *Quartiles* The prices of the 20 top-rated dishwashers, as rated in the February 2006 issue of *Consumer Reports*, are as follows:

$380	$430	$435	$460	$500
$500	$550	$580	$600	$600
$700	$800	$800	$800	$800
$830	$850	$880	$1100	$1550

Determine

a) Q_2. b) Q_1. c) Q_3.

49. *The 50th Percentile* Give the names of two other statistics that have the same value as the 50th percentile.

50. *College Admissions* Jonathan Burd took an admission test for the University of California and scored in the 85th percentile. The following year, Jonathan's sister Kendra took a similar admission test for the University of California and scored in the 90th percentile.

a) Is it possible to determine which of the two answered the higher percent of questions correctly on their respective exams? Explain your answer.

b) Is it possible to determine which of the two was in a better relative position with regard to their respective populations? Explain.

51. *Employee Salaries* The following statistics represent weekly salaries at the Midtown Construction Company:

Mean	$550	First quartile	$510
Median	$540	Third quartile	$575
Mode	$530	83rd percentile	$615

a) What is the most common salary?

b) What salary did half the employees' salaries surpass?

c) About what percent of employees' salaries surpassed $575?

d) About what percent of employees' salaries were less than $510?

e) About what percent of employees' salaries surpassed $615?

f) If the company has 100 employees, what is the total weekly salary of all employees?

CHALLENGE PROBLEMS/GROUP ACTIVITIES

52. *The Mean of the Means* Consider the following five sets of values.

i) 5 6 7 7 8 9 14
ii) 3 6 8 9
iii) 1 1 1 2 5
iv) 6 8 9 12 15
v) 50 51 55 60 80 100

a) Compute the mean of each of the five sets of data.

b) Compute the mean of the five means in part (a).

c) Find the mean of the 27 pieces of data.

d) Compare your answer in part (b) to your answer in part (c). Are the values the same? Does your answer make sense? Explain.

53. ***Ruth versus Mantle*** The tables below compare the batting performances for selected years for two well-known former baseball players, Babe Ruth and Mickey Mantle.

Babe Ruth
Boston Red Sox 1914–1919
New York Yankees 1920–1934

Year	At Bats	Hits	Pct.
1925	359	104	
1930	518	186	
1933	459	138	
1916	136	37	
1922	406	128	
Total	1878	593	

Mickey Mantle
New York Yankees 1951–1968

Year	At Bats	Hits	Pct.
1954	543	163	
1957	474	173	
1958	519	158	
1960	527	145	
1962	377	121	
Total	2440	760	

a) For each player, compute the batting average percent (pct.) for each year by dividing the number of hits by the number of at bats. Round to the nearest thousandth. Place the answers in the pct. column.

b) Going across each of the five horizontal lines (for example Ruth, 1925, vs. Mantle, 1954), compare the percents (pct.) and determine which is greater in each case.

c) For each player, compute the mean batting average percent for the 5 given years by dividing the total hits by the total at bats. Which is greater, Ruth's or Mantle's?

d) Based on your answer in part (b), does your answer in part (c) make sense? Explain.

e) Find the mean percent for each player by adding the five pcts. and dividing by 5.Which is greater, Ruth's or Mantle's?

f) Why do the answers obtained in parts (c) and (e) differ? Explain.

g) Who would you say has the better batting average percent for the 5 years selected? Explain.

54. ***Employee Salaries*** The following table gives the annual salary distribution for employees at Kulzer's Home Improvement.

Annual Salary	Number Receiving Salary
$100,000	1
85,000	2
24,000	6
21,000	4
18,000	5
17,000	7

Using the information provided in the table, determine the

a) mean annual salary.

b) median annual salary.

c) mode annual salary.

d) midrange annual salary.

e) Which is the best measure of central tendency for this set of data? Explain your answer.

Weighted Average *Sometimes when we wish to find an average, we may wish to assign more importance, or weight, to some of the pieces of data. To calculate a* ***weighted average****, we use the formula: weighted average* $= \dfrac{\Sigma xw}{\Sigma w}$, *where w is the weight of the piece of data, x; Σxw is the sum of the products of each piece of data multiplied by its weight; and Σw is the sum of the weights. For example, suppose that* *students in a class need to submit a report that counts for 20% of their grade, they need to take a midterm exam that counts for 30% of their grade, and they need to take a final exam that counts for 50% of their grade. Suppose that a student got a 72 on the report, an 85 on the midterm exam, and a 93 on the final exam. To determine this student's weighted average, first find* Σxw: $\Sigma xw = 72(0.20) + 85(0.30) + 93(0.50) = 86.4$. *Next find* Σw, *the sum of the weights:* $\Sigma w = 0.20 + 0.30 + 0.50 = 1.00$. *Now determine the weighted average as follows.*

$$\text{Weighted average} = \frac{\Sigma xw}{\Sigma w} = \frac{86.4}{1.00} = 86.4$$

Thus, the weighted average is 86.4. Note that Σw *does not always have to be 1.00. In Exercises 55 and 56, use the weighted average formula.*

55. ***Course Average*** Suppose that your final grade for a course is determined by a midterm exam and a final exam. The midterm exam is worth 40% of your grade, and the final exam is worth 60%. If your midterm exam grade is 84 and your final exam grade is 94, calculate your final average.

56. ***Grade Point Average*** In a four-point grade system, an A corresponds to 4.0 points, a B corresponds to 3.0 points, a C corresponds to 2.0 points, and a D corresponds to 1.0 points. No points are awarded for an F. Last semester, Tanya Reeves received a B in a four-credit hour course, an A in a three-credit hour course, a C in a three-credit hour course, and an A in another three-credit hour course. Grade point average (GPA) is calculated as a weighted average using the credit hours as weights and the number of points corresponding to the grade as pieces of data. Calculate Tanya's GPA for the previous semester. (Round your answer to the nearest hundredth.)

RECREATIONAL MATHEMATICS

57. ***Your Exam Average*** **a)** Calculate the mean, median, mode, and midrange of your exam grades in your mathematics course.

b) Which measure of central tendency best represents your average grade?

c) Which measure of central tendency would you rather use as your average grade?

58. ***Purchases*** Matthew Riveria purchased some items at Staples each day for five days. The mode of the number of items Matthew purchased is higher than the median of the number of items he purchased. The median of the number of items Matthew purchased is higher than the mean of the number of items he purchased. He purchased at least two items but no more than seven items each day.

a) How many items did Matthew purchase each day? (*Note:* There is more than one correct answer.)

b) Determine the mean, median, and mode for your answer to part (a).

INTERNET/RESEARCH ACTIVITY

59. Two other measures of location that we did not mention in this section are *stanines* and *deciles*. Use statistics books, books on educational testing and measurements, and Internet websites to write a report on what stanines and deciles are and when percentiles, quartiles, stanines, and deciles are used.

13.6 MEASURES OF DISPERSION

▲ The average life span of an airplane engine may not be enough information to make a sound purchasing decision.

The measures of central tendency by themselves do not always give sufficient information to analyze a situation and make decisions. For example, two manufacturers of airplane engines are being considered for a contract. Manufacturer A's engines have an average (mean) life of 1000 hours of flying time before they fail. Manufacture B's engines have an average life of 950 hours of flying time before they fail. If you assume that both engines cost the same, which ones should be purchased? The average engine life may not be the most important factor. The fact that manufacturer A's engines have an average life of 1000 hours could mean that half will last about 500 hours and the other half will last about 1500 hours. If in fact all manufacturer B's engines have a life span of between 925 and 975 hours, then all of manufacturer B's engines are more consistent and reliable. This example illustrates the importance of knowing something about the *spread*, or *variability*, of the data. In this section, we will discuss two measures of variability or dispersion.

Measures of dispersion are used to indicate the *spread of the data*. The range and standard deviation* are the measures of dispersion that will be discussed in this book.

Range and Standard Deviation

The *range* is the difference between the highest and lowest values; it indicates the total spread of the data.

Range = highest value − lowest value

EXAMPLE 1 *Determine the Range*

The amount of carbohydrates, in grams, of 12 different soft drinks is given below. Determine the range of these data.

26, 27, 31, 35, 31, 29, 24, 26, 27, 25, 30, 31

SOLUTION Range = highest value − lowest value = 35 − 24 = 11. The range of the amounts of carbohydrates is 11 grams. ●

The second measure of dispersion we discuss in this section, the *standard deviation*, measures how much the data *differ from the mean*. It is symbolized either by the letter s or by the Greek letter sigma, σ.† The s is used when the standard deviation of a *sample* is calculated. The σ is used when the standard deviation of the entire *population* is calculated. Since we are assuming that all data presented in this section are for samples, we use s to represent the standard deviation (note, however, that on the height and weight charts on page 879, σ is used. Also, we will use σ in the next section when we determine standard scores.) The larger the spread of the data about the mean, the larger the standard deviation is. Consider the following two sets of data.

5, 8, 9, 10, 12, 13   8, 9, 9, 10, 10, 11

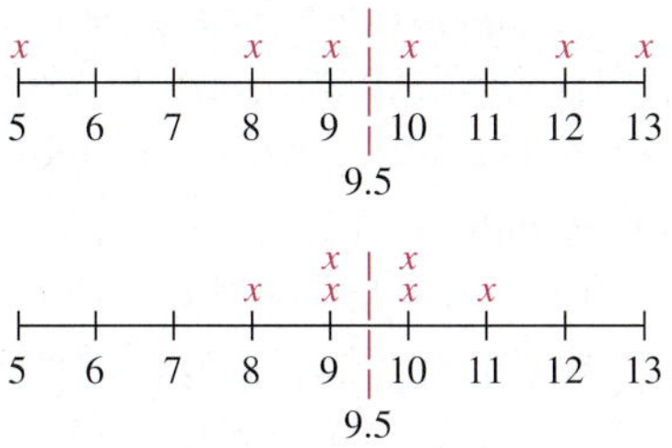

Figure 13.16

Both have a mean of 9.5. Which set of values on the whole do you believe differs less from the mean of 9.5? Figure 13.16 may make the answer more apparent. The scores in the second set of data are closer to the mean and therefore have a smaller standard deviation. You will soon be able to verify such relationships yourself.

Sometimes only a very small standard deviation is desirable or acceptable. Consider a cereal box that is to contain 8 oz of cereal. If the amount of cereal put into the boxes varies too much—sometimes underfilling, sometimes overfilling—the manufacturer will soon be in trouble with consumer groups and government agencies.

At other times, a larger spread of data is desirable or expected. For example, intelligence quotients (IQs) are expected to exhibit a considerable spread about the

**Variance*, another measure of dispersion, is the square of the standard deviation.

†Our alphabet uses both uppercase and lowercase letters, for example, A and a. The Greek alphabet also uses both uppercase and lowercase letters. The symbol Σ is the capital Greek letter sigma, and σ is the lowercase Greek letter sigma.

mean because everyone is different. The following procedure explains how we determine the standard deviation of a set of data.

TO DETERMINE THE STANDARD DEVIATION OF A SET OF DATA

1. Determine the mean of the set of data.
2. Make a chart having three columns:

   Data Data − Mean $(\text{Data} - \text{Mean})^2$
3. List the data vertically under the column marked Data.
4. Subtract the mean from each piece of data and place the difference in the Data − Mean column.
5. Square the values obtained in the Data − Mean column and record these values in the $(\text{Data} - \text{Mean})^2$ column.
6. Determine the sum of the values in the $(\text{Data} - \text{Mean})^2$ column.
7. Divide the sum obtained in step 6 by $n - 1$, where n is the number of pieces of data.*
8. Determine the square root of the number obtained in step 7. This number is the standard deviation of the set of data.

Example 2 illustrates the procedure to follow to determine the standard deviation of a set of data.

EXAMPLE 2 ***Determine the Standard Deviation***

A veterinarian in an animal hospital recorded the following life spans of selected Labrador retrievers (to the nearest year):

7, 9, 11, 15, 18, 12

Determine the standard deviation of the life spans.

SOLUTION First determine the mean:

$$\bar{x} = \frac{\Sigma x}{n} = \frac{7 + 9 + 11 + 15 + 18 + 12}{6} = \frac{72}{6} = 12$$

Next construct a table with three columns, as illustrated in Table 13.4, and list the data in the first column (it is often helpful to list the data in ascending or descending order). Complete the second column by subtracting the mean, 12 in this case, from each piece of data in the first column.

*To determine the standard deviation of a sample, divide the sum of $(\text{Data} - \text{Mean})^2$ column by $n - 1$. To find the standard deviation of a population, divide the sum by n. In this book, we assume that the set of data represents a sample and divide by $n - 1$. The quotient obtained in step 7 represents a measure of dispersion called the *variance*.

DID YOU KNOW?

Statistics and Opera Houses

Sydney Opera House in Sydney, Australia

Architects have developed a mathematical rule based on statistics to help them construct opera houses with exceptional acoustics. The rule was first developed by having conductors rate the overall sound quality in 23 opera houses. Then acoustical engineers measured several acoustical properties in those 23 buildings. By using statistical analysis, the engineers were able to determine which combination of properties produced exceptional sound and which of the acoustic characteristics were most important. This mathematical rule is now used in the development of new opera houses.

Table 13.4

Data	Data − Mean	(Data − Mean)^2
7	$7 - 12 = -5$	
9	$9 - 12 = -3$	
11	$11 - 12 = -1$	
12	$12 - 12 = 0$	
15	$15 - 12 = 3$	
18	$18 - 12 = 6$	
	0	

The sum of the values in the Data − Mean column should always be zero; if not, you have made an error. (If a rounded value of $\bar{x}$ is used, the sum of the values in the Data − Mean column will not always be exactly zero; however, the sum will be very close to zero.)

Next square the values in the second column and place the squares in the third column (Table 13.5).

Table 13.5

Data	Data − Mean	(Data − Mean)^2
7	−5	$(-5)^2 = (-5)(-5) = 25$
9	−3	$(-3)^2 = (-3)(-3) = 9$
11	−1	$(-1)^2 = (-1)(-1) = 1$
12	0	$(0)^2 = (0)(0) = 0$
15	3	$(3)^2 = (3)(3) = 9$
18	6	$(6)^2 = (6)(6) = 36$
	0	80

Add the squares in the third column. In this case, the sum is 80. Divide this sum by one less than the number of pieces of data $(n - 1)$. In this case, the number of pieces of data is 6. Therefore, we divide by 5 and get

$$\frac{80}{5} = 16^*$$

Finally, take the square root of this number. Since $\sqrt{16} = 4$, the standard deviation, symbolized s, is 4.

Now we will develop a formula for determining the standard deviation of a set of data. If we call the individual data x and the mean $\bar{x}$, we could write the three column heads Data, Data − Mean, and (Data − Mean)^2 in Table 13.4 as

$$x \qquad x - \bar{x} \qquad (x - \bar{x})^2$$

Let's follow the procedure we used to obtain the standard deviation in Example 2. We found the sum of the (Data − Mean)^2 column, which is the same as the sum of the $(x - \bar{x})^2$ column. We can represent the sum of the $(x - \bar{x})^2$ column by using the summation notation, $\Sigma(x - \bar{x})^2$. Thus, in Table 13.5, $\Sigma(x - \bar{x})^2 = 80$. We then divided this number by 1 less than the number of pieces of data, $n - 1$. Thus, we have

$$\frac{\Sigma(x - \bar{x})^2}{n - 1}$$

*16 is the variance, symbolized s^2, of this set of data.

Finally, we took the square root of this value to obtain the standard deviation.

Standard Deviation

$$s = \sqrt{\frac{\Sigma(x - \bar{x})^2}{n - 1}}$$

EXAMPLE 3 *Determine the Standard Deviation of Stock Prices*

The following are the prices of nine stocks on the New York Stock Exchange. Determine the standard deviation of the prices.

$17, $28, $32, $36, $50, $52, $66, $74, $104

SOLUTION The mean, $\bar{x}$, is

$$\bar{x} = \frac{\Sigma x}{n} = \frac{17 + 28 + 32 + 36 + 50 + 52 + 66 + 74 + 104}{9} = \frac{459}{9} = 51$$

The mean is $51.

Table 13.6

x	$x - \bar{x}$	$(x - \bar{x})^2$
17	−34	1156
28	−23	529
32	−19	361
36	−15	225
50	−1	1
52	1	1
66	15	225
74	23	529
104	53	2809
	0	5836

Table 13.6 shows us that $\Sigma(x - \bar{x})^2 = 5836$. Since there are nine pieces of data, $n - 1 = 9 - 1$, or 8.

$$s = \sqrt{\frac{\Sigma(x - \bar{x})^2}{n - 1}} = \sqrt{\frac{5836}{8}} = \sqrt{729.5} \approx 27.01$$

The standard deviation, to the nearest tenth, is $27.01.

Standard deviation will be used in Section 13.7 to find the percent of data between any two values in a normal curve. Standard deviations are also often used in determining norms for a population (see Exercise 31).

TECHNOLOGY TIP In this Technology Tip, we will explain how to find the standard deviation using Excel as well as with the TI-83 Plus and TI-84 Plus graphing calculators. In our illustration, we will use the data from Example 3 on page 876, which represent the prices of nine stocks on the New York Stock Exchange.

EXCEL

The instructions used to determine the standard deviation are very similar to those used to determine the mean in the Technology Tip on pages 866–867 in Section 13.5. Please read that material now. Then enter the nine pieces of data in columns A1–A9 and press the Enter key. Now select the following:

Insert > Function . . . > Statistical > STDEV

Then click the [OK] box. The program will then generate a gray box where you need to enter the data. In the area to the right of **Number1** you need to enter the data for which you want to find the standard deviation. Since you have already entered the data in column A, rows 1 to 9, if A1:A9 is not already listed, you can enter A1:A9 in the area to the right of **Number1**. At the bottom of the gray box, *Formula Results* = 27.00925767, which is the standard deviation, is displayed. If you click OK, Excel will place the standard deviation in cell A10.

TI-83 PLUS AND TI-84 PLUS GRAPHING CALCULATORS

To find the standard deviation on Texas Instruments graphing calculators, follow the instructions for finding the mean in the Technology Tip on page 867 in Section 13.5. As explained there, press [STAT] [>] [EDIT] [>] [ENTER]. Remove existing data by highlighting **L1** and then pressing [CLEAR] [>] [ENTER]. Then enter the nine pieces of data, pressing the Enter key after each entry. Then press [STAT] [>] [CALC] [>] [ENTER] [>] [ENTER]. The fourth statistic down is $S_x = 27.00925767$. This value is the standard deviation.

SECTION 13.6 EXERCISES

CONCEPT/WRITING EXERCISES

1. Explain how to find the range of a set of data.
2. What does the standard deviation of a set of data measure?
3. Explain how to find the standard deviation of a set of data.
4. Why is measuring dispersion in observed data important?
5. What is the standard deviation of a set of data in which all the data values are the same? Explain.
6. What symbol is used to represent the sample standard deviation?
7. What symbol is used to represent the population standard deviation?
8. Can you think of any situations in which a large standard deviation may be desirable? Explain.
9. Can you think of any situations in which a small standard deviation may be desirable? Explain.
10. Without actually doing the calculations, decide which, if either, of the following two sets of data will have the greater standard deviation. Explain why.

    10, 13, 14, 15, 17, 21 16, 17, 17, 18, 18, 19

11. Without actually doing the calculations, decide which, if either, of the following two sets of data will have the greater standard deviation. Explain why.

    2, 4, 6, 8, 10 102, 104, 106, 108, 110

12. By studying the standard deviation formula, explain why the standard deviation of a set of data will always be greater than or equal to 0.
13. Patricia Wolff teaches two statistics classes, one in the morning and the other in the evening. On the midterm exam,

the morning class had a mean of 75.2 and a standard deviation of 5.7. The evening class had a mean of 75.2 and a standard deviation of 12.5.

a) How do the means compare?

b) If we compare the set of scores from the first class with those in the second class, how will the distributions of the two sets of scores compare? Explain.

14. Explain why the standard deviation is usually a better measure of dispersion than the range.

PRACTICE THE SKILLS

In Exercises 15–22, determine the range and standard deviation of the set of data. When appropriate, round standard deviations to the nearest hundredth.

15. 11, 9, 6, 12, 17

16. 15, 15, 19, 21, 13, 13

17. 130, 131, 132, 133, 134, 135, 136

18. 3, 7, 8, 12, 0, 9, 11, 12, 6, 2

19. 4, 8, 9, 11, 13, 15

20. 9, 9, 9, 9, 9, 9, 9

21. 7, 9, 7, 9, 9, 10, 12

22. 60, 58, 62, 67, 48, 51, 72, 70

23. ***Toaster Ovens*** Determine the range and standard deviation of the following prices of selected toaster ovens: \$58, \$58, \$80, \$75, \$60, \$75, \$78, \$48, \$75, \$53.

24. ***Years until Retirement*** Seven employees at a large company were asked the number of additional years they planned to work before retirement. Their responses were 10, 23, 28, 4, 1, 6, 12. Determine the range and standard deviation of the number of years.

25. ***Fishing Poles*** Determine the range and standard deviation of the following prices of selected fishing poles: \$50, \$120, \$130, \$60, \$55, \$75, \$200, \$110, \$125, \$175.

26. ***Prescription Prices*** The amount of money seven people spent on prescription medication in a year are as follows: \$600, \$100, \$850, \$350, \$250, \$140, \$300. Determine the range and standard deviation of the amounts.

PROBLEM SOLVING

27. ***Count Your Money*** Six people were asked to determine the amount of money they were carrying, to the nearest dollar. The results were

\$32, \$60, \$14, \$25, \$5, \$68

a) Determine the range and standard deviation of the amounts.

b) Add \$10 to each of the six amounts. How do you expect the range and standard deviation of the new set of data to change? Explain your answer.

c) Determine the range and standard deviation of the new set of data. Do the results agree with your answer to part (b)? If not, explain why.

28. **a)** ***Adding to or Subtracting from Each Number*** Pick any five numbers. Compute the mean and the standard deviation of this distribution.

b) Add 20 to each of the numbers in your original distribution and compute the mean and the standard deviation of this new distribution.

c) Subtract 5 from each number in your original distribution and compute the mean and standard deviation of this new distribution.

d) What conclusions can you draw about changes in the mean and the standard deviation when the same number is added to or subtracted from each piece of data in a distribution?

e) How will the mean and standard deviation of the numbers 8, 9, 10, 11, 12, 13, 14 differ from the mean and standard deviation of the numbers 648, 649, 650, 651, 652, 653, 654? Determine the mean and standard deviation of both sets of numbers.

29. **a)** ***Multiplying Each Number*** Pick any five numbers. Compute the mean and standard deviation of this distribution.

b) Multiply each number in your distribution by 3 and compute the mean and the standard deviation of this new distribution.

c) Multiply each number in your original distribution by 9 and compute the mean and the standard deviation of this new distribution.

d) What conclusions can you draw about changes in the mean and the standard deviation when each value in a distribution is multiplied by the same number?

e) The mean and standard deviation of the distribution 1, 3, 4, 4, 5, 7 are 4 and 2, respectively. Use the conclusion drawn in part (d) to determine the mean and standard deviation of the distribution

5, 15, 20, 20, 25, 35

30. ***Waiting in Line*** Consider the following illustrations of two bank-customer waiting systems.

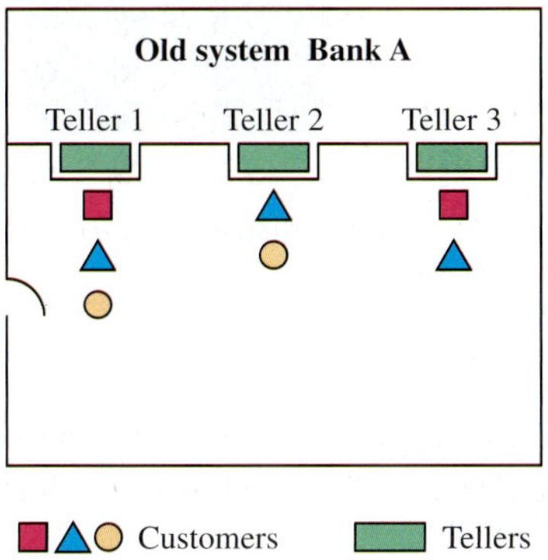

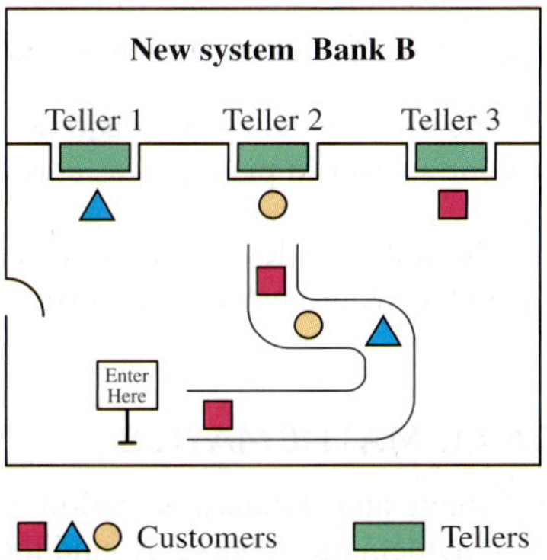

a) How would you expect the mean waiting time in Bank A to compare with the mean waiting time in Bank B? Explain your answer.

b) How would you expect the standard deviation of waiting times in Bank A to compare with the standard deviation of waiting times in Bank B? Explain your answer.

31. ***Height and Weight Distribution*** The chart shown on the right uses the symbol σ to represent the standard deviation. Note that 2σ represents the value that is two standard deviations above the mean; -2σ represents the value that is two standard deviations below the mean. The unshaded areas, from two standard deviations below the mean to two standard deviations above the mean, are considered the normal range. For example, the average (mean) 8-year-old boy has a height of about 50 inches, but any heights between approximately 45 inches and 55 inches are considered normal for 8-year-old boys. Refer to the chart below to answer the following questions.

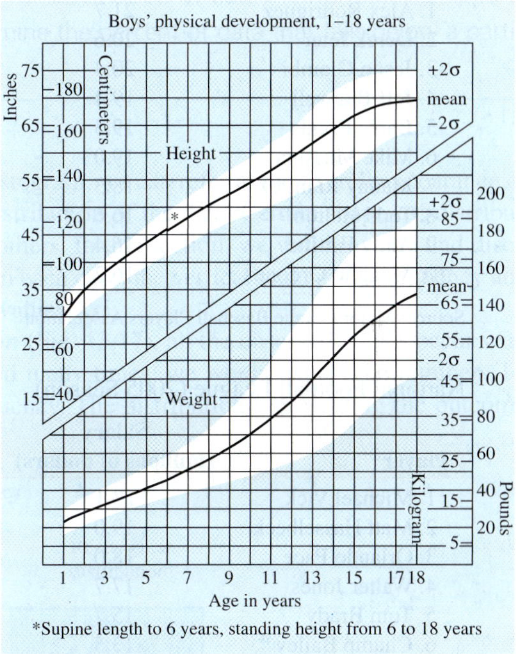

a) What happens to the standard deviation for weights of boys as the age of boys increases? What is the significance of this fact?

b) At age 16, what is the mean weight, in pounds, of boys?

c) What is the approximate standard deviation of boys' weights at age 16?

d) Determine the mean weight and normal range for boys at age 13.

e) Determine the mean height and normal range for boys at age 13.

f) Assuming that this chart was constructed so that approximately 95% of all boys are always in the normal range, determine what percentage of boys are not in the normal range.

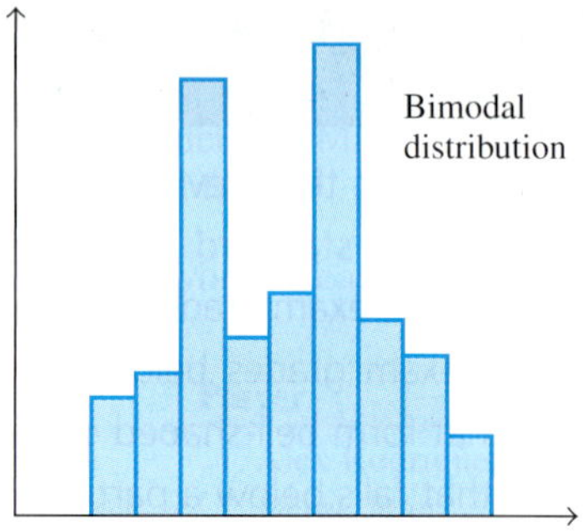

Figure 13.19

A ***bimodal distribution*** (Fig. 13.19) is one in which two nonadjacent values occur more frequently than any other values in a set of data. For example, if an equal number of men and women were weighed, the distribution of their weights would probably be bimodal, with one mode for the women's weights and the second for the men's weights. For a distribution to be considered bimodal, both modes need not have the same frequency but they must both have a frequency greater than the frequency of each of the other values in the distribution.

The life expectancy of lightbulbs has a bimodal distribution: a small peak very near 0 hours of life, resulting from the bulbs that burned out very quickly because of a manufacturing defect, and a much higher peak representing the nondefective bulbs. A bimodal frequency distribution generally means that you are dealing with two distinct populations, in this case, defective and nondefective lightbulbs.

Another distribution, called a ***skewed distribution***, has more of a "tail" on one side than the other. A skewed distribution with a tail on the right (Fig. 13.20a) is said to be skewed to the right. If the tail is on the left (Fig. 13.20b), the distribution is referred to as skewed to the left.

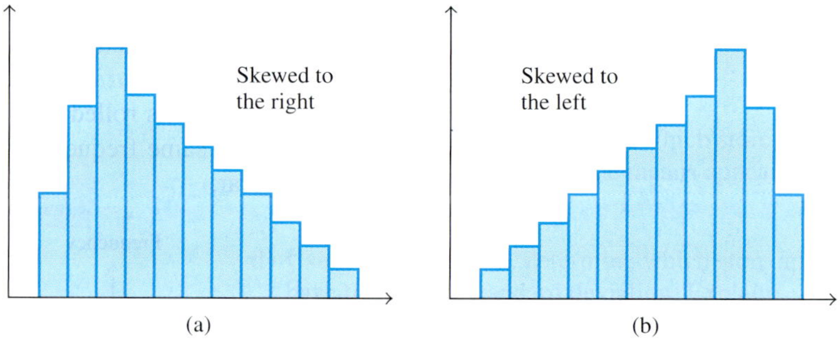

Figure 13.20

The number of children per family might be a distribution skewed to the right. Some families have no children, more families may have one child, the greatest percentage may have two children, fewer may have three children, still fewer may have four children, and so on.

Since few families have high incomes, distributions of family incomes might be skewed to the right.

Smoothing the histograms of the skewed distributions shown in Fig. 13.20 to form curves gives the curves illustrated in Fig. 13.21.

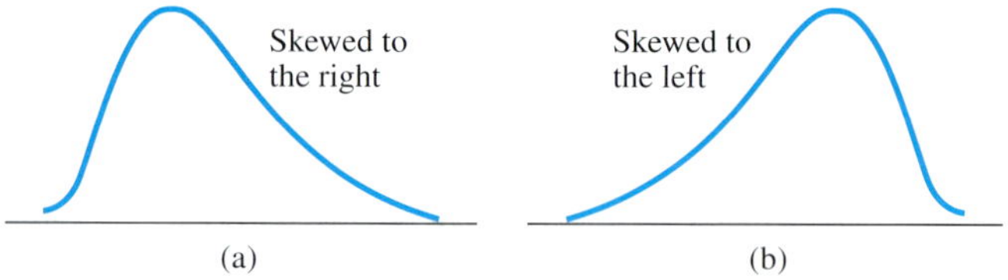

Figure 13.21

In Fig. 13.21(a), the greatest frequency appears on the left side of the curve and the frequency decreases from left to right. Since the mode is the value with the greatest frequency, the mode would appear on the left side of the curve.

Every value in the set of data is considered in determining the mean. The values on the far right side of the curve in Fig. 13.21(a) would tend to increase the value of the mean. Thus, the value of the mean would be farther to the right than the mode. The median

DID YOU KNOW?

What Conclusions Can You Draw?

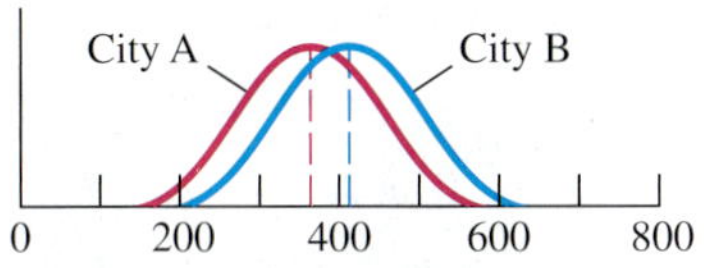

Based on the figure, which shows the distribution of scores on the mathematics part of the SAT test for two different cities, can we say that any given person selected at random from city B has outperformed any given person selected at random from city A? Both distributions appear normal, and the mean of city A is slightly smaller than the mean of city B. Consider, however, two randomly selected students who took this test: Sally from city A and Kendra from city B. The graph shows that many students in city A outperformed students from city B, so we cannot conclude that Sally scored higher than Kendra or that Kendra scored higher than Sally.

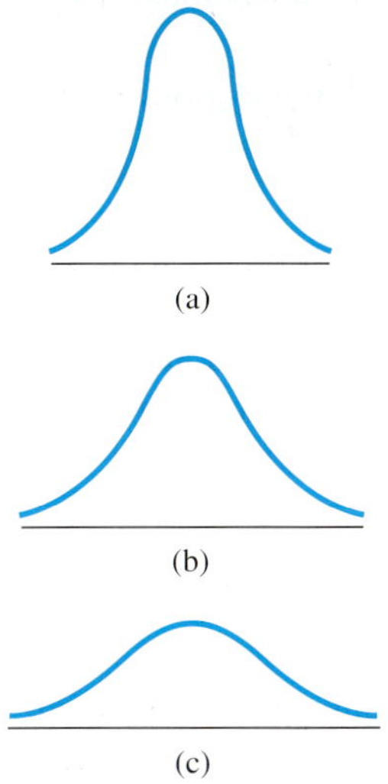

Figure 13.24

would be between the mode and the mean. The rel
and mode for curves that are skewed to the right ar

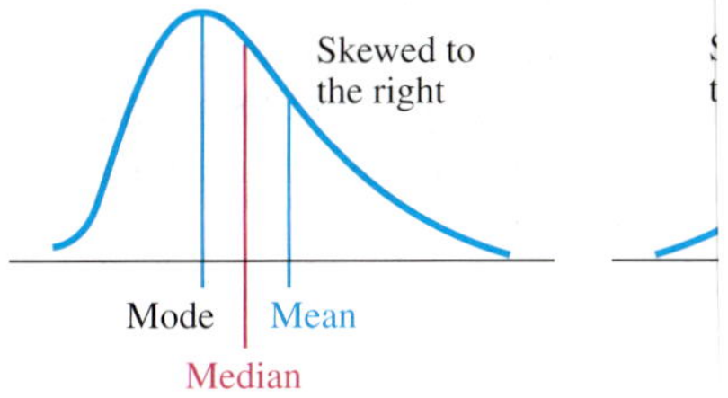

(a)

Figure 13.22

Normal Distributions

Each of these distributions is useful in describing
portant distribution is the *normal* or *Gaussian dis*
ematician Carl Friedrich Gauss. The histogram o
in Fig. 13.23.

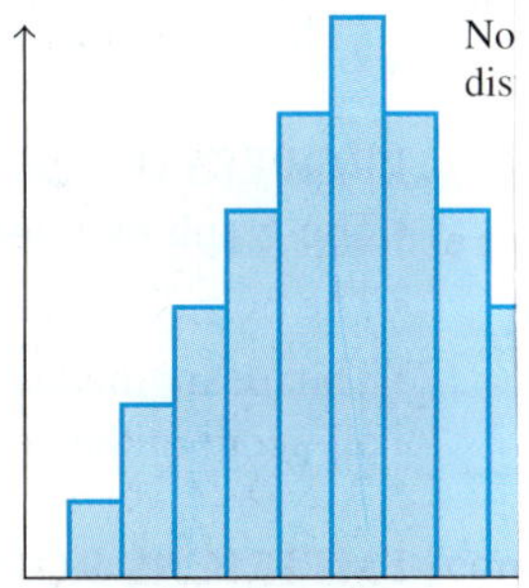

Figure 13.23

The normal distribution is important becaus
distributed or closely resemble a normal distri
intelligence quotients, heights and weights of mal
lengths of full-grown boa constrictors, weights o
automobile brakes, and life spans of refrigerators,

The normal distribution is symmetric about th
togram of a normal distribution down the middle, th
actly. **In a normal distribution, the mean, median,**

When the histogram of a normal distributior
curve is bell-shaped. The bell may be high and na
three curves in Fig. 13.24 represents a normal cur
standard deviation (spread from the mean); curve

When we work with a distribution, we are
Therefore, when we discuss the normal distributio
the standard deviation.

Since the curve representing the normal distrib
always falls above (to the right of) the mean and 50
of) the mean. In addition, every normal distribution
between the value that is one standard deviation bel
that is one standard deviation above the mean, μ

d) What conclusions can you draw about changes in the mean and the standard deviation when each value in a distribution is multiplied by the same number?

e) The mean and standard deviation of the distribution 1, 3, 4, 4, 5, 7 are 4 and 2, respectively. Use the conclusion drawn in part (d) to determine the mean and standard deviation of the distribution

5, 15, 20, 20, 25, 35

30. ***Waiting in Line*** Consider the following illustrations of two bank-customer waiting systems.

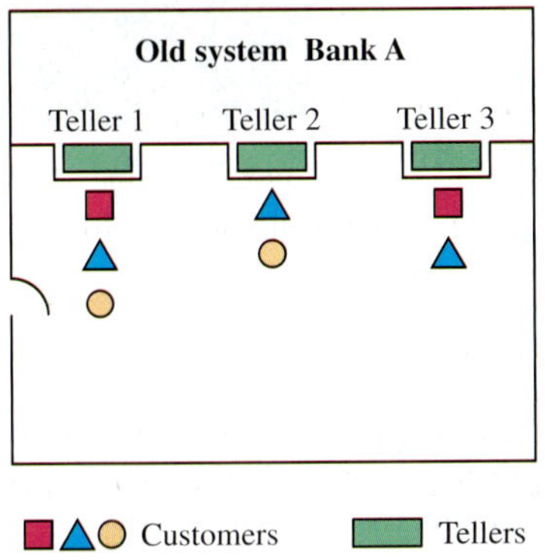

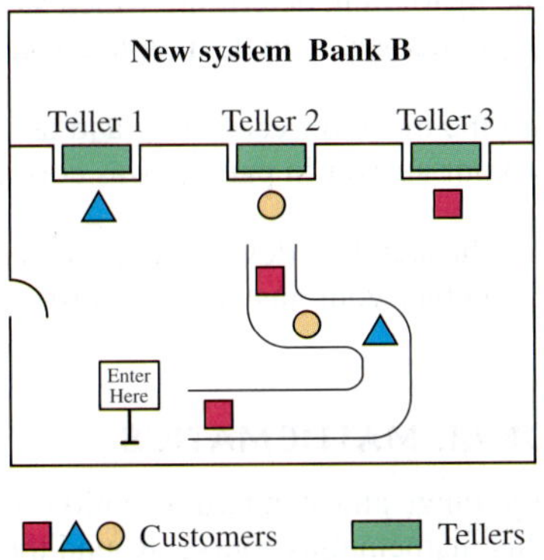

a) How would you expect the mean waiting time in Bank A to compare with the mean waiting time in Bank B? Explain your answer.

b) How would you expect the standard deviation of waiting times in Bank A to compare with the standard deviation of waiting times in Bank B? Explain your answer.

31. ***Height and Weight Distribution*** The chart shown on the right uses the symbol σ to represent the standard deviation. Note that 2σ represents the value that is two standard deviations above the mean; -2σ represents the value that is two standard deviations below the mean. The unshaded areas, from two standard deviations below the mean to two standard deviations above the mean, are considered the normal range. For example, the average (mean) 8-year-old boy has a height of about 50 inches, but any heights between approximately 45 inches and 55 inches are considered normal for 8-year-old boys. Refer to the chart below to answer the following questions.

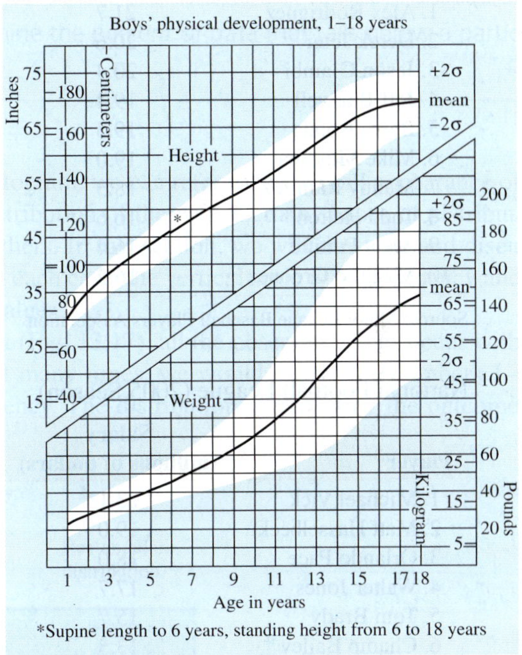

a) What happens to the standard deviation for weights of boys as the age of boys increases? What is the significance of this fact?

b) At age 16, what is the mean weight, in pounds, of boys?

c) What is the approximate standard deviation of boys' weights at age 16?

d) Determine the mean weight and normal range for boys at age 13.

e) Determine the mean height and normal range for boys at age 13.

f) Assuming that this chart was constructed so that approximately 95% of all boys are always in the normal range, determine what percentage of boys are not in the normal range.

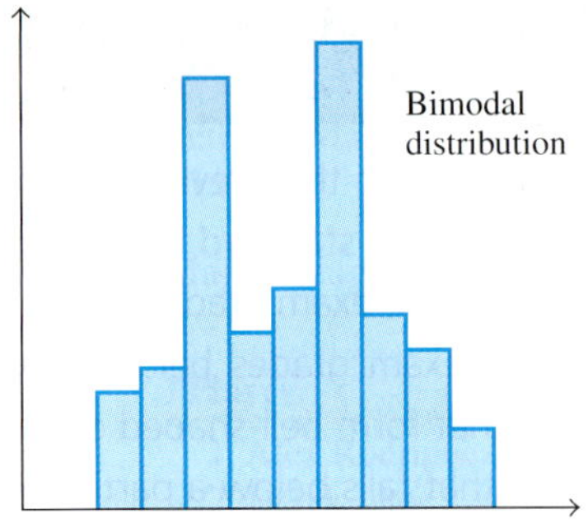

Figure 13.19

A *bimodal distribution* (Fig. 13.19) is one
occur more frequently than any other values in a s
number of men and women were weighed, the d
probably be bimodal, with one mode for the wome
men's weights. For a distribution to be considered l
the same frequency but they must both have a frequ
each of the other values in the distribution.

The life expectancy of lightbulbs has a bimo
near 0 hours of life, resulting from the bulbs that bu
manufacturing defect, and a much higher peak repr
bimodal frequency distribution generally means tha
populations, in this case, defective and nondefectiv

Another distribution, called a *skewed distribu*
side than the other. A skewed distribution with a ta
to be skewed to the right. If the tail is on the left (
ferred to as skewed to the left.

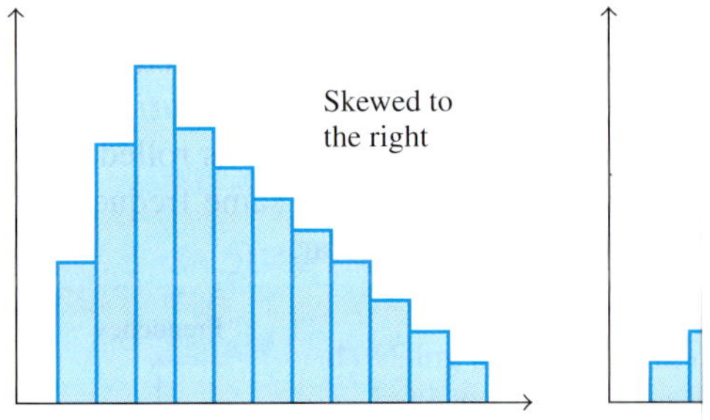

(a)

Figure 13.20

The number of children per family might be a
Some families have no children, more families ma
centage may have two children, fewer may have th
four children, and so on.

Since few families have high incomes, distribu
skewed to the right.

Smoothing the histograms of the skewed dist
form curves gives the curves illustrated in Fig. 13.2

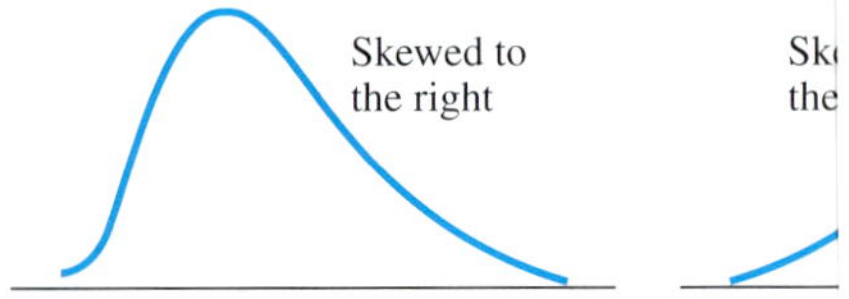

(a)

Figure 13.21

In Fig. 13.21(a), the greatest frequency appear
the frequency decreases from left to right. Since the
est frequency, the mode would appear on the left sic

Every value in the set of data is considered in de
the far right side of the curve in Fig. 13.21(a) would te
Thus, the value of the mean would be farther to the

b) $$z_{115} = \frac{115 - 100}{10} = \frac{15}{10} = 1.5$$

A score of 115 is 1.5 standard deviations above the mean.

c) $$z_{100} = \frac{100 - 100}{10} = \frac{0}{10} = 0$$

The mean always has a z-score of 0.

d) $$z_{84} = \frac{84 - 100}{10} = \frac{-16}{10} = -1.6$$

A score of 84 is 1.6 standard deviations below the mean. ●

If we are given any normal distribution with a known mean and standard deviation, it is possible through the use of Table 13.7 on pages 888 and 889 (the z-table) to determine the percent of data between any two given values. The total area under any normal curve is 1.00. Table 13.7 will be used to determine the cumulative area under the normal curve that lies to the *left of a specified z-value*. We will use Table 13.7(a) when we wish to determine area to the left of a *negative z-value*, and we will use Table 13.7(b) when we wish to determine area to the left of a *positive z-value*.

Example 3 illustrates the procedure to follow when using Table 13.7 to determine the area under the normal curve. When you are determining the area under the normal curve, it is often helpful to draw a picture and shade the area to be determined.

EXAMPLE 3 *Determining the Area under the Normal Curve*

Determine the area under the normal curve

a) to the left of $z = -1.00$.
b) to the left of $z = 1.19$.
c) to the right of $z = 1.19$.
d) between $z = -1.62$ and $z = 2.57$.

SOLUTION

a) To determine the area under the normal curve to the left of $z = -1.00$, as illustrated in Fig. 13.26, we use Table 13.7(a) since we are looking for an area to the left of a negative z-score. In the upper-left corner of the table, we see the letter z. The column under z gives the units and the tenths value for z. To locate the hundredths value of z, we use the column headings to the right of z. In this case, the hundredths value of $z = -1.00$ is 0, so we use the first column labeled .00. To determine the area to the left of $z = -1.00$, we use the row labeled -1.0 and move to the column labeled .00. The table entry, .1587, is circled in blue. Therefore, the total area to the left of $z = -1.00$ is 0.1587.

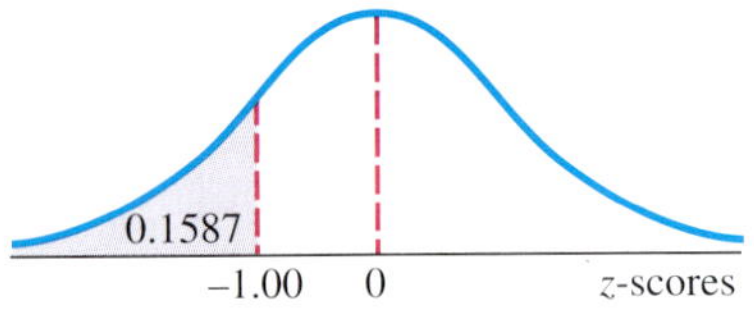

Figure 13.26

b) To determine the area under the normal curve to the left of $z = 1.19$ (Fig. 13.27), we use Table 13.7(b) since we are looking for an area to the left of a positive z-score. We first look for 1.1 in the column under z. Since the hundredths value of $z = 1.19$ is 9, we move to the column labeled .09. Using the row labeled 1.1

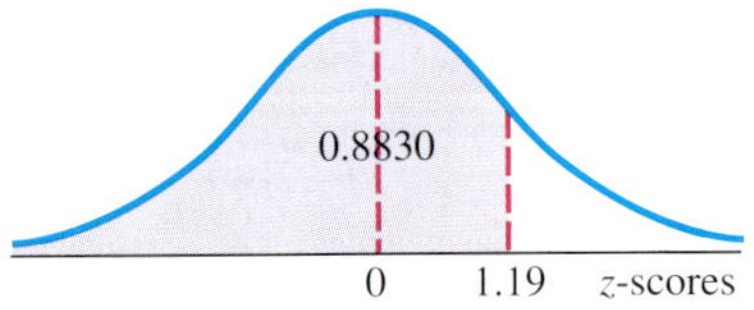

Figure 13.27

d) What conclusions can you draw about changes in the mean and the standard deviation when each value in a distribution is multiplied by the same number?

e) The mean and standard deviation of the distribution 1, 3, 4, 4, 5, 7 are 4 and 2, respectively. Use the conclusion drawn in part (d) to determine the mean and standard deviation of the distribution

5, 15, 20, 20, 25, 35

30. ***Waiting in Line*** Consider the following illustrations of two bank-customer waiting systems.

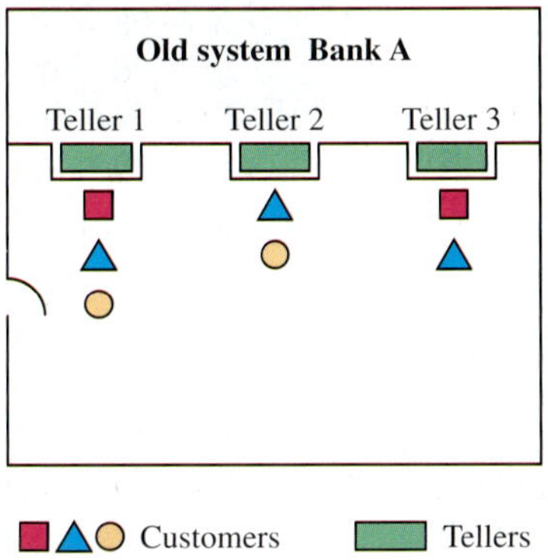

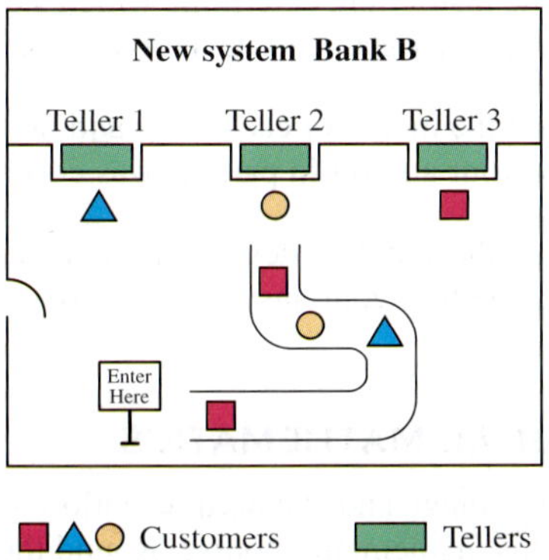

a) How would you expect the mean waiting time in Bank A to compare with the mean waiting time in Bank B? Explain your answer.

b) How would you expect the standard deviation of waiting times in Bank A to compare with the standard deviation of waiting times in Bank B? Explain your answer.

31. ***Height and Weight Distribution*** The chart shown on the right uses the symbol σ to represent the standard deviation. Note that 2σ represents the value that is two standard deviations above the mean; -2σ represents the value that is two standard deviations below the mean. The unshaded areas, from two standard deviations below the mean to two standard deviations above the mean, are considered the normal range. For example, the average (mean) 8-year-old boy has a height of about 50 inches, but any heights between approximately 45 inches and 55 inches are considered normal for 8-year-old boys. Refer to the chart below to answer the following questions.

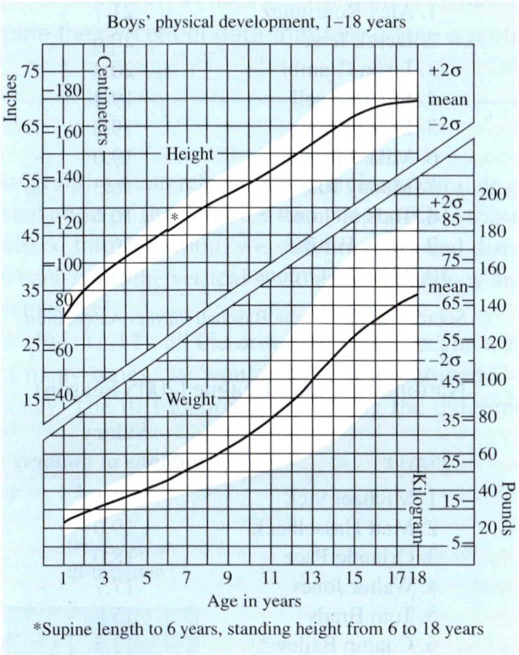

a) What happens to the standard deviation for weights of boys as the age of boys increases? What is the significance of this fact?

b) At age 16, what is the mean weight, in pounds, of boys?

c) What is the approximate standard deviation of boys' weights at age 16?

d) Determine the mean weight and normal range for boys at age 13.

e) Determine the mean height and normal range for boys at age 13.

f) Assuming that this chart was constructed so that approximately 95% of all boys are always in the normal range, determine what percentage of boys are not in the normal range.

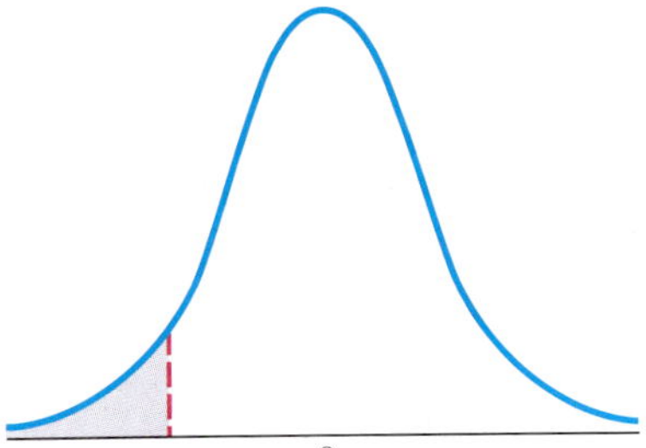

Table entry for z is the area to the left of z.

Table 13.7 Areas of a Standard Normal Distribution

(a) Table of Areas to the Left of z When z Is Negative

z	.00	.01	.02	.03	.04	.05	.06	.07	.08	.09
−3.4	.0003	.0003	.0003	.0003	.0003	.0003	.0003	.0003	.0003	.0002
−3.3	.0005	.0005	.0005	.0004	.0004	.0004	.0004	.0004	.0004	.0003
−3.2	.0007	.0007	.0006	.0006	.0006	.0006	.0006	.0005	.0005	.0005
−3.1	.0010	.0009	.0009	.0009	.0008	.0008	.0008	.0008	.0007	.0007
−3.0	.0013	.0013	.0013	.0012	.0012	.0011	.0011	.0011	.0010	.0010
−2.9	.0019	.0018	.0018	.0017	.0016	.0016	.0015	.0015	.0014	.0014
−2.8	.0026	.0025	.0024	.0023	.0023	.0022	.0021	.0021	.0020	.0019
−2.7	.0035	.0034	.0033	.0032	.0031	.0030	.0029	.0028	.0027	.0026
−2.6	.0047	.0045	.0044	.0043	.0041	.0040	.0039	.0038	.0037	.0036
−2.5	.0062	.0060	.0059	.0057	.0055	.0054	.0052	.0051	.0049	.0048
−2.4	.0082	.0080	.0078	.0075	.0073	.0071	.0069	.0068	.0066	.0064
−2.3	.0107	.0104	.0102	.0099	.0096	.0094	.0091	.0089	.0087	.0084
−2.2	.0139	.0136	.0132	.0129	.0125	.0122	.0119	.0116	.0113	.0110
−2.1	.0179	.0174	.0170	.0166	.0162	.0158	.0154	.0150	.0146	.0143
−2.0	.0228	.0222	.0217	.0212	.0207	.0202	.0197	.0192	.0188	.0183
−1.9	.0287	.0281	.0274	.0268	.0262	.0256	.0250	.0244	.0239	.0233
−1.8	.0359	.0351	.0344	.0336	.0329	.0322	.0314	.0307	.0301	.0294
−1.7	.0446	.0436	.0427	.0418	.0409	.0401	.0392	.0384	.0375	.0367
−1.6	.0548	.0537	.0526	.0516	.0505	.0495	.0485	.0475	.0465	.0455
−1.5	.0668	.0655	.0643	.0630	.0618	.0606	.0594	.0582	.0571	.0559
−1.4	.0808	.0793	.0778	.0764	.0749	.0735	.0721	.0708	.0694	.0681
−1.3	.0968	.0951	.0934	.0918	.0901	.0885	.0869	.0853	.0838	.0823
−1.2	.1151	.1131	.1112	.1093	.1075	.1056	.1038	.1020	.1003	.0985
−1.1	.1357	.1335	.1314	.1292	.1271	.1251	.1230	.1210	.1190	.1170
−1.0	.1587	.1562	.1539	.1515	.1492	.1469	.1446	.1423	.1401	.1379
−0.9	.1841	.1814	.1788	.1762	.1736	.1711	.1685	.1660	.1635	.1611
−0.8	.2119	.2090	.2061	.2033	.2005	.1977	.1949	.1922	.1894	.1867
−0.7	.2420	.2389	.2358	.2327	.2296	.2266	.2236	.2206	.2177	.2148
−0.6	.2743	.2709	.2676	.2643	.2611	.2578	.2546	.2514	.2483	.2451
−0.5	.3085	.3050	.3015	.2981	.2947	.2912	.2877	.2843	.2810	.2776
−0.4	.3446	.3409	.3372	.3336	.3300	.3264	.3228	.3192	.3156	.3121
−0.3	.3821	.3783	.3745	.3707	.3669	.3632	.3594	.3557	.3520	.3483
−0.2	.4207	.4168	.4129	.4090	.4052	.4013	.3974	.3936	.3897	.3859
−0.1	.4602	.4562	.4522	.4483	.4443	.4404	.4364	.4325	.4286	.4247
−0.0	.5000	.4960	.4920	.4880	.4840	.4801	.4761	.4721	.4681	.4641

For values of z less than −3.49, use 0.000 to approximate the area.

d) What conclusions can you draw about changes in the mean and the standard deviation when each value in a distribution is multiplied by the same number?

e) The mean and standard deviation of the distribution 1, 3, 4, 4, 5, 7 are 4 and 2, respectively. Use the conclusion drawn in part (d) to determine the mean and standard deviation of the distribution

5, 15, 20, 20, 25, 35

30. ***Waiting in Line*** Consider the following illustrations of two bank-customer waiting systems.

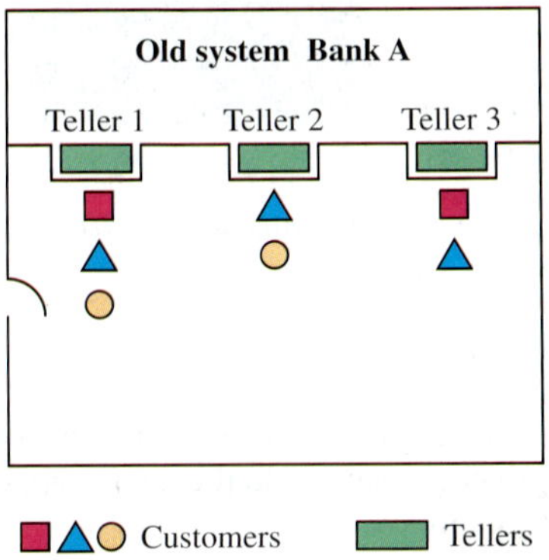

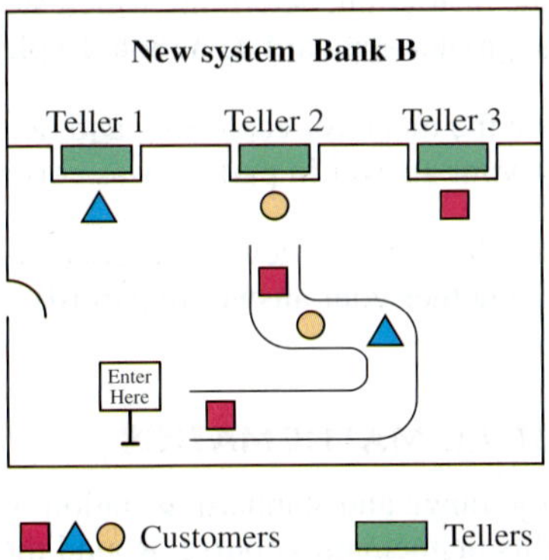

a) How would you expect the mean waiting time in Bank A to compare with the mean waiting time in Bank B? Explain your answer.

b) How would you expect the standard deviation of waiting times in Bank A to compare with the standard deviation of waiting times in Bank B? Explain your answer.

31. ***Height and Weight Distribution*** The chart shown on the right uses the symbol σ to represent the standard deviation. Note that 2σ represents the value that is two standard deviations above the mean; -2σ represents the value that is two standard deviations below the mean. The unshaded areas, from two standard deviations below the mean to two standard deviations above the mean, are considered the normal range. For example, the average (mean) 8-year-old boy has a height of about 50 inches, but any heights between approximately 45 inches and 55 inches are considered normal for 8-year-old boys. Refer to the chart below to answer the following questions.

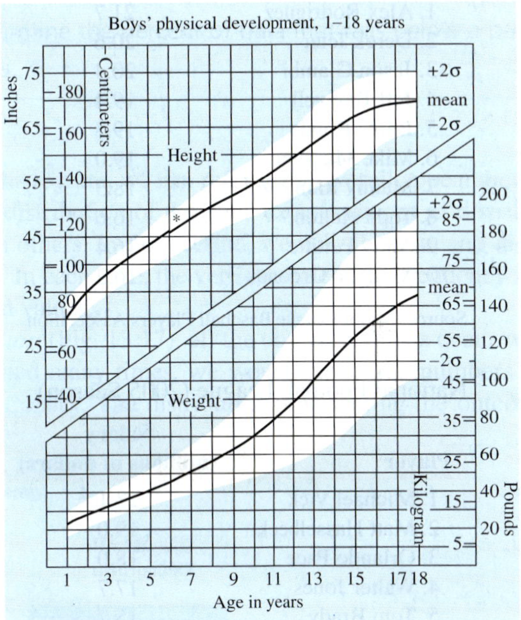

*Supine length to 6 years, standing height from 6 to 18 years

a) What happens to the standard deviation for weights of boys as the age of boys increases? What is the significance of this fact?

b) At age 16, what is the mean weight, in pounds, of boys?

c) What is the approximate standard deviation of boys' weights at age 16?

d) Determine the mean weight and normal range for boys at age 13.

e) Determine the mean height and normal range for boys at age 13.

f) Assuming that this chart was constructed so that approximately 95% of all boys are always in the normal range, determine what percentage of boys are not in the normal range.

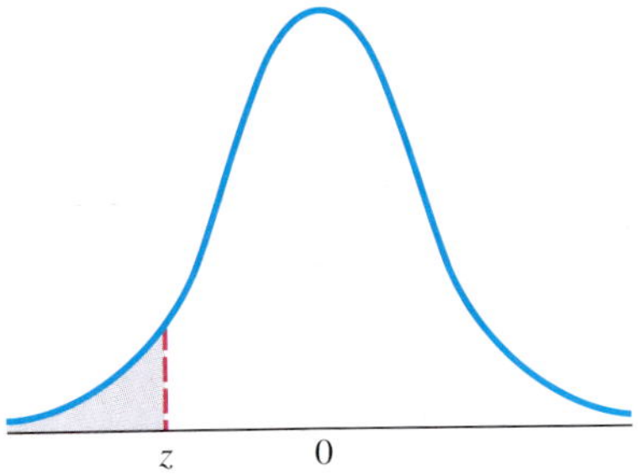

Table entry for z is the area to the left of z.

Table 13.7 Areas of a Standard Normal Distribution

(a) Table of Areas to the Left of z When z Is Negative

z	.00	.01	.02	.03	.04	.05	.06	.07	.08	.09
−3.4	.0003	.0003	.0003	.0003	.0003	.0003	.0003	.0003	.0003	.0002
−3.3	.0005	.0005	.0005	.0004	.0004	.0004	.0004	.0004	.0004	.0003
−3.2	.0007	.0007	.0006	.0006	.0006	.0006	.0006	.0005	.0005	.0005
−3.1	.0010	.0009	.0009	.0009	.0008	.0008	.0008	.0008	.0007	.0007
−3.0	.0013	.0013	.0013	.0012	.0012	.0011	.0011	.0011	.0010	.0010
−2.9	.0019	.0018	.0018	.0017	.0016	.0016	.0015	.0015	.0014	.0014
−2.8	.0026	.0025	.0024	.0023	.0023	.0022	.0021	.0021	.0020	.0019
−2.7	.0035	.0034	.0033	.0032	.0031	.0030	.0029	.0028	.0027	.0026
−2.6	.0047	.0045	.0044	.0043	.0041	.0040	.0039	.0038	.0037	.0036
−2.5	.0062	.0060	.0059	.0057	.0055	.0054	.0052	.0051	.0049	.0048
−2.4	.0082	.0080	.0078	.0075	.0073	.0071	.0069	.0068	.0066	.0064
−2.3	.0107	.0104	.0102	.0099	.0096	.0094	.0091	.0089	.0087	.0084
−2.2	.0139	.0136	.0132	.0129	.0125	.0122	.0119	.0116	.0113	.0110
−2.1	.0179	.0174	.0170	.0166	.0162	.0158	.0154	.0150	.0146	.0143
−2.0	.0228	.0222	.0217	.0212	.0207	.0202	.0197	.0192	.0188	.0183
−1.9	.0287	.0281	.0274	.0268	.0262	.0256	.0250	.0244	.0239	.0233
−1.8	.0359	.0351	.0344	.0336	.0329	.0322	.0314	.0307	.0301	.0294
−1.7	.0446	.0436	.0427	.0418	.0409	.0401	.0392	.0384	.0375	.0367
−1.6	.0548	.0537	.0526	.0516	.0505	.0495	.0485	.0475	.0465	.0455
−1.5	.0668	.0655	.0643	.0630	.0618	.0606	.0594	.0582	.0571	.0559
−1.4	.0808	.0793	.0778	.0764	.0749	.0735	.0721	.0708	.0694	.0681
−1.3	.0968	.0951	.0934	.0918	.0901	.0885	.0869	.0853	.0838	.0823
−1.2	.1151	.1131	.1112	.1093	.1075	.1056	.1038	.1020	.1003	.0985
−1.1	.1357	.1335	.1314	.1292	.1271	.1251	.1230	.1210	.1190	.1170
−1.0	.1587	.1562	.1539	.1515	.1492	.1469	.1446	.1423	.1401	.1379
−0.9	.1841	.1814	.1788	.1762	.1736	.1711	.1685	.1660	.1635	.1611
−0.8	.2119	.2090	.2061	.2033	.2005	.1977	.1949	.1922	.1894	.1867
−0.7	.2420	.2389	.2358	.2327	.2296	.2266	.2236	.2206	.2177	.2148
−0.6	.2743	.2709	.2676	.2643	.2611	.2578	.2546	.2514	.2483	.2451
−0.5	.3085	.3050	.3015	.2981	.2947	.2912	.2877	.2843	.2810	.2776
−0.4	.3446	.3409	.3372	.3336	.3300	.3264	.3228	.3192	.3156	.3121
−0.3	.3821	.3783	.3745	.3707	.3669	.3632	.3594	.3557	.3520	.3483
−0.2	.4207	.4168	.4129	.4090	.4052	.4013	.3974	.3936	.3897	.3859
−0.1	.4602	.4562	.4522	.4483	.4443	.4404	.4364	.4325	.4286	.4247
−0.0	.5000	.4960	.4920	.4880	.4840	.4801	.4761	.4721	.4681	.4641

For values of z less than −3.49, use 0.000 to approximate the area.

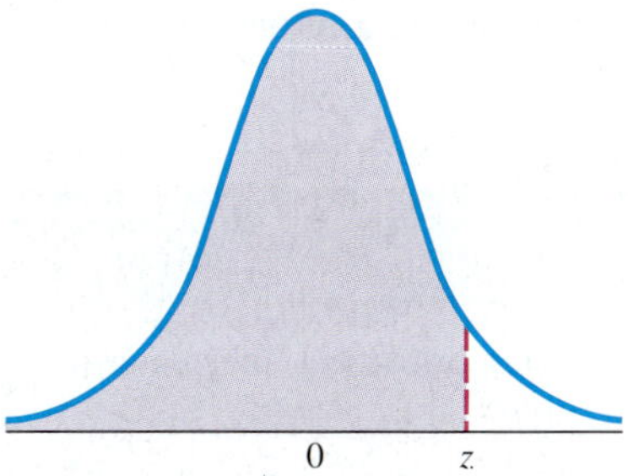

Table entry for z is the area to the left of z.

Areas of a Standard Normal Distribution (*continued*)

(b) Table of Areas to the Left of z When z Is Positive

z	.00	.01	.02	.03	.04	.05	.06	.07	.08	.09
0.0	.5000	.5040	.5080	.5120	.5160	.5199	.5239	.5279	.5319	.5359
0.1	.5398	.5438	.5478	.5517	.5557	.5596	.5636	.5675	.5714	.5753
0.2	.5793	.5832	.5871	.5910	.5948	.5987	.6026	.6064	.6103	.6141
0.3	.6179	.6217	.6255	.6293	.6331	.6368	.6406	.6443	.6480	.6517
0.4	.6554	.6591	.6628	.6664	.6700	.6736	.6772	.6808	.6844	.6879
0.5	.6915	.6950	.6985	.7019	.7054	.7088	.7123	.7157	.7190	.7224
0.6	.7257	.7291	.7324	.7357	.7389	.7422	.7454	.7486	.7517	.7549
0.7	.7580	.7611	.7642	.7673	.7704	.7734	.7764	.7794	.7823	.7852
0.8	.7881	.7910	.7939	.7967	.7995	.8023	.8051	.8078	.8106	.8133
0.9	.8159	.8186	.8212	.8238	.8264	.8289	.8315	.8340	.8365	.8389
1.0	.8413	.8438	.8461	.8485	.8508	.8531	.8554	.8577	.8599	.8621
1.1	.8643	.8665	.8686	.8708	.8729	.8749	.8770	.8790	.8810	.8830
1.2	.8849	.8869	.8888	.8907	.8925	.8944	.8962	.8980	.8997	.9015
1.3	.9032	.9049	.9066	.9082	.9099	.9115	.9131	.9147	.9162	.9177
1.4	.9192	.9207	.9222	.9236	.9251	.9265	.9279	.9292	.9306	.9319
1.5	.9332	.9345	.9357	.9370	.9382	.9394	.9406	.9418	.9429	.9441
1.6	.9452	.9463	.9474	.9484	.9495	.9505	.9515	.9525	.9535	.9545
1.7	.9554	.9564	.9573	.9582	.9591	.9599	.9608	.9616	.9625	.9633
1.8	.9641	.9649	.9656	.9664	.9671	.9678	.9686	.9693	.9699	.9706
1.9	.9713	.9719	.9726	.9732	.9738	.9744	.9750	.9756	.9761	.9767
2.0	.9772	.9778	.9783	.9788	.9793	.9798	.9803	.9808	.9812	.9817
2.1	.9821	.9826	.9830	.9834	.9838	.9842	.9846	.9850	.9854	.9857
2.2	.9861	.9864	.9868	.9871	.9875	.9878	.9881	.9884	.9887	.9890
2.3	.9893	.9896	.9898	.9901	.9904	.9906	.9909	.9911	.9913	.9916
2.4	.9918	.9920	.9922	.9925	.9927	.9929	.9931	.9932	.9934	.9936
2.5	.9938	.9940	.9941	.9943	.9945	.9946	.9948	.9949	.9951	.9952
2.6	.9953	.9955	.9956	.9957	.9959	.9960	.9961	.9962	.9963	.9964
2.7	.9965	.9966	.9967	.9968	.9969	.9970	.9971	.9972	.9973	.9974
2.8	.9974	.9975	.9976	.9977	.9977	.9978	.9979	.9979	.9980	.9981
2.9	.9981	.9982	.9982	.9983	.9984	.9984	.9985	.9985	.9986	.9986
3.0	.9987	.9987	.9987	.9988	.9988	.9989	.9989	.9989	.9990	.9990
3.1	.9990	.9991	.9991	.9991	.9992	.9992	.9992	.9992	.9993	.9993
3.2	.9993	.9993	.9994	.9994	.9994	.9994	.9994	.9995	.9995	.9995
3.3	.9995	.9995	.9995	.9996	.9996	.9996	.9996	.9996	.9996	.9997
3.4	.9997	.9997	.9997	.9997	.9997	.9997	.9997	.9997	.9997	.9998

For z values greater than 3.49, use 1.000 to approximate the area.

Below, we summarize the procedure to determine the percent of data for any interval under the normal curve.

TO DETERMINE THE PERCENT OF DATA BETWEEN ANY TWO VALUES

1. Draw a diagram of the normal curve, indicating the area or percent to be determined.

2. Use the formula $z = \dfrac{x - \mu}{\sigma}$ to convert the given values to z-scores. Indicate these z-scores on the diagram.

3. Look up the areas that correspond to the specified z-scores in Table 13.7.

   a) When determining the area to the left of a negative z-score, use Table 13.7(a).

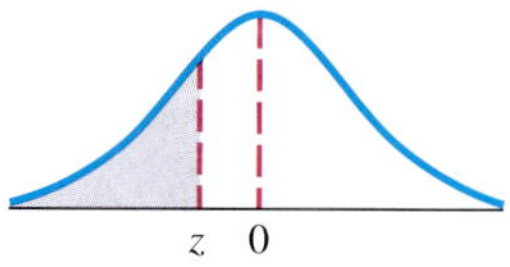

   b) When determining the area to the left of a positive z-score, use Table 13.7(b).

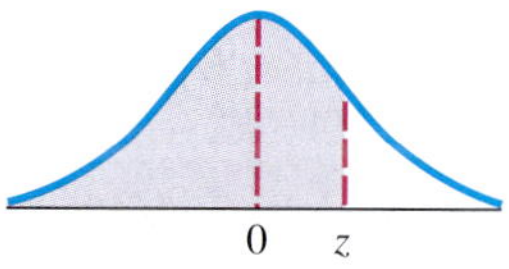

   c) When determining the area to the right of a z-score, subtract the percent of data to the left of the specified z-score from 100%.

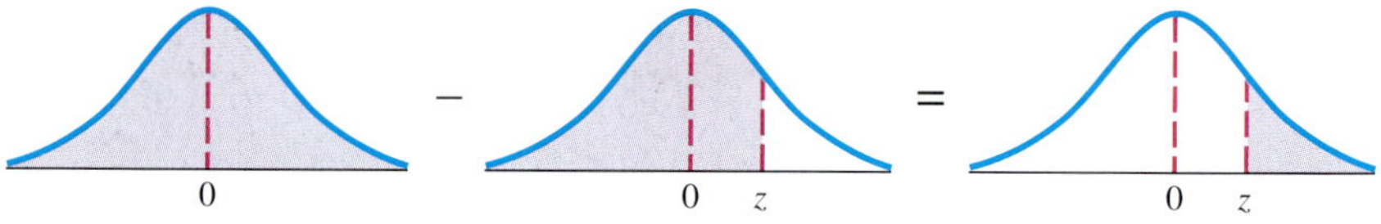

Area under entire curve = 1.0000 | Area to the left of z | Area to the right of z

   Or, use the symmetry of a normal distribution.

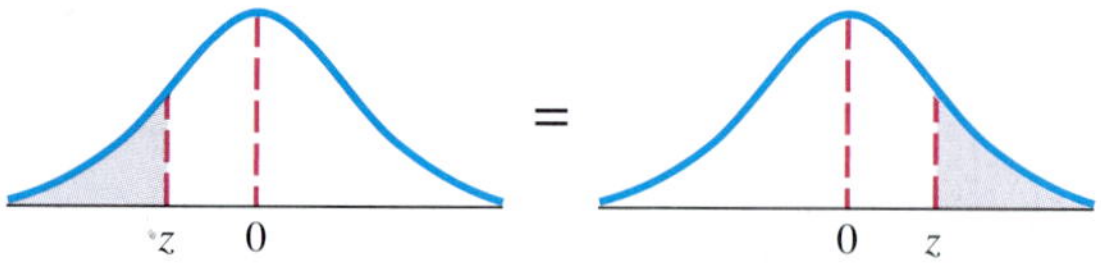

Area to the left of z ($z < 0$) | Area to the right of z ($z > 0$)

   d) When determining the area between two z-scores, subtract the smaller area from the larger area.

   In the figure below, we let z_1 represent the smaller z-score and z_2 represent the larger z-score.

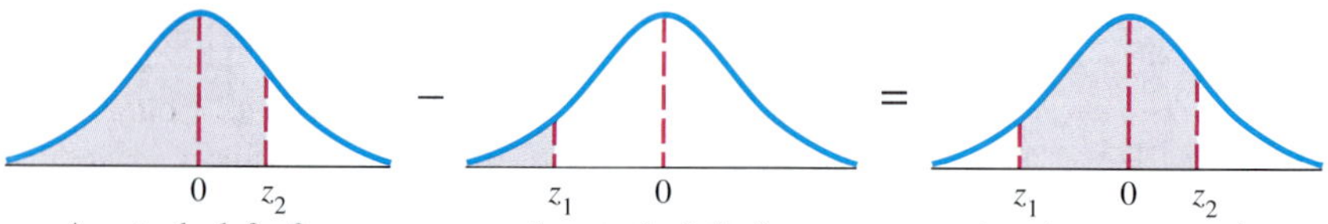

Area to the left of z_2 | Area to the left of z_1 | Area between z_1 and z_2

4. Change the areas you determined in step 3 to percents as explained on page 887.

EXAMPLE 4 *IQ Scores*

Intelligence quotients (IQ scores) are normally distributed with a mean of 100 and a standard deviation of 15. Determine the percent of individuals with IQ scores

a) below 115. b) below 130.

c) below 70. d) between 70 and 115.

e) between 115 and 130. f) above 122.5.

SOLUTION

a) We want to determine the area under the normal curve below the value of 115, as illustrated in Fig. 13.30(a). Converting 115 to a z-score yields a z-score of 1.00.

$$z_{115} = \frac{115 - 100}{15} = \frac{15}{15} = 1.00$$

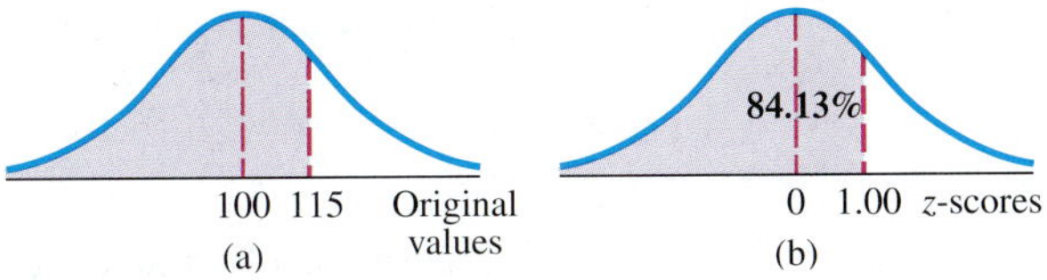

Figure 13.30

The percent of individuals with IQ scores below 115 is the same as the percent of data below a z-score of 1.00 (Fig. 13.30b). Since our z-score is positive, we use Table 13.7(b). From Table 13.7(b), we determine that the area to the left of a z-score of 1.00 is .8413. Therefore, 84.13% of all the IQ scores are below a z-score of 1.00. Thus, 84.13% of individuals have IQ scores below 115.

b) Begin by finding the z-score for 130.

$$z_{130} = \frac{130 - 100}{115} = \frac{30}{15} = 2.00$$

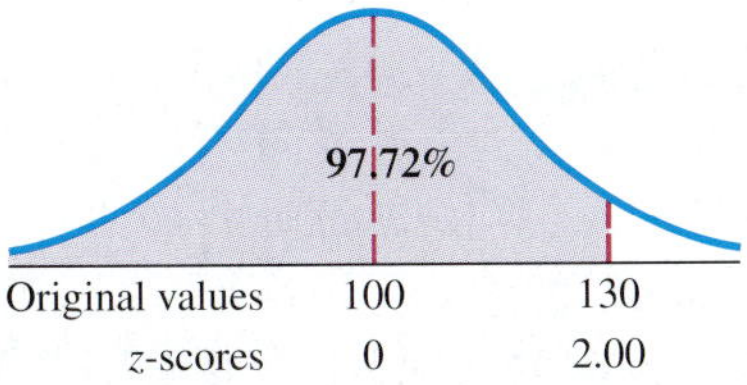

Figure 13.31

The percent of data below a z-score of 130 is the same as the percent of data below a z-score of 2.00 (Fig. 13.31). Using Table 13.7(b), we determine that the area to the left of a z-score of 2.00 is .9772. Therefore, 97.72% of the IQ scores are below a z-score of 2.00. Thus, 97.72% of all individuals have IQ scores below 130.

c) Begin by finding the z-score for 70.

$$z_{70} = \frac{70 - 100}{15} = \frac{-30}{15} = -2.00$$

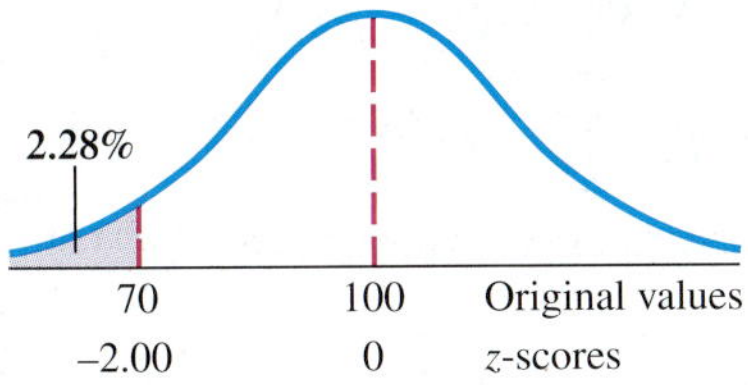

Figure 13.32

The percent of data below a score of 70 is the same as the percent of data below a z-score of -2.00 (Fig. 13.32). Since our z-score is negative, we use Table 13.7(a). Using the table, we determine that the area to the left of $z = -2.00$ is .0228. Therefore, 2.28% of the data is below a z-score of -2.00. Thus, 2.28% of all individuals have IQ scores below 70.

d) In part (a), we determined that $z_{115} = 1.00$, and in part (c), we determined that $z_{70} = -2.00$. The percent of data below a z-score of 1.00 is 84.13% (Fig. 13.33a on page 892). The percent of data below a z-score of -2.00 is 2.28% (Fig. 13.33b). Since we want to find the percent of data between two z-scores, we subtract the smaller percent from the larger percent: 84.13% − 2.28% = 81.85% (Fig. 13.33c). Thus, 81.85% of all individuals have IQ scores between 70 and 115.

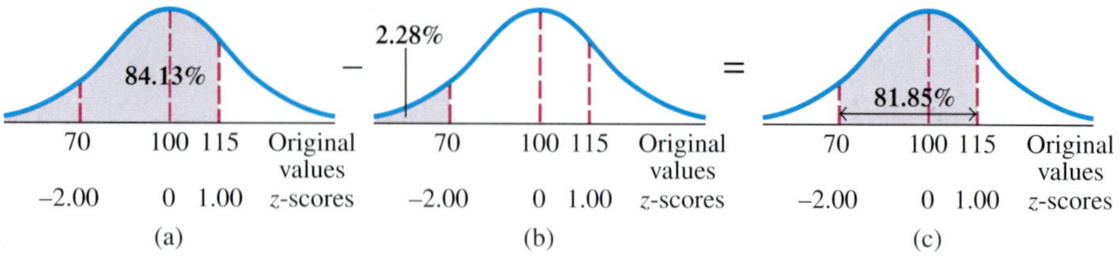

Figure 13.33

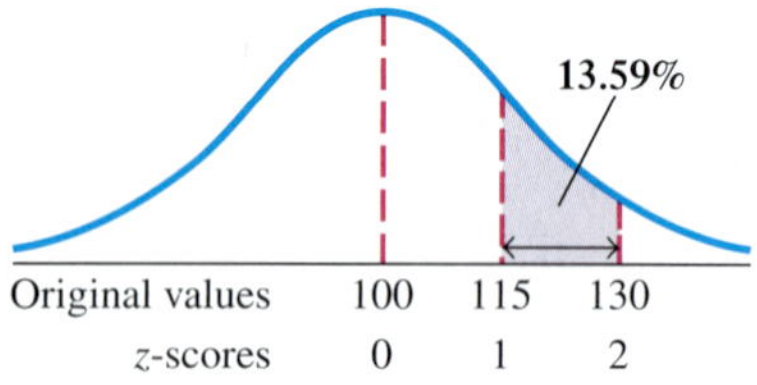

Figure 13.34

e) In part (a), we determined that $z_{115} = 1.00$, and in part (b), we determined $z_{130} = 2.00$. The percent of data below a z-score of 1.00 is 84.13%. The percent of data below a z-score of 2.00 is 97.72%. Since we want to find the percent of data between two z-scores, we subtract the smaller percent from the larger percent: 97.72% − 84.13% = 13.59% (Fig. 13.34). Thus, 13.59% of all individuals have IQ scores between 115 and 130.

f) Begin by determining a z-score for 122.5.

$$z_{122.5} = \frac{122.5 - 100}{15} = \frac{22.5}{15} = 1.50$$

The percent of IQ scores above 122.5 is the same as the percent of data above $z = 1.50$ (Fig. 13.35). To determine the percent of data above $z = 1.50$, we can determine the percent of data below $z = 1.50$ and subtract this percent from 100%. In Table 13.7(b), we see that the area to the left of $z = 1.50$ is .9332. Therefore, 93.32% of the IQ scores are below $z = 1.50$. The percent of IQ scores above $z = 1.50$ are 100% − 93.32%, or 6.68%. Thus, 6.68% of all IQ scores are greater than 122.5.

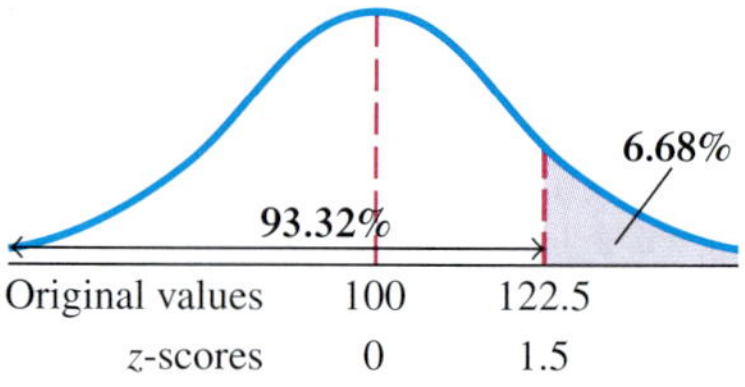

Figure 13.35

EXAMPLE 5 *Horseback Rides*

Assume that the length of time for a horseback ride on the trail at Triple R Ranch is normally distributed with a mean of 3.2 hours and a standard deviation of 0.4 hour.

a) What percent of horseback rides last at least 3.2 hours?

b) What percent of horseback rides last less than 2.8 hours?

c) What percent of horseback rides are at least 3.7 hours?

d) What percent of horseback rides are between 2.8 hours and 4.0 hours?

e) In a random sample of 500 horseback rides at Triple R Ranch, how many are at least 3.7 hours?

SOLUTION

a) In a normal distribution, half the data are always above the mean. Since 3.2 hours is the mean, half, or 50%, of the horseback rides last at least 3.2 hours.

b) Convert 2.8 hours to a z-score.

$$z_{2.8} = \frac{2.8 - 3.2}{0.4} = -1.00$$

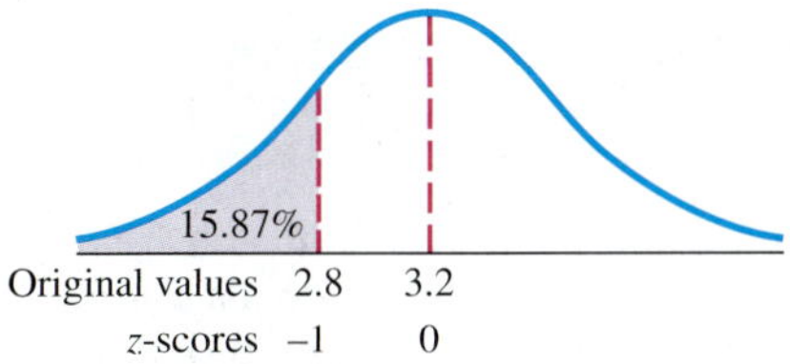

Figure 13.36

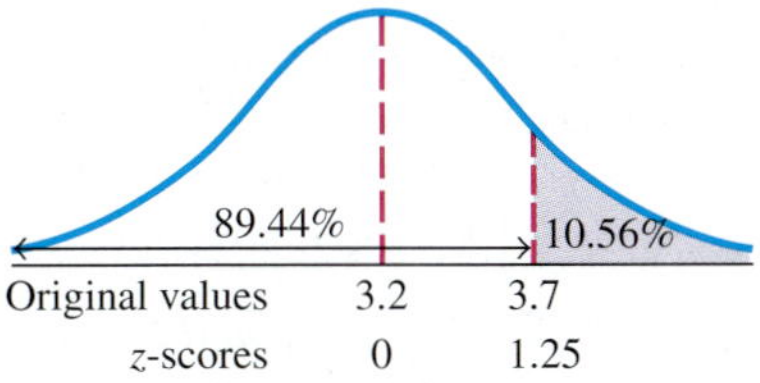

Figure 13.37

Use Table 13.7(a) to find the area of the normal curve that lies below a z-score of -1.00. The area to the left of $z = -1.00$ is 0.1587. Therefore, the percent of horseback rides that last less than 2.8 hours is 15.87% (Fig. 13.36).

c) At least 3.7 hours means greater than or equal to 3.7 hours. Therefore, we are seeking to find the percent of data to the right of 3.7 hours. Convert 3.7 hours to a z-score.

$$z_{3.7} = \frac{3.7 - 3.2}{0.4} = 1.25$$

From Table 13.7(b), we determine that the area to the left of $z = 1.25$ is .8944. Therefore, 89.44% of the data are below $z = 1.25$. The percent of data above $z = 1.25$ (or to the right of $z = 1.25$) is 100% − 89.44%, or 10.56% (Fig. 13.37). Thus, 10.56% of horseback rides last at least 3.7 hours.

d) Convert 4.0 to a z-score.

$$z_{4.0} = \frac{4.0 - 3.2}{0.4} = 2.00$$

From Table 13.7(b), we determine that the area to the left of $z = 2.00$ is .9772 (Fig. 13.38a). Therefore the percent of data below a z-score of 2.00 is 97.72%. From part (b), we determined that $z_{28} = -1.00$ and that the percent of data below a z-score of -1.00 is 15.87% (Fig. 13.38b). To find the percent of data between a z-score of -1.00 and a z-score of 2.00, we subtract the smaller percent from the larger percent. Thus, the percent of horseback rides that last between 2.8 hours and 4.0 hours is 97.72% − 15.87%, or 81.85% (Fig. 13.38c).

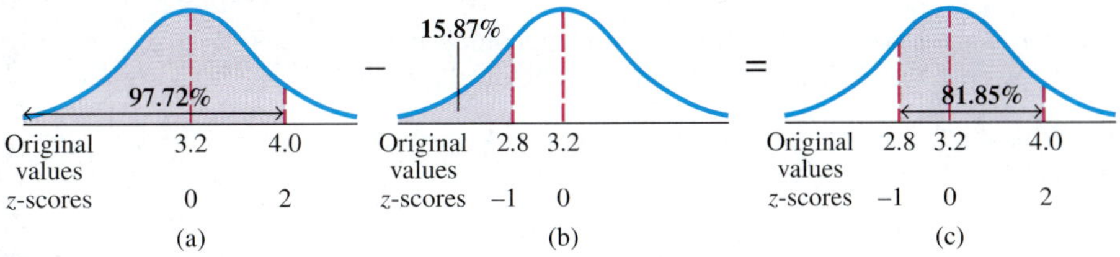

Figure 13.38

e) In part (c), we determined that 10.56% of all horseback rides last at least 3.7 hours. We now multiply 0.1056 times the number in the random sample, 500, to determine the number of horseback rides that last at least 3.7 hours. There are $0.1056 \times 500 = 52.8$, or approximately 53, horseback rides that last at least 3.7 hours. ●

TIMELY TIP Following is a summary of some important items presented in this section.

- The normal curve is symmetric about the mean.
- The area under the normal curve cannot be negative.
- A negative z-score indicates that the corresponding value in the original distribution is less than the mean.
- A positive z-score indicates that the corresponding value in the original distribution is greater than the mean.
- A z-score of 0 indicates that the corresponding value in the original distribution is the mean.
- Table 13.7 provides the area to the left of a specified z-score.
- When using Table 13.7 to determine the area to the left of a specified z-score, locate the units value and tenths value of your specified z-score under the column labeled z. Then move to the column containing the hundredths value of your specified z-score to obtain the area.

SECTION 13.7 EXERCISES

CONCEPT/WRITING EXERCISES

In Exercises 1–6, describe a

1. rectangular distribution.

2. J-shaped distribution.

3. bimodal distribution.

4. distribution that is skewed to the right.

5. distribution that is skewed to the left.

6. normal distribution.

7. What does a z-score measure?

8. Explain how to determine a z-score of a particular piece of data.

9. a) If a given piece of data has a negative z-score, is the given piece of data above the mean or below the mean?

   b) If a given piece of data has a positive z-score, is the given piece of data above the mean or below the mean?

10. What is the z-score of a value that is the mean of a set of data?

11. Consider the following normal curve, representing a normal distribution, with points A, B, and C. One of these points corresponds to μ, one point corresponds to $\mu + \sigma$, and one point corresponds to $\mu - 2\sigma$.

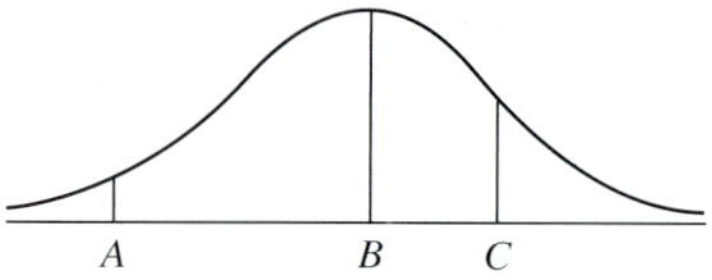

   a) Which point corresponds to μ?

   b) Which point corresponds to $\mu + \sigma$?

   c) Which point corresponds to $\mu - 2\sigma$?

12. Consider the following two normal curves.

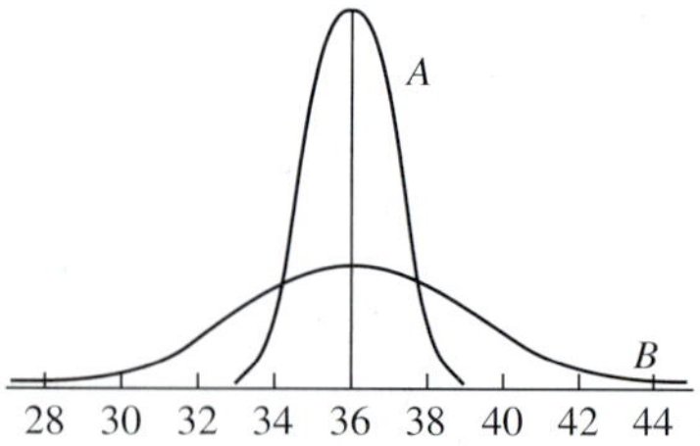

   a) Do these distributions have the same mean? If so, what is the mean?

   b) One of these curves corresponds to a normal distribution with $\sigma = 1$. The other curve corresponds to a normal distribution with $\sigma = 3$. Which curve, A or B, has $\sigma = 3$? Explain.

In Exercises 13–16, give an example of the type of distribution.

13. Rectangular

14. Skewed

15. J-shaped

16. Bimodal

For the distributions in Exercises 17–20, state whether you think the distribution would be normal, J-shaped, bimodal, rectangular, skewed left, or skewed right. Explain your answers.

17. The wearout mileage of automobile tires

18. The numbers resulting from tossing a die many times

19. The number of people per household in the United States

20. The heights of a sample of high school seniors, where there are an equal number of males and females

21. In a distribution that is skewed to the right, which has the greatest value: the mean, median, or mode? Which has the smallest value? Explain.

22. In a distribution skewed to the left, which has the greatest value: the mean, median, or mode? Which has the smallest value? Explain.

23. List three populations other than those given in the text that may be normally distributed.

24. List three populations other than those given in the text that may not be normally distributed.

25. In a normal distribution, what is the relationship between the mean, median, and mode?

26. In a normal distribution, approximately what percent of the data is between

a) one standard deviation below the mean to one standard deviation above the mean?

b) two standard deviations below the mean to two standard deviations above the mean?

c) three standard deviations below the mean to three standard deviations above the mean?

PRACTICE THE SKILLS

In Exercises 27–38, use Table 13.7 on pages 888 and 889 to find the specified area.

27. Above the mean

28. Below the mean

29. Between two standard deviations below the mean and one standard deviation above the mean

30. Between 1.10 and 1.70 standard deviations above the mean

31. To the right of $z = 1.34$

32. To the left of $z = 1.62$

33. To the left of $z = -1.78$

34. To the right of $z = -1.78$

35. Between $z = -1.32$ and $z = -1.64$

36. To the left of $z = 1.84$

37. To the left of $z = -1.62$

38. To the left of $z = -0.92$

In Exercises 39–48, use Table 13.7 on pages 888 and 889 to determine the percent of data specified.

39. Less than $z = 0.71$

40. Less than $z = -0.82$

41. Between $z = -1.34$ and $z = 2.24$

42. Between $z = -2.18$ and $z = -1.90$

43. Greater than $z = -1.90$

44. Greater than $z = 2.66$

45. Less than $z = 1.96$

46. Between $z = -1.53$ and $z = -1.82$

47. Between $z = 0.72$ and $z = 2.14$

48. Between $z = -2.15$ and $z = 3.31$

PROBLEM SOLVING

Fitness Test Scores *In Exercises 49 and 50, assume that the heights of 7-year-old children are normally distributed. The heights of 8 children are given in z-scores below.*

Emily	0.9	Jason	0.0	Heather	−1.3	Juan	0.0
Sarah	1.7	Omar	−0.2	Carol	0.8	Kim	−1.2

49. a) Which of these children are taller than the mean?

b) Which of these children are at the mean?

c) Which of these children are shorter than the mean?

50. a) Which child is the tallest?

b) Which child is the shortest?

Hours Worked by College Students *In Exercises 51–54, assume that the number of hours college students spend working per week is normally distributed with a mean of 18 hours and standard deviation of 4 hours.*

51. Determine the percent of college students who work at least 18 hours per week.

52. Determine the percent of college students who work between 14 and 26 hours per week.

53. Determine the percent of college students who work at least 23 hours per week.

54. In a random sample of 500 college students, how many work at least 23 hours per week?

SAT Scores *In Exercises 55–60, assume that the mathematics scores on the Scholastic Aptitude Test (SAT) are normally distributed with a mean of 500 and a standard deviation of 100.*

55. What percent of students who took the test have a mathematics score below 550?

56. What percent of students who took the test have a mathematics score above 650?

57. What percent of students who took the test have a mathematics score between 550 and 650?

58. What percent of students who took the test have a mathematics score below 300?

59. What percent of students who took the test have a mathematics score between 400 and 525?

60. What percent of students who took the test have a mathematics score above 380?

Vending Machine *In Exercises 61–64, a vending machine is designed to dispense a mean of 7.6 oz of coffee into an 8 oz cup. If the standard deviation of the amount of coffee dispensed is 0.4 oz and the amount is normally distributed, find the percent of times the machine will*

61. dispense from 7.4 oz to 7.7 oz.

62. dispense less than 7.0 oz.

63. dispense less than 7.7 oz.

64. result in the cup overflowing (therefore dispense more than 8 oz).

Automobile Speed In Exercises 65–70, assume that the speed of automobiles on an expressway during rush hour is normally distributed with a mean of 62 mph and a standard deviation of 5 mph.

65. What percent of cars are traveling faster than 62 mph?

66. What percent of cars are traveling between 58 mph and 66 mph?

67. What percent of cars are traveling slower than 56 mph?

68. What percent of cars are traveling faster than 70 mph?

69. If 200 cars are selected at random, how many will be traveling slower than 56 mph?

70. If 200 cars are selected at random, how many will be traveling faster than 70 mph?

Corn Flakes In Exercises 71–74, assume that the amount of corn flakes in a box is normally distributed with a mean of 16 oz and a standard deviation of 0.1 oz.

71. Determine the percent of boxes that will contain between 15.83 oz and 16.32 oz of corn flakes.

72. Determine the percent of boxes that will contain more than 16.16 oz of corn flakes.

73. If the manufacturer produces 300,000 boxes, how many of them will contain less than 15.83 oz of corn flakes?

74. If the manufacturer produces 300,000 boxes, how many of them will contain more than 16.16 oz of corn flakes?

Ages of Children at Day Care In Exercises 75–80, assume that the ages of children at Happy Times Day Care are normally distributed with a mean of 3.7 years and a standard deviation of 1.2 years.

75. What percent of the children are older than 3.1 years?

76. What percent of the children are between 2.5 and 4.3 years?

77. What percent of the children are older than 6.7 years?

78. What percent of the children are younger than 6.7 years?

79. If 120 children are enrolled at Happy Times Day Care, how many of them are older than 3.1 years?

80. If 120 children are enrolled at Happy Times Day Care, how many of them are between 2.5 and 4.3 years?

81. ***Weight Loss*** A weight-loss clinic guarantees that its new customers will lose at least 5 lb by the end of their first month of participation or their money will be refunded. If the weight loss of customers at the end of their first month is normally distributed, with a mean of 6.7 lb and a standard deviation of 0.81 lb, determine the percent of customers who will be able to claim a refund.

82. ***Battery Warranty*** The warranty on a car battery is 36 months. If the breakdown times of this battery are normally distributed with a mean of 46 months and a standard deviation of 8 months, determine the percent of batteries that can be expected to require repair or replacement under warranty.

83. ***Coffee Machine*** A vending machine that dispenses coffee does not appear to be working correctly. The machine rarely gives the proper amount of coffee. Some of the time the cup is underfilled, and some of the time the cup overflows. Does this variation indicate that the mean number of ounces dispensed has to be adjusted, or does it indicate that the standard deviation of the amount of coffee dispensed by the machine is too large? Explain your answer.

84. ***Grading on a Normal Curve*** Mr. Sanderson marks his class on a normal curve. Those with z-scores above 1.8 will receive an A, those between 1.8 and 1.1 will receive a B, those between 1.1 and -1.2 will receive a C, those between -1.2 and -1.9 will receive a D, and those under -1.9 will receive an F. Determine the percent of grades that will be A, B, C, D, and F.

▲ See Exercise 84

CHALLENGE PROBLEMS/GROUP ACTIVITIES

85. ***Salesperson Promotion*** The owner at Kim's Home Interiors is reviewing the sales records of two managers who are up for promotion, Katie and Stella, who work in different stores. At Katie's store, the mean sales have been \$23,200 per month, with a standard deviation of \$2170. At Stella's store, the mean sales have been \$25,600 per month, with a standard deviation of \$2300. Last month, Katie's store sales were \$28,408 and Stella's store sales were \$29,510. At both stores, the distribution of monthly sales is normal.

    a) Convert last month's sales for Katie's store and for Stella's store to z-scores.

    b) If one of the two were to be promoted based solely on the increase in sales last month, who should be promoted? Explain.

86. ***Chebyshev's Theorem*** How can you determine whether a distribution is approximately normal? A statistical theorem called ***Chebyshev's theorem*** states that the *minimum percent* of data between plus and minus K standard deviations from the mean $(K > 1)$ in *any distribution* can be found by the formula

$$\text{Minimum percent} = 1 - \frac{1}{K^2}$$

Thus, for example, between ± 2 standard deviations from the mean there will always be a minimum of 75% of data. This minimum percent applies to any distribution. For $K = 2$,

$$\text{Minimum percent} = 1 - \frac{1}{2^2}$$
$$= 1 - \frac{1}{4} = \frac{3}{4}, \quad \text{or} \quad 75\%$$

Likewise, between ± 3 standard deviations from the mean there will always be a minimum of 89% of the data. For $K = 3$,

$$\text{Minimum percent} = 1 - \frac{1}{3^2}$$
$$= 1 - \frac{1}{9} = \frac{8}{9}, \quad \text{or} \quad 89\%$$

The following table lists the minimum percent of data in *any distribution* and the actual percent of data in *the normal distribution* between ± 1.1, ± 1.5, ± 2.0, and ± 2.5 standard deviations from the mean. The minimum percents of data in any distribution were calculated by using Chebyshev's theorem. The actual percents of data for the normal distribution were calculated by using the area given in the standard normal, or z, table.

	$K = 1.1$	$K = 1.5$	$K = 2$	$K = 2.5$
Minimum (for any distribution)	17.4%	55.6%	75%	84%
Normal distribution	72.9%	86.6%	95.4%	98.8%
Given distribution				

The third row of the chart has been left blank for you to fill in the percents when you reach part (e).

Consider the following 30 pieces of data obtained from a quiz.

1, 1, 1, 1, 2, 2, 2, 2, 3, 3, 4, 4, 4, 5, 6,
6, 6, 7, 7, 7, 7, 8, 8, 8, 8, 9, 9, 9, 10, 10

a) Determine the mean of the set of scores.

b) Determine the standard deviation of the set of scores.

c) Determine the values that correspond to 1.1, 1.5, 2, and 2.5 standard deviations above the mean. (For example, the value that corresponds to 1.5 standard deviations above the mean is $\mu + 1.5\sigma$.)

Then determine the values that correspond to 1.1, 1.5, 2, and 2.5 standard deviations below the mean. (For example, the value that corresponds to 1.5 standard deviations below the mean is $\mu - 1.5\sigma$.)

d) By observing the 30 pieces of data, determine the actual percent of quiz scores between

± 1.1 standard deviations from the mean.
± 1.5 standard deviations from the mean.
± 2 standard deviations from the mean.
± 2.5 standard deviations from the mean.

e) Place the percents found in part (d) in the third row of the chart.

f) Compare the percents in the third row of the chart with the minimum percents in the first row and the normal percents in the second row, and then make a judgment as to whether this set of 30 scores is approximately normally distributed. Explain your answer.

87. *Test Scores* Obtain a set of test scores from your instructor.

a) Determine the mean, median, mode, and midrange of the test scores.

b) Determine the range and standard deviation of the set of scores. (You may round the mean to the nearest tenth when finding the standard deviation.)

c) Construct a frequency distribution of the set of scores. Select your first class so that there will be between 5 and 12 classes.

d) Construct a histogram and frequency polygon of the frequency distribution in part (c).

e) Does the histogram in part (d) appear to represent a normal distribution? Explain.

f) Use the procedure explained in Exercise 86 to determine whether the set of scores approximates a normal distribution. Explain.

88. Determine a value of z such that $z \geq 0$ and 47.5% of the standard normal curve lies between 0 and the z-value.

89. Determine a value of z such that $z \leq 0$ and 38.1% of the standard normal curve lies between 0 and the z-value.

RECREATIONAL MATHEMATICS

90. Ask your instructor for the class mean and class standard deviation for one of the exams taken by your class. For that exam, calculate the z-score for your exam grade. How many standard deviations is your exam grade away from the mean?

91. If the mean score on a math quiz is 12.0 and 77% of the students in your class scored between 9.6 and 14.4, determine the standard deviation of the quiz scores.

INTERNET/RESEARCH ACTIVITY

92. In this project, you actually become the statistician.

a) Select a project of interest to you in which data must be collected.

b) Write a proposal and submit it to your instructor for approval. In the proposal, discuss the aims of your project and how you plan to gather the data to make your sample unbiased.

c) After your proposal has been approved, gather 50 pieces of data by the method you proposed.

d) Rank the data from smallest to largest.

e) Compute the mean, median, mode, and midrange.

f) Determine the range and standard deviation of the data. You may round the mean to the nearest tenth when computing the standard deviation.

g) Construct a frequency distribution, histogram, frequency polygon, and stem-and-leaf display of your data. Select your first class so that there will be between 5 and 12 classes. Be sure to label your histogram and frequency polygon.

h) Does your distribution appear to be normal? Explain your answer. Does it appear to be another type of distribution discussed? Explain.

i) Determine whether your distribution is approximately normal by using the technique discussed in Exercise 86.

13.8 LINEAR CORRELATION AND REGRESSION

▲ You can predict the value of a car based on the age of the car.

Do you believe that there is a relationship between the time a person studied for an exam and the exam grade received? Is there a relationship between the age of a car and the value of the car? Can we predict the value of a car based on the age of the car? In this section, we will learn how to determine whether there is a relationship between two quantities and, if so, how strong that relationship is. We will also learn how to determine the equation of the line that best describes the relationship between two quantities.

In this section, we discuss two important statistical topics: correlation and regression. *Correlation* is used to determine whether there is a relationship between two quantities and, if so, how strong the relationship is. *Regression* is used to determine the equation that relates the two quantities. Although there are other types of correlation and regression, in this section we discuss only linear correlation and linear regression. We begin by discussing linear correlation.

Linear Correlation

The ***linear correlation coefficient***, *r*, is a unitless measure that describes the strength of the linear relationship between two variables. A positive value of *r*, or a positive correlation, means that as one variable increases, the other variable also increases. A negative value of *r*, or a negative correlation, means that as one variable increases, the other variable decreases. The correlation coefficient, *r*, will always be a value between -1 and 1 inclusive. A value of 1 indicates the strongest possible positive correlation, a value of -1 indicates the strongest possible negative correlation, and a value of 0 indicates no correlation (Fig. 13.39).

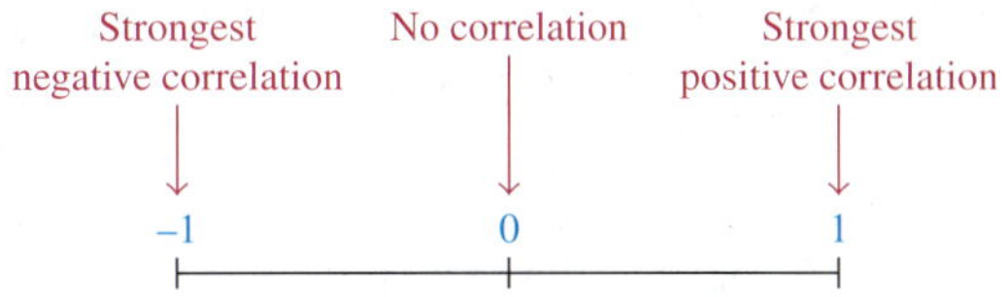

Figure 13.39

A visual aid used with correlation is the ***scatter diagram***, a plot of data points. To help understand how to construct a scatter diagram, consider the following data from Egan Electronics. During a 6-day period, Egan Electronics kept daily records of the number of assembly line workers absent and the number of defective parts produced. The information is provided in the following chart.

Day	1	2	3	4	5	6
Number of workers absent	3	5	0	1	2	6
Number of defective parts	15	22	7	12	20	30

For each of the 6 days, two pieces of data are provided: number of workers absent and number of defective parts. We call the set of data ***bivariate data***. Often when we have a set of bivariate data, we can control one of the quantities. We generally denote the quantity that can be controlled, the ***independent variable***, *x*. The other variable, the ***dependent variable***, is denoted as *y*. In this problem, we will assume that the number of defective parts produced is affected by the number of workers absent. Therefore, we will call the number of workers absent *x* and the number of defective parts produced *y*. When we plot bivariate data, the independent variable is marked on the horizontal axis and the dependent variable is marked on the vertical axis. Therefore, for this problem, number of workers absent is marked on the horizontal axis and number of defective parts is marked on the vertical axis. If we plot the six pieces of bivariate data in the Cartesian coordinate system, we get a scatter diagram, as shown in Fig. 13.40.

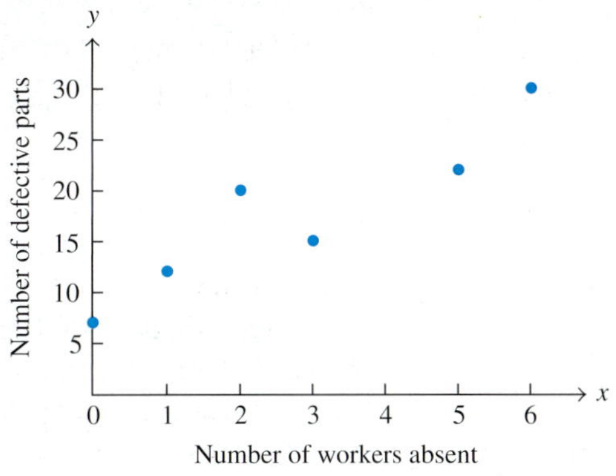

Figure 13.40

The figure shows that, generally, the more workers that are absent, the more defective parts are produced.

In Fig. 13.41, we show some scatter diagrams and indicate the corresponding strength of correlation between the quantities on the horizontal and vertical axes.

Earlier, we mentioned that r will always be a value between -1 and 1 inclusive. A value of $r = 1$ is obtained only when every point of the bivariate data on a scatter diagram lies in a straight line and the line is increasing from left to right (see Fig. 13.41a). In other words, the line has a positive slope, as discussed in Section 6.6.

A value of $r = -1$ will be obtained only when every point of the bivariate data on a scatter diagram lies in a straight line and the line is decreasing from left to right (see Fig. 13.41e). In other words, the line has a negative slope.

The value of r is a measure of how far a set of points varies from a straight line. The greater the spread, the weaker the correlation and the closer the value of r is to 0. Figure 13.41 shows that the more the points diverge from a straight line, the weaker the correlation becomes.

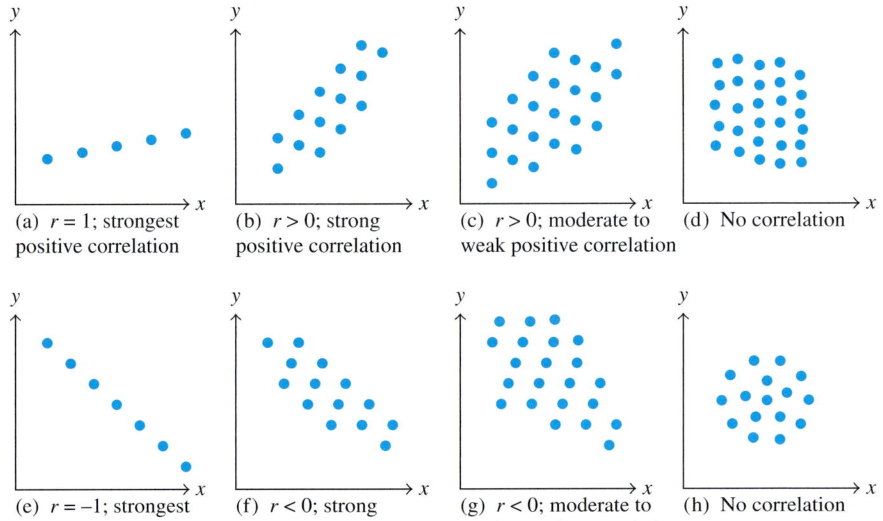

(a) $r = 1$; strongest positive correlation
(b) $r > 0$; strong positive correlation
(c) $r > 0$; moderate to weak positive correlation
(d) No correlation
(e) $r = -1$; strongest negative correlation
(f) $r < 0$; strong negative correlation
(g) $r < 0$; moderate to weak negative correlation
(h) No correlation

Figure 13.41

The following formula is used to calculate r.

LINEAR CORRELATION COEFFICIENT

The formula to calculate the **correlation coefficient, *r***, is as follows.

$$r = \frac{n(\Sigma xy) - (\Sigma x)(\Sigma y)}{\sqrt{n(\Sigma x^2) - (\Sigma x)^2}\sqrt{n(\Sigma y^2) - (\Sigma y)^2}}$$

To determine the correlation coefficient, r, and the equation of the line of best fit (to be discussed shortly), a statistical calculator may be used. On the statistical calculator, you enter the ordered pairs, (x, y) and press the appropriate keys. At the end of this section, we indicate the procedure to follow to use the computer software spreadsheet program, Microsoft Excel, and the TI-83 Plus or the TI-84 Plus calculators to determine the correlation coefficient.

In Example 1, we show how to determine r for a set of bivariate data without the use of a statistical calculator. We will use the same set of bivariate data given on page 899 that was used to make the scatter diagram in Fig. 13.40.

EXAMPLE 1 *Number of Absences versus Number of Defective Parts*

Egan Electronics provided the following daily records about the number of assembly line workers absent and the number of defective parts produced for 6 days. Determine the correlation coefficient between the number of workers absent and the number of defective parts produced.

Day	**1**	**2**	**3**	**4**	**5**	**6**
Number of workers absent	3	5	0	1	2	6
Number of defective parts	15	22	7	12	20	30

SOLUTION We plotted this set of data on the scatter diagram in Figure 13.40. We will call the number of workers absent x. We will call the number of defective parts produced y. We list the values of x and y and calculate the necessary sums: Σx, Σy, Σxy, Σx^2, Σy^2. We determine the values in the column labeled x^2 by squaring the x's (multiplying the x's by themselves). We determine the values in the column labeled y^2 by squaring the y's. We determine the values in the column labeled xy by multiplying each x value by its corresponding y value.

Number of Workers Absent	**Number of Defective Parts**			
x	y	x^2	y^2	xy
3	15	9	225	45
5	22	25	484	110
0	7	0	49	0
1	12	1	144	12
2	20	4	400	40
6	30	36	900	180
17	106	75	2202	387

Thus, $\Sigma x = 17$, $\Sigma y = 106$, $\Sigma x^2 = 75$, $\Sigma y^2 = 2202$, and $\Sigma xy = 387$. In the formula for r, we use both $(\Sigma x)^2$ and Σx^2. Note that $(\Sigma x)^2 = (17)^2 = 289$ and that $\Sigma x^2 = 75$. Similarly, $(\Sigma y)^2 = (106)^2 = 11{,}236$ and $\Sigma y^2 = 2202$.

The n in the formula represents the number of pieces of bivariate data. Here $n = 6$. Now let's determine r.

$$r = \frac{n(\Sigma xy) - (\Sigma x)(\Sigma y)}{\sqrt{n(\Sigma x^2) - (\Sigma x)^2}\sqrt{n(\Sigma y^2) - (\Sigma y)^2}}$$

$$= \frac{6(387) - (17)(106)}{\sqrt{6(75) - (17)^2}\sqrt{6(2202) - (106)^2}}$$

$$= \frac{2322 - 1802}{\sqrt{6(75) - 289}\sqrt{6(2202) - 11{,}236}}$$

$$= \frac{520}{\sqrt{450 - 289}\sqrt{13{,}212 - 11{,}236}}$$

$$= \frac{520}{\sqrt{161}\sqrt{1976}} \approx 0.922$$

Table 13.8 Correlation Coefficient, r

n	$\alpha = 0.05$	$\alpha = 0.01$
4	0.950	0.990
5	0.878	0.959
6	0.811	0.917
7	0.754	0.875
8	0.707	0.834
9	0.666	0.798
10	0.632	0.765
11	0.602	0.735
12	0.576	0.708
13	0.553	0.684
14	0.532	0.661
15	0.514	0.641
16	0.497	0.623
17	0.482	0.606
18	0.468	0.590
19	0.456	0.575
20	0.444	0.561
22	0.423	0.537
27	0.381	0.487
32	0.349	0.449
37	0.325	0.418
42	0.304	0.393
47	0.288	0.372
52	0.273	0.354
62	0.250	0.325
72	0.232	0.302
82	0.217	0.283
92	0.205	0.267
102	0.195	0.254

The derivation of this table is beyond the scope of this text. It shows the critical values of the Pearson correlation coefficient.

Since the maximum possible value for r is 1.00, a correlation coefficient of 0.922 is a strong, positive correlation. This result implies that, generally, the more assembly line workers absent, the more defective parts produced. ●

In Example 1, had we found r to be a value greater than 1 or less than -1, it would have indicated that we had made an error. Also, from the scatter diagram, we should realize that r should be a positive value and not negative.

In Example 1, there appears to be a cause–effect relationship. That is, the more assembly line workers who are absent, the more defective parts are produced. *However, a correlation does not necessarily indicate a cause–effect relationship.* For example, there is a positive correlation between police officers' salaries and the cost of medical insurance over the past 10 years (both have increased), but that does not mean that the increase in police officers' salaries caused the increase in the cost of medical insurance.

Suppose in Example 1 that r had been 0.53. Would this value have indicated a correlation? What is the minimum value of r needed to assume that a correlation exists between the variables? To answer this question, we introduce the term *level of significance*. The ***level of significance***, denoted α (alpha), is used to identify the cutoff between results attributed to chance and results attributed to an actual relationship between the two variables. Table 13.8 gives ***critical values**** (or cutoff values) that are sometimes used for determining whether two variables are related. The table indicates two different levels of significance: $\alpha = 0.05$ and $\alpha = 0.01$. A level of significance of 5%, written $\alpha = 0.05$, means that there is a 5% chance that, when you say the variables are related, they actually are *not* related. Similarly, a level of significance of 1%, or $\alpha = 0.01$, means that there is a 1% chance that, when you say the variables are related, they actually are *not* related. More complete critical value tables are available in statistics books.

To explain the use of the table, we use ***absolute value***, symbolized $|\ |$. The absolute value of a nonzero number is the positive value of the number, and the absolute value of 0 is 0. Therefore,

$$|3| = 3, \qquad |-3| = 3, \qquad |5| = 5, \qquad |-5| = 5, \qquad \text{and} \qquad |0| = 0$$

If the absolute value of r, written $|r|$, is *greater than* the value given in the table under the specified α and appropriate sample size n, we assume that a correlation does exist between the variables. If $|r|$ is less than the table value, we assume that no correlation exists.

Returning to Example 1, if we want to determine whether there is a correlation at a 5% level of significance, we find the critical value (or cutoff value) that corresponds to $n = 6$ (there are 6 pieces of bivariate data) and $\alpha = 0.05$. The value to the right of $n = 6$ and under the $\alpha = 0.05$ column is the critical value 0.811. From the formula, we had obtained $r = 0.922$. Since $|0.922| > 0.811$, or $0.922 > 0.811$, we assume that a correlation between the variables exists.

Note in Table 13.8 that the larger the sample size, the smaller is the value of r needed for a significant correlation.

EXAMPLE 2 *Amount of Drug Remaining in the Bloodstream*

To test the length of time that an infection-fighting drug stays in a person's bloodstream, a doctor gives 300 milligrams of the drug to 10 patients, labeled 1–10 in the table on page 903. Once each hour, for 10 hours, one of the 10 patients is selected

*This table of values may be used only under certain conditions. If you take a statistics course, you will learn more about which critical values to use to determine whether a linear correlation exists.

at random and that person's blood is tested to determine the amount of the drug remaining in the bloodstream. The results are as follows.

Patient	**1**	**2**	**3**	**4**	**5**	**6**	**7**	**8**	**9**	**10**
Time (hr)	1	2	3	4	5	6	7	8	9	10
Drug remaining (mg)	250	230	200	210	140	120	210	100	90	85

Determine at a level of significance of 5% whether a correlation exists between the time elapsed and the amount of drug remaining.

SOLUTION Let time be represented by x and the amount of drug remaining by y. We first draw a scatter diagram (Fig. 13.42).

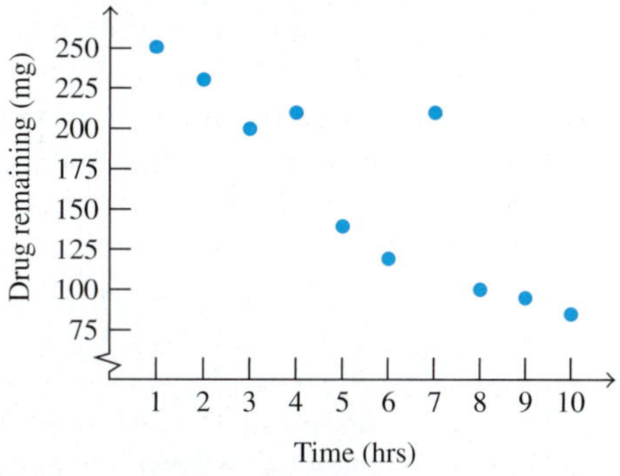

Figure 13.42

The scatter diagram suggests that, if a correlation exists, it will be negative. We now construct a table of values and calculate r.

x	y	x^2	y^2	xy
1	250	1	62,500	250
2	230	4	52,900	460
3	200	9	40,000	600
4	210	16	44,100	840
5	140	25	19,600	700
6	120	36	14,400	720
7	210	49	44,100	1470
8	100	64	10,000	800
9	90	81	8100	810
10	85	100	7225	850
55	1635	385	302,925	7500

$$r = \frac{n(\Sigma xy) - (\Sigma x)(\Sigma y)}{\sqrt{n(\Sigma x^2) - (\Sigma x)^2}\sqrt{n(\Sigma y^2) - (\Sigma y)^2}}$$

$$= \frac{10(7500) - (55)(1635)}{\sqrt{10(385) - (55)^2}\sqrt{10(302{,}925) - (1635)^2}}$$

$$= \frac{-14{,}925}{\sqrt{825}\sqrt{356{,}025}} \approx \frac{-14{,}925}{17{,}138.28} \approx -0.871$$

From Table 13.8, for $n = 10$ and $\alpha = 0.05$, we get 0.632. Since $|-0.871| = 0.871$ and $0.871 > 0.632$, a correlation exists. The correlation is negative, which indicates that the longer the time period, the smaller is the amount of drug remaining. ●

Linear Regression

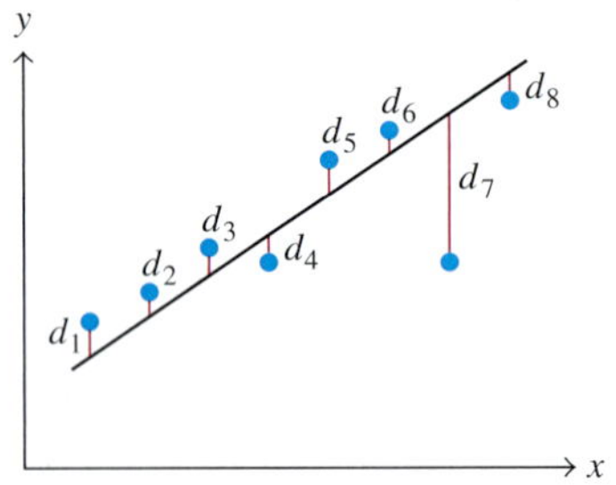

Figure 13.43

Let's now turn to regression. ***Linear regression*** is the process of determining the linear relationship between two variables. Recall from Section 6.6 that the slope–intercept form of a straight line is $y = mx + b$, where m is the slope and b is the y-intercept.

Using the set of bivariate data, we will determine the equation of ***the line of best fit***. The line of best fit is also called ***the regression line***, or ***the least squares line***. The *line of best fit* is the line such that the sum of the squares of the vertical distances from the line to the data points (on the scatter diagram) is a minimum, as shown in Fig. 13.43. In Fig. 13.43, the line of best fit minimizes the sum of d_1 through d_8. To determine the equation of the line of best fit, $y = mx + b$,* we must find m and then b. The formulas for finding m and b are as follows.

The equation of the line of best fit is

$$y = mx + b,$$

$$\text{where} \quad m = \frac{n(\Sigma xy) - (\Sigma x)(\Sigma y)}{n(\Sigma x^2) - (\Sigma x)^2} \quad \text{and} \quad b = \frac{\Sigma y - m(\Sigma x)}{n}$$

Note that the numerator of the fraction used to find m is identical to the numerator used to find r. Therefore, if you have previously found r, you do not need to repeat this calculation. Also, the denominator of the fraction used to find m is identical to the radicand of the first square root in the denominator of the fraction used to find r.

EXAMPLE 3 *The Line of Best Fit*

a) Use the data in Example 1 on page 901 to find the equation of the line of best fit that relates the number of workers absent on an assembly line and the number of defective parts produced.

b) Graph the equation of the line of best fit on a scatter diagram that illustrates the set of bivariate points.

SOLUTION

a) In Example 1, we found $n(\Sigma xy) - (\Sigma x)(\Sigma y) = 520$ and $n(\Sigma x^2) - (\Sigma x)^2 = 161$. Thus,

$$m = \frac{n(\Sigma xy) - (\Sigma x)(\Sigma y)}{n(\Sigma x^2) - (\Sigma x)^2} = \frac{520}{161} \approx 3.23$$

*Some statistics books use $y = ax + b$, $y = b_0 + b_1x$, or something similar for the equation of the line of best fit. In any case, the letter next to the variable x represents the slope of the line of best fit and the other letter represents the y-intercept of the graph.

Now we find the y-intercept, b. In Example 1, we found $n = 6$, $\Sigma x = 17$, and $\Sigma y = 106$.

$$b = \frac{\Sigma y - m(\Sigma x)}{n}$$

$$\approx \frac{106 - 3.23(17)}{6} \approx \frac{51.09}{6} \approx 8.52$$

Therefore, the equation of the line of best fit is

$$y = mx + b$$
$$y = 3.23x + 8.52$$

where x represents the number of workers absent and y represents the predicted number of defective parts produced.

b) To graph $y = 3.23x + 8.52$, we need to plot at least two points. We will plot three points and then draw the graph.

$$y = 3.23x + 8.52$$

$x = 2$ $\quad y = 3.23(2) + 8.52 = 14.98$

$x = 4$ $\quad y = 3.23(4) + 8.52 = 21.44$

$x = 6$ $\quad y = 3.23(6) + 8.52 = 27.90$

x	y
2	14.98
4	21.44
6	27.90

These three calculations indicate that if 2 assembly line workers are absent on the assembly line, the predicted number of defective parts produced is about 15. If 4 assembly line workers are absent, the predicted number of defective parts produced is about 21, and if 6 assembly line workers are absent, the predicted number of defective parts produced is about 28. Plot the three points (the three black points in Figure 13.44) and then draw a straight line through the three points. The scatter diagram and graph of the equation of the line of best fit are plotted in Fig. 13.44.

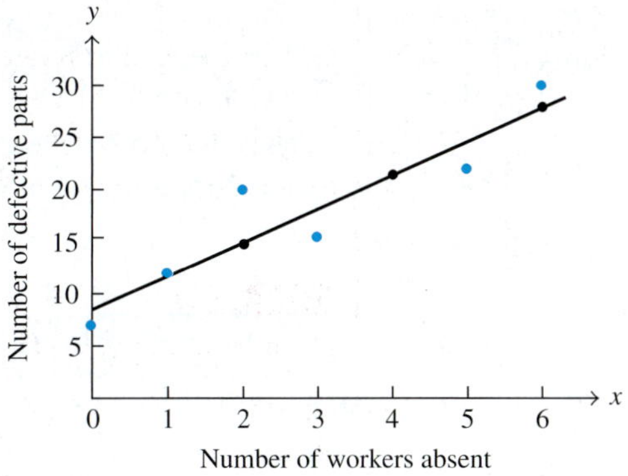

Figure 13.44

In Example 3, the line of best fit intersects the y-axis at 8.52, the value we determined for the y-intercept, b, in part (a).

EXAMPLE 4 *Line of Best Fit for Example 2*

a) Determine the equation of the line of best fit between the time elapsed and the amount of drug remaining in a person's bloodstream in Example 2 on pages 902–904.

b) If the average person is given 300 mg of the drug, how much will remain in the person's bloodstream after 5 hr?

SOLUTION

a) From the scatter diagram on page 903, we see that the slope of the line of best fit, m, will be negative. In Example 2, we found that $n(\Sigma xy) - (\Sigma x)(\Sigma y) = -14{,}925$ and that $n(\Sigma x^2) - (\Sigma x)^2 = 825$. Thus,

$$\begin{aligned} m &= \frac{n(\Sigma xy) - (\Sigma x)(\Sigma y)}{n(\Sigma x^2) - (\Sigma x)^2} \\ &= \frac{-14{,}925}{825} \\ &\approx -18.09 \end{aligned}$$

From Example 2, $n = 10$, $\Sigma x = 55$, and $\Sigma y = 1635$.

$$\begin{aligned} b &= \frac{\Sigma y - m\Sigma x}{n} \\ &= \frac{1635 - (-18.09)(55)}{10} \\ &\approx 263.00 \end{aligned}$$

Thus, the equation of the line of best fit is

$$y = mx + b$$
$$y = -18.09x + 263.00$$

where x is the elapsed time and y is the amount of drug remaining.

b) We evaluate $y = -18.09x + 263.00$ at $x = 5$.

$$y = -18.09x + 263.00$$
$$y = -18.09(5) + 263.00 = 172.55$$

Thus, after 5 hr, about 173 mg of the drug remains in the average person's bloodstream.

TECHNOLOGY TIP We can use EXCEL and both the TI-83 Plus and the TI-84 Plus graphing calculators to determine the correlation coefficient, r, and the equation of the line of best fit. Below we use the data from Example 1 on page 901, where x represents the number of workers absent and y represents the number of defective parts for 6 selected days.

EXCEL

Begin by reading the Technology Tip on page 866. The instructions for determining the correlation coefficient and equation of the line of best fit are similar to the instructions given on page 866. In column A, rows 1 through 6, enter the 6 values representing the number of workers absent. In column B, rows 1 through 6, enter the 6 values that represent the number of defective parts. In general, the items listed on the horizontal axis in the scatter diagram (the independent variable if there is one) are placed in column A and the items listed on the vertical axis are placed in column B. Then select the following:

Insert > Function . . . > Statistical

Then to determine the value of the correlation coefficient, select **CORREL** and click OK. Excel will then generate a gray box and ask you for information about your data. In the box to the right of **Array1**, type in A1:A6, and in the box to the right of **Array2**, type in B1:B6. Then at the bottom of the box you will see *Formula results* = *.921927852*, which is the correlation coefficient. If you then wish to find the equation of the line of best fit, press CANCEL on the bottom of the gray box. Then select

Insert > Function . . . > Statistical > SLOPE > OK

Excel will then generate a gray box. In the box to the right of **Known_y's**, type in B1:B6. In the box to the right of **Known_x's**, type in A1:A6. At that point, at the bottom of the box you will see *Formula results* = *3.229813665*, which is the slope of the equation of the line of best fit. To find the y-intercept, press cancel to remove the gray box. Then select

Insert > Function . . . > Statistical > INTERCEPT > OK

Excel will then generate a gray box. In the box to the right of **Known_y's**, type in B1:B6. In the box to the right of **Known_x's**, type in A1:A6. At that point, at the bottom of the box you will see *Formula results* = *8.51552795*, which is the y-intercept of the equation of the line of best fit.

TI-83 PLUS AND TI-84 PLUS GRAPHING CALCULATORS

To determine the correlation coefficient and the equation of the line of best fit on a graphing calculator, press STAT. Then select Edit. If you have any data in **L1** or **L2**, delete the data by going to **L1** and pressing CLEAR and then ENTER. Follow a similar procedure to clear column **L2**. Now enter the 6 pieces of data representing the number of workers absent in column **L1** and enter the 6 pieces of data representing the number of defective parts in **L2**. Then press STAT and highlight **CALC**. At this point, **1: 1−Var Stats** should be highlighted. Scroll down to highlight **4: LinReg (ax+b)**. Then press the Enter key twice. You will get r = .921927852, which represents the correlation coefficient. On the screen, you will also see values for a and b. Note that a represents the slope and b represents the y-intercept of the equation of the line of best fit. Thus, the slope is 3.229813665 and the y-intercept is 8.51552795.

SECTION 13.8 EXERCISES

CONCEPT/WRITING EXERCISES

1. What does the correlation coefficient measure?
2. What is the purpose of linear regression?
3. What value of r represents the maximum positive correlation?
4. What value of r represents the maximum negative correlation?
5. What value of r represents no correlation between the variables?
6. **a)** What does a negative correlation between two variables indicate?

   b) Give an example of two variables that have a negative correlation.

44. *Selling Popcorn at the Movies* The number of movie tickets sold and the number of units of popcorn sold at AMC Cinema for 8 days is shown below.

Ticket sales	89	110	125	92	100	95	108	97
Units of popcorn	22	28	30	26	22	21	28	25

a) Determine the correlation coefficient between ticket sales and units of popcorn sold.

b) Determine whether a correlation exists at $\alpha = 0.05$.

c) Determine the equation of the line of best fit for tickets sold and units of popcorn sold.

d) Use the equation in part (c) to estimate the units of popcorn sold if 115 tickets are sold.

45. *Fuel Efficiency of Cars* The following table shows the weights, in hundreds of pounds, for six selected cars. Also shown is the corresponding fuel efficiency, in miles per gallon (mpg), for the car in city driving.

Weight (hundreds of pounds)	27	31	35	32	30	30
Fuel efficiency (mpg)	23	22	20	21	24	22

a) Determine the correlation coefficient between the weight of a car and the fuel efficiency.

b) Determine whether a correlation exists at $\alpha = 0.01$.

c) Determine the equation of the line of best fit for the weight of a car and the fuel efficiency of a car.

d) Use the equation in part (c) to estimate the fuel efficiency of a car that weighs 33 hundred pounds.

46. *City Muggings* In a certain section of a city, muggings have been a problem. The number of police officers patrolling that section of the city has varied. The following chart shows the number of police officers and the number of muggings for 8 successive days.

Police officers	20	12	18	15	22	10	20	12
Muggings	8	10	12	9	6	15	7	18

a) Determine the correlation coefficient for number of police officers and number of muggings.

b) Determine whether a correlation exists at $\alpha = 0.05$.

c) Find the equation of the line of best fit for number of police officers and number of muggings.

d) Use the equation in part (c) to estimate the average number of muggings when 14 police officers are patrolling that section of the city.

47. *Chlorine in a Swimming Pool* A gallon of chlorine is put into a swimming pool. Each hour later for the following 6 hr the percent of chlorine that remains in the pool is measured. The following information is obtained.

Time	1	2	3	4	5	6
Chlorine remaining (percent)	80.0	76.2	68.7	50.1	30.2	20.8

a) Determine the correlation coefficient for time and percent of chlorine remaining.

b) Determine whether a correlation exists at $\alpha = 0.01$.

c) Determine the equation of the line of best fit for time and amount of chlorine remaining.

d) Use the equation in part (c) to estimate the average amount of chlorine remaining after 4.5 hr.

48. *Social Security Numbers* **a)** Match the first 9 digits of your phone number (including area code) with the 9 digits in your social security number. To do so, match the first digit in your phone number with the first digit in your social security number to get one ordered pair. Match the second digits to get a second ordered pair. Continue this process until you get a total of nine ordered pairs.

b) Do you believe that this set of bivariate data has a positive correlation, a negative correlation, or no correlation? Explain your answer.

c) Construct a scatter diagram for the nine ordered pairs.

d) Calculate the correlation coefficient, r.

e) Is there a correlation at $\alpha = 0.05$? Explain.

f) Calculate the equation of the line of best fit.

g) Use the equation in part (f) to estimate the digit in a social security number that corresponds with a 7 in a telephone number.

49. ***Hitting the Brakes*** **a)** Examine the art below. Do you believe that there is a positive correlation, a negative correlation, or no correlation between speed of a car and stopping distance when the brakes are applied? Explain.

b) Do you believe that there is a stronger correlation between speed of a car and stopping distance on wet or dry roads? Explain.

c) Use the figure to construct two scatter diagrams, one for dry pavement and the other for wet pavement. Place the speed of the car on the horizontal axis.

d) Compute the correlation coefficient for speed of the car and stopping distance for dry pavement.

e) Repeat part (d) for wet pavement.

f) Were your answers to parts (a) and (b) correct? Explain.

g) Determine the equation of the line of best fit for dry pavement.

h) Repeat part (g) for wet pavement.

i) Use the equations in parts (g) and (h) to estimate the stopping distance of a car going 77 mph on both dry and wet pavements.

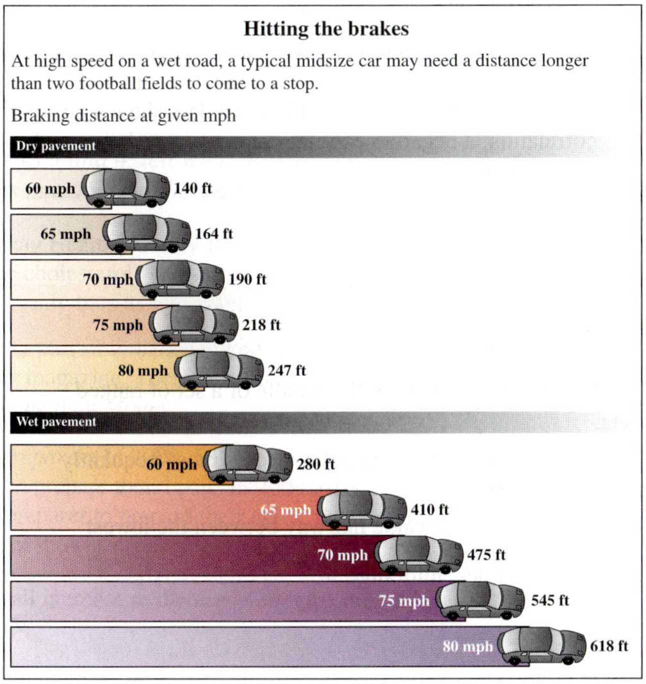

Source: Data from *Car and Driver*, American Automobile Association

CHALLENGE PROBLEMS/GROUP ACTIVITIES

50. ***Interchanging Variables*** **a)** Assume that a set of bivariate data yields a specific correlation coefficient. If the x and y values are interchanged and the correlation coefficient is recalculated, will the correlation coefficient change? Explain.

b) Make up a table of five pieces of bivariate data and determine r using the data. Then switch the values of the x's and y's and recompute the correlation coefficient. Has the value of r changed?

51. ***Height vs. Length*** **a)** Do you believe that a correlation exists between a person's height and the length of a person's arm? Explain.

b) Select 10 people from your class and measure (in inches) their heights and the lengths of their arms.

c) Plot the 10 ordered pairs on a scatter diagram.

d) Calculate the correlation coefficient, r.

e) Determine the equation of the line of best fit.

f) Estimate the length of the arm of a person who is 58 in. tall.

52. ***Calculating a Correlation Coefficient*** **a)** Have your group select a category of bivariate data that it thinks has a strong positive correlation. Designate the independent variable and the dependent variable. Explain why your group believes that the bivariate data have a strong positive correlation.

b) Collect at least 10 pieces of bivariate data that can be used to determine the correlation coefficient. Explain how your group chose these data.

c) Plot a scatter diagram.

d) Calculate the correlation coefficient.

b) Construct a histogram of the frequency distribution.

c) Construct a frequency polygon of the frequency distribution.

d) Construct a stem-and-leaf display.

13.5, 13.6

In Exercises 8–13, for the following test scores 65, 76, 79, 83, 84, 93, determine the

8. mean.

9. median.

10. mode.

11. midrange.

12. range.

13. standard deviation.

In Exercises 14–19, for the set of data 4, 5, 12, 14, 19, 7, 12, 23, 7, 17, 15, 21, determine the

14. mean.

15. median.

16. mode.

17. midrange.

18. range.

19. standard deviation.

13.7

Police Response Time *In Exercises 20–24, assume that police response time to emergency calls is normally distributed with a mean of 9 minutes and a standard deviation of 2 minutes. Determine the percent of emergency calls with a police response time*

20. between 7 and 11 minutes.

21. between 5 and 13 minutes.

22. less than 12.2 minutes.

23. more than 12.2 minutes.

24. more than 7.8 minutes.

Pizza Delivery *In Exercises 25–28, assume that the amount of time to prepare and deliver a pizza from Pepe's Pizza is normally distributed with a mean of 20 min and standard deviation of 5 min. Determine the percent of pizzas that were prepared and delivered*

25. between 20 and 25 min.

26. in less than 18 min.

27. between 22 and 28 min.

28. If Pepe's Pizza advertises that the pizza is free if it takes more than 30 min to deliver, what percent of the pizza will be free?

13.8

29. ***Advertising Rates*** The following table shows the cost, in millions of dollars, for a 30-second television advertisement during the Super Bowl. The column labeled Year refers to the number of years since 1997. Thus, 0 would correspond to 1997 and 8 would correspond to 2005.

Year	0	1	2	3	4	5	6	7	8
Cost ($ millions)	1.2	1.3	1.6	2.0	2.0	2.0	2.3	2.3	2.4

a) Construct a scatter diagram with year on the horizontal axis.

b) Use the scatter diagram in part (a) to determine whether you believe that a correlation exists between the year and the cost of a Super Bowl advertisement. If so, is it a positive or negative correlation? Explain.

c) Calculate the correlation coefficient between the year and the cost of an advertisement.

d) Determine whether a correlation exists at $\alpha = 0.05$.

e) Determine the equation of the line of best fit between the year and the cost of an advertisement. Round both the slope and y-intercept to the nearest hundredth.

f) Assuming that this trend continues, use the equation of the line of best fit to estimate the cost of a 30-second commercial in 2010.

30. ***Daily Sales*** Ace Hardware recorded the number of a particular item sold per week for 6 weeks and the corresponding weekly price, in dollars, of the item as shown in the table below.

Price ($)	0.75	1.00	1.25	1.50	1.75	2.00
Number sold	200	160	140	120	110	95

a) Construct a scatter diagram with price on the horizontal axis.

b) Use the scatter diagram in part (a) to determine whether you believe that a correlation exists between the price of the item and number sold. If so, it is a positive or a negative correlation? Explain.

c) Determine the correlation coefficient between the price and the number sold.

d) Determine whether a correlation exists at $\alpha = 0.05$. Explain how you arrived at your answer.

e) Determine the equation of the line of best fit for the price and the number sold.

f) Use the equation in part (e) to estimate the number sold if the price is $1.60.

13.5–13.7

Men's Weight *In Exercises 31–38, use the following data obtained from a study of the weights of adult men.*

Mean	192 lb	First quartile	178 lb
Median	185 lb	Third quartile	232 lb
Mode	180 lb	86th percentile	239 lb
Standard deviation	23 lb		

31. What is the most common weight?

32. What weight did half of those surveyed exceed?

33. About what percent of those surveyed weighed more than 232 lb?

34. About what percent of those surveyed weighed less than 178 lb?

35. About what percent of those surveyed weighed more than 239 lb?

36. If 100 men were surveyed, what is the total weight of all men?

37. What weight represents two standard deviations above the mean?

38. What weight represents 1.8 standard deviations below the mean?

13.2–13.7

Presidential Children *The following list shows the names of the 42 U.S. presidents and the number of children in their families.*

Washington	0	Cleveland	5
J. Adams	5	B. Harrison	3
Jefferson	6	McKinley	2
Madison	0	T. Roosevelt	6
Monroe	2	Taft	3
J. Q. Adams	4	Wilson	3
Jackson	0	Harding	0
Van Buren	4	Coolidge	2
W. H. Harrison	10	Hoover	2
Tyler	14	F. D. Roosevelt	6
Polk	0	Truman	1
Taylor	6	Eisenhower	2
Fillmore	2	Kennedy	3
Pierce	3	L. B. Johnson	2
Buchanan	0	Nixon	2
Lincoln	4	Ford	4
A. Johnson	5	Carter	4
Grant	4	Reagan	4
Hayes	8	G. Bush	6
Garfield	7	Clinton	1
Arthur	3	G. W. Bush	2

In Exercises 39–50, use the data to determine the following.

39. Mean

40. Mode

41. Median

42. Midrange

43. Range

44. Standard deviation (round the mean to the nearest tenth)

45. Construct a frequency distribution; let the first class be 0–1.

46. Construct a histogram.

47. Construct a frequency polygon.

48. Does this distribution appear to be normal? Explain.

49. On the basis of this sample, do you think the number of children per family in the United States is a normal distribution? Explain.

50. Do you believe that this sample is representative of the population? Explain.

CHAPTER 13 TEST

In Exercises 1–6, for the set of data 27, 43, 43, 45, 52, determine the

1. mean.

2. median.

3. mode.

4. midrange.

5. range.

6. standard deviation.

In Exercises 7–9, use the set of data

26	28	35	46	49	56
26	30	36	46	49	58
26	32	40	47	50	58
26	32	44	47	52	62
27	35	46	47	54	66

CHAPTER 14

Graph Theory

▲ The map shows the ten top-rated metropolitan areas in the United States in 2006 as determined by www.cityrating.com. Given the airfare from each area shown to every other area shown, graph theory can be used to determine the least expensive way for a person to fly to all the areas shown.

WHAT YOU WILL LEARN

- Graphs, paths, and circuits
- The Königsberg bridge problem
- Euler paths and Euler circuits
- Hamilton paths and Hamilton circuits
- Traveling salesman problems
- Brute force method
- Nearest neighbor method
- Trees, spanning trees, and minimum-cost spanning trees

WHY IT IS IMPORTANT

In our everyday lives, we often face problems that require us to carry out several tasks to reach a desired solution. *Graph theory* is an important branch of mathematics used to represent and find solutions to such problems. Businesses, governments, and other organizations make use of graph theory to determine the most cost-effective means of solving problems and providing products and services. For example, graph theory can be used to determine the least expensive route for someone to visit the cities shown on the map.

13.5–13.7

Men's Weight *In Exercises 31–38, use the following data obtained from a study of the weights of adult men.*

Mean	192 lb	First quartile	178 lb
Median	185 lb	Third quartile	232 lb
Mode	180 lb	86th percentile	239 lb
Standard deviation	23 lb		

31. What is the most common weight?

32. What weight did half of those surveyed exceed?

33. About what percent of those surveyed weighed more than 232 lb?

34. About what percent of those surveyed weighed less than 178 lb?

35. About what percent of those surveyed weighed more than 239 lb?

36. If 100 men were surveyed, what is the total weight of all men?

37. What weight represents two standard deviations above the mean?

38. What weight represents 1.8 standard deviations below the mean?

13.2–13.7

Presidential Children *The following list shows the names of the 42 U.S. presidents and the number of children in their families.*

Washington	0	Cleveland	5
J. Adams	5	B. Harrison	3
Jefferson	6	McKinley	2
Madison	0	T. Roosevelt	6
Monroe	2	Taft	3
J. Q. Adams	4	Wilson	3
Jackson	0	Harding	0
Van Buren	4	Coolidge	2
W. H. Harrison	10	Hoover	2
Tyler	14	F. D. Roosevelt	6
Polk	0	Truman	1
Taylor	6	Eisenhower	2
Fillmore	2	Kennedy	3
Pierce	3	L. B. Johnson	2
Buchanan	0	Nixon	2
Lincoln	4	Ford	4
A. Johnson	5	Carter	4
Grant	4	Reagan	4
Hayes	8	G. Bush	6
Garfield	7	Clinton	1
Arthur	3	G. W. Bush	2

In Exercises 39–50, use the data to determine the following.

39. Mean

40. Mode

41. Median

42. Midrange

43. Range

44. Standard deviation (round the mean to the nearest tenth)

45. Construct a frequency distribution; let the first class be 0–1.

46. Construct a histogram.

47. Construct a frequency polygon.

48. Does this distribution appear to be normal? Explain.

49. On the basis of this sample, do you think the number of children per family in the United States is a normal distribution? Explain.

50. Do you believe that this sample is representative of the population? Explain.

CHAPTER 13 TEST

In Exercises 1–6, for the set of data 27, 43, 43, 45, 52, determine the

1. mean.

2. median.

3. mode.

4. midrange.

5. range.

6. standard deviation.

In Exercises 7–9, use the set of data

26	28	35	46	49	56
26	30	36	46	49	58
26	32	40	47	50	58
26	32	44	47	52	62
27	35	46	47	54	66

CHAPTER 14

Graph Theory

▲ The map shows the ten top-rated metropolitan areas in the United States in 2006 as determined by www.cityrating.com. Given the airfare from each area shown to every other area shown, graph theory can be used to determine the least expensive way for a person to fly to all the areas shown.

WHAT YOU WILL LEARN

- Graphs, paths, and circuits
- The Königsberg bridge problem
- Euler paths and Euler circuits
- Hamilton paths and Hamilton circuits
- Traveling salesman problems
- Brute force method
- Nearest neighbor method
- Trees, spanning trees, and minimum-cost spanning trees

WHY IT IS IMPORTANT

In our everyday lives, we often face problems that require us to carry out several tasks to reach a desired solution. *Graph theory* is an important branch of mathematics used to represent and find solutions to such problems. Businesses, governments, and other organizations make use of graph theory to determine the most cost-effective means of solving problems and providing products and services. For example, graph theory can be used to determine the least expensive route for someone to visit the cities shown on the map.

13.5–13.7

Men's Weight *In Exercises 31–38, use the following data obtained from a study of the weights of adult men.*

Mean	192 lb	First quartile	178 lb
Median	185 lb	Third quartile	232 lb
Mode	180 lb	86th percentile	239 lb
Standard deviation	23 lb		

31. What is the most common weight?

32. What weight did half of those surveyed exceed?

33. About what percent of those surveyed weighed more than 232 lb?

34. About what percent of those surveyed weighed less than 178 lb?

35. About what percent of those surveyed weighed more than 239 lb?

36. If 100 men were surveyed, what is the total weight of all men?

37. What weight represents two standard deviations above the mean?

38. What weight represents 1.8 standard deviations below the mean?

13.2–13.7

Presidential Children *The following list shows the names of the 42 U.S. presidents and the number of children in their families.*

Washington	0	Cleveland	5
J. Adams	5	B. Harrison	3
Jefferson	6	McKinley	2
Madison	0	T. Roosevelt	6
Monroe	2	Taft	3
J. Q. Adams	4	Wilson	3
Jackson	0	Harding	0
Van Buren	4	Coolidge	2
W. H. Harrison	10	Hoover	2
Tyler	14	F. D. Roosevelt	6
Polk	0	Truman	1
Taylor	6	Eisenhower	2
Fillmore	2	Kennedy	3
Pierce	3	L. B. Johnson	2
Buchanan	0	Nixon	2
Lincoln	4	Ford	4
A. Johnson	5	Carter	4
Grant	4	Reagan	4
Hayes	8	G. Bush	6
Garfield	7	Clinton	1
Arthur	3	G. W. Bush	2

In Exercises 39–50, use the data to determine the following.

39. Mean

40. Mode

41. Median

42. Midrange

43. Range

44. Standard deviation (round the mean to the nearest tenth)

45. Construct a frequency distribution; let the first class be 0–1.

46. Construct a histogram.

47. Construct a frequency polygon.

48. Does this distribution appear to be normal? Explain.

49. On the basis of this sample, do you think the number of children per family in the United States is a normal distribution? Explain.

50. Do you believe that this sample is representative of the population? Explain.

CHAPTER 13 TEST

In Exercises 1–6, for the set of data 27, 43, 43, 45, 52, determine the

1. mean.

2. median.

3. mode.

4. midrange.

5. range.

6. standard deviation.

In Exercises 7–9, use the set of data

26	28	35	46	49	56
26	30	36	46	49	58
26	32	40	47	50	58
26	32	44	47	52	62
27	35	46	47	54	66

to construct the following.

7. a frequency distribution; let the first class be 25–30

8. a histogram of the frequency distribution

9. a frequency polygon of the frequency distribution

Statistics on Salaries *In Exercises 10–16, use the following data on weekly salaries at Donovan's Construction Company.*

Mean	\$740	First quartile	\$690
Median	\$710	Third quartile	\$745
Mode	\$735	79th percentile	\$752
Standard deviation	\$40		

10. What is the most common salary?

11. What salary did half the employees exceed?

12. About what percent of employees' salaries exceeded \$690?

13. About what percent of employees' salaries was less than \$752?

14. If the company has 100 employees, what is the total weekly salary of all employees?

15. What salary represents one standard deviation above the mean?

16. What salary represents 1.5 standard deviations below the mean?

Anthropology *In Exercises 17–20, assume that anthropologists have determined that the akidolestes, a small primitive mammal believed to have lived with the dinosaurs, has a head circumference that was normally distributed with a mean of 42 cm and a standard deviation of 5 cm.*

17. What percent of head circumferences were between 36 and 53 cm?

18. What percent of head circumferences were greater than 35.75 cm?

19. What percent of head circumferences were greater than 48.25 cm?

20. What percent of head circumferences were less than 50 cm?

21. ***Average Earnings*** The following chart shows the average hourly earnings, to the nearest 10 cents, for U.S. production workers for 2001–2005, where the column labeled Year refers to the numbers of years since 2001.

Average Earnings for U.S. Production Workers

Year	Average Hourly Earnings
0	\$14.5
1	\$15.0
2	\$15.4
3	\$15.7
4	\$16.1

Source: Bureau of Labor Statistics

a) Construct a scatter diagram placing the year on the horizontal axis.

b) Use the scatter diagram in part (a) to determine whether you believe that a correlation exists between the year and the average hourly earnings for a U.S. production worker. Explain.

c) Determine the correlation coefficient between the year and the average hourly earnings for a U.S. production worker.

d) Determine whether a correlation exists at $\alpha = 0.05$.

e) Determine the equation of the line of best fit between the year and the average hourly earnings for a U.S. production worker. Round both the slope and y-intercept to the nearest hundredth.

f) Use the equation in part (e) to predict the average hourly earnings for a U.S. production worker in 2030, or 29 years after 2001.

GROUP PROJECTS

WATCHING TV

1. Do you think that men or women, aged 17–20, watch more hours of TV weekly, or do you think that they watch the same number of hours?

   a) Write a procedure to use to determine the answer to that question. In your procedure, use a sample of 30 men and 30 women. State how you will obtain an unbiased sample.

   b) Collect 30 pieces of data from men aged 17–20 and 30 pieces of data from women aged 17–20. Round answers to the nearest 0.5 hr. Follow the procedure developed in part (a) to obtain your unbiased sample.

   c) Compute the mean for your two groups of data to the nearest tenth.

   d) Using the means obtained in part (c), answer the question asked at the beginning of the problem.

   e) Is it possible that your conclusion in part (d) is wrong? Explain.

   f) Compute the standard deviation for each group to the nearest tenth. How do the standard deviations compare?

   g) Do you believe that the distribution of data from either or both groups resembles a normal distribution? Explain.

   h) Add the two groups of data to get one group of 60 pieces of data. If these 60 pieces of data are added and divided by 60, will you obtain the same mean as when you add the two means from part (c) and divide the sum by 2? Explain.

   i) Compute the mean of the 60 pieces of data by using both methods mentioned in part (h). Are they the same? If so, why? If not, why not?

   j) Do you believe that this group of 60 pieces of data represents a normal distribution? Explain.

BIVARIATE DATA EXPERIMENT

2. a) Have your group select a category of bivariate data that it thinks has a strong negative correlation. Indicate the variable that you will designate as the independent variable and the variable that you will designate as the dependent variable. Explain why your group believes that the bivariate data have a strong negative correlation.

   b) Collect at least 10 pieces of bivariate data that can be used to determine the correlation coefficient. Explain how your group chose these data.

   c) Plot a scatter diagram.

   d) Calculate the correlation coefficient.

   e) Is there a negative correlation at $\alpha = 0.05$? Explain your answer.

   f) Calculate the equation of the line of best fit.

   g) Explain how the equation in part (f) may be used.

C H A P T E R 14

Graph Theory

▲ The map shows the ten top-rated metropolitan areas in the United States in 2006 as determined by www.cityrating.com. Given the airfare from each area shown to every other area shown, graph theory can be used to determine the least expensive way for a person to fly to all the areas shown.

WHAT YOU WILL LEARN

- Graphs, paths, and circuits
- The Königsberg bridge problem
- Euler paths and Euler circuits
- Hamilton paths and Hamilton circuits
- Traveling salesman problems
- Brute force method
- Nearest neighbor method
- Trees, spanning trees, and minimum-cost spanning trees

WHY IT IS IMPORTANT

In our everyday lives, we often face problems that require us to carry out several tasks to reach a desired solution. *Graph theory* is an important branch of mathematics used to represent and find solutions to such problems. Businesses, governments, and other organizations make use of graph theory to determine the most cost-effective means of solving problems and providing products and services. For example, graph theory can be used to determine the least expensive route for someone to visit the cities shown on the map.

14.1 GRAPHS, PATHS, AND CIRCUITS

▲ What is the shortest route to deliver Girl Scout cookies to 15 different homes?

Suppose that you are a Girl Scout leader and you must deliver cookies to 15 different homes. You will use your car to drive from home to home to deliver all the cookies. You realize that you have many possible routes to accomplish the task, but how can you determine the shortest route? In this section, we will introduce a type of diagram that can help you analyze such problems.

The study of graph theory can be traced back to the eighteenth century when people of the East Prussian town of Königsberg sought the solution to a popular problem. Königsberg was situated on both banks and two islands of the Prigel River. From Fig. 14.1, we see that the sections of town were connected with a series of seven bridges. The townspeople wondered if one could walk through town and cross all seven bridges without crossing any of the bridges twice. This question was presented to Swiss mathematician Leonhard Euler (pronounced "oiler," 1707–1783). To study this problem, Euler reduced the problem to one that could be represented with a series of dots and lines. This problem came to be known as the Königsberg bridge problem.

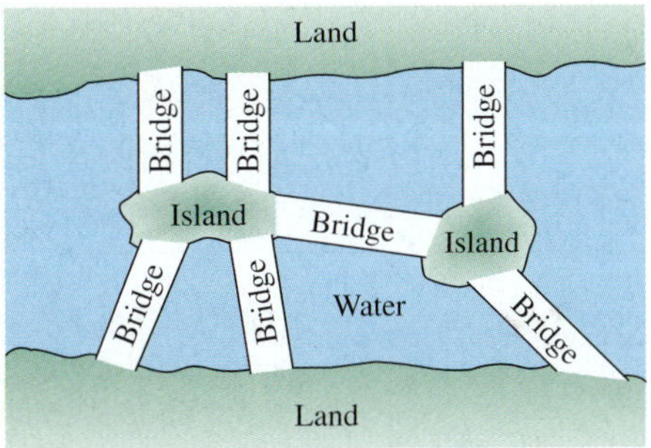

Figure 14.1

We will revisit the Königsberg bridge problem several times in this section and in Section 14.2. Our main focuses in this section are to introduce definitions of graph theory and to explain how these definitions can help represent problems from the physical world. In the next section, we will introduce some basic graph theory used to solve these problems.

Graphs

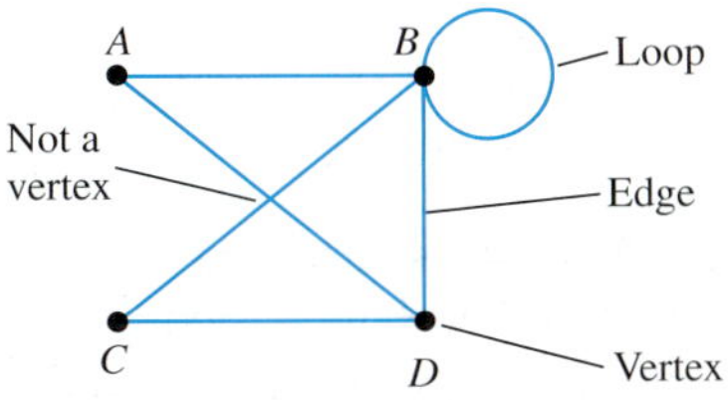

Figure 14.2

A ***graph*** is a finite set of points called ***vertices*** (singular form is ***vertex***) connected by line segments (not necessarily straight) called ***edges***. An edge that connects a vertex to itself is called a ***loop***. Vertices, edges, and a loop are displayed in the graph in Fig. 14.2. We generally refer to vertices with capitalized letters. The edge between two vertices will be referred to using the two vertices. For example, in Fig. 14.2, the edge that connects vertex A to vertex B is referred to as edge AB or as edge BA. The loop in Fig. 14.2 is referred to as edge BB or loop BB. Not every place where two edges cross is a vertex. A dot must be present to represent a vertex. For instance, in Fig. 14.2, even though edges AD and BC cross, since no dot is indicated, the place where these lines cross is not a vertex.

With these basic definitions, we can begin using graphs to represent physical settings.

EXAMPLE 1 *Representing the Königsberg Bridge Problem*

Using the definitions of vertex and edge, represent the Königsberg bridge problem with a graph.

SOLUTION Note that each bridge in Fig. 14.1 on page 919 connects two pieces of land in a manner similar to an edge connecting two vertices. To begin representing the Königsberg bridge problem as a graph, we will label each piece of land with a capital letter, as shown in Fig. 14.3.

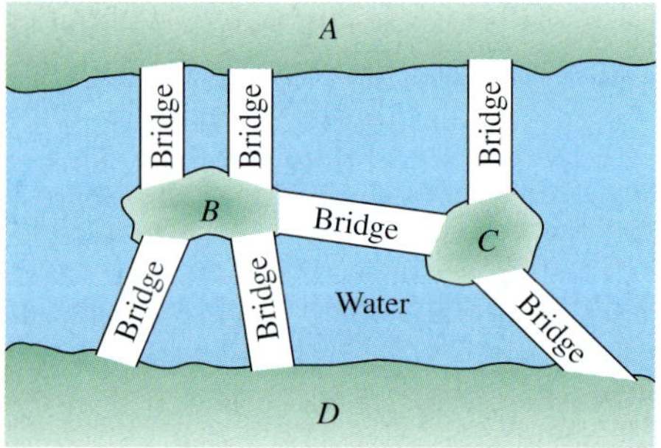

Figure 14.3

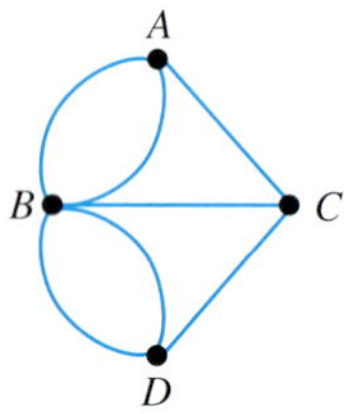

Figure 14.4

Next, we draw edges to represent the bridges. Notice that there are two bridges connecting land A to land B in Fig. 14.3. Therefore, we need two edges to connect vertex A to vertex B in our graph (see Fig. 14.4). Notice that there is one bridge connecting land A to land C in Fig. 14.3. Therefore, we need one edge to connect vertex A to vertex C in our graph. We continue to let edges represent bridges, and the resulting graph is given in Fig. 14.4.

Although the picture and the graph do not look very similar, the graph represents the key aspects of the problem: the land is represented with vertices, and the bridges are represented with edges.

Graphs are powerful tools in problem solving because they allow us to focus on the key aspects of problems without getting distracted with unnecessary details. Our next three examples show other ways that graphs can be used to represent settings from the physical world.

EXAMPLE 2 *Tenth U.S. Court of Appeals*

The Tenth United States Court of Appeals hears cases from the states shown in Fig. 14.5 on page 921. Construct a graph to show the states that share a common border. States whose borders only touch at a single corner point will not be considered to share a common border.

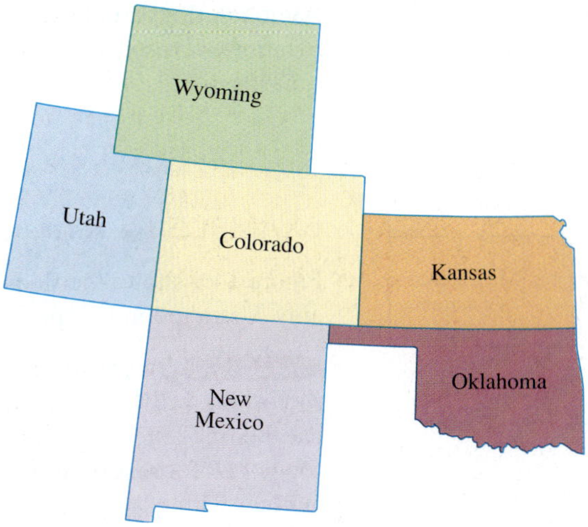

Figure 14.5

SOLUTION In our graph, each vertex will represent one of the six states shown in Fig. 14.5. We begin our graph by placing the six vertices in the same relative positions as the six states on the map and then labeling each vertex with a two-letter state abbreviation (Fig. 14.6). The exact placement of the vertices is not critical.

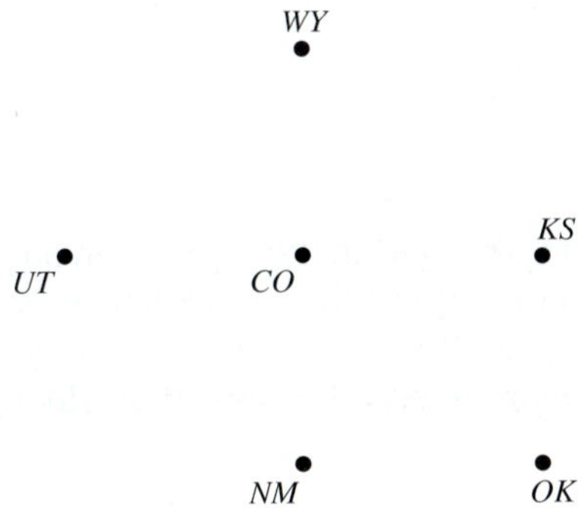

Figure 14.6

Next, if two states share a common border, connect the respective vertices with an edge. For example, Utah shares a common border with Wyoming and Colorado, so there will be an edge that connects vertex *UT* to vertex *WY* and an edge that connects vertex *UT* to vertex *CO*, as shown in Fig. 14.7. Utah and New Mexico have borders that touch only at a corner point and thus are not considered to share a common border. Therefore, in the graph shown in Fig. 14.7, no edge connects *UT* to *NM*.

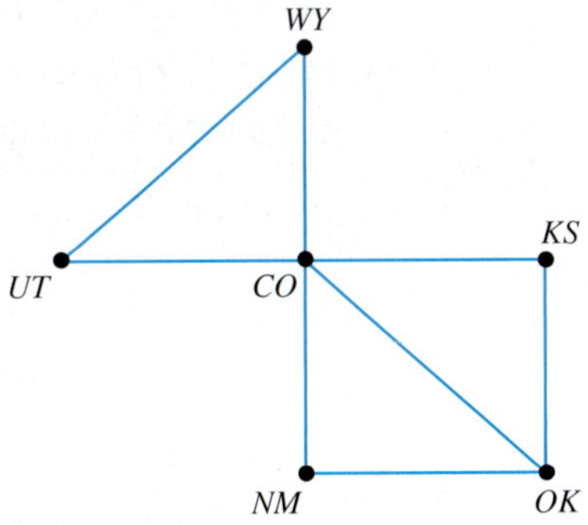

Figure 14.7

Next, note that Colorado shares a common border with Utah, Wyoming, Kansas, Oklahoma, and New Mexico. Therefore, the graph in Fig. 14.7 will have edges that connect *CO* to *UT*, *WY*, *KS*, *OK*, and *NM*, respectively. We continue adding edges until for every two states in Fig. 14.5 that share a common border there is an edge between their respective vertices in Fig. 14.7.

The graph shown in Fig. 14.7 is only one possible arrangement of vertices and edges that can show which states share a common border. Many other graphs that display the same relationship are equally valid. Figure 14.8 shows one such graph.

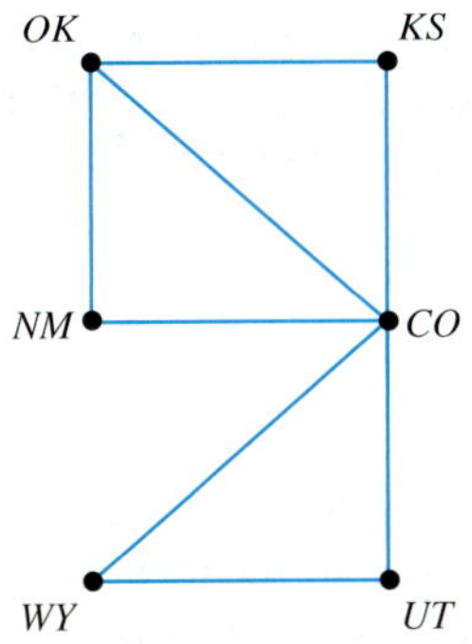

Figure 14.8

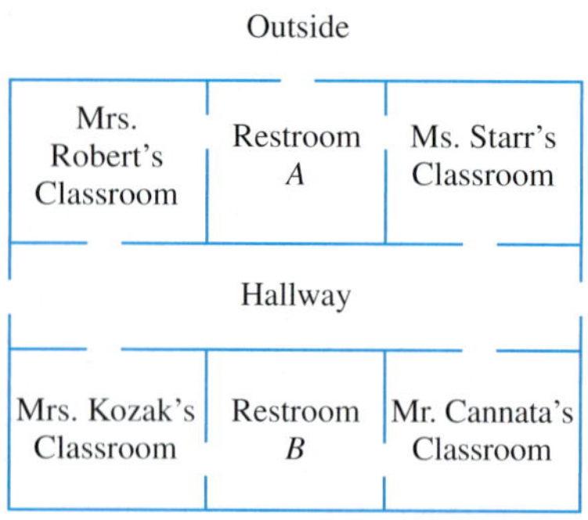

Figure 14.9

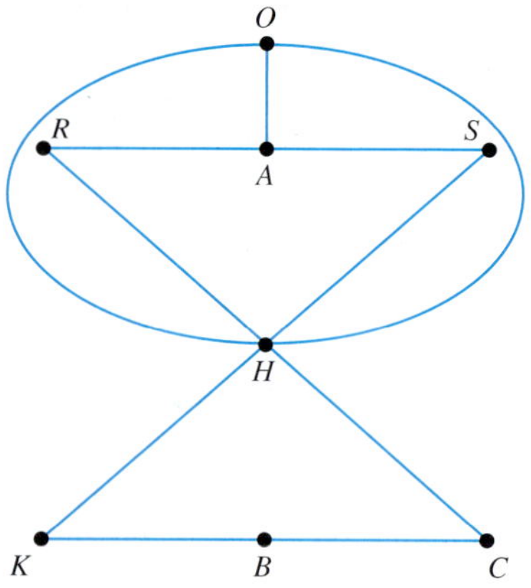

Figure 14.10

Although the vertices in the graph in Fig. 14.8 do not resemble the respective locations of the states they represent, the relationship between the vertices is the same as in Fig. 14.7.

EXAMPLE 3 *Representing a Floor Plan*

Figure 14.9 shows the floor plan of the kindergarten building at the Pullen Academy. Use a graph to represent the floor plan.

SOLUTION To create a graph, we start by letting vertices represent each of the six rooms, the hallway, and the outside of the building. To label the vertices, we will use the first letter of the teacher's last names along with H for hallway, A for restroom A, B for restroom B, and O for the outside of the building. Next, connect the vertices with edges. To determine placement of the edges within the graph, visualize walking through the various doorways in the building. For instance, since a person can walk from Mrs. Kozak's room directly into the hallway or into restroom B, there is an edge between K and H and an edge between K and B. The resulting graph is shown in Fig. 14.10. Notice that since the hallway has two separate doorways that lead to the outside of the building, there are two separate edges that connect vertices H and O.

In Example 3, we placed the vertex O, for outside, near the top of the graph. However, vertex O could have been placed in another location either above, below, or on either side of the vertices that represent the floor plan. If vertex O had been placed elsewhere, the graph would look different, but the graph would still represent the floor plan shown in Fig. 14.9.

EXAMPLE 4 *Representing a Neighborhood*

Figure 14.11 shows a sketch of the Country Oaks subdivision of homes. Use a graph to represent the streets of this neighborhood.

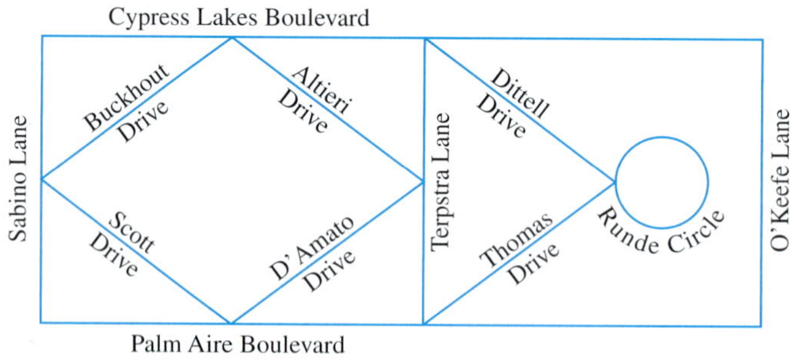

Figure 14.11

SOLUTION In this problem, the vertices will represent the street intersections and the edges will represent the street blocks. We will use the letters $A, B, C, \ldots, K$ to name the vertices. The resulting graph is shown in Fig. 14.12 on page 923.

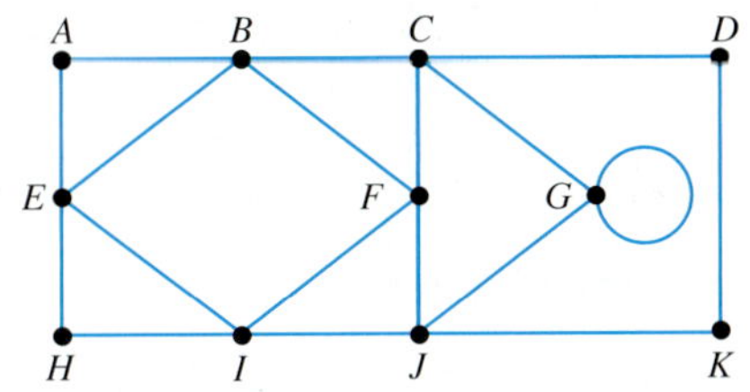

Figure 14.12

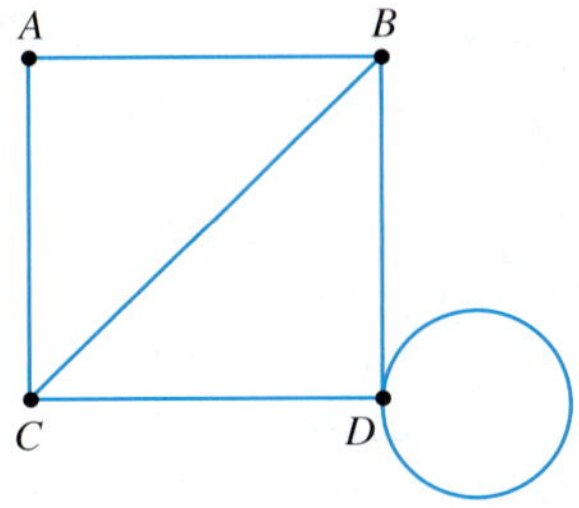

Figure 14.13

Now that we have some experience using graphs to represent real-life settings, we will introduce some additional vocabulary used in graph theory. We will begin with a method of classifying vertices. The ***degree*** of a vertex is the number of edges that connect to that vertex. For instance, in Fig. 14.13, vertex *A* has two edges connected to it. Therefore, the degree of vertex *A* is two. We also see from Fig. 14.13 that vertex *B* has three edges connected to it, as does vertex *C*. Therefore, vertices *B* and *C* each have degree three. Vertex *D* has a loop connected to it. A question arises: Should loop *DD* count as one edge or two edges when determining the degree of vertex *D*? We will agree to count each end of a loop when determining the degree of a vertex. Thus, vertex *D* from Fig. 14.13 will have degree four. A vertex with an even number of edges connected to it is an ***even vertex***, and a vertex with an odd number of edges connected to it is an ***odd vertex***. In Fig. 14.13, vertices *A* and *D* are even and vertices *B* and *C* are odd.

Paths, Circuits, and Bridges

We next introduce some definitions of graph theory that can be used to describe "movement."

A **path** is a sequence of adjacent vertices and the edges connecting them.

By adjacent vertices, we mean two vertices that are connected by a common edge. For example, in Fig. 14.13, vertices *A* and *B* are adjacent vertices since there is a common edge between vertex *A* and vertex *B*. Vertices *A* and *D* are not adjacent vertices since there is not a common edge between vertex *A* and vertex *D*. An example of a path is given in Fig. 14.14. It is important to recognize the difference between a graph and a path. The *graph* is the set of black vertices and the blue edges connecting them. The *path*, shown in red, can be thought of as movement from vertex *C* to vertex *D* to vertex *A* to vertex *B*. For convenience, and to help in explanations, paths will be referred to with the sequence of vertices separated by commas. For example, the path in Fig. 14.14 will be referred to as path *C*, *D*, *A*, *B*.

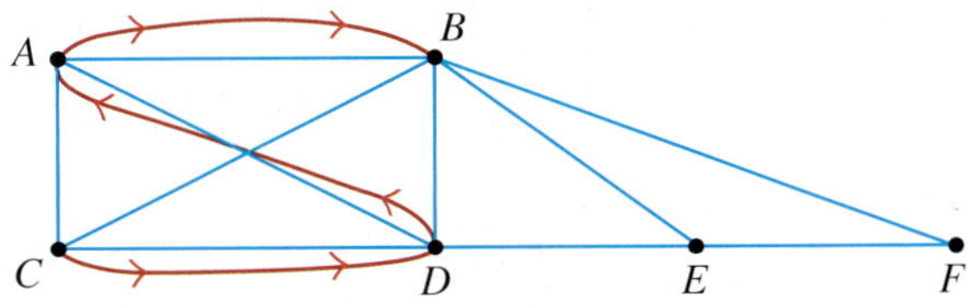

Figure 14.14

DID YOU KNOW?

Graph Theory Representations

Boston Area Subway Map

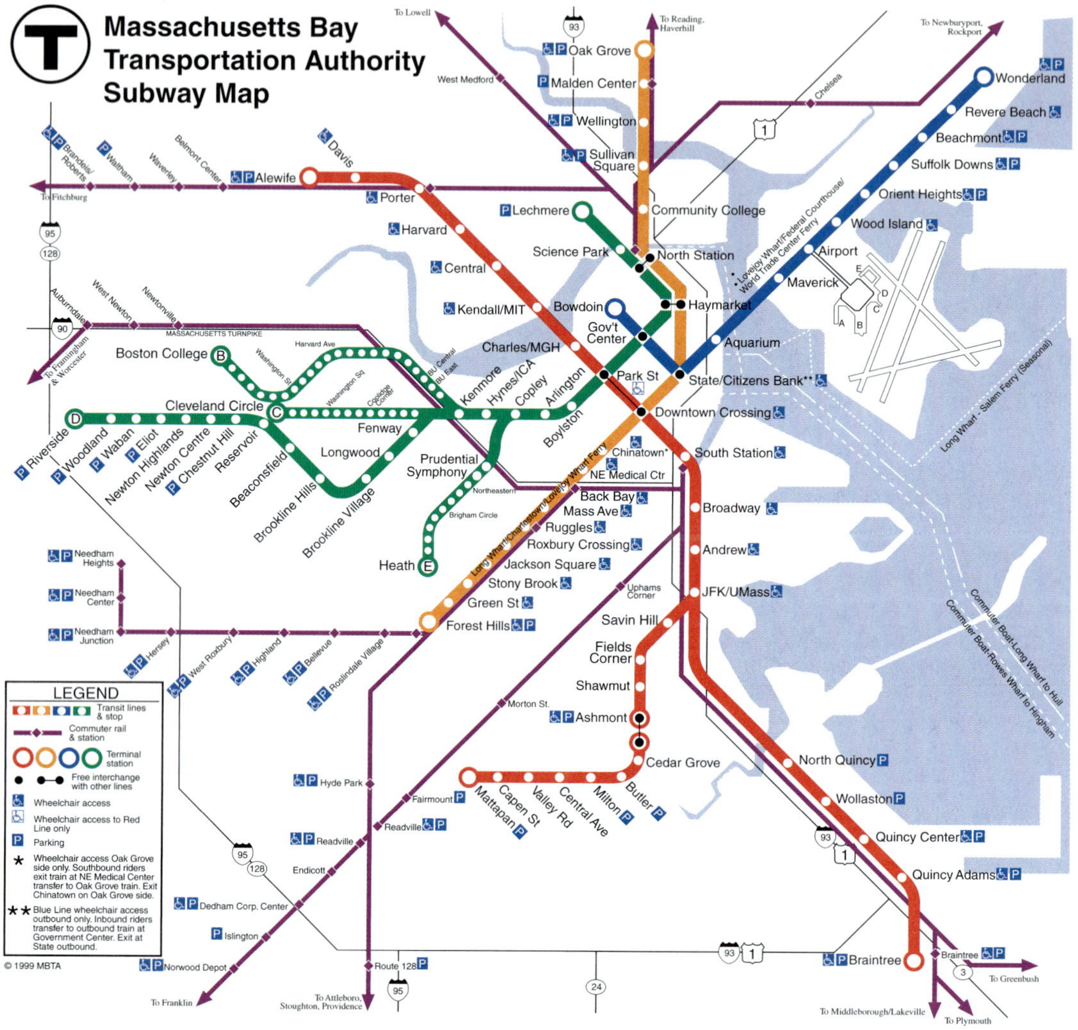

raphs consisting of points called vertices and lines called edges can represent numerous applications used in many professions. Chemists use graphs to sketch atoms in a molecule. Engineers use graphs in schematic drawings of electrical networks. Genealogists use graphs to represent family trees. Transportation officials use graphs to represent airline, railway, and ground transportation routes. Industrial scientists use graphs to represent human and machine interaction times in factories. All these applications can be represented and studied with the branch of mathematics called graph theory, the origins of which can be traced back to Leonhard Euler (1707–1783). The subway map above for the city and surrounding area of Boston, Massachusetts, is a graph of vertices and edges.

A path does not need to include every edge and every vertex of a graph. In addition, a path could include the same vertices and the same edges several times. For example, in Fig. 14.15, we see a graph with four vertices. The path *A, B, C, D, A, B, C, D, A, B, C, D, A, B, C* starts at vertex *A*, "circles" the graph three times, and then goes through vertex *B* to vertex *C*.

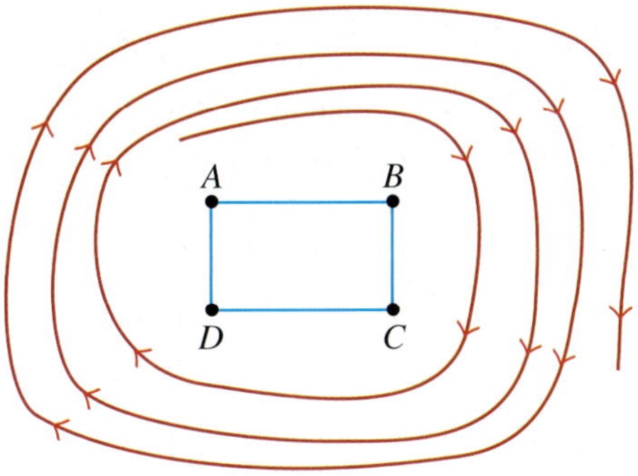

Figure 14.15

A **circuit** is a path that begins and ends at the same vertex.

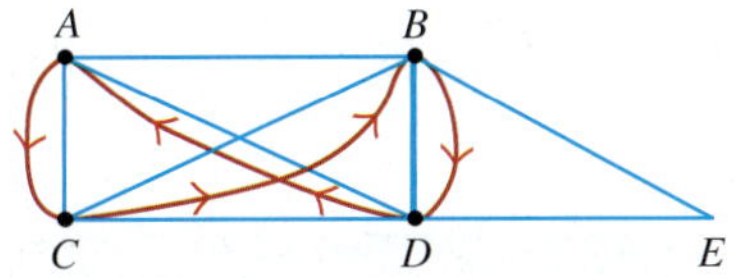

Figure 14.16

Examine Fig. 14.16. The path given by *A, C, B, D, A* forms a circuit. The path given by *B, D, E, B* forms another circuit.

Note that every circuit is, by definition, a path; however, not every path is a circuit. A circuit needs to start and stop at the same vertex. In Sections 14.2 and 14.3, we will discuss paths and circuits in more depth.

Our next definitions classify graphs themselves. A graph is ***connected*** if, for any two vertices in the graph, there is a path that connects them. All the graphs we have studied thus far have been connected because a path existed between each pair of vertices in each graph. If a graph is not connected, it is ***disconnected***. Figure 14.17 shows three examples of disconnected graphs. In Fig. 14.17(c), although edge *AD* and edge *BC* cross, the graph is not connected because there is no vertex indicated where the edges cross.

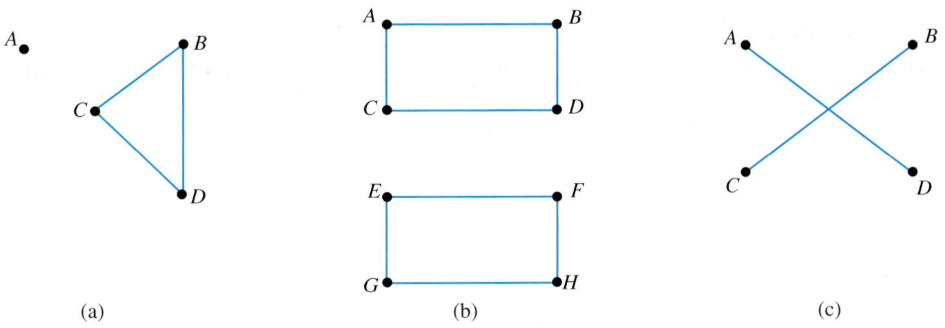

Figure 14.17

A **bridge** is an edge that if removed from a connected graph would create a disconnected graph.

Figure 14.18 shows graphs with bridges indicated. Compare these graphs with those in Fig. 14.17. The graphs in Fig. 14.18 are the same graphs as those in Fig. 14.17 with a bridge added. Note in Fig. 14.18(c) that, in addition to edge *AB* that we added, edges *AD* and *BC* are also bridges. If edge *AD* were removed, vertex *D* would be isolated from the rest of the graph. If edge *BC* were removed, vertex *C* would be isolated from the rest of the graph. Note that our graph theory definition of *bridge* is different from bridges as they are used in the Königsberg bridge problem. In the Königsberg bridge problem, none of the edges that represents real bridges is a bridge in the graph theory sense.

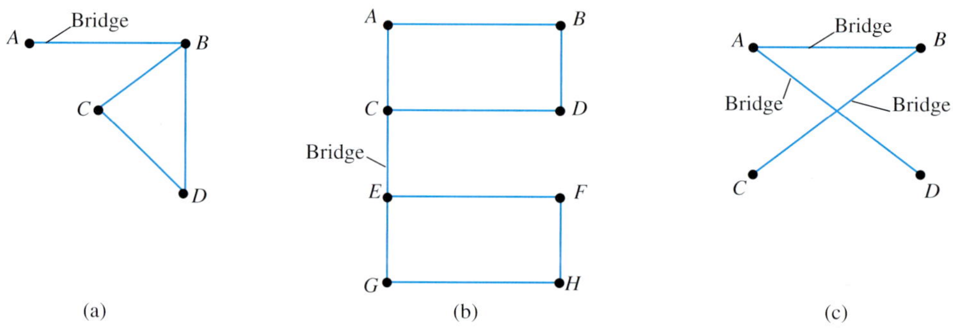

Figure 14.18

SECTION 14.1 EXERCISES

CONCEPT/WRITING EXERCISES

1. Using one sentence, give the definition of a graph. Use the words *vertices* and *edges* in your definition.

2. Draw an example of a graph that contains a loop.

3. Draw an example of a graph that contains four vertices and six edges.

4. Explain how to determine the degree of a vertex.

5. Explain how to determine if a vertex is odd or even.

6. Draw an example of a graph that contains a bridge. Identify the bridge and explain why this edge is a bridge.

7. a) What is a path?

   b) What is a circuit?

   c) Draw a graph and give an example of a path that is not a circuit and an example of a path that is a circuit.

8. Draw an example of a graph that is connected and draw an example of a graph that is disconnected. Explain why your second graph is disconnected.

PRACTICE THE SKILLS

In Exercises 9–14, create a graph with the given properties. There are many possible answers for each problem.

9. Two odd vertices

10. Four even vertices

11. Two even and two odd vertices

12. Four odd vertices and one even vertex

13. Seven vertices with one bridge

14. Seven vertices with two bridges

In Exercises 15–20, use the graph below to answer the following questions.

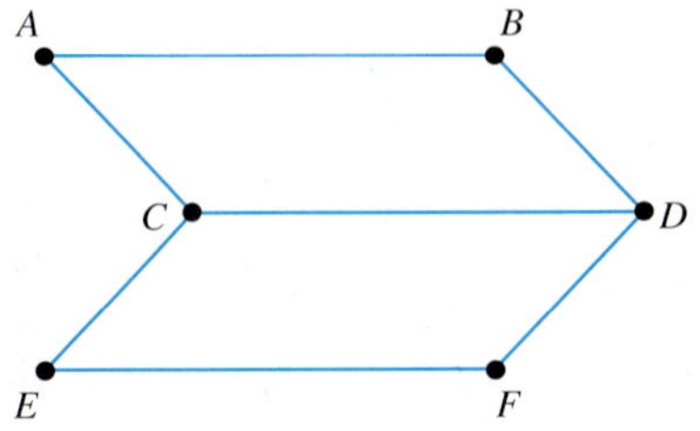

15. Is A, B, C, D, F, E a path? Explain.

16. Which edge(s) shown on the graph are not included in the following path: A, B, D, F, E, C?

17. On the graph, is it possible to determine a path that begins with vertex A, contains all the edges exactly once, and ends with vertex B? If so, determine one such path.

18. On the graph, is it possible to determine a circuit that includes only four distinct vertices? If so, determine one such circuit.

19. On the graph, can you determine a path that includes all the edges without using any edge twice? If so, determine one such path.

20. On the graph, can you determine a circuit that includes all the vertices without using any vertex (other than the beginning and ending vertex) twice? If so, determine one such circuit.

PROBLEM SOLVING

Modified Königsberg Bridge Problems *In Exercises 21 and 22, suppose that the people of Königsberg decide to add several bridges to their city. Two such possibilities are shown. Create graphs that would represent the Königsberg bridge problem with these new bridges.*

21.

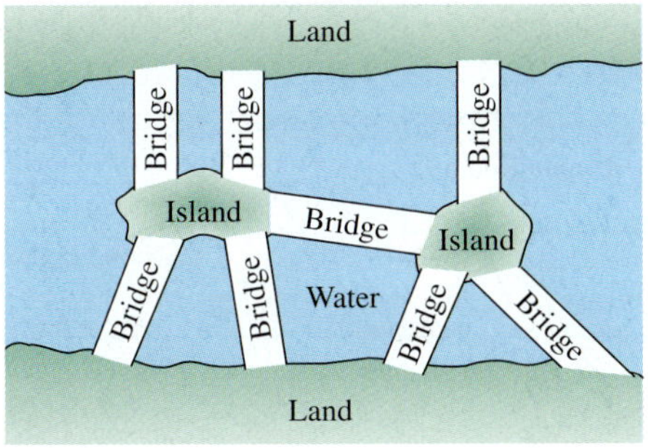

22.

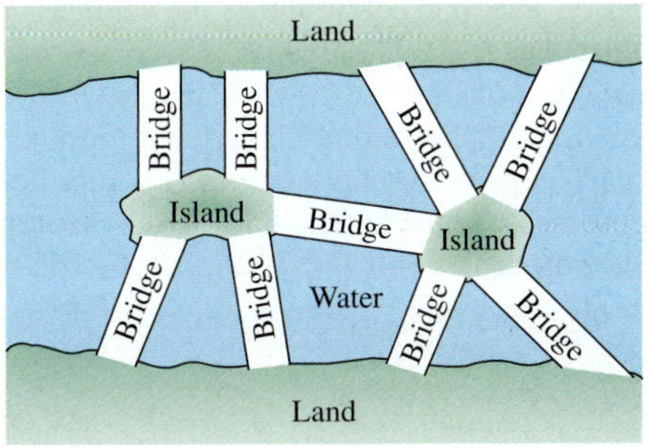

Other U.S. Courts of Appeal *In Exercises 23 and 24, maps of states whose cases are heard in other U.S. Courts of Appeal (see Example 2) are shown. Represent each map as a graph where each vertex represents a state and each edge represents a common border between the states. The two-letter state abbreviation is shown in parentheses.*

23. The Eighth U.S. Court of Appeals

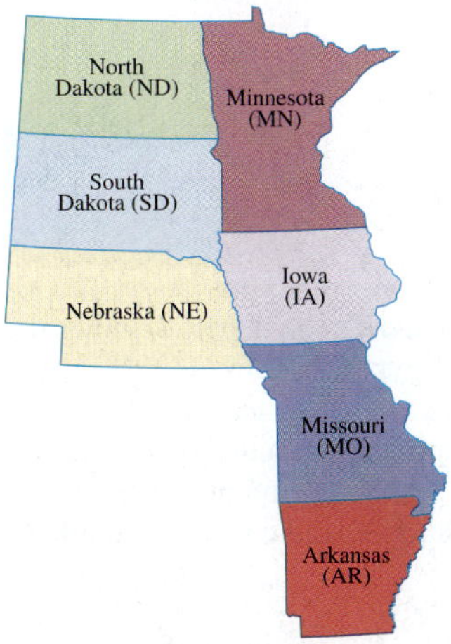

24. The Ninth U.S. Court of Appeals

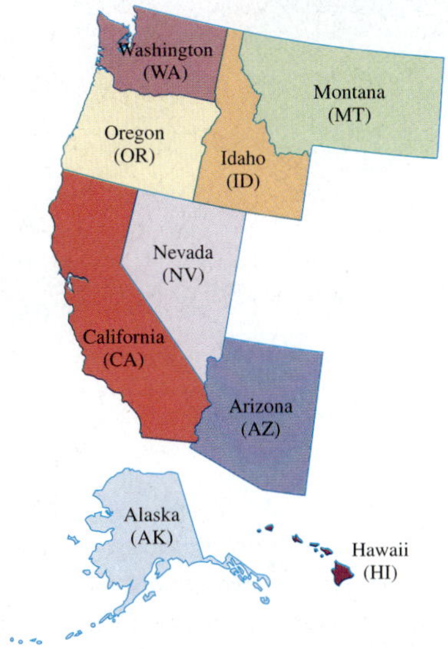

25. ***Central America*** The map below shows the countries of Belize (B), Costa Rica (C), El Salvador (E), Guatemala (G), Honduras (H), Nicaragua (N), and Panama (P). Represent the map as a graph where each vertex represents a country and each edge represents a common border between the countries. Use the letters indicated in parentheses to label the vertices.

26. ***Northern Africa*** The map below shows the countries of Algeria (A), Egypt (E), Libya (L), Morocco (M), Sudan (S), Tunisia (T), and Western Sahara (W). Represent the map as a graph where each vertex represents a country and each edge represents a common border between the countries. Use the letters indicated in parentheses to label the vertices.

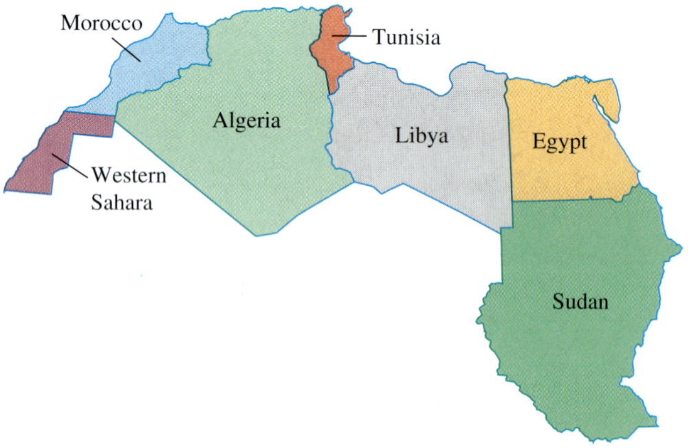

For Exercises 27–30, use a graph to represent the floor plan shown. In each graph, use the letter O to represent the outside of the building. See Example 3.

27. ***Floor Plan*** The drawing below shows the first-floor plan of the Aloha model home offered by the Kim Ghiselin Family Construction Company of Woonsocket, Rhode Island. Place vertex O near the top of the graph.

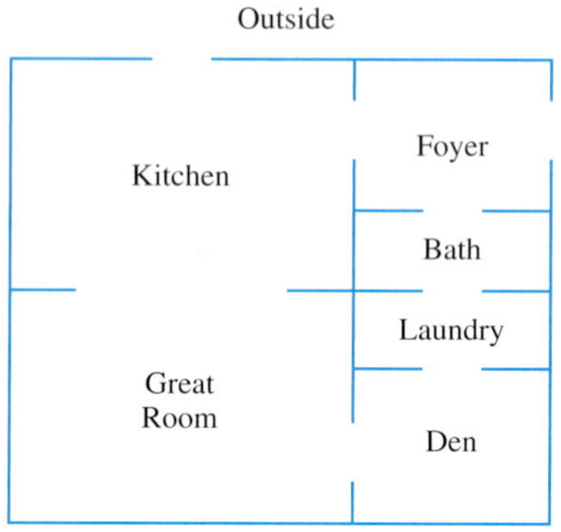

28. ***Floor Plan*** The drawing below shows the floor plan of the Sun Valley model home offered by Wubben Builders of Pocatello, Idaho. Place vertex O near the bottom of the graph.

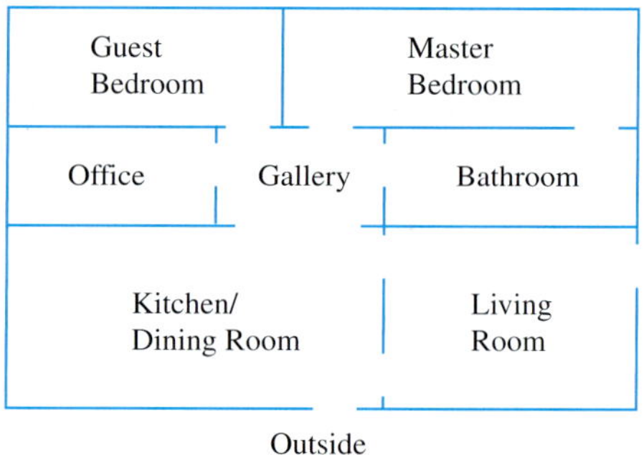

29. ***Floor Plan*** The drawing below shows the floor plan of the first floor of the Oleander model home offered by Nisely Builders of Albuquerque, New Mexico. Place vertex O near the bottom of the graph.

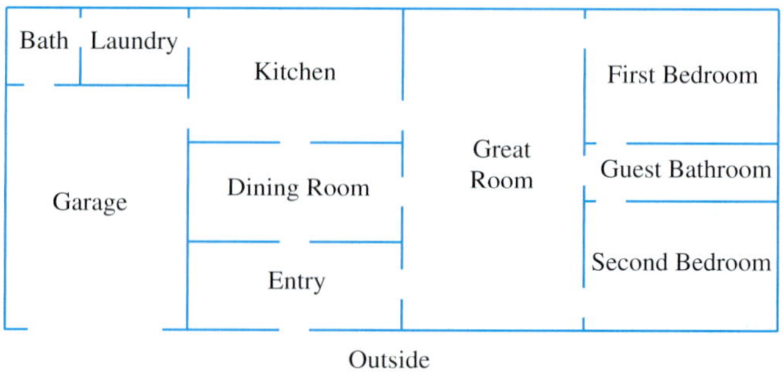

30. ***Floor Plan*** The drawing below shows the floor plan of the Bougainvillea model home offered by Alstrom Builders of Waco, Texas. Place vertex O near the bottom of the graph.

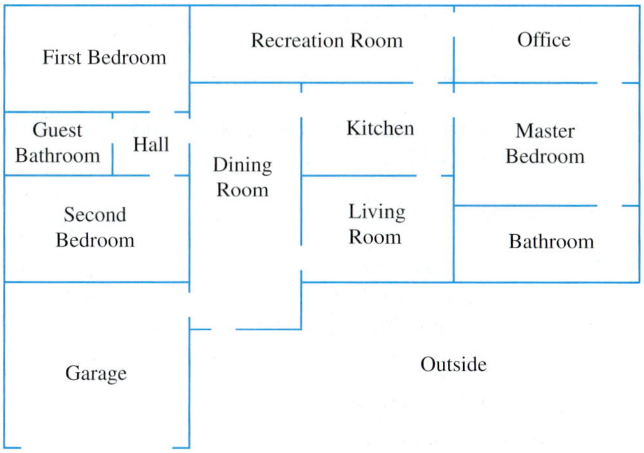

31. ***Representing a Neighborhood*** The map of the Tree Tops subdivision in Prince George County, Virginia, is shown below. Use a graph to represent this subdivision. Use letters to label the vertices.

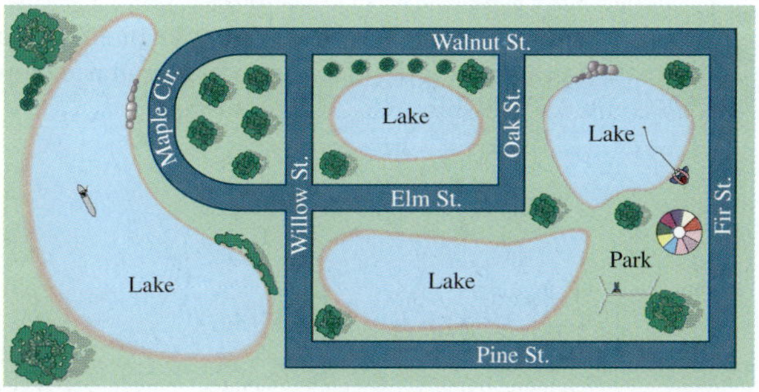

32. ***Representing a Neighborhood*** The map of the Crescent Lakes subdivision in Essex County, Maryland, is shown below. Use a graph to represent this subdivision. Use letters to label the vertices.

In Exercises 33–36, determine whether the graph shown is connected or disconnected.

33.

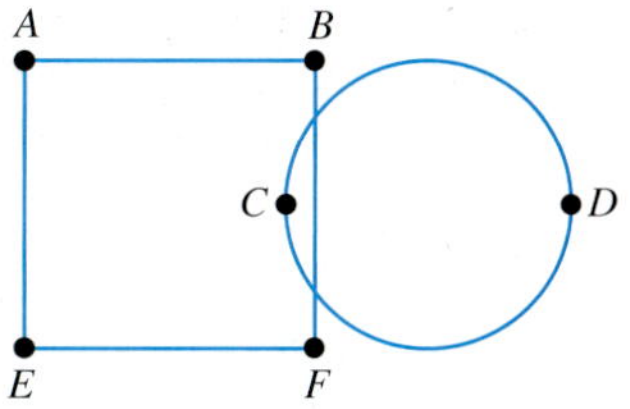

34.

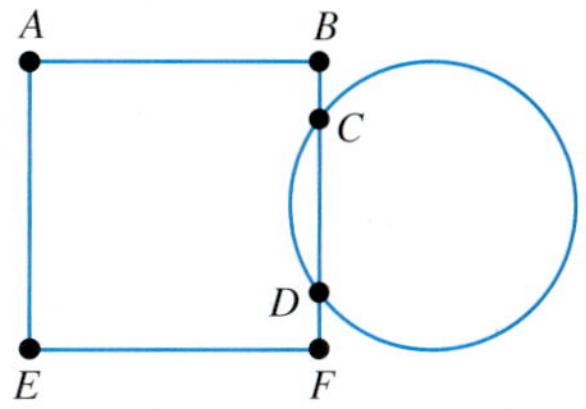

35.

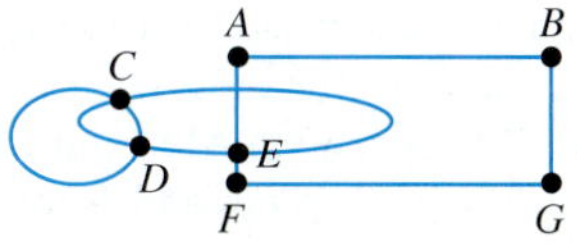

36.

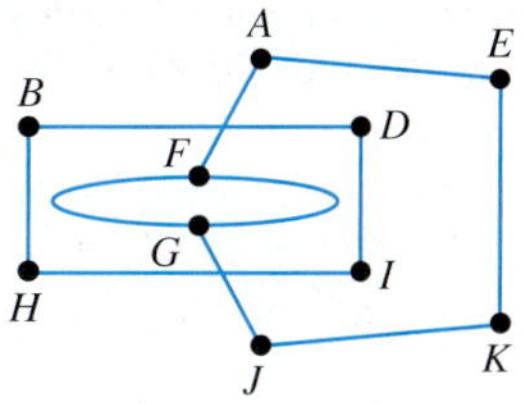

In Exercises 37–40, a connected graph is shown. Identify any bridges in each graph.

37.

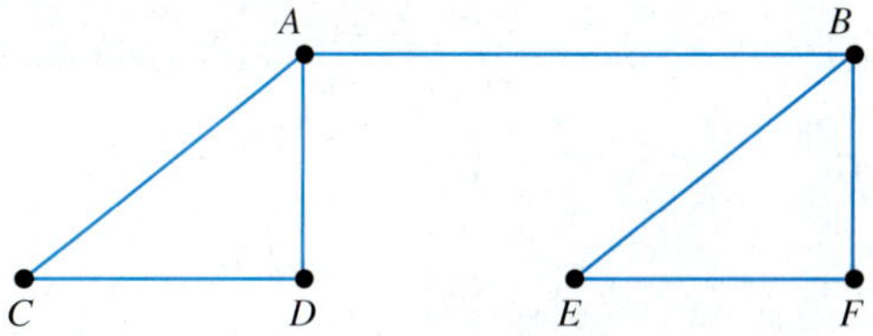

38.

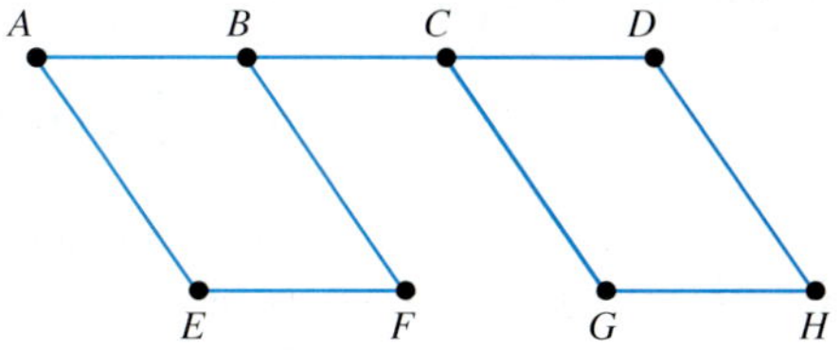

39.

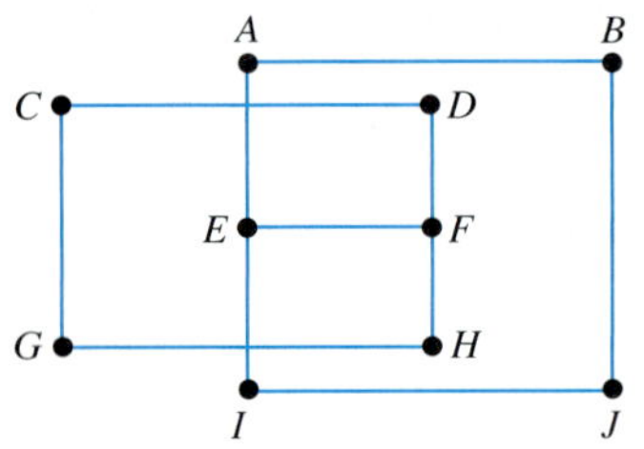

40.

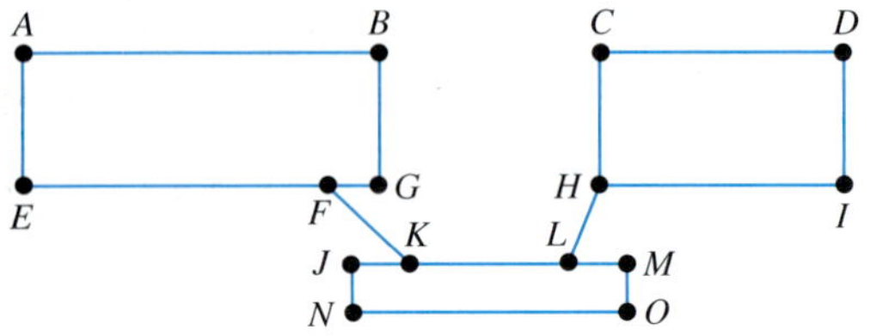

CHALLENGE PROBLEMS/GROUP ACTIVITIES

41. Draw a graph that conveys the following friendships among five children in a kindergarten class. Alex is friends with Drew, Hayden, Joshua, and Tony. Drew is friends with Alex and Tony. Hayden is friends with Alex and Joshua. Joshua is friends with Alex, Hayden, and Tony. Tony is friends with Alex, Drew, and Joshua.

42. Poll your entire class to determine which students knew each other prior to taking this course. Draw a graph that shows this relationship.

43. Attempt to draw a graph that has an odd number of odd vertices. What conclusion can you draw from this exercise?

44. Draw four different graphs and then for each graph:

a) determine the degree of each vertex.

b) determine the sum of the degrees from all vertices from this same graph.

c) determine the number of edges on this same graph.

d) What conclusion relating the sum of the degrees of vertices to the number of edges can you draw from this exercise? Explain why this conclusion is true.

RECREATIONAL MATHEMATICS

45. Use a graph to represent

a) the floor plan of your home.

b) the streets in your neighborhood within $\frac{1}{2}$ mile from your home in all directions.

INTERNET/RESEARCH ACTIVITY

46. Choose a continent other than North America or Australia and create a graph showing the common boundaries between countries. Be sure to use a current map.

14.2 EULER PATHS AND EULER CIRCUITS

▲ Graph theory can be used to determine the best route for a snowplow to travel.

A snowplow driver must plow every road in the township. To minimize the time and cost involved, the driver would like to establish a route that would allow the plow to travel over each road exactly one time. In this section, we will study paths and circuits that can be used to determine the best route for the snowplow driver.

Section 14.1 provided us with some of the basic definitions of graph theory. Although we were able to represent physical settings with graphs, we have not yet discussed solving problems related to these settings. In this section, we will provide more details of graph theory, which allows us to reach solutions to the Königsberg bridge problem and other real-world problems. Before we do so, we need to give two more definitions.

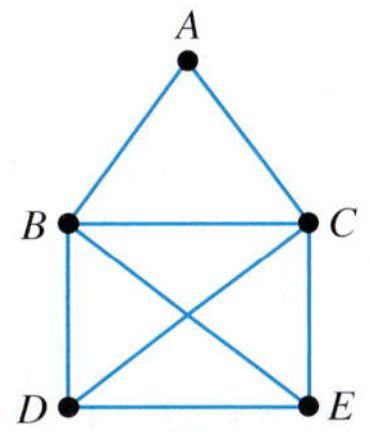

(a)

An Euler path
D, E, B, C, A, B, D, C, E

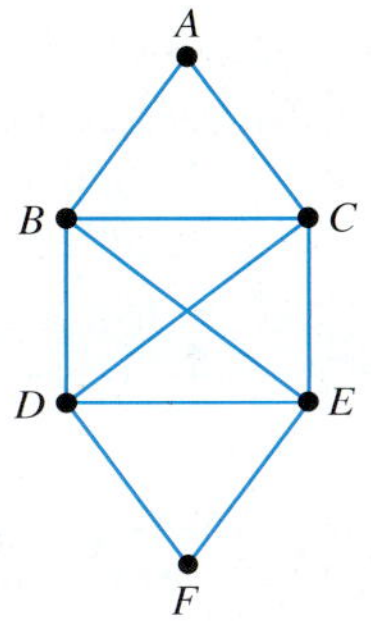

(b)

An Euler circuit
D, E, B, C, A, B, D, C, E, F, D

Figure 14.19

An **Euler path** is a path that passes through each edge of a graph exactly one time.

Recall that a circuit is a path that begins and ends at the same vertex.

An **Euler circuit** is a circuit that passes through each edge of a graph exactly one time.

The difference between an Euler path and an Euler circuit is that an Euler circuit must start and end at the same vertex. An alternate definition for an Euler circuit is that it is an Euler path that begins and ends at the same vertex. To become familiar with Euler paths and Euler circuits, examine the graphs in Fig. 14.19(a) and (b). As we discuss each graph, trace each path or circuit with a pencil or with your finger, or draw the path or circuit on a separate sheet of paper.

In Fig. 14.19(a), the path *D, E, B, C, A, B, D, C, E* is an Euler path since each edge was traced only one time. However, since the path begins at vertex *D* and ends at a different vertex, *E*, the path is not a circuit and therefore is not an Euler circuit. In Fig. 14.19(b), the path *D, E, B, C, A, B, D, C, E, F, D* is an Euler path since each edge was traced only one time. Furthermore, since the path begins and ends with the same vertex, *D*, it is also an Euler circuit. Note that every Euler circuit is an Euler path, but not every Euler path is an Euler circuit.

Our next task is to determine whether a given graph has an Euler path, an Euler circuit, neither, or both. In the next example, we will use trial and error to find Euler paths and Euler circuits. Our discussion will help us introduce *Euler's theorem*, which will be used to solve the Königsberg bridge problem and other problems involving graph theory.

EXAMPLE 1 *Finding Euler Paths and Circuits*

For the graphs shown in Fig. 14.20, determine if an Euler path, an Euler circuit, neither, or both exist.

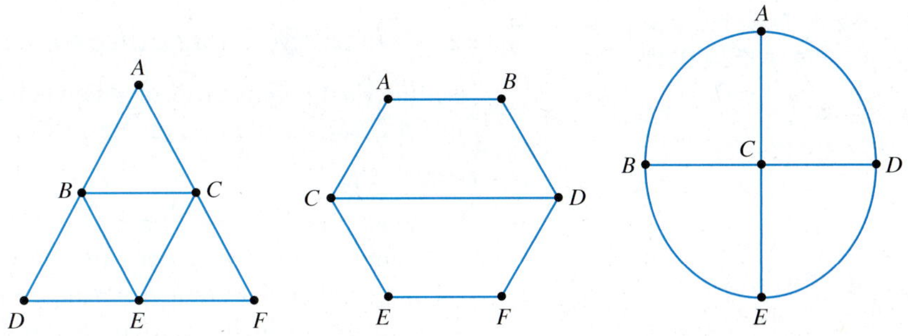

(a) (b) (c)

Figure 14.20

SOLUTION

a) The graph in Fig. 14.20(a) has many Euler circuits, each of which is also an Euler path. One Euler circuit is *A, B, D, E, B, C, E, F, C, A*. Trace this circuit now. Notice that you traced each edge exactly one time and that you started and finished with vertex *A*. Notice that this graph has no odd vertices (and therefore has all even vertices). This graph has both Euler paths and Euler circuits.

b) Using trial and error, we can see that the graph in Fig. 14.20(b) has an Euler path but it does not have an Euler circuit. One Euler path is *C*, *A*, *B*, *D*, *C*, *E*, *F*, *D*. Another Euler path is *D*, *B*, *A*, *C*, *D*, *F*, *E*, *C*. There are actually several Euler paths for this graph, but each one must begin at either vertex *C* or vertex *D*. If the Euler path begins at vertex *C*, it must end at vertex *D*, and vice versa (you should try to find a few more to confirm this observation). Notice that there are exactly two odd vertices, vertex *C* and vertex *D*.

c) Our attempts to trace the graph in Fig. 14.20(c) lead to either tracing at least one edge twice or to omitting at least one of the edges. We must therefore conclude that this graph has neither an Euler path nor an Euler circuit. Notice that this graph has more than two odd vertices.

We are now ready to introduce Euler's theorem, which is used to determine if a graph contains Euler paths and Euler circuits. As you will see, the number of *odd* vertices of a graph determines whether the graph has Euler paths and Euler circuits.

EULER'S THEOREM

For a connected graph, the following statements are true.

1. A graph with *no odd vertices* (all even vertices) has at least one Euler path, which is also an Euler circuit. An Euler circuit can be started at any vertex, and it will end at the same vertex.
2. A graph with *exactly two odd vertices* has at least one Euler path but no Euler circuits. Each Euler path must begin at one of the two odd vertices, and it will end at the other odd vertex.
3. A graph with *more than two odd vertices* has neither an Euler path nor an Euler circuit.

The proof of this theorem is beyond the scope of this book. In Example 2, we will reexamine the graphs used in Example 1. However, this time we will use Euler's theorem to determine whether the graph has Euler paths and Euler circuits.

EXAMPLE 2 *Using Euler's Theorem*

Use Euler's theorem to determine whether an Euler path or an Euler circuit exists in Fig. 14.20(a), (b), and (c) on page 931.

SOLUTION

a) The graph in Fig. 14.20(a) has no odd vertices (all the vertices are even). According to item 1 in Euler's theorem, at least one Euler circuit exists. An Euler circuit can be determined starting at any vertex. The Euler circuit will end at the vertex from which it started. Recall that each Euler circuit is also an Euler path.

b) We see that the graph in Fig. 14.20(b) has four even vertices (*A*, *B*, *E*, and *F*) and two odd vertices (*C* and *D*). Based on item 2 in Euler's theorem, we conclude that since there are exactly two odd vertices, at least one Euler path exists but no Euler circuit exists. Each Euler path must begin at one of the odd vertices and end at the other odd vertex.

c) The graph in Fig. 14.20(c) has four odd vertices (*A*, *B*, *D*, and *E*) and one even vertex (*C*). According to item 3 in Euler's theorem, since this graph has more than two odd vertices, the graph has neither an Euler path nor an Euler circuit.

DID YOU KNOW?

Topology

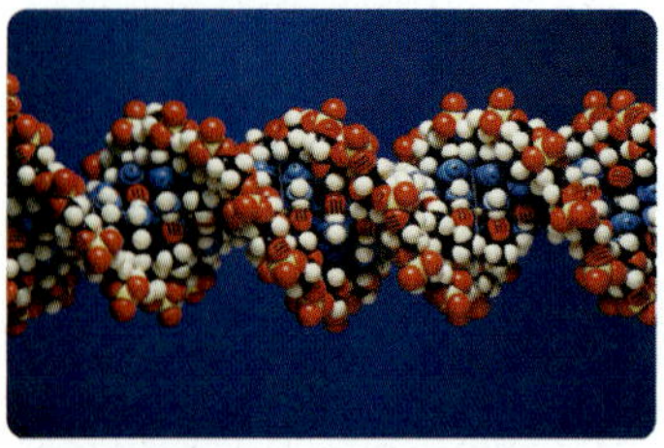

DNA Molecule

Graph theory is part of a larger branch of mathematics known as *topology*. Although topology involves concepts from other older branches of mathematics such as geometry, algebra, and analysis, the name *topology* wasn't used until 1930 in a paper by Solomon Lefschetz (see Profile in Mathematics on page 961). Topology is frequently referred to as rubber-sheet geometry. This descriptive name refers to topology's focus on geometric properties that remain consistent even when an object is bent, twisted, or stretched. One of the more interesting areas within topology is knot theory. Once a very obscure area of mathematics, knot theory is increasingly important because of its applications to science. Among other things, knot theory is used to study DNA molecules, to find new drugs, and to study how infectious diseases spread.

The results found in Example 2 using Euler's theorem are the same as the results found in Example 1 using trial and error. Although both methods led to the same result, Euler's theorem is superior to trial and error because it gives us a tool to examine more complicated graphs.

Next, we will use Euler's theorem to solve questions regarding the Königsberg bridge problem as well as problems relating to the graphs we developed in Section 14.1.

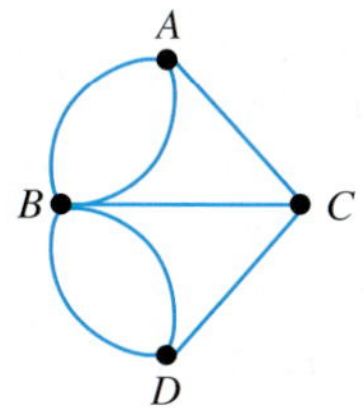

Figure 14.21

EXAMPLE 3 *Solving the Königsberg Bridge Problem*

In Section 14.1, Example 1 on page 920, we discussed the Königsberg bridge problem and drew a graph to represent the situation. The graph is repeated in Fig. 14.21. Could a walk be taken through Königsberg during which each bridge is crossed exactly one time?

SOLUTION The vertices of the graph given in Fig. 14.21 represent the land areas, and the edges of the graph represent the bridges in the Königsberg bridge problem. The original problem posed by the townspeople of Königsberg, in graph theory terms, was whether or not an Euler path exists for the graph in Fig. 14.21. From the figure, we see that the graph has four odd vertices (A, B, C, and D). Thus, according to item 3 of Euler's theorem, no Euler path exists. Therefore, no walk could be taken through Königsberg in which each bridge was crossed exactly one time. •

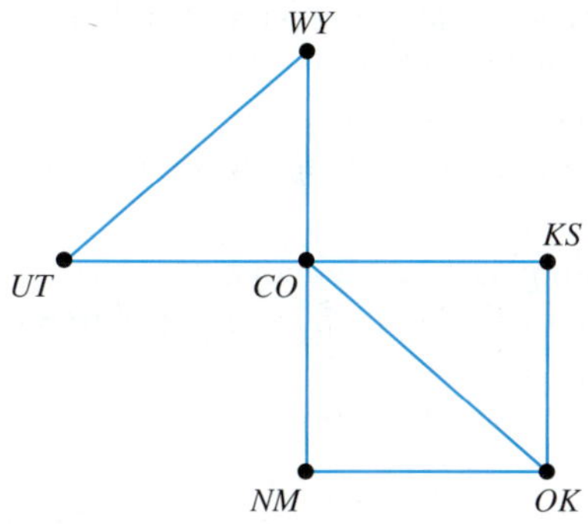

Figure 14.22

EXAMPLE 4 *Tenth U.S. Court of Appeals Visits*

In Section 14.1, Example 2 on page 920, we discussed the states whose cases are heard in the Tenth United States Court of Appeals. We drew a graph that showed the states that share a common border. The graph is repeated in Fig. 14.22. Chief Justice John Roberts wishes to visit each of these states. He wishes to travel between these states and cross each common state border (or use each edge) exactly one time.

a) Is it possible for him to travel among these six states and cross each common state border (or use each edge) exactly one time?
b) If yes, determine a path that he could take to travel between these states and cross each common state border exactly one time.

SOLUTION

a) Since Justice Roberts must use each edge (or common border) exactly one time, he is seeking an Euler path for the graph in Fig. 14.22. Notice that this graph has exactly two odd vertices, *CO* and *OK*. According to item 2 of Euler's theorem, at least one Euler path, but no Euler circuit, exists. Therefore, it *is* possible for him to travel among these states and cross each common state border exactly one time. However, since there is no Euler circuit, he must start his travel and end his travel in different states.
b) Many Euler paths exist, but each one must either start with *CO* and end with *OK* or start with *OK* and end with *CO*. One such path is *CO*, *UT*, *WY*, *CO*, *NM*, *OK*, *KS*, *CO*, *OK*. There are many other Euler paths on this graph. You should try to find at least two now. •

Euler's theorem can be used to determine whether a graph has an Euler path or an Euler circuit. If an Euler path or Euler circuit exists, how can we determine it? We can always attempt to determine Euler paths and Euler circuits using trial and error. This strategy might work well for simpler graphs. However, we would like to attempt this task with a more systematic approach. Our final topic in this section, Fleury's algorithm, will give us such an approach. An ***algorithm*** is a procedure for accomplishing some task.

FLEURY'S ALGORITHM

To determine an Euler path or an Euler circuit:

1. Use Euler's theorem to determine whether an Euler path or an Euler circuit exists. If one exists, proceed with steps 2–5.
2. If the graph has no odd vertices (therefore has an Euler circuit, which is also an Euler path), choose any vertex as the starting point. If the graph has exactly two odd vertices (therefore has only an Euler path), choose one of the two odd vertices as the starting point.
3. Begin to trace edges as you move through the graph. Number the edges as you trace them. Since you can't trace any edges twice in Euler paths and Euler circuits, once an edge is traced consider it "invisible."
4. When faced with a choice of edges to trace, if possible, choose an edge that is not a bridge (i.e., don't create a disconnected graph with your choice of edges).
5. Continue until each edge of the entire graph has been traced once.

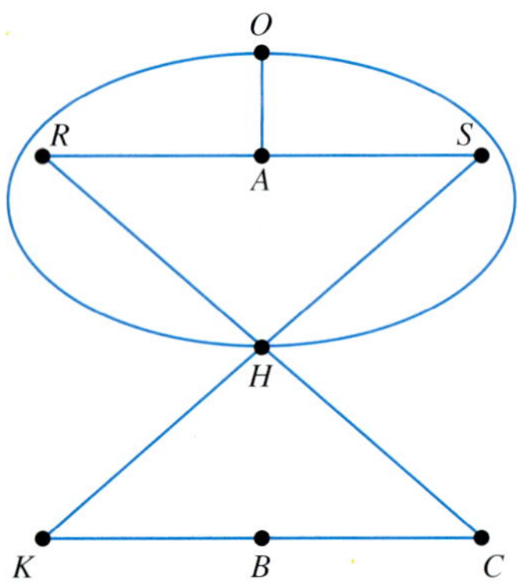

Figure 14.23

EXAMPLE 5 *Locking a Kindergarten Building*

In Section 14.1, Example 3 on page 922, we introduced the kindergarten building at Pullen Academy and drew the graph that represented its floor plan. The graph is repeated in Fig. 14.23. Joe Mays, the school's custodian, is responsible for locking all the doors at the end of the day. Joe wishes to determine a way to move through the building using each doorway exactly once.

a) Is it possible for Joe to move through the building using each doorway of the building exactly one time? (In other words, does an Euler path exist in the graph in Fig. 14.23?)

b) If so, determine one such Euler path.

SOLUTION

a) To determine if there is an Euler path, we will use Euler's theorem. The graph in Fig. 14.23 contains exactly two odd vertices, O and A. By item 2 of Euler's theorem, the graph has at least one Euler path (but no Euler circuit). By Euler's theorem, the Euler path would have to begin at one of the two odd vertices, either O or A, and end at the other odd vertex. Therefore, Joe could move through the building and use each doorway of the building exactly one time by starting at the outside (vertex O) or at restroom A (vertex A).

b) In part (a), we determined that an Euler path exists. Now we use Fleury's algorithm to determine one such path. We can start at either odd vertex, O or A. Let us choose to start at vertex O (see Fig. 14.23). As a matter of convention, once we trace an edge, we will redraw it using a dashed, instead of solid, edge. The dashes will indicate that the edge has already been traced and so it cannot be traced again. At vertex O, there are three choices of edges: the left-side edge OH, the right-side edge OH, or edge OA. None of these edges is a bridge. Thus, if we were to remove any one of these edges, we would not create a disconnected graph. Since none of these edges is a bridge, we can choose to trace any one of these edges. Let us choose the left-side edge OH as our first edge and number it 1 (see Fig. 14.24).

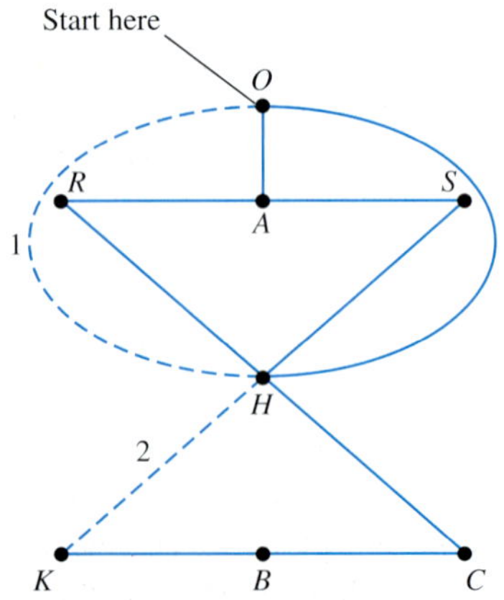

Figure 14.24

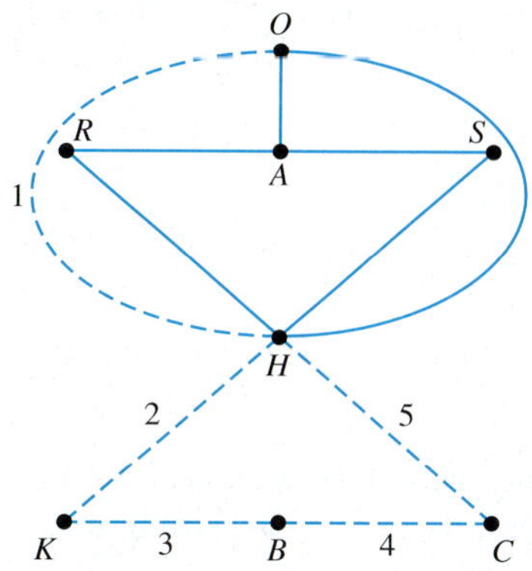

Figure 14.25

Now closely examine our position at vertex *H*. We now must make a choice among the following edges: *HR*, *HS*, *HK*, *HC*, and the right-side edge *HO*. None of these edges is a bridge. Thus, if we were to remove any of these edges, we would not create a disconnected graph. Since none of these edges is a bridge, we can choose to trace any one of these edges. Let us choose edge *HK* and label it edge 2 as shown in Fig. 14.24.

Now we are at vertex *K*. At vertex *K*, we have no choice but to choose edge *KB*, followed by edge *BC*, followed by edge *CH*. We will choose these edges and label them edges 3, 4, and 5, respectively, as in Fig. 14.25. We are now back at vertex *H*.

Examine our position at vertex *H* in Fig. 14.25. There are three choices of edges: *HR*, *HS*, and right-side edge *HO*. None of these edges is a bridge, so we may choose to trace any one of these edges. Let us choose edge *HR* and label it 6. That will put us at vertex *R*, where our only choice is to trace edge *RA* and label it 7. We are now at vertex *A*. At vertex *A*, there is a choice between edges *AO* and *AS*. Because neither edge is a bridge, we may choose either one of these edges. Let us choose edge *AS* and label it 8. Our choices up to this point are shown in Fig. 14.26.

We are now at vertex *S*, where our only choice is to trace edge *SH* followed by edge *HO*, followed by edge *OA*. These edges are labeled 9, 10, and 11, respectively, and the completed Euler path is given in Fig. 14.27.

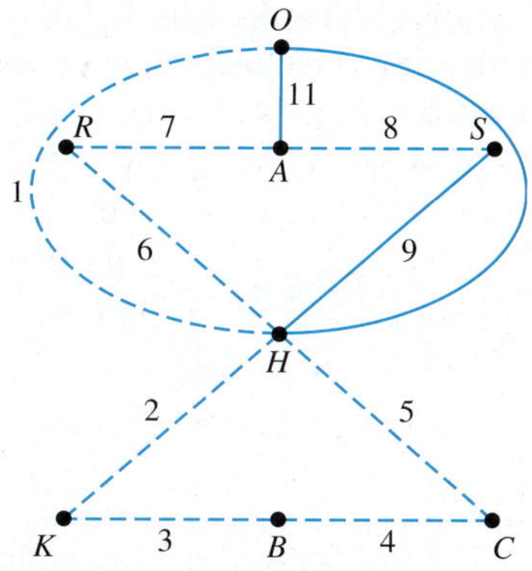

Figure 14.26

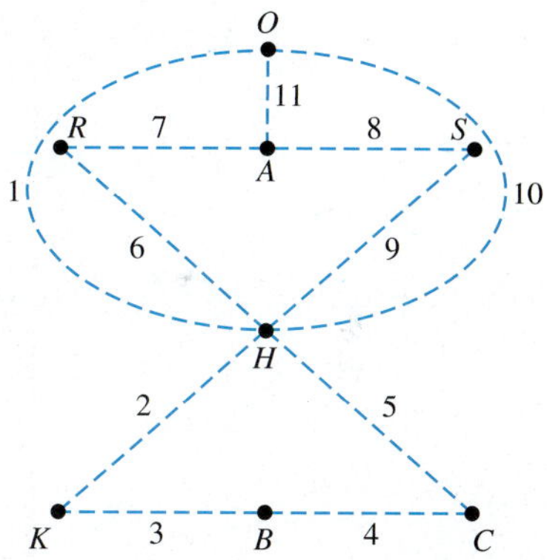

Figure 14.27

Notice that we started at one odd vertex, *O*, and ended at the other odd vertex, *A*. Therefore, if Joe followed the path *O*, *H*, *K*, *B*, *C*, *H*, *R*, *A*, *S*, *H*, *O*, *A*, he would travel through each doorway exactly one time. Many other Euler paths also exist for this problem. You will be asked to find another one in Exercise 50. ●

EXAMPLE 6 *Crime Stoppers Problem*

In Section 14.1, Example 4 on page 922, we introduced the Country Oaks subdivision of homes and drew a graph to represent it. The graph is repeated in Fig. 14.28 on page 936. The Country Oaks Neighborhood Association is planning to organize a crime stopper group in which residents take turns walking through the neighborhood with cell phones to report any suspicious activity to the police.

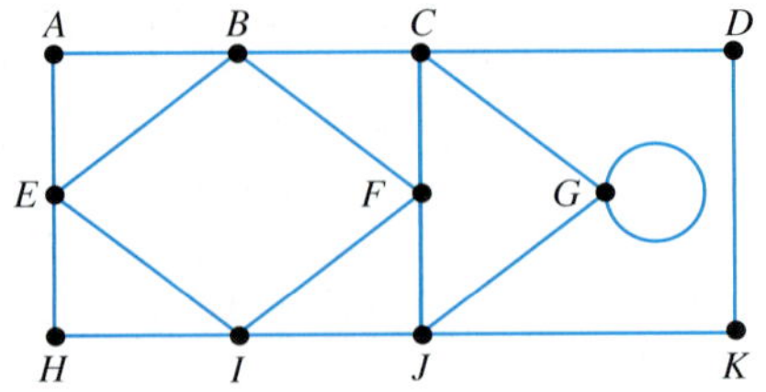

Figure 14.28

a) Can the residents of Country Oaks start at one intersection (or vertex) and walk each street block (or edge) in the neighborhood exactly once and return to the intersection where they started?

b) If yes, determine a circuit that could be followed to accomplish their walk.

SOLUTION

a) Since the residents wish to start at one intersection (or vertex) and return to the same intersection (or vertex), we need to determine if this graph has an Euler circuit. From Fig. 14.28, we see that there are no odd vertices. Therefore, by item 1 of Euler's theorem, we determine that there is at least one Euler circuit.

b) In part (a), we determined that an Euler circuit exists. Now we use Fleury's algorithm to determine one such circuit. Euler's theorem states that we can start at any vertex to determine an Euler circuit. Let us choose to begin at vertex *A*. We face a choice of tracing edge *AB* or edge *AE*. Since neither edge *AB* nor edge *AE* is a bridge, we can choose either edge. Let us choose edge *AB*. Our first choice is indicated in Fig. 14.29.

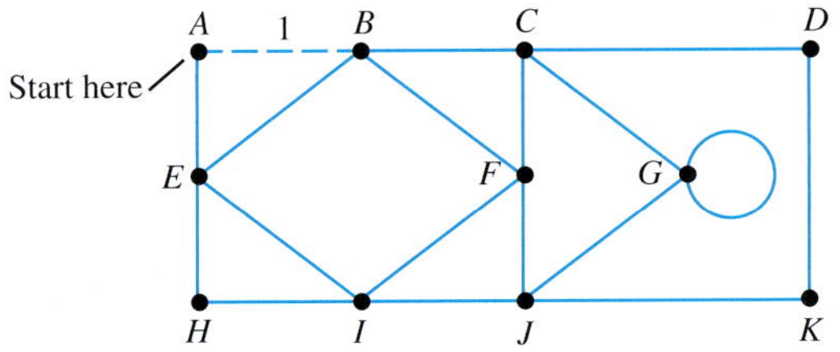

Figure 14.29

We will continue to trace from vertex to vertex around the outside of the graph, as indicated in Fig. 14.30. Notice that no edge chosen is a bridge.

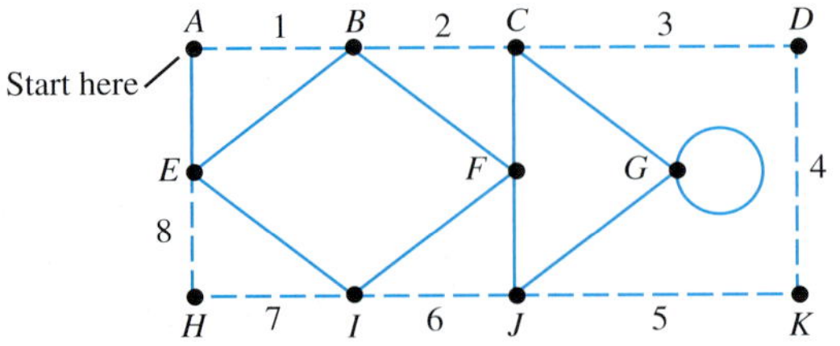

Figure 14.30

We are now at vertex *E*. Our choices are edge *EA*, edge *EB*, or edge *EI*. Edge *EA* is a bridge because removing edge *EA* would leave vertex *A* disconnected from the rest of the untraced graph. Therefore, we need to choose either edge *EB* or edge *EI*. Let us choose edge *EB*. Now we are at vertex *B* and must choose edge *BF*. Our choices so far are shown in Fig. 14.31.

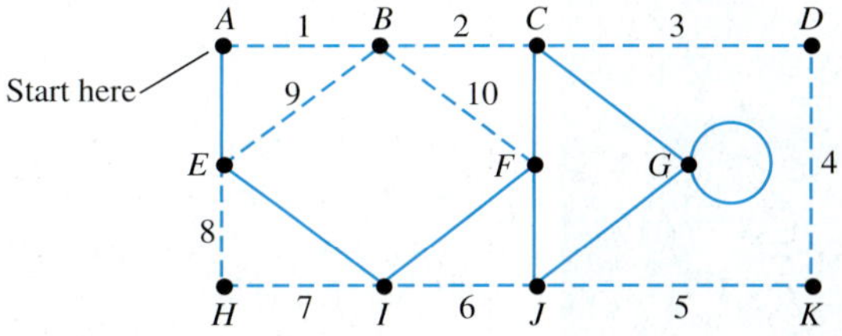

Figure 14.31

We are now at vertex *F* with choices of edge *FC*, edge *FJ*, or edge *FI*. Notice that edge *FI* is a bridge (it connects vertices *A*, *E*, and *I* to the rest of the untraced graph). We can therefore choose either edge *FC* or edge *FJ*. Let us choose edge *FC*, which will put us at vertex *C* (see Fig. 14.32). At vertex *C*, our only choice is edge *CG*.

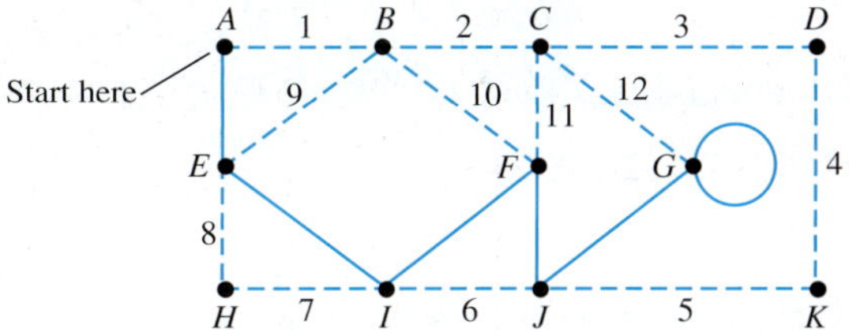

Figure 14.32

We are now at vertex *G* and have to choose between edge *GJ* and the edge *GG* (which is a loop): see Fig. 14.32. Since *GJ* is a bridge, we choose the edge *GG*. We then are back at vertex *G*. Here and throughout the rest of our circuit we only have one choice at each of the remaining vertices. We trace each of these remaining edges to get the result given in Fig. 14.33.

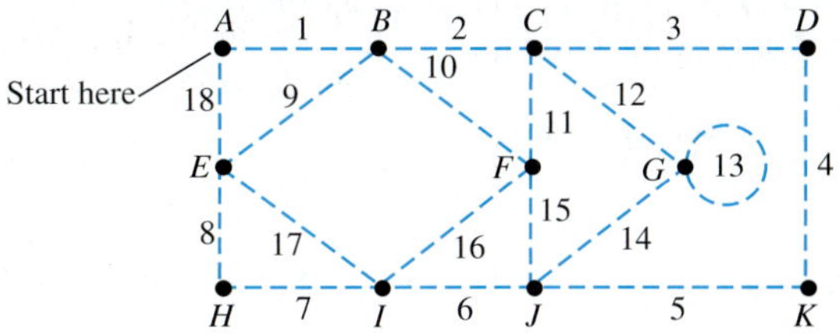

Figure 14.33

The circuit given in Fig. 14.33 gives one possible Euler circuit that would provide the residents of Country Oaks with one way to cover each street block exactly one time starting and ending at vertex *A*. Fleury's algorithm could be used to find many alternative circuits. You will be asked to find several such alternative circuits in Exercises 51 through 54.

▲ **Figure 14.34**
The game Pac-Man

In our first two sections on graph theory, we introduced key concepts and theory that allowed us to represent physical settings as graphs and to solve some problems relating to these settings. We solved problems dealing with the bridges of Königsberg, the states whose cases are heard in the Tenth U.S. Court of Appeals, the doorways of a kindergarten building, and the street blocks involved in a neighborhood crime stopper program. Although these problems are very different, they are all related because in each case a graph is used to represent a physical setting.

There are many other examples of how Euler paths and Euler circuits can be applied to real-life problems, including problems that deal with the most efficient route for a snowplow or a street sweeper. Another example of a path problem is found in the classic arcade games Pac-Man and Ms. Pac-Man (Fig. 14.34). In both these games, the ideal path (disregarding the ghosts) is an Euler path.

SECTION 14.2 EXERCISES

CONCEPT/WRITING EXERCISES

1. a) What is an Euler path?

   b) Draw a graph that contains an Euler path and give an example of a path that is an Euler path.

   c) Using the same graph as in part (b), give an example of a path that is not an Euler path.

2. a) What is an Euler circuit?

   b) Draw a graph that contains an Euler circuit and give an example of a circuit that is an Euler circuit.

   c) Using the same graph as in part (b), give an example of a circuit that is not an Euler circuit.

3. Does a graph have an Euler path if the graph has

   a) no odd vertices?

   b) exactly two odd vertices?

   c) more than two odd vertices?

4. Does a graph have an Euler circuit if the graph has

   a) no odd vertices?

   b) exactly two odd vertices?

   c) more than two odd vertices?

5. Explain how to tell if a connected graph has an Euler circuit.

6. a) Explain how to determine an Euler path provided one exists for a given graph.

   b) Explain how to determine an Euler circuit provided one exists for a given graph.

PRACTICE THE SKILLS

For Exercises 7–10, use the following graph.

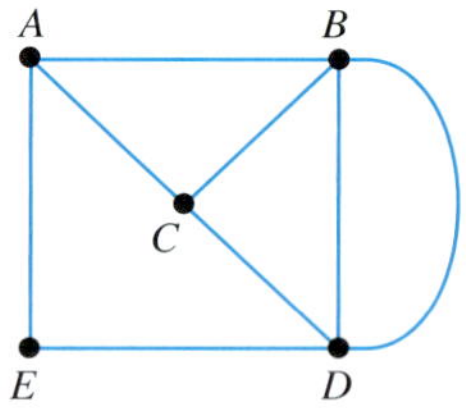

7. Determine an Euler path that begins with vertex A.

8. Determine an Euler path that begins with vertex C.

9. Is it possible to determine an Euler path that begins with vertex B? Explain.

10. Is it possible to determine an Euler circuit for this graph? Explain.

For Exercises 11–14, use the following graph.

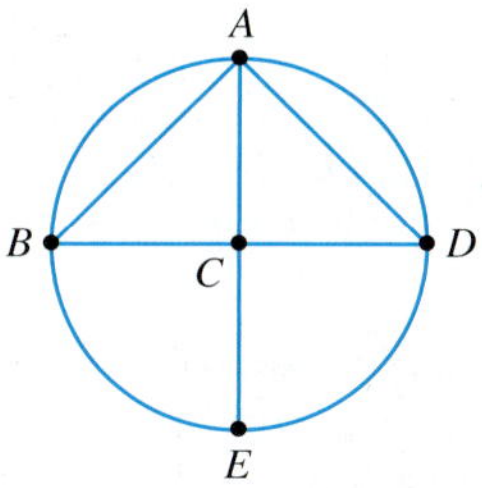

11. Determine an Euler path that begins with vertex *A*.

12. Determine an Euler path that begins with vertex *E*.

13. Is it possible to determine an Euler circuit for this graph? Explain.

14. Is it possible to determine an Euler path that begins with vertex *C*? Explain.

For Exercises 15–20, use the following graph.

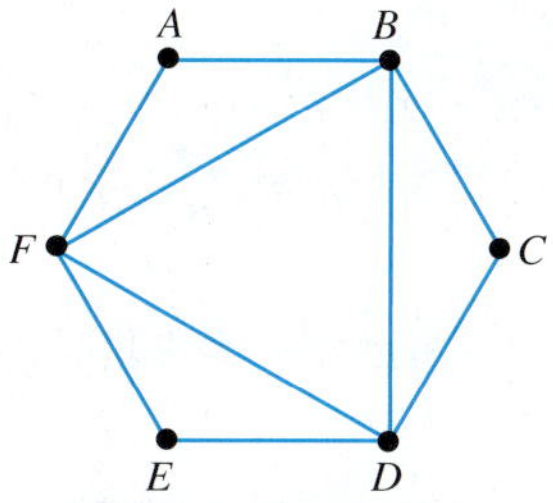

15. Determine an Euler circuit that begins and ends with vertex *A*.

16. Determine an Euler circuit that begins and ends with vertex *B*.

17. Determine an Euler circuit that begins and ends with vertex *C*.

18. Determine an Euler circuit that begins and ends with vertex *D*.

19. Determine an Euler circuit that begins and ends with vertex *E*.

20. Determine an Euler circuit that begins and ends with vertex *F*.

21. Imagine a very large connected graph that has 400 even vertices and no odd vertices.

a) Does an Euler path exist for this graph? Explain.

b) Does an Euler circuit exist for this graph? Explain.

22. Imagine a very large connected graph that has two odd vertices and 398 even vertices.

a) Does an Euler path exist for this graph? Explain.

b) Does an Euler circuit exist for this graph? Explain.

23. Imagine a very large connected graph that has 400 odd vertices and no even vertices.

a) Does an Euler path exist for this graph? Explain.

b) Does an Euler circuit exist for this graph? Explain.

24. Imagine a very large connected graph that has 200 odd vertices and 200 even vertices.

a) Does an Euler path exist for this graph? Explain.

b) Does an Euler circuit exist for this graph? Explain.

PROBLEM SOLVING

Revisiting the Königsberg Bridge Problem *In Exercises 25 and 26, suppose that the people of Königsberg decide to add several bridges to their city. Two such possibilities are shown.*

a) *Would the townspeople be able to walk across all the bridges without crossing the same bridge twice?*

b) *If so, where should they begin and where would they end?*

25.

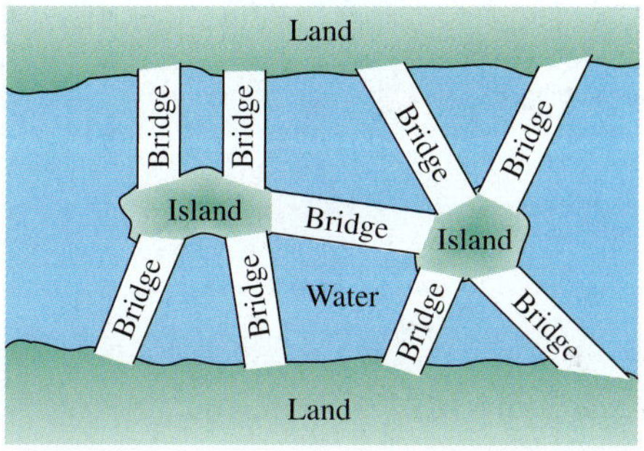

26.

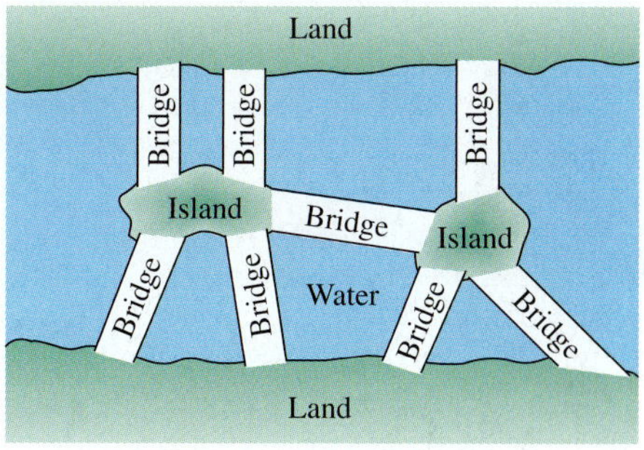

Areas of the World *In Exercises 27–32 on page 940, use each map shown.*

a) *Represent the map as a graph. Use the letter indicated in parentheses to label vertices of the graph.*

b) *Determine (state yes or no) whether the graph in part (a) has an Euler path. If yes, give one such Euler path.*

c) *Determine (state yes or no) whether the graph in part (a) has an Euler circuit. If yes, give one such Euler circuit.*

27. ***Southwest Asia*** Afghanistan (A), Iran (N), Iraq (Q), Jordan (J), Pakistan (P), and Turkey (T)

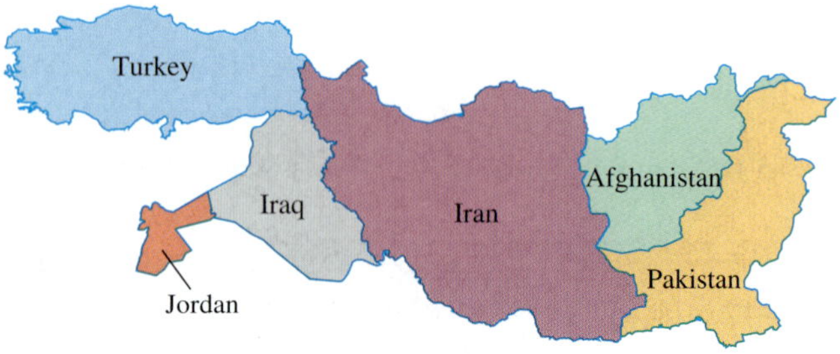

28. ***Northern Africa*** Algeria (A), Chad (C), Libya (L), Niger (N), and Tunisia (T)

29. ***Southeast Asia*** Burma (B), Cambodia (C), Laos (L), Thailand (T), and Vietnam (V)

30. ***Central Europe*** Austria (A), Czech Republic (C), Germany (G), Italy (I), Poland (P), and Switzerland (S)

31. ***Western Canada*** Alberta (A), British Columbia (B), Manitoba (M), Northwest Territories (T), Nunavut (N), Saskatchewan (S), and Yukon (Y)

32. ***South America*** Argentina (A), Bolivia (B), Brazil (Z), Chile (C), Paraguay (P), and Uruguay (U)

Locking Doors *Joe Mays, the custodian at the Pullen Academy (see Example 5 on page 934), is also responsible for locking the doors of four other buildings. The floor plans for these buildings are shown in Exercises 33–36.*

a) *For each floor plan, represent the floor plan as a graph. Use the letters shown to label the vertices of the graph and use the letter O to label the vertex that represents the outside of the building. Place vertex O near where the word* Outside *is placed in the diagram.*

b) *Determine (state yes or no) whether it is possible for Joe to move through the building using each doorway of the building exactly one time. (In other words, does an Euler path exist in the graph?)*

c) *If your answer to part (b) is yes, determine one such Euler path.*

33. Administration Building

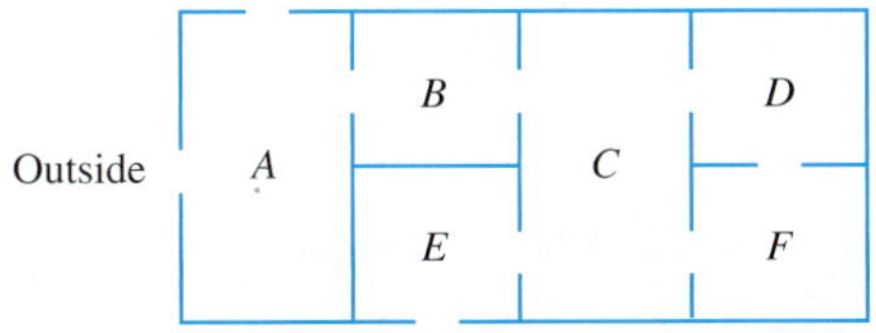

34. Library

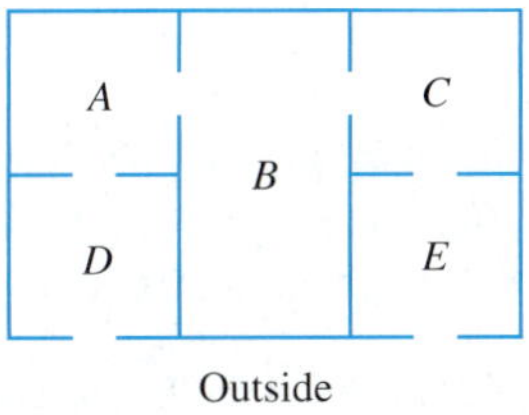

35. Computer Center

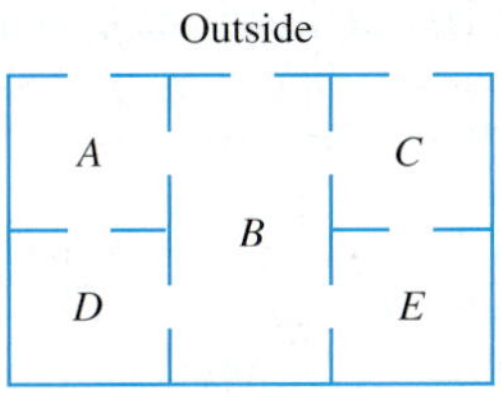

36. Fitness Center

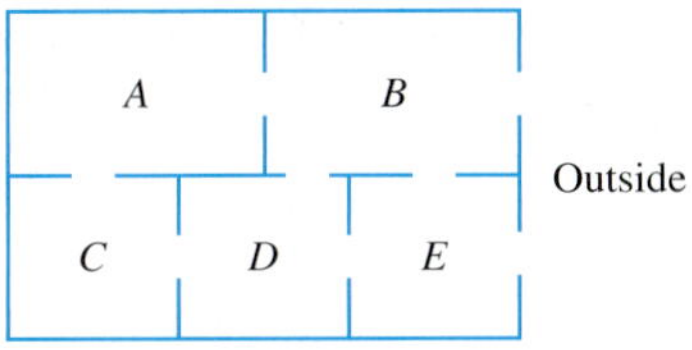

Crime Stopper Routes *The Country Oaks crime stopper organization (see Example 6 on page 935) was so successful that the residents shared their strategies with friends living in the subdivisions of Crescent Lakes and Tree Tops. In Exercises 37 and 38, the respective maps of these communities are given.*

a) *Determine whether the residents in each subdivision will be able to establish a path through their communities so that each street block is walked exactly one time.*

b) *If yes, where would the residents need to start their walk?*

37. Tree Tops

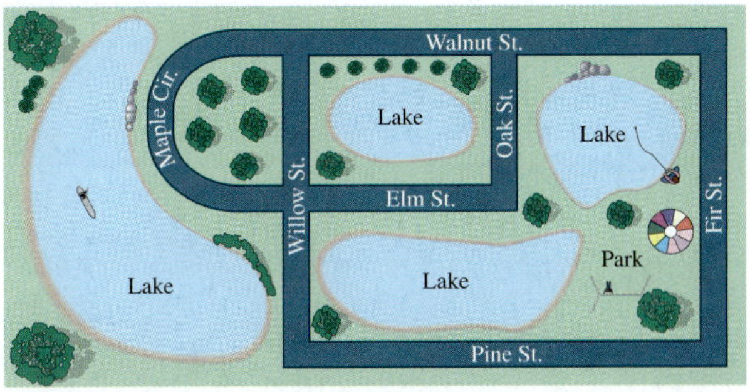

38. Crescent Lakes

In Exercises 39–42, use Fleury's algorithm to determine an Euler path.

39.

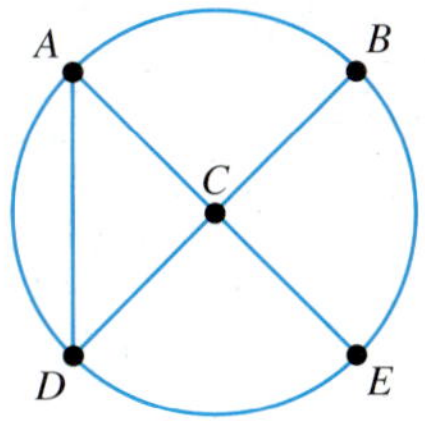

40.

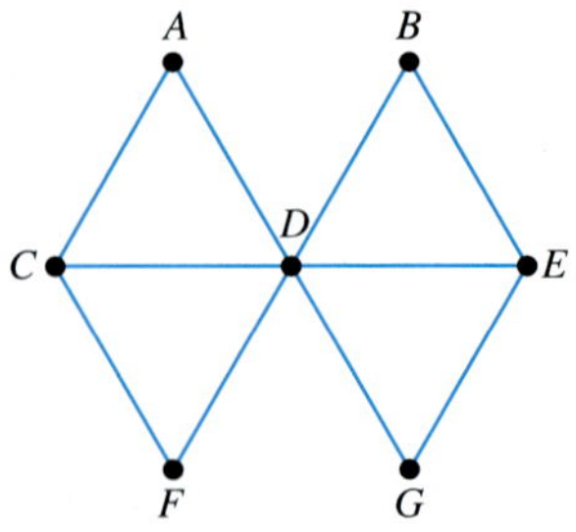

41.

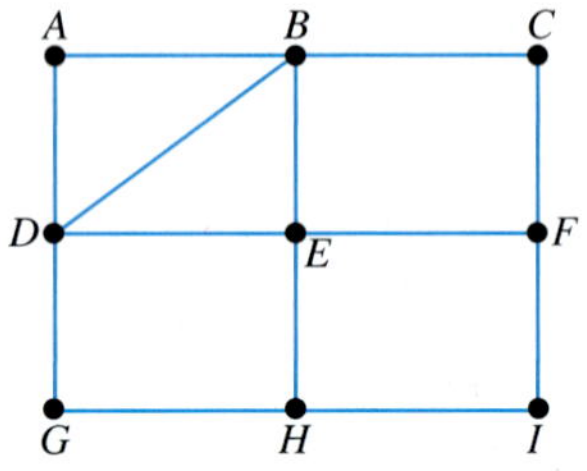

42.

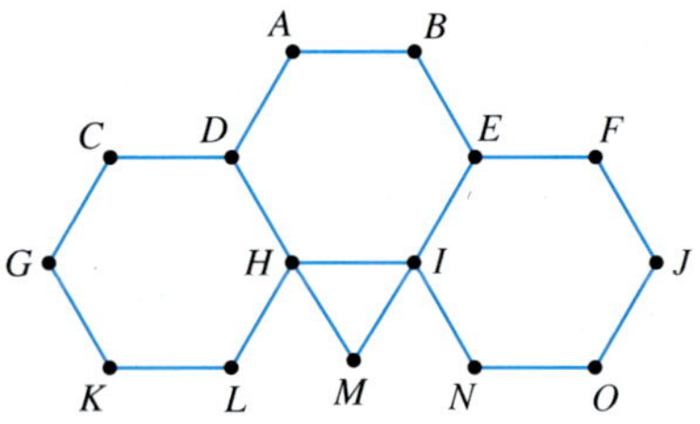

In Exercises 43–48, use Fleury's algorithm to find an Euler circuit.

43.

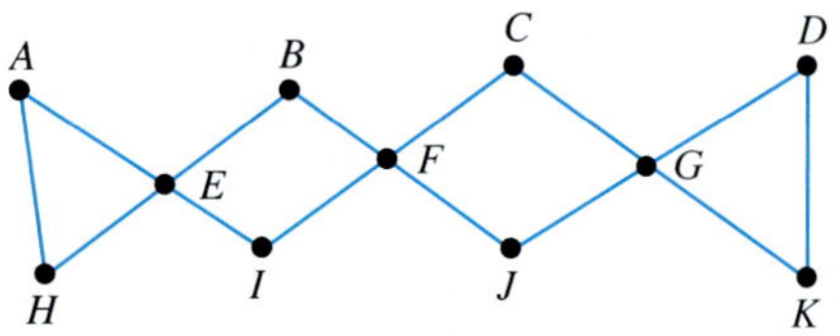

44.

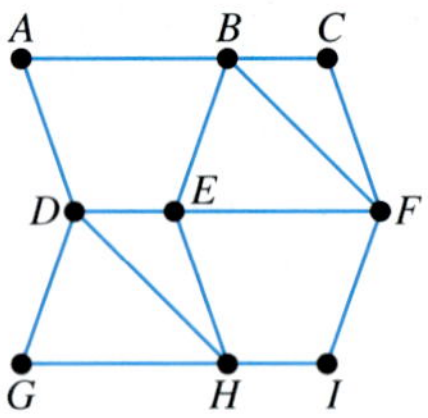

45.

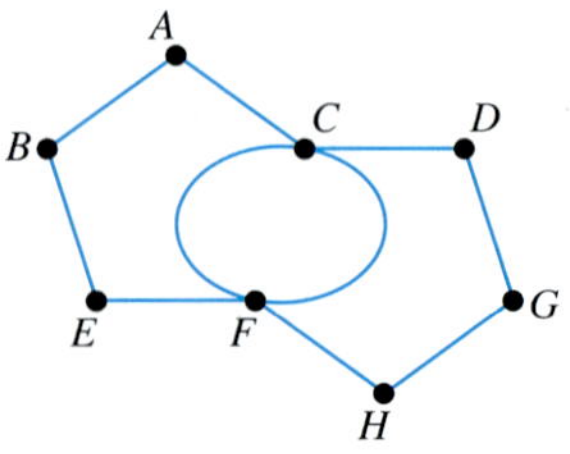

46.

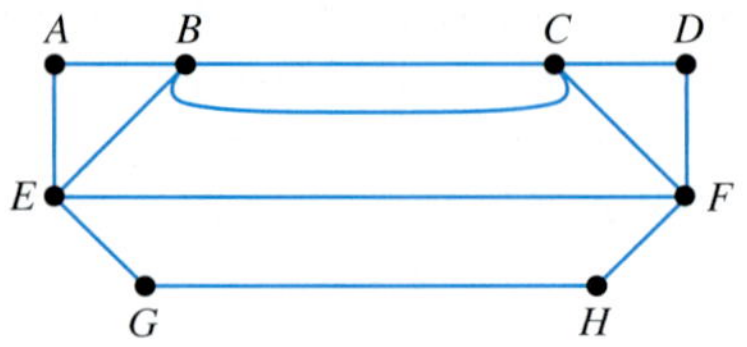

47.

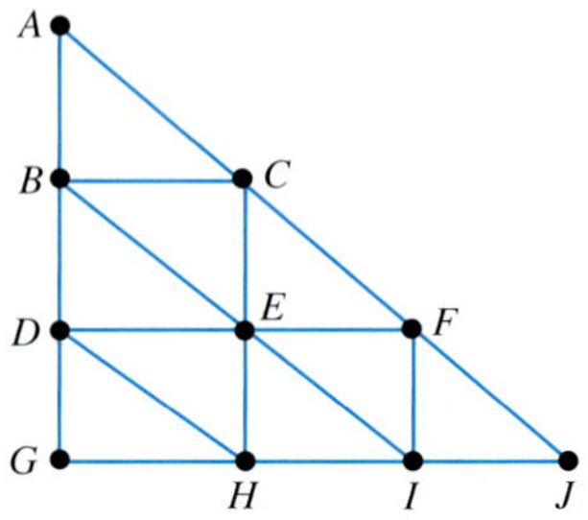

48.

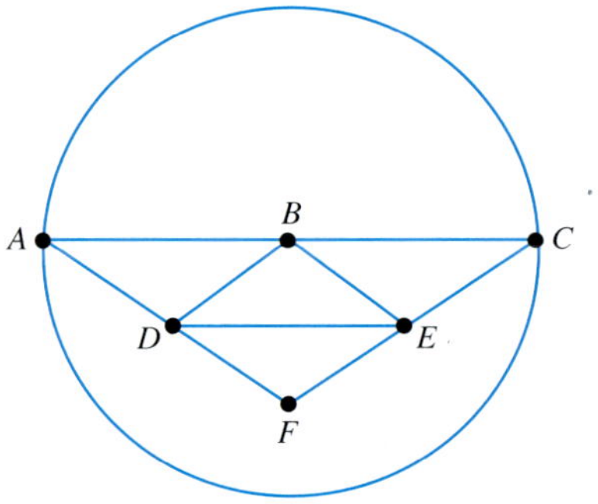

49. Determine an Euler path through the states whose cases are heard in the Tenth U.S. Court of Appeals (see Example 4 on page 933) that begins with the state of Oklahoma.

50. Determine an Euler path for the kindergarten building at the Pullen Academy (see Example 5 on page 934) that begins in restroom (vertex) *A*.

51. Determine an Euler circuit for the Country Oaks crime stopper group (see Example 6 on page 935) that begins with vertex *B* followed by vertex *A*.

52. Determine an Euler circuit for the Country Oaks crime stopper group (see Example 6 on page 935) that begins with vertex *B* followed by vertex *E*.

53. Determine an Euler circuit for the Country Oaks crime stopper group that begins with vertex *J* followed by vertex *G*.

54. Determine an Euler circuit for the Country Oaks crime stopper group that begins with vertex *J* followed by vertex *F*.

CHALLENGE PROBLEMS/GROUP ACTIVITIES

55. Look at the map of the contiguous United States on page 918. Imagine a graph with 48 vertices in which each vertex represents one of the contiguous states. Each edge would represent a common border between states.

    a) Would this graph have an Euler path?

    b) Explain why or why not.

56. Attempt to draw a graph with an Euler circuit that has a bridge. What conclusion can you develop from this exercise?

RECREATIONAL MATHEMATICS

57. a) Draw a graph with one vertex that has both an Euler path and an Euler circuit.

b) Draw a graph with two vertices that has an Euler path but no Euler circuit.

c) Draw a graph with two vertices that has both an Euler path and an Euler circuit.

INTERNET/RESEARCH ACTIVITY

58. Write a paper on the history and development of the branch of mathematics known as graph theory.

14.3 HAMILTON PATHS AND HAMILTON CIRCUITS

▲ Graph theory can be used to determine the shortest route for a mail carrier.

Paula Kunkel delivers mail in a rural area for the United States Postal Service. Paula wishes to find the shortest route that begins at the post office, goes to each mailbox in her delivery area exactly one time, and ends back at the post office. In this section, we will study a special type of problem like the one facing Paula.

In this section, we continue our study of graph theory by studying Hamilton paths and Hamilton circuits. These paths and circuits are named for Irish mathematician and astronomer William Rowan Hamilton (1805–1865); see the Profile in Mathematics on page 948. Before formally defining Hamilton paths and Hamilton circuits, we will introduce an example that will be examined at several points throughout this section.

EXAMPLE 1 *A Transportation Problem*

Julienne Ward has just been promoted to Southeast District Sales Director for the Addison Wesley Publishing Company. She lives in Orlando, Florida, and oversees regional sales offices in Orlando; Atlanta, Georgia; Memphis, Tennessee; and New Orleans, Louisiana. She frequently needs to fly to each regional office for sales meetings. She would like to determine the least expensive route to visit each city one time and then return to Orlando. To help analyze this problem, Julienne used the Internet to find the least expensive one-way fares offered between each of the four cities (see Table 14.1).

Table 14.1

	Orlando	Atlanta	Memphis	New Orleans
Orlando (*O*)	*	$67	$95	$69
Atlanta (*A*)	$67	*	$57	$68
Memphis (*M*)	$95	$57	*	$99
New Orleans (*N*)	$69	$68	$99	*

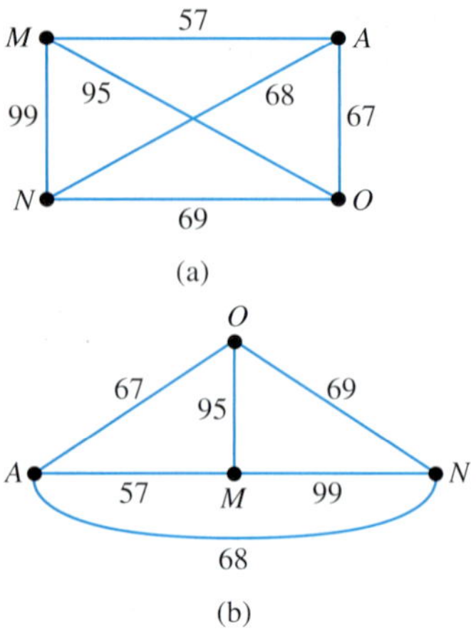

Figure 14.35

Use this table to create a graph. Let the vertices represent the cities, then connect each pair of cities with an edge. List the airfare between each two cities on the respective edges.

SOLUTION First we note that the cost of a one-way flight between two given cities in the table is the same regardless from which city Julienne starts. For example, the one-way fare from New Orleans to Atlanta is \$68 and the one-way fare from Atlanta to New Orleans is also \$68. This information will allow us to list the flight cost along an edge with a single number. Next we draw a graph with four vertices that represent the four cities and six edges that represent the flights between each city. We also include the price of one-way flights along the appropriate edge. Two such graphs are shown in Fig. 14.35. Many other graphs could also display this information.

Problems like the one in Example 1 are called ***traveling salesman problems***. There are usually variations to the situation, but these problems generally involve seeking the least expensive or shortest way to travel among several locations. To help analyze traveling salesman problems, the costs or distances to travel between locations are indicated along each edge of a graph. Such a graph, as in Fig. 14.35(a) or (b), is called a ***weighted graph***. We will use weighted graphs to solve traveling salesman problems, but first we need to introduce Hamilton paths and Hamilton circuits.

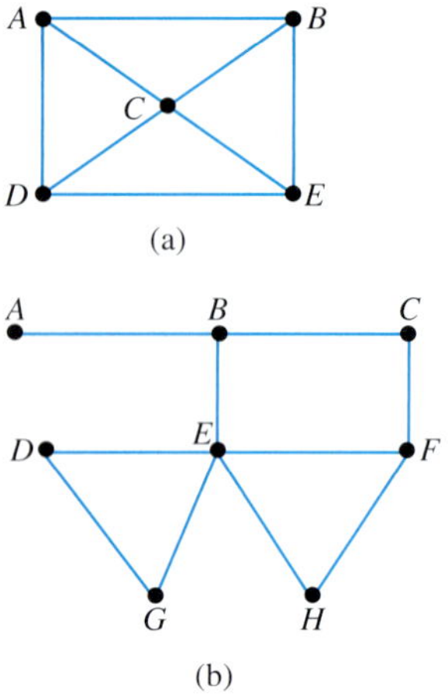

Figure 14.36

Hamilton Paths, Hamilton Circuits, and Complete Graphs

Now we introduce another important definition of graph theory.

A **Hamilton path** is a path that contains each *vertex* of a graph exactly one time.

The graph in Fig. 14.36(a) has Hamilton path *A, B, C, E, D*. The graph in Fig. 14.36(a) also has Hamilton path *C, B, A, D, E*. The graph in Fig. 14.36(b) has a Hamilton path *A, B, C, F, H, E, G, D*. The graph in Fig. 14.36(b) also has Hamilton path *G, D, E, H, F, C, B, A*. Both of the graphs in Fig. 14.36 have other Hamilton paths. Can you find some of them?

Notice that, unlike Euler paths, not every edge needs to be traced in a Hamilton path. To help distinguish between Euler paths and Hamilton paths, remember that *Euler paths are concerned with visiting all the edges, whereas Hamilton paths are concerned with visiting all the vertices.*

Now we introduce Hamilton circuits.

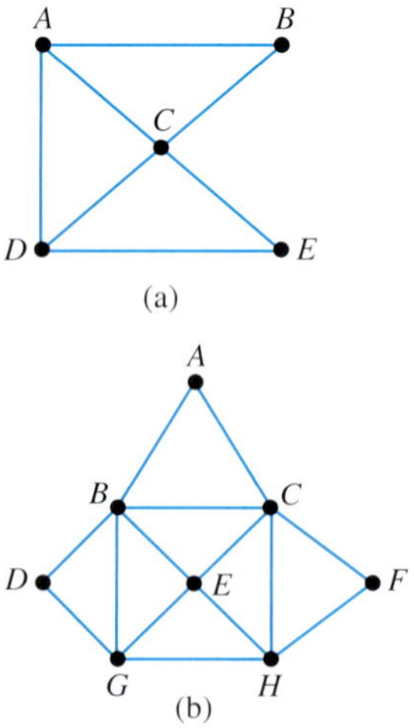

Figure 14.37

A **Hamilton circuit** is a path that begins and ends at the same vertex and passes through all other vertices of a graph exactly one time.

An alternate definition for a Hamilton circuit is that a Hamilton circuit is a Hamilton path that starts and ends at the same vertex. For example, the graph in Fig. 14.37(a) has Hamilton circuit *A, B, C, E, D, A*. The graph in 14.37(a) also has Hamilton circuit *B, A, D, E, C, B*. The graph in Fig. 14.37(b) has Hamilton circuit *A, B, D, G, E, H, F, C, A*. The graph in Fig. 14.37(b) also has Hamilton circuit *E, G, D, B, A, C, F, H, E*. Note that in an ***Euler** circuit*, the path followed must include every edge and must begin and end at the same vertex. In a ***Hamilton circuit***, the path followed must include every vertex and must begin and end at the same vertex. Unlike the Euler circuit, however, a Hamilton circuit does not have to include every edge.

In Section 14.2, we were fortunate that Euler's theorem could tell us under what conditions an Euler path or an Euler circuit would exist. We are not as fortunate with Hamilton paths and Hamilton circuits; no such general theorem exists. We now shift our focus to graphs that are guaranteed to have Hamilton circuits.

A ***complete graph*** is a graph that has an edge between each pair of its vertices. Figure 14.38 shows complete graphs with three, four, and five vertices, respectively.

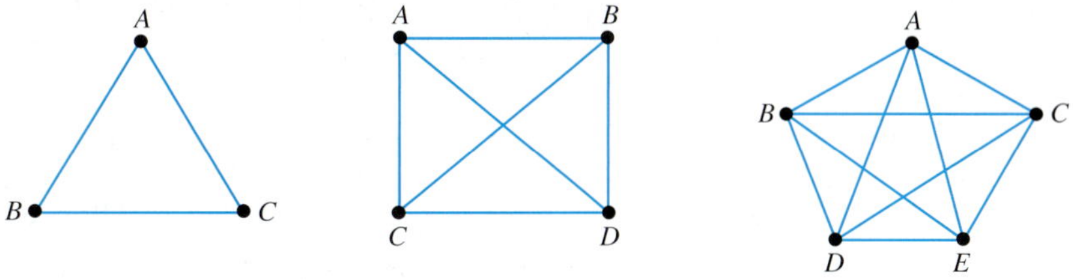

Figure 14.38

Complete graphs are important because *every complete graph has a Hamilton circuit* (but not necessarily an Euler circuit), and traveling salesman problems can be represented by complete graphs. Since each pair of vertices on a complete graph is connected, we can create a Hamilton circuit by simply starting at any vertex and moving from vertex to vertex until we have passed through each vertex exactly one time. Then to complete the circuit we simply move back to the vertex from which we started. Notice in Fig. 14.38 that although all the graphs have Hamilton circuits, only the first and third graphs have Euler circuits.

EXAMPLE 2 *Finding Hamilton Circuits*

Determine a Hamilton circuit for the complete graph shown in Fig. 14.39.

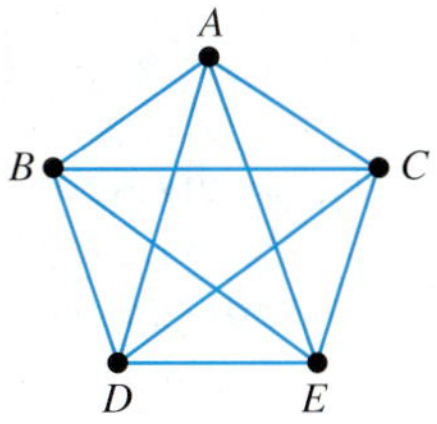

Figure 14.39

SOLUTION We can determine many Hamilton circuits in this complete graph. For example, one is *A*, *B*, *C*, *D*, *E*, *A*. Another is *B*, *A*, *D*, *C*, *E*, *B*. Another is *E*, *B*, *C*, *D*, *A*, *E*. To build a Hamilton circuit from this complete graph, we can list all five vertices in any order and then return to the first vertex. •

A question that naturally arises from Example 2 is, "How many different Hamilton circuits are there in a complete graph?" Before we can give a formula that answers this question, we need to discuss ***factorials***, which were presented in

DID YOU KNOW?

Six Degrees of Kevin Bacon

The Six Degrees of Kevin Bacon is a trivia game started by Albright College students Craig Fass, Brian Turtle, and Mike Ginelli. The object is to link any actor to the actor Kevin Bacon through associations from different films. For example, Queen Latifah has a Bacon number of 1 since she was in the movie *Beauty Shop* (2005) with Kevin Bacon. Owen Wilson has a Bacon number of 2 since he was in *Meet the Parents* (2000) with John Elsen and John Elsen was in *Loverboy* (2005) with Kevin Bacon.

What does this game have to do with graph theory? A graph with actors as vertices and common movies as edges could be used to represent these associations. Such a graph would be very large and difficult to draw. However, computerized versions of the game can manage the associations quite nicely. The web site The Oracle of Bacon (www.oracleofbacon.org) at the University of Virginia computes the Bacon number for any one of 815,811 actors, ranging in Bacon number of 0 (Bacon himself) to 8 (13 different individuals).

Section 12.8. We will review that material here. The symbol 5! is read "five factorial" and means to multiply 5 by each natural number less than 5. So $5! = 5 \cdot 4 \cdot 3 \cdot 2 \cdot 1 = 120$. Also $3! = 3 \cdot 2 \cdot 1 = 6$ and $7! = 7 \cdot 6 \cdot 5 \cdot 4 \cdot 3 \cdot 2 \cdot 1 = 5040$. Note that 0! is defined to be 1.

NUMBER OF UNIQUE HAMILTON CIRCUITS IN A COMPLETE GRAPH

The number of unique Hamilton circuits in a complete graph with n vertices is $(n - 1)!$, where

$$(n-1)! = (n-1)(n-2)(n-3)\cdots(3)(2)(1)$$

EXAMPLE 3 *Number of Hamilton Circuits*

How many unique Hamilton circuits are there in a complete graph with the following number of vertices?

a) Three b) Eight c) Ten d) Eleven

SOLUTION

a) According to our formula, a complete graph with three vertices has $(3 - 1)! = 2! = 2(1) = 2$ unique Hamilton circuits.

b) A complete graph with eight vertices has $(8 - 1)! = 7! = 7 \cdot 6 \cdot 5 \cdot 4 \cdot 3 \cdot 2 \cdot 1 = 5040$ unique Hamilton circuits.

c) A complete graph with ten vertices has $(10 - 1)! = 9! = 9 \cdot 8 \cdot 7 \cdot 6 \cdot 5 \cdot 4 \cdot 3 \cdot 2 \cdot 1 = 362{,}880$ unique Hamilton circuits.

d) A complete graph with eleven vertices has $(11 - 1)! = 10! = 10 \cdot 9 \cdot 8 \cdot 7 \cdot 6 \cdot 5 \cdot 4 \cdot 3 \cdot 2 \cdot 1 = 3{,}628{,}800$ unique Hamilton circuits.

As we see from Example 3, as the number of vertices in a complete graph increases the number of Hamilton circuits increases rapidly.

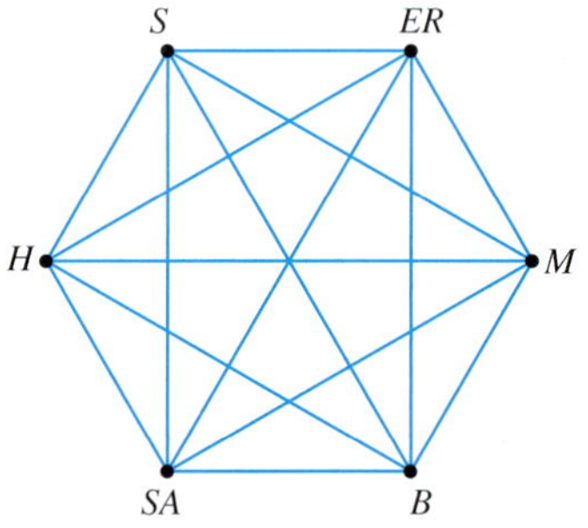

Figure 14.40

EXAMPLE 4 *American Idol Travel*

Paula Abdul, Simon Cowell, and Randy Jackson, the judges for the television show *American Idol*, are in Hollywood (H). They need to travel to the following cities to judge contestants' auditions: San Antonio (SA), East Rutherford (ER), Birmingham (B), Memphis (M), and Seattle (S). In how many different ways can Paula, Simon, and Randy, traveling together, visit each of these cities and return to Hollywood?

SOLUTION We can represent this problem with the complete graph in Fig. 14.40. In the graph, the six vertices represent the six locations and the edges represent the one-way flights between these locations.

To determine the number of possible routes, we need to determine the number of Hamilton circuits within this graph. We know that there are $(6 - 1)! = 5! = 5 \cdot 4 \cdot 3 \cdot 2 \cdot 1 = 120$ different Hamilton circuits within this graph. So the judges have 120 different ways to start in Hollywood, visit each of these five cities, and return to Hollywood.

▲ Randy Jackson, Paula Abdul, and Simon Cowell

Now that we have been introduced to Hamilton circuits, we will apply this knowledge to solving traveling salesman problems.

Traveling Salesman Problems

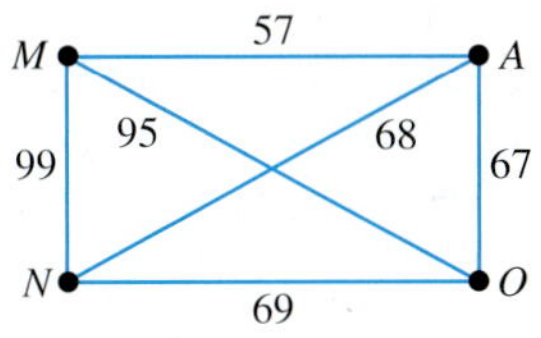

Figure 14.41

In Examples 1 and 4, we saw how complete graphs can represent cities and the process of traveling between these cities. Our goal in a traveling salesman problem is to find the least expensive or shortest way to visit each city once and return home. In terms of graph theory, our goal is to find the Hamilton circuit with the lowest associated cost or distance. The Hamilton circuit with the lowest associated cost (or shortest distance, etc.) is called the ***optimal solution***. We will discuss two methods, the *brute force method* and the *nearest neighbor method*, for determining the optimal solution. In both methods, we will use the term *complete, weighted graph*. A ***complete, weighted graph*** is a complete graph with the weights (or numbers) listed on the edges, as illustrated in Fig. 14.41. We now introduce the brute force method.

THE BRUTE FORCE METHOD OF SOLVING TRAVELING SALESMAN PROBLEMS

To determine the optimal solution:

1. Represent the problem with a complete, weighted graph.
2. List all possible Hamilton circuits for this graph.
3. Determine the cost (or distance) associated with each of these Hamilton circuits.
4. The Hamilton circuit with the lowest cost (or shortest distance) is the optimal solution.

EXAMPLE 5 *Using the Brute Force Method*

In Example 1 on page 943, we introduced Julienne Ward, the Southeast District Sales Director for Addison Wesley. We now want to use the brute force method to determine the optimal solution for Julienne to visit her regional sales offices. She will start in Orlando (O); visit offices in Atlanta (A), Memphis (M), and New Orleans (N); and then return to Orlando.

SOLUTION We illustrated two complete, weighted graphs for the example in Fig. 14.35 on page 944. The graph in Fig. 14.35(a) is repeated in Fig. 14.41. The numbers shown represent the one-way fares, in dollars, between the two cities.

We know that since there are 4 cities that must be visited, there are $(4 - 1)! = 3! = 6$ possible unique Hamilton circuits we need to examine for cost. These Hamilton circuits are listed in the first column of Table 14.2 on page 948.

The Hamilton circuits are listed in the column on the far left. In the next four columns, we place the cost associated with each leg of the given circuit. In the last column, we place the total cost of travel using the given circuit. From this last column, we see that two circuits have the lowest cost of \$289 (circled). Julienne has two choices for an optimal solution. She can either fly from Orlando to Memphis to Atlanta to New Orleans then back to Orlando, or she can fly from Orlando to New Orleans to Atlanta to Memphis then back to Orlando. These two Hamilton circuits

PROFILE IN MATHEMATICS

William Rowan Hamilton

Born in Dublin, Ireland, Sir William Rowan Hamilton (1805–1865) made many contributions to mathematics and the sciences. His work in mathematics included abstract algebra, calculus, geometry, graph theory, logic, and number theory. Perhaps his most important work involved the use of *quaternions* in abstract algebra. Quaternions are elements in a noncommutative group (see Chapter 10 for a discussion of group theory) that Hamilton labeled with the letters *i*, *j*, and *k*. For years, Hamilton puzzled over how to multiply the quaternions. Finally, one day while walking with his wife, he realized the solution. In his excitement, Hamilton carved the solution in the stone of Brougham (or Broom) Bridge as he and his wife passed by on their walk: $i^2 = j^2 = k^2 = ijk = -1$. To this day, there is a plaque on this bridge commemorating Hamilton's discovery.

Hamilton's interests also included philosophy, religion, foreign languages, and poetry. Hamilton also invented a game called the *Icosian Game* (see Exercise 35) that includes many aspects of graph theory you will study in this section. For more information on Hamilton, see the Profiles in Mathematics on page 439.

Table 14.2

Hamilton Circuit	First Leg/Cost	Second Leg/Cost	Third Leg/Cost	Fourth Leg/Cost	Total Cost
O, *A*, *M*, *N*, *O*	*O* to *A* $67	*A* to *M* $57	*M* to *N* $99	*N* to *O* $69	$292
O, *A*, *N*, *M*, *O*	*O* to *A* $67	*A* to *N* $68	*N* to *M* $99	*M* to *O* $95	$329
O, *M*, *A*, *N*, *O*	*O* to *M* $95	*M* to *A* $57	*A* to *N* $68	*N* to *O* $69	$289
O, *M*, *N*, *A*, *O*	*O* to *M* $95	*M* to *N* $99	*N* to *A* $68	*A* to *O* $67	$329
O, *N*, *A*, *M*, *O*	*O* to *N* $69	*N* to *A* $68	*A* to *M* $57	*M* to *O* $95	$289
O, *N*, *M*, *A*, *O*	*O* to *N* $69	*N* to *M* $99	*M* to *A* $57	*A* to *O* $67	$292

are shown on the maps in Fig. 14.42. Notice that the second circuit involves visiting the cities in the exact reverse order of the first circuit. Although these are different circuits, the cost is the same regardless of the direction flown. Either circuit provides Julienne with the least expensive way to visit each of her regional sales offices.

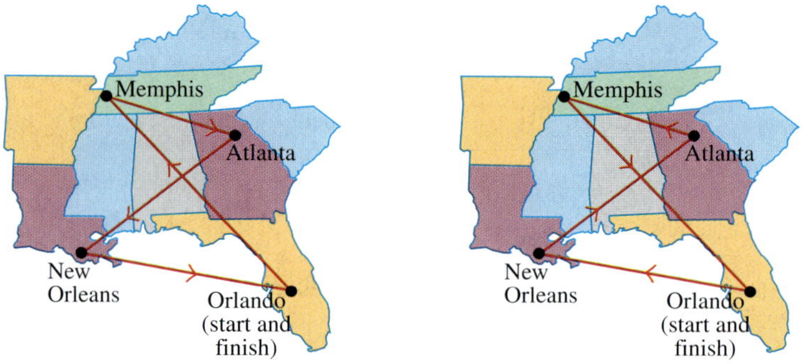

Figure 14.42

Although we used the brute force method in Example 5, it becomes impractical as the number of vertices increases. In fact, on more complex problems, the brute force method is impractical on even the world's fastest supercomputers. For example, suppose that a presidential candidate wished to visit each of the 31 largest cities in the United States. The candidate would have 30! or about 3×10^{32} choices to accomplish this trip. To understand the magnitude of this number, consider that it would take the world's fastest supercomputer (as of September 2006) about 930,000,000,000 (930 *billion*) *years* to perform the brute force method to find the optimal solution. Computer scientists and mathematicians have developed much more efficient algorithms for seeking the optimal solution to traveling salesman problems.

Now we introduce a method for finding an ***approximate solution*** to a traveling salesman problem. Approximate solutions can be used in cases where determining the optimal solution is not reasonable. One method for finding an approximate solution is the *nearest neighbor method*. In this method, the salesperson begins at a given location. If the salesperson wishes to minimize the distance traveled, the salesperson first visits the city (or location) closest to his or her starting location. Then the salesperson visits the next city (or location) closest to his or her present location. This process continues until all the cities (or locations) are visited. Thus, the salesperson moves to the *nearest neighbor*. Sometimes the salesperson wishes to minimize the cost of travel, such as airfare. In this case, from the starting location the salesperson first visits the city (or location) in which the cost of travel is a minimum. Then from the present location, the salesperson visits the city (or location) where the cost of travel is a minimum. This process continues until all the cities (or locations) are visited. Approximate solutions will always be Hamilton circuits.

NEAREST NEIGHBOR METHOD OF DETERMINING AN APPROXIMATE SOLUTION TO A TRAVELING SALESMAN PROBLEM

To approximate the optimal solution:

1. Represent the problem with a complete, weighted graph.
2. Identify the starting vertex.
3. Of all the edges attached to the starting vertex, choose the edge that has the smallest weight. This edge is generally either the shortest distance or the lowest cost. Travel along this edge to the second vertex.
4. At the second vertex, choose the edge that has the smallest weight that does not lead to a vertex already visited. Travel along this edge to the third vertex.
5. Continue this process, each time moving along the edge with the smallest weight until all vertices are visited.
6. Travel back to the original vertex.

EXAMPLE 6 *Using the Nearest Neighbor Method*

In Example 4 on page 946, we discussed the *American Idol* judges' plan to visit five cities in which auditions would take place and then return to Hollywood. Use the nearest neighbor method to determine an approximate solution for the judges' visits. The one-way per person flight prices between cities are given in Table 14.3.

Table 14.3

	Birmingham	**East Rutherford**	**Hollywood**	**Memphis**	**San Antonio**	**Seattle**
Birmingham (*B*)	*	$258	$114	$324	$274	$155
East Rutherford (*ER*)	$258	*	$134	$104	$355	$154
Hollywood (*H*)	$114	$134	*	$144	$129	$108
Memphis (*M*)	$324	$104	$144	*	$634	$299
San Antonio (*SA*)	$274	$355	$129	$634	*	$119
Seattle (*S*)	$155	$154	$108	$299	$119	*

SOLUTION We will use the nearest neighbor method to determine the least expensive per person cost for the judges to complete the trip. We begin by modifying the graph from Example 4 by putting the one-way per person prices into place, making it a weighted graph (see Fig. 14.43). The judges start in Hollywood and choose the least expensive flight, which is to Seattle ($108). In Seattle, the judges choose the least expensive flight to a location they have not already visited, San Antonio ($119). Next, in San Antonio, they choose the least expensive flight to a location they have not already visited, Birmingham ($274). In Birmingham, they choose the least expensive flight to a location they have not already visited, East Rutherford ($258). Once in East Rutherford, the judges have no choice but to fly to the only location not yet visited, Memphis ($104). Since they began in Hollywood, are now in Memphis, and have visited all five audition cities, the judges now return to Hollywood ($144) to complete their trip. The nearest neighbor method would produce a Hamilton circuit, of *H*, *S*, *SA*, *B*, *ER*, *M*, *H*. The cost per person is $108 + $119 + $274 + $258 + $104 + $144 = $1007. The total cost for all three judges is $1007 × 3, or $3021.

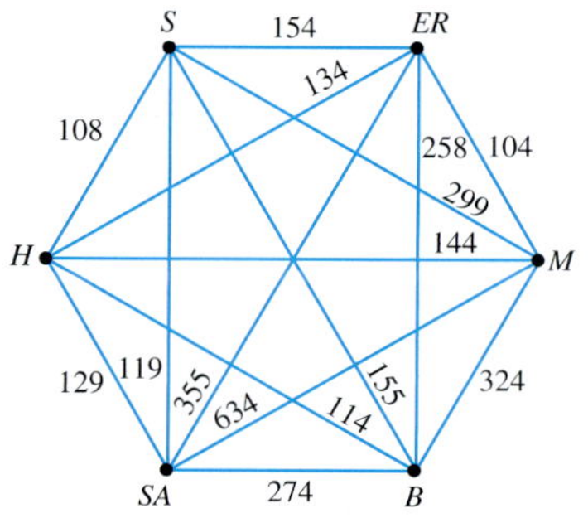

Figure 14.43

For the sake of comparison, let us look at four other randomly chosen Hamilton circuits from Example 6 and the per person costs associated with them (see Table 14.4).

Table 14.4

Randomly Chosen Hamilton Circuit	Cost Calculation	Per Person Cost
H, *SA*, *ER*, *B*, *M*, *S*, *H*	$129 + $355 + $258 + $324 + $299 + $108	$1473
H, *M*, *S*, *B*, *ER*, *SA*, *H*	$144 + $299 + $155 + $258 + $355 + $129	$1340
H, *ER*, *S*, *M*, *SA*, *B*, *H*	$134 + $154 + $299 + $634 + $274 + $114	$1609
H, *ER*, *M*, *B*, *S*, *SA*, *H*	$134 + $104 + $324 + $155 + $119 + $129	$ 965

From Table 14.4, we see that the Hamilton circuit *H*, *ER*, *M*, *B*, *S*, *SA*, *H* results in a per person cost of $965, which is less than the $1007 we obtained in Example 6. Thus, we see that the nearest neighbor method does not always produce the optimal solution. Without using the brute force method, we cannot determine if the Hamilton circuit *H*, *ER*, *M*, *B*, *S*, *SA*, *H* is the optimal solution. The nearest neighbor method produces an *approximation* for the optimal solution. Note also that the nearest neighbor method did produce a Hamilton circuit that was less expensive than three of the four randomly chosen Hamilton circuits.

SECTION 14.3 EXERCISES

CONCEPT/WRITING EXERCISES

1. **a)** What is a Hamilton circuit?

   b) What is the difference between a Hamilton circuit and an Euler circuit?

2. **a)** What is a Hamilton path?

   b) What is the difference between a Hamilton path and an Euler path?

3. **a)** What is a weighted graph?

   b) What is a complete graph?

   c) What is a complete, weighted graph?

4. **a)** Describe how to compute the factorial of a given number.

   b) Determine 4!.

   c) Determine 6!.

   d) Determine 8!.

5. **a)** Describe how to determine the number of unique Hamilton circuits that exist in a complete graph with n vertices.

   b) How many unique Hamilton circuits exist in a complete graph with seven vertices?

   c) How many unique Hamilton circuits exist in a complete graph with ten vertices?

6. **a)** What is meant by the optimal solution to a traveling salesman problem?

   b) What is meant by an approximate solution to a traveling salesman problem?

7. Describe the brute force method for determining the optimal solution to a traveling salesman problem.

8. Describe the nearest neighbor method for approximating the optimal solution to a traveling salesman problem.

PRACTICE THE SKILLS

In Exercises 9–14, determine two different Hamilton paths in each of the following graphs.

9.

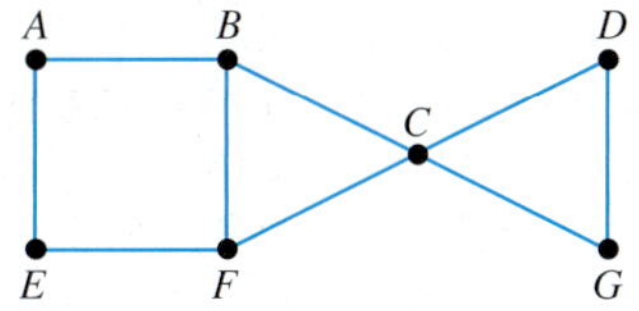

10.

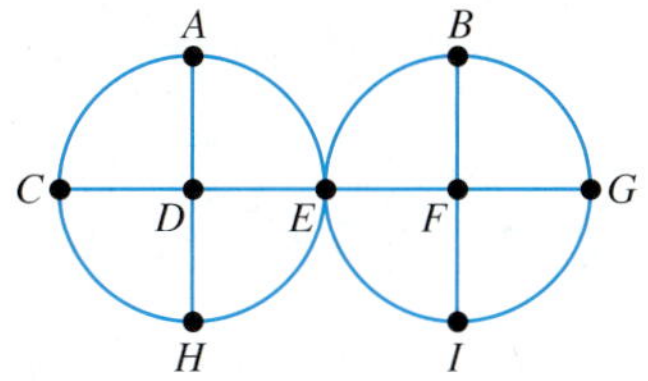

11.

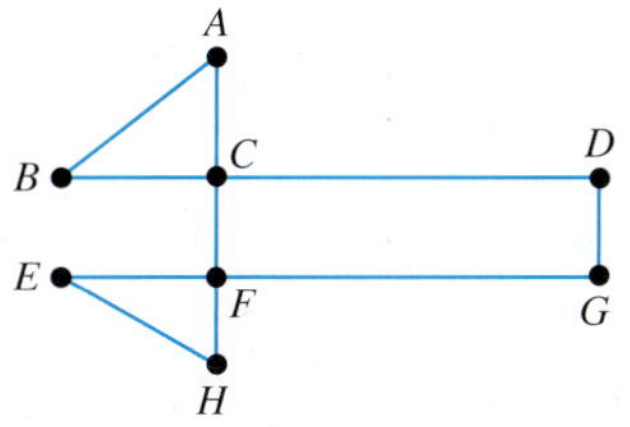

12.

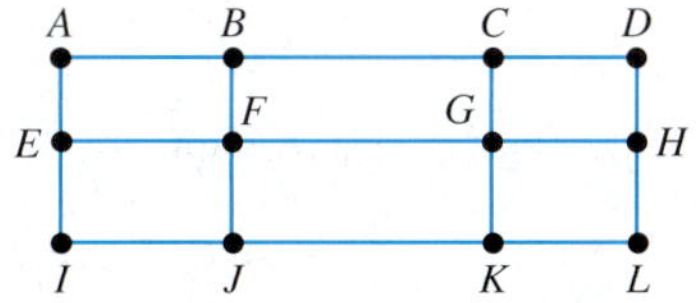

13.

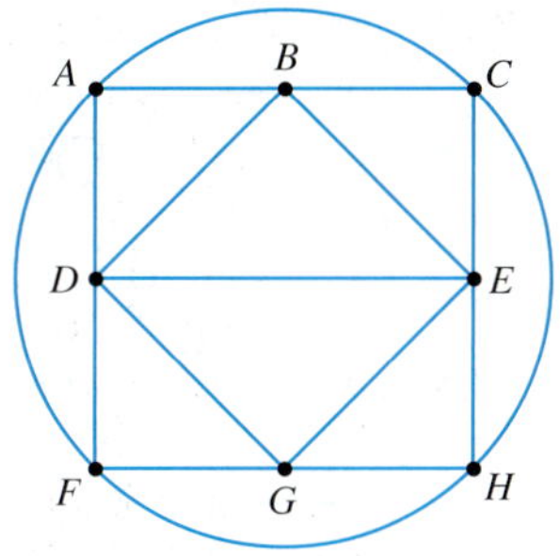

14.

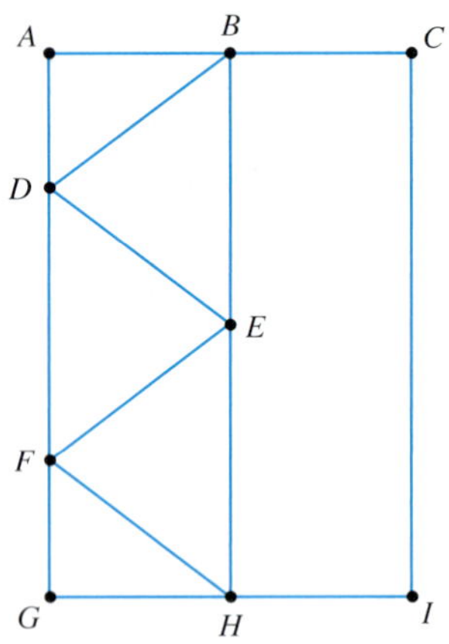

In Exercises 15–18, determine two different Hamilton circuits in each of the following graphs.

15.

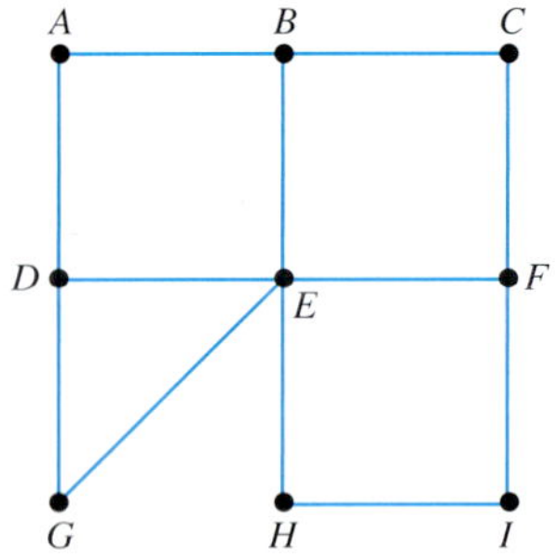

16.

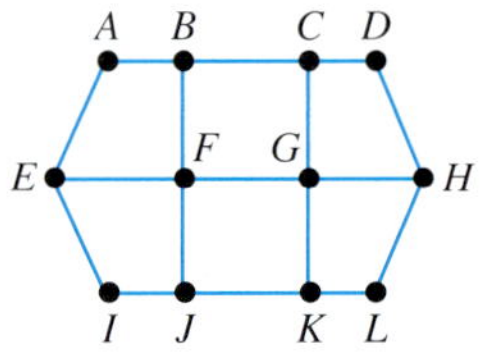

17.

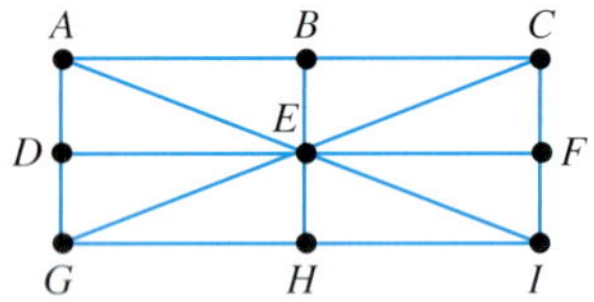

18.

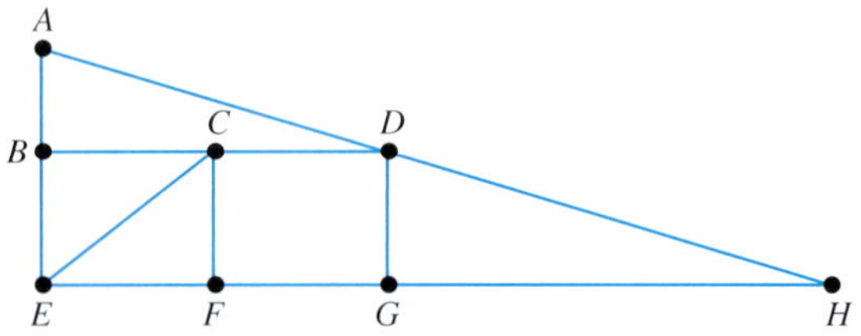

19. Draw a complete graph with four vertices.

20. Draw a complete graph with five vertices.

PROBLEM SOLVING

21. ***John Deere Factories*** Robert W. Lane, the chief executive officer of Deere and Company, is at corporate headquarters in Moline, Illinois, and wishes to visit John Deere factories in the following locations: Coon Rapids, Minnesota; Valley City, North Dakota; Ottumwa, Iowa; Thibodaux, Louisiana; Horicon, Wisconsin; and Augusta, Georgia. In how many ways can Robert visit each of these factories and return to his office in Moline?

22. ***Visiting Relatives*** Rita Hinderman lives in Sumter, South Carolina, and wishes to visit relatives in the following South Carolina cities: Anderson, Charleston, Columbia, Florence, Greenville, Rock Hill, and Spartanburg. In how many ways can she visit each of these cities and return to her home in Sumter?

23. ***Inspecting Weigh Stations*** Sally Ivan lives in Nashville, Tennessee, and works for the Department of Transportation. She wishes to inspect weigh stations in the following Tennessee cities: Caruthersville, Clarksville, Cleveland, Dyersburg, Jackson, Johnson City, Kingsport, Knoxville, Memphis, Murfreesboro, Oak Ridge, and Pulaski. In how many ways can she visit each of these cities and return to her home in Nashville?

▲ A Weigh Station

24. ***A Milk Truck Route*** Dale Klitzke is a milk truck driver for Swiss Valley Farms Cooperative in eastern Iowa. Dale has to start at the processing plant and pick up milk on 10 different farms. In how many ways can Dale visit each farm and return to the processing plant?

25. ***A Vacationing Family*** John and Robyn Pearse promised their three sons that they will each get to pick one spot within 500 miles from their home in Chicago to visit on their vacation. Scott chooses to visit St. Louis so that he can go up in the Gateway Arch. Jacob chooses to visit Green Bay so that he can visit the National Railroad Museum. Jevon chooses to visit Ottawa, Illinois, so that he can visit his grandparents. The approximate distances between these cities are as follows: Chicago to St. Louis is 300 miles, Chicago to Green Bay is 205 miles, Chicago to Ottawa is 80 miles, St. Louis to Green Bay is 500 miles,

St. Louis to Ottawa is 245 miles, and Green Bay to Ottawa is 280 miles.

▲ The Gateway Arch in St. Louis, Missouri

a) Represent this traveling salesman problem with a complete, weighted graph showing the distances on the appropriate edges.

b) Use the brute force method to determine the shortest route for the Pearses to complete their vacation.

c) What is the minimum distance the Pearses can travel?

26. ***Job Interviews*** Christina Dwyer is searching for a new job. She lives in Shreveport, Louisiana, and has interviews in Barrow, Alaska; Tucson, Arizona; and Rochester, New York. The costs of the one-way flights between these four cities are as follows: Shreveport to Barrow costs \$855, Shreveport to Tucson costs \$803, Shreveport to Rochester costs \$113, Barrow to Tucson costs \$393, Barrow to Rochester costs \$337, and Tucson to Rochester costs \$841.

a) Represent this traveling salesman problem with a complete, weighted graph showing the prices of the flights on the appropriate edges.

b) Use the brute force method to determine the least expensive route for Christina to travel to each city and return home to Shreveport.

c) What is the minimum cost she can pay?

27. ***Picking up Children*** Kristin Greenhalgh has to pick up her children, who all attend different schools. She is at home (H) and needs to go to Children First Preschool (C), Braden River Elementary School (B), and Stephens Middle School (S). Kristin estimates the distance among these locations as follows: H to C is 7 miles, H to B is 5 miles, H to S is 13 miles, C to B is 11 miles, C to S is 18 miles, and B to S is 9 miles.

a) Represent this traveling salesman problem with a complete, weighted graph showing the distances on the appropriate edges.

b) Use the brute force method to determine the shortest route for Kristin to start from home, pick up her children, and return to home.

c) What is the minimum distance Kristin can travel?

28. ***Running Errands on Campus*** Mary Mahan needs to run errands on the campus of Clarke College. She is in her office and needs to go to the duplicating center, student center, and library. She estimates the walking distances as follows: from her office to the duplicating center is 150 feet, from her office to the student center is 100 feet, from her office to the library is 400 feet, from the duplicating center to the student center is 125 feet, from the duplicating center to the library is 450 feet, and from the student center to the library is 250 feet.

▲ Clarke College in Dubuque, Iowa

a) Represent this traveling salesman problem with a complete, weighted graph showing the distances on the appropriate edges.

b) Use the brute force method to determine the shortest route for Mary to accomplish her errands and then return back to her office.

c) What is the minimum distance Mary can walk?

29. ***Bass Pro Shops*** Jim Hagale, the chief executive officer of Bass Pro Shops, is at his office in Springfield, Missouri, and is considering adding new stores in the following locations: Bismarck, North Dakota; Carson City, Nevada; Helena, Montana; and Knoxville, Tennessee. He wishes to visit each of these locations and return to his office in Springfield. The one-way flight prices are given in the following table.

	Bismarck	Carson City	Helena	Knoxville	Springfield
Bismarck	*	$431	$483	$476	$378
Carson City	$431	*	$144	$159	$505
Helena	$483	$144	*	$542	$492
Knoxville	$476	$159	$542	*	$459
Springfield	$378	$505	$492	$459	*

a) Represent this traveling salesman problem with a complete, weighted graph showing the prices of flights on the appropriate edges.

b) Use the nearest neighbor method to approximate the optimal route for Jim to travel to each city and return to Springfield. Give the cost of the route determined.

c) Randomly select another route for Jim to travel from Springfield to each of the other cities and return to Springfield and then compute the cost of this route. Compare this cost with the cost determined in part (b).

30. ***Cranberry Plants*** Altay Özgener lives in Boston, Massachusetts, and works for Ocean Spray Cranberries, Inc. Altay wishes to visit cranberry farms in the following locations: Madison, Wisconsin; Princeton, New Jersey; Salem, Oregon; and Walla Walla, Washington. The one-way flight prices are given in the following table.

	Boston	Madison	Princeton	Salem	Walla Walla
Boston	*	$131	$ 256	$298	$ 576
Madison	$131	*	$ 154	$356	$ 970
Princeton	$256	$154	*	$353	$1164
Salem	$298	$356	$ 353	*	$ 179
Walla Walla	$576	$970	$1164	$179	*

a) Represent this traveling salesman problem with a complete, weighted graph showing the prices of flights on the appropriate edges.

b) Use the nearest neighbor method to approximate the optimal route for Altay to travel to each city and return to Boston. Give the cost of the route determined.

c) Randomly select another route for Altay to travel from Boston to each of the other cities and return to Boston and then compute the cost of this route. Compare this cost with the cost found in part (b).

31. ***Presenting Awards*** Donna DeSimone lives in Cleveland and works for the Tupperware Corporation. Donna's job requires her to award prizes to the top five sales representatives in the country. In 2008, the top five sales representatives were in Cleveland, Ohio; Detroit, Michigan; Grand Forks, North Dakota; Morgantown, West Virginia; and Tupelo, Mississippi. The costs of one-way flights between these cities are given in the table below.

	Cleveland	Detroit	Grand Forks	Morgantown	Tupelo
Cleveland	*	$ 39	$418	$119	$105
Detroit	$ 39	*	$319	$109	$128
Grand Forks	$418	$319	*	$271	$520
Morgantown	$119	$109	$271	*	$701
Tupelo	$105	$128	$520	$701	*

a) Represent this traveling salesman problem with a complete, weighted graph showing the prices of the flights on the appropriate edges.

b) Use the nearest neighbor method to approximate the optimal route for Donna to travel to each city and return to Cleveland. Give the cost of the route determined.

c) Randomly select another route for Donna to travel from Cleveland to each of the other cities and return to Cleveland and then compute the cost of this route. Compare this cost with the cost determined in part (b).

32. ***Locations for a New Factory*** Darren Magot works for US Paper Products in New York City. His company is investigating locations for a new factory that will produce a line of disposable clothing. The cities Darren needs to visit are Atlanta, Dallas, Philadelphia, and Washington, D.C. The one-way flight prices between these cities are given in the table below.

	Atlanta	Dallas	New York	Philadelphia	Washington, D.C.
Atlanta	*	$182	$197	$159	$180
Dallas	$182	*	$115	$115	$110
New York	$197	$115	*	$ 55	$156
Philadelphia	$159	$115	$ 55	*	$205
Washington, D.C.	$180	$110	$156	$205	*

a) Represent this traveling salesman problem with a complete, weighted graph showing the prices of the flights on the appropriate edges.

b) Use the nearest neighbor method to approximate the optimal route for Darren to travel to each city and return to New York City. Give the cost of the route determined.

c) Randomly select another route for Darren to travel from New York City to the other cities and return to New York City and then compute the cost of this route. Compare this cost with the cost determined in part (b).

CHALLENGE PROBLEMS/GROUP ACTIVITIES

33. *Visiting Five Cities* Come up with a list of five cities you would like to visit. Use the Internet to search for airline prices between these five cities (make sure to include the city with the airport nearest your home from which you would start and end your trip).

a) Draw a complete, weighted graph that represents these cities and the costs associated with flying between each pair of cities.

b) Use the brute force method to determine the optimal solution to visiting each city and returning home.

c) Use the nearest neighbor method to approximate the optimal solution.

d) How much money does the optimal solution, obtained in part (b), save you over the approximation obtained in part (c)?

34. *Number of Circuits* The following tasks are intended to help you understand the formula for finding the number of unique Hamilton circuits in a complete graph.

a) Draw a complete graph with three vertices labeled A, B, and C. Assume that you are starting at vertex A and wish to move to another vertex. How many choices do you have for moving to the second vertex? Once you choose the second vertex, how many choices do you have for moving to a third vertex? Multiply the number of choices you had from vertex A by the number of choices you had from the second vertex. Compare the number you obtained with the number of Hamilton circuits found by using $(n - 1)!$.

b) Draw a complete graph with four vertices labeled A, B, C, and D. Assume that you are starting at vertex A and wish to move to a second vertex. How many choices do you have for moving to this second vertex? Once you choose the second vertex, how many choices do you have for moving to the third vertex? Once you choose the third vertex, how many choices do you have for the fourth vertex? Multiply the number of choices you had from each vertex together. Compare the number you obtained with the number of Hamilton circuits found by using $(n - 1)!$.

c) Repeat this process for complete graphs with five and six vertices.

d) Explain why $(n - 1)!$ gives the number of Hamilton circuits in a complete graph with n vertices.

RECREATIONAL MATHEMATICS

35. ***The Icosian Game*** In 1857, William Rowan Hamilton invented the game called the Icosian Game. The game consisted of a round board with 20 holes and 20 numbered pegs. On the surface of the board was a pattern similar to the graphs we have studied in this chapter (see photo and artwork below).

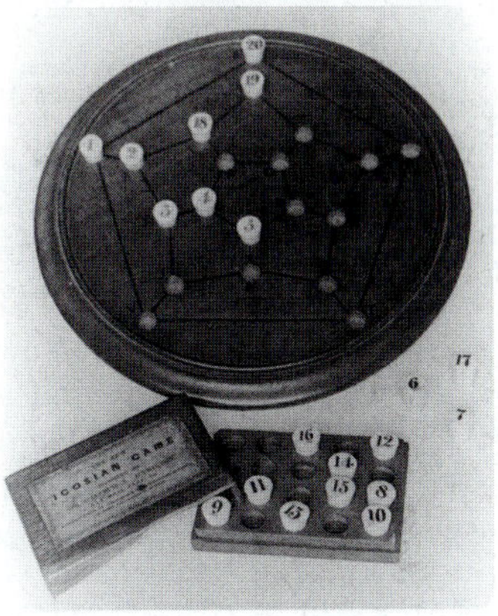

In essence, the object of the game was to use the pegs to form a Hamilton circuit on the graph. The graph below is taken from the Icosian Game. Determine a Hamilton circuit on this graph.

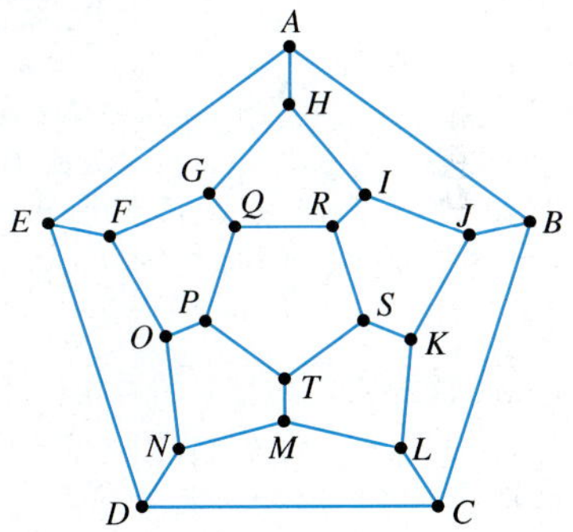

INTERNET/RESEARCH ACTIVITY

36. Write a paper on traveling salesman problems. Include a brief history of the problems. Also include advances made toward reducing the computational time of determining the optimal solution as well as advances made toward determining better approximate solutions.

14.4 TREES

▲ Graph theory can be used to determine the irrigation system with the lowest cost.

Jennifer Duncan wishes to install an irrigation system to water her five flower gardens. How can Jennifer determine the irrigation system that reaches each of her five flower gardens and has the lowest cost? In this section, we will introduce a type of graph that can help solve the problem facing Jennifer.

Introduction

In this chapter, we have introduced graphs as a means to represent problems from everyday life (Section 14.1). We used graph theory to solve a variety of problems by finding Euler paths and Euler circuits (Section 14.2). We also used graph theory to find the optimal and approximate solutions to traveling salesman problems using Hamilton circuits (Section 14.3). We now turn our attention to another type of graph, called a *tree*, which is also frequently used to represent problems from everyday life. Before we define a *tree*, let's look at Example 1, in which we create a family tree.

EXAMPLE 1 *A Family Tree*

Wendy has three children: Adan, Konnor, and Norah. Adan has two children: Justine and Seth. Konnor has no children. Norah has three children: Miya, Brianna, and Reuben. Use a graph to represent this family.

SOLUTION We will construct this graph with three layers, one for each generation. The vertices represent the people, and the edges represent the parent–child relationship. Start with Wendy and make edges to each of her three children, Adan, Konnor, and Norah. Then, make edges from Adan to his two children and from Norah to her three children. Since Konnor has no children, there are no edges below his name.

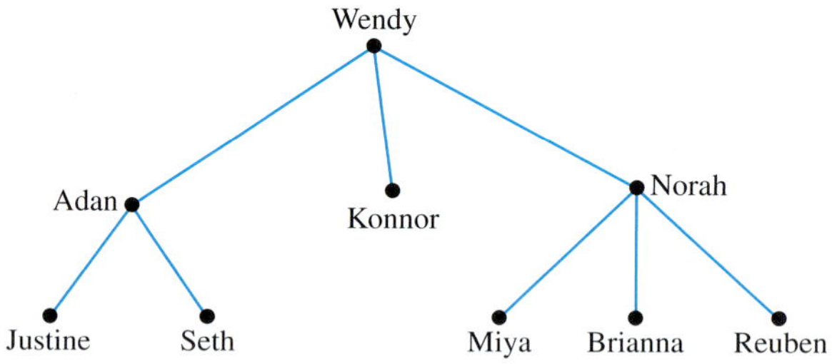

Figure 14.44

The graph shown in Fig. 14.44 is an example of a tree.

A **tree** is a connected graph in which each edge is a bridge.

Recall from Section 14.1 that a bridge is an edge that if removed from a connected graph would create a disconnected graph. Thus, if you remove *any edge* in a tree, it creates a disconnected graph. Since each edge would create a disconnected graph if removed, a tree cannot have any Euler circuits or Hamilton circuits. Can you explain why? Figure 14.45(a) gives four examples of graphs that are trees, and Fig. 14.45(b) gives four examples of graphs that are not trees.

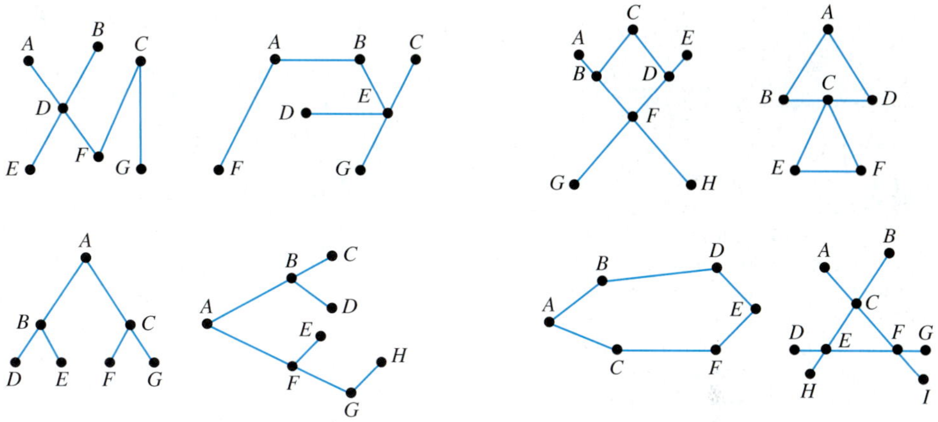

(a) *Graphs that are trees* (b) *Graphs that are not trees*

Figure 14.45

Spanning Trees

Some applications, like the family tree in Example 1, can be represented with a graph that is a tree. In other applications, the problem may initially be represented with a graph that is not a tree. In this section, to solve certain problems, we will need to remove edges from a graph that is not a tree to form a tree known as a *spanning tree.* We now define spanning tree.

A **spanning tree** is a tree that is created from another graph by removing edges while still maintaining a path to each vertex.

Since a spanning tree is a tree, and since a tree cannot have any circuits, a spanning tree cannot have circuits.

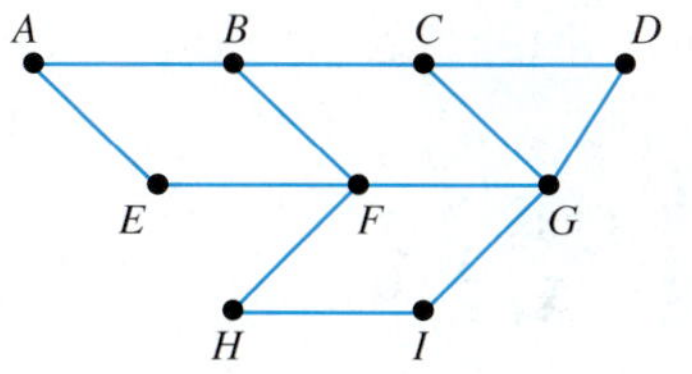

(a)

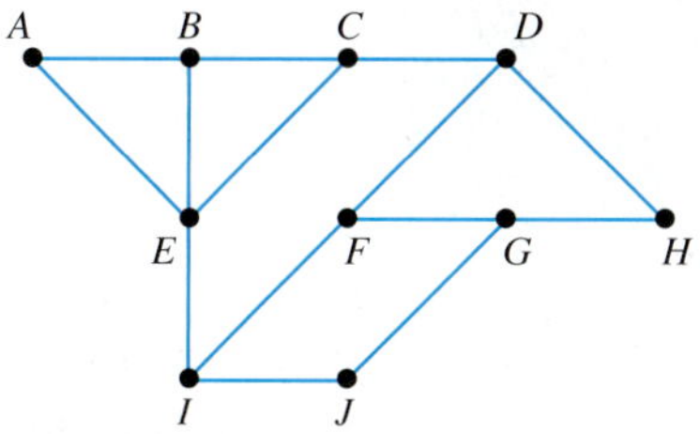

(b)

Figure 14.46

EXAMPLE 2 *Determining Spanning Trees*

Determine two different spanning trees for each graph shown in Fig. 14.46.

SOLUTION Each of the spanning trees we create will need to have a path connecting all vertices, but cannot have any circuits. To create a spanning tree from a graph, we remove edges one at a time while leaving all the vertices in place. When we remove an edge, we must make sure the edge is not a bridge. Removing a bridge would create a disconnected graph. We need to reduce our original graph to one that is still connected. Keeping these guidelines in mind, we continue removing edges until the remaining graph is a tree. Two spanning trees formed from each graph in Fig. 14.46(a) and (b) are given in Fig. 14.47(a) and (b), respectively on page 958. Other spanning trees are possible in each case.

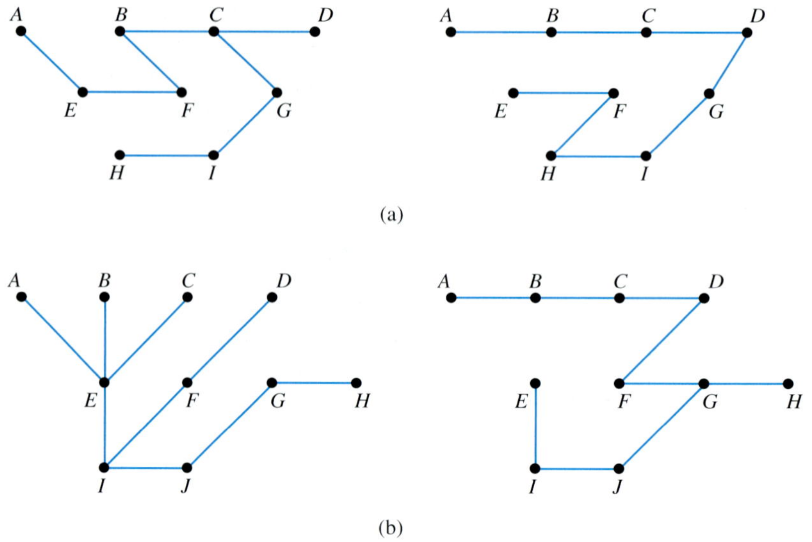

Figure 14.47

Notice in all four spanning trees in Example 2 that each edge is a bridge and that no circuits are present. Our next example provides an application of spanning trees.

EXAMPLE 3 *A Spanning Tree Problem*

Schoolcraft College is considering adding awnings above its sidewalks to help shelter students from the snow and rain while they walk between some of the buildings on campus. A diagram of the buildings and the connecting sidewalks where the awnings are to be added is given in Fig. 14.48.

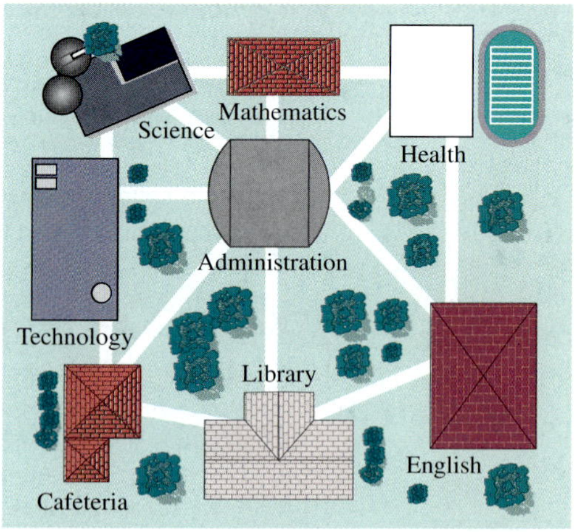

Figure 14.48

Originally, the president of the college wished to have awnings placed over all the sidewalks shown in Fig. 14.48, but that was found to be too costly. Instead, the president has proposed to place just enough awnings over a select number of

DID YOU KNOW?

Data Networks

In our current "information age," data networks support so many daily conveniences that we probably don't even realize a network is involved. Each time we place a telephone call, fly in an airplane, send an e-mail, or use the Internet, we are using some form of modern data network. Each of these networks can be represented as a graph with many vertices and edges. It might surprise you that data networks predate the twentieth century. The most well known example of a primitive data network is the electromagnetic telegraph used in the nineteenth century in the United States and around the world prior to the invention of the telephone. An even older example of a data network is the optical telegraph (with nearly 1000 stations) used in the eighteenth century all across Europe. Perhaps the oldest, yet maybe the slowest, form of data network is one still in use today: the writing and sending of letters!

sidewalks so that, by moving from building to building, students would still be able to reach any location shown without being exposed to the elements.

a) Represent all the buildings and sidewalks shown with a graph.

b) Create three different spanning trees from this graph that would satisfy the president's proposal.

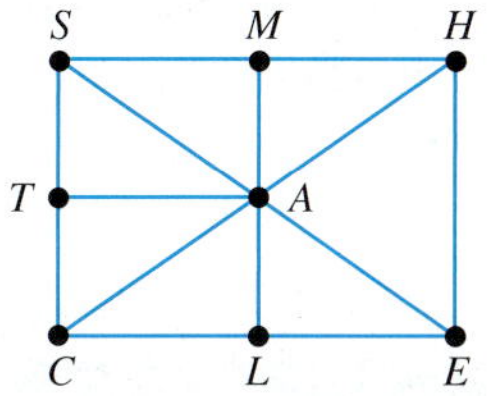

Figure 14.49

SOLUTION

a) Using letters to represent the building names, vertices to represent the buildings, and edges to represent the sidewalks between buildings, we generate the graph in Fig. 14.49.

b) To create a spanning tree we remove nonbridge edges until a tree is created. Three possible spanning trees are given in Fig. 14.50; however, many others are possible.

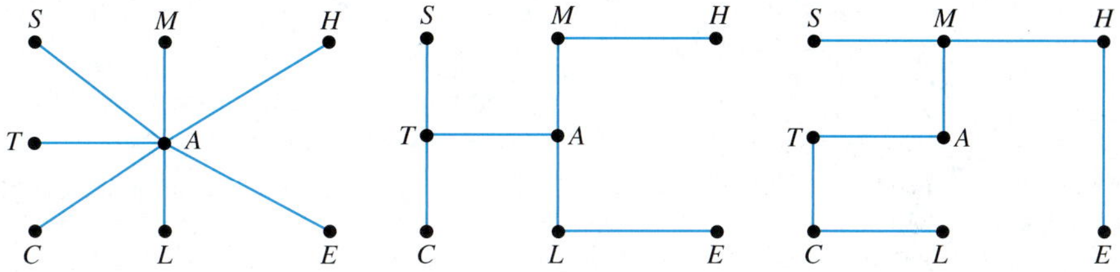

Figure 14.50

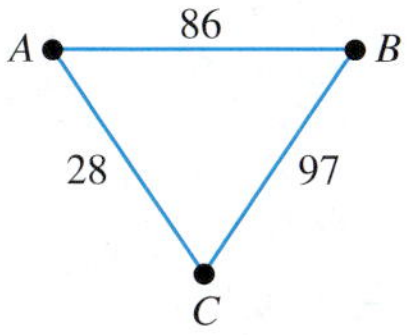

Figure 14.51

Example 3 shows us how spanning trees can be used to represent a real-life problem. However, the administrators at Schoolcraft College still have a problem: Which sidewalks should be chosen to cover with awnings (or which spanning tree should be selected)? Problems like the one faced by Schoolcraft College usually have additional considerations that need to be included in the decision-making process. The most common consideration is the cost of the project. To help analyze these problems, weighted graphs—graphs with costs or distances associated with each edge—and minimum-cost spanning trees are used.

A **minimum-cost spanning tree** is the least expensive spanning tree of all spanning trees under consideration.

Example 4 will explain how we determine a minimum-cost spanning tree for a graph with three edges.

EXAMPLE 4 *Finding a Minimum-Cost Spanning Tree*

Examine the weighted graph in Fig. 14.51. This graph shows the costs, in dollars, associated with each edge. Determine the minimum-cost spanning tree for this graph.

SOLUTION There are three spanning trees associated with this graph; they are shown in Fig. 14.52.

The spanning tree in Fig. 14.52(a) has a cost of $86 + $97 = $183. The spanning tree in Fig. 14.52(b) has a cost of $28 + $97 = $125. The spanning tree in Fig. 14.52(c) has a cost of $86 + $28 = $114. Therefore, the minimum-cost spanning tree is shown in Fig. 14.52(c). It has a cost of $114.

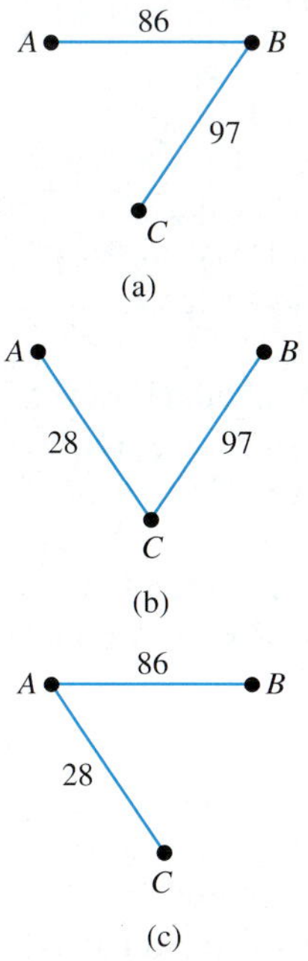

Figure 14.52

The process used in Example 4 is similar to the brute force method used to find the optimal solution to traveling salesman problems in Section 14.3. Although this process is easy to carry out for a small graph, such a process would be impossible for graphs with a large number of vertices. Fortunately, we have an algorithm, called Kruskal's algorithm, to determine the minimum-cost spanning tree.

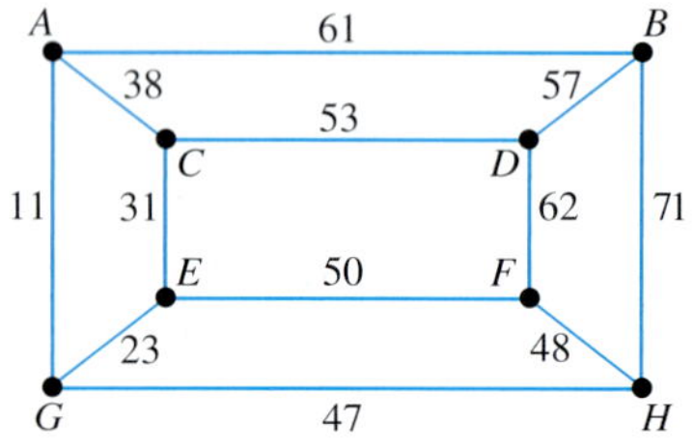

Figure 14.53

KRUSKAL'S ALGORITHM

To construct the minimum-cost spanning tree from a weighted graph:

1. Select the lowest-cost edge on the graph.
2. Select the next lowest-cost edge that does not form a circuit with the first edge.
3. Select the next lowest-cost edge that does not form a circuit with previously selected edges.
4. Continue selecting the lowest-cost edges that do not form circuits with the previously selected edges.
5. When a spanning tree is complete, you have the minimum-cost spanning tree.

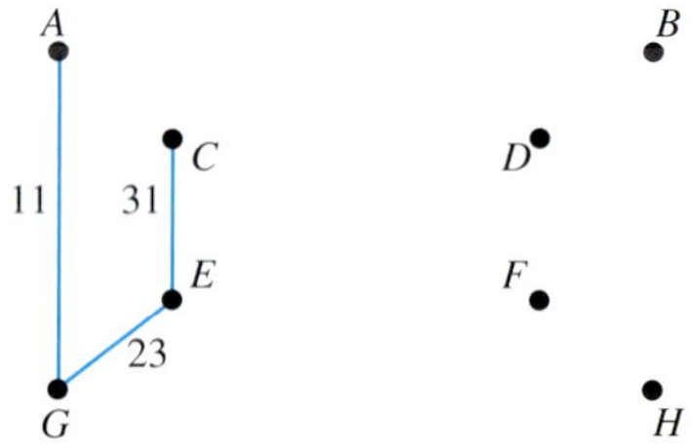

Figure 14.54

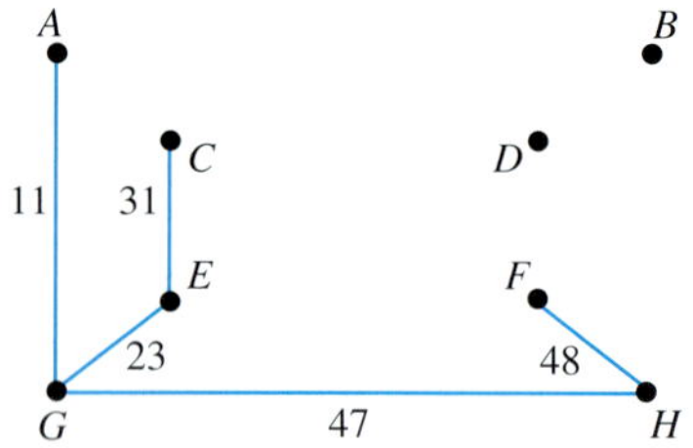

Figure 14.55

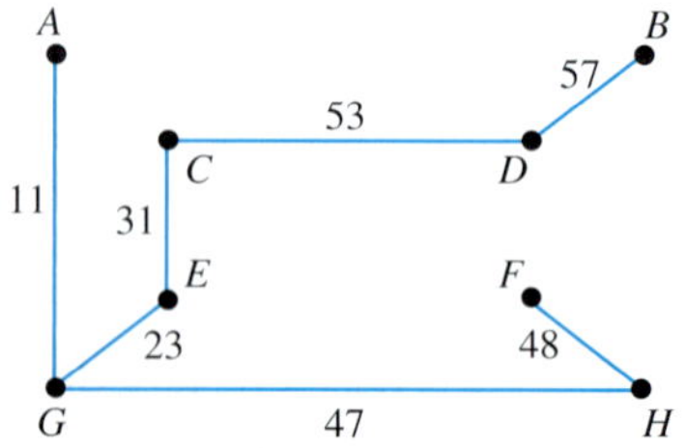

Figure 14.56

EXAMPLE 5 *Using Kruskal's Algorithm*

Use Kruskal's algorithm to determine the minimum-cost spanning tree for the weighted graph shown in Fig. 14.53. The numbers along the edges in Fig. 14.53 represent dollars.

SOLUTION We begin by selecting the lowest-cost edge of the graph, edge *AG*, which is \$11 (see Fig. 14.54). Next we select the next lowest-cost edge that does not form a circuit, edge *GE*, which is \$23. Once again, looking for the lowest-cost edge that does not form a circuit, we select edge *CE*, which is \$31. The result of our first three selections is shown in Fig. 14.54.

Looking at the original weighted graph in Fig. 14.53, we see that the next lowest-cost edge is edge *AC*, which is \$38. However, selecting edge *AC* would create a circuit among vertices *A*, *C*, *E*, and *G*, so we must not select edge *AC*. Instead, we select edge *GH*, which is \$47, since it is the next lowest-cost edge and its selection does not lead to a circuit; see Fig. 14.55. Next, we select edge *FH*, which is \$48.

The next lowest-cost edge not yet selected is edge *EF*, which is \$50. However, this edge would create a circuit among vertices *E*, *F*, *H*, and *G*. Instead, we select edge *CD*, which is \$53, followed by selecting edge *BD*, which is \$57. The result of our selections thus far is shown in Fig. 14.56.

Notice that the graph in Fig. 14.56 is a spanning tree. According to Kruskal's algorithm, the tree in Fig. 14.56 is the minimum-cost spanning tree for the original weighted graph.

Note that in Example 5, all the edges selected were connected to edges we had already selected. However, that will not always be the case. When we select the edges in Example 6, some will be disconnected at the time we select them.

We now will apply our knowledge of minimum-cost spanning trees. Note that in examples and exercises, the graphs need not be drawn to scale.

PROFILE IN MATHEMATICS

Solomon Lefschetz

Solomon Lefschetz (1884–1972) was born in Moscow, studied engineering in Paris, and moved to the United States in 1907. After a laboratory accident claimed both his hands and forearms, Lefschetz turned to mathematics. Lefschetz's work gained him worldwide recognition in algebraic geometry, topology, and nonlinear differential equations. In fact, topology took its name from the title of one of Lefschetz's papers. Lefschetz probably will be remembered most for his reaction to the Soviet Union's launching of the satellite *Sputnik*. Lefschetz believed that the gap between U.S. and Soviet technology was a result of the Soviet superior knowledge of nonlinear differential equations. To help close this gap, Lefschetz came out of retirement to establish and direct a group of 17 research scientists at the Martin Company's Research Institute for Advanced Study in Baltimore. At the age of 81, Lefschetz was awarded the national Medal of Science for his role in mathematics and science development in the United States.

EXAMPLE 6 *An Application Using Kruskal's Algorithm*

Recall from Example 3 that Schoolcraft College is considering the construction of awnings over a select number of sidewalks on campus. The distances, in feet, between buildings are shown in the weighted graph in Fig. 14.57.

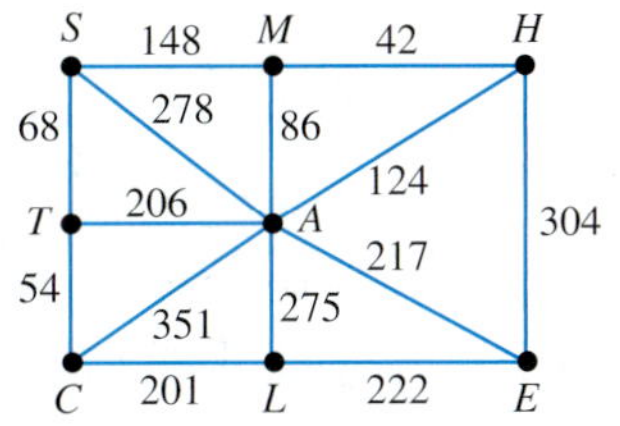

Figure 14.57

a) Determine the shortest series of sidewalks to cover so that students would be able to move between any buildings shown without being exposed to the elements.
b) The cost of the awnings is $23 per foot regardless of where they are placed on campus. What will it cost to cover the sidewalks found in part (a)?

SOLUTION

a) Schoolcraft College is seeking a minimum-cost spanning tree for the weighted graph shown in Fig. 14.57. To determine the minimum-cost spanning tree, we use Kruskal's algorithm. First, we select the edge *MH* since it has the shortest distance and therefore has the lowest cost. We next select the edge *TC* since it has the second shortest distance and it doesn't form a circuit with our previously selected edge (see Fig. 14.58). Our third and fourth choices will be edge *ST* and edge *MA*, respectively, since these edges have the next shortest distances and they do not form a circuit with previously selected edges. The result of our first four selections is shown in Fig. 14.58.

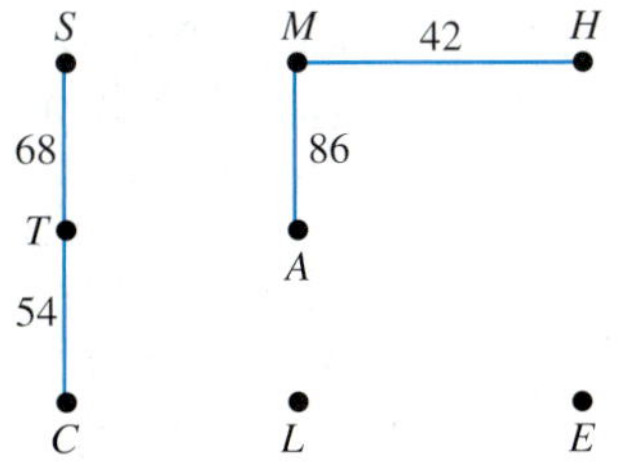

Figure 14.58

The next lowest-cost edge is edge *AH*. However, selection of edge *AH* would form a circuit with vertices *A*, *H*, and *M*. Instead, we choose the next lowest-cost edge—edge *SM*—since it does not form a circuit with the previously selected edges (see Fig. 14.59 on page 962). Our next lowest-cost edge will be edge CL. Note that edge CL does not form a circuit with the previously selected edges. The result of all our selections so far is shown in Fig. 14.59.

We note that the next lowest-cost edge is edge *TA*, but selecting this edge would create a circuit with vertices *S*, *M*, *A*, and *T*. So we select edge *AE*. The result of our selections so far is shown in Fig. 14.60 on page 962.

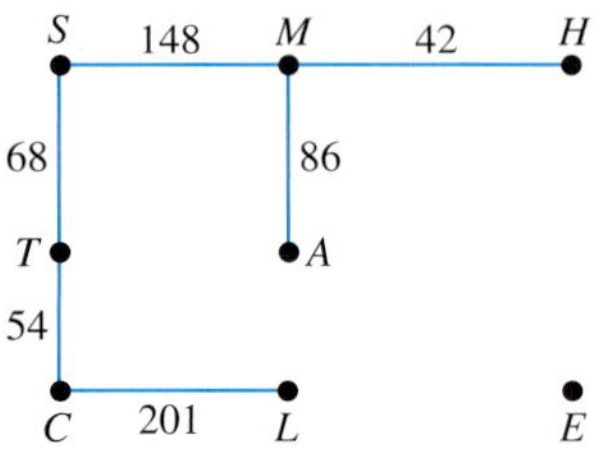

Figure 14.59

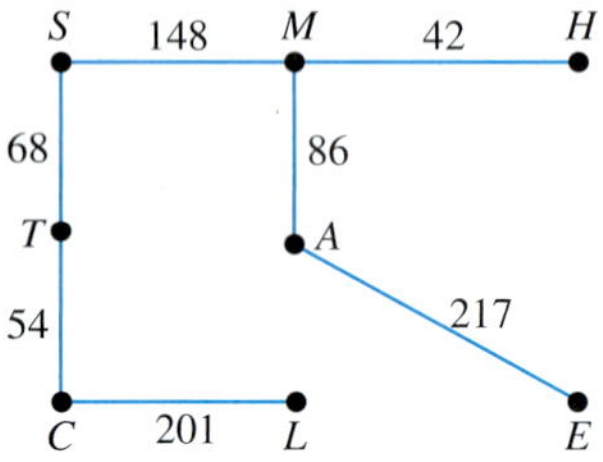

Figure 14.60

From Fig. 14.60, we can see that we have formed a spanning tree. According to Kruskal's algorithm, this graph is the minimum-cost spanning tree. Therefore, Fig. 14.60 shows which sidewalks should be covered with the awnings. Covering these sidewalks will provide protection to students at the lowest cost to Schoolcraft College.

b) From Fig. 14.60, we see that there are 148 + 42 + 68 + 86 + 54 + 217 + 201 = 816 feet of sidewalks that need to be covered. At \$23 per foot, the cost to cover these sidewalks is \$23 × 816 = \$18,768.

EXAMPLE 7

Schools in Budville, Fairplay, Happy Corners, Kieler, Louisburg, and Sinsinawa, Wisconsin, all wish to establish a fiber-optic computer network to share information and to obtain Internet access. The most efficient method of establishing such a network would be to install fiber-optic cable along roadsides. The weighted graph in Fig. 14.61 shows the distance in miles between schools along existing roads.

a) Determine the shortest distance to link these six schools.

b) The cost to install fiber-optic cable is \$1257 per mile. What is the minimum cost to install the fiber-optic cable along the roadsides determined in part (a)?

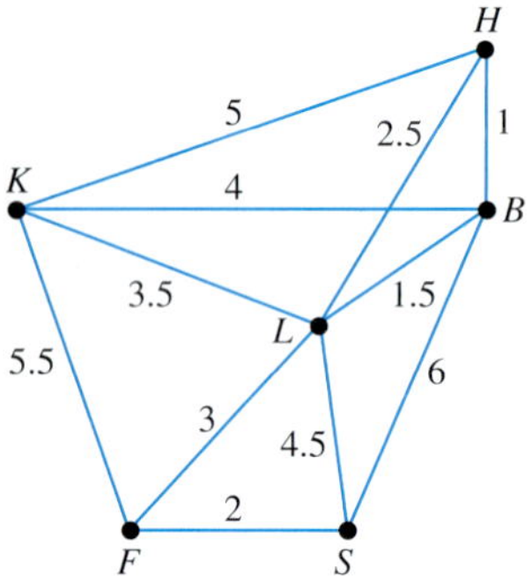

Figure 14.61

SOLUTION

a) We are seeking a minimum-cost spanning tree. We will carry out Kruskal's algorithm to find it. The first three edges selected will be edge *HB* (1 mile), edge *BL* (1.5 miles), and edge *FS* (2 miles). Note that these are the three lowest-cost edges, and their selection does not lead to a circuit. The result of our first three selections is shown in Fig. 14.62.

The next lowest-cost edge is edge *HL* (2.5 miles), but selecting this edge would lead to a circuit between vertices *H*, *B*, and *L*. Instead, we select the next lowest-cost edge, edge *LF* (3 miles). This edge is followed by the next lowest-cost edge, edge *KL* (3.5 miles). The result of all our selections so far is shown in Fig. 14.63. We now have a spanning tree.

According to Kruskal's algorithm, Fig. 14.63 shows the minimum-cost spanning tree. Therefore, Fig. 14.63 gives the paths along which fiber-optic cable should be placed to connect the schools for the minimum cost.

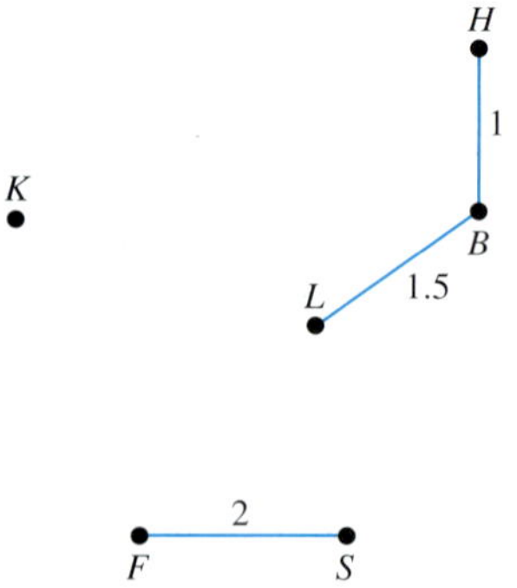

Figure 14.62

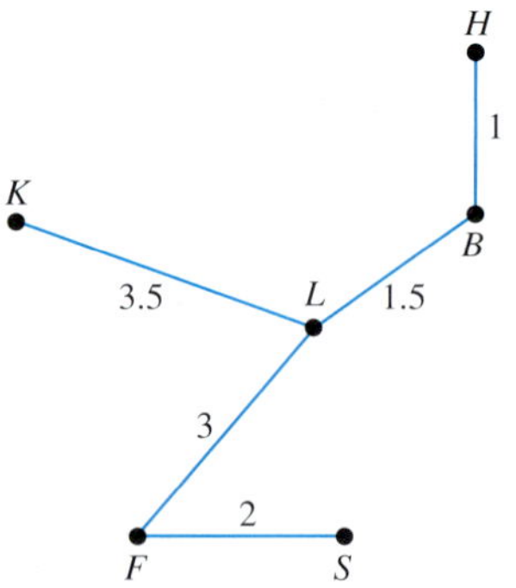

Figure 14.63

b) From Fig. 14.63, we can see that there are 1 + 1.5 + 3.5 + 3 + 2 = 11 miles of fiber-optic cable needed for this network. At \$1257 per mile, the cost to install this fiber-optic network is \$1257 × 11 = \$13,827.

SECTION 14.4 EXERCISES

CONCEPT/WRITING EXERCISES

1. What is a tree?

2. How does a tree differ from a graph that is not a tree?

3. Consider two distinct vertices connected by one edge. Is this graph a tree? Explain.

4. How is a spanning tree obtained from a graph that is not a tree?

5. What is a minimum-cost spanning tree?

6. Describe how to find a minimum-cost spanning tree from a weighted graph.

PRACTICE THE SKILLS

7. ***A Family Tree*** Use a tree to show the parent–child relationships in the following family. Pat has two daughters: Sue and Kris. Sue has four children: John, Thomas, Lee, and Rhiannon. Kris has three children: Alex, Nicholas, and Max.

8. ***A Family Tree*** Use a tree to show the parent–child relationships in the following family. Joe has two children: Allan and Rosemary. Allan has three children: Christopher, Donetta, and AJ. Rosemary has three children: Peter, Paula, and Martin.

9. ***Corporate Structure*** Use a tree to show the following employee relationships at DTX Industries. Kroger is president and has three vice presidents: Dorfman, Blutarsky, and Hoover. Dorfman has three managers: Stratton, Day, and Stork. Blutarsky has three managers: Schoenstein, De Pasto, and Liebowitz. Hoover has two managers: Jennings and Wormer.

10. ***Employment Structure*** Use a tree to show the employee relationships at Stephens College. President Pullen has three division directors working for her: Anderson, Wolcott, and Lisk. Anderson has two department heads working for her: Pomeroy and Thomas. Wolcott has three department heads working for her: Hoey, Jones, and Kent. Lisk has four department heads working for him: Teske, Watts, Groell, and Tuff.

In Exercises 11–18, determine two different spanning trees for the given graph. There are many possible answers.

11.
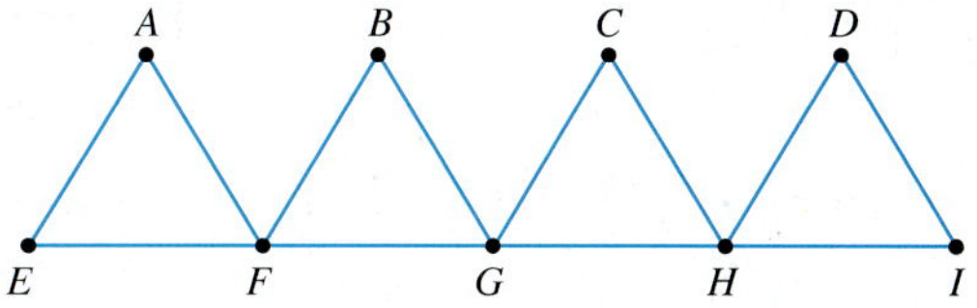

12.
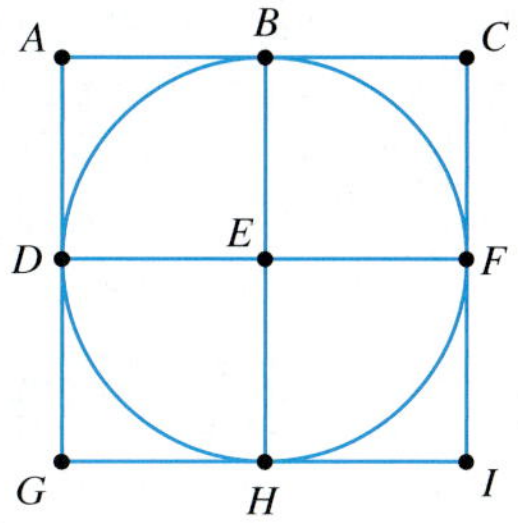

13.
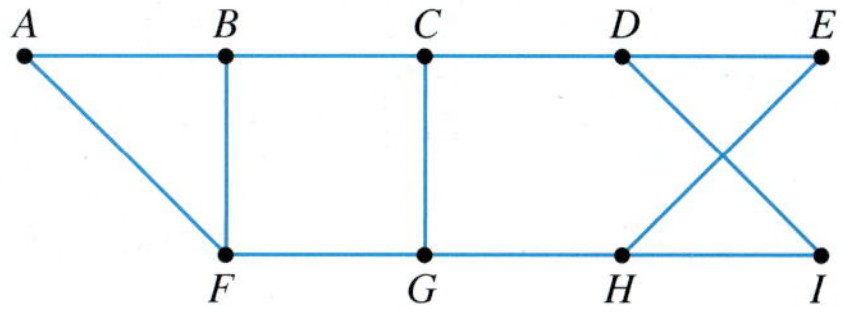

14.
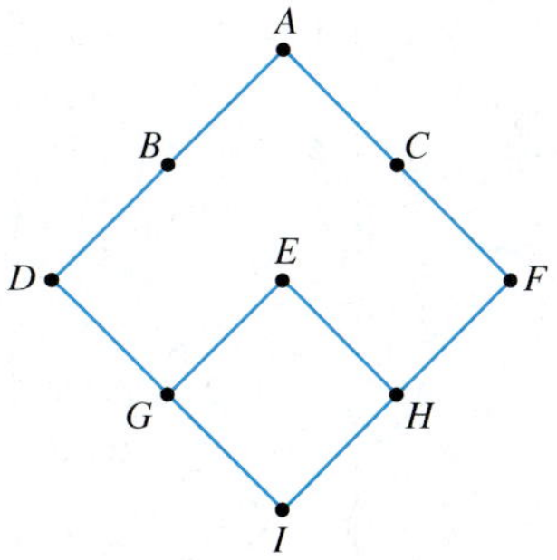

15.
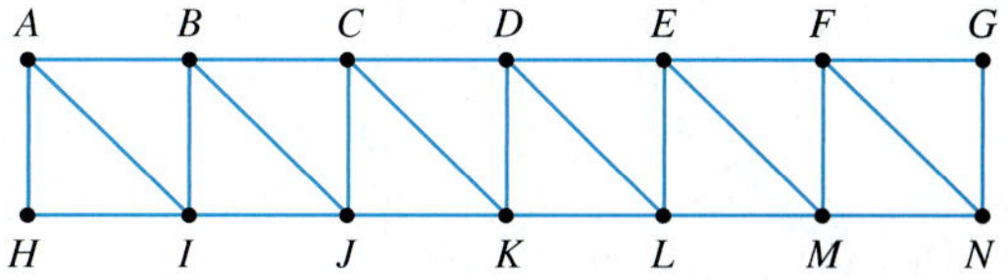

16.
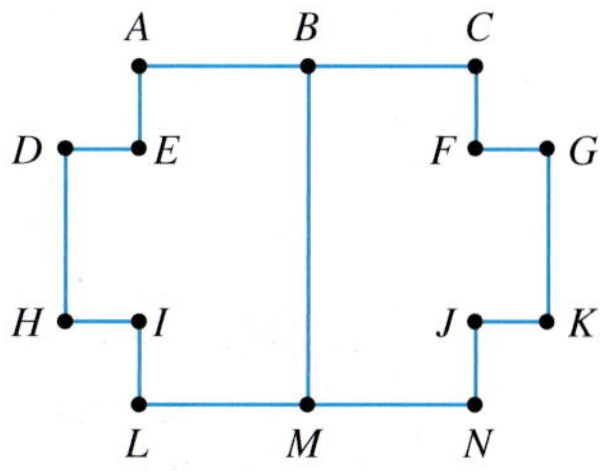

17.

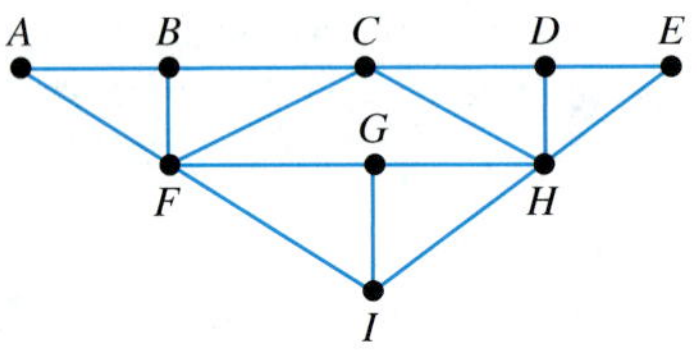

18.

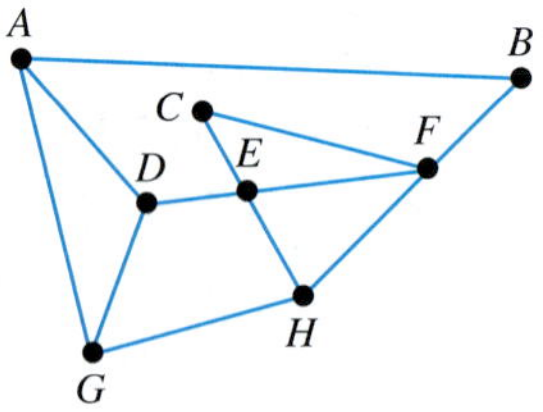

In Exercises 19–26, determine the minimum-cost spanning trees for the given graph.

19.

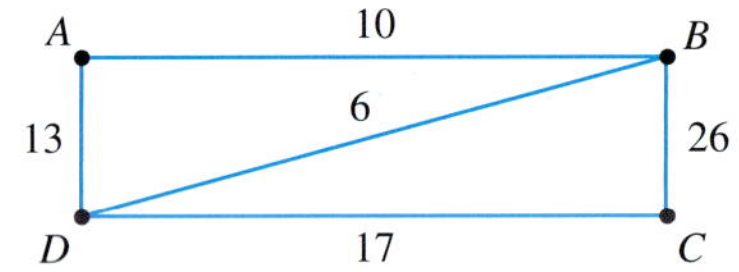

20.

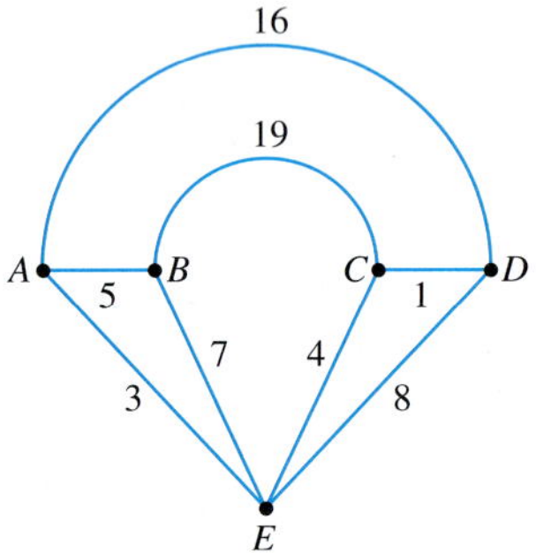

21.

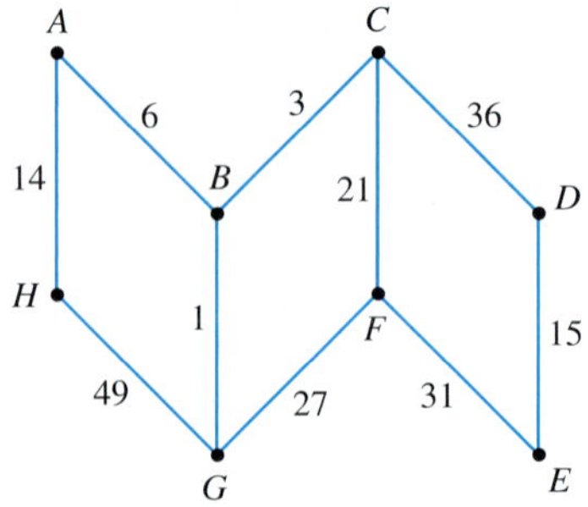

22.

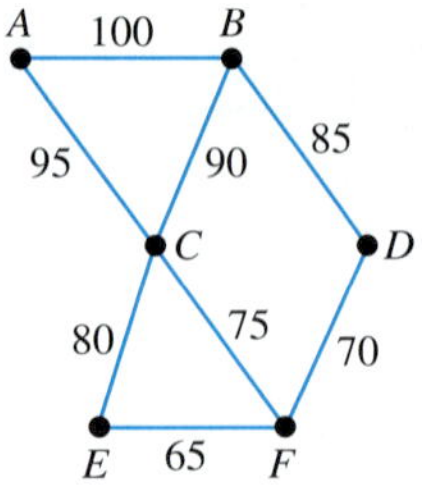

23.

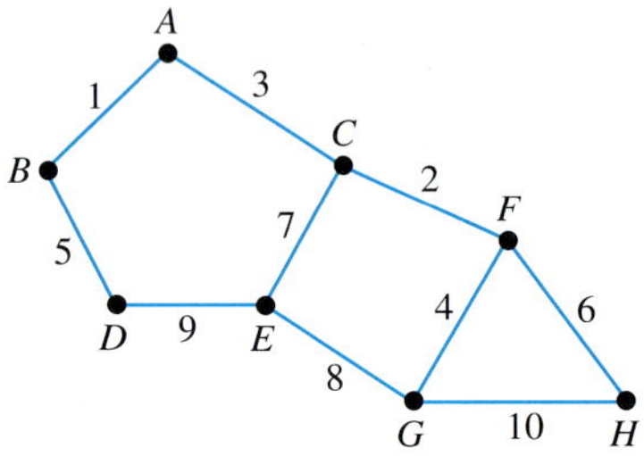

24.

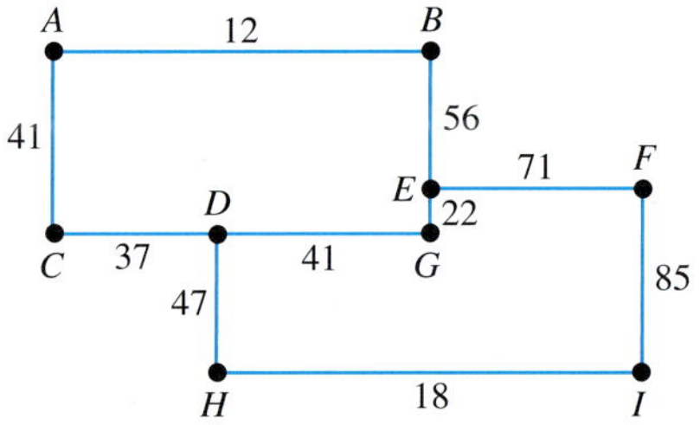

25.

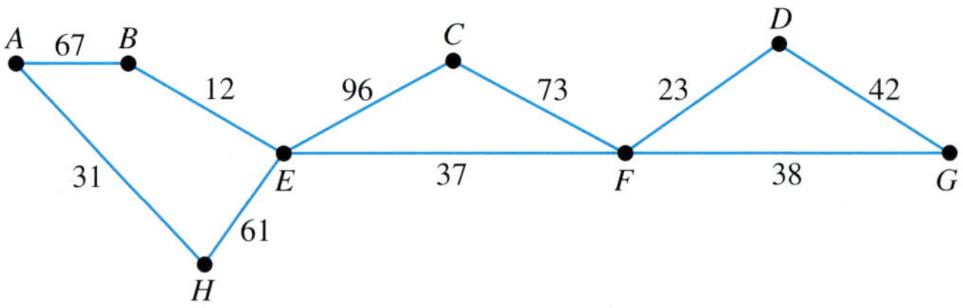

26.

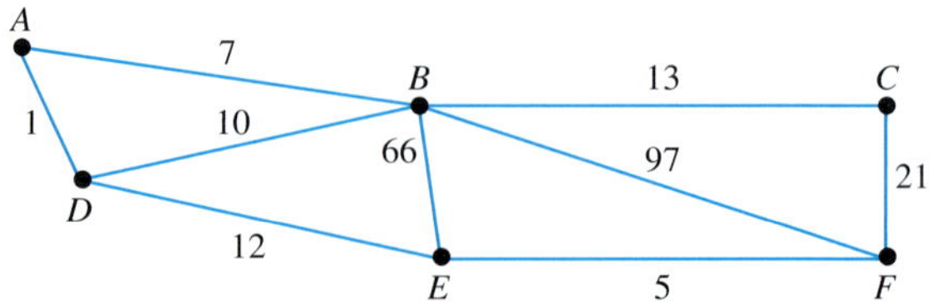

PROBLEM SOLVING

27. ***An Irrigation System*** Jennifer Duncan wishes to install an irrigation system to water her five flower gardens. The distances in feet between her gardens are shown in the figure below.

a) Represent this information with a weighted graph.

b) Use Kruskal's algorithm to determine the minimum-cost spanning tree.

c) If the cost of installing irrigation pipe is $25 per foot, determine the minimum cost of installing the irrigation system from part (b).

28. ***Art Sculpture*** Joe Loccisano is creating a modern art sculpture that needs to have electricity available at five different sites. The figure below shows the items included in the sculpture. The distances shown are in inches.

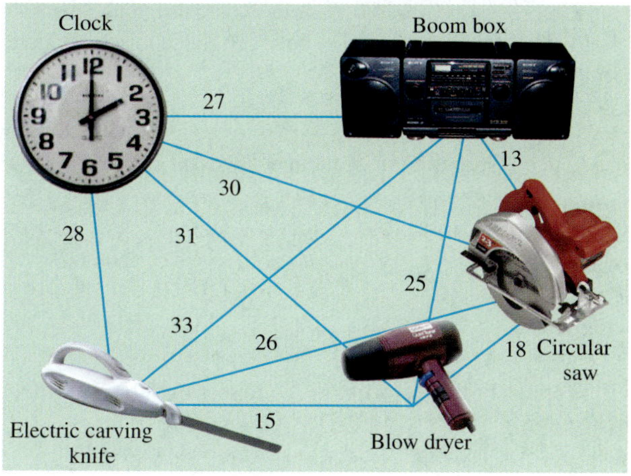

a) Represent this sculpture with a weighted graph.

b) Use Kruskal's algorithm to determine the minimum-cost spanning tree that would provide electricity to each site on the sculpture.

c) If the cost of installing wiring on the sculpture is $0.75 per inch, determine the minimum cost for wiring this sculpture.

29. ***Commuter Train System*** Several communities in eastern Pennsylvania wish to establish a commuter rail train system between the cities shown in the map below (distances are in miles).

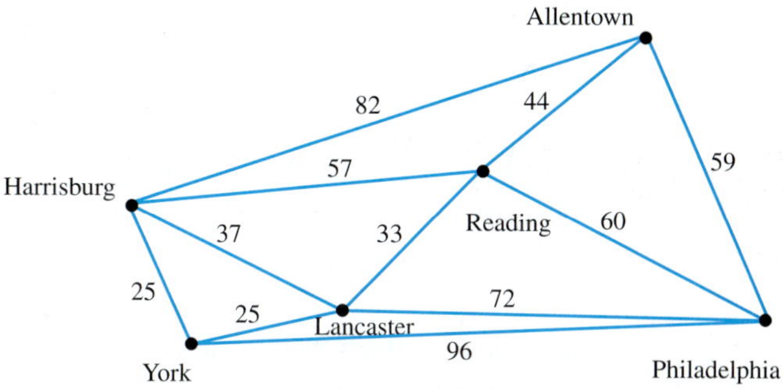

a) Use Kruskal's algorithm to determine the minimum-cost spanning tree that would link the cities using the shortest distance.

b) If it costs $6800 per mile of railroad track, how much does the commuter rail system determined in part (a) cost?

30. ***Linking Campuses*** Five of the campuses in the University of Texas system would like to establish a high-speed telephone and data network. The campuses are located in the following cities: Brownsville, Dallas, El Paso, San Antonio, and Tyler. The map below gives the approximate distances in miles between these five campuses.

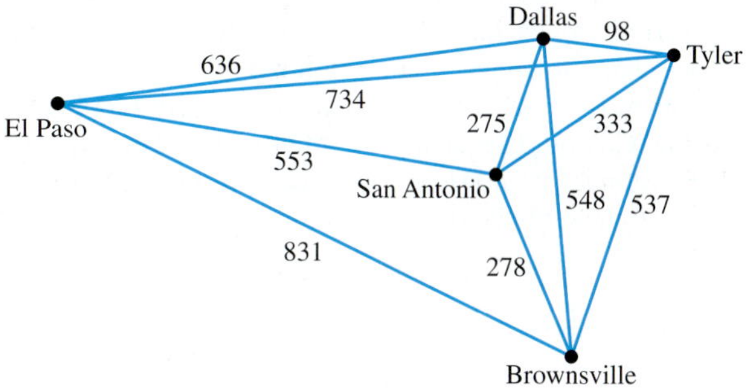

a) Use Kruskal's algorithm to determine the minimum-cost spanning tree that would link each university to create the telephone and data network with the shortest distance.

b) If it costs $1700 per mile to install the network, how much will it cost to produce the network determined in part (a)?

▲ University of Texas at El Paso

31. ***Horse Trails*** The Darlington County, South Carolina, tourism office wishes to build a horse trail that connects the towns of Darlington, Hartsville, Lamar, and Society Hill. The distances, in miles, between these cities are given in the table below.

	Darlington	Hartsville	Lamar	Society Hill
Darlington	*	12	14	15
Hartsville	12	*	13	16
Lamar	14	13	*	26
Society Hill	15	16	26	*

a) Represent this information with a complete, weighted graph (see Section 14.3 for the definition of a complete, weighted graph).

b) Use Kruskal's algorithm to determine the minimum-cost spanning tree that would link each city to create the least expensive horse trail.

c) If the cost of building such a trail is $3500 per mile, what is the cost of building the trail determined in part (b)?

32. ***Electrical Power Lines*** The Eastern Ohio Electric Cooperative wishes to connect cities within its district with new, higher-quality electrical power lines. The cities include Akron, Canton, Cleveland, Columbus, and Youngstown. The distances in miles between the cities involved are given in the table below.

	Akron	Canton	Cleveland	Columbus	Youngstown
Akron	*	23	45	127	48
Canton	23	*	68	125	53
Cleveland	45	68	*	146	75
Columbus	127	125	146	*	178
Youngstown	48	53	75	178	*

a) Represent this information with a complete, weighted graph.

b) Use Kruskal's algorithm to determine the minimum-cost spanning tree that would link each city to create the least expensive power line network.

c) If the cost to establish electrical power lines is $2300 per mile, determine the cost of creating the electrical power lines determined in part (b).

33. ***Recreation Trail*** The Missouri Park and Recreation Association would like to build a recreation trail for biking and hiking that connects the cities of Jefferson City, Kansas City, Springfield, St. Joseph, and St. Louis. The distances in miles between these cities are given in the following table.

	Jefferson City	Kansas City	Springfield	St. Joseph	St. Louis
Jefferson City	*	158	138	215	134
Kansas City	158	*	191	56	249
Springfield	138	191	*	227	218
St. Joseph	215	56	227	*	307
St. Louis	134	249	218	307	*

a) Represent this information with a complete, weighted graph.

b) Use Kruskal's algorithm to determine the minimum-cost spanning tree that would link each city to create the least expensive recreation trail.

c) If the cost of building such a trail is $3700 per mile, what is the cost of building the trail determined in part (b)?

34. ***Light-Rail Transit System*** The state of Illinois is applying for a grant to connect several metropolitan areas of the state with a light-rail transport system. The cities involved are Champaign, Chicago, Peoria, Rockford, and Springfield. The distances in miles between these cities are given in the following table.

	Champaign	Chicago	Peoria	Rockford	Springfield
Champaign	*	135	89	181	86
Chicago	135	*	170	85	202
Peoria	89	170	*	129	74
Rockford	181	85	129	*	193
Springfield	86	202	74	193	*

a) Represent this information with a complete, weighted graph.

b) Use Kruskal's algorithm to determine the minimum-cost spanning tree that would link each city using the shortest distance.

c) What would be the total distance of the light rail transportation system determined in part (b)?

CHALLENGE PROBLEMS/GROUP ACTIVITIES

35. *Computer Network* Create a minimum-cost spanning tree that would serve as the least expensive way to create a computer network among the five largest cities in your state. Consult maps, atlases, almanacs, or the Internet for distances. Assume that the cost per mile is the same regardless of the path taken.

36. *College Structure* Create a tree that shows the administrative structure at your college or university. Start with the highest-ranking officer (that is, president, chancellor, provost, headmaster) and work down to the department chair level.

37. *Sidewalk Covers* Create a minimum-cost spanning tree that would serve as the least expensive way to provide sheltered sidewalks between major buildings on your college campus (see Example 6 on page 961). Assume that the cost per foot is the same regardless of the path taken.

INTERNET/RESEARCH ACTIVITY

38. Write a research paper on the life and work of Joseph Kruskal, who developed Kruskal's algorithm.

CHAPTER 14 SUMMARY

IMPORTANT FACTS

PATHS AND CIRCUITS

A **path** is a sequence of adjacent vertices and the edges connecting them.

A **circuit** is a path that begins and ends at the same vertex.

An **Euler path** is a path that passes through each *edge* of a graph exactly one time.

An **Euler circuit** is a circuit that passes through each *edge* of a graph exactly one time.

A **Hamilton path** is a path that contains each *vertex* of a graph exactly once.

A **Hamilton circuit** is a path that begins and ends at the same vertex and passes through all *vertices* exactly one time.

EULER'S THEOREM

For a connected graph, the following statements are true:

1. A graph with *no odd vertices* (all even vertices) has at least one Euler path, which is also an Euler circuit. An Euler circuit can be started at any vertex, and it will end at the same vertex.
2. A graph with *exactly two odd vertices* has at least one Euler path but no Euler circuits. Each Euler path must begin at one of the two odd vertices and end at the other odd vertex.
3. A graph with *more than two odd vertices* has no Euler paths or Euler circuits.

FLEURY'S ALGORITHM

To determine an Euler path or an Euler circuit:

1. Use Euler's theorem to determine whether an Euler path or an Euler circuit exists. If one exists, proceed with steps 2–5.
2. If the graph has no odd vertices (therefore has an Euler circuit which is also an Euler path), choose any vertex as the starting point. If the graph has exactly two odd vertices (therefore has only an Euler path), choose one of the two odd vertices as the starting point.
3. Begin to trace edges as you move through the graph. Number the edges as you trace them. Since you can't trace any edges twice in Euler paths and Euler circuits, once an edge is traced consider it "invisible."
4. When faced with a choice of edges to trace, if possible, choose an edge that is not a bridge (i.e., don't create a disconnected graph with your choice of edges).
5. Continue until each edge of the entire graph has been traced once.

NUMBER OF UNIQUE HAMILTON CIRCUITS IN A COMPLETE GRAPH

The number of unique Hamilton circuits in a complete graph with n vertices is $(n-1)!$, where

$$(n-1)! = (n-1)(n-2)(n-3)\cdots(3)(2)(1)$$

THE BRUTE FORCE METHOD OF SOLVING TRAVELING SALESMAN PROBLEMS

To determine the optimal solution:

1. Represent the problem with a complete, weighted graph.
2. List all possible Hamilton circuits in this graph.
3. Determine the cost associated with each of these Hamilton circuits.
4. The Hamilton circuit with the lowest cost is the optimal solution.

THE NEAREST NEIGHBOR METHOD FOR DETERMINING AN APPROXIMATE SOLUTION TO A TRAVELING SALESMAN PROBLEM

To approximate the optimal solution:

1. Represent the problem with a complete, weighted graph.
2. Identify the starting vertex.
3. Of all the edges attached to the starting vertex, choose the edge that has the smallest weight. This edge is the shortest distance or the lowest cost. Travel along this edge to the second vertex.
4. At the second vertex, choose the edge that has the smallest weight that does not lead to a vertex already visited. Travel along this edge to the third vertex.
5. Continue this process, each time moving along the edge with the smallest weight until all vertices are visited.
6. Travel back to the original vertex.

KRUSKAL'S ALGORITHM

To construct the minimum-cost spanning tree from a weighted graph:

1. Select the lowest-cost edge on the graph.
2. Select the next lowest-cost edge that does not form a circuit with the first edge.
3. Select the next lowest-cost edge that does not form a circuit with previously selected edges.
4. Continue selecting the lowest-cost edges that do not form circuits with the previously selected edges.
5. When a spanning tree is complete, you have the minimum-cost spanning tree.

CHAPTER 14 REVIEW EXERCISES

14.1

In Exercises 1 and 2, create a graph with the given properties.

1. Create a graph with two even vertices, two odd vertices, and a bridge.
2. Create a graph with four even vertices and two loops.

In Exercises 3 and 4, use the following graph.

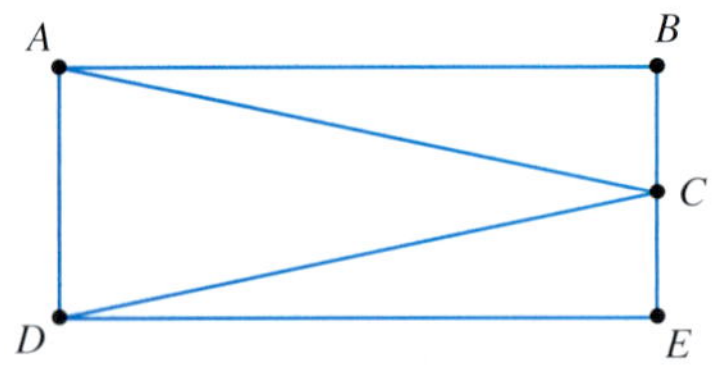

3. Determine a path that contains each edge exactly one time.
4. Is it possible to have a path that begins at vertex A, includes all edges exactly once, and ends at vertex B? Explain.
5. Represent the map as a graph where each vertex represents a state and each edge represents a common border between the states.

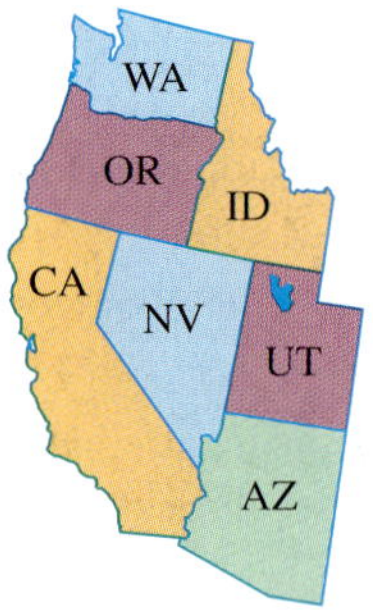

6. The drawing below shows the first-floor plan of Raintree Montessori School. Construct a graph to represent the school using the letters shown to label the vertices. Use the letter O to label the vertex that represents the outside of the school. Place vertex O near the bottom of the graph.

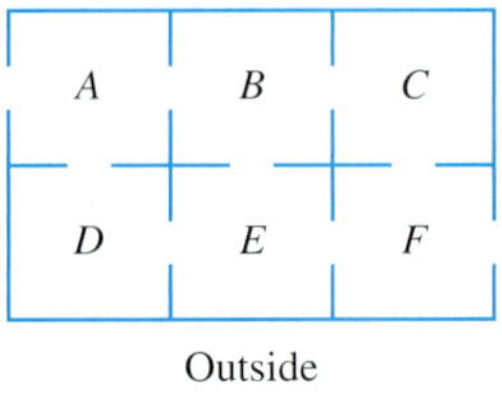

In Exercises 7 and 8, determine whether the graph shown is connected or disconnected.

7.

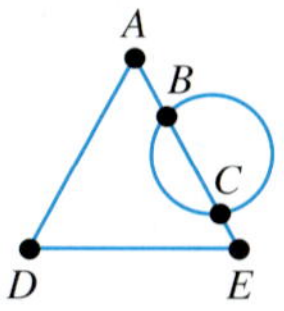

8.

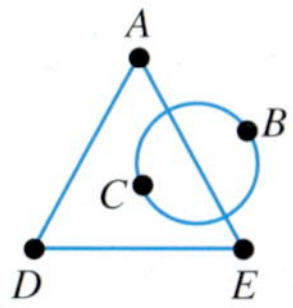

9. Identify all bridges in the graph below.

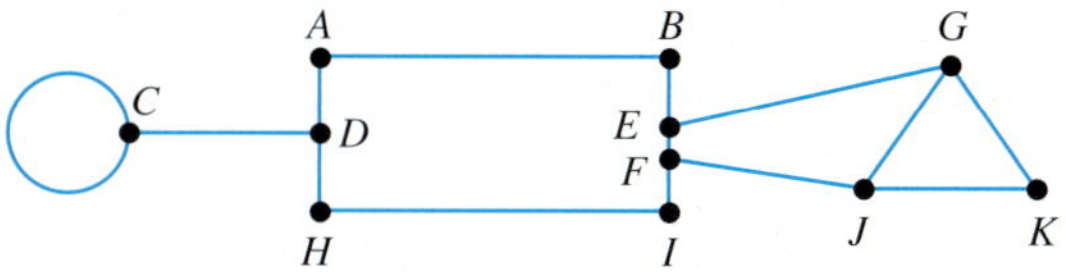

14.2

Use the following graph for Exercises 10 and 11.

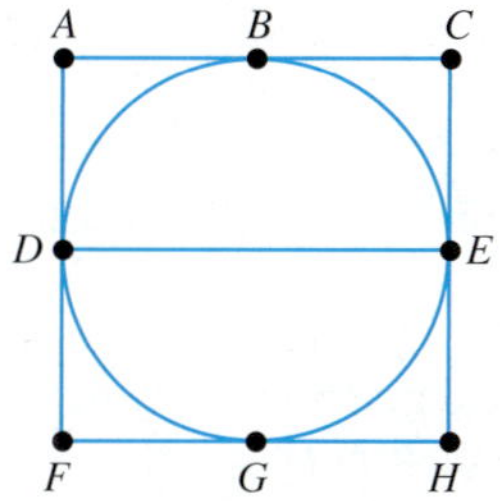

10. Determine an Euler path that begins with vertex *D*.

11. Determine an Euler path that begins with vertex *E*.

Use the following graph for Exercises 12 and 13.

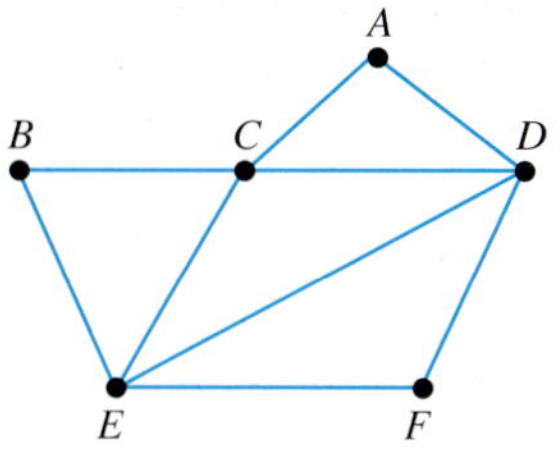

12. Determine an Euler circuit that begins with vertex *B*.

13. Determine an Euler circuit that begins with vertex *E*.

14. Consider the following map.

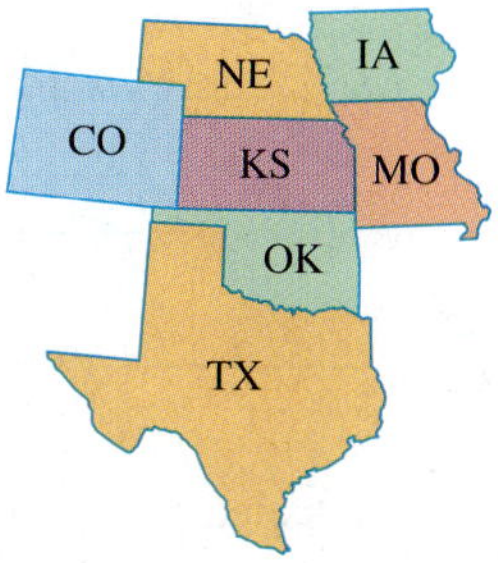

a) Represent the map as a graph.

b) Determine (state yes or no) whether the graph has an Euler path. If yes, give one such Euler path.

c) Determine (state yes or no) whether the graph has an Euler circuit. If yes, give one such Euler circuit.

15. a) Construct a graph that represents the floor plan below.

b) Is it possible for a person to walk through each doorway in the house whose floor plan is shown below without using any of the doorways twice? Explain.

c) If so, where can the person start and where will the person finish? Explain.

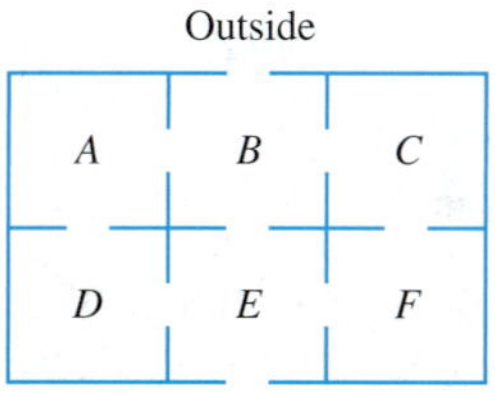

16. a) Can a police officer walk each street shown in the figure below without walking any street more than once? Explain.

b) If yes, where would the police officer have to start the walk?

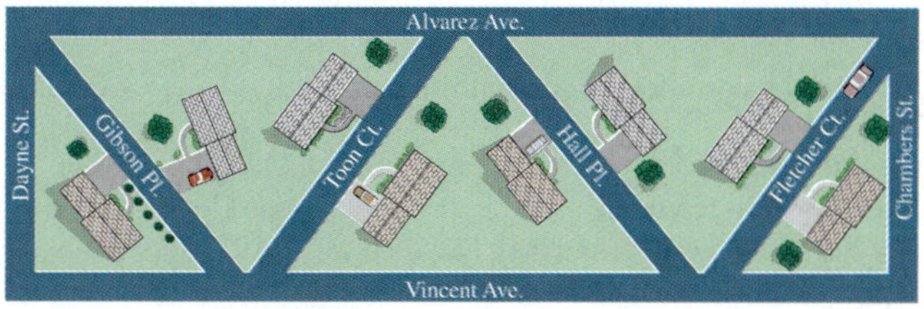

17. Use Fleury's algorithm to determine an Euler path in the following graph.

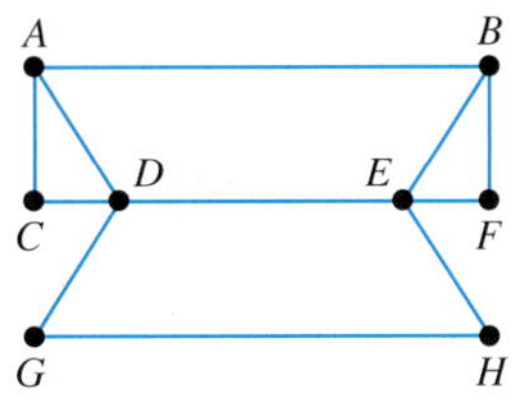

18. Use Fleury's algorithm to determine an Euler circuit in the following graph.

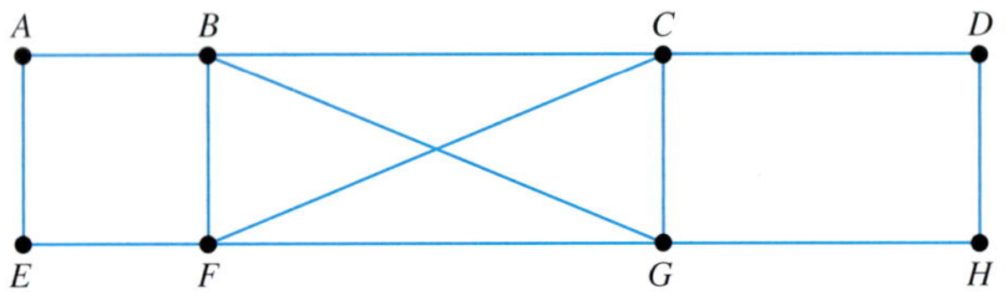

14.3

19. Determine two different Hamilton paths in the following graph.

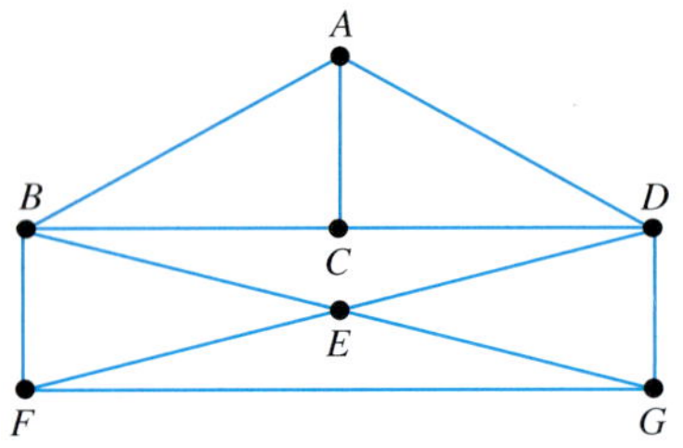

20. Determine two different Hamilton circuits in the following graph.

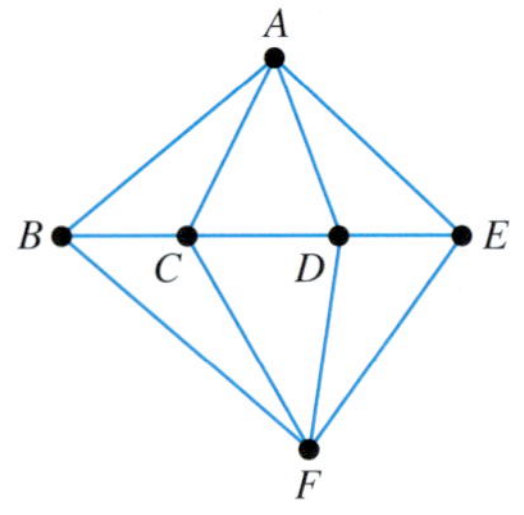

21. Draw a complete graph with five vertices.

22. ***Baseball*** The American League Eastern Division has the following teams: Baltimore Orioles, Boston Red Sox, New York Yankees, Tampa Bay Devil Rays, and Toronto Blue Jays. How many different possible ways could a Boston Red Sox fan visit each team in the American League Eastern Division one time and return home to Boston?

23. ***Job Interviews*** Lance Lopez is searching for a new job. He lives in Portland, Oregon, and has interviews in Des Moines, Iowa; Charleston, West Virginia; and Montgomery, Alabama. The costs of the one-way flights between these four cities are as follows: Portland to Des Moines is \$428, Portland to Charleston is \$787, Portland to Montgomery is \$902, Des Moines to Charleston is \$449, Des Moines to Montgomery is \$458, and Charleston to Montgomery is \$415.

a) Represent this traveling salesman problem with a complete, weighted graph showing the prices of the flights on the appropriate edges.

b) Use the brute force method to determine the least expensive route for Lance Lopez to travel to each city and return home to Portland.

c) What is the total cost of air fare for Lance's trip in part (b)?

24. ***Visiting Sales Offices*** Jennifer Adams is the sales manager for AT&T for the state of Missouri. There are major sales offices in Columbia, Kansas City, St. Joseph, St. Louis, and Springfield. The distances in miles between these cities are given in the following table.

	Columbia	Kansas City	St. Joseph	St. Louis	Springfield
Columbia	*	130	177	127	168
Kansas City	130	*	54	256	192
St. Joseph	177	54	*	304	224
St. Louis	127	256	304	*	210
Springfield	168	192	224	210	*

a) Represent this traveling salesman problem with a complete, weighted graph showing the prices of the flights on the appropriate edges.

b) Use the nearest neighbor method to approximate the optimal route for Jennifer to begin in St. Joseph, visit each city once, and return to St. Joseph.

c) Suppose that Jennifer starts in Springfield. Use the nearest neighbor method to approximate the optimal route for Jennifer to visit each city once and return to Springfield.

14.4

25. ***Employee Relationships*** Use a tree to show the employees' relationships in the Hulka Consulting Corporation. Hulka is president. Winger and Ziskey are vice presidents who work for Hulka. Markowicz, Oxburger, and Soyer are district managers who work for Winger. Elmo, Hector, and Jenesky are district managers who work for Ziskey.

26. Determine two different spanning trees for the following graph.

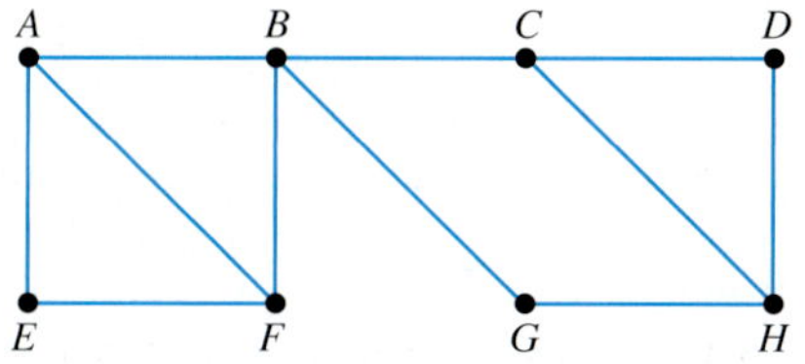

27. Determine the minimum-cost spanning tree for the following graph.

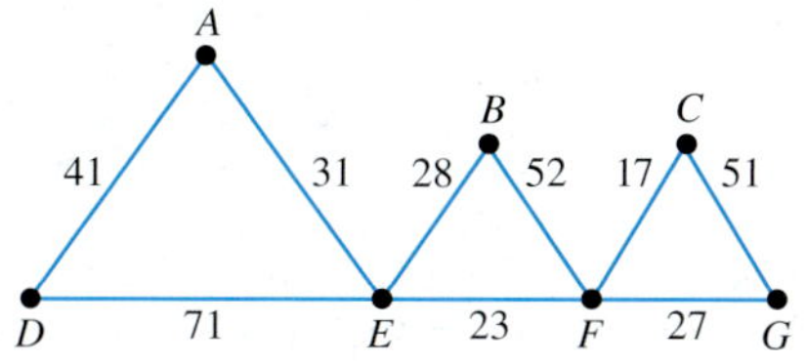

28. ***An Irrigation System*** Lucille Groenke wants to install an irrigation system to water all six of her flowerbeds in her back yard. The drawing shows the location of each flowerbed and the distance between them in feet.

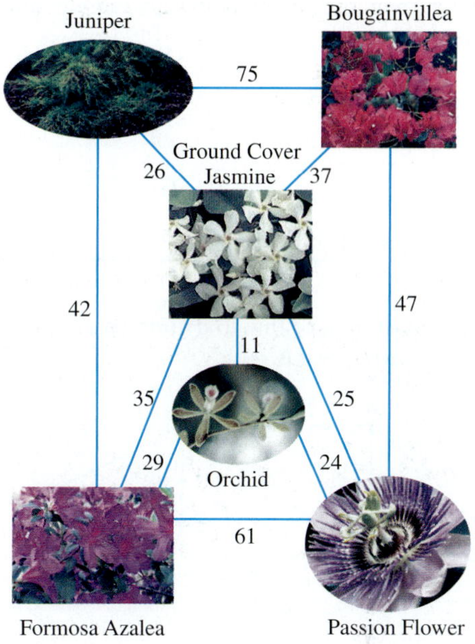

a) Represent this information with a weighted graph.

b) Use Kruskal's algorithm to determine the minimum-cost spanning tree.

c) The irrigation materials cost $2.50 per foot. Determine the cost of installing the irrigation system found in part (b).

CHAPTER 14 TEST

1. Create a graph with six vertices, a bridge, and two loops.

2. ***Part of Africa*** Represent the map below as a graph where each vertex represents a country and each edge represents a common border between the countries.

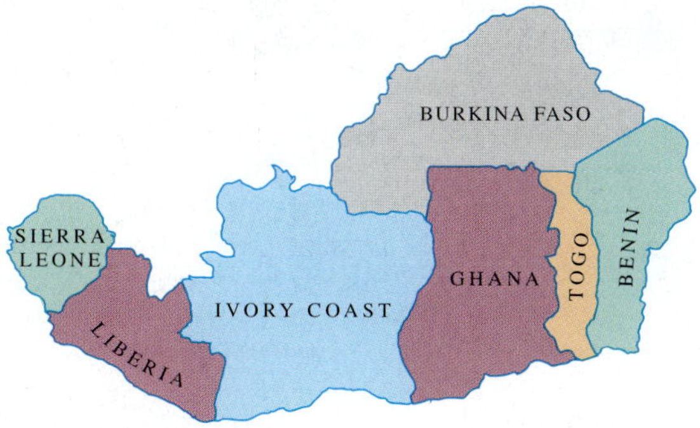

3. Draw a disconnected graph.

4. In the following graph, determine an Euler circuit.

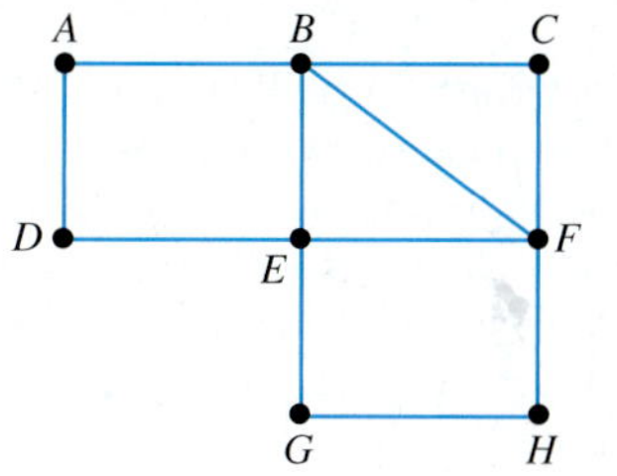

5. Is it possible for a person to walk through each doorway in the house, whose floor plan is on page 972, without using any of the doorways twice? If so, indicate in which room the person may start and where he or she will end.

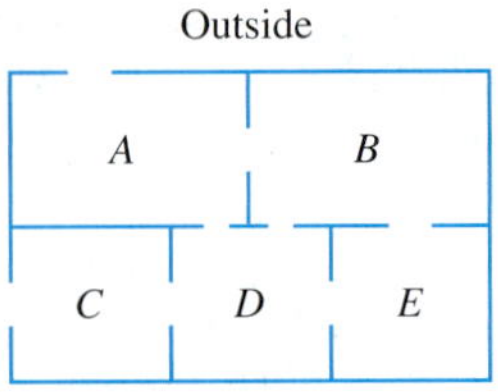

6. Use Fleury's algorithm to determine an Euler circuit in the following graph.

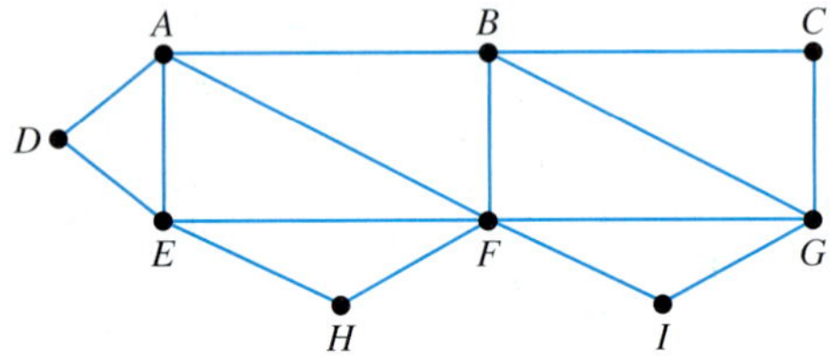

7. In the following graph, determine a Hamilton circuit.

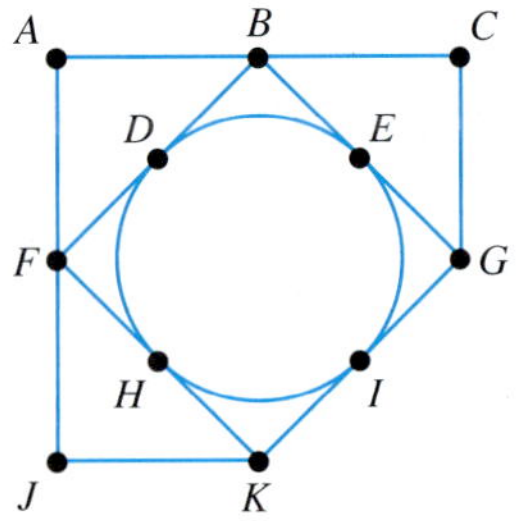

8. ***Conference Sites*** Rachel Goodwin, president of the English Teachers Association (ETA), is trying to determine the site of the next ETA conference. Rachel lives in Concord, Ohio, and wishes to travel to the following cities: Washington, D.C.; Las Vegas, Nevada; Boston, Massachusetts; and Austin, Texas. How many different ways can Rachel visit each city and return to her home in Concord?

9. ***Job Interviews*** Kate Kozak lives in Indianapolis, Indiana, and has job interviews in Pensacola, Florida; El Paso, Texas; and Albuquerque, New Mexico. The costs of one-way flights between these four cities are as follows: Indianapolis to Pensacola is $449, Indianapolis to El Paso is $201, Indianapolis to Albuquerque is $203, Pensacola to El Paso is $728, Pensacola to Albuquerque is $677, and El Paso to Albuquerque is $49.

   a) Represent this traveling salesman problem with a complete, weighted graph showing the prices of flights on the appropriate edges.

   b) Use the brute force method to determine the least expensive route for Kate to visit each city once and return home to Indianapolis. What is the cost when using this route?

   c) Use the nearest neighbor method to approximate the optimal route for Kate to visit each city once and return home to Indianapolis. What is the cost when using this route?

10. Determine a spanning tree for the graph shown below.

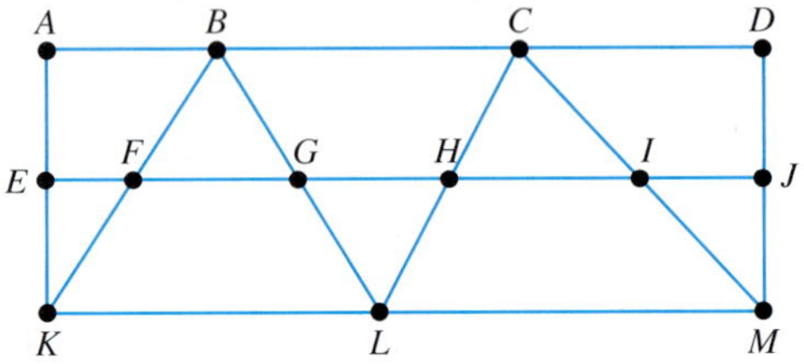

11. Determine the minimum-cost spanning tree for the following weighted graph.

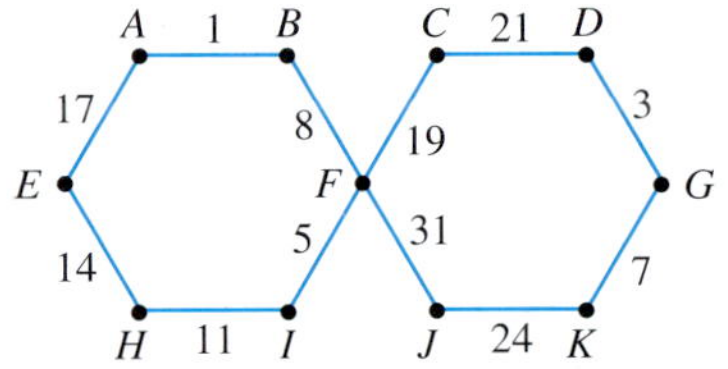

12. ***Irrigation System*** Daniel Kimborowicz is planning a new irrigation system for his yard. His current system has valves already in place as shown in the figure below. The numbers shown are in feet.

   a) Determine the minimum-cost spanning tree that reaches each valve.

   b) If the new irrigation materials cost $1.25 per foot, what is the cost of installing the system determined in part (a)?

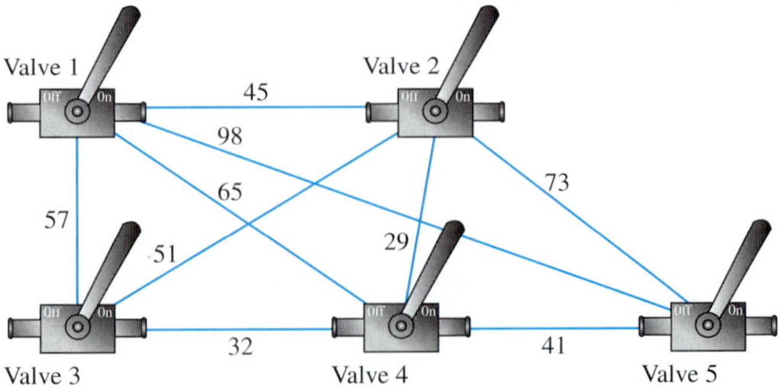

GROUP PROJECTS

1. STREET SWEEPER ROUTE

Make a map of all streets or roads within 2 miles of your college campus. Now add roads as needed so that a street sweeper would be able to sweep all streets on this map, starting at the school and returning to the school without sweeping any street twice.

2. HOMETOWN ROAD TRIP

Select five students in your class so that no two have the same hometown. Use the Internet to find the distances between each pair of towns (MapQuest is a good source to check).

a) Draw a complete, weighted graph that represents the five towns and the distances between them.

b) Use the nearest neighbor method to determine an approximation for the optimal solution when starting at your college, visiting each hometown, and then returning to your college.

c) Use Kruskal's algorithm to determine the minimum-cost spanning tree for the complete, weighted graph in part (a).

d) Determine the total distance one would travel along the minimum-cost spanning tree found in part (c).

3. STATE CAPITAL TOUR

Identify the five state capitals closest to where you live. Use the Internet or consult an almanac to find the driving distances between these five capitals.

a) Draw a complete, weighted graph that represents the capitals and the distances between them.

b) Use the nearest neighbor method to determine an approximation for the optimal solution of starting at your state capital, visiting the other four capitals, and then returning to your state capital.

c) Use Kruskal's algorithm to determine the minimum-cost spanning tree for the complete, weighted graph determined in part (a).

d) Determine the total distance a person would travel along the minimum-cost spanning tree found in part (c).

4. COMPUTER NETWORK ON CAMPUS

Measure the distance between the five busiest buildings on your campus. Create a weighted graph showing the buildings and the distances between buildings. Determine the minimum-cost spanning tree that could be used to install a new computer/telephone network on your campus. Assume that the cost per foot is the same regardless of the path used.

Before we discuss any of these voting methods, though, we will illustrate how the results of an election might be indicated in a table called a ***preference table***. A preference table summarizes the results of an election and shows how often each particular outcome was selected, as illustrated in Example 1.

▲ A husky

EXAMPLE 1 *Balloting by Committee Members*

The planning committee of Erie, Pennsylvania, is voting on a mascot for an upcoming winter festival. The choices are a snowman (S), a polar bear (P), and a husky (H). There are 15 planning committee members (labeled CM1, CM2, CM3, . . . , CM15). Their ballots are shown in Fig. 15.1.

CM1 1ST S 2ND P 3RD H	CM2 1ST S 2ND H 3RD P	CM3 1ST S 2ND P 3RD H	CM4 1ST H 2ND S 3RD P	CM5 1ST P 2ND H 3RD S	CM6 1ST P 2ND S 3RD H	CM7 1ST P 2ND S 3RD H	CM8 1ST S 2ND H 3RD P
CM9 1ST P 2ND S 3RD H	CM10 1ST H 2ND S 3RD P	CM11 1ST S 2ND H 3RD P	CM12 1ST H 2ND S 3RD P	CM13 1ST H 2ND S 3RD P	CM14 1ST H 2ND S 3RD P	CM15 1ST H 2ND P 3RD S	

Figure 15.1

Construct a preference table to illustrate the results of the election.

SOLUTION Notice that committee member 1 (CM1) listed S, P, H as his or her choice. Also, notice that committee member 3 (CM3) also listed S, P, H, in that order. None of the other committee members listed this order of selection. Therefore, there are two committee members who listed S, P, H in that order. This information is indicated in boldface in column one of Table 15.2. Next look at CM2. He or she listed S, H, P. Two other committee members, CM8 and CM11, listed S, H, P in that order. Therefore, three committee members list the order S, H, P. This information is indicated in column two of Table 15.2. Continue this process to obtain the preference table.

Table 15.2 Winter Festival Mascot Preference Table

Number of Votes	2	3	5	1	3	1
First	**S**	S	H	P	P	H
Second	**P**	H	S	H	S	P
Third	**H**	P	P	S	H	S

Notice that the total number of votes indicated in the preference table is 15, which checks with the individual number of votes given. ●

When an election has many voters, a preference table is an efficient way to report the results.

The initial purpose of Example 1 was to introduce the preference table. We will revisit the winter festival mascot election several times in this section. Example 2 introduces another election that will be used frequently in this section.

EXAMPLE 2 *Voting for Performing Arts Club President*

Four students are running for president of the Performing Arts Club: Antoine (A), Betty (B), Camille (C), and Don (D). The club members were asked to rank all candidates. The resulting preference table for this election is given in Table 15.3.

Table 15.3 Performing Arts Club Preference Table

Number of Votes	17	13	10	6	2
First	B	C	D	A	C
Second	A	A	C	D	D
Third	C	D	A	C	A
Fourth	D	B	B	B	B

a) How many students voted in the election?

b) How many students selected the candidates in this order: C, A, D, B?

c) How many students selected C as their first choice?

SOLUTION

a) To find the number of students who voted, we add the numbers in the row labeled Number of Votes.

$$17 + 13 + 10 + 6 + 2 = 48$$

Therefore, 48 students voted in the election.

b) Thirteen students voted in the given order (the second column of letters).

c) To determine the number who voted for C as their first choice, read across the row that says First. When you see a C in the row, record the number above it. Then find the sum of the numbers. Thus,

$$13 + 2 = 15 \text{ students selected C as their first choice.}$$

Now we will begin to examine the results of the elections in Examples 1 and 2. Each election will be analyzed using each of the four different voting methods discussed in this section: plurality method, Borda count method, plurality with elimination method, and pairwise comparison method.

Plurality Method

When electing a candidate using the ***plurality method***, each voter votes for only one candidate. The candidate receiving the most votes is declared the winner. With the plurality method, a preference table does not have to be used since the candidates are not ranked. In this section, we will often use a preference table since the data may often be given in that form from a previous example. The plurality method is the most commonly used method, and it is the easiest method to use when there are more than two candidates.

PLURALITY METHOD
Each voter votes for one candidate. The candidate receiving the most votes is declared the winner.

> **DID YOU KNOW?**
>
> ***Plurality Presidents***
>
> 
>
> Harry Truman
>
> Although it may seem strange, the United States has often elected presidents whom a majority of the voters opposed. When viable third-party candidates are involved in elections, winning candidates may gain a majority of electoral votes, but not a majority of popular votes. In the twentieth century alone, Woodrow Wilson in 1912 and in 1916, Harry Truman in 1948, Richard Nixon in 1968, and Bill Clinton in 1992 and in 1996 all won the election with less than 50% of the popular vote. Although each of these presidents failed to get a majority of popular votes, each did obtain a plurality of popular votes. An even stranger scenario occurred in 1888 when Grover Cleveland received 5,540,365 votes and Benjamin Harrison received 5,445,269 votes. Yet, since Harrison had a majority of electoral votes, he was elected president. Once again, this scenario occurred in 2000 when Al Gore received 49,921,267 votes and George W. Bush received 49,658,276 votes. However, since Bush received a majority of electoral votes (not without some controversy), he was elected president.

If a candidate has a majority of votes, he or she will win an election using the plurality method. However, it is possible for a candidate to win an election using the plurality method without the candidate having a majority of votes, as we will see in the next example.

EXAMPLE 3 *Selecting a Mascot Using the Plurality Method*

Consider the winter festival mascot election given in Example 1. Which mascot is selected using the plurality method?

SOLUTION Using the plurality method, the candidate with the most votes is selected as mascot. In Example 1, the planning committee members ranked the candidates. We will assume that each planning committee member would vote for his or her first choice as listed in Table 15.2 on page 976. From Table 15.2, we see that the snowman received $2 + 3 = 5$ votes, the polar bear received $1 + 3 = 4$ votes, and the husky received $5 + 1 = 6$ votes. These results indicate that by using the plurality method, the husky is chosen as the mascot for the winter festival. Note that the husky received only $\frac{6}{15}$, or 40%, of the first-place votes, which is less than a majority.

EXAMPLE 4 *Electing the Performing Arts Club President by the Plurality Method*

Consider the Performing Arts Club election given in Example 2. Who is elected president using the plurality method?

SOLUTION We will assume that each member would vote for the person he or she listed in first place in Table 15.3 on page 977. From Table 15.3, we can see that Antoine received 6 votes, Betty received 17 votes, Camille received $13 + 2 = 15$ votes, and Don received 10 votes. Therefore, using the plurality method, Betty received the most votes and is elected president of the Performing Arts Club. Note that Betty received only $\frac{17}{48}$, or about 35%, of the first-place votes, which is less than a majority.

As simple as the plurality method is, this voting method ignores voters' second and subsequent choices. The rest of the voting methods we will discuss take all voters' choices into consideration.

Borda Count Method

The *Borda count* method, developed by Jean-Charles de Borda, requires that voters rank candidates in order from most favorable to least favorable. The Borda count method then quantifies this ranking to determine a winner.

> **BORDA COUNT METHOD**
>
> Voters rank the candidates from the most favorable to the least favorable. Each last-place vote is awarded one point, each next-to-last-place vote is awarded two points, each third-from-last-place vote is awarded three points, and so forth. The candidate receiving the most points is the winner of the election.

PROFILE IN MATHEMATICS

Jean-Charles de Borda

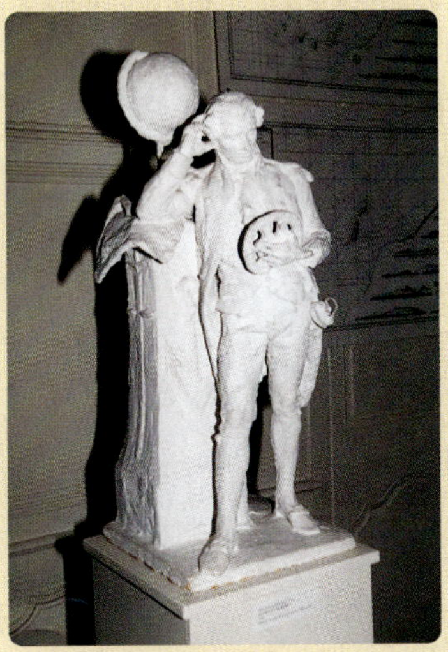

Jean-Charles de Borda (1733–1799), a French calvary officer and naval captain for whom the Borda count method is named, wrote on a variety of subjects including mathematics, physics, design of scientific experiments, and voting theory.

The primary advantage of the Borda count method is that voters are able to provide more information than in the plurality method. Many well-known ranking systems use variations of the Borda count method. The Heisman trophy, given annually to the best college football player, is awarded by having the voters (sportswriters and former Heisman trophy winners) rank their first, second, and third choices out of possibly thousands of college football players. A player receives 3 points for each first-place vote, 2 points for each second-place vote, and 1 point for each third-place vote.

EXAMPLE 5 *Selecting the Winter Festival Mascot Using the Borda Count Method*

Let's reexamine the Erie, Pennsylvania, election for a winter festival mascot. For convenience, the preference table is reproduced in Table 15.4. Recall that S represents snowman, P represents polar bear, and H represents husky. Which mascot is selected using the Borda count method?

Table 15.4 Winter Festival Mascot Preference Table

Number of Votes	2	3	5	1	3	1
First	S	S	H	P	P	H
Second	P	H	S	H	S	P
Third	H	P	P	S	H	S

SOLUTION In this problem, there are three candidates, so a first-place vote is worth 3 points, a second-place vote is worth 2 points, and a third-place vote is worth 1 point. To award points, we examine each candidate starting with the snowman. From Table 15.4, we see that the **snowman receives**

$2 + 3 = \mathbf{5}$ first-place votes producing $\mathbf{5} \times 3 = 15$ points

$5 + 3 = \mathbf{8}$ second-place votes producing $\mathbf{8} \times 2 = 16$ points, and

$1 + 1 = \mathbf{2}$ third-place votes producing $\mathbf{2} \times 1 = 2$ points

Therefore, the snowman receives $15 + 16 + 2 = 33$ points.

Next we see that the **husky receives**

$5 + 1 = \mathbf{6}$ first-place votes producing $\mathbf{6} \times 3 = 18$ points

$3 + 1 = \mathbf{4}$ second-place votes producing $\mathbf{4} \times 2 = 8$ points, and

$2 + 3 = \mathbf{5}$ third-place votes producing $\mathbf{5} \times 1 = 5$ points

Therefore, the husky receives $18 + 8 + 5 = 31$ points.

Finally, we see that the **polar bear receives**

$1 + 3 = \mathbf{4}$ first-place votes producing $\mathbf{4} \times 3 = 12$ points

$2 + 1 = \mathbf{3}$ second-place votes producing $\mathbf{3} \times 2 = 6$ points, and

$3 + 5 = \mathbf{8}$ third-place votes producing $\mathbf{8} \times 1 = 8$ points

Therefore, the polar bear receives $12 + 6 + 8 = 26$ points.

The snowman receives 33 points, the husky receives 31 points, and the polar bear receives 26 points. By using the Borda count method, the snowman is chosen as the winter festival mascot.

We will now reexamine the election for president of the Performing Arts Club by using the Borda count method.

EXAMPLE 6 *Electing the Performing Arts Club President Using the Borda Count Method*

Use the Borda count method to determine the winner of the election for president of the Performing Arts Club discussed in Example 2. Recall that the candidates are Antoine (A), Betty (B), Camille (C), and Don (D). For convenience, the preference table is reproduced below in Table 15.5.

Table 15.5 Performing Arts Club Preference Table

Number of Votes	17	13	10	6	2
First	B	C	D	A	C
Second	A	A	C	D	D
Third	C	D	A	C	A
Fourth	D	B	B	B	B

SOLUTION In this problem, there are four candidates, so a first-place vote is worth 4 points, a second-place vote is worth 3 points, a third-place vote is worth 2 points, and a fourth-place vote is worth 1 point.

Using Table 15.5, we see that **Antoine receives**

6 first-place votes producing **6** $\times$ 4 = 24 points

17 + 13 = **30** second-place votes producing **30** $\times$ 3 = 90 points, and

10 + 2 = **12** third-place votes producing **12** $\times$ 2 = 24 points

Antoine has no fourth-place votes. Therefore, Antoine receives 24 + 90 + 24 = 138 points.

Betty receives

17 first-place votes producing **17** $\times$ 4 = 68 points, and

13 + 10 + 6 + 2 = **31** fourth-place votes producing **31** $\times$ 1 = 31 points

Betty has no second- or third-place votes. Therefore, Betty receives 68 + 31 = 99 points.

Camille receives

13 + 2 = **15** first-place votes producing **15** $\times$ 4 = 60 points

10 second-place votes producing **10** $\times$ 3 = 30 points, and

17 + 6 = **23** third-place votes producing **23** $\times$ 2 = 46 points

Camille has no fourth-place votes. Therefore, Camille receives 60 + 30 + 46 = 136 points.

Don receives

10 first-place votes producing **10** $\times$ 4 = 40 points

6 + 2 = **8** second-place votes producing **8** $\times$ 3 = 24 points

13 third-place votes producing **13** $\times$ 2 = 26 points, and

17 fourth-place votes producing **17** $\times$ 1 = 17 points

Therefore, Don receives 40 + 24 + 26 + 17 = 107 points.

Antoine, with 138 points, receives the most points using the Borda count method and is declared the winner.

Even though Antoine received the fewest number of first-place votes, he is declared the winner under the Borda count method.

Notice that with both the winter festival mascot election (Examples 3 and 5) and the Performing Arts Club president election (Examples 4 and 6), the plurality method and the Borda count method produced different winners. This result indicates the importance of choosing a voting method *prior to the election*. This will become even more evident when we examine the remaining voting methods. Next we will examine the plurality with elimination method, which also takes voters' second and subsequent preferences into account.

Plurality with Elimination

The next method we will discuss, ***plurality with elimination***, may involve a series of elections. When using the plurality with elimination method each voter votes for only one candidate. If a candidate initially receives a majority of votes, that candidate is declared the winner. If no candidate receives a majority of votes, the candidate with the fewest number of votes is eliminated and a second election is held. This process continues until a candidate receives a majority of votes.

PLURALITY WITH ELIMINATION METHOD

Each voter votes for one candidate. If a candidate receives a majority of votes, that candidate is declared the winner. If no candidate receives a majority, eliminate the candidate with the fewest votes and hold another election. (If there is a tie for the fewest votes, eliminate all candidates tied for the fewest votes.) Repeat this process until a candidate receives a majority.

One of the immediate disadvantages of the plurality with elimination method is the possible need for two or more elections, depending on the number of candidates. One way to limit the number of elections held is to begin with a ***runoff election*** in which voters choose among a field of candidates. In the second election, the voters only select among the top two candidates from the runoff election. Another way to avoid multiple elections entirely is to have voters rank candidates from most to least favorable just as we did with the Borda count method. We assume that in the first round, the voters would have selected their first choice. We must also assume that after a candidate is eliminated, the order of preference is not changed. For instance, let's say that a voter's original ranking order was E, G, B, D, F. Suppose that B is eliminated. This voter's ranking order now becomes E, G, D, F. The next two examples will demonstrate the elimination process when used with the plurality with elimination method.

EXAMPLE 7 ***Selecting the Winter Festival Mascot Using the Plurality with Elimination Method***

Once again, we will consider the winter festival mascot election in Erie, Pennsylvania. Recall that S represents snowman, P represents polar bear, and H represents husky. Which mascot is selected using the plurality with elimination method?

SOLUTION Reexamine the preference table shown in Table 15.6.

Table 15.6 Winter Festival Mascot Preference Table

Number of Votes	2	3	5	1	3	1
First	S	S	H	P	P	H
Second	P	H	S	H	S	P
Third	H	P	P	S	H	S

We first count the number of first-place votes for each mascot:

Snowman: $2 + 3 = 5$ Husky: $5 + 1 = 6$ Polar bear: $1 + 3 = 4$

To receive a majority, a candidate must have more than half of the votes. Since there are 15 people voting, 8 or more votes are needed to receive a majority. In this election, no candidate received a majority. Since the polar bear received the fewest number of first-place votes, it is eliminated in the first round. Next, we assume that all preference orders remain the same. For example, the three committee members who had their preference order of P, S, and H will now have a preference order of S and H. Table 15.7 shows the new preference table after the polar bear is eliminated.

Table 15.7 Winter Festival Mascot Preference Table After the Polar Bear Is Eliminated

Number of Votes	2	3	5	1	3	1
First	S	S	H	H	S	H
Second	H	H	S	S	H	S

We see now that the snowman has $2 + 3 + 3 = 8$ first-place votes and the husky has $5 + 1 + 1 = 7$ first-place votes. The snowman has eight of 15 first-place votes, a majority. Using the plurality with elimination method, the snowman is chosen as the winter festival mascot.

EXAMPLE 8 *Electing the Performing Arts Club President Using the Plurality with Elimination Method*

Use the plurality with elimination method to determine the winner of the election for president of the Performing Arts Club from Example 2. The preference table is shown again in Table 15.8. Recall that A represents Antoine, B represents Betty, C represents Camille, and D represents Don.

Table 15.8 Performing Arts Club Preference Table

Number of Votes	17	13	10	6	2
First	B	C	D	A	C
Second	A	A	C	D	D
Third	C	D	A	C	A
Fourth	D	B	B	B	B

SOLUTION First we count the number of first-place votes for each candidate:

Antoine: 6 Betty: 17 Camille: 15 Don: 10

Since 48 votes were cast, a candidate must have at least 25 first-place votes to receive a majority. In this election, no candidate received a majority. Antoine has the fewest number of first-place votes, so he is eliminated. From column one of Table 15.8 on page 982, we see that 17 voters ranked their preference as B, A, C, D. With A eliminated, we assume that those same 17 voters would now rank their preference as B, C, D. From column two in Table 15.8, 13 voters ranked their preference as C, A, D, B. With A eliminated, we assume that those same 13 voters would now rank their preference as C, D, B. We eliminate A from the other columns of the table in a similar manner.

The new preference table with A eliminated is given in Table 15.9.

Table 15.9 Performing Arts Club Preference Table After Antoine Is Eliminated

Number of Votes	17	13	10	6	2
First	B	C	D	D	C
Second	C	D	C	C	D
Third	D	B	B	B	B

The number of first-place votes is now

Betty: 17 Camille: 15 Don: 16

Still, no candidate received a majority. Camille now has the fewest number of first-place votes, so she is eliminated. With Camille eliminated, the new preference table is given in Table 15.10.

Table 15.10 Performing Arts Club Preference Table After Camille Is Eliminated

Number of Votes	17	13	10	6	2
First	B	D	D	D	D
Second	D	B	B	B	B

The number of first-place votes is now

Betty: 17 Don: 31

Don has a majority of first-place votes and is declared the winner using the plurality with elimination method. ●

The Performing Arts club elections in the previous examples are quite interesting. When the plurality method was used, Betty was declared the winner. When the Borda count method was used, Antoine was declared the winner. When the plurality with elimination method was used, Don was declared the winner. This dilemma, once again, underscores the importance of publicly stating the voting method to be used *before* the election takes place.

The selection of the city to host the Olympic Games occurs by using a type of plurality with elimination. The election process for the Academy Awards involves several stages. One stage involves a variation of the plurality with elimination method in which the motion picture nominee with the fewest number of first-place votes is eliminated.

Now we will discuss our next voting method, the pairwise comparison method.

Pairwise Comparison Method

To use the pairwise comparison method, each voter first ranks all the candidates. Then each candidate is compared with each of the other candidates using the rankings. It is treated like a round-robin tournament. For instance, if the candidates for an election were Bert, Ernie, and Louis, the following comparisons would be made: Bert versus Ernie, Bert versus Louis, and Ernie versus Louis. The winner of each individual comparison is awarded one point. If there is a tie, each candidate receives $\frac{1}{2}$ point. The candidate who obtains the most points in the one-to-one comparisons is declared the winner.

PAIRWISE COMPARISON METHOD

Voters rank the candidates. A series of comparisons in which each candidate is compared with each of the other candidates follows. If candidate A is preferred to candidate B, A receives 1 point. If candidate B is preferred to candidate A, B receives 1 point. If the candidates tie, each receives $\frac{1}{2}$ point. After making all comparisons among the candidates, the candidate receiving the most points is declared the winner.

EXAMPLE 9 *Selecting the Winter Festival Mascot Using the Pairwise Comparison Method*

We will again examine the Erie, Pennsylvania, winter festival mascot election. Which mascot is selected if the pairwise comparison method is used?

SOLUTION The preference table is shown again in Table 15.11. Recall that S stands for snowman, P stands for polar bear, and H stands for husky.

Table 15.11 Winter Festival Mascot Preference Table

Number of Votes	**2**	**3**	**5**	**1**	**3**	**1**
First	S	S	H	P	P	H
Second	P	H	S	H	S	P
Third	H	P	P	S	H	S

To determine the winner, it is necessary to make comparisons between the snowman and the polar bear, between the snowman and the husky, and between the polar bear and the husky.

We will begin by comparing the snowman and the polar bear. We see from Table 15.11 that the 2 voters in the first column, the 3 voters in the second column, and the 5 voters in the third column all prefer the snowman to the polar bear. We also see that the 1 voter in the fourth column, the 3 voters in the fifth column, and the 1 voter in the sixth column all prefer the polar bear to the snowman.

The pairwise comparison of the **snowman versus the polar bear** is

Snowman: $2 + 3 + 5 = 10$ votes   Polar bear: $1 + 3 + 1 = 5$ votes

Thus, the snowman wins the pairwise comparison between the snowman and the polar bear and is awarded 1 point.

The pairwise comparison of the **snowman versus the husky** is

Snowman: $2 + 3 + 3 = 8$ votes   Husky: $5 + 1 + 1 = 7$ votes

The snowman wins the pairwise comparison between the snowman and the husky and is awarded another point.

The pairwise comparison of the **polar bear versus the husky** is

Polar bear: $2 + 1 + 3 = 6$ votes   Husky: $3 + 5 + 1 = 9$ votes

The husky wins the pairwise comparison between the polar bear and the husky and is awarded 1 point. Since the snowman received 2 points, the husky received 1 point, and the polar bear received 0 points, using the pairwise comparison method the snowman is declared the winter festival mascot.

TIMELY TIP When we compared the snowman and the polar bear in Example 9, we found the sum of the snowman's votes to be 10. Then we found the sum of the polar bear's votes to be 5. Notice that $10 + 5 = 15$, the number of votes in the preference table. If the snowman had 10 votes, the polar bear must have $15 - 10$ or 5 votes. When we use the pairwise comparison method, if you determine the number of votes of one member in the comparison, the number of votes of the second member can be found by subtracting the first number of votes found from the total number of votes.

Let's look at how we arrived at the comparisons. Each candidate was compared one-to-one with all the other candidates. In Example 9, we started with the snowman. Since the snowman had to be compared with each of the other candidates, we compared the snowman with the polar bear and then we compared the snowman with the husky. Next, we chose the polar bear. Since the polar bear had already been compared with the snowman, we only needed to compare the polar bear with the husky. The last candidate was the husky. Since the husky was already compared with each of the other candidates, we have made all the necessary comparisons. This process is used regardless of the number of candidates.

In Example 9, there were 3 candidates and we made 3 comparisons. The formula to determine the number of comparisons that are needed when n candidates are being considered follows.

THE NUMBER OF COMPARISONS NEEDED

The number of comparisons, c, needed when there are n candidates is

$$c = \frac{n(n-1)}{2}$$

For example, when there are 5 candidates, $n = 5$ and the number of comparisons is

$$c = \frac{n(n-1)}{2} = \frac{5(4)}{2} = \frac{20}{2} = 10$$

In this book, we will not work with more than 5 candidates.

EXAMPLE 10 *Electing the Performing Arts Club President Using the Pairwise Comparison Method*

Use the pairwise comparison method to determine the winner of the election for president of the Performing Arts Club that was originally discussed in Example 2 on page 977.

SOLUTION The preference table is shown again in Table 15.12. Recall that A represents Antoine, B represents Betty, C represents Camille, and D represents Don.

Table 15.12 Performing Arts Club Preference Table

Number of Votes	17	13	10	6	2
First	B	C	D	A	C
Second	A	A	C	D	D
Third	C	D	A	C	A
Fourth	D	B	B	B	B

Since we have 4 candidates, $n = 4$ and the number of comparisons needed is

$$c = \frac{n(n-1)}{2} = \frac{4(3)}{2} = 6$$

The comparisons we will make are A versus B, A versus C, A versus D, B versus C, B versus D, and C versus D.

The pairwise comparison of **Antoine versus Betty** is

Antoine: $13 + 10 + 6 + 2 = 31$ votes   Betty: 17 votes

Antoine wins this comparison and is awarded 1 point.

The pairwise comparison of **Antoine versus Camille** is

Antoine: $17 + 6 = 23$ votes   Camille: $13 + 10 + 2 = 25$ votes

Camille wins this comparison and is awarded 1 point.

The pairwise comparison of **Antoine versus Don** is

Antoine: $17 + 13 + 6 = 36$ votes   Don: $10 + 2 = 12$ votes

Antoine wins this comparison and is awarded a second point.

The pairwise comparison of **Betty versus Camille** is

Betty: 17 votes   Camille: $13 + 10 + 6 + 2 = 31$ votes

Camille wins this comparison and is awarded a second point.

The pairwise comparison of **Betty versus Don** is

Betty: 17 votes   Don: $13 + 10 + 6 + 2 = 31$ votes

Don wins this comparison and is awarded 1 point.

The pairwise comparison of **Camille versus Don** is

Camille: $17 + 13 + 2 = 32$ votes   Don: $10 + 6 = 16$ votes

Camille wins this comparison and is awarded a third point.

DID YOU KNOW?

Presidential Elections

Hayes lost the popular vote, but won the electoral college and the election.

Since the United States was founded in 1776, Americans and their elected representatives have had many disputes regarding election procedures. As you read through this chapter, you will see various voting methods that were used to try to make the voting procedures as fair as possible. You will also learn that there is no ideal voting method, that is, one that has no flaws. Although there may be flaws in the voting procedures, the American election procedure is still a model to democracies all around the world. One current debate is whether we should continue using our present electoral system or change to use the popular vote to decide presidential elections. As of this writing, four presidents were elected even though they lost the popular vote: John Quincy Adams (1824), Rutherford B. Hayes (1876), Benjamin Harrison (1888), and George W. Bush (2000).

We have made all possible one-to-one comparisons. Antoine received 2 points, Betty received 0 points, Camille received 3 points, and Don received 1 point. Since Camille received 3 points, the most points from the pairwise comparison method, Camille wins the election.

As we have illustrated, using four different voting methods resulted in four different winners for the president of the Performing Arts Club. Each candidate could claim that he or she should be declared the winner. Thus, the voting method we choose to use could greatly affect the election results and therefore should be chosen prior to an election.

Tie Breaking

Just as the four methods we have examined can produce four different winners, each method could also produce a tie between two or more candidates. In each of the previous examples, the voting method used led to one winner, but that is not always the case. With each method described—plurality method, Borda count method, plurality with elimination method, and pairwise comparison method—the possibility of the method producing a tie is very real. A tie could be achieved simply by having only two candidates each with the same number of supporters.

Breaking a tie can be achieved by either making an arbitrary choice, such as flipping a coin, or by bringing in an additional voter. For example, under Robert's Rules of Order, the president of any group votes only when there is a tie or to create a tie. There are other ways of breaking a tie that are less arbitrary than flipping a coin. If a tie results from using the Borda count method, it could be broken by choosing the candidate with the most first-place votes. If a tie results from the pairwise comparison method, it could be broken by choosing the winner of the one-to-one comparison between the candidates involved in the tie. Different tie-breaking methods could produce different winners. Therefore, to remain fair, the tie-breaking method should be decided upon in advance, as is done in the next example.

EXAMPLE 11 *Choosing a Commencement Speaker*

The University of North Carolina–Chapel Hill Student Government Association (SGA) is trying to decide whom to invite as the keynote speaker for commencement exercises. The choices are Michael Jordan (J), Oprah Winfrey (W), and Tom Brokaw (B). To make this difficult decision, the SGA decides to hold an election among graduating seniors. It also decides that the plurality method will be used to determine the winner and that the Borda count method will decide the winner in the event that the plurality method leads to a tie. The results of the election are represented in Table 15.13. Who will be asked to speak at the commencement ceremony?

Table 15.13 University of North Carolina–Chapel Hill Commencement Speaker Election

Number of Votes	400	250	150
First	J	B	B
Second	W	W	J
Third	B	J	W

SOLUTION Michael Jordan and Tom Brokaw each receive 400 first-place votes; therefore, the plurality method leads to a tie. Thus, the Borda count method is used to break the tie between Jordan and Brokaw. From Table 15.13 on page 987, we award the following points:

$$\text{Michael Jordan: } 400(3) + 150(2) + 250(1) = 1750$$
$$\text{Tom Brokaw: } 400(3) + 0(2) + 400(1) = 1600$$

Since the Borda count method breaks the tie, Michael Jordan is asked to present the commencement address. ●

Table 15.14 summarizes the voting methods we discussed in this section.

Table 15.14 Summary of Voting Methods

Voting Method	Description
Plurality method	Each voter votes for one candidate. The candidate receiving the most votes is declared the winner.
Borda count method	Voters rank candidates from the most favorable to the least favorable. Each last-place vote is awarded 1 point, each next-to-last-place vote is awarded 2 points, each third-from-last-place vote is awarded 3 points, and so forth. The candidate receiving the most points is the winner.
Plurality with elimination method	Each voter votes for one candidate. If a candidate receives a majority of first-place votes, that candidate is declared the winner. If no candidate receives a majority of first-place votes, eliminate the candidate with the fewest number of first-place votes and hold another election. (If there is a tie for the fewest number of first-place votes, eliminate all candidates tied for the fewest number of first-place votes.) Repeat this process until a candidate receives a majority of first-place votes.
Pairwise comparison method	Voters rank the candidates. A series of comparisons in which each candidate is compared with each of the other candidates follows. If candidate A is preferred to candidate B, A receives 1 point. If candidate B is preferred to candidate A, B receives 1 point. If the candidates tie, each receives $\frac{1}{2}$ point. The candidate receiving the most points is declared the winner.

SECTION 15.1 EXERCISES

CONCEPT/WRITING EXERCISES

1. What does it mean for a candidate to have a majority of votes?

2. Describe how the plurality method is used to determine the winner of an election.

3. Describe how the Borda count method is used to determine the winner of an election.

4. Describe how the plurality with elimination method is used to determine the winner of an election.

5. Describe how the pairwise comparison method is used to determine the winner of an election.

6. Why is it important for election officials to specify prior to an election which voting method is to be used?

7. What is a preference table?

8. a) Explain how to determine with the pairwise comparison method the number of comparisons, c, needed if there are n candidates.

   b) If there are four candidates, how many comparisons will there be with the pairwise comparison method?

   c) If there are six candidates, how many comparisons will there be with the pairwise comparison method?

9. Describe a benefit of the Borda count method over the plurality method.

10. Describe one way other than flipping a coin to settle a tied election.

PRACTICE THE SKILLS

11. ***Plurality Method*** Three candidates are running for mayor of Jersey City and receive the following votes:

Bryant	192,827	Swoopes	210,361
Jeter	265,128		

a) Determine the winner using the plurality method.

b) Did this candidate receive a majority of votes?

12. ***Plurality Method*** Five candidates are running for president of the Student Senate and receive the following votes:

Michael	2192	Gerry	2562	Karen	1671
Felicia	2863	Stephen	1959		

a) Determine the winner using the plurality method.

b) Did this candidate receive a majority of votes?

13. ***Preference Table for Hot Fudge Sauce*** Nine voters are asked to rank three brands of hot fudge sauce: Hershey's (H), Ghiradelli (G), and Smucker's (S). The nine voters turn in the following ballots showing their preferences in order:

H	G	H	S	S	S	H	G	S
G	H	G	H	H	G	G	S	G
S	S	S	G	G	H	S	H	H

Make a preference table for these ballots.

14. ***Preference Table for Yogurt*** Eight voters are asked to rank three brands of yogurt: Dannon (D), Breyers (B), and Columbo (C). The eight voters turn in the following ballots showing their preferences in order:

D	C	C	B	C	B	C	D
B	D	D	D	B	D	D	B
C	B	B	C	D	C	B	C

Make a preference table for these ballots.

PROBLEM SOLVING

Logo Choice *In Exercises 15–20, employees at Paul's Plumbing are choosing a new company logo. The choices are a wrench (W), a faucet (F), and a water heater (H). Each employee ranks the three choices from first to third. The preference table below shows the results of the ballots.*

Number of Votes	10	5	4	2
First	W	H	H	F
Second	H	W	F	H
Third	F	F	W	W

15. How many employees voted?

16. Is there a majority winner?

17. Determine the winner using the plurality method.

18. Determine the winner using the Borda count method.

19. Determine the winner using the plurality with elimination method.

20. Determine the winner using the pairwise comparison method.

Choosing a Vacation Destination *In Exercises 21–24, the Zellner family is trying to decide whether to go to Yellowstone National Park (Y), Rocky Mountain National Park (R), or Great Smoky Mountains National Park (S) on their summer vacation. The two parents, five children, and two grandparents rank their three choices according to the following preference table.*

Number of Votes	3	2	1	2	1
First	Y	S	Y	R	S
Second	R	Y	S	S	R
Third	S	R	R	Y	Y

▲ Yellowstone National Park

21. Determine the winner using the Borda count method.

22. Determine the winner using the plurality method.

23. Determine the winner using the pairwise comparison method.

24. Determine the winner using the plurality with elimination method.

NFL Expansion *In Exercises 25–28, the National Football League (NFL) is considering four cities for expansion: San Antonio (S), Los Angeles (L), Honolulu (H), and Toronto (T). The 32 current owners rank the four candidates according to the following preference table.*

Number of Votes	8	6	3	4	3	3	2	1	2
First	S	L	L	H	S	H	S	T	H
Second	L	H	S	L	L	L	H	H	S
Third	H	T	H	T	T	S	L	S	L
Fourth	T	S	T	S	H	T	T	L	T

25. Determine the winner using the plurality method.

26. Determine the winner using the Borda count method.

27. Determine the winner using the plurality with elimination method.

28. Determine the winner using the pairwise comparison method.

▲ Honolulu, Hawaii

Board of Trustees Election *In Exercises 29–33, 12 members of the executive committee of the Student Senate must vote for a student representative for the college board of trustees from among three candidates: Williams (W), Diaz (D), and Johnson (J). The preference table follows.*

Number of Votes	5	1	4	2
First	W	D	J	J
Second	D	J	W	D
Third	J	W	D	W

29. Determine the winner using the Borda count method.

30. Determine the winner using the plurality method.

31. Determine the winner using the pairwise comparison method.

32. Determine the winner using the plurality with elimination method.

33. Another way to determine the winner if the plurality with elimination method is used is to eliminate the candidate with the most *last*-place votes at each step. Using the preference table given above, determine the winner if the plurality with elimination method is used and the candidate with the most *last*-place votes is eliminated at each step.

Post Office Sites *In Exercises 34–38, the 11 members of the Henrietta Town Board must decide where to build a new post office. Their three choices are Lehigh Road (L), Erie Road (E), and Ontario Road (O). The preference table follows.*

Number of Votes	5	2	4
First	L	E	O
Second	E	O	E
Third	O	L	L

34. Determine the winner using the plurality method.

35. Determine the winner using the Borda count method.

36. Determine the winner using the plurality with elimination method.

37. Determine the winner using the pairwise comparison method.

38. Determine the winner using the plurality with elimination method and if the candidate with the most *last*-place votes is eliminated at each step.

39. ***Choosing a Contractor*** The board of directors of Birds Eye Foods is holding an election to decide which company should be awarded the contract to build an expansion to its corporate headquarters. The 15 members of Birds Eye's board of directors are considering a proposal from each of the following companies: Becker (B), DiMarco (D), LaChase (L), and Merendez (M). The members rank the four choices according to the preference table below.

Number of Votes	8	4	2	1
First	B	B	L	D
Second	L	D	B	B
Third	M	L	M	L
Fourth	D	M	D	M

a) Determine which company is selected using the plurality method.

b) Determine which company is selected using the Borda count method.

c) Determine which company is selected using the plurality with elimination method.

d) Determine which company is selected using the pairwise comparison method.

40. ***Prime-Time Programming*** The programmers at the Retro-Vision cable television station are deciding which classic TV show should lead their prime-time programming. Their choices are *Happy Days* (H), *Seinfeld* (S), *I Love Lucy* (L), and *Cheers* (C). The 17 programmers rank their choices according to the following table.

Number of Votes	8	4	3	2
First	L	H	S	C
Second	C	C	C	S
Third	S	S	H	H
Fourth	H	L	L	L

a) Determine which TV show is selected using the plurality method.

b) Determine which TV show is selected using the Borda count method.

c) Determine which TV show is selected using the plurality with elimination method.

d) Determine which TV show is selected using the pairwise comparison method.

41. ***Flowers in a Garden*** The flowers in a garden at a resort need to be replaced. The choices for the flowers are geraniums (G), impatiens (I), petunias (P), and zinnias (Z). The head gardener holds an election in which all employees get to vote. The results of the election are given in the table below.

Number of Votes	8	11	3	4	3
First	G	P	P	Z	I
Second	P	Z	I	I	G
Third	I	I	Z	G	P
Fourth	Z	G	G	P	Z

a) Determine which flowers are selected using the Borda count method.

b) Determine which flowers are selected using the plurality method.

c) Determine which flowers are selected using the plurality with elimination method.

d) Determine which flowers are selected using the pairwise comparison method.

42. ***Choosing a Computer*** The Wizards Computer Club is ordering five new computers. To obtain a discount price, the club has decided that all five should come from the same manufacturer. The choices they are considering are Mac (M), Compaq (C), Dell (D), and Gateway (G). The 142 members' choices are reflected below.

Number of Votes	43	30	29	26	14
First	G	M	C	D	D
Second	M	D	D	C	C
Third	C	C	M	G	M
Fourth	D	G	G	M	G

In this section, we will examine four criteria, known as the *fairness criteria*, that mathematicians and political scientists have agreed that a voting method should meet to be considered fair. The fairness criteria include the *majority criterion*, the *head-to-head criterion*, the *monotonicity criterion*, and the *irrelevant alternatives criterion*. The first fairness criterion we will discuss is the majority criterion.

Majority Criterion

If a single candidate is the first choice of a majority (more than 50%) of voters, most voters would agree that that candidate should be declared the winner. If that does not happen with a voting method, that voting method violates the majority criterion.

MAJORITY CRITERION
If a candidate receives a majority (more than 50%) of first-place votes, that candidate should be declared the winner.

EXAMPLE 1 *Pool Maintenance Company Selection*

The members of the Fairport Town Board are holding an election to select a company to maintain the pool located at the town's community center. The choices are Allen Pool Maintenance (A), B & N Pool Maintenance (B), and Mayzon Pool Maintenance (M). The 13 board members rank the three choices and then use the Borda count method to make their selection. The preference table is shown in Table 15.15.

a) Which company is chosen using the Borda count method?

b) Does the winner from part (a) have a majority of votes?

Table 15.15 Pool Maintenance Company Selection Preference Table

Number of Votes	7	4	2
First	A	B	B
Second	B	M	A
Third	M	A	M

SOLUTION

a) Using the Borda count method, B & N Pool Maintenance is chosen. Verify this outcome yourself.

b) No. B & N has 6 out of 13 first place votes which is less than 50%. In fact a majority of the town board members, 7 out of 13, chose Allen Pool Maintenance. ●

In Example 1, although a majority of town board members preferred one choice, Allen Pool Maintenance, the Borda count method yielded a different choice, B & N Pool Maintenance. This example demonstrates that **the Borda count method has the potential to violate the majority criterion.**

EXAMPLE 2 *Applying the Majority Criterion*

Preference Table 15.16 shows the outcomes of 28 votes.

Decide which candidate would be chosen with each of the following voting methods. Also discuss whether or not the method violates the majority criterion.

a) The plurality method

b) The Borda count method

c) The plurality with elimination method

d) The pairwise comparison method

Table 15.16

Number of Votes	15	10	3
First	A	B	C
Second	B	C	B
Third	C	A	A

SOLUTION

a) The plurality method awards the election to A. Notice that A also has a majority—15 out of 28—of first-place votes. The plurality method does not violate the majority criterion. In general, a candidate who holds a majority of first-place votes also holds a plurality of first-place votes. Therefore, **the plurality method never violates the majority criterion**. However, that does not mean that the plurality method always produces a winner that has a majority of the votes. It does mean that if a candidate has a majority of first-place votes, that candidate will also have a plurality of votes and will be selected with the plurality method.

b) The Borda count method awards the election to candidate B. Since candidate A holds a majority of first-place votes, this method violates the majority criterion. This example is the second example in which the Borda count method violates the majority criterion. The Borda count method does not always violate the majority criterion; however, from our examples, we can conclude that **the Borda count method has the potential for violating the majority criterion**.

c) The plurality with elimination method awards the election to candidate A. Therefore, the majority criterion is not violated. In general, a candidate who holds a majority of first-place votes is awarded the election without having to hold a second election, or without having to consider eliminating a candidate and then realigning voters' choices. Therefore, **the plurality with elimination method never violates the majority criterion**.

d) The pairwise comparison method awards the election to candidate A. In general, if a candidate holds a majority of first-place votes, this candidate always wins every pairwise comparison, thereby granting this candidate the victory. Thus, **the pairwise comparison method never violates the majority criterion**.

Of the four methods we discussed, only the Borda count method can violate the majority criterion. Therefore, the Borda count method has a flaw that none of the other voting methods has. An advantage of the Borda count method is that it takes into account voters' preferences by having all candidates ranked. However, a candidate with a majority of first-place votes can lose an election with the Borda count method. Next we will introduce a second fairness criterion and use it to uncover flaws in other voting methods.

Head-to-Head Criterion

Suppose that it is found that one candidate is preferred over each of the other candidates using head-to-head comparisons. It would seem reasonable that this candidate should be chosen as the winner. That, however, is not always the case, and the head-to-head criterion will be used to uncover the next voting method flaw.

HEAD-TO-HEAD CRITERION

If a candidate is favored when compared head-to-head with every other candidate, that candidate should be declared the winner.

The head-to-head criterion is also known as the *Condorcet criterion*. This criterion is named after the Marquis de Condorcet; see the Profile in Mathematics on page 996.

DID YOU KNOW?

The "Flawed" Electoral College System

When citizens of the United States vote for president, they are actually voting to decide which presidential candidate receives all the electoral votes from their state. The Electoral College is a group from each state who are selected by the voters to represent that state in presidential elections. This system was reached by compromise at the constitutional convention in 1787 between those politicians who wanted a direct election of a president and those who wanted Congress to select a president. The resulting system has the potential for some very interesting results that some people consider flaws. For example, it is possible for a candidate to receive a majority of the popular vote and not get elected. This situation occurred only once in U.S. history. In 1876, Samuel J. Tilden received a majority of the popular vote but lost the election to Rutherford B. Hayes. Three other times a candidate won a plurality of the vote but was not elected president. They were Andrew Jackson (1824), Grover Cleveland (1888), and Al Gore (2000).

Americans have been dissatisfied with the Electoral College almost since its inception. In fact, there have been more constitutional amendments proposed to restructure the procedure for electing presidents than for any other purpose. Nonetheless, despite all the dissatisfaction with the Electoral College, the rules regarding it remain identical to when it was first adopted.

Our next example demonstrates this unusual occurrence.

EXAMPLE 4 *Choosing a New Product*

The members of the new product development committee at Albertson's Appliances are considering three products for their next new product to develop and manufacture. The choices under consideration are a microwave (M), a refrigerator (R), and an oven (O). To obtain an initial measure of the intentions of the committee, a straw vote is taken. The results of this first election, the straw vote, are given in Table 15.18.

Table 15.18 New Product Development Committee Preference Table: First Election

Number of Votes	8	9	5	11
First	M	R	M	O
Second	R	O	O	M
Third	O	M	R	R

After several hours of debate and discussion, a second vote is taken. In the second election, only five people change their initial vote. The five people who previously voted M, O, R now vote O, M, R. That is, they moved oven into first place ahead of microwave. This change has the effect of eliminating the M, O, R column of Table 15.18 and increasing the number voting O, M, R from 11 to 16. The results of the second election are given in Table 15.19.

Table 15.19 New Product Development Committee Preference Table: Second Election

Number of Votes	8	9	16
First	M	R	O
Second	R	O	M
Third	O	M	R

a) Using the plurality with elimination method, which product wins the first election?

b) Using the plurality with elimination method, which product wins the second election?

c) Do these results violate the monotonicity criterion?

SOLUTION

a) Using the plurality with elimination method, the first election (the straw vote) results in the refrigerator being eliminated and in the oven gaining a majority vote over the microwave by a vote of 20 to 13. Thus, the oven would be chosen as the next new product to develop and manufacture.

b) Using the plurality with elimination method, the second election results in the microwave being eliminated and in the refrigerator gaining a majority over the oven by a vote of 17 to 16. Thus, the refrigerator would be chosen as the next new product to develop and manufacture.

c) Although the five voters who changed their ballots did so in a way to add support to the oven, the oven's victory in the first election was not repeated in the second election. This occurrence shows that **the plurality with elimination method has the potential to violate the monotonicity criterion**.

SOLUTION

a) The plurality method awards the election to A. Notice that A also has a majority—15 out of 28—of first-place votes. The plurality method does not violate the majority criterion. In general, a candidate who holds a majority of first-place votes also holds a plurality of first-place votes. Therefore, **the plurality method never violates the majority criterion**. However, that does not mean that the plurality method always produces a winner that has a majority of the votes. It does mean that if a candidate has a majority of first-place votes, that candidate will also have a plurality of votes and will be selected with the plurality method.

b) The Borda count method awards the election to candidate B. Since candidate A holds a majority of first-place votes, this method violates the majority criterion. This example is the second example in which the Borda count method violates the majority criterion. The Borda count method does not always violate the majority criterion; however, from our examples, we can conclude that **the Borda count method has the potential for violating the majority criterion**.

c) The plurality with elimination method awards the election to candidate A. Therefore, the majority criterion is not violated. In general, a candidate who holds a majority of first-place votes is awarded the election without having to hold a second election, or without having to consider eliminating a candidate and then realigning voters' choices. Therefore, **the plurality with elimination method never violates the majority criterion**.

d) The pairwise comparison method awards the election to candidate A. In general, if a candidate holds a majority of first-place votes, this candidate always wins every pairwise comparison, thereby granting this candidate the victory. Thus, **the pairwise comparison method never violates the majority criterion**.

Of the four methods we discussed, only the Borda count method can violate the majority criterion. Therefore, the Borda count method has a flaw that none of the other voting methods has. An advantage of the Borda count method is that it takes into account voters' preferences by having all candidates ranked. However, a candidate with a majority of first-place votes can lose an election with the Borda count method. Next we will introduce a second fairness criterion and use it to uncover flaws in other voting methods.

Head-to-Head Criterion

Suppose that it is found that one candidate is preferred over each of the other candidates using head-to-head comparisons. It would seem reasonable that this candidate should be chosen as the winner. That, however, is not always the case, and the head-to-head criterion will be used to uncover the next voting method flaw.

HEAD-TO-HEAD CRITERION

If a candidate is favored when compared head-to-head with every other candidate, that candidate should be declared the winner.

The head-to-head criterion is also known as the *Condorcet criterion*. This criterion is named after the Marquis de Condorcet; see the Profile in Mathematics on page 996.

DID YOU KNOW?

The "Flawed" Electoral College System

When citizens of the United States vote for president, they are actually voting to decide which presidential candidate receives all the electoral votes from their state. The Electoral College is a group from each state who are selected by the voters to represent that state in presidential elections. This system was reached by compromise at the constitutional convention in 1787 between those politicians who wanted a direct election of a president and those who wanted Congress to select a president. The resulting system has the potential for some very interesting results that some people consider flaws. For example, it is possible for a candidate to receive a majority of the popular vote and not get elected. This situation occurred only once in U.S. history. In 1876, Samuel J. Tilden received a majority of the popular vote but lost the election to Rutherford B. Hayes. Three other times a candidate won a plurality of the vote but was not elected president. They were Andrew Jackson (1824), Grover Cleveland (1888), and Al Gore (2000).

Americans have been dissatisfied with the Electoral College almost since its inception. In fact, there have been more constitutional amendments proposed to restructure the procedure for electing presidents than for any other purpose. Nonetheless, despite all the dissatisfaction with the Electoral College, the rules regarding it remain identical to when it was first adopted.

PROFILE IN MATHEMATICS

Marquis de Condorcet

The Marquis de Condorcet (1743–1794) was one of the most influential mathematicians, economists, political scientists, and sociologists during the American and French revolutions. Condorcet believed that principles of fair government could be discovered mathematically. By analyzing voting methods, he discovered, to his disappointment, that occasionally there can be no clear and fair way to declare the winner of an election. The pairwise comparison method is also referred to as the Condorcet method. One of Condorcet's discoveries was that in an election with three candidates A, B, and C, it was possible for the voters to prefer A to B and B to C but also C to A.

EXAMPLE 3 *Applying the Head-to-Head Criterion*

Suppose that four candidates are running for mayor of Springwater: Alvarez (A), Buchannon (B), Czechanski (C), and Davis (D). The election involves voters ranking the candidates with the results shown in Table 15.17.

Table 15.17 Springwater Mayor Preference Table

Number of Votes	129	90	87	78	42
First	D	A	B	C	C
Second	A	C	C	B	B
Third	B	B	A	D	A
Fourth	C	D	D	A	D

a) Is there one candidate who is favored over all others using a head-to-head comparison?

b) Who wins this election if the plurality method is used? Does that result violate the head-to-head criterion?

c) Who wins this election if the Borda count method is used? Does that result violate the head-to-head criterion?

d) Who wins this election if the plurality with elimination method is used? Does that result violate the head-to-head criterion?

e) Who wins this election if the pairwise comparison method is used? Does that result violate the head-to-head criterion?

SOLUTION

a) Using Table 15.17, we can determine that Alvarez is favored, using head-to-head comparison, over Buchannon. Alvarez is favored over Buchannon in columns 1 and 2, giving Alvarez 129 + 90 or 219 votes. Buchannon is favored over Alvarez in columns 3, 4, and 5, giving Buchannon 87 + 78 + 42 or 207 votes. Alvarez is favored over Buchannon by a margin of 219 votes to 207 votes. Using this procedure, we can determine that Alvarez is also favored over Czechanski by a margin of 219 votes to 207 votes and Alvarez is also favored over Davis by a margin of 219 votes to 207 (you should verify these numbers). Therefore, Alvarez is favored over all the candidates, and according to the head-to-head criterion, Alvarez should be declared the winner.

b) Using the plurality method, Davis is elected by virtue of the 129 first-place votes. Therefore, this example demonstrates that **the plurality method has the potential to violate the head-to-head criterion**.

c) Using the Borda count method, Buchannon receives 1146 points, Czechanski receives 1140 points, Alvarez receives 1083 points, and Davis receives 891 points (you should verify these tallies). Therefore, Buchannon is elected using the Borda count method. This example also demonstrates that **the Borda count method has the potential to violate the head-to-head criterion**.

d) Using plurality with elimination, we see that Buchannon is eliminated in the first round, Alvarez is eliminated in the second round, and in the third round Czechanski defeats Davis by a count of 297 to 129 (you should verify these results). Therefore, Czechanski is elected using the plurality with elimination method. This example demonstrates that **the plurality with elimination method has the potential to violate the head-to-head criterion**.

e) Using the pairwise comparison method, we see that Alvarez is favored over Buchannon, Czechanski, and Davis, thereby giving Alvarez 3 points. Buchannon is favored using head-to-head comparison over Czechanski and Davis, thereby giving Buchannon 2 points. Czechanski is favored over Davis, thereby giving Czechanski 1 point. Davis is not favored over any of the other candidates, thereby giving Davis 0 points. (You should verify these results.) Therefore, Alvarez is elected using the pairwise comparison method. It should be clear that if a certain candidate were favored to all other candidates, then this candidate would be elected using the pairwise comparison method. Therefore, **the pairwise comparison method never violates the head-to-head criterion**.

By looking at Example 3, we can conclude that the **only voting method that does not have the potential to violate the head-to-head criterion is the pairwise comparison method**.

At this point in our discussion, we should note the differences between the *head-to-head fairness criterion* and the *pairwise comparison voting method*. The head-to-head criterion is one of the fairness criteria used to determine the fairness of a voting method and therefore is not used to determine the winner of an election. The pairwise comparison method is a voting method and is used to determine the winner of an election. When applying the head-to head criterion, we are trying to determine if there is one candidate who is favored when compared head-to-head with every other candidate. If there is such a candidate, this candidate, according to the head-to-head criterion, should win an election, regardless of which voting method is used to determine the winner. As we have seen in Example 3, it is possible for a candidate to be favored when compared head-to-head with every other candidate yet lose an election. When a candidate is favored when compared with all other candidates and does not win the election, as in Example 3, we say that the voting method used violated the head-to-head criterion.

Example 3 highlighted how the head-to-head criterion uncovered flaws in the plurality method, Borda count method, and the plurality with elimination method. Now, we will introduce a third criterion for fair elections, the monotonicity criterion, and use it to uncover an additional flaw in the plurality with elimination method.

Monotonicity Criterion

The third criterion for fair elections is only relevant when an election is repeated. A repeated election refers to the same number of voters choosing among the same candidates. One example of a repeated election is a ***straw vote*** followed by another vote. Frequently, preceding an election, voters discuss the candidates' strengths and weaknesses. Prior to holding an official election, voters may take a straw vote to gain a preliminary measure of the candidates' strength. After discussions, some voters may decide to change their preferences. If a candidate gains votes at the expense of the other candidates, this candidate's chances of winning should increase. If that candidate was already leading, this candidate should still win with additional votes in his or her favor. The repeated election leads us to our next criterion, the monotonicity criterion. The monotonicity criterion uncovers a strange flaw in the plurality with elimination voting method. As we soon shall see, it is possible for the winner of the first election to gain additional support before the second election and then lose the second election!

MONOTONICITY CRITERION

A candidate who wins a first election and then gains additional support without losing any of the original support should also win a second election.

Our next example demonstrates this unusual occurrence.

EXAMPLE 4 *Choosing a New Product*

The members of the new product development committee at Albertson's Appliances are considering three products for their next new product to develop and manufacture. The choices under consideration are a microwave (M), a refrigerator (R), and an oven (O). To obtain an initial measure of the intentions of the committee, a straw vote is taken. The results of this first election, the straw vote, are given in Table 15.18.

Table 15.18 New Product Development Committee Preference Table: First Election

Number of Votes	8	9	5	11
First	M	R	M	O
Second	R	O	O	M
Third	O	M	R	R

After several hours of debate and discussion, a second vote is taken. In the second election, only five people change their initial vote. The five people who previously voted M, O, R now vote O, M, R. That is, they moved oven into first place ahead of microwave. This change has the effect of eliminating the M, O, R column of Table 15.18 and increasing the number voting O, M, R from 11 to 16. The results of the second election are given in Table 15.19.

Table 15.19 New Product Development Committee Preference Table: Second Election

Number of Votes	8	9	16
First	M	R	O
Second	R	O	M
Third	O	M	R

a) Using the plurality with elimination method, which product wins the first election?

b) Using the plurality with elimination method, which product wins the second election?

c) Do these results violate the monotonicity criterion?

SOLUTION

a) Using the plurality with elimination method, the first election (the straw vote) results in the refrigerator being eliminated and in the oven gaining a majority vote over the microwave by a vote of 20 to 13. Thus, the oven would be chosen as the next new product to develop and manufacture.

b) Using the plurality with elimination method, the second election results in the microwave being eliminated and in the refrigerator gaining a majority over the oven by a vote of 17 to 16. Thus, the refrigerator would be chosen as the next new product to develop and manufacture.

c) Although the five voters who changed their ballots did so in a way to add support to the oven, the oven's victory in the first election was not repeated in the second election. This occurrence shows that **the plurality with elimination method has the potential to violate the monotonicity criterion**.

In Example 4, we showed that the plurality with elimination method has the potential to violate the monotonicity criterion. **Note that the Borda count method and the pairwise comparison method also have the potential to violate the monotonicity criterion**.

We have thus far shown how the majority criterion, the head-to-head criterion, and the monotonicity criterion have all uncovered flaws in our voting methods. You may have noticed that we have yet to find a flaw in the pairwise comparison method. Our next criterion, the irrelevant alternatives criterion, uncovers a flaw in the pairwise comparison method.

Irrelevant Alternatives Criterion

The last criterion we will discuss, the irrelevant alternatives criterion, will address the result of removing a candidate from an election who has no chance of winning. Suppose that four candidates, A, B, C, and D, are involved in an election. The voting takes place and it is found that candidate B is the winner. However, it is discovered that just prior to the election, candidate C had dropped out of the election. Since candidate C had no chance of winning, we might conclude that this action should not have any effect on the outcome of the election. We shall soon see that such action can negatively affect the winner. We now have our fourth fairness criterion.

IRRELEVANT ALTERNATIVES CRITERION
If a candidate is the winner of an election and in a second election one or more of the other candidates is removed, the previous winner should still be the winner.

EXAMPLE 5 *Selecting a Co-op President*

The members of the Tennessee Farmers Co-op are holding an election to select the president of the co-op's board of directors. The choices are Anderson (A), Bacari (B), Mendolson (M), Sanchez (S), and Tompkins (T). The election results are shown in Table 15.20.

Table 15.20 Tennessee Farmer's Co-op President Preference Table

Number of Votes	24	16	16	16	8	4	4
First	B	B	T	A	S	M	M
Second	S	S	S	M	T	B	T
Third	M	T	A	T	M	S	S
Fourth	T	A	M	B	B	T	B
Fifth	A	M	B	S	A	A	A

Prior to the announcement of the election results, it is discovered that Mendolson had dropped out of the election. The election officials then decide to eliminate Mendolson's name from the election results.

a) Using the pairwise comparison method, which candidate wins this election if Mendolson is included?

b) Using the pairwise comparison method, which candidate wins the election if Mendolson is eliminated from the preference table?

c) Does this method violate the irrelevant alternatives criterion?

SOLUTION

a) Using the pairwise comparison method, the following results can be verified from Table 15.20 on page 999. Since we have five candidates, $n = 5$ and the number of comparisons needed is

$$c = \frac{n(n-1)}{2} = \frac{5(4)}{2} = 10$$

We will do the comparisons in the following order: Anderson versus Bacari, Anderson versus Mendolson, Anderson versus Sanchez, Anderson versus Tompkins, Bacari versus Mendolson, Bacari versus Sanchez, Bacari versus Tompkins, Mendolson versus Sanchez, Mendolson versus Tompkins, Sanchez versus Tompkins. These are the 10 possible comparisons.

Bacari defeats Anderson (56 to 32), so Bacari gets 1 point.
Anderson defeats Mendolson (48 to 40), so Anderson gets 1 point.
Sanchez defeats Anderson (72 to 16), so Sanchez gets 1 point.
Tompkins defeats Anderson (72 to 16), so Tompkins gets 1 point.
Mendolson defeats Bacari (48 to 40), so Mendolson gets 1 point.
Bacari defeats Sanchez (60 to 28), so Bacari gets 1 point.
Bacari ties Tompkins (44 to 44), so Bacari gets $\frac{1}{2}$ point and Tompkins gets $\frac{1}{2}$ point.
Sanchez defeats Mendolson (64 to 24), so Sanchez gets 1 point.
Mendolson defeats Tompkins (48 to 40), so Mendolson gets 1 point.
Sanchez defeats Tompkins (52 to 36), so Sanchez gets 1 point.

We have made all possible one-to-one comparisons. Sanchez has 3 points, Bacari has $2\frac{1}{2}$ points, Mendolson has 2 points, Tompkins has $1\frac{1}{2}$ points, and Anderson has 1 point. Therefore, Sanchez is the winner when Mendolson is included.

b) Once Mendolson is eliminated from consideration, there is a new preference table. This new table is obtained by eliminating Mendolson from Table 15.20 and keeping the remaining candidates in the same order. See Table 15.21. Since we now have four candidates, $n = 4$ and the number of comparisons needed is

$$c = \frac{n(n-1)}{2} = \frac{4(3)}{2} = 6$$

Table 15.21 Tennessee Farmer's Co-op President Preference Table

Number of Votes	24	16	16	16	8	4	4
First	B	B	T	A	S	B	T
Second	S	S	S	T	T	S	S
Third	T	T	A	B	B	T	B
Fourth	A	A	B	S	A	A	A

The pairwise comparison method, using Table 15.21, produces the following results:

Bacari defeats Anderson (56 to 32), so Bacari gets 1 point.
Sanchez defeats Anderson (72 to 16), so Sanchez gets 1 point.
Tompkins defeats Anderson (72 to 16), so Tompkins gets 1 point.
Bacari defeats Sanchez (60 to 28), so Sanchez gets 1 point.

Bacari ties Tompkins (44 to 44), so Bacari gets $\frac{1}{2}$ point and Tompkins gets $\frac{1}{2}$ point.

Sanchez defeats Tompkins (52 to 36), so Sanchez gets 1 point.

We have made all possible one-to-one comparisons. Now Bacari has $2\frac{1}{2}$ points, Sanchez has 2 points, Tompkins has $1\frac{1}{2}$ points, and Anderson has 0 points. Thus, Bacari is selected.

c) The first election produced Sanchez as the winner. Then, even though Mendolson was not the winning candidate, Mendolson's removal from the election produced Bacari as the winner. This example demonstrates that the **pairwise comparison method has the potential for violating the irrelevant alternatives criterion.** •

In Example 5, we showed that the pairwise comparison method has the potential to violate the irrelevant alternatives criterion. **Note that each voting method we have discussed—the plurality method, the Borda count method, the plurality with elimination method, and the pairwise comparison method—has the potential to violate the irrelevant alternatives criterion**.

Table 15.22

Number of Votes	18	8	6
First	B	A	C
Second	A	C	A
Third	C	B	B

EXAMPLE 6 *Satisfying the Four Fairness Criteria*

Suppose that the plurality method is used to determine the winner of the election given in Table 15.22. Does this method satisfy the four fairness criteria, that is, the majority criterion, the head-to-head criterion, the monotonicity criterion, and the irrelevant alternatives criterion?

SOLUTION By the plurality method, B is the winner with 18 votes. Since a total of 18 votes is also a majority of the 32 votes cast (56%), the majority criterion is satisfied. Examining pairwise comparison matchups, we see that B is favored to A (18 to 14) and that B is also favored to C (18 to 14). Therefore, the head-to-head criterion is satisfied. Since B holds a majority of first-place votes, the monotonicity criterion is satisfied when a second election is held in which B picks up additional votes; B still wins by plurality. Last, if A or C drops out, B still wins by plurality and the irrelevant alternatives criterion is satisfied. Therefore, this election satisfies all four fairness criteria. •

We saw in Example 6 that in a particular election, it is possible that all of the fairness criteria are satisfied. However in a different election, with a different preference table, it is quite possible for the same voting method to violate one or more of the fairness criteria.

Table 15.23 summarizes the four fairness criteria, and Table 15.24 on page 1002 summarizes which voting methods always satisfy the criteria.

Table 15.23 Summary of Fairness Criteria

Majority criterion	If a candidate receives a majority of first-place votes, that candidate should be declared the winner.
Head-to-head criterion	If a candidate is favored when compared individually with every other candidate, that candidate should be declared the winner.
Monotonicity criterion	A candidate who wins a first election and then gains additional support without losing any of the original support should also win a second election.
Irrelevant alternatives criterion	If a candidate is the winner of an election and in a second election one or more of the other candidates is removed, the previous winner should still be the winner.

PROFILE IN MATHEMATICS

Kenneth Arrow

Kenneth Arrow (1921–), an economist, was interested in the theory of corporations and how corporations make decisions when their leaders have different opinions. His particular interest in stockholders led to his discovery that majority voting can be flawed. While working at the RAND Corporation, he studied how mathematical concepts, game theory in particular, apply to decision making in the military and in diplomatic affairs. As a result, he developed a set of properties for any fair and reasonable voting method. After experimenting for days, he realized that there was no voting method that would satisfy all reasonable conditions. This conclusion is called the Arrow impossibility theorem, which he proved in 1951. In 1972, Arrow received the Nobel Prize in Economics for his exceptional work in the theory of general economic equilibrium.

Table 15.24 Summary of the Voting Methods and Whether They Satisfy the Fairness Criteria

	Plurality Method	Borda Count Method	Plurality with Elimination Method	Pairwise Comparison Method
Majority criterion	Always satisfies	May not satisfy	Always satisfies	Always satisfies
Head-to-head criterion	May not satisfy	May not satisfy	May not satisfy	Always satisfies
Monotonicity criterion	Always satisfies	May not satisfy	May not satisfy	May not satisfy
Irrelevant alternatives criterion	May not satisfy	May not satisfy	May not satisfy	May not satisfy

As we have shown, each of the voting methods introduced in Section 15.1 has the potential to violate at least one of the fairness criteria. Is there another method that does satisfy all four criteria? Mathematicians and political scientists struggled with this question for two centuries. Finally, in 1951, an economist, Kenneth Arrow, proved mathematically that there is no voting method that can satisfy all four fairness criteria. This result is called *Arrow's impossibility theorem*. See the Profile in Mathematics on page 1001.

ARROW'S IMPOSSIBILITY THEOREM

It is mathematically impossible for any democratic voting method to simultaneously satisfy each of the fairness criteria:

The majority criterion

The head-to-head criterion

The monotonicity criterion

The irrelevant alternatives criterion

SECTION 15.2 EXERCISES

CONCEPT/WRITING EXERCISES

1. Describe the majority criterion.
2. Describe the monotonicity criterion.
3. Describe the head-to-head criterion.
4. Describe the irrelevant alternatives criterion.
5. Explain why the pairwise comparison method always satisfies the head-to-head criterion.
6. Explain why the pairwise comparison method always satisfies the majority criterion.
7. Explain why the plurality method always satisfies the majority criterion.
8. Explain why the plurality with elimination method always satisfies the majority criterion.

PRACTICE THE SKILLS/PROBLEM SOLVING

9. ***Annual Meeting*** Members of the board of directors of the American Psychological Association are voting to select a city to host the association's annual meeting. The board is considering the following cities: San Antonio (S), Tampa (T), and Portland (P). The preference table for the 19 members of the board of directors is shown below. Show that the Borda count method violates the majority criterion.

Number of Votes	10	6	3
First	T	P	P
Second	P	S	T
Third	S	T	S

▲ Tampa, Florida

10. ***Restructuring a Company*** The board of directors at The Limited is considering three different administrative restructuring plans, A, B, and C. The 11 members of the board of directors rank the plans according to the preference table below.

 a) Which restructuring plan is preferred to all the others in a head-to-head comparison?

 b) Suppose that the plurality method is used to determine the winner. Is the head-to-head criterion satisfied? Explain.

Number of Votes	2	4	2	3
First	B	A	C	C
Second	A	B	A	B
Third	C	C	B	A

11. ***Hiring a New Director*** Irvine Valley College is hiring a new director of counseling and advisement. The hiring committee ranks the four candidates, Alvarez (A), Brown (B), Singleton (S), and Yu (Y), according to the preference table at the top of the next column. Suppose that the Borda count method is used to determine the winner. Is the majority criterion satisfied? Explain.

Number of Votes	4	3	1	1
First	A	B	B	A
Second	B	S	Y	B
Third	S	Y	S	Y
Fourth	Y	A	A	S

12. ***Party Theme*** The children in Ms. Cohn's seventh-grade class are voting on a theme for their class party. Their choices are beach (B), western (W), superheroes (S), or rock and roll (R). The results are shown in the preference table below.

 a) Which theme is preferred to all the other themes in a head-to-head comparison?

 b) Suppose that the plurality method is used to determine the winner. Is the head-to-head criterion satisfied? Explain.

Number of Votes	12	6	4	3
First	B	S	R	W
Second	W	B	S	R
Third	S	W	B	B
Fourth	R	R	W	S

13. ***Residence Hall Improvements*** The administration at St. Cloud State University is considering three possible areas of improvement to the school's residence halls: parking (P), security (S), and lounge areas (L). The nine members of the executive committee of the Student Senate rank the choices according to the preference table shown below. Suppose that the Borda count method is used to determine the winner. Is the head-to-head criterion satisfied? Explain.

Number of Votes	4	1	1	1	2
First	P	L	S	S	L
Second	L	P	L	P	S
Third	S	S	P	L	P

14. ***Design of a Hospital Wing*** The planning committee of General Hospital is considering three proposals, A, B, and C, for the design of a new children's wing in the hospital. The nine members of the planning committee rank the proposals according to the preference table shown below. Suppose that the Borda count method is used to determine the winner. Is the head-to-head criterion satisfied? Explain.

Number of Votes	5	2	2
First	C	A	B
Second	A	B	A
Third	B	C	C

15. ***Preference for Grape Jelly*** Twenty-one people are asked to taste test and rank three different brands of grape jelly. The preference table below shows the rankings of the 21 voters. Suppose that the plurality with elimination method is used to determine the winner. Is the head-to-head criterion satisfied? Explain.

Number of Votes	8	7	6
First	B	C	A
Second	A	A	C
Third	C	B	B

16. ***A Taste Test*** Twenty-five people are surveyed in the Mall of America and asked to taste and rank four different brands of vanilla ice cream. The preference table below shows the rankings of the 25 voters. Suppose that the plurality with elimination method is used to determine the winner. Is the head-to-head criterion satisfied? Explain.

Number of Votes	10	2	5	8
First	A	B	D	C
Second	B	C	B	B
Third	D	A	C	A
Fourth	C	D	A	D

▲ The Mall of America, Minneapolis, MN

17. ***Plurality: Irrelevant Alternatives Criterion*** Suppose that the plurality method is used on the following preference table. If B drops out, is the irrelevant alternatives criterion satisfied? Explain.

Number of Votes	12	9	4
First	A	C	B
Second	C	B	C
Third	B	A	A

18. ***Plurality: Irrelevant Alternatives Criterion*** Suppose that the plurality method is used on the following preference table. If C drops out, is the irrelevant alternatives criterion satisfied? Explain.

Number of Votes	6	5	3
First	B	C	A
Second	C	A	C
Third	A	B	B

19. ***Borda Count: Irrelevant Alternatives Criterion*** Suppose that the Borda count method is used on the following preference table. If C drops out, is the irrelevant alternatives criterion satisfied? Explain.

Number of Votes	10	9	8
First	B	C	A
Second	A	B	C
Third	C	A	B

20. ***Borda Count: Irrelevant Alternatives Criterion*** Suppose that the Borda count method is used on the following preference table. If B drops out, is the irrelevant alternatives criterion satisfied? Explain.

Number of Votes	7	6	5
First	A	C	B
Second	B	A	C
Third	C	B	A

21. ***Plurality with Elimination: Monotonicity Criterion*** Suppose that the plurality with elimination method is used on the following preference table. If the three voters who voted A, C, B, in that order, change their vote to C, A, B, is the monotonicity criterion satisfied? Explain.

Number of Votes	8	9	3	12
First	A	B	A	C
Second	B	C	C	A
Third	C	A	B	B

22. ***Plurality with Elimination: Monotonicity Criterion*** Suppose that the plurality with elimination method is used on the following preference table. If the three voters who voted C, B, A, in that order, change their vote to B, C, A, is the monotonicity criterion satisfied? Explain.

Number of Votes	5	6	3	7
First	C	A	C	B
Second	A	B	B	C
Third	B	C	A	A

23. ***Pairwise Comparison Method: Monotonicity Criterion*** Suppose that the pairwise comparison method is used on the following preference table. If the five voters who voted C, D, A, B, in that order, change their votes to D, B, C, A, is the monotonicity criterion satisfied? Explain.

Number of Votes	8	7	6	5
First	D	B	B	C
Second	C	A	A	D
Third	A	C	D	A
Fourth	B	D	C	B

24. ***Borda Count: Monotonicity Criterion*** Suppose that the Borda count method is used on the following preference table. If the five voters who voted D, A, C, B, in that order, change their votes to A, B, D, C and the three voters who voted C, A, D, B, in that order, change their votes to A, B, C, D, is the monotonicity criterion satisfied? Explain.

Number of Votes	8	7	5	3
First	A	B	D	C
Second	B	D	A	A
Third	C	C	C	D
Fourth	D	A	B	B

25. ***Pairwise Comparison: Irrelevant Alternatives Criterion*** Suppose that the pairwise comparison method is used on the following preference table. If A, C, and E drop out, is the irrelevant alternatives criterion satisfied? Explain.

Number of Votes	1	1	1	1	1
First	B	B	A	C	E
Second	A	C	D	D	D
Third	C	E	B	B	B
Fourth	E	A	C	E	A
Fifth	D	D	E	A	C

26. ***Pairwise Comparison: Irrelevant Alternatives Criterion*** Suppose that the pairwise comparison method is used on the preference table above to the right. If A, B, and E drop out, is the irrelevant alternatives criterion satisfied? Explain.

Number of Votes	2	1	1	1	1	1
First	C	C	D	A	E	B
Second	E	B	C	D	A	E
Third	A	A	B	C	D	D
Fourth	B	E	A	B	C	C
Fifth	D	D	E	E	B	A

27. ***Borda Count: Majority Criterion*** Suppose that the Borda count method is used on the following preference table. Is the majority criterion satisfied?

Number of Votes	4	2	1
First	A	B	C
Second	B	C	B
Third	C	A	A

28. ***Borda Count: Majority Criterion*** Suppose that the Borda count method is used on the following preference table. Is the majority criterion satisfied?

Number of Votes	5	1	4	1
First	B	B	C	A
Second	C	A	A	C
Third	A	C	B	B

29. ***Spring Trip*** The History Club of St. Louis is voting on which city to visit for its spring trip. The choices are New York City (N), Washington, D.C. (W), Philadelphia (P), and Boston (B). The preference table for the 23 members' choices is given below.

▲ The Jefferson Memorial in Washington, D.C.

Number of Votes	12	8	3
First	W	P	B
Second	N	N	W
Third	P	B	P
Fourth	B	W	N

a) Which city, if any, holds a majority of first-place votes?

b) Which city is selected using the plurality method?

c) Which city is selected using the Borda count method?

d) Which city is selected using the plurality with elimination method?

e) Which city is selected using the pairwise comparison method?

f) For this preference table, which voting method(s), if any, violate the majority criterion?

30. ***Investment Choices*** The Motley Crew Investing Club wishes to purchase stock in a pharmaceutical company. The choices are American Pharmacy (A), Burrows-Welcome (B), Chicago Labs (C), Dow Chemical (D), and Emerging Medicines (E). The choice rankings are shown in the following preference table.

Number of Votes	12	8	8	8	4	2	2
First	B	B	E	A	D	C	C
Second	D	D	D	C	E	B	E
Third	C	E	A	E	C	D	D
Fourth	E	A	C	B	B	E	B
Fifth	A	C	B	D	A	A	A

Before calculating the results of this election, the club learns that Chicago Labs has filed for bankruptcy.

a) Using the pairwise comparison method, which company is chosen if Chicago Labs is included?

b) Using the pairwise comparison method, which company is chosen if Chicago Labs is eliminated from the preference table?

c) Does this second result violate the irrelevant alternatives criterion?

31. ***Selecting a Spokesperson*** The Campbell Soup Company is selecting a new spokesperson for its advertisements. The choices are Jennifer Aniston (A), Denzel Washington (W), and Lance Armstrong (L). At the annual meeting of its managers, a straw vote is taken to gauge the managers' initial reaction to the candidates. The results of this straw vote are given in the following preference table.

Number of Votes	28	24	20	10
First	A	W	L	L
Second	L	A	W	A
Third	W	L	A	W

▲ Lance Armstrong

After discussion about the candidates, eight people who had supported Lance Armstrong over Jennifer Aniston now switch their preferences to Jennifer Aniston over Lance Armstrong. This switch leads to the following preference table.

Number of Votes	36	24	20	2
First	A	W	L	L
Second	L	A	W	A
Third	W	L	A	W

a) Using the plurality with elimination method, which person is selected from the first (or straw) vote?

b) Using the plurality with elimination method, which person is selected from the second vote?

c) Does this second result violate the monotonicity criterion?

32. ***Electing a Parks Director*** The City of Birmingham is electing a new parks director. The choices for the position are Diane Allen (A), Brac Brady (B), Faye Conrad (C), Stacy Davis (D), and Bill Evers (E). The election is conducted, and the preference table is shown below.

Number of Votes	80	50	50	50	30	20	20
First	B	B	E	A	D	C	C
Second	D	D	D	C	E	B	E
Third	C	E	A	E	C	D	D
Fourth	E	A	C	B	B	E	B
Fifth	A	C	B	D	A	A	A

Before calculating the results of this election, it is learned that Faye Conrad has dropped out of the race for personal reasons.

a) Using the pairwise comparison method, which candidate is chosen if Faye Conrad is included?

b) Using the pairwise comparison method, which candidate is chosen if Faye Conrad is eliminated from the preference table?

c) Does this second result violate the irrelevant alternatives criterion?

CHALLENGE PROBLEMS/GROUP ACTIVITIES

33. Explain why the plurality method always satisfies the monotonicity criterion.

34. Construct a preference table with four candidates and four rankings of the candidates, similar to the preference table in Exercise 11 on page 1003, such that the plurality method violates the head-to-head criterion. Have the total number of votes be 25.

35. Construct a preference table with three candidates and four rankings of the candidates, similar to the preference table in Exercise 10 on page 1003, such that the plurality with elimination method violates the monotonicity criterion. Have the total number of votes be 21.

36. Construct a preference table with three candidates and three rankings of the candidates, similar to the preference table in Exercise 9 on page 1003, such that the Borda count method violates the majority criterion. Have the total number of votes be 19.

37. Construct a preference table with four candidates and three rankings of the candidates, similar to the preference table in Exercise 29 on page 1005, such that the pairwise comparison method violates the irrelevant alternatives criterion. Have the total number of votes be 6.

Not voting according to a voter's true preference, called ***insincere voting****, is frequently practiced to affect the outcome of an election. Insincere voting is used in Exercise 38.*

38. ***Voting Strategy*** Consider the preference table below. Assume that a majority is needed to win the election. Since no candidate has a majority but C has the most first-place votes, a runoff election is held between A and B. The winner of the runoff election will run against C.

Number of Votes	5	4	2
First	C	A	B
Second	A	C	A
Third	B	B	C

a) Using the plurality method, which candidate will win the runoff election, A or B?

b) Will the candidate determined in part (a) win the election against candidate C?

c) Suppose that the voters who support candidate C, the candidate with the most first-place votes, find out prior to the vote that C will not win if B is eliminated and C runs against A. How can the voters who support C vote insincerely to enable C to win?

INTERNET/RESEARCH ACTIVITIES

39. Choose a country other than the United States and write a research paper on how the president or prime minister is chosen. Describe in detail the voting method used. Include the strengths and weaknesses of the method used.

40. Write a research paper on the history of voting methods. Include how the flaws in the voting methods discussed in this section were uncovered.

15.3 APPORTIONMENT METHODS

▲ Apportionment methods can be used to distribute printers to computer labs.

Suppose that your college receives a grant to purchase 50 new laser printers to be distributed to five different computer labs on campus. Since the computer labs do not all have the same number of computers, the director of computing decides to apportion the laser printers based on the number of computers in each lab. In this section, we will discuss some apportionment methods the college can use to distribute the printers to the five computer labs.

When the delegates for the original 13 states met in 1787 to draft a constitution, their most important discussion concerned the representation of the states in the legislature. Some states, particularly the small states, preferred that each state have the same number of representatives. The large states wanted a proportional representation based on population. The delegates resolved this issue by creating a Senate, in which each state received two Senators, and a House of Representatives, in which each state received a number of representatives based on its population. Article 1, Section 2, of the Constitution of the United States includes the statement "Representatives . . . shall be apportioned among the several states . . . according to their respective numbers." What the constitution did not say is how the apportionments are determined. In this section, we will discuss four different apportionment methods.

The goal of ***apportionment*** is to determine a method to allocate the total number of items to be apportioned in a fair manner. One of the most important examples of apportionment is in determining the representation of governing bodies. However, apportionment problems occur in many other places, from determining how many police officers should be assigned to each precinct in a city to determining how many nurses should be assigned to each shift at a hospital.

In this section, we will discuss four different apportionment methods:

1. Hamilton's method
2. Jefferson's method
3. Webster's method
4. Adams's method

DID YOU KNOW?

Too Young to Run for Congress?

According to the 17th Amendment of the Constitution, "A senator must be at least 30 years of age, a U.S. citizen for at least 9 years, and a resident of the state in which he or she is elected." If you do not want to wait until you are 30 years old to become a member of Congress, you may run for office in the House of Representatives. Members of the House must be at least 25 years of age, a U.S. citizen for at least 7 years, and a resident of the state in which he or she is elected, but not necessarily a resident of the congressional district he or she hopes to represent.

Apportionment Problems

Let's begin by considering a problem facing the Shanahan Law Firm. The firm needs to apportion 60 new fax machines to be distributed among the firm's five offices. Since the offices do not all have the same number of employees, the firm's managing partner decides to apportion the fax machines based on the number of employees at each office. Table 15.25 shows the number of employees at each office.

Table 15.25 Shanahan Law Firm Employees

Office	A	B	C	D	E	Total
Employees	246	201	196	211	226	1080

If the firm wants to distribute the fax machines based on the number of employees at each office, it would take each office's number of employees and divide it by the total number of employees. For example, office A would be entitled to $\frac{246}{1080}$, or $22\frac{7}{9}\%$, of the fax machines. Similarly, office B would be entitled to $\frac{201}{1080}$, or $18\frac{11}{18}\%$, of the fax machines. Dividing the number of employees in each office by the total number of employees leaves each office with a fractional part of a fax machine. Since an office cannot receive a fraction of a fax machine, the law firm has a problem. How can it apportion the fax machines so that each office receives its fair allotment?

Before we discuss how to apply apportionment methods, we will introduce some important terms used in this section, *standard divisor* and *standard quota*. The ***standard divisor*** is found by dividing the total population under consideration by the number of items to be allocated.

$$\text{Standard divisor} = \frac{\text{total population}}{\text{number of items to be allocated}}$$

For our fax machine problem, the standard divisor is found as follows:

$$\text{Standard divisor} = \frac{\text{total number of employees}}{\text{number of fax machines to be allocated}} = \frac{1080}{60} = 18$$

The ***standard quota*** for a particular group is found by dividing the population of the group by the standard divisor.

$$\text{Standard quota} = \frac{\text{population for the particular group}}{\text{standard divisor}}$$

For our fax machine problem, the standard quota for office A is found as follows:

$$\text{Standard quota for office A} = \frac{\text{number of employees at office A}}{\text{standard divisor}} = \frac{246}{18} \approx 13.67$$

We will round all standard divisors and standard quotas to the nearest hundredth.

EXAMPLE 1 *Determining Standard Quotas*

Determine the standard quotas for offices B, C, D, and E of the Shanahan Law Firm and complete Table 15.26. Use 18 as the standard divisor, as was determined above.

Table 15.26 Shanahan Law Firm Employees

Office	A	B	C	D	E	Total
Employees	246	201	196	211	226	1080
Standard quota	13.67					

SOLUTION The standard quota for office B is $\frac{201}{18} = 11.17$, rounded to the nearest hundredth. The other standard quotas are found in a similar manner. Table 15.27 shows the completed table.

Table 15.27 Shanahan Law Firm Employees

Office	A	B	C	D	E	Total
Employees	246	201	196	211	226	1080
Standard quota	13.67	11.17	10.89	11.72	12.56	60.01

The standard quota represents the exact number of fax machines each office would receive if we were able to divide the fax machines into fractional parts. Notice that the sum of the standard quotas is slightly above 60, the total number of fax machines, due to rounding.

Now we introduce two more important definitions, the *lower quota* and the *upper quota.* The ***lower quota*** is the standard quota rounded down to the nearest integer. The ***upper quota*** is the standard quota rounded up to the nearest integer. For example, office A has a lower quota of 13 and an upper quota of 14, office B has a lower quota of 11 and an upper quota of 12, and so on.

DID YOU KNOW?

Mathematical Contributions from Political Leaders

Alexander Hamilton (1757–1804)

As a result of his distinctive service in the Revolutionary War, Hamilton became George Washington's aide-de-camp and personal secretary in 1777. In 1789, Washington appointed Hamilton as the first secretary of the treasury. Hamilton and Thomas Jefferson were strong political adversaries. In 1804, Hamilton was killed in a duel with Aaron Burr, who was serving as Jefferson's vice president.

How should we round the standard quotas so that each office receives its fair share? If we were to use conventional rounding and round each office's standard quota to the nearest whole number, office A would receive 14 fax machines and office B would receive 11 fax machines. Offices C, D, and E would receive 11, 12, and 13 fax machines, respectively. The total number of fax machines needed would be 61. Since there are only 60 fax machines to distribute, conventional rounding of standard quotas can be problematic. We need a more sophisticated apportionment method to avoid the problems that can occur with rounding standard quotas. We will now discuss the first of our four apportionment methods, Hamilton's method.

Hamilton's Method

Mathematically, Hamilton's method is the easiest to apply.

HAMILTON'S METHOD

1. Calculate the standard divisor for the set of data.
2. Calculate each group's* standard quota.
3. Round each standard quota **down** to the nearest integer (the lower quota). Initially, each group receives its lower quota.
4. Distribute any leftover items to the groups with the largest fractional parts until all items are distributed.

EXAMPLE 2 *Using Hamilton's Method for Apportioning Fax Machines*

Use Hamilton's method to distribute the 60 fax machines for the Shanahan Law Firm discussed in Example 1 on page 1009.

SOLUTION Table 15.28 includes the standard quotas (row 2) calculated from Example 1, along with the lower quotas (row 3) and the apportionment using Hamilton's method (row 4).

Table 15.28 Shanahan Law Firm Employees

Office	A	B	C	D	E	Total
Employees	246	201	196	211	226	1080
Standard quota	13.67	11.17	10.89	11.72	12.56	60.01
Lower quota	13	11	10	11	12	57
Hamilton's apportionment	14	11	11	12	12	60

The sum of the lower quotas is 57, leaving three additional fax machines to distribute. Since offices C, D, and A, in this order, have the three highest fractional parts (0.89, 0.72, and 0.67, respectively) in the standard quota, each receives one of the additional fax machines using Hamilton's method. Note that with Hamilton's method, each office receives either its lower quota or its upper quota of laser printers. Also note that the total number of fax machines apportioned by Hamilton's method is 60, which is what we expected. ●

*We use the word *group* in steps 1, 2, and 3 as a general term. However, *group* could refer to a state, a school, or even an individual.

Let's look at another example using Hamilton's method.

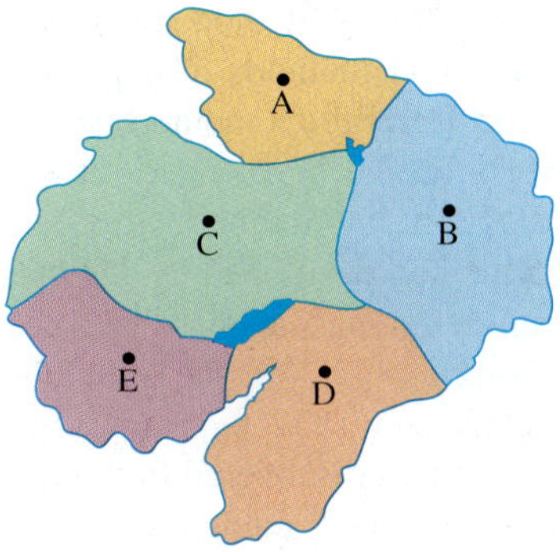

Republic of Geranium population 8,800,000

EXAMPLE 3 *Using Hamilton's Method for Apportioning Legislative Seats*

The Republic of Geranium needs to apportion 275 seats in the legislature. Suppose that the population is 8,800,000 and that there are five states, A, B, C, D, and E. The 275 seats are to be divided among the five states according to their respective populations, given in Table 15.29.

Table 15.29 Republic of Geranium Population

State	A	B	C	D	E	Total
Population	1,003,200	1,228,600	4,990,700	813,000	764,500	8,800,000

Use Hamilton's method to apportion the seats.

SOLUTION The standard divisor is determined by dividing the total population by the number of seats in the legislature.

$$\text{Standard divisor} = \frac{8{,}800{,}000}{275} = 32{,}000$$

The standard quotas are determined by dividing the population of each state by the standard divisor. The results are shown in the second row in Table 15.30.

Table 15.30 Republic of Geranium Population

State	A	B	C	D	E	Total
Population	1,003,200	1,228,600	4,990,700	813,000	764,500	8,800,000
Standard quota	31.35	38.39	155.96	25.41	23.89	
Lower quota	31	38	155	25	23	272
Hamilton's apportionment	31	38	156	26	24	275

The sum of the lower quotas is 272, leaving three additional seats to distribute. Since states C, E, and D, in this order, have the three highest fractional parts in the standard quota (0.96, 0.89 and 0.41, respectively), they each receive the additional seats using Hamilton's method.

As we can see from both examples, an advantage of Hamilton's method is every apportionment is either the lower quota or the upper quota. If an apportionment method assigns either the lower or upper quota to every group under consideration, it is said to satisfy the ***quota rule***.

THE QUOTA RULE

An apportionment for every group under consideration should always be either the upper quota or the lower quota.

DID YOU KNOW?

Mathematical Contributions from Political Leaders

Thomas Jefferson (1743–1826)

Thomas Jefferson, third president of the United States, authored the Declaration of Independence. He excelled in many careers, including philosopher, educator, politician, inventor, scientist, and writer. Under his administration, the United States doubled in area with the Louisiana Purchase in 1803.

We have just discussed Alexander Hamilton's method for apportionment. In this section, we will also discuss Thomas Jefferson's method for apportionment, Daniel Webster's method for apportionment, and John Quincy Adams's method for apportionment. Here is a brief history of the use of these different methods to apportion members of the U.S. House of Representatives. Although Hamilton's method for apportionment was the first method approved by Congress after the 1790 census, it was not used until 1852. Shortly after Congress approved Hamilton's method, President George Washington vetoed it. Congress then adopted a method developed by Thomas Jefferson. As there were apportionment problems with Jefferson's method in 1822 and 1832, Webster's method replaced Jefferson's method in 1842. Webster's method had similar flaws as Jefferson's method and was replaced in 1852 by Hamilton's method, the first apportionment method proposed. When it was discovered in the late 1800s that there were also flaws with Hamilton's method, Webster's method was once again used to apportion the number of representatives for each state. Webster's method was used from 1900 until it was replaced by the current method in 1941, the Hill–Huntington method (see Exercise 51). Adams's method, although considered, was never actually used by Congress to apportion members.

Jefferson's Method

To determine the standard quota with Hamilton's method, the population of each group was divided by the standard divisor. Then each lower quota was used. Any leftover items were distributed to the groups with the highest fractional part. As we saw with Examples 2 and 3, each group received either its lower quota or its upper quota. A weakness that can occur with Hamilton's method is that the groups that received their upper quota gained an advantage over the groups that received their lower quota. Is there a way to modify the quotas to overcome this weakness in Hamilton's method? Jefferson's method uses a divisor other than the standard divisor, called the ***modified divisor***, to obtain quotas. A modified divisor is a divisor that approximates the standard divisor. We will explain how to determine modified divisors shortly. **A modified divisor is used in Jefferson's method, Webster's method, and Adams's method, each of which will be discussed shortly**.

In each case, a modified divisor is used to obtain ***modified quotas***. Modified quotas are obtained by dividing a group's population by a modified divisor. **The differences in Jefferson's method, Webster's method, and Adams's method are in the modified divisor used to obtain the modified quotas and in the rounding process used to apportion the items. Hamilton's method is the only method that uses the standard quota rather than a modified quota in determining apportionments**.

Now we will discuss Jefferson's method.

JEFFERSON'S METHOD

1. Determine a modified divisor, d, such that when each group's modified quota is *rounded **down** to the nearest integer*, the total of the integers is the exact number of items to be apportioned. We will refer to the modified quotas that are rounded down as **modified lower quotas**.
2. Apportion to each group its modified lower quota.

With Jefferson's method, we use a modified divisor that is *less than* the standard divisor. This results in the modified quotas being slightly greater than the standard quotas. When using Hamilton's method, the standard quotas are *rounded down* and then leftover items are distributed. With Jefferson's method, the modified quotas are also *rounded down*. The modified divisor used with Jefferson's method must result in all items being distributed, with no items left over, when the modified quotas are rounded down. We will explain Jefferson's method in Example 4.

EXAMPLE 4 *Using Jefferson's Method for Apportioning Legislative Seats*

Consider the Republic of Geranium from Example 3. Use Jefferson's method to apportion the 275 legislature seats among the five states. Use the population from Table 15.31.

Table 15.31 Republic of Geranium Population

State	A	B	C	D	E	Total
Population	1,003,200	1,228,600	4,990,700	813,000	764,500	8,800,000

SOLUTION In Example 3, the standard divisor was calculated to be 32,000. With Jefferson's method, the modified quota for each group needs to be slightly greater than the standard quota. To accomplish this we use a *modified divisor, which is slightly less than the standard divisor*. When you divide a quantity by a smaller number, the quotient becomes greater. There is no formula to determine this new modified divisor, and usually more than one divisor will work. We will use trial and error until we find a divisor that apportions exactly 275 legislative seats using modified lower quotas.

Suppose that we try 31,500 as our modified divisor, d. To determine each modified quota, we will divide each state's population by 31,500. For example, the modified quota for state A is

$$\frac{1,003,200}{31,500} \approx 31.85$$

We will round modified quotas to the nearest hundredth as we did with standard quotas. Since the modified quota for state A is 31.85, the modified *lower* quota for state A is 31. This result is indicated in column 1, row 4, of Table 15.32.

Table 15.32 shows the results for all the states. In the table, we give the standard quotas as references, but they are not used in obtaining the apportionment using Jefferson's method.

Table 15.32 Republic of Geranium Population Using a Modified Divisor, $d = 31,500$

State	A	B	C	D	E	Total
Population	1,003,200	1,228,600	4,990,700	813,000	764,500	8,800,000
Standard quota	31.35	38.39	155.96	25.41	23.89	
Modified quota	31.85	39.00	158.43	25.81	24.27	
Modified lower quota	31	39	158	25	24	277

The sum of the modified lower quotas, 277, exceeds the number of available seats, 275. If the sum of the modified lower quotas is too high (as it is in this case), use a higher modified divisor. If the sum of the modified lower quotas is too low, use a lower modified divisor. Since our sum is too high, we will try a higher modified divisor. Let's see what happens if we use 31,600 as our modified divisor. This modified divisor is a little higher than the modified divisor of 31,500 previously used.

Using 31,600 as the modified divisor, the modified quota for state A is

$$\frac{1{,}003{,}200}{31{,}600} \approx 31.75$$

Table 15.33 shows the modified quotas for all the states.

Table 15.33 Republic of Geranium Population Using a Modified Divisor, $d = 31{,}600$

State	A	B	C	D	E	Total
Population	1,003,200	1,228,600	4,990,700	813,000	764,500	8,800,000
Standard quota	31.35	38.39	155.96	25.41	23.89	
Modified quota	31.75	38.88	157.93	25.73	24.19	
Modified lower quota	31	38	157	25	24	275

The sum of the modified lower quotas is now 275, our desired sum. Each state is awarded the number of legislative seats listed in Table 15.33 under the category of modified lower quota.

In Example 4, we used a modified divisor of 31,600 to obtain the desired sum of 275. There are an infinite number of modified divisors (in a small range) that could be used to obtain the modified lower quotas that we obtained. In fact, any modified divisor from about 31,587 to about 31,787 would result in the modified lower quotas we obtained in Table 15.33. If you selected a modified divisor lower than about 31,587, you would obtain modified lower quotas for which the sum of the seats is too high. For example, if you selected a modified divisor of 31,586, you would obtain modified lower quotas of 31, 38, 158, 25, and 24. The sum of these seats is 276, which is above the number of seats to be allocated. If you selected a modified divisor greater than about 31,787, you would obtain modified lower quotas for which the sum of the seats is too low. For example, if you selected a modified divisor of 31,788, you would obtain modified lower quotas of 31, 38, 156, 25, and 24. The sum of these seats is 274, which is below the number of seats to be allocated.

With Jefferson's method, we will always use a modified divisor that is less than the standard divisor used with Hamilton's method.

Recall from the discussion of Hamilton's method that if an apportionment method assigns either the lower or upper quota, it is said to satisfy the quota rule. Hamilton's method satisfies the quota rule since it uses either the lower quota or upper quota of the standard quota. Notice that the standard quota for state C from Example 3 is 155.96. According to the quota rule, state C should receive either 155 or 156 seats. Using Jefferson's method, state C received 157 seats (see Table 15.33 above). Therefore, Jefferson's method violates the quota rule.

The first case in the House of Representatives in which Jefferson's method led to a violation of the quota rule occurred in 1832. New York had a standard quota of 38.59 but was awarded 40 seats using Jefferson's method. Daniel Webster argued that this result was unconstitutional and suggested a compromise between Hamilton's method and Jefferson's method, leading to Webster's method. Let's discuss how Webster proposed to apportion the seats in the House of Representatives.

DID YOU KNOW?

Mathematical Contributions from Political Leaders

Daniel Webster (1782–1852)

Daniel Webster, a native of New Hampshire and educated at Dartmouth College, was a member of the Senate for many years. As a result of his successful defense of several famous constitutional cases argued before the Supreme Court, he was regarded as one of the leading lawyers of the country. He was also regarded as one of the best-known orators of his time.

Webster's Method

Hamilton's method rounded ***standard*** *quotas* ***down*** to the nearest integer. Jefferson's method rounded ***modified*** *quotas* ***down*** to the nearest integer. With Webster's method, we round ***modified*** *quotas* ***to*** the nearest integer such that the rounding gives the exact sum to be apportioned.

WEBSTER'S METHOD

1. Determine a modified divisor, d, such that when each group's modified quota is *rounded* ***to the nearest*** *integer*, the total of the integers is the exact number of items to be apportioned. We will refer to modified quotas that are rounded to the nearest integer as **modified rounded quotas**.
2. Apportion to each group its modified rounded quota.

With Webster's method, we use a modified divisor that may be *less than*, *equal to*, or *greater than* the standard divisor and we use *modified rounded quotas*. With Webster's method, as with Jefferson's method, the modified divisor used must result in all the items being distributed, with no items left over. We will explain Webster's method in Example 5.

EXAMPLE 5 *Using Webster's Method for Apportioning Legislative Seats*

Once again, let's consider the Republic of Geranium and apportion the 275 seats among the five states using Webster's method. Use the population from Table 15.34.

Table 15.34 Republic of Geranium Population

State	A	B	C	D	E	Total
Population	1,003,200	1,228,600	4,990,700	813,000	764,500	8,800,000

SOLUTION Our first task is to determine the modified divisor, d. We could choose a number either less than, greater than, or equal to the standard divisor. As with Jefferson's method, there is no formula to determine this new divisor and usually there is more than one divisor that works. Again, we must use trial and error until we find a divisor that apportions the 275 legislative seats.

Let's use the results from Example 3 on page 1011 as a guide. If we round the standard quotas from Example 3 to the nearest integer, the sum is $31 + 38 + 156 + 25 + 24$, or 274. This number is too low since we want a sum of 275. If the sum is too low, as it is in this case, we need to use a smaller divisor. If the sum is too high, we need to use a larger divisor. In Example 3, we used the standard divisor of 32,000. Let's try using 31,950 as our modified divisor, d. Using 31,950 as the modified divisor, the modified quota for state A is

$$\frac{1,003,200}{31,950} \approx 31.40$$

Table 15.35 shows the results using $d = 31,950$.

Table 15.35 Republic of Geranium Population Using a Modified Divisor, $d = 31{,}950$

State	A	B	C	D	E	Total
Population	1,003,200	1,228,600	4,990,700	813,000	764,500	8,800,000
Standard quota	31.35	38.39	155.96	25.41	23.89	
Modified quota	31.40	38.45	156.20	25.45	23.93	
Modified rounded quota	31	38	156	25	24	274

Now round each modified quota to the nearest integer to get the modified rounded quotas, as shown in Table 15.35. Our sum of the modified rounded quotas, 274, is still too low, so we will try a lower modified divisor. Let us try 31,900. Using 31,900 as the modified divisor, the modified quota for state A is

$$\frac{1{,}003{,}200}{31{,}900} \approx 31.45$$

Table 15.36 shows the results using $d = 31{,}900$.

Table 15.36 Republic of Geranium Population Using a Modified Divisor, $d = 31{,}900$

State	A	B	C	D	E	Total
Population	1,003,200	1,228,600	4,990,700	813,000	764,500	8,800,000
Standard quota	31.35	38.39	155.96	25.41	23.89	
Modified quota	31.45	38.51	156.45	25.49	23.97	
Modified rounded quota	31	39	156	25	24	275

Now the sum of the modified rounded quotas is 275, as desired. Therefore, each state is awarded the number of legislative seats listed in Table 15.36 under the category of modified rounded quota. ●

In Example 5, we used a modified divisor of 31,900 to obtain the desired sum of 275. There are an infinite number of modified divisors (in a small range) that could be used to obtain the modified rounded quotas we obtained in Table 15.36. In fact, any modified divisor from about 31,891 to about 31,911 would result in the modified rounded quotas we obtained in Table 15.36. If you selected a modified divisor lower than about 31,891, you would obtain modified rounded quotas for which the sum of the seats is too high. If you selected a modified divisor greater than about 31,911, you would obtain modified rounded quotas for which the sum of the seats is too low.

Occasionally, the results of Webster's method agree with the results from Hamilton's method. With Webster's method, we can use a modified divisor that is less than, greater than, or equal to the standard divisor used with Hamilton's method.

Webster's method is similar to Jefferson's method since they both make use of a modified divisor. Webster's method, however, is a little more difficult in practice to apply than Jefferson's method since the modified divisor could be less than, equal to, or greater than the standard divisor. Webster's method may seem the most reasonable since the modified quotas are rounded in the conventional way. Webster's method,

DID YOU KNOW?

Mathematical Contributions from Political Leaders

John Quincy Adams (1767–1848)

John Quincy Adams, sixth president of the United States, was the son of John Adams, second president of the United States. Adams's varied career included the positions of diplomat, senator, secretary of state, president, and member of the House of Representatives. Adams was instrumental in the expansion of U.S. borders southward and westward. He convinced Spain to cede Florida to the United States and also persuaded Spain to agree that the western border of the Louisiana Purchase extended all the way to the Pacific Ocean.

however, does have a flaw in that it can violate the quota rule. Even though Example 5 did not uncover this flaw, there are examples in which Webster's method violates the quota rule. This violation happens more in theory than in practice. As a result, many experts consider Webster's method the best overall apportionment method.

Let's discuss one more apportionment method that was proposed (but never used by the House of Representatives) by John Quincy Adams around the same time that Webster proposed his method.

Adams's Method

When Jefferson's method was discovered to be problematic because it violated the quota rule, John Quincy Adams proposed a method that is exactly the opposite of Jefferson's method. Instead of using modified lower quotas as Jefferson proposed, Adams suggested using modified upper quotas.

ADAMS'S METHOD

1. Determine a modified divisor, d, such that when each group's modified quota is *rounded* ***up*** *to the nearest integer* the total of the integers is the exact number of items to be apportioned. We will refer to the modified quotas that are rounded up as **modified upper quotas**.
2. Apportion to each group its modified upper quota.

With Adams's method, we use a modified divisor that is *greater than* the standard divisor and we use quotas that are *rounded up*, or *modified upper quotas*. As with Jefferson's and Webster's methods, the modified divisor used must result in all items being distributed, with no items left over.

Let's see what happens when we apply Adams's method to the Republic of Geranium.

EXAMPLE 6 *Using Adams's Method for Apportioning Legislative Seats*

Once again, consider the Republic of Geranium. Apportion the 275 seats among the five states using Adams's method. Use the population from Table 15.37.

Table 15.37 Republic of Geranium Population

State	**A**	**B**	**C**	**D**	**E**	**Total**
Population	1,003,200	1,228,600	4,990,700	813,000	764,500	8,800,000

SOLUTION With Adams's method, the modified quota for each group needs to be slightly smaller than the standard quota. To accomplish that, we use a modified divisor that is slightly greater than the standard divisor. When you divide a quantity by a larger number, the quotient becomes smaller. Since the standard divisor is 32,000, let's try letting $d = 32{,}350$. Using 32,350 as the modified divisor, the modified quota for state A is

$$\frac{1{,}003{,}200}{32{,}350} \approx 31.01$$

Table 15.38 shows the results.

Table 15.38 Republic of Geranium Population Using a Modified Divisor, $d = 32{,}350$

State	A	B	C	D	E	Total
Population	1,003,200	1,228,600	4,990,700	813,000	764,500	8,800,000
Standard quota	31.35	38.39	155.96	25.41	23.89	
Modified quota	31.01	37.98	154.27	25.13	23.63	
Modified upper quota	32	38	155	26	24	275

Using $d = 32{,}350$ and rounding up to the modified quotas, we have a sum of 275 seats, as desired. Therefore, each state will be awarded the number of legislative seats listed in Table 15.38 under the category of modified upper quota. •

In Example 6, we used a modified divisor of 32,350 to obtain the desired sum of 275 seats. There are an infinite number of modified divisors (in a small range) that could be used to obtain the modified upper quotas that we obtained. In fact, any modified divisor from about 32,336 to about 32,356 would result in the modified upper quotas we obtained in Table 15.38. If you selected a modified divisor lower than about 32,336, you would obtain modified upper quotas for which the sum of the seats is too high. If you selected a modified divisor greater than about 32,356, you would obtain modified upper quotas for which the sum of the seats is too low.

With Adams's method, we will always use a modified divisor that is greater than the standard divisor used with Hamilton's method.

In some cases, the range of numbers that can be used as a modified divisor is very narrow and you may need to use a decimal. For example, 34 may be too small for a modified divisor and 35 may be too large for a modified divisor. In that case, you will need to use a decimal number between 34 and 35 as the modified divisor.

Looking at the results from Example 6, we may want to conclude that Adams's method is the best since the quota rule was not violated in this example. Unfortunately, that is not always the case. We have seen that Jefferson's method can result in a state receiving more than its upper quota of seats. Adams's method can lead to just the opposite: a state receiving fewer than its lower quota of seats. Although the Jefferson method favors large states, such as his state of Virginia, Adams's method favors small states, such as those in his area of New England.

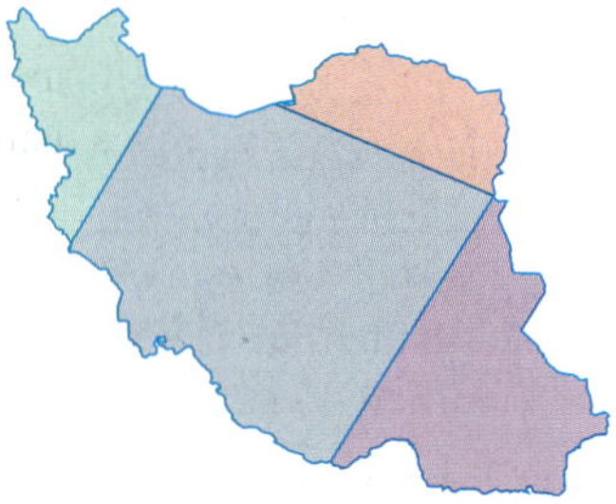

EXAMPLE 7 *Comparing Apportionment Methods*

Suppose that the Panda Republic is a small country with a population of 10,400,000 that consists of four states. There are 260 seats in the legislature that need to be apportioned among the four states. The population of each state is shown in Table 15.39.

Table 15.39

State	A	B	C	D	Total
Population	3,350,000	1,850,000	2,365,000	2,835,000	10,400,000

a) Find each state's apportionment using Hamilton's method.

b) Find each state's apportionment using Jefferson's method.

c) Find each state's apportionment using Adams's method.

d) Find each state's apportionment using Webster's method.

SOLUTION

a) First we must find the standard divisor. The standard divisor is found by dividing the total population by the number of seats to be apportioned. The standard divisor is

$$\frac{10{,}400{,}000}{260} = 40{,}000$$

Next we find the standard quotas by dividing each state's population by the standard divisor, 40,000. Recall that we round standard quotas to the nearest hundredth. With Hamilton's method, we round each standard quota down to obtain the lower quota. Then we distribute any leftover seats to the states with the largest fractional parts until all seats are distributed. Table 15.40 shows the apportionment using Hamilton's method.

Table 15.40 Hamilton's Method, Standard Divisor of 40,000

State	A	B	C	D	Total
Population	3,350,000	1,850,000	2,365,000	2,835,000	10,400,000
Standard quota	83.75	46.25	59.13	70.88	
Lower quota	83	46	59	70	258
Hamilton's apportionment	84	46	59	71	260

Since states D and A, in that order, had the highest fractional parts (0.88 and 0.75, respectively), each receives one additional seat.

b) With Jefferson's method, we use a modified divisor that is *less than* the standard divisor and use modified *lower quotas*. Table 15.41 shows the results using a modified divisor of 39,700.

Table 15.41 Jefferson's Method, Modified Divisor, $d = 39{,}700$

State	A	B	C	D	Total
Population	3,350,000	1,850,000	2,365,000	2,835,000	10,400,000
Modified quota	84.38	46.60	59.57	71.41	
Modified lower quota	84	46	59	71	260

Using $d = 39{,}700$, we have a sum of 260 seats, as desired. Therefore, each state receives the number of seats listed under the modified lower quotas in Table 15.41.

c) With Adams's method, we use a modified divisor that is *greater than* the standard divisor and use modified *upper quotas*. Table 15.42 shows the results using a modified divisor of 40,300.

Table 15.42 Adams's Method, Modified Divisor, $d = 40{,}300$

State	A	B	C	D	Total
Population	3,350,000	1,850,000	2,365,000	2,835,000	10,400,000
Modified quota	83.13	45.91	58.68	70.35	
Modified upper quota	84	46	59	71	260

Using $d = 40{,}300$, we have a sum of 260 seats, as desired. Therefore, each state receives the number of seats listed under the modified upper quotas in Table 15.42.

d) With Webster's method, we need to find a modified divisor such that when each state's modified quota is rounded to the nearest integer, the total of the integers is 260. The modified divisor may be *less than*, *equal to*, or *greater than* the standard divisor. If we round the standard quotas from part (a) to the nearest integer, the sum of the rounded quotas is 260. Therefore, we will use the standard divisor as our modified divisor. Table 15.43 shows the results when we round the modified quotas to the nearest integer.

Table 15.43 Webster's Method, Modified Divisor, $d = 40{,}000$

State	A	B	C	D	Total
Population	3,350,000	1,850,000	2,365,000	2,835,000	10,400,000
Standard quota	83.75	46.25	59.13	70.88	
Modified quota	83.75	46.25	59.13	70.88	
Modified rounded quota	84	46	59	71	260

Each state receives the number of seats listed under the modified rounded quotas in Table 15.43. Note that in this example, all four methods led to the same apportionment. However, that is not always the case.

Of the four methods we have discussed in this section, Hamilton's method uses standard quotas. Jefferson's method, Webster's method, and Adams's method all make use of a modified quota and can all lead to violations of the quota rule. As we will see in the next section, Hamilton's method can also be problematic by producing paradoxes. Table 15.44 summarizes the four apportionment methods.

Table 15.44 Summary of Apportionment Methods

Method	Divisor	Apportionment
Hamilton's	$\text{Standard divisor} = \dfrac{\text{total population}}{\text{number of items to be allocated}}$	Round each standard quota down. Distribute any leftover items to the groups with the largest fractional parts until all items are distributed.
Jefferson's	The modified divisor is *less than* the standard divisor.	Each group's modified quota is *rounded* ***down*** *to the nearest integer*. Apportion to each group its modified lower quota.
Webster's	The modified divisor is *less than*, *greater than*, or *equal to*, the standard divisor.	Each group's modified quota is *rounded* ***to the nearest*** *integer*. Apportion to each group its modified rounded quota.
Adams's	The modified divisor is *greater than* the standard divisor.	Each group's modified quota is *rounded* ***up*** *to the nearest integer*. Apportion to each group its modified upper quota.

SECTION 15.3 EXERCISES

CONCEPT/WRITING EXERCISES

1. How do you determine a standard divisor?
2. How do you determine a standard quota?
3. What is a lower quota?
4. What is an upper quota?
5. What is the quota rule?
6. Which apportionment method requires rounding the standard quota to the lower quota?
7. Which apportionment method(s) require using modified quotas?
8. a) Which apportionment method uses a modified divisor that is less than the standard divisor?

   b) Which apportionment method uses a modified divisor that is greater than the standard divisor?

   c) Which apportionment method uses a modified divisor that could be either less than, greater than, or equal to the standard divisor?
9. a) Which apportionment method uses a modified quota that is always rounded to the nearest integer?

   b) Which apportionment method uses a modified quota that is always rounded up to the nearest integer?

   c) Which apportionment method uses a modified quota that is always rounded down to the nearest integer?
10. Which apportionment method(s) can violate the quota rule?

PRACTICE THE SKILLS/PROBLEM SOLVING

In Exercises 11–50, round quotas to the nearest hundredth.

Legislative Seats *In Exercises 11–18, suppose that Turtlestan is a small country with a population of 8,000,000 that consists of four states, A, B, C, and D. There are 160 seats in the legislature that need to be apportioned among the four states. The population of each state is shown in the table.*

State	A	B	C	D	Total
Population	1,345,000	2,855,000	982,000	2,818,000	8,000,000

11. a) Determine the standard divisor.

    b) Determine each state's standard quota.
12. Determine each state's apportionment using Hamilton's method.
13. a) Determine each state's modified quota using the divisor 49,300.

    b) Determine each state's apportionment using Jefferson's method.
14. a) Determine each state's modified quota using the divisor 49,250.

    b) Determine each state's apportionment using Jefferson's method.
15. a) Determine each state's modified quota using the divisor 50,700.

    b) Determine each state's apportionment using Adams's method.
16. a) Determine each state's modified quota using the divisor 50,600.

    b) Determine each state's apportionment using Adams's method.
17. Determine each state's apportionment using Webster's method using the standard divisor.
18. a) Determine each state's modified quota using the divisor 50,100.

    b) Determine each state's apportionment using Webster's method.

Hotel Staff *In Exercises 19–26, a large hotel chain needs to apportion 25 new staff members among three hotels based on the numbers of rooms in each hotel as shown below.*

Hotel	A	B	C	Total
Number of rooms	306	214	155	675

19. a) Determine the standard divisor.

b) Determine each hotel's standard quota.

20. Determine each hotel's apportionment using Hamilton's method.

21. a) Determine each hotel's modified quota using the divisor 25.8.

b) Determine each hotel's apportionment using Jefferson's method.

22. a) Determine each hotel's modified quota using the divisor 25.7.

b) Determine each hotel's apportionment using Jefferson's method.

23. a) Determine each hotel's modified quota using the divisor 29.

b) Determine each hotel's apportionment using Adams's method.

24. a) Determine each hotel's modified quota using the divisor 28.

b) Determine each hotel's apportionment using Adams's method.

25. Determine each hotel's apportionment using Webster's method using the standard divisor.

26. a) Determine each hotel's modified quota using the divisor 28.1.

b) Determine each hotel's apportionment using Webster's method.

New Skis *In Exercises 27–30, Mountain View Resorts operates four ski resorts in the Rocky Mountains. Mountain View Resorts plans to apportion 50 pairs of new skis among the four resorts based on the number of rooms in each resort as shown below.*

Resort	A	B	C	D	Total
Number of rooms	86	102	130	232	550

27. a) Determine the standard divisor.

b) Determine each resort's standard quota.

c) Determine each resort's apportionment using Hamilton's method.

28. Determine each resort's apportionment using Jefferson's method. (*Hint:* Some divisors between 10 and 11 will work.)

29. Determine each resort's apportionment using Adam's method. (*Hint:* Some divisors between 11 and 12 will work.)

30. Determine each resort's apportionment using Webster's method. (*Hint:* Some divisors between 10.5 and 11.5 will work.)

New Computers *In Exercises 31–34, a university is made up of five schools: Liberal Arts, Sciences, Engineering, Business, and Humanities. There are 250 new computers to be apportioned among the five schools based on their enrollment shown below. The total enrollment is 13,000.*

School	Liberal Arts	Sciences	Engineering	Business	Humanities
Enrollment	1746	7095	2131	937	1091

▲ Princeton University

31. a) Determine the standard divisor.

b) Determine each school's standard quota.

c) Determine each school's apportionment using Hamilton's method.

32. Determine each school's apportionment using Adams's method. (*Hint:* Some divisors between 52 and 53 will work.)

33. Determine each school's apportionment using Jefferson's method. (*Hint:* Some divisors between 51 and 52 will work.)

34. Determine each school's apportionment using Webster's method.

New Cars *In Exercises 35–38, a car manufacturer has 150 cars of a new model to be apportioned to four dealerships. The manufacturer decides to apportion the cars based on the number of cars each dealership sold in the previous year. The number of cars sold by each dealership is shown below.*

Dealership	A	B	C	D	Total
Annual Sales	4800	3608	2990	2102	13,500

35. a) Determine the standard divisor.

b) Determine each dealership's standard quota.

c) Determine each dealership's apportionment using Hamilton's method.

36. Determine each dealership's apportionment using Jefferson's method. (*Hint:* Some divisors between 88 and 89 will work.)

37. Determine each dealership's apportionment using Webster's method. (*Hint:* Some divisors between 89 and 90 will work.)

38. Determine each dealership's apportionment using Adams's method.

New Buses *In Exercises 39–42, the Transit Department in the city of Houston has 100 new buses to be apportioned among six routes. The department decides to apportion the buses based on the average number of daily passengers per route, as shown in the table below.*

Route	A	B	C	D	E	F	Total
Passengers	9070	15,275	12,810	5720	25,250	6875	75,000

39. a) Determine the standard divisor.

b) Determine each route's standard quota.

c) Determine each route's apportionment using Hamilton's method.

40. Determine each route's apportionment using Jefferson's method.

41. Determine each route's apportionment using Adam's method.

42. Determine each route's apportionment using Webster's method.

Nursing Shifts *In Exercises 43–46, a hospital has 200 nurses to be apportioned among four shifts: Shifts A, B, C, and D. The hospital decides to apportion the nurses based on the average number of room calls reported during each shift. Room calls are shown in the table below.*

Shift	A	B	C	D	Total
Room calls	751	980	503	166	2400

43. a) Determine the standard divisor.

b) Determine each shift's standard quota.

c) Determine each shift's apportionment using Hamilton's method.

44. Determine each shift's apportionment using Adams's method. (*Note:* Divisors do not have to be whole numbers.)

45. Determine each shift's apportionment using Jefferson's method. (*Note:* Divisors do not have to be whole numbers.)

46. Determine each shift's apportionment using Webster's method. (*Hint:* Some divisors between 12 and 12.05 will work.)

CHALLENGE PROBLEMS/GROUP ACTIVITIES

47. ***The First Census*** The first census taken in the United States after the adoption of the U.S. Constitution was in 1790. The following table shows the results of this census.

State	Population
Connecticut	236,841
Delaware	55,540
Georgia	70,835
Kentucky	68,705
Maryland	278,514
Massachusetts	475,327
New Hampshire	141,822
New Jersey	179,570
New York	331,589
North Carolina	353,523
Pennsylvania	432,879
Rhode Island	68,446
South Carolina	206,236
Vermont	85,533
Virginia	630,560
Total	3,615,920

a) Determine the apportionment that would have resulted if Hamilton's method had been used as the method originally approved by Congress. One hundred five seats were to be distributed.

b) Determine the apportionment that was used with Jefferson's method.

c) Compare the apportionments from parts (a) and (b). Which state(s) benefited from Jefferson's method? Which state(s) were at a disadvantage from Jefferson's method?

48. ***Legislative Seats*** Suppose that a country with a population of 10,000,000 has 250 legislative seats to be apportioned among four states, where each state has a different population. Determine a population for each state in which Hamilton's method, Jefferson's method, Webster's method, and Adams's method all lead to the same apportionment of the 250 legislative seats. Many answers are possible.

49. ***Police Officers*** Suppose that a police department has 210 new officers to apportion among six precincts. The department plans to apportion the officers based on the number of crimes committed during the previous year in each precinct. Suppose that the number of crimes committed in each precinct is different and that the total number of crimes committed in all six precincts was 2940. Determine the number of crimes committed in each precinct such that Hamilton's method, Jefferson's method, Webster's method, and Adams's method all lead to the same apportionment of the 210 new officers. Many answers are possible.

INTERNET/RESEARCH ACTIVITIES

50. Do research and write a report on the apportionment method used in the House of Representatives in 1872. Include in your report a description of the method used.

51. Do research and write a report on the Huntington–Hill method, the current method used to apportion the representatives in the House of Representatives. Include historical background and describe how the method works.

15.4 FLAWS OF THE APPORTIONMENT METHODS

▲ It is possible for a computer lab to receive fewer printers when 51 printers are to be apportioned than when 50 printers are to be apportioned.

Recall from Section 15.3 the situation in which your college receives a grant to purchase 50 laser printers that will be placed into five different computer labs on campus. Suppose that before the printers are purchased the price of the printers decreases enough so that 51 printers can now be purchased and apportioned to the five labs. Because of a flaw with the apportionment method, it is possible that one of the labs would receive fewer printers with 51 printers being apportioned than with 50 printers being apportioned. In this section, we will discuss this flaw and other flaws of apportionment methods.

Just as we have discovered that voting methods can have flaws, apportionment methods can also have flaws. In this section, as we did in Section 15.2 with voting methods, we will consider several reasonable properties that an apportionment method should have. Then we will see examples in which these properties are violated. Problems with apportionment can occur in a variety of ways. Population changes, changes in the number of items to be distributed, and the addition of one or more groups can lead to problems with apportionment.

Recall from Section 15.3 that the quota rule states that an apportionment for every group under consideration should always be either the upper quota or the lower quota. Hamilton's method, which satisfies the quota rule, would appear to be a reasonable and fair apportionment method. As we will discuss in this section, though, Hamilton's method can result in some serious flaws. In this section, we will discuss three flaws of Hamilton's method: the ***Alabama paradox***, the ***population paradox***, and the ***new-states paradox***. **These flaws only apply to Hamilton's method and not to Jefferson's method, Webster's method, or Adams's method.**

The Alabama Paradox

The first, and perhaps most serious, flaw of Hamilton's method occurs when an increase in the total number of items to be apportioned results in a loss of an item for one of the groups. This flaw first occurred in the apportionment of the House of Representatives in 1880 when a discussion occurred on whether to have 299 or 300 members in the House. Using Hamilton's method with 299 members, Alabama would receive eight seats. But if the total number of representatives were increased to 300, Alabama would only receive seven seats. As a result, this situation became known as the ***Alabama paradox.***

ALABAMA PARADOX

The **Alabama paradox** occurs when an increase in the total number of items to be apportioned results in a loss of an item for a group.

Example 1 illustrates the Alabama paradox.

EXAMPLE 1 *Demonstrating the Alabama Paradox*

Consider Silverstar, a small country with a population of 18,000 people and three states A, B, and C (Table 15.45). There are 150 seats in the legislature that must be apportioned among the three states, according to their population. Show that the Alabama paradox occurs if the number of seats is increased to 151.

Round standard divisors and standard quotas to the nearest hundredth.

Table 15.45

State	A	B	C	Total
Population	888	8076	9036	18,000

SOLUTION With 150 seats in the legislature, the standard divisor is

$$\frac{18{,}000}{150} = 120$$

The standard quotas for each state and the apportionment for each state are shown in Table 15.46.

Table 15.46

State	A	B	C	Total
Population	888	8076	9036	18,000
Standard quota	7.40	67.30	75.30	
Lower quota	7	67	75	149
Hamilton's apportionment	8	67	75	150

With 151 seats in the legislature, the standard divisor is

$$\frac{18{,}000}{151} \approx 119.21$$

The standard quotas for each state and the apportionment for each state with 151 total seats are shown in Table 15.47.

Table 15.47

State	A	B	C	Total
Population	888	8076	9036	18,000
Standard quota	7.45	67.75	75.80	
Lower quota	7	67	75	149
Hamilton's apportionment	7	68	76	151

When the number of seats increased from 150 to 151, state A's apportionment actually decreased, from 8 to 7. This example illustrates the Alabama paradox.

When the total number of items increases, each group's standard quota increases, but not by the same amount. Therefore, it is possible that the order of assignment of the items can change. As a result, some groups can lose items they already had. When the total number of items increases, usually the larger groups benefit at the expense of the smaller groups.

The next paradox we will discuss is the population paradox.

Population Paradox

Another paradox that can occur with Hamilton's method may occur when the population of one or more states changes. It was discovered in the early 1900s that under Hamilton's method, one state could lose a seat to another state even though its population is growing at a faster rate. At the time, Virginia lost a seat in the House of Representatives while Maine gained a seat, although Virginia's population was growing at a much faster rate than Maine's population. Thus, this paradox became known as the *population paradox*.

POPULATION PARADOX

The **population paradox** occurs when group A loses items to group B even though group A's population grew at a faster rate than group B's.

Example 2 illustrates the population paradox.

EXAMPLE 2 *Demonstrating the Population Paradox*

Consider Alexandria, a small country with a population of 100,000 and three states A, B, and C. There are 100 seats in the legislature that must be apportioned among the three states. Using Hamilton's method, the apportionment is shown in Table 15.48.

Table 15.48

State	A	B	C	Total
Population	23,527	5548	70,925	100,000
Standard quota	23.53	5.55	70.93	
Lower quota	23	5	70	98
Hamilton's apportionment	23	6	71	100

DID YOU KNOW?

Redistricting or Gerrymandering?

In response to population changes, state legislatures may revise the boundaries of congressional districts. In 2006, the U.S. Supreme Court ruled that a state legislature can redraw district lines anytime. Prior to the 2006 ruling, states were only allowed to redraw district lines every 10 years. The intention of "redistricting" is to make certain that congressional seats are fairly apportioned among the residents of the state. Some legislatures, however, deliberately manipulate the boundaries of congressional districts to influence election outcomes. A legislator may design a boundary so that the district is full of his or her supporters. This process is called *gerrymandering*. Gerrymandering was first recognized in 1812 when Massachusetts' governor Elbridge Gerry convinced the state legislature to create an oddly shaped district, that some thought resembled a salamander, so as to give his political party an advantage in elections. Thus, the term *gerrymander* was born.

Suppose that the population increases according to Table 15.49 and that the 100 seats are reapportioned. Show that the population paradox occurs when Hamilton's method is used.

Table 15.49

State	A	B	C	Total
Population	23,926	5648	71,110	100,684

SOLUTION To calculate the percent increase, we use the procedure discussed in Section 11.1. State A has an increase of 399 people. Therefore, state A has a percent increase of

$$\frac{399}{23{,}527} \approx 0.01696 \approx 1.696\%$$

State B has an increase of 100 people. Therefore, state B has a percent increase of

$$\frac{100}{5548} \approx 0.01802 \approx 1.802\%$$

State C has an increase of 185 people. Therefore, state C has a percent increase of

$$\frac{185}{70{,}925} \approx 0.00261 \approx 0.261\%$$

All three states had an increase in their population, but state B is increasing at a faster rate than state A and state C.

The standard divisor using the new population is

$$\frac{100{,}684}{100} = 1006.84$$

Table 15.50 shows the reapportionment, using Hamilton's method using the standard divisor 1006.84.

Table 15.50

State	A	B	C	Total
Population	23,926	5,648	71,110	100,684
Standard quota	23.76	5.61	70.63	
Lower quota	23	5	70	98
Hamilton's apportionment	24	5	71	100

State B has lost a seat to state A even though state B's population grew at a faster rate than state A's. As a result, we have an example of the population paradox. ●

TIMELY TIP In Example 2, the population paradox occurred because state B lost a seat to state A even though state B's population grew at a faster rate than state A's population. When checking for the population paradox, it is possible for a state to lose a seat to another state without the population paradox occurring. Remember that for the population paradox to occur, the state that loses the seat must be growing at a faster rate than the state that gains the seat.

The next and final paradox we will discuss is the new-states paradox.

The New-States Paradox

The new-states paradox was discovered in 1907 when Oklahoma was added as a state. When a new state is added, new seats must be added to the legislature. How do we determine the number of new seats to add? A reasonable answer would be to add the number of seats the new state would be entitled to based on its population. When reapportioning the House of Representatives with the additional five seats Oklahoma was entitled to, Maine's apportionment increased from three to four seats and New York's apportionment decreased from 38 to 37. By adding a new state and its fair share of seats, New York was required to give a seat to Maine. This paradox became known as the *new-states paradox*.

NEW-STATES PARADOX

The **new-states paradox** occurs when the addition of a new group reduces the apportionment of another group.

EXAMPLE 3 *Demonstrating the New-States Paradox*

The Oklahoma Public Library System has received a grant to purchase 100 laptop computers to be distributed between two libraries A and B. The 100 laptops will be apportioned based on the population served by each library. The apportionment using Hamilton's method is shown in Table 15.51. The standard divisor is

$$\frac{10{,}000}{100} = 100$$

Table 15.51

Library	A	B	Total
Population	2145	7855	10,000
Standard quota	21.45	78.55	
Lower quota	21	78	99
Hamilton's apportionment	21	79	100

Suppose that an anonymous donor decides to donate money to purchase six more laptops for a third library that serves a population of 625. Show that the new-states paradox occurs when the laptops are reapportioned.

SOLUTION When the third library is added, the new standard divisor is

$$\frac{10{,}625}{106} = 100.24$$

The new standard quotas and Hamilton's apportionment are shown in Table 15.52 on page 1029.

Table 15.52 Using a New Standard Divisor of 100.24

Library	A	B	C	Total
Population	2145	7855	625	10,625
Standard quota	21.40	78.36	6.24	
Lower quota	21	78	6	105
Hamilton's apportionment	22	78	6	106

Before library C was added, library B would receive 79 laptops. By adding a new library and increasing the total number of laptops to be apportioned, library B ended up losing a laptop to library A. Thus, we have a case of the new-states paradox.

When a new group is added to the apportionment, we must determine how many additional items should be added to the total to be apportioned. In Example 3, we were told that six additional laptops were to be added to the total apportioned when a new library was included in the apportionment. If we are given the new group's population but are not given the number of additional items to be added to the total to be apportioned, we calculate the number of items to be apportioned to the new group as follows. First determine the new group's standard quota. The number of additional items would be the new group's standard quota *rounded down* to the nearest integer. For example, in Example 3, the new library, library C, served a population of 625. The standard quota for library C would be

$$\frac{625}{100} = 6.25$$

Rounding 6.25 down to the nearest integer gives us six laptops to be added to the total to be apportioned, or 106 laptops.

As we have discovered in Section 15.3, Hamilton's method appears to be a fair and reasonable apportionment method since it satisfies the quota rule. In this section, however, we discovered that Hamilton's method can produce paradoxes. Jefferson's, Adams's, and Webster's methods can all violate the quota rule but do not produce paradoxes. Hamilton's and Jefferson's methods can favor large states, whereas Adams's and Webster's methods can favor small states. Is there a perfect apportionment method that satisfies the quota rule, does not produce any paradoxes, and favors neither large nor small states? In 1980, mathematicians Michel Balinski and H. Payton Young proved that there is no apportionment method that satisfies the quota rule while also avoiding any paradoxes. Their theorem is called Balinski and Young's impossibility theorem.

BALINSKI AND YOUNG'S IMPOSSIBILITY THEOREM

There is no perfect apportionment method that satisfies the quota rule and avoids any paradoxes.

Table 15.53 on page 1030 summarizes the four apportionment methods we have discussed in this chapter and indicates which methods may violate the quota rule and which methods may produce the paradoxes we have discussed.

Table 15.53 Comparison of Apportionment Methods

	Apportionment Method			
	Hamilton	**Jefferson**	**Adams**	**Webster**
May violate the quota rule (apportionment should always be either upper or lower quota)	No	Yes	Yes	Yes
May produce the Alabama paradox (an increase in the total number of items results in a loss of an item for a group)	Yes	No	No	No
May produce the population paradox (group A loses an item to group B although group A's population grew faster than group B's population)	Yes	No	No	No
May produce the new-states paradox (the addition of a new group reduces the apportionment of another group)	Yes	No	No	No
Apportionment method favors	Large states	Large states	Small states	Small states

Just as there is no perfect voting method, there is also no perfect apportionment method.

SECTION 15.4 EXERCISES

CONCEPT/WRITING EXERCISES

1. What is the Alabama paradox?
2. What is the new-states paradox?
3. What is the population paradox?
4. Does Hamilton's method produce any paradoxes? If so, name them.
5. Which apportionment method(s) favors small states?
6. Which apportionment method(s) favors large states?

PRACTICE THE SKILLS/PROBLEM SOLVING

In Exercises 7–18, round quotas to the nearest hundredth.

7. ***Fax Machines*** Consider the apportionment of 60 fax machines for Shanahan Law Firm given in Example 2 on page 1010 of Section 15.3. The apportionment using Hamilton's method is shown in the table below.

Office	A	B	C	D	E	Total
Employees	246	201	196	211	226	1080
Standard quota	13.67	11.17	10.89	11.72	12.56	60.01
Lower quota	13	11	10	11	12	57
Hamilton's apportionment	14	11	11	12	12	60

Does the Alabama paradox occur using Hamilton's method if the number of fax machines is increased from 60 to 61? Explain your answer.

8. ***Ergonomic Chairs*** A large company with offices in four cities must distribute 144 new ergonomic chairs to the four offices. The chairs will be apportioned based on the number of employees in each office as shown in the table below.

Office	A	B	C	D	Total
Employees	739	277	618	958	2592

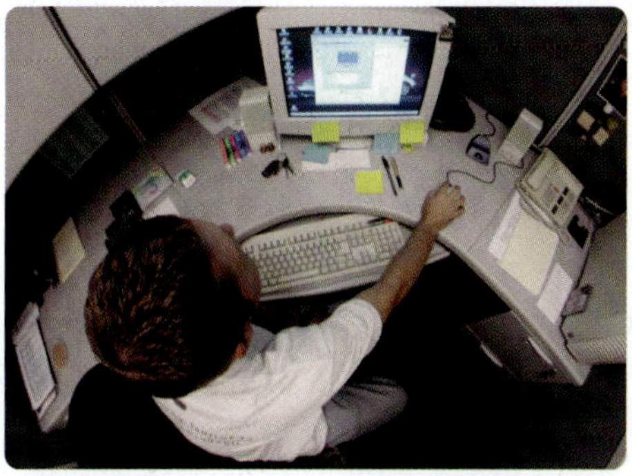

a) Apportion the chairs using Hamilton's method.

b) Does the Alabama paradox occur using Hamilton's method if the number of new chairs is increased from 144 to 145? Explain your answer.

9. ***Legislative Seats*** A country with three states has 30 seats in the legislature. The population of each state is shown in the table below.

State	A	B	C	Total
Population	161	250	489	900

a) Apportion the seats using Hamilton's method.

b) Does the Alabama paradox occur using Hamilton's method if the number of seats is increased from 30 to 31? Explain your answer.

10. ***Legislative Seats*** A country with three states has 200 seats in the legislature. The population of each state is shown in the table below.

State	A	B	C	Total
Population	247,100	481,900	271,000	1,000,000

a) Apportion the seats using Hamilton's method.

b) Does the Alabama paradox occur using Hamilton's method if the number of seats is increased from 200 to 201? Explain your answer.

In Exercises 11–14, assume that the number of items to be apportioned does not change.

11. ***Apportioning Promotions*** Eastman Kodak Company has 25,000 employees in three cities as shown in the table below. It wishes to give promotions to 200 employees.

Cities	A	B	C	Total
Employees	8130	4030	12,840	25,000

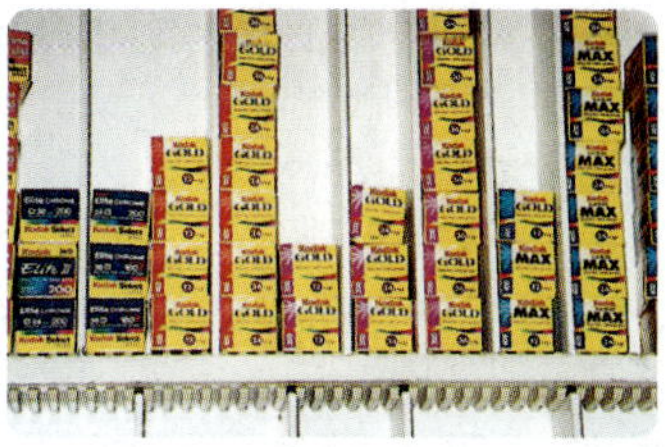

a) Apportion the promotions using Hamilton's method.

b) Suppose that in 10 years the cities have the following number of employees and the company wishes to again give promotions to 200 employees. Does the population paradox occur using Hamilton's method?

Cities	A	B	C	Total
Employees	8150	4030	12,945	25,125

12. ***Apportioning Doctors*** First Physicians Group has 30 doctors to apportion among 3 clinics. The doctors are to be apportioned based on the weekly average number of patients at each clinic as shown below.

Clinic	A	B	C	Total
Patients	217	436	337	990

a) Apportion the doctors using Hamilton's method.

b) Suppose that 1 year later the clinics have the following weekly average number of patients. If the 30 doctors are reapportioned to the clinics, does the population paradox occur using Hamilton's method?

Clinic	A	B	C	Total
Patients	230	450	362	1042

13. ***College Internships*** A college with five divisions has funds for 54 internships. The student population for each division is shown in the table below.

Division	A	B	C	D	E	Total
Population	733	1538	933	1133	1063	5400

a) Apportion the internships using Hamilton's method.

b) Suppose that 1 year later the divisions have the following populations. If the college wishes to apportion 54 internships, does the population paradox occur using Hamilton's method?

Division	A	B	C	D	E	Total
Population	733	1539	933	1133	1116	5454

14. ***Legislative Seats*** A country with three states has 250 seats in the legislature. The population of each state is shown in the table below.

State	A	B	C	Total
Population	459	10,551	18,990	30,000

a) Apportion the seats using Hamilton's method.

b) Suppose that in 10 years the states have the following population. If the country reapportions 250 seats in the legislature, does the population paradox occur using Hamilton's method?

State	A	B	C	Total
Population	464	10,551	19,100	30,115

15. ***Additional Employees*** Cynergy Telecommunications has employees in Europe, labeled A, and in the United States, labeled B. The number of employees in each group is shown in the table below. There are 48 managers to be apportioned between the two groups.

Cynergy Telecommunications	A	B	Total
Employees	844	3956	4800

a) Apportion the managers using Hamilton's method.

b) Suppose that additional employees in Asia, labeled C, with the number of employees shown in the table below, are added with seven new managers. Does the new-states paradox occur using Hamilton's method?

Cynergy Telecommunications	A	B	C	Total
Employees	844	3956	724	5524

16. ***Adding a Park*** The town of Manlius purchased 25 new picnic tables to be apportioned between two parks. The picnic tables are to be apportioned based on the annual number of visitors to each park as shown below.

Park	A	B	Total
Visitors	3750	6250	10,000

a) Apportion the picnic tables using Hamilton's method.

b) Suppose that the town decides to purchase five additional picnic tables and include a third park with an annual number of visitors as shown in the table above on the right. The town will now apportion 30 picnic tables among the three parks. Does the new-states paradox occur using Hamilton's method?

Park	A	B	C	Total
Visitors	3750	6250	2100	12,100

17. ***Adding a State*** A country with three states has 66 seats in the legislature. The population of each state is shown in the table below.

State	A	B	C	Total
Population	68,970	253,770	667,260	990,000

a) Apportion the seats using Hamilton's method.

b) Suppose that a fourth state with the population shown in the table below is added, with five additional seats. Does the new-states paradox occur using Hamilton's method?

State	A	B	C	D	Total
Population	68,970	253,770	667,260	85,800	1,075,800

18. ***Adding a State*** A country with two states has 33 seats in the legislature. The population of each state is shown in the table below.

State	A	B	Total
Population	744	2556	3300

a) Apportion the states using Hamilton's method.

b) Suppose that a third state with the population shown in the table below is added, with seven additional seats. Does the new-states paradox occur using Hamilton's method?

State	A	B	C	Total
Population	744	2556	710	4010

INTERNET/RESEARCH ACTIVITY

19. Write a paper on which apportionment method you think is the best. Include reasons to support your choice including the advantages and disadvantages of the method you have selected.

CHAPTER 15 SUMMARY

IMPORTANT FACTS

VOTING METHODS

- Plurality method
- Borda count method
- Plurality with elimination method
- Pairwise comparison method

FLAWS OF VOTING METHODS

- Majority criterion
- Head-to-head criterion
- Monotonicity criterion
- Irrelevant alternatives criterion

APPORTIONMENT METHODS

- Hamilton's method
- Jefferson's method
- Webster's method
- Adams's method

FLAWS OF APPORTIONMENT METHODS

- Alabama paradox
- Population paradox
- New-states paradox

CHAPTER 15 REVIEW EXERCISES

15.1

1. ***Electing the Club President*** The town of Riverside's chess club is holding an election to choose a club president. The 34 votes were cast as follows: Robert Rivera, 15 votes; Charlotte Kaiser, 12 votes; and Rhonda Timmerman, 7 votes.

   a) Using the plurality method, which candidate is elected president?

   b) Did this candidate receive a majority of the votes?

2. ***Electing a Chairperson*** Suffolk Community College is holding an election to appoint a chairperson of the board of trustees. The 413 faculty members vote as follows: Michelle MacDougal, 231 votes; Jeffrey Stuart, 155 votes; and Donald Campbell, 27 votes.

   a) Using the plurality method, who is elected?

   b) Did this candidate receive a majority of votes?

3. ***Ranking Candidates*** Ten voters are asked to rank four candidates. The 10 voters turn in the following ballots showing their preferences in order:

B	A	D	B	A	C	B	C	D	C
A	C	C	A	C	B	A	B	A	B
C	D	A	C	D	A	C	A	B	A
D	B	B	D	B	D	D	D	C	D

Make a preference table for these ballots.

4. ***Ranking Candidates*** Seven voters are asked to rank three candidates. The seven voters turn in the following ballots showing their preferences in order:

C	C	A	C	A	B	B
A	A	C	B	C	A	A
B	B	B	A	B	C	C

Make a preference table for these ballots.

In Exercises 5–10, the members of the Student Council at Manatee Community College must decide which item to use in a raffle to raise money for the Salvation Army. Their choices are an iPod (I), a portable DVD player (D), a digital camera (C), or a Nintendo Wii (N). The members rank their choices according to the following preference table.

Number of Votes	8	5	4	2	1
First	D	C	I	C	N
Second	I	I	N	N	D
Third	C	N	C	D	I
Fourth	N	D	D	I	C

5. How many members voted?

6. Using the plurality method, which item is chosen?

7. Using the Borda count method, which item is chosen?

8. Using the plurality with elimination method, which item is chosen?

9. Using the pairwise comparison method, which item is chosen?

10. Which item is selected if the plurality with elimination method is used and the item with the most *last*-place votes is eliminated at each step?

Sports Preferences In Exercises 11–16, the employees at Delphi Engineering must decide whether to play baseball (B), soccer (S), or volleyball (V) at their year-end picnic. The preference table follows.

Number of Votes	38	30	25	7	10
First	S	V	B	B	V
Second	B	S	V	S	B
Third	V	B	S	V	S

11. How many employees voted?

12. Determine the winner using the plurality method.

13. Determine the winner using the Borda count method.

14. Determine the winner using the plurality with elimination method.

15. Determine the winner using the pairwise comparison method.

16. Determine the winner if the plurality with elimination method is used and the candidate with the most *last*-place votes is eliminated at each step.

17. ***Choosing a License Plate Style*** Park Forest retirement community in Florida is purchasing a van to be used by the residents. The 372 residents are unable to agree on a style of license plate for the van and decide to hold an election in which the residents rank their choices from among the following: American Association of Retired Persons (A), standard issue Florida license plate (F), save the manatee (M), and protect the panther (P). The results of this election are given in the preference table below.

Number of Votes	161	134	65	12
First	A	A	F	M
Second	P	M	A	P
Third	F	P	M	F
Fourth	M	F	P	A

a) Does any plate receive a majority of first-place votes? If so, which plate received a majority?

b) Using the plurality method, which plate is selected?

c) Using the Borda count method, which plate is selected?

d) Using the plurality with elimination method, which plate is selected?

e) Using the pairwise comparison method, which plate is selected?

18. ***Architects' Convention*** The National Association of Architects held an election among its delegates to decide on the 2014 conference site. The 200 delegates ranked their choices among Chicago (C), Seattle (S), Dallas (D), and Las Vegas (L). The preference table giving the results of the election is shown below.

Number of Votes	65	50	45	25	15
First	S	L	D	C	S
Second	L	D	C	L	D
Third	C	C	L	D	L
Fourth	D	S	S	S	C

▲ Dallas, Texas

a) Does any city receive a majority of first-place votes? If so, which city received a majority?

b) Using the plurality method, which city is selected?

c) Using the Borda count method, which city is selected?

d) Using the plurality with elimination method, which city is selected?

e) Using the pairwise comparison method, which city is selected?

19. ***Selecting an Encyclopedia*** Park Street Library is planning to invest in a set of encyclopedias for the library, but the staff members differ regarding which set to buy. To settle the debate, the 16 staff members decide to rank the following choices: *Encyclopedia Britannica* (EB), *Funk and Wagnall's* (FW), *Grolier* (G), and *World Book* (WB). Aware that sentiments seem to be evenly divided, the staff members agree to use the plurality with elimination method, with the Borda count method used in case of a tie. The results of this election are given in the following preference table.

Number of Votes	6	4	3	1	1	1
First	WB	EB	EB	FW	FW	WB
Second	G	G	WB	WB	EB	FW
Third	EB	WB	FW	G	WB	EB
Fourth	FW	FW	G	EB	G	G

a) Which encyclopedia is selected if the plurality with elimination method is used?

b) According to the rules described above, which encyclopedia is chosen?

c) Because the staff members had agreed to break a tie using the pairwise comparison method, which encyclopedia is chosen?

15.2

Hiring a New Paralegal *In Exercises 20 and 21, a law firm is hiring one new paralegal from among four candidates, A, B, C, and D. The search committee decides to use the Borda count method to determine the winner. The preference table follows.*

Number of Votes	9	6	2
First	A	C	B
Second	B	B	D
Third	C	D	C
Fourth	D	A	A

20. Is the majority criterion satisfied?

21. Is the head-to-head criterion satisfied?

22. ***Plurality with Elimination*** Consider the following preference table.

Number of Votes	10	14	6	12
First	B	C	B	A
Second	A	B	C	C
Third	C	A	A	B

a) Who wins the election if the plurality with elimination method is used?

b) Assume that in a second election the six voters who voted for B, C, A, in that order, all change their preference to C, B, A, in that order. If the plurality with elimination method is used, is the monotonicity criterion satisfied?

c) Using the preference table from part (a), assume that B drops out. Does the plurality with elimination method satisfy the irrelevant alternatives criterion?

23. ***A Taste Test*** In a taste test, 114 people are asked to taste and rank four different brands of hot dogs. The choices are Ball Park (B), Oscar Mayer (O), Nathan's (N), and Hebrew National (H). The preference table is shown below.

Number of Votes	34	24	23	21	12
First	O	B	N	H	H
Second	B	H	H	N	N
Third	N	N	B	O	B
Fourth	H	O	O	B	O

a) Which brand is favored over all others in a head-to-head comparison?

b) Which brand wins if the plurality method is used?

c) Which brand wins if the Borda count method is used?

d) Which brand wins if the plurality with elimination method is used?

e) Which brand wins if the pairwise comparison method is used?

f) Which voting method(s) in parts (b) through (e) violate the head-to-head criterion?

24. ***Selecting a Band*** The Southwestern High School Class of 1979 is organizing its 30-year class reunion and is trying to pick out a band to play at the reunion. The choices are REO Speedwagon (R), Boston (B), Journey (J), and Fleetwood Mac (F). The class members have ranked their choices as indicated in the following preference table.

Number of Votes	34	25	15	9	4
First	B	F	R	J	J
Second	F	J	J	R	R
Third	R	R	F	B	F
Fourth	J	B	B	F	B

▲ REO Speedwagon

a) Is one band favored over all others in a head-to-head comparison?

b) Which band is chosen if the plurality method is used?

c) Which band is chosen if the Borda count method is used?

d) Which band is chosen if the plurality with elimination method is used?

e) Which band is chosen if the pairwise comparison method is used?

f) Which voting methods in parts (a) through (e) violate the head-to-head criterion?

25. ***Violating the Majority Criterion*** Which voting method(s) violate the majority criterion in the following election data?

Number of Votes	36	20	8	6
First	A	B	C	D
Second	B	C	B	B
Third	D	A	D	A
Fourth	C	D	A	C

26. ***Violating the Monotonicity Criterion*** Using the following tables, determine which voting method(s) violate the monotonicity criterion.

First Election

Number of Votes	28	24	20	10
First	A	B	C	C
Second	C	A	B	A
Third	B	C	A	B

Second Election

Number of Votes	36	24	20	2
First	A	B	C	C
Second	C	A	B	A
Third	B	C	A	B

27. ***Violating the Irrelevant Alternatives Criterion*** Using the following table, determine which voting method(s) violates the irrelevant alternatives criterion. Assume that candidate D drops out prior to the conclusion of the election.

Number of Votes	24	16	16	16	8	4	5
First	B	B	E	C	A	D	D
Second	A	A	A	D	E	B	E
Third	D	E	C	E	D	A	A
Fourth	E	C	D	B	B	E	B
Fifth	C	D	B	A	C	C	C

15.3, 15.4

Postal Service Apportionment *In Exercises 28–32, a post office in the city of Riverside has three regions in which to distribute the mail and 10 mail trucks. The trucks are to be apportioned based on the number of houses in each region, as shown in the table below.*

Region	A	B	C	Total
Number of houses	2592	1428	1980	6000

28. Determine each region's apportionment using Hamilton's method.

29. Determine each region's apportionment using Jefferson's method.

30. Determine each region's apportionment using Adams's method.

31. Determine each region's apportionment using Webster's method.

32. Suppose that the post office purchases one new truck. Using Hamilton's method, does the Alabama paradox occur if the number of trucks is increased from 10 to 11?

Apportioning Sociology Sections *In Exercises 33–37, Miami Dade College plans to offer 29 sections of three different sociology courses: Introduction to Sociology (A), Advanced Sociology (B), and Social Problems (C). The sections will be apportioned based on preregistration as shown in the table below.*

Course	A	B	C	Total
Number of students	371	279	220	870

33. Determine the apportionment for each course using Hamilton's method.

34. Determine the apportionment for each course using Jefferson's method.

35. Determine the apportionment for each course using Adams's method.

36. Determine the apportionment for each course using Webster's method.

37. Suppose that the following table shows the final registration for each course. If the sections are reapportioned using Hamilton's method, does the population paradox occur?

Course	A	B	C	Total
Number of students	376	279	222	877

Apportioning Seats *In Exercises 38–42, a country has two states, A and B, and 55 seats in the legislature. The population of each state is shown in the table below.*

State	A	B	Total
Population	4862	50,138	55,000

38. Determine each state's apportionment using Hamilton's method.

39. Determine each state's apportionment using Jefferson's method.

40. Determine each state's apportionment using Adams's method.

41. Determine each state's apportionment using Webster's method.

42. Suppose that a third state, C, with the population shown in the table below, is added along with five new seats. Does the new-states paradox occur using Hamilton's method?

State	A	B	C	Total
Population	4862	50,138	5940	60,940

CHAPTER 15 TEST

Lunch Choices In Exercises 1–6, the employees of an accounting firm are planning to have lunch delivered to a meeting. Their choices are tacos (T), pizza (P), and burgers (B). The preference table follows.

Number of Votes	7	6	6	5
First	T	P	B	P
Second	P	B	T	T
Third	B	T	P	B

1. How many members voted?

2. Does any choice have a majority of votes?

3. Determine the winner using the plurality method.

4. Determine the winner using the Borda count method.

5. Determine the winner using the plurality with elimination method.

6. Determine the winner using the pairwise comparison method.

7. ***Favorite Animal*** The children of Happy Faces Preschool are voting on their favorite classroom animal. The choices are hamster (H), iguana (I), lemming (L), and salamander (S). The results of the election are given in the following preference table.

Number of Votes	43	30	29	26	14
First	S	L	I	H	H
Second	L	H	H	I	I
Third	I	I	L	S	L
Fourth	H	S	S	L	S

   a) Which animal wins this election if the plurality method is used?

   b) Which animal wins this election if the Borda count method is used?

   c) Which animal wins this election if the plurality with elimination method is used?

   d) Which animal wins this election if the pairwise comparison method is used?

8. ***Violating the Head-to-Head Criterion*** Which voting method(s)—plurality, Borda count, plurality with elimination, or pairwise comparison—violate the head-to-head criterion using the following election data?

Number of Votes	86	60	58	52	28
First	W	Y	Z	X	X
Second	Y	X	X	Z	Z
Third	Z	Z	Y	W	Y
Fourth	X	W	W	Y	W

9. ***Voting for a Logo Design*** The park rangers in Yosemite National Park are holding a contest to choose a new logo design. The four logo choices for their stationery are El Capitan (E), a sequoia tree (S), a mule deer (M), and a waterfall (W). The 35 rangers rank their choices according to the following preference table.

Number of Votes	18	10	4	3
First	E	M	W	S
Second	M	W	M	M
Third	S	E	S	E
Fourth	W	S	E	W

   Using the data provided, does the Borda count method violate the majority criterion?

10. ***Apportioning Legislative Seats*** A country has three states and 30 seats in the legislature. The population of each state is shown below.

State	A	B	C	Total
Population	6933	9533	16,534	33,000

   a) Determine each state's apportionment using Hamilton's method.

   b) Determine each state's apportionment using Jefferson's method.

c) Determine each state's apportionment using Adams's method.

d) Determine each state's apportionment using Webster's method.

e) If the number of seats in the legislature increases to 31, does the Alabama paradox occur using Hamilton's method?

f) Suppose that in 10 years the states have the following population and 30 seats are apportioned. If the seats are reapportioned, does the population paradox occur using Hamilton's method?

State	A	B	C	Total
Population	7072	9724	17,030	33,826

g) Suppose that a fourth state with the population shown in the table below is added, with five additional seats for a total of 35 seats. Does the new-states paradox occur using Hamilton's method?

State	A	B	C	D	Total
Population	6933	9533	16,534	5100	38,100

GROUP PROJECT

1. ***Voting Project*** In this project, each group will select a topic of their choice that students in your class can vote on. Have each student in the class cast a ballot where they rank the candidates or items on which they are voting.

   a) After the ballots are cast, construct a preference table.

   b) Determine the winner of the election using each of the four voting methods discussed in Section 15.1.

   c) Check to see if any of the fairness criteria are violated as described in Section 15.2.

   d) After you complete your project, write a final report of how the entire process was conducted and the results obtained and submit it to your instructor.

APPENDIX A Using Excel®

Microsoft Excel is a program that can be used to perform a multitude of tasks, including making spreadsheets and graphs. In this appendix, we will present a brief introduction to Excel. Many publications are available that offer detailed and complete explanations on using Excel. We will give instructions for a PC. The screens and instructions for a Mac are a little different.

Figure A.1 shows an Excel screen. Depending on the version of Excel you are using, the Excel screen on your computer may be a little different.

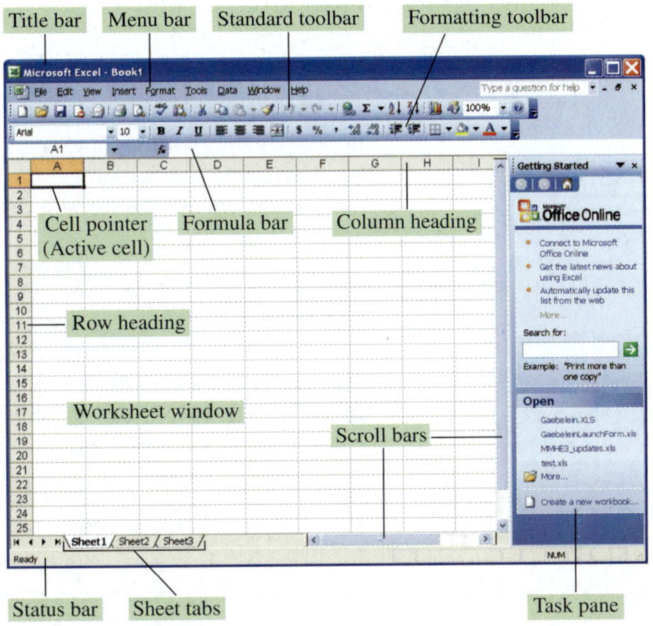

Figure A.1

In Excel, columns are referred to with letters and rows are referred to with numbers. The highlighted, or active cell, in Fig. A.1 is A1. When entering information in Excel, you can enter **labels**, used to identify information, **values** (or numbers), or **formulas** (calculations written in a special notation). When you enter a formula in a cell, Excel displays the result of the formula in the active cell, not the formula itself, after you press the Enter key. Formulas always begin with an equal (=) sign. After you enter a label, a value, or a formula in a cell, you press the Enter key to enter the information. Do not be concerned if, when listing labels, the labels overflow into a second cell. After you finish inserting all the labels, you can extend the column width by placing the mouse pointer on the line between the two columns at the top of the column and then double clicking the mouse button. The column widths will then automatically adjust in size.

Consider the Excel spreadsheet in Fig. A.2 on page AA-2. Labels, shaded in orange, are given across row 1 and down column A. Values, shaded in green, are given in cells B2 through D2 and B3 through D3. Calculations from formulas, shaded in yellow, are given in cells E2 and E3.

In Fig. A.2, we are using Excel to calculate simple interest on a loan, using the simple interest formula, as will be explained shortly. Under column A, we list the dates on which the loans were made.

	A	B	C	D	E
1	**Date**	**Principal**	**Rate**	**Time**	**Interest**
2	**May 22, 2006**	$ 2,000.00	0.06	1	$ 120.00
3	**June 5, 2007**	$ 5,000.00	0.05	3	$ 750.00

Figure A.2

To get the dollar signs in column B, highlight column B by placing the mouse pointer on the B at the top of column B and then click on the $ in the formatting toolbar. Repeat the process to get the dollar signs in column E. After you insert the labels and values, we use the formula

$$\text{Interest} = \text{principal} \times \text{rate} \times \text{time}$$

to determine the simple interest in column E. To use an Excel formula to calculate the interest for May 22, 2006, move the cursor to cell E2 and then click the mouse to activate that cell. Type in the formula =B2*C2*D2. After you press the Enter key, the answer, $120, is displayed. To obtain the answer for cell E3, use =B3*C3*D3. Notice that * is used to indicate multiplication and that there are no blank spaces in the formula.

Now consider the spreadsheet in Fig. A.3, which shows a budget.

	A	B	C	D
1		**Budgeted**	**Actual**	**Difference**
2	**Income Items**			
3	Sales income	$ 35,600.00	$ 39,200.00	$ 3,600.00
4	CD interest	$ 1,620.00	$ 1,490.00	$ (130.00)
5	Other income	$ 5,080.00	$ 6,580.00	$ 1,500.00
6	*Total Income*	$ 42,300.00	$ 47,270.00	
7				
8	**Expense Items**			
9	Rent	$ 10,230.00	$ 10,150.00	$ (80.00)
10	Food	$ 8,573.00	$ 8,450.00	$ (123.00)
11	Clothing	$ 2,510.00	$ 2,370.00	$ (140.00)
12	Medical	$ 1,080.00	$ 960.00	$ (120.00)
13	Entertainment	$ 1,280.00	$ 1,280.00	$ –
14	Insurance	$ 1,200.00	$ 1,420.00	$ 220.00
15	Other	$ 6,110.00	$ 7,190.00	$ 1,080.00
16	*Total Expenses*	$ 30,983.00	$ 31,820.00	
17				
18	**Net Income**	$ 11,317.00	$ 15,450.00	

Figure A.3

To obtain the spreadsheet in Fig. A.3, do the following. First input the labels in cells B1, C1, and D1 and also under column A. Then input the values in cells B3 through B5, B9 through B15, C3 through C5, and C9 through C15. Next we can determine the difference between the actual amounts and budgeted amounts, column D, by subtracting the values in column B from the values in column C. Begin by activating cell D3. Then type =C3-B3 and press the Enter key. At this point, the difference, $39,200 − $35,600, or $3600, fills in cell D3. If we wanted to determine the amount to go in cell D4, we could type in =C4-B4 and continue this process for all amounts in column D. Excel, however, has a number of procedures that can be used to copy formulas from one cell to another cell that needs a similar formula. Copying can save time when building a spreadsheet with multiple columns or rows that need similar formulas. When copying formulas, Excel automatically rewrites the cell reference so that the formula refers to the appropriate cells. You can see the formula for a particular cell at any time by activating that cell and looking in the formula bar near the top of the screen. You can copy formulas using the **Copy** and **Paste** commands in the Edit menu. To copy a formula from one cell to another, click on the cell containing the formula you wish to copy, click on the Edit menu, and then click on Copy. Click on the cell where you want the copied formula to go. Then click on the Edit menu and click Paste. This method will copy the formula to the new cell, and the cell reference used in the formula will automatically change to the appropriate cells for the new calculation.

A quick way to copy a formula to one or more adjacent cells is with the ***fill handle***. To copy this way, position the mouse pointer on the selection's *fill handle* (a tiny square in the bottom right-hand corner of the cell selected). The mouse pointer turns into a black cross. Press and hold the mouse button while you drag the cross down the column, or to the right across the row, until you reach the desired cell. A gray or black border stretches over the cells you pass, highlighting the desired cells. When you release the mouse button, the formula is copied to all the cells that were highlighted.

You can complete cells D4 and D5 using the fill handle. Then, in cell D9, we enter the formula =C9-B9. We then use the fill handle to complete cells D10 through D15.

To obtain the sum of the budgeted income, cell B6, we can use the auto sum button on the standard toolbar or we can enter a formula. To determine the sum of the values in cells B3, B4, and B5, we can enter the formula =SUM(B3:B5) in cell B6. After you press enter, the sum $42,300 is displayed. Then you can obtain the sum of cells C3 through C5, which will go in cell C6, by using =SUM(C3:C5) or by using the fill handle as explained earlier. We obtain the sum of the total budgeted expenses, B16, and total actual expenses, C16, in a similar manner. For example, in cell B16, we use the formula =SUM(B9:B15) to obtain the sum of columns B9 through B15.

Finally, to obtain the budgeted net income, cell B18, we subtract the total budgeted expenses from the total budgeted income. Thus, in cell B18 we use the formula =B6-B16. Similarly, in cell C18 we use =C6-C16.

The preceding explanation should give you a basic idea of how Excel works. Note that when you change a value in a cell in the spreadsheet, the corresponding formulas that use that value automatically recalculate the answer and display the resulting answer.

TIMELY TIP

In Microsoft Excel formulas, use

* for multiplication
/ for division
^ for exponents

Suppose that cell B3 of a spreadsheet has a value of 5. If in cell B4 you had the formula =B3^2, you would get an answer of 25 in cell B4 after pressing the Enter key. Suppose that you wanted to find the square root of the number 16, which is in cell D7 of a spreadsheet. In cell D8, you could use the formula =D7^(1/2) or =D7^0.5. After the Enter key is pressed, the answer 4 is displayed in cell D8.

If you find you left something out and wish to add another row or column to your spreadsheet, place the cursor where you want to add the new row or column. Then go to the Insert menu and select either Row or Column. If you want to delete either a row or column, highlight that row or column and then go to the Edit menu and click on Delete.

If you want to find the sum of the numbers in consecutive cells and another number in a non-consecutive cell, you can use a comma in the formula. For example, =SUM(C5:C10, C:13) adds the numbers in cells C5 through C10 and the number in cell C13.

If items such as dates are automatically converted to a format different than what you want, use the Format menu and select Cells. Here you will see many formats that can be adjusted. To change the format of dates, for example, select the Format menu, then select Cells, select Numbers, and then select Dates.

The only way to get to understand Excel is to work and experiment with it. When doing so, you may want to go to the Insert menu and select Function. Here you can select from many different types of functions in many different areas of mathematics and business.

ANSWERS

CHAPTER 1

SECTION 1.1, PAGE 5

1. a) 1, 2, 3, 4, 5, . . .
b) Counting numbers

3. Inductive reasoning is the process of reasoning to a general conclusion through observation of specific cases.

5. A counterexample is a specific case that satisfies the conditions of the conjecture but shows the conjecture is false.

7. Inductive reasoning

9. Inductive reasoning, a general conclusion is obtained from observation of specific cases.

11. $5 \times 7 = 35$

13. 1 5 10 10 5 1

15. 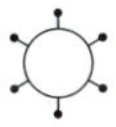 **17.** 

19. 25, 30, 35 **21.** $-1, 1, -1$

23. $\frac{1}{16}, \frac{1}{64}, \frac{1}{256}$ **25.** 36, 49, 64

27. 34, 55, 89 **29.** Y

31. a) 36, 49, 64
b) square 6, 7, 8, 9 and 10
c) No, 72 is between 8^2 and 9^2, so it is not a square number.

33. Blue: 1, 5, 7, 10, 12 Purple: 2, 4, 6, 9, 11 Yellow: 3, 8

35. a) $\approx$ \$6 billion
b) We are using specific cases to make a prediction.

37. 

39. a) You should obtain the original number.
b) You should obtain the original number.
c) The result is the original number.
d) $n, 8n, 8n + 16, \frac{8n + 16}{8} = n + 2, n + 2 - 2 = n$

41. a) 5
b) You should obtain the number 5.
c) The result is always the number 5.
d) $n, n + 1, \frac{n + (n + 1) + 9}{2} = \frac{2n + 10}{2} = n + 5,$
$n + 5 - n = 5$

43. $3 + 5 = 8$, which is not an odd number.

45. $(3 + 2)/2 = 5/2$, which is not an even number.

47. $1 - 2 = -1$, which is not a counting number.

49. a) The sum of the measures of the interior angles should be 180°.
b) Yes, the sum of the measures of the interior angles should be 180°.
c) The sum of the measures of the interior angles of a triangle is 180°.

51. 129, the numbers in positions are found as follows:

$a \quad b$
$c \quad a + b + c$

53. (c)

SECTION 1.2, PAGE 14

Answers in this section will vary depending on how you round your numbers. All answers are approximate.

1. 2210 **3.** 1,200,000,000 **5.** 8000 **7.** 900

9. 200 **11.** 1,200,000,000 **13.** \$240 **15.** \$32.80

17. 180 miles **19.** 13,200 lb **21.** \$3.90 **23.** 16

25. \$37 **27.** \$120 **29.** \$41 **31.** $\approx$ 20 mi

33. a) 100
b) 50
c) 125

35. a) 5 million
b) 98 million
c) 65 million
d) 280 million

37. a) 85%
b) 15%
c) 59,500,000 acres
d) No, since we are not given the area of each state.

39. 25 **41.** $\approx$ 160 bananas **43.** 150° **45.** 10%

47. 9 square units **49.** 150 feet

51.–59. Answers will vary.

61. There are 336 dimples on a regulation golf ball.

SECTION 1.3, PAGE 29

1. 76.5 mi **3.** 19.36 ft **5.** \$29,026

7. a) \$196,800
b) \$26,600
c) \$220,320

9. \$12.50 **11.** \$70 **13.** \$71,989.20

15. a) $\approx$ 122
b) Answers will vary. A close approximation can be obtained by multiplying the U.S. sizes by 2.54.

17. a) 9.2 min
b) 62 min
c) 40 min
d) 47 min

19. a) \$1.5 trillion
b) \$22,000

21. \$82.08

23. a) \$74.40
b) \$264
c) \$64

25. a) 3,153,600 cm^3
b) ≈ 1.4 days

27. a) \$75
b) \$15
c) Long term by \$3

29. a) Divide the Total Emissions by the Emissions per Capita.
b) ≈ 298.6 million
c) ≈ 1307.6 million or 1.3076 billion

31. \$990, less than initial investment

33. a) 48 rolls
b) \$198 if she purchases four 10 packs and two 4 packs

35. a) Water/milk: 3 cups; salt: $\frac{3}{8}$ tsp; Cream of Wheat: 9 tbsp (or $\frac{9}{16}$ cup)
b) Water/milk: $2\frac{7}{8}$ cups; salt: $\frac{3}{8}$ tsp; Cream of Wheat: $\frac{5}{8}$ cup (or 10 tbsp)
c) Water/milk: $2\frac{3}{4}$ cups; salt: $\frac{3}{8}$ tsp; Cream of Wheat: $\frac{9}{16}$ cup (or 9 tbsp)
d) Differences exist in water/milk because the amount for 4 servings is not twice that for 2 servings. Differences also exist in Cream of Wheat because $\frac{1}{2}$ cup is not twice 3 tbsp.

37. 144 square inches

39. The area is 4 times as large.

41. 66 ft **43.** at −1 **45.** \$60,000

47. a) 30
b) 140

49.

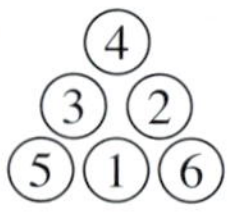

51.

8	6	16
18	10	2
4	14	12

53. The sum of the four corners is 4 times the number in the center.

55. Multiply the center number by 9.

57. 6 ways

59.

	7	
3	1	4
5	8	6
	2	

Other answers are possible, but 1 and 8 must appear in the center.

61.

1	2	3	4	5
2	3	4	5	1
3	4	5	1	2
4	5	1	2	3
5	1	2	3	4

Other answers are possible.

63. Mary is the skier. **65.** 714 square units **67.** \$120

REVIEW EXERCISES, PAGE 37

1. 31, 36, 41 **2.** 25, 36, 49 **3.** −48, 96, −192

4. 25, 32, 40 **5.** 10, 4, −3 **6.** $\frac{3}{8}, \frac{3}{16}, \frac{3}{32}$

7.  **8.** 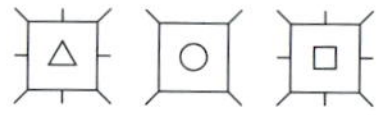 **9.** (c)

10. a) The final number is twice the original number.
b) The final number is twice the original number.
c) The final number is twice the original number.
d) n, $10n$, $10n + 5$, $\frac{10n + 5}{5} = 2n + 1$,
$2n + 1 - 1 = 2n$

11. This process will always result in an answer of 3.

12. $1^2 + 2^2 = 5$

Answers to Exercises 13–25 will vary depending on how you round your numbers. All answers are approximate.

13. 420,000,000 **14.** 2000 **15.** 200

16. Answers will vary. **17.** \$300 **18.** \$36

19. 3 mph **20.** \$14.00 **21.** 2 mi **22.** 0.15 million

23. 0.8 million **24.** 13 square units

25. Length ≈ 22 ft; height ≈ 8 ft

26. \$7.50 **27.** \$1.16

28. Berkman's is cheaper by \$20.00.

29. \$32,996

30. a) 288 lb
b) 12,500 ft^2

31. \$311 **32.** 7.05 mg

33. \$882 **34.** 6 hr 45 min **35.** July 26, 11:00 A.M.

36. a) 6.45 cm^2
b) 16.39 cm^3
c) 1 cm ≈ 0.39 in.

37. 201

38.

21	7	8	18
10	16	15	13
14	12	11	17
9	19	20	6

39.

23	25	15
13	21	29
27	17	19

40. 59 min 59 sec **41.** 6

42. \$25 Room
\$ 3 Men
\$ 2 Clerk
\$30

43. 140 lb

44. Yes; 3 quarters and 4 dimes, or 1 half dollar, 1 quarter and 4 dimes, or 1 quarter and 9 dimes. Other answers are possible.

45. 216 cm^3

46. Place six coins in each pan with one coin off to the side. If it balances, the heavier coin is the one on the side. If the pan does not balance, take the six coins on the heavier side and split them into two groups of three. Select the three heavier coins and weigh two coins. If the pan balances, it is the third coin. If the pan does not balance, you can identify the heavier coin.

47. 125,250 **48.** 16 blue **49.** 90

50. The fifth figure will be an octagon with sides of equal length. Inside the octagon will be a seven sided figure with each side of equal length. The figure will have one antenna.

51. 61

52. Some possible answers are shown. Others are possible.

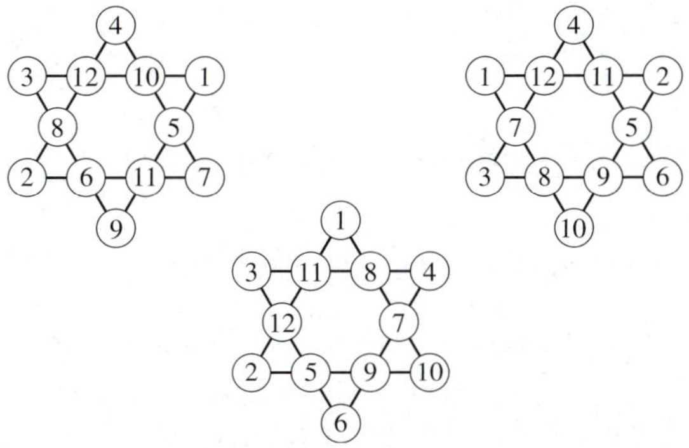

53. a) 2 **b)** 6 **c)** 24 **d)** 120
e) $n(n-1)(n-2)\ldots1$, (or $n!$), where n = the number of people in line

CHAPTER TEST, PAGE 41

1. 19, 23, 27

2. $\frac{1}{16}, \frac{1}{32}, \frac{1}{64}$

3. a) The result is the original number plus 1.
b) The result is the original number plus 1.
c) The result will always be the original number plus 1.
d) $n, 5n, 5n+10, \dfrac{5n+10}{5} = n+2, n+2-1 = n+1$

The answers for Exercises 4–6 are approximate.

4. 12,000 **5.** 2,100,000 **6.** 9 square units

7. a) ≈ 23.03
b) He is in the at risk range.

8. 36 min **9.** 32 cans **10.** $7\frac{1}{2}$ min

11. $\approx$ 39.5 in. by 29.6 in. (The actual dimensions are 100.5 cm by 76.5 cm.)

12. \$49.00

13.

40	15	20
5	25	45
30	35	10

14. Less time if she had driven at 45 mph for the entire trip

15. $2 \cdot 6 \cdot 8 \cdot 9 \cdot 13$; 11 does not divide 11,232.

16. 243 jelly beans

17. a) \$11.97
b) \$11.81
c) Save 16 cents by using the 25% off coupon.

18. 24

CHAPTER 2

SECTION 2.1, PAGE 50

1. A set is a collection of objects.

3. Description, roster form, and set-builder notation; the set of counting numbers less than 7, {1, 2, 3, 4, 5, 6}, and $\{x \mid x \in N \text{ and } x < 7\}$

5. A set is finite if it either contains no elements or the number of elements in the set is a natural number.

7. Two sets are equivalent if they contain the same number of elements.

9. The empty set is a set that contains no elements.

11. A universal set is a set that contains all the elements for any specific discussion.

13. Not well defined

15. Well defined **17.** Well defined

19. Infinite **21.** Infinite **23.** Infinite

25. {Maine, Maryland, Massachusetts, Michigan, Minnesota, Mississippi, Missouri, Montana}

27. {11, 12, 13, 14, . . . , 177}

29. $B = \{2, 4, 6, 8, \ldots\}$

31. { } or $\emptyset$

33. $E = \{14, 15, 16, 17, \ldots, 84\}$

35. {Switzerland, Denmark, Sweden, United Kingdom, Germany, New Zealand}

37. {Switzerland, Denmark, Sweden, United Kingdom, Germany}

39. {2004, 2005}

41. {1998, 1999, 2000}

43. $B = \{x \mid x \in N \text{ and } 4 < x < 13\}$ or
$B = \{x \mid x \in N \text{ and } 5 \le x \le 12\}$

45. $C = \{x \mid x \in N \text{ and } x \text{ is a multiple of } 3\}$

47. $E = \{x \mid x \in N \text{ and } x \text{ is odd}\}$

49. $C = \{x \mid x \text{ is February}\}$

51. Set A is the set of natural numbers less than or equal to 7.

53. Set V is the set of vowels in the English alphabet.

55. Set T is the set of species of trees.

57. Set S is the set of seasons.

59. {Johnson & Johnson, Google, Home Depot}

61. {United Airlines}

63. {1996, 1997, 1998, 1999}

65. {1998, 1999, 2000, 2001, 2002, 2003, 2004}

67. False; $\{e\}$ is a set, and not an element of the set.

69. False; h is not an element of the set.

71. False; 3 is an element of the set.

73. True **75.** 4 **77.** 0

79. Both **81.** Neither

83. Equivalent

85. a) Set A is the set of natural numbers greater than 2. Set B is the set of all numbers greater than 2.
b) Set A contains only natural numbers. Set B contains other types of numbers, including fractions and decimal numbers.
c) $A = \{3, 4, 5, 6, \dots\}$
d) No; because there are an infinite number of elements between any two elements in set B, we cannot write set B in roster form.

87. Cardinal **89.** Ordinal

91. Answers will vary.

93. Answers will vary.

SECTION 2.2 PAGE 58

1. Set A is a subset of set B, symbolized $A \subseteq B$, if and only if all the elements of set A are also elements of set B.

3. If $A \subseteq B$, then every element of set A is an element of set B. If $A \subset B$, then every element of set A is an element of set B and set $A \neq$ set B.

5. The number of proper subsets is determined by the formula $2^n - 1$, where n is the number of elements in the set.

7. False; Spanish is an element of the set, not a subset.

9. True **11.** True

13. False; the set $\{\emptyset\}$ contains the element $\emptyset$.

15. True

17. False; the set $\{0\}$ contains the element 0.

19. False; {swimming} is a set not an element.

21. True **23.** False; no set is a proper subset of itself.

25. $B \subseteq A$, $B \subset A$ **27.** $A \subseteq B$, $A \subset B$ **29.** $B \subseteq A$, $B \subset A$

31. $A = B$, $A \subseteq B$, $B \subseteq A$ **33.** $\{\ \}$

35. { }, {pen}, {pencil}, {pen, pencil}

37. a) $\{\ \}, \{a\}, \{b\}, \{c\}, \{d\}, \{a, b\}, \{a, c\}, \{a, d\}, \{b, c\}, \{b, d\}, \{c, d\}, \{a, b, c\}, \{a, b, d\}, \{a, c, d\}, \{b, c, d\}, \{a, b, c, d\}$
b) $\{a, b, c, d\}$

39. False **41.** True **43.** True **45.** True **47.** True

49. True **51.** 2^4 or 16 **53.** 2^6 or 64 **55.** $E = F$

57. a) Yes **b)** No **c)** Yes **59.** 1 **61.** Yes

SECTION 2.3, PAGE 68

1.

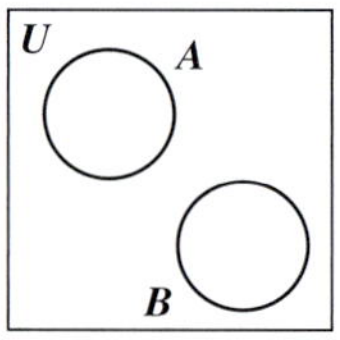

3.

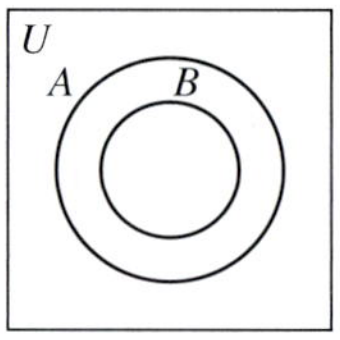

5.

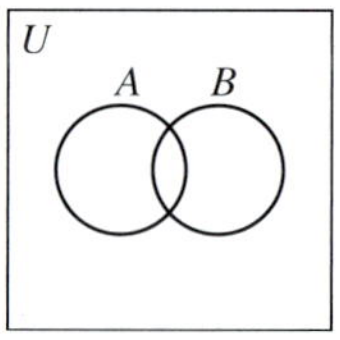

7. Determine the elements that are in the universal set that are not in set A.

9. Select the elements common to both set A and set B.

11. a) *Or* is generally interpreted to mean *union*.
b) *And* is generally interpreted to mean *intersection*.

13. The difference of two sets A and B is the set of elements that belong to set A but not to set B.

15.

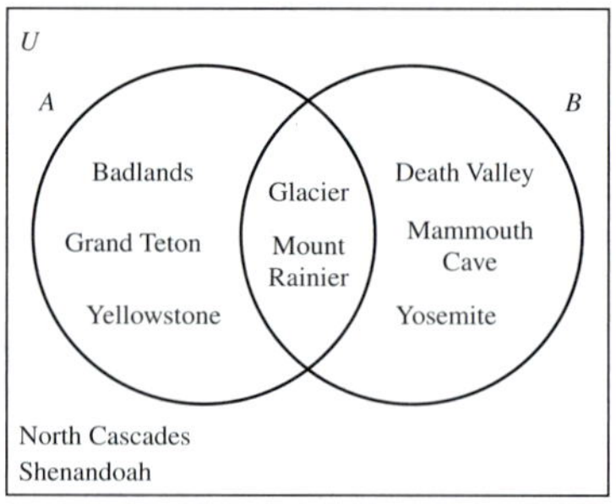

17.

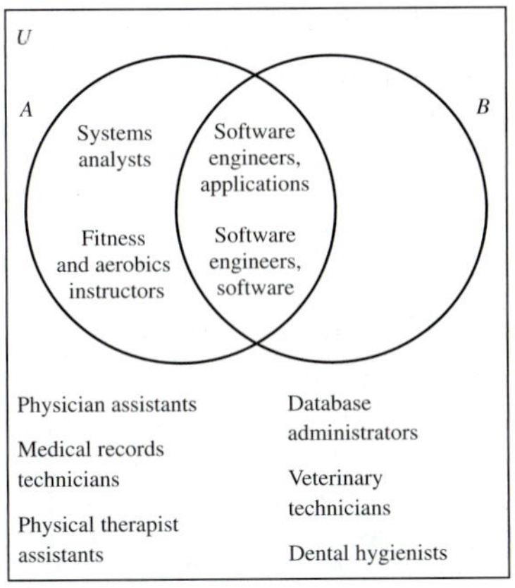

19. The set of animals in U.S. zoos that are not in the San Diego Zoo

21. The set of insurance companies in the United States that do not offer life insurance

23. The set of insurance companies in the United States that offer life insurance or car insurance

25. The set of insurance companies in the United States that offer life insurance and do not offer car insurance

27. The set of furniture stores that sell mattresses and outdoor furniture

29. The set of furniture stores that do not sell outdoor furniture and sell leather furniture

31. The set of furniture stores that sell mattresses or outdoor furniture or leather furniture

33. $\{b, c, t, w, a, h\}$ **35.** $\{a, h\}$

37. $\{c, w, b, t, a, h, f, r, d, g\}$ **39.** $\{p, m, z\}$

41. $\{\text{L}, \Delta, @, \$, *\}$

43. $\{\text{L}, \Delta, @, *, \$, \text{R}, \square, \alpha, \infty, \Sigma, \text{Z}\}$

45. $\{\text{R}, \square, \alpha, *, \$, \infty, Z, \Sigma\}$ **47.** $\{\text{R}, \square, \alpha\}$

49. $\{1, 2, 3, 4, 5, 6, 8\}$ **51.** $\{1, 5, 7, 8\}$ **53.** $\{7\}$

55. $\{\ \}$ **57.** $\{7\}$ **59.** $\{a, e, h, i, j, k\}$

61. $\{a, f, i\}$ **63.** $\{b, c, d, e, g, h, j, k\}$

65. $\{a, c, d, e, f, g, h, i, j, k\}$

67. $\{a, b, c, d, e, f, g, h, i, j, k\}$, or U

69. $\{2, 6, 9\}$ **71.** $\{1, 4\}$ **73.** $\{1, 3, 4, 5, 7, 8, 10\}$

75. $\{4, 9\}$ **77.** $\{(a, 1), (a, 2), (b, 1), (b, 2), (c, 1), (c, 2)\}$

79. No. The ordered pairs are not the same. For example, $(a, 1) \neq (1, a)$.

81. 6 **83.** $\{\ \}$ **85.** $\{2, 4, 6, 8\}$, or B **87.** $\{7, 9\}$

89. $\{1, 3, 5, 6, 7, 8, 9\}$ **91.** $\{6, 8\}$

93. $\{1, 2, 3, 4, 5, 6, 7, 8, 9\}$, or U

95. $\{1, 2, 3, 4, 5\}$, or C

97. A set and its complement will always be disjoint. For example, if $U = \{1, 2, 3\}$ and $A = \{1, 2\}$, then $A' = \{3\}$, and $A \cap A' = \{\ \}$.

99. 49

101. a) $8 = 4 + 6 - 2$ **b)** and **c)** Answers will vary.

103. $\{1, 2, 3, 4, \ldots\}$, or A **105.** $\{4, 8, 12, 16, \ldots\}$, or B

107. $\{2, 4, 6, 8, \ldots\}$, or C **109.** $\{2, 6, 10, 14, 18, \ldots\}$

111. $\{2, 6, 10, 14, 18, \ldots\}$ **113.** U **115.** A

117. U **119.** U **121.** $B \subseteq A$

123. A and B are disjoint sets. **125.** $A \subseteq B$

SECTION 2.4, PAGE 77

1. 8 **3.** II, IV, VI **5.** 5

7. a) Yes
b) No, one specific case cannot be used as proof.
c) No

9.

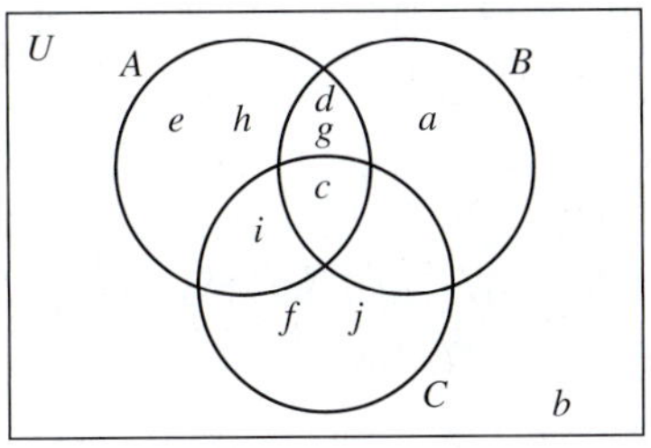

11.

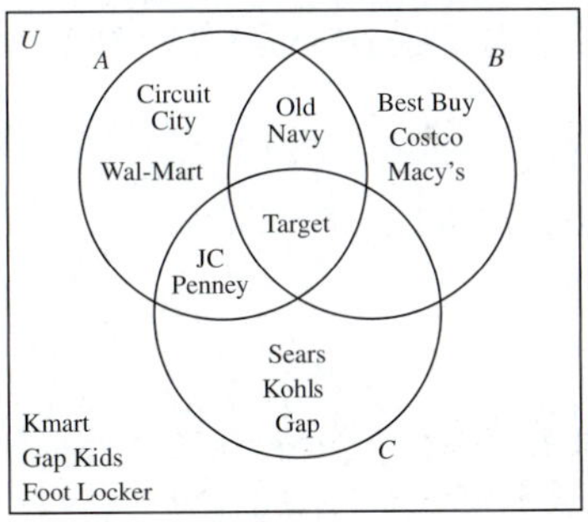

13.

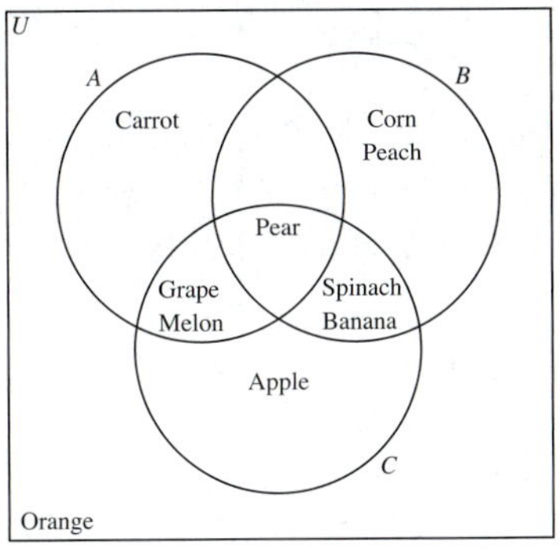

15.

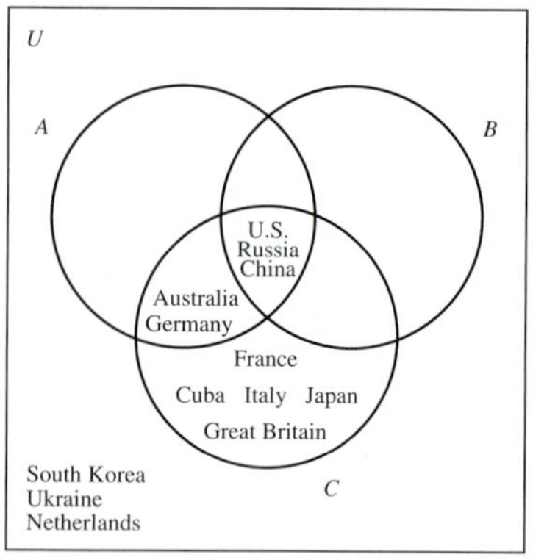

17. V **19.** I **21.** III **23.** II **25.** VIII **27.** III
29. VI **31.** III **33.** III **35.** V **37.** II **39.** VII
41. I **43.** VIII **45.** VI **47.** $\{1, 3, 4, 5, 7, 9\}$
49. $\{2, 3, 4, 5, 6, 8, 12, 14\}$ **51.** $\{3, 4, 5\}$
53. $\{1, 2, 3, 7, 9, 10, 11, 12, 13, 14\}$
55. $\{1, 2, 3, 4, 5, 6, 7, 8, 9, 12, 14\}$
57. $\{2, 11, 12, 13, 14\}$
59. $\{2, 6, 8, 10, 11, 12, 13, 14\}$
61. Yes **63.** No **65.** No **67.** Yes **69.** No
71. Yes **73.** Yes **75.** Yes **77.** No
79. $(A \cup B)'$ **81.** $(A \cup B) \cap C'$
83. a) Both equal $\{6, 7\}$. **b)** Answers will vary.
c) Both are represented by the regions IV, V, VI.

85.

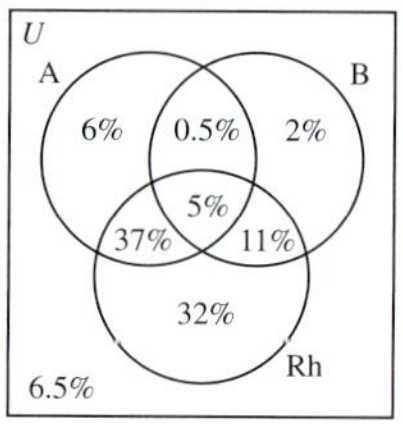

87. a)

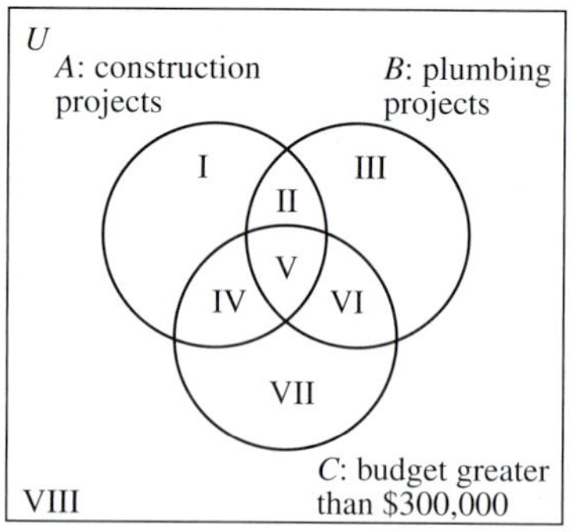

b) V; $A \cap B \cap C$
c) VI; $A' \cap B \cap C$
d) I; $A \cap B' \cap C'$

89. a)

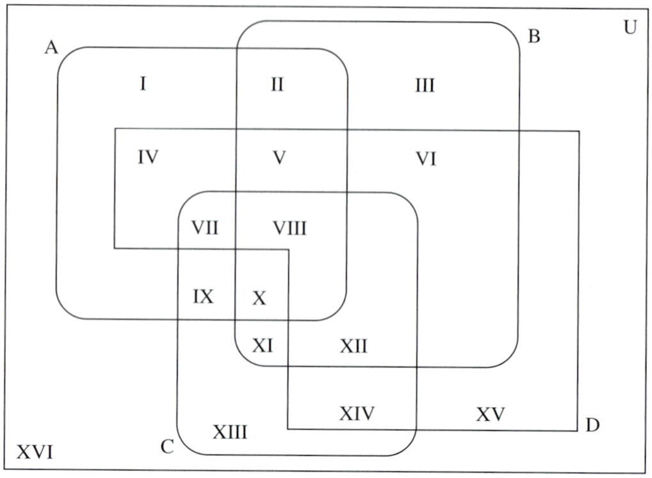

b)

Region	*Set*	*Region*	*Set*
I	$A \cap B' \cap C' \cap D'$	IX	$A \cap B' \cap C \cap D'$
II	$A \cap B \cap C' \cap D'$	X	$A \cap B \cap C \cap D'$
III	$A' \cap B \cap C' \cap D'$	XI	$A' \cap B \cap C \cap D'$
IV	$A \cap B' \cap C' \cap D$	XII	$A' \cap B \cap C \cap D$
V	$A \cap B \cap C' \cap D$	XIII	$A' \cap B' \cap C \cap D'$
VI	$A' \cap B \cap C' \cap D$	XIV	$A' \cap B' \cap C \cap D$
VII	$A \cap B' \cap C \cap D$	XV	$A' \cap B' \cap C' \cap D$
VIII	$A \cap B \cap C \cap D$	XVI	$A' \cap B' \cap C' \cap D'$

SECTION 2.5, PAGE 86

1.

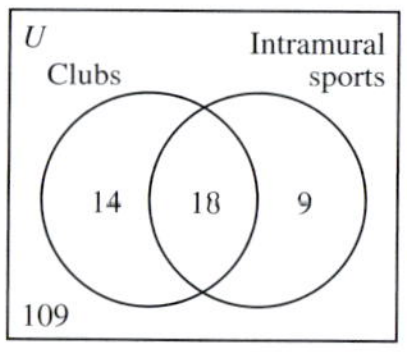

a) 14
b) 9
c) 109

3.

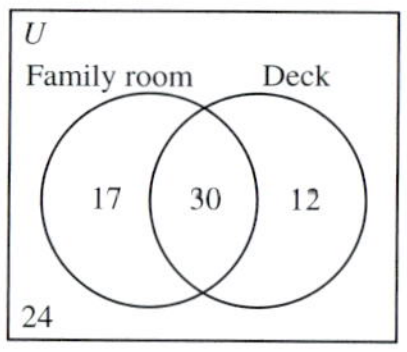

a) 17
b) 12
c) 59

5.

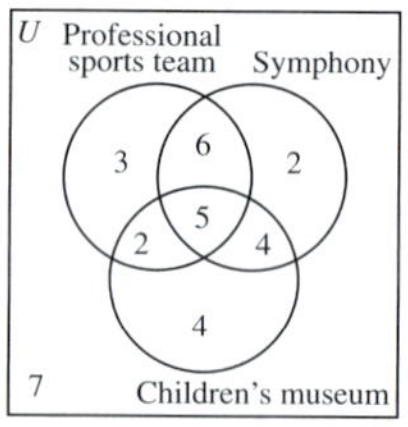

a) 3
b) 6
c) 22
d) 11
e) 12

7.

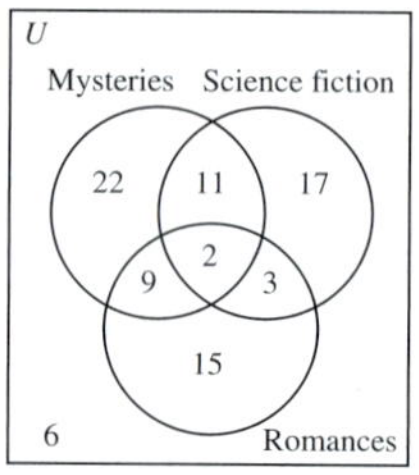

a) 22
b) 11
c) 64
d) 50
e) 23

9.

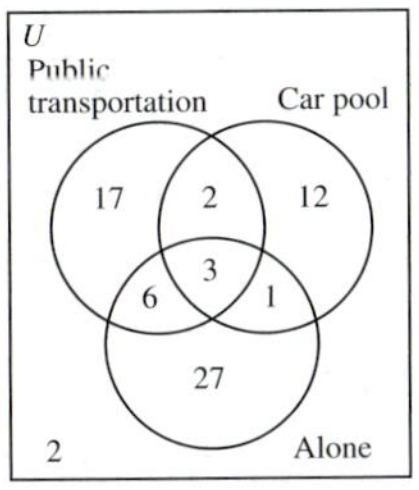

a) 17
b) 27
c) 2
d) 31
e) 2

11.

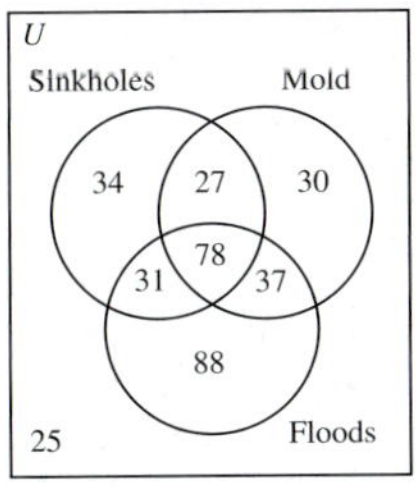

a) 67
b) 237
c) 37
d) 25

13. In a Venn diagram, regions II, IV, and V contain a total of 37 cars driven by women. This total is greater than the 35 cars driven by women, as given in the exercise.

15.

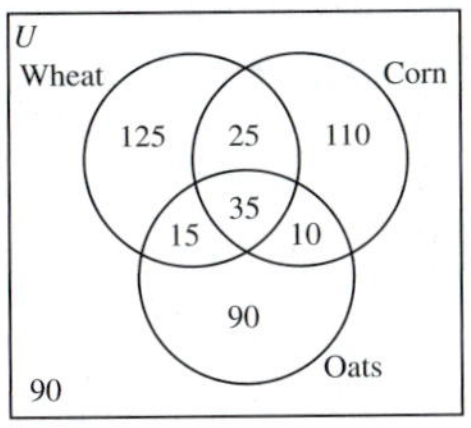

a) 410
b) 35
c) 90
d) 50

SECTION 2.6, PAGE 93

1. An infinite set is a set that can be placed in a one-to-one correspondence with a proper subset of itself.

3. $\{5, 6, 7, 8, 9, \ldots, n + 4, \ldots\}$
$\downarrow\ \downarrow\ \downarrow\ \downarrow\ \downarrow\ \qquad \downarrow$
$\{6, 7, 8, 9, 10, \ldots, n + 5, \ldots\}$

5. $\{3, 5, 7, 9, 11, \ldots, 2n + 1, \ldots\}$
$\downarrow\ \downarrow\ \downarrow\ \downarrow\ \downarrow\ \qquad \downarrow$
$\{5, 7, 9, 11, 13, \ldots, 2n + 3, \ldots\}$

7. $\{3, 7, 11, 15, 19, \ldots, 4n - 1, \ldots\}$
$\downarrow\ \downarrow\ \downarrow\ \downarrow\ \downarrow\ \qquad \downarrow$
$\{7, 11, 15, 19, 23, \ldots, 4n + 3, \ldots\}$

9. $\{6, 11, 16, 21, 26, \ldots, 5n + 1, \ldots\}$
$\downarrow\ \downarrow\ \downarrow\ \downarrow\ \downarrow\ \qquad \downarrow$
$\{11, 16, 21, 26, 31, \ldots, 5n + 6, \ldots\}$

11. $\left\{\frac{1}{2}, \frac{1}{4}, \frac{1}{6}, \frac{1}{8}, \frac{1}{10}, \ldots, \frac{1}{2n}, \ldots\right\}$
$\downarrow\ \downarrow\ \downarrow\ \downarrow\ \downarrow\ \qquad \downarrow$
$\left\{\frac{1}{4}, \frac{1}{6}, \frac{1}{8}, \frac{1}{10}, \frac{1}{12}, \ldots, \frac{1}{2n + 2}, \ldots\right\}$

13. $\{1, 2, 3, 4, 5, \ldots, n, \ldots\}$
$\downarrow\ \downarrow\ \downarrow\ \downarrow\ \downarrow\ \qquad \downarrow$
$\{3, 6, 9, 12, 15, \ldots, 3n, \ldots\}$

15. $\{1, 2, 3, 4, 5, \ldots, n, \ldots\}$
$\downarrow\ \downarrow\ \downarrow\ \downarrow\ \downarrow\ \qquad \downarrow$
$\{4, 6, 8, 10, 12, \ldots, 2n + 2, \ldots\}$

17. $\{1, 2, 3, 4, 5, \ldots, n, \ldots\}$
$\downarrow\ \downarrow\ \downarrow\ \downarrow\ \downarrow\ \qquad \downarrow$
$\{2, 5, 8, 11, 14, \ldots, 3n - 1, \ldots\}$

19. $\{1, 2, 3, 4, 5, \ldots, n, \ldots\}$
$\downarrow\ \downarrow\ \downarrow\ \downarrow\ \downarrow\ \qquad \downarrow$
$\{5, 9, 13, 17, 21, \ldots, 4n + 1, \ldots\}$

21. $\{1, 2, 3, 4, 5, \ldots, n, \ldots\}$
$\downarrow\ \downarrow\ \downarrow\ \downarrow\ \downarrow\ \qquad \downarrow$
$\left\{\frac{1}{3}, \frac{1}{4}, \frac{1}{5}, \frac{1}{6}, \frac{1}{7}, \ldots, \frac{1}{n + 2}, \ldots\right\}$

23. $\{1, 2, 3, 4, 5, \ldots, n, \ldots\}$
$\downarrow\ \downarrow\ \downarrow\ \downarrow\ \downarrow\ \qquad \downarrow$
$\{1, 4, 9, 16, 25, \ldots, n^2, \ldots\}$

25. $\{1, 2, 3, 4, 5, \ldots, n, \ldots\}$
$\downarrow\ \downarrow\ \downarrow\ \downarrow\ \downarrow\ \qquad \downarrow$
$\{3, 9, 27, 81, 243, \ldots, 3^n, \ldots\}$

27. = 29. = 31. =

REVIEW EXERCISES, PAGE 94

1. True
2. False; the word *best* makes the statement not well defined.
3. True
4. False; no set is a proper subset of itself.
5. False; the elements 6, 12, 18, 24, ... are members of both sets.
6. True
7. False; both sets do not contain exactly the same elements.
8. True 9. True 10. True 11. True 12. True
13. True 14. True 15. $A = \{7, 9, 11, 13, 15\}$
16. {Colorado, Nebraska, Missouri, Oklahoma}
17. $C = \{1, 2, 3, 4, \ldots, 161\}$
18. $D = \{9, 10, 11, 12, \ldots, 96\}$
19. $A = \{x \mid x \in N \text{ and } 52 < x < 100\}$
20. $B = \{x \mid x \in N \text{ and } x > 42\}$
21. $C = \{x \mid x \in N \text{ and } x < 5\}$
22. $D = \{x \mid x \in N \text{ and } 27 \leq x \leq 51\}$
23. *A* is the set of capital letters in the English alphabet from E through M, inclusive.

24. *B* is the set of U.S. coins with a value of less than a dollar.

25. *C* is the set of the last three lowercase letters in the English alphabet.

26. *D* is the set of numbers greater than or equal to 3 and less than 9.

27. {5, 7} **28.** {1, 2, 3, 4, 5, 6, 7, 8} **29.** {9, 10}

30. {1, 2, 4, 6, 7, 8, 10} **31.** {1, 3} **32.** {1, 7}

33. {(1, 1), (1, 7), (1, 10), (3, 1), (3, 7), (3, 10), (5, 1), (5, 7), (5, 10), (7, 1), (7, 7), (7, 10)}

34. {(5, 1), (5, 3), (5, 5), (5, 7), (7, 1), (7, 3), (7, 5), (7, 7), (9, 1), (9, 3), (9, 5), (9, 7), (10, 1), (10, 3), (10, 5), (10, 7)}

35. 16 **36.** 15

37.

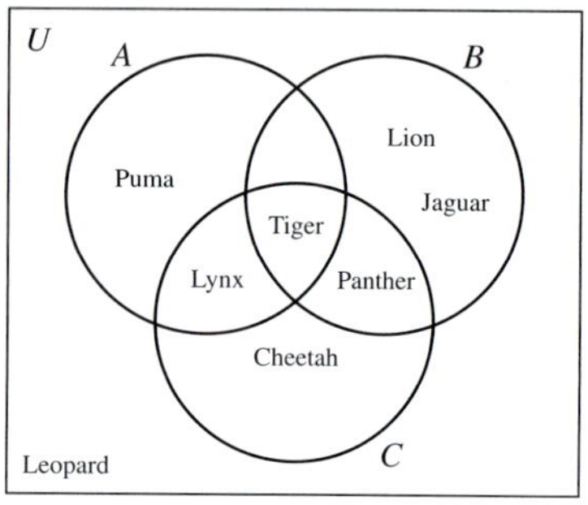

38. {a, c, d, f, g, i, k, l} **39.** {i, k}

40. {a, b, c, d, f, g, h, i, k, l} **41.** {f} **42.** {a, f, i}

43. {a, b, d, f, h, i, l} **44.** True **45.** True **46.** II

47. III **48.** I **49.** IV **50.** IV **51.** II **52.** \$450

53.

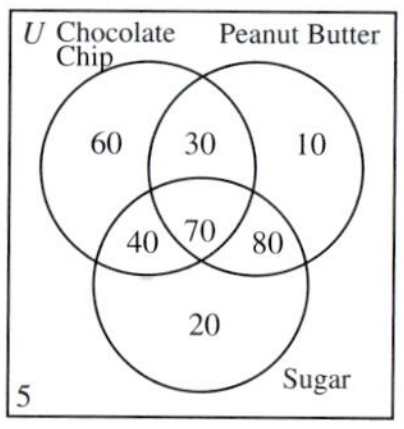

a) 315
b) 10
c) 30
d) 110

54.

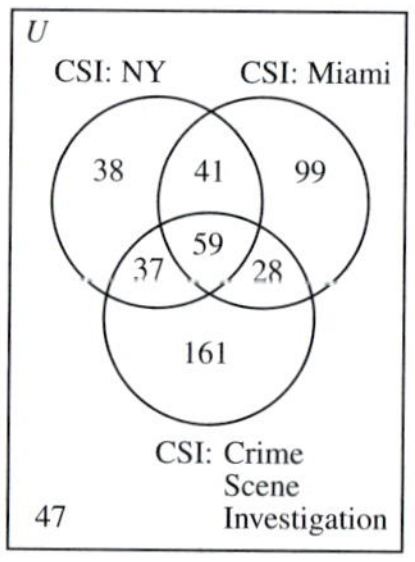

a) 38
b) 298
c) 28
d) 236
e) 106

55. $\{2, 4, 6, 8, 10, \ldots, 2n, \ldots\}$
$\downarrow\ \downarrow\ \downarrow\ \downarrow\ \downarrow\ \quad \downarrow$
$\{4, 6, 8, 10, 12, \ldots, 2n + 2, \ldots\}$

56. $\{3, 5, 7, 9, 11, \ldots, 2n + 1, \ldots\}$
$\downarrow\ \downarrow\ \downarrow\ \downarrow\ \downarrow\ \quad \downarrow$
$\{5, 7, 9, 11, 13, \ldots, 2n + 3, \ldots\}$

57. $\{1, 2, 3, 4, 5, \ldots, n, \ldots\}$
$\downarrow\ \downarrow\ \downarrow\ \downarrow\ \downarrow\ \quad \downarrow$
$\{5, 8, 11, 14, 17, \ldots, 3n + 2, \ldots\}$

58. $\{1, 2, 3, 4, 5, \ldots, n, \ldots\}$
$\downarrow\ \downarrow\ \downarrow\ \downarrow\ \downarrow\ \quad \downarrow$
$\{4, 9, 14, 19, 24, \ldots, 5n - 1, \ldots\}$

CHAPTER TEST, PAGE 97

1. True

2. False; the sets do not contain exactly the same elements.

3. True

4. False; the second set has no subset that contains the element 7.

5. False; the empty set is a subset of every set.

6. False; the set has 2^4, or 16 subsets. **7.** True

8. False; for any set *A*, $A \cup A' = U$, not { }. **9.** True

10. $A = \{1, 2, 3, 4, 5, 6, 7, 8\}$

11. Set *A* is the set of natural numbers less than 9.

12. {7, 9} **13.** {3, 5, 7, 9, 13}

14. {3, 5, 7, 9}, or *A* **15.** 2 **16.** {3, 5}

17. {(3, 3), (3, 11), (3, 15), (5, 3), (5, 11), (5, 15), (7, 3), (7, 11), (7, 15), (9, 3), (9, 11), (9, 15)}

18.

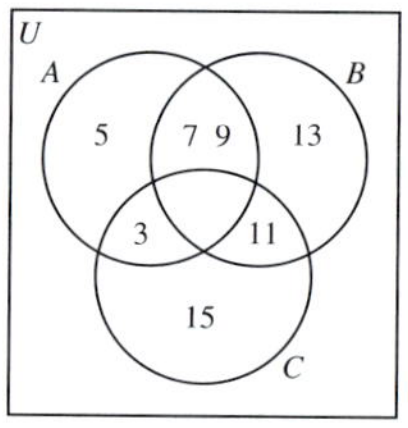

19. Equal

20.

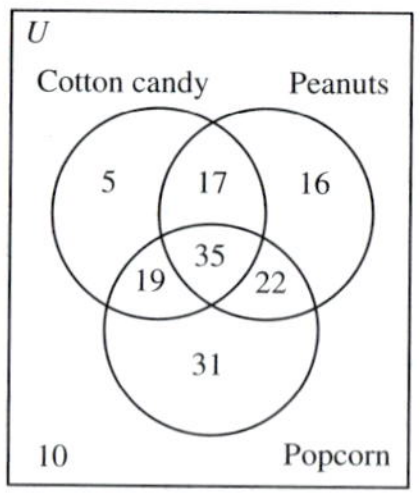

a) 52 **b)** 10 **c)** 93 **d)** 17 **e)** 38 **f)** 31

21. $\{7, 8, 9, 10, 11, \ldots, n + 6, \ldots\}$
$\downarrow\ \downarrow\ \downarrow\ \downarrow\ \downarrow\ \quad \downarrow$
$\{8, 9, 10, 11, 12, \ldots, n + 7, \ldots\}$

22. $\{1, 2, 3, 4, 5, \ldots, n, \ldots\}$
$\downarrow\ \downarrow\ \downarrow\ \downarrow\ \downarrow\ \quad \downarrow$
$\{1, 3, 5, 7, 9, \ldots, 2n - 1, \ldots\}$

CHAPTER 3

SECTION 3.1, PAGE 109

1. a) A sentence that can be judged either true or false is called a statement.
b) A simple statement is a statement that conveys only one idea.
c) Compound statements are statements consisting of two or more simple statements.

3. *All, none,* and *some* are quantifiers.

5. a) Some are. **b)** All are.
c) Some are not. **d)** None are.

7. a) No, the *exclusive or* is used when one or the other of the events can take place, but not both.
b) Yes, the *inclusive or* is used when one or the other, or both events can take place.
c) *Inclusive or*

9. Compound; disjunction, $\vee$

11. Compound; biconditional, $\leftrightarrow$

13. Compound; conjunction, $\wedge$

15. Simple statement

17. Compound; negation, $\sim$

19. Compound; conjunction, $\wedge$

21. Compound; negation, $\sim$

23. Some butterflies are not insects.

25. Some aldermen are running for mayor.

27. All turtles have claws.

29. Some bicycles have three wheels.

31. All pine trees produce pinecones.

33. No pedestrians are in the crosswalk.

35. $\sim p$ **37.** $\sim q \vee \sim p$ **39.** $\sim p \rightarrow \sim q$

41. $\sim p \wedge q$ **43.** $\sim q \leftrightarrow p$ **45.** $\sim(p \vee q)$

47. Ken Jennings did not win more than $3 million.

49. Ken Jennings won 74 games of *Jeopardy!* and Ken Jennings won more than $3 million.

51. If Ken Jennings did not win 74 games of *Jeopardy!* then Ken Jennings won more than $3 million.

53. Ken Jennings did not win 74 games of *Jeopardy!* or Ken Jennings did not win more than $3 million.

55. It is false that Ken Jennings won 74 games of *Jeopardy!* and Ken Jennings won more than $3 million.

57. $(p \wedge \sim q) \wedge r$ **59.** $(p \wedge q) \vee r$ **61.** $p \rightarrow (q \vee \sim r)$

63. $(r \leftrightarrow q) \wedge p$ **65.** $q \rightarrow (p \leftrightarrow r)$

67. The water is 70° or the sun is shining, and we do not go swimming.

69. The water is not 70°, and the sun is shining or we go swimming.

71. If we do not go swimming, then the sun is shining and the water is 70°.

73. If the sun is shining then we go swimming, and the water is 70°.

75. The sun is shining if and only if the water is 70°, and we go swimming.

77. Not permissible, you cannot have both soup and salad. The *or* used on menus is the *exclusive or.*

79. Not permissible, you cannot have both potatoes and pasta. The *or* used on menus is the *exclusive or.*

81. a) $w \wedge \sim p$ **b)** Conjunction

83. a) $\sim(b \rightarrow \sim p)$ **b)** Negation

85. a) $(f \vee v) \rightarrow h$ **b)** Conditional

87. a) $c \leftrightarrow (\sim f \vee p)$ **b)** Biconditional

89. a) $(c \leftrightarrow w) \vee s$ **b)** Disjunction

91. a) Answers will vary.
b) Answers will vary.

SECTION 3.2, PAGE 122

1. a) 4 **b)**

p	q
T	T
T	F
F	T
F	F

3. a)

p	q	$p \wedge q$
T	T	T
T	F	F
F	T	F
F	F	F

b) Only when both p and q are true

c)

p	q	$p \vee q$
T	T	T
T	F	T
F	T	T
F	F	F

d) Only when both p and q are false

5.	7.	9.	11.
F	T	F	T
F	F	T	F
	T	T	T
	T	T	T

13.	15.	17.	19.
T	T	F	T
F	F	F	T
T	T	T	T
T	T	F	T
F	T	F	F
F	F	F	F
T	T	T	T
T	F	T	F

21. $p \wedge q$	23. $p \wedge \sim q$	25. $\sim(p \wedge q)$
T	F	F
F	T	T
F	F	T
F	F	T

27. $p \vee (q \vee r)$	29. $p \wedge (q \vee \sim q)$
T	T
T	T
T	F
T	F
T	
T	
T	
F	

31. a) False **b)** True

33. a) True **b)** True

35. a) True **b)** False

37. a) False **b)** True

39. a) True **b)** True

41. a) True **b)** True

43. True **45.** True **47.** False **49.** False

51. False **53.** True **55.** True **57.** False

59. $p \wedge \sim q$	61. $p \vee \sim q$
F	T
T	T
F	F
F	T

59. True in case 2

61. True in cases 1, 2, and 4, when p is true, or when p and q are both false.

63. $(r \vee q) \wedge p$	65. $q \vee (p \wedge \sim r)$
T	T
T	T
T	F
F	T
F	T
F	T
F	F
F	F

63. True in cases 1, 2, and 3

65. True in cases 1, 2, 4, 5, and 6. True except when p, q, r have truth values TFT, FFT, or FFF.

67. a) Mr. Duncan and Mrs. Tuttle qualify.
b) Mrs. Rusinek does not qualify, since their combined income is less than \$46,000.

69. a) Wing Park qualifies; the other four do not.
b) Gina Vela is returning on April 2. Kara Sharo is returning on a Monday. Christos Supernaw is not staying over on a Saturday. Alex Chang is returning on a Monday.

71. T
T
T
T
F
T
F
T

73. Yes

SECTION 3.3, PAGE 133

1. a)

p	q	$p \rightarrow q$
T	T	T
T	F	F
F	T	T
F	F	T

b) The conditional is false only when the antecedent is true and the consequent is false.

c)

p	q	$p \leftrightarrow q$
T	T	T
T	F	F
F	T	F
F	F	T

d) The biconditional is true only when both p and q are true or when p and q are both false.

3. a) Substitute the truth values for the simple statements. Then evaluate the compound statement, using the assigned truth values.
b) True

5. A self-contradiction is a compound statement that is always false.

7.	9.	11.	13.	15.
T	T	F	T	F
T	F	T	T	T
T	F	T	F	T
F	F	F	T	T

17.	19.	21.	23.	25.
T	T	F	T	T
T	T	F	T	T
T	T	T	T	T
T	F	F	T	F
T	F	T	T	T
F	F	T	F	T
F	F	F	F	F
F	T	T	F	T

27. $p \rightarrow (q \wedge r)$	29. $(p \leftrightarrow q) \vee r$
T	T
F	T
F	T
F	F
T	T
T	F
T	T
T	T

31. $\underline{(\sim p \rightarrow q) \vee r}$
T
T
T
T
T
T
T
F

33. Neither **35.** Self-contradiction **37.** Tautology

39. Not an implication **41.** Implication **43.** Implication

45. True **47.** False **49.** False **51.** True

53. True **55.** True **57.** True **59.** False

61. True **63.** True **65.** True **67.** False

69. False **71.** True **73.** True **75.** True

77. No, the statement only states what will occur if your sister gets straight A's. If your sister does not get straight A's, your parents may still get her a computer.

79. F
F
T
T
T
F
F
F

81. It is a tautology. The statement may be expressed as $(p \rightarrow q) \vee (\sim p \rightarrow q)$, where p: It is a head and q: I win. This statement is a tautology.

83.

Tiger	Boots	Sam	Sue
Blue	Yellow	Red	Green
Nine Lives	Whiskas	Friskies	Meow Mix

SECTION 3.4, PAGE 146

1. a) Statements that have exactly the same truth values

b) Construct truth tables for each statement. If both have the same truth values in the answer columns of the truth tables, the statements are equivalent.

3. $\sim(p \wedge q) \Leftrightarrow \sim p \vee \sim q$
$\sim(p \vee q) \Leftrightarrow \sim p \wedge \sim q$

5. a and c, b and d **7.** $p \wedge \sim q$ **9.** Not equivalent

11. Not equivalent **13.** Equivalent **15.** Equivalent

17. Equivalent **19.** Equivalent **21.** Equivalent

23. Equivalent **25.** Not equivalent

27. Not equivalent **29.** Equivalent

31. The Rocky Mountains are not in the East or the Appalachian Mountains are not in the West.

33. It is false that the watch was a Swatch or the watch was a Swiss Army watch.

35. It is false that the hotel has a weight room and the conference center has an auditorium.

37. If Ashley Tabai takes the new job, then it is false that she will move and she will not buy a new house in town.

39. Ena Salter does not select a new textbook or she will have to write a new syllabus.

41. If Bob the Tomato didn't visit the nursing home then he did not visit the Cub Scout meeting.

43. The plumbers do not meet in Kansas City or the Rainmakers will provide the entertainment.

45. If Chase is hiding, then the pitcher is broken.

47. We go to Cincinnati and we will not go to the zoo.

49. It is false that if I am cold then the heater is working.

51. Borders has a sale and we will not buy $100 worth of books.

53. It is false that if John Deere will hire new workers then the city of Dubuque will not retrain the workers.

55. *Converse:* If we can finish the quilt in 1 week, then we work every night.
Inverse: If we do not work every night, then we cannot finish the quilt in 1 week.
Contrapositive: If we cannot finish the quilt in 1 week, then we do not work every night.

57. *Converse:* If I buy silver jewelry, then I go to Mexico.
Inverse: If I do not go to Mexico, then I do not buy silver jewelry.
Contrapositive: If I do not buy silver jewelry, then I do not go to Mexico.

59. *Converse:* If I scream, then that annoying paper clip shows up on my computer screen.
Inverse: If that annoying paper clip does not show up on my computer screen, then I will not scream.
Contrapositive: If I do not scream, then that annoying paper clip does not show up on my screen.

61. If a natural number is divisible by 10, then the natural number is divisible by 5. True.

63. If a natural number is not divisible by 6, then the natural number is not divisible by 3. False.

65. If two lines are not parallel, then the two lines intersect in at least one point. True.

67. b) and **c)** are equivalent.

69. a) and **c)** are equivalent.

71. b) and **c)** are equivalent.

73. b) and **c)** are equivalent.

75. None are equivalent.

77. None are equivalent.

79. a) and **c)** are equivalent.

81. a) and **b)** are equivalent.

83. True. If $p \rightarrow q$ is false, it must be of the form $T \rightarrow F$. Therefore, the converse must be of the form $F \rightarrow T$, which is true.

85. False. A conditional statement and its contrapositive always have the same truth values.

87. Answers will vary.

89. Answers will vary.

SECTION 3.5, PAGE 159

1. a) The conclusion necessarily follows from the given set of premises.
b) The conclusion does not necessarily follow from the given set of premises.

3. Yes, if the conclusion does not necessarily follow from the premises, the argument is invalid, even if the conclusion is a true statement.

5. Yes, if the conclusion does not follow from the set of premises.

7. a) $p \rightarrow q$
$\underline{p}$
$\therefore q$
b) Answers will vary.

9. a) $p \rightarrow q$
$\underline{\sim q}$
$\therefore \sim p$
b) Answers will vary.

11. a) $p \rightarrow q$
$\underline{q}$
$\therefore p$
b) Answers will vary.

13. Invalid **15.** Valid **17.** Valid **19.** Invalid
21. Valid **23.** Valid **25.** Valid **27.** Invalid
29. Invalid **31.** Valid

33. a) $p \rightarrow q$
$\underline{\sim p}$
$\therefore \sim q$
b) Invalid

35. a) $p \rightarrow q$
$\underline{p}$
$\therefore q$
b) Valid

37. a) $p \rightarrow q$
$\underline{\sim q}$
$\therefore \sim p$
b) Valid

39. a) $p \rightarrow q$
$\underline{q}$
$\therefore p$
b) Invalid

41. a) $p \vee q$
$\underline{\sim p}$
$\therefore q$
b) Valid

43. a) $p \rightarrow q$
$\underline{q \rightarrow r}$
$\therefore p \rightarrow r$
b) Valid

45. a) $p \wedge q$
$\underline{q \rightarrow r}$
$\therefore r \rightarrow p$
b) Valid

47. a) $s \wedge g$
$\underline{g \rightarrow c}$
$\therefore s \rightarrow c$
b) Valid

49. a) $h \rightarrow b$
$\underline{\sim p \rightarrow \sim b}$
$\therefore h \rightarrow p$
b) Valid

51. a) $p \rightarrow q$
$\underline{\sim q}$
$\therefore \sim p$
b) Valid

53. a) $p \vee q$
$\underline{\sim p}$
$\therefore q$
b) Valid

55. a) $t \wedge g$
$\underline{\sim t \vee \sim g}$
$\therefore \sim t$
b) Valid

57. a) $c \wedge \sim h$
$\underline{h \rightarrow c}$
$\therefore h$
b) Invalid

59. a) $f \rightarrow d$
$\underline{d \rightarrow \sim s}$
$\therefore f \rightarrow s$
b) Invalid

61. Therefore, your face will break out.

63. Therefore, I am stressed out.

65. Therefore, you did not close the deal.

67. Therefore, if you do not pay off your credit card bills, then the bank makes money.

69. No. The conditional statement will always be true, and therefore it will be a tautology, and a valid argument.

SECTION 3.6, PAGE 167

1. a) It is a valid argument.
b) It is an invalid argument.

3. a)

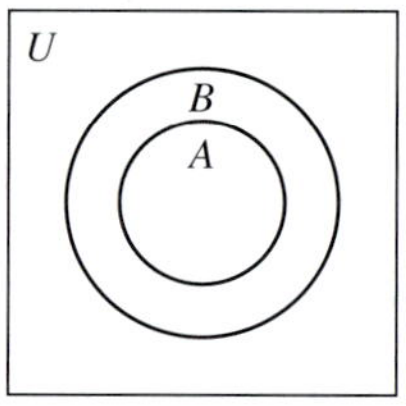

b)

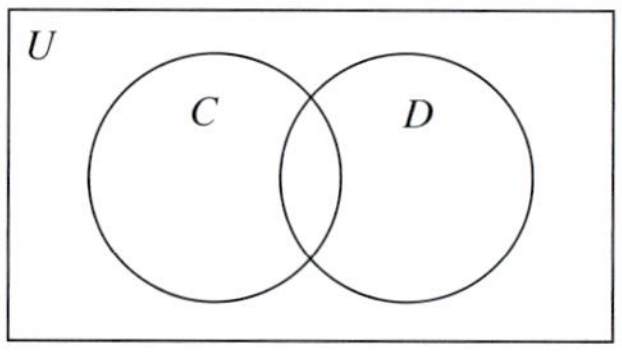

c)

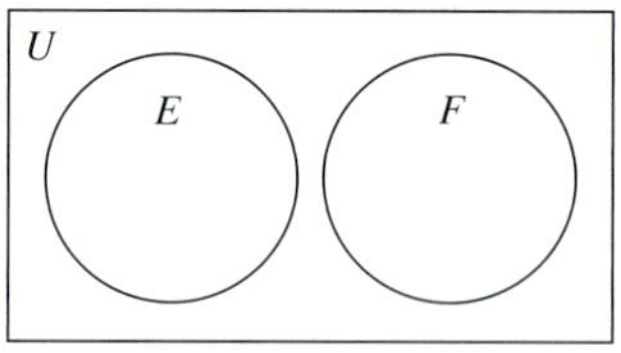

5. Yes, if the conclusion necessarily follows from the premises, the argument is valid.

7. Valid **9.** Valid **11.** Invalid
13. Valid **15.** Invalid **17.** Invalid
19. Invalid **21.** Valid **23.** Invalid **25.** Invalid
27. Valid **29.** Invalid

SECTION 3.7, PAGE 174

1. a) Answers will vary.
b) $\wedge$ (and)

3. It is a series circuit; therefore, both switches must be closed for current to flow and the lightbulb to go on. When the p switch is closed, the $\bar{p}$ switch is open and no current will flow through the circuit. When the $\bar{p}$ switch is closed, the p switch is open and no current will flow through the circuit.

5. a) $p \wedge q$
b) The lightbulb will be on when both p and q are closed.

7. a) $(p \vee q) \wedge \sim q$
b) The lightbulb will be on when p is closed and q is open.

9. a) $(p \wedge q) \wedge [(p \wedge \sim q) \vee r]$
b) The lightbulb will be on when p, q, and r are all closed.

11. a) $p \vee q \vee (r \wedge \sim p)$
b) The lightbulb will be on in all cases except when p, q, and r are all open.

13.

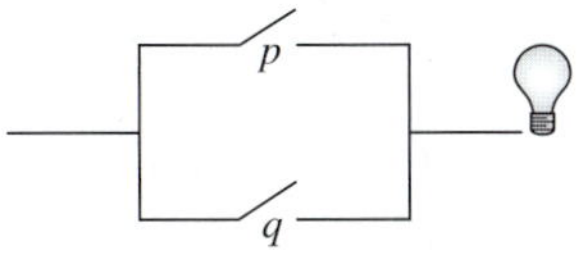

15.

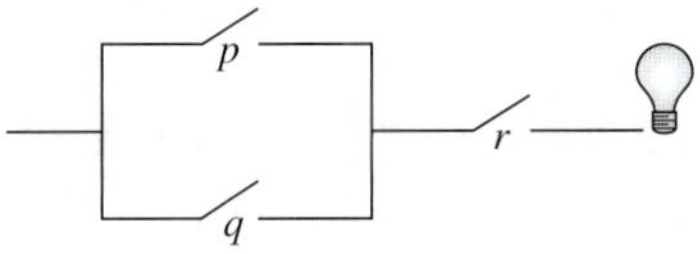

17.

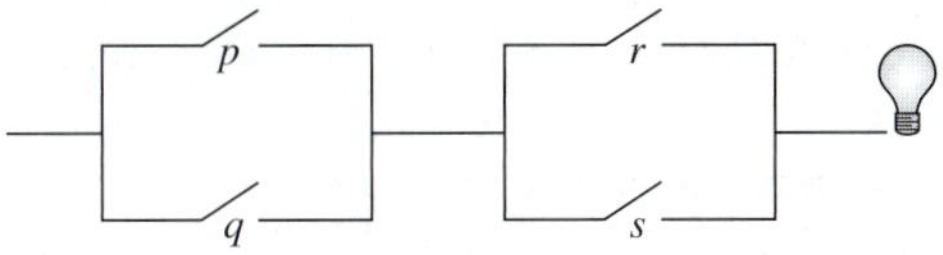

19.

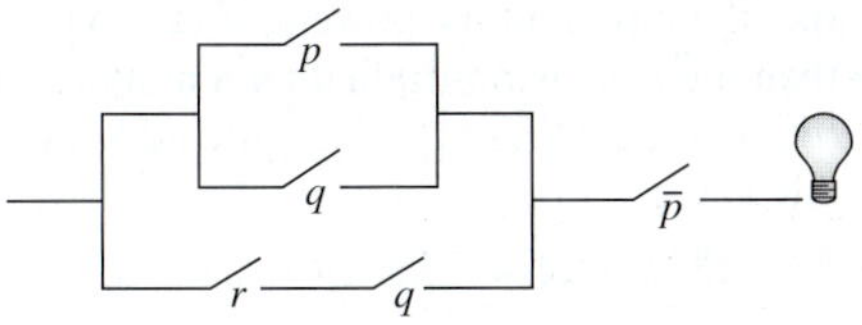

21. $p \vee q$; $\sim p \wedge \sim q$; not equivalent

23. $[(p \wedge q) \vee r] \wedge p$; $(q \vee r) \wedge p$; equivalent

25. $(p \vee \sim p) \wedge q \wedge r$; $p \wedge q \wedge r$; not equivalent

27. a)

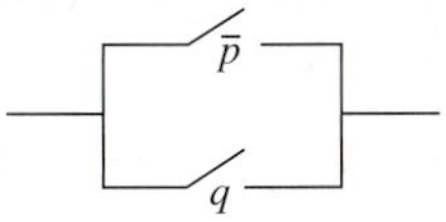

b)

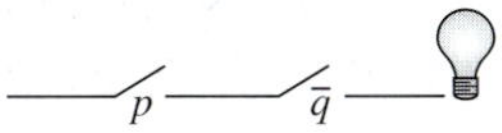

REVIEW EXERCISES, PAGE 176

1. Some gift cards are exchangeable.

2. Some bears are not mammals.

3. No women are presidents.

4. All pine trees are green.

5. The coffee is Maxwell House or the coffee is hot.

6. The coffee is not hot and the coffee is strong.

7. If the coffee is hot, then the coffee is strong and the coffee is not Maxwell House.

8. The coffee is Maxwell House if and only if the coffee is not strong.

9. The coffee is not Maxwell House, if and only if the coffee is strong and the coffee is not hot.

10. The coffee is Maxwell House or the coffee is not hot, and the coffee is not strong.

11. $r \wedge q$ **12.** $p \rightarrow r$ **13.** $(r \rightarrow q) \vee \sim p$

14. $(q \leftrightarrow p) \wedge \sim r$ **15.** $(r \wedge q) \vee \sim p$ **16.** $\sim(r \wedge q)$

17.	18.	19.	20.	21.	22.
F	T	T	T	F	F
F	F	T	F	T	T
T	F	T	T	F	T
F	F	T	T	F	T
		T	F	T	T
		F	F	T	T
		F	F	T	T
		T	F	T	T

23. False **24.** True **25.** False **26.** True **27.** True

28. True **29.** False **30.** False **31.** Not equivalent

32. Equivalent **33.** Equivalent **34.** Not equivalent

35. It is false that if Bobby Darin sang *Mack the Knife* then Elvis wrote *Memphis*.

36. If Lynn Swann did not play for the Steelers, then Jack Tatum played for the Raiders.

37. Altec Lansing does not produce only speakers and Harman Kardon does not produce only stereo receivers.

38. It is false that Travis Tritt won an Academy Award or Randy Jackson does commercials for Milk Bone Dog Biscuits.

39. The temperature is above 32° or we will go ice fishing at O'Leary's Lake.

40. a) If you soften your opinion, then you hear a new voice today.
b) If you do not hear a new voice today, then you do not soften your opinion.
c) If you do not soften your opinion, then you do not hear a new voice today.

41. a) If we will learn the table's value, then we take the table to *Antiques Roadshow*.
b) If we do not take the table to *Antiques Roadshow*, then we will not learn the table's value.
c) If we will not learn the table's value, then we do not take the table to *Antiques Roadshow*.

42. a) If Maureen Gerald is helping at the school, then she is not in attendance.
b) If Maureen Gerald is in attendance, then she is not helping at the school.
c) If Maureen Gerald is not helping at the school, then she is in attendance.

43. a) If we will not buy a desk at Miller's Furniture, then the desk is made by Winner's Only and the desk is in the Rose catalog.
b) If the desk is not made by Winner's Only or the desk is not in the Rose catalog, then we will buy a desk at Miller's Furniture.
c) If we will buy a desk at Miller's Furniture, then the desk is not made by Winner's Only or the desk is not in the Rose catalog.

44. a) If I let you attend the prom, then you will get straight A's on your report card.
b) If you do not get straight A's on your report card, then I will not let you attend the prom.
c) If I will not let you attend the prom, then you did not get straight A's on your report card.

45. a), b), and **c)** are equivalent. **46.** None are equivalent.

47. a) and **c)** are equivalent. **48.** None are equivalent.

49. Invalid **50.** Valid **51.** Valid **52.** Invalid

53. Invalid **54.** Invalid **55.** Invalid **56.** Invalid

57. a) $p \wedge [(q \wedge r) \vee \sim p]$
b) The lightbulb will be on when p, q, and r are all closed.

58.

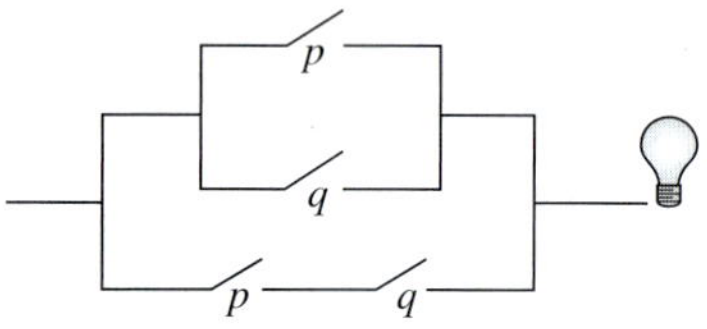

59. Equivalent

CHAPTER TEST, PAGE 179

1. $(p \wedge r) \vee \sim q$ **2.** $(r \rightarrow q) \vee \sim p$ **3.** $\sim(r \leftrightarrow \sim q)$

4. Ann is not the secretary and Elaine is the president, if and only if Dick is not the vice president.

5. If Ann is the secretary or Dick is not the vice president, then Elaine is the president.

6.	**7.**
F	T
T	T
F	T
F	T
F	F
F	T
F	T
F	F

8. True **9.** True **10.** True **11.** True **12.** Equivalent

13. a) and **b)** are equivalent.

14. a) and **b)** are equivalent.

15. $s \rightarrow f$
$f \rightarrow p$
$\therefore\ s \rightarrow p$
Valid

16. Invalid

17. Some highways are not roads.

18. Nick did not play football or Max did not play baseball.

19. *Converse:* If today is Saturday, then the garbage truck comes.
Inverse: If the garbage truck does not come, then today is not Saturday.
Contrapositive: If today is not Saturday, then the garbage truck does not come.

20.

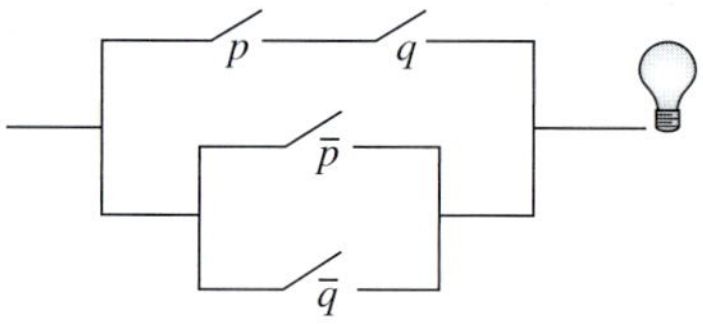

CHAPTER 4

SECTION 4.1, PAGE 188

1. A number is a quantity, and it answers the question "How many?" A numeral is a symbol used to represent the number.

3. A system of numeration consists of a set of numerals and a scheme or rule for combining the numerals to represent numbers.

5. The Hindu–Arabic numeration system

7. In a multiplicative system, there are numerals for each number less than the base and for powers of the base. Each numeral less than the base is multiplied by a numeral for the power of the base, and these products are added to obtain the number.

9. 345 **11.** 2423 **13.** 334,214 **15.** ∩∩∩∩∩|||

17.  **19.** 

21. 8 **23.** 43 **25.** 1236 **27.** 2946 **29.** 12,666

31. 9464 **33.** XXVII **35.** CCCXLI **37.** MMV

39. $\overline{\text{IV}}$DCCXCIII **41.** $\overline{\text{IX}}$CMXCIX **43.** $\overline{\text{XX}}$DCXLIV

45. 74 **47.** 4081 **49.** 8550 **51.** 4003

53. 五十三 **55.** 三百七十八 **57.** 四千二百六十 **59.** 七千零五十六

61. 26 **63.** 279 **65.** 2883
67. $\nu\theta$ **69.** $\psi\kappa$f **71.** $^{\iota}\varepsilon\varepsilon$
73. 1021, MXXI, 一千零二十一, $^{\iota}\alpha\kappa\alpha$
75. 527, ,
DXXVII, $\phi\kappa\zeta$
77. $\overline{\text{CMXCIX}}$CMXCIX
79. Advantage: You can write some numerals more compactly.
Disadvantage: There are more numerals to memorize.
81. Advantage: You can write some numerals more compactly.
Disadvantage: There are more numerals to memorize.
83. MM

SECTION 4.2, PAGE 196

1. Positional value system
3. The 4 in 40 represents four tens. The 4 in 400 represents 4 hundreds.
5. a) 10 **b)** 0, 1, 2, 3, 4, 5, 6, 7, 8, 9
7. Write the sum of the products of each digit times its corresponding positional value.
9. a) There may be confusion because numerals could be interpreted in different ways. For example, ▼ could be interpreted to be either 1 or 60.
b)  for both numerals.
11. 1, 20, 18×20, $18 \times (20)^2$, $18 \times (20)^3$
13. $(2 \times 10) + (3 \times 1)$
15. $(3 \times 100) + (5 \times 10) + (9 \times 1)$
17. $(8 \times 100) + (9 \times 10) + (7 \times 1)$
19. $(4 \times 1000) + (3 \times 100) + (8 \times 10) + (7 \times 1)$
21. $(1 \times 10{,}000) + (6 \times 1000) + (4 \times 100) + (0 \times 10) + (2 \times 1)$
23. $(3 \times 100{,}000) + (4 \times 10{,}000) + (6 \times 1000) + (8 \times 100) + (6 \times 10) + (1 \times 1)$
25. 14 **27.** 784 **29.** 4868 **31.** 
33.  **35.**  **37.** 57
39. 4682 **41.** 4000 **43.**  **45.**  **47.** 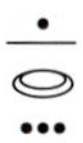
49. 1944, 
51. 

53. a) No largest numeral
b) 
55.  **57.** 
59. a) Answers will vary. **b)** Answers will vary.
c) Answers will vary.

SECTION 4.3, PAGE 202

1. a) Answers will vary. **b)** Answers will vary.
3. a) There is no digit 4 in base 4.
b) There is no digit 3 in base 2.
c) There is no digit L in base 12.
d) There is no digit G in base 16.
5. 1 **7.** 13 **9.** 184 **11.** 441 **13.** 1395 **15.** 53
17. 4012 **19.** 1549 **21.** 50,809 **23.** 110_2 **25.** 110212_3
27. 311_4 **29.** 402_5 **31.** 2112_8 **33.** 1861_9 **35.** 5264_{12}
37. $1ED_{16}$ **39.** $24EF_{16}$ **41.** 2202102_3 **43.** 31014_5
45. 5600_7 **47.** 1567_{11} **49.** $8DE_{15}$ **51.** 23 **53.** 78
55. 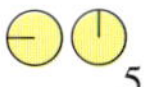$_5$ **57.** $_5$ **59.** 14 **61.** 36 **63.** $_4$
65. $_4$
67. a) Answers will vary. **b)** 10213_5 **c)** 1373_8
69. Answers will vary. **71.** $b = 6$
73. a) 876 **b)** $_4$

SECTION 4.4, PAGE 212

1. a) $1, b, b^2, b^3, b^4$ **b)** $1, 2, 2^2, 2^3, 2^4$
3. No, cannot have a 3 in base 3.
5. Answers will vary. **7.** 111_3 **9.** 2130_5
11. $9B5_{12}$ **13.** 2200_3 **15.** 24001_7 **17.** 10100_2
19. 11_3 **21.** 831_9 **23.** 110_2 **25.** 11_2 **27.** 3616_7
29. 1011_3 **31.** 121_3 **33.** 4103_8 **35.** 21020_6
37. 77676_{12} **39.** 100011_2 **41.** 6031_7 **43.** 101_2
45. 223_6 R1_6 **47.** 123_4 **49.** 103_4 R1_4 **51.** 41_5 R1_5
53. 45_7 R2_7 **55.** $_5$ **57.** $_5$ **59.** $_4$
61. $_4$ **63.** $_4$ **65.** $_4$

33. $2\frac{3}{5}$ **35.** $-12\frac{1}{6}$ **37.** $-58\frac{8}{15}$ **39.** 0.7 **41.** $0.\overline{2}$

43. 0.375 **45.** $2.1\overline{6}$ **47.** $5.\overline{6}$ **49.** $\frac{75}{100} = \frac{3}{4}$

51. $\frac{45}{1000} = \frac{9}{200}$ **53.** $\frac{2}{10} = \frac{1}{5}$ **55.** $\frac{131}{10{,}000}$ **57.** $\frac{1}{10{,}000}$ **59.** $\frac{1}{9}$

61. $\frac{2}{1}$ **63.** $\frac{15}{11}$ **65.** $\frac{37}{18}$ **67.** $\frac{574}{165}$ **69.** $\frac{2}{5}$ **71.** $\frac{2}{5}$ **73.** $\frac{49}{64}$

75. $\frac{36}{35}$ **77.** $\frac{20}{21}$ **79.** $\frac{11}{12}$ **81.** $\frac{9}{22}$ **83.** $\frac{23}{54}$ **85.** $\frac{17}{144}$

87. $-\frac{109}{600}$ **89.** $\frac{19}{24}$ **91.** $-\frac{1}{24}$ **93.** $\frac{19}{24}$ **95.** 1 **97.** $\frac{11}{10}$

99. $\frac{23}{42}$ **101.** $1\frac{3}{4}$ in. **103.** $120\frac{3}{4}$ in. **105.** $9\frac{15}{16}$ in.

107. $\frac{1}{10}$ **109.** $58\frac{7}{8}$ in.

111. a) $1\frac{49}{60}, 2\frac{48}{60}, 9\frac{6}{60}, 6\frac{3}{60}, 2\frac{9}{60}, \frac{22}{60}$

b) $22\frac{17}{60}$; 22 hours 17 minutes

113. $26\frac{5}{32}$ in. **115. a)** $29\frac{3}{16}$ in. **b)** 33 in. **c)** $32\frac{3}{4}$ in.

In Exercises 117–121, an infinite number of answers are possible. We give one answer.

117. 0.105 **119.** −2.1755 **121.** 4.8725

123. $\frac{1}{2}$ **125.** $\frac{11}{200}$ **127.** $\frac{11}{200}$

129. a) $1\frac{3}{8}$ cup water (or milk) and $\frac{3}{4}$ cup oatmeal

b) $1\frac{1}{2}$ cup water (or milk) and $\frac{3}{4}$ cup oatmeal

131. a) $\frac{1}{8}$ **b)** $\frac{1}{16}$ **c)** 5 **d)** 6

SECTION 5.4, PAGE 266

1. A rational number can be written as a ratio of two integers. Real numbers that cannot be written as a ratio of two integers are irrational numbers.

3. A perfect square is any number that is the square of a natural number.

5. a) To add or subtract two or more square roots with the same radicand, add or subtract their coefficients and then multiply the sum or difference by the common radical.

b) $5\sqrt{5}$

7. a) Multiply both the numerator and denominator by a radical that will result in the radicand in the denominator becoming a perfect square.

b) $\dfrac{2\sqrt{3}}{3}$

9. Rational **11.** Rational **13.** Irrational

15. Rational **17.** Irrational **19.** 4 **21.** 10

23. −13 **25.** −9 **27.** −10

29. Rational number, integer, natural number

31. Rational number, integer, natural number

33. Rational number **35.** Rational number

37. Rational number

39. $2\sqrt{3}$ **41.** $4\sqrt{3}$ **43.** $3\sqrt{7}$ **45.** $2\sqrt{21}$ **47.** $9\sqrt{2}$

49. $5\sqrt{5}$ **51.** $\sqrt{2}$ **53.** $-13\sqrt{3}$ **55.** $4\sqrt{3}$ **57.** $23\sqrt{2}$

59. 9 **61.** $2\sqrt{15}$ **63.** $10\sqrt{2}$ **65.** 2 **67.** 3

69. $\dfrac{\sqrt{5}}{5}$ **71.** $\dfrac{\sqrt{21}}{7}$ **73.** $\dfrac{2\sqrt{15}}{3}$ **75.** $\dfrac{\sqrt{15}}{3}$ **77.** $\dfrac{\sqrt{15}}{3}$

79. $\sqrt{5}$ is between 2 and 3 since 5 is between 4 and 9. $\sqrt{5}$ is between 2 and 2.5 since 5 is closer to 4 than to 9. $\sqrt{5} \approx 2.24$.

81. $\sqrt{107}$ is between 10 and 11 since 107 is between 100 and 121. $\sqrt{107}$ is between 10 and 10.5 since 107 is closer to 100 than to 121. $\sqrt{107} \approx 10.34$.

83. $\sqrt{170}$ is between 13 and 14 since 170 is between 169 and 196. $\sqrt{170}$ is between 13 and 13.5 since 170 is closer to 169 than to 196. $\sqrt{170} \approx 13.04$.

In Exercises 85–89, false answers can be modified in a variety of ways. We give one possible answer.

85. False. $\sqrt{c}$ may be a rational number or an irrational number for a composite number c. (For example, $\sqrt{25}$ is a rational number; $\sqrt{8}$ is an irrational number.)

87. True

89. False. The product of a rational number and an irrational number may be a rational number or an irrational number.

91. $3\sqrt{2} + 5\sqrt{2} = 8\sqrt{2}$ **93.** $\sqrt{2} \cdot \sqrt{3} = \sqrt{6}$

95. $\sqrt{2} \neq 1.414$ since $\sqrt{2}$ is irrational and 1.414 is rational.

97. No. π is irrational; therefore, it cannot equal $\frac{22}{7}$ or 3.14, both of which are rational.

99. $\sqrt{4 \cdot 9} = \sqrt{4} \cdot \sqrt{9}$, $\sqrt{36} = 2 \cdot 3$, $6 = 6$

101. a) 10 mph **b)** 20 mph **c)** 40 mph **d)** 80 mph

103. a) Rational. $\sqrt{0.04} = 0.2$, which is a rational number.

b) Irrational. $\sqrt{0.7} = 0.8366600265\ldots$. Since the decimal number is not a terminating or a repeating decimal number, this number is an irrational number.

105. a) $(44 \div \sqrt{4}) \div \sqrt{4} = 11$ **b)** $(44 \div 4) + \sqrt{4} = 13$

c) $4 + 4 + 4 + \sqrt{4} = 14$

d) $\sqrt{4}(4 + 4) + \sqrt{4} = 18$

Other answers are possible.

SECTION 5.5, PAGE 273

1. The real numbers are the union of the rational numbers and the irrational numbers.

3. If whenever the operation is performed on two elements of a set the result is also an element of the set, then the set is closed under that operation.

5. $a \cdot b = b \cdot a$, the order in which two numbers are multiplied is immaterial. One example is $4(5) = 5(4)$.

7. $(a + b) + c = a + (b + c)$, when adding three numbers, you may place parentheses around any two adjacent numbers. One example is $(1 + 2) + 3 = 1 + (2 + 3)$.

9. No **11.** Yes **13.** Yes **15.** Yes **17.** Yes

19. No **21.** No **23.** No **25.** Yes **27.** Yes

61. 26 **63.** 279 **65.** 2883
67. νθ **69.** ψκϛ **71.** ͵εε
73. 1021, MXXI, 一千零二十一, ͵ακα
75. 527, 99999∩∩|||||||, DXXVII, φκζ
77. $\overline{\text{CMXCIX}}$CMXCIX
79. Advantage: You can write some numerals more compactly. Disadvantage: There are more numerals to memorize.
81. Advantage: You can write some numerals more compactly. Disadvantage: There are more numerals to memorize.
83. MM

SECTION 4.2, PAGE 196

1. Positional value system
3. The 4 in 40 represents four tens. The 4 in 400 represents 4 hundreds.
5. a) 10 **b)** 0, 1, 2, 3, 4, 5, 6, 7, 8, 9
7. Write the sum of the products of each digit times its corresponding positional value.
9. a) There may be confusion because numerals could be interpreted in different ways. For example, ▼ could be interpreted to be either 1 or 60.
b) ▼▼ <▼▼▼ for both numerals.
11. 1, 20, 18 × 20, 18 × $(20)^2$, 18 × $(20)^3$
13. (2 × 10) + (3 × 1)
15. (3 × 100) + (5 × 10) + (9 × 1)
17. (8 × 100) + (9 × 10) + (7 × 1)
19. (4 × 1000) + (3 × 100) + (8 × 10) + (7 × 1)
21. (1 × 10,000) + (6 × 1000) + (4 × 100) + (0 × 10) + (2 × 1)
23. (3 × 100,000) + (4 × 10,000) + (6 × 1000) + (8 × 100) + (6 × 10) + (1 × 1)
25. 14 **27.** 784 **29.** 4868 **31.** <<<▼▼▼▼▼▼
33. <▼▼▼▼ <<<<<<▼ **35.** ▼ ▼ <<▼▼▼▼▼ **37.** 57
39. 4682 **41.** 4000 **43.**  **45.**  **47.** 
49. 1944, <<<▼▼ <<▼▼▼▼
51. 

53. a) No largest numeral
b) 
55. ▼▼ <<<<<▼▼▼▼▼ **57.** 
59. a) Answers will vary. **b)** Answers will vary. **c)** Answers will vary.

SECTION 4.3, PAGE 202

1. a) Answers will vary. **b)** Answers will vary.
3. a) There is no digit 4 in base 4.
b) There is no digit 3 in base 2.
c) There is no digit L in base 12.
d) There is no digit G in base 16.
5. 1 **7.** 13 **9.** 184 **11.** 441 **13.** 1395 **15.** 53
17. 4012 **19.** 1549 **21.** 50,809 **23.** 110_2 **25.** 110212_3
27. 311_4 **29.** 402_5 **31.** 2112_8 **33.** 1861_9 **35.** 5264_{12}
37. $1ED_{16}$ **39.** $24EF_{16}$ **41.** 2202102_3 **43.** 31014_5
45. 5600_7 **47.** 1567_{11} **49.** $8DE_{15}$ **51.** 23 **53.** 78
55. 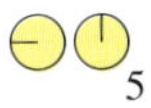 **57.**  **59.** 14 **61.** 36 **63.** 
65. 
67. a) Answers will vary. **b)** 10213_5 **c)** 1373_8
69. Answers will vary. **71.** $b = 6$
73. a) 876 **b)** 

SECTION 4.4, PAGE 212

1. a) $1, b, b^2, b^3, b^4$ **b)** $1, 2, 2^2, 2^3, 2^4$
3. No, cannot have a 3 in base 3.
5. Answers will vary. **7.** 111_3 **9.** 2130_5
11. $9B5_{12}$ **13.** 2200_3 **15.** 24001_7 **17.** 10100_2
19. 11_3 **21.** 831_9 **23.** 110_2 **25.** 11_2 **27.** 3616_7
29. 1011_3 **31.** 121_3 **33.** 4103_8 **35.** 21020_6
37. 77676_{12} **39.** 100011_2 **41.** 6031_7 **43.** 101_2
45. 223_6 $R1_6$ **47.** 123_4 **49.** 103_4 $R1_4$ **51.** 41_5 $R1_5$
53. 45_7 $R2_7$ **55.** 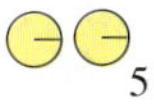 **57.**  **59.** 
61.  **63.**  **65.** 

33. $2\frac{3}{5}$ **35.** $-12\frac{1}{6}$ **37.** $-58\frac{8}{15}$ **39.** $0.\overline{7}$ **41.** $0.\overline{2}$
43. 0.375 **45.** $2.1\overline{6}$ **47.** $5.\overline{6}$ **49.** $\frac{75}{100} = \frac{3}{4}$
51. $\frac{45}{1000} = \frac{9}{200}$ **53.** $\frac{2}{10} = \frac{1}{5}$ **55.** $\frac{131}{10{,}000}$ **57.** $\frac{1}{10{,}000}$ **59.** $\frac{1}{9}$
61. $\frac{2}{1}$ **63.** $\frac{15}{11}$ **65.** $\frac{37}{18}$ **67.** $\frac{574}{165}$ **69.** $\frac{2}{5}$ **71.** $\frac{2}{5}$ **73.** $\frac{49}{64}$
75. $\frac{36}{35}$ **77.** $\frac{20}{21}$ **79.** $\frac{11}{12}$ **81.** $\frac{9}{22}$ **83.** $\frac{23}{54}$ **85.** $\frac{17}{144}$
87. $-\frac{109}{600}$ **89.** $\frac{19}{24}$ **91.** $-\frac{1}{24}$ **93.** $\frac{19}{24}$ **95.** 1 **97.** $\frac{11}{10}$
99. $\frac{23}{42}$ **101.** $1\frac{3}{4}$ in. **103.** $120\frac{3}{4}$ in. **105.** $9\frac{15}{16}$ in.
107. $\frac{1}{10}$ **109.** $58\frac{7}{8}$ in.
111. a) $1\frac{49}{60}, 2\frac{48}{60}, 9\frac{6}{60}, 6\frac{3}{60}, 2\frac{9}{60}, \frac{22}{60}$
b) $22\frac{17}{60}$; 22 hours 17 minutes
113. $26\frac{5}{32}$ in. **115. a)** $29\frac{3}{16}$ in. **b)** 33 in. **c)** $32\frac{3}{4}$ in.

In Exercises 117–121, an infinite number of answers are possible. We give one answer.

117. 0.105 **119.** −2.1755 **121.** 4.8725
123. $\frac{1}{2}$ **125.** $\frac{11}{200}$ **127.** $\frac{11}{200}$
129. a) $1\frac{3}{8}$ cup water (or milk) and $\frac{3}{4}$ cup oatmeal
b) $1\frac{1}{2}$ cup water (or milk) and $\frac{3}{4}$ cup oatmeal
131. a) $\frac{1}{8}$ **b)** $\frac{1}{16}$ **c)** 5 **d)** 6

SECTION 5.4, PAGE 266

1. A rational number can be written as a ratio of two integers. Real numbers that cannot be written as a ratio of two integers are irrational numbers.

3. A perfect square is any number that is the square of a natural number.

5. a) To add or subtract two or more square roots with the same radicand, add or subtract their coefficients and then multiply the sum or difference by the common radical.
b) $5\sqrt{5}$

7. a) Multiply both the numerator and denominator by a radical that will result in the radicand in the denominator becoming a perfect square.
b) $\dfrac{2\sqrt{3}}{3}$

9. Rational **11.** Rational **13.** Irrational
15. Rational **17.** Irrational **19.** 4 **21.** 10
23. −13 **25.** −9 **27.** −10
29. Rational number, integer, natural number
31. Rational number, integer, natural number
33. Rational number **35.** Rational number
37. Rational number
39. $2\sqrt{3}$ **41.** $4\sqrt{3}$ **43.** $3\sqrt{7}$ **45.** $2\sqrt{21}$ **47.** $9\sqrt{2}$
49. $5\sqrt{5}$ **51.** $\sqrt{2}$ **53.** $-13\sqrt{3}$ **55.** $4\sqrt{3}$ **57.** $23\sqrt{2}$
59. 9 **61.** $2\sqrt{15}$ **63.** $10\sqrt{2}$ **65.** 2 **67.** 3

69. $\dfrac{\sqrt{5}}{5}$ **71.** $\dfrac{\sqrt{21}}{7}$ **73.** $\dfrac{2\sqrt{15}}{3}$ **75.** $\dfrac{\sqrt{15}}{3}$ **77.** $\dfrac{\sqrt{15}}{3}$

79. $\sqrt{5}$ is between 2 and 3 since 5 is between 4 and 9. $\sqrt{5}$ is between 2 and 2.5 since 5 is closer to 4 than to 9. $\sqrt{5} \approx 2.24$.

81. $\sqrt{107}$ is between 10 and 11 since 107 is between 100 and 121. $\sqrt{107}$ is between 10 and 10.5 since 107 is closer to 100 than to 121. $\sqrt{107} \approx 10.34$.

83. $\sqrt{170}$ is between 13 and 14 since 170 is between 169 and 196. $\sqrt{170}$ is between 13 and 13.5 since 170 is closer to 169 than to 196. $\sqrt{170} \approx 13.04$.

In Exercises 85–89, false answers can be modified in a variety of ways. We give one possible answer.

85. False. $\sqrt{c}$ may be a rational number or an irrational number for a composite number c. (For example, $\sqrt{25}$ is a rational number; $\sqrt{8}$ is an irrational number.)

87. True

89. False. The product of a rational number and an irrational number may be a rational number or an irrational number.

91. $3\sqrt{2} + 5\sqrt{2} = 8\sqrt{2}$ **93.** $\sqrt{2} \cdot \sqrt{3} = \sqrt{6}$

95. $\sqrt{2} \neq 1.414$ since $\sqrt{2}$ is irrational and 1.414 is rational.

97. No. π is irrational; therefore, it cannot equal $\frac{22}{7}$ or 3.14, both of which are rational.

99. $\sqrt{4 \cdot 9} = \sqrt{4} \cdot \sqrt{9}$, $\sqrt{36} = 2 \cdot 3$, $6 = 6$

101. a) 10 mph **b)** 20 mph **c)** 40 mph **d)** 80 mph

103. a) Rational. $\sqrt{0.04} = 0.2$, which is a rational number.
b) Irrational. $\sqrt{0.7} = 0.8366600265\ldots$. Since the decimal number is not a terminating or a repeating decimal number, this number is an irrational number.

105. a) $(44 \div \sqrt{4}) \div \sqrt{4} = 11$ **b)** $(44 \div 4) + \sqrt{4} = 13$
c) $4 + 4 + 4 + \sqrt{4} = 14$
d) $\sqrt{4}(4 + 4) + \sqrt{4} = 18$
Other answers are possible.

SECTION 5.5, PAGE 273

1. The real numbers are the union of the rational numbers and the irrational numbers.

3. If whenever the operation is performed on two elements of a set the result is also an element of the set, then the set is closed under that operation.

5. $a \cdot b = b \cdot a$, the order in which two numbers are multiplied is immaterial. One example is $4(5) = 5(4)$.

7. $(a + b) + c = a + (b + c)$, when adding three numbers, you may place parentheses around any two adjacent numbers. One example is $(1 + 2) + 3 = 1 + (2 + 3)$.

9. No **11.** Yes **13.** Yes **15.** Yes **17.** Yes
19. No **21.** No **23.** No **25.** Yes **27.** Yes

29. Commutative property of addition. The only difference between the expressions on both sides of the equal sign is the order of 5 and x.

31. $(-3) + (-4) = (-4) + (-3) = -7$

33. No. $4 - 3 \neq 3 - 4$.

35. $[(-2) + (-3)] + (-4) = (-2) + [(-3) + (-4)] = -9$

37. No. $(16 \div 8) \div 2 \neq 16 \div (8 \div 2)$.

39. No. $(81 \div 9) \div 3 \neq 81 \div (9 \div 3)$.

41. Distributive property

43. Associative property of multiplication

45. Associative property of addition

47. Commutative property of multiplication

49. Distributive property

51. Commutative property of multiplication

53. Distributive property

55. Commutative property of multiplication

57. $4z + 4$ **59.** $-\frac{3}{4}x + 9$ **61.** $3x + 4$

63. $2x - 1$ **65.** 2 **67.** $5\sqrt{2} + 5\sqrt{3}$

69. a) Distributive property
b) Associative property of addition

71. a) Distributive property
b) Associative property of addition
c) Commutative property of addition
d) Associative property of addition

73. a) Distributive property
b) Commutative property of addition
c) Associative property of addition
d) Commutative property of addition

75. Yes **77.** No **79.** Yes **81.** No **83.** Yes

85. Yes **87.** Yes **89.** Answers will vary.

91. No. $0 \div a = 0$ (when $a \neq 0$), but $a \div 0$ is undefined.

SECTION 5.6, PAGE 284

1. The 2 is the base and the 3 is the exponent or power.

3. a) To multiply two exponential expressions with the same base, add the exponents and use this sum as the exponent on the common base.
b) $2^3 \cdot 2^4 = 2^{3+4} = 2^7 = 128$

5. a) Any nonzero expression raised to the power of 0 equals 1.
b) $7^0 = 1$

7. a) Any base with an exponent raised to another exponent is equal to the base raised to the product of the exponents.
b) $(3^2)^4 = 3^{2 \cdot 4} = 3^8 = 6561$

9. a) -1^{500} means $-(1)^{500}$ or $-1 \cdot 1^{500}$. Since 1 raised to any power equals 1, $-1^{500} = -1 \cdot 1^{500} = -1 \cdot 1 = -1$.
b) $(-1)^{500}$ means (-1) multiplied by itself 500 times. Since 500 is even, $(-1)^{500} = 1$.
c) -1^{501} means $-(1)^{501}$ or $-1 \cdot 1^{501}$. Since 1 raised to any power equals 1, $-1^{501} = -1 \cdot 1^{501} = -1 \cdot 1 = -1$.
d) $(-1)^{501}$ means (-1) multiplied by itself 501 times. Since 501 is odd, $(-1)^{501} = -1$.

11. a) If the exponent is positive, move the decimal point in the number to the right the same number of places as the exponent, adding zeros where necessary. If the exponent is negative, move the decimal point in the number to the left the same number of places as the exponent, adding zeros where necessary.
b) 0.0000291 **c)** 7,020,000

13. a) 9 **b)** 8 **15. a)** 25 **b)** -25

17. a) -16 **b)** 16 **19. a)** -64 **b)** -64

21. a) $\frac{1}{64}$ **b)** $\frac{9}{16}$ **23. a)** 1000 **b)** 1

25. a) 243 **b)** -243 **27. a)** 25 **b)** 25

29. a) 1 **b)** -1 **31. a)** 1 **b)** 6

33. a) $\frac{1}{27}$ **b)** $\frac{1}{49}$ **35. a)** $-\frac{1}{81}$ **b)** $\frac{1}{81}$

37. a) 64 **b)** 64 **39. a)** 4 **b)** $\frac{1}{16}$

41. 1.75×10^5 **43.** 2.3×10^{-4} **45.** 5.6×10^{-1}

47. 1.9×10^4 **49.** 1.86×10^{-4} **51.** 4.23×10^{-6}

53. 7.11×10^2 **55.** 1.53×10^{-1} **57.** 170

59. 0.0001097 **61.** 0.0000862 **63.** 0.312

65. 9,000,000 **67.** 231 **69.** 35,000 **71.** 10,000

73. 750,000 **75.** 0.0153 **77.** 250 **79.** 0.0021

81. 20 **83.** 1.0×10^{11} **85.** 4.5×10^{-7}

87. 7.0×10^1 **89.** 2.0×10^{-7} **91.** 3.0×10^8

93. 3.6×10^{-3}; 1.7; 9.8×10^2; 1.03×10^4

95. 8.3×10^{-5}; 0.00079; 4.1×10^3; 40,000

97. 0.046 **99.** 0.168

101. 2.708×10^{51} sec **103.** \$6779

105. 11.95 hours **107.** 2.9×10^8 cells

109. 8.64×10^9 ft^3

111. a) \$720,000,000 **b)** \$300,000,000 **c)** \$120,000,000
d) \$60,000,000

113. 1000 **115.** 333,333 times

117. a) About 5.87×10^{12} (5.87 trillion) mi
b) About 500 sec or 8 min 20 sec

SECTION 5.7, PAGE 294

1. A sequence is a list of numbers that are related to each other by a given rule. One example is 1, 3, 5, 7, 9, ...

3. a) An arithmetic sequence is one in which each term differs from the preceding term by a constant amount. One example is 4, 7, 10, 13, 16, ...
b) A geometric sequence is one in which the ratio of any two successive terms is a constant amount. One example is 3, 6, 12, 24, ...

5. a) a_n is the nth term or the general term.
b) a_1 is the first term.
c) d is the common difference.
d) s_n is the sum of the first n terms.

7. 5, 6, 7, 8, 9 **9.** 12, 10, 8, 6, 4

11. 5, 3, 1, −1, −3 **13.** $\frac{3}{4}, 1, \frac{5}{4}, \frac{3}{2}, \frac{7}{4}$ **15.** 8 **17.** 13

19. $-\frac{91}{5}$ **21.** −8 **23.** $a_n = n$ **25.** $a_n = 2n$

27. $a_n = \frac{3}{4}n - \frac{1}{2}$ **29.** $a_n = \frac{3}{2}n - \frac{9}{2}$ **31.** $s_{50} = 1275$

33. $s_{50} = 2500$ **35.** $s_8 = -52$ **37.** $s_{24} = 60$

39. 1, 5, 25, 125, 625 **41.** 2, −4, 8, −16, 32

43. −3, 3, −3, 3, −3 **45.** 81, −27, 9, −3, 1

47. 160 **49.** $\frac{3}{4}$ **51.** −3645 **53.** $a_{10} = -39{,}366$

55. $a_n = 2^{n-1}$ **57.** $a_n = (-1)(-1)^{n-1}$

59. $a_n = 2 \cdot \left(\frac{1}{2}\right)^{n-1}$ **61.** $a_n = 9 \cdot \left(\frac{1}{3}\right)^{n-1}$

63. 186 **65.** −4095 **67.** −620,011

69. −10,923 **71.** 5050 **73.** 10,100

75. a) 63 in. **b)** 954 in.

77. a) \$44,800 **b)** \$319,200 **79.** 496 pinecones

81. 52.4288 g **83.** ≈ \$70,088 **85.** \$486,000

87. 161.4375 **89.** 267 **91.** 191.3568 ft

SECTION 5.8, PAGE 302

1. The first and second terms are 1. Each term thereafter is the sum of the previous two terms.

3. a) The golden number is $\dfrac{\sqrt{5} + 1}{2}$.
b) When a line segment AB is divided at a point C, such that the ratio of the whole, AB, to the larger part, AC, is equal to the ratio of the larger part, AC, to the smaller part, CB, then each of the two ratios AB/AC and AC/CB is known as the golden ratio.
c) The proportion made by using the two golden ratios, $AB/AC = AC/CB$, is known as the golden proportion.
d) A golden rectangle is one where the ratio of the length to the width is equal to the golden number.

5. Answers will vary.

7. a) 1.618 **b)** 0.618 **c)** 1

9. $\frac{1}{1} = 1, \frac{2}{1} = 2, \frac{3}{2} = 1.5, \frac{5}{3} \approx 1.667, \frac{8}{5} = 1.6, \frac{13}{8} = 1.625, \frac{21}{13} \approx 1.615, \frac{34}{21} \approx 1.619, \frac{55}{34} \approx 1.6176, \frac{89}{55} \approx 1.6182$. The consecutive ratios alternate, increasing and decreasing about the golden ratio.

11. Answers will vary. **13.** Answers will vary.

15. Answers will vary. **17.** Answers will vary.

19. Answers will vary. **21.** Answers will vary.

23. No **25.** Yes; 3, 5 **27.** Yes; 105, 170

29. Yes; −1, −1 **31.** Answers will vary.

33. Answers will vary.

35. a) 1, 3, 4, 7, 11, 18, 29, 47
b) 8 + 21 = 29, 13 + 34 = 47
c) It is the Fibonacci sequence.

37. Answers will vary.

39. Answers will vary.

REVIEW EXERCISES, PAGE 306

1. 2, 3, 4, 5, 6, 10 **2.** 2, 3, 4, 6, 9 **3.** $2^2 \cdot 3^3 \cdot 5$

4. $3^2 \cdot 7 \cdot 11$ **5.** $2^3 \cdot 3 \cdot 5 \cdot 7$ **6.** $2 \cdot 3^2 \cdot 7^2$

7. $2^2 \cdot 3 \cdot 11^2$ **8.** 15; 210 **9.** 9; 756 **10.** 5; 2250

11. 30; 900 **12.** 4; 480 **13.** 36; 432 **14.** 45 days

15. −3 **16.** 3 **17.** −6 **18.** −4 **19.** −9 **20.** 3

21. 0 **22.** 4 **23.** −15 **24.** 24 **25.** −56 **26.** 5

27. −2 **28.** 6 **29.** 4 **30.** 3 **31.** 0.3 **32.** 0.44

33. 0.375 **34.** 3.25 **35.** $0.\overline{857142}$ **36.** $0.58\overline{3}$

37. 0.375 **38.** 0.6875 **39.** $0.\overline{714285}$ **40.** $\frac{9}{40}$

41. $\frac{14}{10} = \frac{7}{5}$ **42.** $\frac{2}{3}$ **43.** $\frac{51}{99}$ **44.** $\frac{83}{1000}$ **45.** $\frac{73}{10{,}000}$ **46.** $\frac{211}{90}$

47. $\frac{7}{4}$ **48.** $\frac{25}{6}$ **49.** $-\frac{13}{4}$ **50.** $-\frac{283}{8}$ **51.** $2\frac{1}{5}$ **52.** $9\frac{3}{8}$

53. $-1\frac{5}{7}$ **54.** $-27\frac{1}{5}$ **55.** $\frac{13}{12}$ **56.** $\frac{1}{4}$ **57.** $\frac{17}{12}$ **58.** $\frac{1}{4}$

59. $\frac{35}{54}$ **60.** $\frac{53}{28}$ **61.** $\frac{1}{6}$ **62.** $\frac{13}{40}$ **63.** $\frac{8}{15}$ **64.** $2\frac{7}{32}$ tsp

65. $3\sqrt{5}$ **66.** $10\sqrt{2}$ **67.** $8\sqrt{5}$ **68.** $-3\sqrt{2}$

69. $8\sqrt{2}$ **70.** $-20\sqrt{3}$ **71.** $5\sqrt{7}$ **72.** $3\sqrt{2}$

73. $4\sqrt{3}$ **74.** 10 **75.** $2\sqrt{7}$ **76.** $\dfrac{4\sqrt{3}}{3}$ **77.** $\dfrac{\sqrt{35}}{5}$

78. $6 + 3\sqrt{7}$ **79.** $4\sqrt{3} + 3\sqrt{2}$ **80.** $3\sqrt{2} + 3\sqrt{5}$

81. Commutative property of addition

82. Commutative property of multiplication

83. Associative property of addition

84. Distributive property

85. Associative property of addition

86. Commutative property of addition

87. Associative property of multiplication

88. Commutative property of multiplication

89. Distributive property

90. Commutative property of multiplication

91. No **92.** Yes **93.** No **94.** Yes **95.** No **96.** No

97. 25 **98.** $\frac{1}{25}$ **99.** 81 **100.** 125 **101.** 1 **102.** $\frac{1}{64}$

103. 64 **104.** 81 **105.** 8.2×10^9 **106.** 1.58×10^{-5}

107. 2.309×10^{-2} **108.** 4.95×10^6 **109.** 280,000

110. 0.000139 **111.** 0.000175 **112.** 10,000,000

113. 6.0×10^{-5} **114.** 3.75×10^{12} **115.** 2.1×10^1

116. 3.0×10^0 **117.** 1,100,000,000,000 **118.** 0.7

119. 120 **120.** 5 **121.** ≈388 times **122.** ≈ \$5555.56

123. Arithmetic; 17, 21 **124.** Geometric; 8, 16
125. Arithmetic; −15, −18 **126.** Geometric; $\frac{1}{32}, \frac{1}{64}$
127. Arithmetic; 16, 19 **128.** Geometric; $\frac{1}{2}, -\frac{1}{2}$
129. 27 **130.** 10 **131.** 25 **132.** 48 **133.** $\frac{1}{4}$
134. −48 **135.** 3825 **136.** −25 **137.** 632
138. 57.5 **139.** 682 **140.** 45 **141.** 33 **142.** −21
143. Arithmetic; $a_n = 3n$
144. Arithmetic; $a_n = 3n - 2$
145. Arithmetic; $a_n = -\frac{3}{2}n + \frac{11}{2}$
146. Geometric; $a_n = 3(2)^{n-1}$
147. Geometric; $a_n = 2(-1)^{n-1}$
148. Geometric; $a_n = 5\left(\frac{1}{3}\right)^{n-1}$ **149.** No
150. Yes; −8, −13 **151.** No **152.** No

CHAPTER TEST, PAGE 309

1. 5 **2.** $2 \cdot 3^2 \cdot 23$ **3.** 8 **4.** −20
5. −175 **6.** $\frac{37}{8}$ **7.** $19\frac{5}{9}$ **8.** 0.625 **9.** $\frac{129}{20}$
10. $\frac{121}{240}$ **11.** $\frac{1}{8}$ **12.** $9\sqrt{3}$ **13.** $\frac{\sqrt{14}}{7}$
14. Yes; the product of any two integers is an integer.
15. Associative property of addition
16. Distributive property **17.** 64 **18.** 1024 **19.** $\frac{1}{81}$
20. 8.0×10^{11} **21.** $a_n = -4n + 2$ **22.** −187
23. 486 **24.** −510 **25.** $a_n = 3(2)^{n-1}$
26. 1, 1, 2, 3, 5, 8, 13, 21, 34, 55

CHAPTER 6

SECTION 6.1, PAGE 315

1. Letters of the alphabet used to represent numbers are called variables.
3. The solution to an equation is the number or numbers that replace the variable to make the equation a true statement.
5. a) The 4 is the base and the 5 is the exponent.
b) Answers will vary.
7. a) 12 **b)** 27 **9.** 25 **11.** −4 **13.** 686 **15.** −3
17. 12 **19.** −20 **21.** $-\frac{10}{9}$ **23.** 7 **25.** 9 **27.** 39
29. No **31.** No **33.** No **35.** Yes **37.** Yes
39. \$62.93 **41.** 96.8 million **43.** 88.72 min **45.** 1.71 in.
47. The two expressions are not equal.

SECTION 6.2, PAGE 327

1. The parts that are added or subtracted in an algebraic expression are called terms. In $3x - 2y$, the $3x$ and $-2y$ are terms.
3. The numerical part of a term is called its numerical coefficient. For the term $3x$, 3 is the numerical coefficient.
5. To simplify an expression means to combine like terms by using the commutative, associative, and distributive properties.
7. If $a = b$, then $a \cdot c = b \cdot c$ for all real numbers a, b, and c, where $c \neq 0$. If $\frac{x}{3} = 2$, then $3\left(\frac{x}{3}\right) = 3(2)$.
9. If $a = b$, then $a/c = b/c$ for all real numbers a, b, and c, where $c \neq 0$. If $4x = 8$, then $\frac{4x}{4} = \frac{8}{4}$.
11. A ratio is a quotient of two quantities. An example is $\frac{7}{9}$.
13. Yes. They have the same variable and the same exponent on the variable.
15. $10x$ **17.** $2x + 12$ **19.** $3x + 11y$ **21.** $-8x + 2$
23. $-5x + 3$ **25.** $13.3x - 8.3$ **27.** $-\frac{2}{15}x - 4$
29. $10x - 9y + 3$ **31.** $8s - 17$ **33.** $1.4x - 2.8$
35. $-\frac{5}{12}x + \frac{4}{5}$ **37.** $4.52x - 13.5$ **39.** 17 **41.** −1 **43.** $\frac{24}{7}$
45. $\frac{2}{3}$ **47.** 4 **49.** 3 **51.** 17 **53.** −8 **55.** No solution
57. All real numbers **59.** 3 **61.** −3 **63.** 5
65. \$209.25 **67.** 4 gallons
69. ≈ 15,868,800 households
71. a) 1.6 kph **b)** 56.25 mph **73.** 0.375 cc
75. a) Answers will vary. **b)** −1
77. a) An equation that has no solution.
b) You will obtain a false statement.
79. a) 2 : 5 **b)** $m : m + n$

SECTION 6.3, PAGE 337

1. A formula is an equation that typically has a real-life application.
3. Subscripts are numbers (or letters) placed below and to the right of variables. They are used to help clarify a formula.
5. An exponential equation is of the form $y = a^x$, $a > 0$, $a \neq 1$.
7. 56 **9.** 56 **11.** 10 **13.** 251.2 **15.** 37.1
17. 2 **19.** 3000 **21.** 12 **23.** 25 **25.** 6
27. 200 **29.** 7.2 **31.** 14 **33.** 3240
35. 66.67 **37.** 39 **39.** 7609.81
41. $y = \frac{4x - 14}{9}$ or $y = \frac{4}{9}x - \frac{14}{9}$
43. $y = \frac{-8x + 21}{7}$ or $y = -\frac{8}{7}x + 3$
45. $y = \frac{2x + 6}{3}$ or $y = \frac{2}{3}x + 2$

47. $y = \dfrac{2x - z + 15}{3}$ or $y = \dfrac{2}{3}x - \dfrac{1}{3}z + 5$

49. $y = \dfrac{9x + 4z - 7}{8}$ or $y = \dfrac{9}{8}x + \dfrac{1}{2}z - \dfrac{7}{8}$

51. $r = \dfrac{d}{t}$ **53.** $a = p - b - c$

55. $B = \dfrac{3V}{h}$ **57.** $r = \dfrac{C}{2\pi}$

59. $b = y - mx$ **61.** $w = \dfrac{P - 2l}{2}$

63. $c = 3A - a - b$ **65.** $T = \dfrac{PV}{K}$

67. $C = \frac{5}{9}(F - 32)$ **69.** $h = \dfrac{S - 2\pi r^2}{2\pi r}$

71. a) \$112.50 **b)** \$4612.50 **73.** ≈ 4.19 in.^{3}
75. 486,000 bacteria **77.** $\approx \$4.49 \times 10^{14}$
79. ≈ 1051.47 in.^{3}

SECTION 6.4, PAGE 343

1. A mathematical expression is a collection of variables, numbers, parentheses, and operation symbols. An equation is two algebraic expressions joined by an equal sign.

3. 4 more than x **5.** 2 times x, decreased by 3

7. $8 + x$ **9.** $3 + 2z$ **11.** $6w + 9$ **13.** $4x + 6$

15. $\dfrac{18 - s}{4}$ **17.** $3(x + 7)$ **19.** $x + 5 = 11$; 6

21. $x - 4 = 20$; 24 **23.** $4x - 10 = 42$; 13

25. $4x + 12 = 32$; 5 **27.** $x + 6 = 2x - 3$; 9

29. $x + 10 = 2(x + 3)$; 4

31. $150 + 0.42x = 207.54$; 137 miles

33. $x + 0.05x = 42$; \$40 per half hour

35. $0.08x = 250$; 3125 copies

37. $x + 3x = 1000$; \$250 for business, \$750 for liberal arts

39. $2w + 2(w + 3) = 54$; width: 12 ft, length: 15 ft

41. $x + 6x + 2515 = 21{,}730$; Utah: 2745 mustangs; Nevada: 18,985 mustangs

43. $x + 14x + 11 = 341$; December: 22 tornados; May: 319 tornados

45. $3w + 2(2w) = 140$; width: 20 ft, length: 40 ft

47. $70x = 760$; ≈ 11 months

49. $\dfrac{r}{2} + 0.07r = 257$; \$450.88

51. Deduct \$720 from Mr. McAdams's income and \$2920 from Mrs. McAdams's income.

53.
$$\begin{aligned} x + (x + 1) + (x + 2) &= 3(x + 2) - 3 \\ 3x + 3 &= 3x + 6 - 3 \\ 3x + 3 &= 3x + 3 \end{aligned}$$

55. $-40°$

SECTION 6.5, PAGE 352

1. Direct variation: As one variable increases, so does the other, and as one variable decreases, so does the other.

3. Joint variation: One quantity varies directly as the product of two or more other quantities.

5. Direct **7.** Inverse **9.** Direct **11.** Inverse
13. Inverse **15.** Inverse **17.** Direct **19.** Direct
21. Answers will vary. **23. a)** $y = kx$ **b)** 120

25. a) $m = \dfrac{k}{n^2}$ **b)** 0.25 **27. a)** $A = \dfrac{kB}{C}$ **b)** 2.5

29. a) $F = kDE$ **b)** 210 **31. a)** $t = \dfrac{kd^2}{f}$ **b)** 200

33. a) $Z = kWY$ **b)** 100 **35. a)** $H = kL$ **b)** 3

37. a) $A = kB^2$ **b)** 720 **39. a)** $F = \dfrac{kq_1q_2}{d^2}$ **b)** 1200

41. a) $t = kv$ **b)** \$2700 **43. a)** $l = \dfrac{k}{d^2}$ **b)** 80 dB

45. a) $R = \dfrac{kA}{P}$ **b)** 4800 tapes

47. a) $v = \dfrac{k\sqrt{t}}{l}$ **b)** 4 vibrations per second

49. a) $N = \dfrac{kp_1p_2}{d}$ **b)** $\approx 121{,}528$ calls

51. a) Inversely **b)** Stays 0.3 **53.** \$132.27

SECTION 6.6, PAGE 360

1. $a < b$ means that a is less than b, $a \le b$ means that a is less than or equal to b, $a > b$ means that a is greater than b, $a \ge b$ means that a is greater than or equal to b.

3. When both sides of an inequality are multiplied or divided by a negative number, the direction of the inequality symbol must be reversed.

5. Yes, the inequality symbol points to the -3 in both cases.

7. a) An inequality of the form $a < x < b$ is called a compound inequality.
b) $-5 < x < 3$

9. 

11. 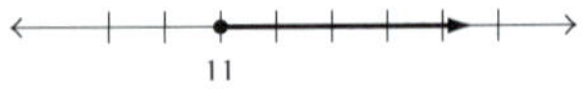

13. 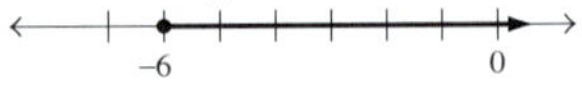

15. 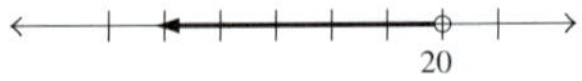

17. 

19. 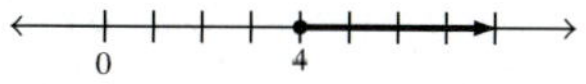

21. 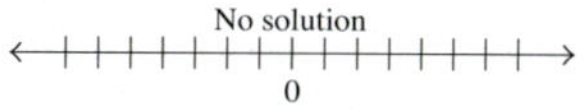

23. 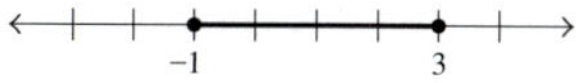

25. 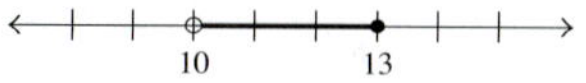

27. 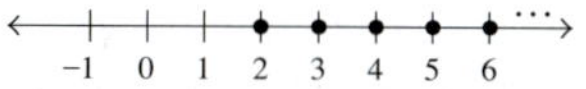

29. 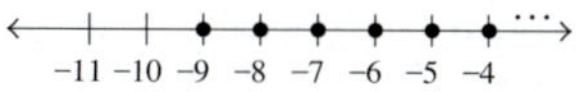

31. 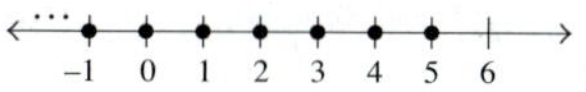

33. 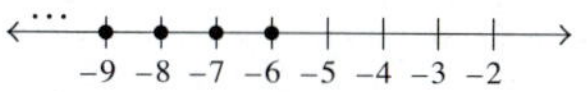

35. 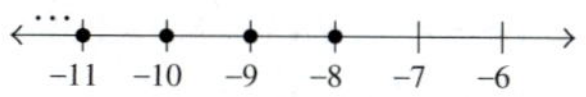

37. 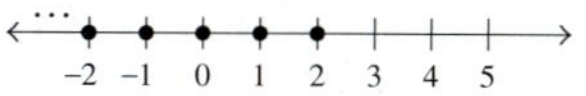

39. 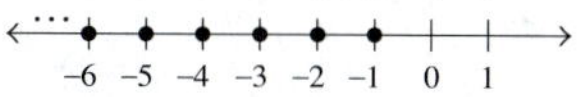

41. 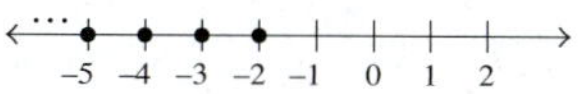

43. 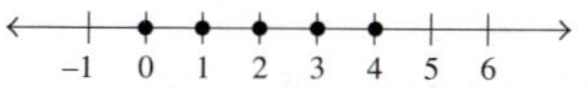

45. 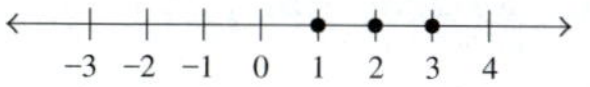

47. a) 2009–2015 **b)** 2005–2008 **c)** 2006–2015 **d)** 2005–2006

49. 19 videos **51.** 350 hours

53. 6 hours **55.** $0.5 < t < 1.5$

57. $94 \leq x \leq 100$, assuming 100 is the highest grade possible

59. $6.875 \leq x \leq 11$

61. The student's answer is $x \leq -12$, whereas the correct answer is $x \geq -12$. Yes, -12 is in both solution sets.

SECTION 6.7, PAGE 372

1. A graph is an illustration of all the points whose coordinates satisfy an equation.

3. To find the y-intercept, set $x = 0$ and solve the equation for y.

5. a) Answers will vary. **b)** $-\frac{1}{3}$

7. a) First **b)** Third

For Exercises 9–15, see the following figure.

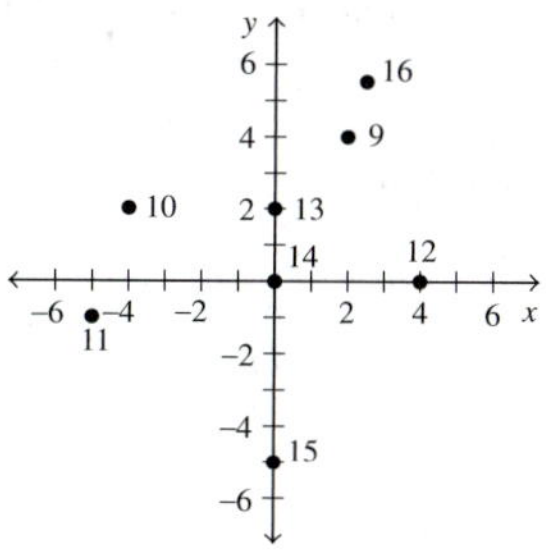

For Exercises 17–23, see the following figure.

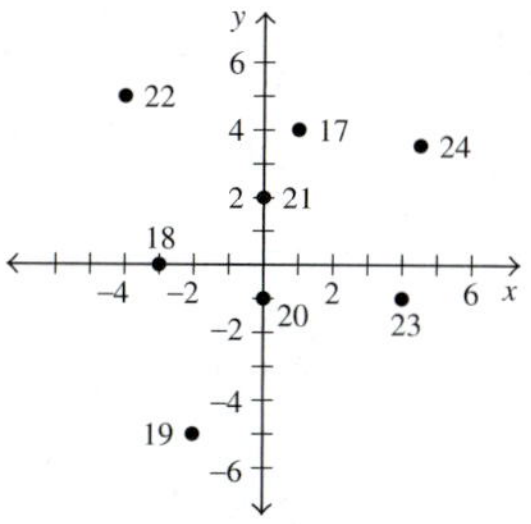

25. (0, 2) **27.** $(-2, 0)$ **29.** $(-5, -3)$ **31.** $(2, -3)$

33. (2, 2) **35.** (5, 2), (1, 4) **37.** (8, 2), $(0, -\frac{10}{3})$

39. $(-3, -2)$ **41.** (8, 0), (0, 3)

43.

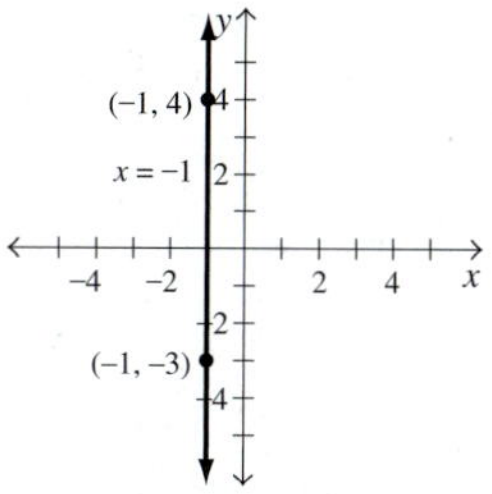

Slope: undefined

45.

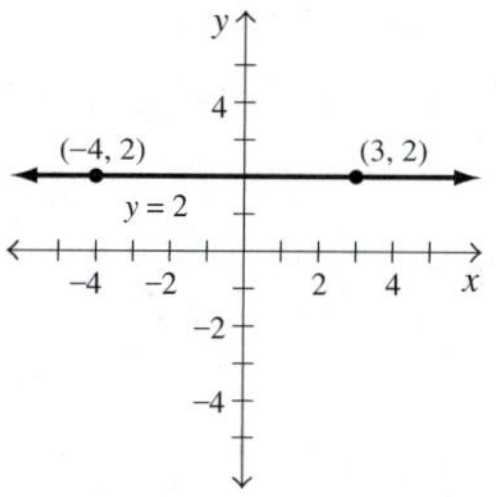

Slope is 0.

47.

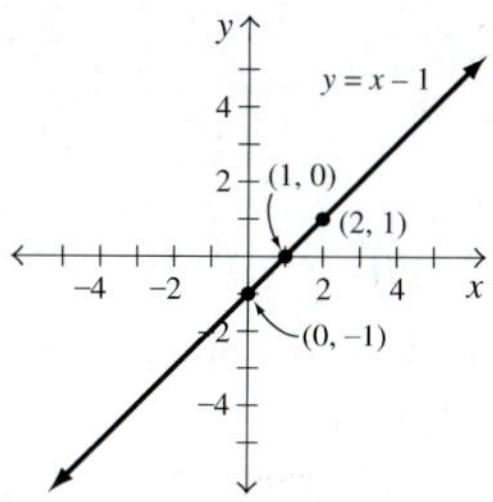

49.

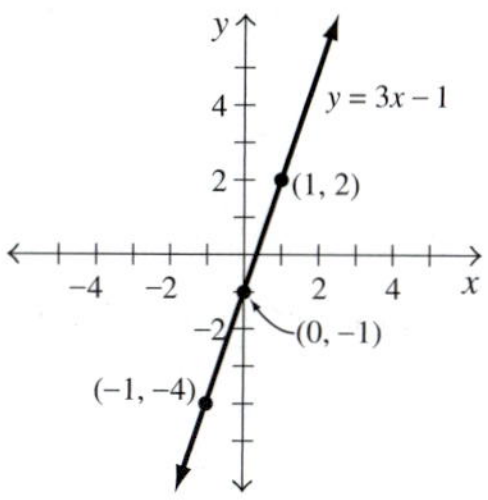

51.

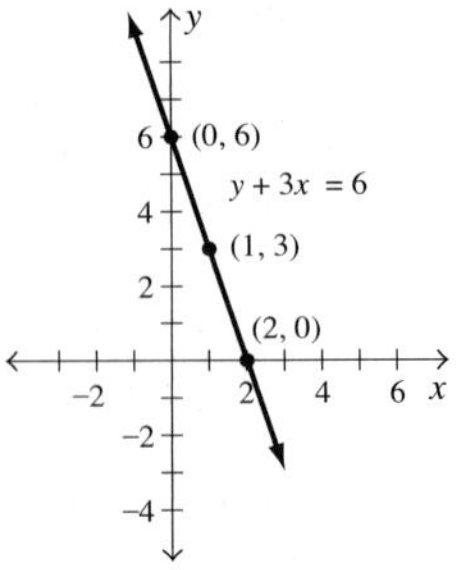

53.

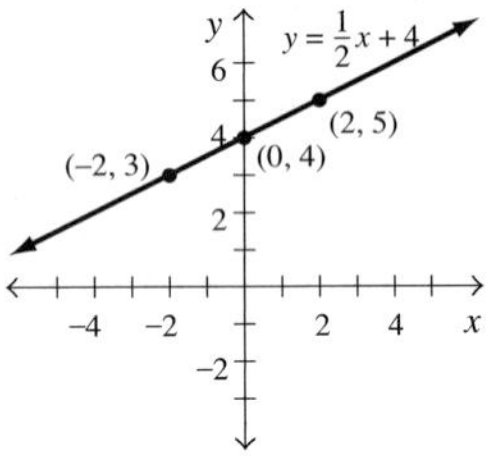

55.

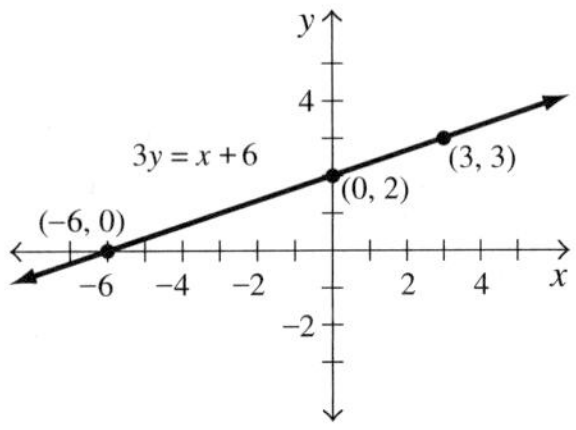

57.

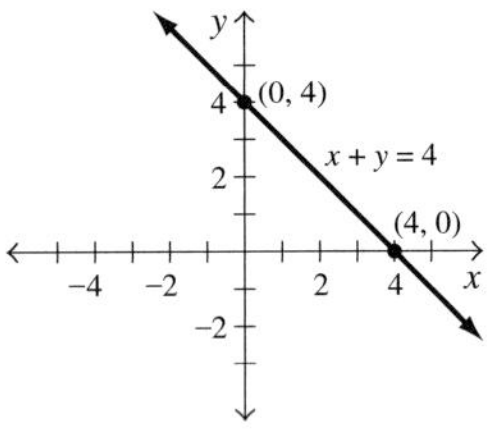

59.

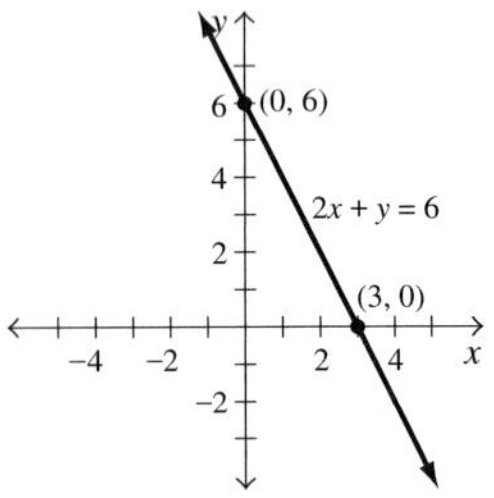

61.

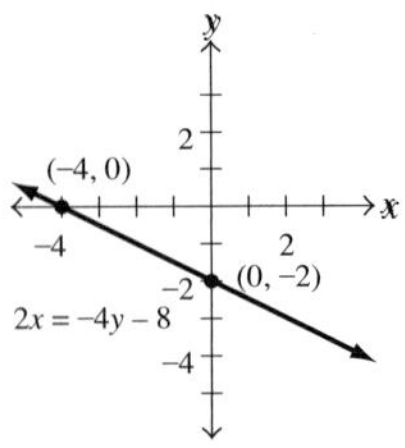

63.

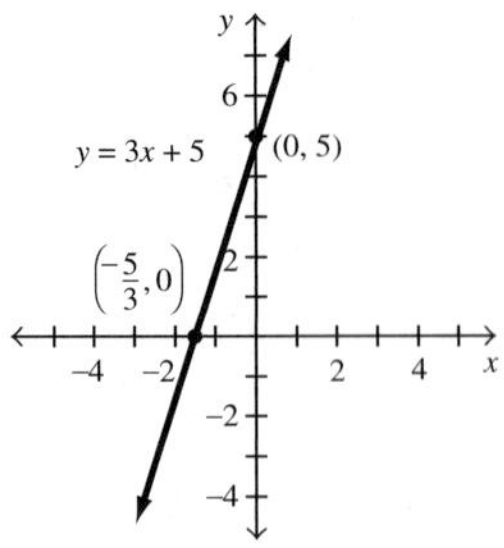

65.

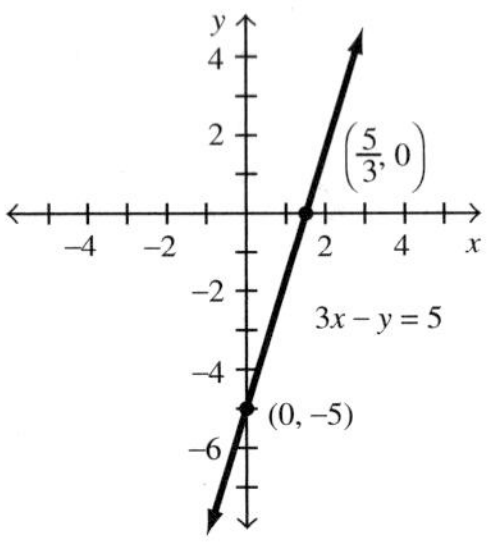

67. 2 **69.** $-\frac{5}{4}$ **71.** 0 **73.** Undefined **75.** $-\frac{8}{11}$

77.

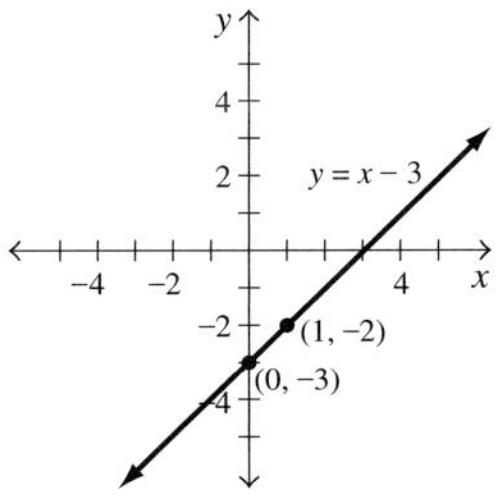

79.

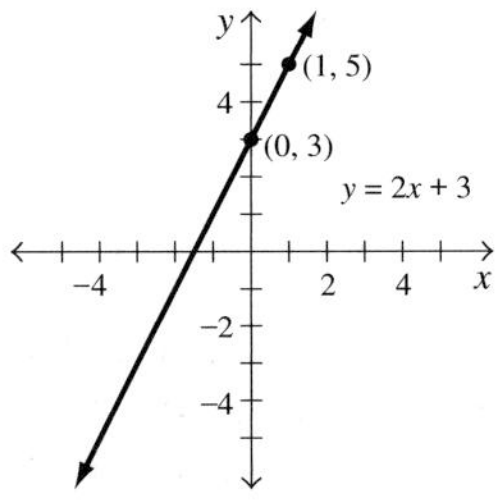

81.

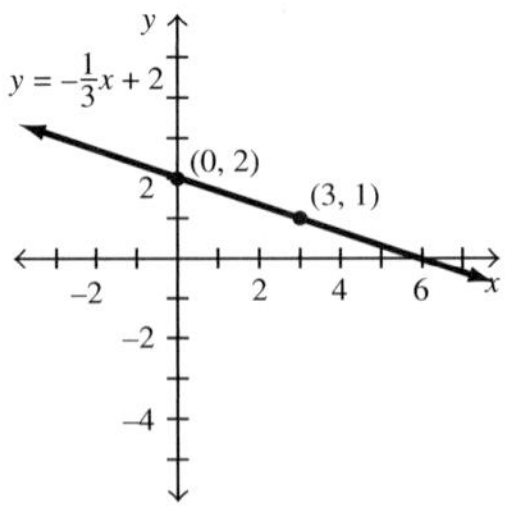

83.

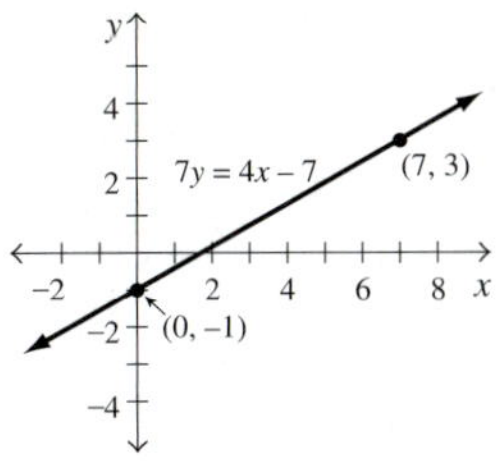

85.

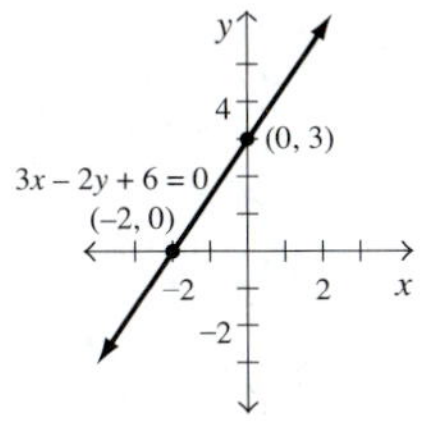

87. $y = -2x + 4$ **89.** $y = 3x + 2$

91. **a)** $D(3, -2)$ **b)** $A = 20$ square units

93. $(7, 2)$ or $(-1, 2)$ **95.** -3 **97.** 3

99. **a)**

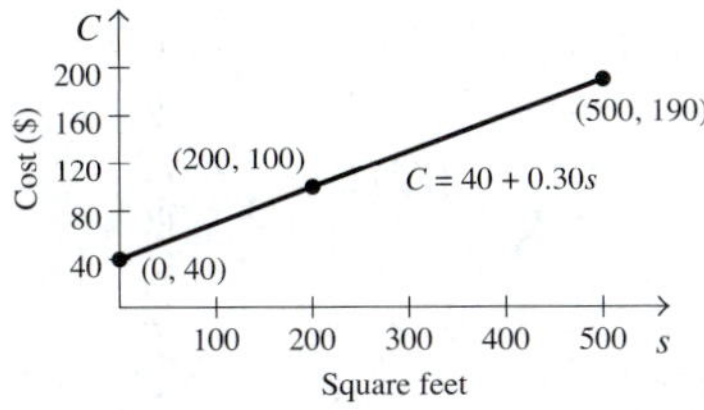

b) \$130 **c)** 100 square feet

101. **a)**

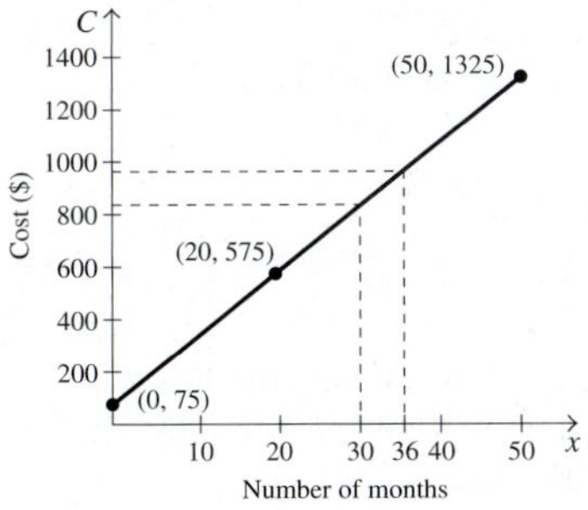

b) \$825 **c)** 36 months

103. **a)** 2 **b)** $y = 2x + 9$ **c)** 15 defects **d)** 4 workers

105. **a)** ≈ -0.05 **b)** $y = -0.05x + 2.60$
c) $\approx$ \$2.50 billion **d)** 4 years after 2000, or in 2004

107. **a)** Solve the equations for y to put them in slope–intercept form. Then compare the slopes and y-intercepts. If the slopes are equal but the y-intercepts are different, then the lines are parallel.
b) The lines are parallel.

SECTION 6.8, PAGE 378

1. (1) Mentally substitute the equal sign for the inequality sign and plot points as if you were graphing the equation. (2) If the inequality is $<$ or $>$, draw a dashed line through the points. If the inequality is $\leq$ or $\geq$, draw a solid line through the points. (3) Select a test point not on the line and substitute the x- and y-coordinates into the inequality. If the substitution results in a true statement, shade in the area on the same side of the line as the test point. If the test point results in a false statement, shade in the area on the opposite side of the line as the test point.

3. A half plane is the set of all the points in a plane on one side of a line.

5. **a)** No **b)** Yes **c)** Yes **d)** No

7.

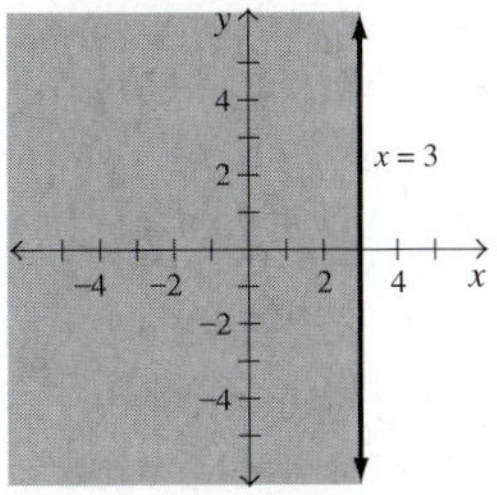

9.

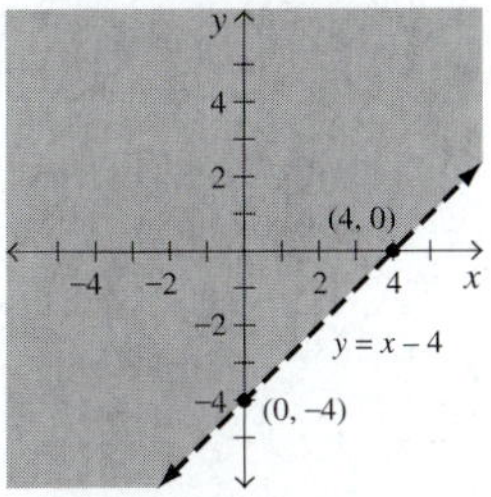

11.

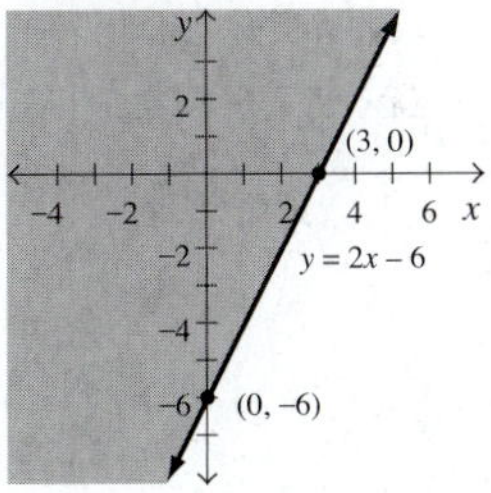

13.

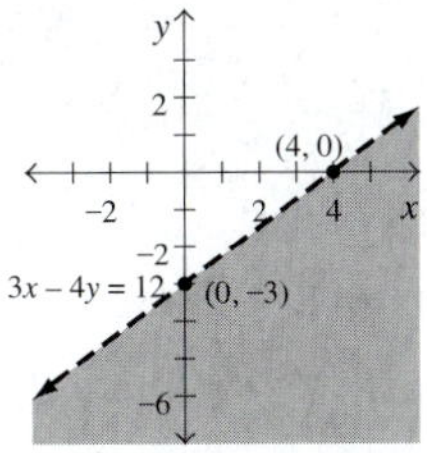

15.

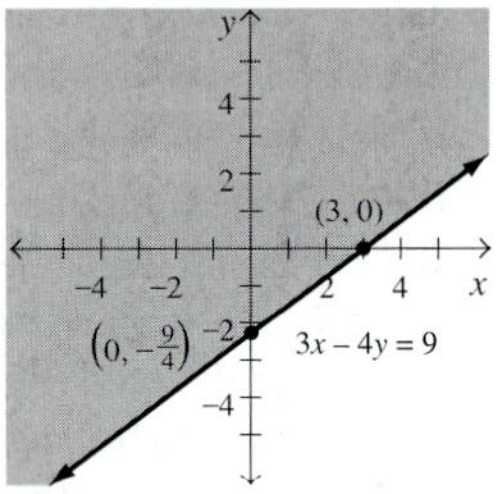

17.

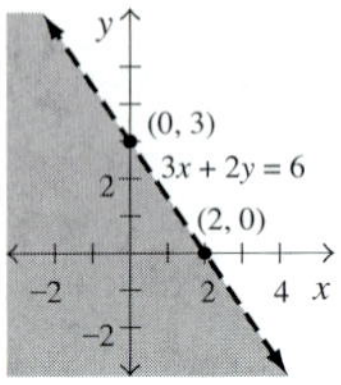

19.

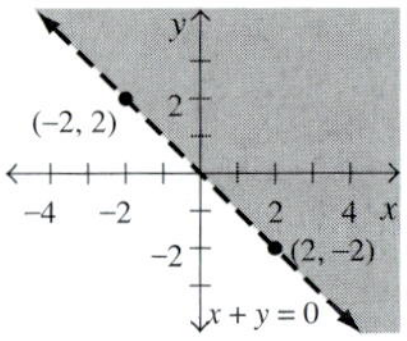

21.

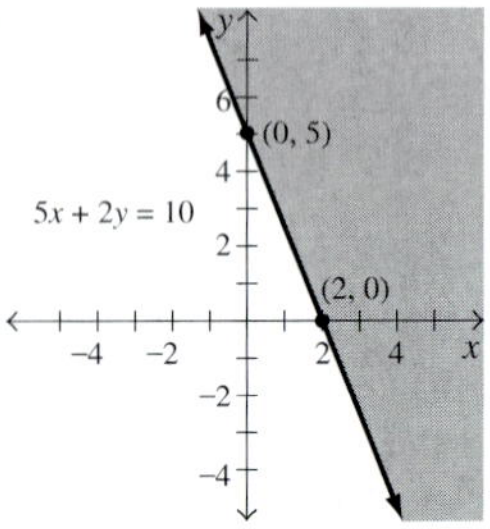

23.

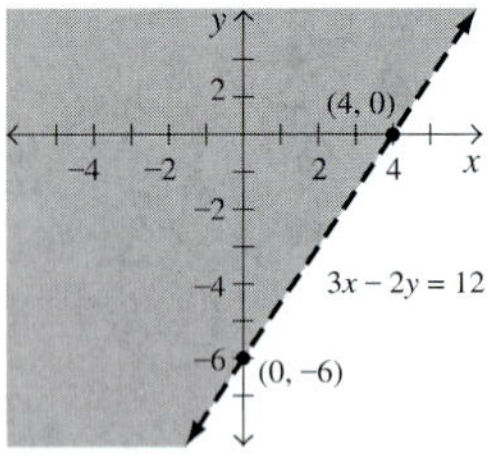

25.

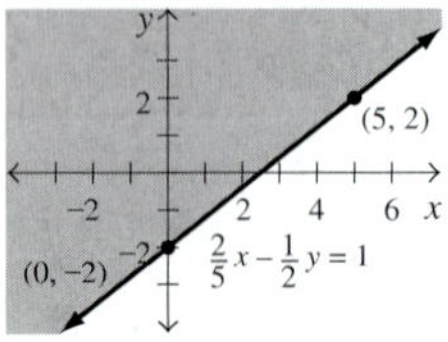

27.

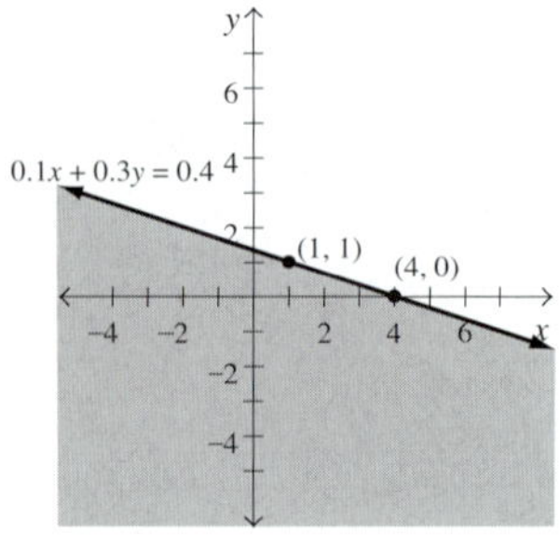

29. a) $x + y \leq 300$

b)

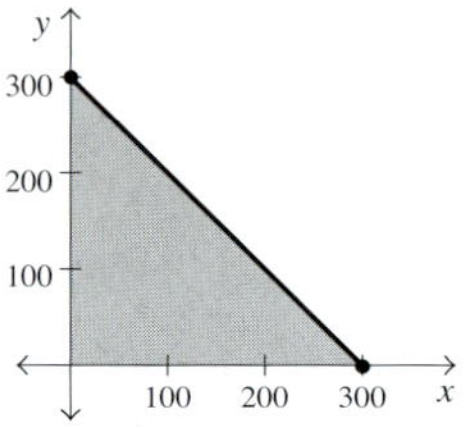

31. a) No, you cannot have a negative number of shirts.

b)

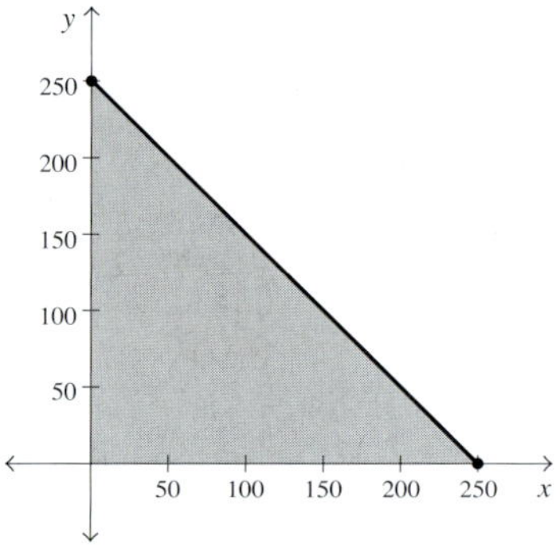

c) Answers will vary.

SECTION 6.9, PAGE 388

1. A binomial is an expression that contains two terms in which each exponent that appears on the variable is a whole number. $2x + 3$, $x - 7$, $x^2 - 9$

3. Answers will vary. **5.** $ax^2 + bx + c = 0, a \neq 0$

7. $(x + 5)(x + 3)$ **9.** $(x - 2)(x + 1)$

11. $(x + 6)(x - 4)$ **13.** $(x + 1)(x - 3)$

15. $(x - 7)(x - 3)$ **17.** $(x - 7)(x + 7)$

19. $(x + 7)(x - 4)$ **21.** $(x + 9)(x - 7)$

23. $(2x + 3)(x - 2)$ **25.** $(5x + 1)(x + 3)$

27. $(5x + 2)(x + 2)$ **29.** $(4x + 3)(x + 2)$

31. $(4x - 3)(x - 2)$ **33.** $(4x + 1)(2x - 3)$

35. $3, -6$ **37.** $-\frac{4}{3}, \frac{1}{2}$ **39.** $-5, -2$ **41.** $6, 1$ **43.** $5, -3$

45. $3, 1$ **47.** $9, -9$ **49.** $-9, 4$ **51.** $\frac{2}{3}, -4$ **53.** $-\frac{1}{5}, -2$

55. $\frac{1}{3}, 1$ **57.** $\frac{1}{3}, \frac{3}{2}$ **59.** $3, -5$ **61.** $6, -3$ **63.** $9, -1$

65. No real solution **67.** $2 \pm \sqrt{2}$

69. $\dfrac{4 \pm \sqrt{13}}{3}$ **71.** $\dfrac{3 \pm 2\sqrt{6}}{3}$ **73.** $-1, -\frac{5}{2}$ **75.** $\frac{7}{3}, 1$

77. No real solution

79. Width = 12 m, length = 22 m

81. a) The zero-factor property cannot be used.
b) $\approx 8.37, \approx 2.63$

83. $x^2 - 2x - 3 = 0$

SECTION 6.10, PAGE 401

1. A function is a special type of relation in which each value of the independent variable corresponds to a unique value of the dependent variable.

3. The domain of a function is the set of values that can be used for the independent variable.

5. If a vertical line touches more than one point on the graph, then for each value of x there is not a unique value for y and the graph does not represent a function.

7. $x = -\dfrac{b}{2a}$

9. Not a function

11. Function, domain: $\mathbb{R}$; range: $\mathbb{R}$

13. Not a function

15. Function, domain: $\mathbb{R}$; range: $y \geq -4$

17. Not a function

19. Function, domain: $0 \leq x < 12$; range: $y = 1, 2, 3$

21. Not a function

23. Function, domain: $\mathbb{R}$; range: $y > 0$

25. Not a function

27. Yes **29.** No **31.** Yes **33.** 8 **35.** 1

37. −6 **39.** 5 **41.** −23 **43.** 45 **45.** −17

47.

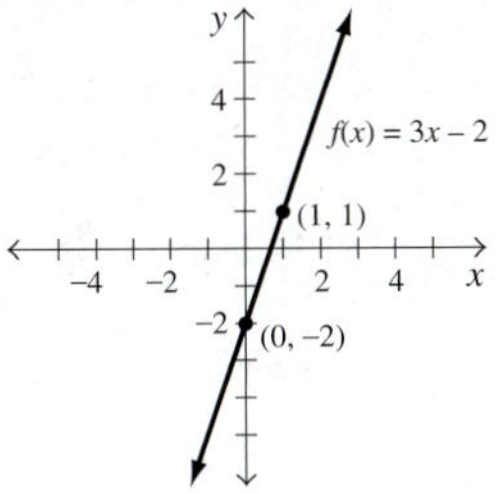

49.

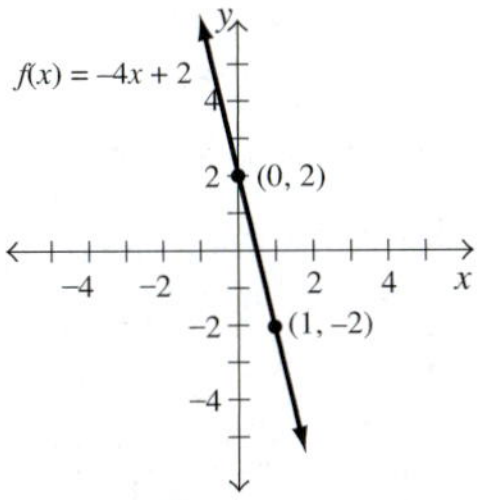

51.

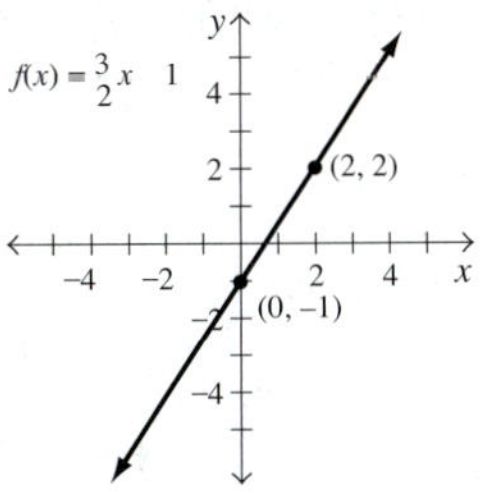

53. a) Upward **b)** $x = 0$ **c)** $(0, -9)$ **d)** $(0, -9)$
e) $(3, 0), (-3, 0)$
f)

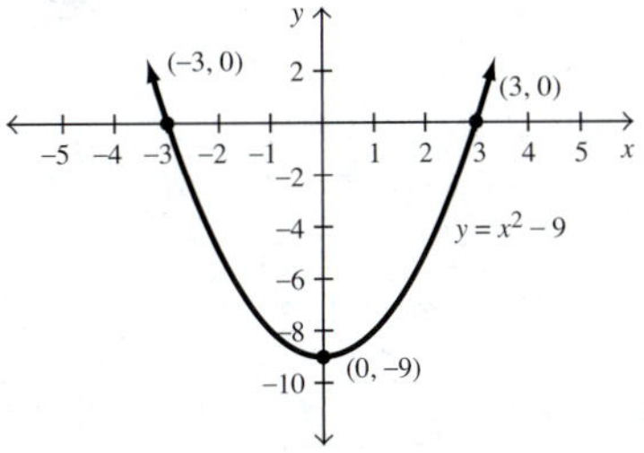

g) Domain: $\mathbb{R}$; range: $y \geq -9$

55. a) Downward **b)** $x = 0$ **c)** $(0, 4)$ **d)** $(0, 4)$
e) $(-2, 0), (2, 0)$
f)

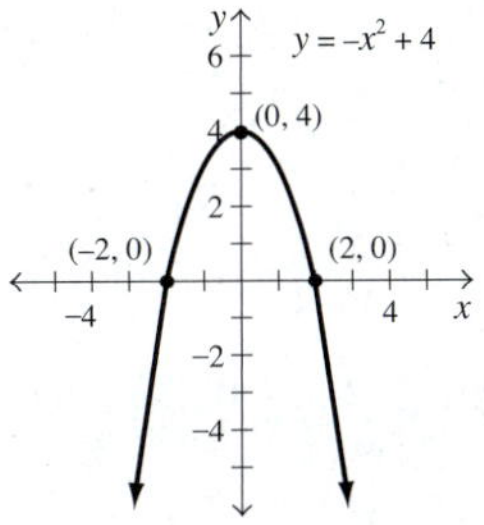

g) Domain: $\mathbb{R}$; range: $y \leq 4$

57. a) Downward **b)** $x = 0$ **c)** $(0, -8)$ **d)** $(0, -8)$
e) No x-intercepts
f)

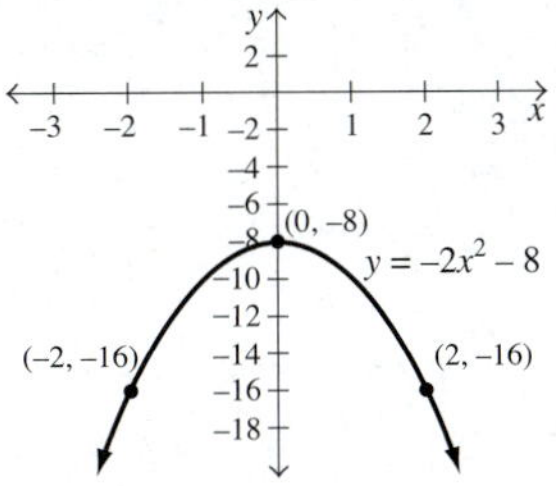

g) Domain: $\mathbb{R}$; range: $y \leq -8$

59. a) Upward **b)** $x = 0$ **c)** $(0, -3)$ **d)** $(0, -3)$
e) $(-1.22, 0), (1.22, 0)$
f)

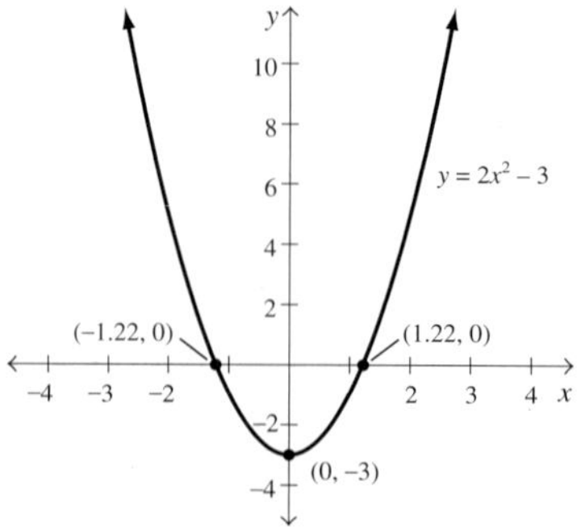

g) Domain: $\mathbb{R}$; range: $y \geq -3$

61. a) Upward **b)** $x = -1$ **c)** $(-1, 5)$ **d)** $(0, 6)$
e) No x-intercepts
f)

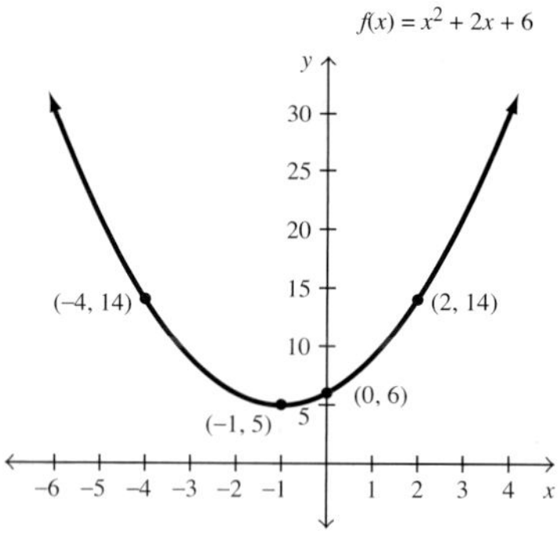

g) Domain: $\mathbb{R}$; range: $y \geq 5$

63. a) Upward **b)** $x = -\frac{5}{2}$ **c)** $(-2.5, -0.25)$ **d)** $(0, 6)$
e) $(-3, 0), (-2, 0)$
f)

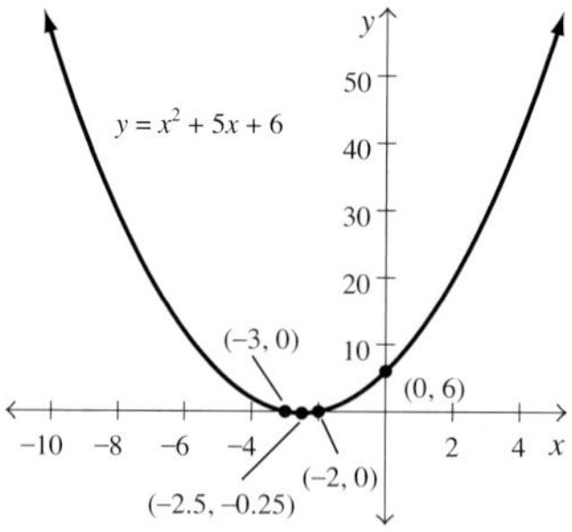

g) Domain: $\mathbb{R}$; range: $y \geq -0.25$

65. a) Downward **b)** $x = 2$ **c)** $(2, -2)$ **d)** $(0, -6)$
e) No x-intercepts
f)

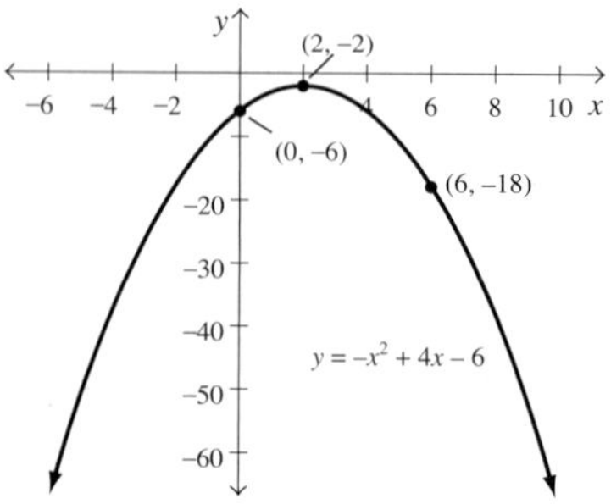

g) Domain: $\mathbb{R}$; range: $y \leq -2$

67. a) Downward **b)** $x = \frac{3}{4}$ **c)** $(\frac{3}{4}, -\frac{7}{8})$ **d)** $(0, -2)$
e) No x-intercepts
f)

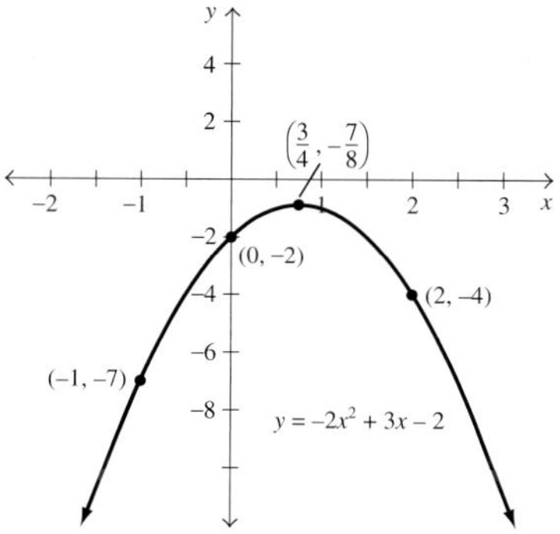

g) Domain: $\mathbb{R}$; range: $y \leq -\frac{7}{8}$

69. Domain: $\mathbb{R}$; range: $y > 0$

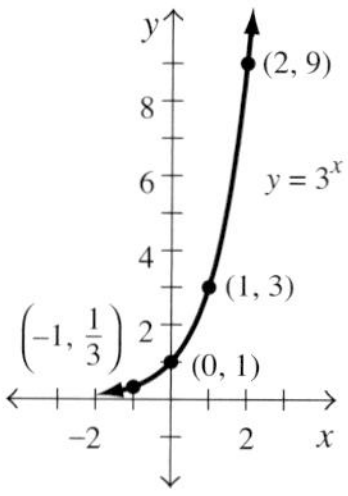

71. Domain: $\mathbb{R}$; range: $y > 0$

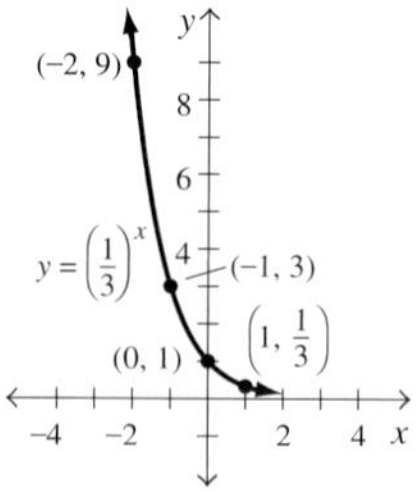

73. Domain: $\mathbb{R}$; range: $y > 1$

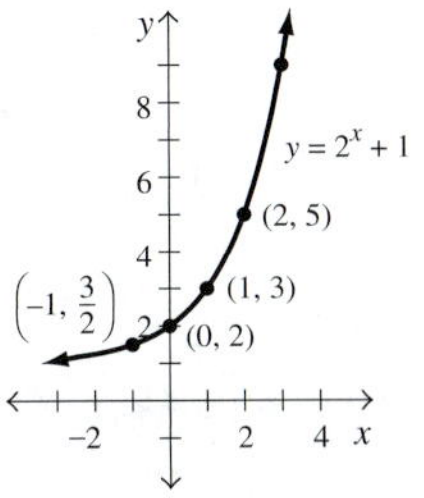

75. Domain: $\mathbb{R}$; range: $y > 1$

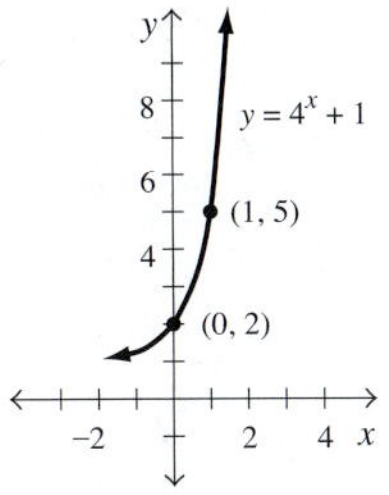

77. Domain: $\mathbb{R}$; range: $y > 0$

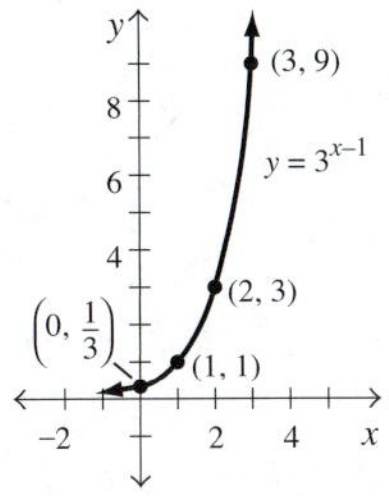

79. Domain: $\mathbb{R}$; range: $y > 0$

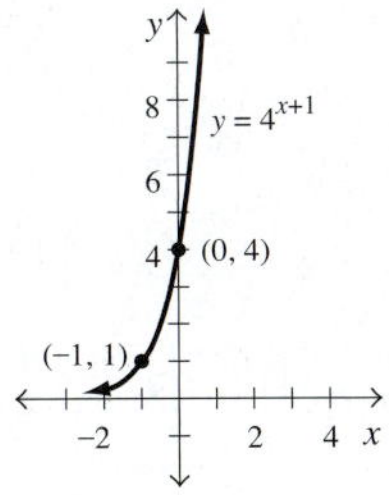

81. \$41,000 **83. a)** $\approx 13\%$ **b)** 1970 **c)** $x \approx 36$; $\approx 5\%$
85. a) 5200 people **b)** $\approx$ 14,852 people
87. a) Yes **b)** $\approx$ \$59
89. a) 23.2 cm **b)** 55.2 cm **c)** 69.5 cm
91. a) 170 beats per minute
b) $\approx$ 162 beats per minute
c) $\approx$ 145 beats per minute
d) 136 beats per minute
e) 120 years of age

REVIEW EXERCISES, PAGE 407

1. 19 **2.** −8 **3.** 17 **4.** $\frac{1}{4}$ **5.** −13 **6.** 13
7. $4x + 5$ **8.** $15x - 10$ **9.** $7x - 3$ **10.** −10
11. −3 **12.** −13 **13.** −31 **14.** $\frac{128}{5}$ **15.** $\frac{1}{2}$ cup
16. 250 min, or 4 hr 10 min **17.** 48 **18.** ≈ 173.1
19. 101.5 **20.** 25
21. $y = 2x - 6$
22. $y = \dfrac{-2x + 15}{7}$ or $y = -\dfrac{2}{7}x + \dfrac{15}{7}$
23. $y = \dfrac{2x + 22}{3}$ or $y = \dfrac{2}{3}x + \dfrac{22}{3}$
24. $y = \dfrac{-3x + 5z - 4}{4}$ or $y = -\dfrac{3}{4}x + \dfrac{5}{4}z - 1$
25. $w = \dfrac{A}{l}$ **26.** $w = \dfrac{P - 2l}{2}$
27. $l = \dfrac{L - 2wh}{2h}$ or $l = \dfrac{L}{2h} - w$
28. $d = \dfrac{a_n - a_1}{n - 1}$
29. $7 - 4x$ **30.** $2y + 7$ **31.** $10 + 3r$ **32.** $\dfrac{9}{q} - 15$
33. $3 + 7x = 17$; $x = 2$ **34.** $3x + 8 = x - 6$; $x = -7$
35. $5(x - 4) = 45$; $x = 13$
36. $10x + 14 = 8(x + 12)$; $x = 41$
37. $x + 2x = 15{,}000$; bonds: \$5000, mutual funds: \$10,000
38. $9.50x + 15{,}000 = 95{,}000$; $\approx$ 8421 chairs
39. $x + 6x + 2 = 79$; 11 species of threatened mammals, 68 species of endangered mammals
40. $x + (x + 12{,}000) = 68{,}000$; \$28,000 for B and \$40,000 for A
41. 64 **42.** 2 **43.** 20 **44.** ≈ 426.7
45. a) 150 lb **b)** 5 bags **46.** 4 in.
47. \$119.88 **48.** 1.75 in.
49. 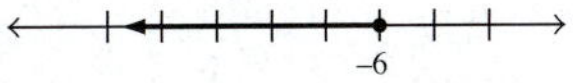

50. 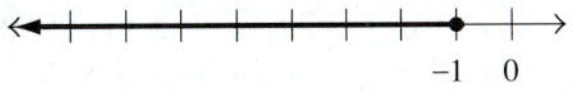

51. 

52. 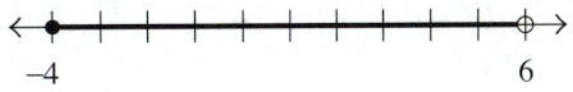

53. 

54. 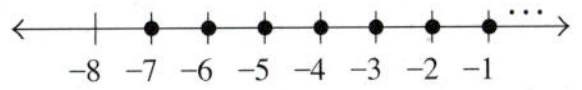

55. 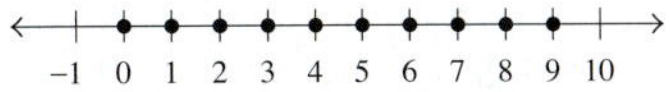

56. 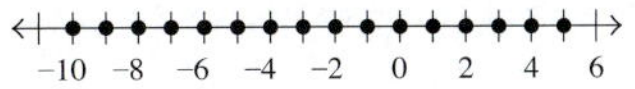

For Exercises 57–60, see the following figure.

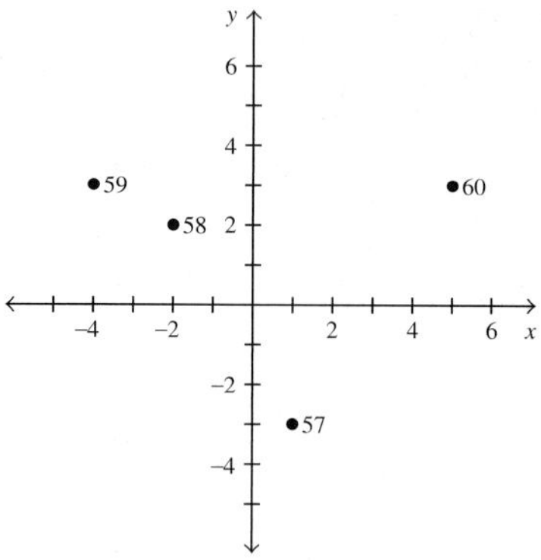

61. $D(-3, -1)$; area $= 20$ square units

62. $D(4, 1)$; area $= 21$ square units

63.

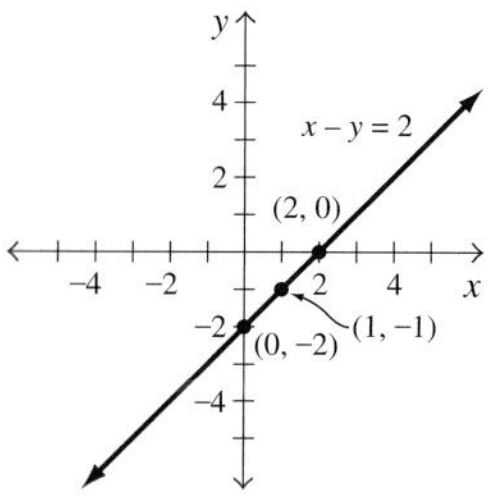

64.

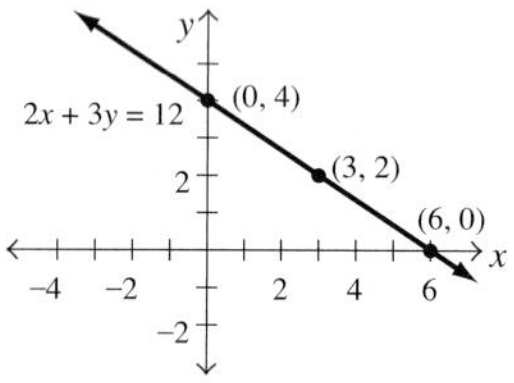

65.

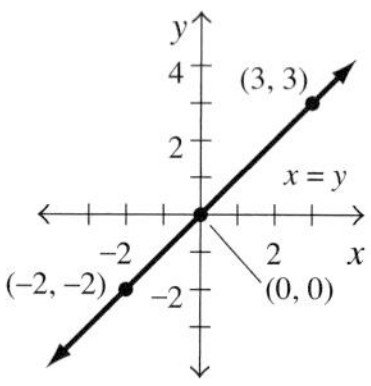

66.

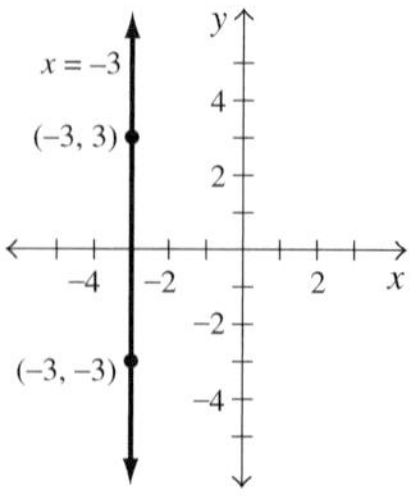

67.

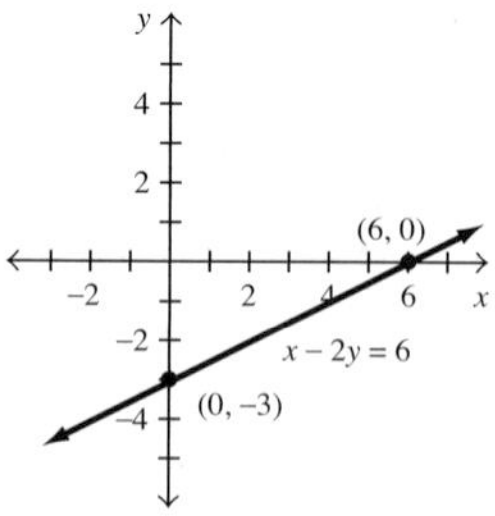

68.

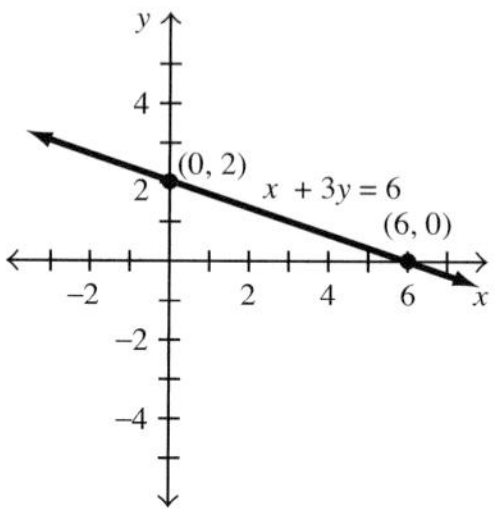

69.

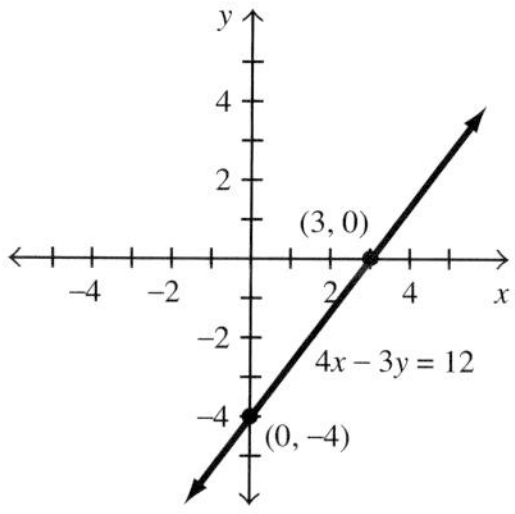

70.

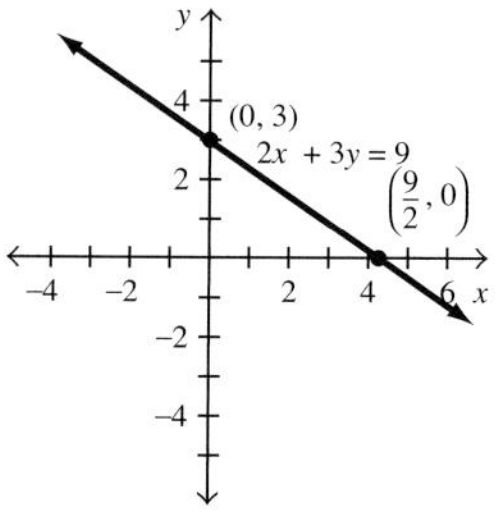

71. $\frac{2}{5}$ **72.** $-\frac{3}{2}$ **73.** $\frac{7}{3}$ **74.** Undefined

75.

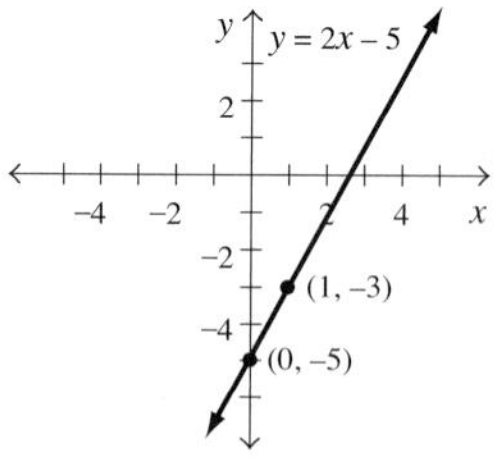

76.

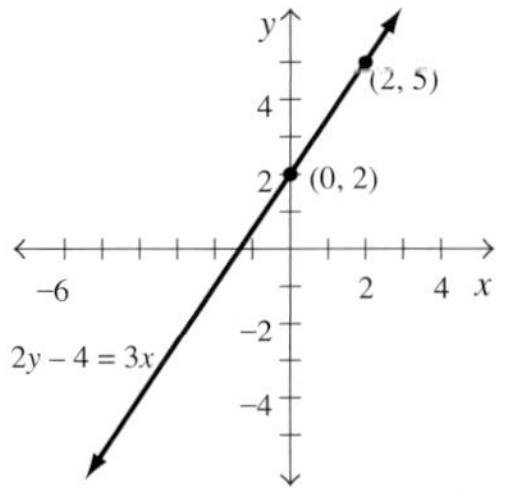

77.

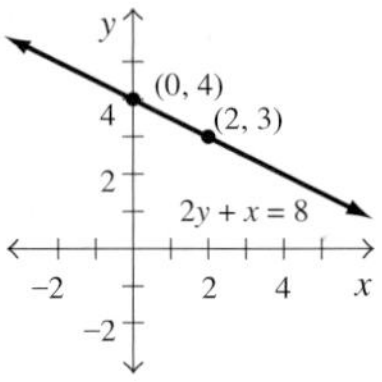

78.

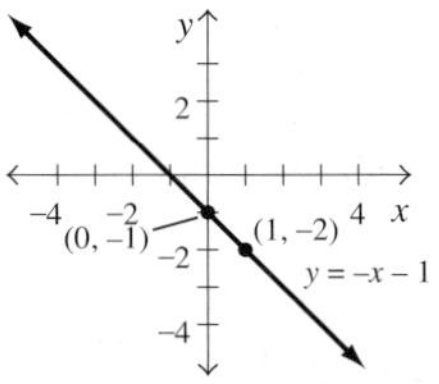

79. $y = 2x + 4$ **80.** $y = -x + 1$

81. a)

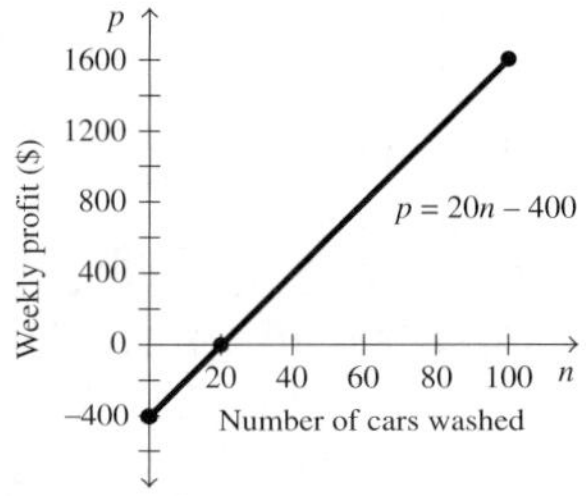

b) \$600 **c)** 70 cars

82. a)

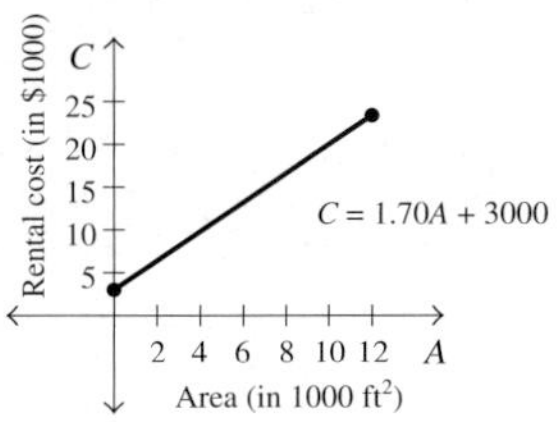

b) $\approx$ \$6400 **c)** ≈ 4120 ft^2

83.

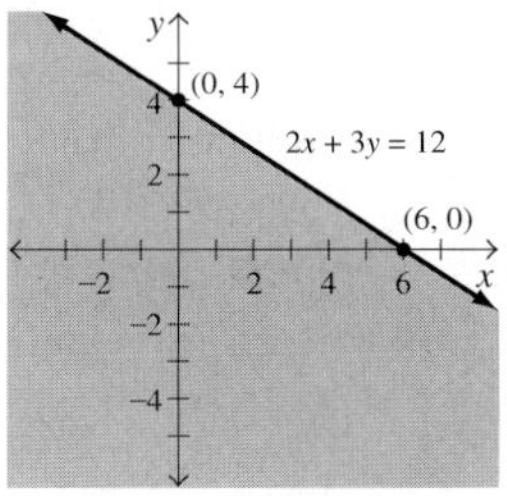

84.

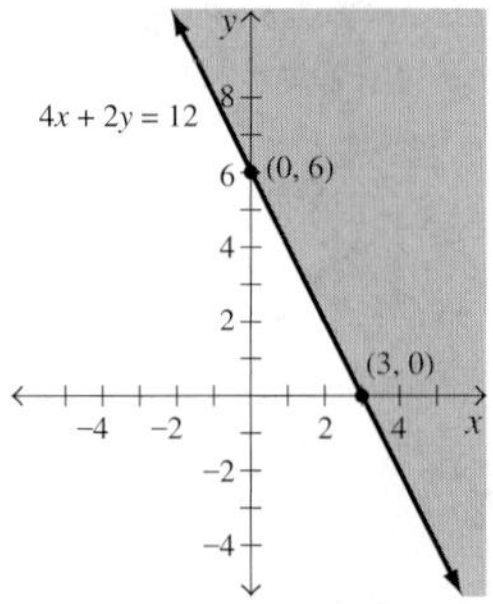

85.

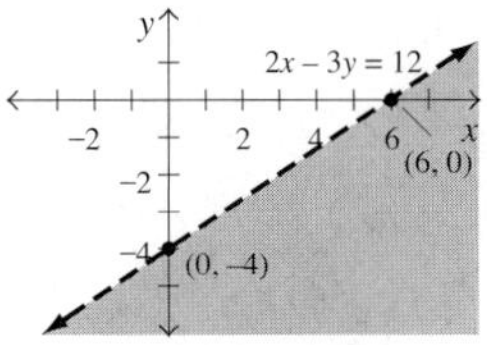

86.

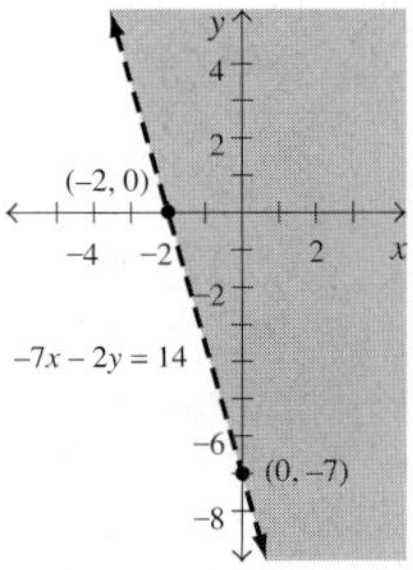

87. $(x + 2)(x + 9)$ **88.** $(x + 6)(x - 5)$
89. $(x - 6)(x - 4)$ **90.** $(x - 5)(x - 4)$
91. $(3x - 1)(2x + 3)$ **92.** $(2x - 1)(x + 7)$
93. $-1, -3$ **94.** $-6, 3$ **95.** $\frac{2}{3}, 5$
96. $-2, -\frac{1}{3}$ **97.** $2 \pm \sqrt{5}$ **98.** $-2, 8$
99. No real solution **100.** $-1, \frac{3}{2}$
101. Function, domain: $x = -2, -1, 2, 3$; range: $y = -1, 0, 2$
102. Not a function **103.** Not a function
104. Function, domain: $\mathbb{R}$; range: $\mathbb{R}$
105. 5 **106.** 15 **107.** 39 **108.** -21
109. **a)** Downward **b)** $x = -2$ **c)** $(-2, 25)$ **d)** $(0, 21)$
e) $(-7, 0), (3, 0)$
f)

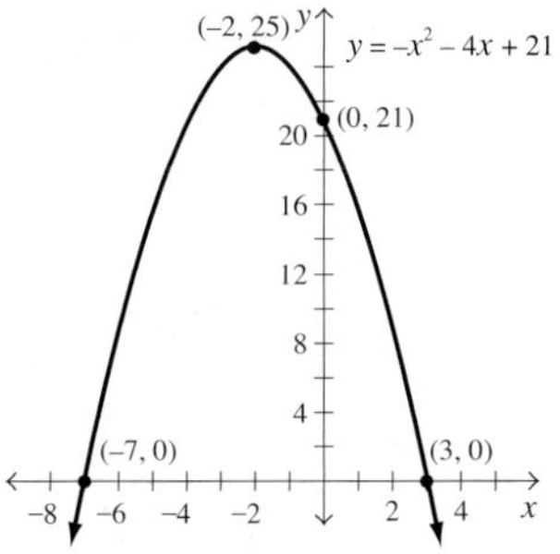

g) Domain: $\mathbb{R}$; range: $y \leq 25$

110. **a)** Upward **b)** $x = -2$ **c)** $(-2, -2)$ **d)** $(0, 6)$
e) $(-3, 0), (-1, 0)$
f)

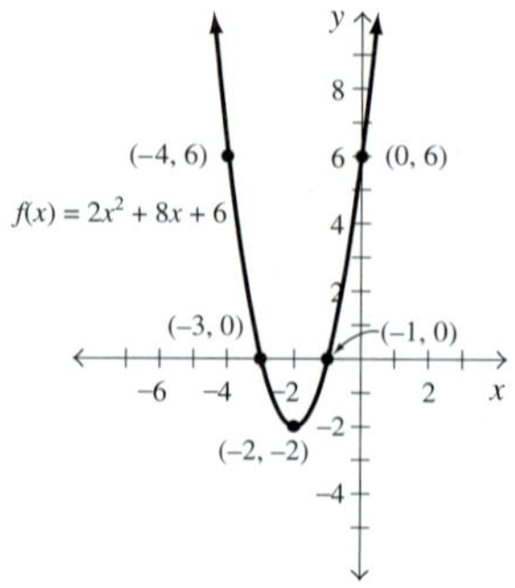

g) Domain: $\mathbb{R}$; range: $y \geq -2$

111. Domain: $\mathbb{R}$; range: $y > 0$

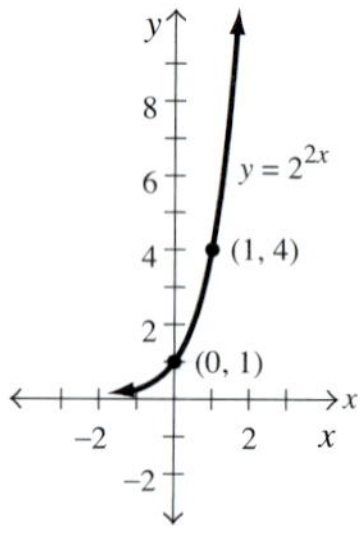

112. Domain: $\mathbb{R}$; range: $y > 0$

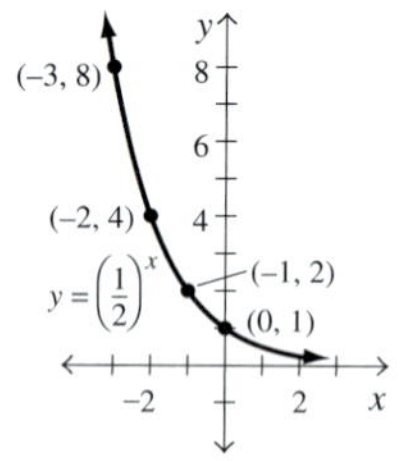

113. 22.8 mpg

114. **a)** 4208 **b)** 4250

115. $\approx 68.7\%$

12.

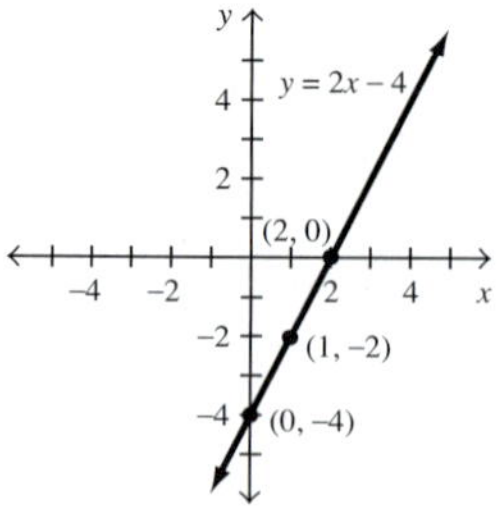

13.

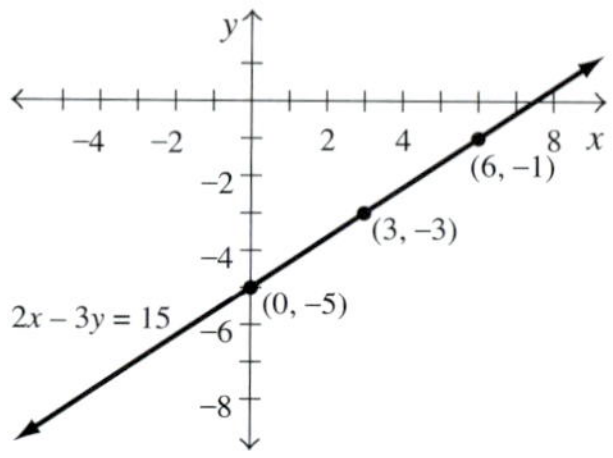

14.

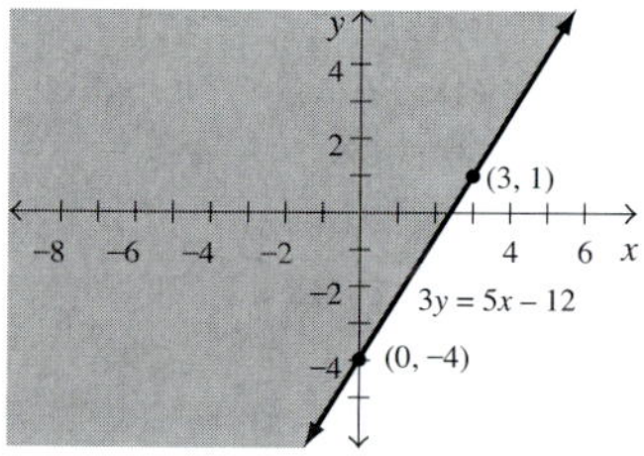

15. $-1, -4$ **16.** $\frac{4}{3}, -2$

17. It is a function. **18.** 17

19. **a)** Upward **b)** $x = 1$ **c)** $(1, 3)$ **d)** $(0, 4)$
e) No x-intercepts
f)

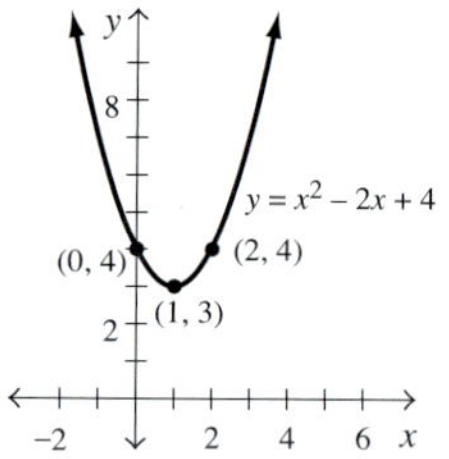

g) Domain: $\mathbb{R}$; range: $y \geq 3$

CHAPTER TEST, PAGE 411

1. 9 **2.** $\frac{19}{5}$ **3.** 18 **4.** $3x + 5 = 17; 4$

5. $350 + 0.06x = 710$; \$6000 **6.** 77

7. $y = \dfrac{-3x + 11}{5}$ or $y = -\dfrac{3}{5}x + \dfrac{11}{5}$

8. $3\frac{1}{3}$ **9.** 6.75 ft

10.

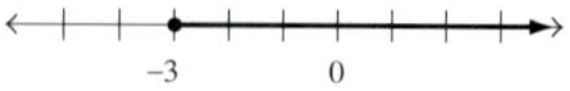

11. 2

CHAPTER 7

SECTION 7.1, PAGE 419

1. Two or more linear equations form a system of linear equations.

3. An inconsistent system of equations is one that has no solution.

5. A dependent system of equations is one that has an infinite number of solutions.

7. The graphs of the equations will be parallel lines.
9. The graphs of the equations will be the same line.
11. (3, 5)

13.

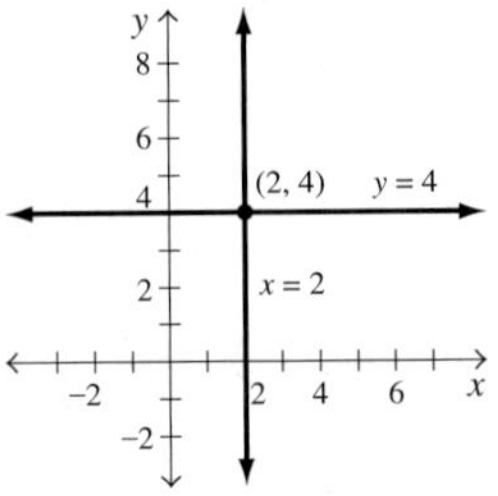

15.

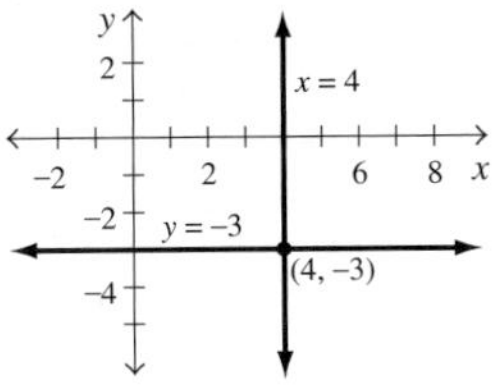

17.

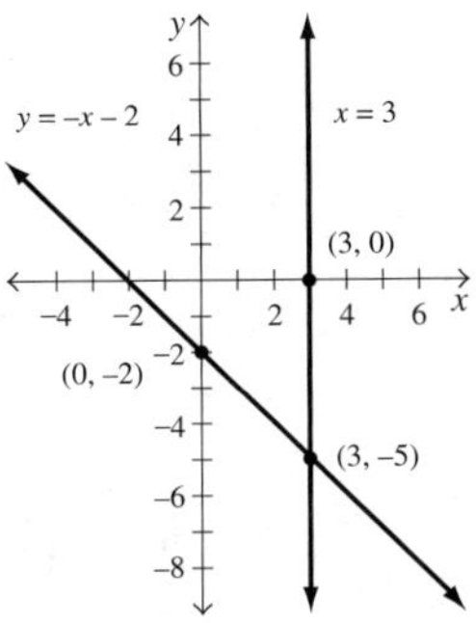

19.

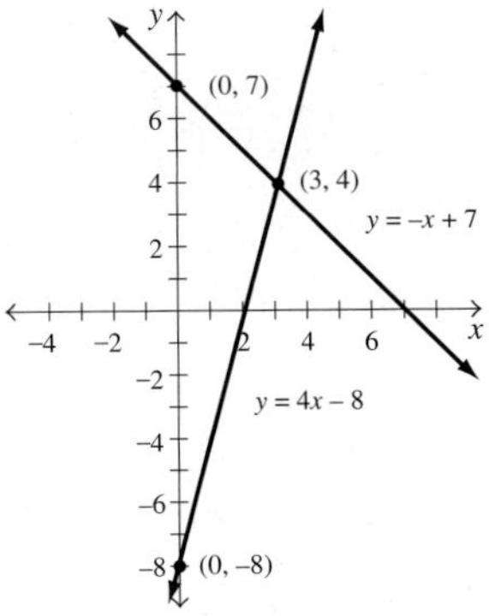

21.

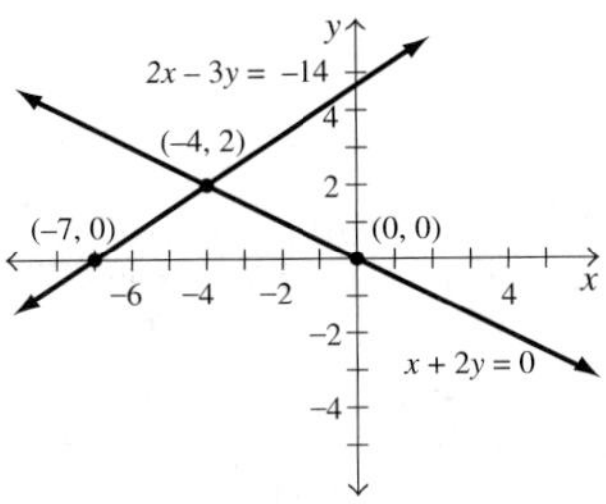

23.

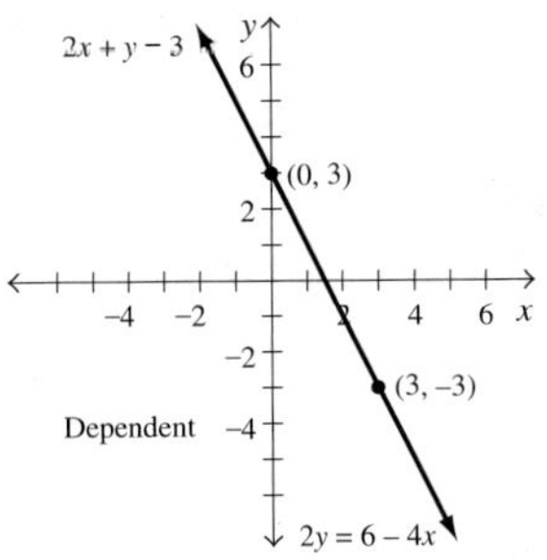

25.

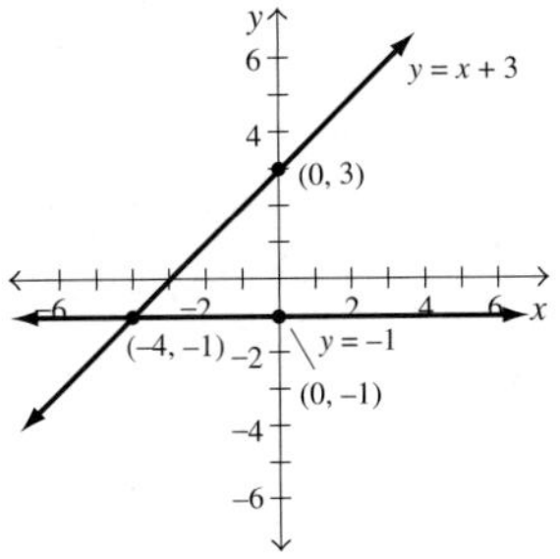

27.

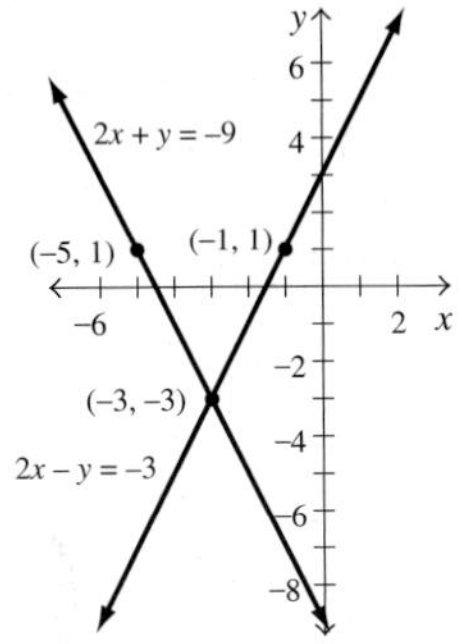

29.

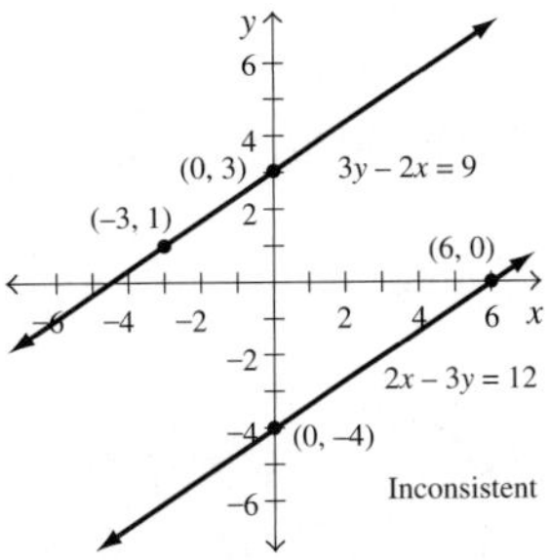

31.

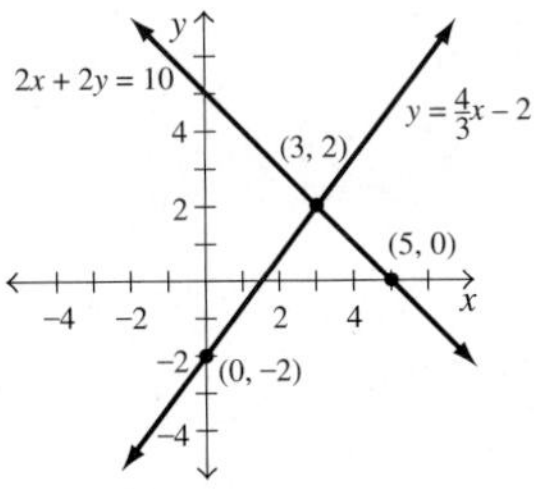

33. a) One unique solution; the lines intersect at one and only one point.
b) No solution; the lines do not intersect.
c) Infinitely many solutions; the lines coincide.
35. An infinite number of solutions
37. One solution **39.** No solution
41. An infinite number of solutions
43. No solution **45.** One solution
47. Not perpendicular **49.** Perpendicular
51. a) Cost to repair: $C = 375x + 250$
Cost to replace: $C = 225x + 700$

b)

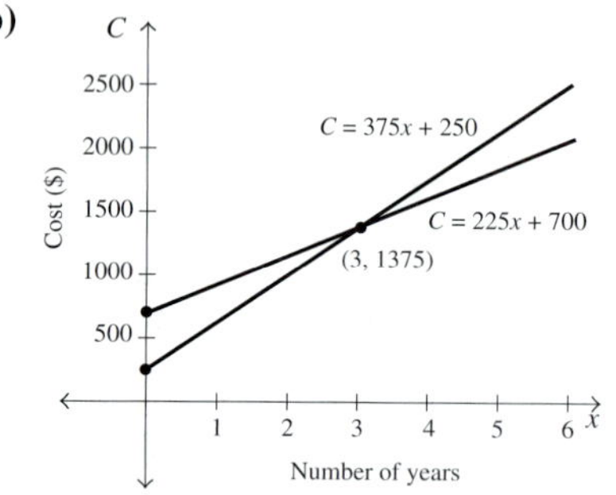

c) 3 years

53. a) $C = 15x + 400$
$R = 25x$

b)

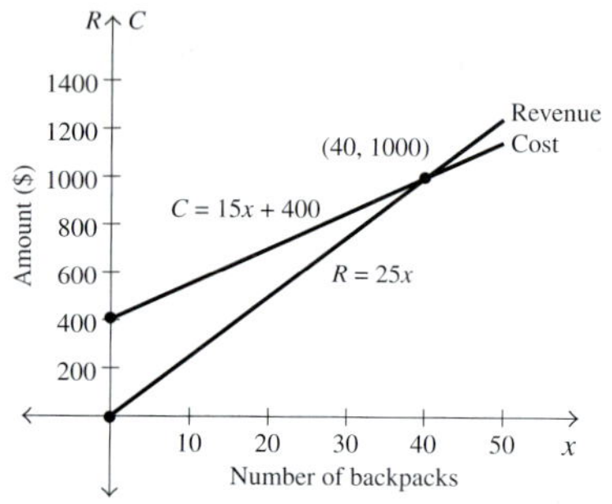

c) 40 backpacks **d)** $P = 10x - 400$
e) Loss of $100 **f)** 140 backpacks

55. a) $C = 230x + 8400$, $R = 300x$

b)

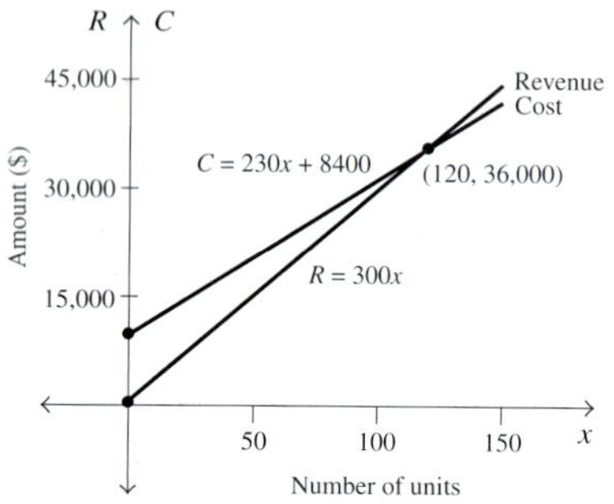

c) 120 units **d)** $P = 70x - 8400$
e) Loss of $1400 **f)** 138 units

57. a) Job 1: $s = 0.15x + 500$
Job 2: $s = 650$

b)

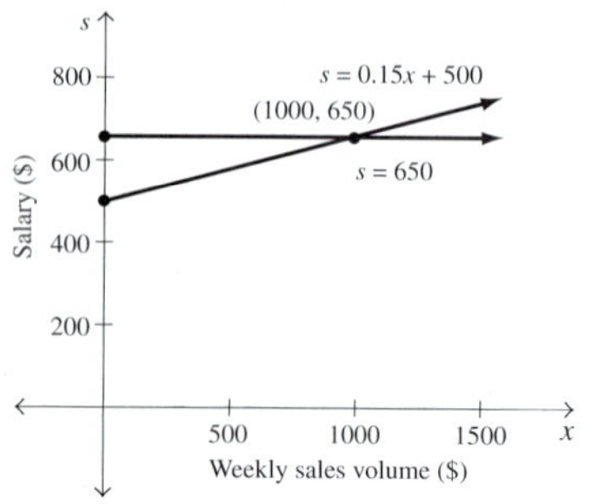

c) $1000

59. a) One **b)** Three **c)** Six **d)** Ten
e) To determine the number of points of intersection for n lines, add $n - 1$ to the number of points of intersection for $n - 1$ lines. For example, for 5 lines there were 10 points of intersection. Therefore, for 6 lines ($n = 6$) there are $10 + (6 - 1) = 15$ points of intersection.

SECTION 7.2, PAGE 431

1. Answers will vary.

3. The system is dependent if the same value is obtained on both sides of the equal sign.

5. x, in the first equation

7. $(-2, 6)$ **9.** $(1, -1)$

11. No solution; inconsistent system **13.** $(-2, \frac{8}{3})$

15. An infinite number of solutions; dependent system

17. $(-3, -6)$ **19.** $(\frac{11}{5}, -\frac{13}{5})$ **21.** $(-\frac{1}{5}, -\frac{8}{5})$

23. No solution; inconsistent system

25. $(3, 0)$ **27.** $(7, 5)$ **29.** $(-2, 0)$

31. $(-4, 5)$ **33.** $(3, 5)$ **35.** $(1, -2)$

37. No solution; inconsistent system **39.** $(2, 1)$

41. $s = 0.15p + 12{,}000$
$s = 0.05p + 27{,}000$
$150,000 annual profit

43. $m + l = 50$
$10.95m + 14.95l = 663.50$
21 medium, 29 large

45. $x + y = 10$
$0.25x + 0.50y = 0.40(10)$
4ℓ of 25%, 6ℓ of 50%

47. $y = 0.02x + 18$
$y = 0.015x + 24$
1200 copies

49. $x + y = 20$
$3x + y = 30$
Mix 5 lb of nuts with 15 lb of pretzels.

51. $x + y = 4600$
$27x + 14y = 104{,}700$
3100 amphitheatre tickets and 1500 lawn tickets

53. $\approx$ 11.2 years after 2000 or in 2011

55. $(\frac{1}{2}, \frac{1}{3})$ **57.–61.** Answers will vary.

SECTION 7.3, PAGE 441

1. A matrix is a rectangular array of elements.

3. A square matrix contains the same number of rows and columns.

5. 4

7. a) Answers will vary. **b)** $\begin{bmatrix} 7 & 9 & -7 \\ 2 & 16 & 9 \end{bmatrix}$

9. a) The number of columns of the first matrix must be the same as the number of rows of the second matrix.
b) 2×3

11. a) $I = \begin{bmatrix} 1 & 0 \\ 0 & 1 \end{bmatrix}$ **b)** $I = \begin{bmatrix} 1 & 0 & 0 \\ 0 & 1 & 0 \\ 0 & 0 & 1 \end{bmatrix}$

13. $\begin{bmatrix} -3 & 9 \\ 9 & 9 \end{bmatrix}$ **15.** $\begin{bmatrix} -1 & 4 \\ -5 & 4 \\ 7 & 6 \end{bmatrix}$

17. $\begin{bmatrix} 6 & -7 \\ -12 & 4 \end{bmatrix}$ **19.** $\begin{bmatrix} 1 & 9 \\ 18 & 17 \\ -2 & 2 \end{bmatrix}$

21. $\begin{bmatrix} 6 & 4 \\ 10 & 0 \end{bmatrix}$ **23.** $\begin{bmatrix} -2 & 16 \\ 26 & 0 \end{bmatrix}$

25. $\begin{bmatrix} 16 & 2 \\ 12 & 0 \end{bmatrix}$ **27.** $\begin{bmatrix} 26 & 18 \\ 48 & 24 \end{bmatrix}$

29. $\begin{bmatrix} 15 \\ 22 \end{bmatrix}$ **31.** $\begin{bmatrix} 4 & 7 & 6 \\ -2 & 3 & 1 \\ 5 & 1 & 2 \end{bmatrix}$

33. $A + B = \begin{bmatrix} 9 & 1 & 8 \\ 6 & -1 & 4 \end{bmatrix}$; cannot be multiplied

35. Cannot be added; $A \times B = \begin{bmatrix} 26 & 38 \\ 24 & 24 \end{bmatrix}$

37. Cannot be added; $\begin{bmatrix} 1 \\ -1 \end{bmatrix}$

39. $A + B = B + A = \begin{bmatrix} 7 & 10 \\ 4 & 4 \end{bmatrix}$

41. $A + B = B + A = \begin{bmatrix} 8 & 0 \\ 6 & -8 \end{bmatrix}$

43. $(A + B) + C = A + (B + C) = \begin{bmatrix} 7 & 10 \\ 6 & 13 \end{bmatrix}$

45. $(A + B) + C = A + (B + C) = \begin{bmatrix} 5 & 5 \\ 7 & -37 \end{bmatrix}$

47. No **49.** No **51.** Yes

53. $(A \times B) \times C = A \times (B \times C) = \begin{bmatrix} 41 & 13 \\ 56 & 16 \end{bmatrix}$

55. $(A \times B) \times C = A \times (B \times C) = \begin{bmatrix} 16 & -10 \\ -24 & 2 \end{bmatrix}$

57. $(A \times B) \times C = A \times (B \times C) = \begin{bmatrix} 17 & 0 \\ -7 & 0 \end{bmatrix}$

59.

Tomatoes	Onions	Carrots	
88	58	70	Chase's Farm
78	54	71	Gro-More Farms

61. Total cost

53	Java's Coffee Shop
55	Spot Coffee Shop

63.

Large	Small	
38	50	Sugar
56	72	Flour
17	26	Milk
10	14	Eggs

65. [36.04 47.52]

67. Answers will vary.

69. Yes **71.** False

73. a) \$28.70 **b)** \$60.10 **c)** $\begin{bmatrix} 28.7 & 24.6 \\ 41.3 & 35.7 \\ 69.3 & 60.1 \end{bmatrix}$

75. Yes. Answers will vary. One example is

$$A = \begin{bmatrix} 2 & 7 & 6 \\ -3 & 0 & 8 \end{bmatrix}, \quad B = \begin{bmatrix} 1 & 2 \\ 3 & 4 \\ 5 & 6 \end{bmatrix}.$$

SECTION 7.4, PAGE 452

1. a) An augmented matrix is a matrix formed with the coefficients of the variables and the constants. The coefficients of the variables are separated from the constants by a vertical bar.
b) $\left[\begin{array}{cc|c} 1 & 3 & 7 \\ 2 & -1 & 4 \end{array}\right]$

3. If you obtain an augmented matrix in which a 0 appears across an entire row, the system of equations is dependent.

5. Change the -2 to a 1 by multiplying the numbers in the second row by $-\frac{1}{2}$.

7. (1, 2) **9.** (3, 2)

11. An infinite number of solutions; dependent system

13. $(\frac{7}{2}, -1)$ **15.** (1, 1)

17. No solution; inconsistent system

19. (3, 4)

21. Fitted caps: 18, stretch-fit caps: 14

23. Truck driver: $7\frac{1}{9}$ hours; laborer: $9\frac{1}{9}$ hours

25. Premium paper: 150 reams, paper for color printers: 50 reams

SECTION 7.5, PAGE 456

1. The solution set of a system of linear inequalities is the set of points that satisfy all inequalities in the system.

3. Yes. A point of intersection satisfies both inequalities and is, therefore, a solution.

5.

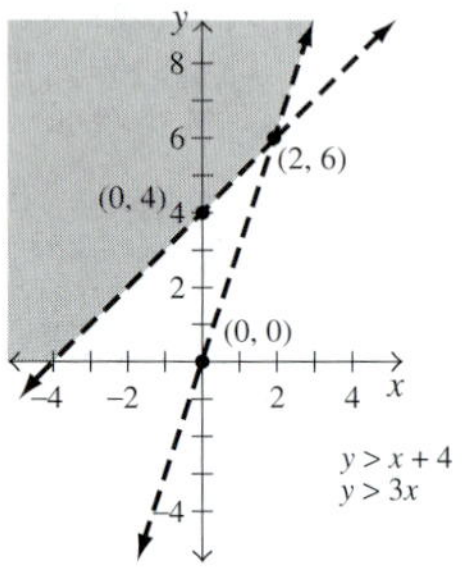

7.

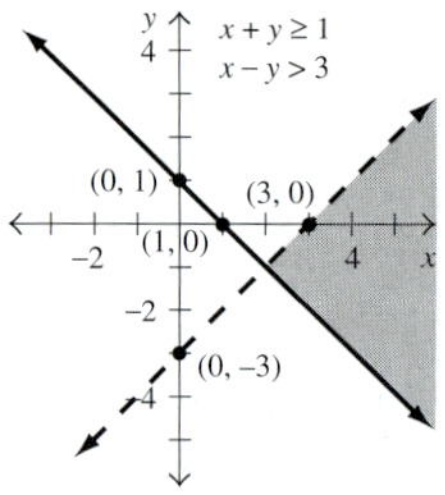

9.

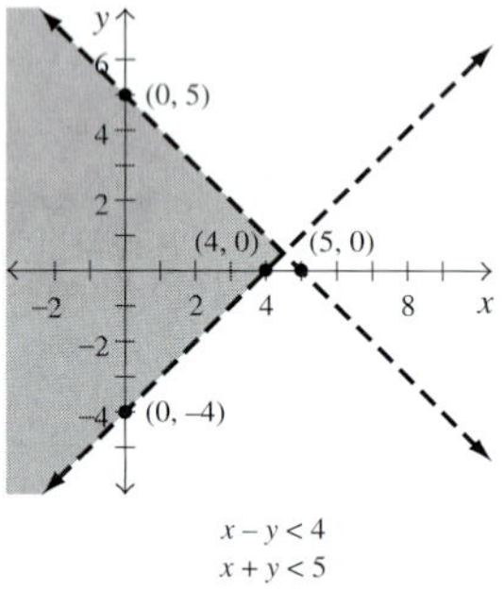

11.

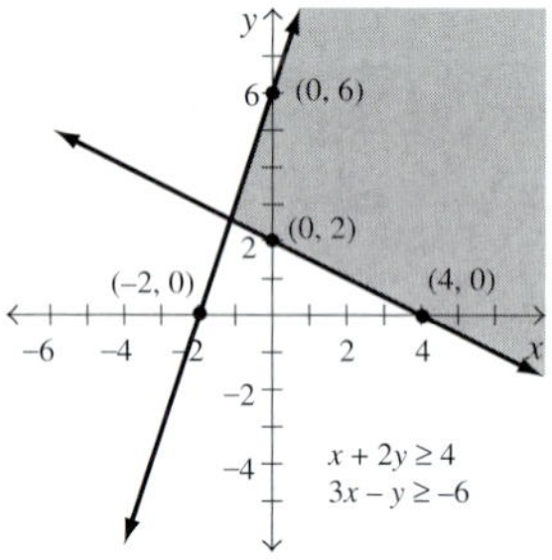

13.

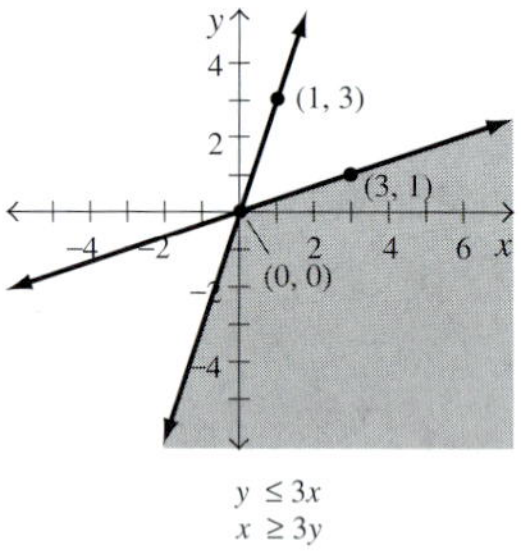

15.

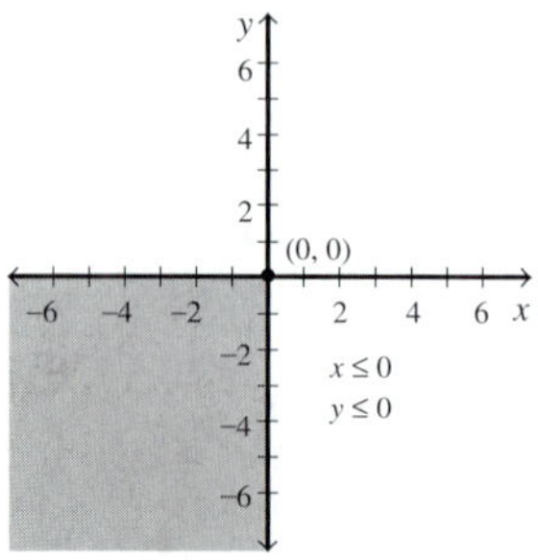

17.

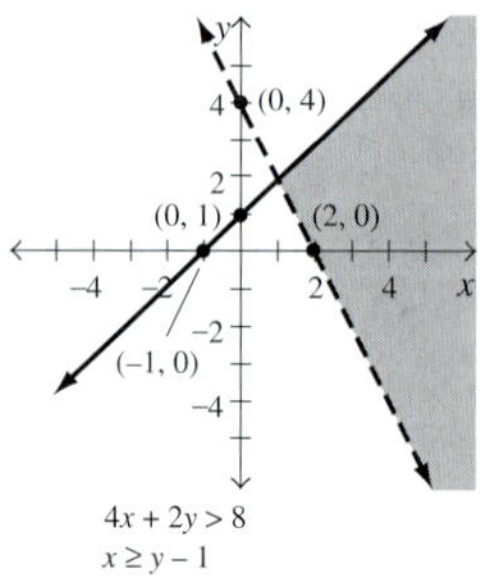

19.

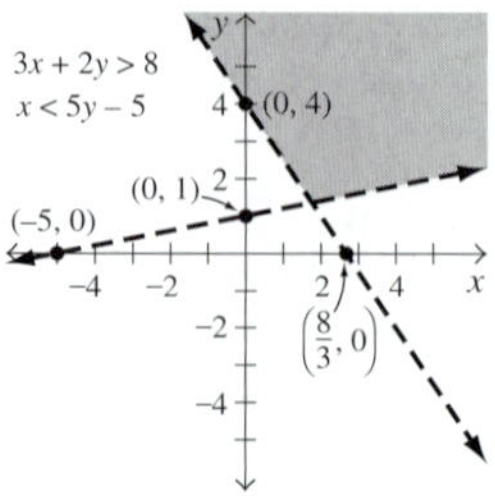

21. a) $20x + 30y \leq 600, x \geq 2y, x \geq 10, y \geq 5$

b)

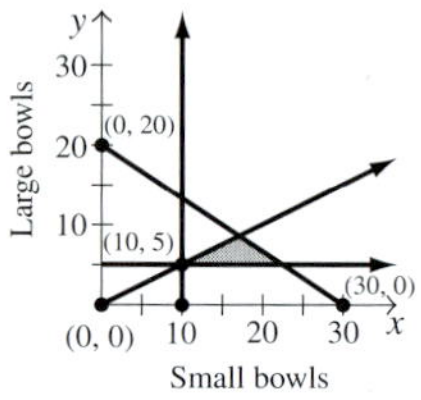

c) One example is 15 small bowls, 7 large bowls

23. $x \leq 0, y \geq 0$

25. Yes. One example is $x \geq 0, y \geq 0, x \leq 0, y \leq 0$ which has solution (0, 0).

27. Answers will vary.

SECTION 7.6, PAGE 461

1. Constraints are restrictions that are represented as linear inequalities.

3. Vertices

5. Answers will vary.

7. Maximum is 30 at (5, 0), minimum is 0 at (0, 0).

9. Maximum is 190 at (50, 30), minimum is 70 at (20, 10).

11. a)

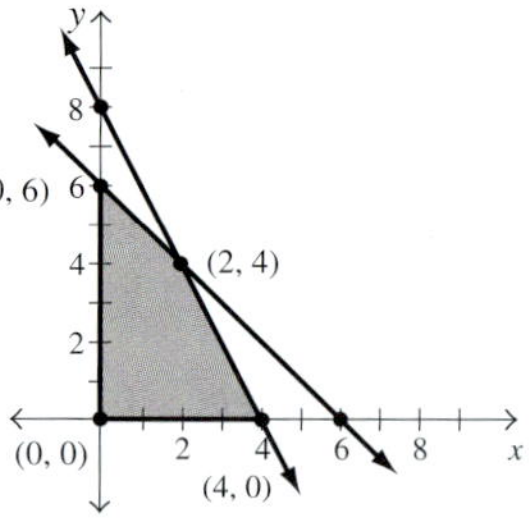

b) Maximum is 30 at (0, 6), minimum is 0 at (0, 0).

13. a)

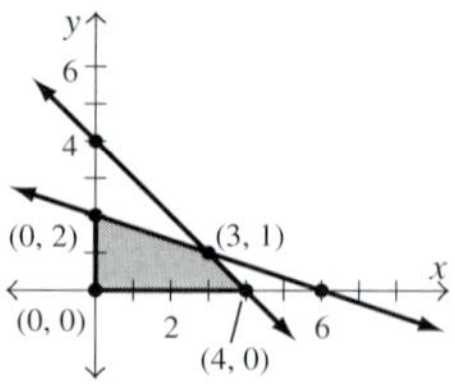

b) Maximum is 28 at (4, 0), minimum is 0 at (0, 0).

15. a)

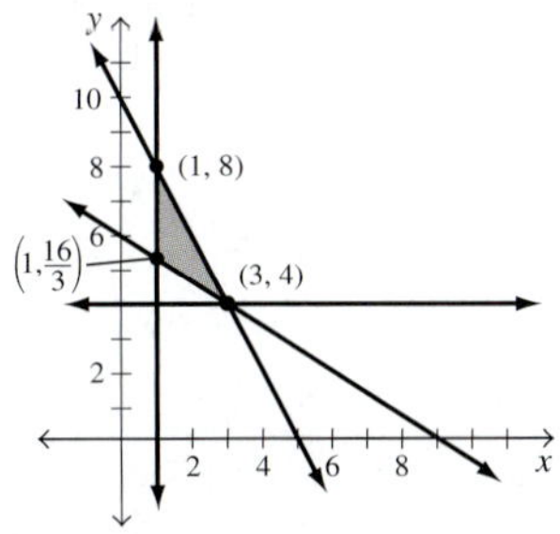

b) Maximum is 15.4 at (1, 8), minimum is 11 at $(1, \frac{16}{3})$.

17. a) $x + y \leq 24, x \geq 2y, y \geq 4, x \geq 0, y \geq 0$

b) $P = 40x + 55y$

c)

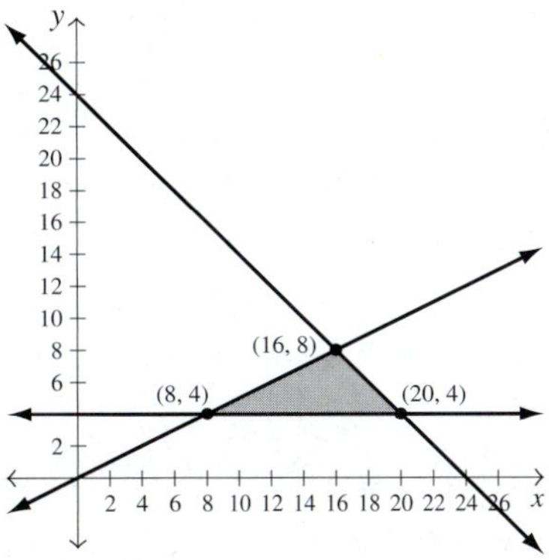

d) (8, 4), (16, 8), (20, 4)

e) 16 Kodak cameras and 8 Canon cameras

f) $1080

19. a) $3x + 4y \geq 60, 10x + 5y \geq 100, x \geq 0, y \geq 0$

b) $C = 28x + 33y$

c)

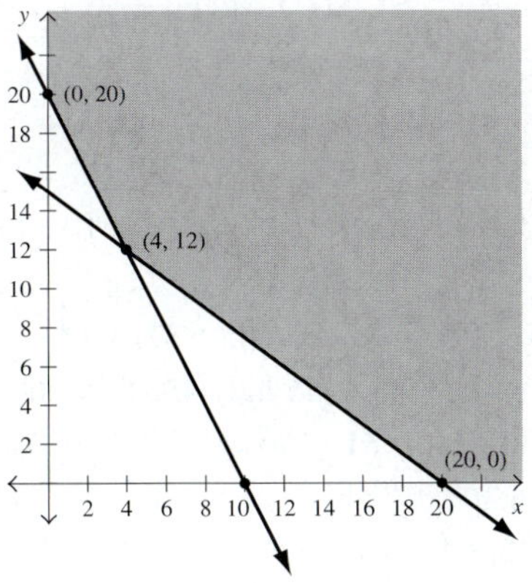

d) (0, 20), (4, 12), (20, 0)

e) 4 hours for Machine I and 12 hours for Machine II

f) $508

21. Three car seats and seven strollers, $320

REVIEW EXERCISES, PAGE 464

1.

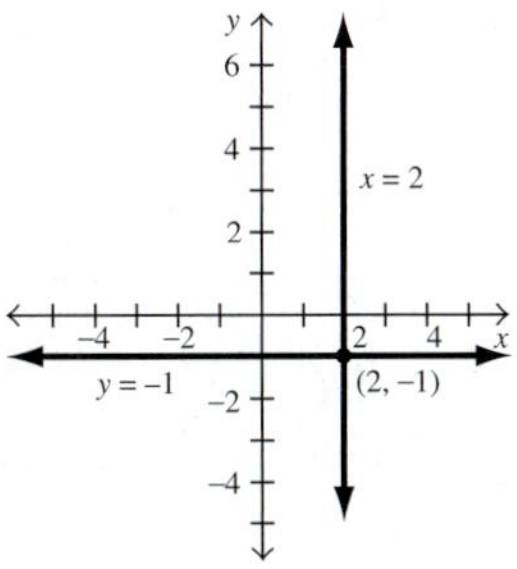

2.

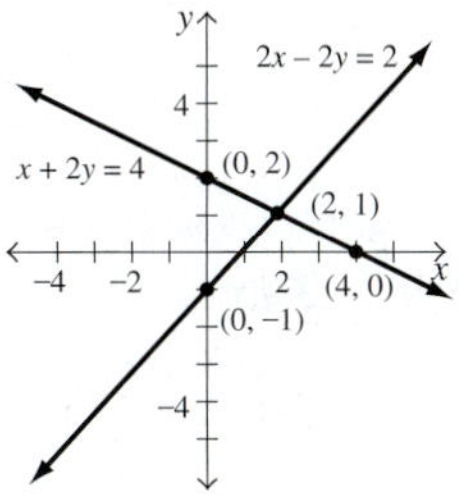

3.

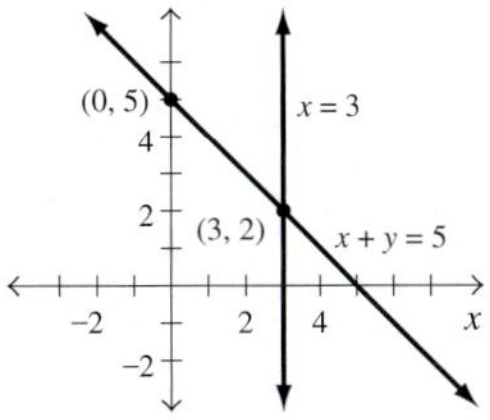

4.

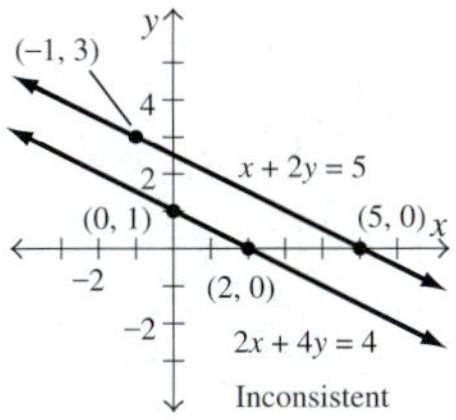

5. An infinite number of solutions **6.** No solution

7. One solution **8.** One solution **9.** (3, 1)

10. $(-1, -5)$ **11.** $(-2, -8)$

12. No solution; inconsistent **13.** $(4, -2)$ **14.** $(-7, 16)$

15. $(4, -2)$ **16.** An infinite number of solutions; dependent

17. $(2, -1)$ **18.** $(2, 0)$

19. $\begin{bmatrix} -1 & -8 \\ 8 & 7 \end{bmatrix}$ **20.** $\begin{bmatrix} 3 & 2 \\ -4 & 1 \end{bmatrix}$

21. $\begin{bmatrix} 2 & -6 \\ 4 & 8 \end{bmatrix}$ **22.** $\begin{bmatrix} 8 & 9 \\ -14 & -1 \end{bmatrix}$

23. $\begin{bmatrix} -20 & -14 \\ 20 & 2 \end{bmatrix}$ **24.** $\begin{bmatrix} -12 & -14 \\ 12 & -6 \end{bmatrix}$

25. (2, 2) **26.** $(-2, 2)$ **27.** $(3, -3)$

28. (1, 0) **29.** $(\frac{12}{11}, \frac{7}{11})$ **30.** (1, 2)

31. $350,000 at 4%, $250,000 at 6%

32. Mix $83\frac{1}{3}\ell$ of 80% acid solution with $16\frac{2}{3}\ell$ of 50% acid solution.

33. $245 per ton for topsoil, $183 per ton for mulch

34. a) 32.5 months **b)** Model 6070B

35. a) 3 hr **b)** All-Day parking lot

36.

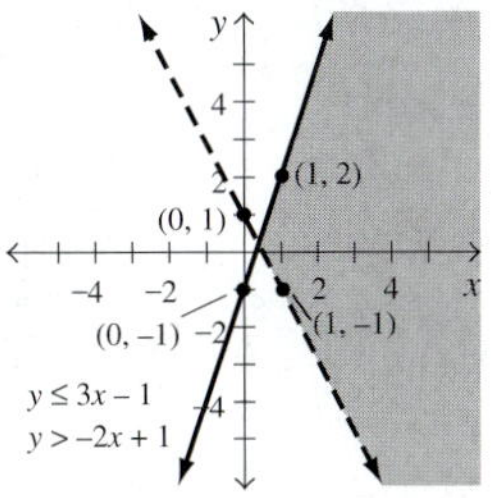

37.

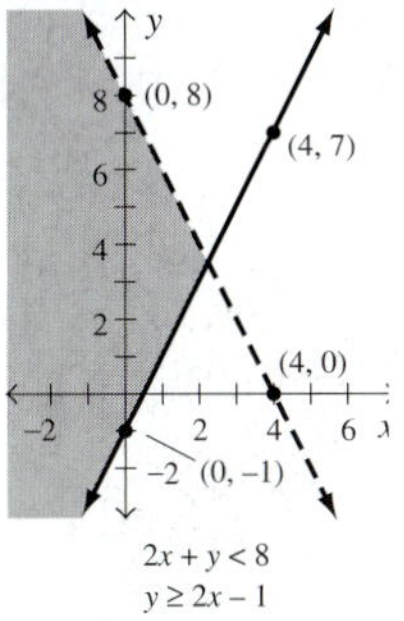

38.

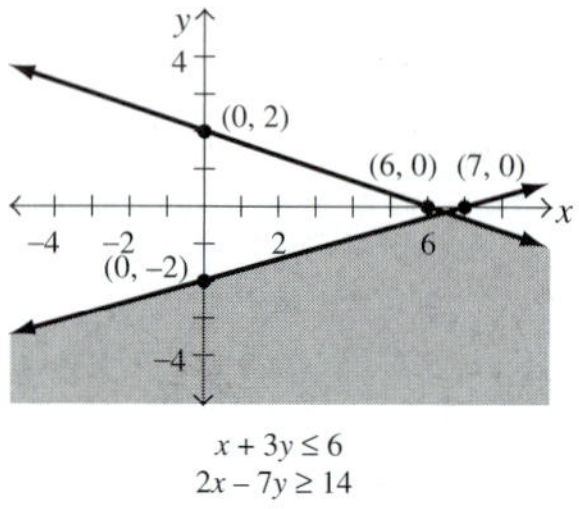

39.

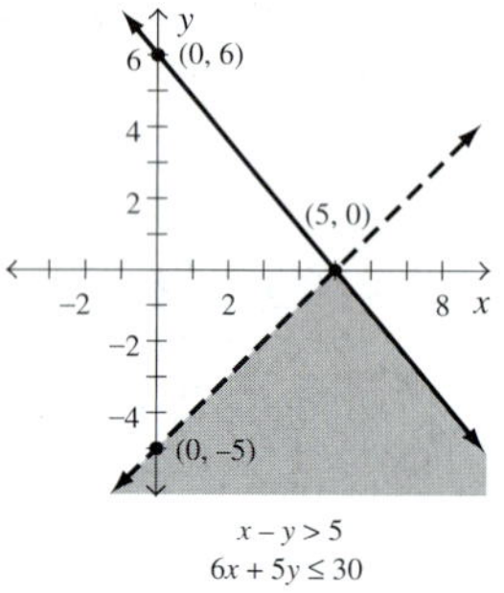

40. a)

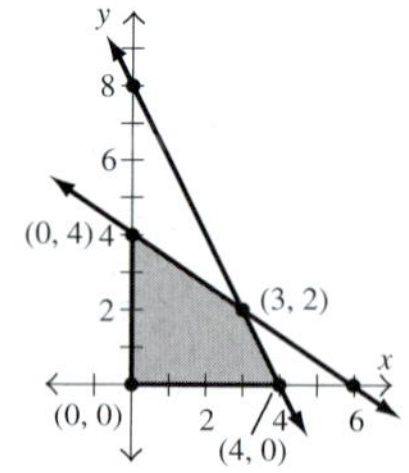

b) Maximum is 21 at (3, 2), minimum is 0 at (0, 0).

CHAPTER TEST, PAGE 466

1. If the lines do not intersect (are parallel), the system of equations is inconsistent. The system of equations is consistent if the lines intersect. If both equations represent the same line, the system of equations is dependent.

2.

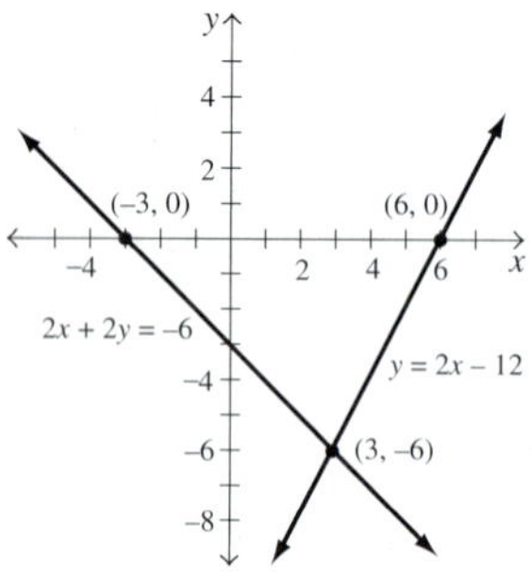

3. One solution **4.** (2, −3) **5.** (−2, −13)

6. (−2, −6) **7.** (−1, 3) **8.** (2, 0)

9. (−2, 2) **10.** $\begin{bmatrix} 1 & -8 \\ 6 & 5 \end{bmatrix}$

11. $\begin{bmatrix} 4 & 1 \\ -9 & -1 \end{bmatrix}$ **12.** $\begin{bmatrix} -27 & -16 \\ 14 & 3 \end{bmatrix}$

13.

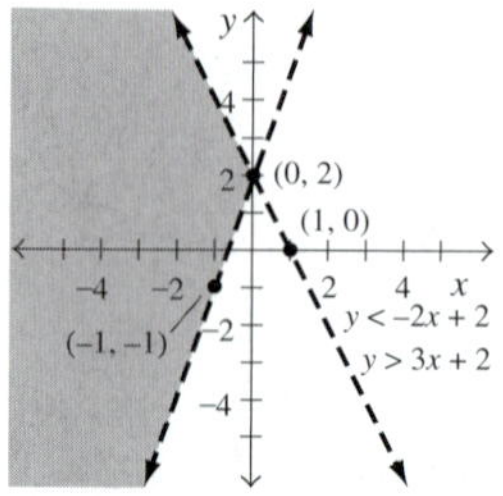

14. Daily fee: \$35, mileage charge: \$0.18

15. a) 40 checks **b)** Citrus Bank

16. a)

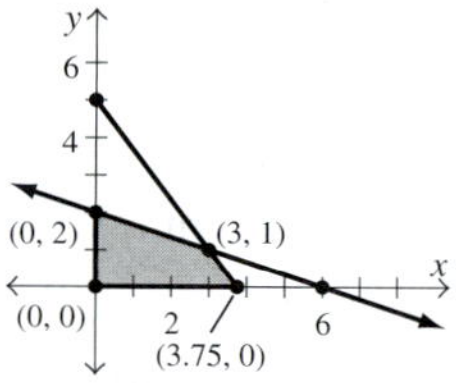

b) Maximum is 22.5 at (3.75, 0), minimum is 0 at (0, 0).

CHAPTER 8

SECTION 8.1, PAGE 474

1. The metric system

3. It is the standard of measurement accepted worldwide. There is only one basic unit of measurement for each quantity. It is based on the number 10, which makes many calculations easier than the U.S. customary system.

5. a) Answers will vary. **b)** 0.002 146 km **c)** 60 800 dm

7. Answers will vary.

9. a) 10,000 times **b)** 10 000 cm **c)** 0.0001 hm

11. Yard **13.** 5 **15.** 22 **17.** (b) **19.** (c) **21.** (f)

23. a) 100 **b)** 0.001 **c)** 1000 **d)** 0.01 **e)** 10 **f)** 0.1

25. cg; $\frac{1}{100}$ g **27.** dg; $\frac{1}{10}$ g **29.** hg; 100 g

31. 3 000 000 g **33.** 5000 **35.** 0.0085 **37.** 0.024 26

39. 0.024 35 **41.** 13 400 **43.** 325 hg **45.** 895 000 mℓ

47. 1.40 g **49.** 4.0302 daℓ **51.** 590 cm, 2.3 dam, 0.47 km

53. 1.4 kg, 1600 g, 16 300 dg

55. 203 000 mm, 2.6 km, 52.6 hm

57. Jim, 1 m > 1 yd

59. The pump that removes 1 daℓ per minute

61. a) 346 cm **b)** 3460 mm

63. a) 6.417 km/ℓ **b)** 6417 m/ℓ

65. a) 2160 mℓ **b)** 2.16 ℓ **c)** \$1.13 per liter

67. a) 108 m **b)** 0.108 km **c)** 108 000 mm
69. a) 11 120 km **b)** 11 120 000 m **71.** 1000
73. $1 \times 10^{24} = 1\ 000\ 000\ 000\ 000\ 000\ 000\ 000\ 000$
75. ≈ 30 eggs **77.** ≈ 4.1 cups **79.** 7 dam
81. 6 mg **83.** 2 daℓ **85.** gram **87.** liter **89.** meter
91. kilometer **93.** degrees celsius

SECTION 8.2, PAGE 485

1. Volume **3.** Area **5.** Volume **7.** Volume
9. Area **11.** Length **13.–17.** Answers will vary.
19. A cubic decimeter **21.** A cubic centimeter
23. Area **25.** Centimeters **27.** Centimeters or millimeters
29. Centimeters **31.** Kilometers **33.** Centimeters
35. Kilometers **37.** (c) **39.** (c) **41.** (b) **43.** (c)
45.–49. Answers will vary. **51.** Centimeter, kilometer
53. Meter **55.** Centimeter **57.** Square centimeters
59. Square centimeters or square meters
61. Square meters or hectares **63.** Square centimeters
65. Square meters **67.** (a) **69.** (b) **71.** (a)
73. (c) **75.–79.** Answers will vary.
81. Liters **83.** Kiloliters
85. Cubic meters or cubic centimeters
87. Liters **89.** Cubic meters **91.** (c) **93.** (c)
95. (c) **97.** (a) **99. b)** 152 561 cm^3
101. ≈ 0.20 m^3
103. Longer side ≈ 4 cm, shorter side ≈ 2.2 cm, area ≈ 8.8 cm^2
105. ≈ 326.85 m^2 **107. a)** 3869 m^2 **b)** 369 m^2
109. a) 3151.2 m^2 **b)** 0.315 12 ha
111. $304 **113. a)** 56 000 cm^3 **b)** 56 000 mℓ **c)** 56 ℓ
115. 100 times larger **117.** 1000 times larger
119. 100 **121.** 100 **123.** 0.000 001
125. 1 000 000 **127.** 218 **129.** 76 **131.** 60 kℓ
133. Answers will vary. **135.** 6700
137. a) 4,014,489,600 sq in. **b)** Answers will vary.
139. a) Answers will vary. The average use is 5150.7 ℓ/day. **b)** Answers will vary. The average use is 493.2 ℓ/day.

SECTION 8.3, PAGE 495

1. Kilogram **3.** 2
5. Answers will vary. One possible answer is 32°C.
7. Answers will vary. **9.** Kilograms
11. Grams **13.** Grams **15.** Kilograms
17. Kilograms or metric tonnes **19.** (b)
21. (b) **23.** (b) **25.–27.** Answers will vary.
29. (c) **31.** (b) **33.** (c) **35.** (c) **37.** (c) **39.** 77°F
41. ≈ 33.3°C **43.** ≈ −17.8°C **45.** 98.6°F
47. ≈ −10.6°C **49.** 32°F **51.** ≈ −28.9°C
53. ≈ 73.9 by the formula **55.** 71.6°F
57. 95.18°F **59.** 64.04°F–74.30°F **61.** $20.25
63. 444 g **65. a)** 2304 m^3 **b)** 2304 kℓ **c)** 2304 t
67. Yes: 78°F is about 25.6°C.
69. 0.0042 **71.** 17 400 000 g
73. a) 1200 g **b)** 1200 cm^3
75. a) 5.625 ft^3 **b)** ≈ 351.6 lb **c)** ≈ 42.4 gal
77. a) −79.8°F **b)** 36.5°F **c)** 510 000 000°C

SECTION 8.4, PAGE 505

1. Dimensional analysis is a procedure used to convert from one unit of measurement to a different unit of measurement.
3. $\frac{60\text{ seconds}}{1\text{ minute}}$ or $\frac{1\text{ minute}}{60\text{ seconds}}$
5. $\frac{1\text{ lb}}{0.45\text{ kg}}$ **7.** $\frac{3.8\ \ell}{1\text{ gal}}$ **9.** 157.48 cm **11.** 1.26 m
13. 266.67 lb **15.** 62.4 km **17.** 1687.5 acres
19. 33.19 pints **21.** 1.46 mi^2 **23.** 54 kg **25.** 28 grams
27. 0.45 kilogram **29.** 2.54 centimeters, 1.6 kilometers
31. 0.9 meter **33.** ≈ 561.11 yd **35.** ≈ 1146.67 ft
37. 37.5 mph **39.** 43.2 m^2 **41.** 50 mph **43.** 0.21 oz
45. 360 m^3 **47.** $0.495 per pound **49.** ≈ 9078.95 gal
51. 6 qt **53. a)** −8460 cm **b)** −84.6 m
55. a) 10.89 ft^2 **b)** 35.937 ft^3 **57.** 25.2 mg
59. 6840 mg, or 6.84 g
61. a) 25 mg **b)** 900 mg
63. a) 289.2 m **b)** 76 500 t **c)** 44.8 kph
65. a) 0.9 € per pound **b)** $1.17 per pound
67. 1.0 cc, or b)
69. a) 4000 cc **b)** ≈ 244.09 $in.^3$ **71.** A kilogram
73. A liter **75.** A decimeter **77.** wonton **79.** 1 kilohurtz
81. 1 megaphone **83.** 2 kilomockingbird **85.** 1 decoration

REVIEW EXERCISES, PAGE 509

1. $\frac{1}{100}$ of base unit **2.** 1000 × base unit
3. $\frac{1}{1000}$ of base unit **4.** 100 × base unit
5. 10 × base unit **6.** $\frac{1}{10}$ of base unit **7.** 0.040 g
8. 320 cℓ **9.** 0.016 mm **10.** 1 kg **11.** 4620 ℓ
12. 19 260 dg **13.** 3000 mℓ, 14 630 cℓ, 2.67 kℓ
14. 0.047 km, 47 000 cm, 4700 m **15.** Centimeters
16. Grams **17.** Degrees Celsius
18. Millimeters or centimeters **19.** Square meters

20. Milliliters or cubic centimeters **21.** Millimeters
22. Kilograms or tonnes **23.** Kilometers
24. Meters or centimeters
25. a) Answers will vary. **b)** Answers will vary.
26. a) Answers will vary. **b)** Answers will vary.
27. (c) **28.** (b) **29.** (c) **30.** (a) **31.** (a) **32.** (b)
33. 3.6 t **34.** 4 300 000 g **35.** 75.2°F **36.** 20°C
37. ≈ -21.1°C **38.** 102.2°F
39. $l = 4$ cm, $w = 1.6$ cm, $A = 6.4 \text{ cm}^2$
40. $r = 1.5$ cm, $A \approx 7.07 \text{ cm}^2$
41. a) 80 m^3 **b)** 80 000 kg
42. a) 899.79 cm^2 **b)** $0.089\,979 \text{ m}^2$
43. a) $96\,000 \text{ cm}^3$ **b)** 0.096 m^3
c) 96 000 mℓ **d)** 0.096 kℓ
44. 10,000 times larger **45.** ≈ 9.84 in. **46.** ≈ 233.33 lb
47. 74.7 m **48.** ≈ 111.11 yd **49.** 72 kph **50.** ≈ 63.16 qt
51. 76 ℓ **52.** $\approx 78.95 \text{ yd}^3$ **53.** $\approx 12.77 \text{ in.}^2$ **54.** 3.8 ℓ
55. 11.4 m^3 **56.** 99.2 km **57.** 0.9 ft **58.** 82.55 mm
59. a) 1050 kg **b)** ≈ 2333.33 lb **60.** 32.4 m^2
61. a) 190 kℓ **b)** 190 000 kg
62. a) 104 kph **b)** 104 000 meters per hour
63. a) 252 ℓ **b)** 252 kg
64. $1.58 per pound

CHAPTER TEST, PAGE 512

1. 0.497 daℓ **2.** 2 730 000 cm **3.** 100 times greater
4. 2.4 km **5.** (b) **6.** (a) **7.** (c) **8.** (c) **9.** (b)
10. 10,000 times greater **11.** 1,000,000,000 times greater
12. 6300 g **13.** ≈ 28.13 mi **14.** ≈ -26.11°C
15. 68°F **16. a)** 300 000 g **b)** ≈ 666.67 lb
17. a) 3200 m^3 **b)** 3 200 000 ℓ (or 3200 kℓ)
c) 3 200 000 kg
18. $245

CHAPTER 9

SECTION 9.1, PAGE 523

1. a) Undefined terms, definitions, postulates (axioms), and theorems
b) First, Euclid introduced undefined terms. Second, he introduced certain definitions. Third, he stated primitive propositions called postulates about the undefined terms and definitions. Fourth, he proved, using deductive reasoning, other propositions called theorems.

3. Two lines that do not lie in the same plane and do not intersect are called skew lines.

5. Two angles in the same plane are adjacent angles when they have a common vertex and a common side but no common interior points.

7. Two angles, the sum of whose measure is 180°, are called supplementary angles.

9. An angle whose measure is greater than 90° but less than 180° is an obtuse angle.

11. An angle whose measure is less than 90° is an acute angle.

13. Ray, $\overrightarrow{BA}$ **15.** Half line, $\overset{\circ\!\rightarrow}{BA}$ **17.** Ray, $\overrightarrow{AB}$
19. Half open line segment, $\overset{\circ\!-\!\bullet}{AB}$ **21.** $\overrightarrow{BG}$ **23.** $\overset{\bullet\!-\!\circ}{BD}$
25. $\{B, F\}$ **27.** $\{C\}$ **29.** $\measuredangle CFG$ **31.** $\overleftrightarrow{BC}$ **33.** $\{B\}$
35. $\overrightarrow{BC}$ **37.** $\measuredangle ABE$ **39.** $\overrightarrow{BF}$ **41.** $\overset{\circ\!-\!\circ}{AC}$ **43.** $\overset{\bullet\!-\!\circ}{BE}$
45. Obtuse **47.** Straight **49.** Right
51. None of these angles **53.** 64° **55.** $57\frac{1}{4}°$ **57.** 25.3°
59. 91° **61.** 159.5° **63.** $136\frac{2}{7}°$ **65.** (b) **67.** (f)
69. (a) **71.** $m\measuredangle 1 = 49°, m\measuredangle 2 = 41°$ **73.** 134° and 46°
75. Angles 3, 4, and 7 each measure 125°; angles 1, 2, 5, and 6 each measure 55°.
77. Angles 2, 4, and 5 each measure 120°; angles 1, 3, 6, and 7 each measure 60°.
79. $m\measuredangle 1 = 64°, m\measuredangle 2 = 26°$
81. $m\measuredangle 1 = 33°, m\measuredangle 2 = 57°$
83. $m\measuredangle 1 = 134°, m\measuredangle 2 = 46°$
85. $m\measuredangle 1 = 29°, m\measuredangle 2 = 151°$
87. a) An infinite number **b)** An infinite number
89. An infinite number

For Exercises 91–97, each answer given is one of many possible answers.

91. Plane ABG and plane JCD **93.** $\overleftrightarrow{BG}$ and $\overleftrightarrow{DG}$
95. Plane $AGB \cap$ plane $ABC \cap$ plane $BCD = \{B\}$
97. $\overleftrightarrow{BC} \cap$ plane $ABG = \{B\}$
99. Always true. If any two lines are parallel to a third line, then they must be parallel to each other.
101. Sometimes true. Vertical angles are only complementary when each is equal to 45°.
103. Sometimes true. Alternate interior angles are only complementary when each is equal to 45°.
105. Answers will vary.
107. No. Line l and line n may be parallel or skew.
109. 360°

SECTION 9.2, PAGE 533

1. A polygon is a closed figure in a plane determined by three or more straight line segments.

3. The different types of triangles are acute, obtuse, right, isosceles, equilateral, and scalene. Descriptions will vary.

5. If the corresponding sides of two similar figures are the same length, the figures are congruent figures.

7. a) Triangle **b)** Regular

9. a) Octagon **b)** Regular

11. a) Rhombus **b)** Not regular

13. a) Octagon **b)** Not regular

15. a) Isosceles **b)** Acute

17. a) Isosceles **b)** Right

19. a) Scalene **b)** Acute

21. a) Scalene **b)** Right

23. Rectangle **25.** Square **27.** Rhombus

29. 17° **31.** 150°

33. $m\measuredangle 1 = 90°$, $m\measuredangle 2 = 50°$, $m\measuredangle 3 = 130°$, $m\measuredangle 4 = 50°$, $m\measuredangle 5 = 50°$, $m\measuredangle 6 = 40°$, $m\measuredangle 7 = 90°$, $m\measuredangle 8 = 130°$, $m\measuredangle 9 = 140°$, $m\measuredangle 10 = 40°$, $m\measuredangle 11 = 140°$, $m\measuredangle 12 = 40°$

35. 720° **37.** 1080° **39.** 3240°

41. a) 60° **b)** 120° **43. a)** 108° **b)** 72°

45. a) 144° **b)** 36° **47.** $x = 2, y = 1.6$

49. $x = 2\frac{2}{5}, y = 7\frac{1}{2}$ **51.** $x = 20, y = 21\frac{1}{4}$

53. 3 **55.** $3\frac{1}{3}$ **57.** 28 **59.** 30 **61.** 28° **63.** 9

65. 10 **67.** 70° **69.** 55° **71.** 35° **73.** 70 ft

75. a) ≈ 113.14 mi **b)** ≈ 75.43 mi

77. $D'E' = 4, E'F' = 5, D'F' = 3$

79. a) $m\measuredangle HMF = m\measuredangle TMB$, $m\measuredangle HFM = m\measuredangle TBM$, $m\measuredangle MHF = m\measuredangle MTB$
b) 44 ft

SECTION 9.3, PAGE 545

Throughout this section, we used the [π] *key on a scientific calculator to determine answers in calculations involving* π*. If you use 3.14 for* π*, your answers may vary slightly.*

1. a) Answers will vary.
b) Answers will vary.
c)

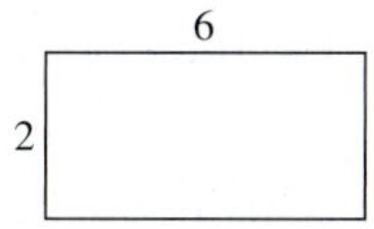

The area of this rectangle is 12 square units. The perimeter of this rectangle is 16 units.

3. a) To determine the number of square feet, multiply the number of square yards by 9.
b) To determine the number of square yards, divide the number of square feet by 9.

5. 17.5 in.^{2} **7.** 17.5 cm^2 **9. a)** 210 ft^2 **b)** 62 ft

11. a) 6000 cm^2 **b)** 654 cm

13. a) 288 in.^{2} **b)** 74 in.

15. a) 50.27 m^2 **b)** 25.13 m

17. a) 132.73 ft^2 **b)** 40.84 ft

19. a) 17 yd **b)** 40 yd **c)** 60 yd^2

21. a) 12 km **b)** 30 km **c)** 30 km^2

23. ≈ 21.46 m^2 **25.** 8 in.^{2} **27.** 90 yd^2

29. ≈ 65.73 in.^{2} **31.** ≈ 307.88 cm^2 **33.** 23 yd^2

35. 132.3 ft^2 **37.** 234,000 cm^2 **39.** 0.8625 m^2

41. a) $5973 **b)** $7623 **43.** $2890

45. $2908.80 **47.** $234.21 **49. a)** 288 ft **b)** 4700 tiles

51. 21 ft **53.** ≈ 312 ft

55. a) $A = s^2$ **b)** $A = 4s^2$ **c)** Four times larger

57. 24 cm^2 **59.** Answers will vary.

SECTION 9.4, PAGE 561

Throughout this section, we used the [π] *key on a scientific calculator to determine answers in calculations involving* π*. If you use 3.14 for* π*, your answers may vary slightly.*

1. a) Volume is a measure of the capacity of a figure.
b) Surface area is the sum of the areas of the surfaces of a three-dimensional figure.

3. A polyhedron is a closed surface formed by the union of polygonal regions. A regular polyhedron is one whose faces are all regular polygons of the same size and shape.

5. Answers will vary. **7. a)** 8 m^3 **b)** 28 m^2

9. a) 125 ft^3 **b)** 150 ft^2

11. a) 150.80 in.^{3} **b)** 175.93 in.^{2}

13. a) 131.95 cm^3 **b)** 163.22 cm^2

15. a) 381.70 cm^3 **b)** 254.47 cm^2

17. 384 in.^{3} **19.** 524.33 cm^3 **21.** 324 mm^3 **23.** 392 ft^3

25. 284.46 cm^3 **27.** 24 ft^3 **29.** 243 ft^3 **31.** ≈ 5.67 yd^3

33. 3,700,000 cm^3 **35.** 7.5 m^3

37. Tubs ≈ 141.37 in.^{3}; boxes = 125 in.^{3}

39. 40,840 mm^2 **41.** ≈ 2.50 qt

43. a) 120,000 cm^3 **b)** 120,000 mℓ **c)** 120 ℓ

45. ≈ 283.04 in.^{3}

47. a) ≈ 323.98 in.^{3} **b)** ≈ 0.19 ft^3

49. ≈ 14.14 in.^{3} **51.** Ten edges

53. Six vertices **55.** Fourteen edges

57. a) ≈ 5.11×10^8 km^2 **b)** ≈ 3.79×10^7 km^2
c) ≈ 13 times larger **d)** ≈ 1.09×10^{12} km^3
e) ≈ 2.20×10^{10} km^3 **f)** ≈ 50 times larger

59. a)–e) Answers will vary.
f) If we double the length of each edge of a cube, the new volume will be eight times the original volume.

61. a) Answers will vary.

b) $V_1 = a^3; V_2 = a^2b; V_3 = a^2b; V_4 = ab^2;$
$V_5 = a^2b; V_6 = ab^2; V_7 = b^3$

c) ab^2

SECTION 9.5, PAGE 579

1. The act of moving a geometric figure from some starting position to some ending position without altering its shape or size is called rigid motion. The four main rigid motions studied in this section are reflections, translations, rotations, and glide reflections.

3. A reflection is a rigid motion that moves a figure to a new position that is a mirror image of the figure in the starting position.

5. A rotation is a rigid motion performed by rotating a figure in the plane about a specific point.

7. A translation is a rigid motion that moves a figure by sliding it along a straight line segment in the plane.

9. A glide reflection is a rigid motion formed by performing a translation (or glide) followed by a reflection.

11. A geometric figure is said to have reflective symmetry if the positions of a figure before and after a reflection are identical (except for vertex labels).

13. A tessellation is a pattern consisting of the repeated use of the same geometric figures to entirely cover a plane, leaving no gaps.

This figure contains the answers for Exercises 15 and 16.

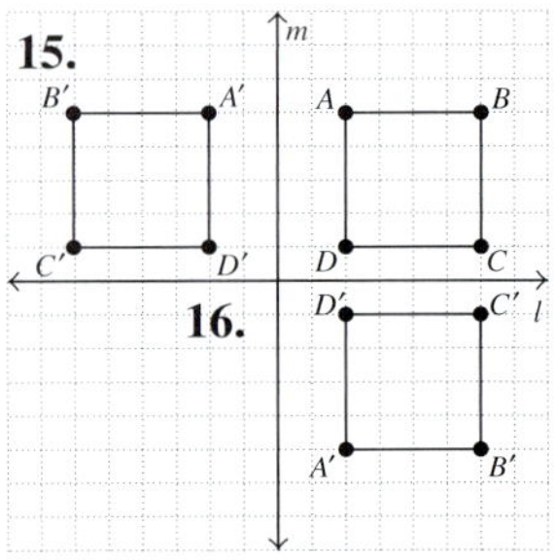

This figure contains the answers for Exercises 17 and 18.

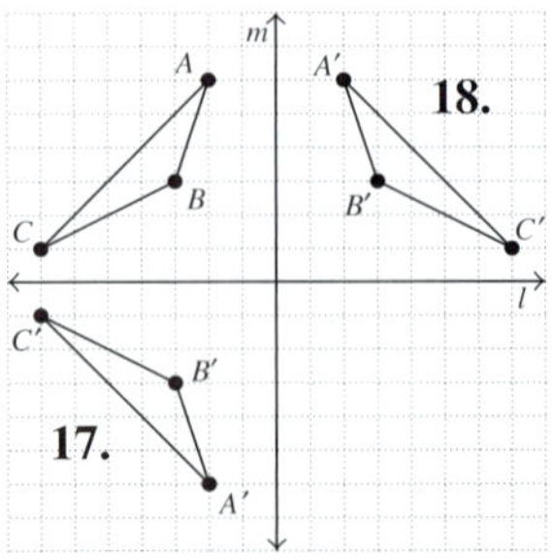

This figure contains the answers for Exercises 19 and 20.

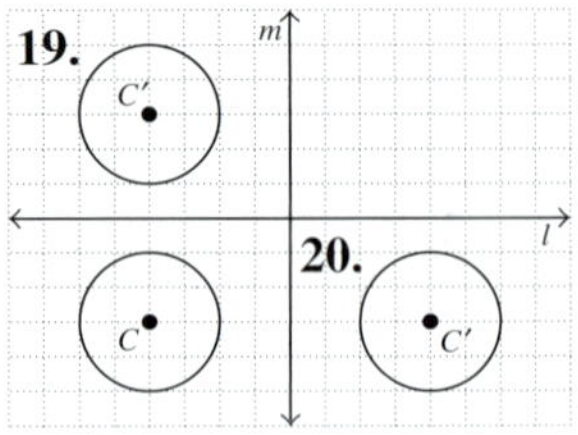

This figure contains the answers for Exercises 21 and 22.

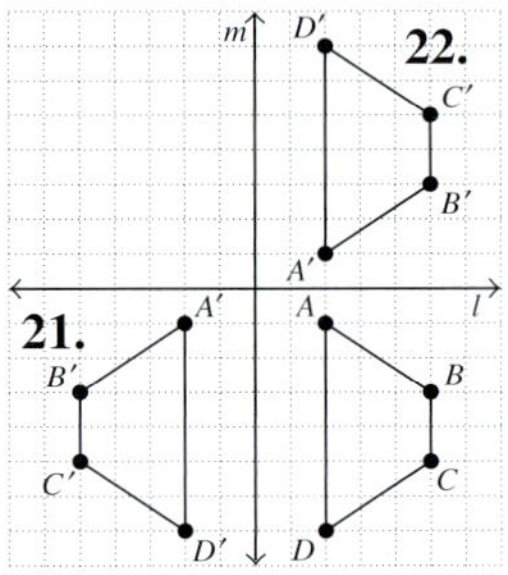

This figure contains the answers to Exercises 23 and 24.

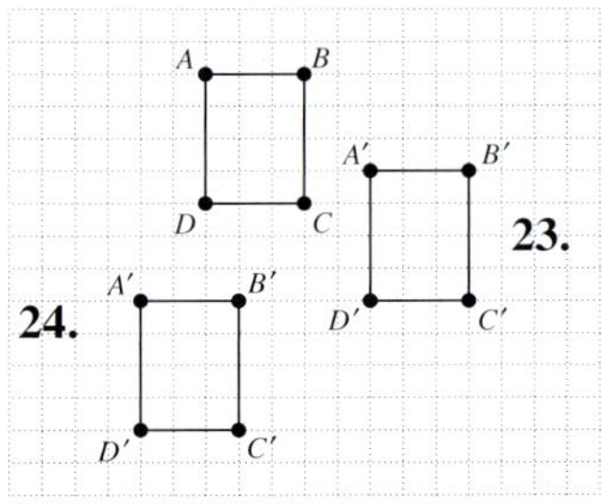

This figure contains the answers to Exercises 25 and 26.

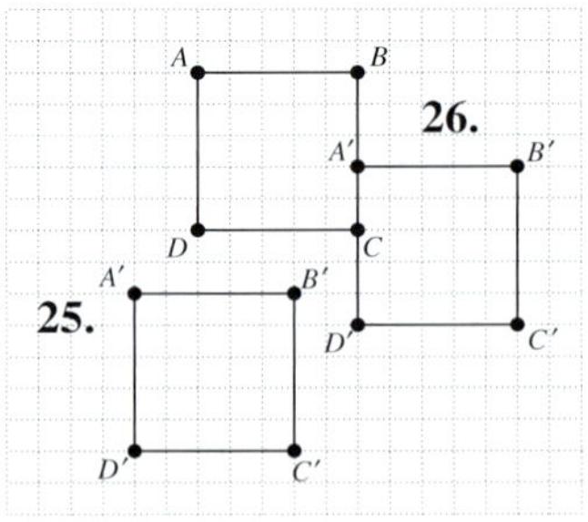

This figure contains the answers to Exercises 27 and 28.

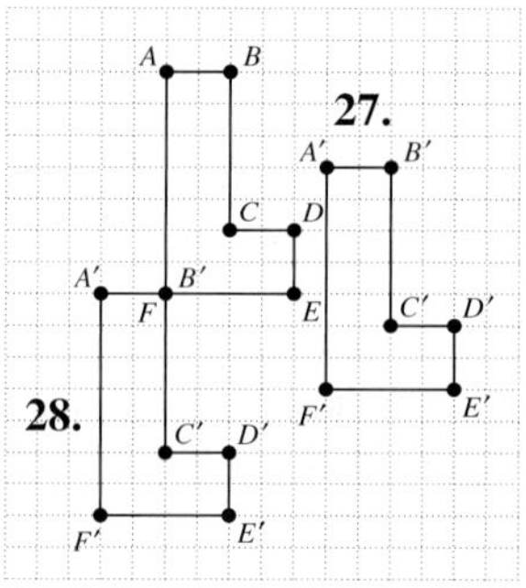

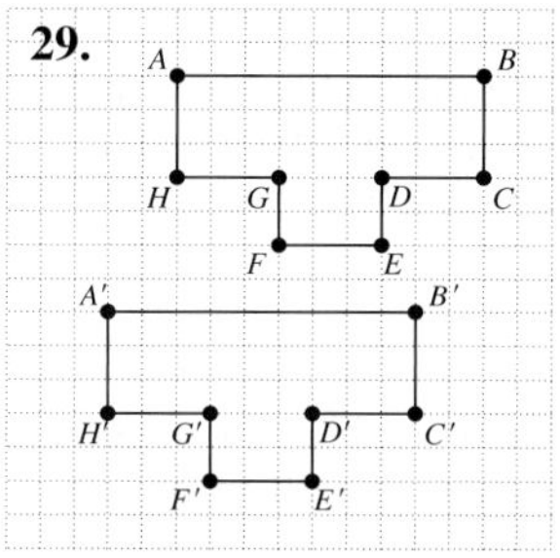

This figure contains the answers to Exercises 31 and 32.

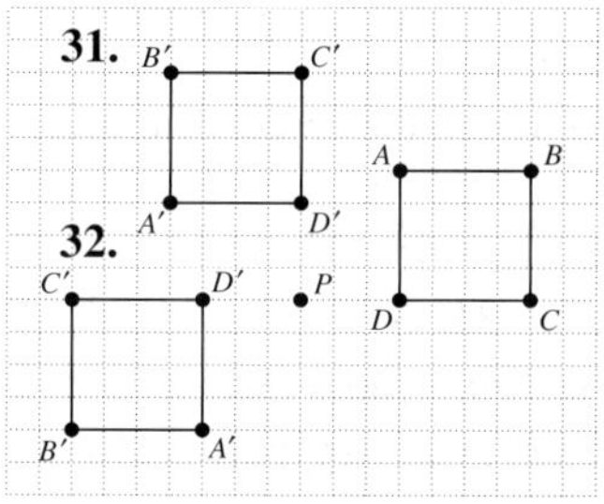

This figure contains the answers to Exercises 33 and 34.

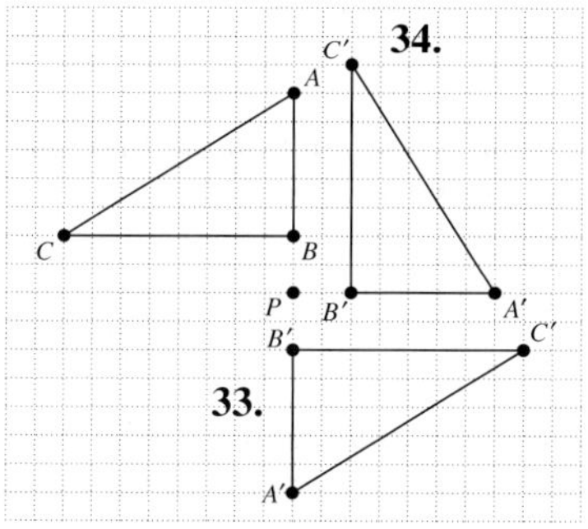

This figure contains the answers to Exercises 35 and 36.

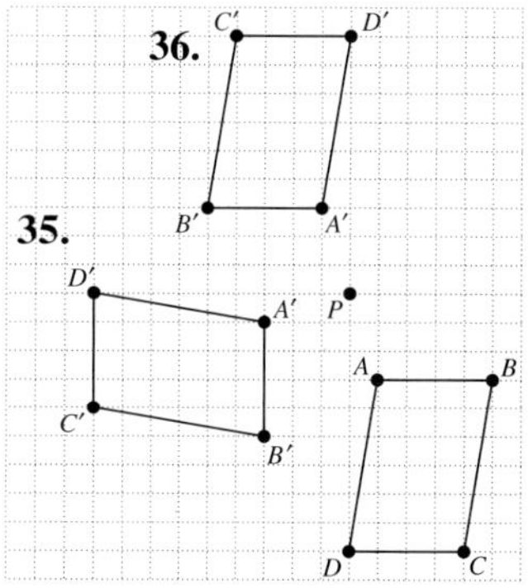

This figure contains the answers to Exercises 37 and 38.

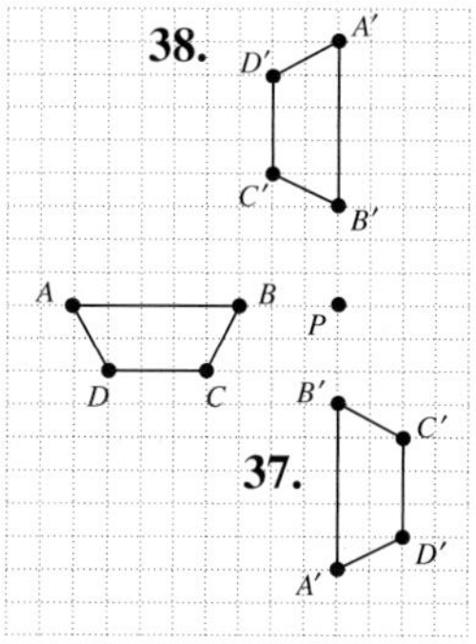

This figure contains the answers to Exercises 39 and 40.

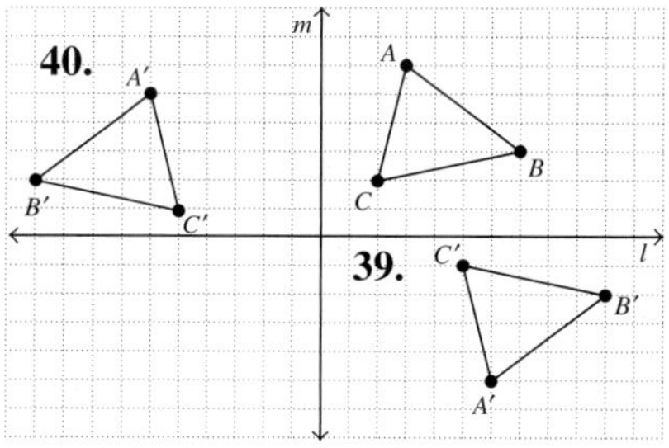

This figure contains the answers to Exercises 41 and 42.

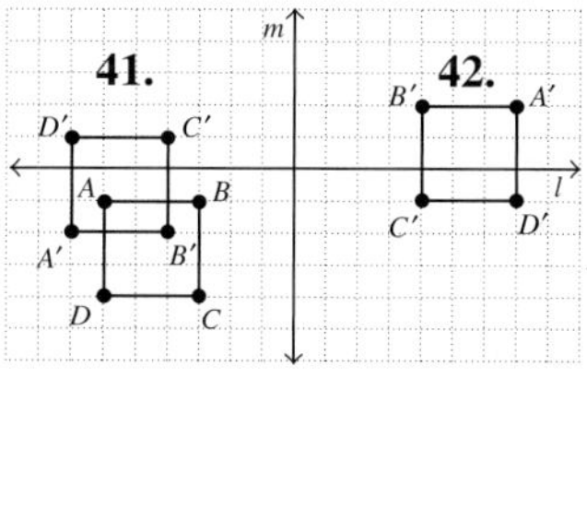

41. Yes **42.** No **43.** No **44.** Yes

45. a)–d) Answers will vary. **46.** Answers will vary.

47. Outside

48. Euclidean: Given a line and a point not on the line, one and only one line can be drawn parallel to the given line through the given point. Elliptical: Given a line and a point not on the line, no line can be drawn through the given point parallel to the given line. Hyperbolic: Given a line and a point not on the line, two or more lines can be drawn through the given point parallel to the given line.

49.

50.

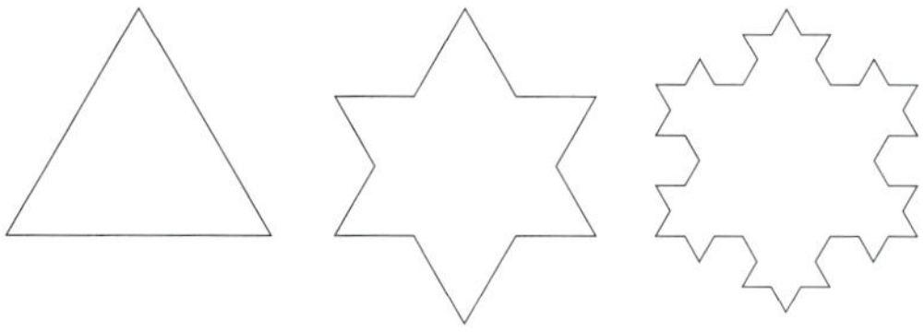

CHAPTER TEST, PAGE 607

1. $\overleftrightarrow{EF}$ **2.** ΔBCD **3.** $\{D\}$ **4.** $\overleftrightarrow{AC}$ **5.** 77.6°
6. 128.3° **7.** 64° **8.** 1440° **9.** $\approx$ 2.69 cm

10. a) 12 in. **b)** 30 in. **c)** 30 in.^{2}

11. a) 1436.76 cm^3 **b)** 615.75 cm^2

12. 59.43 m^3 **13.** 112 ft^3

14.

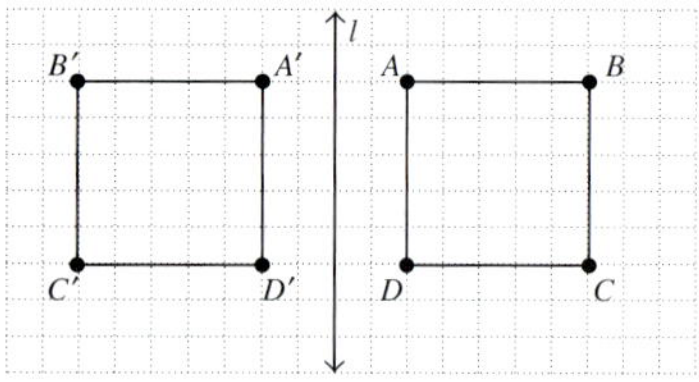

15.

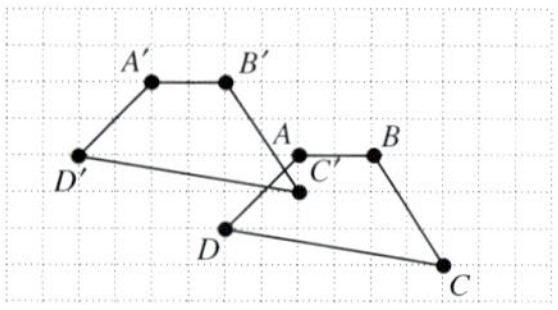

16.

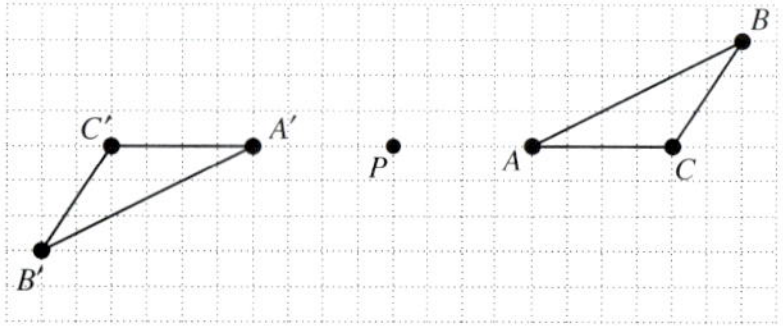

17.

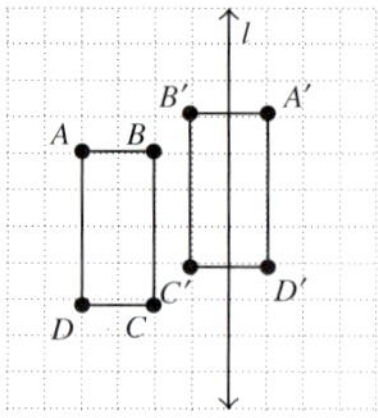

18. a) No **b)** Yes

19. A surface with one side and one edge

20. a)–b) Answers will vary.

21. Euclidean: Given a line and a point not on the line, one and only one line can be drawn parallel to the given line through the given point. Elliptical: Given a line and a point not on the line, no line can be drawn through the given point parallel to the given line. Hyperbolic: Given a line and a point not on the line, two or more lines can be drawn through the given point parallel to the given line.

CHAPTER 10

SECTION 10.1, PAGE 617

1. A binary operation is an operation, or rule, that can be performed on two and only two elements of a set. The result is a single element.

3. a) When we add two numbers, the sum is one number: $4 + 5 = 9$.
b) When we subtract two numbers, the difference is one number: $5 - 4 = 1$.
c) When we multiply two numbers, the product is one number: $5 \times 4 = 20$.
d) When we divide two numbers, the quotient is one number: $20 \div 5 = 4$.

5. A mathematical system is a commutative group if all five of the following conditions hold.
1. The set of elements is closed under the given operation.
2. An identity element exists for the set.
3. Every element in the set has an inverse.
4. The set of elements is associative under the given operation.
5. The set of elements is commutative under the given operation.

7. An identity element is an element in a set such that when a binary operation is performed on it and any given element in the set, the result is the given element. For the set of integers, the additive identity element is 0 and the multiplicative identity element is 1. Examples: $5 + 0 = 5, 5 \times 1 = 5$.

9. When a binary operation is performed on two elements in a set and the result is the identity element for the binary operation, each element is said to be the *inverse* of the other. For the set of rational numbers, the additive inverse of 2 is -2 since $2 + (-2) = 0$ and the multiplicative inverse of 2 is $\frac{1}{2}$ since $2 \times \frac{1}{2} = 1$.

11. Yes; for a group the commutative property need not apply.

13. (d); the commutative property need not apply.

15. $a + b = b + a$ for any elements a and b; $3 + 4 = 4 + 3$

17. $(a \cdot b) \cdot c = a \cdot (b \cdot c)$ for any elements a, b, and c; $(2 \cdot 3) \cdot 4 = 2 \cdot (3 \cdot 4)$

19. $4 \div 2 \neq 2 \div 4$

21. $(6 - 4) - 1 \neq 6 - (4 - 1)$
$2 - 1 \neq 6 - 3$
$1 \neq 3$

23. Yes; it satisfies the five properties needed.

25. No; there is no identity element.

27. No; the system is not closed.

29. No; there is no identity element.

31. No; not all elements have inverses.

33. Yes; it satisfies the four properties needed.

35. No; the system is not closed. For example, $\frac{1}{0}$ is undefined.

37. No; it does not have an identity element and does not satisfy the associative property.

39. Yes

41. No; the system is not closed. For example, $\sqrt{2} + (-\sqrt{2}) = 0$, which is rational. There is also no identity element.

43. Answers will vary.

SECTION 10.2, PAGE 625

1. Numbers are obtained by starting at the first addend (on a clock face) and then moving clockwise the number of hours equal to the second addend.

3. a) Add $2 + 11$ to get 1, then add $1 + 5$. **b)** 6

5. a) Add 12 to 2 to get $14 - 6$. **b)** 8
c) Since 12 is the identity element, you can add 12 to any number without changing its value.

7. Yes, 12

9. Yes; 1:11, 2:10, 3:9, 4:8, 5:7, 6:6, 7:5, 8:4, 9:3, 10:2, 11:1, and 12:12

11. Yes, $6 + 9 = 9 + 6$ since both equal 3

13. a) 7 **b)** 5. The additive inverse of 2 is 5 since $2 + 5 = 7$.

15. No; the elements are not symmetric about the main diagonal.

17. Yes; C is the identity element since the row next to C is identical to the top row, and the column under C is identical to the left-hand column.

19. a) The inverse of A is B since A ☮ B = B ☮ A = C.
b) The inverse of B is A since B ☮ A = A ☮ B = C.
c) The inverse of C is C since C ☮ C = C.

21. 7 **23.** 4 **25.** 4 **27.** 8 **29.** 6 **31.** 6

33. 5 **35.** 11 **37.** 3 **39.** 1 **41.** 12 **43.** 12

45.

+	1	2	3	4	5	6
1	2	3	4	5	6	1
2	3	4	5	6	1	2
3	4	5	6	1	2	3
4	5	6	1	2	3	4
5	6	1	2	3	4	5
6	1	2	3	4	5	6

47. 3 **49.** 3 **51.** 2 **53.** 4

55.

+	1	2	3	4	5	6	7
1	2	3	4	5	6	7	1
2	3	4	5	6	7	1	2
3	4	5	6	7	1	2	3
4	5	6	7	1	2	3	4
5	6	7	1	2	3	4	5
6	7	1	2	3	4	5	6
7	1	2	3	4	5	6	7

57. 4 **59.** 6 **61.** 4 **63.** 7

65. Yes; it satisfies the five required properties.

67. a) {0, 2, 4, 6} **b)** £ **c)** Yes
d) Yes; 0 **e)** Yes; 0–0, 2–6, 4–4, 6–2
f) (2 £ 4) £ 6 = 2 £ (4 £ 6)
g) Yes; 2 £ 6 = 6 £ 2 **h)** Yes

69. a) $\{*, 5, L\}$
b) 
c) Yes
d) Yes, L
e) Yes, ∗–5, 5–∗, L–L
f) (∗  5)  5 = ∗  (5  5)
g) Yes, L  ∗ = ∗  L
h) Yes

71. a) $\{f, r, o, m\}$ **b)**  **c)** Closed
d) m **e)** f **f)** f **g)** m **h)** r

73. Not closed

75. No inverse for ⊙ or for ∗, not associative

77. No identity element, no inverses, not associative, not commutative

79. a)

+	E	O
E	E	O
O	O	E

b) Yes, it is a commutative group; it satisfies the five properties.

81. Answers will vary.

83. a) Is closed; identity element is 6; inverses: 1–5, 2–2, 3–3, 4–4, 5–1, 6–6; is associative; for example,

$$(2 \infty 5) \infty 3 = 2 \infty (5 \infty 3)$$
$$3 \infty 3 = 2 \infty 2$$
$$6 = 6$$

b) $3 \infty 1 \neq 1 \infty 3$
$2 \neq 4$

85. a)

*	*R*	*S*	*T*	*U*	*V*	*I*
R	*V*	*T*	*U*	*S*	*I*	*R*
S	*U*	*I*	*V*	*R*	*T*	*S*
T	*S*	*R*	*I*	*V*	*U*	*T*
U	*T*	*V*	*R*	*I*	*S*	*U*
V	*I*	*U*	*S*	*T*	*R*	*V*
I	*R*	*S*	*T*	*U*	*V*	*I*

b) Yes. It is closed; identity element is *I*; inverses $R-V$, $S-S$, $T-T$, $U-U$, $V-R$, $I-I$; and the associative property will hold.

c) No, it is not commutative. For example, $R * S \neq S * R$

87.

+	0	1	2	3	4
0	0	1	2	3	4
1	1	2	3	4	0
2	2	3	4	0	1
3	3	4	0	1	2
4	4	0	1	2	3

89. Add the number in the top row and the number in the left-hand column and divide the sum by 4. The remainder is placed in the table.

SECTION 10.3, PAGE 638

1. A modulo *m* system consists of *m* elements, 0 through $m - 1$, and a binary operation.

3. 3;

0	1	2
0	1	2
3	4	5
6	7	8
.	.	.
.	.	.

5. 16 classes **7.** (b), (c) or (d) **9.** Sunday **11.** Friday **13.** Saturday **15.** Friday **17.** February **19.** May **21.** September **23.** July **25.** 1 **27.** 2 **29.** 3 **31.** 2 **33.** 1 **35.** 2 **37.** 0 **39.** 0 **41.** 5 **43.** 2 **45.** 6 **47.** 9 **49.** 5 **51.** 5 **53.** 3 **55.** 5 **57.** 5 **59.** { } **61.** 1 and 6 **63.** 5 **65.** 0

67. a) 2020, 2024, 2028, 2032, 2036 **b)** 3004
c) 2552, 2556, 2560, 2564, 2568, 2572

69. a) Resting (for the second of two days)
b) Resting (for the second of two days)
c) Morning and afternoon lessons
d) No

71. a) 5 **b)** No **c)** 54 weeks from this week

73. a) Evening **b)** Day **c)** Day

75. a)

+	0	1	2
0	0	1	2
1	1	2	0
2	2	0	1

b) Yes **c)** Yes, 0 **d)** Yes; 0–0, 1–2, 2–1
e) $(1 + 2) + 2 = 1 + (2 + 2)$
f) Yes; $2 + 1 = 1 + 2$ **g)** Yes **h)** Yes

77. a)

×	0	1	2	3
0	0	0	0	0
1	0	1	2	3
2	0	2	0	2
3	0	3	2	1

b) Yes **c)** Yes, 1
d) No; no inverse for 0 or for 2, inverse of 1 is 1, inverse of 3 is 3
e) $(1 \times 2) \times 3 = 1 \times (2 \times 3)$
f) Yes, $2 \times 3 = 3 \times 2$ **g)** No

79. 7 **81.** 1, 2, 3 **83.** 0 **85.** 2

87. a) Tuesday **b)** Wednesday **89.** 0

REVIEW EXERCISES, PAGE 641

1. A mathematical system consists of a set of elements and at least one binary operation.

2. A binary operation is an operation that can be performed on two and only two elements of a set. The result is a single element.

3. No; for example, $2 - 5 = -3$, and -3 is not a whole number.

4. Yes; the difference of any two real numbers is a real number.

5. 3 **6.** 5 **7.** 9 **8.** 8 **9.** 9 **10.** 11

11. Closure, identity element, inverses, and associative property

12. A commutative group **13.** Yes

14. No; no inverse for any integer except 1 and -1

15. No; no identity element **16.** No; no inverse for 0

17. No identity element; no inverses

18. Not every element has an inverse; not associative. For example, $(P ? P) ? 4 \neq P ? (P ? 4)$.

19. Not associative. For example, $(! \square p) \square ? \neq ! \square (p \square ?)$

20. a) {☺, ●, ♀, ♂} **b)** ⌂ **c)** Yes **d)** Yes; ☺
e) Yes; ☺ – ☺, ● – ♂, ♀ – ♀, ♂ – ●
f) (● ⌂ ♀) ⌂ ♂ = ● ⌂ (♀ ⌂ ♂)
g) Yes; ● ⌂ ♂ = ♂ ⌂ ● **h)** Yes

21. 0 **22.** 5 **23.** 1 **24.** 3 **25.** 11 **26.** 2 **27.** 4
28. 12 **29.** 9 **30.** 11 **31.** 1 **32.** 1 **33.** 6 **34.** 1
35. 0, 2, 4, 6 **36.** 4 **37.** 5 **38.** 9 **39.** 7 **40.** 8

41.

+	0	1	2	3	4	5
0	0	1	2	3	4	5
1	1	2	3	4	5	0
2	2	3	4	5	0	1
3	3	4	5	0	1	2
4	4	5	0	1	2	3
5	5	0	1	2	3	4

Yes, it is a commutative group.

42.

×	0	1	2	3
0	0	0	0	0
1	0	1	2	3
2	0	2	0	2
3	0	3	2	1

No; no inverse for 0 or 2

43. a) Yes **b)** Yes

CHAPTER TEST, PAGE 643

1. A set of elements and a binary operation

2. Closure, identity element, inverses, associative property, commutative property

3. Yes

4.

+	1	2	3	4	5
1	2	3	4	5	1
2	3	4	5	1	2
3	4	5	1	2	3
4	5	1	2	3	4
5	1	2	3	4	5

5. Yes, it is a commutative group.

6. 3 **7.** 2

8. a) □ **b)** Yes **c)** Yes, T **d)** S **e)** S

9. No, not closed.

10. Yes, it is a commutative group.

11. Yes, it is a commutative group.

12. 2 **13.** 0 **14.** 6 **15.** 2 **16.** 5

17. 2 **18.** { } **19.** 7

20. a)

×	0	1	2	3	4
0	0	0	0	0	0
1	0	1	2	3	4
2	0	2	4	1	3
3	0	3	1	4	2
4	0	4	3	2	1

b) No; no inverse for 0

CHAPTER 11

SECTION 11.1, PAGE 653

1. A percent is a ratio of some number to 100.

3. Divide the numerator by the denominator, multiply the quotient by 100, and add a percent sign.

5. $$\text{Percent change} = \frac{\left(\begin{array}{c}\text{amount in}\\\text{latest period}\end{array}\right) - \left(\begin{array}{c}\text{amount in}\\\text{previous period}\end{array}\right)}{\text{amount in the previous period}} \times 100$$

7. 75.0% **9.** 55.0% **11.** 0.8% **13.** 378.0%

15. 0.05 **17.** 0.0515 **19.** 0.0025 **21.** 0.002

23. 0.01 **25.** 25% **27.** 4.7 grams **29.** $208,341.12

31. $651,066.00 **33.** $1550 **35.** $2100 **37.** 41.3%

39. 10.0% **41.** 0.9% decrease

43. a) 98.8% increase **b)** 224.9% increase **c)** 145.9% increase **d)** 58.5% increase

45. a) 78.3% increase **b)** 27.4% decrease **c)** 28.5% increase **d)** 66.2% increase

47. $6.75 **49.** 25% **51.** 300

53. a) $2.61 **b)** $46.11 **c)** $6.92 **d)** $53.03

55. 12 students **57.** $39,055 **59.** 5.3% decrease

61. 12.5% increase **63.** 18.6% decrease **65.** $3750

67. He will have a loss of $10. **69.** $21.95

SECTION 11.2, PAGE 665

1. Interest is the money the borrower pays for the use of the lender's money.

3. Security or collateral is anything of value pledged by the borrower that the lender may sell or keep if the borrower does not repay the loan.

5. i is the *interest*, p is the *principal*, r is the interest *rate* expressed as a percent, and t is the *time*.

7. The United States rule states that if a partial payment is made on a loan, interest is computed on the principal from the first day of the loan until the date of the partial payment.

9. $15 **11.** $24.06 **13.** $15.85 **15.** $80.06

17. $113.20 **19.** 5% **21.** $600 **23.** 2 years **25.** $6630

27. a) $131.25 **b)** $3631.25

29. a) $182.50 **b)** $3467.50 **c)** ≈ 7.9%

31. $23,793.75 **33.** 284 days **35.** 272 days **37.** 264 days

39. June 13 **41.** March 24 **43.** $2017.03 **45.** $1459.33

47. $5278.99 **49.** $850.64 **51.** $6086.82 **53.** $2646.24

55. a) November 3, 2007 **b)** $978.06 **c)** $21.94 **d)** ≈ 4.44%

57. a) ≈ 409.0% **b)** ≈ 204.5% **c)** ≈ 102.3%

59. a) 6.663% **b)** ≈ 7.139% **c)** $6663 **d)** $6996.15

SECTION 11.3, PAGE 675

1. An investment is the use of money or capital for income or profit.

3. A variable investment is one in which neither the principal nor the interest is guaranteed.

5. a) The effective annual yield is the simple interest rate that gives the same amount of interest as a compound rate over the same period of time.
b) Another name for effective annual yield is annual percentage yield.

In the remainder of this exercise set, your answers may vary in the last digit, depending on the calculator used.

7. $A = \$1216.65$; $i = \$216.65$

9. $A = \$2253.98$; $i = \$253.98$

11. $A = \$8252.64$; $i = \$1252.64$

13. $A = \$8666.26$; $i = \$666.26$

15. $30,695.66 **17.** $85,282.13 **19.** $54,142.84

21. $1653.36 **23.** $2341.82

25. a) $4195.14 **b)** $4214.36

27. $3106.62 **29.** $7609.45

31. ≈ 3.53% **33.** Yes, the APY should be 2.43%.

35. He will earn more interest in the account that pays the 5% simple interest

37. a) $555,000 **b)** $58,907.61 **c)** $98.51

39. $212,687.10 **41.** $12,015.94

43. a) $1040.60, $40.60 **b)** $1082.43, $82.43 **c)** $1169.86, $169.86 **d)** No

45. $1.53

47. a) 24 years **b)** 12 years **c)** 9 years **d)** 6 years **e)** 3.27%

49. The simple interest is $20,000. The compound interest is $22,138.58. The compound interest is greater by $2,138.58.

SECTION 11.4, PAGE 688

1. An open-end installment loan is one with which you can make different payments each month. A fixed installment loan is one in which you pay a fixed amount each month for a set number of months.

3. The APR is the true rate of interest charged on a loan.

5. The total installment price is the sum of all the monthly payments and the down payment, if any.

7. The unpaid balance method and the average daily balance method

9. a) $1719.55 **b)** $170.33

11. a) $1834.57 **b)** $223.85

13. a) $628.40 **b)** 9.0%

15. a) $1752 **b)** 9%

17. a) 6% **b)** $726.00 **c)** $7858.00

19. a) $2818.20 **b)** $689.39 **c)** $347.90 **d)** $8614.17

21. a) $26 **b)** $46.01, which rounds up to $47

23. a) $21.14, which rounds up to $22
b) $34.39, which rounds up to $35

25. a) $19.76 **b)** $743.41

27. a) $1.56 **b)** $133.11

29. a) $512.00 **b)** $6.66 **c)** $638.43

31. a) $121.78 **b)** $1.52 **c)** $133.07
d) The interest charged using the average daily balance method is $0.04 less than the interest charged using the unpaid balance method.

33. a) $11.96 **b)** $886.96

35. a) $25 **b)** $35.60 **c)** 8.5% **d)** 6.5%

37. $14,077.97

39. a) $6872.25 **b)** $610.37 **c)** $2637.42 **d)** $2501.05

41. $201.48

SECTION 11.5, PAGE 702

1. A mortgage is a long-term loan in which the property is pledged as security for payment of the difference between the down payment and the sale price.

3. The major difference is that the interest rate for a conventional loan is fixed for the duration of the loan, whereas the interest rate for a variable-rate loan may change every period, as specified in the loan agreement.

5. A buyer's adjusted monthly income is found by subtracting any fixed monthly payments with more than 10 months remaining from the gross monthly income.

7. An amortization schedule lists payment dates and payment numbers. For each payment, it lists the amount that goes to pay the interest and the principal. It also gives the balance remaining on the loan after each payment.

9. Equity is the difference between the appraised value of your home and the loan balance.

11. a) $70,000 **b)** $2438.80

13. a) $262,500 **b)** $9579.50

15. a) $39,000 **b)** $156,000 **c)** $3120

17. a) $802.20 **b)** $1411.50 **c)** No

19. a) $187,736.40 **b)** $112,736.40 **c)** $38.68

21. a) $17,025 **b)** $2894.25 **c)** $1212.40 **d)** $930.98 **e)** $1057.65 **f)** Yes **g)** $127.02

23. Bank B

25. a) $664,704 **b)** $927,360 **c)** $1,231,488

27. a) 805

b)

Payment Number	Interest	Principal	Balance of Loan
1	$750.00	$55.00	$99,945.00
2	$749.59	$55.41	$99,889.59
3	$749.17	$55.83	$99,833.76

c) 9.38%

d)

Payment Number	Interest	Principal	Balance of Loan
4	$780.37	$24.63	$99,809.13
5	$780.17	$24.83	$99,784.30
6	$779.98	$25.02	$99,759.28

e) 9.46%

29. a) The variable-rate loan **b)** $2160

SECTION 11.6, PAGE 714

1. An annuity is an account into which, or out of which, a sequence of scheduled payments is made.

3. a) An annuity into which equal payments are made at regular intervals, with interest compounded at the end of each interval and with a fixed interest rate for each compounding period, is called an ordinary annuity.
b) A sinking fund is a type of annuity in which the goal is to save a specific amount of money in a specified amount of time.

5. An immediate annuity is an annuity that is established with a lump sum of money for the purpose of providing the investor with regular, usually monthly, payments for the rest of the investor's life.

7. With a 401k plan, the money invested is not subject to taxes; however, when money is withdrawn, it is subject to income taxes. With a Roth 401k plan, the investor has already paid taxes on the money invested, and money withdrawn is not subject to income taxes.

9. $199,316.54 **11.** $299,929.32 **13.** $493.00
15. $175.48 **17.** $37,078.18 **19.** $56,559.36
21. $331.46 **23.** $5160.71
25. a) $23,003.87 **b)** $826,980.88 **c)** $349,496.41 **d)** $12,000 **e)** $36,000 **f)** Alberto

REVIEW EXERCISES, PAGE 717

1. 75.0% **2.** 83.3% **3.** 62.5% **4.** 4.1%
5. 0.98% ≈ 1.0% **6.** 314.1% **7.** 0.09 **8.** 0.141
9. 1.23 **10.** 0.0025 **11.** $0.008\overline{3}$ **12.** 0.0000045
13. ≈17.6% **14.** ≈11.0% **15.** 18.75% **16.** 80
17. 91.8 **18.** $6.42 **19.** 40 people **20.** 26.7%
21. $10 **22.** 5% **23.** $450 **24.** 0.5 year
25. $6214.25 **26. a)** $162 **b)** $3162
27. a) $1380 **b)** $4620 **c)** ≈14.9%
28. a) $7\frac{1}{2}\%$ **b)** $830 **c)** at least $941.18
29. a) $6691.13; $1691.13 **b)** $6719.58; $1719.58 **c)** $6734.28; $1734.28 **d)** $6744.25; $1744.25 **e)** $6749.13; $1749.13
30. $5076.35 **31.** 5.76% **32.** $13,415.00
33. a) 6.0% **b)** 253.16 **c)** $4150.34
34. a) 6.5% **b)** $64.32 **c)** $1962.51
35. a) 4.5% **b)** $32.06 **c)** $1420.43
36. a) $6.31 **b)** $847.61 **c)** $508.99 **d)** $6.62 **e)** $847.92
37. a) $2.60 **b)** $546.92 **c)** $382.68 **d)** $5.36 **e)** $549.68
38. a) $14,400 **b)** $57,600 **c)** $8415.60 **d)** 5.5%
39. a) $10.44 **b)** 8.5%
40. a) $33,925 **b)** $1345.49 **c)** $855.93 **d)** $1172.60 **e)** Yes
41. a) $13,485 **b)** $756.51 **c)** $24.20 **d)** $285,828.60 **e)** $195,928.60
42. a) $550.46 **b)** 8% **c)** 7.75%
43. $48,378.57 **44.** $955.34

CHAPTER TEST, PAGE 720

1. a) $40 **b)** 3 years **2.** $819.38 **3.** $6519.38
4. $2523.20 **5.** $123.20
6. a) $7961.99, $461.99 **b)** $3036.68, $536.68
7. $1997.50 **8.** $181.46 **9.** 8.5%
10. a) 4.5% **b)** $64.02 **c)** $2836.28
11. a) $12.30 **b)** $1146.57 **c)** $765.67 **d)** $10.72 **e)** $1144.99
12. $21,675 **13.** $2456.50 **14.** $1848.93 **15.** $1123.85
16. $1428.02 **17.** Yes **18. a)** $428,717.50 **b)** $284,217.50
19. $231,020.45 **20.** $277.28

CHAPTER 12

SECTION 12.1, PAGE 729

1. An experiment is a controlled operation that yields a set of results.

3. Empirical probability is the relative frequency of occurrence of an event. It is determined by actual observation of an experiment.

$$P(E) = \frac{\text{number of times event } E \text{ has occurred}}{\text{number of times experiment was performed}}$$

5. Answers will vary.

7. No, it means that if a coin was flipped many times, about $\frac{1}{2}$ of the tosses would land heads up.

9. a) Roll a die many times and then find the relative frequency of 5's to the total number of rolls.
b) Answers will vary. **c)** Answers will vary.

11.–13. Answers will vary.

15. a) $\frac{7}{15}$ **b)** $\frac{1}{3}$ **c)** $\frac{1}{5}$ **17. a)** $\frac{3}{7}$ **b)** $\frac{8}{21}$ **c)** $\frac{1}{21}$

19. a) $\frac{4737}{129{,}098}$, or ≈ 0.0367 **b)** $\frac{21{,}922}{129{,}098}$, or ≈ 0.1698
c) $\frac{11{,}418}{129{,}098}$, or ≈ 0.0884

21. a) $\frac{29}{186}$ **b)** $\frac{1}{2}$ **c)** $\frac{2}{31}$

23. a) $\frac{11}{40}$ **b)** $\frac{9}{40}$ **c)** $\frac{1}{4}$ **d)** $\frac{7}{40}$ **e)** $\frac{3}{40}$

25. a) $\frac{6}{20} = \frac{3}{10}$ **b)** $\frac{14}{20} = \frac{7}{10}$ **c)** $\frac{14}{20} = \frac{7}{10}$
d) $\frac{2}{20} = \frac{1}{10}$

27. a) 0 **b)** $\frac{50}{250} = 0.2$ **c)** 1

29. a) $\frac{224}{929} \approx 0.2411$ **b)** $\frac{705}{929} \approx 0.7589$

31. Answers will vary.

SECTION 12.2, PAGE 737

1. If each outcome of an experiment has the same chance of occurring as any other outcome, they are said to be equally likely outcomes.

3. $P(A) + P(\text{not } A) = 1$

5. 0.3 **7.** $\frac{7}{12}$

9. Answers will vary. **11.** 0 and 1

13. a) $\frac{1}{5}$ **b)** $\frac{1}{4}$

15. $\frac{1}{40}$ **17.** $\frac{1}{13}$ **19.** $\frac{12}{13}$ **21.** $\frac{1}{2}$ **23.** 1 **25.** $\frac{4}{13}$

27. a) $\frac{1}{2}$ **b)** $\frac{1}{4}$ **c)** $\frac{1}{4}$ **d)** 0

29. a) $\frac{1}{3}$ **b)** $\frac{1}{6}$ **c)** $\frac{1}{3}$ **d)** $\frac{1}{6}$ **31.** $\frac{1}{5}$ **33.** $\frac{9}{10}$ **35.** $\frac{1}{12}$

37. $\frac{1}{6}$ **39.** $\frac{1}{2}$ **41.** $\frac{2}{3}$ **43.** $\frac{11}{17}$ **45.** $\frac{12}{17}$ **47.** $\frac{2}{9}$

49. $\frac{5}{9}$ **51.** 0 **53.** $\frac{3}{51}$ **55.** $\frac{27}{51}$ **57.** $\frac{1}{26}$ **59.** $\frac{5}{26}$

61. $\frac{17}{26}$ **63.** $\frac{83}{130}$ **65.** $\frac{11}{26}$ **67.** $\frac{6}{19}$ **69.** $\frac{15}{38}$ **71.** $\frac{5}{19}$

73. $\frac{13}{36}$ **75.** $\frac{1}{3}$ **77.** $\frac{23}{36}$ **79. a)** 0 **b)** 1

81. a) $\frac{1}{4}$ **b)** $\frac{1}{4}$ **c)** $\frac{1}{4}$ **83.** 29 dots

SECTION 12.3, PAGE 745

1. a) Answers will vary. **b)** Answers will vary.

3. Odds against

5. 7:2 **7. a)** $\frac{1}{2}$ **b)** $\frac{1}{2}$

9. a) $\frac{8}{15}$ **b)** $\frac{7}{15}$ **c)** 7:8 **d)** 8:7

11. 6:1, 1:6 **13.** 5:1 **15.** 4:2 or 2:1 **17.** 12:1, 1:12
19. 10:3, 3:10 **21.** 1:1 **23.** 5:3 **25. a)** 8:7 **b)** 7:8

27. 8:7 **29.** 7:8 **31.** 8:7 **33. a)** $\frac{3}{10}$ **b)** 7:3

35. 1:18 **37. a)** $\frac{7}{11}$ **b)** $\frac{4}{11}$ **39.** $\frac{11}{15}$ **41.** $\frac{1}{5}$ **43.** 1:4

45. 74:1 **47.** 0.34 **49.** 33:17 **51.** 43:57 **53.** 1:9

55. 7:1 **57. a)** $\frac{1}{33}$ **b)** 32:1

59. Horse 2, $\frac{2}{9}$; Horse 3, $\frac{1}{3}$; Horse 4, $\frac{1}{16}$; Horse 5, $\frac{5}{12}$; Horse 6, $\frac{1}{2}$

61. $\approx 97:3$

SECTION 12.4, PAGE 755

1. The expected value is the expected gain or loss of an experiment over the long run.

3. The fair price is the amount that should be charged for the game to be fair and result in an expectation of 0.

5. To obtain the fair price, add the cost to the expected value.

7. \$0.50. Since you would lose \$1.00 on average for each game you played, the price of the game should be \$1.00 less than the actual cost. Then the expectation would be \$0, and the game would be fair. The results could also be obtained from the fair price formula, fair price = expectation + cost to play.

9. −\$1.20 **11.** 10,080 people **13.** 70 points

15. 1.44 million viewers **17.** \$3840 **19.** \$1.60 off

21. a) $\approx -\$0.17$ **b)** $\approx \$0.17$

23. a) Yes, because you have a positive expectation of $\frac{1}{5}$
b) Yes, because you have a positive expectation of $\frac{1}{2}$

25. a) \$−1.20 **b)** \$0.80 **27. a)** \$−2.00 **b)** \$1.00

29. \$3.00 **31.** \$−1.25 **33.** \$800 **35.** \$400

37. a) \$5.50 **b)** \$7.50 **39. a)** \$2.25 **b)** \$4.25

41. a) \$2.50 **b)** \$12.50 **43. a)** −\$6.00 **b)** \$4.00

45. 0.75 base **47.** 2.9 points **49.** 56 employees

51. \$7200 **53.** 3.5 **55.** ≈141.51 service calls

57. a) $\frac{9}{16}, \frac{1}{4}, \frac{1}{8}, \frac{1}{16}$ **b)** \$11.81 **c)** \$11.81

59. An amount greater than \$1200

61. −\$0.053 or −5.3¢

63. a) \$458.33 **b)** \$308.33

SECTION 12.5, PAGE 766

1. If a first experiment can be performed in *M* distinct ways and a second experiment can be performed in *N* distinct ways, then the two experiments in that specific order can be performed in $M \cdot N$ distinct ways.

3. 14

5. The first selection is made. Then the second selection is made without the first selection being returned to the group of items being selected.

7. a) 2500 **b)** 2450

9. a) 343 **b)** 210

11. a) 4
b)

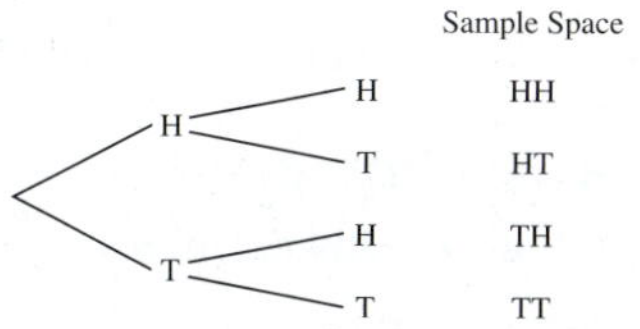

c) $\frac{1}{4}$ **d)** $\frac{1}{2}$ **e)** $\frac{1}{4}$

13. a) 9
b)

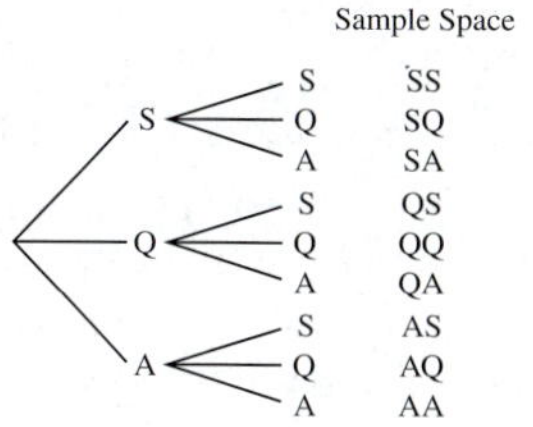

c) $\frac{1}{9}$ **d)** $\frac{1}{9}$ **e)** $\frac{5}{9}$

15. a) 12
b)

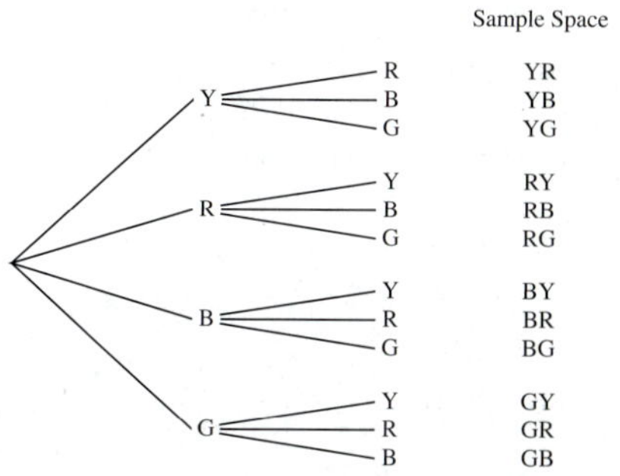

c) $\frac{1}{2}$ **d)** 1 **e)** $\frac{1}{2}$

17. a) 9
b)

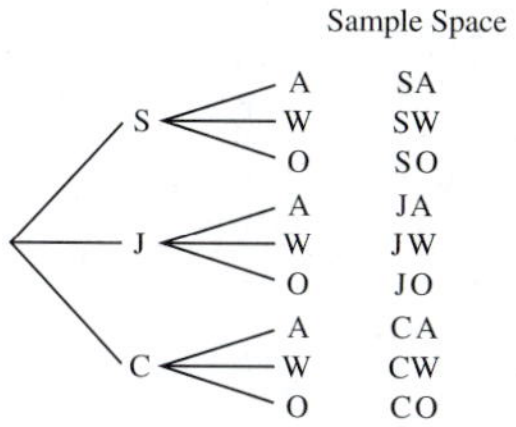

c) $\frac{1}{3}$ **d)** $\frac{1}{9}$ **e)** $\frac{2}{3}$

19. a) 36
b)

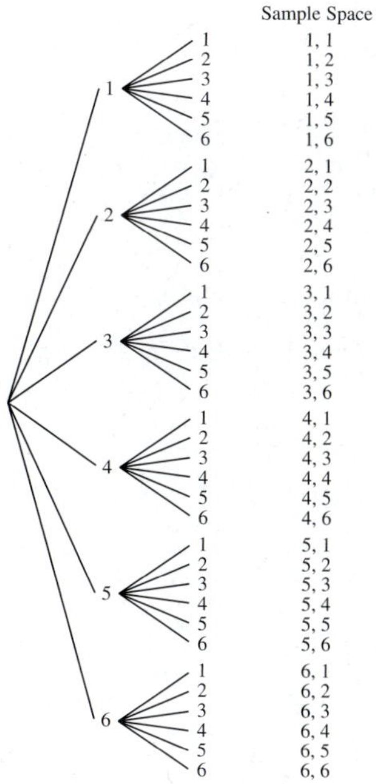

c) $\frac{1}{6}$ **d)** $\frac{1}{6}$ **e)** $\frac{1}{36}$ **f)** No

21. a) 6

b)

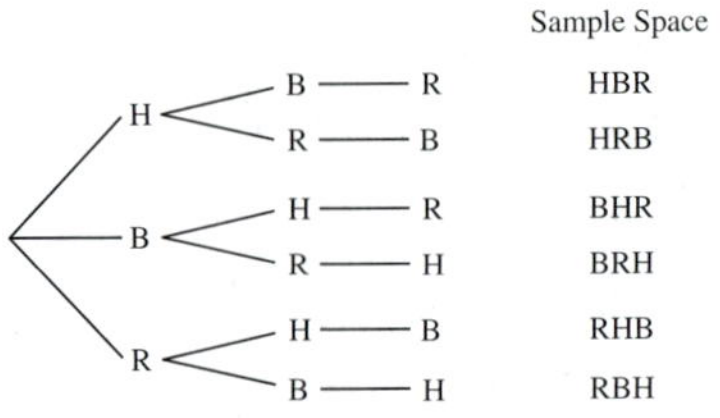

c) $\frac{1}{3}$ **d)** $\frac{1}{6}$ **e)** $\frac{1}{6}$

23. a) 18

b)

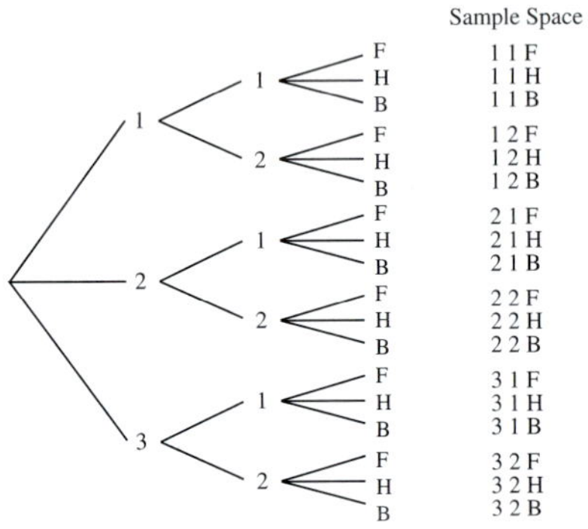

c) $\frac{1}{3}$ **d)** $\frac{1}{9}$ **e)** $\frac{2}{3}$

25. a) 27

b)

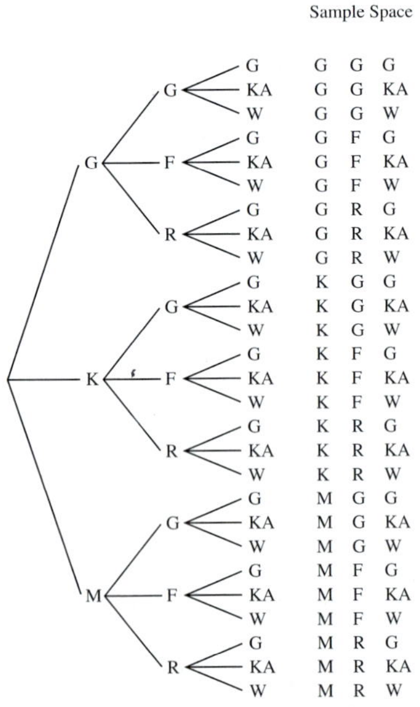

c) $\frac{1}{27}$ **d)** $\frac{8}{27}$ **e)** $\frac{19}{27}$

27. a) 24

b)

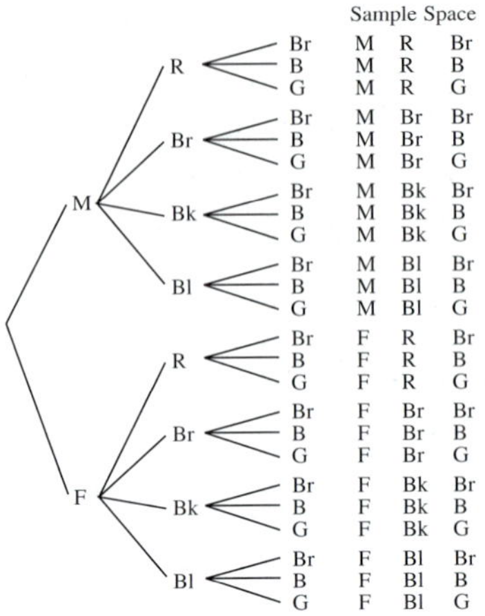

c) $\frac{1}{24}$ **d)** $\frac{1}{8}$

29. a) $\frac{1}{3}$ **b)** $\frac{2}{3}$ **c)** No, the probability of selecting a red chip is not the same as the probability of selecting a white chip. **d)** Answers will vary.

31. 3; 1 red, 1 blue, and 1 brown

SECTION 12.6, PAGE 778

1. a) At least one event, A *or* B, must occur.
b) Both events, A *and* B, must occur.

3. a) Events that cannot happen simultaneously are mutually exclusive events.
b) $P(A \text{ or } B) = P(A) + P(B)$

5. We assume that event A has already occurred.

7. a) No, both can like classical music.
b) No, the daughter may be more or less likely to like classical music because the mother likes it.

9. Events A and B cannot occur at the same time. Therefore, $P(A \text{ and } B) = 0$.

11. 0.7 **13.** 0.4 **15.** 0.75 **17.** $\frac{1}{3}$ **19.** $\frac{1}{2}$ **21.** $\frac{2}{13}$

23. $\frac{8}{13}$ **25.** $\frac{17}{26}$ **27. a)** $\frac{1}{16}$ **b)** $\frac{1}{19}$ **29. a)** $\frac{1}{16}$ **b)** $\frac{5}{76}$

31. a) $\frac{3}{80}$ **b)** $\frac{3}{76}$ **33. a)** $\frac{9}{25}$ **b)** $\frac{33}{95}$ **35.** $\frac{11}{20}$ **37.** $\frac{2}{5}$

39. $\frac{1}{4}$ **41.** $\frac{1}{8}$ **43.** $\frac{9}{64}$ **45.** $\frac{1}{8}$ **47.** $\frac{3}{8}$ **49.** $\frac{4}{7}$ **51.** $\frac{9}{49}$

53. $\frac{1}{35}$ **55.** $\frac{1}{35}$ **57.** $\frac{1}{8}$ **59.** $\frac{1}{8}$ **61. a)** $\frac{1}{32}$ **b)** $\frac{1}{2}$

63. a) $\frac{4}{49}$ **b)** $\frac{2}{21}$ **65. a)** $\frac{24}{49}$ **b)** $\frac{11}{21}$ **67.** $\frac{93}{245}$

69. $\frac{152}{245}$ 71. $\frac{1771}{9880}$ 73. $\frac{7}{988}$ 75. 0.7 77. $\frac{343}{1000}$

79. $\frac{1}{4}$ 81. $\frac{27}{1024}$ 83. $\frac{243}{1024}$ 85. $\frac{5}{22}$ 87. $\frac{1050}{1331}$

89. $\frac{1}{24}$ 91. $\frac{5}{12}$ 93. 0.36 95. 0.36

97. a) No b) 0.001 c) 0.00004 d) 0.00096 e) 0.000999 f) 0.998001

99. $\frac{9}{250}$ 101. $\frac{2169}{62{,}500}$ 103. $\frac{14}{45}$

105. Favors dealer. The probability of at least one diamond is ≈ 0.44, which is less than 0.5.

107. $\frac{1}{9}$ 109. $\frac{1}{2}$

SECTION 12.7, PAGE 787

1. The probability of E_2 given that E_1 has occurred

3. $\frac{1}{3}$ 5. $\frac{1}{3}$ 7. $\frac{2}{3}$ 9. $\frac{2}{3}$ 11. $\frac{3}{4}$ 13. $\frac{2}{3}$ 15. $\frac{2}{3}$ 17. $\frac{1}{5}$

19. $\frac{1}{3}$ 21. $\frac{3}{5}$ 23. $\frac{1}{7}$ 25. $\frac{1}{16}$ 27. $\frac{1}{7}$ 29. $\frac{5}{36}$ 31. $\frac{1}{6}$

33. $\frac{2}{3}$ 35. $\frac{1}{3}$ 37. $\frac{1}{3}$ 39. $\frac{1}{2}$ 41. 0.5941 43. 0.3605

45. 0.6251 47. $\frac{7}{13}$ 49. $\frac{4}{15}$ 51. $\frac{11}{15}$ 53. $\frac{29}{59}$ 55. $\frac{33}{50}$

57. $\frac{30}{61}$ 59. $\frac{10}{11}$ 61. $\frac{3}{19}$ 63. $\frac{44}{47}$ 65. $\frac{11}{27}$ 67. $\frac{10}{29}$

69. $\frac{11}{29}$ 71. $\frac{93}{200}$ 73. $\frac{15}{52}$

75. a) 140 b) 120 c) $\frac{7}{10}$ d) $\frac{3}{5}$ e) $\frac{2}{3}$ f) $\frac{4}{7}$ g) Because A and B are not independent events

77. a) 0.3 b) 0.5 c) Yes; $P(A \mid B) = P(A) \cdot P(B)$

79. $\frac{2}{3}$ 81. $\frac{1}{3}$ 83. $\frac{1}{3}$

SECTION 12.8, PAGE 799

1. Answers will vary.

3. a) $n! = n(n-1)(n-2)\cdots(3)(2)(1)$

b) $\frac{n!}{n_1!n_2!\cdots n_r!}$

5. The number of permutations of n items taken r at a time.

7. ${}_nP_r = \frac{n!}{(n-r)!}$

9. 24 11. 42 13. 1 15. 1 17. 3024 19. 336

21. 10,000 23. a) 1,404,000 b) 1,757,600

25. a) $5^5 = 3125$ b) $\frac{1}{3125} = 0.00032$

27. 57,106,944 29. 336

31. a) 720 b) 120 c) 24 d) 600

33. 720 35. a) 479,001,600 b) 3,628,800 c) 14,400

37. 78,624,000 39. 131,040 41. 676,000 43. 104,000

45. a) 8,000,000 b) 6,400,000,000 c) 64,000,000,000,000

47. 3,603,600 49. 5040 51. 280 53. 362,880

55. 302,400 57. 630 59. 15,120

61. a) 40,320 b) 362,880

63. a) 3125 b) ≈ 128 c) 0.00032

65. 12,600 sec, or 3.5 hr 67. No 69. 600

71. a) 2520 b) SCROOGE

SECTION 12.9, PAGE 806

1. Answers will vary. 3. ${}_nC_r = \frac{n!}{(n-r)!r!}$

5. Answers will vary. 7. 6 9. a) 15 b) 360

11. a) 1 b) 1 13. a) 120 b) 720 15. $\frac{1}{6}$ 17. $\frac{7}{2}$

19. 72 21. 28 23. 5 25. 56 27. 35 29. 495

31. 70 33. 28 35. 6160 37. 5880 39. 8820

41. 700 43. 1200 45. a) 45 b) 56

47. a) and b)

```
                1
              1   1
            1   2   1
          1   3   3   1
        1   4   6   4   1
      1   5   10  10  5   1
```

49. a) 24 b) 24 51. 60,060

SECTION 12.10, PAGE 813

1. $\frac{{}_8C_4}{{}_{12}C_4}$ 3. $\frac{{}_5C_3}{{}_{26}C_3}$ 5. $\frac{{}_8C_5}{{}_{15}C_5}$ 7. $\frac{{}_{28}C_8}{{}_{50}C_8}$ 9. $\frac{5}{42}$ 11. $\frac{4}{143}$

13. $\frac{1}{12}$ 15. $\frac{2}{11}$ 17. $\frac{1}{9{,}366{,}819}$ 19. $\frac{3}{10}$ 21. $\frac{7}{10}$

23. $\frac{1}{115}$ 25. $\frac{27}{230}$ 27. 0.0012 29. 0.3083 31. $\frac{2}{55}$

33. $\frac{4}{33}$ 35. $\frac{1}{77}$ 37. $\frac{5}{77}$ 39. $\frac{15}{253}$

41. a) $\frac{1}{123{,}760}$ b) $\frac{1}{30{,}940}$

43. **a)** $\frac{33}{54,145}$ **b)** $\frac{1}{2,598,960}$

45. **a)** $\frac{1}{2,162,160}$ **b)** $\frac{1}{6435}$

47. 1; Since there are more hairs than people, two or more people must have the same number of hairs on their head.

SECTION 12.11, PAGE 822

1. A probability distribution shows the probability associated with each specific outcome of an experiment. In a probability distribution, every possible outcome must be listed and the sum of all the probabilities must be 1.

3. $P(x) = ({}_nC_x)p^xq^{n-x}$ **5.** 0.0512

7. 0.3456 **9.** 0.015625

11. **a)** $P(x) = ({}_nC_x)(0.14)^x(0.86)^{n-x}$
b) $P(2) = ({}_{12}C_2)(0.14)^2(0.86)^{10}$

13. 0.24576 **15.** 0.25412 **17.** 0.06877 **19.** 0.4096

21. **a)** 0.00098 **b)** 0.08789 **c)** 0.10352

23. **a)** ≈ 0.1119 **b)** ≈ 0.2966 **25.** 0; it will be midnight.

REVIEW EXERCISES, PAGE 825

1. Answers will vary. **2.** Answers will vary.

3. $\frac{1}{5}$ **4.** Answers will vary. **5.** $\frac{2}{5}$ **6.** $\frac{1}{2}$ **7.** $\frac{7}{10}$

8. 1 **9.** $\frac{1}{10}$ **10.** $\frac{7}{25}$ **11.** $\frac{11}{50}$ **12.** $\frac{16}{25}$ **13.** $\frac{43}{50}$

14. **a)** 69:31 **b)** 31:69 **15.** 5:3 **16.** $\frac{3}{85}$ **17.** 7:3

18. **a)** −$1.20 **b)** −$3.60 **c)** $0.80

19. **a)** −$0.23 **b)** $0.23 **c)** Lose $23.08

20. 660 people

21. **a)**

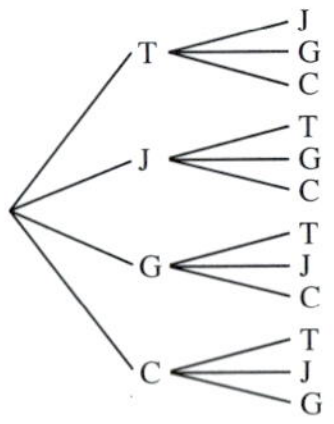

b) Sample Space

TJ
TG
TC
JT
JG
JC
GT
GJ
GC
CT
CJ
CG

c) $\frac{1}{12}$

22. **a)**

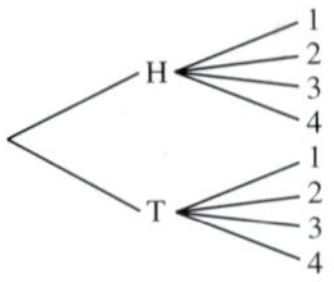

b) Sample Space

H1
H2
H3
H4
T1
T2
T3
T4

c) $\frac{1}{4}$ **d)** $\frac{3}{4}$

23. $\frac{1}{4}$ **24.** $\frac{9}{64}$ **25.** $\frac{5}{16}$ **26.** $\frac{7}{8}$ **27.** 1 **28.** $\frac{3}{16}$

29. $\frac{1}{22}$ **30.** $\frac{14}{55}$ **31.** $\frac{41}{55}$ **32.** $\frac{1}{22}$ **33.** $\frac{1}{4}$

34. Against, 3:1; in favor, 1:3 **35.** $13.75

36. $\frac{1}{8}$ **37.** $\frac{5}{8}$ **38.** In favor, 3:5; against, 5:3 **39.** $3.75

40. $\frac{7}{8}$ **41.** $\frac{15}{17}$ **42.** $\frac{17}{19}$ **43.** $\frac{2}{15}$ **44.** $\frac{1}{2}$ **45.** $\frac{23}{40}$ **46.** $\frac{3}{17}$

47. $\frac{3}{4}$ **48.** $\frac{12}{17}$ **49.** **a)** 24 **b)** $4500 **50.** 30 **51.** 720

52. 504 **53.** 20 **54.** **a)** 3003 **b)** 3,628,800

55. **a)** $\frac{1}{3,819,816}$ **b)** $\frac{1}{175,711,536}$ **56.** 35,640 **57.** 560

58. $\frac{1}{221}$ **59.** $\frac{1}{12}$ **60.** $\frac{1}{18}$ **61.** $\frac{1}{24}$ **62.** $\frac{11}{12}$ **63.** $\frac{5}{182}$

64. $\frac{45}{364}$ **65.** $\frac{2}{13}$ **66.** $\frac{11}{13}$

67. **a)** $P(x) = ({}_nC_x)(0.6)^x(0.4)^{n-x}$
b) $P(75) = ({}_{100}C_{75})(0.6)^{75}(0.4)^{25}$

68. 0.0512 **69.** **a)** 0.01024 **b)** 0.98976

CHAPTER TEST, PAGE 829

1. $\frac{11}{20}$ **2.** $\frac{2}{9}$ **3.** $\frac{5}{9}$ **4.** $\frac{7}{9}$ **5.** $\frac{1}{3}$ **6.** $\frac{1}{6}$ **7.** $\frac{1}{6}$

8. $\frac{5}{18}$ **9.** $\frac{5}{12}$ **10.** $\frac{8}{13}$ **11.** 18

12.

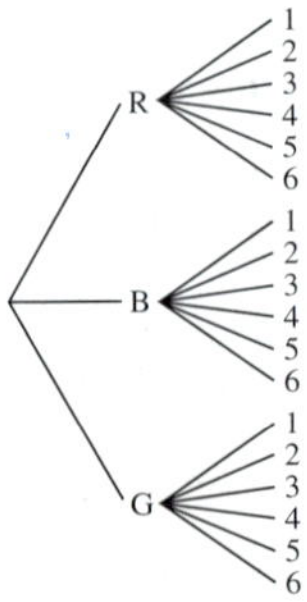

Sample Space

R1
R2
R3
R4
R5
R6
B1
B2
B3
B4
B5
B6
G1
G2
G3
G4
G5
G6

13. $\frac{1}{18}$ **14.** $\frac{4}{9}$ **15.** $\frac{5}{6}$ **16.** 1,581,840

17. a) 5:4 **b)** 5:4 **18.** $\frac{2}{9}$ **19.** \$0

20. a) $\frac{107}{228}$ **b)** $\frac{115}{228}$ **c)** $\frac{68}{115}$ **d)** $\frac{60}{107}$

21. 120 **22.** $\frac{3}{38}$ **23.** $\frac{35}{38}$ **24.** $\frac{175}{396}$ **25.** 0.1323

CHAPTER 13

SECTION 13.1, PAGE 837

1. Answers will vary.

3.–5. Answers will vary.

7. a) A population is all items or people of interest.
b) A sample is a subset of the population.

9. a) A systematic sample is a sample obtained by selecting every *n*th item on a list or production line.
b) Use a random number table to select the first item and then select every *n*th item after that.

11. a) A cluster sample is a random selection of groups of units.
b) Divide a geographical area into sections. Randomly select sections or clusters. Either each member of the selected cluster is included in the sample or a random sample of the members of each selected cluster is used.

13. An unbiased sample is one that is a small replica of the entire population with regard to income, education, gender, race, religion, political affiliation, age, and so forth.

15. Stratified sample **17.** Cluster sample

19. Systematic sample **21.** Convenience sample

23. Random sample **25. a)–c)** Answers will vary.

27. President; four out of 42 U.S. presidents have been assassinated (Lincoln, Garfield, McKinley, Kennedy).

SECTION 13.2, PAGE 841

1. Answers will vary.

3. Not all people who request a brochure will purchase a travel package.

5. Although the cookies are fat free, they still contain calories. Eating many of them may still cause you to gain weight.

7. The fact that Morgan's is the largest department store does not imply that it is inexpensive.

9. People with asthma may move to Arizona because of its climate. Therefore, more people with asthma may live in Arizona.

11. The quality of a steak does not necessarily depend on the price of the steak.

13. There may be deep sections in the pond, so it may not be safe to go wading.

15. Half the students in a population are expected to be below average.

17. a)

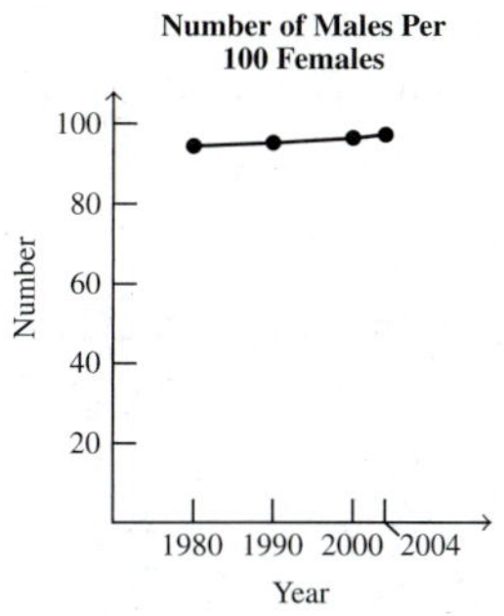

b)

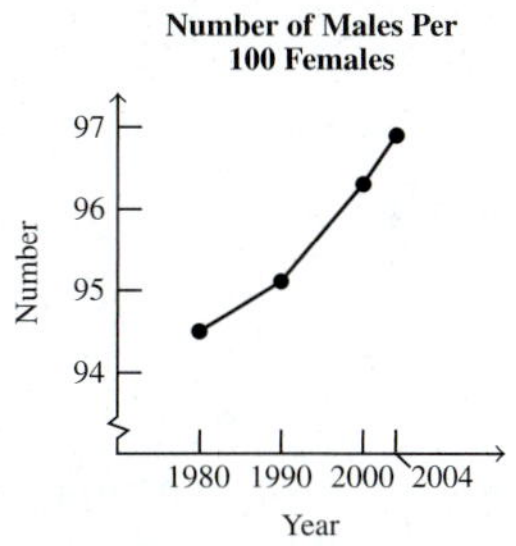

19. a)

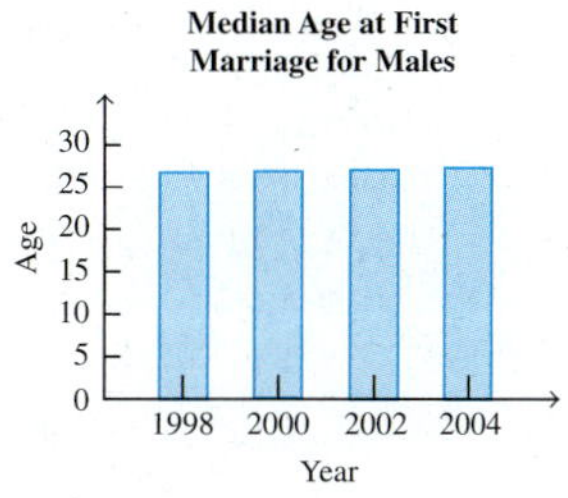

b)

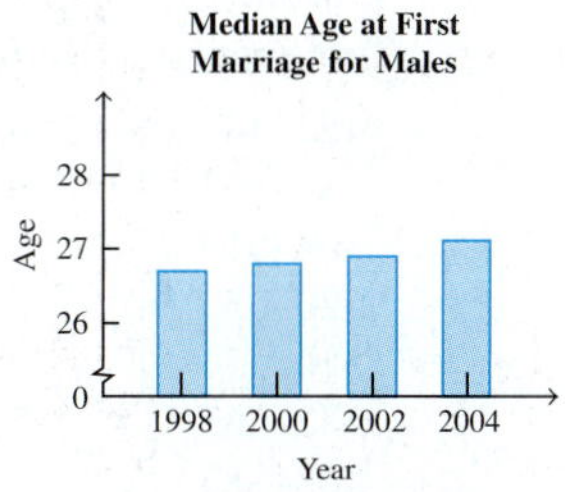

21. a)

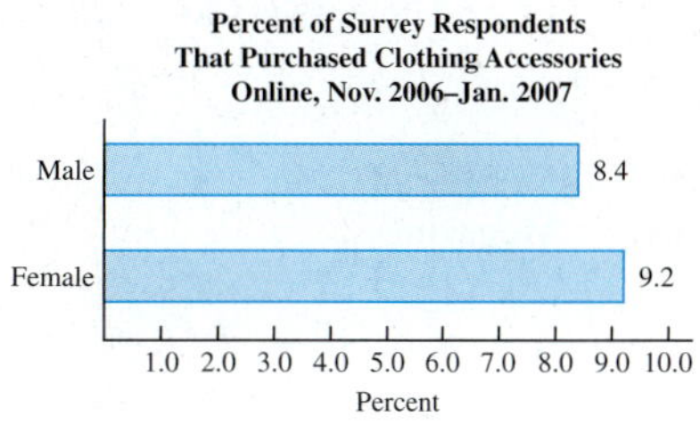

b) Answers will vary.

23. A decimal point

SECTION 13.3, PAGE 847

1. A frequency distribution is a listing of observed values and the corresponding frequency of occurrence of each value.

3. a) 7 **b)** 16–22 **c)** 16 **d)** 22

5. The modal class is the class with the greatest frequency.

7. a) 20 **b)** 7 **c)** 19 **d)** 16–22 **e)** 51–57

9.

Number of Visits	Number of Students
0	3
1	8
2	3
3	5
4	2
5	7
6	2
7	3
8	4
9	1
10	2

11.

IQ	Number of Students
78–86	2
87–95	15
96–104	18
105–113	7
114–122	6
123–131	1
132–140	1

13.

IQ	Number of Students
80–90	8
91–101	22
102–112	11
113–123	7
124–134	1
135–145	1

15.

Placement Test Scores	Number of Students
472–492	9
493–513	9
514–534	5
535–555	2
556–576	3
577–597	2

17.

Placement Test Scores	Number of Students
472–487	4
488–503	9
504–519	7
520–535	3
536–551	2
552–567	2
568–583	2
584–599	1

19.

Circulation (ten thousands)	Number of Magazines
157–306	34
307–456	9
457–606	2
607–756	1
757–906	1
907–1056	1

21.

Circulation (ten thousands)	Number of Magazines
157–256	29
257–356	8
357–456	6
457–556	2
557–656	0
657–756	1
757–856	1
857–956	0
957–1056	1

23.

Population (millions)	Number of Cities
6.0–6.9	2
7.0–7.9	5
8.0–8.9	6
9.0–9.9	2
10.0–10.9	3
11.0–11.9	2

25.

Population (millions)	Number of Cities
5.5–6.5	2
6.6–7.6	4
7.7–8.7	6
8.8–9.8	3
9.9–10.9	3
11.0–12.0	2

27.

Percent	Number of States
5.6–7.5	2
7.6–9.5	8
9.6–11.5	16
11.6–13.5	10
13.6–15.5	6
15.6–17.5	8

29.

Percent	Number of States
5.6–7.0	1
7.1–8.5	4
8.6–10.0	13
10.1–11.5	8
11.6–13.0	9
13.1–14.5	3
14.6–16.0	7
16.1–17.5	5

31. February, since it has the fewest number of days

32. a) Did You Know (page 845) There are 6 F's.

b) Answers will vary.

SECTION 13.4, PAGE 855

1. Answers will vary.

3. Answers will vary.

5. a) Answers will vary.

b)

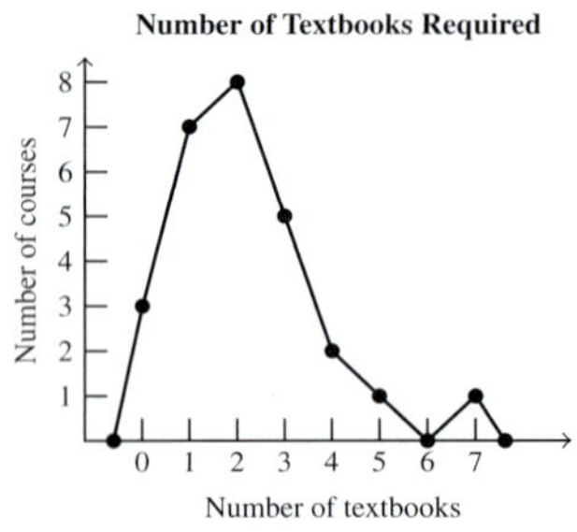

7. a) Answers will vary.

b)

Observed Values	Frequency
45	3
46	0
47	1
48	0
49	1
50	1
51	2

9. Tuition: $23,241.44; room: $4867.49; board: $3578.09; fees: $548.00

11.

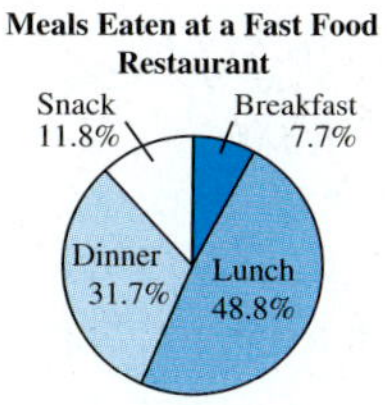

13. a) and **b)**

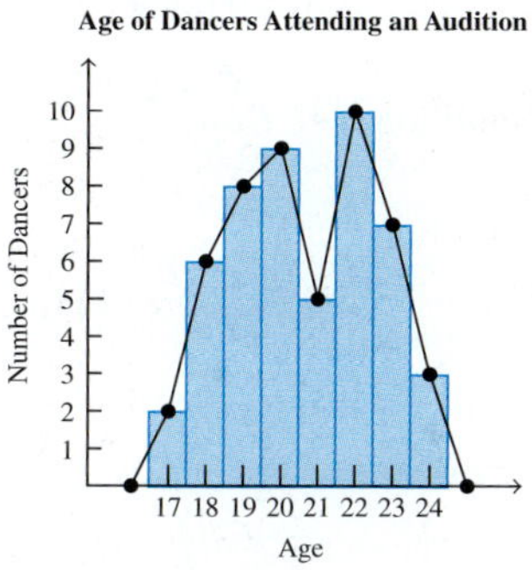

15. a) and **b)**

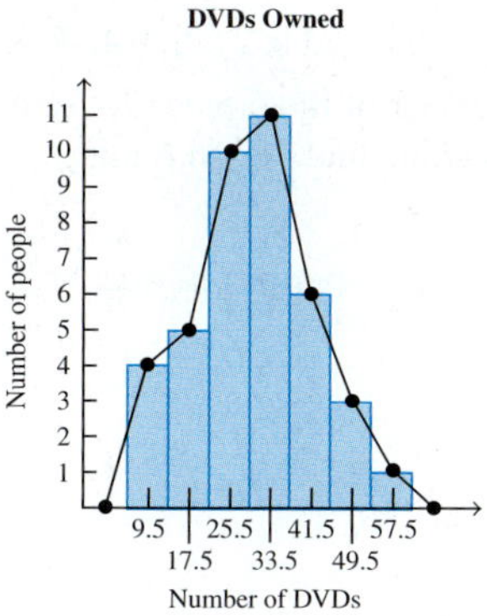

17. a) 28 **b)** 4 **c)** 2 **d)** 75

e)

Number of Televisions	Number of Homes
0	2
1	4
2	8
3	6
4	4
5	3
6	1

19. a) 7 **b)** 16 **c)** 36

d)

Number of Messages	Number of People	Number of Messages	Number of People
3	2	7	3
4	3	8	8
5	7	9	6
6	4	10	3

e)

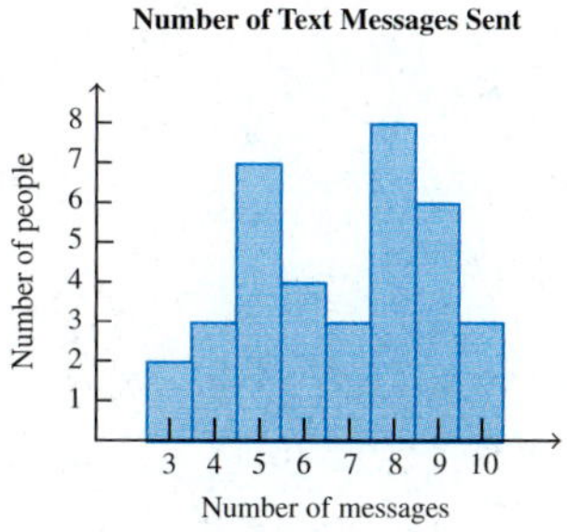

21.

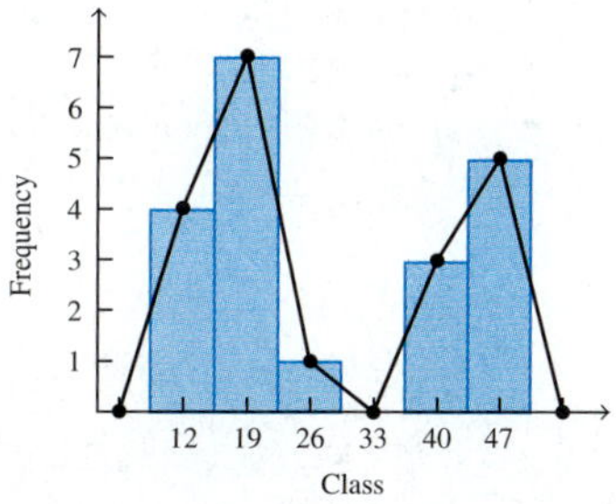

23. 1 | 2 represents 12

0	4	6	7	8				
1	2	2	3	5	6	7	8	9
2	1	2	3	5	7			
3	3	4						
4	0							

25. a)

Salaries (1000s of dollars)	Number of Social Workers
27	1
28	7
29	4
30	3
31	2
32	3
33	3
34	2

b) and **c)**

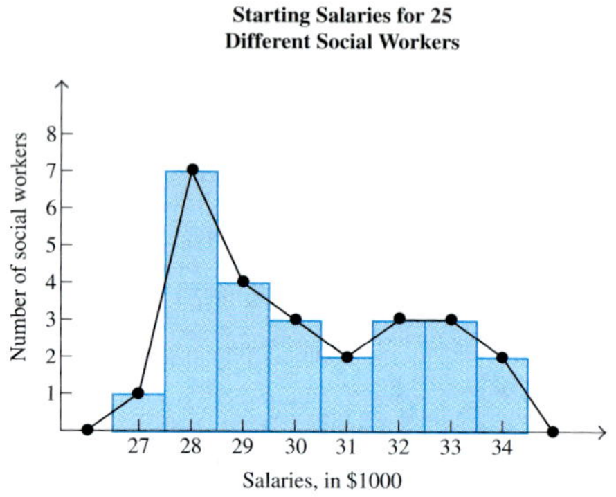

d) 2|8 represents 28

2|7 8 8 8 8 8 8 8 9 9 9 9
3|0 0 0 1 1 2 2 2 3 3 3 4 4

27. a)

Number of Performances	Number of Shows
1508–2548	33
2549–3589	7
3590–4630	4
4631–5671	1
5672–6712	3
6713–7753	1
7754–8794	1

b) and **c)**

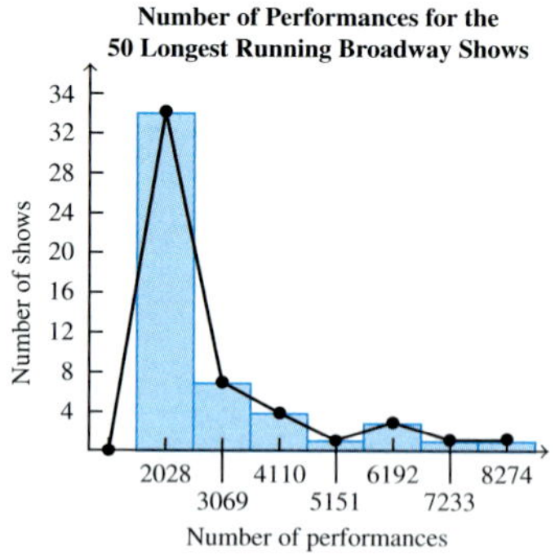

29. a)–e) Answers will vary.

SECTION 13.5, PAGE 867

1. a) The mean is the balancing point of a set of data. It is the sum of the data divided by the number of pieces of data.
b) The median is the value in the middle of a set of ranked data. To find the median, rank the data and select the value in the middle.

3. The median might be used when there are some values that differ greatly from the rest of the values in the set, for example, salaries.

5. The midrange might be used when the item being studied is constantly fluctuating, for example, daily temperature.

7. a) $\bar{x}$ **b)** μ **9.** Answers will vary.

11. 13, 12, 12, 16 **13.** 79.3, 82, none, 76

15. 8, 8, none, 8 **17.** 13.1, 11, 1, 18.5

19. 11.9, 12.5, 13, 11.5 **21.** 3.5, 3, 3, 4

23. a) 4.9, 5, 5, 6 **b)** 5.3, 5, 5, 6
c) Only the mean **d)** The mean and the midrange

25. A 79 mean average on 10 quizzes gives a total of 790 points. An 80 mean average on 10 quizzes requires a total of 800 points. Thus, Jim missed a B by 10 points, not 1 point.

27. a) 52.9 million **b)** 43.8 million **c)** None
d) 61.8 million

29. a) 10.8 million **b)** 10.1 million
c) None **d)** 11.5 million

31. 552

33. One example is 72, 73, 74, 76, 77, 78.

35. a) Yes **b)** No **c)** No **d)** Yes

37. a) 29 or greater
b) Yes, 99 or greater
c) 20 or greater **d)** 80 or greater

39. One example: 1, 2, 3, 3, 4, 5 changed to 1, 2, 3, 4, 4, 5.

41. No. By changing only one piece of the 6 pieces of data, you cannot alter both the median and the midrange.

43. The data must be ranked.

45. He is taller than approximately 35 percent of all kindergarten children.

47. a) $25 **b)** $22 **c)** $34

49. Second quartile, median

51. a) $530 **b)** $540 **c)** 25%
d) 25% **e)** 17% **f)** $55,000

53. a)

Ruth	Mantle
0.290	0.300
0.359	0.365
0.301	0.304
0.272	0.275
0.315	0.321

b) Mantle's is greater in every case.
c) Ruth: 0.316; Mantle: 0.311; Ruth's is greater.
d) Answers will vary.
e) Ruth: 0.307; Mantle: 0.313; Mantle's is greater.
f) Answers will vary. **g)** Answers will vary.

55. 90 **57. a)–c)** Answers will vary.

SECTION 13.6, PAGE 877

1. Range = highest value − lowest value

3. Answers will vary.

5. Zero **7.** σ

9. Answers will vary.

11. They would be the same since the spread of data about each mean is the same.

13. a) The mean is the same for both classes.
b) The spread of the data about the mean is greater for the evening class since the standard deviation is greater for the evening class.

15. 11, $\sqrt{16.5} \approx 4.06$ **17.** 6, $\sqrt{4.67} \approx 2.16$

19. 11, $\sqrt{15.2} \approx 3.90$ **21.** 5, $\sqrt{3} \approx 1.73$

23. \$32, $\sqrt{137.78} \approx \$11.74$

25. \$150, $\sqrt{2600} \approx \$50.99$

27. a) \$63, $\sqrt{631.6} \approx \$25.13$ **b)** Answers will vary.
c) Answers remain the same, range: \$63, standard deviation ≈ \$25.13.

29. a)–c) Answers will vary.
d) If each number in a distribution is multiplied by n, the mean and standard deviation of the new distribution will be n times that of the original distribution.
e) The mean of the second set is $4 \times 5 = 20$, and the standard deviation of the second set is $2 \times 5 = 10$.

31. a) The standard deviation increases. There is a greater spread from the mean as they get older.
b) ≈133 lb **c)** ≈21 lb
d) Mean: ≈100 lb; normal range: ≈60 to 140 lb
e) Mean: ≈62 in.; normal range: ≈53 to 68 in.
f) 5%

33. a)

East		West	
Number of Oil Changes Made	Number of Days	Number of Oil Changes Made	Number of Days
15–20	2	15–20	0
21–26	2	21–26	0
27–32	5	27–32	6
33–38	4	33–38	9
39–44	7	39–44	4
45–50	1	45–50	6
51–56	1	51–56	0
57–62	2	57–62	0
63–68	1	63–68	0

b)

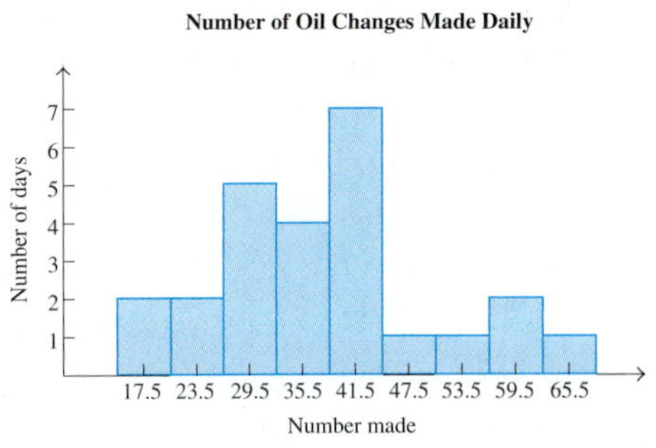

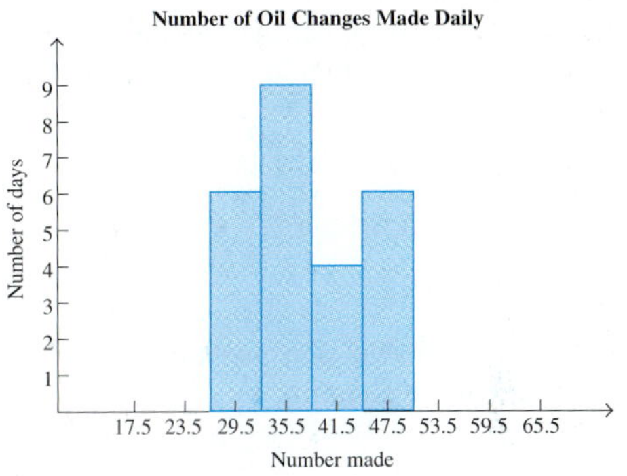

c) They appear to have about the same mean since they are both centered around 38.
d) The distribution for East is more spread out. Therefore, East has a greater standard deviation.
e) East: 38, West: 38 **f)** East: ≈12.64, West: ≈5.98

35. 6, 6, 6, 6, 6

SECTION 13.7, PAGE 894

1. A rectangular distribution is one in which all the values have the same frequency.

3. A bimodal distribution is one in which two nonadjacent values occur more frequently than any other values in a set of data.

5. A distribution skewed to the left is one that has a "tail" on its left.

7. A z-score measures how far, in terms of standard deviations, a given score is from the mean.

9. a) Below the mean
b) Above the mean

11. a) *B* **b)** *C* **c)** *A*

13.–15. Answers will vary.

17. Normal **19.** Skewed right

21. The mean is the greatest value. The median is lower than the mean. The mode is the lowest value.

23. Answers will vary. **25.** They all have the same value.

27. 0.5000 **29.** 0.8185 **31.** 0.0901 **33.** 0.0375

35. 0.0429 **37.** 0.0526 **39.** 76.11% **41.** 89.74%

43. 97.13% **45.** 97.50% **47.** 21.96%

49. a) Emily, Sarah, Carol **b)** Jason, Juan
c) Omar, Heather, Kim

51. 50% **53.** 10.56% **55.** 69.15% **57.** 24.17%

59. 44.00% **61.** 29.02% **63.** 59.87% **65.** 50.00%

67. 11.51% **69.** ≈ 23 cars **71.** 95.47%

73. 13,380 boxes **75.** 69.15% **77.** 0.62%

79. ≈ 83 children **81.** 1.79%

83. The standard deviation is too large.

85. a) Katie: $z = 2.4$; Stella: $z = 1.7$
b) Katie. Her z-score is higher than Stella's z-score, which means her sales are further above the mean than Stella's sales.

87. a)–f) Answers will vary.

89. −1.18 **91.** 2

SECTION 13.8, PAGE 907

1. The correlation coefficient measures the strength of the relationship between the quantities.

3. 1 **5.** 0

7. a) A positive correlation indicates that as one quantity increases, the other increases.
b) Answers will vary.

9. The level of significance is used to identify the cutoff between results attributed to chance and results attributed to an actual relationship between the two variables.

11.–13. Answers will vary.

15. Yes **17.** Yes

19. No **21.** No

The answers in the remainder of this section may differ slightly from your answers, depending on how your answers are rounded and which calculator you used. The answers given here were obtained from a Texas Instruments TI-36x solar calculator.

23. a)

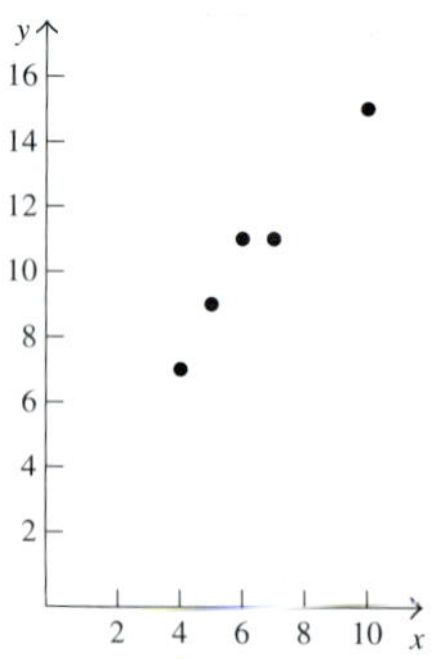

b) 0.981 **c)** Yes **d)** Yes

25. a)

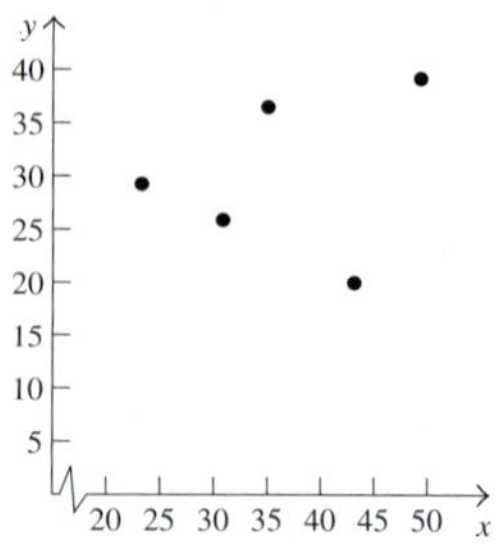

b) 0.228 **c)** No **d)** No

27. a)

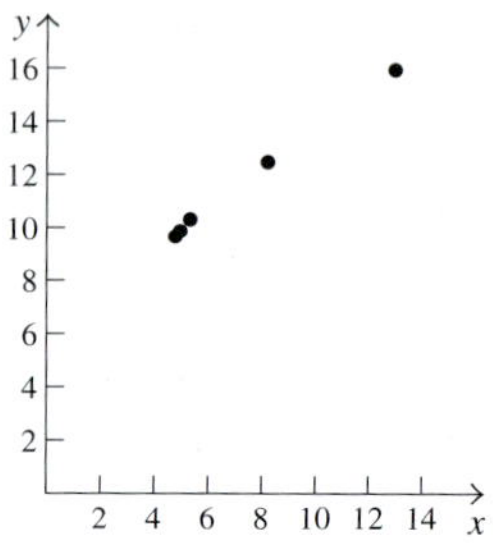

b) 0.999 **c)** Yes **d)** Yes

29. a)

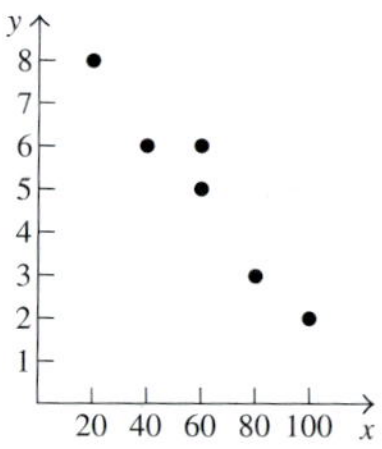

b) −0.968 **c)** Yes **d)** Yes

31. $y = 1.26x + 2.51$

33. $y = 0.18x + 23.82$

35. $y = 0.81x + 5.84$

37. $y = -0.08x + 9.50$

39. a) 0.974 **b)** Yes **c)** $y = 0.74x + 1.26$

41. a) 0.950 **b)** Yes **c)** $y = 0.77x + 24.86$

43. a) 0.993 **b)** Yes **c)** $y = 3.90x + 1.94$
d) 21.4 kilocalories

45. a) −0.804 **b)** No **c)** $y = -0.43x + 35.28$
d) 21.1 mpg

47. a) −0.977 **b)** Yes **c)** $y = -12.93x + 99.59$
d) 41.4%

49. a) and **b)** Answers will vary.
c)

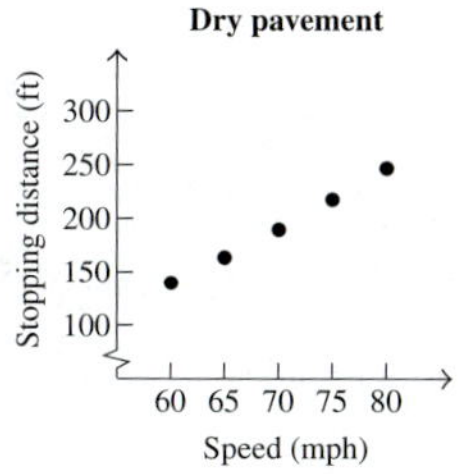

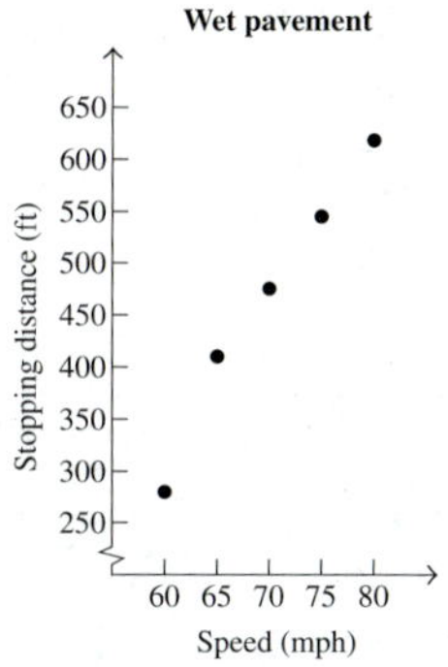

d) 0.999 **e)** 0.990 **f)** Answers will vary.
g) $y = 5.36x - 183.40$ **h)** $y = 16.22x - 669.80$
i) Dry, 229.3 ft; wet, 579.1 ft

51. a)–f) Answers will vary.

53. a) 0.993 **b)** Should be the same.
c) 0.993, the values are the same.

REVIEW EXERCISES, PAGE 913

1. a) A population consists of all items or people of interest.
b) A sample is a subset of the population.

2. A random sample is one where every item in the population has the same chance of being selected.

3. The candy bars may have lots of calories, or fat, or sodium. Therefore, it may not be healthy to eat them.

4. Sales may not necessarily be a good indicator of profit. Expenses must also be considered.

5. a)

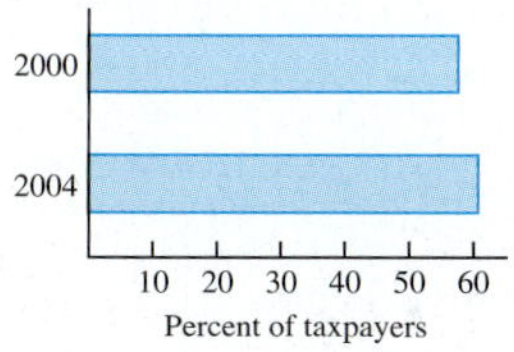

b)

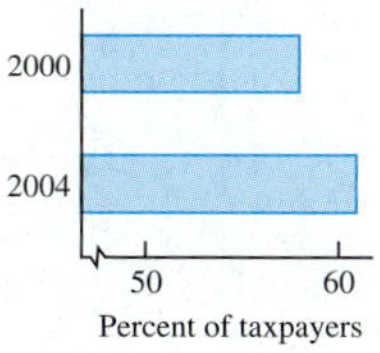

6. a)

Class	Frequency
35	1
36	3
37	6
38	2
39	3
40	0
41	4
42	1
43	3
44	1
45	1

b) and **c)**

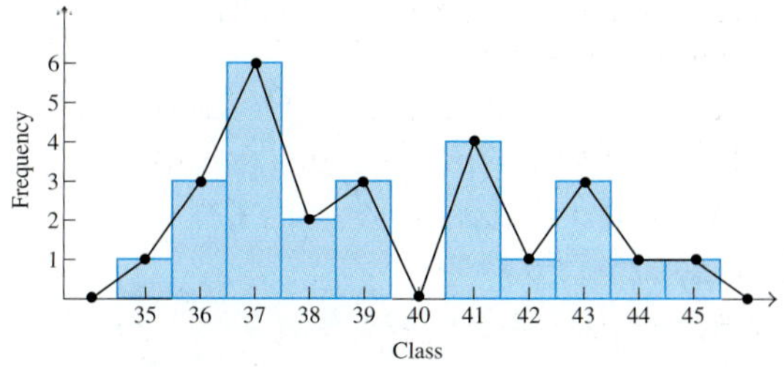

7. a)

High Temperature	Number of Cities
58–62	1
63–67	4
68–72	9
73–77	10
78–82	11
83–87	4
88–92	1

b) and **c)**

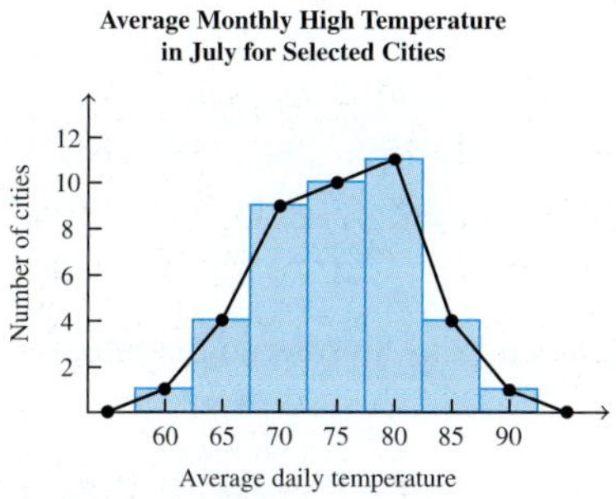

d) 5|8 represents 58

5	8
6	3 6 6 7 8 8 9
7	0 1 1 1 2 2 3 3 3 4 5 5 5 6 6 7 9 9 9
8	0 0 0 0 1 2 2 2 3 4 4 7
9	1

8. 80 **9.** 81 **10.** None **11.** 79 **12.** 28
13. $\sqrt{87.2} \approx 9.34$ **14.** 13 **15.** 13
16. 7 and 12 **17.** 13.5 **18.** 19
19. $\sqrt{40} \approx 6.32$ **20.** 68.26% **21.** 95.44%
22. 94.52% **23.** 5.48% **24.** 72.57% **25.** 34.1%
26. 34.5% **27.** 29.0% **28.** 2.3%

29. a)

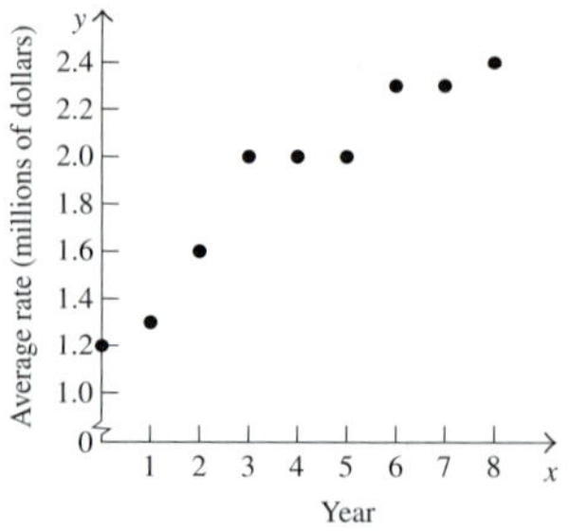

b) Yes; positive **c)** 0.957 **d)** Yes
e) $y = 0.15x + 1.29$ **f)** \$3.2 million

30. a)

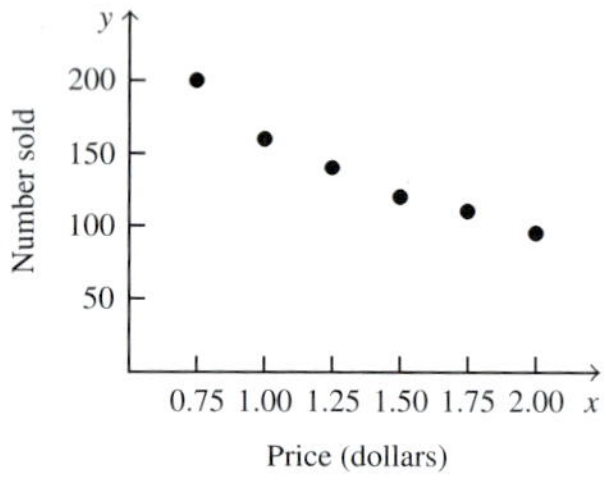

b) Yes; negative **c)** −0.973 **d)** Yes
e) $y = -79.4x + 246.7$ **f)** ≈120 sold

31. 180 lb **32.** 185 lb **33.** 25% **34.** 25% **35.** 14%
36. 19,200 lb **37.** 238 lb **38.** 150.6 lb **39.** ≈3.57
40. 2 **41.** 3 **42.** 7 **43.** 14 **44.** $\sqrt{8.105} \approx 2.85$

45.

Number of Children	Number of Presidents
0–1	8
2–3	15
4–5	10
6–7	6
8–9	1
10–11	1
12–13	0
14–15	1

46. and **47.**

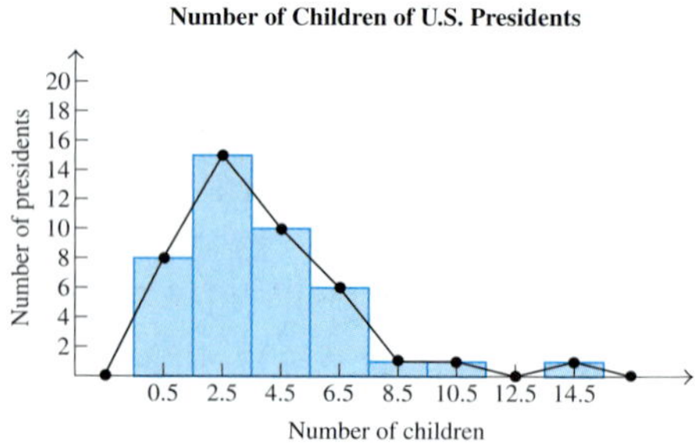

48. No, it is skewed to the right.
49. Answers will vary.
50. Answers will vary.

CHAPTER TEST, PAGE 915

1. 42 **2.** 43 **3.** 43 **4.** 39.5 **5.** 25 **6.** $\sqrt{84} \approx 9.17$

7.

Class	Frequency
25–30	7
31–36	5
37–42	1
43–48	7
49–54	5
55–60	3
61–66	2

8.

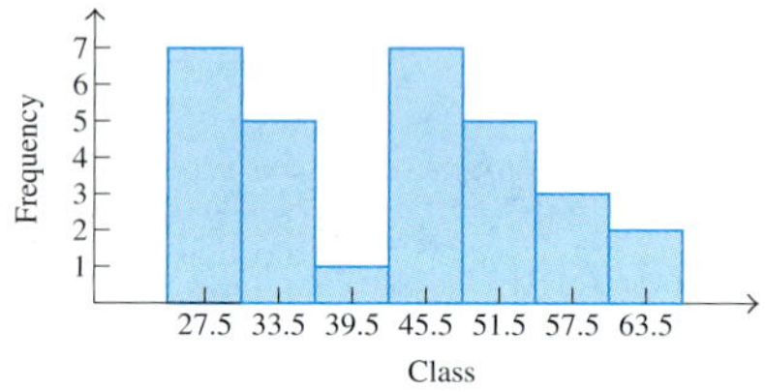

9.

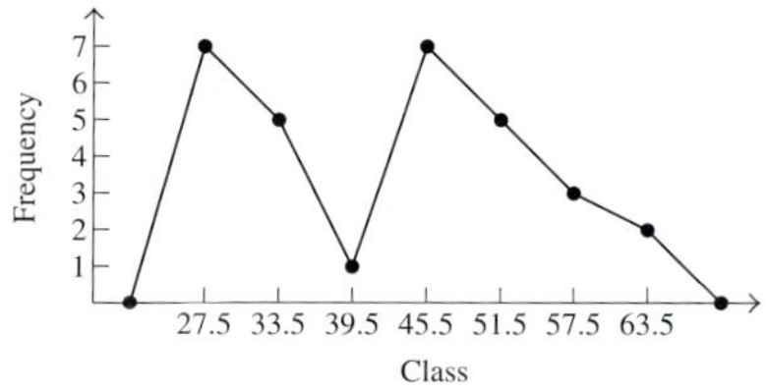

10. \$735 **11.** \$710 **12.** 75% **13.** 79%
14. \$74,000 **15.** \$780 **16.** \$680 **17.** 87.10%
18. 89.44% **19.** 10.56% **20.** 94.52%

21. a)

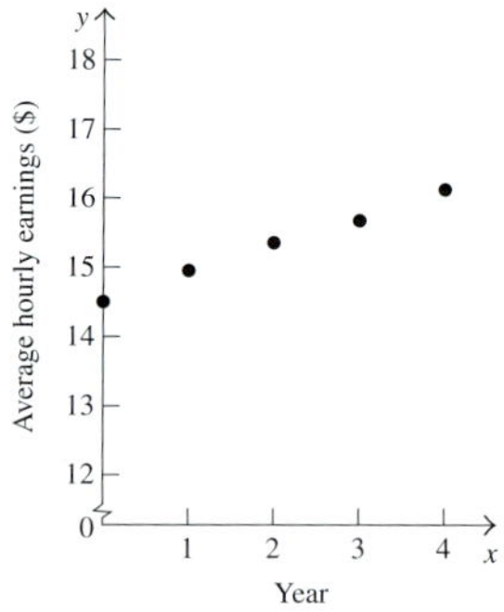

b) Yes **c)** 0.996 **d)** Yes **e)** $y = 0.39x + 14.56$
f) \$25.87

CHAPTER 14

SECTION 14.1, PAGE 926

1. A graph is a finite set of points, called vertices, that are connected with line segments, called edges.

3. One example:

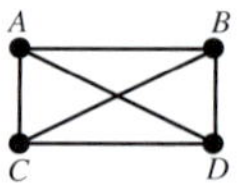

5. If the number of edges connected to the vertex is even, the vertex is even. If the number of edges connected to the vertex is odd, the vertex is odd.

7. a) A path is a sequence of adjacent vertices and the edges connecting them.

b) A circuit is a path that begins and ends at the same vertex.

c) In the following graph:

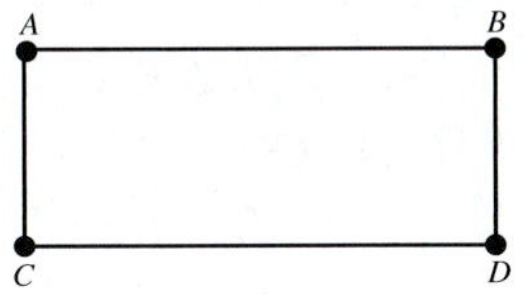

The path *A*, *B*, *D*, *C* is a path that is not a circuit. The path *A*, *B*, *D*, *C*, *A* is a path that is also a circuit.

9.

11.

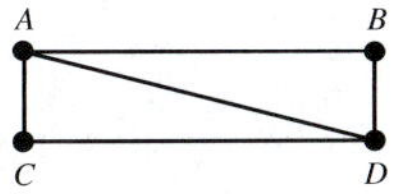

13.

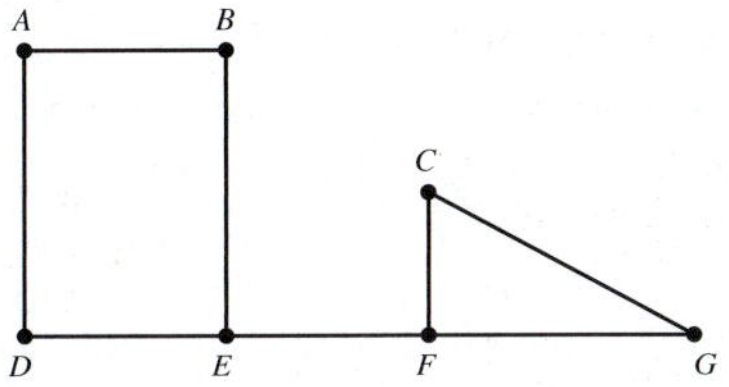

15. No. There is no edge connecting vertices *B* and *C*.

17. No

19. Yes. One example is *C*, *A*, *B*, *D*, *F*, *E*, *C*, *D*

In Exercises 21–31, one graph is shown. Other graphs are possible.

21.

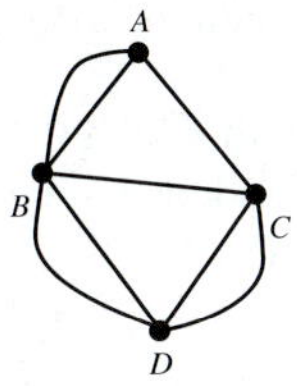

23.

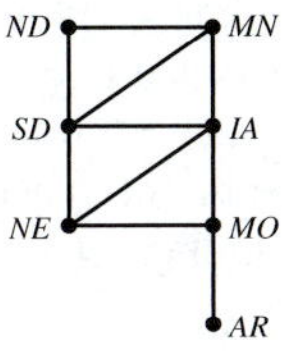

25.

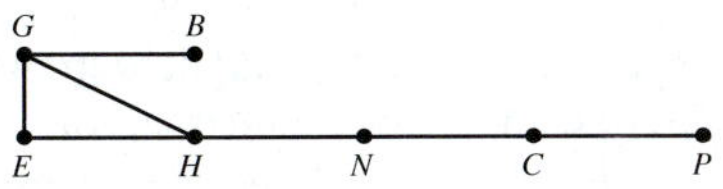

27.

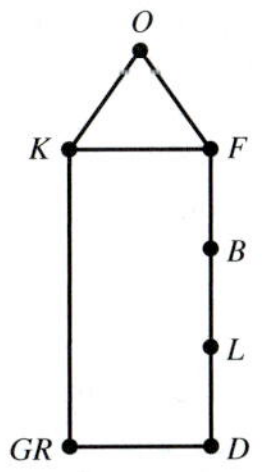

29.

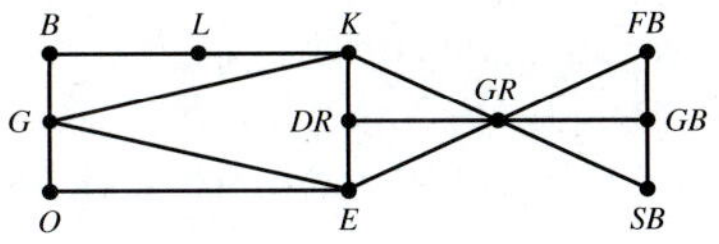

31.

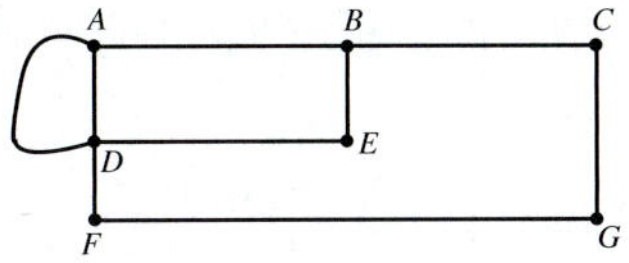

33. Disconnected **35.** Connected

37. Edge *AB* **39.** Edge *EF*

41.

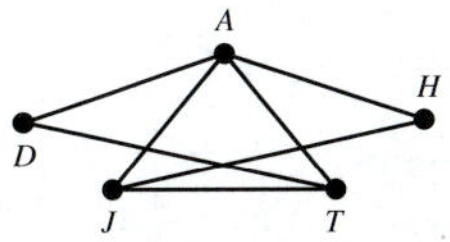

Other answers are possible.

43. It is impossible to have a graph with an odd number of odd vertices.

45. a) and **b)** Answers will vary.

SECTION 14.2, PAGE 938

In this section, when asked to determine an Euler path or Euler circuit, we will provide one answer here. Often, other answers are possible.

1. a) An Euler path is a path that must include each edge of a graph exactly one time.

b) Answers will vary. One example is shown. The path *A*, *B*, *E*, *D*, *C*, *A*, *D*, *B* is an Euler path.

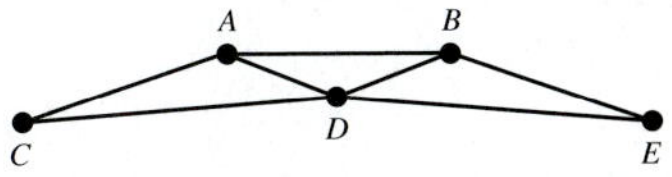

c) In the graph shown, path *A*, *B*, *E*, *D*, *C* is a path that is not an Euler path.

3. a) Yes **b)** Yes **c)** No

5. If all the vertices are even, the graph has an Euler circuit.

7. *A*, *B*, *D*, *B*, *C*, *D*, *E*, *A*, *C*.

9. No. This graph has exactly two odd vertices. Each Euler path must begin with an odd vertex. *B* is an even vertex.

11. *A*, *B*, *A*, *C*, *B*, *E*, *C*, *D*, *A*, *D*, *E*

13. No. A graph with exactly two odd vertices has no Euler circuits.

15. *A, B, C, D, E, F, B, D, F, A*

17. *C, D, E, F, A, B, D, F, B, C*

19. *E, F, A, B, C, D, F, B, D, E*

21. a) Yes. There are no odd vertices.
b) Yes. There are no odd vertices.

23. a) No. There are more than two odd vertices.
b) No. There is at least one odd vertex.

25. a) Yes
b) They could start on either island and finish at the other.

In Exercises 27–31, one graph is shown. Other graphs are possible.

27. a)

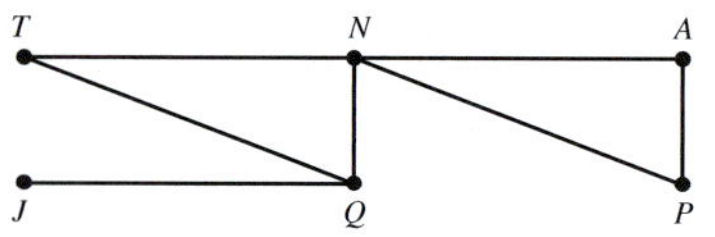

b) Yes; *J, Q, T, N, A, P, N, Q*
c) No

29. a)

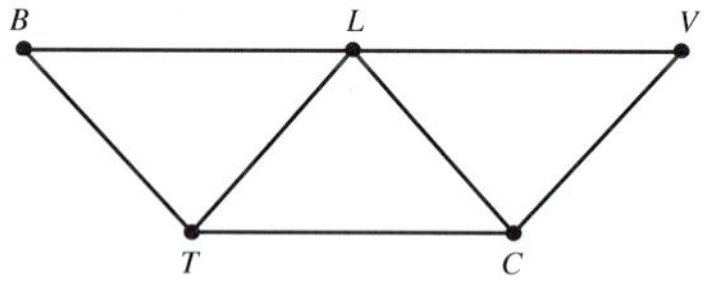

b) Yes; *T, B, L, V, C, L, T, C*
c) No

31. a)

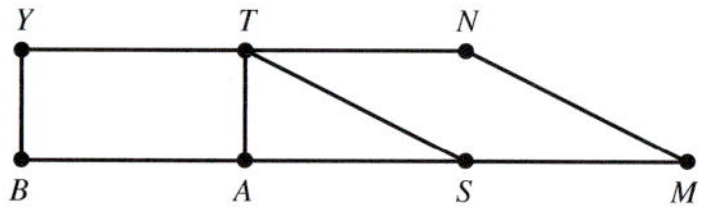

b) Yes; *A, S, M, N, T, Y, B, A, T, S*
c) No

33. a)

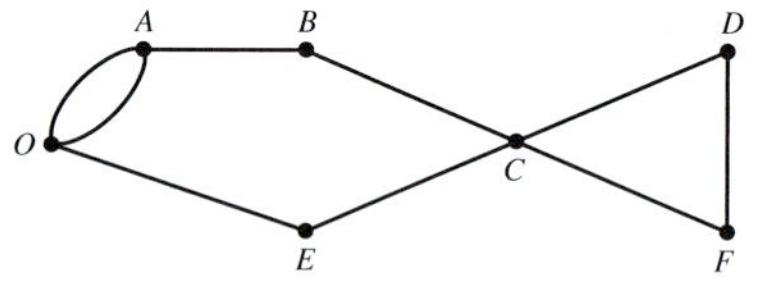

b) Yes
c) *O, A, B, C, D, F, C, E, O, A*

35. a)

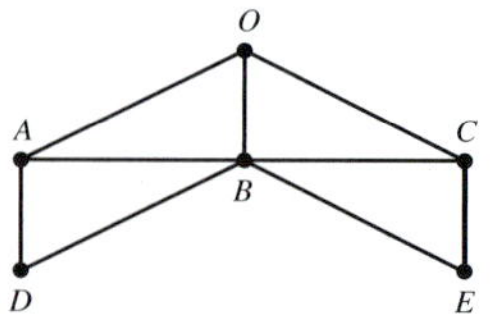

b) No

37. a) Yes
b) The residents would need to start at the intersection of Maple Cir., Walnut St., and Willow St. or at the intersection of Walnut St. and Oak St.

39. *B, E, D, A, B, C, A, D, C, E*

41. *H, I, F, C, B, D, G, H, E, D, A, B, E, F*

43. *A, E, B, F, C, G, D, K, G, J, F, I, E, H, A*

45. *A, C, D, G, H, F, C, F, E, B, A*

47. *A, B, C, E, B, D, E, F, I, E, H, D, G, H, I, J, F, C, A*

49. *OK, KS, CO, WY, UT, CO, NM, OK, CO*

51. *B, A, E, H, I, J, K, D, C, G, G, J, F, C, B, F, I, E, B*

53. *J, G, G, C, F, J, K, D, C, B, F, I, E, B, A, E, H, I, J*

55. a) No
b) California, Nevada, and Louisiana (and others) have an odd number of states bordering them. Since a graph of the United States would have more than two odd vertices, no Euler path and no Euler circuit exist.

57. a)

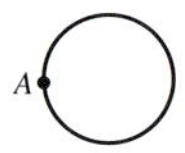

b)

c)

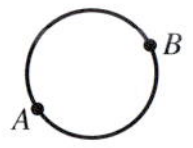

SECTION 14.3, PAGE 951

In this section, when asked to determine a Hamilton path or Hamilton circuit, we will provide one answer. Often, other answers are possible.

1. a) A Hamilton circuit is a path that begins and ends at the same vertex and passes through all other vertices exactly one time.
b) Both Hamilton and Euler circuits begin and end at the same vertex. A Hamilton circuit passes through all other *vertices* exactly once, whereas an Euler circuit passes through each *edge* exactly once.

3. a) A weighted graph is a graph with a number, or weight, assigned to each edge.
b) A complete graph is a graph in which there is an edge between each pair of vertices.
c) A complete, weighted graph is a graph in which there is an edge between each pair of vertices and each edge has a number, or weight, assigned to it.

5. a) The number of unique Hamilton circuits in a complete graph with n vertices is found by computing $(n - 1)!$
b) $6! = 720$ **c)** $9! = 362,880$

7. Represent the problem with a complete, weighted graph and determine the cost or distance for each Hamilton circuit for the graph. The Hamilton circuit with the lowest cost or shortest distance is the optimal solution.

9. *A, E, F, B, C, D, G* and *G, D, C, F, E, A, B*

11. *A, B, C, D, G, F, E, H* and *E, H, F, G, D, C, A, B*

13. *A, B, C, E, D, F, G, H* and *F, G, H, E, D, A, B, C*

15. *A, B, C, F, I, H, E, G, D, A* and *E, H, I, F, C, B, A, D, G, E*

17. *A, B, C, F, I, E, H, G, D, A* and *A, E, B, C, F, I, H, G, D, A*

19.

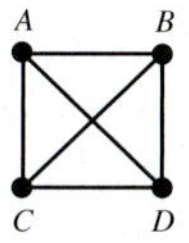

21. $6! = 720$

23. $12! = 479{,}001{,}600$ ways

In Exercises 25–31, other graphs are possible.

25. a)

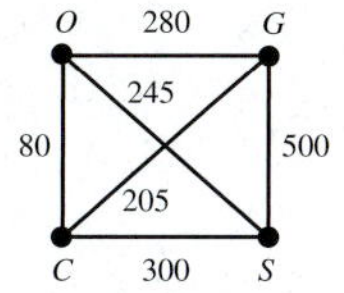

b) *C, G, S, O, C* or *C, G, O, S, C* or *C, S, O, G, C* or *C, O, S, G, C*

c) 1030 miles

27. a)

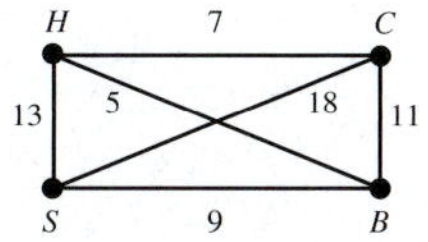

b) *H, C, S, B, H* or *H, B, S, C, H*

c) 39 miles

29. a)

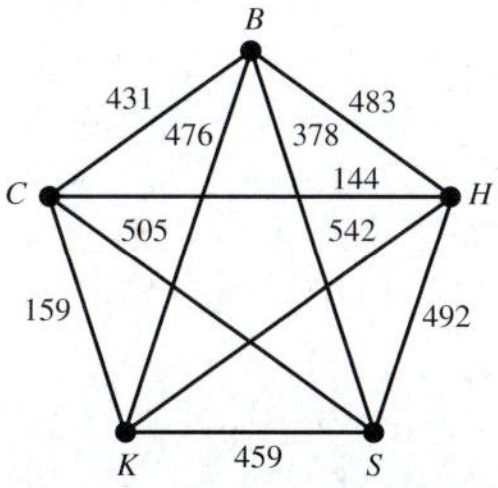

b) *S, B, C, H, K, S* for $1954

c) Answers will vary.

31. a)

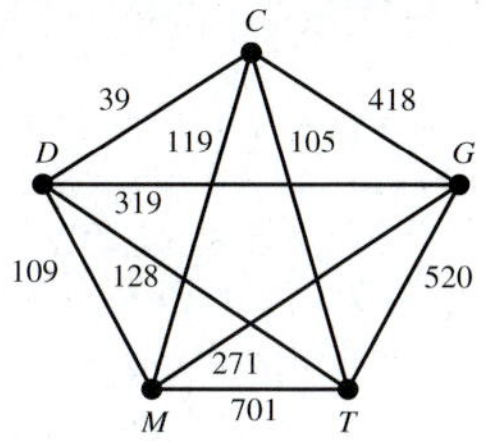

b) *C, D, M, G, T, C* for $1044

c) Answers will vary.

33. a–d) Answers will vary.

35. *A, E, D, N, O, F, G, Q, P, T, M, L, C, B, J, K, S, R, I, H, A*

SECTION 14.4, PAGE 963

1. A tree is a connected graph in which each edge is a bridge.

3. Yes. Removing the edge would create a disconnected graph.

5. A minimum-cost spanning tree is a spanning tree that has the lowest cost or shortest distance of all spanning trees for a given graph.

7.

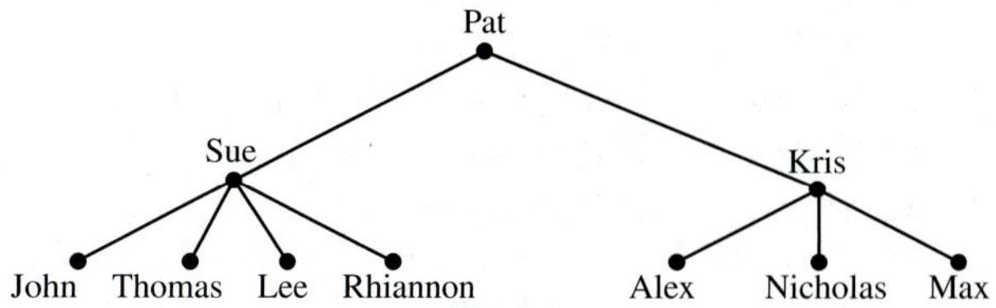

9.

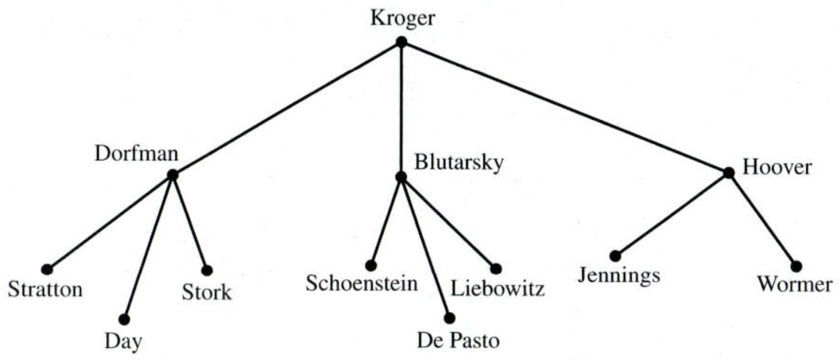

11.

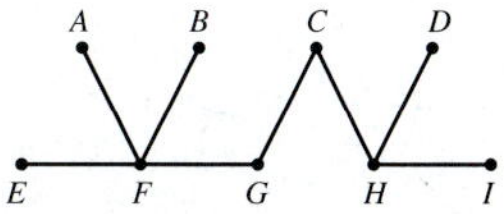

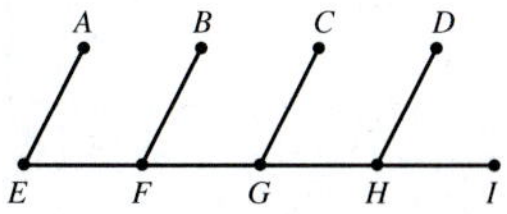

Other answers are possible.

13.

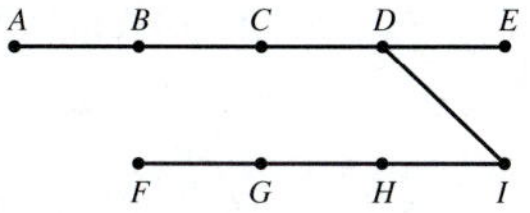

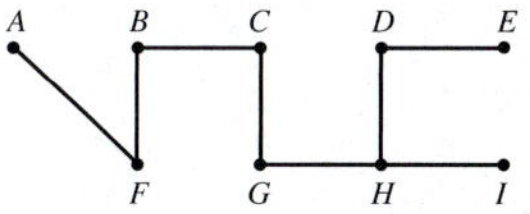

Other answers are possible.

15.

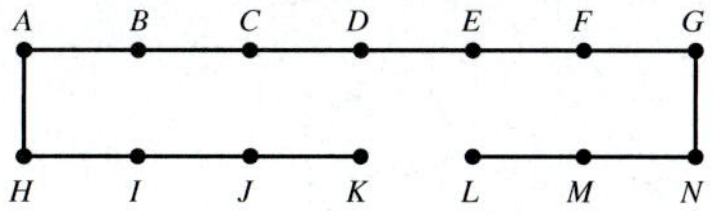

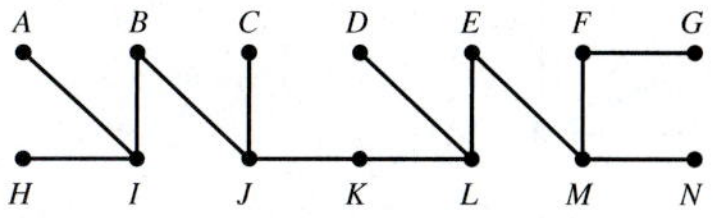

Other answers are possible.

17. 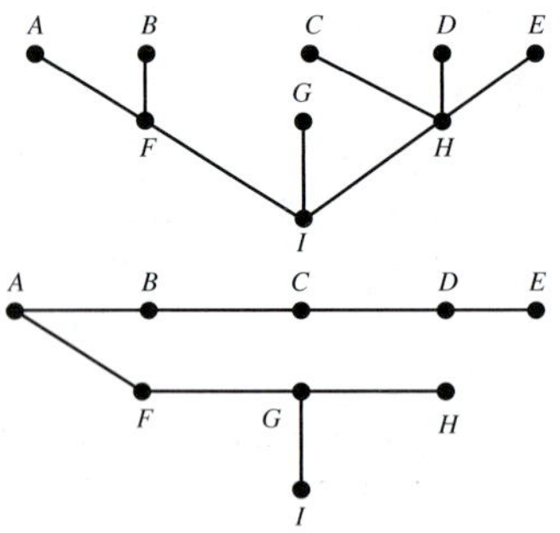

Other answers are possible.

19. 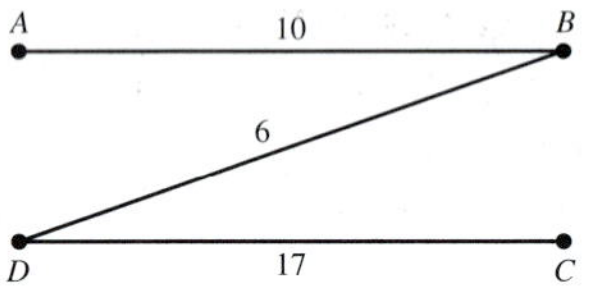

21. 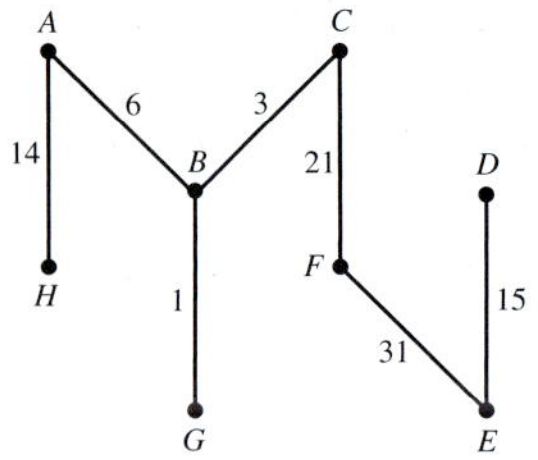

23. 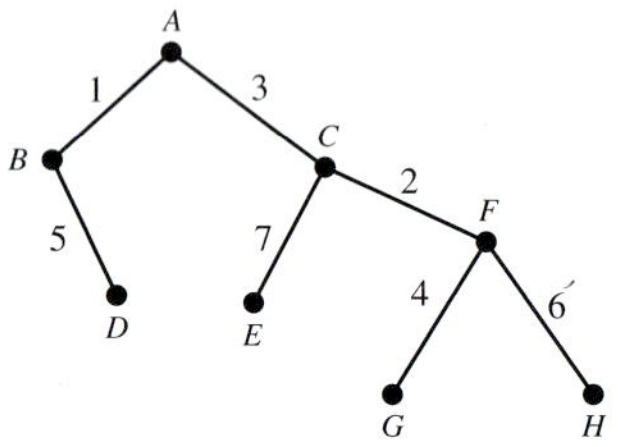

25. 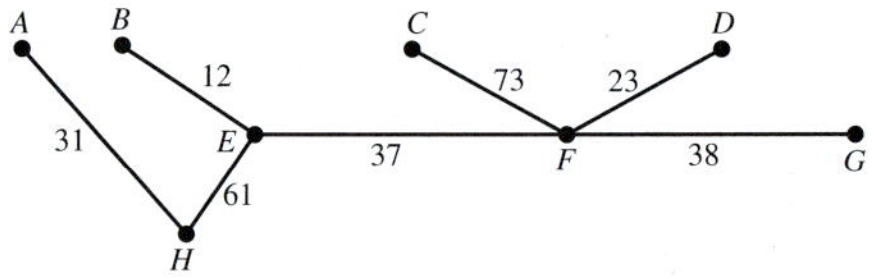

27. a) 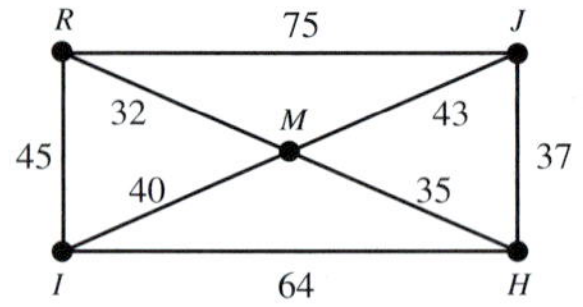

Other answers are possible.

b) 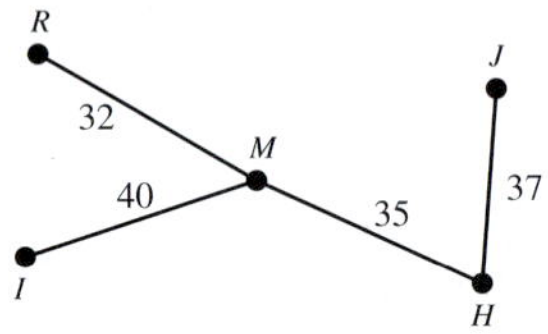

c) $3600

29. a) 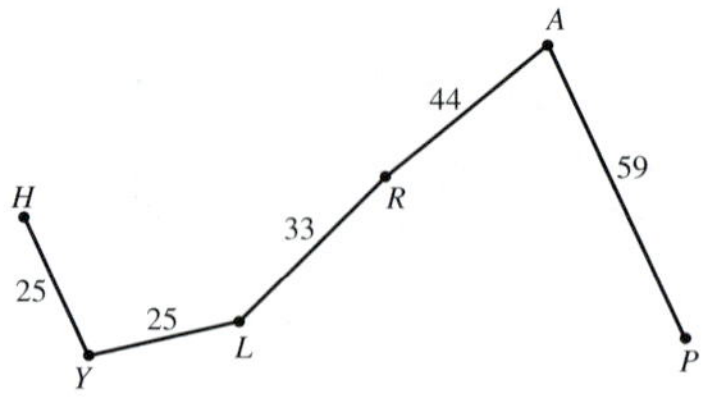

b) $1,264,800

31. a) 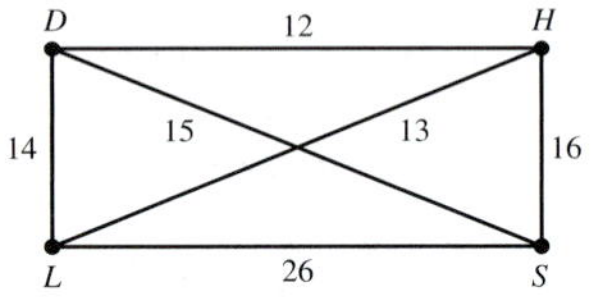

Other answers are possible.

b) 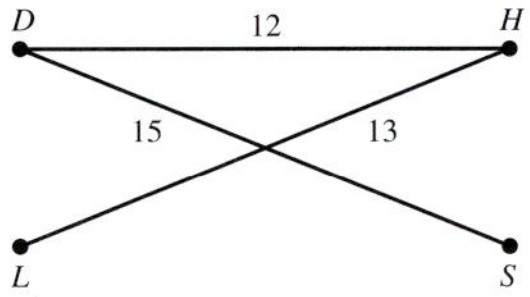

c) $140,000

33. a) 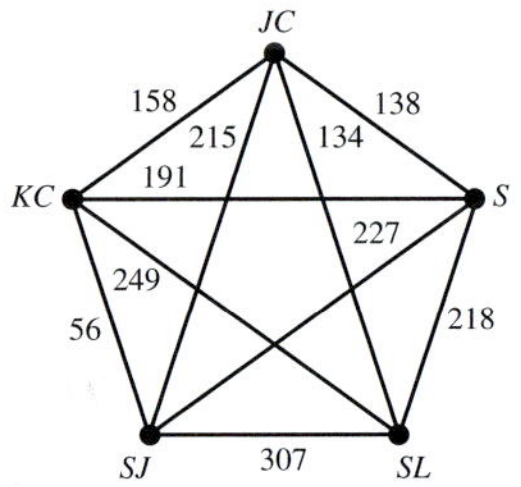

b) 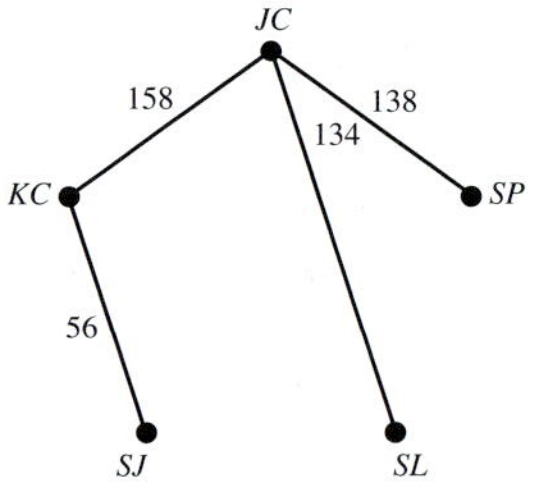

c) $1,798,200

35.–37. Answers will vary.

REVIEW EXERCISES, PAGE 968

1. 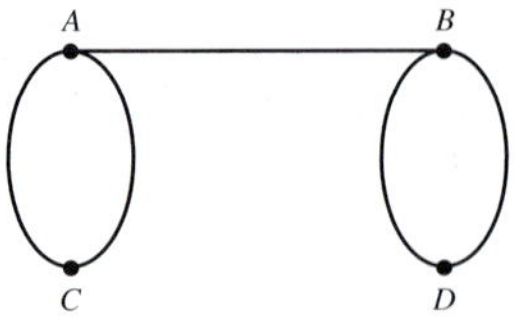

Other answers are possible.

2.

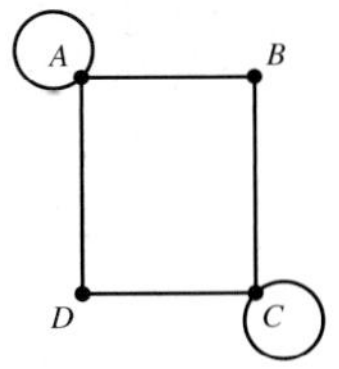

Other answers are possible.

3. A, B, C, A, D, C, E, D; other answers are possible.

4. No. A path that includes each edge exactly one time would start at vertex A and end at vertex D, or vice versa.

5.

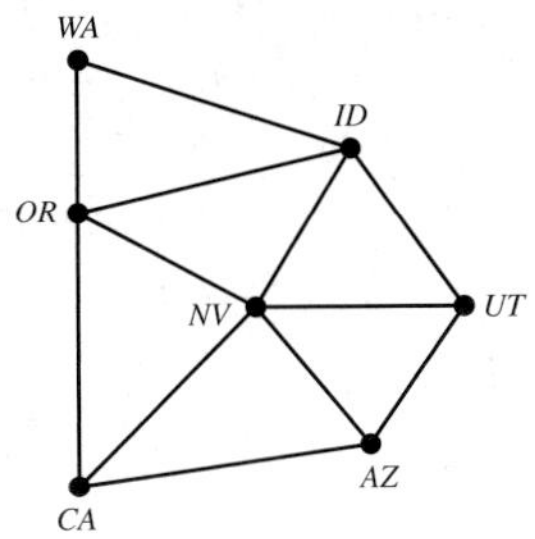

6.

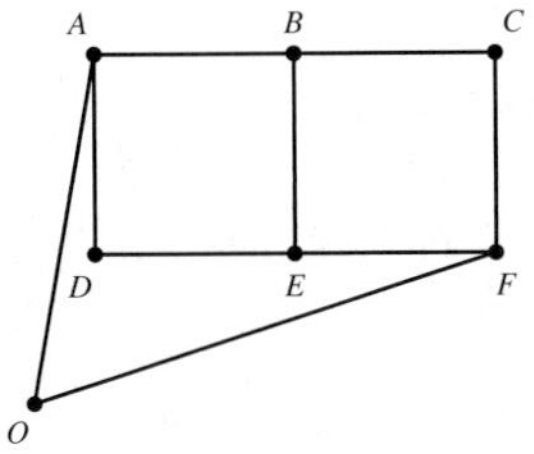

7. Connected **8.** Disconnected **9.** Edge CD

10. $D, A, B, C, E, H, G, F, D, B, E, G, D, E$; other answers are possible.

11. $E, D, B, E, G, D, A, B, C, E, H, G, F, D$; other answers are possible.

12. $B, C, A, D, F, E, C, D, E, B$; other answers are possible.

13. $E, F, D, E, C, D, A, C, B, E$; other answers are possible.

14. a)

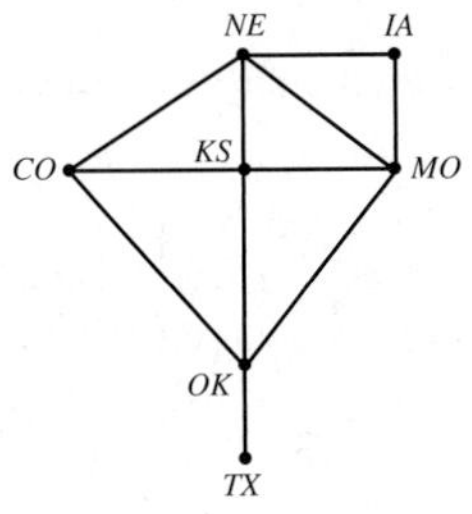

b) Yes. $CO, NE, IA, MO, NE, KS, MO, OK, CO, KS, OK, TX$; other answers are possible.

c) No

15. a)

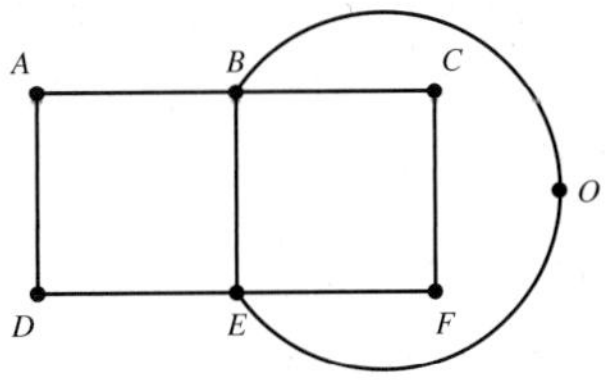

b) Yes

c) The person may start in any room or outside and will finish in the location from which he or she started.

16. a) Yes

b) The officer would have to start at either the intersection of Dayne St., Gibson Pl., and Alvarez Ave. or at the intersection of Chambers St., Fletcher Ct., and Alvarez Ave.

17. $A, B, F, E, H, G, D, C, A, D, E, B$; other answers are possible.

18. $A, B, C, D, H, G, C, F, G, B, F, E, A$; other answers are possible.

19. A, C, B, F, E, D, G and A, C, D, G, F, B, E; other answers are possible.

20. A, B, C, D, F, E, A and A, E, F, B, C, D, A; other answers are possible.

21.

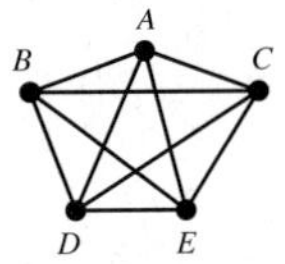

22. $4! = 24$ ways

23. a)

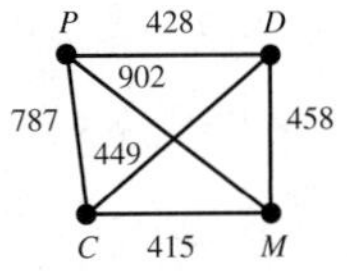

b) P, D, M, C, P or P, C, M, D, P

c) $2088

24. a)

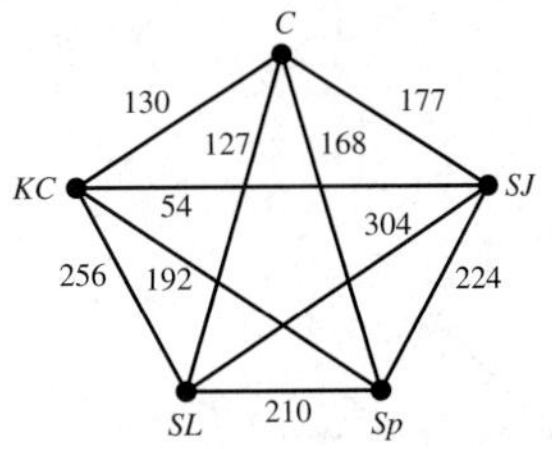

b) SJ, KC, C, SL, Sp, SJ traveling a total of 745 miles

c) Sp, C, SL, KC, SJ, Sp traveling a total of 829 miles

25.

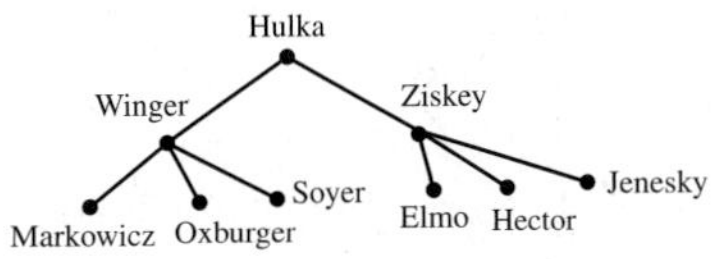

26.

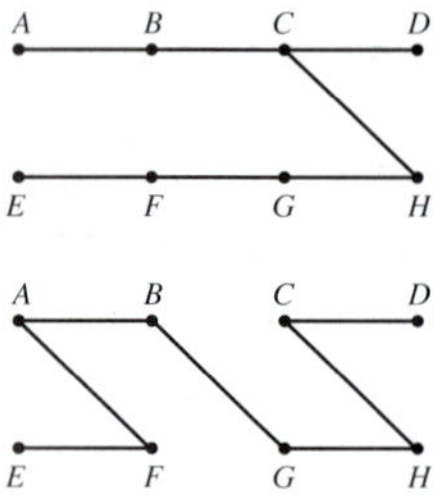

Other answers are possible.

27.

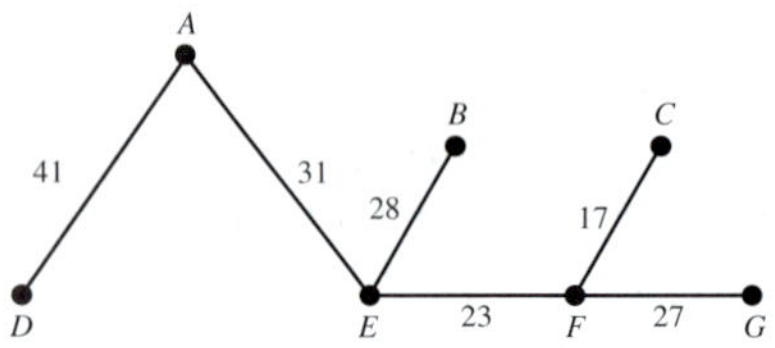

28. a)

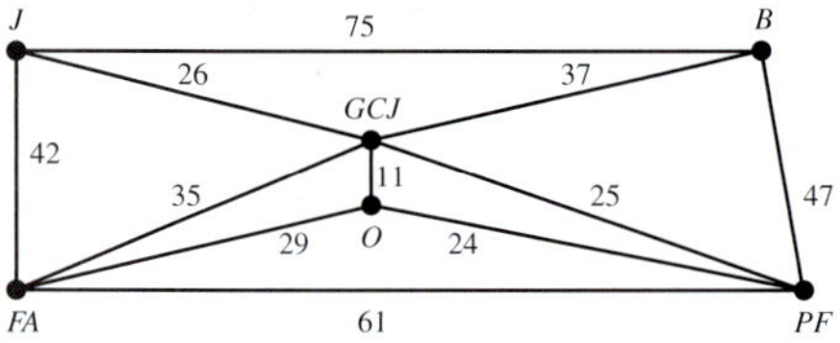

b)

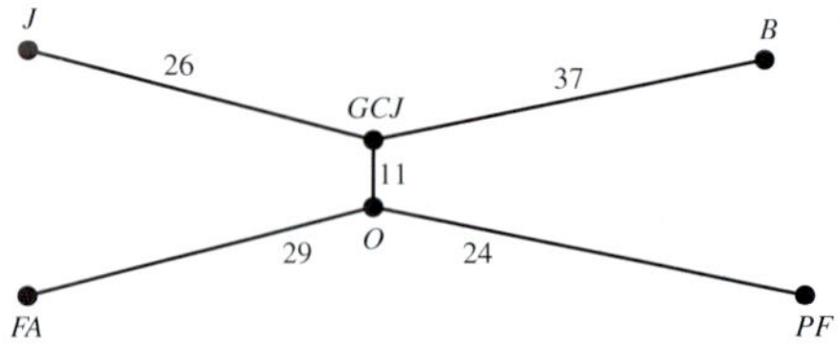

c) $317.50

CHAPTER TEST, PAGE 971

1.

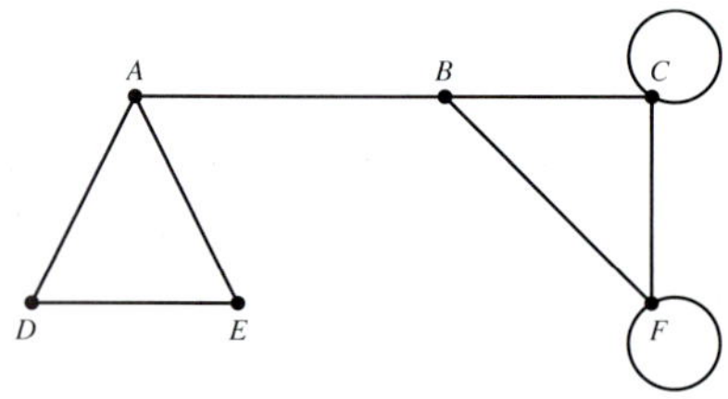

Other answers are possible.

2.

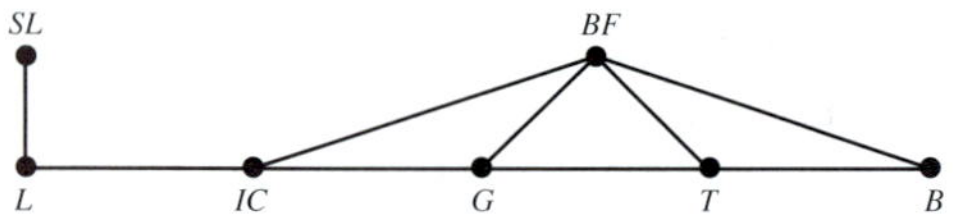

3. One example:

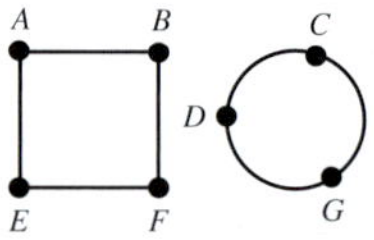

4. *A, B, C, F, H, G, E, F, B, E, D, A*; other answers are possible.

5. Yes. The person may start in room *A* and end at room *B*, or vice versa.

6. *A, D, E, A, F, E, H, F, I, G, F, B, G, C, B, A*; other answers are possible.

7. *A, B, C, G, E, D, H, I, K, J, F, A*; other answers are possible.

8. $4! = 24$

9. a)

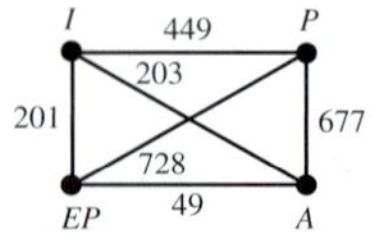

b) *I, P, A, EP, I* or *I, EP, A, P, I* for $1376

c) *I, EP, A, P, I* for $1376

10.

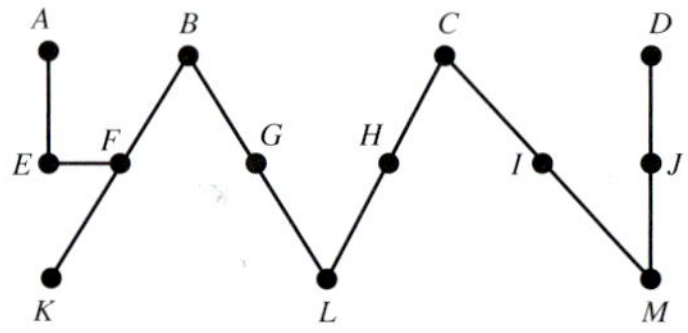

Other answers are possible.

11.

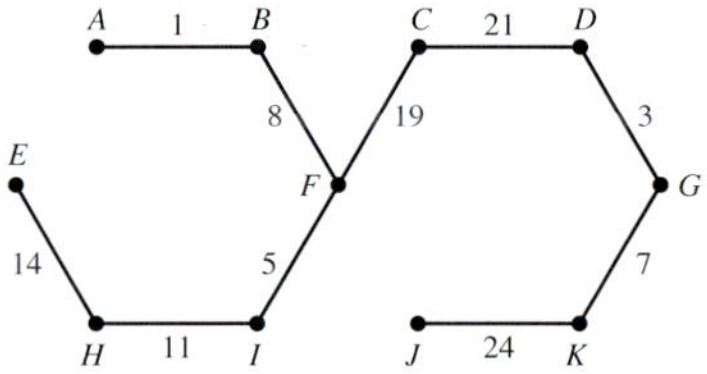

12. a)

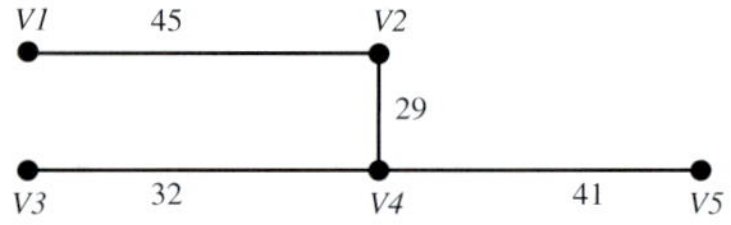

b) $183.75

CHAPTER 15

SECTION 15.1, PAGE 988

1. When a candidate receives more than 50% of the votes, he or she receives a majority.

3. Voters rank candidates from most favorable to least favorable. Each last-place vote is awarded 1 point, each next-to-last-place vote is awarded 2 points, each third-from-last-place vote is awarded 3 points, and so forth. The candidate receiving the most points is the winner.

5. Voters rank the candidates. A series of comparisons is conducted in which each candidate is compared with each of the other candidates. If candidate A is preferred to candidate B, A receives 1 point. If candidate B is preferred to candidate A, B receives 1 point. If the candidates tie, each receives $\frac{1}{2}$ point. The candidate receiving the most points is declared the winner.

7. A preference table summarizes the results of an election.

9. Voters are able to provide more information, such as ranking choices, with the Borda count method.

11. a) Jeter **b)** No

13.

Number of Votes	3	1	2	2	1
First	H	G	S	S	G
Second	G	H	H	G	S
Third	S	S	G	H	H

15. 21 **17.** Wrench **19.** Water heater

21. Yellowstone National Park

23. Three-way tie, no winner

25. San Antonio

27. No winner, a tie between Honolulu and San Antonio

29. Williams **31.** Johnson **33.** Williams **35.** Erie

37. Erie **39. a)** Becker **b)** Becker **c)** Becker **d)** Becker

41. a) Petunias **b)** Petunias **c)** Geraniums
d) Tie between petunias and zinnias

43. a) If there were only two columns, only two of the candidates were the first choice of the voters. If each of the 15 voters cast a ballot, one of the voters must have received a majority of votes because 15 cannot be split evenly.
b) An odd number cannot be divided evenly, so one of the two first-choice candidates must receive more than half of the votes.

45. a) Warriors, Rams and Tigers tied, Comets
b) Rams, Tigers, Warriors, Comets

47. a) 150 **b)** 50 **c)** Yes

49. One possible answer is:

Number of Votes	5	5	2
First	A	B	C
Second	B	A	B
Third	C	C	A

SECTION 15.2, PAGE 1002

1. If a candidate receives a majority of first-place votes, that candidate should be declared the winner.

3. If a candidate is favored when compared individually with every other candidate, that candidate should be declared the winner.

5. A candidate who is preferred to all others will win each head-to-head comparison and is selected with the pairwise comparison method.

7. If a candidate receives a majority of votes, that candidate also has the most votes and wins by the plurality method.

9. Portland wins with the Borda count method, but Tampa has a majority of first-place votes.

11. No. Brown is the Borda count winner, but Alvarez has a majority of first-place votes.

13. No. Lounge areas wins by the Borda count method, but Parking is preferred to Security and to Lounge areas.

15. No. C wins by plurality with elimination, but A is preferred over each of the other candidates using head-to-head comparisons.

17. No. A wins the first election by plurality. If B drops out, C wins by plurality.

19. Yes. B wins by the Borda count method. If C drops out, B still wins.

21. No. C wins by plurality with elimination. If the voters change their preference, B wins by plurality with elimination.

23. No. D wins by pairwise comparison. If the voters change their preference, B wins by pairwise comparison.

25. No. B wins by pairwise comparison. If A, C, and E drop out, D wins by pairwise comparison.

27. No. B wins by the Borda count method, but A has a majority of first-place votes.

29. a) Washington, D.C.
b) Washington, D.C.
c) Washington, D.C.
d) Washington, D.C.
e) Washington, D.C.
f) None of them

31. a) Jennifer Aniston **b)** Denzel Washington **c)** Yes

33. A candidate who holds a plurality only gains strength and holds an even larger lead if more favorable votes are added.

35.–37. Answers will vary.

SECTION 15.3, PAGE 1021

1. If we divide the total population by the number of items to be apportioned, we obtain a number called the standard divisor.

3. A lower quota is the standard quota rounded down to the nearest whole number.

5. An apportionment should always be either the upper quota or the lower quota.

7. Jefferson's method, Webster's method, and Adams's method

9. a) Webster's method **b)** Adams's method
c) Jefferson's method

11. a) 50,000
b) 26.9, 57.1, 19.64, 56.36

13. a) 27.28, 57.91, 19.92, 57.16 **b)** 27, 57, 19, 57

15. a) 26.53, 56.31, 19.37, 55.58 **b)** 27, 57, 20, 56

17. 27, 57, 20, 56

19. a) 27 **b)** 11.33, 7.93, 5.74

21. a) 11.86, 8.29, 6.01 **b)** 11, 8, 6

23. a) 10.55, 7.38, 5.34 **b)** 11, 8, 6

25. 11, 8, 6

27. a) 11 **b)** 7.82, 9.27, 11.82, 21.09 **c)** 8, 9, 12, 21

29. 8, 9, 12, 21

31. a) 52 **b)** 33.58, 136.44, 40.98, 18.02, 20.98
c) 34, 136, 41, 18, 21

33. 33, 137, 41, 18, 21

35. a) 90 **b)** 53.33, 40.09, 33.22, 23.36 **c)** 53, 40, 33, 24

37. 54, 40, 33, 23

39. a) 750 **b)** 12.09, 20.37, 17.08, 7.63, 33.67, 9.17
c) 12, 20, 17, 8, 34, 9

41. 12, 20, 17, 8, 34, 9

43. a) 12 **b)** 62.58, 81.67, 41.92, 13.83 **c)** 62, 82, 42, 14

45. 63, 82, 42, 13

47. a) 7, 2, 2, 2, 8, 14, 4, 5, 10, 10, 13, 2, 6, 2, 18
b) 7, 1, 2, 2, 8, 14, 4, 5, 10, 10, 13, 2, 6, 2, 19
c) Virginia, Delaware

49. Answers will vary. One possible answer is A: 743, B: 367, C: 432, D: 491, E: 519, F: 388

SECTION 15.4, PAGE 1030

1. The Alabama paradox occurs when an increase in the total number of items to be apportioned results in a loss of items for a group.

3. The population paradox occurs when group A loses items to group B, although group A's population grew at a higher rate than group B's.

5. Adams's method, Webster's method

7. No. The new apportionment is 14, 11, 11, 12, 13. No office suffers a loss, so the Alabama paradox doesn't occur.

9. a) 6, 8, 16
b) Yes. When the number of seats increases, states B and C gain a seat and state A loses a seat.

11. a) 65, 32, 103 **b)** No

13. a) 7, 16, 9, 11, 11
b) Yes. Division B loses an internship to division A even though the population of division B grew faster than the population of division A.

15. a) 8, 40
b) Yes. The apportionment is now 9, 39, 7, and group B loses a manager.

17. a) 5, 17, 44
b) Yes. The apportionment is now 4, 17, 44, 6, and state A loses a seat.

REVIEW EXERCISES, PAGE 1033

1. a) Robert Rivera **b)** No

2. a) Michelle MacDougal **b)** Yes

3.

Number of Votes	3	2	1	1	3
First	B	A	D	D	C
Second	A	C	C	A	B
Third	C	D	A	B	A
Fourth	D	B	B	C	D

4.

Number of Votes	2	2	2	1
First	C	A	B	C
Second	A	C	A	B
Third	B	B	C	A

5. 20 **6.** Portable DVD player **7.** iPod **8.** Digital camera

9. Tie between iPod and digital camera **10.** iPod **11.** 110

12. Volleyball **13.** Soccer **14.** Volleyball

15. None, a three-way tie **16.** Soccer

17. a) Yes, American Association of Retired Persons
b) American Association of Retired Persons
c) American Association of Retired Persons
d) American Association of Retired Persons
e) American Association of Retired Persons

18. a) No **b)** Seattle **c)** Las Vegas
d) Las Vegas **e)** Las Vegas

19. a) A tie between *Encyclopedia Britannica* and *World Book*
b) *World Book*
c) A tie between *Encyclopedia Britannica* and *World Book*

20. No. A has a majority, but B wins by the Borda count method.

21. No. B wins by the Borda count method, but A is preferred over each of the other candidates using head-to-head comparisons.

22. a) C wins the election by plurality with elimination.
b) When the order is changed A wins. Therefore, the monotonicity criterion is not satisfied.
c) No. If B drops out, A is the winner by plurality with elimination. Therefore, the irrelevant alternatives criterion is not satisfied.

23. a) Ball Park **b)** Oscar Mayer
c) Tie between Nathan's and Hebrew National
d) Hebrew National **e)** Ball Park
f) Plurality method, Borda count method, and plurality with elimination method all violate the head-to-head criterion.

24. a) Fleetwood Mac **b)** Boston **c)** Fleetwood Mac
d) REO Speedwagon **e)** Fleetwood Mac
f) The plurality method and the plurality with elimination method

25. The Borda count method

26. The plurality with elimination method; note the plurality method does not violate the monotonicity criterion because C does not gain support in the second election.

27. The pairwise comparison method and the Borda count method

28. 4, 3, 3 **29.** 5, 2, 3 **30.** 4, 3, 3 **31.** 5, 2, 3

32. Yes. The apportionment is 5, 2, 4. Region B loses one truck.

33. 13, 9, 7 **34.** 13, 9, 7 **35.** 12, 9, 8 **36.** 13, 9, 7

37. No. The apportionment with the additional population is still 13, 9, 7.

38. 5, 50 **39.** 4, 51 **40.** 5, 50 **41.** 5, 50

42. Yes. The new apportionment is 5, 49, 6. State B loses a seat.

CHAPTER TEST, PAGE 1038

1. 24 **2.** No **3.** Pizza **4.** Pizza **5.** Tacos

6. Tacos

7. a) Salamander **b)** Iguana **c)** Hamster **d)** Lemming

8. Plurality method, Borda count method, and plurality with elimination method

9. Yes, El Capitan has a majority of first-place votes but mule deer wins.

10. a) 6, 9, 15 **b)** 6, 9, 15 **c)** 6, 9, 15 **d)** 6, 9, 15
e) No. The apportionment is 6, 9, 16.
f) No. The apportionment is 6, 9, 15.
g) No. The apportionment is 6, 9, 15, 5.

CREDITS

Page 1, Couple working on budgets, Digital Vision; Page 2, DisneyWorld finger scanner, Apani Networks; Page 3, ATM eye scan, International Business Machines Corporation; Page 4, Astronaut on moon, NASA; Page 9, Jury, Corbis; Page 9, Buying stamps in post office, Beth Anderson; Page 10, Bushels of grapes in vineyard, PhotoDisc Red; Page 13, Counting birds, Art Wolfe/Stone/Getty Images; Page 14, Repeat detail of birds, Art Wolfe/Stone/Getty Images; Page 15, Picking strawberries, PhotoDisc Blue; Page 18, Bananas, Allen R. Angel; Page 18, Blueberries, PhotoDisc; Page 18, Statue of Liberty, Allen R. Angel; Page 19, Three men and palm tree, Corbis Royalty Free; Page 20, Buying CDs in office supply store, PhotoDisc Red; Page 22, George Polya at blackboard, Palo Alto Weekly; Page 23, Archimedes, The Bridgeman Art Library International; Page 25, Restaurant waitstaff with customers, PhotoDisc Red; Page 26, Vacation in the mountains, Allen R. Angel; Page 27, Spraying weed killer, Corbis; Page 30, Large fishing boat, Digital Vision; Page 33, African safari, PhotoDisc Red; Page 42, Van Gogh's *Sunflowers*, Christie's Images/Corbis; Page 44, Child playing with abacus, Stockbyte Platinum Getty RF; Page 45, Horses, Allen R. Angel; Page 45, Georg Cantor, The Granger Collection; Page 46, Planets in Earth's solar system, Bennett/Donahue/Schneider/Voit, *The Essential Cosmic Perspective, Fourth Edition*, Pearson Addison-Wesley, © 2007; Page 50, Astronauts on the moon, NASA; Page 51, Snow White and the Seven Dwarfs, Walt Disney/The Kobal Collection; Page 52, The Beatles, The Everett Collection; Page 53, J.K. Rowling, Corbis; Page 54, Fireman, PhotoDisc Red; Page 57, Pizza with toppings, Stockbyte Platinum Getty RF; Page 58, Exercise class at gym, Purestock Getty RF; Page 60, Koala and Mrs. Angel, Allen R. Angel; Page 60, Buying a soft drink, Image Source Getty RF; Page 69, San Diego Zoo, PhotoDisc Red; Page 68, Yosemite National Park, PhotoDisc; Page 71, Person with dog, Ingram Publishing Getty RF; Page 71, High school band, Beth Anderson; Page 72, Doctors examining x-rays, PhotoDisc Blue; Page 77, Walmart, Beth Anderson; Page 78, Athlete from summer Olympics, Digital Vision; Page 82, Using treadmill or Stairmaster, PhotoDisc; Page 85, Hawaii, Beth Anderson; Page 87, Movie theatre attendees, PhotoDisc Red; Page 87, Water slide at amusement park, Digital Vision; Page 88, Women at restaurant, Digital Vision; Page 89, Farmer and wheatfield, PhotoDisc; Page 89, Georg Cantor, The Granger Collection; Page 91, Leopold Kronecker, The Granger Collection; Page 95, Tiger, PhotoDisc; Page 97, Actors from *CSI:NY*, Getty Editorial; Page 98, Cotton candy snack, PhotoDisc Red; Page 100, Lawyer arguing case, Digital Vision; Page 101, Morton salt ad, Morton International, Inc.; Page 102, Brooklyn Bridge, Corbis RF; Page 103, Abbott and Costello, Bettmann/Corbis; Page 105, Billie Joe Armstrong/Green Day, AP Wideworld Photos; Page 106, George Boole, The Granger Collection; Page 108, Painting by Beth Anderson, Beth Anderson; Page 109, Martin St. Louis, Getty Images Sport; Page 110, Louis Armstrong playing trumpet, Time & Life Pictures/Getty Images; Page 110, Butterflies and flowers, Allen R. Angel; Page 111, Ken Jennings, Jeopardy champ, Getty Images News; Page 113, Idaho potatoes, Digital Vision; Page 116, Guitarists, PhotoDisc Red; Page 123, Girl with cookies and milk, Corbis Royalty Free; Page 123, Collie, PhotoDisc Blue; Page 123, John Madden, sports announcer, AP Wideworld Photos; Page 125, Airplane travel, Digital Vision; Page 126, Car for A student, Digital Vision; Page 133, Soccer, Corbis Royalty Free; Page 134, Nike shoes, Corbis; Page 136, Sunny beach, Dennis Runde; Page 137, Charles Dodgson, The Granger Collection; Page 137, Alice in Wonderland, The Granger Collection; Page 141, Augustus de Morgan, The Granger Collection; Page 144, Sitar, PhotoDisc; Page 146, Digital photographer, ThinkStock RF; Page 147, Timex watch, Beth Anderson; Page 148, *Jimmy Neutron*, Paramount Pictures/Courtesy: Everett Collection; Page 148, Sailboat, Beth Anderson; Page 149, Yellow lab, Christine Abbott; Page 149, People fishing, PhotoDisc Blue; Page 150, FedEx truck, Beth Anderson; Page 150, Person taking a picture, PhotoDisc Red; Page 151, John Mellencamp, Major League Baseball/Getty Images; Page 153, Alaska glacier, PhotoDisc Blue; Page 156, U.S. Constitution photo, Corbis RF; Page 158, Prices at gas station, Getty Images News; Page 160, Les Paul and guitar, Corbis; Page 160, Graduation, StockDisc; Page 161, Football quarterback about to throw, Stockbyte Platinum/Getty RF; Page 162, Meg Ryan in *In the Land of Women*, Warner Independent/Courtesy: Everett Collection; Page 165, Barcode scanner, PhotoDisc; Page 167, Gingerbread man, National Geographic/Getty Royalty Free; Page 168, Morel mushrooms, PhotoDisc; Page 168, Caterpillar, PhotoDisc Red; Page 169, Reading lamp, Digital Vision; Page 176, Ellen Johnson-Sirleaf, President of Liberia, Jason Szenes/epa/Corbis; Page 177, Bobby Darin and Sandra Dee, Hulton Archive/Getty Images; Page 180, Ontario Science Center logic game, Courtesy Ontario Science Center; Page 181, Kids using math manipulatives, Richard Hutchings/PhotoEdit Inc.; Page 182, Hines Ward and Super Bowl XL logo, Getty Images Sport; Page 182, Denise Schmandt-Besserat, Courtesy Denise Schmandt-Besserat; Page 183, Rhind papyrus, Public domain; Page 184, Clock with Roman numerals, Allen R. Angel; Page 186, Sign showing Hindu-Arabic numerals, Mark Gibson; Page 190, Great pyramid at Khufu, Giza Egypt, Digital Vision (PP); Page 190, Bob Barker game in *Price is Right*, CBS/Photofest; Page 192, Counting board, Taxi/Getty Images; Page 193, Globe, PhotoDisc; Page 194, Photo of Jan Fleck's *Numbers*, Allen R. Angel; Page 197, Boys with iPods, Christine Abbott; Page 198, Students in computer lab, Blend Images/Getty RF; Page 200, Arabic and Chinese numerals, PhotoAlto/Getty RF; Page 204, Middle school kids in computer lab, PhotoDisc; Page 204, Arithmetic teacher and students, Purestock/Getty RF; Page 213, Internet researcher, Stockbyte Platinum/Getty RF; Page 214, Pay check/time clock, PhotoDisc; Page 215, Gradeschool class—new old math, PhotoDisc; Page 216, John Napier, The Granger Collection; Page 223, Teenagers on cell phones. PhotoDisc Red (PP); Page 224, Box of chocolates, Beth Anderson; Page 227, Factoring machine 1914, Jeffrey Shallit; Page 228, Srinivasa Ramanujan, The Granger Collection; Page 230, Bank vault, Digital Vision; Page 233, Barbie dolls, Beth Anderson; Page 233, School band, Merrill Education; Page 235, Playing in snow, PhotoDisc; Page 238, Mount Everest, Corbis RF; Page 243, NYSE floor traders, PhotoDisc; Page 243, Reggie Bush, New Orleans Saints, Getty Editorial; Page 245, Preparing cake mix, Pixland/Corbis Royalty Free; Page 253, Waves of string vibration, Fundamental Photos; Page 257, Measuring boy's height, Corbis RF; Page 258, Farmer and silo, Digital Vision; Page 259, San Antonio, Texas, Allen R. Angel; Page 261, Superbowl champs playing, Getty Editorial; Page 262, Pythagoras of Samos, Corbis; Page 268, Car in motion, Allen R. Angel; Page 269, Elementary school students doing arithmetic, Digital Vision; Page 271, Daniel Tammet, Colin McPherson/Corbis; Page 275, English bulldog, PhotoDisc Red; Page 276, Breaking an egg into a bowl, PhotoDisc; Page 276, Galaxy, NASA headquarters; Page 280, Another galaxy, NASA Media Services; Page 281, Preda Mihailescu, Used with permission of Preda Mihailescu; Page 283, WWI poster, U.S. bonds, Library of Congress; Page 286, Jupiter, U.S. Geological Survey/U.S. Department of the Interior; Page 286, IBM BlueGene/L system computer, Kim Kulish/Corbis; Page 287, Niagara Falls, PhotoDisc; Page 288, U.S. mint/uncut sheets of currency, Charles O'Rear/Corbis; Page 289, Scientists in lab with microscope, Digital Vision; Page 290, Carl Frederich Gauss, Public domain; Page 292, Sun and planets, PhotoDisc; Page 296, Squirrel, Digital Vision; Page 296, Cuckoo clock, PhotoDisc Red; Page 296, Samurai sword production, Bettmann/Corbis; Page 297, Black Jack game, PhotoDisc; Page 298, Beehive, Digital Vision; Page 298, Leonardo of Pisa, Fibonacci, Corbis;

Page 943, USPS mail truck rural, Corbis RM; Page 945, Kevin Bacon, Reuters NewMedia Inc./Corbis; Page 946, *American Idol* judges (Paula Abdul, Randy Jackson, Simon Cowell), Getty Images Entertainment; Page 952, Truck weigh station, Allen R. Angel; Page 953, Gateway Arch, St. Louis, PhotoDisc; Page 953, Child getting into minivan, PhotoDisc; Page 953, Clarke College, Courtesy Clarke College; Page 955, Icosian Game, Royal Irish Academy; Page 956, Woman and large flower garden, Digital Vision; Page 961, Solomon Lefschetz, Archive Photos; Page 965, Hibiscus, Beth Anderson; Page 966, University of Texas at El Paso, Venegas Engineering Management and Construction; Page 971, Six flowers, Allen R. Angel; Page 974, Voting, Digital Vision; Page 975, People at a party, Stockbyte/Getty Royalty Free; Page 975, Home garden, PhotoDisc; Page 976, Husky, PhotoDisc; Page 977, Violin players, Digital Vision; Page 978, Harry S. Truman, Bettmann/Corbis; Page 979, Jean-Charles de Borda, Courtesy Musée de Borda; Page 982, Ice sculpture, Beth Anderson; Page 984, Snowman, PhotoDisc Blue; Page 987, Rutherford B. Hayes, Library of Congress; Page 987, UNC Chapel Hill, Getty Images News; Page 989, Three brands of yogurt, Beth Anderson; Page 989, Yellowstone National Park, Allen R. Angel; Page 990, Honolulu, PhotoDisc Blue; Page 991, Cast of *Cheers*, NBC/Photofest; Page 991, Flower garden, Allen R. Angel; Page 992, Kittens, PhotoDisc; Page 993, Mary J. Blige, 2007 Grammy awards, AFP/Getty Images; Page 993, Party guests eating pizza, PhotoDisc Red; Page 995, Voting, Mark Richards; Page 996, Marquis de Condorcet, The Granger Collection; Page 999, Farmer and tractor, PhotoDisc; Page 1001, Kenneth Arrow, AP Wideworld Photos; Page 1003, Tampa, Florida, PhotoDisc Red; Page 1003, Teacher and students, Digital Vision; Page 1004, Mall of America, Owen Franken/Corbis; Page 1005, Jefferson Memorial, PhotoDisc; Page 1006, Lance Armstrong, AFP/Getty Images; Page 1007 and 1024 Laser printer in computer lab, iStockPhoto; Page 1008, Congress, Corbis; Page 1010, Alexander Hamilton, Public domain; Page 1012, Thomas Jefferson, Public domain; Page 1015, Daniel Webster, Public domain; Page 1017, John Quincy Adams, Public domain; Page 1021, Hotel staff, PhotoDisc; Page 1022, Ski resort, Digital Vision; Page 1022, Princeton University, Courtesy Princeton University; Page 1028, Oklahoma Public Library, Hola Images/Getty Royalty Free; Page 1030, Fax machine, iStockPhoto; Page 1031, Office worker at a workstation, iStockPhoto; Page 1034, Student with an iPod, iStockPhoto; Page 1034, Florida manatee license plate, Florida Dept. of Highway Safety and Motor Vehicles; Page 1035, Dallas, TX, iStockPhoto; Page 1036, REO Speedwagon, Getty Images Entertainment; Page 1036, Girl eating hot dogs, iStockPhoto.

INDEX of Applications

CHEMISTRY AND PHYSICS

COMPUTERS

CONSTRUCTION

CONSUMER

DOMESTIC

ECONOMICS

EDUCATION

ENVIRONMENT

FINANCE

GENERAL INTEREST

NUMBERS

PROBABILITY

SPORTS AND ENTERTAINMENT

STATISTICS AND DEMOGRAPHICS

TRAVEL

INDEX

Note: Page numbers preceded by AA indicate appendix.